Physics Galaxy

19 Years'

2002-2020 (Offline & Online)

JEE MAIN

Chapter-Wise Solutions

Physics

Ashish Arora

G K Publications (P) Ltd

CAREER LAUNCHER INFRASTRUCTURE (P) LTD.

First Edition 2000

Revised Edition : 2020

ISBN : 978-93-90187-74-4

Typeset by: *CLIP DTP Unit*

Administrative and Production Offices

Published by : CLIP

A-45, Mohan Cooperative Industrial Area,
Near Mohan Estate Metro Station,
New Delhi - 110044

Marketed by : G.K. Publications (P) Ltd.

A-45, Mohan Cooperative Industrial Area,
Near Mohan Estate Metro Station,
New Delhi - 110044

For product information :

Visit *www.gkpublications.com* or email to *gkp@gkpublications.com*

Dedicated

to

My Parents, Son, Daughter

and

My beloved wife

In his teaching career since 1992 Ashish Arora personally mentored more than 10000 IITians and students who reached global heights in various career and profession chosen. It is his helping attitude toward students with which all his students remember him in life for his contribution in their success and keep connections with him live. Below is the list of some of the successful students in International Olympiad personally taught by him.

NAVNEET LOIWAL	*International GOLD Medal in IPhO-2000 at LONDON*, Also secured **AIR-4** in **IIT JEE 2000** PROUD FOR INDIA : Navneet Loiwal was the first Indian Student who won first International GOLD Medal for our country in International Physics Olympiad.
DUNGRA RAM CHOUDHARY	**AIR-1** in **IIT JEE 2002**
HARSHIT CHOPRA	*National Gold Medal in INPhO-2002* and got **AIR-2** in **IIT JEE-2002**
KUNTAL LOYA	A Girl Student got position **AIR-8** in **IIT JEE 2002**
LUV KUMAR	*National Gold Medal in INPhO-2003* and got **AIR-3** in **IIT JEE-2003**
RAJHANS SAMDANI	*National Gold Medal in INPhO-2003* and got **AIR-5** in **IIT JEE-2003**
SHANTANU BHARDWAJ	*International SILVER Medal in IPhO-2002 at INDONESIA*
SHALEEN HARLALKA	*International GOLD Medal in IPhO-2003 at CHINA* and got **AIR-46** in **IIT JEE-2003**
TARUN GUPTA	*National GOLD Medal in INPhO-2005*
APEKSHA KHANDELWAL	*National GOLD Medal in INPhO-2005*
ABHINAV SINHA	*Hon'ble Mension Award in APhO-2006* at *KAZAKHSTAN*
RAMAN SHARMA	*International GOLD Medal in IPhO-2007* at *IRAN* and got **AIR-20** in **IIT JEE-2007**
PRATYUSH PANDEY	*International SILVER Medal in IPhO-2007* at *IRAN* and got **AIR-85 in IIT JEE-2007**
GARVIT JUNIWAL	*International GOLD Medal in IPhO-2008* at *VIETNAM* and got **AIR-10** in **IIT JEE-2008**
ANKIT PARASHAR	*National GOLD Medal in INPhO-2008*
HEMANT NOVAL	*National GOLD Medal in INPhO-2008* and got **AIR-25** in **IIT JEE-2008**
ABHISHEK MITRUKA	*National GOLD Medal in INPhO-2009*
SARTHAK KALANI	*National GOLD Medal in INPhO-2009*
ASTHA AGARWAL	*International SILVER Medal in IJSO-2009* at *AZERBAIJAN*
RAHUL GURNANI	*International SILVER Medal in IJSO-2009* at *AZERBAIJAN*
AYUSH SINGHAL	*International SILVER Medal in IJSO-2009* at *AZERBAIJAN*
MEHUL KUMAR	*International SILVER Medal in IPhO-2010* at *CROATIA* and got **AIR-19** in **IIT JEE-2010**
ABHIROOP BHATNAGAR	*National GOLD Medal in INPhO-2010*
AYUSH SHARMA	*International Double GOLD Medal in IJSO-2010* at *NIGERIA*
AASTHA AGRAWAL	*Hon'ble Mension Award in APhO-2011* at *ISRAEL* and got **AIR-93** in **IIT JEE 2011**
ABHISHEK BANSAL	*National GOLD Medal in INPhO-2011*
SAMYAK DAGA	*National GOLD Medal in INPhO-2011*
SHREY GOYAL	*National GOLD Medal in INPhO-2012* and secured **AIR-24** in **IIT JEE 2012**
RAHUL GURNANI	*National GOLD Medal in INPhO-2012*
JASPREET SINGH JHEETA	*National GOLD Medal in INPhO-2012*
DIVYANSHU MUND	*National GOLD Medal in INPhO-2012*
SHESHANSH AGARWAL	*International SILVER Medal in IAO-2012 at KOREA*
SWATI GUPTA	*International SILVER Medal in IJSO-2012 at IRAN*
PRATYUSH RAJPUT	*International SILVER Medal in IJSO-2012 at IRAN*
SHESHANSH AGARWAL	*International BRONZE Medal in IOAA-2013 at GREECE*
SHESHANSH AGARWAL	*International GOLD Medal in IOAA-2014 at ROMANIA*
SHESHANSH AGARWAL	*International SILVER Medal in IPhO-2015 at INDIA* and secured **AIR-58** in **JEE(Advanced)-2015**
VIDUSHI VARSHNEY	*International SILVER Medal in IJSO-2015 at SOUTH KOREA*
AMAN BANSAL	**AIR-1 in JEE Advanced 2016**
KUNAL GOYAL	**AIR-3 in JEE Advanced 2016**
GOURAV DIDWANIA	**AIR-9 in JEE Advanced 2016**
DIVYANSH GARG	*International SILVER Medal in IPhO-2016 at SWITZERLAND*

ABOUT THE AUTHOR

The complexities of Physics have given nightmares to many, but the homegrown genius of Jaipur-Ashish Arora has helped several students to live their dreams by decoding it.

Newton Law of Gravitation and Faraday's Magnetic force of attraction apply perfectly well with this unassuming genius. A Pied Piper of students, his webportal https://www.physicsgalaxy.com, The world's largest encyclopedia of video lectures on high school Physics possesses strong gravitational pull and magnetic attraction for students who want to make it big in life.

Ashish Arora, gifted with rare ability to train masterminds, has mentored over 10,000 IITians in his past 24 years of teaching sojourn including lots of students made it to Top 100 in IIT-JEE/JEE(Advance) including AIR-1 and many in Top-10. Apart from that, he has also groomed hundreds of students for cracking International Physics Olympiad. No wonder his student Navneet Loiwal brought laurel to the country by becoming the first Indian to win a Gold medal at the 2000 - International Physics Olympiad in London (UK).

His special ability to simplify the toughest of the Physics theorems and applications rates him as one among the best Physics teachers in the world. With this, Arora simply defies the logic that perfection comes with age. Even at 18 when he started teaching Physics while pursuing engineering, he was as engaging as he is now. Experience, besides graying his hair, has just widened his horizon.

Now after encountering all tribes of students - some brilliant and some not-so-intelligent - this celebrated teacher has embarked upon a noble mission to make the entire galaxy of Physics inform of his webportal PHYSICSGALAXY.COM to serve and help global students in the subject. Today students from 183 countries are connected with this webportal. On any topic of physics students can post their queries in comment section in Physics Galaxy Youtube channel on which many Physics experts with Ashish Arora reply to several queries posted online by students.

Dedicated to global students of middle and high school level, his website *www.physicsgalaxy.com* also has teaching sessions dubbed in American accent and subtitles in 87 languages. For students in India preparing for JEE & NEET, his online courses will be available soon on PHYSICSGALAXY.COM.

In 2007 he started authoring Physics Galaxy Youtube Channel which has now become World's Largest Youtube Channel for online physics lectures on high school physics and preparation of JEE & NEET. Daily about 35000+ video lectures are being watched on the channel by students preparing for JEE, NEET & High school Physics.

FOREWORD

I am extremely happy to share my views about this book written by Er. Ashish Arora. He has contributed immensely to the teaching-learning of Physics though many innovative ways. His approach has always been student centric and this contribution is yet another example of the same.

As the title states "Physics Galaxy JEE Main Solutions" the book provides valuable information and covers the necessary components by offering solution to all questions asked in old JEE Main (all Online & Offline 2002-2020) with detailed explanations. All questions are categorized into 30 topics based on weightage of questions asked in the past 19 years. Most of the questions are linked to its video solution available on YouTube channel of Physics Galaxy. This is purely to indicate and promote understanding of the methods or reasons for specific/particular solutions. Aspirants can also scan the respective QR code given in the book to see its video explanation as and when required. All solutions of the questions are based on the conceptual analysis covered in the popular Physics Galaxy book series (5 volumes) so that aspirants following the book series will be able to relate and recall all the explanations already done.

I am sure the book promotes the concept of deeper learning in the domain of Physics. My best wishes to the many aspirants who will certainly find this book a very valuable tool to enhance thorough understanding of the subject.

September, 2020

Prof. Sandeep Sancheti
Ph. D. (U.K.), B.Tech. FIETE, MIEEE
VC, SRMIST, Chennai &
President, AIU, New Delhi

PREFACE

For a science student, preparing for JEE Main or NEET, Physics is the most important subject, unlike to other subjects it requires logical reasoning and high imagination of brain. Without improving the level of physics it is very difficult to achieve a goal in the present age of competitions. To score better, one does not require hard working at least in physics. It just requires a simple understanding and approach to think a physical situation. Actually physics is the surrounding of our everyday life. If you wish to make the concepts of physics strong, you should try to understand core concepts of physics in practical approach rather than theoretical.

This book "*PHYSICS GALAXY JEE Main Solutions*" of past 19years is designed in a totally different and friendly approach to develop the physics concepts psychologically. The book covers detailed solutions to all the questions previously asked in JEE Main all online and offline papers from 2002 to 2020. From the analysis of 2020 first attempt in January papers it is clearly seen that all six papers held were designed on the basis of past 19year papers. More than 60% questions are based on concepts asked in previous years which were matching in style with the questions of these old papers.

All the questions of these online and offline papers are divided in 30 topics for topic level understanding of questions asked in JEE Main. In my classroom teaching also I found that topic level understanding of previous year papers of exam helps students a lot in concept visualization from the point of view of preparation of final exam as well as for the purpose of last moment revision of all topics.

For every question asked in these old JEE Main papers, video solutions are recorded and made available for students on YouTube channel of Physics Galaxy as well as these video solutions can be accessed through Physics Galaxy mobile app available on iOS and Android playstore. While following the book if students face any doubt in understanding the solution of the specific question then they can scan the QR codes given at the end of the book for all the questions to directly start the video solution of that question.

I don't have words for my best friend my wife Anuja for always being together with me to complete this book in the unique style and format along with the whole serires of Physics Galaxy book series and lecture notes series.

I would like to pay my gratitude to Sh. Dayashankar Prajapati in assisting me to complete the task in Design Labs of PHYSICSGALAXY.COM and presenting the book in totally new format of second edition.

At last but the most important person, my father who has devoted his valuable time to finally present the book in such a format and a simple language, thanks is a very small word for his dedication in this book.

In this first edition of book I have tried my best to make this book error free but owing to the nature of work, inadvertently, there is possibility of errors left untouched. I shall be grateful to the readers, if they point out me regarding errors and oblige me by giving their valuable and constructive suggestions via emails for further improvement of the book.

Date : September, 2020

Ashish Arora

PHYSICSGALAXY.COM

B-80, Model Town, Malviya Nagar, Jaipur-302017

e-mails: ashisharora@physicsgalaxy.com

CONTENTS

* * * * *

CHAPTER-1 UNITS AND DIMENSIONS

1.1 Measurements and Dimensional Analysis

1. Identify the pair whose dimensions are equal : **[2002]**
(A) Torque and work (B) Stress and energy
(C) Force and stress (D) Force and work

2. The physical quantities not having same dimensions are : **[2003]**
(A) torque and work
(B) momentum and planck's constant
(C) stress and young's modulus
(D) speed and $(\mu_0\varepsilon_0)^{-1/2}$

3. Dimensions of $\frac{1}{\mu_0\varepsilon_0}$, where symbols have their usual meaning, are : **[2003]**
(A) $[L^{-1}T]$ (B) $[L^{-2}T^2]$
(C) $[L^2T^{-2}]$ (D) $[LT^{-1}]$

4. Which one of the following represents the correct dimensions of the coefficient of viscosity ? **[2004]**
(A) $[ML^{-1}T^{-1}]$ (B) $[MLT^{-1}]$
(C) $[ML^{-1}T^{-2}]$ (D) $[ML^{-2}T^{-2}]$

5. Out of the following pair, which one does NOT have identical dimensions : **[2005]**
(A) Impulse and momentum
(B) Angular momentum and planck's constant
(C) Work and torque
(D) Moment of inertia and moment of a force

6. Which of the following units denotes the dimensions $\frac{ML^2}{Q^2}$, where Q denotes the electric charge ? **[2006]**
(A) Wb/m^2 (B) Henry (H)
(C) H/m^2 (D) Weber (Wb)

7. The dimensions of magnetic field in M, L, T and C (coulomb) is given as : **[2008]**
(A) $[MLT^{-1}C^{-1}]$ (B) $[MT^2C^{-2}]$
(C) $[MT^{-1}C^{-1}]$ (D) $[MT^{-2}C^{-1}]$

8. Given that K = energy, V = velocity, T = time. If they are chosen as the fundamental units, then what is dimensional formula for surface tension ? **[2012 (Online)]**
(A) $[KV^{-2}T^{-2}]$ (B) $[K^2V^2T^{-2}]$
(C) $[K^2V^{-2}T^{-2}]$ (D) $[KV^2T^2]$

9. Let $[\epsilon_0]$ denote the dimensional formula of the permittivity of vacuum. If M = mass, L = length, T = time and A = electric current, then : **[2013]**
(A) $\epsilon_0 = [M^{-1}L^{-3}T^2A]$ (B) $\epsilon_0 = [M^{-1}L^{-3}T^4A^2]$
(C) $\epsilon_0 = [M^1L^2T^1A^2]$ (D) $\epsilon_0 = [M^1L^2T^1A]$

10. The dimensions of angular momentum, latent heat and capacitance are, respectively : **[2013 (Online)]**
(A) $ML^2T^1A^2, L^2T^{-2}, M^{-1}L^{-2}T^2$
(B) $ML^2T^{-2}, L^2T^2, M^{-1}L^{-2}T^4A^2$
(C) $ML^2T^{-1}, L^2T^{-2}, ML^2TA^2$
(D) $ML^2T^{-1}, L^2T^{-2}, M^{-1}L^{-2}T^4A^2$

11. If the time period t of the oscillation of a drop of liquid of density d, radius r, vibrating under surface tension s is given by the formula $t = \sqrt{r^{2b}s^cd^{a/2}}$. It is observed that the time period is directly proportional to $\sqrt{\frac{d}{s}}$. The value of b should therefore be : **[2013 (Online)]**
(A) $\frac{3}{4}$ (B) $\sqrt{3}$
(C) $\frac{3}{2}$ (D) $\frac{2}{3}$

12. In terms of resistance R and time T, the dimensions of ratio $\frac{\mu}{\varepsilon}$ of the permeability μ and permittivity ε is : **[2014 (Online)]**
(A) $[RT^{-2}]$ (B) $[R^2T^{-1}]$
(C) $[R^2]$ (D) $[R^2T^2]$

13. From the following combinations of physical constants (expressed through their usual symbols) the only combination, that would have the same value in different systems of units, is : **[2014 (Online)]**
(A) $\frac{ch}{2\pi\epsilon_e^2}$
(B) $\frac{e^2}{2\pi\epsilon_0 Gm_e^2}$ (m_e = mass of electron)
(C) $\frac{\mu_0\epsilon_0}{c^2}\frac{G}{he^2}$
(D) $\frac{2\pi\sqrt{\mu_0\epsilon_0}}{ce^2}\frac{h}{G}$

14. If the capacitance of a nanocapacitor is measured in terms of a unit 'u' made by combining the electric charge 'e', Bohr radius 'a_0', Planck's constant 'h' and speed of light 'c' then : **[2015 (Online)]**

(A) $u = \frac{e^2 h}{a_0}$ (B) $u = \frac{hc}{e^2 a_0}$

(C) $u = \frac{e^2 c}{h a_0}$ (D) $u = \frac{e^2 a_0}{hc}$

15. If electronic charge e, electron mass m, speed of light in vacuum c and Planck's constant h are taken as fundamental quantities, the permeability of vacuum μ_0 can be expressed in units of : **[2015 (Online)]**

(A) $\left(\frac{h}{me^2}\right)$ (B) $\left(\frac{hc}{me^2}\right)$

(C) $\left(\frac{h}{ce^2}\right)$ (D) $\left(\frac{mc^2}{he^2}\right)$

16. In the following 'I' refers to current and other symbols have their usual meaning, Choose the option that corresponds to the dimensions of electrical conductivity : **[2016 (Online)]**

(A) $M^{-1} L^{-3} T^3 I$ (B) $M^{-1} L^{-3} T^3 I^2$

(C) $M^{-1} L^3 T^3 I$ (D) $M L^{-3} T^3 I^2$

17. A, B, C and D are four different physical quantities having different dimensions. None of them is dimensionless. But we know that the equation $AD = C \ln (BD)$ holds true. Then which of the combination is not a meaningful quantity ? **[2016 (Online)]**

(A) $\frac{C}{BD} - \frac{AD^2}{C}$ (B) $A^2 - B^2C^2$

(C) $\frac{A}{B} - C$ (D) $\frac{(A-C)}{D}$

18. Time (T), velocity (C) and angular momentum (h) are chosen as fundamental quantities instead of mass, length and time. In terms of these, the dimensions of mass would be : **[2017 (Online)]**

(A) $[M] = [T^{-1} C^{-2} h]$ (B) $[M] = [T^{-1} C^2 h]$

(C) $[M] = [T^{-1} C^{-2} h^{-1}]$ (D) $[M] = [T C^{-2} h]$

19. The characteristic distance at which quantum gravitational effects are significant, the Planck length, can be determined from a suitable combination of the fundamental physical constants G, h and c. Which of the following correctly gives the Planck length ? **[2018 (Online)]**

(A) G^2hc (B) $\left(\frac{Gh}{c^3}\right)^{\frac{1}{2}}$

(C) $G^{\frac{1}{2}} h^2 c$ (D) Gh^2c^3

20. Expression for time in terms of G (universal gravitational constant), h (Planck constant) and c (speed of light) is proportional to : **[9 Jan. 2019 (Evening)]**

(A) $\sqrt{\frac{Gh}{c^3}}$ (B) $\sqrt{\frac{hc^5}{G}}$

(C) $\sqrt{\frac{c^3}{Gh}}$ (D) $\sqrt{\frac{Gh}{c^5}}$

21. The density of a material in SI units is 128 kg m^{-3}. In certain units in which the unit of length is 25 cm and the unit of mass is 50 g, the numerical value of density of the material is : **[10 Jan. 2019 (Morning)]**

(A) 410 (B) 640

(C) 16 (D) 40

22. The force of interaction between two atoms is given by $F = \alpha\beta \exp\left(-\frac{x^2}{\alpha kt}\right)$; where x is the distance, k is the Boltzmann constant and T is temperature and α and β are two constants. The dimension of β is : **[11 Jan. 2019 (Morning)]**

(A) $M^2L^2T^{-2}$ (B) M^2LT^{-4}

(C) $M^0L^2T^{-4}$ (D) MLT^{-2}

23. If speed (V), acceleration (A) and force (F) are considered as fundamental units, the dimension of Young's modulus will be : **[11 Jan. 2019 (Evening)]**

(A) $V^{-2} A^2F^2$ (B) $V^{-4}A^2F$

(C) $V^{-4}A^{-2}F$ (D) $V^{-2}A^2F^{-2}$

24. Let l, r, c and v represent inductance, resistance, capacitance and voltage, respectively. The dimension of $\frac{l}{rcv}$ in SI units will be : **[12 Jan. 2019 (Evening)]**

(A) [LTA] (B) $[LA^{-2}]$

(C) $[A^{-1}]$ (D) $[LT^2]$

25. If surface tension (S), Moment of inertia (I) and Planck's constant (h), were to be taken as the fundamental units, the dimensional formula for linear momentum would be : **[8 April 2019 (Evening)]**

(A) $S^{3/2}I^{1/2}h^0$ (B) $S^{1/2}I^{1/2}h^0$

(C) $S^{1/2}I^{1/2}h^{-1}$ (D) $S^{1/2}I^{3/2}h^{-1}$

26. In the formula $X = 5YZ^2$, X and Z have dimensions of capacitance and magnetic field, respectively. What are the dimensions of Y in SI units ? **[10 April 2019 (Evening)]**

(A) $[M^{-2} L^{-2} T^6 A^3]$ (B) $[M^{-1} L^{-2} T^4 A^2]$

(C) $[M^{-3} L^{-2} T^8 A^4]$ (D) $[M^{-2} L^0 T^{-4} A^{-2}]$

* * * * *

CHAPTER-2 MOTION IN A STRAIGHT LINE

2.1 Motion in One Dimension

1. If a body loses half of its velocity on penetrating 3 cm in a wooden block, then how much will it penetrate more before coming to rest? **[2002]**

(A) 1 cm (B) 2 cm
(C) 3 cm (D) 4 cm

2. From a building two balls A and B are thrown with equal speeds such that A is thrown upwards and B downwards (both vertically). If v_A and v_B are their respective velocities on reaching the ground, then : **[2002]**

(A) $v_B > v_A$
(B) $v_A = v_B$
(C) $v_A > v_B$
(D) their velocities depends on their masses

3. Speeds of two identical cars are u and $4u$ at a specific instant. At constant retardation the ratio of the respective distances at which the two cars are stopped from that instant is : **[2002]**

(A) 1 : 1 (B) 1 : 4
(C) 1 : 8 (D) 1 : 16

4. A car moving with a speed of 50 km/hr, can be stopped by brakes after at least 6 m. If the same car is moving at a speed of 100 km/hr, the minimum stopping distance is : **[2003]**

(A) 12m (B) 18m
(C) 24m (D) 6m

5. A ball is released from the top of a tower of height h metres. It takes T seconds to reach the ground. What is the position of the ball after $T/3$ seconds? **[2004]**

(A) $h/9$ metre from the ground
(B) $7h/9$ metre from the ground
(C) $8h/9$ metre from the ground
(D) $17h/9$ metre from the ground

6. An automobile travelling with a speed of 60 km/h, can brake to stop within a distance of 20 m. If the car is going twice as fast, i.e. 120 km/h, the stopping distance will be : **[2004]**

(A) 20m (B) 40m
(C) 60m (D) 80m

7. The relation between time t and distance x is $t = ax^2 + bx$, where a and b are constants. The acceleration is : **[2005]**

(A) $-2abv^2$ (B) $2bv^2$
(C) $-2av^3$ (D) $2av^3$

8. A car, starting from rest, accelerates at the rate f through a distance S, then continues at constant speed for time t and then decelerates at the rate $f/2$ to come to rest. If the total distance travelled is 15S, then : **[2005]**

(A) $S = ft$ (B) $S = \frac{1}{6}ft^2$
(C) $S = \frac{1}{72}ft^2$ (D) $S = \frac{1}{4}ft^2$

9. A particle is moving eastwards with a velocity of 5 ms^{-1}. In 10 second the velocity changes to 5 ms^{-1} northwards. The average acceleration in this time is : **[2005]**

(A) $\frac{1}{\sqrt{2}}$ ms^{-1} towards north-west
(B) $\frac{1}{2}$ ms^{-2} towards north
(C) Zero
(D) $\frac{1}{2}$ ms^{-2} towards north-west

10. A parachutist after bailing out falls 50 m without friction. When parachute opens, it decelerates at 2 m/s^2. He reaches the ground with a speed of 3 m/s. At what height, did he bail out? **[2005]**

(A) 91 m (B) 182 m
(C) 293 m (D) 111 m

11. A particle located at $x = 0$ at time $t = 0$, starts moving along the positive x-direction with a velocity v that varies as $v = \alpha\sqrt{x}$. The displacement of the particle varies with time as : **[2006]**

(A) $t^{1/2}$ (B) t^3
(C) t^2 (D) t

12. The velocity of a particle is $v = v_0 + gt + ft^2$. If its position is $x = 0$ at $t = 0$, then its displacement after unit time ($t = 1$) is : **[2007]**

(A) $v_0 + 2g + 3f$ (B) $v_0 + \frac{g}{2} + \frac{f}{3}$
(C) $v_0 + g + f$ (D) $v_0 + \frac{g}{2} + f$

13. An object, moving with a speed of 6.25 m/s, is decelerated at a rate given by $\frac{dv}{dt} = -2.5\sqrt{v}$ where v is the instantaneous speed. The time taken by the object, to come to rest, would be: **[2011]**

(A) 1s (B) 2s
(C) 4s (D) 8s

14. A good train accelerating uniformly on a straight railway track, approaches an electric pole standing on the side of track. Its engine passes the pole with velocity u and the guard's room passes with velocity v. The middle wagon of the train passes the pole with a velocity : **[2012 (Online)]**

(A) $\dfrac{u+v}{2}$ (B) $\dfrac{1}{2}\sqrt{u^2+v^2}$

(C) $\sqrt{uv}$ (D) $\sqrt{\left(\dfrac{u^2+v^2}{2}\right)}$

15. The graph of an object's motion (along the x-axis) is shown in the figure. The instantaneous velocity of the object at point A and B are v_A and v_B respectively. Then : **[2012 (Online)]**

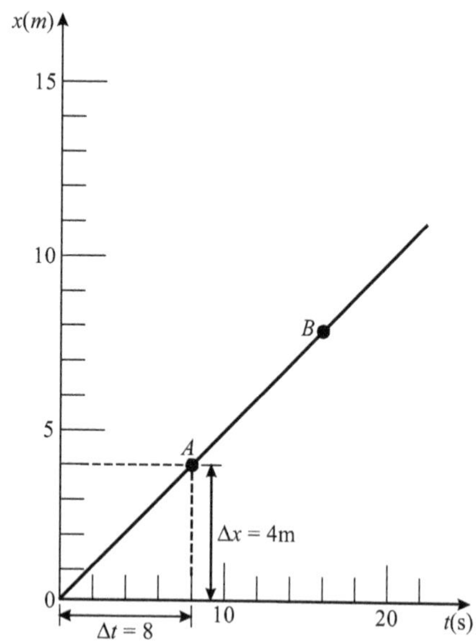

(A) $v_A = v_B = 0.5$ m/s (B) $v_A = 0.5$ m/s $< v_B$

(C) $v_A = 0.5$ m/s $> v_B$ (D) $v_A = v_B = 2$ m/s

16. The distance travelled by a body moving along a line in time t is proportional to t^3. The acceleration-time (a, t) graph for the motion of the body will be : **[2012 (Online)]**

(A) 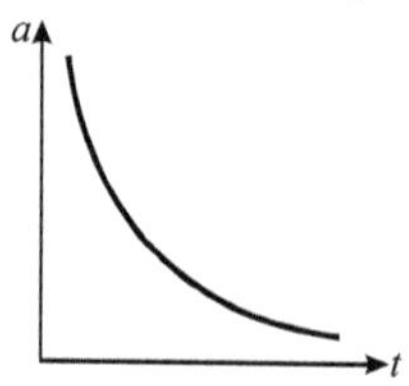

(B) 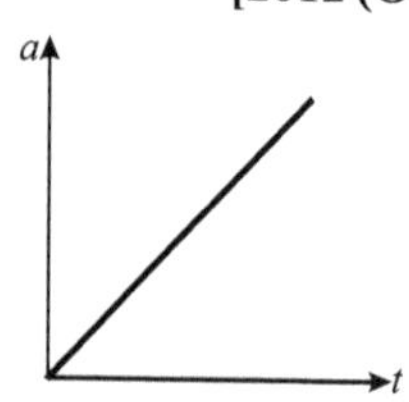

(C) 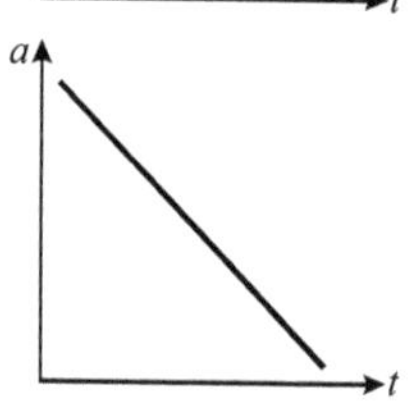

(D) 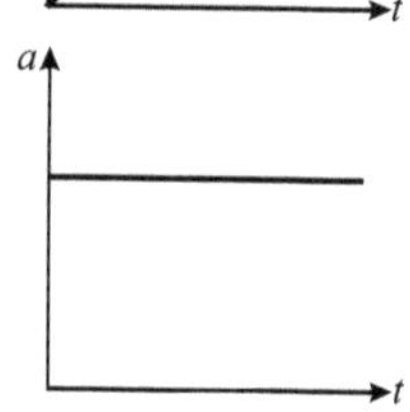

17. A car covers the first half of the distance between two places at 40 km/h and other half at 60 km/h. The average speed of the car is : **[2012 (Online)]**

(A) 40 km/h (B) 45 km/h

(C) 48 km/h (D) 60 km/h

18. From a tower of height H, a particle is thrown vertically upwards with a speed u. The time taken by the particle, to hit the ground, is n times that taken by it to reach the highest point of its path. The relation between H, u and n is : **[2014]**

(A) $2gH = n^2u^2$ (B) $gH = (n-2)^2u^2$

(C) $2gH = nu^2(n-2)$ (D) $gH = (n-2)u^2$

19. A person climbs up a stalled escalator in 60 s. If standing on the same but escalator running with constant velocity he takes 40 s. How much time is taken by the person to walk up the moving escalator ? **[2014 (Online)]**

(A) 37 s (B) 27 s

(C) 24 s (D) 45 s

20. A body is thrown vertically upwards. Which one of the following graphs correctly represent the velocity *vs* time ? **[2017]**

(A) 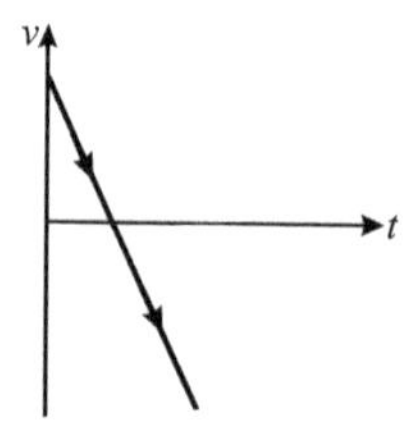

(B) 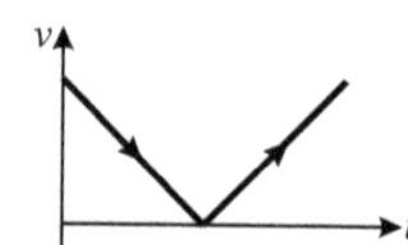

(C) 

(D) 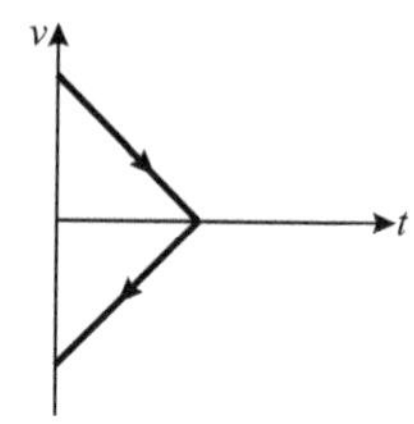

21. Which graph corresponds to an object moving with a constant negative acceleration and a positive velocity ? **[2017 (Online)]**

(A) 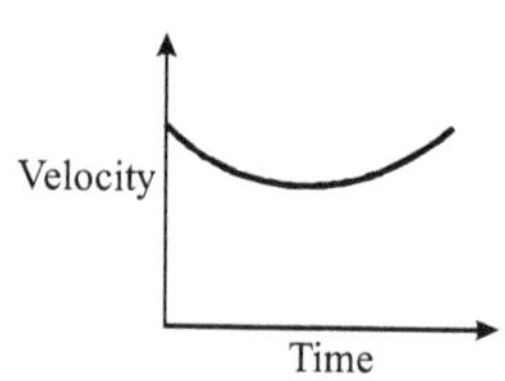

(B) 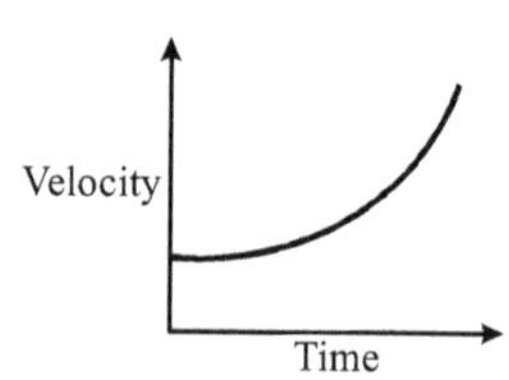

(C) 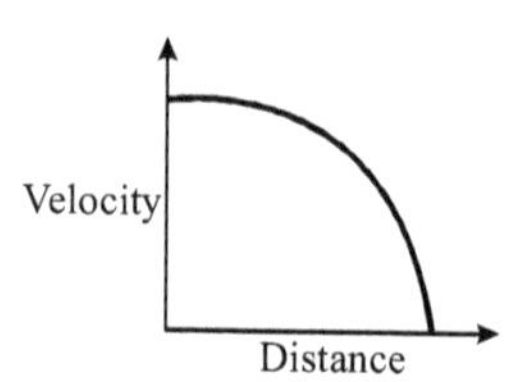

(D) 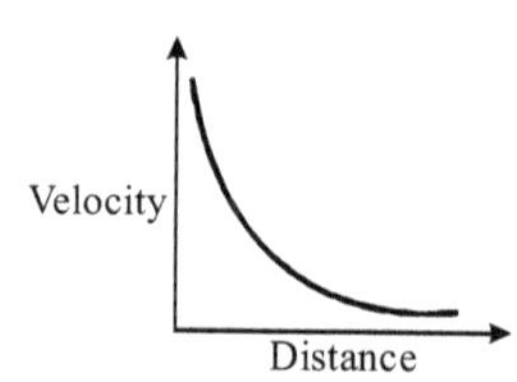

22. All the graphs below are intended to represent the same motion. One of them does it incorrectly. Pick it up : **[2018]**

(A) 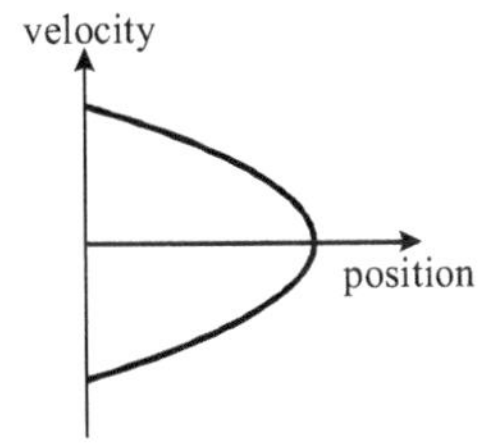

(B) 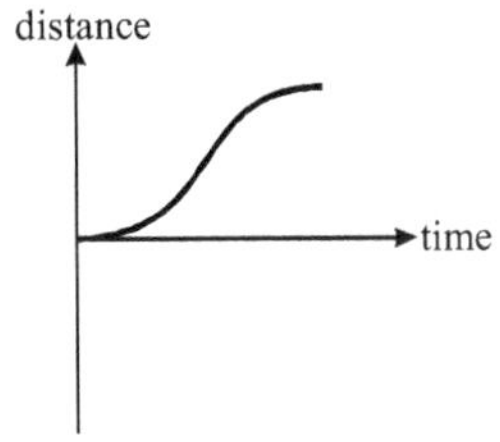

(C) 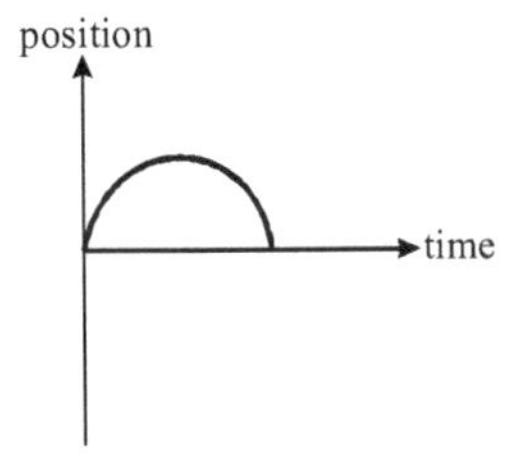

(D) 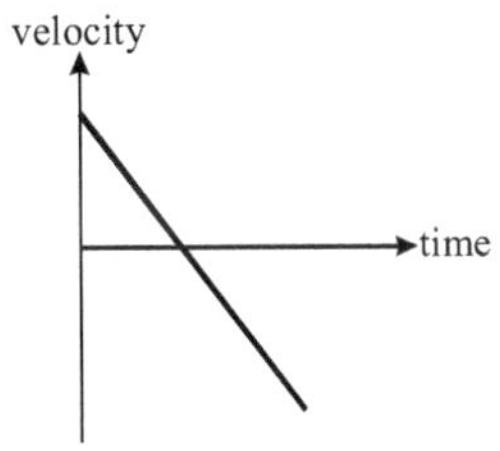

23. An automobile, travelling at 40 km/h, can be stopped at a distance of 40 m by applying brakes. If the same automobile is travelling at 80 km/h, the minimum stopping distance, in metres, is (assume no skidding) : **[2018 (Online)]**

(A) 75m (B) 160m

(C) 100m (D) 150m

24. The velocity-time graph of a car and a scooter are shown in the figure-(i) the difference between the distance travelled by the car and the scooter in 15 s and (ii) the time at which the car will catch up with the scooter are, respectively : **[2018 (Online)]**

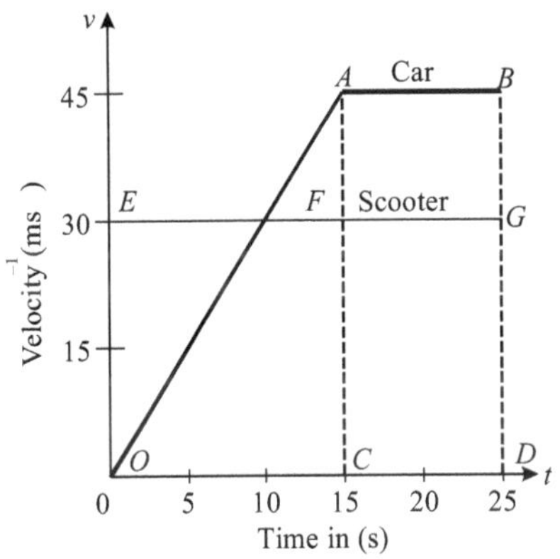

(A) 337.5 m and 25 s (B) 225.5 m and 10 s

(C) 112.5 m and 22.5 s (D) 112.5 m and 15 s

25. A man in a car at location Q on a straight highway is moving with speed v. He decides to reach a point P in a field at a distance d from highway (point M) as shown in the figure. Speed of the car in the field is half to that on the highway. What should be the distance RM, so that the time taken to reach P is minimum? **[2018 (Online)]**

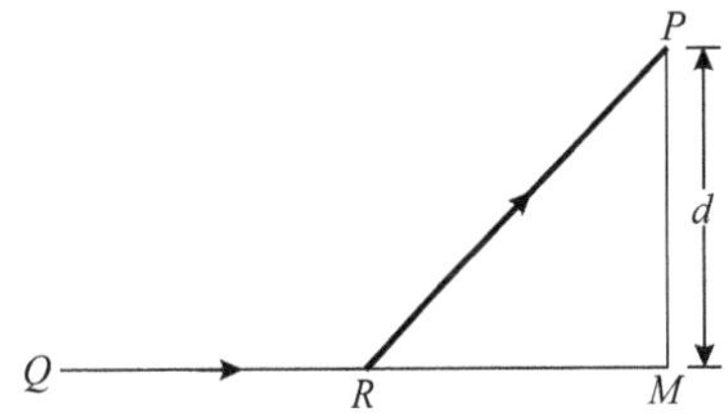

(A) $\frac{d}{\sqrt{3}}$ (B) $\frac{d}{2}$

(C) $\frac{d}{\sqrt{2}}$ (D) d

26. In a car race on straight road, car A takes a time t less than car B at the finish and passes finishing point with a speed 'v' more than that of car B. Both the cars start from rest and travel with constant acceleration a_1 and a_2 respectively. Then 'v' is equal to : **[9 Jan. 2019 (Evening)]**

(A) $\frac{a_1 + a_2}{2}t$ (B) $\sqrt{2a_1a_2}t$

(C) $\frac{2a_1a_2}{a_1 + a_2}t$ (D) $\sqrt{a_1a_2}t$

27. A particle starts from the origin at time $t = 0$ and moves along the positive x-axis. The graph of velocity with respect to time is shown in figure. What is the position of the particle at time $t = 5$s ? **[10 Jan. 2019 (Evening)]**

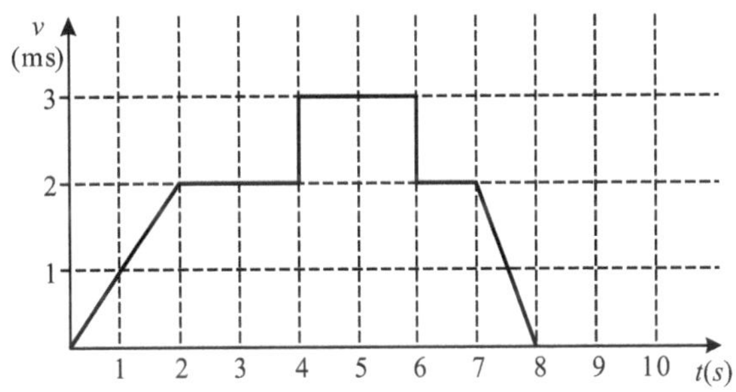

(A) 6m (B) 9m

(C) 3m (D) 10m

28. A particle moves from the point $(2.0\hat{i} + 4.0\hat{j})\,m$, at $t = 0$, with an initial velocity $(5.0\hat{i} + 4.0\hat{j})\,\text{ms}^{-1}$. It is acted upon by a constant force which produces a constant acceleration $(4.0\hat{i} + 4.0\hat{j})\,\text{ms}^{-2}$. What is the distance of the particle from the origin at time 2s ? **[11 Jan. 2019 (Evening)]**

(A) $20\sqrt{2}$ m (B) $10\sqrt{2}$ m

(C) 5m (D) 15m

29. A passenger train of length 60 m travels at a speed of 80 km/hr. Another freight train of length 120 m travels at a speed of 30 km/hr. The ratio of times taken by the passenger train to completely cross the freight train when : (i) they are moving in the same direction, and (ii) in the opposite directions is :

[12 Jan. 2019 (Morning)]

(A) $\frac{5}{2}$ (B) $\frac{25}{11}$

(C) $\frac{3}{2}$ (D) $\frac{11}{5}$

30. A particle starts from origin O from rest and moves with a uniform acceleration along the positive x-axis. Identify all figures that correctly represent the motion qualitatively. (a = acceleration, v = velocity, x = displacement, t = time) :

[8 April 2019 (Evening)]

(1)
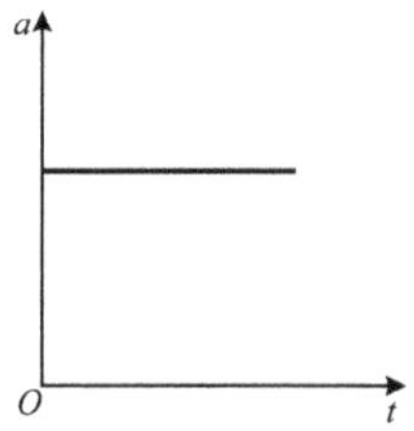

(2)
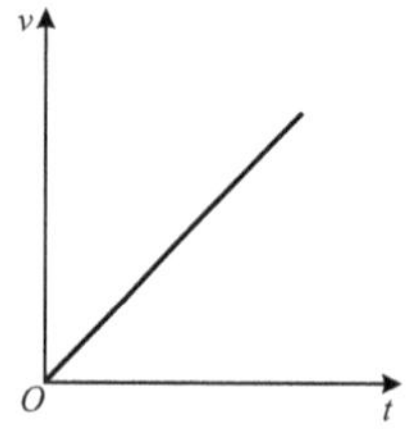

(3)
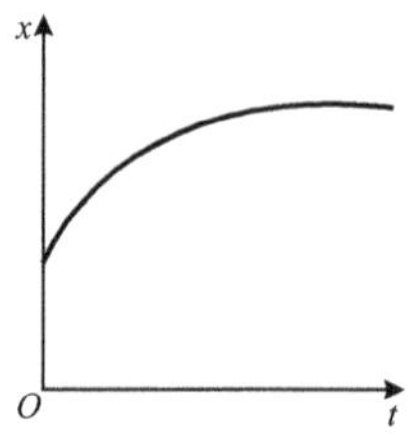

(4)
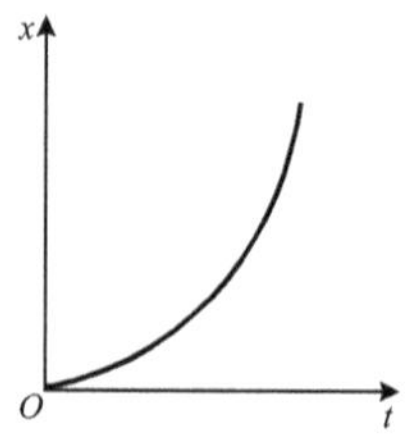

(A) (1), (2), (3) (B) (1)

(C) (1), (2), (4) (D) (2), (3)

31. Let $|\vec{A}_1| = 3$, $|\vec{A}_2| = 5$ and $|\vec{A}_1 + \vec{A}_2| = 5$. The value of $(2\vec{A}_1 + 3\vec{A}_2)\cdot(3\vec{A}_1 - 2\vec{A}_2)$ is : **[8 April 2019 (Evening)]**

(A) -112.5 (B) -106.5

(C) -118.5 (D) -99.5

32. A ball is thrown vertically up (taken as $+z$-axis) from the ground. The correct momentum-height (p-h) diagram is :

[9 April 2019 (Morning)]

(A)
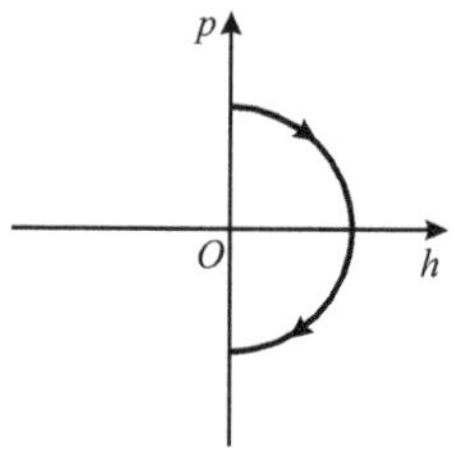

(B)
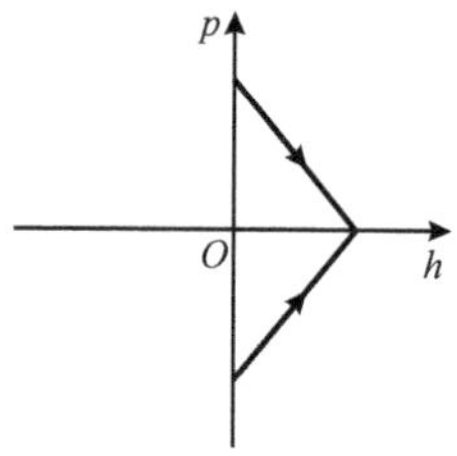

(C)
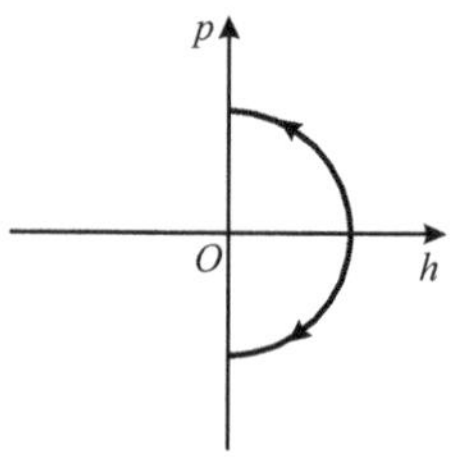

(D)
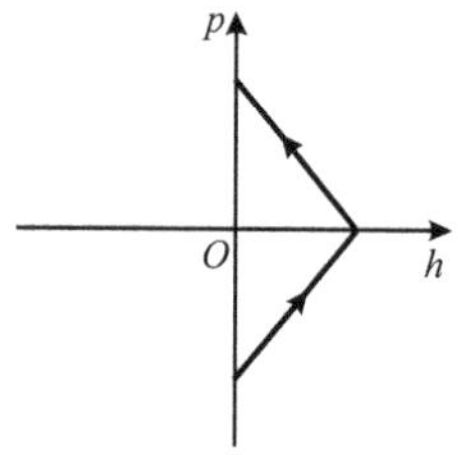

33. The position of a particle as a function of time t, is given by $x(t) = at + bt^2 - ct^3$ where a, b and c are constants. When the particle attains zero acceleration, then its velocity will be :

[9 April 2019 (Evening)]

(A) $a+\frac{b^2}{4c}$ (B) $a+\frac{b^2}{c}$

(C) $a+\frac{b^2}{2c}$ (D) $a+\frac{b^2}{3c}$

34. A particle is moving with speed $v = b\sqrt{x}$ along positive x-axis. Calculate the speed of the particle at time $t = \tau$ (assume that the particle is at origin at $t = 0$) : **[12 April 2019 (Evening)]**

(A) $\frac{b^2\tau}{4}$ (B) $\frac{b^2\tau}{2}$

(C) $b^2\tau$ (D) $\frac{b^2\tau}{\sqrt{2}}$

* * * * *

CHAPTER-3 MOTION IN A PLANE

3.1 Motion in Two Dimensions

1. A ball whose kinetic energy is E, is projected at an angle of 45° to the horizontal. The kinetic energy of the ball at the highest point of its flight will be : **[2002]**

(A) E (B) $E/\sqrt{2}$
(C) $E/2$ (D) Zero

2. A boy playing on the roof of a 10 m high building throws a ball with a speed of 10 m/s at an angle of 30° with the horizontal. How far from the throwing point will the ball be at the height of 10 m from the ground? **[2002]**

[$g = 10$ m/s², $\sin 30° = 1/2$, $\cos 30° = \frac{\sqrt{3}}{2}$]

(A) 5.20m (B) 4.33m
(C) 2.60m (D) 8.66m

3. The coordinates of a moving particle at any time t are given by $x = \alpha t^3$ and $y = \beta t^3$. The speed of the particle at time t is given by : **[2003]**

(A) $\sqrt{\alpha^2+\beta^2}$ (B) $3t^2\sqrt{\alpha^2+\beta^2}$
(C) $t^2\sqrt{\alpha^2+\beta^2}$ (D) $\sqrt{\alpha^2+\beta^2}$

4. A ball is thrown from a point with a speed v_0 at angle of projection θ. From the same point and at the same instant, a person starts running with a constant speed $v_0/2$ to catch the ball? If yes, what should be the angle of projection? **[2004]**

(A) Yes, 60° (B) Yes, 30°
(C) No (D) Yes, 45°

5. If $\vec{A}\times\vec{B} = \vec{B}\times\vec{A}$, then the angle between A and B is : **[2004]**

(A) $\frac{\pi}{2}$ (B) $\frac{\pi}{3}$
(C) π (D) $\frac{\pi}{4}$

6. A projectile can have the same range 'R' for two angles of projection. If 'T_1' and 'T_2' to be time of flights in the two cases, then the product of the two time of flights is directly proportional to : **[2004]**

(A) R (B) $\frac{1}{R}$
(C) $\frac{1}{R^2}$ (D) R^2

7. A particle is projected at 60° to the horizontal with a kinetic energy K. The kinetic energy at the highest point is : **[2007]**

(A) K (B) Zero
(C) $K/4$ (D) $K/2$

8. A particle has an initial velocity of $3\hat{i}+4\hat{j}$ and an acceleration of $0.4\hat{i}+0.3\hat{j}$. Its speed after 10 s is : **[2009]**

(A) $7\sqrt{2}$ units (B) 7 units
(C) 8.5 units (D) 10 units

9. A particle is moving with velocity $\vec{v} = K(y\hat{i}+x\hat{j})$, where K is a constant. The general equation for its path is : **[2010]**

(A) $y = x^2$ + constant (B) $y^2 = x$ + constant
(C) xy = constant (D) $y^2 = x^2$ + constant

10. A water fountain on the ground sprinkles water all around it. If the speed of water coming out of the fountain is v, the total area around the fountain that gets wet is : **[2011]**

(A) $\pi\frac{v^4}{g^2}$ (B) $\frac{\pi}{2}\frac{v^4}{g^2}$
(C) $\pi\frac{v^2}{g^2}$ (D) $\pi\frac{v^4}{g}$

11. A boy can throw a stone up to a maximum height of 10 m. The maximum horizontal distance that the boy can throw the same stone up to, will be : **[2012]**

(A) $20\sqrt{2}$m (B) 10m
(C) $10\sqrt{2}$m (D) 20m

12. A projectile is given an initial velocity of $(\hat{i}+2\hat{j})$ m/s, where $\hat{i}$ is along the ground and $\hat{j}$ is along the vertical. If $g = 10$ m/s², the equation of its trajectory is : **[2013]**

(A) $y = 2x - 5x^2$ (B) $4y = 2x - 5x^2$
(C) $4y = 2x - 25x^2$ (D) $y = x - 5x^2$

13. The maximum range of a bullet fired from a toy pistol mounted on a car at rest is $R_0 = 40$ m. What will be the acute angle of inclination of the pistol for maximum range when the car is moving in the direction of firing with uniform velocity $v = 20$ m/s, on a horizontal surface ? ($g = 10$ m/s²) : **[2013 (Online)]**

(A) 30° (B) 60°
(C) 75° (D) 45°

14. A ball projected from ground at an angle of 45° just clears a wall in front. If point of projection is 4m from the foot of wall and ball strikes the ground at a distance of 6 m on the other side of the wall, the height of the wall is : **[2013 (Online)]**

(A) 4.4m (B) 2.4m
(C) 3.6m (D) 1.6m

15. The initial speed of a bullet fired from a rifle is 630 m/s. The rifle is fired at the centre of a target 700 m away at the same level as the target. How far above the centre of the target ? **[2014 (Online)]**

(A) 1.0m (B) 4.2m
(C) 6.1m (D) 9.8m

16. The position of a projectile launched from the origin at $t = 0$ is given by $\vec{r} = (40\hat{i} + 50\hat{j})$ m at $t = 2$ s. If the projectile was launched at an angle θ from the horizontal, then θ is : (take $g = 10\ \text{ms}^{-2}$) : **[2014 (Online)]**

(A) $\tan^{-1}\frac{2}{3}$ (B) $\tan^{-1}\frac{3}{2}$
(C) $\tan^{-1}\frac{7}{4}$ (D) $\tan^{-1}\frac{4}{5}$

17. A vector $\vec{A}$ is rotated by a small angle $\Delta\theta$ radian ($\Delta\theta << 1$) to get a new vector $\vec{B}$. In that case $|\vec{B} - \vec{A}|$ is : **[2015 (Online)]**

(A) $|\vec{A}|\Delta\theta$ (B) $|\vec{B}|\Delta\theta - |\vec{A}|$
(C) $|\vec{A}|\left(1 - \frac{\Delta\theta^2}{2}\right)$ (D) 0

18. Let $\vec{A} = (\hat{i} + \hat{j})$ and $\vec{B} = (2\hat{i} - \hat{j})$. The magnitude of a coplanar vector $\vec{C}$ such that $\vec{A}\cdot\vec{C} = \vec{B}\cdot\vec{C} = \vec{A}\cdot\vec{B}$ is given by: **[2018 (Online)]**

(A) $\sqrt{\frac{5}{9}}$ (B) $\sqrt{\frac{10}{9}}$
(C) $\sqrt{\frac{20}{9}}$ (D) $\sqrt{\frac{9}{12}}$

19. A particle is moving with a velocity $\vec{v} = K(y\hat{i} + x\hat{j})$, where K is a constant. The general equation for its path is : **[9 Jan. 2019 (Morning)]**

(A) xy = constant (B) $y^2 = x^2$ + constant
(C) $y = x^2$ + constant (D) $y^2 = x$ + constant

20. The position co-ordinates of a particle moving in a 3-D coordinate system is given by

$x = a\cos\omega t$

$y = a\sin\omega t$

and $z = a\omega t$

The speed of the particle is : **[9 Jan. 2019 (Evening)]**

(A) $a\omega$ (B) $\sqrt{3}\,a\omega$
(C) $\sqrt{2}\,a\omega$ (D) $2a\omega$

21. Two guns A and B can fire bullets at speeds 1 km/s and 2 km/s respectively. From a point on a horizontal ground, they are fired in all possible directions. The ratio of maximum areas covered by the bullets fired by the two guns, on the ground is : **[10 Jan. 2019 (Morning)]**

(A) 1 : 2 (B) 1 : 4
(C) 1 : 8 (D) 1 : 16

22. In the cube of side 'a' shown in the figure, the vector from the central point of the face $ABOD$ to the central point of the face $BEFO$ will be : **[10 Jan. 2019 (Morning)]**

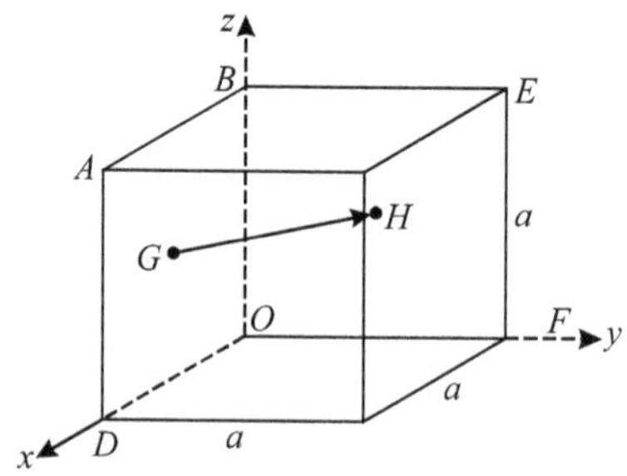

(A) $\frac{1}{2}a(\hat{i} - \hat{k})$ (B) $\frac{1}{2}a(\hat{j} - \hat{i})$
(C) $\frac{1}{2}a(\hat{k} - \hat{i})$ (D) $\frac{1}{2}a(\hat{j} - \hat{k})$

23. Two forces P and Q of magnitude $2F$ and $3F$, respectively, are at an angle θ with each other. If the force Q is doubled, then their resultant also gets doubled. Then, the angle is : **[10 Jan. 2019 (Evening)]**

(A) 30° (B) 60°
(C) 90° (D) 120°

24. Two vectors $\vec{A}$ and $\vec{B}$ have equal magnitudes. The magnitude of $(\vec{A} + \vec{B})$ is 'n' times the magnitude of $(\vec{A} - \vec{B})$. The angle between $\vec{A}$ and $\vec{B}$ is : **[10 Jan. 2019 (Evening)]**

(A) $\sin^{-1}\left[\frac{n^2 - 1}{n^2 + 1}\right]$ (B) $\cos^{-1}\left[\frac{n - 1}{n + 1}\right]$
(C) $\cos^{-1}\left[\frac{n^2 - 1}{n^2 + 1}\right]$ (D) $\sin^{-1}\left[\frac{n - 1}{n + 1}\right]$

25. A body is projected at $t = 0$ with a velocity 10 ms^{-1} at an angle of 60° with the horizontal. The radius of curvature of its trajectory at $t = 1$s is R. Neglecting air resistance and taking acceleration due to gravity $g = 10$ ms^{-2}, the value of R is : **[11 Jan. 2019 (Morning)]**

(A) 2.5m (B) 10.3m
(C) 2.8m (D) 5.1m

26. The position vector of a particle changes with time according to the relation $\vec{r}(t) = 15t^2\hat{i} + (4 - 20t^2)\hat{j}$. What is the magnitude of the acceleration at $t = 1$?

[9 April 2019 (Evening)]

(A) 40 (B) 100
(C) 25 (D) 50

27. A plane is inclined at an angle $\alpha = 30°$ with a respect to the horizontal. A particle is projected with a speed $u = 2\ \text{ms}^{-1}$ from the base of the plane, making an angle $\theta = 15°$ with respect to the plane as shown in the figure. The distance from the base, at which the particle hits the plane is close to :
(Take $g = 10\ \text{ms}^{-2}$) **[10 April 2019 (Evening)]**

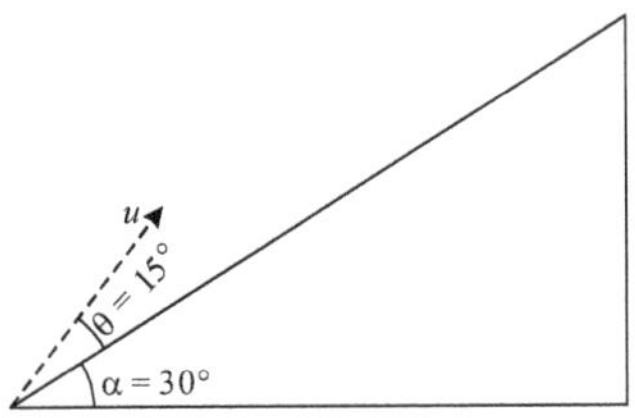

(A) 14 cm (B) 20 cm
(C) 18 cm (D) 26 cm

28. The trajectory of a projectile near the surface of the earth is given as $y = 2x - 9x^2$. If it were launched at an angle θ_0 with speed v_0 then ($g = 10\ \text{ms}^{-2}$) : **[12 April 2019 (Morning)]**

(A) $\theta_0 = \cos^{-1}\left(\frac{1}{\sqrt{5}}\right)$ and $v_0 = \frac{5}{3}\text{ms}^{-1}$

(B) $\theta_0 = \sin^{-1}\left(\frac{1}{\sqrt{5}}\right)$ and $v_0 = \frac{5}{3}\text{ms}^{-1}$

(C) $\theta_0 = \sin^{-1}\left(\frac{2}{\sqrt{5}}\right)$ and $v_0 = \frac{3}{5}\text{ms}^{-1}$

(D) $\theta_0 = \cos^{-1}\left(\frac{2}{\sqrt{5}}\right)$ and $v_0 = \frac{3}{5}\text{ms}^{-1}$

29. A shell is fired from a fixed artillery gun with an initial speed u such that it hits the target on the ground at a distance R from it. If t_1 and t_2 are the values of the time taken by it to hit the target in two possible ways, the product $t_1 t_2$ is :

[12 April 2019 (Morning)]

(A) R/g (B) $R/4g$
(C) $2R/g$ (D) $R/2g$

30. Two particles are projected from the same point with the same speed u such that they have the same range R, but different maximum heights, h_1 and h_2. Which of the following is correct?

[12 April 2019 (Evening)]

(A) $R^2 = 2h_1h_2$ (B) $R^2 = 16h_1h_2$
(C) $R^2 = 4h_1h_2$ (D) $R^2 = h_1h_2$

* * * * *

CHAPTER-3 MOTION IN A PLANE

3.2 Relative Motion

1. A body is at rest at $x = 0$. At $t = 0$, it starts moving in the positive x-direction with a constant acceleration. At the same instant another body passes through $x = 0$ moving in the positive x-direction with a constant speed. The position of the first body is given by $x_1(t)$ after time 't' and that of second body by $x_2(t)$ after the same time interval. Which of the following graphs correctly describes $(x_1 - x_2)$ as a function of time 't' ? **[2008]**

(A) 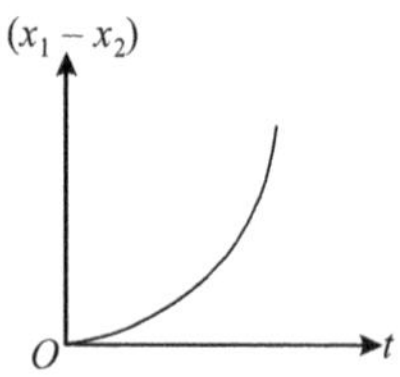

(B) 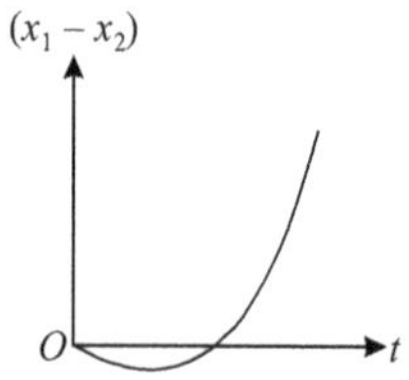

(C) 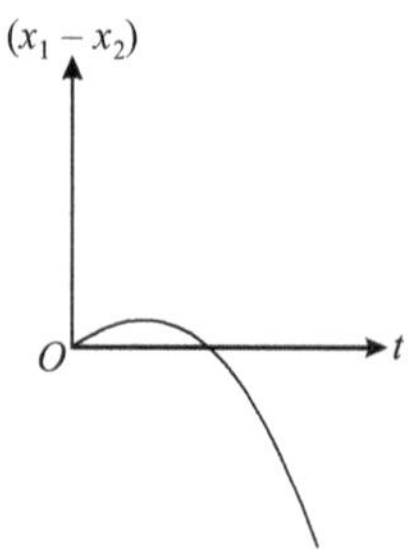

(D) 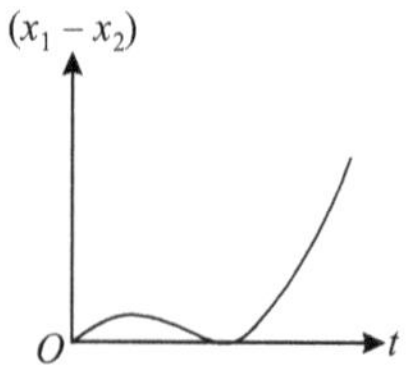

2. Two fixed frictionless inclined planes making an angle 30° and 60° with the vertical are shown in the figure. Two blocks A and B are placed on the two planes. What is the relative vertical acceleration of A with respect to B? **[2010]**

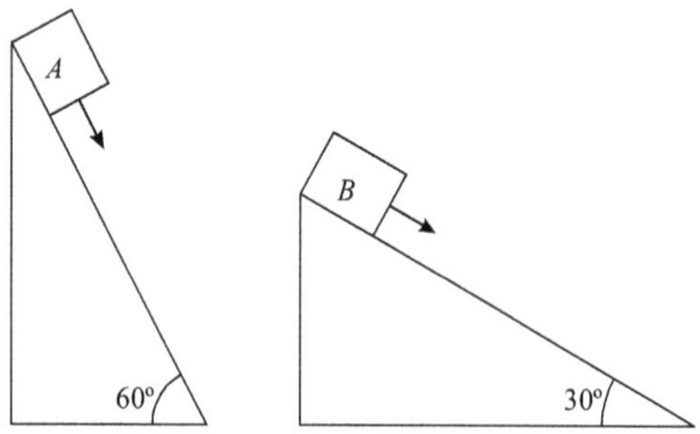

(A) 4.9 ms^{-2} in horizontal direction
(B) 9.8 ms^{-2} in vertical direction
(C) Zero
(D) 4.9 ms^{-2} in vertical direction

3. Two stones are thrown up simultaneously from the edge of a cliff 240 m high with initial speed of 10 m/s and 40 m/s respectively. Which of the following graph best represents the time variation of relative position of the second stone with respect of the first ?

(Assume stones do not rebound after hitting the ground and neglect air resistance, take $g = 10$ m/s^2)

(The figure are schematic and not drawn to scale) **[2015]**

(A) 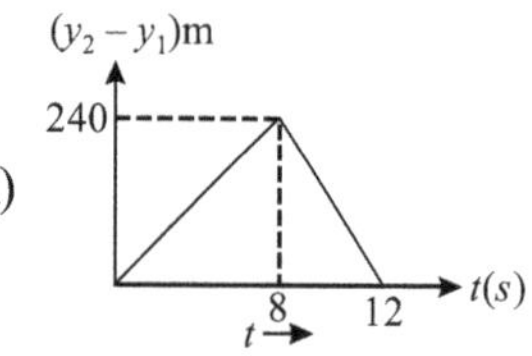

(B) 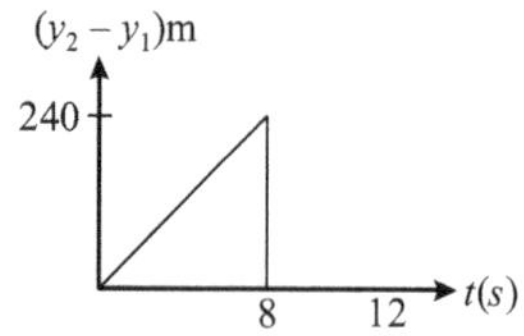

(C) 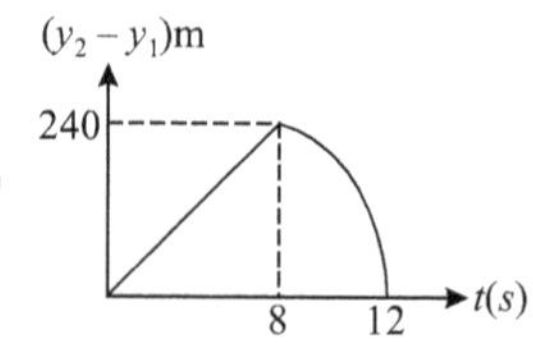

(D) 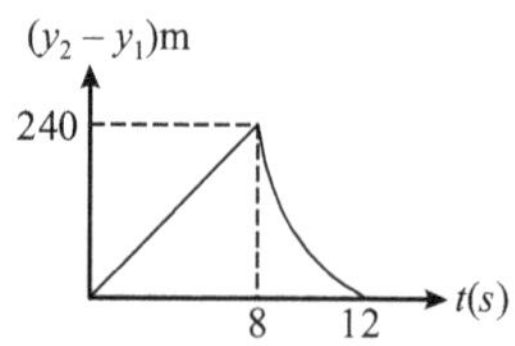

4. A car is standing 200 m behind a bus, which is also at rest. The two start moving at the same instant but with different forward accelerations. The bus has acceleration 2 m/s^2 and the car has acceleration 4 m/s^2. The car will catch up with the bus after a time of: **[2017 (Online)]**

(A) $\sqrt{100}$ s (B) $\sqrt{120}$ s
(C) $10\sqrt{2}$ s (D) 15 s

5. Ship A is sailing towards north-east with velocity $\vec{v} = 30\hat{i} + 50\hat{j}$ km/hr where $\hat{i}$ points east and $\hat{j}$, north. Ship B is at a distance of 80 km east and 150 km north of Ship A and is sailing towards west at 10 km/hr. A will be at minimum distance from B in : **[8 April 2019 (Morning)]**

(A) 4.2 hrs. (B) 2.2 hrs.
(C) 3.2 hrs. (D) 2.6 hrs.

6. The stream of a river is flowing with a speed of 2 km/h. A swimmer can swim at a speed of 4 km/h. What should be the direction of the swimmer with respect to the flow of the river to cross the river straight ? **[9 April 2019 (Morning)]**

(A) 60° (B) 150°
(C) 90° (D) 120°

* * * * *

CHAPTER-4 LAWS OF MOTION

4.1 Newton's Laws of Motion

1. When forces F_1, F_2, F_3 are acting on a particle of mass m such that F_2 and F_3 are mutually perpendicular, then the particle remains stationary. If the force F_1 is now removed then the acceleration of the particle is : **[2002]**

(A) F_1/m (B) F_2F_3/mF_1
(C) $(F_2 - F_3)/m$ (D) F_2/m

2. Three identical blocks of masses $m = 2$ kg are drawn by a force F with an acceleration of $0.6\ ms^{-2}$ on a frictionless surface, then what is the tension (in N) in the string between the blocks B and C : **[2002]**

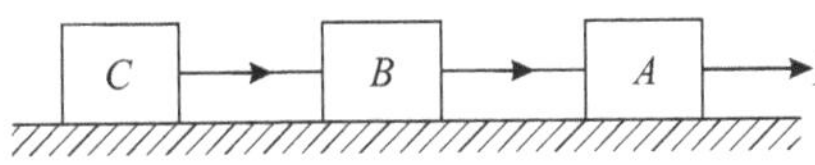

(A) 9.2 (B) 1.2
(C) 4 (D) 9.8

3. One end of massless rope, which passes over a massless and frictionless pulley P is tied to a hook C while the other end is free. Maximum tension that the rope can bear is 840 N. With what value of maximum safe acceleration (in ms^{-2}) can a man of 60 kg climb on the rope? **[2002]**

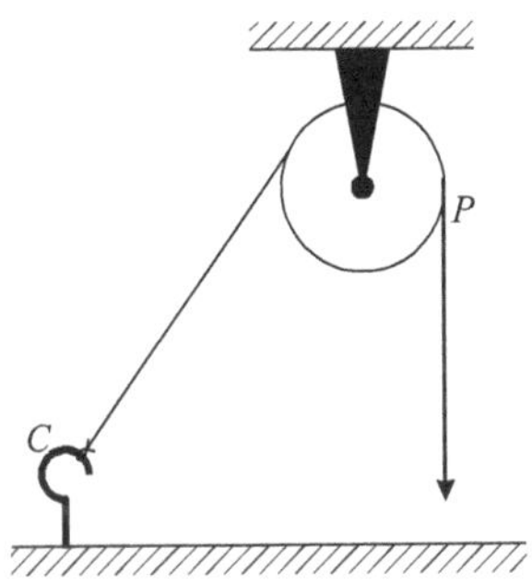

(A) 16 (B) 6
(C) 4 (D) 8

4. A solid sphere, a hollow sphere and a ring are released from top of an inclined plane (frictionless) so that they slide down the plane. Then maximum acceleration down the plane is for (no rolling) : **[2002]**

(A) Solid sphere (B) Hollows sphere
(C) Ring (D) All same

5. A lift is moving down with acceleration a. A man in the lift drops a ball inside the lift. The acceleration of the ball as observed by the man in the lift and a man standing stationary on the ground are respectively : **[2002]**

(A) g, g (B) $g - a, g - a$
(C) $g - a, g$ (D) a, g

6. Two forces are such that the sum of their magnitudes is 18 N and their resultant is 12 N which is perpendicular to the smaller force. Then the magnitudes of the forces are : **[2002]**

(A) 12 N, 6 N (B) 13 N, 5 N
(C) 10 N, 8 N (D) 16 N, 2 N

7. A light string passing over a smooth light pulley connects two blocks of masses m_1 and m_2 (vertically). If the acceleration of the system is $g/8$, then the ratio of the masses is : **[2002]**

(A) 8 : 1 (B) 9 : 7
(C) 4 : 3 (D) 5 : 3

8. A light spring balance hangs from the hook of the other light spring balance and a block of mass M kg hangs from the former one. Then the true statement about the scale reading is: **[2003]**

(A) Both the scales read M kg each
(B) The scale of the lower one reads M kg and of the upper one zero
(C) The reading of the two scales can be anything but the sum of the readings will be M kg
(D) Both the scales read $M/2$ kg

9. A spring balance is attached to the ceiling of a lift. A man hangs his bag on the spring and the spring reads 49 N, when the lift is stationary. If the lift moves downward with an acceleration of $5\ m/s^2$, the reading of the spring balance will be : **[2003]**

(A) 24 N (B) 74 N
(C) 15 N (D) 49 N

10. A rocket which has a mass of 3.5×10^4 kg is blasted upwards with an initial acceleration of $10\ m/s^2$. Then the initial thrust of the blast is : **[2003]**

(A) 3.5×10^5 N (B) 7.0×10^5 N
(C) 14.0×10^5 N (D) 1.75×10^5 N

11. Three forces start acting simultaneously on a particle moving with velocity $\vec{v}$. These forces are represented in magnitude and direction by the three sides of a triangle ABC (as shown). The particle will now move with velocity : **[2003]**

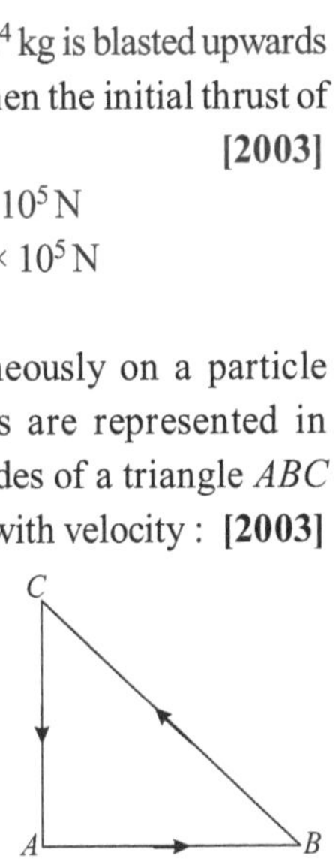

(A) Less than $\vec{v}$
(B) Greater than $\vec{v}$
(C) $|v|$ in the direction of largest force BC
(D) $\vec{v}$, remaining unchanged

12. A block of mass M is pulled along a horizontal frictionless surface by a rope of mass m. If a force P is applied at the free end of the rope, the force exerted by the rope on the block is : **[2003]**

(A) $\frac{Pm}{M+m}$ (B) $\frac{Pm}{M-m}$

(C) P (D) $\frac{PM}{M+m}$

13. Two masses $m_1 = 5$ kg and $m_2 = 4.8$ kg tied to a string are hanging over a light frictionless pulley. What is the acceleration of the masses when they are free to move ? ($g = 9.8\ m/s^2$) : **[2004]**

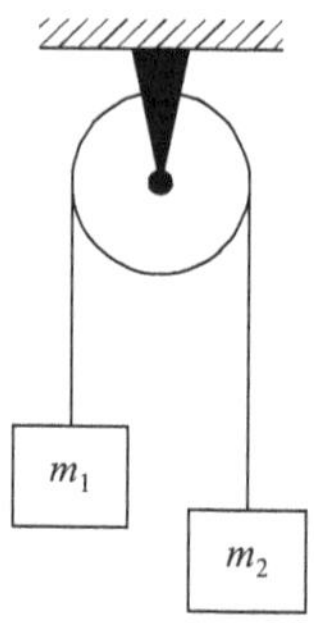

(A) $0.2\ m/s^2$ (B) $9.8\ m/s^2$
(C) $5\ m/s^2$ (D) $4.8\ m/s^2$

14. A machine gun fires a bullet of mass 40 g with a velocity $1200\ ms^{-1}$. The man holding it, can exert maximum force of 144 N on the gun. How many bullets can he fire per second at the most? **[2004]**

(A) One (B) Four
(C) Two (D) Three

15. A block is kept on a frictionless inclined surface with angle of inclination α. The incline is given an acceleration a to keep the block stationary. Then a is equal to : **[2005]**

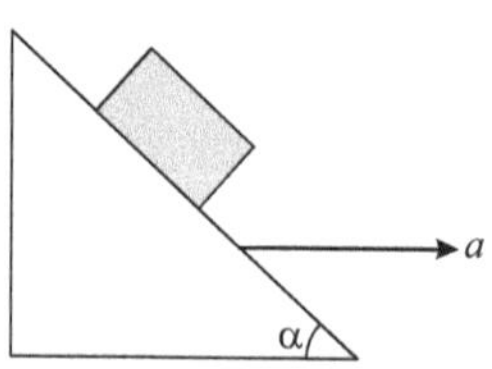

(A) $g/\tan\alpha$ (B) $g \operatorname{cosec} \alpha$
(C) g (D) $g \tan \alpha$

16. A player caught a cricket ball of mass 150 g moving at a rate of 20 m/s. If the catching process is completed in 0.1 s, the force of the blow exerted by the ball on the hand of the player is equal to : **[2006]**

(A) 150 N (B) 3 N
(C) 30 N (D) 300 N

17. A ball of mass 0.2 kg is thrown vertically upwards by applying a force by hand. If the hand moves 0.2 m while applying the force and the ball goes upto 2 m height further, find the magnitude of the force. (Consider $g = 10\ m/s^2$) : **[2006]**

(A) 4 N (B) 16 N
(C) 20 N (D) 22 N

18. A block of mass m is connected to another block of mass M by a spring (massless) of spring constant k. The blocks are kept on a smooth horizontal plane. Initially the blocks are at rest and the spring is unstretched. Then a constant force F starts acting on the block of mass M to pull it. Find the force on the block of mass m : **[2007]**

(A) $\frac{mF}{M}$ (B) $\frac{(M+m)F}{m}$

(C) $\frac{mF}{(m+M)}$ (D) $\frac{MF}{(m+M)}$

19. If a spring of stiffness 'k' is cut into two parts 'A' and 'B' of length $l_A : l_B = 2 : 3$, then the stiffness of spring 'A' is given by : **[2011]**

(A) $\frac{5}{2}k$ (B) $\frac{3k}{5}$

(C) $\frac{2k}{5}$ (D) k

20. A particle of mass m is at rest at the origin at time $t = 0$. It is subjected to a force $F(t) = F_0 e^{-bt}$ in the x-direction. Its speed $v(t)$ is depicted by which of the following curves ? **[2012]**

(A)
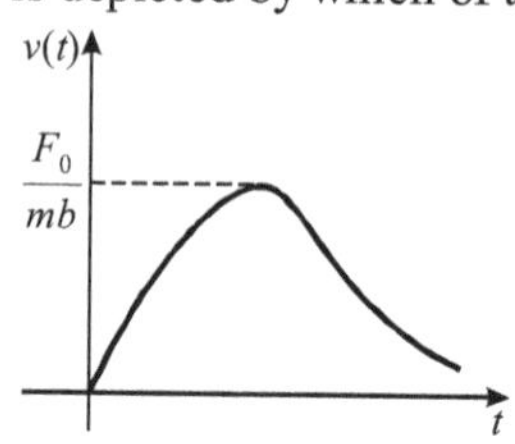

(B)
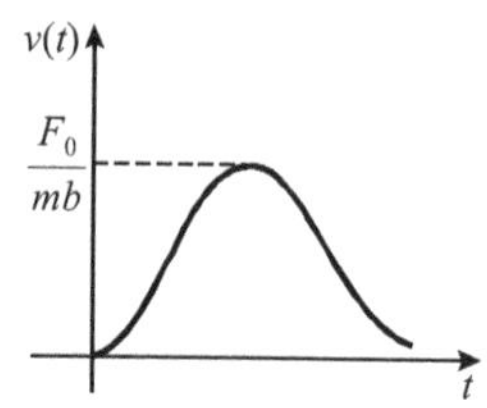

(C)
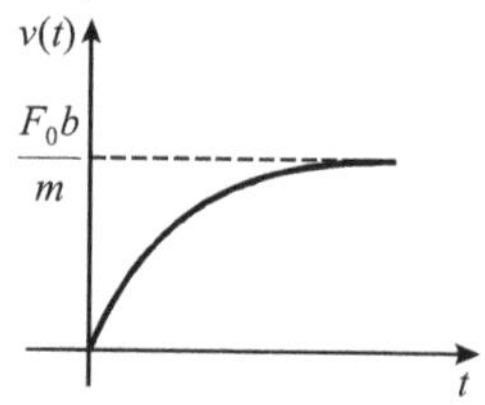

(D)
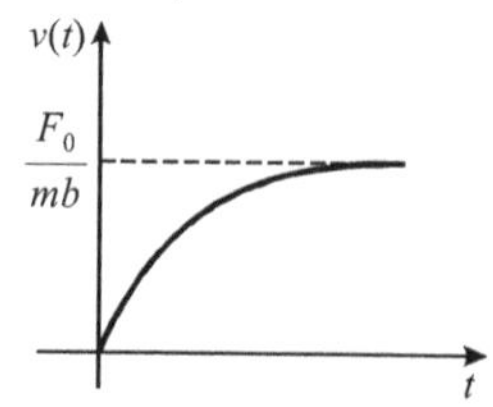

21. This question has Statement-1 and Statement-2. Of the four choices given after the Statements, choose the one that best describes the two Statements. **[2012 (Online)]**

Statement-1 : If you push on a cart being pulled by a horse so that it does not move, the cart pushes you back with an equal and opposite force :

Statement-2 : The cart does not move because the force described in statement-1 cancel each other.

(A) Statement-1 is true, Statement-2 is true, Statement-2 is the correct explanation of Statement-1

(B) Statement-1 is false, Statement-2 is true

(C) Statement-1 is true, Statement-2 is false

(D) Statement-1 is true, Statement-2 is true, Statement-2 is not the correct explanation of Statement-1.

22. Two blocks of mass $M_1 = 20$ kg and $M_2 = 12$ kg are connected by a metal rod of mass 8 kg. The system is pulled vertically up by applying a force of 480 N as shown. The tension at the mid-point of the rod is : **[2013 (Online)]**

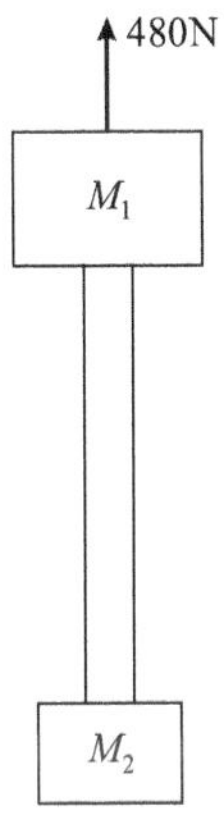

(A) 144 N (B) 96 N
(C) 240 N (D) 192 N

23. A uniform sphere of weight W and radius 5 cm is being held by a string as shown in the figure. The tension in the string will be : **[2013 (Online)]**

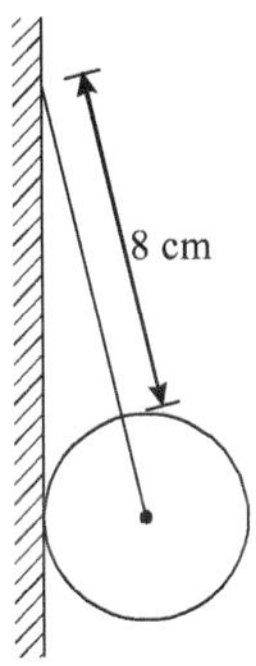

(A) $12\frac{W}{5}$ (B) $5\frac{W}{12}$

(C) $13\frac{W}{5}$ (D) $13\frac{W}{12}$

24. A body of mass 5 kg under the action of constant force $\vec{F} = F_x\hat{i} + F_y\hat{j}$ has velocity at $t = 0$ s as $\vec{v} = (6\hat{i} - 2\hat{j})$ m/s and at $t = 10$ s as $\vec{v} = +6\hat{j}$ m/s. The force $\vec{F}$ is: **[2014 (Online)]**

(A) $(-3\hat{j} + 4\hat{j})$ N (B) $\left(-\frac{3}{5}\hat{i} + \frac{4}{5}\hat{j}\right)$N

(C) $(3\hat{i} - 4\hat{j})$ N (D) $\left(\frac{3}{5}\hat{i} - \frac{4}{5}\hat{j}\right)$N

25. A heavy box is to dragged along a rough horizontal floor. To do so, person A pushes it at an angle 30° from the horizontal and requires a minimum force F_A, while person B pulls the box at an angle 60° from the horizontal and needs minimum force F_B. If the coefficient of friction between the box and the floor is $\frac{\sqrt{3}}{5}$, the ratio $\frac{F_A}{F_B}$ is : **[2014 (Online)]**

(A) $\sqrt{3}$ (B) $\frac{5}{\sqrt{3}}$

(C) $\sqrt{\frac{3}{2}}$ (D) $\frac{2}{\sqrt{3}}$

26. A particle of mass m is acted upon by a force F given by the empirical law $F = \frac{R}{t^2} v(t)$. If this law is to be tested experimentally by observing the motion starting from rest, the best way is to plot : **[2016 (Online)]**

(A) $\log v(t)$ against $\frac{1}{t}$ (B) $v(t)$ against t^2

(C) $\log v(t)$ against $\frac{1}{t^2}$ (D) $\log v(t)$ against t

27. A rocket is fired vertically from the earth with an acceleration of $2g$, where g is the gravitational acceleration. On an inclined plane inside the rocket, making an angle θ with the horizontal, a point object of mass m is kept. The minimum coefficient of friction μ_{min} between the mass and the inclined surface such that the mass does not move is : **[2016 (Online)]**

(A) $\tan 2\theta$ (B) $\tan \theta$
(C) $3 \tan \theta$ (D) $2 \tan \theta$

28. A mass of 10 kg is suspended vertically by a rope from the roof. When a horizontal force is applied on the rope at some point, the rope deviated at an angle of 45° at the roof point. If the suspended mass is at equilibrium, the magnitude of the force applied is ($g = 10$ ms^{-2}) : **[9 Jan. 2019 (Evening)]**

(A) 200 N (B) 100 N
(C) 140 N (D) 70 N

29. A particle of mass m is moving in a straight line with momentum p. Starting at time $t = 0$, a force $F = kt$ acts in the same direction on the moving particle during time interval T so that its momentum changes from p to $3p$. Here k is a constant. The value of T is : **[11 Jan. 2019 (Evening)]**

(A) $2\sqrt{\frac{p}{k}}$ (B) $\sqrt{\frac{2p}{k}}$

(C) $\sqrt{\frac{2k}{p}}$ (D) $2\sqrt{\frac{k}{p}}$

* * * * *

CHAPTER-4 LAWS OF MOTION

4.2 Friction

1. A horizontal force of 10 N is necessary to just hold a block stationary against a wall. The coefficient of friction between the block and the wall is 0.2. The weight of the block is : **[2003]**

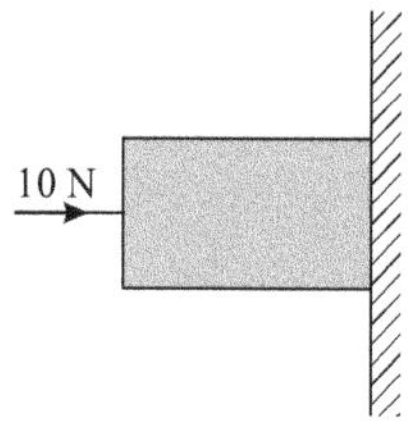

(A) 20 N (B) 50 N
(C) 100 N (D) 2 N

2. A marble block of mass 2 kg lying on ice when given a velocity of 6 m/s is stopped by friction in 10 s. Then the coefficient of friction is : **[2003]**
(A) 0.02 (B) 0.03
(C) 0.06 (D) 0.01

3. A block rests on a rough inclined plane making an angle of 30° with the horizontal. The coefficient of static friction between the block and the plane is 0.8. If the frictional force on the block is 10 N, the mass of the block (in kg) is : (taken $g = 10\ m/s^2$) : **[2004]**
(A) 2.0 (B) 4.0
(C) 1.6 (D) 2.5

4. A smooth block is released at rest on a 45° incline and then slides a distance d. The time taken to slide is n times as much to slide on rough incline than on a smooth incline. The coefficient of friction is : **[2005]**

(A) $\mu_k = 1 - \frac{1}{n^2}$ (B) $\mu_k = \sqrt{1 - \frac{1}{n^2}}$

(C) $\mu_s = 1 - \frac{1}{n^2}$ (D) $\mu_s = \sqrt{1 - \frac{1}{n^2}}$

5. The upper half of an inclined plane with inclination ϕ is perfectly smooth, while the lower half is rough. A body starting from rest at the top will again come to rest at the bottom, if the coefficient of friction for the lower half is given by : **[2005]**
(A) $2 \sin \phi$ (B) $2 \cos \phi$
(C) $2 \tan \phi$ (D) $\tan \phi$

6. Consider a car moving on a straight road with a speed of 100 m/s. The distance at which car can be stopped, is : $[\mu_k = 0.5]$ **[2005]**
(A) 800 m (B) 1000 m
(C) 100 m (D) 400 m

7. The minimum force required to start pushing a body up a rough (frictional coefficient μ) inclined plane is F_1 while the minimum force needed to prevent it from sliding down is F_2. If the inclined plane makes an angle θ from the horizontal such that $\tan \theta = 2\mu$ then the ratio $\frac{F_1}{F_2}$ is : **[2011]**
(A) 4 (B) 1
(C) 2 (D) 3

8. An insect crawls up a hemispherical surface very slowly. The coefficient of friction between the insect and the surface is 1/3. If the line joining the centre of the hemispherical surface to the insect makes an angle α with the vertical, the maximum possible value of α so that the insect does not slip is given by: **[2012 (Online)]**

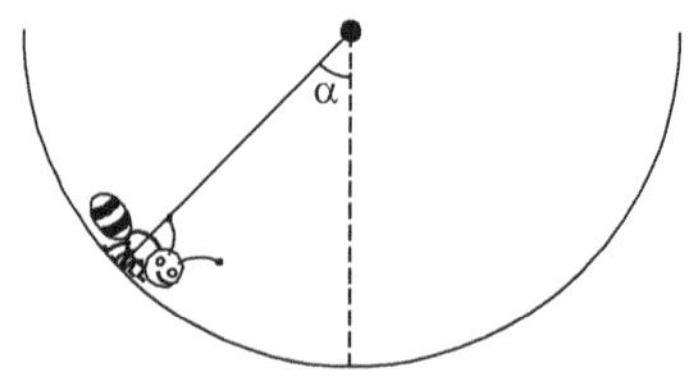

(A) $\cot \alpha = 3$ (B) $\sec \alpha = 3$
(C) $\text{cosec}\ \alpha = 3$ (D) $\cos \alpha = 3$

9. A block of weight W rest on a horizontal floor with coefficient of static friction μ. It is desired to make the block move by applying minimum amount of force. The angle θ from the horizontal at which the force should be applied and magnitude of the force F are respectively : **[2012 (Online)]**

(A) $\theta = \tan^{-1}(\mu), F = \frac{\mu W}{\sqrt{1+\mu^2}}$

(B) $\theta = \tan^{-1}\left(\frac{1}{\mu}\right), F = \frac{\mu W}{\sqrt{1+\mu^2}}$

(C) $\theta = 0, F = \mu W$

(D) $\theta = \tan^{-1}\left(\frac{\mu}{1+\mu}\right), F = \frac{\mu W}{1+\mu}$

10. A body starts from rest on a long inclined plane of slope 45°. The coefficient of friction between the body and the plane varies as $\mu = 0.3\,x$, where x is distance travelled down the plane. The body will have maximum speed : **[2013 (Online)]**
(A) 9.8 m (B) 27 m
(C) 12 m (D) 3.33 m

11. A block is placed on a rough horizontal plane. A time dependent horizontal force $F = kt$ acts on the block, where k is a positive constant. The acceleration-time graph of the block is: **[2013 (Online)]**

(A) 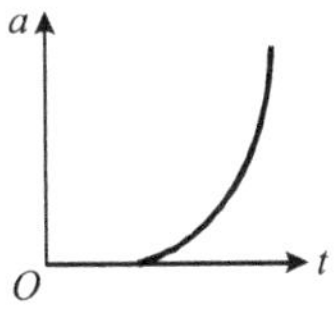

(B) 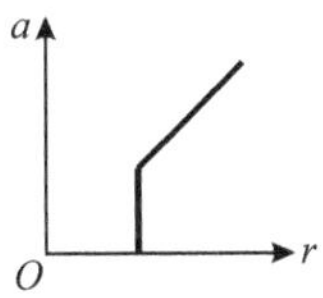

(C) 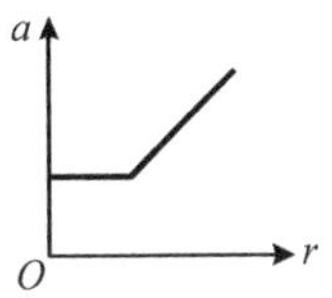

(D) 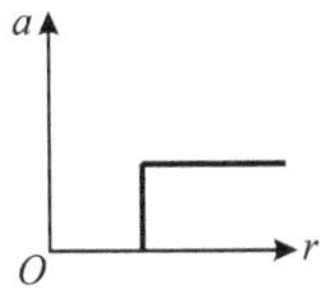

12. A block of mass m is placed on a surface with a vertical cross section given by $y = \frac{x^3}{6}$. If the coefficient of friction is 0.5, the maximum height above the ground at which the block can be placed without slipping is : **[2014]**

(A) $\frac{1}{6}m$ (B) $\frac{2}{3}m$

(C) $\frac{1}{3}m$ (D) $\frac{1}{2}m$

13. Given in the figure are two blocks A and B of weight 20 N and 100 N, respectively. These are being pressed against a wall by a force F as shown. If the coefficient of friction between the blocks is 0.1 and between block B and the wall is 0.15, the frictional force applied by the wall on block B is : **[2015]**

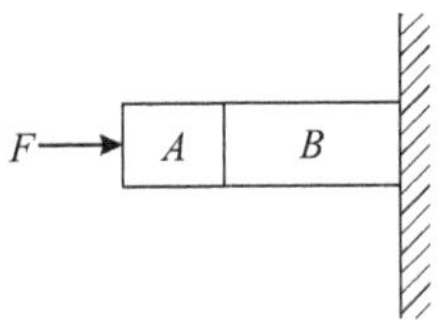

(A) 100 N (B) 80 N

(C) 120 N (D) 150 N

14. The masses m_1 = 5 kg and m_2 = 10 kg, connected by an inextensible string over a friction less pulley, are moving as shown in the figure. The coefficient of friction of horizontal surface is 0.15. The minimum weight m that should be put on top of m_2 to stop the motion is : **[2018]**

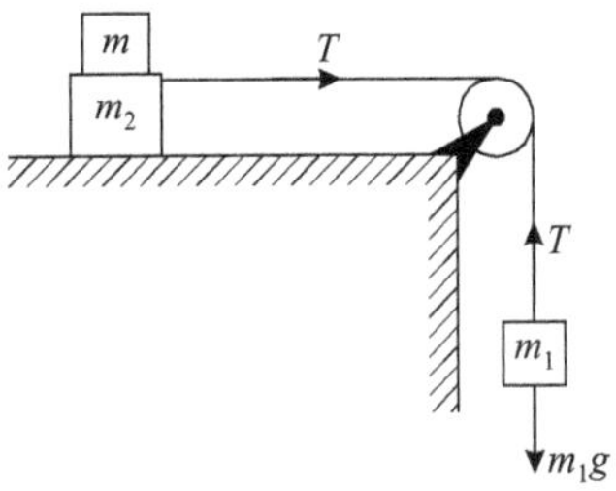

(A) 18.3 kg (B) 27.3 kg

(C) 43.3 kg (D) 10.3 kg

15. A given object taken n times more time to slide down a 45° rough inclined plane as it takes to slide down a perfectly smooth 45° incline. The coefficient of kinetic friction between the object and the incline is : **[2018 (Online)]**

(A) $\sqrt{1-\frac{1}{n^2}}$ (B) $1-\frac{1}{n^2}$

(C) $\frac{1}{2-n^2}$ (D) $\sqrt{\frac{1}{1-n^2}}$

16. A body of mass 2 kg slide down with an acceleration of 3 m/s^2 on a rough inclined plane having a slope of 30°. The external force required to take the same body up the plane with the same acceleration will be : (g = 10 m/s^2) : **[2018 (Online)]**

(A) 4 N (B) 14 N

(C) 6 N (D) 20 N

17. A block of mass 10 kg is kept on a rough inclined plane as shown in the figure. A force of 3 N is applied on the block. The coefficient of static friction between the plane and the block is 0.6. What should be the minimum value of force P, such that the block does not move downward ? [Take g = 10 ms^{-2}] **[9 Jan. 2019 (Morning)]**

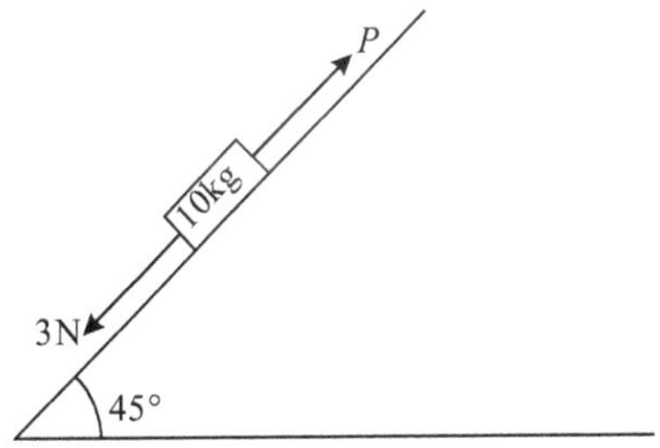

(A) 32 N (B) 25 N

(C) 23 N (D) 18 N

18. A block kept on a rough inclined plane, as shown in the figure, remains at rest upto a maximum force 2 N down the inclined plane. The maximum external force up the inclined plane that does not move the block is 10 N. The coefficient of static friction between the block and the plane is : [Take g = 10 m/s^2] **[12 Jan. 2019 (Evening)]**

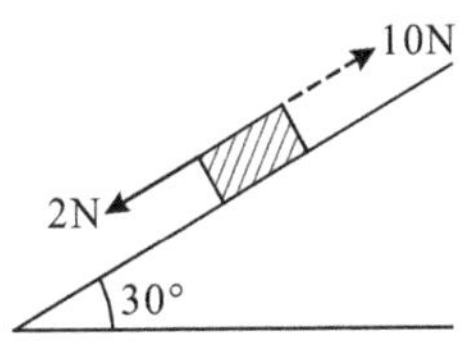

(A) $\frac{2}{3}$ (B) $\frac{\sqrt{3}}{2}$

(C) $\frac{\sqrt{3}}{4}$ (D) $\frac{1}{2}$

19. Two blocks A and B of masses $m_A = 1$ kg and $m_B = 3$ kg are kept on the table as shown in figure. The coefficient of friction between A and B is 0.2 and between B and the surface of the table is also 0.2. The maximum force F that can be applied on B horizontally, so that the block A does not slide over the block B is : (Take $g = 10$ m/s^2) : **[10 April 2019 (Evening)]**

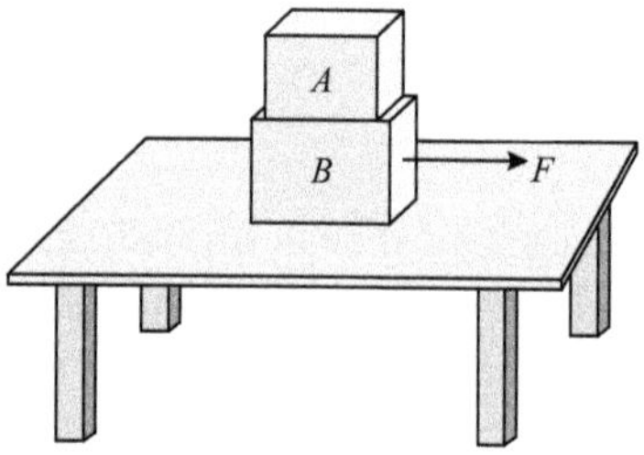

(A) 16 N (B) 40 N
(C) 12 N (D) 8 N

20. A block of mass 5 kg is (i) pushed in case (A) and (ii) pulled in case (B), by a force $F = 20$ N, making an angle of 30° with the horizontal, as shown in the figures. The coefficient of friction between the block and floor is $\mu = 0.2$. The difference between the accelerations of the block, in case (B) and case (A) will be : ($g = 10$ ms^{-2}) : **[12 April 2019 (Evening)]**

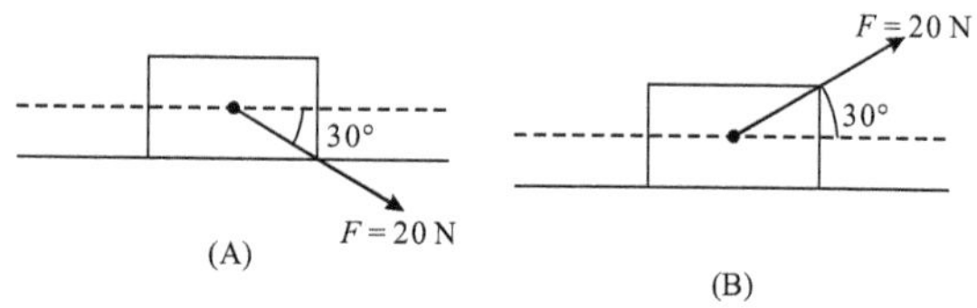

(A) 0 ms^{-2} (B) 0.8 ms^{-2}
(C) 0.4 ms^{-2} (D) 3.2 ms^{-2}

* * * * *

CHAPTER-5 WORK ENERGY POWER & APPLLICATIONS

5.1 Work Energy Power

1. A spring of force constant 800 N/m has an extension of 5 cm. The work done in extending it from 5 cm to 15 cm is : **[2002]**

(A) 16 J (B) 8 J
(C) 32 J (D) 24 J

2. A spring of spring constant 5×10^3 N/m is stretched initially by 5 cm from the unstretched position. Then the work required to stretch it further by another 5 cm is : **[2003]**

(A) 12.50 N-m (B) 18.75 N-m
(C) 25.00 N-m (D) 6.25 N-m

3. A body is moved along a straight line by a machine delivering a constant power. The distance moved by the body in time t is proportional to : **[2003]**

(A) $t^{3/4}$ (B) $t^{3/2}$
(C) $t^{1/4}$ (D) $t^{1/2}$

4. A particle moves in a straight line with retardation proportional to its displacement. Its loss of kinetic energy for any displacement x is proportional to : **[2004]**

(A) x^2 (B) e^x
(C) x (D) $\log_e x$

5. A body of mass m accelerates uniformly from rest to v_1 in time t_1. The instantaneous power delivered to the body as a function of time t is : **[2004]**

(A) $\frac{mv_1 t}{t_1}$ (B) $\frac{mv_1^2 t}{t_1^2}$
(C) $\frac{mv_1 t^2}{t_1}$ (D) $\frac{mv_1^2 t}{t_1}$

6. A uniform chain of length 2 m is kept on a table such that a length of 60 cm hangs freely from the edge of the table. The total mass of the chain is 4 kg. What is the work done in pulling the entire chain on the table : **[2004]**

(A) 7.2 J (B) 3.6 J
(C) 120 J (D) 1200 J

7. A force $\vec{F} = (5\vec{i} + 3\vec{j} + 2\vec{k})N$ is applied over a particle which displaces it from its origin to the point $\vec{r} = (2\vec{i} - \vec{j})m$. The work done on the particle in joule is : **[2004]**

(A) +10 (B) +7
(C) −7 (D) +13

8. A bullet fired into a fixed target loses half of its velocity after penetrating 3 cm. How much further it will penetrate before coming to rest, assuming that it faces constant resistance to motion ? **[2005]**

(A) 3.0 m (B) 2.0 m
(C) 1.5 cm (D) 1.0 cm

9. The block of mass M moving on the frictionless horizontal surface collides with the spring of spring constant k and compresses it by length L. The maximum momentum of the block after collision is : **[2005]**

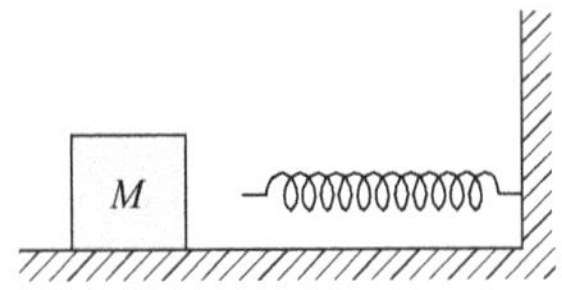

(A) $\sqrt{Mk}\,L$ (B) $\frac{kL^2}{2M}$
(C) Zero (D) $\frac{ML^2}{k}$

10. A spherical ball of mass 20 kg is stationary at the top of a hill of height 100 m. It rolls down a smooth surface to the ground, then climbs up another hill of height 30 m and finally rolls down to a horizontal base at a height of 20 m above the ground. The velocity attained by the ball is : **[2005]**

(A) 20 m/s (B) 40 m/s
(C) $10\sqrt{30}$ m/s (D) 10 m/s

11. A particle of mass 100 g is thrown vertically upwards with a speed of 5 m/s. The work done by the force of gravity during the time the particle goes up is : **[2006]**

(A) −0.5 J (B) −1.25 J
(C) 1.25 J (D) 0.5 J

12. The potential energy of a 1 kg particle free to move along the x-axis is given by $V(x) = \left(\frac{x^4}{4} - \frac{x^2}{2}\right)$ J. The total mechanical energy of the particle is 2 J. Then, the maximum speed (in m/s) is: **[2006]**

(A) $3/\sqrt{2}$ (B) $\sqrt{2}$
(C) $1/\sqrt{2}$ (D) 2

13. A mass of M kg is suspended by a weightless string. The horizontal force that is required to displace it until the string makes an angle of 45° with the initial vertical direction is : **[2006]**

(A) Mg $(\sqrt{2} + 1)$ (B) Mg $\sqrt{2}$
(C) $\frac{\text{Mg}}{\sqrt{2}}$ (D) Mg $(\sqrt{2} - 1)$

14. A 2 kg block slides on a horizontal floor with a speed of 4 m/s. It strikes an uncompressed spring, and compresses it till the block is motionless. The kinetic friction force is 15 N and spring constant is 10,000 N/m. The spring compresses by : **[2007]**

(A) 5.5 cm (B) 2.5 cm
(C) 11.0 cm (D) 8.5 cm

15. An athlete in the Olympic games covers distance of 100 m in 10 s. His kinetic energy can be estimated to be in the range : **[2008]**

(A) 200 J – 500 J (B) 2×10^5 J – 3×10^5 J
(C) 2×10^4 J – 3×10^4 J (D) 2×10^3 J – 5×10^3 J

16. The potential energy function for the force between two atoms in a diatomic molecule is approximately given by $U(x) = \frac{a}{x^{12}} - \frac{b}{x^6}$, where a and b are constant and x is the distance between the atoms. If the dissociation energy of the molecule is $D = [U(x = \infty) - U_{\text{at equilibrium}}]$, D is : **[2010]**

(A) $\frac{b^2}{6a}$ (B) $\frac{b^2}{2a}$
(C) $\frac{b^2}{12a}$ (D) $\frac{b^2}{4a}$

17. At time $t = 0$s particle starts moving along the x-axis. If its kinetic energy increases uniformly with time 't', the net force acting on it must be proportional to : **[2011]**

(A) $\sqrt{t}$ (B) Constant
(C) t (D) $\frac{1}{\sqrt{t}}$

18. This question has Statement-1 and Statement-2. Of the four choices given after the statements, choose the one that best describes the two statements.

If two springs S_1 and S_2 of force constants k_1 and k_2, respectively, are stretched by the same force, it is found that more work is done on spring S_1 than on spring S_2.

Statement-1 : If stretched by the same amount, work done on S_1, will be more than that on S_2

Statement-2 : $k_1 < k_2$. **[2012]**

(A) Statement-1 is true, Statement-2 is true and Statement-2 is not the correct explanation of Statement-1.
(B) Statement-1 is false, Statement-2 is true
(C) Statement-1 is true, Statement-2 is false
(D) Statement-1 is true, Statement-2 is true and Statement-2 is the correct explanation of statement-1.

19. The force $\vec{F} = F\hat{i}$ on a particle of mass 2 kg, moving along the x-axis is given in the figure as a function of its position x. The particle is moving with a velocity of 5 m/s along the x-axis at $x = 0$. What is the kinetic energy of the particle at $x = 8$ m ? **[2012 (Online)]**

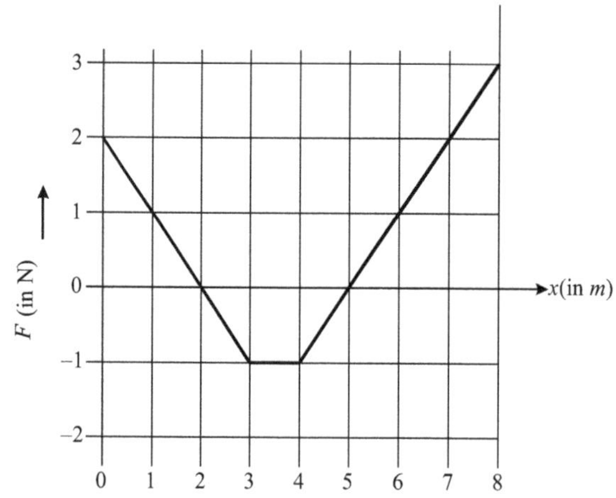

(A) 34 J (B) 34.5 J
(C) 4.5 J (D) 29.4 J

20. A particle gets displaced by $\Delta\vec{r} = (2\hat{i} + 3\hat{j} + 4\hat{k})$ m under the action of a force $\vec{F}$ $(7\hat{i} + 4\hat{j} + 3\hat{k})$. The change in its kinetic energy is : **[2012 (Online)]**

(A) 38 J (B) 70 J
(C) 52.5 J (D) 126 J

21. The springs of force constants 300 N/m (Spring A) and 400 N/m (Spring B) are joined together in series. The combination is compressed by 8.75 cm. The ratio of energy stored in A and B is $\frac{E_A}{E_B}$. Then $\frac{E_A}{E_B}$ is equal : **[2013 (Online)]**

(A) $\frac{4}{3}$ (B) $\frac{16}{9}$
(C) $\frac{3}{4}$ (D) $\frac{9}{16}$

22. A 70 kg man leaps vertically into the air from a crouching position. To take the leap the man pushes the ground with a constant force F to raise himself. The center of gravity rises by 0.5 m before he leaps. After the leap the c.g. rises by another 1 m. The maximum power delivered by the muscles is : [Take $g = 10$ ms^{-2}] **[2013 (Online)]**

(A) 6.26×10^3 Watts at the start
(B) 6.26×10^3 Watts at take off
(C) 6.26×10^4 Watts at the start
(D) 6.26×10^4 Watts at take off

23. When a rubber-band is stretched by a distance x, it exerts a restoring force of magnitude $F = ax + bx^2$ where a and b are constants. The work done in stretching the unstretched rubber-band by L is : **[2014]**

(A) $aL^2 + bL^3$ (B) $\frac{1}{2}(aL^2 + bL^3)$

(C) $\frac{aL^2}{2} + \frac{bL^3}{3}$ (D) $\frac{1}{2}\left(\frac{aL^2}{2} + \frac{bL^3}{3}\right)$

24. A spring of unstretched length 1 has a mass m with one end fixed to a rigid support. Assuming spring to be made of a uniform wire, the kinetic energy possessed by it if its free end is pulled with uniform velocity v is : **[2014 (Online)]**

(A) $\frac{1}{2}mv^2$ (B) mv^2

(C) $\frac{1}{3}mv^2$ (D) $\frac{1}{6}mv^2$

25. A wind-powered generator converts wind energy into electrical energy. Assume that the generator converts a fixed fraction of the wind energy intercepted by its blades into electrical energy. For wind speed v, the electrical power output will be most likely proportional to : **[2014 (Online)]**

(A) v^4 (B) v^2

(C) v (D) v^3

26. A bullet looses $\left(\frac{1}{n}\right)^{th}$ of its velocity passing through one plank. The number of such planks that are required to stop the bullet can be : **[2014 (Online)]**

(A) $\frac{n^2}{2n-1}$ (B) $\frac{2n^2}{n-1}$

(C) Infinite (D) n

27. A small ball of mass m starts at a point A with speed v_0 and moves along a frictionless track AB as shown. The track BC has coefficient of friction μ. The ball comes to stop at C after travelling a distance L which is : **[2014 (Online)]**

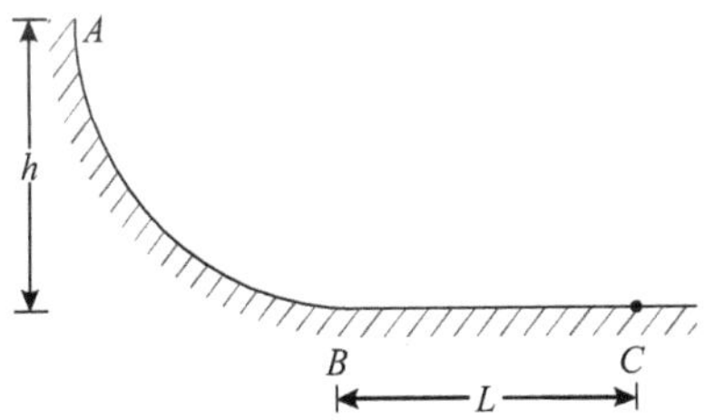

(A) $\frac{2h}{\mu} + \frac{v_0^2}{2\mu g}$ (B) $\frac{h}{\mu} + \frac{v_0^2}{2\mu g}$

(C) $\frac{h}{2\mu} + \frac{v_0^2}{\mu g}$ (D) $\frac{h}{2\mu} + \frac{v_0^2}{2\mu g}$

28. A point particle of mass m, moves along the uniformly rough track PQR as shown in the figure. The coefficient of friction, between the particle and the rough track equals m. The particle released, from rest, from the point P and it comes to rest at a point R. The energies, lost by the ball, over the parts, PQ and QR, of the track, are equal to each other, and no energy is lost when particle changes direction from PQ to QR The values of the coefficient of friction m and the distance $x(= QR)$, are, respectively close to : **[2016]**

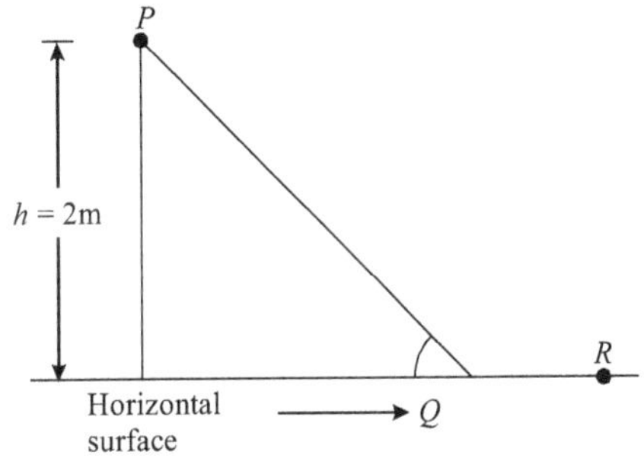

(A) 0.2 and 6.5 m (B) 0.2 and 3.5 m

(C) 0.29 and 3.5 m (D) 0.29 and 6.5 m

29. A person trying to lose weight by burning fat lifts a mass of 10 kg upto a height of 1 m 1000 times. Assume that the potential energy lost each time he lowers the mass is dissipated. How much fat will he use up considering the work done only when the weight is lifted up? Fat supplies 3.8×10^7 J of energy per kg which is converted to mechanical energy with a 20% efficiency rate. Take $g = 9.8\ ms^{-2}$: **[2016]**

(A) 2.45×10^{-3} kg (B) 6.45×10^{-3} kg

(C) 9.89×10^{-3} kg (D) 12.89×10^{-3} kg

30. Velocity-time graph for a body of mass 10 kg is shown in figure. Work-done on the body in first two seconds of the motion is : **[2016 (Online)]**

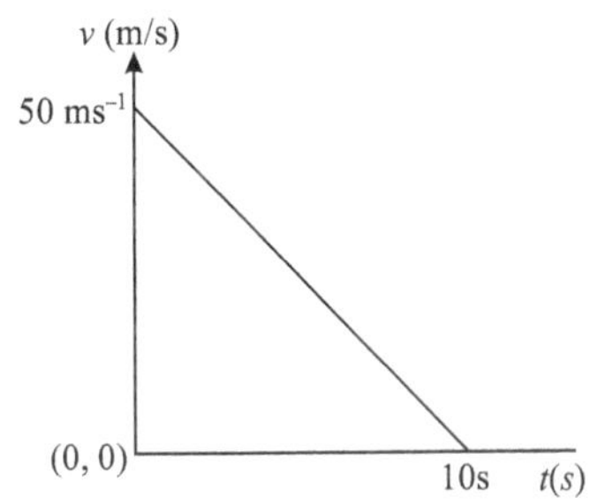

(A) −9300 J (B) 12000 J

(C) −4500 J (D) −12000 J

31. A car of weight W is on an inclined road that rises by 100 m over a distance of 1 Km and applies a constant frictional force $\frac{W}{20}$ on the car. While moving uphill on the road at a speed of $10\ ms^{-1}$, the car needs power P. If it needs power $\frac{P}{2}$ while moving downhill at speed v then value of v is : **[2016 (Online)]**

(A) $20\ ms^{-1}$ (B) $5\ ms^{-1}$

(C) $15\ ms^{-1}$ (D) $10\ ms^{-1}$

32. A time dependent force $F = 6t$ acts on a particle of mass 1 kg. If the particle starts from rest, the work done by the force during the first 1 second will be : **[2017]**

(A) 9 J (B) 18 J

(C) 4.5 J (D) 22 J

33. A body of mass $m = 10^{-2}$ kg is moving in a medium and experiences a frictional force $F = -kv^2$. Its initial speed is $v_0 = 10\ \text{ms}^{-1}$. If, after 10s, its energy is $\frac{1}{8}mv_0^2$, the value of k will be : **[2017]**

(A) 10^{-4} kg m^{-1} (B) 10^{-1} kg m^{-1} s^{-1}

(C) 10^{-3} kg m^{-1} (D) 10^{-3} kg s^{-1}

34. A particle is moving in a circular path of radius a under the action of an attractive potential $U = -\frac{k}{2r^2}$. Its total energy is : **[2018]**

(A) $-\frac{k}{4a^2}$ (B) $\frac{k}{2a^2}$

(C) Zero (D) $-\frac{3}{2}\frac{k}{a^2}$

35. Two particles of the same mass m are moving in circular orbits because of force, given by $F(r) = \frac{-16}{r} - r^3$.
The first particle is at a distance $r = 1$, and the second, at $r = 4$. The best estimate for the ratio of kinetic energies of the first and the second particle is closest to : **[2018 (Online)]**

(A) 10^{-1} (B) 6×10^{-2}

(C) 6×10^2 (D) 3×10^{-3}

36. A body of mass m starts moving from rest along x-axis so that its velocity varies as $v = a\sqrt{s}$ where a is a constant and s is the distance covered by the body. The total work done by all the forces acting on the body in the first t second after the start of the motion is : **[2018 (Online)]**

(A) $\frac{1}{8}ma^4t^2$ (B) $4\,ma^4t^2$

(C) $8\,ma^4t^2$ (D) $\frac{1}{4}ma^4t^2$

37. A block of mass m, lying on a smooth horizontal surface, is attached to a spring (of negligible mass) of spring constant k. The other end of the spring is fixed, as shown in the figure. The block is initially at rest in its equilibrium position. If now the block is pulled with a constant force F, the maximum speed of the block is : **[9 Jan. 2019 (Morning)]**

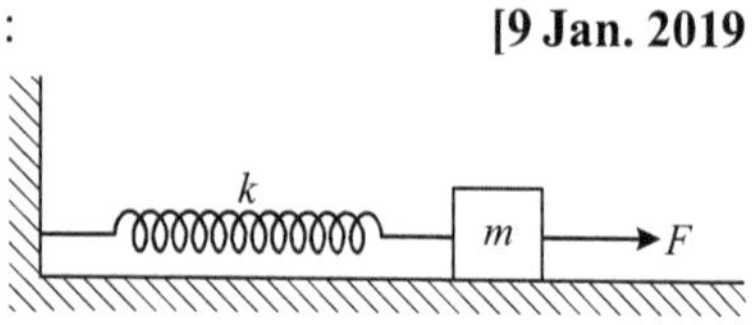

(A) $\frac{\pi F}{\sqrt{mk}}$ (B) $\frac{2F}{\sqrt{mk}}$

(C) $\frac{F}{\sqrt{mk}}$ (D) $\frac{F}{\pi\sqrt{mk}}$

38. A force acts on a 2 kg object so that its position is given as a function of time as $x = 3t^2 + 5$. What is the work done by this force in first 5 seconds ? **[9 Jan. 2019 (Evening)]**

(A) 850 J (B) 900 J

(C) 950 J (D) 875 J

39. A block of mass m is kept on a platform which starts from rest with constant acceleration $g/2$ upward, as shown in figure. Work done by normal reaction on block in time t is : **[10 Jan. 2019 (Morning)]**

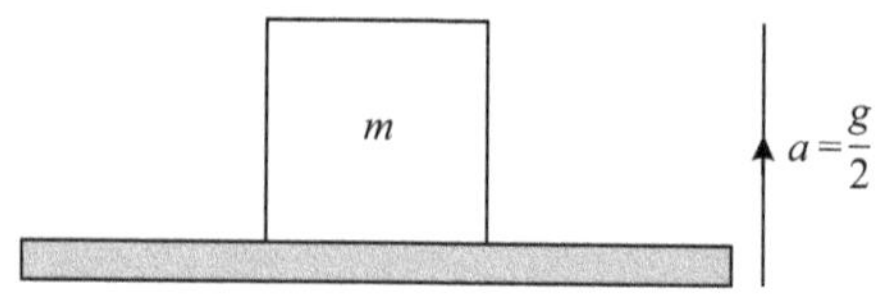

(A) 0 (B) $\frac{3mg^2t^2}{8}$

(C) $-\frac{mg^2t^2}{8}$ (D) $\frac{mg^2t^2}{8}$

40. A particle which is experiencing a force, given by $\vec{F} = 3\vec{i} - 12\vec{j}$, undergoes a displacement of $\vec{d} = 4\vec{i}$. If the particle had a kinetic energy of 3 J at the beginning of the displacement, what is its kinetic energy at the end of the displacement ? **[10 Jan. 2019 (Evening)]**

(A) 15 J (B) 10 J

(C) 12 J (D) 9 J

41. A body of mass 1 kg falls freely from a height of 100 m on a platform of mass 3 kg which is mounted on a spring having spring constant $k = 1.25 \times 10^6$ N/m. The body sticks to the platform and the spring's maximum compression is found to be x. Given that $g = 10\ \text{ms}^{-2}$, the value of x will be close to : **[11 Jan. 2019 (Morning)]**

(A) 4 cm (B) 8 cm

(C) 80 cm (D) 40 cm

42. A particle moves in one dimension from rest under the influence of a force that varies with the distance travelled by the particle as shown in the figure. The kinetic energy of the particle after it has travelled 3 m is : **[8 April 2019 (Morning)]**

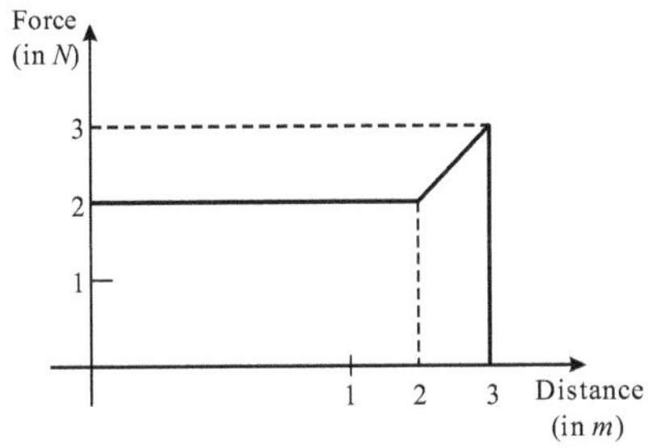

(A) 6.5 J (B) 2.5 J
(C) 4 J (D) 5 J

43. A uniform cable of mass 'M' and length 'L' is placed on a horizontal surface such that its $\left(\frac{1}{n}\right)^{th}$ part is hanging below the edge of the surface. To lift the hanging part of the cable upto the surface, the work done should be : **[9 April 2019 (Morning)]**

(A) $\frac{MgL}{n^2}$ (B) $\frac{MgL}{2n^2}$

(C) $\frac{2MgL}{n^2}$ (D) $nMgL$

44. A ball is thrown upward with an initial velocity V_0 from the surface of the earth. The motion of the ball is affected by a drag force equal to $m\gamma\upsilon^2$ (where m is mass of the ball, υ is its instantaneous velocity and γ is a constant). Time taken by the ball to rise to its zenith is : **[10 April 2019 (Morning)]**

(A) $\frac{1}{\sqrt{\gamma g}}\sin^{-1}\left(\sqrt{\frac{\gamma}{g}}V_0\right)$ (B) $\frac{1}{\sqrt{\gamma g}}\tan^{-1}\left(\sqrt{\frac{\gamma}{g}}V_0\right)$

(C) $\frac{1}{\sqrt{2\gamma g}}\tan^{-1}\left(\sqrt{\frac{2\gamma}{g}}V_0\right)$ (D) $\frac{1}{\sqrt{\gamma g}}\ln\left(1+\sqrt{\frac{\gamma}{g}}V_0\right)$

45. A spring whose unstretched length is l has a force constant k. The spring is cut into two pieces of unstretched lengths l_1 and l_2 where, $l_1 = nl_2$ and n is an integer. The ratio k_1/k_2 of the corresponding force constants, will be : **[12 April 2019 (Evening)]**

(A) $\frac{1}{n^2}$ (B) n^2

(C) $\frac{1}{n}$ (D) n

* * * * *

CHAPTER-5 WORK ENERGY POWER & APPLLICATIONS

5.2 Circular Motion

1. The minimum velocity (in ms^{-1}) with which a car driver must traverse a flat curve of radius 150 m and coefficient of friction 0.6 to avoid skidding is : **[2002]**

(A) 60 (B) 30
(C) 15 (D) 25

2. A particle is acted upon by a force of constant magnitude which is always perpendicular to the velocity of the particle. The motion of the particle takes place in a plane, it follows that: **[2004]**

(A) Its velocity is constant
(B) Its acceleration is constant
(C) Its kinetic energy is constant
(D) It moves in a straight line

3. Which of the following statements is false for a particle moving in a circle with a constant angular speed? **[2004]**

(A) The velocity vector is tangent to the circle
(B) The acceleration vector is tangent to the circle
(C) The acceleration vector point to the center of the circle
(D) The velocity and acceleration vectors are perpendicular to each other

4. For a particle in uniform circular motion in anticlockwise sense the acceleration at $\vec{a}$ point $P(R, \theta)$ on the circle of radius R is (Here θ is measured from the x-axis) : **[2010]**

(A) $-\frac{v^2}{R}\cos\theta\,\hat{i}+\frac{v^2}{R}\sin\theta\,\hat{j}$ (B) $-\frac{v^2}{R}\sin\theta\,\hat{i}+\frac{v^2}{R}\cos\theta\,\hat{j}$

(C) $-\frac{v^2}{R}\cos\theta\,\hat{i}-\frac{v^2}{R}\sin\theta\,\hat{j}$ (D) $\frac{v^2}{R}\hat{i}+\frac{v^2}{R}\hat{j}$

5. A point P moves in counter-clockwise direction on a circular path as shown in the figure. The movement of 'P' is such that it sweeps out a length $s = t^3 + 5$, where s is in metres and t is in seconds. The radius of the path is 20 m. The acceleration of 'P' when $t = 2$ s is nearly : **[2010]**

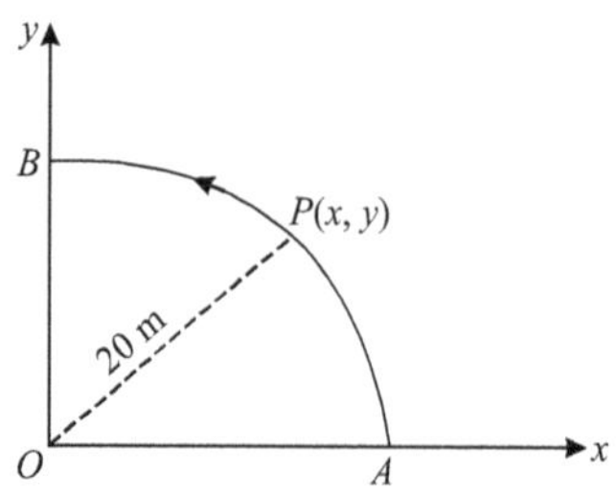

(A) 13 m/s^2 (B) 12 m/s^2
(C) 7.2 m/s^2 (D) 14 m/s^2

6. Two cars of masses m_1 and m_2 are moving in circles of radii r_1 and r_2, respectively. Their speeds are such that they make complete circles in the same time t. The ratio of their centripetal acceleration is : **[2012]**

(A) $m_1 r_1 : m_2 r_2$ (B) $m_1 : m_2$
(C) $r_1 : r_2$ (D) $1 : 1$

7. A body of mass 'm' is tied to one end of a spring and whirled round in a horizontal plane with a constant angular velocity. The elongation in the spring is 1 cm. If the angular velocity is doubled, the elongation in the spring is 5 cm. The original length of the spring is : **[2013 (Online)]**

(A) 15 cm (B) 12 cm
(C) 16 cm (D) 10 cm

8. A particle is released on a vertical smooth semicircular track from point X so that OX makes angle θ from the vertical (see figure). The normal reaction of the track on the particle vanishes at point Y where OY makes angle ϕ with the horizontal. Then : **[2014 (Online)]**

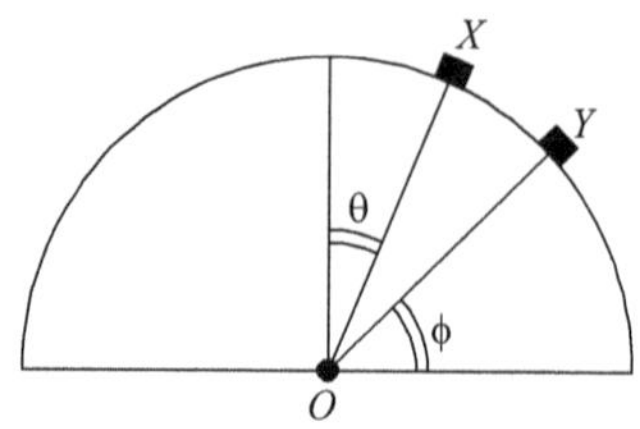

(A) $\sin\phi = \cos\phi$ (B) $\sin\phi = \frac{1}{2}\cos\theta$

(C) $\sin\phi = \frac{2}{3}\cos\theta$ (D) $\sin\phi = \frac{3}{4}\cos\theta$

9. If a body moving in circular path maintains constant speed of 10 ms^{-1}, then which of the following correctly describes relation between acceleration and radius ? **[2015 (Online)]**

(A) 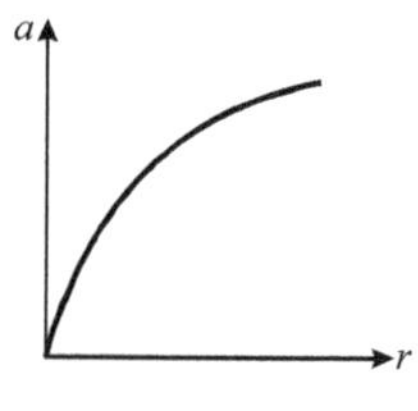

(B) 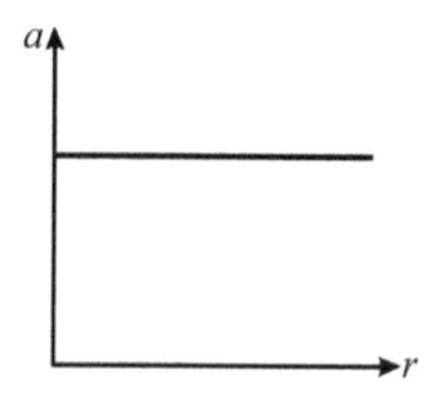

(C) 

(D) 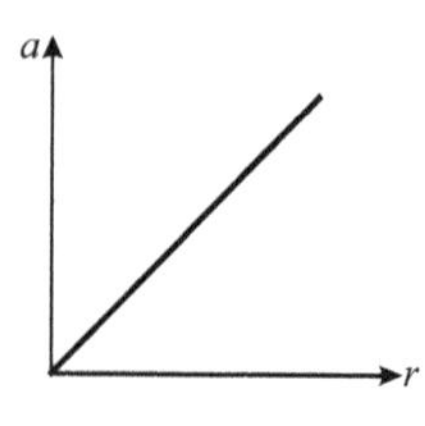

10. A particle is moving in a circle of radius r under the action of a force $F = \alpha r^2$ which is directed towards centre of the circle. Total mechanical energy (kinetic energy + potential energy) of the particle is (take potential energy = 0 for $r = 0$) : **[2015 (Online)]**

(A) $\frac{1}{2}\alpha r^3$ (B) $\frac{5}{6}\alpha r^3$

(C) $\frac{4}{3}\alpha r^3$ (D) αr^3

11. A particle of mass M is moving in a circle of fixed radius R in such a way that its centripetal acceleration at time t is given by $n^2 R t^2$ where n is a constant. The power delivered to the particle by the force acting on it, is : **[2016 (Online)]**

(A) $\frac{1}{2} M n^2 R^2 t^2$ (B) $M n^2 R^2 t$

(C) $M n R^2 t^2$ (D) $M n R^2 t$

12. Concrete mixture is made by mixing cement, stone and sand in a rotating cylinder drum. If the drum rotates too fast, the ingredients remain stuck to the wall of the drum and proper mixing of ingredients does not take place. The maximum rotational speed of the drum in revolutions per minute (rpm) to ensure proper mixing is close to : [Take the radius of the drum to be 1.25 m and its axle to be horizontal] **[2016 (Online)]**

(A) 27.0 (B) 0.4

(C) 1.3 (D) 8.0

13. A conical pendulum of length 1 m making an angle $\theta = 45°$ w.r.t. Z-axis and moves in a circle in the XY plane. The radius of the circle is 0.4 m and its centre is vertically below O. The speed of the pendulum, in its circular path, will be : [Take $g = 10\ \text{ms}^{-2}$] **[2017 (Online)]**

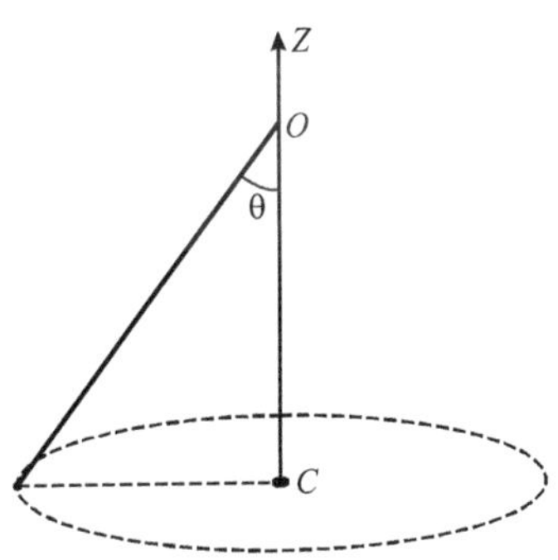

(A) 0.4 m/s (B) 4 m/s

(C) 0.2 m/s (D) 2 m/s

14. A disc rotates about its axis of symmetry in a horizontal plane at a steady rate of 3.5 revolutions per second. A coin placed at a distance of 1.25 cm from the axis of rotation remains at rest on the disc. The coefficient of friction between the coin and the disc is ($g = 10\ \text{m/s}^2$) : **[2018 (Online)]**

(A) 0.5 (B) 0.7

(C) 0.3 (D) 0.6

15. A particle is moving along a circular path with a constant speed of 10 ms^{-1}. What is the magnitude of the change in velocity of the particle, when it moves through an angle of 60° around the centre of the circle? **[11 Jan. 2019 (Morning)]**

(A) Zero (B) 10 m/s

(C) $10\sqrt{3}$ m/s (D) $10\sqrt{2}$ m/s

16. Two particles A, B are moving on two concentric circles of radii R_1 and R_2 with equal angular speed ω. At $t = 0$, their positions and direction of motion are shown in the figure. The relative velocity $\vec{\upsilon}_A - \vec{\upsilon}_B$ at $t = \pi/2\omega$ is given by : **[12 Jan. 2019 (Evening)]**

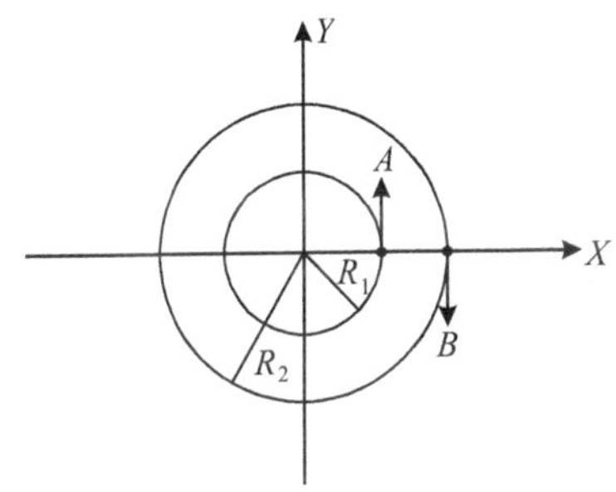

(A) $-\omega(R_1 + R_2)\hat{i}$ (B) $\omega(R_1 + R_2)\hat{i}$

(C) $\omega(R_1 - R_2)\hat{i}$ (D) $\omega(R_2 - R_1)\hat{i}$

17. A smooth wire of length $2\pi r$ is bent into a circle and kept in a vertical plane. A bead can slide smoothly on the wire. When the circle is rotating with angular speed ω about the vertical diameter AB, as shown in figure, the bead is at rest with respect to the circular ring at position P as shown. Then the value of ω^2 is equal to : **[12 April 2019 (Evening)]**

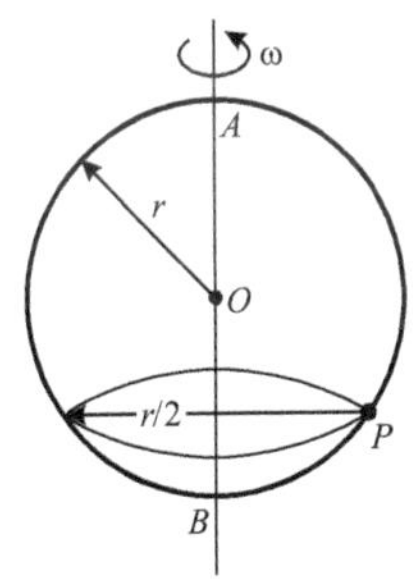

(A) $(g\sqrt{3})/r$ (B) $\frac{\sqrt{3}g}{2r}$

(C) $2g/r$ (D) $2g/(r\sqrt{3})$

* * * * *

CHAPTER-6 SYSTEM OF PARTICLES & ROTATIONAL MOTION

6.1 System of Particles

1. Two identical particles move towards each other with velocity $2v$ and v respectively. The velocity of centre of mass is: **[2002]**

(A) v (B) $v/3$
(C) $v/2$ (D) Zero

2. Consider the following two statements :
A : Linear momentum of a system of particles is zero
B : Kinetic energy of a system of particles is zero
Then : **[2003]**
(A) A does not imply B and B does not imply A
(B) A implies B but B does not imply A
(C) A does not imply B but B implies A
(D) A implies B and B implies A

3. Two spherical bodies of mass M and $5M$ and radii R and $2R$ respectively are released in free space with initial separation between their centres equal to $12R$. It they attract each other due to gravitational force only, then the distance covered by the smaller body just before collision is : **[2003]**
(A) $2.5R$ (B) $4.5R$
(C) $7.5R$ (D) $1.5R$

4. A body A of mass M while falling vertically downwards under gravity breaks into two parts; a body B of mass $1/3M$ and, a body C of mass $2/3M$. The centre of mass of bodies B and C taken together shifts compared to that of body A towards : **[2005]**
(A) Depends on height of breaking
(B) Does not shift
(C) Body C
(D) Body B

5. A mass m moves with a velocity v and collides inelastically with another identical mass. After collision the I[st] mass moves with velocity $v/\sqrt{3}$ in a direction perpendicular to the initial direction of motion. Find the speed of the second mass after collision : **[2005]**
(A) v (B) $\sqrt{3}\,v$
(C) $\frac{2}{\sqrt{3}}v$ (D) $\frac{v}{\sqrt{3}}$

6. A T shaped object with dimensions shown in the figure, is lying on a smooth floor. A force $\vec{F}$ is applied at the point P parallel to AB, such that the object has only the translational motion without rotation. Find the location of P with respect to C: **[2005]**

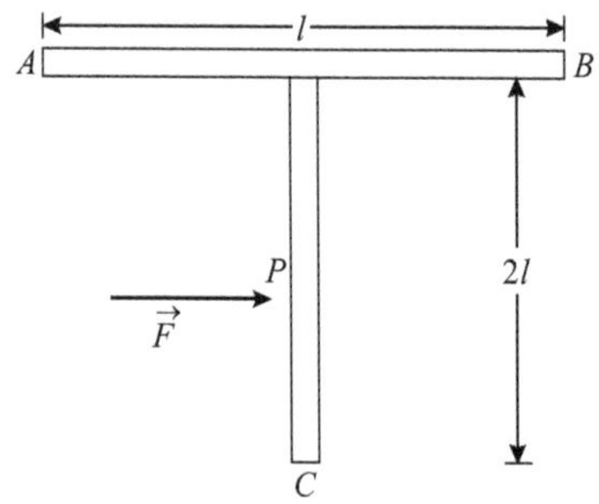

(A) $\frac{2}{3}l$ (B) $\frac{3}{2}l$
(C) $\frac{4}{3}l$ (D) l

7. A bomb of mass 16 kg at rest explodes into two pieces of masses 4 kg and 12 kg. The velocity of the 12 kg mass is 4 ms^{-1}. The kinetic energy of the other mass is : **[2006]**
(A) 144 J (B) 288 J
(C) 192 J (D) 96 J

8. Consider a two particle system with particles having masses m_1 and m_2. If the first particle is pushed towards the centre of mass through a distance d, by what distance should the second particle be moved, so as to keep the centre of mass at the same position ? **[2006]**
(A) $\frac{m_2}{m_1}d$ (B) $\frac{m_1}{m_1+m_2}d$
(C) $\frac{m_1}{m_2}d$ (D) d

9. A block of mass 0.50 kg is moving with a speed of 2.00 ms^{-1} on a smooth surface. It strikes another mass of 1.00 kg and then they move together as a single body. The energy loss during the collision is : **[2008]**
(A) 0.16 J (B) 1.00 J
(C) 0.67 J (D) 0.34 J

10. A thin rod of length 'L' is lying along the x-axis with its ends at $x = 0$ and $x = L$. It linear density (mass/length) varies with x as $k\left(\frac{x}{L}\right)^n$ where n can be zero or any positive number. If the position x_{CM} of the centre of mass of the rod is plotted against 'n', which of the following graphs best approximates the dependence of x_{CM} on n ? **[2008]**

(A) 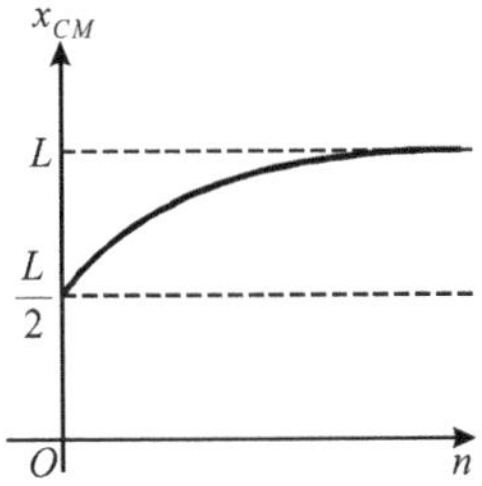

(B) 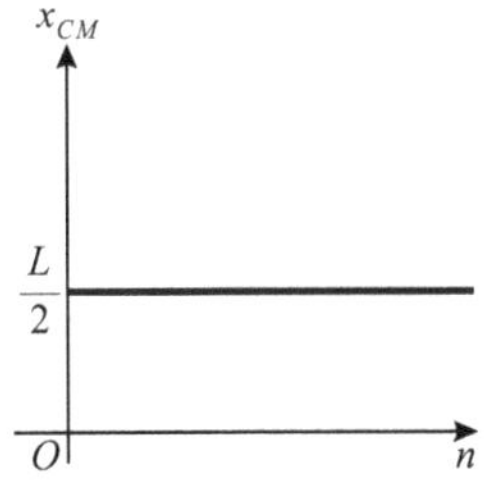

(C) 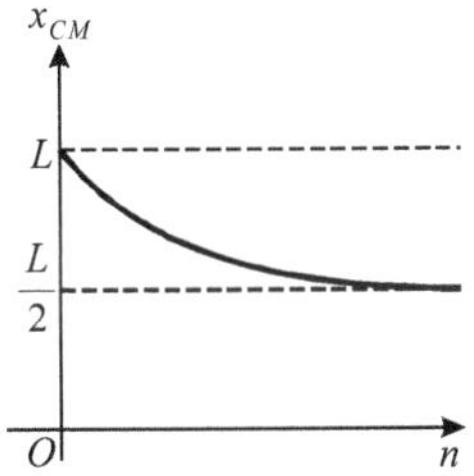

(D) 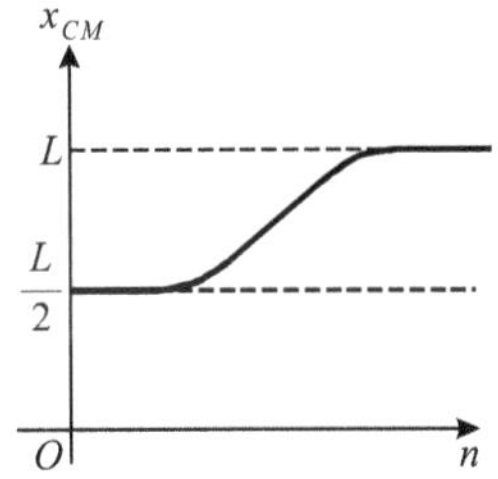

11. Consider a rubber ball freely falling from a height $h = 4.9$ m onto a horizontal elastic plate. Assume that the duration of collision is negligible and the collision with the plate is totally elastic. Then the velocity as a function of time and the height as a function of time will be : **[2009]**

(A) 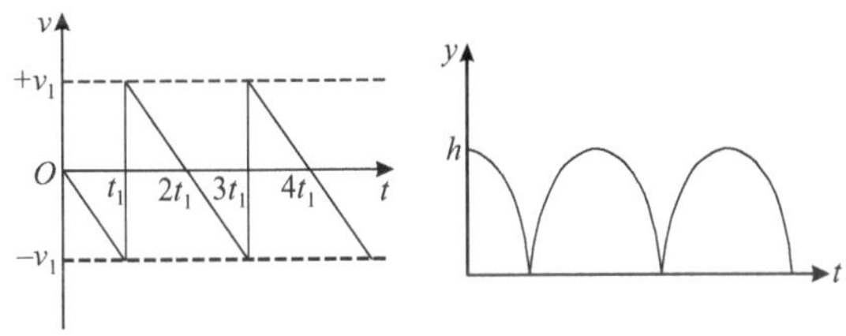

(B) 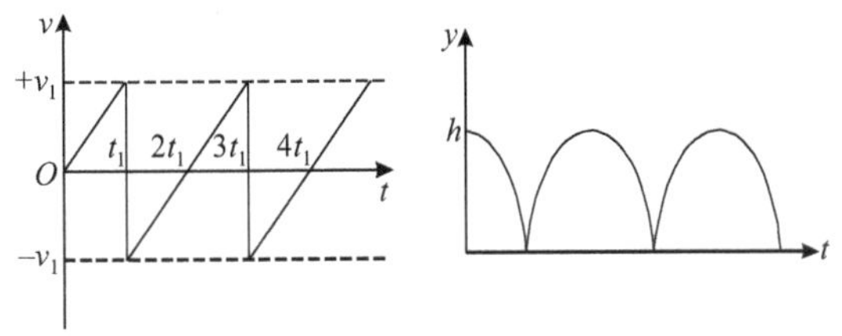

(C) 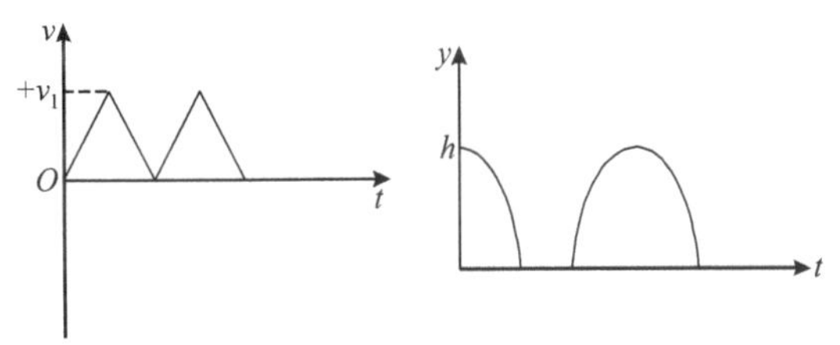

(D) 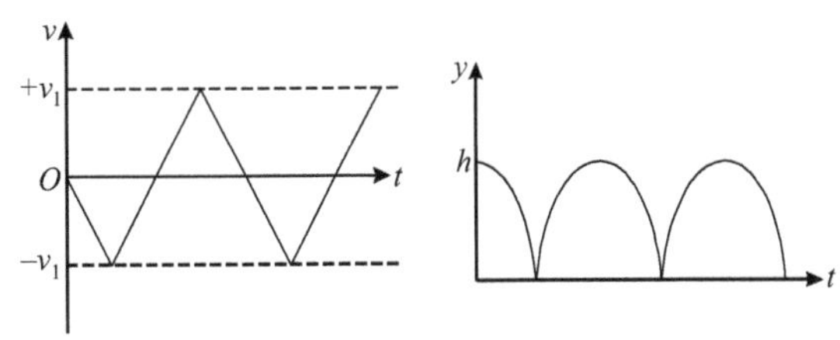

12. The figure shows the position-time (x-t) graph of one dimensional motion of a body of mass 0.4 kg. The magnitude of each impulse is : **[2010]**

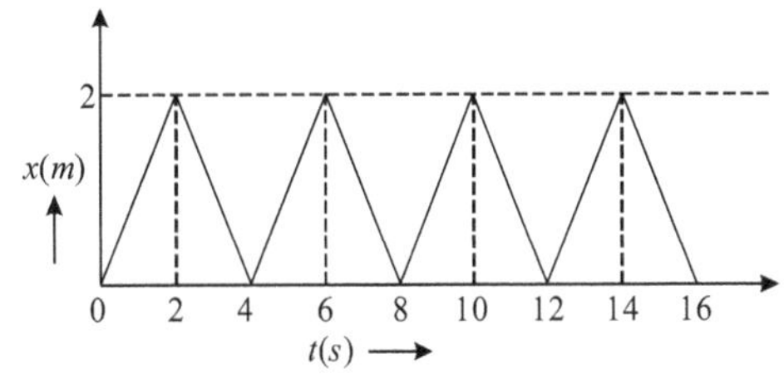

(A) 0.4 Ns (B) 0.8 Ns
(C) 1.6 Ns (D) 0.2 Ns

Directions : This Questions contain Statement-1 and Statement-2.
Of the four choices given after the statements, choose the one that best describes the two statements.

13. Statement-1 : Two particles moving in the same direction do not lose all their energy in a completely inelastic collision.
Statement-2 : Principle of conservation of momentum holds true for all kinds of collisions. **[2010]**

(A) Statement-1 is true, Statement-2 is false
(B) Statement-1 is true, Statement-2 is true; Statement-2 is the correct explanation of Statement-1
(C) Statement-1 is true, Statement-2 is true; Statement-2 is not the correct explanation of Statement-1
(D) Statement-1 is false, Statement-2 is true

14. A particle of mass 'm' is projected with a velocity v making an angle of 30° with the horizontal. The magnitude of angular momentum of the projectile about the point of projection when the particle is at its maximum height 'h' is: **[2011]**

(A) $\frac{\sqrt{3}}{2}\frac{mv^2}{g}$ (B) Zero

(C) $\frac{mv^3}{\sqrt{2}g}$ (D) $\frac{\sqrt{3}}{16}\frac{mv^3}{g}$

15. A spring is compressed between two blocks of masses m_1 and m_2 placed on a horizontal frictionless surface as shown in the figure. When the blocks are released, they have initial velocity of v_1 and v_2 as shown. The block travel distances x_1 and x_2 respectively before coming to rest. The ratio (x_1/x_2) is : **[2012 (Online)]**

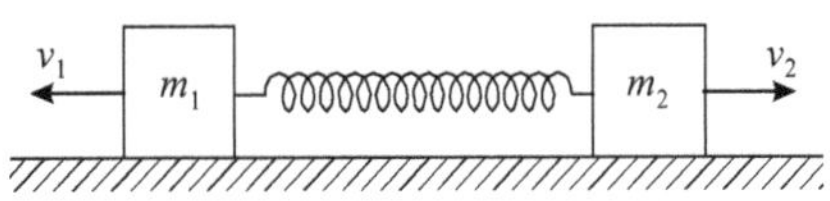

(A) $\frac{m_2}{m_1}$ (B) $\frac{m_1}{m_2}$

(C) $\sqrt{\frac{m_2}{m_1}}$ (D) $\sqrt{\frac{m_1}{m_2}}$

16. A projectile moving vertically upwards with a velocity of 200 ms^{-1} breaks into two equal parts at a height of 490 m. One part starts moving vertically upwards with a velocity of 400 ms^{-1}. How much time it will take, after the break up with the other part to hit the ground ? **[2012 (Online)]**

(A) $2\sqrt{2}$ s (B) 5s
(C) 10s (D) $\sqrt{10}$ s

17. Two bodies A and B of mass m and $2m$ respectively are placed on a smooth floor. They are connected by a spring of negligible mass. A third body C of mass m is placed on the floor. The body C moves with a velocity v_0 along the line joining A and B and collides elastically with A. At a certain time after the collision it is found that the instantaneous velocities of A and B are same and the compression of the spring is x_0. The spring constant k will be : **[2012 (Online)]**

(A) $m\dfrac{v_0^2}{x_0^2}$ (B) $m\dfrac{v_0}{2x_0}$

(C) $2m\dfrac{v_0}{x_0}$ (D) $\dfrac{2}{3}m\left(\dfrac{v_0}{x_0}\right)^2$

18. A moving particle of mass m, makes a head on elastic collision with another particle of mass $2m$, which is initially at rest. The percentage loss in energy of the colliding particle on collision, is close to : **[2012 (Online)]**

(A) 33% (B) 67%
(C) 90% (D) 10%

Directions : This Questions contain Statement-1 and Statement-2.

Of the four choices given after the statements, choose the one that best describes the two statements.

19. This question has statement-1 and statement-2. Of the four choices given after the statements, choose the one that best describes the two statements.

Statement-1 : A point particle of mass m moving with speed v collides with stationary point particle of mass M. If the maximum energy loss possible is given as $f\left(\frac{1}{2}mv^2\right)$ then $f=\left(\dfrac{m}{M+m}\right)$.

Statement-2 : Maximum energy loss occurs when the particles get stuck together as a result of the collision. **[2013]**

(A) Statement-1 is true, Statement-2 is false; Statement-2 is the correct explanation of Statement-1.
(B) Statement-1 is true, Statement-2 is true; Statement-2 is not the correct explanation of Statement-2
(C) Statement-1 is true, Statement-2 is false
(D) Statement-1 is false, Statement-2 is true

20. A projectile of mass M is fired so that the horizontal range is 4 km. At the highest point the projectile explodes in two parts of masses $M/4$ and $3M/4$ respectively and the heavier part starts falling down vertically with zero initial speed. The horizontal range (distance from point of firing) of the lighter part is : **[2013 (Online)]**

(A) 16km (B) 1km
(C) 10km (D) 2km

21. A boy of mass 20 kg is standing on a 80 kg free to move on a long cart. There is negligible friction between cart and ground. Initially, the boy is standing 25 m from a wall. If he walks 10 m on the cart towards the wall, then the final distance of the boy from the wall will be : **[2013 (Online)]**

(A) 15m (B) 12.5m
(C) 15.5m (D) 17m

22. Three masses m, $2m$ and $3m$ are moving in x-y plane with speed $3u$, $2u$ and u respectively as shown in figure. The three masses collide at the same point at P and stick together. The velocity of resulting mass will be : **[2014 (Online)]**

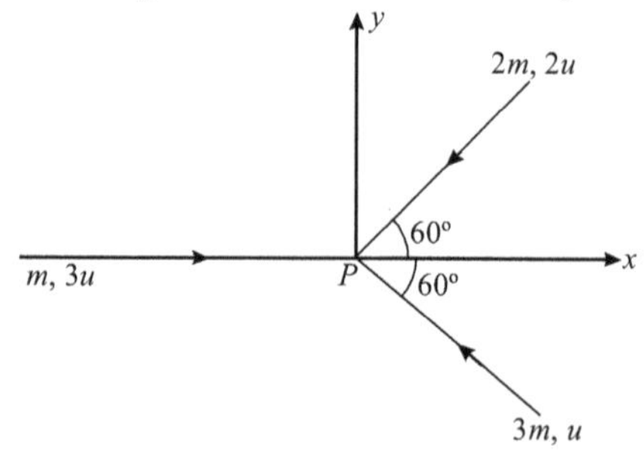

(A) $\frac{u}{12}(\hat{i}+\sqrt{3}\hat{j})$ (B) $\frac{u}{12}(\hat{i}-\sqrt{3}\hat{j})$

(C) $\frac{u}{12}(-\hat{i}+\sqrt{3}\hat{j})$ (D) $\frac{u}{12}(-\hat{i}-\sqrt{3}\hat{j})$

23. A thin bar of length L has a mass per unit length λ, that increases linearly with distance from one end. If its total mass is M and its mass per unit length at the lighter end is λ_0, then the distance of the centre of mass from the lighter end is : **[2014 (Online)]**

(A) $\dfrac{L}{2}-\dfrac{\lambda_0 L^2}{4M}$ (B) $\dfrac{L}{3}+\dfrac{\lambda_0 L^2}{8M}$

(C) $\dfrac{L}{3}+\dfrac{\lambda_0 L^2}{4M}$ (D) $\dfrac{2L}{3}-\dfrac{\lambda_0 L^2}{6M}$

24. Distance of the centre of mass of a solid uniform cone from its vertex is z_0. If the radius of its base is R and its height is h then z_0 is equal to : **[2015]**

(A) $\dfrac{h^2}{4R}$ (B) $\dfrac{3h}{4}$

(C) $\dfrac{5h}{8}$ (D) $\dfrac{3h^2}{8R}$

25. A particle of mass m moving in the x direction with speed $2v$ is hit by another particle of mass $2m$ moving in the y direction with speed v. If the collision is perfectly inelastic, the percentage loss in the energy during the collision is close to : **[2015]**

(A) 44% (B) 50%
(C) 56% (D) 62%

26. A bullet of mass 4 g is fired horizontally with a speed of 300 m/s into 0.8 kg block of wood at rest on a table. If the coefficient of friction between the block and the table is 0.3, how far will the block slide approximately? **[2015 (Online)]**

(A) 0.19 m (B) 0.379 m
(C) 0.569 m (D) 0.758 m

27. A block of mass m = 0.1 kg is connected to a spring of unknown spring constant k. It is compressed to a distance x from its equilibrium position and released from rest. After approaching half the distance $\left(\frac{x}{2}\right)$ from equilibrium position, it hits another block and comes to rest momentarily, while the other block moves with a velocity 3 ms^{-1}. The total initial energy of the spring is : **[2015 (Online)]**

(A) 0.3 J (B) 0.6 J
(C) 0.8 J (D) 1.5 J

28. A block of mass m = 10 kg rest on a horizontal table. The coefficient of friction between the block and the table is 0.05. When hit by a bullet of mass 50 g moving with speed v, that gets embedded in it, the block moves and comes to stop after moving a distance of 2 m on the table. If a freely falling object were to acquire speed $\frac{v}{10}$ after being dropped from height H, then neglecting energy losses and taking g = 10 ms^{-2}, the value of H is close to : **[2015 (Online)]**

(A) 0.05 km (B) 0.02 km
(C) 0.03 km (D) 0.04 km

29. A large number (n) of identical beads, each of mass m and radius r are strung on a thin smooth rigid horizontal rod of length L ($L >> r$) and are at rest at random positions. The rod is mounted between two rigid supports (see figure). If one of the beads is now given a speed v, the average force experienced by each support after a long time is (assume all collisions are elastic): **[2015 (Online)]**

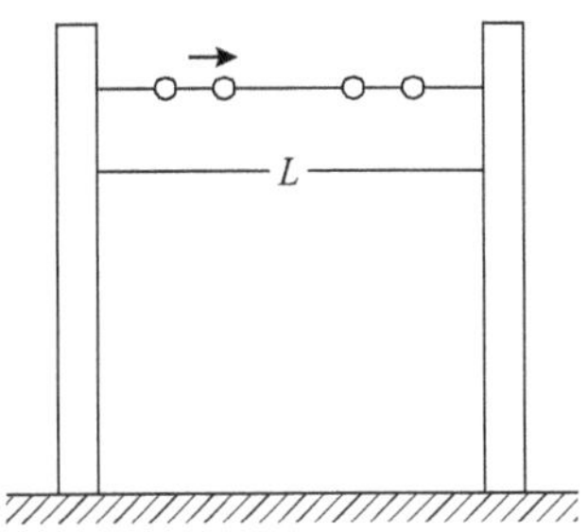

(A) $\frac{mv^2}{2(L-nr)}$ (B) $\frac{mv^2}{L-2nr}$
(C) $\frac{mv^2}{L-nr}$ (D) Zero

30. A uniform thin rod AB of length L has linear mass density $\mu(x) = a + \frac{bx}{L}$, where x is measured from A. If the CM of the rod lies at a distance of $\left(\frac{7}{12}\right)L$ from A, then a and b are related as: **[2015 (Online)]**

(A) $a = 2b$ (B) $2a = b$
(C) $a = b$ (D) $3a = 2b$

31. In the figure shown ABC is a uniform wire. If centre of mass of wire lies vertically below point A, then $\frac{BC}{AB}$ is close to : **[2016 (Online)]**

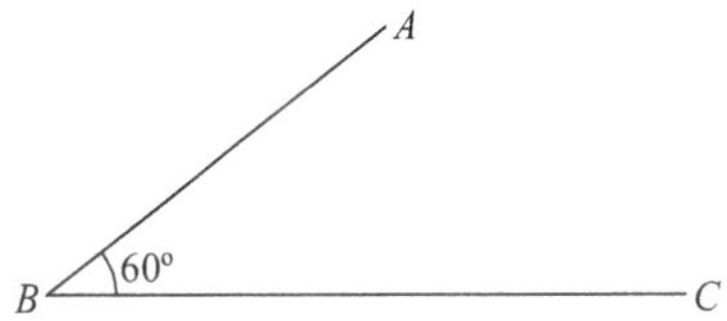

(A) 1.85 (B) 1.5
(C) 1.37 (D) 3

32. An object is dropped from a height h from the ground. Every time it hits the ground it looses 50% of its kinetic energy. The total distance covered as $t \to \infty$ is : **[2017 (Online)]**

(A) $3h$ (B) ∞
(C) $\frac{5}{3}h$ (D) $\frac{8}{3}h$

33. Two particles A and B of equal mass M are moving with the same speed v as shown in the figure. They collide completely inelastically and move as a single particle C. The angle θ that the path of C makes with the X-axis is given by : **[2017 (Online)]**

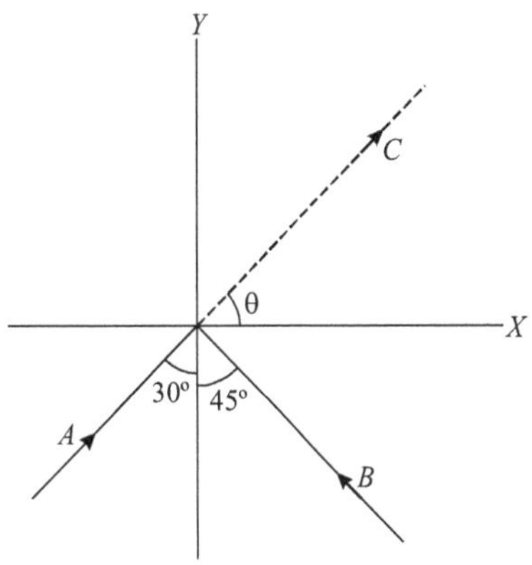

(A) $\tan\theta = \dfrac{\sqrt{3}+\sqrt{2}}{1-\sqrt{2}}$ (B) $\tan\theta = \dfrac{\sqrt{3}-\sqrt{2}}{1-\sqrt{2}}$

(C) $\tan\theta = \dfrac{1-\sqrt{2}}{\sqrt{2}(1+\sqrt{3})}$ (D) $\tan\theta = \dfrac{1-\sqrt{3}}{1+\sqrt{2}}$

34. In a collinear collision, a particle with an initial speed v_0 strikes a stationary particle of the same mass. If the final total kinetic energy is 50% greater than the original kinetic energy, the magnitude of the relative velocity between the two particles, after collision, is : **[2018]**

(A) $\dfrac{v_0}{4}$ (B) $\sqrt{2}v_0$

(C) $\dfrac{v_0}{2}$ (D) $\dfrac{v_0}{\sqrt{2}}$

35. The mass of a hydrogen molecule is 3.32×10^{-27} kg. If 10^{23} hydrogen molecules strike, per second, a fixed wall of area 2 cm^2 at an angle of 45° to the normal, and rebound elastically with a speed of 10^3 m/s, then the pressure on the wall is nearly : **[2018]**

(A) 2.35×10^3 N/m^2 (B) 4.70×10^3 N/m^2
(C) 2.35×10^2 N/m^2 (D) 4.70×10^2 N/m^2

36. It is found that if a neutron suffers an elastic collinear collision with deuterium at rest, fractional loss of its energy is P_d; while for its similar collision with carbon nucleus at rest, fractional loss of energy is P_c. The values of P_d and P_c are respectively : **[2018]**

(A) (·89, ·28) (B) (·28, ·89)
(C) (0, 0) (D) (0, 1)

37. A proton of mass m collides elastically with a particle of unknown mass at rest. After the collision, the proton and the unknown particle are seen moving at an angle of 90° with respect to each other. The mass of unknown particle is : **[2018 (Online)]**

(A) $\dfrac{m}{\sqrt{3}}$ (B) $\dfrac{m}{2}$

(C) 2m (D) m

38. Three blocks A, B and C are lying on a smooth horizontal surface, as shown in the figure. A and B have equal masses, m while C has mass M. Block A is given an initial speed v towards B due to which it collides with B perfectly inelastically. The combined mass collides with C, also perfectly inelastically $\dfrac{5}{6}$ th of the initial kinetic energy is lost in whole process. What is value of M/m ? **[9 Jan. 2019 (Morning)]**

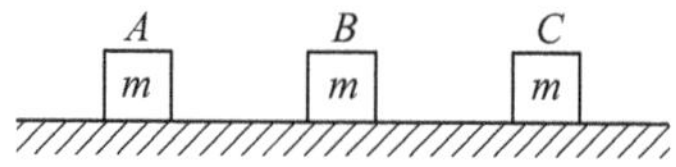

(A) 4 (B) 5
(C) 3 (D) 2

39. A piece of wood of mass 0.03 kg is dropped from the top of a 100 m height building. At the same time, a bullet of mass 0.02 kg is fired vertically upward, with a velocity 100 ms^{-1}, from the ground. The bullet gets embedded in the wood. Then the maximum height to which the combined system reaches above the top of the building before falling below is : (g = 10 ms^{-2}) : **[10 Jan. 2019 (Morning)]**

(A) 30m (B) 10m
(C) 40m (D) 20m

40. A simple pendulum, made of a string of length l and a bob of mass m, is released from a small angle θ_0. It strikes a block of mass M, kept on a horizontal surface at its lowest point of oscillations, elastically. It bounces back and goes up to an angle θ_1. Then M is given by : **[12 Jan. 2019 (Morning)]**

(A) $\dfrac{m}{2}\left(\dfrac{\theta_0-\theta_1}{\theta_0+\theta_1}\right)$ (B) $\dfrac{m}{2}\left(\dfrac{\theta_0+\theta_1}{\theta_0-\theta_1}\right)$

(C) $m\left(\dfrac{\theta_0+\theta_1}{\theta_0-\theta_1}\right)$ (D) $m\left(\dfrac{\theta_0-\theta_1}{\theta_0+\theta_1}\right)$

41. The position vector of the centre of mass $\vec{r}$ cm of an symmetric uniform bar of negligible area of cross-section as shown in figure is : **[12 Jan. 2019 (Morning)]**

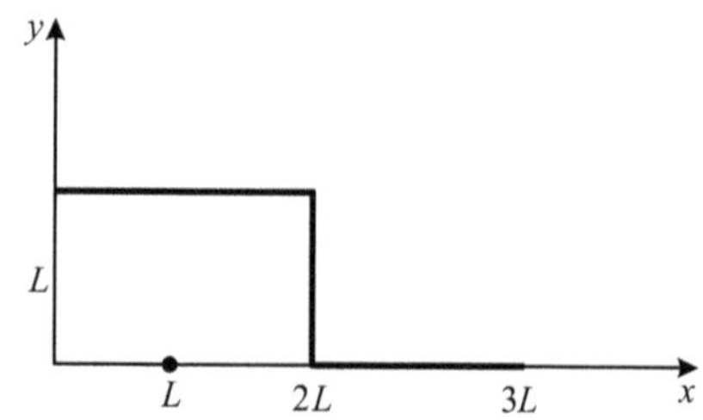

(A) $\vec{r}$ cm $= \dfrac{13}{8}L\hat{x}+\dfrac{5}{8}L\hat{y}$ (B) $\vec{r}$ cm $= \dfrac{11}{8}L\,\hat{x}+\dfrac{3}{8}L\hat{y}$

(C) $\vec{r}$ cm $= \dfrac{3}{8}L\,\hat{x}+\dfrac{11}{8}L\hat{y}$ (D) $\vec{r}$ cm $= \dfrac{5}{8}L\,\hat{x}+\dfrac{13}{8}L\hat{y}$

42. An alpha-particle of mass m suffers 1-dimensional elastic collision with a nucleus at rest of unknown mass. It is scattered directly backwards losing, 64% of its initial kinetic energy. The mass of the nucleus is : **[12 Jan. 2019 (Evening)]**

(A) 4m (B) 3.5m
(C) 2m (D) 1.5m

43. Four particles A, B, C and D with masses $m_A = m$, $m_B = 2m$, $m_C = 3m$ and $m_D = 4m$ are at the corners of a square. They have accelerations of equal magnitude with directions as shown. The acceleration of the centre of mass of the particles is : **[8 April 2019 (Morning)]**

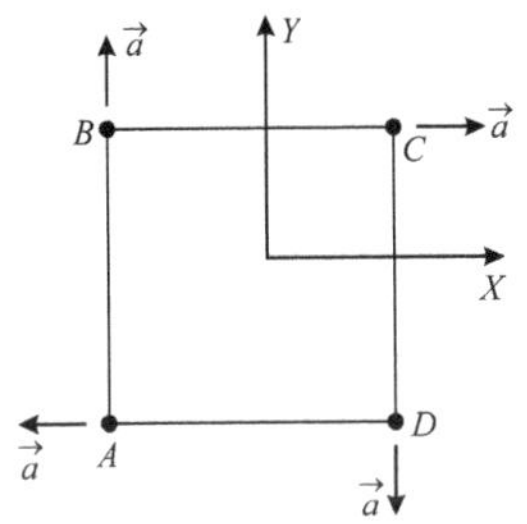

(A) $\frac{a}{5}(\hat{i} - \hat{j})$ (B) $\frac{a}{5}(\hat{i} + \hat{j})$

(C) Zero (D) $a(\hat{i} + \hat{j})$

44. A uniform rectangular thin sheet $ABCD$ of mass M has length a and breadth b, as shown in the figure. If the shaded portion $HBGO$ is cut-off, the coordinates of the centre of mass of the remaining portion will be : **[8 April 2019 (Evening)]**

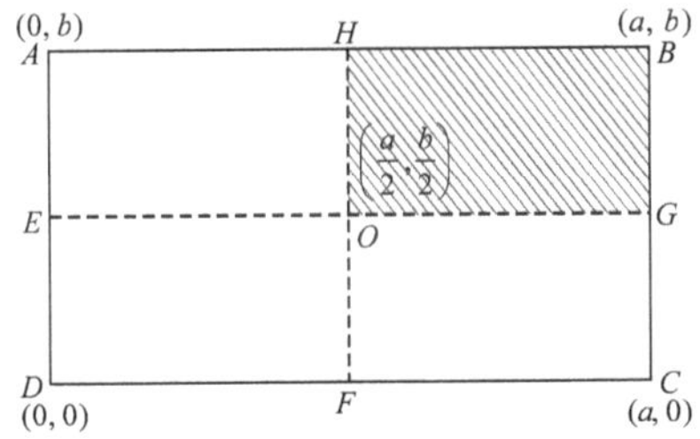

(A) $\left(\frac{2a}{3}, \frac{2b}{3}\right)$ (B) $\left(\frac{5a}{3}, \frac{5b}{3}\right)$

(C) $\left(\frac{3a}{4}, \frac{3b}{4}\right)$ (D) $\left(\frac{5a}{12}, \frac{5b}{12}\right)$

45. A body of mass m_1 moving with an unknown velocity of $v_1\hat{i}$, undergoes a collinear collision with a body of mass m_2 moving with a velocity $v_2\hat{i}$. After collision, m_1 and m_2 move with velocities of $v_3\hat{i}$ and $v_4\hat{i}$, respectively. If $m_2 = 0.5\, m_1$ and $v_3 = 0.5v_1$, then v_1 is : **[8 April 2019 (Evening)]**

(A) $v_4 - \frac{v_2}{4}$ (B) $v_4 - \frac{v_2}{2}$

(C) $v_4 - v_2$ (D) $v_4 + v_2$

46. A body of mass 2 kg makes an eleastic collision with a second body at rest and continues to move in the original direction but with one fourth of its original speed. What is the mass of the second body ? **[9 April 2019 (Morning)]**

(A) 1.8 kg (B) 1.2 kg

(C) 1.5 kg (D) 1.0 kg

47. A particle of mass 'm' is moving with speed '$2v$' and collides with a mass '$2m$' moving with speed 'v' in the same direction. After collision, the first mass is stopped completely while the second one splits into two particles each of mass 'm', which move at angle 45° with respect to the original direction. The speed of each of the moving particle will be : **[9 April 2019 (Evening)]**

(A) $v/(2\sqrt{2})$ (B) $2\sqrt{2}v$

(C) $\sqrt{2}v$ (D) $v/\sqrt{2}$

48. A wedge of mass $M = 4m$ lies on a frictionless plane. A particle of mass m approaches the wedge with speed v. There is no friction between the particle and the plane or between the particle and the wedge. The maximum height climbed by the particle on the wedge is given by : **[9 April 2019 (Evening)]**

(A) $\frac{2v^2}{7g}$ (B) $\frac{v^2}{g}$

(C) $\frac{2v^2}{5g}$ (D) $\frac{v^2}{2g}$

49. Two particles, of masses M and $2M$, moving, as shown, with speeds of 10 m/s and 5 m/s, collide elastically at the origin. After the collision, they move along the indicated directions with speeds υ_1 and υ_2, respectively. The values of υ_1 and υ_2 are nearly : **[10 April 2019 (Morning)]**

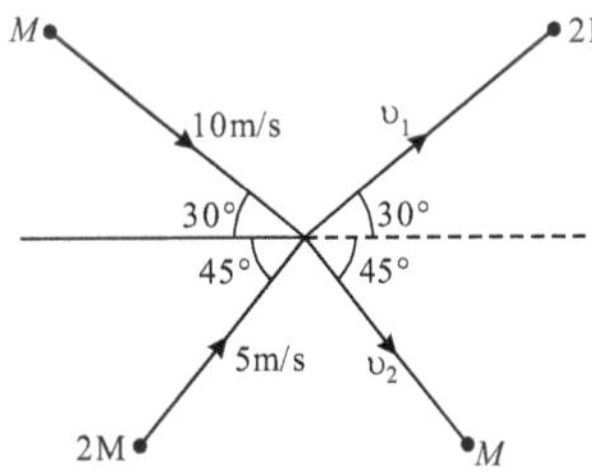

(A) 3.2 m/s and 6.3 m/s (B) 3.2 m/s and 12.6 m/s

(C) 6.5 m/s and 6.3 m/s (D) 6.5 m/s and 3.2 m/s

50. A bullet of mass 20 g has an initial speed of 1 ms^{-1}, just before it starts penetrating a mud wall of thickness 20 cm. If the wall offers a mean resistance of 2.5×10^{-2} N, the speed of the bullet after emerging from the other side of the wall is close to: **[10 April 2019 (Evening)]**

(A) 0.4 ms^{-1} (B) 0.1 ms^{-1}

(3) 0.3 ms^{-1} (D) 0.7 ms^{-1}

51. A person of mass M is, sitting on a swing of length L and swinging with an angular amplitude θ_0. If the person stands up when the swing passes through its lowest point, the work done by him, assuming that his centre of mass moves by a distance $l(l << L)$, is close to : **[12 April 2019 (Morning)]**

(A) Mgl (B) $Mgl(1 + \theta_0^2)$

(C) $Mgl(1 - \theta_0^2)$ (D) $Mgl\left(1 + \frac{\theta_0^2}{2}\right)$

52. A man (mass = 50 kg) and his son (mass = 20 kg) are standing on a frictionless surface facing each other. The man pushes his son so that he starts moving at a speed of 0.70 ms^{-1} with respect to the man. The speed of the man with respect to the surface is:

[12 April 2019 (Morning)]

(A) 0.20 ms^{-1} (B) 0.14 ms^{-1}

(C) 0.47 ms^{-1} (D) 0.28 ms^{-1}

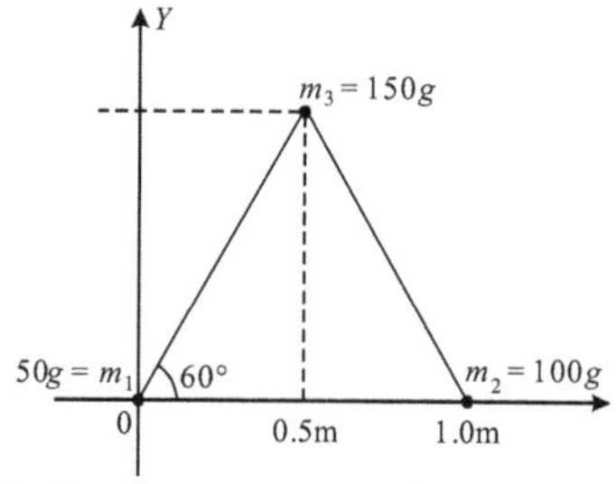

53. Three particles of masses 50 g, 100 g and 150 g are placed at the vertices of an equilateral triangle of side 1m (as shown in the figure). The (x, y) coordinates of the centre of mass will be:

[12 April 2019 (Evening)]

(A) $\left(\frac{7}{12}m, \frac{\sqrt{3}}{8}m\right)$ (B) $\left(\frac{\sqrt{3}}{4}m, \frac{5}{12}m\right)$

(C) $\left(\frac{7}{12}m, \frac{\sqrt{3}}{4}m\right)$ (D) $\left(\frac{\sqrt{3}}{8}m, \frac{7}{12}m\right)$

* * * * *

CHAPTER-6 SYSTEM OF PARTICLES & ROTATIONAL MOTION

6.2 Rigid Body Dynamics

1. Initial angular velocity of a circular disc of mass M is ω_1. Then two small spheres of mass m are attached gently to two diametrically opposite points on the edge of the disc. What is the final angular velocity of the disc ? **[2002]**

(A) $\left(\frac{M+m}{M}\right)\omega_1$ (B) $\left(\frac{M+m}{m}\right)\omega_1$

(C) $\left(\frac{M}{M+4m}\right)\omega_1$ (D) $\left(\frac{M}{M+2m}\right)\omega_1$

2. Moment of inertia of a circular wire of mass M and radius R about its diameter is : **[2002]**

(A) $MR^2/2$ (B) MR^2

(C) $2MR^2$ (D) $MR^2/4$

3. A particle of mass m moves along line PC with velocity v as shown. What is the angular momentum of the particle about O? **[2002]**

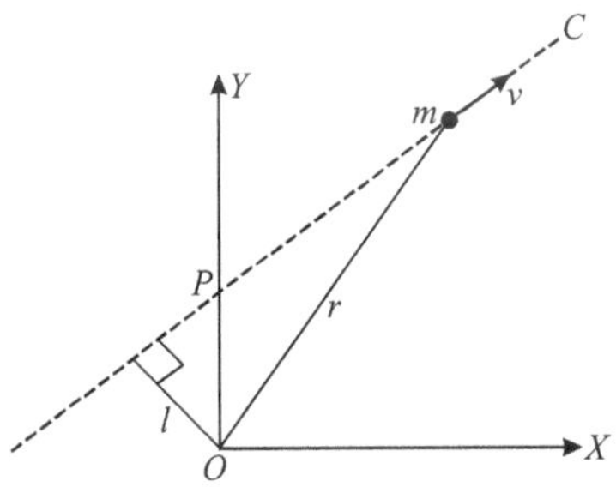

(A) mvL (B) mvl

(C) mvr (D) Zero

4. A circular disc X of radius R is made from an iron plate of thickness t and another disc Y of radius $4R$ is made from an iron plate of thickness $t/4$. Then the relation between the moment of inertia I_X and I_Y is : **[2003]**

(A) $I_Y = 32 I_X$ (B) $I_Y = 16 I_X$

(C) $I_Y = I_X$ (D) $I_Y = 64 I_X$

5. A particle performing uniform circular motion has angular momentum L. If its angular frequency is doubled and its kinetic energy halved, then the new angular momentum is : **[2003]**

(A) $\frac{L}{4}$ (B) $2L$

(C) $4L$ (D) $\frac{L}{2}$

6. Let $\vec{F}$ be the force acting on particle having position vector $\vec{r}$ and $\vec{\tau}$ be the torque of this force about the origin. Then : **[2003]**

(A) $\vec{r}\cdot\vec{\tau}=0$ and $\vec{F}\cdot\vec{\tau}\neq 0$ (B) $\vec{r}\cdot\vec{\tau}\neq 0$ and $\vec{F}\cdot\vec{\tau}=0$

(C) $\vec{r}\cdot\vec{\tau}\neq 0$ and $\vec{F}\cdot\vec{\tau}\neq 0$ (D) $\vec{r}\cdot\vec{\tau}=0$ and $\vec{F}\cdot\vec{\tau}=0$

7. A solid sphere is rotating in free space. If the radius of the sphere is increased keeping mass same which one of the following will not be affected ? **[2004]**

(A) Moment of inertia (B) Angular momentum

(C) Angular velocity (D) Rotational kinetic energy

8. One solid sphere A and another hollow sphere B are of same mass and same outer radii. Their moment of inertia about their diameters are respectively I_A and I_B such that : **[2004]**

(A) $I_A = I_B$ (B) $I_A > I_B$

(C) $I_A < I_B$ (D) $\frac{I_A}{I_B} = \frac{d_A}{d_B}$

Where d_A and d_B are their densities.

9. An annular ring with inner and outer radii R_1 and R_2 is rolling without slipping with a uniform angular speed. The ratio of the forces experienced by the two particles situated on the inner and outer parts of the ring F_1/F_2 is : **[2005]**

(A) $\frac{R_2}{R_1}$ (B) $\left(\frac{R_1}{R_2}\right)^2$

(C) 1 (D) $\frac{R_1}{R_2}$

10. The moment of inertia of uniform semicircular disc of mass M and radius r about a line perpendicular to the plane of the disc through the centre is : **[2005]**

(A) $\frac{1}{4}Mr^2$ (B) $\frac{2}{5}Mr^2$

(C) Mr^2 (D) $\frac{1}{2}Mr^2$

11. Four point masses, each of value m, are placed at the corners of a square $ABCD$ of side l. The moment of inertia of this system about an axis passing through A and parallel to BD is : **[2006]**

(A) $2ml^2$ (B) $\sqrt{3}\,ml^2$

(C) $3ml^2$ (D) ml^2

12. A force of $-F\hat{k}$ acts on O, the origin of the co-ordinate system. The torque about the point $(1, -1)$ is : **[2006]**

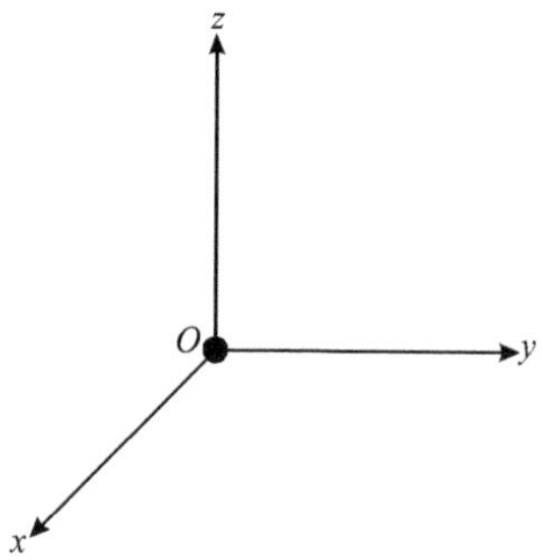

(A) $F(\hat{i}-\hat{j})$ (B) $-F(\hat{i}+\hat{j})$

(C) $F(\hat{i}+\hat{j})$ (D) $-F(\hat{i}-\hat{j})$

13. A thin circular ring of mass m and radius R is rotating about its axis with a constant angular velocity ω. Two objects each of mass M are attached gently to the opposite ends of a diameter of the ring. The ring now rotates with an angular velocity $\omega' =$: **[2006]**

(A) $\frac{\omega(m+2m)}{M}$ (B) $\frac{\omega(m-2m)}{(m+2M)}$

(C) $\frac{\omega m}{(m+M)}$ (D) $\frac{\omega m}{(m+2M)}$

14. For the given uniform square lamina $ABCD$, whose centre is O: **[2007]**

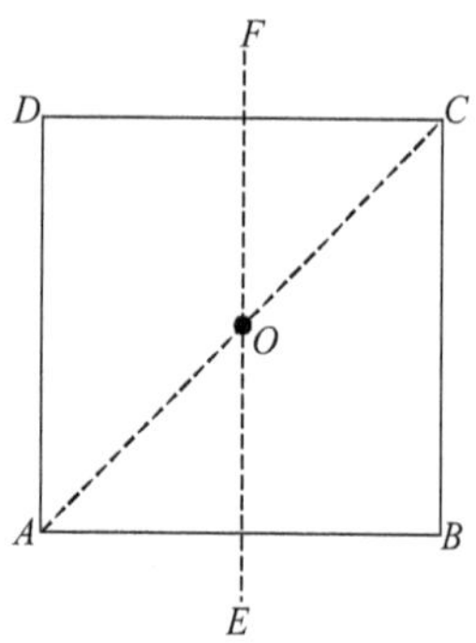

(A) $\sqrt{2}I_{AC}=I_{EF}$ (B) $I_{AD}=3I_{EF}$

(C) $I_{AC}=I_{EF}$ (D) $I_{AC}=\sqrt{2}I_{EF}$

15. A circular disc of radius R is removed from a bigger circular disc of radius 2 R such that the circumferences of the discs coincide. The centre of mass of the new disc is αR from the centre of the bigger disc. The value of α is : **[2007]**

(A) 1/3 (B) 1/2

(C) 1/6 (D) 1/4

16. A round uniform body of radius R, mass M and moment of inertia I, rolls down (without slipping) an inclined plane making an angle θ with the horizontal. Then its acceleration is : **[2007]**

(A) $\frac{g\sin\theta}{1+I/MR^2}$ (B) $\frac{g\sin\theta}{1+MR^2/I}$

(C) $\frac{g\sin\theta}{1-I/MR^2}$ (D) $\frac{g\sin\theta}{1-MR^2/I}$

17. Angular momentum of the particle rotating with a central force is constant due to : **[2007]**

(A) Constant force (B) Constant linear momentum

(C) Zero torque (D) Constant torque

18. Consider a uniform square plate of side 'a' and mass 'm' the moment of inertia of this plate about an axis perpendicular to its plane and passing through one of its corners is : **[2008]**

(A) $\frac{5}{6}ma^2$ (B) $\frac{1}{12}ma^2$

(C) $\frac{7}{12}ma^2$ (D) $\frac{2}{3}ma^2$

19. A thin uniform rod of length l and mass m is swinging freely about a horizontal axis passing through its end. Its maximum angular speed is ω. Its centre of mass rises to a maximum height of : **[2009]**

(A) $\frac{1}{6}\frac{l\omega}{g}$ (B) $\frac{1}{2}\frac{l^2\omega^2}{g}$

(C) $\frac{1}{6}\frac{l^2\omega^2}{g}$ (D) $\frac{1}{3}\frac{l^2\omega^2}{g}$

20. A small particle of mass m is projected at an angle θ with the x-axis with an initial velocity v_0 in the x-y plane as shown in the figure. At a time $t < \frac{v_0\sin\theta}{g}$, the angular momentum of the particle is : **[2010]**

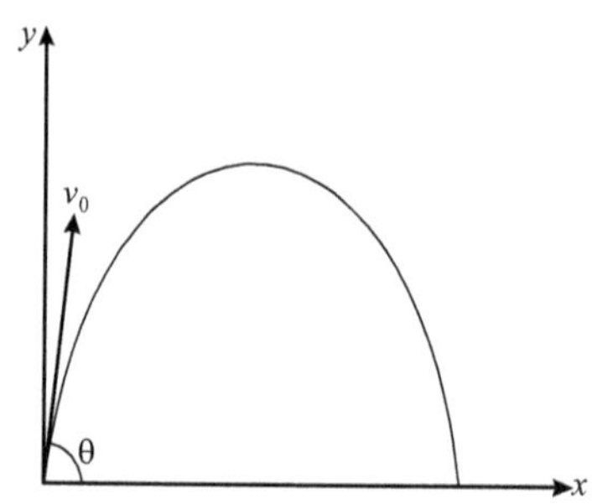

(A) $\frac{1}{2}mg\,v_0\,t^2\cos\theta\,\hat{i}$ (B) $-mg\,v_0\,t^2\cos\theta\,\hat{j}$

(C) $mg\,v_0\,t\cos\theta\,\hat{k}$ (D $-\frac{1}{2}mg\,v_0\,t^2\cos\theta\,\hat{k}$

Where $\hat{i}, \hat{j}$ and $\hat{k}$ are unit vectors along x, y and z-axis respectively.

21. A mass m hangs with the help of a string wrapped around a pulley on a frictionless bearing. The pulley has mass m and radius R. Assuming pulley to be a perfect uniform circular disc, the acceleration of the mass m, if the string does not slip on the pulley, is : **[2011]**

(A) g (B) $\frac{2}{3}g$

(C) $\frac{g}{3}$ (D) $\frac{3}{2}g$

22. A thin horizontal circular disc is rotating about a vertical axis passing through its centre. An insect is at rest at a point near the rim of the disc. The insect now moves along a diameter of the disc to reach its other end. During the journey of the insect, then angular speed of the disc : **[2011]**

(A) Continuously increases
(B) First increases and then decreases
(C) Remains unchanged
(D) Continuously decreases

23. A pulley of radius $2m$ is rotated about its axis by a force $F = (20t - 5t^2)$ newton (where t is measured in seconds) applied tangentially. If the moment of inertia of the pulley about its axis of rotation is 10 kg-m^2, the number of rotations made by the pulley before its direction of motion it reversed, is : **[2011]**

(A) More than 6 but less than 9
(B) More than 9
(C) Less than 3
(D) More than 3 but less than 6

24. A diameter R is cut from a disc of mass M and radius R; the circumference of the cut passes through the centre of the disc. The moment of inertia of the remaining portion of the disc about an axis perpendicular to the disc and passing through its centre is: **[2012 (Online)]**

(A) $\left(\frac{15}{32}\right)MR^2$ (B) $\left(\frac{1}{8}\right)MR^2$

(C) $\left(\frac{3}{8}\right)MR^2$ (D) $\left(\frac{13}{32}\right)MR^2$

25. A solid sphere having mass m and radius r rolls down an inclined plane. Then its kinetic energy is : **[2012 (Online)]**

(A) $\frac{5}{7}$ rotational and $\frac{2}{7}$ translational

(B) $\frac{2}{7}$ rotational and $\frac{5}{7}$ translational

(C) $\frac{2}{5}$ rotational and $\frac{3}{5}$ translational

(D) $\frac{1}{2}$ rotational and $\frac{1}{2}$ translational

26. This question has Statement-1 and Statement-2. Of the four choices given after the Statements, choose the one the best describes the two Statements. **[2012 (Online)]**

Statement-1 : When moment of inertia I of a body rotating about an axis with angular speed ω increases, its angular momentum L is unchanged but the kinetic energy K increases if there is no torque applied on it.

Statement-2 : $L = I\omega$, kinetic energy of rotation $= \frac{1}{2}I\omega^2$:

(A) Statement-1 is true, Statement-2 is true, Statement-2 is not the correct explanation of Statement-1
(B) Statement-1 is false, Statement-2 is true
(C) Statement-1 is true, Statement-2 is true, Statement-2 is correct explanation of the Statement-1
(D) Statement-1 is true, Statement-2 is false

27. A stone of mass m, tied to the end of a string, is whirled around in a circle on a horizontal frictionless table. The length of the string is reduced gradually keeping the angular momentum of the stone about the centre of the circle constant. Then, the tension in the string is given by $T = Ar^n$, where A is a constant, r is the instantaneous radius of the circle. The value of n is equal to : **[2012 (Online)]**

(A) -1 (B) -2
(C) -4 (D) -3

28. A solid sphere is rolling on a surface as shown in figure, with a translational velocity v ms^{-1}. If it is to climb the inclined surface continuing to roll without slipping, then minimum velocity for this to happen is : **[2012 (Online)]**

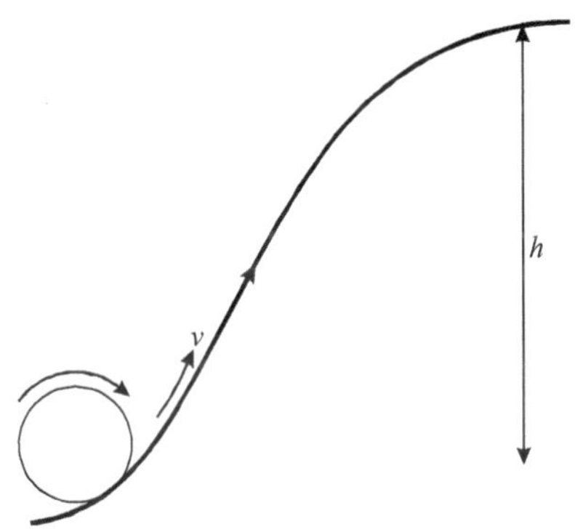

(A) $\sqrt{2gh}$ (B) $\sqrt{\frac{7}{5}gh}$

(C) $\sqrt{\frac{7}{2}gh}$ (D) $\sqrt{\frac{10}{7}gh}$

29. A thick-walled hollow sphere has outside radius R_0. It rolls down an incline without slipping and its speed at the bottom is v_0. Now the incline is waxed, so that it is practically frictionless and the sphere is observed to slide down (without any rolling). Its speed at the bottom is observed to be $5v_0/4$. The radius of gyration of the hollow sphere about an axis through its centre is : **[2012 (Online)]**

(A) $3R_0/2$ (B) $3R_0/4$
(C) $9R_0/16$ (D) $3R_0$

30. Two point masses of mass $m_1 = fM$ and $m_2 = (1-f)M$ $(f<1)$ are in outer space (far from gravitational influence of other objects) at a distance R from each other. They move in circular orbits about their centre of mass with angular velocities ω_1 for m_1 and ω_2 for m_2. In that case : **[2012 (Online)]**

(A) $(1-f)\,\omega_1 = f\omega$
(B) $\omega_1 = \omega_2$ and independent of f
(C) $f\omega_1 = (1-f)\,\omega_2$
(D) $\omega_1 = \omega_2$ and depend on f

31. A hoop of radius r and mass m rotating with an angular velocity ω_0 is placed on a rough horizontal surface. The initial velocity of the centre of the hoop is zero. What will be the velocity of the centre of the hoop when it ceases to slip ? **[2013]**

(A) $\frac{r\omega_0}{4}$ (B) $\frac{r\omega_0}{3}$

(C) $\frac{r\omega_0}{2}$ (D) $r\omega_0$

32. A bullet of mass 10 g and speed 500 m/s is fired into a door and gets embedded exactly at the centre of the door. The door is 1.0 m wide and weighs 12 kg. It is hinged at one end and rotates about a vertical axis practically without friction. The angular speed of the door just after the bullet embeds into it will be : **[2013 (Online)]**

(A) 6.25 rad/sec (B) 0.625 rad/sec
(C) 3.35 rad/sec (D) 0.335 rad/sec

33. A particle of mass 2 kg is moving such that at time t, its position, in meter, is given by $\vec{r}\,(t) = 5\hat{i} - 2t^2\hat{j}$. The angular momentum of the particle at $t = 2$s about the origin in kg m^{-2} s^{-1} is : **[2013 (Online)]**

(A) $-80\,\hat{k}$ (B) $(10\hat{i} - 16\hat{j})$

(C) $-40\,\hat{k}$ (D) $40\,\hat{k}$

34. A tennis ball (treated as hollow spherical shell) starting from O rolls down a hill. At point A the ball becomes air borne leaving at an angle of 30° with the horizontal. The ball strikes the ground at B. What is the value of the distance AB ? (Moment of inertia of a spherical shell of mass m and radius R about its diameter $= 2/3mR^2$) : **[2013 (Online)]**

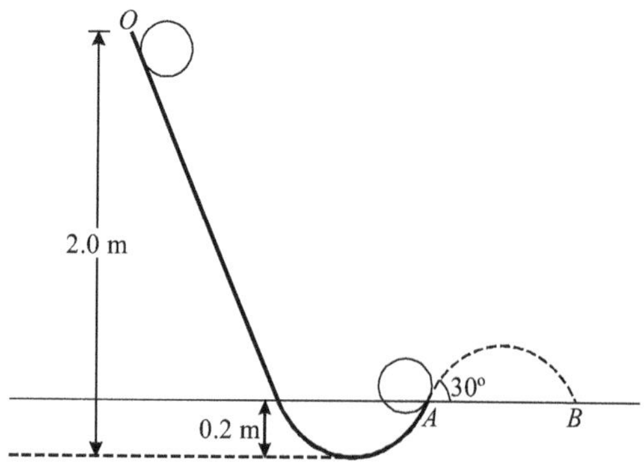

(A) 1.87m (B) 2.08m
(C) 1.57m (D) 1.77m

35. A ring of mass M and radius R is rotating about its axis with angular velocity ω. Two identical bodies each of mass m are now gently attached at the two ends of a diameter of the ring. Because of this, the kinetic energy loss will be : **[2013 (Online)]**

(A) $\frac{m(M+2m)}{M}\omega^2 R^2$ (B) $\frac{Mm}{(M+m)}\omega^2 R^2$

(C) $\frac{Mm}{(M+2m)}\omega^2 R^2$ (D) $\frac{(M+m)M}{(M+2m)}\omega^2 R^2$

36. A mass 'm' is supported by a massless string wound around a uniform hollow cylinder of mass m and radius R. If the string does not slip on the cylinder, with what acceleration will the mass fall on release ? **[2014]**

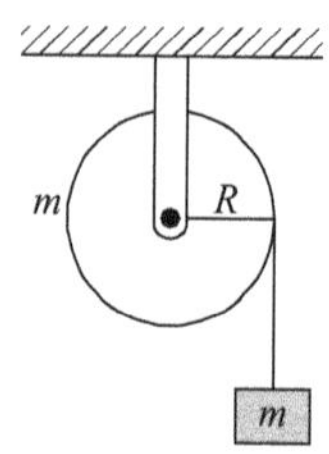

(A) $\frac{2g}{3}$

(B) $\frac{g}{2}$

(C) $\frac{5g}{6}$

(D) g

37. A bob of mass m attached to an inextensible string of length l is suspended from a vertical support. The bob rotates in a horizontal circle with an angular speed ω rad/s about the vertical. About, the point of suspension : **[2014]**

(A) Angular momentum is conserved
(B) Angular momentum changes in magnitude but not in direction
(C) Angular momentum changes in direction but not in magnitude
(D) Angular momentum changes both in direction and magnitude

38. Consider a cylinder of mass M resting on a rough horizontal rug that is pulled out from under it with acceleration 'a' perpendicular to the axis of the cylinder. What is $F_{friction}$ at point P ? It is assumed that the cylinder does not slip : **[2014 (Online)]**

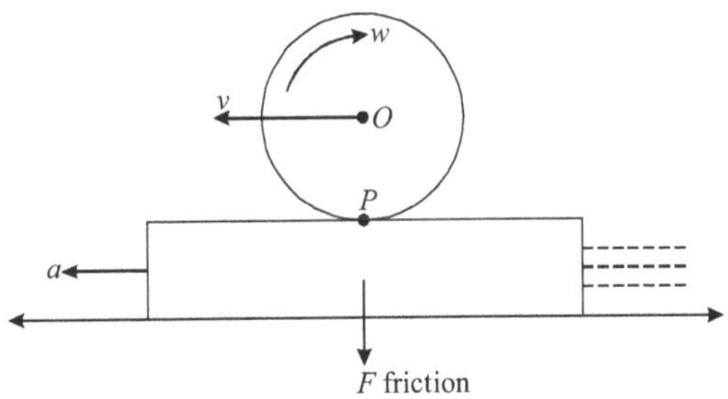

(A) Mg (B) Ma

(C) $\frac{Ma}{2}$ (D) $\frac{Ma}{3}$

39. A cylinder of mass M_C and sphere of mass M_S are placed at point A and B of two inclines, respectively (See figure). If they roll on the incline without slipping such that their acceleration are the same, then the ratio $\frac{\sin\theta_C}{\sin\theta_S}$ is : **[2014 (Online)]**

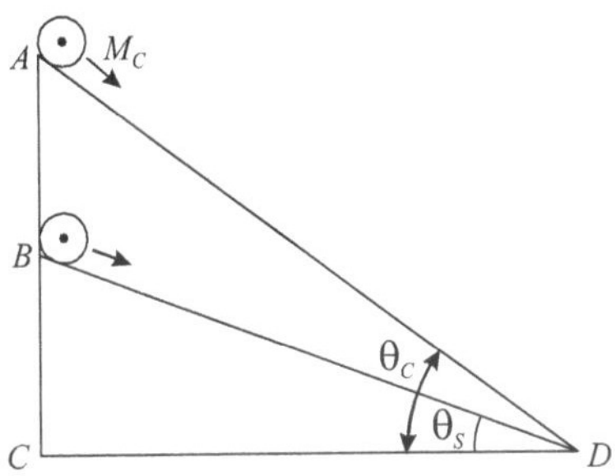

(A) $\sqrt{\frac{8}{7}}$ (B) $\sqrt{\frac{15}{14}}$

(C) $\frac{8}{7}$ (D) $\frac{15}{14}$

40. A ball of mass 160 g is thrown up at an angle of 60° to the horizontal at a speed of 10 ms^{-1}. The angular momentum of the ball at the highest point of the trajectory with respect to the point from which the ball is thrown is nearly ($g = 10\ ms^{-2}$) : **[2014 (Online)]**

(A) 1.73 kg m^2/s (B) 3.0 kg m^2/s

(C) 3.46 kg m^2/s (D) 6.9 kg m^2/s

41. A particle of mass m is moving in a circular path of radius a, with a constant velocity v as shown in the figure. The centre of circle is marked by 'C'. The angular momentum from the origin O can be written as : **[2014 (Online)]**

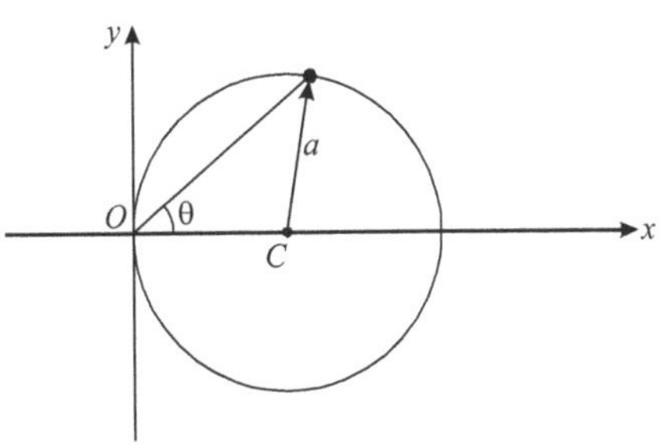

(A) $mva\,(1 + \cos 2\theta)$ (B) $mva\,(1 + \cos\theta)$

(C) $mva \cos 2\theta$ (D) mva

42. From a solid sphere of mass M and radius R a cube of maximum possible volume is cut. Moment of inertia of cube about an axis passing through its center and perpendicular to one of its faces is : **[2015]**

(A) $\frac{MR^2}{32\sqrt{2}\pi}$ (B) $\frac{MR^2}{16\sqrt{2}\pi}$

(C) $\frac{4MR^2}{9\sqrt{3}\pi}$ (D) $\frac{4MR^2}{3\sqrt{3}\pi}$

43. A uniform solid cylindrical roller of mass 'm' is being pulled on a horizontal surface with force F parallel to the surface and applied at its centre. If the acceleration of the cylinder is 'a' and it rolling without slipping then the value of 'F' is : **[2015 (Online)]**

(A) ma (B) $\frac{5}{3}ma$

(C) $\frac{3}{2}ma$ (D) $2ma$

44. A particle of mass 2 kg is on a smooth horizontal table and moves in a circular path of radius 0.6 m. The height of the table from the ground is 0.8 m. If the angular speed of the particle is 12 rad s^{-1}, the magnitude of its angular momentum about a point on the ground right under the centre of the circle is : **[2015 (Online)]**

(A) 14.4 kg $m^2 s^{-1}$ (B) 8.64 kg $m^2 s^{-1}$

(C) 0.16 kg $m^2 s^{-1}$ (D) 11.52 kg $m^2 s^{-1}$

45. Consider a thin uniform square sheet made of a rigid material. If its side is 'a' mass m and moment of inertia I about one of its diagonals, then : **[2015 (Online)]**

(A) $I > \frac{ma^2}{12}$ (B) $\frac{ma^2}{24} < I < \frac{ma^2}{12}$

(C) $I = \frac{ma^2}{24}$ (D) $I = \frac{ma^2}{12}$

46. A particle of mass m is moving along the side of a square of side 'a', with a uniform speed v in x-y plane as shown in the figure : **[2016]**

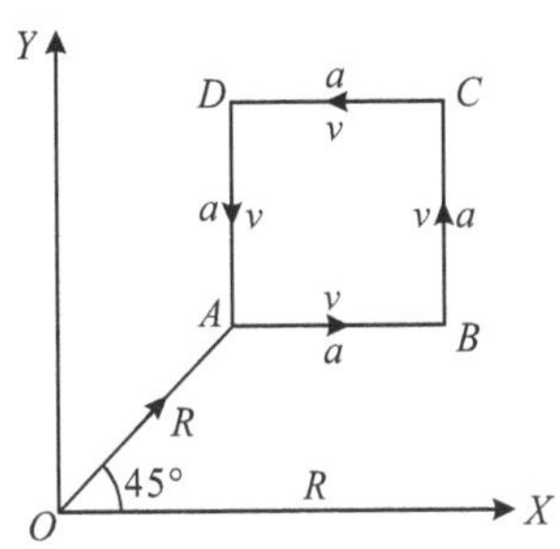

Which of the following statements is false for the angular momentum $\vec{L}$ about the origin ?

(A) $\vec{L} = mv\left[\frac{R}{\sqrt{2}} + a\right]\hat{k}$ when the particle is moving from B to C

(B) $\vec{L} = \frac{mv}{\sqrt{2}}R\hat{k}$ when the particle is moving from D to A

(C) $\vec{L} = -\frac{mv}{\sqrt{2}}R\hat{k}$ when the particle is moving from A to B

(D) $\vec{L} = mv\left[\frac{R}{\sqrt{2}} - a\right]\hat{k}$ when the particle is moving from C to D

47. A roller is made by joining together two cones at their vertices O. It is kept on two rails AB and CD, which are placed asymmetrically (see figure), with its axis perpendicular to CD and its centre O at the centre of line joining AB and CD (see figure). It is given a light push so that its starts rolling with its centre O moving parallel to CD in the direction shown. As it moves, the roller will tend to : **[2016]**

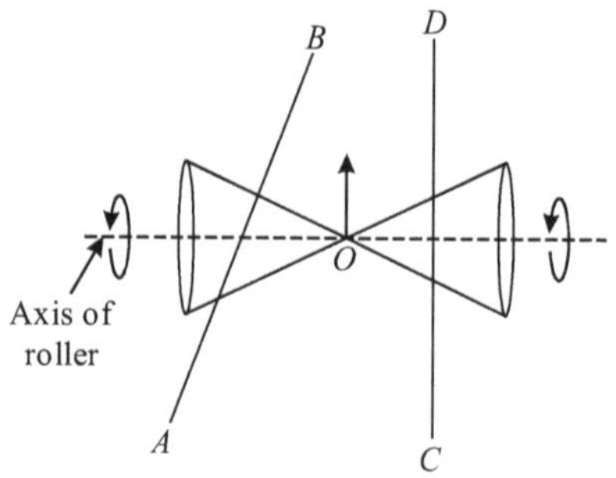

(A) Go straight
(B) Turn left and right alternately
(C) Turn left
(D) Turn right

48. A cubical block of side 30 cm is moving with velocity 2 ms^{-1} on a smooth horizontal surface. The surface has a bump at a point O as shown in figure. The angular velocities (in rad/s) of the block immediately after it hits the bump, is : **[2016 (Online)]**

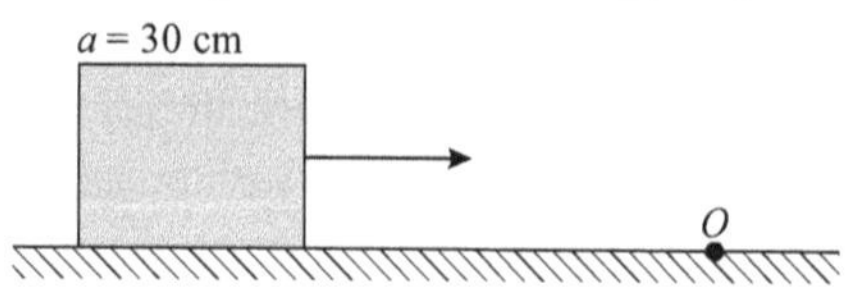

(A) 13.3 (B) 5.0
(C) 9.4 (D) 6.7

49. The moment of inertia of a uniform cylinder of length l and radius R about its perpendicular bisector is I. What is the ratio l/R such that the moment of inertia is minimum ? **[2017]**

(A) $\frac{\sqrt{3}}{2}$ (B) 1

(C) $\frac{3}{\sqrt{2}}$ (D) $\sqrt{\frac{3}{2}}$

50. A slender uniform rod of mass M and length l is pivoted at one end so that it can rotate in a vertical plane (see figure.) There is negligible friction at the pivot. The free end is held vertically above the pivot and then released. The angular acceleration of the rod when it make an angle θ with the vertical is : **[2017]**

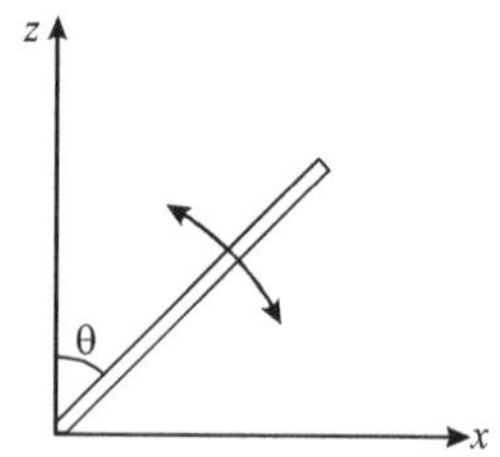

(A) $\frac{3g}{3l}\sin\theta$ (B) $\frac{3g}{3l}\cos\theta$

(C) $\frac{2g}{3l}\cos\theta$ (D) $\frac{3g}{2l}\sin\theta$

51. A circular hole of radius $R/4$ is made in a thin uniform disc having mass M and radius R, as shown in figure. The moment of inertia of the remaining portion of the disc about an axis passing through the point O and perpendicular to the plane of the disc is : **[2017 (Online)]**

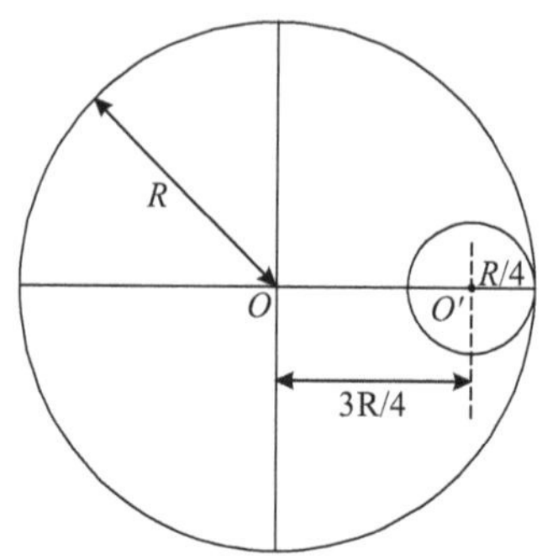

(A) $\frac{219MR^2}{256}$ (B) $\frac{237MR^2}{512}$

(C) $\frac{19MR^2}{512}$ (D) $\frac{197MR^2}{256}$

52. Moment of inertia of an equilateral triangular lamina ABC, about the axis passing through its centre O and perpendicular to its plane is I_0 as shown in the figure. A cavity DEF is cut out from the lamina, where D, E, F are the mid points of the sides. Moment of inertia of the remaining part of lamina about the same axis is : **[2017 (Online)]**

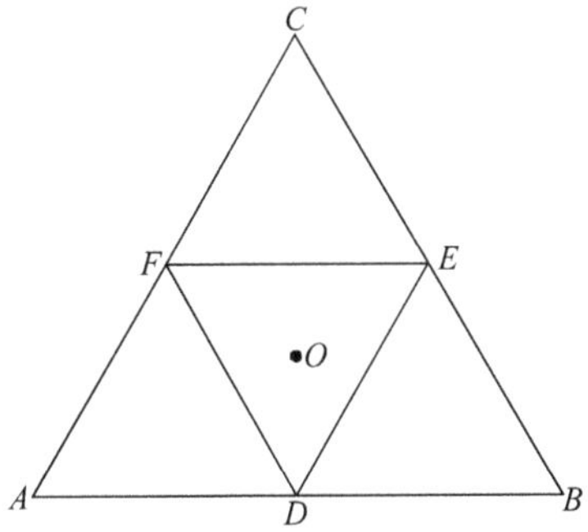

(A) $\frac{7}{8}I_0$ (B) $\frac{15}{16}I_0$

(C) $\frac{3I_0}{4}$ (D) $\frac{31I_0}{32}$

53. The machine as shown has 2 rods of length 1 m connected by a pivot at the top. The end of one rod is connected to the floor by a stationary pivot and the end of the other rod has a roller that rolls along the floor in a slot.
As the roller goes back and forth, a 2 kg weight moves up and down. If the roller is moving towards right at a constant speed, the weight moves up with a : **[2017 (Online)]**

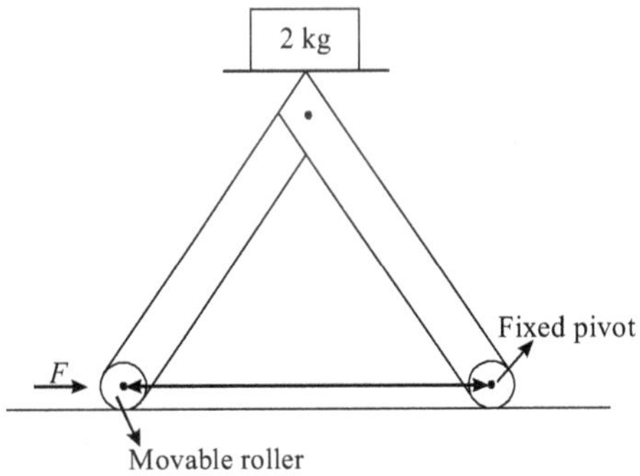

(A) Constant speed
(B) Decreasing speed
(C) Increasing speed
(D) Speed which is 3/4th of that of the roller when the weight is 0.4 m above the ground

54. In a physical balance working on the principle of moments, when 5 mg weight is placed on the left pan, the beam becomes horizontal. Both the empty pans of the balance are of equal mass. Which of the following statements is correct ? **[2017 (Online)]**

(A) Left arm is longer than the right arm
(B) Both the arms are of same length
(C) Left arm is shorter than the right arm
(D) Every object that is weighed using this balance appears lighter than its actual weight

55. A particle is moving with a uniform speed in a circular orbit of radius R in a central force inversely proportional to the nth power of R. If the period of rotation of the particle is T, then : **[2018]**

(A) $T \propto R^{3/2}$ for any n (B) $T \propto R^{n/2+1}$
(C) $T \propto R^{(n+1)/2}$ (D) $T \propto R^{n/2}$

56. Seven identical circular planar disks, each of mass M and radius R are welded symmetrically as shown. The moment of inertia of the arrangement about the axis normal to the plane and passing through the point P is : **[2018]**

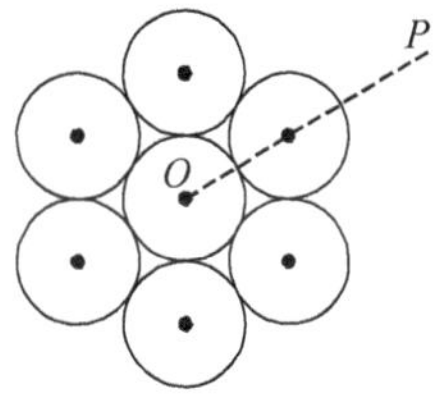

(A) $\frac{19}{2}MR^2$ (B) $\frac{55}{2}MR^2$

(C) $\frac{73}{2}MR^2$ (D) $\frac{181}{2}MR^2$

57. From a uniform circular disc of radius R and mass $9M$, a small disc of radius $R/3$ is removed as shown in the figure. The moment of inertia of the remaining disc about an axis perpendicular to the plane of the disc and passing, through centre of disc is : **[2018]**

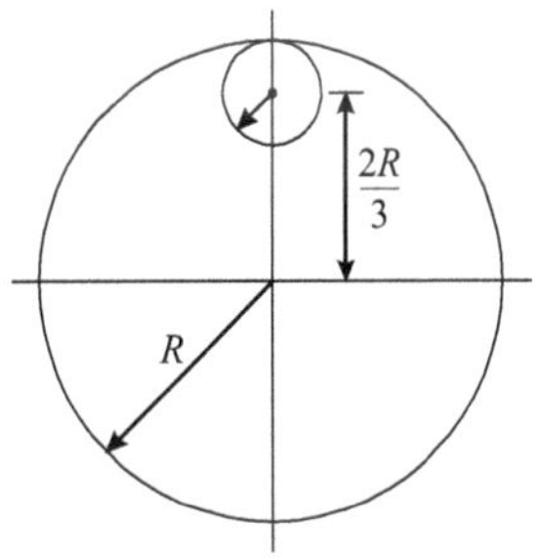

(A) $4MR^2$ (B) $\frac{40}{9}MR^2$

(C) $10MR^2$ (D) $\frac{37}{9}MR^2$

58. A force of 40N acts on a point B a the end of an L-shaped object, as shown in the figure. The angle θ that will produce maximum moment of the force about point A is given by : **[2018 (Online)]**

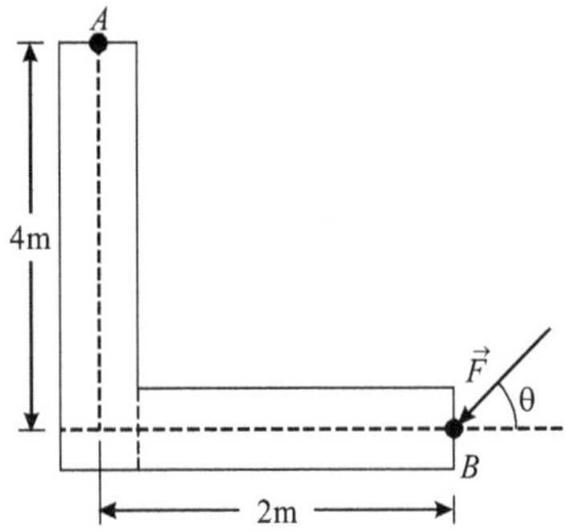

(A) $\tan\theta = \frac{1}{4}$ (B) $\tan\theta = 2$

(C) $\tan\theta = \frac{1}{2}$ (D) $\tan\theta = 4$

59. A uniform rod AB is suspended from a point X, at a variable distance from x from A, as shown. To make the rod horizontal, a mass m is suspended from its end A. A set of (m, x) values is recorded. The appropriate variable that give a straight line, when plotted, are : **[2018 (Online)]**

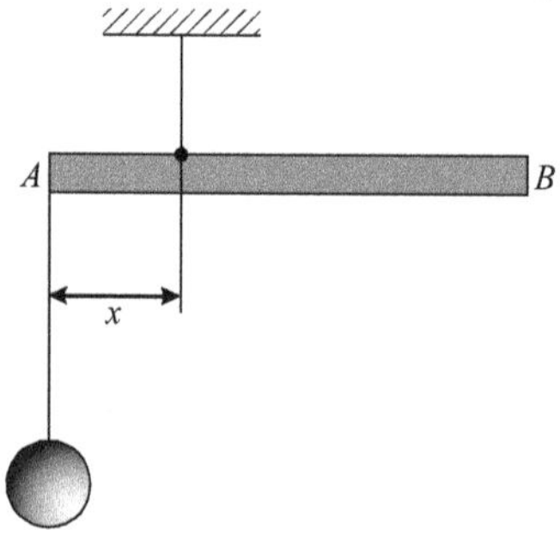

(A) $m, \frac{1}{x}$ (B) $m, \frac{1}{x^2}$

(C) m, x (D) m, x^2

60. A thin uniform bar of length L and mass $8m$ lies on a smooth horizontal table. Two point masses m and $2m$ moving in the same horizontal plane from opposite sides of the bar with speeds $2v$ and v respectively. The masses stick to the bar after collision at a distance $L/3$ and $L/6$ respectively from the centre of the bar. If the bar starts rotating about its center of mass as a result of collision, the angular speed of the bar will be : **[2018 (Online)]**

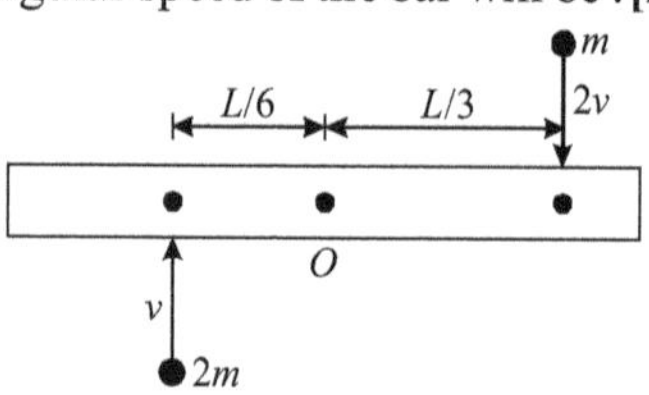

(A) $\frac{v}{6L}$ (B) $\frac{6v}{5L}$

(C) $\frac{3v}{5L}$ (D) $\frac{v}{5L}$

61. A thin circular disc is in the xy plane as shown in the figure. The ratio of its moment of inertia about z and z' axes will be : **[2018 (Online)]**

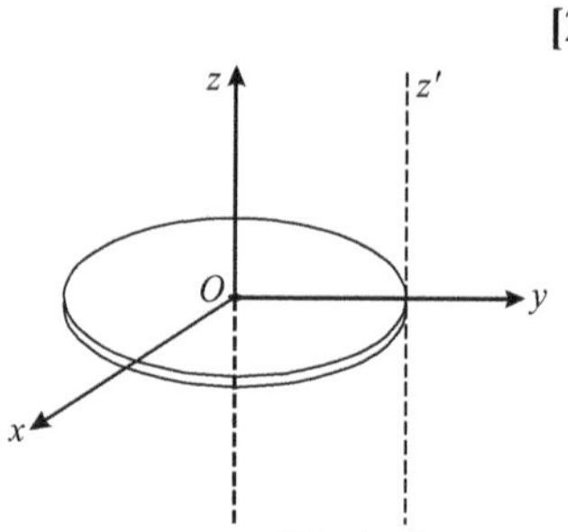

(A) 1 : 2 (B) 1 : 4

(C) 1 : 3 (D) 1 : 5

62. A thin rod MN, free the rotate in the vertical plane about the fixed end N, is held horizontal. When the end M is released the speed of this end, when the rod makes an angle α with the horizontal, will be proportional to (see figure) : **[2018 (Online)]**

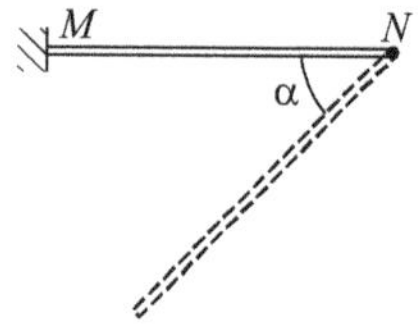

(A) $\sqrt{\cos\alpha}$ (B) $\cos\alpha$

(C) $\sin\alpha$ (D) $\sqrt{\sin\alpha}$

63. An L-shaped object, made of thin rods of uniform mass density, is suspended with a string as shown in figure. If $AB = BC$, and the angle made by AB with downward vertical is θ, then : **[9 Jan. 2019 (Morning)]**

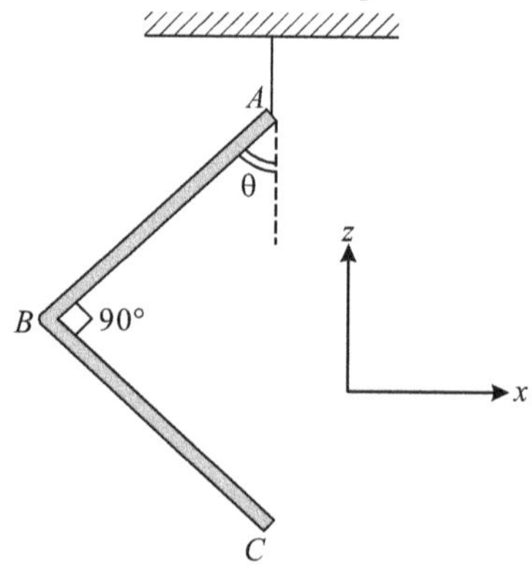

(A) $\tan\theta = \frac{2}{\sqrt{3}}$ (B) $\tan\theta = \frac{1}{3}$

(C) $\tan\theta = \frac{1}{2}$ (D) $\tan\theta = \frac{1}{2\sqrt{3}}$

64. A rod of length 50 cm is pivoted at one end. It is raised such that it makes an angle of 30° from the horizontal as shown and released from rest. Its angular speed when it passes through the horizontal (in rad s^{-1}) will be ($g = 10\ \text{ms}^{-2}$) : **[9 Jan. 2019 (Evening)]**

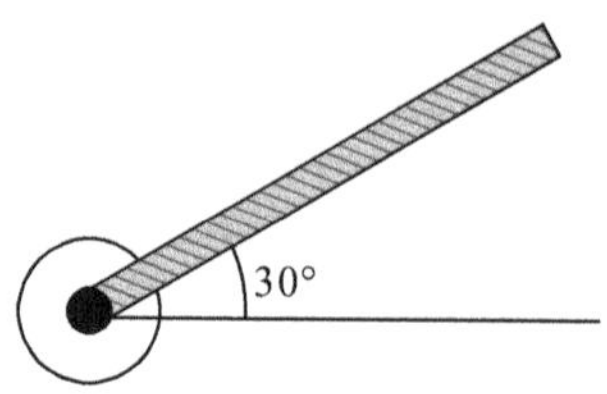

(A) $\sqrt{30}$ (B) $\sqrt{\frac{30}{2}}$

(C) $\frac{\sqrt{30}}{2}$ (D) $\frac{\sqrt{20}}{3}$

65. To mop-clean a floor, a cleaning machine presses a circular mop of radius R vertically down with a total force F and rotates it with a constant angular speed about its axis. If the force F

is distributed uniformly over the mop and if coefficient of friction between the mop and the floor is μ, the torque, applied by the machine on the mop is : **[10 Jan. 2019 (Morning)]**

(A) $\frac{2}{3}\mu FR$ (B) $\mu FR/3$
(C) $\mu FR/2$ (D) $\mu FR/6$

66. A homogeneous solid cylindrical roller of radius R and mass M is pulled on a cricket pitch by a horizontal force. Assuming rolling without slipping, angular acceleration of the cylinder is: **[10 Jan. 2019 (Morning)]**

(A) $\frac{3F}{2mR}$ (B) $\frac{F}{3mR}$
(C) $\frac{2F}{3mR}$ (D) $\frac{F}{2mR}$

67. A rigid massless rod of length $3l$ has two masses attached at each end as shown in the figure. The rod is pivoted at point P on the horizontal axis (see figure). When released from initial horizontal position, its instantaneous angular acceleration will be : **[10 Jan. 2019 (Evening)]**

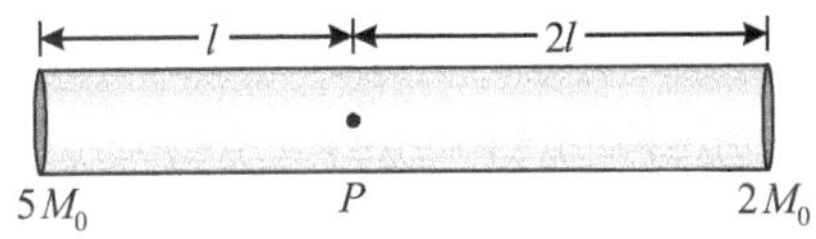

(A) $\frac{g}{2l}$ (B) $\frac{7g}{3l}$
(C) $\frac{g}{13l}$ (D) $\frac{g}{3l}$

68. Two identical spherical balls of mass M and radius R each are stuck on two ends of a rod of length $2R$ and mass M (see figure). The moment of inertia of the system about the axis passing perpendicularly through the centre of the rod is : **[10 Jan. 2019 (Evening)]**

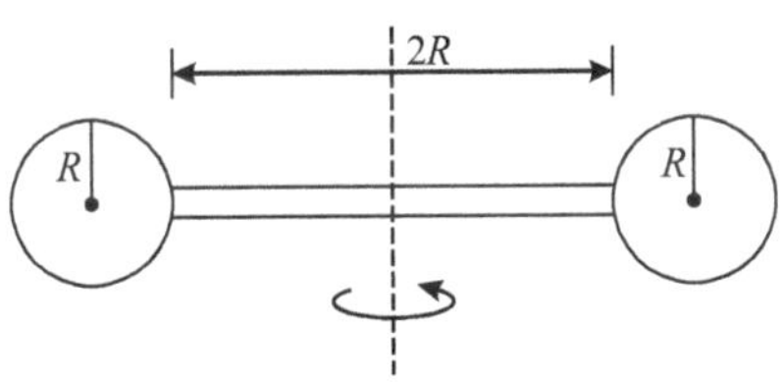

(A) $\frac{152}{15}MR^2$ (B) $\frac{17}{15}MR^2$
(C) $\frac{137}{15}MR^2$ (D) $\frac{209}{15}MR^2$

69. An equilateral triangle ABC is cut from a thin solid sheet of wood. (see figure) D, E and F are the mid-points of its sides as shown and G is the centre of the triangle. The moment of inertia of the triangle about an axis passing through G and perpendicular to the plane of the triangle is I_0. It the smaller triangle DEF is removed from ABC, the moment of inertia of the remaining figure about the same axis is I. Then : **[11 Jan. 2019 (Morning)]**

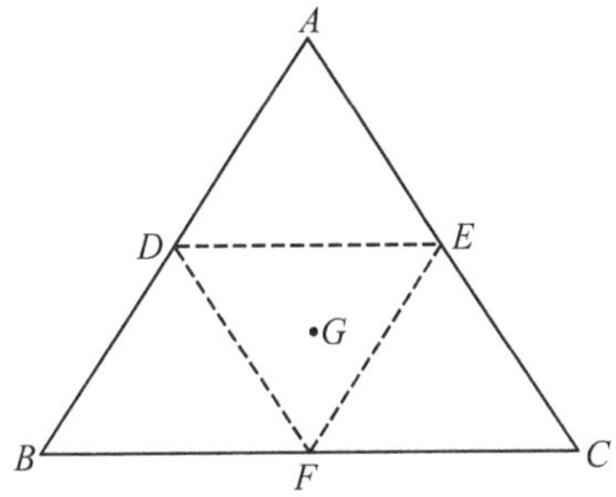

(A) $I = \frac{9}{16}I_0$ (B) $I = \frac{3}{4}I_0$
(C) $I = \frac{I_0}{4}$ (D) $I = \frac{15}{16}I_0$

70. A slab is subjected to two forces $\vec{F_1}$ and $\vec{F_2}$ of same magnitude F as shown in the figure. Force $\vec{F_2}$ is in XY-plane while force F_1 acts along z-axis at the point $(2\vec{i} + 3\vec{j})$. The moment of these forces about point O will be : **[11 Jan. 2019 (Morning)]**

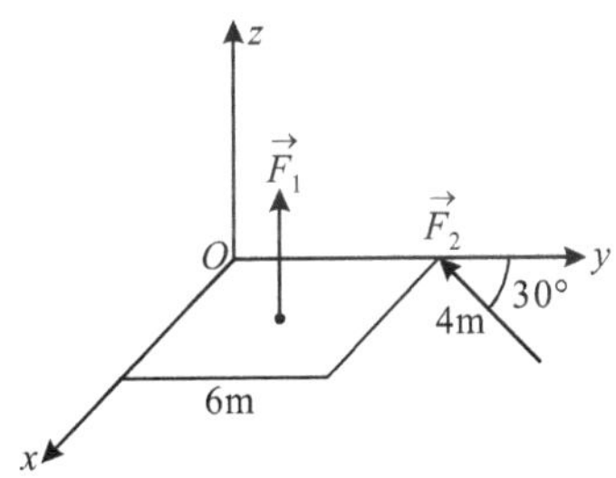

(A) $(3\hat{i} - 2\hat{j} - 3\hat{k})F$ (B) $(3\hat{i} + 2\hat{j} + 3\hat{k})F$
(3) $(3\hat{i} + 2\hat{j} - 3\hat{k})F$ (D) $(3\hat{i} - 2\hat{j} + 3\hat{k})F$

71. A string is wound around a hollow cylinder of mass 5 kg and radius 0.5 m. If the string is now pulled with a horizontal force of 40 N, and the cylinder is rolling without slipping on a horizontal surface (see figure), then the angular acceleration of the cylinder will be (Neglect the mass and thickness of the string) : **[11 Jan. 2019 (Evening)]**

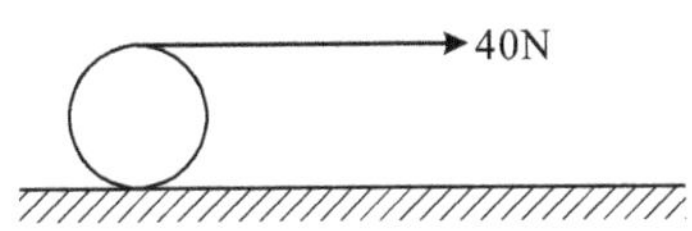

(A) 12 rad/s^2 (B) 16 rad/s^2
(C) 10 rad/s^2 (D) 20 rad/s^2

72. The magnitude of torque on a particle of mass 1 kg is 2.5 Nm about the origin. If the force acting on it is 1 N, and the distance of the particle from the origin is 5 m, the angle between the force and the position vector is (in radians) : **[11 Jan. 2019 (Evening)]**

(A) $\frac{\pi}{8}$ (B) $\frac{\pi}{6}$

(C) $\frac{\pi}{4}$ (D) $\frac{\pi}{3}$

73. A circular disc D_1 of mass M and radius R has two identical discs D_2 and D_3 of the same mass M and radius R attached rigidly at its opposite ends (see figure). The moment of inertia of the system about the axis OO', passing through the centre of D_1, as shown in the figure, will be: **[11 Jan. 2019 (Evening)]**

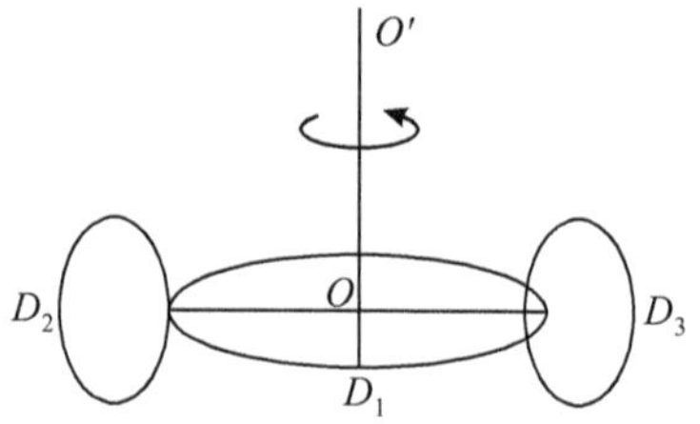

(A) $3MR^2$ (B) $\frac{2}{3}MR^2$

(C) MR^2 (D) $\frac{4}{5}MR^2$

74. Let the moment of inertia of a hollow cylinder of length 30 cm (inner radius 10 cm and outer radius 20 cm), about its axis be I. The radius of a thin cylinder of the same mass such that its moment of inertia about its axis is also I, is : **[12 Jan. 2019 (Morning)]**

(A) 12 cm (B) 18 cm

(C) 16 cm (D) 14 cm

75. A particle of mass 20g is released with an initial velocity 5 m/s along the curve from the point A, as shown in the figure. The point A is at height h from point B. The particle slides along the frictionless surface. When the particle reaches point B, its angular momentum about O will be : [Take $g = 10$ m/s^2] **[12 Jan. 2019 (Evening)]**

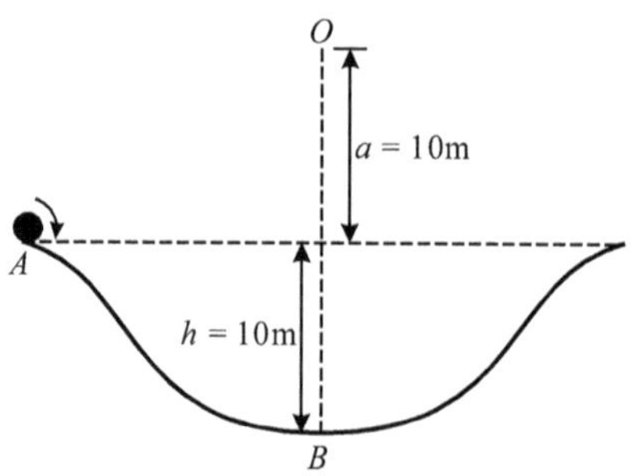

(A) 8 kg-m^2/s (B) 6 kg-m^2/s

(C) 3 kg-m^2/s (D) 2 kg-m^2/s

76. The moment of inertia of a solid sphere, about an axis parallel to its diameter and at a distance of x from it, is $I(x)'$. Which one of the graphs represents the variation of $I(x)$ with x correctly? **[12 Jan. 2019 (Evening)]**

(A) 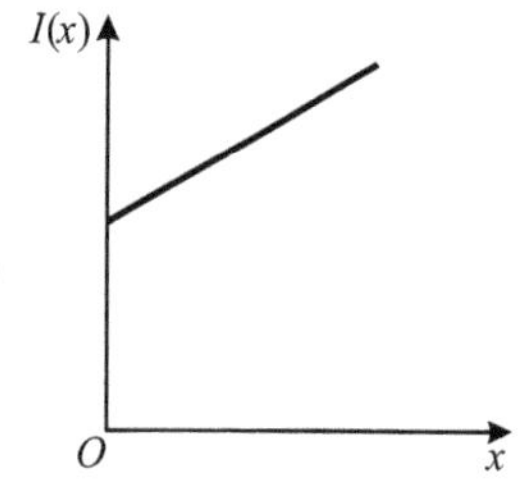

(B) 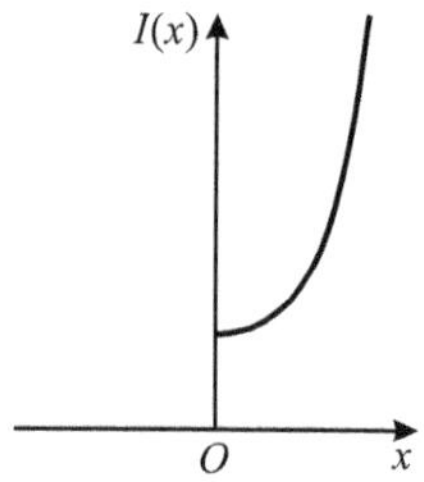

(C) 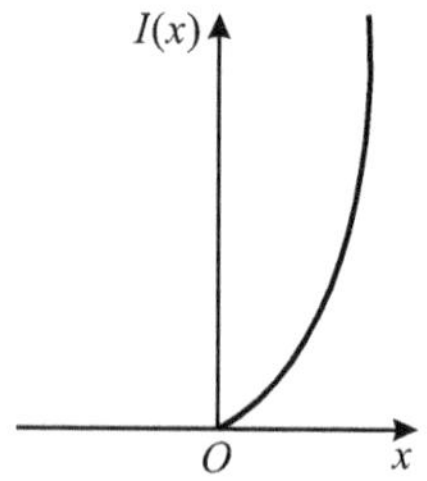

(D) 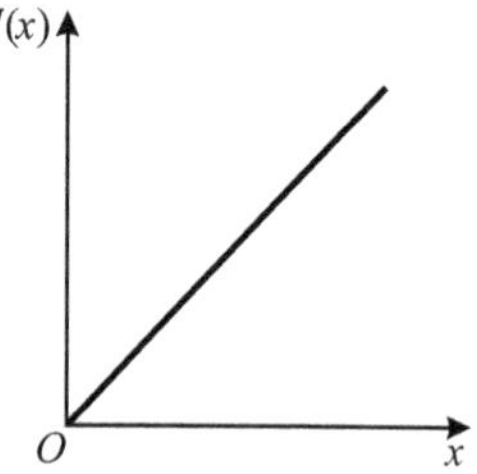

77. A thin circular plate of mass M and radius R has its density varying as $\rho(r) = \rho_0 r$ with ρ_0 as constant and r is the distance from its centre. The moment of Inertia of the circular plate about an axis perpendicular to the plate and passing through its edge is $I = aMR^2$. The value of the coefficient a is : **[8 April 2019 (Morning)]**

(A) $\frac{3}{2}$ (B) $\frac{1}{2}$

(C) $\frac{3}{5}$ (D) $\frac{8}{5}$

78. A solid sphere and solid cylinder of identical radii approach an incline with the same linear velocity (see figure). Both roll without slipping all throughout. The two climb maximum heights h_{sph} and h_{cyl} on the incline. The ratio $\frac{h_{sph}}{h_{cyl}}$ is given by : **[8 April 2019 (Evening)]**

(A) $\frac{14}{15}$ (B) $\frac{4}{5}$

(C) 1 (D) $\frac{2}{\sqrt{5}}$

79. A rectangular solid box of length 0.3 m is held horizontally, with one of its sides on the edge of a platform of height 5 m. When released, it slips off the table in a very short time $\tau = 0.01$ s, remaining essentially horizontal. The angle by which it would rotate when it hits the ground will be (in radians) close to : **[8 April 2019 (Evening)]**

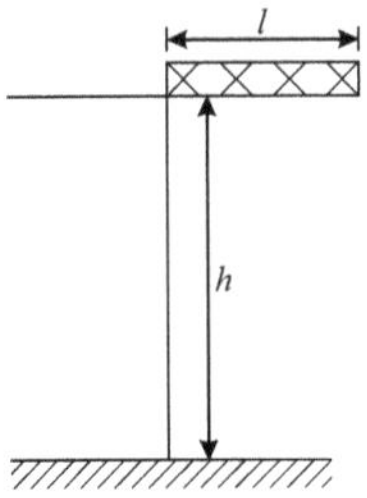

(A) 0.02
(B) 0.28
(C) 0.5
(D) 0.3

80. The following bodies are made to roll up (without slipping) the same inclined plane from a horizontal plane : (i) a ring of radius R, (ii) a solid cylinder of radius $\frac{R}{2}$ and (iii) a solid sphere of radius $\frac{R}{4}$. If in each case, the speed of the centre of mass at the bottom of the incline is same, the ratio of the maximum heights they climb is : **[9 April 2019 (Morning)]**

(A) $4:3:2$
(B) $14:15:20$
(C) $10:15:7$
(D) $2:3:4$

81. A stationary horizontal disc is free to rotate about its axis. When a torque is applied on it, its kinetic energy as a function of θ, where θ is the angle by which it has rotated, is given as $k\theta^2$. If its moment of inertia is I then the angular acceleration of the disc is : **[9 April 2019 (Morning)]**

(A) $\frac{k}{2I}\theta$
(B) $\frac{k}{I}\theta$
(C) $\frac{k}{4I}\theta$
(D) $\frac{2k}{I}\theta$

82. Moment of inertia of a body about a given axis is 1.5 kg m^2. Initially the body is at rest. In order to produce a rotational kinetic energy of 1200 J, the angular accleration of 20 rad/s^2 must be applied about the axis for a duration of : **[9 April 2019 (Evening)]**

(A) 2s
(B) 5s
(C) 2.5s
(D) 3s

83. A thin smooth rod of length L and mass M is rotating freely with angular speed ω_0 about an axis perpendicular to the rod and passing through its center. Two beads of mass m and negligible size are at the center of the rod initially. The beads are free to slide along the rod. The angular speed of the system, when the beads reach the opposite ends of the rod, will be : **[9 April 2019 (Evening)]**

(A) $\frac{M\omega_0}{M+3m}$
(B) $\frac{M\omega_0}{M+m}$
(C) $\frac{M\omega_0}{M+2m}$
(D) $\frac{M\omega_0}{M+6m}$

84. A thin disc of mass M and radius R has mass per unit area $\sigma(r) = kr^2$ where r is the distance from its centre. Its moment of inertia about an axis going through its centre of mass and perpendicular to its plane is : **[10 April 2019 (Morning)]**

(A) $\frac{MR^2}{6}$
(B) $\frac{MR^2}{3}$
(C) $\frac{2MR^2}{3}$
(D) $\frac{MR^2}{2}$

85. Two coaxial discs, having moments of inertia I_1 and $\frac{I_1}{2}$, are rotating with respective angular velocities ω_1 and $\frac{\omega_1}{2}$, about their common axis. They are brought in contact with each other and thereafter they rotate with a common angular velocity. If E_f and E_i are the final and initial total energies, then $(E_f - E_i)$ is : **[10 April 2019 (Morning)]**

(A) $\frac{I_1\omega_1^2}{12}$
(B) $\frac{3}{8}I_1\omega_1^2$
(C) $\frac{I_1\omega_1^2}{6}$
(D) $\frac{I_1\omega_1^2}{24}$

86. A particle of mass m is moving along a trajectory given by
$x = x_0 + a\cos\omega_1 t$
$y = y_0 + b\sin\omega_2 t$
The torque, acting on the particle about the origin, at $t = 0$ is : **[10 April 2019 (Morning)]**

(A) $m(-x_0 b + y_0 a)\,\omega_1^2\hat{k}$
(B) $+my_0 a\,\omega_1^2\hat{k}$
(C) $-m(x_0 b\omega_2^2 - y_0 a\omega_1^2)\hat{k}$
(D) Zero

87. A metal coin of mass $5g$ and radius 1cm is fixed to a thin stick AB of negligible mass as shown in the figure. The system is initially at rest. The constant torque, that will make the system rotate about AB at 25 rotations per second in 5 s, is close to : **[10 April 2019 (Evening)]**

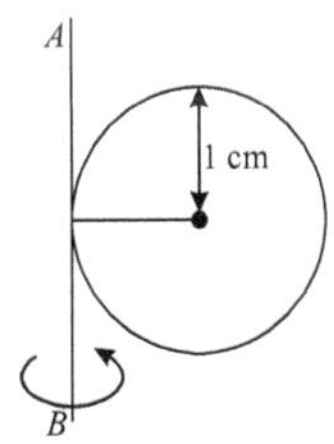

(A) 4.0×10^{-6}Nm
(B) 2.0×10^{-5}Nm
(C) 1.6×10^{-5}Nm
(D) 7.9×10^{-6}Nm

88. The time dependence of the position of a particle of mass $m = 2$ is given by $\vec{r}(t) = 2t\hat{i} - 3t^2\hat{j}$. Its angular momentum, with respect to the origin, at time $t = 2$ is : **[10 April 2019 (Evening)]**

(A) $36\hat{k}$ (B) $-34(\hat{k}-\hat{i})$

(C) $48(\hat{i}+\hat{j})$ (D) $-48\hat{k}$

89. A solid sphere of mass M and radius R is divided into two unequal parts. The first part has a mass of $\frac{7M}{8}$ and is converted into a uniform disc of radius $2R$. The second part is converted into a uniform solid sphere. Let I_1 be the moment of inertia of the disc about its axis and I_2 be the moment of inertia of the new sphere about its axis. The ratio I_1/I_2 is given by :

[10 April 2019 (Evening)]

(A) 185 (B) 65

(C) 285 (D) 140

90. A circular disc of radius b has a hole of radius a at its centre (see figure). If the mass per unit area of the disc varies as $\left(\frac{\sigma_0}{r}\right)$, then the radius of gyration of the disc about its axis passing through the centre is : **[12 April 2019 (Morning)]**

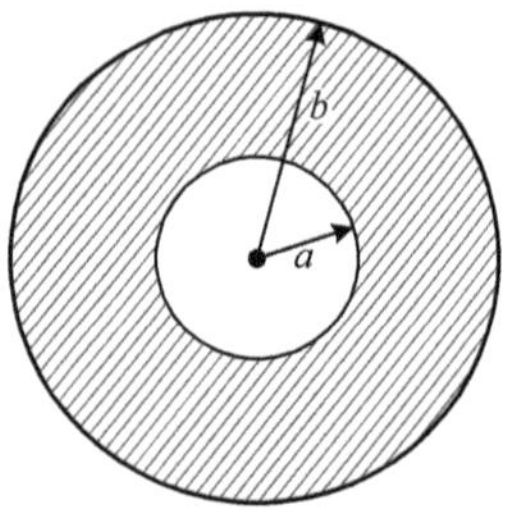

(A) $\frac{a+b}{2}$ (B) $\frac{a+b}{3}$

(C) $\sqrt{\frac{a^2+b^2+ab}{2}}$ (D) $\sqrt{\frac{a^2+b^2+ab}{3}}$

91. A uniform rod of length l is being rotated in a horizontal plane with a constant angular speed about an axis passing through one of its ends. If the tension generated in the rod due to rotation is $T(x)$ at a distance x from the axis, then which of the following graphs depicts it most closely ?

[12 April 2019 (Morning)]

(A)

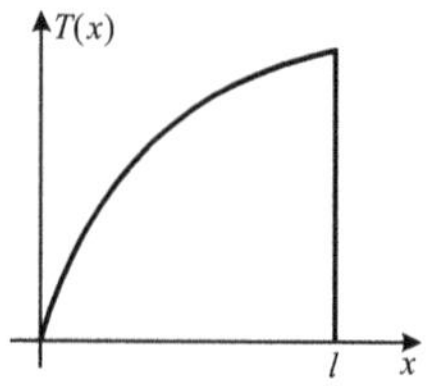

(B)

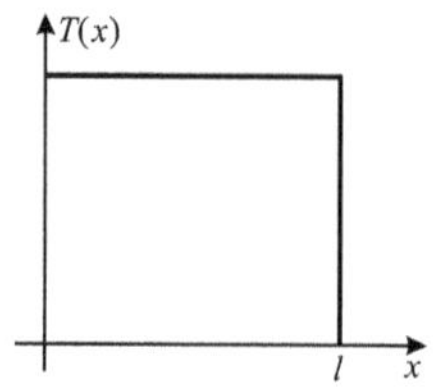

(C)

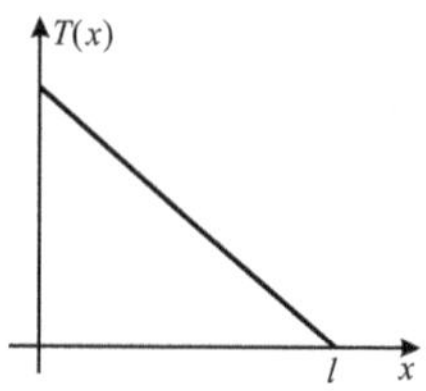

(D)

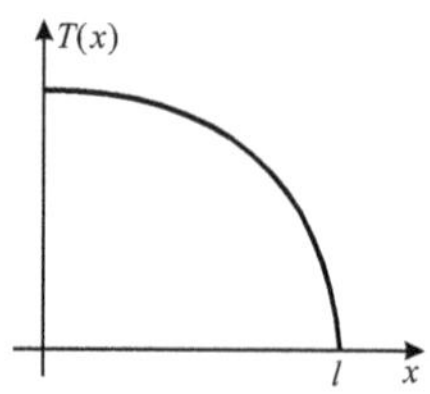

* * * * *

CHAPTER-7 GRAVITATION

7.1 Gravitational Field

1. The kinetic energy needed to project a body of mass m from the earth's surface (radius R) to infinity is : **[2002]**

(A) $\frac{mgR}{2}$ (B) $2mgR$

(C) mgR (D) $\frac{mgR}{4}$

2. If g is the acceleration due to gravity on the earth's surface, the gain in the potential energy of an object of mass m raised from the surface of the earth to a height equal to the radius R of the earth, is : **[2004]**

(A) $2mgR$ (B) $\frac{1}{2}mgR$

(C) $\frac{1}{4}mgR$ (D) mgR

3. A particle of mass 10 g is kept on the surface of a uniform sphere of mass 100 kg and radius 10 cm. Find the work to be done against the gravitational force between them, to take the particle far away from the sphere : [you may take $G = 6.67 \times 10^{-11}$ Nm^2/kg^2] **[2005]**

(A) 13.34×10^{-10} J (B) 3.33×10^{-10} J

(C) 6.67×10^{-9} J (D) 6.67×10^{-10} J

4. The change in the value of g at a height h above the surface of the earth is the same as at a depth d below the surface of earth. When both d and h are much smaller than the radius of earth, then which one of the following is correct ? **[2005]**

(A) $d = \frac{h}{2}$ (B) $d = \frac{3h}{2}$

(C) $d = 2h$ (D) $d = h$

5. Average density of the earth : **[2005]**

(A) Does not depend on g

(B) Is a complex function of g

(C) Is directly proportional to g

(D) Is inversely proportional to g

6. If g_E and g_M are the accelerations due to gravity on the surfaces of the earth and the moon respectively and if Millikan's oil drop experiment could be performed on the two surfaces, one will find the ratio $\frac{\text{electronic charge on the moon}}{\text{electronic charge on the earth}}$ to be : **[2007]**

(A) 1 (B) Zero

(C) g_E/g_M (D) g_M/g_E

7. This question contains statement-1 and statement-2 of the four choices given after the statements, choose the one that best describes the two statements. **[2008]**

Statement 1 : For a mass M kept at the centre of a cube of side 'a', the flux of gravitational field passing through its sides is $4\pi GM$.

Statement 2 : If the direction of a field due to a point source is radial and its dependence on the distance 'r' from the source is given as $\frac{I}{r^2}$, its flux through a closed surface depends only on the strength of the source enclosed by the surface and not on the size or shape of the surface.

(A) Statement-1 is false, Statement-2 is true

(B) Statement-1 is true, Statement-2 is true; Statement-2 is a correct explanation for Statement-1

(C) Statement-1 is true, Statement-2 is true; Statement-2 is not a correct explanation for Statement-1

(D) Statement-1 is true, Statement-2 is false

8. The height at which the acceleration due to gravity becomes $\frac{g}{9}$ (where g = the acceleration due to gravity on the surface of the earth) in terms of R, the radius of the earth, is : **[2009]**

(A) $\frac{R}{2}$ (B) $\sqrt{2}R$

(C) $2R$ (D) $\frac{R}{\sqrt{2}}$

9. Two bodies of masses m and $4m$ are placed at a distance r. The gravitational potential at a point on the line joining them where the gravitational field is zero is : **[2011]**

(A) $-\frac{6Gm}{r}$ (B) $-\frac{9Gm}{r}$

(C) Zero (D) $-\frac{4Gm}{r}$

10. The mass of a spaceship is 1000 kg. It is to be launched from the earth's surface out into free space. The value of 'g' and 'R' (radius of earth) are 10 m/s^2 and 6400 km respectively. The required energy for this work will be : **[2012]**

(A) 6.4×10^{10} J (B) 6.4×10^{11} J

(C) 6.4×10^{8} J (D) 6.4×10^{9} J

11. Assuming the earth to be a sphere of uniform density, the acceleration due to gravity inside the earth at a distance of r from the centre is proportional to : **[2012 (Online)]**

(A) r (B) r^{-1}

(C) r^2 (D) r^{-2}

12. The variation of acceleration due to gravity g with distance d from centre of the earth is best represented by (R = Earth's radius) : **[2012 (Online)]**

(A)

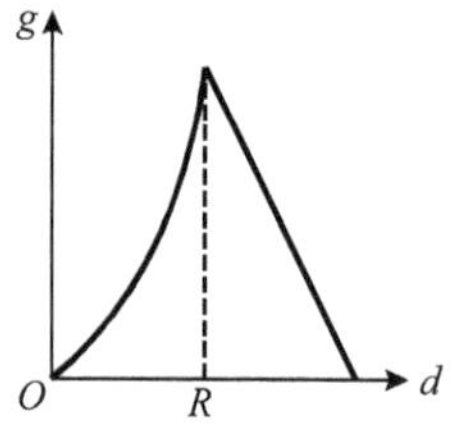

(B)

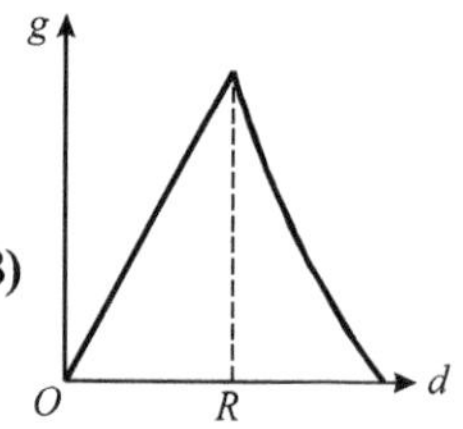

(C)

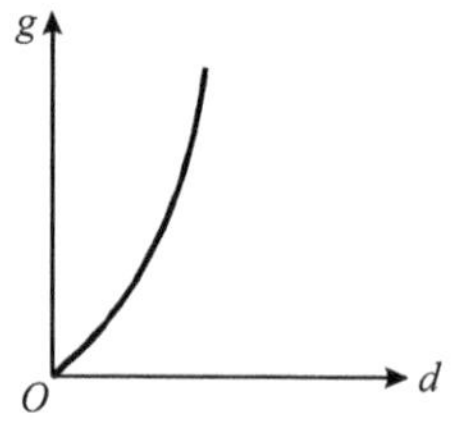

(D)

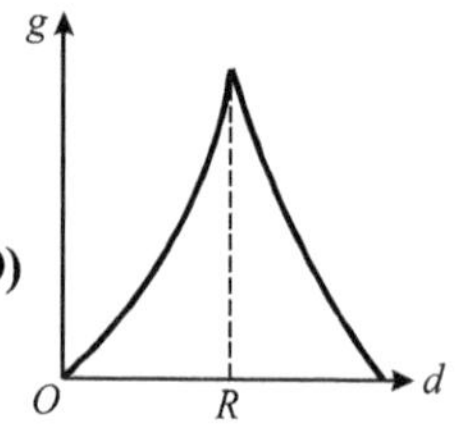

13. The gravitational field, due to the 'left over part' of a uniform sphere (from which a part as shown, has been 'removed out'), at a very far off point P, located as shown, would be (nearly) : **[2013 (Online)]**

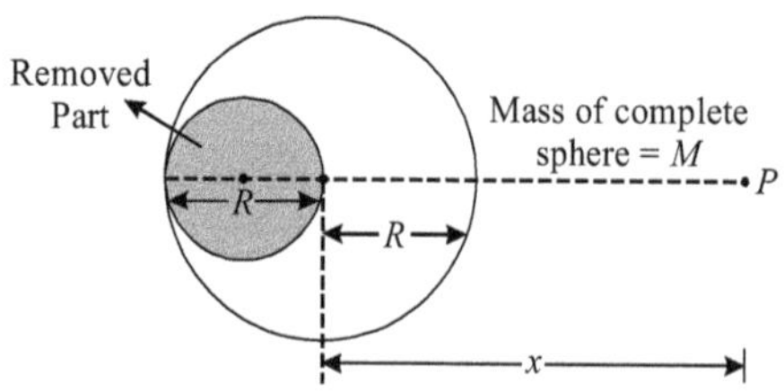

(A) $\frac{5}{6}\frac{GM}{x^2}$ (B) $\frac{8}{9}\frac{GM}{x^2}$

(C) $\frac{7}{8}\frac{GM}{x^2}$ (D) $\frac{6}{7}\frac{GM}{x^2}$

14. The change in the value of acceleration of earth towards sun, when the moon comes from the position of solar eclipse to the position on the other side of earth in line with sun is : (mass of the moon = 7.36×10^{22} kg, radius of the moon's orbit $= 3.8 \times 10^8$ m) **[2013 (Online)]**

(A) 6.73×10^{-5} m/s^2 (B) 6.73×10^{-3} m/s^2

(C) 6.73×10^{-2} m/s^2 (D) 6.73×10^{-4} m/s^2

15. Four particles, each of mass M and equidistant from each other, move along a circle of radius R under the action of their mutual gravitational attraction. The speed of each particle is : **[2014]**

(A) $\sqrt{\frac{GM}{R}}$ (B) $\sqrt{2\sqrt{2}\frac{GM}{R}}$

(C) $\sqrt{\frac{GM}{R}(1+2\sqrt{2})}$ (D) $\frac{1}{2}\sqrt{\frac{GM}{R}(1+2\sqrt{2})}$

16. A capillary tube is immersed vertically in water and the height of the water column is x. When this arrangement is taken into a mine of depth d, the height of the water column is y. If R is the radius of earth, the ratio $\frac{x}{y}$ is : **[2014 (Online)]**

(A) $\left(1-\frac{d}{R}\right)$ (B) $\left(1-\frac{2d}{R}\right)$

(C) $\left(\frac{R-d}{R+d}\right)$ (D) $\left(\frac{R+d}{R-d}\right)$

17. The gravitational field in a region is given by $\vec{g} = 5$ N/kg $\hat{i}$ + 12 N/kg $\hat{j}$. The change in the gravitational potential energy of a particle of mass 1 kg when it is taken from the origin to a point (7 m, –3 m) is : **[2014 (Online)]**

(A) 71 J (B) $13\sqrt{58}$ J

(C) –71 J (D) 1 J

18. From a sphere of mass M and radius R, a smaller sphere of radius $\frac{R}{2}$ is carved out such that the cavity made in the original sphere is between its centre and the periphery (See figure). For the configuration in the figure where the distance between the centre of the original sphere and the removed sphere is $3R$, the gravitational force between the two sphere is : **[2014 (Online)]**

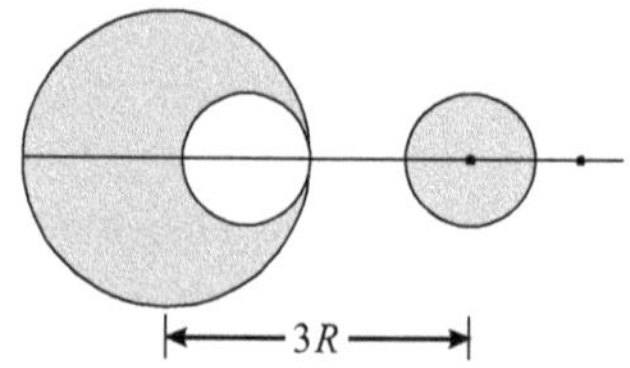

(A) $\frac{41GM^2}{3600R^2}$ (B) $\frac{41GM^2}{450R^2}$

(C) $\frac{59GM^2}{450R^2}$ (D) $\frac{GM^2}{225R^2}$

19. Two hypothetical planets of masses m_1 and m_2 are at rest when they are infinite distance apart. Because of the gravitational force they move towards each other along the line joining their centres. What is their speed when their separation is 'd' ? (Speed of m_1 is v_1 and that of m_2 is v_2) **[2014 (Online)]**

(A) $v_1 = v_2$

(B) $v_1 = m_2\sqrt{\frac{2G}{(d(m_1+m_2)}}$ $v_2 = m_1\sqrt{\frac{2G}{(d(m_1+m_2)}}$

(C) $v_1 = m_1\sqrt{\frac{2G}{(d(m_1+m_2)}}$ $v_2 = m_2\sqrt{\frac{2G}{(d(m_1+m_2)}}$

(D) $v_1 = m_2\sqrt{\frac{2G}{m_1}}$ $v_2 = m_2\sqrt{\frac{2G}{m_2}}$

20. From a solid sphere of mass M and radius R, a spherical portion of radius $\frac{R}{2}$ is removed, as shown in figure. Taking gravitational potential $V = 0$ at $r = \infty$, the potential at the centre of the cavity thus formed is : **[2015]**
(G = gravitational constant)

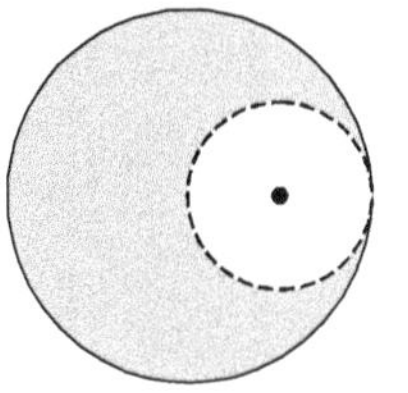

(A) $\frac{-GM}{2R}$ (B) $\frac{-GM}{R}$
(C) $\frac{-2GM}{3R}$ (D) $\frac{-2GM}{R}$

21. Which of the following most closely depicts the correct variation of the gravitational potential $V(r)$ due to a large planet of radius R and uniform mass density ? (Figures are not drawn to scale) **[2015 (Online)]**

(A)
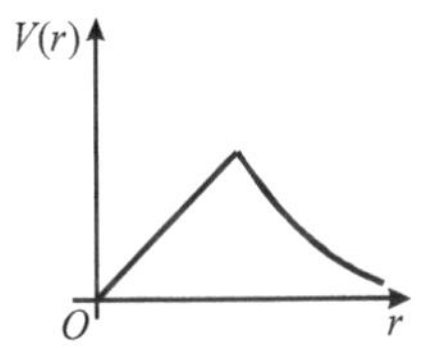

(B)
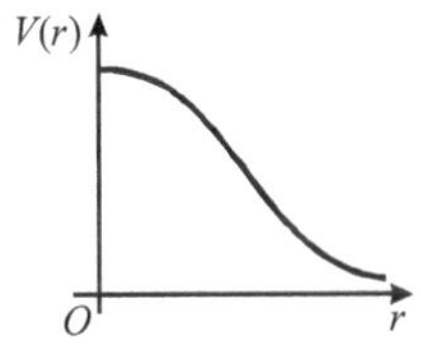

(C)
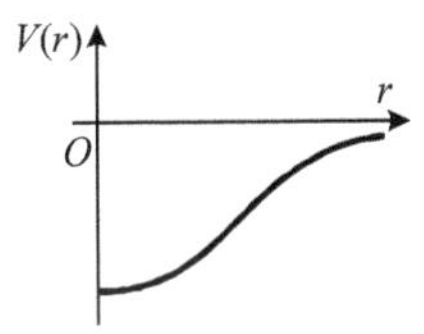

(D)
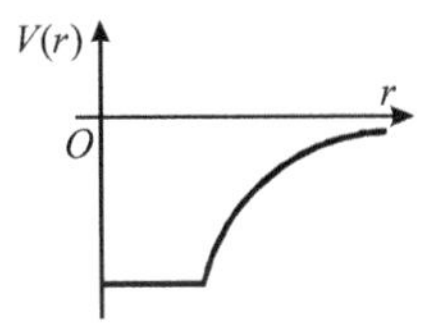

22. A point particle is held on the axis of a ring of mass m and radius r at a distance r from it's centre C. When released, it reaches C under the gravitational attraction of the ring. Its speed at C will be : **[2016 (Online)]**

(A) $\sqrt{\frac{2Gm}{r}(\sqrt{2}-1)}$ (B) $\sqrt{\frac{Gm}{r}}$
(C) $\sqrt{\frac{2Gm}{r}\left(1-\frac{1}{\sqrt{2}}\right)}$ (D) $\sqrt{\frac{2Gm}{r}}$

23. If the Earth has no rotational motion, the weight of a person on the equator is W. Determine the speed with which the earth would have to rotate about its axis so that the person at the equator will weight $\frac{3}{4}W$. Radius of the Earth is 6400 km and $g = 10$ m\s^2 : **[2017 (Online)]**
(A) 1.1×10^{-3} rad/s (B) 0.83×10^{-3} rad/s
(C) 0.63×10^{-3} rad/s (D) 0.28×10^{-3} rad/s

24. The mass density of a spherical body is given by $\rho(r) = \frac{k}{r}$ for $r \le R$ and $\rho(r) = 0$ for $r > R$, where r is the distance from the centre. The correct graph that describes qualitatively the acceleration a, of a test particle as a function of r is : **[2017 (Online)]**

(A)
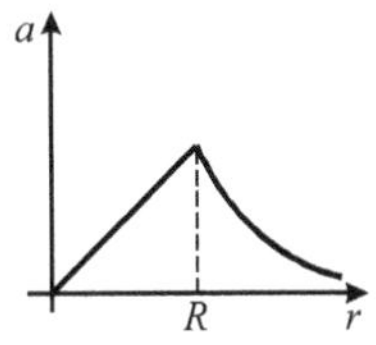

(B)
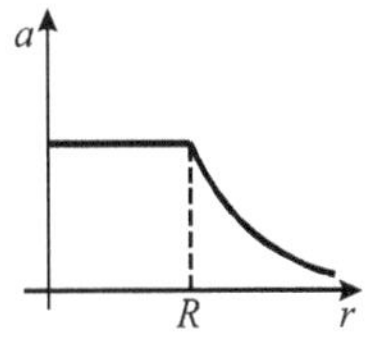

(C)
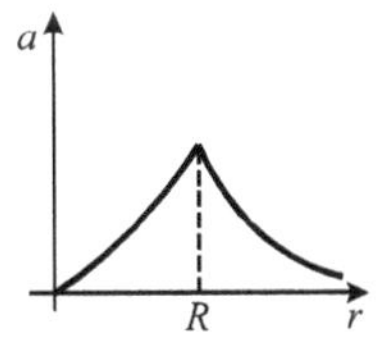

(D)
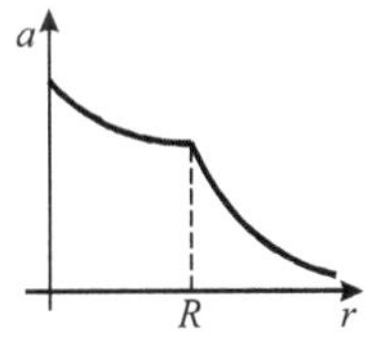

25 Take the mean distance of the moon and the sun from the earth to be 0.4×10^6 km and 150×10^6 km respectively. Their masses are 8×10^{22} kg and 2×10^{30} kg respectively. The radius of the earth is 6400 km. Let ΔF_1 be the difference in the forces exerted by the moon at the nearest and farthest points on the earth and ΔF_2 be the difference in the force exerted by the sun at the nearest and farthest points on the earth. Then, the number closest to $\frac{\Delta F_1}{\Delta F_2}$ is : **[2018 (Online)]**
(A) 2 (B) 6
(C) 10^{-2} (D) 0.6

26. Suppose that the angular velocity of rotation of earth is increased. Then, as a consequence : **[2018 (Online)]**
(A) There will be no change in weight anywhere on the earth
(B) Weight of the object, everywhere on the earth, will decrease
(C) Weight of the object, everywhere on the earth, will increase
(D) Except at poles, weight of the object on the earth will decrease

27. The mass and the diameter of a planet are three times the respective values for the Earth. The period of oscillation of a simple pendulum on the Earth is 2 s. The period of oscillation of the same pendulum on the planet would be : **[11 Jan. 2019 (Evening)]**

(A) $\frac{2}{\sqrt{3}}$ s (B) $2\sqrt{3}$ s
(C) $\frac{\sqrt{3}}{2}$ s (D) $\frac{3}{2}$ s

28. A straight rod of length L extends from $x = a$ to $x = L + a$. The gravitational force exerted on a point mass 'm' at $x = 0$, if the mass per unit length of the rod is $A + Bx^2$, is given by : **[12 Jan. 2019 (Morning)]**

(A) $Gm\left[A\left(\frac{1}{a+L}-\frac{1}{a}\right)-BL\right]$

(B) $Gm\left[A\left(\frac{1}{a}-\frac{1}{a+L}\right)+BL\right]$

(C) $Gm\left[A\left(\frac{1}{a+L}-\frac{1}{a}\right)+BL\right]$

(D) $Gm\left[A\left(\frac{1}{a}-\frac{1}{a+L}\right)-BL\right]$

29. Four identical particles of mass M are located at the corners of a square of side 'a'. What should be their speed if each of them revolves under the influence of other's gravitational field in a circular orbit circumscribing the square ? **[8 April 2019 (Morning)]**

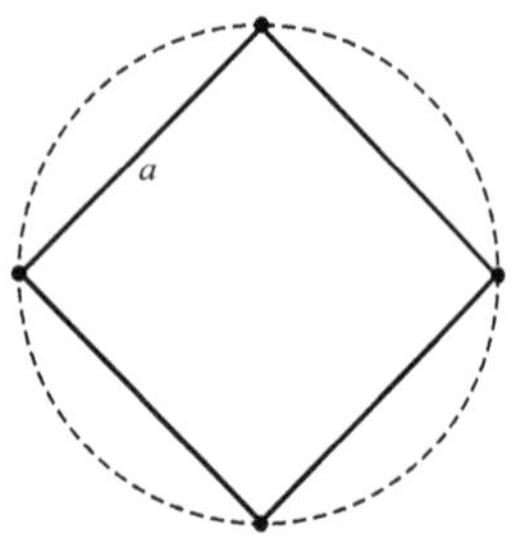

(A) $1.21\sqrt{\frac{GM}{a}}$ (B) $1.41\sqrt{\frac{GM}{a}}$

(C) $1.16\sqrt{\frac{GM}{a}}$ (D) $1.35\sqrt{\frac{GM}{a}}$

30. A rocket has to be launched from earth in such a way that it never returns. If E is the minimum energy delivered by the rocket launcher, what should be the minimum energy that the launcher should have if the same rocket is to be launched from the surface of the moon ? Assume that the density of the earth and the moon are equal and that the earth's volume is 64 times the volume of the moon : **[8 April 2019 (Evening)]**

(A) $\frac{E}{4}$ (B) $\frac{E}{16}$

(C) $\frac{E}{32}$ (D) $\frac{E}{64}$

31. A solid sphere of mass 'M' and radius 'a' is surrounded by a uniform concentric spherical shell of thickness $2a$ and mass $2M$. The gravitational field at distance '$3a$' from the centre will be : **[9 April 2019 (Morning)]**

(A) $\frac{2GM}{9a^2}$ (B) $\frac{GM}{3a^2}$

(C) $\frac{GM}{9a^2}$ (D) $\frac{2GM}{3a^2}$

32. The value of acceleration due to gravity at Earth's surface is 9.8 ms^{-2}. The altitude above its surface at which the acceleration due to gravity decreases to 4.9 ms^{-2}, is close to : (Radius of earth = 6.4×10^6 m) **[10 April 2019 (Morning)]**

(A) 1.6×10^6 m (B) 6.4×10^6 m

(C) 9.0×10^6 m (D) 2.6×10^6 m

* * * * *

CHAPTER-7 GRAVITATION

7.2 Satellite Motion

1. If suddenly the gravitational force of attraction between earth and a satellite revolving around it becomes zero, then the satellite will : **[2002]**

(A) Continue to move in its orbit with same velocity
(B) Move tangentially to the original orbit with same velocity
(C) Become stationary in its orbit
(D) Move towards the earth

2. Energy required to move a body of mass m from an orbit of radius $2R$ to $3R$ is : **[2002]**

(A) $GMm/12R$ (B) $GMm/3R$
(C) $GMm/8R$ (D) $GMm/6R$

3. The escape velocity of a body depends upon mass as : **[2002]**

(A) m^0 (B) m^1
(C) m^2 (D) m^3

4. The time period of a satellite of earth is 5 hours. If the separation between the centre of earth and the satellite is increased to 4 times the previous value, the new time period will become : **[2003]**

(A) 10 h (B) 80 h
(C) 40 h (D) 20 h

5. The escape velocity for a body projected vertically upwards from the surface of earth is 11 km/h. If the body is projected at an angle of 45° with the vertical, the escape velocity will be : **[2003]**

(A) $11\sqrt{2}$ km/s (B) 22 km/s
(C) 11 km/s (D) $\frac{11}{\sqrt{2}}$ km/s

6. Suppose the gravitational force varies inversely as the nth power of distance. Then the time period of a planet in circular orbit of radius R around the sun will be proportional to : **[2004]**

(A) $R^{\left(\frac{n+1}{2}\right)}$ (B) $R^{\left(\frac{n-1}{2}\right)}$
(C) R^n (D) $R^{\left(\frac{n-2}{2}\right)}$

7. The time period of an earth satellite in circular orbit is independent of : **[2004]**

(A) The mass of the satellite
(B) Radius of its orbit
(C) Both the mass and radius of the orbit
(D) Neither the mass of the satellite nor the radius of its orbit

8. A satellite of mass m revolves around the earth of radius R at a height x from its surface. If g is the acceleration due to gravity on the surface of the earth, the orbital speed of the satellite is : **[2004]**

(A) gx (B) $\frac{gR}{R-x}$
(C) $\frac{gR^2}{R+x}$ (D) $\left(\frac{gR^2}{R+x}\right)^{1/2}$

9. A planet in a distant solar system is 10 times more massive than the earth and its radius is 10 times smaller. Given that the escape velocity from the earth is 11 km s^{-1}, the escape velocity from the surface of the planet would be : **[2008]**

(A) 1.1 km s^{-1} (B) 11 km s^{-1}
(C) 110 km s^{-1} (D) 0.11 km s^{-1}

10. Two particles of equal mass 'm' go around a circle of radius R under the action of their mutual gravitational attraction. The speed of each particle with respect to their centre of mass is : **[2011]**

(A) $\sqrt{\frac{Gm}{R}}$ (B) $\sqrt{\frac{Gm}{4R}}$
(C) $\sqrt{\frac{Gm}{3R}}$ (D) $\sqrt{\frac{Gm}{2R}}$

11. India's Mangalyan was sent to the Mars by launching it into a transfer orbit EOM around the sun. It leaves the earth at E and meets Mars at M. If the semi-major axis of Earth's orbit is $a_e = 1.5 \times 10^{11}$ m, that of Mars orbit $a_m = 2.28 \times 10^{11}$ m, taken Kepler's laws give the estimate of time for Mangalyan to reach Mars from Earth to be close to : **[2014 (Online)]**

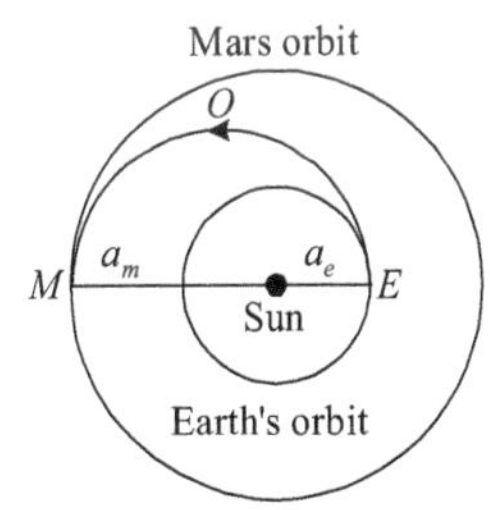

(A) 500 days (B) 320 days
(C) 260 days (D) 220 days

12. A very long (length L) cylindrical galaxy is made of uniformly distributed mass and has radius R ($R << L$). A star outside the galaxy is orbiting the galaxy in a plane perpendicular to the galaxy and passing through its centre. If the time period of star is T and its distance from the galaxy's axis is r, then : **[2015 (Online)]**

(A) $T \propto r$ (B) $T \propto \sqrt{r}$
(C) $T \propto r^2$ (D) $T^2 \propto r^3$

13. A satellite is revolving in a circular orbit at a height 'h' from the earth's surface (radius of earth R; $h << R$). The minimum increase in its orbital velocity required, so that the satellite could escape from the earth's gravitational field, is close to : (Neglect the effect of atmosphere.) **[2016]**

(A) $\sqrt{2gR}$ (B) $\sqrt{gR}$
(C) $\sqrt{gR/2}$ (D) $\sqrt{gR}(\sqrt{2}-1)$

14. Figure shows elliptical path $abcd$ of a planet around the sun S such that the area of triangle csa is 1/4 the area of the ellipse. (See figure) With db as the semimajor axis and ca as the semiminor axis. If t_1 is the time taken for planet to go over path abc and t_2 for path taken over cda then : **[2016 (Online)]**

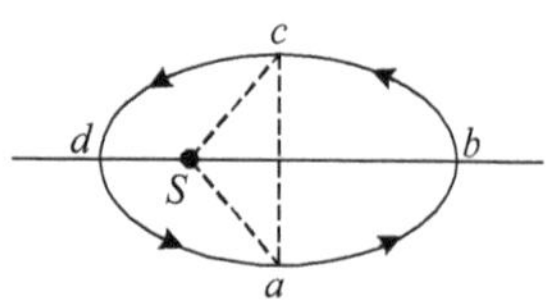

(A) $t_1 = 4t_2$ (B) $t_1 = 2t_2$
(C) $t_1 = 3t_2$ (D) $t_1 = t_2$

15. An astronaut of mass m is working on a satellite orbiting the earth at a distance h from the earth's surface. The radius of the earth is R, while its mass is M. The gravitational pull F_G on the astronaut is : **[2016 (Online)]**

(A) Zero since astronaut feels weightless

(B) $\frac{GMm}{(R+h)^2} < F_G < \frac{GMm}{R^2}$

(C) $F_G = \frac{GMm}{(R+h)^2}$ (D) $0 < F_G < \frac{GMm}{R^2}$

16. A body of mass m is moving in a circular orbit of radius R about a planet of mass M. At some instant, it splits into two equal masses. The first mass moves in a circular orbit of radius $\frac{R}{2}$, and the other mass, in a circular orbit of radius $\frac{3R}{2}$. The difference between the final and initial total energies is : **[2018 (Online)]**

(A) $-\frac{GMm}{2R}$ (B) $+\frac{GMm}{6R}$

(C) $-\frac{GMm}{6R}$ (D) $\frac{GMm}{2R}$

17. If the angular momentum of a planet of mass m, moving around the Sun in a circular orbit is L, about the center of the Sun, its areal velocity is : **[9 Jan. 2019 (Morning)]**

(A) $\frac{4L}{m}$ (B) $\frac{L}{m}$

(C) $\frac{L}{2m}$ (D) $\frac{2L}{m}$

18. The energy required to take a satellite to a height 'h' above Earth surface (radius of Earth $= 6.4 \times 10^3$ km) is E_1 and kinetic energy required for the satellite to be in a circular orbit at this height is E_2. The value of h for which E_1 and E_2 are equal, is : **[9 Jan. 2019 (Evening)]**

(A) 1.28×10^4 km (B) 6.4×10^3 km
(C) 3.2×10^3 km (D) 1.6×10^3 km

19. A satellite is moving with a constant speed v in circular orbit around the earth. An object of mass 'm' is ejected from the satellite such that it just escapes from the gravitational pull of the earth. At the time of ejection, the kinetic energy of the object is : **[10 Jan. 2019 (Morning)]**

(A) $\frac{3}{2}mv^2$ (B) mv^2

(C) $2mv^2$ (D) $\frac{1}{2}mv^2$

20. Two stars of masses 3×10^{31} kg each, and at distance 2×10^{11} m rotate in a plane about their common centre of mass O. A meteorite passes through O moving perpendicular to the star's rotation plane. In order to escape from the gravitational field of this double star, the minimum speed that meteorite should have at O is :
(Take Gravitational constant $G = 6.67 \times 10^{-11}$ Nm^2 kg^{-2}) **[10 Jan. 2019 (Evening)]**

(A) 1.4×10^5 m/s (B) 24×10^4 m/s
(C) 3.8×10^4 m/s (D) 2.8×10^5 m/s

21. A satellite is revolving in a circular orbit at a height h from the earth surface, such that $h << R$ where R is the radius of the earth. Assuming that the effect of earth's atmosphere can be neglected then minimum increase in the speed required so that the satellite could escape from the gravitational field of earth is : **[11 Jan. 2019 (Morning)]**

(A) $\sqrt{gR}(\sqrt{2}-1)$ (B) $\sqrt{2gR}$

(C) $\sqrt{gR}$ (D) $\sqrt{\frac{gR}{2}}$

22. A satellite of mass M is in a circular orbit of radius R about the centre of the earth. A meteorite of the same mass, falling towards the earth, collides with the satellite completely inelastically. The speeds of the satellite and the meteorite are the same, just before the collision. The subsequent motion of the combined body will be : **[12 Jan. 2019 (Morning)]**

(A) In a circular orbit of a different radius
(B) In the same circular orbit of radius R
(C) In an elliptical orbit
(D) Such that it escapes to infinity

23. Two satellites, A and B, have masses m and $2m$ respectively. A is in a circular orbit of radius R, and B is in a circular orbit of radius $2R$ around the earth. The ratio of their kinetic energies, T_A/T_B, is : **[12 Jan. 2019 (Evening)]**

(A) 2 (B) $\sqrt{\frac{1}{2}}$
(C) 1 (D) $\frac{1}{2}$

24. A test particle is moving in a circular orbit in the gravitational field produced by a mass density $\rho(r) = \frac{K}{r^2}$. Identify the correct relation between the radius R of the particle's orbit and its period T: **[9 April 2019 (Evening)]**

(A) T/R^2 is a constant (B) TR is a constant
(C) T^2/R^3 is a constant (D) T/R is a constant

25. A spaceship orbits around a planet at a height of 20 km from its surface. Assuming that only gravitational field of the planet acts on the spaceship, what will be the number of complete revolutions made by the spaceship in 24 hours around the planet ?
[Given : Mass of planet = 8×10^{22} kg;
Radius of planet = 2×10^6 m,
Gravitational constant $G = 6.67 \times 10^{-11}$ Nm^2/kg^2] **[10 April 2019 (Evening)]**

(A) 9 (B) 11
(C) 13 (D) 17

26. The ratio of the weights of a body on the Earth's surface to that on the surface of a planet is 9 : 4. The mass of the planet is $\frac{1}{9}$th of that of the Earth. If 'R' is the radius of the Earth, what is the radius of the planet ? (Take the planets to have the same mass density) : **[12 April 2019 (Evening)]**

(A) $\frac{R}{3}$ (B) $\frac{R}{2}$
(C) $\frac{R}{4}$ (D) $\frac{R}{9}$

* * * * *

CHAPTER-8 MECHANICAL PROPERTIES OF SOLIDES

8.1 Elasticity

1. A wire suspended vertically from one of its ends is stretched by attaching a weight of 200 N to the lower ends. The weight stretches the wire by 1 mm. Then the elastic energy stored in the wire is : **[2003]**

(A) 0.2 J (B) 10 J
(C) 20 J (D) 0.1 J

2. A wire fixed at the upper end stretches by length l by applying a force F. The work done in stretching is : **[2004]**

(A) $\frac{F}{2l}$ (B) Fl
(C) $2Fl$ (D) $\frac{Fl}{2}$

3. If S is stress and Y is Young's modulus of material of a wire, the energy stored in the wire per unit volume is : **[2005]**

(A) $2S^2Y$ (B) $\frac{S^2}{2Y}$
(C) $\frac{2Y}{S^2}$ (D) $\frac{S}{2Y}$

4. A wire elongates by l mm when a load W is hanged from it. If the wire goes over a pulley and two weights W each are hung at the two ends, the elongation of the wire will be (in mm) : **[2006]**

(A) l (B) $2l$
(C) Zero (D) $l/2$

5. Two wires are made of the same material and have the same volume. However wire 1 has cross-sectional area A and wire 2 has cross-sectional area $3A$. If the length of wire 1 increases by Δx on applying force F, how much force is needed to stretch wire 2 by the same amount ? **[2009]**

(A) $6F$ (B) $9F$
(C) F (D) $4F$

6. The load versus elongation graphs for four wires of same length and made of the same material are shown in the figure. The thinnest wire is represented by the line : **[2012 (Online)]**

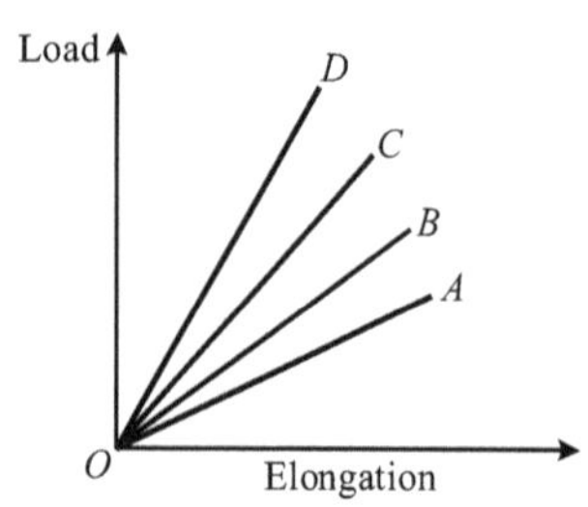

(A) OA (B) OC
(C) OD (D) OB

7. A structural steel rod has a radius of 10 mm and length of 1.0 m. A 100 kN for stretches it along its length. Young's modulus of structural steel is 2×10^{11} Nm^{-2}. The percentage strain is about : **[2012 (Online)]**

(A) 0.16% (B) 0.32%
(C) 0.08% (D) 0.24%

8. A steel wire can sustain 100 kg weight without breaking. If the wire is cut into two equal parts, each can sustain a weight of : **[2012 (Online)]**

(A) 50 kg (B) 400 kg
(C) 100 kg (D) 200 kg

9. Two blocks of masses m and M are connected by means of a metal wire of cross-sectional area A passing over a frictionless fixed pulley as shown in the figure. The system is then released. If $M = 2$m, then the stress produced in the wire is : **[2013 (Online)]**

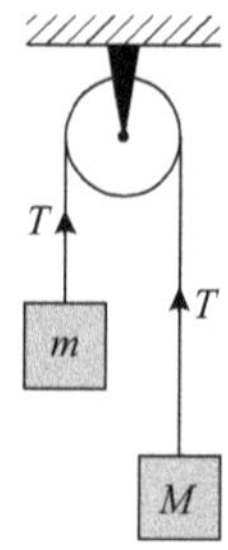

(A) $\frac{2mg}{3A}$ (B) $\frac{4mg}{3A}$
(C) $\frac{mg}{A}$ (D) $\frac{3mg}{4A}$

10. If the ratio of lengths, radii and Young's moduli of steel and brass wires in the figure are a, b and c respectively, then the corresponding ratio of increase in their lengths is : **[2013 (Online)]**

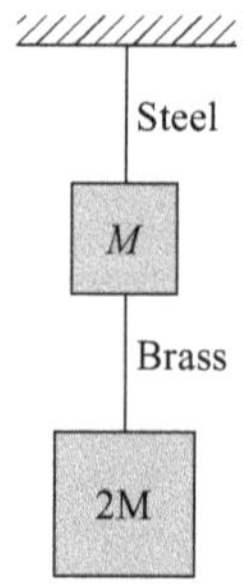

(A) $\frac{3c}{2ab^2}$ (B) $\frac{2a^2c}{b}$

(C) $\frac{3a}{2b^2c}$ (D) $\frac{2ac}{b^2}$

11. A uniform wire (Young's modulus 2×10^{11} Nm^{-2}) is subjected to longitudinal tensile stress of 5×10^7 Nm^{-2}. If the overall volume change in the wire is 0.02%, the fractional decrease in the radius of the wire is close to : **[2013 (Online)]**

(A) 1.0×10^{-4} (B) 1.5×10^{-4}

(C) 0.25×10^{-4} (D) 5×10^{-4}

12. A copper wire of length 1.0 m and a steel wire of length 0.5 m having equal cross-sectional areas are joined end to end. The composite wire is stretched by a certain load which stretches the copper wire by 1 mm. If Young's moduli of copper and steel are respectively 1.0×10^{11} Nm^{-2} and 2.0×10^{11} Nm^{-2}, the total extension of the composite wire is : **[2013 (Online)]**

(A) 1.75 mm (B) 2.0 mm

(C) 1.50 mm (D) 1.25 mm

13. In materials like aluminium and copper, the correct order of magnitude of various elastic moduli is : **[2014 (Online)]**

(A) Young's modulus < shear modulus < bulk modulus

(B) Bulk modulus < shear modulus < Young's modulus

(C) Shear modulus < Young's modulus < bulk modulus

(D) Bulk modulus < Young's modulus < shear modulus

14. The bulk moduli of ethanol, mercury and water are given as 0.9, 25 and 2.2 respectively in units of 10^9 Nm^{-2}. For a given value of pressure, the fractional compression in volume is $\frac{\Delta V}{V}$. Which of the following statements about $\frac{\Delta V}{V}$ for these three liquids is correct ? **[2014 (Online)]**

(A) Ethanol > Water > Mercury

(B) Water > Ethanol > Mercury

(C) Mercury > Ethanol > Water

(D) Ethanol > Mercury > Water

15. Steel ruptures when a shear of 3.5×10^8 Nm^{-2} is applied. The force needed to punch a 1 cm diameter hole in a steel sheet 0.3 cm thick is nearly: **[2014 (Online)]**

(A) 1.4×10^4 N (B) 2.7×10^4 N

(C) 3.3×10^4 N (D) 1.1×10^4 N

16. A uniform tapering conical wire is made from a material of Young's modulus Y and has a normal, unextended length L. The radii, at the upper and lower ends of this conical wire, have values R and $3R$, respectively. The upper end of the wire is fixed to a rigid support and a mass M is suspended from its lower end. The equilibrium extended length, of this wire, would equal: **[2016 (Online)]**

(A) $L\left(1+\frac{2}{9}\frac{Mg}{\pi YR^2}\right)$ (B) $L\left(1+\frac{1}{9}\frac{Mg}{\pi YR^2}\right)$

(C) $L\left(1+\frac{1}{3}\frac{Mg}{\pi YR^2}\right)$ (D) $L\left(1+\frac{2}{3}\frac{Mg}{\pi YR^2}\right)$

17. A bottle has an opening of radius a and length b. A cork of length b and radius $(a+\Delta a)$ where $(\Delta a << a)$ is compressed to fit into the opening completely (see figure). If the bulk modulus of cork is B and frictional coefficient between the bottle and cork is μ then the force needed to push the cork into the bottle is : **[2016 (Online)]**

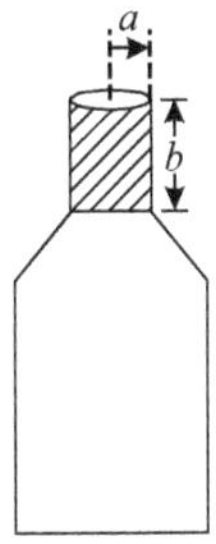

(A) $(\pi\mu Bb)\,a$ (B) $(2\pi\mu Bb)\,\Delta a$

(C) $(\mu\pi Bb)\,\Delta a$ (D) $(4\pi\mu Bb)\,\Delta a$

18. A load of mass M kg is suspended from a steel wire of length 2 m and radius 1.0 mm in Searle's apparatus experiment. The increase in length produced in the wire is 4.0 mm. Now the load is fully immersed in a liquid of relative density 2. The relative density of the material of load is 8. The new value of increase in length of the steel wire is : **[12 Jan. 2019 (Evening)]**

(A) 4.0 mm (B) 3.0 mm

(C) 5.0 mm (D) Zero

19. Water from a pipe is coming at a rate of 100 litres per minute. If the radius of the pipe is 5 cm, the Reynolds number for the flow is of the order of: (density of water = 1000 kg/m^3, coefficient of viscosity of water = 1 mPas) : **[8 April 2019 (Morning)]**

(A) 10^6 (B) 10^3

(C) 10^4 (D) 10^2

20. A boy's catapult is made of rubber cord which is 42 cm long, with 6 mm diameter of cross-section and of negligible mass. The boy keeps a stone weighing 0.02 kg on it and stretches the cord by 20 cm by applying a constant force. When released, the stone flies off with a velocity of 20 ms^{-1}. Neglect the change in the area of cross-section of the cord while stretched. The Young's modulus of rubber is closest to : **[8 April 2019 (Morning)]**

(A) 10^4 Nm^{-2} (B) 10^8 Nm^{-2}

(C) 10^6 Nm^{-2} (D) 10^3 Nm^{-2}

21. A steel wire having a radius of 2.0 mm, carrying a load of 4 kg, is hanging from a ceiling. Given that $g = 3.1\,\pi$ ms^{-2}, what will be the tensile stress that would be developed in the wire ? **[8 April 2019 (Morning)]**

(A) 4.8×10^6 Nm^{-2} (B) 5.2×10^6 Nm^{-2}

(C) 6.2×10^6 Nm^{-2} (D) 3.1×10^6 Nm^{-2}

22. Young's moduli of two wires A and B are in the ratio 7 : 4. Wire A is 2 m long and has radius R. Wire B is 1.5 m long and has radius 2 mm. If the two wires stretch by the same length for a given load, then the value of R is close to : **[8 April 2019 (Evening)]**

(A) 1.9 mm (B) 1.7 mm
(C) 1.5 mm (D) 1.3 mm

23. The elastic limit of brass is 379 MPa. What should be the minimum diameter of a brass rod if it is to support a 400 N load without exceeding its elastic limit ? **[10 April 2019 (Evening)]**

(A) 1.16 mm (B) 0.90 mm
(C) 1.36 mm (D) 1.00 mm

24. In an experiment, bras and steel wires of length 1 m each with areas of cross section 1 mm^2 are used. The wires are connected in series and one end of the combined wire is connected to a rigid support and other end is subjected to elongation. The stress required to produce a net elongation of 0.2 mm is :
(Given, the Young's Modulus for steel and brass are respectively, 120×10^9 N/m^2 and 60×10^9 N/m^2) **[10 April 2019 (Evening)]**

(A) 0.2×10^6 N/m^2 (B) 4.0×10^6 N/m^2
(C) 1.8×10^6 N/m^2 (D) 1.2×10^6 N/m^2

25. At 40°C, a brass wire of 1 mm radius is hung from the ceiling. A small mass, M is hung from the free end of the wire. When the wire is cooled down from 40°C to 20°C it regains its original length of 0.2 m. The value of M is close to : (Coefficient of linear expansion and Young's modulus of brass are 10^{-5}/°C and 10^{11} N/m^2, respectively; $g = 10$ ms^{-2}): **[12 April 2019 (Morning)]**

(A) 1.5 kg (B) 9 kg
(C) 0.9 kg (D) 0.5 kg

26. A uniform cylindrical rod of length L and radius r, is made from a material whose Young's modulus of Elasticity equals Y. When this rod is heated by temperature T and simultaneously subjected to a net longitudinal compressional force F, its length remains unchanged. The coefficient of volume expansion, of the material of the rod, is (nearly) equals to : **[12 April 2019 (Evening)]**

(A) $F/(3\pi r^2 YT)$ (B) $3F/(\pi r^2 YT)$
(C) $6F/(\pi r^2 YT)$ (D) $9F/(\pi r^2 YT)$

* * * * *

CHAPTER-9 MECHANICAL PROPERTIES OF FLUIDS

9.1 Fluid Statics

1. A jar is filled with two non-mixing liquids 1 and 2 having densities ρ_1 and ρ_2, respectively. A solid ball, made of a material of density ρ_3, is dropped in the jar. It comes to equilibrium in the position shown in the figure. Which of the following is true for ρ_1, ρ_2 and ρ_3 : **[2008]**

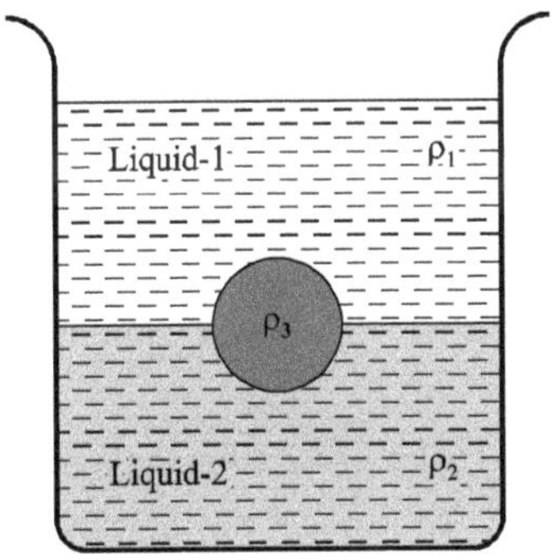

(A) $\rho_3 < \rho_1 < \rho_2$ (B) $\rho_3 < \rho_2 < \rho_1$
(C) $\rho_1 < \rho_2 < \rho_3$ (D) $\rho_1 < \rho_3 < \rho_2$

2. A ball is made of a material of density ρ where $\rho_{oil} < \rho < \rho_{water}$ with ρ_{oil} and ρ_{water} representing the densities of oil and water, respectively. The oil and water are immiscible. If the above ball is in equilibrium in a mixture of this oil and water, which of the following pictures represents its equilibrium position ? **[2010]**

(A) 

(B) 

(C) 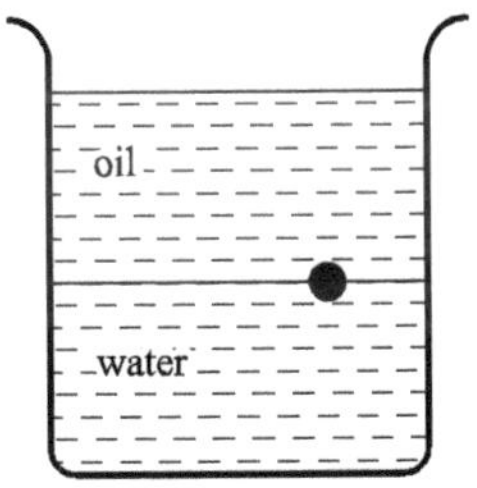

(D) 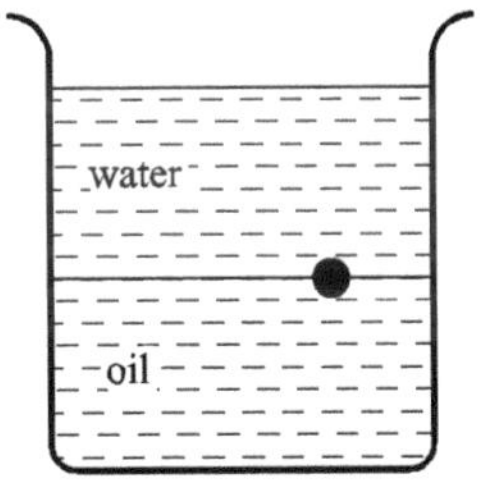

3. A uniform cylinder of length L and mass M having cross-sectional area A is suspended, with its length vertical, from a fixed point by a massless spring, such that it is half submerged in a liquid of density σ at equilibrium position. The extension x_0 of the spring when it is in equilibrium is : **[2013]**

(A) $\frac{Mg}{k}\left(1-\frac{LA\sigma}{M}\right)$ (B) $\frac{Mg}{k}\left(1-\frac{LA\sigma}{2M}\right)$

(C) $\frac{Mg}{k}\left(1+\frac{LA\sigma}{M}\right)$ (D) $\frac{Mg}{k}$

4. There is a circular tube in a vertical plane. Two liquids which do not mix and of densities d_1 and d_2 are filled in the tube. Each liquid subtends 90° angle at centre. Radius joining their interface makes an angle α with vertical. Ratio d_1/d_2 is : **[2014]**

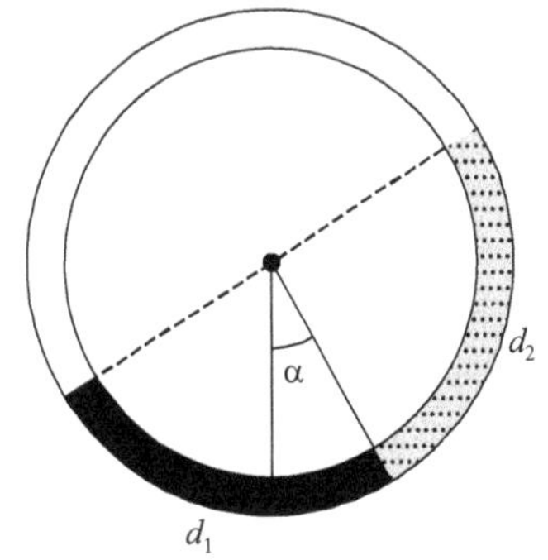

(A) $\frac{1+\sin\alpha}{1-\sin\alpha}$ (B) $\frac{1+\cos\alpha}{1-\cos\alpha}$

(C) $\frac{1+\tan\alpha}{1-\tan\alpha}$ (D) $\frac{1+\sin\alpha}{1-\cos\alpha}$

5. A wooden block floating in a bucket of water has 4/5 of its volume submerged. When certain amount of an oil is poured into the bucket, it is found that the block is just under the oil surface with half of its volume under water and half in oil. The density of oil relative to that of water is :

[9 April 2019 (Evening)]

(A) 0.5 (B) 0.7
(C) 0.6 (D) 0.8

6. A submarine experiences a pressure of 5.05×10^6 Pa at a depth of d_1 in a sea. When it goes further to a depth of d_2, it experiences a pressure of 8.08×10^6 Pa. Then $d_2 - d_1$ is approximately (density of water = 10^3 kg/m^3 and acceleration due to gravity = 10 ms^{-2}) : **[10 April 2019 (Evening)]**

(A) 500m (B) 400m
(C) 300m (D) 600m

7. A cubical block of side 0.5 m floats on water with 30% of its volume under water. What is the maximum weight that can be put on the block without fully submerging it under water? (Take density of water = 10^3 kg/m^3) :**[10 April 2019 (Evening)]**

(A) 65.4 kg (B) 87.5 kg
(C) 30.1 kg (D) 46.3 kg

* * * * *

CHAPTER-9 MECHANICAL PROPERTIES OF FLUIDS

9.2 Fluid Dynamics

1. A cylinder of height 20 m is completely filled with water. The velocity of efflux of water (in ms^{-1}) through a small hole on the side wall of the cylinder near its bottom, is : **[2002]**

(A) 10 (B) 20

(C) 25.5 (D) 5

2. Water is flowing continuously from a tap having an internal diameter 8×10^{-3} m. The water velocity as it leaves the tap is 0.4 ms^{-1}. The diameter of the water stream at a distance 2×10^{-1} m below the tap is close to : **[2011]**

(A) 9.6×10^{-3} m (B) 3.6×10^{-3} m

(C) 5.0×10^{-3} m (D) 7.5×10^{-3} m

3. A square hole of side length l is made at a depth of h and a circular hole of radius r is made at a depth of 4h from the surface of water in a water tank kept on a horizontal surface. If $l << h$, $r << h$ and the rate of water flow from the holes is the same, then r is equal to : **[2012]**

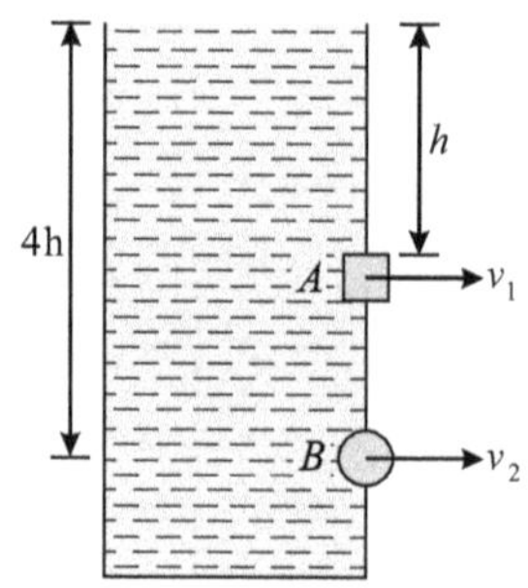

(A) $\frac{l}{\sqrt{2\pi}}$ (B) $\frac{l}{\sqrt{3\pi}}$

(C) $\frac{l}{3\pi}$ (D) $\frac{l}{2\pi}$

4. Water is flowing through a horizontal tube having cross-sectional area of its two ends being A and A' such that the ratio A/A' is 5. If the pressure difference of water between the two ends is 3×10^5 Nm^{-2}, the velocity of water with which it enters the tube will be (neglect gravity effects) : **[2012 (Online)]**

(A) 5 ms^{-1} (B) 10 ms^{-1}

(C) 25 ms^{-1} (D) $50\sqrt{10}$ ms^{-1}

5. In a cylindrical water tank, there are two small holes A and B on the wall at a depth of h_1, from the surface of water and at a height of h_2 from the bottom of water tank. Surface of water is at height H from the bottom of water tank. Water coming out from both holes strikes the ground at the same point S. Find the ratio of h_1 and h_2 : **[2012 (Online)]**

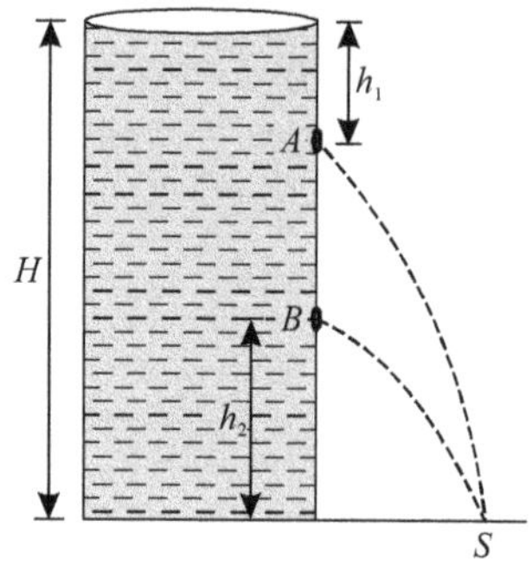

(A) Depends on H (B) 1 : 1

(C) 2 : 2 (D) 1 : 2

6. Air of density 1.2 kg m^{-3} is blowing across the horizontal wings of an aeroplane in such a way that its speeds above and below the wings are 150 ms^{-1} and 100 ms^{-1}, respectively. The pressure difference between the upper and lower sides of the wings, is : **[2013 (Online)]**

(A) 60 Nm^{-2} (B) 180 Nm^{-2}

(C) 7500 Nm^{-2} (D) 12500 Nm^{-2}

7. Water is flowing at a speed of 1.5 ms^{-1} through horizontal tube of cross-sectional area 10^{-2} m^2 and you are trying to stop the flow by your palm. Assuming that the water stops immediately after hitting the palm, the minimum force that you must exert should be : (density of water = 10^3 kgm^{-3}) **[2014 (Online)]**

(A) 15 N (B) 22.5 N

(C) 33.7 N (D) 45 N

8. A cylindrical vessel of cross-section A contains water to a height h. There is a hole in the bottom of radius 'a'. The time in which it will be emptied is : **[2014 (Online)]**

(A) $\frac{2A}{\pi a^2}\sqrt{\frac{h}{g}}$ (B) $\frac{\sqrt{2}A}{\pi a^2}\sqrt{\frac{h}{g}}$

(C) $\frac{2\sqrt{2}A}{\pi a^2}\sqrt{\frac{h}{g}}$ (D) $\frac{A}{\sqrt{2}\pi a^2}\sqrt{\frac{h}{g}}$

9. A tank with a small hole at the bottom has been filled with water and kerosene (specific gravity 0.8). The height of water is 3 m and that of kerosene 2 m. When the hole is opened the velocity of fluid coming out from it is nearly : (take $g = 10$ ms^{-2} and density of water = 10^3 kg m^{-3}) : **[2014 (Online)]**

(A) 10.7 ms^{-1} (B) 9.6 ms^{-1}

(C) 8.5 ms^{-1} (D) 7.6 ms^{-1}

10. In the diagram shown, the difference in the two tubes of the manometer is 5 cm, the cross section of the tube at A and B is 6 mm^2 and 10 mm^2 respectively. The rate at which water flows through the tube is ($g = 10\ ms^{-2}$) : **[2014 (Online)]**

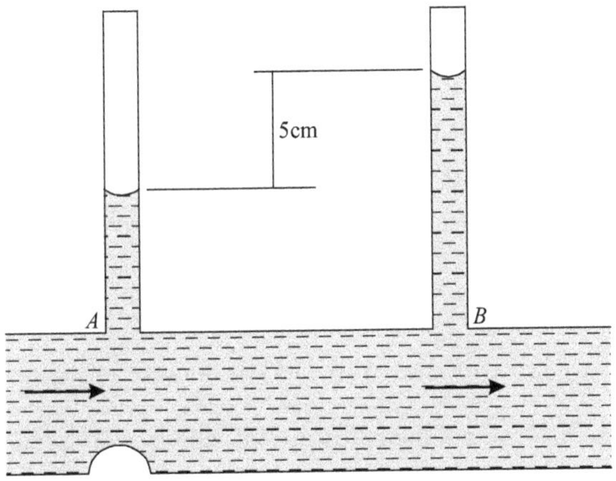

(A) 7.5 cc/s (B) 8.0 cc/s
(C) 10.0 cc/s (D) 12.5 cc/s

11. Consider a water jar of radius R that has water filled up to height H and is kept on a stand of height h (see figure). Through a hole of radius r ($r << R$) at its bottom, the water leaks out and the stream of water coming down towards the ground has a shape like a funnel as shown in the figure. If the radius of the cross-section of water stream when it hits the ground is x. Then: **[2016 (Online)]**

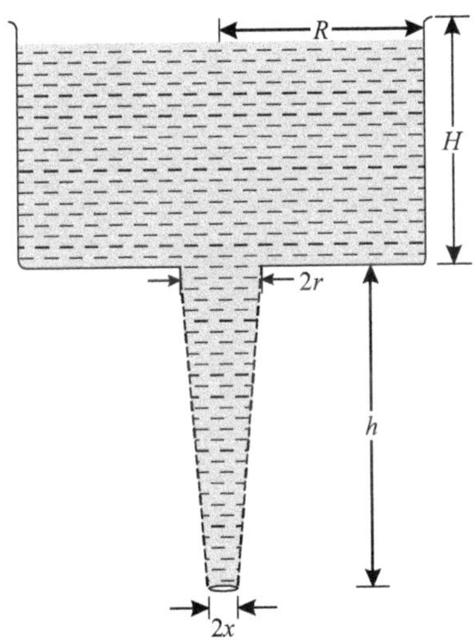

(A) $x = r\left(\frac{H}{H+h}\right)^{\frac{1}{4}}$ (B) $x = r\left(\frac{H}{H+h}\right)$

(C) $x = r\left(\frac{H}{H+h}\right)^{2}$ (D) $x = r\left(\frac{H}{H+h}\right)^{\frac{1}{2}}$

12. The top of a water tank is open to air and its water level is maintained. It is giving out 0.74 m^3 water per minute through a circular opening of 2 cm radius in its wall. The depth of the centre of the opening from the level of water in the tank is close to : **[9 Jan. 2019 (Evening)]**

(A) 9.6m (B) 4.8m
(C) 2.9m (D) 6.0m

13. Water flows into a large tank with flat bottom at the rate of $10^{-4}\ m^3s^{-1}$. Water is also leaking out of a hole of area 1 cm^2 at its bottom. If the height of the water in the tank remains steady, then this height is : **[10 Jan. 2019 (Morning)]**

(A) 4 cm (B) 2.9 cm
(C) 1.7 cm (D) 5.1 cm

14. A liquid of density ρ is coming out of a hose pipe of radius a with horizontal speed v and hits a mesh. 50% of the liquid passes through the mesh unaffected. 25% looses all of its momentum and 25% comes back with the same speed. The resultant pressure on the mesh will be : **[11 Jan. 2019 (Morning)]**

(A) ρv^2 (B) $\frac{3}{4}\rho v^2$

(C) $\frac{1}{2}\rho v^2$ (D) $\frac{1}{4}\rho v^2$

15. A long cylindrical vessel is half filled with a liquid. When the vessel is rotated about its own vertical axis, the liquid rises up near the wall. If the radius of vessel is 5 cm and its rotational speed is 2 rotations per second, then the difference in the heights between the centre and the sides, in cm, will be : **[12 Jan. 2019 (Evening)]**

(A) 1.2 (B) 0.1
(C) 2.0 (D) 0.4

16. Water from a tap emerges vertically downwards with an initial speed of 1.0 ms^{-1}. The cross-sectional area of the tap is $10^{-4}\ m^2$. Assume that the pressure is constant throughout the stream of water and that the flow is streamlined. The cross-sectional area of the stream, 0.15 m below the tap would be : (Take $g = 10\ ms^{-2}$) **[10 April 2019 (Evening)]**

(A) $1 \times 10^{-5}\ m^2$ (B) $5 \times 10^{-5}\ m^2$
(C) $2 \times 10^{-5}\ m^2$ (D) $5 \times 10^{-4}\ m^2$

* * * * *

CHAPTER-9 MECHANICAL PROPERTIES OF FLUIDS

9.3 Viscosity

1. Spherical balls of radius R are falling in a viscous fluid of viscosity η with a velocity v. The retarding viscous force acting on the spherical ball is : **[2004]**

(A) Directly proportional to R but inversely proportional to v
(B) Directly proportional to both radius R and velocity v
(C) Inversely proportional to both radius R and velocity v
(D) Inversely proportional to R but directly proportional to velocity v

2. If the terminal speed of a sphere of gold (density = 19.5 kg/m^3) is 0.2 m/s in a viscous liquid (density = 1.5 kg/m^3), find the terminal speed of a sphere of silver (density = 10.5 kg/m^3) of the same size in the same liquid : **[2006]**

(A) 0.4 m/s (B) 0.133 m/s
(C) 0.1 m/s (D) 0.2 m/s

3. A spherical solid ball of volume V is made of a material of density ρ_1. It is falling through a liquid of density ρ_2 ($\rho_2 < \rho_1$). Assume that the liquid applies a viscous force on the ball that is proportional to the square of its speed v, i.e., $F_{viscous} = -kv^2$ ($k > 0$). Then terminal speed of the ball is : **[2008]**

(A) $\sqrt{\frac{Vg(\rho_1 - \rho_2)}{k}}$ (B) $\frac{Vg\rho_1}{k}$

(C) $\sqrt{\frac{Vg\rho_1}{k}}$ (D) $\frac{Vg(\rho_1 - \rho_2)}{k}$

4. If a ball of steel (density $\rho = 7.8$ g cm^{-3}) attains a terminal velocity of 10 cm s^{-1} when falling in a tank of water (coefficient of viscosity $\eta_{water} = 8.5 \times 10^{-4}$ Pa.s) then its terminal velocity in glycerine : **[2011]**
($\rho = 12$ g cm^{-3}, $\eta = 13.2$ Pa.s) would be nearly :

(A) 1.6×10^{-5} cm s^{-1} (B) 6.25×10^{-4} cm s^{-1}
(C) 6.45×10^{-4} cm s^{-1} (D) 1.5×10^{-5} cm s^{-1}

5. The terminal velocity of a small sphere of radius a in a viscous liquid is proportional to : **[2012 (Online)]**

(A) a^2 (B) a^3
(C) a (D) a^{-1}

6. In an experiment, a small steel ball falls through a liquid at a constant speed of 10 cm/s. If the steel ball is pulled upward with a force equal to twice its effective weight, how fast will it move upward ? **[2013 (Online)]**

(A) 5 cm/s (B) Zero
(C) 10 cm/s (D) 20 cm/s

7. The average mass of rain drops is 3.0×10^{-5} kg and their average terminal velocity is 9 m/s. Calculate the energy transferred by rain to each square metre of the surface at a place which receives 100 cm of rain in a year : **[2014 (Online)]**

(A) 3.5×10^5 J (B) 4.05×10^4 J
(C) 3.0×10^5 J (D) 9.0×10^4 J

8. The velocity of water in a river is 18 km/hr near the surface. If the river is 5 m deep, find the shearing stress between the horizontal layers of water. The co-efficient of viscosity of water = 10^{-2} poise : **[2014 (Online)]**

(A) 10^{-1} N/m^2 (B) 10^{-2} N/m^2
(C) 10^{-3} N/m^2 (D) 10^{-4} N/m^2

9. If it takes 5 minutes to fill a 15 litre bucket from a water tap of diameter $\frac{2}{\sqrt{\pi}}$ cm then the Reynolds number for the flows is (density of water = 10^3 kg/m^3) and viscosity of water = 10^{-3} Pa.s) close to : **[2015 (Online)]**

(A) 1100 (B) 11,000
(C) 550 (D) 5500

10. Which of the following option correctly describes the variation of the speed v and acceleration 'a' of a point mass falling vertically in a viscous medium that applies a force $F = -kv$, where 'k' is a constant, on the body ? (Graphs are schematic and not drawn to scale) : **[2016 (Online)]**

(A) 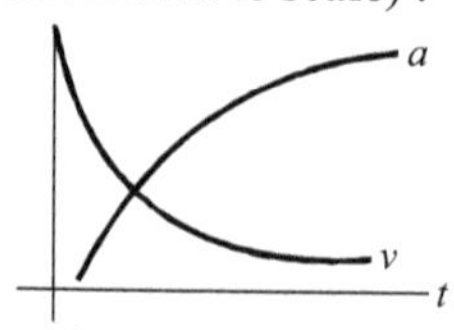

(B) 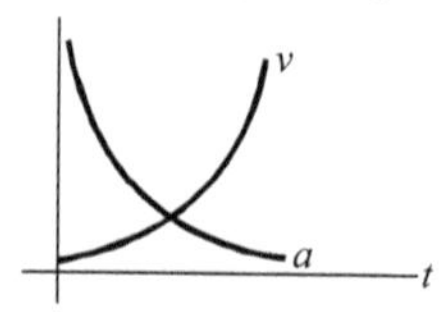

(C) 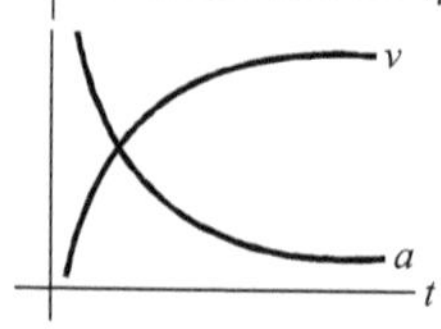

(D) 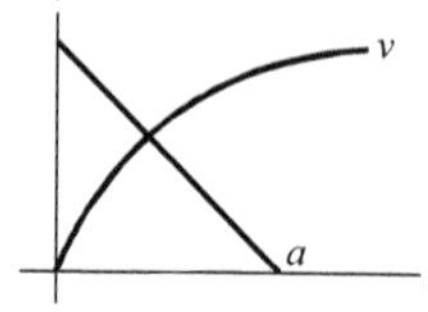

11. Two tubes of radii r_1 and r_2, and lengths l_1 and l_2, respectively, are connected in series and a liquid flows through each of them in streamline conditions. P_1 and P_2 are pressure differences across the two tubes. If P_2 is $4P_1$ and l_2 is $l_1/4$, then the radius r_2 will be equal to : **[2017 (Online)]**

(A) r_1 (B) $2r_1$
(C) $4r_1$ (D) $\frac{r_1}{2}$

12. A solid sphere, of radius R acquires a terminal velocity v_1 when falling (due to gravity) through a viscous fluid having a coefficient of viscosity η. The sphere is broken into 27 identical solid spheres. If each of these spheres acquires a terminal velocity v_2, when falling through the same fluid, the ratio (v_1/v_2) equals : **[12 April 2019 (Evening)]**

(A) 1/27 (B) 1/9
(C) 27 (D) 9

* * * * *

CHAPTER-9 MECHANICAL PROPERTIES OF FLUIDS

9.4 Surface Tension

1. If two soap bubbles of different radii are connected by a tube : **[2004]**

(A) Air flows from the bigger bubble to the smaller bubble till the sizes become equal

(B) Air flows from bigger bubble to the smaller bubble till the sizes are interchanged

(C) Air flows from the smaller bubble to the bigger bubble

(D) There is no flow of air

2. A 20 cm long capillary tube is dipped in water. The water rises upto 8 cm. If the entire arrangement is put in a freely falling elevator, the length of water column in the capillary tube will be: **[2005]**

(A) 8 cm (B) 10 cm
(C) 4 cm (D) 20 cm

3. A capillary tube (A) is dipped in water. Another identical tube (B) is dipped in a soap-water solution. Which of the following shows the relative nature of the liquid columns in the two tubes : **[2008]**

(A)
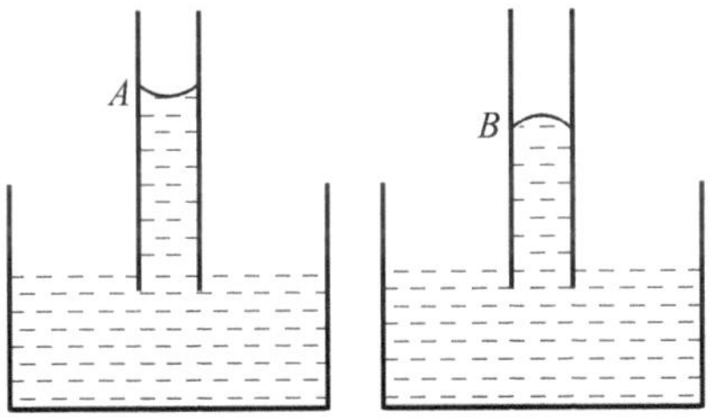

(B)
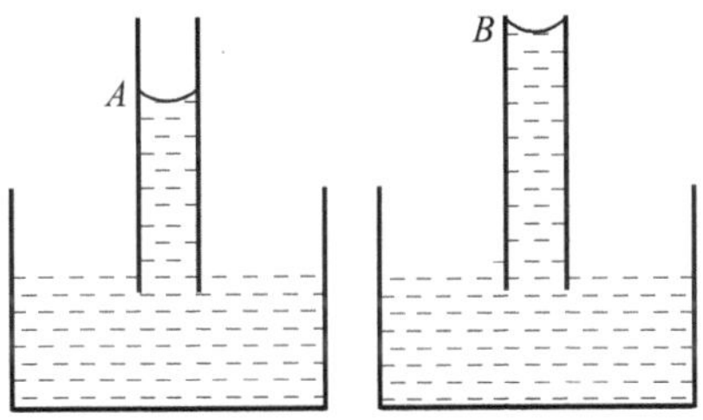

(C)
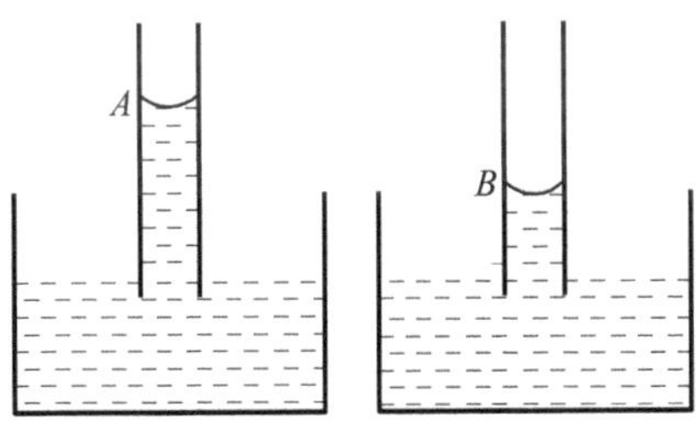

(D)
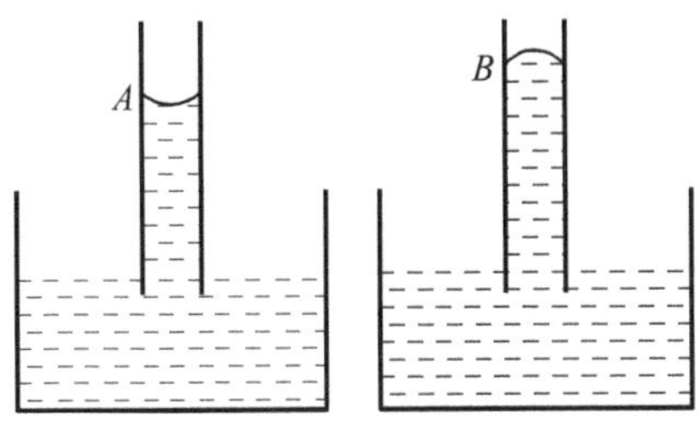

4. Two mercury drops (each of radius 'r') merge to form a bigger drop. The surface energy of the bigger drop, if T is the surface tension, is : **[2011]**

(A) $2^{5/3}\pi r^2 T$ (B) $4\pi r^2 T$
(C) $2\pi r^2 T$ (D) $2^{8/3}\pi r^2 T$

5. Work done in increasing the size of a soap bubble from a radius of 3 cm to 5 cm is nearly (Surface tension of soap solution $= 0.03\ \text{Nm}^{-1}$) : **[2011]**

(A) 2πmJ (B) 0.4πmJ
(C) 4πmJ (D) 0.2πmJ

6. A thin liquid film formed between a U-shaped wire and a light slider supports a weight of 1.5×10^{-2} N (see figure). The length of the slider is 30 cm and its weight negligible. The surface tension of the liquid film is : **[2012]**

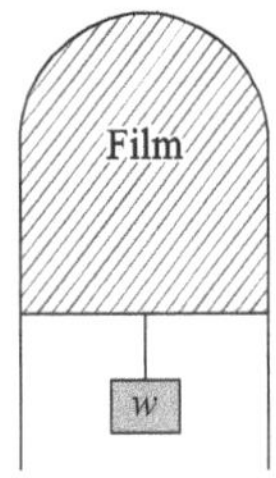

(A) 0.025 Nm^{-1} (B) 0.012525 Nm^{-1}
(C) 0.125 Nm^{-1} (D) 0.0525 Nm^{-1}

7. A large number of liquid drops each of radius r coalesce to from a single drop of radius R. The energy released in the process is converted into kinetic energy of the big drop so formed. The speed of the big drop is (given, surface tension of liquid T, density ρ) : **[2012 (Online)]**

(A) $\sqrt{\dfrac{T}{\rho}\left(\dfrac{1}{r}-\dfrac{1}{R}\right)}$ (B) $\sqrt{\dfrac{2T}{\rho}\left(\dfrac{1}{r}-\dfrac{1}{R}\right)}$

(C) $\sqrt{\dfrac{4T}{\rho}\left(\dfrac{1}{r}-\dfrac{1}{R}\right)}$ (D) $\sqrt{\dfrac{6T}{\rho}\left(\dfrac{1}{r}-\dfrac{1}{R}\right)}$

8. Assume that a drop of liquid evaporates by decrease in its surface energy, so that its temperature remains unchanged. What should be the minimum radius of the drop for this to be possible ? The surface tension is T, density of liquid is ρ and L is its latent heat of vaporization : **[2013]**

(A) $\sqrt{T/\rho L}$ (B) $T/\rho L$
(C) $2T/\rho L$ (D) $\rho L/T$

9. This question has Statement-1 and Statement-2. Of the four choices given after the Statements, choose the one that best describes the two Statements. **[2013 (Online)]**

Statement-1 : A capillary is dipped in a liquid and liquid rises to a height h in it. As the temperature of the liquid is raised, the height h increases (if the density of the liquid and the angle of contact remain the same).

Statement-2 : Surface tension of a liquid decreases with the rise in its temperature.

(A) Statement-1 is true, Statement-2 is true; Statement-2 is not the correct explanation for Statement-1

(B) Statement-1 is false, Statement-2 is true

(C) Statement-1 is true, Statement-2 is false

(D) Statement-1 is true, Statement-2 is true; Statement-2 is the correct explanation for Statement-1

10. Wax is coated on the inner wall of a capillary tube and the tube is then dipped in water. Then, compared to the unwaxed capillary, the angle of contact θ and the height h upto which water rises change. These changes are : **[2013 (Online)]**

(A) θ increases and h also increases

(B) θ decreases h also decreases

(C) θ increases and h decreases

(D) θ decreases and h increases

11. On heating water, bubbles being formed at the bottom of the vessel detach and rise. Take the bubbles to be spheres of radius R and making a circular contact of radius r with the bottom of the vessel. If $r << R$ and the surface tension of water is T, value of r just before bubbles detach is : (density of water is ρ_w) : **[2014]**

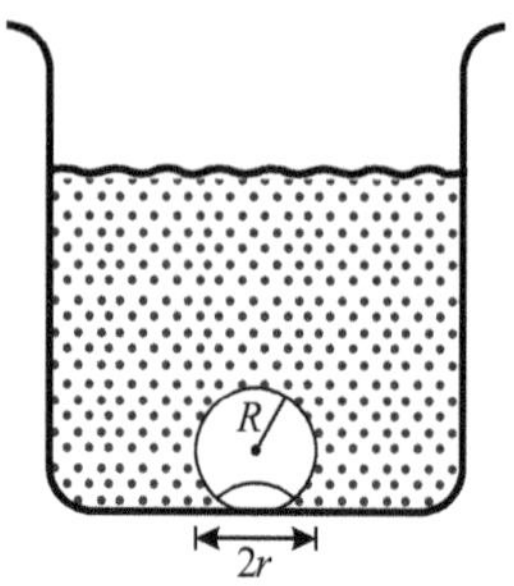

(A) $R^2\sqrt{\dfrac{2\rho_w g}{3T}}$ (B) $R^2\sqrt{\dfrac{\rho_w g}{6T}}$

(C) $R^2\sqrt{\dfrac{\rho_w g}{T}}$ (D) $R^2\sqrt{\dfrac{3\rho_w g}{T}}$

12. An air bubble of radius 0.1 cm is in a liquid having surface tension 0.06 N/m and density 10^3 kg/m^3. The pressure inside the bubble is 1100 Nm^{-2} greater than the atmospheric pressure. At what depth is the bubble below the surface of the liquid ? ($g = 9.8$ ms^{-2}) : **[2014 (Online)]**

(A) 0.1 m (B) 0.15 m

(C) 0.20 m (D) 0.25 m

13. Two soap bubbles coalesce to form a single bubble. If V is the subsequent change in volume of contained air and S change in total surface are, T is the surface tension and P atmospheric pressure, then which of the following relation is correct ? **[2014 (Online)]**

(A) $4PV + 3ST = 0$ (B) $3PV + 4ST = 0$

(C) $2PV + 3ST = 0$ (D) $3PV + 2ST = 0$

14. If two glass plates have water between them and are separated by very small distance (see figure), it is very difficult to pull them apart. It is because the water in between form cylindrical surface on the side that gives rise to lower pressure in the water in comparison to atmosphere. If the radius of the cylindrical surface is R and surface tension of water is T then the pressure in water between the plates is lower by : **[2015 (Online)]**

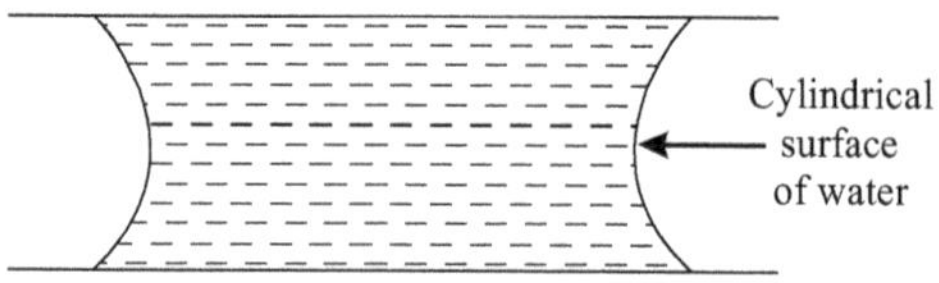

(A) $\dfrac{2T}{R}$ (B) $\dfrac{4T}{R}$

(C) $\dfrac{T}{4R}$ (D) $\dfrac{T}{R}$

15. A soap bubble, blown by a mechanical pump at the mouth of a tube, increases in volume, with time, at a constant rate. The graph that correctly depicts the time dependence of pressure inside the bubble is given by : **[12 Jan. 2019 (Evening)]**

(A)

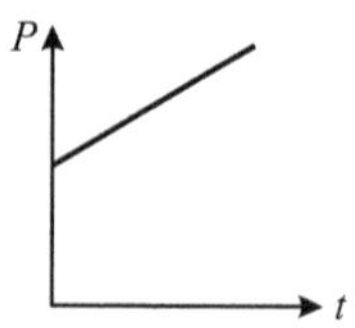

(B)

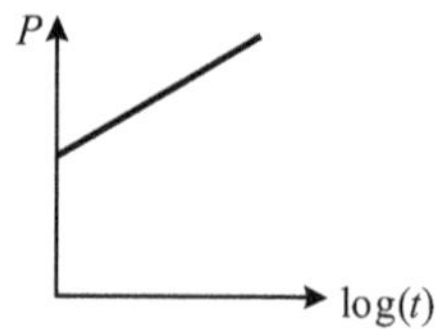

(C)

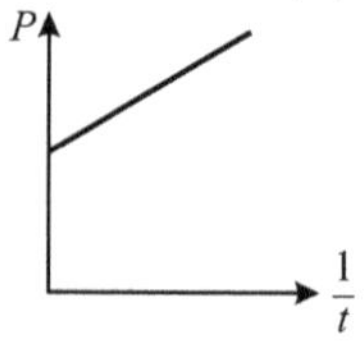

(D)

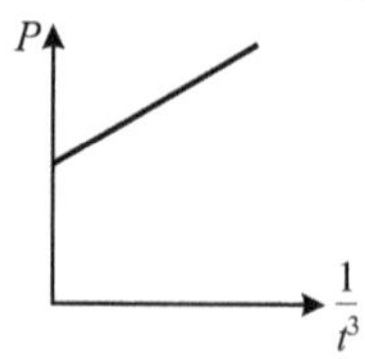

16. If 'M' is the mass of water that rises in a capillary tube of radius 'r', then mass of water which will rise in a capillary tube of radius '$2r$' is : **[9 April 2019 (Morning)]**

(A) $4M$ (B) M

(C) $2M$ (D) $\dfrac{M}{2}$

17. The ratio of surface tensions of mercury and water is given to be 7.5 while the ratio of thier densities is 13.6. Their contact angles, with glass, are close to 135° and 0°, respectively. It is observed that mercury gets depressed by an amount h in a capillary tube of radius r_1, while water rises by the same amount h in a capillary tube of radius r_2. The ratio, (r_1/r_2), is then close to : **[10 April 2019 (Morning)]**

(A) 2/3 (B) 3/5

(C) 2/5 (D) 4/5

* * * * *

CHAPTER-10 THERMAL PROPERTIES OF MATTER

10.1 Thermal Expansion

1. A metal rod of Young's modulus Y and coefficient of thermal expansion α is held at its two ends such that its length remains invariant. If its temperature is raised by t°C, the linear stress developed in it is : **[2011]**

(A) $\frac{\alpha t}{Y}$ (B) $\frac{Y}{\alpha t}$

(C) $Y\alpha t$ (D) $\frac{1}{(Y\alpha t)}$

2. An aluminium sphere of 20 cm diameter is heated from 0°C to 100°C. Its volume changes by (given that coefficient of linear expansion for aluminium $\alpha_{Al} = 23 \times 10^{-6}$/°C) : **[2011]**

(A) 28.9 cc (B) 2.89 cc
(C) 9.28 cc (D) 49.8 cc

3. A wooden wheel of radius R is made of two semicircular parts (see figure). The two parts are held together by a ring made of a metal strip of cross sectional area S and Length L. L is slightly less than $2\pi R$. To fit the ring on the wheel, it is heated so that its temperature rises by ΔT and it just steps over the wheel. As it cools down to surrounding temperature, it presses the semicircular parts together. If the coefficient of linear expansion of the metal is α, and its Young's modulus is Y, the force that one part of the wheel applies on the other part is : **[2012]**

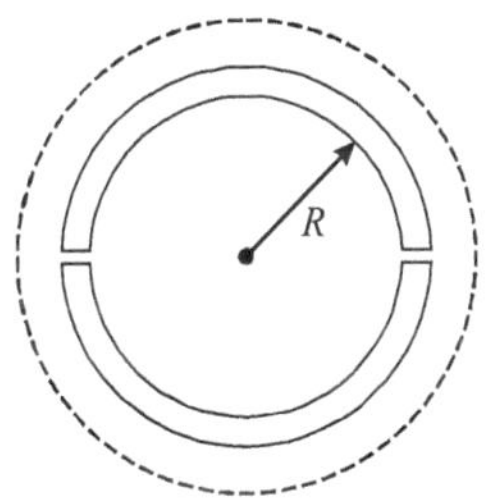

(A) $2SY\alpha\Delta T$ (B) $2\pi SY\alpha\Delta T$
(C) $SY\alpha\Delta T$ (D) $\pi SY\alpha\Delta T$

4. On a linear temperature scale Y, water freezes at – 160°Y and boils at – 50°Y. On this Y scale, a temperature of 340 K would be read as : (water freezes at 273 K and boils at 373 K) : **[2013 (Online)]**

(A) –73.7°Y (B) –233.7°Y
(C) –86.3°Y (D) –106.3°Y

5. The ratio of the coefficient of volume expansion of a glass container to that of a viscous liquid kept inside the container is 1 : 4. What fraction of the inner volume of the container should the liquid occupy so that the volume of the remaining vacant space will be same at all temperatures ? **[2013 (Online)]**

(A) 2 : 5 (B) 1 : 4
(C) 1 : 64 (D) 1 : 8

6. The pressure that has to be applied to the ends of a steel wire of length 10 cm to keep its length constant when its temperature is raised by 100°C is (For steel Young's modulus is 2×10^{11} Nm^{-2} and coefficient of thermal expansion is 1.1×10^{-5} K^{-1}) : **[2014]**

(A) 2.2×10^8 Pa (B) 2.2×10^9 Pa
(C) 2.2×10^7 Pa (D) 2.2×10^6 Pa

7. A pendulum clock loses 12 s a day if the temperature is 40°C and gains 4 s a day if the temperature is 20°C. The temperature at which the clock will show correct time, and the co-efficient of linear expansion (α) of the metal of the pendulum shaft are respectively : **[2016]**

(A) 25°C; $\alpha = 1.85 \times 10^{-5}$/°C (B) 60°C; $\alpha = 1.85 \times 10^{-4}$/°C
(C) 30°C; $\alpha = 1.85 \times 10^{-3}$/°C (D) 55°C; $\alpha = 1.85 \times 10^{-2}$/°C

8. An external pressure P is applied on a cube at 0°C so that it is equally compressed from all sides. K is the bulk modulus of the material of the cube and α is its coefficient of linear expansion. Suppose we want to bring the cube to its original size by heating. The temperature should be raised by : **[2017]**

(A) $\frac{3\alpha}{PK}$ (B) $3PK\alpha$

(C) $\frac{P}{3\alpha K}$ (D) $\frac{P}{\alpha K}$

9. A compressive force, F is applied at the two ends of a long thin steel rod. It is heated, simultaneously, such that its temperature increases by ΔT. The net change in its length is zero. Let l be the length of the rod, A its area of cross-section, Y its Young's modulus, and α its coefficient of linear expansion. Then, F is equal to : **[2017 (Online)]**

(A) $l^2 Y\alpha\Delta T$ (B) $lA\, Y\alpha\Delta T$

(C) $AY\alpha\Delta T$ (D) $\frac{AY}{\alpha\Delta T}$

10. A steel rail of length 5 m and area of cross-section 40 cm^2 is prevented from expanding along its length while the temperature rises by 10°C. If coefficient of linear expansion and Young's modulus of steel are 1.2×10^{-5} K^{-1} and 2×10^{11} Nm^{-2} respectively, the force developed in the rail is approximately : **[2017 (Online)]**

(A) 2×10^7 N (B) 1×10^5 N
(C) 2×10^9 N (D) 3×10^{-5} N

11. A rod, of length L at room temperature and uniform area of cross section A, is made of a metal having coefficient of linear expansion α/ °C^{-1}. It is observed that an external compressive force F, is applied on each of its ends, prevents any change in the length of the rod, when its temperature rises by $\Delta T\ K$. Young's modulus Y, for this metal is : **[9 Jan. 2019 (Morning)]**

(A) $\frac{F}{2A\alpha\Delta T}$ (B) $\frac{F}{A\alpha(\Delta T - 273)}$

(C) $\frac{F}{A\alpha\Delta T}$ (D) $\frac{2F}{A\alpha\Delta T}$

12. Two rods A and B of identical dimensions are at temperature 30°C. If A is heated upto 180°C and B upto T°C, then the new lengths are the same. If the ratio of the coefficients of linear expansion of A and B is 4 : 3, then the value of T is : **[11 Jan. 2019 (Evening)]**

(A) 270°C (B) 230°C

(C) 250°C (D) 200°C

13. A thermometer graduated according to a linear scale reads a value x_0 when in contact with boiling water, and $x_0/3$ when in contact with ice. What is the temperature of an object in °C, if this thermometer is in the contact with the object reads $x_0/2$? **[11 Jan. 2019 (Evening)]**

(A) 35 (B) 25

(C) 60 (D) 40

* * * * *

CHAPTER-10 THERMAL PROPERTIES OF MATTER

10.2 Calorimetry

1. Heat given to a body which raises its temperature by 1°C is: **[2002]**

(A) Water equivalent (B) Thermal capacity
(C) Specific heat (D) Temperature gradient

2. If mass-energy equivalence is taken into account, when water is cooled to form ice, the mass of water should : **[2002]**

(A) Increase (B) Remain unchanged
(C) Decrease (D) First increase then decrease

3. Given that 1 g of water in liquid phase has volume 1 cm^3 and in vapour phase 1671 cm^3 at atmospheric pressure and the latent heat of vaporization of water is 2256 J/g; the change in the internal energy in joules for 1 g of water at 373 K when it changes from liquid phase to vapour phase at the same temperature is : **[2013 (Online)]**

(A) 2256 (B) 167
(C) 2089 (D) 1

4. 500 g of water and 100 g of ice at 0°C are in a calorimeter whose water equivalent is 40 g. 10 g of steam at 100°C is added to it. Then water in the calorimeter is : (Latent heat of ice = 80 cal/g, Latent heat of steam = 540 cal/g) : **[2013 (Online)]**

(A) 580 g (B) 590 g
(C) 600 g (D) 610 g

5. Water of volume 2*L* in a closed container is heated with a coil of 1 kW. While water is heated, the container loses energy at a rate of 160 J/s. In how much time will the temperature of water rise from 27°C to 77°C ? (Specific heat of water is 4.2 kJ/kg and that of the container is negligible) : **[2014 (Online)]**

(A) 8 min 20 s (B) 6 min 2 s
(C) 7 min (D) 14 min

6. An experiment takes 10 minutes to raise the temperature of water in a container from 0°C to 100°C and another 55 minutes to convert it totally into steam by a heater supplying heat at a uniform rate. Neglecting the specific heat of the container and taking specific heat of water to be 1 cal/g °C, the heat of vapourization according to this experiment will come out to be: **[2015 (Online)]**

(A) 560 cal/g (B) 550 cal/g
(C) 540 cal/g (D) 530 cal/g

7. A copper ball of mass 100 gm is at a temperature *T*. It is dropped in a copper calorimeter of mass 100 gm, filled with 170 gm of water at room temperature. Subsequently, the temperature of the system is found to be 75°C. *T* is given by (Given : room temperature = 30°C, specific heat of copper = 0.1 cal/gm°C) : **[2017]**

(A) 1250°C` (B) 825°C
(C) 800°C (D) 885°C

8. In an experiment a sphere of aluminium of mass 0.20 kg is heated upto 150°C. Immediately, it is put into water of volume 150 cc at 27°C kept in a calorimeter of water equivalent to 0.025 kg. Final temperature of the system is 40°C. The specific heat of aluminium is : [Take 4.2 Joule = 1 calorie] **[2017 (Online)]**

(A) 378 J/kg – °C (B) 315 J/kg – °C
(C) 476 J/kg – °C (D) 434 J/kg – °C

9. An unknown metal of mass 192 g heated to a temperature of 100°C was immersed into a brass calorimeter of mass 128 g containing 240 g of water a temperature of 8.4°C Calculate the specific heat of the unknown metal if water temperature stabilizes at 21.5°C (Specific heat of brass is 394 J kg^{-1} K^{-1}) : **[10 Jan. 2019 (Evening)]**

(A) 1232 J kg^{-1} K^{-1} (B) 458 J kg^{-1} K^{-1}
(C) 654 J kg^{-1} K^{-1} (D) 916 J kg^{-1} K^{-1}

10. Ice at –20°C is added to 50 g of water at 40°C. When the temperature of the mixture reaches 0°C, it is found that 20 g of ice is still unmelted. The amount of ice added to the water was close to **[11 Jan. 2019 (Morning)]**
(Specific heat of water = 4.2 J/g/°C)
Specific heat of Ice = 2.1 J/g/°C
Heat of fusion of water at 0°C = 334 J/g) :

(A) 50 g (B) 40 g
(C) 60 g (D) 100 g

11. When 100 g of a liquid *A* at 100°C is added to 50 g of a liquid *B* at temperature 75°C, the temperature of the mixture becomes 90°C. The temperature of the mixture, if 100 g of liquid *A* at 100°C is added to 50 g of liquid *B* at 50°C, will be : **[11 Jan. 2019 (Evening)]**

(A) 80°C (B) 60°C
(C) 70°C (D) 85°C

12. A metal ball of mass 0.1 kg is heated upto 500°C and dropped into a vessel of heat capacity 800 JK^{-1} and containing 0.5 kg water. The initial temperature of water and vessel is 30°C. What is the approximate percentage increment in the temperature of the water ? [Specific Heat Capacities of water and metal are, respectively, 4200 $Jkg^{-1}K^{-1}$ and 400 $JKg^{-1}K^{-1}$] : **[11 Jan. 2019 (Evening)]**

(A) 30% (B) 20%
(C) 25% (D) 15%

13. A thermally insulated vessel contains 150 g of water at 0°C. Then the air from the vessel is pumped out adiabatically. A fraction of water turns into ice and the rest evaporates at 0°C itself. The mass of evaporated water will be closest to :
(Latent heat of vaporization of water = 2.10×10^6 J kg^{-1} and Latent heat of Fusion of water = 3.36×10^5 J kg^{-1}) :

[8 April 2019 (Morning)]

(A) 130 g (B) 35 g
(C) 20 g (D) 150 g

14. When M_1 gram of ice at −10°C (specific heat = 0.5 cal g^{-1} °C^{-1}) is added to M_2 gram of water at 50°C, finally no ice is left and the water is at 0°C. The value of latent heat of ice, in cal g^{-1} is :

[12 April 2019 (Morning)]

(A) $\frac{5M_1}{M_2} - 50$ (B) $\frac{50M_2}{M_1}$
(C) $\frac{50M_2}{M_1} - 5$ (D) $\frac{5M_2}{M_1} - 5$

* * * * *

CHAPTER-10 THERMAL PROPERTIES OF MATTER

10.3 Heat Transfer

1. Two spheres of the same material have radii 1 m and 4 m and temperatures 4000 K and 2000 K respectively. The ratio of the energy radiated per second by the first sphere to that by the second is : **[2002]**

(A) 1 : 1 (B) 16 : 1
(C) 4 : 1 (D) 1 : 9

2. Infrared radiations are detected by : **[2002]**

(A) Spectrometer (B) Pyrometer
(C) Nanometer (D) Photometer

3. Which of the following is more close to a black body : **[2002]**

(A) Black board paint (B) Green leaves
(C) Black holes (D) Red roses

4. The earth radiates in the infra-red region of the spectrum. The spectrum is correctly given by : **[2003]**

(A) Rayleigh Jeans law (B) Plank's law of radiation
(C) Stefan's law of radiation (D) Wien's law

5. According to Newton's law of cooling, the rate of cooling of a body is proportional to $(\Delta\theta)^n$, where $\Delta\theta$ is the difference of the temperature of the body and the surroundings, and n is equal to : **[2003]**

(A) Two (B) Three
(C) Four (D) One

6. If the temperature of the sun were to increase from T to $2T$ and its radius from R to $2R$, then the ratio of the radiant energy received on earth to what it was previously, will be : **[2004]**

(A) 4 (B) 16
(C) 32 (D) 64

7. The temperature of the two outer surfaces of a composite slab, consisting of two materials having coefficients of thermal conductivity K and $2K$ and thickness x and $4x$, respectively are T_2 and T_1 ($T_2 > T_1$). The rate of heat transfer through the slab, in a steady state is $\left(\frac{A(T_2 - T_1)K}{x}\right) f$, with f equals to (A = cross sectional area) : **[2004]**

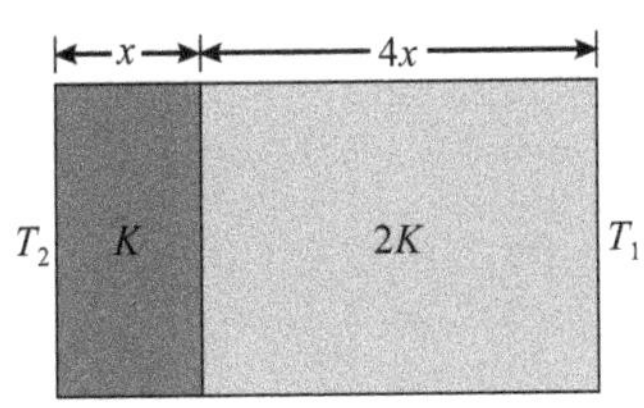

(A) 1 (B) 1/2
(C) 2/3 (D) 1/3

8. The figure shows a system of two concentric spheres of radii r_1 and r_2 and kept at temperatures T_1 and T_2, respectively. The radial rate of flow of heat in a substance between the two concentric spheres, is proportional to : **[2005]**

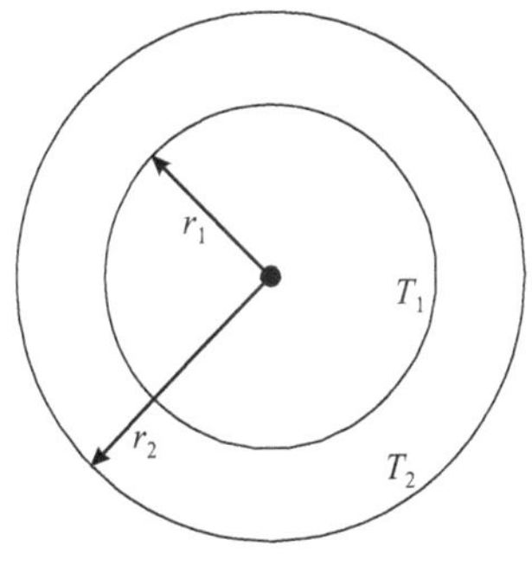

(A) $\frac{(r_2 - r_1)}{(r_1 r_2)}$ (B) $\ln\left(\frac{r_2}{r_1}\right)$

(C) $\frac{r_1 r_2}{(r_2 - r_1)}$ (D) $(r_2 - r_1)$

9. Assuming the sun to be a spherical body of radius R at a temperature of T K, evaluate the total radiant power, incident on earth, at a distance r from the sun-(when radius of earth is r_0) : **[2006]**

(A) $4\pi r_0^2 R^2 \sigma T^4/r^2$ (B) $\pi r_0^2 R^2 \sigma T^4/r^2$
(C) $r_0^2 R^2 \sigma T^4/4\pi r^2$ (D) $R^2 \sigma T^4/r^2$

10. One end of a thermally insulated rod is kept at a temperature T_1 and the other at T_2. The rod is composed of two sections of lengths l_1 and l_2 and thermal conductivities K_1 and K_2 respectively. The temperature at the interface of the two sections is : **[2007]**

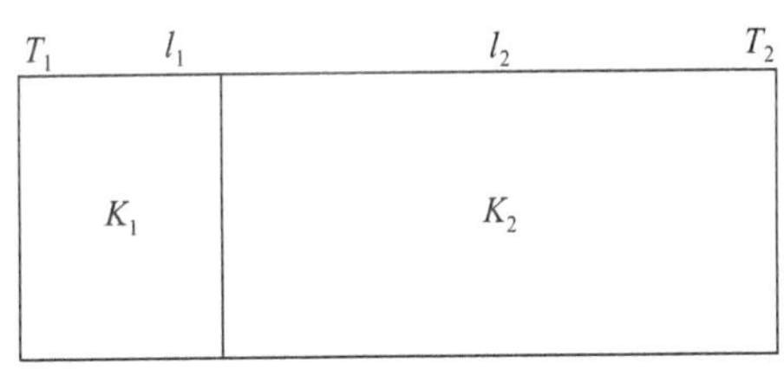

(A) $(K_2 l_2 T_1 + K_1 l_1 T_2)/(K_1 l_1 + K_2 l_2)$
(B) $(K_2 l_1 T_1 + K_1 l_2 T_2)/(K_2 l_1 + K_1 l_2)$
(C) $(K_1 l_2 T_1 + K_2 l_1 T_2)/(K_1 l_2 + K_2 l_1)$
(D) $(K_1 l_1 T_1 + K_2 l_2 T_2)/(K_1 l_1 + K_2 l_2)$

11. A long metallic bar is carrying heat from one of its ends to the other end under steady-state. The variation of temperature θ along the length x of the bar from its hot end is best described by which of the following figures? **[2009]**

(A) 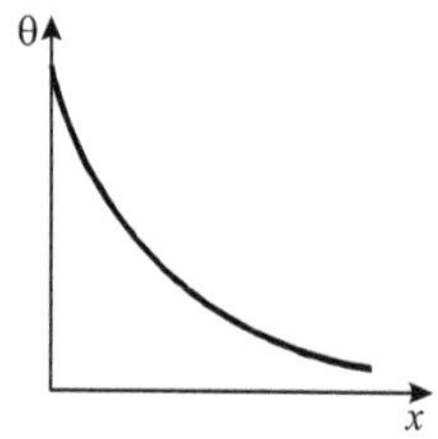 (B) 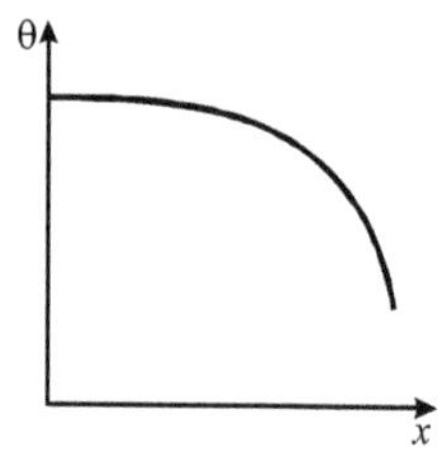

(C) 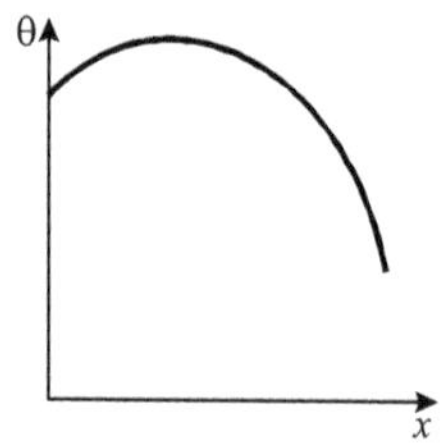 (D) 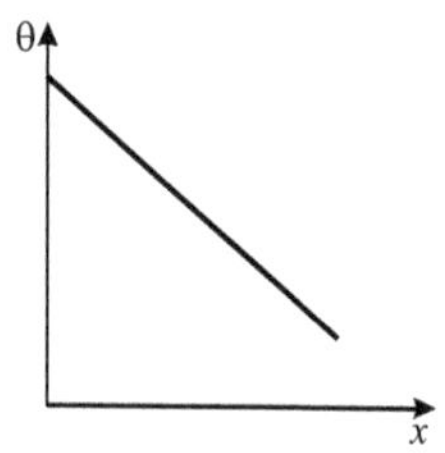

12. A liquid in a beaker has temperature $\theta(t)$ at time t and θ_0 is temperature of surroundings, then according to Newton's law of cooling the correct graph between $\log_e(\theta - \theta_0)$ and t is : **[2012]**

(A) 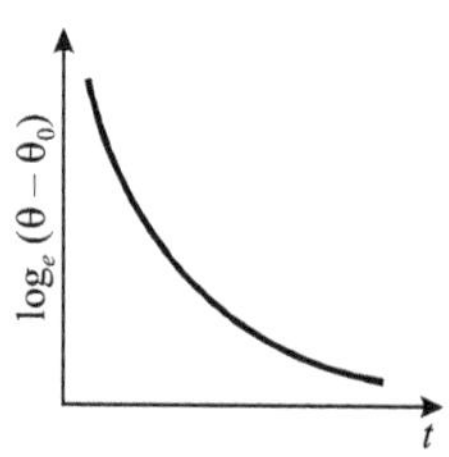 (B) 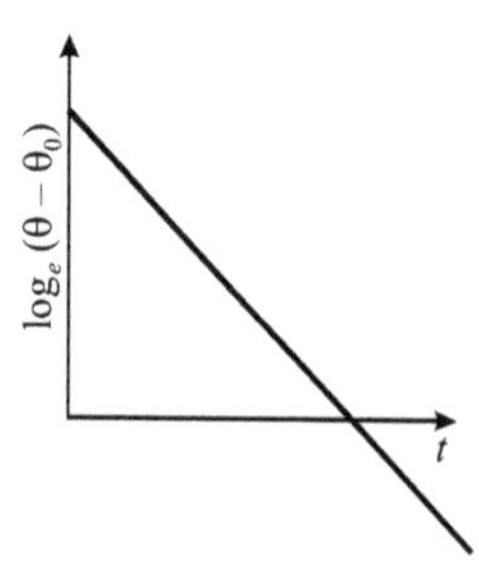

(C) 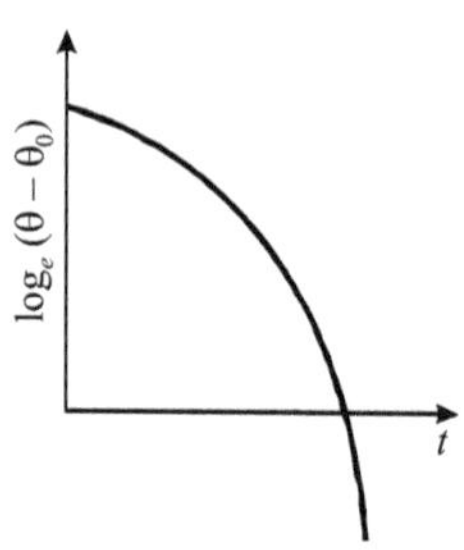 (D) 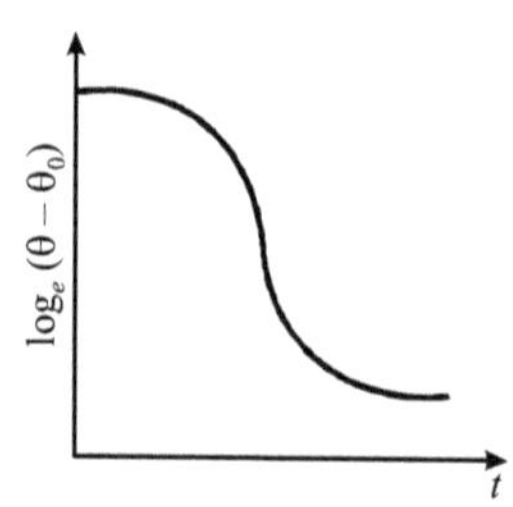

13. The heat radiated per unit area in 1 hour by a furnace whose temperature is 3000 K is ($\sigma = 5.7 \times 10^{-8}$ Wm^{-2} K^{-4}) : **[2012 (Online)]**

(A) 1.7×10^{10} J (B) 1.1×10^{12} J
(C) 2.8×10^{8} J (D) 4.6×10^{6} J

14. A large cylindrical rod of length L is made by joining two identical rods of copper and steel of length $\left(\frac{L}{2}\right)$ each. The rods are completely insulated from the surroundings. If the free end of copper rod is maintained at 100°C and that of steel at 0°C then the temperature of junction is (Thermal conductivity of copper is 9 times that of steel) : **[2012 (Online)]**

(A) 90°C (B) 50°C
(C) 10°C (D) 67°C

15. If a piece of metal is heated to temperature θ and then allowed to cool in a room which is at temperature θ_0 the graph between the temperature T of the metal and time t will be closed to : **[2013]**

(A) 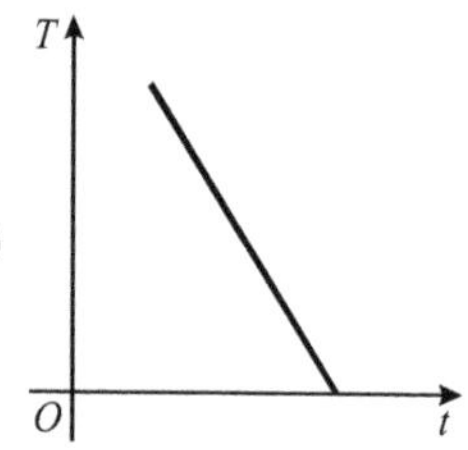 (B) 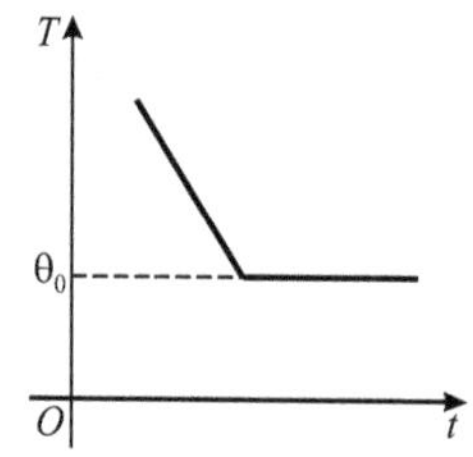

(C) 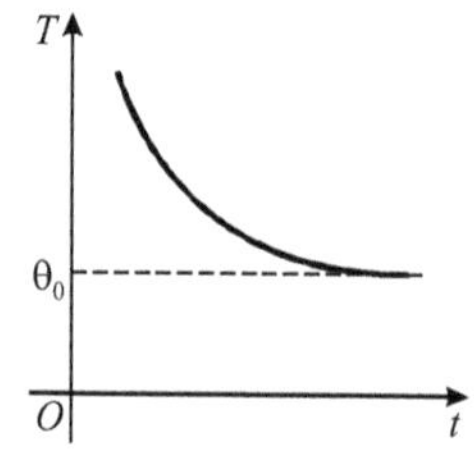 (D) 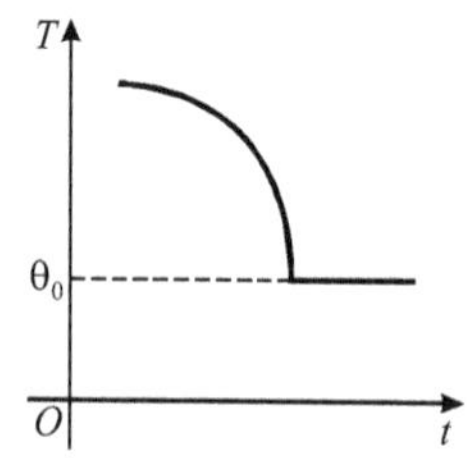

16. A mass of 50g of water in a closed vessel, with surroundings at a constant temperature takes 2 minutes to cool from 30°C to 25°C. A mass of 100 g of another liquid in an identical vessel with identical surroundings takes the same time to cool from 30° C to 25°C. The specific heat of the liquid is :
(The water equivalent of the vessel is 30 g) : **[2013 (Online)]**

(A) 2.0 kcal/kg (B) 7 kcal/kg
(C) 3 kcal/kg (D) 0.5 kcal/kg

17. Three rods of Copper, Brass and Steel are welded together to form a Y-shaped structure. Area of cross-section of each rod = 4 cm^2. End of copper rod is maintained at 100°C where as ends of brass and steel are kept at 0°C. Lengths of the copper, brass and steel rods are 46, 13 and 12 cms respectively. The rods are thermally insulated from surroundings except at ends. Thermal conductivities of copper, brass and steel are 0.92, 0.26 and 0.12 CGS units respectively. Rate of heat flow through copper rod is : **[2014]**

(A) 1.2 cal/s (B) 2.4 cal/s
(C) 4.8 cal/s (D) 6.0 cal/s

18. A black coloured solid sphere of radius R and mass M is inside a cavity with vacuum inside. The walls of the cavity are maintained at temperature T_0. The initial temperature of the sphere is $3T_0$. If the specific heat of the material of the sphere varies as αT^3 per unit mass with the temperature T of the sphere, where α is a constant, then the time taken for the sphere to cool down to temperature $2T_0$ will be (σ is Stefan Boltzmann constant): **[2014 (Online)]**

(A) $\frac{M\alpha}{4\pi R^2\sigma} In\left(\frac{3}{2}\right)$ (B) $\frac{M\alpha}{4\pi R^2\sigma} In\left(\frac{16}{3}\right)$

(C) $\frac{M\alpha}{16\pi R^2\sigma} In\left(\frac{16}{3}\right)$ (D) $\frac{M\alpha}{16\pi R^2\sigma} In\left(\frac{3}{2}\right)$

19. A hot body, obeying Newton's law of cooling is colling down from its peak value 80°C to an ambient temperature of 30°C. it takes 5 minutes in colling down from 80°C to 40°C. How much time will it take to cool down from 62°C to 32°C? (Given In 2 = 0.693, In 5 = 1.609) : **[2014 (Online)]**

(A) 3.75 minutes (B) 8.6 minutes
(C) 9.6 minutes (D) 6.5 minutes

20. Hot water cools from 60°C to 50°C in the first 10 minutes and to 42°C in the next 10 minutes. The temperature of the surroundings is : **[2014 (Online)]**

(A) 25°C (B) 10°C
(C) 15°C (D) 20°C

21. Temperature difference of 120°C is maintained between two ends of a uniform rod AB of length $2L$. Another bent rod PQ, of same cross-section as AB and length $\frac{3L}{2}$, is connected across AB (See figure). In steady state, temperature difference between P and Q will be close to : **[9 Jan. 2019 (Morning)]**

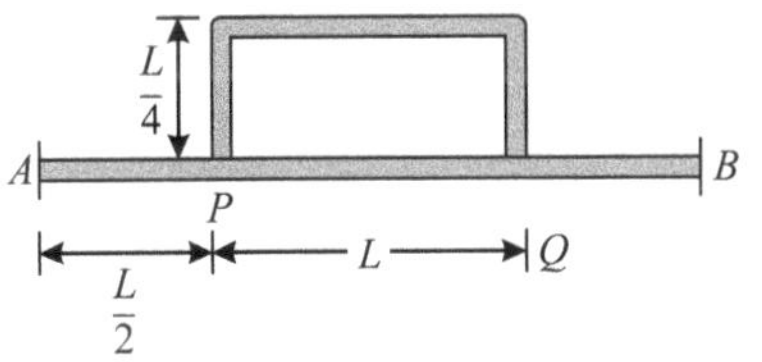

(A) 60°C (B) 75°C
(C) 35°C (D) 45°C

22. A heat source at $T = 10^3$ K is connected to another heat reservoir at $T = 10^2$ K by a copper slab which is 1 m thick. Given that the thermal conductivity of copper is 0.1 WK^{-1} m^{-1}, the energy flux through it in the steady state is : **[10 Jan. 2019 (Morning)]**

(A) 90 Wm^{-2} (B) 200 Wm^{-2}
(C) 65 Wm^{-2} (D) 120 Wm^{-2}

23. A cylinder of radius R is surrounded by a cylindrical shell of inner radius R and outer radius $2R$. The thermal conductivity of the material of the inner cylinder is K_1 and that of the outer cylinder is K_2. Assuming no loss of heat, the effective thermal conductivity of the system for heat flowing along the length of the cylinder is : **[12 Jan. 2019 (Morning)]**

(A) $K_1 + K_2$ (B) $\frac{K_1 + K_2}{2}$
(C) $\frac{2K_1 + 3K_2}{5}$ (D) $\frac{K_1 + 3K_2}{4}$

24. Two identical beakers A and B contain equal volumes of two different liquids at 60°C each and left to cool down. Liquid in A has density of 8×10^2 kg/m^3 and specific heat of 2000 J kg^{-1} K^{-1} while liquid in B has density of 10^3 kg m^{-3} and specific heat of 4000 J kg^{-1} K^{-1}. Which of the following best describes their temperature versus time graph schematically? (assume the emissivity of both the beakers to be the same) : **[8 April 2019 (Morning)]**

(A) 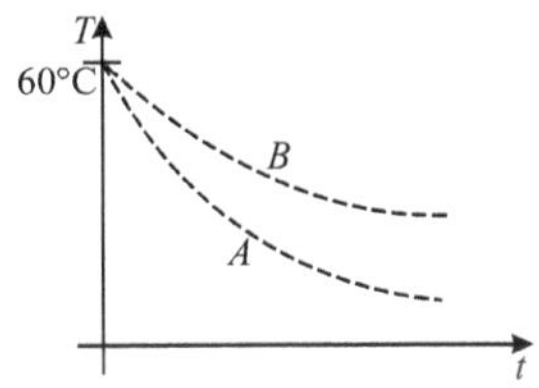

(B) 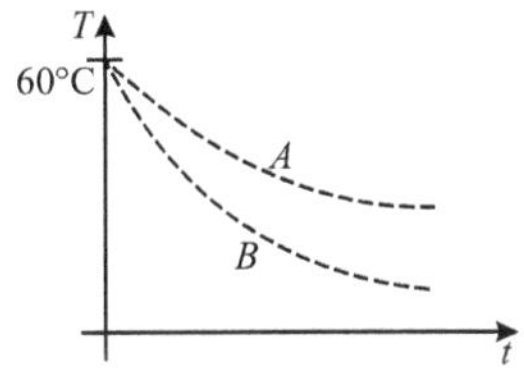

(C) 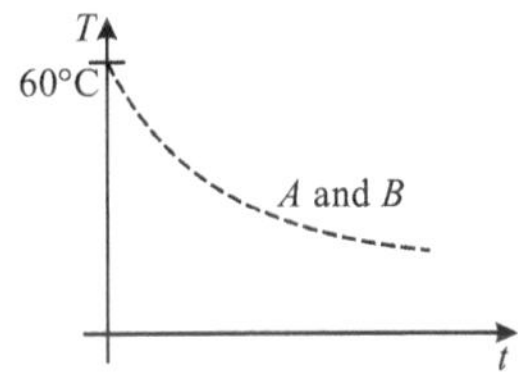

(D) 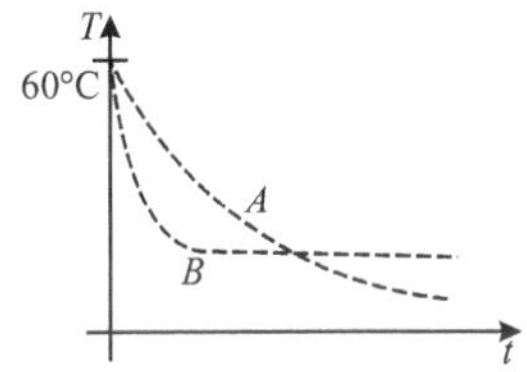

25. Two materials having coefficients of thermal conductivity '$3K$' and 'K' and thickness 'd' and '$3d$', respectively, are joined to form a slab as shown in the figure. The temperatures of the outer surfaces are 'θ_2' and 'θ_1' respectively, ($\theta_2 > \theta_1$). The temperature at the interface is : **[9 April 2019 (Evening)]**

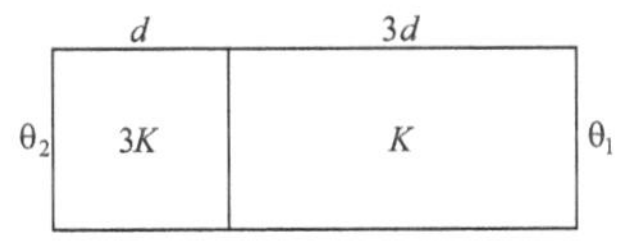

(A) $\frac{\theta_2 + \theta_1}{2}$ (B) $\frac{\theta_1}{10} + \frac{9\theta_2}{10}$
(C) $\frac{\theta_1}{3} + \frac{2\theta_2}{3}$ (D) $\frac{\theta_1}{6} + \frac{5\theta_2}{6}$

* * * * *

CHAPTER-11 THERMODYNAMICS

11.1 Laws of Thermodynamics

1. Which statement is incorrect ? **[2002]**
(A) All reversible cycles have same efficiency
(B) Reversible cycle has more efficiency than an irreversible one
(C) Carnot cycle is a reversible one
(D) Carnot cycle has the maximum efficiency in all cycles

2. Even carnot engine cannot give 100% efficiency because we cannot : **[2002]**
(A) Prevent radiation
(B) Find ideal sources
(C) Reach absolute zero temperature
(D) Eliminate friction

3. A carnot engine takes 3×10^6 cal of heat from a reservoir at 627 °C and gives it to a sink at 27 °C. The work done by the engine is : **[2003]**
(A) 4.2×10^6 J (B) 8.4×10^6 J
(C) 16.8×10^6 J (D) Zero

4. During an adiabatic process, the pressure of a gas is found to be proportional to the cube of its absolute temperature. The ratio C_P/C_V for the gas is : **[2003]**
(A) 4/3 (B) 2
(C) 5/3 (D) 3/2

5. Which of the following parameters does not characterise the thermodynamic state of matter ? **[2003]**
(A) Temperature (B) Pressure
(C) Work (D) Volume

6. "Heat cannot be itself flow from a body at lower temperature to a body at higher temperature" is a statement or consequence of: **[2003]**
(A) Second law of thermodynamics
(B) Conservation of momentum
(C) Conservation of mass
(D) First law of thermodynamics

7. Which of the following statements is correct for any thermodynamic system ? **[2004]**
(A) The internal energy changes in all processes
(B) Internal energy and entropy are state functions
(C) The change in entropy can never be zero
(D) The work done in an adiabatic process is always zero

8. A system goes from A to B via two processes I and II as shown in figure. If ΔU_1 and ΔU_2 are the changes in internal energies in the processes I and II respectively then : **[2005]**

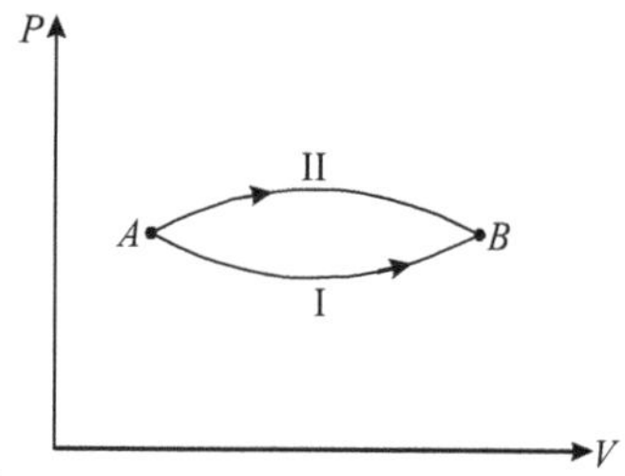

(A) $\Delta U_1 = \Delta U_2$
(B) Relation between ΔU_1 and ΔU_2 cannot be determined
(C) $\Delta U_2 > \Delta U_1$ (D) $\Delta U_2 < \Delta U_1$

9. The temperature-entropy diagram of a reversible engine cycle is given in the figure. Its efficiency is : **[2005]**

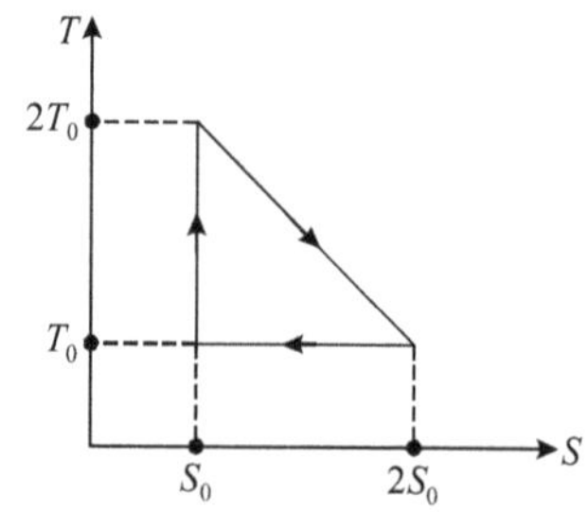

(A) 1/2 (B) 1/4
(C) 1/3 (D) 2/3

10. Which of the following is incorrect regarding the first law of thermodynamics ? **[2005]**
(A) It is applicable to any cyclic process
(B) It is a restatement of the principle of conservation of energy
(C) It introduces the concept of the internal energy
(D) It introduced the concept of the entropy

11. The work of 146 kJ is performed in order to compress one kilo mole of a gas adiabatically and in this process the temperature of the gas increases by 7°C. The gas is : $[R = 8.3 \text{ J mol}^{-1} \text{ K}^{-1}]$ **[2006]**
(A) Diatomic
(B) Triatomic
(C) A mixture of monoatomic and diatomic
(D) Monoatomic

12. When a system is taken from state i to state f along the path iaf, it is found that $Q = 50$ cal and $W = 20$ cal. Along the path ibf $Q = 36$ cal. W along the path ibf is : **[2007]**

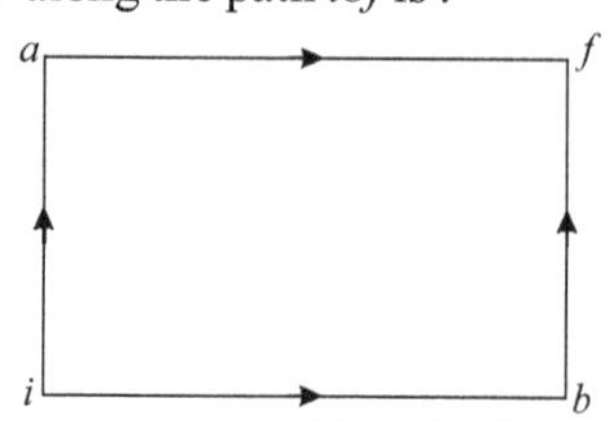

(A) 6 cal (B) 16 cal
(C) 66 cal (D) 14 cal

13. A carnot engine, having an efficiency of $\eta = 1/10$ as heat engine, is used as a refrigerator. If the work done on the system is 10 J, the amount of energy absorbed from the reservoir at lower temperature is : **[2007]**

(A) 99 J (B) 90 J
(C) 1 J (D) 100 J

14. Directions : This question is based on the following paragraph. Two moles of helium gas are taken over the cycle *ABCDA*, as shown in the *P-T* diagram.

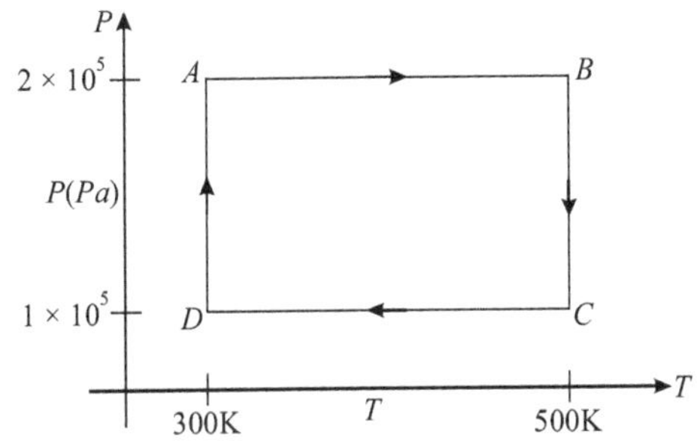

Assuming the gas to be ideal, the work done on the gas in taking it from *A* to *B* is : **[2009]**

(A) 400 R (B) 500 R
(C) 200 R (D) 300 R

15. Directions : This question is based on the following paragraph. Two moles of helium gas are taken over the cycle *ABCDA*, as shown in the *P-T* diagram. **[2009]**

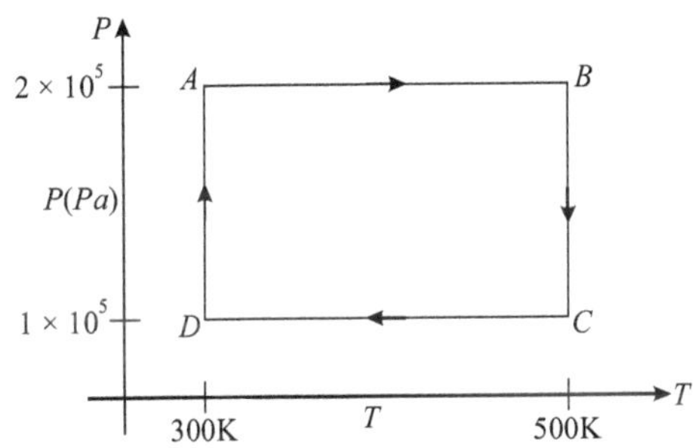

The work done on the gas in taking it from *D* to *A* is :

(A) −690 R (B) +690 R
(C) −414 R (D) +414 R

16. Directions : This question is based on the following paragraph.
Two moles of helium gas are taken over the cycle *ABCDA*, as shown in the *P-T* diagram.

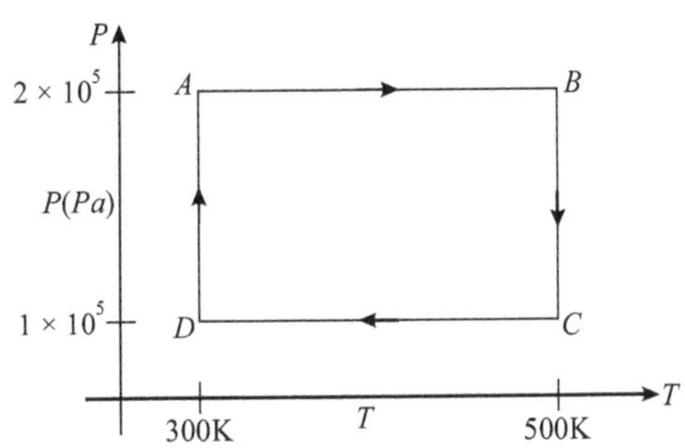

The net work done on the gas in the cycle *ABCDA* is : **[2009]**

(A) 1076R (B) 1904R
(C) Zero (D) 276R

17. A diatomic ideal gas is used in a carnot engine as the working substance. If during the adiabatic expansion part of the cycle the volume of the gas increases from V to $32V$, the efficiency of the engine is : **[2010]**

(A) 0.25 (B) 0.5
(C) 0.75 (D) 0.99

18. 100 g of water is heated from 30°C to 50°C Ignoring the slight expansion of the water, the change in its internal energy is (specific heat of water is 4184 J/kg/K) : **[2011]**

(A) 84 kJ (B) 2.1 kJ
(C) 4.2 kJ (D) 8.4 kJ

19. A container with insulating walls is divided into two equal parts by a partition fitted with a valve. One part is filled with an ideal gas at a pressure P and temperature T, whereas the other part is completely evacuated. If the valve is suddenly opened, the pressure and temperature of the gas will be : **[2011]**

(A) $\frac{P}{2}, T$ (B) $\frac{P}{2}, \frac{T}{2}$

(C) P, T (D) $P, \frac{T}{2}$

20. The specific heat capacity of a metal at low temperature (T) is given as $C_p(\text{kJK}^{-1}\,\text{kg}^{-1}) = 32\left(\frac{T}{400}\right)^3$. A 100 gram vessel of this metal is to be cooled from 20 K to 4 K by a special refrigerator operating at room temperature (27°C). The amount of work required to cool the vessel is : **[2011]**

(A) Equal to 0.002 kJ
(B) Greater than 0.148 kJ
(C) Between 0.148 kJ and 0.028 kJ
(D) Less than 0.028 kJ

21. A Carnot engine operating between temperatures T_1 and T_2 has efficiency has efficiency 1/6. When T_2 is lowered by 62 K, its efficiency increases to 1/3. Then T_1 and T_2 are, respectively : **[2011]**

(A) 330 K and 268 K (B) 310 K and 248 K
(C) 372 K and 310 K (D) 372 K and 330 K

22. A thermally insulated vessel contains an ideal gas of molecular mass M and ratio of specific heats γ. It is moving with speed v and is suddenly brought to rest. Assuming no heat is lost to the surroundings, its temperature increases by : **[2011]**

(A) $\frac{\gamma M v^2}{2R} K$ (B) $\frac{(\gamma-1)}{2R} M v^2 K$

(C) $\frac{(\gamma-1)}{2(\gamma+1)R} M v^2 K$ (D) $\frac{(\gamma-1)}{2\gamma R} M v^2 K$

23. A Carnot engine, whose efficiency is 40% takes in heat from a source maintained at a temperature of 500 K. It is desired to have an engine of efficiency 60%. Then, the intake temperature for the same exhaust (sink) temperature must be : **[2012]**

(A) 600 K

(B) Efficiency of Carnot engine cannot be made larger than 50%

(C) 1200 K

(D) 750 K

24. Helium gas goes through a cycle *ABCDA* (consisting of two isochoric and two isobaric lines) as shown in figure. Efficiency of this cycle is nearly (Assume the gas to be close to ideal gas) **[2012]**

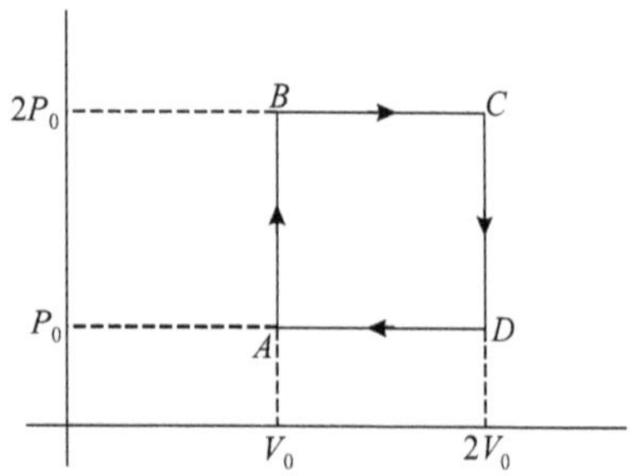

(A) 12.5% (B) 15.4%

(C) 9.1% (D) 10.5%

25. This question has Statement-1 and Statement-2. Of the four choices given after the Statements, choose the one that best describes the two Statements. **[2012 (Online)]**

Statement-1 : An inventor claims to have constructed an engine that has an efficiency of 30% when operated between the boiling and freezing points of water. This is not possible.

Statement-2 : The efficiency of a real engine is always less than the efficiency of a carnot engine operating between the same two temperatures.

(A) Statement-1 is true; Statement-2 is true, Statement-2 not the correct explanation of Statement-1

(B) Statement-1 is true, Statement-2 is false

(C) Statement-1 is false, Statement-2 is true

(D) Statement-1 is true, Statement-2 is true, Statement-2 is the correct explanation of Statement-1

26. This question has Statement-1 and Statement-2. Of the four choices given after the Statements, choose the one that best describes the two Statements. **[2012 (Online)]**

Statement-1 : In an adiabatic process, change in internal energy of a gas is equal to work done on/by the gas in the process.

Statement-2 : The temperature of a gas remains constant in an adiabatic process.

(A) Statement-1 is true; Statement-2 is true, Statement-2 is a correct explanation of Statement-1

(B) Statement-1 is true, Statement-2 is true

(C) Statement-1 is false, Statement-2 is true

(D) Statement-1 is false, Statement-2 is true, Statement-2 is not a correct explanation of Statement-1

27. *n* moles of an ideal gas undergo a process $A \to B$ as shown in the figure. Maximum temperature of the gas during the process is : **[2012 (Online)]**

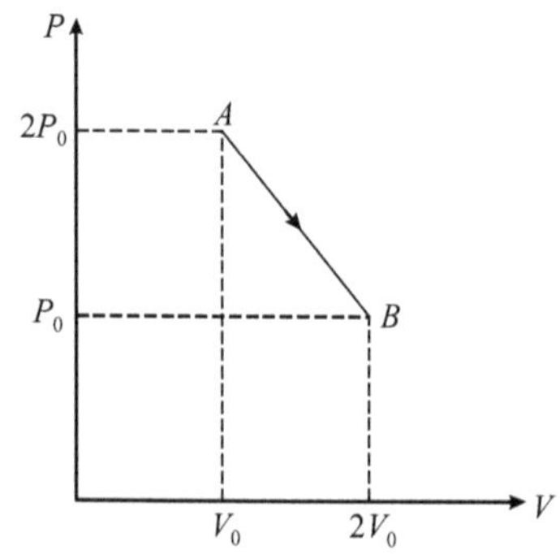

(A) $\frac{9P_0V_0}{nR}$ (B) $\frac{3P_0V_0}{2nR}$

(C) $\frac{9P_0V_0}{2nR}$ (D) $\frac{9P_0V_0}{4nR}$

28. The pressure of an ideal gas varies with volume as $P = \alpha V$, where α is a constant. One mole of the gas is allowed to undergo expansion such that its volume becomes '*m*' times its initial volume. The work done by the gas in the process is : **[2012 (Online)]**

(A) $\frac{\alpha V}{2}(m^2-1)$ (B) $\frac{\alpha^2 V^2}{2}(m^2-1)$

(C) $\frac{\alpha}{2}(m^2-1)$ (D) $\frac{\alpha V^2}{2}(m^2-1)$

29. An ideal monatomic gas with pressure *P*, volume *V* and temperature *T* is expanded isothermally to a volume 2*V* and a final pressure P_i. If the same gas is expanded adiabatically to a volume 2*V*, the final pressure is P_a. The ratio $\frac{P_a}{P_i}$ is : **[2012 (Online)]**

(A) $2^{-1/3}$ (B) $2^{1/3}$

(C) $2^{2/3}$ (D) $2^{-2/3}$

30. A given ideal gas with $\gamma = \frac{C_p}{C_v} = 1.5$ at a temperature *T*. If the gas is compressed adiabatically to one-fourth of its initial volume, the final temperature will be : **[2012 (Online)]**

(A) $2\sqrt{2}T$ (B) $4T$

(C) $2T$ (D) $8T$

31. The door of a working refrigerator is left open in a well insulated room. The temperature of air in the room will : **[2012 (Online)]**

(A) Decrease

(B) Increase in winters and decrease in summers

(C) Remains the same

(D) Increase

32. The above *p-v* diagram represents the thermodynamic cycle of an engine, operating with an ideal monoatomic gas. The amount of heat, extracted from the source in a single cycle is : **[2013]**

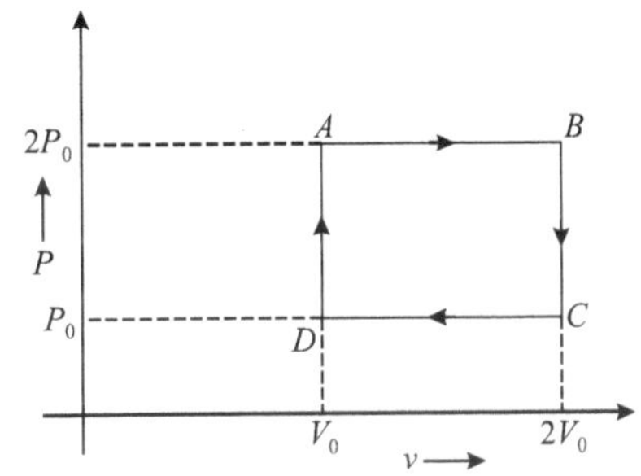

(A) p_0v_0 (B) $\left(\frac{13}{2}\right)p_0v_0$

(C) $\left(\frac{11}{2}\right)p_0v_0$ (D) $4p_0v_0$

33. An ideal gas at atmospheric pressure is adiabatically compressed so that its density becomes 32 times of its initial value. If the final pressure of gas is 128 atmospheres, the value of 'γ' of the gas is : **[2013 (Online)]**

(A) 1.5 (B) 1.4

(C) 1.3 (D) 1.6

34. A certain amount of gas is taken through a cyclic process (*A, B, C, D, A*) that has two isobars, one isochore and one isothermal. The cycle can be represented on a *P-V* indicator diagram as : **[2013 (Online)]**

(A)

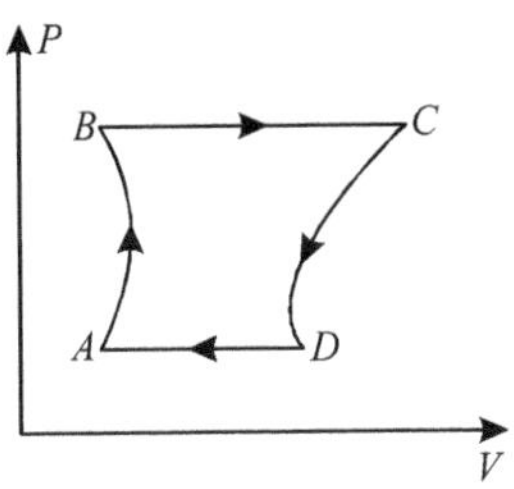

(B)

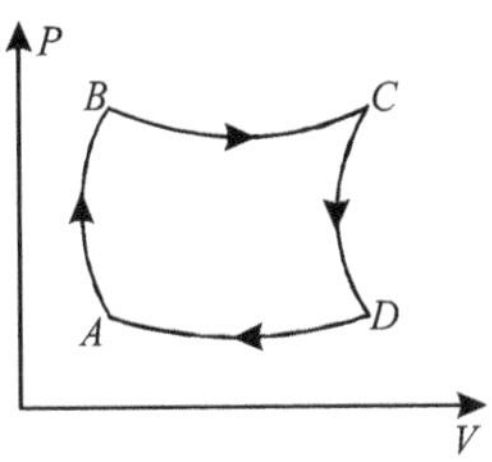

(C)

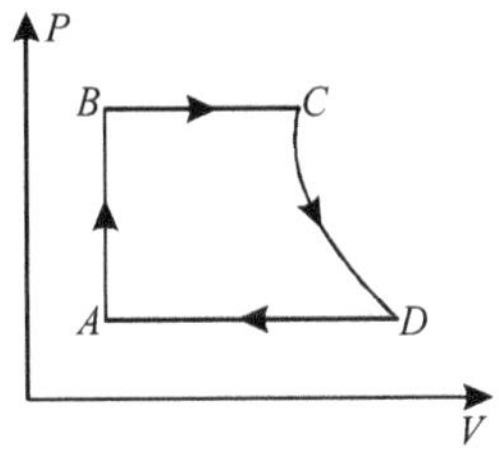

(D)

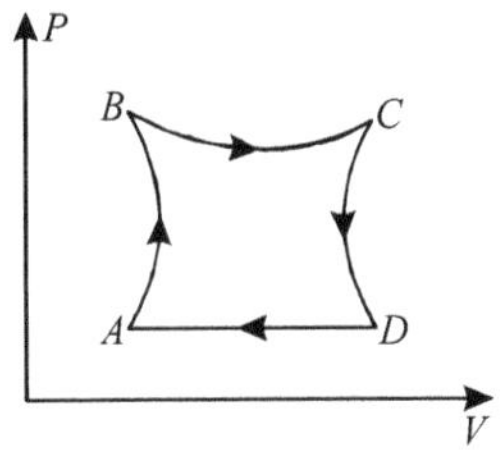

35. In the isothermal expansion of 10 g of gas from volume *V* to 2 V the work done by the gas is 575 J. What is the root mean square speed of the molecules of the gas at that temperature ? **[2013 (Online)]**

(A) 398 m/s (B) 520 m/s

(C) 499 m/s (D) 532 m/s

36. One mole of diatomic ideal gas undergoes a cyclic process *ABC* as shown in figure. The process *BC* is adiabatic. The temperatures at *A*, *B* and *C* are 400 K, 800 K and 600 K respectively. Choose the correct statement : **[2014]**

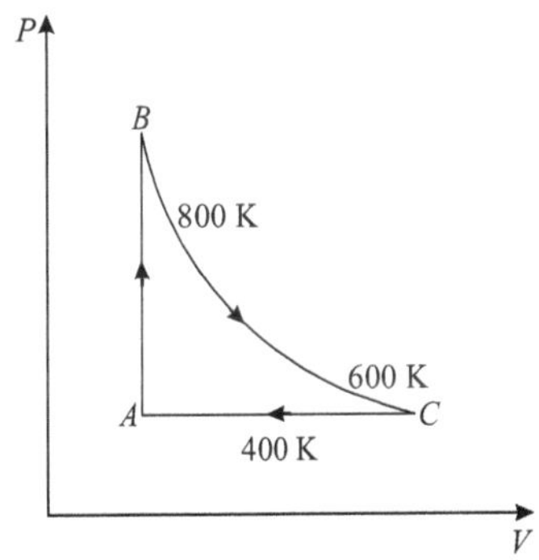

(A) The change in internal energy in whole cyclic process is 250 R

(B) The change in internal energy in the process *CA* is 700 R

(C) The change in internal energy in the process *AB* is –350 R

(D) The change in internal energy in the process *BC* is –500 R

37. The equation of state for a gas is given by $PV = nRT + \alpha V$, where *n* is the number of moles and α is a positive constant. The initial temperature and pressure of one mole of the gas contained in a cylinder are T_0 and P_0 respectively. The work done by the gas when its temperature doubles isobarically will be : **[2014 (Online)]**

(A) $\frac{P_0T_0R}{P_0 - \alpha}$ (B) $\frac{P_0T_0R}{P_0 + \alpha}$

(C) $P_0T_0R \ln 2$ (D) P_0T_0R

38. An ideal monoatomic gas is confined in a cylinder by a spring loaded piston of cross section 8.0×10^{-3} m^2. Initially the gas is at 300 K and occupies a volume of 2.4×10^{-3} m^3 and the spring is in its relaxed state as shown in figure. The gas is heated by a small heater until the piston moves out slowly by

0.1 m. The force constant of the spring is 8000 N/m and the atmospheric pressure is 1.0×10^5 N/m^2. The cylinder and the piston are thermally insulated. The piston and the spring are massless and there is no friction between the piston and the cylinder. The final temperature of the gas will be :
(Neglect the heat loss through the lead wires of the heater. The heat capacity of the heater coil is also negligible) : **[2014 (Online)]**

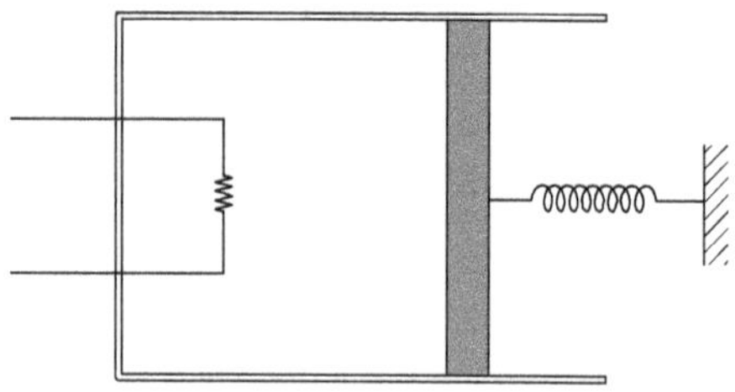

(A) 300 K (B) 800 K
(C) 500 K (D) 1000 K

39. During an adiabatic compression, 830 J of work is done on 2 moles of a diatomic ideal gas to reduce its volume by 50%. The change in its temperature is nearly: $[R = 8.3 \text{ JK}^{-1} \text{mol}^{-1}]$ **[2014 (Online)]**

(A) 40 K (B) 33 K
(C) 20 K (D) 14 K

40. A gas is compressed from a volume of 2 m^3 to a volume of 1 m^3 at a constant pressure of 100 N/m^2. Then it is heated at constant volume by supplying 150 J of energy. As a result, the internal energy of the gas : **[2014 (Online)]**

(A) Increases by 250 J (B) Decreases by 250 J
(C) Increases by 50 J (D) Decreases by 50 J

41. A Carnot engine absorbs 1000 J of heat energy from a reservoir at 127°C and rejects 600 J of heat energy during each cycle. The efficiency of engine and temperature of sink will be : **[2014 (Online)]**

(A) 20% and – 43°C (B) 40% and – 33°C
(C) 50% and – 20°C (D) 70% and – 10°C

42. Consider an ideal gas confined in an isolated closed chamber. As the gas undergoes an adiabatic expansion, the average time of collision between molecules increases as V^q, where V is the volume of the gas. The value of q is $\left(\gamma = \dfrac{C_p}{C_v}\right)$: **[2015]**

(A) $\dfrac{3\gamma+5}{6}$ (B) $\dfrac{3\gamma-5}{6}$

(C) $\dfrac{\gamma+1}{2}$ (D) $\dfrac{\gamma-1}{2}$

43. Consider a spherical shell of radius R at temperature T. The black body radiation inside it can be considered as an ideal gas of photons with internal energy per unit volume $u = \dfrac{U}{V} \propto T^4$ and pressure $p = \dfrac{1}{3}\left(\dfrac{U}{V}\right)$. If the shell now undergoes an adiabatic expansion the relation between T and R is : **[2015]**

(A) $T \propto \dfrac{1}{R}$ (B) $T \propto \dfrac{1}{R^3}$

(C) $T \propto e^{-R}$ (D) $T \propto e^{-3R}$

44. A solid body of constant heat capacity 1 J/°C is being heated by keeping it in contact with reservoirs in two ways : **[2015]**

(i) Sequentially keeping in contact with 2 reservoirs such that each reservoir supplied same amount of heat
(ii) Sequentially keeping in contact with 8 reservoirs such that each reservoir supplies same amount of heat

In both the cases body is brought from initial temperature 100°C to final temperature 200°C. Entropy change of the body in the two cases respectively is :

(A) ln2, 2 ln2 (B) 2 ln2, 8 ln2
(C) ln2, 4 ln2 (D) ln2, ln2

45. An ideal gas goes through a reversible cycle $a \to b \to c \to d$ has the V-T diagram shown below. Process $d \to a$ and $b \to c$ are adiabatic.

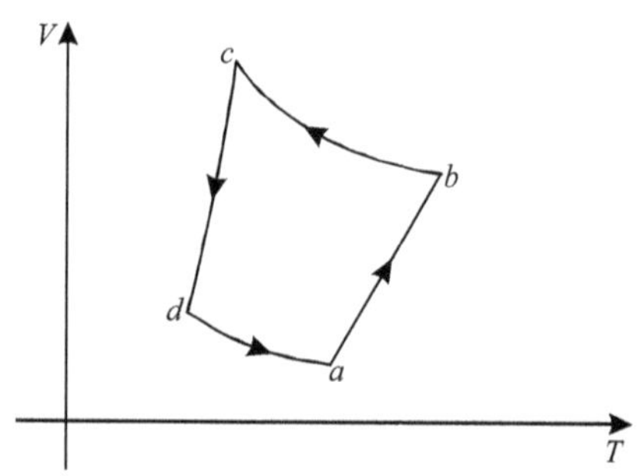

The corresponding P-V diagram for the process is (all figures are schematic and not drawn to scale) : **[2015 (Online)]**

(A)
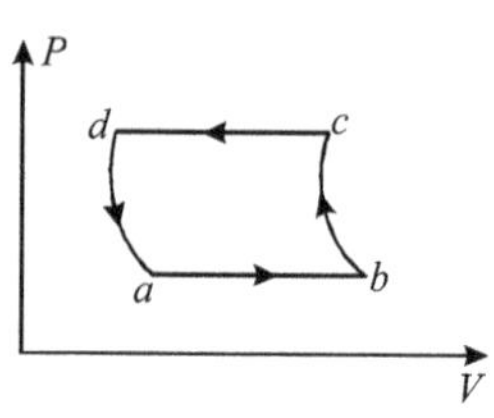

(B)
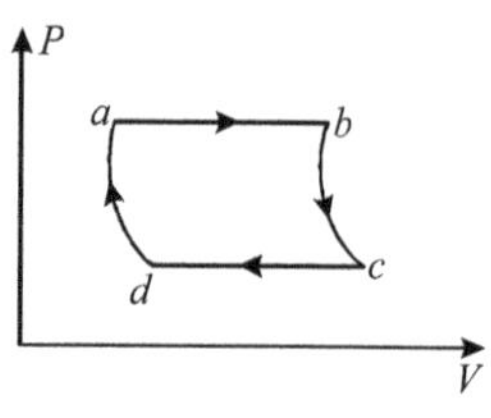

(C)
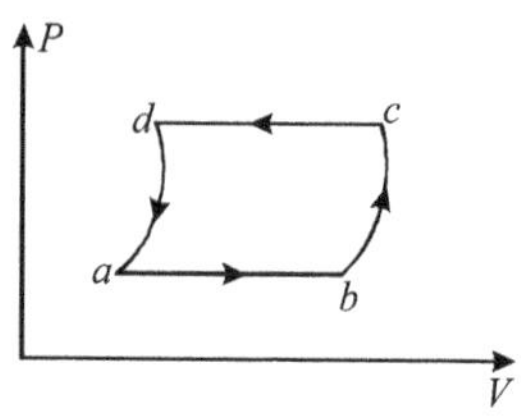

(D)
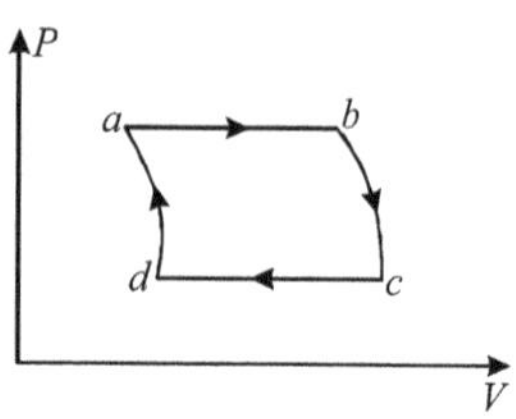

46. 'n' moles of an ideal gas undergoes a process $A \rightarrow B$ as shown in the figure. The maximum temperature of the gas during the process will be : **[2016]**

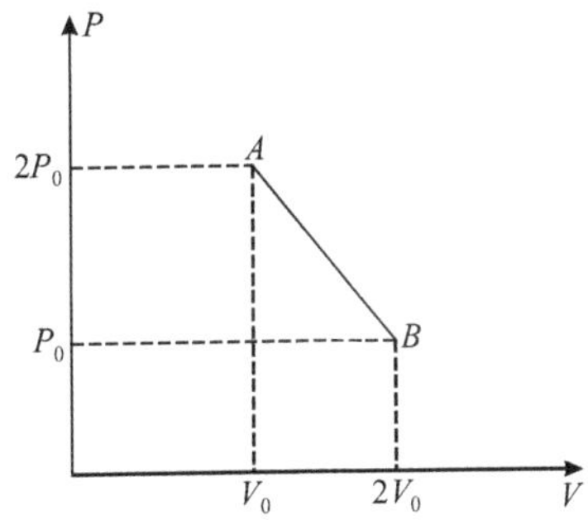

(A) $\frac{9P_0V_0}{4nR}$ (B) $\frac{3P_0V_0}{2nR}$

(C) $\frac{9P_0V_0}{2nR}$ (D) $\frac{9P_0V_0}{nR}$

47. An ideal gas undergoes a quasi static, reversible process in which its molar heat capacity C remains constant. If during this process the relation of pressure P and volume V is given by PV^n = constant, then n is given by (Here C_P and C_V are molar specific heat at constant pressure and constant volume, respectively) : **[2016]**

(A) $n = \frac{C_P}{C_V}$ (B) $n = \frac{C - C_P}{C - C_V}$

(C) $n = \frac{C_P - C}{C - C_V}$ (D) $n = \frac{C - C_V}{C - C_P}$

48. A Carnot freezer takes heat from water at 0°C inside it and rejects it to the room at a temperature of 27°C. The latent heat of ice is 336×10^3 J kg^{-1}. If 5 kg of water at 0°C is converted into ice at 0°C by the freezer, then the energy consumed by the freezer is closed to : **[2016 (Online)]**

(A) 1.51×10^5 J (B) 1.68×10^6 J

(C) 1.71×10^7 J (D) 1.67×10^5 J

49. The ratio of work done by an ideal monoatomic gas to the heat supplied to it in an isobaric process is : **[2016 (Online)]**

(A) $\frac{2}{5}$ (B) $\frac{3}{2}$

(C) $\frac{3}{5}$ (D) $\frac{2}{3}$

50. 200 g water is heated from 40°C to 60°C. Ignoring the slight expansion of water, the change in its internal energy is close to (Given specific heat of water = 4184 J/kg K) : **[2016 (Online)]**

(A) 167.4 kJ (B) 8.4 kJ

(C) 4.2 kJ (D) 16.7 kJ

51. An engine operates by taking n moles of an ideal gas through the cycle $ABCDA$ shown in figure. The thermal efficiency of the engine is : (Take $C_v = 1.5$ R, where R is gas constant) : **[2017 (Online)]**

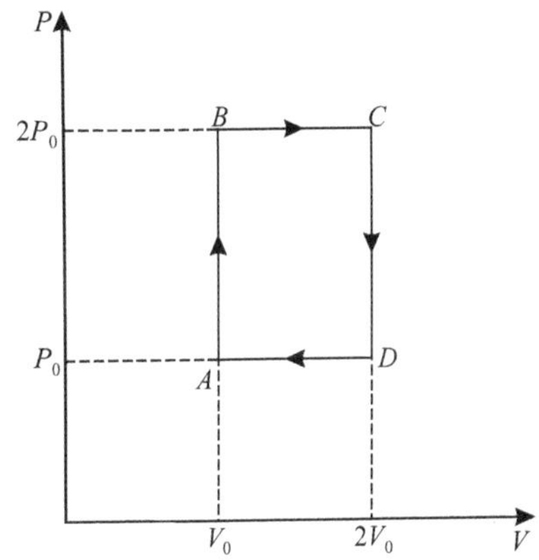

(A) 0.24 (B) 0.15

(C) 0.32 (D) 0.08

52. A gas can be taken from A to B via two different processes ACB and ADB. When path ACB is used 60 J of heat flows into the system and 30 J of work is done by the system. If path ADB is used work done by the system is 10 J. The heat Flow into the system in path ADB is : **[9 Jan. 2019 (Morning)]**

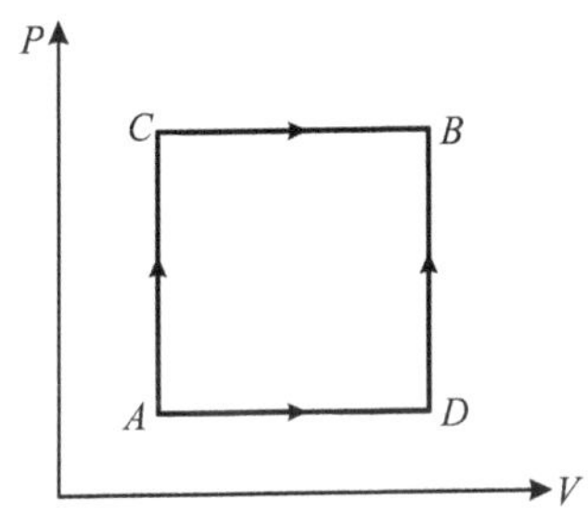

(A) 80 J (B) 20 J

(C) 100 J (D) 40 J

53. A 15g mass of nitrogen gas is enclosed in a vessel at a temperature 27°C. Amount of heat transferred to the gas, so that rms velocity of molecules is doubled, is about :

[Take R = 8.3 J/K mole] **[9 Jan. 2019 (Evening)]**

(A) 10 kJ (B) 0.9 kJ

(C) 6 kJ (D) 14 kJ

54. Two Carrnot engines A and B are operated in series. The first one, A, receives heat at T_1 (= 600 K) and rejects to a reservoir at temperature T_2. The second engine B receives heat rejected by the first engine and, in turn, rejects to a heat reservoir at T_3 (= 400 K). Calculate the temperature T_2 if the work outputs of the two engines are equal : **[9 Jan. 2019 (Evening)]**

(A) 400 K (B) 600 K

(C) 500 K (D) 300 K

55. Three Carnot engines operate in series between a heat source at a temperature T_1 and a heat sink at temperature T_4 (see figure). There are two other reservoirs at temperature T_2, and T_3, as shown, with $T_2 > T_2 > T_3 > T_4$. The three engines are equally efficient if: **[10 Jan. 2019 (Morning)]**

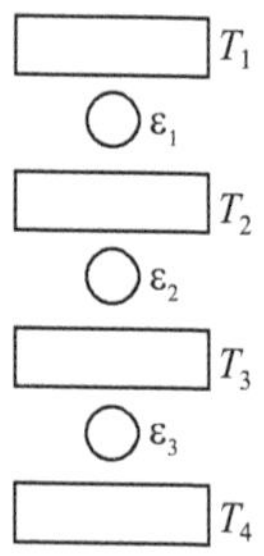

(A) $T_2 = (T_1^2 T_4)^{1/3}; T_3 = (T_1 T_4^2)^{1/3}$

(B) $T_2 = (T_1 T_4^2)^{1/3}; T_3 = (T_1^2 T_4)^{1/3}$

(C) $T_2 = (T_1^3 T_4)^{1/4}; T_3 = (T_1 T_4^3)^{1/4}$

(D) $T_2 = (T_1 T_4)^{1/2}; T_3 = (T_1^2 T_4)^{1/3}$

56. Half mole of an ideal monatomic gas is heated at constant pressure of 1 atm from 20 °C to 90°C. Work done by gas is close to : (Gas constant $R = 8.31$ J/mol.K) **[10 Jan. 2019 (Evening)]**

(A) 73 J
(B) 291 J
(C) 581 J
(D) 146 J

57. A rigid diatomic ideal gas undergoes an adiabatic process at room temperature. The relation between temperature and volume of this process is TV^x = constant, then x is : **[11 Jan. 2019 (Morning)]**

(A) $\frac{5}{3}$
(B) $\frac{2}{5}$
(C) $\frac{2}{3}$
(D) $\frac{3}{5}$

58. In a process, temperature and volume of one mole of an ideal monoatomic gas are varied according to the relation $VT = K$, where K is a constant. In this process the temperature of the gas is increased by ΔT. The amount of heat absorbed by gas is (R is gas constant) : **[11 Jan. 2019 (Evening)]**

(A) $\frac{1}{2}R\Delta T$
(B) $\frac{3}{2}R\Delta T$
(C) $\frac{1}{2}KR\Delta T$
(D) $\frac{2K}{3}\Delta T$

59. For the given cyclic process CAB as shown for a gas, the work done is : **[12 Jan. 2019 (Morning)]**

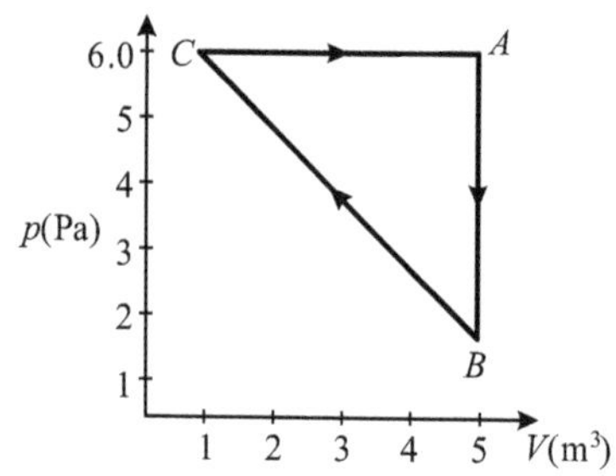

(A) 1 J
(B) 5 J
(C) 10 J
(D) 30 J

60. The given diagram shows four processes i.e., isochoric, isobaric, isothermal and adiabatic. The correct assignment of the processes, in the same order is given by : **[8 April 2019 (Evening)]**

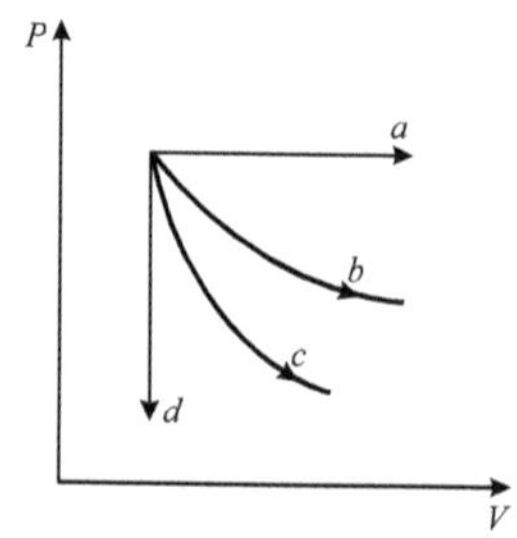

(A) $d\ a\ c\ b$
(B) $a\ d\ c\ b$
(C) $a\ d\ b\ c$
(D) $d\ a\ b\ c$

61. Following figure shows two processes A and B for a gas. If ΔQ_A and ΔQ_B are the amount of heat absorbed by the system in two cases, and ΔU_A and ΔU_B are changes in internal energies, respectively, then : **[9 April 2019 (Morning)]**

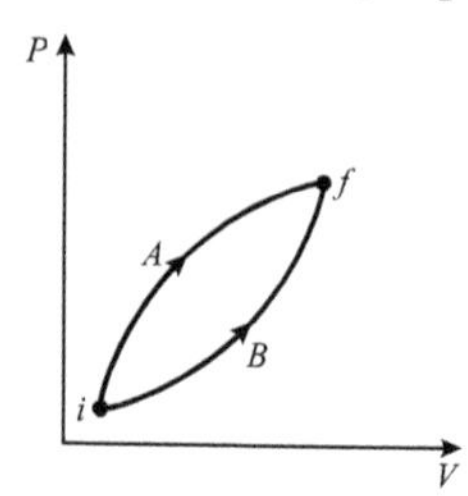

(A) $\Delta Q_A = \Delta Q_B; \Delta U_A = \Delta U_B$
(B) $\Delta Q_A > \Delta Q_B; \Delta U_A = \Delta U_B$
(C) $\Delta Q_A > \Delta Q_B; \Delta U_A > \Delta U_B$
(D) $\Delta Q_A < \Delta Q_B; \Delta U_A < \Delta U_B$

62. A cylinder with fixed capacity of 67.2 lit contains helium gas at STP. The amount of heat needed to raise the temperature of the gas by 20°C is : [Given that $R = 8.31$ J mol^{-1} K^{-1}] : **[10 April 2019 (Morning)]**

(A) 748 J
(B) 374 J
(C) 350 J
(D) 700 J

63. n moles of an ideal gas with constant volume heat capcity C_V undergo an isobaric expansion by certain volume. The ratio of the work done in the process, to the heat supplied is : **[10 April 2019 (Morning)]**

(A) $\frac{4nR}{C_V - nR}$ (B) $\frac{nR}{C_V - nR}$

(C) $\frac{nR}{C_V + nR}$ (D) $\frac{4nR}{C_V + nR}$

64. When heat Q is supplied to a diatomic gas of rigid molecules, at constant volume its temperature increases by ΔT. The heat required to produce the same change in temperature, at a constant pressure is : **[10 April 2019 (Evening)]**

(A) $\frac{7}{5}Q$ (B) $\frac{3}{2}Q$

(C) $\frac{5}{3}Q$ (D) $\frac{2}{3}Q$

65. One mole of an ideal gas passes through a process where pressure and volume obey the relation $P = P_o\left[1 - \frac{1}{2}\left(\frac{V_0}{V}\right)^2\right]$. Here P_0 and V_0 are constants. Calculate the change in the temperature of the gas if its volume changes from V_0 to $2V_0$: **[10 April 2019 (Evening)]**

(A) $\frac{1}{2}\frac{P_oV_o}{R}$ (B) $\frac{3}{4}\frac{P_oV_o}{R}$

(C) $\frac{5}{4}\frac{P_oV_o}{R}$ (D) $\frac{1}{4}\frac{P_oV_o}{R}$

66. A sample of an ideal gas is taken through the cyclic process abca as shown in the figure. The change in the internal energy of the gas along the path *ca* is –180 J. The gas absorbs 250 J of heat along the path ab and 60 J along the path *bc*. The work done by the gas along the path *abc* is : **[12 April 2019 (Morning)]**

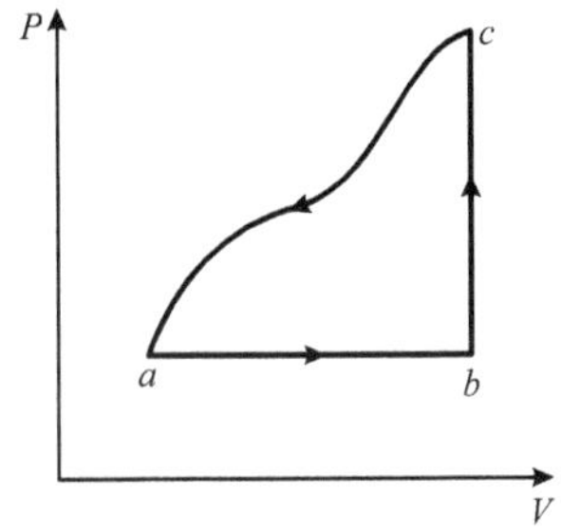

(A) 100 J (B) 120 J

(C) 140 J (D) 130 J

67. A Carnot engine has an efficiency of 1/6. When the temperature of the sink is reduced by 62°C, its efficiency is doubled. The temperatures of the source and the sink are, respectively : **[12 April 2019 (Evening)]**

(A) 124°C, 62°C (B) 37°C, 99°C

(C) 62°C, 124°C (D) 99°C, 37°C

68. A diatomic gas with rigid molecules does 10 J of work when expanded at constant pressure. What would be the heat energy absorbed by the gas, in this process ? **[12 April 2019 (Evening)]**

(A) 35 J (B) 40 J

(C) 25 J (D) 30 J

* * * * *

CHAPTER-12 KINETIC THEORY

12.1 Kinetic Theory of Gases

1. Cooking gas containers are kept in a lorry moving with uniform speed. The temperature of the gas molecules inside will : **[2002]**

(A) Increase
(B) Decrease
(C) Remains same
(D) Decrease for some, while increase for others

2. At what temperature is the rms velocity of a hydrogen molecule equal to that of an oxygen molecules at 47° C? **[2002]**

(A) 80 K (B) −73 K
(C) 3 K (D) 20 K

3. 1 mole of a gas with $\gamma = 7/5$ is mixed with 1 mole of a gas with $\gamma = 5/3$, then the value of γ for the resulting mixture is : **[2002]**

(A) 7/5 (B) 2/5
(C) 24/16 (D) 12/7

4. Two thermally insulated vessels 1 and 2 are filled with air at temperatures (T_1, T_2), volume (V_1, V_2) and pressure (P_1, P_2) respectively. If the valve joining the two vessels is opened, the temperature inside the vessel at equilibrium will be : **[2004]**

(A) $T_1 + T_2$ (B) $\frac{(T_1 + T_2)}{2}$

(C) $\frac{T_1T_2(P_1V_1 + P_2V_2)}{P_1V_1T_2 + P_2V_2T_1}$ (D) $\frac{T_1T_2(P_1V_1 + P_2V_2)}{P_1V_1T_1 + P_2V_2T_2}$

5. One mole of ideal monoatomic gas ($\gamma = 5/3$) is mixed with one mole of diatomic gas ($\gamma = 7/5$). What is γ for the mixture ? γ denotes the ratio of specific heat at constant pressure, to that at constant volume : **[2004]**

(A) 3/2 (B) 23/15
(C) 35/23 (D) 4/3

6. A gaseous mixture consists of 16 g of helium and 16 g of oxygen. The ratio C_p/C_v of the mixture is : **[2005]**

(A) 1.59 (B) 1.62
(C) 1.4 (D) 1.54

7. Two rigid boxes containing different ideal gases are placed on a table. Box A contains one mole of nitrogen at temperature T_0, while box B contains one mole of helium at temperature (7/3) T_0. The boxes are then put into thermal contact with each other, and heat flows between them until the gases reach a common final temperature (Ignore the heat capacity of boxes). Then, the final temperature of gases, T_f in terms of T_0 is : **[2006]**

(A) $T_f = \frac{3}{7}T_0$ (B) $T_f = \frac{7}{3}T_0$

(C) $T_f = \frac{3}{2}T_0$ (D) $T_f = \frac{5}{2}T_0$

8. If C_P and C_V denote the specific heats of nitrogen per unit mass at constant pressure and constant volume respectively, then : **[2007]**

(A) $C_P - C_V = R/28$ (B) $C_P - C_V = R/14$
(C) $C_P - C_V = R$ (D) $C_P - C_V = 28R$

9. An insulated container of gas has two chambers separated by an insulating partition. One of the chambers has volume V_1 and contains ideal gas at pressure P_1 and temperature T_1. The other chamber has volume V_2 and contains ideal gas at pressure P_2 and temperature T_2. If the partition is removed without doing any work on the gas, the final equilibrium temperature of the gas in the container will be : **[2008]**

(A) $\frac{T_1T_2(P_1V_1 + P_2V_2)}{P_1V_1T_2 + P_2V_2T_1}$ (B) $\frac{P_1V_1T_1 + P_2V_2T_2}{P_1V_1 + P_2V_2}$

(C) $\frac{P_1V_1T_2 + P_2V_2T_1}{P_1V_1 + P_2V_2}$ (D) $\frac{T_1T_2(P_1V_1 + P_2V_2)}{P_1V_1T_1 + P_2V_2T_2}$

10. One kg of a diatomic gas is at a pressure of 8×10^4 N/m^2. The density of the gas is 4 kg/m^3. What is the energy of the gas due to its thermal motion ? **[2009]**

(A) 6×10^4 J (B) 7×10^4 J
(C) 3×10^4 J (D) 5×10^4 J

11. Three perfect gases at absolute temperatures T_1, T_2 and T_3 are mixed. The masses of molecules are m_1, m_2, and m_3 and the number of molecules are n_1, n_2 and n_3 respectively. Assuming no loss of energy, then final temperature of the mixture is : **[2011]**

(A) $\frac{n_1T_1^2 + n_2T_2^2 + n_3T_3^2}{n_1T_1 + n_2T_2 + n_3T_3}$ (B) $\frac{n_1^2T_1^2 + n_2^2T_2^2 + n_3^2T_3^2}{n_1T_1 + n_2T_2 + n_3T_3}$

(C) $\frac{T_1 + T_2 + T_3}{3}$ (D) $\frac{n_1T_1 + n_2T_2 + n_3T_3}{n_1 + n_2 + n_3}$

12. A perfect gas at 27°C is heated at constant pressure so as to double its volume. The final temperature of the gas will be, close to : **[2012 (Online)]**

(A) 327°C (B) 200°C
(C) 54°C (D) 300°C

13. A thin tube sealed at both ends in 100 cm long. It lies horizontally, the middle 20 cm containing mercury and two equal ends containing air at standard atmospheric pressure. If the tube is now turned to a vertical position, by what amount will the mercury be displaced ? (Given : cross-section of the tube can be assumed to be uniform): **[2013 (Online)]**

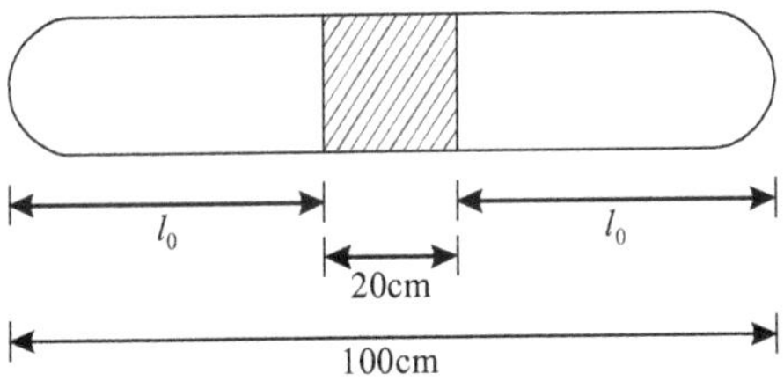

(A) 2.95 cm (B) 5.18 cm
(C) 8.65 cm (D) 0.0 cm

14. Figure shows the variation in temperature (ΔT) with the amount of heat supplied (Q) in an isobaric process corresponding to a monoatomic (M), diatomic (D) and a polyatomic (P) gas. The initial state of all the gases are the same and the scales for the two axes coincide. Ignoring vibrational degrees of freedom, the lines a, b and c respectively correspond to : **[2013 (Online)]**

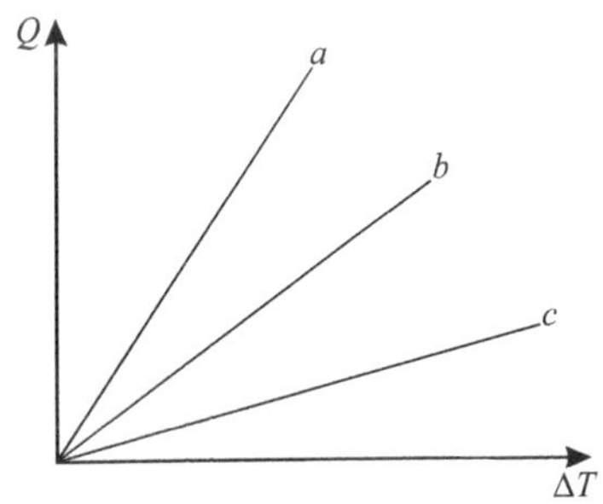

(A) P, M and D (B) M, D and P
(C) P, D and M (D) D, M and P

15.

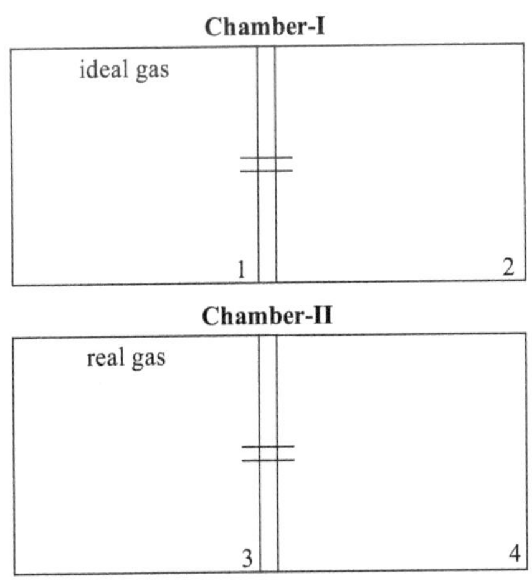

There are two identical chambers, completely thermally insulated from surroundings. Both chambers have a partition wall dividing the chambers in two compartments. Compartment 1 is filled with an ideal gas and Compartment 3 is filled with a real gas. Compartments 2 and 4 are vacuum. A small hole (orifice) is made in the partition walls and the gases are allowed to expand in vacuum. **[2013 (Online)]**

Statement-1 : No change in temperature of the gas takes place when ideal gas expands in vacuum. However, the temperature of real gas goes down (cooling) when it expands in vacuum.

Statement-2 : The internal energy of an ideal gas in only kinetic. The internal energy of a real gas is kinetic as well as potential.

(A) Statement-1 is false and Statement-2 is true
(B) Statement-1 and Statement-2 both are true. Statement-2 is the correct explanation of Statement-1
(C) Statement-1 is true and Statement-2 is false
(D) Statement-1 and Statement-2 both are true. Statement-2 is not correct explanation of Statement-1

16. An open glass tube is immersed in mercury in such a way that a length of 8 cm extends above the mercury level. The open end of the tube is then closed and sealed and the tube is raised vertically up by additional 46 cm. What will be length of the air column above mercury in the tube now ?
(Atmospheric pressure = 76 cm of Hg) **[2014]**

(A) 16 cm (B) 22 cm
(C) 38 cm (D) 6 cm

17. Modern vacuum pumps can evacuate a vessel down to a pressure of 4.0×10^{-15} atom. At room temperature (300 K). Taking $R = 8.0\ \text{JK}^{-1}\ \text{mole}^{-1}$, atom $= 10^5\ Pa$ and $N_{\text{Avogadro}} = 6 \times 10^{23}\ \text{mole}^{-1}$, the mean distance between molecules of gas in an evacuated vessel will be of the order of : **[2014 (Online)]**

(A) 0.2 μm (B) 0.2 mm
(C) 0.2 cm (D) 0.2 nm

18. At room temperature a diatomic gas is found to have an r.m.s. speed of 1930 ms^{-1}. The gas is : **[2014 (Online)]**

(A) H_2 (B) Cl_2
(C) O_2 (D) F_2

19. A gas molecule of mass M at the surface of the earth has kinetic energy equivalent to 0°C. If it were to go up straight without colliding with any other molecules, how high it would rise ? Assume that the height attained in much less than radius of the earth. (k_B is Boltzmann constant) : **[2014 (Online)]**

(A) 0 (B) $\dfrac{273k_B}{2Mg}$

(C) $\dfrac{546k_B}{3Mg}$ (D) $\dfrac{819k_B}{2Mg}$

20. Using equipartition of energy, the specific heat (in $\text{J kg}^{-1}\ \text{K}^{-1}$) of aluminium at room temperature can be estimated to be (atomic weight of aluminium = 27) : **[2015 (Online)]**

(A) 410 (B) 25
(C) 1850 (D) 925

21. In an ideal gas at temperature T, the average force that a molecule applies on the walls of a closed container depends on T as T^q. A good estimate for q is : **[2015 (Online)]**

(A) $\frac{1}{2}$ (B) 2

(C) 1 (D) $\frac{1}{4}$

22. C_p and C_v are specific heats at constant pressure and constant volume respectively. It is observed that

$C_p - C_v$ = a for hydrogen gas

$C_p - C_v = b$ for nitrogen gas

The correct relation between a and b is : **[2017]**

(A) $a = 14b$ (B) $a = 28b$

(C) $a = \frac{1}{14}b$ (D) $a = b$

23. The temperature of an open room of volume 30 m^3 increases from 17°C to 27°C due to sunshine. The atmospheric pressure in the room remains 1×10^5 *Pa*. If n_i and n_f are the number of molecules in the room before and after heating, the $n_f - n_i$ will be : **[2017]**

(A) 2.5×10^{25} (B) -2.5×10^{25}

(C) -1.61×10^{23} (D) 1.38×10^{23}

24. An ideal gas has molecules with 5 degrees of freedom. The ratio of specific heats at constant pressure (C_p) and at constant volume (C_v) is : **[2017 (Online)]**

(A) 6 (B) $\frac{7}{2}$

(C) $\frac{5}{2}$ (D) $\frac{7}{5}$

25. N moles of a diatomic gas in a cylinder are at a temperature T. Heat is supplied to the cylinder such that the temperature remains constant but n moles of the diatomic gas get converted into monoatomic gas. What is the change in the total kinetic energy of the gas ? **[2017 (Online)]**

(A) $\frac{1}{2}nRT$ (B) 0

(C) $\frac{3}{2}nRT$ (D) $\frac{5}{2}nRT$

26. For the *P-V* diagram given for an ideal gas, out of the following which one correctly represents the *T-P* diagram? **[2017 (Online)]**

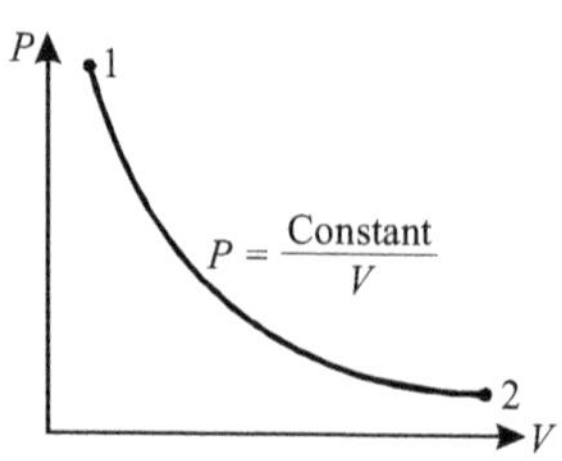

(A)

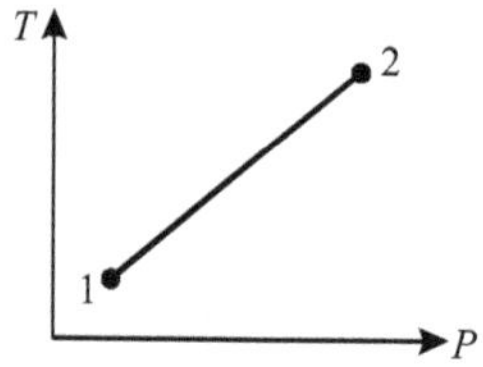

(B)

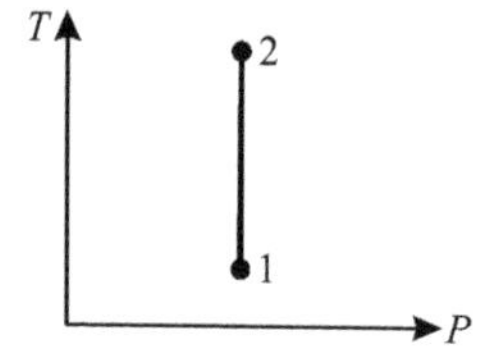

(C)

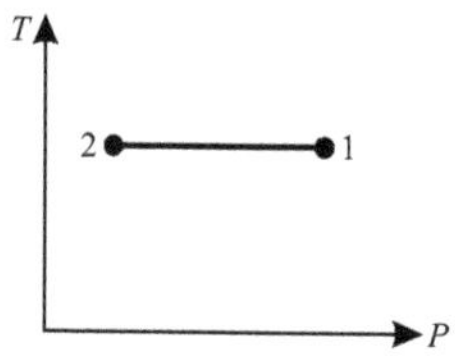

(D)

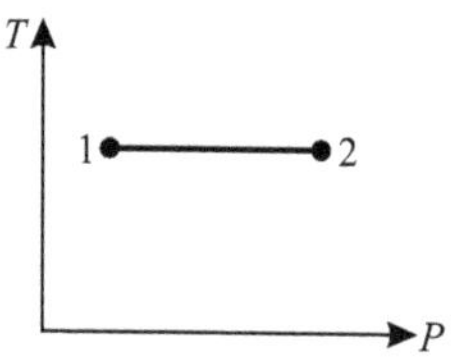

27. A mixture of 2 moles of helium gas (atomic mass = 4 u), and 1 mole of argon gas (atomic mass = 40u) is kept at 300 K in a container. The ratio of their rms speeds $\left[\frac{V_{rms}(\text{helium})}{V_{rms}(\text{argon})}\right]$, is close to : **[9 Jan. 2019 (Morning)]**

(A) 2.24 (B) 0.45

(C) 0.32 (D) 3.16

28. Two kg of a monoatomic gas is at a pressure of 4×10^4 N/m^2. The density of the gas is 8 kg/m^3. What is the order of energy of the gas due to its thermal motion ? **[10 Jan. 2019 (Evening)]**

(A) 10^3 J (B) 10^5 J

(C) 10^6 J (D) 10^4 J

29. The gas mixture consists of 3 moles of oxygen and 5 moles of argon at temperature T. Considering only translational and rotational modes, the total internal energy of the system is : **[11 Jan. 2019 (Morning)]**

(A) 12 RT (B) 20 RT

(C) 15 RT (D) 4 RT

30. An ideal gas occupies a volume of 2 m^3 at a pressure of 3×10^6 Pa. The energy of the gas is : **[12 Jan. 2019 (Morning)]**

(A) 3×10^2 (B) 10^8 J

(C) 6×10^4 J (D) 9×10^6 J

31. A vertical closed cylinder is separated into two parts by a frictionless piston of mass m and of negligible thickness. The piston is free to move along the length of the cylinder. The length of the cylinder above the piston is l_1, and that below the piston is l_2, such that $l_1 > l_2$. Each part of the cylinder contains n moles of an ideal gas at equal temperature T. If the piston is stationary, its mass, m, will be given by : (R is universal gas constant and g is the acceleration due to gravity) : **[12 Jan. 2019 (Evening)]**

(A) $\frac{nRT}{g}\left[\frac{1}{l_2}+\frac{1}{l_1}\right]$ (B) $\frac{nRT}{g}\left[\frac{l_1-l_2}{l_1l_2}\right]$

(C) $\frac{RT}{g}\left[\frac{2l_1+l_2}{l_1l_2}\right]$ (D) $\frac{RT}{ng}\left[\frac{l_1-3l_2}{l_1l_2}\right]$

32. An ideal gas is enclosed in a cylinder at pressure of 2 atm and temperature, 300 K. The mean time between two successive collisions is 6×10^{-8} s. If the pressure is doubled and temperature is increased to 500 K, the mean time between two successive collisions will be close to : **[12 Jan. 2019 (Evening)]**

(A) 4×10^{-8} s (B) 3×10^{-6} s

(C) 2×10^{-7} s (D) 0.5×10^{-8} s

33. If 10^{22} gas molecules each of mass 10^{-26} kg collide with a surface (perpendicular to it) elastically per second over an area 1 m^2 with a speed 10^4 m/s, the pressure exerted by the gas molecules will be of the order of: **[8 April 2019 (Morning)]**

(A) 10^8 N/m^2 (B) 10^4 N/m^2

(C) 10^3 N/m^2 (D) 10^{16} N/m^2

34. The temperature, at which the root mean square velocity of hydrogen molecules equals their escape velocity from the earth, is closest to :

[Boltzmann Constant $k_B = 1.38 \times 10^{-23}$ J/K

Avogadro Number $N_A = 6.02 \times 10^{26}$ /kg

Radius of Earth : 6.4×10^6 m

Gravitational acceleration on Earth = 10ms^{-2}]

[8 April 2019 (Evening)]

(A) 650 K (B) 3×10^5 K

(C) 10^4 K (D) 800 K

35. An HCl molecule has rotational, translational and vibrational motions. If the rms velocity of HCl molecules in its gaseous phase is $\bar{v}$, m is its mass and k_B is Boltzmann constant, then its temperature will be : **[9 April 2019 (Morning)]**

(A) $\frac{m\bar{v}^2}{6k_B}$ (B) $\frac{m\bar{v}^2}{5k_B}$

(C) $\frac{m\bar{v}^2}{3k_B}$ (D) $\frac{m\bar{v}^2}{7k_B}$

36. For a given gas at 1 atm pressure, rms speed of the molecule is 200 m/s at 127°C. At 2 atm pressure and at 227°C, the rms speed of the molecules will be : **[9 April 2019 (Morning)]**

(A) 80 m/s (B) $100\sqrt{5}$ m/s

(C) $80\sqrt{5}$ m/s (D) 100 m/s

37. The specific heats, C_P and C_V of a gas of diatomic molecules, A, are given (in units of J mol^{-1} K^{-1}) by 29 and 22, respectively. Another gas of diatomic molecules, B, has the corresponding values 30 and 21. If they are treated as ideal gases, then : **[9 April 2019 (Evening)]**

(A) A has one vibrational mode and B has two

(B) Both A and B have a vibrational mode each

(C) A is rigid but B has a vibrational mode

(D) A has a vibrational mode but B has none

38. A 25×10^{-3} m^3 volume cylinder is filled with 1 mol of O_2 gas at room temperature (300K). The molecular diameter of O_2, and its root mean square speed, are found to be 0.3nm, and 200 m/s, respectively. What is the average collision rate (per second) for an O_2 molecule ? **[10 April 2019 (Morning)]**

(A) $\sim 10^{11}$ (B) $\sim 10^{13}$

(C) $\sim 10^{10}$ (D) $\sim 10^{12}$

39. Two moles of helium gas is mixed with three moles of hydrogen molecules (taken to be rigid). What is the molar specific heat of mixture at constant volume ? (R = 8.3 J/mol K): **[12 April 2019 (Morning)]**

(A) 21.6 J/mol K (B) 19.7 J/mol K

(C) 17.4 J/mol K (D) 15.7 J/mol K

40. The number density of molecules of a gas depends on their distance r from the origin as, $n(r) = n_0 e^{-\alpha r^4}$. Then the total number of molecules is proportional to : **[12 April 2019 (Evening)]**

(A) $n_0\alpha^{1/4}$ (B) $n_0\alpha^{-3}$

(C) $n_0\alpha^{-3/4}$ (D) $\sqrt{n_0}\,\alpha^{1/2}$

* * * * *

CHAPTER-13 OSCILLATIONS

13.1 Simple Harmonic Motion

1. If a spring has time period T, and is cut into n equal parts, then the time period of each part will be : **[2002]**

(A) $T\sqrt{n}$ (B) $\frac{T}{\sqrt{n}}$

(C) nT (D) T

2. In a simple harmonic oscillator, at the mean position : **[2002]**

(A) Kinetic energy is minimum, potential energy is maximum
(B) Both kinetic and potential energies are maximum
(C) Kinetic energy is maximum, potential energy is minimum
(D) Both kinetic and potential energies are minimum

3. A child swinging on a swing in sitting position, stands up, then the time period of the swing will : **[2002]**

(A) Increase
(B) Decrease
(C) Remain same
(D) Increase if the child is long and decrease if the child is short

4. A mass M is suspended from a spring of negligible mass. The spring is pulled a little and then released so that the mass executes SHM of time period T. If the mass is increased by m, the time period becomes $5T/3$, then the ratio of $\frac{m}{M}$ is : **[2003]**

(A) $\frac{3}{5}$ (B) $\frac{25}{9}$

(C) $\frac{16}{9}$ (D) $\frac{5}{3}$

5. The displacement of a particle varies according to the relation $x = 4(\cos\pi t + \sin\pi t)$. The amplitude of the particle is : **[2003]**

(A) -4 (B) 4

(C) $4\sqrt{2}$ (D) 8

6. A body executes simple harmonic motion. The potential energy (PE), the kinetic energy (KE) and total energy (TE) are measured as function of displacement x. Which of the following statements is true ? **[2003]**

(A) KE is maximum when $x = 0$
(B) TE is zero when $x = 0$
(C) KE is maximum when x is maximum
(D) PE is maximum when $x = 0$

7. Two particles A and B of equal masses are suspended from two massless springs of spring constants k_1 and k_2, respectively. If the maximum velocities, during oscillations are equal, the ratio of amplitudes of A and B is : **[2003]**

(A) $\sqrt{k_1/k_2}$ (B) k_1/k_2

(C) $\sqrt{k_2/k_1}$ (D) k_2/k_1

8. The length of a simple pendulum executing simple harmonic motion is increased by 21%. The percentage increase in the time period of the pendulum of increased length is : **[2003]**

(A) 11% (B) 21%

(C) 42% (D) 10%

9. The bob of a simple pendulum executes simple harmonic motion in water with a period t, while the period of oscillation of the bob is t_0 in air. Neglecting frictional force of water and given that the density of the bob is $(4/3) \times 1000$ kg/m^3. What relationship between t and t_0 is true? **[2004]**

(A) $t = t_0$ (B) $t = t_0/2$

(C) $t = 2t_0$ (D) $t = 4t_0$

10. A particle at the end of a spring executes simple harmonic motion with a period t_1, while the corresponding period for another spring is t_2. If the period of oscillation with the two springs in series is T, then : **[2004]**

(A) $T = t_1 + t_2$ (B) $T^2 = t_1^2 + t_2^2$

(C) $T^{-1} = t_1^{-1} + t_2^{-1}$ (D) $T^{-2} = t_1^{-2} + t_1^{-2}$

11. The total energy of a particle, executing simple harmonic motion is : **[2004]**

(A) $\propto x$ (B) $\propto x^2$

(C) independent of x (D) $\propto x^{1/2}$

12. A particle of mass m is attached to a spring (of spring constant k) and has a natural angular frequency ω_0. An external force $F(t)$ proportional to $\cos\omega t$ $(\omega \neq \omega_0)$ is applied to the oscillator. The time displacement of the oscillator will be proportional to : **[2004]**

(A) $\frac{m}{\omega_0^2 - \omega^2}$ (B) $\frac{1}{m(\omega_0^2 - \omega^2)}$

(C) $\frac{1}{m(\omega_0^2 + \omega^2)}$ (D) $\frac{m}{(\omega_0^2 + \omega^2)}$

13. In forced oscillation of a particle the amplitude is maximum for a frequency ω_1 of the force while the energy is maximum for a frequency ω_2 of the force; then : **[2004]**

(A) $\omega_1 < \omega_2$ when damping is small and $\omega_1 > \omega_2$ when damping is large
(B) $\omega_1 > \omega_2$
(C) $\omega_1 = \omega_2$
(D) $\omega_1 < \omega_2$

14. A particle of mass 0.3 kg is subjected to a force $F = -kx$ with $k = 15$ N/m. What will be its initial acceleration, if it is released from a point 20 cm away from the origin ? **[2005]**

(A) 3 m/s^2 (B) 15 m/s^2
(C) 5 m/s^2 (D) 10 m/s^2

15. The function $\sin^2(\omega t)$ represents : **[2005]**

(A) A periodic, but not simple harmonic, motion with a period $2\pi/\omega$
(B) A periodic, but not simple harmonic, motion with a period π/ω
(C) A simple harmonic motion with a period $2\pi/\omega$
(D) A simple harmonic motion with a period π/ω

16. If a simple harmonic motion is represented by $\frac{d^2x}{dt^2} + \alpha x = 0$, its time period is : **[2005]**

(A) $\frac{2\pi}{\alpha}$ (B) $\frac{2\pi}{\sqrt{\alpha}}$
(C) $2\pi\alpha$ (D) $2\pi\sqrt{\alpha}$

17. The bob of a simple pendulum is a spherical hollow ball filled with water. A plugged hole near the bottom of the oscillating bob gets suddenly unplugged. During observation, till water is coming out, the time period of oscillation would : **[2005]**

(A) First increase and then decrease to the original value
(B) First decrease and then increase to the original value
(C) Remain unchanged
(D) Increase towards a saturation value

18. Two simple harmonic motions are represented by the equations $y_1 = 0.1 \sin\left(100\pi t + \frac{\pi}{3}\right)$ and $y_2 = 0.1 \cos 100\pi t$. The phase difference of the velocity of particle 1, with respect to the velocity of particle 2 is : **[2005]**

(A) $\frac{-\pi}{6}$ (B) $\frac{\pi}{3}$
(C) $\frac{-\pi}{3}$ (D) $\frac{\pi}{6}$

19. The maximum velocity of a particle, executing simple harmonic motion with an amplitude 7 mm, is 4.4 m/s. The period of oscillation is : **[2006]**

(A) 0.01 s (B) 10 s
(C) 0.1 s (D) 100 s

20. Starting from the origin a body oscillates simple harmonically with a period of 2s. After what time will its kinetic energy be 75% of the total energy ? **[2006]**

(A) $\frac{1}{6}$ s (B) $\frac{1}{4}$ s
(C) $\frac{1}{3}$ s (D) $\frac{1}{12}$ s

21. A coin is placed on a horizontal platform which undergoes vertical simple harmonic motion of angular frequency ω. The amplitude of oscillation is gradually increased. The coin will leave contact with the platform for the first time : **[2006]**

(A) At the mean position of the platform
(B) For an amplitude of $\frac{g}{\omega^2}$
(C) For an amplitude of $\frac{g^2}{\omega^2}$
(D) At the highest position on the platform

22. The displacement of an object attached to a spring and executing simple harmonic motion is given by $x = 2 \times 10^{-2} \cos \pi t$ metre. The time at which the maximum speed first occurs is : **[2007]**

(A) 0.5 s (B) 0.75 s
(C) 0.125 s (D) 0.25 s

23. A point mass oscillates along the x-axis according to the law $x = x_0 \cos(\omega t - \pi/4)$ If the acceleration of the particle is written as- $a = A \cos(\omega t + \delta)$, then : **[2007]**

(A) $A = x_0, \delta = -\pi/4$ (B) $A = x_0\omega^2, \delta = \pi/4$
(C) $A = x_0\omega^2, \delta = -\pi/4$ (D) $A = x_0\omega^2, \delta = 3\pi/4$

24. Two springs, of force constants k_1 and k_2, are connected to a mass m as shown. The frequency of oscillation of the mass is f. If both k_1 and k_2 are made four times their original values, the frequency of oscillation becomes : **[2007]**

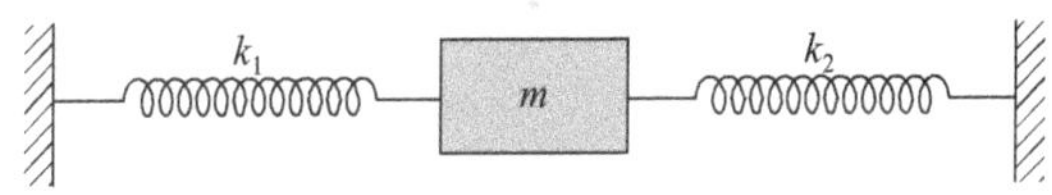

(A) $f/2$ (B) $f/4$
(C) $4f$ (D) $2f$

25. A particle of mass m executes simple harmonic motion with amplitude a and frequency ν. The average kinetic energy during its motion from the position of equilibrium to the end is : **[2007]**

(A) $\pi^2 ma^2\nu^2$ (B) $\frac{1}{4}ma^2\nu^2$
(C) $4\pi^2 ma^2\nu^2$ (D) $2\pi^2 ma^2\nu^2$

26. If x, v and a denote the displacement, the velocity and the acceleration of a particle executing simple harmonic motion of time period T, then, which of the following does not change with time ? **[2009]**

(A) $aT + 2\pi v$ (B) $\frac{aT}{v}$

(C) $a^2T^2 + 4\pi^2v^2$ (D) $\frac{aT}{x}$

27. A mass M, attached to a horizontal spring, executes S.H.M. with amplitude A_1. When the mass M passes through its mean position then a smaller mass m is placed over it and both of them move together with amplitude A_2. The ratio of $\left(\frac{A_1}{A_2}\right)$ is : **[2011]**

(A) $\left(\frac{M}{M+m}\right)^{1/2}$ (B) $\left(\frac{M+m}{M}\right)^{1/2}$

(C) $\frac{M}{M+m}$ (D) $\frac{M+m}{M}$

28. Two particles are executing simple harmonic motion of the same amplitude A and frequency ω along the x-axis. Their mean position is separated by distance X_0 ($X_0 > A$). If the maximum separation between them is ($X_0 + A$), the phase difference between their motion is : **[2011]**

(A) $\frac{\pi}{4}$ (B) $\frac{\pi}{6}$

(C) $\frac{\pi}{2}$ (D) $\frac{\pi}{3}$

29. A wooden cube (density of wood 'd') of side 'l' floats in a liquid of density 'ρ' with its upper and lower surfaces horizontal. If the cube is pushed slightly down and released, it performs simple harmonic motion of period 'T'. Then, 'T' is equal to : **[2011]**

(A) $2\pi\sqrt{\frac{l\pi}{(\rho-d)g}}$ (B) $2\pi\sqrt{\frac{ld}{\rho g}}$

(C) $2\pi\sqrt{\frac{l\rho}{dg}}$ (D) $2\pi\sqrt{\frac{ld}{(\rho-d)g}}$

30. If a simple pendulum has significant amplitude (up to a factor of $1/e$ of original) only in the period between $t = 0s$ to $t = \tau s$, then τ may be called the average life of the pendulum. When the spherical bob of the pendulum suffers a retardation (due to viscous drag) proportional to its velocity, with 'b' as the constant of proportionality, the average life time of the pendulum is (assuming damping is small) in seconds : **[2012]**

(A) $\frac{2}{b}$ (B) $\frac{0.693}{b}$

(C) b (D) $\frac{1}{b}$

31. A ring is suspended from a point S on its rim as shown in the figure. When displaced from equilibrium, it oscillates with time period of 1 second. The radius of the ring is : (take $g = \pi^2$): **[2012 (Online)]**

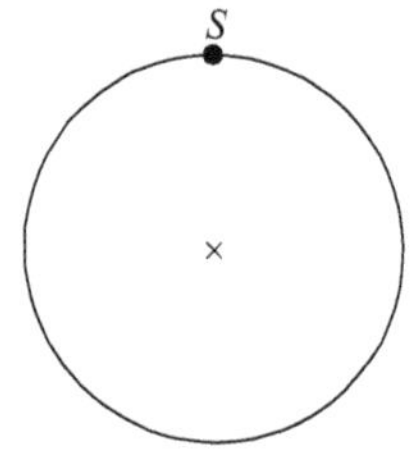

(A) 0.125 m (B) 1.5 m

(C) 1.0 m (D) 0.5 m

32. The displacement $y(t) = A\sin(\omega t + \phi)$ of a pendulum for $\phi = \frac{2\pi}{3}$ is correctly represented by : **[2012 (Online)]**

(A) 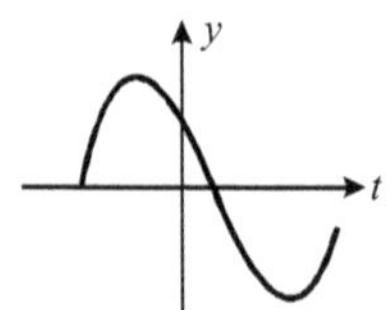

(B) 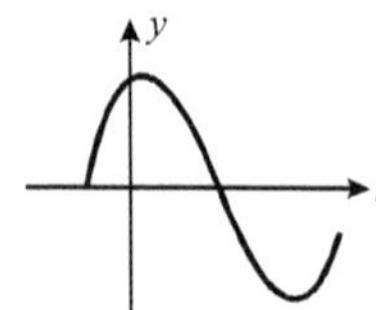

(C) 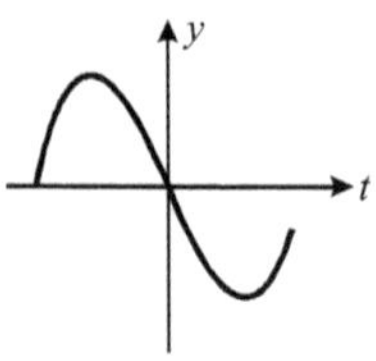

(D) 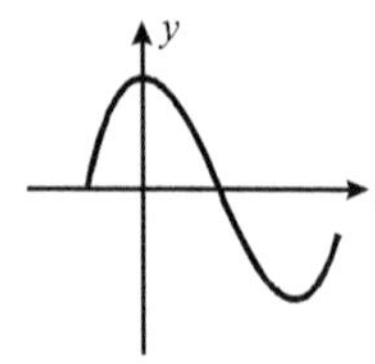

33. An ideal gas enclosed in a vertical cylindrical container supports a freely moving piston of mass M. The piston and the cylinder have equal cross sectional area A. When the piston is in equilibrium, the volume of the gas is V_0 and its pressure is P_0. The piston is slightly displaced from the equilibrium position and released. Assuming that the system is completely isolated from its surrounding, the piston executes a simple harmonic motion with frequency : **[2013]**

(A) $\frac{1}{2\pi}\frac{A\gamma P_0}{V_0 M}$ (B) $\frac{1}{2\pi}\frac{V_0 M P_0}{A^2\gamma}$

(C) $\frac{1}{2\pi}\sqrt{\frac{A^2\gamma P_0}{MV_0}}$ (D) $\frac{1}{2\pi}\sqrt{\frac{MV_0}{A\gamma P_0}}$

34. The amplitude of a damped oscillator decreases to 0.9 times its original magnitude in 5s. In another 10s it will decrease to α times its original magnitude, where α equals : **[2013]**

(A) 0.7 (B) 0.81

(C) 0.729 (D) 0.6

35. Two simple pendulums of length 1 m and 4 m respectively are both given small displacement in the same direction at the same instant. They will be again in phase after the shorter pendulum has completed number of oscillations equal to : **[2013 (Online)]**

(A) 2 (B) 7
(C) 5 (D) 3

(D) $T < 2\pi\sqrt{\frac{l}{g}}$ and damping is larger than in air alone

36. A mass $m = 1.0$ kg is put on a flat pan attached to a vertical spring fixed on the ground. The mass of the spring and the pan is negligible. When pressed slightly and released, the mass executes simple harmonic motion. The spring constant is 500 N/m. What is the amplitude A of the motion, so that the mass m tends to get detached from the pan? (Take $g = 10$ m/s^2) The spring is stiff enough so that it does not get distorted during the motion : **[2013 (Online)]**

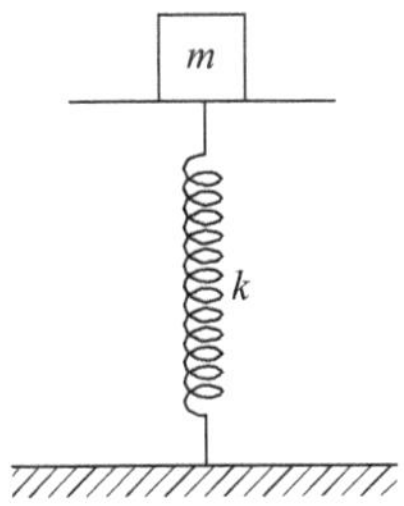

(A) $A > 2.0$ cm (B) $A = 2.0$ cm
(C) $A < 2.0$ cm (D) $A = 1.5$ cm

37. A uniform cylinder of length L and mass M having cross-sectional area A is suspended, with its length vertical, from a fixed point by a massless spring, such that it is half submerged in a liquid of density σ at equilibrium position. When the cylinder is given a downward push and released, it starts oscillating vertically with a small amplitude. The time period T of the oscillations of the cylinder will be : **[2013 (Online)]**

(A) Smaller than $2\pi\left[\frac{M}{(k + A\sigma g)}\right]^{1/2}$

(B) $2\pi\sqrt{\frac{M}{k}}$

(C) Larger than $2\pi\left[\frac{M}{(k + A\sigma g)}\right]^{1/2}$

(D) $2\pi\left[\frac{M}{(k + A\sigma g)}\right]^{1/2}$

38. Bob of a simple pendulum of length l is made of iron. The pendulum is oscillating over a horizontal coil carrying direct current. If the time period of the pendulum is T then : **[2013 (Online)]**

(A) $T < 2\pi\sqrt{\frac{l}{g}}$ and damping is smaller than in air alone

(B) $T = 2\pi\sqrt{\frac{l}{g}}$ and damping is larger than in air alone

(C) $T > 2\pi\sqrt{\frac{l}{g}}$ and damping is smaller than in air alone

39. A particle moves with simple harmonic motion in a straight line. In first τ, s after starting from rest it travels a distance a, and in next τ s it travels $2a$, in same direction, then : **[2014]**

(A) Amplitude of motion is $3a$
(B) Time period of oscillations is 8τ
(C) Amplitude of motion is $4a$
(D) Time period of oscillations is 6τ

40. The amplitude of a simple pendulum, oscillating in air with a small spherical bob, decreases from 10 cm to 8 cm in 40 seconds. Assuming that Stokes law is valid, and ratio of the coefficient of viscosity of air to that of carbon dioxide is 1.3. The time in which amplitude of this pendulum will reduce from 10 cm to 5 cm in carbon dioxide will be close to :
(In 5 = 1.601, In 2 = 0.693) **[2014 (Online)]**

(A) 231 s (B) 208 s
(C) 161 s (D) 142 s

41. In an experiment for determining the gravitational acceleration g of a place with the help of a simple pendulum, the measured time period square is plotted against the string length of the pendulum in the figure. What is the value of g at the place ? **[2014 (Online)]**

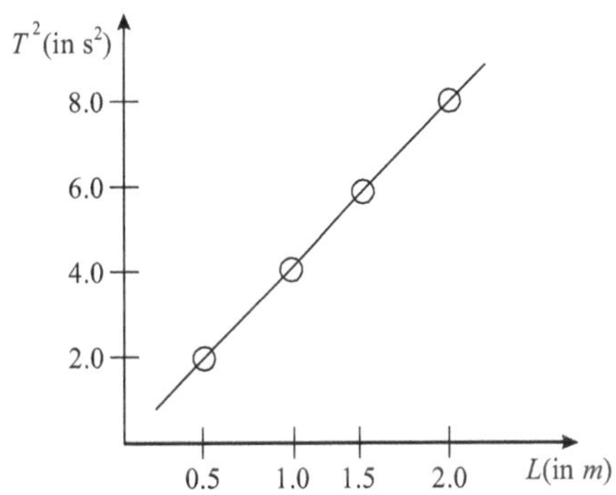

(A) 9.81 m/s^2 (B) 9.87 m/s^2
(C) 9.91 m/s^2 (D) 10.0 m/s^2

42. A particle which is simultaneously subjected to two perpendicular simple harmonic motions represented by; $x = a_1 \cos \omega t$ and $y = a_2 \cos 2\omega t$ traces a curve given by : **[2014 (Online)]**

(A) 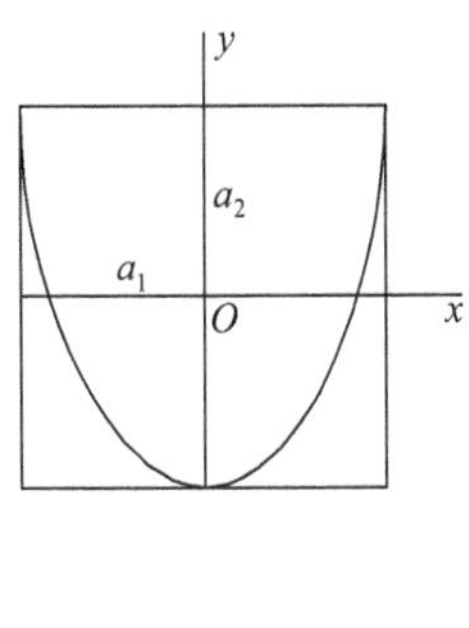

(B) 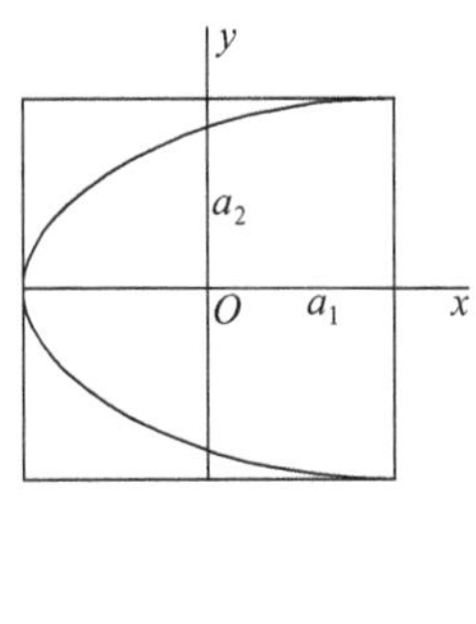

(C) 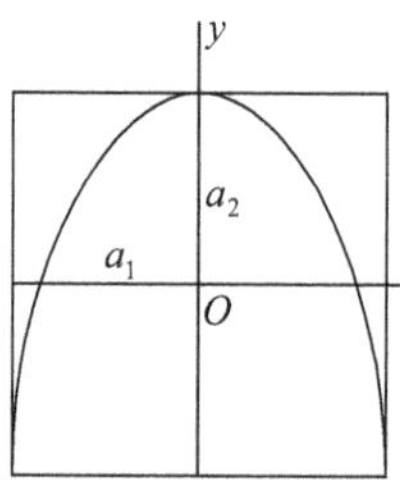

(D) 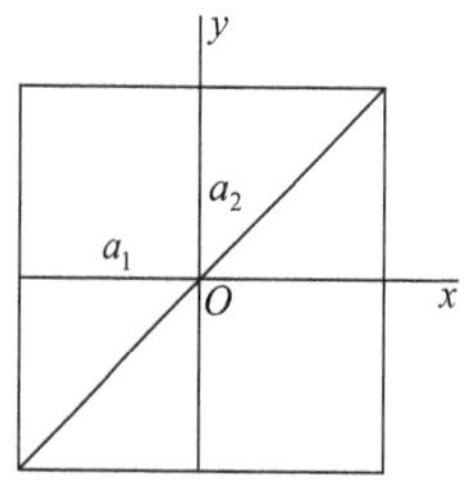

43. Which of the following expressions corresponds to simple harmonic motion along a straight line, where x is the displacement and a, b, c are positive constants ? **[2014 (Online)]**

(A) $a + bx - cx^2$ (B) bx^2
(C) $a - bx + cx^2$ (D) $-bx$

44. A body is in simple harmonic motion with time period half second ($T = 0.5$ s) and amplitude one cm ($A = 1$ cm). Find the average velocity in the interval in which it moves form equilibrium position to half of its amplitude : **[2014 (Online)]**

(A) 4 cm/s (B) 6 cm/s
(C) 12 cm/s (D) 16 cm/s

45. Two bodies of masses 1 kg and 4 kg are connected to a vertical spring, as shown in the figure. The smaller mass executes simple harmonic motion of angular frequency 25 rad/s, and amplitude 1.6 cm while the bigger mass remains stationary on the ground. The maximum force exerted by the system on the floor is (take $g = 10\ ms^{-2}$) : **[2014 (Online)]**

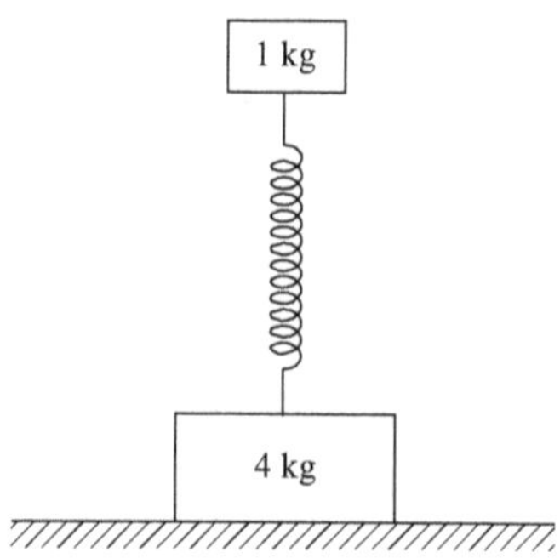

(A) 20 N (B) 10 N
(C) 60 N (D) 40 N

46. The angular frequency of the damped oscillator is given by, $\omega = \sqrt{\frac{k}{m} - \frac{r^2}{4m^2}}$ where k is the spring constant, m is the mass of the oscillator and r is the damping constant. If the ratio $\frac{r^2}{mk}$ is 8%, the change in time period compared to the undamped oscillator is approximately as follows : **[2014 (Online)]**

(A) Increases by 1% (B) Increases by 8%
(C) Decreases by 1% (D) Decreases by 8%

47. A pendulum made of a uniform wire of cross sectional area A has time period T. When an additional mass M is added to its bob, the time period changes to T_M. If the Young's modulus of the material of the wire is Y then $\frac{1}{Y}$ is equal to (g = gravitational acceleration) : **[2015]**

(A) $\left[\left(\frac{T_M}{T}\right)^2 - 1\right]\frac{A}{Mg}$ (B) $\left[\left(\frac{T_M}{T}\right)^2 - 1\right]\frac{Mg}{A}$

(C) $\left[1 - \left(\frac{T_M}{T}\right)^2\right]\frac{A}{Mg}$ (D) $\left[1 - \left(\frac{T}{T_M}\right)^2\right]\frac{A}{Mg}$

48. For a simple pendulum, a graph is plotted between its kinetic energy (KE) and potential energy (PE) against its displacement d. Which one of the following represents these correctly ? (Graph are schematic and not drawn to scale) **[2015]**

(A) 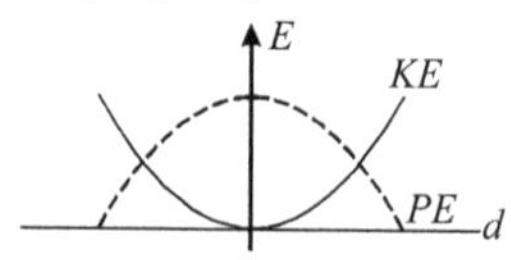

(B) 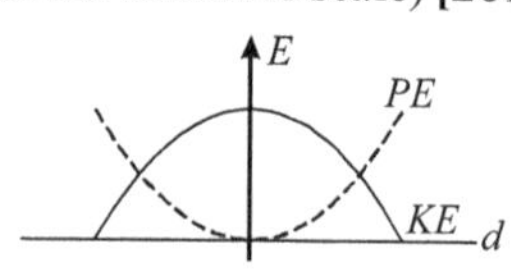

(C) 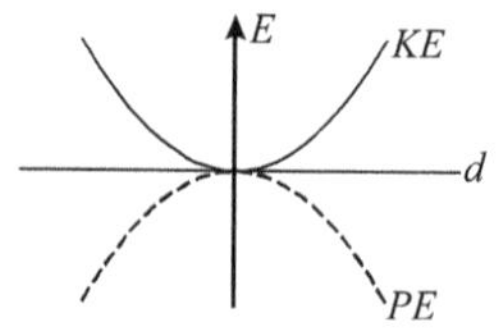

(D) 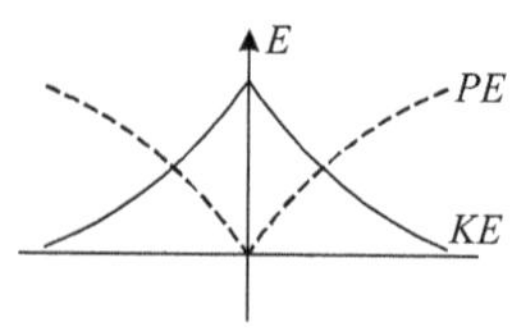

49. A pendulum with time period of 1s is losing energy. At certain time its energy is 45 J. If after completing 15 oscillations, its energy has become 15 J, its damping constant (in s^{-1}) is : **[2015 (Online)]**

(A) $\frac{1}{2}$ (B) $\frac{1}{30}\ln 3$

(C) 2 (D) $\frac{1}{15}\ln 3$

50. x and y displacements of a particle are given as $x(t) = a \sin \omega t$ and $y(t) = a \sin 2\omega t$. Its trajectory will look like : **[2015 (Online)]**

(A) 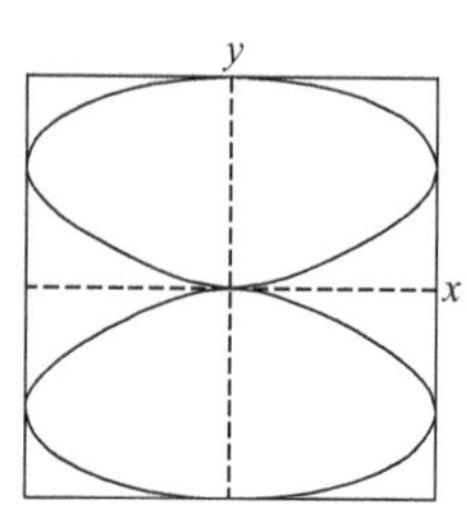

(B) 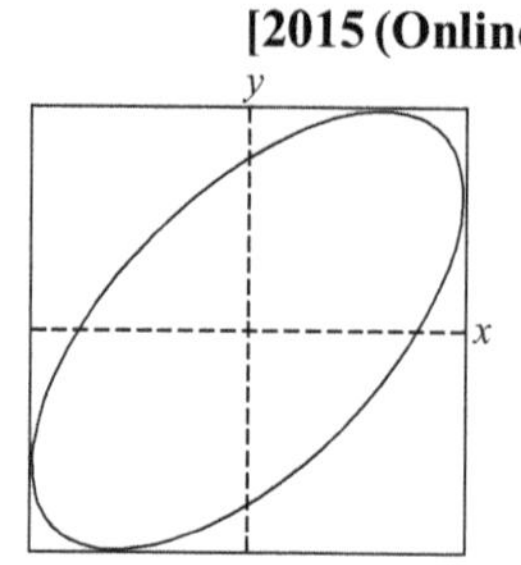

(C) 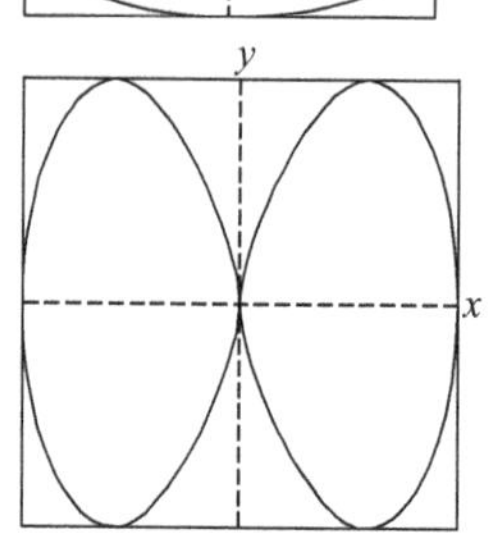

(D) 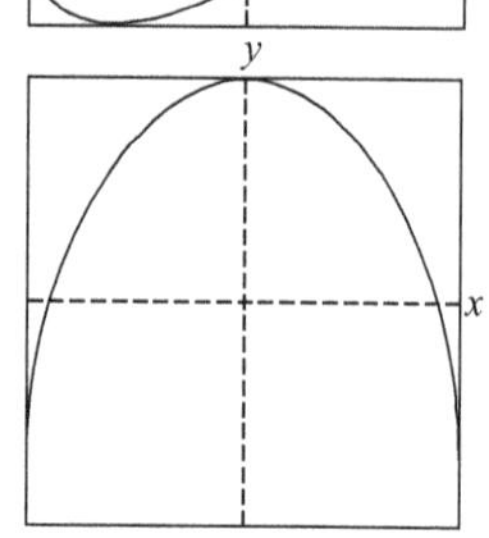

51. A simple harmonic oscillator of angular frequency 2 rad s^{-1} is acted upon by an external force $F = \sin t$ N. If the oscillator is at rest in its equilibrium position at $t = 0$, its position at later times is proportional to : **[2015 (Online)]**

(A) $\sin t + \frac{1}{2}\cos 2t$ (B) $\cot t - \frac{1}{2}\sin 2t$

(C) $\sin t - \frac{1}{2}\sin 2t$ (D) $\sin t + \frac{1}{2}\sin 2t$

52. A particle performs simple harmonic motion with amplitude A. Its speed is trebled at the instant that it is at a distance $\frac{2A}{3}$ from equilibrium position. The new amplitude of the motion is : **[2016]**

(A) $\frac{A}{3}\sqrt{41}$ (B) $3A$

(C) $A\sqrt{3}$ (D) $\frac{7A}{3}$

53. In an engine the piston undergoes vertical simple harmonic motion with amplitude 7 cm. A washer rests on top of the piston and moves with it. The motor speed is slowly increased. The frequency of the piston at which the washer no longer stays in contact with the piston, is close to : **[2016 (Online)]**

(A) 0.7 Hz (B) 1.9 Hz
(C) 1.2 Hz (D) 0.1 Hz

54. Two particles are performing simple harmonic motion in a straight line about the same equilibrium point. The amplitude and time period for both particles are same and equal to A and T, respectively. At time $t = 0$ one particle has displacement A while the other one has displacement $\frac{-A}{2}$ and they are moving towards each other. If they cross each other at time t, then t is : **[2016 (Online)]**

(A) $\frac{5T}{6}$ (B) $\frac{T}{3}$

(C) $\frac{T}{4}$ (D) $\frac{T}{6}$

55. A particle is executing simple harmonic motion with a time period T. At time $t = 0$, it is at its position of equilibrium. The kinetic energy-time graph of the particle will look like : **[2017]**

(A)
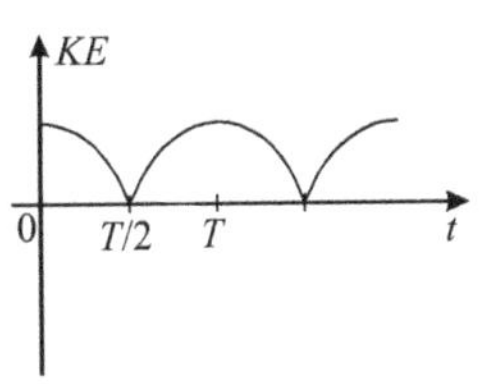

(B)
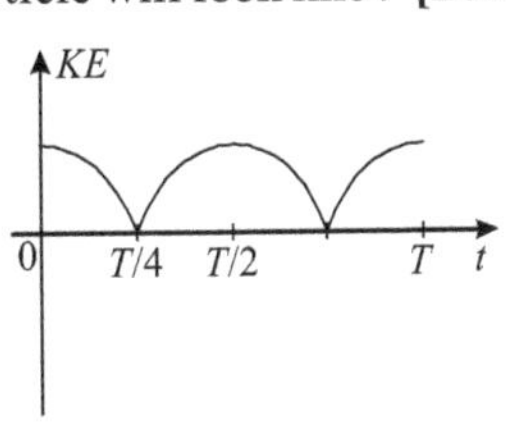

(C)
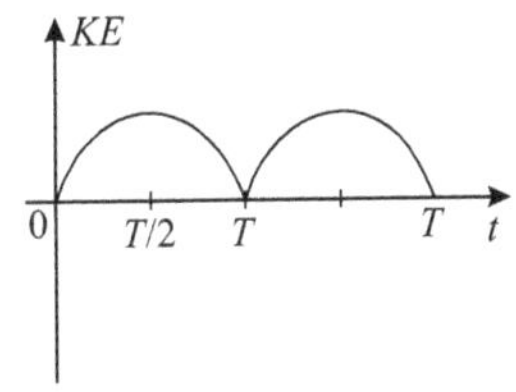

(D)
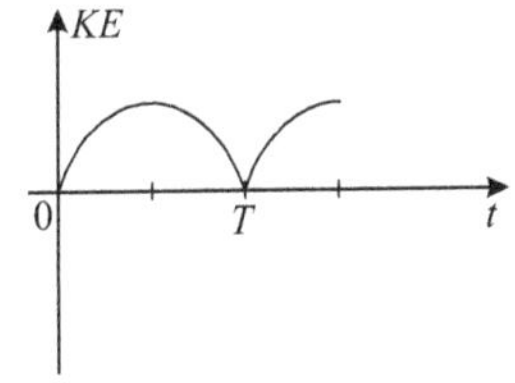

56. A 1 kg block attached to a spring vibrates with a frequency of 1 Hz on a frictionless horizontal table. Two springs identical to the original spring are attached in parallel to an 8 kg block placed on the same table. So, the frequency of vibration of the 8 kg block is : **[2017 (Online)]**

(A) $\frac{1}{4}$ Hz (B) $\frac{1}{2\sqrt{2}}$ Hz

(C) $\frac{1}{2}$ Hz (D) 2 Hz

57. In an experiment to determine the period of a simple pendulum of length 1m, it is attached to different spherical bobs of radii r_1 and r_2. The two spherical bobs have uniform mass distribution. If the relative difference in the periods, is found to be 5×10^{-4} s, the difference in radii, $|r_1 - r_2|$ is best given : **[2017 (Online)]**

(A) 1 cm (B) 0.1 cm
(C) 0.5 cm (D) 0.01 cm

58. A block of mass 0.1 kg is connected to an elastic spring of spring constant 640 Nm^{-1} and oscillates in a medium of constant 10^{-2} kg s^{-1}. The system dissipates its energy gradually. The time taken for its mechanical energy of vibration to drop to half of its initial value, is closest to : **[2017 (Online)]**

(A) 2 s (B) 3.5 s
(C) 5 s (D) 7 s

59. The ratio of maximum acceleration to maximum velocity in a simple harmonic motion is 10 s^{-1}. At, $t = 0$ the displacement is 5 m. What is the maximum acceleration ? The initial phase is $\frac{\pi}{4}$: **[2017 (Online)]**

(A) 500 m/s^2 (B) $500\sqrt{2}$ m/s^2

(C) 750 m/s^2 (D) $750\sqrt{2}$ m/s^2

60. A silver atom in a solid oscillates in simple harmonic motion in some direction with a frequency of 10^{12}/sec. What is the force constant of the bonds connecting one atom with the other? (Mole wt. of silver = 108 and Avagadro number = 6.02×10^{23} gm $mole^{-1}$) : **[2018]**

(A) 6.4 N/m (B) 7.1 N/m
(C) 2.2 N/m (D) 5.5 N/m

61. Two simple harmonic motions, as shown, are at right angles. They are combined to form Lissajous figures.

$x(t) = A\sin(at + \delta)$

$y(t) = B\sin(bt)$

Identify the correct match below **[2018 (Online)]**

(A) Parameters : $A = B, a = 2b; \delta = \frac{\pi}{2}$; Curve: Circle

(B) Parameters : $A = B, a = b; \delta = \frac{\pi}{2}$; Curve: Line

(C) Parameters : $A \neq B, a = b; \delta = \frac{\pi}{2}$; Curve: Ellipse

(D) Parameters : $A \neq B, a = b; \delta = 0$; Curve: Parabola

62. A particle executes simple harmonic motion and is located at $x = a$, b and c at times t_0, $2t_0$ and $3t_0$ respectively. The frequency of the oscillation is : **[2018 (Online)]**

(A) $\frac{1}{2\pi t_0}\cos^{-1}\left(\frac{a+b}{2c}\right)$ (B) $\frac{1}{2\pi t_0}\cos^{-1}\left(\frac{a+b}{3c}\right)$

(C) $\frac{1}{2\pi t_0}\cos^{-1}\left(\frac{2a+3c}{b}\right)$ (D) $\frac{1}{2\pi t_0}\cos^{-1}\left(\frac{a+c}{2b}\right)$

63. An oscillator of mass M is at rest in its equilibrium position in a potential $V = \frac{1}{2}k(x - X)^2$. A particle of mass m comes from right with speed u and collides completely inelastically with M and sticks to it. This process repeats every time the oscillator crosses its equilibrium position. The amplitude of oscillations after 13 collisions is : $[M = 10, m = 5, u = 1, k = 1]$ **[2018 (Online)]**

(A) $\frac{1}{2}$ (B) $\frac{1}{\sqrt{3}}$

(C) $\frac{2}{3}$ (D) $\sqrt{\frac{3}{5}}$

64. Two masses m and $\frac{m}{2}$ are connected at the two ends of a massless rigid rod of length l. The rod is suspended by a thin wire of torsional constant k at the centre of mass of the rod-mass system (see figure). Because of torsional constant k, the restoring torque is $\tau = k\theta$ for angular displacement θ. If the rod is rotated by θ_0 and released, the tension in it when it passes through its mean position will be : **[9 Jan. 2019 (Morning)]**

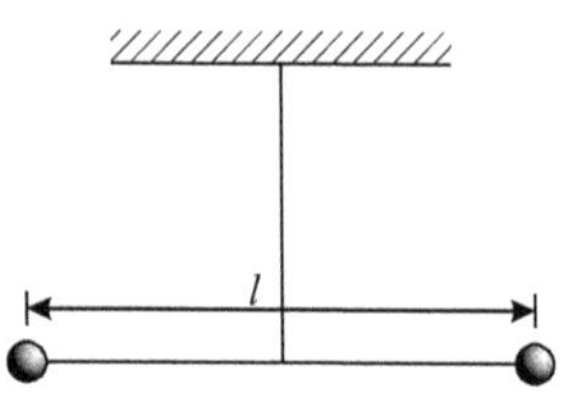

(A) $\frac{3k\theta_0^2}{l}$ (B) $\frac{k\theta_0^2}{2l}$

(C) $\frac{2k\theta_0^2}{l}$ (D) $\frac{k\theta_0^2}{l}$

65. A rod of mass 'M' and length '$2L$' is suspended at its middle by a wire. It exhibits torsional oscillations; If two masses each of 'm' are attached at distance '$L/2$' from its centre on both sides, it reduces the oscillation frequency by 20%. The value of ratio m/M is close to : **[9 Jan. 2019 (Evening)]**

(A) 0.17 (B) 0.37

(C) 0.57 (D) 0.77

66. A particle is executing simple harmonic motion (SHM) of amplitude A, along the x-axis, about $x = 0$. When its potential Energy (PE) equals kinetic energy (KE), the position of the particle will be : **[9 Jan. 2019 (Evening)]**

(A) $\frac{A}{2}$ (B) $\frac{A}{2\sqrt{2}}$

(C) $\frac{A}{\sqrt{2}}$ (D) A

67. A particle executes simple harmonic motion with an amplitude of 5 cm. When the particle is at 4 cm from the mean position, the magnitude of its velocity in SI units is equal to that of its acceleration. Then, its periodic time in seconds is : **[10 Jan. 2019 (Evening)]**

(A) $\frac{7}{3}\pi$ (B) $\frac{3}{8}\pi$

(C) $\frac{4\pi}{3}$ (D) $\frac{8\pi}{3}$

68. A cylindrical plastic bottle of negligible mass is filled with 310 ml of water and left floating in a pond with still water. If pressed downward slightly and released, it starts performing simple harmonic motion at angular frequency ω. If the radius of the bottle is 2.5 cm then ω close to :

(Density of water $= 10^3$ kg/m^3) : **[10 Jan. 2019 (Evening)]**

(A) 5.00 rad s^{-1} (B) 1.25 rad s^{-1}

(C) 3.75 rad s^{-1} (D) 2.50 rad s^{-1}

69. A particle undergoing simple harmonic motion has time dependent displacement given by $x(t)\ A\sin\frac{\pi t}{90}$. The ratio of potential to kinetic energy of this particle at $t = 210$s will be : **[11 Jan. 2019 (Morning)]**

(A) 2 (B) $\frac{1}{9}$

(C) 3 (D) 1

70. A simple pendulum of length 1 m is oscillating with an angular frequency 10 rad/s. The support of the pendulum starts oscillating up and down with a small angular frequency of 1 rad/s and an amplitude of 10^{-2} m. The relative change in the angular frequency of the pendulum is best given by :

[11 Jan. 2019 (Evening)]

(A) 10^{-3} rad/s (B) 10^{-1} rad/s

(C) 1 rad/s (D) 10^{-5} rad/s

71. A pendulum is executing simple harmonic motion and its maximum kinetic energy is K_1. If the length of the pendulum is doubled and it performs simple harmonic motion with the same amplitude as in the first case, its maximum kinetic energy is K_2. Then : **[11 Jan. 2019 (Evening)]**

(A) $K_2 = \frac{K_1}{4}$ (B) $K_2 = \frac{K_1}{2}$

(C) $K_2 = 2K_1$ (D) $K_2 = K_1$

72. Two light identical springs of spring constant k are attached horizontally at the two ends of a uniform horizontal rod AB of length l and mass m. The rod is pivoted at its centre 'O' and can rotate freely in horizontal plane. The other ends of the two springs are fixed to rigid supports as shown in figure. The rod is gently pushed through a small angle and released. The frequency of resulting oscillation is : **[12 Jan. 2019 (Morning)]**

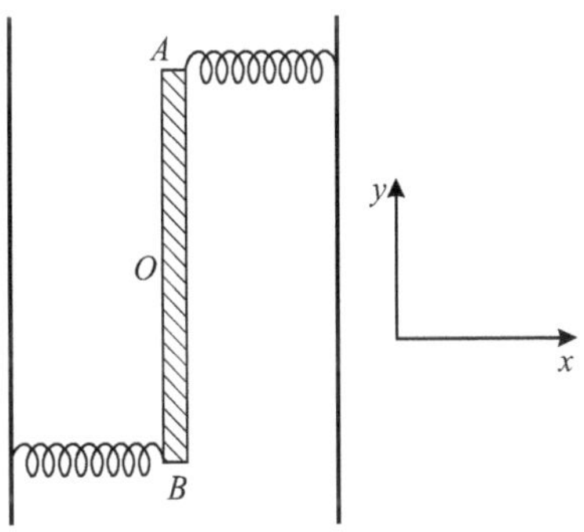

(A) $\frac{1}{2\pi}\sqrt{\frac{6k}{m}}$ (B) $\frac{1}{2\pi}\sqrt{\frac{2k}{m}}$

(C) $\frac{1}{2\pi}\sqrt{\frac{k}{m}}$ (D) $\frac{1}{2\pi}\sqrt{\frac{3k}{m}}$

73. A simple harmonic motion is represented by :

$y = 5(\sin 3\pi t + \sqrt{3}\cos 3\pi t)$ cm

The amplitude and time period of the motion are :

[12 Jan. 2019 (Evening)]

(A) 5cm, $\frac{3}{2}s$ (B) 5cm, $\frac{2}{3}s$

(C) 10cm, $\frac{3}{2}s$ (D) 10cm, $\frac{2}{3}s$

74. A damped harmonic oscillator has a frequency of 5 oscillations per second. The amplitude drops to half its value for every 10 oscillations. The time it will take to drop to $\frac{1}{1000}$ of the original amplitude is close to : **[8 April 2019 (Evening)]**

(A) 100 s (B) 20 s

(C) 10 s (D) 50 s

75. A simple pendulum oscillating in air has period T. The bob of the pendulum is completely immersed in a non-viscous liquid. The density of the liquid is $\frac{1}{16}$ th of the material of the bob. If the bob is inside liquid all the time, its period of oscillation in this liquid is : **[9 April 2019 (Morning)]**

(A) $4T\sqrt{\frac{1}{15}}$ (B) $2T\sqrt{\frac{1}{10}}$

(C) $4T\sqrt{\frac{1}{14}}$ (D) $2T\sqrt{\frac{1}{14}}$

76. A massless spring (k = 800 N/m), attached with a mass (500 g) is completely immersed in 1kg of water. The spring is stretched by 2cm and released so that it starts vibrating. What would be the order of magnitude of the change in the temperature of water when the vibrations stop completely ? (Assume that the water container and spring receive negligible heat and specific heat of mass = 400 J/kg K, specific heat of water = 4184 J/kg K) : **[9 April 2019 (Evening)]**

(A) 10^{-3}K (B) 10^{-4}K

(C) 10^{-1}K (D) 10^{-5}K

77. The displacement of a damped harmonic oscillator is given by $x(t) = e^{-01.1t}\cos(10\pi t + \phi)$. Here t is in seconds. The time taken for its amplitude of vibration to drop to half of its initial value is close to : **[10 April 2019 (Morning)]**

(A) 13 s (B) 7 s

(C) 27 s (D) 4 s

* * * * *

CHAPTER-14 WAVES

14.1 Wave Motion

1. The displacement y of a wave travelling in the x-direction is given by $y = 10^{-4} \sin\left(600t - 2x + \frac{\pi}{3}\right)$ metre, where, x is expressed in metres and t in seconds. The speed of the wave-motion, in ms^{-1} is : **[2003]**

(A) 300 (B) 600
(C) 1200 (D) 200

2. A sound absorber attenuates the sound level by 20dB. The intensity decreases by a factor of : **[2007]**

(A) 1000 (B) 10000
(C) 10 (D) 100

3. A wave travelling along the x-axis is described by the equation $y(x, t) = 0.005 \cos(\alpha x - \beta t)$. If the wavelength and the time period of the wave in 0.08 m and 2.0 s respectively then α and β in appropriate units are : **[2008]**

(A) $\alpha = 25.00\pi, \beta = \pi$ (B) $\alpha = \frac{0.08}{\pi}, \beta = \frac{2.0}{\pi}$

(C) $\alpha = \frac{0.04}{\pi}, \beta = \frac{1.0}{\pi}$ (D) $\alpha = 12.50\pi, \beta = \frac{\pi}{2.0}$

4. The speed of sound in oxygen (O_2) at a certain temperature is 460 ms^{-1}. The speed of sound in helium (He) at the same temperature will be (assume both gases to be ideal) : **[2008]**

(A) $460\sqrt{\frac{200}{21}}\ ms^{-1}$ (B) $500\sqrt{\frac{200}{21}}\ ms^{-1}$

(C) $650\sqrt{2}\ ms^{-1}$ (D) $330\sqrt{2}\ ms^{-1}$

5. The equation of a wave on a string of linear mass density 0.04 kg m^{-1} is given by $y = 0.02\ (m) \sin\left[2\pi\left(\frac{t}{0.04(s)} - \frac{x}{0.50(m)}\right)\right]$. The tension in the string is : **[2010]**

(A) 6.25 N (B) 4.0 N
(C) 12.5 N (D) 0.5 N

6. **Statement-1 :** Two longitudinal waves given by equations: $y_1(x, t) = 2a \sin(\omega t - kx)$ and $y_2(x, t) = a \sin(2\omega t - 2kx)$ will have equal intensity.

Statement-2 : Intensity of waves of given frequency in same medium is proportional to square of amplitude only. **[2011]**

(A) Statement-1 is false, statement-2 is true
(B) Statement-1 is true, statement-2 is false
(C) Statement-1 is true, statement-2 true; statement-2 is the correct explanation of statement-1
(D) Statement-1 is true, statement-2 is true; statement-2 is not correct explanation of statement-1

7. The transverse displacement $y(x, t)$ of a wave on a string is given by $y(x, t) = e^{-(ax^2 + bt^2 + 2\sqrt{ab}xt)}$. This represents a : **[2011]**

(A) Standing wave of frequency $\sqrt{b}$

(B) Standing wave of frequency $\frac{1}{\sqrt{b}}$

(C) Wave moving in $+x$ direction with speed $\sqrt{\frac{a}{b}}$

(D) Wave moving in $-x$ direction with speed $\sqrt{\frac{b}{a}}$

8. The disturbance $y(x, t)$ of a wave propagating in the positive x-direction is given by $y = \frac{1}{1+x^2}$ at time $t = 0$ and by $y = \frac{1}{[1+(x-1)^2]}$ at $t = 2$s, where x and y are in meters. The shape of the wave disturbance does not change during the propagation. The velocity of wave in m/s is : **[2012 (Online)]**

(A) 2.0 (B) 4.0
(C) 0.5 (D) 1.0

9. When two sound waves travel in the same direction in a medium, the displacements of a particle located at 'x' at time 't' is given by :

$y_1 = 0.05 \cos(0.50\pi x - 100\pi t)$
$y_2 = 0.05 \cos(0.45\pi x - 92\pi t)$

where y_1, y_2 and x are in meters and t in seconds. The speed of sound in the medium is : **[2013 (Online)]**

(A) 92 m/s (B) 200 m/s
(C) 100 m/s (D) 332 m/s

10. In a transverse wave the distance between a crest and neighbouring trough at the same instant is 0.4 cm and the distance between a crest and trough at the same place is 1.0 cm. The next crest appears at the same place after a time interval of 0.4s. The maximum speed of the vibrating particles in the medium is : **[2013 (Online)]**

(A) $\frac{3\pi}{2}$ cm/s (B) $\frac{5\pi}{2}$ cm/s

(C) $\frac{\pi}{2}$ cm/s (D) 2π cm/s

11. An engine approaches a hill with a constant speed. When it is at a distance of 0.9 km, it blows a whistle whose echo is heard by the driver after 5 seconds. If the speed of sound in air is 330 m/s, then the speed of the engine is : **[2013 (Online)]**

(A) 32 m/s (B) 27.5 m/s
(C) 60 m/s (D) 30 m/s

12. A transverse wave is represented by $y = \frac{10}{\pi}\sin\left(\frac{2\pi}{T}t - \frac{2\pi}{\lambda}x\right)$. For what value of the wavelength the wave velocity is twice the maximum particle velocity ? **[2014 (Online)]**

(A) 40 cm (B) 20 cm
(C) 10 cm (D) 60 cm

13. A uniform string of length 20 m is suspended from a rigid support. A short wave pulse is introduced at its lowest end. It starts moving up the string. The time taken to reach the support is : (take $g = 10\ ms^{-2}$) : **[2016]**

(A) $2\pi\sqrt{2}$ s (B) 2s
(C) $2\sqrt{2}$ s (D) $\sqrt{2}$ s

14. A heavy ball of mass M is suspended from the ceiling of a car by a light string of mass m ($m << M$). When the car is at rest, the speed of transverse waves in the string is 60 ms^{-1}. When the car has acceleration a, the wave-speed increases to 60.5 ms^{-1}. The value of a, in terms of gravitational acceleration g, is closest to : **[9 Jan. 2019 (Morning)]**

(A) $\frac{g}{5}$ (B) $\frac{g}{20}$

(C) $\frac{g}{10}$ (D) $\frac{g}{30}$

15. Equation of travelling wave on a stretched string of linear density 5 g/m is $y = 0.03 \sin(450t - 9x)$ where distance and time are measured is SI units. The tension in the string is : **[11 Jan. 2019 (Morning)]**

(A) 10 N (B) 12.5 N
(C) 7.5 N (D) 5 N

16. A travelling harmonic wave is represented by the equation $y(x, t) = 10^{-3} \sin(50t + 2x)$, where x and y are in meter and t is in seconds. Which of the following is a correct statement about the wave? The wave is propagating along the : **[12 Jan. 2019 (Morning)]**

(A) Negative x-axis with speed 25 ms^{-1}
(B) The wave is propagating along the positive x-axis with speed 25 ms^{-1}
(C) The wave is propagating along the positive x-axis with speed 100 ms^{-1}
(D) The wave is propagating along the negative x-axis with speed 100 ms^{-1}

17. The pressure wave, $P = 0.01 \sin[1000t - 3x]Nm^{-2}$, corresponds to the sound produced by a vibrating blade on a day when atmospheric temperature is 0°C. On some other day, when temperature is T, the speed of sound produced by the same blade and at the same frequency is found to be 336 ms^{-1}. Approximate value of T is : **[9 April 2019 (Morning)]**

(A) 15°C (B) 12°C
(C) 4°C (D) 11°C

18. A progressive wave travelling along the positive x-direction is represented by $y(x, t) = A \sin(kx - \omega t + \phi)$. Its snapshot at $t = 0$ is given in the figure. For this wave, the phase ϕ is : **[12 April 2019 (Morning)]**

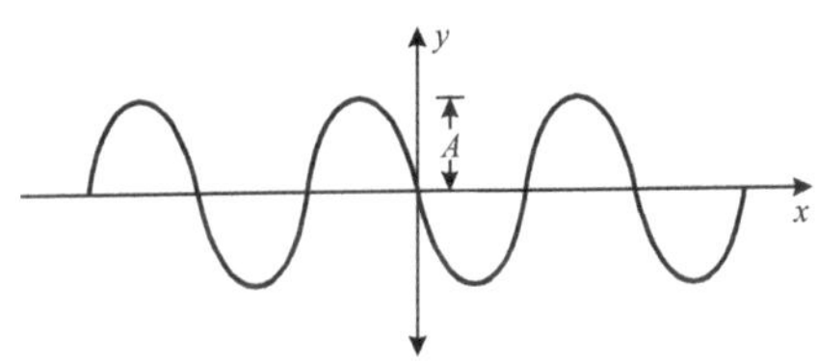

(A) 0 (B) $-\frac{\pi}{2}$

(C) π (D) $\frac{\pi}{2}$

19. A small speaker delivers $2W$ of audio output. At what distance from the speaker will one detect 120 dB intensity sound? [Given reference intensity of sound as $10^{-12}\ W/m^2$] : **[12 April 2019 (Evening)]**

(A) 10 cm (B) 30 cm
(C) 40 cm (D) 20 cm

* * * * *

CHAPTER-14 WAVES

14.2 Stationary Waves and Beats

1. Tube A has both ends open while tube B has one end closed, otherwise they are identical. The ratio of fundamental frequency of tubes A and B is : **[2002]**

(A) $1:2$ (B) $1:4$

(C) $2:1$ (D) $4:1$

2. A tuning fork arrangement (pair) produces 4 beats/s with one fork of frequency 288 cps. A little wax is placed on the unknown fork and it then produces 2 beats/s. The frequency of the unknown fork is : **[2002]**

(A) 286 cps (B) 292 cps

(C) 294 cps (D) 288 cps

3. Length of a string tied to two rigid supports is 40 cm. Maximum length (wavelength in cm) of a stationary wave produced on it, is : **[2002]**

(A) 20 (B) 80

(C) 40 (D) 120

4. A wave $y = a \sin(\omega t - kx)$ on a string meets with another wave producing a node at $x = 0$. Then the equation of the unknown wave is : **[2002]**

(A) $y = a \sin(\omega t + kx)$ (B) $y = -a \sin(\omega t + kx)$

(C) $y = a \sin(\omega t - kx)$ (D) $y = -a \sin(\omega t - kx)$

5. A tuning fork of known frequency 256 Hz makes 5 beats per second with the vibrating string of a piano. The beat frequency decreases to 2 beats per second when the tension in the piano string is slightly increased. The frequency of the piano string before increasing the tension was : **[2003]**

(A) $(256+2)$ Hz (B) $(256-2)$ Hz

(C) $(256-5)$ Hz (D) $(256+5)$ Hz

6. When two tuning forks (fork 1 and fork 2) are sounded simultaneously, 4 beats per second are heard. Now, some tape is attached on the prong of the fork 2. When the tuning forks are sounded again, 6 beats per second are heard. If the frequency of fork 1 is 200 Hz, then what was the original frequency of fork 2? **[2005]**

(A) 200 Hz (B) 202 Hz

(C) 196 Hz (D) 204 Hz

7. A string is stretched between fixed points separated by 75.0 cm. It is observed to have resonant frequencies of 420 Hz and 315 Hz. Their are no other resonant frequencies between these two. Then, the lowest resonant frequency for this string is : **[2006]**

(A) 105 Hz (B) 1.05 Hz

(C) 1050 Hz (D) 10.5 Hz

8. While measuring the speed of sound by performing a resonance column experiment, a student gets the first resonance condition at column length of 18 cm during winter. Repeating the same experiment during summer, student measures the column length to be x cm for the second resonance. Then : **[2008]**

(A) $18 > x$ (B) $x > 54$

(C) $54 > x > 36$ (D) $36 > x > 18$

9. Three sound waves of equal amplitudes have frequencies $(\nu - 1)$, ν, $(\nu + 1)$. They superpose to give beats. The number of beats produced per second will be : **[2009]**

(A) 2 (B) 1

(C) 4 (D) 3

10. A travelling wave represented by $y = A \sin(\omega t - kx)$ is superimposed on another wave represented by $y = A \sin(\omega t + kx)$. The resultant is : **[2011]**

(A) A standing wave having nodes at $x = \left(n + \frac{1}{2}\right)\frac{\lambda}{2}, n = 0, 1, 2$

(B) A wave travelling along $+x$ direction

(C) A wave travelling along $-x$ direction

(D) A standing wave having nodes at $x = \frac{n\lambda}{2}; n = 0, 1, 2$

11. A cylindrical tube, open at both ends, has a fundamental frequency f, in air. The tube is dipped vertically in water so that half of it is in water. The fundamental frequency of the air-column is now : **[2012]**

(A) $2f$ (B) f

(C) $f/2$ (D) $3f/4$

12. Following are expressions for four plane simple harmonic waves : **[2012 (Online)]**

(i) $y_1 = A \cos 2\pi\left(n_1 t + \frac{x}{\lambda_1}\right)$

(ii) $y_2 = A \cos 2\pi\left(n_1 t + \frac{x}{\lambda_1} + \pi\right)$

(iii) $y_3 = A \cos 2\pi\left(n_2 t + \frac{x}{\lambda_2}\right)$

(iv) $y_4 = A \cos 2\pi\left(n_2 t - \frac{x}{\lambda_2}\right)$

The pair of waves which will produce destructive interference and stationary wave respectively in a medium, are :

(A) (iii, iv), (i, ii) (B) (i, iii), (ii, iv)

(C) (i, iv), (ii, iii) (D) (i, ii), (iii, iv)

13. A wave represented by the equation $y_1 = a\cos(kx - \omega t)$ is superimposed with another wave to form a stationary wave such that the point $x - 0$ is node. The equation for the other wave is : **[2012 (Online)]**

(A) $a\cos(kx - \omega t - \pi)$ (B) $a\cos(kx + \omega t + \pi)$

(C) $a\cos\left(kx + \omega t + \frac{\pi}{2}\right)$ (D) $a\cos\left(kx - \omega t + \frac{\pi}{2}\right)$

14. A uniform tube of length 60.5 cm is held vertically with its lower end dipped in water. A sound source of frequency 500 Hz sends sound waves into the tube. When the length of tube above water is 16 cm and again when it is 50 cm, the tube resonates with the source of sound. Two lowest frequencies (in Hz), to which tube will resonate when it is taken out of water, are (approximately) : **[2012 (Online)]**

(A) 281, 562 (B) 281, 843

(C) 276, 552 (D) 272, 544

15. An air column in a pipe, which is closed at one end, will be in resonance with a vibrating tuning fork of frequency 264 Hz if the length of the column in cm is (velocity of sound = 330 m/s): **[2012 (Online)]**

(A) 125.00 (B) 93.75

(C) 62.50 (D) 187.50

16. A sonometer wire of length 1.5 m is made of steel. The tension in it produces an elastic strain of 1%. What is the fundamental frequency of steel if density and elasticity of steel are 7.7×10^3 kg/m^3 and 2.2×10^{11} N/m^2 respectively? **[2013]**

(A) 188.5 Hz (B) 178.2 Hz

(C) 200.5 Hz (D) 770 Hz

17. A sonometer wire of length 114 cm is fixed at both the ends. Where should the two bridges be placed so as to divide the wire into three segments whose fundamental frequencies are in the ratio 1 : 3 : 4 ? **[2013 (Online)]**

(A) At 36 cm and 84 cm from one end

(B) At 24 cm and 72 cm from one end

(C) At 48 cm and 96 cm from one end

(D) At 72 cm and 96 cm from one end

18. A pipe of length 85 cm is closed from one end. Find the number of possible natural oscillations of air column in the pipe whose frequencies lie below 1250 Hz. The velocity of sound in air is 340 m/s : **[2014]**

(A) 12 (B) 8

(C) 6 (D) 4

19. The total length of a sonometer wire between fixed ends is 110 cm. Two bridges are placed to divide the length of wire in ratio 6 : 3 : 2. The tension in the wire is 400 N and the mass per unit length is 0.01 kg/m. What is the minimum common frequency with which three parts can vibrate ? **[2014 (Online)]**

(A) 1100 Hz (B) 1000 Hz

(C) 166 Hz (D) 100 Hz

20. A pipe open at both ends has a fundamental frequency f in air. The pipe is dipped vertically in water so that half of it is in water. The fundamental frequency of the air column is now : **[2016]**

(A) $\frac{f}{2}$ (B) $\frac{3f}{4}$

(C) $2f$ (D) f

21. A standing wave is formed by the superposition of two waves travelling in opposite directions. The transverse displacement is given by $y(x, t) = 0.5\sin\left(\frac{5\pi}{4}x\right)\cos(200\pi t)$. What is the speed of the travelling wave moving in the positive x direction ? (x and t are in meter and second, respectively) : **[2017 (Online)]**

(A) 160 m/s (B) 90 m/s

(C) 180 m/s (D) 120 m/s

22. Two wires W_1 and W_2 have the same radius r and respective densities ρ_1 and ρ_2 such that $\rho_2 = 4\rho_1$. They are joined together at the point O, as shown in the figure. The combination is used as a sonometer wire and kept under tension T. The point O is midway between the two bridges. When a stationary waves is set up in the composite wire, the joint is found to be a node. The ratio of the number of antinodes formed in W_1 to W_2 is : **[2017 (Online)]**

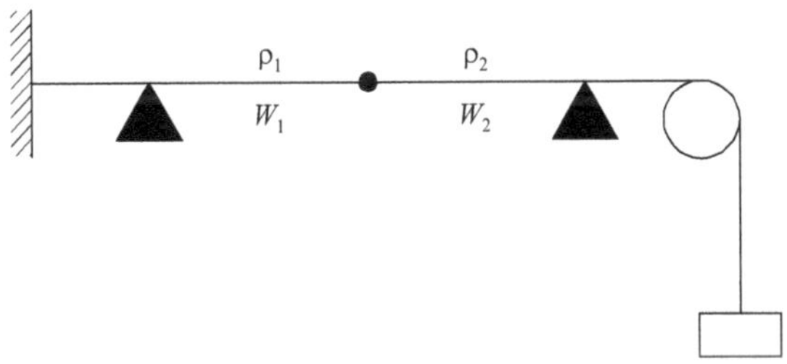

(A) 1 : 1 (B) 1 : 2

(C) 1 : 3 (D) 4 : 1

23. A granite rod of 60 cm length is clamped at its middle point and is set into longitudinal vibrations. The density of granite is 2.7×10^3 kg/m^3 and its Young's modulus is 9.27×10^{10} Pa. What will be the fundamental frequency of the longitudinal vibrations? **[2018]**

(A) 5 kHz (B) 2.5 kHz

(C) 10 kHz (D) 7.5 kHz

24. Two sitar strings, A and B, playing the note 'Dha' are slightly out of tune and produce beats and frequency 5 Hz. The tension of the string B is slightly increased and the beat frequency is found to decrease by 3 Hz. If the frequency of A is 425 Hz, the original frequency of B is : **[2018 (Online)]**

(A) 430 Hz (B) 428 Hz
(C) 422 Hz (D) 420 Hz

25. 5 beats/second are heard when a turning fork is sounded with a sonometer wire under tension, when the length of the sonometer wire is either 0.95 m or 1m. The frequency of the fork will be : **[2018 (Online)]**

(A) 195 Hz (B) 251 Hz
(C) 150 Hz (D) 300 Hz

26. A tuning fork vibrates with frequency 256 Hz and gives one beat per second with the third normal mode of vibration of an open pipe. What is the length of the pipe ? (Speed of sound of air is 340 s^{-1}) : **[2018 (Online)]**

(A) 190 cm (B) 180 cm
(C) 220 cm (D) 200 cm

27. The end correction of a resonance column is 1 cm. If the shortest length resonating with the tuning fork is 10 cm, the next resonating length should be : **[2018 (Online)]**

(A) 32 cm (B) 40 cm
(C) 28 cm (D) 36 cm

28. A musician using an open flute of length 50 cm produces second harmonic sound waves. A person runs towards the musician from another end of a hall at a speed of 10 km/h. If the wave speed is 330 m/s, the frequency heard by the running person shall be close to : **[9 Jan. 2019 (Evening)]**

(A) 753 Hz (B) 500 Hz
(3) 333 Hz (D) 666 Hz

29. A string of length 1 m and mass 5 g is fixed at both ends. The tension in the string is 8.0 N. The string is set into vibration using an external vibrator of frequency 100 Hz. The separation between successive nodes on the string is close to : **[10 Jan. 2019 (Morning)]**

(A) 16.6 cm (B) 20.0 cm
(C) 10.0 cm (D) 33.3 cm

30. A closed organ pipe has a fundamental frequency of 1.5 kHz. The number of overtones that can be distinctly heard by a person with this organ pipe will be : (Assume that the highest frequency a person can hear is 20,000 Hz) : **[10 Jan. 2019 (Evening)]**

(A) 7 (B) 5
(C) 6 (D) 4

31. A resonance tube is old and has jagged end. It is still used in the laboratory to determine velocity of sound in air. A tuning fork of frequency 512 Hz produces first resonance when the tube is filled with water to a mark 11 cm below a reference mark, near the open end of the tube. The experiment is repeated with another fork of frequency 256 Hz which produces first resonance when water reaches a mark 27 cm below the reference mark. The velocity of sound in air, obtained in the experiment, is close to : **[12 Jan. 2019 (Evening)]**

(A) 328 ms^{-1} (B) 322 ms^{-1}
(C) 341 ms^{-1} (D) 335 ms^{-1}

32. A wire of length $2L$, is made by joining two wires A and B of same length but different radii r and $2r$ and made of the same material. It is vibrating at a frequency such that the joint of the two wires forms a node. If the number of antinodes in wire A is p and that in B is q then the ratio $p : q$ is : **[8 April 2019 (Morning)]**

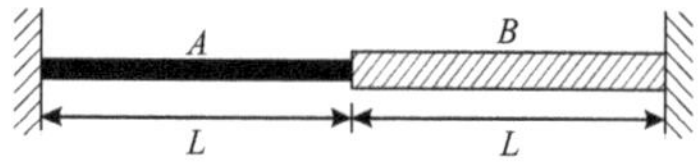

(A) 4 : 9 (B) 3 : 5
(C) 1 : 4 (D) 1 : 2

33. A string is clamped at both the ends and it is vibrating in its 4th harmonic. The equation of the stationary wave is $Y = 0.3 \sin(0.157x)\cos(200\pi t)$. The length of the string is : (All quantities are in SI units.) **[9 April 2019 (Morning)]**

(A) 20 m (B) 80 m
(C) 60 m (D) 40 m

34. A string 2.0 m long and fixed at its ends is driven by a 240 Hz vibrator. The string vibrates in its third harmonic mode. The speed of the wave and its fundamental frequency is : **[9 April 2019 (Evening)]**

(A) 320 m/s, 120 Hz (B) 180 m/s, 80 Hz
(C) 180 m/s, 120 Hz (D) 320 m/s, 80 Hz

35. The correct figure that shows, schematically, the wave pattern produced by superposition of two waves of frequencies 9 Hz and 11 Hz is : **[10 April 2019 (Evening)]**

(A)
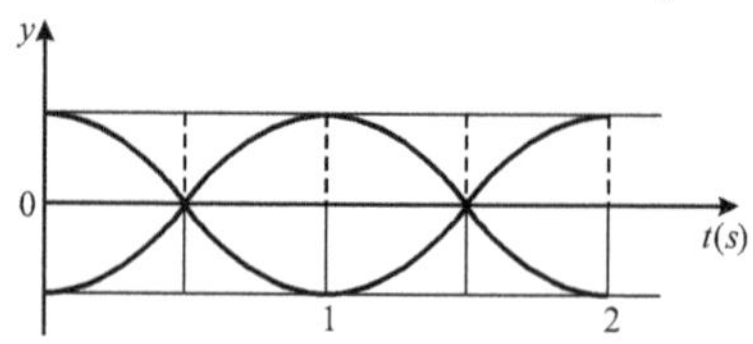

(B)
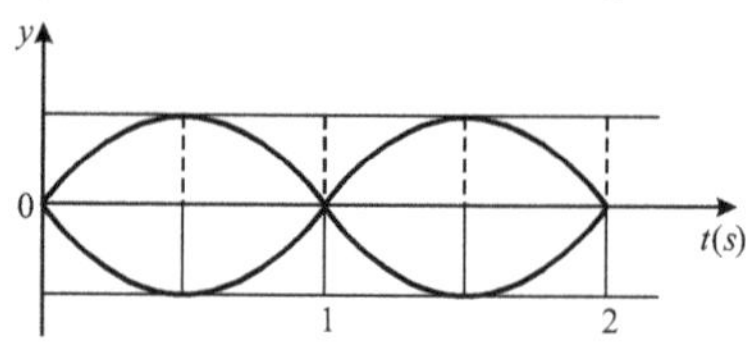

(C)
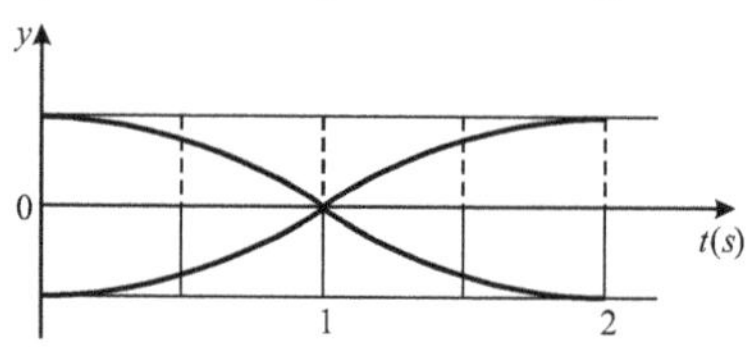

(D) 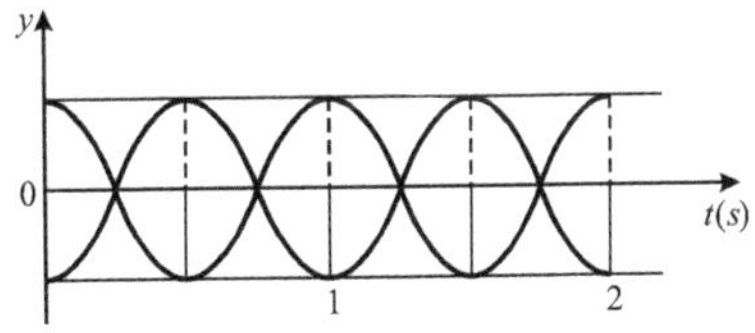

36. A tuning fork of frequency 480 Hz is used in an experiment for measuring speed of sound (v) in air by resonance tube method. Resonance is observed to occur at two successive lengths of the air column, $l_1 = 30$ cm and $l_2 = 70$ cm. Then v is equal to : **[12 April 2019 (Evening)]**

(A) 332 ms^{-1} (B) 379 ms^{-1}
(C) 384 ms^{-1} (D) 338 ms^{-1}

* * * * *

CHAPTER-14 WAVES

14.3 Doppler's Effect

1. An observer moves towards a stationary source of sound, with a velocity one-fifth of the velocity of sound. What is the percentage increase in the apparent frequency ? **[2005]**

(A) Zero (B) 0.5%

(C) 5% (D) 20%

2. A whistle producing sound waves of frequencies 9500 Hz and above is approaching a stationary person with speed v ms^{-1}. The velocity of sound in air is 300 ms^{-1}. If the person can hear frequencies upto a maximum of 10,000 Hz, the maximum value of v upto which he can hear the whistle is : **[2006]**

(A) $15\sqrt{2}$ ms^{-1} (B) $15/\sqrt{2}$ ms^{-1}

(C) 15 ms^{-1} (D) 30 ms^{-1}

3. A motor cycle starts from rest and accelerates along a straight path at 2 m/s^2. At the starting point of the motor cycle there is a stationary electric siren. How far has the motor cycle gone when the driver hears the frequency of the siren at 94% of its value when the motor cycle was at rest ? (Speed of sound = 330 ms^{-1}) : **[2009]**

(A) 147m (B) 196m

(C) 49m (D) 98m

4. This question has Statement-1 and Statement-2. Of the four choices given after the Statements, choose the one that best describes the two Statements. **[2012 (Online)]**

Statement-1 : Bats emitting ultrasonic waves can detect the location of a prey by hearing the waves reflected from it.

Statement-2 : When the source and the detector are moving, the frequency of reflected waves is changed.

(A) Statement-1 is false, Statement-2 is true

(B) Statement-1 is true, Statement-2 is false

(C) Statement-1 is true, Statement-2 is true, Statement-2 is not the correct explanation of Statement-1

(D) Statement-1 is true, Statement-2 is true, Statement-2 is the correct explanation of Statement-1

5. A and B are two sources generating sound waves. A listener is situated at C. The frequency of the source at A is 500 Hz. A, now, moves towards C with a speed 4 m/s. The number of beats heard at C is 6. When A moves away from C with speed 4 m/s, the number of beats heard at C is 18. The speed of sound is 340 m/s. The frequency of the source at B is : **[2013 (Online)]**

(A) 500 Hz (B) 506 Hz

(C) 512 Hz (D) 494 Hz

6. Two factories are sounding their sirens at 800 Hz. A man goes from one factory to other at a speed of 2 m/s. The velocity of sound is 320 m/s. The number of beats heard by the person in one second will be : **[2014 (Online)]**

(A) 2 (B) 4

(C) 8 (D) 10

7. A source of sound A emitting waves of frequency 1800 Hz is falling towards ground with a terminal speed v. The observer B on the ground directly beneath the source receives waves of frequency 2150 Hz. The source A receives waves, reflected from ground of frequency nearly : (Speed of sound = 343 m/s) : **[2014 (Online)]**

(A) 2150 Hz (B) 2500 Hz

(C) 1800 Hz (D) 2400 Hz

8. A train is moving on a straight track with speed 20 ms^{-1}. It is blowing its whistle at the frequency of 1000 Hz. The percentage change in the frequency heard by a person standing near the track as the train passes him is (speed of sound = 320 ms^{-1}) close to : **[2015]**

(A) 6% (B) 12%

(C) 18% (D) 24%

9. A bat moving at 10 ms^{-1} towards a wall sends a sound signal of 8000 Hz towards it. On reflection it hears a sound of frequency f. The value of f in *Hz* is close to (speed of sound = 320 ms^{-1}) : **[2015 (Online)]**

(A) 8516 (B) 8258

(C) 8424 (D) 800

10. A source of sound emits sound waves at frequency f_0. It is moving towards an observer with fixed speed vs ($vs < v$, where v is the speed of sound in air). If the observer were to move towards the source with speed v_0, one of the following two graphs (A and B) will given the correct variation of the frequency f heard by the observer as v_0 is changed : **[2015 (Online)]**

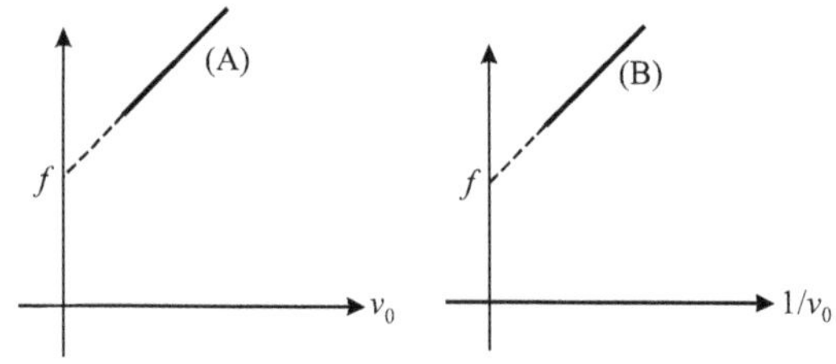

The variation of f with v_0 is given correctly by :

(A) Graph A with slope = $\dfrac{f_0}{(v+v_s)}$

(B) Graph B with slope $= \dfrac{f_0}{(v - v_s)}$

(C) Graph A with slope $= \dfrac{f_0}{(v - v_s)}$

(D) Graph B with slope $= \dfrac{f_0}{(v + v_s)}$

11. Two engines pass each other moving in opposite directions with uniform speed of 30 m/s. One of them is blowing a whistle of frequency 540 Hz. Calculate the frequency heard by driver of second engine before they pass each other. Speed of sound is 330 m/sec : **[2016 (Online)]**

(A) 450 Hz (B) 540 Hz
(C) 270 Hz (D) 648 Hz

12. A toy-car, blowing its horn, is moving with a steady speed of 5 m/s, away from a wall. An observer, towards whom the toy car is moving, is able to hear 5 beats per second. If the velocity of sound in air is 340 m/s, the frequency of the horn of the toy car is close to : **[2016 (Online)]**

(A) 680 Hz (B) 510 Hz
(C) 340 Hz (D) 170 Hz

13. An observer is moving with half the speed of light towards a stationary microwave source emitting waves at frequency 10 GHz. What is the frequency of the microwave measured by the observer ? (Speed of light $= 3 \times 10^8$ ms^{-1}) : **[2017]**

(A) 17.3 GHz (B) 15.3 GHz
(C) 10.1 GHz (D) 12.1 GHz

14. A train moves towards a stationary observer with speed 34 m/s. The train sounds a whistle and its frequency registered by the observer is f_1. If the speed of the train is reduced to 17 m/s, the frequency registered is f_2. If speed of sound is 340 m/s, then the ratio f_1/f_2 is : **[10 Jan. 2019 (Morning)]**

(A) 18/17 (B) 19/18
(C) 20/19 (D) 21/20

15. A person standing on an open ground hears the sound of a jet aeroplane, coming from north at an angle 60° with ground level. But he finds the aeroplane right vertically above his position. If v is the speed of sound, speed of the plane is : **[12 Jan. 2019 (Morning)]**

(A) $\dfrac{2v}{\sqrt{3}}$ (B) v
(C) $\dfrac{v}{2}$ (D) $\dfrac{\sqrt{3}}{2}v$

16. Two cars A and B are moving away from each other in opposite directions. Both the cars are moving with a speed of 20 ms^{-1} with respect to the ground. If an observer in car A detects a frequency 2000 Hz of the sound coming from car B, what is the natural frequency of the sound source in car B ? (speed of sound in air = 340 ms^{-1}) : **[9 April 2019 (Evening)]**

(A) 2250 Hz (B) 2060 Hz
(C) 2150 Hz (D) 2300 Hz

17. A stationary source emits sound waves of frequency 500 Hz. Two observers moving along a line passing through the source detect sound to be of frequencies 480 Hz and 530 Hz. Their respective speeds are, in ms^{-1}, (Given speed of sound = 300 m/s) : **[10 April 2019 (Morning)]**

(A) 16, 14 (B) 12, 18
(C) 12, 16 (D) 8, 18

18. A source of sound S is moving with a velocity of 50 m/s towards a stationary observer. The observer measures the frequency of the source as 1000 Hz. What will be the apparent frequency of the source when it is moving away from the observer after crossing him ? (Take velocity of sound in air is 350 m/s) **[10 April 2019 (Evening)]**

(A) 857 Hz (B) 807 Hz
(C) 750 Hz (D) 1143 Hz

19. A submarine (A) travelling at 18 km/hr is being chased along the line of its velocity by another submarine (B) travelling at 27 km/hr. B sends a sonar signal of 500 Hz to detect A and receives a reflected sound of frequency ν. The value of ν is close to : (Speed of sound in water = 1500 ms^{-1}) : **[12 April 2019 (Morning)]**

(A) 499 Hz (B) 502 Hz
(C) 507 Hz (D) 504 Hz

20. Two sources of sound S_1 and S_2 produce sound waves of same frequency 660 Hz. A listener is moving from source S_1 towards S_2 with a constant speed u m/s and he hears 10 beats/s. The velocity of sound is 330 m/s. Then, u equals : **[12 April 2019 (Evening)]**

(A) 2.5 m/s (B) 15.0 m/s
(C) 5.5 m/s (D) 10.0 m/s

* * * * *

CHAPTER-15 ELECTRIC CHARGES AND FIELDS

15.1 Electric Force and Electric Field

1. If a charge q is placed at the centre of the line joining two equal charges Q such that the system is in equilibrium then the value of q is : **[2002]**

(A) $Q/2$ (B) $-Q/2$
(C) $Q/2$ (D) $-Q/4$

2. Three charges $-q_1$, $+q_2$ and $-q_3$ are place as shown in the figure. The x-component of the force on $-q_1$ is proportional to: **[2003]**

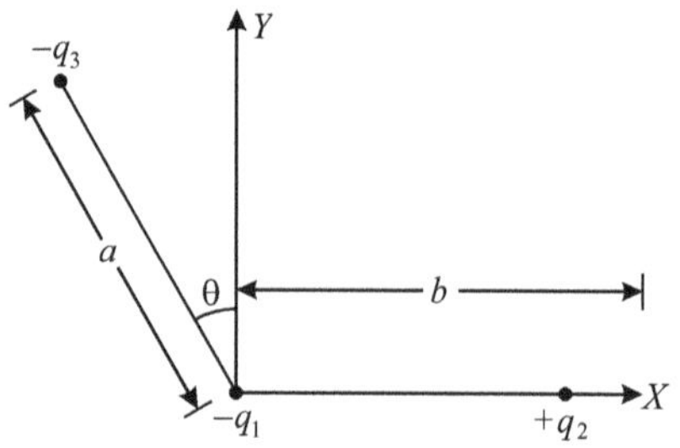

(A) $\frac{q_2}{b^2}-\frac{q_3}{a^2}\cos\theta$ (B) $\frac{q_2}{b^2}+\frac{q_3}{a^2}\sin\theta$

(C) $\frac{q_2}{b^2}+\frac{q_3}{a^2}\cos\theta$ (D) $\frac{q_2}{b^2}-\frac{q_3}{a^2}\sin\theta$

3. Four charges equal to $-Q$ are placed at the four corners of a square and a charge q is at its centre. If the system is in equilibrium, the value of q is : **[2004]**

(A) $-\frac{Q}{4}(1+2\sqrt{2})$ (B) $\frac{Q}{4}(1+2\sqrt{2})$

(C) $-\frac{Q}{2}(1+2\sqrt{2})$ (D) $\frac{Q}{2}(1+2\sqrt{2})$

4. Two spherical conductors B and C having equal radii and carrying equal charges in them, repel each other with a force F when kept apart at some distance. A third spherical conductor having same radius as that of B but uncharged, is brought in contact with B, then brought in contact with C and finally removed away from both. The new force of repulsion between B and C is : **[2004]**

(A) $\frac{F}{4}$ (B) $\frac{3F}{4}$

(C) $\frac{F}{8}$ (D) $\frac{3F}{8}$

5. A charged oil drop is suspended in a uniform field of 3×10^4 v/m so that it neither falls nor rises. The charge on the drop will be (Take the mass of the charge $= 9.9 \times 10^{-15}$ kg and $g = 10$ m/s^2) : **[2004]**

(A) 1.6×10^{-18} C (B) 3.2×10^{-18} C
(C) 3.3×10^{-18} C (D) 4.8×10^{-18} C

6. Two point charges $+8q$ and $-2q$ are located at $x = 0$ and $x = L$ respectively. The location of a point on the x-axis at which the net electric field due to these two point charges is zero is : **[2005]**

(A) $2L$ (B) $L/4$
(C) $8L$ (D) $4L$

7. A charged ball B hangs from a silk thread S, which makes an angle θ with a large charged conducting sheet P, as shown in the figure. The surface charge density σ of the sheet is proportional to : **[2005]**

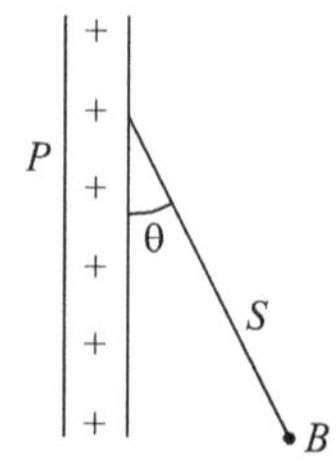

(A) cos θ (B) cot θ
(C) sin θ (D) tan θ

8. A thin spherical shell of radius R has a charge Q spread uniformly over its surface. Which of the following graphs most closely represents the electric field $E(r)$ produced by the shell in the range $0 \le r < \infty$, where r is the distance from the centre of the shell : **[2008]**

(A)
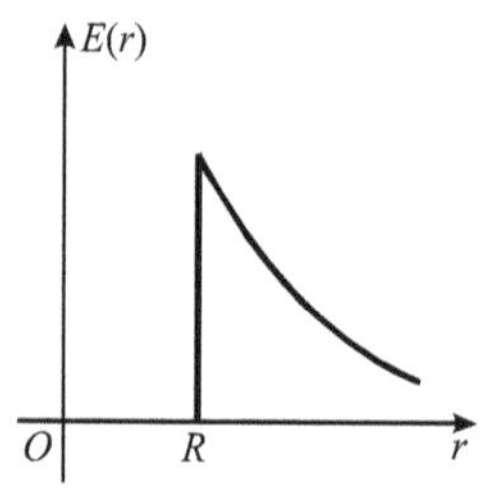

(B)
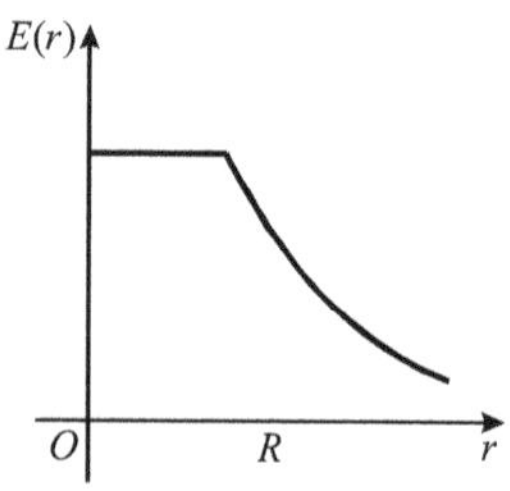

(C)
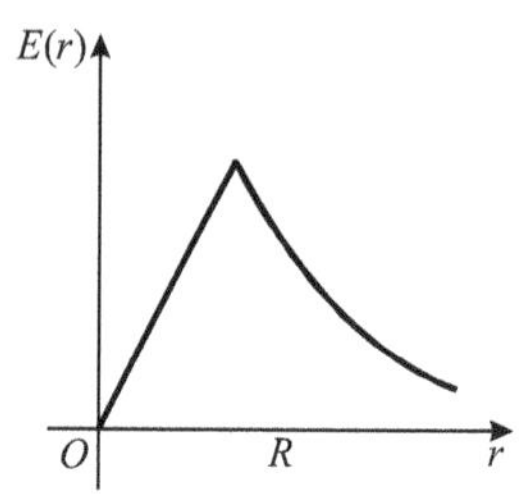

(D)
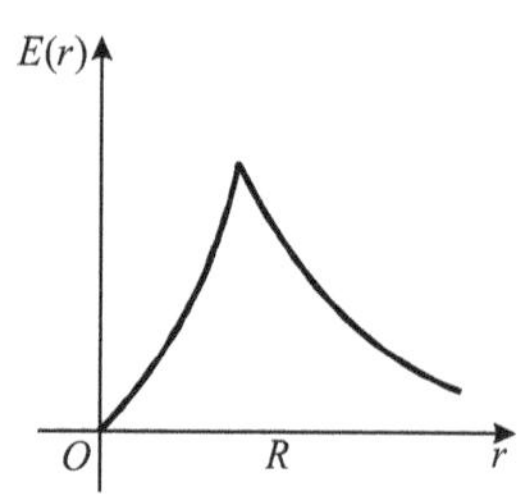

9. A charge Q is placed at each of the opposite corners of a square. A charge q is placed at each of the other two corners. If the net electrical force on Q is zero, then $\frac{Q}{q}$ equals : **[2009]**

(A) 1 (B) $-\frac{1}{\sqrt{2}}$

(C) $-2\sqrt{2}$ (D) -1

10. Let $\rho(r) = \frac{Q}{\pi R^4} r$ be the charge density distribution for a solid sphere of radius R and total charge Q. For a point 'p' inside the sphere at distance r_1 from the centre of the sphere, the magnitude of electric field is : **[2009]**

(A) $\frac{Qr_1^2}{4\pi \epsilon_0 R^4}$ (B) $\frac{Qr_1^2}{3\pi \epsilon_0 R^4}$

(C) 0 (D) $\frac{Q}{4\pi \epsilon_0 r_1^2}$

11. Two identical charged spheres are suspended by strings of equal lengths. The strings make an angle of 30° with each other. When suspended in a liquid of density 0.8 g cm^{-3}, the angle remains the same. If density of the material of the sphere is 1.6 g cm^{-3}, the dielectric constant of the liquid is : **[2010]**

(A) 1 (B) 4

(C) 3 (D) 2

12. Let there be a spherically symmetric charge distribution with charge density varying as $\rho(r) = \rho_0\left(\frac{5}{4} - \frac{r}{R}\right)$ upto $r = R$, and $\rho(r) = 0$ for $r > R$, where r is the distance from the origin. The electric field at a distance r $(r < R)$ from the origin is given by : **[2010]**

(A) $\frac{\rho_0 r}{3\epsilon_0}\left(\frac{5}{4} - \frac{r}{R}\right)$ (B) $\frac{4\pi\rho_0 r}{3\epsilon_0}\left(\frac{5}{3} - \frac{r}{R}\right)$

(C) $\frac{\rho_0 r}{4\epsilon_0}\left(\frac{5}{3} - \frac{r}{R}\right)$ (D) $\frac{4\rho_0 r}{3\epsilon_0}\left(\frac{5}{4} - \frac{r}{R}\right)$

13. A thin semi-circular ring of radius r has a positive charge q distributed uniformly over it. The net field $\vec{E}$ at the centre O is: **[2010]**

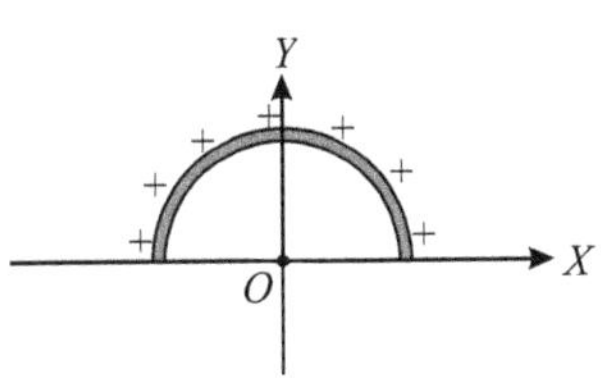

(A) $\frac{q}{2\pi^2 \epsilon_0 r^2}\hat{j}$ (B) $\frac{q}{4\pi^2 \epsilon_0 r^2}\hat{j}$

(C) $-\frac{q}{4\pi^2 \epsilon_0 r^2}\hat{j}$ (D) $-\frac{q}{2\pi^2 \epsilon_0 r^2}\hat{j}$

14. Two identical charged spheres suspended from a common point by two massless string of length l are initially a distance $d(d << l)$ apart because of their mutual repulsion. The charge begins to leak from both the spheres at a constant rate. As a result the charges approach each other with a velocity v. Then as a function of distance x between them : **[2011]**

(A) $v \propto x^{1/x}$ (B) $v \propto x$

(C) $v \propto x^{-1/2}$ (D) $v \propto x^{-1}$

15. In a uniformly charged sphere of total charge Q and radius R, the electric field E is plotted as a function of distance from the centre. The graph which would correspond to the above will be : **[2012]**

(A)

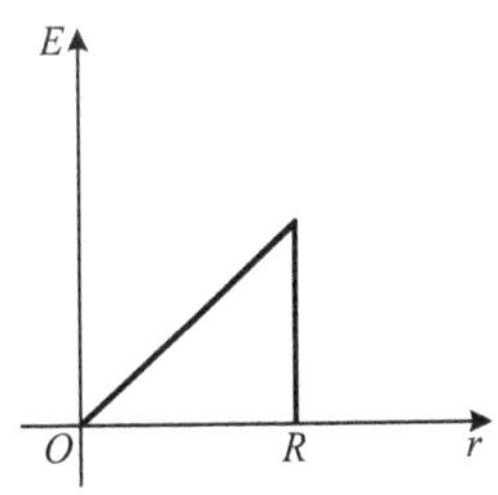

(B)

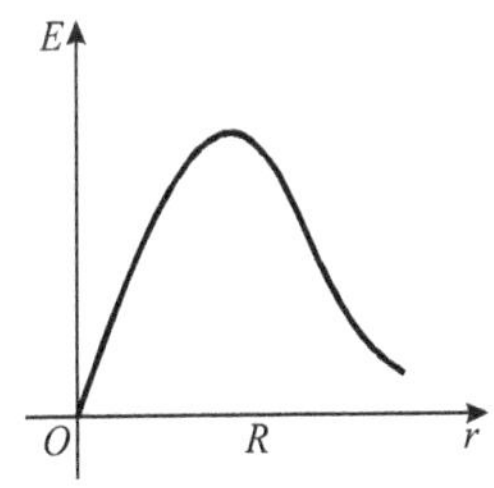

(C)

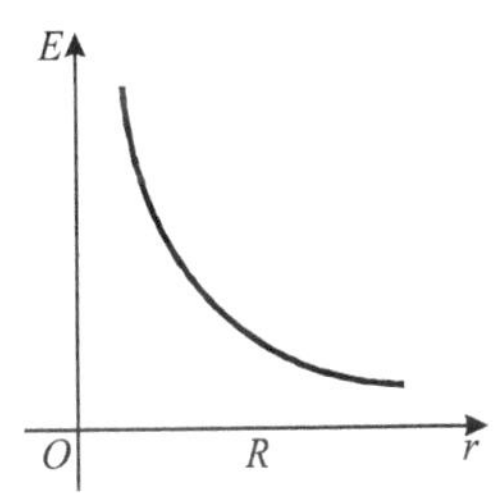

(D)

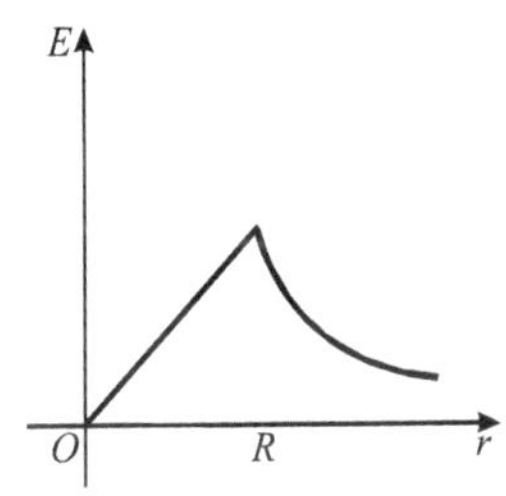

16. Two charges, each equal to q, are kept at $x = -a$ and $x = a$ on the x-axis. A particle of mass m and charge $q_0 = \frac{q}{2}$ is placed at the origin. If charge q_0 is given a small displacement $(y << a)$ along the y-axis, the net force acting on the particle is proportional to : **[2013]**

(A) y (B) $-y$

(C) $\frac{1}{y}$ (D) $-\frac{1}{y}$

17. A liquid drop having 6 excess electrons is kept stationary under a uniform electric field of 25.5 kVm^{-1}. The density of liquid is 1.26×10^3 kg m^{-3}. The radius of the drop is (neglect buoyancy) : **[2013 (Online)]**

(A) 4.3×10^{-7} m (B) 7.8×10^{-7} m

(C) 0.078×10^{-7} m (D) 3.4×10^{-7} m

18. The surface charge density of a thin charged disc of radius R is σ. The value of the electric field at the centre of the disc is $\frac{\sigma}{2\epsilon_0}$. With respect to the field at the centre, the electric field along the axis at a distance R from the centre of the disc : **[2013 (Online)]**

(A) Reduces by 70.7% (B) Reduces by 29.3%

(C) Reduces by 9.7% (D) Reduces by 14.6%

19. Two balls of same mass and carrying equal charge are hung from a fixed support of length l. At electrostatic equilibrium, assuming that angles made by each thread is small, the separation, x between the balls is proportional to : **[2013 (Online)]**

(A) l (B) l^2
(C) $l^{2/3}$ (D) $l^{1/3}$

20. The magnitude of the average electric field normally present in the atmosphere just above the surface of the Earth is about 150 N/C, directed inward towards the center of the Earth. This gives the total net surface charge carried by the Earth to be : [Given $\varepsilon_0 = 8.85 \times 10^{-12}$ C^2/N-m^2, $R_E = 6.37 \times 10^6$ m] **[2014 (Online)]**

(A) +670 kC (B) −670 kC
(C) −680 kC (D) +680 kC

21. A spherically symmetric charge distribution is characterised by a charge density having the following variations :

$$\rho(r) = \rho_0\left(1 - \frac{r}{R}\right) \text{ for } r < R$$
$$\rho(r) = 0 \text{ for } r \geq R$$

Where r is the distance from the centre of the charge distribution ρ_0 is a constant. The electric field at an internal point ($r < R$) is: **[2014 (Online)]**

(A) $\frac{\rho_0}{4\varepsilon_0}\left(\frac{r}{3} - \frac{r^2}{4R}\right)$ (B) $\frac{\rho_0}{\varepsilon_0}\left(\frac{r}{3} - \frac{r^2}{4R}\right)$

(C) $\frac{\rho_0}{3\varepsilon_0}\left(\frac{r}{3} - \frac{r^2}{4R}\right)$ (D) $\frac{\rho_0}{12\varepsilon_0}\left(\frac{r}{3} - \frac{r^2}{4R}\right)$

22. A thin disc of radius $b = 2a$ has a concentric hole of radius 'a' in it (see figure). It carries uniform surface charge 'σ' on it. If the electric field on its axis at height 'h' ($h << a$) from its centre is given as 'Ch' then value of 'C' is : **[2015 (Online)]**

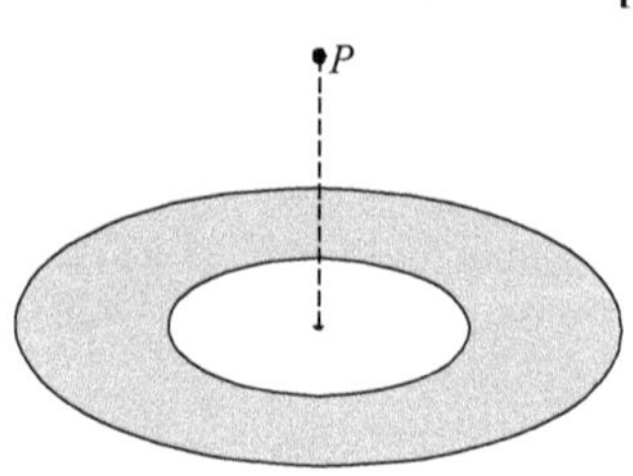

(A) $\frac{\sigma}{4a \epsilon_0}$ (B) $\frac{\sigma}{8a \epsilon_0}$

(C) $\frac{\sigma}{a \epsilon_0}$ (D) $\frac{\sigma}{2a \epsilon_0}$

23. A wire of length L (= 20 cm), is bent into a semicircular arc. If the two equal halves of the arc were each to be uniformly charged with charges $\pm Q$, [$|Q| = 10^3\varepsilon_0$ Coulomb where ε_0 is the permittivity (in SI units) of free space] the net electric field at the centre O of the semicircular arc would be : **[2015 (Online)]**

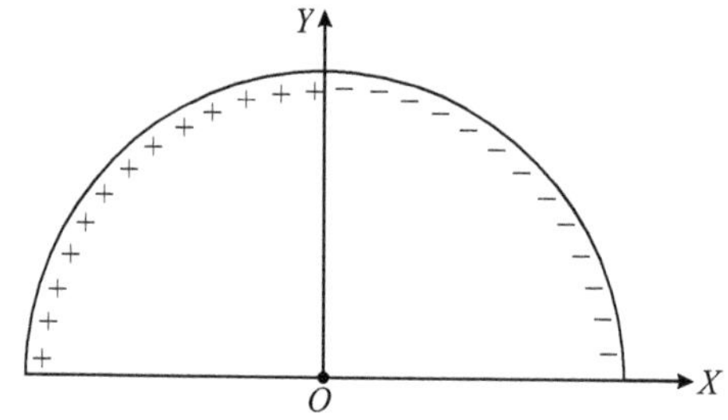

(A) $(50 \times 10^3 \text{ N/C})\,\hat{j}$ (B) $(50 \times 10^3 \text{ N/C})\,\hat{i}$

(C) $(25 \times 10^3 \text{ N/C})\,\hat{j}$ (D) $(25 \times 10^3 \text{ N/C})\,\hat{i}$

24. Shown in the figure are two point charges $+Q$ and $-Q$ inside the cavity of a spherical shell. The charges are kept near the surface of the cavity on opposite sides of the centre of the shell. If σ_1 is the surface charge on the inner surface and Q_1 net charge on it and σ_2 the surface charge on the outer surface and Q_2 net charge on it then : **[2015 (Online)]**

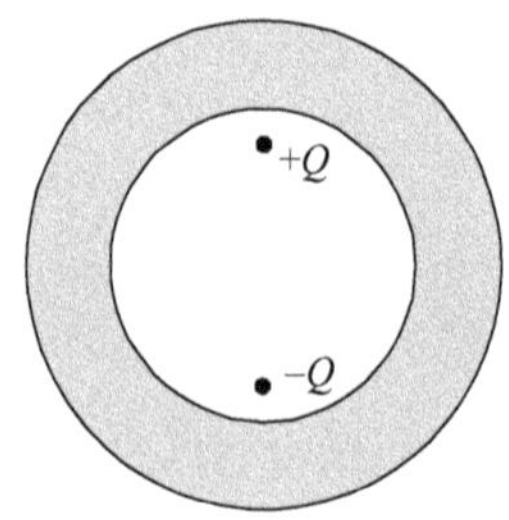

(A) $\sigma_1 \neq 0, Q_1 = 0$
$\sigma_2 \neq 0, Q_2 = 0$
(B) $\sigma_1 \neq 0, Q_1 = 0$
$\sigma_2 = 0, Q_2 = 0$
(C) $\sigma_1 = 0, Q_1 = 0$
$\sigma_2 = 0, Q_2 = 0$
(D) $\sigma_1 \neq 0, Q_1 \neq 0$
$\sigma_2 \neq 0, Q_2 \neq 0$

25. The region between two concentric spheres of radii 'a' and 'b' respectively (see figure), has volume charge density $\rho = \frac{A}{r}$, where A is a constant and r is the distance from the centre. At the centre of the spheres is a point charge Q. The value of A such that the electric field in the region between the spheres will be constant, is : **[2016]**

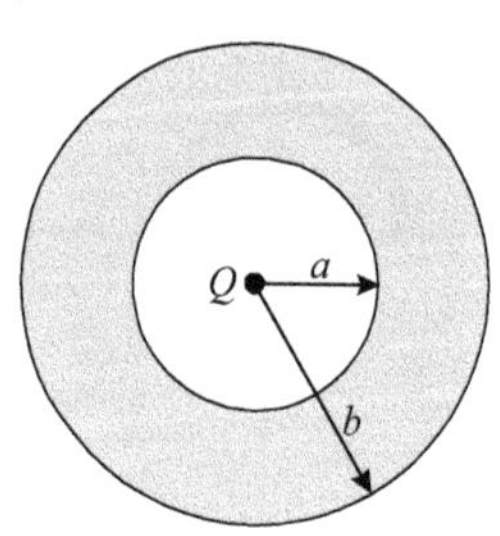

(A) $\frac{Q}{2\pi a^2}$ (B) $\frac{Q}{2\pi(b^2 - a^2)}$

(C) $\frac{2Q}{\pi(a^2 - b^2)}$ (D) $\frac{2Q}{\pi a^2}$

26. A solid ball of radius R has a charge density ρ given by $\rho = \rho_0\left(1-\frac{r}{R}\right)$ for $0 \le r \le R$. The electric field outside the ball is: **[2018 (Online)]**

(A) $\frac{\rho_0 R^3}{\epsilon_0 r^2}$ (B) $\frac{4\rho_0 R^3}{3\epsilon_0 r^2}$

(C) $\frac{3\rho_0 R^3}{4\epsilon_0 r^2}$ (D) $\frac{\rho_0 R^3}{12\epsilon_0 r^2}$

27. A body of mass M and charge q is connected to a spring of spring constant k. It is oscillating along x-direction about its equilibrium position, taken to be at $x = 0$, with an amplitude A. An electric field E is applied along the x-direction. Which of the following statements is correct ? **[2018 (Online)]**

(A) The total energy of the system is $\frac{1}{2}m\omega^2A^2 + \frac{1}{2}\frac{q^2E^2}{k}$

(B) The new equilibrium position is at a distance : $\frac{2qE}{k}$ from $x = 0$

(C) The new equilibrium position is at a distance : $\frac{qE}{2k}$ from $x = 0$

(D) The total energy of the system is $\frac{1}{2}m\omega^2A^2 - \frac{1}{2}\frac{q^2E^2}{k}$

28. Two identical conducting sphere A and B, carry equal charge. They are separated by a distance much larger than their diameter, and the force between them is F. A third identical conducting sphere, C, is uncharged. Sphere C is first touched to A, then to B, and then removed. As a result, the force between A and B would be equal to : **[2018 (Online)]**

(A) $\frac{3F}{4}$ (B) $\frac{F}{2}$

(C) F (D) $\frac{3F}{8}$

29. Three charges $+Q$, q, $+Q$ are placed respectively, at distance 0, $d/2$ and d from the origin, on the x-axis. If the net force experienced by $+Q$, placed at $x = 0$, is zero, then value of q is : **[9 Jan. 2019 (Morning)]**

(A) $+Q/2$ (B) $-Q/2$

(C) $-Q/4$ (D) $+Q/4$

30. For a uniformly charged ring of radius R, the electric field on its axis has the largest magnitude at a distance h from its centre. Then value of h is : **[9 Jan. 2019 (Morning)]**

(A) $\frac{R}{\sqrt{5}}$ (B) R

(C) $\frac{R}{\sqrt{2}}$ (D) $R\sqrt{2}$

31. Two point charges $q_1(\sqrt{10}\ \mu C)$ and $q_2(-25\ \mu C)$ are placed on the x-axis at $x = 1$ m and $x = 4$ m respectively. The electric field (in V/m) at a point $y = 3$ m on y-axis is,

$\left[\text{take } \frac{1}{4\pi\varepsilon_0} = 9\times10^9\ \text{Nm}^2\text{C}^{-2}\right]$: **[9 Jan. 2019 (Evening)]**

(A) $(-63\hat{i}+27\hat{j})\times10^2$ (B) $(81\hat{i}-81\hat{j})\times10^2$

(C) $(63\hat{i}-27\hat{j})\times10^2$ (D) $(-81\hat{i}+81\hat{j})\times10^2$

32. Charge is distributed within a sphere of radius R with a volume charge density $\rho(r) = \frac{A}{r^2}e^{-2r/a}$, where A and a are constants. If Q is the total charge of this charge distribution, the radius R is : **[9 Jan. 2019 (Evening)]**

(A) $\frac{a}{2}\log\left(1-\frac{Q}{2\pi aA}\right)$ (B) $a\log\left(1-\frac{Q}{2\pi aA}\right)$

(C) $a\log\left(\frac{1}{1-\frac{Q}{2\pi aA}}\right)$ (D) $\frac{a}{2}\log\left(\frac{1}{1-\frac{Q}{2\pi aA}}\right)$

33. The bob of a simple pendulum has mass $2g$ and a charge of 5.0 μC. It is at rest in a uniform horizontal electric field of intensity 2000 V/m. At equilibrium, the angle that the pendulum makes with the vertical is : (take $g = 10\ \text{m/s}^2$) : **[8 April 2019 (Morning)]**

(A) $\tan^{-1}(5.0)$ (B) $\tan^{-1}(2.0)$

(C) $\tan^{-1}(0.5)$ (D) $\tan^{-1}(0.2)$

34. Four point charges $-q$, $+q$, $+q$ and $-q$ are placed on y-axis at $y = -2d$, $y = -d$, $y = +d$ and $y = +2d$, respectively. The magnitude of the electric field E at a point on the x-axis at $x = D$, with $D >> d$, will behave as : **[9 April 2019 (Evening)]**

(A) $E \propto \frac{1}{D}$ (B) $E \propto \frac{1}{D^3}$

(C) $E \propto \frac{1}{D^2}$ (D) $E \propto \frac{1}{D^4}$

35. Let a total charge $2Q$ be distributed in a sphere of radius R, with the charge density given by $\rho(r) = kr$, where r is the distance from the centre. Two charges A and B, of $-Q$ each, are placed on diametrically opposite points, at equal distance a from the centre. If A and B do not experience any force, then : **[12 April 2019 (Evening)]**

(A) $a = \frac{3R}{2^{1/4}}$ (B) $a = R/\sqrt{3}$

(C) $a = 8^{-1/4}R$ (D) $a = 2^{-1/4}R$

* * * * *

CHAPTER-15 ELECTRIC CHARGES AND FIELDS

15.2 Electric Flux and Gauss's Law

1. A charged particle q is placed at the centre O of cube of length L ($ABCDEFGH$). Another same charge q is placed at a distance L from O. Then the electric flux through $ABCD$ is : **[2002]**

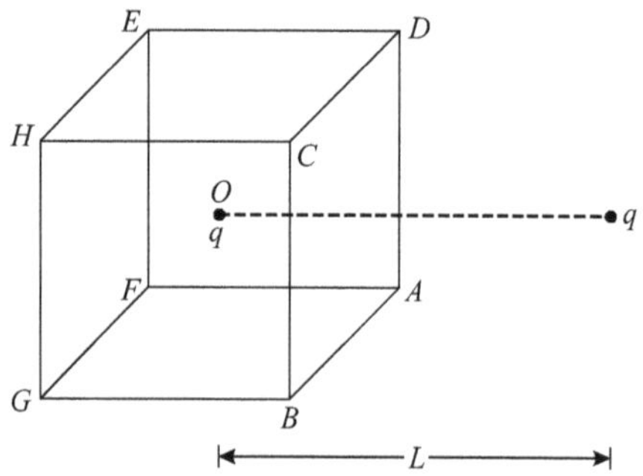

(A) $q/4\pi\epsilon_0 L$ (B) Zero
(C) $q/2\pi\epsilon_0 L$ (D) $q/3\pi\epsilon_0 L$

2. If the electric flux entering and leaving an enclosed surface respectively is ϕ_1 and ϕ_2, the electric charge inside the surface will be : **[2003]**

(A) $(\phi_2 - \phi_1)\varepsilon_0$ (B) $(\phi_1 - \phi_2)/\varepsilon_0$
(C) $(\phi_2 - \phi_1)/\varepsilon_0$ (D) $(\phi_1 - \phi_2)\varepsilon_0$

3. The electrostatic potential inside a charged spherical ball is given by $\phi = ar^2 + b$ where r is the distance from the centre; a, b are constant. Then the charge density inside the ball is : **[2011]**

(A) $-24\pi a \epsilon_0$ (B) $-6 a \epsilon_0$
(C) $-24\pi a \epsilon_0 r$ (D) $-6 a \epsilon_0 r$

4. The flat base of a hemisphere of radius a with no charge inside it lies in a horizontal plane. A uniform electric field $\vec{E}$ is applied at an angle $\pi/4$ with the vertical direction. The electric flux through the curved surface of the hemisphere is : **[2012 (Online)]**

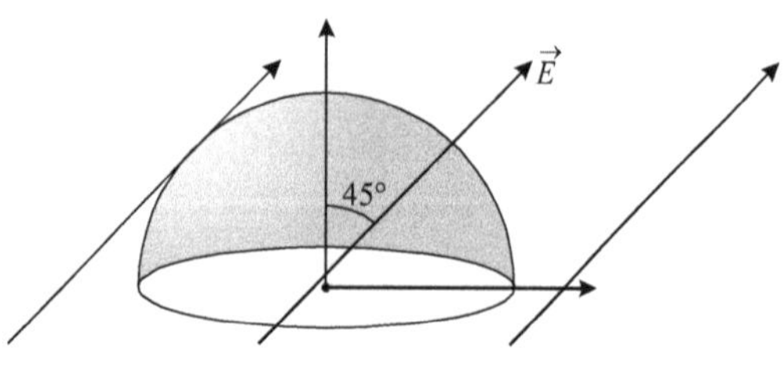

(A) $\pi a^2 E$ (B) $\dfrac{\pi a^2 E}{\sqrt{2}}$

(C) $\dfrac{\pi a^2 E}{2\sqrt{2}}$ (D) $\dfrac{(\pi + 2a\pi)^2 E}{(2\sqrt{2}^2)}$

5. Three positive charges of equal value q are placed at vertices of an equilateral triangle. The resulting lines of force should be sketched as in : **[2012 (Online)]**

(A)
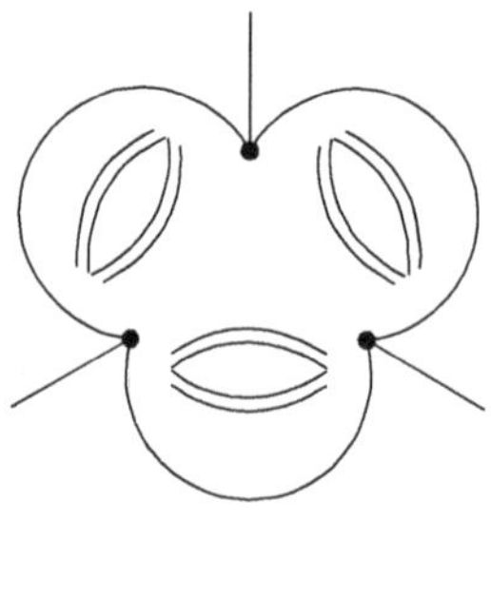

(B)
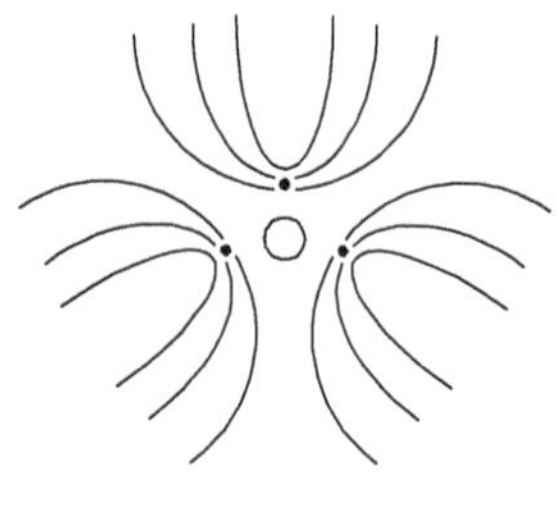

(C)
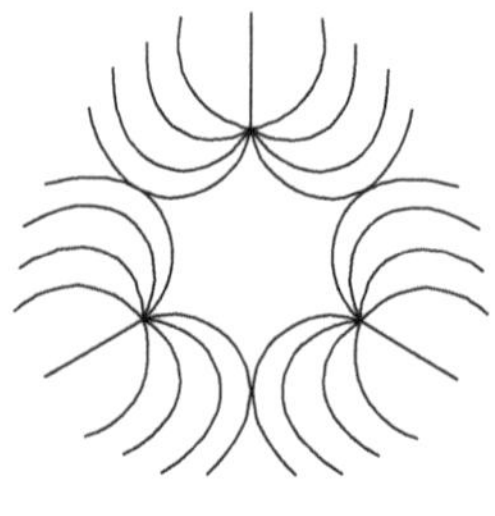

(D)
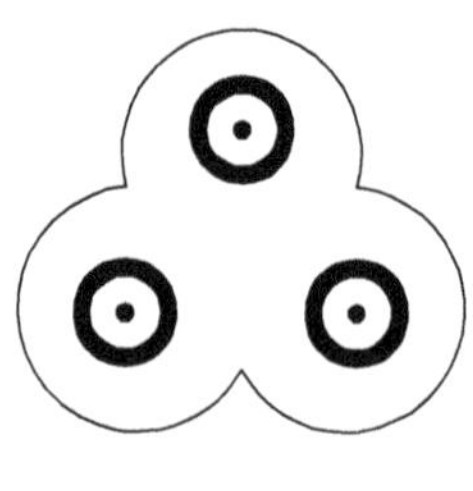

6. The electric potential $V(x)$ in a region around the origin is given by $V(x) = 4x^2$ volts. The electric charge enclosed in a cube of 1m side with its centre at the origin is (in coulomb) : **[2012 (Online)]**

(A) $8\varepsilon_0$ (B) $-4\varepsilon_0$
(C) 0 (D) $-8\varepsilon_0$

7. The electric field in a region of space is given by, $\vec{E} = E_0\hat{i} + 2E_0\hat{j}$ where $E_0 = 100$ N/C. The flux of the field through a circular surface of radius 0.02 m parallel to the Y-Z plane is nearly : **[2014 (Online)]**

(A) 0.125 Nm^2/C (B) 0.02 Nm^2/C
(C) 0.005 Nm^2/C (D) 3.14 Nm^2/C

8. A long cylindrical shell carries positive surface charge ρ in the upper half and negative surface charge $-\sigma$ in the lower half. The electric field lines around the cylinder will look like figure given in (figures are schematic and not drawn to scale) : **[2015]**

(A)
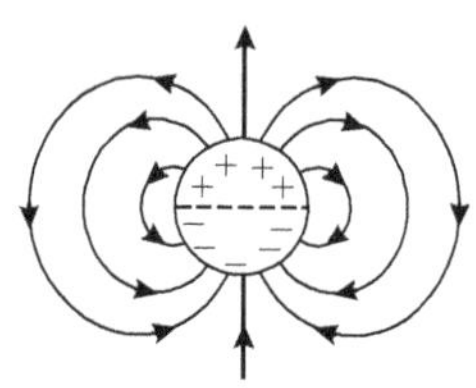

(B)
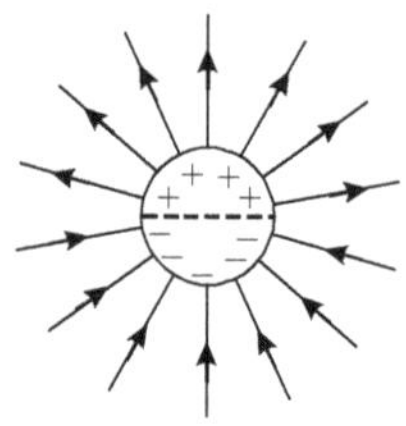

(C)

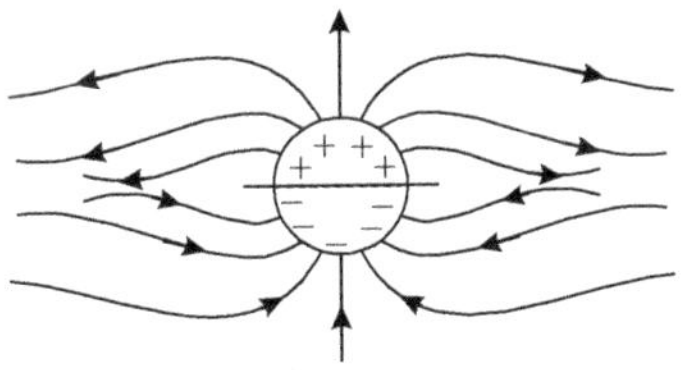

(D)

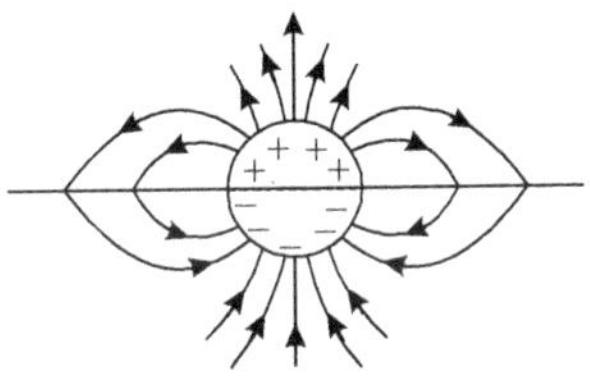

9. Four closed surfaces and corresponding charge distributions are shown below :

Let the respective electric fluxes through the surfaces be Φ_1, Φ_2, Φ_3 and Φ_4. Then : **[2017 (Online)]**

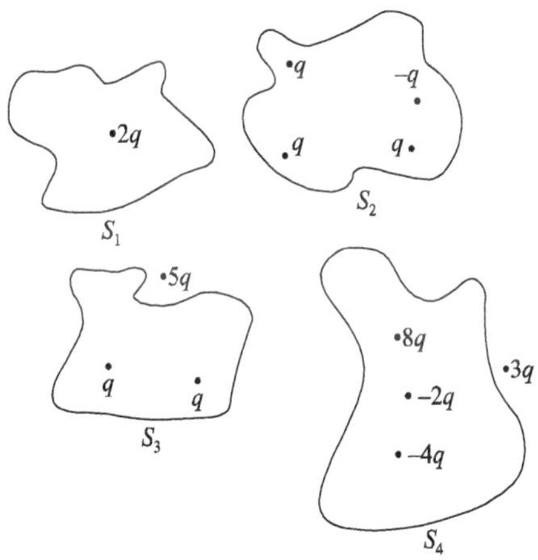

(A) $\Phi_1 < \Phi_2 = \Phi_3 > \Phi_4$ (B) $\Phi_1 > \Phi_2 > \Phi_3 > \Phi_4$

(C) $\Phi_1 = \Phi_2 = \Phi_3 = \Phi_4$ (D) $\Phi_1 > \Phi_3 ; \Phi_2 < \Phi_4$

10. A charge Q is placed at a distance $a/2$ above the centre of the square surface of edge a as shown in the figure. The electric flux through the square surface is : **[2018 (Online)]**

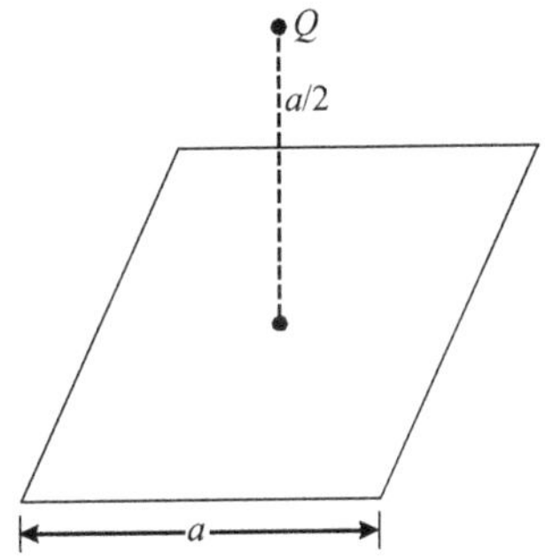

(A) $\dfrac{Q}{3\epsilon_0}$ (B) $\dfrac{Q}{6\epsilon_0}$

(C) $\dfrac{Q}{2\epsilon_0}$ (D) $\dfrac{Q}{\epsilon_0}$

* * * * *

CHAPTER-16 ELECTRIC POTENTIAL AND CAPACITANCE

16.1 Electric Potential and Dipole

1. On moving a charge of 20 C by 2 cm, 2 J of work is done, then the potential difference between the points is : **[2002]**

(A) 0.1 V (B) 8V
(C) 2V (D) 0.5 V

2. A thin spherical conducting shell of radius R has a charge q. Another charge Q is placed at the centre of the shell. The electrostatic potential at a point P at a distance $R/2$ from the centre of the shell is : **[2003]**

(A) $\frac{2Q}{4\pi \epsilon_0 R}$ (B) $\frac{2Q}{4\pi \epsilon_0 R} - \frac{2q}{4\pi \epsilon_0 R}$

(C) $\frac{2Q}{4\pi \epsilon_0 R} + \frac{q}{4\pi \epsilon_0 R}$ (D) $\frac{(q+Q)}{4\pi \epsilon_0} \frac{2}{R}$

3. A charged particle q is shot towards another charged particle Q which is fixed, with a speed v. It approaches Q upto a closest distance r and then returns. If q was given a speed $2v$, the closest distance of approach would be : **[2004]**

(A) r (B) $2r$
(C) $r/2$ (D) $r/4$

4. Two thin wire rings each having a radius R are placed at a distance d apart with their axes coinciding. The charges on the two rings are $+q$ and $-q$. The potential difference between the centres of the two rings is : **[2005]**

(A) $\frac{qR}{4\pi \epsilon_0 d^2}$ (B) $\frac{q}{2\pi \epsilon_0}\left[\frac{1}{R} - \frac{1}{\sqrt{R^2+d^2}}\right]$

(C) Zero (D) $\frac{q}{4\pi \epsilon_0}\left[\frac{1}{R} - \frac{1}{\sqrt{R^2+d^2}}\right]$

5. Two spherical conductors A and B of radii 1 mm and 2 mm are separated by a distance of 5 cm and are uniformly charged. If the spheres are connected by a conducting wire then in equilibrium condition, the ratio of the magnitude of the electric fields at the surfaces of spheres A and B is : **[2006]**

(A) 4 : 1 (B) 1 : 2
(C) 2 : 1 (D) 1 : 4

6. An electric dipole is placed at an angle of 30° to a non-uniform electric field. The dipole will experience : **[2006]**

(A) A translational force only in the direction of the field
(B) A translational force only in a direction normal to the direction of the field
(C) A torque as well as a translational force
(D) A torque only

7. The potential at a point x (measured in μm) due to some charges situated on the x-axis is given by : $V(x) = 20/(x^2 - 4)$ volt. The electric field E at $x = 4$ μm is given by: **[2007]**

(A) $\frac{5}{3}$ V/μm and in the $-ve$ x direction

(B) $\frac{5}{3}$ V/μm and in the $+ve$ x direction

(C) $\frac{10}{9}$ V/μm and in the $-ve$ x direction

(D) $\frac{10}{9}$ V/μm and in the $+ve$ x direction

8. Charges are placed on the vertices of a square as shown. Let $\vec{E}$ be the electric field and V the potential at the centre. If the charges on A and B are interchanged with those on D and C respectively, then : **[2007]**

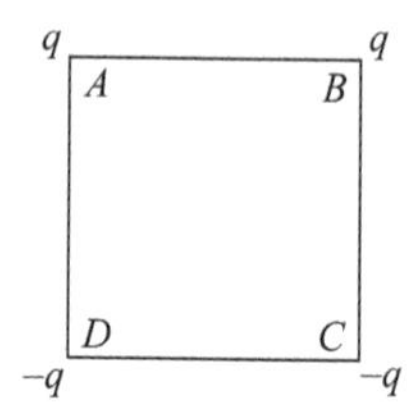

(A) $\vec{E}$ remains unchanged, V changes
(B) both $\vec{E}$ and V change
(C) $\vec{E}$ and V remain unchanged
(D) $\vec{E}$ changes and V remains unchanged

9. An electric charge 10^{-3} μC is placed at the origin (0, 0) of X-Y coordinate system. Two points A and B are situated at $(\sqrt{2}, \sqrt{2})$ and (2, 0) respectively. The potential difference between the points A and B will be : **[2007]**

(A) 9V (B) Zero
(C) 2V (D) 4.5 V

10. Statement-1 : For a charged particle moving from point P to point Q the net work done by an electrostatic field on the particle is independent of the path connecting point P to point Q.
Statement-2 : The net work done by a conservative force on an object moving along closed loop is zero. **[2009]**

(A) Statement-1 is true, Statement-2 is true; Statement-2 is not the correct explanation of Statement-1

(B) Statement-1 is false, Statement-2 is true
(C) Statement-1 is true, Statement-2 is false
(D) Statement-1 is true, Statement-2 is true; Statement-2 is the correct explanation of Statement-1

11. Two points P and Q are maintained at the potential of 10V and –4V, respectively. The work done in moving 100 electrons from P to Q is : **[2009]**

(A) -2.24×10^{-16} J (B) 2.24×10^{-16} J
(C) -9.60×10^{-17} J (D) 9.60×10^{-17} J

12. Two positive charges of magnitude 'q' are placed at the ends of a side (side-1) of a square of side '$2a$'. Two negative charges of the same magnitude are kept at the other corners. Starting from rest, if a charge Q moves from the middle of side-1 to the centre of square, its kinetic energy at the centre of square is : **[2011]**

(A) $\frac{1}{4\pi\epsilon_0}\frac{2qQ}{a}\left(1-\frac{1}{\sqrt{5}}\right)$ (B) Zero

(C) $\frac{1}{4\pi\epsilon_0}\frac{2qQ}{a}\left(1+\frac{1}{\sqrt{5}}\right)$ (D) $\frac{1}{4\pi\epsilon_0}\frac{2qQ}{a}\left(1-\frac{2}{\sqrt{5}}\right)$

13. An insulating solid sphere of radius R has a uniformly positive charge density ρ. As a result of this uniform charge distribution there is a finite value of electric potential at the centre of the sphere, at the surface of the sphere and also at a point out side the sphere. The electric potential at infinity is zero.

Statement-1 : When a charge 'q' is taken from the centre to the surface of the sphere, its potential energy changes by $\frac{q\rho}{3\epsilon_0}$.

Statement-2 : The electric field at a distance $r (r < R)$ from the centre of the sphere is $\frac{\rho r}{3\epsilon_0}$. **[2012]**

(A) Statement-1 is true, Statement-2 is true and Statement-2 is the correct explanation of Statement-1
(B) Statement-1 is true, Statement-2 is true and Statement-2 is not the correct explanation of Statement-1
(C) Statement-1 is true, Statement-2 is false
(D) Statement-1 is false, Statement-2 is true

14. A charge of total amount Q is distributed over two concentric hollow spheres of radii r and $R (R > r)$ such that the surface charge densities on the two spheres are equal. The electric potential at the common centre is : **[2012 (Online)]**

(A) $\frac{1}{4\pi\varepsilon_0}\frac{(R-r)Q}{(R^2+r^2)}$ (B) $\frac{1}{4\pi\varepsilon_0}\frac{(R+r)Q}{2(R^2+r^2)}$

(C) $\frac{1}{4\pi\varepsilon_0}\frac{(R+r)Q}{(R^2+r^2)}$ (D) $\frac{1}{4\pi\varepsilon_0}\frac{(R-r)Q}{(R^2+r^2)}$

15. A charge Q is uniformly distributed over a long rod AB of length L as shown in the figure. The electric potential at the point O lying at a distance L from the end A is : **[2013]**

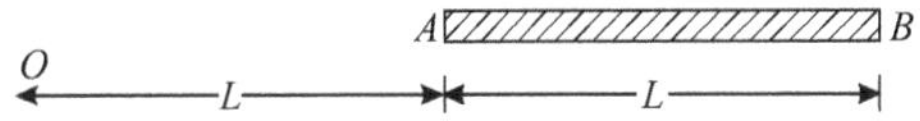

(A) $\frac{Q}{8\pi\epsilon_0 L}$ (B) $\frac{3Q}{4\pi\epsilon_0 L}$

(C) $\frac{Q}{4\pi\epsilon_0 L\ln 2}$ (D) $\frac{Q\ln 2}{4\pi\epsilon_0 L}$

16. Two point dipoles of dipole moment $\vec{P}_1$ and $\vec{P}_2$ are at a distance x from each other and $\vec{P}_1 \parallel \vec{P}_2$. The force between the dipoles is : **[2013 (Online)]**

(A) $\frac{1}{4\pi\varepsilon_0}\frac{4p_1p_2}{x^4}$ (B) $\frac{1}{4\pi\varepsilon_0}\frac{3p_1p_2}{x^3}$

(C) $\frac{1}{4\pi\varepsilon_0}\frac{6p_1p_2}{x^4}$ (D) $\frac{1}{4\pi\varepsilon_0}\frac{8p_1p_2}{x^4}$

17. This question has Statement-1 and Statement-2. Of the four choices given after the Statements, choose the one that best describes the two Statements. **[2013 (Online)]**

Statement-1 : No work is required to be done to move a test charge between any two points on an equipotential surface.

Statement-2 : Electric lines of force at the equipotential surfaces are mutually perpendicular to each other.

(A) Statement-1 is true, Statement-2 is true, Statement-2 is the correct explanation of Statement-1
(B) Statement-1 is true, Statement-2 is true, Statement-2 is not the correct explanation of Statement-1
(C) Statement-1 is true, Statement-2 is false
(D) Statement-1 is false, Statement-2 is true

18. A point charge of magnitude + 1 μC is fixed at (0, 0, 0). An isolated uncharged spherical conductor, is fixed with its center at (4, 0, 0). The potential and the induced electric field at the centre of the sphere is : **[2013 (Online)]**

(A) 1.8×10^5 V and -5.625×10^6 V/m
(B) 0 V and 0 V/m
(C) 2.25×10^5 V and -5.625×10^6 V/m
(D) 2.25×10^5 V and 0 V/m

19. Consider a finite insulated, uncharged conductor placed near a finite positively charged conductor. The uncharged body must have a potential : **[2013 (Online)]**

(A) Less than the charged conductor and more than at infinity
(B) More than the charged conductor and less than at infinity
(C) More than the charged conductor and more than at infinity
(D) Less than the charged conductor and less than at infinity

20. Two small equal point charges of magnitude q are suspended from a common point on the ceiling by insulating mass less strings of equal lengths. They come to equilibrium with each string making angle θ from the vertical. If the mass of each charge is m, then the electrostatic potential at the centre of line joining them will be $\left(\frac{1}{4\pi\epsilon_0}=k\right)$: **[2013 (Online)]**

(A) $2\sqrt{kmg\tan\theta}$ (B) $\sqrt{kmg\tan\theta}$

(C) $4\sqrt{kmg/\tan\theta}$ (D) $4\sqrt{kmg\tan\theta}$

21. Assume that an electric field $\vec{E} = 30x^2\hat{i}$ exists in space. Then the potential difference $V_A - V_O$, where V_O is the potential at the origin and V_A the potential at $x = 2$ m is : **[2014]**

(A) 120 V (B) -120 V

(C) -80 V (D) 80 V

22. A uniformly charged solid sphere of radius R has potential V_0 (measured with respect to ∞) on its surface. For this sphere the equipotential surfaces with potentials $\frac{3V_0}{2}$, $\frac{5V_0}{4}$, $\frac{3V_0}{4}$ and $\frac{V_0}{4}$ have radius R_1, R_2, R_3, and R_4 respectively. Then : **[2015]**

(A) $R_1 = 0$ and $R_2 > (R_4 - R_3)$

(B) $R_1 \neq 0$ and $(R_2 - R_1) > (R_4 - R_3)$

(C) $R_1 = 0$ and $R_2 < (R_4 - R_3)$

(D) $2R < R_4$

23. An electric field $\vec{E} = (25\hat{i} + 30\hat{j})NC^{-1}$ exists in a region of space. If the potential at the origin is taken to be zero then the potential at $x = 2$m, $y = 2$m is : **[2015 (Online)]**

(A) -110 J/C (B) -140 J/C

(C) -120 J/C (D) -130 J/C

24. The potential (in volts) of a charge distribution is given by

$V(z) = 30 - 5z^2$ for $|z| \leq 1$m

$V(z) = 35 - 10|z|$ for $|z| \geq 1$m

$V(z)$ does not depend on x and y. If this potential is generated by a constant charge per unit volume ρ_0 (in units of ε_0) which is spread over a certain region, then choose the correct statement: **[2016 (Online)]**

(A) $\rho_0 = 20\,\varepsilon_0$ in the entire region

(B) $\rho_0 = 10\,\varepsilon_0$ for $|z| \leq 1$ m and $p_0 = 0$ elsewhere

(C) $\rho_0 = 20\,\varepsilon_0$ for $|z| \leq 1$ m and $p_0 = 0$ elsewhere

(D) $\rho_0 = 40\,\varepsilon_0$ in the entire region

25. Within a spherical charge distribution of charge density $\rho(r)$, N equipotential surfaces of potential V_0, $V_0 + \Delta V$, $V_0 + 2\Delta V$, ... $V_0 + N\Delta V (\Delta V > 0)$, are drawn and have increasing radii r_0, r_1, r_2, ... r_N, respectively. If the difference in the radii of the surfaces is constant for all values of V_0 and ΔV then : **[2016 (Online)]**

(A) $\rho(r) =$ constant (B) $\rho(r) \propto \frac{1}{r^2}$

(C) $\rho(r) \propto \frac{1}{r}$ (D) $\rho(r) \propto r$

26. There is a uniform electrostatic field in a region. The potential at various points on a small sphere centred at P, in the region, is found to vary between in the limits 589.0 V to 589.8 V. What is the potential at a point on the sphere whose radius vector makes an angle of 60° with the direction of the field ? **[2017 (Online)]**

(A) 589.5 V (B) 589.2 V

(C) 589.4 V (D) 589.6 V

27. Three concentric metal shells A, B and C of respective radii a, b and c ($a < b < c$) have surface charge densities $+\sigma$, $-\sigma$ and $+\sigma$ respectively. The potential of shell B is : **[2018]**

(A) $\frac{\sigma}{\epsilon_0}\left[\frac{a^2-b^2}{a}+c\right]$ (B) $\frac{\sigma}{\epsilon_0}\left[\frac{a^2-b^2}{b}+c\right]$

(C) $\frac{\sigma}{\epsilon_0}\left[\frac{b^2-c^2}{b}+a\right]$ (D) $\frac{\sigma}{\epsilon_0}\left[\frac{b^2-c^2}{c}+a\right]$

28. A charge Q is distributed over three concentric spherical shells of radii a, b, c ($a < b < c$) such that their surface charge densities are equal to one another. The total potential at a point at distance r from their common centre, where $r < a$, would be: **[10 Jan. 2019 (Morning)]**

(A) $\frac{Q}{4\pi\varepsilon_0(a+b+c)}$ (B) $\frac{Q(a+b+c)}{4\pi\varepsilon_0(a^2+b^2+c^2)}$

(C) $\frac{Q}{12\pi\varepsilon_0}\frac{ab+bc+ca}{abc}$ (D) $\frac{Q}{4\pi\varepsilon_0}\frac{(a^2+b^2+c^2)}{(a^3+b^3+c^3)}$

29. Two electric dipoles, A, B with respective dipole moments $\vec{d}_A = -4qa\hat{i}$ and $\vec{d}_B = -2qa\hat{i}$ placed on the x-axis with a separation R, as shown in the figure : **[10 Jan. 2019 (Morning)]**

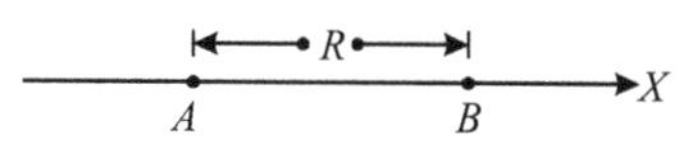

The distance from A at which both of them produce the same potential is :

(A) $\frac{\sqrt{2}R}{\sqrt{2}+1}$ (B) $\frac{R}{\sqrt{2}+1}$

(C) $\frac{\sqrt{2}\,R}{\sqrt{2}-1}$ (D) $\frac{R}{\sqrt{2}-1}$

30. Charges $-q$ and $+q$ located at A and B, respectively, constitute an electric dipole. Distance $AB = 2a$, O is the mid point of the dipole and OP is perpendicular to AB. A charge Q is placed at P where $OP = y$ and $y >> 2a$. The charge Q experiences and electrostatic force F. If Q is now moved along the equatorial line to P' such that $OP' = \left(\frac{y}{3}\right)$, the force on Q will be close to. $\left(\frac{y}{3} >> 2a\right)$: **[10 Jan. 2019 (Evening)]**

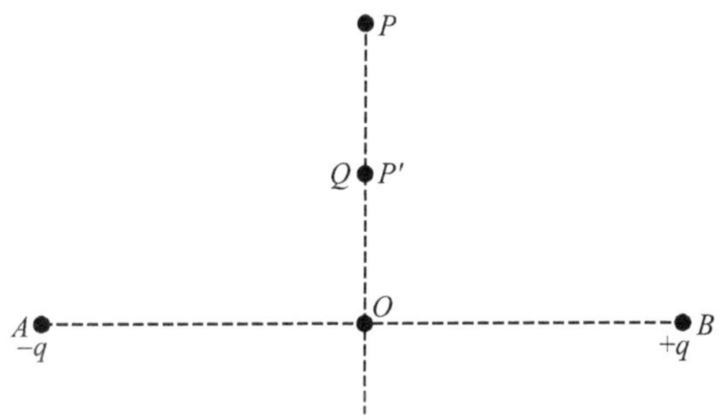

(A) $\frac{F}{3}$ (B) $3F$

(C) $9F$ (D) $27F$

31. Four equal point charges Q each are placed in the xy plane at (0, 2), (4, 2), (4, –2) and (0, –2). The work required to put a fifth charge Q at the origin of the coordinate system will be : **[10 Jan. 2019 (Evening)]**

(A) $\frac{Q^2}{2\sqrt{2}\pi\varepsilon_0}$ (B) $\frac{Q^2}{4\pi\varepsilon_0}\left(1+\frac{1}{\sqrt{5}}\right)$

(C) $\frac{Q^2}{4\pi\varepsilon_0}\left(1+\frac{1}{\sqrt{3}}\right)$ (D) $\frac{Q^2}{4\pi\varepsilon_0}$

32. The charges Q, $+q$ and $+q$ are placed at the vertices of a right-angle isosceles triangle as shown below. The net electrostatic energy of the configuration is zero, if the value of Q is: **[11 Jan. 2019 (Morning)]**

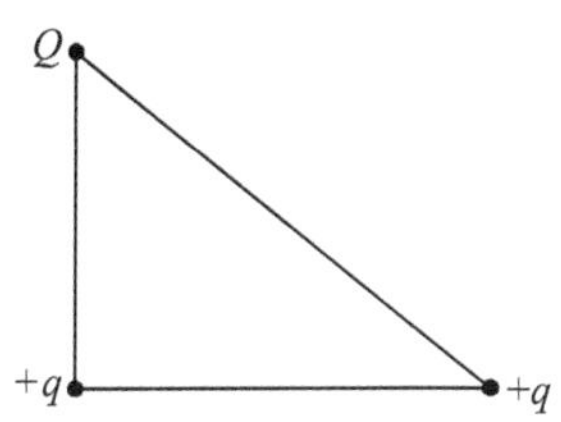

(A) $\frac{-\sqrt{2}q}{\sqrt{2}+1}$ (B) $-2q$

(C) $\frac{-q}{1+\sqrt{2}}$ (D) $+q$

33. The given graph shows variation (with distance r from centre) of : **[11 Jan. 2019 (Morning)]**

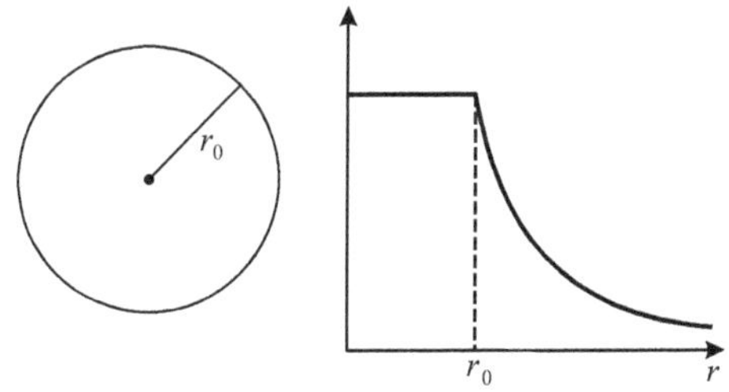

(A) Potential of a uniformly charged sphere
(B) Potential of a uniformly charged spherical shell
(C) Electric field of uniformly charged spherical shell
(D) Electric field of uniformly charged sphere

34. An electric field of 1000 V/m is applied to an electric dipole at angle of 45°. The value of electric dipole moment is 10^{-29} Cm. What is the potential energy of the electric dipole ? **[11 Jan. 2019 (Evening)]**

(A) -9×10^{-20} J (B) -7×10^{-27} J

(C) -10×10^{-29} J (D) -20×10^{-18} J

35. Determine the electric dipole moment of the system of three charges, placed on the vertices of an equilateral triangle, as shown in the figure : **[12 Jan. 2019 (Morning)]**

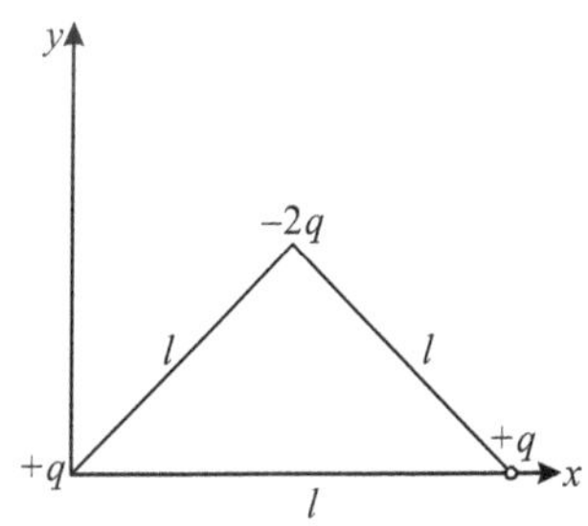

(A) $(ql)\frac{\hat{i}+\hat{j}}{\sqrt{2}}$ (B) $\sqrt{3}ql\frac{\hat{j}-\hat{i}}{\sqrt{2}}$

(C) $-\sqrt{3}ql\,\hat{j}$ (D) $2ql\,\hat{j}$

36. There is a uniform spherically symmetric surface charge density at a distance R_0 from the origin. The charge distribution is initially at rest and starts expanding because of mutual repulsion. The figure that represents best the speed $V(R(t))$ of the distribution as a function of its instantaneous radius $R(t)$ is : **[12 Jan. 2019 (Morning)]**

(A)
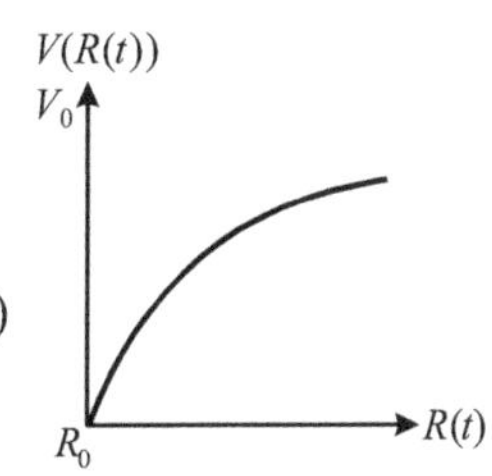

(B)
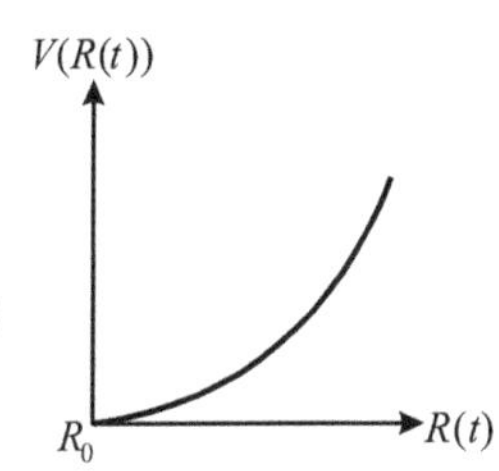

(C) 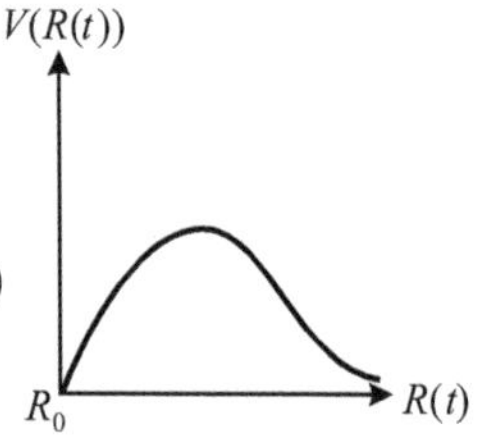

(D) 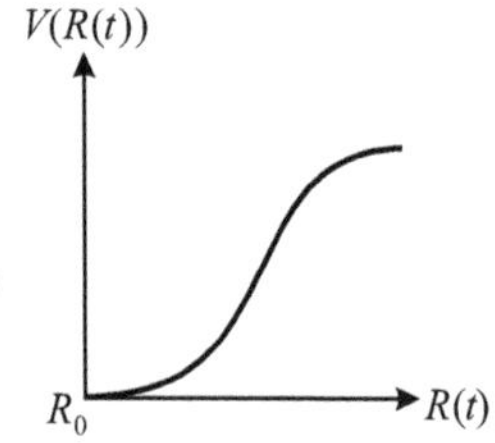

37. A solid conducting sphere, having a charge Q, is surrounded by an uncharged conducting hollow spherical shell. Let the potential difference between the surface of the solid sphere and that of the outer surface of the hollow shell be V. If the shell is now given a charge of $-4Q$, the new potential difference between the same two surfaces is :

[8 April 2019 (Morning)]

(A) V
(B) 2V
(C) –2 V
(D) 4V

38. An electric dipole is formed by two equal and opposite charges q with separation d. The charges have same mass m. It is kept in a uniform electric field E. If it is slightly rotated from its equilibrium orientation, then its angular frequency ω is :

[8 April 2019 (Evening)]

(A) $\sqrt{\frac{qE}{2md}}$
(B) $2\sqrt{\frac{qE}{md}}$
(C) $\sqrt{\frac{2qE}{md}}$
(D) $\sqrt{\frac{qE}{md}}$

39. A positive point charge is released from rest at a distance r_0 from a positive line charge with uniform density. The speed (v) of the point charge, as a function of instantaneous distance r from line charge, is proportional to : **[8 April 2019 (Evening)]**

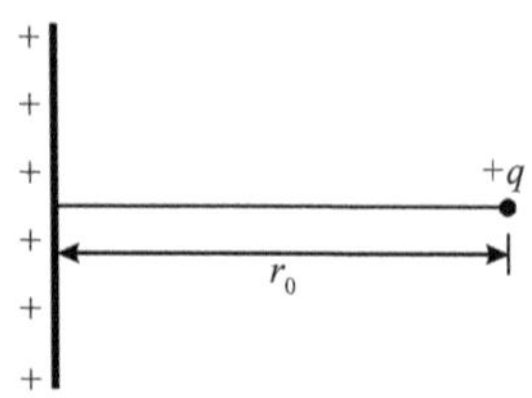

(A) $v \propto e^{+r/r_0}$
(B) $v \propto \ln\left(\frac{r}{r_0}\right)$
(C) $v \propto \left(\frac{r}{r_0}\right)$
(D) $v \propto \sqrt{\ln\left(\frac{r}{r_0}\right)}$

40. The electric field in a region is given by $\vec{E} = (Ax + B)\hat{i}$, where E is in NC^{-1} and x is in metres. The values of constants are $A = 20$ SI unit and $B = 10$ SI unit. If the potential at $x = 1$ is V_1 and that at $x = -5$ is V_2, then $V_1 - V_2$ is :

[8 April 2019 (Evening)]

(A) –48 V
(B) –520 V
(C) 180 V
(D) 320 V

41. A system of three charges are placed as shown in the figure.

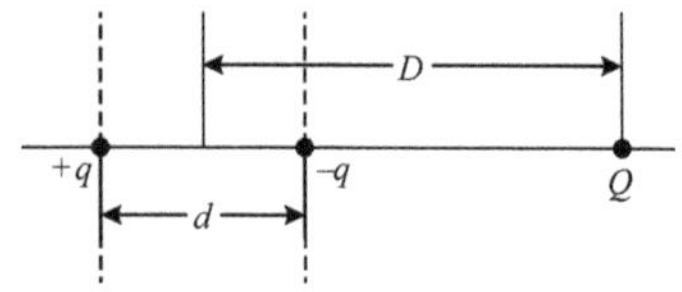

If $D >> d$, the potential energy of the system is best given by :

[9 April 2019 (Morning)]

(A) $\frac{1}{4\pi\varepsilon_0}\left[-\frac{q^2}{d} - \frac{qQd}{2D^2}\right]$
(B) $\frac{1}{4\pi\varepsilon_0}\left[+\frac{q^2}{d} + \frac{qQd}{D^2}\right]$
(C) $\frac{1}{4\pi\varepsilon_0}\left[-\frac{q^2}{d} + \frac{2qQd}{D^2}\right]$
(D) $\frac{1}{4\pi\varepsilon_0}\left[-\frac{q^2}{d} - \frac{qQd}{D^2}\right]$

42. A uniformly charged ring of radius $3a$ and total charge q is placed in xy-plane centred at origin. A point charge q is moving towards the ring along the z-axis and has speed u at $z = 4a$. The minimum value of u such that it crosses the origin is :

[10 April 2019 (Morning)]

(A) $\sqrt{\frac{2}{m}}\left(\frac{1}{15}\frac{q^2}{4\pi\varepsilon_0 a}\right)^{1/2}$
(B) $\sqrt{\frac{2}{m}}\left(\frac{2}{15}\frac{q^2}{4\pi\varepsilon_0 a}\right)^{1/2}$
(C) $\sqrt{\frac{2}{m}}\left(\frac{4}{15}\frac{q^2}{4\pi\varepsilon_0 a}\right)^{1/2}$
(D) $\sqrt{\frac{2}{m}}\left(\frac{1}{5}\frac{q^2}{4\pi\varepsilon_0 a}\right)^{1/2}$

43. In free space, a particle A of charge 1 μC is held fixed at a point P. Another particle B of the same charge and mass 4 μg is kept at a distance of 1 mm from P. If B is released, then its velocity at a distance of 9 mm from P is :

$\left[\text{Take } \frac{1}{4\pi\varepsilon_0} = 9 \times 10^9\ \text{Nm}^2\text{C}^{-2}\right]$ **[10 April 2019 (Evening)]**

(A) 2.0×10^3 m/s
(B) 3.0×10^4 m/s
(C) 1.5×10^2 m/s
(D) 1.0 m/s

44. Shown in the figure is a shell made of a conductor. It has inner radius a and outer radius b, and carries charge Q. At its centre is a dipole $\vec{p}$ as shown. In this case :

[12 April 2019 (Morning)]

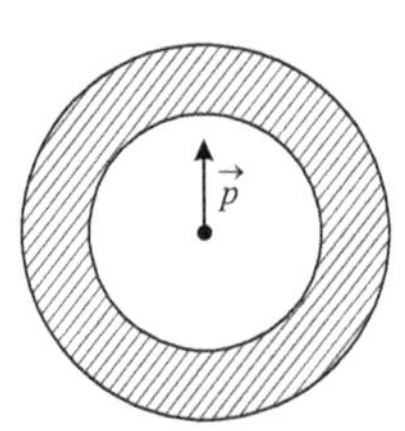

(A) Electric field outside the shell is the same as that of a point charge at the centre of the shell
(B) Surface charge density on the inner surface of the shell is zero everywhere

(C) Surface charge density on the inner surface is uniform and equal to $\frac{(Q/2)}{4\pi a^2}$

(D) Surface charge density on the outer surface depends on $|\vec{p}|$

45. A point dipole $\vec{p} = -p_0\hat{x}$ is kept at the origin. The potential and electric field due to this dipole on the y-axis at a distance d are, respectively : (Take $V = 0$ at infinity) :

[12 April 2019 (Morning)]

(A) $\frac{|\vec{p}|}{4\pi\varepsilon_0 d^2}, \frac{-\vec{p}}{4\pi\varepsilon_0 d^3}$

(B) $0, \frac{\vec{p}}{4\pi\varepsilon_0 d^3}$

(C) $\frac{|\vec{p}|}{4\pi\varepsilon_0 d^2}, \frac{\vec{p}}{4\pi\varepsilon_0 d^3}$

(D) $0, \frac{-\vec{p}}{4\pi\varepsilon_0 d^3}$

* * * * *

CHAPTER-16 ELECTRIC POTENTIAL AND CAPACITANCE

16.2 Capacitance

1. Capacitance (in F) of a spherical conductor having radius 1 m, is : **[2002]**

(A) 1.1×10^{-10} (B) 10^{-6}

(C) 9×10^{-9} (D) 10^{-3}

2. If there are n capacitors in parallel connected to V volt source, then the energy stored is equal to : **[2002]**

(A) CV (B) $\frac{1}{2}nCV^2$

(C) CV^2 (D) $\frac{1}{2n}CV^2$

3. The work done in placing a charge of 8×10^{-18} C on a condenser of capacity 100 micro-farad is : **[2003]**

(A) 16×10^{-32} J (B) 3.1×10^{-26} J

(C) 4×10^{-10} J (D) 32×10^{-32} J

4. A sheet of aluminium foil of negligible thickness is introduced between the plates of a capacitor. The capacitance of the capacitor : **[2003]**

(A) Decreases (B) Remains unchanged

(C) Becomes infinite (D) Increases

5. A parallel plate capacitor is made by stacking n equally spaced plates connected alternatively. If the capacitance between any two adjacent plates is C, then the resultant capacitance is : **[2005]**

(A) $(n-1)C$ (B) $(n+1)C$

(C) C (D) nC

6. A fully charged capacitor has a capacitance C. It is discharged through a small coil of resistance wire embedded in a thermally insulated block of specific heat capacity s and mass m. If the temperature of the block is raised by ΔT, the potential difference V across the capacitance is : **[2005]**

(A) $\sqrt{\frac{2mC\Delta T}{s}}$ (B) $\frac{mC\Delta T}{s}$

(C) $\frac{ms\Delta T}{C}$ (D) $\sqrt{\frac{2ms\Delta T}{C}}$

7. Two insulating plates are both uniformly charged in such a way that the potential difference between them is $V_2 - V_1 = 20$ V. (i.e., plate 2 is at a higher potential). The plates are separated by $d = 0.1$ m and can be treated as infinitely large. An electron is released from rest on the inner surface of plate 1. What is its speed when it hits plate 2? **[2006]**

$(e = 1.6 \times 10^{-19}$ C, $m_0 = 9.11 \times 10^{-31}$ kg)

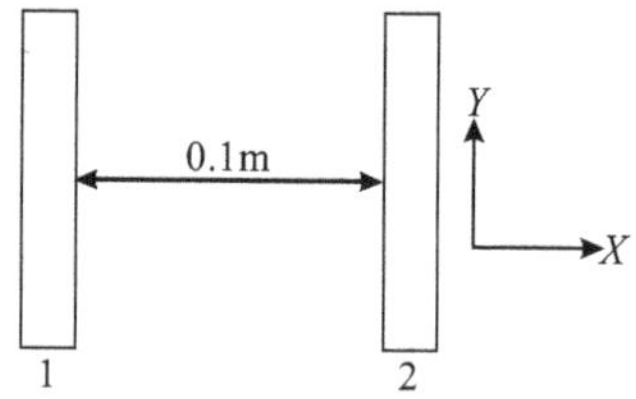

(A) 2.65×10^6 m/s (B) 7.02×10^{12} m/s

(C) 1.87×10^6 m/s (D) 32×10^{-19} m/s

8. A parallel plate condenser with a dielectric of dielectric constant K between the plates has a capacity C and is charged to a potential V volts. The dielectric slab is slowly removed from between the plates and then re-inserted. The net work done by the system in this process is : **[2007]**

(A) $\frac{1}{2}(K-1)CV^2$ (B) $CV^2(K-1)/K$

(C) $(K-1)CV^2$ (D) Zero

9. A battery is used to charge a parallel plate capacitor till the potential difference between the plates becomes equal to the electromotive force of the battery. The ratio of the energy stored in the capacitor and the work done by the battery will be : **[2007]**

(A) 1 (B) 2

(C) 1/4 (D) 1/2

10. A parallel plate capacitor with air between the plates has a capacitance of 9 pF. The separation between its plates is 'd'. The space between the plates is now filled with two dielectrics. One of the dielectric has dielectric constant $K_1 = 3$ and thickness $\frac{d}{3}$ while the other one has dielectric constant $K_2 = 6$ and thickness $\frac{2d}{3}$. Capacitance of the capacitor is now : **[2008]**

(A) 1.8 pF (B) 45 pF

(C) 40.5 pF (D) 20.25 pF

11. Two circuits (a) and (b) have charged capacitors of capacitance C, 2C and 3C with open switches. Charges on each of the capacitor are as shown in the figures. On closing the switches : **[2012 (Online)]**

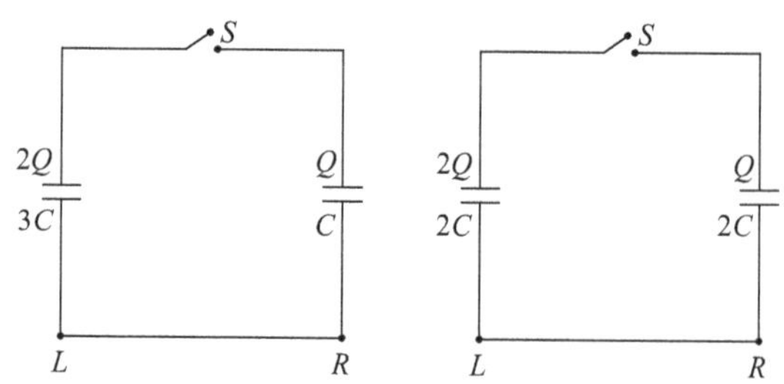

(A) No charge flows in (a) but charge flows from R to L in (b)
(B) Charges flow from L to R in both (a) and (b)
(C) Charges flows from R to L in (a) and from L to R in (b)
(D) No charge flows in (a) but charge flows from L to R in (b)

12. A series combination of n_1 capacitors, each of capacity C_1 is charged by source of potential difference 4V. When another parallel combination of n_2 capacitors each of capacity C_2 is charged by a source of potential difference V, it has the same total energy stored in it as the first combination has. The value of C_2 in terms of C_1 is then : **[2012 (Online)]**

(A) $16\frac{n_2}{n_1}C_1$ (B) $\frac{2C_1}{n_1 n_2}$
(C) $2\frac{n_2}{n_1}C_1$ (D) $\frac{16C_1}{n_1 n_2}$

13. This question has Statement-1 and Statement-2. Of the four choices given after the Statements, choose the one that best describes the two Statements. **[2012 (Online)]**
Statement-1 : It is not possible to make a sphere of capacity 1 farad using a conducting material.
Statement-2 : It is possible for earth as its radius is 6.4×10^6 m.
(A) Statement-1 is true, Statement-2 is true, Statement-2 is the correct explanation of Statement-1
(B) Statement-1 is false, Statement-2 is true
(C) Statement-1 is true, Statement-2 is true, Statement-2 is not the correct explanation of Statement-1
(D) Statement-1 is true, Statement-2 is false

14. Two capacitors C_1 and C_2 are charged to 120V and 200V respectively. It is found that by connecting them together the potential on each one can be made zero. Then : **[2013]**
(A) $5C_1 = 3C_2$ (B) $3C_1 = 5C_2$
(C) $3C_1 + 5C_2 = 0$ (D) $9C_1 = 4C_2$

15. A uniform electric field $\vec{E}$ exists between the plates of a charged condenser. A charged particle enters the space between the plates and perpendicular to $\vec{E}$. The path of the particle between the plates is a : **[2013 (Online)]**
(A) Straight line (B) Hyperbola
(C) Parabola (D) Circle

16. To establish an instantaneous current of 2 A through a 1 μF capacitor; the potential difference across the capacitor plates should be changed at the rate of : **[2013 (Online)]**
(A) 2×10^4 V/s (B) 4×10^6 V/s
(C) 2×10^6 V/s (D) 4×10^4 V/s

17. A parallel plate capacitor having a separation between the plates d, plate are A and material with dielectric constant K has capacitance C_0. Now one-third of the material is replaced by another material with dielectric constant 2 K, so that effectively there are two capacitors one with area $\frac{1}{3}A$, dielectric constant 2K and another with area $\frac{2}{3}A$ and dielectric constant K. If the capacitance of this new capacitor is C then $\frac{C}{C_0}$ is : **[2013 (Online)]**

(A) 1 (B) $\frac{4}{3}$
(C) $\frac{2}{3}$ (D) $\frac{1}{3}$

18. A parallel plate capacitor is made of two circular plates separated by a distance of 5 mm and with a dielectric of dielectric constant 2.2 between them. When the electric field in the dielectric is 3×10^4 V/m, the charge density of the positive plate will be closed to : **[2014]**
(A) 6×10^{-7} C/m^2 (B) 3×10^{-7} C/m^2
(C) 3×10^4 C/m^2 (D) 6×10^4 C/m^2

19. The gap between the plates of a parallel plate capacitor of area A and distance between plates d, is filled with a dielectric whose permittivity varies linearly from $\in_1$ at one plate to $\in_2$ at the other. The capacitance of capacitor is: **[2014 (Online)]**
(A) $\in_0(\in_1 + \in_2)A/d$ (B) $\in_0(\in_2 + \in_1)A/2d$
(C) $\in_0 A/[d\ln(\in_2/\in_1)]$ (D) $\in_0(\in_2 - \in_1)A/[d\ln(\in_2/\in_1)]$

20. Three capacitors, each of 3μF, are provided these cannot be combined to provide the resultant capacitance of : **[2014 (Online)]**
(A) 1 μF (B) 2 μF
(C) 4.5 μF (D) 6 μF

21. A parallel plate capacitor is made of two plates of length , width w and separated by distance d. A dielectric slab (dielectric constant K) that fits exactly between the plates is held near the edge of the plates. It is pulled into the capacitor by a force $F = -\frac{\partial U}{\partial x}$ where U is the energy of the capacitor when dielectric is inside the capacitor up to distance x (See figure). If the charge on the capacitor is Q then the force on the dielectric when it is near the edge is : **[2014 (Online)]**

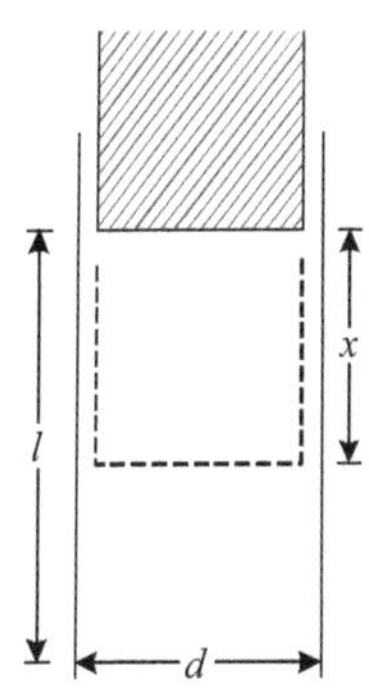

(A) $\frac{Q^2 d}{2\omega l^2 \varepsilon_0}K$ (B) $\frac{Q^2 w}{2\omega l^2 \varepsilon_0}(K-1)$

(C) $\dfrac{Q^2 d}{2wl^2\varepsilon_0}(K-1)$ (D) $\dfrac{Q^2 w}{2dl^2\varepsilon_0}K$

22. The space between the plates of a parallel plate capacitor is filled with a 'dielectric' whose 'dielectric constant' varies with distance as per the relation :

$K(x) = K_0 + \lambda x (\lambda = a \text{ constant})$

The capacitance C, of the capacitor, would be related to its vacuum capacitance C_0 for the relation : **[2014 (Online)]**

(A) $C = \dfrac{\lambda d}{\ln(1+K_0\lambda d)}C_0$ (B) $C = \dfrac{\lambda}{d\cdot\ln(1+K_0\lambda d)}C_0$

(C) $C = \dfrac{\lambda d}{\ln(1+\lambda d/K_0)}C_0$ (D) $C = \dfrac{\lambda}{d\cdot\ln(1+K_0/\lambda d)}C_0$

23. In the given circuit, charge Q_2 on the 2 μF capacitor changes as C is varied from 1 μF to 3 μF. Q_2 as a function of 'C' is given properly by : (figures are drawn schematically and are not to scale) : **[2015]**

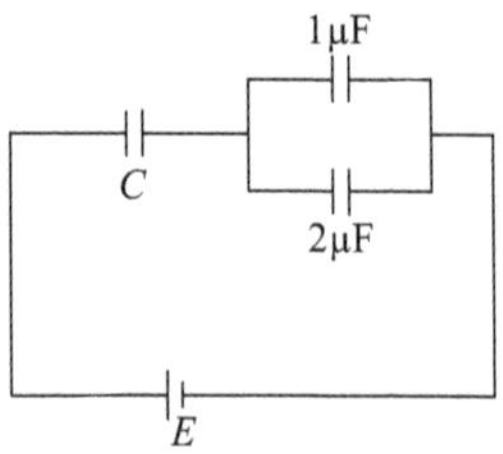

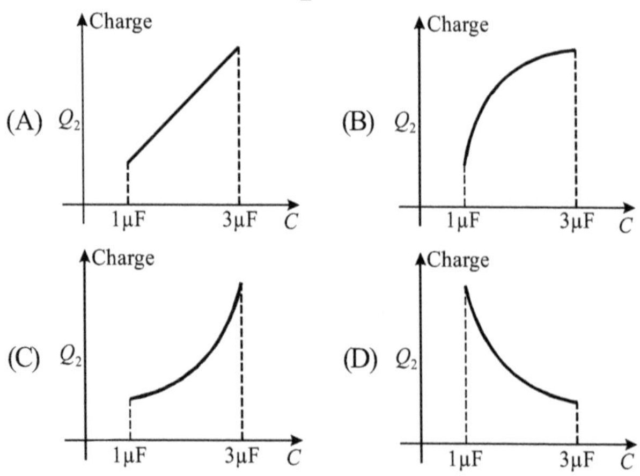

24. In figure a system of four capacitors connected across a 10 V battery is shown. Charge that will flow from switch S when it is closed is : **[2015 (Online)]**

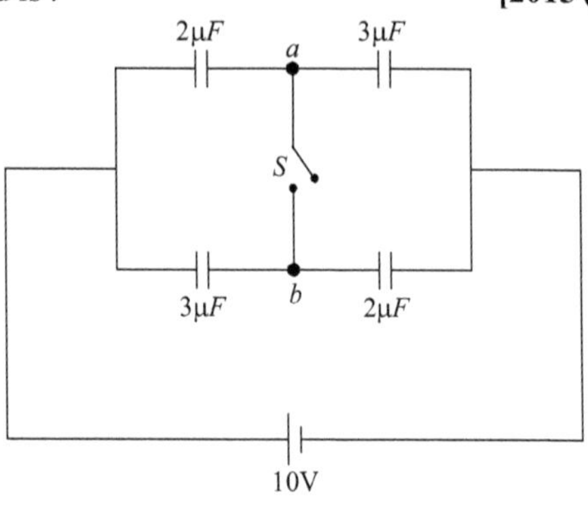

(A) 5 μC from b to a (B) 20 μC from a to b
(C) Zero (D) 5 μC from a to b

25. A combination of capacitors is set up as shown in the figure. The magnitude of the electric field, due to a point charge Q (having a charge equal to the sum of the charges on the 4 μF and 9 μF capacitors), at a point distant 30 m from it, would equal : **[2016]**

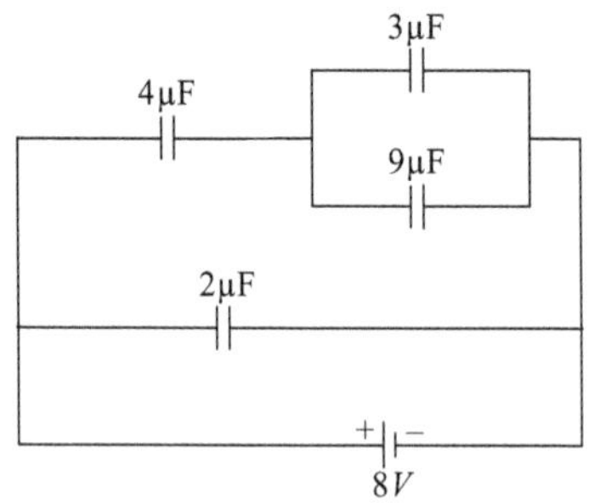

(A) 240 N/C (B) 360 N/C
(C) 420 N/C (D) 480 N/C

26. Three capacitors each of 4 μF are to be connected in such a way that the effective capacitance is 6 μF. This can be done by connecting them : **[2016 (Online)]**

(A) All in series
(B) All in parallel
(C) Two in parallel and one in series
(D) Two in series and one in parallel

27. Figure shows a network of capacitors where the numbers indicates capacitances in micro Farad. The value of capacitance C if the equivalent capacitance between point A and B is to be 1 μF is : **[2016 (Online)]**

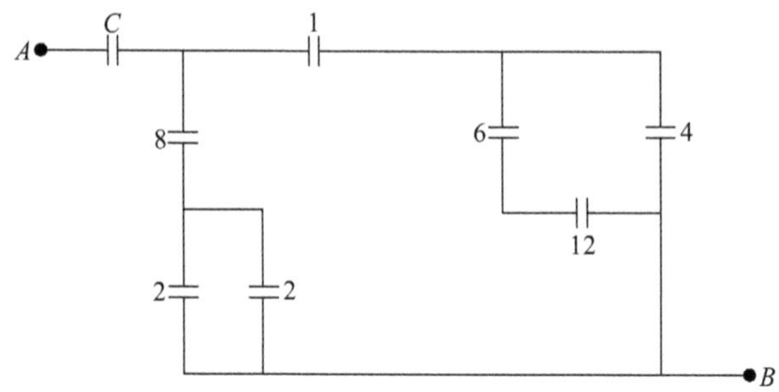

(A) $\dfrac{32}{23}\mu F$ (B) $\dfrac{31}{23}\mu F$

(C) $\dfrac{33}{23}\mu F$ (D) $\dfrac{34}{23}\mu F$

28. A capacitance of 2 μF is required in an electrical circuit across a potential difference of 1.0 kV. A large number of 1 μF capacitors are available which can withstand a potential difference of not more than 300 V. The minimum number of capacitors required to achieve this is : **[2017]**

(A) 24 (B) 32
(C) 2 (D) 16

29. The energy stored in the electric field produced by a metal sphere is 4.5 J. If the sphere contains 4 μC charge, its radius will be : [Take : $\frac{1}{4\pi\varepsilon_0} = 9 \times 10^9$ N – m²/C²] **[2017 (Online)]**

(A) 20mm (B) 32mm
(C) 28mm (D) 16mm

30. A combination of parallel plate capacitors is maintained at a certain potential difference. When a 3 mm thick slab is introduced between all the plates, in order to maintain the same potential difference, the distance between the plates in increased by 2.4 mm. Find the dielectric constant of the slab : **[2017 (Online)]**

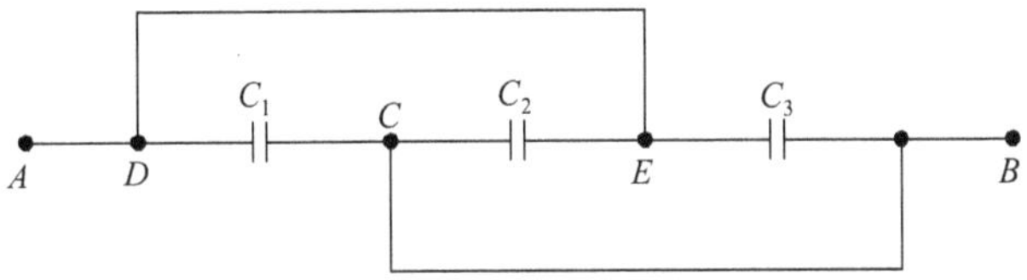

(A) 3 (B) 4
(C) 5 (D) 6

31. A parallel plate capacitor of capacitance 90 pF is connected to a battery of emf 20 V. If a dielectric material of dielectric constant $k = \frac{5}{3}$ is inserted between the plates, the magnitude of the induced charge will be : **[2018]**

(A) 1.2 nC (B) 0.3 nC
(C) 2.4 nC (D) 0.9 nC

32. A parallel plate capacitor with area 200 cm² and separation between the plates 1.5 cm, is connected across a battery of emf V. If the force of attraction between the plates is 25×10^{-6} N, the value of V is approximately : **[2018 (Online)]**

$$\left(\epsilon_0 = 8.85 \times 10^{-12} \frac{C^2}{N.m^2}\right)$$

(A) 150 V (B) 100 V
(C) 250 V (D) 300 V

33. The equivalent capacitance between A and B in the circuit given below is : **[2018 (Online)]**

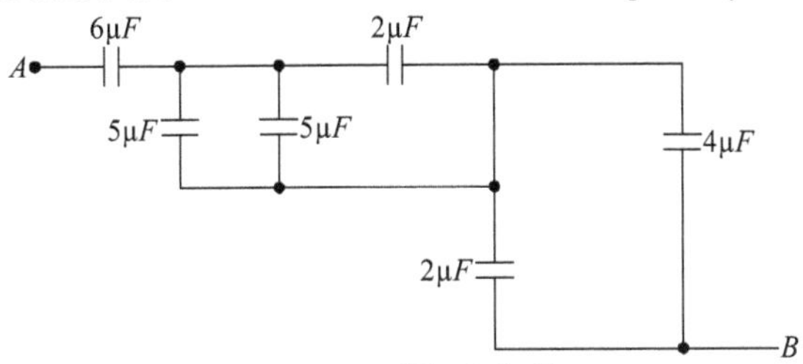

(A) 4.9 μF (B) 3.6 μF
(C) 5.4 μF (D) 2.4 μF

34. A capacitor $C_1 = 1\ \mu$F is charged up to a voltage $V = 60$ V by connecting it to battery B through switch (1), Now C_1 is disconnected from battery and connected to a circuit consisting of two uncharged capacitors $C_2 = 3.0\ \mu$F and $C_3 = 6.0\ \mu$F through a switch (2) as shown in the figure. The sum of final charges on C_2 and C_3 is : **[2018 (Online)]**

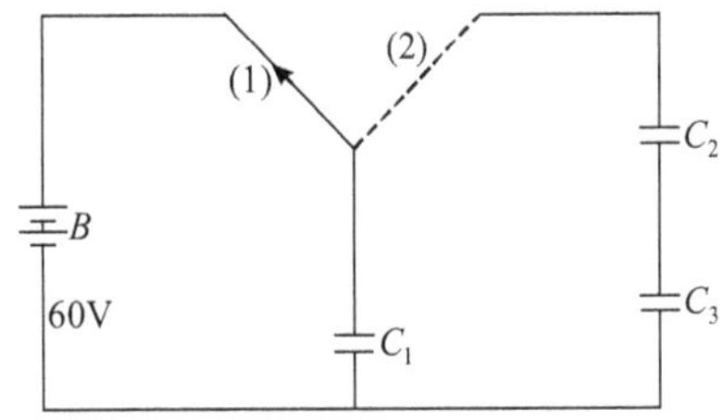

(A) 36 μC (B) 20 μC
(C) 54 μC (D) 40 μC

35. A parallel plate capacitor is made of two square plates of side 'a', separated by a distance d ($d << a$). The lower triangular portion is filled with a dielectric of dielectric constant K, as shown in the figure. Capacitance of this capacitor is : **[9 Jan. 2019 (Morning)]**

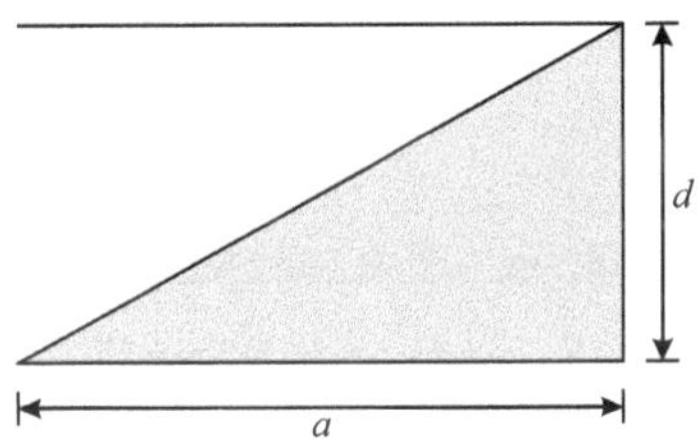

(A) $\frac{1}{2}\frac{k \in_0 a^2}{d}$ (B) $\frac{k \in_0 a^2}{d} \ln K$

(C) $\frac{k \in_0 a^2}{d(K-1)} \ln K$ (D) $\frac{k \in_0 a^2}{2d(K+1)}$

36. A parallel plate capacitor with square plates is filled with four dielectrics of dielectric constants K_1, K_2, K_3, K_4 arranged as shown in the figure. The effective dielectric constant K will be : **[9 Jan. 2019 (Evening)]**

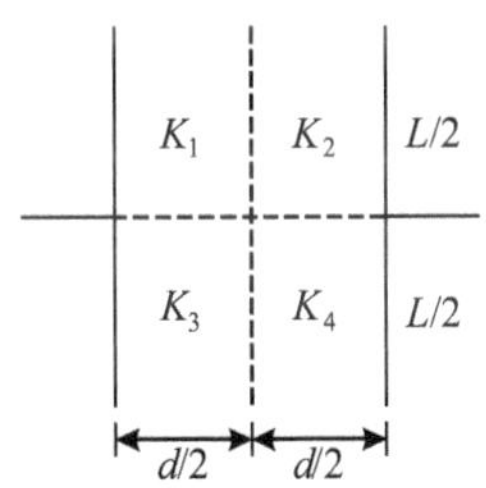

(A) $K = \frac{(K_1 + K_2)(K_3 + K_4)}{2(K_1 + K_2 + K_3 + K_4)}$

(B) $K = \frac{(K_1 + K_2)(K_3 + K_4)}{(K_1 + K_2 + K_3 + K_4)}$

(C) $K = \frac{(K_1 + K_4)(K_2 + K_3)}{2(K_1 + K_2 + K_3 + K_4)}$

(D) $K = \frac{(K_1 + K_3)(K_2 + K_4)}{K_1 + K_2 + K_3 + K_4}$

37. A parallel plate capacitor is of area 6 cm^2 and a separation 3 mm. The gap is filled with three dielectric materials of equal thickness (see figure) with dielectric constants $K_1 = 10$, $K_2 = 12$ and $K_3 = 14$. The dielectric constant of a material which when fully inserted in above capacitor, gives same capacitance would be : **[10 Jan. 2019 (Morning)]**

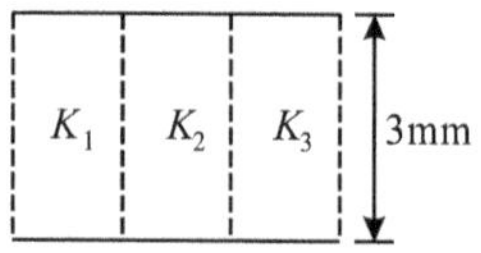

(A) 12 (B) 4
(C) 36 (D) 14

38. A parallel plate capacitor having capacitance 12 pF is charged by a battery to a potential difference of 10 V between its plates. The charging battery is now disconnected and a porcelain slab of dielectric constant 6.5 is slipped between the plates, the work done by the capacitor on the slab is : **[10 Jan. 2019 (Evening)]**

(A) 692 pJ (B) 60 pJ
(C) 508 pJ (D) 560 pJ

39. In the figure shown below, the charge on the left plate of the 10 μF capacitor is –30 μC. The charge on the right plate of the 6 μF capacitor is : **[11 Jan. 2019 (Morning)]**

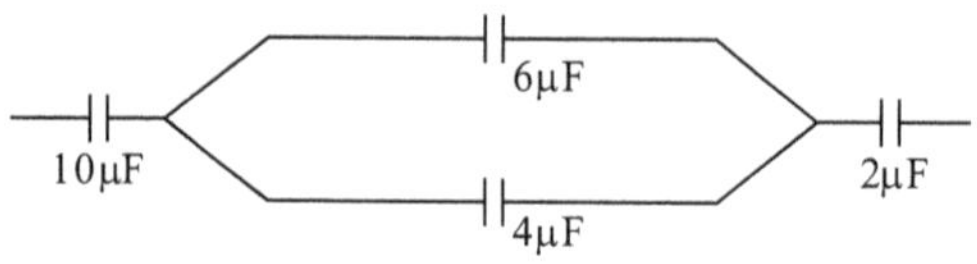

(A) –18 μC (B) –12 μC
(C) +12 μC (D) +18 μC

40. Seven capacitors, each of capacitance 2 μF, are to be connected in a configuration to obtain an effective capacitance of $\left(\frac{6}{13}\right)$ μF. Which of the combinations, shown in figures below, will achieve the desired value ? **[11 Jan. 2019 (Evening)]**

(A) 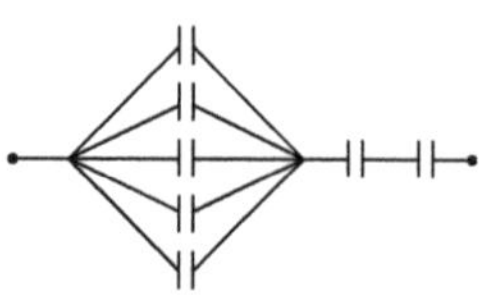 (B) 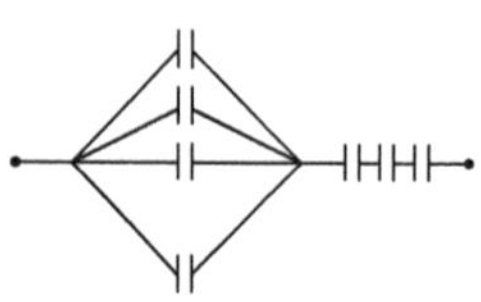

(C) 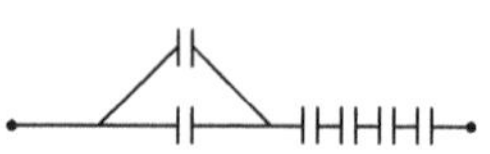 (D) 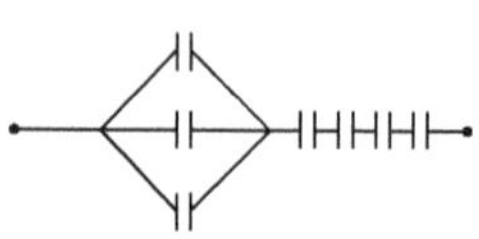

41. In the figure shown, after the switch 'S' is turned from position 'A' to position 'B', the energy dissipated in the circuit in terms of capacitance 'C' and total charge 'Q' is : **[12 Jan. 2019 (Morning)]**

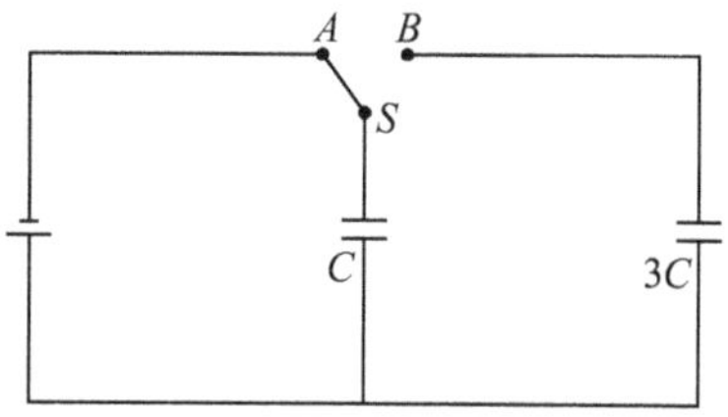

(A) $\frac{3}{8}\frac{Q^2}{C}$ (B) $\frac{3}{4}\frac{Q^2}{C}$

(C) $\frac{1}{8}\frac{Q^2}{C}$ (D) $\frac{5}{8}\frac{Q^2}{C}$

42. In the circuit shown, find C if the effective capacitance of the whole circuit is to be 0.5 μF. All values in the circuit are in μF. **[12 Jan. 2019 (Evening)]**

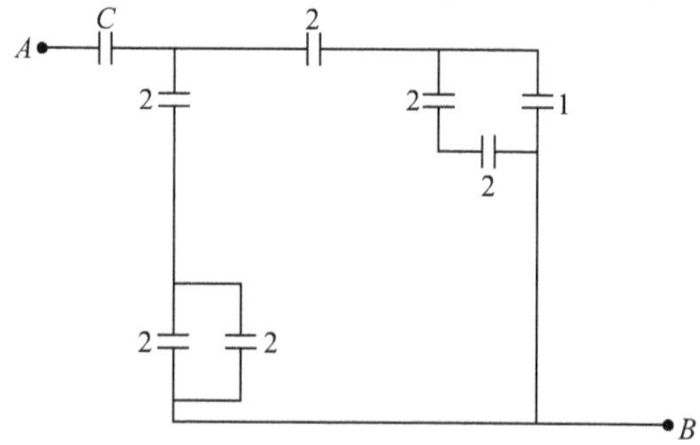

(A) $\frac{7}{10}\mu F$ (B) $\frac{7}{11}\mu F$

(C) $\frac{6}{5}\mu F$ (D) $4\mu F$

43. The charge on a capacitor plate in a circuit, as a function of time, is shown in the figure. What is the value of current at $t = 4$ s? **[12 Jan. 2019 (Evening)]**

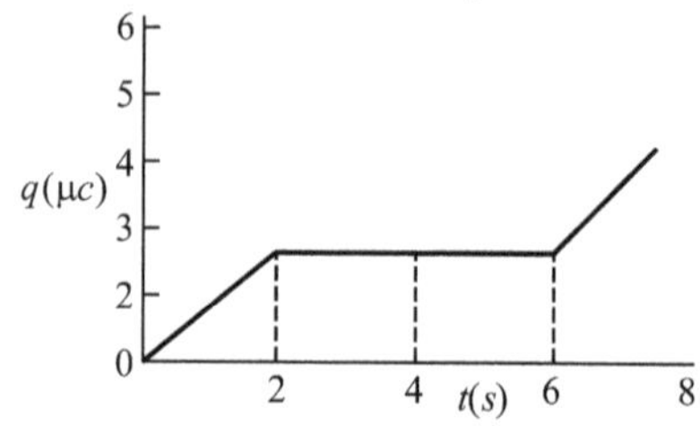

(A) 3 μA (B) 2 μA
(C) Zero (D) 1.5 μA

44. A parallel plate capacitor with plates of area 1 m^2 each, and at a separation of 0.1 m. If the electric field between the plates is 100 N/C, the magnitude of charge each plate is.

[Take $\varepsilon_0 = 8.85 \times 10^{-12} \frac{C^2}{N - m^2}$]: **[12 Jan. 2019 (Evening)]**

(A) 7.85×10^{-10} C (B) 6.85×10^{-10} C
(C) 9.85×10^{-10} C (D) 8.85×10^{-10} C

45. Voltage rating of a parallel plate capacitor is 500 V. Its dielectric can withstand a maximum electric field of 10^6 V/m. The plate area is 10^{-4} m^2. What is the dielectric constant of the

capacitance is 15 pF ?
[Given $\in_0 = 8.86 \times 10^{-12}\ C^2/Nm^2$] **[8 April 2019 (Morning)]**

(A) 3.8 (B) 4.5

(C) 6.2 (D) 8.5

46. A parallel plate capacitor has 1 μF capacitance. One of its two plates is given +2μC charge and the other plate, +4μC charge. The potential difference developed across the capacitor is : **[8 April 2019 (Evening)]**

(A) 5V (B) 2V

(C) 3V (D) 1V

47. A capacitor with capacitance 5μF is charged to 5μC. If the plates are pulled apart to reduce the capacitance to 2μF, how much work is done ? **[9 April 2019 (Morning)]**

(A) 3.75×10^{-6} J (B) 2.55×10^{-6} J

(C) 2.16×10^{-6} J (D) 6.25×10^{-6} J

48. The parallel combination of two air filled parallel plate capacitors of capacitance C and nC is connected to a battery of voltage, V. When the capacitors are fully charged, the battery is removed and after that a dielectric material of dielectric constant K is placed between the two plates of the first capacitor. The new potential difference of the combined system is : **[9 April 2019 (Evening)]**

(A) $\dfrac{V}{K+n}$ (B) V

(C) $\dfrac{(n+1)V}{(K+n)}$ (D) $\dfrac{nV}{K+n}$

49. Figure shows charge (q) versus voltage (V) graph for series and parallel combination of two given capacitors. The capacitances are : **[10 April 2019 (Morning)]**

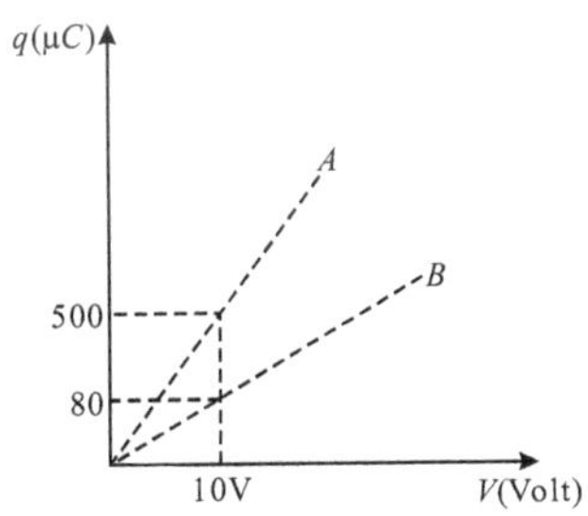

(A) 50 μF and 30 μF (B) 20 μF and 30 μF

(C) 60 μF and 40 μF (D) 40 μF and 10 μF

50. A simple pendulum of length L is placed between the plates of a parallel plate capacitor having electric field E, as shown in figure. Its bob has mass m and charge q. The time period of the pendulum is given by : **[10 April 2019 (Evening)]**

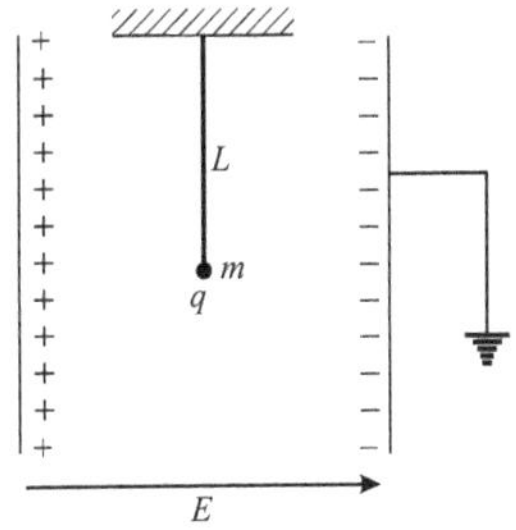

(A) $2\pi\sqrt{\dfrac{L}{\sqrt{g^2+\left(\dfrac{qE}{m}\right)^2}}}$ (B) $2\pi\sqrt{\dfrac{L}{\left(g+\dfrac{qE}{m}\right)}}$

(C) $2\pi\sqrt{\dfrac{L}{\left(g-\dfrac{qE}{m}\right)}}$ (D) $2\pi\sqrt{\dfrac{L}{\sqrt{g^2-\dfrac{q^2E^2}{m^2}}}}$

51. Two identical parallel plate capacitors, of capacitance C each, have plates of area A, separated by a distance d. The space between the plates of the two capacitors, is filled with three dielectrics, of equal thickness and dielectric constants K_1, K_2 and K_3. The first capacitor is filled as shown in figure-I, and the second one is filled as shown in figure-II. If these two modified capacitors are charged by the same potential V, the ratio of the energy stored in the two, would be (E_1 refers to capacitor (I) and E_2 to capacitor (II)) : **[12 April 2019 (Morning)]**

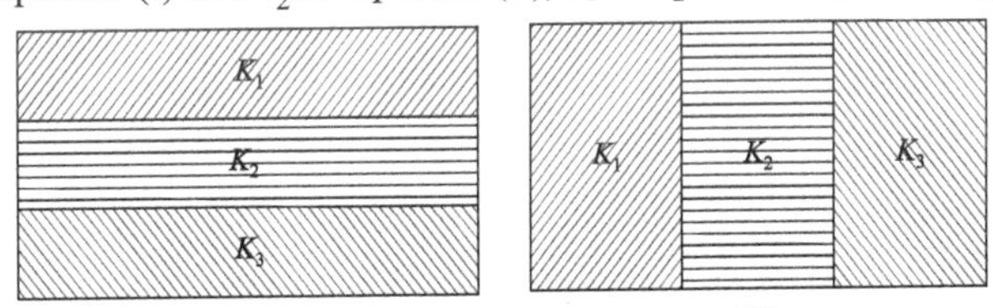

(I) (II)

(A) $\dfrac{E_1}{E_2}=\dfrac{9K_1K_2K_3}{(K_1+K_2+K_3)(K_2K_3+K_3K_1+K_1K_2)}$

(B) $\dfrac{E_1}{E_2}=\dfrac{K_1K_2K_3}{(K_1+K_2+K_3)(K_2K_3+K_3K_1+K_1K_2)}$

(C) $\dfrac{E_1}{E_2}=\dfrac{(K_1+K_2+K_3)(K_2K_3+K_3K_1+K_1K_2)}{K_1K_2K_3}$

(D) $\dfrac{E_1}{E_2}=\dfrac{(K_1+K_2+K_3)(K_2K_3+K_3K_1+K_1K_2)}{9K_1K_2K_3}$

52. In the given circuit, the charge on 4μF capacitor will be : **[12 April 2019 (Evening)]**

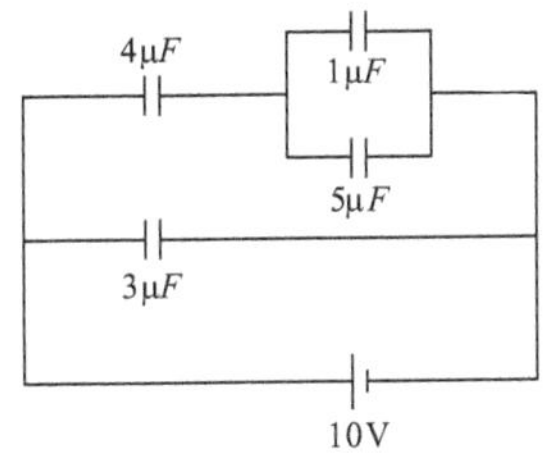

(A) 5.4μC (B) 24μC

(C) 13.4μC (D) 9.6μC

* * * * *

CHAPTER-17 CURRENT ELECTRICITY

17.1 Current Electricity

1. The length of a given cylindrical wire is increased by 100%. Due to the consequent decrease in diameter the change in the resistance of the wire will be : **[2003]**

(A) 200 % (B) 100 %
(C) 50 % (D) 300 %

2. The thermistor are usually made of : **[2004]**

(A) metals with low temperature coefficient of resistivity
(B) metals with high temperature coefficient of resistivity
(C) metal oxides with high temperature coefficient of resistivity
(D) semiconducting materials having low temperature coefficient of resistivity

3. An electric current is passed through a circuit containing two wires of the same material, connected in parallel. If the lengths and radii of the wires are in the ratio of 4/3 and 2/3, then the ratio of the currents passing through the wire will be : **[2004]**

(A) 3 (B) 1/3
(C) 8/9 (D) 2

4. The resistance of the series combination of two resistance is S. When they are joined in parallel, the total resistance is P. If $S = nP$, then the minimum possible value of n is : **[2004]**

(A) 4 (B) 3
(C) 2 (D) 1

5. The resistance of bulb filament is 100 Ω at a temperature of 100°C. If its temperature coefficient of resistance be 0.005 /°C, its resistance will become 200 Ω at a temperature of : **[2006]**

(A) 300 °C (B) 400 °C
(C) 500 °C (D) 200 °C

6. A material 'B' has twice the specific resistance of 'A'. A circular wire made of 'B' has twice the diameter of a wire made of 'A'. Then for the two wires to have the same resistance, the ratio l_B/l_A of their respective lengths must be: **[2006]**

(A) 1 (B) $\frac{1}{2}$
(C) $\frac{1}{4}$ (D) 2

7. The resistance of a wire is 5 ohm at 50°C and 6 ohm at 100°C. The resistance of the wire at 0°C will be : **[2007]**

(A) 3 ohm (B) 2 ohm
(C) 1 ohm (D) 4 ohm

8. Consider a block of conducting material of resistivity 'ρ' shown in the figure. Current 'I' enters at 'A' and leaves from 'D'. We apply superposition principle to find voltage 'ΔV' developed between 'B' and 'C'. The calculation is done in the following steps :

(i) Take current 'I' entering from 'A' and assume it to spread over a hemispherical surface in the block.
(ii) Calculate field $E(r)$ at distance 'r' from A by using Ohm's law $E = \rho j$, where j is the current per unit area at 'r'
(iii) From the 'r' dependence of $E(r)$, obtain the potential $V(r)$ at r.
(iv) Repeat (i), (ii) and (iii) for current 'I' leaving 'D' and superpose results for 'A' and 'D'

For current entering at A, the electric field at a distance 'r' from A is : **[2008]**

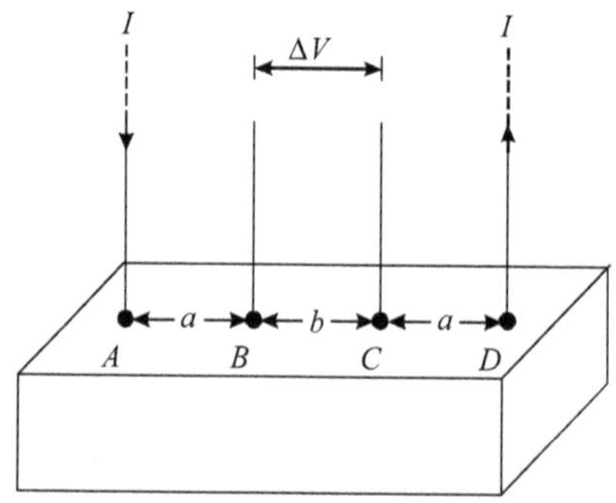

(A) $\frac{\rho I}{8\pi r^2}$ (B) $\frac{\rho I}{r^2}$
(C) $\frac{\rho I}{2\pi r^2}$ (D) $\frac{\rho I}{4\pi r^2}$

9. Consider a block of conducting material of resistivity 'ρ' shown in the figure. Current 'I' enters at 'A' and leaves from 'D'. We apply superposition principle to find voltage 'ΔV' developed between 'B' and 'C'. The calculation is done in the following steps :

(i) Take current 'I' entering from 'A' and assume it to spread over a hemispherical surface in the block.
(ii) Calculate field $E(r)$ at distance 'r' from A by using Ohm's law $E = \rho j$, where j is the current per unit area at 'r'
(iii) From the 'r' dependence of $E(r)$, obtain the potential $V(r)$ at r.
(iv) Repeat (i), (ii) and (iii) for current 'I' leaving 'D' and superpose results for 'A' and 'D'

ΔV measured between B and C is : **[2008]**

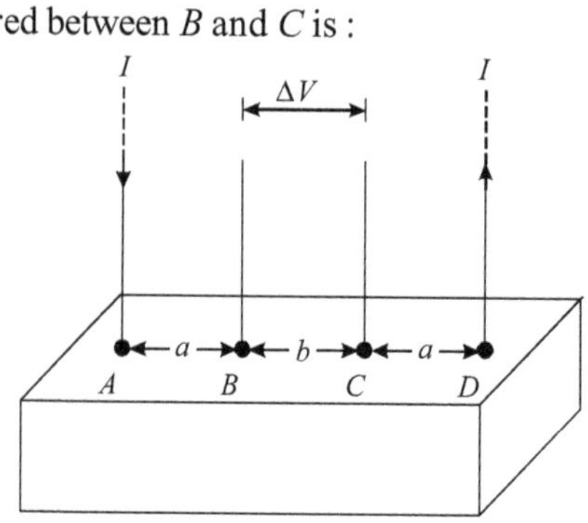

(A) $\frac{\rho I}{\pi a}-\frac{\rho I}{\pi(a+b)}$ (B) $\frac{\rho I}{a}-\frac{\rho I}{(a+b)}$

(C) $\frac{\rho I}{2\pi a}-\frac{\rho I}{2\pi(a+b)}$ (D) $\frac{\rho I}{2\pi(a+b)}$

10. Statement-1 : The temperature dependence of resistance is usually given as $R = R_0(1 + \alpha\Delta t)$. The resistance of a wire changes from 100 Ω to 150 Ω when its temperature is increased from 27°C to 227°C. This implies that $\alpha = 2.5 \times 10^{-3}$/°C.

Statement-2 : $R = R_0(1 + \alpha\Delta t)$ is valid only when the change in the temperature ΔT is small and $\Delta R = (R - R_0) << R_0$. **[2009]**

(A) Statement-1 is true, Statement-2 is true; Statement-2 is not the correct explanation of Statement-1

(B) Statement-1 is false, Statement-2 is true

(C) Statement-1 is true, Statement-2 is false

(D) Statement-1 is true, Statement-2 is true; Statement-2 is the correct explanation of Statement-1

11. Two conductors have the same resistance at 0°C but their temperature coefficients of resistance are α_1 and α_2. The respective temperature coefficients of their series and parallel combinations are nearly : **[2010]**

(A) $\frac{\alpha_1+\alpha_2}{2}, \frac{\alpha_1+\alpha_2}{2}$ (B) $\frac{\alpha_1+\alpha_2}{2}, \alpha_1+\alpha_2$

(C) $\alpha_1+\alpha_2, \frac{\alpha_1+\alpha_2}{2}$ (D) $\alpha_1+\alpha_2, \frac{\alpha_1\alpha_2}{\alpha_1+\alpha_2}$

12. If 400 Ω of resistance is made by adding four 100 Ω resistance of tolerance 5%, then the tolerance of the combination is : **[2011]**

(A) 20% (B) 5%

(C) 10% (D) 15%

13. If a wire is stretched to make it 0.1 % longer, its resistance will : **[2011]**

(A) Decrease by 0.2% (B) Decrease by 0.05%

(C) Increase by 0.05% (D) Increase by 0.2%

14. The resistance of a wire is R. It is bent at the middle by 180° and both the ends are twisted together to make a shorter wire. The resistance of the new wire is : **[2012 (Online)]**

(A) $2R$ (B) $R/2$

(C) $R/4$ (D) $R/8$

15. When 5 V potential difference is applied across a wire of length 0.1 m, the drift speed of electrons is 2.5×10^{-4} ms^{1}. If the electron density in the wire is 8×10^{28} m^{-3}, the resistivity of the material is close to : **[2015]**

(A) 1.6×10^{-8} Ωm (B) 1.6×10^{-7} Ωm

(C) 1.6×10^{-6} Ωm (D) 1.6×10^{-5} Ωm

16. Suppose the drift velocity v_d in a material varied with the applied electric field E as $v_d \propto \sqrt{E}$. The $V-I$ graph for a wire made of such a material is best given by : **[2015 (Online)]**

(A) 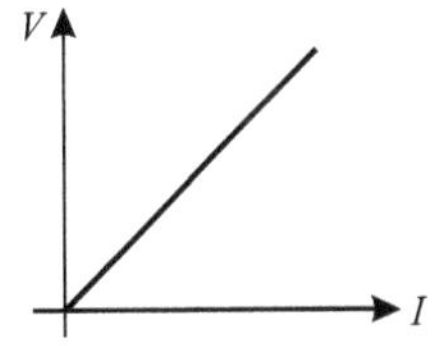

(B) 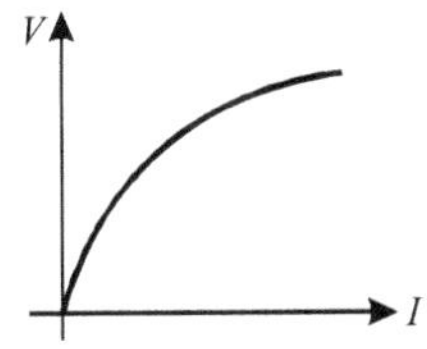

(C) 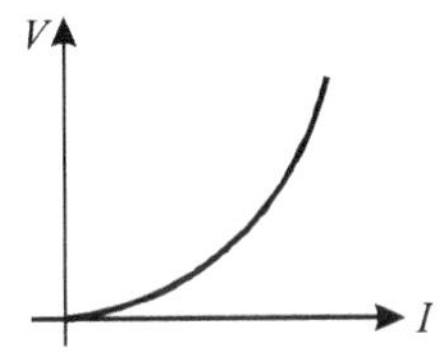

(D) 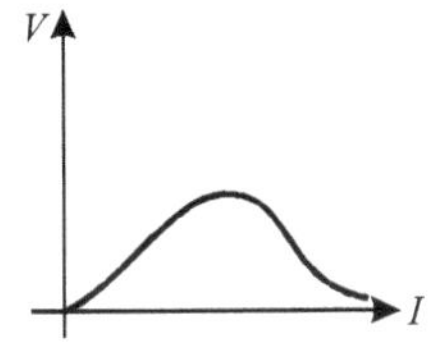

17. A uniform wire of length l and radius r has a resistance of 100 Ω. It is recast into a wire of radius $r/2$. The resistance of new wire will be : **[2017 (Online)]**

(A) 1600 Ω (B) 400 Ω

(C) 200 Ω (D) 100 Ω

18. A copper rod of cross-sectional area A carries a uniform current I through it. At temperature T, if the volume charge density of the rod is ρ, how long will the charges take to travel a distance d? **[2018 (Online)]**

(A) $\frac{2\rho dA}{IT}$ (B) $\frac{2\rho dA}{I}$

(C) $\frac{\rho dA}{I}$ (D) $\frac{\rho dA}{IT}$

19. A heating element has a resistance of 100 Ω at room temperature. When it is connected to a supply of 220 V, a steady current of 2 A passes in it and temperature is 500°C more than room temperature. What is the temperature coefficient of resistance of the heating element ? **[2018 (Online)]**

(A) 1×10^{-4} °C^{-1} (B) 5×10^{-4} °C^{-1}

(C) 2×10^{-4} °C^{-1} (D) 0.5×10^{-4} °C^{-1}

20. A resistance is shown in the figure. Its value and tolerance are given respectively by : **[9 Jan. 2019 (Morning)]**

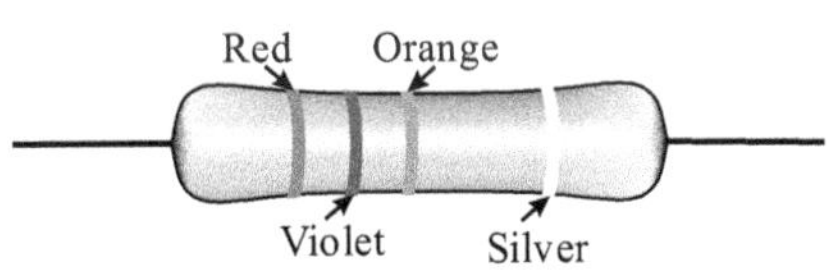

(A) 27 KΩ, 20% (B) 270 KΩ, 5%

(C) 270 KΩ, 10% (D) 27 KΩ, 10%

21. Drift speed of electrons, when 1.5 A of current flows in a copper wire of cross section 5 mm^2, is v. If the electron density in copper is 9×10^{28} /m^3 the value of v in mm/s is close to (Take charge of electron to be $=1.6 \times 10^{-19}$ C) :

[9 Jan. 2019 (Morning)]

(A) 0.2 (B) 3
(C) 2 (D) 0.02

22. A carbon resistance has a following colour code. What is the value of the resistance ? **[9 Jan. 2019 (Evening)]**

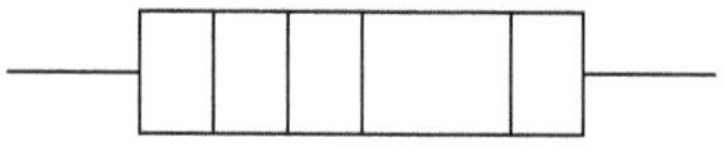

(A) $1.64\,M\Omega \pm 5\%$ (B) $530\,k\Omega \pm 5\%$
(C) $64\,k\Omega \pm 10\%$ (D) $5.3\,M\Omega \pm 5\%$

23. A 2 W carbon resistor is color coded with green, black, red and brown respectively. The maximum current which can be passed through this resistor is : **[10 Jan. 2019 (Morning)]**

(A) 63 mA (B) 0.4 mA
(C) 100 mA (D) 20 mA

24. A 200 Ω resistor has a certain color code. If one replaces the red color by green in the code, the new resistance will be : **[8 April 2019 (Morning)]**

(A) 100 Ω (B) 400 Ω
(C) 500 Ω (D) 300 Ω

25. A metal wire of resistance 3Ω is elongated to make a uniform wire of double its previous length. This new wire is now bent and the ends joined to make a circle. If two points on this circle make an angle 60° at the centre, the equivalent resistance between these two points will be : **[9 April 2019 (Evening)]**

(A) $\frac{12}{5}\Omega$ (B) $\frac{5}{3}\Omega$
(C) $\frac{5}{2}\Omega$ (D) $\frac{7}{2}\Omega$

26. In a conductor, if the number of conduction electrons per unit volume is $8.5 \times 10^{28}\,m^{-3}$ and mean free time is 25 *fs* (femto second), it's approximate resistivity is : ($m_e = 9.1 \times 10^{-31}$ kg) : **[9 April 2019 (Evening)]**

(A) $10^{-5}\,\Omega m$ (B) $10^{-6}\,\Omega m$
(C) $10^{-7}\,\Omega m$ (D) $10^{-8}\,\Omega m$

27. A current of 5A passes through a copper conductor (resistivity $= 1.7 \times 10^{-8}\,\Omega m$) of radius of cross-section 5 mm. Find the mobility of the charges if their drift velocity is 1.1×10^{-3} m/s: **[10 April 2019 (Morning)]**

(A) $1.3\,m^2/Vs$ (B) $1.5\,m^2/Vs$
(C) $1.8\,m^2/Vs$ (D) $1.0\,m^2/Vs$

28. In an experiment, the resistance of a material is plotted as a function of temperature (in some range). As shown in the figure, it is a straight line. One may conclude that : **[10 April 2019 (Morning)]**

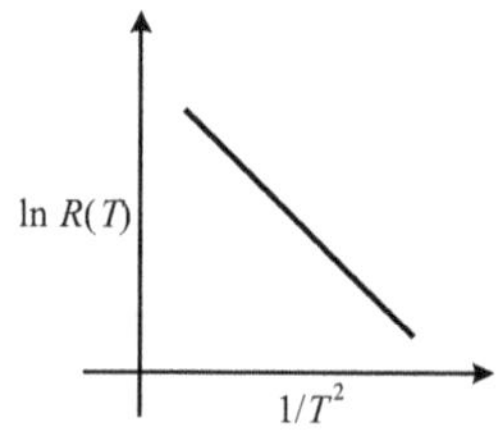

(A) $R(T) = \frac{R_0}{T^2}$ (B) $R(T) = R_0 e^{-T^2/T_0^2}$
(C) $R(T) = R_0 e^{-T_0^2/T^2}$ (D) $R(T) = R_0 e^{T^2/T_0^2}$

29. Space between two concentric conducting spheres of radii a and $b (b > a)$ is filled with a medium of resistivity ρ. The resistance between the two spheres will be : **[10 April 2019 (Evening)]**

(A) $\frac{\rho}{4\pi}\left(\frac{1}{a} - \frac{1}{b}\right)$ (B) $\frac{\rho}{2\pi}\left(\frac{1}{a} - \frac{1}{b}\right)$
(C) $\frac{\rho}{2\pi}\left(\frac{1}{a} + \frac{1}{b}\right)$ (D) $\frac{\rho}{4\pi}\left(\frac{1}{a} + \frac{1}{b}\right)$

* * * * *

CHAPTER-17 CURRENT ELECTRICITY

17.2 Circuit Analysis

1. A 3 volt battery with negligible internal resistance is connected in a circuit as shown in the figure. The current I, in the circuit will be : **[2003]**

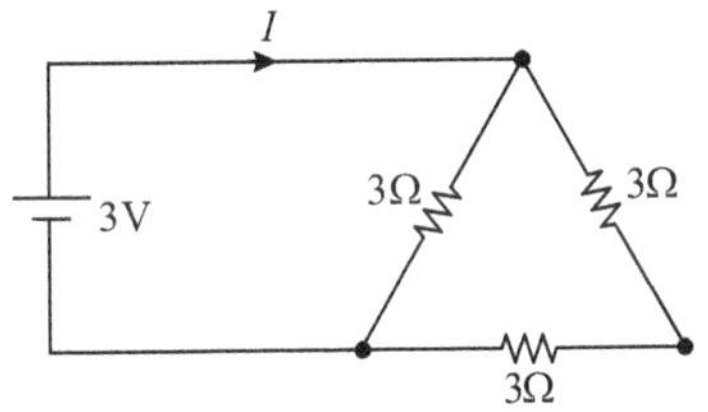

(A) 1 A (B) 1.5 A
(C) 2 A (D) 1/3 A

2. Two sources of equal emf are connected to an external resistance R. The internal resistances of the two sources are R_1 and R_2 ($R_2 > R_1$). If the potential difference across the source having internal resistance R_2, is zero, then : **[2005]**

(A) $R = \dfrac{R_2 \times (R_1 + R_2)}{(R_2 - R_1)}$ (B) $R = R_2 - R_1$

(C) $R = \dfrac{R_1 R_2}{(R_1 + R_2)}$ (D) $R = \dfrac{R_1 R_2}{(R_2 - R_1)}$

3. The current I drawn from the 5 volt source will be : **[2006]**

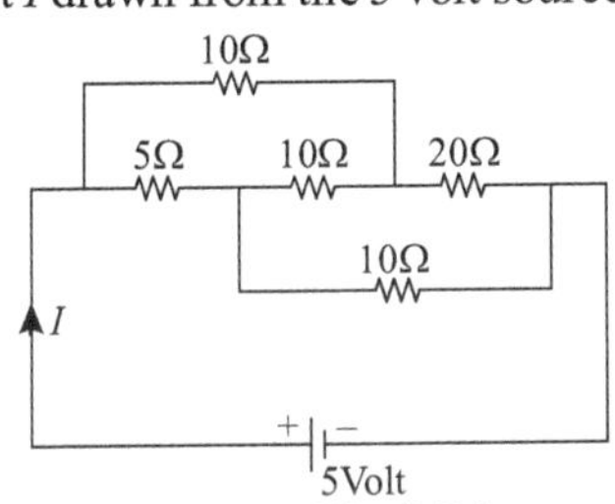

(A) 0.33 A (B) 0.5 A
(C) 0.67 A (D) 0.17 A

4. In a Wheatstone's bridge, three resistances P, Q and R are connected in the three arms and the fourth arm is formed by two resistances S_1 and S_2 connected in parallel. The condition for the bridge to be balanced will be : **[2006]**

(A) $\dfrac{P}{Q} = \dfrac{2R}{S_1 + S_2}$ (B) $\dfrac{P}{Q} = \dfrac{R(S_1 + S_2)}{S_1 S_2}$

(C) $\dfrac{P}{Q} = \dfrac{R(S_1 + S_2)}{2S_1 S_2}$ (D) $\dfrac{P}{Q} = \dfrac{R}{S_1 + S_2}$

5. The kirchhoff's first law ($\Sigma i = 0$) and second law ($\Sigma iR = \Sigma E$), Where the symbols have their usual meanings, are respectively based on : **[2006]**

(A) conservation of charge, conservation of momentum
(B) conservation of energy, conservation of charge
(C) conservation of momentum, conservation of charge
(D) conservation of charge, conservation of energy

6. A 5 V battery with internal resistance 2 Ω and a 2 V battery with internal resistance 1 Ω are connected to a 10 Ω resistor as shown in the figure. The current in the 10 Ω resistor is : **[2008]**

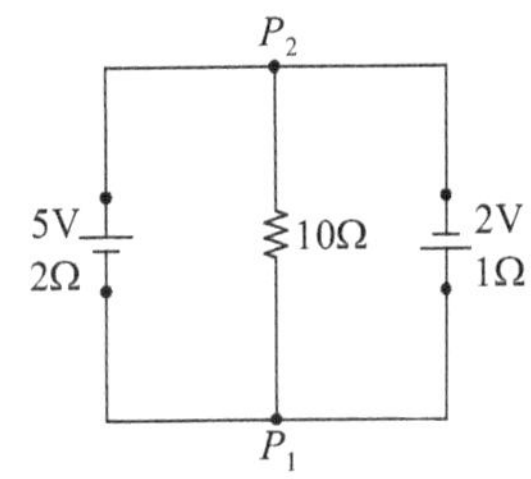

(A) 0.27 A P_2 to P_1 (B) 0.03 A P_1 to P_2
(C) 0.03 A P_2 to P_1 (D) 0.27 A P_1 to P_2

7. A d.c. source of emf $E_1 = 100$ V and internal resistance $r = 0.5\ \Omega$, a storage battery of emf $E_2 = 90$ V and an external resistance R are connected as shown in figure. For what value of R no current will pass through the battery ? **[2013 (Online)]**

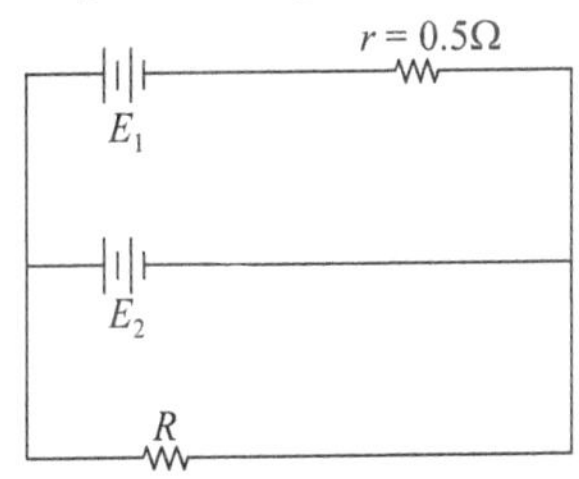

(A) 5.5 Ω (B) 3.5 Ω
(C) 4.5 Ω (D) 2.5 Ω

8. A letter 'A' is constructed of a uniform wire with resistance 1.0 Ω per cm. The sides of the letter are 20 cm and the cross piece in the middle is 10 cm long. The apex angle is 60°. The resistance between the ends of the legs is closed to : **[2013 (Online)]**

(A) 50.0 Ω (B) 10 Ω
(C) 36.7 Ω (D) 26.7 Ω

9. Six equal resistances are connected between points P, Q and R as shown in figure. Then net resistance will be maximum between : **[2013 (Online)]**

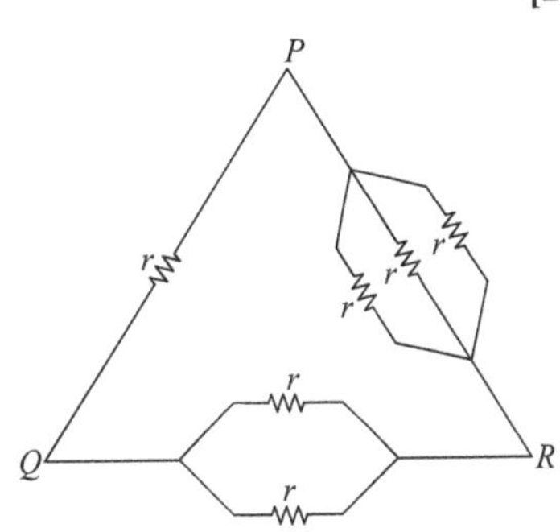

(A) P and R (B) P and Q
(C) Q and R (D) Any two points

10. A d.c. main supply of e.m.f. 220 V is connected across a storage battery of e.m.f. 200 V through a resistance of 1 Ω. The battery terminals are connected to an external resistance 'R'. The minimum value of 'R', so that a current passes through the battery to charge it is : **[2014 (Online)]**

(A) 7 Ω (B) 9 Ω
(C) 11 Ω (D) Zero

11. In the circuit shown, current (in A) through 50 V and 30 V batteries are, respectively : **[2014 (Online)]**

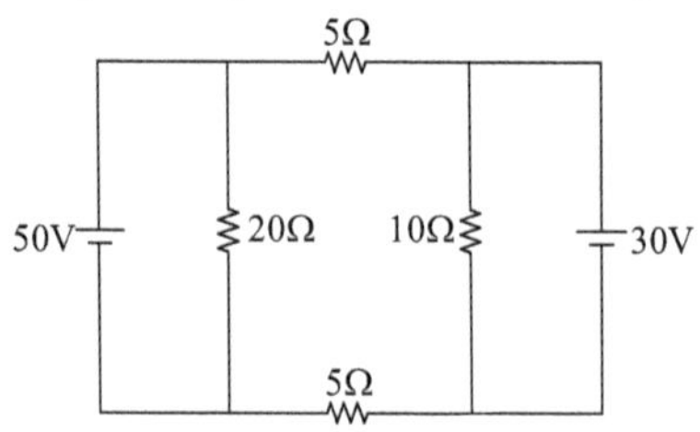

(A) 2.5 and 3 (B) 3.5 and 2
(C) 4.5 and 1 (D) 3 and 2.5

12. In the circuit shown, the current in the 1 Ω resistor is : **[2015]**

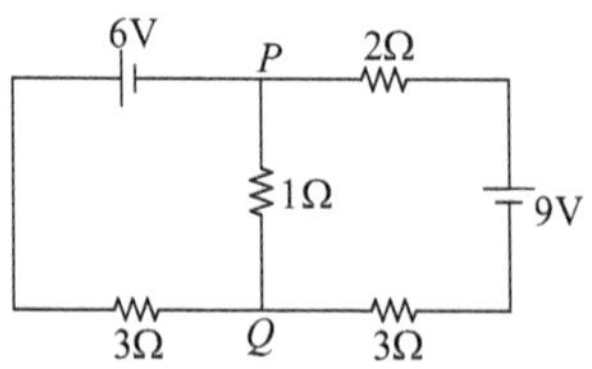

(A) 1.3 A, from P to Q (B) 0 A
(C) 0.13 A, from Q to P (D) 0.13 A, from P to Q

13. In the electric network shown, when no current flows through the 4 Ω resistor in the arm EB, the potential difference between the points A and D will be : **[2015 (Online)]**

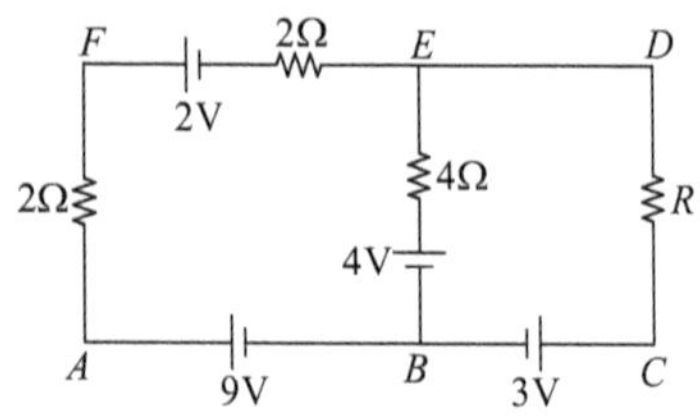

(A) 6V (B) 3V
(C) 5V (D) 4V

14.

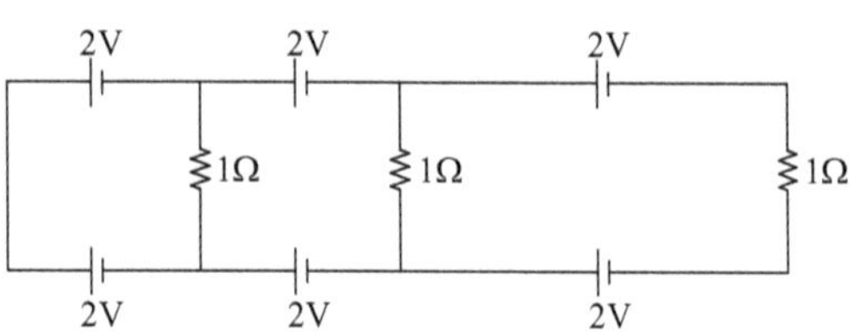

In the above circuit the current in each resistance is : **[2017]**

(A) 0.5 A (B) 0 A
(C) 1 A (D) 0.25 A

15. A 9 V battery with internal resistance of 0.5 Ω is connected across an infinite network as shown in the figure. All ammeters A_1, A_2, A_3 and voltmeter V are ideal : **[2017 (Online)]**

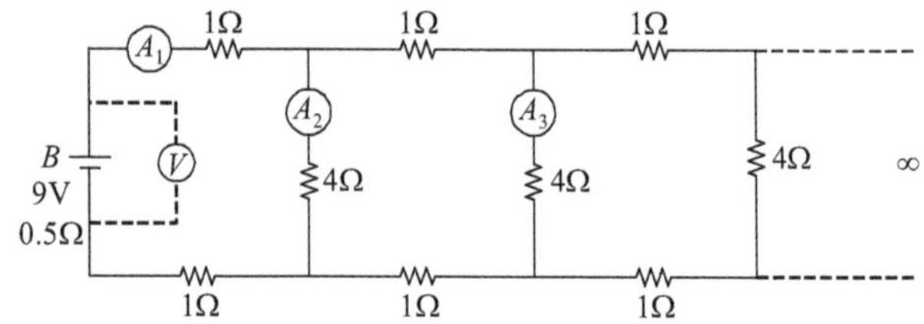

(A) Reading of A_1 is 2 A (B) Reading of A_1 is 18 A
(C) Reading of V is 9 V (D) Reading of V is 7 V

16. Two batteries with e.m.f. 12 V and 13 V are connected in parallel across a load resistor of 10 Ω. The internal resistance of the two batteries are 1 Ω and 2 Ω respectively. The voltage across the load lies between : **[2018]**

(A) 11.6 V and 11.7 V (B) 11.5 V and 11.6 V
(C) 11.4 and 11.5 V (D) 11.7 V and 11.8 V

17. In the given circuit all resistances are of volume R ohm each. The equivalent resistance between A and B is : **[2018 (Online)]**

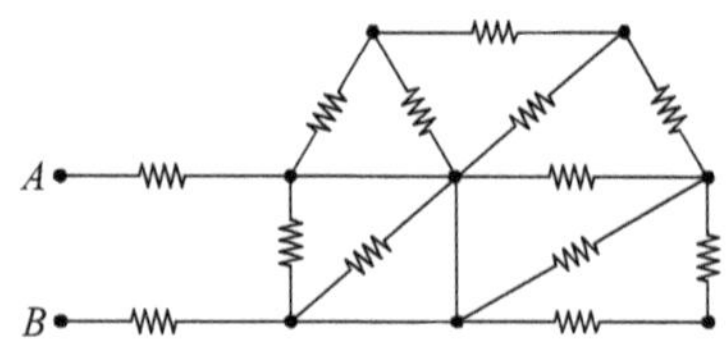

(A) $2R$ (B) $\frac{5R}{2}$
(C) $\frac{5R}{3}$ (D) $3R$

18. When the switch S, in the circuit shown, is closed, then the value of current i will be : **[9 Jan. 2019 (Morning)]**

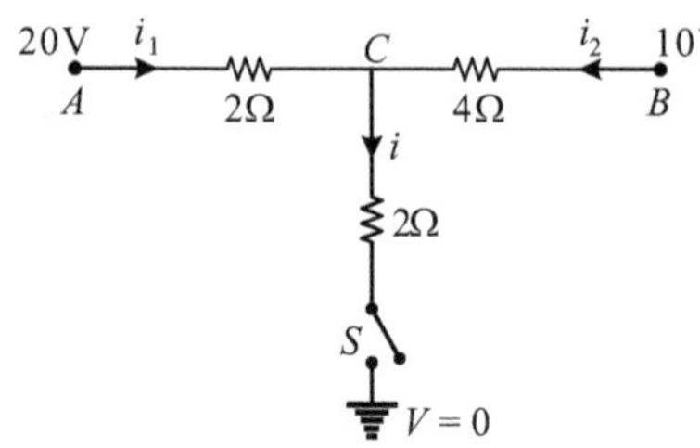

(A) 3 A (B) 5 A
(C) 4 A (D) 2 A

19. A uniform metallic wire has a resistance of 18 Ω and is bent into an equilateral triangle. Then, the resistance between any two vertices of the triangle is : **[10 Jan. 2019 (Morning)]**

(A) 8 Ω (B) 12 Ω
(C) 4 Ω (D) 2 Ω

20. In the given circuit the cells have zero internal resistance.

The currents (in Amperes) passing through resistance R_1, and R_2 respectively, are : **[10 Jan. 2019 (Morning)]**

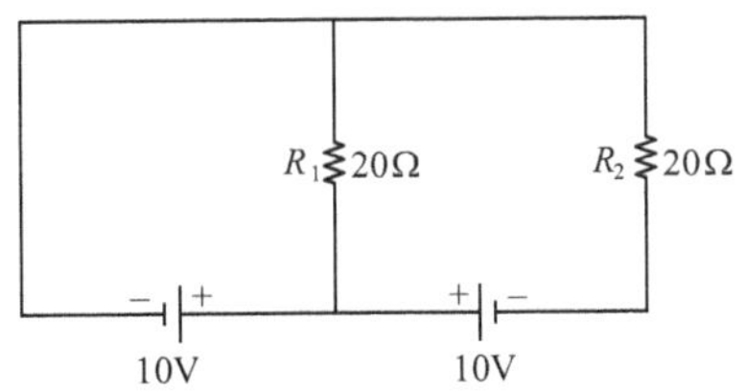

(A) 2, 2 (B) 0,1
(C) 1, 2 (D) 0.5, 0

21. In the circuit, the potential difference between A and B is: **[11 Jan. 2019 (Evening)]**

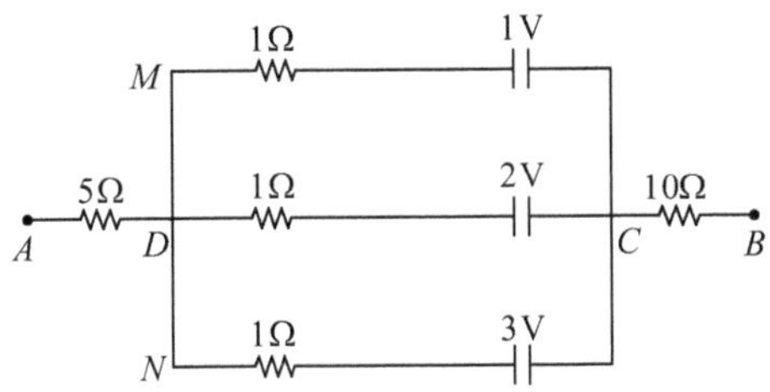

(A) 6 V (B) 1 V
(C) 3 V (D) 2 V

22. In the given circuit diagram, the currents, $I_1 = -0.3$A, $I_4 = 0.8$ A and $I_5 = 0.4$ A, are flowing as shown. The currents I_2, I_3 and I_6, respectively, are : **[12 Jan. 2019 (Evening)]**

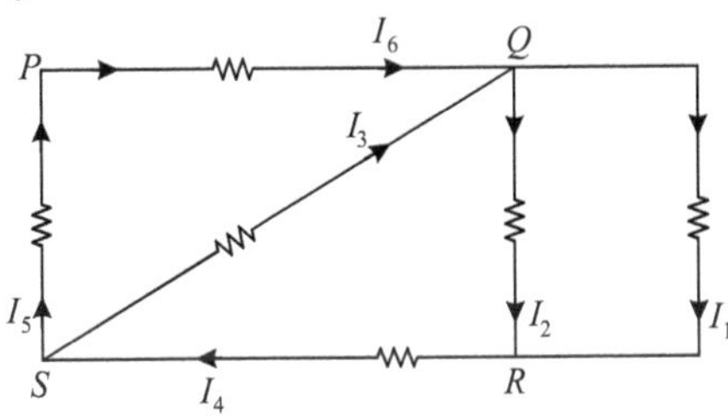

(A) 1.1 A, 0.4 A, 0.4 A (B) –0.4 A, 0.4 A, 1.1 A
(C) 0.4 A, 1.1 A, 0.4 A (D) 1.1 A,–0.4 A, 0.4 A

23. For the circuit shown, with $R_1 = 1.0\,\Omega$, $R_2 = 2.0\,\Omega$, $E_1 = 2$V and $E_2 = E_3 = 4$V, the potential difference between the points 'a' and 'b' is approximately (in V) : **[8 April 2019 (Morning)]**

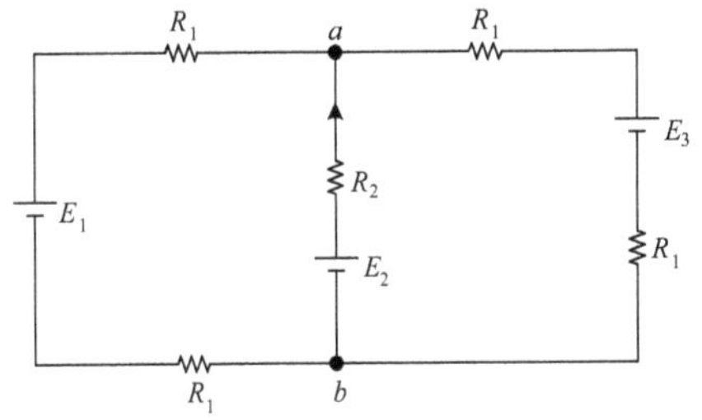

(A) 2.7 (B) 3.3
(C) 2.3 (D) 3.7

24. In the figure shown, what is the current (in Ampere) drawn from the battery ? You are given.
$R_1 = 15\Omega, R_2 = 10\,\Omega, R_3 = 20\,\Omega, R_4 = 5\Omega, R_5 = 25\Omega,$
$R_6 = 30\,\Omega, E = 15$ V **[8 April 2019 (Evening)]**

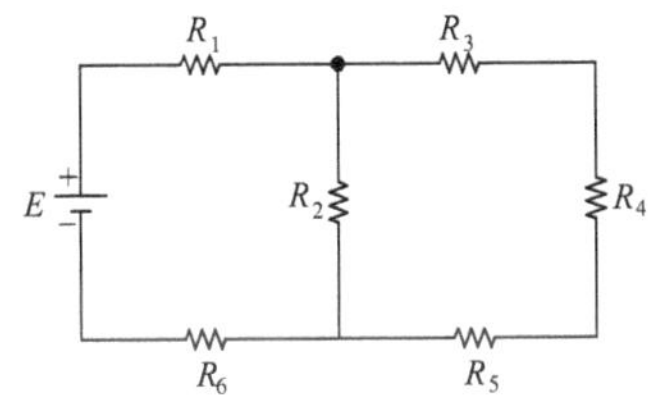

(A) 7/18 (B) 13/24
(C) 9/32 (D) 20/3

25. A wire of resistance R is bent to form a square $ABCD$ as shown in the figure. The effective resistance between E and C is : (E is mid-point of arm CD) : **[9 April 2019 (Morning)]**

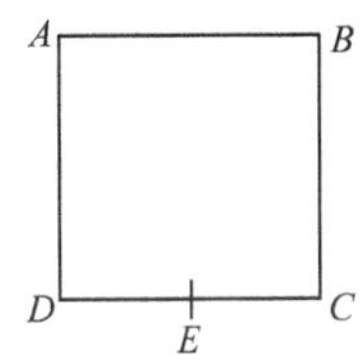

(A) R (B) $\frac{1}{16}R$

(C) $\frac{7}{64}R$ (D) $\frac{3}{4}R$

* * * * *

CHAPTER-17 CURRENT ELECTRICITY
17.3 Thermal Effects of Current & RC Circuits

1. The mass of product liberated on anode in an electrochemical cell depends on : **[2002]**

(A) $(It)^{1/2}$ (B) It
(C) I/t (D) I^2t

2. A wire when connected to 220 V mains supply has power dissipation P_1. Now the wire is cut into two equal pieces which are connected in parallel to the same supply. Power dissipation in this case is P_2. Then $P_2 : P_1$ is: **[2002]**

(A) 1 (B) 4
(C) 2 (D) 3

3. If in the circuit, power dissipation is 150 W, then R is : **[2002]**

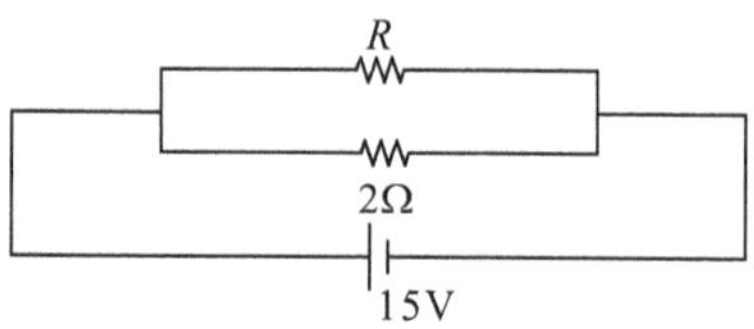

(A) 2 Ω (B) 6 Ω
(C) 5 Ω (D) 4 Ω

4. The thermo e.m.f. of a thermo-couple is 25μ V/°C at room temperature. A galvanometer of 40 ohm resistance, capable of detecting current as low as 10^{-5} A, is connected with the thermo couple. The smallest temperature difference that can be detected by this system is : **[2003]**

(A) 16°C (B) 12°C
(C) 8°C (D) 20°C

5. The negative *Zn* pole of a Daniell cell, sending a constant current through a circuit, decreases in mass by 0.13 g in 30 minutes. If the electrochemical equivalent of *Zn* and *Cu* are 32.5 and 31.5 respectively, the increase in the mass of the positive *Cu* pole in this time is : **[2003]**

(A) 0.180 g (B) 0.141 g
(C) 0.126 g (D) 0.242 g

6. A 220 volt, 1000 watt bulb is connected across a 110 volt mains supply. The power consumed will be : **[2003]**

(A) 750 watt (B) 500 watt
(C) 250 watt (D) 1000 watt

7. The thermo emf of a thermocouple varies with the temperature θ of the hot junction as $E = a\theta + b\theta^2$ in volts where the ratio a/b is 700°C. If the cold junction is kept at 0°C, then the neutral temperature is : **[2004]**

(A) 1400 °C
(B) 350 °C
(C) 700 °C
(D) No neutral temperature is possible for this thermocouple

8. The electrochemical equivalent of a metal is 3.35×10^{-7} kg per Coulomb. The mass of the metal liberated at the cathode when a 3A current is passed for 2 seconds will be : **[2004]**

(A) 6.6×10^{57} kg (B) 9.9×10^{-7} kg
(C) 19.8×10^{-7} kg (D) 1.1×10^{-7} kg

9. Time taken by a 836 W heater to heat one litre of water from 10°C to 40°C is : **[2004]**

(A) 150 s (B) 100 s
(C) 50 s (D) 200 s

10. An energy source will supply a constant current into the load, if its internal resistance is : **[2005]**

(A) Equal to the resistance of the load
(B) Very large as compared to the load resistance
(C) Zero
(D) Non-zero but less than the resistance of the load

11. Two voltameters, one of copper and another of silver, are joined in parallel. When a total charge q flows through the voltameters, equal amount of metals are deposited. If the electrochemical equivalents of copper and silver are Z_1 and Z_2 respectively the charge which flows through the silver voltameter is : **[2005]**

(A) $\dfrac{q}{1+\dfrac{Z_2}{Z_1}}$ (B) $\dfrac{q}{1+\dfrac{Z_1}{Z_2}}$

(C) $q\dfrac{Z_2}{Z_1}$ (D) $q\dfrac{Z_1}{Z_2}$

12. A heater coil is cut into two equal parts and only one part is now used in the heater. The heat generated will now be : **[2005]**

(A) Four times (B) Doubled
(C) Halved (D) One fourth

13. The resistance of hot tungsten filament is about 10 times the cold resistance. What will be the resistance of 100 W and 200 V lamp when not in use ? **[2005]**

(A) 20 Ω (B) 40 Ω
(C) 200 Ω (D) 400 Ω

14. A thermocouple is made from two metals, Antimony and Bismuth. If one junction of the couple is kept hot and the other is kept cold, then, an electric current will : **[2006]**
(A) Flow from Antimony to Bismuth at the hot junction
(B) Flow from Bismuth to Antimony at the cold junction
(C) Now flow through the thermocouple
(D) Flow from Antimony to Bismuth at the cold junction

15. An electric bulb is rated 220 V-100 W. The power consumed by it when operated on 110 V will be : **[2006]**
(A) 75 W (B) 40 W
(C) 25 W (D) 50 W

16. Let C be the capacitance of a capacitor discharging through a resistor R. Suppose t_1 is the time taken for the energy stored in the capacitor to reduce to half its initial value and t_2 is the time taken for the charge to reduce to one-fourth its initial value. Then the ratio t_1/t_2 will be : **[2010]**
(A) 2 (B) 1
(C) 1/2 (D) 1/4

17. Combination of two identical capacitors, a resistor R and a *dc* voltage source of voltage 6 V is used in an experiment on $(C-R)$ circuit. It is found that for a parallel combination of the capacitor the time in which the voltage of the fully charged combination reduces to half its original voltage is 10 second. For series combination the time needed for reducing the voltage of the fully charged series combination by half is : **[2011]**
(A) 20 second (B) 10 second
(C) 5 second (D) 2.5 second

18. A resistor 'R' and 2μF capacitor in series is connected through a switch to 200 V direct supply. Across the capacitor is a neon bulb that lights up at 120 V. Calculate the value of R to make the bulb light up 5s after the switch has been closed. $(\log_{10} 2.5 = 0.4)$: **[2011]**
(A) $2.7 \times 10^6 \Omega$ (B) $3.3 \times 10^7 \Omega$
(C) $1.3 \times 10^4 \Omega$ (D) $1.7 \times 10^5 \Omega$

19. Two electric bulbs marked 25W-220 V and 100 W-220 V are connected in series to a 440 V supply. Which of the bulbs will fuse ? **[2012]**
(A) Neither (B) Both
(C) 100 W (D) 25 W

20. The figure shows an experimental plot for discharging of a capacitor in an R–C circuit. The time constant τ of this circuit lies between : **[2012]**

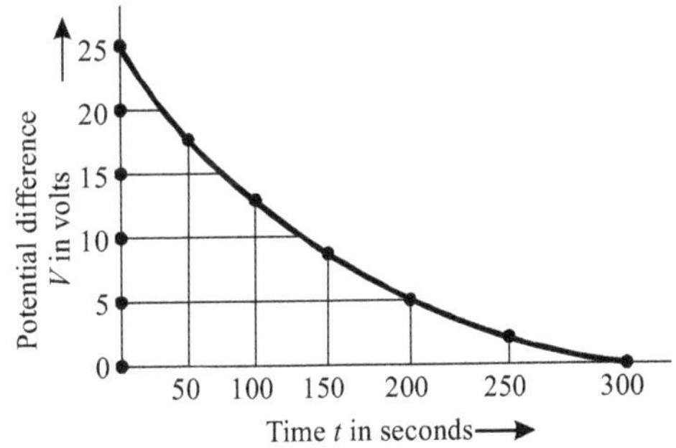

(A) 100 sec and 150 sec (B) 150 sec and 200 sec
(C) 0 and 50 sec (D) 50 sec and 100 sec

21. This question has Statement-1 and Statement-2. Of the four choice given after the Statements, choose the one that best describes the two Statements.
Statement-1 : The possibility of an electric bulb fusing is higher at the time of switching ON.
Statement-2 : Resistance of an electric bulb when it is not lit up is much smaller than when it is lit up. **[2012 (Online)]**
(A) Statement-1 is true, Statement-2 is false
(B) Statement-1 is false, Statement-2 is true, Statement-2 is not a correct explanation of Statement-1
(C) Statement-1 is true, Statement-2 is true, Statement-2 is a correct explanation of Statement-1
(D) Statement-1 is false, Statement-2 is true

22. Three resistors of 4 Ω, 6 Ω and 12 Ω are connected in parallel and the combination is connected in series with a 1.5 V battery of 1 Ω internal resistance. The rate of Joule heating in the 4 Ω resistor is : **[2012 (Online)]**
(A) 0.55 W (B) 0.33 W
(C) 0.25 W (D) 0.86 W

23. A 6.0 volt battery is connected to two light bulbs as shown in figure. Light bulb-1 has resistance 3 ohm while light bulb-2 has resistance 6 ohm. Battery has negligible internal resistance. Which bulb will glow brighter ? **[2012 (Online)]**

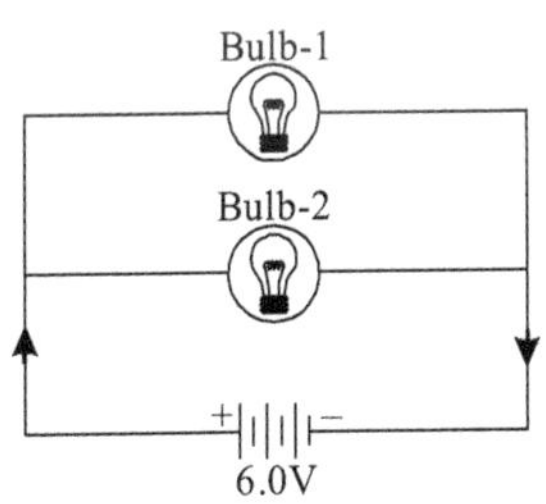

(A) Bulb-1 will glow more first and then its brightness will become less than bulb-2
(B) Bulb-1
(C) Bulb-2
(D) Both glow equally

24. A resistance R and a capacitance C are connected in series to a battery of negligible internal resistance through a key. The key is closed at $t = 0$. If after t sec the voltage across the capacitance was seven times the voltage across R, the value of t is : **[2012 (Online)]**

(A) 3 RC ln 2 (B) 2 RC ln 2
(C) 2 RC ln 7 (D) 3 RC ln 7

25. The supply voltage to a room is 120 V. The resistance of the lead wires is 6 Ω. A 60 W bulb is already switched on. What is the decrease of voltage across the bulb, when a 240 W heater is switched on in parallel to the bulb? **[2013]**

(A) Zero Volt (B) 2.9 Volt
(C) 13.3 Volt (D) 10.04 Volt

26. Which of the four resistances P, Q, R and S generate the greatest amount of heat when a current flows from A to B? **[2013 (Online)]**

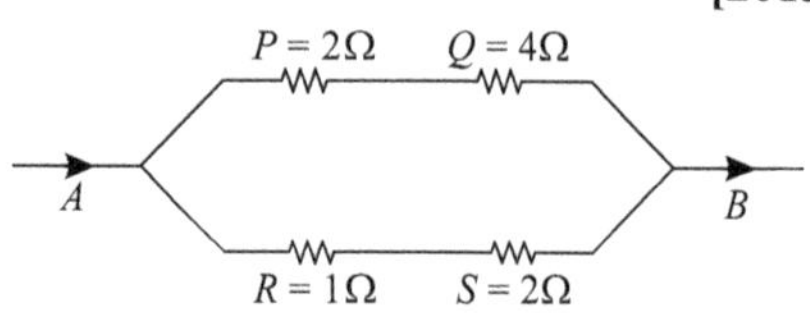

(A) Q (B) S
(C) P (D) R

27. In a large building, there are 15 bulbs of 40 W, 5 bulbs of 100 W, 5 fans of 80 W and 1 heater of 1 kW. The voltage of the electric mains is 220 V. The minimum capacity of the main fuse of the building will be : **[2014]**

(A) 8 A (B) 10 A
(C) 12 A (D) 14 A

28. Four bulbs B_1, B_2, B_3 and B_4 of 100 W each are connected to 220 V main as shown in the figure. The reading in an ideal ammeter will be : **[2014 (Online)]**

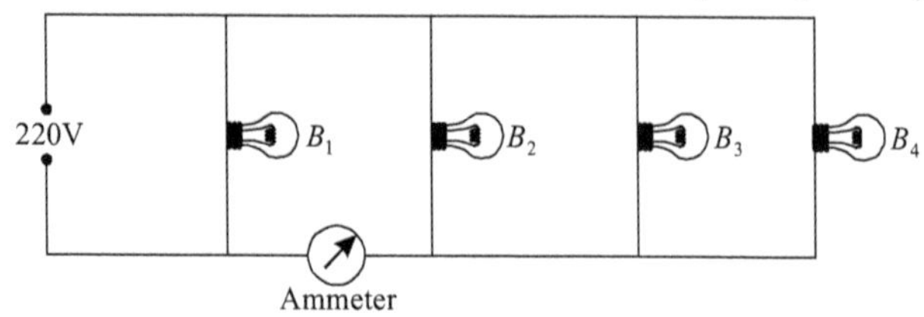

(A) 0.45 A (B) 0.90 A
(C) 1.35 A (D) 1.80 A

29. The circuit shown here has two batteries of 8.0 V and 16.0 V and three resistors 3 Ω, 9 Ω and 9 Ω and a capacitor of 5.0 μF. How much is the current I in the circuit in steady state ? **[2014 (Online)]**

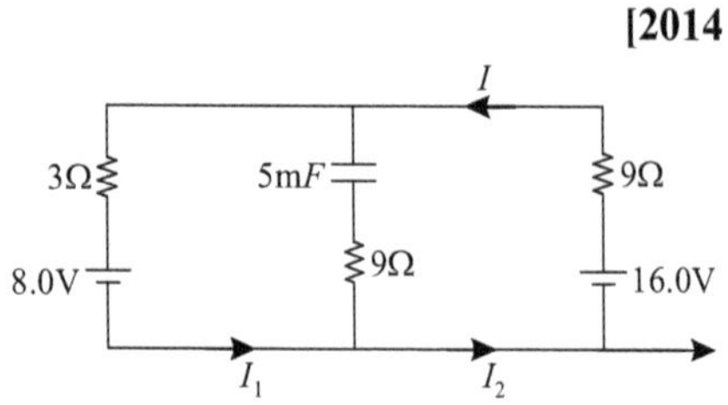

(A) 1.6 A (B) 0.67 A
(C) 2.5 A (D) 0.25 A

30. In the circuit shown, the resistance r is a variable resistance. If for $r = fR$, the heat generation in r is maximum then the value of f is : **[2016 (Online)]**

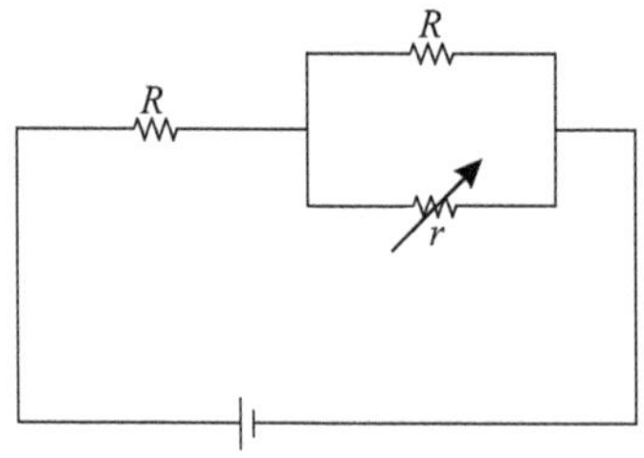

(A) $\frac{1}{2}$ (B) 1
(C) $\frac{1}{4}$ (D) $\frac{3}{4}$

31. The resistance of an electrical toaster has a temperature dependence given by $R(T) = R_0[1 + \alpha(T - T_0)]$ in its range of operation. At $T_0 = 300$ K, $R = 100\,\Omega$ and at $T = 500$ K, $R = 120\,\Omega$. The toaster is connected to a voltage source at 200 V and its temperature is raised at a constant rate from 300 to 500 K in 30 s. The total work done in raising the temperature is : **[2016 (Online)]**

(A) $400 \ln \frac{5}{6}$ J (B) $200 \ln \frac{2}{3}$ J
(C) 300 J (D) $400 \ln \frac{1.5}{1.3}$ J

32. In the given circuit diagram when the current reaches steady state in the circuit, the charge on the capacitor of capacitance C will be : **[2017]**

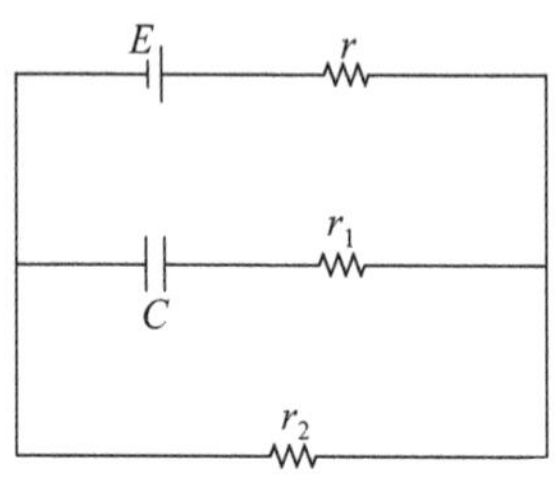

(A) $CE\frac{r_2}{(r + r_2)}$ (B) $CE\frac{r_1}{(r_1 + r)}$
(C) CE (D) $CE\frac{r_1}{(r_2 + r)}$

33. The figure shows three circuits I, II, and III which are connected to a 3 V battery. If the powers dissipated by the configurations I, II, and III are P_1, P_2 and P_3 respectively, then : **[2017 (Online)]**

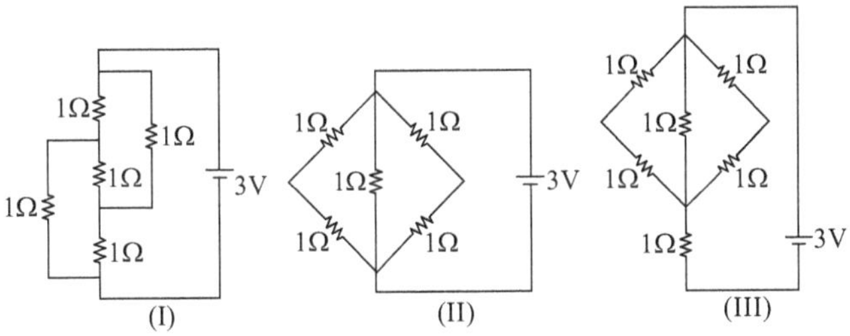

(A) $P_1 > P_2 > P_3$ (B) $P_1 > P_3 > P_2$
(C) $P_2 > P_1 > P_3$ (D) $P_3 > P_2 > P_1$

34. In the following circuit, the switch S is closed at $t = 0$. The charge on the capacitor C_1 as a function of time will be given by $\left(C_{eq} = \frac{C_2 C_2}{C_1 + C_2}\right)$. **[2018 (Online)]**

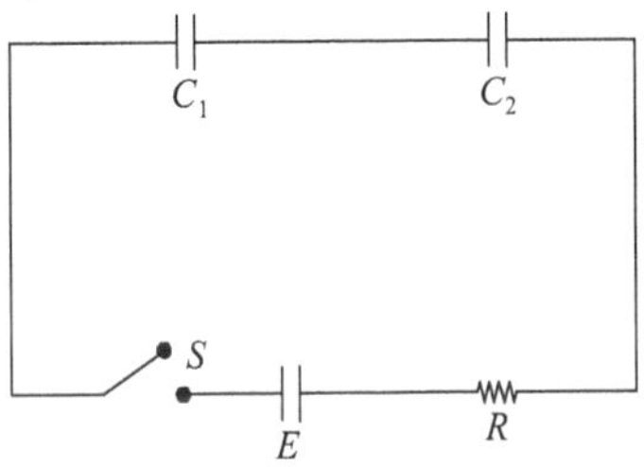

(A) $C_{eq} E[1 - \exp(-t/RC_{eq})]$ (B) $C_1 E[1 - \exp(-tR/C_1)]$
(C) $C_2 E[1 - \exp(-t/RC_2)]$ (D) $C_{eq} E \exp(-t/RC_{eq})$

35. A constant voltage is applied between two ends of a metallic wire. If the length is halved and the radius of the wire is doubled, the rate of heat developed in the wire will be : **[2018 (Online)]**
(A) Increased 8 times (B) Doubled
(C) Halved (D) Unchanged

36. A current of 2 mA was passed through an unknown resistor which dissipated a power of 4.4 W. Dissipated power when an ideal power supply of 11 V is connected across it is :
[10 Jan. 2019 (Evening)]
(A) 11×10^{-5} W (B) 11×10^{-4} W
(C) 11×10^{5} W (D) 11×10^{-3} W

37. Two equal resistance when connected in series to a battery, consume electric power of 60 W. If these resistances are now connected in parallel combination to the same battery, the electric power consumed will be : **[11 Jan. 2019 (Morning)]**
(A) 60 W (B) 240 W
(C) 30 W (D) 120 W

38. Two electric bulbs, rated at (25 W, 220 V) and (100 W, 220 V), are connected in series across a 220 V voltage source. If the 25 W and 100 W bulbs draw powers P_1 and P_2 respectively, then :
[12 Jan. 2019 (Morning)]
(A) $P_1 = 9$ W, $P_2 = 16$ W (B) $P_1 = 4$ W, $P_2 = 16$ W
(C) $P_1 = 16$ W, $P_2 = 4$ W (D) P_1 16 W, $P_2 = 9$ W

39. A 20 Henry inductor coil is connected to a 10 ohm resistance in series as shown in figure. The time at which rate of dissipation of energy (joule's heat) across resistance is equal to the rate at which magnetic energy is stored in the inductor is :
[8 April 2019 (Morning)]

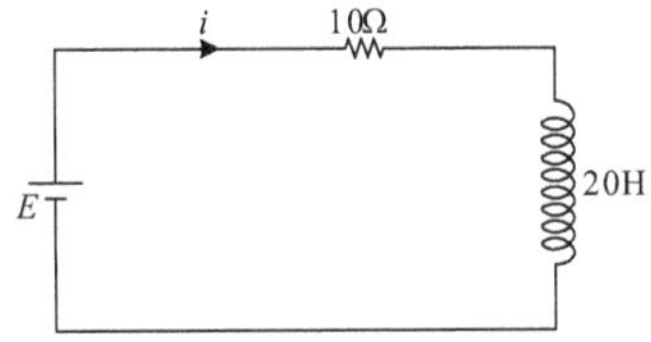

(A) $\frac{2}{\ln 2}$ (B) $\ln 2$
(C) $2 \ln 2$ (D) $\frac{1}{2} \ln 2$

40. A cell of internal resistance r drives current through an external resistance R. The power delivered by the cell to the external resistance will be maximum when :
[8 April 2019 (Evening)]
(A) $R = 1000\,r$ (B) $R = 0.001\,r$
(C) $R = 2r$ (D) $R = r$

41. Determine the charge on the capacitor in the following circuit : **[9 April 2019 (Morning)]**

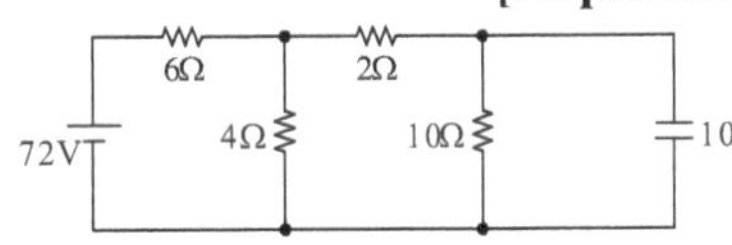

(A) 2μC (B) 60μC
(C) 200μC (D) 10μC

42. The resistive network shown below is connected to a D.C. source of 16V. The power consumed by the network is 4 Watt. The value of R is : **[12 April 2019 (Morning)]**

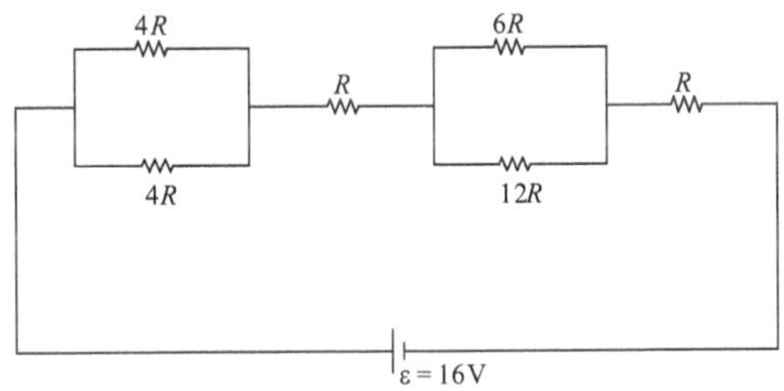

(A) 8 Ω (B) 6 Ω
(C) 1 Ω (D) 16 Ω

43. One kg of water, at 20°C, is heated in an electric kettle whose heating element has a mean (temperature averaged) resistance of 20Ω. The rms voltage in the mains is 200V. Ignoring heat loss from the kettle, time taken for water to evaporate fully, is close to : [Specific heat of water = 4200J/kg°C). Latent heat of water = 2260 kJ/kg] **[12 April 2019 (Evening)]**
(A) 3 minutes (B) 22 minutes
(C) 10 minutes (D) 16 minutes

* * * * *

CHAPTER-17 CURRENT ELECTRICITY
17.4 Electrical Measuring Instruments

1. If an ammeter is to be used in place of a voltmeter, then we must connect with the ammeter a : **[2002]**
(A) low resistance in parallel
(B) high resistance in parallel
(C) high resistance in series
(D) low resistance in series

2. An ammeter reads upto 1 A. Its internal resistance is 0.81 Ω. To increase the range to 10 A, the value of the required shunt is : **[2003]**
(A) 0.03 Ω (B) 0.3 Ω
(C) 0.9 Ω (D) 0.09 Ω

3. The length of a wire of a potentiometer is 100 cm, and the emf of its stand and cell is E volt. It is employed to measure the emf of a battery whose internal resistance is 0.5 Ω. If the balance point is obtained at $l = 30$ cm from the positive end, the emf of the battery is : **[2003]**
(A) $\frac{30E}{100.5}$ (B) $\frac{30E}{100-0.5}$
(C) $\frac{30(E-0.5i)}{100}$, where i is the current in the potentiometer wire
(D) $\frac{30E}{100}$

4. In a metre bridge experiment, null point is obtained at 20 cm from one end of the wire when resistance X is balanced against another resistance Y. If $X < Y$, then where will be the new position of the null point from the same end, if one decides to balance a resistance of $4X$ against Y? **[2004]**
(A) 50 cm (B) 80 cm
(C) 40 cm (D) 70 cm

5. In a potentiometer experiment the balancing with a cell is at length 240 cm. On shunting the cell with a resistance of 2 Ω, the balancing length becomes 120 cm. The internal resistance of the cell is : **[2005]**
(A) 1 Ω (B) 0.5 Ω
(C) 4 Ω (D) 2 Ω

6. In the circuit, the galvanometer G shows zero deflection. If the batteries A and B have negligible internal resistance, the value of the resistor R will be : **[2005]**

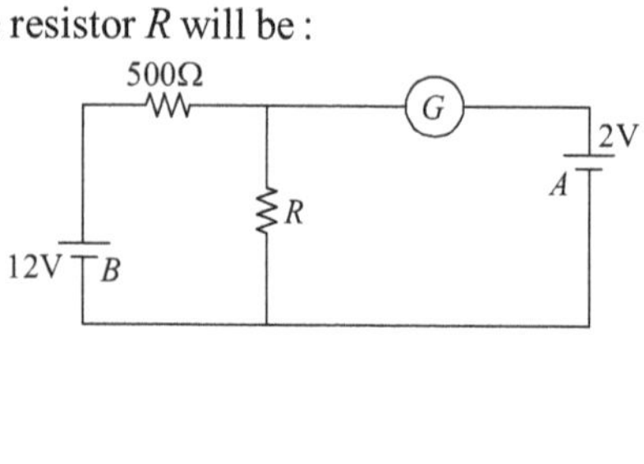

(A) 200 Ω (B) 100 Ω
(C) 500 Ω (D) 1000 Ω

7. A moving coil galvanometer has 150 equal divisions. Its current sensitivity is 10 divisions per milli ampere and voltage sensitivity is 2 divisions per millivolt. In order that each division reads 1 V, the resistance in Ohm's needed to be connected in series with the coil will be : **[2005]**
(A) 10^3 (B) 10^5
(C) 99995 (D) 9995

8. Shown in the figure below is a meter-bridge set up with null deflection in the galvanometer. The value of the unknown resistor R is : **[2008]**

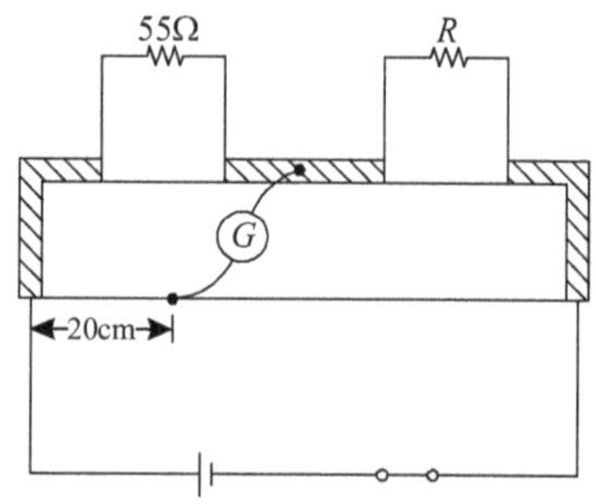

(A) 13.75 Ω (B) 220 Ω
(C) 110 Ω (D) 55 Ω

9. The current in the primary circuit of a potentiometer is 0.2 A. The specific resistance and cross-section of the potentiometer wire are 4×10^{-7} ohm metre and 8×10^{-7} m^2 respectively. The potential gradient will be equal to : **[2011]**
(A) 0.2 V/m (B) 1 V/m
(C) 0.5 V/m (D) 0.1 V/m

10. It is preferable to measure the e.m.f. of a cell by potentiometer than by a voltmeter because of the following possible reasons : **[2012 (Online)]**
(i) In case of potentiometer, no current flows through the cell
(ii) The length of the potentiometer allows greater precision
(iii) Measurement by the potentiometer is quicker
(iv) The sensitivity of the galvanometer, when using a potentiometer is not relevant
(A) (i), (iii), (iv) (B) (i), (iii), (iv)
(C) (i), (ii) (D) (i), (ii), (iii), (iv)

11. In an experiment of potentiometer for measuring the internal resistance of primary cell a balancing length l is obtained on the potentiometer wire when the cell is open circuit. Now the cell is short circuited by a resistance R. If R is to be equal to the internal resistance of the cell the balancing length on the potentiometer wire will be : **[2012 (Online)]**

(A) l (B) $2l$

(C) $l/2$ (D) $l/4$

12. This question has Statement-1 and Statement-2. Of the four choice given after the Statements, choose the one that best describes the two Statements.

Statement-1 : Higher the range, greater is the resistance of ammeter.

Statement-2 : To increase the range of ammeter, additional shunt needs to be used across it **[2013]**

(A) Statement-1 is true, Statement-2 is true, Statement-2 is the correct explanation of Statement-1

(B) Statement-1 is true, Statement-2 is true, Statement-2 is not the correct explanation of Statement-1

(C) Statement-1 is true, Statement-2 is false

(D) Statement-1 is false, Statement-2 is true

13. Correct set up to verify Ohm's law is : **[2013 (Online)]**

(A)

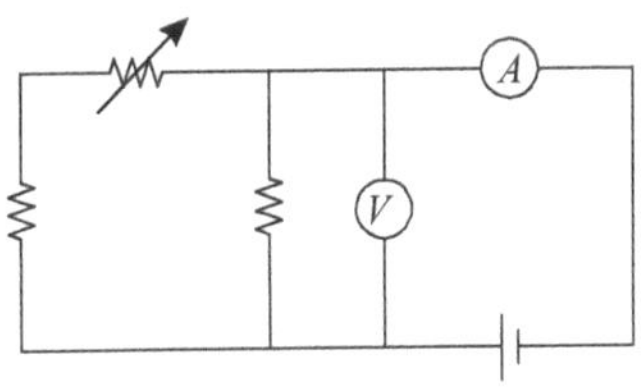

(B)

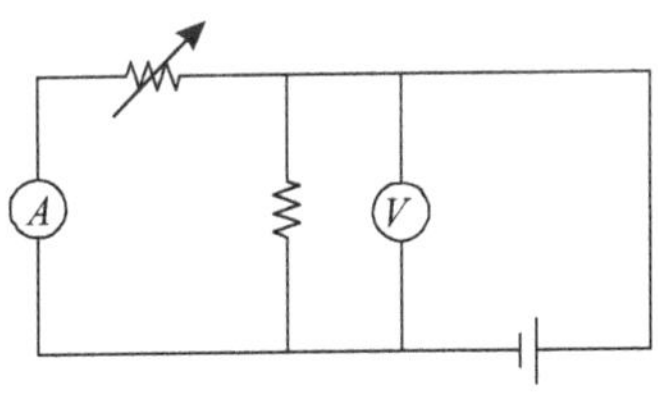

(C)

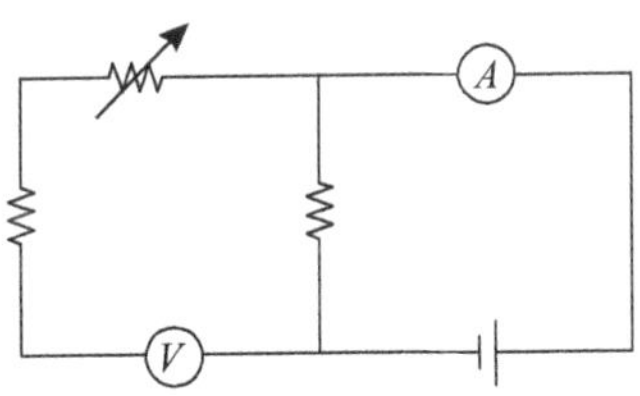

(D)

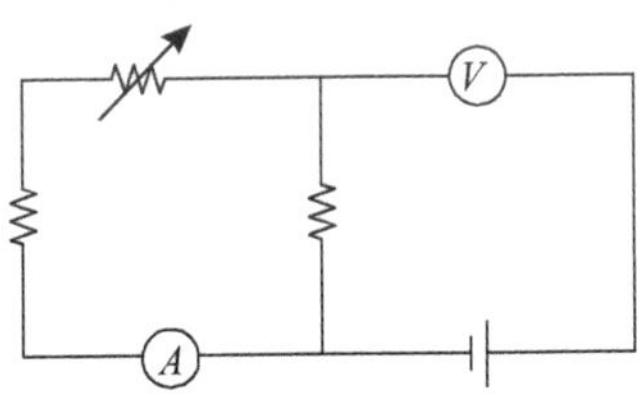

14. In a metre bridge experiment null point is obtained at 40 cm from one end of the wire when resistance X is balanced against another resistance Y. If $X < Y$, then the new position of the null point from the same end, if one decides to balance a resistance of $3X$ against Y, will be close to : **[2013 (Online)]**

(A) 80 cm (B) 75 cm

(C) 67 cm (D) 50 cm

15. A shunt of resistance 1 Ω is connected across a galvanometer of 120 Ω resistance. A current of 5.5 ampere gives full scale deflection in the galvanometer. The current that will give full scale deflection in the absence of the shunt is nearly : **[2013 (Online)]**

(A) 5.5 ampere (B) 0.5 ampere

(C) 0.004 ampere (D) 0.045 ampere

16. To find the resistance of a galvanometer by the half deflection method the following circuit is used with resistances $R_1 = 9970\ \Omega$, $R_2 = 30\ \Omega$ and $R_3 = 0$. The deflection in the galvanometer is d. With $R_3 = 107\ \Omega$ the deflection changed to $\frac{d}{2}$. The galvanometer resistance is approximately : **[2013 (Online)]**

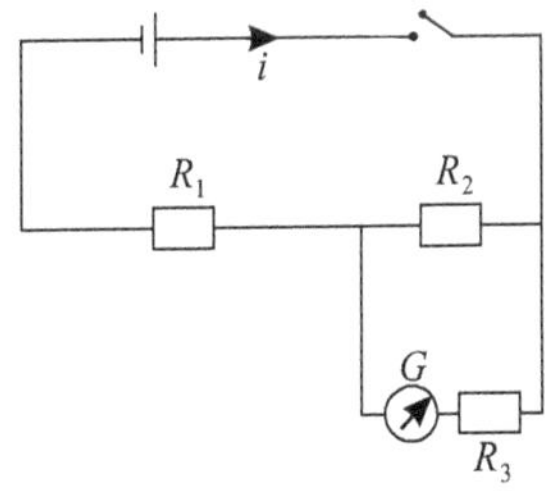

(A) 107 Ω (B) 137 Ω

(C) 107/2 Ω (D) 77 Ω

17. In the circuit diagrams (A, B, C and D) shown below, R is a high resistance and S is a resistance of the order of galvanometer resistance G. The correct circuit, corresponding to the half deflection method for finding the resistance and figure of merit of the galvanometer, is the circuit labelled as : **[2014 (Online)]**

(A)

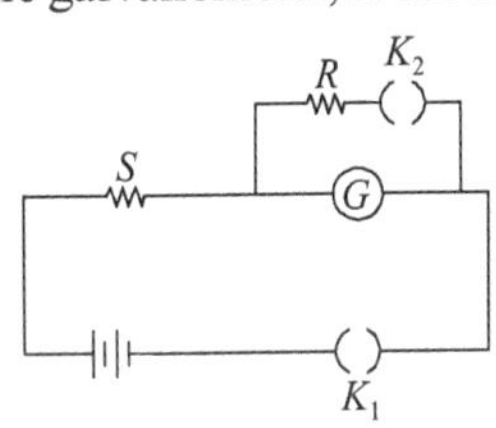

(B)

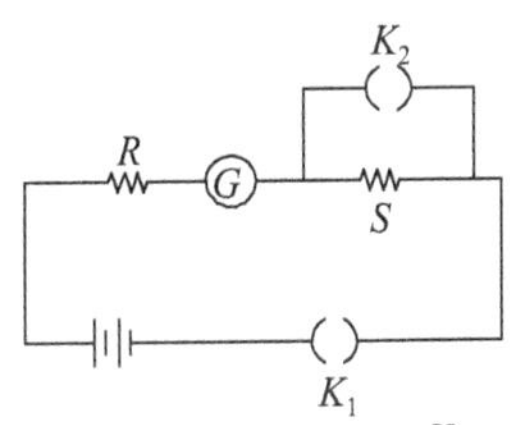

(C)

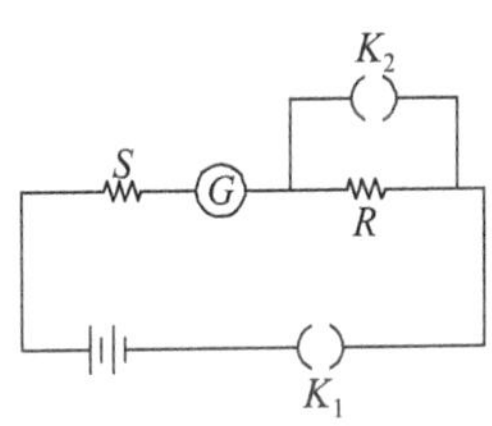

(D)

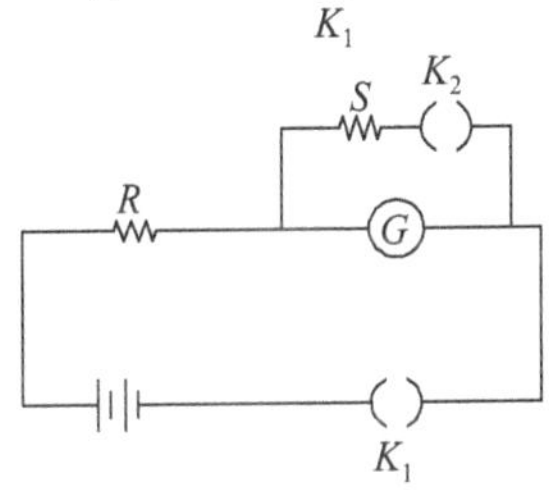

(A) Circuit A with $G = \frac{RS}{(R-S)}$

(B) Circuit B with $G = S$

(C) Circuit C with $G = S$

(D) Circuit D with $G = \frac{RS}{(R-S)}$

18. A 10 V battery with internal resistance 1 Ω and a 15 V battery with internal resistance 0.6 Ω are connected in parallel to a voltmeter (see figure). The reading in the voltmeter will be close to : **[2015 (Online)]**

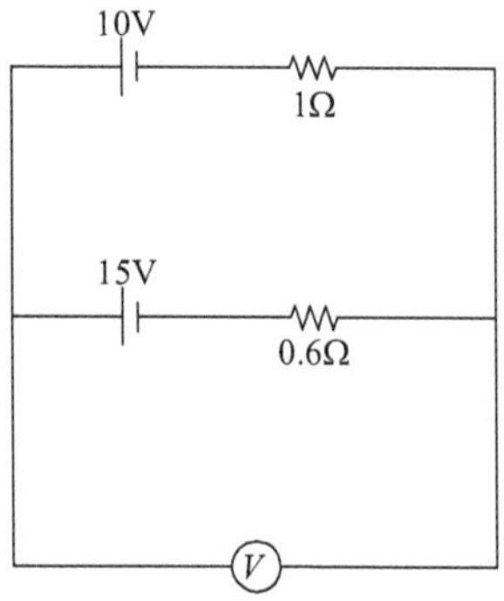

(A) 12.5 V (B) 24.5 V
(C) 13.1 V (D) 11.9 V

19. A galvanometer having a coil resistance of 100 Ω gives a full scale deflection, when a current of 1 mA is passed through it. The value of the resistance, which can convert this galvanometer into ammeter giving a full scale deflection for a current of 10 A, is : **[2016]**

(A) 0.01 Ω (B) 2 Ω
(C) 0.1 Ω (D) 3 Ω

20. To know the resistance G of a galvanometer by half deflection method, a battery of emf V_E and resistance R is used to deflect the galvanometer by angle θ. If a shunt of resistance S is needed to get half deflection then G, R and S related by the equation : **[2016 (Online)]**

(A) $S(R+G)=RG$ (B) $2S(R+G)=RG$
(C) $2G=S$ (D) $2S=G$

21. A 50 Ω resistance is connected to a battery of 5 V. A galvanometer of resistance 100 Ω is to be used as an ammeter to measure current through the resistance, for this a resistance r_s is connected to the galvanometer. Which of the following connections should be employed if the measured current is within 1% of the current without the ammeter in the circuit ? **[2016 (Online)]**

(A) $r_s = 0.5\ \Omega$ in series with the galvanometer
(B) $r_s = 1\ \Omega$ in series with galvanometer
(C) $r_s = 1\ \Omega$ in parallel with galvanometer
(D) $r_s = 0.5\ \Omega$ in parallel with the galvanometer

22. A galvanometer has a 50 division scale. Battery has no internal resistance. It is found that there is deflection of 40 divisions when $R = 2400\ \Omega$. Deflection becomes 20 divisions when resistance taken from resistance box is 4900 Ω. Then we can conclude : **[2016 (Online)]**

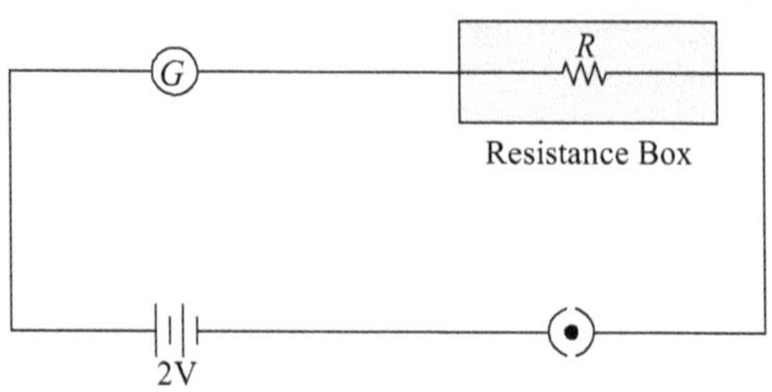

(A) Current sensitivity of galvanometer is 20 μA/division
(B) Resistance of galvanometer is 200 Ω
(C) Resistance required on $R.B.$ for a deflection of 10 divisions is 9800 Ω
(D) Full scale deflection current is 2 mA

23. Which of the following statements is false ? **[2017]**

(A) A rheostat can be used as a potential divider
(B) Kirchhoff's second law represents energy conservation
(C) Wheatstone bridge is the most sensitive when all the four resistances are of the same order of magnitude
(D) In a balanced wheatstone bridge if the cell and the galvanometer are exchanged, the null point is disturbed

24. When a current of 5 mA is passed through a galvanometer having a coil of resistance 15 Ω, it shows full scale deflection. The value of the resistance to be put in series with the galvanometer to convert it into to voltmeter of range 0 – 10 V is: **[2017]**

(A) $2.535 \times 10^3\ \Omega$ (B) $4.005 \times 10^3\ \Omega$
(C) $1.985 \times 10^3\ \Omega$ (D) $2.045 \times 10^3\ \Omega$

25. A potentiometer PQ is set up to compare two resistances as shown in the figure. The ammeter A in the circuit reads 1.0 A when two way key K_3 is open. The balance point is at a length l_1 cm from P when two way key K_3 is plugged in between 2 and 1, while the balance point is at a length l_2 cm from P when key K_3 is plugged in between 3 and 1. The ratio of two resistances $\frac{R_1}{R_2}$, is found to be : **[2017 (Online)]**

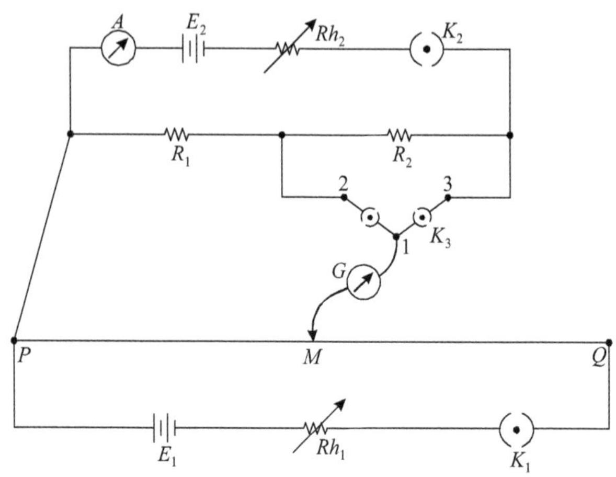

(A) $\frac{l_1}{l_1+l_2}$ (B) $\frac{l_2}{l_2-l_1}$
(C) $\frac{l_1}{l_1-l_2}$ (D) $\frac{l_1}{l_2-l_1}$

26. In a meter bridge experiment resistances are connected as shown in the figure. Initially resistance $P = 4\ \Omega$ and the neutral point N is at 60 cm from A. Now an unknown resistance R is connected in series to P and the new position of the neutral point is at 80 cm from A. The value of unknown resistance R is: **[2017 (Online)]**

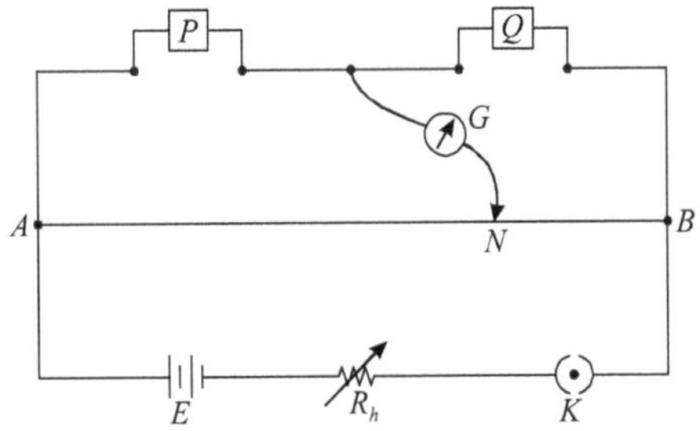

(A) $\frac{33}{5}\Omega$ (B) $6\,\Omega$

(C) $7\,\Omega$ (D) $\frac{20}{3}\Omega$

27. In a potentiometer experiment, it is found that no current passes through the galvanometer when the terminals of the cell are connected across 52 cm of the potentiometer wire. If the cell is shunted by a resistance of 5 Ω, a balance is found when the cell is connected across 40 cm of the wire. Find the internal resistance of the cell : **[2018]**

(A) 1 Ω (B) 1.5 Ω

(C) 2 Ω (D) 2.5 Ω

28. On interchanging the resistances, the balance point of a meter bridge shifts to the left by 10 cm. The resistance of their series combination is 1 kΩ. How much was the resistance on the left slot before interchanging the resistances? **[2018]**

(A) 990 Ω (B) 505 Ω

(C) 550 Ω (D) 910 Ω

29. In a meter bridge, as shown in the figure, it is given that resistance $Y = 12.5\ \Omega$ and that the balance is obtained at a distance 39.5 cm from end A (by jockey J). After interchanging the resistances X and Y, a new balance point is found at a distance l_1 from end A. What are the values of X and l_1 ? **[2018 (Online)]**

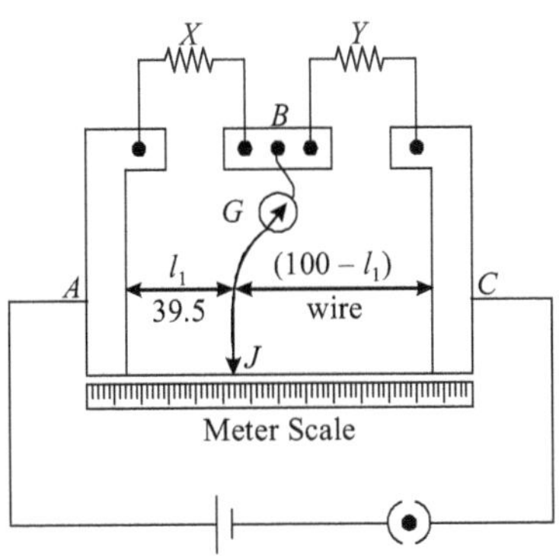

(A) 19.15 Ω and 39.5 cm (B) 8.16 Ω and 60.5 cm

(C) 19.15 Ω and 60.5 cm (D) 8.16 Ω and 39.5 cm

30. In a circuit for finding the resistance of a galvanometer by half deflection method, a 6 V battery and a high resistance of 11 kΩ are used. The figure of merit of the galvanometer 60 μA/division. In the absence of shunt resistance, the galvanometer produces a deflection of $\theta = 9$ divisions when current flows in the circuit. The value of the shunt resistance that can cause the deflection of $\theta/2$, is closest to : **[2018 (Online)]**

(A) 55 Ω (B) 110 Ω

(C) 220 Ω (D) 550 Ω

31. A galvanometer with its coil resistance 25 Ω requires a current of 1 mA for its full deflection. In order to construct an ammeter to read up to a current of 2 A, the approximate value of the shunt resistance should be : **[2018 (Online)]**

(A) $2.5 \times 10^{-2}\,\Omega$ (B) $1.25 \times 10^{-3}\,\Omega$

(C) $2.5 \times 10^{-3}\,\Omega$ (D) $1.25 \times 10^{-2}\,\Omega$

32. In the given circuit the internal resistance of the 18 V cell is negligible. If $R_1 = 400\ \Omega$, $R_3 = 100\ \Omega$ and $R_4 = 500\ \Omega$ and the reading of an ideal voltmeter across R_4 is 5 V, then the value R_2 will be : **[9 Jan. 2019 (Evening)]**

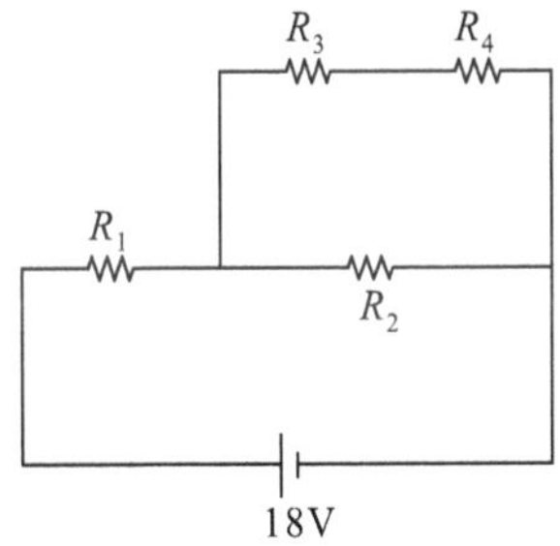

(A) 300 Ω (B) 230 Ω

(C) 450 Ω (D) 550 Ω

33. A potentiometer wire AB having length L and resistance $12r$ is joined to a cell D of emf ε and internal resistance r. A cell C having emf ε/2 and internal resistance $3r$ is connected. The length AJ at which the galvanometer as shown in figure shows no deflection is : **[10 Jan. 2019 (Morning)]**

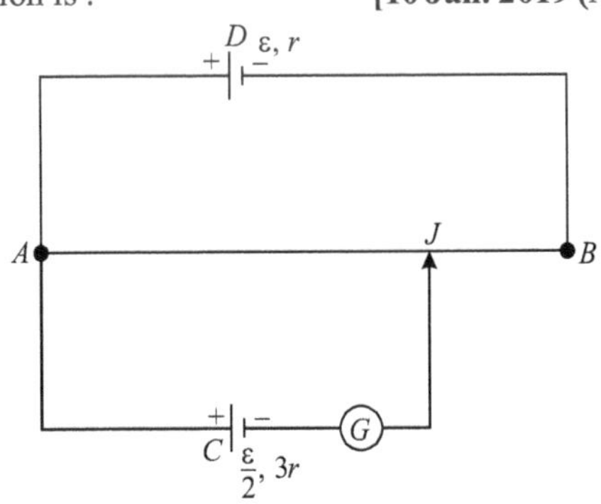

(A) $\frac{5}{12}L$ (B) $\frac{11}{24}L$

(C) $\frac{11}{12}L$ (D) $\frac{13}{24}L$

34. The actual value of resistance R, shown in the figure is 30 Ω. This is measured in an experiment as shown using the standard formula $R = \frac{V}{I}$, where V and I are the readings of the voltmeter and ammeter, respectively. If the measured value of R is 5% less, then the internal resistance of the voltmeter is : **[10 Jan. 2019 (Evening)]**

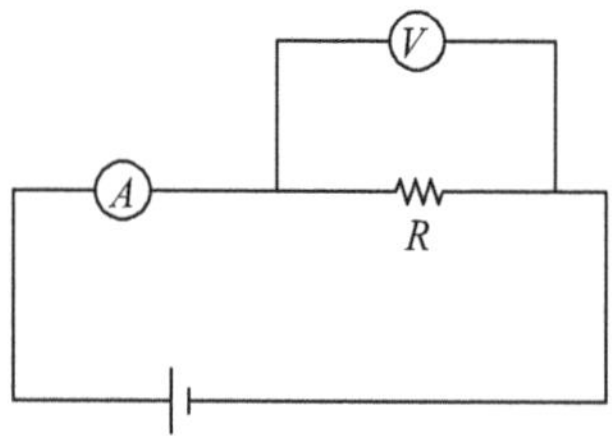

(A) 350 Ω (B) 570 Ω
(C) 35 Ω (D) 600 Ω

35. The Wheatstone bridge shown in Figure here, gets balanced when the carbon resistor used as R_1 has the colour code (Orange, Red, Brown). The resistors R_2 and R_4 are 80 Ω and 40 Ω, respectively.
Assuming that the colour code for the carbon resistors gives their accurate values, the colour code for the carbon resistor, used as R_3, would be : **[10 Jan. 2019 (Evening)]**

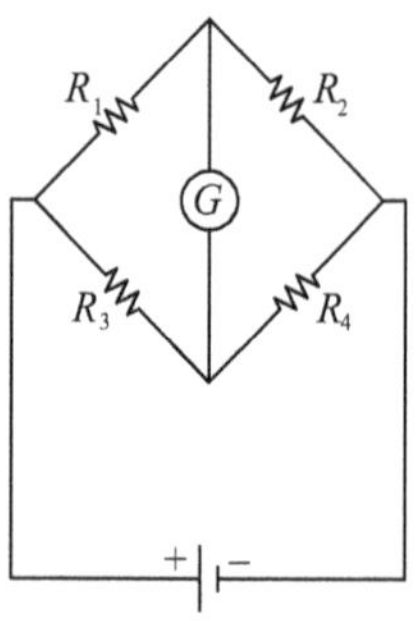

(A) Red, Green, Brown (B) Brown, Blue, Brown
(C) Grey, Black, Brown (D) Brown, Blue, Black

36. The resistance of the meter bridge AB in given figure is 4 Ω. With a cell of emf $\varepsilon = 0.5$ V and rheostat resistance $R_h = 2\,\Omega$ the null point is obtained at some point J. When the cell is replaced by another one of emf $\varepsilon = \varepsilon_2$ the same null point J is found for $R_h = 6\,\Omega$. The emf ε_2 is : **[11 Jan. 2019 (Morning)]**

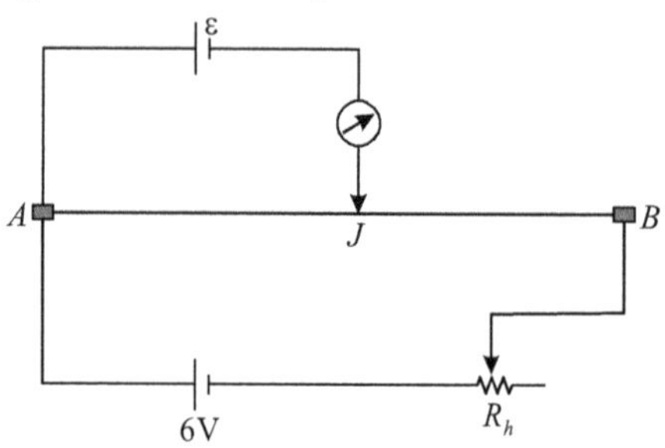

(A) 0.6 V (B) 0.5 V
(C) 0.3 V (D) 0.4 V

37. In a Wheatstone bridge (see figure). Resistances P and Q are approximately equal. When $R = 400\,\Omega$, the bridge is balanced. On inter-changing P and Q, the value of R, for balance, is 405 Ω. The value of X is close to : **[11 Jan. 2019 (Morning)]**

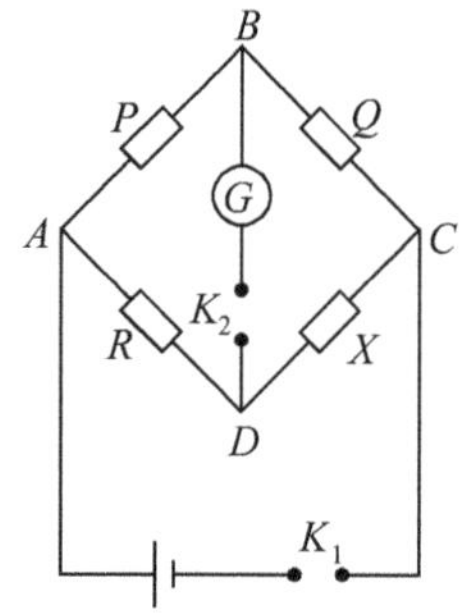

(A) 403.5 ohm (B) 404.5 ohm
(C) 401.5 ohm (D) 402.5 ohm

38. A galvanometer having a resistance of 20 Ω and 30 divisions on both sides has figure of merit 0.005 ampere/division. The resistance that should be connected in series such that it can be used as a voltmeter upto 15 volt, is : **[11 Jan. 2019 (Evening)]**

(A) 80 Ω (B) 120 Ω
(C) 125 Ω (D) 100 Ω

39. In the experimental set up of metre bridge shown in the figure, the null point is obtained at a distance of 40 cm from A. If a 10 Ω resistor is connected in series with R_1, the null point shifts by 10 cm. The resistance that should be connected in parallel with $(R_1 + 10)\,\Omega$ such that the null point shifts back to its initial position is : **[11 Jan. 2019 (Evening)]**

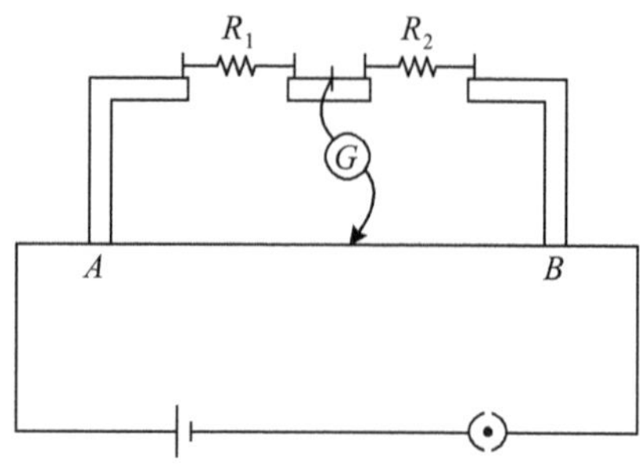

(A) 40 Ω (B) 60 Ω
(C) 20 Ω (D) 30 Ω

40. The galvanometer deflection, when key K_1 is closed but K_2 is open, equals θ_0 (see figure). On closing K_2 also and adjusting R_2 to 5 Ω, the deflection in galvanometer becomes $\frac{\theta_0}{5}$. The resistance of the galvanometer is, then, given by [Neglect the internal resistance of battery] : **[12 Jan. 2019 (Morning)]**

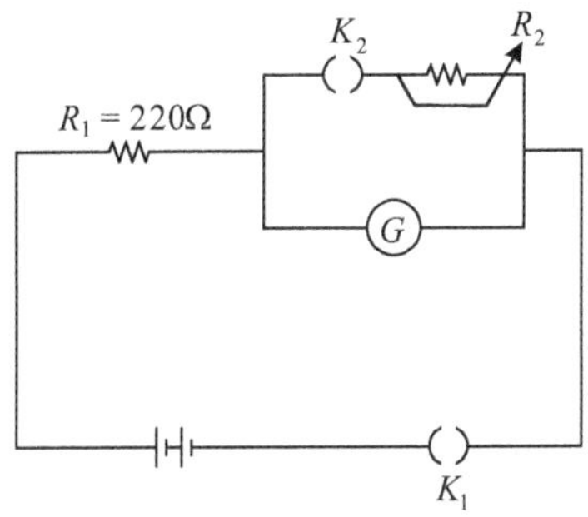

(A) $12\,\Omega$ (B) $25\,\Omega$
(C) $5\,\Omega$ (D) $22\,\Omega$

41. In a meter bridge, the wire of length 1 m has a non-uniform cross-section such that, the variation $\frac{dR}{dl}$ of its resistance R with length l is $\frac{dR}{dl} \propto \frac{1}{\sqrt{l}}$. Two equal resistances are connected as shown in the figure. The galvanometer has zero deflection when the jockey is at point P. What is the length AP?

[12 Jan. 2019 (Morning)]

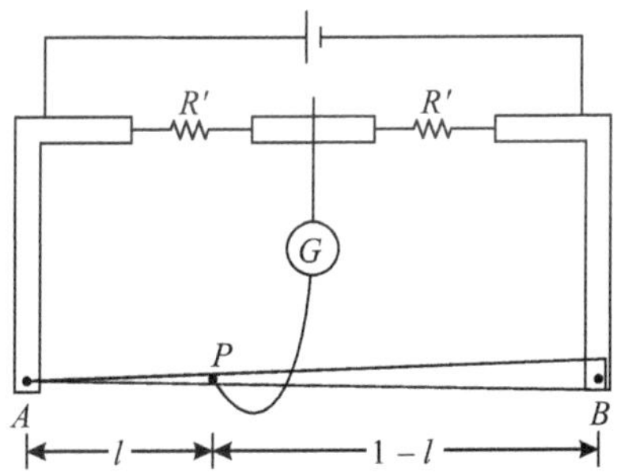

(A) 0.25 m (B) 0.3 m
(C) 0.35 m (D) 0.2 m

42. An ideal battery of 4 V and resistance R are connected in series in the primary circuit of a potentiometer of length 1 m and resistance $5\,\Omega$. The value of R, to give a potential difference of 5 mV across 10 cm of potentiometer wire, is :

[12 Jan. 2019 (Morning)]

(A) $490\,\Omega$ (B) $480\,\Omega$
(C) $395\,\Omega$ (D) $495\,\Omega$

43. A galvanometer, whose resistance is 50 ohm, has 25 divisions in it. When a current of 4×10^{-4} A passes through it, its needle (pointer) deflects by one division. To use this galvanometer as a voltmeter of range 2.5 V, it should be connected to a resistance of : **[12 Jan. 2019 (Evening)]**

(A) 6250 ohm (B) 250 ohm
(C) 200 ohm (D) 6200 ohm

44. In the circuit shown, a four-wire potentiometer is made of a 400 cm long wire, which extends between A and B. The resistance per unit length of the potentiometer wire is $r = 0.01\ \Omega/\text{cm}$. If an ideal voltmeter is connected as shown with jockey J at 50 cm from end A, the expected reading of the voltmeter will be :

[8 April 2019 (Evening)]

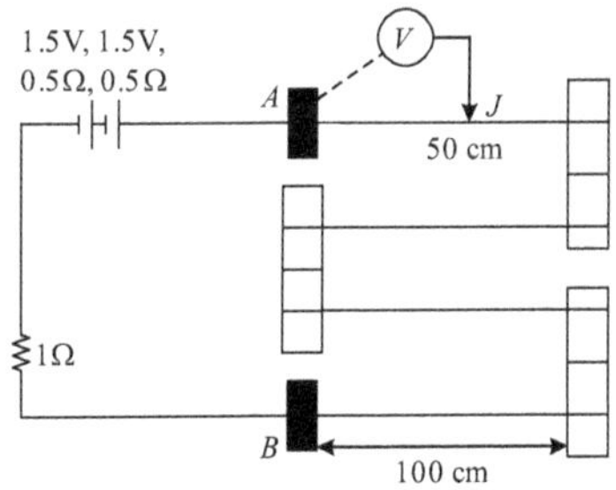

(A) 0.20 V (B) 0.25 V
(C) 0.75 V (D) 0.50 V

45. The resistance of a galvanometer is 50 ohm and the maximum current which can be passed through it is 0.002 A. What resistance must be connected to it in order to convert it into an ammeter of range 0 – 0.5 A? **[9 April 2019 (Evening)]**

(A) 0.2 ohm (B) 0.002 ohm
(C) 0.02 ohm (D) 0.5 ohm

46. In a meter bridge experiment, the circuit diagram and the corresponding observation table are shown in figure

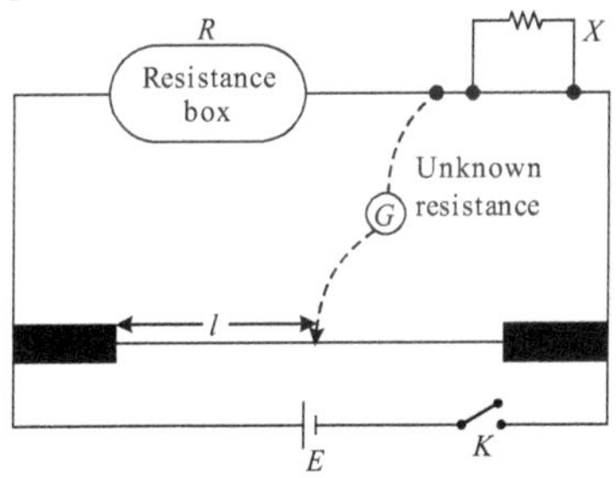

Sl.No.	$R(\Omega)l$	l(cm)
1.	1000	60
2.	100	13
3.	10	1.5
4.	1	1.0

Which of the readings is inconsistent ?

[10 April 2019 (Morning)]

(A) 4 (B) 1
(C) 2 (D) 3

47. In the given circuit, an ideal voltmeter connected across the $10\,\Omega$ resistance reads 2 V. The internal resistance r, of each cell is : **[10 April 2019 (Morning)]**

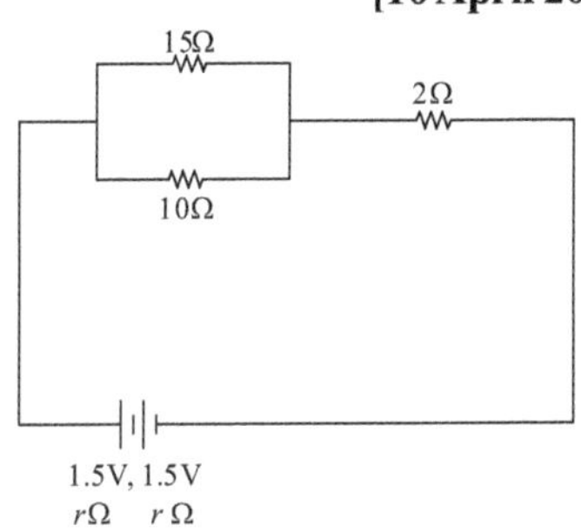

(A) 1 Ω (B) 1.5 Ω
(C) 0 Ω (D) 0.5 Ω

48. A galvanometer of resistance 100 Ω has 50 divisions on its scale and has sensitivity of 20 μA/division. It is to be converted to a voltmeter with three ranges, of 0 – 2V, 0 – 10V and 0 – 20V. The appropriate circuit to do so is : **[12 April 2019 (Morning)]**

(A) 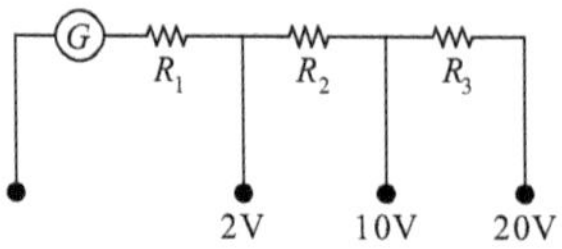

$R_1 = 1900\,\Omega$
$R_2 = 9900\,\Omega$
$R_3 = 19900\,\Omega$

(B) 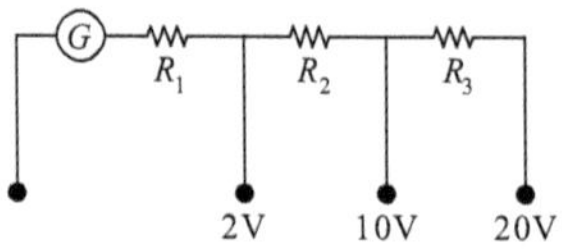

$R_1 = 2000\,\Omega$
$R_2 = 8000\,\Omega$
$R_3 = 10000\,\Omega$

(C) 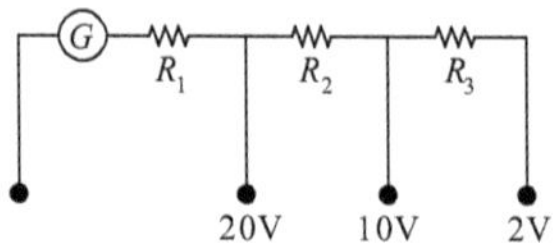

$R_1 = 19900\,\Omega$
$R_2 = 9900\,\Omega$
$R_3 = 1900\,\Omega$

(D) 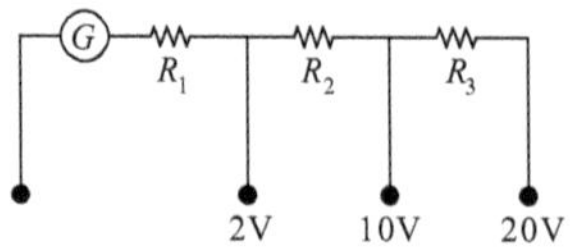

$R_1 = 1900\,\Omega$
$R_2 = 8000\,\Omega$
$R_3 = 10000\,\Omega$

49. To verify Ohm's law, a student connects the voltmeter across the battery as, shown in the figure. The measured voltage is plotted as a function of the current, and the following graph is obtained. If V_0 is almost zero, identify the correct statement : **[12 April 2019 (Morning)]**

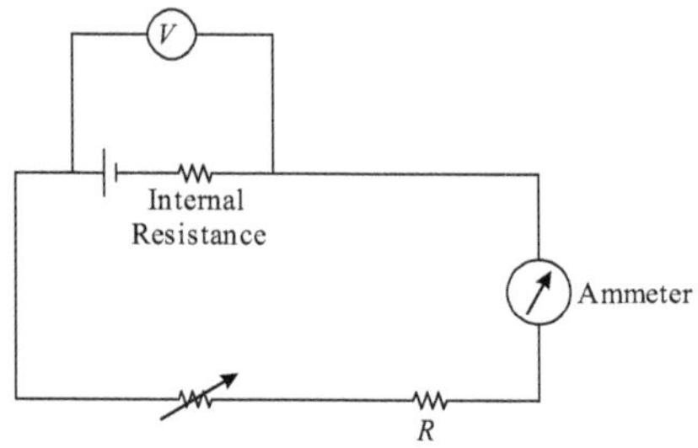

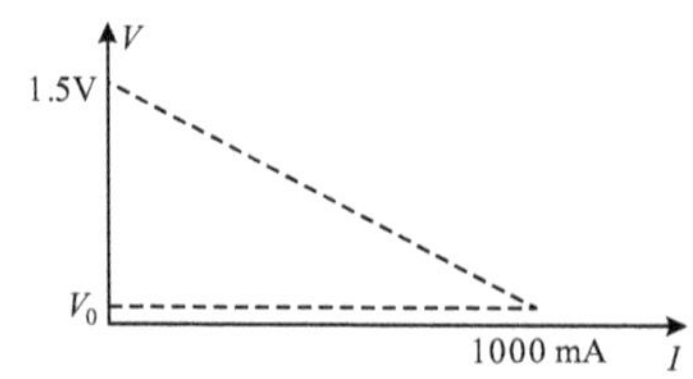

(A) The value of the resistance R is 1.5 Ω
(B) The emf of the battery is 1.5 V and the value of R is 1.5 Ω
(C) The emf of the battery is 1.5 V and its internal resistance is 1.5 Ω
(D) The potential difference across the battery is 1.5 V when it sends a current of 1000 mA

* * * * *

CHAPTER-18 MOVING CHARGES & MAGNETISM

18.1 Magnetic Effects of Current

1. If in a circular coil A of radius R, current i is flowing and in another coil B of radius $2R$ a current $2i$ is flowing, then the ratio of the magnetic fields, B_A and B_B produced by them will be : **[2002]**

(A) 1 (B) 2
(C) 1/2 (D) 4

2. The magnetic field due to a current carrying circular loop of radius 3 cm at a point on the axis at a distance of 4 cm from the centre is 54 μT. What will be its value at the centre of the loop ? **[2004]**

(A) 250 μT (B) 150 μT
(C) 125 μT (D) 75 μT

3. A long wire carries a steady current. It is bent into a circle of one turn and the magnetic field at the centre of the coil is B. It is then bent into a circular loop of n turns. The magnetic field at the centre of the coil will be : **[2004]**

(A) nB (B) n^2B
(C) $2nB$ (D) $2n^2B$

4. A current i ampere flows along an infinitely long straight thin walled tube, then the magnetic induction at any point inside the tube is : **[2004]**

(A) $\frac{\mu_0}{4\pi}\cdot\frac{2i}{r}$ tesla (B) Zero
(C) Infinite (D) $\frac{2i}{r}$ tesla

5. Two concentric coils each of radius equal to 2π cm are placed at right angles to each other. 3 A and 4 A are the currents flowing in each coil respectively. The magnetic induction in Wb/m^2 at the centre of the coils will be : $[\mu_0 = 4\pi \times 10^{-7}$ Wb/Am$]$ **[2005]**

(A) 12×10^{-5} (B) 10^{-5}
(C) 5×10^{-5} (D) 7×10^{-5}

6. A long solenoid has 200 turns per cm and carries a current i. The magnetic field at its centre is 6.28×10^{-2} weber/m^2. Another long solenoid has 100 turns per cm and it carries a current $i/3$. The value of the magnetic field at its centre is : **[2006]**

(A) 1.05×10^{-2} weber/m^2 (B) 1.05×10^{-5} weber/m^2
(C) 1.05×10^{-3} weber/m^2 (D) 1.05×10^{-4} weber/m^2

7. Two identical conducting wires AOB and COD are placed at right angles to each other. The wire AOB carries an electric current I_1 and COD carries a current I_2. The magnetic field on a point lying at a distance d from O, in a direction perpendicular to the plane of the wires AOB and COD, will be given by : **[2007]**

(A) $\frac{\mu_0}{2\pi}\left(\frac{I_1+I_2}{d}\right)^{1/2}$ (B) $\frac{\mu_0}{2\pi d}(I_1^2+I_2^2)^{1/2}$
(C) $\frac{\mu_0}{2\pi d}(I_1+I_2)$ (D) $\frac{\mu_0}{2\pi d}(I_1^2+I_2^2)$

8. A long straight wire of radius a carries a steady current i. The current is uniformly distributed across its cross section. The ratio of the magnetic field at $a/2$ and $2a$ is : **[2007]**

(A) 1/2 (B) 1/4
(C) 4 (D) 1

9. A current I flows along the length of an infinitely long, straight, thin walled pipe. Then : **[2007]**

(A) The magnetic field at all points inside the pipe is the same, but not zero
(B) The magnetic field is zero only on the axis of the pipe
(C) The magnetic field is different at different points inside the pipe
(D) The magnetic field at any point inside the pipe is zero

10. A horizontal overhead powerline is at a height of 4 m from the ground and carries a current of 100 A from east to west. The magnetic field directly below it on the ground is : $[\mu_0 = 4\pi \times 10^{-7}$ TmA$^{-1}]$: **[2008]**

(A) 2.5×10^{-7} T southward (B) 5×10^{-6} T northward
(C) 5×10^{-6} T southward (D) 2.5×10^{-7} T northward

11. The magnitude of the magnetic field (B) due to the loop $ABCD$ at the origin (O) is : **[2009]**

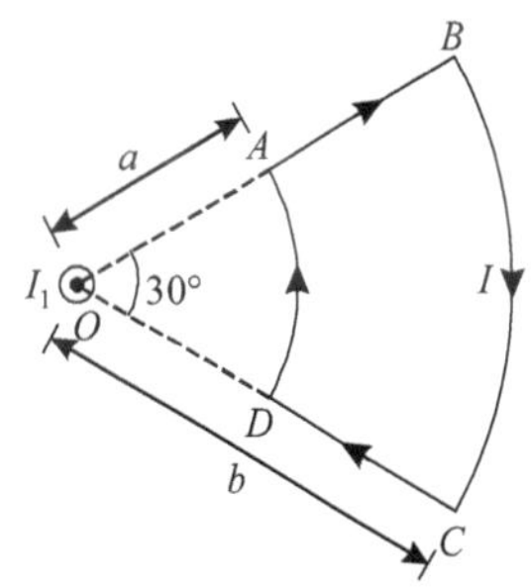

(A) $\frac{\mu_0 I}{4\pi}\left[\frac{b-a}{ab}\right]$ (B) $\frac{\mu_0 I}{4\pi}\left[2(b-a)+\frac{\pi}{3}(a+b)\right]$
(C) Zero (D) $\frac{\mu_0 I(b-a)}{24ab}$

12. Due to the presence of the current I_1 at the origin : **[2009]**

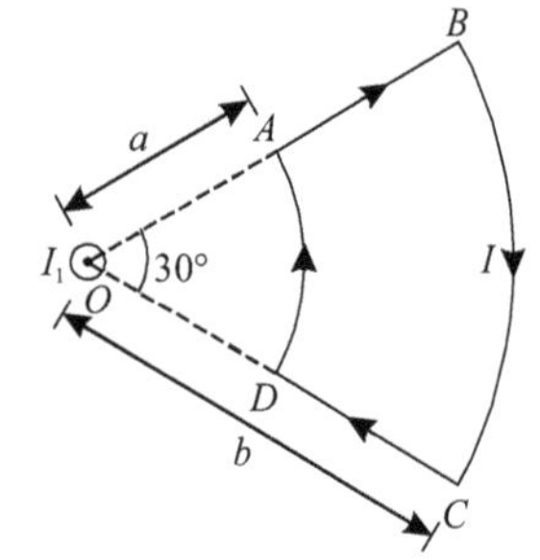

(A) The magnitude of the net force on the loop is given by

$$\frac{I_1 I}{4\pi}\mu_0\left[2(b-a)+\frac{\pi}{3}(a+b)\right]$$

(B) The magnitude of the net force on the loop is given by

$$\frac{\mu_0 I\, I_1}{24ab}(b-a)$$

(C) The forces on AB and DC are zero

(D) The forces on AD and BC are zero

13. Two long parallel wire are at a distance 2d apart. They carry steady equal currents flowing out of the plane of the paper as shown. The variation of the magnetic field B along the line XX' is given by : **[2010]**

(A) 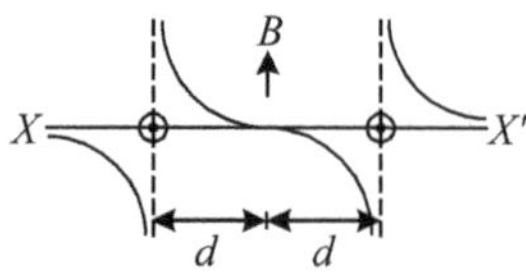

(B) 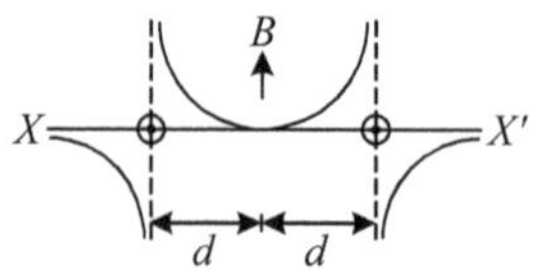

(C) 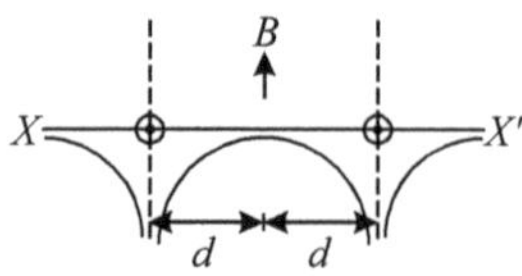

(D) 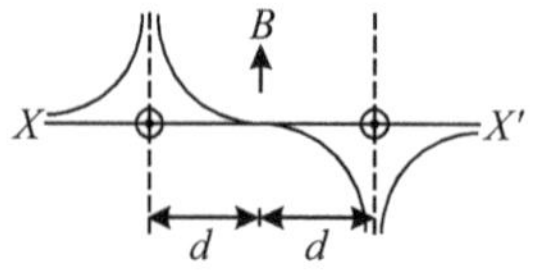

14. A current I flows in an infinity long wire with cross section in the from of a semicircular ring of radius R. The magnitude of the magnetic induction along its axis is : **[2011]**

(A) $\frac{\mu_0 I}{2\pi R}$ (B) $\frac{\mu_0 I}{4\pi R}$

(C) $\frac{\mu_0 I}{\pi^2 R}$ (D) $\frac{\mu_0 I}{2\pi^2 R}$

15. A charge Q is uniformly distributed over the surface of non-conducting disc of radius R. The disc rotates about an axis perpendicular to its plane and passing through its centre with an angular velocity ω. As a result of this rotation a magnetic field of induction B is obtained at the centre of the disc. If we keep both the amount of charge placed on the disc and its angular velocity to be constant and vary the radius of the disc then the variation of the magnetic induction at the centre of the disc will be represented by the figure : **[2012]**

(A) 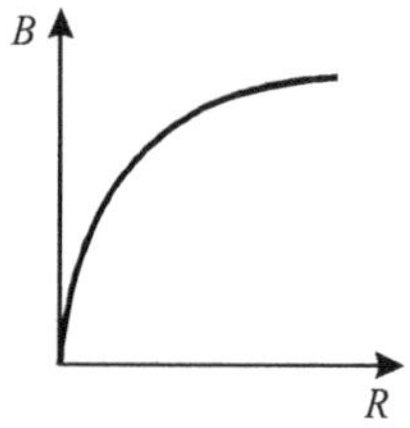

(B) 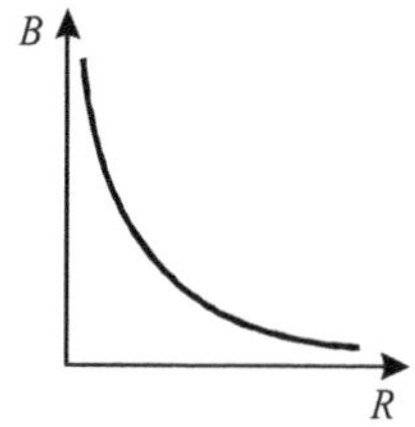

(C) 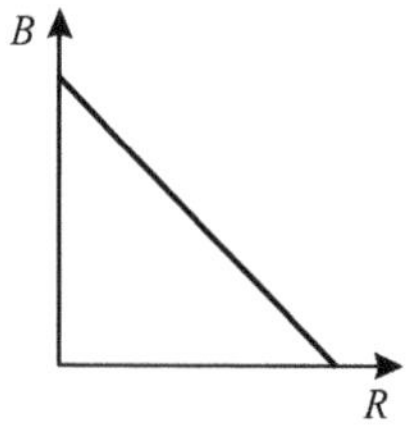

(D) 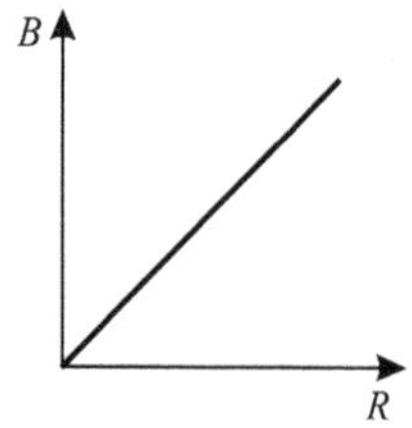

16. An electric current is flowing through a circular coil of radius R. The ratio of the magnetic field at the centre of the coil and that at a distance $2\sqrt{2}R$ from the centre of the coil and on its axis is : **[2013 (Online)]**

(A) $2\sqrt{2}$ (B) 27

(C) 36 (D) 8

17. A current i is flowing in a straight conductor of length L. The magnetic induction at a point on its axis at a distance $L/4$ from its centre will be : **[2013 (Online)]**

(A) Zero (B) $\frac{\mu_0 i}{2\pi L}$

(C) $\frac{\mu_0 i}{\sqrt{2}L}$ (D) $\frac{4\mu_0 i}{\sqrt{5}\mu L}$

18. Choose the correct sketch of the magnetic field lines of a circular current loop shown by the dot ⊙ and the cross ⊗: **[2013 (Online)]**

(A) 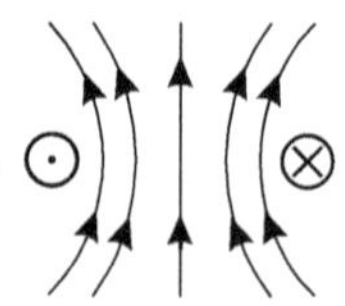

(B) 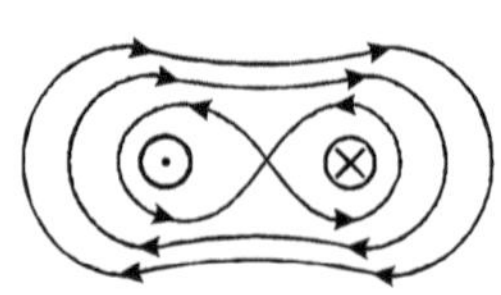

(C) 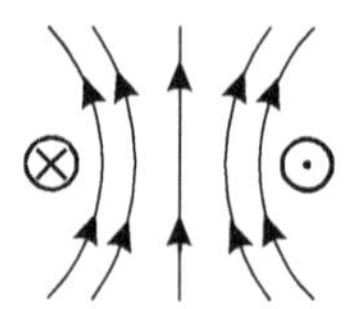

(D) 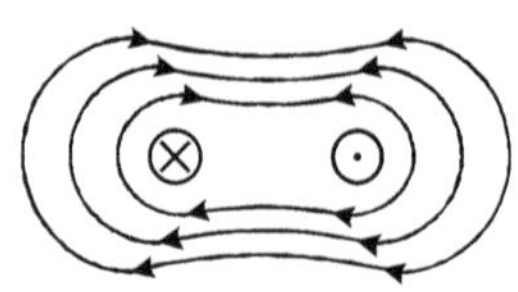

19. Consider two thin identical conducting wires covered with very thin insulating material. One of the wires is bent into a loop and produces magnetic field B_1, at its centre when a current I passes through it. The second wire is bent into a coil with three identical loops adjacent to each other and produces magnetic field B_2 at the centre of the loops when a current $I/3$ passes through it. Then ratio $B_1 : B_2$ is : **[2014 (Online)]**

(A) 1 : 1 (B) 1 : 3

(C) 1 : 9 (D) 9 : 1

20. Two identical wires A and B, each of length 'l', carry the same current I. Wire A is bent into a circle of radius R and wire B is bent to form a square of side 'a'. If B_A and B_B are the values of magnetic field at the centres of the circle and square respectively, then the ratio $\frac{B_A}{B_B}$ is : **[2016]**

(A) $\frac{\pi^2}{8}$ (B) $\frac{\pi^2}{16\sqrt{2}}$

(C) $\frac{\pi^2}{16}$ (D) $\frac{\pi^2}{8\sqrt{2}}$

21. A current of 1 A is flowing on the sides of an equilateral triangles of side 4.5×10^{-2} m. The magnetic field at the centre of the triangle will be : **[2018 (Online)]**

(A) 4×10^{-5} Wb/m^2 (B) Zero

(C) 2×10^{-5} Wb/m^2 (D) 8×10^{-5} Wb/m^2

22. A Helmholtz coil has pair of loops, each with N turns and radius R. They are placed coaxially at distance R and the same current I flows through the loops in the same direction. The magnitude of magnetic field at P, midway between the centres A and C, is given by (Refer to figure) : **[2018 (Online)]**

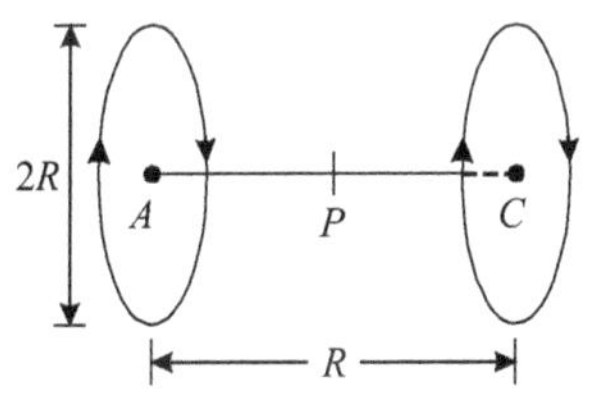

(A) $\frac{4N\mu_0 I}{5^{3/2} R}$ (B) $\frac{8N\mu_0 I}{5^{3/2} R}$

(C) $\frac{4N\mu_0 I}{5^{1/2} R}$ (D) $\frac{8N\mu_0 I}{5^{1/2} R}$

23. A current loop, having two circular arcs joined by two radial lines is shown in the figure. It carries a current of 10 A. The magnetic field at point O will be close to : **[9 Jan. 2019 (Morning)]**

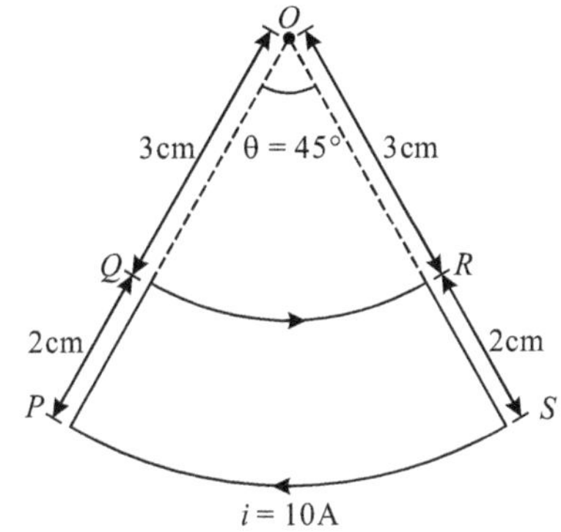

(A) 1.0×10^{-5} T (B) 1.5×10^{-5} T

(C) 1.0×10^{-7} T (D) 1.0×10^{-7} T

24. One of the two identical conducting wires of length L is bent in the form of a circular loop and the other one into a circular coil of N identical turns. If the same current is passed in both, the ratio of the magnetic field at the center of the loop (B_L) to that at the centre of the coil (B_C), then $\frac{B_L}{B_C}$ will be : **[9 Jan. 2019 (Evening)]**

(A) $\frac{1}{N}$ (B) N^2

(C) $\frac{1}{N^2}$ (D) N

25. As shown in the figure, two infinitely long, identical wires are bent by 90° and placed in such a way that the segments LP and QM are along the x-axis, while segments PS and QN are parallel to the y-axis. If $OP = OQ = 4$ cm, and the magnitude of the magnetic field at O is 10^{-4} T, and the two wires carry equal currents (see figure), the magnitude of the current in each wire and the direction of the magnetic field at O will be ($\mu_0 = 4\pi \times 10^{-7}$ NA^{-2}) : **[12 Jan. 2019 (Morning)]**

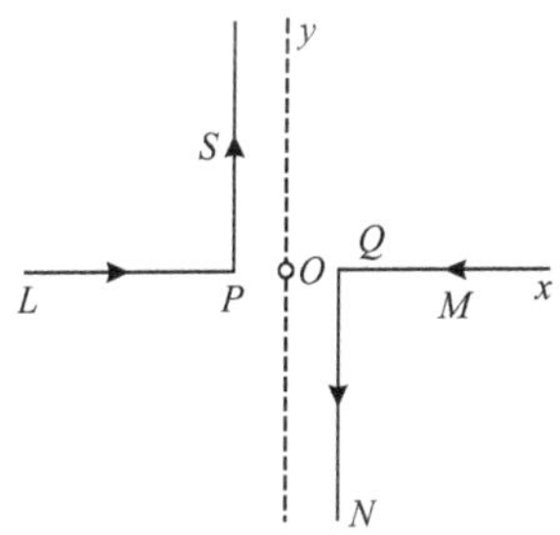

(A) 40 A, perpendicular into the page

(B) 40 A, perpendicular out of the page

(C) 20 A, perpendicular out of the page

(D) 20 A, perpendicular into the page

26. In SI units, the dimesions of $\sqrt{\frac{\epsilon_0}{\mu_0}}$ is :

[8 April 2019 (Morning)]

(A) $A^{-1} TML^3$ (B) $A^2T^3M^{-1}L^{-2}$

(C) $AT^2M^{-1}L^{-1}$ (D) $AT^{-3}ML^{3/2}$

27. Two very long, straight, and insulated wires are kept at 90° angle from each other in xy-plane as shown in the figure. These wires carry currents of equal magnitude I, whose directions are shown in the figure. The net magnetic field at point P will be : **[8 April 2019 (Evening)]**

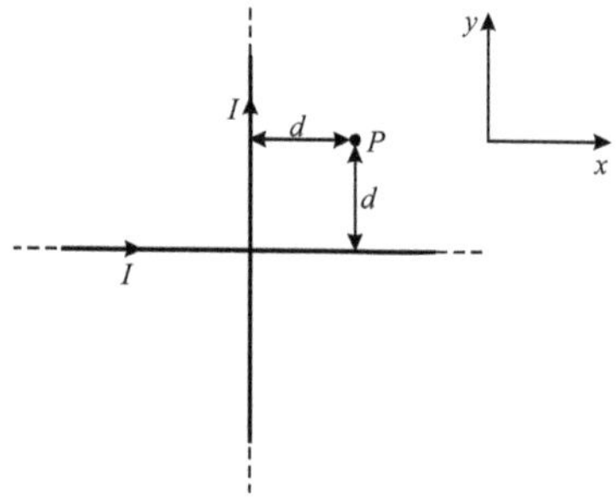

(A) Zero (B) $\frac{+\mu_0 I}{\pi d}(\hat{z})$

(C) $-\frac{\mu_0 I}{2\pi d}(\hat{x}+\hat{y})$ (D) $\frac{\mu_0 I}{2\pi d}(\hat{x}+\hat{y})$

28. The magnitude of the magnetic field at the center of an equilateral triangular loop of side 1 m which is carrying a current of 10 A is : [Take $\mu_0 = 4\pi \times 10^{-7}$ NA^{-2}] **[10 April 2019 (Evening)]**

(A) 18 μT (B) 3 μT
(C) 1 μT (D) 9 μT

29. Which of the following combinations has the dimension of electrical resistance (ε_0 is the permittivity of vacuum and μ_0 is the permeability of vacuum) ? **[12 April 2019 (Morning)]**

(A) $\sqrt{\frac{\varepsilon_0}{\mu_0}}$ (B) $\frac{\mu_0}{\varepsilon_0}$

(C) $\sqrt{\frac{\mu_0}{\varepsilon_0}}$ (D) $\frac{\varepsilon_0}{\mu_0}$

30. A thin ring of 10 cm radius carries a uniformly distributed charge. The ring rotates at a constant angular speed of 40π rad s^{-1} about its axis, perpendicular to its plane. If the magnetic field at its centre is 3.8×10^{-9} T, then the charge carried by the ring is close to ($\mu_0 = 4\pi \times 10^{-7}$ N/A^2) : **[12 April 2019 (Morning)]**

(A) 2×10^{-6} C (B) 3×10^{-5} C
(C) 4×10^{-5} C (D) 7×10^{-6} C

31. Find the magnetic field at point *P* due to a straight line segment *AB* of length 6 cm carrying a current of 5A. (See figure) ($\mu_0 = 4\pi \times 10^{-7}$ N-A^{-2}) : **[12 April 2019 (Evening)]**

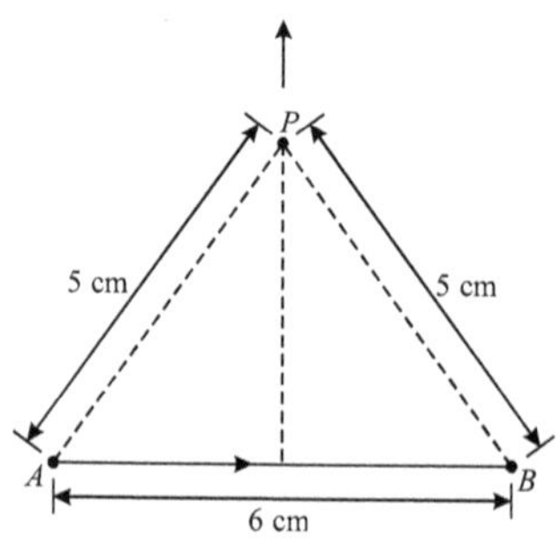

(A) 3.0×10^{-5} T (B) 2.5×10^{-5} T
(C) 2.0×10^{-5} T (D) 1.5×10^{-5} T

* * * * *

CHAPTER-18 MOVING CHARGES & MAGNETISM

18.2 Magnetic Force on Moving Charges

1. Wires 1 and 2 carrying currents i_1 and i_2 respectively are inclined at an angle θ to each other. What is the force on a small element dl of wire 2 at a distance r from wire 1 (as shown in figure) due to the magnetic field of wire 1 ? **[2002]**

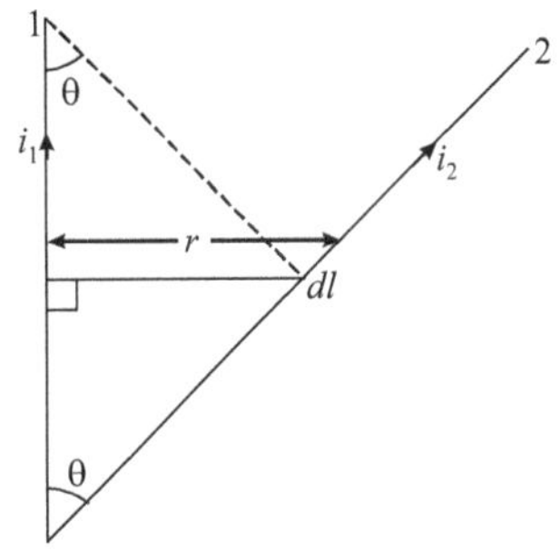

(A) $\frac{\mu_0}{2\pi r} i_1 i_2 \, dl \tan\theta$ (B) $\frac{\mu_0}{2\pi r} i_1 i_2 \, dl \sin\theta$

(C) $\frac{\mu_0}{2\pi r} i_1 i_2 \, dl \cos\theta$ (D) $\frac{\mu_0}{4\pi r} i_1 i_2 \, dl \sin\theta$

2. The time period of a charged particle undergoing a circular motion in a uniform magnetic field is independent of its : **[2002]**

(A) Speed (B) Mass
(C) Charge (D) Magnetic induction

3. If a current is passed through a spring then the spring will : **[2002]**
(A) Expand (B) Compress
(C) Remain same (D) None of these

4. If an electron and a proton having same momentum enter perpendicularly to a magnetic field, then : **[2002]**
(A) Curved path of electron and proton will be same (ignoring the sense of revolution)
(B) They will move undeflected
(C) Curved path of electron is more curved than that of proton
(D) Path of proton is more curved

5. A particle of charge -16×10^{-18} C moving with velocity 10 ms^{-1} along the x-axis enters a region where a magnetic field of induction B is along the y-axis and an electric field of magnitude 10^4 V/m is along the negative z-axis If the charged particle continues moving along the x-axis, the magnitude of B is : **[2003]**
(A) 10^3 Wb/m^2 (B) 10^5 Wb/m^2
(C) 10^{16} Wb/m^2 (D) 10^{-3} Wb/m^2

6. A magnetic needle laying parallel to a magnetic field requires W unit of work to turn it through 60°. The torque needed to maintain the needle in this position will be : **[2003]**

(A) $\sqrt{3}\,W$ (B) W

(C) $(\sqrt{3}/2)\,W$ (D) $2W$

7. A particle of mass M and charge Q moving with velocity $\vec{v}$ describes a circular path of radius R when subjected to a uniform transverse magnetic field of induction B. The work done by the field when the particle completes one full circle is : **[2003]**

(A) $\left(\frac{Mv^2}{R}\right) 2\pi R$ (B) Zero

(C) $BQ2\pi R$ (D) $BQv2\pi R$

8. Two long conductors, separated by a distance d carry currents I_1 and I_2 in the same direction. They exert a force F on each other. Now the current in one of them is increased to two times and its direction is reversed. The distance is also increased to $3d$. The new value of the force between them is : **[2004]**
(A) $-2F$ (B) $F/3$
(C) $-2F/3$ (D) $-F/3$

9. A charged particle of mass m and charge q travels on a circular path of radius r that is perpendicular to a magnetic field B. The time taken by the particle to complete one revolution is : **[2005]**

(A) $\frac{2\pi mq}{B}$ (B) $\frac{2\pi q^2 B}{m}$

(C) $\frac{2\pi qB}{m}$ (D) $\frac{2\pi m}{qB}$

10. Two thin, long, parallel wires, separated by a distance d carry a current of 'i' in the same direction. They will : **[2005]**

(A) attract each other with a force of $\frac{\mu_0 i^2}{(2\pi d)}$

(B) repel each other with a force of $\frac{\mu_0 i^2}{(2\pi d)}$

(C) attract each other with a force of $\frac{\mu_0 i^2}{(2\pi d^2)}$

(D) repel each other with a force of $\frac{\mu_0 i^2}{(2\pi d^2)}$

11. A uniform electric field and a uniform magnetic field are acting along the same direction in a certain region. If an electron is projected along the direction of the fields with a certain velocity then : **[2005]**

(A) Its velocity will increase
(B) Its velocity will decrease
(C) It will turn towards left of direction of motion
(D) It will turn towards right of direction of motion

12. In a region, steady and uniform electric and magnetic fields are present. These two fields are parallel to each other. A charged particle is released from rest in this region. The path of the particle will be a : **[2006]**

(A) helix (B) straight line
(C) ellipse (D) circle

13. A charged particle moves through a magnetic field perpendicular to its direction. Then : **[2007]**

(A) the momentum changes but the kinetic energy is constant
(B) both momentum and kinetic energy of the particle are not constant
(C) both momentum and kinetic energy of the particle are constant
(D) kinetic energy changes but the momentum is constant

14. A charged particle with charge q enters a region of constant, uniform and mutually orthogonal fields $\vec{E}$ and $\vec{B}$ with a velocity $\vec{V}$ perpendicular to both $\vec{E}$ and $\vec{B}$, and comes out without any change in magnitude or direction of $\vec{V}$. Then : **[2007]**

(A) $\vec{V} = \frac{\vec{E}\times\vec{B}}{B^2}$ (B) $\vec{V} = \frac{\vec{B}\times\vec{E}}{B^2}$

(C) $\vec{V} = \frac{\vec{E}\times\vec{B}}{E^2}$ (D) $\vec{V} = \frac{\vec{B}\times\vec{E}}{E^2}$

15. A thin circular disk of radius R is uniformly charged with density $\sigma > 0$ per unit area. The disk rotates about its axis with a uniform angular speed ω. The magnetic moment of the disk is : **[2011]**

(A) $2\pi R^4\sigma\omega$ (B) $\pi R^4\sigma\omega$

(C) $\frac{\pi R^4}{2}\sigma\omega$ (D) $\frac{\pi R^4}{4}\sigma\omega$

16. An electric charge $+q$ moves with velocity $\vec{V} = 3\hat{i} + 4\hat{j} + \hat{k}$, in an electromagnetic field given by : $\vec{E} = 3\hat{i} + \hat{j} + 2\hat{k}$. $\vec{B} = \hat{i} + \hat{j} - 3\hat{k}$. The y-component of the force experienced by $+q$ is : **[2011]**

(A) $2q$ (B) $11q$
(C) $5q$ (D) $3q$

17. Proton, Deuteron and alpha particle of the same kinetic energy are moving in circular trajectories in a constant magnetic field. The radii of proton, deuteron and alpha particle are respectively r_p, r_d and r_α. Which one of the following relations is correct ? **[2012]**

(A) $r_\alpha = r_d > r_p$ (B) $r_\alpha = r_d = r_p$
(C) $r_\alpha = r_p < r_d$ (D) $r_\alpha > r_d > r_p$

18. The circuit in figure consists of wires at the top and bottom and identical springs as the left and right sides. The wire at the bottom has a mass of 10 g and is 5 cm long. The wire is hanging as shown in the figure. The springs stretch 0.5 cm under the weight of the wire and the circuit has a total resistance of 12 Ω. When the lower wire is subjected to a static magnetic field, the springs, stretch an additional 0.3 cm. The magnetic field is : **[2012 (Online)]**

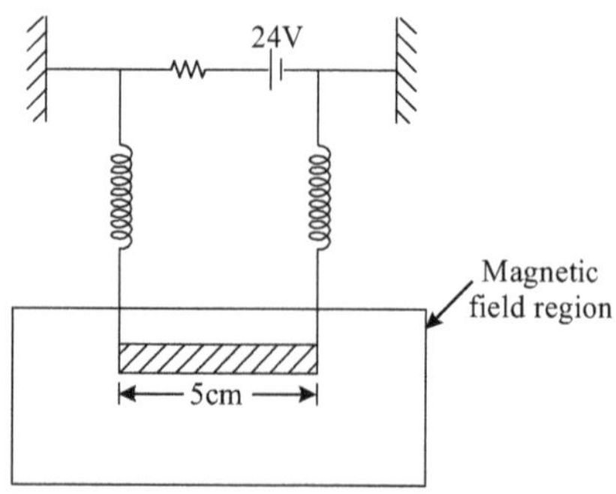

(A) 0.6 T and directed out of page
(B) 1.2 T and directed into the plane of page
(C) 0.6 T and directed into the plane of page
(D) 1.2 T and directed out of page

19. Currents of a 10 ampere and 2 ampere are passed through two parallel thin wires A and B respectively in opposite directions. Wire A is infinitely long and the length of the wire B is 2 m. The force acting on the conductor B, which is situated at 10 cm distance from A will be : **[2012 (Online)]**

(A) 8×10^{-5} N (B) 5×10^{-5} N
(C) $8\pi \times 10^{-7}$ N (D) $4\pi \times 10^{-7}$ N

20. The velocity of certain ions that pass undeflected through crossed electric field $E = 7.7$ k V/m and magnetic field $B = 0.14$ T is : **[2012 (Online)]**

(A) 18 km/s (B) 77 km/s
(C) 55 km/s (D) 1078 km/s.

21. The magnetic force acting on charged particle of charge 2 μC in magnetic field of 2T acting in y-direction, when the particle velocity is $(2\hat{i} + 3\hat{j}) \times 10^6$ ms^{-1} is : **[2012 (Online)]**

(A) 8 N in z-direction (B) 8 N in y-direction
(C) 4 N in y-direction (D) 4 N in z-direction

22. A proton and a deuteron are both accelerated through the same potential difference and enter in a magnetic field perpendicular to the direction of the field. If the deuteron follows a path of radius R, assuming the neutron and proton masses are nearly equal, the radius of the proton's path will be : **[2012 (Online)]**

(A) $\sqrt{2}R$ (B) $\frac{R}{\sqrt{2}}$

(C) $\frac{R}{2}$ (D) R

23. This question has Statement-1 and Statement-2. Of the four choices given after the Statements, choose the one that best describes the two Statements.

Statement-1 : A charged particle is moving at right angle to a static magnetic field. During the motion the kinetic energy of the charge remains unchanged.

Statement-2 : Static magnetic field exert force on a moving charge in the direction perpendicular to the magnetic field : **[2012 (Online)]**

(A) Statement-1 is false, Statement-2 is true

(B) Statement-1 is true, Statement-2 is true, Statement-2 is not the correct explanation of Statement-1

(C) Statement-1 is true, Statement-2 is false

(D) Statement-1 is true, Statement-2 is true, Statement-2 is the correct explanation of Statement-1

24. A rectangular loop of wire, supporting a mass m, hangs with one end in a uniform magnetic field B pointing out of the plane of the paper. A clockwise current is set up such that $i > mg/Ba$, where a is the width of the loop. Then : **[2013 (Online)]**

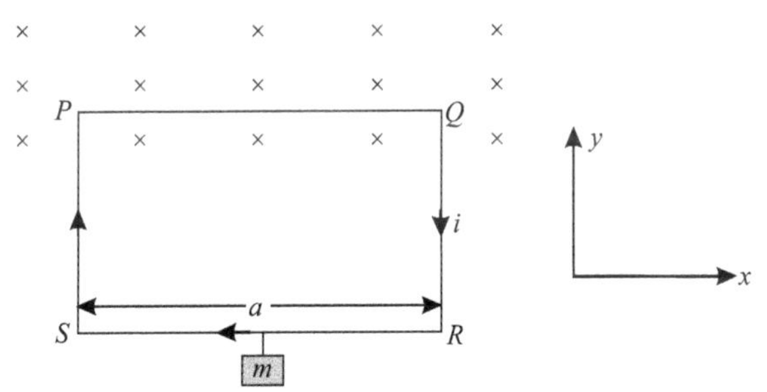

(A) The weight rises due to a vertical force caused by the magnetic field and work is done on the system

(B) The weight do not rise due to vertical for caused by the magnetic field and work is done on the system

(C) The weight rises due to a vertical force caused by the magnetic field but no work is done on the system

(D) The weight rises due to a vertical force caused by the magnetic field and work is extracted from the magnetic field

25. A particle of charge 16×10^{-16} C moving with velocity 10 ms^{-1} along x-axis enters a region where magnetic field of induction $\vec{B}$ is along the y-axis and an electric field of magnitude 10^4 Vm^{-1} is along the negative z-axis. If the charged particle continues moving along x-axis, the magnitude of $\vec{B}$ is : **[2013 (Online)]**

(A) $16 \times 10^3 \text{ Wb m}^{-2}$ (B) $2 \times 10^3 \text{ Wb m}^{-2}$

(C) $1 \times 10^3 \text{ Wb m}^{-2}$ (D) $4 \times 10^3 \text{ Wb m}^{-2}$

26. A metal sample carrying a current along X-axis with density J_X is subjected to magnetic field B_Z (along Z-axis). The electric field E_Y developed along Y-axis is directly proportional to J_X as well as B_Z. The constant of proportionality has SI unit : **[2013 (Online)]**

(A) $\frac{\text{m}^2}{\text{A}}$ (B) $\frac{\text{m}^3}{\text{As}}$

(C) $\frac{\text{m}^2}{\text{As}}$ (D) $\frac{\text{As}}{\text{m}^3}$

27. A conductor lies along the z-axis at $-1.5 \le z < 1.5$ m and carries a fixed current of 10.0 A in $-\hat{a}_Z$ direction (see figure). For a field $\vec{B} = 3.0 \times 10^{-4} e^{-0.2x} \hat{a}_y$ T, find the power required to move the conductor at constant speed to $x = 2.0$ m, $y = 0$ m in 5×10^{-3} s. Assume parallel motion along the x-axis: **[2014]**

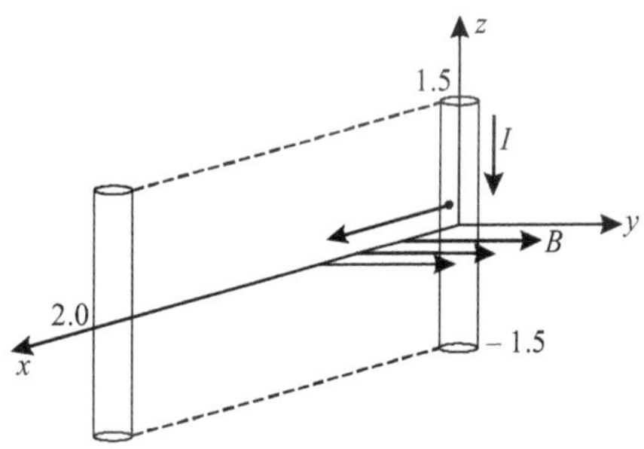

(A) 1.57 W (B) 2.97 W

(C) 14.85 W (D) 29.7 W

28. Three straight parallel current carrying conductors are shown in the figure. The force experienced by the middle conductor of length 25 cm is : **[2014 (Online)]**

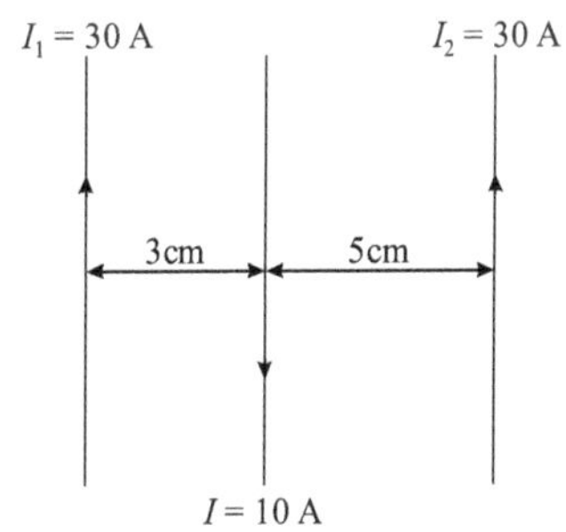

(A) 3×10^{-4} N toward right (B) 6×10^{-4} N toward right

(C) 9×10^{-4} N toward right (D) Zero

29. A positive charge 'q' of mass 'm' is moving along the $+x$ axis.

We wish to apply a uniform magnetic field B for time Δt so that the charge reverses its direction crossing the y-axis at a distance d. Then : **[2014 (Online)]**

(A) $B=\dfrac{mv}{qd}$ and $\Delta t=\dfrac{\pi d}{v}$ (B) $B=\dfrac{mv}{2qd}$ and $\Delta t=\dfrac{\pi d}{2v}$

(C) $B=\dfrac{2mv}{qd}$ and $\Delta t=\dfrac{\pi d}{2v}$ (D) $B=\dfrac{2mv}{qd}$ and $\Delta t=\dfrac{\pi d}{v}$

30. A rectangular loop of sides 10 cm and 5 cm carrying a current I of 12 A is placed in different orientations as shown in the figures below :

(a)

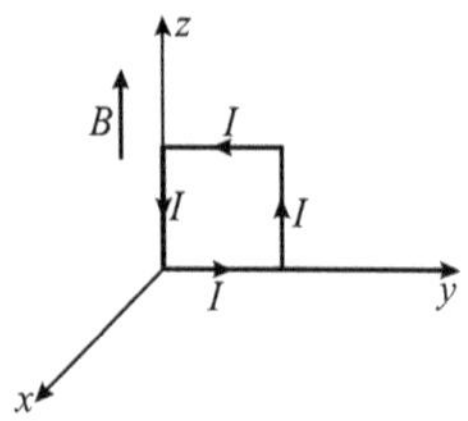

(b)

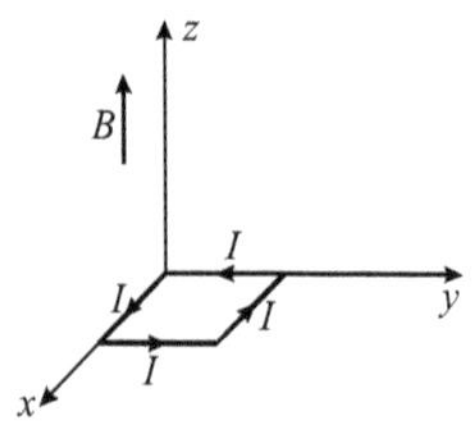

(c)

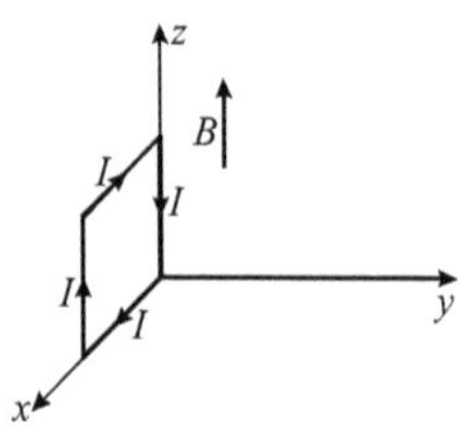

(d)

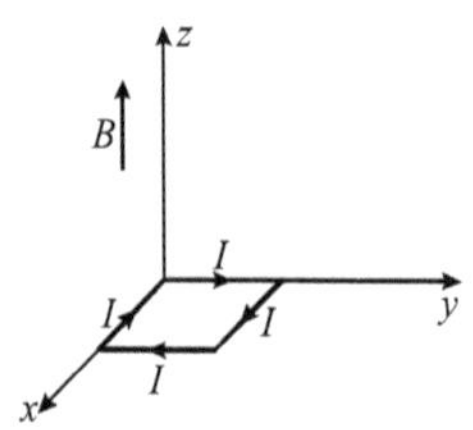

If there is a uniform magnetic field of 0.3 T in the positive z direction, in which orientations the loop would be in (i) stable equilibrium and (ii) unstable equilibrium? **[2015]**

(A) (a) and (b), respectively (B) (a) and (c), respectively

(C) (b) and (d), respectively (D) (b) and (c), respectively

31. Two long current carrying thin wires, both with current I, are held by insulating threads of length L and are in equilibrium as shown in the figure, with threads making an angle 'θ' with the vertical. If wires have mass λ per unit length then the value of I is : (g = gravitational acceleration) : **[2015]**

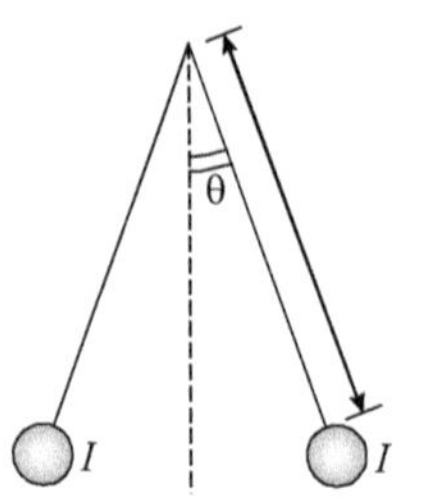

(A) $\sin\theta\sqrt{\dfrac{\pi\lambda gL}{\mu_0\cos\theta}}$ (B) $2\sin\theta\sqrt{\dfrac{\pi\lambda gL}{\mu_0\cos\theta}}$

(C) $2\sqrt{\dfrac{\pi gL}{\mu_0}\tan\theta}$ (D) $\sqrt{\dfrac{\pi\lambda gL}{\mu_0}\tan\theta}$

32. Two coaxial solenoids of different radii carry current I in the same direction. Let $\vec{F}_1$ be the magnetic force on the inner solenoid due to the outer one and $\vec{F}_2$ be the magnetic force on the outer solenoid due to the inner one. Then : **[2015]**

(A) $\vec{F}_1=\vec{F}_2=0$

(B) $\vec{F}_1$ is radially inwards and $\vec{F}_2$ is radially outwards

(C) $\vec{F}_1$ is radially inwards and $\vec{F}_2=0$

(D) $\vec{F}_1$ is radially outwards and $\vec{F}_2=0$

33. A wire carrying current I is tied between points P and Q and is in the shape of a circular arc of radius R due to a uniform magnetic field B (perpendicular to the plane of the paper, shown by ×××) in the vicinity of the wire. If the wire subtends an angle $2\theta_0$ at the centre of the circle (of which it forms an arc) then the tension in the wire is : **[2015 (Online)]**

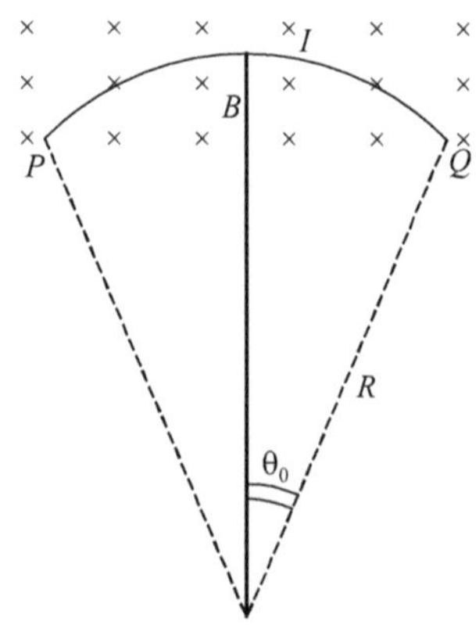

(A) $\dfrac{IBR}{2\sin\theta_0}$ (B) $\dfrac{IBR\theta_0}{\sin\theta_0}$

(C) IBR (D) $\dfrac{IBR}{\sin\theta_0}$

34. Two long straight parallel wires, carrying (adjustable) current I_1 and I_2, are kept at a distance d apart. If the force 'F' between the two wires is taken as 'positive' when the wires repel each other and 'negative' when the wires attract each other, the graph showing the dependence of 'F', on the product $I_1 I_2$, would be : **[2015 (Online)]**

(A)

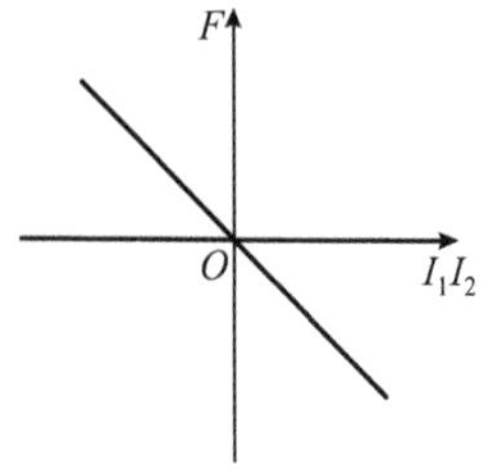

(B)

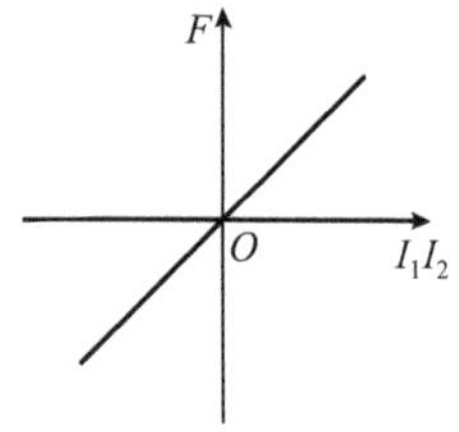

(C)

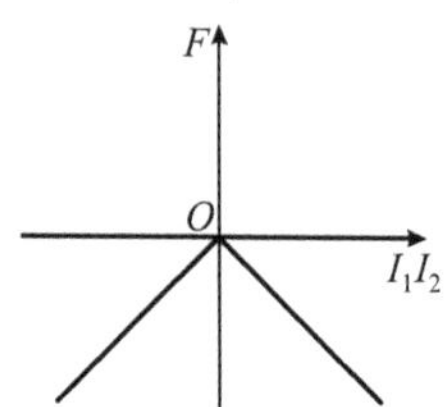

(D)

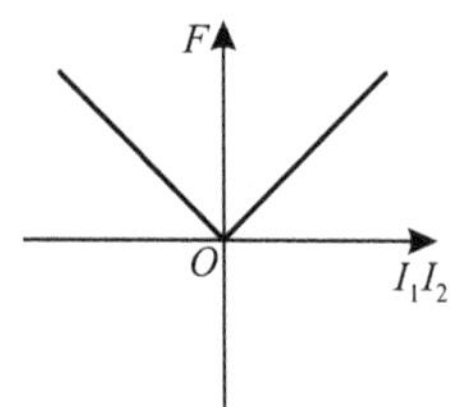

35. A proton (mass m) accelerated by a potential difference V flies through a uniform transverse magnetic field B. The field occupies a region of space by width 'd'. If α be the angle of deviation of proton from initial direction of motion (see figure), the value of $\sin\alpha$ will be : **[2015 (Online)]**

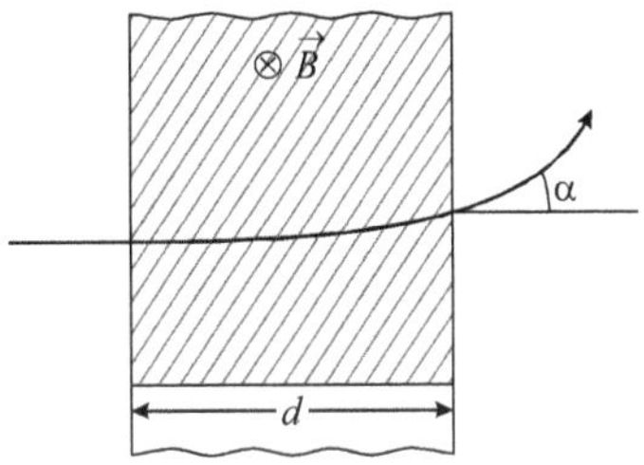

(A) $qV\sqrt{\frac{Bd}{2m}}$ (B) $\frac{B}{2}\sqrt{\frac{qd}{mV}}$

(C) $\frac{B}{d}\sqrt{\frac{q}{2mV}}$ (D) $Bd\sqrt{\frac{q}{2mV}}$

36. A uniform magnetic field B of 0.3 T is along the positive Z-direction. A rectangular loop ($abcd$) of sides 10 cm × 5 cm carries a current I of 12 A. Out of the following different orientations which one corresponds to stable equilibrium? **[2017 (Online)]**

(A) 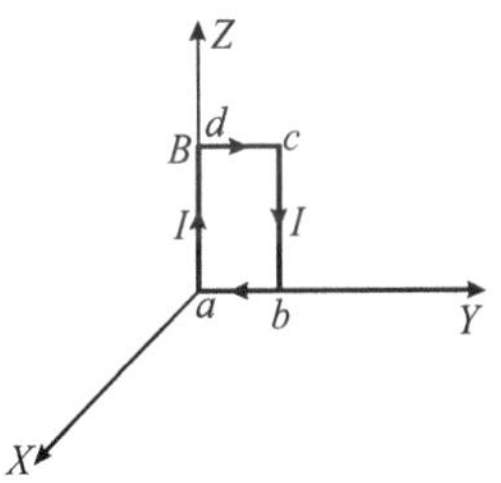

(B) 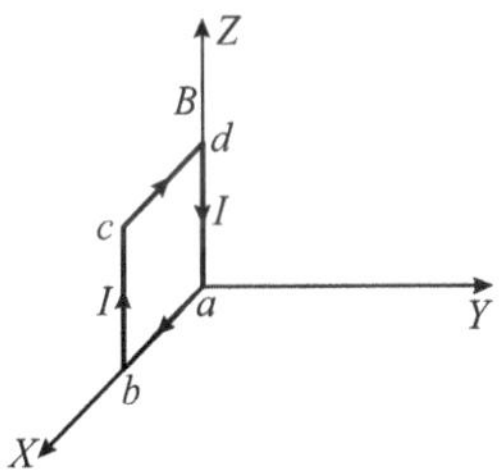

(C) 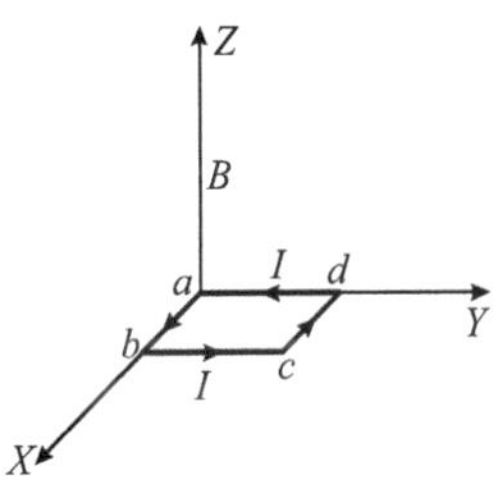

(D) 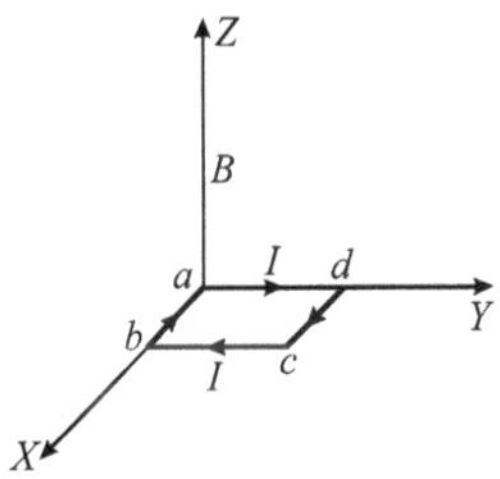

37. In a certain region static electric and magnetic fields exist. The magnetic field is given by $\vec{B} = B_0(\hat{i} + 2\hat{j} - 4\hat{k})$. If a test charge moving with a velocity $\vec{v} = v_0(3\hat{i} - \hat{j} + 2\hat{k})$ experiences no force in that region, then the electric field in the region, in SI units, is : **[2017 (Online)]**

(A) $\vec{E} = -v_0B_0(3\hat{i} - 2\hat{j} - 4\hat{k})$

(B) $\vec{E} = -v_0B_0(\hat{i} + \hat{j} + 7\hat{k})$

(C) $\vec{E} = v_0B_0(14\hat{j} + 7\hat{k})$

(D) $\vec{E} = -v_0B_0(14\hat{j} + 7\hat{k})$

38. A negative test charge is moving near a long straight wire. Carrying a current. The force acting on the test charge is parallel to the direction of the current. The motion of the charge is : **[2017 (Online)]**

(A) Away from the wire

(B) Towards the wire

(C) Parallel to the wire along the current

(D) Parallel to the wire opposite to the current

39. The dipole moment of a circular loop carrying a current I, is m and the magnetic field at the centre of the loop is B_1. When the dipole moment is doubled by keeping the current constant, the magnetic field at the centre of the loop is B_2. The ratio $\frac{B_1}{B_2}$ is : **[2018]**

(A) 2 (B) $\sqrt{3}$

(C) $\sqrt{2}$ (D) $\frac{1}{\sqrt{2}}$

40. An electron, a proton and an alpha particle having the same kinetic energy are moving in circular orbits of radii r_e, r_p, r_α respectively in a uniform magnetic field B. The relation between r_e, r_p, r_α is : **[2018]**

(A) $r_e > r_p = r_\alpha$ (B) $r_e < r_p = r_\alpha$

(C) $r_e < r_p < r_\alpha$ (D) $r_e < r_\alpha < r_p$

41. A charge q is spread uniformly over an insulated loop of radius r. If it is rotated with an angular velocity ω with respect to normal axis then the magnetic moment of the loop is : **[2018 (Online)]**

(A) $\frac{1}{2}q\omega r^2$ (B) $\frac{4}{3}q\omega r^2$

(C) $\frac{3}{2}q\omega r^2$ (D) $q\omega r^2$

42. An infinitely long current carrying wire and a small current carrying loop are in the plane of the paper as shown. The radius of the loop is 'a' and distance of its centre from the wire is d ($d >> a$). If the loop applies a force F on the wire then : **[9 Jan. 2019 (Morning)]**

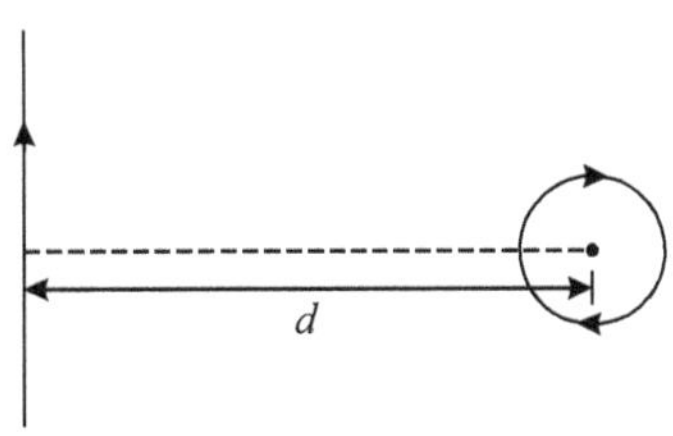

(A) $F \propto \left(\frac{a^2}{d^3}\right)$ (B) $F \propto \left(\frac{a}{d}\right)$

(C) $F \propto \left(\frac{a}{d}\right)^2$ (D) $F = 0$

43. A particle having the same charge as of electron moves in a circular path of radius 0.5 cm under the influence of a magnetic field of 0.5 T. If an electric field of 100 V/m makes it to move in a straight path, then the mass of the particle is (Given charge of electron $=1.6 \times 10^{-19}$ C) : **[9 Jan. 2019 (Evening)]**

(A) 2.0×10^{-24} kg (B) 1.6×10^{-19} kg
(C) 1.6×10^{-27} kg (D) 9.1×10^{-31} kg

44. An insulating thin rod of length l has a linear charge density $\rho(x) = \rho_0 \frac{x}{l}$ on it. The rod is rotated about an axis passing through the origin ($x = 0$) and perpendicular to the rod. If the rod makes n rotation per second, then the time averaged magnetic moment of the rod is : **[10 Jan. 2019 (Morning)]**

(A) $\frac{\pi}{4} n\rho l^3$ (B) $n\rho l^3$
(C) $\pi n\rho l^3$ (D) $\frac{\pi}{3} n\rho l^3$

45. In an experiment electrons are accelerated, from rest, by applying a voltage of 500 V. Calculate the radius of the path if a magnetic field 100 mT is then applied.
[Charge of the electron $= 1.6 \times 10^{-19}$ C
Mass of the electron $= 9.1 \times 10^{-31}$ kg] : **[11 Jan. 2019 (Morning)]**

(A) 7.5×10^{-4} m (B) 7.5×10^{-3} m
(C) 7.5 m (D) 7.5×10^{-2} m

46. The region between $y = 0$ and $y = d$ contains a magnetic field $\vec{B} = B\hat{z}$. A particle of mass m and charge q enters the region with a velocity $\vec{v} = v\hat{i}$. If $d = \frac{mv}{2qB}$, the acceleration of the charged particle at the point of its emergence at the other side is : **[11 Jan. 2019 (Evening)]**

(A) $\frac{qvB}{m}\left(\frac{\hat{i}+\hat{j}}{\sqrt{2}}\right)$ (B) $\frac{qvB}{m}\left(\frac{1}{2}\hat{i}-\frac{\sqrt{3}}{\sqrt{2}}\hat{j}\right)$

(C) $\frac{qvB}{m}\left(\frac{-\hat{j}+\hat{i}}{\sqrt{2}}\right)$ (D) $\frac{qvB}{m}\left(\frac{\sqrt{3}}{2}\hat{i}+\frac{1}{2}\hat{j}\right)$

47. A particle of mass m and charge q is in an electric and magnetic field given by

$$\vec{E} = 2\hat{i} + 3\hat{j} \; ; \; \vec{B} = 4\hat{j} + 6\hat{k} .$$

The charged particle is shifted from the origin to the point $P(x = 1; y = 1)$ along a straight path. The magnitude of the total work done is : **[11 Jan. 2019 (Evening)]**

(A) $(0.35)q$ (B) $(0.15)q$
(C) $(2.5)q$ (D) $5q$

48. A proton and an α-particle (with their masses in the ratio of 1 : 4 and charges in the ratio of 1 : 2) are accelerated from rest through a potential difference V. If a uniform magnetic field (B) is set up perpendicular to their velocities, the ratio of the radii $r_p : r_\alpha$ of the circular paths described by them will be : **[12 Jan. 2019 (Morning)]**

(A) $1:\sqrt{2}$ (B) $1:2$
(C) $1:3$ (D) $1:\sqrt{3}$

49. A circular coil having N turns and radius r carries a current I. It is held in the XZ plane in a magnetic field $B\hat{i}$. The torque on the coil due to the magnetic field is : **[8 April 2019 (Morning)]**

(A) $B\pi r^2 IN$ (B) $\frac{Br^2 I}{\pi N}$

(C) Zero (D) $\frac{B\pi r^2 I}{N}$

50. A moving coil galvanometer has resistance 50 Ω and it indicates full deflection at 4 mA current. A voltmeter is made using this galvanometer and a 5 kΩ resistance. The maximum voltage, that can be measured using this voltmeter, will be close to : **[9 April 2019 (Morning)]**

(A) 10 V (B) 20 V
(C) 40 V (D) 15 V

51. A rigid square loop of side 'a' and carrying current I_2 is lying on a horizontal surface near a long current I_1 carrying wire in the same plane as shown in figure. The net force on the loop due to wire will be : **[9 April 2019 (Morning)]**

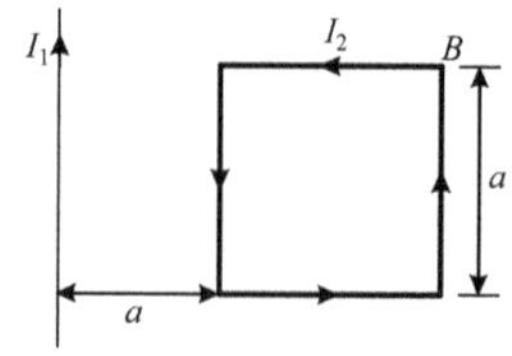

(A) Attractive and equal to $\frac{\mu_0 I_1 I_2}{3\pi}$

(B) Repulsive and equal to $\frac{\mu_0 I_1 I_2}{4\pi}$

(C) Repulsive and equal to $\frac{\mu_0 I_1 I_2}{2\pi}$

(D) Zero

52. A rectangular coil (Dimension 5 cm × 2.5 cm) with 100 turns, carrying a current of 3A in the clock-wise direction is kept centered at the origin and in the X-Z plane. A magnetic field of 1T is applied along X-axis. If the coil is tilted through 45° about Z-axis, then the torque on the coil is : **[9 April 2019 (Morning)]**

(A) 0.55 Nm (B) 0.27 Nm
(C) 0.38 Nm (D) 0.42 Nm

53. A moving coil galvanometer has a coil with 175 turns and area 1 cm^2. It uses a torsion band of torsion constant 10^{-6} N-m/rad. The coil is placed in a maganetic field B parallel to its plane. The coil deflects by 1° for a current of 1mA. The value of B (in Tesla) is approximately: **[9 April 2019 (Evening)]**

(A) 10^{-3} (B) 10^{-1}
(C) 10^{-4} (D) 10^{-2}

54. A proton, an electron, and a Helium nucleus, have the same energy. They are in circular orbits in a plane due to magnetic field perpendicualr to the plane. Let r_p, r_e and r_{He} be their respective radii, then : **[10 April 2019 (Morning)]**

(A) $r_e > r_p > r_{He}$ (B) $r_e < r_p < r_{He}$
(C) $r_e < r_p = r_{He}$ (D) $r_e > r_p = r_{He}$

55. Two wires A & B are carrying currents I_1 & I_2 as shown in the figure. The separation between them is d. A third wire C carrying a current I is to be kept parallel to them at a distance x from A such that the net force acting on it is zero. The possible values of x are : **[10 April 2019 (Morning)]**

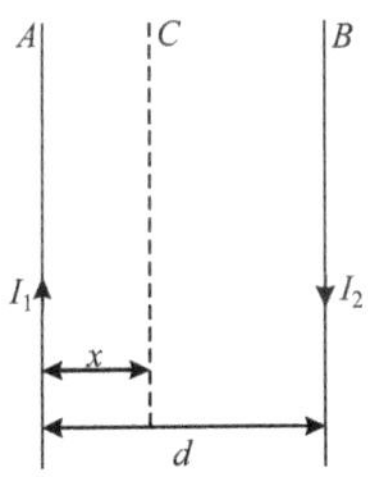

(A) $x = \left(\frac{I_1}{I_1 - I_2}\right)d$ and $x = \frac{I_2}{(I_1 + I_2)}d$

(B) $x = \pm\frac{I_1 d}{(I_1 - I_2)}$

(C) $x = \left(\frac{I_1}{I_1 + I_2}\right)d$ and $x = \frac{I_2}{(I_1 - I_2)}d$

(D) $x = \left(\frac{I_2}{I_1 + I_2}\right)d$ and $x = \left(\frac{I_2}{I_1 - I_2}\right)d$

56. A moving coil galvanometer allows a full scale current of $10^{-4}A$. A series resistance of $2M\Omega$ is required to convert the above galvanometer into a voltmeter of range 0-5V. Therefore the value of shunt resistance required to convert the above galvanometer into an ammeter of range 0-10 mA is : **[10 April 2019 (Morning)]**

(A) 200 Ω (B) 100 Ω
(C) 10 Ω (D) 500 Ω

57. A moving coil galvanometer, having a resistance G, produces full scale deflection when a current I_g flows through it. This galvanometer can be converted into (i) an ammeter of range 0 to I_0 ($I_0 > I_g$) by connecting a shunt resistance R_A to it and (ii) into a voltmeter of range 0 to $V(V = GI_0)$ by connecting a series resistance R_V to it. Then : **[12 April 2019 (Evening)]**

(A) $R_A R_V = G^2\left(\frac{I_g}{I_0 - I_g}\right)$ and $\frac{R_A}{R_V} = \left(\frac{I_0 - I_g}{I_g}\right)^2$

(B) $R_A R_V = G^2$ and $\frac{R_A}{R_V} = \left(\frac{I_g}{I_0 - I_g}\right)^2$

(C) $R_A R_V = G^2$ and $\frac{R_A}{R_V} = \frac{I_g}{(I_0 - I_g)}$

(D) $R_A R_V = G^2\left(\frac{I_0 - I_g}{I_g}\right)$ and $\frac{R_A}{R_V} = \left(\frac{I_g}{I_0 - I_g}\right)^2$

58. An electron, moving along the x-axis with an initial energy of 100 eV, enters a region of magnetic field $\vec{B} = (1.5\times10^{-3}T)\hat{k}$ at S (See figure). The field extends between $x = 0$ and $x = 2$ cm. The electron is detected at the point Q on a screen placed 8 cm away from the point S. The distance d between P and Q (on the screen) is : (electron's charge = 1.6 × 10^{-19} C, mass of electron = 9.1 × 10^{-31} kg) : **[12 April 2019 (Evening)]**

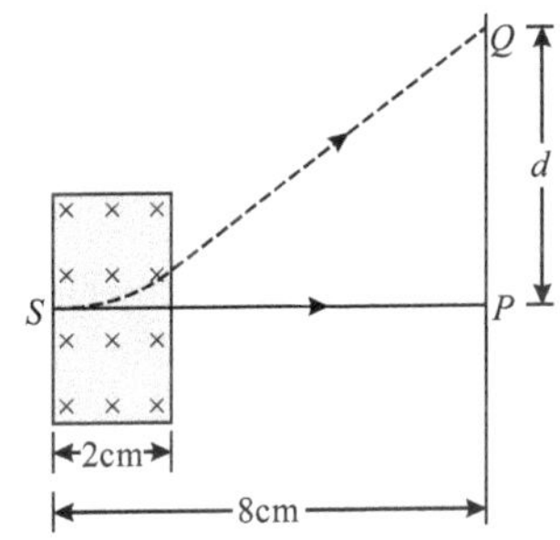

(A) 12.87 cm (B) 1.22 cm
(C) 11.65 cm (D) 2.25 cm

* * * * *

CHAPTER-19 MAGNETISM AND MATTER

19.1 Classical and Terrestrial Magnetism

1. The magnetic lines of force inside a bar magnet : **[2003]**
(A) are from north-pole to south-pole of the magnet
(B) do not exist
(C) depend upon the area of cross-section of the bar magnet
(D) are from south-pole to north-pole of the magnet

2. A thin rectangular magnet suspended freely has a period of oscillation equal to T. Now it is broken into two equal halves (each having half of the original length) and one piece is made to oscillate freely in the same field. If its period of oscillation is T', the ratio $\frac{T'}{T}$ is : **[2003]**

(A) $\frac{1}{2\sqrt{2}}$ (B) $\frac{1}{2}$
(C) 2 (D) $\frac{1}{4}$

3. The length of a magnet is large compared to its width and breadth. The time period of its oscillation in a vibration magnetometer is 2 s. The magnet is cut along its length into three equal parts and these parts are then placed on each other with their like poles together. The time period of this combination will be : **[2004]**

(A) $2\sqrt{3}$ s (B) $\frac{2}{3}$ s
(C) 2 s (D) $\frac{2}{\sqrt{3}}$ s

4. A magnetic needle is kept in a non-uniform magnetic field. It experiences : **[2005]**
(A) a torque but not a force (B) neither a force nor a torque
(C) a force and a torque (D) a force but not a torque

5. A bar magnet of length 6 cm has a magnetic moment of 4 JT^{-1}. Find the strength of magnetic field at a distance of 200 cm from the centre of the magnet along its equatorial line : **[2012 (Online)]**
(A) 4×10^{-8} tesla (B) 3.5×10^{-8} tesla
(C) 5×10^{-8} tesla (D) 3×10^{-8} tesla

6. Two short bar magnets of length 1 cm each have magnetic moments 1.20 Am^2 and 1.00 Am^2 respectively. They are placed on a horizontal table parallel to each other with their N poles pointing towards the South. They have a common magnetic equator and are separated by a distance of 20.0 cm. The value of the resultant horizontal magnetic induction at the mid-point O of the line joining their centres is close to : (Horizontal component of earth's magnetic induction is 3.6×10^{-5} Wb/m^2) : **[2013]**
(A) 3.6×10^{-5} Wb/m^2 (B) 2.56×10^{-4} Wb/m^2
(C) 3.50×10^{-4} Wb/m^2 (D) 5.80×10^{-4} Wb/m^2

7. The earth's magnetic field lines resemble that of a dipole at the centre of the earth. If the magnetic moment of this dipole at the centre of the earth. If the magnetic moment of this dipole is close to 8×10^{22} Am^2, the value of earth's magnetic field near the equator is close to (radius of the earth = 6.4×10^6 m) : **[2013 (Online)]**
(A) 0.6 Gauss (B) 1.2 Gauss
(C) 1.8 Gauss (D) 0.32 Gauss

8. The mid points of two small magnetic dipoles of length d in end-on positions, are separated by a distance x, ($x >> d$). The force between them is proportional to x^{-n} where n is : **[2014 (Online)]**

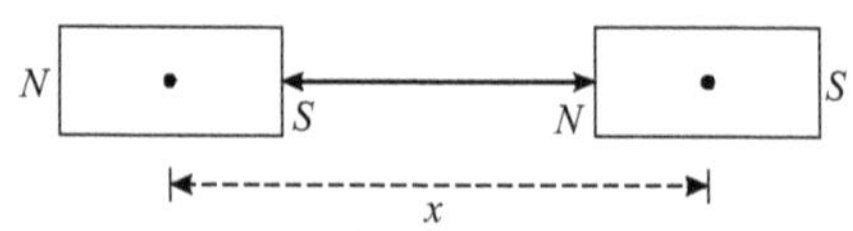

(A) 1 (B) 2
(C) 3 (D) 4

9. The magnetic field of earth at the equator is approximately 4×10^{-5} T. The radius of earth is 6.4×10^6 m. Then the dipole moment of the earth will be nearly of the order of : **[2014 (Online)]**
(A) 10^{23} A m^2 (B) 10^{20} A m^2
(C) 10^{16} A m^2 (D) 10^{10} A m^2

10. A short bar magnet is placed in the magnetic meridian of the earth with north pole pointing north. Neutral points are found at a distance of 30 cm from the magnet on the East-West line, drawn through the middle point of the magnet. The magnetic moment of the magnet in Am^2 is close to : (Given $\frac{\mu_0}{4\pi} = 10^{-7}$ in SI units and B_H = Horizontal component of earth's magnetic field = 3.6×10^{-5} tesla) : **[2015 (Online)]**
(A) 14.6 (B) 19.4
(C) 9.7 (D) 4.9

11. A 25 cm long solenoid has radius 2 cm and 500 total number of turns. It carries a current of 15 A. If it is equivalent to a magnet of the same size and magnetization $\vec{M}$ (magnetic moment/volume), then $|\vec{M}|$ is : **[2015 (Online)]**
(A) $30000\,\pi$ Am^{-1} (B) $3\,\pi$ Am^{-1}
(C) 30000 Am^{-1} (D) 300 Am^{-1}

12. A magnetic dipole is acted upon by two magnetic fields which are inclined to each other at an angle of 75°. One of the fields has a magnitude of 15 mT. The dipole attains stable equilibrium at an angle of 30° with this field. The magnitude of the other field (in *mT*) is close to : **[2016 (Online)]**

(A) 1 (B) 11
(C) 36 (D) 1060

13. A magnetic needle of magnetic moment 6.7×10^{-2} Am^2 and moment of inertia 7.5×10^{-6} kg m^2 is performing simple harmonic oscillations in a magnetic field of 0.01 T. Time taken for 10 complete oscillations is : **[2017]**

(A) 6.98 s (B) 8.76 s
(C) 6.65 s (D) 8.89 s

14. A Magnetic dipole in a constant magnetic field has : **[2017 (Online)]**

(A) Maximum potential energy when the torque is maximum
(B) Zero potential energy when the torque is minimum
(C) Zero potential energy when the torque is maximum
(D) Minimum potential energy when the torque is maximum

15. At some location on earth the horizontal component of earth's magnetic field is 18×10^{-6} T. At this location, magnetic needle of length 0.12 m and pole strength 1.8 Am is suspended from its mid-point using a thread, it makes 45° angle with horizontal in equilibrium. To keep this needle horizontal, the vertical force that should be applied at one of its ends is : **[10 Jan. 2019 (Evening)]**

(A) 3.6×10^{-5} N (B) 6.5×10^{-5} N
(C) 1.3×10^{-5} N (D) 1.8×10^{-5} N

16. A hoop and a solid cylinder of same mass and radius are made of a permanent magnetic material with their magnetic moment parallel to their respective axes. But the magnetic moment of hoop is twice of solid cylinder. They are placed in a uniform magnetic field in such a manner that their magnetic moments make a small angle with the field. If the oscillation periods of hoop and cylinder are T_h and T_c respectively, then: **[10 Jan. 2019 (Evening)]**

(A) $T_h = 0.5\,T_c$ (B) $T_h = 2\,T_c$
(C) $T_h = 1.5\,T_c$ (D) $T_h = T_c$

17. A square loop is carrying a steady current *I* and the magnitude of its magnetic dipole moment is *m*. If this square loop is changed to a circular loop and it carries the same current, the magnitude of the magnetic dipole moment of circular loop will be : **[10 April 2019 (Evening)]**

(A) $\frac{3m}{\pi}$ (B) $\frac{4m}{\pi}$
(C) $\frac{2m}{\pi}$ (D) $\frac{m}{\pi}$

18. A magnetic compass needle oscillates 30 times per minute at a place where the dip is 45°, and 40 times per minute where the dip is 30°. If B_1 and B_2 are respectively the total magnetic field due to the earth at the two places, then the ratio B_1/B_2 is best given by : **[12 April 2019 (Morning)]**

(A) 2.2 (B) 1.8
(C) 0.7 (D) 3.6

* * * * *

CHAPTER-19 MAGNETISM AND MATTER

19.2 Magnetic Properties of Materials

1. Curie temperature is the temperature above which : **[2003]**

(A) A ferromagnetic material becomes paramagnetic
(B) A paramagnetic material becomes diamagnetic
(C) A ferromagnetic material becomes diamagnetic
(D) A paramagnetic material becomes ferromagnetic

2. The materials suitable for making electromagnets should have : **[2004]**

(A) High retentivity and low coercivity
(B) Low retentivity and low coercivity
(C) High retentivity and high coercivity
(D) Low retentivity and high coercivity

3. Needles N_1, N_2 and N_3 are made of a ferromagnetic, a paramagnetic and a diamagnetic substance respectively. A magnet when brought close to them will : **[2006]**

(A) Attract N_1 and N_2 strongly but repel N_3
(B) Attract N_1 strongly, N_2 weakly and repel N_3 weakly
(C) Attract N_1 strongly, but repel N_2 and N_3 weakly
(D) Attract all three of them

4. Relative permittivity and permeability of a material ε_r and μ_r, respectively. Which of the following values of these quantities are allowed for a diamagnetic material ? **[2008]**

(A) $\varepsilon_r = 0.5, \mu_r = 1.5$ (B) $\varepsilon_r = 1.5, \mu_r = 0.5$
(C) $\varepsilon_r = 0.5, \mu_r = 0.5$ (D) $\varepsilon_r = 1.5, \mu_r = 1.5$

5. The coercivity of a small magnet where the ferromagnet gets demagnetized is 3×10^3 A m^{-1}. The current required to be passed in a solenoid of length 10 cm and number of turns 100, so that the magnet gets demagnetized when inside the solenoid, is : **[2014]**

(A) 30 mA (B) 60 mA
(C) 3 A (D) 6 A

6. Three identical bars A, B and C are made of different magnetic materials. When kept in a uniform magnetic field, the field lines around them look as follows :

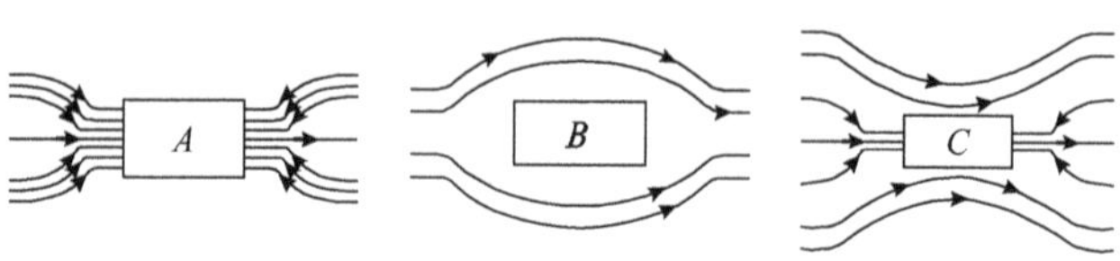

Make the correspondence of these bars with their material being diamagnetic (D), ferromagnetic (F) and paramagnetic (P) : **[2014 (Online)]**

(A) $A \leftrightarrow D, B \leftrightarrow P, C \leftrightarrow F$ (B) $A \leftrightarrow F, B \leftrightarrow D, C \leftrightarrow P$
(C) $A \leftrightarrow P, B \leftrightarrow F, C \leftrightarrow D$ (D) $A \leftrightarrow F, B \leftrightarrow P, C \leftrightarrow D$

7. An example of a perfect diamagnet is a superconductor. This implies that when a superconductor is put in a magnetic field of intensity B, the magnetic field B_s inside the superconductor will be such that : **[2014 (Online)]**

(A) $B_s = -B$ (B) $B_s = 0$
(C) $B_s = B$ (D) $B_s < B$ but $Bs \neq 0$

8. Hysteresis loops for two magnetic materials A and B are given below :

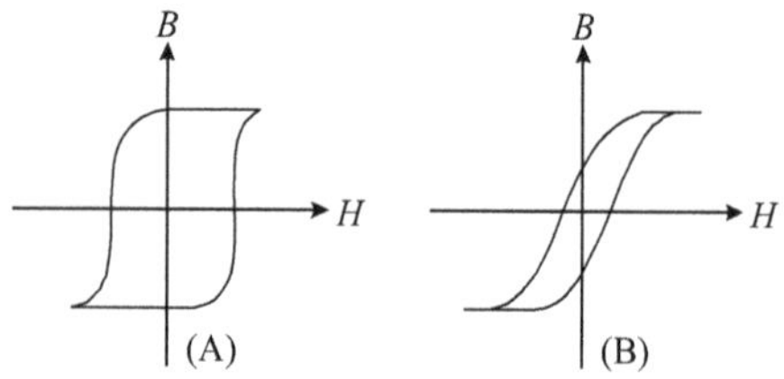

These materials are used to make magnets for electric generators, transformer core and electromagnet core. Then it is proper to use : **[2016]**

(A) A for electric generators and transformers
(B) A for electromagnets and B for electric generators
(C) A for transformers and B for electric generators
(D) B for electromagnets and transformers

9. The B-H curve for a ferromagnet is shown in the figure. The ferromagnet is placed inside a long solenoid with 1000 turns/cm. The current that should be passed in the solenoid to demagnetize the ferromagnet completely is : **[2018 (Online)]**

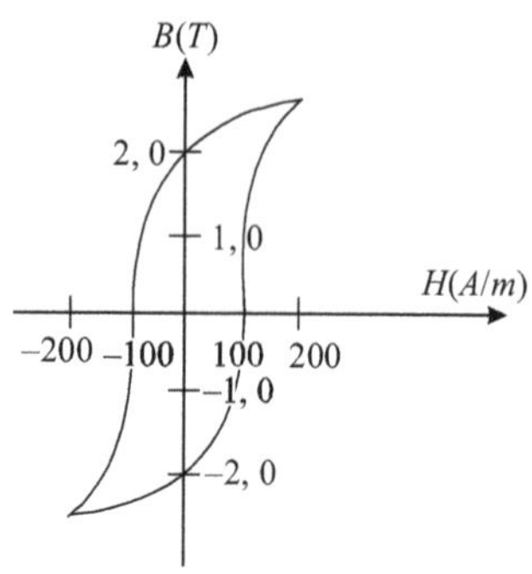

(A) 2 mA (B) 1 mA
(C) 40 μA (D) 20 μA

10. A bar magnet is demagnetized by inserting it inside a solenoid of length 0.2 m, 100 turns, and carrying a current of 5.2 A. The coercivity of the bar magnet is : **[9 Jan. 2019 (Morning)]**

(A) 1200 A/m (B) 2600 A/m
(C) 520 A/m (D) 285 A/m

11. A paramagnetic substance in the form of a cube with sides 1 cm has a magnetic dipole moment of 20×10^{-6} J/T when a magnetic intensity of 60×10^{3} A/m is applied. Its magnetic susceptibility is : **[11 Jan. 2019 (Evening)]**

(A) 2.3×10^{-2} (B) 3.3×10^{-2}

(C) 3.3×10^{-4} (D) 4.3×10^{-2}

12. A paramagnetic material has 10^{28} atoms/m^3. Its magnetic susceptibility at temperature 350 K is 2.8×10^{-4}. Its susceptibility at 300 K is : **[12 Jan. 2019 (Evening)]**

(A) 3.672×10^{-4} (B) 3.726×10^{-4}

(C) 3.267×10^{-4} (D) 2.672×10^{-4}

13. Two magnetic dipoles X and Y are placed at a separation d, with their axes perpendicular to each other. The dipole moment of Y is twice that of X. A particle of charge q is passing, through their midpoint P, at angle $\theta = 45°$ with the horizontal line, as shown in figure. What would be the magnitude of force on the particle at that instant ? (d is much larger than the dimensions of the dipole) : **[8 April 2019 (Evening)]**

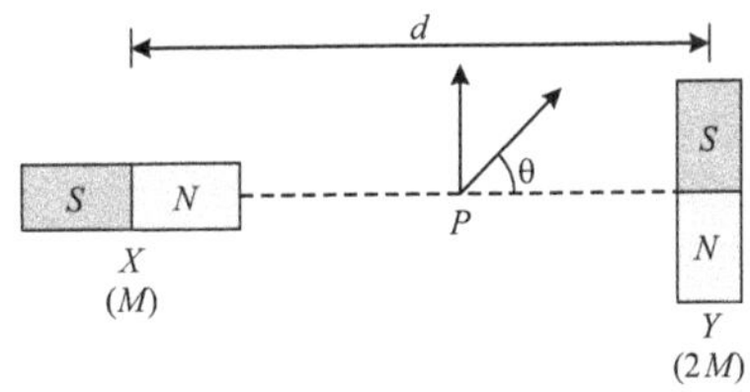

(A) $\sqrt{2}\left(\frac{\mu_0}{4\pi}\right)\frac{M}{(d/2)^3}\times qv$ (B) $\left(\frac{\mu_0}{4\pi}\right)\frac{2M}{(d/2)^3}\times qv$

(C) $\left(\frac{\mu_0}{4\pi}\right)\frac{M}{(d/2)^3}\times qv$ (D) 0

* * * * *

CHAPTER-20 ELECTROMAGNETIC INDUCTION

20.1 Electromagnetic Induction

1. A conducting square loop of side L and resistance R moves in its plane with a uniform velocity v perpendicular to one of its sides. A magnetic induction B constant in time and space, pointing perpendicular and into the plane at the loop exists everywhere with half the loop outside the field, as shown in figure. The induced emf is : **[2002]**

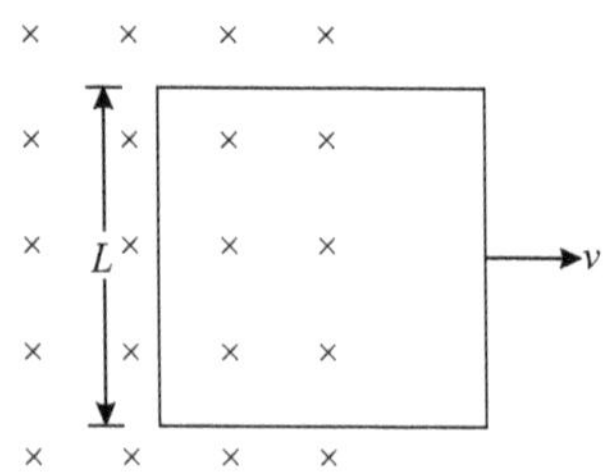

(A) Zero (B) RvB

(C) $\frac{vBL}{R}$ (D) vBL

2. A metal conductor of length 1 m rotates vertically about one of its ends at angular velocity 5 radians per second. If the horizontal component of earth's magnetic field is 0.2×10^{-4} T, then the emf developed between the two ends of the conductor is : **[2004]**

(A) 5 μV (B) 50 μV

(C) 5mV (D) 50mV

3. In a uniform magnetic field of induction B, a wire in the form of semicircle of radius r rotates about the diameter of the circle with angular frequency ω. If the total resistance of the circuit is R, the mean power generated per period of rotation is: **[2004]**

(A) $\frac{B\pi r^2 \omega}{2R}$ (B) $\frac{(B\pi r^2 \omega)^2}{8R}$

(C) $\frac{(B\pi r\omega)^2}{2R}$ (D) $\frac{(B\pi r\omega^2)^2}{8R}$

4. A coil having n turns and resistance $R\ \Omega$ is connected with a galvanometer of resistance $4\ R\Omega$. This combination is moved in time t seconds from a magnetic field W_1 weber to W_2 weber. The induced current in the circuit is : **[2004]**

(A) $\frac{W_2 - W_1}{5\,Rnt}$ (B) $-\frac{n(W_2 - W_1)}{5\,Rt}$

(C) $-\frac{(W_2 - W_1)}{Rnt}$ (D) $-\frac{n(W_2 - W_1)}{Rt}$

5. One conducting U-tube can slide inside another as shown in figure, maintaining electrical contacts between the tubes. The magnetic field B is perpendicular to the plane of the figure. If each tube moves towards the other at a constant speed v, then the emf induced in the circuit in terms of B, l and v, where l is the width of each tube, will be : **[2005]**

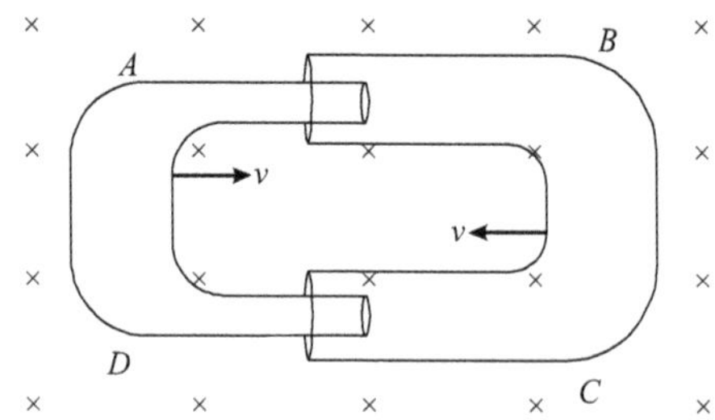

(A) Blv (B) $-Blv$

(C) zero (D) $2Blv$

6. A rectangular loop has a sliding connector PQ of length l and resistance $R\Omega$ and it is moving with a speed v as shown. The set-up is placed in a uniform magnetic field going into the plane of the paper. The three currents I_1, I_2 and I are : **[2010]**

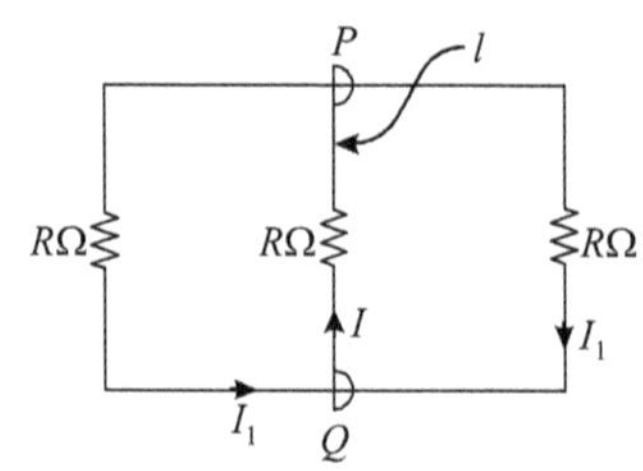

(A) $I_1 = I_2 = \frac{Blv}{6R}, I = \frac{Blv}{3R}$ (B) $I_1 = -I_2 = \frac{Blv}{R}, I = \frac{2Blv}{R}$

(C) $I_1 = I_2 = \frac{Blv}{3R}, I = \frac{2Blv}{3R}$ (D) $I_1 = I_2 = , I = \frac{Blv}{R}$

7. A horizontal straight wire 20 m long extending from east to west is falling with a speed of 5.0 m/s, at right angles to the horizontal component of the earth's magnetic field 0.30×10^{-4} Wb/m^2. The instantaneous value of the e.m.f. induced in the wire will be : **[2011]**

(A) 6.0mV (B) 3mV

(C) 4.5mV (D) 1.5mV

8. A boat is moving due east in a region where the earth's magnetic field is 5.0×10^{-5} NA^{-1}m^{-1} due north and horizontal. The boat carries a vertical aerial 2 m long. If the speed of the boat is 1.50 ms^{-1}, the magnitude of the induced emf in the wire of aerial is : **[2011]**

(A) 0.50mV (B) 0.15mV

(C) 1mV (D) 0.75mV

9. A coil is suspended in a uniform magnetic field, with the plane of the coil parallel to the magnetic lines of force. When a current is passed through the coil it starts oscillating; it is very difficult to stop. But if an aluminium plate is placed near to the coil, it stops. This is due to : **[2012]**

(A) Electromagnetic induction in the aluminium plate giving rise to electromagnetic damping
(B) Development of air current when the plate is placed
(C) Induction of electrical charge on the plate
(D) Shielding of magnetic lines of force as aluminium is a paramagnetic material

10. Magnetic flux through a coil of resistance 10 Ω is changed by $\Delta\phi$ in 0.1 s. The resulting current in the coil varies with time as shown in the figure. Then $|\Delta\phi|$ is equal to (in weber) : **[2012 (Online)]**

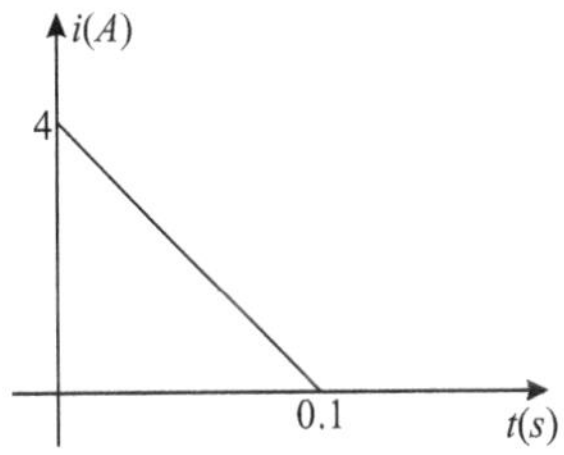

(A) 6 (B) 4
(C) 2 (D) 8

11. A circular loop of radius 0.3 cm lies parallel to a much bigger circular loop of radius 20 cm. The centre of the small loop is on the axis of the bigger loop. The distance between their centres is 15 cm. If a current of 2.0 A flows through the smaller loop, then the flux linked with bigger loop is : **[2013]**

(A) 9.1×10^{-11} weber (B) 6×10^{-11} weber
(C) 3.3×10^{-11} weber (D) 6.6×10^{-9} weber

12. A metallic rod of length 'l' is tied to a string of length $2l$ and made to rotate with angular speed ω on a horizontal table with one end of the string fixed. If there is a vertical magnetic field 'B' in the region, the e.m.f. induced across the ends of the rod is : **[2013]**

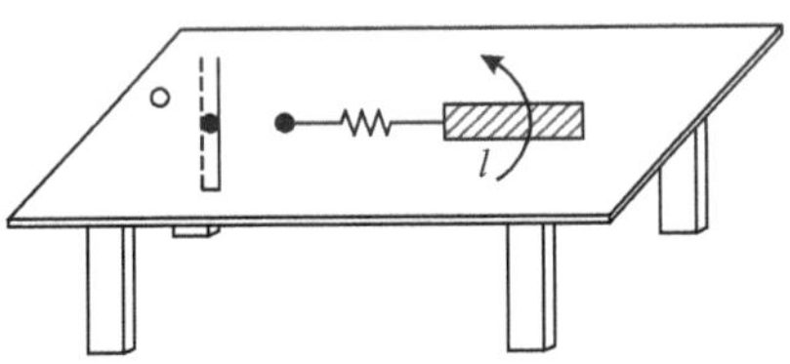

(A) $\frac{2B\omega l^2}{2}$ (B) $\frac{3B\omega l^2}{2}$

(C) $\frac{4B\omega l^2}{2}$ (D) $\frac{5B\omega l^2}{2}$

13. A square frame of side 10 cm and a long straight wire carrying current 1 A are in the plate of the paper. Starting from close to the wire, the frame moves towards the right with a constant speed of 10 ms^{-1} (see figure).
The e.m.f. induced at the time the left arm of the frame is at x = 10 cm from the wire is: **[2014 (Online)]**

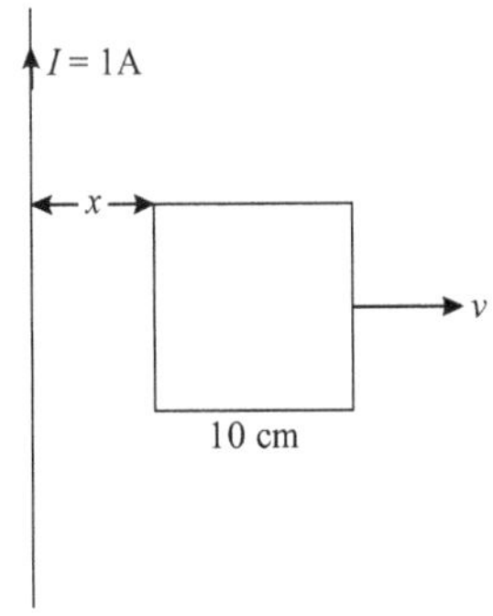

(A) 2 μV (B) 1 μV
(C) 0.75 μV (D) 0.5 μV

14. A coil of circular cross-section having 1000 turns and 4 cm^2 face area is placed with its axis parallel to a magnetic field which decreases by 10^{-2} Wb m^{-2} in 0.01 s. The e.m.f. induced in the coil is : **[2014 (Online)]**

(A) 400mV (B) 200mV
(C) 4mV (D) 0.4mV

15. Figure shows a circular area of radius R where a uniform magnetic field $\vec{B}$ is going into the plane of paper and increasing in magnitude at a constant rate.
In that case, which of the following graphs, drawn schematically, correctly shows the variation of the induced electric field $E(r)$? **[2014 (Online)]**

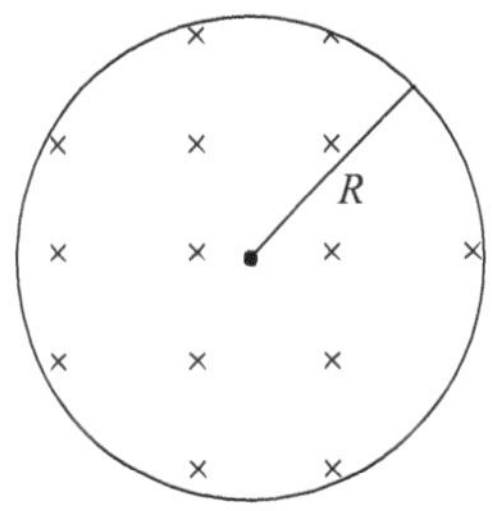

(A) 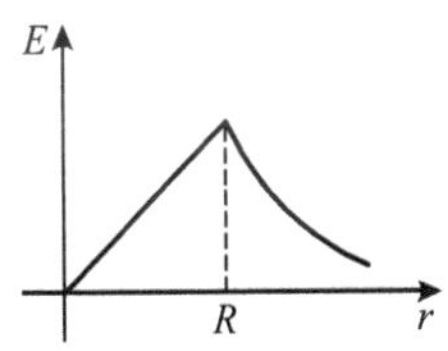

(B) 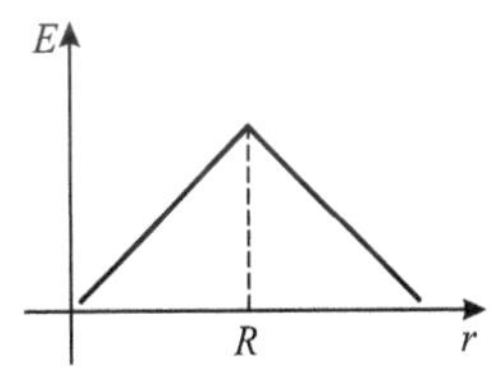

(C) 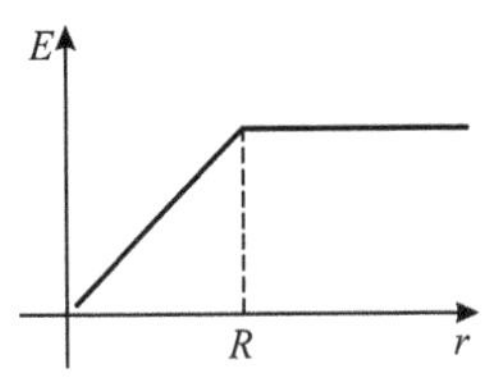

(D) 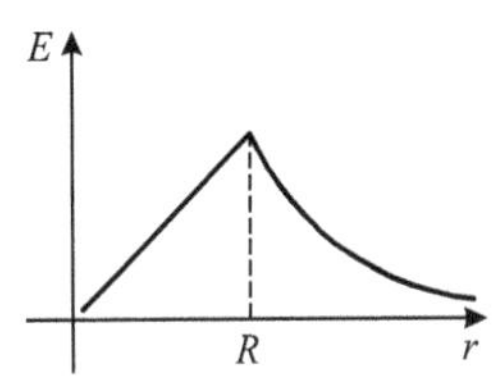

16. A fighter plane of length 20 m, wing span (distance from tip of one wing to the tip of the other wing) of 15 m and height 5 m is lying towards east over Delhi. Its speed is 240 ms^{-1}. The earth's magnetic field over Delhi is 5×10^{-5} T with the declination angle ~ 0° and dip of θ such that sin θ = 2/3. If the voltage developed is V_B between the lower and upper side of the plane of V_W between the tips of the wings then V_B and V_W are close to: **[2016 (Online)]**

(A) $V_B = 40$ mV; $V_w = 120$ mV with left side of pilot at higher voltage
(B) $V_B = 45$ mV; $V_W = 120$ mV with right side of pilot at higher voltage
(C) $V_B = 40$ mV; $V_W = 135$ mV with right side of pilot at higher voltage
(D) $V_B = 45$ mV; $V_W = 120$ mV with left side of pilot at higher voltage

17. Consider a thin metallic sheet perpendicular to the plane of the paper moving with speed 'v' in a uniform magnetic field B going into the plane of the paper (See figure). If charge densities σ_1 and σ_2 are induced on the left and right surfaces, respectively, of the sheet then (ignore fringe effects) : **[2016 (Online)]**

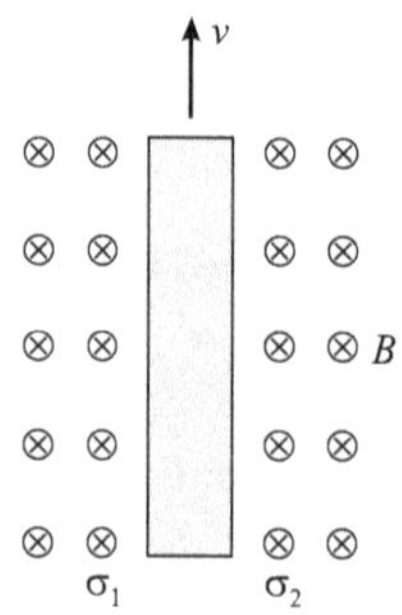

(A) $\sigma_1 = \dfrac{-\epsilon_0 vB}{2}, \sigma_2 = \dfrac{\epsilon_0 vB}{2}$

(B) $\sigma_1 = \epsilon_0 vB, \sigma_2 = -\epsilon_0 vB$

(C) $\sigma_1 = \dfrac{\epsilon_0 vB}{2}, \sigma_2 = \dfrac{-\epsilon_0 vB}{2}$

(D) $\sigma_1 = \sigma_2 = \epsilon_0 vB$

18. A conducting metal circular-wire-loop of radius r is placed perpendicular to a magnetic field which varies with time as $B = B_0 e^{-t/\tau}$, where B_0 and τ are constants, at time $t = 0$. If the resistance of the loop is R then the heat generated in the loop after a long time ($t \to \infty$) is : **[2016 (Online)]**

(A) $\dfrac{\pi^2 r^2 B_0^4}{2\tau R}$ (B) $\dfrac{\pi^2 r^4 B_0^2}{2\tau R}$

(C) $\dfrac{\pi^2 r^4 B_0^2 R}{\tau}$ (D) $\dfrac{\pi^2 r^4 B_0^2}{\tau R}$

19. In a coil of resistance 100 Ω, a current is induced by changing the magnetic flux through it as shown in the figure. The magnitude of change in flux through the coil is : **[2017]**

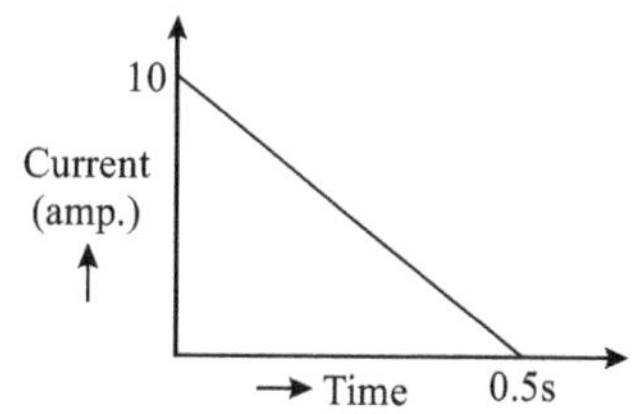

(A) 250 Wb (B) 275 Wb
(C) 200 Wb (D) 225 Wb

20. At the centre of a fixed large circular coil of radius R, a much smaller circular coil of radius r is placed. The two coils are concentric and are in the same plane. The larger coil carries a current I. The smaller coil is set to rotate with a constant angular velocity ω about an axis along their common diameter. Calculate the emf induced in the smaller coil after a time t of its start of rotation : **[2018 (Online)]**

(A) $\dfrac{\mu_0 I}{2R} \omega r^2 \sin \omega t$ (B) $\dfrac{\mu_0 I}{4R} \omega \pi r^2 \sin \omega t$

(C) $\dfrac{\mu_0 I}{2R} \omega \pi r^2 \sin \omega t$ (D) $\dfrac{\mu_0 I}{4R} \omega r^2 \sin \omega t$

21. A copper rod of mass m slides under gravity on two smooth parallel rails, with separation 1 and set at angle of θ with the horizontal. At the bottom, rails are joined by a resistance R. There is a uniform magnetic field B normal to the plane of the rails, as shown in the figure. The terminal speed of the copper rod is : **[2018 (Online)]**

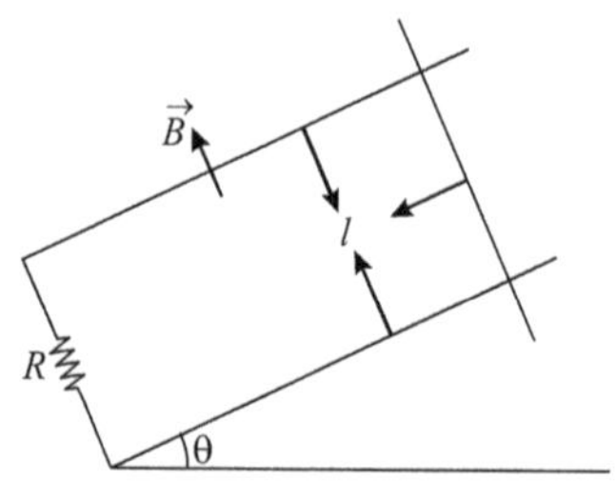

(A) $\dfrac{mgR \cos \theta}{B^2 l^2}$ (B) $\dfrac{mgR \sin \theta}{B^2 l^2}$

(C) $\dfrac{mgR \tan \theta}{B^2 l^2}$ (D) $\dfrac{mgR \cot \theta}{B^2 l^2}$

22. An ideal capacitor of capacitance 0.2 μF is charged to a potential difference of 10V. The charging battery is then disconnected. The capacitor is then connected to an ideal inductor of self inductance 0.5 mH. The current at a time when the potential difference across the capacitor is 5 V, is : **[2018 (Online)]**

(A) 0.17 A (B) 0.15 A
(C) 0.34 A (D) 0.25 A

23. A coil of cross-sectional area A having n turns is placed in a uniform magnetic field B. When it is rotated with an angular velocity ω, the maximum e.m.f. induced in the coil will be : **[2018 (Online)]**

(A) $nBA\,\omega$
(B) $\frac{3}{2}\,nBA\,\omega$
(C) $3\,nBA\,\omega$
(D) $\frac{1}{2}\,nBA\,\omega$

24. A conducting circular loop made of a thin wire, has area 3.5×10^{-3} m^2 and resistance 10 Ω. It is placed perpendicular to a time dependent magnetic field $B(t) = (0.4\text{T}) \sin(50\pi t)$. The field is uniform in space. Then the net charge flowing through the loop during $t = 0$ s and $t = 10$ ms is close to : **[9 Jan. 2019 (Morning)]**

(A) 14mC
(B) 21 mC
(C) 6mC
(D) 7mC

25. A solid metal cube of edge length 2 cm is moving in a positive y-direction at a constant speed of 6 m/s. There is a uniform magnetic field of 0.1 T in the positive z-direction. The potential difference between the two faces of the cube perpendicular to the x-axis, is : **[10 Jan. 2019 (Morning)]**

(A) 6mV
(B) 1mV
(C) 12mV
(D) 2mV

26. A magnet of total magnetic moment $10^{-2}\,\hat{i}$ A-m^2 is placed in a time varying magnetic field, $B\,\hat{i}$ (cost ωt) where $B = 1$ Tesla and $\omega = 0.125$ rad/s. The work done for reversing the direction of the magnetic moment at $t = 1$ second, is : **[10 Jan. 2019 (Morning)]**

(A) 0.007 J
(B) 0.014 J
(C) 0.01 J
(D) 0.028 J

27. A 10 m long horizontal wire extends from North East to South West. It is falling with a speed of 5.0 ms^{-1}, at right angles to the horizontal component of the earth's magnetic field, of 0.3×10^{-4} Wb/m^2. The value of the induced emf in wire is : **[12 Jan. 2019 (Evening)]**

(A) 2.5×10^{-3} V
(B) 1.1×10^{-3} V
(C) 0.3×10^{-3} V
(D) 1.5×10^{-3} V

28. A thin strip 10 cm long is on a U shaped wire of negligible resistance and it is connected to a spring of spring constant 0.5 Nm^{-1} (see figure). The assembly is kept in a uniform magnetic field of 0.1 T. If the strip is pulled from its equilibrium position and released, the number of oscillation it performs before its amplitude decreases by a factor of e is N. If the mass of the strip is 50 grams, its resistance 10 Ω and air drag negligible, N will be close to : **[8 April 2019 (Morning)]**

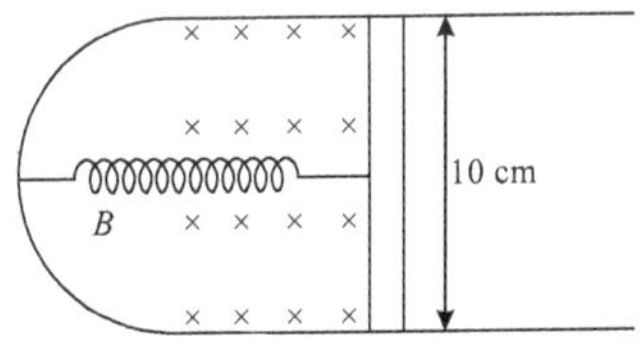

(A) 50000
(B) 5000
(C) 10000
(D) 1000

29. The figure shows a square loop L of side 5 cm which is connected to a network of resistances. The whole setup is moving towards right with a constant speed of 1 cms^{-1}. At some instant, a part of L is in a uniform magnetic field of 1T, perpendicular to the plane of the loop. If the resistance of L is 1.7 Ω, the current in the loop at that instant will be close to : **[12 April 2019 (Morning)]**

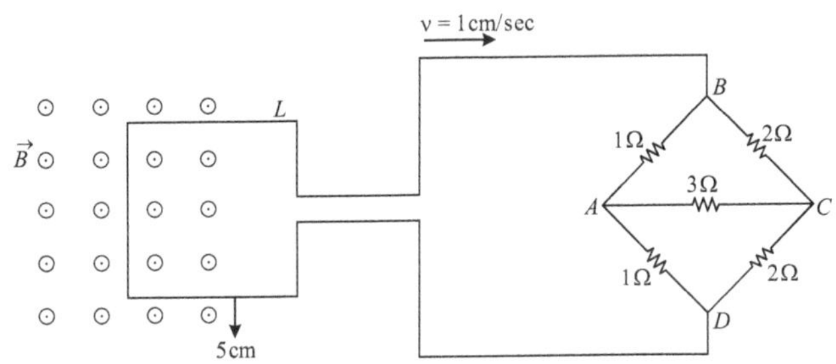

(A) 115 μA
(B) 170 μA
(C) 60 μA
(D) 150 μA

* * * * *

CHAPTER-20 ELECTROMAGNETIC INDUCTION

20.2 Self and Mutual Induction

1. The inductance between A and D is : **[2002]**

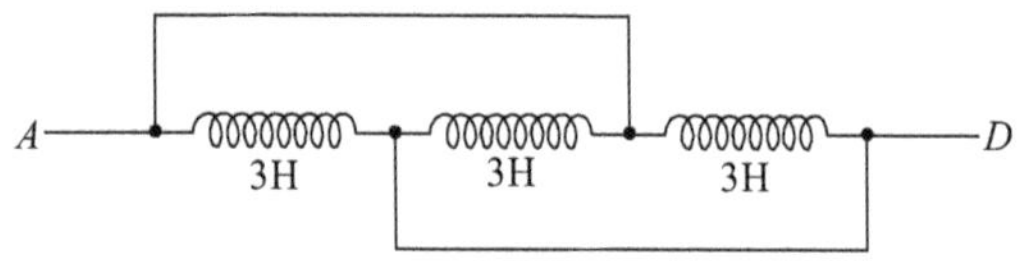

(A) 3.66 H (B) 9 H
(C) 0.66 H (D) 1 H

2. When the current changes from +2 A to – 2 A in 0.05 s, an emf of 8 V is induced in a coil. The coefficient of self-induction of the coil is : **[2003]**

(A) 0.2 H (B) 0.4 H
(C) 0.8 H (D) 0.1 H

3. Two coils are placed close to each other. The mutual inductance of the pair of coils depends upon : **[2003]**

(A) the rates at which currents are changing in the two coils
(B) relative position and orientation of the two coils
(C) the materials of the wires of the coils
(D) the currents in the two coils

4. A coil of inductance 300 mH and resistance 2 Ω is connected to a source of voltage 2 V. The current reaches half of its steady state value in : **[2005]**

(A) 0.05 s (B) 0.1 s
(C) 0.15 s (D) 0.3 s

5. The flux linked with a coil at any instant 't' is given by : $\phi = 10t^2 - 50t + 250$. The induced emf at $t = 3$s is : **[2006]**

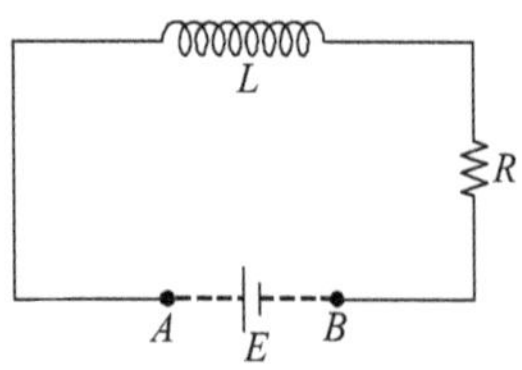

(A) – 190 V (B) – 10 V
(C) 10V (D) 190V

6. An inductor (L = 100 mH), a resistor (R = 100 Ω) and a battery (E = 100 V) are initially connected in series as shown in the figure. After a long time the battery is disconnected after short circuiting the points A and B. The current in the circuit 1 ms after the short circuit is : **[2006]**

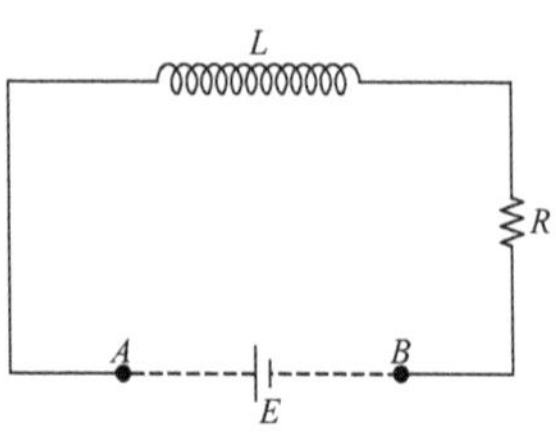

(A) $1/e$ A (B) e A
(C) 0.1 A (D) 1 A

7. An ideal coil of 10 H is connected in series with a resistance of 5 Ω and a battery of 5 V. 2 s after the connection is made, the current flowing (in ampere) in the circuit is : **[2007]**

(A) $(1 - e)$ (B) e
(C) e^{-1} (D) $(1 - e^{-1})$

8. Two coaxial solenoids are made by winding thin insulated wire over a pipe of cross-sectional area A = 10 cm^2 and length = 20 cm. If one of the solenoids has 300 turns and the other 400 turns, their mutual inductance is : $[\mu_0 = 4\pi \times 10^{-7}\ \text{TmA}^{-1}]$: **[2008]**

(A) $2.4\,\pi \times 10^{-5}$ H (B) $4.8\,\pi \times 10^{-4}$ H
(C) $4.8\,\pi \times 10^{-5}$ H (D) $2.4\,\pi \times 10^{-4}$ H

9. An inductor of inductance L = 400 mH and resistors of resistances R_1 = 2 Ω and R_2 = 2 Ω are connected to a battery of emf 12 V as shown in the figure. The internal resistance of the battery is negligible. The switch S is closed at t = 0. The potential drop across L as a function of time is : **[2009]**

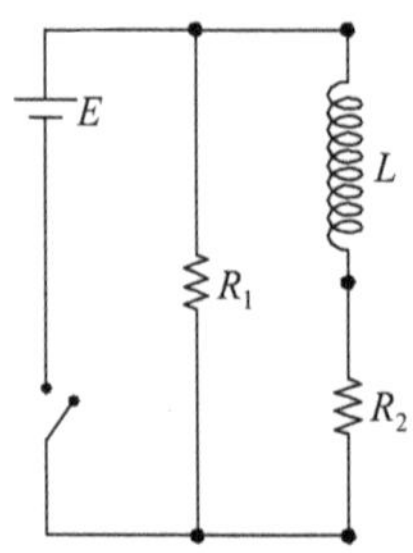

(A) $6(1 - e^{-t/0.2})$ V (B) $12e^{-5t}$ V
(C) $6e^{-5t}$ V (D) $\frac{12}{t}e^{-3t}$ V

10. In the circuit show below, the key K is closed at t = 0. The current through the battery is : **[2010]**

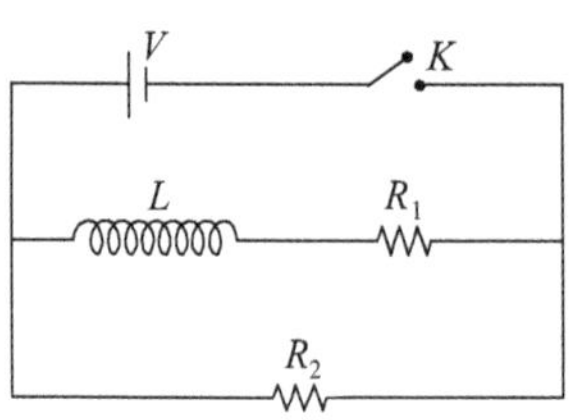

(A) $\frac{V(R_1 + R_2)}{R_1 R_2}$ at $t = 0$ and $\frac{V}{R_2}$ at $t = \infty$

(B) $\frac{VR_1R_2}{\sqrt{R_1^2 + R_2^2}}$ at $t = 0$ and $\frac{V}{R_2}$ at $t = \infty$

(C) $\frac{V}{R_2}$ at $t = 0$ and $\frac{V(R_1 + R_2)}{R_1 R_2}$ at $t = \infty$

(D) $\frac{V}{R_2}$ at $t = 0$ and $\frac{V R_1 R_2}{\sqrt{R_1^2 + R_2^2}}$ at $t = \infty$

11. This question has Statement-1 and Statement-2. Of the four choices given after the Statements, choose the one that best describes the two Statements.

Statement-1 : Self inductance of a long solenoid of length L, total number of turns N and radius r is less than $\frac{\pi\mu_0 N^2 r^2}{L}$.

Statement-2 : The magnetic induction in the solenoid in Statement-1 carrying current I is $\frac{\mu_0 N I}{L}$ in the middle of the solenoid but becomes less as we move towards its ends : **[2012 (Online)]**

(A) Statement-1 is true, Statement-2 is false

(B) Statement-1 is true, Statement-2 is true, Statement-2 is the correct explanation of Statement-1

(C) Statement-1 is false, Statement-2 is true

(D) Statement-1 is true, Statement-2 is true, Statement-2 is not the correct explanation of Statement-1

12. A coil of self inductance L is connected at one end of two rails as shown in figure. A connector of length l, mass m can slide freely over the two parallel rails. The entire set up is placed in a magnetic field of induction B going into the page. At an instant $t = 0$ an initial velocity v_0 is imparted to it and as a result of that it starts moving along x-axis. The displacement of the connector is represented by the figure : **[2012 (Online)]**

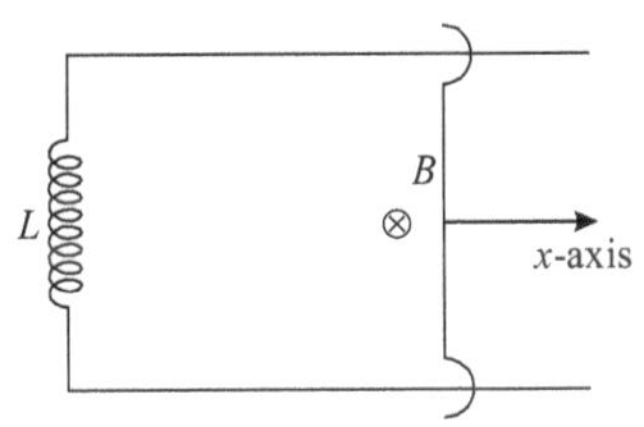

(A)
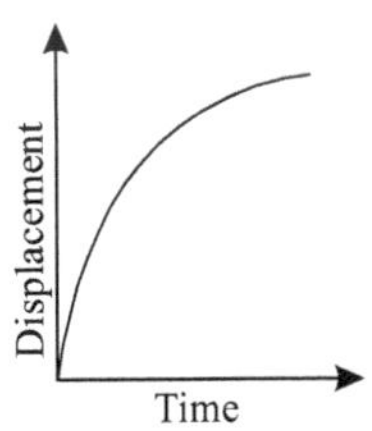

(B)
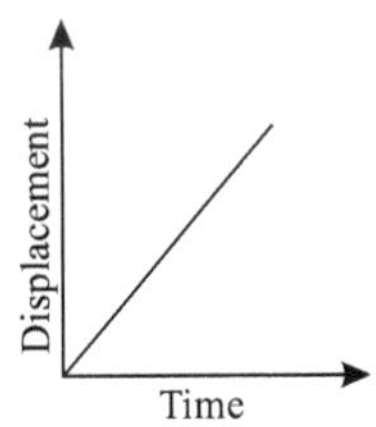

(C)
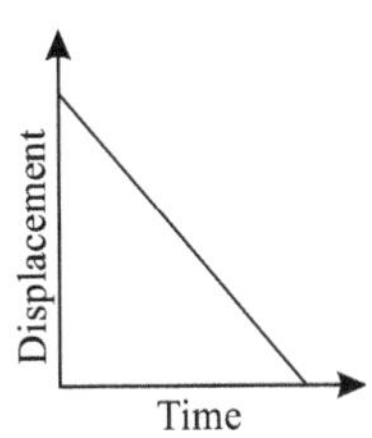

(D)
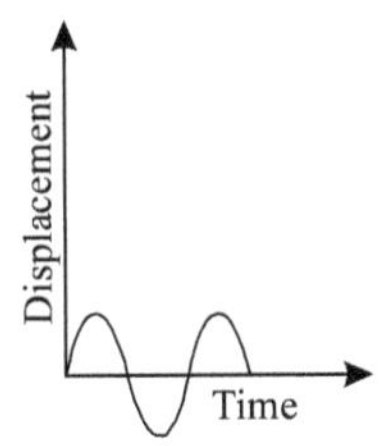

13. In an LCR circuit as shown below both switches are open initially. Now switch S_1 is closed, S_2 kept open. (q is charge on the capacitor and $\tau = RC$ is Capacitive time constant). Which of the following statement is correct? **[2013]**

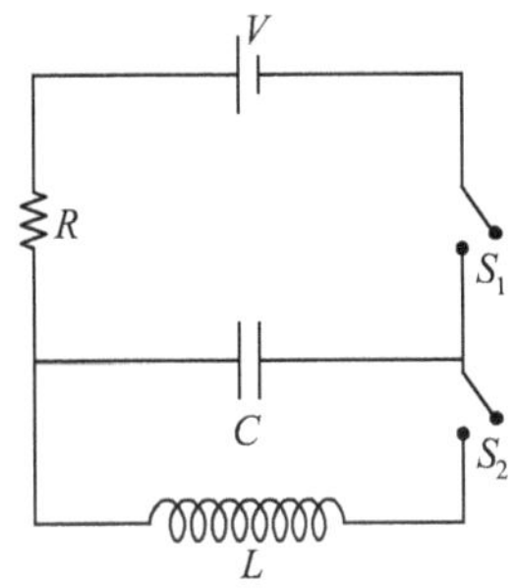

(A) Work done by the battery is half of the energy dissipated in the resistor

(B) At, $t = \tau$, $q = CV/2$

(C) At, $t = 2\tau$, $q = CV(1 - e^{-2})$

(D) At, $t = 2\tau$, $q = CV(1 - e^{-1})$

14. Two coils, X and Y, are kept in close vicinity of each other. When a varying current, $I(t)$, flows through coil X, the induced e.m.f ($V(t)$) in coil Y, varies in the manner shown here. The variation of $I(t)$, with time, can then be represented by the graph labelled as graph : **[2013 (Online)]**

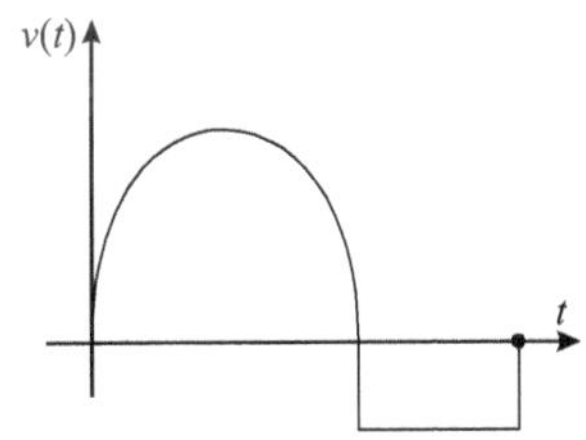

(A)
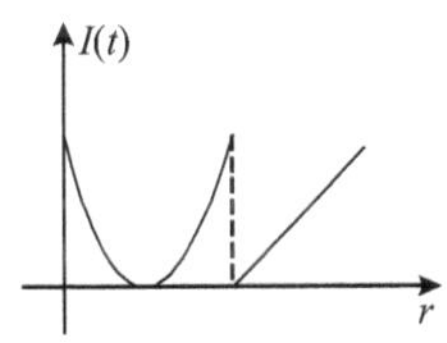

(B)
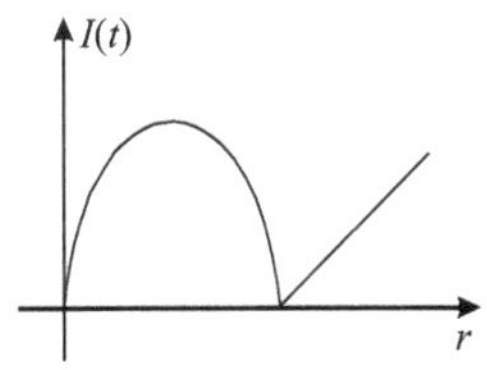

(C)
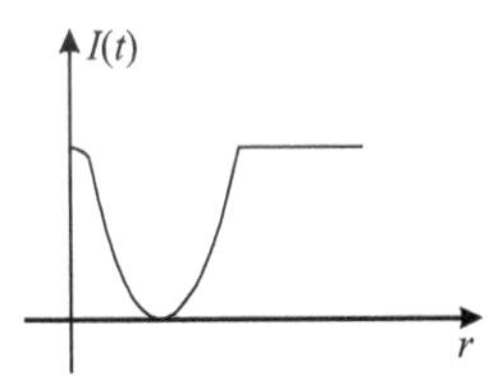

(D)
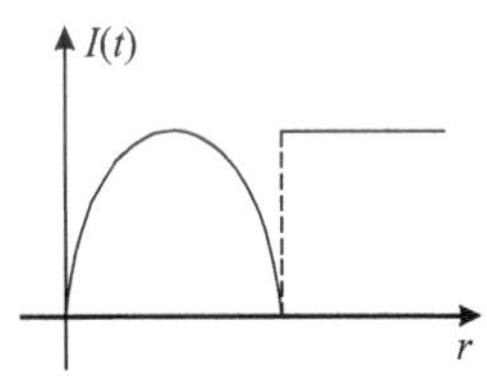

(A) A
(B) C
(C) B
(D) D

15. In the circuit shown here, the point 'C' is kept connected to point 'A' till the current flowing through the circuit becomes constant. Afterward, suddenly, point 'C' is disconnected from point 'A' and connected to point 'B' at time $t = 0$. Ratio of the voltage across resistance and the inductor at $t = L/R$ will be equal to : **[2014]**

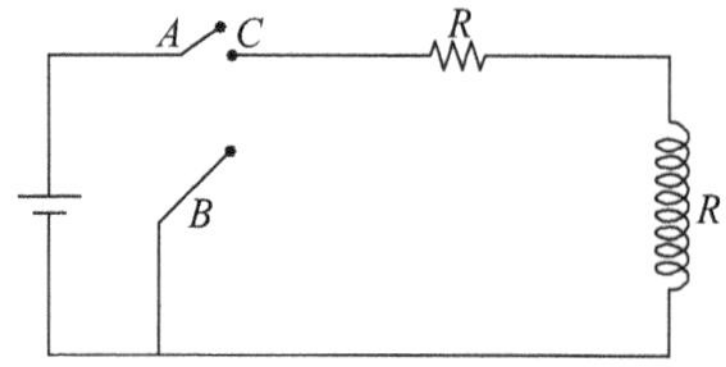

(A) $\dfrac{e}{1-e}$ (B) 1

(C) -1 (D) $\dfrac{1-e}{e}$

16. An LCR circuit is equivalent to a damped pendulum. In an LCR circuit the capacitor is charged to Q_0 and then connected to the L and R as shown below :

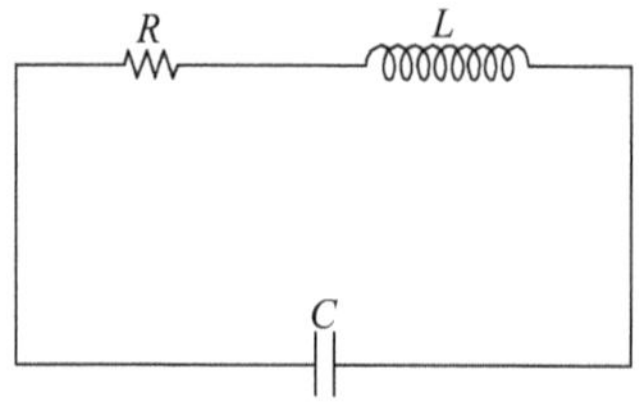

If a student plots graphs of the square of maximum charge (Q^2_{Max}) on the capacitor with time (t) for two different values L_1 and $L_2 (L_1 > L_2)$ of L then which of the following represents this graph correctly? (plots are schematic and not drawn to scale) : **[2015]**

(A) 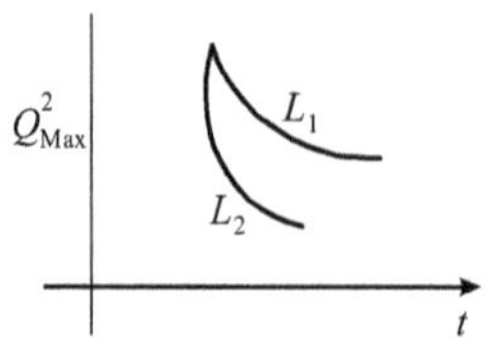

(B) 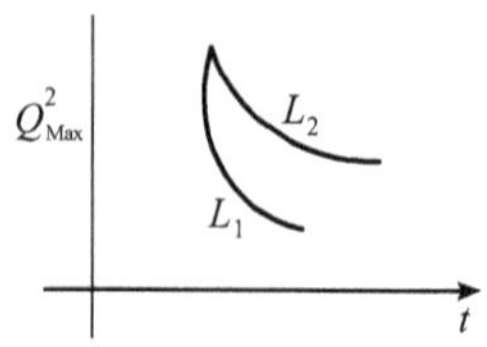

(C) 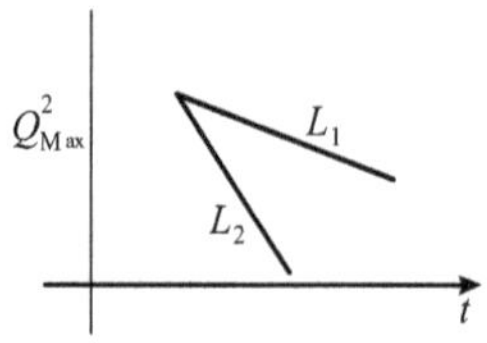

(D) 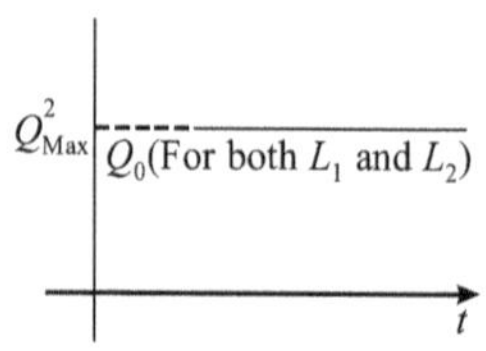

17. An inductor ($L = 0.03$ H) and a resistor ($R = 0.15$ kΩ) are connected in series to a battery of 15V EMF in a circuit shown below. The key K_1 has been kept closed for a long time. Then at $t = 0$, K_1 is opened and key K_2 is closed simultaneously. At $t = 1$ ms, the current in the circuit will be: ($e^5 \cong 150$) : **[2015]**

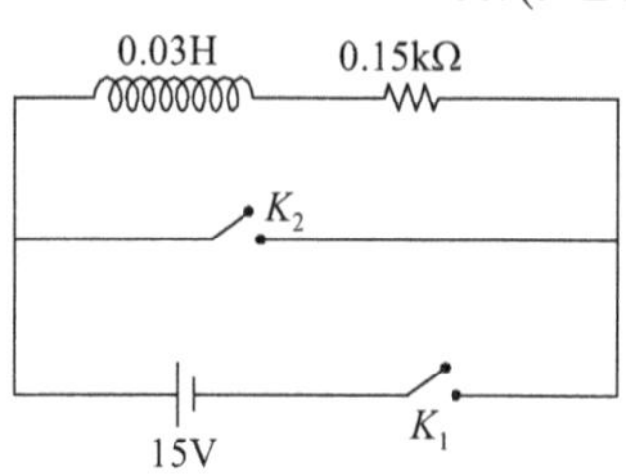

(A) 100 mA (B) 67 mA

(C) 6.7 mA (D) 0.67 mA

18. When current in a coil changes from 5 A to 2A in 0.1 s, average voltage of 50 V is produced. The self-inductance of the coil is : **[2015 (Online)]**

(A) 6 H (B) 0.67 H

(C) 3 H (D) 1.67 H

19. In the circuits (a) and (b) switches S_1 and S_2 are closed at $t = 0$ and are kept closed for a long time. The variation of current in the two circuits for $t \geq 0$ are roughly shown by figure (figures are schematic and not drawn to scale) : **[2015 (Online)]**

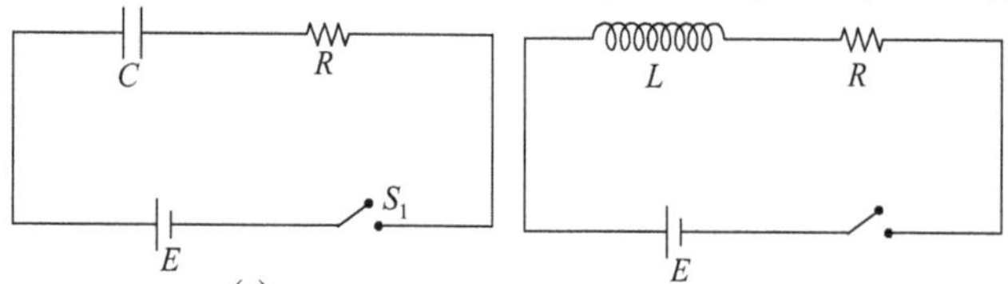

(a) (b)

(A) 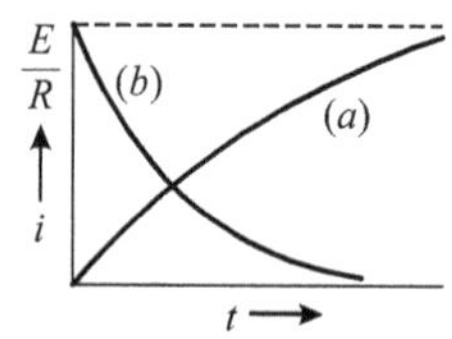

(B) 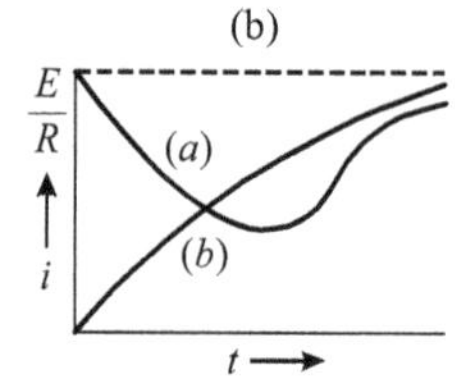

(C) 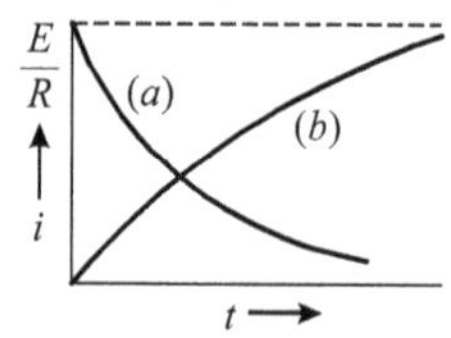

(D) 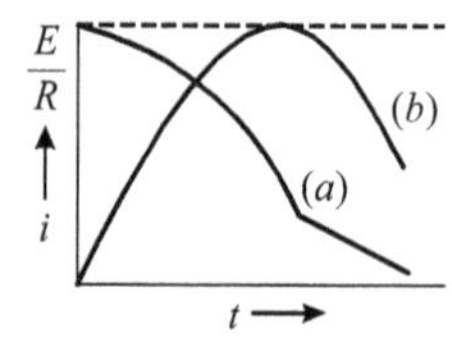

20. A power transmission line feeds input power at 2300 V to a step down transformer with its primary windings having 4000 turns. The output power is delivered at 230 V by the transformer. If the current in the primary of the transformer is 5 A and its efficiency is 90%, the output current would be : **[9 Jan. 2019 (Evening)]**

(A) 25 A (B) 50 A

(C) 35 A (D) 45 A

21. The self induced emf of a coil is 25 volt. When the current in it is changed at uniform rate from 10 A to 25 A in 1 s, the change in the energy of the inductance is : **[10 Jan. 2019 (Evening)]**

(A) 437.5 J (B) 637.5 J

(C) 740 J (D) 540 J

22. In the circuit shown, the switch S_1 is closed at time $t = 0$ and the switch S_2 is kept open. At some later time (t_0), the switch S_1 is opened and S_2 is closed. The behaviour of the current I as a function of time 't' is given by : **[11 Jan. 2019 (Morning)]**

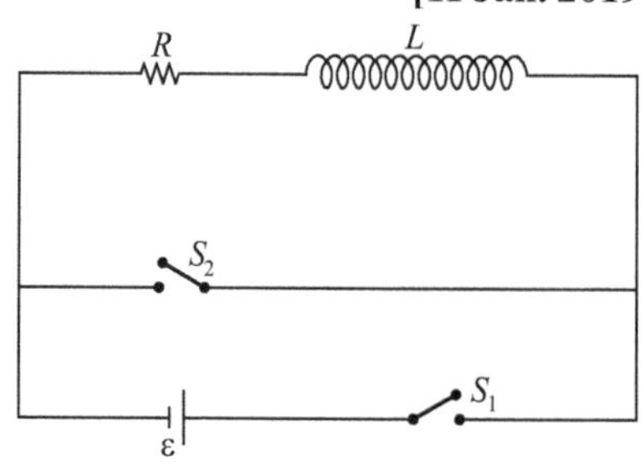

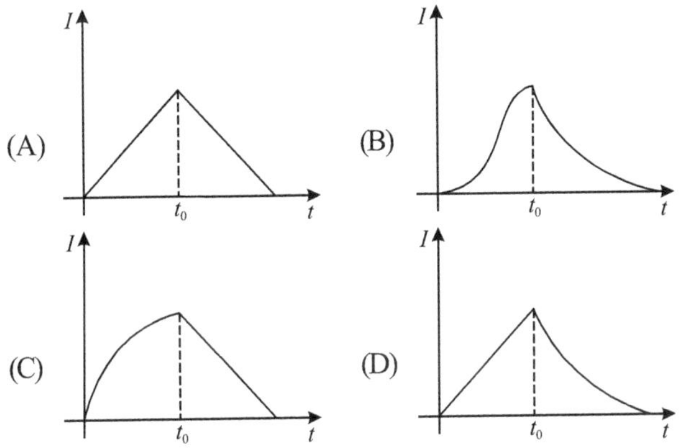

23. There are two long co-axial solenoids of same length l. The inner and outer coils have radii r_1 and r_2 and number of turns per unit length n_1 and n_2 respectively. The ratio of mutual inductance to the self-inductance of the inner-coil is :

[11 Jan. 2019 (Morning)]

(A) $\frac{n_2}{n_1}\cdot\frac{r_2^2}{r_1^2}$ (B) $\frac{n_2}{n_1}\cdot\frac{r_1}{r_2}$

(C) $\frac{n_1}{n_2}$ (D) $\frac{n_2}{n_1}$

24. A copper wire is wound on a wooden frame, whose shape is that of an equilateral triangle. If the linear dimension of each side of the frame is increased by a factor of 3, keeping the number of turns of the coil per unit length of the frame the same, then the self inductance of the coil : **[11 Jan. 2019 (Evening)]**

(A) Decreases by a factor of $9\sqrt{3}$
(B) Increases by a factor of 3
(C) Decreases by a factor of 9
(D) Increases by a factor of 27

25. In the figure shown, a circuit contains two identical resistors with resistance $R = 5\ \Omega$ and an inductance with $L = 2$ mH. An ideal battery of 15 V is connected in the circuit. What will be the current through the battery long after the switch is closed?

[12 Jan. 2019 (Morning)]

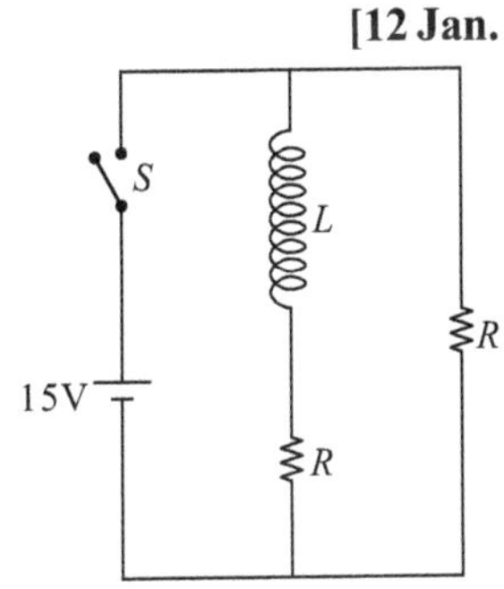

(A) 6 A (B) 7.5 A
(C) 5.5 A (D) 3 A

26. The total number of turns and cross-section area in a solenoid is fixed. However, its length L is varied by adjusting the separation between windings. The inductance of solenoid will be proportional to : **[9 April 2019 (Morning)]**

(A) $1/L^2$ (B) $1/L$
(C) L (D) L^2

27. Two coils 'P' and 'Q' are separated by some distance. When a current of 3 A flows through coil 'P', a magnetic flux of 10^{-3}Wb passes through 'Q'. No current is passed through 'Q'. When no current passes through 'P' and a current of 2A passes through 'Q', the flux through 'P' is : **[9 April 2019 (Evening)]**

(A) 6.67×10^{-3} Wb (B) 6.67×10^{-4} Wb
(C) 3.67×10^{-4} Wb (D) 3.67×10^{-3} Wb

28. A very long solenoid of radius R is carrying current $I(t) = kte^{-\alpha t}(k > 0)$, as a function of time $(t \geq 0)$. Counter clockwise current is taken to be positive. A circular conducting coil of radius $2R$ is placed in the equatorial plane of the solenoid and concentric with the solenoid. The current induced in the outer coil is correctly depicted, as a function of time, by :

[9 April 2019 (Evening)]

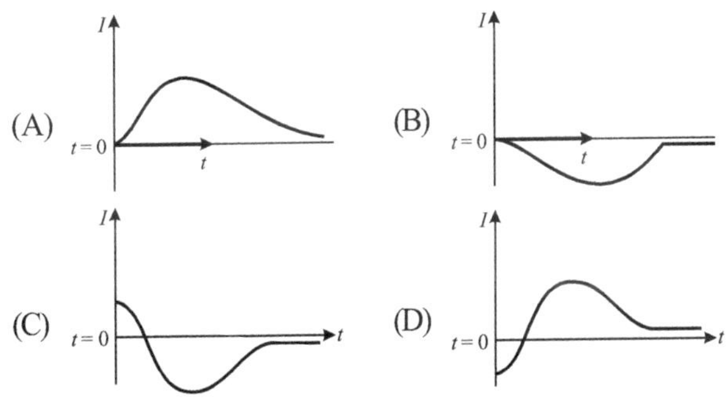

29. A coil of self inductance 10mH and resistance 0.1Ω is connected through a switch to a battery of internal resistance 0.9Ω. After the switch is closed, the time taken for the current to attain 80% of the saturation value is : (Take ln 5 = 1.6) :

[10 April 2019 (Evening)]

(A) 0.103 s (B) 0.016 s
(C) 0.002 s (D) 0.324 s

30. Consider the LR circuit shown in the figure. If the switch S is closed at $t = 0$ then the amount of charge that passes through the battery between $t = 0$ and $t = \frac{L}{R}$ is :

[12 April 2019 (Evening)]

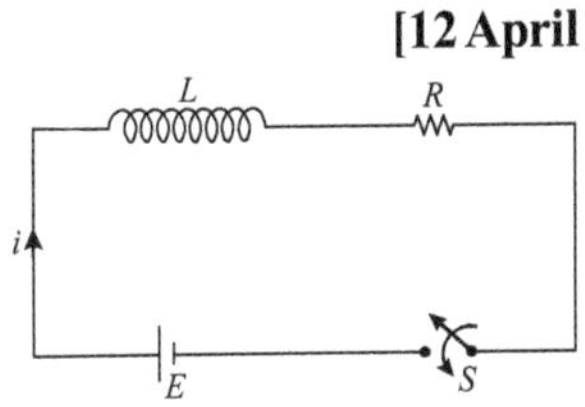

(A) $\frac{EL}{7.3R^2}$ (B) $\frac{EL}{2.7R^2}$

(C) $\frac{7.3EL}{R^2}$ (D) $\frac{2.7EL}{R^2}$

* * * * *

CHAPTER-21 ALTERNATING CURRENT

21.1 Alternating Current

1. In a transformer, number of turns in the primary are 140 and that in the secondary are 280. If current in primary is 4 A, then that in the secondary is : **[2002]**

(A) 4 A (B) 2 A
(C) 6 A (D) 10 A

2. The power factor of an AC circuit having resistance R and inductance L (connected in series) and an angular velocity ω is: **[2002]**

(A) $\frac{R}{\omega L}$ (B) $\frac{R}{(R^2+\omega^2 L^2)^{1/2}}$
(C) $\frac{\omega L}{R}$ (D) $\frac{R}{(R^2-\omega^2 L^2)^{1/2}}$

3. The core of any transformer is laminated so as to : **[2003]**

(A) reduce the energy loss due to eddy currents
(B) make it light weight
(C) make it robust and strong
(D) increase the secondary voltage

4. In an oscillating LC circuit the maximum charge on the capacitor is Q. The charge on the capacitor when the energy is stored equally between the electric and magnetic fields is : **[2003]**

(A) $Q/2$ (B) $Q/\sqrt{3}$
(C) $Q/\sqrt{2}$ (D) Q

5. A metal wire of linear mass density of 9.8 g/m is stretched with a tension of 10 kg-wt between two rigid supports 1 m apart. The wire passes at its middle point between the poles of a permanent magnet and it vibrates in resonance when carrying an alternating current of frequency n. The frequency n of the alternating source is : **[2003]**

(A) 50 Hz (B) 100 Hz
(C) 200 Hz (D) 25 Hz

6. In an LCR circuit, capacitance is changed from C to 2 C. For the resonant frequency to remain unchanged, the inductance should be changed from L to : **[2004]**

(A) $4L$ (B) $2L$
(C) $L/2$ (D) $L/4$

7. In an LCR series ac circuit, the voltage across each of the components. L, C and R is 50 V. The voltage across the LC combination will be : **[2004]**

(A) 50 V (B) $50\sqrt{2}V$
(C) 100 V (D) 0

8. Alternating current can not be measured by DC ammeter because : **[2004]**

(A) AC cannot pass through DC ammeter
(B) AC changes direction
(C) Average value of current for complete cycle is zero
(D) DC ammeter will get damaged

9. The phase difference between the alternating current and emf is $\pi/2$. which of the following cannot be the constituent of the circuit ? **[2005]**

(A) C alone (B) R, L
(C) L, C (D) L alone

10. A circuit has a resistance of 12 Ω and an impedance of 15 Ω. The power factor of circuit will be : **[2005]**

(A) 0.8 (B) 0.4
(C) 1.25 (D) 0.125

11. The self-inductance of the motor of an electric fan is 10 H. In order to impart maximum power at 50 Hz, it should be connected to a capacitance of : **[2005]**

(A) 4 μF (B) 8 μF
(C) 1 μF (D) 2 μF

12. In an AC generator, a coil with N turns, all of the same area A and total resistance R, rotates with frequency ω in a magnetic field B. The maximum value of emf generated in the coil is : **[2006]**

(A) $NABR\omega$ (B) NAB
(C) $NABR$ (D) $NAB\omega$

13. In a series resonant LCR circuit, the voltage across R is 100 volts and R = 1 kΩ with C = 2μF. The resonant frequency ω is 200 rad/s. At resonance the voltage across L is : **[2006]**

(A) 2.5×10^{-2} V (B) 40 V
(C) 250 V (D) 4×10^{-3} V

14. In an AC circuit the voltage applied is $E = E_0 \sin\omega t$. The resulting current in the circuit is $I = I_0 \sin\left(\omega t - \frac{\pi}{2}\right)$. The power consumption in the circuit is given by : **[2007]**

(A) $P = \frac{E_0 I_0}{\sqrt{2}}$ (B) P = zero
(C) $P = \frac{E_0 I_0}{2}$ (D) $P = \sqrt{2}\, E_0 I_0$

15. In a series LCR circuit R = 200 Ω and the voltage and the frequency of the main supply is 220 V and 50 Hz respectively.

On taking out the capacitance from the circuit the current lags behind the voltage by 30°. On taking out the inductor from the circuit the current leads the voltage by 30°. The power dissipated in the LCR circuit is : **[2010]**

(A) 242 W (B) 305 W
(C) 210 W (D) Zero

16. A fully charged capacitor C with initial charge q_0 is connected to a coil of self inductance L at $t = 0$. The time at which the energy is stored equally between the electric and the magnetic fields is : **[2011]**

(A) $2\pi\sqrt{LC}$ (B) $\sqrt{LC}$
(C) $\pi\sqrt{LC}$ (D) $\frac{\pi}{4}\sqrt{LC}$

17. The capacitor of an oscillatory circuit is enclosed in a container. When the container is evacuated, the resonance frequency of the circuit is 10 kHz. When the container is filled with a gas, the resonance frequency changes by 50 Hz. The dielectric constant of the gas is : **[2012 (Online)]**

(A) 1.001 (B) 2.001
(C) 1.01 (D) 3.01

18. In an LCR circuit shown in the following figure, what will be the readings of the voltmeter across the resistor and ammeter if an a.c. source of 220 V and 100 Hz is connected to it as shown ? **[2012 (Online)]**

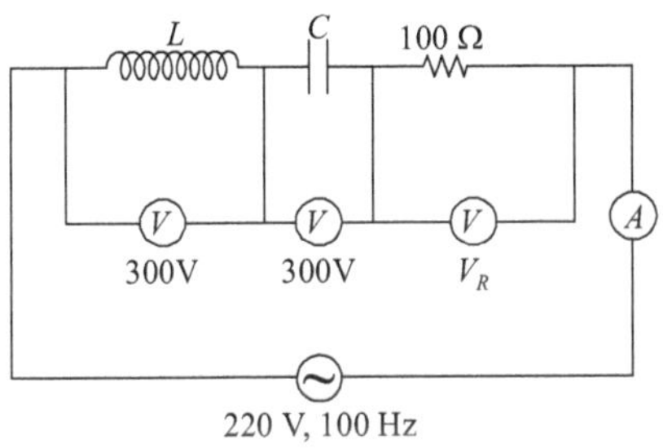

(A) 800 V, 8 A (B) 110 V, 1.1 A
(C) 300 V, 3 A (D) 220 V, 2.2 A

19. An LCR circuit as shown in the figure is connected to a voltage source V_{ac} whose frequency can be varied. The frequency, at which the voltage across the resistor is maximum, is : **[2013 (Online)]**

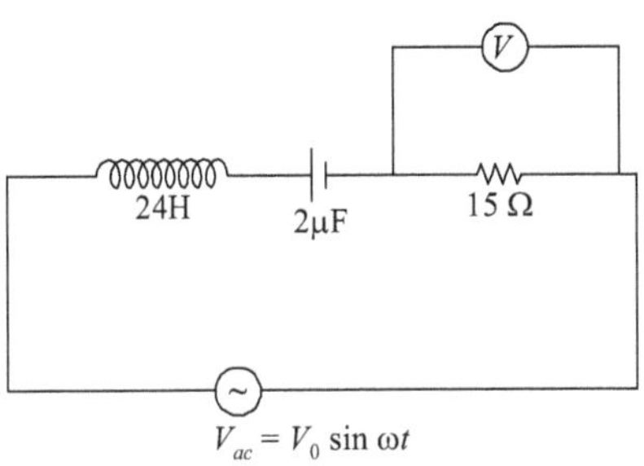

(A) 902 Hz (B) 143 Hz
(C) 23 Hz (D) 345 Hz

20. In a series L-C-R circuit, $C = 10^{-11}$ Farad, $L = 10^{-5}$ Henry and $R = 100$ Ohm, when a constant D.C. voltage E is applied to the circuit, the capacitor acquires a charge 10^{-9} C. The D.C. source is replaced by a sinusoidal voltage source in which the peak voltage E_0 is equal to the constant D.C. voltage E. At resonance the peak value of the charge acquired by the capacitor will be : **[2013 (Online)]**

(A) 10^{-15} C (B) 10^{-6} C
(C) 10^{-10} C (D) 10^{-8} C

21. The plot given below is of the average power delivered to and LRC circuit versus frequency. The quality factor of the circuit is : **[2013 (Online)]**

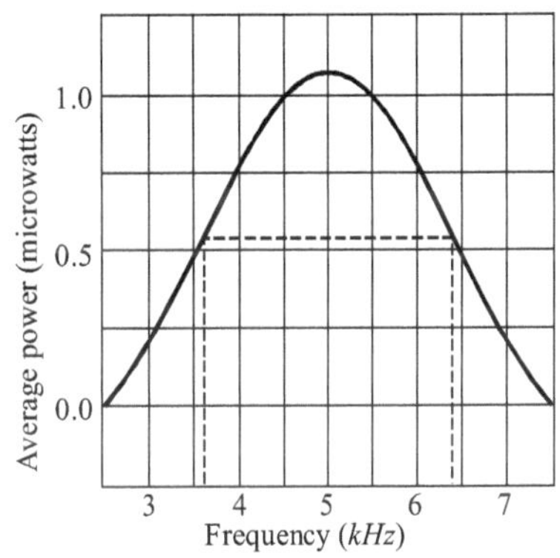

(A) 5.0 (B) 2.0
(C) 2.5 (D) 0.4

22. When resonance is produced in a series LCR circuit, then which of the following is not correct ? **[2013 (Online)]**

(A) Current in the circuit is in phase with the applied voltage
(B) Inductive and capacitive reactances are equal
(C) If R is reduced, the voltage across capacitor will increase
(D) Impedance of the circuit is maximum

23. A series LR circuit is connected to an ac source of frequency ω and the inductive reactance is equal to 2 R. A capacitance of capacitive reactance equal to R is added in series with L and R. The ratio of the new power factor to the old one is : **[2013 (Online)]**

(A) $\sqrt{\frac{2}{3}}$ (B) $\sqrt{\frac{2}{5}}$
(C) $\sqrt{\frac{3}{2}}$ (D) $\sqrt{\frac{5}{2}}$

24. When the rms voltages V_L, V_C and V_R are measured respectively across the inductor L, the capacitor C and the resistor R in a series LCR circuit connected to an AC source, it is found that the ratio $V_L : V_C : V_R = 1 : 2 : 3$. If the rms voltage of the AC sources is 100 V, the V_R is close to : **[2014 (Online)]**

(A) 50 V (B) 70 V
(C) 90 V (D) 100 V

25. A sinusoidal voltage $V(t) = 100 \sin(500t)$ is applied across a pure inductance of $L = 0.02$ H. The current through the coil is: **[2014 (Online)]**

(A) $10 \cos(500\,t)$ (B) $-10 \cos(500t)$
(C) $10 \sin(500t)$ (D) $-10 \sin(500t)$

26. The AC voltage across a resistance can be measured using a : **[2015 (Online)]**

(A) Hot wire voltmeter
(B) Moving coil galvanometer
(C) Potential coil galvanometer
(D) Moving magnet galvanometer

27. For the LCR circuit, shown here, the current is observed to lead the applied voltage. An additional capacitor C', when joined with the capacitor C present in the circuit, makes the power factor of the circuit unity. The capacitor C', must have been connected in : **[2015 (Online)]**

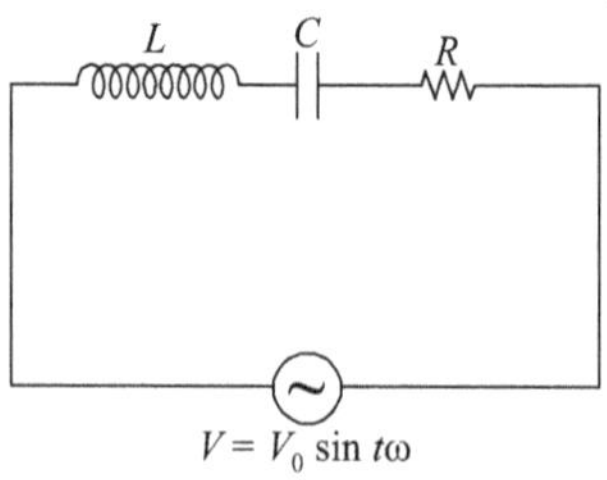

(A) Series with C and has a magnitude $\dfrac{C}{(\omega^2 LC - 1)}$

(B) Series with C and has a magnitude $\dfrac{1-\omega^2 LC}{\omega^2 L}$

(C) Parallel with C and has a magnitude $\dfrac{1-\omega^2 LC}{\omega^2 L}$

(D) Parallel with C and has a magnitude $\dfrac{C}{(\omega^2 LC - 1)}$

28. An arc lamp requires a direct current of 10 A at 80 V to function. If it is connected to a 220 V (rms), 50 Hz AC supply, the series inductor needed for it to work is close to : **[2016]**

(A) 80 H (B) 0.08 H
(C) 0.044 H (D) 0.065 H

29. A series LR circuit is connected to a voltage source with $V(t) = V_0 \sin \omega t$. After very large time, current (t) behaves as $\left(t_0 >> \dfrac{L}{R}\right)$: **[2016 (Online)]**

(A) 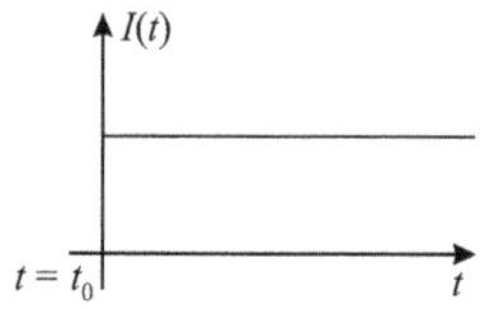

(B) 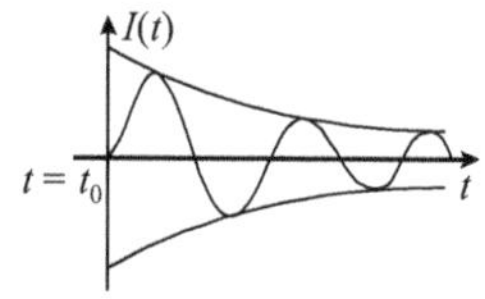

(C) 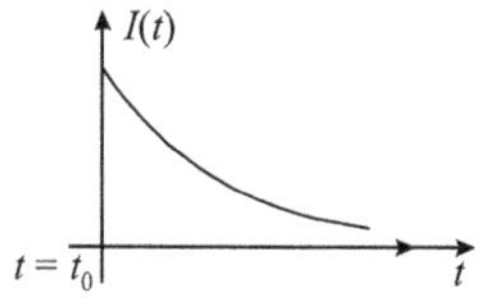

(D) 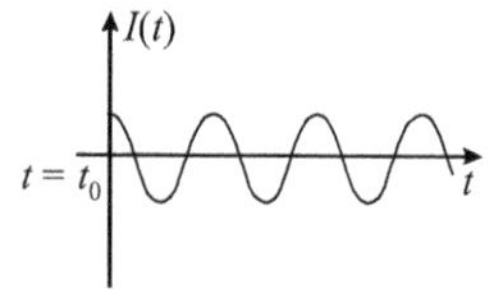

30. A sinusoidal voltage of peak value 283 V and angular frequency 320/s is applied to a series LCR circuit. Given that $R = 5\,\Omega$, $L = 25$ mH and $C = 1000\,\mu F$. The total impedance, and phase difference between the voltage across the source and the current will respectively be : **[2017 (Online)]**

(A) $10\,\Omega$ and $\tan^{-1}\left(\dfrac{5}{3}\right)$ (B) $7\,\Omega$ and 45°

(C) $10\,\Omega$ and $\tan^{-1}\left(\dfrac{8}{3}\right)$ (D) $7\,\Omega$ and $\tan^{-1}\left(\dfrac{5}{3}\right)$

31. A small circular loop of wire of radius a is located at the centre of a much larger circular wire loop of radius b. The two loops are in the same plane. The outer loop of radius b carries an alternating current $I = I_0 \cos(\omega t)$. The emf induced in the smaller inner loop is nearly : **[2017 (Online)]**

(A) $\dfrac{\pi\mu_0 I_0}{2} \cdot \dfrac{a^2}{b}\, \omega \sin(\omega t)$ (B) $\dfrac{\pi\mu_0 I_0}{2} \cdot \dfrac{a^2}{b}\, \omega \cos(\omega t)$

(C) $\pi\mu_0 I_0 \dfrac{a^2}{b}\, \omega \sin(\omega t)$ (D) $\dfrac{\pi\mu_0 I_0 b^2}{a}\, \omega \cos(\omega t)$

32. For an RLC circuit driven with voltage of amplitude v_m and frequency $\omega_0 = \dfrac{1}{\sqrt{LC}}$ the current exhibits resonance. The quality factor, Q is given by : **[2018]**

(A) $\dfrac{\omega_0 L}{R}$ (B) $\dfrac{\omega_0 R}{L}$

(C) $\dfrac{R}{(\omega_0 C)}$ (D) $\dfrac{CR}{\omega_0}$

33. In an AC circuit, the instantaneous e.m.f. and current are given by

$e = 100 \sin 30t$

$i = 20 \sin\left(30t - \dfrac{\pi}{4}\right)$

In one cycle of AC, the average power consumed by the circuit and the wattless current are, respectively : **[2018]**

(A) 50 W, 10 A (B) $\frac{1000}{\sqrt{2}}$ W, 10 A

(C) $\frac{50}{\sqrt{2}}$ W, 0 (D) 50 W, 0

34. A power transmission line feeds input power at 2300 V to a step down transformer with its primary windings having 4000 turns, given the output power at 230 V. If the current in the primary of the transformer is 5 A, and its efficiency is 90%, the output current would be : **[2018 (Online)]**

(A) 20 A (B) 40 A
(C) 45 A (D) 25 A

35. A series *AC* circuit containing an inductor (20 mH), a capacitor (120 μF) and a resistor (60 Ω) is driven by an *AC* source of 24 V/50 Hz. The energy dissipated in the circuit in 60 s is : **[9 Jan. 2019 (Evening)]**

(A) 2.26×10^3 J (B) 3.39×10^3 J
(C) 5.65×10^2 J (D) 5.17×10^2 J

36. 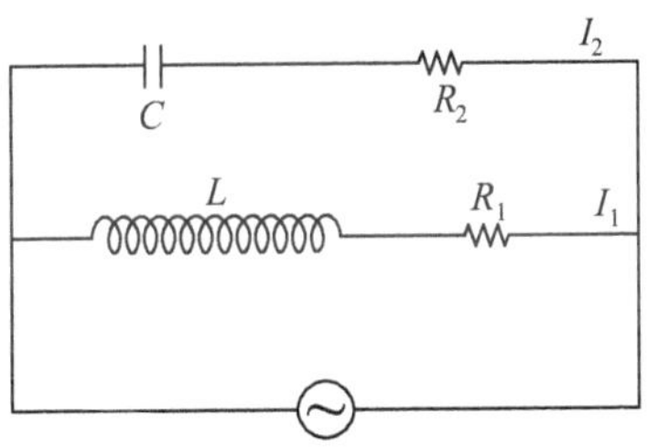

In the above circuit, $C = \frac{\sqrt{3}}{2}\mu F$, $R_2 = 20\ \Omega$, $L = \frac{\sqrt{3}}{10}$ H and $R_1 = 10\ \Omega$. Current in L-R_1 path is I_1 and in C-R_2 path it is I_2. The voltage of A.C source is given by $V = 200\sqrt{2}\sin(100t)$ volts. The phase difference between I_1 and I_2 is : **[12 Jan. 2019 (Evening)]**

(A) 30° (B) 0°
(C) 90° (D) 60°

37. An alternating voltage $v(t) = 220 \sin 100\,\pi t$ volt is applied to a purely resistance load of 50 Ω. The time taken for the current to rise from half of the peak value to the peak value is : **[8 April 2019 (Morning)]**

(A) 2.2 ms (B) 5 ms
(C) 3.3 ms (D) 7.2 ms

38. A circuit connected to an ac source of *emf* $e = e_0 \sin(100t)$ with t in seconds, gives a phase difference of $\pi/4$ between the emf e and current i. Which of the following circuits will exhibit this ? **[8 April 2019 (Evening)]**

(A) *RC* circuit with R = 1 kΩ and *C* = 1 μF
(B) *RL* circuit with R = 1 kΩ and *L* = 1 mH
(C) *RL* circuit with R = 1 kΩ and *L* = 10 mH
(D) *RC* circuit with R = 1 kΩ and *C* = 10 μF

39. A transformer consisting of 300 turns in the primary and 150 turns in the secondary gives output power of 2.2 kW. If the current in the secondary coil is 10A, then the input voltage and current in the primary coil are : **[10 April 2019 (Morning)]**

(A) 220 V and 10A (B) 440 V and 5A
(C) 440 V and 20 A (D) 220 V and 20 A

* * * * *

CHAPTER-22 ELECTROMAGNETIC WAVES

22.1 Electromagnetic Wave

1. Electromagnetic waves are transverse in nature is evident by : **[2002]**

(A) Polarization (B) Interference
(C) Reflection (D) Diffraction

2. An electromagnetic wave of frequency $v = 3.0$ MHz passes from vacuum into a dielectric medium with permittivity $\in = 4.0$. Then : **[2004]**

(A) Wave length is halved and frequency remains unchanged
(B) Wave length is doubled and frequency becomes half
(C) Wave length is doubled and the frequency remains unchanged
(D) Wave length and frequency both remain unchanged

3. If a source of power 4 kW produces 10^{20} photons/second, the radiation belongs to apart of the spectrum called : **[2010]**

(A) γ-rays (B) X-rays
(C) ultraviolet rays (D) microwaves

4. An electromagnetic wave in vacuum has the electric and magnetic field $\vec{E}$ and $\vec{B}$, which are always perpendicular to each other. The direction of polarization is given by $\vec{X}$ and that of wave propagation by $\vec{k}$. Then : **[2012]**

(A) $\vec{X} \parallel \vec{B}$ and $\vec{k} \parallel \vec{B} \times \vec{E}$ (B) $\vec{X} \parallel \vec{E}$ and $\vec{k} \parallel \vec{E} \times \vec{B}$
(C) $\vec{X} \parallel \vec{B}$ and $\vec{k} \parallel \vec{E} \times \vec{B}$ (D) $\vec{X} \parallel \vec{E}$ and $\vec{k} \parallel \vec{B} \times \vec{E}$

5. The frequency of X-rays; γ-rays and ultraviolet rays are respectively a, b and c then : **[2012 (Online)]**

(A) $a < b; b > c$ (B) $a > b; b > c$
(C) $a < b < c$ (D) $a = b = c$

6. An electromagnetic wave with frequency ω and wavelength λ travels in the $+y$ direction. Its magnetic field is along $+x$-axis. The vector equation for the associated electric field (of amplitude E_0) is : **[2012 (Online)]**

(A) $\vec{E} = -E_0 \cos\left(\omega t + \frac{2\pi}{\lambda} y\right)\hat{x}$

(B) $\vec{E} = E_0 \cos\left(\omega t - \frac{2\pi}{\lambda} y\right)\hat{x}$

(C) $\vec{E} = E_0 \cos\left(\omega t - \frac{2\pi}{\lambda} y\right)\hat{z}$

(D) $\vec{E} = -E_0 \cos\left(\omega t + \frac{2\pi}{\lambda} y\right)\hat{z}$

7. The magnetic field in a travelling electromagnetic wave has a peak value of 20 nT. The peak value of electric field strength is : **[2013]**

(A) 3 V/m (B) 6 V/m
(C) 9 V/m (D) 12 V/m

8. Photons of an electromagnetic radiation has an energy 11 keV each. To which region of electromagnetic spectrum does it belong ? **[2013 (Online)]**

(A) X-rays region (B) Ultra violet region
(C) Infrared region (D) Visible region

9. Select the correct statement from the following : **[2013 (Online)]**

(A) Electromagnetic waves cannot travel in vacuum
(B) Electromagnetic waves are longitudinal waves
(C) Electromagnetic waves are produced by charges moving with uniform velocity
(D) Electromagnetic waves carry both energy and momentum as they propagate through space

10. A plane electromagnetic wave in a non-magnetic dielectric medium is given by $\vec{E} = \vec{E}_0 (4 \times 10^{-7} x - 50t)$ with distance being in meter and time in seconds. The dielectric constant of the medium is : **[2013 (Online)]**

(A) 2.4 (B) 5.8
(C) 8.2 (D) 4.8

11. During the propagation of electromagnetic waves in a medium : **[2014]**

(A) Electric energy density is double of the magnetic energy density
(B) Electric energy density is half of the magnetic energy density
(C) Electric energy density is equal to the magnetic energy density
(D) Both electric and magnetic energy densities are zero

12. If microwaves, X-rays, infrared, gamma rays, ultra-violet, radio waves and visible parts of the electromagnetic spectrum are denoted by M, X, I, G, U, R and V then which of the following is the arrangement in ascending order of wavelength ? **[2014 (Online)]**

(A) R, M, I, V, U, X and G (B) M, R, V, X, U, G and I
(C) G, X, U, V, I, M and R (D) I, M, R, U, V, X and G

13. An electromagnetic wave of frequency 1×10^{14} hertz is propagating along z-axis. The amplitude of electric field is 4 V/m. If $\varepsilon_0 = 8.8 \times 10^{-12}$ C^2/N-m^2, then average energy density of electric field will be : **[2014 (Online)]**

(A) 35.2×10^{-10} J/m^3 (B) 35.2×10^{-11} J/m^3
(C) 35.2×10^{-12} J/m^3 (D) 35.2×10^{-13} J/m^3

14. A lamp emits monochromatic green light uniformly in all directions. The lamp is 3% efficient in converting electrical power to electromagnetic waves and consumes 100 W of power. The amplitude of the electric field associated with the electromagnetic radiation at a distance of 5 m from the lamp will be nearly : **[2014 (Online)]**

(A) 1.34 V/m (B) 2.68 V/m
(C) 4.02 V/m (D) 5.36 V/m

15. An electromagnetic wave travelling in the x-direction has frequency of 2×10^{14} Hz and electric field amplitude of 27 Vm^{-1}. From the options given below, which one describes the magnetic field for this wave ? **[2015 (Online)]**

(A) $\vec{B}(x,t) = (3\times10^{-8}T)\hat{j}$; $\sin [2\pi(1.5 \times 10^{-8}x - 2 \times 10^{14}t)]$

(B) $\vec{B}(x,t) = (9\times10^{-8}T)\hat{i}$; $\sin [2\pi(1.5 \times 10^{-8}x - 2 \times 10^{14}t)]$

(C) $\vec{B}(x,t) = (9\times10^{-8}T)\hat{j}$; $\sin [1.5 \times 10^{-6}x - 2 \times 10^{14}t]$

(D) $\vec{B}(x,t) = (9\times10^{-8}T)\hat{k}$; $\sin [2\pi(1.5 \times 10^{-6}x - 2 \times 10^{14}t)]$

16. For plane electromagnetic waves propagating in the z-direction, which one of the following combination gives the correct possible direction for $\vec{E}$ and $\vec{B}$ field respectively ? **[2015 (Online)]**

(A) $(2\hat{i}+3\hat{j})$ and $(\hat{i}+2\hat{j})$

(B) $(-2\hat{i}-3\hat{j})$ and $(3\hat{i}-2\hat{j})$

(C) $(3\hat{i}+4\hat{j})$ and $(4\hat{i}-3\hat{j})$

(D) $(\hat{i}+2\hat{j})$ and $(2\hat{i}-\hat{j})$

17. Arrange the following electromagnetic radiations per quantum in the order of increasing energy : **[2016]**

A : Blue B : Yellow light
C : X-ray D : Radiowave
(A) C, A, B, D (B) B, A, D, C
(C) D, B, A, C (D) A, B, D, C

18. Microwave oven acts on the principle of : **[2016 (Online)]**
(A) Giving rotational energy to water molecules
(B) Giving translational energy to water molecules
(C) Giving vibrational energy to water molecules
(D) Transferring electrons from lower to higher energy levels in water molecule

19. Consider an electromagnetic wave propagating in vacuum. Choose the correct statement : **[2016 (Online)]**
(A) For an electromagnetic wave propagating in $+y$ direction the electric field is $\vec{E} = \frac{1}{\sqrt{2}}E_{yz}(x,t)\hat{z}$ and the magnetic field is $\vec{B} = \frac{1}{\sqrt{2}}B_z(x,t)\hat{y}$
(B) For an electromagnetic wave propagating in $+y$ direction the electric field is $\vec{E} = \frac{1}{\sqrt{2}}E_{yz}(x,t)\hat{y}$ and the magnetic field is $\vec{B} = \frac{1}{\sqrt{2}}B_{yz}(x,t)\hat{z}$
(C) For an electromagnetic wave propagating in $+x$ direction the electric field is $\vec{E} = \frac{1}{\sqrt{2}}E_{yz}(y,z,t)\ (\hat{y}+\hat{z})$ and the magnetic field is $\vec{B} = \frac{1}{\sqrt{2}}B_{yz}(y,z,t)(\hat{y}+\hat{z})$
(D) For an electromagnetic wave propagating in $+x$ direction the electric field is $\vec{E} = \frac{1}{\sqrt{2}}E_{yz}(x,t)(\hat{y}-\hat{z})$ and the magnetic field is $\vec{B} = \frac{1}{\sqrt{2}}B_{yz}(x,t)(\hat{y}+\hat{z})$

20. Magnetic field in a plane electromagnetic wave is given by

$\vec{B} = B_0 \sin (kx + \omega t)\ \hat{j}T$

Expression for corresponding electric field will be :
Where c is speed of light. **[2017 (Online)]**

(A) $\vec{E} = B_0 c \sin (kx + \omega t)\ \hat{k}$ V/m

(B) $\vec{E} = \frac{B_0}{c} \sin (kx + \omega t)\ \hat{k}$ V/m

(C) $\vec{E}\ - B_0 c \sin (kx + \omega t)\ \hat{k}$ V/m

(D) $\vec{E} = B_0 c \sin (kx - \omega t)\ \hat{k}$ V/m

21. The electric field component of a monochromatic radiation is given by

$$\vec{E} = 2E_0\hat{i} \cos kz \cos \omega t$$

Its magnetic field $\vec{B}$ is then given by : **[2017 (Online)]**

(A) $\frac{2E_0}{c}\hat{j}$ sin kz cos ωt (B) $\frac{2E_0}{c}\hat{j}$ cos kz sin ωt

(C) $\frac{2E_0}{c}\hat{j}$ sin kz sin ωt (D) $\frac{2E_0}{c}\hat{j}$ cos kz cos ωt

22. An EM wave from air enters a medium. The electric fields are $\vec{E}_1 = E_{01}\hat{x} \cos\left[2\pi v\left(\frac{z}{c}-t\right)\right]$ in air and $\vec{E}_2 = E_{02}\hat{x} \cos [k(2z - ct)]$ in medium, where the wave number k and frequency v refer to their values in air. The medium is nonmagnetic. If $\in_{r_1}$ and $\in_{r_2}$ refer to relative permittivities of air and medium respectively, which of the following options is correct ? **[2018]**

(A) $\frac{\in_{r_1}}{\in_{r_2}} = 4$ (B) $\frac{\in_{r_1}}{\in_{r_2}} = 2$

(C) $\frac{\in_{r_1}}{\in_{r_2}} = \frac{1}{4}$ (D) $\frac{\in_{r_1}}{\in_{r_2}} = \frac{1}{2}$

23. A monochromatic beam of light has a frequency $\nu = \frac{3}{2\pi} \times 10^{12}$ Hz and is propagating along the direction $\frac{\hat{i}+\hat{j}}{\sqrt{2}}$. It is polarized along the $\hat{k}$ direction. The acceptable form for the magnetic field is : **[2018 (Online)]**

(A) $k\frac{E_0}{C}\left(\frac{\hat{i}-\hat{j}}{\sqrt{2}}\right)\cos\left[10^4\left(\frac{\hat{i}-\hat{j}}{\sqrt{2}}\right)\cdot\vec{r}-(3\times10^{12})t\right]$

(B) $\frac{E_0}{C}\left(\frac{\hat{i}-\hat{j}}{\sqrt{2}}\right)\cos\left[10^4\left(\frac{\hat{i}+\hat{j}}{\sqrt{2}}\right)\cdot\vec{r}-(3\times10^{12})t\right]$

(C) $\frac{E_0}{C}\hat{k}\cos\left[10^4\left(\frac{\hat{i}+\hat{j}}{\sqrt{2}}\right)\cdot\vec{r}+(3\times10^{12})t\right]$

(D) $\frac{E_0}{C}\frac{(\hat{i}+\hat{j}+\hat{k})}{\sqrt{3}}\cos\left[10^4\left(\frac{\hat{i}+\hat{j}}{\sqrt{2}}\right)\cdot\vec{r}+(3\times10^{12})t\right]$

24. A plane electromagnetic wave of wavelength λ has an intensity I. It is propagating along the positive Y-direction. The allowed expressions for the electric and magnetic fields are given by : **[2018 (Online)]**

(A) $\vec{E}=\sqrt{\frac{I}{\epsilon_0 c}}\cos\left[\frac{2\pi}{\lambda}(y-ct)\right]\hat{i}\ ;\ \vec{B}=\frac{1}{c}E\hat{k}$

(B) $\vec{E}=\sqrt{\frac{I}{\epsilon_0 c}}\cos\left[\frac{2\pi}{\lambda}(y-ct)\right]\hat{k}\ ;\ \vec{B}=-\frac{1}{c}E\hat{k}$

(C) $\vec{E}=\sqrt{\frac{2I}{\epsilon_0 c}}\cos\left[\frac{2\pi}{\lambda}(y-ct)\right]\hat{k}\ ;\ \vec{B}=+\frac{1}{c}E\hat{i}$

(D) $\vec{E}=\sqrt{\frac{2I}{\epsilon_0 c}}\cos\left[\frac{2\pi}{\lambda}(y+ct)\right]\hat{k}\ ;\ \vec{B}=\frac{1}{c}E\hat{i}$

25. A plane polarized monochromatic *EM* wave is travelling a vacuum along *z*-direction such that at $t = t_1$ it is found that the electric field is zero at a spatial point z_1. The next zero that occurs in its neighbourhood is at z_2. The frequency of the electromagnetic wave is : **[2018 (Online)]**

(A) $\frac{3\times10^8}{|z_2-z_1|}$ (B) $\frac{6\times10^8}{|z_2-z_1|}$

(C) $\frac{1.5\times10^8}{|z_2-z_1|}$ (D) $\frac{1}{t_1+\frac{|z_2-z_1|}{3\times10^8}}$

26. A plane electromagnetic wave of frequency 50 MHz travels in free space along the positive *x*-direction. At a particular point in space and time, $\vec{E}=6.3\hat{j}$ V/m. The corresponding magnetic field $\vec{B}$, at that point will be : **[9 Jan. 2019 (Morning)]**

(A) $18.9\times10^{-8}\ \hat{k}T$ (B) $6.6\times10^{-8}\ \hat{k}T$

(C) $2.1\times10^{-8}\ \hat{k}T$ (D) $18.9\times10^{8}\ \hat{k}T$

27. The energy associated with electric field is (U_E) and with magnetic field is (U_B) for an electromagnetic wave in free space. Then : **[9 Jan. 2019 (Evening)]**

(A) $U_E=\frac{U_B}{2}$ (B) $U_E<U_B$

(C) $U_E=U_B$ (D) $U_E>U_B$

28. If the magnetic field of a plane electromagnetic wave is given by (The speed of light = 3×10^8/m/s)

$$B=100\times10^{-6}\sin\left[2\pi\times2\times10^{15}\left(t-\frac{x}{c}\right)\right]$$

then the maximum electric field associated with it is : **[10 Jan. 2019 (Morning)]**

(A) 4×10^4 N/C (B) 4.5×10^4 N/C

(C) 6×10^4 N/C (D) 3×10^4 N/C

29. The electric field of a plane polarized electromagnetic wave in free space at time $t = 0$ is given by an expression $\vec{E}(x, y) = 10\hat{j}\cos[(6x+8z)]$.

The magnetic field $\vec{B}(x, z, t)$ is given by (*c* is the velocity of light) : **[10 Jan. 2019 (Evening)]**

(A) $\frac{1}{c}(6\hat{k}+8\hat{i})\cos[6x-8z+10ct]$

(B) $\frac{1}{c}(6\hat{k}-8\hat{i})\cos[6x+8z-10ct]$

(C) $\frac{1}{c}(6\hat{k}+8\hat{i})\cos[6x+8z-10ct]$

(D) $\frac{1}{c}(6\hat{k}-8\hat{i})\cos[6x+8z+10ct]$

30. An electromagnetic wave of intensity 50 Wm^{-2} enters in a medium of refractive index '*n*' without any loss. The ratio of the magnitudes of electric fields, and the ratio of the magnitudes of magnetic fields of the wave before and after entering into the medium are respectively, given by : **[11 Jan. 2019 (Morning)]**

(A) $\left(\frac{1}{\sqrt{n}},\frac{1}{\sqrt{n}}\right)$ (B) $\left(\sqrt{n},\frac{1}{\sqrt{n}}\right)$

(C) $(\sqrt{n},\sqrt{n})$ (D) $\left(\frac{1}{\sqrt{n}},\sqrt{n}\right)$

31. A 27 mW laser beam has a cross-sectional area of 10 mm^2. The magnitude of the maximum electric field in this electromagnetic wave is given by [Given permittivity of space $\epsilon_0 = 9\times10^{-12}$ SI units, Speed of light $c = 3\times10^8$ m/s]: **[11 Jan. 2019 (Evening)]**

(A) 1 kV/m (B) 2 kV/m

(C) 1.4 kV/m (D) 0.7 kV/m

32. A light wave is incident normally on a glass slab of refractive index 1.5. If 4% of light gets reflected and the amplitude of the electric field of the incident light is 30 V/m, then the amplitude of the electric field for the wave propagating in the glass medium will be : **[12 Jan. 2019 (Morning)]**

(A) 10 V/m (B) 24 V/m
(C) 30 V/m (D) 6 V/m

33. The mean intensity of radiation on the surface of the Sun is about 10^8 W/m². The rms value of the corresponding magnetic field is closest to : **[12 Jan. 2019 (Evening)]**

(A) 10^2 T (B) 10^{-4} T
(C) 1 T (D) 10^{-2} T

34. A plane electromagnetic wave travels in free space along the x-direction. The electric field component of the wave at a particular point of space and time is $E = 6\ \text{V m}^{-1}$ along y-direction. Its corresponding magnetic field component, B would be : **[8 April 2019 (Morning)]**

(A) 6×10^{-8} T along z-direction
(B) 6×10^{-8} T along x-direction
(C) 2×10^{-8} T along z-direction
(D) 2×10^{-8} T along y-direction

35. The magnetic field of an electromagnetic wave is given by.

$$\vec{B} = 1.6 \times 10^{-6} \cos(2 \times 10^7 z + 6 \times 10^{15} t)(2\hat{i} + \hat{j})\frac{\text{Wb}}{\text{m}^2}$$

The associated electric field will be : **[8 April 2019 (Evening)]**

(A) $\vec{E} = 4.8 \times 10^2 \cos(2 \times 10^7 z + 6 \times 10^{15} t)\ (\hat{i} - 2\hat{j})\frac{V}{m}$

(B) $\vec{E} = 4.8 \times 10^2 \cos(2 \times 10^7 z - 6 \times 10^{15} t)\ (2\hat{i} + \hat{j})\frac{V}{m}$

(C) $\vec{E} = 4.8 \times 10^2 \cos(2 \times 10^7 z - 6 \times 10^{15} t)\ (-2\hat{j} + \hat{i})\frac{V}{m}$

(D) $\vec{E} = 4.8 \times 10^2 \cos(2 \times 10^7 z + 6 \times 10^{15} t)\ (-\hat{i} + 2\hat{j})\frac{V}{m}$

36. The magnetic field of a plane electromagnetic wave is given by $\vec{B} = B_0 \hat{i}\,[\cos(kz - \omega t)] + B_1 \hat{j} \cos(kz + \omega t)$ where $B_0 = 3 \times 10^{-5}$ T and $B_1 = 2 \times 10^{-6}$ T. The rms value of the force experienced by a stationary charge $Q = 10^{-4}$ C at $z = 0$ is closest to : **[9 April 2019 (Morning)]**

(A) 0.9 N (B) 0.1 N
(C) 3×10^{-2} N (D) 0.6 N

37. The electric field of a plane electromagnetic wave is given by

$$\vec{E} = E_0 \hat{i} \cos(kz)\cos(\omega t)$$

The corresponding magnetic field $\vec{B}$ is then given by : **[10 April 2019 (Morning)]**

(A) $\vec{B} = \frac{E_0}{C}\hat{j}\sin(kz)\cos(\omega t)$ (B) $\vec{B} = \frac{E_0}{C}\hat{j}\sin(kz)\sin(\omega t)$

(C) $\vec{B} = \frac{E_0}{C}\hat{k}\sin(kz)\cos(\omega t)$ (D) $\vec{B} = \frac{E_0}{C}\hat{j}\cos(kz)\sin(\omega t)$

38. An electromagnetic wave is represented by the electric field $\vec{E} = E_0 \hat{n} \sin[\omega t + (6y - 8z)]$. Taking unit vectors in x, y and z directions to be $\hat{i}, \hat{j}, \hat{k}$, the direction of propogation $\hat{s}$, is: **[12 April 2019 (Morning)]**

(A) $\hat{s} = \frac{4\hat{j} - 3\hat{k}}{5}$ (B) $\hat{s} = \frac{3\hat{i} - 4\hat{j}}{5}$

(C) $\hat{s} = \left(\frac{-3\hat{j} + 4\hat{k}}{5}\right)$ (D) $\hat{s} = \frac{-4\hat{k} + 3\hat{j}}{5}$

39. A plane electromagnetic wave having a frequency $\nu = 23.9$ GHz propagates along the positive z-direction in free space. The peak value of the electric field is 60V/m. Which among the following is the acceptable magnetic field component in the electromagnetic wave? **[12 April 2019 (Evening)]**

(A) $\vec{B} = 2 \times 10^7 \sin(0.5 \times 10^3 z + 1.5 \times 10^{11} t)\hat{i}$

(B) $\vec{B} = 2 \times 10^{-7} \sin(1.5 \times 10^2 x + 0.5 \times 10^{11} t)\hat{j}$

(C) $\vec{B} = 2 \times 10^{-7} \sin(0.5 \times 10^3 z - 1.5 \times 10^{11} t)\hat{i}$

(D) $\vec{B} = 60 \sin(0.5 \times 10^3 x + 1.5 \times 10^{11} t)\hat{k}$

* * * * *

CHAPTER-23 RAY OPTICS AND OPTICAL INSTRUMENTS

23.1 Geometrical Optics I - Reflection of Light

1. If two mirrors are kept at 60° to each other, then the number of images formed by them is : **[2002]**

(A) 5 (B) 6

(C) 7 (D) 8

2. To get three images of a single object, one should have two plane mirrors at an angle of : **[2003]**

(A) 60° (B) 90°

(C) 120° (D) 30°

3. A car is fitted with a convex side-view mirror of focal length 20 cm. A second car 2.8 m behind the first car is overtaking the first car at a relative speed of 15 m/s. The speed of the image of the second car as seen in the mirror of the first one is : **[2011]**

(A) 10 m/s (B) 15 m/s

(C) $\frac{1}{10}$ m/s (D) $\frac{1}{15}$ m/s

4. You are asked to design a shaving mirror assuming that a person keeps it 10 cm from his face and views the magnified image of the face at the closest comfortable distance of 25 cm. The radius of curvature of the mirror would then be : **[2015 (Online)]**

(A) 60 cm (B) –24 cm

(C) –60 cm (D) 24 cm

5. A particle is oscillating on the X-axis with an amplitude 2 cm about the point $x_0 = 10$ cm with a frequency ω. A concave mirror of focal length 5 cm is placed at the origin (see figure) Identify the correct statements : **[2018 (Online)]**

(a) The image executes periodic motion

(b) The image executes non-periodic motion

(c) The turning points of the image are asymmetric w.r.t. the image of the point at $x = 10$ cm

(d) The distance between the turning points of the oscillation of the image is $\frac{100}{21}$ cm.

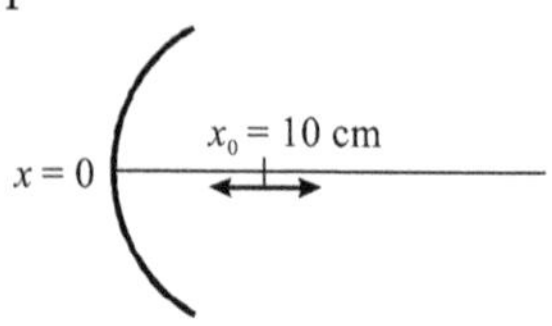

(A) (b), (d) (B) (b), (c)

(C) (a), (c), (d) (D) (a), (d)

6. Two plane mirrors are inclined to each other such that a ray of light incident on the first mirror (M_1) and parallel to the second mirror (M_2) is finally reflected from the second mirror (M_2) parallel to the first mirror (M_1). The angle between the two mirrors will be : **[9 Jan. 2019 (Evening)]**

(A) 90° (B) 45°

(C) 75° (D) 60°

7. A point source of light, S is placed at a distance L in front of the centre of plane mirror of width d which is hanging vertically on a wall. A man walks in front of the mirror along a line parallel to the mirror, at a distance $2L$ as shown below. The distance over which the man can see the image of the light source in the mirror is : **[12 Jan. 2019 (Morning)]**

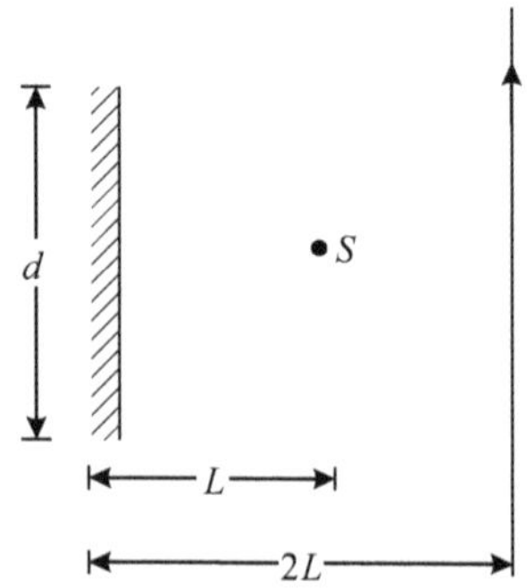

(A) $3d$ (B) $\frac{d}{2}$

(C) d (D) $2d$

8. A concave mirror for face viewing has focal length of 0.4 m. The distance at which you hold the mirror from your face in order to see your image upright with a magnification of 5 is : **[9 April 2019 (Morning)]**

(A) 1.60m (B) 0.24m

(C) 0.16m (D) 0.32m

9. A concave mirror has radius of curvature of 40 cm. It is at the bottom of a glass that has water filled up to 5 cm (see figure). If a small particle is floating on the surface of water, its image as seen, from directly above the glass, is at a distance d from the surface of water. The value of d is close to : (Refractive index of water = 1.33) : **[12 April 2019 (Morning)]**

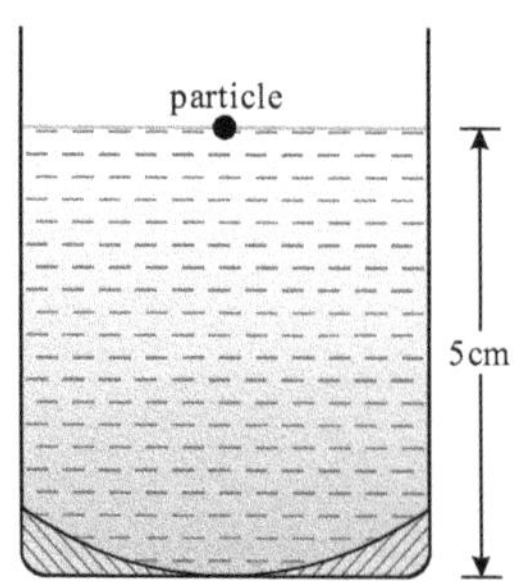

(A) 8.8 cm (B) 11.7 cm

(C) 6.7 cm (D) 13.4 cm

* * * * *

CHAPTER-23 RAY OPTICS AND OPTICAL INSTRUMENTS

23.2 Geometrical Optics II - Refraction of Light

1. Which of the following is used in optical fibres ? **[2002]**

(A) Total internal reflection (B) Scattering

(C) Diffraction (D) Refraction

2. A light ray is incident perpendicular to one face of a 90° prism and is totally internally reflected at the glass-air interface. If the angle of reflection is 45°, we conclude that the refractive index n: **[2004]**

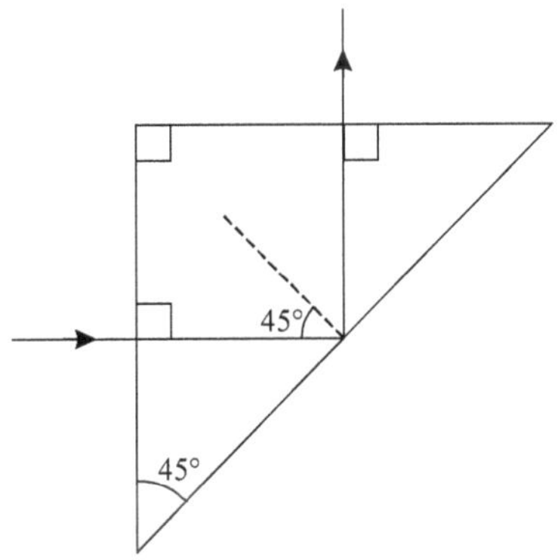

(A) $n < \frac{1}{\sqrt{2}}$ (B) $n > \sqrt{2}$

(C) $n > \frac{1}{\sqrt{2}}$ (D) $n < \sqrt{2}$

3. A fish looking up through the water sees the outside world, contained in a circular horizon. If the refractive index of water is 4/3 and the fish is 12 cm below the water surface, the radius of this circle in cm is : **[2005]**

(A) $36\sqrt{7}$ (B) $\frac{36}{\sqrt{7}}$

(C) $36\sqrt{5}$ (D) $4\sqrt{5}$

4. A transparent solid cylindrical rod has a refractive index of $\frac{2}{\sqrt{3}}$. It is surrounded by air. A light ray is incident at the mid-point of one end of the rod as shown in the figure :

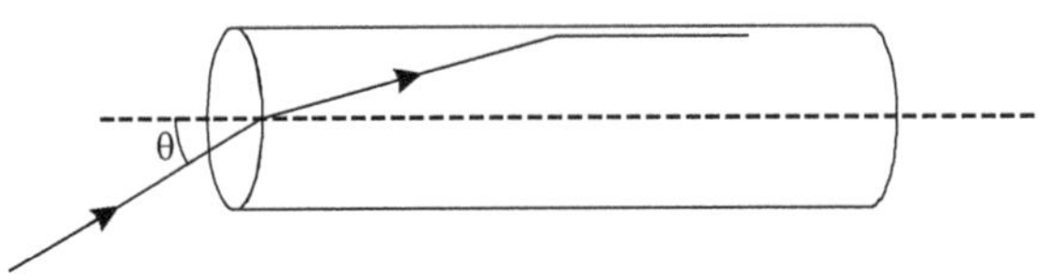

The incident angle θ for which the light ray grazes along the wall of the rod is : **[2009]**

(A) $\sin^{-1}\left(\frac{2}{\sqrt{3}}\right)$ (B) $\sin^{-1}\left(\frac{1}{\sqrt{3}}\right)$

(C) $\sin^{-1}\left(\frac{1}{2}\right)$ (D) $\sin^{-1}\left(\frac{\sqrt{3}}{2}\right)$

5. Let the x-y plane be the boundary between two transparent media. Medium 1 in $z \geq 0$ has a refractive index of $\sqrt{2}$ and medium 2 with $z < 0$ has a refractive index of $\sqrt{3}$. A ray of light in medium 1 given by the vector $\vec{A} = 6\sqrt{3}\hat{i} + 8\sqrt{3}\hat{j} - 10\hat{k}$ is incident on the plane of separation. The angle of refraction in medium 2 is : **[2011]**

(A) 60° (B) 75°

(C) 30° (D) 45°

6. A beaker contains water up to a height h_1 and kerosene of height h_2 above water so that the total height of (water + kerosene) is $(h_1 + h_2)$. Refractive index of water is μ_1 and that of kerosene is μ_2. The apparent shift in the position of the bottom of the beaker when viewed from above is : **[2011]**

(A) $\left(1-\frac{1}{\mu_1}\right)h_2 + \left(1-\frac{1}{\mu_2}\right)h_1$ (B) $\left(1+\frac{1}{\mu_1}\right)h_1 - \left(1+\frac{1}{\mu_2}\right)h_2$

(C) $\left(1-\frac{1}{\mu_1}\right)h_1 + \left(1-\frac{1}{\mu_2}\right)h_2$ (D) $\left(1+\frac{1}{\mu_1}\right)h_2 - \left(1+\frac{1}{\mu_2}\right)h_1$

7. A glass prism of refractive index 1.5 is immersed in water (refractive index 4/3) as shown in figure. A light beam incident normally on the face AB is totally reflected to reach the face BC, if: **[2012 (Online)]**

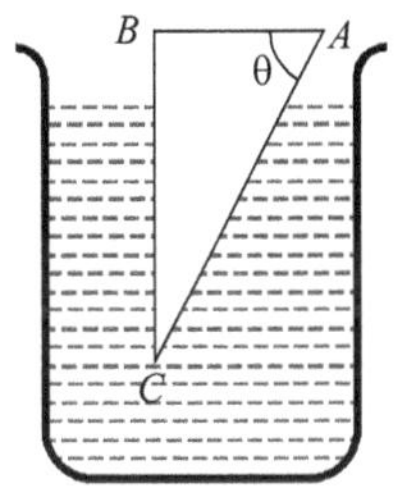

(A) $\sin\theta > \frac{5}{9}$ (B) $\sin\theta > \frac{2}{3}$

(C) $\sin\theta > \frac{8}{9}$ (D) $\sin\theta > \frac{1}{3}$

8. The graph between angle of deviation (δ) and angle of incidence (i) for a triangular prism is represented by : **[2013]**

(A) 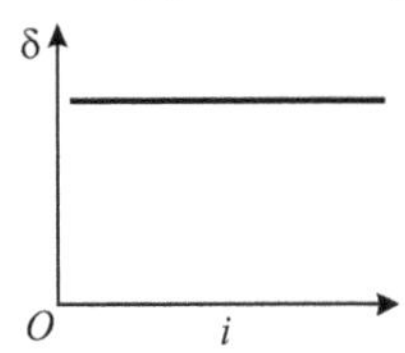

(B) 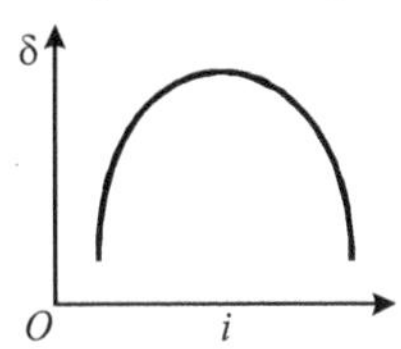

(C) 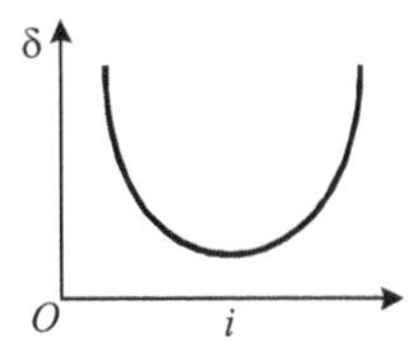

(D) 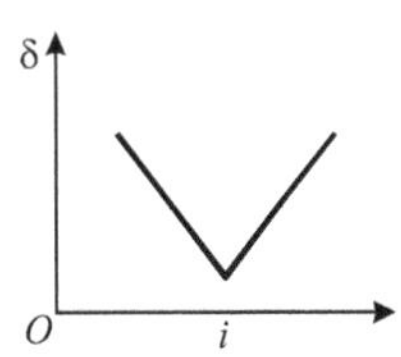

9. Light is incident from a medium into air at two possible angles of incidence (*A*) 20° and (*B*) 40°. In the medium light travels 3.0 cm in 0.2 ns. The ray will : **[2013 (Online)]**

(A) Suffer total internal reflection in both cases (*A*) and (*B*)
(B) Suffer total internal reflection in case (*B*) only
(C) Have partial reflection and partial transmission in case (*B*)
(D) Have 100% transmission in case (*A*)

10. A light ray falls on a square glass slab as shown in the diagram. The index of refraction of the glass, if total internal reflection is to occur at the vertical face, is equal to : **[2013 (Online)]**

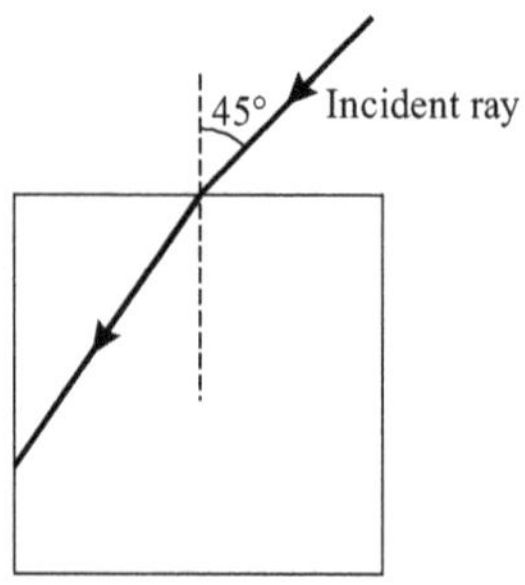

(A) $\frac{(\sqrt{2}+1)}{2}$ (B) $\sqrt{\frac{5}{2}}$

(C) $\frac{3}{2}$ (D) $\sqrt{\frac{3}{2}}$

11. A printed page is pressed by a glass of water. The refractive index of the glass and water is 1.5 and 1.33, respectively. If the thickness of the bottom of glass is 1 cm and depth of water is 5 cm, how much the page will appear to be shifted if viewed from the top ? **[2013 (Online)]**

(A) 1.033 cm (B) 3.581 cm
(C) 1.3533 cm (D) 1.90 cm

12. A diver looking up through the water sees the outside world contained in a circular horizontal. The refractive index of water is 4/3, and the diver's eyes are 15 cm below the surface of water. Then the radius of the circle is : **[2014 (Online)]**

(A) $15\times3\times\sqrt{5}$ cm (B) $15\times3\sqrt{5}$ cm

(C) $\frac{15\times\sqrt{7}}{3}$ cm (D) $\frac{15\times3}{\sqrt{7}}$ cm

13. Monochromatic light is incident of a glass prism of angle *A*. If the refractive index of the material of the prism is μ, a ray, incident at an angle θ, on the face *AB* would get transmitted through the face *AC* of the prism provided : **[2015]**

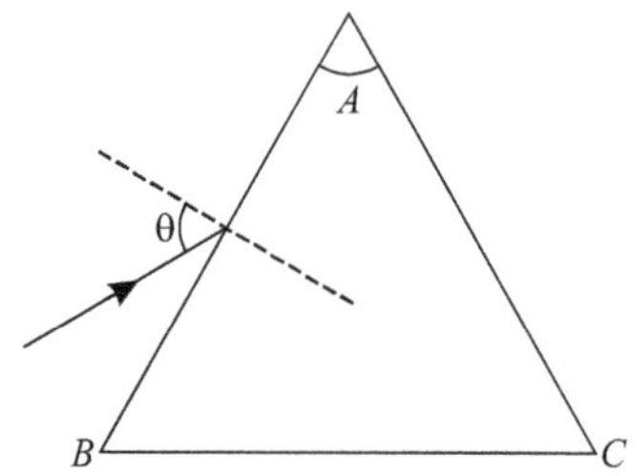

(A) $\theta > \sin^{-1}\left[\mu\sin\left(A-\sin^{-1}\left(\frac{1}{\mu}\right)\right)\right]$

(B) $\theta < \sin^{-1}\left[\mu\sin\left(A-\sin^{-1}\left(\frac{1}{\mu}\right)\right)\right]$

(C) $\theta > \cos^{-1}\left[\mu\sin\left(A+\sin^{-1}\left(\frac{1}{\mu}\right)\right)\right]$

(D) $\theta < \cos^{-1}\left[\mu\sin\left(A+\sin^{-1}\left(\frac{1}{\mu}\right)\right)\right]$

14. In an experiment for determination of refractive index of glass of a prism by $i-\delta$, plot, it was found that a ray incident at angle 35°, suffers a deviation of 40° and that it emerges at angle 79°. In that case which of the following is closest to the maximum possible value of the refractive index ? **[2016]**

(A) 1.5 (B) 1.6
(C) 1.7 (D) 1.8

15. A hemispherical glass body of radius 10 cm and refractive index 1.5 is silvered on its curved surface. A small air bubble is 6 cm below the flat surface inside it along the axis. The position of the image of the air bubble made by the mirror is seen : **[2016 (Online)]**

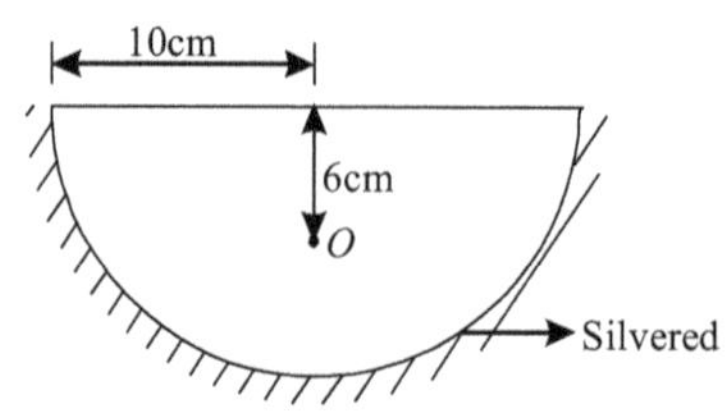

(A) 14 cm below flat surface (B) 20 cm below flat surface
(C) 16 cm below flat surface (D) 30 cm below flat surface

16. Let the refractive index of a denser medium with respect to a rarer medium be n_{12} and its critical angle be θ_C. At an angle of incidence *A* when light is travelling from denser medium to rarer medium, a part of the light is reflected and the rest is refracted and the angle between reflected and refracted rays is 90°. Angle *A* is given by : **[2017 (Online)]**

(A) $\frac{1}{\cos^{-1}(\sin\theta_C)}$ (B) $\frac{1}{\tan^{-1}(\sin\theta_C)}$

(C) $\cos^{-1}(\sin\theta_C)$ (D) $\tan^{-1}(\sin\theta_C)$

17. A ray of light is incident at an angle of 60° on one face of a prism of angle 30°. The emergent ray of light makes an angle of 30° with incident ray. The angle made by the emergent ray with second face of prism will be : **[2018 (Online)]**

(A) 30° (B) 90°
(C) 0° (D) 45°

18. The eye can be regarded as a single refracting surface. The radius of curvature of this surface is equal to that of cornea (7.8 mm). This surface separates two media of refractive indices 1 and 1.34. Calculate the distance from the refracting surface at which a parallel beam of light will come to focus :
[10 Jan. 2019 (Evening)]

(A) 2 cm (B) 1 cm
(C) 3.1 cm (D) 4.0 cm

19. A monochromatic light is incident at a certain angle on an equilateral triangular prism and suffers minimum deviation. If the refractive index of the material of the prism is $\sqrt{3}$, then the angle of incidence is : **[11 Jan. 2019 (Evening)]**

(A) 30° (B) 45°
(C) 90° (D) 60°

20. In figure, the optical fiber is $l = 2$m long and has a diameter of $d = 20$ μm. If a ray of light is incident on one end of the fiber at angle $\theta_1 = 40°$, the number of reflection it makes before emerging from the other end is close to :
(refractive index of fibre is 1.31 and sin 40° = 0.64)
[8 April 2019 (Morning)]

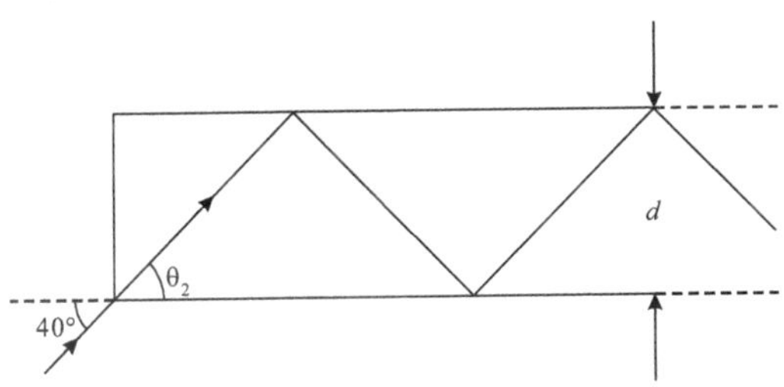

(A) 55000 (B) 57000
(C) 66000 (D) 45000

21. A ray of light AO in vacuum is incident on a glass slab at angle 60° and refracted at angle 30° along OB as shown in the figure. The optical path length of light ray from A to B is :
[10 April 2019 (Morning)]

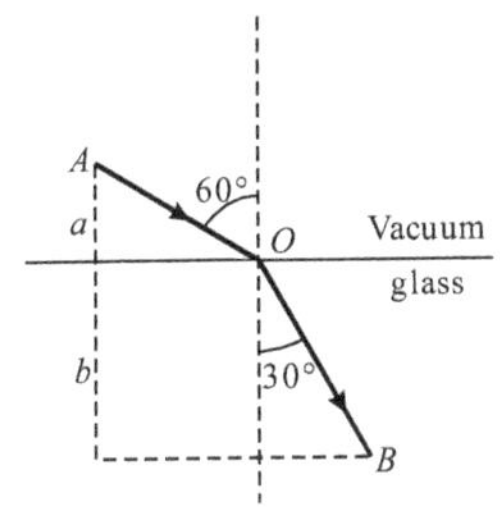

(A) $2a + 2b$ (B) $2a + \frac{2b}{3}$

(C) $\frac{2\sqrt{3}}{a} + 2b$ (D) $2a + \frac{2b}{\sqrt{3}}$

22. A transparent cube of side d, made of a material of refractive index μ_2, is immersed in a liquid of refractive index $\mu_1 (\mu_1 < \mu_2)$. A ray is incident on the face AB at an angle θ (shown in the figure). Total internal reflection takes place at point E on the face BC. The θ must satisfy : **[12 April 2019 (Evening)]**

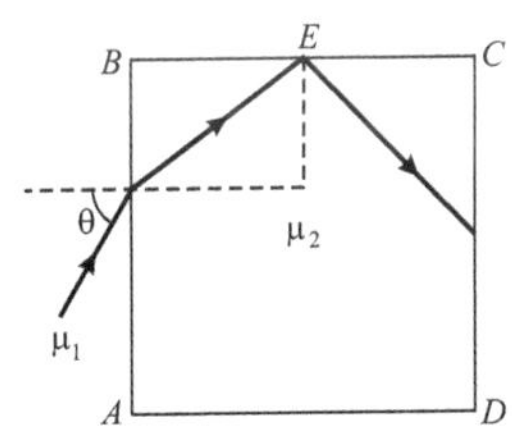

(A) $\theta < \sin^{-1}\frac{\mu_1}{\mu_2}$ (B) $\theta < \sin^{-1}\sqrt{\frac{\mu_2^2}{\mu_1^2} - 1}$

(C) $\theta > \sin^{-1}\frac{\mu_1}{\mu_2}$ (D) $\theta > \sin^{-1}\sqrt{\frac{\mu_2^2}{\mu_1^2} - 1}$

* * * * *

CHAPTER-23 RAY OPTICS AND OPTICAL INSTRUMENTS

23.3 Geometrical Optics III - Thin Lenses

1. A plano-convex lens of refractive index 1.5 and radius of curvature 30 cm is silvered at the curved surface. Now, this lens has been used to form the image of an object. At what distance from this lens, an object be placed in order to have a real image of the size of the object ? **[2004]**

(A) 20 cm (B) 30 cm
(C) 60 cm (D) 80 cm

2. A thin glass (refractive index 1.5) lens has optical power of – 5 D in air. Its optical power in a liquid medium with refractive index 1.6 will be : **[2005]**

(A) 1 D (B) – 1 D
(C) 25 D (D) – 25 D

3. Two lenses of power – 15 D and + 5 D are in contact with each other. The focal length of the combination is : **[2007]**

(A) – 20 cm (B) – 10 cm
(C) + 20 cm (D) + 10 cm

4. A student measures the focal length of a convex lens by putting an object pin at a distance 'u' from the lens and measuring the distance 'v' of the image pin. The graph between 'u' and 'v' plotted by the student should look like : **[2008]**

(A)
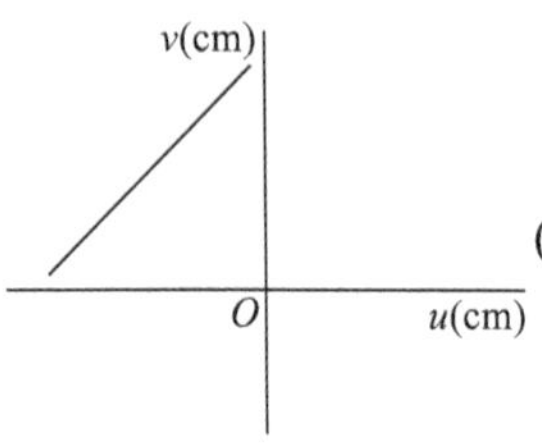

(B)
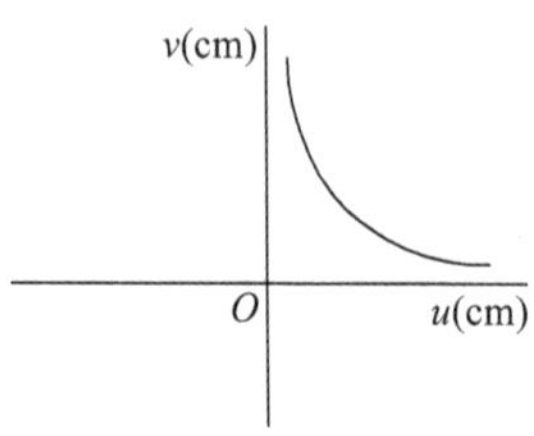

(C)
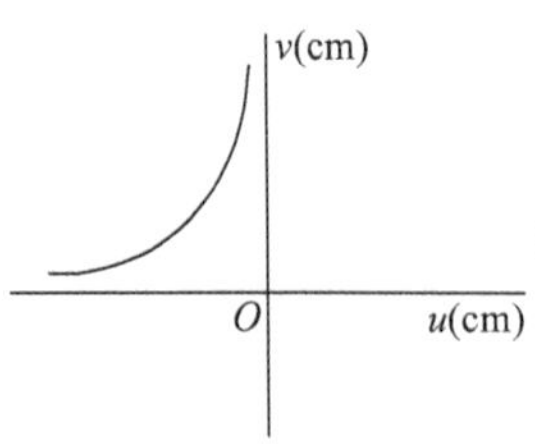

(D)
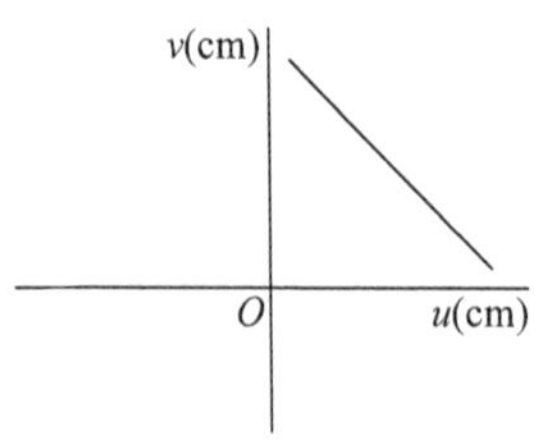

5. In an optics experiment, with the position of the object fixed, a student varies the position of a convex lens and for each position, the screen is adjusted to get a clear image of the object. A graph between the object distance u and the image distance v, from the lens, is plotted using the same scale for the two axes. A straight line passing through the origin and making an angle of 45° with the x-axis meets the experimental curve at P. The coordinates of P will be : **[2009]**

(A) $\left(\frac{f}{2}, \frac{f}{2}\right)$ (B) (f, f)
(C) $(4f, 4f)$ (D) $(2f, 2f)$

6. An object 2.4 m in front of a lens forms a sharp image on a film 12 cm behind the lens. A glass plate 1 cm thick, of refractive index 1.50 is interposed between lens and film with its plane faces parallel to film. At what distance (from lens) should object be shifted to be in sharp focus on film ? **[2012]**

(A) 5.6 m (B) 7.2 m
(C) 2.4 m (D) 3.2 m

7. Diameter of a plano-convex lens is 6 cm and thickness at the centre is 3 mm. If speed of light in material of lens is 2×10^8 m/s, the focal length of the lens is : **[2013]**

(A) 15 cm (B) 20 cm
(C) 30 cm (D) 10 cm

8. The image of an illuminated square is obtained on a screen with the help of a converging lens. The distance of the square from the lens is 40 cm. The area of the image is 9 times that of the square. The focal length of the lens is : **[2013 (Online)]**

(A) 36 cm (B) 27 cm
(C) 60 cm (D) 30 cm

9. A thin convex lens made from crown glass $\left(\mu = \frac{3}{2}\right)$ has focal length f. When it is measured in two different liquids having refractive indices 4/3 and 5/3, it has the focal lengths f_1 and f_2 respectively. The correct relation between the focal length is : **[2014]**

(A) $f_1 = f_2 < f$
(B) $f_1 > f$ and f_2 becomes negative
(C) $f_2 > f$ and f_1 becomes negative
(D) f_1 and f_2 both becomes negative

10. An object is located in a fixed position in front of a screen. Sharp image is obtained on the screen for two positions of a thin lens separated by 10 cm. The size of the images in two situations are in the ratio 3 : 3. What is the distance between the screen and the object ? **[2014 (Online)]**

(A) 124.5 cm (B) 144.5 cm
(C) 65.0 cm (D) 99.0 cm

11. The refractive index of the material of a concave lens is μ. It is immersed in a medium of refractive index μ_1. A parallel beam of light is incident on the lens. The path of the emergent rays when $\mu_1 > \mu$ is : **[2014 (Online)]**

(A) 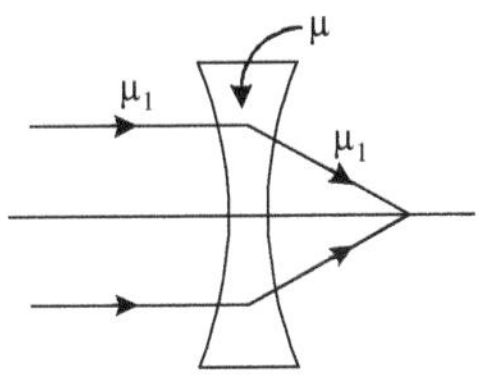

(B) 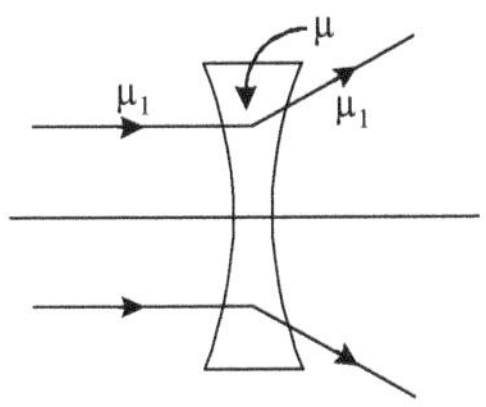

(C) 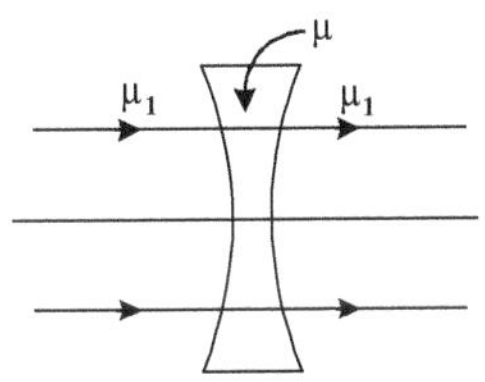

(D) 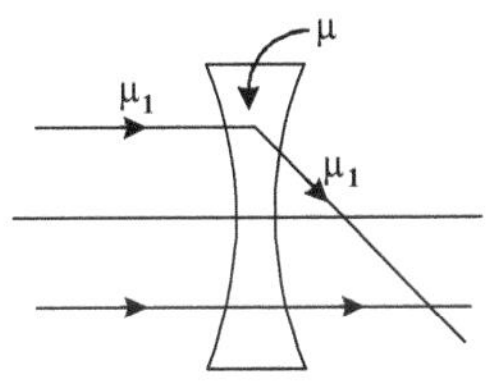

12. A thin convex lens of focal length 'f' is put on a plane mirror as shown in the figure. When an object is kept at a distance 'a' from the lens-mirror combination, its image is formed at a distance $\frac{a}{3}$ in front of the combination. The value of 'a' is : **[2015 (Online)]**

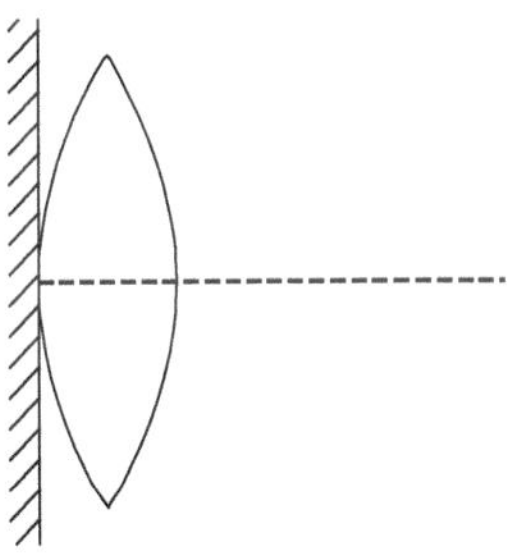

(A) $3f$ (B) $\frac{3}{2}f$

(C) f (D) $2f$

13. To find the focal length of a convex mirror, a student records the following data :

Object Pin	Convex Lens	Convex Mirror	Image Pin
22.2 cm	32.2 cm	45.8 cm	71.2 cm

The focal length of the convex lens is f_1 and that of mirror is f_2. Then taking index correction to be negligibly small, f_1 and f_2 are close to : **[2016 (Online)]**

(A) $f_1 = 7.8$ cm $f_2 = 12.7$ cm

(B) $f_1 = 12.7$ cm $f_2 = 7.8$ cm

(C) $f_1 = 15.6$ cm $f_2 = 25.4$ cm

(D) $f_1 = 7.8$ cm $f_2 = 25.4$ cm

14. A convex lens, of focal length 30 cm, a concave lens of focal length 120 cm, and a plane mirror are arranged as shown. For an object kept at a distance of 60 cm from the convex lens, the final image, formed by the combination, is a real image, at a distance of: **[2016 (Online)]**

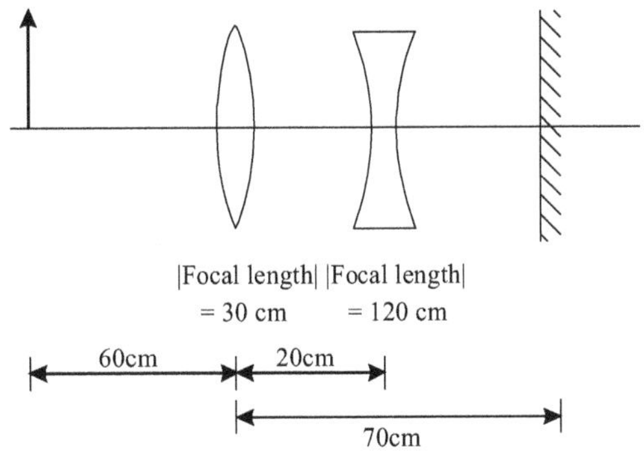

(A) 60 cm from the convex lens

(B) 60 cm from the concave lens

(C) 70 cm from the convex lens

(D) 70 cm from the concave lens

15. In an experiment a convex lens of focal length 15 cm is placed coaxially on an optical bench in front of a convex mirror at a distance of 5 cm from it. It is found that an object and its image coincide, if the object is placed at a distance of 20 cm from the lens. The focal length of the convex mirror is : **[2017 (Online)]**

(A) 27.5 cm (B) 20.0 cm

(C) 25.0 cm (D) 30.5 cm

16. A plano-convex lens becomes an optical system of 28 cm focal length when its plane surface is silvered and illuminated from left to right as shown in figure-A. If the same lens is instead silvered on the curved surface and illuminated from other side as in Figure-B, it acts like an optical system of focal length 10 cm. The refractive index of the material of lens if: **[2018 (Online)]**

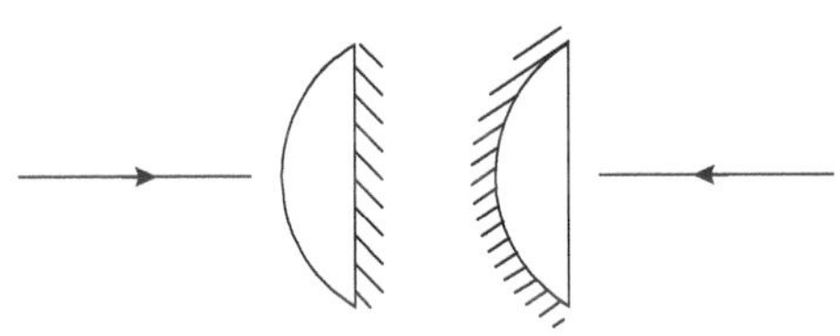

Figure-A Figure-B

(A) 1.50 (B) 1.55

(C) 1.75 (D) 1.51

17. A convex lens is put 10 cm from a light source and it makes a sharp image on a screen, kept 10 cm from the lens. Now a glass block (refractive index 1.5) of 1.5 cm thickness is placed in contact with the light source. To get the sharp image again, the screen is shifted by a distance d. Then d is : **[9 Jan. 2019 (Morning)]**

(A) 0.55 cm away from the lens

(B) 1.1 cm away from the lens

(C) 0.55 cm towards the lens

(D) 0

18. A plano-convex lens of refractive index μ_1 and focal length f_1 is kept in contact with another plano-concave lens of refractive index μ_2 and focal length f_2. If the radius of curvature of their spherical faces is R each and $f_1 = 2f_2$, then μ_1 and μ_2 are related as : **[10 Jan. 2019 (Morning)]**

(A) $\mu_1 + \mu_2 = 3$ (B) $2\mu_1 - \mu_2 = 1$
(C) $2\mu_2 - \mu_1 = 1$ (D) $3\mu_2 - 2\mu_1 = 1$

19. An object is at a distance of 20 m from a convex lens of focal length 0.3 m. The lens forms an image of the object. If the object moves away from the lens at a speed of 5 m/s, the speed and direction of the image will be : **[11 Jan. 2019 (Morning)]**

(A) 0.92×10^{-3} m/s away from the lens
(B) 2.26×10^{-3} m/s away from the lens
(C) 1.16×10^{-3} m/s towards the lens
(D) 3.22×10^{-3} m/s towards the lens

20. What is the position and nature of image formed by lens combination shown in figure? (f_1, f_2 are focal lengths) : **[12 Jan. 2019 (Morning)]**

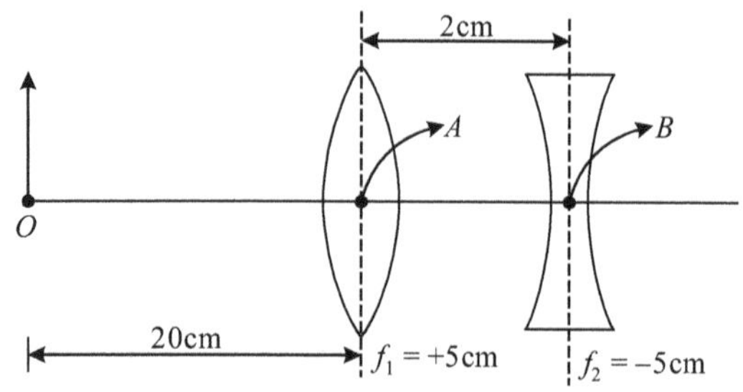

(A) 70 cm from point B at left, virtual
(B) 40 cm from point B at right, real
(C) $\frac{20}{3}$ cm from point B at right, real
(D) 70 cm from point B at right, real

21. Formation of real image using a biconvex lens is shown below.
If the whole set up is immersed in water without disturbing the object and the screen position, what will one observe on the screen ? **[12 Jan. 2019 (Evening)]**

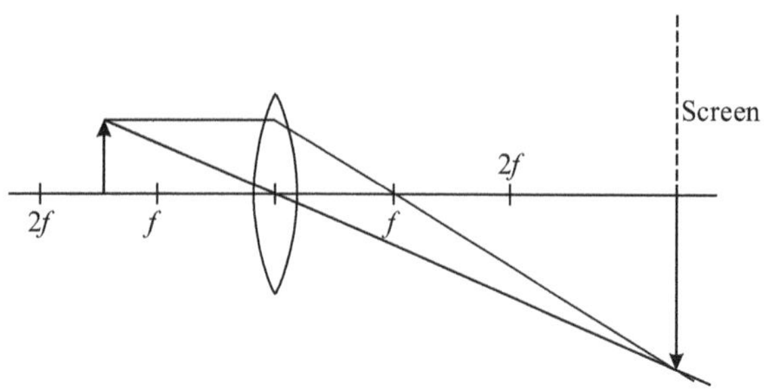

(A) Image disappears (B) No change
(C) Erect real image (D) Magnified image

22. A plano-convex lens (focal length f_2, refractive index μ_2, radius of curvature R) fits exactly into a plano-concave lens (focal length f_1, refractive index μ_1, radius of curvature R). Their plane surfaces are parallel to each other. Then, the focal length of the combination will be : **[12 Jan. 2019 (Evening)]**

(A) $f_1 - f_2$ (B) $f_1 + f_2$
(C) $\frac{R}{\mu_2 - \mu_1}$ (D) $\frac{2f_1f_2}{f_1 + f_2}$

23. An upright object is placed at a distance of 40 cm in front of a convergent lens of focal length 20 cm. A convergent mirror of focal length 10 cm is placed at a distance of 60 cm on the other side of the lens. The position and size of the final image will be : **[8 April 2019 (Morning)]**

(A) 40 cm from the convergent mirror, same size as the object
(B) 20 cm from the convergent mirror, same size as the object
(C) 20 cm from the convergent mirror, twice the size of the object
(D) 40 cm from the convergent lens, twice the size of the object

24. A convex lens (of focal length 20 cm) and a concave mirror, having their principal axes along the same lines, are kept 80 cm apart from each other. The concave mirror is to the right of the convex lens. When an object is kept at a distance of 30 cm to the left of the convex lens, its image remains at the same position even if the concave mirror is removed. The maximum distance of the object for which this concave mirror, by itself would produce a virtual image would be : **[8 April 2019 (Evening)]**

(A) 20 cm (B) 10 cm
(C) 25 cm (D) 30 cm

25. A thin convex lens L (refractive index = 1.5) is placed on a plane mirror M. When a pin is placed at A, such that $OA = 18$ cm, its real inverted image is formed at A itself, as shown in figure. When a liquid of refractive index μ_1 is put between the lens and the mirror, The pin has to be moved to A', such that $OA' = 27$ cm, to get its inverted real image at A' itself. The value of μ_1 will be: **[9 April 2019 (Evening)]**

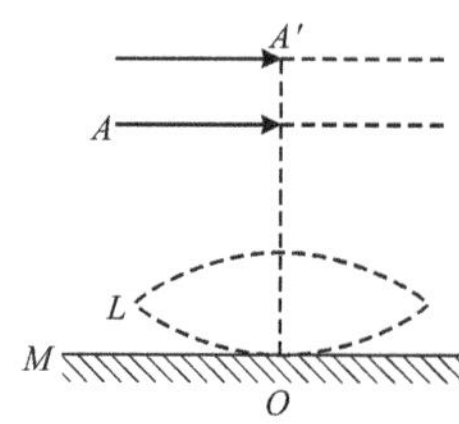

(A) $\sqrt{2}$ (B) $\frac{4}{3}$
(C) $\sqrt{3}$ (D) $\frac{3}{2}$

26. A convex lens of focal length 20 cm produces images of the same magnification 2 when an object is kept at two distances x_1 and $x_2 (x_1 > x_2)$ from the lens. The ratio of x_1 and x_2 is : **[9 April 2019 (Evening)]**

(A) 5 : 3 (B) 2 : 1

(C) 4 : 3 (D) 3 : 1

27. One plano-convex and one plano-concave lens of same radius of curvature 'R' but of different materials are joined side by side as shown in the figure. If the refractive index of the material of 1 is μ_1 and that of 2 is μ_2, Then the focal length of the combination is : **[10 April 2019 (Morning)]**

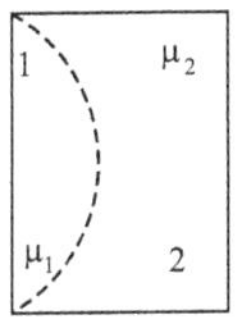

(A) $\dfrac{R}{2-(\mu_1-\mu_2)}$ (B) $\dfrac{2R}{\mu_1-\mu_2}$

(C) $\dfrac{R}{2(\mu_1-\mu_2)}$ (D) $\dfrac{R}{\mu_1-\mu_2}$

28. The graph shows how the magnification m produced by a thin lens varies with image distance v. What is the focal length of the lens used ? **[10 April 2019 (Evening)]**

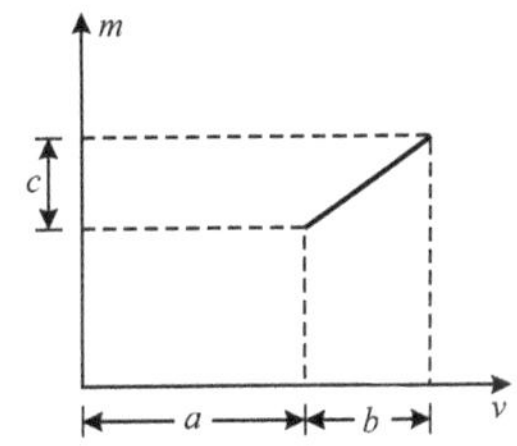

(A) $\dfrac{b^2c}{a}$ (B) $\dfrac{b^2}{ac}$

(C) $\dfrac{a}{c}$ (D) $\dfrac{b}{c}$

* * * * *

CHAPTER-23 RAY OPTICS AND OPTICAL INSTRUMENTS

23.4 Dispersion of Light

1. The refractive index of glass is 1.520 for red light and 1.525 for blue light. Let D_1 and D_2 be angles of minimum deviation for red and blue light respectively in a prism of this glass. Then : **[2006]**

(A) $D_1 < D_2$
(B) $D_1 = D_2$
(C) D_1 can be less than or greater than D_2 depending upon the angle of prism
(D) $D_1 > D_2$

2. When monochromatic red light is used instead of blue light in a convex lens, its focal length will : **[2011]**

(A) Does not depend on colour of light
(B) Increase
(C) Decrease
(D) Remain same

3. Which of the following processes play a part in the formation of a rainbow ? **[2012 (Online)]**

(i) Refraction (ii) Total internal reflection
(iii) Dispersion (iv) Interference
(A) (i), (ii) and (iii) (B) (i) and (ii)
(C) (i), (ii) and (iv) (D) (iii) and (iv)

4. A beam of light consisting of red, green and blue colours is incident on a right-angled prism on face AB. The refractive indices of the material for the above red, green and blue colours are 1.39, 1.44 and 1.47 respectively. A person looking on surface AC of the prism will see : **[2012 (Online)]**

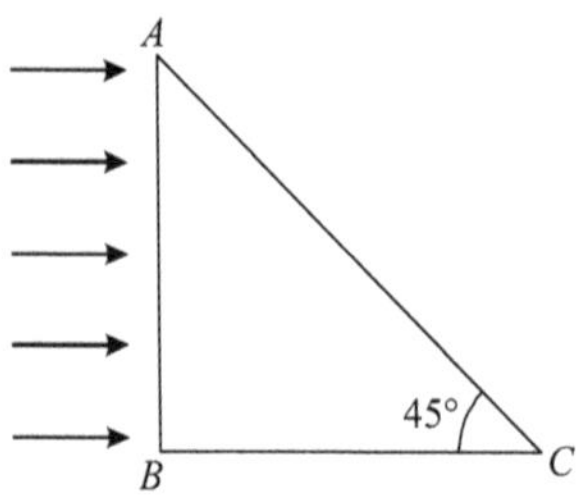

(A) No light (B) Green and blue colours
(C) Red and green colours (D) Red colour only

5. A convergent doublet of separated lenses, corrected for spherical aberration, has resultant focal length of 10 cm. The separation between the two lenses is 2 cm. The focal lengths of the component lenses : **[2018 (Online)]**

(A) 18 cm, 20 cm (B) 10 cm, 12 cm
(C) 12 cm, 14 cm (D) 16 cm, 18 cm

6. The variation of refractive index of a crown glass thin prism with wavelength of the incident light is shown. Which of the following graphs is the correct one, if D_m is the angle of minimum deviation? **[11 Jan. 2019 (Morning)]**

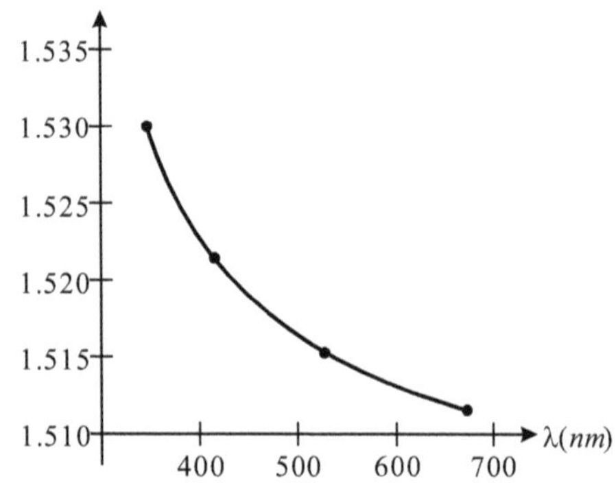

(A)
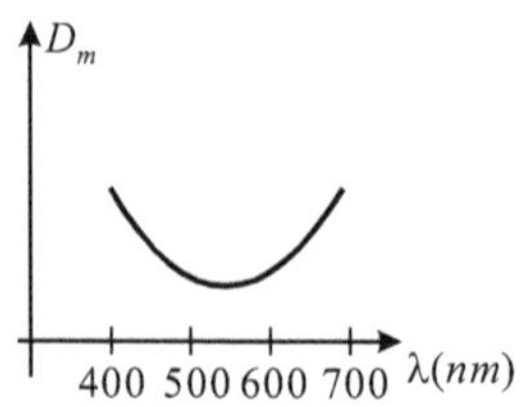

(B)
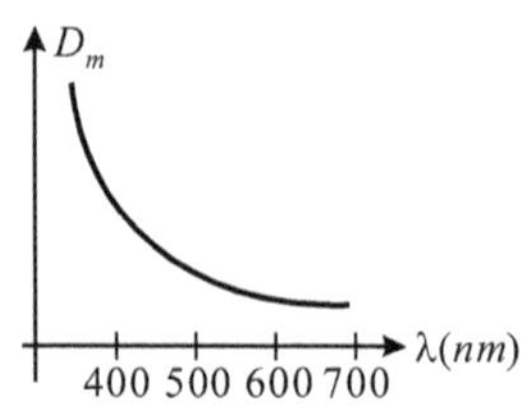

(C)
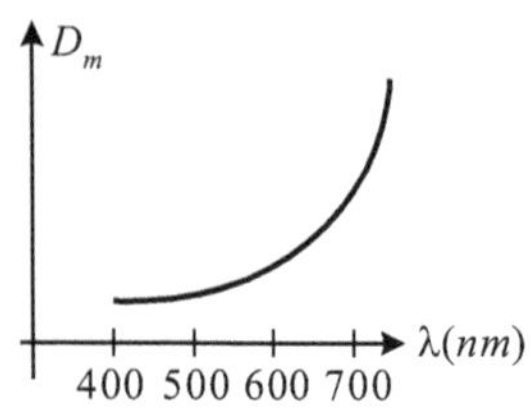

(D)
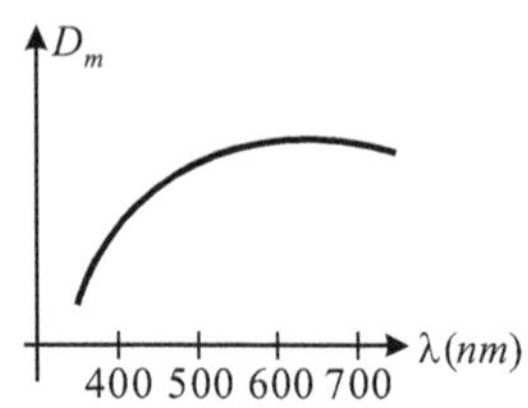

* * * * *

CHAPTER-23 RAY OPTICS AND OPTICAL INSTRUMENTS

23.5 Optical Instruments

1. An astronomical telescope has a large aperture to : **[2002]**
(A) Reduced spherical aberration
(B) Have high resolution
(C) Increase span of observation
(D) Have low dispersion

2. Wavelength of light used in a optical instrument are $\lambda_1 = 4000$ Å and $\lambda_2 = 5000$ Å, then ratio of their respective resolving powers (corresponding to λ_1 and λ_2) is : **[2002]**
(A) 16 : 25 (B) 9 : 1
(C) 4 : 5 (D) 5 : 4

3. The image formed by an objective of a compound microscope is : **[2003]**
(A) Virtual and diminished (B) Real and diminished
(C) Real and enlarged (D) Virtual and enlarged

4. Two point white dots are 1 mm apart on a black paper. They are viewed by eye of pupil diameter 3 mm. Approximately, what is the maximum distance at which these dots can be resolved by the eye ? [Take $\lambda = 500$ nm] : **[2005]**
(A) 5m (B) 1m
(C) 6m (D) 3m

5. An experiment is performed to find the refractive index of glass using a travelling microscope. In this experiment distances are measured by : **[2008]**
(A) A vernier scale provided on the microscope
(B) A standard laboratory scale
(C) A meter scale provided on the microscope
(D) A screws gauge provided on the microscope

6. We wish to make a microscope with the help of two positive lenses both with a focal length of 20 mm each and the object is positioned 25 mm from the objective lens. How far apart the lenses should be so that the final image is formed at infinity ? **[2012 (Online)]**
(A) 20mm (B) 100mm
(C) 120mm (D) 80mm

7. A telescope of aperture 3×10^{-2} m diameter is focused on a window at 80 m distance fitted with a wire mesh of spacing 2×10^{-3} m. Given : $\lambda = 5.5 \times 10^{-7}$ m, which of the following is true for observing the mesh through the telescope? **[2012 (Online)]**
(A) Yes, it is possible with the same aperture size
(B) Possible also with an aperture half the present diameter
(C) No, it is not possible
(D) Given data is not sufficient

8. The focal length of the objective and the eyepiece of a telescope are 50 cm and 5 cm respectively. If the telescope is focussed for distinct vision on a scale distant 2 m from its objective, then its magnifying power will be : **[2013 (Online)]**
(A) – 4 (B) – 8
(C) +8 (D) – 2

9. This question has Statement-1 and Statement-2. Of the four choices given after the Statements, choose the one that best describes the two Statements. **[2013 (Online)]**
Statement-1 : Very large size telescopes are reflecting telescopes instead of refracting telescopes :
Statement-2 : It is easier to provide mechanical support to large size mirrors than large size lenses :
(A) Statement-1 is true and Statement-2 is false
(B) Statement-1 is false and Statement-2 is true
(C) Statement-1 and Statement-2 are true and Statement-2 is correct explanation for Statement-1
(D) Statement-1 and Statement-2 are true and Statement-2 is not the correct explanation for Statement-1

10. The focal lengths of objective lens and eye lens of a Galilean telescope are respectively 30 cm and 3.0 cm. Telescope produces virtual, erect image of an object situated far away from it at least distance of distinct vision from the eye lens. In this condition, the magnifying power of the Galilean telescope should be : **[2014 (Online)]**
(A) +11.2 (B) – 11.2
(C) – 8.8 (D) +8.8

11. In a compound microscope, the focal length of objective lens is 1.2 cm and focal length of eye piece is 3.0 cm. When object is kept at 1.25 cm in front of objective, final image is formed at infinity. Magnifying power of the compound microscope should be : **[2014 (Online)]**
(A) 200 (B) 100
(C) 400 (D) 150

12. Assuming human pupil to have a radius of 0.25 cm and a comfortable viewing distance of 25 cm, the minimum separation between two objects that human eye can resolve at 500 nm wavelength is : **[2015]**
(A) 1 μm (B) 30 μm
(C) 100 μm (D) 300 μm

13. A telescope has an objective lens of focal length 150 cm and an eyepiece of focal length 5 cm. If a 50 m tall tower at a distance of 1 km is observed through this telescope in normal setting, the angle formed by the image of the tower is θ, then θ is close to : [2015 (Online)]

(A) 30° (B) 15°
(C) 60° (D) 1°

14. An observer looks at a distant tree of height 10 m with a telescope of magnifying power of 20. To the observer the tree appears : [2016]

(A) 10 times taller (B) 10 times nearer
(C) 20 times taller (D) 20 times nearer

15. To determine refractive index of glass slab using a travelling microscope, minimum number of readings required are : [2016 (Online)]

(A) Two (B) Four
(C) Three (D) Five

16. Calculate the limit of resolution of a telescope objective having a diameter of 200 cm, if it has to detect light of wavelength 500 nm coming from a star : [8 April 2019 (Evening)]

(A) 305×10^{-9} radian (B) 152.5×10^{-9} radian
(C) 610×10^{-9} radian (D) 457.5×10^{-9} radian

17. Diameter of the objective lens of a telescope is 250 cm. For light of wavelength 600 nm. coming from a distant object, the limit of resolution of the telescope is close to : [9 April 2019 (Evening)]

(A) 1.5×10^{-7} rad (B) 2.0×10^{-7} rad
(C) 3.0×10^{-7} rad (D) 4.5×10^{-7} rad

18. The value of numerical aperature of the objective lens of a microscope is 1.25. If light of wavelength 5000 Å is used, the minimum separation between two points, to be seen as distinct, will be : [12 April 2019 (Morning)]

(A) 0.24 μm (B) 0.48 μm
(C) 0.12 μm (D) 0.38 μm

* * * * *

CHAPTER-24 WAVE OPTICS

24.1 Interference of Light

1. To demonstrate the phenomenon of interference we require two sources which emit radiations of : **[2003]**
(A) nearly the same frequency
(B) the same frequency
(C) different wavelength
(D) the same frequency and having a definite phase relationship

2. The maximum number of possible interference maxima for slit-separation equal to twice the wavelength in Young's double-slit experiment, is : **[2004]**
(A) infinite (B) five
(C) three (D) zero

3. A Young's double slit experiment uses a monochromatic source. The shape of the interference fringes formed on a screen is : **[2005]**
(A) hyperbola (B) circle
(C) straight line (D) parabola

4. In a Young's double slit experiment the intensity at a point where the path difference is $\lambda/6$ (λ being the wavelength of the light used) is I. If I_0 denotes the maximum intensity, I/I_0 is equal to : **[2007]**

(A) $\frac{1}{\sqrt{2}}$ (B) $\frac{\sqrt{3}}{2}$

(C) $\frac{1}{2}$ (D) $\frac{3}{4}$

5. A mixture of light, consisting of wavelength 590 nm and an unknown wavelength, illuminates Young's double slit and gives rise to two overlapping interference patterns on the screen. The central maximum of both lights coincide. Further, it is observed that the third bright fringe of known light coincides with the 4th bright fringe of the unknown light. From this data, the wavelength of the unknown light is : **[2009]**
(A) 442.5 nm (B) 776.8 nm
(C) 393.4 nm (D) 885.0 nm

Direction : Below questions from 6 to 8 are based on the following paragraph.
An initially parallel cylindrical beam travels in a medium of refractive index $\mu(I) = \mu_0 + \mu_2 I$, where μ_0 and μ_2 are positive constants and I is the intensity of the light beam. The intensity of the beam is decreasing with increasing radius :

6. The initial shape of the wavefront of the beam is : **[2010]**
(A) planar
(B) convex
(C) concave
(D) convex near the axis and concave near the periphery

7. The speed of the light in the medium is : **[2010]**
(A) maximum on the axis of the beam
(B) minimum on the axis of the beam
(C) the same everywhere in the beam
(D) directly proportional to the intensity I

8. As the beam enters the medium, it will : **[2010]**
(A) travel as a cylindrical beam
(B) diverge
(C) converge
(D) diverge near the axis and converge near the periphery

9. Direction : The question has a paragraph followed by two statement, Statement-1 and statement-2. Of the given four alternatives after the statements, choose the one that describes the statements.
A thin air film is formed by putting the convex surface of a plane-convex lens over a plane glass plate. With monochromatic light, this film gives an interference pattern due to light reflected from the top (convex) surface and the bottom (glass plate) surface of the film
Statement-1 : When light reflects from the air-glass plate interface, the reflected wave suffers a phase change of π.
Statement-2 : The centre of the interference pattern is dark. **[2011]**
(A) Statement-1 is true, Statement-2 is true and Statement-2 is not the correct explanation of Statement-1
(B) Statement-1 is false, Statement-2 is true
(C) Statement-1 is true, Statement-2 is false
(D) Statement-1 is true, Statement-2 is true and Statement-2 is the correct explanation of statement-1

10. In a Young's double slit experiment, the two slits act as coherent sources of waves of equal amplitude A and wavelength λ. In another experiment with the same arrangement the two slits are made to act as incoherent sources of waves of same amplitude and wavelength. If the intensity at the middle point of the screen in the first case is I_1 and in the second case I_2, then the ratio I_1/I_2 is : **[2011]**
(A) 4 (B) 2
(C) 1 (D) 0.5

11. At two points P and Q on screen in Young's double slit experiment, waves from slits S_1 and S_2 have a path difference of 0 and $\lambda/4$ respectively. The ratio of intensities at P and Q will be : **[2011]**
(A) 3 : 2 (B) 2 : 1
(C) $\sqrt{2}:1$ (D) 4 : 1

12. In Young's double slit experiment, one of the slit is wider than other, so that the amplitude of the light from one slit is double of that from other slit. If I_m be the maximum intensity, the resultant intensity I when they interfere at phase difference ϕ is given by : **[2012]**

(A) $\frac{I_m}{9}\left(1+8\cos^2\frac{\phi}{2}\right)$ (B) $\frac{I_m}{9}(4+5\cos\phi)$

(C) $\frac{I_m}{9}\left(1+2\cos^2\frac{\phi}{2}\right)$ (D) $\frac{I_m}{5}\left(1+4\cos^2\frac{\phi}{2}\right)$

13. In a Young's double slit experiment with light of wavelength λ, fringe pattern on the screen has fringe width β. When two thin transparent glass (refractive index μ) plates of thickness t_1 and t_2 ($t_1 > t_2$) are placed in the path of the two beams respectively, the fringe pattern will shift by a distance : **[2012 (Online)]**

(A) $\frac{\beta(\mu-1)}{\lambda}\left(\frac{t_1}{t_2}\right)$ (B) $\frac{\mu\beta}{\lambda}\frac{t_1}{t_2}$

(C) $\frac{\beta(\mu-1)}{\lambda}(t_1-t_2)$ (D) $(\mu-1)\frac{\lambda}{\beta}(t_1+t_2)$

14. The maximum number of possible interference maxima for slit separation equal to 1.8 λ, where λ is the wavelength of light used, in a Young's double slit experiment is : **[2012 (Online)]**

(A) Zero (B) 3
(C) Infinite (D) 5

15. In Young's double slit interference experiment, the slit widths are in the ratio 1 : 25. Then the ratio of intensity at the maxima and minima in the interference pattern is : **[2012 (Online)]**

(A) 3:2 (B) 1:25
(C) 9:4 (D) 1:5

16. Two coherent plane light waves of equal amplitude makes a small angle $\alpha(<< 1)$ with each other. The fall almost normally on a screen. If λ is the wavelength of light waves, the fringe width Δx of interference patterns of the two sets of waves on the screen is : **[2012 (Online)]**

(A) $\frac{2\lambda}{\alpha}$ (B) $\frac{\lambda}{\alpha}$

(C) $\frac{\lambda}{(2\alpha)}$ (D) $\frac{\lambda}{\sqrt{\alpha}}$

17. Two coherent point sources S_1 and S_2 are separated by a small distance 'd' as shown. The fringes obtained on the screen will be : **[2013]**

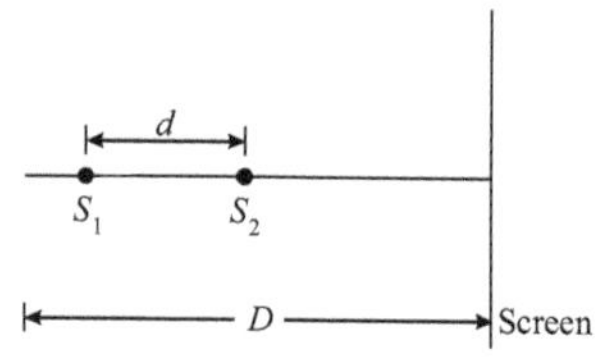

(A) points (B) straight lines
(C) semi-circles (D) concentric circles

18. This question has Statement-1 and Statement-2. Of the four choices given after the Statements, choose the one that best describes the two Statements.

Statement-1 : In young's double slit experiment, the number of fringes observed in the field of view is small with longer wavelength of light and is large with shorter wavelength of light.

Statement-2 : In the double slit experiment the fringe width depends directly on the wavelength of light. **[2013 (Online)]**

(A) Statement-1 is true, Statement-2 is true and the Statement-2 is correct explanation of the Statement-1
(B) Statement-1 is false and the Statement-2 is true
(C) Statement-1 is true Statement-2 is true and the Statement-2 is not correct explanation of the Statement-1
(D) Statement-1 is true and the Statement-2 is false

19. The source that illuminates the double-slit in 'double-slit interference experiment' emits two distinct monochromatic waves of wavelength 500 nm and 600 nm, each of them producing its own pattern on the screen. At the central point of the pattern when path difference is zero, maxima of both the patterns coincide and the resulting interference pattern is most distinct at the region of zero path difference. But as one moves out of this central region, the two fringe systems are gradually out of step such that maximum due to on wavelength coincides with the minimum due to the other and the combined fringe system becomes completely indistinct. This may happen when path difference in *nm* is : **[2013 (Online)]**

(A) 2000 (B) 3000
(C) 1000 (D) 1500

20. A ray of light of intensity I is incident on a parallel glass slab at point A as shown in diagram. It undergoes partial reflection and refraction. At each reflection, 25% of incident energy is reflected. The rays AB and A 'B' undergo interference. The ratio of I_{max} and I_{min} is : **[2013 (Online)]**

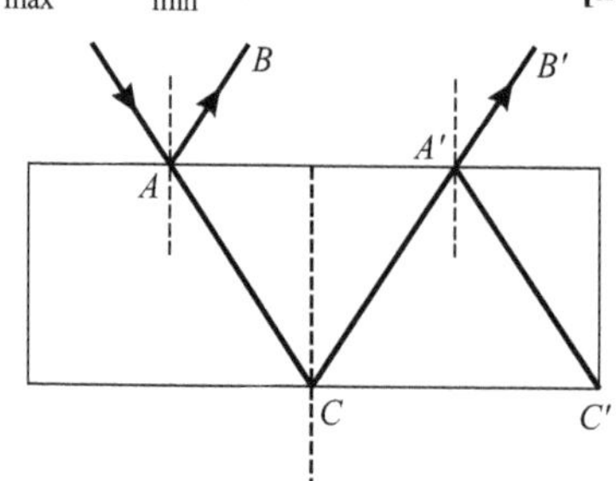

(A) 49 : 1 (B) 7 : 1
(C) 4 : 1 (D) 8 : 1

21. n identical waves each of intensity I_0 interfere with each other. The ratio of maximum intensities if the interference is (i) coherent and (ii) incoherent is : **[2013 (Online)]**

(A) n^2 (B) $\frac{1}{n}$

(C) $\frac{1}{n^2}$ (D) n

22. Two monochromatic light beams of intensity 16 and 9 units are interfering. The ratio of intensities of bright and dark parts of the resultant pattern is : **[2014 (Online)]**

(A) $\frac{16}{9}$ (B) $\frac{4}{3}$

(C) $\frac{7}{1}$ (D) $\frac{49}{1}$

23. On a hot summer night, the refractive index of air is smallest near the ground and increase with height from the ground. When a light beam is directed horizontally, the Huygen's principle leads us to conclude that as it travels, the light beam : **[2015]**

(A) becomes narrower
(B) goes horizontally without any deflection
(C) bends downwards
(D) bends upwards

24. In a Young's double slit experiment with light of wavelength λ the separation of slits is d and distance of screen is D such that $D >> d >> \lambda$. If the fringe width is β, the distance from point of maximum intensity to the point where intensity falls to half of maximum intensity on either side is : **[2015 (Online)]**

(A) $\frac{\beta}{6}$ (B) $\frac{\beta}{3}$

(C) $\frac{\beta}{4}$ (D) $\frac{\beta}{2}$

25. In a Young's double slit experiment, slits are separated by 0.5 mm, and the screen is placed 150 cm away. A beam of light consisting of two wavelengths, 650 nm and 520 nm, is used to obtain interference fringes on the screen. The least distance from the common central maximum to the point where the bright fringes due to both the wavelengths coincide is : **[2017]**

(A) 9.75 mm (B) 15.6 mm
(C) 1.56 mm (D) 7.8 mm

26. Two coherent sources produce waves of different intensities which interfere. After interference, the ratio of the maximum intensity to the minimum intensity is 16. The intensity of the waves are in the ratio : **[9 Jan. 2019 (Morning)]**

(A) 4 : 1 (B) 25 : 9
(C) 16 : 9 (D) 5 : 3

27. In a Young's double slit experiment, the slits are placed 0.320 mm apart. Light of wavelength $\lambda = 500$ nm is incident on the slits. The total number of bright fringes that are observed in the angular range $-30° \le \theta \le 30°$ is : **[9 Jan. 2019 (Evening)]**

(A) 320 (B) 641
(C) 321 (D) 640

28. In a Young's double slit experiment with slit separation 0.1 mm, one observes a bright fringe at angle 1/40 rad by using light of wavelength λ_1. When the light of wavelength λ_2 is used a bright fringe is seen at the same angle in the same set up. Given that λ_1 and λ_2 are in visible range (380 nm to 740 nm), their values are : **[10 Jan. 2019 (Morning)]**

(A) 380 nm, 500 nm (B) 625 nm, 500 nm
(C) 380 nm, 525 nm (D) 400 nm, 500 nm

29. Consider a Young's double slit experiment as shown in figure. What should be the slit separation d in terms of wavelength λ such that the first minima occurs directly in front of the slit (S_1) ? **[10 Jan. 2019 (Evening)]**

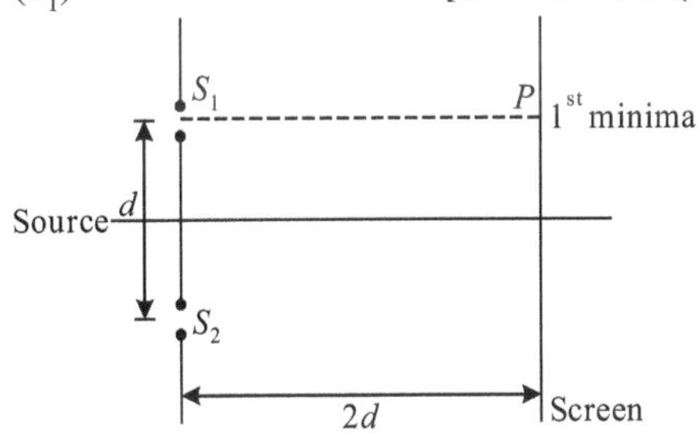

(A) $\frac{\lambda}{2(5-\sqrt{2})}$ (B) $\frac{\lambda}{(5-\sqrt{2})}$

(C) $\frac{\lambda}{(\sqrt{5}-2)}$ (D) $\frac{\lambda}{2(\sqrt{5}-2)}$

30. In a Young's double slit experiment, the path difference, at a certain point on the screen, between two interfering waves is 1/8th of wavelength. The ratio of the intensity at this point to that at the centre of a bright fringe is close to : **[11 Jan. 2019 (Morning)]**

(A) 0.94 (B) 0.74
(C) 0.85 (D) 0.80

31. In a double-slit experiment, green light (5303 Å) falls on a double slit having a separation of 19.44 μm and a width of 4.05 μm. The number of bright fringes between the first and the second diffraction minima is : **[11 Jan. 2019 (Evening)]**

(A) 09 (B) 10
(C) 04 (D) 05

32. In an interference experiment the ratio of amplitudes of coherent waves is $\frac{a_1}{a_2} = \frac{1}{3}$. The ratio of maximum and minimum intensities of fringes will be : **[8 April 2019 (Morning)]**

(A) 4 (B) 2
(C) 9 (D) 18

33. The figure shows a Young's double slit experimental setup. It is observed that when a thin transparent sheet of thickness t and refractive index μ is put in front of one of the slits, the central maximum gest shifted by a distance equal to n fringe widths. If the wavelength of light used is λ, t will be : **[9 April 2019 (Morning)]**

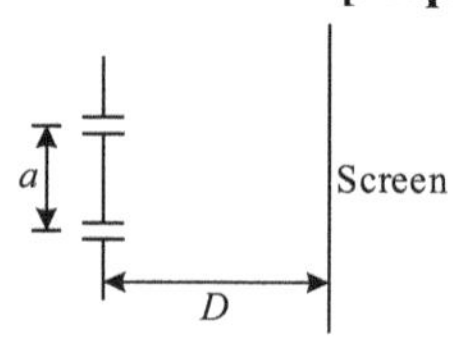

(A) $\frac{2D\lambda}{a(\mu-1)}$ (B) $\frac{D\lambda}{a(\mu-1)}$

(C) $\frac{2nD\lambda}{a(\mu-1)}$ (D) $\frac{nD\lambda}{a(\mu-1)}$

34. In a Young's doubble slit experiment, the ratio of the slit's width is 4 : 1. The ratio of the intensity of maxima to minima, close to the central fringe on the screen, will be : **[10 April 2019 (Evening)]**

(A) $(\sqrt{3}+1)^4 : 16$ (B) $9:1$
(C) $4:1$ (D) $25:9$

35. In a double slit experiment, when a thin film of thickness t having refractive index μ is introduced in front of one of the slits, the maximum at the centre of the fringe pattern shifts by one fringe width. The value of t is (λ is the wavelength of the light used) : **[12 April 2019 (Morning)]**

(A) $\frac{\lambda}{2(\mu-1)}$ (B) $\frac{\lambda}{(2\mu-1)}$

(C) $\frac{2\lambda}{(\mu-1)}$ (D) $\frac{\lambda}{(\mu-1)}$

* * * * *

CHAPTER-24 WAVE OPTICS

24.2 Diffraction of Light

1. If I_0 is the intensity of the principle maximum in the single slit diffraction pattern, then what will be its intensity when the slit width is doubled : **[2005]**

(A) $2I_0$ (B) $4I_0$

(C) I_0 (D) $I_0/2$

2. The first diffraction minimum due to the single slit diffraction is seen at $\theta = 30°$ for a light of wavelength 5000 Å falling perpendicularly on the slit. The width of the slit is : **[2012 (Online)]**

(A) 2.5×10^{-5} cm (B) 1.25×10^{-5} cm

(C) 10×10^{-5} cm (D) 5×10^{-5} cm

3. This question has Statement-1 and Statement-2. Of the four choices given after the Statements, choose the one that best describes the two Statements. **[2013 (Online)]**

Statement-1 : Out of radio waves and microwaves, the radio waves undergo more diffraction.

Statement-2 : Radio waves have greater frequency compared to microwaves.

(A) Statement-1 is true, Statement-2 is true and Statement-2 is the correct explanation of Statement-1

(B) Statement-1 is false, Statement-2 is true

(C) Statement-1 is true, Statement-2 is false

(D) Statement-1 is true, Statement-2 is true but Statement-2 is not the correct explanation of Statement-1

4. The diameter of the objective lens of microscope makes an angle β at the focus of the microscope. Further, the medium between the object and the lens is an oil of refractive index n. Then the resolving power of the microscope: **[2014 (Online)]**

(A) Increases with decreasing value of n

(B) Increases with decreasing value of β

(C) Increases with increasing value of $n \sin 2\beta$

(D) Increases with increasing value of $\frac{1}{n \sin 2\beta}$

5. In a Young's double slit experiment, the distance between the two identical slits is 6.1 times larger than the slit width. Then the number of intensity maxima observed within the central maximum of the single slit diffraction pattern is : **[2014 (Online)]**

(A) 3 (B) 6

(C) 12 (D) 24

6. In an experiment of single slit diffraction pattern, first minimum for red light coincides with first maximum of some other wavelength. If wavelength of red light is 6600 Å, then wavelength of first maximum will be : **[2014 (Online)]**

(A) 3300 Å (B) 4400 Å

(C) 5500 Å (D) 6600 Å

7. The box of a pin hole camera, of length L, has a hole of radius a. It is assumed that when the hole is illuminated by a parallel beam of light of wavelength λ the spread of the spot (obtained on the opposite wall of the camera) is the sum of its geometrical spread and the spread due to diffraction. The spot would then have its minimum size (say b_{min}) when : **[2016]**

(A) $a = \frac{\lambda^2}{L}$ and $b_{min} = \left(\frac{2\lambda^2}{L}\right)$

(B) $a = \sqrt{\lambda L}$ and $b_{min} = \left(\frac{2\lambda^2}{L}\right)$

(C) $a = \sqrt{\lambda L}$ and $b_{min} = \sqrt{4\lambda L}$

(D) $a = \frac{\lambda^2}{L}$ and $b_{min} = \sqrt{4\lambda L}$

8. In Young's double slit experiment, the distance between slits and the screen is 1.0 m and monochromatic light of 600 nm is being used. A person standing near the slits is looking at the fringe pattern. When the separation between the slits is varied, the interference pattern disappears for a particular distance d_0 between the slits. If the angular resolution of the eye is $\frac{1°}{60}$, the value of d_0 is close to : **[2016 (Online)]**

(A) 1mm (B) 3mm

(C) 2mm (D) 4mm

9. Two stars are 10 light years away from the earth. They are seen through a telescope of objective diameter 30 cm. The wavelength of light is 600 nm. To see the stars just resolved by the telescope, the minimum distance between them should be (1 light year = 9.46×10^{15} m) of the order of: **[2016 (Online)]**

(A) 10^8 km (B) 10^{10} km

(C) 10^{11} km (D) 10^6 km

10. A single slit of width 0.1 mm is illuminated by a parallel beam of light of wavelength 6000 Å and diffraction bands are observed on a screen 0.5 m from the slit. The distance of the third dark band from the central bright band is : **[2017 (Online)]**

(A) 3mm (B) 9mm

(C) 4.5mm (D) 1.5mm

11. A single slit of width b is illuminated by a coherent monochromatic light of wavelength λ. If the second and fourth minima in the diffraction pattern at a distance 1m from the slit are at 3 cm and 6 cm respectively from the central maximum, wheat is the width of the central maximum? (i.e. distance between first minimum on either side of the central maximum) : **[2017 (Online)]**

(A) 1.5 cm (B) 3.0 cm
(C) 4.5 cm (D) 6.0 cm

12. The angular width of the central maximum in a single slit diffraction pattern is 60°. The width of the slit is 1 μm. The slit is illuminated by monochromatic plane waves. If another slit of same width is made near it, Young's fringes can be observed on a screen placed at a distance 50 cm from the slits. If the observed fringe width is 1 cm, what is slit separation distance ? (i.e. distance between the centres of each slit) **[2018]**

(A) 25 μm (B) 50 μm
(C) 75 μm (D) 100 μm

13. Light of wavelength 550 nm falls normally on a slit of width 22.0×10^{-5} cm. The angular position of the second minima from the central maximum will be (in radians) : **[2018 (Online)]**

(A) $\frac{\pi}{8}$ (B) $\frac{\pi}{12}$

(C) $\frac{\pi}{4}$ (D) $\frac{\pi}{6}$

* * * * *

CHAPTER-24 WAVE OPTICS

24.3 Polarization of Light

1. Electromagnetic waves are transverse in nature is evident by : **[2002]**
(A) polarization (B) interference
(C) reflection (D) diffraction

2. The angle of incidence at which reflected light is totally polarized for reflection from air to glass (refractive index n), is : **[2004]**
(A) $\tan^{-1}(1/n)$ (B) $\sin^{-1}(1/n)$
(C) $\sin^{-1}(n)$ (D) $\tan^{-1}(n)$

3. When an unpolarized light of intensity I_0 is incident on a polarizing sheet, the intensity of the light which does not get transmitted is : **[2005]**
(A) $1/2I_0$ (B) $1/4I_0$
(C) Zero (D) I_0

4. **Statement-1 :** On viewing the clear blue portion of the sky through a Calcite Crystal, the intensity of transmitted light varies as the crystal is rotated. **[2011]**
Statement-2 : The light coming from the sky is polarized due to scattering of sun light by particles in the atmosphere. The scattering is largest for blue light.
(A) Statement-1 is false, statement-2 is true
(B) Statement-1 is true, statement-2 is false
(C) Statement-1 is true, statement-2 true; statement-2 is the correct explanation of statement-1
(D) Statement-1 is true, statement-2 is true; statement -2 is not correct explanation of statement-1

5. Two polaroids have their polarizing directions parallel so that the intensity of a transmitted light is maximum. The angle through which either polaroid must be turned if the intensity is to drop by one-half is : **[2012 (Online)]**
(A) 135° (B) 90°
(C) 120° (D) 180°

6. A beam of unpolarised light of intensity I_0 is passed through a polaroid A and then through another polaroid B which is oriented so that its principal plane makes an angle of 45° relative to that of A. The intensity of the emergent light is : **[2013]**
(A) I_0 (B) $I_0/2$
(C) $I_0/4$ (D) $I_0/8$

7. A person lives in a high-rise building on the bank of a river 50 m wide. Across the river is a well lit tower of height 40 m. When the person, who is at a height of 10 m, looks through a polarizer at an appropriate angle at light of the tower reflecting from the river surface, he notes that intensity of light coming from distance X from his building is the least and this corresponds to the light coming from light bulbs at height 'Y' on the tower. The values of X and Y are respectively close to (refractive index of water = 4/3) : **[2013 (Online)]**

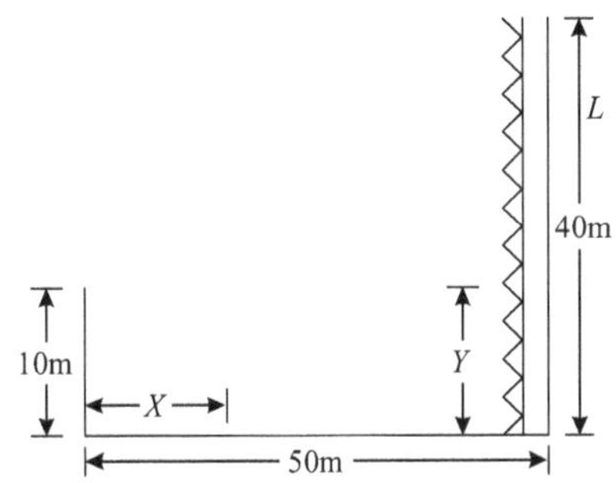

(A) 25 m, 10 m (B) 13 m, 27 m
(C) 22 m, 13 m (D) 17 m, 20 m

8. Two beams, A and B, of plane polarized light with mutually perpendicular planes of polarization are seen through a polaroid. From the position when the beam A has maximum intensity (and beam B has zero intensity), a rotation of polaroid through 30° makes the two beams appear equally bright. If the initial intensities of the two beams are I_A and I_B respectively, then I_A/I_B equals : **[2014]**
(A) 3 (B) 3/2
(C) 1 (D) 1/3

9. A ray of light is incident from a denser to a rarer medium. The critical angle for total internal reflection is θ_{iC} and Brewster's angle of incidence is θ_{iB}, such that $\sin\theta_{iC}/\sin\theta_{iB} = \eta = 1.28$. The relative refractive index of the two media is : **[2014 (Online)]**
(A) 0.2 (B) 0.4
(C) 0.8 (D) 0.9

10. Unpolarized light of intensity I_0 is incident on surface of a block of glass at Brewster's angle. In that case, which one of the following statements is true ? **[2015 (Online)]**
(A) Reflected light is completely polarized with intensity less than $\frac{I_0}{2}$
(B) Transmitted light is completely polarized with intensity less than $\frac{I_0}{2}$
(C) Transmitted light is partially polarized with intensity $\frac{I_0}{2}$
(D) Reflected light is partially polarized with intensity $\frac{I_0}{2}$

11. Unpolarized light of intensity I passes through an ideal polarizer A. Another identical polarizer B is placed behind A. The intensity of light beyond B is found to be $I/2$. Now another identical polarizer C is placed between A and B. The intensity beyond B is now found to be $I/8$. The angle between polarizer A and C is : **[2018]**

(A) 0° (B) 30°
(C) 45° (D) 60°

12. A plane polarized light is incident on a polarizer with its pass axis making angle θ with x-axis, as shown in the figure. At four different values of θ, θ = 8°, 38°, 188° and 218°, the observed intensities are same. What is the angle between the direction of polarization and x-axis : **[2018 (Online)]**

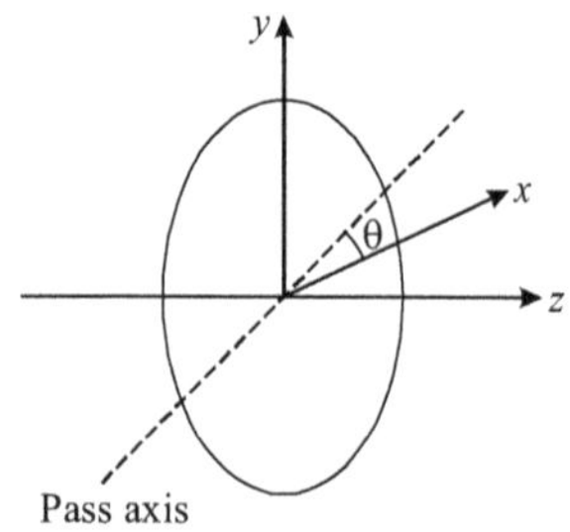

(A) 203° (B) 45°
(C) 98° (D) 128°

13. Unpolarized light of intensity I is incident on a system of two polarizers, A followed by B. The intensity of emergent light is $I/2$. If a third polarizer C is placed between A and B, the intensity of emergent light is reduced to $I/3$. The angle between the polarizers A and C is θ. Then : **[2018 (Online)]**

(A) $\cos\theta = \left(\frac{2}{3}\right)^{1/4}$ (B) $\cos\theta = \left(\frac{1}{3}\right)^{1/4}$

(C) $\cos\theta = \left(\frac{1}{3}\right)^{1/2}$ (D) $\cos\theta = \left(\frac{2}{3}\right)^{1/2}$

14. Consider a tank made of glass (refractive index 1.5) with a thick bottom. It is filled with a liquid of refractive index μ. A student finds that, irrespective of what the incident angle i (see figure) is for a beam of light entering the liquid, the light reflected from the liquid glass interface is never completely polarized. For this to happen, the minimum value of μ is : **[9 Jan. 2019 (Morning)]**

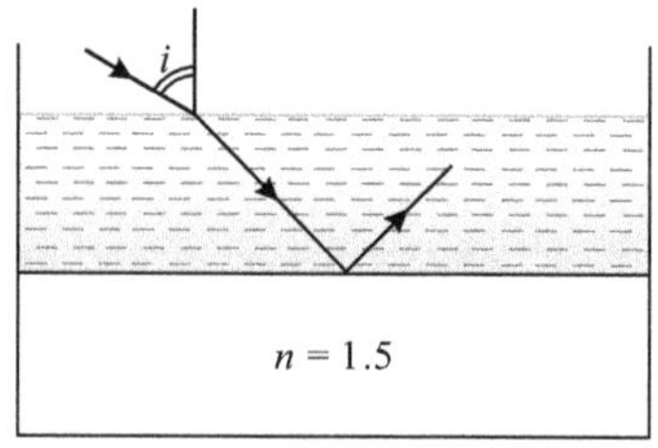

(A) $\frac{3}{\sqrt{5}}$ (B) $\frac{5}{\sqrt{3}}$

(C) $\sqrt{\frac{5}{3}}$ (D) $\frac{4}{3}$

15. A system of three polarizers P_1, P_2, P_3 is set up such that the pass axis of P_3 is crossed with respect to that of P_1. The pass axis of P_2 is inclined at 60° to the pass axis of P_3. When a beam of unpolarized light of intensity I_0 is incident on P_1, the intensity of light transmitted by the three polarizers is I. The ratio (I_0/I) equals (nearly) : **[12 April 2019 (Evening)]**

(A) 16.00 (B) 1.80
(C) 5.33 (D) 10.67

* * * * *

CHAPTER-25 DUAL NATURE OF RADIATION, MATTER & X-RAYS

25.1 Photoelectric Effect

1. Sodium and copper have work functions 2.3 eV and 4.5 eV respectively. Then the ratio of their threshold wavelengths is nearest to : **[2002]**

(A) 1 : 2 (B) 4 : 1
(C) 2 : 1 (D) 1 : 4

2. Two identical, photocathodes receive light of frequencies f_1 and f_2. If the velocities of the photoelectrons (of mass m) coming out are respectively v_1 and v_2, then : **[2003]**

(A) $v_1^2 - v_2^2 = \frac{2h}{m}(f_1 - f_2)$

(B) $v_1 + v_2 = \left[\frac{2h}{m}(f_1 + f_2)\right]^{1/2}$

(C) $v_1^2 + v_2^2 = \frac{2h}{m}(f_1 + f_2)$

(D) $v_1 + v_2 = \frac{2h}{m}\left[\frac{2h}{m}(f_1 - f_2)\right]^{1/2}$

3. The work function of a substance is 4.0 eV. The longest wavelength of light that can cause photoelectron emission from this substance is approximately : **[2004]**

(A) 540 nm (B) 400 nm
(C) 310 nm (D) 220 nm

4. According to Einstein's photoelectric equation, the plot of the kinetic energy of the emitted photoelectrons from a metal *Vs* the frequency, of the incident radiation gives a straight line whose slope : **[2004]**

(A) depends on the nature of the metal used
(B) depends on the intensity of the radiation
(C) depends both on the intensity of the radiation and the metal used
(D) is the same for all metals and independent of the intensity of the radiation

5. A photocell is illuminated by a small bright source placed 1 m away. When the same source of light is place 0.5 m away, the number of electrons emitted by photocathode would : **[2005]**

(A) decrease by a factor of 4
(B) increase by a factor of 4
(C) decrease by a factor of 2
(D) increase by a factor of 2

6. The anode voltage of a photocell is kept fixed. The wavelength λ of the light falling on the cathode is gradually changed. The plate current I of the photocell varies as follows : **[2006]**

(A)
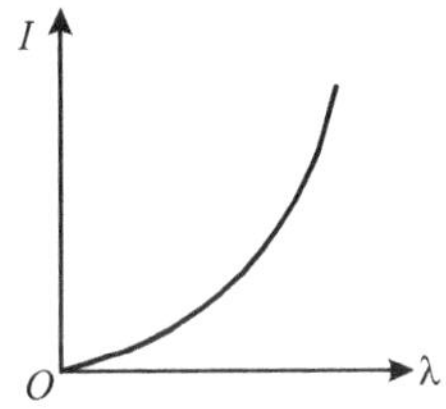

(B)
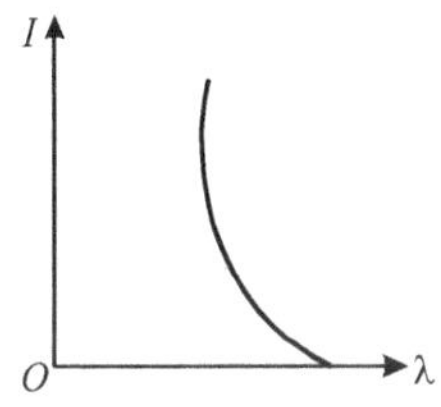

(C)
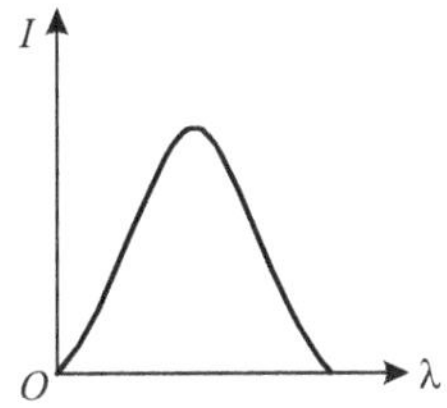

(D)
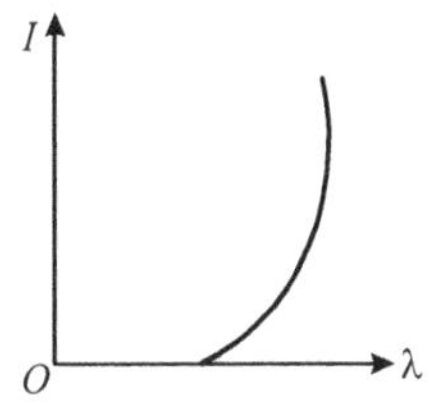

7. The threshold frequency for a metallic surface corresponds to an energy of 6.2 eV and the stopping potential for a radiation incident on this surface is 5 V. The incident radiation lies in : **[2006]**

(A) ultra-violet region (B) infra-red region
(C) visible region (D) X-ray region

8. The surface of a metal is illuminated with the light of 400 nm. The kinetic energy of the ejected photoelectrons was found to be 1.68 eV. The work function of the metal is : [hc = 1240 eV-nm] **[2009]**

(A) 1.51 eV (B) 1.68 eV
(C) 3.09 eV (D) 1.42 eV

9. Directions : Question contain Statement-1 and Statement-2. Of the four choices given after the statements, choose the one that best discribes the two statements. **[2010]**

Statement-1 : When ultraviolet light is incident on a photocell, its stopping potential is V_0 and the maximum kinetic energy of the photoelectrons is K_{max}. When the ultraviolet light is replaced by X-rays, both V_0 and K_{max} increase.

Statement-2 : Photoelectrons are emitted with speeds ranging from zero to a maximum value because of the range of frequencies present in the incident light.

(A) Statement-1 is true, Statement-2 is false
(B) Statement-1 is true, Statement-2 is true; Statement-2 is the correct explanation of Statement-1
(C) Statement-1 is true, Statement-2 is true; Statement-2 is not the correct explanation of Statement-1
(D) Statement-1 is false, Statement-2 is true

10. This question has Statement-1 and Statement-2. Of the four choices given after the statements, choose the one that best describes the two statements. **[2011]**

Statement-1 : A metallic surface is irradiated by a monochromatic light of frequency $v > v_0$ (the threshold frequency). The maximum kinetic energy and the stopping potential are K_{max} and V_0 respectively. If the frequency incident on the surface is doubled, both the K_{max} and V_0 are also doubled.

Statement-2 : The maximum kinetic energy and the stopping potential of photoelectrons emitted from a surface are linearly dependent on the frequency of incident light.

(A) Statement-1 is true, Statement-2 is true, Statement-2 is not the correct explanation of Statement-1

(B) Statement-1 is false, Statement-2 is true

(C) Statement-1 is true, Statement-2 is false

(C) Statement-1 is true, Statement-2 is true, Statement-2 is the correct explanation of Statement-1

11. This question has Statement-1 and Statement-2. Of the four choices given after the Statements, choose the one that best describes the two Statements. **[2012 (Online)]**

Statement-1 : A metallic surface is irradiated by a monochromatic light of frequency $\upsilon > \upsilon_0$ (the threshold frequency). If the incident frequency is now doubled, the photocurrent and the maximum kinetic energy are also doubled.

Statement-2 : The maximum kinetic energy of photoelectrons emitted from a surface is linearly dependent on the frequency of the incident light. The photocurrent depends only on the intensity of the incident light :

(A) Statement-1 is true, Statement-2 is true, Statement-2 is the correct explanation of Statement-1

(B) Statement-1 is false, Statement-2 is true

(C) Statement-1 is true, Statement-2 is false

(D) Statement-1 is true, Statement-2 is true, Statement-2 is not the correct explanation of Statement-1

12. Photoelectrons are ejected from a metal when light of frequency ν falls on it. Pick out the wrong statement from the following : **[2012 (Online)]**

(A) No electrons are emitted if ν is less than W/h, where W is the work function of the metal

(B) The ejection of the photoelectrons is instantaneous

(C) The maximum energy of the photoelectrons is $h\nu$

(D) The maximum energy of the photoelectrons is independent of the intensity of the light

13. The anode voltage of photocell is kept fixed. The wavelength λ of the light falling on the cathode is gradually changed. The plate current I of the photocell varies as follows: **[2013]**

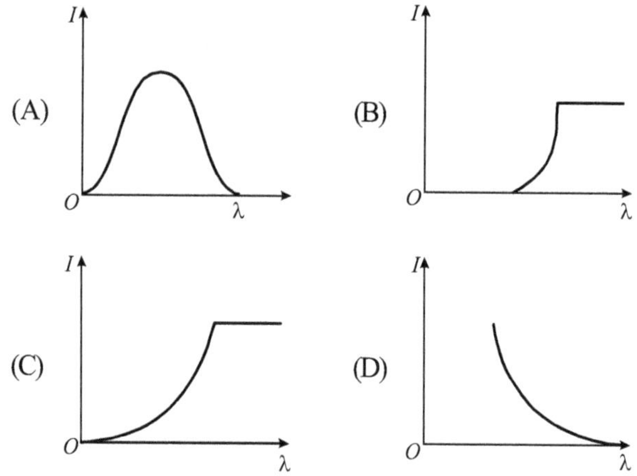

14. A copper ball of radius 1 cm and work function 4.47 eV is irradiated with ultraviolet radiation of wavelength 2500 Å. The effect of irradiation results in the emission of electrons from the ball. Further the ball will acquire charge and due to this there will be a finite value of the potential on the ball. The charge acquired by the ball is : **[2013 (Online)]**

(A) 5.5×10^{-13} C (B) 7.5×10^{-13} C

(C) 4.5×10^{-12} C (D) 2.5×10^{-11} C

15. In an experiment on photoelectric effect, a student plots stopping potential V_0 against reciprocal of the wavelength λ of the incident light for two different metals A and B. These are shown in the figure.

Looking at the graphs, you can most appropriately say that : **[2013 (Online)]**

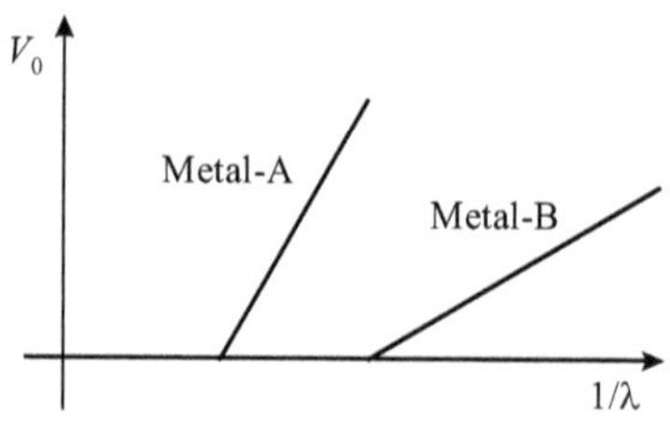

(A) Work function of metal B is greater than that of metal A

(B) For light of certain wavelength falling on both metal, maximum kinetic energy of electrons emitted from A will be greater than those emitted from B

(C) Work function of metal A is greater than that of metal B

(D) Students data is not correct

16. The radiation corresponding to $3 \rightarrow 2$ transition of hydrogen atom falls on a metal surface to produce photoelectrons. These electrons are made to enter a magnetic field of 3×10^{-4} T. If the radius of the largest circular path followed by these electrons is 10.0 mm, the work function of the metal is close to : **[2014]**

(A) 1.8 eV (B) 1.1 eV

(C) 0.8 eV (D) 1.6 eV

17. A beam of light has two wavelengths of 4972 Å and 6216 Å with a total intensity of 3.6×10^{-3} Wm^{-2} equally distributed among the two wavelengths. The beam falls normally on an area of 1 cm^2 of a clean metallic surface of work function 2.3 eV. Assume that there is no loss of light by reflection and that each capable photon ejects one electron. The number of photoelectrons liberated in 2s is approximately : **[2014 (Online)]**

(A) 6×10^{11} (B) 9×10^{11}
(C) 11×10^{11} (D) 15×10^{11}

18. Radiation of wavelength λ, is incident on a photocell. The fastest emitted electron has speed v. If the wavelength is changed to $\frac{3\lambda}{4}$, the speed of the fastest emitted electron will be : **[2016]**

(A) $> v\left(\frac{4}{3}\right)^{\frac{1}{2}}$ (B) $< v\left(\frac{4}{3}\right)^{\frac{1}{2}}$
(C) $= v\left(\frac{4}{3}\right)^{\frac{1}{2}}$ (D) $= v\left(\frac{3}{4}\right)^{\frac{1}{2}}$

19. When photons of wavelength λ_1 are incident on an isolated sphere, the corresponding stopping potential is found to be V. When photons of wavelength λ_2 are used, the corresponding stopping potential was thrice that of the above value. If light of wavelength λ_3 is used then find the stopping potential for this case : **[2016 (Online)]**

(A) $\frac{hc}{e}\left[\frac{1}{\lambda_3}+\frac{1}{\lambda_2}-\frac{1}{\lambda_1}\right]$ (B) $\frac{hc}{e}\left[\frac{1}{\lambda_3}+\frac{1}{2\lambda_2}-\frac{1}{\lambda_1}\right]$
(C) $\frac{hc}{e}\left[\frac{1}{\lambda_3}-\frac{1}{\lambda_2}-\frac{1}{\lambda_1}\right]$ (D) $\frac{hc}{e}\left[\frac{1}{\lambda_3}+\frac{1}{2\lambda_2}-\frac{3}{2\lambda_1}\right]$

20. A photoelectric surface is illuminated successively by monochromatic light of wavelengths λ and $\frac{\lambda}{2}$. If the maximum kinetic energy of the emitted photoelectrons in the second case is 3 times that in the first case, the work function of the surface is : **[2016 (Online)]**

(A) $\frac{hc}{2\lambda}$ (B) $\frac{hc}{\lambda}$
(C) $\frac{hc}{3\lambda}$ (D) $\frac{3hc}{\lambda}$

21. The maximum velocity of the photoelectrons emitted from the surface is v when light of frequency n falls on a metal surface. If the incident frequency is increased to $3n$, the maximum velocity of the ejected photoelectrons will be : **[2017 (Online)]**

(A) Less than $\sqrt{3}v$ (B) v
(C) More than $\sqrt{3}v$ (D) Equal to $\sqrt{3}v$

22. A Laser light of wavelength 600 nm is used to weld retina detachment. If a Laser pulse of width 60 ms and power 0.5 kW is used the approximate number of photons in the pulse are : [Take Plank's constant $h = 6.62 \times 10^{-34}$ Js] : **[2017 (Online)]**

(A) 10^{20} (B) 10^{18}
(C) 10^{22} (D) 10^{19}

23. Surface of certain metal is first illuminated with light of wavelength $\lambda_1 = 350$ nm and then, by light of wavelength $\lambda_2 = 540$ nm. It is found that the maximum speed of the photo electrons in the two cases differ by a factor of 2. The work function of the metal (in eV) is close to :

(Energy of photon $= \frac{1240}{\lambda(\text{in } nm)}$ eV) **[9 Jan. 2019 (Morning)]**

(A) 1.8 (B) 1.4
(C) 2.5 (D) 5.6

24. The magnetic field associated with a light wave is given, at the origin, by

$B = B_0\,[\sin(3.14 \times 10^7)\,ct + \sin(6.28 \times 10^7)\,ct]$.

If this light falls on a silver plate having a work function of 4.7 eV, what will be the maximum kinetic energy of the photo electrons? ($c = 3 \times 10^8$ ms^{-1}, $h = 6.6 \times 10^{-34}$ J-s) : **[9 Jan. 2019 (Evening)]**

(A) 7.72 eV (B) 8.52 eV
(C) 12.5 eV (D) 6.82 eV

25. A metal plate of area 1×10^{-4} m^2 is illuminated by a radiation of intensity 16 mW/m^2. The work function of the metal is 5 eV. The energy of the incident photons is 10 eV and only 10% of it produces photo electrons. The number of emitted photo electrons per second and their maximum energy, respectively, will be. [1 eV = 1.6×10^{-19} J] : **[10 Jan. 2019 (Evening)]**

(A) 10^{10} and 5 eV (B) 10^{14} and 10 eV
(C) 10^{12} and 5 eV (D) 10^{11} and 5 eV

26. In a photoelectric experiment, the wavelength of the light incident on a metal is changed from 300 nm to 400 nm. The decrease in the stopping potential is close to :

$\left(\frac{hc}{e} = 1240\, nm - V\right)$ **[11 Jan. 2019 (Evening)]**

(A) 0.5 V (B) 1.0 V
(C) 2.0 V (D) 1.5 V

27. When a certain photosensitive surface is illuminated with monochromatic light of frequency ν, the stopping potential for the photo current is $-V_0/2$. When the surface is illuminated by monochromatic light of frequency ν/2, the stopping potential is $-V_0$. The threshold frequency for photoelectric emission is: **[12 Jan. 2019 (Evening)]**

(A) $\frac{3\nu}{2}$ (B) 2ν
(C) $\frac{4}{3}\nu$ (D) $\frac{5\nu}{3}$

28. The electric field of light wave is given as $\vec{E}=10^{-3}\cos\left(\frac{2\pi x}{5\times10^{-7}}-2\pi\times6\times10^{14}t\right)\hat{x}\frac{N}{C}$. This light falls on a metal plate of work function $2eV$. The stopping potential of the photo-electrons is.

Given, $E\,(\text{in } eV)=\frac{12375}{\lambda(\text{in Å})}$: **[9 April 2019 (Morning)]**

(A) 0.48 V (B) 2.0 V
(C) 2.48 V (D) 0.72 V

29. In a photoelectric effect experiment the threshold wavelength of the light is 380 nm. If the wavelentgh of incident light is 260 nm, the maximum kinetic energy of emitted electrons will be : Given $E\,(\text{in } eV)=\frac{1237}{\lambda(\text{in } nm)}$ **[10 April 2019 (Morning)]**

(A) 1.5 eV (B) 4.5 eV
(C) 15.1 eV (D) 3.0 eV

30. The stopping potential V_0(in volt) as a function of frequency (ν) for a sodium emitter, is shown in the figure. The work function of sodium, from the data plotted in the figure, will be : (Given : Planck's constant $(h) = 6.63\times10^{-34}$ Js, electron charge $e = 1.6\times10^{-19}$ C) **[12 April 2019 (Morning)]**

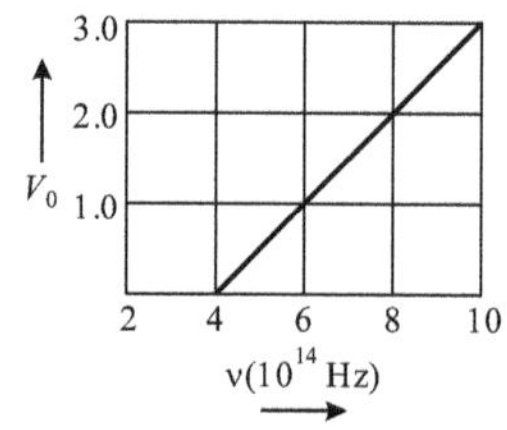

(A) 1.95 eV (B) 1.82 eV
(C) 1.66 eV (D) 2.12 eV

* * * * *

CHAPTER-25 DUAL NATURE OF RADIATION, MATTER & X-RAYS

25.2 Matter Waves & X-Ray

1. Formation of covalent bonds in compounds exhibits : **[2002]**

(A) wave nature of electron
(B) particle nature of electron
(C) both wave and particle nature of electron
(D) none of the above

2. A radiation of energy E falls normally on a perfectly reflecting surface. The momentum transferred to the surface is: **[2004]**

(A) E/c (B) $2E/c$
(C) Ec (D) E/c^2

3. If the kinetic energy of a free electron doubles, its de-Broglie wavelength changes by the factor : **[2005]**

(A) $\frac{1}{2}$ (B) 2
(C) $\frac{1}{\sqrt{2}}$ (D) $\sqrt{2}$

4. Photon of frequency ν has a momentum associated with it. If c is the velocity of light, the momentum is : **[2007]**

(A) ν/c (B) $h\nu c$
(C) $h\nu/c^2$ (D) $h\nu/c$

5. Directions For question 5 and 6 : These questions are based on the following paragraph.

Wave property of electrons implies that they will show diffraction effects. Davisson and Germer demonstrated this by diffracting electrons from crystals. The law governing the diffraction from a crystal is obtained by requiring that electron waves reflected from the planes of atoms in a crystal interfere constructively (see figure).

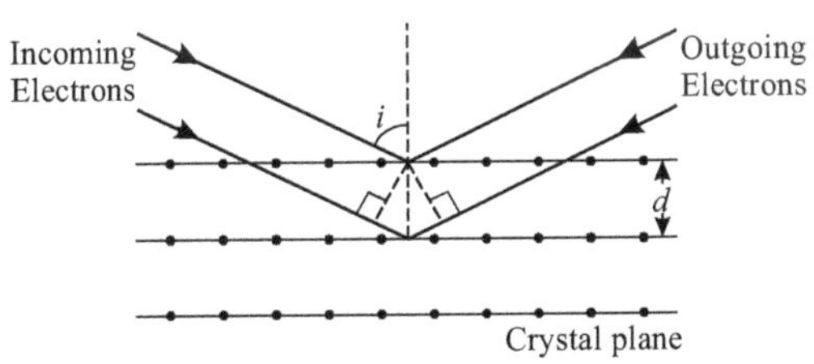

Electrons accelerated by potential V are diffracted from a crystal. If $d = 1$ Å and $i = 30°$ V should be about ($h = 6.6 \times 10^{-34}$ Js, $m_e = 9.1 \times 10^{-31}$ kg, $e = 1.6 \times 10^{-19}$ C) : **[2008]**

(A) 2000 V (B) 50 V
(C) 500 V (D) 1000 V

6. If a strong diffraction peak is observed when electrons are incident at an angle 'i' from the normal to the crystal planes with distance 'd' between them (see figure), de-Broglie wavelength λ_{dB} of electrons can be calculated by the relationship (n is an integer): **[2008]**

(A) $d \sin i = n\lambda_{dB}$ (B) $2d \cos i = n\lambda_{dB}$
(C) $2d \sin i = n\lambda_{dB}$ (D) $d \cos i = n\lambda_{dB}$

7. This question has Statement-1 and Statement-2. Of the four choices given after the statements, choose the one that best describes the two statements. **[2012]**

Statement-1 : Davisson-Germer experiment established the wave nature of electrons.

Statement-2 : If electrons have wave nature, they can interfere and show diffraction.

(A) Statement-1 is true, Statement-2 is true and Statement-2 is not the correct explanation of Statement-1
(B) Statement-1 is false, Statement-2 is true
(C) Statement-1 is true, Statement-2 is false
(D) Statement-1 is true, Statement-2 is true and Statement-2 is the correct explanation of statement-1

8. Electrons are accelerated through a potential difference V and protons are accelerated through a potential difference 4 V. The de-Broglie wavelengths are λ_e and λ_p for electrons and protons respectively. The ratio of $\frac{\lambda_e}{\lambda_p}$ is given by: (given m_e is mass of electron and m_p is mass of proton) : **[2013 (Online)]**

(A) $\frac{\lambda_e}{\lambda_p} = \sqrt{\frac{m_p}{m_e}}$ (B) $\frac{\lambda_e}{\lambda_p} = \sqrt{\frac{m_e}{m_p}}$

(C) $\frac{\lambda_e}{\lambda_p} = \frac{1}{2}\sqrt{\frac{m_e}{m_p}}$ (D) $\frac{\lambda_e}{\lambda_p} = 2\sqrt{\frac{m_p}{m_e}}$

9. Match List-I (Experiment performed) with List-II (Phenomena discovered/associated) and select the correct option from the options given below the lists : **[2014 (Online)]**

List-I	**List-II**
(1) Davisson and germer experiment	(i) Wave nature of electrons
(2) Millikan's oil drop experiment	(ii) Charge of an electron
(3) Rutherford experiment	(iii) Quantisation of energy levels
(4) Franck-Hertz experiment	(iv) Existence of nucleus

(A) (1)→(i); (2)→(ii); (3)→(iii); (4)→(iv)
(B) (1)→(i); (2)→(ii); (3)→(iv); (4)→(iii)
(C) (1)→(iii); (2)→(iv); (3)→(i); (4)→(ii)
(D) (1)→(iv); (2)→(iii); (3)→(ii); (4)→(i)

10. A photon of wavelength λ is scattered from an electron, which was at rest. The wavelength shift $\Delta\lambda$ is three times of λ and the angle of scattering θ is 60°. The angle at which the electron recoiled is ϕ. The value of $\tan\phi$ is : (electron speed is much smaller than the speed of light) : **[2014 (Online)]**

(A) 0.16 (B) 0.22
(C) 0.25 (D) 0.28

11. For which of the following particles will it be most difficult to experimentally verify the de-Broglie relationship ? **[2014 (Online)]**

(A) An electron (B) A proton
(C) An α-particle (D) A dust particle

12. Match List-I (Fundamental Experiment) with List-II (its conclusion) and select the correct option from the choices given below the list : **[2015]**

List-I	List-II
(A) Franck-Hertz Experiment	(p) Particle nature of light
(B) Photo-electric experiment	(q) Discrete energy levels of atom
(C) Davisson-Germer Experiment	(r) Wave nature of electron
	(s) Structure of atom

(A) (A)$\rightarrow$(i); (B)$\rightarrow$(iv); (C)$\rightarrow$(iii)
(B) (A)$\rightarrow$(ii); (B)$\rightarrow$(iv); (C)$\rightarrow$(iii)
(C) (A)$\rightarrow$(ii); (B)$\rightarrow$(i); (C)$\rightarrow$(iii)
(D) (A)$\rightarrow$(iv); (B)$\rightarrow$(iii); (C)$\rightarrow$(ii)

13. If one were to apply Bohr model to a particle of mass 'm' and charge 'q' moving in a plane under the influence of a magnetic field 'B', the energy of the charged particle in the nth level will be : **[2015 (Online)]**

(A) $n\left(\frac{hqB}{2\pi m}\right)$ (B) $n\left(\frac{hqB}{8\pi m}\right)$
(C) $n\left(\frac{hqB}{4\pi m}\right)$ (D) $n\left(\frac{hqB}{\pi m}\right)$

14. The de-Broglie wavelength associated with the electron in the $n = 4$ level is : **[2015 (Online)]**

(A) $1/4^{th}$ of the de-Broglie wavelength of the electron in the ground state
(B) Four times the de-Broglie wavelength of the electron in the ground state
(C) Two times the de-Broglie wavelength of the electron in the ground state
(D) Half of the de-Broglie wavelength of the electron in the ground state

15. de-Broglie wavelength of an electron accelerated by a voltage of 50 V is close to ($|e| = 1.6 \times 10^{-19}$ C, $m_e = 9.1 \times 10^{31}$ kg, $h = 6.6 \times 10^{-34}$ Js) : **[2015 (Online)]**

(A) 2.4 Å (B) 0.5 Å
(C) 1.7 Å (D) 1.2 Å

16. A parallel beam of electrons travelling in x-direction falls on a slit of width d (see figure). If after passing the slit, an electron acquires momentum P_y in the y-direction then for a majority of electrons passing through the slit (h is Plank's constant) : **[2015 (Online)]**

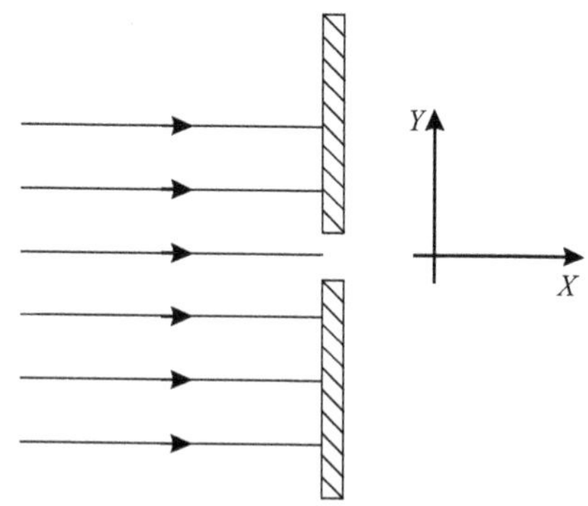

(A) $|P_y| d > h$ (B) $|P_y| d < h$
(C) $|P_y| d = h$ (D) $|P_y| d >> h$

17. A particle A of mass m and initial velocity v collides with a particle B of mass $m/2$ which is at rest. The collision is head on, and elastic. The ratio of the de-Broglie wavelengths λ_A to λ_B after the collision is : **[2017]**

(A) $\frac{\lambda_A}{\lambda_B} = \frac{2}{3}$ (B) $\frac{\lambda_A}{\lambda_B} = \frac{1}{2}$
(C) $\frac{\lambda_A}{\lambda_B} = \frac{1}{3}$ (D) $\frac{\lambda_A}{\lambda_B} = 2$

18. An electron beam is accelerated by a potential difference V to hit a metallic target to produce X-rays. It produces continuous as well as characteristic X-rays. If λ_{min} is the smallest possible wavelength of X-ray in the spectrum, the variation of $\log \lambda_{min}$ with $\log V$ is correctly represented in : **[2017]**

(A)
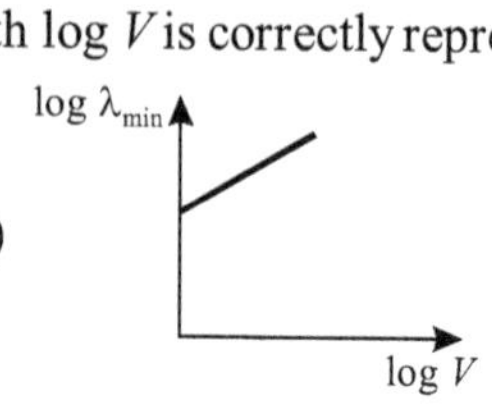

(B)
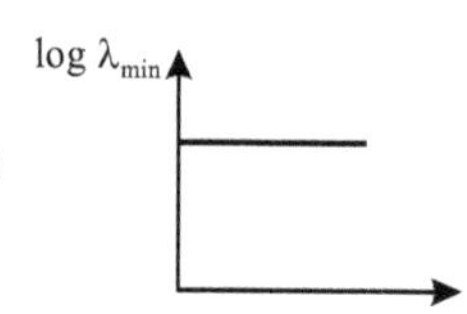

(C)
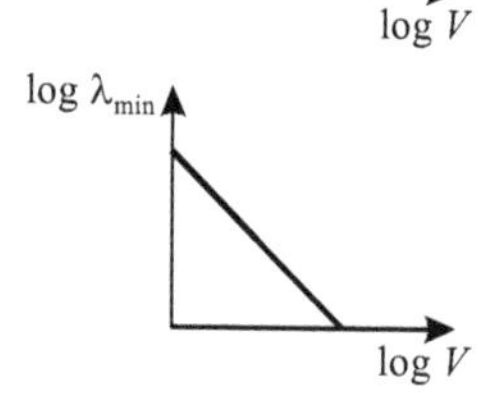

(D)
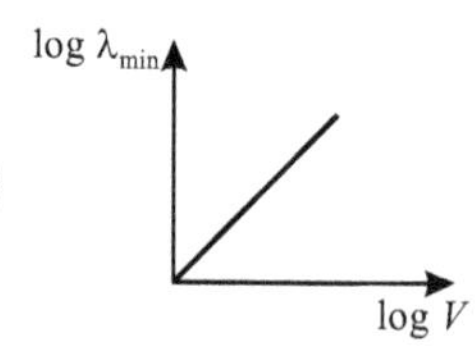

19. If the de Broglie wavelengths associated with a proton and an α-particle are equal, then the ratio of velocities of the proton and the α- particle will be : **[2018 (Online)]**

(A) 1 : 4 (B) 1 : 2
(C) 4 : 1 (D) 2 : 1

20. Two electrons are moving with non-relativistic speeds perpendicular to each other. If corresponding de Broglie wavelengths are λ_1 and λ_2, their de Broglie wavelength in the frame of reference attached to their centre of mass is : **[2018 (Online)]**

(A) $\lambda_{CM}=\lambda_1=\lambda_2$ (B) $\frac{1}{\lambda_1}=\frac{1}{\lambda_1}+\frac{1}{\lambda_2}$

(C) $\lambda_{CM}=\frac{2\lambda_1\lambda_2}{\sqrt{\lambda_1^2+\lambda_2^2}}$ (D) $\lambda_{CM}=\left(\frac{\lambda_1+\lambda_2}{2}\right)$

21. The de-Broglie wavelength (λ_B) associated with the electron orbiting in the second excited state of hydrogen atom is related to that in the ground state (λ_G) by : **[2018 (Online)]**

(A) $\lambda_B=\lambda_G/3$ (B) $\lambda_B=\lambda_G/2$
(C) $\lambda_B=2\lambda_G$ (D) $\lambda_B=3\lambda_G$

22. In an electron microscope, the resolution that can be achieved is of the order of the wavelength of electrons used. To resolve a width of 7.5×10^{-12} m, the minimum electron energy required is close to : **[10 Jan. 2019 (Morning)]**

(A) 100 keV (B) 500 keV
(C) 25 keV (D) 1 keV

23. If the de-Broglie wavelength of an electron is equal to 10^{-3} times the wavelength of a photon of frequency 6×10^{14} Hz, then the speed of electron is equal to : **[11 Jan. 2019 (Morning)]**
(Speed of light = 3×10^8 m/s
Planck's constant = 6.63×10^{-34} J.s
Mass of electron = 9.1×10^{-31} kg)

(A) 1.45×10^6 m/s (B) 1.7×10^6 m/s
(C) 1.8×10^6 m/s (D) 1.1×10^6 m/s

24. A particle A of mass 'm' and charge 'q' is accelerated by a potential difference of 50 V. Another particle B of mass '4 m' and charge 'q' is accelerated by a potential difference of 2500 V.

The ratio of de-Broglie wavelengths $\frac{\lambda_A}{\lambda_B}$ is close to :
[12 Jan. 2019 (Morning)]

(A) 10.00 (B) 14.14
(C) 4.47 (D) 0.07

25. Two particles move at right angle to each other. Their de-Broglie wavelengths are λ_1 and λ_2 respectively. The particles suffer perfectly inelastic collision. The de-Broglie wavelength λ, of the final particle, is given by : **[8 April 2019 (Morning)]**

(A) $\lambda=\frac{\lambda_1+\lambda_2}{2}$ (B) $\frac{2}{\lambda}=\frac{1}{\lambda_1}+\frac{1}{\lambda_2}$

(C) $\lambda=\sqrt{\lambda_1\lambda_2}$ (D) $\frac{1}{\lambda^2}=\frac{1}{\lambda_1^2}+\frac{1}{\lambda_2^2}$

26. A nucleus A, with a finite de-broglie wavelength λ_A, undergoes spontaneous fission into two nuclei B and C of equal mass. B flies in the same direction as that of A, while C flies in the opposite direction with a velocity equal to half of that of B. The de-Broglie wavelengths λ_B and λ_C of B and C are respectively : **[8 April 2019 (Evening)]**

(A) $2\lambda_A, \lambda_A$ (B) $\lambda_A, 2\lambda_A$

(C) $\lambda_A, \frac{\lambda_A}{2}$ (D) $\frac{\lambda_A}{2}, \lambda_A$

27. A particle 'P' is formed due to a completely inelastic collision of particles 'x' and 'y' having de-Broglie wavelengths 'λ_x' and 'λ_y' respectively. If x and y were moving in opposite directions, then the de-Broglie wavelength of 'P' is :
[9 April 2019 (Evening)]

(A) $\lambda_x+\lambda_y$ (B) $\frac{\lambda_x\lambda_y}{\lambda_x+\lambda_y}$

(C) $\frac{\lambda_x\lambda_y}{|\lambda_x-\lambda_y|}$ (D) $\lambda_x-\lambda_y$

28. 50 W/m^2 energy density of sunlight is normally incident on the surface of a solar panel. Some part of incident energy (25%) is reflected from the surface and the rest is absorbed. The force exerted on 1m^2 surface area will be close to ($c = 3 \times 10^8$ m/s) : **[9 April 2019 (Evening)]**

(A) 15×10^{-8} N (B) 35×10^{-8} N
(C) 10×10^{-8} N (D) 20×10^{-8} N

29. Light is incident normally on a completely absorbing surface with an energy flux of 25 Wcm^{-2}. If the surface has an area of 25 cm^{-2}, the momentum transferred to the surface in 40 min time duration will be : **[10 April 2019 (Evening)]**

(A) 5.0×10^{-3} Ns (B) 3.5×10^{-6} Ns
(C) 1.4×10^{-6} Ns (D) 6.3×10^{-4} Ns

30. A 2 mW laser operates at a wavelength of 500 nm. The number of photons that will be emitted per second is :
[Given Planck's constant $h = 6.6 \times 10^{-34}$ Js, speed of light $c = 3.0 \times 10^8$ m/s] **[10 April 2019 (Evening)]**

(A) 2×10^{16} (B) 1.5×10^{16}
(C) 5×10^{15} (D) 1×10^{16}

* * * * *

CHAPTER-26 ATOMS

26.1 Atomic Structure

1. If the binding energy of the electron in a hydrogen atom is 13.6 eV, the energy required to remove the electron from the first excited state of Li^{2+} is : **[2002]**

(A) 30.6 eV (B) 13.6 eV
(C) 3.4 eV (D) 122.4 eV

2. The wavelengths involved in the spectrum of deuterium (^2_1D) are slightly different from that of hydrogen spectrum, because : **[2003]**

(A) sizes of the two nuclei are different
(B) nuclear forces are different in the two cases
(C) masses of the two nuclei are different
(D) attraction between the electron and the nucleus is different in the two cases

3. If 13.6 eV energy is required to ionize the hydrogen atom, then the energy required to remove an electron from $n = 2$ is : **[2003]**

(A) 10.2 eV (B) 0 eV
(C) 3.4 eV (D) 6.8 eV

4. The diagram shows the energy levels for an electron in a certain atom. Which transition shown represents the emission of a photon with the most energy? **[2005]**

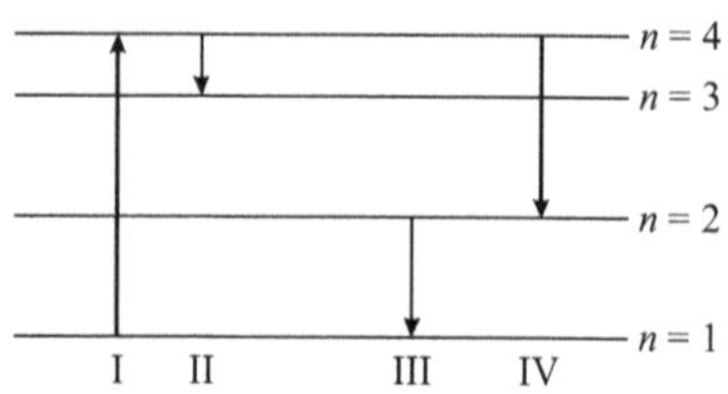

(A) III (B) IV
(C) I (D) II

5. Which of the following transitions in hydrogen atoms emit photons of highest frequency ? **[2007]**

(A) $n = 2$ to $n = 6$ (B) $n = 6$ to $n = 2$
(C) $n = 2$ to $n = 1$ (D) $n = 1$ to $n = 2$

6. Suppose an electron is attracted towards the origin by a force k/r where 'k' is a constant and 'r' is the distance of the electron from the origin. By applying Bohr model to this system, the radius of the nth orbital of the electron is found to be 'r_n' and the kinetic energy of the electron to be 'T_n'. Then which of the following is true? **[2008]**

(A) $T_n \propto \frac{1}{n^2}, r_n \propto n^2$
(B) T_n independent of n, $r_n \propto n$
(C) $T_n \propto \frac{1}{n}, r_n \propto n$
(D) $T_n \propto \frac{1}{n}, r_n \propto n^2$

7. The transition from the state $n = 4$ to $n = 3$ in a hydrogen like atom results in ultraviolet radiation. Infrared radiation will be obtained in the transition from : **[2009]**

(A) $4 \to 2$ (B) $5 \to 4$
(C) $2 \to 1$ (D) $3 \to 2$

8. Energy required for the electron excitation in Li^{++} from the first to the third Bohr orbit is : **[2011]**

(A) 108.8 eV (B) 122.4 eV
(C) 12.1 eV (D) 36.3 eV

9. A diatomic molecule is made of two masses m_1 and m_2 which are separated by a distance r. If we calculate its rotational energy by applying Bohr's rule of angular momentum quantization, its energy will be given by : (n is an integer) : **[2012]**

(A) $\frac{(m_1 + m_2)n^2h^2}{2m_1m_2r^2}$ (B) $\frac{(m_1 + m_2)^2 n^2h^2}{2m_1^2m_2^2r^2}$

(C) $\frac{n^2h^2}{2(m_1 + m_2)r^2}$ (D) $\frac{2n^2h^2}{(m_1 + m_2)r^2}$

10. Hydrogen atom is excited from ground state to another state with principal quantum number equal to 4. Then the number of spectral lines in the emission spectra will be : **[2012]**

(A) 6 (B) 2
(C) 3 (D) 5

11. The electron of a hydrogen atom makes a transition from the $(n + 1)^{th}$ orbit to the nth orbit. For large n the wavelength of the emitted radiation is proportional to : **[2012 (Online)]**

(A) n (B) n^3
(C) n^4 (D) n^2

12. A hypothetical atom has only three energy levels. The ground level has energy, $E_1 = -8$ eV. The two excited states have energies, $E_2 = -6$ eV and $E_3 = -2$ eV. Then which of the following wavelengths will not be present in the emission spectrum of this atom? **[2012 (Online)]**

(A) 207 nm (B) 465 nm
(C) 310 nm (D) 620 nm

13. A doubly ionised Li atom is excited from its ground state ($n = 1$) to $n = 3$ state. The wavelengths of the spectral lines are given by λ_{32}, λ_{31} and λ_{21}. The ratio $\lambda_{32}/\lambda_{31}$ and $\lambda_{21}/\lambda_{31}$ are, respectively : **[2012 (Online)]**

(A) 8.1, 0.67 (B) 8.1, 1.2
(C) 6.4, 1.2 (D) 6.4, 0.67

14. Which of the plots shown in the figure represents speed (v_n) of the electron in a hydrogen atom as a function of the principal quantum number (n) ? **[2012 (Online)]**

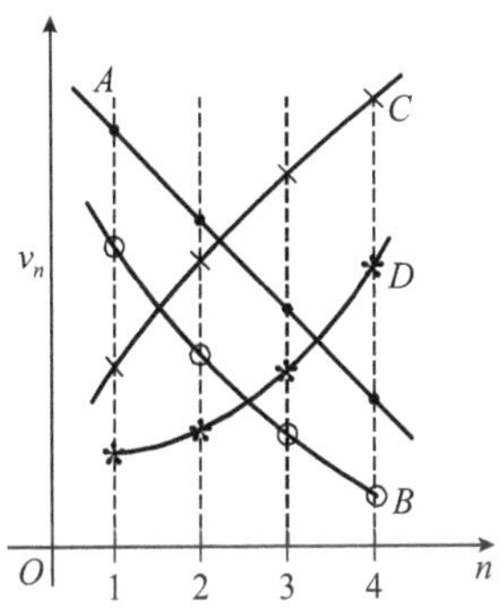

(A) B (B) D
(C) C (D) A

15. In the Rutherford experiment, α-particles are scattered from a nucleus as shown. Out of the four paths, which path is not possible ? **[2012 (Online)]**

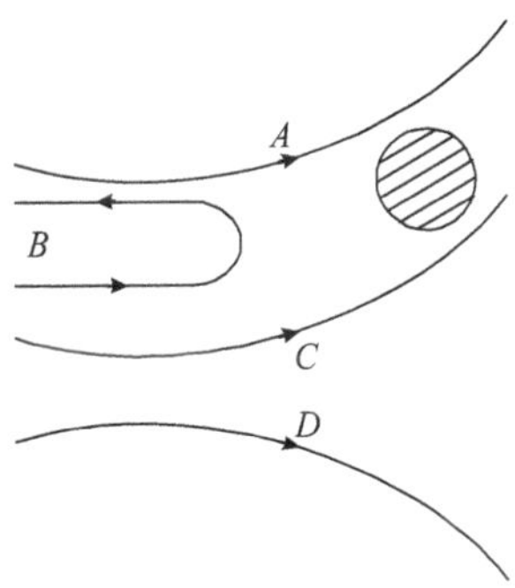

(A) D (B) B
(C) C (D) A

16. Ionisation energy of Li (Lithium) atom in ground state is 5.4 eV. Binding energy of an electron in Li^+ ion in ground state is 75.6 eV. Energy required to remove all three electrons of Lithium (Li) atom is : **[2012 (Online)]**

(A) 81.0 eV (B) 135.4 eV
(C) 203.4 eV (D) 156.6 eV

17. In a hydrogen like atom electron makes transition from an energy level with quantum number n to another with quantum number $(n - 1)$. If $n >> 1$, the frequency of radiation emitted is proportional to : **[2013]**

(A) $\frac{1}{n}$ (B) $\frac{1}{n^2}$
(C) $\frac{1}{n^{3/2}}$ (D) $\frac{1}{n^3}$

18. In the Bohr model an electron moves in a circular orbit around the proton. Considering the orbiting electron to be a circular current loop, the magnetic moment of the hydrogen atom, when the electron is in n^{th} excited state, is : **[2013 (Online)]**

(A) $\left(\frac{e}{2m}\frac{n^2h}{2\pi}\right)$ (B) $\left(\frac{e}{m}\right)\frac{nh}{2\pi}$
(C) $\left(\frac{e}{2m}\right)\frac{nh}{2\pi}$ (D) $\left(\frac{e}{m}\right)\frac{n^2h}{2\pi}$

19. Orbits of a particle moving in a circle are such that the perimeter of the orbit equals an integer number of de-Broglie wavelengths of the particle. For a charged particle moving in a plane perpendicular to a magnetic field, the radius of the nth orbital will therefore be proportional to : **[2013 (Online)]**

(A) n^2 (B) n
(C) $n^{1/2}$ (D) $n^{1/4}$

20. A 12.5 eV electron beam is used to bombard gaseous hydrogen at room temperature. It will emit : **[2013 (Online)]**

(A) 2 lines in the Lyman series and 1 line in the Balmer series
(B) 3 lines in the Lyman series
(C) 1 line in the Lyman series and 2 lines in the Balmer series
(D) 3 lines in the Balmer series

21. In the Bohr's model of hydrogen-like atom the force between the nucleus and the electron is modified as $F = \frac{e^2}{4\pi\varepsilon_0}\left(\frac{1}{r^2}+\frac{\beta}{r^3}\right)$, where β is a constant. For this atom, the radius of the nth orbit in terms of the Bohr radius $\left(a_0 = \frac{\varepsilon_0 h^2}{m\pi e^2}\right)$ is : **[2013 (Online)]**

(A) $r_n = a_0 n - \beta$ (B) $r_n = a_0 n^2 + \beta$
(C) $r_n = a_0 n^2 - \beta$ (D) $r_n = a_0 n + \beta$

22. Hydrogen ($_1H^1$), Deuterium ($_1H^2$), singly ionised Helium $(_2He^4)^+$ and dubly ionised lithium $(_3Li^6)^{++}$ all have one electron around the nucleus. Consider an electron transition from $n = 2$ to $n = 1$. If the wave lengths of emitted radiation are λ_1, λ_2, λ_3 and λ_4 respectively then approximately which one of the following is correct ? **[2014]**

(A) $4\lambda_1 = 2\lambda_2 = 2\lambda_3 = \lambda_4$ (B) $\lambda_1 = 2\lambda_2 = 2\lambda_3 = \lambda_4$
(C) $\lambda_1 = \lambda_2 = 4\lambda_3 = 9\lambda_4$ (D) $\lambda_1 = 2\lambda_2 = 3\lambda_3 = 4\lambda_4$

23. The binding energy of the electron in a hydrogen atom is 13.6 eV, the energy required to remove the electron from the first excited state of Li^{++} is : **[2014 (Online)]**

(A) 122.4 eV (B) 30.6 eV

(C) 13.6 eV (D) 3.4 eV

24. As an electron makes a transition from an excited state to the ground state of a hydrogen-like atom/ion : **[2015]**

(A) its kinetic energy increases but potential energy and total energy decrease

(B) kinetic energy, potential energy and total energy decrease

(C) kinetic energy decreases, potential energy increases but total energy remains same

(D) kinetic energy and total energy decrease but potential energy increases

25. A neutron moving with a speed 'v' makes a head on collision with a stationary hydrogen atom in ground state. The minimum kinetic energy of the neutron for which inelastic collision will take place is : **[2016 (Online)]**

(A) 20.4 eV (B) 10.2 eV

(C) 12.1 eV (D) 16.8 eV

26. A hydrogen atom makes a transition from $n = 2$ to $n = 1$ and emits a photon. This photon strikes a doubly ionized lithium atom ($Z = 3$) in excited state and completely removes the orbiting electron. The least quantum number for the excited state of the ion for the process is : **[2016 (Online)]**

(A) 2 (B) 4

(C) 5 (D) 3

27. Some energy levels of a molecules are shown in the figure. The ratio of the wavelengths $r = \lambda_1/\lambda_2$, is given by: **[2017]**

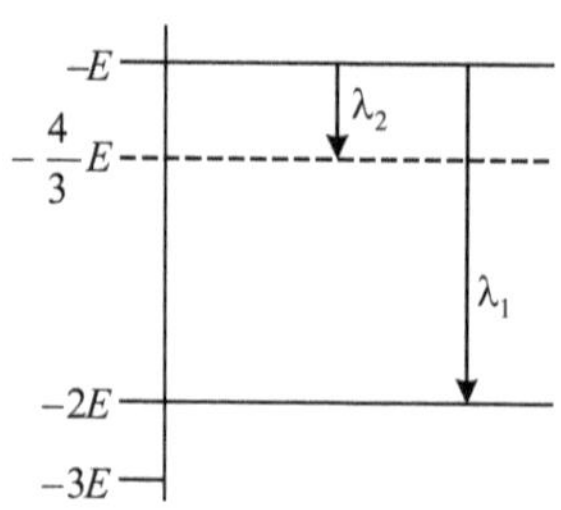

(A) $r = \frac{3}{4}$ (B) $r = \frac{1}{3}$

(C) $r = \frac{4}{3}$ (D) $r = \frac{2}{3}$

28. According to Bohr's theory, the time averaged magnetic field at the centre (i.e. nucleus) of a hydrogen atom due to the motion of electrons in the nth orbit is proportional to : (n = principal quantum number) **[2017 (Online)]**

(A) n^{-4} (B) n^{-5}

(C) n^{-3} (D) n^{-2}

29. The acceleration of an electron in the first orbit of the hydrogen atom ($z = 1$) is : **[2017 (Online)]**

(A) $\frac{h^2}{\pi^2 m^2 r^3}$ (B) $\frac{h^2}{8\pi^2 m^2 r^3}$

(C) $\frac{h^2}{4\pi^2 m^2 r^3}$ (D) $\frac{h^2}{4\pi m^2 r^3}$

30. If the series limit frequency of the Lyman series is v_1, then the series limit frequency of the P-fund series at : **[2018]**

(A) $25\, v_L$ (B) $16\, v_L$

(C) $v_L / 16$ (D) $v_L / 25$

31. An electron from various excited states of hydrogen atom emit radiation to come to the ground state. Let λ_n, λ_g be the de Broglie wavelength of the electron in the nth state and the ground state respectively. Let Λ_A be the wavelength of the emitted photon in the transition from the nth state to the ground state. For large n, (A, B are constants) **[2018]**

(A) $\Lambda_n \approx A + \frac{B}{\lambda_n^2}$ (B) $\Lambda_n \approx A + B\lambda_n$

(C) $\Lambda_n^2 \approx A + B\lambda_n^2$ (D) $\Lambda_n^2 \approx \lambda$

32. Muon (μ^{-1}) is negatively charged ($|q| = |e|$) with a mass $m_\mu = 200\, m_e$, where m_e is the mass of the electron and e is the electronic charge. If μ^{-1} is bound to a proton to form a hydrogen like atom, identify the correct statements : **[2018 (Online)]**

(a) Radius of the muonic orbit is 200 times smaller than that of the electron

(b) the speed of the μ^{-1} in the nth orbit is 1/200 times that of the electron in the nth orbit

(c) The ionization energy of muonic atom is 200 times more than that of an hydrogen atom

(d) The momentum of the muon in the nth orbit is 200 times more than that of the electron

(A) (a), (b), (d) (B) (b), (d)

(C) (c), (d) (D) (a), (c), (d)

33. The energy required to remove the electron from a singly ionized Helium atom is 2.2 times the energy required to remove an electron from Helium atom. The total energy required to ionize the Helium atom completely is : **[2018 (Online)]**

(A) 20 eV (B) 79 eV

(C) 109 eV (D) 34 eV

34. A hydrogen atom, initially in the ground state is excited by absorbing a photon of wavelength 980 Å. The radius of the atom in the excited state, in terms of Bohr radius a_0, will be : (h_c = 12500 eV – Å) **[11 Jan. 2019 (Morning)]**

(A) $9\, a_0$ (B) $25\, a_0$

(C) $4\, a_0$ (D) $16\, a_0$

35. In a hydrogen like atom, when an electron jumps from the M-shell to the L-shell, the wavelength of emitted radiation is λ. If an electron jumps from N-shell to the L-shell, the wavelength of emitted radiation will be : **[11 Jan. 2019 (Evening)]**

(A) $\frac{27}{20}\lambda$ (B) $\frac{16}{25}\lambda$

(C) $\frac{20}{27}\lambda$ (D) $\frac{25}{16}\lambda$

36. A particle of mass m moves in a circular orbit in a central potential field $U(r) = 1/2kr^2$. If Bohr's quantization conditions are applied, radii of possible orbitals and energy levels vary with quantum number n as : **[12 Jan. 2019 (Morning)]**

(A) $r_n \propto n^2, E_n \propto \frac{1}{n^2}$ (B) $r_n \propto \sqrt{n}, E_n \propto \frac{1}{n}$

(C) $r_n \propto n, E_n \propto n$ (D) $r_n \propto \sqrt{n}, E_n \propto n$

37. In a Frank-Hertz experiment, an electron of energy 5.6 eV passes through mercury vapour and emerges with an energy 0.7 eV. The minimum wavelength of photons emitted by mercury atoms is close to : **[12 Jan. 2019 (Evening)]**

(A) 2020 nm (B) 220 nm
(C) 250 nm (D) 1700 nm

38. Radiation coming from transitions $n = 2$ to $n = 1$ of hydrogen atoms fall on He^+ ions in $n = 1$ and $n = 2$ states. The possible transition of helium ions as they absorb energy from the radiation is : **[8 April 2019 (Morning)]**

(A) $n = 1 \rightarrow n = 4$ (B) $n = 2 \rightarrow n = 4$
(C) $n = 2 \rightarrow n = 5$ (D) $n = 2 \rightarrow n = 3$

39. Taking the wavelength of first Balmer line in hydrogen spectrum ($n = 3$ to $n = 2$) as 660 nm, the wavelength of the 2nd Balmer line ($n = 4$ to $n = 2$) will be : **[9 April 2019 (Morning)]**

(A) 889.2 nm (B) 642.7 nm
(C) 488.9 nm (D) 388.9 nm

40. A He^+ ion is in its first excited state. Its ionization energy is: **[9 April 2019 (Evening)]**

(A) 6.04 eV (B) 13.60 eV
(C) 54.40 eV (D) 48.36 eV

41. In Li^{++}, electron in first Bohr orbit is excited to a level by a radiation of wavelength λ. When the ion gets deexcited to the ground state in all possible ways (including intermediate emissions), a total of six spectral lines are observed. What is the value of λ? (Given : $h = 6.63 \times 10^{-34}$ Js ; $c = 3 \times 10^8$ ms^{-1}) **[10 April 2019 (Evening)]**

(A) 9.4 nm (B) 12.3 nm
(C) 10.8 nm (D) 11.4 nm

42. An excited He^+ ion emits two photons in succession, with wavelengths 108.5 nm and 30.4 nm, in making a transition to ground state. The quantum number n, corresponding to its initial excited state is (for photon of wavelength λ, energy $E = \frac{1240\text{eV}}{\lambda(\text{in } nm)}$): **[12 April 2019 (Morning)]**

(A) $n = 5$ (B) $n = 4$
(C) $n = 6$ (D) $n = 7$

43. The electron in a hydrogen atom first jumps from the third excited state to the second excited state and subsequently to the first excited state. The ratio of the respective wavelengths, λ_1/λ_2, of the photons emitted in this process is : **[12 April 2019 (Evening)]**

(A) 9/7 (B) 7/5
(C) 27/5 (D) 20/7

44. Consider an electron in a hydrogen atom, revolving in its second excited state (having radius 4.65Å). The de-Broglie wavelength of this electron is : **[12 April 2019 (Evening)]**

(A) 12.9 Å (B) 3.5 Å
(C) 9.7 Å (D) 6.6 Å

* * * * *

CHAPTER-27 NUCLEI

27.1 Radioactivity

1. At a specific instant emission of radioactive compound is deflected in a magnetic field. The compound can emit : **[2002]**

(i) electrons (ii) protons

(iii) He^{2+} (iv) neutrons

The emission at the instant can be :

(A) i, ii, iii (B) i, ii, iii, iv

(C) iv (D) ii, iii

2. If N_0 is the original mass of the substance of half-life period $t_{1/2} = 5$ years, then the amount of substance left after 15 years is : **[2002]**

(A) $\frac{N_0}{8}$ (B) $\frac{N_0}{16}$

(C) $\frac{N_0}{2}$ (D) $\frac{N_0}{4}$

3. A radioactive sample at any instant has its disintegration rate 5000 disintegration per minute. After 5 minutes, the rate is 1250 disintegrations per minute. Then, the decay constant (per minute) is : **[2003]**

(A) 0.4 ln 2 (B) 0.2 ln 2

(C) 0.1 ln 2 (D) 0.8 ln 2

4. Which of the following radiations has the least wavelength ? **[2003]**

(A) γ-rays (B) β-rays

(C) α-rays (D) X-rays

5. A nucleus disintegrates into two nuclear parts which have their velocities in the ratio 2 : 1. The ratio of their nuclear sizes will be : **[2004]**

(A) $2^{1/3} : 1$ (B) $1 : 3^{1/2}$

(C) $3^{1/2} : 1$ (D) $1 : 2^{1/3}$

6. If radius of the $^{27}_{13}Al$ nucleus is estimated to be 3.6 fermi, then the radius of $^{125}_{52}Te$ nucleus be nearly : **[2005]**

(A) 6 fermi (B) 8 fermi

(C) 4 fermi (D) 5 fermi

7. Starting with a sample of pure ^{66}Cu, 7/8 of it decays into Zn in 15 min. The corresponding half-life is : **[2005]**

(A) 10 min (B) 15 min

(C) 5 min (D) $7\frac{1}{2}$ min

8. The half-life period of a radioactive element X is same as the mean life time of another radioactive element Y. Initially they have the same number of atoms. Then : **[2007]**

(A) X will decay faster than Y

(B) Y will decay faster than X

(C) Y and X have same decay rate initially

(D) X and Y decay at same rate always

9. The half life of a radioactive substance is 20 minutes. The approximate time interval $(t_2 - t_1)$ between the time t_2 when 2/3 of it has decayed and time t_1 when 1/3 of it had decayed is :

(A) 20 min (B) 28 min **[2011]**

(C) 7 min (D) 14 min

10. Half-lives of two radioactive elements A and B are 20 minutes and 40 minutes, respectively. Initially, the samples have equal number of nuclei. After 80 minutes, the ratio of decayed numbers of A and B nuclei will be : **[2012]**

(A) 1 : 16 (B) 4 : 1

(C) 1 : 4 (D) 5 : 4

11. The counting rate observed from a radioactive source at $t = 0$ was 1600 counts s^{-1}, and $t = 8$ s, it was 100 counts s^{-1}. The counting rate observed as counts s^{-1} at $t = 6$ s will be :

[2012 (Online)]

(A) 250 (B) 400

(C) 300 (D) 200

12. A sample originally contained 10^{20} radioactive atoms, which emit α-particles. The ratio of α-particles emitted in the third year to that emitted during the second year is 0.3. How many α-particles were emitted in the first year ? **[2012 (Online)]**

(A) 3×10^{18} (B) 3×10^{19}

(C) 5×10^{18} (D) 7×10^{19}

13. Which of the following Statements is correct ?

[2012 (Online)]

(A) The rate of radioactive decay cannot be controlled but that of nuclear fission can be controlled

(B) Nuclear forces are short range, attractive and charge dependent

(C) Nuclei of atoms having same number of neutrons are known as isobars

(D) Wavelength of matter waves is given by de-Broglie formula but that of photons is not given by the same formula

14. The decay constants of a radioactive substance for α and β emission are λ_α and λ_β respectively. If the substance emits α and β simultaneously, then the average half life of the material will be : **[2012 (Online)]**

(A) $\frac{2T_\alpha T_\beta}{T_\alpha + T_\beta}$ (B) $T_\alpha + T_\beta$

(C) $\frac{T_\alpha T_\beta}{T_\alpha + T_\beta}$ (D) $\frac{1}{2}(T_\alpha + T_\beta)$

15. The half-life of a radioactive element A is the same as the mean-life of another radioactive element B. Initially both substances have the same number of atoms, then : **[2013 (Online)]**

(A) A and B decay at the same rate always
(B) A and B decay at the same rate initially
(C) A will decay at a faster rate than B
(D) B will decay at a faster rate than A

16. A radioactive nuclei with decay constant 0.5/s is being produced at a constant rate of 100 nuclei/s. If at $t = 0$ there were no nuclei, the time when there are 50 nuclei is : **[2014 (Online)]**

(A) 1 s (B) $2\ln\left(\frac{4}{3}\right)$ s
(C) ln 2 s (D) $\ln\left(\frac{4}{3}\right)$ s

17. A piece of bone of an animal from a ruin is found to have ^{14}C activity of 12 disintegrations per minute per *gm* of its carbon content. The ^{14}C activity of a living animal is 16 disintegrations per minute per *gm*. How long ago nearly did the animal die? (Given half life of ^{14}C is $t_{1/2} = 5760$ years) : **[2014 (Online)]**

(A) 1672 years (B) 2391 years
(C) 3291 years (D) 4453 years

18. A piece of wood from a recently cut tree shows 20 decays per minute. A wooden piece of same size placed in a museum (obtained from a tree cut many years back) shows 2 decays per minute. If half life of C^{14} is 5730 years, then age of the wooden piece placed in the museum is approximately : **[2014 (Online)]**

(A) 10439 years (B) 13094 years
(C) 19039 years (D) 39049 years

19. Let N_β be the number of β particles emitted by 1 gram of Na^{24} radioactive nuclei (half life = 15 hrs) in 7.5 hours, N_β is close to (Avogadro number = 6.023×10^{23}/g. mole) : **[2015 (Online)]**

(A) 6.2×10^{21} (B) 7.5×10^{21}
(C) 1.25×10^{22} (D) 1.75×10^{22}

20. A radioactive nucleus A with a half life T, decays into a nucleus B. At $t = 0$, there is no nucleus B. At sometime t, the ratio of the number of B to that of A is 0.3. Then, t is given by : **[2017]**

(A) $t = T\log(1.3)$ (B) $t = \frac{T}{\log(1.3)}$
(C) $t = T\frac{\log(2)}{\log(1.3)}$ (D) $t = T\frac{\log(1.3)}{\log(2)}$

21. A solution containing active cobalt $^{60}_{27}Co$ having activity of 0.8 μCi and decay constant λ is injected in an animal's body. If 1 cm^3 of blood is drawn from the animal's body after 10 hrs of injection, the activity found was 300 decays per minute. What is the volume of blood that is flowing in the body ? (1Ci = 3.7×10^{10} decay per second and at $t = 10$ hrs $e^{-\lambda t} = 0.84$) **[2018 (Online)]**

(A) 6 litres (B) 7 litres
(C) 4 litres (D) 5 litres

22. At some instant, a radioactive sample S_1 having an activity 5 μCi has twice the number of nuclei as another sample S_2 which has an activity of 10 μCi. The half lives of S_1 and S_2 are : **[2018 (Online)]**

(A) 10 years and 20 years, respectively
(B) 5 years and 20 years, respectively
(C) 20 years and 10 years, respectively
(D) 20 years and 5 years, respectively

23. A sample of radioactive material A, that has an activity of 10 mCi (1 Ci = 3.7×10^{10} decays/s), has twice the number of nuclei as another sample of a different radioactive material B which has an activity of 20 mCi. The correct choices for half-lives of A and B would then be respectively : **[9 Jan. 2019 (Morning)]**

(A) 20 days and 5 days (B) 20 days and 10 days
(C) 5 days and 10 days (D) 10 days and 40 days

24. At a given instant, say $t = 0$, two radioactive substances A and B have equal activities. The ratio R_B/R_A of their activities after time t itself decays with time t as e^{-3t}. If the half-life of A is ln2, the half-life of B is : **[9 Jan. 2019 (Evening)]**

(A) $\frac{\ln 2}{2}$ (B) 2 ln2
(C) $\frac{\ln 2}{4}$ (D) 4 ln2

25. Using a nuclear counter the count rate of emitted particles from a radioactive source is measured. At $t = 0$ it was 1600 counts per second and $t = 8$ seconds it was 100 counts per second. The count rate observed, as counts per second, at $t = 6$ seconds is close to : **[10 Jan. 2019 (Morning)]**

(A) 150 (B) 360
(C) 200 (D) 400

26. In a radioactive decay chain, the initial nucleus is $^{232}_{90}Th$. At the end there are 6 α-particles and 4 β-particles which are emitted. If the end nucleus, If $^{A}_{Z}X$, A and Z are given by : **[12 Jan. 2019 (Evening)]**

(A) $A = 208; Z = 80$ (B) $A = 202; Z = 80$
(C) $A = 200; Z = 81$ (D) $A = 208; Z = 82$

27. Two radioactive materials A and B have decay constants 10λ and λ, respectively. It initially they have the same number of nuclei, then the ratio of the number of nuclei of A to that of B will be $1/e$ after a time : **[10 April 2019 (Morning)]**

(A) $\frac{11}{10\lambda}$ (B) $\frac{1}{9\lambda}$

(C) $\frac{1}{10\lambda}$ (D) $\frac{1}{11\lambda}$

28. Two radioactive substances A and B have decay constants 5λ and λ respectively. At $t = 0$, a sample has the same number of the two nuclei. The time taken for the ratio of the number of nuclei to become $\left(\frac{1}{e}\right)^2$ will be : **[10 April 2019 (Evening)]**

(A) $1/4\lambda$ (B) $1/\lambda$

(C) $1/2\lambda$ (D) $2/\lambda$

29. Half lives of two radioactive nuclei A and B are 10 minutes and 20 minutes, respectively. If, initially a sample has equal number of nuclei, then after 60 minutes, the ratio of decayed numbers of nuclei A and B will be : **[12 April 2019 (Evening)]**

(A) 9 : 8 (B) 1 : 8

(C) 8 : 1 (D) 3 : 8

* * * * *

CHAPTER-27 NUCLEI

27.2 Nuclear Physics

1. In the nuclear fusion reaction, ${}_{1}^{2}H + {}_{1}^{3}H \rightarrow {}_{2}^{4}He + n$ given that the repulsive potential energy between the two nuclei is 7.7×10^{-14} J, the temperature at which the gases must be heated to initiate the reaction is nearly [Boltzmann's constant $k = 1.38 \times 10^{-23}$ J/K] : **[2003]**

(A) 10^7 K (B) 10^5 K
(C) 10^3 K (D) 10^9 K

2. Which of the following cannot be emitted by radioactive substances during their decay ? **[2003]**

(A) Protons (B) Neutrinos
(C) Helium nuclei (D) Electrons

3. A nucleus with $Z = 92$ emits the following in a sequence : $\alpha, \alpha, \beta^-, \beta^-, \alpha, \alpha, \alpha, \alpha, \beta^-, \beta^-, \alpha, \beta^+, \beta^+, \alpha$. The Z of the resulting nucleus is : **[2003]**

(A) 76 (B) 78
(C) 82 (D) 74

4. When U^{238} nucleus originally at rest, decays by emitting an alpha particle having a speed u, the recoil speed of the residual nucleus is : **[2003]**

(A) $\frac{4u}{238}$ (B) $-\frac{4u}{234}$

(C) $\frac{4u}{234}$ (D) $-\frac{4u}{238}$

5. An α-particle of energy 5 MeV is scattered through 180° by a fixed uranium nucleus. The distance of the closest approach is of the order of : **[2004]**

(A) 1 Å (B) 10^{-10} cm
(C) 10^{-12} cm (D) 10^{-15} cm

6. The binding energy per nucleon of deuteron $\left({}_{1}^{2}H\right)$ and helium nucleus $\left({}_{2}^{4}He\right)$ is 1.1 MeV and 7 MeV respectively. If two deuteron nuclei react to form a single helium nucleus, then the energy released is : **[2004]**

(A) 13.9 MeV (B) 26.9 MeV
(C) 23.6 MeV (D) 19.2 MeV

7. A nuclear transformation is denoted by $X(n, \alpha) \rightarrow {}_{3}^{7}Li$. Which of the following is the nucleus of element X ? **[2005]**

(A) ${}_{6}^{12}C$ (B) ${}_{5}^{10}B$

(C) ${}_{5}^{9}B$ (D) ${}_{4}^{11}Be$

8. The intensity of gamma radiation from a given source is I. On passing through 36 mm of lead, it is reduced to $I/8$. The thickness of lead, which will reduce the intensity to $I/2$ will be : **[2005]**

(A) 6mm (B) 9mm
(C) 18mm (D) 12mm

9. The energy spectrum of β-particles [number $N(E)$ as a function of β-energy E] emitted from a radioactive source is: **[2006]**

(A)
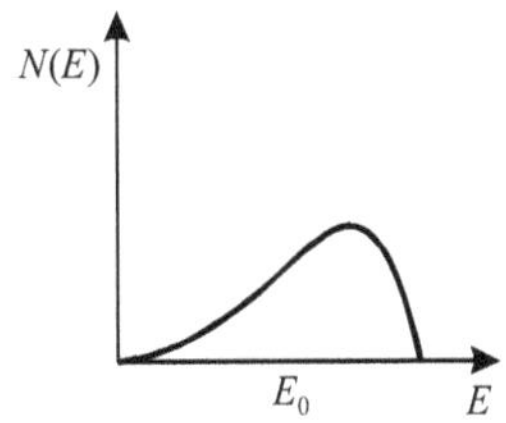

(B)
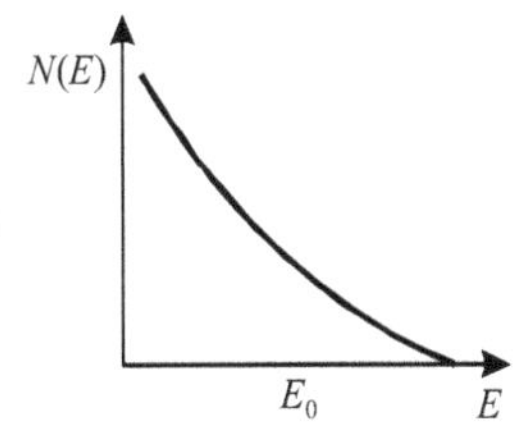

(C)
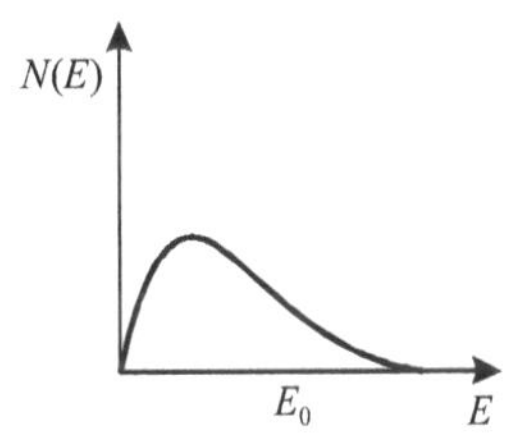

(D)
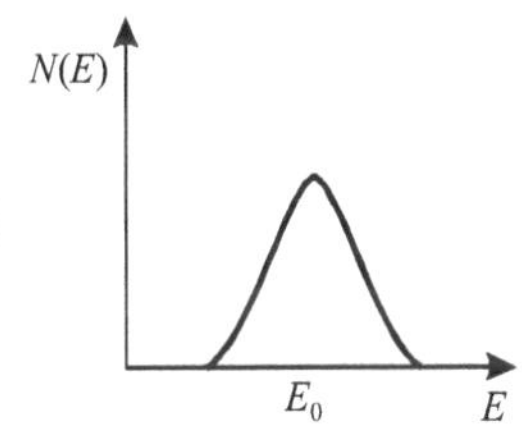

10. When $3Li^7$ nuclei are bombarded by protons and the resultant nuclei are ${}_{4}Be^8$, the emitted particles will be : **[2006]**

(A) alpha particles (B) beta particles
(C) gamma photons (D) neutrons

11. An alpha nucleus of energy $1/2\ mv^2$ bombards a heavy nuclear target of charge Ze. Then the distance of closest approach for the alpha nucleus will be proportional to : **[2006]**

(A) v^2 (B) $1/m$
(C) $1/v^4$ (D) $1/Ze$

12. If the binding energy per nucleon in ${}_{3}^{7}Li$ and ${}_{2}^{4}He$ nuclei are 5.60 MeV and 7.06 MeV respectively, then in the reaction: $p + {}_{3}^{7}Li \rightarrow 2\,{}_{2}^{4}He$ energy of proton must be : **[2006]**

(A) 28.24 MeV (B) 17.28 MeV
(C) 1.46 MeV (D) 39.2 MeV

13. In gamma ray emission from a nucleus : **[2007]**

(A) both the neutron number and the proton number change
(B) there is no change in the proton number and the neutron number
(C) only the neutron number changes
(D) only the proton number changes

14. If M_0 is the mass of an oxygen isotope $_8O^{17}$, M_p and M_n are the masses of a proton and a neutron, respectively, the nuclear binding energy of the isotope is : **[2007]**

(A) $(M_o - 8M_p)c^2$ (B) $(M_o - 8M_p - 9M_n)c^2$
(C) $M_o c^2$ (D) $(M_o - 17M_n)c^2$

15. Which of the following are the constituents of the nucleus? **[2007]**

(A) Electrons and protons (B) Neutrons and protons
(C) Electrons and neutrons (D) Neutrons and positrons

16. This question contains Statement-1 and Statement-2. Out of the four choices given after the statements, choose the one that best describes the two statements. **[2008]**

Statement-1 : Energy is released when heavy nuclei undergo fission or light nuclei undergo fusion.

Statement-2 : For heavy nuclei, binding energy per nucleon increases with increasing Z while for light nuclei it decreases with increasing Z.

(A) Statement-1 is false, Statement-2 is true
(B) Statement-1 is true, Statement-2 is true; Statement-2 is a correct explanation for Statement-1
(C) Statement-1 is true, Statement-2 is true; Statement-2 is not a correct explanation for Statement-1
(D) Statement-1 is true, Statement-2 is false

17. The above is a plot of binding energy per nucleon E_b, against the nuclear mass M; A, B, C, D, E, F correspond to different nuclei. Consider four reactions : **[2009]**

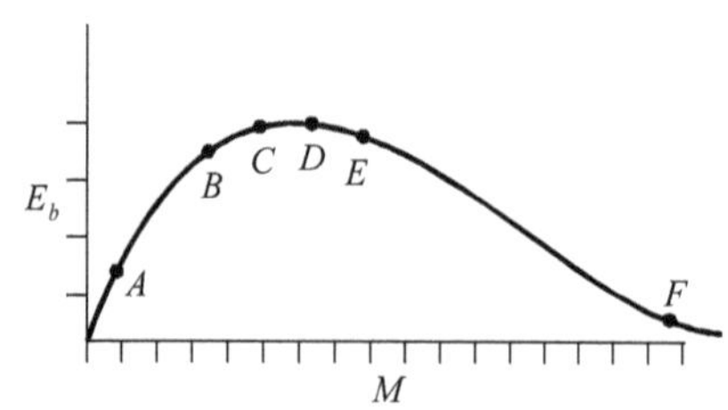

(i) $A + B \rightarrow C + \varepsilon$ (ii) $C \rightarrow A + B + \varepsilon$
(iii) $D + E \rightarrow F + \varepsilon$ (iv) $F \rightarrow D + E + \varepsilon$

where ε is the energy released ? In which reactions is ε positive ?

(A) (ii) and (iv) (B) (ii) and (iii)
(C) (i) and (iv) (D) (i) and (iii)

18. A radioactive nucleus (initial mass number A and atomic number Z) emits 3 α-particles and 2 positrons. The ratio of number of neutrons to that of protons in the final nucleus will be : **[2010]**

(A) $\frac{A-Z-4}{Z-2}$ (B) $\frac{A-Z-8}{Z-4}$

(C) $\frac{A-Z-4}{Z-8}$ (D) $\frac{A-Z-12}{Z-4}$

19. A nucleus of mass $M + \Delta m$ is at rest and decays into two daughter nuclei of equal mass $M/2$ each. Speed of light is c. The binding energy per nucleon for the parent nucleus is E_1 an that for the daughter nuclei is E_2. Then : **[2010]**

(A) $E_1 = 2E_2$ (B) $E_2 = 2E_1$
(C) $E_1 > E_2$ (D) $E_2 > E_1$

20. A nucleus of mass $M + \Delta m$ is at rest and decays into two daughter nuclei of equal mass $M/2$ each. Speed of light is c. The speed of daughter nuclei is : **[2010]**

(A) $c\sqrt{\frac{\Delta m}{M+\Delta m}}$ (B) $c\frac{\Delta m}{M+\Delta m}$

(C) $c\sqrt{\frac{2\Delta m}{M}}$ (D) $c\sqrt{\frac{\Delta m}{M}}$

21. Statement-1 : A nucleus having energy E_1 decays β^- emission to daughter nucleus having energy E_2, but the β^- rays are emitted with a continuous energy spectrum having end point energy $E_1 - E_2$. **[2011]**

Statement-2 : To conserve energy and momentum in β-decay at least three particles must take part in the transformation.

(A) Statement-1 is incorrect, statement-2 is correct
(B) Statement-1 is correct, statement-2 is incorrect
(C) Statement-1 is correct, statement-2 correct; statement-2 is the correct explanation of statement-1
(D) Statement-1 is correct, statement-2 is correct; statement-2 is not the correct explanation of statement-1

22. After absorbing a slowly moving neutron of mass m_N (momentum ~ 0) a nucleus of mass M breaks into two nuclei of masses m_1 and $5\ m_1$ ($6\ m_1 = M + m_N$), respectively. If the de Broglie wavelength of the nucleus with mass m_1 is λ, then de Broglie wavelength of the other nucleus will be : **[2011]**

(A) 25λ (B) 5λ

(C) $\frac{\lambda}{5}$ (D) λ

23. When Uranium is bombarded with neutrons, it undergoes fission. The fission reaction can be written as :

$$_{92}U^{235} + {_0n^1} \rightarrow {_{56}Ba^{141}} + {_{36}Kr^{92}} + 3x + Q \text{ (energy)}$$

where three particles named x are produced and energy Q is released. What is the name of the particle x ? **[2013 (Online)]**

(A) Electron (B) α-particle
(C) Neutron (D) Neutrino

24. Two deuterons undergo nuclear fusion to form a Helium nucleus. Energy released in this process is : (given binding energy per nucleon for deuteron = 1.1 MeV and for helium = 7.0 MeV) : **[2017 (Online)]**

(A) 30.2 MeV (B) 32.4 MeV
(C) 23.6 MeV (D) 25.8 MeV

25. Imagine that a reactor converts all given mass into energy and that it operates at a power level of 10^9 watt. The mass of the fuel consumed per hour in the reactor will be : (velocity of light, c is 3×10^8 m/s) : **[2017 (Online)]**

(A) 0.96 gm (B) 0.8 gm
(C) 4×10^{-2} gm (D) 6.6×10^{-5} gm

26. Both the nucleus and the atom of some element are in their respectively first excited states. They get de-excited by emitting photons of wavelengths λ_N, λ_A respectively. The ratio $\frac{\lambda_N}{\lambda_A}$ is closet to : **[2018 (Online)]**

(A) 10^{-6} (B) 10
(C) 10^{-1} (D) 10^{-10}

27. An unstable heavy nucleus at rest breaks into two nuclei which move away with velocities in the ratio of 8 : 27. The ratio of the radii of the nuclei (assumed to be spherical) is : **[2018 (Online)]**

(A) 8 : 27 (B) 2 : 3
(C) 3 : 2 (D) 4 : 9

28. Consider the nuclear fission

$$Ne^{20} \rightarrow 2He^4 + C^{12}$$

Given that the binding energy/nucleon of Ne^{20}, He^4 and C^{12} are, respectively, 8.03 MeV, 7.07 MeV and 7.86 MeV, identify the correct statement : **[10 Jan. 2019 (Evening)]**

(A) 8.3 MeV energy will be released
(B) Energy of 12.4 MeV will be supplied
(C) Energy of 11.9 MeV has to be supplied
(D) Energy of 3.6 MeV will be released

29. The ratio of mass densities of nuclei of ^{40}Ca and ^{16}O is close to : **[8 April 2019 (Evening)]**

(A) 1 (B) 2
(C) 0.1 (D) 5

* * * * *

CHAPTER-28 SEMICONDUCTORS ELECTRONICS

28.1 Semiconductor

1. The energy band gap is maximum in : [2002]
(A) Metals (B) Superconductors
(C) Insulators (D) Semiconductors

2. By increasing the temperature, the specific resistance of a conductor and a semiconductor : [2002]
(A) Increases for both (B) Decreases for both
(C) Increases, decreases (D) Decreases, increases

3. At absolute zero, *Si* acts as : [2002]
(A) Non-metal (B) Metal
(C) Insulator (D) None of these

4. A strip of copper and another of germanium are cooled from room temperature to 80 K. The resistance of: [2003]
(A) Each of these decreases
(B) Copper strip increases and that of germanium decreases
(C) Copper strip decreases and that of germanium increases
(D) Each of these increases

5. The difference in the variation of resistance with temperature in a metal and a semiconductor arises essentially due to the difference in the : [2003]
(A) Crystal structure
(B) Variation of the number of charge carriers with temperature
(C) Type of bonding
(D) Variation of scattering mechanism with temperature

6. When *p-n* junction diode is forward biased then : [2004]
(A) Both the depletion region and barrier height are reduced
(B) The depletion region is widened and barrier height is reduced
(C) The depletion region is reduced and barrier height is increased
(D) Both the depletion region and barrier height are increased

7. The electrical conductivity of a semiconductor increases when electromagnetic radiation of wavelength shorter than 2480 nm is incident on it. The band gap in (*eV*) for the semiconductor is : [2005]
(A) 2.5 eV (B) 1.1 eV
(C) 0.7 eV (D) 0.5 eV

8. If the ratio of the concentration of electrons to that of holes in a semiconductor is 7/5, and the ratio of currents is 7/4, then what is the ratio of their drift velocities ? [2006]
(A) $\frac{5}{8}$ (B) $\frac{4}{5}$
(C) $\frac{5}{4}$ (D) $\frac{4}{7}$

9. A solid which is not transparent to visible light and whose conductivity increases with temperature is formed by : [2006]
(A) Ionic bonding (B) Covalent bonding
(C) Vander Waals bonding (D) Metallic bonding

10. If the lattice constant of this semiconductor is decreased, then which of the following is correct ? [2006]

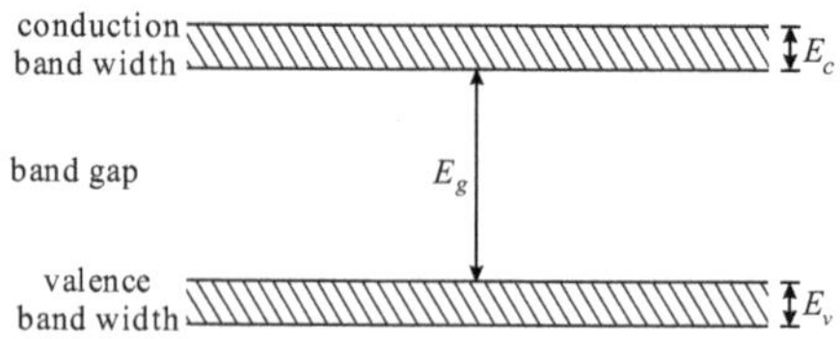

(A) All E_C, E_g, E_v increase
(B) E_c and E_v increase, but E_g decrease
(C) E_c and E_v decrease, but E_g increases
(D) All E_c, E_g, E_v decrease

11. Carbon, silicon and germanium have four valence electrons each. At room temperature which one of the following statements is most appropriate ? [2007]
(A) The number of free electrons for conduction is significant only in *Si* and *Ge* but small in *C*
(B) The number of free conduction electrons is significant in *C* but small in *Si* and *Ge*
(C) The number of free conduction electrons is negligibly small in all the three
(D) The number of free electrons for conduction is significant in all the three

12. This question has Statement-1 and Statement-2. Of the four choices given after the Statements, choose the one that best describes the two Statements. [2012 (Online)]
Statement-1 : A pure semiconductor has negative temperature coefficient of resistance.
Statement-2 : On raising the temperature, more charge carriers are released into the conduction band
(A) Statement-1 is false, Statement-2 is true
(B) Statement-1 is true, Statement-2 is false
(C) Statement-1 is true, Statement-2 is true, Statement-2 is not a correct explanation of Statement-1
(D) Statement-1 is true, Statement-2 is true, Statement-2 is the correct explanation of Statement-1

13. The temperature dependence of resistances of Cu and undoped Si in the temperature range 300 – 400K, is best described by : [2016]

(A) Linear increase for Cu, exponential decrease of Si
(B) Linear decrease for Cu, linear decrease for Si
(C) Linear increase for Cu, linear increase for Si
(D) Linear increase for Cu, exponential increase for Si

14. What is the conductivity of a semiconductor sample having electron concentration of $5 \times 10^{18}\ m^{-3}$, hole concentration of $5 \times 10^{19}\ m^{-3}$, electron mobility of $2.0\ m^2\ V^{-1}\ s^{-1}$ and hole mobility of $0.01\ m^2\ V^{-1}\ s^{-1}$? **[2017 (Online)]**
(Take charge of electron as 1.6×10^{-19} C) :

(A) $1.68\ (\Omega - m)^{-1}$ (B) $1.83\ (\Omega - m)^{-1}$
(C) $0.59\ (\Omega - m)^{-1}$ (D) $1.20\ (\Omega - m)^{-1}$

15. Mobility of electrons in a semiconductor is defined as the ratio of their drift velocity to the applied electric field. If, for an *n*-type semiconductor, the density of electrons is $10^{19}\ m^{-3}$ and their mobility is $1.6\ m^2/(V.s)$ then the resistivity of the semiconductor (since it is an *n*-type semiconductor contribution of holes is ignored) is close to : **[9 Jan. 2019 (Morning)]**

(A) $2\ \Omega m$ (B) $0.4\ \Omega m$
(C) $4\ \Omega m$ (D) $0.2\ \Omega m$

* * * * *

CHAPTER-28 SEMICONDUCTORS ELECTRONICS

28.2 PN Junction and Diodes

1. In the middle of the depletion layer of a reverse-biased *p-n* junction, the : **[2003]**

(A) Electric field is zero (B) Potential is maximum

(C) Electric field is maximum (D) Potential is zero

2. A strip of copper and another of germanium are cooled from room temperature to 80 K. The resistance of : **[2004]**

(A) Each of these decreases

(B) Copper strip increases and that of germanium decreases

(C) Copper strip decreases and that of germanium increases

(D) Each of these increases

3. The circuit has two oppositely connected ideal diodes in parallel. What is the current flowing in the circuit ? **[2006]**

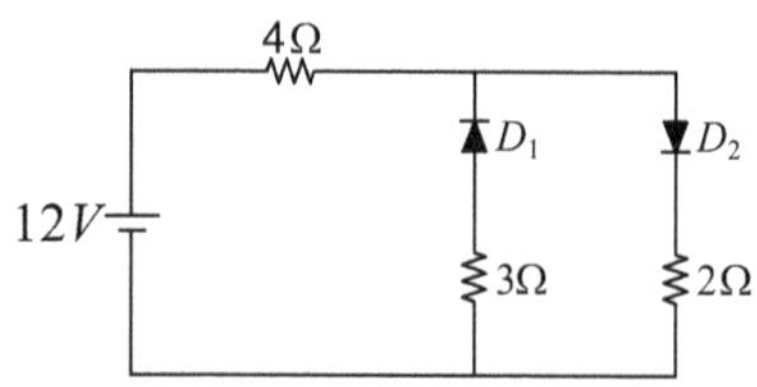

(A) 1.71 A (B) 2.00 A

(C) 2.31 A (D) 1.33 A

4. In the following, which one of the diodes reverse biased ? **[2006]**

(A) 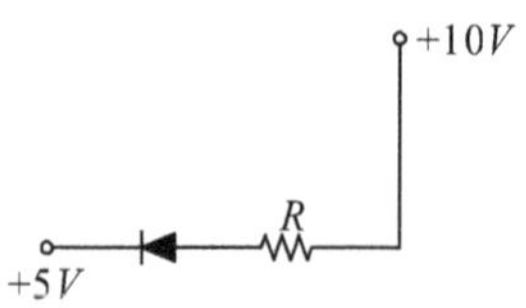

(B) 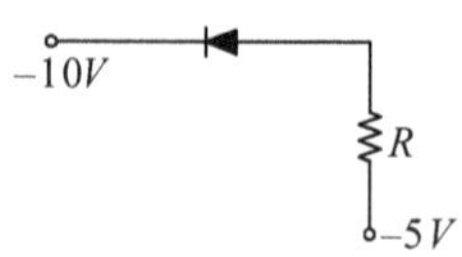

(C) 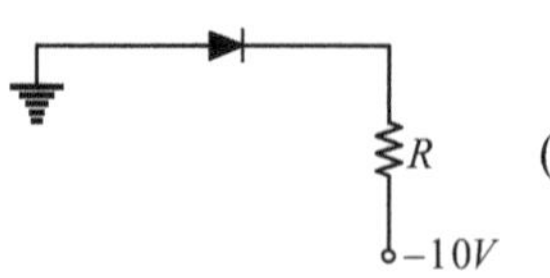

(D) 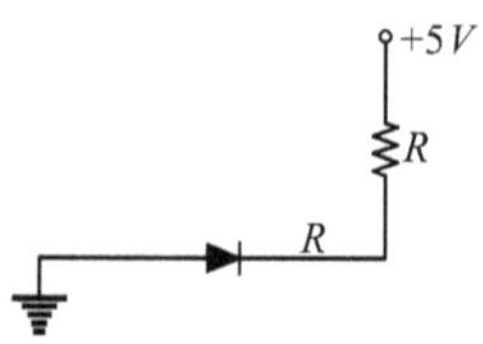

5. If in a *p-n* junction diode, a square input signal of 10 V is applied as shown **[2007]**

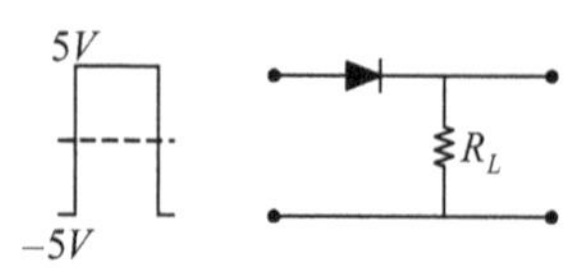

Then the output signal across R_L will be :

(A) 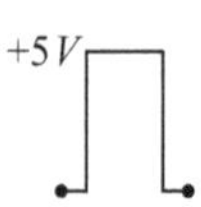

(B) 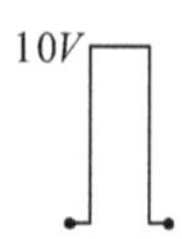

(C) 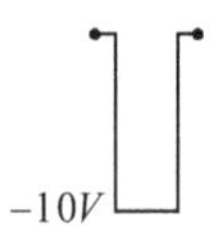

(D) 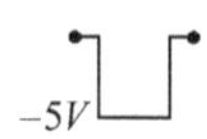

6. A p-n junction (D) shown in the figure can act as a rectifier. An alternating current source (V) is connected in the circuit. **[2009]**

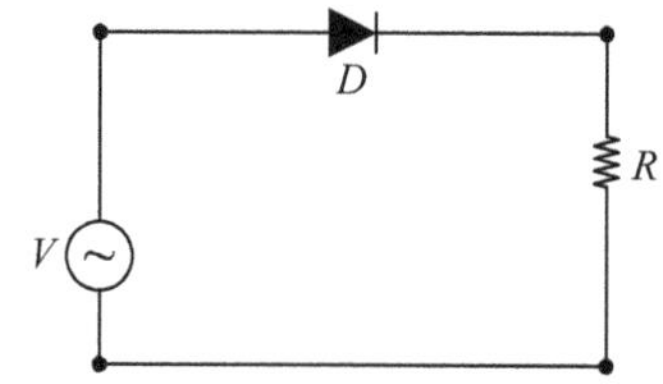

The current (*I*) in the resistor (*R*) can be shown by

(A) 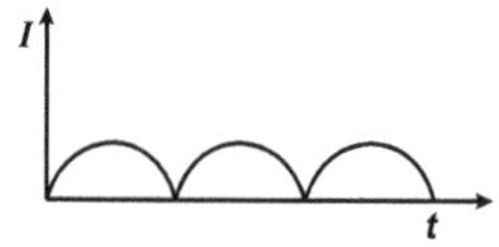

(B) 

(C) 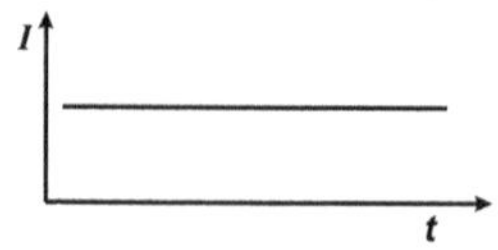

(D) 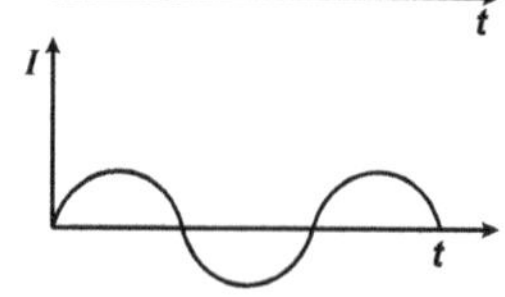

7. The I-V characteristic of an LED is : **[2013]**

(A) 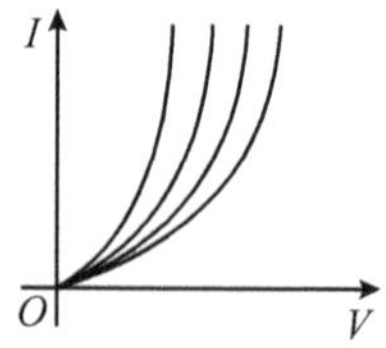

(B) 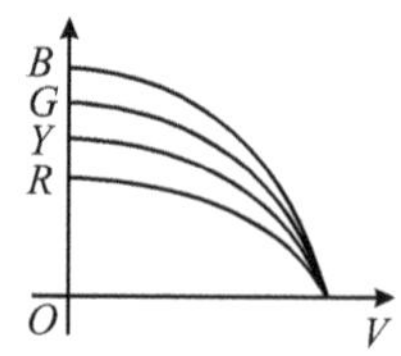

(C) 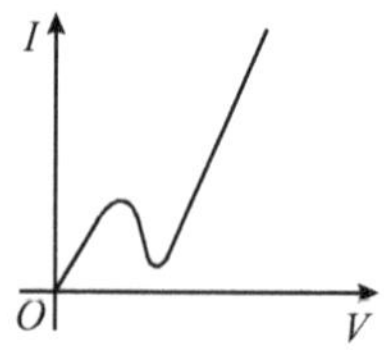

(D) 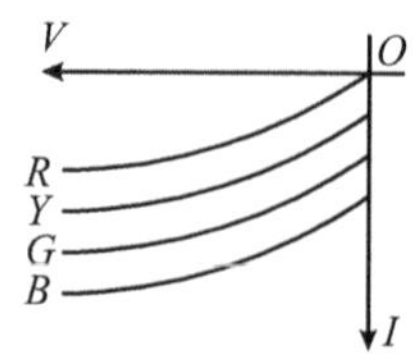

8. Figure shows a circuit in which three identical diodes are used. Each diode has forward resistance of 20 Ω and infinite backward resistance. Resistors $R_1 = R_2 = R_3 = 50\ \Omega$. Battery voltage is 6 V. The current through R_3 is : **[2013 (Online)]**

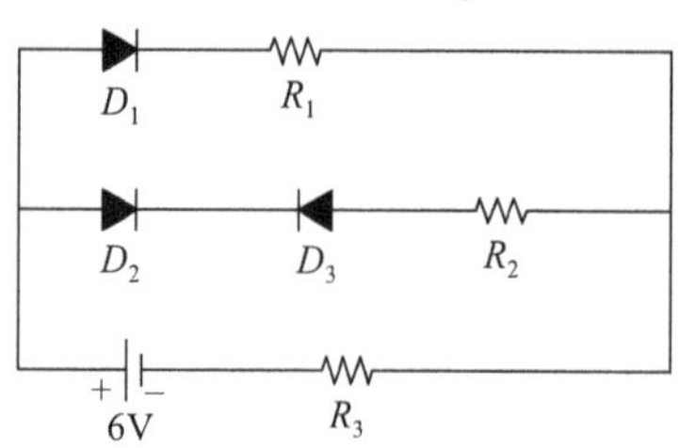

(A) 50 mA (B) 100 mA

(C) 60 mA (D) 25 mA

9. The forward biased diode connection is : **[2014]**

(A) 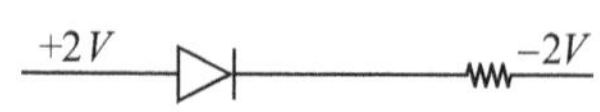

(B) 

(C) 

(D) 

10. A Zener diode is connected to a battery and a load as shown below : **[2014 (Online)]**

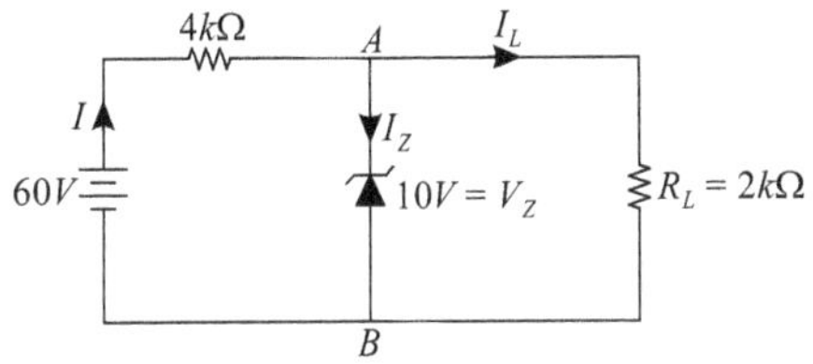

The currents, I, I_Z and I_L are respectively :

(A) 15 mA, 5 mA, 10 mA (B) 15 mA, 7.5 mA, 7.5 mA
(C) 12.5 mA, 5 mA, 7.5 mA (D) 12.5 mA, 7.5 mA, 5 mA

11. For LED's to emit light in visible region of electromagnetic light, it should have energy band gap in the range of : **[2014 (Online)]**

(A) 0.1 eV to 0.4 eV (B) 0.5 eV to 0.8 eV
(C) 0.9 eV to 1.6 eV (D) 1.7 eV to 3.0 eV

12. A red LED emits light at 0.1 watt uniformly around it. The amplitude of the electric field of the light at a distance of 1 m from the diode is : **[2015]**

(A) 5.48 V/m (B) 7.75 V/m
(C) 1.73 V/m (D) 2.45 V/m

13. In an unbiased *n-p* junction electrons diffuse from *n*-region to *p*-region because : **[2015 (Online)]**

(A) Hole in *p*-region attract them
(B) Electrons travel across the junction due to potential difference
(C) Only electrons move from *n* to *p* region and not the vice-versa
(D) Electron concentration in *n*-region is more compared to that in *p*-region

14. A 2 V battery is connected across *AB* as shown in the figure. The value of the current supplied by the battery when in one case battery's positive terminal is connected to *A* and in other case when positive terminal of battery is connected to *B* will respectively be : **[2015 (Online)]**

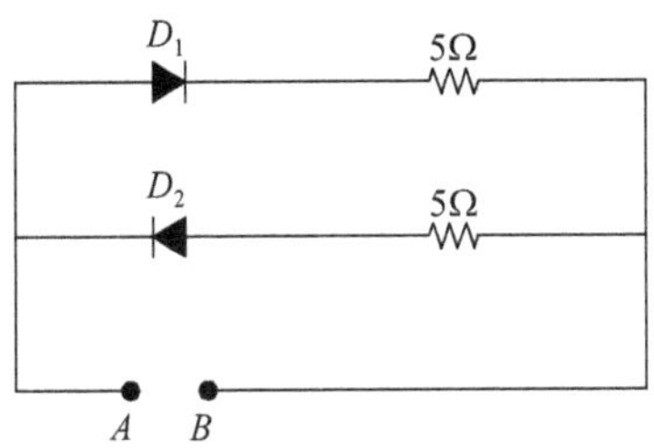

(A) 0.4 A and 0.2 A (B) 0.2 A and 0.4 A
(C) 0.1 A and 0.2 A (D) 0.2 A and 0.1 A

15. Identify the semiconductor devices whose characteristics are given below, in the order (i), (ii), (iii), (iv) : **[2016]**

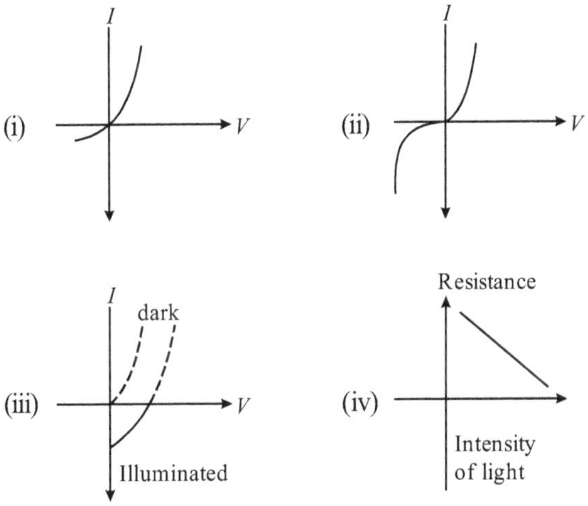

(A) Solar cell, Light dependent resistance, Zener diode, simple diode
(B) Zener diode, Solar cell, simple diode, Light dependent resistance
(C) Simple diode, Zener diode, Solar cell, Light dependent resistance
(D) Zener diode, Simple diode, Light dependent resistance, Solar cell

16. An experiment is performed to determine the I-V characteristic of a Zener diode, which has a protective resistance of $R = 100\ \Omega$, and a maximum power of dissipation rating of 1 W. The minimum voltage range of the *DC* source in the circuit is : **[2016 (Online)]**

(A) 0 – 5 V (B) 0 – 24 V
(C) 0 – 12 V (D) 0 – 8 V

17. The V-I characteristic of a diode is shown in the figure. The ratio of forward to reverse bias resistance is : **[2017 (Online)]**

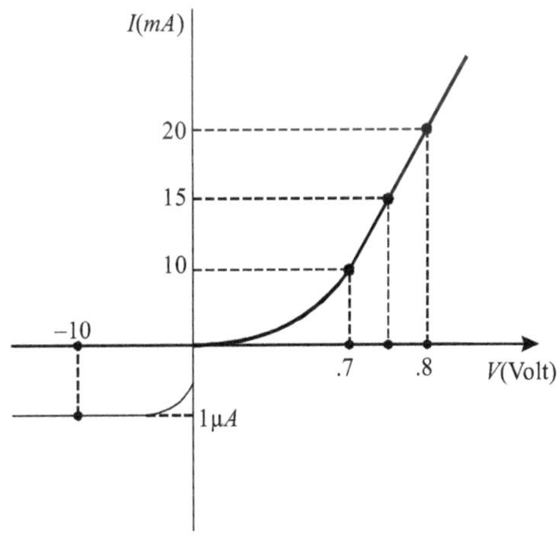

(A) 10 (B) 10^{-6}
(C) 10^{6} (D) 100

18. The reading of the ammeter for a silicon diode in the given circuit is : **[2018]**

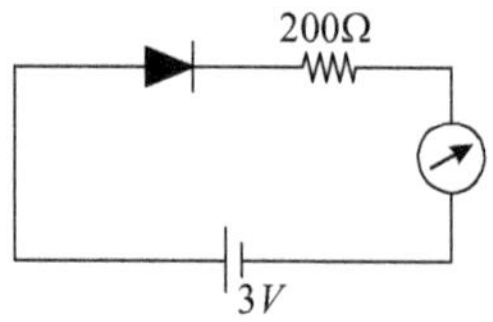

(A) 0 (B) 15 mA
(C) 11.5 mA (D) 13.5 mA

19. In the given circuit, the current through zener diode is : **[2018 (Online)]**

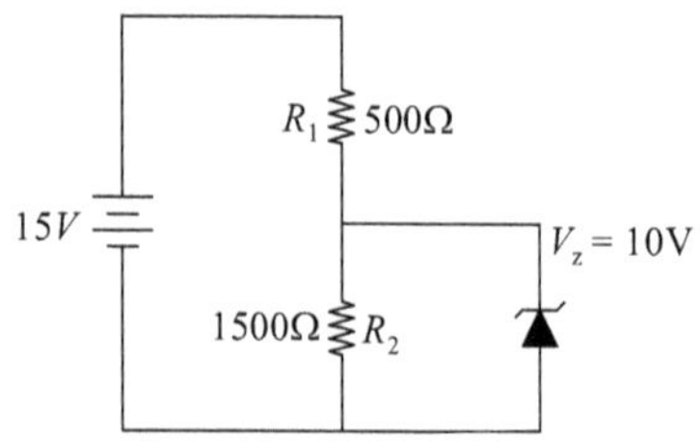

(A) 2.5 mA (B) 3.3 mA
(C) 5.5 mA (D) 6.7 mA

20. *Ge* and *Si* diodes start conducting at 0.3 V and 0.7 V respectively. In the following figure if *Ge* diode connection are reversed, the value of V_0 changes by : (assume that the *Ge* diode has large breakdown voltage) : **[9 Jan. 2019 (Evening)]**

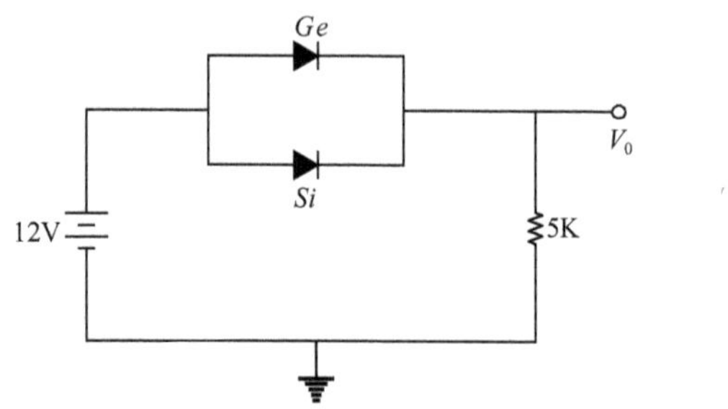

(A) 0.6 V (B) 0.8 V
(C) 0.4 V (D) 0.2 V

21. For the circuit shown below, the current through the Zener diode is : **[10 Jan. 2019 (Evening)]**

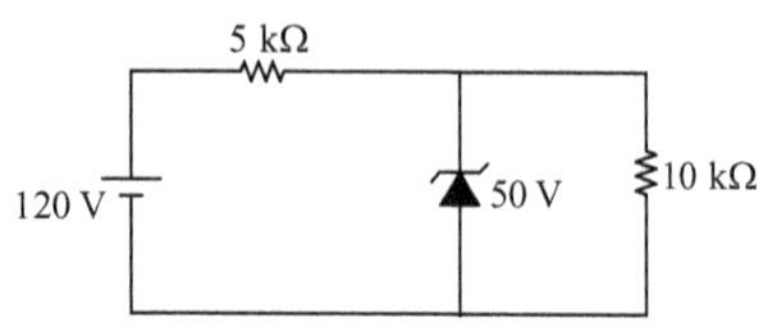

(A) 5 mA (B) Zero
(C) 14 mA (D) 9 mA

22. In the given circuit the current through Zener Diode is close to : **[11 Jan. 2019 (Morning)]**

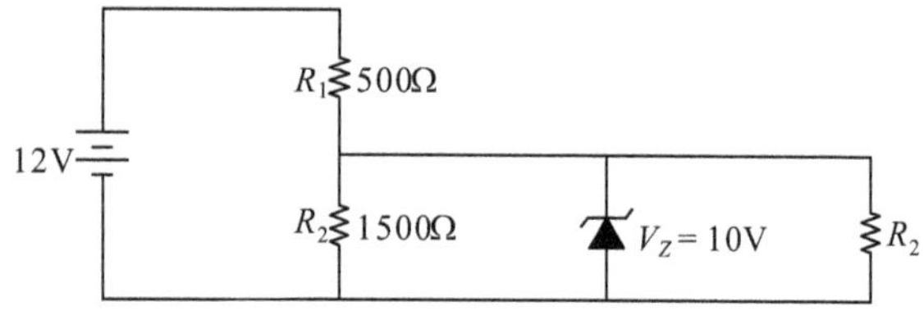

(A) 6.0 mA (B) 4.0 mA
(C) 6.7 mA (D) 0.0 mA

23. The circuit shown below contains two ideal diodes, each with a forward resistance of 50 Ω. If the battery voltage is 6 V, the current through the 100 Ω resistance (in Amperes) is : **[11 Jan. 2019 (Evening)]**

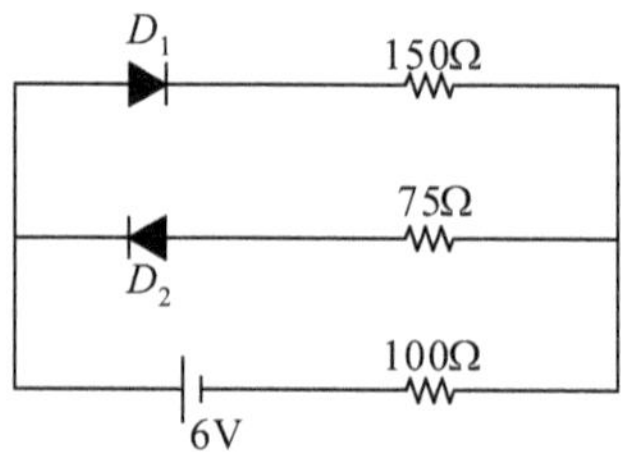

(A) 0.027 (B) 0.020
(C) 0.030 (D) 0.036

24. The reverse breakdown voltage of a Zener diode is 5.6 V in the given circuit. The current I_Z through the Zener is : **[8 April 2019 (Morning)]**

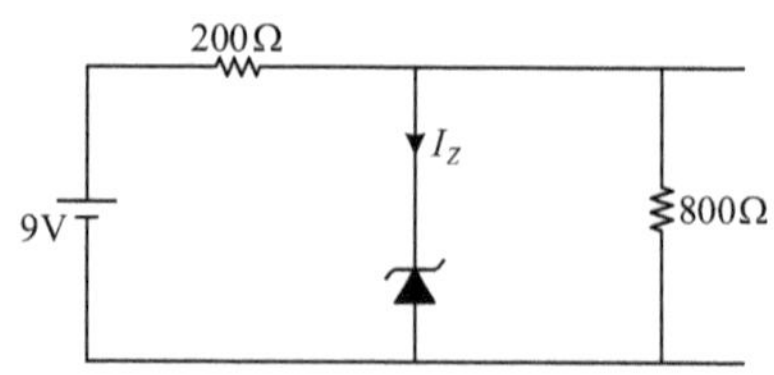

(A) 7 mA (B) 17 mA
(C) 10 mA (D) 15 mA

25. The figure represents a voltage regulator circuit using a Zener diode. The breakdown voltage of the Zener diode is 6 V and the load resistance is $R_L = 4\ k\Omega$. The series resistance of the circuit is $R_i = 1\ k\Omega$. If the battery voltage V_B varies from 8 V to 16 V, what are the minimum and maximum values of the current through Zener diode ? **[10 April 2019 (Evening)]**

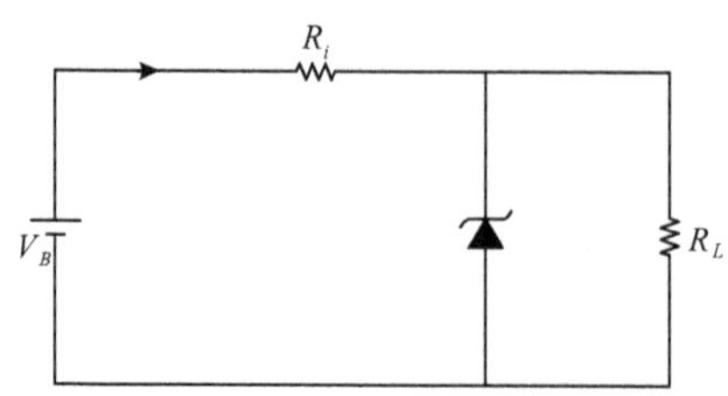

(A) 0.5 mA ; 6 mA (B) 0.5 mA ; 8.5 mA
(C) 1.5 mA ; 8.5 mA (D) 1 mA ; 8.5 mA

26. Figure shown a *DC* voltage regulator circuit, with a Zener diode of breakdown voltage = 6V. If the unregulated input voltage varies between 10V to 16V, then what is the maximum Zener current ? **[12 April 2019 (Evening)]**

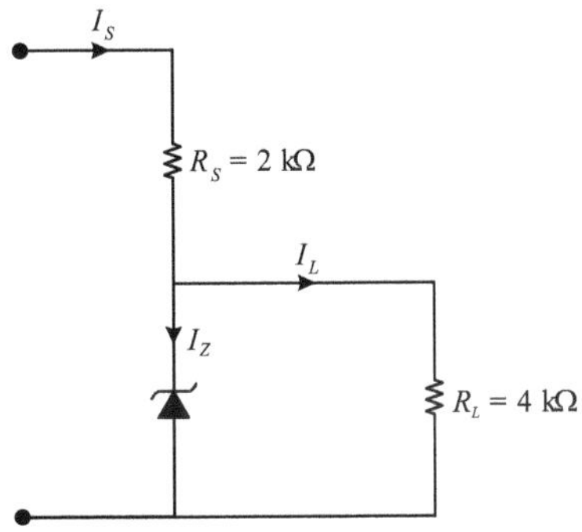

(A) 2.5 mA
(B) 3.5 mA
(C) 7.5 mA
(D) 1.5 mA

* * * * *

CHAPTER-28 SEMICONDUCTORS ELECTRONICS

28.3 Transistors

1. The part of a transistor which is most heavily doped to produce large number of majority carriers is : **[2002]**

(A) Emmiter

(B) Base

(C) Collector

(D) Can be any of the above three

2. When *npn* transistor is used as an amplifier : **[2004]**

(A) Electrons move from collector to base

(B) Holes move from emitter to base

(C) Electrons move from base to collector

(D) Holes move from base to emitter

3. In a common base amplifier, the phase difference between the input signal voltage and output voltage is : **[2005]**

(A) π (B) $\frac{\pi}{4}$

(C) $\frac{\pi}{2}$ (D) 0

4. In a common base mode of a transistor, the collector current is 5.488 mA for an emitter current of 5.60 mA. The value of the base current amplification factor (β) will be : **[2006]**

(A) 49 (B) 50

(C) 51 (D) 48

5. A working transistor with its three legs marked *P*, *Q* and *R* is tested using a multimeter. No conduction is found between *P* and *Q*. By connecting the common (negative) terminal of the multimeter to *R* and the other (positive) terminal to *P* or *Q*, some resistance is seen on the multimeter. Which of the following is true for the transistor ? **[2008]**

(A) It is an *npn* transistor with *R* as base

(B) It is a *pnp* transistor with *R* as base

(C) It is a *pnp* transistor with *R* as emitter

(D) It is an *npn* transistor with *R* as emitter

6. An *n-p-n* transistor has three leads *A*, *B* and *C*. Connecting *B* and *C* by moist fingers, *A* to the positive lead of an ammeter, and *C* to the negative lead of the ammeter, one finds large deflection. Then, *A*, *B* and *C* refer respectively to : **[2014 (Online)]**

(A) Emitter, base and collector

(B) Base, emitter and collector

(C) Base, collector and emitter

(D) Collector, emitter and base

7. For a common emitter configuration, if α and β have their usual meanings, the incorrect relationship between α and β is : **[2016]**

(A) $\alpha = \frac{\beta}{1+\beta}$ (B) $\alpha = \frac{\beta^2}{1+\beta^2}$

(C) $\frac{1}{\alpha} = \frac{1}{\beta} + 1$ (D) None of these

8. An unknown transistor needs to be identified as a *npn* or *pnp* type. A multimeter, with +*ve* and –*ve* terminals, is used to measure resistance between different terminals of transistor. If terminal 2 is the base of the transistor then which of the following is correct for a *pnp* transistor? **[2016 (Online)]**

(A) +*ve* terminals 2, –*ve* terminal 3, resistance low

(B) +*ve* terminals 2, –*ve* terminal 1, resistance high

(C) +*ve* terminals 1, –*ve* terminal 2, resistance high

(D) +*ve* terminals 3, –*ve* terminal 2, resistance high

9. A realistic graph depicting the variation of the reciprocal of input resistance in an input characteristics measurement in a common emitter transistor configuration is : **[2016 (Online)]**

(A) 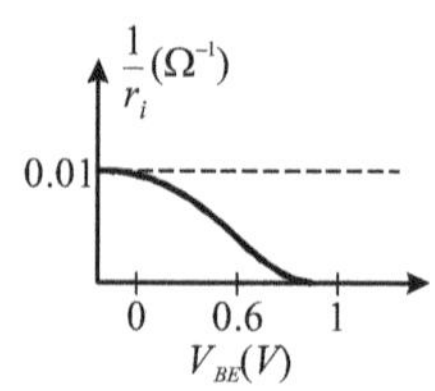

(B) 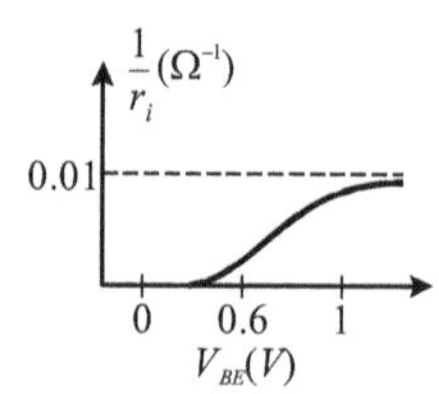

(C) 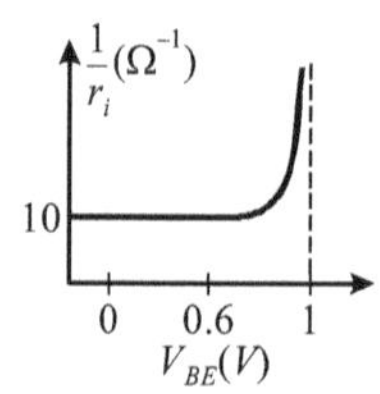

(D) 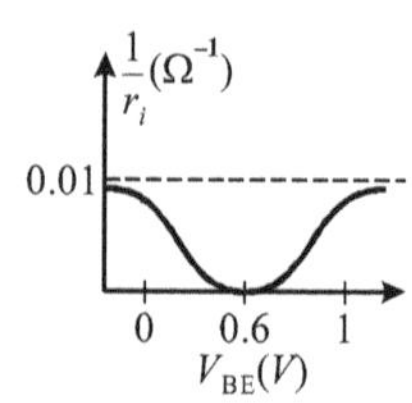

10. The current gain of a common emitter amplifier is 69. If the emitter current is 7.0 mA, collector current is : **[2017 (Online)]**

(A) 9.6 mA (B) 6.9 mA

(C) 0.69 mA (D) 69 mA

11. In a common emitter amplifier circuit using an *n-p-n* transistor, the phase difference between the input and the output voltages will be : **[2017 (Online)]**

(A) 135° (B) 180°

(C) 45° (D) 90°

12. In a common emitter configuration with suitable bias, it is given than R_L is the load resistance and R_{BE} is small signal dynamic resistance (input side). Then, voltage gain, current gain and power gain are given, respectively, by :

(β is current gain, I_B, I_C, I_E are respectively base, collector and emitter currents :) **[2018 (Online)]**

(A) $\beta\frac{R_L}{R_{BE}}, \frac{\Delta I_E}{\Delta I_B}, \beta^2\frac{R_L}{R_{BE}}$ (B) $\beta^2\frac{R_L}{R_{BE}}, \frac{\Delta I_C}{\Delta I_B}, \beta\frac{R_L}{R_{BE}}$

(C) $\beta^2\frac{R_L}{R_{BE}}, \frac{\Delta I_C}{\Delta I_E}, \beta^2\frac{R_L}{R_{BE}}$ (D) $\beta\frac{R_L}{R_{BE}}, \frac{\Delta I_C}{\Delta I_B}, \beta^2\frac{R_L}{R_{BE}}$

13. In the figure, given that V_{BB} supply can vary from 0 to 5.0 V, V_{CC} = 5 V, β_{dc} = 200, R_B = 100 kΩ, R_C = 1 kΩ and V_{BE} = 1.0 V, The minimum base current and the input voltage at which the transistor will go to saturation, will be, respectively :

[12 Jan. 2019 (Evening)]

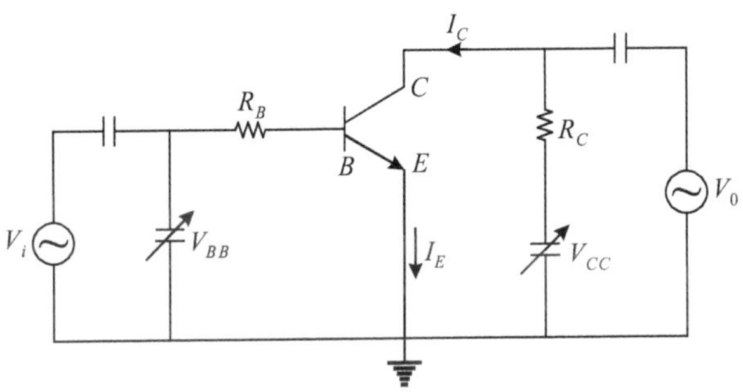

(A) 20 μA and 3.5 V (B) 25 μA and 3.5 V
(C) 25 μA and 2.5 V (D) 20 μA and 2.8 V

14. A common emitter amplifier circuit, built using an npn transistor, is shown in the figure. Its *dc* current gain is 250, R_C = 1 kΩ and V_{CC} = 10 V. What is the minimum base current for V_{CE} to reach saturation ? **[8 April 2019 (Evening)]**

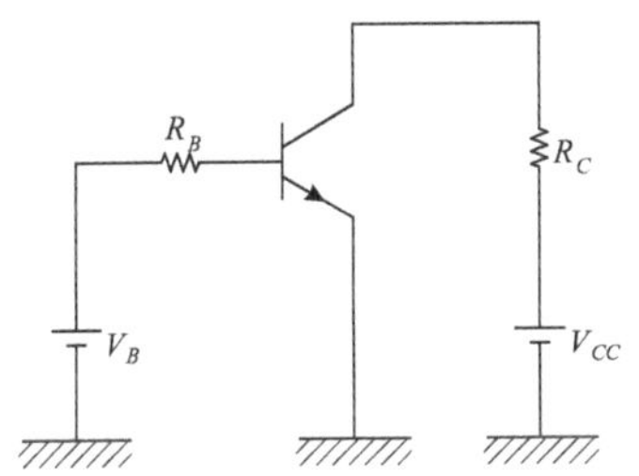

(A) 100 μA (B) 7 μA
(C) 40 μA (D) 10 μA

15. An *NPN* transistor is used in common emitter configuration as an amplifier with 1 kΩ load resistance. Signal voltage of 10 mV is applied across the base-emitter. This produces a 3 mA change in the collector current and 15*μA* change in the base current of the amplifier. The input resistance and voltage gain are : **[9 April 2019 (Morning)]**

(A) 0.33 kΩ, 1.5 (B) 0.67 kΩ, 200
(C) 0.33 kΩ, 300 (D) 0.67 kΩ, 300

16. An *npn* transistor operates as a common emitter amplifier, with a power gain of 60 dB. The input circuit resistance is 100 Ω and the output load resistance is 10 kΩ. The common emitter current gain β is : **[10 April 2019 (Morning)]**

(A) 60 (B) 10^4
(C) 6×10^2 (D) 10^2

17. The transfer characteristic curve of a transistor, having input and output resistance 100 Ω and 100 kΩ respectively, is shown in the figure. The Voltage and Power gain, are respectively : **[12 April 2019 (Morning)]**

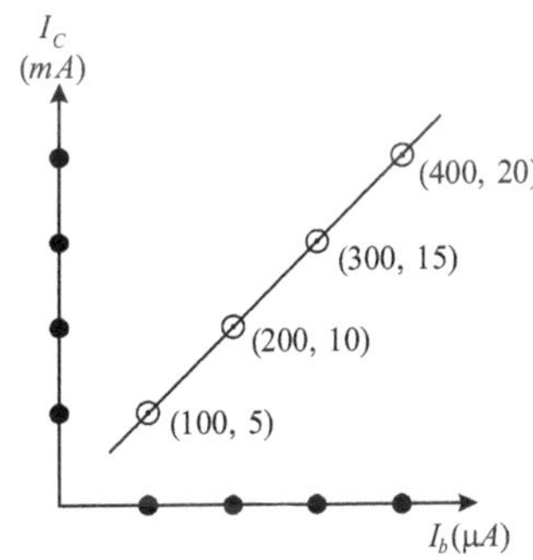

(A) $5 \times 10^4, 5 \times 10^5$ (B) $5 \times 10^4, 5 \times 10^6$
(C) $5 \times 10^4, 2.5 \times 10^6$ (D) $2.5 \times 10^4, 2.5 \times 10^6$

* * * * *

CHAPTER-28 SEMICONDUCTORS ELECTRONICS

28.4 Logic Gates

1. In the circuit below, A and B represent two inputs and C represents the output. **[2008]**

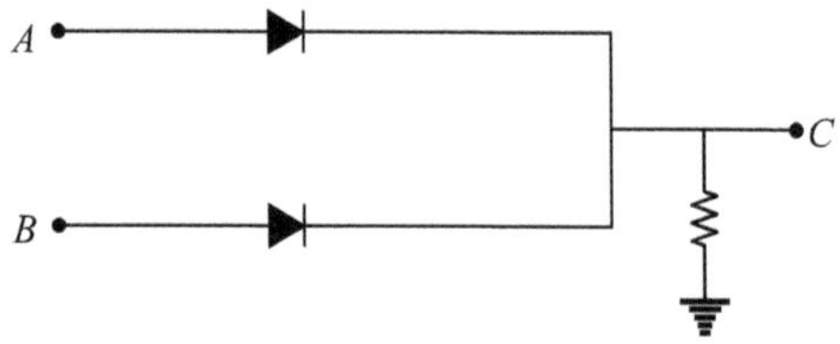

The circuit represents :

(A) NOR gate (B) AND gate

(C) NAND gate (D) OR gate

2. The logic circuit shown below has the input waveforms 'A' and 'B' as shown. Pick out the correct output waveform : **[2009]**

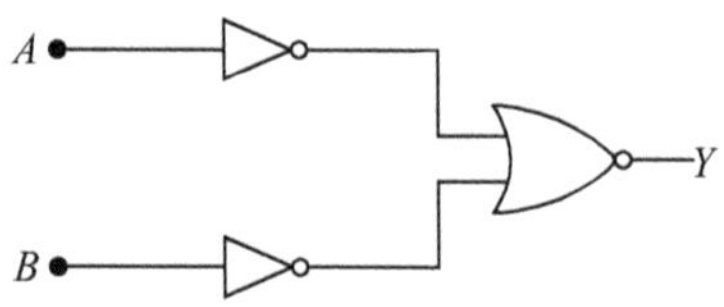

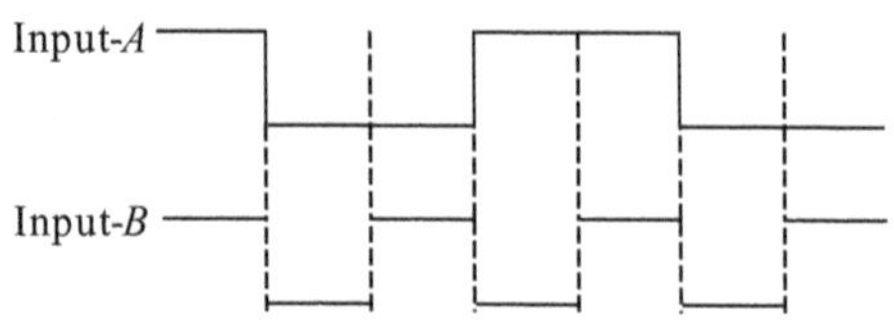

Output is :

(A)

(B)

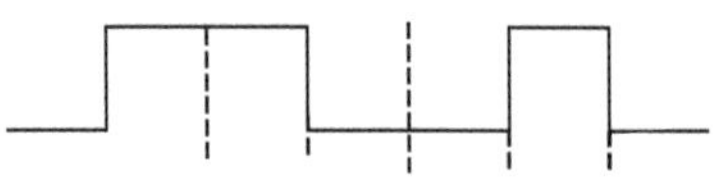

(C)

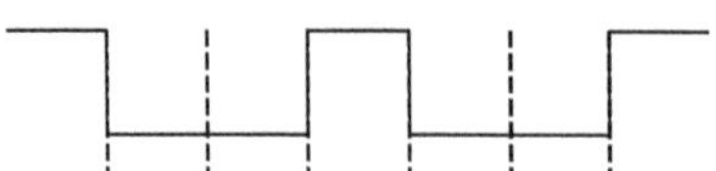

(D)

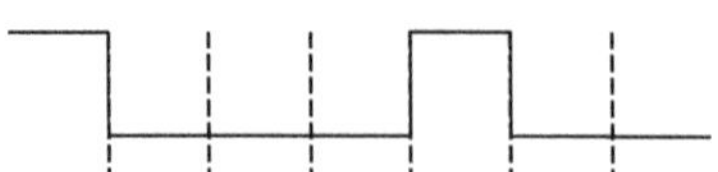

3. The combination of gates shown below yields : **[2010]**

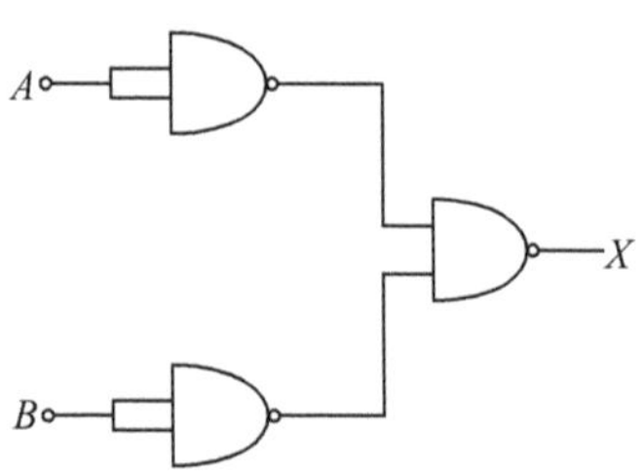

(A) OR gate (B) NOT gate

(C) XOR gate (D) NAND gate

4. The output of an OR gate is connected to both the inputs of a NAND gate. The combination will serve as : **[2011]**

(A) NOT gate (B) NOR gate

(C) AND gate (D) OR gate

5. Truth table for system of four NAND gates as shown in figure is : **[2012]**

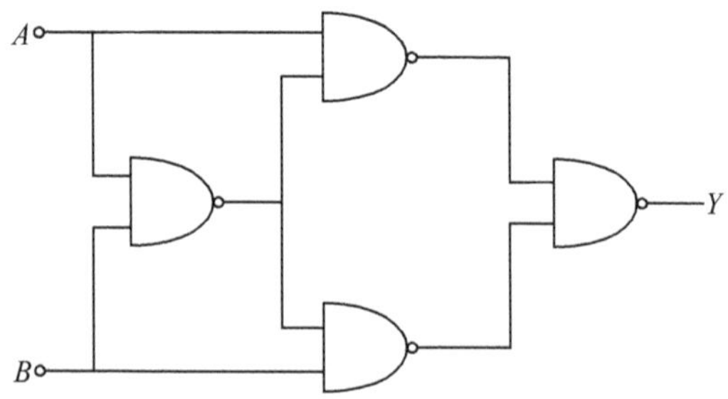

(A)

A	B	Y
0	0	0
0	1	1
1	0	1
1	1	0

(B)

A	B	Y
0	0	0
0	1	0
1	0	1
1	1	1

(C)

A	B	Y
0	0	1
0	1	1
1	0	0
1	1	0

(D)

A	B	Y
0	0	1
0	1	0
1	0	1
1	1	1

6. Which logic gate with inputs A and B performs the same operation as that performed by the following circuit ? **[2012 (Online)]**

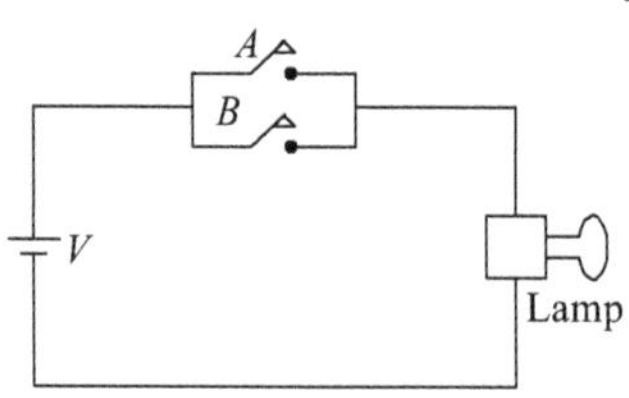

(A) NAND gate (B) OR gate

(C) NOR gate (D) AND gate

7. Which one of the following is the Boolean expression for NOR gate ? **[2012 (Online)]**

(A) $Y = \overline{A+B}$ (B) $Y = \overline{A \cdot B}$

(C) $Y = A \cdot B$ (D) $Y = \overline{A}$

8. The figure shows a combination of two NOT gates and a NOR gate.
The combination is equivalent to a : **[2012 (Online)]**

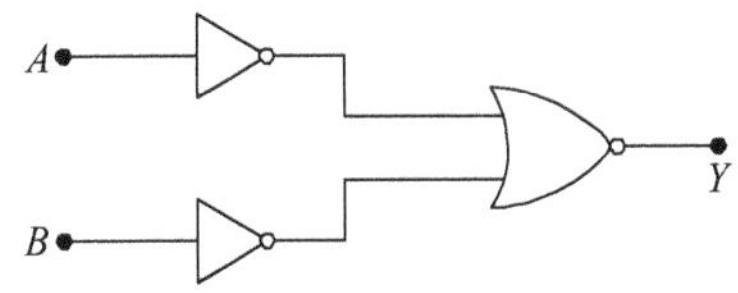

(A) NAND gate (B) NOR gate
(C) AND gate (D) OR gate

9. A system of four gates is set up as shown. The 'truth table' corresponding to this system is : **[2013 (Online)]**

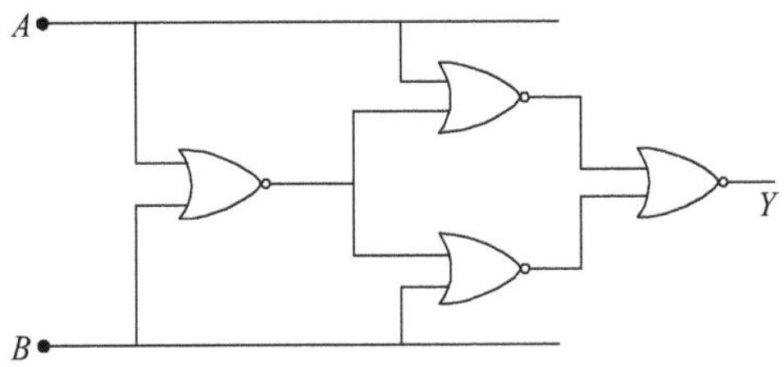

(A)

A	B	Y
0	0	1
0	1	0
1	0	0
1	1	1

(B)

A	B	Y
0	0	0
0	1	0
1	0	1
1	1	0

(C)

A	B	Y
0	0	1
0	1	0
1	0	1
1	1	0

(D)

A	B	Y
0	0	1
0	1	1
1	0	0
1	1	0

10. Which of the following circuits correctly represents the following truth table ? **[2013 (Online)]**

A	B	C
0	0	0
0	1	0
1	0	1
1	1	0

(A) 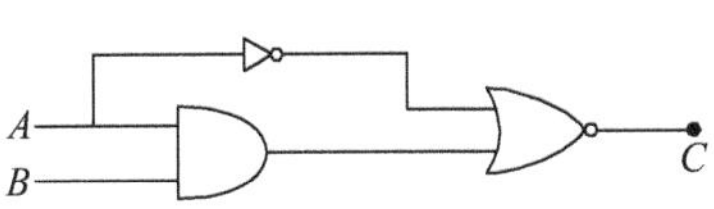

(B) 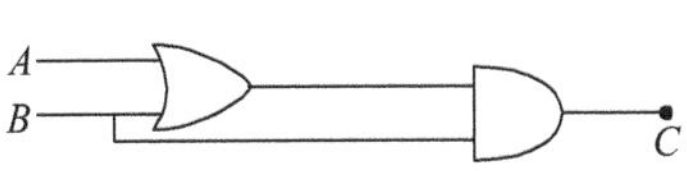

(C) 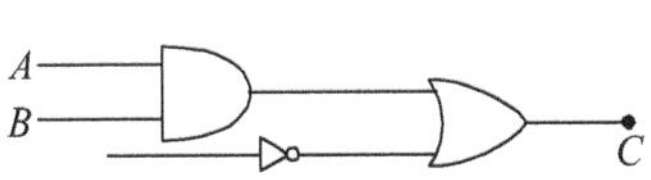

(D) 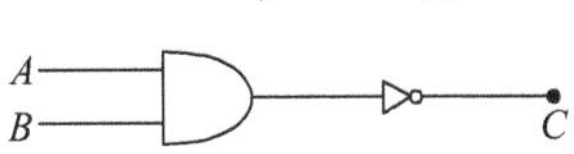

11. Consider two *npn* transistors as shown in figure. If 0 Volts corresponds to false and 5 Volts correspond to true then the output at C corresponds to : **[2013 (Online)]**

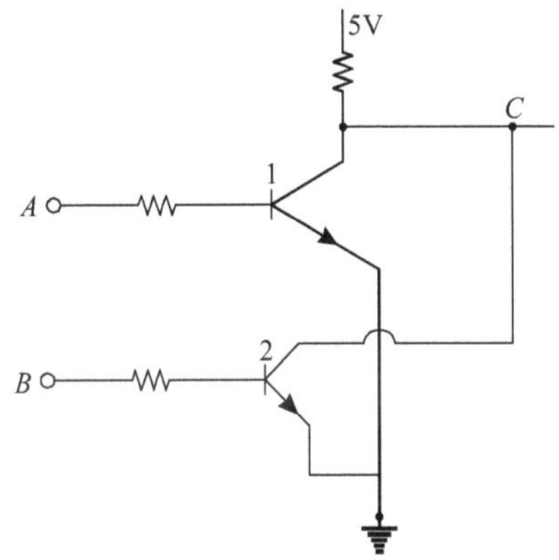

(A) A NAND B (B) A OR B
(C) A AND B (D) A NOR B

12. Identify the gate and match A, B, Y in bracket to check : **[2014 (Online)]**

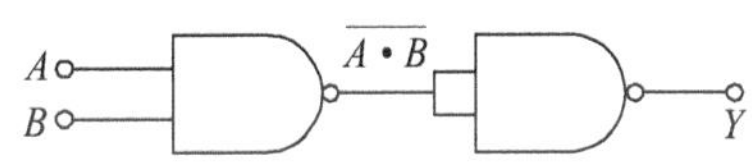

(A) AND $(A=1, B=1, Y=1)$ (B) OR $(A=1, B=1, Y=0)$
(C) NOT $(A=1, B=1, Y=1)$ (D) XOR $(A=0, B=0, Y=0)$

13. Given : A and B are input terminals
Logic 1 = > 5 V
Logic 0 = < 1 V
Which logic gate operation, the above circuit does? **[2014 (Online)]**

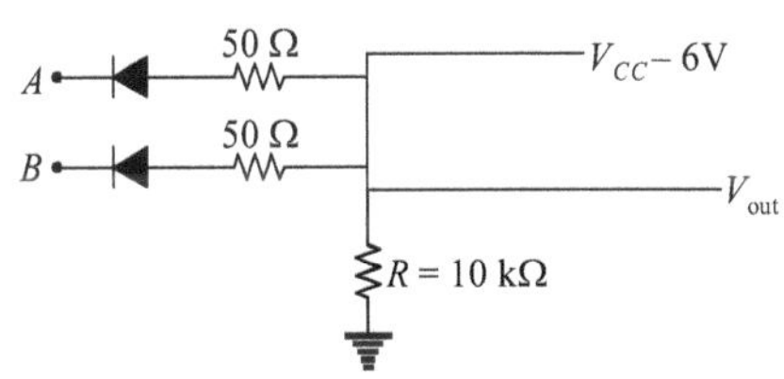

(A) AND Gate (B) OR Gate
(C) XOR Gate (D) NOR Gate

14. The truth table given in figure represents : **[2016 (Online)]**

A	B	Y
0	0	0
0	1	1
1	0	1
1	1	1

(A) OR – Gate (B) NAND – Gate
(C) AND – Gate (D) NOR – Gate

15. To get an output of 1 from the circuit shown in figure the input must be : **[2016 (Online)]**

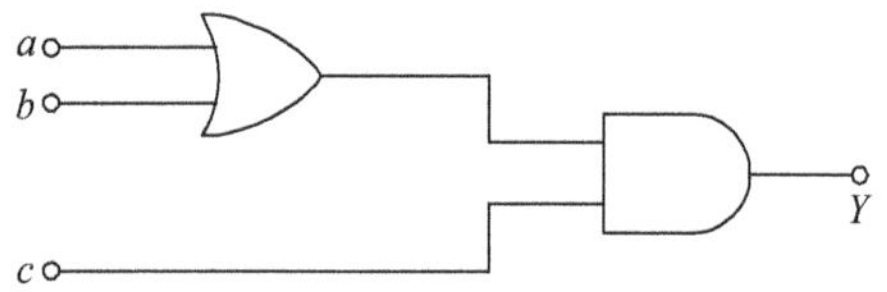

(A) $a=0, b=0, c=1$ (B) $a=1, b=0, c=0$
(C) $a=1, b=0, c=1$ (D) $a=0, b=1, c=0$

16. Truth table for the given circuit will be : **[2018 (Online)]**

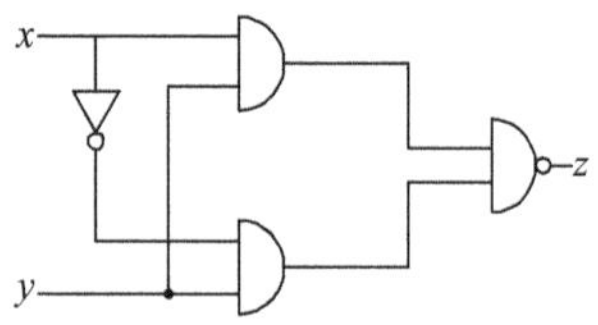

(A)

x	y	z
0	0	1
0	1	1
1	0	1
1	1	0

(B)

x	y	z
0	0	0
0	1	0
1	0	0
1	1	1

(C)

x	y	z
0	0	1
0	1	1
1	0	1
1	1	1

(D)

x	y	z
0	0	0
0	1	1
1	0	1
1	1	1

17. To get output '1' at R, for the given logic gate circuit the input values must be : **[10 Jan. 2019 (Morning)]**

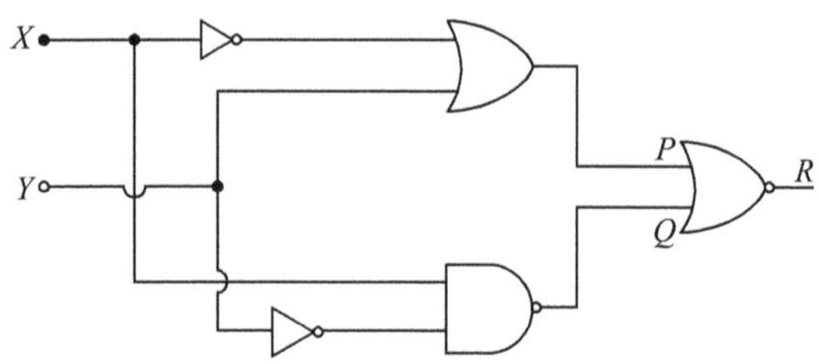

(A) $X=0, Y=1$ (B) $X=1, Y=1$
(C) $X=0, Y=0$ (D) $X=1, Y=0$

18. The output of the given logic circuit is : **[12 Jan. 2019 (Morning)]**

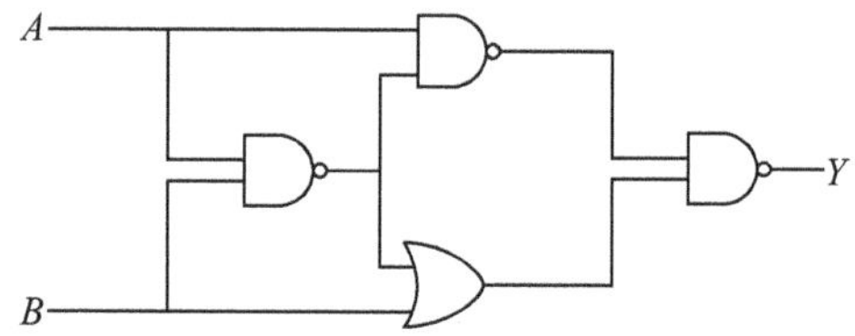

(A) $\bar{A}B$ (B) $A\bar{B}$
(C) $AB+\overline{AB}$ (D) $A\bar{B}+\bar{A}B$

19. The logic gate equivalent to the given logic circuit is : **[9 April 2019 (Evening)]**

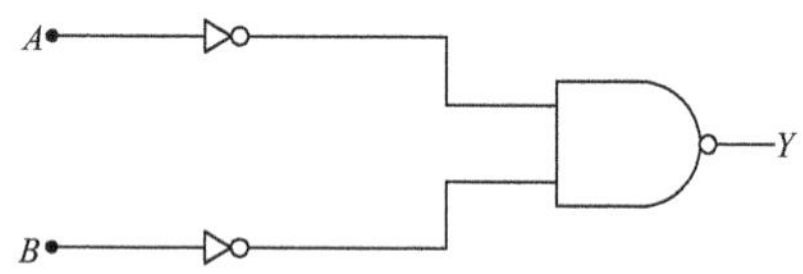

(A) OR (B) AND
(C) NOR (D) NAND

20. The truth table for the circuit given in the figure is: **[12 April 2019 (Morning)]**

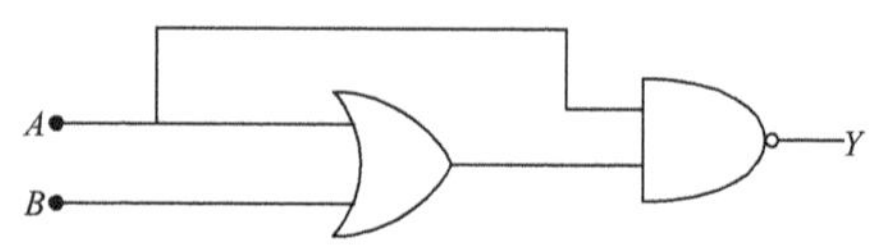

(A)

A	B	Y
0	0	1
0	1	1
1	0	0
1	1	0

(B)

A	B	Y
0	0	0
0	1	0
1	0	1
1	1	1

(C)

A	B	Y
0	0	1
0	1	0
1	0	0
1	1	0

(D)

A	B	Y
0	0	1
0	1	1
1	0	1
1	1	1

* * * * *

CHAPTER-29 COMMUNICATION SYSTEMS

29.1 Communication Systems

1. Consider telecommunication through optical fibres. Which of the following statements is not true? **[2003]**

(A) Optical fibres can be of graded refractive index

(B) Optical fibres are subject to electromagnetic interference from outside

(C) Optical fibres have extremely low transmission loss

(D) Optical fibres may have homogeneous core with a suitable cladding

2. This question has Statement-1 and Statement-2. Of the four choices given after the statements, choose the one that best describes the two statements. **[2011]**

Statement-1 : Sky wave signals are used for long distance radio communication. These signals are in general, less stable than ground wave signals.

Statement-2 : The state of ionosphere varies from hour to hour, day to day and season to season.

(A) Statement-1 is true, Statement-2 is true, Statement-2 is the correct explanation of Statement-1

(B) Statement-1 is true, Statement-2 is true, Statement-2 is not the correct explanation of Statement-1

(C) Statement-1 is false, Statement-2 is true

(D) Statement-1 is true, Statement-2 is false

3. A radar has a power of 1 kW and is operating at a frequency of 10 GHz. It is located on a mountain top of height 500 m. The maximum distance upto which it can detect object located on the surface of the earth **[2012]**

(Radius of earth = 6.4×10^6 m) is :

(A) 80 km (B) 16 km

(C) 40 km (D) 64 km

4. A 10 kW transmitter emits radio waves of wavelength 500 m. The number of photons emitted per second by the transmitter is of the order of : **[2012 (Online)]**

(A) 10^{37} (B) 10^{31}

(C) 10^{25} (D) 10^{43}

5. A radio transmitter transmits at 830 kHz. At a certain distance from the transmitter magnetic field has amplitude 4.82×10^{-11} T. The electric field and the wavelength are respectively : **[2012 (Online)]**

(A) 0.014 N/C, 36 m (B) 0.14 N/C, 36 m

(C) 0.14 N/C, 360 m (D) 0.014 N/C, 360 m

6. This question has Statement-1 and Statement-2. Of the four choices given after the Statements, choose the one that best describes the two Statements. **[2013]**

Statement-1 : Short wave transmission is achieved due to the total internal reflection of the *e-m* wave from an appropriate height in the ionosphere.

Statement-2 : Refractive index of a plasma is independent of the frequency of *e-m* waves.

(A) Statement-1 is true, Statement-2 is false

(B) Statement-1 is false, Statement-2 is true

(C) Statement-1 is true, Statement-2 is true but Statement-2 is not the correct explanation of Statement-1

(D) Statement-1 is true, Statement-2 is true and Statement-2 is the correct explanation of Statement-1

7. Which of the following statement is NOT correct ?

(A) Ground wave signals are more stable than the sky wave signals **[2013]**

(B) The critical frequency of an ionospheric layer is the highest frequency that will be reflected back by the layer when it is vertically incident

(C) Electromagnetic waves of frequencies higher than about 30 MHz cannot penetrate the ionosphere

(D) Sky wave signals in the broadcast frequency range are stronger at night than in the day time

8. A transmitting antenna at the top of a tower has height 32 m and height of the receiving antenna is 50 m. What is the maximum distance between them for satisfactory communication in line of sight (LOS) mode ? **[2014 (Online)]**

(A) 55.4 km (B) 45.5 km

(C) 54.5 km (D) 455 km

9. For sky wave propagation, the radio waves must have a frequency range in between : **[2014 (Online)]**

(A) 1 MHz to 2 MHz (B) 5 MHz to 25 MHz

(C) 35 MHz to 40 MHz (D) 45 MHz to 50 MHz

10. Long range radio transmission is possible when the radio waves are reflected from the ionosphere. For this to happen the frequency of the radio waves must be in the range : **[2014 (Online)]**

(A) 80 - 150 MHz (B) 8 - 25 MHz

(C) 1 - 3 MHz (D) 150 - 1500 kHz

11. A signal is to be transmitted through a wave of wavelength λ, using a linear antenna. The length l of the antenna and effective power radiated P_{eff} will be given respectively as : (K is a constant of proportionality) : **[2017 (Online)]**

(A) $\lambda, P_{eff} = K\left(\frac{1}{\lambda}\right)^2$ (B) $\frac{\lambda}{8}, P_{eff} = K\left(\frac{1}{\lambda}\right)$

(C) $\frac{\lambda}{16}, P_{eff} = K\left(\frac{1}{\lambda}\right)^3$ (D) $\frac{5}{\lambda}, P_{eff} = K\left(\frac{1}{\lambda}\right)^{\frac{1}{2}}$

12. A *TV* transmission tower has a height of 140 m and the height of the receiving antenna is 40 m. What is the maximum distance upto which signals can be broadcasted from this tower in *LOS* (Line of Sight) mode ? (Given : radius of earth = 6.4×10^6m) : **[10 Jan. 2019 (Morning)]**

(A) 80 km (B) 48 km

(C) 40 km (D) 65 km

13. To double the covering range of a *TV* transmission tower, its height should be multiplied by : **[12 Jan. 2019 (Evening)]**

(A) $\frac{1}{\sqrt{2}}$ (B) 4

(C) $\sqrt{2}$ (D) 2

14. In a line of sight radio communication, a distance of about 50 km is kept between the transmitting and receiving antennas. If the height of the receiving antenna is 70 m, then the minimum height of the transmitting antenna should be.

(Radius of the Earth = 6.4×10^6 m) : **[8 April 2019 (Evening)]**

(A) 40 m (B) 51 m

(C) 32 m (D) 20 m

15. Given below in the the left column are different modes of communication using the kinds of waves given the right column: **[10 April 2019 (Morning)]**

A.	Optical Fibre communication	P.	Ultrasound
B.	Radar	Q.	Infrared Light
C.	Sonar	R.	Microwaves
D.	Mobile Phones	S.	Radio Waves

(A) A-S, B-Q, C-R, D-P (B) A-R, B-P, C-S, D-Q

(C) A-Q, B-S, C-R, D-P (D) A-Q, B-S, C-P, D-R

* * * * *

CHAPTER-29 COMMUNICATION SYSTEMS

29.2 Modulation and Demodulation

1. Which of the following four alternatives is not correct ? We need modulation : **[2011]**

(A) To reduce the time lag between transmission and reception of the information signal

(B) To reduce the size of antenna

(C) To reduce the fractional band width, that is the ratio of the signal band width to the centre frequency

(D) To increase the selectivity

2. Given the electric field of a complete amplitude modulated wave as

$$\vec{E} = \hat{i}E_c\left(1+\frac{E_m}{E_c}\cos\omega_m t\right)\cos\omega_c t$$

Where the subscript stands for the carrier wave and m for the modulating signal. The frequencies present in the modulated wave are : **[2012 (Online)]**

(A) ω_c and $\sqrt{\omega_c^2+\omega_m^2}$ (B) ω_c, $\omega_c+\omega_m$ and $\omega_c-\omega_m$

(C) ω_c and ω_m (D) ω_c and $\sqrt{\omega_c\omega_m}$

3. Which of the following modulated signal has the best noise-tolerance ? **[2013]**

(A) Long-wave (B) Short-wave

(C) Medium-wave (D) Amplitude-modulated

4. A diode detector is used to detect an amplitude modulated wave of 60% modulation using a condenser of capacity 250 picofarad in parallel with a load resistance 100 kilo ohm. Find the maximum modulated frequency which could be detected by it : **[2013]**

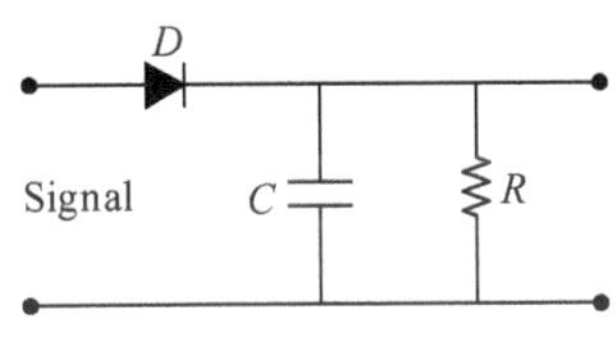

(A) 10.62 MHz (B) 10.62 kHz

(C) 5.31 MHz (D) 5.31 kHz

5. If a carrier wave $c(t) = A\sin\omega_c t$ is amplitude modulated by a modulator signal $m(t) = A\sin\omega_m t$ then the equation of modulated signal $[C_m(t)]$ and its modulation index are respectively : **[2013 (Online)]**

(A) $C_m(t) = A(1+\sin\omega_m t)\sin\omega_c t$ and 2

(B) $C_m(t) = A(1+\sin\omega_m t)\sin\omega_m t$ and 1

(C) $C_m(t) = A(1+\sin\omega_m t)\sin\omega_c t$ and 1

(D) $C_m(t) = A(1+\sin\omega_c t)\sin\omega_m t$ and 2

6. A signal of 5 kHz frequency is amplitude modulated on a carrier wave of frequency 2 MHz. The frequencies of the resultant signal is/are : **[2015]**

(A) 2005 kHz, 2000 kHz and 1995 kHz

(B) 2000 kHz and 1995 kHz

(C) 2 MHz only

(D) 2005 kHz and 1995 kHz

7. Choose the correct statement : **[2016]**

(A) In frequency modulation the amplitude of the high frequency carrier wave is made to vary in proportion to the amplitude of the audio signal

(B) In frequency modulation the amplitude of the high frequency carrier wave is made to vary in proportional to the frequency of the audio signal

(C) In amplitude modulation the amplitude of the high frequency carrier wave is made to vary in proportion to the amplitude of the audio signal

(D) In amplitude modulation the frequency of the high frequency carrier wave is made to vary in proportion to the amplitude of the audio signal

8. A modulated signal $C_m(t)$ has the form $C_m(t) = 30\sin 300\pi t + 10(\cos 200\pi t - \cos 400\pi t)$. The carrier frequency f_c, the modulating frequency (message frequency) f_w and the modulation index μ are respectively given by : **[2016 (Online)]**

(A) $f_c = 200$ Hz; $f_w = 50$ Hz; $\mu = 1/2$

(B) $f_c = 150$ Hz; $f_w = 50$ Hz; $\mu = 2/3$

(C) $f_c = 150$ Hz; $f_w = 30$ Hz; $\mu = 1/3$

(D) $f_c = 200$ Hz; $f_w = 30$ Hz; $\mu = 1/2$

9. An audio signal consists of two distinct sounds : one a human speech signal in the frequency band of 200 Hz to 2700 Hz, while the other is a high frequency music signal in the frequency band of 10200 Hz to 15200 Hz. The ratio of the AM signal band width required to send both the signals together to the *AM* signal bandwidth required to send just the human speech is : **[2016 (Online)]**

(A) 2 (B) 5

(C) 6 (D) 3

10. In amplitude modulations, sinusoidal carrier frequency used is denoted by ω_c and the signal frequency is denoted by ω_m. The bandwidth $(\Delta\omega_m)$ of the signal is such that $\Delta\omega_m < \omega_c$. Which of the following frequencies is not contained in the modulated wave ? **[2017]**

(A) $\omega_m+\omega_c$ (B) $\omega_c-\omega_m$

(C) ω_m (D) ω_c

11. A signal of frequency 20 kHz and peak voltage of 5 Volt is used to modulate a carrier wave of frequency 1.2 MHz and peak voltage 25 Volts. Choose the correct statement :
(A) Modulation index = 5, side frequency bands are at 1400 kHz and 1000 kHz
(B) Modulation index = 5, side frequency bands are at 21.2 kHz and 18.8 kHz
(C) Modulation index = 0.8, side frequency bands are at 1180 kHz and 1220 kHz
(D) Modulation index = 0.2, side frequency bands are at 1220 kHz and 1180 kHz **[2017 (Online)]**

12. A carrier wave of peak voltage 14 V is used for transmitting a message signal. The peak voltage of modulating signal given to achieve a modulation index of 80% will be : **[2018]**
(A) 11.2 V (B) 7 V
(C) 22.4 V (D) 28 V

13. A telephonic communication service is working at carrier frequency of 10 GHz. Only 10% of it is utilized for transmission. How many telephonic channels can be transmitted simultaneously if each channel requires a bandwidth of 5 kHz ? **[2018]**
(A) 2×10^3 (B) 2×10^4
(C) 2×10^5 (D) 2×10^6

14. The carrier frequency of a transmitter is provided by a tank circuit of a coil of inductance 49 μH and a capacitance of 2.5 nF. It is modulated by an audio signal of 12 kHz. The frequency range occupied by the side bands is : **[2018 (Online)]**
(A) 18 kHz – 30 kHz (B) 63 kHz – 75 kHz
(C) 442 kHz – 466 kHz (D) 13482 kHz – 13494 kHz

15. The number of amplitude modulated broadcast stations that can be accommodated in a 300 kHz band width for the highest modulating frequency 15 kHz will be : **[2018 (Online)]**
(A) 20 (B) 10
(C) 8 (D) 15

16. In a communication system operating at wavelength 800 nm, only one percent of source frequency is available as signal bandwidth. The number of channels accomodated for transmitting *TV* signals of band width 6 MHz are (Take velocity of light $c = 3 \times 10^8$ m/s, $h = 6.6 \times 10^{-34}$ J-s) : **[9 Jan. 2019 (Evening)]**
(A) 3.75×10^6 (B) 4.87×10^5
(C) 3.86×10^6 (D) 6.25×10^5

17. The modulation frequency of an *AM* radio station is 250 kHz, which is 10% of the carrier wave. If another *AM* station approaches you for license what broadcast frequency will you allot ? **[10 Jan. 2019 (Evening)]**
(A) 2750 kHz (B) 2000 kHz
(C) 2250 kHz (D) 2900 kHz

18. An amplitude modulated signal is given by $V(t) = 10[1 + 0.3 \cos(2.2 \times 10^4)] \sin(5.5 \times 10^5 t)$. Here t is in seconds. The sideband frequencies (in kHz) are, [Given $\pi = 22/7$] : **[11 Jan. 2019 (Morning)]**
(A) 1785 and 1715 (B) 892.5 and 857.5
(C) 89.25 and 85.75 (D) 178.5 smf 171.5

19. An amplitude modulated signal is plotted below : **[11 Jan. 2019 (Evening)]**

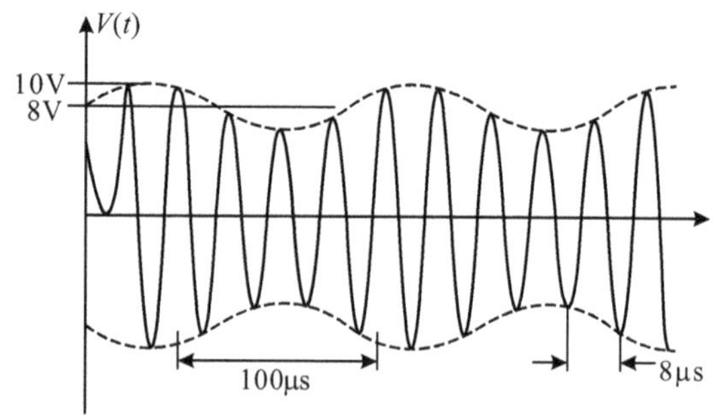

Which one of the following best describes the above signal ?
(A) $(9 + \sin(2.5\pi \times 10^5 t)) \sin(2\pi \times 10^4 t)V$
(B) $(9 + \sin(4\pi \times 10^4 t)) \sin(5\pi \times 10^5 t)V$
(C) $(1 + 9\sin(2\pi \times 10^4 t)) \sin(2.5\pi \times 10^5 t)V$
(D) $(9 + \sin(2\pi \times 10^4 t)) \sin(2.5\pi \times 10^5 t)V$

20. A 100 V carrier wave is made to vary between 160 V and 40 V by a modulating signal. What is the modulation index? **[12 Jan. 2019 (Morning)]**
(A) 0.6 (B) 0.5
(C) 0.3 (D) 0.4

21. The wavelength of the carrier waves in a modern optical fiber communication network is close to : **[8 April 2019 (Morning)]**
(A) 600 nm (B) 900 nm
(C) 2400 nm (D) 1500 nm

22. A signal $A \cos \omega t$ is transmitted using $v_0 \sin \omega_0 t$ as carrier wave. The correct amplitude modulated (*AM*) signal is : **[9 April 2019 (Morning)]**
(A) $v_0 \sin \omega_0 t + A \cos \omega t$
(B) $v_0 \sin \omega_0 t + \frac{A}{2}\sin(\omega_0 - \omega)t + \frac{A}{2}\sin(\omega_0 + \omega)t$
(C) $(v_0 + A) \cos \omega t \sin \omega_0 t$
(D) $v_0 \sin[\omega_0 (1 + 0.01 A \sin \omega t) t]$

23. The physical sizes of the transmitter and receiver antenna in a communication system are : **[9 April 2019 (Evening)]**
(A) Proportional to carrier frequency
(B) Inversely proportional to modulation frequency
(C) Inversely proportional to carrier frequency
(D) Independent of both carrier and modulation frequency

24. A message signal of frequency 100 MHz and peak voltage 100V is used to execute amplitude modulation on a carrier wave of frequency 300 GHz and peak voltage 400 V. The modulation index and difference between the two side band frequencies are : **[10 April 2019 (Morning)]**

(A) 4; 1×10^8 Hz
(B) 0.25; 1×10^8 Hz
(C) 4; 2×10^8 Hz
(D) 0.25; 2×10^8 Hz

25. In an amplitude modulator circuit, the carrier wave is given by, $C(t) = 4 \sin(20000\pi t)$ while modulating signal is given by, $m(t) = 2 \sin(2000\pi t)$. The values of modulation index and lower side band frequency are : **[12 April 2019 (Evening)]**

(A) 0.5 and 9 kHz
(B) 0.5 and 10 kHz
(C) 0.3 and 9 kHz
(D) 0.4 and 10 kHz

* * * * *

CHAPTER-30 ERRORS AND EXPERIMENTS

30.1 Errors and Experiments

1. Two full turns of the circular scale of a screw gauge cover a distance of 1 mm on its main scale. The total number of divisions on the circular scale is 50. Further, it is found that the screw gauge has a zero error of – 0.03 mm. While measuring the diameter of a thin wire, a student notes the main scale reading of 3 mm and the number of circular scale divisions in line with the main scale as 35. The diameter of the wire is : **[2008]**

(A) 3.32 mm (B) 3.73 mm

(C) 3.67 mm (D) 3.38 mm

2. A body of mass $m = 3.513$ kg is moving along the x-axis with a speed of 5.00 ms^{-1}. The magnitude of its momentum is recorded as : **[2008]**

(A) 17.6 kg ms^{-1} (B) 17.565 kg ms^{-1}

(C) 17.56 kg ms^{-1} (D) 17.57 kg ms^{-1}

3. In an experiment the angles are required to be measured using an instrument, 29 divisions of the main scale exactly coincide with the 30 divisions of the vernier scale. If the smallest division of the main scale is half a degree (= 0.5°), then the least count of the instrument is : **[2009]**

(A) half minute (B) one degree

(C) half degree (D) one minute

4. The respective number of significant figures for the numbers 23.023, 0.0003 and 2.1×10^{-3} are : **[2009]**

(A) 5, 1, 2 (B) 5, 1, 5

(C) 5, 5, 2 (D) 4, 4, 2

5. A screw gauge gives the following reading when used to measure the diameter of a wire.

Main scale reading : 0 mm

Circular scale reading : 52 divisions **[2011]**

Given that 1 mm on main scale corresponds to 100 divisions of the circular scale. The diameter of wire from the above data is :

(A) 0.052 cm (B) 0.026 cm

(C) 0.005 cm (D) 0.52 cm

6. A spectrometer gives the following reading when used to measure the angle of a prism.

Main scale reading : 58.5 degree

Vernier scale reading : 09 divisions

Given that 1 division on main scale corresponds to 0.5 degree. Total divisions on the Vernier scale is 30 and match with 29 divisions of the main scale. The angle of the prism from the above data is : **[2012]**

(A) 58.59 degree (B) 58.77 degree

(C) 58.65 degree (D) 59 degree

7. Resistance of a given wire is obtained by measuring the current flowing in it and the voltage difference applied across it. If the percentage errors in the measurement of the current and the voltage difference are 3% each, then error in the value of resistance of the wire is : **[2012]**

(A) 6% (B) Zero

(C) 1% (D) 3%

8. A student measured the diameter of a wire using a screw gauge with the least count 0.001 cm and listed the measurements. The measured value should be recorded as : **[2012 (Online)]**

(A) 5.3200 cm (B) 5.3 cm

(C) 5.32 cm (D) 5.320 cm

9. N divisions on the main scale of a vernier calliper coincide with $(N + 1)$ divisions of the vernier scale. If each division of main scale is 'a' units, then the least count of the instrument is : **[2012 (Online)]**

(A) a (B) $\frac{a}{N}$

(C) $\frac{N}{N+1} \times a$ (D) $\frac{a}{N+1}$

10. A student measured the length of a rod and wrote it as 3.50 cm. Which instrument did he use to measure it ? **[2014]**

(A) A meter scale

(B) A vernier calliper where the 10 divisions in vernier scale matches with 9 division in main scale and main scale has 10 divisions in 1 cm

(C) A screw gauge having 100 divisions in the circular scale and pitch as 1 mm

(D) A screw gauge having 50 divisions in the circular scale and pitch as 1 mm

11. The current voltage relation of a diode is given by $I = (e^{1000 V/T} - 1)$ mA, where the applied voltage V is in volt and the temperature T is in degree kelvin. If a student makes an error measuring ± 0.01 V while measuring the current of 5 mA at 300 K, what will be the error in the value of current in mA ? **[2014]**

(A) 0.2 mA (B) 0.02 mA

(C) 0.5 mA (D) 0.05 mA

12. An experiment is performed to obtain the value of acceleration due to gravity g by using a simple pendulum of length L. In this experiment time for 100 oscillations is measured by using a watch of 1 second least count and the value is 90.0 seconds. The length L is measured by using a meter scale of least count 1 mm and the value is 20.0 cm. The error in the determination of g would be : **[2014 (Online)]**

(A) 1.7% (B) 2.7%
(C) 4.4% (D) 2.27%

13. In the experiment of calibration of voltmeter, a standard cell of e.m.f. 1.1 volt is balanced against 440 cm of potential wire. The potential difference across the ends of resistance is found to balance against 220 cm of the wire. The corresponding reading of voltmeter is 0.5 volt. The error in the reading of voltmeter will be : **[2014 (Online)]**

(A) -0.15 volt (B) 0.15 volt
(C) 0.5 volt (D) -0.05 volt

14. Match List-I (Event) with List-II (Order of the time interval for happening of the event) and select the correct option from the options given below the lists : **[2014 (Online)]**

List-I	List-II
(1) Rotation period of earth	(i) 10^5 s
(2) Revolution period of earth	(ii) 10^7 s
(3) Period of light wave	(iii) 10^{-15} s
(4) Period of sound wave	(iv) 10^{-3} s

(A) (1)-(i), (2)-(ii), (3)-(iii), (4)-(iv)
(B) (1)-(ii), (2)-(i), (3)-(iv), (4)-(iii)
(C) (1)-(i), (2)-(ii), (3)-(iv), (4)-(iii)
(D) (1)-(ii), (2)-(i), (3)-(iii), (4)-(iv)

15. The period of oscillation of a simple pendulum is $T = 2\pi\sqrt{\frac{L}{g}}$. Measured value of L is 20.0 cm know to 1 mm accuracy and time for 100 oscillations of the pendulum is found to be 90 s using a wrist watch of 1 s resolution. The accuracy in the determination of g is : **[2015]**

(A) 1% (B) 5%
(C) 2% (D) 3%

16. Diameter of a steel ball is measured using a Vernier callipers which has divisions of 0.1 cm on its main scale (MS) and 10 divisions of its vernier scale (VS) match 9 divisions on the main scale. Three such measurements for a ball are given as : **[2015 (Online)]**

S.No.	MS(cm)	VS divisions
1.	0.5	8
2.	0.5	4
3.	0.5	6

If the zero error is -0.03 cm, then mean corrected diameter is :

(A) 0.52 cm (B) 0.59 cm
(C) 0.56 cm (D) 0.53 cm

17. A student measures the time period of 100 oscillations of a simple pendulum four times. The data set is 90 s, 91 s, 95 s, and 92 s. If the minimum division in the measuring clock is 1 s, then the reported mean time should be : **[2016]**

(A) 92 ± 1.8 s (B) 92 ± 3 s
(C) 92 ± 1.5 s (D) 92 ± 5.0 s

18. A screw gauge with a pitch of 0.5 mm and a circular scale with 50 divisions is used to measure the thickness of a thin sheet of Aluminium. Before starting the measurement, it is found that wen the two jaws of the screw gauge are brought in contact, the 45th division coincides with the main scale line and the zero of the main scale is barely visible. What is the thickness of the sheet if the main scale reading is 0.5 mm and the 25th division coincides with the main scale line ? **[2016]**

(A) 0.70 mm (B) 0.50 mm
(C) 0.75 mm (D) 0.80 mm

19. The following observations were taken for determining surface tension T of water by capillary method :
Diameter of capillary, $D = 1.25 \times 10^{-2}$ m, rise of water, $h = 1.45 \times 10^{-2}$ m. Using $g = 9.80$ m/s^2 and the simplified relation $T = \frac{rhg}{2} \times 10^3$ N/m, the possible error in surface tension is closest to : **[2017]**

(A) 2.4% (B) 10%
(C) 0.15% (D) 1.5%

20. A physical quantity P is described by the relation $P = a^{1/2} b^2 c^3 d^{-4}$. If the relative errors in the measurement of a, b, c, and d respectively, are 2%, 1%, 3% and 5%, then the relative error in P will be : **[2017 (Online)]**

(A) 8% (B) 12%
(C) 32% (D) 25%

21. The percentage errors in quantities P, Q, R and S are 0.5%, 1%, 3% and 1.5% respectively in the measurement of a physical quantity $A = \frac{P^3Q^2}{\sqrt{RS}}$.
The maximum percentage error in the value of A will be : **[2018 (Online)]**

(A) 8.5% (B) 6.0%
(C) 7.5% (D) 6.5%

22. In a screw gauge, 5 complete rotations of the screw cause it to move a linear distance of 0.25 cm. There are 100 circular scale divisions. The thickness of a wire measured by this screw gauge gives a reading of 4 main scale divisions and 30 circular scale divisions. Assuming negligible zero error, the thickness of the wire is : **[2018 (Online)]**

(A) 0.0430 cm (B) 0.3150 cm
(C) 0.4300 cm (D) 0.2150 cm

23. The relative error in the determination of the surface area of a sphere is α. Then the relative error in the determination of its volume is : **[2018 (Online)]**

(A) $\frac{2}{3}\alpha$ (B) $\frac{1}{3}\alpha$

(C) $\frac{3}{2}\alpha$ (D) α

24. The relative uncertainty in the period of a satellite orbiting around the earth is 10^{-2}. If the relative uncertainty in the radius of the orbit is negligible, the relative uncertainty in the mass of the earth is : **[2018 (Online)]**

(A) 3×10^{-2} (B) 10^{-2}

(C) 2×10^{-2} (D) 6×10^{-2}

25. The density of a material in the shape of a cube is determined by measuring three sides of the cube and its mass. If the relative errors in measuring the mass and length are respectively 1.5% and 1%, the maximum error in determining the density is : **[2018 (Online)]**

(A) 2.5% (B) 3.5%

(C) 4.5% (D) 6%

26. A copper wire is stretched to make it 0.5% longer. The percentage change in its electrical resistance, if its volume remains unchanged is : **[9 Jan. 2019 (Morning)]**

(A) 2.5% (B) 0.5%

(C) 1.0% (D) 2.0%

27. The pitch and the number of divisions, on the circular scale, for a given screw gauge are 0.5 mm and 100 respectively. When the screw gauge is fully tightened without any object, the zero of its circular scale lies 3 divisions below the reference line. The readings of the main scale and the circular scale, for a thin sheet, are 5.5 mm and 48 respectively, the thickness of this sheet is: **[9 Jan. 2019 (Evening)]**

(A) 5.755 m (B) 5.725 mm

(C) 5.740 m (D) 5.950 mm

28. The diameter and height of a cylinder are measured by a meter scale to be 12.6 ± 0.1 cm and 34.2 ± 0.1 cm, respectively. What will be the value of its volume in appropriate significant figures ? **[10 Jan. 2019 (Evening)]**

(A) 4260 ± 80 cm^3 (B) 4300 ± 80 cm^3

(C) 4264.4 ± 81.0 cm^3 (D) 4264 ± 81 cm^3

29. The least count of the main scale of a screw gauge is 1 mm. The minimum number of divisions on its circular scale required to measure 5 μm diameter of wire is : **[12 Jan. 2019 (Morning)]**

(A) 50 (B) 100

(C) 200 (D) 500

30. In a simple pendulum experiment for determination of acceleration due to gravity (g), time taken for 20 oscillations is measured by using a watch of 1 second least count. The mean value of time taken comes out to be 30 s. The length of pendulum is measured by using a meter scale of least count 1 mm and the value obtained is 55.0 cm. The percentage error in the determination of g is close to : **[8 April 2019 (Evening)]**

(A) 0.7% (B) 0.2%

(C) 3.5% (D) 6.8%

31. In the density measurement of a cube, the mass and edge length are measured as (10.00 ± 0.10) kg and (0.10 ± 0.01) m, respectively. The error in the measurement of density is : **[9 April 2019 (Morning)]**

(A) 0.10 kg/m^3 (B) 0.31 kg/m^3

(C) 0.07 kg/m^3 (D) 0.01 kg/m^3

32. The area of a square is 5.29 cm^2. The area of 7 such squares taking into account the significant figures is : **[9 April 2019 (Evening)]**

(A) 37 cm^2 (B) 37.0 cm^2

(C) 37.03 cm^2 (D) 37.030 cm^2

* * * * *

ANSWER KEY

1.1 Measurements and Dimensional Analysis

1. (A) 2. (B) 3. (C) 4. (A) 5. (D)
6. (B) 7. (C) 8. (A) 9. (B) 10. (D)
11. (C) 12. (C) 13. (B) 14. (D) 15. (C)
16. (B) 17. (A, D) 18. (A) 19. (B) 20. (D)
21. (D) 22. (B) 23. (B) 24. (C) 25. (B)
26. (C)

2.1 Motion in One Dimension

1. (A) 2. (B) 3. (D) 4. (C) 5. (C)
6. (D) 7. (C) 8. (C) 9. (A) 10. (C)
11. (C) 12. (B) 13. (B) 14. (D) 15. (A)
16. (B) 17. (C) 18. (C) 19. (C) 20. (A)
21. (C) 22. (B) 23. (B) 24. (C) 25. (A)
26. (D) 27. (B) 28. (A) 29. (D) 30. (C)
31. (C) 32. (A) 33. (D) 34. (B)

3.1 Motion in Two Dimensions

1. (C) 2. (D) 3. (B) 4. (A) 5. (C)
6. (A) 7. (C) 8. (A) 9. (D) 10. (A)
11. (D) 12. (A) 13. (B) 14. (B) 15. (C)
16. (C) 17. (A) 18. (A) 19. (B) 20. (C)
21. (D) 22. (B) 23. (D) 24. (C) 25. (C)
26. (D) 27. (B) 28. (A) 29. (C) 30. (B)

3.2 Relative Motion

1. (B) 2. (D) 3. (C) 4. (C) 5. (D)
6. (D)

4.1 Newton's Laws of Motion

1. (A) 2. (B) 3. (C) 4. (D) 5. (C)
6. (B) 7. (B) 8. (A) 9. (A) 10. (B)
11. (D) 12. (D) 13. (A) 14. (D) 15. (D)
16. (C) 17. (D) 18. (C) 19. (A) 20. (D)
21. (A) 22. (D) 23. (D) 24. (A) 25. (D)
26. (A) 27. (B) 28. (B) 29. (A)

4.2 Friction

1. (D) 2. (C) 3. (A) 4. (A) 5. (C)
6. (B) 7. (D) 8. (A) 9. (A) 10. (D)
11. (B) 12. (A) 13. (C) 14. (B) 15. (B)
16. (D) 17. (A) 18. (B) 19. (A) 20. (B)

5.1 Work Energy Power

1. (B) 2. (B) 3. (B) 4. (A) 5. (B)
6. (B) 7. (B) 8. (D) 9. (A) 10. (B)
11. (B) 12. (A) 13. (D) 14. (A) 15. (D)
16. (D) 17. (D) 18. (B) 19. (D) 20. (A)
21. (A) 22. (B) 23. (C) 24. (D) 25. (D)
26. (A) 27. (B) 28. (C) 29. (D) 30. (C)
31. (C) 32. (C) 33. (A) 34. (C) 35. (B)
36. (A) 37. (C) 38. (B) 39. (B) 40. (A)
41. (A) 42. (A) 43. (B) 44. (B) 45. (C)

5.2 Circular Motion

1. (B) 2. (C) 3. (B) 4. (C) 5. (D)
6. (C) 7. (A) 8. (C) 9. (C) 10. (B)
11. (B) 12. (A) 13. (D) 14. (D) 15. (B)
16. (D) 17. (D)

6.1 System of Particles

1. (C) 2. (C) 3. (C) 4. (B) 5. (C)
6. (C) 7. (B) 8. (C) 9. (C) 10. (A)
11. (A) 12. (B) 13. (B) 14. (D) 15. (A)
16. (C) 17. (D) 18. (C) 19. (D) 20. (C)
21. (D) 22. (D) 23. (D) 24. (B) 25. (C)
26. (B) 27. (B) 28. (D) 29. (B) 30. (B)
31. (C) 32. (A) 33. (A) 34. (B) 35. (A)
36. (A) 37. (D) 38. (A) 39. (C) 40. (C)
41. (A) 42. (A) 43. (A) 44. (D) 45. (C)
46. (B) 47. (B) 48. (C) 49. (C) 50. (D)
51. (B) 52. (A) 53. (C)

6.2 Rigid Body Dynamics

1. (C) 2. (A) 3. (B) 4. (D) 5. (A)
6. (D) 7. (B) 8. (C) 9. (D) 10. (D)
11. (C) 12. (B) 13. (D) 14. (C) 15. (A)
16. (A) 17. (C) 18. (D) 19. (C) 20. (D)
21. (B) 22. (B) 23. (D) 24. (D) 25. (B)
26. (B) 27. (D) 28. (D) 29. (B) 30. (B)
31. (C) 32. (B) 33. (A) 34. (A) 35. (C)
36. (B) 37. (C) 38. (D) 39. (D) 40. (B)
41. (A) 42. (C) 43 (C) 44. (A) 45. (D)
46. (A, C) 47. (C) 48. (B) 49. (D) 50. (D)
51. (B) 52. (B) 53. (B) 54. (C) 55. (C)
56. (D) 57. (A) 58. (C) 59. (A) 60. (B)
61. (C) 62. (D) 63. (B) 64. (A) 65. (A)
66. (C) 67. (C) 68. (C) 69. (D) 70. (D)
71. (B) 72. (B) 73. (A) 74. (None) 75. (B)
76. (B) 77. (D) 78. (A) 79. (C) 80. (C)
81. (D) 82. (A) 83. (D) 84. (C) 85. (D)
86. (B) 87. (B) 88. (D) 89. (D) 90. (D)
91. (D)

7.1 Gravitational Field

1. (C) 2. (B) 3. (D) 4. (C) 5. (A)
6. (A) 7. (B) 8. (C) 9. (B) 10. (A)
11. (A) 12. (B) 13. (C) 14. (A) 15. (D)
16. (A) 17. (D) 18. (A) 19. (B) 20. (B)
21. (C) 22. (C) 23. (C) 24. (B) 25. (A)
26. (D) 27. (B) 28. (B) 29. (C) 30. (B)
31. (B) 32. (D)

7.2 Satellite Motion

1. (B) 2. (A) 3. (A) 4. (C) 5. (C)
6. (A) 7. (A) 8. (D) 9. (C) 10. (B)
11. (C) 12. (A) 13. (D) 14. (C) 15. (C)
16. (C) 17. (C) 18. (C) 19. (B) 20. (D)
21. (A) 22. (C) 23. (C) 24. (D) 25. (B)
26. (B)

8.1 Elasticity

1. (D) 2. (D) 3. (B) 4. (A) 5. (B)
6. (A) 7. (A) 8. (C) 9. (B) 10. (C)
11. (C) 12. (D) 13. (C) 14. (A) 15. (C)
16. (C) 17. (D) 18. (B) 19. (C) 20. (C)
21. (D) 22. (B) 23. (A) 24. (None) 25. (None)
26. (B)

9.1 Fluid Statics

1. (D) 2. (C) 3. (B) 4. (C) 5. (C)
6. (C) 7. (B)

9.2 Fluid Dynamics

1. (B) 2. (B) 3. (A) 4. (A) 5. (B)
6. (C) 7. (B) 8. (B) 9. (B) 10. (A)
11. (A) 12. (B) 13. (D) 14. (B) 15. (C)
16. (B)

9.3 Viscosity

1. (B) 2. (C) 3. (A) 4. (B) 5. (A)
6. (C) 7. (B) 8. (C) 9. (D) 10. (C)
11. (D) 12. (D)

9.4 Surface Tension

1. (C) 2. (D) 3. (C) 4. (D) 5. (B)
6. (A) 7. (D) 8. (C) 9. (B) 10. (C)
11. (A) 12. (A) 13. (B) 14. (D) 15. (D)
16. (C) 17. (C)

10.1 Thermal Expansion

1. (C) 2. (A) 3. (A) 4. (C) 5. (B)
6. (A) 7. (A) 8. (C) 9. (C) 10. (B)
11. (C) 12. (B) 13. (B)

10.2 Calorimetry

1. (B) 2. (C) 3. (C) 4. (B) 5. (A)
6. (B) 7. (D) 8. (D) 9. (D) 10. (B)
11. (A) 12. (B) 13. (C) 14. (C)

10.3 Heat Transfer

1. (A) 2. (B) 3. (A) 4. (D) 5. (D)
6. (D) 7. (D) 8. (C) 9. (B) 10. (C)
11. (D) 12. (B) 13. (A) 14. (A) 15. (C)
16. (D) 17. (C) 18. (C) 19. (B) 20. (B)
21. (D) 22. (A) 23. (D) 24. (A) 25. (B)

11.1 Laws of Thermodynamics

1. (A) 2. (C) 3. (B) 4. (D) 5. (C)
6. (A) 7. (B) 8. (A) 9. (C) 10. (D)
11. (A) 12. (A) 13. (A) 14. (A) 15. (D)
16. (D) 17. (C) 18. (D) 19. (A) 20. (C)
21. (C) 22. (B) 23. (D) 24. (B) 25. (D)
26. (B) 27. (D) 28. (D) 29. (D) 30. (C)
31. (D) 32. (C) 33. (B) 34. (C) 35. (C)
36. (D) 37. (A) 38. (B) 39. (C) 40. (A)
41. (B) 42. (C) 43. (A) 44. (D) 45. (B)
46. (A) 47. (B) 48. (D) 49. (A) 50. (D)
51. (B) 52. (D) 53. (A) 54. (C) 55. (A)
56. (B) 57. (B) 58. (A) 59. (C) 60. (D)
61. (B) 62. (A) 63. (C) 64. (A) 65. (C)
66. (D) 67. (B) 68. (A)

12.1 Kinetic Theory of Gases

1. (C) 2. (D) 3. (C) 4. (C) 5. (A)
6. (B) 7. (C) 8. (A) 9. (A) 10. (D)
11. (D) 12. (A) 13. (B) 14. (C) 15. (A)
16. (A) 17. (B) 18. (A) 19. (D) 20. (D)
21. (C) 22. (A) 23. (B) 24. (D) 25. (A)
26. (C) 27. (D) 28. (D) 29. (C) 30. (D)
31. (B) 32. (A) 33. (None) 34. (C) 35. (C)
36. (B) 37. (D) 38. (C) 39. (C) 40. (C)

13.1 Simple Harmonic Motion

1. (B) 2. (C) 3. (B) 4. (C) 5. (C)
6. (A) 7. (C) 8. (D) 9. (C) 10. (B)
11. (C) 12. (B) 13. (C) 14. (D) 15. (D)
16. (B) 17. (A) 18. (A) 19. (A) 20. (A)

21. (B)	22. (A)	23. (D)	24. (D)	25. (A)
26. (D)	27. (B)	28. (D)	29. (B)	30. (A)
31. (A)	32. (A)	33. (C)	34. (C)	35. (A)
36. (A)	37. (D)	38. (D)	39. (D)	40. (C)
41. (B)	42. (A)	43. (D)	44. (C)	45. (C)
46. (A)	47. (A)	48. (B)	49. (D)	50. (C)
51. (C)	52. (D)	53. (B)	54. (D)	55. (B)
56. (C)	57. (B)	58. (D)	59. (B)	60. (B)
61. (C)	62. (D)	63. (B)	64. (D)	65. (B)
66. (C)	67. (D)	68. (None)	69. (C)	70. (A)
71. (C)	72. (A)	73. (D)	74. (D)	75. (A)
76. (D)	77. (B)			

14.1 Wave Motion

1. (A)	2. (D)	3. (A)	4. (A)	5. (A)
6. (B)	7. (D)	8. (C)	9. (B)	10. (B)
11. (D)	12. (A)	13. (C)	14. (A)	15. (B)
16. (A)	17. (C)	18. (C)	19. (C)	

14.2 Stationary Waves and Beats

1. (C)	2. (B)	3. (B)	4. (B)	5. (C)
6. (C)	7. (A)	8. (B)	9. (B)	10. (A)
11. (B)	12. (D)	13. (B)	14. (D)	15. (B)
16. (B)	17. (D)	18. (C)	19. (B)	20. (D)
21. (A)	22. (B)	23. (A)	24. (D)	25. (A)
26. (D)	27. (A)	28. (D)	29. (B)	30. (A)
31. (A)	32. (D)	33. (B)	34. (D)	35. (D)
36. (C)				

14.3 Doppler's Effect

1. (D)	2. (C)	3. (D)	4. (C)	5. (C)
6. (D)	7. (B)	8. (B)	9. (A)	10. (C)
11. (D)	12. (D)	13. (A)	14. (B)	15. (C)
16. (A)	17. (B)	18. (C)	19. (B)	20. (A)

15.1 Electric Force and Electric Field

1. (D)	2. (B)	3. (B)	4. (D)	5. (C)
6. (A)	7. (D)	8. (A)	9. (C)	10. (A)
11. (D)	12. (C)	13. (A)	14. (C)	15. (D)
16. (A)	17. (B)	18. (A)	19. (D)	20. (C)
21. (B)	22. (A)	23. (D)	24. (B)	25. (A)
26. (D)	27. (A)	28. (D)	29. (C)	30. (C)
31. (C)	32. (D)	33. (C)	34. (D)	35. (C)

15.2 Electric Flux and Gauss Law

1. (B)	2. (A)	3. (B)	4. (B)	5. (C)
6. (D)	7. (A)	8. (A)	9. (C)	10. (B)

16.1 Electric Potential and Dipole

1. (A)	2. (C)	3. (D)	4. (B)	5. (C)
6. (C)	7. (D)	8. (D)	9. (B)	10. (D)
11. (B)	12. (A)	13. (D)	14. (C)	15. (D)
16. (C)	17. (C)	18. (C)	19. (A)	20. (D)
21. (C)	22. (D,C)	23. (A)	24. (B)	25. (C)
26. (B)	27. (B)	28. (B)	29. (C)	30. (D)
31. (B)	32. (A)	33. (B)	34. (B)	35. (C)
36. (A)	37. (A)	38. (C)	39. (D)	40. (C)
41. (D)	42. (B)	43. (A)	44. (A)	45. (D)

16.2 Capacitance

1. (A)	2. (B)	3. (D)	4. (B)	5. (A)
6. (D)	7. (A)	8. (D)	9. (D)	10. (C)
11. (C)	12. (D)	13. (D)	14. (B)	15. (C)
16. (C)	17. (B)	18. (A)	19. (D)	20. (D)
21. (C)	22. (C)	23. (B)	24. (A)	25. (C)
26. (D)	27. (A)	28. (B)	29. (D)	30. (C)
31. (A)	32. (C)	33. (D)	34. (D)	35. (C)
36. (None)	37. (A)	38. (C)	39. (D)	40. (D)
41. (A)	42. (B)	43. (C)	44. (D)	45. (D)
46. (D)	47. (A)	48. (C)	49. (D)	50. (A)
51. (A)	52. (B)			

17.1 Current Electricity

1. (D)	2. (C)	3. (B)	4. (A)	5. (B)
6. (D)	7. (D)	8. (C)	9. (A)	10. (B)
11. (A)	12. (B)	13. (D)	14. (C)	15. (D)
16. (C)	17. (A)	18. (C)	19. (C)	20. (D)
21. (D)	22. (B)	23. (D)	24. (C)	25. (B)
26. (D)	27. (D)	28. (C)	29. (B)	

17.2 Circuit Analysis

1. (B)	2. (B)	3. (B)	4. (B)	5. (D)
6. (C)	7. (C)	8. (D)	9. (B)	10. (C)
11. (C)	12. (C)	13. (C)	14. (B)	15. (A)
16. (B)	17. (A)	18. (B)	19. (C)	20. (D)
21. (D)	22. (A)	23. (B)	24. (C)	25. (C)

17.3 Thermal Effects of Current & RC Circuits

1. (B)	2. (B)	3. (B)	4. (A)	5. (C)
6. (C)	7. (D)	8. (C)	9. (A)	10. (B)
11. (A)	12. (B)	13. (B)	14. (D)	15. (C)
16. (D)	17. (D)	18. (A)	19. (D)	20. (A)
21. (C)	22. (C)	23. (B)	24. (A)	25. (D)
26. (B)	27. (C)	28. (C)	29. (B)	30. (A)
31. (None)	32. (A)	33. (C)	34. (A)	35. (A)
36. (A)	37. (B)	38. (C)	39. (C)	40. (D)
41. (C)	42. (A)	43. (B)		

17.4 Electrical Measuring Instruments

1. (C)	2. (D)	3. (D)	4. (A)	5. (D)
6. (B)	7. (D)	8. (B)	9. (D)	10. (C)
11. (C)	12. (D)	13. (A)	14. (C)	15. (D)
16. (D)	17. (D)	18. (C)	19. (A)	20. (A)
21. (D)	22. (A)	23. (D)	24. (C)	25. (D)
26. (D)	27. (B)	28. (C)	29. (B)	30. (B)
31. (D)	32. (A)	33. (D)	34. (B)	35. (B)
36. (C)	37. (D)	38. (A)	39. (B)	40. (D)
41. (A)	42. (C)	43. (C)	44. (B)	45. (A)
46. (A)	47. (D)	48. (D)	49. (C)	

18.1 Magnetic Effect of Current

1. (A)	2. (A)	3. (B)	4. (B)	5. (C)
6. (A)	7. (B)	8. (D)	9. (D)	10. (C)
11. (D)	12. (D)	13. (A)	14. (B)	15. (B)
16. (B)	17. (A)	18. (A)	19. (A)	20. (D)
21. (A)	22. (B)	23. (A)	24. (C)	25. (D)
26. (B)	27. (A)	28. (A)	29. (C)	30. (B)
31. (D)				

18.2 Magnetic Force on Moving Charges

1. (C)	2. (A)	3. (B)	4. (A)	5. (A)
6. (A)	7. (B)	8. (C)	9. (D)	10. (A)
11. (B)	12. (B)	13. (A)	14. (A)	15. (D)
16. (B)	17. (C)	18. (A)	19. (A)	20. (C)
21. (A)	22. (B)	23. (D)	24. (C)	25. (C)
26. (B)	27. (B)	28. (A)	29. (C)	30. (C)
31. (B)	32. (A)	33. (C)	34. (A)	35. (D)
36. (C)	37. (D)	38. (B)	39. (C)	40. (B)
41. (A)	42. (C)	43. (A)	44. (A)	45. (A)
46. (None)	47. (D)	48. (A)	49. (A)	50. (B)
51. (B)	52. (B)	53. (A)	54. (C)	55. (B)
56. (None)	57. (B)	58. (A)		

19.1 Classical and Terrestrial Magnetism

1. (D)	2. (B)	3. (B)	4. (C)	5. (C)
6. (B)	7. (D)	8. (D)	9. (A)	10. (C)
11. (C)	12. (B)	13. (C)	14. (C)	15. (B)
16. (D)	17. (B)	18. (C)		

19.2 Magnetic Properties of Materials

1. (A)	2. (B)	3. (B)	4. (B)	5. (C)
6. (B)	7. (B)	8. (D)	9. (B)	10. (B)
11. (C)	12. (C)	13. (D)		

20.1 Electromagnetic Induction

1. (D)	2. (B)	3. (B)	4. (B)	5. (D)
6. (C)	7. (B)	8. (B)	9. (A)	10. (C)
11. (A)	12. (D)	13. (B)	14. (A)	15. (A)
16. (D)	17. (B)	18. (B)	19. (A)	20. (C)
21. (B)	22. (A)	23. (A)	24. (A)	25. (C)
26. (B)	27. (D)	28. (B)	29. (B)	

20.2 Self and Mutual Induction

1. (D)	2. (D)	3. (B)	4. (B)	5. (B)
6. (A)	7. (D)	8. (D)	9. (B)	10. (C)
11. (B)	12. (D)	13. (C)	14. (None)	15. (C)
16. (A)	17. (D)	18. (D)	19. (C)	20. (D)
21. (A)	22. (B)	23. (D)	24. (B)	25. (A)
26. (B)	27. (B)	28. (B)	29. (B)	30. (B)

21.1 Alternating Current

1. (B)	2. (B)	3. (A)	4. (C)	5. (A)
6. (C)	7. (D)	8. (C)	9. (B)	10. (A)
11. (C)	12. (D)	13. (C)	14. (B)	15. (A)
16. (D)	17. (C)	18. (D)	19. (B)	20. (D)
21. (B)	22. (D)	23. (D)	24. (C)	25. (B)
26. (A)	27. (C)	28. (D)	29. (D)	30. (B)
31. (A)	32. (A)	33. (B)	34. (C)	35. (D)
36. (None)	37. (C)	38. (D)	39. (B)	

22.1 Electromagnetic Wave

1. (A)	2. (A)	3. (B)	4. (B)	5. (A)
6. (C)	7. (B)	8. (A)	9. (D)	10. (B)
11. (C)	12. (C)	13. (C)	14. (B)	15. (D)
16. (B)	17. (C)	18. (C)	19. (D)	20. (A)
21. (C)	22. (C)	23. (B)	24. (C)	25. (C)
26. (C)	27. (C)	28. (D)	29. (B)	30. (B)
31. (C)	32. (B)	33. (B)	34. (C)	35. (D)
36. (D)	37. (B)	38. (C)	39. (C)	

23.1 Geometrical Optics I - Reflection of Light

1. (A)	2. (B)	3. (D)	4. (C)	5. (C)
6. (D)	7. (A)	8. (D)	9. (A)	

23.2 Geometrical Optics II - Refraction of Light

1. (A)	2. (B)	3. (B)	4. (B)	5. (D)
6. (C)	7. (C)	8. (C)	9. (B)	10. (D)
11. (None)	12. (D)	13. (A)	14. (A)	15. (B)
16. (D)	17. (C)	18. (C)	19. (D)	20. (B)
21. (A)	22. (B)			

23.3 Geometrical Optics III - Thin Lenses

1. (A)	2. (A)	3. (B)	4. (C)	5. (D)
6. (A)	7. (C)	8. (D)	9. (B)	10. (D)
11. (A)	12. (D)	13. (A)	14. (A)	15. (A)

16. (B) 17. (A) 18. (B) 19. (C) 20. (D)
21. (A) 22. (C) 23. (B) 24. (B) 25. (B)
26. (D) 27. (D) 28. (D)

23.4 Dispersion of Light

1. (A) 2. (B) 3. (A) 4. (D) 5. (A)
6. (B)

23.5 Optical Instruments

1. (B) 2. (D) 3. (C) 4. (A) 5. (A)
6. (C) 7. (A) 8. (D) 9. (C) 10. (D)
11. (A) 12. (B) 13. (C) 14. (C) 15. (C)
16. (A) 17. (C) 18. (A)

24.1 Interference of Light

1. (D) 2. (C) 3. (A) 4. (D) 5. (A)
6. (A) 7. (B) 8. (C) 9. (D) 10. (B)
11. (B) 12. (A) 13. (C) 14. (B) 15. (C)
16. (B) 17. (D) 18. (A) 19. (D) 20. (A)
21. (D) 22. (D) 23. (D) 24. (C) 25. (D)
26. (B) 27. (B) 28. (B) 29. (D) 30. (C)
31. (D) 32. (A) 33. (None) 34. (B) 35. (D)

24.2 Diffraction of Light

1. (B) 2. (C) 3. (C) 4. (C) 5. (C)
6. (B) 7. (C) 8. (C) 9. (A) 10. (B)
11. (B) 12. (B) 13. (D)

24.3 Polarization of Light

1. (A) 2. (D) 3. (A) 4. (C) 5. (A)
6. (C) 7. (B) 8. (D) 9. (C) 10. (A)
11. (C) 12. (A) 13. (A) 14. (A) 15. (D)

25.1 Photoelectric Effect

1. (C) 2. (A) 3. (C) 4. (D) 5. (B)
6. (B) 7. (A) 8. (D) 9. (A) 10. (B)
11. (B) 12. (C) 13. (D) 14. (A) 15. (D)
16. (B) 17. (B) 18. (A) 19. (None) 20. (A)
21. (C) 22. (A) 23. (A) 24. (A) 25. (D)
26. (B) 27. (None) 28. (A) 29. (A) 30. (C)

25.2 Matter Waves and X-Ray

1. (A) 2. (B) 3. (C) 4. (D) 5. (B)
6. (B) 7. (D) 8. (D) 9. (B) 10. (C)
11. (D) 12. (C) 13. (C) 14. (B) 15. (C)
16. (A) 17. (D) 18. (C) 19. (C) 20. (C)
21. (D) 22. (C) 23. (A) 24. (B) 25. (D)
26. (D) 27. (C) 28. (D) 29. (A) 30. (B)

26.1 Atomic Structure

1. (A) 2. (C) 3. (C) 4. (A) 5. (C)
6. (B) 7. (B) 8. (A) 9. (A) 10. (A)
11. (B) 12. (B) 13. (C) 14. (A) 15. (C)
16. (C) 17. (D) 18. (C) 19. (C) 20. (A)
21. (C) 22. (C) 23. (B) 24. (A) 25. (A)
26. (B) 27. (B) 28. (B) 29. (C) 30. (D)
31. (A) 32. (D) 33. (B) 34. (D) 35. (C)
36. (D) 37. (C) 38. (B) 39. (C) 40. (B)
41. (C) 42. (A) 43. (D) 44. (C)

27.1 Radioactivity

1. (A) 2. (A) 3. (A) 4. (A) 5. (D)
6. (A) 7. (C) 8. (B) 9. (A) 10. (D)
11. (D) 12. (D) 13. (A) 14. (C) 15. (D)
16. (B) 17. (B) 18. (C) 19. (B) 20. (D)
21. (D) 22. (D) 23. (A) 24. (C) 25. (C)
26. (D) 27. (B) 28. (C) 29. (A)

27.2 Nuclear Physics

1. (D) 2. (A) 3. (B) 4. (C) 5. (C)
6. (C) 7. (B) 8. (D) 9. (C) 10. (C)
11. (B) 12. (B) 13. (B) 14. (B) 15. (B)
16. (D) 17. (C) 18. (C) 19. (C) 20. (C)
21. (C) 22. (D) 23. (C) 24. (C) 25. (C)
26. (A) 27. (C) 28. (C) 29. (A)

28.1 Semiconductor

1. (C) 2. (C) 3. (C) 4. (C) 5. (B)
6. (A) 7. (D) 8. (C) 9. (B) 10. (C)
11. (A) 12. (D) 13. (A) 14. (A) 15. (B)

28.2 PN Junction and Diodes

1. (C) 2. (C) 3. (B) 4. (D) 5. (A)
6. (B) 7. (A) 8. (A) 9. (A) 10. (C)
11. (D) 12. (D) 13. (D) 14. (A) 15. (C)
16. (B) 17. (B) 18. (C) 19. (B) 20. (C)
21. (D) 22. (D) 23. (B) 24. (C) 25. (B)
26. (B)

28.3 Transistors

1. (A) 2. (D) 3. (D) 4. (A) 5. (B)
6. (C) 7. (B) 8. (B) 9. (C) 10. (B)
11. (B) 12. (D) 13. (B) 14. (C) 15. (D)
16. (D) 17. (C)

28.4 Logic Gates

1. (D)	2. (D)	3. (A)	4. (B)	5. (A)
6. (B)	7. (A)	8. (C)	9. (A)	10. (A)
11. (A)	12. (A)	13. (None)	14. (A)	15. (C)
16. (C)	17. (D)	18. (B)	19. (A)	20. (A)

29.1 Communication Systems

1. (B)	2. (A)	3. (A)	4. (B)	5. (D)
6. (A)	7. (C)	8. (B)	9. (B)	10. (B)
11. (A)	12. (D)	13. (B)	14. (C)	15. (D)

29.2 Modulation and Demodulation

1. (A)	2. (B)	3. (B)	4. (B)	5. (C)
6. (A)	7. (C)	8. (B)	9. (C)	10. (C)
11. (D)	12. (A)	13. (C)	14. (C)	15. (B)
16. (D)	17. (B)	18. (C)	19. (D)	20. (A)
21. (D)	22. (B)	23. (C)	24. (D)	25. (A)

30.1 Errors and Experiments

1. (D)	2. (A)	3. (D)	4. (A)	5. (A)
6. (C)	7. (A)	8. (D)	9. (D)	10. (B)
11. (A)	12. (B)	13. (D)	14. (A)	15. (D)
16. (B)	17. (C)	18. (D)	19. (D)	20. (C)
21. (D)	22. (D)	23. (C)	24. (C)	25. (C)
26. (C)	27. (B)	28. (A)	29. (C)	30. (D)
31. (None)	32. (C)			

* * * * *

SOLUTION

1.1 Measurements and Dimensional Analysis

Sol. 1 (A) Torque $\tau = Fr \rightarrow [MLT^{-2}][L] = [ML^2T^{-2}]$
Work $W = Fs \rightarrow [MLT^{-2}][L] = [ML^2T^{-2}]$ } same

(B) Stress $= \frac{F}{A} \rightarrow \frac{[MLT^{-2}]}{[L^2]} = [ML^{-1}T^{-2}]$
Energy $\rightarrow$ [Work] $\rightarrow [ML^2T^{-2}]$ } different

(C) Force $\rightarrow [MLT^{-2}]$
Stress $\rightarrow [ML^{-1}T^{-2}]$ } different

(D) Force $\rightarrow [MLT^{-2}]$
Work $\rightarrow [ML^2T^{-2}]$ } different

Ans. (A)

Sol. 2 (A) Torque $\tau = Fr \rightarrow [MLT^{-2}][L] = [ML^2T^{-2}]$
Work $W = Fs \rightarrow [ML^2T^{-2}]$ } same

(B) Momentum $P = mv \rightarrow [M][LT^{-2}] = [MLT^{-1}]$
Plank's constant $h = \frac{E}{\nu} \rightarrow \frac{[ML^2T^{-2}]}{[T^{-1}]} = [ML^2T^{-1}]$ } different

(C) Stresss $= \frac{F}{A} \rightarrow \frac{[MLT^{-2}]}{[L^2]} = [ML^{-1}T^{-2}]$
Young's modulus $Y = \frac{\text{Stress}}{\text{strain}} = [\text{stress}] = ML^{-1}T^{-2}]$ } same

(D) speed (m/s) and $\frac{1}{\sqrt{\mu_0 \in_0}}$ = speed of light (m/s) } same

Ans. (B)

Sol. 3 Speed of light in free space is given as

$$c = \frac{1}{\sqrt{\mu_0 \in_0}}$$

$$\Rightarrow \quad \frac{1}{\mu_0 \in_0} = c^2$$

$$\Rightarrow \quad D\left(\frac{1}{\mu_0 \in_0}\right) = D(c^2) = [LT^{-1}]^2 = [L^2T^{-2}]$$

Ans. (C)

Sol. 4 From stokes rule we have viscous force on a sphere of radius r moving at a speed v in a medium of coefficient of viscosity η is given as

$$F = 6\pi\eta r v$$

$$\Rightarrow \quad \eta = \frac{F}{6\pi r v}$$

$$\Rightarrow \quad D(\eta) = \frac{[F]}{[r][v]} = \frac{[MLT^{-2}]}{[L][LT^{-1}]} = [ML^{-1}T^{-1}]$$

Ans. (A)

Sol. 5 (A) Impulse $J = Ft \rightarrow [MLT^{-2}][T] = [MLT^{-1}]$
Momentum $P = mv \rightarrow [M][LT^{-1}] = [MLT^{-1}]$ } same

(B) Angular momentum $L = mvr \rightarrow [M][LT^{-1}][L] = [ML^2T^{-1}]$
Planck's constant $h = \frac{E}{\nu} \rightarrow \frac{[ML^2T^{-2}]}{[T^{-1}]} = [ML^2T^{-1}]$ } same

(C) Work $W \rightarrow [ML^2T^{-2}]$
Torque $\tau = Fr \rightarrow [MLT^{-2}][L] = [ML^2T^{-2}]$ } same

(D) Moment of inertia $I = Mr^2 \rightarrow [ML^2]$
Moment of a force $\tau = Fr \rightarrow [ML^2T^{-2}]$ } Different

Ans. (D)

Sol. 6 Dimensions of $\frac{ML^2}{Q^2}$ is given as

$$D\left(\frac{ML^2}{Q^2}\right) = \frac{[ML^2]}{[AT]^2} = [ML^2A^{-2}T^{-2}]$$

(A) Wb/m^2 $\rightarrow$ unit of magnetic induction (B)

$$D(B) = [MT^{-2}A^{-1}]$$

(B) Henry $\rightarrow$ unit for self Induction (L)

We use energy in inductance $E = \frac{1}{2}LI^2 \Rightarrow L = \frac{2E}{I^2}$

$$\Rightarrow \quad D(L) = \frac{[ML^{-2}T^{-2}]}{[A^2]} = [ML^2T^{-2}A^{-2}]$$

(C) $\text{H/m}^2 \rightarrow D\left(\frac{\text{H}}{\text{m}^2}\right) = \frac{[ML^2T^{-2}A^{-2}]}{[L^2]} = [MT^2A^{-2}]$

(D) Wb $\rightarrow$ unit for magnetic flux ϕ

$$\Rightarrow \quad D(\phi) = D(\text{B.S}) = [MT^{-2}A^{-1}][L^2] = [ML^2T^{-2}A^{-1}]$$

Ans. (B)

Sol. 7 The force on moving charge in magnetic field is given as

$$F = qvB$$

$$\Rightarrow \quad B = \frac{F}{qv}$$

$$\Rightarrow \quad D[B] = \frac{[F]}{[q][v]} = \frac{[MLT^{-2}]}{[C][LT^{-1}]} = [MT^{-1}C^{-1}]$$ **Ans. (C)**

Sol. 8 If surface tension S is expressed in terms of K, V and T with dimensions a, b and c then we use

$$[S] = [K]^a [V]^b [T]^c$$

$$\Rightarrow \quad [MT^{-2}] = [ML^2T^{-2}]^a [LT^{-1}]^b [T]^c$$

Comparing dimensions on LHS & RHS we have

$$\Rightarrow \quad a = 1$$

$$2a + b = 0 \Rightarrow b = -2$$

$$-2a - b + c = -2 \Rightarrow c = 2a + b - 2 \Rightarrow c = -2$$

$$\Rightarrow \quad [S] = [K^1 V^{-2} T^{-2}]$$ **Ans. (A)**

Sol. 9 By Coulomb's law we use the force between two point charges is given as

$$F = \frac{1}{4\pi \in_0} \frac{q_1 q_2}{r^2}$$

$$\Rightarrow \quad \in_0 = \frac{1}{4\pi} \frac{q_1 q_2}{Fr^2}$$

$$\Rightarrow \quad D(\in_0) = \frac{[\text{charge}]^2}{[F][L^2]} = \frac{[IT]^2}{[MLT^{-2}][L^2]}$$

$$\Rightarrow \quad D(\in_0) = [M^{-1}L^{-3}T^4I^2] = [M^{-1}L^{-3}T^4A^2]$$ **Ans. (B)**

Sol. 10 Angular momentum is expressed as

$$L = mvr$$

$$\Rightarrow \quad D[L] = [M][LT^{-1}][L] = [ML^2T^{-1}]$$

Latent heat of a substance is expressed as

$$L_f = \frac{Q}{m}$$

$$\Rightarrow \quad D(L_f) = \frac{[ML^2T^{-2}]}{[M]} = [L^2T^{-2}]$$

Capacitance is expressed as

$$C = \frac{q}{V}$$

Potential difference V is expressed as

$$V = \frac{W}{q}$$

$$\Rightarrow \quad D(V) = \frac{[ML^2T^{-2}]}{[AT]}$$

$$\Rightarrow \quad D(V) = [ML^2T^{-3}A^{-1}]$$

$$\Rightarrow \quad D(C) = \frac{D(q)}{D(V)} = \frac{[AT]}{[ML^2T^{-3}A^{-1}]}$$

$$= [M^{-1}L^{-2}T^4]$$ **Ans. (D)**

Sol. 11 Time period $t = r^b s^{c/2} d^{a/4}$ ...(1)

and it is also given that $t \propto \sqrt{\frac{d}{s}} = d^{1/2}s^{-1/2}$ ...(2)

$$t = kd^{1/2}s^{-1/2} \quad \ldots(3)$$

Here k involves the terms of r

From (1) and (3) we have

$$\frac{c}{2} = -\frac{1}{2} \Rightarrow c = -1$$

and $$\frac{a}{4} = \frac{1}{2} \Rightarrow a = 2$$

Writing dimensions of LHS & RHS in equation (1), gives

$$[T] = [L]^b [MT^{-2}]^{c/2} [ML^{-3}]^{a/4}$$

in above equation on RHS dimensions of L and M must be zero

$$\Rightarrow \quad b - \frac{3a}{4} = 0 \Rightarrow b = \frac{3a}{4} = \frac{3 \times 2}{4} = \frac{3}{2}$$ **Ans. (C)**

Sol. 12 Dimensions of R, h, $\in$ are

$$[R] = [ML^2T^{-3}I^{-2}]$$

$$[\in] = [M^{-1}L^{-3}T^4I^2]$$

$$[\mu] = [MLT^{-2}I^{-2}]$$

$$\Rightarrow \quad D\left(\frac{\mu}{\in}\right) = \frac{[MLT^{-2}I^{-2}]}{[M^{-1}L^{-3}T^4I^2]} = [M^2L^4T^{-6}I^{-4}]$$

$$= [R^2]$$ **Ans. (C)**

Sol. 13 Only a dimensionless constants (without units) have same values in different system of unit.

$$[e] = [IT]$$

$$[\in_0] = [M^{-1}L^{-3}T^4I^2]$$

$$[h] = [ML^2T^{-1}]$$

$$[c] = [LT^{-1}]$$

$$[G] = [M^{-1}L^3T^{-2}]$$

We can also use $$c = \frac{1}{\sqrt{\mu_0 \in_0}}$$

(A) $$D\left(\frac{ch}{2\pi \in_0^2}\right) = \frac{[LT^{-1}][ML^2T^{-1}]}{[M^{-1}L^{-3}T^4I^2]}$$

$\rightarrow$ Not dimensionless

(B) $$D\left(\frac{e^2}{2\pi \in_0 Gm_e^2}\right) = \frac{[IT]^2}{[M^{-1}L^{-3}T^4I^2][M^{-1}L^3T^{-2}][M]^2}$$

$\rightarrow$ Dimensionless

So this value does not change with system of units.

(C) $$D\left(\frac{\mu_0 \in_0 G}{c^2 he^2}\right) = D\left(\frac{G}{c^4 he^2}\right)$$

$$= \frac{[M^{-1}L^3T^{-2}]}{[LT^{-1}]^2[ML^2T^{-1}][IT]^2}$$

→ Not dimensionless

(D) $$D\left(\frac{2\pi\sqrt{\mu_0 \in_0}}{ce^2}\frac{h}{G}\right) = D\left(\frac{h}{c^2e^2G}\right)$$

$$= \frac{[M^1L^2T^{-1}]}{[LT^{-1}]^2[IT]^2[M^{-1}L^{-3}T^{-2}]}$$

→ Not dimensionless **Ans. (B)**

Sol. 14 By dimensional analysis we use

$$D(u) = [e]^a[a_0]^b[h]^c[c]^d$$

Comparing dimensions on LHS & RHS we have

$\Rightarrow$ $c = -1$

$b + 2c + d = -2 \Rightarrow b = 1$

$a - c - d = 4 \Rightarrow d = -1$

$a = 2$

$\Rightarrow$ $u = e^2 a_0^1 h^{-1} c^{-1}$

$\Rightarrow$ $u = \dfrac{e^2 a_0}{hc}$ **Ans. (D)**

Sol. 15 Here we use

$$\mu_0 = ke^am^bc^ch^d$$

As speed of light $c = \dfrac{1}{\sqrt{\mu_0 \in_0}}$

$\Rightarrow$ $D(\mu_0) = \dfrac{1}{D(c)^2 . D(\in_0)}$

Then to find dimensions of e, m, c and h we compare the dimensions on LHS & RHS

$\Rightarrow$ $D(\mu_0) = [IT]^a[M]^b[LT^{-1}]^c[ML^2T^{-1}]^d$

$\Rightarrow$ $[MLT^{-2}I^{-2}] = [IT]^a[M]^b[LT^{-1}]^c[ML^2T^{-1}]^d$

$\Rightarrow$ $b + d = 1 \Rightarrow b = 0$

$c + 2d = 1 \Rightarrow d = 1$

$a - c - d = -2 \Rightarrow c + d = 0 \Rightarrow c = -1$

$a = -2$

$\Rightarrow$ $\mu_0 = ke^{-2}m^0c^{-1}h^1$

$\Rightarrow$ $\mu_0 = k\left(\dfrac{h}{ce^2}\right)$ **Ans. (C)**

Sol. 16 Resistance of a conductor is given as

$$R = \rho\frac{l}{A}$$

Conductivity $\sigma = \dfrac{1}{\rho} = \dfrac{l}{RA}$

by Ohm's law $V = IR \Rightarrow R = \dfrac{V}{I}$

$\Rightarrow$ $\sigma = \dfrac{lI}{VA}$ (for dimensions)

Electric potential $V = \dfrac{W}{q}$

$\Rightarrow$ $D(V) = \dfrac{[ML^2T^{-2}]}{[I][T]} = [ML^2T^{-3}I^{-1}]$

and $D(\sigma) = \dfrac{[L][I]}{[V][L^2]} = \dfrac{[L][I]}{[ML^2T^{-3}I^{-1}][L^2]}$

$= [M^{-1}L^{-3}I^2T^3]$ **Ans. (B)**

Sol. 17 From given equation $[BD] \to$ dimensionless

and $[AD] \equiv [C]$

(A) $D\left[\dfrac{C}{BD}\right] \equiv D[C]$ and $D\left[\dfrac{AD^2}{C}\right] = D[D]$

$\Rightarrow$ $\underbrace{\dfrac{C}{BD}}_{\downarrow\,[C]} - \underbrace{\dfrac{AD^2}{C}}_{\downarrow\,[D]}$

$\Rightarrow$ Two physical quantities having different dimensions are subtracted here so this is not a meaningful quantity.

(B) $D(A^2) \equiv D\left(\dfrac{C^2}{D^2}\right)$ and $D(B^2C^2) \equiv D\left(\dfrac{C^2}{D^2}\right)$

$\Rightarrow$ $\underbrace{A^2}_{\downarrow\,\left[\frac{C^2}{D^2}\right]} - \underbrace{B^2C^2}_{\downarrow\,\left[\frac{C^2}{D^2}\right]}$

$\Rightarrow$ This equation is dimensionally homogeneous, so it is meaningful

(C) $D\left(\dfrac{A}{B}\right) \equiv D(AD)$ and $D(C) \equiv D(AD)$

$\Rightarrow$ $\underbrace{\dfrac{A}{B}}_{\downarrow\,[AD]} - \underbrace{C}_{\downarrow\,[AD]}$

$\Rightarrow$ This equation is dimensionally homogeneous.

This is a meaningful quantity.

(D) $D(A) \neq D(C)$

$\Rightarrow$ A-C will not give a meaningful quantity. **Ans. (A) & (D)**

Sol. 18 Dimensions of mass in terms of T, C and h are given as

$$D[M] = D(T)^a . D(C)^b . D(h)^c$$

$$[M] = [T]^a[LT^{-1}]^b[ML^2T^{-1}]^c$$

Comparing LHS & RHS dimensions, we get

$$c = 1$$

$$a-b-c=0 \Rightarrow a=-1$$
$$b+2c=0 \Rightarrow b=-2c=-2$$
$$\Rightarrow \quad D[M]=[T^{-1}C^{-2}h^{1}]$$ **Ans. (A)**

Sol. 19 Let the plank length (l_p) in terms of G, h and c is given as

$$l_p = KG^a h^b c^c \quad ...(1)$$
$$\Rightarrow \quad D(l_p) = D(G^a)\cdot D(h^b)\cdot D(c^c)$$
$$\Rightarrow \quad [L] = [M^{-1}L^3T^{-2}]^a[ML^2T^{-1}]^b[LT^{-1}]^c$$

Comparing dimensions of L, M & T on LHS & RHS gives

$$1 = 3a+2b+c$$
$$0 = -a+b \Rightarrow a=b$$
$$0=-2a-b-c \Rightarrow c=-3a$$
$$\Rightarrow \quad 1=5a-3a \Rightarrow a=1/2$$
$$\Rightarrow \quad b=1/2$$
$$c=-3/2$$

From equation-(1) $l_p = KG^{1/2}h^{1/2}c^{-3/2} = K\sqrt{\frac{Gh}{c^3}}$

For all given options we take $K = 1$ **Ans. (B)**

Sol. 20 We use gravitational force as

$$F = \frac{GM^2}{R^2}$$
$$\Rightarrow \quad [G] = [M^{-1}L^3T^{-2}]$$

Also as $E = h\nu$

$$\Rightarrow \quad [h] = [ML^2T^{-1}]$$

and $[c] = [LT^{-1}]$

Now we use

$$t \propto G^x h^y C^z$$
$$[T] = [M^{-1}L^3T^{-2}]^x[ML^2T^{-1}]^y[LT^{-1}]^z$$
$$[M^0L^0T^1] = [M^{-x+y}L^{3x+2y+z}T^{-2x-y-z}]$$

Comparing the powers of M, L, T in LHS & RHS, we have

$$-x+y=0$$
$$\Rightarrow \quad x=y$$

and $3x+2y+z=0$

$$\Rightarrow \quad 5x+z=0 \quad ...(1)$$

and $-2x-y-z=1$

$$\Rightarrow \quad 3x+z=-1 \quad ...(2)$$

On solving (1) & (2) $x=y=\frac{1}{2}, z=-\frac{5}{2}$

$$\Rightarrow \quad t \propto \sqrt{\frac{Gh}{c^5}}$$ **Ans. (D)**

Sol. 21 Density of material is

$$\rho = \frac{128\text{kg}}{\text{m}^3} = \frac{128\times(50\times 20\text{g})}{(25\times 4\text{cm})}$$
$$\Rightarrow \quad \rho = \frac{128}{64}(20)\text{ units}$$
$$\Rightarrow \quad \rho = 40\text{ units}$$ **Ans. (D)**

Sol. 22 Force is given as

$$F = \alpha\beta e^{\left(\frac{-x^2}{\alpha KT}\right)}$$

Here the exponential power must be dimensionless, so we have

$$D\left(\frac{x^2}{\alpha KT}\right) = M^0L^0T^0$$
$$\Rightarrow \quad \frac{L^2}{[\alpha]ML^2T^{-2}} = M^0L^0T^0$$
$$\Rightarrow \quad [\alpha] = M^{-1}T^2$$

And we use

$$[F] = [\alpha]\,[\beta]$$
$$ML^1T^{-2} = M^{-1}T^2[\beta]$$
$$\Rightarrow \quad [\beta] = M^2L^1T^{-4}$$ **Ans. (B)**

Sol. 23 We use young modulus $= \frac{\text{stress}}{\text{strain}}$

$$\Rightarrow \quad \frac{F}{S} = Y\cdot\frac{\Delta l}{l}$$
$$\Rightarrow \quad [Y] = \frac{[F]}{[S]}$$

If dimensions of Y in unit of V, A and F are x, y, z then we use

$$[Y] = \frac{[F]}{[S]}$$
$$\Rightarrow \quad [V]^x[A]^y[F]^z = \frac{[MLT^{-2}]}{[L^2]}$$
$$[LT^{-1}]^x[LT^{-2}]^y[MLT^{-2}]^z = [ML^{-1}T^{-2}]$$

Comparing dimensions in LHS & RHS, we get

$$z=1$$
$$x+y+z=-1$$
$$\Rightarrow \quad x+y=-2$$
$$-x-2y-2z=-2$$
$$\Rightarrow \quad x+2y=0$$
$$\Rightarrow \quad y=2 \text{ and } x=-4$$
$$\Rightarrow \quad [Y] = V^{-4}A^2F^1$$ **Ans. (B)**

Sol. 24 $\left[\frac{l}{r}\right] = T$

and $[CV] = AT$

$$\Rightarrow \quad \left[\frac{l}{rCV}\right] = \frac{T}{AT} = A^{-1}$$ **Ans. (C)**

Sol. 25 $p = kS^a I^b h^c$

Where k is dimensionless constant

$$MLT^{-1} = (MT^{-2})^a (ML^2)^b (ML^2T^{-1})^c$$

$$a + b + c = 1$$

$$2b + 2c = 1$$

$$-2a - c = -1$$

Solving, we get

$$a = \frac{1}{2}$$

$$b = \frac{1}{2}$$

$$c = 0$$

$$[P] = S^{1/2} I^{1/2} h^0$$ **Ans. (B)**

Sol. 26 Given that $X = 5YZ^2$

$$\Rightarrow \quad Y = \frac{X}{5Z^2}$$

Dimensionally we use

$$[Y] = \frac{[X]}{[Z^2]}$$

$$\Rightarrow \quad [Y] = \frac{A^2 \cdot M^{-1}L^{-2} \cdot T^4}{(MA^{-1}T^{-2})^2}$$

$$\Rightarrow \quad [Y] = M^{-3} \cdot L^{-2} \cdot T^8 \cdot A^4$$ **Ans. (C)**

2.1 Motion in One Dimension

Sol. 1 Considering uniform retardation

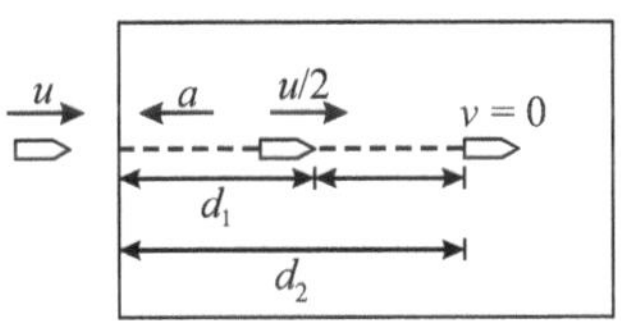

Wooden block

We use $v^2 = u^2 - 2ad$

Initially $\left(\frac{u}{2}\right)^2 = u^2 - 2ad_1$

$$\Rightarrow \quad \frac{3}{4}u^2 = 2a(3)$$

$$\Rightarrow \quad a = \frac{u^2}{8}$$

Finally $(0)^2 = u^2 - 2ad_2$

$$\Rightarrow \quad d_2 = 4 \text{ cm}$$

$\Rightarrow$ more distance is $d_2 - d_1 = 4 - 3 = 1$ cm **Ans. (A)**

Sol. 2

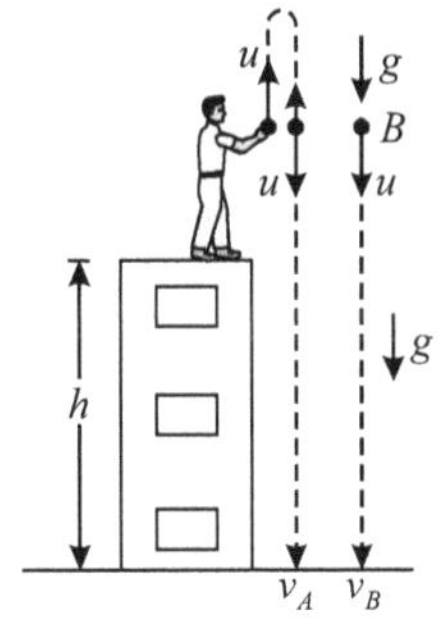

for both balls A & B, we use

$$v^2 = u^2 + 2gh$$

$$\Rightarrow \quad v_A = v_B$$ **Ans. (B)**

Sol. 3

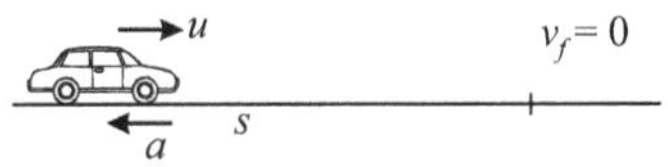

$$s_1 = \frac{u^2}{2a}$$

for car having initial speed $4u$

$$s_2 = \frac{(4u)^2}{2a} = \frac{8u^2}{a}$$

$$\frac{s_1}{s_2} = \frac{1}{16}$$ **Ans. (D)**

Sol. 4 We use $v^2 = u^2 - 2as$; a = constant

For stopping $v = 0$

$\Rightarrow$ stopping distance

$$s = \frac{u^2}{2a}$$

$$\Rightarrow \quad s \propto u^2$$

if u becomes doubled

$$\Rightarrow \quad s_f = 4s_i = 4 \times 6 = 24 \text{ m}$$ **Ans. (C)**

Sol. 5

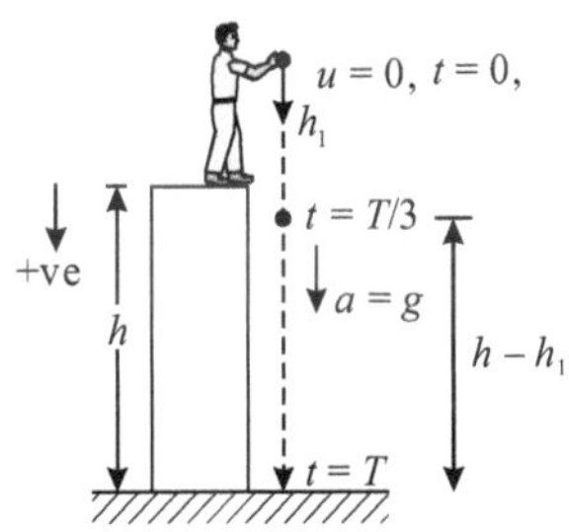

We use $h = ut + \frac{1}{2}gt^2 = \frac{1}{2}gt^2$

$$h = \frac{1}{2}gT^2$$

at $t = T/3, h = h_1$

$$\Rightarrow \quad h_1 = \frac{1}{2}g\left(\frac{T}{3}\right)^2 = \frac{1}{2}g\left(\frac{T}{9}\right)^2 = \frac{h}{9}$$

distance from ground $h - h_1 = h - \frac{h}{9} = \frac{8h}{9}$ **Ans. (C)**

Sol. 6 With reference to solution of question 4 came in 2003, we use

stopping distance $s \propto u^2$

As u is getting doubled

$$\Rightarrow \quad s_f = 4s_i$$

$$s_f = 4 \times 20$$

$$= 80\,\text{m}$$

Ans. (D)

Sol. 7 We are given $t = ax^2 + bx$

differentiate w.r. to time gives

$$1 = 2ax\left(\frac{dx}{dt}\right) + b\left(\frac{dx}{dt}\right) \qquad \ldots(1)$$

$$\Rightarrow \quad x = \frac{1 - b\left(\frac{dx}{dt}\right)}{2a\left(\frac{dx}{dt}\right)}$$

again differentiating equation-(1) w.r. to time gives

$$0 = 2a\left[\left(\frac{dx}{dt}\right)^2 + x\left(\frac{d^2x}{dt^2}\right)\right] + b\left[\frac{d^2x}{dt^2}\right]$$

Acceleration is given as

$$\Rightarrow \quad a_c = \frac{d^2x}{dt^2} = \frac{-2a\left(\frac{dx}{dt}\right)^2}{2ax + b}$$

[considering $v = dx/dt$]

$$a_c = \frac{-2av^2}{2a\left(\frac{1 - bv}{2av}\right) + b} = -2av^3$$

Ans. (C)

Sol. 8 We draw v-t curve for motion of car as shown in figure and for first part of journey, we use

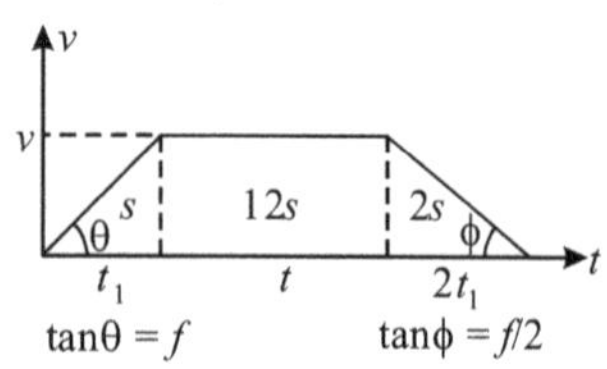

$$v = \sqrt{2fs}$$

$$\Rightarrow \quad 12s = vt$$

$$\Rightarrow \quad 12s = \sqrt{2fs}\,.\,t$$

$$\Rightarrow \quad 144s^2 = 2fs.\,t^2$$

$$\Rightarrow \quad s = \frac{1}{72}ft^2$$

Ans. (C)

Sol. 9 Average acceleration is given as

$$\bar{a} = \left|\frac{\vec{v}_f - \vec{v}_i}{\Delta t}\right| = \frac{|\Delta\vec{v}|}{\Delta t} = \frac{5\sqrt{2}}{10}\,\text{m/s}^2$$

$$\bar{a} = \frac{1}{\sqrt{2}}\,\text{m/s}^2$$

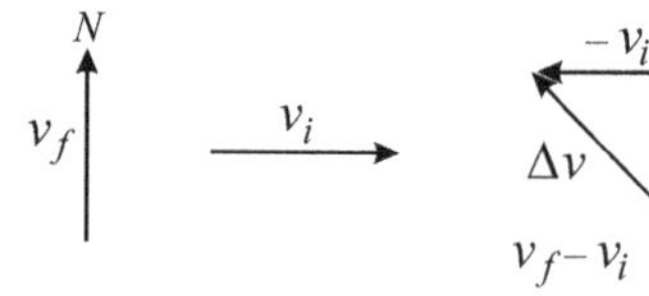

here $\quad |\overrightarrow{\Delta v}| = \sqrt{5^2 + 5^2} = 5\sqrt{2}$ m/s

along N-W direction.

Ans. (A)

Sol. 10 We use $\quad v^2 = u^2 - 2ah$

(for motion after parachute opens)

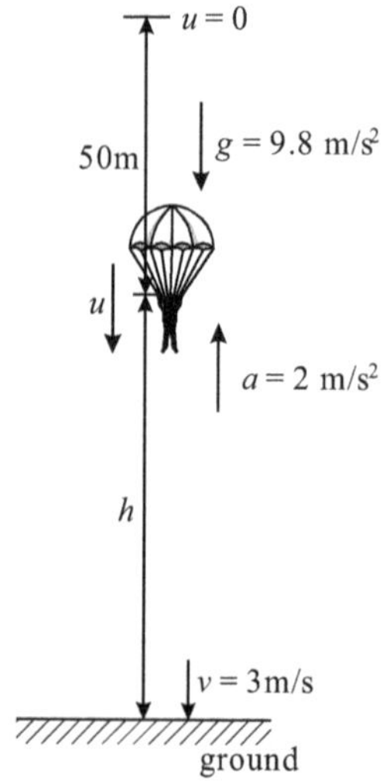

Here we use $\quad u = \sqrt{2gs} = \sqrt{2 \times 9.8 \times 50} = \sqrt{980}$ m/s

$$\Rightarrow \quad (3)^2 = 980 - 2 \times 2 \times h$$

$$\Rightarrow \quad h = \frac{971}{4} = 242.7\,\text{m}$$

Total height $\quad h_T = 242.7 + 50 = 292.7$ m

Ans. (C)

Sol. 11 We are given $v = \alpha\sqrt{x}$

$$\Rightarrow \quad \frac{dx}{dt} = \alpha\sqrt{x}$$

$$\Rightarrow \quad \frac{dx}{\sqrt{x}} = \alpha dt$$

Integrating above relation gives

$$\int_{x=0}^{x=x} \frac{dx}{\sqrt{x}} = \int_{t=0}^{t=t} \alpha dt$$

$$\Rightarrow \quad [2\sqrt{x}]_0^x = \alpha[t]_0^t$$

$$\Rightarrow \quad 2\sqrt{x} = \alpha t$$

$$\Rightarrow \quad x = \frac{1}{4}\alpha^2 t^2$$

$$\Rightarrow \quad x \propto t^2$$

Ans. (C)

Sol. 12 We are given $v = v_0 + gt + ft^2$

$$\Rightarrow \quad \frac{dx}{dt} = v_0 + gt + ft^2$$

$$\Rightarrow \quad dx = (v_0 + gt + ft^2)\, dt$$

integrating above relation gives

$$\int_{x=0}^{x=x} dx = \int_{t=0}^{t=1} (v_0 + gt + ft^2)dt$$

$$\Rightarrow \quad x = \left[v_0 t + \frac{1}{2}gt^2 + \frac{1}{3}ft^3\right]_0^1$$

$$\Rightarrow \quad x = \left(v_0 + \frac{1}{2}g + \frac{1}{3}f\right) - 0$$

Ans. (B)

Sol. 13 Deceleration of particle is

$$a = \frac{dv}{dt} = -2.5\sqrt{v}$$

$$\Rightarrow \quad \frac{dv}{\sqrt{v}} = -2.5\, dt$$

Integrating above relation gives

$$\int_{6.25}^{0} \frac{dv}{\sqrt{v}} = \int_{t=0}^{t=t} -2.5dt$$

$$\Rightarrow \quad \left[2\sqrt{v}\right]_{6.25}^{0} = -2.5[t]_0^t$$

$$\Rightarrow \quad -2 \times 2.5 = -2.5t \Rightarrow t = 2\text{s}$$

Ans. (B)

Sol. 14 As train is moving at constant acceleration, we use

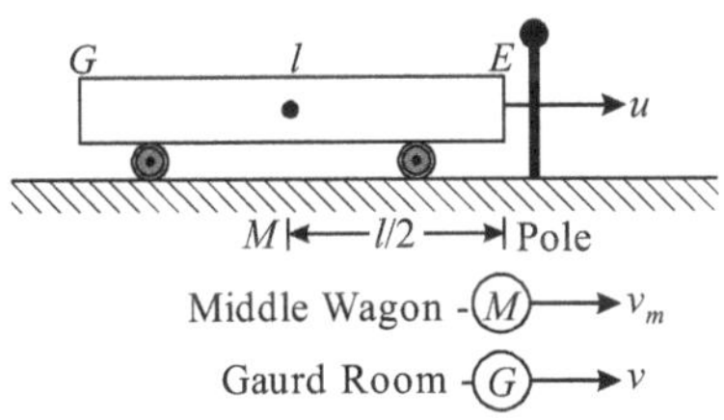

$$v^2 = u^2 + 2al \quad \ldots(1)$$

and
$$v_m^2 = u^2 + 2a(l/2) \quad \ldots(2)$$

from (1)
$$al = \frac{v^2 - u^2}{2}$$

from (2)
$$v_m^2 = u^2 + \frac{v^2 - u^2}{2}$$

$$\Rightarrow \quad v_m = \sqrt{\frac{u^2 + v^2}{2}}$$

Ans. (D)

Sol. 15 As graph is a straight line the instantaneous velocity is equal to the average velocity.

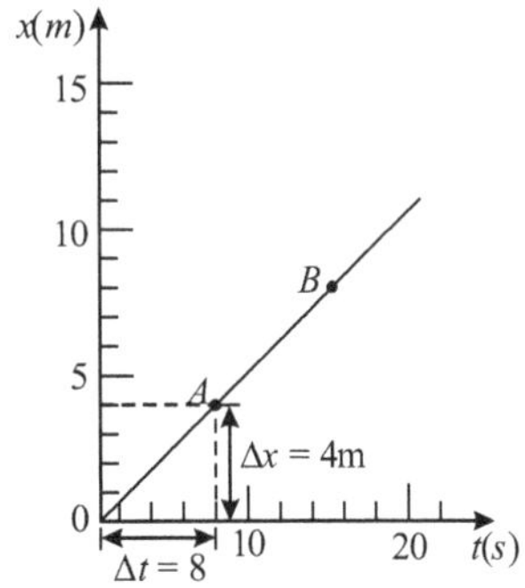

$$\Rightarrow \quad V_A = V_B = \frac{\Delta x}{\Delta t} = \frac{4}{8} = \frac{1}{2} = 0.5 \text{ m/s}$$

Ans. (A)

Sol. 16 We are given that

$$x = Ct^3$$

Velocity of particle is given as

$$v = \frac{dx}{dt} = 3Ct^2$$

$\Rightarrow$ Acceleration of particle

$$a = \frac{dv}{dt} = 6Ct \Rightarrow a \propto t$$

Ans. (B)

Sol. 17 Total time taken

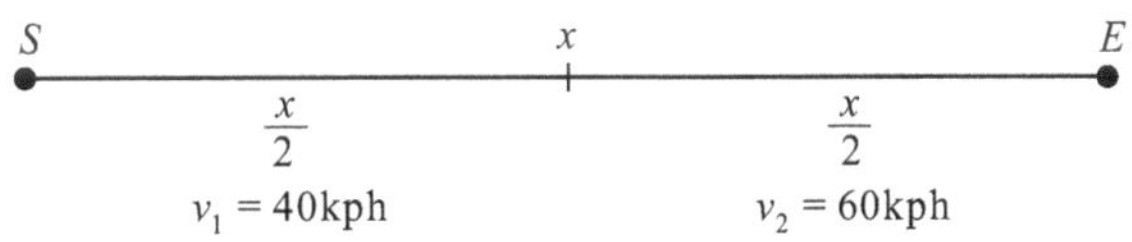

$$T = \frac{x/2}{40} + \frac{x/2}{60}$$

$$\text{Average speed} = \frac{\text{distance}}{\text{time}} = \frac{x}{\frac{x}{80} + \frac{x}{120}}$$

$$= \frac{80 \times 120}{80 + 120}$$

$$= 48 \text{ kph}$$

Ans. (C)

Sol. 18

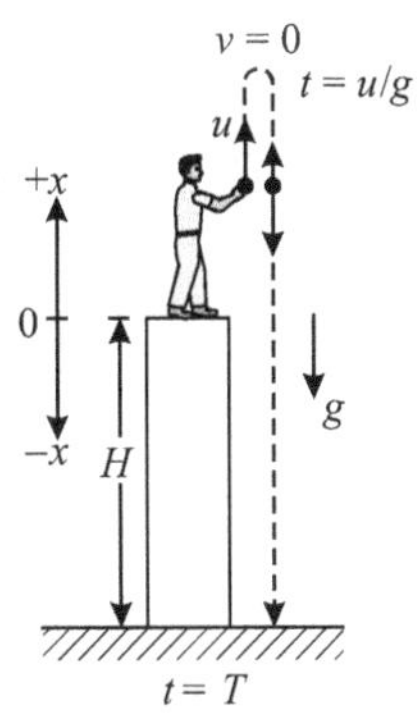

We use $\quad s = ut + \frac{1}{2}at^2$

$$\Rightarrow \quad -H = uT + \frac{1}{2}(-g)T^2$$

Given that $\quad T = nt = \dfrac{nu}{g}$

$$\Rightarrow \quad -H = u\left(\frac{nu}{g}\right) - \frac{1}{2}g\left(\frac{nu}{g}\right)^2$$

$$\Rightarrow \quad H = -\frac{nu^2}{g} + \frac{n^2u^2}{2g}$$

$$\Rightarrow \quad 2gH = nu^2(n-2)$$

Ans. (C)

Sol. 19 If length of escalator is l and velocity of man's walks is v_1 and velocity escalator's motion is v_2.

Then we have $\quad 60 = \dfrac{l}{v_1} \Rightarrow v_1 = \dfrac{l}{60}$

and $\quad 40 = \dfrac{l}{v_2} \Rightarrow v_2 = \dfrac{l}{40}$

When man is walking up on moving escalator then time taken is

$$t = \frac{l}{v_1 + v_2} = \frac{l}{l/60 + l/40} = 24 \text{ s}$$

Ans. (C)

Sol. 20

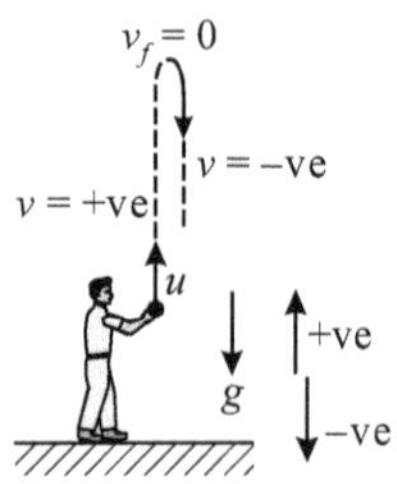

Acceleration $a = \dfrac{dv}{dt} = -g$ (constant)

Option (D) can be discarded as at one time instant velocity in two directions is NOT possible and options (B) can also be discarded as after topmost point direction (sign) of velocity is not changing in this case and option (C) can be discarded as velocity increases from beginning in this case. **Ans. (A)**

Sol. 21

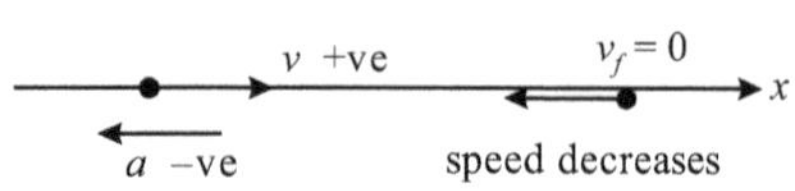

Acceleration of object is taken as

$$a = -v\frac{dv}{dx}$$

slope of v-x curve $\dfrac{dv}{dx} = -\dfrac{a}{v}$

with time initially v decreases so

$$\left|\frac{dv}{dx}\right| = \frac{a}{v} \text{ increases}$$

This is only shown in option (C). **Ans. (C)**

Sol. 22 (C)

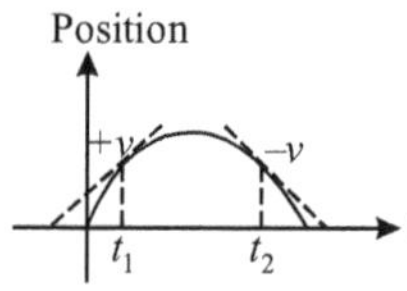

(A)

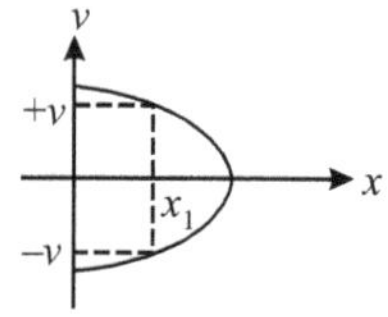

(D)

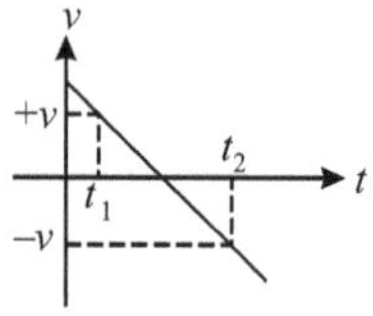

(B)

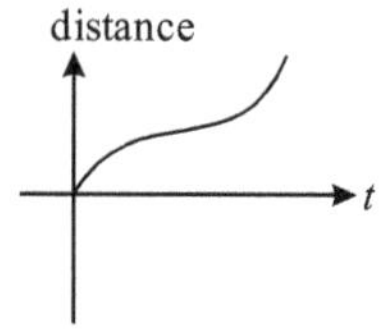

Actual distance-time curve for the given situation is shown above which speed is first increasing then decreasing as in the option (B) curve. **Ans. (B)**

Sol. 23 In first case $u_1 = 40$ kph

Stopping distance $\quad s_1 = 40$ m

$\Rightarrow$ We use $\quad s_1 = \dfrac{u_1^2}{2a}$

$$\Rightarrow \quad a = \frac{u_1^2}{2s_1} = \frac{1600}{2\times 40} = 20 \text{ units}$$

In second case $\quad u_2 = 80$ km/h

Stopping distance $\quad s_2 = \dfrac{u_2^2}{2a} = \dfrac{6400}{2\times 20} = 160$ m **Ans. (B)**

Sol. 24 Distance travelled by car in 15s is

$$S_{car} = \frac{1}{2}\left(\frac{45}{15}\right)(15)^2 = \frac{1}{2}\times 45\times 15 = \frac{675}{2}\text{ m}$$

Distance travelled by scooter in 15s is

$$S_{scooter} = vt = 30\times 15 = 450 \text{ m}$$

difference $S_{scooter} - S_{car} = 450 - \dfrac{675}{2} = \mathbf{112.5\ m}$

To verify: If car catches the scooter at time t then we use

$$S_{car\ t} = S_{scooter\ t} \qquad [t > 15\text{s}]$$

$$\frac{675}{2} + 45(t-15) = 30\,t$$

$$\Rightarrow \quad 15t = \frac{675}{2} \Rightarrow t = \mathbf{22.5\ s}$$

Ans. (C)

Sol. 25 The time taken by car to reach point P is

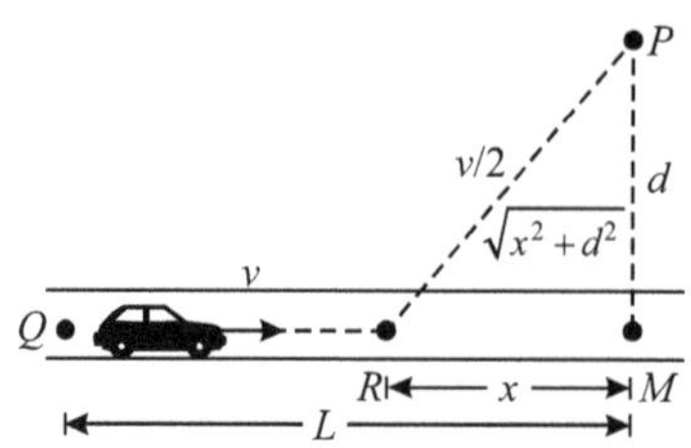

$$t = \frac{L-x}{v} + \frac{\sqrt{x^2+d^2}}{v/2}$$

t is minimum when $\frac{dt}{dx}=0$

$$\Rightarrow \quad \frac{dt}{dx}=\frac{1}{v}\left[-1+2\frac{1}{2\sqrt{x^2+d^2}}\cdot 2x\right]=0$$

$$\Rightarrow \quad x=\frac{d}{\sqrt{4-1}}$$

$$\Rightarrow \quad x=\frac{d}{\sqrt{3}}$$

Ans. (A)

Sol. 26 For A & B let time taken by A is t_0

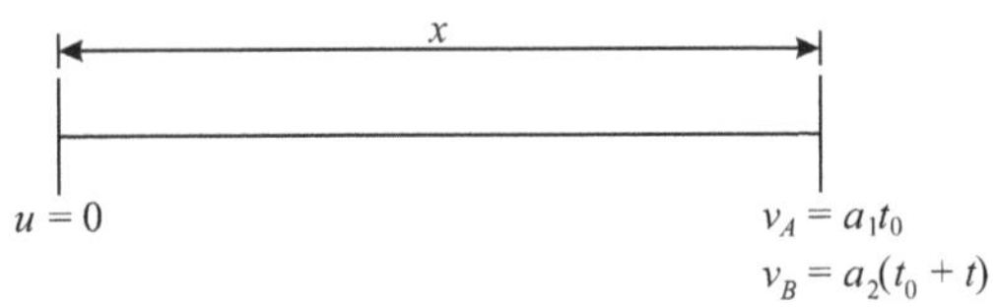

Given that

$$v_A - v_B = v = (a_1 - a_2)t_0 - a_2 t \quad \ldots(1)$$

$$\Rightarrow \quad x_B = x_A = \frac{1}{2}a_1t_0^2 = \frac{1}{2}a_2(t_0+t)^2$$

$$\Rightarrow \quad \sqrt{a_1}\,t_0 = \sqrt{a_2}\,(t_0+t)$$

$$\Rightarrow \quad (\sqrt{a_2}-\sqrt{a_2})t_0 = \sqrt{a_2}\,t \quad \ldots(2)$$

Putting t_0 in equation, we have

$$\Rightarrow \quad v=(a_1-a_2)\left(\frac{\sqrt{a_2}\,t}{\sqrt{a_1}-\sqrt{a_2}}\right)-a_2t$$

$$\Rightarrow \quad v=(\sqrt{a_1}+\sqrt{a_2})\sqrt{a_2}\,t-a_2t$$

$$\Rightarrow \quad v=\sqrt{a_1a_2}\,t+a_2t-a_2t$$

$$\Rightarrow \quad v=\sqrt{a_1a_2}\,t$$

Ans. (D)

Sol. 27 Area under graph upto $t = 5$ s is

$$s=\frac{1}{2}\times 2\times 2+2\times 2+3\times 1=9\text{ m}$$ **Ans. (B)**

Sol. 28 Displacement of particle is

$$\vec{s}=\vec{u}t+\frac{1}{2}at^2$$

$$\Rightarrow \quad \vec{s}=(5\hat{i}+4\hat{j})2+\frac{1}{2}(4\hat{i}+4\hat{j})4$$

$$\Rightarrow \quad \vec{s}=10\hat{i}+8\hat{j}+8\hat{i}+8\hat{j}$$

$$\Rightarrow \quad \vec{r}_f-\vec{r}_i=18\hat{i}+16\hat{j}$$

$$\Rightarrow \quad \vec{r}_f=20\hat{i}+20\hat{j} \qquad (\text{As } \vec{r}_i=2\hat{i}+4\hat{j})$$

$$\Rightarrow \quad \left|\vec{r}_f\right|=20\sqrt{2}$$

Ans. (A)

Sol. 29 When trains travel in same direction, time taken is

$$t_1=\frac{180\times 18}{(80-30)\times 5}=\frac{180\times 18}{50\times 5}\text{ s} \quad \ldots(1)$$

When trains travel in opposite direction, time taken is

$$t_2=\frac{180\times 18}{(80+30)\times 5}=\frac{180\times 18}{110\times 5}\text{ s} \quad \ldots(2)$$

From (1) and (2), we have

$$\frac{t_1}{t_2}=\frac{11}{5}$$

Ans. (D)

Sol. 30 Given initial velocity $u = 0$ and acceleration is constant

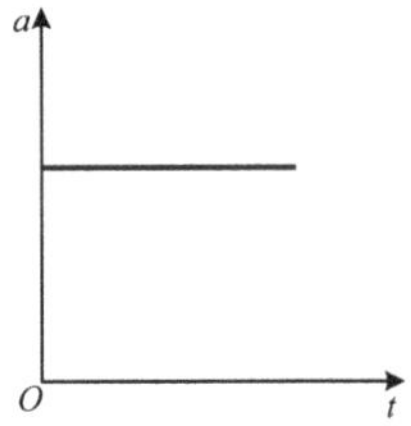

At time t, velocity is given as

$$v=0+at$$

$$\Rightarrow \quad v=at$$

and displacement is given as

$$x=0(t)+\frac{1}{2}at^2$$

$$\Rightarrow \quad x=\frac{1}{2}at^2$$

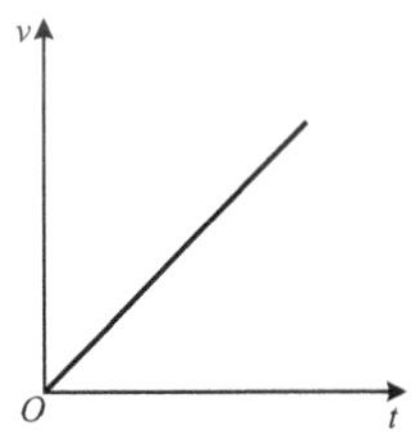

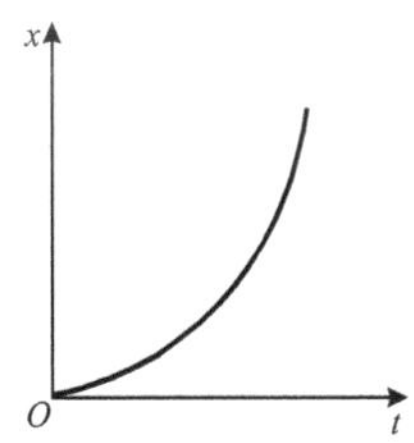

Graph (1) ; (2) and (4) are correct. **Ans. (C)**

Sol. 31 Given that

$$|\vec{A}_1|=3$$

$$|\vec{A}_2|=5$$

$$|\vec{A}_1+\vec{A}_2|=5$$

$$\Rightarrow \quad |\vec{A}_1+\vec{A}_2|=\sqrt{|\vec{A}_1|^2+|\vec{A}_2|^2+2|\vec{A}_1||\vec{A}_2|\cos\theta}$$

$$\Rightarrow \quad 5=\sqrt{9+25+2\times 3\times 5\cos\theta}$$

$$\Rightarrow \quad \cos\theta=-\frac{9}{2\times 3\times 5}=-\frac{3}{10}$$

$$\Rightarrow \quad (2\vec{A}_1+3\vec{A}_2)\cdot(3\vec{A}_1-2\vec{A}_2)$$

$$=6|\vec{A}_1|^2+9\vec{A}_1\cdot\vec{A}_2-4\vec{A}_1\vec{A}_2-6|\vec{A}_2|^2$$

$$=54+5\times 3\times 5\left(-\frac{3}{10}\right)-6\times 25$$

$$=54-150-\frac{45}{2}=-118.5$$

Ans. (C)

Sol. 32 Momentum

$$p = mv \quad \ldots(1)$$

For motion under gravity, we use

$$h = \frac{u^2 - v^2}{2g} \quad \ldots(2)$$

$$h = \frac{u^2 - p^2/m^2}{2g}$$

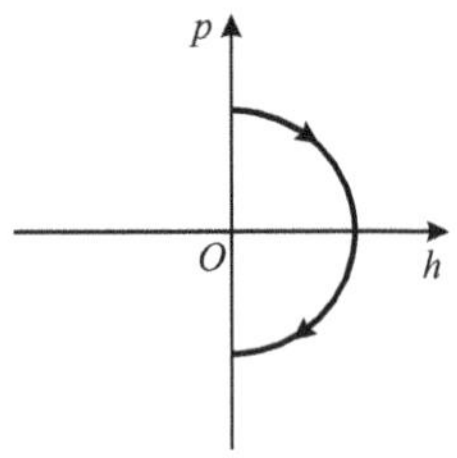

Ans. (A)

Sol. 33 Given that

$$x = at + bt^2 - ct^3$$

$$\Rightarrow \quad v = \frac{dx}{dt} = a + 2bt - 3ct^2$$

$$\Rightarrow \quad a = \frac{dv}{dt} = 2b - 6ct = 0$$

$$\Rightarrow \text{At} \quad t = \frac{b}{3c}, a = 0$$

$$\Rightarrow \quad v_{\left(at\ t=\frac{b}{3c}\right)} = a + 2b\left(\frac{b}{3c}\right) - 3c\left(\frac{b}{3c}\right)$$

$$= a + \frac{b^2}{3c}$$

Ans. (D)

Sol. 34 Given that

$$v = b\sqrt{x}$$

$$\Rightarrow \quad \frac{dv}{dt} = \frac{b}{2\sqrt{x}}\frac{dx}{dt}$$

$$\Rightarrow \quad a = \frac{bv}{2\sqrt{x}}$$

$$\Rightarrow \quad a = \frac{b(b\sqrt{x})}{2\sqrt{x}} = \frac{b^2}{2}$$

For constant acceleration velocity after time $t = \tau$ is given as

$$v = 0 + at$$

$$\Rightarrow \quad v = \frac{b^2}{2}\tau$$

Ans. (B)

3.1 Motion in Two Dimensions

Sol. 1

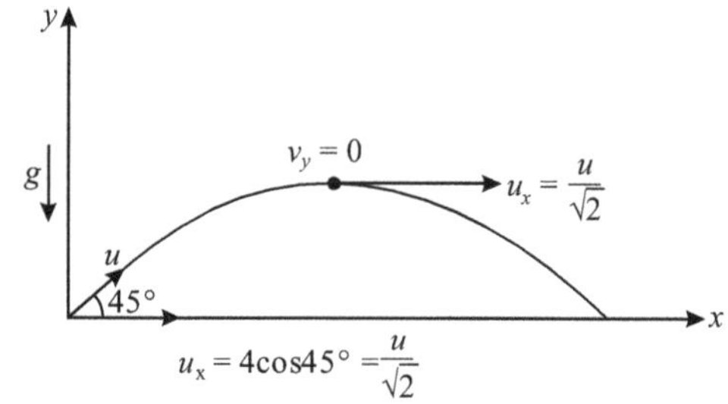

Initial KE of ball $E = \frac{1}{2}mu^2$

KE of ball at highest point of flight

$$E_H = \frac{1}{2}m\left(\frac{u}{\sqrt{2}}\right)^2 = \frac{1}{4}mu^2 = E/2$$ **Ans. (C)**

Sol. 2

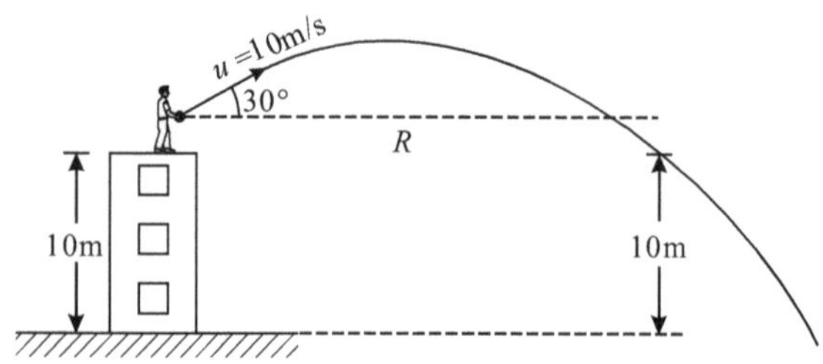

Horizontal Range of ball

$$R = \frac{u^2 \sin 2\theta}{g}$$

$$\Rightarrow \quad R = \frac{(10)^2 \sin(60°)}{10} = 10 \times \frac{\sqrt{3}}{2} = 5\sqrt{3}\text{ m}$$

$$\Rightarrow \quad R = 8.66\text{ m}$$

Ans. (D)

Sol. 3 As $\quad x = \alpha t^3; \quad y = \beta t^3$

$$v_x = \frac{dx}{dt} = 3\alpha t^2; \quad v_y = \frac{dy}{dt} = 3\beta t^2$$

velocity of particle $\quad v = \sqrt{v_x^2 + v_y^2}$

$$v = \sqrt{(3\alpha t^2)^2 + (3\beta t^2)^2}$$

$$v = 3t^2\sqrt{\alpha^2 + \beta^2}$$

Ans. (B)

Sol. 4

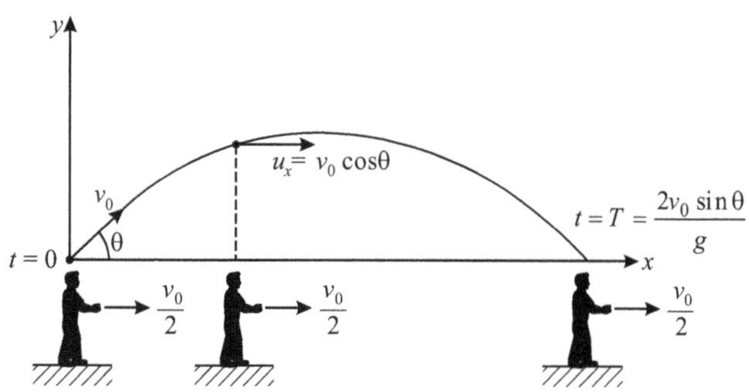

$$R = \frac{v_0}{2} T$$

$$\Rightarrow \quad \frac{v_0^2 \sin 2\theta}{g} = \frac{v_0}{2} \times \frac{2v_0 \sin\theta}{g}$$

$$2\sin\theta \cos\theta = \sin\theta \Rightarrow \cos\theta = 1/2$$

Person will be able to catch the ball if

$$v_0 \cos\theta = \frac{v_0}{2}$$

$$\Rightarrow \quad \cos\theta = \frac{1}{2}$$

$$\Rightarrow \quad \theta = 60°$$

Ans. (A)

Sol. 5 $\vec{A} \times \vec{B} - \vec{B} \times \vec{A} = 0$

$$\Rightarrow \quad \vec{A} \times \vec{B} + \vec{A} \times \vec{B} = 0$$

$$\Rightarrow \quad 2(\vec{A} \times \vec{B}) = 0$$

$$\Rightarrow \quad \vec{A} \times \vec{B} = 0$$

$$\Rightarrow \quad AB \sin\theta = 0$$

$\Rightarrow$ Either $A = 0$ or $B = 0$ or $\theta = 0, \pi, 2\pi$ **Ans. (C)**

Sol. 6 Same range of projectile angles are complementary θ, $90 - \theta$ so time of flight in the two cases are

$$T_1 = \frac{2u \sin\theta}{g}$$

and

$$T_2 = \frac{2u \cos\theta}{g}$$

here

$$T_1 T_2 = \frac{4u^2 \sin\theta \cos\theta}{g^2} = \frac{2u^2 \sin 2\theta}{g^2} = \frac{2R}{g}$$

$$\Rightarrow \quad T_1 T_2 \propto R$$

Ans. (A)

Sol. 7

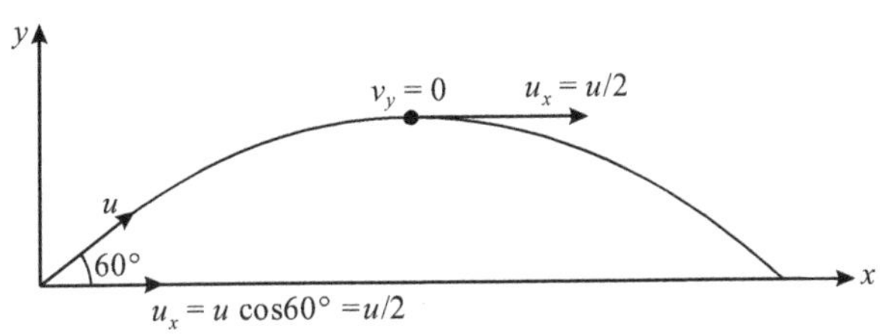

Initial kE $\quad K = \frac{1}{2} m u^2$

KE of particle at highest point

$$K_H = \frac{1}{2} m u_x^2 = \frac{1}{2} m \left(\frac{4}{2}\right)^2 = \frac{1}{8} m u^2$$

$$\Rightarrow \quad K_H = \frac{K}{4}$$

Ans. (C)

Sol. 8 We can use vector speed equation

$$\vec{v} = \vec{u} + \vec{a}\, t$$

After 10 s, find velocity is

$$\vec{v} = (3\hat{i} + 4\hat{j}) + (0.4\hat{i} + 0.3\hat{j}) \times 10$$

$$\vec{v} = 7\hat{i} + 7\hat{j}$$

$$|\vec{v}| = \sqrt{7^2 + 7^2}$$

$$= 7\sqrt{2} \text{ units}$$

Ans. (A)

Sol. 9 We are given velocity components of particle as

$$v_x = \frac{dx}{dt} = Ky \qquad \ldots (1)$$

and

$$v_y = \frac{dy}{dt} = Kx \qquad \ldots (2)$$

By (1)/(2) $\Rightarrow \quad \frac{dx}{dy} = \frac{y}{x}$

$$x\, dx = y\, dy$$

integrating above relation gives

$$\int x\, dx = \int y\, dx$$

$$\Rightarrow \quad \frac{x^2}{2} = \frac{y^2}{2} + C$$

$$\Rightarrow \quad x^2 = y^2 + C$$

$$\Rightarrow \quad y^2 = x^2 + C_1$$

Ans. (D)

Sol. 10

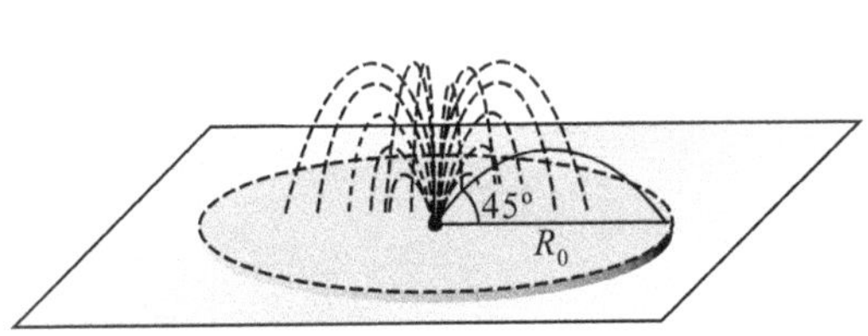

Maximum horizontal range

$$R = \frac{u^2 \sin 2\theta}{g}$$

at $\theta = 45° \rightarrow \quad R = R_0 \text{ (max.)}$

$$\Rightarrow \quad R_0 = \frac{u^2}{g}$$

Area gets wet $= \pi R_0^2 = x\left(\frac{u^2}{g}\right)^2 = \frac{xu^4}{g^2}$ **Ans. (A)**

Sol. 11

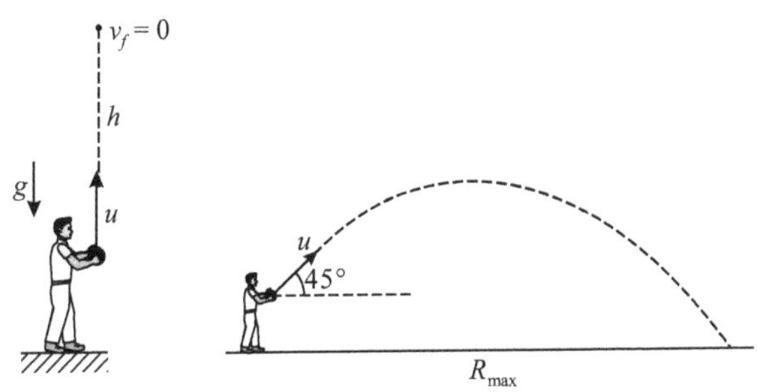

When stories thrown up, maximum height is given as

$$h = \frac{u^2}{2g} = 10\,\text{m}$$

$$\Rightarrow \quad u^2 = 200$$

For projectile maximum range is given for same projection speed, is

$$R_{max} = \frac{u^2}{g} = \frac{200}{10} = 20\,\text{m}$$

Ans. (D)

Sol. 12

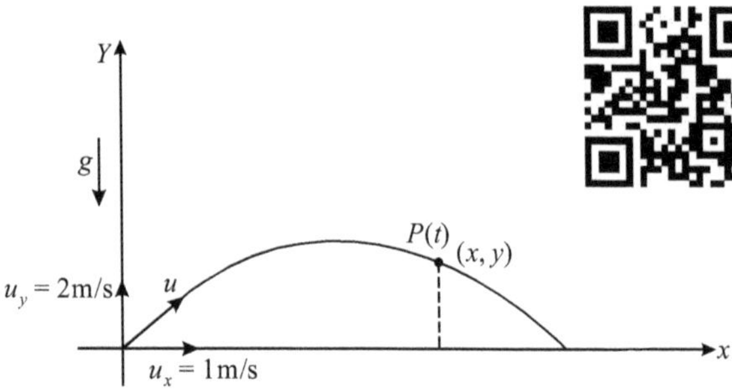

Position coordinates of particle at a time t are

$$x = u_x t = t \qquad \ldots(1)$$

and

$$y = u_y t - \frac{1}{2} gt^2 - 5t^2 \qquad \ldots(2)$$

eliminating t we get

$$\Rightarrow \quad y = 2x - 5x^2$$

Ans. (A)

Sol. 13 When car is at rest

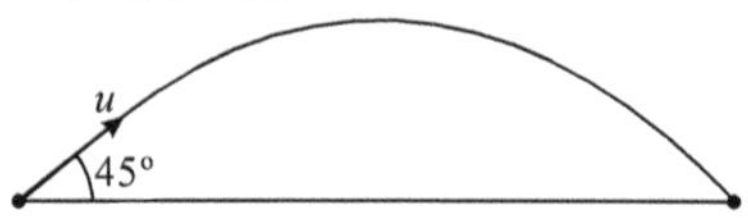

We use (max) $R_{max} = \dfrac{u^2}{g} = 40$

$$\Rightarrow \quad u = \sqrt{400} = 20\,\text{m/s}$$

When car is moving

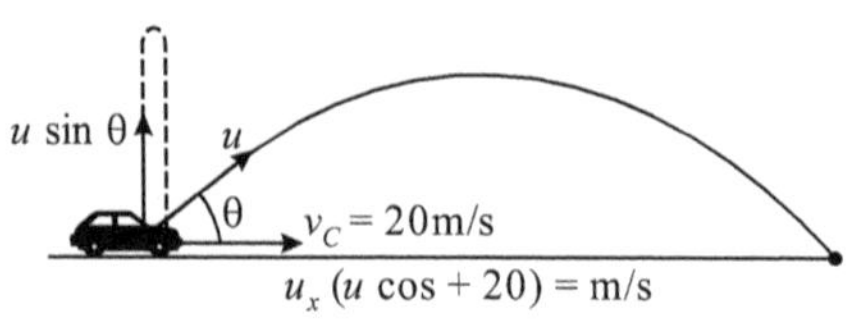

time of flight $t = \dfrac{2u\sin\theta}{g}$

Range of projectile $R_x = u_x t$

$$\Rightarrow \quad R_x = (u\cos\theta + 20) \times \frac{2u\sin\theta}{g}$$

$$= 80(\cos\theta\sin\theta + \sin\theta)$$

$$\Rightarrow \quad R_x = 80\left[\frac{1}{2}\sin 2\theta + \sin\theta\right]$$

R_x is max when $\dfrac{dR_x}{d\theta} = 0$

$$\frac{dR_x}{d\theta} = 80[\cos 2\theta + \cos\theta]$$

$$\Rightarrow \quad \cos 2\theta = -\cos\theta$$

$$\Rightarrow \quad \cos 2\theta = \cos(180 - \theta) \Rightarrow 2\theta = 180 - \theta$$

$$\Rightarrow \quad \theta = 60°$$

Ans. (B)

Sol. 14

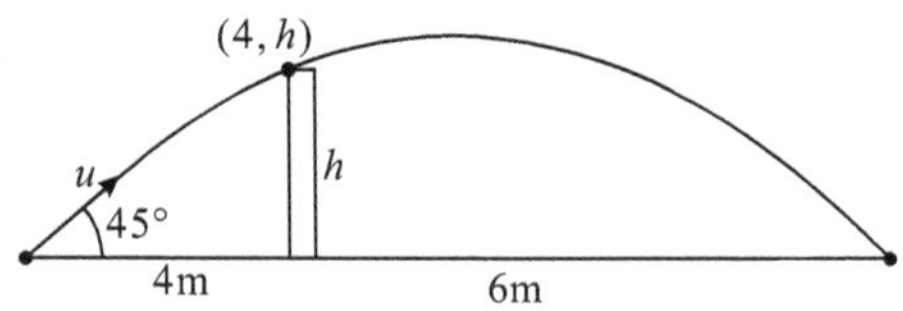

Maximum Range of projectile is

$$\frac{u^2}{g} = 10$$

$$\Rightarrow \quad u = \sqrt{100} = 10\,\text{m/s}$$

Equation of trajectory of projectile

$$y = x\tan\theta - \frac{gx^2}{2u^2\cos^2\theta}$$

$$\Rightarrow \quad y = x - \frac{10x^2}{2(10)^2\left(\frac{1}{\sqrt{2}}\right)^2}$$

$$\Rightarrow \quad y = x - \frac{x^2}{10}$$

at $x = 4$ m, $y = h$

$$\Rightarrow \quad h = 4 - \frac{16}{10} = 2.4\,\text{m}$$

Ans. (B)

Sol. 15

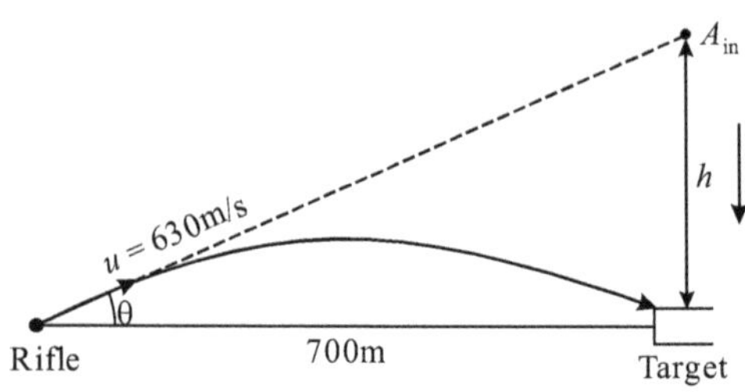

As θ small the time taken by bullet to hit the target is

$$t = \frac{700}{4\cos\theta} = \frac{700}{630} = \frac{10}{9}\,\text{s}$$

($\cos\theta \sim 1$ for small θ)

Distance fallen by bullet in time t is

$$h = \frac{1}{2}gt^2$$

$$= \frac{1}{2} \times 10 \times \left(\frac{10}{9}\right)^2 = 6.1\,\text{m}$$

Ans. (C)

Alternative :

Considering projectile motion we use

$$\frac{u^2 \sin 2\theta}{g} = 700\,\text{m}$$

$$\frac{u^2(2\sin\theta\cos\theta)}{g} = 700$$

for small θ, we take $\cos\theta \sim 1$ and $\sin\theta \sim \tan\theta$

$$\Rightarrow \quad 2\sin\theta = \frac{700\times 10}{(630)^2}$$

here $\quad \tan\theta = \dfrac{h}{700}$

$$\Rightarrow \quad h = 700\tan\theta\ \text{m}$$

$$\Rightarrow \quad h \simeq 700\sin\theta$$

$$\Rightarrow \quad h = 700\times\frac{700\times 10}{2\times(630)^2} = 6.1\,\text{m}$$ **Ans. (C)**

Sol. 16

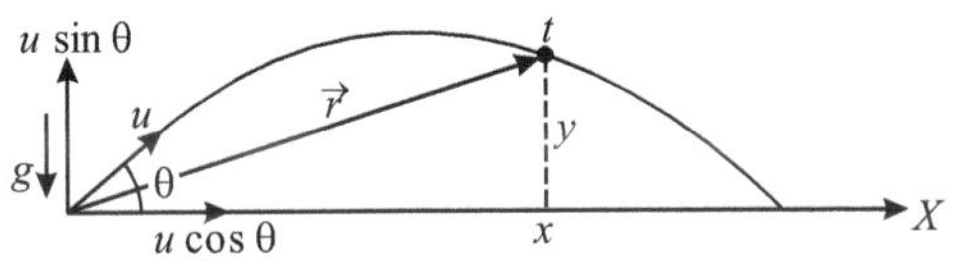

For a given projectile position vector at time t is given as

$$\vec{r} = (u\cos\theta.t)\,\hat{i} + \left(u\sin\theta\cdot t - \frac{1}{2}gt^2\right)\hat{i}$$

$$\vec{r} = 40\hat{i} + 50\hat{j} \text{ (given at } 2s)$$

By Comparing with given values

$$u\cos\theta(2) = 40$$

$$\Rightarrow \quad u\cos\theta = 20 \qquad \ldots(1)$$

$$u\sin\theta\,(2) - 20 = 50 \qquad \ldots(2)$$

By (2)/(1) $\Rightarrow \quad \tan\theta = \dfrac{35}{20} = \dfrac{7}{4}$ **Ans. (C)**

Sol. 17 $|\vec{A}| = |\vec{B}| = A$

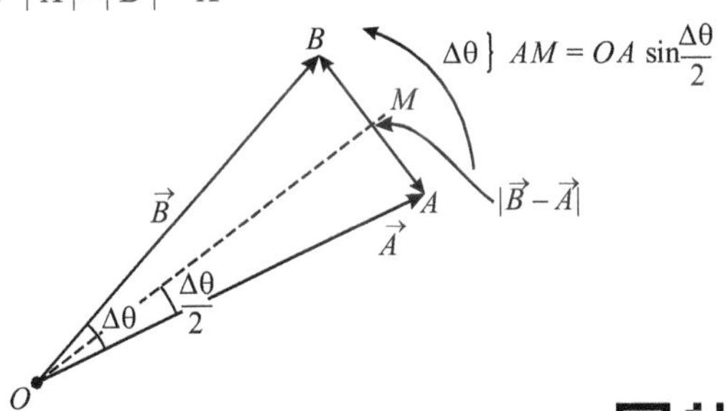

$$|\vec{A}| = |\vec{B}| = A$$

The magnitude of $\vec{B}-\vec{A}$ is given as

$$|\vec{B}-\vec{A}| = AB = 2A\sin\left(\frac{\Delta\theta}{2}\right)$$

$$\Rightarrow \quad |\vec{B}-\vec{A}| = 2A\times\frac{\Delta\theta}{2}$$ [for θ << 1, we use sinθ ~ θ]

$$\Rightarrow \quad |\vec{B}-\vec{A}| = A\,\Delta\theta$$ **Ans. (A)**

Alternatively : From below figure it can be directly seen the magnitude of $|\vec{B}-\vec{A}| = A\,\Delta\theta$.

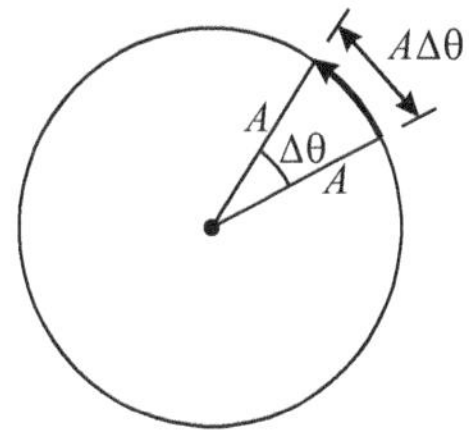

Sol. 18 If we consider

$$\vec{C} = a\hat{i} + b\hat{j}$$

$$\vec{A}.\vec{C} = a + b$$

$$\vec{B}.\vec{C} = 2a - b$$

$$\vec{A}.\vec{B} = 2 - 1 = 1$$

$$a + b = 1 \qquad \ldots(1)$$

$$\Rightarrow \quad 2a - b = 1 \qquad \ldots(2)$$

Solving-(1) and (2) gives

$$a = \frac{1}{3};\ b = \frac{2}{3}$$

$$\Rightarrow \quad |\vec{C}| = \sqrt{a^2 + b^2}$$

$$= \sqrt{\left(\frac{1}{3}\right)^2\left(\frac{2}{3}\right)^2} = \sqrt{\frac{5}{9}}$$ **Ans. (A)**

Sol. 19 From the given velocity function, we use

$$\frac{dx}{dt} = ky \quad \text{and} \quad \frac{dy}{dt} = kx$$

$$\Rightarrow \quad \frac{dy}{dx} = \frac{\frac{dy}{dt}}{\frac{dx}{dt}} = \frac{x}{y}$$

$$\Rightarrow \quad ydy = xdx$$

Integrating both side, we get

$$y^2 = x^2 + C$$ **Ans. (B)**

Sol. 20 Velocity components of particle are

$$v_x = -a\omega\sin\omega t$$

and $\quad v_y = a\omega\cos\omega t$

and $\quad v_z = a\omega$

$$\Rightarrow \quad v = \sqrt{v_x^2 + v_y^2 + v_z^2}$$

$$\Rightarrow \quad v = \sqrt{2}a\omega$$ **Ans. (C)**

Sol. 21 Range of projectile is

$$R = \frac{u^2\sin 2\theta}{g}$$

Area covered on ground is

$$A = \pi R^2$$

$$\Rightarrow \quad A \propto R^2$$

$$\Rightarrow \quad A \propto u^4$$

$$\Rightarrow \quad \frac{A_1}{A_2} = \frac{u_1^4}{u_2^4} = \left[\frac{1}{2}\right]^4 = \frac{1}{16}$$ **Ans. (D)**

Sol. 22 We have position vectors of points G and H are

$$\vec{r}_g = \frac{a}{2}\hat{i} + \frac{a}{2}\hat{k}$$

$$\vec{r}_H = \frac{a}{2}\hat{j} + \frac{a}{2}\hat{k}$$

$$\vec{r}_H - \vec{r}_g = \frac{a}{2}(\hat{j} - \hat{i})$$ **Ans. (B)**

Sol. 23 Resultant of $2F$ and $3F$ is given as

$$4F^2 + 9F^2 + 12F^2 \cos\theta = R^2 \quad \ldots(1)$$

When second force becomes $6F$ then resultant becomes $2R$, so we use

$$4F^2 + 36F^2 + 24F^2 \cos\theta = 4R^2 \quad \ldots(2)$$

From equation-(1) and (2), we have

$$4F^2 + 36F^2 + 24F^2\cos\theta = 4(13F^2 + 12F^2\cos\theta) = 52F^2 + 48F^2\cos\theta$$

$$\Rightarrow \quad \cos\theta = -\frac{12F^2}{24F^2} = -\frac{1}{2}$$

$$\Rightarrow \quad \theta = 120°$$ **Ans. (D)**

Sol. 24 If θ is angle between $\vec{A}$ and $\vec{B}$, and $|\vec{A}| = |\vec{B}| = a$ then we have

$$|\vec{A} + \vec{B}| = 2a\cos\theta/2 \quad \ldots(1)$$

$$|\vec{A} - \vec{B}| = 2a\cos\frac{(\pi - \theta)}{2} = 2a\sin\theta/2 \quad \ldots(2)$$

$$\Rightarrow \quad n\left(2a\cos\frac{\theta}{2}\right) = 2a\frac{\sin\theta}{2}$$

$$\Rightarrow \quad \tan\frac{\theta}{2} = n$$

$$\Rightarrow \quad \theta = 2\tan^{-1} n = \cos^{-1}\left(\frac{n^2 - 1}{n^2 + 1}\right)$$ **Ans. (C)**

Sol. 25 Projection velocity components are

$$u_x = 10\cos 60° = 5 \text{ m/s}$$

$$u_y = 10\cos 30° = 5\sqrt{3} \text{ m/s}$$

velocity after $t = 1$s is given as

$$v_x = 5 \text{ m/s}$$

$$v_y = (5\sqrt{3} - 10) \text{ m/s} (-\text{ve} \Rightarrow \text{downward motion})$$

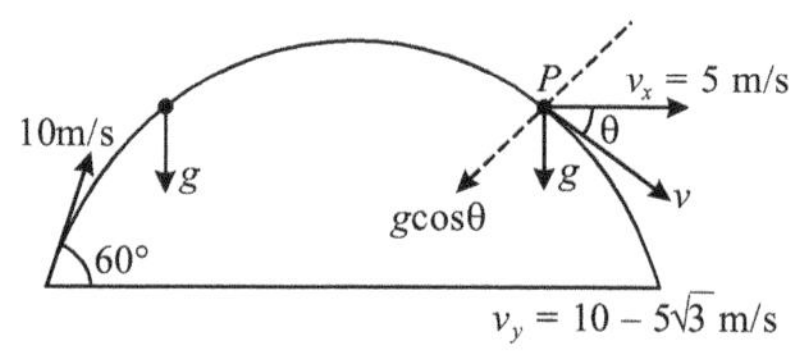

At point P, normal acceleration of body is

$$a_n = \frac{v^2}{R}$$

$$\Rightarrow \quad R = \frac{v_x^2 + v_y^2}{a_n} = \frac{25 + 100 + 75 - 100\sqrt{3}}{10\cos\theta}$$

Where $\tan\theta = \left|\frac{v_y}{v_x}\right| = \frac{10 - 5\sqrt{3}}{5} = 2 - \sqrt{3}$

$$\Rightarrow \quad \theta = 15°$$

$$\Rightarrow \quad R = \frac{100(2 - \sqrt{3})}{10\cos 15°} = 2.8 \text{ m}$$ **Ans. (C)**

Sol. 26 Given $\vec{r} = 15t^2\hat{i} + (4 - 20t^2)\hat{j}$

$$\Rightarrow \quad \vec{v} = \frac{d\vec{r}}{dt} = 30t\hat{i} + (-40t)\hat{j}$$

$$\Rightarrow \quad \vec{a} = \frac{d\vec{v}}{dt} = 30\hat{i} - 40\hat{j}$$

$$\Rightarrow \quad |\vec{a}| = 50 \text{ m/s}^2$$ **Ans. (D)**

Sol. 27 Time of flight is

$$t = \frac{2 \times 2 \times \sin 15°}{g\cos 30°}$$

Range on inclined plane is

$$R = 2\cos 15° \times t - \frac{1}{2} g\sin 30° t^2$$

Solving, we get

$$R \simeq 20 \text{ cm}$$ **Ans. (B)**

Sol. 28 Equation of trajectory is given as

$$y = 2x - 9x^2 \quad \ldots(1)$$

Comparing with general equation of projectile

$$y = x\tan\theta - \frac{g}{2u^2\cos^2\theta}\cdot x^2 \quad \ldots(2)$$

We get $\tan\theta = 2$

$$\Rightarrow \quad \cos\theta = \frac{1}{\sqrt{5}}$$

Also, $\frac{g}{2u^2\cos^2\theta} = 9$

$$\Rightarrow \quad \frac{10}{2 \times 9 \times \left(\frac{1}{\sqrt{5}}\right)^2} = u^2$$

$$\Rightarrow \quad u^2 = \frac{25}{9}$$

$$\Rightarrow \quad u = \frac{5}{3}\ \text{m/s}$$ **Ans. (A)**

Sol. 29 Range will be same for time t_1 & t_2, so angles of projection will be 'θ' & '90° – θ'

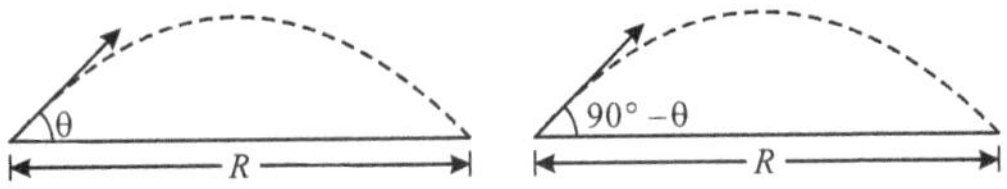

$$t_1 = \frac{2u\sin\theta}{g},\ t_2 = \frac{2u\sin(90^\circ-\theta)}{g}$$

and $$R = \frac{u^2\sin 2\theta}{g}$$

$$t_1 t_2 = \frac{4u^2\sin\theta\cos\theta}{g^2} = \frac{2}{g}\left[\frac{2u^2\sin\theta\cos\theta}{g}\right]$$

$$= \frac{2R}{g}$$ **Ans. (C)**

Sol. 30

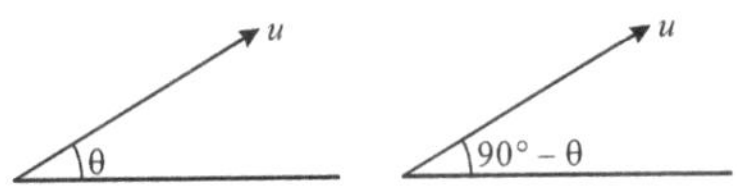

For same range angle of projection will be θ & 90 – θ, so we use

$$R = \frac{u^2\, 2\sin\theta\cos\theta}{g}$$

and time of flights are

$$h_1 = \frac{u^2\sin^2\theta}{g}$$

$$h_2 = \frac{u^2\sin^2(90-\theta)}{g}$$

$$\Rightarrow \quad \frac{R^2}{h_1 h_2} = 16$$ **Ans. (B)**

3.2 Relative Motion

Sol. 1

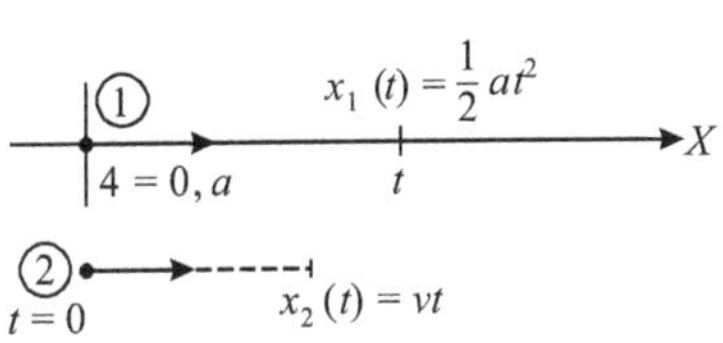

$t = 0 \qquad x_2(t) = vt$

$$\Rightarrow \quad x_1 - x_2 = \frac{1}{2}at^2 - vt$$

at $t = 0+$ we have $x_1 - x_2 \rightarrow$ –ve

Initially $x_1 - x_2$ is negative in option-(B) only. **Ans. (B)**

Sol. 2 Vertical acceleration of A is

$$a_{Ay} = g\sin 60^\circ = \frac{\sqrt{3}g}{2}$$

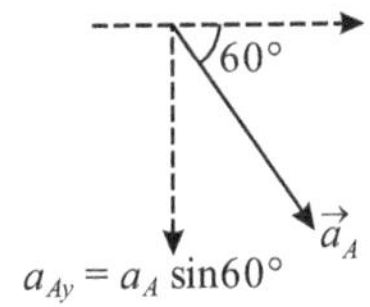

$$\Rightarrow \quad a_{Ay} = \frac{\sqrt{3}g}{2}\left(\frac{\sqrt{3}}{2}\right) = \frac{3g}{4}$$

Vertical acceleration of B is

$$a_{By} = g\sin 30^\circ = \frac{g}{2}$$

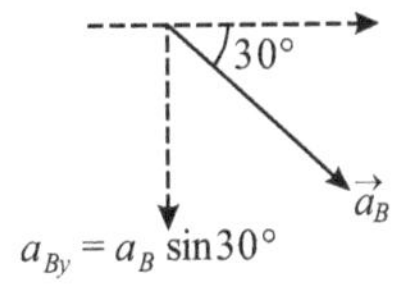

$$\Rightarrow \quad a_{By} = \frac{g}{2}\left(\frac{1}{2}\right) = \frac{g}{4}$$

Acceleration of B w.r.t. A is vertical direction is

$$a_{yAB} = a_{Ay} - a_{By} = \frac{3g}{u} - \frac{g}{u} = \frac{g}{2}$$

$$= 4.9\ \text{m/s}^2$$ **Ans. (D)**

Sol. 3 For $u = 10$ m/s, $h_1 = \frac{u_1^2}{2g} = \frac{100}{2\times 10} = 5$ m

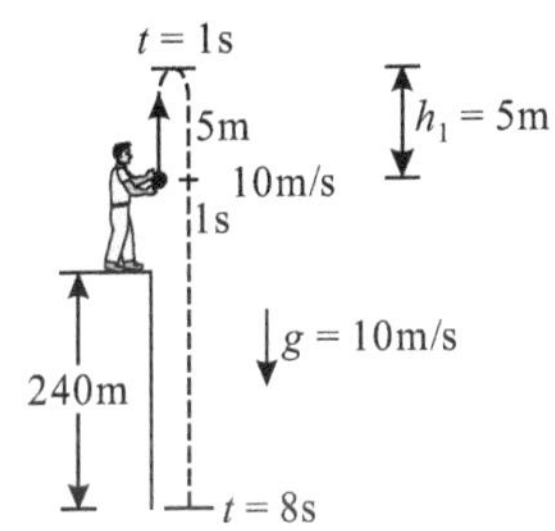

Time of downward journey

$$t_2 = \sqrt{\frac{2h}{g}}$$

$$t_2 = \sqrt{\frac{2\times 245}{10}} = 7\ \text{s}$$

For $u = 40$ m/s, $h_2 = \frac{u^2}{2g} = \frac{(40)^2}{2\times 10} = 80$ m

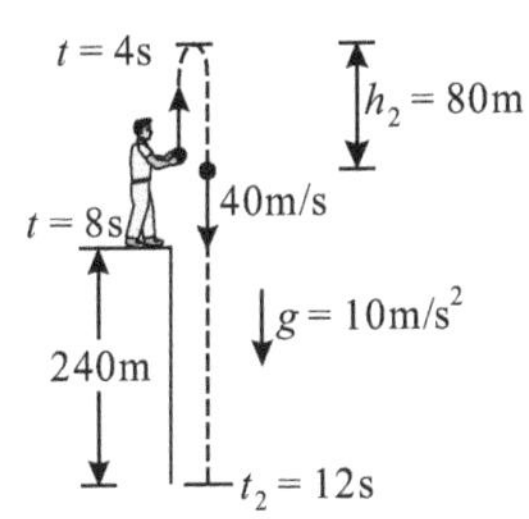

Time for downward journey of stone

$$t_2 = \sqrt{\frac{2h}{g}}$$

$$t_2 = \sqrt{\frac{2\times 320}{10}} = 8\text{s}$$

If y_1 and y_2 are height above ground for the two stones, we have

$$y_1 = 240 + (10t - 5t^2)$$
$$y_1 = 240 + 10t - 5t^2$$
$$y_2 = 240 + (40t - 5t^2)$$
$$y_2 = 240 + 40t - 5t^2$$

$\Rightarrow$ from $t = 0$ to $t = 8$ s when first stone reaches ground

$$y_2 - y_1 = 30t$$

at $t = 8$s $\quad y_2 - y_1 = 240\,\text{m}$

after $t = 8$s, first stone is at rest $\Rightarrow y_1 = 0$

$\Rightarrow \quad \Delta y = y_2 - y_1 = 240 + 40t - 5t^2$

slope of graph

$$\frac{d(\Delta y)}{dt} = 40 - 10t$$

As for $t > 8$ s slope of graph is increasing **Ans. (C)**

Sol. 4

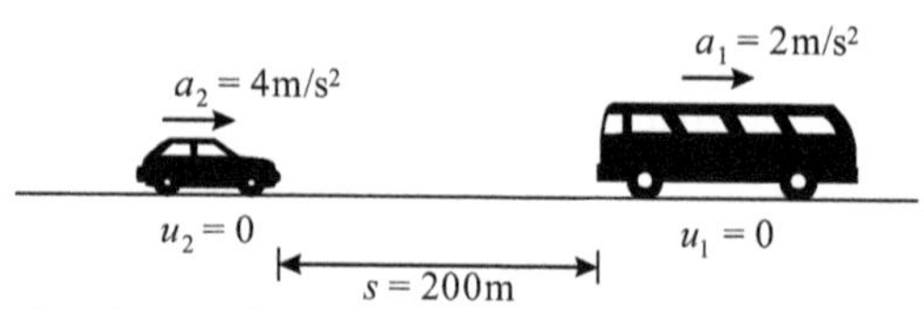

Considering bus to be at rest and relative to bus, speed and acceleration of car are

$$u_{CB} = 0 = u_2 - 4,$$
$$a_{CB} = a_2 - a_1 = 4 - 2 = 2\text{m/s}^2$$

time to bus by car (rel. to bus) a distance of 200 m is

$$t = \sqrt{\frac{25}{a}} = \sqrt{\frac{2\times 200}{2}}$$

$$= 10\sqrt{2}\ \text{s}$$ **Ans. (C)**

Sol. 5 If we take the position of ship 'A' as origin then positions and velocities of both ships can be given as :

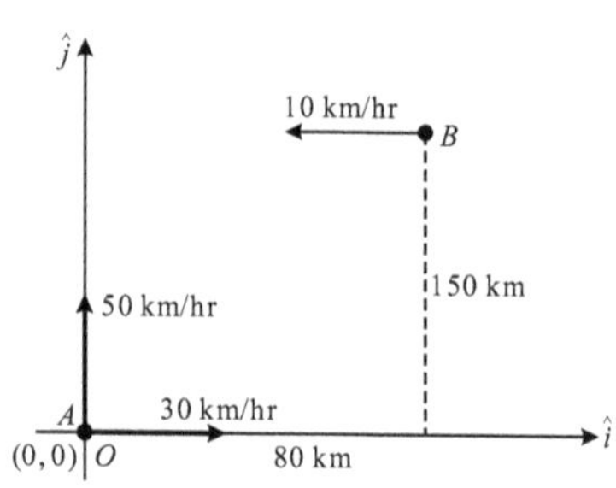

$$\vec{v}_A = (30\hat{i} + 50\hat{j})\,\text{km/hr}$$
$$\vec{v}_B = -10\hat{i}\ \text{km/hr}$$
$$\vec{r}_A = 0\hat{i} + 0\hat{j}$$
$$\vec{r}_B = (80\hat{i} + 150\hat{j})\,\text{km}$$

Time after which distance between them will be minimum

$$t = -\frac{\vec{r}_{BA}\cdot\vec{v}_{BA}}{|\vec{v}_{BA}|^2};$$

where $\vec{r}_{BA} = (80\hat{i} + 150\hat{j})$ km

$$\vec{v}_{BA} = -10\hat{i} - (30\hat{i} + 50\hat{j})$$
$$(-40\hat{i} - 50\hat{j})\ \text{km/hr}$$

$$\therefore \quad t = -\frac{(80\hat{i} + 150\hat{j})\cdot(-40\hat{i} - 50\hat{j})}{|-40\hat{i} - 50\hat{j}|^2}$$

$$= \frac{3200 + 7500}{4100}\,\text{hr} = \frac{10700}{4100}\,\text{hr} = 2.6\,\text{hrs}$$ **Ans. (D)**

Sol. 6

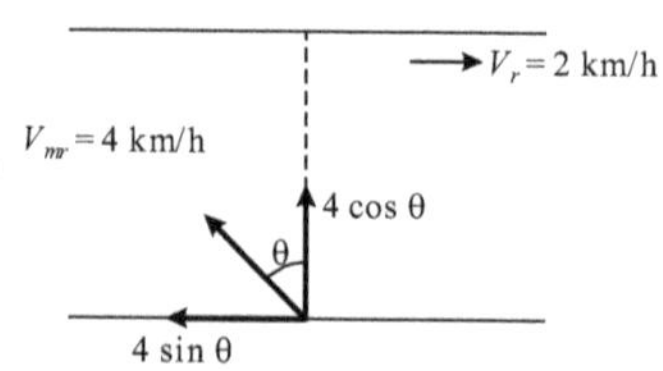

For swimmer to cross the river straight

$\Rightarrow \quad 4\sin\theta = 2$

$\Rightarrow \quad \sin\theta = \frac{1}{2}$

$\Rightarrow \quad \theta = 30°$

So, angle with direction of river flow

$$= 90° + \theta = 120°$$ **Ans. (D)**

4.1 Newton's Laws of Motion

Sol. 1 As particle remain stationary

$$\Rightarrow \quad \vec{F_1} + \vec{F_2} + \vec{F_3} = 0$$
$$\vec{F_1} + \vec{F_R} = 0$$
$$\vec{F_1} = -\vec{F_R}$$

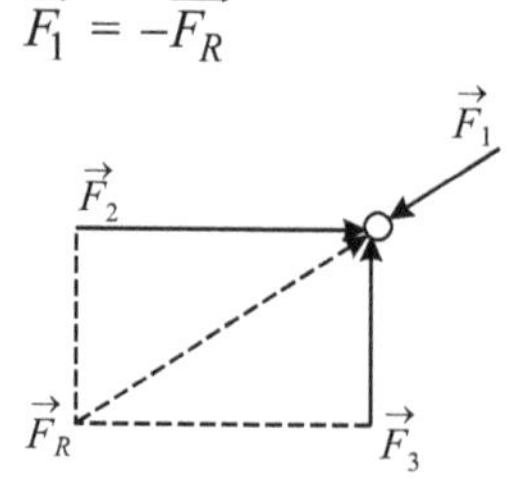

when $\vec{F_1}$ is removed, particle will move in under the influence of $\vec{F_R}$

$\Rightarrow$ acceleration

$$\vec{a} = \frac{\vec{F_R}}{m} \Rightarrow |\vec{a}| = \frac{F_1}{m}$$ **Ans. (A)**

Sol. 2

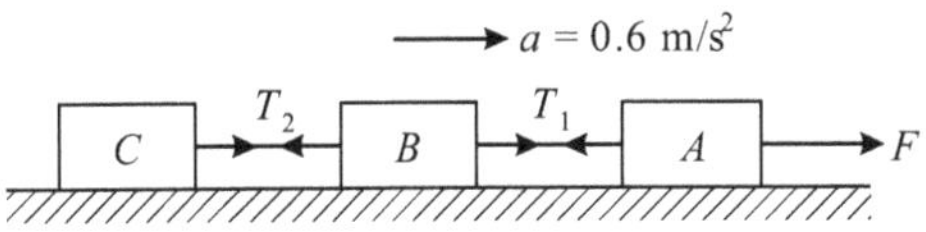

For block C, its equation of motion is

$$T_2 = m_C a$$
$$= 2 \times 0.6$$
$$= 1.2\,\text{N}$$

Ans. (B)

Sol. 3 When man is at rest

$$T = 60g = 600\,\text{N}$$

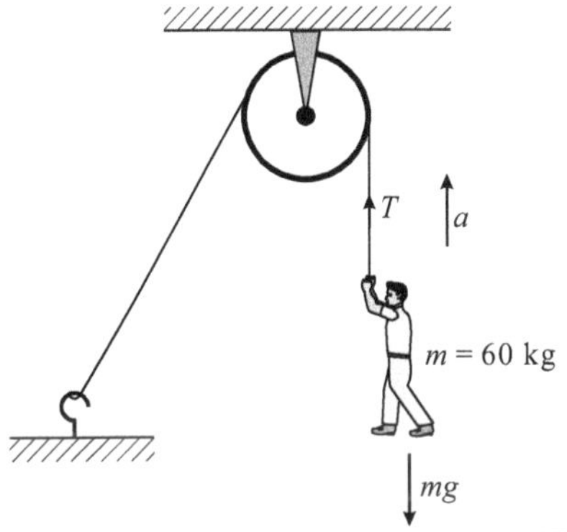

When man starts climbing up the rope at acceleration a, its equation of motion is

$$T - mg = ma$$
$$\Rightarrow \quad 840 - 600 = 60a$$
$$\Rightarrow \quad a = 4\,\text{m/s}^2$$

Ans. (C)

Sol. 4

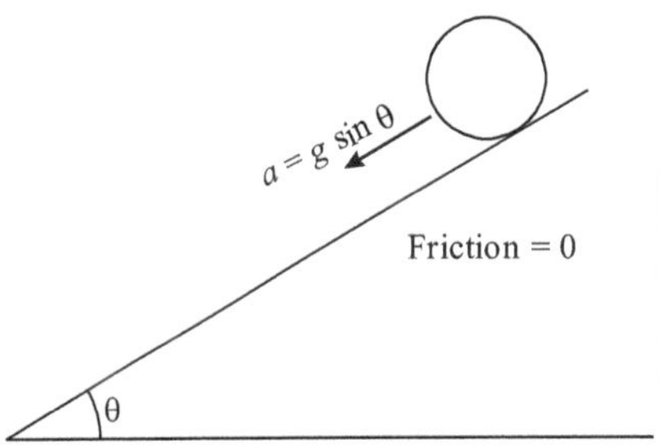

As no friction is there, all bodies slide with acceleration $g \sin\theta$.

Ans. (D)

Sol. 5 Acceleration of ball w.r.t. lift

$$\vec{a}_{BL} = \vec{a}_B - \vec{a}_L$$
$$a_{BL} = g - a$$

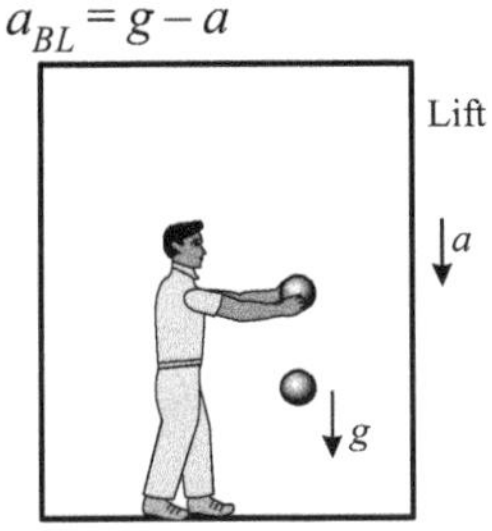

Acceleration of ball w.r.t. ground

$$a_B = g$$

Ans. (C)

Sol. 6 Given that

$$F_1 + F_2 = 18\,\text{N} \qquad \ldots(1)$$

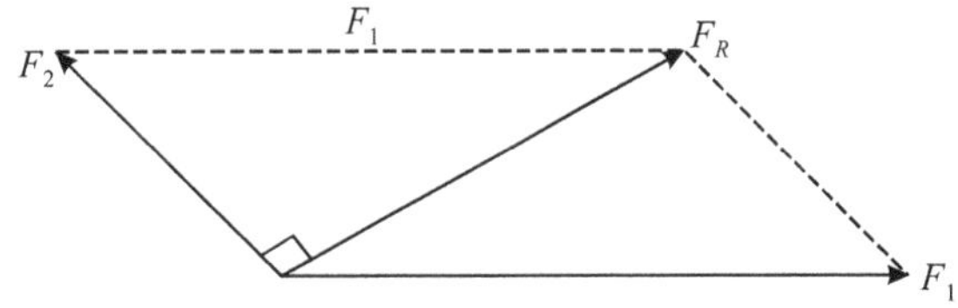

From the given vector triangle, we use

$$F_2^2 + F_R^2 = F_1^2$$
$$\Rightarrow \quad F_1^2 - F_2^2 = F_R^2 = 144$$
$$\Rightarrow \quad (F_1 + F_2)(F_1 - F_2) = 144$$

from (1) $\Rightarrow$ $$F_1 - F_2 = \frac{144}{18} = 8 \qquad \ldots(2)$$

from (1) and (2) $$2F_1 = 26 \Rightarrow F_1 = 13\,\text{N}$$
$$2F_2 = 10 \Rightarrow F_2 = 5\,\text{N}$$

Ans. (B)

Sol. 7 Acceleration of Atwood's machine is given as

$$a = \frac{m_2 - m_1}{m_2 + m_1} \cdot g$$

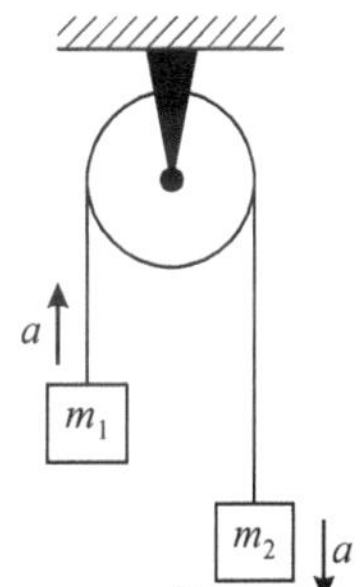

$$a = g/8$$

$$m_2 + m_1 = 8m_2 - 8m_1$$
$$9m_1 = 7m_2$$
$$\Rightarrow \quad \frac{m_2}{m_1} = \frac{9}{7}$$

Ans. (B)

Sol. 8 As spring balances are light, throughout the string length tension remain same.

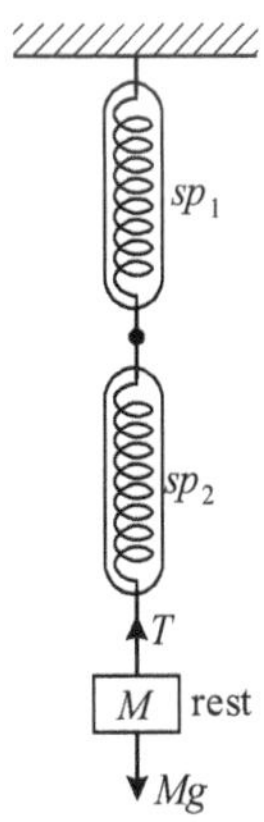

$$T = T_{sp_1} = T_{sp_2} = Mg$$

$\Rightarrow$ Readings of both sp_1 and sp_2 are same

Ans. (A)

Sol. 9 At rest $T = 49\text{N} = mg$

$$\Rightarrow \quad m = \frac{49}{9.8} = 5\text{ kg}$$

When lift moves downward at acceleration a

Then its equation of motion is (for bag)

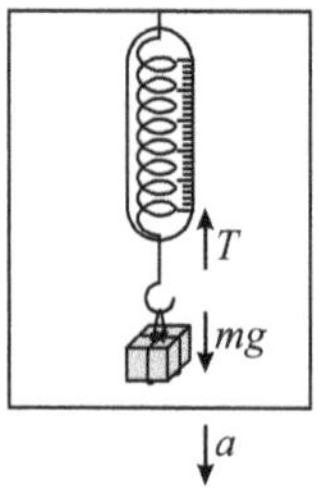

$$mg - T_1 = ma$$

$$\Rightarrow \quad T_1 = mg - ma = m(g - a)$$
$$= 5(9.8 - 5)$$
$$= 5 \times 4.8 = 24\text{N}$$

Ans. (A)

Sol. 10

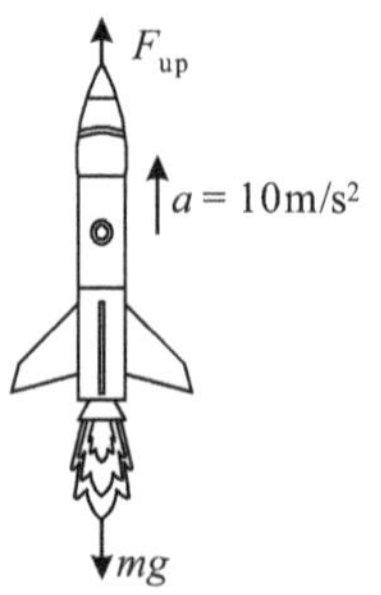

For upward acceleration of rocket

$$a = 10\text{ m/s}^2$$

We use

$$F_{up} - mg = ma$$

$$\Rightarrow \quad F_{up} = mg + ma$$

$$\Rightarrow \quad F_{up} = m(g + a) = 3.5 \times 10^4[10 + 10]$$
$$= 7 \times 10^5\text{ N}$$

Ans. (B)

Sol. 11 As vector sum of the three forces is equal to zero

$$\Rightarrow \quad \vec{a} = 0$$

$\Rightarrow \quad \vec{v} \rightarrow$ remain uncharged. **Ans. (D)**

Sol. 12 *FBD* of block is shown in figure

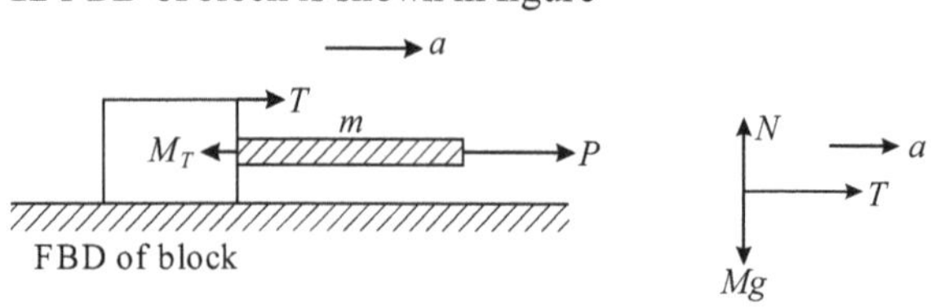

Acceleration of system (block + rope) is given as

$$a = \left(\frac{P}{M + m}\right)$$

Equation of motion of block

$$T = Ma = \frac{MP}{M + m}$$

Ans. (D)

Sol. 13 Acceleration of masses in Atwood's machine is given as

$$a = \left(\frac{m_1 - m_2}{m_1 + m_2}\right)g$$
$$= \left(\frac{5 - 4.8}{5 + 4.8}\right) \times 9.8\text{ m/s}^2$$
$$= 0.2\text{ m/s}^2$$

Ans. (A)

Sol. 14

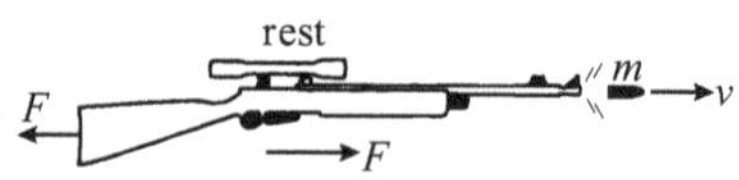

Man has to apply an equal force to keep the gun at rest

F = rate of charge of linear momentum

If N bullets are fired per second

$$\Rightarrow \quad F = N \cdot mv$$
$$144 = N \times 40 \times 10^{-3} \times 1200$$
$$\Rightarrow \quad N = 3$$

Ans. (D)

Sol. 15

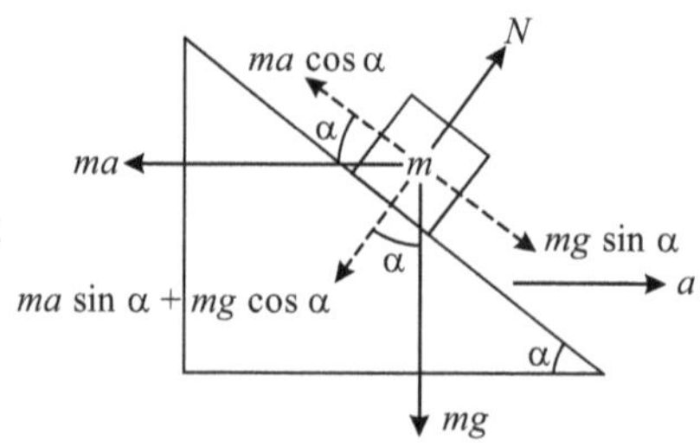

To keep the block at rest, we use

$$mg \sin \alpha = ma \cos \alpha$$
$$a = g \tan \alpha$$

Ans. (D)

Sol. 16

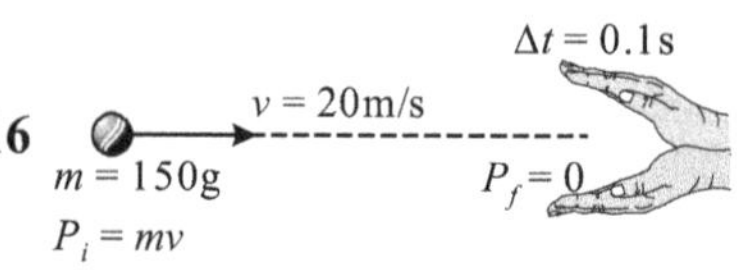

In the process of catching the ball, force on hand is

$$F = \frac{\Delta P}{\Delta t} = \frac{P_i - P_f}{\Delta t} = \frac{150 \times 10^{-3} \times 20}{0.1}$$

$\Rightarrow \quad F = 30\text{ N}$ **Ans. (C)**

Sol. 17 In free fall motion of ball we use

$$v^2 = u^2 - 2gh$$
$$\Rightarrow \quad 0 = u^2 - 2 \times 10 \times 2$$
$$\Rightarrow \quad u = \sqrt{40}\text{ m/s}$$

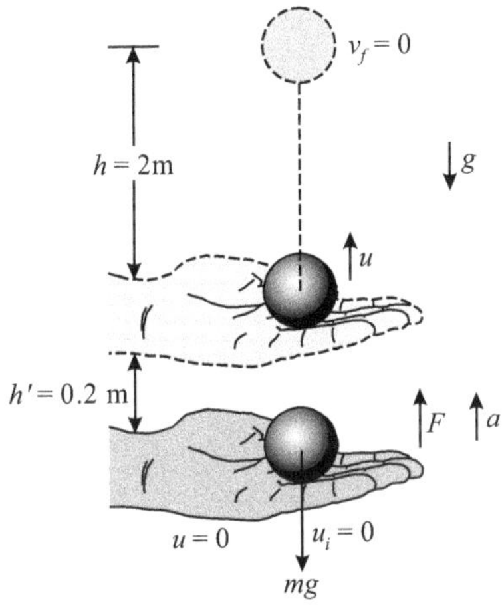

When ball was being pushed by hand, we use

$$u^2 = u_i^2 + 2ah'$$

$$40 = 0 + 2a(0.2)$$

$$\Rightarrow \quad a = 100 \text{ m/s}^2$$

equation of motion of ball is given as

$$F - mg = ma$$

$$\Rightarrow \quad F = m(g + a) = 0.2(10 + 100)$$

$$\Rightarrow \quad F = 0.2 \times 110 = 22 \text{ N}$$

Ans. (D)

Sol. 18

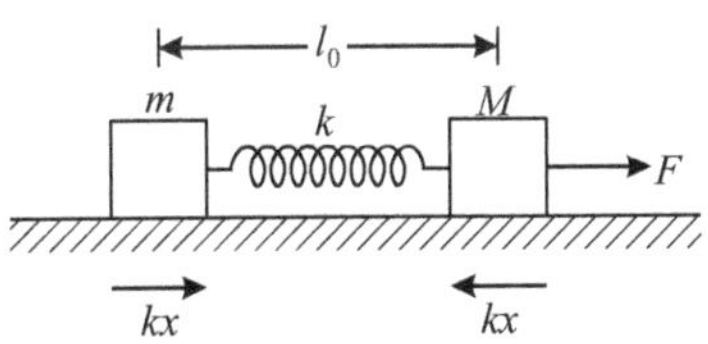

When both the blocks will start moving at same acceleration, say spring has an extension x_0

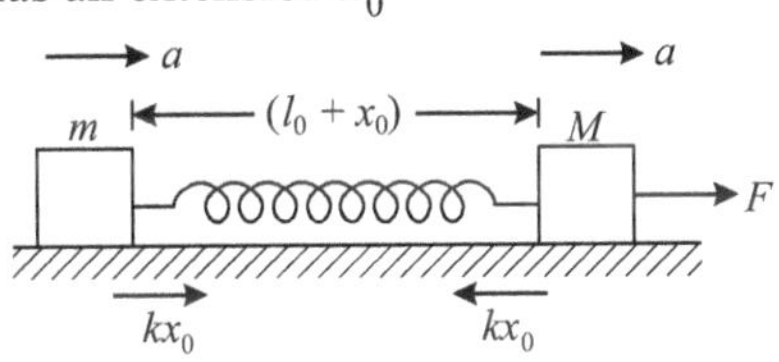

Common acceleration

$$a = \frac{F}{m + M}$$

equation of motion of blocks are

and $\quad F - kx_0 = Ma$

$$kx_0 = ma$$

$$kx_0 = \frac{mF}{m + M}$$

Ans. (C)

Sol. 19 If a spring of length l is cut in different parts then force constant of a part of length l_1 is given as

$$k_1 = \frac{kl}{l_1} \quad [k = \text{force constant of original spring}]$$

For spring A and B we have

$$l_A = 2l_0; \; l_B = 3l_0; \; l = l_A + l_B = 5l_0$$

Thus force constant of spring A is

$$k_A = \frac{kl}{l_A} = k\left(\frac{5l_0}{2l_0}\right) = \frac{5}{2}k$$

Ans. (A)

Sol. 20 Acceleration of particle

$$a = \frac{F}{m} = \frac{F_0}{m}e^{-bt}$$

$$\Rightarrow \quad a = \frac{dv}{dt} = \frac{F_0}{m}e^{-bt}$$

$$\Rightarrow \quad dv = \frac{F_0}{m}e^{-bt}\,dt$$

Integrating above relation gives

$$\int_{v=0}^{v=v} dv = \int_{t=0}^{t=t} \frac{F_0}{m}e^{-bt}\,dt$$

$$\Rightarrow \quad v = \frac{F_0}{m}\left[\frac{e^{-bt}}{-b}\right]_0^t$$

$$\Rightarrow \quad v_{(t)} = \frac{F_0}{mb}[1 - e^{-bt}]$$

Ans. (D)

Sol. 21 Statement-I :

As cart does not move

$\Rightarrow \quad F_H = F_B$ (ignoring friction on cart)

Acceleration to Newton's III law statement-I is true

Statement-II : In cart as F_H is balanced by F_B it does not move

$\Rightarrow$ Statement-II is also true and it is explaining statement-I also. **Ans. (A)**

Sol. 22

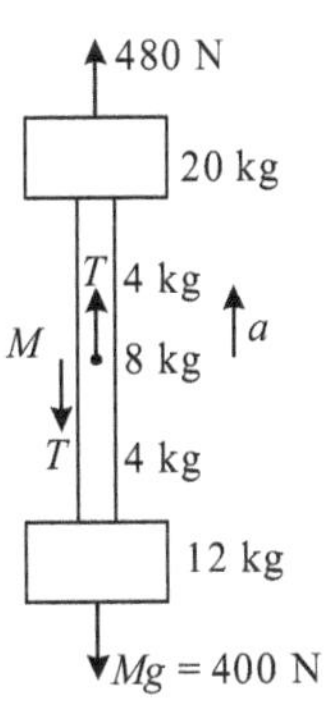

Equation of motion of system is

$$480 - 400 = 40a$$

$$\Rightarrow \quad a = \frac{80}{40} = 2\text{m/s}^2$$

for mass below point M, we use below FBD as shown

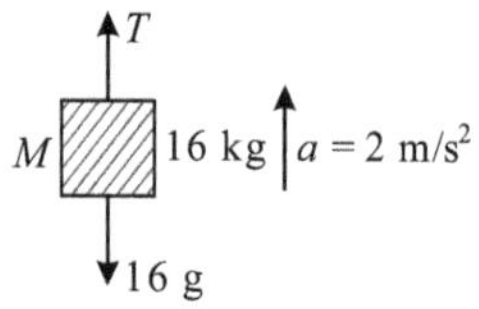

Equation of motion for above FBD is

$$T - 160 = 16 \times 2 = 32$$

$$\Rightarrow \quad T = 192\text{ N}$$ **Ans. (D)**

Sol. 23

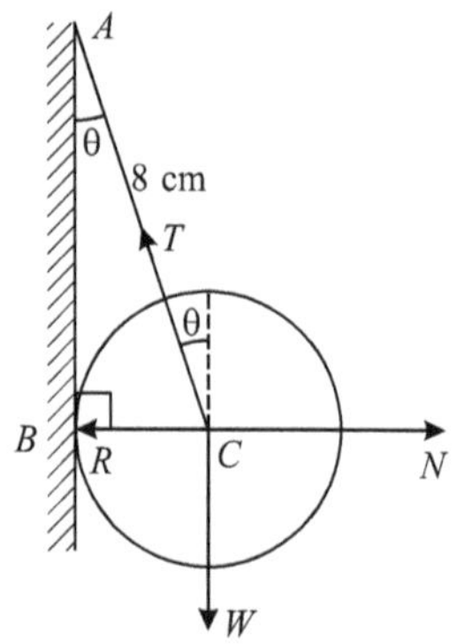

$$R = 5\text{ cm}$$

$$AC = 8 + 5 = 13\text{ cm}$$

and $$AB = \sqrt{AC^2 - BC^2} = \sqrt{13^2 - 5^2}$$

$$= \sqrt{144} = 12\text{ cm}$$

For equilibrium of sphere, we use

$$T\cos\theta = W$$

$$T\sin\theta = N$$

$$\Rightarrow \quad T = \frac{W}{\cos\theta}$$

Here $$\cos\theta = \frac{AB}{AC} = \frac{12}{13}$$

$$\Rightarrow \quad T = \frac{13W}{12}$$ **Ans. (D)**

Sol. 24 As $\vec{F}$ = constant

$\Rightarrow \quad \vec{a}$ = constant

We can use $\vec{a} = \dfrac{\vec{v}_2 - \vec{v}_1}{\Delta t}$

$$\Rightarrow \quad \vec{a} = \frac{6\hat{i} - (6\hat{i} - 2\hat{j})}{10} = \frac{-6\hat{i} + 8\hat{j}}{10}$$

$$\Rightarrow \quad \vec{a} = \frac{-3\hat{i} + 4\hat{j}}{5}\text{ m/s}^2$$

Force on particle is

$$\vec{F} = m\vec{a} = 5\left(\frac{-3\hat{i} + 4\hat{j}}{5}\right) = -3\hat{i} + 4\hat{j}\text{ N}$$ **Ans. (A)**

Sol. 25

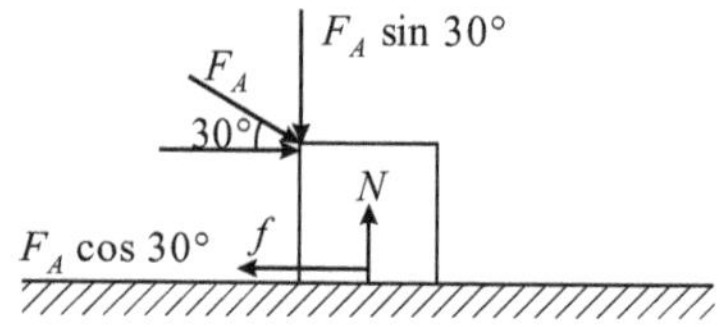

$$N = Mg + F_A \sin 30°$$

$$f = \mu N$$

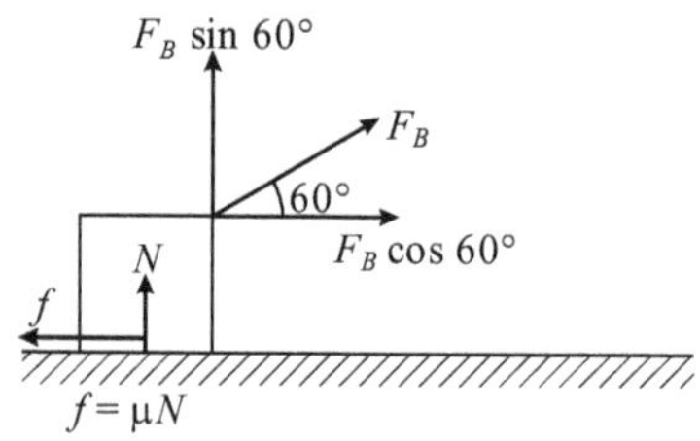

$$N = Mg - F_B \sin 60°$$

to drag the box by person A, we use

$$F_A \cos 30° \geq \mu(Mg + F_A \sin 30°)$$

$$F_A \geq \frac{2\mu Mg}{\sqrt{3} - \mu} \quad \ldots(1)$$

$$\Rightarrow \quad F_A\left(\frac{\sqrt{3}}{2} - \frac{\mu}{2}\right) \geq \mu\, Mg$$

to drag the box by person B, we use

$$F_B \cos 60° \geq \mu(Mg - F_B \sin 60°)$$

$$\Rightarrow \quad F_B \geq \frac{2\mu Mg}{1 + \mu\sqrt{3}} \quad \ldots(2)$$

$$\Rightarrow \quad F_B\left(\frac{1}{2} + \frac{\mu\sqrt{3}}{2}\right) \geq \mu Mg$$

The ratio $\dfrac{F_A}{F_B}$ can be obtained by equation-(1) & (2)

$$\Rightarrow \quad \frac{F_A}{F_B} = \frac{1 + \mu\sqrt{3}}{\sqrt{3} - \mu}$$

$$\Rightarrow \quad \frac{F_A}{F_B} = \frac{1 + \left(\frac{\sqrt{3}}{5}\right)\sqrt{3}}{\sqrt{3} - \frac{\sqrt{3}}{5}} = \frac{1 + 3/5}{\sqrt{3}(4/5)} = \frac{2}{\sqrt{3}}$$

Ans. (D)

Sol. 26 Given that

$$F = ma = \frac{R}{t^2}v$$

$$\Rightarrow \quad \frac{dv}{dt} = \frac{R}{mt^2}v$$

$$\Rightarrow \quad \int \frac{dv}{v} = \int \frac{R}{mt^2}dt$$

$$\Rightarrow \qquad \ln(v) = -\frac{R}{mt} + C$$

$$\Rightarrow \qquad \ln(v) = -\left(\frac{R}{m}\right)\frac{1}{t} + C$$

here above expression stations that graph of ln v and $\frac{1}{t}$ is a straight line with slope $-\frac{R}{m}$ **Ans. (A)**

Sol. 27

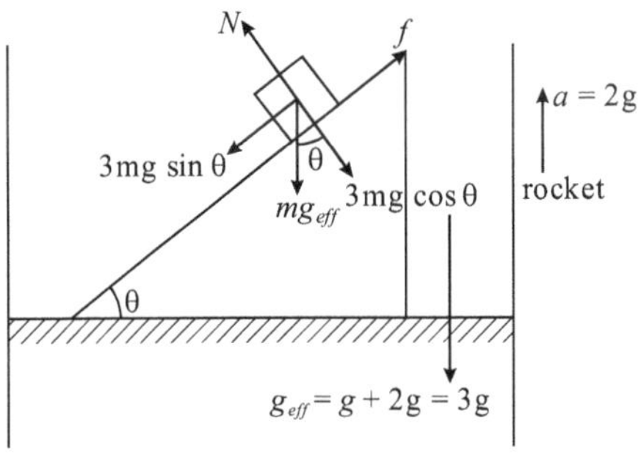

Limiting friction on the block is

$$f_L = \mu N = 3\mu \text{ mg} \cos\theta$$

for block not to slide (f_L)

$$3 \text{ mg} \sin\theta < 3\text{m } mg \cos\theta$$

$$\mu > \tan\theta$$

$$\Rightarrow \qquad \mu_{min} = \tan\theta$$ **Ans. (B)**

Sol. 28

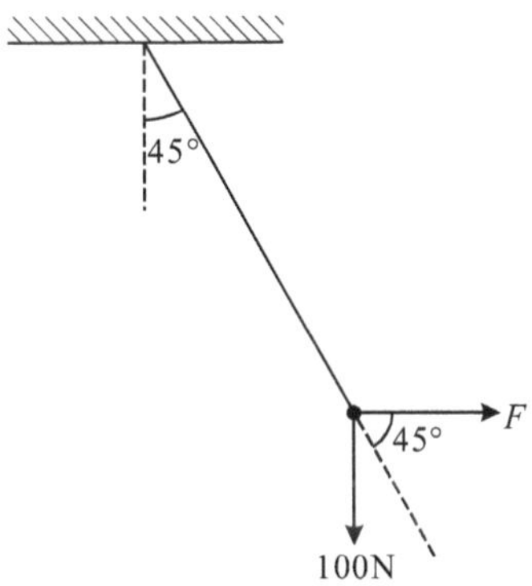

Equilibrium, we use

$$\tan 45° = \frac{100}{F}$$

$$\Rightarrow \qquad F = 100 \text{ N}$$ **Ans. (B)**

Sol. 29 By Newton's second law, we use

$$\frac{dp}{dt} = F = kt$$

$$\Rightarrow \qquad \int_p^{3p} dp = \int_O^T kt\, dt$$

$$\Rightarrow \qquad 2p = \frac{kT^2}{2}$$

$$\Rightarrow \qquad T = 2\sqrt{\frac{p}{k}}$$ **Ans. (A)**

4.2 Friction

Sol. 1

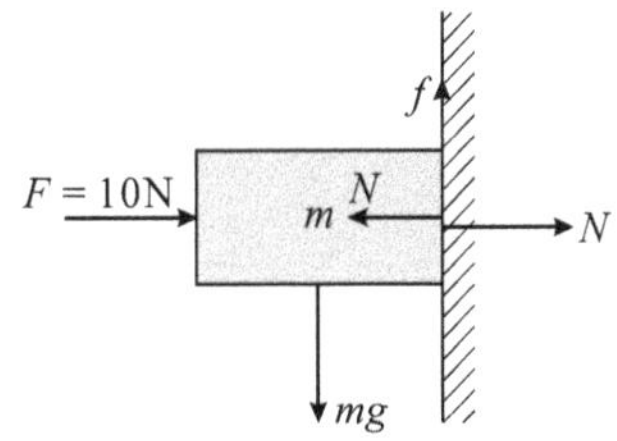

For equilibrium of block, we have

$$N = F = 10 \text{ N}$$

$$mg = f = \mu N$$

$$\Rightarrow \qquad mg = 0.2 \times 10 = 2 \text{ N}$$ **Ans. (D)**

Sol. 2

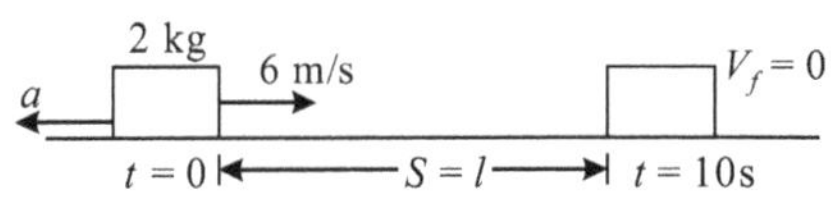

Retardation due to friction on block is

$$a = -\mu g$$

we use $\quad V = u + at$

$$\Rightarrow \qquad 0 = 6 - \mu(10)(10)$$

$$\Rightarrow \qquad \mu = 0.06$$ **Ans. (C)**

Sol. 3

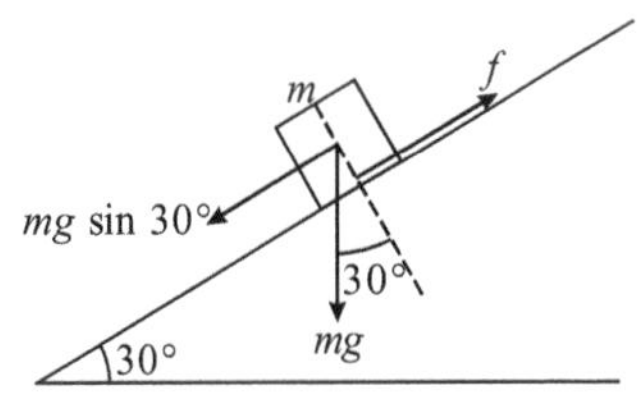

As block is rest on incline plane we use

$$f = mg \sin 30°$$

$$\Rightarrow \qquad 10 = m(10)\left(\frac{1}{2}\right)$$

$$\Rightarrow \qquad m = 2 \text{ kg}$$ **Ans. (A)**

Sol. 4

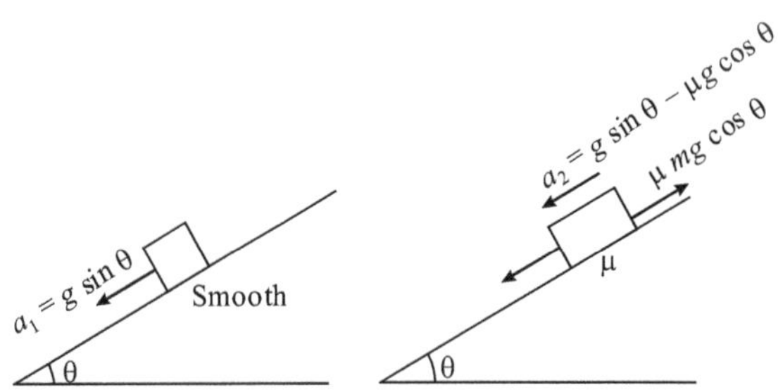

To slide a distance l along incline, time required for smooth incline

$$t_1 = \sqrt{\frac{2l}{a_1}} = \sqrt{\frac{2l}{g\sin\theta}}$$

For rough incline

$$t_2 = \sqrt{\frac{2l}{a_2}} = \sqrt{\frac{2l}{g\sin\theta - \mu g\cos\theta}}$$

Given that $t_2 = nt_1$

$$\Rightarrow \quad \sqrt{\frac{2l\sqrt{2}}{g(1-\mu)}} = n\sqrt{\frac{2l\sqrt{2}}{g}}$$

$$\Rightarrow \quad \sqrt{\frac{1}{1-\mu}} = n$$

$$\Rightarrow \quad 1-\mu = \frac{1}{n^2}$$

$$\Rightarrow \quad \mu_k = 1 - \frac{1}{n^2}$$

Ans. (A)

Sol. 5

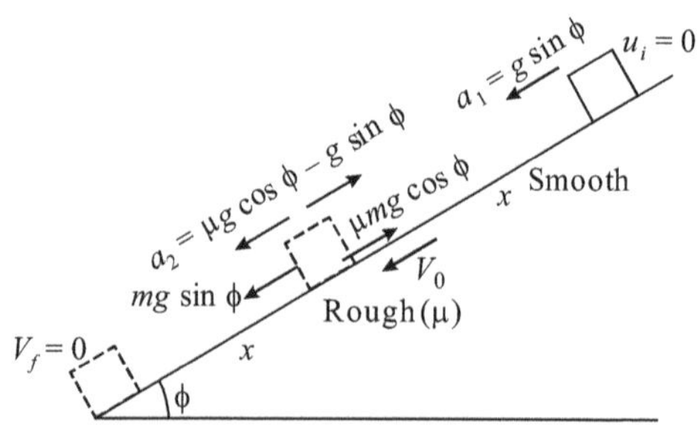

for upper half of motion, we use

$$V_0 = \sqrt{2a_1x} = \sqrt{2g\sin\phi \cdot x} \qquad \ldots(1)$$

for lower half of motion, we use

$$V_0 = \sqrt{2a_2x} = \sqrt{2g(\mu\cos\phi - \sin\phi)x} \qquad \ldots(2)$$

from (1) & (2), we have

$$2xg\sin\phi = 2xg(\mu\cos\phi - \sin\phi)$$

$$\Rightarrow \quad 2\sin\phi = \mu\cos\phi$$

$$\Rightarrow \quad \mu = 2\tan\phi$$

Ans. (C)

Sol. 6

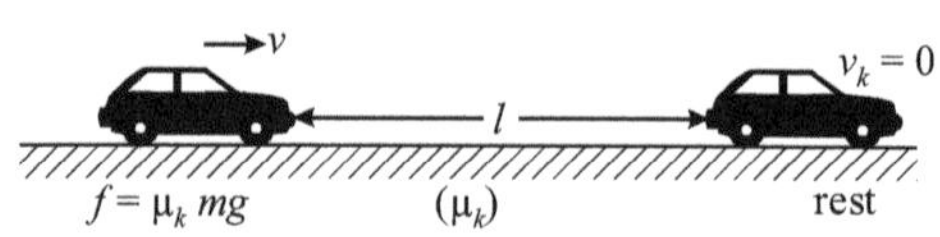

Due to friction maximum retardation in car is

$$a_{max} = \mu_x g$$

Using $\quad v_f^2 = v^2 - 2al$

$$\Rightarrow \quad l = \frac{v^2}{2a} \quad [\text{As } v_f = 0]$$

$$\Rightarrow \quad l = \frac{(100)^2}{2\times 0.5\times 10} = \frac{10000}{10}$$

$$\Rightarrow \quad l = 1000\,\text{m}$$

Ans. (B)

Sol. 7

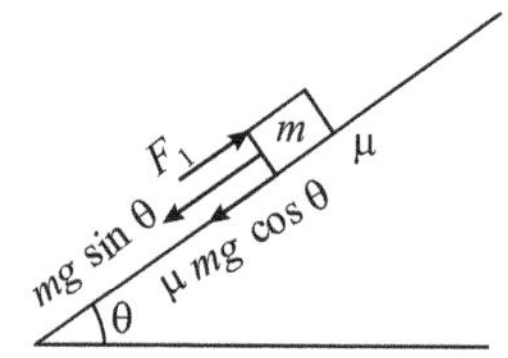

To push the block up on incline plane, minimum, force required is

$$F_1 = mg\sin\theta + \mu mg\cos\theta$$

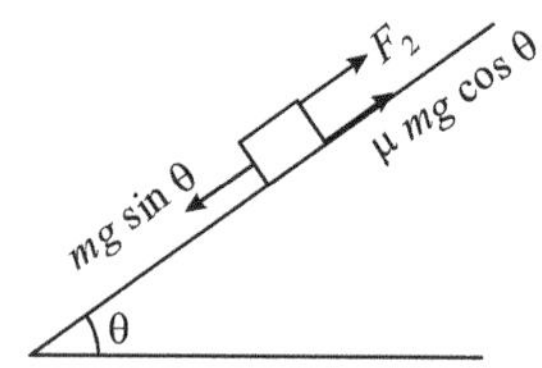

To prevent the body sliding down

$$F_2 = mg\sin\theta - \mu mg\cos\theta.$$

$$\Rightarrow \quad \frac{F_1}{F_2} = \frac{\sin\theta + \mu\cos\theta}{\sin\theta - \mu\cos\theta} = \frac{\tan\theta + \mu}{\tan\theta - \mu} = \frac{3\mu}{\mu} = 3$$

Ans. (D)

Sol. 8 For insect not to slip α is the angle of repose

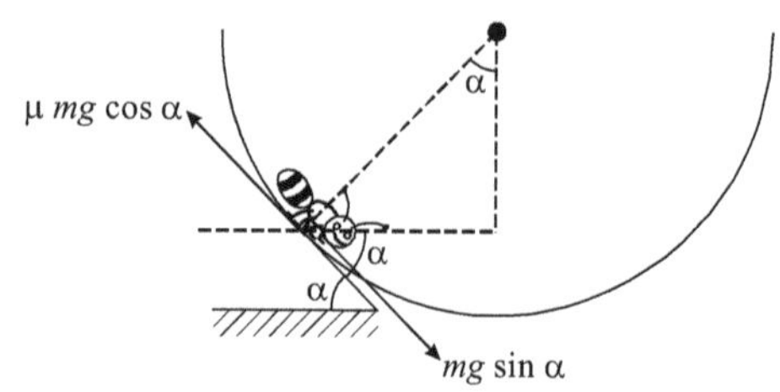

$$mg\sin\alpha \le \mu mg\cos\alpha$$

$$\Rightarrow \quad \mu \ge \tan\alpha \quad \Rightarrow \quad \tan\alpha = \frac{1}{3}$$

$$\Rightarrow \quad \cot\alpha = 3$$

Ans. (A)

Sol. 9

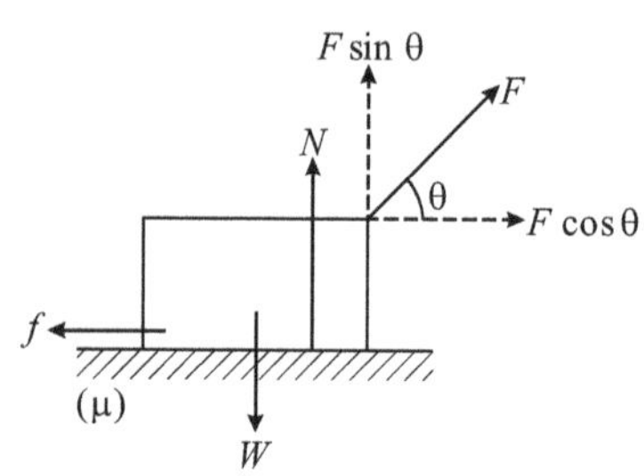

$$N = W - F\sin\theta$$

To slide the block we use

$$F\cos\theta = \mu(W - F\sin\theta)$$

$$\Rightarrow \quad F = \frac{\mu W}{\cos\theta + \mu\sin\theta} \qquad \ldots(1)$$

F is minimum when

$$\frac{dF}{d\theta} = 0$$

$$\Rightarrow \quad -\sin\theta + \mu\cos\theta = 0$$

$$\Rightarrow \quad \mu = \tan\theta$$

$$\Rightarrow \quad \theta = \tan^{-1}(\mu)$$

$$\Rightarrow \quad \sin\theta = \frac{\mu}{\sqrt{1+\mu^2}};\ \cos\theta = \frac{1}{\sqrt{1+\mu^2}}$$

Minimum value of force can be given as

$$F_{\min} = \frac{\mu W}{\frac{1}{\sqrt{1+\mu^2}} + \frac{\mu^2}{\sqrt{1+\mu^2}}} = \frac{\mu W}{\sqrt{1+\mu^2}}$$

Ans. (A)

Sol. 10

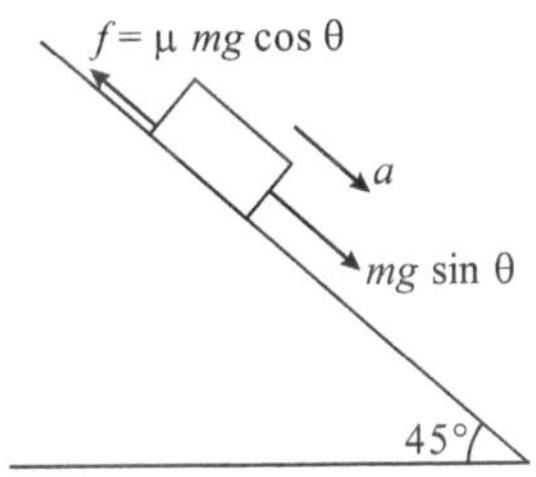

Acceleration body down the plane is

$$a = g\sin\theta - mg\cos\theta \qquad (\mu = 0.3x)$$

Speed of body increases until $a = 0$

After some time at a specific distance

$$a = 0$$

$$\Rightarrow \quad g\left(\frac{1}{\sqrt{2}}\right) - 0.3x\, g\cos 45° = 0$$

$$\Rightarrow \quad x = \frac{1}{0.3} = 3.33\ \text{m}$$

Ans. (D)

Sol. 11

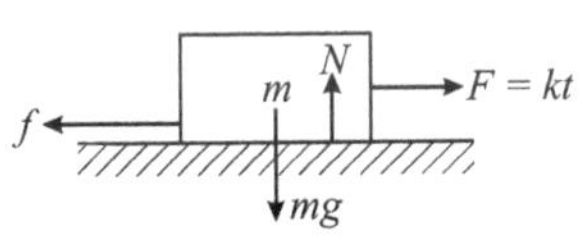

Limiting value of friction $= \mu\, mg$

for small values of F,

$$f = F$$

static friction and block static sliding at time $t = t_0$ when

$$kt_0 = \mu\, mg$$

and for $t > t_0$, equation of motion, we use

$$kt - \mu\, mg = ma$$

$$\Rightarrow \quad a = \frac{kt}{m} - mg$$

The graph will be a line with $+ve$ slope after

$$t = t_0$$

Ans. (B)

Sol. 12

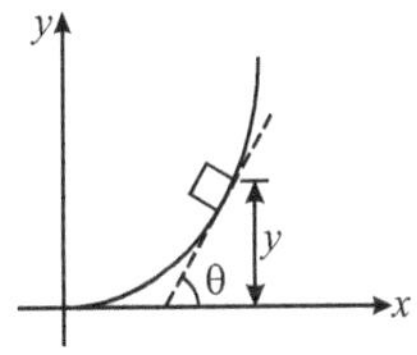

Block will remain at rest at a height y where

if $\quad \tan\theta = \mu$

$\theta \rightarrow$ Angle of Repose

here $\quad \tan\theta = \frac{dy}{dx} = \frac{d}{dx}\left(\frac{x^3}{6}\right) = \frac{x^2}{2} = 0.5$

$$\Rightarrow \quad x^2 = 0.5 \times 2 = 1$$

$$\Rightarrow \quad x = \pm 1$$

$$\Rightarrow \quad y = \frac{1}{6}m$$

Ans. (A)

Sol. 13

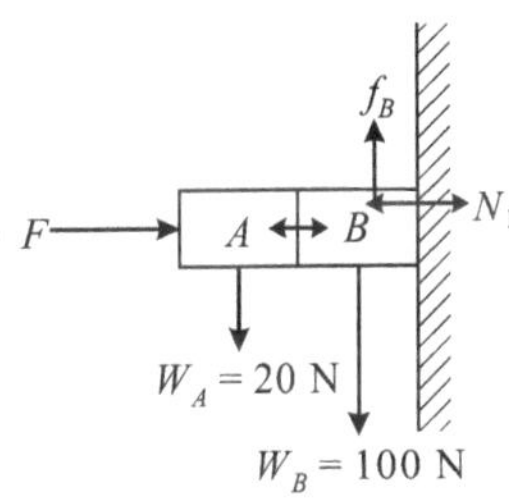

We consider blocks to be at rest and combined *FBD* of blocks with static friction is shown as

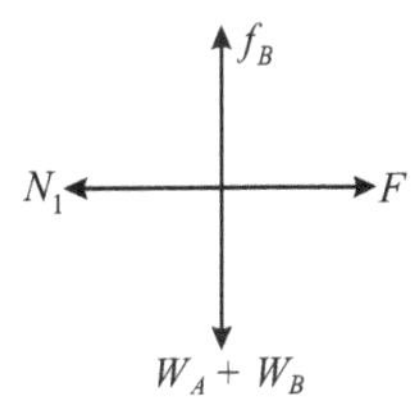

For equilibrium of blocks we have

$$f_B = W_A + W_B = 120\ \text{N}$$

Ans. (C)

Sol. 14

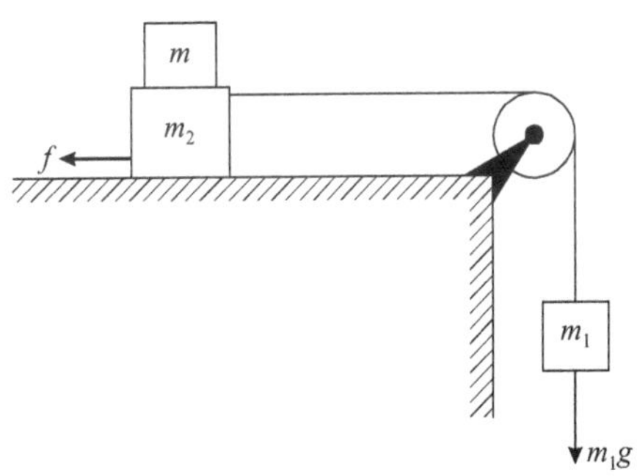

$$f_{max} = \mu(m + m_2)g$$

To stop the motion, we use

$$m_1 g \le \mu(m + m_2)g$$

$$\Rightarrow \quad 5 \le 0.15\,(m + 10)$$

$$\Rightarrow \quad \frac{100}{3} \le m + 10 \quad \Rightarrow \quad m \ge 23.33\ \text{kg}$$

Ans. (B)

Sol. 15

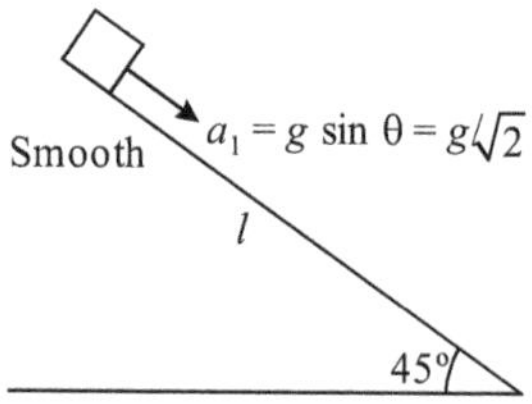

Time taken to slide is

$$t_1 = \sqrt{\frac{2l}{a_1}} = \sqrt{\frac{2\sqrt{2}l}{g}}$$

Given that $\qquad t_2 = nt_1$

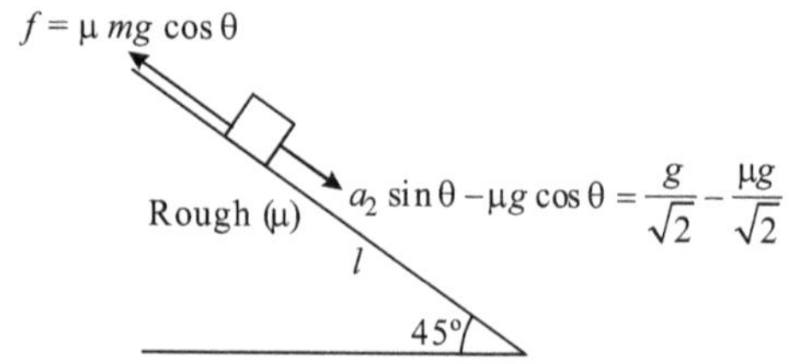

Time taken to slide is

$$t_2 = \sqrt{\frac{2l}{a_2}} = \sqrt{\frac{2\sqrt{2}l}{g(1-\mu)}}$$

$$\Rightarrow \quad \sqrt{\frac{2\sqrt{2}l}{g(1-\mu)}} = n\sqrt{\frac{2\sqrt{2}l}{g}}$$

$$\Rightarrow \quad \frac{1}{1-\mu} = n^2 \Rightarrow \mu = 1 - \frac{1}{n^2}$$

Ans. (B)

Sol. 16

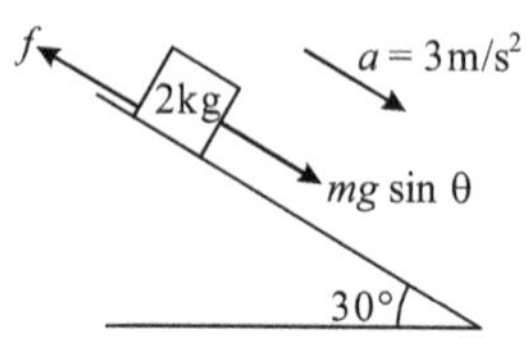

Equation motion of block

$$mg\sin\theta - f = ma$$

$$\Rightarrow \quad 2\times 10 \times \frac{1}{2} - f = 2\times 3$$

$$\Rightarrow \quad f = 10 - 6 = 4\text{ N}$$

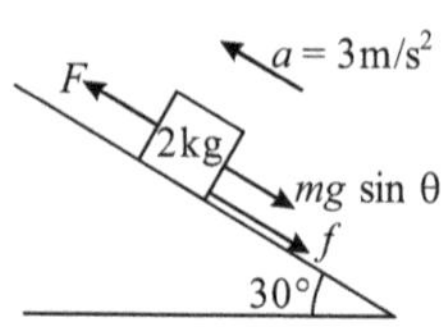

Equation of motion of block

$$F - mg\sin\theta - f = ma$$

$$\Rightarrow \quad F - 10 - 4 = 6$$

$$\Rightarrow \quad F = 20\text{ N}$$

Ans. (D)

Sol. 17 To keep the block at rest *FBD* is shown in figure. At equilibrium of block we use

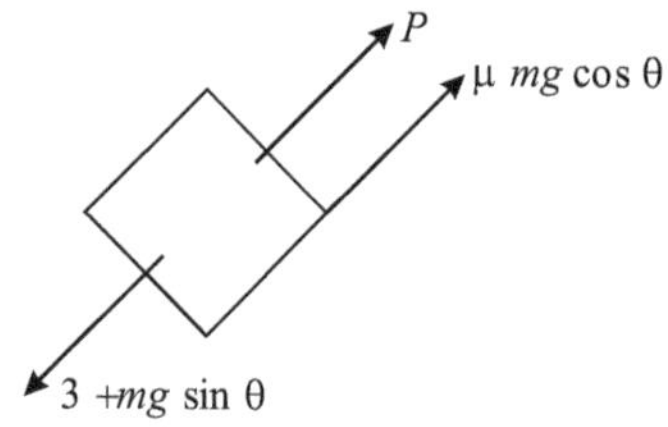

$$P + \mu mg\cos\theta = 3 + mg\sin\theta$$

$$\Rightarrow \quad P = 3 + 10\times 10\times\frac{1}{\sqrt{2}} - 0.6\times 10\times 10\times\frac{1}{\sqrt{2}}$$

$$\Rightarrow \quad P = 3 + 20\sqrt{2} = 31.28\text{ N}$$

Ans. (A)

Sol. 18 For limiting equilibrium down the plane, we use

$$2 + mg\sin 30 = \mu mg\cos 30° \qquad \ldots(1)$$

for limiting equilibrium up the plane, we use

$$10 = mg\sin 30 + \mu mg\cos 30° \qquad \ldots(2)$$

from (1) and (2)

$$\Rightarrow \quad 6 = \mu mg\cos 30°$$

and $\quad 4 = mg\sin 30°$

$$\Rightarrow \quad \frac{3}{2} = \mu\times\sqrt{3}$$

$$\Rightarrow \quad \mu = \frac{\sqrt{3}}{2}$$

Ans. (B)

Sol. 19

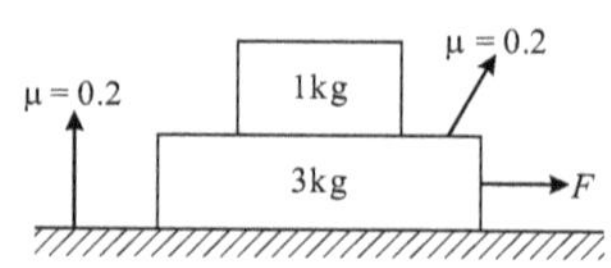

$$a_{A\max} = \mu g = 2\text{ m/s}^2$$

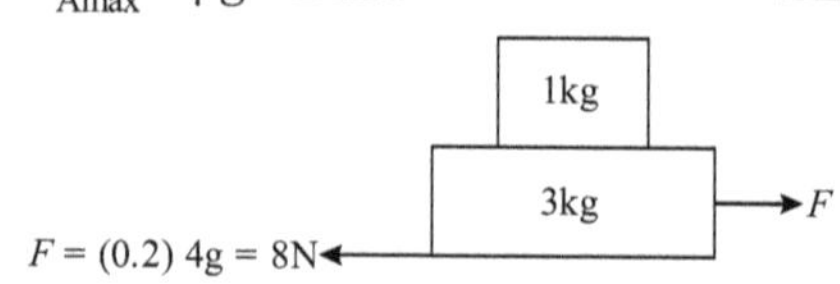

Equation of motion of blocks for combined motion is

$$F - 8 = 4\times 2$$

$$F = 16\text{ N}$$

Ans. (A)

Sol. 20 FBD of the two cases are shown below

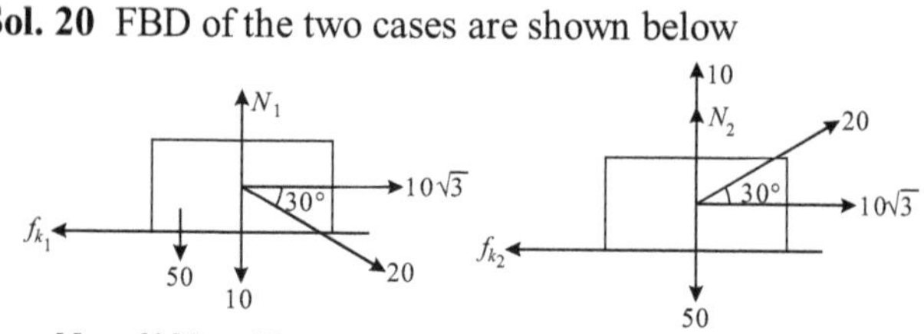

$$N_1 = 60 \; N_2 = 40$$

$$a_1 = \frac{10\sqrt{3} - 0.2\times 60}{5}$$

$$a_2 = \frac{10\sqrt{3} - 0.2\times 40}{5}$$

$$a_1 - a_2 = 0.8$$

Ans. (B)

5.1 Work Energy Power

Sol. 1 The energy stored in a spring having extension x is

$$U = \frac{1}{2}kx^2$$

Work done in extending the extension in spring from x_1 to x_2 is

$$W = U_f - U_i$$

$$\Rightarrow \quad W = \frac{1}{2}k(x_2^2 - x_1^2)$$

$$\Rightarrow \quad W = \frac{1}{2} \times 800[(0.15)^2 - (0.05)^2]$$

$$\Rightarrow \quad W = 400[0.0225 - 0.0025] = 8\text{J}$$

Ans. (B)

Sol. 2 Work done in extending spring extension from x_1 to x_2 is

$$W = U_f - U_i = \frac{1}{2}kx_2^2 - \frac{1}{2}kx_1^2$$

$$\Rightarrow \quad W = \frac{1}{2}(5\times10^3)\times[(0.1)^2 - (0.05)^2]$$

$$\Rightarrow \quad W = 18.75\text{ J}$$

Ans. (B)

Sol. 3 Along a straight line motion, power is given as

$$P = F.v = mav$$

$$\Rightarrow \quad m\left(\frac{dv}{dt}\right)v = P \rightarrow \text{constant}$$

$$\Rightarrow \quad vdv = \frac{P}{m}dt$$

Integrating this relation gives

$$\int_0^v vdv = \frac{P}{m}\int_0^t dt$$

$$\Rightarrow \quad \frac{v^2}{2} = \frac{P}{m}t$$

$$\Rightarrow \quad v = \sqrt{\frac{2P}{m}t}$$

$$\Rightarrow \quad \frac{dx}{dt} = \sqrt{\frac{2P}{m}}\sqrt{t}$$

$$\Rightarrow \quad dx = \sqrt{\frac{2P}{m}}\sqrt{t}\,dt$$

$$\Rightarrow \quad \int_0^x dx = \sqrt{\frac{x}{m}}\int_0^t \sqrt{t}dt$$

$$\Rightarrow \quad x = \sqrt{\frac{x}{m}}\frac{t^{3/2}}{(3/2)}$$

$$\Rightarrow \quad x \propto t^{3/2}$$

Ans. (B)

Sol. 4 Given that

$$a \propto -x$$

$$\Rightarrow \quad v\frac{dv}{dx} = a = -Cx$$

$$\Rightarrow \quad \int_{v_i}^{v_f} vdv = -\int_{x=0}^{x} Cx\,dx$$

$$\Rightarrow \quad \left[\frac{v^2}{2}\right]_{v_i}^{v_f} = -\frac{Cx^2}{2}$$

$$\Rightarrow \quad \frac{v_i^2}{2} - \frac{v_f^2}{2} = -\frac{Cx^2}{2}$$

If equation is multiplied with mass of particle m

Here $v_i \rightarrow$ initial speed at $x = 0$

$v_f \rightarrow$ final speed after covering displacement x

$$\frac{1}{2}mv_i^2 - \frac{1}{2}mv_f^2 = \frac{1}{2}mCx^2$$

$$\Rightarrow \quad \Delta K = \frac{1}{2}mCx^2$$

$$\Rightarrow \quad \Delta K \propto x^2$$

Ans. (A)

Sol. 5

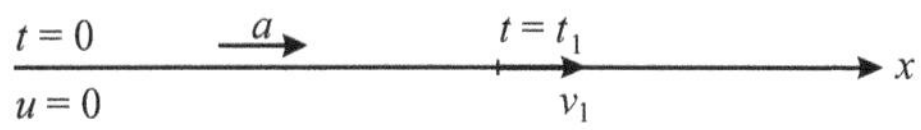

Using $\quad v = u + at$

$$v_1 = at_1$$

$$\Rightarrow \quad a = \frac{v_1}{t_1}$$

Instantaneous power

$$P = F.v = m\,a\,v$$

After time t_1 speed of particle is $v = at$

$$\Rightarrow \quad P = ma^2t = \frac{mv_1^2}{t_1^2}\cdot t$$

Ans. (B)

Sol. 6

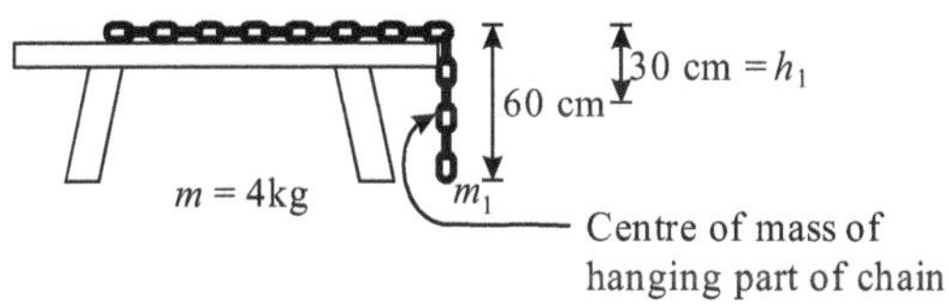

Mass of hanging part of chain is

$$m_1 = \frac{4}{2}\times0.6 = 1.2\text{ kg}$$

Work done in lifting hanging part of chain is

$$W = m_1 g h_1 = 1.2\times10\times0.3$$

$$= 3.6\text{ J}$$

Ans. (B)

Sol. 7 Work done in displacing a particle is given as

$$w = \vec{F}\cdot\vec{r}$$

$$\Rightarrow \quad w = (5\hat{i} + 3\hat{j} + 2\hat{k})\cdot(2\hat{i} - \hat{j})$$

$$\Rightarrow \quad w = 10 - 3 = 7\text{ J}$$ **Ans. (B)**

Sol. 8

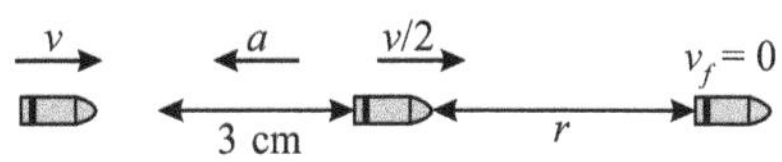

We use $\quad v^2 = u^2 + 2as$

For first part of motion

$$\left(\frac{v}{2}\right)^2 = (v)^2 - 2a(3)$$

$$\Rightarrow \quad \frac{3v^2}{4} = 6a$$

$$\Rightarrow \quad a = \frac{v^2}{8}$$

For second part of motion

$$(0)^2 = \left(\frac{v}{2}\right)^2 - 2as$$

$$\Rightarrow \quad s = \frac{v^2}{8a} = 1 \text{ cm}$$

Ans. (D)

Sol. 9 By conservation of energy

$$\frac{1}{2}Mv^2 = \frac{1}{2}kL^2$$

$$\Rightarrow \quad v = \sqrt{\frac{k}{M}}L$$

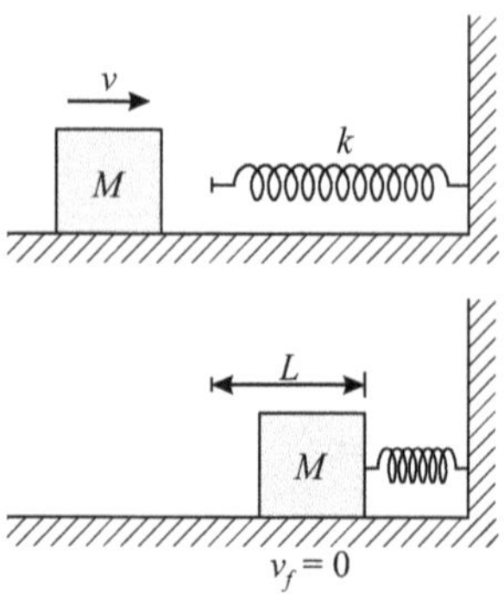

Maximum momentum of the block is

$$P_{max} = Mv = M\sqrt{\frac{k}{M}}\cdot L = \sqrt{kM}\cdot L$$

Ans. (A)

Sol. 10 All surface are smooth

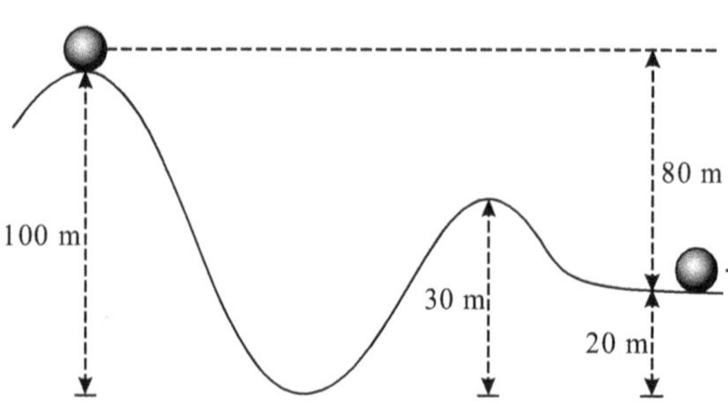

Using work energy theorem, we have

$$0 + mg(80) = \frac{1}{2}mv^2$$

$$\Rightarrow \quad v = \sqrt{1600} = 40 \text{ m/s}$$

Ans. (B)

Sol. 11

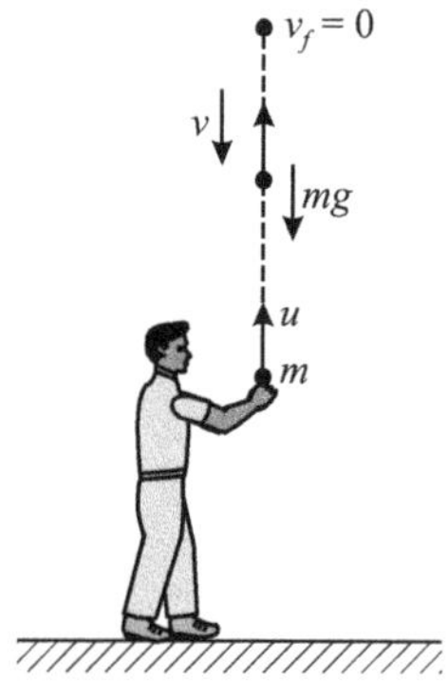

By applying work-energy theorem, we have

$$\frac{1}{2}mu^2 + W_g = 0$$

$$\Rightarrow \quad W_g = \frac{1}{2}mu^2 = -\frac{1}{2} \times 100 \times 10^{-3} \times (5)^2$$

$$\Rightarrow \quad W_g = -1.25 \text{ J}$$

Ans. (B)

Sol. 12 At any instant total mechanical energy is given as

$$E = K + U$$

$$\Rightarrow \quad K = E - U = 2 - \left(\frac{x^4}{4} - \frac{x^2}{2}\right)$$

K is maximum when

$$\frac{dK}{dx} = 0 = -\frac{4x^3}{4} + \frac{2x}{2}$$

$$\Rightarrow \quad -x^3 + x = 0$$

$$\Rightarrow \quad x = 0, -1, +1$$

$$K_{x=0} = 2\text{J}$$

and

$$K_{x=\pm 1} = 2 - \frac{1}{4} + \frac{1}{2} = \frac{9}{4}\text{J (max)}$$

$$\Rightarrow \quad \frac{1}{2}mv_{max}^2 = \frac{9}{4}$$

$$\Rightarrow \quad v_{max} = \sqrt{\frac{9}{2}} = \frac{3}{\sqrt{2}} \text{ m/s}$$

Ans. (A)

Sol. 13 As initial and final kinetic energy of mass is zero, using work-energy theorem, we have

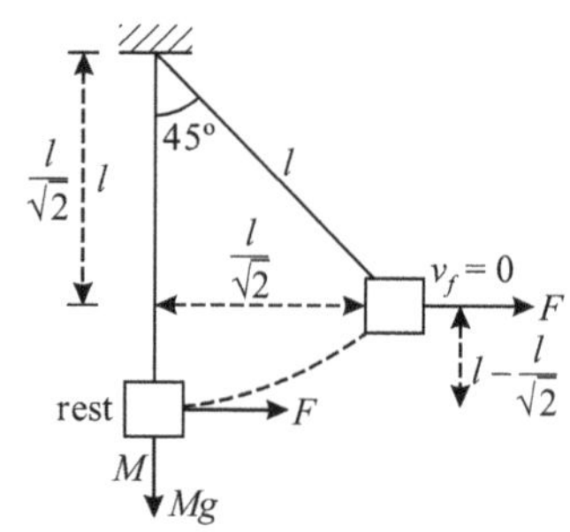

$$K_i(=0)\, F\left(\frac{l}{\sqrt{2}}\right) - Mg\left(l - \frac{l}{\sqrt{2}}\right) = K_f(=0)$$

$$\Rightarrow \quad F = Mg(\sqrt{2} - 1)$$

Ans. (D)

Sol. 14 Using work-energy theorem for the compression of spring, we have

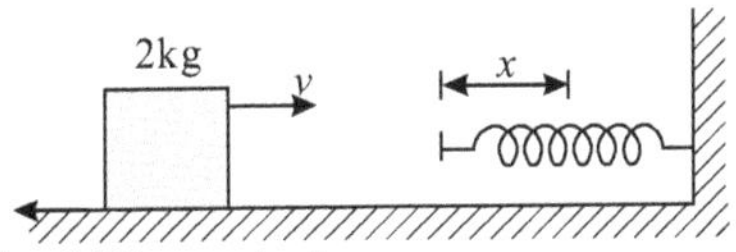

$f_k = 15\text{N} \quad v = 4\text{m/s}$

$$\Rightarrow \quad \frac{1}{2}mv^2 - f_k x - \frac{1}{2}kx^2 = 0$$

$$\Rightarrow \quad \frac{1}{2}\times 2\times (4)^2 - 15x - \frac{1}{2}\times 10000\,x^2 = 0$$

$$\Rightarrow \quad 5000x^2 + 15x - 16 = 0$$

$$\Rightarrow \quad x = \frac{-15 \pm \sqrt{225 + 4\times 5000\times 16}}{10000} = 0.055\text{ m}$$

$$\Rightarrow \quad x = 5.5\text{ cm}$$

Ans. (A)

Sol. 15 Speed of athlete can be taken as

$$\bar{v} = \frac{100}{10} = 10\text{m/s}$$

KE of athlete $\quad K = \frac{1}{2}m\bar{v}^2$

$$= \frac{1}{2}\times 60\times (10)^2 = 3000\text{ J}$$

mass of an athlete can be estimated ~ 60 kg **Ans. (D)**

Sol. 16 At $x = \infty$, $U(x) = 0$

at equilibrium, force on atoms = 0

we use $\quad F = -\frac{dU}{dx} = \frac{12a}{x^{13}} - \frac{6b}{x^7} = 0$

$$\Rightarrow \quad x^6 = \frac{2a}{b}$$

$$\Rightarrow \quad x = \left(\frac{2a}{b}\right)^{1/6}$$

PE at equilibrium is given as

$$U_{\text{at eqm}} = \frac{a}{\left(\frac{2a}{b}\right)^{12/6}} - \frac{b}{\left(\frac{2a}{b}\right)^{6/6}}$$

$$= \frac{a}{\left(\frac{4a^2}{b^2}\right)} - \frac{b}{\frac{2a}{b}}$$

$$= \frac{b^2}{4a} - \frac{b^2}{2a} = -\frac{b^2}{4a}$$

$$\Rightarrow \quad D = U_\infty - U_{\text{at eqm}} = 0 - \left(-\frac{b^2}{4a}\right) = \frac{b^2}{4a}$$

Ans. (D)

Sol. 17 We are given that

$$K = \frac{1}{2}mv^2 = Ct$$

$$\Rightarrow \quad v = \sqrt{\frac{2Ct}{m}}$$

$\Rightarrow$ Acceleration of particle

$$a = \frac{dv}{dx} = \sqrt{\frac{2C}{m}}\cdot\frac{1}{2\sqrt{t}}$$

$\Rightarrow$ Force on particle is given as

$$a = \frac{dv}{dx} = \sqrt{\frac{2C}{m}}\cdot\frac{1}{2\sqrt{t}} = \sqrt{\frac{Cm}{2t}}$$

$$\Rightarrow \quad F \propto \frac{1}{\sqrt{t}}$$

Ans. (D)

Sol. 18 We are given that applied force $F = k_1x_1 = k_2x_2$

and $\quad W_{s_1} > W_{s_2}$

$$\frac{1}{2}k_1x_1^2 > \frac{1}{2}k_1x_2^2$$

$$\Rightarrow \quad \frac{1}{2}Fx_1 > \frac{1}{2}Fx_2$$

$$\Rightarrow \quad x_1 > x_2$$

$$\Rightarrow \quad k_1 < k_2$$

$\Rightarrow$ S_{II} is correct

If in the two springs extension is equal, x

$$\Rightarrow \quad W_{s_1} = \frac{1}{2}k_1x^2 \text{ and } W_{s_2} = \frac{1}{2}k_2x^2$$

$$\Rightarrow \quad W_{s_1} < W_{s_2}$$

$\Rightarrow$ S_{I} is false **Ans. (B)**

Sol. 19 Using work-energy theorem we have

$$\frac{1}{2}mv_i^2 + W_F = \frac{1}{2}mv_f^2$$

By area under F-x curve

$$W_{0.2s} = \frac{1}{2}\times 2\times 2 = 2\text{ J}$$

$$W_{2.5s} = -\frac{1}{2}\times 4\times 1 = -2\text{ J}$$

$$W_{5.8s} = \frac{1}{2}\times 3\times 3 = 4.5\text{ J}$$

$$\Rightarrow \quad k_f = \frac{1}{2}\times 2\times (5)^2 + [W_{0.2} + W_{2.5} + W_{5.8}]$$

$$\Rightarrow \quad k_f = 25 + [2 - 2 + 4.5] = 29.5\text{ J}$$

Ans. (D)

Sol. 20 By work-energy theorem

Gain (change) in $\quad KE$ = Work done on particle

$$\Delta k = \vec{F}\cdot\overrightarrow{\Delta r}$$

$$\Rightarrow \quad \Delta k = (7\hat{i} + 4\hat{j} + 3\hat{k})\cdot(2\hat{i} + 3\hat{j} + 4\hat{k})$$

$$\Rightarrow \quad \Delta k = 14 + 12 + 12 = 38\text{ J}$$

Ans. (A)

Sol. 21 For springs in series, tension remain same

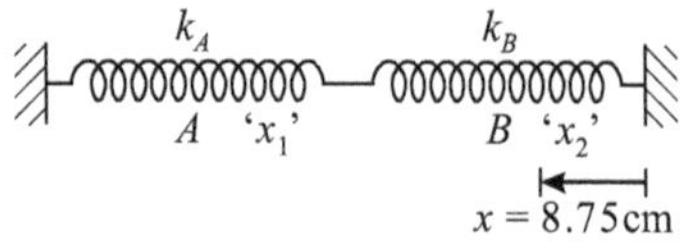

If compression in A and B are x_1 and x_2

$$\Rightarrow \quad k_A x_1 = k_B x_2$$

and $$3x_1 = 4x_2$$

$$x_1 + x_2 = 8.75$$

$$\Rightarrow \quad x_1 + \frac{3}{4}x_1 = 8.75$$

$$\Rightarrow \quad x_1 = \frac{8.75 \times 4}{7} = 5 \text{ cm}$$

$$\Rightarrow \quad x_2 = 3.75 \text{ cm}$$

Ratio of energy stored in A & B is

$$\frac{E_A}{E_B} = \frac{\frac{1}{2}k_A x_1^2}{\frac{1}{2}k_B x_2^2} = \frac{300 \times (5)^2}{400 \times (3.75)^2}$$

$$\Rightarrow \quad \frac{E_A}{E_B} = \frac{4}{3}$$

Ans. (A)

Sol. 22 If a is upward acceleration of man, we can use

$$v = \sqrt{2ah_1} = \sqrt{2gh_2}$$

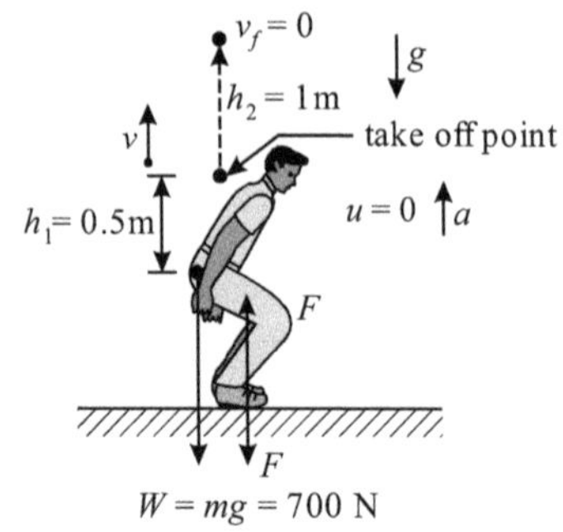

$$\Rightarrow \quad a = \frac{gh_2}{h_1} = \frac{10 \times 1}{0.5} = 20 \text{ m/s}^2$$

$\Rightarrow$ Velocity at take off is

$$v = \sqrt{2 \times 10 \times 1} = \sqrt{20} \text{ m/s}$$

Power delivered by muscles is

$$P = ma.v$$

$$\Rightarrow \quad P = 70 \times 20 \times \sqrt{20}$$

$$\Rightarrow \quad P = 6.26 \times 10^3 \text{ W}$$

Ans. (B)

Sol. 23 When the rubber band is stretched by a distance x than work done in further stretching it by dx is given as

$$dW = Fdx = (ax + bx^2)\,dx$$

$$\Rightarrow \quad W = \int dW = \int_0^L (ax + bx^2)dx$$

$$\Rightarrow \quad W = \left[\frac{ax^2}{2} + \frac{bx^3}{3}\right]_0^L$$

$$\Rightarrow \quad W = \frac{aL^2}{2} + \frac{bL^3}{3}$$

Ans. (C)

Sol. 24 To find KE of spring, we consider an elemental part of spring as shown mass of element is $dm = \frac{m}{l}dx$

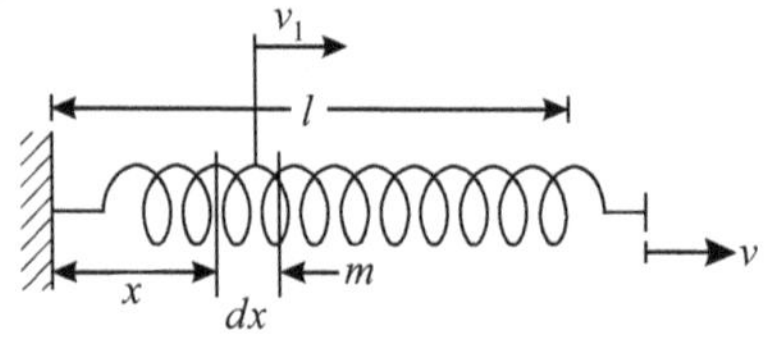

As spring is uniformly stretched, the speed of element is

$$v_1 = \frac{v}{l} \cdot x$$

KE of element is $dK = \frac{1}{2}dm v_1^2$

$$\Rightarrow \quad dK = \frac{1}{2}\left(\frac{m}{l}dx\right)\left(\frac{v}{l}x\right)^2$$

$$\Rightarrow \quad \int dK = \frac{1}{2}\frac{mv^2}{l^3}\int_0^l x^2 dx$$

Total KE of spring is given as

$$K = \frac{1}{2}\frac{mv^2}{l^3}\left(\frac{l^3}{3}\right) = \frac{1}{6}mv^2$$

Ans. (D)

Sol. 25 Wind volume flowing through the cross sectional area of blades (rotating) of wind mill per unit time is given as

$$V = Av$$

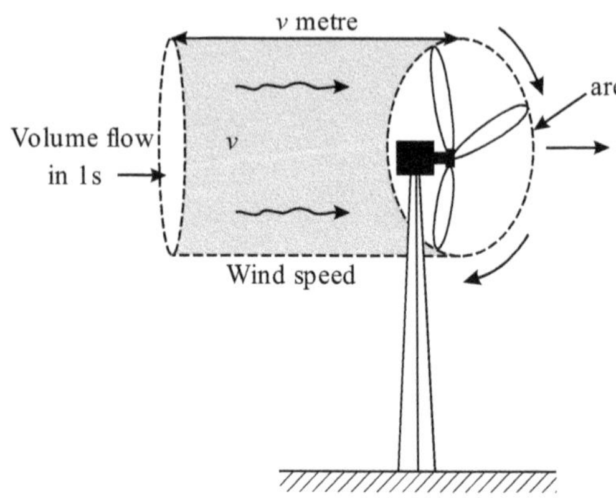

Mass of wind intercepting the blades per second is

$$m = \rho Av \quad [\rho \rightarrow \text{wind density of air}]$$

Kinetic energy of wind mass intercepting the section of blades per second is

$$\text{KE/sec} = \frac{1}{2}mv^2 = \frac{1}{2}\rho Av^3$$

$\Rightarrow$ Electrical power (J/s) $\propto v^3$ **Ans. (D)**

Sol. 26 If plank thickness is x, we have

$$u^2\left(1-\frac{1}{n}\right)^2 = u^2 - 2ax$$

$$\Rightarrow \quad 2ax = u^2\left[1-\left(\frac{n-1}{n}\right)^2\right]$$

$$\Rightarrow \quad 2ax = u^2\left[\frac{2n-1}{n^2}\right] \quad \dots(1)$$

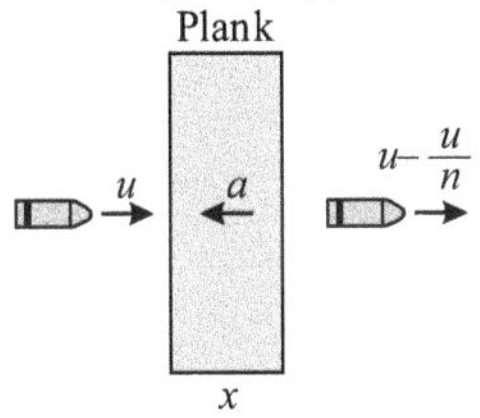

Using constant retardation, we have

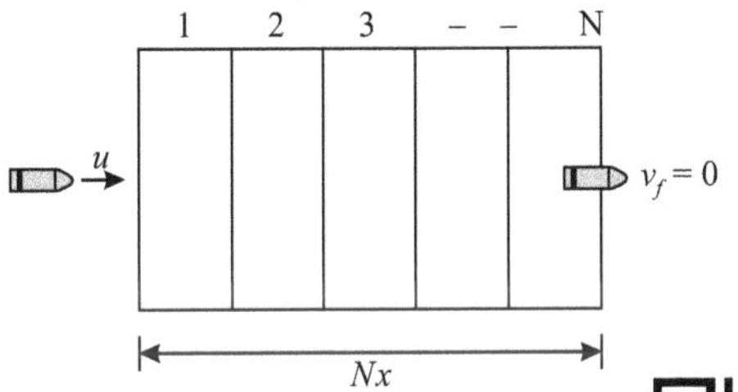

$$0 = u^2 - 2a(Nx)$$

$$\Rightarrow \quad u^2 = Nu^2\left[\frac{2n-1}{n^2}\right]$$

$$\Rightarrow \quad N = \frac{n^2}{2n-1}$$

Ans. (A)

Sol. 27 Using work energy theorem from point A to C, we have

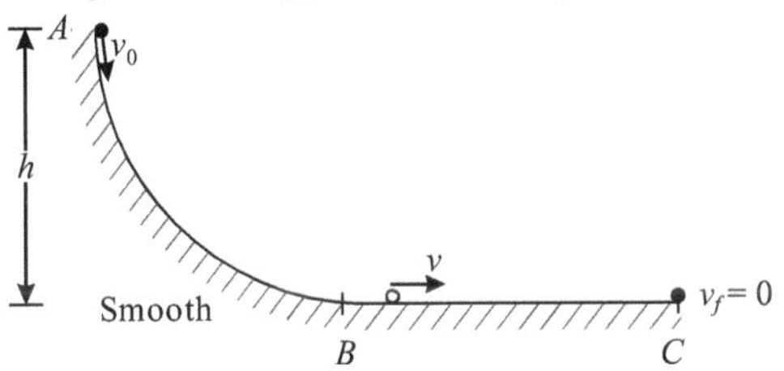

$$\frac{1}{2}mv_0^2 + mgh - \mu mg\,L = 0$$

$$\Rightarrow \quad L = \frac{v_0^2}{2\mu g} + \frac{h}{\mu} \quad \textbf{Ans. (B)}$$

Sol. 28

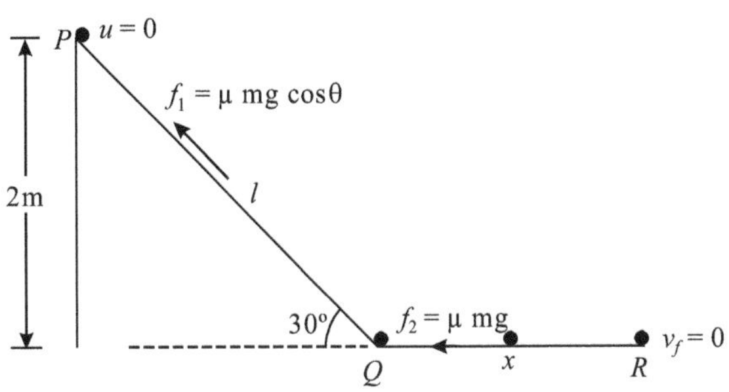

$$l = \frac{2}{\sin 30^\circ} = 4\,\text{m}$$

As work done by friction is equal in part PQ & QR

$$\mu mg \cos 30^\circ\,(\mu) = \mu mg\,x$$

$$\Rightarrow \quad x = 2\sqrt{3}\ \text{m} = 3.464\ \text{m} \simeq 3.5\ \text{m}$$

We use work energy theorem from P to R as

$$K_i(=0) + \text{mg}\,(2) - 2\mu\,mg\,x = K_f(=0)$$

$$\Rightarrow \quad x = 3.5 = \frac{1}{\mu}$$

$$\Rightarrow \quad \mu = \frac{1}{3.5} = 0.29$$

Ans. (C)

Sol. 29 Work done in lifting 10 kg mass by 1 m, 1000 times is

$$W = mgh \times 1000 = 10 \times 9.8 \times 1 \times 1000\ \text{J}$$

If m in kg mass of fat will be used in doing this work then energy released is

$$E = m \times 3.8 \times 10^7\,\text{J} \times \frac{20}{100}$$

By conservation of energy

$$m \times 3.8 \times 10^7 \times \frac{20}{100} = 10 \times 9.8 \times 1 \times 1000$$

$$\Rightarrow \quad m = 12.89 \times 10^{-3}\ \text{kg}$$

Ans. (D)

Sol. 30 Time function of body's speed

$$v = -5t + 50\ \text{m/s}$$

at $t = 0$; $v_1 = 50$ m/s

at $t = 2s$; $v_2 = -5 \times 2 + 50 = 40$ m/s

Work done on body is equal to gain in kinetic energy

$$\Rightarrow \quad W = \frac{1}{2}m(v_2^2 - v_1^2) = \frac{1}{2} \times 10 \times ((40)^2 - (50)^2) = -4500\ \text{J}$$

Ans. (C)

Sol. 31 To rise up at constant speed v_0, we have driving force

$$f_D = \left(\frac{W}{20} + \frac{W}{10}\right)$$

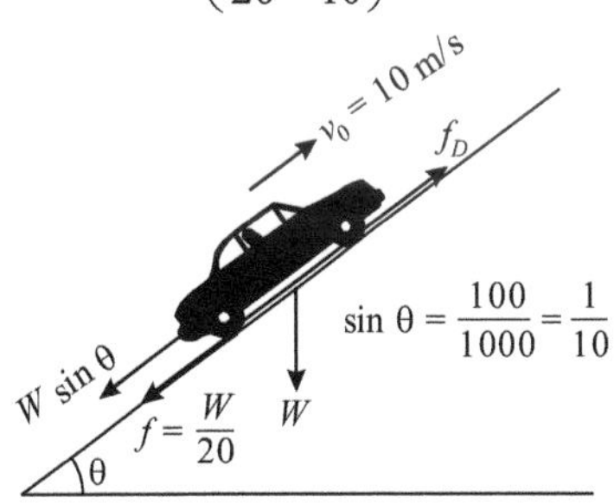

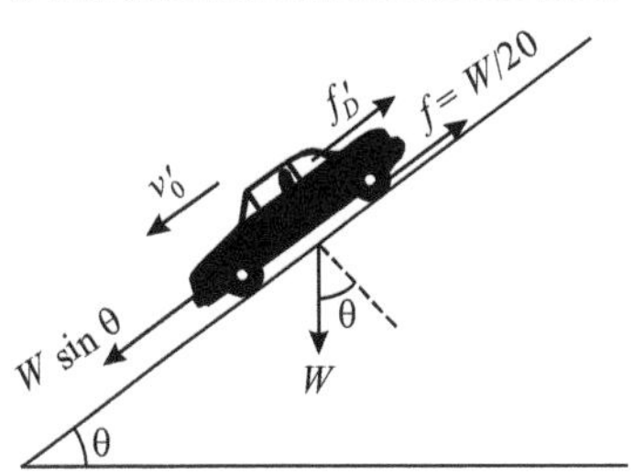

Given that power needed is

$$P = f_D v_0 = W\left(\frac{1}{20}+\frac{1}{10}\right)\times 10 \quad \ldots(1)$$

To move down at constant speed v, we have

$$f_D' = \left(\frac{W}{10}-\frac{W}{20}\right) \text{ and power is } \frac{P}{2}$$

$$\Rightarrow \quad \frac{P}{2} = f_D' v_0' = W\left(\frac{1}{10}-\frac{1}{20}\right)v_0' \quad \ldots(2)$$

$$\text{By (1)/(2)} \Rightarrow \quad 2 = \frac{3/2}{1/20v_0'}$$

$$\Rightarrow \quad v_0' = \frac{30}{2} = 15 \text{ m/s}$$ **Ans. (C)**

Sol. 32 Work done by force = Gain in KE = $\frac{1}{2}mv^2$

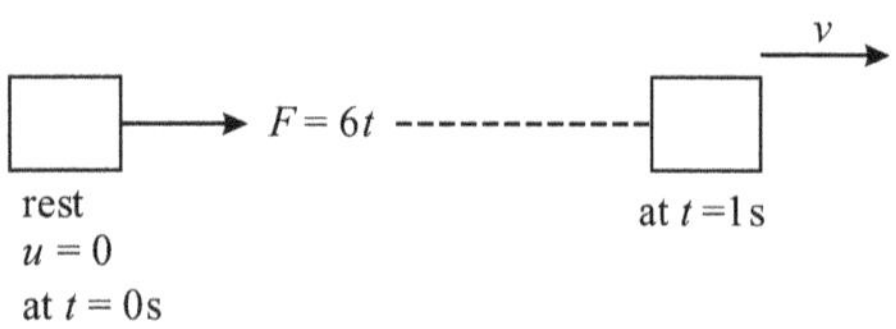

Acceleration of particle $a = \frac{F}{m} = \frac{6t}{1} = 6t \text{ m/s}^2$

$$\Rightarrow \quad \frac{dv}{dt} = 6t$$

$$\Rightarrow \quad \int_0^v dv = \int_0^1 6t\,dt$$

$$\Rightarrow \quad v = 6\left(\frac{t^2}{2}\right)_0^1 = 6\times\frac{1}{2} = 3 \text{ m/s}$$

$$= \frac{1}{2}\times 1\times(3)^2 = \frac{9}{2} = 4.5 \text{ J}$$ **Ans. (C)**

Sol. 33 At $t = 0$, $v = 10$ m/s

and at $t = 10$s, $\frac{1}{2}mv_f^2 = \frac{1}{8}m(10)^2$

$$\Rightarrow \quad v_f^2 = \frac{100}{4} = 25 \quad \Rightarrow v_f = 5 \text{ m/s}$$

$$F = -kv^2$$

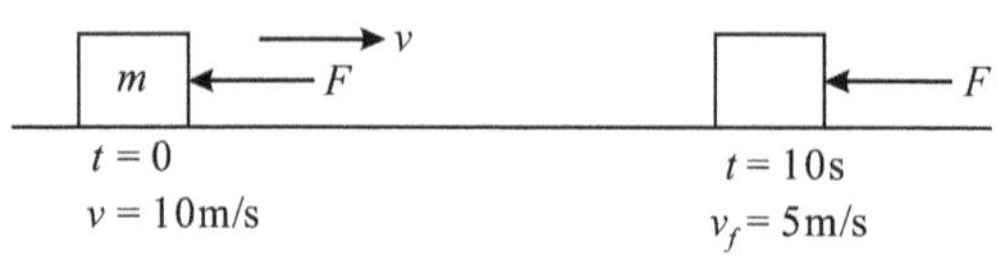

$$\Rightarrow \quad \text{Acceleration } a = \frac{F}{m} = -\frac{k}{m}v^2$$

$$\Rightarrow \quad \frac{dv}{dt} = -\frac{k}{m}v^2$$

$$\Rightarrow \quad \int_{10}^{5}\frac{dv}{v^2} = \int_0^{10}\frac{k}{m}dt$$

$$\Rightarrow \quad \left[+\frac{1}{v}\right]_{10}^{5} = +\frac{k}{m}[t]_0^{10}$$

$$\Rightarrow \quad \frac{1}{5}-\frac{1}{10} = \frac{k}{m}\times 10$$

$$\Rightarrow \quad k = 10^{-3}\left[\frac{1}{5}-\frac{1}{10}\right] = 10^{-4} \text{ kg/m}$$ **Ans. (A)**

Sol. 34 Force on particle in the given potential field is given as

$$\vec{F} = -\frac{dU}{dr}\hat{r} = \frac{k}{r^3}\hat{r}$$

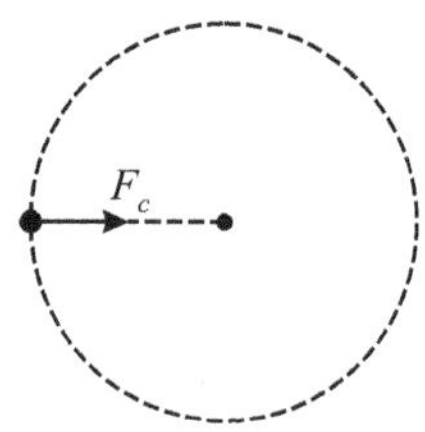

For stable circular motion we use

$$\frac{k}{r^3} = \frac{mv^2}{r}$$

$$\Rightarrow \quad mv^2 = \frac{k}{r^2}$$

$\Rightarrow$ The kinetic energy of particle is

$$\text{KE} = \frac{1}{2}\text{mv}^2 = \frac{k}{2r^2}$$

Total energy of particle TE = KE + PE

$$= \frac{k}{2r^2} - \frac{k}{2r^2} = 0$$ **Ans. (C)**

Sol. 35 If particle moves at speed v in circular orbit then

$$\frac{16}{r}+r^3 = \frac{mv^2}{r} \Rightarrow mv^2 = \left(\frac{16}{r}+r^3\right)\times r = 16 + r^4$$

Kinetic energy of particle $\frac{1}{2}mv^2 = \frac{1}{2}(16+r^4)$

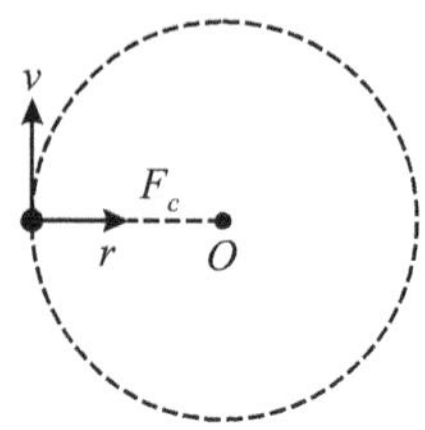

Ratio of kinetic energy of the two particles is given as

$$\frac{k_1}{k_2} = \frac{\frac{1}{2}(16+(1)^4)}{\frac{1}{2}(16+(4)^4)} = \frac{16+1}{16+256} = \frac{17}{272} = 0.06$$ **Ans. (B)**

Sol. 36 Velocity of body is

$$v = a\sqrt{s}$$

$$\Rightarrow \qquad \frac{ds}{dt} = a\sqrt{s}$$

$$\Rightarrow \qquad \int_0^s \frac{ds}{\sqrt{s}} = \int_0^t a\,dt$$

$$\Rightarrow \qquad 2\sqrt{s} = at$$

$$\Rightarrow \qquad s = \frac{1}{4}a^2t^2$$

$$\Rightarrow \qquad v = \frac{ds}{dt} = \frac{1}{2}a^2t$$

By work-energy theorem we have work done by all forces on body is

$$W = \Delta K = \frac{1}{2}mv^2 - 0$$

$$\Rightarrow \qquad W = \frac{1}{2}m\left(\frac{1}{2}a^2t\right)^2 = \frac{1}{8}ma^4t^2 \qquad \textbf{Ans. (A)}$$

Alternative :

$$W = F\cdot s \quad \left[\text{As } F = m\left(\frac{1}{2}a^2\right) \rightarrow \text{constant}\right]$$

$$\Rightarrow \quad W = \frac{1}{2}ma^2\left(\frac{1}{4}a^2t^2\right) = \frac{1}{8}ma^4t^2 \quad \textbf{Ans. (A)}$$

Sol. 37 Maximum speed is at mean position (equilibrium) where

$$x = \frac{F}{k}$$

Using work energy theorem, we have

$$F(x) - \frac{1}{2}kx^2 = \frac{1}{2}mv^2 - 0$$

$$\Rightarrow \quad F\left(\frac{F}{k}\right) - \frac{1}{2}k\left(\frac{F}{k}\right)^2 = \frac{1}{2}mv^2$$

$$\Rightarrow \qquad v_{max} = \frac{F}{\sqrt{mk}} \qquad \textbf{Ans. (C)}$$

Sol. 38 Given that

$$x = 3t^2 + 5$$

$$\Rightarrow \qquad v = \frac{dx}{dt} = 6t + 0$$

at $t = 0$; $v = 0$

and at $t = 5$ sec ; $v = 30$ m/s

By work energy theorem, we use

$$\text{Work done} = \frac{1}{2}mv^2 - 0$$

$$= \frac{1}{2}(2)(30)^2 = 900 \text{ J} \qquad \textbf{Ans. (B)}$$

Sol. 39 For block we use

$$N - mg = \frac{mg}{2}$$

$$\Rightarrow \qquad N = \frac{3mg}{2}$$

Work done by normal reaction is

$$W = Ns = \left(\frac{3mg}{2}\right)\left[\frac{1}{2}\left(\frac{g}{2}\right)t^2\right]$$

$$\Rightarrow \qquad W = \frac{3mg^2t^2}{8} \qquad \textbf{Ans. (B)}$$

Sol. 40 Work done

$$W = \vec{F}\cdot\vec{d} = (3\hat{i} - 12\hat{j})\cdot(4\hat{i})$$

$$\Rightarrow \qquad W = 12 \text{ J}$$

By work energy theorem, we use

$$K_i + W_{net} = K_f$$

$$\Rightarrow \qquad 3 + 12 = K_f$$

$$\Rightarrow \qquad K_f = 15 \text{ J} \qquad \textbf{Ans. (A)}$$

Sol. 41 Velocity of 1 kg block just before it collides with 3kg block

$$= \sqrt{2gh} = \sqrt{2000} \text{ m/s}$$

Applying momentum conversation just before and just after collision

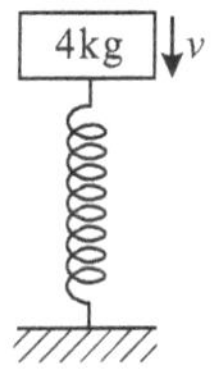

$$1\times\sqrt{2000} = 4v$$

$$\Rightarrow \qquad v = \frac{\sqrt{2000}}{4} \text{ m/s}$$

Initial compression of spring

$$1.25 \times 10^6 x_0 = 30$$

$$\Rightarrow \qquad x_0 \simeq 0$$

Applying work energy theorem, we have

$$W_g + W_{sp} = \Delta KE$$

$$\frac{1}{2}(m_1 + m_2)v^2 + (m_1 + m_2)gx + \frac{1}{2}kx^2 = 0$$

$$\frac{1}{2} \times 4 \times v^2 + 40 \times x - \frac{1}{2} \times 1.25 \times 10^6 x^2 = 0$$

Solving, we get $x \simeq 4$ cm **Ans. (A)**

Sol. 42 According to work energy theorem

Work done by force on the particle = Change in KE

Work done = Area under F-x graph

$$\Rightarrow \qquad W = \int F.dx$$

$$\Rightarrow \qquad W = 2\times 2 + \frac{(2+3)\times 1}{2}$$

$$\Rightarrow \qquad W = KE_{final} - KE_{initial} = 6.5$$

As $KE_{initial} = 0$

$$\Rightarrow \qquad KE_{final} = 6.5 \text{ J} \qquad \textbf{Ans. (A)}$$

Sol. 43 Mass of the hanging part $= \frac{M}{n}$

$$h_{COM} = \frac{L}{2n}$$

Work done by gravity in lifting the cable is

$$W = mgh_{COM} = \left(\frac{M}{n}\right)g\left(\frac{L}{2n}\right) = \frac{MgL}{2n^2}$$ **Ans. (B)**

Sol. 44 Considering gravity, acceleration of ball is

$$\frac{dv}{dt} = -(g + \gamma v^2)$$

$$\Rightarrow \quad \frac{g}{\gamma}\left(\frac{dv}{\frac{g}{\gamma} + v^2}\right) = -gdt$$

Integrating in limits from

$t = 0$ to $t = t$ and $v = V_0$ to $v = 0$

$$\int_{V_0}^{0} \frac{g}{\gamma}\left(\frac{dv}{\frac{g}{\gamma} + v^2}\right) = -\int_0^t gdt$$

$$\Rightarrow \quad -gt = -\sqrt{\frac{g}{\gamma}}\tan^{-1}\left(\frac{V_0}{\sqrt{\frac{g}{\gamma}}}\right)$$

$$\Rightarrow \quad t = \frac{1}{\sqrt{\gamma g}}\tan^{-1}\left(\sqrt{\frac{\gamma}{g}}V_0\right)$$ **Ans. (B)**

Sol. 45 For a spring of length l and force constant k when cut in length l_1 and l_2 then force constants of these parts are given as

$$k_1 = \frac{kl}{l_1} \quad \ldots(1)$$

$$k_2 = \frac{kl}{l_2}$$

$$\Rightarrow \quad \frac{k_1}{k_2} = \frac{l_2}{l_1}$$

$$\Rightarrow \quad \frac{k_1}{k_2} = \frac{l_2}{nl_2} = \frac{1}{n}$$ **Ans (C)**

5.2 Circular Motion

Sol. 1 Friction provides the necessary centripetal force for circular motion so we use

$$f = \frac{mv^2}{R}$$

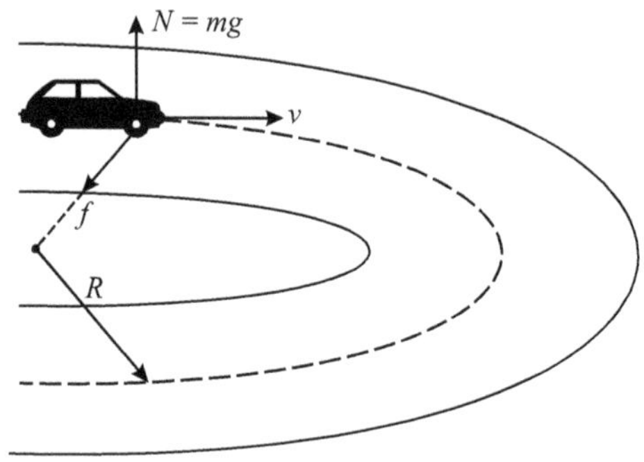

For skidding $\quad f = \frac{mv^2}{R} \geq \mu mg$

to avoid skidding

$$f = \frac{mv^2}{R} \leq \mu mg$$

$$\Rightarrow \quad v_{max} = \sqrt{\mu Rg} = \sqrt{0.6 \times 150 \times 10}$$

$$\Rightarrow \quad v_{max} = \sqrt{900} = 30 \text{ m/s}$$ **Ans. (B)**

Sol. 2

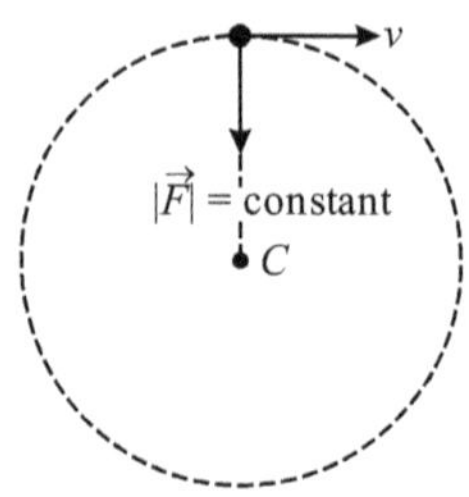

$$|\vec{F}| = \frac{mv^2}{R} \text{ (centripetal force)} = \text{constant}$$

$\Rightarrow \quad |v| \to$ constant

$\Rightarrow \quad$ KE $\to$ constant **Ans. (C)**

Sol. 3

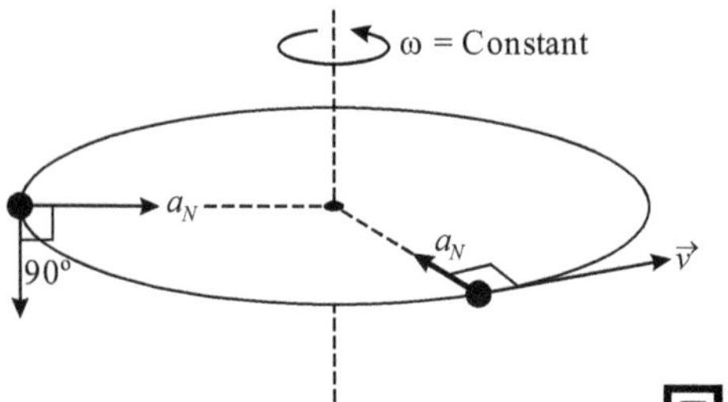

$v = R\omega$ constant

$\Rightarrow$ tangential acceleration $= 0$

Here normal acceleration $a_N = \frac{v^2}{R} = \omega^2 R$ (constant in magnitude)

Option (B) is false here. **Ans. (B)**

Sol. 4

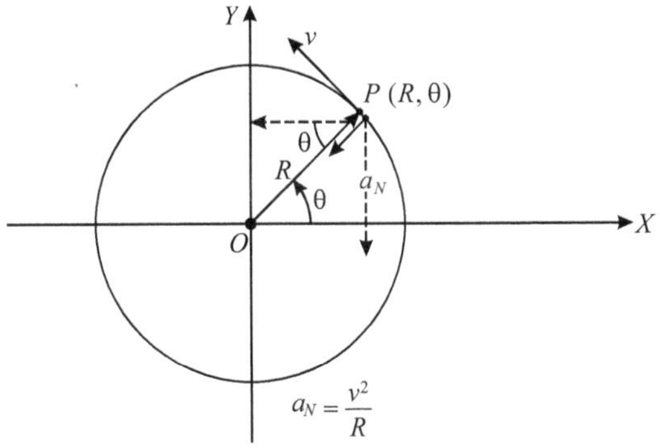

As in uniform circular motion $a_{\text{tangential}} = 0$

and only centripetal acceleration is there so its vector is given as

$$\vec{a} = -a_N \cos\theta\,\hat{i} - a_N \sin\theta\,\hat{j}$$

$$\vec{a} = -\frac{v^2}{R}\cos\theta\,\hat{i} - \frac{v^2}{R}\sin\theta\,\hat{j} \quad \textbf{Ans. (C)}$$

Sol. 5

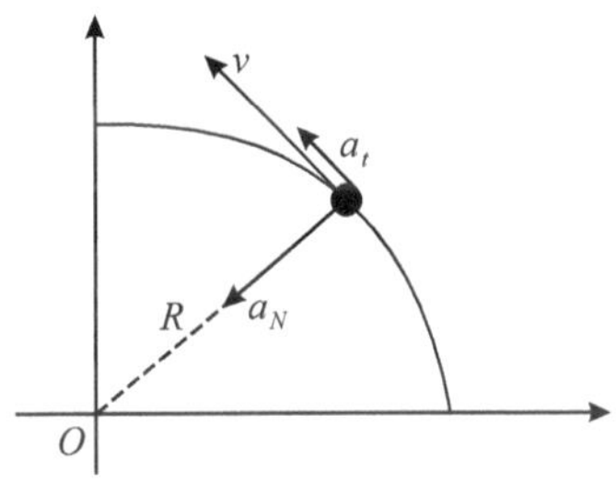

Tangential acceleration of point P is given as

$$a_t = \frac{dv}{dt} = \frac{d^2v}{dt^2} = 6t$$

at $t = 2\text{s}$, $a_t = 6 \times 2 = 12\ \text{m/s}^2$

$$s = t^3 + 5$$

$$\Rightarrow \quad v = \frac{ds}{dt} = 3t^2$$

at $\quad t = 2\text{s}$

$$v = 3 \times 4$$

$$\Rightarrow \quad v = 12\ \text{m/s}$$

Normal acceleration of point P is given as

at $\quad t = 2\text{s},\ a_N = \frac{v^2}{R} = \frac{144}{20} = 7.2\ \text{m/s}^2$

Total acceleration of point P at $t = 2\text{s}$ is

$$a_T = \sqrt{a_t^2 + a_N^2}$$

$$\Rightarrow \quad a_T = \sqrt{(12)^2 + (7.2)^2}$$

$$\Rightarrow \quad a_T = 13.99\ \text{m/s}^2 \quad \textbf{Ans. (D)}$$

Sol. 6 As cars complete the circles in same time $\Rightarrow$ Their angular speeds are equal

Centripetal acceleration of cars

$$a_{N_1} = \omega^2 r_1$$

$$a_{N_2} = \omega^2 r_2$$

$$\Rightarrow \quad \frac{a_{N_1}}{a_{N_2}} = \frac{r_1}{r_2}$$

Ans. (C)

Sol. 7

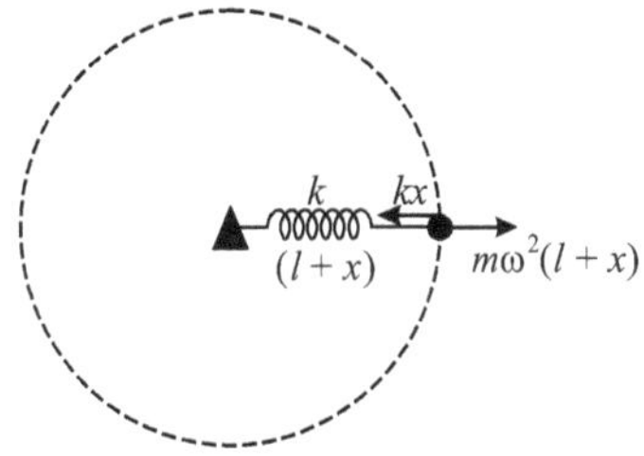

Case-I : $\quad x = 1\ \text{cm}$

If original length of spring is l, we use

for circular motion of body we have along radial direction

$$x = 1\ \text{cm}$$

$$k(1) = m\omega^2(l+1) \quad \ldots(1)$$

Case-II : $\quad x = 5\ \text{cm}$

$$\omega \to 2\omega$$

$$k(5) = m(2\omega)^2(l+5) \quad \ldots(2)$$

$$(2)/(1) \quad \Rightarrow \quad \frac{4(l+5)}{l+1} = \frac{5}{1}$$

$$\Rightarrow \quad 4l + 20 = 5l + 5$$

$$\Rightarrow \quad l = 15\ \text{cm} \quad \textbf{Ans. (A)}$$

Sol. 8

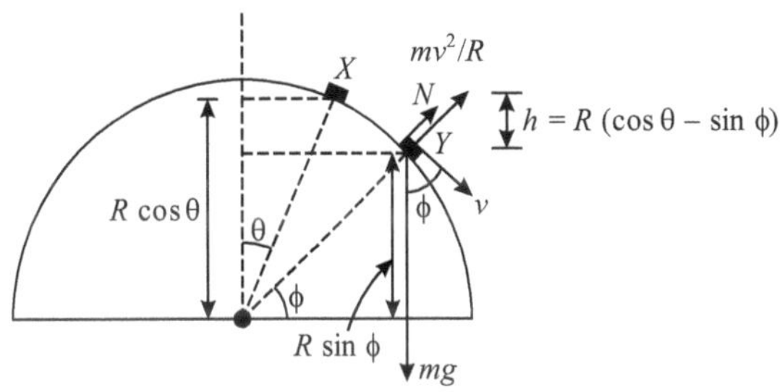

By work energy theorem, we use

$$mgh = \frac{1}{2}mv^2$$

$$\Rightarrow \quad v = \sqrt{2gR(\cos\theta - \sin\phi)}$$

At point Y we have along radial direction in circular motion (in frame of particle)

$$N + \frac{mv^2}{R} = mg\sin\phi$$

As $N = 0$

$\Rightarrow \quad 2\,mg(\cos\theta - \sin\phi) = mg\sin\phi$

$\Rightarrow \quad 2\cos\theta = 3\sin\phi$ **Ans. (C)**

Sol. 9

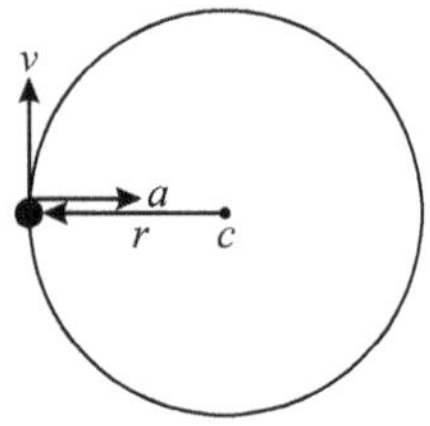

For uniform circular motion

$$a_{\tan} = 0$$

and $$a_N = \frac{v^2}{r} \text{ (Centripetal acceleration)}$$

$$\Rightarrow \quad a_{total} = a_N = \frac{v^2}{r}$$

$$a \propto \frac{1}{r}$$

Ans. (C)

Sol. 10

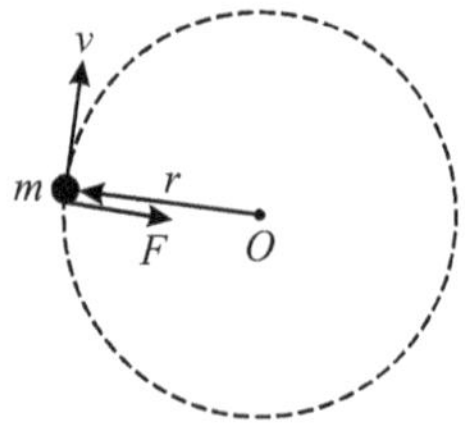

The centripetal force $F = \alpha r^2$

The PE of particle in the field of this force is given as

$$U = -\int_0^r \vec{F}\cdot\vec{dr} = \int_0^r F dr = \int_0^r \alpha r^2 dr = \frac{\alpha r^3}{3} - 0$$

We also use

$$F = \frac{mv^2}{r} = \alpha r^2 \quad \Rightarrow \quad \text{KE} = \frac{1}{2}mv^2 = \frac{1}{2}\alpha r^3$$

Total mechanical energy $= \frac{1}{2}\alpha r^3 + \frac{1}{3}\alpha r^3 = \frac{5}{6}\alpha r^3$ **Ans. (B)**

Sol. 11

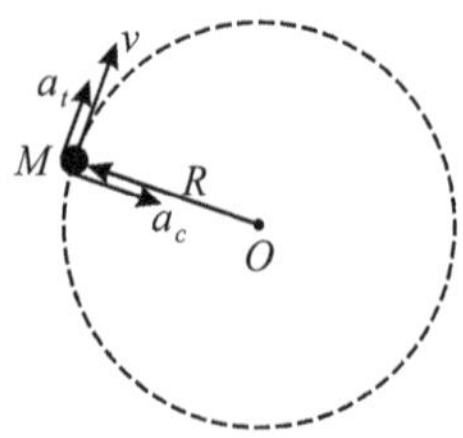

Centripetal acceleration of particle is given as

$$a_c = n^2Rt^2 = \frac{v^2}{R}$$

$$\Rightarrow \quad v^2 = n^2R^2t^2$$

$$\Rightarrow \quad v^2 = nRt \quad \Rightarrow \quad a_t = \frac{dv}{dt} = nR$$

As v increases, tangential acceleration is in the direction of velocity

Power delivered to the particle is given as

$$P = ma_tv = m(nR)(nRt)$$

$$\Rightarrow \quad P = mn^2R^2t$$

Ans. (B)

Sol. 12

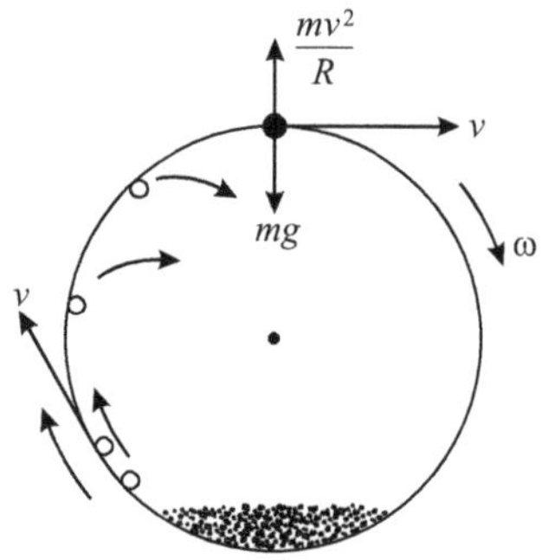

In the frame of rotating drum at the topmost point particle (mixture ingredients) will not loose contact from drum wall if

$$\frac{mv^2}{R} \geq mg \quad \Rightarrow \quad v \geq \sqrt{Rg}$$

If $v \geq \sqrt{Rg}$ than proper mixing will not take place

Thus for proper mixing $v < \sqrt{Rg}$

$$\Rightarrow \quad \omega = \frac{v}{R} = \sqrt{\frac{g}{R}} = \sqrt{\frac{10}{1.25}} \text{ rad/s}$$

ω in rpm is $\omega_{RPM} = \frac{60}{2\pi}\sqrt{\frac{10}{1.25}} \gtrsim 27 \text{ rpm}$

$\Rightarrow$ If $\omega \simeq 27$ rpm then before reaching top, ingredients fall and mix properly **Ans. (A)**

Sol. 13

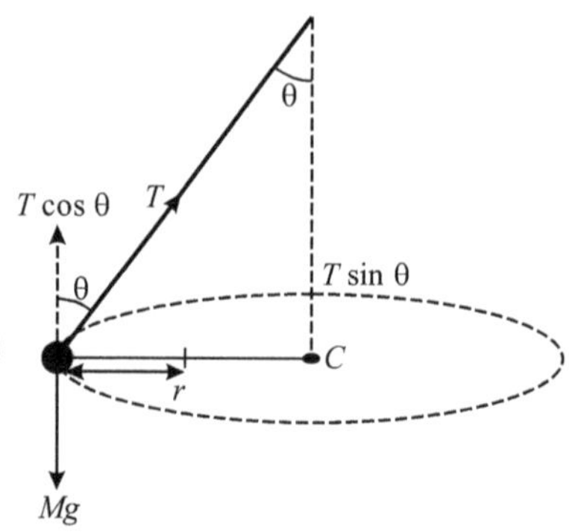

For stable circular motion, we use

$$T\sin\theta = \frac{mv^2}{r} \quad \ldots(1)$$

$$\text{T}\cos\theta = \text{mg} \quad \ldots(2)$$

$$(1)/(2) \quad \Rightarrow \quad \tan\theta = \frac{v^2}{rg}$$

$$\Rightarrow \quad v = \sqrt{rg\tan\theta}$$

$$\Rightarrow \quad v = \sqrt{0.4\times10\times1} = 2 \text{ m/s}$$

Ans. (D)

Sol. 14

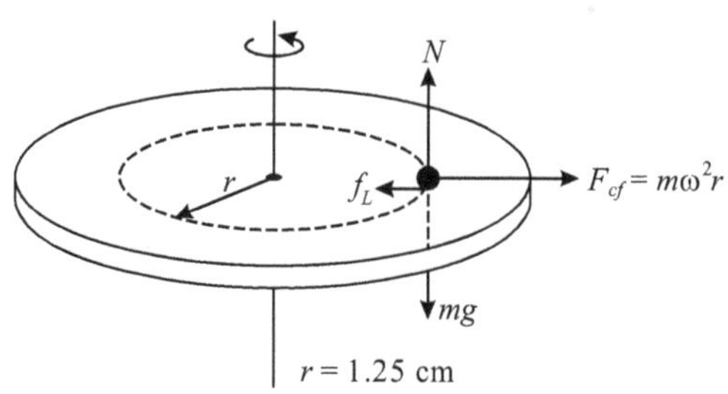

Angular speed $\omega = 2\pi n = 2\pi \times 3.5 = 7\pi$ rad/s

in frame of disc $\quad F_{cf} = m\omega^2 r$

If coin is at rest on disc

$\Rightarrow \quad m\omega^2 r \le f_L$

$\Rightarrow \quad m\omega^2 r \le \mu g m$

$\Rightarrow \quad \mu \ge \dfrac{\omega^2 r}{g} = \dfrac{(7\times 3.14)^2 \times (1.25\times 10^{-2})}{10}$

$\Rightarrow \quad \mu = 0.6$ **Ans. (D)**

Sol. 15

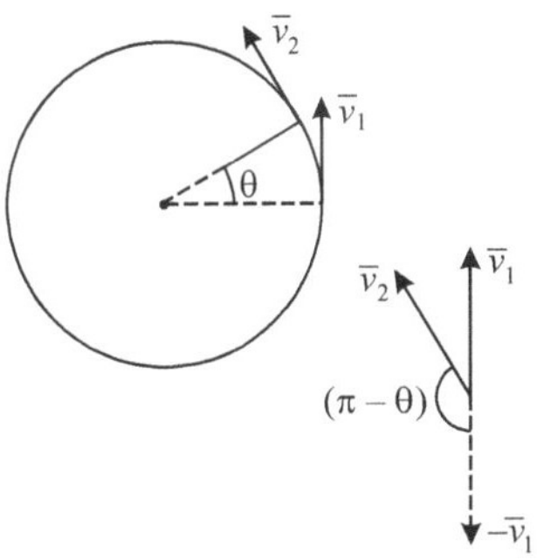

For above circular motion, change in velocity is

$$|\Delta \bar{v}| = \sqrt{v_1^2 + v_2^2 + 2v_1 v_2 \cos(\pi - \theta)}$$

$\Rightarrow \quad |\Delta \bar{v}| = 2v \sin\dfrac{\theta}{2} \qquad$ since $[|\bar{v}_1| = |\bar{v}_2|]$

$\Rightarrow \quad |\Delta \bar{v}| = (2\times 10)\times \sin(30°)$

$\Rightarrow \quad |\Delta \bar{v}| = 10$ m/s **Ans. (B)**

Sol. 16 At $t = \dfrac{\pi}{2\omega}$ angle transversed by particles is

$$\theta = \omega t = \omega\frac{\pi}{2\omega} = \frac{\pi}{2}$$

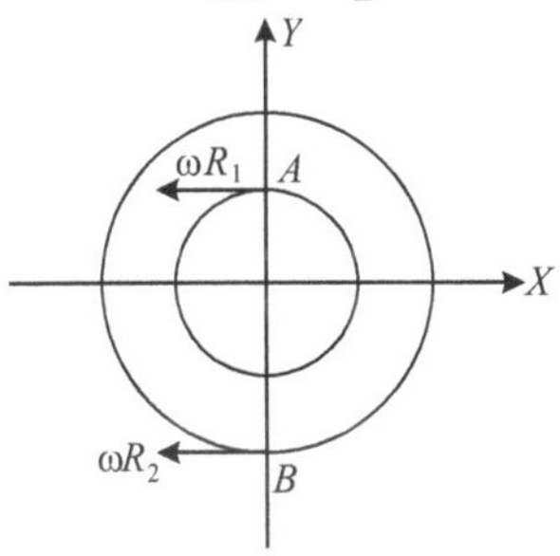

$\Rightarrow \quad \vec{v}_A - \vec{v}_B = \omega R_1(-\hat{i}) - \omega R_2(-i)$

$\qquad \omega = \omega(R_2 - R_1)\hat{i}$ **Ans. (D)**

Sol. 17 FBD of bead is shown in figure below

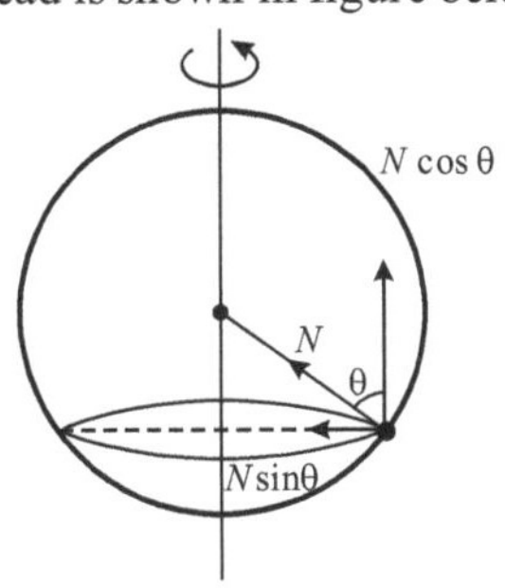

For equilibrium of bead in frame of wire, we have

$$N\sin\theta = m\frac{r}{2}\omega^2 \qquad \ldots(1)$$

$$N\cos\theta = mg \qquad \ldots(2)$$

$\Rightarrow \quad \tan\theta = \dfrac{r\omega^2}{2g}$

$\Rightarrow \quad \dfrac{r}{2\dfrac{\sqrt{3}r}{2}} = \dfrac{r\omega^2}{2g}$

$\Rightarrow \quad \omega^2 = \dfrac{2g}{\sqrt{3}r}$ **Ans. (D)**

6.1 System of Particles

Sol. 1

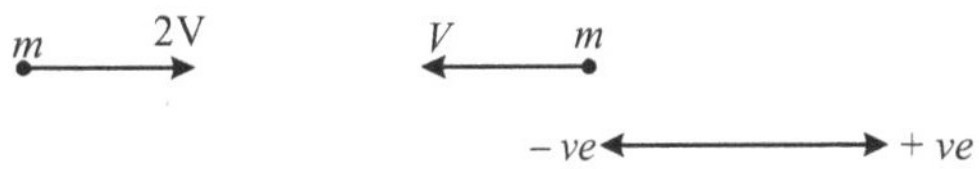

Velocity of centre of mass is given as

$$\vec{v}_{cm} = \frac{m_1\vec{V}_1 + m_2\vec{V}_2}{m_1 + m_2} = \frac{m(2V) + m(-v)}{2m} = \frac{v}{2}$$

Ans. (C)

Sol. 2 Linear Momentum → vector

Kinetic energy → scalar

If KE = 0 $\Rightarrow$ Speeds of all particles of system will be individually zero

$\Rightarrow$ linear momentum = 0 **Ans. (C)**

Sol. 3

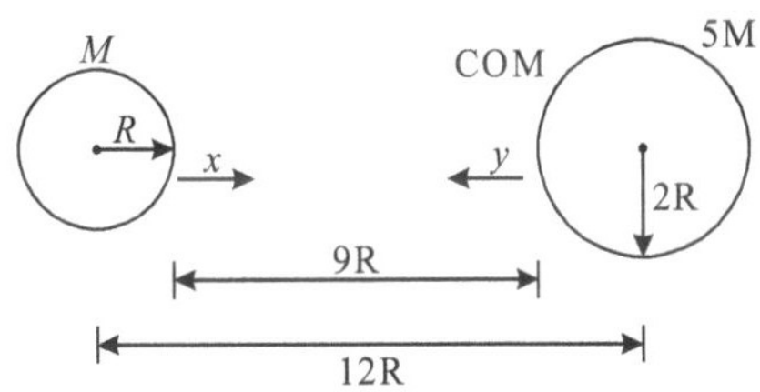

Due to internal gravitational force bodies move such that their centre of mass remain at rest

If M and 5 M travels distances x and y before collision

$\Rightarrow \quad x + y = 9R$

In motion displacement of centre of mass is zero

$\Rightarrow \quad \Delta x_{CM} = \dfrac{M\Delta x_1 + 5M\Delta x_2}{M + 5M} = 0$

$\Rightarrow \quad \Delta x_{CM} = Mx - 5M(9R - x) = 0$

$\Rightarrow \quad x = 7.5R$ **Ans. (C)**

Sol. 4

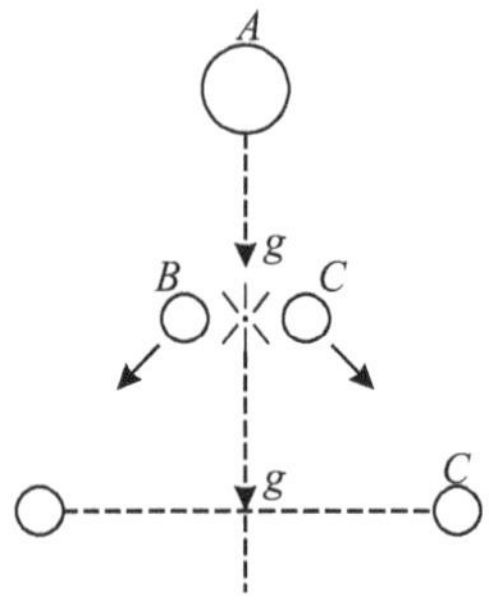

Breaking is due to internal forces which do not affect the motion of centre of mass

$\Rightarrow$ Centre of mass of bodies B & C continue to move as if body A is not broken. **Ans. (B)**

Sol. 5

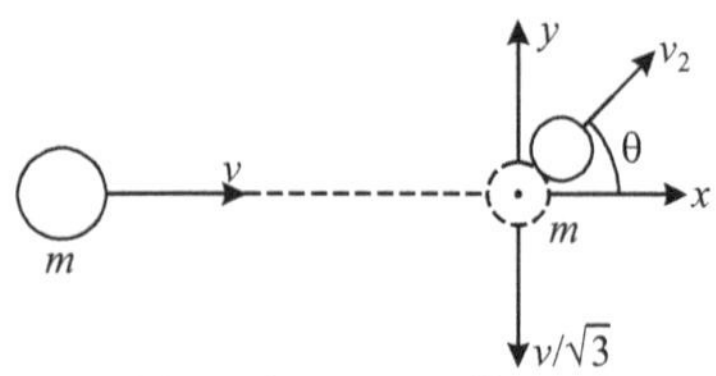

As no external force is present linear momentum of system will remain conserved so we use

in x-dir $\quad mv = mv_2 \cos\theta$

in y-dir $\quad O = mv_2 \sin\theta - \dfrac{mv}{\sqrt{3}}$

$\Rightarrow \quad v_2 \cos\theta = v \quad \ldots(1)$

and $\quad v_2 \sin\theta = \dfrac{v}{\sqrt{3}} \quad \ldots(2)$

Squaring and adding equation-(1) and (2) gives

$$v_2 = \sqrt{v^2 + \frac{v^2}{3}} = \sqrt{\frac{4v^2}{3}} = \frac{2v}{\sqrt{3}}$$ **Ans. (C)**

Sol. 6 For only translational motion force $\vec{F}$ must be applied at COM of system

Distance PC is given as

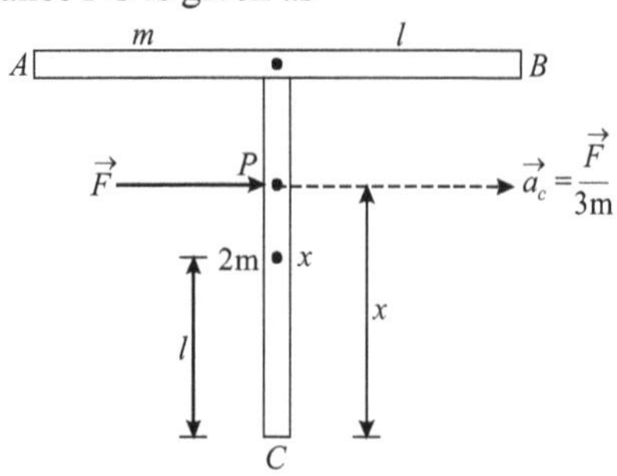

$$x = \frac{m_1x_1 + m_2x_2}{M_1 + m_2} = \frac{2m(l) + m(2l)}{2m + m} = \frac{4l}{3}$$

Ans. (C)

Sol. 7

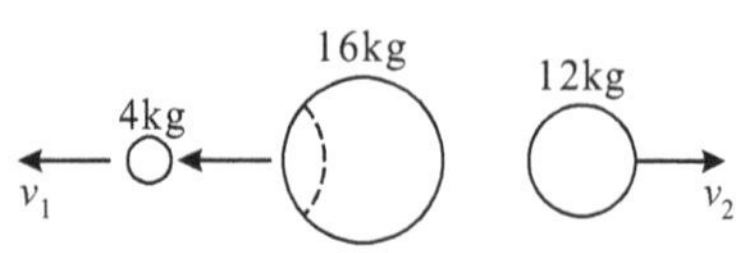

Rest

$$P_i = 0 = P_f$$

$\Rightarrow \quad 0 = 12v_2 - 4v_1$

$\Rightarrow \quad 0 = 12 \times 4 - 4v_1$

$\Rightarrow \quad v_1 = 12$ m/s

KE of 4 kg mass is

$$k = \frac{1}{2} m_1 v_1^2$$

$\Rightarrow \quad k = \dfrac{1}{2} \times 4 \times (12)^2$

$\Rightarrow \quad k = 288$ J **Ans. (B)**

Sol. 8

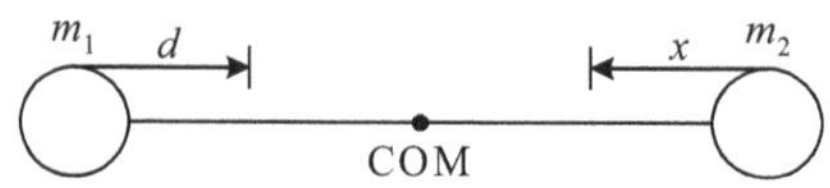

Displacement of COM

$$\Delta x_{COM} = 0$$

We use

$$\vec{\Delta x}_{COM} = \frac{m_1\overrightarrow{\Delta x_1} + m_2\overrightarrow{\Delta x_2}}{m_1 + m_2}$$

$\Rightarrow \quad 0 = \dfrac{m_1 d + m_2 x}{m_1 + m_2}$

$\Rightarrow \quad x = -\dfrac{m_1}{m_2} d$

[$-ve$ sign indicates opposite direction]

Ans. (C)

Sol. 9

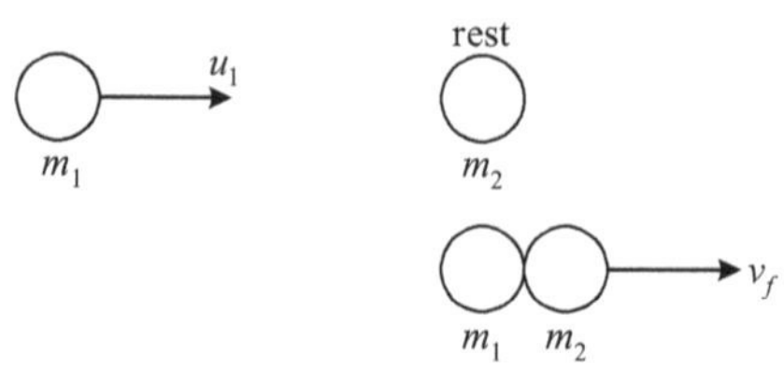

By conservation of linear momentum

$$m_1u_1 = (m_1 + m_2)V_f$$

$\Rightarrow \quad V_f = \dfrac{m_1u_1}{m_1 + m_2}$

loss in energy $\quad \Delta E_L = k_i - k_f$

$\Rightarrow \quad \Delta E_L = \dfrac{1}{2} m_1 u_1^2 - \dfrac{1}{2}(m_1 + m_2) v_f^2$

$\Rightarrow \quad \Delta E_L = \dfrac{1}{2}\left(\dfrac{m_1 m_2}{m_1 + m_2}\right) u_1^2$

$\Rightarrow \quad \Delta E_L = \dfrac{1}{2} \times \dfrac{0.5 \times 1}{0.5 + 1} \times (2)^2 = \dfrac{2}{3}$ J **Ans. (C)**

Sol. 10 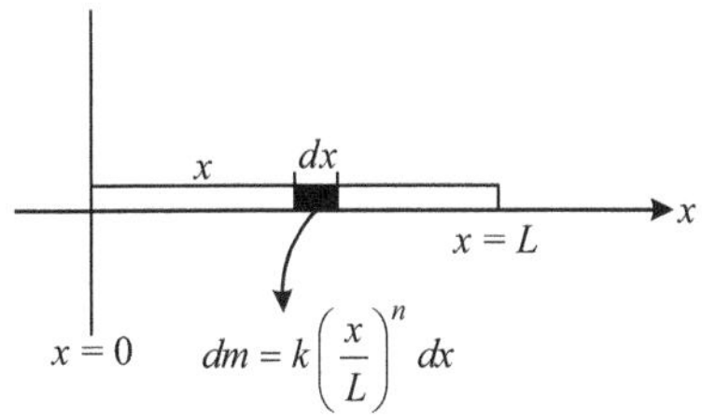

Mass of rod $M = \int dm = \int_0^L \frac{k}{L^n} \cdot \left[\frac{x^{n+1}}{n+1}\right]_0^L = \left(\frac{k}{n+1}\right)L$

Centre of mass position of rod

$$x_{cm} = \frac{1}{M}\int dmx$$

$$\Rightarrow \quad x_{cm} = \frac{1}{\left(\frac{k}{n+1}\right)L} \cdot \int_0^L \frac{k}{L^n} \cdot x^{n+1} dx$$

$$\Rightarrow \quad x_{cm} = \frac{n+1}{L^{n+1}}\left(\frac{x^{n+2}}{n+2}\right)_0^L = \left(\frac{n+1}{n+2}\right)L$$

at $n = 0 \rightarrow x_{cm} = \frac{L}{2}$ and at $n \rightarrow \infty$ we have $x_{cm} = L$ **Ans. (A)**

Sol. 11 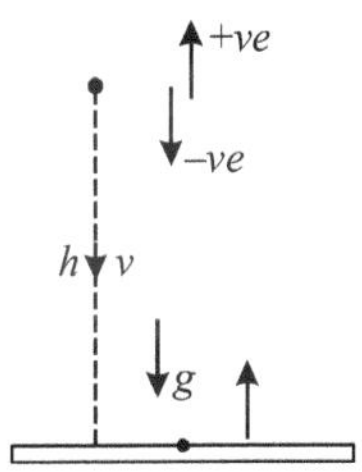

$$v = gt$$ **Ans. (A)**

Sol. 12 Slope of x-t curve from $t = 0$ to $t = 2$ s

$$v_1 = \frac{2}{2} = 1 \text{ m/s} \quad +ve$$

Slope of x-t curve from $t = 2$ s to $t = 4$ s

$$v_2 = \frac{2}{2} = -1 \text{ m/s}$$

Impulse

$$\Rightarrow \quad \vec{J} = \Delta\vec{P} = m\vec{v}_2 - m\vec{v}_1$$

$$\Rightarrow \quad |\vec{J}| = |0.4(-1-1)|$$

$$|\vec{J}| = 0.4 \times 2 = 0.8 \text{ Ns}$$ **Ans. (B)**

Sol. 13 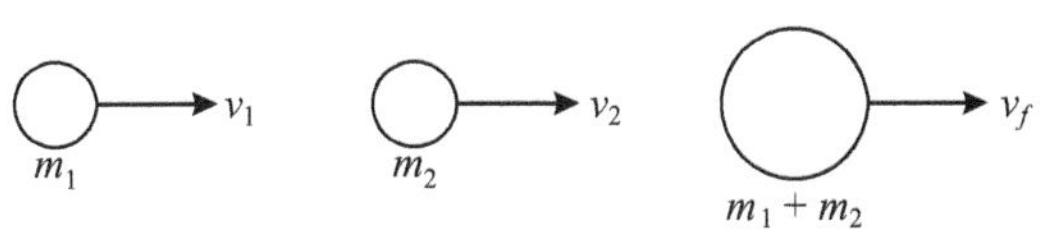

$$m_1v_1 + m_2v_2 = (m_1 + m_2)v_f$$

$$\Rightarrow \quad v_f = \frac{m_1V_1 + m_2V_2}{m_1 + m_2} \neq 0$$

$\Rightarrow$ $k_f \neq 0$ **Ans. (B)**

$\Rightarrow$ S-1 and S-2 are true and S-2 is explaining S-1

Sol. 14 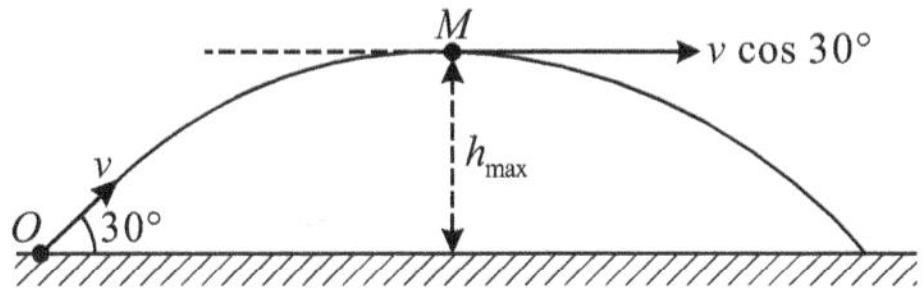

Maximum height of projectile

$$h_{max} = \frac{v^2 \sin^2(30°)}{2g}$$

Angular momentum of particle at M about O is given as

$$L_{MO} = m(V\cos 30°) \times h_{max}$$

$$\Rightarrow \quad L_{MO} = mv\left(\frac{\sqrt{3}}{2}\right) \times \frac{v^2}{2g}\left(\frac{1}{4}\right) = \frac{\sqrt{3}mv^3}{16g}$$

Ans. (D)

Sol. 15 As finally blocks come to rest

$$\Rightarrow \quad m_1V_1 = m_2V_2$$

$$\Rightarrow \quad V_{CM} = 0$$

Thus COM of system will remain at rest

$\Rightarrow$ displacement of $COM = 0$

Thus we use

$$\Delta x_{cm} = \frac{m_2x_2 - m_1x_1}{m_1 + m_2} = 0$$

$$\Rightarrow \quad m_1x_1 = m_2x_2$$

$$\Rightarrow \quad \frac{x_1}{x_2} = \frac{m_2}{m_1}$$ **Ans. (A)**

Sol. 16 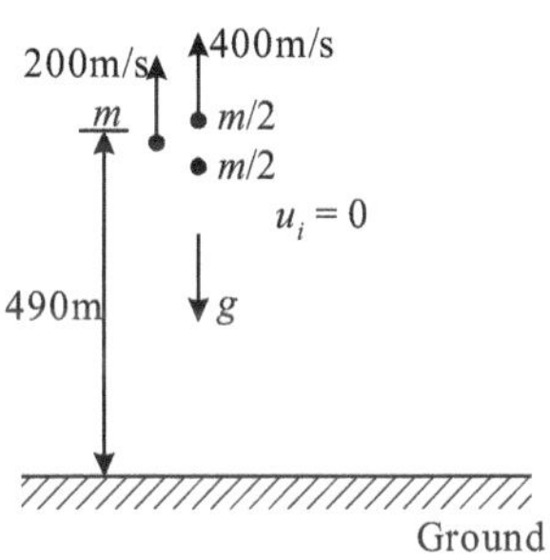

Initial upward momentum of moving body is

$$P_i = m(200)$$

Final upward momentum of $m/2$ (one part) is

$$P_f = \frac{m}{2}(400) = m(200)$$

As the moving part carries the same momentum as that of the initial body, No momentum is left in the other part

$$\Rightarrow \quad v_{\text{other part}} = 0$$

As it is in free fall, time taken to hit the ground is

$$t = \sqrt{\frac{2h}{g}} = \sqrt{\frac{2 \times 490}{9.8}} = \sqrt{100} = 10 \text{ s}$$ **Ans. (C)**

Sol. 17

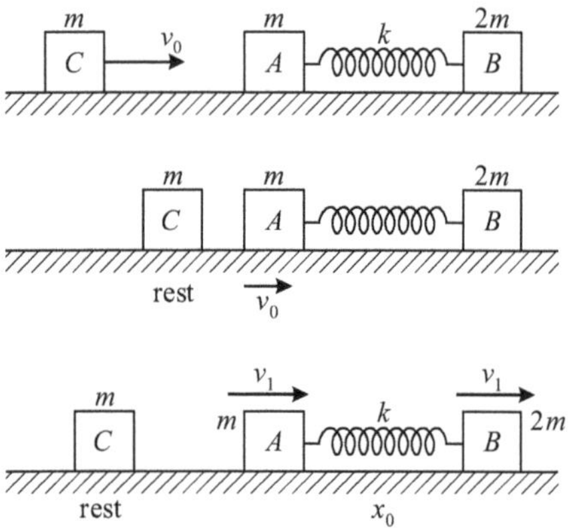

By conservation of linear momentum, we have

$$mv_0 = 3mv_1 \quad \Rightarrow \quad v_1 = \frac{v_0}{3}$$

By conservation of energy, we have

$$\frac{1}{2}mv_0^2 = \frac{1}{2}(3m)\left(\frac{v_0}{3}\right)^2 + \frac{1}{2}kx_0^2$$

$$\Rightarrow \quad \frac{1}{2}mv_0^2\left(1-\frac{1}{3}\right) = \frac{1}{2}kx_0^2$$

$$\Rightarrow \quad k = \frac{2}{3}mv_0^2\left(\frac{1}{x_0^2}\right)$$

Ans. (D)

Sol. 18

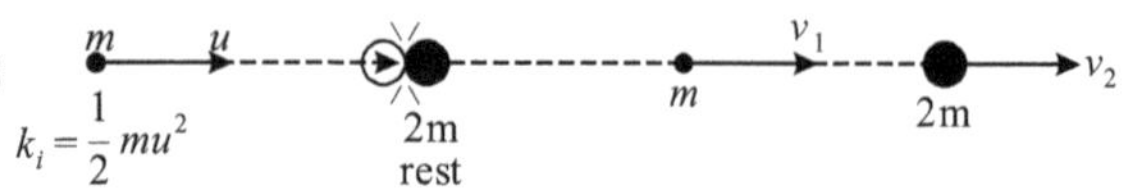

For elastic collision (Head-on), The final velocity of particle m is given as

$$v_1 = \left(\frac{m_1 - m_2}{m_1 + m_2}\right)u_1 + \left(\frac{2m_2}{m_1 + m_2}\right)u_2$$

$$\Rightarrow \quad v_1 = \left(\frac{m-2m}{m+2m}\right)u + 0 = -\frac{1}{3}u$$

Final KE of particle – 1 is

$$k_f = \frac{1}{2}m\left(\frac{u}{3}\right)^2 = \frac{1}{2}mu^2\left(\frac{1}{9}\right)$$

% loss in KE of first particle is % loss

$$\Delta k_L = \frac{k_i - k_f}{k_i}\times 100 = \left(1-\frac{1}{9}\right)\times 100\ \%$$

$$\Rightarrow \quad \Delta k_L \simeq 89\%$$

Ans. (C)

Sol. 19 As maximum energy loss occurs in completely inelastic collision

$\Rightarrow$ Statement-2 is true

by conservation of momentum

$$mv = (m+M)v_f$$

$$\Rightarrow \quad v_f = \frac{mv}{m+M}$$

Energy loss (maximum)

$$\Delta E_{max} = \frac{1}{2}mv^2 - \frac{1}{2}(m+M)\left(\frac{mv}{m+M}\right)^2$$

$$\Delta E_{max} = \frac{1}{2}\left(\frac{mM}{m+M}\right)v^2 = \underbrace{\frac{M}{m+M}}_{f}\left(\frac{1}{2}mv^2\right)$$

Ans. (D)

Sol. 20

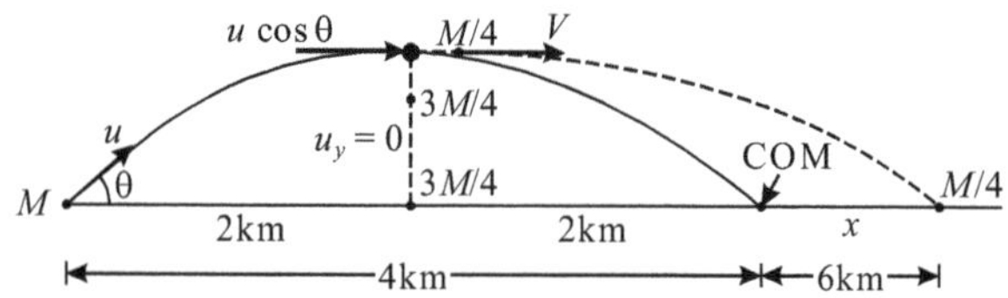

By Conservation of linear momentum

$$M(u\cos\theta) = \frac{M}{u}\cdot v$$

$$\Rightarrow \quad v = 4u\cos\theta$$

$3M/4$ and $M/4$ will fall on ground such that their centre of mass is located at the original path of projectile

$$\Rightarrow \quad \frac{3M}{4}\times 2 = \frac{M}{4}\times x$$

$$\Rightarrow \quad x = 6\text{ km}$$

$$\text{Range} = 4 + 6 = 10\text{ km}$$

Ans. (C)

Sol. 21

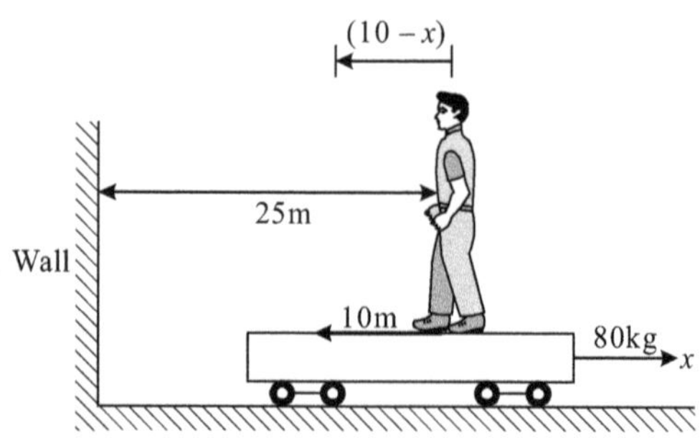

Due to motion of boy, if cart displaces toward right by x such that

$$\Delta x_{COM} = 0 = \frac{M_1x_1 + M_2x_2}{M_1 + M_2}$$

$$\Rightarrow \quad 0 = \frac{80x - 20(10-x)}{100}$$

$$\Rightarrow \quad 80x = 20(10-x)$$

$$\Rightarrow \quad 5x = 10$$

$$\Rightarrow \quad x = 2\text{ m}$$

Final distance of boy from wall is

$$d = 25 - (10-2) = 17\text{ m}$$

Ans. (D)

Sol. 22 After collision if the resulting mass $6m$ has velocity components v_x and v_y then by conservation of linear momentum, we have

Along x-direction

$$+m(3u)-2m(2u)\cos 60° - 3mu\cos 60° = 6mv_x$$

$$\Rightarrow \quad 3u-2u-\frac{3}{2}u = 6v_x$$

$$\Rightarrow \quad v_x = -\frac{u}{12}$$

Along y-direction

$$-(2m)(2u)\sin 60° + 3mu\sin 60° = 6mv_y$$

$$\Rightarrow \quad -4mu\left(\frac{\sqrt{3}}{2}\right)+3mu\left(\frac{\sqrt{3}}{2}\right) = 6mv_y$$

$$\Rightarrow \quad -\frac{\sqrt{3}}{2}u = 6v_y$$

$$\Rightarrow \quad v_y = -\frac{\sqrt{3}}{12}u$$

$$\Rightarrow \quad \vec{v} = v_x\hat{i}+v_y\hat{i}$$

$$\Rightarrow \quad \vec{v}_{6m} = -\frac{u}{12}\hat{i}-\frac{\sqrt{3}}{12}u\hat{j}$$

$$= \frac{u}{12}(-i-\sqrt{3}\hat{j})$$

Ans. (D)

Sol. 23

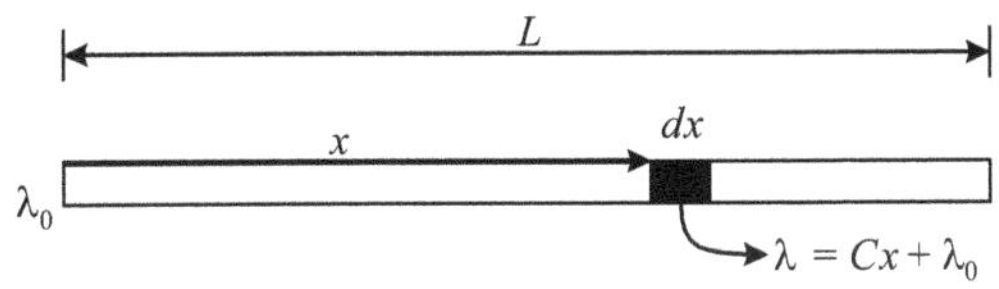

Mass of element is

$$M = \int dm = \int_0^L \lambda dx = \int_0^L (Cx+\lambda_0)dx$$

$$\Rightarrow \quad M = \int_0^L (Cx+\lambda_0)dx$$

$$\Rightarrow \quad M = \frac{CL^2}{2}+\lambda_0 L$$

$$\Rightarrow \quad C = \frac{2}{L^2}(M-\lambda_0 L)$$

Distance of COM from lighter end is

$$x_{COM} = \frac{1}{M}\int dmx = \frac{1}{M}\int_0^L (Cx^2+\lambda_0 x)dx$$

$$\Rightarrow \quad x_{COM} = \frac{1}{M}\left(\frac{LC^3}{3}+\frac{\lambda_0 L^2}{2}\right) = \frac{L^2}{M}\left(\frac{CL}{3}+\frac{\lambda_0}{2}\right)$$

$$\Rightarrow \quad x_{COM} = \frac{L^2}{M}\left(\frac{2}{3L}(M-\lambda_0 L)+\frac{\lambda_0}{2}\right)$$

$$\Rightarrow \quad x_{COM} = \frac{2}{3}L-\frac{L^2}{M}\left(-\frac{2}{3}\lambda_0+\frac{\lambda_0}{2}\right)$$

$$\Rightarrow \quad x_{COM} = \frac{2L}{3}-\frac{\lambda_0 L^2}{6M}$$

Ans. (D)

Sol. 24

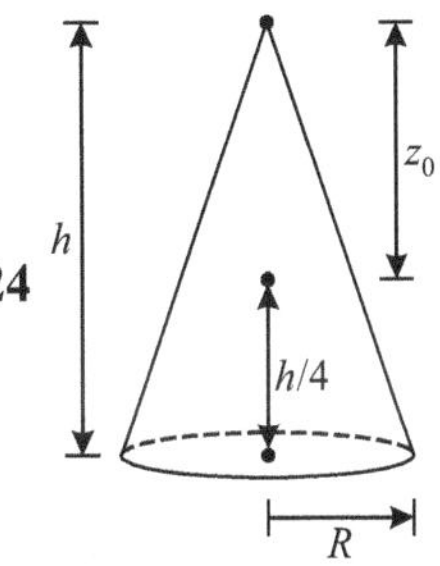

COM distance from vertex

$$z_0 = h-\frac{h}{4} = \frac{3h}{4}$$

Ans. (B)

Sol. 25

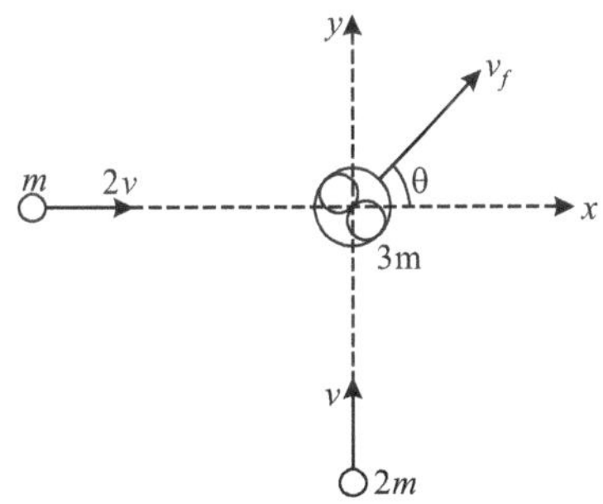

By conservation of linear momentum

in x-direction $\quad 2mv = 3mv_f\cos\theta \quad \ldots(1)$

in y-direction $\quad 2mv = 3mv_f\sin\theta \quad \ldots(2)$

Squaring and adding (1) and (2)

$$2v_f = \sqrt{8}v$$

$$\Rightarrow \quad v_f = \frac{2\sqrt{2}}{3}v$$

Initial KE of system is

$$k_i = \frac{1}{2}m(2V)^2+\frac{1}{2}2m(V)^2 = 3mv^2$$

Final KE of system is

$$k_f = \frac{1}{2}3m\left(\frac{2\sqrt{2}}{3}v\right)^2 = \frac{4}{3}mv^2$$

Loss in energy during collision

$$\Delta E_L = k_i - k_y = \frac{5}{3}mv^2$$

$$\%\text{ loss} = \frac{k_i-k_f}{k_i}\times 100\%$$

$$\Rightarrow \quad \%\text{ loss} = \frac{500}{9} = 55.55\%$$

Ans. (C)

Sol. 26 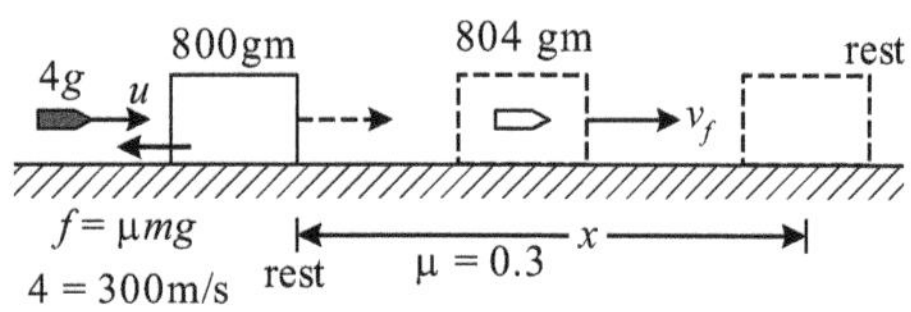

By conservation of momentum we use

$$4u + 0 = 804v_f$$

$$v_f = \frac{4\times 300}{804} = \frac{1200}{804} = 1.49\text{ m/s}$$

using work energy theorem for block sliding on rough surface, we have

$$\frac{1}{2}mv_f^2 - \mu mgx = 0$$

$$\Rightarrow \quad x = \frac{V_f^2}{2\mu g} = \frac{(1.49)^2}{2\times 0.3\times 10} = 0.379\text{ m}$$

Ans. (B)

Sol. 27 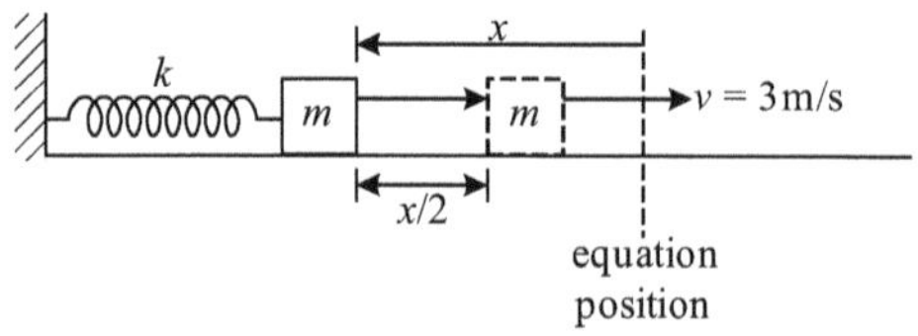

As it is given that on collision moving block comes to rest & other moves, this implies for elastic collision masses are equal. Thus just before collision velocity of mass attached to spring is 3 m/s.

By conservation of energy, we use

$$\frac{1}{2}kx^2 = \frac{1}{2}k\left(\frac{x}{2}\right)^2 + \frac{1}{2}mv^2$$

$$\Rightarrow \quad \frac{1}{2}kx^2\left(1 - \frac{1}{4}\right) = \frac{1}{2}\times 0.1\times (3)^2$$

$$\Rightarrow \quad \frac{1}{2}kx^2 = \frac{0.9}{2}\times\frac{4}{3}$$

$$\Rightarrow \quad \frac{1}{2}kx^2 = 0.6\text{ J}$$

initial stored energy in spring. **Ans. (B)**

Sol. 28 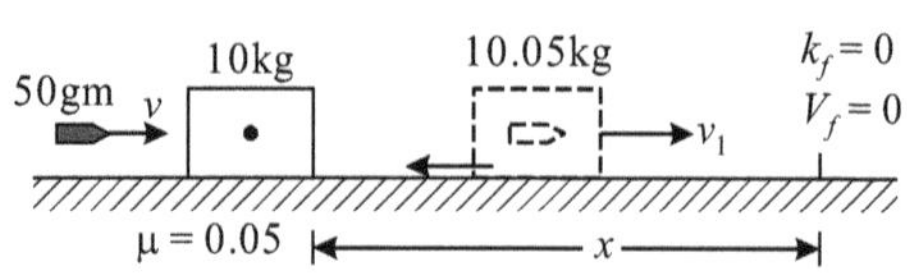

If after collision block moves at speed v_1 we have by momentum conservation

$$0.05V = 10.05v_1$$

$$\Rightarrow \quad v_1 = \frac{0.05v}{10.05}$$

Using work-energy theorem after collision we have

$$\frac{1}{2}(10.05)\,v_1^2 - \mu\,(m+M)gx = 0$$

$$\Rightarrow \quad \frac{1}{2}(10.05)\left(\frac{0.05v}{10.05}\right)^2 - 0.05\times 10.05\times 10\times 2 = 0$$

$$\Rightarrow \quad v = 201\sqrt{2}\text{ m/s}$$

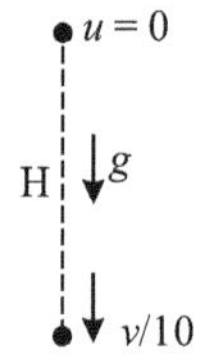

We have

$$\Rightarrow \quad \frac{v}{10} = \sqrt{2gH}$$

$$\Rightarrow \quad H = \frac{v^2}{200g} = \frac{(201\sqrt{2})^2}{2\times 10^3} \cong 40\text{ m}$$ **Ans. (D)**

Sol. 29 When a bead strikes with speed v and rebond the change in momentum is

$$\Delta P = mv - (-me) = 2mv$$

Between two supoort, the distance beads travel on rod is

$$r = L - n(2r)\ \text{(diameter of each bead)}$$

time between two successive collisions on same support is

$$\Delta t = \frac{2r}{v} = \frac{2[L-2nr]}{v}$$

Average force on each support is

$$F = \frac{\Delta P}{\Delta t} = \frac{2mv}{2(L-2nr)}\times v = \frac{mv^2}{L-2nr}$$ **Ans. (B)**

Sol. 30 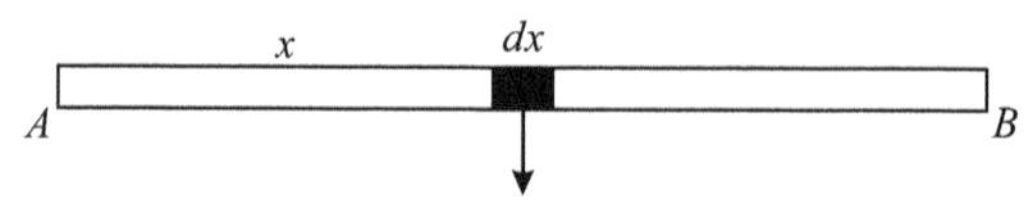

$$dm = \mu\,dx = \left(a + \frac{bx}{L}\right)dx$$

COM of rod is given as

$$x_{cm} = \frac{\int dm\,x}{\int dm} = \frac{\int_0^L\left(a+\frac{bx}{L}\right)x\,dx}{\int_0^L\left(a+\frac{bx}{L}\right)dx}$$

$$\Rightarrow \quad x_{cm} = \frac{\left(\frac{aL^2}{2} + \frac{bL^2}{3L}\right)}{\left(aL + \frac{bL^2}{2L}\right)} = \frac{7L}{12}$$

$$\Rightarrow \quad \frac{7L}{12} = \frac{\frac{a}{2}+\frac{b}{3}}{a+\frac{b}{2}}$$

$$\Rightarrow \quad 7a+\frac{7b}{2}=6a+4b$$

$$\Rightarrow \quad a=\frac{b}{2}$$

$$\Rightarrow \quad b=2a \quad \textbf{Ans. (B)}$$

Sol. 31

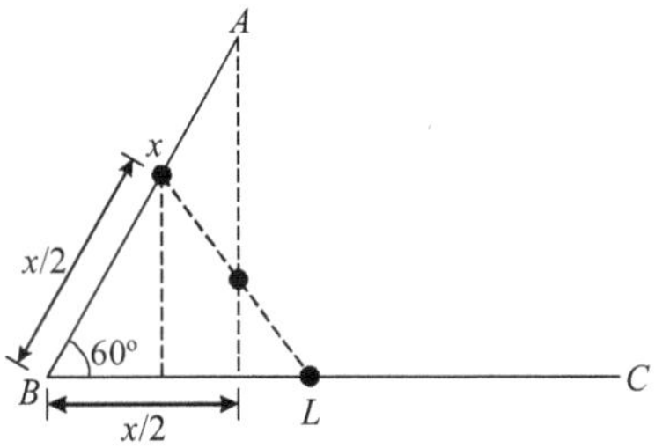

From point B, x coordinate of COM of ABC is located at a distance $x/2$ from B and for BC, COM is at a distance $L/2$ from B.

$$\Rightarrow \quad x_{cm}=\frac{m_1x_1+m_2x_2}{m_1+m_2}$$

$$\Rightarrow \quad \frac{x}{2}=\frac{\lambda x\left(\frac{x}{2}\cos 60^\circ\right)+\lambda L\left(\frac{L}{2}\right)}{\lambda x+\lambda L}$$

If λ kg/m is linear mass density of wire

$\Rightarrow \quad m_1=\lambda x \quad \& \quad m_2=\lambda L$
(AB) (BC)

$$\Rightarrow \quad \frac{x}{2}(x+L)=\frac{x^2}{4}+\frac{L^2}{2}$$

$$\Rightarrow \quad \frac{x^2}{2}+\frac{xL}{2}=\frac{x^2}{4}+\frac{L^2}{2}$$

$$\Rightarrow \quad \frac{x^2}{4}+\frac{xL}{2}=\frac{L^2}{2}$$

$$\Rightarrow \quad \left(\frac{L}{x}\right)^2-\left(\frac{L}{x}\right)-\frac{1}{2}=0$$

$$\Rightarrow \quad \frac{L}{x}=\frac{1+\sqrt{1+4(1/2)}}{2}=\frac{1+\sqrt{3}}{2}\simeq 1.37$$

Ans. (C)

Sol. 32

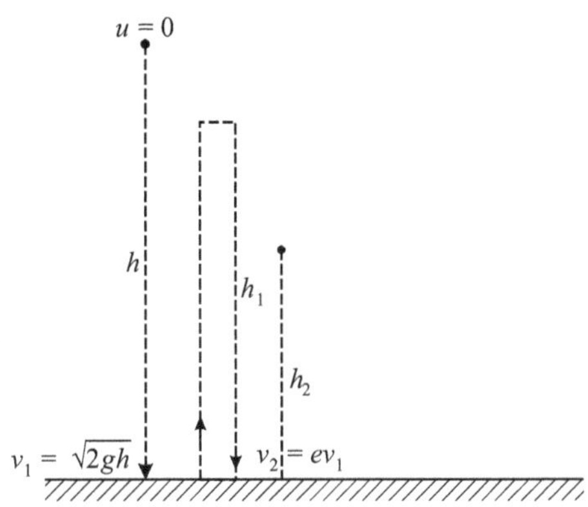

Given that

$$\frac{1}{2}mv_2^2=\frac{1}{2}\left(\frac{1}{2}mv_1^2\right)$$

$$\Rightarrow \quad v_2=\frac{1}{\sqrt{2}}v_1$$

We consider coefficient of restitution for this collision to be

$$e=\frac{1}{\sqrt{2}}$$

height attained by object after first impact

$$h_1=\frac{v_2^2}{2g}=e^2\left(\frac{v_1^2}{2g}\right)=e^2h$$

after second impact

$$h_2=e^2h_1=e^4h$$

at $t\to\infty$ total distance covered is

$$s=h+2h_1+2h_2+\ldots\infty$$

$$\Rightarrow \quad s=h(1+2e^2+2e^4+\ldots\infty)$$

$$\Rightarrow \quad s=h\left[1+2e^2\left(\frac{1}{1-e^2}\right)\right]=h\left(\frac{1+e^2}{1-e^2}\right)$$

$$\Rightarrow \quad s=h\left(\frac{1+\frac{1}{2}}{1-\frac{1}{2}}\right)=3h \quad \textbf{Ans. (A)}$$

Sol. 33 By conservation of linear momentum, we have along x-direction $\rightarrow mv\sin 30^\circ - mv\sin 45^\circ = 2mv_1\cos\theta$

$$\Rightarrow \quad \frac{mv}{2}-\frac{mv}{\sqrt{2}}=2mv_1\cos\theta \quad \ldots(1)$$

along y-direction $\rightarrow$

$$mv\cos 30^\circ + mv\cos 45^\circ = 2mv_1\sin\theta$$

$$\Rightarrow \quad \frac{\sqrt{3}}{2}mv+\frac{mv}{\sqrt{2}}=2mv_1\sin\theta \quad \ldots(2)$$

here $\frac{(2)}{(1)}$ gives

$$\Rightarrow \quad \tan\theta=\frac{\frac{\sqrt{3}}{2}+\frac{1}{\sqrt{2}}}{\frac{1}{2}-\frac{1}{\sqrt{2}}}=\frac{\sqrt{3}+\sqrt{2}}{1-\sqrt{2}}$$

Ans. (A)

Sol. 34

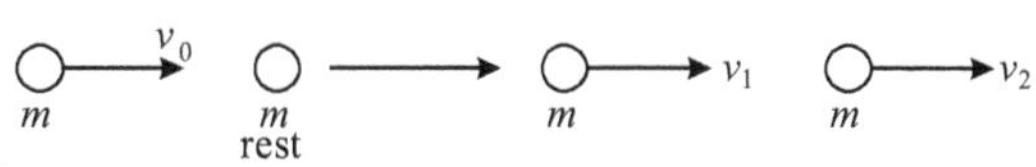

Given that

$$\frac{1}{2}mv_1^2+\frac{1}{2}mv_2^2=\frac{3}{2}\left(\frac{1}{2}mv_0^2\right)$$

$$\Rightarrow \quad v_1^2+v_2^2=\frac{3}{2}v_0^2 \quad \ldots(1)$$

By conservation of linear momentum, we have

$$mv_0=mv_1+mv_2$$

$$\Rightarrow \quad v_1 + v_2 = b_0 \qquad \ldots(2)$$

$$(2)^2 \Rightarrow v_1^2 + v_2^2 + 2v_1v_2 = v_0^2$$

$$(2)^2 - (1) \Rightarrow \quad 2v_1v_2 = -\frac{v_0^2}{2}$$

The relative velocity of separation is

$$(v_2 - v_1)^2 = v_1^2 + v_2^2 - 2v_1v_2$$

$$\Rightarrow \quad (v_2 - v_1)^2 = \frac{3}{2}v_0^2 + \frac{v_0^2}{2} = 2v_0^2$$

$$\Rightarrow \quad v_2 - v_1 = \sqrt{2}v_0$$

Ans. (B)

Sol. 35

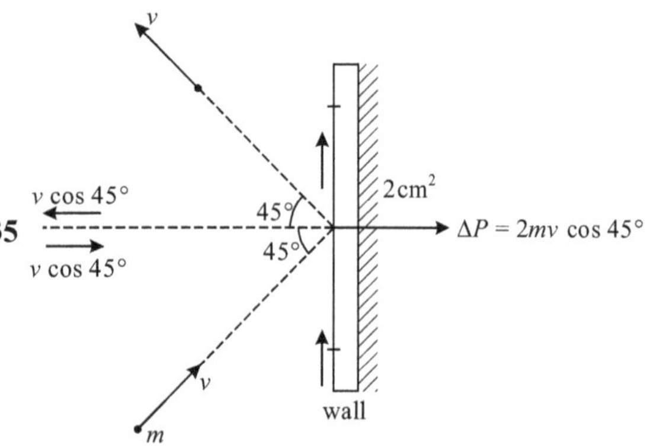

Change in momentum of *H*-molecule in one collision

Force on wall = change in momentum/sec due to collision on wall.

$$= 2mv \cos 45° \times 10^{23}\,\text{N}$$

$$\text{Pressure on wall} = \frac{F}{A}$$

$$\Rightarrow \quad P = \frac{2mv\left(\frac{1}{\sqrt{2}}\right)\times 10^{23}}{2\times 10^{-4}}$$

$$\Rightarrow \quad P = \frac{2\times 3.32\times 10^{-27}\times 10^{3}\times \frac{1}{\sqrt{3}}\times 10^{23}}{2\times 10^{-4}}$$

$$= 2.35 \times 10^3\,\text{N/m}^2$$

Ans. (A)

Sol. 36

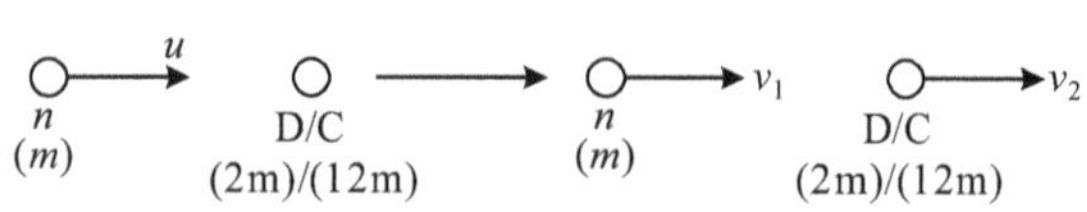

For elastic collision, final speed of neutron is given as

$$v_1 = \frac{m_1 - m_2}{m_1 + m_2}\cdot u_1 + \frac{2m_2}{m_1 + m_2}u_2$$

$$\Rightarrow \quad v_1 = -\frac{1}{3}u$$

Frictional loss in *KE* of neutron

$$P_d = \frac{\frac{1}{2}mu^2 - \frac{1}{2}m\left(\frac{u}{3}\right)^2}{\frac{1}{2}mu^2} = \frac{8}{9} = 0.89$$

For collision with *C*-nucleus we use $m_2 = 12m$ so final speed of neutron is given as

$$v_1' = -\frac{11}{13}u$$

fractional loss in *KE* of neutron

$$P_c = \frac{\frac{1}{2}mu^2 - \frac{1}{2}m\left(\frac{4}{13}u\right)^2}{\frac{1}{2}mu^2} = \frac{48}{469} = 0.28$$

Ans. (A)

Sol. 37

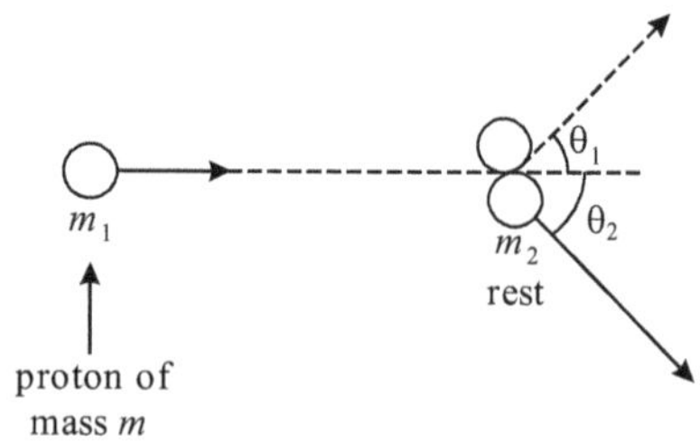

We have discussed that for an elastic collision if $m_1 = m_2$

Then divergence angle is

$$\theta_1 + \theta_2 = 90°$$

here $m_2 = m$ for $\theta_1 + \theta_2 = 90°$

Ans. (D)

Sol. 38 $k_i = \frac{1}{2}mv_0^2$

From linear momentum conservation

$$mv_0 = (2\text{m} + M)\,v_f$$

$$\Rightarrow \quad v_f = \frac{mv_0}{2m + M}$$

Given that

$$\frac{k_i}{k_f} = 6$$

$$\Rightarrow \quad \frac{\frac{1}{2}mv_0^2}{\frac{1}{2}(2m+M)\left(\frac{mv_0}{2m+M}\right)^2} = 6$$

$$\Rightarrow \quad \frac{2m+M}{m} = 6$$

$$\Rightarrow \quad \frac{M}{m} = 4$$

Ans. (A)

Sol. 39

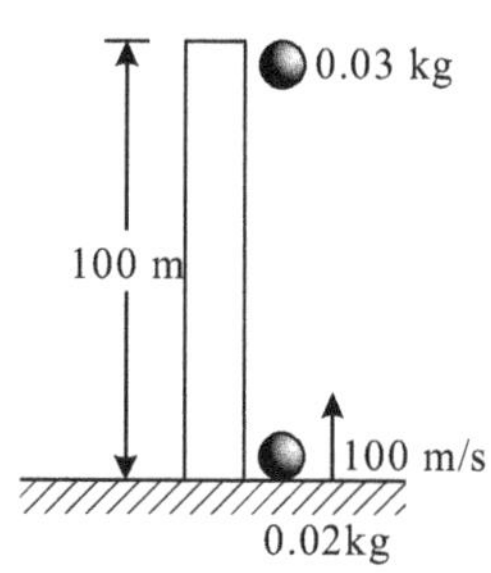

Time taken for the particles to collide,

$$t = \frac{d}{V_{rel}} = \frac{100}{100} = 1 \text{ sec}$$

In 1s distance fallen by wood is

$$h_1 = \frac{1}{2}gt^2 = \frac{1}{2} \times 10 \times (1)^2 = 5 \text{ m}$$

Speed of wood just before collision = gt = 10 m/s & speed of bullet just before collision v-gt

$$= 100 - 10 = 90 \text{ m/s}$$

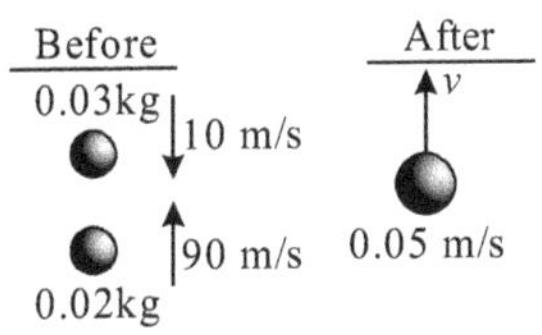

Now, conservation of linear momentum just before and after the collision

$$-(0.03)(1v) + (0.02)(9v) = (0.05)v$$

$$\Rightarrow \quad 150 = 5v$$

$$\Rightarrow \quad v = 30 \text{ m/s}$$

Maximum height reached by body $h = \frac{v^2}{2g}$

$$\Rightarrow \quad h = \frac{30 \times 30}{2 \times 10} = 45 \text{ m}$$

$\Rightarrow$ Height above tower = 45 – 5 = 40 m **Ans. (C)**

Sol. 40

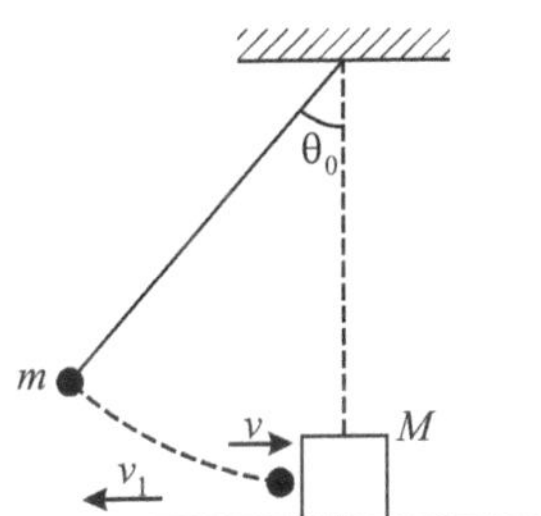

If m strikes M at speed v and rebound at v_1, we have

$$v = \sqrt{2gl(1-\cos\theta_0)}$$

and

$$v_1 = \sqrt{2gl(1-\cos\theta_1)}$$

By momentum conservation, we use

$$m\sqrt{2gl(1-\cos\theta_0)} = MV_m - m\sqrt{2gl(1-\cos\theta)}$$

$$\Rightarrow \quad m\sqrt{2gl}\{\sqrt{1-\cos\theta_0} + \sqrt{1-\cos\theta_1}\} = MV_m \quad \ldots(1)$$

For elastic collision, we use

$$e = 1 = \frac{V_m + \sqrt{2gl(1-\cos\theta_1)}}{\sqrt{2gl(1-\cos\theta_0)}}$$

$$\Rightarrow \quad \sqrt{2gl}(\sqrt{1-\cos\theta_0} - \sqrt{1-\cos\theta_1}) = V_m \quad \ldots(2)$$

(1)/(2) gives

$$\frac{(\sqrt{1-\cos\theta_0} + \sqrt{1-\cos\theta_1})}{(\sqrt{1-\cos\theta_0} - \sqrt{1-\cos\theta_1})} = \frac{M}{m}$$

$$\Rightarrow \quad \frac{m-M}{m+M} = \frac{\sqrt{1-\cos\theta_1}}{\sqrt{1-\cos\theta_0}} = \frac{\sin\left(\frac{\theta_1}{2}\right)}{\sin\left(\frac{\theta_0}{2}\right)}$$

$$\Rightarrow \quad \frac{M}{m} = \frac{\theta_0 - \theta_1}{\theta_0 + \theta_1}$$

$$\Rightarrow \quad M = m\left(\frac{\theta_0 - \theta_1}{\theta_0 + \theta_1}\right)$$

Ans. (C)

Sol. 41

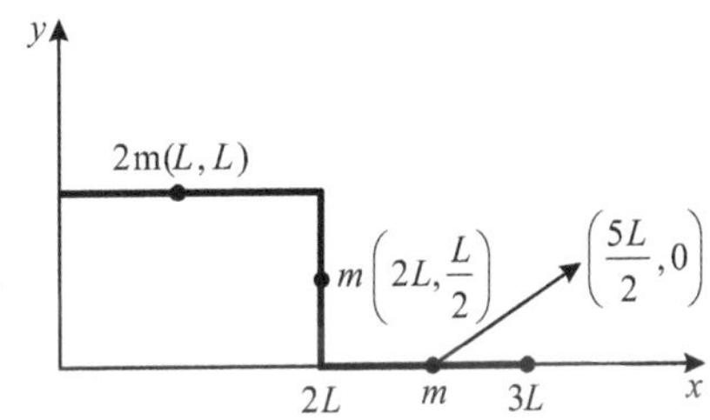

Centre of mass of three segments of bar are shown in figure. So centre of mass coordinates (X_{cm}, Y_{cm}) are given as

$$X_{cm} = \frac{2mL + 2mL + \frac{5mL}{2}}{4m} = \frac{13}{8}L$$

$$Y_{cm} = \frac{2m \times L + m \times \left(\frac{L}{2}\right) + m \times 0}{4m} = \frac{5L}{8}$$

Ans. (A)

Sol. 42

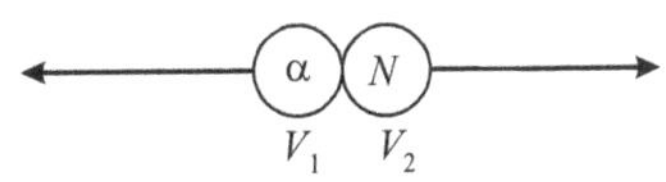

Using momentum conservation

$$mv_0 = Mv_2 - mv_1 \quad \ldots(1)$$

Given that

$$\frac{1}{2}mv_1^2 = 0.36 \times \frac{1}{2}mv_0^2$$

$$\Rightarrow \quad v_1 = 0.6v_0$$

$$\Rightarrow \quad \frac{1}{2}Mv_2^2 = 0.64 \times \frac{1}{2}mv_0^2$$

$$\Rightarrow \quad v_2 = \sqrt{\frac{m}{M}} \times 0.8v_0$$

From equation-(1)

$$mv_0 = \sqrt{mM} \times 0.8v_0 - m \times 0.6v_0$$

$$\Rightarrow \quad 1.6\text{m} = 0.8\sqrt{mM}$$

$$\Rightarrow \quad 4\text{ m}^2 = mM$$

$$\Rightarrow \quad M = 4\text{ m}$$

Ans. (A)

Sol. 43

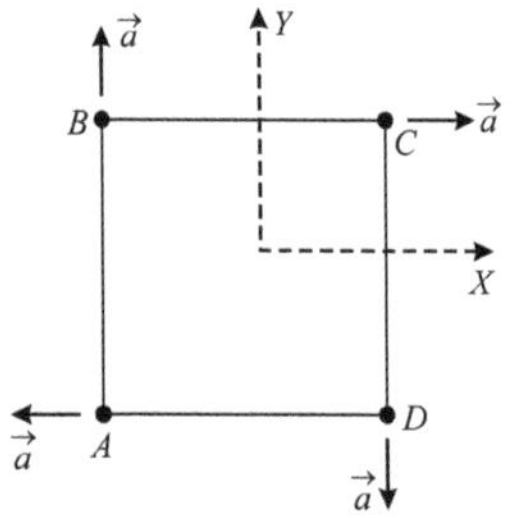

The acceleration vector of four particles are given as

$$\vec{a}_A = -a\hat{i}$$
$$\vec{a}_B = a\hat{j}$$
$$\vec{a}_C = a\hat{i}$$
$$\vec{a}_D = -a\hat{j}$$

Thus acceleration of centre of mass of this system is

$$\vec{a}_{\text{cm}} = \frac{m_a\vec{a}_a + m_b\vec{a}_b + m_c\vec{a}_c + m_d\vec{a}_d}{m_a + m_b + m_c + m_d}$$

$$\Rightarrow \quad \vec{a}_{\text{cm}} = \frac{-ma\hat{i} + 2ma\hat{j} + 3ma\hat{i} - 4ma\hat{j}}{10\text{m}}$$

$$= \frac{2ma\hat{i} - 2ma\hat{j}}{10\text{m}}$$

$$= \frac{a}{5}\hat{i} - \frac{a}{5}\hat{j}$$

$$= \frac{a}{5}(\hat{i} - \hat{j})$$

Ans. (A)

Sol. 44 In figure below O is COM of complete sheet and C is COM of the shaded portion which is cut off

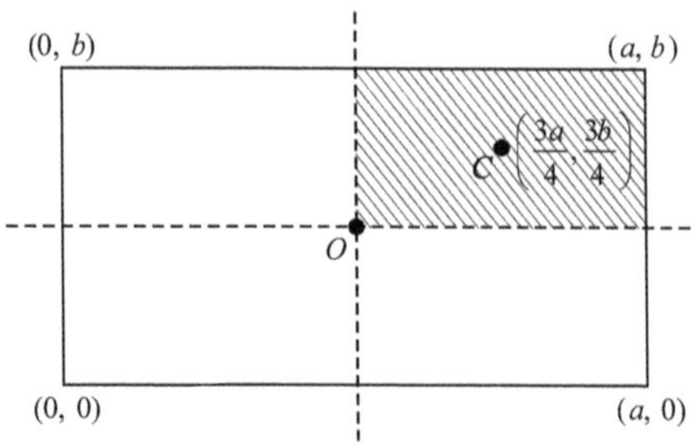

x-coordinate of COM of remaing sheet will be

$$x = \frac{M\frac{a}{2} - \frac{M}{4} \times \frac{3a}{4}}{M - \frac{M}{4}}$$

$$\Rightarrow \quad x = \frac{\frac{a}{2} - \frac{3a}{16}}{\frac{3}{4}} = \frac{\frac{5a}{16}}{\frac{3}{4}} = \frac{5a}{12}$$

y-coordinate of COM of remaining sheet will be

$$y = \frac{M\frac{b}{2} - \frac{M}{4} \times \frac{3b}{4}}{M - \frac{M}{4}} = \frac{5b}{12}$$

Ans. (D)

Sol. 45 Using linear momentum conservation, we have

$$m_1v_1\hat{i} + m_2v_2\hat{i} = m_1v_3\hat{i} + m_2v_4\hat{i}$$

$$m_1v_1 + 0.5m_1v_2 = m_1(0.5v_1) + 0.5m_1v_4$$

$$0.5m_1v_1 = 0.5m_1(v_4 - v_2)$$

$$v_1 = v_4 - v_2$$

Ans. (C)

Sol. 46

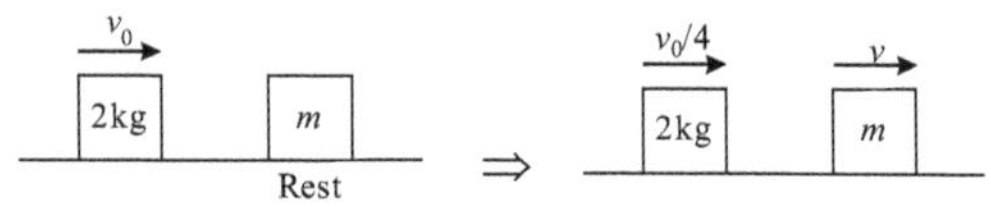

Using conservation of linear momentum, we have

$$2v_0 = 2\left(\frac{v_0}{4}\right) + mv$$

$$\Rightarrow \quad 2v_0 = \frac{v_0}{2} + mv$$

$$\Rightarrow \quad \frac{3v_0}{2} = mv \qquad \ldots(1)$$

As collision is elastic, we use $e = 1$, so we have

$$v_{\text{separation}} = v_{\text{approch}}$$

$$\Rightarrow \quad v - \frac{v_0}{4} = v_0$$

$$\Rightarrow \quad \frac{5v_0}{4} = v \qquad \ldots(2)$$

From (1) and (2), we use

$$\Rightarrow \quad \frac{3v_0}{2} = m\left(\frac{5v_0}{4}\right)$$

$$\Rightarrow \quad m = \frac{6}{5} = 1.2 \text{ kg}$$

Ans. (B)

Sol. 47

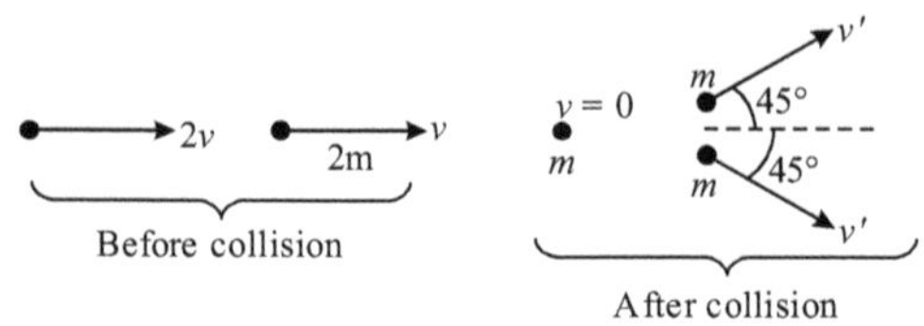

Using linear momentum conservation, we have

$$m(2v) + 2m(v) = m \times 0 + m\left(\frac{v'}{\sqrt{2}} \times 2\right)$$

$$\Rightarrow \quad v' = 2\sqrt{2}\, v$$

Ans. (B)

Sol. 48

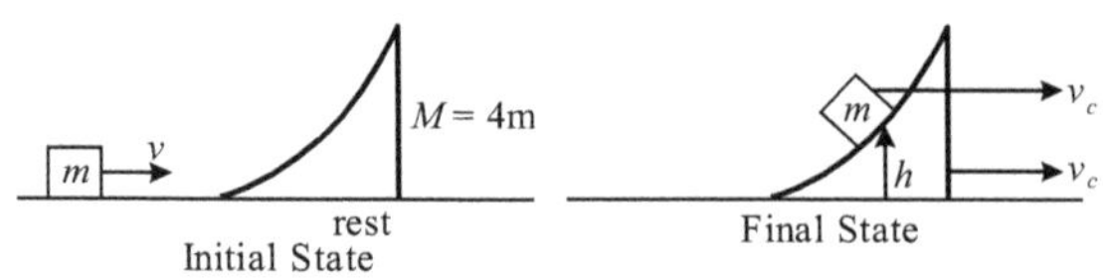

Using linear momentum conservation, we have

$$mv = (m + M)v_c$$

$$\Rightarrow \quad v_c = \frac{v}{5}$$

Using work energy theorem, we have

$$-mgh = \frac{1}{2}(m+M)v_c^2 - \frac{1}{2}mv^2$$

$$\Rightarrow \quad h = \frac{2v^2}{5g} \qquad \textbf{Ans. (C)}$$

Sol. 49 By conservation of momentum, we use

in X-direction $M(10)\cos 30^\circ + 2M(5)\cos 45^\circ = 2Mv_1 \cos 30^\circ + Mv_2 \cos 45^\circ$

$$5\sqrt{3} + 5\sqrt{2} = \sqrt{3}\,v_1 = \frac{v_2}{\sqrt{2}} \qquad \ldots(1)$$

in Y-direction $10M\sin 30^\circ - 2M(5)\sin 45^\circ = Mv_2 \sin 45^\circ - 2Mv_1 \sin 30^\circ$

$$5 - 5\sqrt{2} = \frac{v_2}{\sqrt{2}} - v_1 \qquad \ldots(2)$$

(1)(2) $5 \times 1.732 + 10 \times 1.414 - 5 = (\sqrt{3}+1)v_1$

$$\Rightarrow \quad v_1 = \frac{8.66 + 14.14 - 5}{2.732}$$

$$\Rightarrow \quad v_1 = 6.51 \text{ m/s}$$

From (2) we use

$$v_2 = (5 - 7.07 + 6.51) \times \sqrt{2}$$

$$\Rightarrow \quad v_2 = 4.44 \times 1.414$$

$$\Rightarrow \quad v_2 = 6.28 \text{ m/s} \qquad \textbf{Ans. (C)}$$

Sol. 50 $m = 20g,$

$u = 1$ m/s,

$v = ?$

$S = 20 \times 10^{-2}$ m

$$a = \frac{-2.5\times10^{-2}}{20\times10^{-3}} \text{ m/s}^2$$

$$v^2 = u^2 + 2as$$

$$\Rightarrow \quad v^2 = 1 - 2\times\frac{2.5\times10^{-2}}{20\times10^{-3}}\times\frac{20}{100}$$

$$\Rightarrow \quad v = \frac{1}{\sqrt{2}} \simeq 0.7\text{m/s} \qquad \textbf{Ans. (D)}$$

Sol. 51 By conservation of angular momentum we use

$$Mv_0 L = Mv_1(L-l)$$

$$\Rightarrow \quad v_1 = v_0\left(\frac{L}{L-l}\right) \qquad \ldots(1)$$

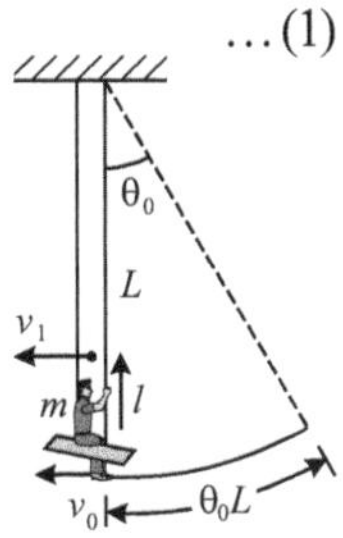

Gain in KE of system in

$$\Delta k = \frac{1}{2}M(v_1^2 - v_0^2)$$

$$= \frac{1}{2}Mv_0^2\left[\left(\frac{L}{L-l}\right)^2 - 1\right]$$

$$= \frac{1}{2}Mv_0^2\left[\left(1-\frac{l}{L}\right)^{-2} - 1\right]$$

$$= Mv_0^2 \cdot \frac{l}{L}$$

Total W/D by person is $W = Mgl + \frac{Mv_0^2 l}{L}$

$$= Mgl + \frac{Ml}{L}\cdot gL\theta_0^2$$

$$= mgl(1+\theta_0^2) \qquad \textbf{Ans. (B)}$$

Sol. 52

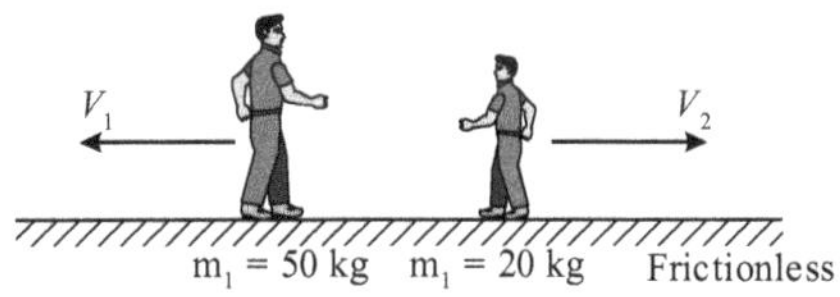

By conservation of momentum, we use

$$m_1 v_1 = m_2 v_2$$

$$50v_1 = 20v_2$$

and $\quad v_1 + v_2 = 0.7$

$$\Rightarrow \quad v_1 + 2.5v_1 = 0.7$$

$$\Rightarrow \quad v_1 = \frac{0.7}{3.5} = 0.2 \text{ m/a} \qquad \textbf{Ans. (A)}$$

Sol. 53

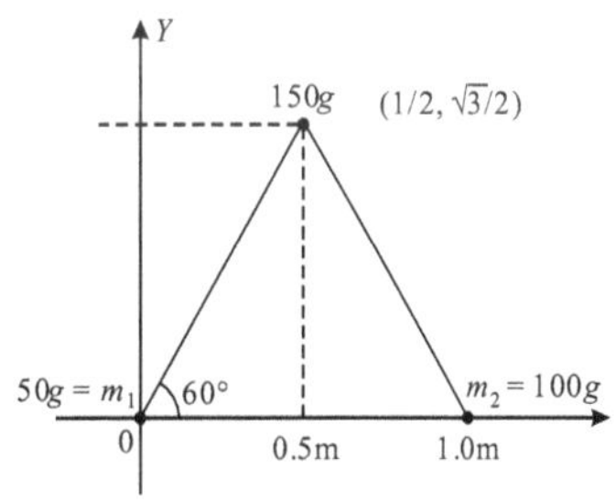

Co-ordinates of COM are

$$x_c = \frac{m_1x_1 + m_2x_2 + m_3x_3}{m_1+m_2+m_3} = \frac{50\times0 + 100\times1 + 150\times0.5}{300}$$

$$= \frac{175}{300} = \frac{7}{12}\text{m}$$

$$y_c = \frac{m_1y_1 + m_2y_2 + m_3y_3}{m_1+m_2+m_3} = \frac{50\times0\times100\times0 + 150\times\sqrt{3}/2}{300}$$

$$= \frac{\sqrt{3}}{4}\text{m} \qquad \textbf{Ans. (C)}$$

6.2 Rigid Body Dynamics

Sol. 1

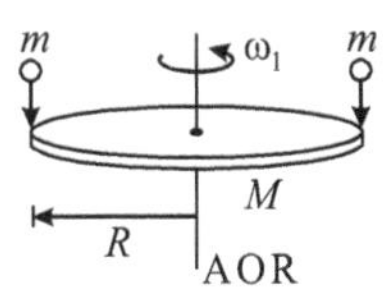

Initial MI of disc

$$I_1 = \frac{1}{2}MR^2$$

Initial MI of disc with small spheres

$$I_2 = \frac{1}{2}MR^2 + 2mR^2$$

As no external torque is applied on disc, its angular momentum remain conserved.

$\Rightarrow$ $I_1\omega_1 = I_2\omega_2$ (where ω_2 is final angular speed)

$\Rightarrow$ $\omega_2 = \frac{I_1\omega_1}{I_2} = \left(\frac{M}{M+4m}\right)\omega_1$ **Ans. (C)**

Sol. 2

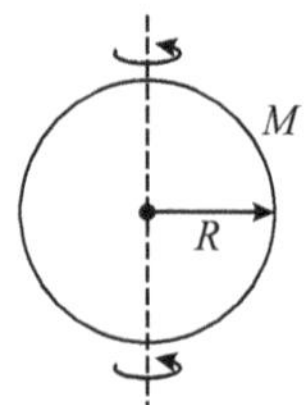

MI of a circular ring about its diametrical axis is given as

$$I = \frac{1}{2}MR^2$$ **Ans. (A)**

Sol. 3

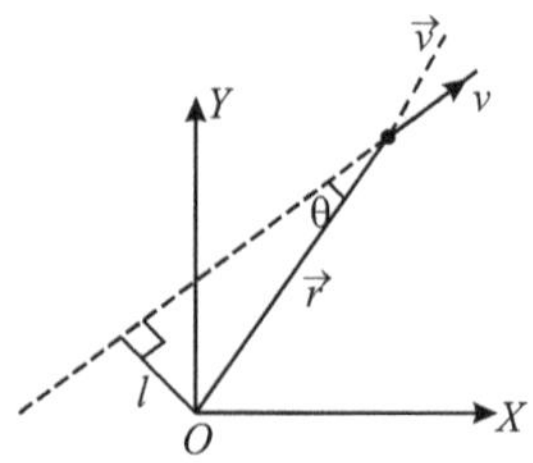

Angular momentum

$\vec{L} = m(\vec{r}\times\vec{v}) = mvr\sin\theta = mvl$ **Ans. (B)**

Sol. 4

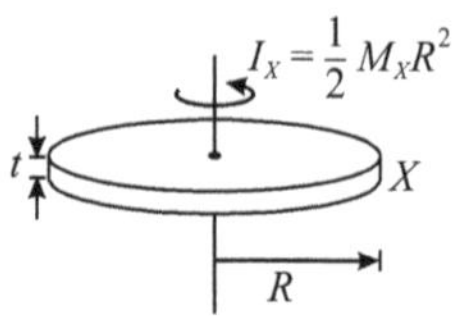

Mass of disc X is

$$M_x = \rho(\pi R^2 t)$$

$\Rightarrow$ $I_x = \frac{1}{2}\rho\pi R^4 t$

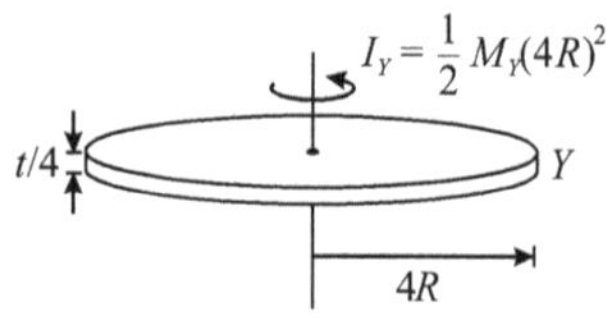

Mass of disc Y is

$$M_y = \rho(\pi(4R)^2(t/4)) = 4\rho\pi R^2 t$$

$\Rightarrow$ $I_Y = \frac{1}{2}(64\rho\pi R^4 t)$

$\Rightarrow$ $I_y = 64\, I_X$ **Ans. (D)**

Sol. 5 For a particle in circular motion, we have

Angular momentum $L = I\omega \Rightarrow I = \frac{L}{\omega}$

Kinetic Energy $K = \frac{1}{2}I\omega^2 = \frac{1}{2}\left(\frac{L}{\omega}\right)\omega^2$

$\Rightarrow$ $L = \frac{2K}{\omega}$

If $K \to K/2$ and $\omega \to 2\omega$

Then final angular momentum is

$$L_f = \frac{2(K/2)}{2\omega} = \frac{K}{2\omega} = \frac{L}{4}$$ **Ans. (A)**

Sol. 6 For the given particle, we know

$$\vec{\tau} = \vec{r}\times\vec{F}$$

$\Rightarrow$ $\vec{\tau}$ is $\perp$ to $\vec{r}$ and also $\vec{\tau}$ is $\perp$ to $\vec{F}$

$\Rightarrow$ $\vec{r}\,.\,\vec{\tau} = 0$ and $\vec{F}\,.\,\vec{\tau} = 0$ **Ans. (D)**

Sol. 7

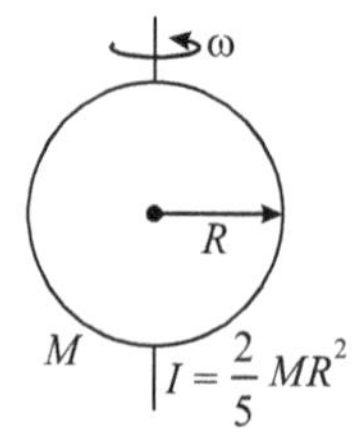

If R increases $\Rightarrow$ I increases

As no external torque is applied

$\Rightarrow$ Angular momentum remain conserved

So as I increases, ω decreases

Rotational KE $K = \frac{1}{2}I\omega^2 = \frac{1}{2}L\omega$ **Ans. (B)**

Sol. 8

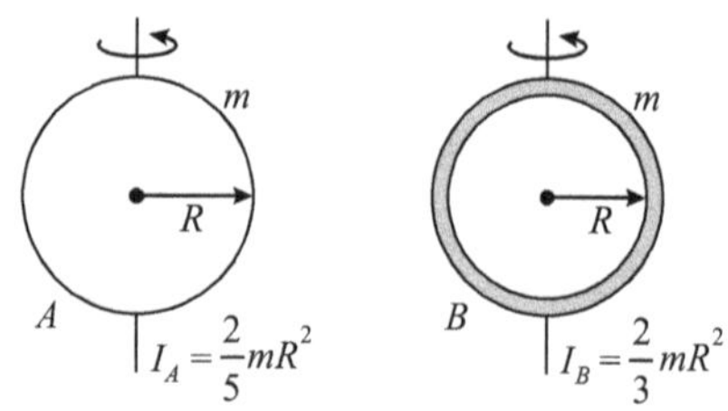

$\Rightarrow$ $I_B > I_A$ **Ans. (C)**

Sol. 9

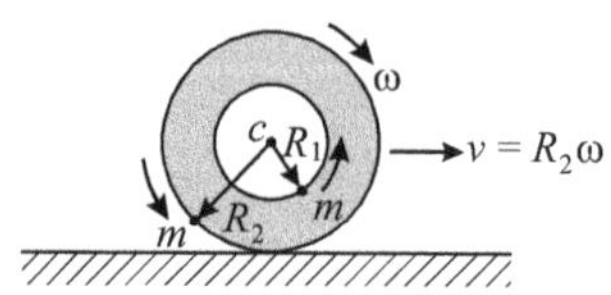

With respect to ring centre the two particles on inner and outer parts of ring are in uniform circular motion at speed ω.

$\Rightarrow$ Forces on the two particles are

$$F_1 = m\omega^2 R_1 \text{ and } F_2 = m\omega^2 R_2$$

$\Rightarrow$ $\frac{F_1}{F_2} = \frac{R_1}{R_2}$ **Ans. (D)**

Sol. 10

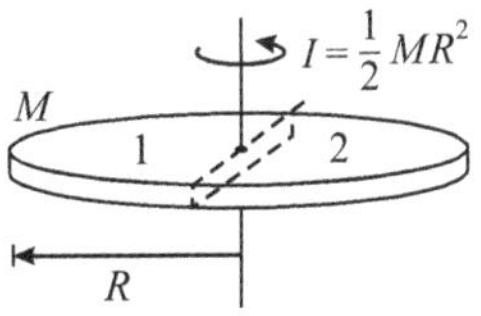

By property of mass distribution for all identical parts of a body, expression of moment of inertia remain same.

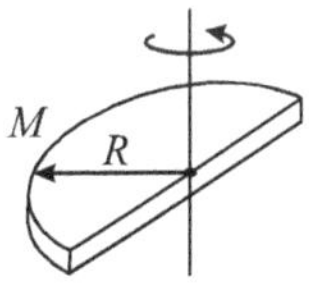

$\Rightarrow$ $$I=\frac{1}{2}MR^2$$ **Ans. (D)**

Sol. 11

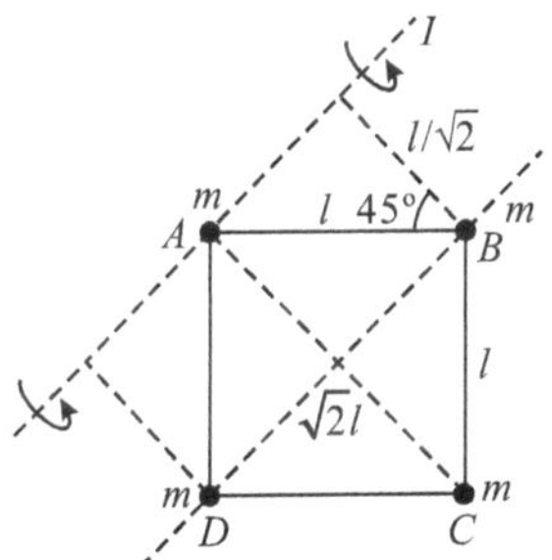

MI of all the masses about given AOR is

$$I=I_A+I_B+I_C+I_D$$

$\Rightarrow$ $$I=0+m\left(\frac{l}{\sqrt{2}}\right)^2+m(\sqrt{2}l)^2+m\left(\frac{l}{\sqrt{2}}\right)^2$$

$\Rightarrow$ $$I=\frac{ml^2}{2}+2ml^2+\frac{ml^2}{2}=3ml^2$$ **Ans. (C)**

Sol. 12

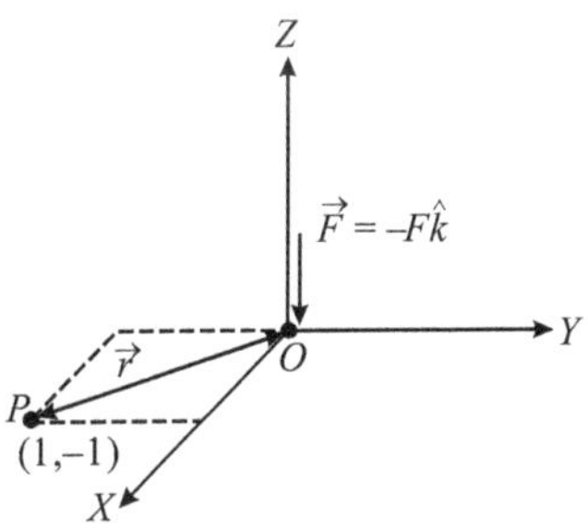

Position vector of origin about the given point is

$$\vec{r}=-\hat{i}+\hat{j}$$

Torque of the given force about point P is

$$\vec{\tau}=\vec{r}\times\vec{F}=(-\hat{i}+\hat{j})\times(-F\hat{k})$$

$\Rightarrow$ $$\vec{\tau}=F(-\hat{j}-\hat{i})=-F(\hat{i}+\hat{j})$$ **Ans. (B)**

Sol. 13

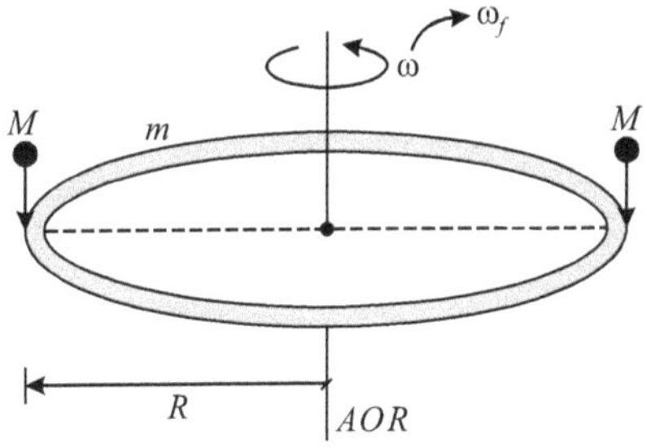

As no external torque is applied on ring, its angular momentum remain conserved

$$L=I\omega=I_f\omega_f$$

here $$I=mR^2 \text{ and } I_f=mR^2+2MR^2$$

$\Rightarrow$ $$mR^2\omega=(m+2M)R^2\omega_f$$

$\Rightarrow$ $$\omega_f=\frac{m\omega}{m+2M}$$ **Ans. (D)**

Sol. 14

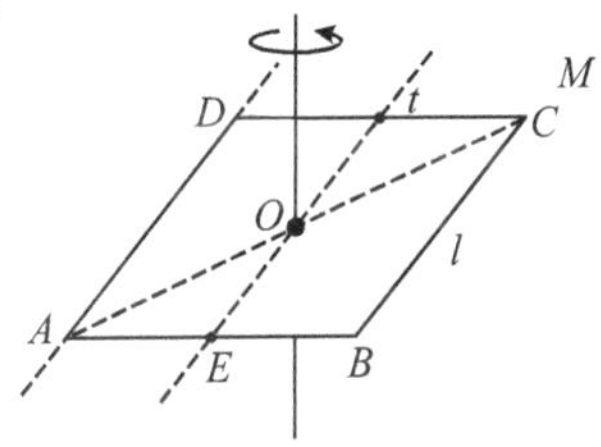

MI of a square plate about AOR passing through its center and ⊥ to its plane is given as

$$I_0=\frac{1}{6}Ml^2$$

For any AOR in the plane of lamina passing through its center MI can be given as

$$2I_P=\frac{1}{6}Ml^2$$

$\Rightarrow$ $$I_P=\frac{1}{12}Ml^2=I_{EF}=I_{AC}$$

$\Rightarrow$ $$I_{AD}=I_{EF}+\frac{Ml^2}{4}=\frac{Ml^2}{3}$$ **Ans. (C)**

Sol. 15

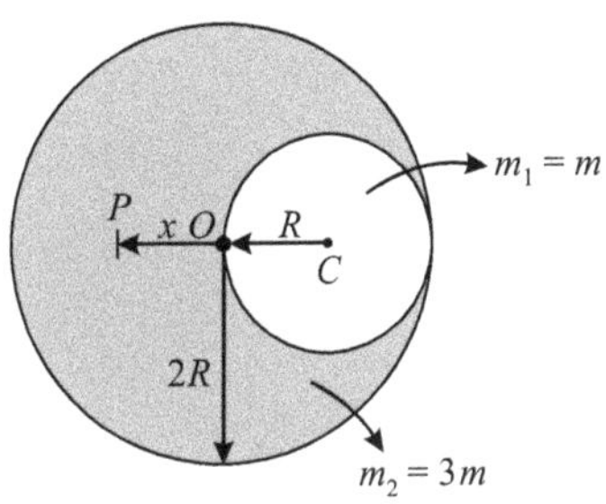

As disc is uniform, its mass is proportional to its surface area

$\Rightarrow$ $$m_1=C\cdot\pi R^2=m$$

and $$m_2=C[\pi(2R)^2-\pi R^2]=C(3\pi R^2)=3m$$

As point O is the COM of bigger disc of radius $2R$, we use

$$3mx = mR$$

$$\Rightarrow \quad x = \frac{R}{3} = \alpha R$$

$$\Rightarrow \quad \alpha = \frac{1}{3}$$ **Ans. (A)**

Sol. 16

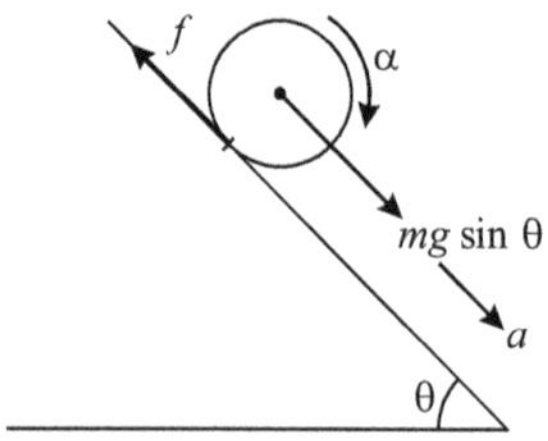

Equation of motion of body

$$mg\sin\theta - f = ma \qquad \ldots(1)$$

$$\Rightarrow \quad fR = I\left(\frac{a}{R}\right) \qquad \ldots(2)$$

[For pure rolling $a = R\alpha$]

$$(1)+(2) \Rightarrow \quad mg\sin\theta = a\left(M + \frac{I}{R^2}\right)$$

$$\Rightarrow \quad a = \frac{g\sin\theta}{\left(1+\frac{I}{MR^2}\right)}$$ **Ans. (A)**

Sol. 17

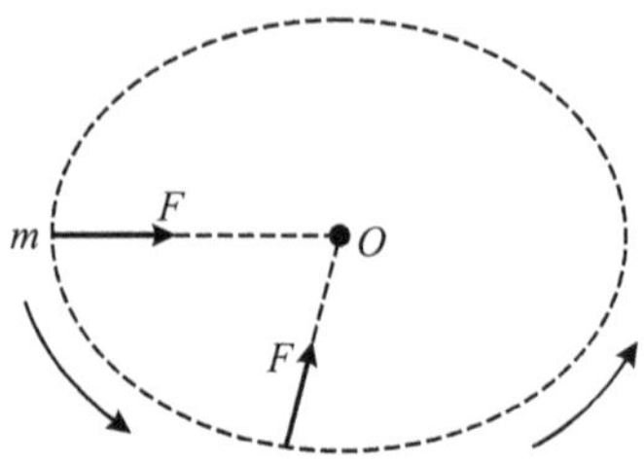

Due to central force which always passes through centre so the torque due to this force about O is always zero during motion.

As $\quad \vec{\tau}_0 = 0$

$\Rightarrow$ Angular momentum about O remain conserved. **Ans. (C)**

Sol. 18

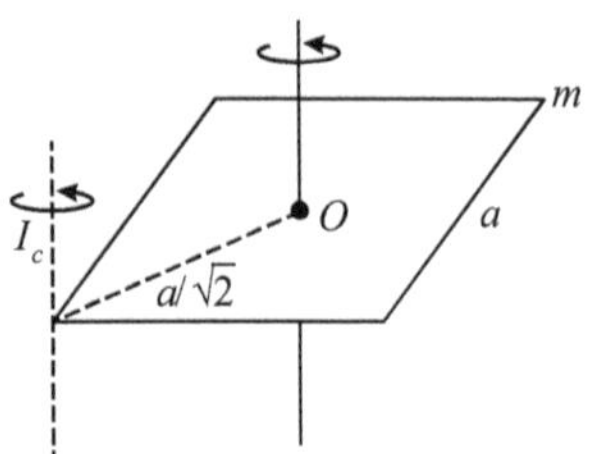

MI of square plate about its central AOR is

$$I_0 = \frac{1}{6}Ma^2$$

From parallel AOR, we have

$$I_c = I_0 + Md^2$$

$$\Rightarrow \quad I_c = \frac{1}{6}Ma^2 + \frac{1}{2}Ma^2 = \frac{2}{3}Ma^2$$ **Ans. (D)**

Sol. 19 By conservation of energy, we have

$$\frac{1}{2}I\omega^2 = mgh$$

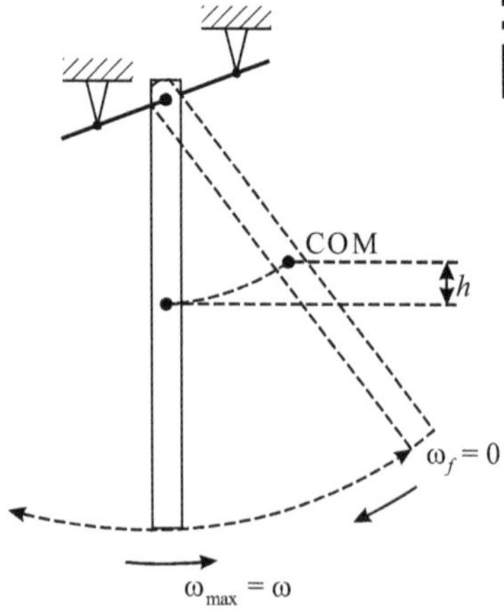

$$\Rightarrow \quad h = \frac{I\omega^2}{2mg} = \frac{\omega^2}{2mg}\left(\frac{Ml^2}{3}\right) = \frac{\omega^2 l^2}{6g}$$ **Ans. (C)**

Sol. 20 Angular momentum of particle is given as

$$\vec{L} = m(\vec{r}\times\vec{v})$$

where position vector of particle at time t is given as

$$\vec{r} = (v_0\cos\theta t)\,\hat{i} + (v_0\sin\theta t - \frac{1}{2}gt^2)\,\hat{j}$$

velocity vector of particle at time t is given as

$$\vec{v} = (v_0\cos\theta)\,\hat{i} + (v_0\sin\theta - gt)\,\hat{j}$$

$$\Rightarrow \quad \vec{L} = -\frac{1}{2}mgv_0t^2\cos\theta\,\hat{k}$$ **Ans. (D)**

Sol. 21 Equation of motion for mass m

$$mg - T = ma \qquad \ldots(1)$$

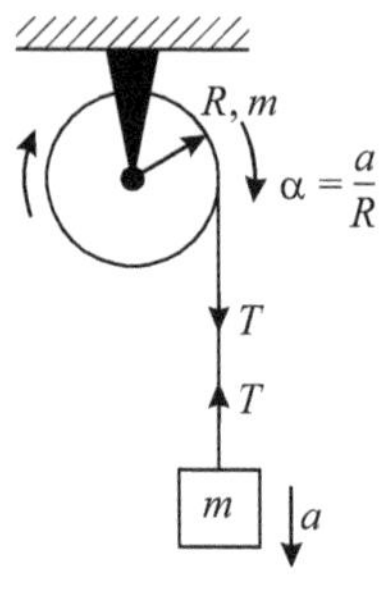

For pulley

$$TR = \left(\frac{1}{2}mR^2\right)\left(\frac{a}{R}\right) \qquad \ldots(2)$$

$$(1)+(2) \Rightarrow \quad mg = \frac{3}{2}ma$$

$$\Rightarrow \quad a = \frac{2g}{3}$$ **Ans. (B)**

Sol. 22

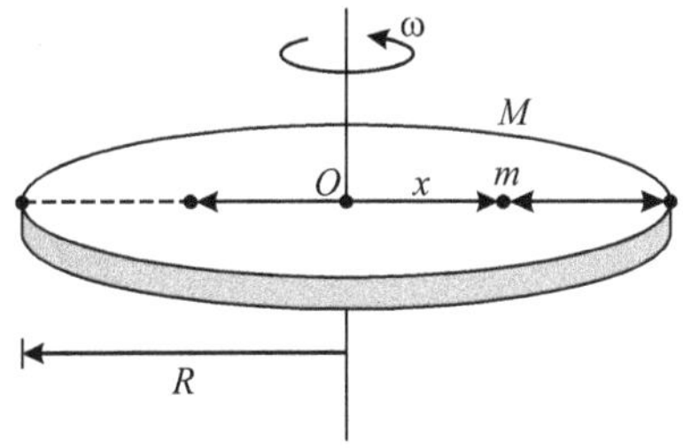

$$I = \frac{1}{2}MR^2 + mx^2$$

An no external torque is acting on disc, its angular momentum remain conserved

$\Rightarrow$ $$L = I\omega = \text{constant}$$

$\Rightarrow$ $$\left(\frac{1}{2}MR^2 + mx^2\right)\omega = \text{constant}$$

$\Rightarrow$ $$\omega = \frac{\text{constant}}{\frac{1}{2}MR^2 + mx^2}$$

As $x\downarrow \Rightarrow \omega\uparrow$ and after passing centre O $x\uparrow \Rightarrow \omega\downarrow$ **Ans (B)**

Sol. 23 Torque on pulley

$$\tau = rF = 40t - 10t^2 = I\alpha$$

Angular acceleration of pulley

$$\alpha = 4t - t^2$$

$\Rightarrow$ $$\frac{d\omega}{dt} = 4t - t^2$$

$\Rightarrow$ $$\int_0^\omega d\omega = \int_0^t (4t - t^2)dt$$

$\Rightarrow$ $$\omega = 2t^2 - \frac{t^3}{3}$$

here ω first ↑ then ↓ and it becomes zero when

$$2t^2 - \frac{t^3}{3} = 0$$

$\Rightarrow$ $$t = 6\text{s}$$

at $t = 6$s direction of pulley rotation is reversed

$$\frac{d\theta}{dt} = 2t^2 - \frac{t^3}{3}$$

$$\int_0^\theta d\theta = \int_0^6 \left(2t^2 - \frac{t^3}{3}\right)dt$$

$\Rightarrow$ $$\theta = \left[\frac{2t^3}{3} - \frac{t^4}{12}\right]_0^6 = \frac{2}{3}\times 216 - \frac{(6)^4}{12} - 0$$

$\Rightarrow$ $$\theta = 36 \text{ rad}$$

$$\text{No of turns} = \frac{\theta}{2\pi} = \frac{36}{2\pi} = 5.73$$ **Ans. (D)**

Sol. 24

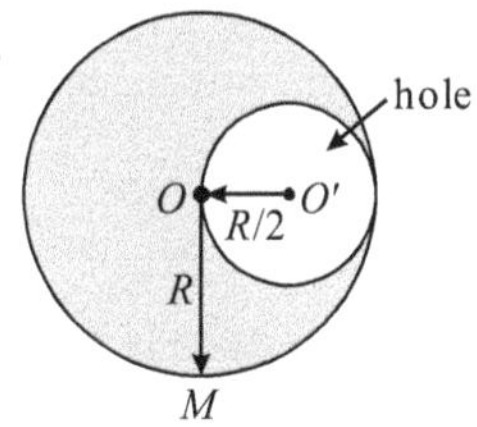

MI of whole disc is

$$I_0 = \frac{1}{2}MR^2$$

Mass of small (hole) disc is

$$m = \frac{M}{\pi R^2}\times\frac{\pi R^2}{4} = \frac{M}{4}$$

MI of small disc about axis at O is

$$I_{\text{hole}} = I_{O'} + \left(\frac{M}{4}\right)\left(\frac{R}{2}\right)^2$$

$\Rightarrow$ $$I_{\text{hole}} = \frac{1}{2}\left(\frac{M}{4}\right)\left(\frac{R}{2}\right)^2 + \left(\frac{M}{4}\right)\left(\frac{R}{2}\right)^2$$

$\Rightarrow$ $$I_{\text{hole}} = \frac{3}{2}\times\frac{M}{4}\times\frac{R^2}{4} = \frac{3}{32}MR^2$$

MI of remaining disc is

$$I = I_0 - I_{\text{hole}}$$

$\Rightarrow$ $$I = \frac{1}{2}MR^2 - \frac{3}{32}MR^2 = \frac{13}{32}MR^2$$ **Ans. (D)**

Sol. 25 Translational $KE_T = \frac{1}{2}mv^2$ ...(1)

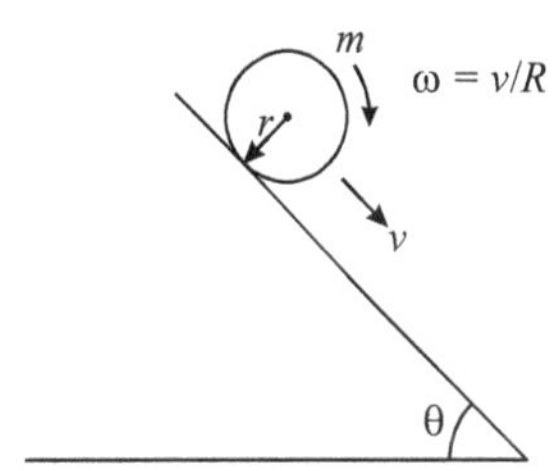

Rotational $KE_R = \frac{1}{2}I\omega^2 = \frac{1}{2}\left(\frac{2}{5}mR^2\right)\left(\frac{v}{R}\right)^2$

$$KE_R = \frac{1}{5}mv^2 \quad \text{...(2)}$$

By (2)/(1) $\Rightarrow$ $$\frac{KE_R}{KE_T} = \frac{2}{5}$$

$\Rightarrow$ $$KE_R = \frac{2}{7} \text{ times of TE}$$

$$KE_T = \frac{5}{7} \text{ times of TE}$$ **Ans. (B)**

Sol. 26 If there is No torque $\Rightarrow L \to$ constant

we use $$L = I\omega$$

and $$K = \frac{1}{2}I\omega^2 = \frac{L^2}{2I}$$

$\Rightarrow$ As $L \to$ constant so on increasing $I \Rightarrow$ KE decreases.

S-1 is false.

S-2 is true. **Ans. (B)**

Sol. 27 Angular momentum of revolving stone is

$$L = mvr = C \text{ (constant)}$$

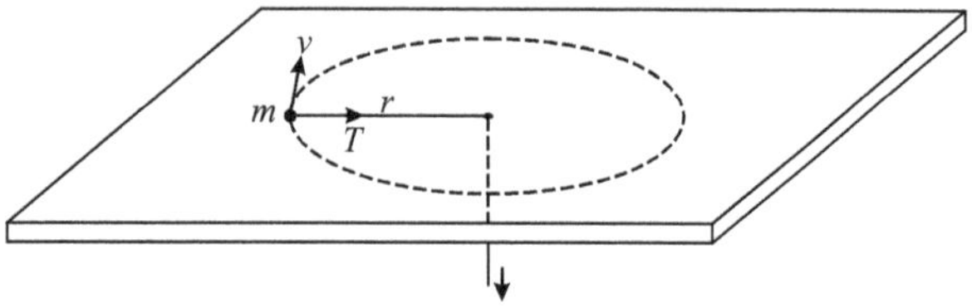

Considering stable AB (r is gradually decreased) circular motion we use

$$T = mv^2/r$$

$$\Rightarrow \quad T = \frac{m}{r}\left(\frac{C}{mr}\right)^2 = \frac{C^2}{m} r^{-3} = Ar^n$$

$$\Rightarrow \quad n = -3$$

Ans. (D)

Sol. 28 Initial speed of sphere should be sufficient to take sphere to top when its $v_f = 0$.

So using work-energy theorem we have

$$\frac{1}{2}mv^2 + \frac{1}{2}\left(\frac{2}{5}mr^2\right)\left(\frac{v}{r}\right)^2 - mgh = 0$$

$$\left(\frac{1}{2}+\frac{1}{5}\right)mv^2 = mgh$$

$$\Rightarrow \quad v = \sqrt{\frac{10}{7}gh}$$

Ans. (D)

Sol. 29 In case of pure rolling, the speed attained by the body at bottom of incline is

$$v_0 = \sqrt{\frac{2gh}{1+\frac{k^2}{R_0^2}}} \quad \ldots(1)$$

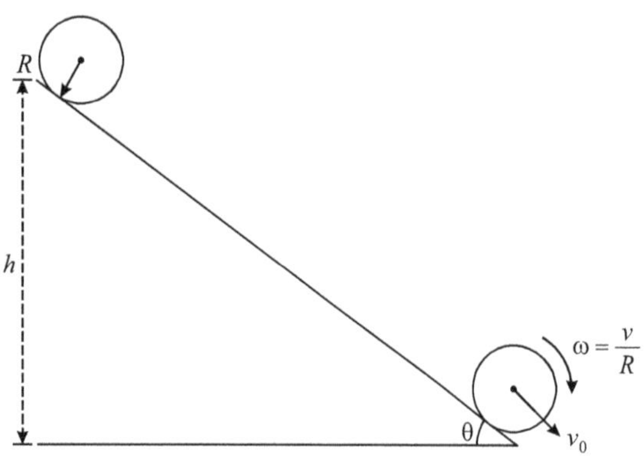

In case of smooth incline no rotational KE is gained by body so speed attained is

$$v = \sqrt{2gh} = \frac{5v_0}{4} \quad \ldots(2)$$

Using (2)/(1) gives $\frac{5}{4} = \sqrt{1+\frac{k^2}{R_0^2}}$

$$\Rightarrow \quad 1+\frac{k^2}{R_0^2} = \frac{25}{16}$$

$$\Rightarrow \quad \frac{k^2}{R_0^2} = \frac{9}{16} \Rightarrow k = \frac{3}{4}R_0$$

Ans. (B)

Sol. 30

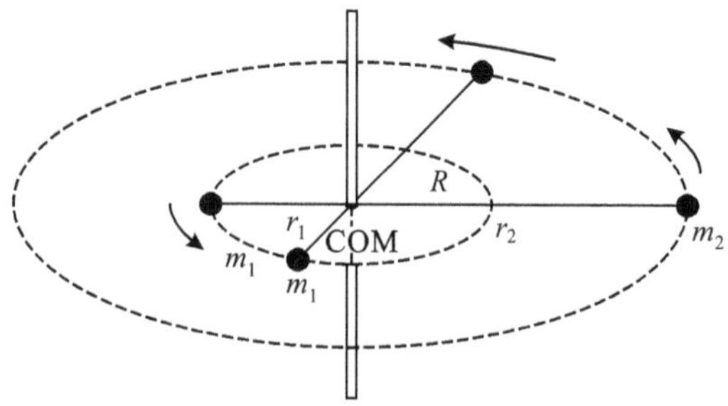

As no external force is acting on the system of these two particles

$\Rightarrow$ Their COM will remain at rest

$\Rightarrow$ During revolution the two masses will always remain opposite to each other as shown on the two sides of COM.

$\Rightarrow$ Their time period will be same for them to be on opposite sides of COM

$\Rightarrow$ $\omega_1 = \omega_2$ (independent of m_1 & m_2) **Ans. (B)**

Sol. 31 At pure rotting we use

$$v_f = r\omega_f$$

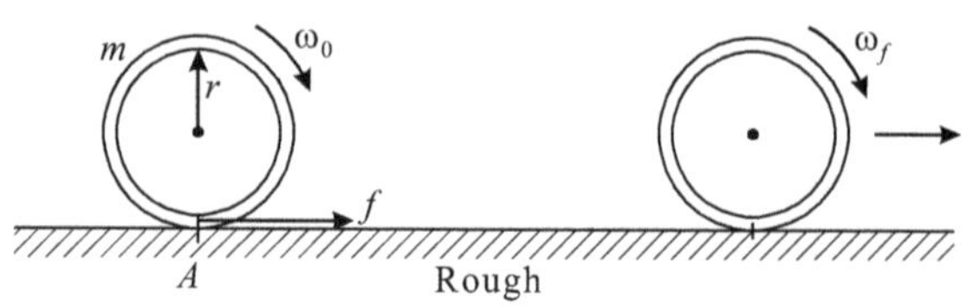

By conversation of angular momentum about point A, we use

$$(mr^2)\omega_0 = (mr^2)\omega_f + mv_f r = 2mr^2\omega_f$$

$$\Rightarrow \quad \omega_f = \frac{\omega_0}{2}$$

$$\Rightarrow \quad v_f = r\omega_f = \frac{r\omega_0}{2}$$

Ans. (C)

Sol. 32 In this process of collision the angular momentum of door + bullet system remain conserved.

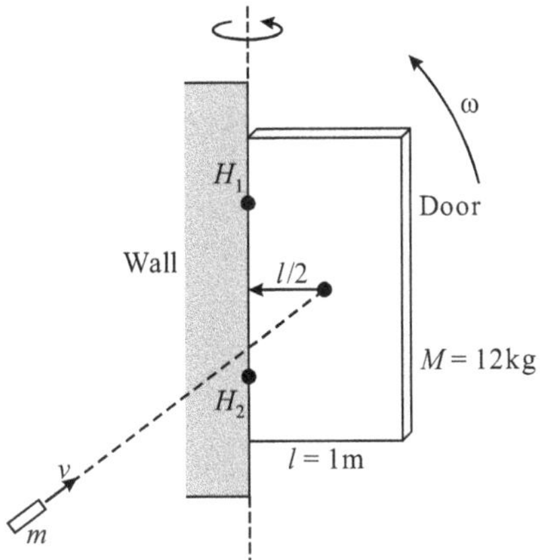

$I = \frac{1}{3}ml^2$

As angular momentum of door + bullet system remain conserved, we use

$$mv\left(\frac{l}{2}\right) = \frac{1}{3}Ml^2\omega$$

$$\Rightarrow \quad \omega = \frac{3mvl}{2Ml^2} = \frac{3\times0.01\times500}{2\times12\times1} = \frac{15}{24}$$

$$\Rightarrow \quad \omega = 0.625 \text{ rad/s}$$ **Ans. (B)**

Sol. 33 Angular momentum

$$\vec{L} = m(\vec{r}\times\vec{v})$$

Where $\vec{r} = 5\hat{i} - 2t^2\hat{j}$

and $\vec{v} = \frac{d\vec{r}}{dt} = -4t\,\hat{j}$

$$\Rightarrow \quad \vec{L} = m[(5\hat{i}-2t^2\hat{j})\times(-4t\hat{j})]$$

$$\Rightarrow \quad \vec{L}_{(t=2s)} = 2[(5\hat{i}-8\hat{j})\times(-8\hat{j})]$$

$$\vec{L}_{(t=2s)} = 2\times[-40\hat{k}] = -80\hat{k}$$ **Ans. (A)**

Sol. 34 By work-energy theorem from 0 to A, we have

$$0 + mg(1.8) = \frac{1}{2}mv^2 + \frac{1}{2}\left(\frac{2}{3}mR^2\right)\left(\frac{v}{R}\right)^2 = \frac{5}{6}mv^2$$

$$\Rightarrow \quad v = \sqrt{\frac{6}{5}\times10\times1.8}$$

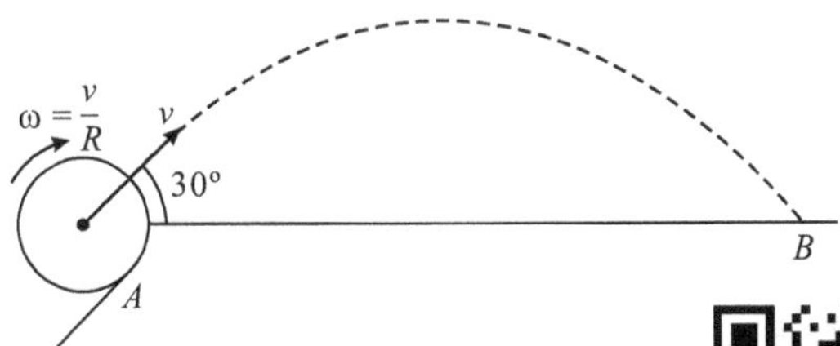

Horizontal Range $\quad R = \frac{v^2\sin2\theta}{g}$

$$\Rightarrow \quad R = AB = \frac{6}{5}\times\frac{10\times1.8}{10}\times\frac{\sqrt{3}}{2}$$

$$\Rightarrow \quad R = 1.87\text{ m}$$ **Ans. (A)**

Sol. 35 In gently attaching the bodies, no external torque acts on system, so angular momentum remain conserved.

$$MR^2\omega = (M+2m)R^2\omega_f$$

$$\Rightarrow \quad \omega_f = \frac{m\omega}{M+2m}$$

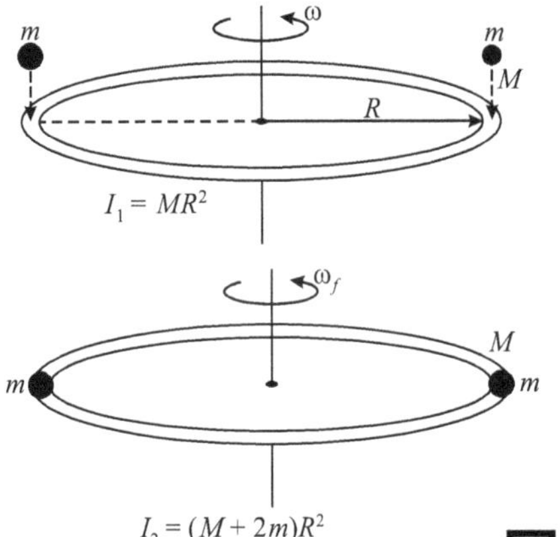

Loss in kinetic energy is given as

$$\Delta K_{\text{Loss}} = K_i - K_f$$

$$\Rightarrow \quad \Delta K_{\text{Loss}} = \frac{1}{2}I_1\omega^2 - \frac{1}{2}I_1\omega_f^2$$

$$\Rightarrow \quad \Delta K_{\text{Loss}} = \frac{1}{2}(MR^2)\omega^2 - \frac{1}{2}(M+2m)R^2\left(\frac{M}{M+2m}\right)^2\omega^2$$

$$\Rightarrow \quad \Delta K_{\text{Loss}} = \frac{1}{2}MR^2\omega^2 - \frac{1}{2}\frac{M^2R^2}{(M+2m)}\omega^2$$

$$\Rightarrow \quad \Delta K_{\text{Loss}} = \frac{1}{2}R^2\omega^2\left[M - \frac{M^2}{M+2m}\right]$$

$$\Rightarrow \quad \Delta K_{\text{Loss}} = \frac{1}{2}R^2\omega^2\left[\frac{2mM}{M+2m}\right] = \frac{mMR^2\omega^2}{(M+2m)}$$

Ans. (C)

Sol. 36

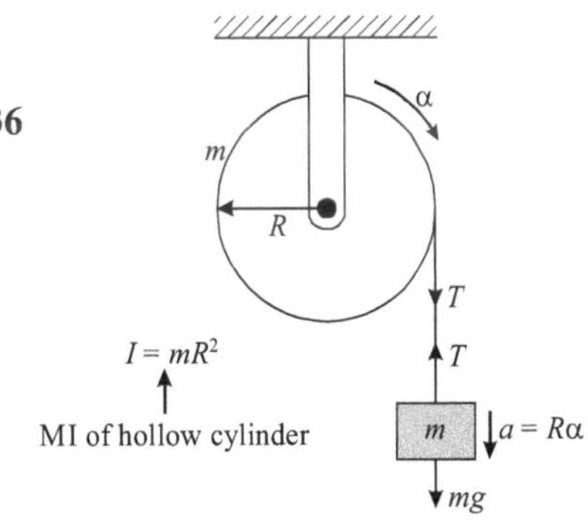

Equation of motion for falling mass

$$mg - T = ma \qquad \ldots(1)$$

For cylinder $\quad TR = (mR^2)\left(\frac{a}{R}\right) \qquad \ldots(2)$

$(1)+(2)\Rightarrow \quad mg = 2ma$

$$\Rightarrow \quad a = g/2$$ **Ans. (B)**

Sol. 37

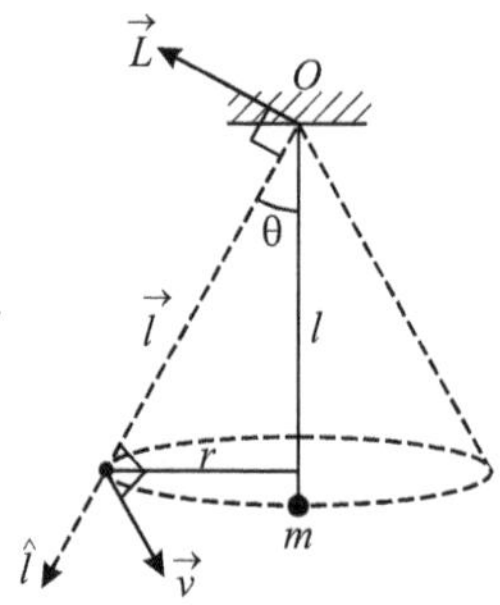

$$|\vec{v}| = \text{constant}$$

Angular momentum of bob about point O is

$$\vec{L} = m(\vec{l} \times \vec{v})$$

Direction of $\vec{L}$ is $\perp$ to both $\vec{l}$ and $\vec{v}$ as shown in figure and its magnitude is

$$L = mlv \rightarrow \text{constant}$$ **Ans. (C)**

Sol. 38 relative to rug we consider cylinder is having acceleration a, for which equations of motion of cylinder are

$$ma - f = Ma_1 \qquad \ldots(1)$$

$$fR = \left(\frac{1}{2}MR^2\right)\left(\frac{a_1}{R}\right) \qquad \ldots(2)$$

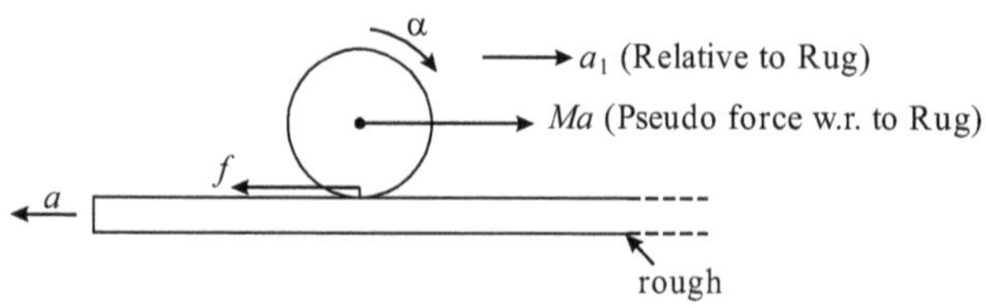

From (1) and (2), we have

$$Ma - f = 2f$$

$$\Rightarrow \qquad 3f = Ma$$

$$\Rightarrow \qquad f = \frac{1}{3}Ma$$ **Ans. (D)**

Sol. 39 For pure rolling of a body on incline plane, acceleration is given as

$$a = \frac{g\sin\theta}{\left(1 + \frac{k^2}{R^2}\right)}$$

for cylinder $\quad k_c = \frac{R}{\sqrt{2}} \Rightarrow a_c = \frac{g\sin\theta}{1 + \frac{1}{2}} = \frac{2}{3}g\sin\theta_c$

for sphere $\quad k_s = \sqrt{\frac{2}{5}}R \Rightarrow a_s = \frac{g\sin\theta_s}{1 + \frac{2}{5}} = \frac{5}{7}g\sin\theta_s$

$$a_c = a_s \Rightarrow \frac{2}{3}g\sin\theta_c = \frac{5}{7}g\sin\theta_s$$

$$\Rightarrow \qquad \frac{\sin\theta_c}{\sin\theta_s} = \frac{1}{2/3 \big/ 5/7} = \frac{15}{14}$$ **Ans. (D)**

Sol. 40 Angular momentum of particle at highest point of trajectory about O is given as

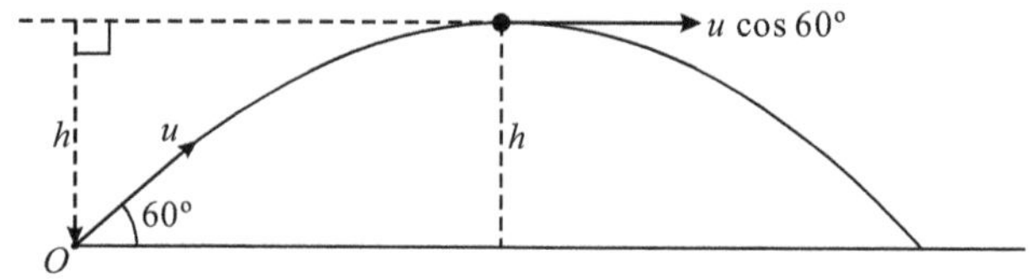

$$L_0 = m(u\cos 60°) \times h$$

$$\Rightarrow \qquad L_0 = mu\cos 60° \times \frac{u^2\sin^2 60°}{2g}$$

$$\Rightarrow \qquad L_0 = 0.16 \times 10 \times \frac{1}{2} \times \frac{(10)^2\left(\frac{\sqrt{3}}{2}\right)^2}{2 \times 10}$$

$$\Rightarrow \qquad L_0 = 3\ \text{kg m}^2/\text{s}$$ **Ans. (B)**

Sol. 41 Angle α is given as in ΔPCA

$$\alpha = 180° - (2\theta + 90 - \theta)$$

$$\Rightarrow \qquad \alpha = 90° - \theta$$

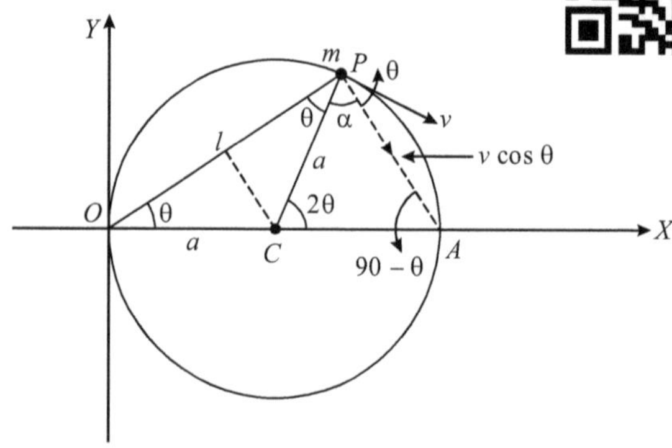

Angular momentum of mass m about O is given as

$$L_0 = m(v\cos\theta)l$$

here $\qquad l = 2a\cos\theta$

$$\Rightarrow \qquad L_0 = mv\cos\theta \times 2a\cos\theta$$

$$\Rightarrow \qquad L_0 = 2mva\cos^2\theta$$

$$\Rightarrow \qquad L_0 = mva(1 + \cos 2\theta)$$ **Ans. (A)**

Sol. 42 For largest cube we use

$$2R = \sqrt{3}l$$

$$\Rightarrow \qquad l = \frac{2R}{\sqrt{3}}$$

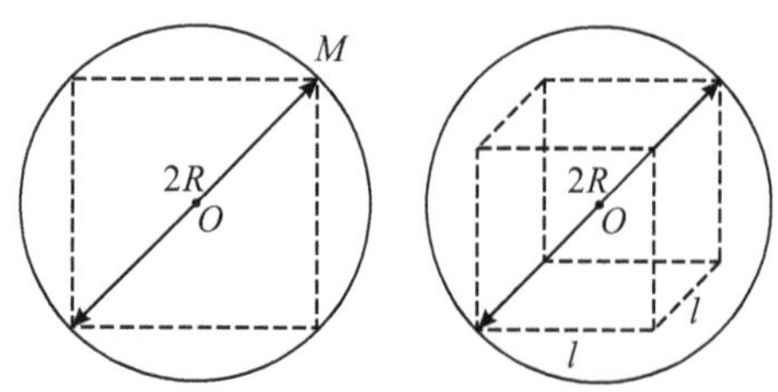

2D view $\qquad$ 3D view

Mass of cube $$m=\frac{M}{\frac{4}{3}\pi R^3}\times\left(\frac{2R}{\sqrt{3}}\right)^3$$

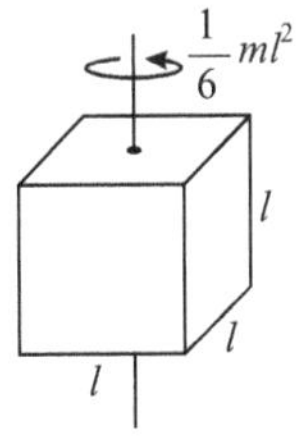

MI of cube about its central axis is

$$I=\frac{1}{6}\left(\frac{3M}{4\pi R^3}\frac{8R^3}{9\sqrt{3}}\right)\left(\frac{2R}{\sqrt{3}}\right)^2$$

$$\Rightarrow \quad I=\frac{4MR^2}{9\sqrt{3}\pi}$$ **Ans. (C)**

Sol. 43 Equations of motion of cylinder are

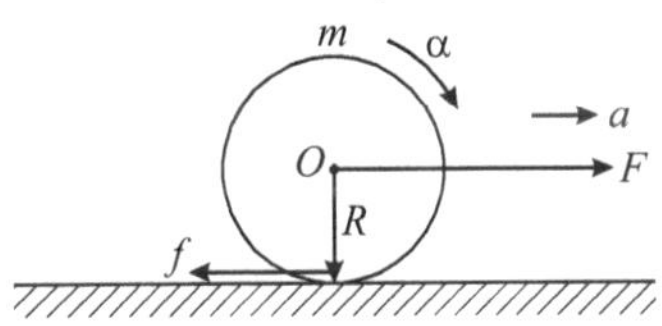

$$F-f=ma \quad \dots(1)$$

$$fR=\left(\frac{1}{2}mR^2\right)\left(\frac{a}{R}\right)$$

$$f=\frac{1}{2}ma \quad \dots(2)$$

equation (1) + (2)

$$F=\frac{3}{2}ma$$ **Ans. (C)**

Sol. 44 $$l=\sqrt{r^2+h^2}$$

$$l=\sqrt{(0.6)^2+(0.8)^2}=1\text{ m}$$

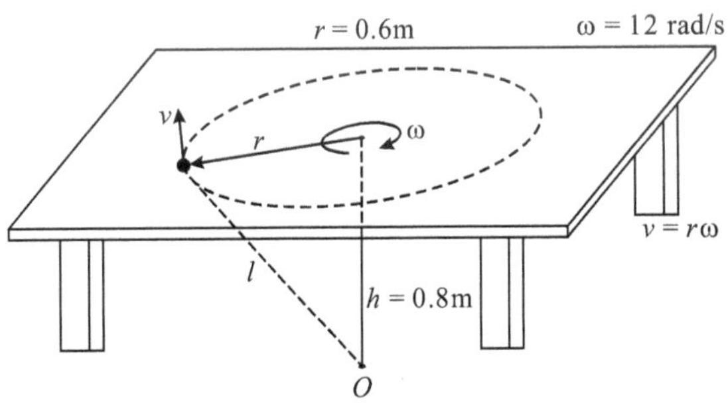

In figure angular momentum of particle about O is given as

$$L=mvl$$

$$\Rightarrow \quad L=2\times(0.6\times 12)\times 1$$

$$\Rightarrow \quad L=14.4\text{ kg m}^2\text{/s}$$ **Ans. (A)**

Sol. 45 MI of a square lamina about its central AOR is given as

$$I_0=\frac{Ma^2}{6}$$

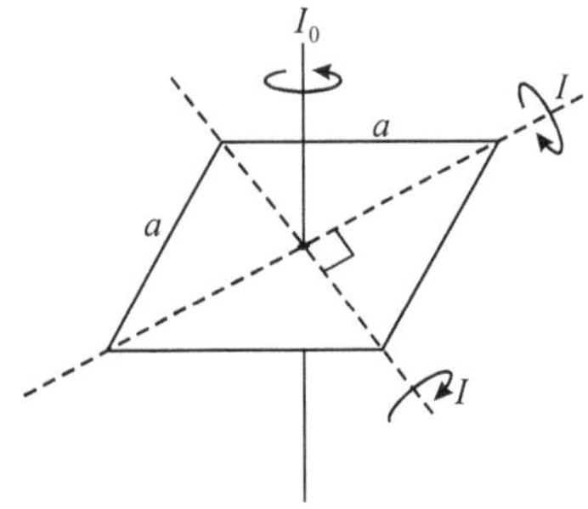

By perpendicular axes theorem we have

$$I+I=I_0$$

$$\Rightarrow \quad I=\frac{1}{2}I_0=\frac{ma^2}{12}$$ **Ans. (D)**

Sol. 46 Angular momentum of particle about origin for its motion

along AB $\quad \vec{L}=mv\left(\frac{R}{\sqrt{2}}\right)(-\hat{k})$

along BC $\quad \vec{L}=mv\left(a+\frac{R}{\sqrt{2}}\right)\hat{k}$

along CD $\quad \vec{L}=mv\left(a+\frac{R}{\sqrt{2}}\right)\hat{k}$

along DA $\quad \vec{L}=mv\left(\frac{R}{\sqrt{2}}\right)(-\hat{k})$ **Ans. (A, C)**

Sol. 47 Here we analyze friction on the bottom contact points of cones and rails

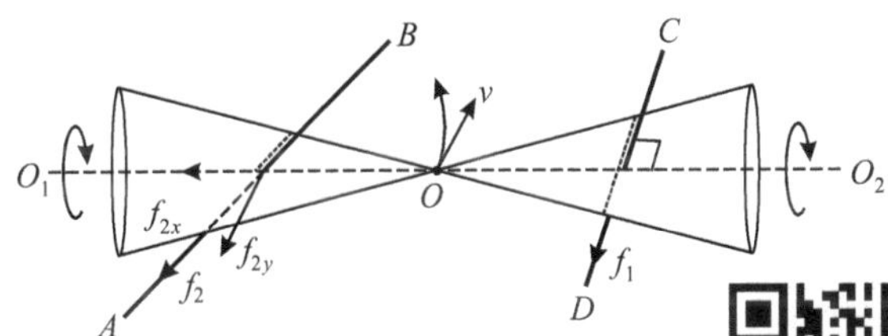

Due to a component of friction f_2 on roller at rail AB it will turn toward left. **Ans. (C)**

Sol. 48 As no torque is acting about O at the instant of impact so angular momentum of cube remain conserved about O.

$$mv\frac{a}{2}=I\omega$$

$$\Rightarrow \quad \omega=\frac{mva}{2I}$$

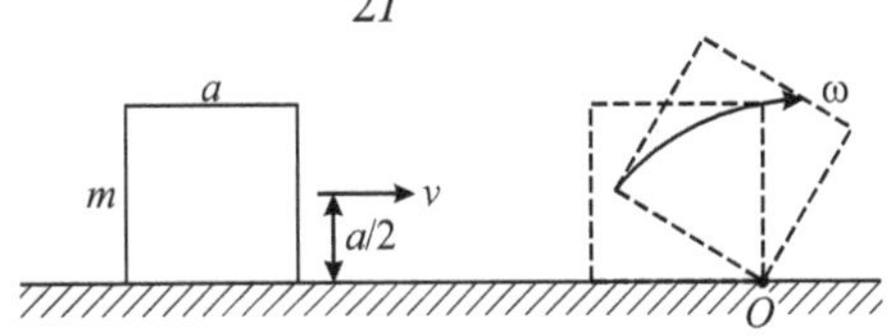

MI of cube about its corner

$$I = I_c + M\left(\frac{a}{\sqrt{2}}\right)^2 = \frac{Ma^2}{6} + \frac{Ma^2}{2}$$

$$\Rightarrow \quad I = \frac{2}{3}Ma^2$$

$$\Rightarrow \quad \omega = \frac{mva}{2\left(\frac{2}{3}Ma^2\right)} = \frac{3v}{4a}$$

$$\Rightarrow \quad \omega = \frac{3\times 2}{4\times 0.3} = 5 \text{ rad/s}$$

Ans. (B)

Sol. 49 MI of the given cylinder about AOR shown is

$$I = \frac{1}{12}Ml^2 + \frac{1}{4}MR^2$$

$$\Rightarrow \quad I = \frac{1}{12}Ml^2 + \frac{1}{4}M\left(\frac{M}{\rho\pi l}\right)$$

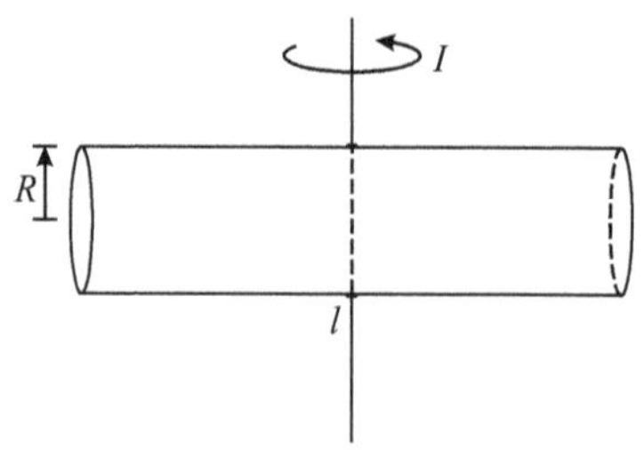

for I to be minimum. We use $\frac{dI}{dl} = 0$

$$\Rightarrow \quad \frac{dI}{dl} = \frac{1}{6}Ml + \frac{M^2}{4\rho\pi}\left(-\frac{1}{l^2}\right) = 0$$

$$\Rightarrow \quad l^3 = \frac{6M}{4\rho\pi} = \frac{3}{2}\frac{\rho\pi R^2 l}{\rho\pi}$$

$$\Rightarrow \quad \frac{l^2}{R^2} = \frac{3}{2} \Rightarrow \frac{l}{R} = \sqrt{\frac{3}{2}}$$

Ans. (D)

Sol. 50 Torque of rod's weight about point O is

$$\tau = Mg \times \frac{l}{2}\sin\theta$$

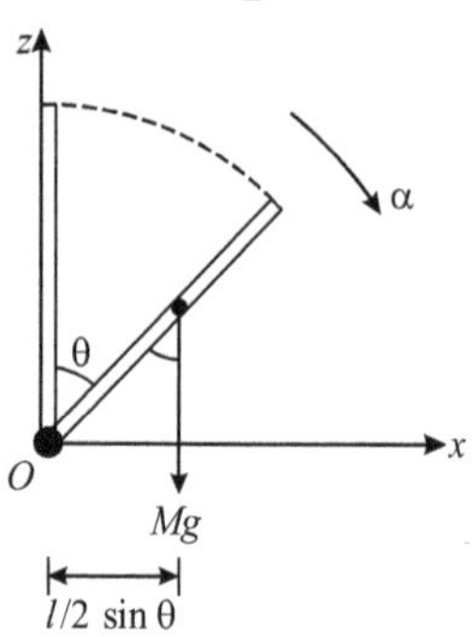

If α be the angular acceleration of rod, we use

$$Mg\frac{l}{2}\sin\theta = \left(\frac{Ml^2}{3}\right)\alpha$$

$$\Rightarrow \quad \alpha = \frac{3g}{2l}\sin\theta$$

Ans. (D)

Sol. 51 MI of whole disc is

$$I_0 = \frac{1}{2}MR^2$$

Mass of small disc is

$$m = \frac{M}{\pi R^2}\pi\left(\frac{R}{4}\right)^2 = \frac{M}{16}$$

MI of small disc about O is

$$I_{\text{hole}} = I_c + md^2$$

$$I_{\text{hole}} = \frac{1}{2}m\left(\frac{R}{4}\right)^2 + m\left(\frac{3R}{4}\right)^2$$

$$\Rightarrow \quad I_{\text{hole}} = \frac{1}{2}\left(\frac{M}{16}\right)\left(\frac{R^2}{16}\right) + \left(\frac{M}{16}\right)\left(\frac{9R^2}{16}\right)$$

$$\Rightarrow \quad I_{\text{hole}} = \frac{19}{512}MR^2$$

MI of given remaining portion is

$$I = I_0 - I_{\text{hole}}$$

$$\Rightarrow \quad I = \left(\frac{1}{2} - \frac{19}{512}\right)MR^2 = \frac{237}{512}MR^2$$

Ans. (B)

Sol. 52 By mass distribution property for same shape of objects of different dimensions MI expression remain same.

For lamina ABC $\quad I_0 = \gamma ML^2$

$$\Rightarrow \quad I_{DEF} = \gamma ml^2; \text{ where } m_{DEF} = \frac{1}{4}M \text{ and } l = \frac{l}{2}$$

$$\Rightarrow \quad I_{DEF} = \gamma\left(\frac{M}{4}\right)\left(\frac{L}{2}\right)^2 = \frac{\gamma}{16}ML^2 = \frac{I_0}{16}$$

MI of remaining part of lamina after removing DEF is

$$I = I_0 - I_{DEF} = \frac{15I_0}{16}$$

Ans. (B)

Sol. 53 In the given situation we have

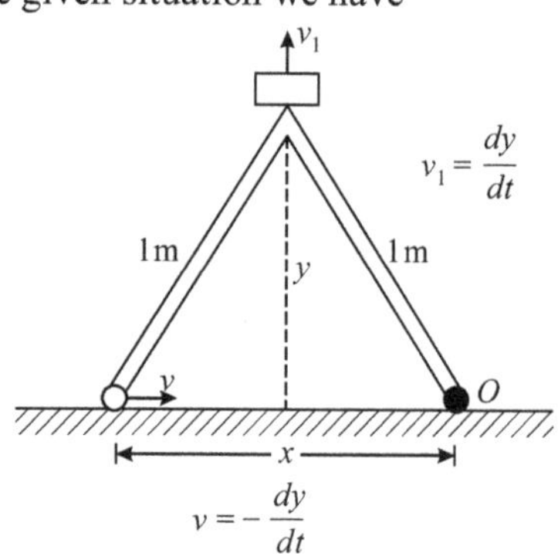

$$\left(\frac{x}{2}\right)^2 + y^2 = 1$$

Differentiate w.r. to time

$$\frac{x}{2}\frac{dx}{dt}+2y\frac{dy}{dt}=0$$

$$\Rightarrow \qquad -\frac{x}{2}v+2yv_1=0 \Rightarrow v_1=\frac{x}{4y}v \qquad [v = \text{constant}]$$

As with time $x\downarrow$ and $y\uparrow \Rightarrow v_1$ decreases **Ans. (B)**

Sol. 54 For equilibrium of beam balance we use

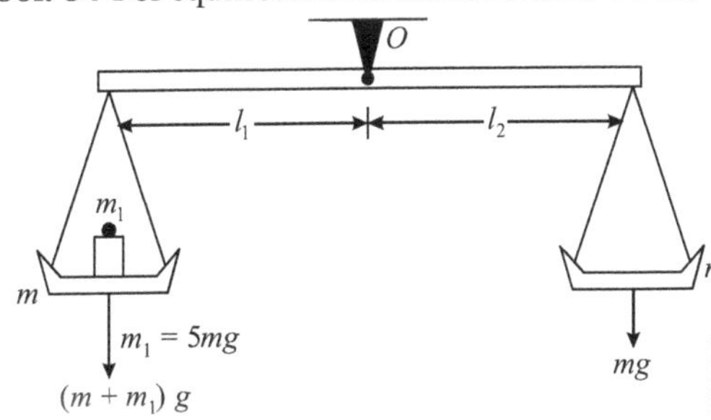

$$(m+m_1)g\times l_1 = mg\times l_2$$

$$\Rightarrow \qquad l_1 < l_2 \qquad \textbf{Ans. (C)}$$

Sol. 55 The central force on particle is given as

$$F=\frac{k}{R^n}=m\omega^2R \qquad \text{(for circular motion)}$$

$$\Rightarrow \qquad \omega=\sqrt{\frac{k}{mR^{n+1}}}$$

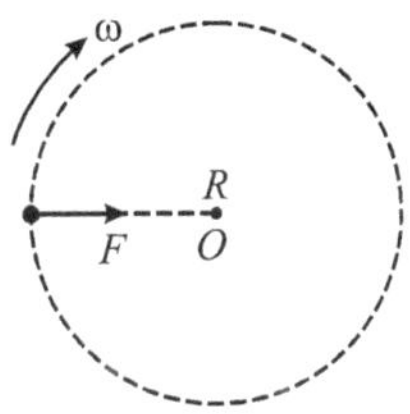

Time period of revolution is

$$T=\frac{2\pi}{\omega}=2\pi\sqrt{\frac{mR^{n+1}}{k}}$$

$$\Rightarrow \qquad T\propto R^{\frac{n+1}{2}} \qquad \textbf{Ans. (C)}$$

Sol. 56 MI of this system about centre O of central disc is

$$I_0=\frac{1}{2}MR^2+6\left(\frac{1}{2}MR^2+M(2R)^2\right)$$

$$I_0=\frac{7}{2}MR^2+24MR^2=\frac{55}{2}MR^2$$

Now using parallel axes theorem for the system, its MI about P is given

$$I_P=I_0+7M(3R)^2=\left(\frac{55}{2}+63\right)MR^2$$

$$\Rightarrow \qquad I_P=\frac{181}{2}MR^2 \qquad \textbf{Ans. (D)}$$

Sol. 57 Mass per unit area of disc $=\dfrac{9M}{\pi R^2}$

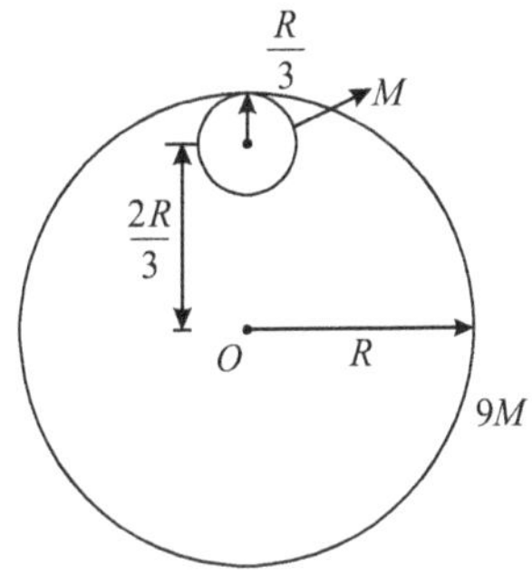

Mass of small disc is

$$m=\frac{9M}{\pi R^2}\times\pi\left(\frac{R}{3}\right)^2=M$$

MI of remaining disc is

$$I_0=(I_{9M}-I_M)_{\text{about }O}$$

$$I_0=\frac{1}{2}(9M)R^2-\left[\frac{1}{2}M\left(\frac{R}{3}\right)^2+M\left(\frac{2R}{3}\right)^2\right]$$

$$\Rightarrow \qquad I_0=\frac{9}{2}MR^2-\frac{1}{2}MR^2=4MR^2 \qquad \textbf{Ans. (A)}$$

Sol. 58 Moment of force (Torque) about point A will be maximum when line AB is $\perp$ direction of $\vec{F}$.

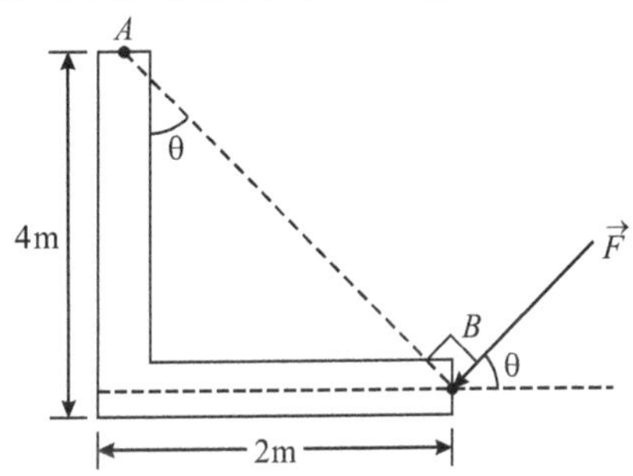

For this we use $\tan\theta=\dfrac{2}{4}=\dfrac{1}{2}$ **Ans. (C)**

Sol. 59 To keep the rod horizontal net force and torque on it must be zero.

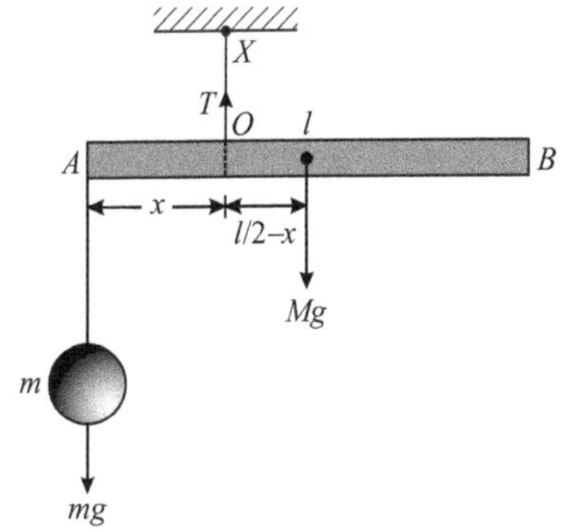

$$mg\,x=Mg\left(\frac{l}{2}-x\right)$$

Where $mgx \to$ Anticlockwise torque about O

and $Mg\left(\frac{l}{2}-x\right) \to$ Clockwise torque about O

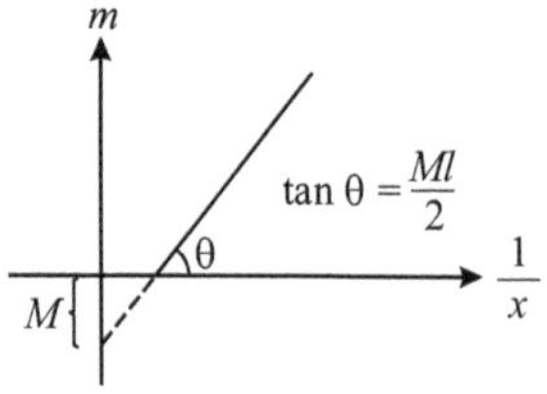

$$m=\left(\frac{Ml}{2}\right)\frac{1}{x}-M$$

where $\frac{Ml}{2} \to$ Slope

and $M \to$ Intercept on Y axis **Ans. (A)**

Sol. 60 After masses stick to rod, final COM of rod remains at mid point of rod as mass moments of m and $2m$ about O are equal & on opposite sides.

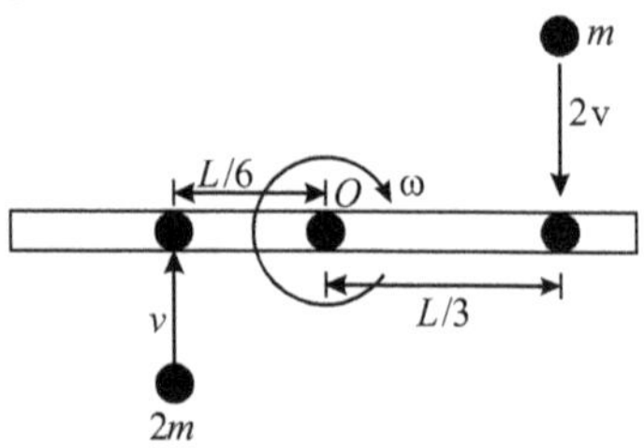

As no external torque is acting on system.

$\Rightarrow$ Total angular momentum will remain conserved about O.

$$L_i = L_f$$

$$2m(v)\left(\frac{L}{6}\right)+m(2v)\left(\frac{L}{3}\right)=I\omega$$

$$\Rightarrow \quad \frac{mvL}{3}+\frac{2mvL}{3}=\left[\frac{1}{12}(8m)L^2+m\left(\frac{L}{3}\right)^2+2m\left(\frac{L}{6}\right)^2\right]\omega$$

$$mvL = mL^2\omega\left[\frac{2}{3}+\frac{1}{9}+\frac{1}{18}\right]=\frac{15}{18}mL^2\omega$$

$$\Rightarrow \quad mvL=\frac{5}{6}mL^2\omega$$

$$\Rightarrow \quad \omega=\frac{6v}{5L}$$ **Ans. (B)**

Sol. 61 Ratio of I_z and $I_{z'}$ is

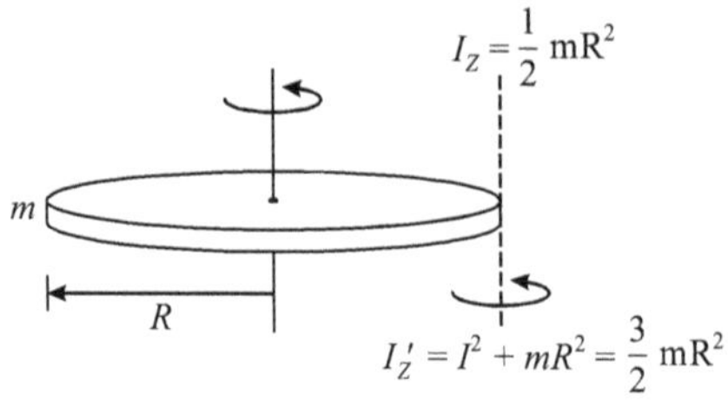

$$\frac{I_z}{I_z'}=\frac{\frac{1}{2}mR^2}{\frac{3}{2}mR^2}=\frac{1}{3}$$ **Ans. (C)**

Sol. 62 The COM of rod has fallen by

$$h=\frac{l}{2}\sin\alpha$$

By work-energy theorem, we have

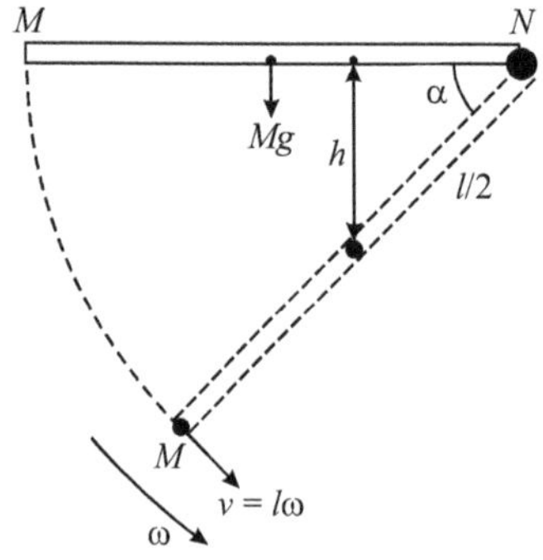

$v_M = \sqrt{3gl \sin \alpha}$

$$mg\left(\frac{l}{2}\sin\alpha\right)=\frac{1}{2}I\omega^2$$

$$\Rightarrow \quad mg\frac{l}{2}\sin\alpha=\frac{1}{2}\left(\frac{ml^2}{3}\right)\omega^2$$

$$\Rightarrow \quad \omega=\sqrt{\frac{3g\sin\alpha}{l}}$$ **Ans. (D)**

Sol. 63 Centre of mass must be vertically below the suspension point.

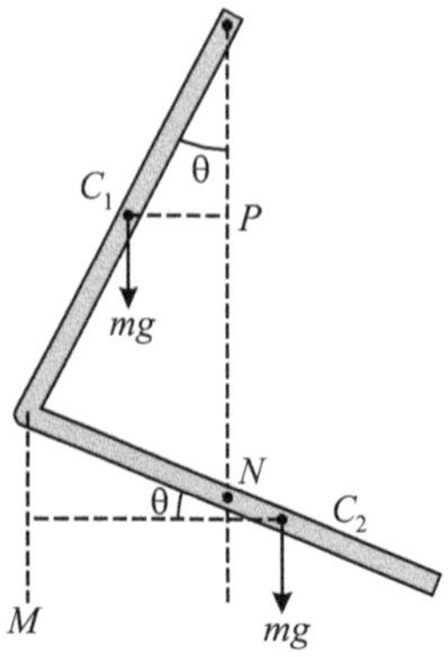

Balancing torque about suspension point, we have

$$mg\left(\frac{L}{2}\sin\theta\right)=mg\left(\frac{L}{2}\cos\theta-L\sin\theta\right)$$

$$\Rightarrow \quad \frac{3}{2}mgL\sin\theta=\frac{mgL}{2}\cos\theta$$

$$\Rightarrow \quad \tan\theta=\frac{1}{3}$$ **Ans. (B)**

Sol. 64

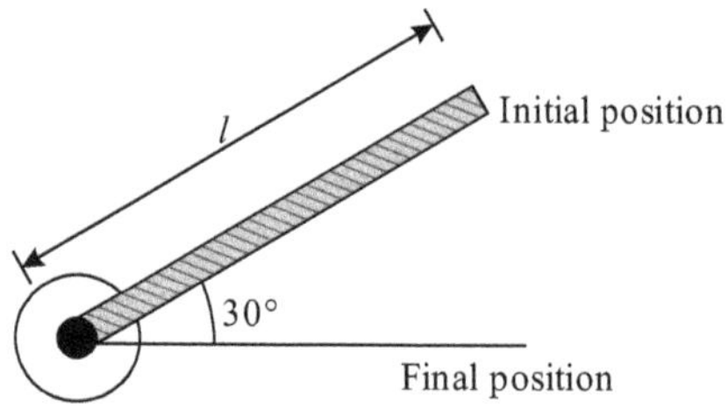

$$mg\left(\frac{l}{2}\right)\sin 30^\circ = \frac{1}{2}I\omega^2$$

$$\Rightarrow \quad \frac{1}{2}\left(\frac{ml^2}{3}\right)\omega^2 = \frac{mgl}{4}$$

$$\Rightarrow \quad \omega = \sqrt{\frac{3g}{2l}} = \sqrt{\frac{3\times 10}{2\times 0.5}}$$

$$\Rightarrow \quad \omega = \sqrt{30}\ \text{rad/s}$$

Ans. (A)

Sol. 65

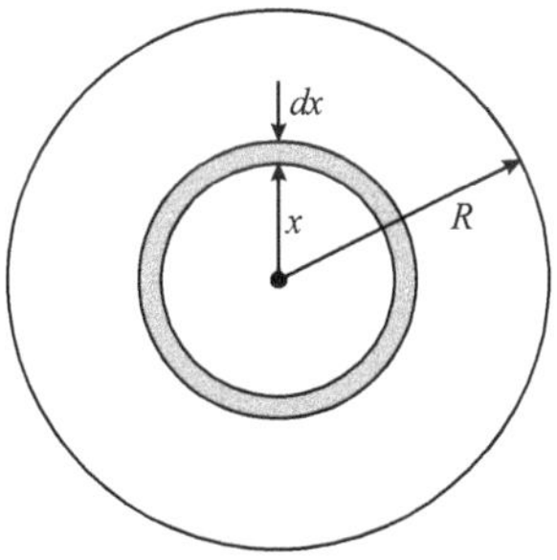

Consider a strip of radius x & thickness dx, Torque due to friction on this strip.

$$\Rightarrow \quad \tau = \int d\tau = \int_0^R \frac{\mu F\cdot 2\pi x dx}{\pi R^2}\times x$$

$$\Rightarrow \quad \tau = \frac{2\mu F}{R^2}\cdot\frac{R^3}{3}$$

$$\Rightarrow \quad \tau = \frac{2\mu FR}{3}$$

Ans. (A)

Sol. 66

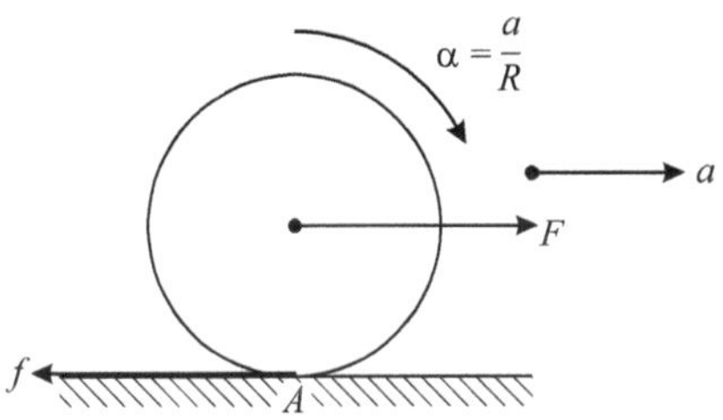

Equation of motion of cylinder about bottom point of contact A is written as

$$FR = \frac{3}{2}MR^2\alpha$$

$$\Rightarrow \quad \alpha = \frac{2F}{3MR}$$

Ans. (C)

Sol. 67

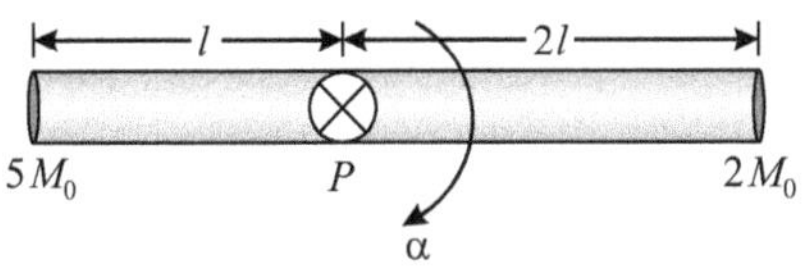

Applying torque equation about point P, we have

$$2M_0g(2l) - 5M_0gl = I\alpha$$

Where $I = 2M_0(2l)^2 + 5M_0$

$$l^2 = 13M_0l^2d$$

$$\Rightarrow \quad \alpha = -\frac{M_0gl}{13M_0l^2}$$

$$\Rightarrow \quad \alpha = -\frac{g}{13l}$$

$$\Rightarrow \quad \alpha = \frac{g}{13l}\ \text{anticlockwise}$$

Ans. (C)

Sol. 68 For Balls (considered solid spheres)

Using parallel axis theorem, the moment of inertia of each balls is

$$I_{ball} = \frac{2}{5}MR^2 + M(2R)^2$$

$$\Rightarrow \quad I_{ball} = \frac{22}{5}MR^2$$

For two balls it is $\frac{44}{5}MR^2$

Moment of inertia of rod is

$$I_{rod} = \frac{ML^2}{12} = \frac{M(2R)^2}{12} = \frac{MR^2}{3}$$

$$I_{system} = I_{Ball} + I_{rod}$$

$$\Rightarrow \quad I_{system} = \frac{44}{5}MR^2 + \frac{MR^2}{3}$$

$$\Rightarrow \quad I_{system} = \frac{137}{15}MR^2$$

Ans. (C)

Sol. 69 Suppose M is mass and a is side of larger triangle, then $\frac{M}{4}$ and $\frac{a}{2}$ will be mass and side length of smaller triangle. By symmetry expression of moment of inertia of a triangle remain same as

$$I = \gamma ML^2$$

So we use

$$\frac{I_{removed}}{I_{original}} = \frac{\frac{M}{4}\cdot\left(\frac{a}{2}\right)^2}{M\ (a)^2}$$

$$I_{removed} = \frac{I_0}{16}$$

Thus moment of inertia of remaining triangle is

$$I = I_0 - \frac{I_0}{16} = \frac{15I_0}{16}$$

Ans. (D)

Sol. 70 Torque for F_2 force

$$\vec{F_2} = \frac{F}{2}(-\hat{i}) + \frac{F\sqrt{3}}{2}(-\hat{j})$$

$$\vec{r_2} = 0\hat{i} + 6\hat{j}$$

$$\vec{\tau}_{F_2} = \vec{r_2}\times\vec{F_2} = 3F\hat{k}$$

Torque for F_1 force

$$\vec{F_2} = F\hat{k}$$
$$\vec{r_1} = 2\hat{i}+3\hat{j}$$
$$\vec{\tau}_{F_1} = \vec{r_1}\times\vec{F_1} = 3F\hat{i}+2F(-\hat{j})$$
$$\vec{\tau}_{net} = \vec{\tau}_{F_1}+\vec{\tau}_{F_2}$$

$\Rightarrow$ $\vec{\tau}_{net} = 3F\hat{i}+2F(-\hat{j})+3F(\hat{k})$ **Ans. (D)**

Sol. 71

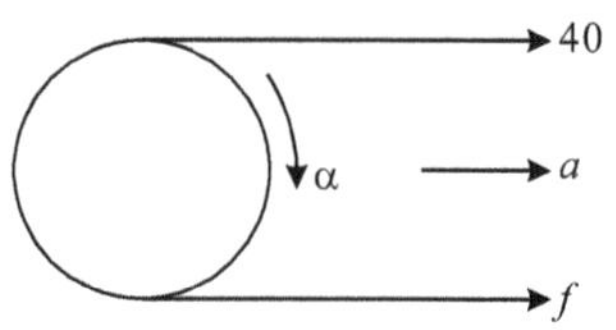

For pure rolling

$$\alpha = \frac{a}{R}$$

Equation of motion of cylinder is

$$40+f = m(R\alpha) \qquad \ldots(1)$$
$$40\times R - f\times R = mR^2\alpha$$

$\Rightarrow$ $40-f = mR\alpha \qquad \ldots(2)$

From (1) and (2)

$\Rightarrow$ $\alpha = \dfrac{40}{mR} = \dfrac{40}{5\times 0.5}$

$= 16\text{ rad/s}^2$ **Ans. (B)**

Sol. 72 Torque on particle is given as

$$\tau = Fr\sin\theta$$

$\Rightarrow$ $2.5 = 1\times 5\sin\theta$

$\Rightarrow$ $\sin\theta = 0.5 = \dfrac{1}{2}$

$\Rightarrow$ $\theta = \dfrac{\pi}{6}$ **Ans. (B)**

Sol. 73 Moment of inertia of system is given as

$$I = \frac{MR^2}{2}+2\left(\frac{MR^2}{4}+MR^2\right)$$

$\Rightarrow$ $I = \dfrac{MR^2}{2}+\dfrac{MR^2}{2}+2MR^2$

$\Rightarrow$ $I = 3MR^2$ **Ans. (A)**

Sol. 74 Moment of inertia of hollow cylinder is

$$I = \frac{1}{2}M(R_1^2+R_2^2) = \frac{1}{2}M(10^2+20^2) = 250\,M$$

Moment of inertia of a solid cylinder is

$$I = \frac{1}{2}MR^2 = 250\,M$$

$\Rightarrow$ $R = \sqrt{5000} = 22.36\text{ cm}$ **Ans. (None)**

Sol. 75 Using Work Energy Theorem from A to B, we have

$$\frac{1}{2}mv_A^2+mgh = \frac{1}{2}mv_B^2$$

$\Rightarrow$ $2gh = v_B^2 - v_A^2$

$\Rightarrow$ $2\times 10\times 10 = v_B^2 - 5^2$

$\Rightarrow$ $v_B = 15\text{ m/s}$

Angular momentum about O is given as

$$L_0 = mvr$$

$\Rightarrow$ $L_0 = 20\times 10^{-3}\times 20$

$\Rightarrow$ $L_0 = 6\text{ kg.m}^2\text{/s}$ **Ans. (B)**

Sol. 76 By parallel axes theorem, we use

$$I = \frac{2}{5}mR^2+mx^2$$ **Ans. (B)**

Sol. 77 By parallel axes theorem, we have

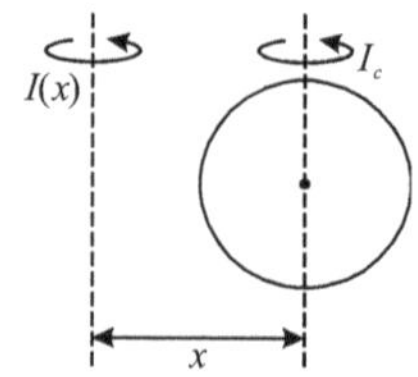

$$I(x) = I_c + Mx^2$$

$\Rightarrow$ $I(x) = \dfrac{2}{5}MR^2+Mx^2$ **Ans. (D)**

Sol. 78 As the box is in contact with the edge for $\Delta t = 0.01$ s. So angular impulse on box is

$$\Delta L = \tau\cdot\Delta t = mg\cdot\frac{l}{2}\times\Delta t = I\omega$$

Gain in angular speed of box

$$\omega = \frac{mgl\Delta t}{2I} = \frac{mgl\Delta t}{2\left(\dfrac{ml^2}{3}\right)} = \frac{3g\Delta t}{2l}$$

$\Rightarrow$ $\omega = \dfrac{3\times 10\times 0.01}{2\times 0.3} = 0.5\text{ rad/s}$

Time taken in reaching ground is

$$t = \sqrt{\frac{2h}{g}} = \sqrt{\frac{2\times 5}{10}} = 1\text{s}$$

In 1s angle rotated by box is

$$\theta = \omega t = 0.5\times 1 = 0.5\text{ rad}$$ **Ans. (A)**

Sol. 79 Angular impulse = change in angular momentum

$$\tau\Delta t = \Delta L$$

$\Rightarrow$ $mg\dfrac{l}{2}\times.01 = \dfrac{ml^2}{3}\omega$

$$\Rightarrow \qquad \omega = \frac{3g \times 0.01}{2l}$$

$$\omega = \frac{3 \times 10 \times .01}{2 \times 0.3}$$

$$\omega = \frac{1}{2} = 0.5 \text{ rad/s}$$

time taken by rod to hit the ground

$$t = \sqrt{\frac{2h}{g}} = \sqrt{\frac{2 \times 5}{10}} = 1 \text{ s}$$

in this time angle rotate by rod

$$\theta = \omega t = 0.5 \times 1 = 0.5 \text{ rad}$$

Ans. (C)

Sol. 80 If moment of inertia of rolling object is taken as I and speed of centre of mass at bottom of incline is v, we use

$$\frac{1}{2}\left(m + \frac{I}{R^2}\right)v^2 = mgh$$

If radius of gyration is k, then, we use $I = mk^2$, so we have

$$h = \frac{\left(1 + \frac{k^2}{R^2}\right)v^2}{2g} \qquad \ldots(1)$$

For given object we use

$$\frac{k_{\text{ring}}}{R_{\text{ring}}} = 1$$

$$\frac{k_{\text{solid cylinder}}}{R_{\text{solid cylinder}}} = \frac{1}{\sqrt{2}}$$

$$\Rightarrow \qquad \frac{k_{\text{solid sphere}}}{R_{\text{solid sphere}}} = \sqrt{\frac{2}{5}}$$

From equation-(1) we have

$$h_1 : h_2 : h_3 :: (1+1) : \left(1 + \frac{1}{2}\right) : \left(1 + \frac{2}{5}\right) :: 20 : 15 : 14$$

Therefor **most appropriate option is (B)** which in not in proper sequence. **Ans. (C)**

Sol. 81 Kinetic energy of disc is given as

$$KE = \frac{1}{2}I\omega^2 = k\theta^2$$

$$\Rightarrow \qquad \omega^2 = \frac{2k\theta^2}{I}$$

$$\Rightarrow \qquad \omega = \sqrt{\frac{2k}{I}}\,\theta \qquad \ldots(1)$$

Differentiating-(1) wrt time, we get angular acceleration

$$\alpha = \frac{d\omega}{dt} = \sqrt{\frac{2k}{I}}\left(\frac{d\theta}{dt}\right)$$

$$\Rightarrow \qquad \alpha = \sqrt{\frac{2k}{I}} \cdot \sqrt{\frac{2k}{I}}\,\theta$$

$$\Rightarrow \qquad \alpha = \frac{2k}{I}\theta$$

Ans. (D)

Sol. 82 Kinetic energy of rotating body is

$$KE = \frac{1}{2}I\omega^2$$

$$\Rightarrow \qquad 1200 = \frac{1}{2}1.5 \times \omega^2$$

$$\Rightarrow \qquad \omega^2 = \frac{1200 \times 2}{1.5} = 1600$$

$$\Rightarrow \qquad \omega = 40 \text{ rad/s}^2$$

If body rotates for time t at angular acceleration $\alpha = 20$ rad/s^2, we use

$$\omega = \omega_0 + \alpha t$$

$$40 = 0 + 20t$$

$$t = 2 \text{ s}$$

Ans. (A)

Sol. 83 Applying angular momentum conservation, about axis of rotation, we have

$$L_i = L_f$$

$$\frac{ML^2}{12}\omega_0 = \left(\frac{ML^2}{12} + m\left(\frac{L}{2}\right)^2 \times 2\right)\omega$$

$$\Rightarrow \qquad \omega = \frac{M\omega_0}{M + 6m}$$

Ans. (D)

Sol. 84 We consider an elemental strip of radius r and width dr as shown in figure mass of elemental strip is

$$dm = \sigma(r) \cdot 2\pi r dr$$

Moment of inertia of elemental strip is

$$dI = dmr^2$$

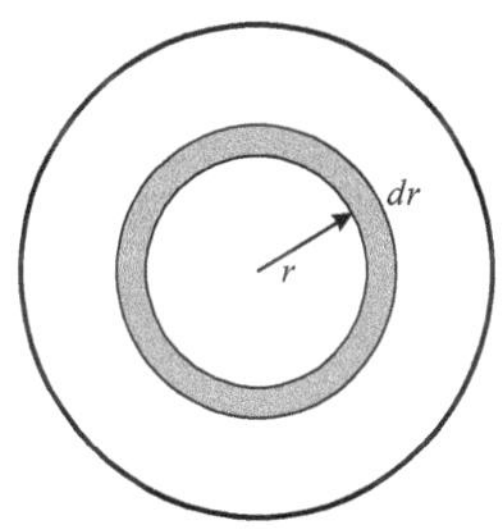

$$\Rightarrow \qquad I_{\text{Disc}} = \int_0^R (dm)r^2$$

$$\Rightarrow \qquad I_{\text{Disc}} = \int_0^R (\sigma 2\pi r dr)r^2$$

$$\Rightarrow \qquad I_{\text{Disc}} = \int_0^R (kr^2\, 2\pi r dr)r^2$$

$$\Rightarrow \qquad I_{\text{Disc}} = 2\pi k\int_0^R r^5 dr = \frac{2\pi k R^6}{6} = \frac{\pi k R^6}{3} \qquad \ldots(1)$$

Total mass of disc is

$$M=\int dm=\int_0^R 2\pi r dr\,(kr^2)$$

$$\Rightarrow \quad M=2\pi k\int_0^R r^3 dr$$

$$\Rightarrow \quad M=2\pi k\left.\frac{r^4}{4}\right|_0^R=\frac{\pi k R^4}{2} \quad \ldots(2)$$

From equation-(1) and (2), we have

$$I_{\text{Disc}}=\frac{\pi k R^6}{3}=\left(\frac{\pi k R^4}{2}\right)\times\frac{2}{3}R^2$$

$$\Rightarrow \quad I_{\text{Disc}}=M\times\frac{2}{3}R^2$$

$$\Rightarrow \quad I_{\text{Disc}}=\frac{2}{3}MR^2$$

Ans. (C)

Sol. 85 Initial *KE* of system is

$$E_i=\frac{1}{2}I_1\times\omega_1^2+\frac{1}{2}\left(\frac{I_1}{2}\right)\left(\frac{\omega_1^2}{4}\right)$$

$$\Rightarrow \quad E_i=\frac{I_1\omega_1^2}{2}\left(\frac{9}{8}\right)=\frac{9}{16}I_1\omega_1^2$$

By conservation of angular momentum, we have

$$I_1\omega_1+\frac{I_1\omega_1}{4}=\frac{3I_1}{2}\omega$$

$$\Rightarrow \quad \frac{5}{4}I_1\omega_1=\frac{3I_1}{2}\omega$$

$$\Rightarrow \quad \omega=\frac{5}{6}\omega_1$$

Final KE of system is

$$E_f=\frac{1}{2}I_f\omega^2=\frac{1}{2}\left(\frac{3}{2}I_1\right)\left(\frac{5}{6}\omega_1\right)^2$$

$$\Rightarrow \quad E_f=\frac{25}{48}I_1\omega_1^2$$

Difference in kinetic energies (E_f-E_i) is given as

$$E_f-E_i=I_1\omega_1^2\left(\frac{25}{48}-\frac{9}{16}\right)=\frac{-2}{48}I_1\omega_1^2$$

$$\Rightarrow \quad E_f-E_i=\frac{-I_1\omega_1^2}{24}$$

Ans. (D)

Sol. 86 Acceleration of particle

$$\overrightarrow{a_x}=\frac{d^2x}{dt^2}=-a\omega_1^2\cos\omega_1 t\,\hat{i}$$

$$\overrightarrow{a_y}=\frac{d^2y}{dt^2}=-b\omega_2^2\sin\omega_2 t\,\hat{i}$$

$$\vec{a}_{t=0}=-a\omega_1^2\,\hat{i}$$

Position vector of particle at $t=0 \rightarrow \vec{r}=(x_0+a)\,\hat{i}+y_0\,\hat{j}$

Torque on particle is $\vec{\tau}=\vec{r}\times\vec{F}=\vec{r}\times m\,\vec{a}_{t=0}$

$$=[(x_0+a)\,\hat{i}+y_0\hat{j}\,]\times m[-a\omega_1^2\,\hat{i}\,]$$

$$=my_0\,a\omega_1^2\,\hat{k}$$

Ans. (B)

Sol. 87 Angular acceleration of coin is

$$\alpha=\frac{\Delta\omega}{\Delta t}=\frac{25\times 2\pi}{5}=10\pi\text{ rad/s}^2$$

Torque on coin is

$$\tau=\left(\frac{5}{4}MR^2\right)\alpha$$

$$\Rightarrow \quad \tau=\frac{5}{4}\times 5\times 10^{-3}\times(10^{-2})^2\times 10\pi$$

$$\Rightarrow \quad \tau=1.9625\times 10^{-5}\text{ Nm}\simeq 2.0\times 10^{-5}\text{ Nm}$$ **Ans. (B)**

Sol. 88 Velocity at $t=2$s is

$$\vec{v}_{(t)}=\frac{\overrightarrow{dr}}{dt}=2\,\hat{i}-6t\,\hat{j}\,|_{t=2s}\,(2\,\hat{i}-12\,\hat{j})$$

Position vector at $t=2$s is

$$\vec{r}_{(t)}=(4\,\hat{i}-12\,\hat{j})$$

Angular momentum at $t=2$s is

$$\vec{L}=m(\vec{r}\times\vec{v})$$

$$=2\,(4\,\hat{i}-12\,\hat{j})\times(2\,\hat{i}-12\,\hat{j})$$

$$=2\,[-48\hat{k}+24\hat{k}]$$

$$=-48\,\hat{k}$$

Ans. (D)

Sol. 89 For disc $m_1=\dfrac{7M}{8}$, $R_1=2R$

MI of disc $I_1=\dfrac{1}{2}m_1R_1^2=\dfrac{1}{2}\left(\dfrac{7M}{8}\right)(2R)^2=\dfrac{7}{4}MR^2 \quad \ldots(1)$

For new solid sphere $m_2=\dfrac{M}{8}$,

$R_2=\dfrac{R}{2}$ [we use $\dfrac{1}{8}\left(\dfrac{4}{3}\pi R^3\right)=\dfrac{4}{3}\pi R_2^3 \Rightarrow R_2=\dfrac{R}{2}$]

MI of new sphere $I_2=\dfrac{2}{5}m_2R_2^2=\dfrac{2}{5}\left(\dfrac{M}{8}\right)\left(\dfrac{R}{2}\right)^2$

$$=\frac{1}{80}MR^2 \quad \ldots(2)$$

From (1) and (2) $\dfrac{I_1}{I_2}=\dfrac{\frac{7}{4}MR^2}{\frac{1}{80}MR^2}=140$

Ans. (D)

Sol. 90 Area of elemental strip = $ds = 2\pi x\, dx$

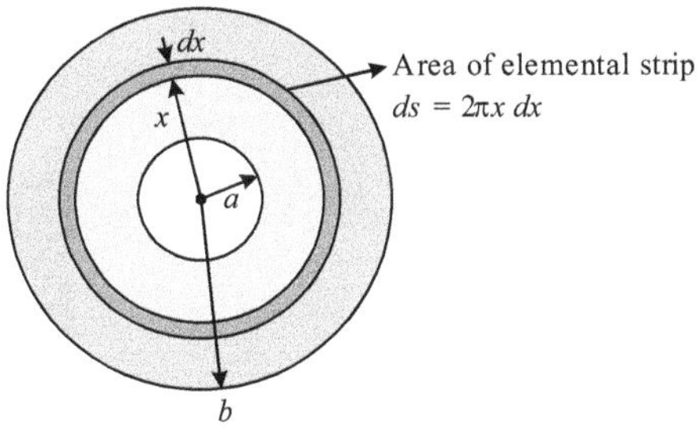

Mass of elemental strip $dm = \dfrac{\sigma_0 ds}{x} = \dfrac{\sigma_0}{x} \cdot 2\pi x\, dx = 2\pi\sigma_0\, dx$

Total mass of disc $m = \int dm = 2\pi\sigma_0 = \int_a^b dx = 2\pi\sigma_0 (b - a)$

MI of elemental strip $dI = dmx^2$

$\Rightarrow \quad dI = 2\pi\sigma_0 x^2 dx$

$\Rightarrow \quad I = \int dI = \int_a^b 2\pi\sigma_0 x^2 dx = \dfrac{2\pi\sigma_0}{3} \cdot (b^3 - a^3)$

Radius of gyration of disc is

$$k = \sqrt{\frac{I}{m}} = \sqrt{\frac{(b^3 - a^3)}{3(b-a)}} = \sqrt{\frac{a^2 + b^2 + ab}{3}}$$ **Ans. (D)**

Sol. 91

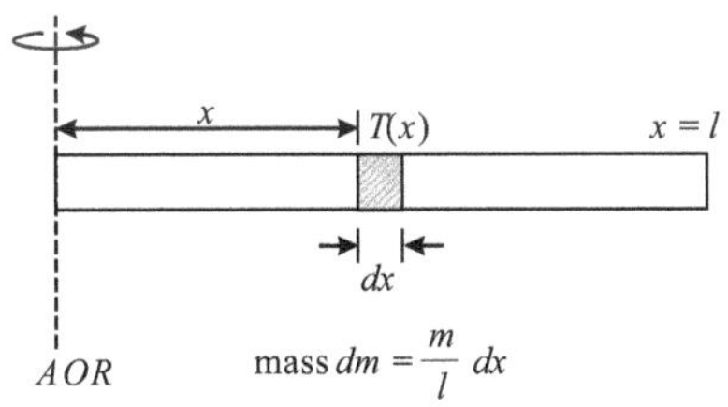

At a distance x from AOR, tension in rod will be due to the centrifugal force on part of rod outside it in frame of rod.
Centrifugal force on dm is

$$dF = dm\omega^2 x$$

Total centrifugal force on outer part of rod i.e. from x to l is

$$F = T_{(x)} = \int dF = \int_x^l \frac{m}{l}\omega^2 x dx = \frac{m\omega^2}{2l}(l^2 - x^2)$$

This is a parabolic function **Ans. (D)**

7.1 Gravitational Field

Sol. 1

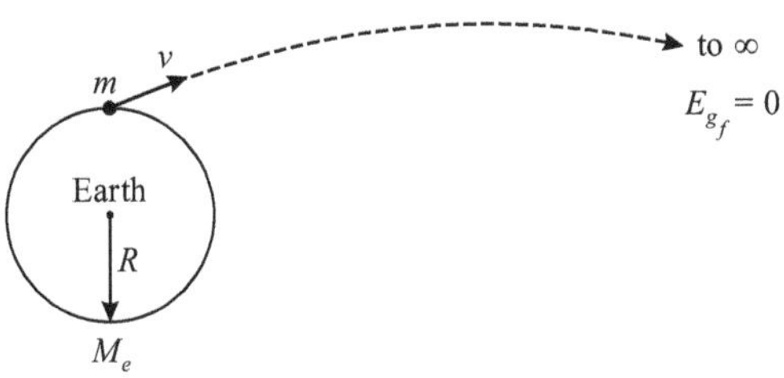

On surface of earth for the body mechanical energy is

$$E_{gi} = -\frac{GM_e m}{R} + \frac{1}{2}mv^2 = E_{gf} = 0$$

$$\Rightarrow \quad KE = \frac{GM_e m}{R} = (\text{gR})\,\text{m} \quad [\text{As } g = \frac{GM_e}{R^2}]$$

$$\Rightarrow \quad KE = mgR$$ **Ans. (C)**

Sol. 2

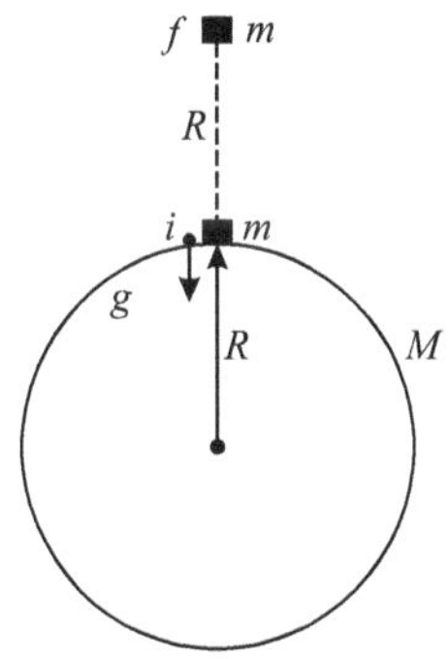

Acceleration due to gravity on

At earth surface $g = \dfrac{GM}{R^2} \Rightarrow \dfrac{GM}{R} = gR$

Gain in PE when body of mass m is raised from position i to f is

$$\Delta U = U_f - U_i$$

$$\Rightarrow \quad \Delta U = \left(-\frac{GMm}{2R}\right) - \left(-\frac{GMm}{R}\right)$$

$$\Rightarrow \quad \Delta U = \frac{GMm}{2R} = \frac{1}{2}mgR$$ **Ans. (B)**

Sol. 3

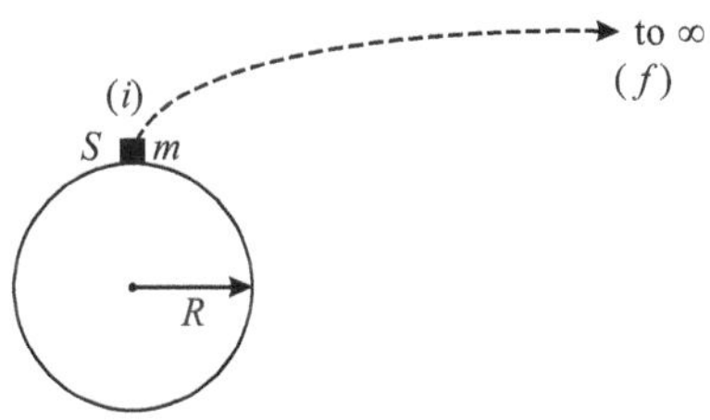

Work done in shifting mass from point S to ∞, is given as

$$W = m(V_f - V_i)$$

$$\Rightarrow \quad W = m\left(0 - \left(-\frac{GM}{R}\right)\right)$$

$$\Rightarrow \quad W = \frac{GMm}{R} = \frac{6.67 \times 10^{-11} \times 100 \times 10 \times 10^{-3}}{0.1}$$

$$\Rightarrow \quad W = 6.67 \times 10^{-10}\,\text{J}$$ **Ans. (D)**

Sol. 4 At a small height h above earth surface, value of g is given as

$$g_h = g_s\left(1 - \frac{2h}{R_e}\right) \qquad \ldots(1)$$

At a depth d below earth surface, value of g is given as

$$g_d = g_s\left(1 - \frac{d}{R_e}\right) \quad \ldots (2)$$

Given that $g_h = g_d$ or $g_s - g_h = g_s - g_d$

$$\Rightarrow \quad \frac{2h}{R_e} = \frac{d}{R_e} \rightarrow 2h = d$$

Ans. (C)

Sol. 5 For earth, its mass M and radius R are constants

$$\Rightarrow \text{ Average density } \rho = \frac{M}{\frac{4}{3}\pi R^3} = \frac{3M}{4\pi R^3} \rightarrow \text{constant}$$

If for a planet its mass M is constant and its radius R is changed

$$\text{Acceleration due to gravity } g = \frac{GM}{R^2} \Rightarrow R = \sqrt{\frac{GM}{g}}$$

$$\text{Average density } \rho = \frac{3M}{4\pi\left(\frac{GM}{g}\right)^{3/2}} = \frac{3g^{3/2}}{4\pi G^{3/2} M^{1/2}}$$

$$\Rightarrow \quad \rho \propto g^{3/2}$$

If for a planet its radius R is constant and its mass M is changed

$$M = \frac{gR^2}{G}$$

$$\text{Average density} \quad \rho = \frac{3}{4\pi R^3} \times \frac{gR^2}{G} = \frac{3}{4\pi RG} \cdot g$$

$$\Rightarrow \quad \rho \propto g$$

Ans. (C)

Sol. 6 Electronic charge is independent of g.

$$\Rightarrow \quad \frac{q_m}{q_e} = 1$$ **Ans. (A)**

Sol. 7 By Gauss's law for gravitational field we use

$$\phi_m = \oint \vec{g} \cdot \vec{ds} = 4\pi GM$$

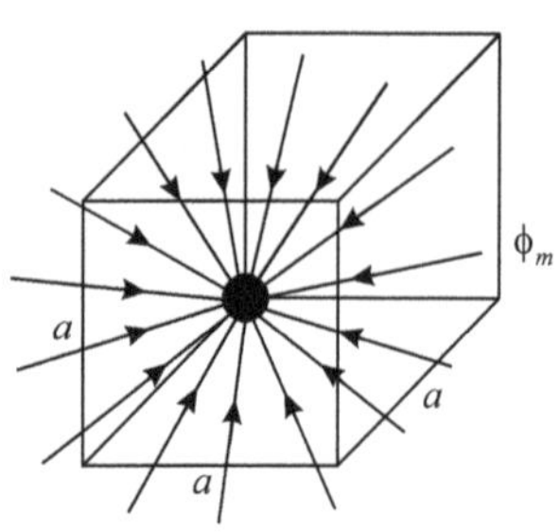

S-1 is True

By Gauss's law the gravitational field due to a point m is radially inward (radial) and for the surface enclosing the mass $\phi = 4\pi GM$.

Ans. (B)

Sol. 8 At a height h above the earth surface the value of g is given as

$$g_h = \frac{g_s}{\left(1 + \frac{h}{R_e}\right)^2} = \frac{g_s}{9}$$

$$\Rightarrow \quad \left(1 + \frac{h}{g}\right)^2 = 9$$

$$\Rightarrow \quad 1 + \frac{h}{R} = 3 \Rightarrow h = 2R$$

Ans. (C)

Sol. 9

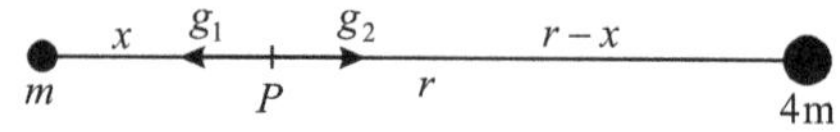

Gravitational field strength due to m and $4m$ at point P are given as

$$g_1 = \frac{Gm}{x^2} = \frac{4Gm}{(r-x)^2}$$

$$\Rightarrow \quad 2x = r - x$$

$$\Rightarrow \quad x = \frac{r}{3}$$

Gravitational potential at point P is given as

$$V_P = -\frac{Gm}{x} - \frac{G(4m)}{r-x}$$

$$\Rightarrow \quad V_P = -\frac{3Gm}{r} - \frac{3(4Gm)}{2r} = -\frac{9Gm}{r}$$

Ans. (B)

Sol. 10

Initial gravitational PE of space ship on surface of earth is

$$U_i = -\frac{GM_e m}{R_e} = -(gR)m \quad \left[\text{As } g = \frac{GM_e}{R_e^2}\right]$$

Work done required to launch space ship to ∞ (free space) is

$$W = U_f - U_i = 0 - (-mgR) = mgR$$

$$= 1000 \times 10 \times 6400 \times 10^3$$

$$= 6.4 \times 10^{10} \text{ J}$$

Ans. (A)

Sol. 11

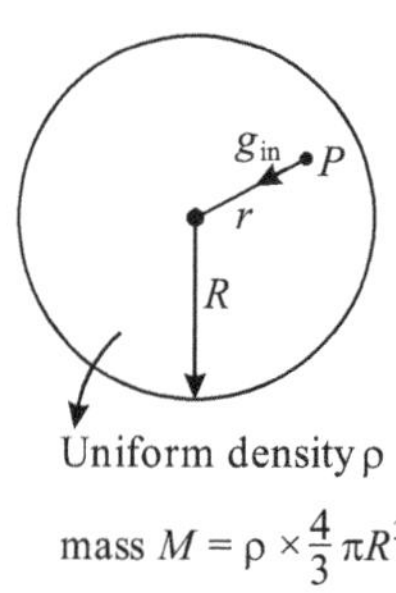

For $r < R$ acceleration due to gravity inside a sphere is given as

$$g_{\text{in}} = \frac{GMr}{R^3} = \frac{G}{R^3}\left(\rho \cdot \frac{4}{3}\pi R^3\right) \cdot r$$

$$\Rightarrow \quad g_{\text{in}} = \frac{4}{3}\pi G\rho r$$

$$\Rightarrow \quad g_{\text{in}} \propto r$$

Ans. (A)

Sol. 12 Inside earth, value of g is given as

For $x < R_e$ $\quad g_{\text{in}} = \dfrac{GM_e x}{R_e^3}$

For $x \geq R_e$ $\quad g_{\text{s/out}} = \dfrac{GM_e}{x^2}$

(Where x is distance from earth centre)

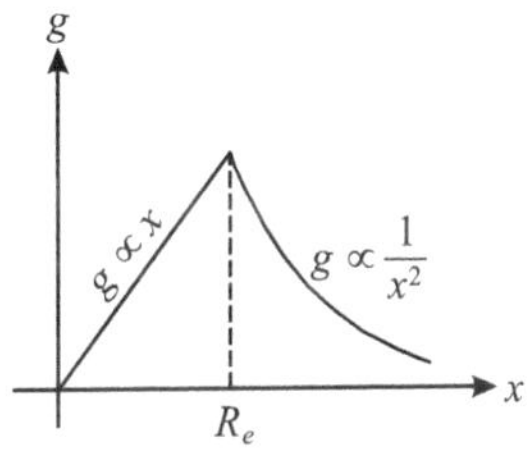

The closest/best possible option given is (B) **Ans. (B)**

Sol. 13 The mass of removed part of sphere is given as

$$m = \left(\frac{M}{\frac{4}{3}\pi R^3}\right) \times \frac{4}{3}\pi\left(\frac{R}{2}\right)^3 = \frac{M}{8}$$

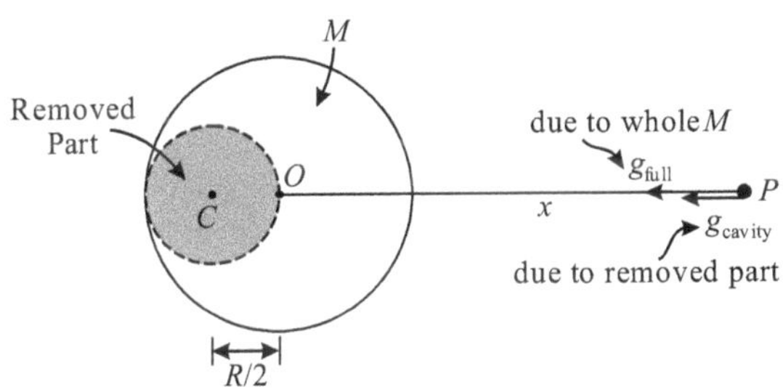

Net gravitational field at point P due to remaining mass is

$$g_P = g_{\text{full}} - g_{\text{cavity}}$$

$$\Rightarrow \quad g_P = \frac{GM}{x^2} - \frac{G\left(\frac{M}{8}\right)}{\left(x + \frac{R}{2}\right)^2}$$

$$\Rightarrow \quad g_P = \frac{GM}{x^2} - \frac{GM}{8x^2}$$

[As $x >> \dfrac{R}{2}$ so $x + \dfrac{R}{2} \simeq x$]

$$\Rightarrow \quad g_P = \frac{7}{8}\frac{GM}{x^2}$$

Ans. (C)

Sol. 14

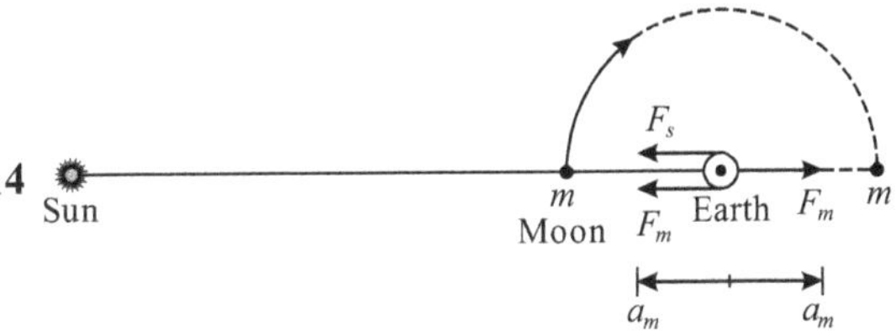

Change in acceleration of earth toward Sun due to change in position of moon is

$$\Delta a = a_m - (-a_m)$$

$$\Rightarrow \quad \Delta a = 2a_m$$

$$\Delta a = 2a_m = 2\left(\frac{Gm}{R^2}\right) = \frac{2 \times 6.67 \times 10^{-11} \times 7.36 \times 10^{22}}{(3.8 \times 10^8)^2}$$

$$\Rightarrow \quad \Delta a = 6.75 \times 10^{-5}\ \text{m/s}^2$$

Ans. (A)

Sol. 15

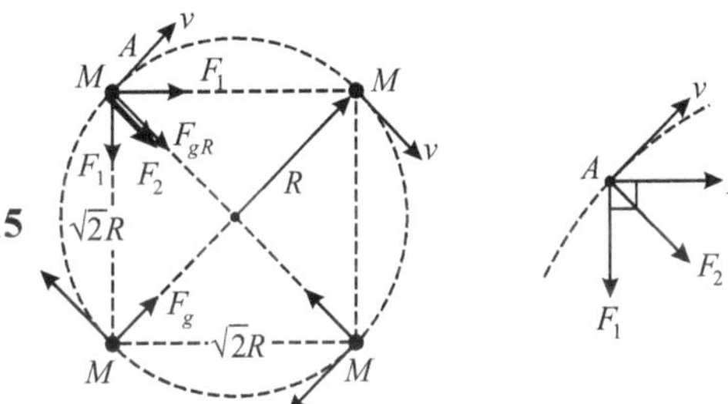

Gravitational forces acting on particle A are

$$F_1 = \frac{GM^2}{(\sqrt{2}R)^2};\ F_2 = \frac{GM^2}{4R^2}$$

Net gravitational force on particle A, is given as

$$F_{gA} = F_2 + \sqrt{2}F_1$$

$$\Rightarrow \quad F_{gA} = \frac{GM^2}{4R^2} + \sqrt{2}\left(\frac{GM^2}{2R^2}\right) = \frac{Mv^2}{R}$$

$$\Rightarrow \quad \frac{GM^2}{2R^2}\left(\frac{1}{2} + \sqrt{2}\right) = \frac{Mv^2}{R}$$

$$\Rightarrow \quad v = \sqrt{\frac{GM}{4R}(2\sqrt{2} + 1)}$$

Ans. (D)

Sol. 16 Capillary height above the water level in tube is given as

$$h = \frac{2T\cos\theta}{r\rho g}$$

at earth surface $\quad x = \dfrac{2T\cos\theta}{r\rho g}$ ...(1)

at a depth d below the surface of earth

$$y = \frac{2T\cos\theta}{r\rho\left[g\left(1 - \frac{d}{R}\right)\right]} \quad \text{...(2)}$$

(1)/(2) $\Rightarrow \quad \dfrac{x}{y} = 1 - \dfrac{d}{R}$ **Ans. (A)**

Sol. 17 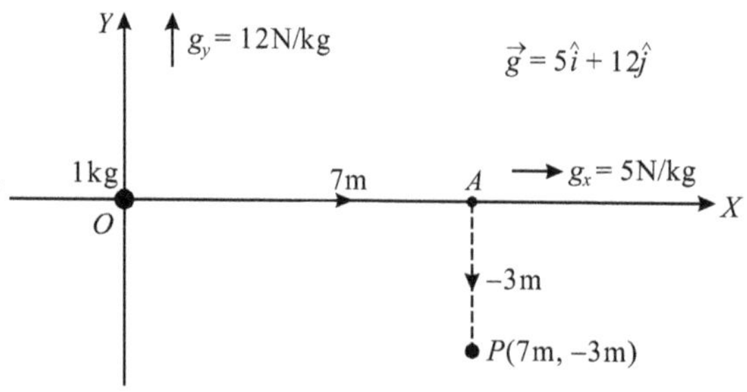

In the given situation work done in taking particle from $O \rightarrow A$ and $A \rightarrow P$ are given as

$$V_0 - V_A = 5 \times 7 = 35 \text{ J}$$

$$V_A - V_P = 12 \times (-3) = -36 \text{ J}$$

Work done or change in PE of 1 kg mass in shifting from O to P is

$$V_P - V_0 = 1 \text{ J}$$

Ans. (D)

Sol. 18 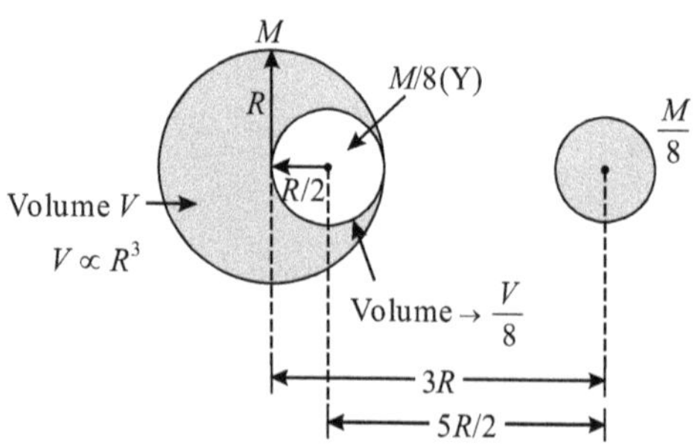

If force F_1 acts between M and $\frac{M}{8}$ than it is given as

$$F_1 = \frac{GM\left(\frac{M}{8}\right)}{(3R)^2} = \frac{GM^2}{72R^2} \quad \ldots(1)$$

If F_2 is the force between M/8 (X) and M/8 (Y) than it is given as

$$F_2 = \frac{G\left(\frac{M}{8}\right)\left(\frac{M}{8}\right)}{\left(5\frac{R}{2}\right)^2} = \frac{GM^2}{64 \times \frac{25}{4}R^2}$$

$$\Rightarrow \quad F_2 = \frac{GM^2}{400R^2} \quad \ldots(2)$$

The net force between the sphere (with cavity) and other part (x) is given as

$$F_{net} = F_1 - F_2$$

$$\Rightarrow \quad F_{net} = \frac{GM^2}{72R^2} - \frac{GM^2}{400R^2}$$

$$\Rightarrow \quad F_{net} = \frac{41GM^2}{3600R^2}$$

Ans. (A)

Sol. 19 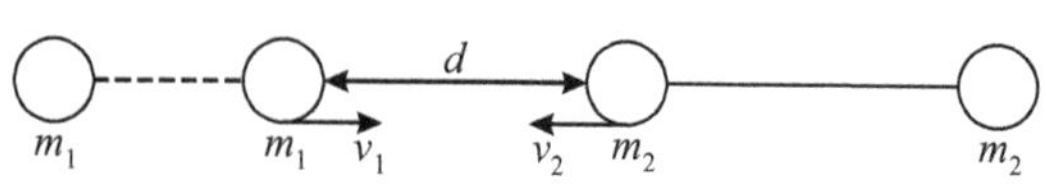

By conservation of energy

$$\Rightarrow \quad E_i = E_f$$

$$\underset{PE}{0} + \underset{KE}{0} = -\frac{Gm_1m_2}{d} + \frac{1}{2}m_1v_1^2 + \frac{1}{2}m_2v_2^2 \quad \ldots(1)$$

By conservation of momentum

$$\Rightarrow \quad m_1v_1 = m_2v_1 \quad \ldots(2)$$

$$\Rightarrow \quad v_2 = \left(\frac{m_1}{m_2}\right)v_1$$

We use v_2 in equation-(1) as

$$\Rightarrow \quad \frac{Gm_1m_2}{d} = \frac{1}{2}m_1v_1^2\left[1 + \frac{m_1}{m_2}\right]$$

$$\Rightarrow \quad v_1 = \sqrt{\frac{2Gm_2^2}{d(m_1+m_2)}} = m_2\sqrt{\frac{2G}{d(m_1+m_2)}}$$

From equation-(2) $v_2 = \frac{m_1}{m_2} \cdot v_1 = m_1\sqrt{\frac{2G}{d(m_1+m_2)}}$ **Ans. (B)**

Sol. 20 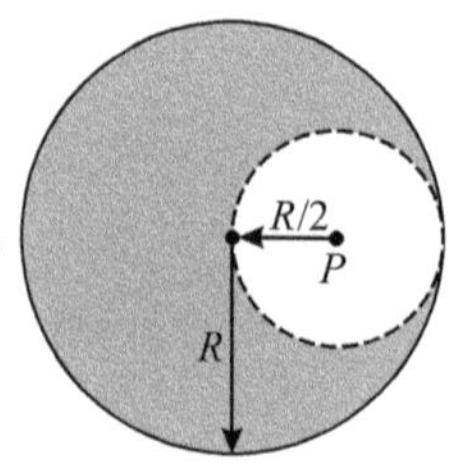

Removed mass

$$m = \frac{M}{\frac{4}{3}\pi R^3} \times \frac{4}{3}\pi\left(\frac{R}{2}\right)^3 = \frac{M}{8}$$

Gravitational potential at centre of cavity is

$$V_P = V_{\text{sphere } P} - V_{mP} \quad \ldots(1)$$

Due to a solid sphere gravitational potential at its interior point is given as

$$V_{in} = -\frac{GM}{2R^3}(3R^2 - x^2)$$

$\Rightarrow$ at $P\left(x = \frac{R}{2}\right)$ so we have

$$V_{in} = -\frac{GM}{2R^3}\left(3R^2 - \frac{R^2}{4}\right)$$

$$\Rightarrow \quad V_{in} = -\frac{11}{8}\frac{GM}{R}$$

Gravitational potential due to removed mass m at centre of cavity is

$$V_{Pm} = -\frac{3}{2}\left(\frac{GM}{(R/2)}\right) = -\frac{3}{2}\frac{G(M/8)}{(R/2)}$$

$$V_{Pm} = -\frac{3}{8}\frac{GM}{R}$$

From equation-(1)

$$V_P = \left(-\frac{11}{8}\frac{GM}{R}\right) - \left(-\frac{3}{8}\frac{GM}{R}\right)$$

$$\Rightarrow \quad V_P = -\frac{GM}{R}$$ **Ans. (B)**

Sol. 21 Gravitational potential due to uniform solid spherical planet is given as

$$V_{in} = -\frac{GM}{2R^3}(3R^2 - r^2) \text{ for } r < R$$

at $r = 0$
$$V_c = -\frac{3}{2}\frac{GM}{R}$$

$$V_s = -\frac{GM}{R} \quad \text{for } r = R$$

$$V_{out} = -\frac{GM}{r} \quad \text{for } r > R$$ **Ans. (C)**

Sol. 22

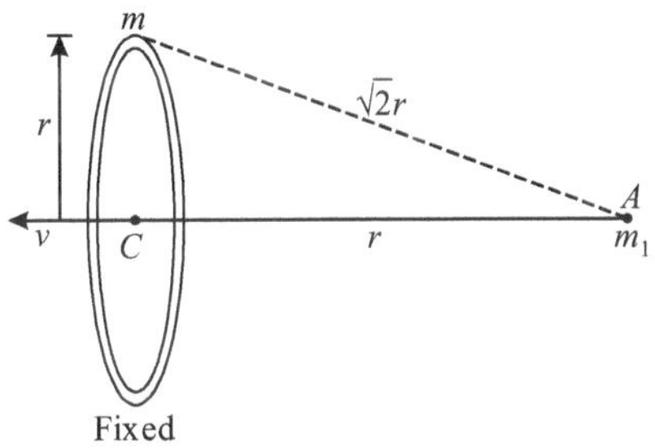

By using energy conservation, we have

$$E_{iA} = E_{fC}$$

$$-\frac{Gmm_1}{\sqrt{2}r} = -\frac{Gmm_1}{r} + \frac{1}{2}m_1v^2$$

$$\Rightarrow \quad v = \sqrt{\frac{2Gm}{r}\left(1 - \frac{1}{\sqrt{2}}\right)}$$ **Ans. (C)**

Sol. 23 Weight of a person of mass m on earth surface is given as $W = mg_s$

If earth is rotating at angular speed ω, value of g on earth's equator is

$$g_{eq} = g_s - \omega^2 R_e$$

For weight to be $\frac{3}{4}W$, we use

$$g_{eq} = \frac{3}{4}g_s = g_s - \omega^2 R_e$$

$$\Rightarrow \quad \omega^2 R_e = \frac{g_s}{4}$$

$$\Rightarrow \quad \omega = \sqrt{\frac{g_s}{4R_e}} = \sqrt{\frac{10}{4 \times 6400 \times 1000}}$$

$$\Rightarrow \quad \omega = \frac{1}{16 \times 10^2} = 0.625 \times 10^{-3} \text{ rad/s}$$ **Ans. (C)**

Sol. 24

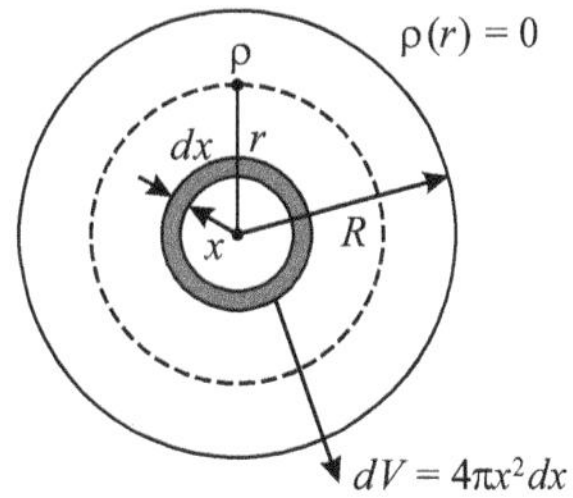

The mass of elemental shell of radius x and thickness dx is given as

$$dm = \rho(x) \times 4\rho x^2 dx$$

Total mass contained within the inner sphere of radius r is given as

$$m = \int dm = \int_0^r \frac{k}{x} \cdot 4\pi x^2 dx$$

$$\Rightarrow \quad m = 4\pi k \int_0^r x\, dx = 2\pi k r^2$$

acceleration of a test particle at point P is given as

$$a = -\frac{Gm}{r^2} = -\frac{G}{r^2} \cdot 2\pi k r^2 = -2\pi k G$$

–ve sign only denotes inward direction.

$|a| \rightarrow$ constant

independent of r for $r < R$ **Ans. (B)**

Sol. 25

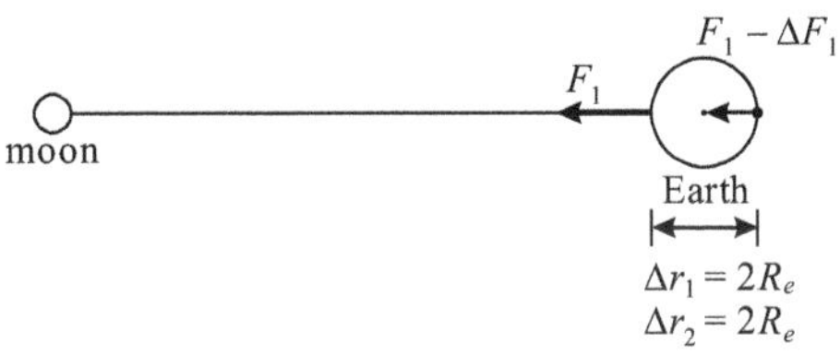

Similarly for Sun and earth this difference of force is given as

Force between earth and moon is

$$F_1 = \frac{GM_eM_m}{r_1^2}$$

$$\Rightarrow \quad \Delta F_1 = -\frac{2GM_eM_m}{r_1^3} \cdot \Delta r_1 \qquad \dots(1)$$

and
$$\Delta F_2 = -\frac{2GM_eM_S}{r_2^3} \cdot \Delta r_2 \qquad \dots(2)$$

From (a) (2) $\Rightarrow$
$$\frac{\Delta F_1}{\Delta F_2} = \frac{M_m}{M_S} \times \left(\frac{r_2}{r_1}\right)^3$$

$$\Rightarrow \quad \frac{\Delta F_1}{\Delta F_2} = \left(\frac{8 \times 10^{22}}{2 \times 10^{30}}\right) \times \left(\frac{150 \times 10^6}{0.4 \times 10^6}\right)^3 = 2.1$$

Ans. (A)

Sol. 26

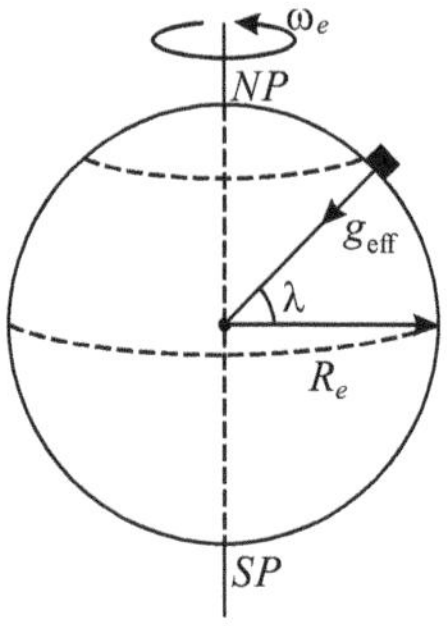

As effective gravity at a point on earth surface at latitude λ is given as

$$g_{eff} = g_s - \omega_e^2 R_e \cos^2\lambda \qquad \ldots(1)$$

at poles $\lambda = \pm 90° \Rightarrow g_{eff} = g_s$

at all points on earth other than poles

$$g_{eff} < g_s$$

If ω_e increases $\Rightarrow g_{eff}$ everywhere decreases except poles

$$\text{weight} = mg_{eff}$$

Ans. (D)

Sol. 27 Acceleration due to gravity is

$$g = \frac{GM}{R^2}$$

$$\frac{g_p}{g_e} = \frac{M_e}{M_e}\left(\frac{R_e}{R_p}\right)^2 = 3\left(\frac{1}{3}\right)^2 = \frac{1}{3}$$

For a simple pendulum, we have

$$T \propto \frac{1}{\sqrt{g}}$$

$$\Rightarrow \quad \frac{T_p}{T_e} = \sqrt{\frac{g_e}{g_p}} = \sqrt{3}$$

$$\Rightarrow \quad T_p = 2\sqrt{3}\text{s}$$

Ans. (B)

Sol. 28

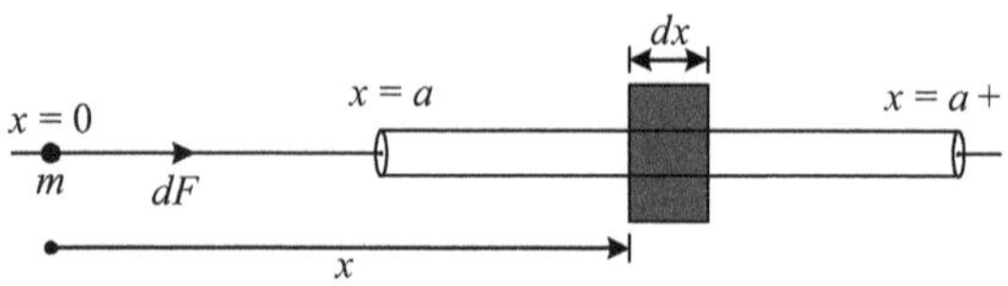

Mass of element of width dx considered is

$$dm = (A + Bx^2)dx$$

Force on this element due to m is given as

$$dF = \frac{Gm\,dm}{x^2} = \frac{Gm}{x^2}(A + Bx^2)dx$$

Net force on rod is

$$F = \int_a^{a+L} \frac{Gm}{x^2}(A + Bx^2)dx$$

$$\Rightarrow \quad F = Gm\left[-\frac{A}{x} + Bx\right]_a^{a+L}$$

$$\Rightarrow \quad F = Gm\left[A\left(\frac{1}{a} - \frac{1}{a+L}\right) + BL\right]$$

Ans. (B)

Sol. 29

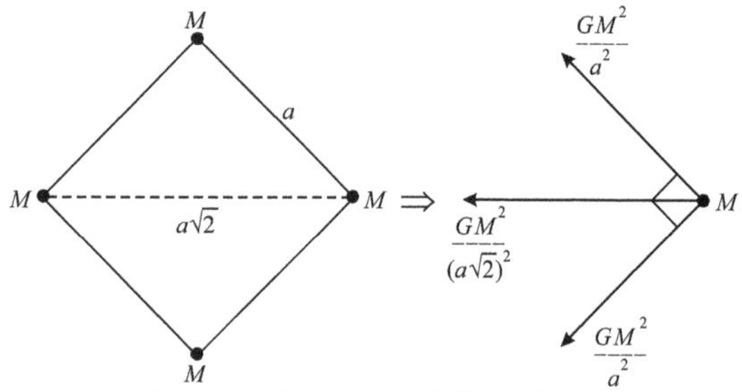

Net force on particle towards centre of circle is

$$F_C = \frac{GM^2}{2a^2} + \frac{GM^2}{a^2}\sqrt{2}$$

$$\Rightarrow \quad F_C = \frac{GM^2}{a^2}\left(\frac{1}{2} + \sqrt{2}\right)$$

This force will act as centripetal force for circular motion in a circle of radius $\frac{a}{\sqrt{2}}$, given as

$$F_C = \frac{mv^2}{r}$$

$$\Rightarrow \quad \frac{mv^2}{\frac{a}{\sqrt{2}}} = \frac{GM^2}{a^2}\left(\frac{1}{2} + \sqrt{2}\right)$$

$$\Rightarrow \quad v^2 = \frac{GM}{a}\left(\frac{1}{2\sqrt{2}} + 1\right)$$

$$\Rightarrow \quad v^2 = \frac{GM}{a}(1.35)$$

$$\Rightarrow \quad v = 1.16\sqrt{\frac{GM}{a}}$$

Ans. (C)

Sol. 30

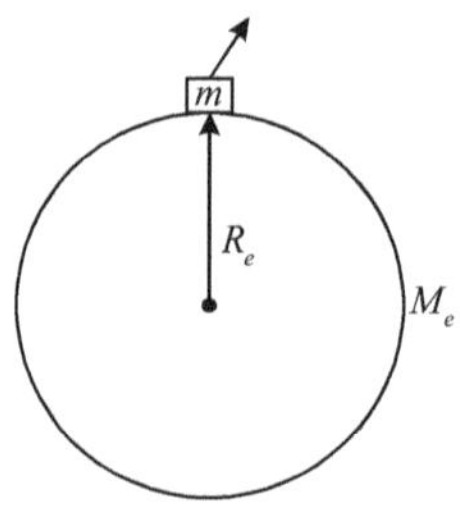

Energy required (minimum) for rocket to escape is

$$E = \frac{GM_e m}{R_e}$$

For moon, we have

$$R_m = \frac{R_e}{4} \Rightarrow V_m = \frac{V_e}{64}$$

$$\Rightarrow \quad M_m = \frac{M_e}{64}$$

Minimum energy required for rocket to escape from moon is

$$E_{moon} = \frac{GM_m m}{R_m} \qquad \left(\because M_m = \frac{M_e}{64}, R_m = \frac{R_e}{4}\right)$$

$$E_{moon} = \frac{E}{16}$$

Ans. (B)

Sol. 31

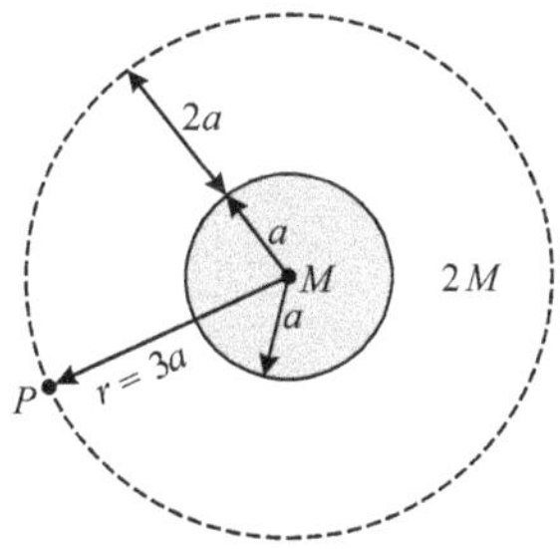

At point P gravitational field is due to the whole spherically symmetric mass $3M$ which will act as concentrated at centre for this point, so

$$g_P = \frac{G(3M)}{r^2} = \frac{G(3M)}{(3a)^2} = \frac{GM}{3a^2}$$ **Ans. (B)**

Sol. 32 Acceleration due to gravity at height h above earth surface is given as

$$g_h = \frac{GM_e}{(R_e+h)^2}$$

Given that $\frac{GM_e}{(R_e+h)^2} = \frac{1}{2}\frac{GM_e}{R_e^2}$

$\Rightarrow$ $\sqrt{2}R_e = R_e + h$

$\Rightarrow$ $h = (\sqrt{2}-1)R_e$

$= 0.414 \times 6.4 \times 10^6 \text{ m}$

$= 2.6 \times 10^6 \text{ m}$ **Ans. (D)**

7.2 Satellite Motion

Sol. 1

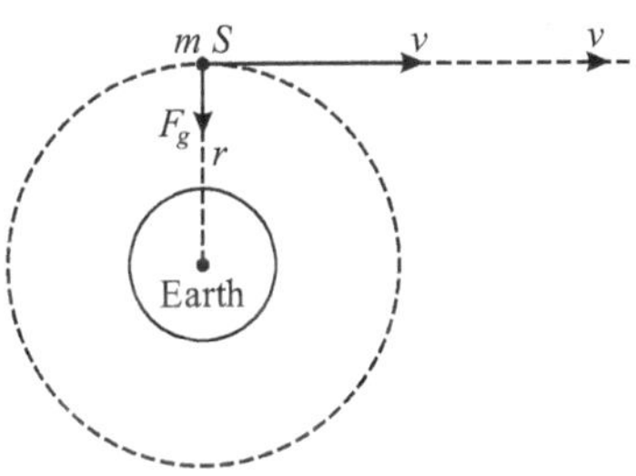

$$F_g = \frac{GM_e m}{r^2} = \frac{mv^2}{r}$$

If gravitational force between earth and satellite becomes zero than there will be no centripetal force. So no circular motion will be continued and particle will move with $\vec{v}$ due to its inertia tangentially. **Ans. (B)**

Sol. 2

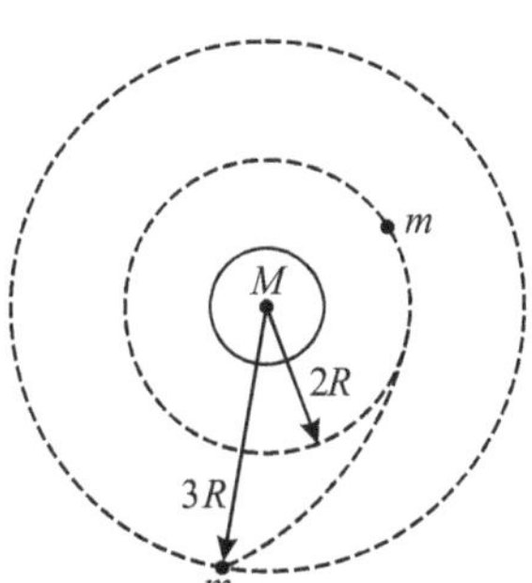

For a satellite of mass m, total energy in an orbit of radius r is given as

$$E = -\frac{GMm}{2r}$$

Initial energy of satellite ($r = 2R$) $E_i = -\frac{GMm}{2(2R)} = -\frac{GMm}{4R}$

Final energy of satellite ($r = 3R$) $E_f = -\frac{GMm}{2(3r)} = -\frac{GMm}{6R}$

Energy required $\Delta E = E_f - E_i = \left(-\frac{GMm}{6R}\right) - \left(-\frac{GMm}{4R}\right)$

$\Rightarrow$ $\Delta E = \frac{GMm}{2R}\left(\frac{1}{2}-\frac{1}{3}\right)$

$\Rightarrow$ $\Delta E = \frac{GMm}{12R}$ **Ans. (A)**

Sol. 3 v_{es} is such that its KE is numerically equal to the magnitude of gravitational PE so that $TE = 0$

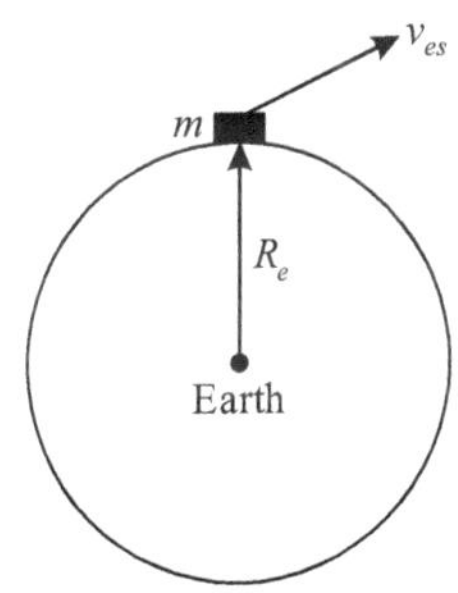

$$\frac{1}{2}mv_{es}^2 + \left(-\frac{GM_e m}{R_e}\right) = 0$$

$\Rightarrow$ $v_{es} = \sqrt{\frac{2GM_e}{R_e}} = \sqrt{2gR_e}$

This is independent of m **Ans. (A)**

Sol. 4 By Kepler's III law we have

$$T^2 \propto R^3$$

$\Rightarrow$ $\left(\frac{T_1}{T_2}\right)^2 = \left(\frac{R_1}{R_2}\right)^3$

Here $T_1 = 5$ hr

$R_2 = 4R_1$

$\Rightarrow$ $\frac{5}{T_2} = \left(\frac{R_1}{4R_1}\right)^{3/2}$

$\Rightarrow$ $T_2 = 5(4)^{3/2} = 40$ hr **Ans. (C)**

Sol. 5 Escape velocity is required to overcome the $-ve$ PE in body due to Earth's gravity

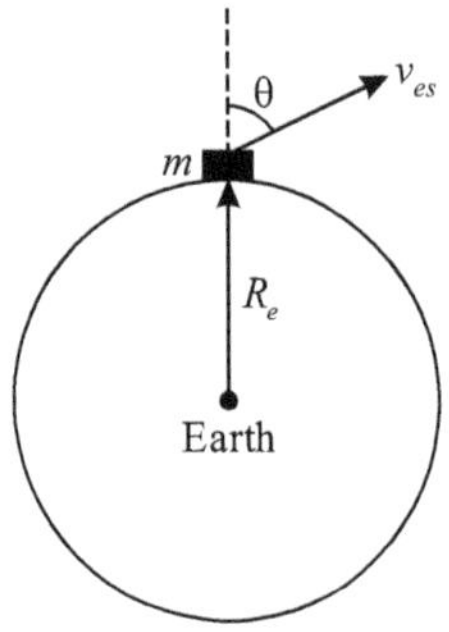

For escape velocity

$$\frac{1}{2}mv_{es}^2 + \left(-\frac{GM_e m}{R_e}\right) = 0$$

$$v_{es} = \sqrt{\frac{2GM_e}{R_e}} \simeq 11 \text{ km/s}$$

This is independent of θ **Ans. (C)**

Sol. 6

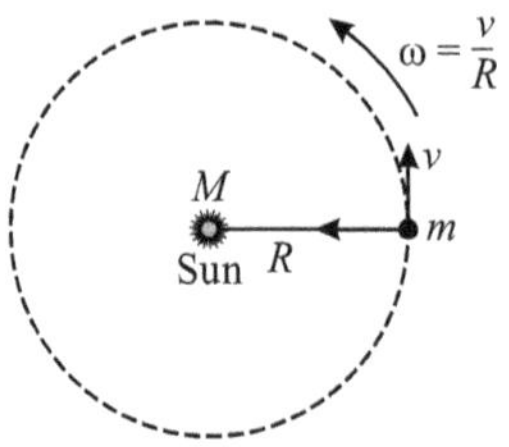

Here gravitational force provides the centripetal force to planet for its circular motion, so we have

$$F_g = \frac{GMm}{R^n} = m\omega^2 R$$

$$\Rightarrow \quad \omega = \sqrt{\frac{GM}{R^{n+1}}}$$

time period of orbital motion is

$$T = \frac{2\pi}{\omega} = \frac{2\pi}{\sqrt{GM}}\sqrt{R^{n+1}}$$

$$\Rightarrow \quad T \propto R^{\left(\frac{n+1}{2}\right)}$$ **Ans. (A)**

Sol. 7

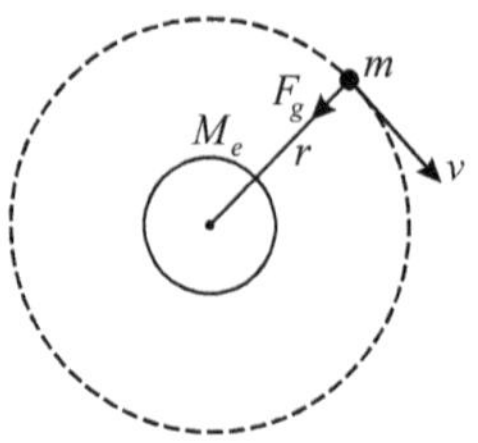

Orbital speed of a satellite in an orbit radius r is given as

$$v = \sqrt{\frac{GM_e}{r}}$$

time period of satellite is

$$T = \frac{2\pi r}{v} = \frac{2\pi}{\sqrt{GM_e}} \cdot r^{3/2}$$

T is independent of mass of satellite **Ans. (A)**

Sol. 8

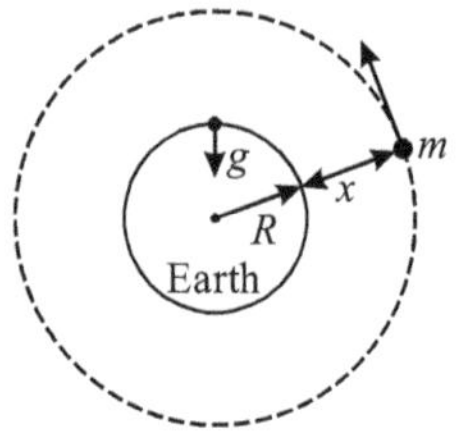

If g is acceleration due to gravity on earth surface then we have

$$g = \frac{GM}{R^2}$$

orbital speed of satellite in the orbit of radius $r = R + x$ is given as

$$v = \sqrt{\frac{GM}{r}} = \sqrt{\frac{GM}{R+x}} = \sqrt{\frac{GM}{R^2} \cdot \frac{R^2}{R+x}}$$

$$= \sqrt{\frac{gR^2}{R+x}}$$ **Ans. (D)**

Sol. 9 From a planet escape velocity is given as

$$v_e = \sqrt{\frac{2GM}{R}} \quad \text{for earth } v_{e_1} = \sqrt{\frac{2GM_e}{R_e}} = 11 \text{ km/s}$$

$$\Rightarrow \quad v_e \propto \sqrt{\frac{M}{R}}$$

For this planet $M = 10\,M_e$ and $R = \frac{R_e}{10}$

$$\Rightarrow \quad v_e = \sqrt{\frac{2G(10M_e)}{(R_e/10)}} = 10\sqrt{\frac{2GM_e}{R_e}}$$

$$\Rightarrow \quad v_e = 110 \text{ km/s}$$ **Ans. (C)**

Sol. 10

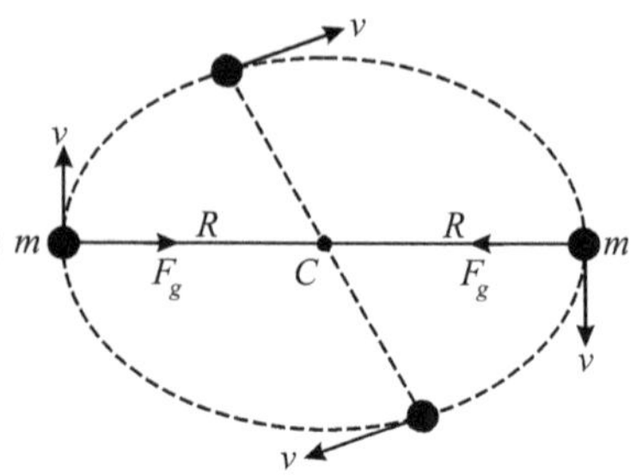

In the given double star system the gravitational force will provide the centripetal force for this circular motion. so we use

$$\frac{Gm^2}{(2R)^2} = \frac{mv^2}{R}$$

$$\Rightarrow \quad v = \sqrt{\frac{Gm}{4R}}$$ **Ans. (B)**

Sol. 11 For EOM elliptical orbit if time period of motion in this orbit is T then we use

$$T^2 \propto \left(\frac{a_m + a_e}{2}\right)^3 \qquad \ldots(1)$$

Semi major axis of elliptical orbit for earth if T_e be the period around sun then we use

$$T_e^2 \propto a_e^3 \qquad \ldots(2)$$

Here (1)/(2) gives

$$\frac{T^2}{T_e^2} = \left[\frac{(a_m + a_e)/2}{a_e}\right]^3$$

$$T = \left[\frac{(a_m + a_e)/2}{a_e}\right]^{3/2} T_e$$

$$T = \left[\frac{(1.5 + 2.28)/2}{1.5}\right]^{3/2} \times 365 \text{ days}$$

$$T = \left(\frac{1.89}{1.5}\right)^{3/2} \times 365 \text{ days}$$

Time of motion in half ellipse as described (path EOM) is given as

$$t = \frac{T}{2} = \left[\frac{1.89}{1.5}\right]^{3/2} \times \frac{365}{2} \cong 258 \text{ days}$$

Ans. (C)

Sol. 12

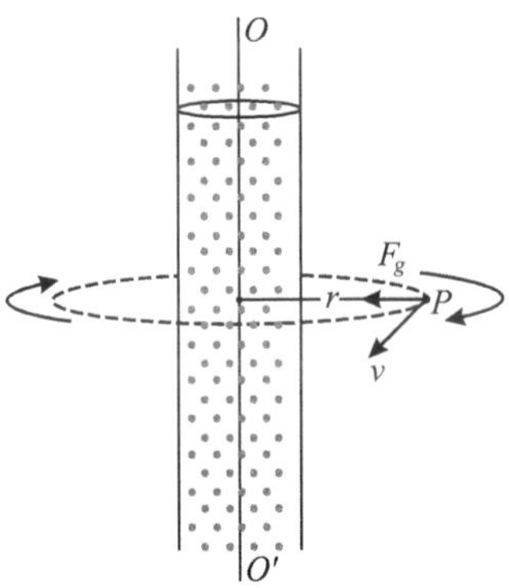

If λ be the linear mass density along the length of galaxy then gravitational field at the location of star is given as

$$g_P = \frac{2G\lambda}{r}$$

For circular motion of star, we use

$$mg_P = \frac{mv^2}{r}$$

$$\Rightarrow \quad \frac{2G\lambda}{r} = \frac{v^2}{r} \quad \Rightarrow \quad v = \sqrt{2G\lambda} \quad \text{orbital speed of star}$$

Period of revolution of star in circular orbit is

$$T = \frac{2\pi r}{v}$$

$$\Rightarrow \quad T = \frac{2\pi r}{\sqrt{2G\lambda}}$$

$$\Rightarrow \quad T \propto r$$

Ans. (A)

Sol. 13

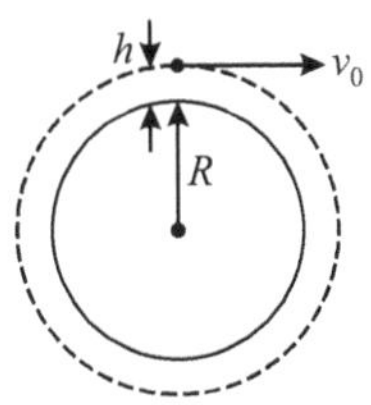

Orbital speed velocity close to earth surface is given as

$$v_0 = \sqrt{gR}$$

Escape velocity very close to earth surface is given as

$$v_e = \sqrt{2gR}$$

Additional velocity required for satellite to escape is

$$\Delta v = v_e - v_0 = \sqrt{gR}\,(\sqrt{2} - 1)$$

Ans. (D)

Sol. 14

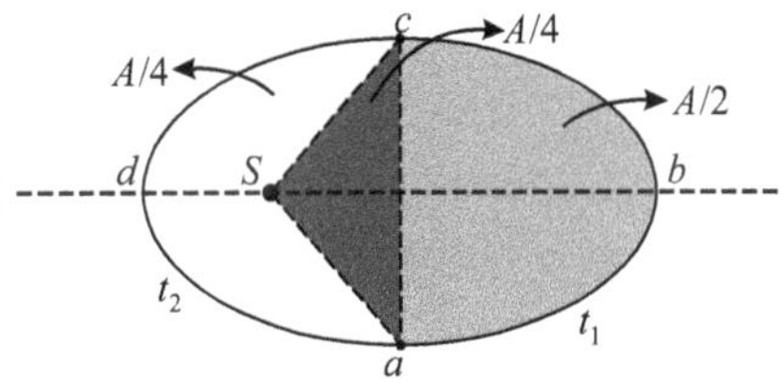

Total area = A

According Kepler's law of areas, rate of sweeping out area by position vector of planet w.r. to sun remain constant.

As t $\propto$ area swept, we have

$$\Rightarrow \qquad t_1 \propto \frac{3A}{4} \qquad \ldots(1)$$

$$\text{and} \qquad t_2 \propto \frac{A}{4} \qquad \ldots(2)$$

$$(1)/(2) \Rightarrow \frac{t_1}{t_2} = 3 \quad \Rightarrow \quad t_1 = 3t_2$$

Ans. (C)

Sol. 15

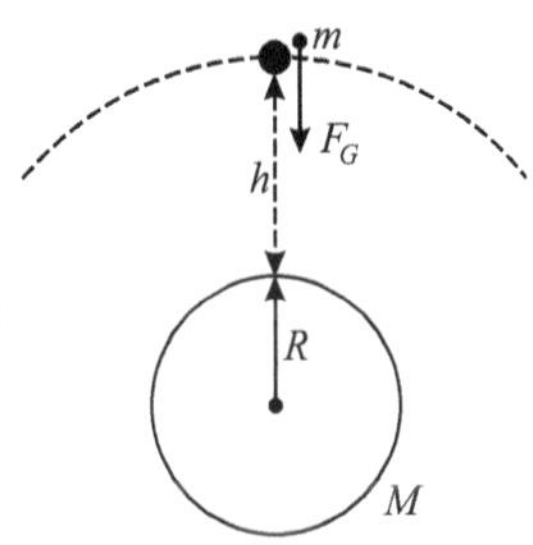

Along with satellite we consider the astronaut is also revolving around earth in orbit of radius $(R + h)$ so gravitational force on astronaut is given as

$$F_G = \frac{GMm}{(R+h)^2}$$

Here we neglect the gravitational force between astronaut and satellite.

Ans. (C)

Sol. 16 Initial State

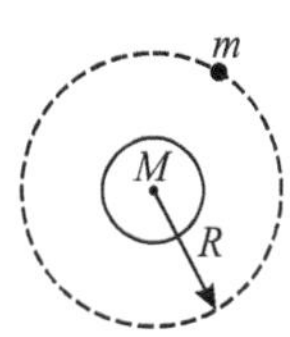

Total energy of body is

$$E_i = -\sqrt{\frac{GMm}{2R}}$$

Final State

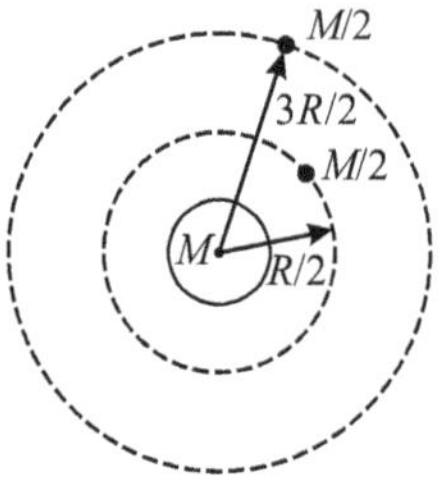

$$E_f = -\frac{GM\left(\frac{m}{2}\right)}{2\left(\frac{R}{2}\right)} - \frac{GM\left(\frac{m}{2}\right)}{2\left(3\frac{R}{2}\right)}$$

$$\Rightarrow \quad E_f = -\frac{2}{3}\frac{GMm}{R}$$

Difference in final and initial energies is

$$\Delta E = E_f - E_i$$

$$\Rightarrow \quad \Delta E = -\frac{GMm}{R}\left(\frac{2}{3}-\frac{1}{2}\right)$$

$$\Rightarrow \quad \Delta E = -\frac{GMm}{6R}$$

Ans. (C)

Sol. 17 By Kepler's II law areal velocity of a planet is constant even in elliptical path and given as

$$\frac{dA}{dt} = \frac{L}{2m}$$ **Ans. (C)**

Sol. 18 $U_{surface} + E_1 = U_h$

KE of satellite is zero at earth surface & at height *h*, we use

$$-\frac{GM_em}{R_e} + E_1 = -\frac{GM_em}{(R_e+h)}$$

$$E_1 = GM_em\left(\frac{1}{R_e} - \frac{1}{R_e+h}\right)$$

$$E_1 = \frac{GM_em}{(R_e+h)} \times \frac{h}{R_e}$$

Gravitational attraction at height *h* will be

$$F_G = mg = \frac{mv^2}{(R_e+h)}$$

$$\Rightarrow \quad \frac{GM_em}{(R_e+h)^2} = \frac{mv^2}{(R_e+h)}$$

$$\Rightarrow \quad mv^2 = \frac{GM_em}{(R_e+h)}$$

$$E_2 = \frac{1}{2}mv^2 = \frac{GM_em}{2(R_e+h)}$$

Given that $\quad E_1 = E_2$

$$\Rightarrow \quad \frac{h}{R_e} = \frac{1}{2}$$

$$\Rightarrow \quad h = \frac{R_e}{2} = 3200\text{ km}$$ **Ans. (C)**

Sol. 19 At height *r* from center of earth orbital velocity is given as

$$v = \sqrt{\frac{GM}{r}}$$

By energy conservation, we have

$$KE \text{ of 'm'} + \left(-\frac{GMm}{r}\right) = 0 + 0$$

$$\Rightarrow \quad KE \text{ of 'm'} = \frac{GMm}{r} = \left(\sqrt{\frac{GM}{r}}\right)^2 m = mv^2$$ **Ans. (B)**

Sol. 20 If *M* is mass of star *m* is mass of meteorite then by energy conservation between *O* & ∞, we use

$$-\frac{GMm}{r} + \frac{-GMm}{r} + \frac{1}{2}mv^2 = 0 + 0$$

$$\Rightarrow \quad v = \sqrt{\frac{4GM}{r}} = 2.8 \times 10^5 \text{ m/s}$$ **Ans. (D)**

Sol. 21 Orbital speed of satellite is

$$v_0 = \sqrt{g(R+h)} \approx \sqrt{gR}$$

Escape velocity is

$$v_e = \sqrt{2g(R+h)} \approx \sqrt{2gR}$$

Increment in speed required is

$$\Delta v = v_e - v_0 = (\sqrt{2}-1)\sqrt{gR}$$ **Ans. (A)**

Sol. 22 As after impact the resulting velocity of the combined body is at some angle to the radial direction so resulting motion will be an elliptical orbit. **Ans. (C)**

Sol. 23 Orbital speed of satellite in orbit of radius P_c is given

$$v = \sqrt{\frac{GM}{r}}$$

Kinetic energies of the two satellites are

$$T_A = \frac{1}{2}m_Av_A^2$$

and $$T_B = \frac{1}{2}m_Bv_B^2$$

$$\Rightarrow \quad \frac{T_A}{T_B} = \frac{m \times \frac{GM}{R}}{2m \times \frac{GM}{2R}}$$

$$\Rightarrow \quad \frac{T_A}{T_B} = 1$$ **Ans (C)**

Sol. 24 Mass contained in the elemental shell is

$$dm = \rho \cdot 4\pi x^2 dx$$

$$M=\int dm=\int_0^R \frac{k}{x^2}\cdot 4\pi x^2\,dx=4\pi kR$$

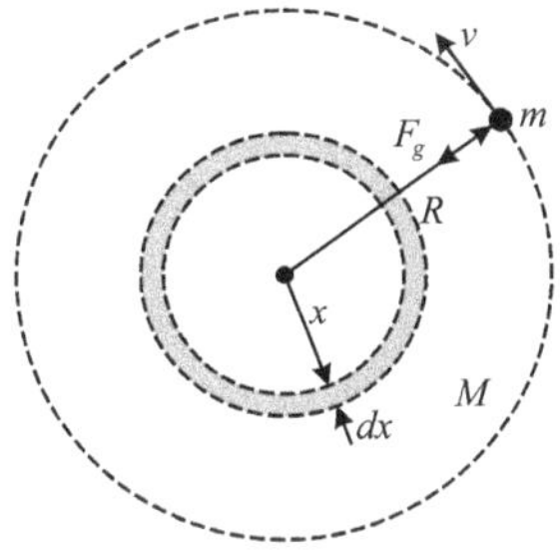

Gravitational force on mass in orbit is

$$F_g=\frac{GMm}{R^2}=\frac{G(4\pi kR)m}{R^2}$$

$$\Rightarrow \qquad F_g=\frac{4\pi kGm}{R}$$

For circular motion, we use

$$F_g=\frac{4\pi kGm}{R}=\frac{mv^2}{R}$$

$$\Rightarrow \qquad v=\sqrt{4\pi kG}$$

Time period $T=\dfrac{2\pi R}{\sqrt{4\pi kG}} \Rightarrow \dfrac{T}{R}=$ constant **Ans. (D)**

Sol. 25 Orbital speed of spaceship is

$$v=\sqrt{\frac{Gm}{r}}$$

Where $r=R_P+h=2.02\times 10^6$ m

$$v=\sqrt{\frac{6.67\times10^{-11}\times8\times10^{22}}{2.02\times10^6}}\simeq 1.625\times10^3 \text{ m/s}$$

Time period of revolutions of spaceship is

$$T=\frac{2\pi r}{v}$$

Number of revolutions in 24 hrs are

$$n=\frac{24\times3600}{T}=\frac{24\times3600\times1.625\times10^3}{2\times3.14\times2.02\times10^6}$$

$$\simeq 11$$ **Ans. (B)**

Sol. 26 Gravity on surface of a planet is given as

$$g_P=\frac{GM_P}{R_P^2}$$

$$\frac{g_e}{g_P}=\frac{9}{4}\,\frac{M_e/R_e^2}{M_P/R_P^2} \qquad \left(\text{here } M_P=\frac{1}{9}M_e\right)$$

$$\Rightarrow \qquad \frac{9}{4}=\frac{9R_P^2}{R_e^2} \Rightarrow R_P=\frac{R_e}{2}=\frac{R}{2}$$ **Ans. (B)**

8.1 Elasticity

Sol. 1

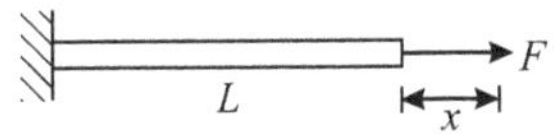

Work done by external force is stretching the wire, so $W=fx$

Half of above work is stored in elastic PE of wire, so we use

and $$U=\frac{1}{2}fx$$

$$\Rightarrow \qquad U=\frac{1}{2}\times200\times1\times10^{-3}=0.1\text{ J}$$ **Ans. (D)**

Sol. 2

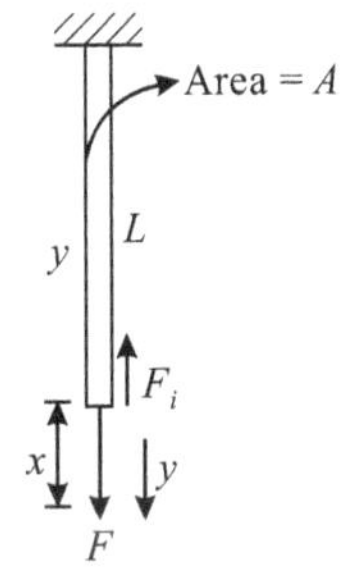

Work done by external force $W=Fx$

Due to stretching by internal force because of elasticity it develops stress in wire, using Young's modulus we have

$$Y=\frac{\text{stress}}{\text{stress}}=\left(\frac{f_i/A}{y/L}\right)\Rightarrow F_i=\frac{YA}{L}\cdot y$$

y increases from 0 to x.

$y\rightarrow$ instantaneous extension is wire

Work done by internal force

$$W_i=\int_0^x F_i day=-\frac{yA}{L}\int_0^x y\,dy$$

$$\Rightarrow \qquad W_i=\frac{1}{-2}\frac{yA}{L}x^2 \qquad \ldots(1)$$

At final extension x, we have for equilibrium of wire

$$F=\frac{YA}{L}\cdot x \Rightarrow \frac{YA}{L}=\frac{F}{x} \qquad \ldots(2)$$

from (1) and (2)

$$W_i=-\frac{1}{2}\frac{YA}{L}x^2=-\frac{1}{2}\left(\frac{F}{x}\right)x^2=-\frac{1}{2}Fx$$

– ve sign denotes absorption of energy

$\Rightarrow \quad \dfrac{1}{2}Fx$ energy is stored in the wire.

Work done in stretching by external force $=Fl$

Work done by elastic forces in stretching $=\dfrac{1}{2}Fl$

Heat dissipation $=Fl-\dfrac{1}{2}Fl=\dfrac{1}{2}Fl$ **Ans. (D)**

Sol. 3 If a wire is stretched by an external force then elastic PE density (J/m^3) in the wire is given as

$$u_E = \frac{1}{2} \times \text{stress} \times \text{strain}$$

By definition of Young's modulus, we use

$$y = \frac{\text{stress}}{\text{strain}} = \frac{S}{\text{strain}}$$

$$\Rightarrow \quad \text{Strain} = \frac{S}{Y}$$

$$u_E = \frac{1}{2} \times S \times \frac{S}{Y} = \frac{S^2}{2Y}$$

Ans. (B)

Sol. 4

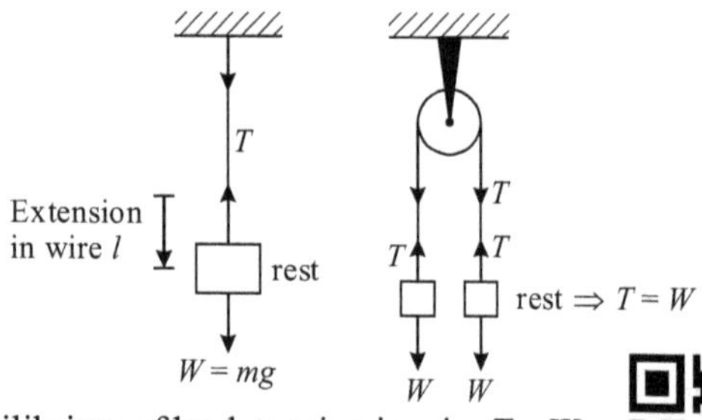

For equilibrium of load, tension in wire $T = W$
As tension in wire in the two cases is same
$\Rightarrow$ extension in these cases must be same.

Ans. (A)

Sol. 5

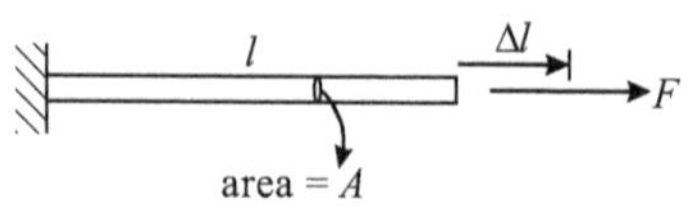

Volume of wire $\quad V = Al$

By definition of young's modulus, we have

$$Y = \frac{\text{Stress}}{\text{strain}} = \frac{(F/A)}{\Delta l/l}$$

$$\Rightarrow \quad F = \frac{YA}{l}\Delta l \times \frac{A}{A}$$

$$\Rightarrow \quad F = \frac{YA^2\Delta l}{V}$$

For same Δl and V we have

$$F \propto A^2$$

$$\Rightarrow \quad \frac{F_1}{F_2} = \left(\frac{A_1}{A_2}\right)^2$$

If $\quad A_2 = 3A \;\&\; A_1 = A$

$\Rightarrow \quad F_2 = F_1(3)^2 = 9F_1 = 9F$

Ans. (B)

Sol. 6

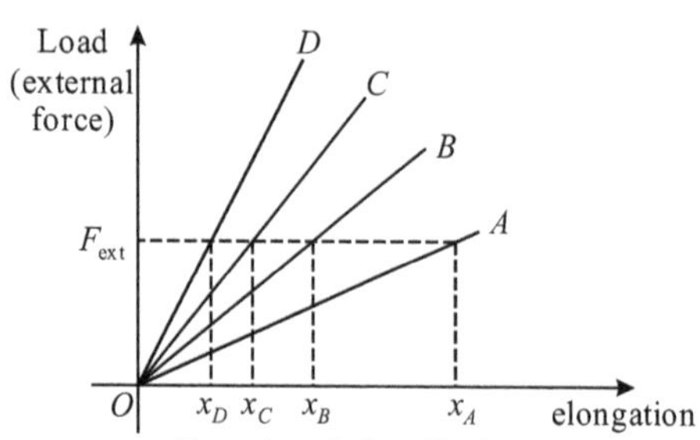

Same length for all wires

For a given value of load (external force)

$$x_A > x_B > x_C > x_D$$

As x_A is maximum

$\Rightarrow$ Strain is maximum in wire OA

$\Rightarrow$ Stress is maximum in wire OA

$$\text{stress} = \frac{F_{ext}}{A}$$

$\Rightarrow$ A is minimum for wire OA

Ans. (A)

Sol. 7 Young's modulus $Y = \dfrac{\text{Stress}}{\text{Strain}}$

$$\Rightarrow \quad \text{Strain} = \frac{\text{Stress}}{Y} = \frac{F/A}{Y} = \frac{F}{\pi r^2 Y}$$

$$\Rightarrow \quad \text{Strain} = \frac{10^5}{3.14 \times (10 \times 10^{-3})^2 \times 2 \times 10^{11}}$$

$$\Rightarrow \quad \text{Strain} = \frac{1}{628}$$

$$\%\text{ strain in wire} = \frac{1}{628} \times 100\% = 0.16\%$$

Ans. (A)

Sol. 8 Breaking force of a wire is the force at which stress in wire becomes equal to rupture strength of the wire material and it is independent of length of wire. **Ans. (C)**

Sol. 9 Stress produced in wire is given as

$$\text{Stress} = \frac{T}{A}$$

In the simple Atwood's machine shown in figure tension in wire is given as

$$T = \frac{2mm}{M+m} \cdot g = \frac{2(2m)m}{2m+m} \cdot g = \frac{4mg}{3}$$

$$\Rightarrow \quad \text{Stress} = \frac{4mg}{3A}$$

Ans. (B)

Sol. 10 Given that $\dfrac{l_{st}}{l_{br}} = a$; $\dfrac{r_{st}}{r_{br}} = b$ and $\dfrac{1/st}{1/br} = C$

We use $\quad Y = \dfrac{\text{Stress}}{\text{Strain}} = \dfrac{(T/\pi r^2)}{\Delta l/l}$

$$\Rightarrow \quad \Delta l = \frac{Tl}{\pi r^2 Y}$$

For steel wire $\quad \Delta l_{st} = \dfrac{3mg.l_{st}}{\pi r_{st}^2 Y_{st}}$

For brass wire $\quad \Delta l_{br} = \dfrac{2mg\, l_{br}}{\pi r_{br}^2 Y_{br}}$

$$\Rightarrow \quad \frac{\Delta l_{st}}{\Delta l_{br}} = \frac{3l_{st}}{2l_{br}} \times \frac{r_{br}^2}{r_{st}^2} \times \frac{Y_{br}}{Y_{st}}$$

$$\Rightarrow \quad \frac{\Delta l_{st}}{\Delta l_{br}} = \frac{3}{2}a \times \frac{1}{b^2} \times \frac{1}{c} = \frac{3a}{2b^2c} \quad \textbf{Ans. (C)}$$

Sol. 11 Young's modulus of wire

$$Y = \frac{\text{Stress}}{\text{Strain}} = \frac{\text{Stress}}{\Delta l / l}$$

$[l \to \text{wire length } \& \ \Delta l \to \text{extension}]$

$$\Rightarrow \quad \frac{\Delta l}{l} = \frac{\text{Stress}}{Y} = \frac{5 \times 10^7}{2 \times 10^{11}} = 2.5 \times 10^{-4}$$

If wire is having a cross-sectional radius r than its value is

$$V = \pi r^2 l$$

$$\Rightarrow \quad \Delta V = \pi r^2 \Delta l + 2\pi r\, \Delta r l$$

$$\Rightarrow \quad \frac{\Delta V}{V} = \frac{\Delta r^2 \Delta l + 2\pi r \Delta r l}{\pi r^2 l} = \frac{0.02}{100}$$

$$\Rightarrow \quad \frac{\Delta l}{l} + \frac{2\Delta r}{r} = \frac{0.02}{100} = 2 \times 10^{-4}$$

$$\Rightarrow \quad \frac{2\Delta r}{r} = (2 - 2.5) \times 10^{-4} = -0.5 \times 10^{-4}$$

$$\Rightarrow \quad \frac{\Delta r}{r} = -0.25 \times 10^{-4} \quad \textbf{Ans. (C)}$$

Sol. 12 As cross-sectional areas of the two wires are equal

$$\Rightarrow \quad (\text{Stress})_{Cu} = (\text{Stress})_{steel}$$

$$Y_{cm} \times \frac{\Delta l_{Cu}}{l_{Cu}} = Y_s \times \frac{\Delta l_{st}}{l_{st}}$$

$$1.0 \times 10^{11} \times \frac{1 \times 10^{-3}}{1} = 2.0 \times 10^{11} \times \frac{\Delta l_{st}}{0.5}$$

$$\Rightarrow \quad \Delta l_s = \frac{1}{4} \times 10^{-3}\,\text{m}$$

$$= 0.25\,\text{mm}$$

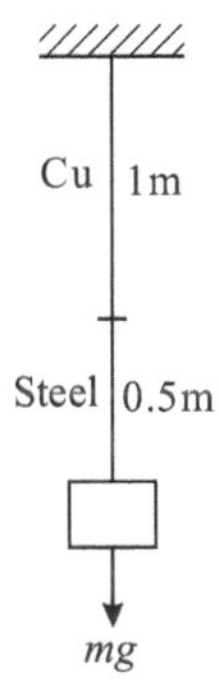

Total extension is composite wire

$$\Delta l = \Delta l_{cu} + \Delta l_{st}$$

$$= 1 + 0.25$$

$$\Rightarrow \quad \Delta l = 1.25\,\text{mm} \quad \textbf{Ans. (D)}$$

Sol. 13

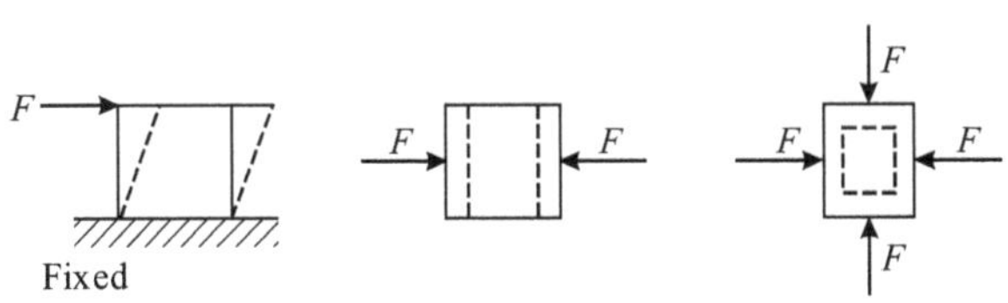

Modulus of elasticity

Shear modulus

$$\eta = \frac{\text{shear strain}}{s.\text{strain}}$$

Young's modulus

$$Y = \frac{\text{long stress}}{\text{long strain}}$$

Bulk modulus

$$B = \frac{\text{Bulk stress}}{\text{Volume strain}}$$

For equal magnitude of strains we have

Bulk stress > longitudinal stress > shear stress

$$\Rightarrow \quad B > Y > \eta \quad \textbf{Ans. (C)}$$

Sol. 14 By definition of Bulk modulus

$$B = \left| \frac{\Delta P_{ex}}{\Delta V / V} \right|$$

Here $\frac{\Delta V}{V}$ is fractional compression in volume due to excess pressure ΔP_{ex}

$\Rightarrow$ For ΔP_{ex} = constant for all liquids

We have $\quad \frac{\Delta V}{V} \propto \frac{1}{B}$

$$\Rightarrow \quad \left|\frac{\Delta V}{V}\right|_{Ethanol} > \left|\frac{\Delta V}{V}\right|_{Water} > \left|\frac{\Delta V}{V}\right|_{Mercury} \quad \textbf{Ans. (A)}$$

Sol. 15

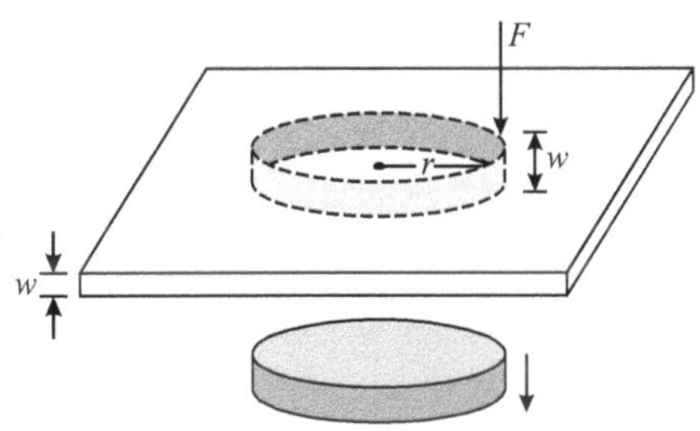

Shear stress by the applied force on the part of steel sheet (disc shaped) which is to be cut/removed is given as

$$\text{Shear stress} \quad \sigma = \frac{F}{2\pi r w}$$

When the shear stress becomes equal to rapture strength of steel than hole is cut

$\Rightarrow$ required force is given as

$$F = 2\pi r w \times \sigma_{max}$$

$$\Rightarrow \quad F = 2 \times 3.14 \times 0.5 \times 10^{-2} \times 0.3 \times 10^{-2} \times 3.5 \times 10^8$$

$$\Rightarrow \quad F = 3.29 \times 10^4\,\text{N} \quad \textbf{Ans. (C)}$$

Sol. 16

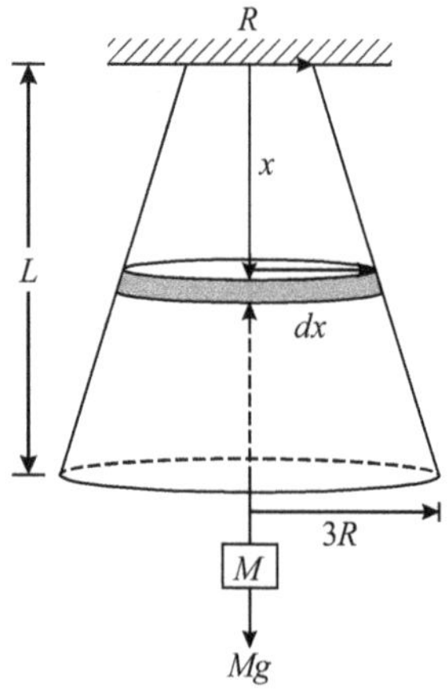

Radius of elemental disc from top end at a distance x is

$$R=\left(\frac{3R-R}{L}\right)x+R=\frac{2R}{L}x+R$$

for whole wire we can use on all such elements

$$Y=\frac{\text{Stress}}{\text{Strain}}$$

If dy is the elongation in width dx

$$\Rightarrow \qquad Y=\frac{Mg/\pi r^2}{dy/dx}$$

$$\Rightarrow \qquad dy=\frac{Mg\,dx}{Y\pi\left(\frac{2R}{L}x+R\right)^2}$$

Total extension in wire is

$$y=\int dy=\frac{Mg}{Y\pi}\int_0^L\frac{dx}{\left(\frac{2R}{L}x+R\right)^2}$$

$$\Rightarrow \qquad y=\frac{Mg}{Y\pi}\left[\frac{L}{2R}\left(-\frac{1}{\left(\frac{2R}{L}x+R\right)}\right)\right]_0^L$$

$$\Rightarrow \qquad y=\frac{Mg\,L}{2Y\pi R}\left[-\frac{1}{3R}-\left(-\frac{1}{R}\right)\right]=\frac{Mg\,L}{3\pi YR^2}$$

Total extended length of wire

$$L_f=L+y=L\left(1+\frac{Mg}{3\pi YR^2}\right)$$ **Ans. (C)**

Sol. 17

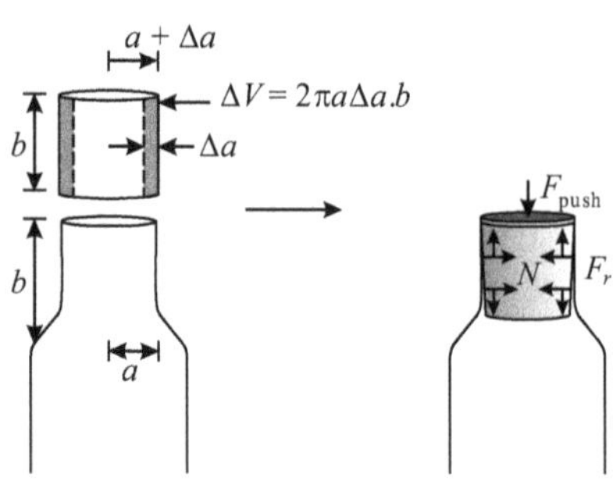

To fit the cork in the bottle opening a contact force compressive in nature is there and stress due to this, contact force on cork volume is

excess pressure $\quad \Delta P=\frac{N}{3}=\frac{N}{2\pi ab}$

Volume compression in cork

$$\Delta V=(2\pi a\Delta a).b$$

Volume of cork

$$V=\pi a^2 b$$

Volume strain in cork

$$\Rightarrow \qquad \frac{\Delta V}{V}=\frac{2\pi a\Delta a.b}{\pi a^2 b}$$

Bulk modulus of cork is given as

$$B=\frac{\Delta P_{ex}}{\left|\frac{\Delta V}{V}\right|}$$

$$\Rightarrow \qquad B=\frac{N/2\pi ab}{(2\pi a\Delta ab)/\pi a^2 b}=\frac{N.a^2 b}{2\pi\Delta ab.2\pi ab}$$

$$\Rightarrow \qquad N=4\pi\Delta abB$$

to push the cork inside the bottle the external force has to overcome the friction on cork

$$\Rightarrow \qquad F_{\text{push}}\geq\mu N=4\pi\mu\Delta abB$$ **Ans. (D)**

Sol. 18

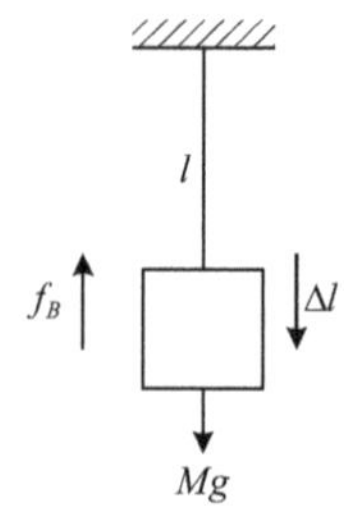

Young's modulus of wire is given as

$$Y=\frac{\text{Stress}}{\text{Strain}}=\frac{Mg/s}{\Delta l/l}$$

$$\Rightarrow \qquad \Delta l \;\alpha\; MG \qquad \ldots(1)$$

If load is immersed in liquid then the new extension in wire can be given as

$$\Delta l' \;\alpha\; Mg_{\text{eff}}=Mg\left(1-\frac{\rho_l}{\rho_s}\right)=Mg\left(1-\frac{2}{8}\right)=\frac{3}{4}Mg \qquad \ldots(2)$$

$$(1)/(2) \Rightarrow \qquad \frac{\Delta l}{\Delta l'}=\frac{4}{3}$$

$$\Rightarrow \qquad \Delta l'=\frac{3}{4}\times\Delta l=\frac{3}{4}\times 4\text{ mm}=3\text{ mm}$$ **Ans (B)**

Sol. 19 Volume flow rate of water is

$$Q=vA=v(\pi r^2)$$

$$\Rightarrow \quad v = \frac{Q}{\pi r} = \frac{100\times10^{-3}}{60} \times \frac{1}{\pi\times25\times10^{-4}}$$

$$\Rightarrow \quad v = \frac{2}{3\pi}\text{ m/s}$$

Reynolds Number is given as

$$R = \frac{\rho vd}{\eta} = \frac{10^3\times2\times10\times10^{-2}}{10^{-3}\times3\pi}$$

$$\simeq 2\times10^4$$

Thus order R is 10^4 **Ans. (C)**

Sol. 20 Due to stretch cord stores elastic *PE* and on releasing, this *PE* is transformed into *KE* of stone.

Energy density of cord

$$= \frac{1}{2}\times\text{stress}\times\text{strain}$$

$$= \frac{1}{2}\times Y\times(\text{strain})^2 = \frac{1}{2}Y\left(\frac{\Delta l}{l}\right)^2$$

PE in cord $E_C = \frac{1}{2}Y\left(\frac{\Delta l}{l}\right)^2\times A\times l$

By energy conservation

$$E_C = \frac{1}{2}mv^2$$

$$\Rightarrow \quad \frac{1}{2}Y\frac{(\Delta l)^2}{l}A = \frac{1}{2}mv^2$$

$$\Rightarrow \quad Y = \frac{mv^2 l}{(\Delta l^2)A} = \frac{0.02\times(20)^2\times42\times10^{-2}}{(20\times10^{-2})^2\times\pi\times(3\times10^{-3})^2}$$

$$\Rightarrow \quad Y \cong 3\times10^6\text{ N/m}^2$$ **Ans. (C)**

Sol. 21 Tensile stresss in wire will be

$$\text{Stress} = \frac{\text{Tensile force}}{\text{Cross section Area}}$$

$$= \frac{mg}{\pi R^2} = \frac{4\times3.1\pi}{\pi\times4\times10^{-6}}\text{Nm}^{-2}$$

$$= 3.1\times10^6\text{ Nm}^{-2}$$ **Ans. (D)**

Sol. 22 Given that

$$\frac{Y_A}{Y_B} = \frac{7}{4};\ L_A = 2\text{ m};\ L_B = 1.5\text{ m};\ r_A = R;\ r_B = 2\text{ mm}$$

Using $$Y = \frac{\text{Sress}}{\text{Strain}} = \frac{F/\pi r^2}{x/L}$$

$$\Rightarrow \quad \frac{F}{x} = \frac{Y\pi r^2}{L} \quad \text{Same for } A \text{ and } B$$

$$\Rightarrow \quad \frac{Y_A\pi r_A^2}{L_A} = \frac{Y_B\pi r_B^2}{L_B}$$

$$\Rightarrow \quad \frac{7}{4}\times\frac{R^2}{2} = \frac{(2\times10^{-3})^2}{1.5}$$

$$\Rightarrow \quad R \simeq 1.74\times10^{-3}\text{ m} \simeq 1.7\text{ mm}$$ **Ans. (B)**

Sol. 23 Stress in rod is

$$\text{Stress} = \frac{F}{A} = \frac{400\times4}{\pi d^2} = 379\times10^6$$

$$\Rightarrow \quad d^2 = \frac{1600}{\pi\times379\times10^6} = 1.34\times10^{-6}$$

$$\Rightarrow \quad d = \sqrt{1.34}\times10^{-3} = 1.15\times10^{-3}\text{m}$$ **Ans. (A)**

Sol. 24 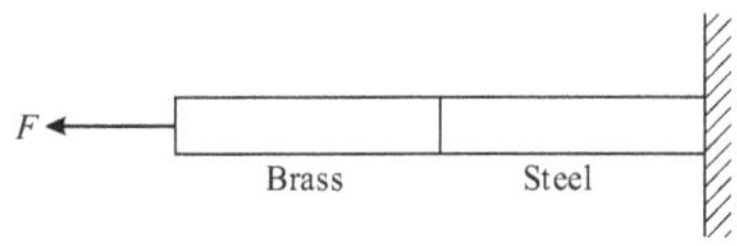

Equivalent force constaints for brass and steel wires are

$$k_1 = \frac{Y_1A_1}{l_1} = \frac{120\times10^9\times A}{1}$$

$$k_2 = \frac{Y_2A_2}{l_2} = \frac{60\times10^9\times A}{1}$$

For series combination, combined force constent is

$$k_{eq} = \frac{k_1k_2}{k_1\times k_2} = \frac{120\times60}{180}\times10^9\times A$$

$$\Rightarrow \quad k_{eq} = 40\times10^9\times\text{A}$$

For applied force F if net elongation is x, we have

$$F = k_{eq}(x)$$

$$\Rightarrow \quad F = (40\times10^9)A\cdot(0.2\times10^{-3})$$

$$\Rightarrow \quad \text{Stress} = \frac{F}{A} = 8\times10^6\text{ N/m}^2$$

None of the given option is correct **Ans. (None)**

Sol. 25 Thermal contraction is given as

$$\Delta l = l\alpha\Delta T$$

If Δl is elengation in wire due to hanging mass, we use Young's Modulus

$$Y = \frac{\text{Sress}}{\text{Strain}} = \frac{Mg/A}{\Delta l/l} = \frac{Mg}{A\alpha\Delta T}$$

$$\Rightarrow \quad M = \frac{YA\alpha\Delta T}{g}$$

$$= \frac{10^4\times\pi(10^{-3})^2\times10^{-5}\times20}{10} = 2\pi = 6.28\text{ kg}$$

None of the given options is correct. **Ans. (None)**

Sol. 26 The expension in rod due to heating is

$$\Delta L = \alpha LT \quad \text{(Coefficient of linear expension)}$$

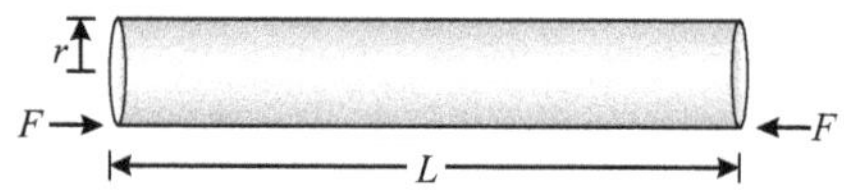

Compressional stress in rod is $= \dfrac{F}{\pi r^2}$

The compression due to F, exactly balances the thermal expension so its length remain unchanged.

Using young's modulus, we have

$$Y = \frac{\text{Sress}}{\text{Strain}} = \frac{F/\pi r^2}{\Delta L/L} = \frac{F}{\pi r^2 dT} \Rightarrow \alpha = \frac{F}{\pi r^2 YT}$$

Coefficient of volume expension

$$\gamma = 3\alpha = \frac{3F}{\pi r^2 YT}$$ **Ans. (B)**

9.1 Fluid Statics

Sol. 1 As ball sinks in liquid-1

$\Rightarrow \quad \rho_3 > \rho_1 \quad \ldots(1)$

and ball is not sinking in liquid-2

$\Rightarrow \quad \rho_3 < \rho_2 \quad \ldots(2)$

Here in jar liquid-1 is above liquid-2

$\Rightarrow \quad \rho_1 < \rho_2 \quad \ldots(3)$

From-(1), (2) and 3

$$\rho_1 < \rho_3 < \rho_2$$ **Ans. (D)**

Sol. 2 As $\quad \rho_{\text{oil}} < \rho_{\text{water}}$

$\Rightarrow$ Oil will be above water in container

$$\rho_{\text{oil}} < \rho$$

$\Rightarrow$ Ball will sink in oil

$$\rho < \rho_{\text{water}}$$

$\Rightarrow$ Ball will float in water **Ans. (C)**

Sol. 3

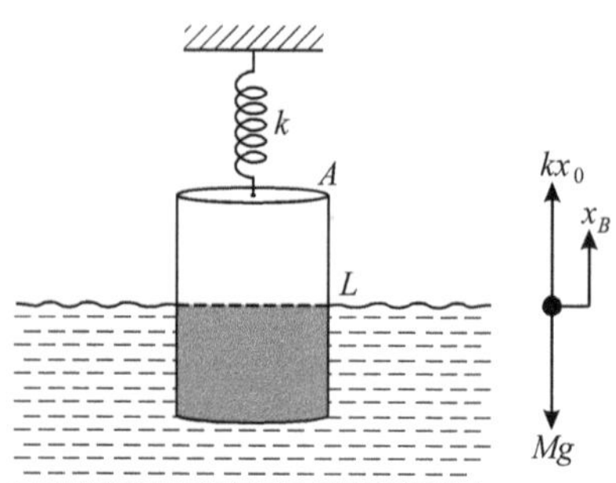

Buoyant force on cylinder due to liquid displaced by submerged part is

$$F_B = V_{\sigma g} = \left(\frac{L}{2}A\right)\sigma g$$

For equilibrium, we use

$$kx_0 + F_B = Mg$$

$$\Rightarrow \quad kx_0 = Mg - \frac{LA\sigma g}{2}$$

$$\Rightarrow \quad x_0 = \frac{1}{k}\left(Mg - \frac{LA\sigma g}{2}\right)$$

$$\Rightarrow \quad x_0 = \frac{Mg}{k}\left(1 - \frac{\sigma LA}{2M}\right)$$ **Ans. (B)**

Sol. 4 For equilibrium of liquids pressure at any point of a liquid is same at one horizontal level.

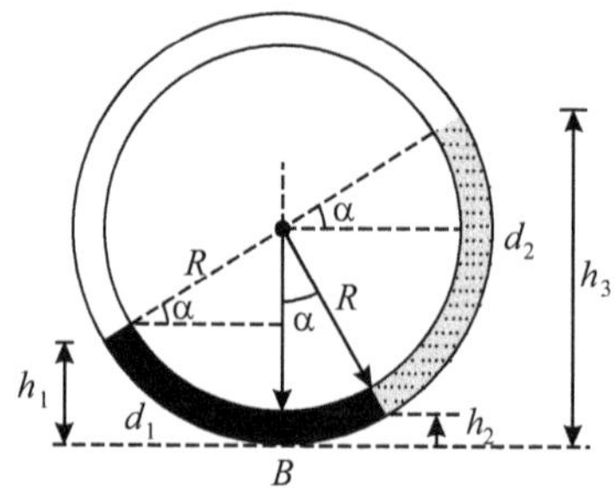

In figure if R is the radius of tube, we have

$$h_1 = R - R\sin\alpha$$

$$h_2 = R - R\cos\alpha$$

$$h_3 = R - R\sin\alpha$$

Pressure due to liquid at point B is given as

$$P_B = h_1 d_1 g$$

and $\quad P_B = (h_3 - h_2)d_2 g + h_2 d_1 g$

$$\Rightarrow \quad h_1 d_1 g = (h_3 - h_2)\, d_2 g + h_2 d_1 g$$

$$\Rightarrow \quad (h_1 - h_2)d_1 = (h_3 - h_2)d_2$$

$$\Rightarrow \quad \frac{d_1}{d_2} = \frac{h_3 - h_2}{h_1 - h_2} = \frac{\sin\alpha + \cos\alpha}{\cos\alpha - \sin\alpha}$$

$$\Rightarrow \quad \frac{d_1}{d_2} = \frac{1 + \tan\alpha}{1 - \tan\alpha}$$ **Ans. (C)**

Sol. 5 In initial state we use for floatation of block of volume v_B

$$V_B \rho_B g = \frac{4}{5} V_B \rho_w g$$

$$\Rightarrow \quad \rho_B = \frac{4}{5}\rho_w \quad \ldots(1)$$

Later after pouring oil over the water, we use

$$V_B \rho_B g = \frac{V_B}{2}\rho_w g + \frac{V_B}{2}\rho_0 g$$

$$\Rightarrow \quad \frac{4}{5}\rho_w = \frac{1}{2}\rho_w + \frac{1}{2}\rho_0$$

$$\Rightarrow \quad \frac{3}{10}\rho_w = \frac{1}{2}\rho_0$$

$$\Rightarrow \quad \frac{P_0}{\rho_w} = \frac{6}{10} = 0.6$$ **Ans. (C)**

Sol. 6 At a depth d_1, we have

$$P_{atm} + d_1\rho g = P_1 = 5.05 \times 10^6 \quad \ldots(1)$$

At a depth d_2, we have

$$P_{atm} + d_2\rho g = P_2 = 8.08 \times 10^6 \quad \ldots(2)$$

$$(2) - (1) \Rightarrow (d_2 - d_1)\,\rho g = 3.03 \times 10^6$$

$$\Rightarrow \quad d_2 - d_1 = \frac{3.03 \times 10^6}{10^3 \times 10}$$

$$= 3.03 \times 10^2 = 303\text{ m} \qquad \textbf{Ans. (C)}$$

Sol. 7

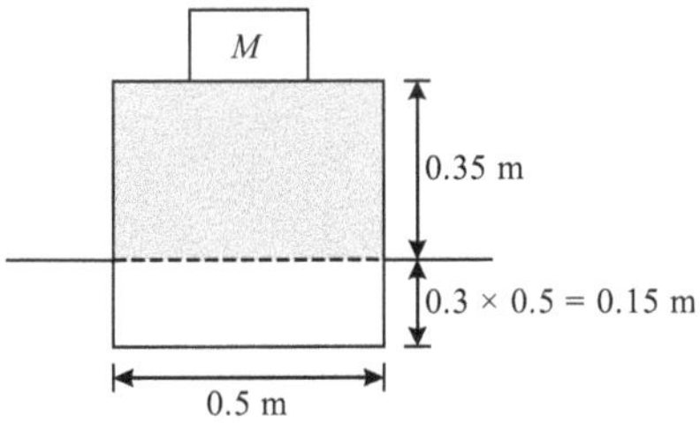

To fully submerge first submerge the complete block weight of mass M is to be balanced by buoyancy due to the 70% part of cubical block outside initially

$$\Rightarrow \quad Mg = (0.5 \times 0.5 \times 0.35) \times \rho_w \cdot g$$

$$\Rightarrow \quad M = (0.25 \times 0.35) \times 10^3 = 87.5\text{ kg} \qquad \textbf{Ans. (B)}$$

9.2 Fluid Dynamics

Sol. 1

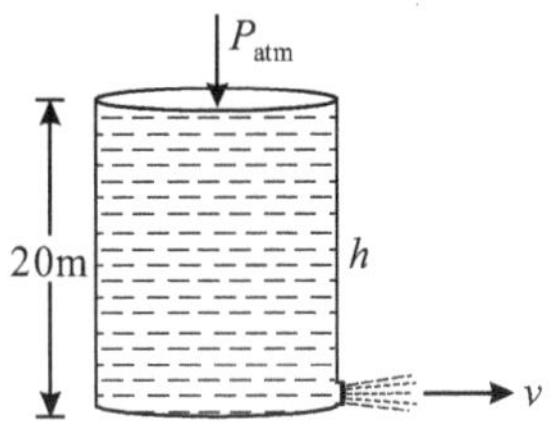

Efflux velocity $\quad v = \sqrt{2gh}$

$$\Rightarrow \quad v = \sqrt{2 \times 10 \times 20} = 20\text{ m/s} \qquad \textbf{Ans. (B)}$$

Sol. 2

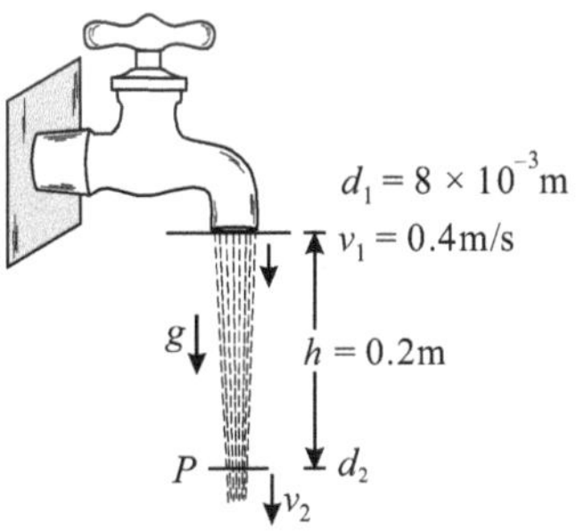

By speed equation for free fall motion of water stream we have

$$v_2^2 = v_1^2 + 2gh$$

$$\Rightarrow \quad v_2^2 = (0.4)^2 + 2 \times 10 \times 0.2 = 4.16$$

$$\Rightarrow \quad v_2 = \sqrt{4.16}\text{ m/s}$$

By continuity equation is stream of water, we use

$$A_1v_1 = A_2v_2$$

$$\left(\frac{\pi d_1^2}{4}\right)v_1 = \left(\frac{\pi d_2^2}{u}\right)v_2$$

$$d_2 = d_1\sqrt{\frac{v_1}{v_2}} = 8 \times 10^{-3} \times \sqrt{\frac{0.4}{4.16}}$$

$$\Rightarrow \quad d_2 \cong 3.6 \times 10^{-3} \qquad \textbf{Ans. (B)}$$

Sol. 3 Efflux velocity of water coming out from holes

$$v_1 = \sqrt{2gh}$$

and $$v_2 = \sqrt{2g(4h)} = 2\sqrt{2gh}$$

As flows rate from holes is same, we use

$$A_1v_1 = A_2v_2$$

$$\Rightarrow \quad l^2\sqrt{2gh} = \pi r^2 \times 2\sqrt{2gh}$$

$$\Rightarrow \quad r^2 = \frac{l^2}{2\pi}$$

$$\Rightarrow \quad r = \frac{l}{\sqrt{2\pi}} \qquad \textbf{Ans. (A)}$$

Sol. 4

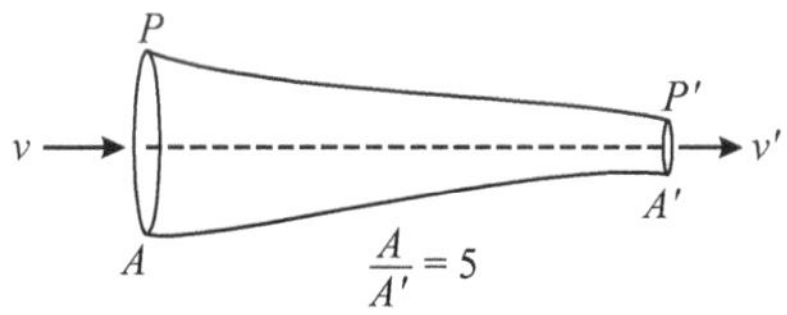

$$P - P' = 3 \times 10^5$$

Using Bernoulli's equation at the ends of tube we have

$$P + \frac{1}{2}\rho v^2 = P' + \frac{1}{2}\rho V'^2$$

$$\Rightarrow \quad P - P' = \frac{1}{2}\rho(v'^2 - v^2) \quad \ldots(1)$$

By continuity equation, we use

$$Av = A'v'$$

$$\Rightarrow \quad v' = \frac{A}{A'}v = 5v$$

From (1), we use

$$3 \times 10^5 = \frac{1}{2} \times 10^3 \times 24\,v^2$$

$$\Rightarrow \quad v = \sqrt{\frac{2 \times 3 \times 10^5}{10^3 \times 24}} = \sqrt{25} = 5\text{ m/s} \qquad \textbf{Ans. (A)}$$

Sol. 5 Efflux velocity of water from hole A is

$$v_1 = \sqrt{2gh}$$

Range of water from hole A is

$$R_A = v_1 \times \sqrt{\frac{2(H-h_1)}{g}}$$

$$R_A = 2\sqrt{h_1(H-h_1)} \quad \ldots(1)$$

Efflux velocity of water from hole B is

$$v_2 = \sqrt{2g(H-h_2)}$$

Range of water from hole B is

$$R_B = v_2 \times \sqrt{\frac{2h_2}{g}} = 2\sqrt{h_2(H-h_2)} \quad \ldots(2)$$

Given that $R_A = R_B$

$\Rightarrow \quad h_1(H-h_1) = h_2(H-h_2)$

$\Rightarrow \quad h_1 = h_2$ **Ans. (B)**

Sol. 6

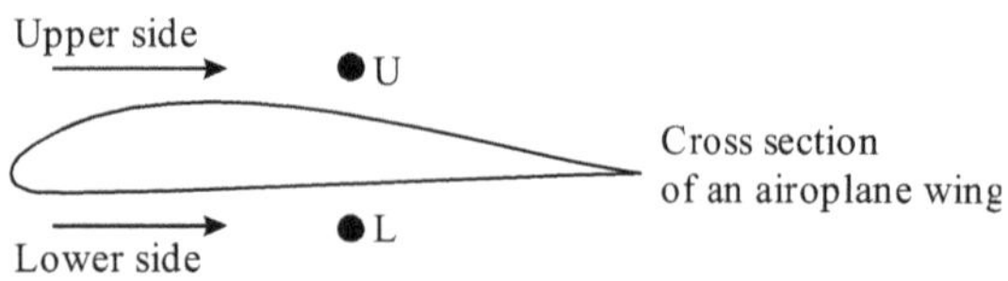

Wind velocity

$$v_U = 150 \text{ m/s}$$

$$v_L = 100 \text{ m/s}$$

If P_U and P_L are the air pressures at upper and lower sides of wings then we use Barnoullie's theorem, we have

$$P_U + \frac{1}{2}\rho v_U^2 = P_L + \rho v_L^2$$

$\Rightarrow \quad P_L - P_U = \frac{1}{2}\rho(v_U^2 - v_L^2)$

$\Rightarrow \quad P_L - P_U = \frac{1}{2} \times 1.2 \times ((150)^2 - (100)^2)$

$\Rightarrow \quad P_L - P_U = 0.6 \times 10^4 \times (2.25 - 1) = 6 \times 1.25 \times 103$

$\Rightarrow \quad P_L - P_U = 7500 \text{ N/m}^2$ **Ans. (C)**

Sol. 7

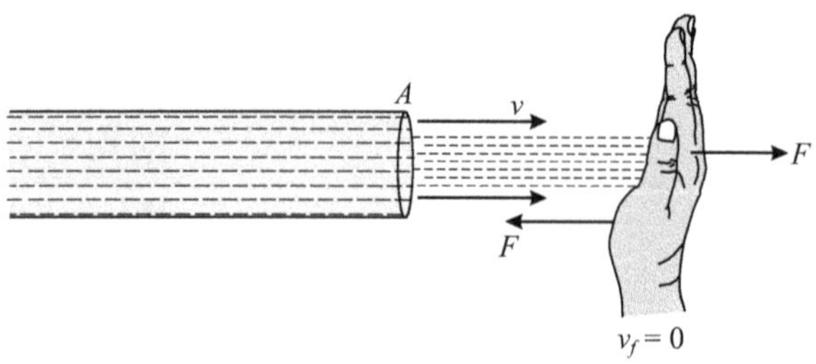

Water stops when hit the palm and it transfers momentum to the hand.

Force exerted = rate of change of momentum

= momentum in hitting water/sec

$$F = \rho\,(Av) \times V = \rho A v^2$$

$$= 10^3 \times 10^{-2} \times (1.5)^2$$

$\Rightarrow \quad F = 22.5 \text{ N}$ **Ans (B)**

$\rho \to$ density; $Av \to$ volume flow rate

Sol. 8

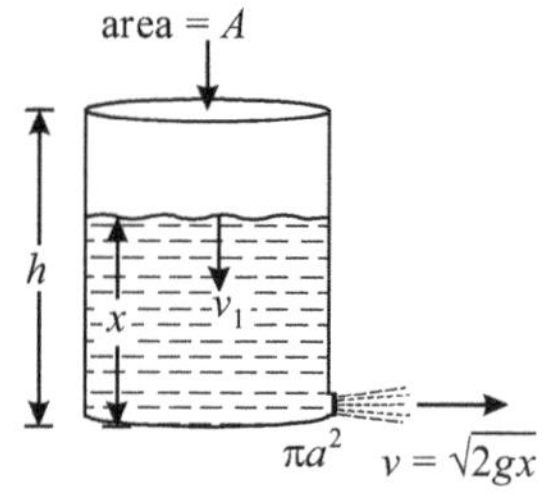

Here we can use $\quad v_1 = -\frac{dx}{dt}$

By equation of contains we use

$$V_1 A = \sqrt{2gx}\,.\,(\pi a^2)$$

$\Rightarrow \quad -A\frac{dx}{dt} = \pi a^2 \sqrt{2g}.\sqrt{x}$

$\Rightarrow \quad \frac{dx}{\sqrt{x}} = -\frac{\pi a^2}{A}\sqrt{2g}\,dt$

with time x is decreasing and if in time T if the vessel is emptied than we use

$$\int_{x=h}^{x=0} \frac{dx}{\sqrt{x}} = -\frac{\pi a^2}{A}\sqrt{2g}\int_{t=0}^{t=T} dt$$

$\Rightarrow \quad \left[2\sqrt{x}\right]_h^0 = -\frac{\pi a^2}{A}\sqrt{2g}\,[t]_0^T$

$\Rightarrow \quad 2\sqrt{h} = \frac{\pi a^2}{A}\sqrt{2g}\,.\,T$

$\Rightarrow \quad T = \frac{A}{\pi a^2}\sqrt{\frac{2h}{g}}$ **Ans. (B)**

Sol. 9

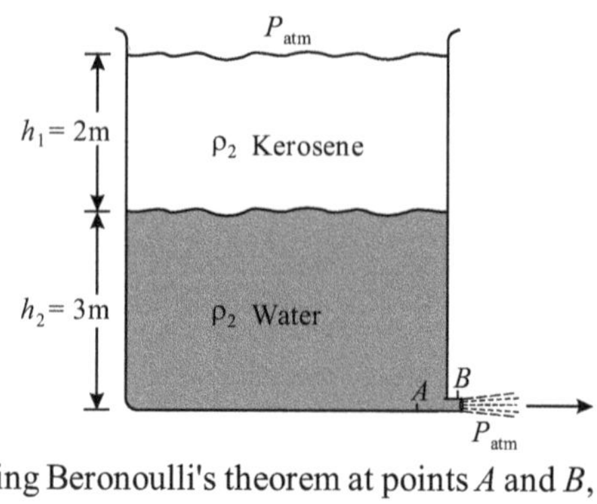

Applying Beronoulli's theorem at points A and B, we have

$$P_{atm} + h_1\rho_1 g + h_2\rho_2 g = P_{atm} + \frac{1}{2}\rho_2 v^2$$

$\Rightarrow \quad v = \sqrt{\frac{2h_1\rho_1 g}{\rho_2} + 2h_2 g}$

$\Rightarrow \quad v = \sqrt{\frac{2\times 2\times 800\times 10}{1000} + 2\times 3\times 10}$

$\Rightarrow \quad v = \sqrt{32+60} = \sqrt{92} = 9.59 \text{ m/s}$ **Ans. (B)**

Sol. 10 We apply Beronoulli's equation at A and B

$$P_A + \frac{1}{2}\rho v_A^2 = P_B + \frac{1}{2}\rho v_B^2$$

$$\Rightarrow \quad P_B - P_A = \rho\left(v_A^2 - v_B^2\right)$$

$$\Rightarrow \quad h\rho g = \frac{1}{2}\rho v_A^2\left(-\frac{A_A^2}{A_B^2}\right)$$

$$\Rightarrow \quad v_A = \sqrt{\frac{2gh}{\left(-\frac{A_A^2}{A_B^2}\right)}} = \sqrt{\frac{2\times 10\times 0.05}{1-(3/5)^2}}$$

$$\Rightarrow \quad v_A = \sqrt{\frac{1}{16/25}} = \frac{5}{4}\text{ m/s}$$

By continuity equation

$$A_A v_B = A_B v_B \quad \ldots(1)$$

We have $\quad P_B - P_A = h\rho g$

From (1) $\quad v_B = \dfrac{A_A v_A}{A_B}$

Fluid flow rate through tube is

$$\frac{dV}{dt} = A_A v_A = 6\times 10^{-6}\times 1.25$$

$$\Rightarrow \quad \frac{dV}{dt} = 7.5\times 10^{-6}\text{m}^3/\text{s}$$

Ans. (A)

Sol. 11 As $r << R$, the efflux velocity of water coming out from hole is given as

$$u = \sqrt{2gh}$$

If v is the speed at which it hits ground

$$\Rightarrow \quad v^2 = u^2 + 2gh = 2g(H+h)$$

$$\Rightarrow \quad v = \sqrt{2g(H+h)}$$

Now we use equation of continuity at hole and where water stream strikes the ground

$$\Rightarrow \quad A_1 v_1 = A_2 v_2$$

$$\Rightarrow \quad \pi r^2\sqrt{2gH} = \pi x^2\sqrt{2g(H+h)}$$

$$\Rightarrow \quad x = r\left(\frac{H}{H+h}\right)^{1/4}$$

Ans. (A)

Sol. 12 By equation of continuity, we use

In flow volume = Out flow volume

$$\Rightarrow \quad \frac{0.74}{60} = (\pi\times 4\times 10^{-4})\times\sqrt{2gh}$$

$$\Rightarrow \quad \sqrt{2gh} = \frac{74\times 100}{240\pi}$$

$$\Rightarrow \quad 2gh = \frac{740\times 740}{24\times 24\times 10}$$

$$\Rightarrow \quad h = \frac{74\times 74}{2\times 24\times 24}$$

$$\Rightarrow \quad h \approx 4.8\text{ m}$$

Ans. (B)

Sol. 13

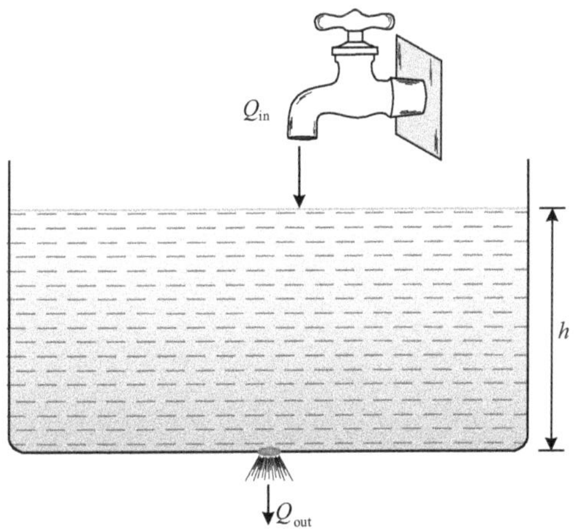

As height of water column is constant, we use water inflow rate (Q_{in}) = water outflow rate (Q_{out})

$$\Rightarrow \quad Q_{in} = 10^{-4}\text{ m}^3/\text{s}$$

$$\Rightarrow \quad Q_{out} = Av = 10^{-4}\times\sqrt{2gh}$$

$$\Rightarrow \quad 10^{-4} = 10^{-4}\sqrt{20\times h}$$

$$\Rightarrow \quad h = \frac{1}{20}\text{m} = 0.05\text{ m} = 5\text{ cm}$$

Ans. (D)

Sol. 14 Momentum per second carried by liquid per second is ρAv^2 where area of hose pipe is $A = \pi a^2$

Net force due to reflected liquid $= 2\times\left[\dfrac{1}{4}\rho Av^2\right]$

Net force due to stopped liquid $= \dfrac{1}{4}\rho Av^2$

Total force $F_{net} = \dfrac{3}{4}\rho Av^2$

Net pressure $= p = \dfrac{F_{net}}{A} = \dfrac{3}{4}\rho v^2$

Ans. (B)

Sol. 15

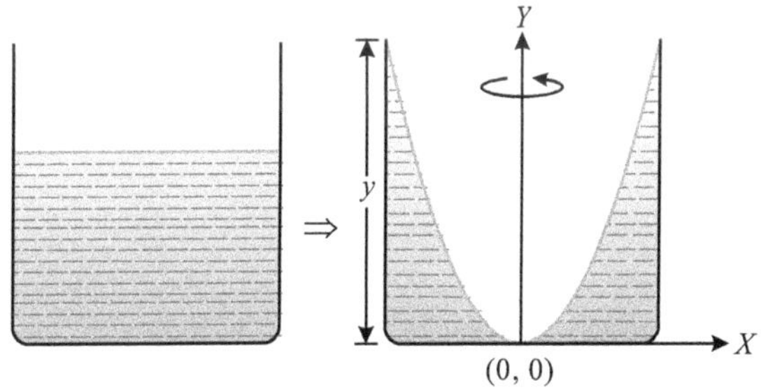

For the figure shown above equation of the liquid surface profile in a rotating container is given as

$$y = \frac{\omega^2 x^2}{2g} = \frac{(2\times 2\pi)^2\times(0.05)^2}{20} \simeq 2\text{ cm}$$

Ans. (C)

Sol. 16

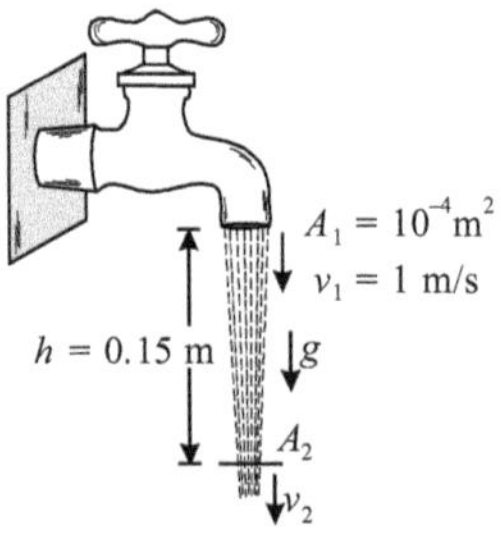

By equation of continuity we use

$$A_1v_1 = A_2v_2 \quad \dots(1)$$

By acceleration due to gravity, we have

$$v_2^2 = v_1^2 + 2gh$$

$$v_2 = \sqrt{(1)^2 + 2\times10\times0.15} = \sqrt{4} = 2 \text{ m/s}$$

From equation-(1)

$$A_2 = \frac{A_1v_1}{v_2} = \frac{10^{-4}\times1}{2} = 5\times10^{-5} \text{ m}^2$$ **Ans. (B)**

9.3 Viscosity

Sol. 1

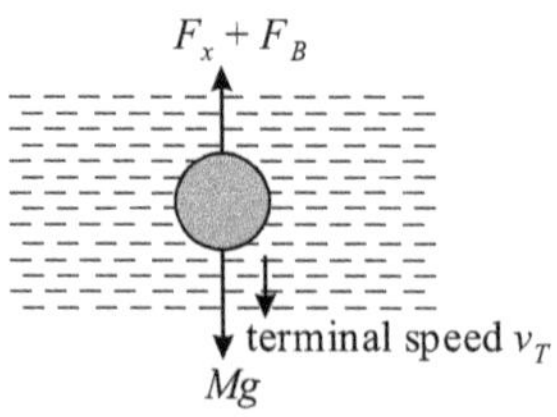

Using stoke's Rule we know before terminal speed is obtained, ball will have an acceleration which decreases because viscous force on ball is given as

$$F_v = 6\pi\eta Rv$$

$$\Rightarrow \quad F_v \propto Rv$$ **Ans. (B)**

Sol. 2 In a liquid, terminal speed of sphere is given as

$$v_T = \frac{2}{9}\frac{gR^2}{\eta}(\rho_s - \rho_l)$$

For gold sphere, we have

$$v_1 = \frac{2}{9}\frac{9R^2}{\eta}(19.5 - 1.5) \quad \dots(1)$$

For silver sphere, we have

$$v_2 = \frac{2}{9}\frac{gR^2}{\eta}(10.5 - 1.5) \quad \dots(2)$$

Using (1)/(2) gives

$$\frac{v_1}{v_2} = \frac{18}{9} = 2$$

$$\Rightarrow \quad v_2 = \frac{v_1}{2} = \frac{0.2}{2} = 0.1 \text{ m/s}$$ **Ans. (C)**

Sol. 3

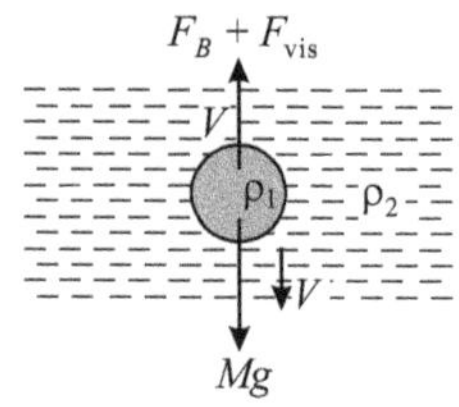

Terminal speed is attained by ball when its acceleration $a = 0$

$$\Rightarrow \quad F_B + F_{vis} = mg$$

[F_{vis} is in opposite direction of v so need to consider-ve sign again]

$$\Rightarrow \quad V\rho_2g + kv^2 = V\rho_1g$$

$$\Rightarrow \quad kv^2 = V(\rho_1 - \rho_2)g$$

$$\Rightarrow \quad v = \sqrt{\frac{V(\rho_1 - \rho_2)g}{k}}$$ **Ans. (A)**

Sol. 4 Terminal velocity of a ball in a viscous liquid is given as

$$v = \frac{2}{9}\frac{gR^2}{\eta}(\rho_S - \rho_L)$$

for ball falling in water

$$10 = v_w = \frac{2}{9}\frac{gR^2}{8.5\times10^{-4}}\times(7.8 - 1) \quad \dots(1)$$

for ball falling in glycerine

$$v_g = \frac{2}{9}\frac{gR^2}{13.2}\times(7.8 - 1.2) \quad \dots(2)$$

From (2)/(1) we have

$$\frac{v_g}{10} = \frac{8.5\times10^{-4}}{13.2}\times\left(\frac{7.8-1.2}{7.8-1}\right)$$

$$\Rightarrow \quad v_g = 6.25\times10^{-4} \text{ cm/s}$$ **Ans. (B)**

Sol. 5

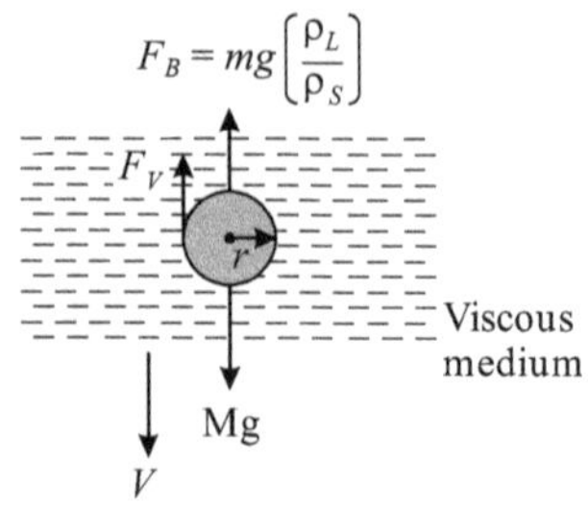

Upward viscous force on ball is given by stoke's rule as

$$F_v = 6\pi\eta rv$$

At terminal speed, all forces on sphere are balanced, so we have

$$mg = mg\left(\frac{\rho_L}{\rho_S}\right) + 6\pi\eta rv$$

$$\Rightarrow \quad \rho_3\left(\frac{4}{3}\pi r^3\right)g\left(1 - \frac{\rho_L}{\rho_S}\right) = 6\pi\eta rv$$

$$\Rightarrow \quad v = \frac{2}{9}\frac{gr^2}{\eta}(\rho_S - \rho_L)$$

$$\Rightarrow \quad v \propto r^2$$ **Ans. (A)**

Sol. 6 Case-1

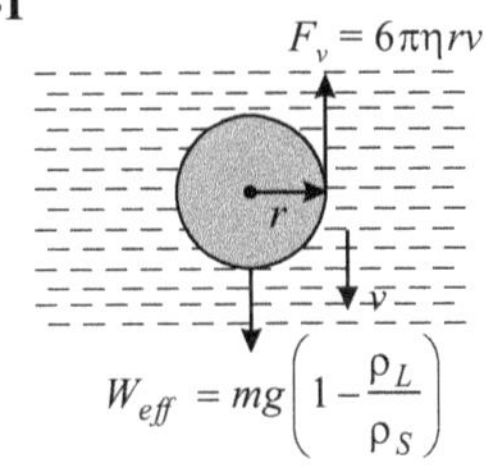

At terminal speed of ball

$$W_{eff} = 6\pi\eta rv \qquad \ldots(1)$$

When ball is pulled upward at $F_{ext} = 2W_{eff}$ then at terminal speed v', we have

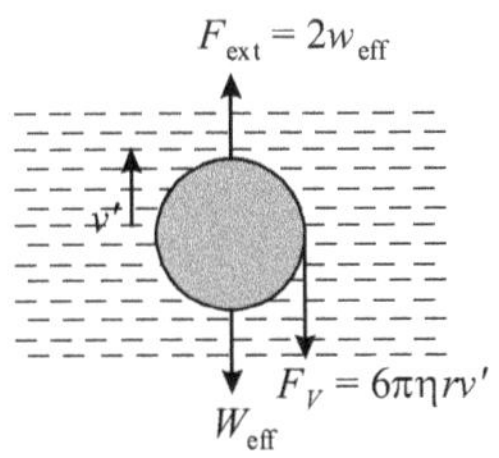

$$2W_{eff} = W_{eff} + 6\pi\eta rv'$$

$$\Rightarrow \quad W_{eff} = 6\pi\eta rv' \qquad \ldots(2)$$

from (1) and (2)

$$\Rightarrow \quad v' = v = 10 \text{ cm/s}$$ **Ans. (C)**

Sol. 7

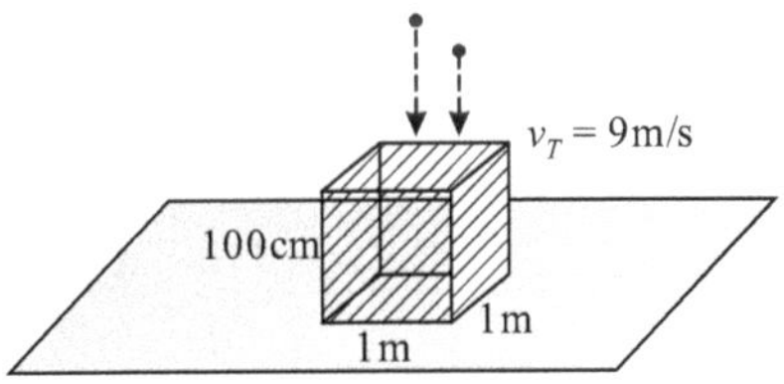

Total calculate volume of rain falling on the place per square meter in one year is

$$V = 1 \times 1 \times 100 \text{ cm} = 1 \text{ m}^3$$

KE of the rain volume which becomes zero on striking at ground is

$$\text{KE} = \frac{1}{2}(PV).\, V^2 = \frac{1}{2} \times 10^3 \times (9)^2$$

$$\Rightarrow \quad \text{KE} = \frac{81}{2} \times 1000$$

$$\Rightarrow \quad \text{KE} = 4.05 \times 10^4 \text{J}$$ **Ans. (B)**

Sol. 8

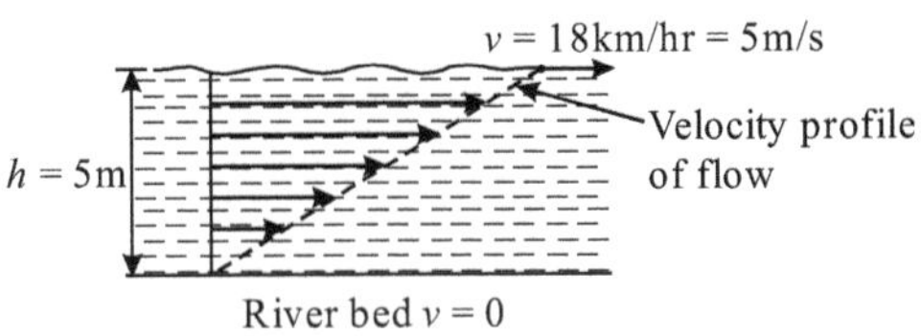

Viscous force between layers of river water is given as

$$F_v = \eta A \frac{dv}{d\pi} = \eta A\left(\frac{\Delta v}{h}\right)$$

(we assume velocity profile to be linear as shown in figure)

Shearing stress between layers of flowing water is given as

$$\text{shearing stress} = \frac{F_V}{A} = \eta\left(\frac{\Delta v}{n}\right) = \eta\left(\frac{v-0}{h}\right)$$

$$\Rightarrow \quad \sigma_s = 10^{-3} \times \frac{5}{5} = 10^{-3} \text{N/m}^2$$

Using 1 poise $= \frac{1}{10}$ poiseuille $= \frac{1}{10}$ N·s/m² **Ans. (C)**

Sol. 9 For a liquid flow in a pipeline, Reynolds number is given as

$$R = \frac{PvD}{\eta}$$

$v \rightarrow$ flow velocity

$P \rightarrow$ density

$D \rightarrow$ internal daintier of pipeline

The volume flow rate of water through tap is given as

$$Av = \frac{V}{t} = \frac{15\times10^3}{5\times60} \text{ cm}^3\text{/s}$$

$$A = \pi r^2$$

Av = Cross-section area of tap

$$\Rightarrow \quad v = \frac{15\times10^3}{300\times\pi\times\left(\frac{1}{\sqrt{\pi}}\right)^2} = 50 \text{ cm/s} = 0.5 \text{ m/s}$$

$\Rightarrow$ Reynolds number is given as

$$R = \frac{pvd}{\eta} = \frac{10^3\times0.5\times\frac{2}{\sqrt{3.14}}\times10^{-2}}{10^{-3}}$$

$$R = \frac{1}{\sqrt{3.14}} \times 10^4 = 5643$$ **Ans. (D)**

Sol. 10

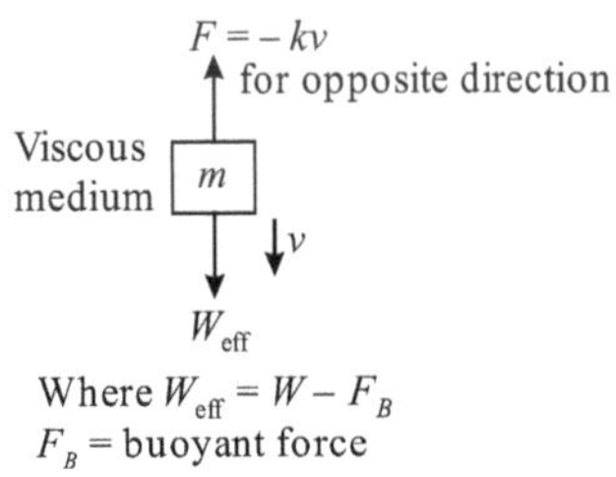

If a be the acceleration of mass, we use

$$W_{eff} - kv = ma$$

$$\Rightarrow \quad a = \frac{W_{eff} - kv}{m}$$

[If we put $a = \frac{dv}{dt}$ and integrative then it given $v(t)$ as exponent]

Here we can see that as $v\uparrow \Rightarrow a\downarrow$ and after some (long time) time when $v=\frac{W_{eff}}{k}$ (constant) $\Rightarrow a=0$ **Ans. (C)**

Sol. 11

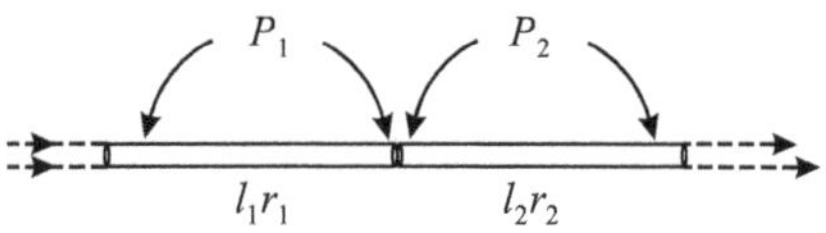

For viscous fluid flowing in a capillary tube we use

flow rate $$\frac{dQ}{dt}=\frac{\pi r^4}{8\eta l}\Delta P$$

As tubes in this case are in series, so we use

$$\left.\frac{dQ}{dt}\right|_1=\left.\frac{dQ}{dt}\right|_2$$

$$\frac{\pi r_1^4}{8\eta l_1}P_1=\frac{\pi r_2^4}{8\eta l_2}.P_2$$

$$\Rightarrow \quad \frac{r_1^4}{l_1}P_1=\frac{r_2^4}{l_1/4}(4P_1)$$

$$\Rightarrow \quad r_1^4=16r_2^4$$

$$\Rightarrow \quad r_2=\left(\frac{r_1^4}{16}\right)^{1/4}=\frac{r_1}{2}$$ **Ans. (D)**

Sol. 12 We have

$$V_T=\frac{2}{9}\left(\frac{r^2g}{\eta}\right)(\rho_0-\rho_l)$$

$$\Rightarrow \quad v_T\propto r^2$$

As mass of sphere will remain same before and after breaking, we have

$$\Rightarrow \quad \rho\left(\frac{4}{3}\pi R^3\right)=\rho\left(\frac{4}{3}\pi R^2\right)\times 27$$

$$\Rightarrow \quad r=\frac{R}{3}$$

$$\Rightarrow \quad \frac{v_1}{v_2}=\frac{R^2}{r^2}=9$$ **An.s (D)**

9.4 Surface Tension

Sol. 1

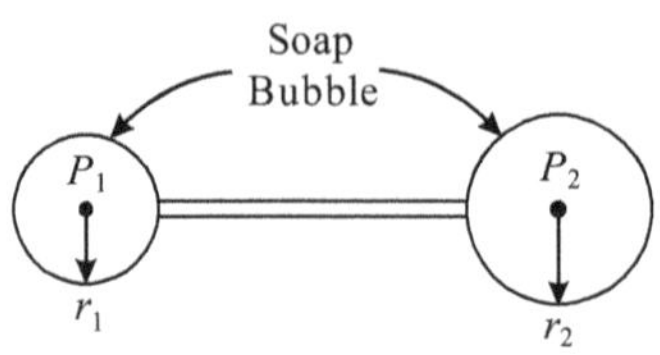

Excess pressure inside a soap bubble of radius r is given as-

$$P_{ex}=\frac{4T}{r}$$

Where $T\rightarrow$ surface tension of soap solution.

Excess pressure inside the two given soap bubble are

$$P_1=\frac{4T}{r_1} \text{ and } P_2=\frac{4T}{r_2}$$

As $r_1<r_2\Rightarrow P_1>P_2$

$\Rightarrow$ Air flows from smaller bubble to bigger bubble **Ans. (C)**

Sol. 2

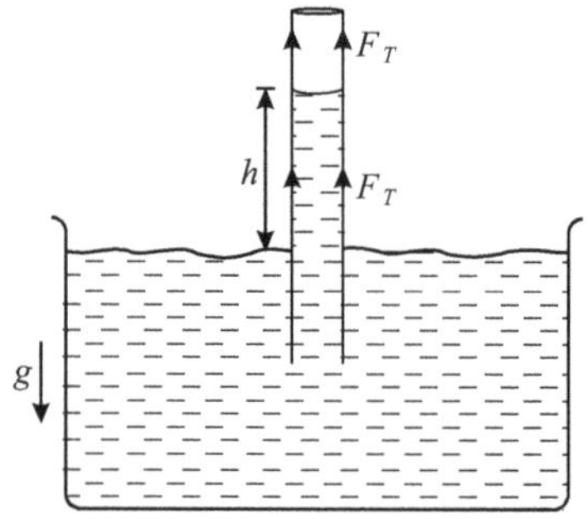

Inside capillary liquid is raised upto the level where force of surface tension balances the weight of liquid raised.

In a free falling state $g_{eff}=0$

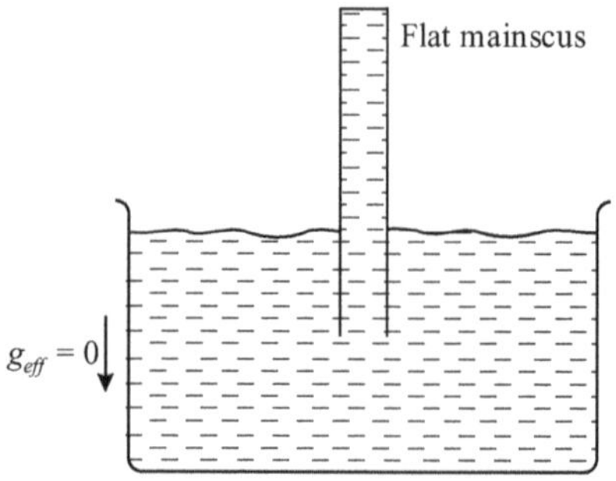

$\Rightarrow$ Effective weight of liquid raised $=0$

$\Rightarrow$ F_T(upward) $=0\Rightarrow$ liquid will fill the complete height of capillary. **Ans. (D)**

Sol. 3 As $T_{\text{soap sol.}}<T_{\text{water}}$

$\Rightarrow$ Capillary height $h=\frac{2T\cos\theta}{r\rho g}$

$\Rightarrow \quad h_B<h_A$ **Ans. (C)**

Sol. 4

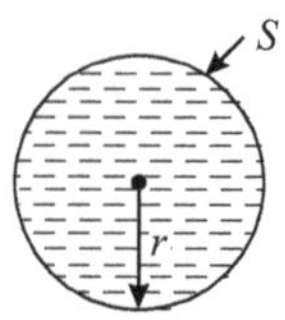

A liquid drop has only one surface of area outside

$$S=4\pi r^2$$

When two mercury drops merge to form a bigger drop radius R then we have

$$2\times\frac{4}{3}\pi r^3=\frac{4}{3}\pi R^3$$

$$\Rightarrow \quad R=2^{1/3}r$$

Surface energy of bigger drop

$$E = T \times S$$

$$\Rightarrow \quad E = T(4\pi R^2)$$

$$\Rightarrow \quad E = 4\pi T(2^{1/3}r)^2$$

$$\Rightarrow \quad E = 2^{8/3}\pi r^2 T$$

Ans. (D)

Sol. 5

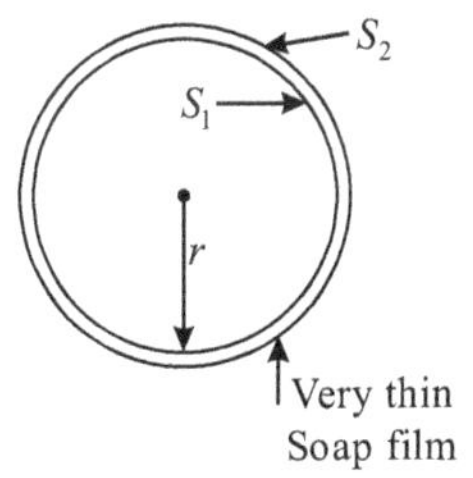

Soap bubble has the surface S_1 and S_2 which carries surface energy, given as

$$E = T \times 2(4\pi r^2) = 8\pi T r^2$$

If radius of bubble is increased from r_1 to r_2 than work done in this process is

$$W = E_f - E_i = 8\pi T\,(r_2^2 - r_1^2)$$

$$\Rightarrow \quad W = 8\pi \times 0.03\,[(0.05)^2 - (0.03)^2]$$

$$\Rightarrow \quad W = 0.384\,\pi\,\text{mJ} \sim 0.4\,\pi\,\text{mJ}$$

Ans. (B)

Sol. 6

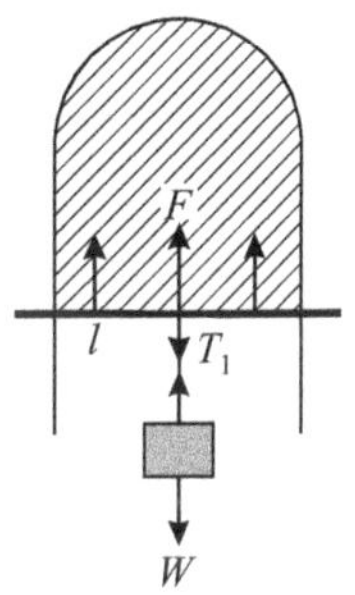

If surface tension of liquid is T than total upward force on slider due to two surfaces of the film is

$$F = 2Tl = w$$

$$\Rightarrow \quad T = \frac{w}{2l} = \frac{1.5\times 10^{-2}}{2\times 0.3} = 0.025\ \text{N/m}$$

Ans. (A)

Sol. 7 If n small drops coalesce to from the big drop of radius R, we have

$$n \times \frac{4}{3}\pi r^3 = \frac{4}{3}\pi R^3 \Rightarrow n = \frac{R^3}{r^3}$$

Amount of surface energy released in process

$$\Delta E = n(4\pi r^2 T) - 4\pi R^2 T$$

$$\Delta E = 4\pi T\left[r^2 \cdot \frac{R^3}{r^2} - R^2\right] = 4\pi T R^3\left[\frac{1}{r} - \frac{1}{R}\right]$$

If v is the speed attained by big drop then we have

$$\Delta E = \frac{1}{2} m v^2$$

$$\frac{1}{2}\left(\rho \times \frac{4}{3}\pi R^3\right)v^2 = 4\pi T R^3\left[\frac{1}{r} - \frac{1}{R}\right]$$

$$\Rightarrow \quad v = \sqrt{\frac{6T}{\rho}\left[\frac{1}{r} - \frac{1}{R}\right]}$$

Ans. (D)

Sol. 8 Due to surface energy the outer layer (very thin) of a liquid drop if evaporates by absorption of liquid heat, whole drop can get evaporated at same temperature.

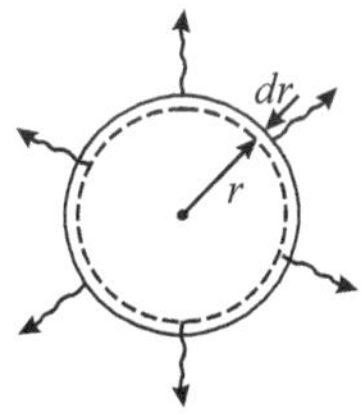

Due to evaporation of thin layer of liquid at the surface of drop the change in surface energy is

$$\Delta E = T(4\pi(r + dr)^2) - T(4\pi r^2)$$

$$\Rightarrow \quad \Delta E = T(8\pi r dr)$$

for evaporation of the thin elemental layer of liquid of width dr and out of heat to be absorbed by it is

$$dQ = dm\,L \qquad dm \to \text{mass of layer}$$

$$\Rightarrow \quad dQ = (\rho dv)L \qquad dv \to \text{volume of layer}$$

$$\Rightarrow \quad dQ = \rho L(4\pi r^2\,dr) \qquad [dv = 4\pi r^2\,dr]$$

The drop gets evaporated only by its surface energy if

$$\Delta E = dQ$$

$$T(8\pi r dr) = \rho L(4\pi r^2 dr)$$

$$\Rightarrow \quad r = \frac{2T}{\rho L}$$

Ans. (C)

Sol. 9 Capillary height in tube is given as

$$h = \frac{2T\cos\theta}{r\rho g}$$

As with temperature surface temperature describes

$\Rightarrow$ as temperature interface, h decreases

S-1 Is false

S-2 Is true as it is a fact studied in theory **Ans. (B)**

Sol. 10

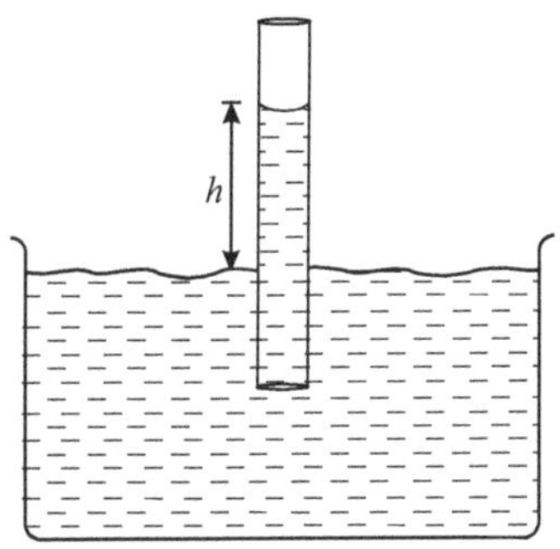

Capillary height is given as

$$h=\frac{2T\cos\theta}{r\rho g}$$

$\theta < 90°$ for glass- water interface as water wets the glass surface. When capillary is coated with wax which is not wet by water so angle of contact for wax - water surface is $\theta > 90°$

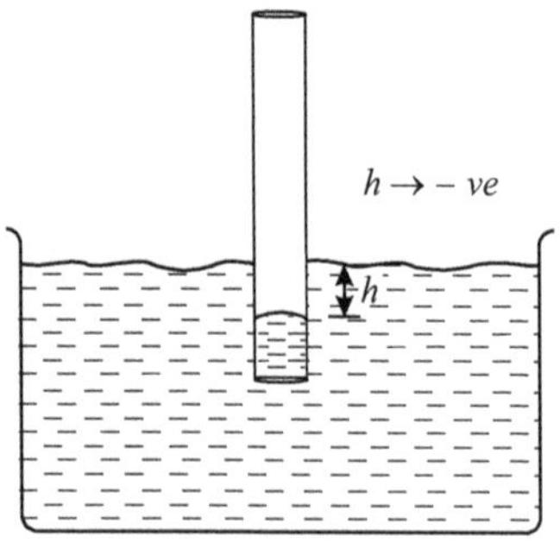

$\Rightarrow$ $\cos\theta = -ve$

$\Rightarrow$ $h = -ve$ **Ans. (C)**

Sol. 11 The excess pressure inside soap bubble is

$$P_{ex}=\frac{2T}{R}$$

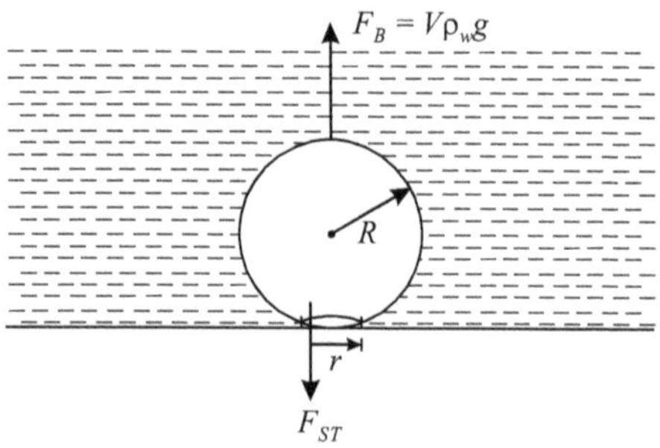

Force of surface tension at bottom on bubble is

$$F_{ST}=P_{ex}\times\pi r^2$$

At the moment spherical bubble detach from base, we use

$$F_B=F_{ST}$$

$\Rightarrow$ $$\frac{4}{3}\pi R^3\rho_\omega g=\frac{2T}{R}\times\pi r^2$$

$\Rightarrow$ $$r=R^2\sqrt{\frac{2\rho_\omega g}{3T}}$$ **Ans. (A)**

Sol. 12

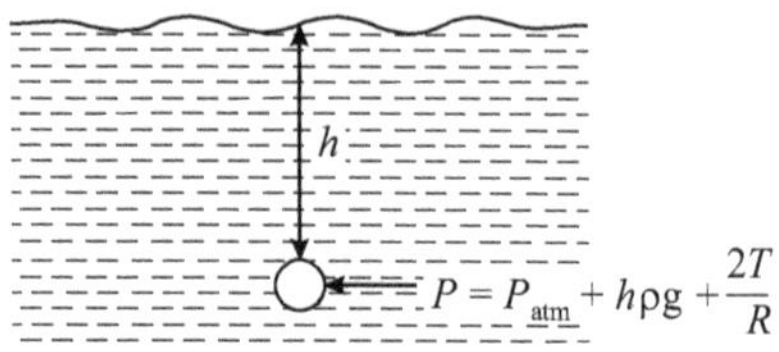

Inside the bubble excess pressure is given as

$$P_{ex}=P-P_{atm}=h\rho g+\frac{2T}{R}=1100$$

$\Rightarrow$ $$h\times1000\times9.8+\frac{2\times0.06}{10^{-3}}=1100$$

$\Rightarrow$ $$9800h+120=1100$$

$\Rightarrow$ $$h=\frac{980}{9800}=0.1\text{ m}$$ **Ans. (A)**

Sol. 13

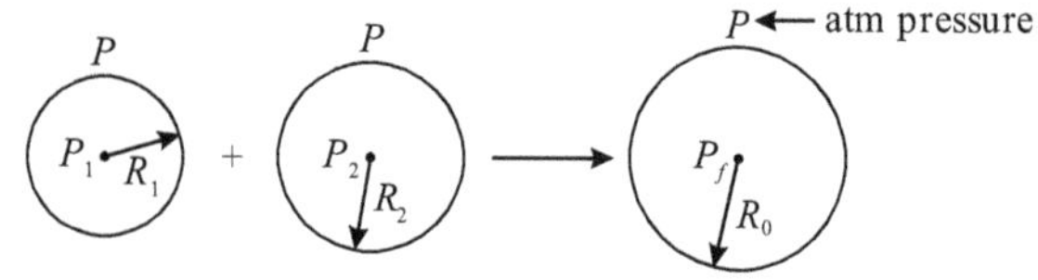

$V_1=\frac{4}{3}\pi R_1^3$ $\quad V_1=\frac{4}{3}\pi R_2^3$ $\quad V_0=\frac{4}{3}\pi R_0^3$

$P_1=P+\frac{4T}{R_1}$ $\quad P_2=P+\frac{4T}{R_2}$ $\quad P_f=P+\frac{4T}{R_0}$

If temperature is contained constant then we use

$$P_1V_1+P_2V_2=P_fV_0$$

$\Rightarrow$ $$\left(P+\frac{4T}{R_1}\right)\left(\frac{4}{3}\pi R_1^3\right)+\left(P+\frac{4T}{R_2}\right)\left(\frac{4}{3}\pi R_2^3\right)=\left(P+\frac{4T}{R_o}\right)\left(\frac{4}{3}\pi R_o^3\right)$$

$$\Rightarrow P\left[\frac{4}{3}\pi R_1^3+\frac{4}{3}\pi R_2^3-\frac{4}{3}\pi R_o^3\right]+\frac{4T}{3}(4\pi R_1^2)+\frac{4T}{3}(4\pi R_2^2)-\frac{4T}{3}(4\pi R_o^2)=0$$

$\Rightarrow$ $$PV+\frac{4T}{3}S=0$$

$\Rightarrow$ $$3PV+4TS=0$$ **Ans. (B)**

Sol. 14

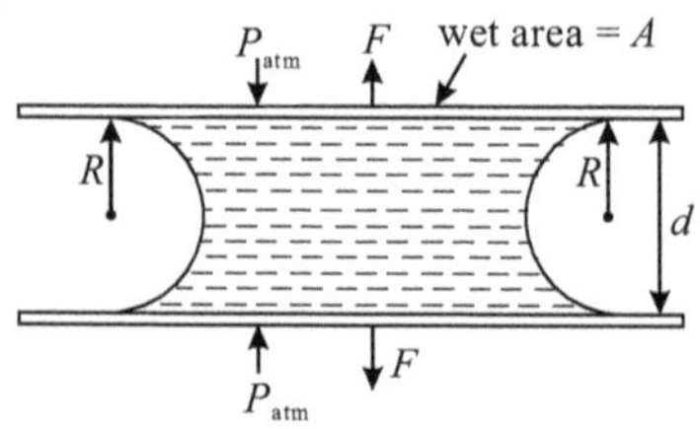

To pull apart the plates, the required force is given as

$$F=\frac{2TA}{d}=\frac{2TA}{2R}=\frac{TA}{R}$$

The pressure in water will be lower by

$$\Delta P_L=\frac{F}{A}=\frac{T}{R}$$ **Ans. (D)**

Sol. 15 Pressure inside a soap bubble of radius r is given as

$$P_{in}=P_{atm}+\frac{4T}{r} \qquad \ldots(1)$$

If volume of soap bubble increases at constant rate, we have

$$V\propto t$$

$\Rightarrow$ $$r\propto t^{1/3}$$

$\Rightarrow$ $$r=kt^{1/3}$$

From equation-(1), we have

$$P_{in} = P_{atm} + \frac{4T}{kt^{1/3}}$$

Thus graph between P and $\frac{1}{t^{1/3}}$ is a straight line with positive slope and positive intercept. **Ans. (D)**

Sol. 16 Height of liquid rise in capillary tube

$$h = \frac{2T\cos\theta_C}{\rho r g}$$

$$\Rightarrow \quad h \propto \frac{1}{r}$$

When radius becomes double height become half

$$\Rightarrow \quad h' = \frac{h}{2}$$

Now, $\quad M = \pi r^2 h \times \rho$

and $\quad M' = \pi(2r)^2(h/2) \times \rho = 2M$ **Ans. (C)**

Sol. 17 Capillary height is given as

$$h = \frac{2T\cos\theta}{r\rho g}$$

For H_g, we use $h = \frac{2T_{Hg}\cos\theta_{Hg}}{r_1\rho_{Hg}g}$

and for water, we use

$$h = \frac{2T_w\cos\theta_w}{r_2\rho_w g}$$

$$\Rightarrow \quad \frac{T_{Hg}\cos\theta_{Hg}}{r_1\rho_{Hg}} = \frac{T_w\cos\theta_w}{r_2\rho_w}$$

$$\Rightarrow \quad \frac{r_1}{r_2} = \left|\frac{T_{Hg}\cos\theta_H}{T_w\cos\theta_w} \times \frac{\rho_w}{\rho_{Hg}}\right|$$

$$\Rightarrow \quad \frac{r_1}{r_2} = \frac{7.5 \times 1/\sqrt{2}}{13.6}$$

$$\Rightarrow \quad \frac{r_1}{r_2} = \frac{5}{9\sqrt{2}} = \frac{5}{12.72} \simeq \frac{2}{5}$$ **Ans. (C)**

10.1 Thermal Expansion

Sol. 1

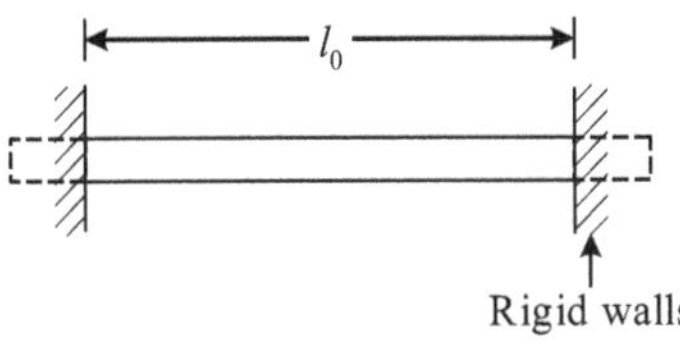

If on rising temperature by t°C rod expands by Δl then due to rigid walls, it is compressed by the same length elastically strain developed in rod due to this is

$$\text{strain} = \frac{\Delta l}{l} = \frac{\alpha l \Delta t}{l} = \alpha\Delta t = \alpha t$$

$\Rightarrow$ By definition of Young's modulus

$$Y = \frac{\text{Stress}}{\text{Strain}}$$

$\Rightarrow \quad$ linear stress $= Y \times \text{Strain} = Y\alpha t$ **Ans. (C)**

Sol. 2 If a body of volume V is heated by ΔT then volume expansion of body is given as

$$\Delta V = \gamma V \Delta T$$

Coefficient of volume expansion $\gamma = 3\alpha$

$$\Rightarrow \quad \Delta V = 3\alpha V\Delta T = 3\alpha\left(\frac{4}{3}\pi r^3\right)\Delta T$$

$$= 3 \times 23 \times 10^{-6} \times \left(\frac{4}{3} \times 3.14 \times (10)^3\right) \times 100$$

$$= 28.88 \text{ cm}^3 \sim 28.9 \text{ cc}$$ **Ans. (A)**

Sol. 3 Here we consider wood is having negligible expansion coefficient compared to metal so on rising temperature by ΔT, strain due to thermal expansion which is compensated by elastic compression on cooling down is given as

$$\text{Strain} = \frac{\Delta l}{l} = \alpha\Delta T$$

$$\Rightarrow \quad \text{Stress} = \frac{F}{S} = Y \times \text{Strain} = Y\alpha\Delta T$$

force between two parts of wheel is

$$2F = 2YS\alpha\Delta T$$

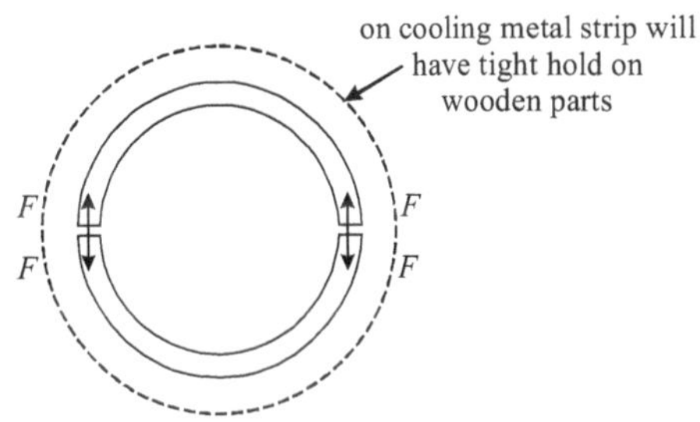

Ans. (A)

Sol. 4 For all linear scales, we use

$$\frac{\text{Reading} - LFP}{UFP - LFP} = \text{Constant}$$

For Kelvin scale & Y scale we use

$$\frac{Y - (-160)}{-50 - (-160)} = \frac{340 - 273}{373 - 273} = \frac{67}{100}$$

$$Y + 160 = \frac{67}{100} \times 110 = 73.7$$

$$\Rightarrow \quad Y = 73.7 - 160 = -86.3°Y$$ **Ans. (C)**

Sol. 5 Given that

$$\frac{\gamma_G}{\gamma_L} = \frac{1}{4}$$

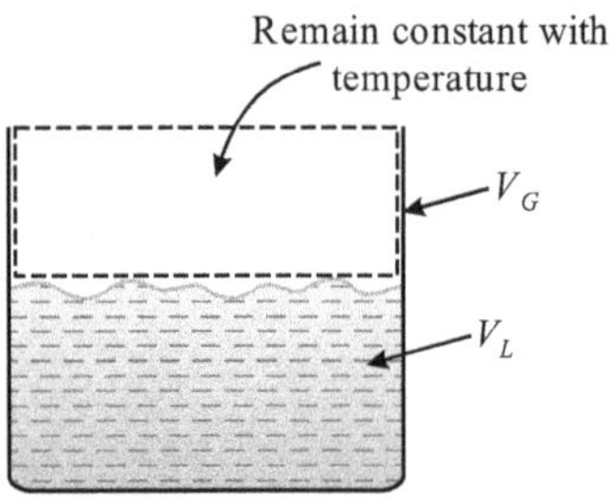

$\Rightarrow$ For any rise in temperature (say ΔT) the expansion in container is equal to expansion in liquid volume.

$$\Rightarrow \quad \Delta V_G = \Delta V_L$$

$$\Rightarrow \quad \gamma_G V_G \Delta T = \gamma_L V_L \Delta T$$

$$\Rightarrow \quad \frac{V_L}{V_G} = \frac{\gamma_G}{\gamma_L} = \frac{1}{4}$$ **Ans. (B)**

Sol. 6 On the ends of steel wire if F is the force needed to keep its length constant then stress in wire is

$$\text{Stress} = \frac{F}{A} = Y \times \text{Strain}$$

$$\text{Strain} = \frac{\Delta l}{l} = \alpha \Delta T$$

$$\Rightarrow \quad \text{Stress} = \frac{F}{A} = Y\alpha\Delta T$$

pressure required

$$p = 2 \times 10^{11} \times 1.1 \times 10^{-5} \times 100$$

$$\Rightarrow \quad p = 2.2 \times 10^8 \text{ Pa}$$ **Ans. (A)**

Sol. 7. Fractional time lost/second by a pendulum clock is given as

$$\frac{\delta t}{t} = \frac{1}{2}\alpha\Delta T$$

Case-1 : When $T = 40°C$ and we consider T_g is graduation temperature of clock when it reads correct time then we have

$$\delta t = 12 \text{ s}$$

$$t = 1 \text{ day} \equiv 86400 \text{ s}$$

$$\Rightarrow \quad \frac{12}{86400} = \frac{1}{2}\alpha(40 - T_g) \quad \ldots(1)$$

Case-2 :

$$\delta t = 4\text{s}$$

$$t = 1 \text{ day}$$

$$\Rightarrow \quad \frac{4}{86400} = \frac{1}{2}\alpha(T_g - 20) \quad \ldots(2)$$

using $\frac{(1)}{(2)}$ gives

$$\frac{12}{4} = \frac{40 - T_g}{T_g - 20}$$

$$\Rightarrow \quad 3T_g - 60 = 40 - T_g$$

$$\Rightarrow \quad 4T_g = 100 \quad \Rightarrow \quad T_g = 25°C$$

From equation-(1)

$$\frac{12}{86400} = \frac{1}{2}\alpha(40 - 25)$$

$$\Rightarrow \quad \alpha = \frac{12 \times 2}{86400 \times 15} = 1.85 \times 10^{-5} \, °C^{-1}$$ **Ans. (A)**

Sol. 8 For a cube of volume V, if on heating its volume increases by ΔV then we use for ΔT rise in temperature

$$\Delta V = V\gamma\Delta T = V(3\alpha)\Delta T \quad \ldots(1)$$

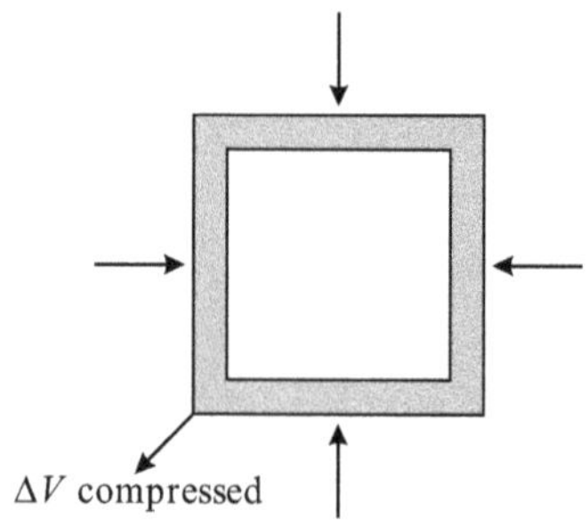

On applying pressure to cube if its volume is compressed by ΔV then we have

Bulk modulus

$$K = \frac{P}{\left|\frac{\Delta V}{V}\right|} = \frac{P}{3\alpha\Delta T}$$

$$\Rightarrow \quad \Delta T = \frac{P}{3\alpha K}$$

for same size ΔV increased (due to thermal expansion).

Ans. (C)

Sol. 9

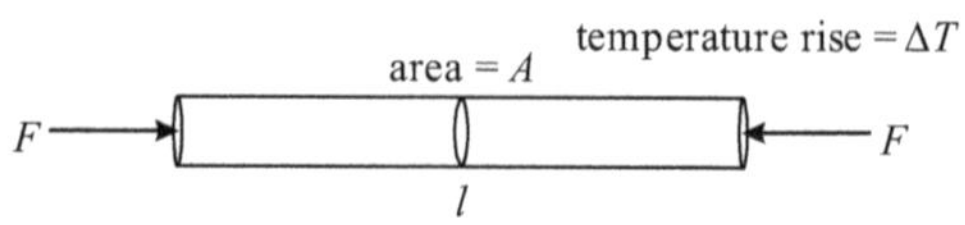

As length of rod $\rightarrow$ constant

$\Rightarrow$ Thermal expansion due to rise in temperature is compensated by elastic compression due to compressive force F.

Here we use

$$Y = \frac{\text{Stress}}{\text{Strain}} = \frac{F/A}{\Delta l / l}$$

Thermal expansion

$$\Delta l = \alpha l \Delta T$$

$$\Rightarrow \quad Y = \frac{F/A}{\alpha\Delta T}$$

$$\Rightarrow \quad F = YA\alpha\Delta T$$ **Ans. (C)**

Sol. 10 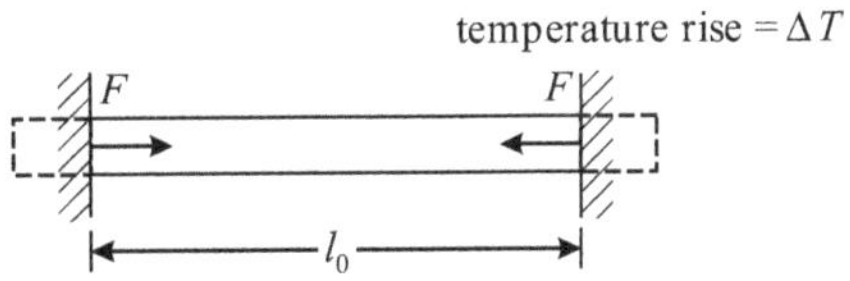

Due to the external force at ends, then thermal expansion in rail due to rise in temperature in compensated by elastic compression due to external force.

$$\Rightarrow \quad Y = \frac{\text{Stress}}{\text{Strain}} = \frac{F/A}{\Delta l / l_0}$$

Thermal expansion

$$\Delta l = \alpha l_0 \Delta T$$

$$\Rightarrow \quad Y = \frac{F/A}{\alpha \Delta T}$$

$$\Rightarrow \quad F = YA\alpha\Delta T$$

$$\Rightarrow \quad F = 2 \times 10^{11} \times 40 \times 10^{-4} \times 1.2 \times 10^{-5} \times 10$$

$$\Rightarrow \quad F = 9.6 \times 10^4 \text{ N}$$

$$\Rightarrow \quad F \simeq 1 \times 10^5 \text{ N}$$

Ans. (B)

Sol. 11 Young's modulus

$$Y = \frac{\text{Stress}}{\text{Strain}}$$

$$\Rightarrow \quad Y = \frac{F/A}{(\Delta l / l)}$$

$$\Rightarrow \quad Y = \frac{F}{A(\alpha \Delta T)}$$

Ans. (C)

Sol. 12 Given that

$$\Delta l_1 = \Delta l_2$$

$$\Rightarrow \quad l\alpha_1 \Delta T_1 = l\alpha_2 \Delta T_2$$

$$\Rightarrow \quad \frac{\alpha_1}{\alpha_2} = \frac{\Delta T_1}{\Delta T_2}$$

$$\Rightarrow \quad \frac{4}{3} = \frac{T-30}{180-30}$$

$$\Rightarrow \quad T = 230°\text{C}$$

Ans. (B)

Sol. 13 For the given temperature and scale, we use

$$\frac{\frac{x_0}{2} - \frac{x_0}{3}}{x_0 - \frac{x_0}{3}} = \frac{R-0}{100-0}$$

$$\Rightarrow \quad R = 100 \times \left(\frac{x_0/6}{2x_0/3} \right)$$

$$\Rightarrow \quad R = 25°\text{C}$$

Ans. (B)

10.2 Calorimetry

Sol. 1 Specific heat → amount of heat which raises the temperature of unit mass of a substance by 1°C

Total heat required to raise the temperature of mass m by temperature t°C is

$$Q = mst \text{ (specific heat)}$$

If $t = 1$°C

$$\Rightarrow \quad Q = ms \rightarrow \text{Thermal capacity}$$

amount of heat to raise a body temperature by 1°C **Ans. (B)**

Sol. 2 By mass energy equivalence we have

$$E = mc^2$$

In case of water forming ice, it releases latent heat to surrounding

As heat is released from substance ⇒ some mass is converted to heat (energy)

⇒ mass of water should decrease. **Ans. (C)**

Sol. 3 Amount of heat required to change the phase of 1g water from liquid to vapour is

$$\Delta Q = mL_v = 1 \times 2256 \text{ J/g}$$

$$\Rightarrow \quad \Delta Q = 2256 \text{ J}$$

volume change due to vaporization is

$$\Delta V = 1671 - 1 = 1670 \text{ cm}^3$$

As expansion take place at constant atmospheric pressure (10^5 Pa), the work done by steam is

$$W = P\Delta V = 10^5 \times 1670 \times 10^{-6} = 167 \text{ J}$$

By first law of thermodynamics we use

$$\Delta Q = \Delta U + W$$

$$\Rightarrow \quad \Delta U = \Delta Q - W$$

$$= 2256 - 167 = 2089 \text{ J}$$

Ans. (C)

Sol. 4 Mixture = 500 gm water at 0°C + 100 gm ice at 0°C + 40 gm water (calorimeter) at 0°C

+ 10 gm steam at 100°C

heat required to melt 100 gm ice is

$$Q_{ice} = 100 \times 80 = 8000 \text{ cal}$$

Max heat which can be supplied by steam is

$$Q_{steam} = 10 \times 540 + 10 \times 1 \times 100$$

$$= 6400 \text{ cal}$$

As $\quad Q_{steam} < Q_{ice}$

⇒ Whole ice will not melt and find temperature mixture is 0°C

$\Rightarrow$ Amount of ice melted is given as

$$m = \frac{Q_{\text{steam}}}{L_{\text{ice}}} = \frac{6400}{80} = 80 \text{ gm}$$

$\Rightarrow$ Mixture= 500 gm water at 0°C + 80 gm water at 0°C + 20 gm ice at 0°C + Calorimeter (40 gm water) + 10 gm water at 0°C

In mixture finally there is 590 gm water at 0°C. **Ans. (B)**

Sol. 5 For heating 2 L water (2 kg) by 50°C (from 27°C to 77°C) amount of heat required is

$$Q = ms\Delta T = 2 \times 4200 \times 50 \text{ J}$$

Net heat supplied (rate) to water is

$$r = 1000 - 160 = 840 \text{ W}$$

(supply by coil is 1000W, loss is 160 J/s)

If in time t it happens then

$$840\, t = 2 \times 4200 \times 50$$

$$\Rightarrow \quad t = 500 \text{ s} = 8 \text{ m } 20 \text{ s}$$ **Ans. (A)**

Sol. 6 Let mass of water is m in gm and the uniform rate at which heat is being supplied is r cal/s, then we use

for first 10 min. $\rightarrow$ Total heat supplied

$$r \times 10 \times 60 = m \times 1 \times 100 \qquad \ldots(1)$$

for next 55 min $\rightarrow$ Total heat supplied

$$r \times 55 \times 60 = m \times L_v \qquad \ldots(2)$$

$$\frac{(2)}{(1)} \Rightarrow \frac{L_v}{100} = \frac{55}{10}$$

$$\Rightarrow \quad L_v = 550 \text{ cal/gm}$$ **Ans. (B)**

Sol. 7

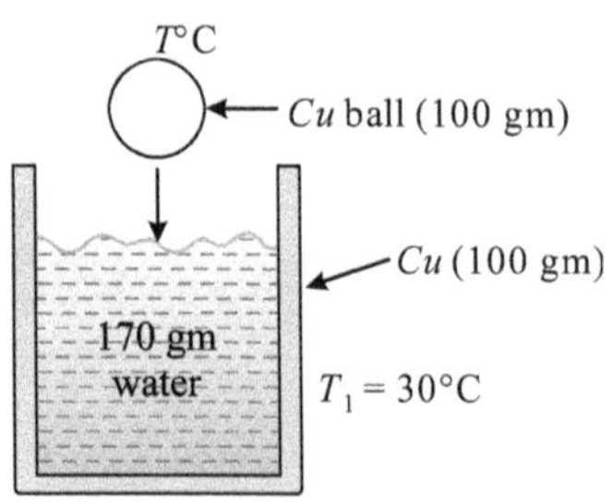

As $T_f = 75°\text{C}$

So rise in temperature of calorimeter + water

$$\Delta T = 75° - 30°$$

$$\Rightarrow \quad \Delta T = 45°\text{C}$$

As final temperature of mixture is 75°C

$\Rightarrow$ Heat lost by *Cu*-ball = Heat gained by *Cu*-calorimeter + water

$\Rightarrow$ $100 \times 0.1 \times (T - 75) = 100 \times 0.1 \times 45 + 170 \times 1 \times 45$

$\Rightarrow$ $10\text{T} - 750 = 450 + 7650$

$\Rightarrow$ $10\text{T} = 8100 + 750 = 8850$

$$\Rightarrow \quad T = \frac{8850}{10} = 885°\text{C}$$ **Ans. (D)**

Sol. 8

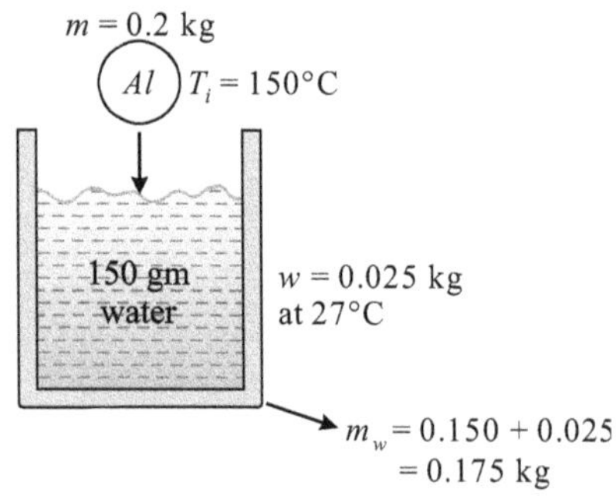

Equilibrium temperature attained = 40°C

Heat lost by *Al* - sphere = heat gained by calorimeter + water

$$0.2 \times s_M \times 110 = 0.175 \times 4200 \times 13$$

$$\Rightarrow \quad s_{Al} = \frac{0.175 \times 4200 \times 13}{0.2 \times 110} = 434.31 \text{ J/kg°C}$$

Ans. (D)

Sol. 9 By energy conservation, we use

$$192 \times s \times (100 - 21.5)$$
$$= 128 \times 394 \times (21.5 - 8.4)$$
$$+ 240 \times 4200 \times (21.5 - 8.4)$$

$$\Rightarrow \quad s = 916 \text{ J/kg-K}$$ **Ans. (D)**

Sol. 10 Let amount of ice is m gm. According to principal of calorimeter heat taken by ice = heat given by water

$$\Rightarrow \quad 20 \times 2.1 \times m + (m - 20) \times 334$$
$$= 50 \times 4.2 \times 40$$
$$376\text{m} = 8400 + 6680$$
$$m = 40.1\text{g}$$ **Ans. (B)**

Sol. 11 By conservation of energy, we use

$$100 \times s_A \times [100 - 90] = 50 \times s_B \times (90 - 75)$$

$$\Rightarrow \quad 2s_A = 1.5s_B$$

$$\Rightarrow \quad s_A = \frac{3}{4}s_B$$

And $\quad 100 \times s_A \times [100 - T] = 50 \times s_B\,(T - 50)$

$$\Rightarrow \quad 2 \times \left(\frac{3}{4}\right)(100 - T) = (T - 50)$$

$$\Rightarrow \quad 300 - 3T = 2T - 100$$

$$400 = 5T$$

$$T = 80°\text{C}$$ **Ans. (A)**

Sol. 12 By conservation of energy, we use

$$0.1 \times 400 \times (500 - T) = 0.5 \times 4200 \times (T - 30) + 800\,(T - 30)$$

$$\Rightarrow \quad 40(500 - T) = (T - 30)\,(2100 + 800)$$

$$\Rightarrow \quad 20000 - 40T = 2900\,T - 30 \times 2900$$

$\Rightarrow \quad 20000 + 30 \times 2900 = T(2940)$

$\Rightarrow \quad T = 36.4°C$

$\Rightarrow \quad \frac{\Delta T}{T} \times 100 = \frac{6.4}{30} \times 100$

$= 21.3\%$ **Ans. (B)**

Sol. 13

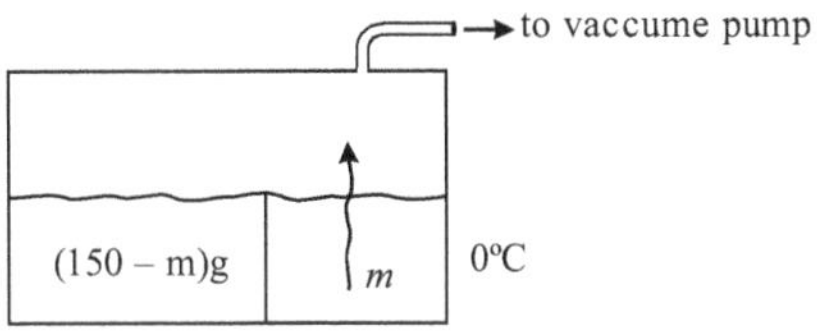

If mass m gets evaporated

$\Rightarrow$ heat required is

$$\Delta Q_{evap} = mL_v$$

This heat is supplied by remaining water in vessel which will transform into ice

$\Rightarrow$ heat supplied is $\Delta Q_{freeze} = (150 - m) L_f$

$\Rightarrow \quad \Delta Q_{evap} = \Delta Q_{freeze}$

$$mL_v = (150 - m) L_f$$

$$m = \frac{150 L_f}{L_v + L_f} = \frac{150 \times 3.36 \times 10^5}{(21 + 3.36) \times 10^5}$$

$m \simeq 20\text{g}$ **Ans. (C)**

Sol. 14 Heat lost by water = Heat gained by ice

$\Rightarrow \quad M_2 S_w \Delta T_w = M_1 S_i \Delta T_i + M_1 L_f$

$\Rightarrow \quad M_2 \times 1 \times 50 = M_1 \times 0.5 \times 10 + M_1 L_f$

$\Rightarrow \quad 50 M_2 = 5M_1 + M_1 L_f$

$\Rightarrow \quad L_f = \frac{50 M_2}{M_1} - 5$ **Ans. (C)**

10.3 Heat Transfer

Sol. 1 By Stefan's-Boltzmann law, the heat energy radiated/sec (Thermal radiation Power) by body of surface area A is given as

$$P_{th} = \frac{dQ}{dt} = \sigma \varepsilon A T^4$$

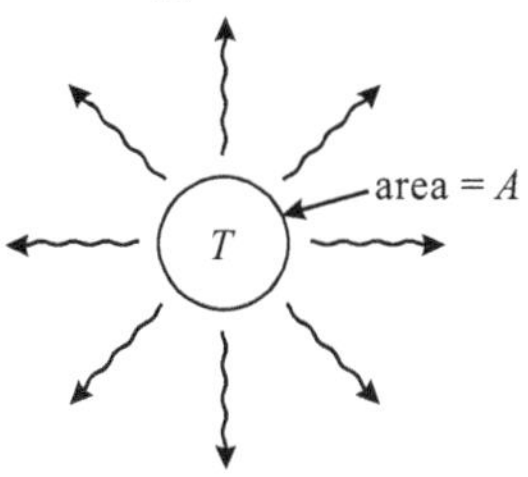

$e \rightarrow$ emissivity of body surface

$e < 1$ for general bodies

$e = 1$ for black body

For the given spheres we can use

$$P_1 = e\sigma (4\pi r_1^2) T_1^4$$

and $\quad P_2 = e\sigma (4\pi r_2^2) T_2^4$

$\Rightarrow \quad \frac{P_1}{P_2} = \frac{r_1^2 T_1^4}{r_2^2 T_2^4} = \frac{(1)^2 (4000)^4}{(4)^2 (2000)^4} = \frac{256}{256} = \frac{1}{1}$

Ans. (A)

Sol. 2 (A) Spectrometer $\rightarrow$ used to analyse EM radiation spectrum

(B) Pyrometer $\rightarrow$ used to measure thermal radiations & radiations from a hot body

(C) Nanometer $\rightarrow$ used to measure small dimensions

(D) Photometer $\rightarrow$ used to measure light intensity **Ans. (B)**

Sol. 3 Black Body $\rightarrow$ which absorbs all the radiation incident on it and emits maximum radiation from it at a specific temperature In above options, options (A) is most suitable for choosing it close to a Black body. **Ans. (A)**

Sol. 4

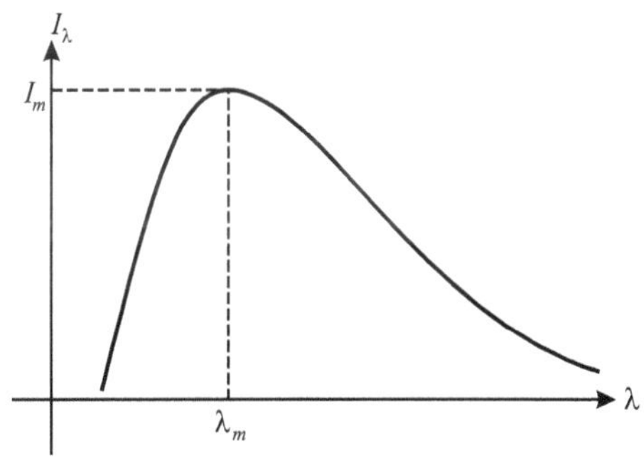

For any Black Body The EM radiation spectrum is given by Wien's Displacement law & Wien's fifth power law

$$\lambda_m \propto \frac{1}{T}$$

$I_m \propto T^5$ **Ans. (D)**

Sol. 5 By Newton's law of cooling the rate of cooling of a body is given as

$$\frac{\Delta\theta}{\Delta t} \propto \Delta\theta$$

($\Delta\theta$ = temperature difference of body with surrounding)

$\Rightarrow \quad n = 1$ **Ans. (D)**

Sol. 6 We consider sum as a black body and by Stefan's law, radiant power emitted by sun is given as

$$P_{Rad} = \sigma A T^4 = \sigma(4\pi R^2) T^4$$

($R \rightarrow$ radius of Sun)

$\Rightarrow \quad P_{rad} \propto R^2 T^4$

If $T \to 2T$ and $R \to 2R$

Then we have $$\frac{P_f}{P_i} = \frac{(2R)^2(2T)^4}{R^2T^4}$$

$$= (2)^2(2)^4 = 64$$

$$\Rightarrow \quad P_f = 64\,P_i$$

Ans. (D)

Sol. 7

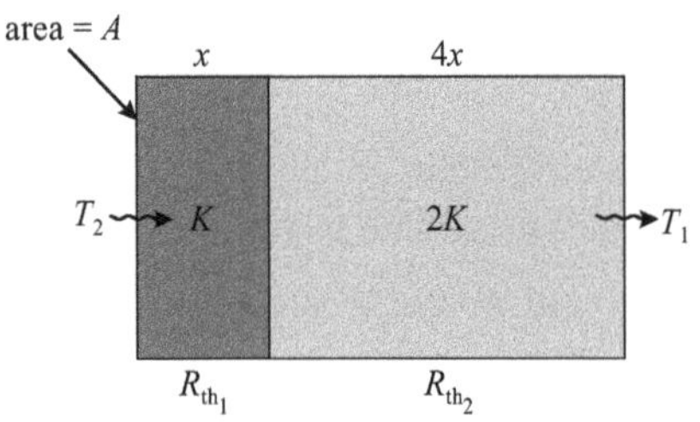

Where $$R_{th_1} = \frac{1}{k}\cdot\frac{x}{A}$$

and $$R_{th_2} = \frac{1}{2k}\cdot\frac{4x}{A}$$

Heat conduction through composite slab in steady state is given as

$$\frac{dQ}{dt} = \frac{T_2 - T_1}{(R_{th_1} + R_{th_2})}$$

$$\Rightarrow \quad \frac{dQ}{dt} = \frac{T_2 - T_1}{\frac{x}{kA} + \frac{2x}{kA}} = \frac{kA(T_2 - T_1)}{3x}$$

$$\Rightarrow \quad f = 1/3$$

Ans. (D)

Sol. 8

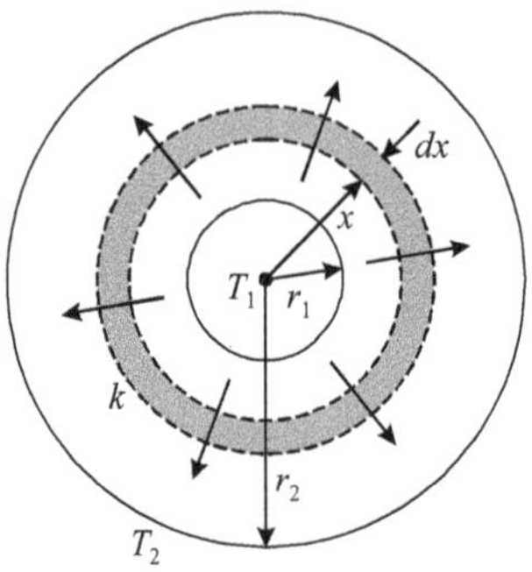

$(T_1 > T_2)$

Rate of heat flow between shells is given as

$$\frac{dQ}{dt} = \frac{T_1 - T_2}{R_{th}} \qquad \ldots(1)$$

Where R_{th} is the thermal resistance for radial heat flow.

The thermal resistance of elemental shell for radial heat flow is given as

$$dR = \frac{1}{k}\cdot\frac{dx}{4\pi x^2}$$

Total thermal resistance between inner and outer shells is

$$R_{th} = \int dR = \frac{1}{4\pi k}\int_{r_1}^{r_2}\frac{dx}{x^2} = \frac{1}{4\pi k}\left[-\frac{1}{x}\right]_{r_1}^{r_2}$$

$$= \frac{1}{4\pi k}\left[\frac{1}{r_1} - \frac{1}{r_2}\right]$$

$$\Rightarrow \quad R_{th} = \frac{(r_2 - r_1)}{4\pi k r_1 r_2}$$

Rate of heat flow in radial direction from equation-(1)

$$\frac{dQ}{dt} = \frac{(T_1 - T_2)}{(r_2 - r_1)} \times 4\pi k r_1 r_2$$

$$\Rightarrow \quad \frac{dQ}{dt} \propto \frac{r_1 r_2}{r_2 - r_1}$$

Ans. (C)

Sol. 9

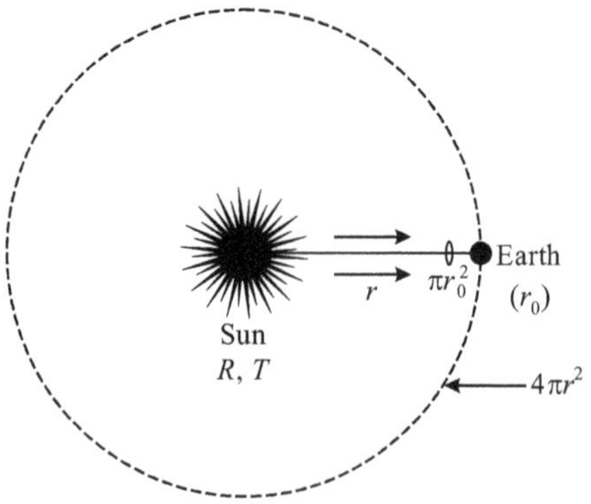

Surface area of spherical shell in which whole power of Sun will be distributed

Radiant power of Sun is

$$P_{rad} = \sigma(4\pi R^2)\,T^4$$

Radiant power of Sun falling per unit area of the earth surface (Normally) is

$$S = \frac{P_{rad}}{4\pi r^2} = \frac{\sigma(4\pi R^2)T^4}{4\pi r^2}$$

Solar constant $$S = \frac{\sigma R^2 T^4}{r^2}$$

Energy incident on Earth is

$$E_e = S.(\pi r_0^2)$$

$$\Rightarrow \quad E_e = \frac{\pi\sigma R^2 r_0^2 T^4}{r^2}$$

Ans. (B)

Sol. 10

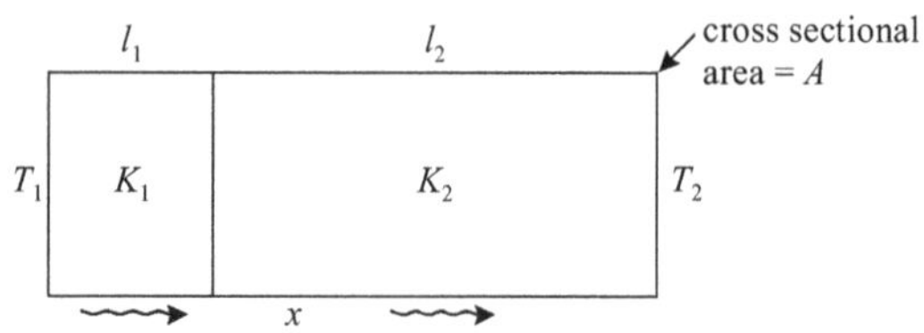

In steady state throughout the rod heat current is constant as heat absorption is zero, so we have

$$\left.\frac{dQ}{dt}\right|_1 = \left.\frac{dQ}{dt}\right|_2$$

$$\frac{T_i - x}{\left(\frac{1}{K_1}\cdot\frac{l_1}{A}\right)} = \frac{x - T_2}{\left(\frac{1}{K_2}\cdot\frac{l_2}{A}\right)}$$

$$\Rightarrow \qquad x = \frac{K_1 l_2 T_1 + K_2 l_1 T_2}{K_1 l_2 + K_2 l_1}$$

Ans. (C)

Sol. 11

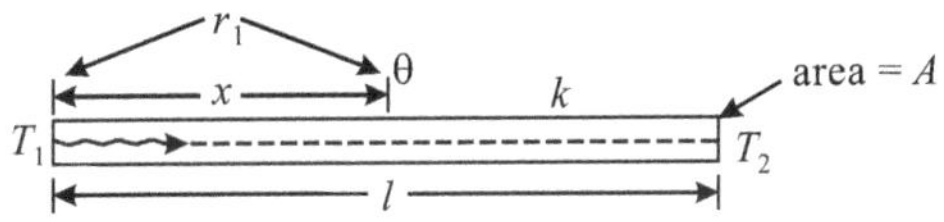

$T_1 > T_2$

Thermal current in rod is given as

$$I_{th} = \frac{dQ}{dt} = \frac{T_1 - T_2}{\left(\frac{1}{k}\cdot\frac{l}{A}\right)}$$

$\Rightarrow \qquad I_{th}$ = constant under steady state

Using Ohm's law we have

$$\theta = T_1 - I_{th} r_1 \text{ where } r_1 = \frac{1}{k}\cdot\frac{x}{A}$$

$$\Rightarrow \qquad \theta = T_1 - \frac{(T_1 - T_2)}{\left(\frac{1}{k}\frac{l}{A}\right)}\left(\frac{1}{k}\frac{x}{A}\right)$$

$$\Rightarrow \qquad \theta = T_1 - \frac{(T_1 - T_2)}{l}\cdot x$$

Above equation has negative slope and positive intercept on y-axis. **Ans. (D)**

Sol. 12 By Newton's law of cooling, the rate of colling is given as

$$\frac{d\theta}{dt} = -k(\theta - \theta_0)\ ^\circ\text{C/s}$$

$$\Rightarrow \qquad \int_{\theta_1}^{\theta} \frac{d\theta}{\theta - \theta_0} = \int_0^t -k dt$$

θ_1 is temperature of liquid at $t = 0$ ($\theta_1 > \theta_0$)

$$[\log_e(\theta - \theta_0)]_{\theta_1}^{\theta} = -kt$$

$$\Rightarrow \qquad \log_e(\theta - \theta_0) = -kt + \log_e(\theta_1 - \theta_0) \text{ (constant)}$$

$$y = mx + C \text{ [equation of a straight line]}$$

$\Rightarrow$ – ve slope straight line **Ans. (B)**

Sol. 13 At such a high temperature (3000 K) we can consider the furnace like a black body so heat radiated per unit area/sec is given by Stefan's law as

$$E = \sigma T^4 \text{ J/m}^2\text{-s}$$

$\Rightarrow$ in 1 Hr. total heat/m^2 is given as

$$\Rightarrow \qquad Q = Et = \sigma T^4.t$$

$$= 5.7 \times 10^{-8} \times (3000)^4 \times 3600$$

$$\Rightarrow \qquad Q = 1.66 \times 10^{10} \text{ J/m}^2$$

$$Q \simeq 1.7 \times 10^{10} \text{ J/m}^2$$

Ans. (A)

Sol. 14 Rate of Heat conducting in steady state is given as

$$\frac{dQ}{dt} = \frac{kA(T_1 - T_2)}{l}$$

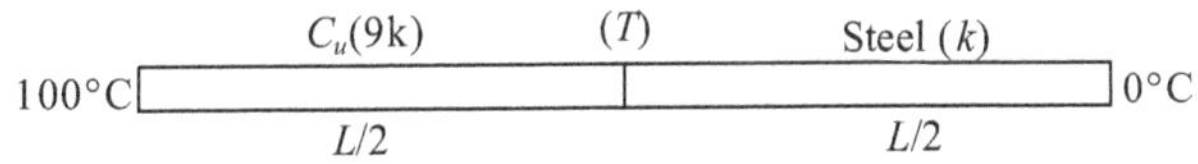

In steady state of conduction heat flow rate through the two rods will be same

$$\Rightarrow \qquad \frac{(9k)A(100 - T)}{L/2} = \frac{kA(T - 0)}{L/2}$$

$$\Rightarrow \qquad 9(100 - T) = T$$

$$\Rightarrow \qquad 900 = T + 9T$$

$$\Rightarrow \qquad T = \frac{900}{10} = 90^\circ\text{C}$$

Ans. (A)

Sol. 15 By Newton's law of cooling, we use rate of colling of metal is given as

$$\frac{dT}{dt} = -k(T - \theta_0)$$

$$\Rightarrow \qquad \int_{\theta}^{T} \frac{dT}{T - \theta_0} = \int_0^t -k dt$$

$$\Rightarrow \qquad [\ln(T - \theta_0)]_{\theta}^{T} = -kt$$

$$\ln\left(\frac{T - \theta_0}{\theta - \theta_0}\right) = -kt$$

$$\Rightarrow \qquad T = \theta_0 + (\theta - \theta_0)e^{-kt}$$

at $t = 0$; $T = \theta$

at $t \to \infty$ (long time), $T = \theta_0$

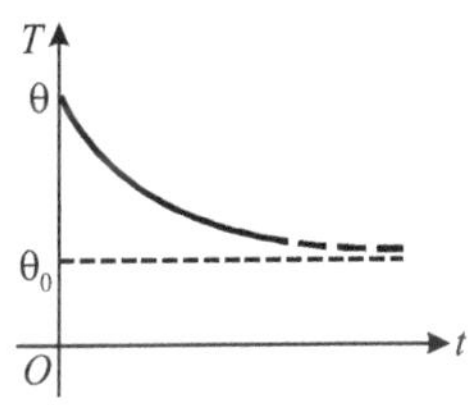

Ans. (C)

Sol. 16 By average from of Newton's law of cooling, we use

$$\frac{\Delta T}{t} = \frac{k}{(ms + \omega)}\left[\frac{T_1 + T_2}{2} - T_s\right]$$

where $\qquad \Delta T = T_2 - T_1$

Initially with water, we use

$$\frac{30 - 25}{2} = \frac{k}{50 \times 1 + 30}\left[\frac{30 + 25}{2} - T_s\right] \qquad \ldots(1)$$

Then with given liquid, we use

$$\frac{30 - 25}{2} = \frac{k}{100s + 30}\left[\frac{30 + 25}{2} - T_s\right] \qquad \ldots(2)$$

$$\frac{(1)}{(2)} \Rightarrow \quad 50+30=100\,s+30$$

$$\Rightarrow \quad s=\frac{50}{100}$$

$$=0.5 \text{ cal/gm}=0.5 \text{ kcal/kg}$$

Ans. (D)

Sol. 17

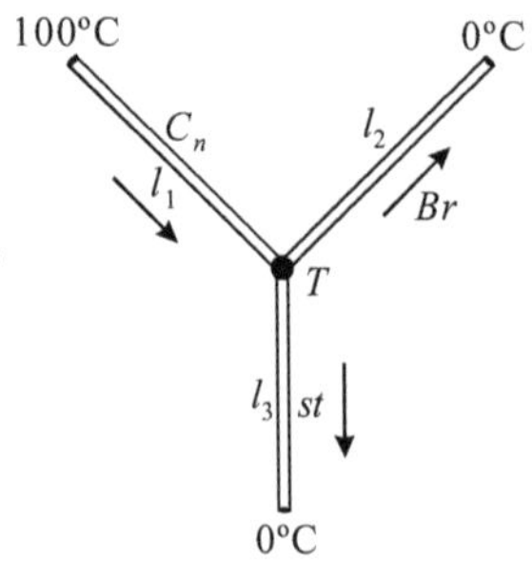

$$A=4 \text{ cm}^2$$

$$l_1=46 \text{ cm}; k_1=0.92 \text{ CGS units}$$

$$l_2=13 \text{ cm}; k_2=0.26 \text{ CGS units}$$

$$l_3=12 \text{ cm}; k_3=0.12 \text{ CGS units}$$

Using Junction rule at temperature T, we have

$$I_{Cu}=I_{Br}+I_{st} \qquad \ldots(1)$$

Thermal resistances of rods are

$$R_{Cn}=\frac{1}{k_1}\cdot\frac{l_1}{A}=\frac{1}{0.92}\times\frac{46}{4}=\frac{100}{8} CGS \text{ Units}$$

$$R_{Br}=\frac{1}{k_2}\frac{l_2}{A}=\frac{1}{0.26}\times\frac{13}{4}=\frac{100}{8} CGS \text{ Units}$$

$$R_{st}=\frac{1}{k_3}\frac{l_3}{A}=\frac{1}{0.12}\times\frac{12}{4}=\frac{100}{4} CGS \text{ Units}$$

$$I_{Cu}=\frac{100-T}{R_{Cu}}; I_{Br}=\frac{T-0}{R_{Br}}; I_{st}=\frac{T-0}{R_{st}}$$

from equation (1) we have

$$\frac{100-T}{(100/8)}=\frac{T-0}{(100/8)}+\frac{T-0}{(100/4)}$$

$$\Rightarrow \quad 100-T=T+\frac{T}{2}=\frac{3T}{2}$$

$$\Rightarrow \quad 200=5T$$

$$\Rightarrow \quad T=40°\text{C}$$

heat current through Cu-rod is

$$\left.\frac{dQ}{dt}\right|_{Cu}=\frac{100-T}{100/8}$$

$$=\frac{100-40}{100/8}=\frac{8\times 60}{100}=4.8 \text{ cal/s}$$

Ans. (C)

Sol. 18 Specific heat of sphere

$$s=\alpha T^3$$

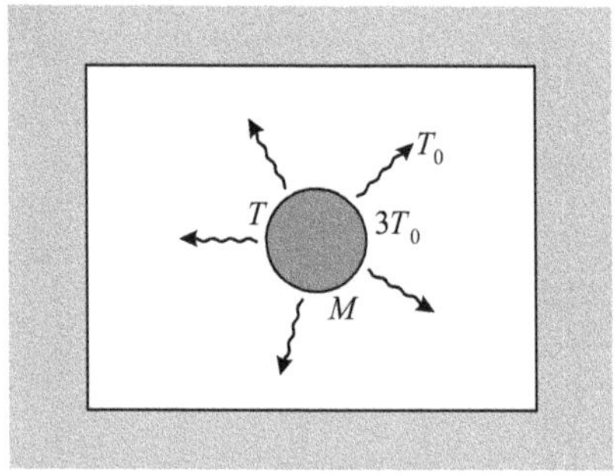

Due to radiation rate of heat loss by sphere is given as, when sphere temperature is T

$$2T_0<T<3T_0$$

$$\frac{dQ}{dt}=\sigma(4\pi R^2)(T^4-T_0^4)$$

$$\Rightarrow \quad -\frac{MsdT}{dt}=4\pi\sigma R^2(T^4-T_0^4)$$

$$\Rightarrow \quad -\frac{M\alpha T^3 dT}{dt}=4\pi\sigma R^2(T^4-T_0^4)$$

$$\Rightarrow \quad \frac{T^3 dT}{T^4-T_0^4}=-\frac{4\pi\sigma R^2}{M\alpha}dt$$

Integrating above expression, gives

$$\int_{T=3T_0}^{T=2T_0}\frac{T^3 dT}{T^4-T_0^4}=-\frac{4\pi\sigma R^2}{M\alpha}\int_{t=0}^{t=t}dt$$

Let $\quad x=T^4-T_0^4$

$$\Rightarrow \quad dx=4T^3 dT$$

at $\quad T=3T_0 \rightarrow x_1=80T_0^4$

and at $\quad T=2T_0 \rightarrow x_2=15T_0^4$

$$\Rightarrow \quad \frac{1}{4}\int_{x_1}^{x_2}\frac{dx}{x}=-\frac{4\pi\sigma R^2}{M\alpha}\cdot t$$

$$\Rightarrow \quad \frac{1}{4}[\ln x]_{80T_0^4}^{15T_0^4}=-\frac{4\pi\sigma R^2}{M\alpha}t$$

$$\Rightarrow \quad [\ln(80T_0^4)-\ln(15T_0^4)]\cdot\frac{M\alpha}{16\pi\sigma R^2}=t$$

$$\Rightarrow \quad t=\frac{M\alpha}{16\pi\sigma R^2}\ln\left(\frac{16}{3}\right)$$

Ans. (C)

Sol. 19 By Newton's law of cooling for a body colling down, its temperature as a function of time is given as

$$T_f=T_0+(T_H-T_0)e^{-kt}$$

Here $T_H \rightarrow$ initial high temperature

$T_0 \rightarrow$ surrounding temperature

For case-1, we use

$$40 = 30 + 50e^{-k(5)}$$

$$\Rightarrow \quad e^{5k} = 5 \Rightarrow 5k = \ln(5) \quad \ldots(1)$$

For case-2, we use

$$32 = 30 + 32e^{-kt}$$

$$\Rightarrow \quad e^{kt} = 16 \Rightarrow kt = \ln(16) \quad \ldots(2)$$

$$\frac{(2)}{(1)} \Rightarrow \quad \frac{t}{5} = \frac{4\ln(2)}{\ln 5}$$

$$\Rightarrow \quad t = \frac{5\times4\times0.693}{1.609} = 8.61 \text{ min}$$

Ans. (B)

Sol. 20 Here we use average form of Newton's law of colling as

$$\frac{T_2 - T_1}{t} = k\left(\frac{T_1 + T_2}{2} - T_s\right)$$

Case-I : First 10 minimum

$$\frac{60-50}{10} = k(55 - T_s) \quad \ldots(1)$$

Case-II : Next 10 minimum

$$\frac{50-42}{10} = k(46 - T_s) \quad \ldots(2)$$

$$\frac{(1)}{(2)} \Rightarrow \quad \frac{10}{8} = \frac{55 - T_s}{46 - T_s}$$

$$\Rightarrow \quad 230\sqrt{T_s} = 220 - 4T_s$$

$$\Rightarrow \quad T_s = 230 - 220 = 10°\text{C}$$

Ans. (B)

Sol. 21 Equivalent circuit of the given system of rods is shown in figure below.

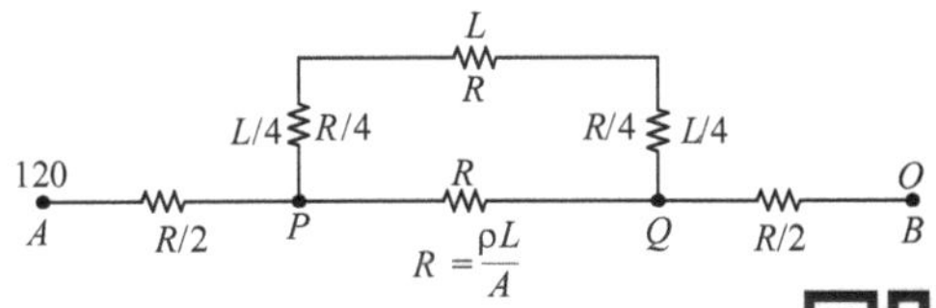

Thermal current is given as

$$I_{th} = \frac{\Delta T}{R_{eq.}} = \frac{(120)5}{8R} = \frac{120\times5}{8R}$$

Temperature difference of P and Q is given as

$$\Delta T_{PQ} = I_{th}R_{PQ} = \left(\frac{120\times5}{8R}\right)\times\frac{3}{5}R = \frac{360}{8} = 45°\text{C}$$

Ans. (D)

Sol. 22

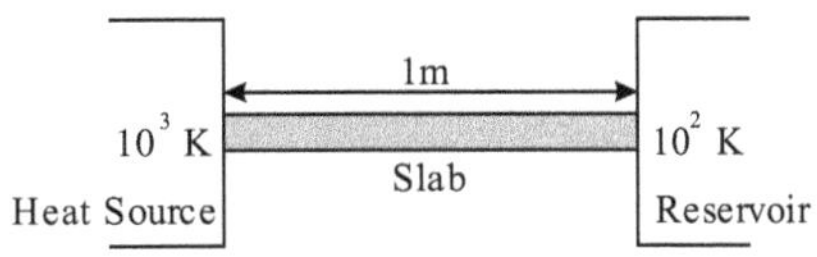

Rate of heat conduction through slab in steady state is given as

$$\left(\frac{dQ}{dt}\right) = \frac{kA\Delta T}{l}$$

Energy flux (w/m²) through slab is given as

$$\frac{1}{A}\left(\frac{dQ}{dt}\right) = \frac{(0.1)(900)}{1} = 90\text{W/m}^2$$

Ans. (A)

Sol. 23

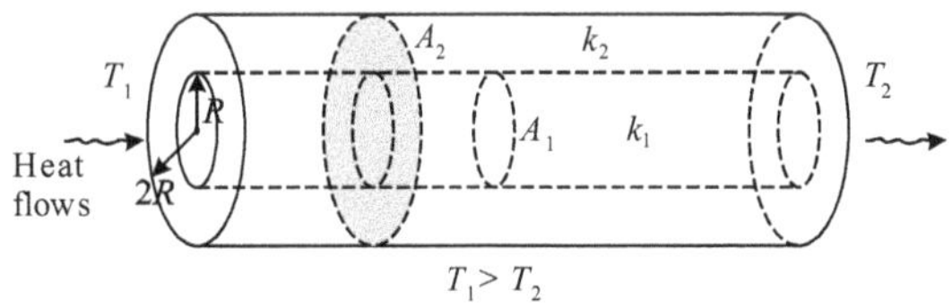

In parallel combination, equivalent thermal conductivity is given as

$$K_{eq} = \frac{k_1A_1 + k_2A_2}{A_1 + A_2}$$

$$K_{eq} = \frac{k_1(\pi R^2) + k_2(\pi(2R)^2 - \pi R^2)}{4\pi R^2}$$

$$K_{eq} = \frac{k_1 + 3k_2}{4}$$

Ans. (D)

Sol. 24 By Newton's law of cooling

$$\frac{dT}{dt} = -\frac{4eA\sigma T_0^3}{ms}\cdot(\Delta T)$$

$$\Rightarrow \quad \left|\frac{dT}{dt}\right| \propto \frac{1}{ms}$$

$$\Rightarrow \quad \left|\frac{dT}{dt}\right| \propto \frac{1}{\rho s}$$

(As for equal volume $m \propto \rho$)

Here $(\rho s)_A = 8\times10^2\times2000 = 16\times15^5$ SI units

$(\rho s)_B = 10^3\times4000 = 4\times10^6$ SI units

$$(\rho s)_B > (\rho s)_A \Rightarrow \left|\frac{dT}{dt}\right|_B < \left|\frac{dT}{dt}\right|_A$$

Here T vs t curve of A is steeper than that of B.

Ans. (A)

Sol. 25 If interface temperature is T, we use in steady state

$$\frac{3kA(\theta_2 - T)}{d} = \frac{kA(T - \theta_1)}{3d}$$

$$\Rightarrow \quad 3\theta_2 - 3T = \frac{T - \theta_1}{3}$$

$$\Rightarrow \quad 10T = 9\theta_2 + \theta_1$$

$$\Rightarrow \quad T = \frac{9\theta_2 + \theta_1}{10}$$

Ans. (B)

11.1 Laws of Thermodynamics

Sol. 1 (A) False for reversible cycles efficiency

$$\eta = \frac{\text{work done}}{\text{heat input}}\times100\%$$

for different thermodynamic processes w/o by gas in different for same heat input.

(B, C, D) True.

Ans. (A)

Sol. 2 Efficiency of a carnot engine is given as

$$\eta = \left(1 - \frac{T_2}{T_1}\right) \times 100\%$$

Where $T_2 \rightarrow$ temperature sink

$T_1 \rightarrow$ temperature of source

To achieve $\eta = 100\%$

either $T_2 = 0 \rightarrow$ Not attainable

or $T_1 = \infty$ [Impossible] **Ans. (C)**

Sol. 3 Efficiency of carnot engine is given as

$$\eta = 1 - \frac{T_{\text{Sink}}}{T_{\text{Source}}}$$

here $T_{\text{Sources}} = 627 + 273 = 900\text{ K}$

and $T_{\text{Sink}} = 27 + 273 = 300\text{ K}$

$\Rightarrow \quad \eta = 1 - \frac{300}{900} = \frac{2}{3}$

For engine efficiency relates work done and heat input as

$$\eta = \frac{\text{work done}}{\text{Heat input}}$$

$\Rightarrow$ Work done $= \eta \times$ heat supplied

$= \frac{2}{3} \times 3 \times 10^6 \times 4.2$

$= 8.4 \times 10^6\text{ J}$ **Ans. (B)**

Sol. 4 For an adiabatic process we use

$PV^\gamma = \text{constant}$ ...(1)

We are given that $P \propto T^3$

$\Rightarrow \quad T = kP^{1/3}$

By Gas law we have

$PV = nRT$

$\Rightarrow \quad PV = nRkP^{1/3}$

$\Rightarrow \quad P^{1-1/3}V = \text{Cost}$

$\Rightarrow \quad P^{2/3}\ V = \text{Cost}$

$\Rightarrow \quad PV^{3/2} = \text{Cost}$ ...(2)

Comparing equation-(1) and (2) we have $\gamma = 3/2$ **Ans. (D)**

Sol. 5 State of a gas is indicate by thermodynamic coordinates (P, V, T) which helps in denoting gas state on indicator diagram and also we can plot process equation of ID curve. **Ans. (C)**

Sol. 6 Above statement is a consequence of "Second law of thermodynamics"

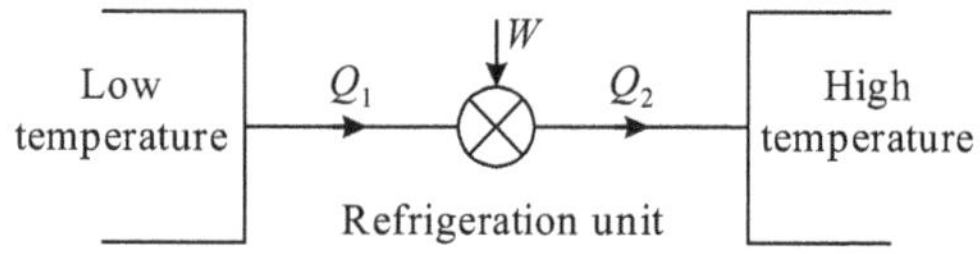

$$W + Q_1 = Q_2$$

Without external work it is impossible to transfer heat from low temperature zone to high temperature zone. **Ans. (A)**

Sol. 7. (A) As internal energy is a state function

$U = \frac{f}{2}nRT$ 'False'

This is for an isothermal process as

$$\Delta T = 0\,;\, \Delta U = 0$$

(B) 'True' Both U & S are state functions

(C) As entropy is a state function so in cyclic process or whenever a system returns to its initial state $\Delta S = 0 \rightarrow$ 'False'

(D) For adiabatic process $\Delta Q = 0 \Rightarrow \Delta W = -\Delta U \rightarrow$ 'False'

Ans. (B)

Sol. 8 As we know internal energy of a gas is a 'State function'

$$U = \frac{f}{2}nRT$$

At state A and B if internal energy of system are U_A and U_B respectively

$\Rightarrow \quad \Delta U_1 = \Delta U_2 = U_B - U_A$

As U is not a 'Path function' it does not depend on paths. **Ans. (A)**

Sol. 9 Efficiency of a heat engine is given as

$$\eta = \frac{\text{Work done}}{\text{Heat input}}$$

Heat $\rightarrow$ Area under T-S diagram

heat supplied in cycle

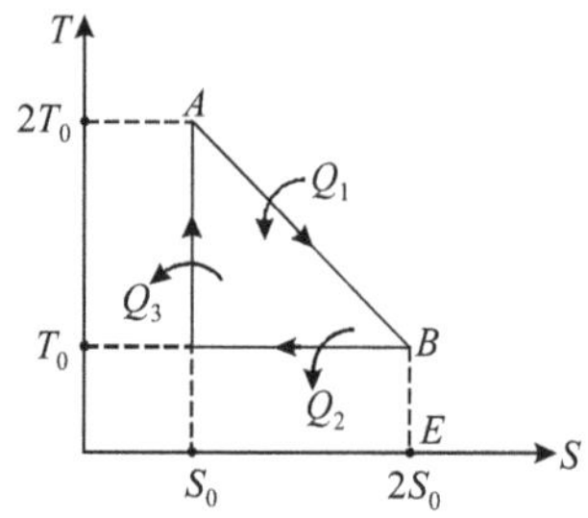

$Q_1 = $ Area $ABEDA$

$= T_0(2S_0 - S_0) + \frac{1}{2}T_0S_0 = \frac{3}{2}T_0S_0$

Work done in cycle $W = $ Area ABC

$= \frac{1}{2}T_0S_0$

$\Rightarrow$ Efficiency of cycle

$$\eta = \frac{W}{Q_1} = \frac{1/2T_0S_0}{3/2T_0S_0} = \frac{1}{3} \qquad \textbf{Ans. (C)}$$

Sol. 10 (A) **True :** for all process (cyclic or not)

$$\Delta Q = \Delta U + \Delta W$$

or for elemental duration *FLT* is valid as

$$dQ = dU + dW$$

(B) **True :** *FLT* is defined by using energy conservation

(C) **True :** For *FLT* we use internal energy of gas/substance along with heat supplied & work done.

(D) **False :** Concept entropy is introduced in second law of thermodynamics

Sol. 11 For an adiabatic process work done is given as

$$W = \frac{nR(T_2 - T_1)}{\gamma - 1}$$

$$\Rightarrow \quad \gamma - 1 = \frac{nR(T_2 - T_1)}{W}$$

$$\Rightarrow \quad \gamma = 1 + \frac{10^3 \times 8.3 \times 7}{146 \times 10^3} = 1 + 0.4 = 1.4$$

$\gamma = 1.4$ for a diatomic gas

$$(f = 5 \Rightarrow \gamma = \frac{f+2}{f} = \frac{7}{5} = 1.4) \qquad \textbf{Ans. (A)}$$

Sol. 12 As we know for same initial and final states of all the thermodynamic process we have

$$\overset{iaf}{\Delta U_1} = \overset{ibf}{\Delta U_2} \quad (\text{as } U \to \text{state function})$$

$$\Rightarrow \quad Q_1 - W_1 = Q_2 - W_L$$

$$\Rightarrow \quad 50 - 20 = 36 - W_2$$

$$\Rightarrow \quad W_2 = 36 - 30 = 6 \text{ cal} \qquad \textbf{Ans. (A)}$$

Sol. 13 If same engine is used as refrigerator then relation in its efficiency (η) and coefficient of performance (k) is given as

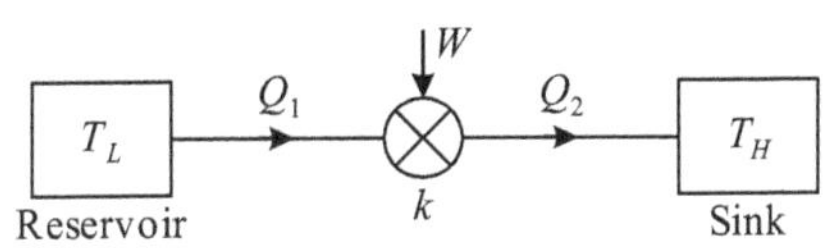

$$k = \frac{1}{\eta} - 1 = \frac{1}{1/10} - 1 = 9$$

COP of a refrigerator is given as

$$k = \frac{Q_1}{W} = \frac{\text{Heat extracted}}{\text{Work input}}$$

$$\Rightarrow \quad 9 = \frac{Q_1}{W} \Rightarrow Q_1 = 9 \times 10 = 90 \text{ J} \qquad \textbf{Ans. (B)}$$

Sol. 14 Work done by gas in Isobaric Process *AB* is

$$W_{AB} = P(V_2 - V_1) = nR(T_2 - T_1)$$

$$W_{AB} = 2R(500 - 300) = 400R$$

Work done on the gas

$$= -W_{AB} = -400R \qquad \textbf{Ans. (A)}$$

Sol. 15 Work done by gas in Isothermal process *DA* is

$$W_{DA} = nRT \ln\left(\frac{P_1}{P_2}\right)$$

$$\Rightarrow \quad W_{DA} = 2.303\, nR\,(300)\, \log_{10}\left(\frac{1}{2}\right)$$

$$\Rightarrow \quad W_{DA} = 2.303 \times 2R \times 300 \times (-0.301) = -414R$$

Work done on the gas is

$$= -W_{DA} = 414R \qquad \textbf{Ans. (D)}$$

Sol. 16 Total work done in complete cycle is given as

$$W_{\text{cycle}} = W_{AB} + W_{BC} + W_{CD} + W_{DA}$$

$$\Rightarrow \quad W_{\text{cycle}} = nR(T_B - T_A) + nRT_B \ln\frac{P_B}{P_C} + nR(T_D - T_C) + nRT_D \ln\frac{P_D}{P_A}$$

$$\Rightarrow \quad W_{\text{cycle}} = 2R(200) + 2.303 \times 2R \times 500 \log_{10}(2) + 2R(-200) + 2.303 \times 2R \times 300 \log_{10}(1/2)$$

$$\Rightarrow \quad W_{\text{cycle}} = 690\,R - 414\,R = 276\,R$$

Work done on the gas $= -W_{\text{cycle}} = -276\,R$ **Ans. (D)**

Sol. 17

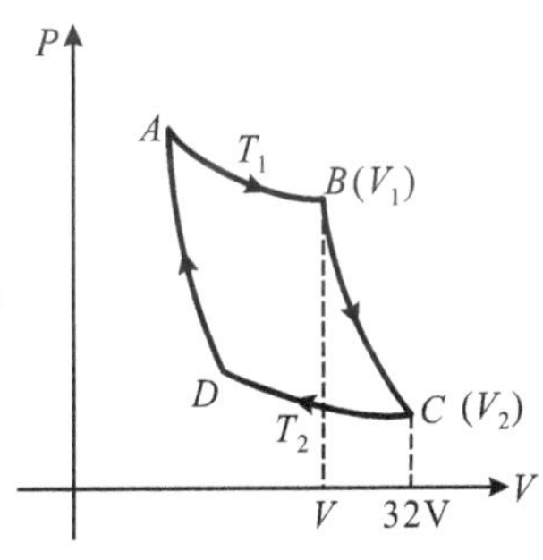

AB and *CD* $\to$ isothermal parts

BC and *DA* $\to$ adiabatic parts

$T_1 \to$ temperature of source

$T_2 \to$ temperature of sink

for adiabatic process *BC*, we use process equation

$$TV^{\gamma-1} = \text{constant} \quad (\text{for diatomic gases } \gamma = 7/5)$$

$$\Rightarrow \quad T_1V_1^{\gamma-1} = T_2V_2^{\gamma-1}$$

$$\Rightarrow \quad \frac{T_2}{T_1} = \left(\frac{V_1}{V_2}\right)^{\gamma-1} = \left(\frac{V_1}{32V}\right)^{7/5-1}$$

$$= \left(\frac{1}{2^5}\right)^{2/5} = \frac{1}{2^2} = \frac{1}{4}$$

Engine efficiency is given as

$$\eta = 1 - \frac{T_2}{T_1} = 1 - \frac{1}{4} = \frac{3}{4} = 0.75$$

in % $\quad \eta = \frac{3}{4} \times 100\% = 75\%$ **Ans. (C)**

Sol. 18 First law of thermodynamics increase in internal energy of gas = Amount of heat supplied

$$\Delta U = Q = mS\Delta T$$

$$\Rightarrow \quad \Delta U = 0.1 \times 4184 \times 20$$

$$\Rightarrow \quad \Delta U \cong 8.4 \text{ kJ}$$ **Ans. (D)**

Sol. 19

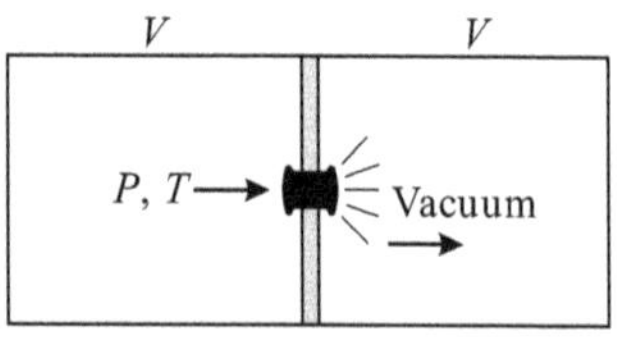

In case of 'Free expansion' (no opposition) no work is done by gas so as $\Delta Q = 0$ (insulating walls)

or $\quad U =$ constant

$\Rightarrow \quad T =$ constant

As gas temperature remain constant

we use $\quad P_1V_1 = P_2V_2$

$$PV = P_2(2V) \Rightarrow P_2 = P/2$$ **Ans. (A)**

Sol. 20 The amount of heat to be last by metal in cooling from 20 K to 4 K is

$$Q = \int dQ = -\int mSdT = -\int_{20}^{4} m\left(32\left[\frac{T}{400}\right]^3\right)dT$$

heat lost

$$\Rightarrow Q = -\frac{0.1\times 32}{(400)^3}\cdot\left[\frac{T^4}{4}\right]_{20}^{4} = \frac{0.1\times 32}{(400)^3\times 4}[(20)^4 - (4)^2]$$

$$Q = 0.002 \text{ kJ}$$

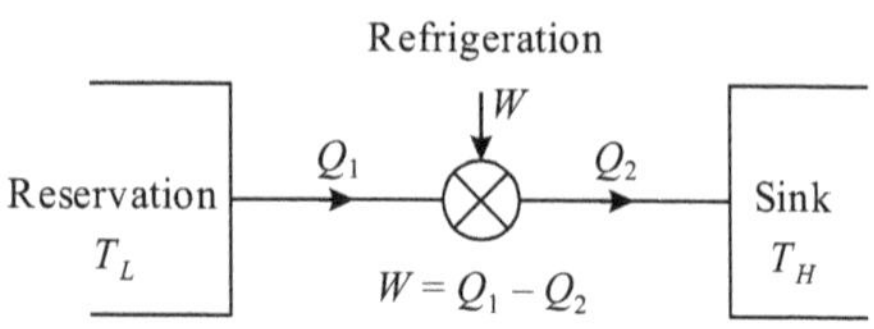

Coefficient of performance

$$K = \frac{Q_1}{W}$$

$$K = \frac{Q_1}{Q_2 - Q_1} = \frac{T_1}{T_2 - T_1}$$

At $\quad T_1 = 20\text{ K} \ \& \ T_S = 27 + 273 = 300\text{ K}$

The amount of work done in refrigeration unit is

$$\frac{Q_1}{W_1} = \frac{T_1}{T_2 - T_1} = \frac{20}{300-20} = \frac{20}{280}$$

$$\Rightarrow \quad W_1 = \frac{0.002\text{kJ}}{20/280} = 0.028 \text{ kJ}$$

As metal is cooled to 4 K, so at T_L = 4 K, work done by refrigeration unit is

$$\frac{Q_1}{W_2} = \frac{T_1}{T_2 - T_1} = \frac{4}{300-4} = \frac{4}{296}$$

$$\Rightarrow \quad W_2 = \frac{0.002\text{kJ}}{(4/296)} = 0.148 \text{ kJ}$$

As metal is cooling from 20 K to 4 K so any temperature of metal during cooling is between 20 K and 4 K so work required by refrigeration unit in between W_1 and W_2. **Ans. (C)**

Sol. 21 If $T_1 \to$ Source temperature

& $\ T_2 \to$ Sink temperature

$\Rightarrow$ Engine efficiency $\eta = 1 - \frac{T_2}{T_1}$

Initially $\quad 1 - \frac{T_2}{T_1} = \frac{1}{6} \Rightarrow \frac{T_2}{T_1} = \frac{5}{6}$ ...(1)

Later $\quad 1 - \frac{T_2 - 62}{T_1} = \frac{1}{3} \Rightarrow \frac{T_2 - 62}{T_1} = \frac{2}{3}$ ...(2)

Solving equation-(1) and (2) we get

$$T_1 = 372 \text{ K and } T_2 = 310 \text{ K}$$ **Ans. (C)**

Sol. 22

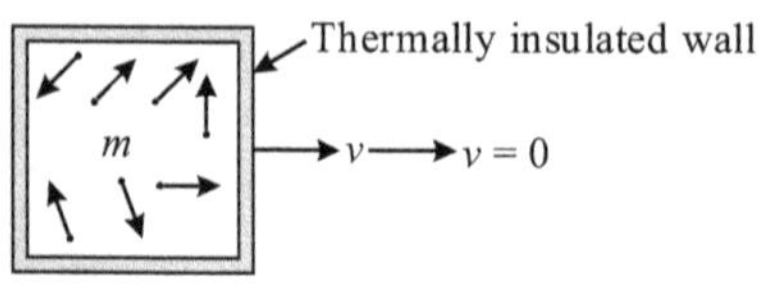

KE of gas particles due to the motion of vessel

$$K = \frac{1}{2}mv^2$$

When suddenly the vessel in brought to rest then above *KE* will increase the energy of random motion (thermal energy) of molecules as no energy can transmit through container walls

$$\Rightarrow \quad \frac{1}{2}mv^2 = \Delta U = n\left(\frac{R}{T-1}\right)\Delta T$$

$$\Rightarrow \quad \frac{1}{2}mv^2 = \frac{m}{M}\left(\frac{R}{\gamma-1}\right)\Delta T$$

$$\Rightarrow \quad \Delta T = \frac{Mv^2}{2R}(\gamma - 1)$$ **Ans. (B)**

Sol. 23 Efficiency of carnot engine is given as

$$\eta = 1 - \frac{T_{sink}}{T_{Source}}$$

for first engine $T_{Source} = 500\text{ K} \;\&\; \eta = 0.4$

$$\Rightarrow \quad 0.4 = 1 - \frac{T_{Sink}}{500}$$

$$\Rightarrow \quad T_{Sink} = 500 \times 0.6$$

$$\Rightarrow \quad T_{Sink} = 300\text{ K}$$

for another engine with

$$\eta = 0.6 \;\&\; T_{Sink} = 300\text{ K}$$

$$\Rightarrow \quad 0.6 = 1 - \frac{300}{T_{Source}}$$

$$\Rightarrow \quad T_{Source} = \frac{300}{0.4} = 750\text{ K}$$ **Ans. (D)**

Sol. 24 Cycle efficiency is given as

$$\eta = \frac{\text{Work Done}}{\text{Heat input}} \times 100\%$$

Work done by gas in cycle W = Area enclosed by $ABCD = P_0V_0$

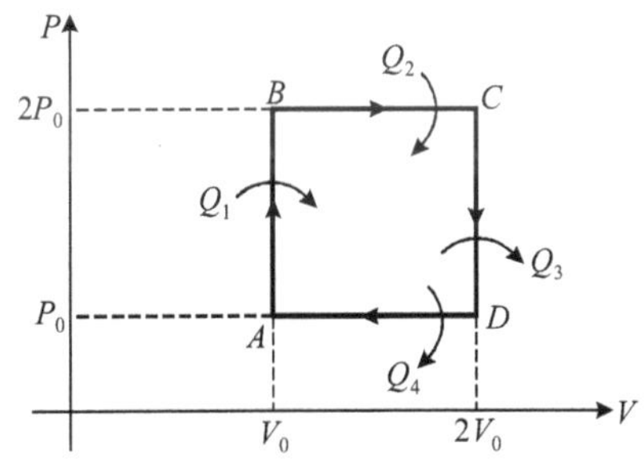

$$W = (Q_1 + Q_2) - (Q_3 - Q_4)$$

Heat input in cycle $Q_{in} = Q_1 + Q_2$

$$\Rightarrow \quad Q_{in} = nC_v\Delta T_{AB} + nC_p\Delta T_{BC}$$

$$\Rightarrow \quad Q_{in} = \frac{3}{2}nR\Delta T_{AB} + \frac{5}{2}nR\Delta T_{BC}$$

$$\Rightarrow \quad Q_{in} = \frac{3}{2}(P_0V_0) + \frac{5}{2}(2P_0V_0) = \frac{13}{2}P_0V_0$$

Cycle efficiency

$$\eta = \frac{W}{Q_{in}} \times 100\%$$

$$\Rightarrow \quad \eta = \frac{P_0V_0}{13/2P_0V_0} \times 100$$

$$\Rightarrow \quad \eta = \frac{200}{13}\% \sim 15.4\%$$ **Ans. (B)**

Sol. 25 Statement-2 : It is true

So carnot efficiency for source and sink at temperature 373 K and is given as

$$\eta = \left(1 - \frac{T_2}{T_1}\right) \times 100 = \left(1 - \frac{273}{373}\right) \times 100$$

$$\Rightarrow \quad \eta = (1 - 0.732) \times 100$$

$$= 26.8\% \text{ max possible}$$

Statement-1 : Is true & it is explained by Statement-2. **Ans. (D)**

Sol. 26 As in an adiabatic process there is no heat exchange

$$\Rightarrow \quad \Delta Q = 0$$

So from first law of thermodynamics, we have

$$\Delta Q = \Delta U + W$$

$$\Rightarrow \quad \Delta U = -W$$

$\Rightarrow$ Statement-1 is true

As T = constant in isothermal process $\Rightarrow$ Statement-2 is false

In adiabatic process we use $TV^{\gamma-1}$ = constant **Ans. (B)**

Sol. 27 On the PV diagram equation of process $A \rightarrow B$ is

$$P = mV + C \text{ where}$$

$$m = -\frac{P_0}{V_0} \text{ and } C = 3P_0$$

$$\Rightarrow \quad P = -\frac{P_0}{V_0} \cdot V + 3P_0$$

By gas law, we use

$$T = \frac{PV}{nR} = -\frac{P_0}{nRV_0} \cdot V^2 + \frac{3P_0}{nR} \cdot V \qquad \ldots(1)$$

T is max when $\frac{dT}{dV} = 0$

$$\Rightarrow \quad -\frac{P_0}{nRV_0}(2V) + \frac{3P_0}{nR} = 0$$

$$\Rightarrow \quad \frac{2V}{V_0} = 3 \quad \Rightarrow \quad V = \frac{3V_0}{2}$$

at maximum temperature

$\Rightarrow$ Maximum temperature gas in given by equation-(1) as

$$T_m = -\frac{P_0}{nRV_0}\left(\frac{3V_0}{2}\right)^2 + \frac{3P_0}{nR}\left(\frac{3V_0}{2}\right)$$

$$\Rightarrow \quad T_m = -\frac{P_0}{nRV_0} \times \frac{9V_0^2}{4} + \frac{9}{2}\frac{P_0V_0}{nR}$$

$$= \frac{P_0V_0}{nR}\left(\frac{9}{2} - \frac{9}{4}\right) = \frac{9}{4}\frac{P_0V_0}{nR}$$ **Ans. (D)**

Sol. 28 Gas expands from volume $V \rightarrow mV$

Work done by gas in the process is given as

$$W = \int_V^{mV} PdV = \int_V^{mV} \alpha VdV$$

$$\Rightarrow \quad W = \alpha\left[\frac{V^2}{2}\right]_V^{mV}$$

$$\Rightarrow \quad W = \frac{\alpha V^2}{2}(m^2 - 1) \quad \textbf{Ans. (D)}$$

Sol. 29 Initial state of gas for ideal monatomic gas

$$\gamma = 5/3$$

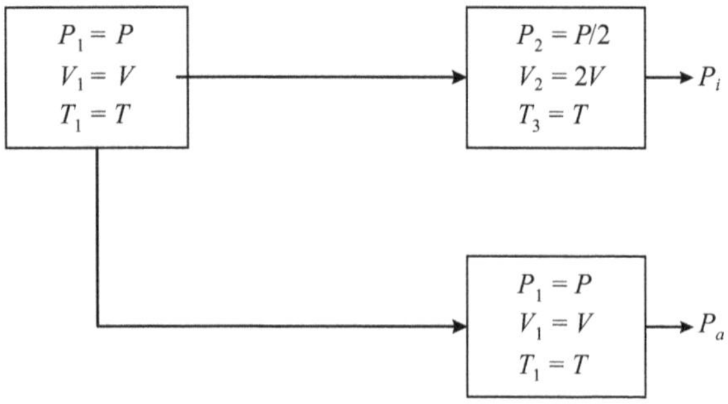

Isothermal expanded

$$T = \text{constant}$$

$$PV = \text{constant}$$

$$P.V = P_2(2V)$$

$$\Rightarrow \quad P_2 = P/2$$

Adiabatic expanded

$$\Delta Q = 0$$

$$PV^{5/3} = \text{constant}$$

$$PV^{5/3} = P_2'(2V)^{5/3}$$

$$\Rightarrow \quad P_2' = \frac{P^{5/3}}{2}$$

$$\text{Ratio} \quad \frac{P_a}{P_i} = \frac{P/2^{5/3}}{P/2} = 2^{1-5/3} = 2^{-2/3} \quad \textbf{Ans. (D)}$$

Sol. 30 For an adiabatic process state parameters V and T are related as

$$TV^{\gamma-1} = \text{constant}$$

$$\Rightarrow \quad T_1V_1^{\gamma-1} = T_2V_2^{\gamma-1}$$

$$\text{As} \quad V_2 = \frac{V_1}{4}$$

$$\Rightarrow \quad T_1V_1^{0.5} = T_2\left(\frac{V_1}{4}\right)^{0.5}$$

$$\Rightarrow \quad T_2 = \sqrt{4}T_1 = 2T_1 = 2T \quad \textbf{Ans. (C)}$$

Sol. 31 Refrigerator

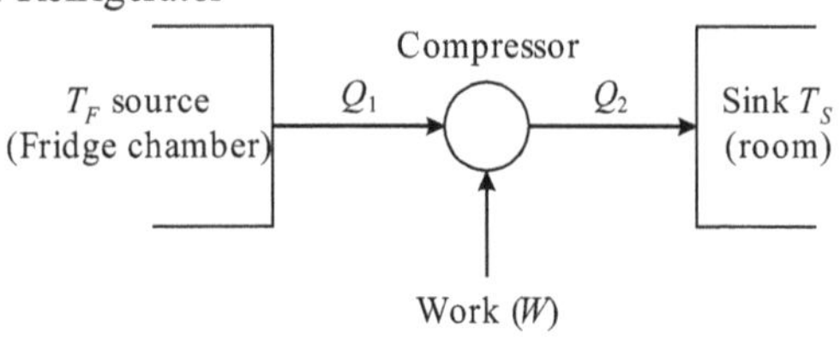

By conservation of energy

$$Q_2 = Q_1 + W$$

If door of refrigerator is kept open

$\Rightarrow$ Source and sink are same.

$\Rightarrow$ Net heat W will be given to room

$\Rightarrow$ Temperature of room increases **Ans. (D)**

Sol. 32 In the given cycle, heat extracted from the source is given as

$$Q_{out} = Q_3 + Q_4$$

$$\Rightarrow \quad Q_{out} = nC_V\Delta T_{BC} + nC_P\Delta T_{CD}$$

$$\Rightarrow \quad Q_{out} = \frac{3}{2}nR\Delta T_{BC} + \frac{5}{2}nR\Delta T_{CD}$$

$$\Rightarrow \quad Q_{out} = \frac{3}{2}\times 2P_0V_0 + \frac{5}{2}P_0V_0 = \frac{11}{2}P_0V_0$$

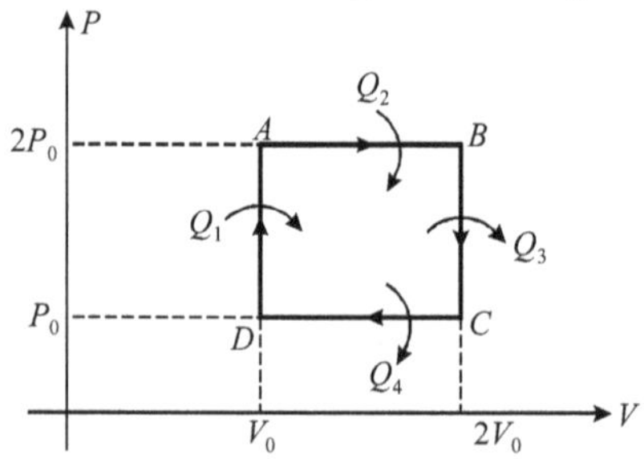

$$nR\Delta T_{BC} = 2V_0P_0$$

$$nR\Delta T_{CD} = P_0V_0. \quad \textbf{Ans. (C)}$$

Sol. 33

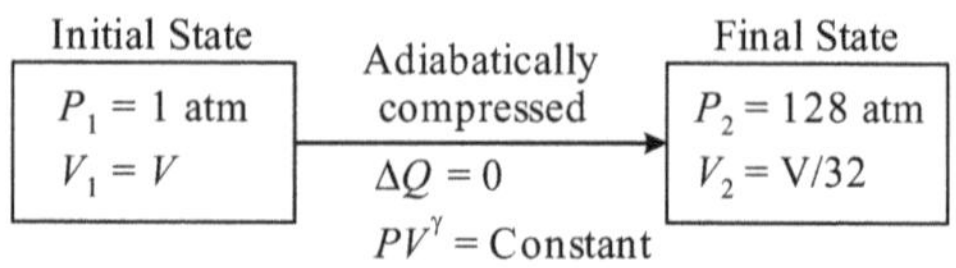

As density becomes 32 times

$$\Rightarrow \quad V \to V/32$$

$$\Rightarrow \quad P_1V_1^\gamma = P_2V_2^\gamma$$

$$\Rightarrow \quad (1)(V)^\gamma = 128\left(\frac{V}{32}\right)^\gamma$$

$$\Rightarrow \quad 32^\gamma = 128$$

$$\Rightarrow \quad 2^{5\gamma} = 2^7$$

$$\Rightarrow \quad 5\gamma = 7$$

$$\Rightarrow \quad \gamma = \frac{7}{5} = 1.4 \quad \textbf{Ans. (B)}$$

Sol. 34 Isochores on PV-diagram

$$P = \text{constant}$$

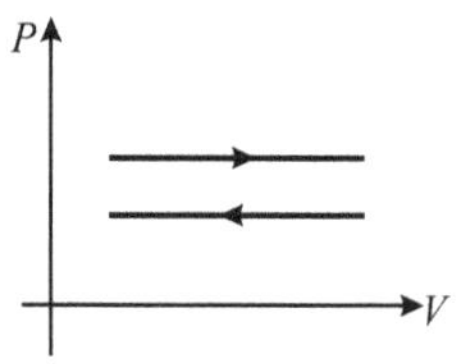

Isocharms on PV-diagram

$$V = \text{constant}$$

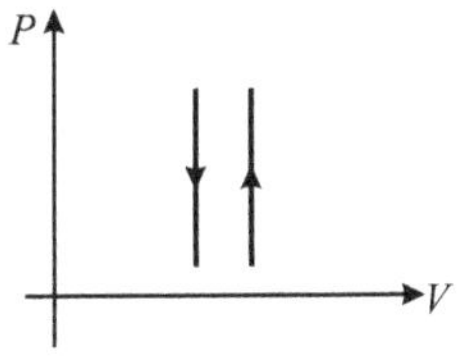

Isotherms on PV-diagram

$$T = \text{constant}$$

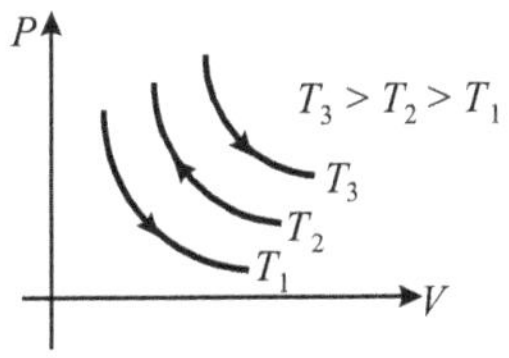

The given process has

$$\begin{bmatrix} \text{2 Isobars} \rightarrow AD \text{ and } BC \\ \text{1 Isotherm} \rightarrow CD \\ \text{1 Isochore} \rightarrow AB \end{bmatrix}$$

Ans. (C)

Sol. 35 Work done by a gas in isothermal expansion to given as

$$W = nRT \ln\left(\frac{V_2}{V_1}\right) = \frac{10 \times 10^{-3}}{M} RT \ln(2)$$

$$\Rightarrow \quad 575 = \frac{10^{-2}}{M} \cdot RT \times 0.693$$

$$\Rightarrow \quad \frac{RT}{M} = \frac{57500}{0.693}$$

$$v_{rms} = \sqrt{\frac{3RT}{M}} = \sqrt{\frac{3 \times 57500}{0.693}}$$

$$\Rightarrow \quad v_{rms} \simeq 498.6 \text{ m/s}$$

Ans. (C)

Sol. 36 Change in internal energy is given as

$$\Delta U = \frac{nR\Delta T}{\gamma - 1} = \frac{nR\Delta T}{0.4}$$

(for diatomic gas $\gamma = 1.4$)

$$\Rightarrow \quad \Delta U_{CA} = \frac{R}{0.4}(-200) = -500\,\text{R}$$

$$\Rightarrow \quad \Delta U_{AB} = \frac{R}{0.4}(400) = 1000\,\text{R}$$

$$\Rightarrow \quad \Delta U_{BC} = \frac{R}{0.4}(-200) = -500\,\text{R}$$

Ans. (D)

Sol. 37 Work done in an isobaric process is given as

$$W = P\Delta V = P(V_f - V_i)$$

as initial process is P_0

$$\Rightarrow \quad W = P_0(V_f - V_i) \quad \ldots(1)$$

From the given equation, we have

$$PV = nRT + \alpha V$$

for $n = 1$ mole $\quad V = \dfrac{RT}{P - \alpha}$

at $T = T_0 \rightarrow V_i = \dfrac{RT_0}{P_0 - \alpha}$

at $T = 2T_0 \rightarrow V_f = \dfrac{2RT_0}{P_0 - \alpha}$

from (1)

$$W = P_0\left(\frac{2RT_0}{P_0 - \alpha} - \frac{RT_0}{P_0 - \alpha}\right) = \frac{RP_0T_0}{P_0 - \alpha}$$

Ans. (A)

Sol. 38 Initial state of gas

$$P_1 = 10^5 \text{ N/m}^2$$
$$V_1 = 2.4 \times 10^{-3} \text{ m}^3$$
$$T_1 = 300 \text{ K}$$

After expansion, gas process is

$$P_f = P_{atm} + \frac{kx}{A}$$

$$\Rightarrow \quad P_f = 10^5 + \frac{8000 \times 0.1}{8.0 \times 10^{-3}}$$

$$\Rightarrow \quad P_f = 10^5 + 10^5 = 2 \times 10^5 \text{ N/m}^2$$

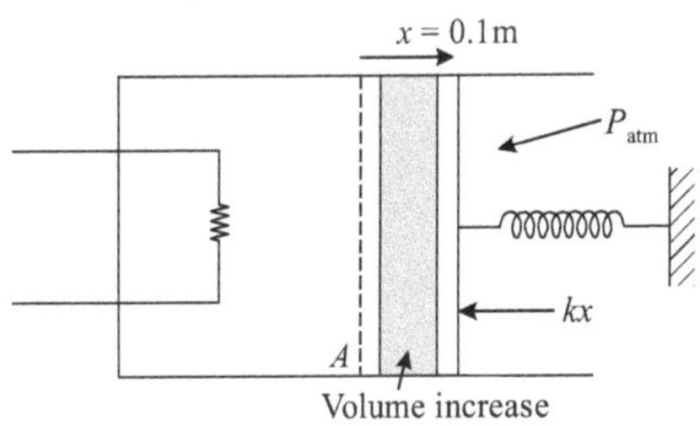

$$\Delta V = Ax$$
$$\Delta V = 8.0 \times 10^{-3} \times 0.1$$
$$\Delta V = 0.8 \times 10^{-3} \text{ m}^3$$

Find state of gas

$$P_f = 2 \times 10^5 \text{ N/m}^2$$
$$V_f = V_i + \Delta V = 2.4 \times 10^{-3}\text{m}^3 + 0.8 \times 10^{-3}\text{m}^3$$
$$= 3.2 \times 10^{-3} \text{ m}^2$$
$$T_f = \rho$$

As no of moles of gas are constant. So we use

$$\frac{P_1V_1}{T_1} = \frac{P_fV_f}{T_f}$$

$$\Rightarrow \quad \frac{10^5 \times 2.4 \times 10^{-3}}{300} = \frac{2 \times 10^5 \times 3.2 \times 10^{-3}}{T_F}$$

$$\Rightarrow \quad T_f = \frac{2 \times 3.2}{2.4} \times 300 = 800 \text{ K}$$

Ans. (B)

Sol. 39 In an adiabatic process, work done by gas is given as

$$W = -\Delta U = -\frac{nR\Delta T}{\gamma - 1}$$

given that $W = -830$ J (work done on gas) for $n = 2$ mole and $\gamma = 1.4$ for a diatomic ideal gas

$$\Rightarrow \quad W = -830 = -\frac{2 \times 8.3 \times \Delta T}{1.4 - 1}$$

$$\Rightarrow \quad \Delta T = \frac{830 \times 0.4}{2 \times 8.3} = \frac{40}{2} = 20\ \text{K}$$

$\Rightarrow$ Temperature gas increases by 20 K **Ans. (C)**

Sol. 40 By first law of thermodynamics we have

$$\Delta Q = \Delta U + \Delta W \quad \ldots(1)$$

In whole process $\Delta Q = 150$ J

and $\Delta W = P\Delta V = P(V_2 - V_1) = 100\,(1 - 2)$

$\Rightarrow \quad \Delta W = -100$ J

As gas is compressed $\Rightarrow$ work is done on the gas ($-ve$)

So from equation-(1), we use

$\Rightarrow \quad 150 = \Delta U - 100$

$\Rightarrow \quad \Delta U = 250$ J

$\Rightarrow$ Internal energy increases by 250 J **Ans. (A)**

Sol. 41

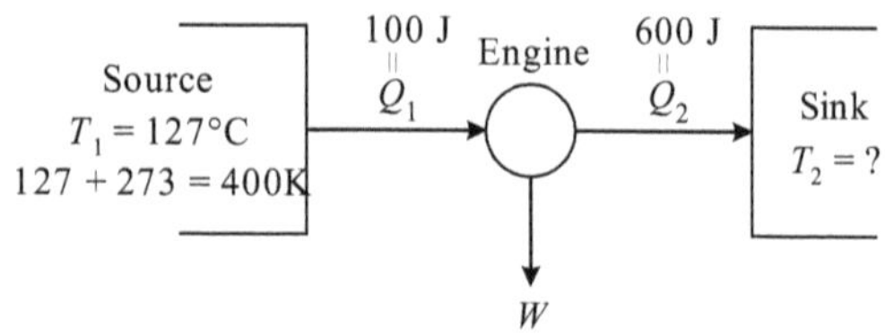

Work done per cycle by engine

$$W = Q_1 - Q_2$$

$$= 1000 - 600 = 400\ \text{J}$$

Efficiency of engine

$$\eta = \frac{W}{Q_1} \times 100 = \frac{400}{1000} \times 100 = 40\%$$

In carnot engine Q_1 and Q_2 are proportional to T_1 and T_2

$$\Rightarrow \quad \frac{Q_1}{Q_2} = \frac{T_1}{T_2}$$

$$\Rightarrow \quad \frac{1000}{600} = \frac{400}{T_2}$$

$$\Rightarrow \quad T_2 = 240\ \text{K} = -33°\text{C}$$ **Ans. (B)**

Sol. 42 Average time between collisions of gas molecules

$$t \propto \frac{V}{\sqrt{T}}$$

In Adiabatic process we use $TV^{\gamma-1}$ = constant

$$\Rightarrow \quad T = CV^{1-\gamma}$$

$$\Rightarrow \quad t \propto \frac{V}{\sqrt{CV^{1-\gamma}}} = C_1 V^{1-\frac{1-\gamma}{2}} = C_1 V^{\frac{1+\gamma}{2}}$$

$$\Rightarrow \quad t \propto V^{\frac{1+\gamma}{2}} \Rightarrow q = \frac{1+\gamma}{2}$$ **Ans. (C)**

Sol. 43 Given that

$$\frac{U}{V} = kT^4$$

$\Rightarrow$ Gas pressure $P = \frac{1}{3}kT^4$

By Gas law $PV = nR_gT$

$R_g \to$ Gas constant

$$\Rightarrow \quad \frac{1}{3}kT^4 \cdot \left(\frac{4}{3}\pi R^3\right) = nR_gT$$

$$\Rightarrow \quad T^3R^3 = \text{constant}$$

$$\Rightarrow \quad T \propto \frac{1}{R}$$ **Ans. (A)**

Sol. 44 Change in Entropy of a solid body is given as

$$\Delta S = \int \frac{dQ}{T}$$

for a solid body we use

$$dQ = CdT$$

$$\Rightarrow \quad \Delta S = \int_{T_1}^{T_2} \frac{CdT}{T} = C\ln\left(\frac{T_2}{T_1}\right)$$

$$\Rightarrow \quad \Delta S = 1\ln\left(\frac{200}{100}\right) = \ln(2)$$

As entropy change depends only on temperature (initial & final) so in both cases ΔS remains same. **Ans. (D)**

Sol. 45 In V-T diagram 'ab' and 'cd' are straight line

$\Rightarrow$ These are isobars on PV diagram

Isobars on VT curve pass through O (assuming this by looking at given options)

$\Rightarrow \quad P$ = constant

from 'bc' and 'da' which are adiabatic

Option (B) is the correct Ans.

(B) $b \to c\ (\to)$ volume is increasing

$d \to a\ (\to)$ volume is decreasing

Corresponding to given VT curve.

Also slope of curve in $-ve$ and decreasing

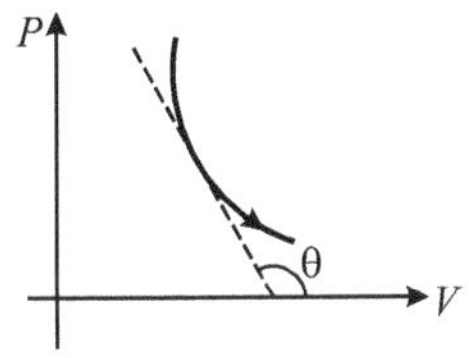

$$m = \tan\theta \rightarrow -ve$$

m is decreasing in magnitude with volume. **Ans. (B)**

Sol. 46 Equation of Straight line shown in indicator diagram is

$$P = -\frac{P_0}{V_0}\cdot V + 3P_0$$

$$\text{slope} = -\frac{P_0}{V_0}$$

$$\text{intercept} = 3P_0$$

$$\text{equation } y = mx + c$$

Temperature of gas during the process is

$$T = \frac{PV}{nR} = -\frac{P_0}{nRV_0}\cdot V^2 + \frac{3P_0}{nR}\cdot V$$

For maximum temperature we use

$$\frac{dT}{dV} = 0 = -\frac{2P_0}{nRV_0}\cdot V + \frac{3P_0}{nR}$$

$$\Rightarrow \quad V = \frac{3}{2}V_0$$

The maximum temperature of gas is given as

$$T_{max} = -\frac{P_0}{nRV_0}\left(\frac{3V_0}{2}\right)^2 + \frac{3P_0}{nR}\left(\frac{3V_0}{2}\right)$$

$$\Rightarrow \quad T_{max} = -\frac{9}{4}\frac{P_0V_0}{nR} + \frac{9}{2}\frac{P_0V_0}{nR} = \frac{9}{4}\frac{P_0V_0}{nR}$$ **Ans. (A)**

Sol. 47 For a polytrophic process here $n \rightarrow$ Polytropic constant

$$PV^n = \text{Constant}$$

& for this process the molar specific heat of gas is given as

$$C = C_V + \frac{R}{1-n}$$

$$\Rightarrow \quad \frac{R}{1-n} = C - C_V$$

$$\Rightarrow \quad 1 - n = \frac{R}{C - C_V} = \frac{C_P - C_v}{C - C_V}$$

$$\Rightarrow \quad n = 1 - \frac{C_P - C_V}{C - C_V} = \frac{C - C_V - C_P + C_V}{C - C_V}$$

$$\Rightarrow \quad n = \frac{C - C_P}{C - C_V}$$ **Ans. (B)**

Sol. 48 Refrigerator

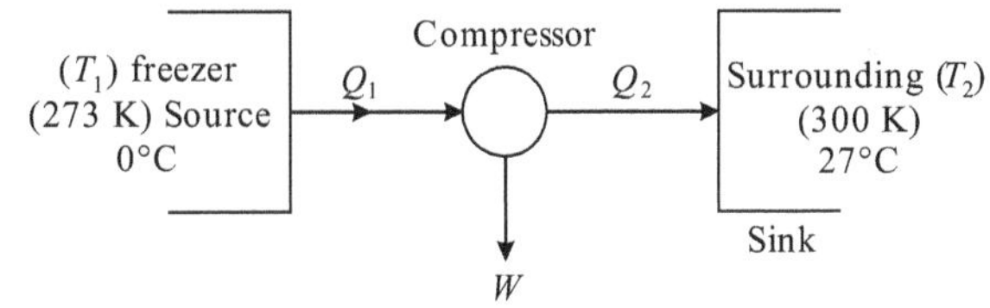

by conservation of energy

$$Q_1 + W_{ext} = Q_2$$

$$\Rightarrow \quad W_{ext} = Q_2 - Q_1$$

For a carnot freezer we use

$$\frac{Q_1}{Q_2} = \frac{T_1}{T_2} \quad \ldots(1)$$

here $Q_1 \rightarrow$ heat released by freezer in freezing ice

Here $\quad Q_1 = mL = 5 \times 336 \times 10^3\,(\text{J/kg}) \times \text{kg}$

$$= 5 \times 336 \times 10^3\,\text{J}$$

We use by equation-(1)

$$Q_2 = \left(\frac{T_2}{T_1}\right) \times Q_1 = \frac{300}{273} \times 5 \times 336 \times 10^3$$

Energy consumed by freezer in

$$W = Q_2 - Q_1$$

$$\Rightarrow \quad W = 5 \times 336 \times 10^3\left(\frac{300}{273} - 1\right)$$

$$\Rightarrow \quad W = 1.67 \times 10^5\,\text{J}$$ **Ans. (D)**

Sol. 49 In an isobaric process (P = constant)

Heat supplied $\quad \Delta Q = nC_P\Delta T$

[for a monoatomic gas $C_P = \frac{J}{2}R$]

$$\Rightarrow \quad \Delta Q = \frac{5}{2}nR\Delta T \quad \ldots(1)$$

Work done $\quad W = P\Delta V = nRT \quad \ldots(2)$

$$\frac{(2)}{(1)} \Rightarrow \frac{W}{\Delta Q} = \frac{nR\Delta T}{5/2nR\Delta T} = \frac{2}{5}$$ **Ans. (A)**

Sol. 50 By first law of thermodynamics

$$\Delta Q = \Delta U + \Delta W$$

If expansion of water is neglected

$$\Rightarrow \quad \Delta W = 0 \text{ (heat supplied)}$$

$$\Delta U = mS\Delta T$$

$$\Rightarrow \quad \Delta U = 0.2 \times 4184 \times 20 = 16.7\,\text{kJ}$$ **Ans. (D)**

Sol. 51 Work done by the gas in the cycle is

$$W = P_0V_0 \text{ (area enclosed)}$$

Heat supplied in cycle can be given as

$$\Delta Q_{supply} = Q_{AB} + Q_{BC} = nC_V(T_B - T_A) + nC_P(T_C - T_B)$$

Here we use $C_V = 1.5R$ and $C_P = C_V + R = 2.5R$

$$\Rightarrow \quad \Delta Q_{supply} = 1.5(nRT_B - nRT_A) + 2.5(nRT_C - nRT_B)$$

By Gas law, we use

$$\Delta Q_{supply} = 1.5(2P_0V_0 - P_0V_0) + 2.5(4P_0V_0 - 2P_0V_0)$$

$$\Rightarrow \quad \Delta Q_{supply} = 1.5P_0V_0 + 5P_0V_0$$

$$\Rightarrow \quad \Delta Q_{supply} = 6.5P_0V_0$$

Engine efficiency

$$\eta = \frac{W}{\Delta Q_{supply}}$$

$$= \frac{P_0V_0}{(13/2)P_0V_0} = \frac{2}{13} = 0.15 \quad \textbf{Ans. (B)}$$

Sol. 52 By First law of thermodynamics, we have for ACB

$$\Delta Q_{ACB} = \Delta U_{ACB} + \Delta W_{ACB}$$

$$\Rightarrow \quad 60 = \Delta U_{ACB} + 30 \text{ J}$$

$$\Rightarrow \quad \Delta U_{ACB} = 30 \text{ J} = \Delta U_{ADB}$$

For process ADB

$$\Delta Q_{ADB} = \Delta U_{ADB} + \Delta W_{ADB}$$

$$\Delta Q_{ADB} = 30 \text{ J} + 10 \text{ J} = 40 \text{ J}$$ **Ans. (D)**

Sol. 53 For closed vessel, heat supplied is given as

$$\Delta Q = nC_v\Delta T$$

$$\Rightarrow \quad \Delta Q = \frac{15}{28} \times \frac{5R}{2} \times (4T - T) \quad [T = 300 \text{ K}]$$

$$\Rightarrow \quad \Delta Q = 10021 \text{ J} \cong 10 \text{ kJ} \quad \textbf{Ans. (A)}$$

Sol. 54 Given system of carnot engines can be drawn as

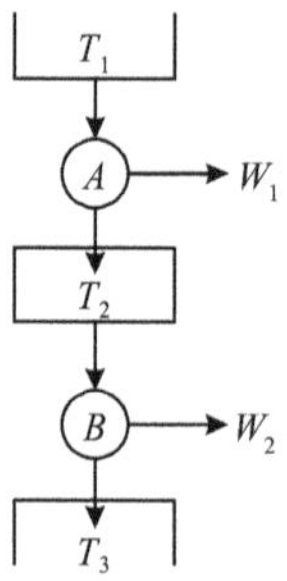

As we use $\quad W_1 = W_2$

$$\Rightarrow \quad \Delta Q_1 = \Delta Q_2$$

$$\Rightarrow \quad T_2 - T_1 = T_3 - T_2$$

$$\Rightarrow \quad T_2 = \frac{T_1 + T_3}{2} + \frac{600 + 400}{2}$$

$$= 500\,K \quad \textbf{Ans. (C)}$$

Sol. 55 As engines are equally efficient, we use

$$\eta = 1 - \frac{T_2}{T_1} = 1 - \frac{T_3}{T_2} = 1 - \frac{T_4}{T_3}$$

$$\Rightarrow \quad \frac{T_2}{T_1} = \frac{T_3}{T_4} = \frac{T_4}{T_3}$$

$$\Rightarrow \quad T_2 = \sqrt{T_1T_3}$$

and $\quad T_3 = \sqrt{T_2T_4}$

$$\Rightarrow \quad T_2^{3/4} = \sqrt{T_1^{1/2}T_4^{1/4}}$$

$$\Rightarrow \quad T_2 = T_1^{2/3}T_4^{1/3}$$

and $\quad T_3 = T_1^{1/3}T_4^{2/3}$ **Ans. (A)**

Sol. 56 Work done by gas is

$$W = P\Delta V = nR\Delta T$$

$$\Rightarrow \quad W = \frac{1}{2} \times 8.31 \times 70$$

$$\Rightarrow \quad W = 291 \text{ J}$$ **Ans. (B)**

Sol. 57 For adiabatic process we use $TV^{\gamma-1}$ = constant and for a diatomic gas we use

$$\gamma - 1 = \frac{7}{5} - 1 = \frac{2}{5}$$

$$\Rightarrow \quad x = \frac{2}{5} \quad \textbf{Ans. (B)}$$

Sol. 58 Given that $VT = K$

$$\Rightarrow \quad V\left(\frac{PV}{nR}\right) = K$$

$$\Rightarrow \quad PV^2 = K$$

For a polytropic process, molar specific heat of gas is given as

$$C = \frac{R}{1-n} + C_V \qquad (n = 2)$$

$$\Rightarrow \quad C = \frac{R}{1-2} + \frac{3R}{2} = \frac{R}{2}$$

Heat supplied to gas is

$$\Delta Q = nC\,\Delta T$$

$$\Rightarrow \quad \Delta Q = 1 \times \frac{R}{2} \times \Delta T = \frac{R\Delta T}{2}$$ **Ans. (A)**

Sol. 59 Since P–V indicator diagram is given, so work done by gas is area enclosed by the cyclic diagram

$$\Rightarrow \quad \text{Work done by gas} = \frac{1}{2} \times 4 \times 5 \text{ J}$$

$$\Rightarrow \quad W = 10 \text{ J} \quad \textbf{Ans. (C)}$$

Sol. 60 Process a → constant ⇒ Isobaric
Process b → Isothermal
Process c → Adiabatic
Process d → Isochoric

Adiabatic curve is steeper them isothermal curve

Order as specified in que → d, a, b, c **Ans. (D)**

Sol. 61

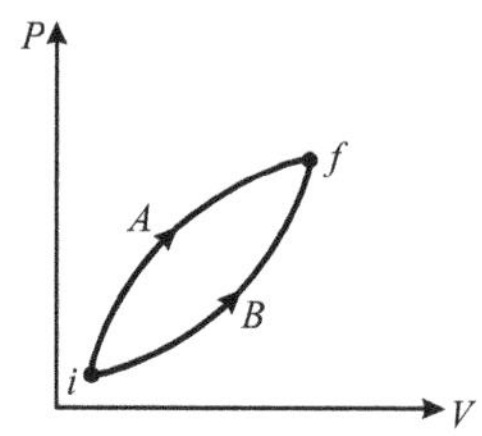

Initial and final states for both the processes are same so we use

$$\Delta U_A = \Delta U_B$$

Work done during process A is greater than in process B so using First Law of thermodynamics, we have

$$\Delta Q = \Delta U + W$$

$$\Rightarrow \quad \Delta Q_A > \Delta Q_B \qquad \textbf{Ans. (B)}$$

Sol. 62 For constant volume heating, we use

$$\Delta Q = nC_V\Delta T = n\frac{3}{2}R\Delta T$$

$$= \left(\frac{67.2}{22.4}\right)\left(\frac{3}{2}\times 8.31\right)(20)$$

$$\approx 748 \text{ J}$$

Ans. (A)

Sol. 63 For isobaric process work done by gas is given as

Heat supplied is $\quad W = nR\Delta T$

$$\Delta Q = (C_v + nR)\Delta T$$

$$\Rightarrow \quad \frac{W}{\Delta Q} = \frac{nR}{C_v + nR} \qquad \textbf{Ans. (C)}$$

Sol. 64 Heat required for constant volume and constant pressure to raise gas temperature by ΔT is

$$Q = nC_v\Delta T$$

$$Q' = nC_p\Delta T$$

$$\Rightarrow \quad \frac{Q'}{Q} = \frac{C_p}{C_v}$$

For diatomic gas we have

$$\frac{C_p}{C_v} = \gamma = \frac{7}{5}$$

$$\Rightarrow \quad Q' = \frac{7}{5}Q$$

Ans. (A)

Sol. 65 Temperature of gas ($n = 1$) is given as

$$PV = RT \Rightarrow T = \frac{PV}{R} = \frac{P_0V}{R}\left(1 - \frac{1}{2}\left(\frac{V_0}{V}\right)^2\right)$$

At $V = V_0$, temperature is

$$T_1 = \frac{P_0V_0}{R}\left[1 - \frac{1}{2}\right] = \frac{P_0V_0}{2R}$$

At $V = 2V_0$, temperature is

$$T_2 = \frac{2P_0V_0}{R}\left[1 - \frac{1}{2}\left(\frac{1}{4}\right)\right] = \frac{7P_0V_0}{4R}$$

Temperature change

$$\Delta T = T_2 - T_1 = \frac{7P_0V_0}{4R} - \frac{P_0V_0}{2R} = \frac{5P_0V_0}{4R} \qquad \textbf{Ans. (C)}$$

Sol. 66 Work done along path abc is

$$W_{abc} = W_{ab}$$

As $\quad W_{bc} = 0 \qquad$ (being an isochoric process)

By First law of thermodynamics, we use

$$W_{ab} = \Delta Q_{ab} - \Delta U_{ab}$$

In a cyclic process net

$$\Delta U_{\text{cycle}} = 0$$

$$\Delta U_{ab} + \Delta U_{bc} + \Delta U_{ca} = 0$$

$$\Rightarrow \quad \Delta U_{ab} + 60 - 180 = 0 \qquad [\Delta U_{bc} = \Delta Q_{bc}]$$

$$\Rightarrow \quad \Delta U_{ab} = 120 \text{ J}$$

$$= 250 - 120 = 130 \text{ J}$$

Ans. (D)

Sol. 67 Efficiency of Carnot engine

$$\eta = 1 - \frac{T_{\text{sink}}}{T_{\text{source}}}$$

$$\Rightarrow \quad \frac{1}{6} = 1 - \frac{T_{\text{sink}}}{T_{\text{source}}}$$

$$\Rightarrow \quad \frac{T_{\text{sink}}}{T_{\text{source}}} = \frac{5}{6} \qquad \ldots(1)$$

Efficiency gets doubled when temperature of sink decreases by 62°C so we use

$$\frac{2}{6} = 1 - \frac{T_{\text{sink}} - 62}{T_{\text{source}}}$$

$$\Rightarrow \quad \frac{62}{T_{\text{source}}} = \frac{1}{6} \qquad \ldots(2)$$

$$\Rightarrow \quad T_{\text{source}} = 372 \text{ K} = 99^\circ\text{C}$$

From equation-(1), we have

$$T_{\text{sink}} = \frac{5}{6}\times 372 = 310 \text{ K} = 37^\circ\text{C} \qquad \textbf{Ans. (B)}$$

Sol. 68 For a diatomic gas, we have

$$C_p = \frac{7}{2}R$$

As gas undergoes isobaric process, heat absorbed by gas is

$$\Rightarrow \quad \Delta Q = nC_P\Delta T$$

In isobaric process, work done is

Also, $\quad \Delta W = nR\Delta T = 10\text{J (given)}$

$$\Rightarrow \quad \Delta Q = n\left(\frac{7}{2}R\Delta T\right) = \frac{7}{2}(nR\Delta T)$$

$$= \frac{7}{2}\times 10 = 35 \text{ J} \qquad \textbf{Ans. (A)}$$

12.1 Kinetic Theory of Gases

Sol. 1

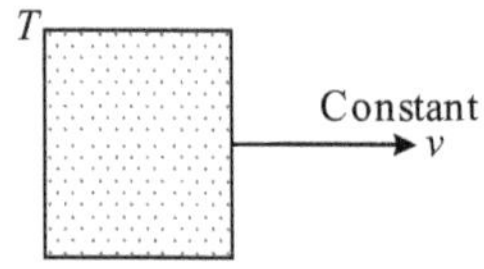

Container with gas

Temperature of gas is a measure of *KE* of random motion of gas molecules

$$E=\frac{f}{2}kT$$

where $f\rightarrow$ degrees of freedom

If container is moving at uniform speed in a specific direction then it does not affect the random motion of gas molecules in container. Due to v, all particles will have an ordered motion which is not affecting its T. **Ans. (C)**

Sol. 2 $T_{O_2}=273+47=320\text{ K}$

RMS velocity of molecules is given as

$$v=\sqrt{\frac{3RT}{M}}$$

If at T_{H_2} the velocity (*RMS*) of H_2 molecule is equal to that of O_2 molecule at 320 K

$$\Rightarrow \quad v_{H_2}=v_{O_2}$$

$$\Rightarrow \quad \sqrt{\frac{3RT_{H_2}}{2\times10^{-3}}}=\sqrt{\frac{3RT_{O_2}}{32\times10^{-3}}}$$

$$\Rightarrow \quad T_{H_2}=\frac{2}{32}\times320\text{K}=20\text{ K}$$ **Ans. (D)**

Sol. 3 The equivalent value of γ for a mixture of gas is given as

$$\frac{n_1+n_2}{\gamma_{eq}-1}=\frac{n_1}{\gamma_1-1}+\frac{n_2}{\gamma_2-1}$$

$$\Rightarrow \quad \frac{1+1}{\gamma_{eq}-1}=\frac{1}{\frac{7}{5}-1}+\frac{1}{\frac{5}{3}-1}$$

$$\Rightarrow \quad \frac{2}{\gamma_{eq}-1}=\frac{5}{2}+\frac{3}{2}=4$$

$$\Rightarrow \quad \gamma_{eq}-1=\frac{1}{2}$$

$$\Rightarrow \quad \gamma_{eq}=\frac{3}{2}=\frac{24}{16}$$ **Ans. (C)**

Sol. 4

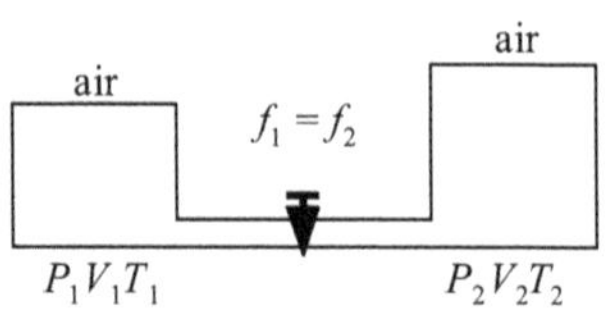

No of moles of gas (air) in the two vessels are

$$n_1=\frac{P_1V_1}{RT_1} \text{ and } n_2=\frac{P_2V_2}{RT_2}$$

By conservation of internal energy final equilibrium temperature of mixture is given as

$$T_f=\frac{f_1n_1T_1+f_2n_2T_2}{f_1n_1+f_2n_2}=\frac{n_1T_1+n_2T_2}{n_1+n_2}$$

$$\Rightarrow \quad T_f=\frac{P_1V_1+P_2V_2}{\frac{P_1V_1}{T_1}+\frac{P_2V_2}{T_2}}=\frac{T_1T_2(P_1V_1+P_2V_2)}{P_1V_1T_2+P_2V_2T_1}$$ **Ans. (C)**

Sol. 5 For equivalent γ of mixture of gases we use

$$\frac{n_1+n_2}{\gamma_{eq}-1}=\frac{n_1}{\gamma_1-1}+\frac{n_2}{\gamma_2-1}$$

$$\Rightarrow \quad \frac{1+1}{\gamma_{eq}-1}=\frac{1}{\frac{5}{3}-1}+\frac{1}{\frac{7}{5}-1}$$

$$\Rightarrow \quad \frac{2}{\gamma_{eq}-1}=\frac{3}{2}+\frac{5}{2}=4$$

$$\Rightarrow \quad \gamma_{eq}-1=\frac{1}{2} \Rightarrow \gamma_{eq}=\frac{3}{2}$$ **Ans. (A)**

Sol. 6 Equivalent γ for a gaseous mixture is given as

$$\frac{n_1+n_2}{\gamma_{eq}-1}=\frac{n_1}{\gamma_1-1}+\frac{n_2}{\gamma_2-1}$$

for H_e gas $\quad n_1=\frac{m_1}{M_1}=\frac{16}{4}=4\text{ mole}$

for O_2 gas $\quad n_2=\frac{m_2}{M_2}=\frac{16}{32}=0.5\text{ mole}$

$$\Rightarrow \quad \frac{4.5}{\gamma_{eq}-1}=\frac{4}{\frac{5}{3}-1}+\frac{0.5}{\frac{7}{5}-1}=6+\frac{2.5}{2}=7.25$$

$$\Rightarrow \quad \gamma_{eq}-1=\frac{4.5}{7.25} \Rightarrow \gamma_{eq}=1.62$$ **Ans. (B)**

Sol. 7 When heat flows between boxes, total internal energy of the two gases will remain constant so for this cases the final equilibrium temperature of gases is given as

$$T_{eq}=\frac{f_1n_1T_1+f_2n_2T_2}{f_1n_1+f_2n_2}$$

for $N_2\rightarrow$ $\begin{bmatrix} n_1=1 \\ f_1=5 \\ T_1=T_0 \end{bmatrix}$ and for $He\rightarrow$ $\begin{bmatrix} n_2=1 \\ f_2=3 \\ T_2=\frac{7}{3}T_0 \end{bmatrix}$

$$\Rightarrow \quad T_{eq}=\frac{5\times1\times T_0+3\times1\times(7/3)T_0}{5\times1+3\times1}$$

$$\Rightarrow \quad T_{eq}=\frac{12T_0}{8}=\frac{3}{2}T_0$$ **Ans. (C)**

Sol. 8 C'_P and C'_V are the specific heats for an ideal gas per unit mole then we use the relation

$$C'_P - C'_V = R \qquad \ldots(1)$$

C_P and C_V per unit mass for N_2 gas can be given as

$$M = 28\text{gm}$$

$$\Rightarrow \quad C_P = \frac{C'_P}{28} \text{ and } C_V = \frac{C'_V}{28}$$

$\Rightarrow$ from equation (1), we have

$$28C_P - 28C_V = R$$

$$\Rightarrow \quad C_P - C_V = \frac{R}{28}$$

Ans. (A)

Sol. 9

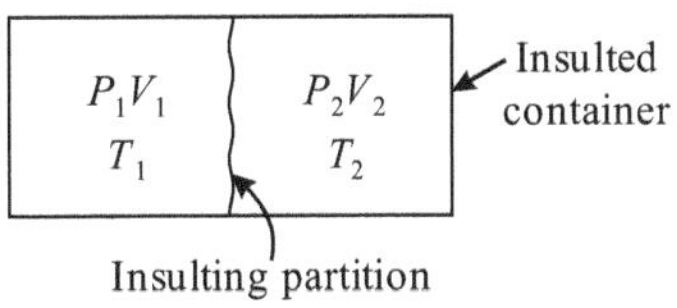

$$f_1 = f_2$$

$$n_1 = \frac{P_1V_1}{RT_1} \quad \text{and} \quad n_2 = \frac{P_2V_2}{RT_2}$$

When gases are mixed after removing partition then by internal energy conservation final equation temperature mixture is given as

$$T_f = \frac{f_1n_1T_1 + f_2n_2T_2}{f_1n_1 + f_2n_2} = \frac{n_1T_1 + n_2T_2}{n_1 + n_2}$$

$$\Rightarrow \quad T_f = \frac{T_1T_2(P_1V_1 + P_2V_2)}{P_1V_1T_2 + P_2V_2T_1}$$

Ans. (A)

Sol. 10 Energy due to thermal motion $\rightarrow$ internal energy of gas (energy of rendom motion)

For a given sample of gas its internal energy is given as

$$U = \frac{f}{2}nRT$$

For a diatomic gas $f = 5$ and here gas volume

$$V = \frac{m}{\rho} = \frac{1}{4}m^3$$

$$\Rightarrow \quad U = \frac{5}{2}nRT = \frac{5}{2}PV = \frac{5}{2} \times 8 \times 10^4 \times \frac{1}{4}$$

$$\Rightarrow \quad U = 5 \times 10^4 \text{ J}$$

Ans. (D)

Sol. 11 For gaseous mixture, final equilibrium temperature is given as

$$T_{eq} = \frac{f_1n_1T_1 + f_2n_2T_2 + f_3n_3T_3}{f_1n_1 + f_2n_2 + f_3n_3}$$

We consider $f_1 = f_2 = f_3$

As no of moles α no of molecules

$$\Rightarrow \quad T_{eq} = \frac{n_1T_1 + n_2T_2 + n_3T_3}{n_1 + n_2 + n_3}$$

Ans. (D)

Sol. 12 As gas pressure is constant so we use by gas law

$$\frac{V_1}{T_1} = \frac{V_2}{T_2}$$

here $\quad V_2 = 2V_1$

$$\Rightarrow \quad T_2 = 2T_1 = 2 \times (273 + 27) = 60 \text{ K}$$

$$\Rightarrow \quad T_2 = 600 - 273 = 327°\text{C}$$

Ans. (A)

Sol. 13 Initial situation

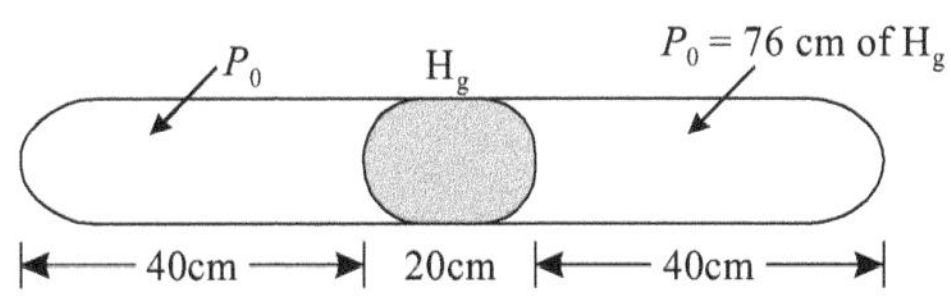

Final situation

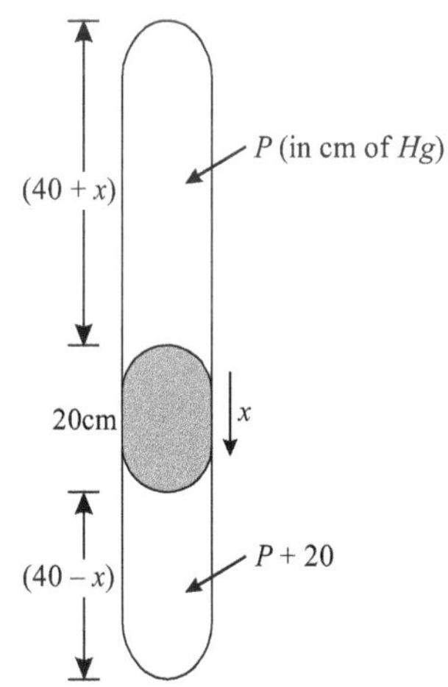

we consider constant temperature, so we have

upper section $\rightarrow P_0 \times 40 = P \times (40 + x) \qquad \ldots(1)$

lower section $\rightarrow P_0 \times 40 = (P + 20) \times (40 - x) \qquad \ldots(2)$

from (1) and (2)

$$40\,P + P_x = 40\,P + 800 - P_x = 20x$$

$$\Rightarrow \quad P_x = 400 - 10x$$

$$\Rightarrow \quad P = \frac{400 - 10x}{x} \qquad \ldots(3)$$

from equation (1)

$$76 \times 40 = \frac{(400 - 10x)}{x}(40 + x)$$

$$\Rightarrow \quad 3040x = 1600 - 400x + 400x - 10x^2$$

$$\Rightarrow \quad x^2 + 304x - 1600 = 0$$

$$\Rightarrow \quad x = \frac{-304 + \sqrt{(304)^2 - 6400}}{2} \simeq 5.17 \text{ cm.}$$

Ans. (B)

Sol. 14 For an isobaric process heat supplied is related to the rise in temperature as

$$Q = nC_P\Delta T$$

Slope of $Q - \Delta T$ curve $= \dfrac{Q}{\Delta T} = nC_P$

Here all gases are having same initial states $(P_1, V_1, T_1) \rightarrow$ same

$\Rightarrow$ Same have equal no of moles.

For a gas $C_P = \dfrac{f+2}{2}$

Where $f = 3$ for monoatomic gas (M)

$f = 5$ for diatomic gas (D)

$f = 6$ for polyatomic gas (P)

Maximum linear molecule

As $C_P(P) > C_P(D) > C_P(M)$ **Ans. (C)**

Sol. 15 Statement-1 : It is false because in free expansion temperature do not change as $W = 0$.

Statement-2 : True as for ideal gas we neglect *PE* as particles have no interaction but in real gas *PE* is there **Ans. (A)**

Sol. 16

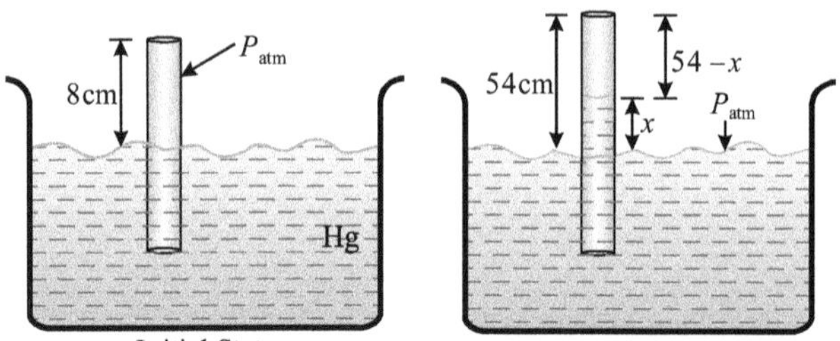

Initial State Final State

As total no of moles of air trapped in tube is constant we use

$$P_1V_1 = P_2V_2$$

$\Rightarrow$ $76 \times 8 = (76-x) \times (54-x)$

$\Rightarrow$ $76 \times 8 = 76 \times 54 - 130x + x^2$

$\Rightarrow$ $x^2 - 130x + 76 \times 46 = 0$

$\Rightarrow$ $x = \dfrac{130 \pm \sqrt{(130)^2 - 4 \times 76 \times 46}}{2}$

$\Rightarrow$ $x = \dfrac{130-54}{2} = \dfrac{76}{2} = 38$ cm

Length of air column in final state

$l = 54 - x = 54 - 38 = 16$ cm **Ans. (A)**

Sol. 17 If there are N molecules in a vessel of volume V then mean distance between molecules in the vessel is given as

$$d_m = \left(\frac{V}{N}\right)^{1/3} \quad \ldots(1)$$

Using gas law, we have

$$PV = nRT$$

$\Rightarrow$ $PV = \dfrac{N}{N_{AV}} \cdot RT$

$\Rightarrow$ $\dfrac{V}{N} = \dfrac{RT}{PN_{AV}} = \dfrac{8.0 \times 300}{4 \times 10^{-10} \times 6 \times 10^{23}}$

$= \dfrac{100}{10^{12}} = 10^{-11}$ m^3/molecule

$\Rightarrow$ from equation-(1)

$d_m = (10^{-11})^{1/2} = (10 \times 10^{-12})^{1/3}$

$\Rightarrow$ $d_m = 2.1 \times 10^{-4}$ m $= 0.21$ mm. **Ans. (B)**

Sol. 18 *RMS* speed of a gas molecule is given as

$$v_{rms} = \sqrt{\frac{3RT}{M}}$$

If we consider room temperature to be at $\sim 27°C = 300$ K.

$\Rightarrow$ $(1930)^2 = \dfrac{3 \times 8.314 \times 300}{M}$

$\Rightarrow$ $M = \dfrac{3 \times 8.314 \times 300}{(1930)^2} \simeq 2 \times 10^{-3}$ kg

Thus gas is H_2 **Ans. (A)**

Sol. 19 The (Translational) *KE* of a gas molecule at temperature T is given as

$$KE = \frac{3}{2}k_BT$$

at $T = 0°C = 273$ K

$\Rightarrow$ $KE = \dfrac{3}{2}k_B \times 273 = \dfrac{819}{2}k_B$

If it rises to a height h above the earth surface

$\Rightarrow$ $KE = Mgh = \dfrac{819}{2}k_B$

[As $h << R_e$ so we consider g = constant]

$\Rightarrow$ $h = \dfrac{819k_B}{2Mg}$ **Ans. (D)**

Sol. 20 For atoms in a solid lattice we take 6 vibrational degrees of freedom $(KE + PE)$

Then molar specific heat of *Al* at constant volume is given as

$$C_v = \frac{f}{2}R = \frac{6}{2}R = 3R \text{ J/mol.K}$$

The specific heat in J/kg-K can be given as

$$s = \frac{C_v}{M} = \frac{3R}{M} = \frac{3 \times 8.314}{27 \times 10^{-3}} \text{ J/kg-K}$$

$\Rightarrow$ $s \simeq 925$ J/kg-K **Ans. (D)**

Sol. 21 The pressure exerted by gas molecules on container walls is

$$P=\frac{1}{3}\rho v_{rms}^2 \qquad \left[\text{We use } v_{\text{rms}}=\sqrt{\frac{3RT}{M}}\right]$$

$$\Rightarrow \quad P=\frac{\rho RT}{M} \Rightarrow P \alpha T$$

[for a closed container $\rho \to$ constant]

$\Rightarrow$ Force on walls α pressure by gas

Force $\alpha\ T^{(1)}$

$\Rightarrow \quad q=1$ **Ans. (C)**

Sol. 22 If C_p and C_v are molar specific heats (J/mol-k) then we use for all gases

$$C_p - C_v = R = 8.314 \text{ J/mol-k}$$

for C_p and C_v taken as specific heats (J/kg-k) we use

for H_2 gas $\quad C_p - C_v = \frac{R}{M_{H_2}} = \frac{R}{2\times10^{-3}} = a$

for N_2 gas $\quad C_p - C_v = \frac{R}{28\times10^{-3}} = b$

$\Rightarrow \quad a=14b$ **Ans. (A)**

Sol. 23 From Gas law we use no of molecules in room are given as

$$N = n.A_v = \left(\frac{PV}{RT}\right)\cdot Av$$

Initial molecules in room

$$n_i = \frac{PVA_v}{RT_i} \quad (\text{where } T_i = 273+17=290 \text{ k})$$

Final molecules in room

$$n_f = \frac{PVA_v}{RT_f} \quad (\text{where } T_f = 273+27=300 \text{ k})$$

$$\Rightarrow \quad \Delta n = n_f - n_i = \frac{PVA_v}{R}\left[\frac{1}{T_f}-\frac{1}{T_i}\right]$$

$$\Rightarrow \quad \Delta n = \frac{10^5\times30\times6.023\times10^{23}}{8.314}\left[\frac{1}{300}-\frac{1}{290}\right]$$

$\Rightarrow \quad \Delta n = -2.5\times10^{25}$ **Ans. (D)**

Sol. 24 For a gas molecule having f degrees of freedom, we use

$$C_v=\frac{f}{2}R \quad \text{and} \quad C_p=\frac{f+2}{2}R$$

for $f=5$ we have

$$C_v=\frac{5}{2}R \text{ and } C_p=\frac{7}{2}R$$

$\Rightarrow \quad \gamma=\frac{C_p}{C_v}=\frac{7}{5}$ **Ans. (D)**

Sol. 25 The total KE of a gas having n moles is given as

$$U=\frac{f}{2}nRT$$

[where $f\to$ degrees of freedom]

for diatomic gas $f=5$

for monoatomic gas $f=3$

Initially N moles of diatomic gas

$$\Rightarrow \quad U_i=\frac{f}{2}NRT=\frac{5}{2}NRT \qquad \ldots(1)$$

Initially there are $(N-n)$ moles of diatomic gas and $2n$ moles of monoatomic gas

$$\Rightarrow \quad U_f=\frac{5}{2}(N\text{-}n)RT+\frac{3}{2}(2n)RT$$

$$\Rightarrow \quad U_f=\frac{5}{2}NRT+3nRT-\frac{5}{2}nRT$$

$$\Rightarrow \quad U_f=\frac{5}{2}NRT+\frac{1}{2}nRT \qquad \ldots(2)$$

from (1) and (2) change in total KE of gas

$$\Delta U = U_f - U_i = \frac{1}{2}nRT$$ **Ans. (A)**

Sol. 26 By Gas law

$$P=\frac{nRT}{V}$$

$\Rightarrow \quad T=$ constant from 1 to 2

$P_1 > P_2$ **Ans. (C)**

Sol. 27 By Gas law, we use

$$\frac{V_{\text{rms}}(He)}{V_{\text{rms}}(Ar)}=\sqrt{\frac{M_{Ar}}{M_{He}}}$$

$$=\sqrt{\frac{40}{4}} \cong 3.16$$ **Ans. (D)**

Sol. 28 For a monoatomic gas, energy of thermal motion

$$U=\frac{3}{2}kT\times N$$

$$\Rightarrow \quad U=\frac{3}{2}nRT=\frac{3}{2}PV \quad \left[\text{Volume } V=\frac{2}{8}=\frac{1}{4}\text{m}^3\right]$$

$$\Rightarrow \quad U=\frac{3}{2}\times4\times10^4\times\frac{1}{4}$$

$$\Rightarrow \quad U=1.5\times10^4 \text{ J}$$

Order is $=10^4$ J **Ans. (D)**

Sol. 29 Total internal energy of a gaseous mixture is given as

$$U=\frac{f_1}{2}n_1RT+\frac{f_2}{2}n_2RT$$

$$\Rightarrow \quad U=\frac{5}{2}(3RT)+\frac{3}{2}\times(5RT)$$

$\Rightarrow \quad U=15RT$ **Ans. (C)**

Sol. 30 Energy $= \frac{f}{2}nRT = \frac{f}{2}PV$

$$\Rightarrow \quad E = \frac{f}{2}(3\times10^6)(2)$$

$$\Rightarrow \quad E = f \times 3 \times 10^6$$

Considering gas is monatomic for ideal case, we use

$$\Rightarrow \quad f = 3$$

$$\Rightarrow \quad E = 9 \times 10^6 \text{ J}$$

Ans. (D)

Sol. 31

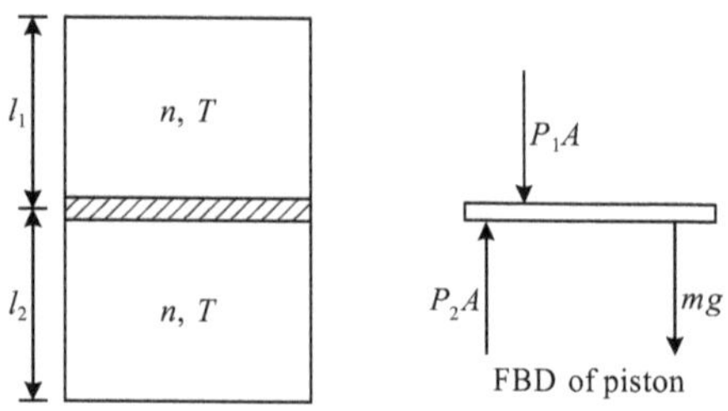

At equilibrium, we use

$$P_2A = P_1A + mg$$

$$\Rightarrow \quad \frac{nRT.A}{Al_2} = \frac{nRT.A}{Al_1} + mg$$

$$\Rightarrow \quad nRT\left(\frac{1}{l_2} - \frac{1}{l_1}\right) = mg$$

$$\Rightarrow \quad m = \frac{nRT}{g}\left(\frac{l_1 - l_2}{l_1 \cdot l_2}\right)$$

Ans (B)

Sol. 32 Mean time between two successive collisions is given as

$$t_m = \frac{\text{average separation}}{\text{mean velocity}}$$

Average separation between molecules is

$$\lambda_m = \frac{(V)^{1/3}}{N}$$

and $\quad v_m = \sqrt{\frac{8RT}{\pi M}}$

$$\Rightarrow \quad t_m = \frac{(V)^{1/3}}{N}\sqrt{\frac{\pi M}{8RT}}$$

$$\Rightarrow \quad t_m \propto \frac{T^{1/3}}{P^{1/3}\sqrt{T}}$$

$$\Rightarrow \quad t_m \propto \frac{1}{P^{1/3}} T^{1/6}$$

$$\Rightarrow \quad \frac{t_1}{t_2} = \left(\frac{P_2}{P_1}\right)^{1/3}\left(\frac{T_2}{T_1}\right)^{1/6}$$

$$\Rightarrow \quad t_2 = 6 \times 10^{-8} \times \left(\frac{1}{2}\right)^{1/3}\left(\frac{3}{5}\right)^{1/6}$$

$$\Rightarrow \quad t_2 = 6 \times 10^{-8}/1.26 \times 1.09 = 4.37 \times 10^{-8} \text{ s}$$

Ans. (A)

Sol. 33

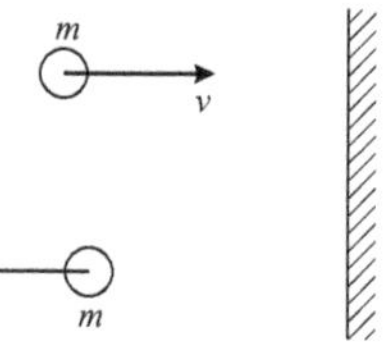

Magnitude of change in momentum per collision $= 2mv$

$$\text{Pressure} = \frac{\text{Force}}{\text{Area}} = \frac{N(2mv)}{1}$$

$$p = \frac{10^{22} \times 2 \times 10^{-26} \times 10^4}{1}$$

$$\Rightarrow \quad p = 2 \text{ N/m}^2$$

Ans. (None)

Sol. 34 RMS velocity of hydrogen molecules is given as

$$v_{rms} = \sqrt{\frac{3RT}{M}}$$

If RMS velocity is equal to escape velocity then we use

$$\sqrt{\frac{3RT}{M}} = \sqrt{2gRe}$$

$$\Rightarrow \quad T = \frac{2gR_eM_e}{3R} = \frac{2\times10\times6.4\times10^6\times2\times10^{-3}}{3\times8.314}$$

$$\Rightarrow \quad T \simeq 10 \times 10^3 = 10^4 \; 6.4 \times 10^6$$

$$\Rightarrow \quad T = \frac{4\times10\times6.4\times10^6}{3\times1.38\times6.02\times10^3}$$

$$\Rightarrow \quad T = 10 \times 10^3 = 10^4 \text{K}$$

Ans. (C)

Sol. 35 According to law of equipartion of energy, we use

$$\frac{1}{2}m(v_{rms}^2) = 3\times\frac{1}{2}K_bT$$

$$\Rightarrow \quad T = \frac{m\bar{v}_{rms}^2}{3k}$$

Ans. (C)

Sol. 36 RMS speed is given as

$$v_{rms} = \sqrt{\frac{3RT}{M}}$$

$$\Rightarrow \quad v_{rms} \propto \sqrt{T}$$

For the given data, we use

$$\frac{v}{200} = \sqrt{\frac{500}{400}}$$

$$\Rightarrow \quad \frac{v}{200} = \frac{\sqrt{5}}{2}$$

$$\Rightarrow \quad v = 100\sqrt{5} \text{ m/s}$$

Ans. (B)

Sol. 37 For A, we use

$$R = C_p - C_v = 7$$

$$C_v = \frac{fR}{2} = 22$$

$$\Rightarrow \quad f = \frac{44}{7} = 6.3$$

$$f \simeq 6 \begin{cases} 5(\text{Rotation + Translational}) \\ 1(\text{Vibration}) \end{cases}$$

For B, we use

$$R = C_p - C_v = 9$$

$$C_v = \frac{fR}{2} = 21$$

$$\Rightarrow \quad f = \frac{42}{9}$$

$$f \simeq 5 \begin{cases} 5(\text{Rotation + Translational}) \\ 0(\text{Vibration}) \end{cases}$$

Ans. (D)

Sol. 38 Mean free path of O_2 gas

$$\lambda = \frac{RT}{\sqrt{2}\pi d^2 N_A P} = \frac{V}{\sqrt{2}\pi d^2 N_A}$$

Average collision rate is given as

$$r_c = \frac{v_{av}}{\lambda} = \frac{\sqrt{8/3\pi} \times v_{ms}}{\lambda}$$

$$\Rightarrow \quad r_c = \frac{\sqrt{8/3\pi} \times 200 \text{ m/s}}{25 \times 10^{-3}} \times \sqrt{2}\pi (2 \times 0.3 \times 10^{-9}) \times 6.023 \times 10^{23}$$

$$\Rightarrow \quad r_c \simeq 0.18 \times 10^{10} \text{ sec}^{-1} \sim 10^{10} \text{ sec}^{-1}$$

Ans. (C)

Sol. 39 C_V of a mixture of gases is given as

$$C_{V_{eq}} = \frac{n_1 C_{V_1} + n_2 C_{V_2}}{n_1 + n_2}$$

For He, $C_{V_1} = \frac{3R}{2}$ and For H_2, $C_{V_2} = \frac{5R}{2}$

$$\Rightarrow \quad C_{V_{eq}} = \frac{2 \times 3R/2 + 3 \times 5R/2}{2 + 3}$$

$$\Rightarrow \quad C_{V_{eq}} = \frac{21}{10} R = \frac{21}{10} \times 8.314 = 17.46 \text{ J/mol K}$$

Ans. (C)

Sol. 40

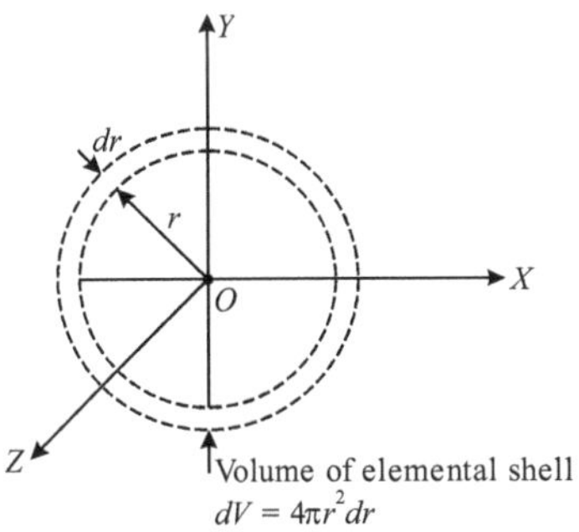

Number of molecule in the elemental shell is

$$dN = n(r) \cdot dV$$

$$\Rightarrow \quad dN = n_0 e^{-\alpha r^4} \cdot 4\pi r^2 dr$$

Total molecules

$$N = \int dN = 4\pi n_0 \int_0^{\infty} e^{-\alpha r^4} \cdot r^2 dr$$

Let $\quad t = \alpha r^4 \quad \Rightarrow \quad dt = 4\pi r^3 dr$

$$\Rightarrow \quad r^2 dr = \frac{dt}{4\alpha r} \quad \Rightarrow \quad r = \frac{t^{1/4}}{\alpha^{1/4}}$$

$$\Rightarrow \quad N = 4\pi n_0 \int_0^{\infty} e^{-t} \cdot \frac{dt\alpha^{1/4}}{4\alpha t^{1/4}}$$

$$\Rightarrow \quad N = \pi n_0 \alpha^{-3/4} \int_0^{\infty} t^{-1/4} \cdot e^{-t} dt$$

Ans. (C)

13.1 Simple Harmonic Motion

Sol. 1

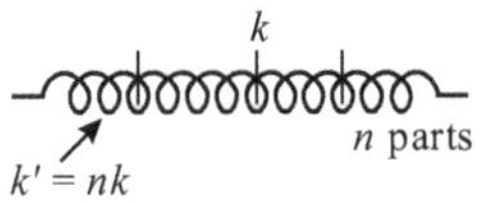

Each part has force constant n times of the original spring if it is divided in n equal parts.

For a force constant k, the time period of oscillations is

$$T = 2\pi\sqrt{\frac{m}{k}}$$

If $\quad k \to k' = nk$

New time period will be

$$T' = 2\pi\sqrt{\frac{m}{nk}} = \frac{T}{\sqrt{n}}$$

Ans. (B)

Sol. 2 In SHM, KE and PE with position are given as

$$K_y = \frac{1}{2} m\omega^2 (A^2 - y^2)$$

and

$$P_y = \frac{1}{2} m\omega^2 y^2$$

At mean position $\quad y = 0$

$$K_m = \frac{1}{2} m\omega^2 A^2 \quad \text{(max)}$$

$$P_m = 0 \quad \text{(min)}$$

At extreme position $\quad y = \pm A$

$$K_e = 0 \quad \text{(min)}$$

$$P_e = \frac{1}{2} m\omega^2 A^2 \quad \text{(max)}$$

Ans. (C)

Sol. 3 Time period of a swing (similar to simple pendulum) is given as

$$T = 2\pi\sqrt{\frac{l}{g}}$$

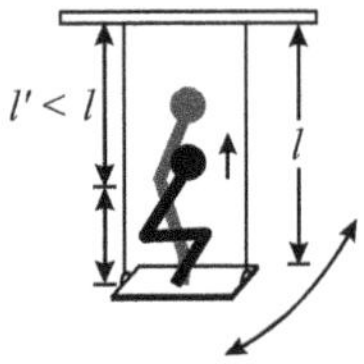

As boy stands up, length of pendulum decreases

$\Rightarrow$ new time period $T' = 2\pi\sqrt{\frac{l'}{g}} \Rightarrow T' < T$ **Ans. (B)**

Sol. 4

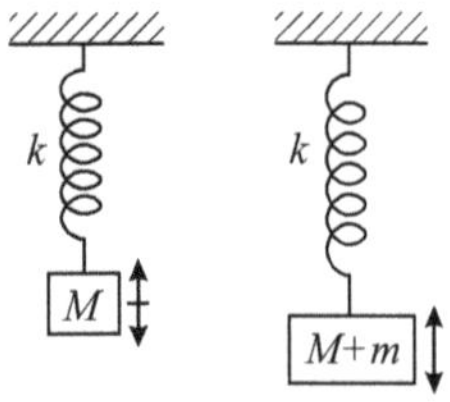

Time period of SHM in first condition

$$T = 2\pi\sqrt{\frac{M}{k}} \quad \ldots(1)$$

Time period of SHM in second condition

$$\frac{5T}{3} = T' = 2\pi\sqrt{\frac{M+m}{k}} \quad \ldots(2)$$

By (2)/(1) $\Rightarrow$ $\frac{5}{3} = \sqrt{\frac{M+m}{M}}$

$\Rightarrow$ $\frac{25}{9} = \frac{M+m}{M}$

$\Rightarrow$ $\frac{m}{M} = \frac{16}{9}$ **Ans. (C)**

Sol. 5 SHM equation is given as

$$x = 4\cos\pi t + 4\sin\pi t = A\sin(\omega t + \theta)$$

Where A is SHM amplitude

here we use $A\cos\theta = 4$

and $A\sin\theta = 4$

square above equations and add

$\Rightarrow$ $A = \sqrt{4^2 + 4^2} = 4\sqrt{2}$ **Ans. (C)**

Sol. 6 Energies as a function of position x in SHM are given as

$$TE = \frac{1}{2}m\omega^2A^2 \rightarrow \text{constant}$$

$$K_x = \frac{1}{2}m\omega^2(A^2 - x^2)$$

$$P_x = \frac{1}{2}m\omega^2x^2$$

At mean position $x = 0$

$$K_x = \frac{1}{2}m\omega^2A^2\,(\text{max})$$

$$P_x = 0\,(\text{min})$$

At extreme position

$$x = \pm A$$

$$K_x = 0$$

$$P_x = \frac{1}{2}m\omega^2A^2\,(\text{max})$$ **Ans. (A)**

Sol. 7 If amplitudes of A and B are A_1 and A_2 respectively then we use $v_{\max A} = v_{\max B}$

for SHM $v_{\max} = \omega A$

$$\omega_A A_1 = \omega_B A_2$$

$$\omega = \sqrt{\frac{k}{m}}$$

$\Rightarrow$ $\frac{A_1}{A_2} = \sqrt{\frac{k_2}{k_1}}$ **Ans. (C)**

Sol. 8 Time period of a simple pendulum is given as

$$T = 2\pi\sqrt{\frac{l}{g}}$$

If $l \rightarrow 1.21l$ $\Rightarrow$ new time period will be

$$T' = 2\pi\sqrt{\frac{1.21l}{g}} = 1.1 \times 2\pi\sqrt{\frac{l}{g}}$$

$\Rightarrow$ $T' = 1.1T$

% increase in time period is

$$= \frac{T'-T}{T} \times 100$$

$$= \frac{(1.1-1)}{1} \times 100 = 10\%$$ **Ans. (D)**

Sol. 9 Time period of a simple pendulum is

$$T = 2\pi\sqrt{\frac{l}{g}} \text{ in air}$$

$$T' = 2\pi\sqrt{\frac{l}{g_{\text{eff}}}} \text{ in water}$$

Where $g_{\text{eff}} = g\left(1 - \frac{\rho_l}{\rho_s}\right)$

$$T' = 2\pi\sqrt{\frac{l}{g\left(\frac{\rho_s - \rho_l}{\rho_s}\right)}} = T\sqrt{\frac{\rho_s}{\rho_s - \rho_l}}$$

$\Rightarrow$ $T' = t_0\sqrt{\frac{\frac{4}{3}\times 10^3}{\left(\frac{4}{3}-1\right)\times 10^3}} = t_0\sqrt{\frac{4}{3}\times\frac{3}{1}} = 2t_0$ **Ans. (C)**

Sol. 10 For first spring (force constant k_1), time period

$$t_1 = 2\pi\sqrt{\frac{m}{k_1}}$$

For second spring (force constant k_2), time period

$$t_2 = 2\pi\sqrt{\frac{m}{k_2}}$$

In series combination of springs, effective force constant k_{eq} is given by

$$\frac{1}{k_{eq}} = \frac{1}{k_1}+\frac{1}{k_2}$$

Time period of SHM using series combination is given as

$$T_{eq} = T = 2\pi\sqrt{\frac{m}{k_{eq}}} = 2\pi\sqrt{\frac{m}{k_1}+\frac{m}{k_2}}$$

Squaring gives

$$T^2 = 4\pi^2\frac{m}{k_1}+4\pi^2\frac{m}{k_2}$$

$$\Rightarrow \quad T^2 = t_1^2 + t_2^2$$ **Ans. (B)**

Sol. 11 Total energy in SHM is given as

$$TE = \frac{1}{2}m\omega^2A^2 + U_{min}$$

Constant minimum PE does not change with position.

with position KE & PE oscillates and

$$K_x + P_x = \text{TE} \rightarrow \text{constant}$$ **Ans. (C)**

Sol. 12 For a forced oscillation in a damping medium if external force is $F(t) = F_0\cos\omega t$

then the oscillation amplitude is given as

$$A = \frac{F_0}{\sqrt{m^2(\omega_0^2-\omega^2)^2+b^2\omega^2}}$$

(where $b \rightarrow$ damping coefficient)

in this question $b = 0 \Rightarrow A = \dfrac{F_0}{m(\omega_0^2-\omega^2)}$ **Ans. (B)**

Sol. 13 Energy of oscillations $\propto$ (amplitude)^2

$\Rightarrow$ When amplitude is maximum, energy will also be maximum and this happens at same frequency of force

$\Rightarrow \quad \omega_1 = \omega_2$ **Ans. (C)**

Sol. 14 Acceleration of particle is given as

$$a = \frac{F}{m} = -\frac{kx}{m}$$

$$\Rightarrow \quad a = -\frac{15\times 0.2}{0.3} = -10 \text{ m/s}^2$$

Negative sign shows opposite direction. **Ans. (D)**

Sol. 15 The position given is

$$y = \sin^2\omega t = \frac{1}{2}(1-\cos 2\omega t)$$

$$\Rightarrow \quad y = \frac{1}{2} - \frac{1}{2}\cos 2\omega t$$

$$\Rightarrow \quad y = y_0 + A\cos(2\omega t + \pi)$$

This represents a function of SHM with mean position at $y = y_0$, Amplitude A and angular frequency 2ω.

Where $A = 1/2$ and $y_0 = 1/2$

$\Rightarrow$ Time period $= \dfrac{2\pi}{2\omega} = \dfrac{\pi}{\omega}$ **Ans. (D)**

Sol. 16 The basic differential equation of SHM is given as

$$\frac{d^2x}{dt^2} + \omega^2 x = 0$$

where $\omega \rightarrow$ Angular frequency of SHM.

comparing both the above equation, we have

$$\omega^2 = \alpha \text{ or } \omega = \sqrt{\alpha}$$

Time period of SHM $T = \dfrac{2\pi}{\omega} = \dfrac{2\pi}{\sqrt{\alpha}}$ **Ans. (B)**

Sol. 17

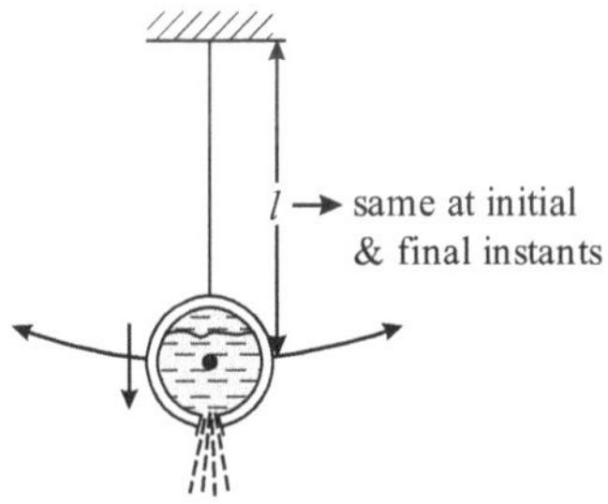

As water flows out, inside the shell water level comes down.

Time period of oscillation of pendulum

$$T = 2\pi\sqrt{\frac{l}{g}}$$

As water level goes down

$\Rightarrow$ l increases as centre of mass is going down.

When shell becomes empty then again COM of shell is at its centre $\Rightarrow$ after some time l decreases.

$\Rightarrow$ T first increases then decreases to same value. **Ans. (A)**

Sol. 18 Velocity of particle-1 is

$$v_1 = \frac{dy_1}{dt} = 0.1\times 100\pi\cos\left(100\pi t + \frac{\pi}{3}\right)$$

$$\Rightarrow \quad v_1 = 10\pi\cos\left(100\pi t + \frac{\pi}{3}\right) \quad \ldots(1)$$

Velocity of particle-2 is

$$v_2 = \frac{dy_2}{dt} = -0.1\times 100\pi\sin(100\pi t)$$

$$\Rightarrow \quad v_2 = 10\pi\cos\left(100\pi t + \frac{\pi}{2}\right) \quad \ldots(2)$$

Phase difference between v_1 and v_2 is

$$\alpha = \left(100\pi t + \frac{\pi}{3}\right) - \left(100\pi t + \frac{\pi}{2}\right)$$

$$\Rightarrow \quad \alpha = -\frac{\pi}{6}$$

Ans. (A)

Sol. 19 In SHM maximum velocity of particle is given as

$$v_{max} = A\omega = A\left(\frac{2\pi}{T}\right)$$

$$\Rightarrow \quad T = \frac{2\pi A}{v_{max}} = \frac{2\times3.14\times7\times10^{-3}}{4.4}$$

$$\Rightarrow \quad T = 0.01\text{s}$$

Ans. (A)

Sol. 20

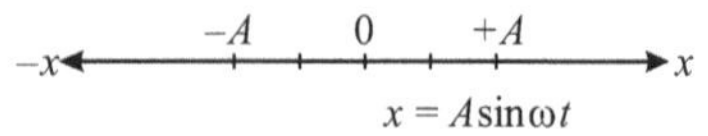

For the particle in SHM, its KE is given as

$$K = \frac{1}{2}m\omega^2(A^2 - x^2) \quad \ldots(1)$$

Given that at time t $K = \frac{3}{4}E = \frac{3}{4}\left(\frac{1}{2}m\omega^2A^2\right)$

From (1) $\frac{1}{2}m\omega^2(A^2 - x^2) = \frac{3}{4}\left(\frac{1}{2}m\omega^2A^2\right)$

$$\Rightarrow \quad A^2 - x^2 = \frac{3}{4}A^2$$

$$\Rightarrow \quad x = \pm A/2$$

As $T = 2\text{s}$

$$\Rightarrow \quad \omega = \frac{2\pi}{2} = \pi \text{ rad/s}$$

$$\Rightarrow \quad x = A\sin\pi t$$

at $\quad x = A/2 = A\sin\pi t$

$$\Rightarrow \quad \sin\pi t = \frac{1}{2} \Rightarrow \pi t = \frac{\pi}{6} \Rightarrow t = \frac{1}{6}\text{s}$$

Ans. (A)

Sol. 21 In SHM

$$a_{max} = \omega^2 A$$

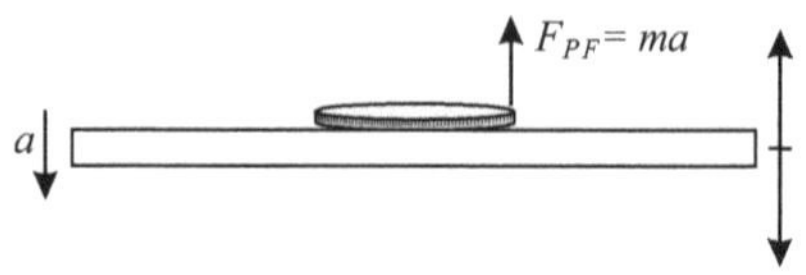

During SHM when platform goes down at acceleration a, coin experiences a pseudo force *ma* in upwards direction, it will leave contact when

$$ma = mg$$

$$\Rightarrow \quad \omega^2 A = g$$

$$\Rightarrow \quad A = \frac{g^2}{\omega^2}$$

Ans. (B)

Sol. 22 The SHM equation given is of the form

$$x = A\cos\omega t \quad [A = 2\times10^{-2}\text{m}, \omega = \pi \text{ rad/s}]$$

here particle starts SHM from external position as $t = 0, x = A$. Particle speed is maximum in SHM when it crosses the mean position which happens at

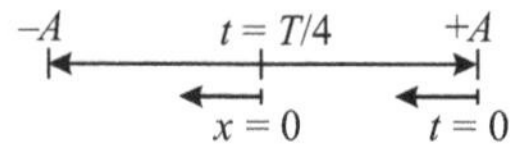

$$t = \frac{T}{4} = \frac{1}{4}\left(\frac{2\pi}{\omega}\right) = \frac{1}{2}\text{s} = 0.5\text{s}$$

Ans.(A)

Sol. 23 SHM equation as

$$x = x_0\cos(\omega t - \pi/4)$$

velocity in SHM is given as

$$v = \frac{dx}{dt} = -\omega x_0\sin(\omega t - \pi/4)$$

Acceleration in SHM is given as

$$a = \frac{dv}{dt} = -\omega^2 x_0\cos(\omega t - \pi/4)$$

$$\Rightarrow \quad a = \omega^2 x_0\cos(\omega t - \frac{\pi}{4} + \pi)$$

$$\Rightarrow \quad a = \omega^2 x_0\cos(\omega t + \frac{3\pi}{4}) \quad \ldots(1)$$

Comparing equation-(1) with the given acceleration we get

$$A = \omega^2 x_0$$

and $\quad \delta = \frac{3\pi}{4}$

Ans. (D)

Sol. 24 The frequency of oscillation of block in this case is given as

$$f = \frac{1}{2\pi}\sqrt{\frac{k_1 + k_2}{m}}$$

if $\quad k_1 \leftarrow 4k_1$

and $k_2 \leftarrow 4k_2$

$$\Rightarrow \quad f' = \frac{1}{2\pi}\sqrt{\frac{4(k_1 + k_2)}{m}} = 2f$$

Ans. (D)

Sol. 25

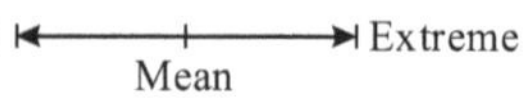

Average KE of a particle in SHM for one period (or *T*/4 period from mean to extreme position) is given as

$$(\text{KE})_{avg} = \frac{1}{2}\text{TE} = \frac{1}{2}\left(\frac{1}{2}m\omega^2A^2\right)$$

$$\Rightarrow \quad (\text{KE})_{avg} = \frac{1}{4}\text{m}(2\pi\nu)^2a^2$$

$$\Rightarrow \quad (\text{KE})_{avg} = m\pi^2\nu^2a^2$$

Ans. (A)

Sol. 26 In SHM $a = -\omega^2 x$ and $v = \omega\sqrt{A^2 - x^2}$

(A) $-\omega^2 xT + 2\pi\omega\sqrt{A^2 - x^2}$

(B) $\dfrac{aT}{v} = -\dfrac{\omega^2 xT}{\omega\sqrt{A^2 - x^2}}$

(C) $a^2T^2 + 4\pi^2v^2 = \omega^2x^2T^2 + 4\pi^2\omega^2(A^2 - x^2)$

above options are dependent on x which varies with time.

(D) $\dfrac{aT}{x} = -\omega^2 T$ independent of x. **Ans. (D)**

Sol. 27

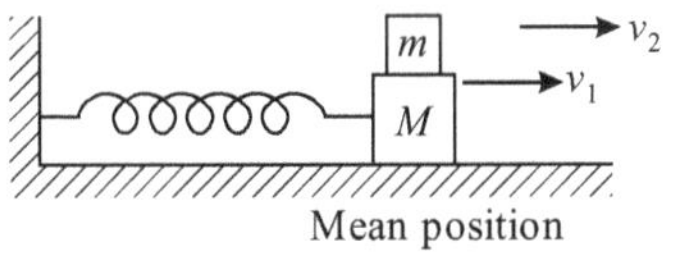

$$v_1 = A_1\omega_1 = A_1\sqrt{\frac{k}{M}}$$

Due to addition of m, the new speed of blocks is given by linear momentum conservation as

$$Mv_1 = (m + M)v_2$$

$$v_2 = \frac{Mv_1}{M+m} = A_2\omega_2 \text{ [here } \omega_2 = \sqrt{\frac{k}{m+M}}\text{]}$$

$$\Rightarrow \quad A_2 = \frac{Mv_1}{(M+m)}\cdot\sqrt{\frac{m+M}{k}}$$

$$\Rightarrow \quad A_2 = \frac{M}{\sqrt{M+m}}\cdot\frac{1}{\sqrt{k}}\cdot A_1\sqrt{\frac{k}{m}}$$

$$\Rightarrow \quad A_2 = A_1\sqrt{\frac{M}{M+m}}$$

$$\Rightarrow \quad \frac{A_1}{A_2} = \sqrt{\frac{M+m}{M}}$$

Ans. (B)

Sol. 28 SHM equation of the two particles are given as

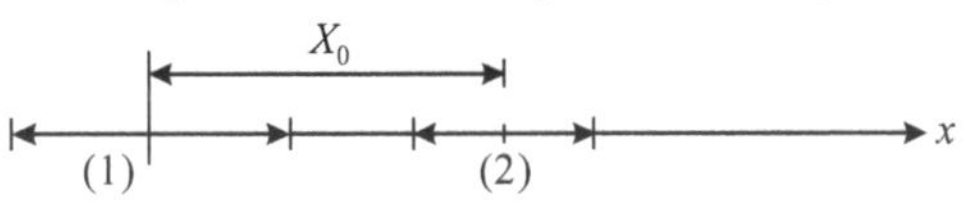

$$x_1 = A\sin\omega t$$

$$x_2 = X_0 + A\sin(\omega t + \phi)$$

separation between the particles is

$$x_2 - x_1 = X_0 + A\sin(\omega t + \phi) - A\sin\omega t$$

given that

$$(x_2 - x_1)_{max} = X_0 + A$$

$$(A\sin(\omega t + \phi) - A\sin\omega t)_{max} = A$$

$$\Rightarrow \quad \left[2A\cos\left(\omega t + \frac{\phi}{2}\right)\sin\frac{\phi}{2}\right]_{max} = A$$

$$\Rightarrow \quad \left[\left(2A\sin\frac{\phi}{2}\right)\cos\left(\omega t + \frac{\phi}{2}\right)\right]_{max} = A$$

Here $\cos\left(\omega t + \dfrac{\phi}{2}\right) = (+1)_{max}$

$$\Rightarrow \quad 2A\sin\frac{\phi}{2} = A$$

$$\Rightarrow \quad \phi = \pi/3$$

Ans. (D)

Sol. 29 Under the state of equilibrium we have

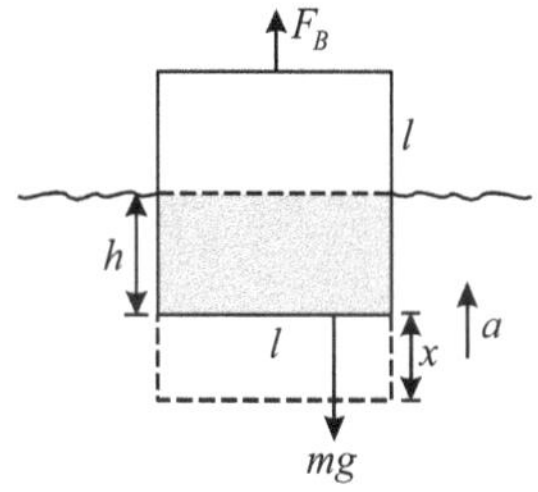

$$F_B = mg$$

$$\Rightarrow \quad l^2h\rho g = l^3dg$$

$$\Rightarrow \quad h\rho = ld \qquad \ldots(1)$$

After pushing the cube by x, restoring force on it upward will be $F_B' - mg = ma$

$$\Rightarrow \quad (h + x)l^2\rho g - mg = ma$$

$$hl^2\rho g + xl^2\rho g - l^3dg = l^3da$$

$$\Rightarrow \quad a = -\frac{\rho g}{ld}x$$

–ve sign denotes opposite direction

Comparing with $a = -\omega^2 x$

we get $\omega = \sqrt{\dfrac{\rho g}{ld}}$

$\Rightarrow$ Time period $T = \dfrac{2\pi}{\omega}$

$$\Rightarrow \quad T = 2\pi\sqrt{\frac{ld}{\rho g}}$$

Ans. (B)

Sol. 30 For a simple pendulum having damped oscillations its equation of SHM in differential form is given as

$$\frac{d^2x}{dt^2} + bv + \omega^2 x = 0$$

Solution to above equation is

$$x = A_0e^{-bt/2}\sin(\omega t + \phi)$$

Where $A_0e^{-bt/2}$ = Amplitude envelope.

Given that at $t = \tau$, $A = \dfrac{A_0}{e}$

we use $A = A_0e^{-bt/2}$

$$\Rightarrow \quad \frac{A_0}{e} = A_0e^{-b\tau/2}$$

$$\Rightarrow \quad \frac{b\tau}{2} = 1$$

$$\Rightarrow \quad \tau = \frac{2}{b}$$

Ans. (A)

Sol. 31 This ring oscillates as a compound pendulum for which oscillation period is given as

$$T = 2\pi\sqrt{\frac{I}{mgl}}$$

here $I = mR^2 + mR^2 = 2mR^2$; $l = R$

$$\Rightarrow \quad 1 = 2\pi\sqrt{\frac{2mR^2}{mgR}} = 2\pi\sqrt{\frac{2R}{g}}$$

squaring, we get

$$\Rightarrow \quad 1 = 4\pi^2 \times \frac{2R}{g} \Rightarrow R = \frac{1}{8}\text{ m} = 0.125\text{m}$$

Ans. (A)

Sol. 32 General SHM equation

$$y(t) = A\sin\omega t$$

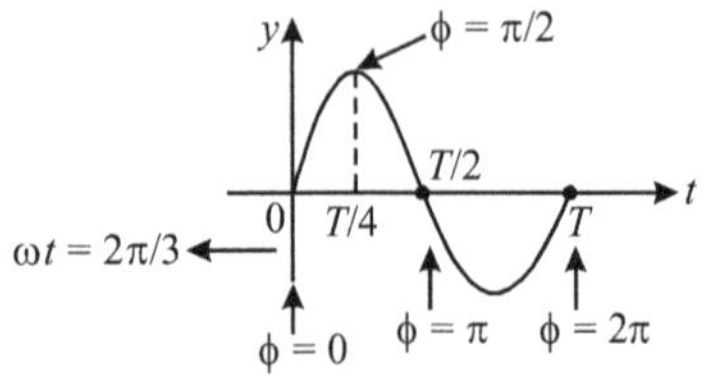

$$y = A\sin(\omega t + \phi) = A\sin\left(\omega t + \frac{2\pi}{3}\right)$$

at $\omega t = -\frac{2\pi}{3}$ the function should start from $y = 0$

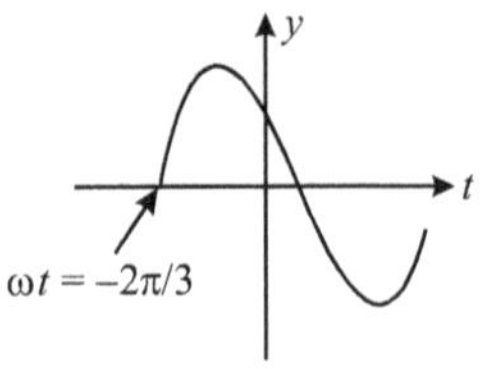

Ans. (A)

Sol. 33

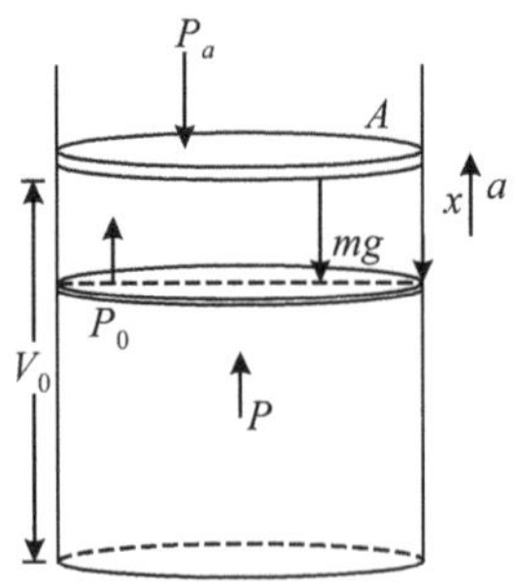

At equilibrium position, we use

$$P_0A = mg + P_aA \qquad \ldots(1)$$

If after displacing piston downward gas pressure becomes P then equation of motion of piston is given as

$$PA - mg - P_aA = ma \qquad \ldots(2)$$

After pushing the piston down, we use

$$P_0V_0^r = P(V_0 - Ax)^\gamma$$

$$\Rightarrow \quad P_0V_0^r = PV_0^r\left(1 - \frac{Ax}{V_0}\right)^\gamma$$

$$\Rightarrow \quad P = P_0\left(1 - \frac{Ax}{V_0}\right)^{-\gamma}$$

Using binomial expansion as $1 >> Ax/V_0$ we have

$$P = P_0\left(1 + \frac{\gamma Ax}{V_0}\right)$$

From equation-(2), we have

$$\Rightarrow \quad P_0A + \frac{P_0\gamma A^2x}{V_0} - mg - P_aA = ma$$

$$\Rightarrow \quad a = -\frac{P_0\gamma A^2}{mV_0} \qquad \ldots(3)$$

–ve sign denotes opposite direction of a compared to x.

Comparing equation-(3) with $a = -\omega^2x$

$$\Rightarrow \quad \omega = \sqrt{\frac{P_0rA^2}{mV_0}}$$

SHM frequency of piston

$$f = \frac{\omega}{2\pi} = \frac{1}{2\pi}\sqrt{\frac{P_0\gamma A^2}{mV_0}}$$

Ans. (C)

Sol. 34 For a damped oscillator amplitude is given

$$A = A_0e^{-bt/2m}$$

Given that at $t = 5$s we use $A = 0.9A_0$

$$\Rightarrow \quad 0.9A_0 = A_0e^{-5b/2m}$$

$$\Rightarrow \quad e^{-5b/2m} = 0.9$$

At $t = 15$s we use

$$A_f = A_0e^{-15b/2m}$$

$$\Rightarrow \quad A_f = A_0(0.9)^3 = \underbrace{0.729}_{\alpha}A_0$$

Ans. (C)

Sol. 35

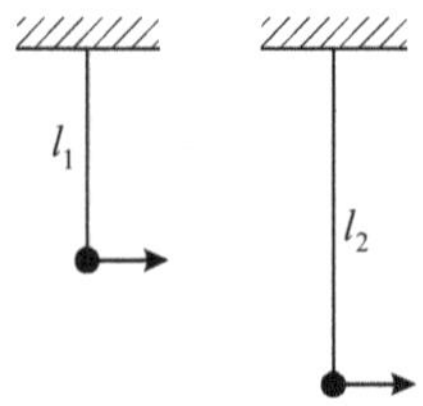

$$T_1 = 2\pi\sqrt{\frac{l}{g}}$$

$$T_2 = 2\pi\sqrt{\frac{4}{g}}$$

As $T_1 < T_2$, shorter pendulum will exceed in phase compared to longer pendulum & these will be in phase again when shorter pendulum will oscillate 1 oscillation more than longer pendulum. If by this time shorter pendulum completes N oscillations.

$$\Rightarrow \quad N.\,2\pi\sqrt{\frac{1}{g}} = (N-1)\,2\pi\sqrt{\frac{4}{g}}$$

$$\Rightarrow \quad N = 2N - 2 \Rightarrow N = 2$$

Ans. (A)

Sol. 36 Angular frequency of vertical SHM of mass is

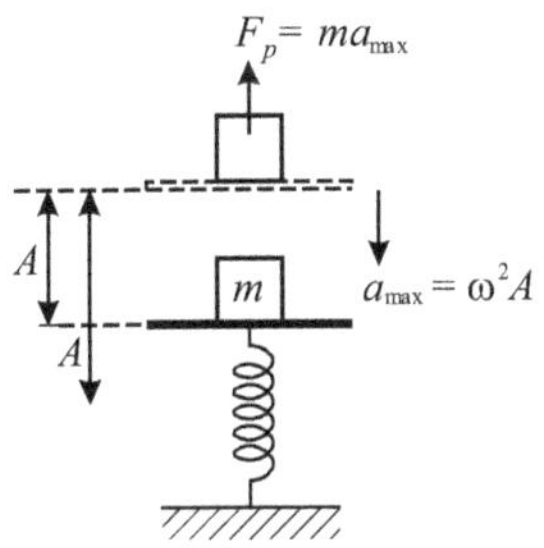

$$\omega = \sqrt{\frac{k}{m}} = \sqrt{500} = 10\sqrt{5}\text{ rad/s}$$

At upper extreme position block will experience a Pseudo force upward w.r. to pan, given as

$$F_P = ma_{max} = m\omega^2 A$$

If $F_P > mg$ then block will have tendency to get detached from pan $\quad m\omega^2 A > mg$

$$\Rightarrow \quad A > \frac{g}{\omega^2}$$

$$\Rightarrow \quad A > \frac{10}{500}$$

$$\Rightarrow \quad A > 0.02\text{ m} = 2\text{ cm}$$

for $A > 2$cm somewhere at position above mean position mass can get detached from pan. **Ans. (A)**

Sol. 37 If at equilibrium spring extension is h then we have

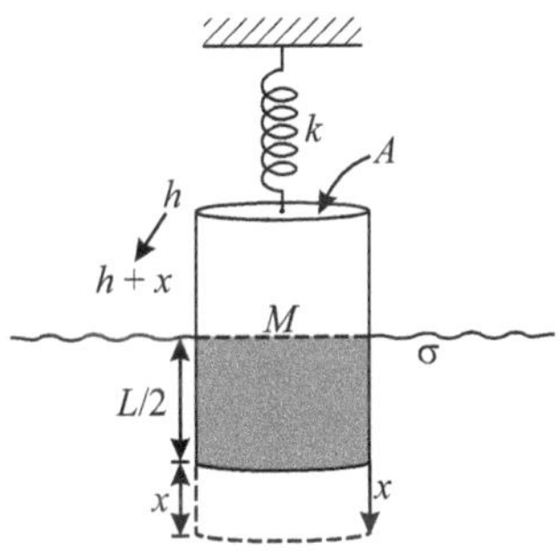

FBD of cylinder

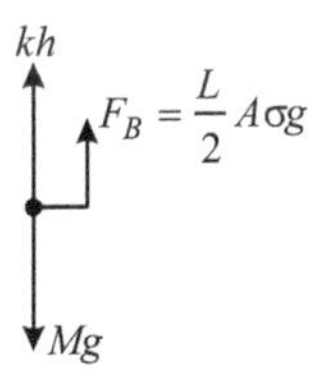

For equation of cylinder, we have

$$kh + \frac{L}{2}A\sigma g = Mg \qquad \dots(1)$$

After pushing down the cylinder by x, its FBD is

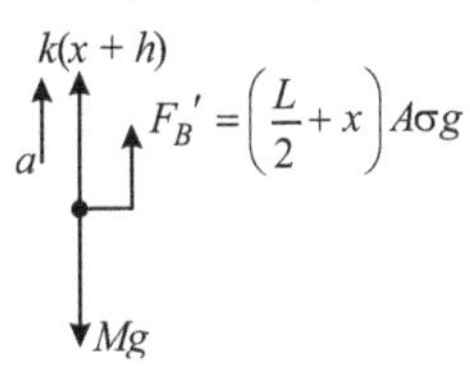

Equation of motion of cylinder

$$k(x+h) + \left(\frac{L}{2}+x\right)A\sigma g - Mg = Ma$$

$$\Rightarrow \quad a = \left(\frac{k + A\sigma g}{M}\right)x$$

As a is in opposite direction to x, we use

$$a = -\left(\frac{k + A\sigma g}{M}\right)x$$

Comparing with SHM equation $a = -\omega^2 x$, we have frequency of its SHM is

$$\Rightarrow \quad \omega = \sqrt{\frac{k + A\sigma g}{M}}$$

Time period of oscillation $T = \dfrac{2\pi}{\omega}$

$$\Rightarrow \quad T = 2\pi\sqrt{\frac{M}{k + A\sigma g}}$$

As nothing is specified in question about damping. **Ans. (D)**

Sol. 38 Time period of oscillation for a simple pendulum

$$T = 2\pi\sqrt{\frac{l}{g}}$$

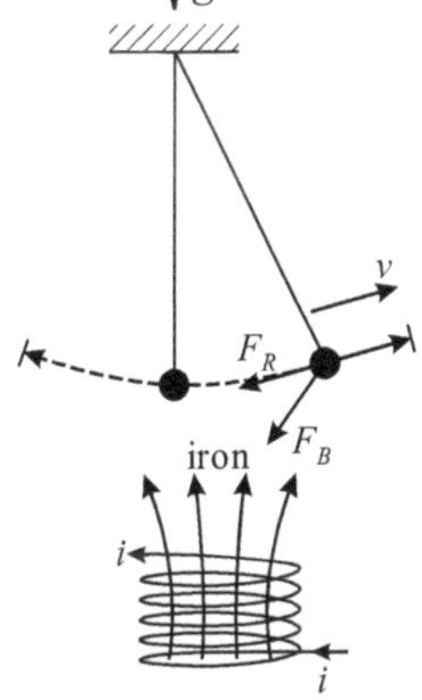

Due to coil the net restoring force on bob during oscillation increases (as shown) so angular frequency of oscillation increases $\Rightarrow T$ decreases.

$$\Rightarrow \quad T < 2\pi\sqrt{\frac{l}{g}}$$

Coil behaves as a magnet and damping is more than in air alone. **Ans. (D)**

Sol. 39

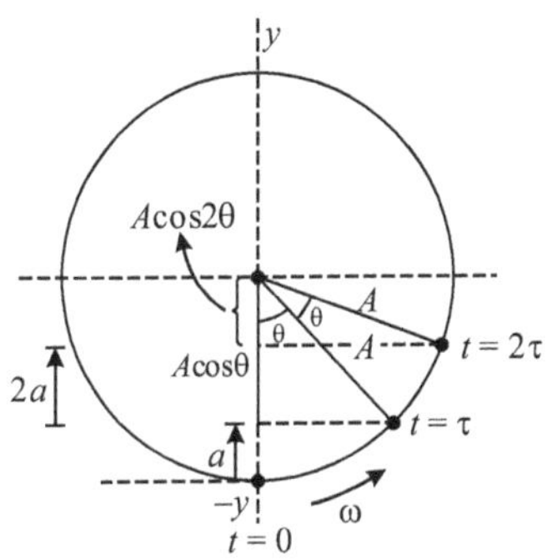

From $t = 0$ to $t = \tau$, displacement of particle in SHM is

$$a = A - A\cos(\omega\tau) \quad \ldots(1)$$

from $t = \tau$ to $t = 2\tau$, displacement of particle is $2a$

$$\Rightarrow \quad 3a = A - A\cos(2\omega\tau) \quad \ldots(2)$$

from equation (1) $\cos(\omega\tau) = \left(\dfrac{A-a}{A}\right)$

from equation (2) $2\cos^2(\omega\tau) - 1 = \dfrac{A-3a}{A}$

$$\Rightarrow \quad 2\left(\frac{A-a}{A}\right)^2 - 1 = \frac{A-3a}{A}$$

$$\Rightarrow \quad \frac{2A^2 + 2a^2 - 4aA}{A^2} = \frac{2A-3a}{A}$$

$$\Rightarrow \quad 2A^2 + 2a^2 - 4aA = 2A^2 - 3aA$$

$$\Rightarrow \quad 2a^2 = aA$$

$$\Rightarrow \quad A = 2a$$

From equation-(1) we have

$$\cos(\omega\tau) = \left(\frac{A-a}{A}\right) = \frac{1}{2}$$

$$\Rightarrow \quad \omega\tau = \frac{\pi}{3}$$

$$\Rightarrow \quad \left(\frac{2\pi}{T}\right)\tau = \frac{\pi}{3}$$

$$\Rightarrow \quad T = 6\tau$$

Ans. (D)

Sol. 40 Stokes's Law

$$F_v = -6\pi\eta rv$$

if $\quad F_v = -bv$

$$\Rightarrow \quad b \propto \eta$$

$$\Rightarrow \quad \frac{b_{air}}{b_{CO_2}} = \frac{\eta_{air}}{\eta_{CO_2}} = 1.3$$

For damped oscillation in air, we use

$$A = A_0 e^{-bt/2m}$$

at $\quad t = 0, A_0 = 10$ cm

and $t = 40$s, $A = 8$ cm

$$\Rightarrow \quad 8 = 10\, e^{\frac{-b_{air} \times 40}{2m}} \quad \ldots(1)$$

For damped oscillations in CO_2, we use

at $\quad t = 0, A_0 = 10$ cm

and $t = t_1, A = 5$ cm

$$\Rightarrow \quad 5 = 10\, e^{\frac{-b_{CO_2} \times t_1}{2m}} \quad \ldots(2)$$

from equation-(1) $\dfrac{40 b_{air}}{2m} = \ln\left(\dfrac{5}{4}\right) \quad \ldots(3)$

from equation-(2) $\dfrac{t_1 b_{CO_2}}{2m} = \ln(2) \quad \ldots(4)$

using (4)/(3) we have

$$\frac{t_1}{40(b_{air}/b_{CO_2})} = \frac{\ln(2)}{\ln(5) - \ln(4)}$$

$$\Rightarrow \quad t_1 = 40 \times 1.3 \times \frac{0.693}{1.601 - 0.693}$$

$$\Rightarrow \quad t_1 \approx 167\text{s}$$

Ans. (C)

Sol. 41 The time period of a simple pendulum is given as

$$T = 2\pi\sqrt{\frac{l}{g}}$$

$$\Rightarrow \quad T^2 = 4\pi^2\left(\frac{l}{g}\right)$$

$$\Rightarrow \quad g = \frac{4\pi^2}{(T^2/l)} = \frac{4\pi^2}{(\text{slope of graph})}$$

$$\Rightarrow \quad g = \frac{4\pi^2}{4.0} = \pi^2 = \left(\frac{22}{7}\right)^2 = 9.87 \text{ m/s}^2$$

Ans. (B)

Sol. 42 As $\quad x = a_1\cos\omega t$

$$y = a_2\cos 2\omega t = a_2(2\cos^2\omega t - 1)$$

$\Rightarrow$ at $t = 0$, $x = a$ and $y = a_2$

$$\Rightarrow \quad y = a_2\left(\frac{2x^2}{a_1^2} - 1\right)$$

$$\Rightarrow \quad x^2 = \frac{a_1^2}{2a_2}y + \frac{a_1^2}{2}$$

At $\quad y = 0, \quad x = \pm a_1/2$

$\Rightarrow$ Trajectory is an upward open parabola. **Ans. (A)**

Sol. 43 In SHM the restoring force $\propto$ displacement from mean position $x = 0$.

If $x \neq 0$ at mean position than force is a linear function of x and direction of force always toward mean position.

In option (A), (B) and (C) function depends upon x^2

$\Rightarrow$ it can not be SHM

In option (D) it is $(-bx)$

$\Rightarrow \quad F = -bx$ (it is SHM) **Ans. (D)**

Sol. 44 $\quad A = 1$ cm

$$\omega = \frac{2\pi}{T} = \frac{2\pi}{0.5} = 4\pi \text{ rad/s}$$

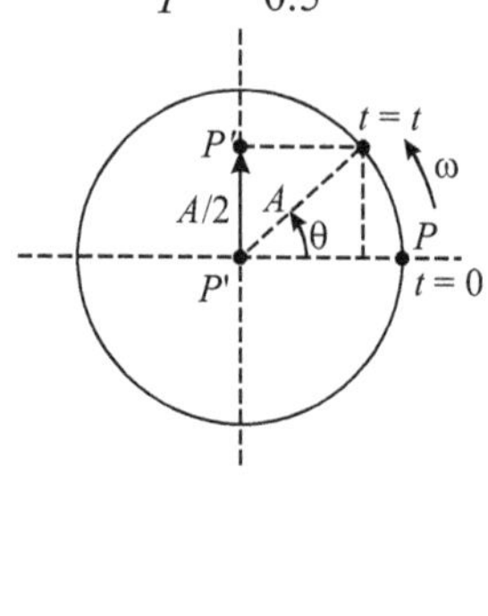

To cover distance from mean position to $A/2$.

Phase change is $\theta = \sin^{-1}\frac{A/2}{A} = \sin^{-1}\frac{1}{2} = \frac{\pi}{6}$

time taken by SHM in going from mean position to $A/2$ is

$$t = \frac{\theta}{\omega} = \frac{\pi/6}{4\pi} = \frac{1}{24}\text{ s}$$

Average velocity in moving from $x = 0$ to $x = A/2$ is

$$v_{avg} = \frac{\Delta x}{\Delta t} = \frac{A/2}{1/24}$$

$$= 12A = 12 \text{ cm/s}$$ **Ans. (C)**

Sol. 45

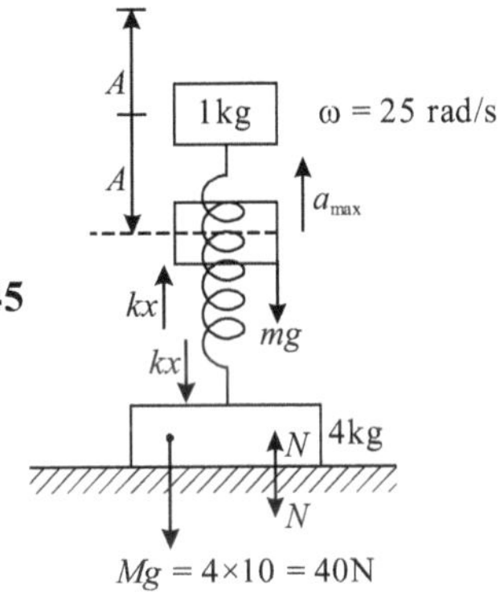

For 4 kg block, force on ground

$$N = Mg + kx|_{max} \quad \ldots(1)$$

Maximum force will be exerted on floor when 1 kg mass is at its lowest extreme so that spring compression is maximum.

Acceleration of 1 kg block at this position is

$$a_{max} = \omega^2 A = (25)^2 \times 0.016 = 10 \text{ m/s}^2$$

Equation of motion of 1 kg block is

$$kx - mg = ma_{max}$$

$$kx = mg + ma_{max} = 1 \times 10 + 1 \times 10$$

$$= 20 \text{ N (max)}$$

From equation-(1)

$$N_{max} = Mg + kx|_{max}$$

$$= 40 \times 10 + 20 = 60 \text{ N}$$ **Ans. (C)**

Sol. 46 Angular frequency of undamped oscillator

$$\omega_0 = \sqrt{\frac{k}{m}}$$

Change in angular frequency $\Delta\omega = \omega_0 - \omega$

$$\Rightarrow \quad \Delta\omega = \sqrt{\frac{k}{m}} - \sqrt{\frac{k}{m} - \frac{r^2}{4m^2}}$$

$$\Rightarrow \quad \Delta\omega = \sqrt{\frac{k}{m}}\left(1 - \sqrt{1 - \frac{r^2}{4mk}}\right)$$

$$\Rightarrow \quad \frac{\Delta\omega}{\omega_0} = \left(1 - \sqrt{1 - \frac{0.08}{4}}\right)$$

$$\Rightarrow \quad \frac{\Delta\omega}{\omega_0} = 1 - \sqrt{1 - 0.02} = 1 - (1 - 0.01) = 0.01$$

As $$T = \frac{2\pi}{\omega}$$

$$\Rightarrow \quad \frac{\Delta T}{T_0} = -\frac{\Delta\omega}{\omega_0}$$

$\Rightarrow$ As ω decreases by 1 % $\Rightarrow$ T increases by 1% **Ans. (A)**

Sol. 47 If for length l of pendulum time period is T then

$$T = 2\pi\sqrt{\frac{l}{g}} \Rightarrow l = \frac{T^2 g}{4\pi^2} \quad \ldots(1)$$

after adding a mass M, if length becomes $l + \Delta l$

then we have $$T_M = 2\pi\sqrt{\frac{l + \Delta l}{g}} \Rightarrow l + \Delta l = \frac{T_M^2 g}{4\pi^2} \quad \ldots(2)$$

If Y is Young's modulus of wire than we use

$$Y = \frac{\text{Stress}}{\text{Strain}} = \frac{Mg/A}{\Delta l / l}$$

$$\Rightarrow \quad \frac{\Delta l}{l} = \frac{Mg}{YA}$$

We use (2)/(1)

$$\Rightarrow \quad 1 + \frac{\Delta l}{l} = \left(\frac{T_M}{T}\right)^2$$

$$\Rightarrow \quad 1 + \frac{Mg}{YA} = \left(\frac{T_M}{T}\right)^2$$

$$\Rightarrow \quad \frac{1}{Y} = \left[\left(\frac{T_M}{T}\right)^2 - 1\right]\frac{A}{Mg}$$ **Ans. (A)**

Sol. 48 For SHM, KE and PE as a function of displacement are given as

$$K_x = \frac{1}{2}m\omega^2(A^2 - x^2)$$

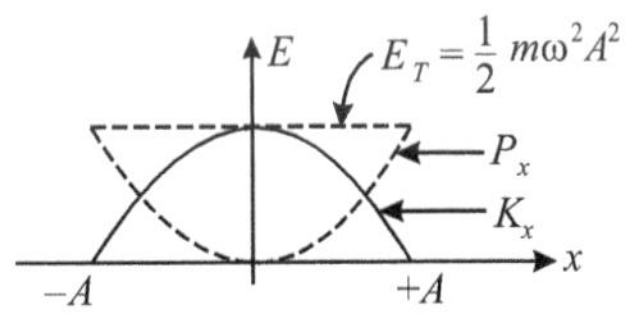

and $$P_x = \frac{1}{2}m\omega^2 x^2$$ **Ans. (B)**

Sol. 49 In damped oscillations, mechanical energy as a function of time is given as

$$E = E_0 e^{-bt/m} \quad \ldots(1)$$

at $t = 0, E_0 = 45$ J

and at $t = 15$s, $E = 15$ J

$$\Rightarrow \quad 15 = 45\, e^{-(b/m)15}$$

$$\Rightarrow \quad \frac{1}{3} = e^{-15(b/m)}$$

$$\Rightarrow \quad \ln(3) = 15(b/m)$$

$$\Rightarrow \quad \frac{b}{m} = \frac{1}{15}\ln(3)\ \text{s}^{-1}$$

Ans. (D)

Sol. 50 As $x = a\sin\omega t$ at $t = 0, x = 0$

and $y = a\sin 2\omega t$ $y = 0$

$\Rightarrow$ Answer could be option (A) or (C), so we check both here

(A) In one half oscillation of y, x completes its one oscillation

$\Rightarrow$ frequency of x is twice that of y.

(C) In one half oscillation of x, y completes its one oscillation

$\Rightarrow$ frequency of y is twice that of x, so this is correct answer for the given equation.

Ans. (C)

Sol. 51 Given that at $t = 0$, $x = 0$ and oscillator starts from rest $v = 0$. Checking given option for that

(A) $x = C(\sin t + \frac{1}{2}\cos t)|_{t=0} = C(0 + \frac{1}{2}) \neq 0$ ×

(B) $x = C(\cot t - \frac{1}{2}\cos t)|_{t=0} \to \infty$ ×

(C) $x = C(\sin t + \frac{1}{2}\cos t)|_{t=0} = 0$ ✓

$v = \frac{dx}{dt} = C(\cos t - \cos 2t)|_{t=0} = 0$ ✓

(D) $x = C(\sin t + \frac{1}{2}\sin 2t)|_{t=0} = 0$ ✓

$v = \frac{dx}{dt} = C(\cos t + \cos 2t)|_{t=0} = 2C$ ×

Ans. (C)

Sol. 52 Speed of particle when it is at a distance $\frac{2A}{3}$ from mean position is

$$v = \omega\sqrt{A^2 - \left(\frac{2A}{3}\right)^2} = \frac{\sqrt{5}}{3}A\omega$$

After increasing the speed 3 times if new amplitude becomes A' then we use

$$3v = \omega\sqrt{(A')^2 - \left(\frac{2A}{3}\right)^2} = \sqrt{5}\,A\text{w}$$

$$\Rightarrow \quad 5A^2\omega^2 = \omega^2\left((A')^2 - \frac{4A^2}{9}\right)$$

$$\Rightarrow \quad (A')^2 = \frac{49}{9}A^2 \Rightarrow A' = \frac{7}{3}A$$

Ans. (D)

Sol. 53

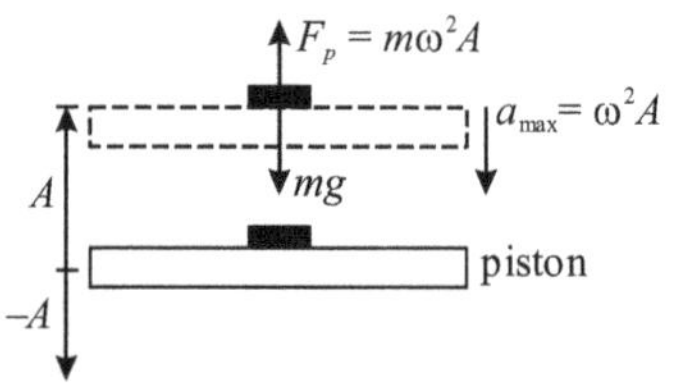

At upper extreme position of piston washer experiences maximum upward Pseudo force w.r. to piston.

If $F_p \geq mg$ then it looses contact from piston

$$\Rightarrow \quad m\omega^2 A = mg$$

$$\Rightarrow \quad \omega = \sqrt{\frac{g}{A}}$$

$$\Rightarrow \quad f = \frac{1}{2\pi}\sqrt{\frac{10}{0.07}} = \frac{1}{2\pi}\sqrt{\frac{100}{7}} = \frac{1}{2}\sqrt{14.2}$$

$$\Rightarrow \quad f \approx \frac{3.8}{2} = 1.9\ \text{Hz}$$

Ans. (B)

Sol. 54

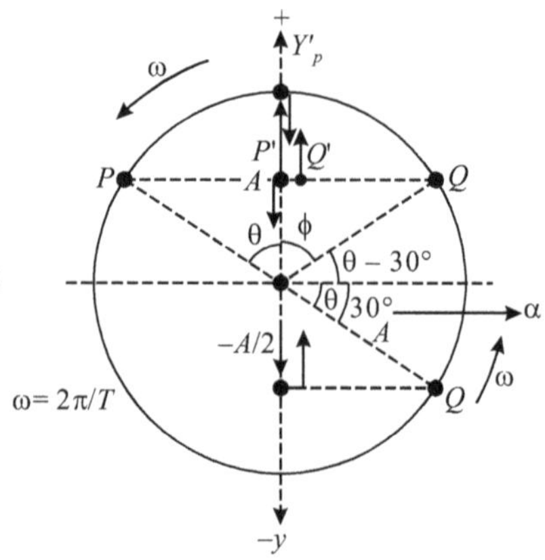

$$A\sin\alpha = A/2$$

$$\Rightarrow \quad \sin\alpha = 1/2$$

$$\Rightarrow \quad \alpha = 30°$$

P' and Q' are projections of P and Q in uniform circular motion along YY' which are in SHM.

In figure shown we have

$$\phi = 90° - (\theta - 30°) = 120° - \theta$$

and we have $\theta = \phi$

$$\Rightarrow \quad \theta = 120 - \theta \Rightarrow \theta = 60° = \pi/3$$

Time taken by P and Q to traverse angle $\pi/3$ is

$$t = \frac{\theta}{\omega} = \frac{\pi/3}{2\pi/T} = \frac{T}{6}$$

Ans. (D)

Sol. 55 For a particle starting SHM from mean position, equation of motion is given as

$$y = A\sin\omega t$$

$\Rightarrow$ particle's velocity $v = \frac{dy}{dt} = A\omega\cos\omega t$

$$\Rightarrow \qquad KE = \frac{1}{2}mv^2 = \frac{1}{2}m\omega^2A^2\cos^2\omega t$$

KE_{max} at $t = 0$

In option (B), KE oscillates at twice the frequency of SHM.

Ans. (B)

Sol. 56 Case-1:

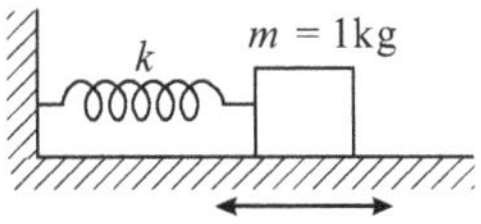

Oscillation frequency of block

$$f = \frac{1}{2\pi}\sqrt{\frac{k}{m}} = 1\text{ Hz}$$

$$\Rightarrow \qquad k = 4\pi^2\text{ N/m}$$

Case-2:

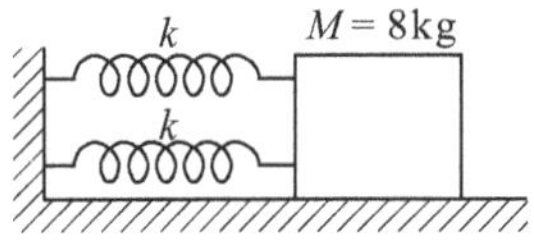

In parallel effective force constant of spring

$$k_{eff} = k_1 + k_2 = 2k = 8\pi^2\text{ N/m}$$

Oscillation frequency $f' = \frac{1}{2\pi}\sqrt{\frac{2k}{M}}$

$$= \frac{1}{2\pi}\sqrt{\frac{8\pi^2}{8}} = \frac{1}{2}\text{Hz}$$

Ans. (C)

Sol. 57 Time period of oscillation of a pendulum is given as

$$T = 2\pi\sqrt{\frac{l}{g}}$$

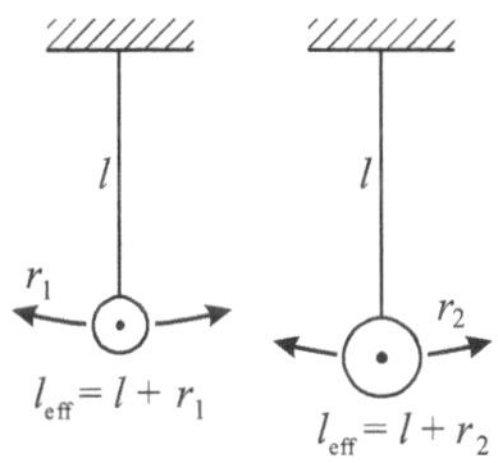

For uniform mass distribution COM is taken at centre of bob.

$$\ln T = \ln\frac{2\pi}{\sqrt{g}} + \ln\sqrt{l}$$

Relative change in period

$$\frac{dT}{T} = \frac{1}{2}\frac{dl}{l} \qquad \text{here } dl = |r_2 - r_1|$$

$$\Rightarrow \qquad 5\times 10^{-4} = \frac{1}{2}\frac{|r_2 - r_1|}{1\text{m}}$$

$$\Rightarrow \qquad |r_2 - r_1| = 2\times 5\times 10^{-4} = 10^{-3}\text{ m} = 1\text{ mm}$$ **Ans.(B)**

Sol. 58 For damped oscillations, the oscillation amplitude is given as

$$A = A_0 e^{\frac{-bt}{2m}} \qquad \ldots(1)$$

$b \rightarrow$ damping coefficient

$m \rightarrow$ mass of block

As we know mechanical energy of oscillation $E \propto A^2$

E will be half when $A \rightarrow \frac{A}{\sqrt{2}}$

from equation-(1), if at $t = t$, $A_0 \rightarrow \frac{A}{\sqrt{2}}$

$$\Rightarrow \qquad \frac{A_0}{\sqrt{2}} = A_0 e^{\frac{-bt}{2m}}$$

$$\Rightarrow \qquad \frac{bt_1}{2m} = \ln\sqrt{2} = \frac{1}{2}\ln 2$$

$$\Rightarrow \qquad t_1 = \frac{m}{b}\ln(2) = \frac{0.1}{0.01}\times 0.693$$

$$\Rightarrow \qquad t_1 = 6.93\text{s}$$

Ans. (D)

Sol. 59 If in SHM amplitude is A and angular frequency is ω, then we have

max. velocity $\quad v_{max} = A\omega$

max. acceleration $\quad a_{max} = \omega^2 A$

Given that $\quad \frac{a_{max}}{v_{max}} = \omega = 10\text{ s}^{-1}$

General equation of SHM can be taken as

$$y = A\sin(\omega t + \alpha)$$

at $\quad t = 0, y = 5\text{m}, \alpha = \frac{\pi}{4}$

$$5 = A\sin\left(\frac{\pi}{4}\right)$$

$$\Rightarrow \qquad 5 = A\left(\frac{1}{\sqrt{2}}\right)$$

$$\Rightarrow \qquad A = 5\sqrt{2}\text{ m}$$

$$\Rightarrow \qquad a_{max} = \omega^2 A = (10)^2\times 5\sqrt{2}$$

$$= 500\sqrt{2}\text{ m/s}^2$$

Ans. (B)

Sol. 60 Mass of one atom of silver is

$$m = \frac{M}{Av} = \frac{108\times 10^{-3}}{6.02\times 10^{23}}$$

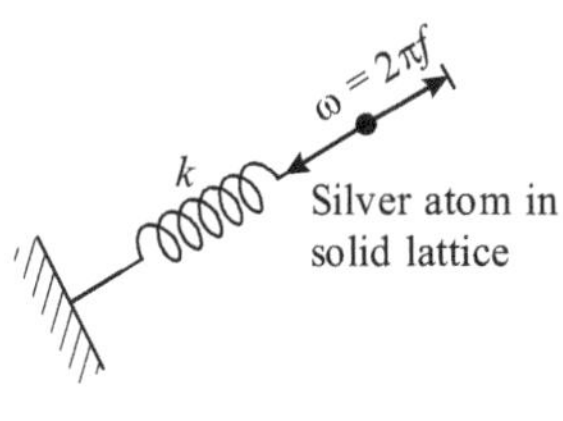

Spring represent the bonds connecting the silver atom with other atoms.

For SHM angular frequency of atom is given as

$$\omega = \sqrt{\frac{k}{m}}$$

$\Rightarrow$ Oscillation frequency $f = \frac{\omega}{2\pi}$

$$\Rightarrow \quad 10^{12} = \frac{1}{2\times 3.14}\sqrt{\frac{k\times 6.02\times 10^{23}}{108\times 10^{-3}}}$$

$$\Rightarrow \quad k = 7.1 \text{ N/m}$$

Ans. (B)

Sol. 61 for $a = b$, general equation of Lissajous's figure is given as

$$\frac{x^2}{A^2}+\frac{y^2}{B^2}-\frac{2xy}{AB}\cos\delta = \sin^2\delta$$

(B) $\frac{x^2}{A^2}+\frac{y^2}{B^2}-\frac{2xy}{A^2}(0) = 1 \Rightarrow x^2+y^2 = A^2 \rightarrow$ circle (✘)

(C) $\frac{x^2}{A^2}+\frac{y^2}{B^2}-\frac{2xy}{AB}(0) = 1 \Rightarrow \frac{x^2}{A^2}+\frac{y^2}{B^2} = 1 \rightarrow$ ellipse (✔)

(D) $\frac{x^2}{A^2}+\frac{y^2}{B^2}-\frac{2xy}{AB}(1) = 0 \rightarrow$ Not a parabola. (✘) **Ans. (C)**

Sol. 62 If particle starts SHM from its mean position (say) then equation of SHM is

$$y = A\sin\omega t$$

with given conditions, we have

$$a = A\sin\omega t_0 \quad \ldots(1)$$

$$b = A\sin 2\omega t_0 \quad \ldots(2)$$

$$c = A\sin 3\omega t_0 \quad \ldots(3)$$

$$(1)+(3) \Rightarrow \quad a+c = 2A\sin 2\omega t_0\cos\omega t_0$$

$$\Rightarrow \quad \cos\omega t_0 = \frac{a+c}{2b}$$

$$\Rightarrow \quad \omega t_0 = \cos^{-1}\left(\frac{a+c}{2b}\right)$$

$$\omega = \frac{1}{t_0}\cos^{-1}\left(\frac{a+c}{2b}\right)$$

$$f = \frac{\omega}{2\pi} = \frac{1}{2\pi t_0}\cos^{-1}\left(\frac{a+c}{2b}\right)$$

Ans. (D)

Sol. 63 Restoring force on mass M is

$$F = \left|\frac{dV}{dx}\right| = k(x-X) = 0 \text{ (a mean position)}$$

$$\Rightarrow \quad x = X$$

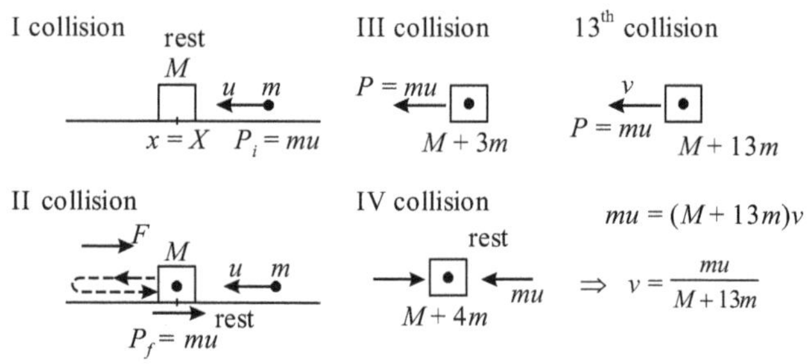

After 13th collision velocity of combined mass $M + 13m$ is given as

$$v = \frac{mu}{M+13m} = \frac{5\times 1}{10+65} = \frac{5}{75} = \frac{1}{15}$$

at mean position

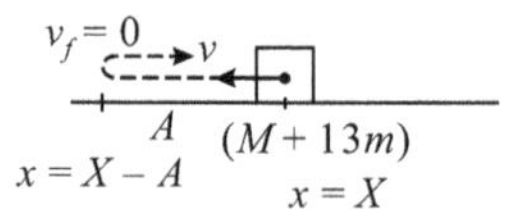

When block reaches extreme position, its KE at mean position is transformed into PE at extreme position.

$\Rightarrow$ By conservation of energy

$$\frac{1}{2}(M+13m)v^2 = \frac{1}{2}k(X-A-x)^2$$

$$\Rightarrow \quad \frac{1}{2}(10+65)\left(\frac{1}{15}\right)^2 = \frac{1}{2}\times 1\times A^2$$

$$\Rightarrow \quad A^2 = \frac{75}{15\times 15} = \frac{1}{3} \Rightarrow A = \frac{1}{\sqrt{3}}$$

Ans. (B)

Sol. 64 Angular frequency of SHM of rod is

$$\omega = \sqrt{\frac{k}{I}}$$

$$\Rightarrow \quad \omega = \sqrt{\frac{3k}{ml^2}} \qquad \left(I = \frac{ml^2}{9}+\frac{m}{2}\cdot\frac{4l^2}{9} = \frac{ml^2}{3}\right)$$

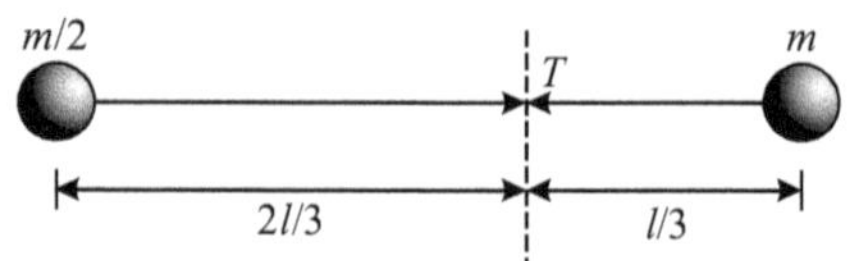

$$\dot{\theta}_{max} = \omega\theta_0 \qquad \text{[maximum angular speed]}$$

tension in rod is given as

$$\Rightarrow \quad T = m\dot{\theta}^2 r_1$$

$$\Rightarrow \quad T = m\dot{\theta}^2\frac{l}{3}$$

$$\Rightarrow \quad T = m\omega^2\theta_0^2\frac{l}{3}$$

$$\Rightarrow \quad T = m\frac{3k}{ml^2}\theta_0^2\frac{l}{3}$$

$$\Rightarrow \quad T = \frac{k\theta_0^2}{l}$$

Ans. (D)

Sol. 65 Angular frequency of torsional oscillations is given by

$$\omega = \sqrt{\frac{k}{I}}$$

$$\omega_1 = \sqrt{\frac{k}{\frac{M(2L)^2}{12}}} \qquad \text{[Oscillations of rod]}$$

$$\omega_2 = \sqrt{\frac{k}{\frac{M(2L)^2}{12} + 2m\left(\frac{L}{2}\right)^2}}$$

[Oscillations of rod with masses]

$$\omega_2 = 0.8\omega_1$$

$$\Rightarrow \quad \frac{m}{M} = 0.375$$

Ans. (B)

Sol. 66 Potential energy $(U) = \frac{1}{2}kx^2$

Kinetic energy $(K) = \frac{1}{2}kA^2 - \frac{1}{2}kx^2$

Given that, $U = k$

$$\Rightarrow \quad \frac{1}{2}kx^2 = \frac{1}{2}kA^2 - \frac{1}{2}kx^2$$

$$\Rightarrow \quad x = \pm\frac{A}{\sqrt{2}}$$

Ans. (C)

Sol. 67 Velocity and acceleration of particle in SHM's

$$v = \omega\sqrt{A^2 - x^2} \quad \ldots(1)$$

$$a = -\omega^2 x \quad \ldots(2)$$

Given that

$$|v| = |a| \quad \ldots(3)$$

$$\omega\sqrt{A^2 - x^2} = \omega^2 x$$

$$\Rightarrow \quad A^2 - x^2 = \omega^2 x^2$$

$$\Rightarrow \quad 5^2 - 4^2 = \omega^2(4^2)$$

$$\Rightarrow \quad 3 = \omega \times 4$$

$$\Rightarrow \quad \omega = 3/4$$

$$T = 2\pi/\omega = \frac{8\pi}{3}$$

Ans. (D)

Sol. 68

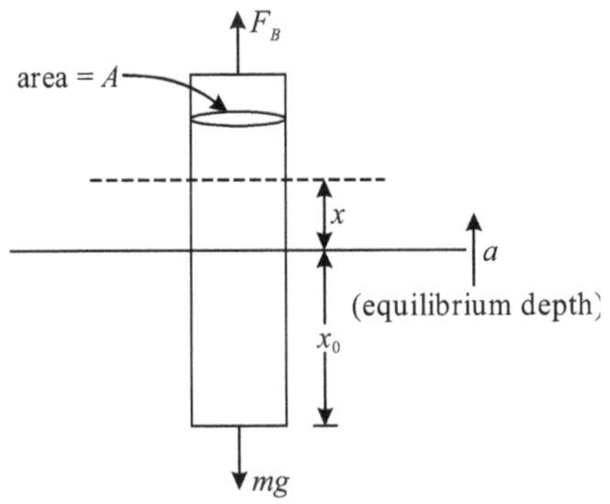

At equilibrium

$$mg = \rho_\omega A x_0 g$$

When further pushed down by x, we use

$$\rho_\omega A(x + x_0)g - mg = ma$$

$$\Rightarrow \quad \rho_\omega Axg = ma$$

$$\Rightarrow \quad a = \frac{\rho_\omega Ag}{m}x$$

Comparing with

$a = \omega^2 x$ we get

$$\omega = \sqrt{\frac{\rho_\omega Ag}{m}}$$

$$\Rightarrow \quad \omega = \sqrt{\frac{10^3 \times \pi(2.5)^2 \times 10^{-4} \times 10}{310 \times 10^{-6} \times 10^3}}$$

$$\Rightarrow \quad \omega = \sqrt{63.30} = 7.95 \text{ rad/s}$$

Ans. (None)

Sol. 69 Kinetic energy of particle in SHM is

$$K = \frac{1}{2}m\omega^2 A^2 \cos^2\omega t \quad \ldots(1)$$

Potential energy of particle in SHM is

$$U = \frac{1}{2}m\omega^2 A^2 \sin^2\omega t \quad \ldots(2)$$

$$\Rightarrow \quad \frac{K}{U} = \cot^2 \omega t = \cot^2 \frac{\pi}{90}(210) = \frac{1}{3}$$

Ans. (C)

Sol. 70 Angular frequency of simple pendulum is given as

$$\omega = \sqrt{\frac{g_{eff}}{l}}$$

$$\Rightarrow \quad \frac{\Delta\omega}{\omega} = \frac{1}{2}\frac{\Delta g_{eff}}{g_{eff}}$$

$$\Rightarrow \quad \Delta\omega = \frac{1}{2}\frac{\Delta g_{eff}}{g_{eff}} \times \omega \quad \ldots(1)$$

Due to vertical oscillations of support change in g_{eff} is given as

$$\Delta g_{eff} = 2\omega^2 A$$

So from equation-(1), we have

$$\Delta\omega = \frac{1}{2} \times \frac{2A\omega_s^2}{10} \times 10$$

$$\Rightarrow \quad \Delta\omega = 10^{-3} \text{ rad/sec}$$

Ans. (A)

Sol. 71 Maximum kinetic energy at lowest point B is given as

$$K = mgl\,(1 - \cos\theta) \quad \ldots(1)$$

Where θ = angular amplitude of pendulum

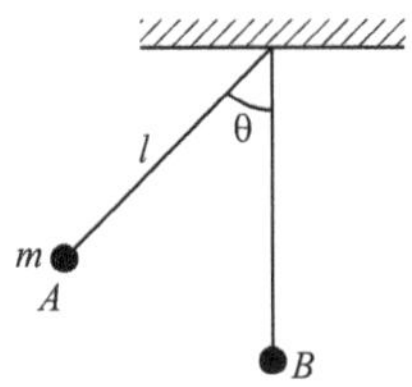

For the given conditions, we use from equation-(1)

$$K_1 = mgl(1 - \cos\theta)$$

and $$K_2 = mgl(2l)(1 - \cos\theta)$$

$\Rightarrow$ $K_2 = 2K_1$ **Ans. (C)**

Sol. 72

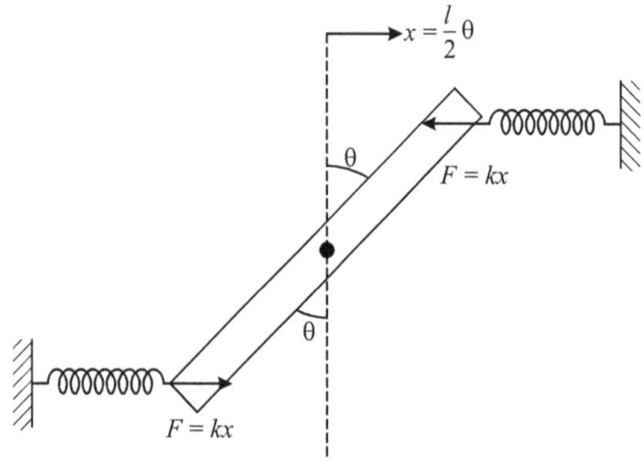

Restoring torque on rod due to its angular displacement is

$$\tau = -2kx + \frac{l}{2}$$

$$\Rightarrow \quad \tau = -\left(\frac{kl^2}{2}\right)\theta$$

Angular acceleration of rod is

$$\alpha = \frac{\tau}{I} = -\frac{\frac{kl^2}{2}\theta}{(Ml^2/12)}$$

$$\Rightarrow \quad \alpha = -\frac{6k}{M}\cdot\theta$$

Comparing with

$$\alpha = -\omega^2\theta, \text{ we get}$$

$$\omega = \sqrt{\frac{6k}{M}}$$

$$\Rightarrow \quad f = \frac{\omega}{2\pi} = \frac{1}{2\pi}\sqrt{\frac{6k}{M}}$$ **Ans. (A)**

Sol. 73 SHM equation given as

$$y = 5\left[\sin(3\pi t) + \sqrt{3}\cos(3\pi t)\right]$$

Above equation cab be rewritten as

$$y = 10\sin\left(3\pi t + \frac{\pi}{3}\right)$$

$\Rightarrow$ Amplitude = 10 cm

$$\Rightarrow \quad T = \frac{2\pi}{\omega} = \frac{2\pi}{3\pi} = \frac{2}{3}\text{ s}$$ **Ans. (D)**

Sol. 74 For a damped oscillator with time amplitude decreases exponentially

$$\Rightarrow \quad A_t = A_0 e^{-\gamma t} \qquad \ldots(1)$$

Given that after 10 oscilations (2 sec)

$$A = \frac{A_0}{2}$$

$$\Rightarrow \text{From (1)} \ \frac{A_0}{2} = A_0 e^{-2\gamma}$$

$$\Rightarrow \ln(2) = 2\gamma \Rightarrow \gamma = \frac{\ln(2)}{2}$$

$$\frac{A_0}{1000} = A_0 e^{-\gamma t}$$

$$\ln(1000) = \frac{\ln 2}{2}\cdot 2$$

$$\Rightarrow \quad t = \frac{6\ln(10)}{\ln 2} = 19.93\text{ s} \simeq 20\text{ s}$$ **Ans. (B)**

Sol. 75 For a simple pendulum

$$T = 2\pi\sqrt{\frac{L}{g_{eff}}}$$

Situation 1 : When pendulum is in air, we use $g_{\text{eff}} = g$

Situation 2 : When pendulum is in liquid, we use

$$g_{\text{eff}} = g\left(1 - \frac{\rho_{\text{liquid}}}{\rho_{\text{body}}}\right) = g\left(1 - \frac{1}{16}\right) = \frac{15g}{16}$$

$$\Rightarrow \quad \frac{T'}{T} = \frac{2\pi\sqrt{\frac{L}{15g/16}}}{2\pi\sqrt{\frac{L}{g}}}$$

$$\Rightarrow \quad T' = \frac{4T}{\sqrt{15}}$$ **Ans. (A)**

Sol. 76 By law of conservation of energy, we use

$$\frac{1}{2}kx^2 = (m_1 s_1 + m_2 s_2)\Delta T$$

$$\Rightarrow \quad \Delta T = \frac{16\times10^{-2}}{4384} = 3.65\times10^{-5}$$ **Ans. (D)**

Sol. 77 For damped oscillation, amplitude is given as

$$A = A_0 e^{-0.1t} = \frac{A_0}{2}$$

$$\Rightarrow \quad \ln 2 = 0.1t$$

$$\Rightarrow \quad t = 10\ln 2 = 6.93$$

$$\Rightarrow \quad \approx 7\text{ sec}$$ **Ans. (B)**

14.1 Wave Motion

Sol. 1 The given wave equation is

$$y = 10^{-4}\sin\left(600t - 2x + \frac{\pi}{3}\right) \qquad \ldots(1)$$

The general wave equation is expressed as

$$y = A\sin(\omega t - kx + \alpha) \quad \ldots(2)$$

Comparing equation-(1) & (2), we have

$$\omega = 600 \text{ rad/s}$$

$$k = 2 \text{ m}^{-1}$$

$$\Rightarrow \text{ Wave speed } \quad v = \frac{\omega}{k}$$

$$= \frac{600}{2} = 300 \text{ m/s}$$

Ans. (A)

Sol. 2 If we use sound level of I_1 w.r. to I_{01}

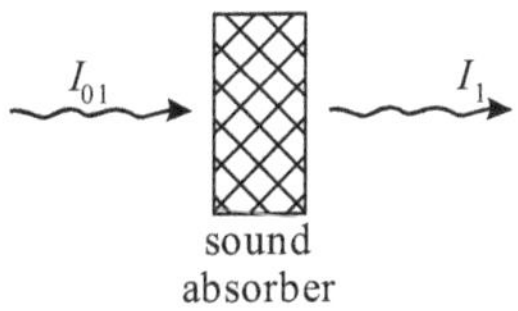

We have

$$L_{I_{01}} - L_{I_1} = 20 \text{ dB}$$

$$\Rightarrow \quad 10\log_{10}\left(\frac{I_{01}}{I_0}\right) - 10\log_{10}\left(\frac{I_1}{I_0}\right) = 20$$

$$\Rightarrow \quad 10\log_{10}\left(\frac{I_{01}}{I_1}\right) = 20$$

$$\Rightarrow \quad \frac{I_{01}}{I_1} = 10^2 = 100 \Rightarrow I_1 = \frac{I_{01}}{100}$$

Ans. (D)

Sol. 3 The given wave equation is

$$y = 0.005\cos(\alpha x - \beta t) \quad \ldots(1)$$

General expression for a wave equation is

$$y = A\cos(kx - \omega t) \quad \ldots(2)$$

$$A\cos(kx - \omega t) = A\sin(\omega t - kx + \phi)$$

Comparing (1) and (2) we have

$$\alpha = k = \frac{2\pi}{\lambda} = \frac{2\pi}{0.08} = 25\pi$$

and

$$\beta = \omega = \frac{2\pi}{T} = \frac{2\pi}{2} = \pi$$

Ans. (A)

Sol. 4 In gaseous medium speed of sound is given as

$$v = \sqrt{\frac{\gamma RT}{M}}$$

$$\Rightarrow \quad \frac{v_{O_2}}{v_{He}} = \sqrt{\frac{\gamma_{O_2}}{M_{O_2}} \times \frac{M_{He}}{\gamma_{He}}}$$

$$\Rightarrow \quad v_{He} = v_{O_2}\sqrt{\frac{\gamma_{He}M_{O_2}}{\gamma_{O_2}M_{He}}} = 460\sqrt{\frac{\frac{5}{3}\times 32}{\frac{7}{5}\times 4}}$$

$$\Rightarrow \quad v_{He} = 460\sqrt{\frac{200}{21}} \text{ m/s}$$

Ans. (A)

Sol. 5 Given wave equation is

$$y = 0.02\sin\left[2\pi\left(\frac{t}{0.04} - \frac{x}{0.05}\right)\right] \quad \ldots(1)$$

General equation of a wave equation is of the form

$$y = A\sin\left[2\pi\left(\frac{t}{T_0} - \frac{x}{A}\right)\right] \quad \ldots(2)$$

Comparing (1) and (2) gives

$$T_0 = 0.04 \text{ s}$$

$$\lambda = 0.5 \text{ m}$$

$$\Rightarrow \text{ Wave speed } \quad v = \frac{\lambda}{T_0} = \frac{0.5}{0.04} = \frac{50}{4} \text{ m/s}$$

Wave speed on a string is given as

$$v = \sqrt{\frac{T}{4}}$$

$$\Rightarrow \quad T = \mu v^2 = 0.04 \times \frac{2500}{16}$$

$$\Rightarrow \quad T = \frac{25}{4} = 6.25 \text{ N}$$

Ans. (A)

Sol. 6 As sound wave intensities are given as

$$I = 2\pi^2 n^2 A^2 \rho v$$

For a given medium $\rho, v \rightarrow$ constant $\Rightarrow I \propto x^2A^2$

in S-I for first wave $n_1 = \frac{\omega}{2\pi}; A_1 = 2a$

For second wave $\quad n_2 = \frac{2\omega}{2\pi}; A_2 = a$

$\Rightarrow \; I_1 = I_2$

S-I is true

As Intensity $\propto n^2A^2$

$\Rightarrow$ S-II is false

Ans. (B)

Sol. 7 The given expression is

$$y = e^{-(ax^2 + bt^2 + 2\sqrt{ab}xt)}$$

$$y = e^{-[\sqrt{a}x + \sqrt{b}t]^2}$$

It is a function of $(c_1x + c_2t)$ thus it represents a propagating wave. +ve sign shows that wave is travelling in $-x$ direction.

having $\omega = \sqrt{b}$ and $k = \sqrt{a}$

$$\Rightarrow \text{ Wave speed } \quad v = \frac{\omega}{k} = \sqrt{\frac{b}{a}}$$

Ans. (D)

Sol. 8 If a wave displacement function is given as

$$y = f(x) \text{ at } t = 0$$

For a wave travelling in $+x$ direction at speed v, then at time $t = t$, the displacement function is given as

$$y = f(x - vt)$$

Given at $t = 0$

$$y = \frac{1}{1 + x^2}$$

So at time t the function becomes

$$y = \frac{1}{1 + (x - vt)^2} \quad \ldots(1)$$

In question at $t = 2$s displacement function is given as

$$y = \frac{1}{1 + (x - 1)^2} \quad \ldots(2)$$

Comparing equation-(2) with equation-(1), we get

$$vt = 1\big|_{t=2\text{s}}$$

$$\Rightarrow \quad v = \frac{1}{2} = 0.5 \text{ m/s} \quad \textbf{Ans. (C)}$$

Sol. 9 Standard wave equation in a medium can be given as

$$y = A\cos(kx - \omega t)$$

Comparing y_1 with above standard equation we get

$$A = 0.05 \text{ m}$$

and $$k_1 = \frac{2\pi}{\lambda_1} = 0.5\pi$$

$$\Rightarrow \quad \lambda_1 = 4 \text{ m}$$

$$\omega_1 = 2\pi n_1 = 100\pi$$

$$\Rightarrow \quad n_1 = 50 \text{ Hz}$$

Wave velocity $v = n_1\lambda_1 = 50 \times 4 = 200$ m/s **Ans. (B)**

Sol. 10 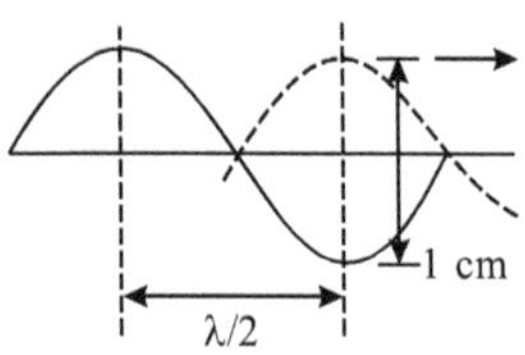

Distance between a crest and neighbouring trough is

$$\frac{\lambda}{2} = 0.4 \text{ cm}$$

$$\Rightarrow \quad \lambda = 0.8 \text{ cm}$$

It is given that $2A = 1$ cm $\Rightarrow A = 0.5$ cm

As distance between two neighbouring crests is λ

$$\Rightarrow \quad t = \frac{\lambda}{v} = 0.4 \text{ s}$$

$$\Rightarrow \quad v = \frac{\lambda}{0.4} = \frac{0.8}{0.4} = 2 \text{ cm/s}$$

$\Rightarrow$ Maximum speed of vibrating particles

$$v_{max} = A\omega = A\left(\frac{2\pi v}{\lambda}\right) = \frac{0.5 \times 2\pi \times 2}{0.8}$$

$$\Rightarrow \quad v_{max} = \frac{5\pi}{2} \text{ cm/s} \quad \textbf{Ans. (B)}$$

Sol. 11 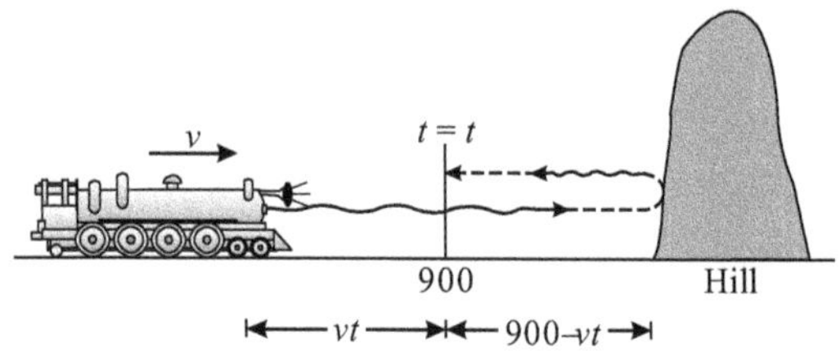

In time t when reflected sound is heard by the driver then distance travelled by sound is

$$900 + (900 - vt) = v_s.t$$

at $$t = 5\text{s}$$

$$\Rightarrow \quad 900 + 900 - 5v = 330 \times 5 = 1650$$

$$\Rightarrow \quad 5v = 1800 - 1650 = 150$$

$$\Rightarrow \quad v = \frac{150}{5} = 30 \text{ m/s} \quad \textbf{Ans. (D)}$$

Sol. 12 Given equation of SHW is

$$y = \frac{10}{\pi}\sin\left(\frac{2\pi}{T}t - \frac{2\pi}{\lambda}x\right)$$

Comparing with $y = A\sin(\omega t - kx)$

Amplitude $A = \frac{10}{\pi}$; $\omega = \frac{2\pi}{T}$; $k = \frac{2\pi}{\lambda}$

As wave velocity $= 2 \times$ maximum particle velocity

$$\frac{\omega}{k} = 2 \times A\omega$$

$$\Rightarrow \quad \frac{\lambda}{2\pi} = 2 \times \frac{10}{\pi}$$

$$\Rightarrow \quad \lambda = 40 \text{ cm} \quad \textbf{Ans. (A)}$$

Sol. 13 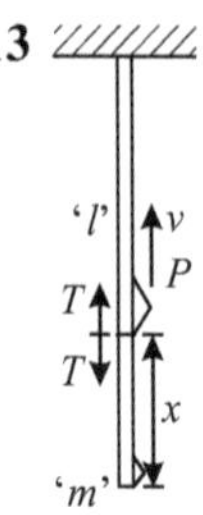

Tension at point P in string is

$$T = \left(\frac{m}{l}x\right)g$$

Wave speed in string at point P is

$$v = \sqrt{\frac{T}{\mu}} = \sqrt{\frac{\left(\frac{m}{l} \times g\right)}{\frac{m}{l}}} = \sqrt{xg}$$

$$\Rightarrow \quad \frac{dx}{dt} = \sqrt{xg}$$

$$\Rightarrow \quad \int_0^l \frac{dx}{\sqrt{x}} = \int_0^t \sqrt{g}\,dt$$

$$\Rightarrow \quad \left[2\sqrt{x}\right]_0^l = \sqrt{g}t$$

$$\Rightarrow \quad t = \frac{2\sqrt{l}}{\sqrt{g}} = 2\sqrt{\frac{20}{10}} = 2\sqrt{2}\,\text{s}$$ **Ans. (C)**

Sol. 14 When car is at rest, wave speed is given as

$$60 = \sqrt{\frac{Mg}{\mu}}$$

When car is accelerating, we use

$$60.5 = \sqrt{\frac{M(g^2+a^2)^{1/2}}{\mu}}$$

$$\Rightarrow \quad \frac{60.5}{60} = \sqrt{\sqrt{\frac{g^2+a^2}{g^2}}}$$

$$\Rightarrow \quad \left(1+\frac{0.5}{60}\right)^4 = \frac{g^2+a^2}{g^2} = 1+\frac{2}{60}$$

$$\Rightarrow \quad g^2 + a^2 = g^2 + g^2 \times \frac{2}{60}$$

$$\Rightarrow \quad a = g\sqrt{\frac{2}{60}} = \frac{g}{\sqrt{30}} = \frac{g}{5.47}$$

$$\Rightarrow \quad a \simeq \frac{g}{5}$$ **Ans. (A)**

Sol. 15 Given equation is

$$y = 0.03\sin(450t - 9x)$$

Comparing with

$y = A\sin(\omega t - kx)$, we get

$$v = \frac{\omega}{k} = \frac{450}{9} = 50 \text{ m/s}$$

Velocity of transverse waves on a string gives

$$v = \sqrt{\frac{T}{\mu}}$$

$$\Rightarrow \quad \frac{T}{\mu} = 2500$$

$$\Rightarrow \quad T = 2500 \times 5 \times 10^{-3}$$

$$\Rightarrow \quad T = 12.5 \text{ N}$$ **Ans. (B)**

Sol. 16 Above wave is moving along $-ve$ x-axis with speed given as

$$v = \frac{\omega}{K}$$

$$\Rightarrow \quad v = \frac{50}{2} = 25 \text{ m/s}$$ **Ans. (A)**

Sol. 17 Speed of wave from wave equation is written as

$$v = -\frac{(\text{coeffecient of } t)}{(\text{coeffecient of } x)}$$

$$v = -\frac{1000}{(-3)} = \frac{1000}{3} \text{ m/s}$$

As speed of wave $v \propto \sqrt{T}$, we have

$$= \frac{1000/3}{336} = \sqrt{\frac{273}{T}}$$

$$\Rightarrow \quad T = 277.41 \text{ K}$$

$$\Rightarrow \quad T = 4.41°\text{C}$$ **Ans. (C)**

Sol. 18

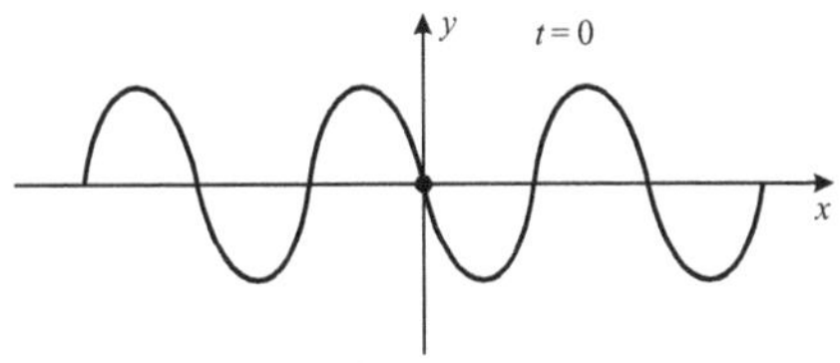

For the given wave equation

$$y = A\sin(kx - wt + \phi)$$

at $x = 0$, $t = 0$, $y = 0$ and slope is negative

$$\Rightarrow \quad \frac{\partial y}{\partial x} = Ak\cos(kx - wt + \phi) = -\text{ve}$$

$$\Rightarrow \quad \phi = x \quad \text{(From the given options)}$$ **Ans. (C)**

Sol. 19 Loudness of sound is given by

$$L(dB) = 10\log\frac{I}{I_0}$$

$$\Rightarrow \quad 120 = 10\log\left(\frac{I}{I_0}\right)$$

$$\Rightarrow \quad I = 1\text{W/m}^2$$

For isotropic point sources, intensity at a distance r is given as

$$I = \frac{P}{4\pi r^2} = \frac{2}{4\pi r^2}$$

$$\Rightarrow \quad r = \sqrt{\frac{2}{4\pi}} = \sqrt{\frac{1}{2\pi}}\,m \quad (\text{Taking } I = 1 \text{ W/m}^2)$$

$$\Rightarrow \quad r = 0.399 \text{ m} \approx 40 \text{ cm}$$ **Ans. (C)**

14.2 Stationary Waves and Beats

Sol. 1 Fundamental frequencies of A and B are

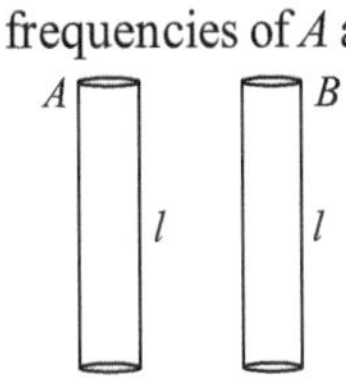

$$v_A = \frac{v}{2l} \text{ and } v_B = \frac{v}{4l}$$

Here we ignore end correction

$$\Rightarrow \qquad \frac{\nu_A}{\nu_B} = \frac{2}{1} \qquad \textbf{Ans. (C)}$$

Sol. 2 If the frequency of unknown fork is ν_0 then we use Initially

$$|\nu_0 - 288| = 4$$

$$\Rightarrow \qquad \nu_0 = 292 \text{ cps or } 284 \text{ cps}$$

On loading ν_0 with some wax its frequency decreases & later it produces 2 beats/s with 288 cps fork.

After loading with wax $\nu_0 \to \nu_0' = 292\downarrow$ or $284\downarrow$

for 292 $\to$ $\quad f_B = 290 \Rightarrow$ 2 beats/sec

for 284 $\to$ $\quad f_B > 4$ beats/s

$\Rightarrow$ 2 beats/s is not possible with 284 Hz. **Ans. (B)**

Sol. 3

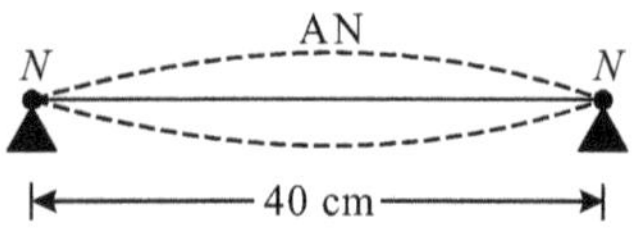

Maximum wavelength of stationary wave in a clamped string is for fundamental frequency given as

$$\nu_0 = \frac{v}{2l}$$

$\Rightarrow$ wavelength $\quad \lambda = 2l = 2 \times 40 = 80$ cm **Ans. (B)**

Sol. 4 General equation of a stationary wave can be

$$y = 2A \sin kx \sin(\omega t + \alpha)$$

$\to$ at $x = 0$; $\quad R = 2A \sin kx = 0$

$\Rightarrow$ Node

or $\quad y = 2A \cos kx \sin(\omega t + \alpha)$

$\to$ at $x = 0$; $\quad R = 2A \cos kx = 2A$

$\Rightarrow$ Antinode

As one wave is $\quad y_1 = a \sin(\omega t - kx)$

with option (A) $\to$ $y_R = a\sin(\omega t - kx) + a\sin(\omega t + kx)$

$$= 2A \sin\omega t \cos kx$$

with option (B) $\to$ $y_R = a\sin(\omega t - kx) + a\sin(\omega t + kx)$

$$= -2A \cos\omega t \sin kx \qquad \textbf{Ans. (B)}$$

Sol. 5 If initial frequency of piano string is ν_0.

then we have $|\nu_0 - 256| = 5 \Rightarrow \nu_0 = 261$ Hz or 251 Hz

On increasing tension in string, its oscillation frequency increases.

$\Rightarrow \nu_0 \to \nu_0' = 261$ Hz $(\uparrow) \to f_B' > 5$ Hz

or $\quad 251$ Hz $(\uparrow) \to f_B' < 5$ Hz

Giving that $f_B' = 2$ Hz

final beat frequency $\Rightarrow \nu_0 = 251$ Hz **Ans. (C)**

Sol. 6 Initially we have $\nu_1 = 200$ Hz and $\nu_2 = ?$

$$f_B = |200 - \nu_2| = 4 \text{ Hz}$$

$$\Rightarrow \qquad \nu_2 = 196 \text{ Hz or } 204 \text{Hz}$$

On loading some tape on fork-2, its frequency decreases

$$\nu_2 \to \nu_2'$$

we use $\nu_2' = 196\downarrow$ or $204\downarrow$

when 196 $\Rightarrow f_B' > 4$ Hz and for 204 $\Rightarrow f_B' < 4$ Hz

later it is given that

$$f_B' = |200 - \nu_2'| = 6 \text{ Hz}$$

$$\Rightarrow \qquad \nu_2 = 196 \text{ Hz} \qquad \textbf{Ans. (C)}$$

Sol. 7 For a clamped string oscillations all the horizontal frequencies of fundamental frequency are available for resonance so for two successive resonant frequency, we use

$$n_2 - n_1 = n_0 \; (n_0 = \text{fundamental frequency})$$

$$\Rightarrow \qquad n_0 = 420 - 315 = 105 \text{ Hz} \qquad \textbf{Ans. (A)}$$

Sol. 8 For resonance column experiment we know that second resonance occurs at 3 times the length of first resonance.

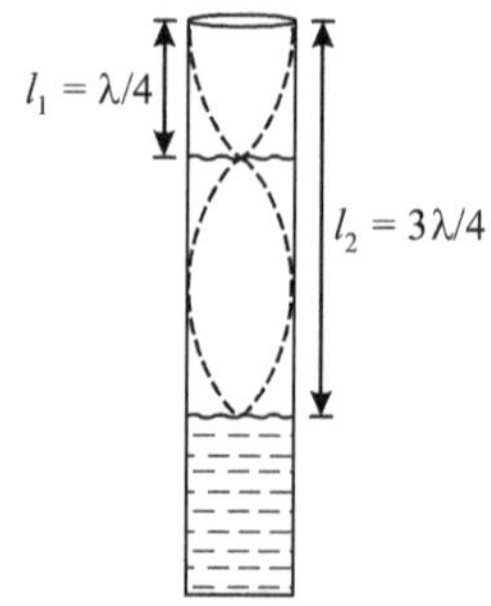

In winter $\qquad l_1 = 18$ cm $\Rightarrow l_2 = 54$ cm

As speed of sound $v \propto \sqrt{T}$ in summer speed of sound increases so frequencies of oscillating column are given as (frequency of tuning fork)

$$\nu_1 = \frac{v}{4l_1} = \frac{3v}{4l_2} \to \text{constant}$$

In summer as v increases $\Rightarrow$ for same experiment l_2 also increases. **Ans. (B)**

Sol. 9 Considering two waves at a time beat frequency can be given as

$$\left.\begin{aligned} \nu_1 &= \nu - 1 \\ \text{and} \quad \nu_2 &= \nu \end{aligned}\right] \to f_B = 1\text{Hz}$$

time of maxima $t = 0, 1, 2, 3, \ldots\ldots$ s

and $\left.\begin{array}{l} \nu_2 = \nu \\ \nu_2 = \nu + 1 \end{array}\right] \rightarrow f_B' = 1\text{Hz}$

$\Rightarrow \quad t_m = 0, 1, 2, 3, \ldots\ldots$ s

and $\left.\begin{array}{l} \nu_2 = \nu - 1 \\ \nu_3 = \nu + 1 \end{array}\right] \rightarrow f_B'' = 2\text{Hz}$

$\Rightarrow \quad t_m = 0, 0.5, 1.5, 2, \ldots$ s

So 0.5, 1.5 .... Very low intensity maxima

and 0, 1, 2, ... Higher maximas are produced with high intensity.

$\Rightarrow$ Time difference = 1s

$$\Rightarrow \quad f_{Bf} = \frac{1}{T} = 1 \text{ Hz}$$ **Ans. (B)**

Sol. 10 Superposition of the two waves is giving resulting displacement as $\quad y = y_1 + y_2$

$$\Rightarrow \quad y = A \sin(\omega t - kx) + A \sin(\omega t + kx)$$

$$\Rightarrow \quad y = 2A \sin \omega t \cos kx$$

$$\Rightarrow \quad y = R \sin\omega t$$

Where $R = 2A \cos kx$ is amplitude of different medium particles.

For nodes $\quad R = 2A \cos kx = 0$

$$\Rightarrow \quad \cos kx = 0$$

$$\Rightarrow \quad kx = (2N+1)\frac{\pi}{2}; N = 0, 1, 2, \ldots\ldots$$

$$\Rightarrow \quad \frac{2\pi}{\lambda}x = (2N+1)\frac{\pi}{2}$$

$$\Rightarrow \quad x = (2N+1)\frac{\lambda}{4} \text{ (position of Nodes)}$$ **Ans. (A)**

Sol. 11 Fundamental oscillations of an open pipe $\Rightarrow$ Antinodes at open ends and 1 node at middle

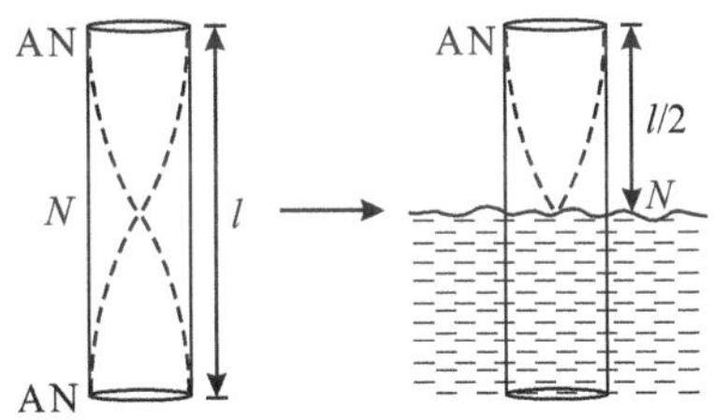

for closed pipe $\rightarrow$ Antinode at open end and node at closed end.

By looking at figures, the wavelength of sound is same in the two cases $\Rightarrow$ frequency is also same

To verify: $\quad f_{\text{open}} = \dfrac{v}{2l} = f$

$$f_{\text{closed}} = \frac{v}{4l'} = \frac{v}{4(l/2)}$$

$$= \frac{v}{2l} = f$$ **Ans. (B)**

Sol. 12 Destructive interference is produced for coherent waves (same frequency) with phase difference of $(2N + 1)\pi$.

In y_1 & y_2 frequency in same & phase difference is π so there SHW will produce Destructive interference.

Stationary waves are produced by superposition of two coherent waves travelling in opposite direction, it happens in y_3 & y_4. **Ans. (D)**

Sol. 13 A general equation of stationary wave can be given in two ways

$$y = A_0 \sin kx \sin(\omega t + \phi) \text{ for a node at } x = 0$$

$$y = A_0 \cos kx \sin(\omega t + \phi) \text{ for an Antinode at } x = 0$$

As one wave is

$$y_1 = a\cos(kx - \omega t)$$

if $\quad y_2 = -a\cos(kx + \omega t) = a\cos(kx + \omega t + \pi)$

by superposition

$$y = y_1 + y_2 = a[\cos(kx - \omega t) - \cos(kx + \omega t)]$$

$$= 2a \sin kx \cos\omega t$$

at $\quad x = 0$ there will be a Node **Ans. (B)**

Sol. 14

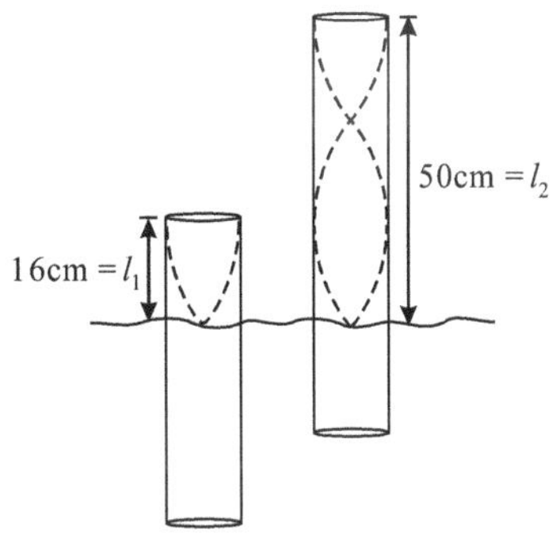

As l_2 is approx 3 times l_1 (but not exact) so we need to account end correction (say 'e'). If v is the velocity of sound than we use

$$500 = \frac{v}{4(l_1 + e)} \qquad \ldots(1)$$

and

$$500 = \frac{3v}{4(l_1 + e)} \qquad \ldots(2)$$

From (1) and (2)

$$\Rightarrow \quad \frac{v}{4(0.16 + e)} = \frac{3v}{4(0.5 + e)}$$

$$\Rightarrow \quad 0.5 + e = 0.48 + 3e$$

$$\Rightarrow \quad 2e = 0.02$$

$$\Rightarrow \quad e = 0.01 \text{ cm}$$

from equation-(1)

$$500 = \frac{v}{4(0.16 + 0.01)} = \frac{v}{4 \times 0.17}$$

$$\Rightarrow \quad v = 500 \times 4 \times 0.17 = 20 \times 17 = 340\ \text{m/s}$$

when tube is pulled out of water, it behaves like an open tube

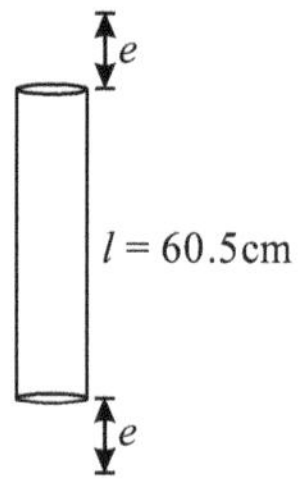

fundamental frequency of the open tube is

$$\nu_0 = \frac{v}{2(l+2e)} = \frac{340}{2(0.605+0.02)}$$

$$\Rightarrow \quad \nu_0 = \frac{340}{2\times 0.625} = \frac{340}{1.25} = 272\ \text{Hz}$$

An open pipe has all restoring frequency which are integral multiple of its fundamental frequency. So lowest two frequencies are ν_0 and $2\nu_0 \Rightarrow 272$ Hz & 544 Hz **Ans. (D)**

Sol. 15 For a closed pipe, it resonates with a given frequency if its length is odd multiple of $\lambda/4$.

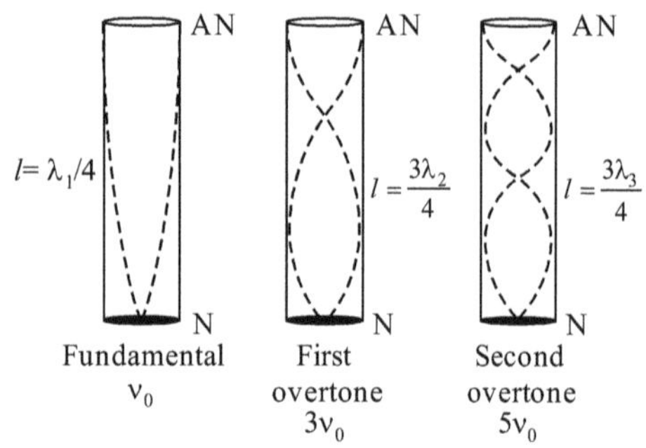

For the given frequency of tuning fork sound wavelength is

$$\lambda = \frac{v}{\nu} = \frac{330}{264} = 1.25\ \text{m} = 125\ \text{cm}$$

The possible length for closed pipe to resonate with above tuning fork are

$$l = \lambda/4, 3\lambda/4, 5\lambda/4, \ldots\ldots$$

$$\Rightarrow \quad l = 31.25\ \text{cm}, 93.75\ \text{cm}, \ldots$$ **Ans. (B)**

Sol. 16 For a stretched string if elastic deformation is there with tension T in it than we have

$$Y = \frac{\text{Stress}}{\text{Strain}} = \frac{T/A}{\text{Strain}}$$

$\Rightarrow$ String tension $T = YA \times$ Strain

Fundamental frequency of a clamped wire is given as

$$\nu_0 = \frac{1}{2l}\sqrt{\frac{T}{\mu}} = \frac{1}{2l}\sqrt{\frac{YA(\text{Strain})}{\rho A}}$$

$$\Rightarrow \quad \nu_0 = \frac{1}{2l}\sqrt{\frac{Y(\text{Strain})}{\rho}}$$

$$\Rightarrow \quad \nu_0 = \frac{1}{2\times 1.5}\sqrt{\frac{2.2\times 10^{11}\times 0.01}{7.7\times 10^3}}$$

$$\Rightarrow \quad \nu_0 = 178.2\ \text{Hz}$$ **Ans. (B)**

Sol. 17 In a damped sonometer wire if l is the length between bridges then

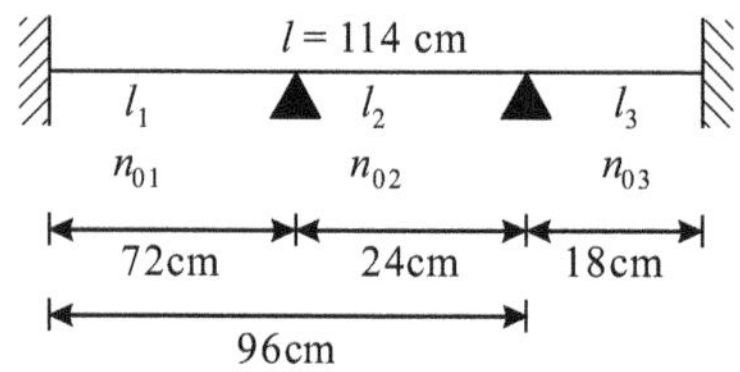

Its fundamental frequency is given as

$$\nu_0 = \frac{v}{2l} = \frac{1}{2l}\sqrt{\frac{T}{\mu}}$$

$$\text{As } n \propto \frac{1}{l} \Rightarrow \frac{1}{l_1}:\frac{1}{l_2}:\frac{1}{l_3} = 1:3:4$$

$$\Rightarrow \quad l_1 : l_2 : l_3 = 12:4:3$$

$$\Rightarrow \quad l_1 = \frac{12}{19}\times 114 = 72\ \text{cm}$$

$$l_2 = \frac{4}{19}\times 114 = 24\ \text{cm}$$

$$l_3 = \frac{3}{19}\times 114 = 18\ \text{cm}$$ **Ans. (D)**

Sol. 18 For a closed pipe, fundamental frequency is given as

$$n_0 = \frac{v}{4L} = \frac{340}{4\times 0.85} = \frac{85}{0.85} = 100\ \text{Hz}$$

For a closed pipe only odd harmonic frequency resonate

$\Rightarrow$ below 1250 Hz the resonating frequency are

100 Hz, 300 Hz, 500 Hz, 700 Hz, 900 Hz, & 1100 Hz

1 2 3 4 5 6 **Ans. (C)**

Sol. 19 The lengths of parts in which 110 cm wire is divided is

$l_1 = 60$ cm; $l_2 = 30$ cm; $l_3 = 20$ cm

fundamental frequencies of the 3 parts of wire are-

$$\nu_{1p} = \frac{1}{2l_1}\sqrt{\frac{T}{\mu}} = \frac{1}{2\times 0.6}\times\sqrt{\frac{400}{0.01}} = \frac{1}{2\times 0.6}\times\frac{20}{0.1}$$

$$= \frac{1000}{6}\ \text{Hz}$$

$$\nu_{2p} = \frac{1}{2l_2}\sqrt{\frac{T}{\mu}} = \frac{1}{2\times 0.3}\times\sqrt{\frac{400}{0.01}} = \frac{1000}{3}\ \text{Hz}$$

$$\nu_{3p} = \frac{1}{2l_3}\sqrt{\frac{T}{\mu}} = \frac{1}{2\times 0.2}\times\sqrt{\frac{400}{0.01}} = \frac{1000}{2}\text{ Hz}$$

As there wires can oscillate at any multiple of their fundamental frequency.

$\Rightarrow$ The lowest common frequency here is

$\nu_0 = 1000$ Hz **Ans. (B)**

Sol. 20 Fundamental frequency of open & closed pipes are shown below

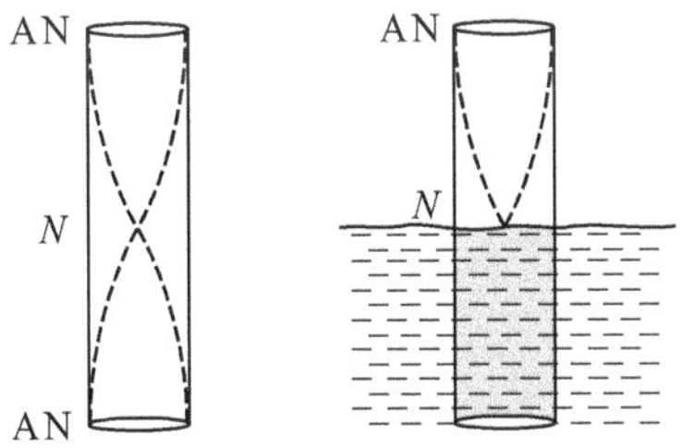

As wavelength in the above two cases is same for fundamental frequencies.

$\Rightarrow$ Their fundamental frequency must be same. **Ans. (D)**

Sol. 21 General equation for stationary wave is given as-

$$y(x, t) = 2A \sin kx \cos \omega t \quad \ldots(1)$$

Comparing equation-(1) with the given equation we have

$$k = \frac{5\pi}{4} \text{ and } \omega = 200\,\pi$$

$\Rightarrow$ Wave speed is given as

$$v = \frac{\omega}{k} = \frac{200\pi}{5\pi/4} = \frac{200\times 4}{5}$$

$\Rightarrow$ $v = 160$ m/s **Ans. (A)**

Sol. 22 If wire W_1 oscillates in p loops then its frequency is

$$\nu_1 = \frac{p}{2l}\sqrt{\frac{T}{\mu_1}} = \frac{p}{2l}\sqrt{\frac{T}{\rho_1 A}}$$

If wire W_2 oscillates in q loops then its frequency is

$$\nu_2 = \frac{q}{2l}\sqrt{\frac{T}{\mu_2}} = \frac{q}{2l}\sqrt{\frac{T}{\rho_2 A}}$$

p and q are no. of Antinodes in wires W_1 and W_2.
For stationary wave in composite wire, we have

$$\nu_1 = \nu_2$$

$$\Rightarrow \frac{p}{2l}\sqrt{\frac{T}{\rho_1 A}} = \frac{q}{2l}\sqrt{\frac{T}{\rho_2 A}}$$

$$\Rightarrow \frac{p}{q} = \frac{1}{2}$$ **Ans. (B)**

Sol. 23

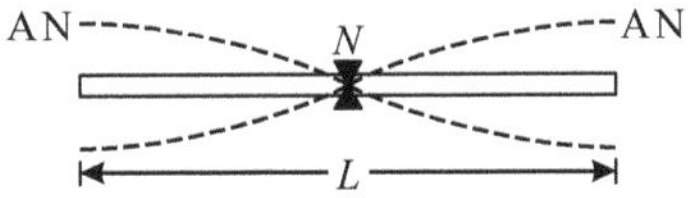

Fundamental frequency oscillations

here $L = \lambda/2$

$\Rightarrow$ $\lambda = 2L$

Frequency of oscillation

$$\nu_0 = \frac{v}{\lambda} = \frac{v}{2L} = \frac{1}{2L}\sqrt{\frac{Y}{\rho}}$$

$$\Rightarrow \nu_0 = \frac{1}{2\times 0.6}\times\sqrt{\frac{9.27\times 10^{10}}{2.7\times 10^3}}$$

$\Rightarrow$ $\nu_0 = 4.88\times 10^3$ Hz ≈ 5 kHz **Ans. (A)**

Sol. 24 Given that $|\nu_A - \nu_B| = 5$

either $\nu_B = 420$ Hz or 430 Hz As $\nu_A = 425$ Hz

As tension in B is increased

$\Rightarrow$ ν_B increases and new beat frequency with A becomes 3 Hz.

If $\nu_B = 420$ Hz then on increasing it can be 422 Hz & $f_B = 3$ Hz with ν_A.

If $\nu_B = 430$ Hz then on increasing always $f_B > 5$ Hz with ν_A.

Ans. (D)

Sol. 25 For a sonometer wire oscillation frequency is given as

$$n_s = \frac{v}{2l}$$

at $l = 0.95$ m, $n_{s_1} = \dfrac{v}{2\times 0.95}$

and at $l = 1$m, $n_{s_2} = \dfrac{v}{2\times 1}$

$\Rightarrow$ $n_{s_1} > n_{s_2}$

As tuning fork produces 5 beats/s with both n_{s_1} and n_{s_2}

$\Rightarrow$ $n_{s_1} > n_f > n_{s_2}$

$$\Rightarrow n_f - 5 = n_{s_2} = \frac{v}{1.9} \quad \ldots(1)$$

and $$n_f - 5 = n_{s_1} = \frac{v}{2} \quad \ldots(2)$$

(2)/(1) gives $$\frac{n_f + 5}{n_f - 5} = \frac{20}{19}$$

$\Rightarrow$ $19n_f + 95 = 20n_f - 100$

$\Rightarrow$ $n_f = 195$ Hz **Ans. (A)**

Sol. 26 Frequency of 3rd Harmonic of open pipe is

$$n_3 = \frac{3v}{2l}$$

As beat frequency is 1 beat/s

$$\Rightarrow \quad \frac{3v}{2l} = 256 \pm 1 = 255 \text{ Hz or } 257 \text{ Hz}$$

Case-I: $\frac{3v}{2l} = 255$

$$\Rightarrow \quad l = \frac{3v}{2\times 255} = \frac{3\times 340}{2\times 255}$$

$$= 2\text{ m} = 200\text{ cm}$$

Case-II: $\frac{3v}{2l} = 257$

$$\Rightarrow \quad l = \frac{3v}{2\times 257} = \frac{3\times 340}{2\times 257}$$

$$\Rightarrow \quad l \simeq 1.97\text{ m} = 197\text{ cm}$$

Ans. (D)

Sol. 27 First resonance length is 10 cm

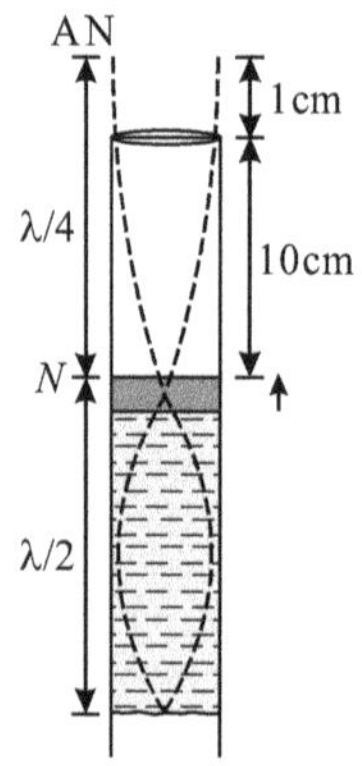

$$\Rightarrow \quad \frac{\lambda}{4} = 10 + 1 = 11\text{ cm}$$

Second resonance occurs when length of column is approx 3λ/4.

$$\frac{3\lambda}{4} = 3\times 11 = 33\text{ cm}$$

Resonance column length now is

$$\frac{3\lambda}{4} - e = 33 - 1 = 32\text{ cm}$$

Ans. (A)

Sol. 28 Frequency of the sound produced by flute is given as

$$f = 2\left(\frac{v}{2l}\right) = \frac{2\times 330}{2\times 0.5} = 660\text{Hz}$$

Velocity of observer,

$$v_0 = 10\times\frac{5}{18} = \frac{25}{9}\text{ m/s}$$

Frequency detected by observer is

$$f_{ap} = \left[\frac{v+v_0}{v}\right]f$$

$$\Rightarrow \quad f' = \left[\frac{\frac{25}{9}+330}{330}\right]660$$

$$\Rightarrow \quad f' = 335.56\times 2 = 671.12\text{ Hz}$$

Ans. (D)

Sol. 29 Velocity of transverse wave on string

$$v = \sqrt{\frac{T}{\mu}} = \sqrt{\frac{8}{5}\times 1000} = 40\text{ m/s}$$

Now, wavelength of wave

$$\lambda = \frac{v}{n} = \frac{40}{100}\text{ m}$$

Separation between successive nodes given as

$$\frac{\lambda}{2} = \frac{20}{100}\text{ m} = 20\text{ cm}$$

Ans. (B)

Sol. 30 For closed organ pipe, resonant frequencies are odd multiple of fundamental frequency. For maximum, we use

$(2n + 1)f_0 \le 20{,}000$

(f_0 is fundamental frequency = 1.5 kHz)

$$\Rightarrow \quad n = 6$$

So total number of overtones that can be heard are 7 (0 to 6)

Ans. (A)

Sol. 31 If antinode is produced at a distance x above the reference mark, with frequency 512 Hz, we have

$$512 = \frac{v}{4(11+x)} \quad \ldots(1)$$

with 256 Hz frequency we use

$$256 = \frac{v}{4(27+x)} \quad \ldots(2)$$

$$\frac{(1)}{(2)} \Rightarrow \quad 2 = \frac{27+x}{11+x}$$

$$\Rightarrow \quad 22 + 2x = 27 + x$$

$$\Rightarrow \quad x = 5\text{ cm}$$

Thus sound speed in air is

$$v = 512\times 4(11+x)$$

$$\Rightarrow \quad v = 512\times 4\times 0.16$$

$$\Rightarrow \quad v = 327.68\text{ m/s}$$

Ans. (A)

Sol. 32

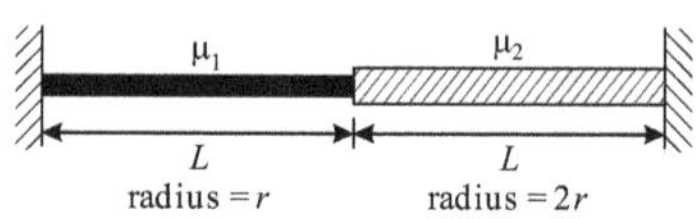

Let mass per unit length of wires are μ_1 and μ_2 respectively

As materials are same, so density ρ is same, we use

$$\mu_1 = \frac{\rho\pi r^2 L}{L} = \mu$$

and $$\mu_2 = \frac{\rho 4\pi r^2 L}{L} = 4\mu$$

Tension in both are same, taken as T, let speed of wave in wires are v_1 and v_2, we have

$$v_1 = \sqrt{\frac{T}{\mu}} = v;$$

$$v_2 = \sqrt{\frac{T}{4\mu}} = \frac{v}{2}$$

So fundamental frequencies in both wires are given as

$$f_{01} = \frac{v_1}{2L} = \frac{v}{2L}$$

and $$f_{02} = \frac{v_2}{2L} = \frac{v}{4L}$$

Minimum frequency at which both wires resonate is L.C.M of both of these frequencies, given as

$$f_0 = \frac{v}{2L}$$

Hence number of loops in wires are 1 and 2 respectively as shown below

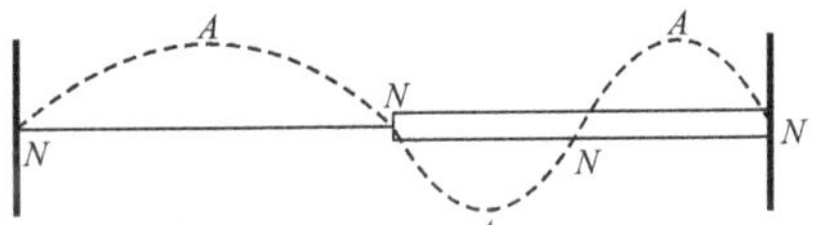

So, ratio of no. of antinodes is 1 : 2 **Ans. (D)**

Sol. 33 4th harmonic oscillations in clamped wire are shown below

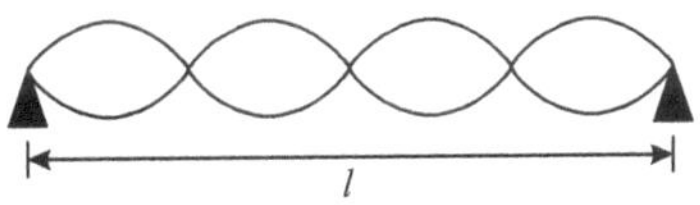

$$\Rightarrow \quad 4\frac{\lambda}{2} = l$$

$$\Rightarrow \quad 2\lambda = l \Rightarrow \lambda = \frac{l}{2}$$

From given equation, we use

$$\frac{2\pi}{\lambda} = 0.157$$

$$\lambda = 40$$

$$l = 2\lambda$$

$$\Rightarrow \quad l = 80\,\text{m}$$

Ans. (B)

Sol. 34 As string vibrates in 3rd harmonic, we use

$$3\left(\frac{v}{2l}\right) = 240$$

$$\Rightarrow \quad 3\frac{v}{2\times 2} = 240$$

$$\Rightarrow \quad v = 320\,\text{m/s}$$

Fundamental frequency of string is given as

$$n_0 = \frac{v}{2l} = \frac{320}{2\times 2} = 80\,\text{Hz}$$

Ans. (D)

Sol. 35 Beat frequency of the given waves is

$$f_{\text{beat}} = 11 - 9 = 2\,\text{Hz}$$

Thus in one second there are two maxima and minima occur. This is corresponding to option (D). **Ans. (D)**

Sol. 36 In resonance tube experiment, wave speed is given as

$$\text{v} = 2f(l_2 - l_1)$$

$$\Rightarrow \quad \text{v} = 2\times 480\times(70-30)\times 10^{-2}$$

$$\Rightarrow \quad \text{v} = 960\times 40\times 10^{-2}$$

$$\Rightarrow \quad \text{v} = 38400\times 10^{-2}\,\text{m/s}$$

$$\Rightarrow \quad \text{v} = 384\,\text{m/s}$$

Ans. (C)

14.3 Doppler's Effect

Sol. 1

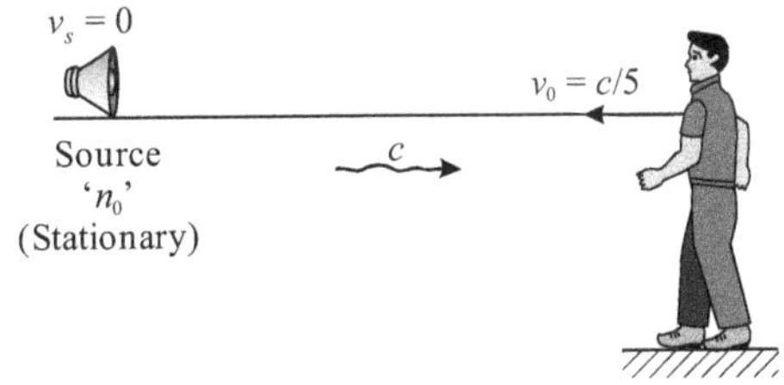

Apparent frequency is

$$\text{v}_{ap} = \text{v}_0\left(\frac{c - v_0}{c - v_s}\right) = \text{v}_0\left(\frac{c - (-c/5)}{c}\right)$$

$$\Rightarrow \quad \text{v}_{ap} = \text{v}_0\left(\frac{6}{5}\right)$$

$$\%\text{ increase in frequency} = \frac{\text{v}_{ap} - \text{v}_0}{\text{v}_0}\times 100$$

$$= \frac{1}{5}\times 100$$

$$= 20\,\%$$

Ans. (D)

Sol. 2

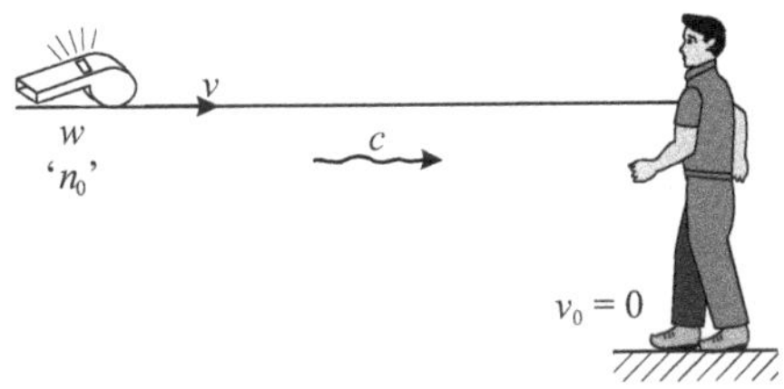

Apparent frequency heard by observer is

$$n_{ap} = n_0\left(\frac{c - v_0}{c - v_s}\right) = n_0\left(\frac{c}{c - v_s}\right)$$

As $n_{ap} = 10000$ Hz and $n_0 = 9500$ Hz

$$\Rightarrow \quad 10000 = 9500\left(\frac{300}{300 - v_s}\right)$$

$$\Rightarrow \quad 300 - v_s = \frac{9500\times 300}{10000} = 285$$

$$\Rightarrow \quad v_s = 300 - 285 = 15\,\text{m/s}$$

Ans. (C)

Sol. 3

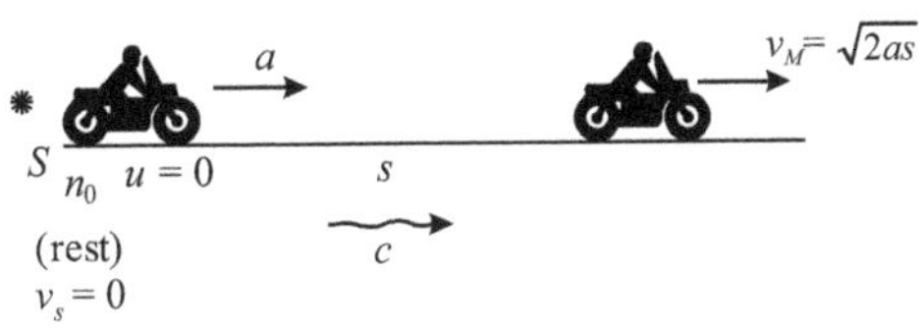

Apparent frequency is $n_{app} = n_0\left(\dfrac{c-v_M}{c}\right)$

$$\Rightarrow \quad \frac{94}{100}n_0 = n_0\left(\frac{c-\sqrt{2as}}{c}\right)$$

$$\Rightarrow \quad \frac{94\times 330}{100} = 330-\sqrt{2\times 2\times s}$$

$$\Rightarrow \quad s = 98\,\text{m}$$

Ans. (D)

Sol. 4 S-1: Echolocation : A phenomenon by which bats detect the objects in their surrounding.

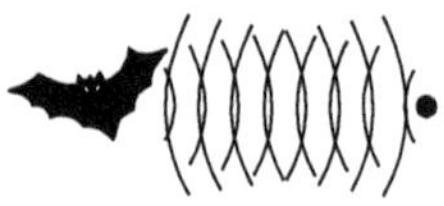

Waves reflected from object through which bats estimate the location of object.

S-1 is true. By Definition of Doppler's effects.

In general **S-2** is True.

As **S-2** is not linked with **S-1** directly. **Ans. (C)**

Sol. 5 Case-I:

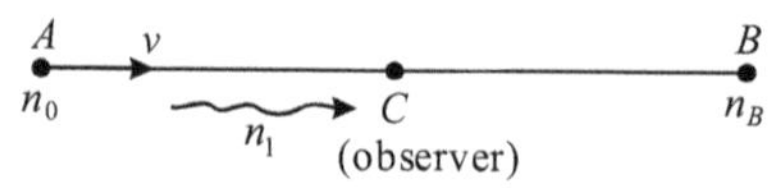

Apparent frequency of A heard by C is

$$n_1 = n_0\left(\frac{c}{c-v}\right)$$

$$\Rightarrow \quad n_1 = 500\times\frac{340}{336} = 506\,\text{Hz}$$

$$f_B = |n_1 - n_B| = 6$$

$$n_B = 500\,\text{Hz} \text{ or } 512\,\text{Hz}$$

Case-2:

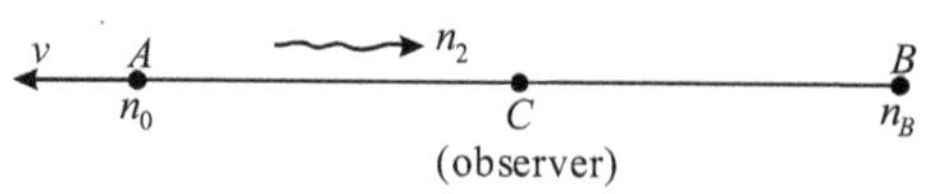

Apparent frequency of A heard by C is

$$n_2 = n_0\left(\frac{c}{c+v}\right)$$

$$\Rightarrow \quad n_2 = 500\times\frac{340}{344} = 494\,\text{Hz}$$

$$f_B' = |n_2 - n_B| = 18$$

$$\Rightarrow \quad n_B = 512\,\text{Hz}$$

Ans. (C)

Sol. 6

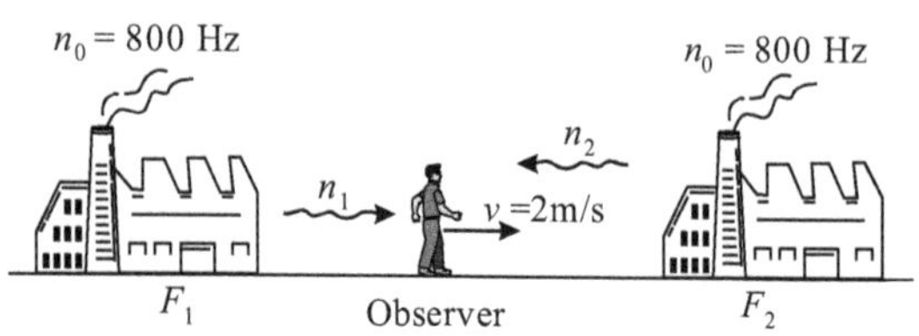

Apparent frequency of sound heard by observer are

from factory-1 it is $n_1 = n_0\left(\dfrac{c-v}{c}\right) = 800\times\left(\dfrac{320-2}{320}\right)$

$$\Rightarrow \quad n_1 = \frac{800\times 318}{320}$$

from factory-2 it is $n_2 = n_0\left(\dfrac{c+v}{c}\right) = 800\times\left(\dfrac{320+2}{320}\right)$

$$\Rightarrow \quad n_2 = \frac{800\times 322}{320}$$

Beat frequency heard $f_B = n_2 - n_1$

$$= \frac{800}{320}(322-318)$$

$$\Rightarrow \quad f_B = \frac{800\times 4}{320} = 10\,\text{Hz}$$

Ans. (D)

Sol. 7

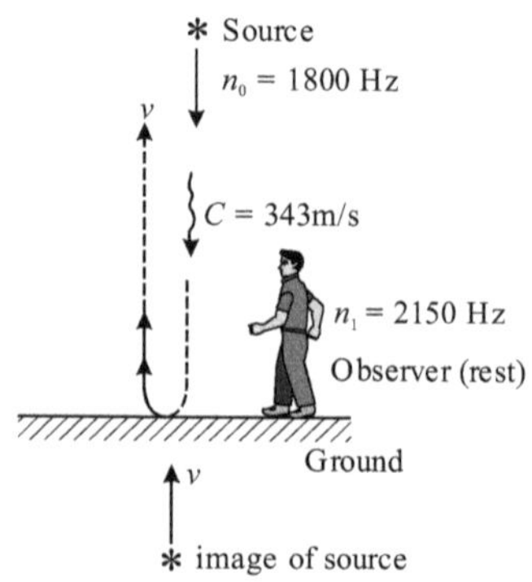

By Doppler's effect apparent frequency of sound received by observer is

$$n_1 = n_0\left(\frac{c}{c-v_s}\right)$$

$$2150 = 1800\left(\frac{343}{343-v}\right)$$

$$\Rightarrow \quad 343-v = \frac{1800\times 340}{2150}$$

$$\Rightarrow \quad v = -\frac{180\times 343}{215} + 343 \simeq 56\,\text{m/s}$$

Reflected waves appear to be coming from the image of source in the ground so the apparent frequency received by source is reflected waves is given as

$$n_{ap} = n_0\left(\frac{c+v}{c-v}\right) = 1800\times\left(\frac{343+56}{343-56}\right)$$

$$\Rightarrow \quad n_{ap} = 1800\times\frac{399}{287} = 2502\,\text{Hz}$$

Ans. (B)

Sol. 8

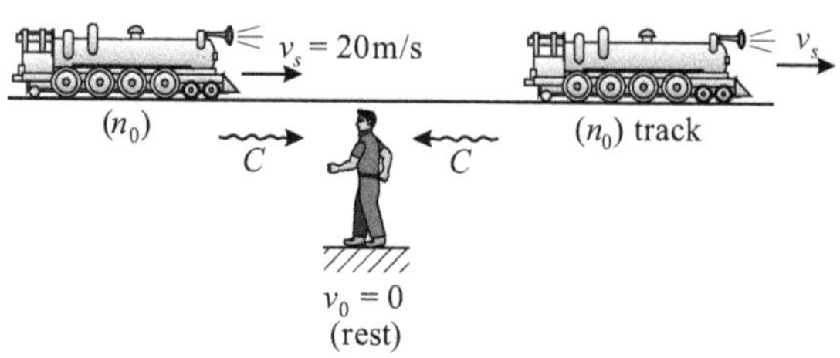

If n_{ap_2} is the frequency observed when train is approaching the person then we have

$$n_{ap_1} = n_0\left(\frac{c}{c-v_s}\right) \qquad \ldots(1)$$

If v_{app_2} is frequency observed when train reaches away from person then we have

$$n_{ap_2} = n_0\left(\frac{c}{c+v_S}\right) \qquad \ldots(2)$$

% change in frequency observed when train passes

$$\frac{\Delta n}{n_0}\times 100 = \frac{n_{ap_1}-n_{ap_2}}{n_0}\times 100\%$$

$$\% \text{ change} = \left[\frac{c}{c-v_S}-\frac{c}{c+v_S}\right]\times 100$$

$$\Rightarrow \quad \% \text{ change} = \left[\frac{320}{300}-\frac{320}{340}\right]\times 100$$

$$\Rightarrow \quad \% \text{ change} = 106.66-94.11 \simeq 12\%$$

Ans. (B)

Sol. 9

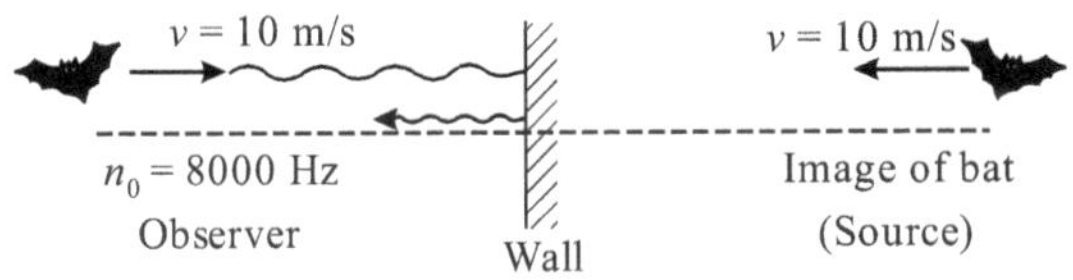

Frequency of reflected sound appear to be coming from image of bat as shown in figure so the apparent frequency heard is given as

$$f_{ap} = n_0\left(\frac{c+v}{c-v}\right)$$

$$= 8000\times\left(\frac{320+10}{320-10}\right)$$

$$\Rightarrow \quad f_{ap} = 8000\times\frac{330}{310}\cong 8516\,\text{Hz}$$

Ans. (A)

Sol. 10 By Doppler's effect apparent frequency of sound heard by observer is given as

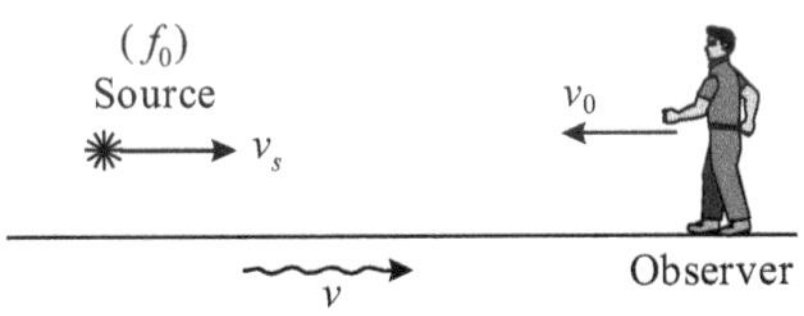

Apparent frequency $f = f_0\left(\dfrac{v+v_0}{v-v_s}\right)$

$$\Rightarrow \quad f = \left(\frac{f_0}{v-v_s}\right)v_0+\left(\frac{f_0 v}{v-v_s}\right)$$

This is of form $\quad y = mx + C$

slope of graph is $\quad m = \dfrac{f_0}{v-v_s}$

Ans. (C)

Sol. 11

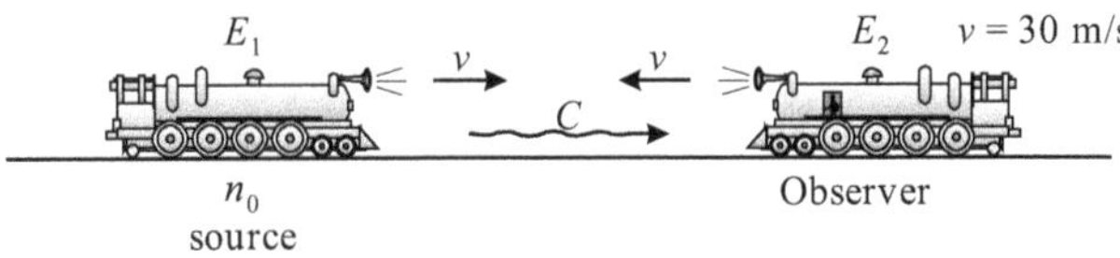

Apparent frequency heard by driver of second engine E_2 is

$$n_{ap} = n_0\left(\frac{c+v}{c-v}\right) = 540\times\left(\frac{330+30}{330-30}\right)$$

$$\Rightarrow \quad n_{ap} = 540\times\frac{360}{300} = 54\times 12 = 648\,\text{Hz}$$

Ans. (D)

Sol. 12

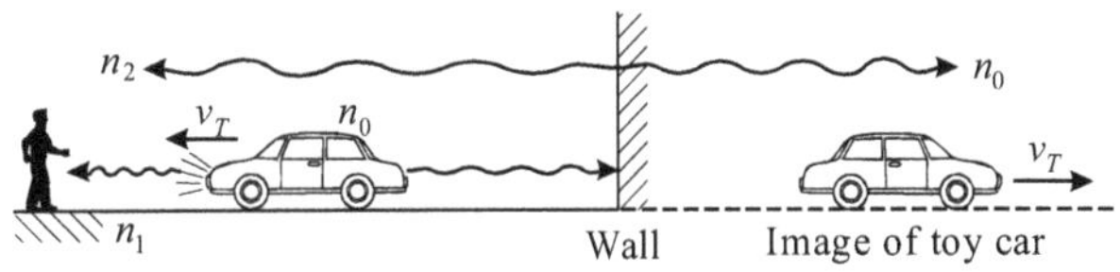

$n_1 \rightarrow$ Apparent frequency of horn from approaching toy car

$$\Rightarrow \quad n_1 = n_0\left(\frac{c}{c-v_T}\right) = n_0\left(\frac{340}{340-5}\right)$$

$$\Rightarrow \quad n_1 = \frac{340}{335}n_0$$

$n_2 \rightarrow$ Apparent frequency of horn from the receding image of toy car due to reflected wave from wall.

$$\Rightarrow \quad n_2 = n_0\left(\frac{c}{c+v_T}\right) = n_0\left(\frac{340}{340+5}\right)$$

$$\Rightarrow \quad n_2 = \frac{340}{345}n_0$$

As beat frequency heard by observer is 5 beats/s

$$\Rightarrow \quad n_1 - n_2 = 5$$

$$\Rightarrow \quad \left(\frac{340}{335}-\frac{340}{345}\right)n_0 = 5$$

$$\Rightarrow \quad n_0 \simeq 170\,\text{Hz}$$

Ans. (D)

Sol. 13

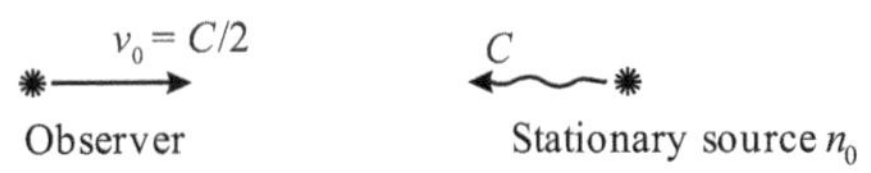

The apparent frequency observed by observer is given by relativistic relation as

$$n_{ap} = n_0\sqrt{\frac{C-V_0}{C+V_0}} = n_0\sqrt{\frac{C+C/2}{C-C/2}} \quad [\because v_0 = -C/2]$$

$$\Rightarrow \quad n_{ap} = n_0\left[\frac{3C/2}{C/2}\right]^{1/2} = 10\times 10^9\times\sqrt{3}$$

$$\Rightarrow \quad n_{ap} = 17.32\,\text{GHz}$$ **Ans. (A)**

Sol. 14 Apparent frequency of sound observed is

$$f_{ap} = f_0\left[\frac{c_2 \pm v_0}{c_2 \mp v_s}\right]$$

when $v_s = 34$ m/s

$$\Rightarrow \quad f_1 = f_0\left[\frac{340}{340-34}\right]$$

when $v_s = 17$ m/s

$$f_2 = f_0\left[\frac{340}{340-17}\right]$$

$$\Rightarrow \quad \frac{f_1}{f_2} = \frac{340-17}{340-34} = \frac{323}{306}$$

$$\Rightarrow \quad \frac{f_1}{f_2} = \frac{19}{18}$$

Ans. (B)

Sol. 15 Sound appears to be coming from location B and plane in at A, so we use

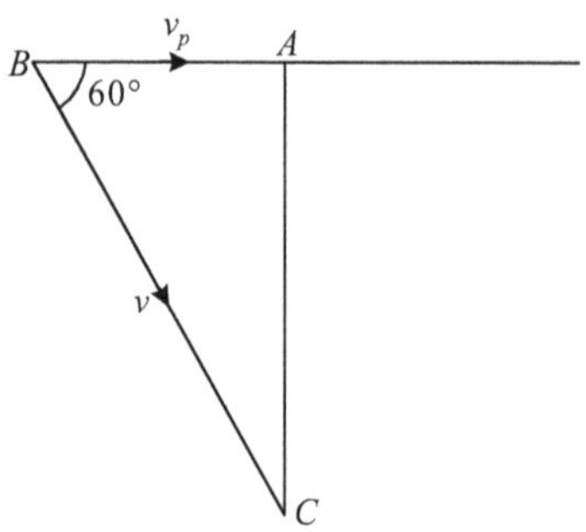

$$AB = v_p \times t \qquad (v_p \to \text{velocity of plane})$$

$$BC = vt$$

From figure

$$\cos 60° = \frac{AB}{BC}$$

$$\Rightarrow \quad \frac{1}{2} = \frac{v_p \times t}{vt}$$

$$\Rightarrow \quad v_p = \frac{v}{2}$$

Ans. (C)

Sol. 16

Applying Doppler effect for sound

$$x = \frac{c - v_0}{c + v_s} n_0$$

(v_0 & v_s is taken ⊕ when approaching each other)

$$\Rightarrow \quad 2000 = \frac{340-20}{340+20} f_0$$

$$\Rightarrow \quad f_0 = 2250 \text{ Hz}$$

Ans. (A)

Sol. 17 Apparent frequency for the recording observer is given as

$$f_1 = 480 = \frac{300 - v_1}{300} \times 500$$

$$\Rightarrow \quad \frac{1440}{5} = 300 - v_1$$

$$\Rightarrow \quad v_1 = \frac{60}{5} = 12 \text{ m/s}$$

Apparent frequency for the approaching observer is given as

$$530 = \frac{300 + v_2}{300} \times 500$$

$$\Rightarrow \quad 1590 = 1500 + 5v_2$$

$$\Rightarrow \quad 5v_2 = 90$$

$$\Rightarrow \quad v_2 = 18 \text{ m/s}$$

Ans. (B)

Sol. 18 When source is approaching as shown below, we use

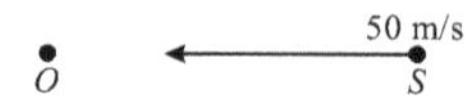

$$f_{app} = \left(\frac{c-0}{c-50}\right) f_{source} \qquad (c \to \text{speed of second})$$

$$\Rightarrow \quad 1000 = \left(\frac{350}{300}\right) f_{source}$$

$$\Rightarrow \quad f_{source} = \frac{1000 \times 300}{350} \text{ Hz}$$

When source is recording as shown below, we use

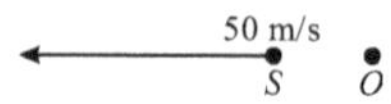

$$f'_{app} = \left(\frac{c}{c+50}\right) \cdot f_{source}$$

$$\Rightarrow \quad f'_{app} = \frac{350}{400} \times 1000 \times \frac{300}{350}$$

$$\Rightarrow \quad f'_{app} = 750 \text{ Hz}$$

Ans. (C)

Sol. 19

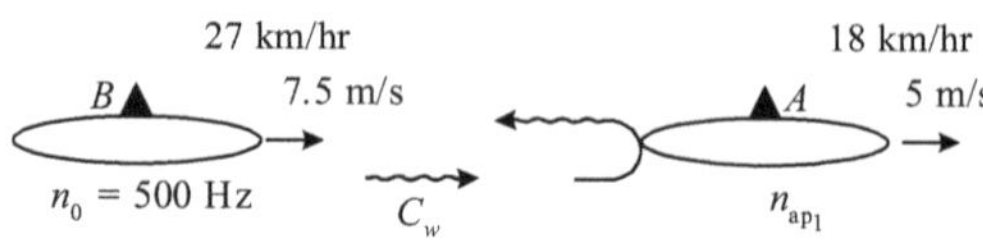

Frequency of signal detected/received by A

$$n_{ap_1} = n_0\left(\frac{C_w - v_{OA}}{C_w - v_{SB}}\right) = 500\left(\frac{1500-5}{1500-7.5}\right)$$

$$\Rightarrow \quad n_{ap_1} = \frac{500 \times 1495}{1492.5} \text{ Hz}$$

Now n_{ap_1} will be reflected by A as source and frequency detected by B for this reflected sound is

$$n_{ap_2} = n_{ap_1}\left(\frac{C_w + v_{OB}}{C_w + v_{SA}}\right) = \frac{500 \times 1495}{1492.5} \times \left(\frac{1500+7.5}{1500+5}\right)$$

$$\Rightarrow \quad n_{ap_2} \simeq 502 \text{ Hz}$$

Ans. (B)

Sol. 20

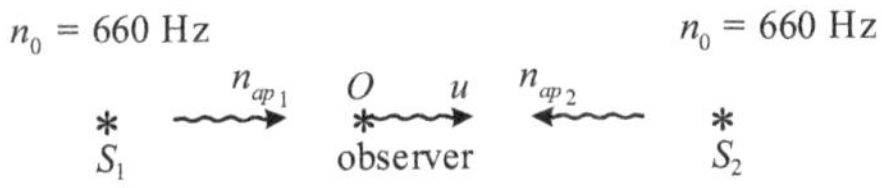

Apparent frequency received from S_2 is

$$n_{ap_2} = n_0\left[\frac{C+4}{C}\right]$$

Apparent frequency received from S_1 is

$$n_{ap_1} = n_0\left[\frac{C-4}{C}\right]$$

Beat frequency heard by observer

$$f_B = n_{ap_2} - n_{ap_1} = n_0\left[\frac{C+4}{C}\right] - n_0\left[\frac{C-4}{C}\right] = n_0\left(\frac{24}{C}\right)$$

$$\Rightarrow \quad 10 = 660 \times \frac{24}{330} \Rightarrow 4 = \frac{10}{4} = 2.5 \text{ m/s}$$ **Ans. (A)**

15.1 Electric Force and Electric Field

Sol. 1

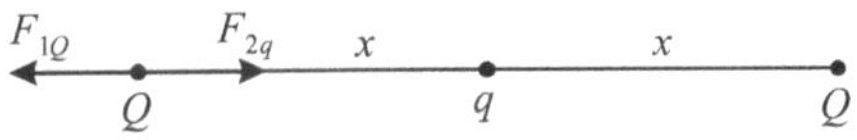

[For equation θ must be of opposite sign to Q]

For equilibrium of Q, we use

$$F_{1Q} = F_{2q}$$

$$\frac{KQ^2}{(2x)^2} = \frac{KQq}{x^2}$$

$$\Rightarrow \quad |q| = \frac{Q}{4} \quad \text{or} \quad q = -\frac{Q}{4}$$ **Ans. (D)**

Sol. 2 By Colomb's law

$$F_{12} = \frac{Kq_1q_2}{b^2}; F_{13} = \frac{Kq_1q_3}{a^2}$$

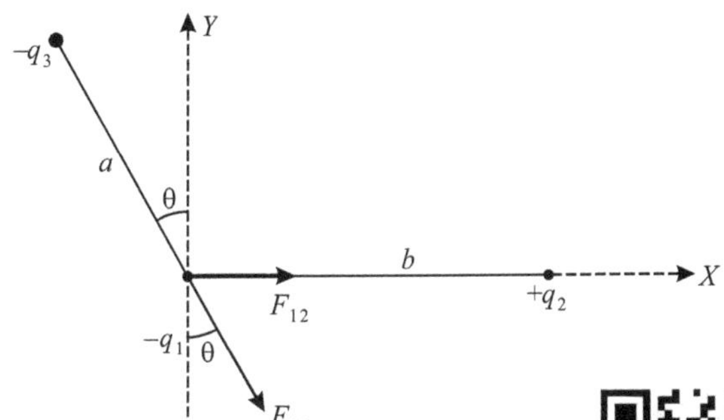

Net force on $-q_1$ along x direction is given as

$$F_x = F_{12} + F_{13}\sin\theta$$

$$\Rightarrow \quad F_x = Kq_1\left[\frac{q_2}{b^2} + \frac{q_3}{a^2}\sin\theta\right]$$ **Ans. (B)**

Sol. 3

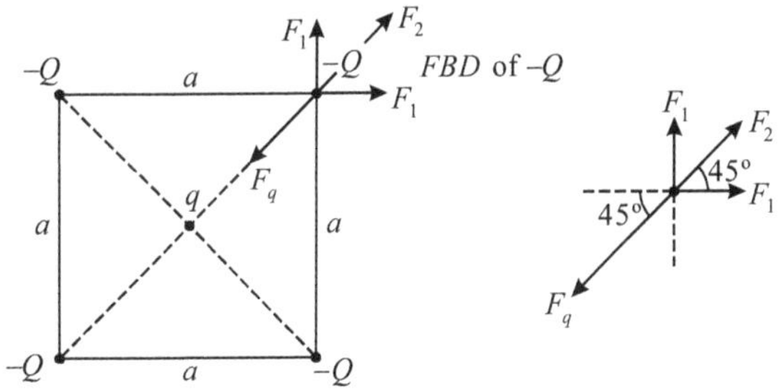

To keep net force on each of $-Q$, q has to be 'Positive'. So that for equilibrium of $-Q$ charge we have

Hare $$F_1 = \frac{KQ^2}{a^2}; F_2 = \frac{KQ^2}{2a^2}$$

$$F_q = F_2 + \sqrt{2}F_1$$

and $$F_q = \frac{KqQ}{(a/\sqrt{2})^2}$$

$$\Rightarrow \quad \frac{KqQ}{a^2 2} = \frac{KQ^2}{2a^2} + \frac{\sqrt{2}KQ^2}{a^2}$$

$$\Rightarrow \quad \frac{2q}{a^2} = \frac{Q}{2a^2} = \frac{\sqrt{2}Q}{a^2}$$

$$\Rightarrow \quad q = \frac{Q}{4}(1+2\sqrt{2})$$ **Ans. (B)**

Sol. 4 Here to solve the case, we ignore induction effects.

$$\Rightarrow \quad r >> R$$

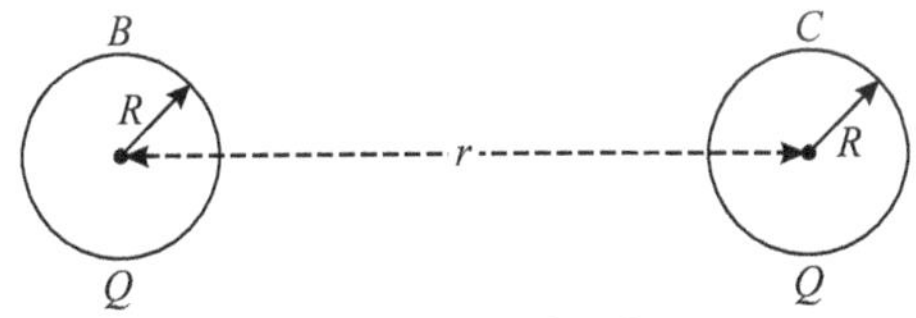

Initially the force between spheres is given as

$$F = \frac{KQ^2}{r^2}$$

In second case when the third identical sphere is located to B

Step-I

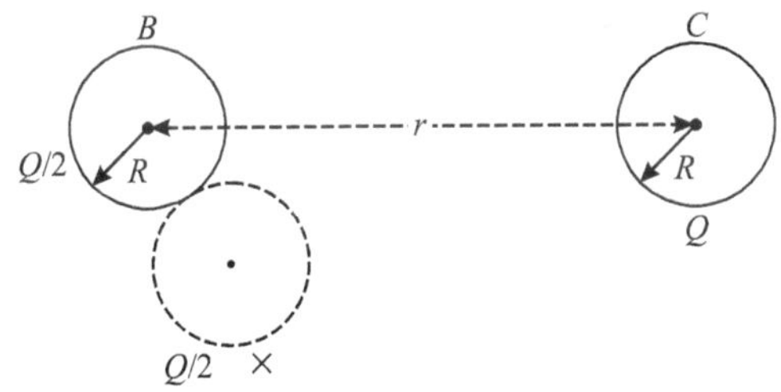

Step-II

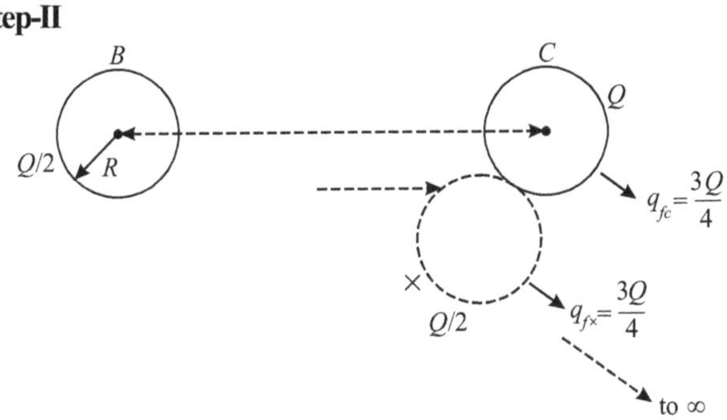

When sphere x is connected to sphere C then again their total charge will be equally distributed in than as these are identical.

$\Rightarrow$ Final charge on B and C are $Q/2$ and $3Q/4$

$\Rightarrow$ Final force between B and C is

$$F' = \frac{K(Q/2)(3Q/4)}{r^2}$$

$$= \frac{3}{8}\frac{KQ^2}{r^2}$$

Ans. (D)

Sol. 5 For equilibrium of drop, we have

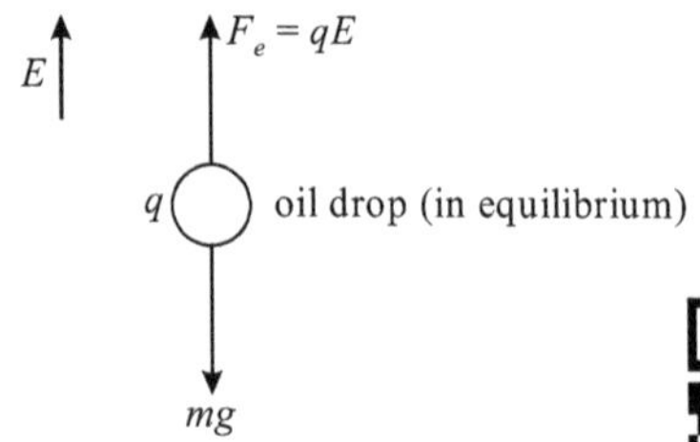

$$qE = mg$$

$$\Rightarrow \quad q = \frac{mg}{E} = \frac{9.9\times10^{-15}\times10}{3\times10^{4}}$$

$$\Rightarrow \quad q = 3.3\times10^{-18}\,\text{C}$$

Ans. (C)

Sol. 6

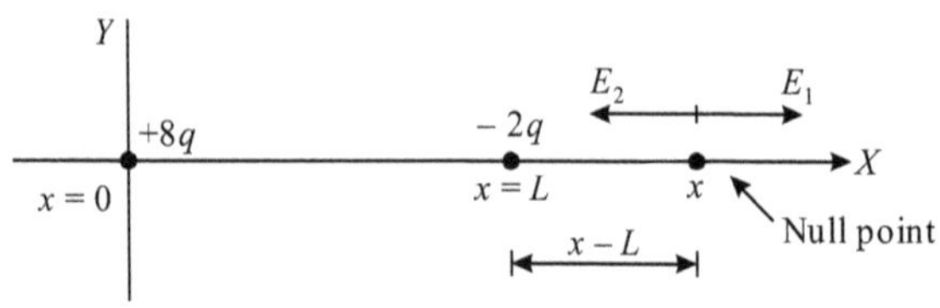

$E_1 \longrightarrow EF$ at x due to $8q$

$E_2 \longrightarrow EF$ at x due to $-2q$

If at x, $EF_{net} = 0$

$$\Rightarrow \quad E_1 = E_2$$

$$\frac{K(8q)}{x^2} = \frac{K(2q)}{(x-L)^2}$$

$$\Rightarrow \quad 8(x-L)^2 = 2x^2$$

$$\Rightarrow \quad x = 2L$$

Ans. (A)

Sol. 7

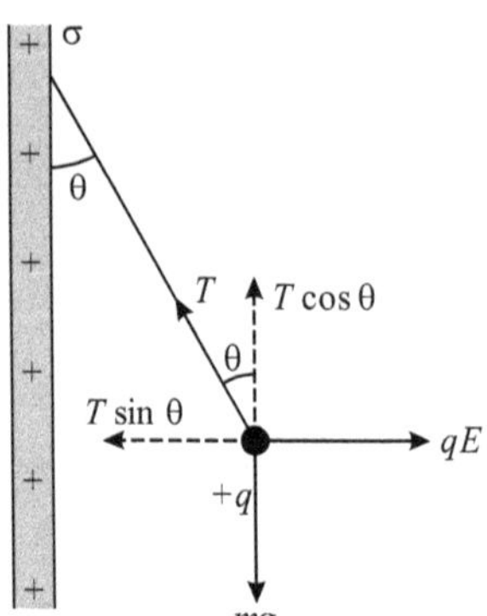

Electric field due to a large charged sheet

$$E = \frac{\sigma}{2\epsilon_0}$$

As ball is in equilibrium, we use

$$T\sin\theta = q\left(\frac{\sigma}{2\epsilon_0}\right) \qquad \ldots(1)$$

$$T\cos\theta = mg \qquad \ldots(2)$$

(1) ÷ (2) gives

$$\Rightarrow \quad \tan\theta = \frac{q\sigma}{2\epsilon_0\, mg}$$

$$\Rightarrow \quad \sigma \propto \tan\theta$$

Ans. (D)

Sol. 8

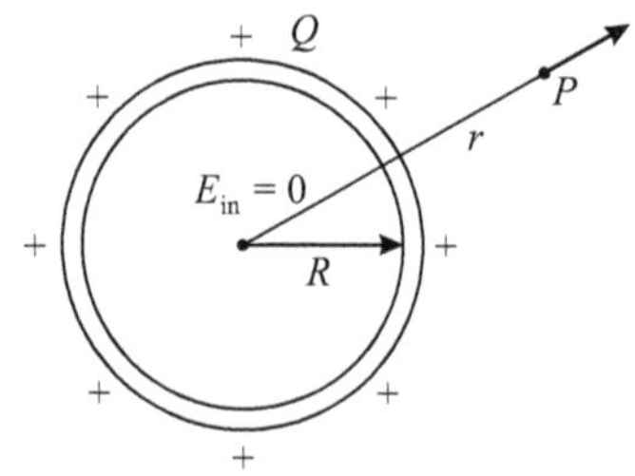

$$E = \frac{KQ}{r^2} \quad (\text{for } r > R)$$

$$E \propto \frac{1}{r^2}$$

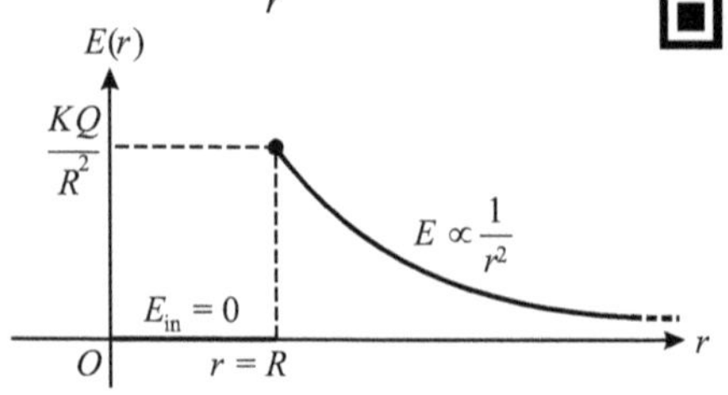

Ans. (A)

Sol. 9

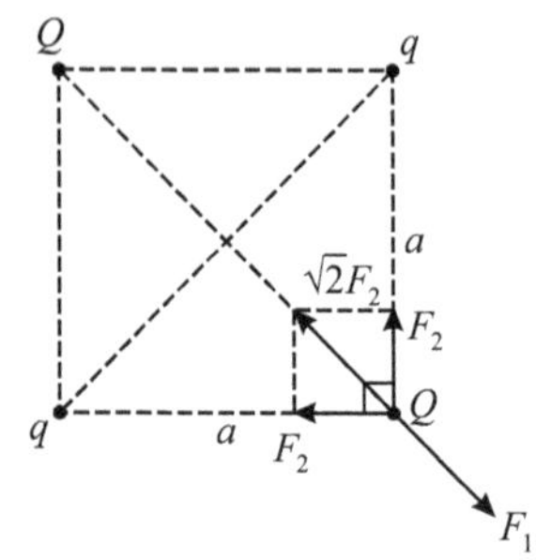

Force on Q due to other Q is $F_1 = \dfrac{KQ^2}{(\sqrt{2}a)^2}$

If not force on Q is to be zero

$\Rightarrow$ q has to be –ve so that resultant of two F_2 on Q balances F_1.

Force on Q due to each of q is $F_2 = \dfrac{KqQ}{a^2}$

As not force on Q is zero

$$\Rightarrow \quad \sqrt{2}F_2 = F_1$$

$$\Rightarrow \qquad \sqrt{2}\,\frac{KqQ}{a^2} = \frac{KQ^2}{2a^2}$$

$$\Rightarrow \qquad Q = 2\sqrt{2}\,q \text{ in magnitude}$$

$$\Rightarrow \qquad Q = -2\sqrt{2}\,q$$ **Ans. (C)**

Sol. 10

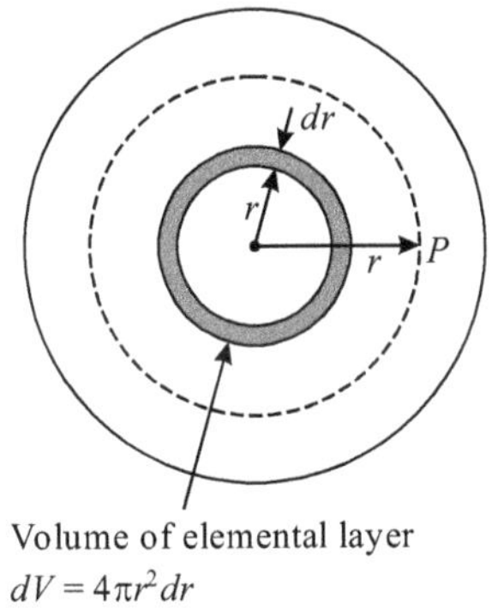

The charge contained inside the thin layer of elemental shell is

$$dq = \rho(r).dV$$

$$\Rightarrow \qquad dq = \frac{Q}{\pi R^4}\cdot r\cdot(4\pi r^2 dr)$$

$$\Rightarrow \qquad dq = \frac{4Q}{R^4}\cdot r^3 dr$$

Total charge inside a sphere of radius r_1 is

$$q = \int dq = \int_0^{r_1}\frac{4Q}{R^4}\cdot r^3 dr$$

$$\Rightarrow \qquad q = \frac{4Q}{R^4}\int_0^{r_1} r^3 dr$$

$$\Rightarrow \qquad q = \frac{4Q}{R^4}\cdot\left[\frac{r^4}{4}\right]_0^{r_1} = \frac{Qr_1^4}{R^4}$$

Electric field at point '*P*' due to the inside charge is given as

$$E = \frac{Kq}{r_1^2} = \frac{1}{4\pi\epsilon_0}\cdot\frac{(Qr_1^4/R^4)}{r_1^2}$$

$$\Rightarrow \qquad E = \frac{1}{4\pi\epsilon_0}\frac{Qr_1^2}{R^4}$$ **Ans. (A)**

Sol. 11 In Air

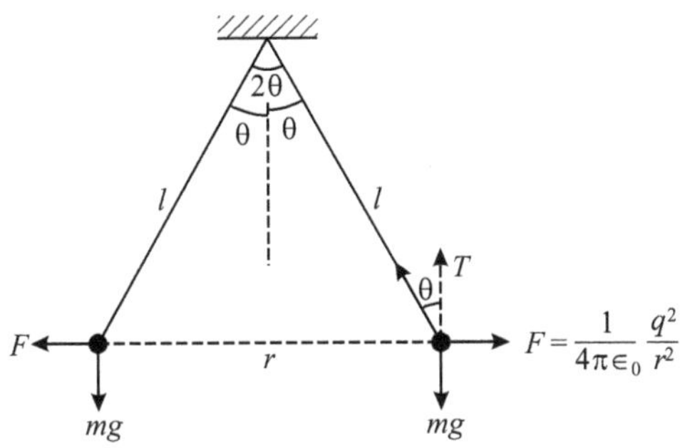

For equilibrium of bob

$$T\sin\theta = \frac{1}{4\pi\epsilon_0}\frac{q^2}{r^2} \qquad \ldots(1)$$

$$T\cos\theta = mg \qquad \ldots(2)$$

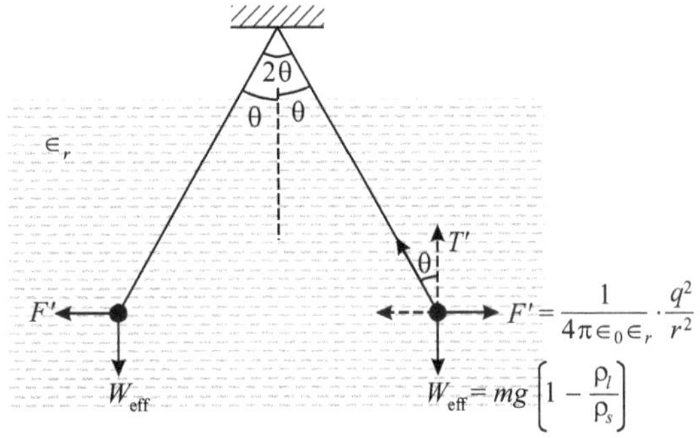

as $\qquad F_B = mg\cdot\frac{\rho_l}{\rho_s}$

force of Buoyancy

$$T'\sin\theta = \frac{1}{4\pi\epsilon_0\epsilon_r}\cdot\frac{q^2}{r^2} \qquad \ldots(3)$$

$$T'\cos\theta = mg\left(1-\frac{\rho_l}{\rho_s}\right) \qquad \ldots(4)$$

$$(1)\div(2) \Rightarrow \qquad \tan\theta = \frac{q^2}{4\pi\epsilon_0\, mgr^2}$$

$$(3)\div(4) \Rightarrow \qquad \tan\theta = \frac{q^2}{4\pi\epsilon_0\epsilon_r\, mg\left(1-\frac{\rho_l}{\rho_s}\right)r^2}$$

Here we use

$$\frac{q^2}{4\pi\epsilon_0\, mgr^2} = \frac{q^2}{4\pi\epsilon_0\epsilon_r\, mg\left(1-\frac{\rho_l}{\rho_s}\right)r^2}$$

$$\Rightarrow \qquad \epsilon_r = \frac{1}{1-\frac{\rho_l}{\rho_s}} = \frac{1}{1-\frac{0.8}{1.6}} = 2$$ **Ans. (D)**

Sol. 12

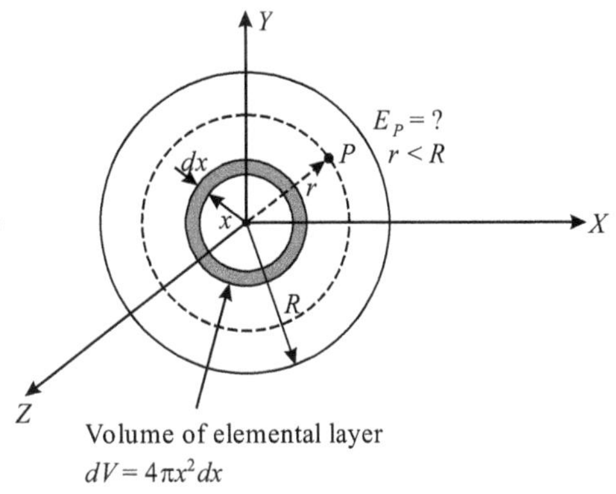

Charge contained in the elemental shell is

$$dq = \rho(r)\cdot dV|_{r=x}$$

$$\Rightarrow \quad dq = \rho_0\left(\frac{5}{4}-\frac{x}{R}\right)\cdot(4\pi x^2 dx)$$

Total charge contained in spherical region of radius r is given as

$$q = \int dq = \int_0^r \rho_0\left(\frac{5}{4}-\frac{x}{R}\right)\cdot 4\pi x^2 dx$$

$$\Rightarrow \quad q = 4\pi\rho_0\int_0^r\left(\frac{5}{4}x^2-\frac{x^3}{R}\right)dx$$

$$\Rightarrow \quad q = 4\pi\rho_0\left[\frac{5x^3}{12}-\frac{x^4}{4R}\right]_0^r$$

$$\Rightarrow \quad q = 4\pi\rho_0\left[\frac{5r^3}{12}-\frac{r^4}{4R}\right]_0^r$$

Electric field at the surface of inside sphere of radius r is

$$E = \frac{Kq}{r^2} = \frac{1}{4\pi\in_0}\cdot\frac{4\pi\rho_0}{r^2}\left[\frac{5r^3}{12}-\frac{r^4}{4R}\right]$$

$$\Rightarrow \quad E = \frac{\rho_0 r}{4\in_0}\left[\frac{5}{3}-\frac{r}{R}\right]$$

Ans. (C)

Sol. 13

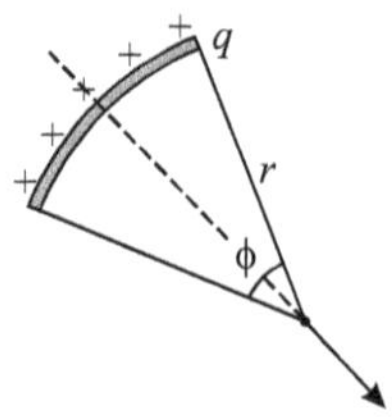

For a half ring

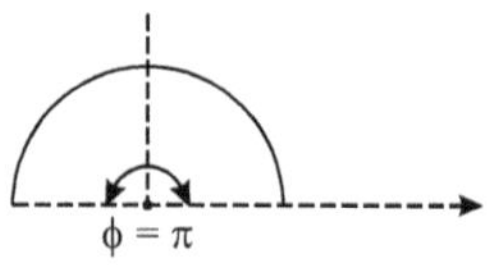

$$E = \frac{2Kq\sin(\phi/2)}{\phi r^2} = \frac{2Kq\sin(\pi/2)}{\pi r^2}$$

$$\Rightarrow \quad E = \frac{2q}{4\pi^2\in_0 r^2} = \frac{q}{2\pi^2\in_0 r^2}$$

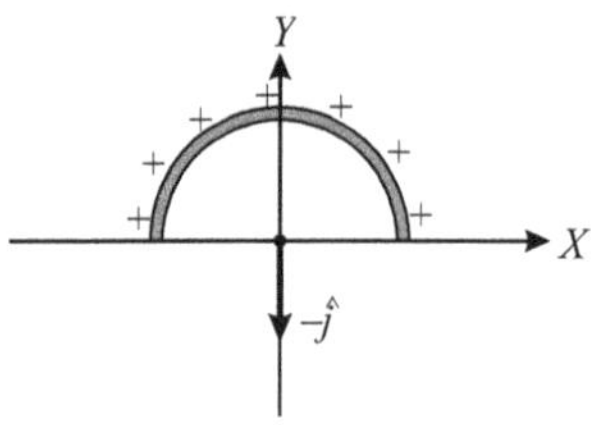

$$\Rightarrow \quad \vec{E} = -\frac{q}{2\pi^2\in_0 r^2}\hat{j}$$

Ans. (D)

Sol. 14

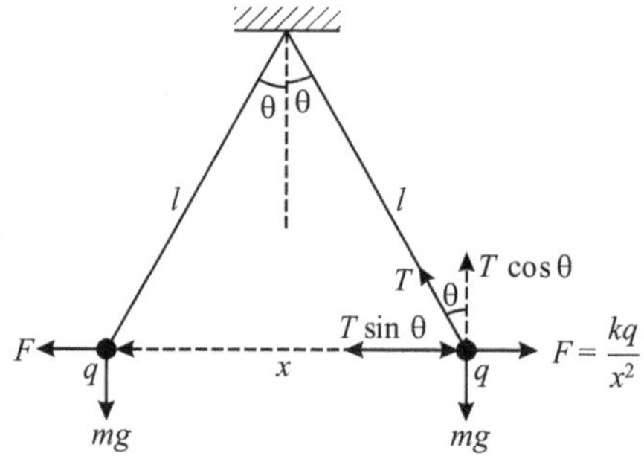

For equilibrium of spheres, we use

$$T\sin\theta = \frac{Kq^2}{x^2} \quad \ldots(1)$$

$$T\cos\theta = mg \quad \ldots(2)$$

$(1)\div(2)$

$$\Rightarrow \quad \tan\theta = \frac{Kq^2}{mgx^2} = \frac{x}{2l}$$

$\Rightarrow$ as θ is small

$$\Rightarrow \quad \tan\theta \simeq \theta = \frac{x}{2l}$$

$$\Rightarrow \quad q = \sqrt{\frac{mg}{2Kl}}\cdot x^{3/2}$$

If q starts leaking (decreasing) than x will decreases and as θ is very small, we can consider very slow variation in x. So we have rate of charge leakage (constant)

$$\frac{dq}{dt} = \sqrt{\frac{mg}{2Kl}}\left(\frac{3}{2}x^{1/2}\right)\left(\frac{dx}{dt}\right)$$

$$\Rightarrow \quad \frac{dx}{dt} = \frac{dq}{dt}\cdot\sqrt{\frac{2Kl}{mg}}\left(\frac{2}{3\sqrt{x}}\right)$$

$$\Rightarrow \quad \frac{dx}{dt} \propto \frac{1}{\sqrt{x}}$$

Ans. (C)

Sol. 15 In a uniformly charged sphere, electric field at a distance r from centre is given as

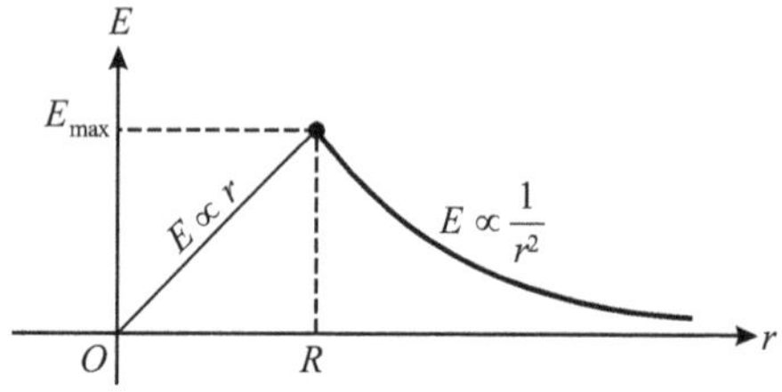

For $r > R$ $\quad E_{out} = \frac{KQ}{r^2} \longrightarrow E \propto \frac{1}{r^2}\downarrow$

For $r = R$ $\quad E_S = \frac{KQ}{R^2} \longrightarrow E_{max}$

For $r < R$ $\quad E_{in} = \frac{KQr}{R^3} \longrightarrow E \propto r\uparrow$

Ans. (D)

Sol. 16

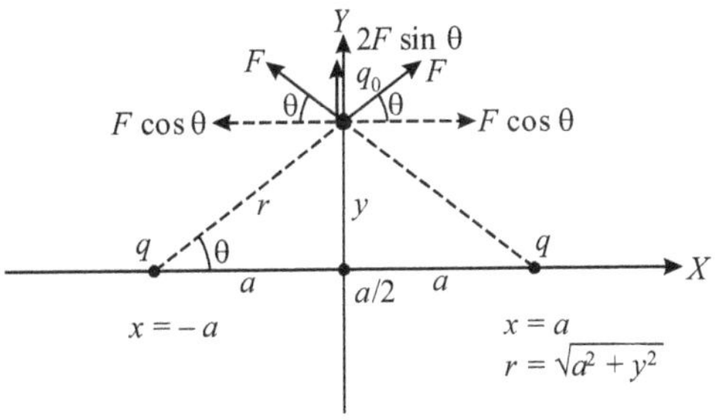

From Figure not force on q_0 due to charges on x-axis (considered fixed) is

$$F_{\text{net}} = 2F\sin\theta$$

$F \longrightarrow$ Coulombic Force

$$F = \frac{Kqq_0}{r^2} = \frac{Kqq_0}{(a^2 + y^2)}$$

$$\Rightarrow \quad F_{\text{net}} = \frac{2Kqq_0}{(a^2+y^2)} \cdot \frac{y}{\sqrt{a^2+y^2}}$$

As $y << a$

$$\Rightarrow \quad a^2 + y^2 \approx a^2$$

$$\Rightarrow \quad F_{\text{net}} = \frac{2Kqq_0}{(a^2+y^2)^{3/2}} \cdot y \approx \frac{2Kqq_0 y}{a^3}$$ **Ans. (A)**

Sol. 17 In space electric force must be in downward direction so that upward electric force on $-6e$ charge balances the weight of drop.

$\Rightarrow$ At equilibrium of drop

$$qE = mg$$

Mass of liquid drop

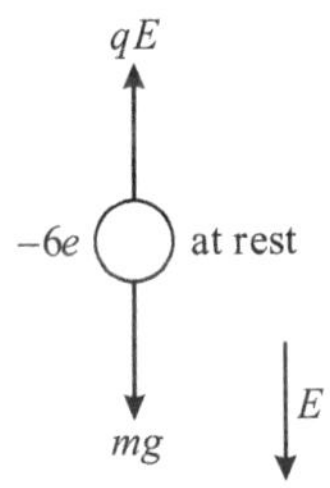

$$m = \frac{4}{3}\pi R^3 \cdot \rho$$

$$\Rightarrow \quad 6eE = \frac{4}{3}\pi R^3 \rho g$$

$$\Rightarrow \quad R = \left(\frac{9eE}{2\pi\rho g}\right)^{1/3}$$

$$\Rightarrow \quad R = \left(\frac{9 \times 1.6 \times 10^{-19} \times 25.5 \times 10^3}{6.28 \times 1.26 \times 10^3 \times 10}\right)^{1/3}$$

$$\Rightarrow \quad R = 7.74 \times 10^{-7}\,\text{m}$$ **Ans. (B)**

Sol. 18 EF at a distance x from the centre of disc is

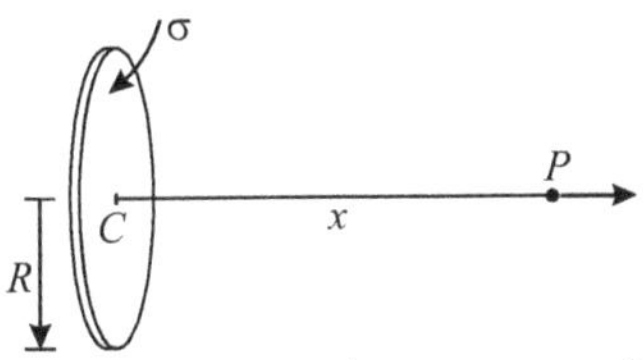

$$E_P = \frac{\rho}{2\epsilon_0}\left(1 - \frac{x}{\sqrt{x^2+R^2}}\right)$$

At $x = 0 \rightarrow$ $$E_C = \frac{\rho}{2\epsilon_0}$$

At $x = R \rightarrow$ $$E_P = \frac{\rho}{2\epsilon_0}\left(1 - \frac{R}{\sqrt{2}R}\right) = \frac{\sigma}{2\epsilon_0}\left(1 - \frac{1}{\sqrt{2}}\right)$$

% reduction in electric force at P w.r.t C is

$$= \frac{E_C - E_P}{E_C} = 100\%$$

$$= \frac{\frac{\sigma}{2\epsilon_0} \cdot \frac{1}{\sqrt{2}}}{\sigma / 2\epsilon_0} \times 100\% = 70.7\%$$ **Ans. (A)**

Sol. 19

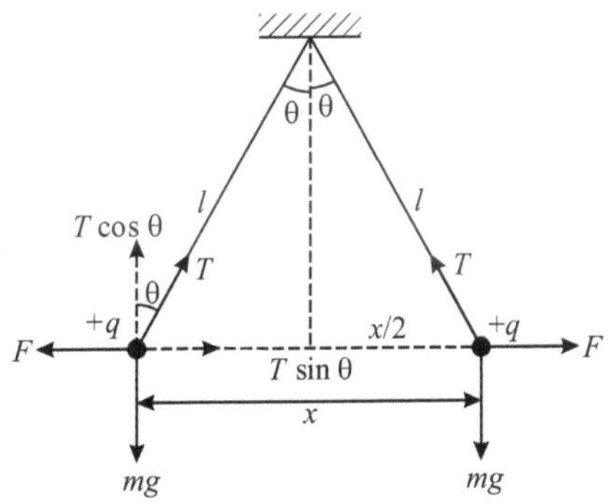

In equilibrium position, for balls, we use

$$T\sin\theta = F = \frac{Kq^2}{x^2} = \frac{1}{4\pi\epsilon_0} \cdot \frac{q^2}{x^2} \quad \ldots(1)$$

and $$T\cos\theta = mg \quad \ldots(2)$$

(1) ÷ (2)

$$\Rightarrow \quad \tan\theta = \frac{F}{mg} = \frac{q^2}{4\pi\epsilon_0 mgx^2}$$

For small θ, we use

$$\tan\theta \simeq \sin\theta = \frac{x}{2l}$$

$$\Rightarrow \quad \frac{x}{2l} = \frac{q^2}{4\pi\epsilon_0 mgx^2}$$

$$\Rightarrow \quad x = \left[\frac{q^2 l}{2\pi\epsilon_0 mg}\right]^{1/3}$$

Ans. (D)

Sol. 20

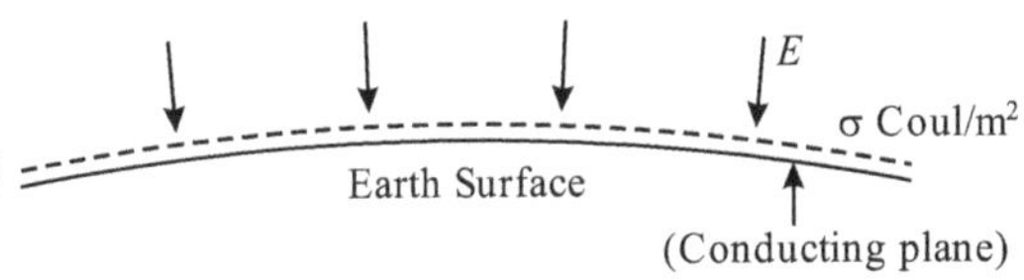

If σ C/m^2 (–ve) is the surface charge density on Earth surface then we use

$$E = \frac{\sigma}{\epsilon_0}$$

$\Rightarrow$ $\sigma = \epsilon_0 E$

The total surface area of earth $= 4\pi R_e^2$

$\Rightarrow$ Total charge on earth surface is

$$q = -\sigma \times 4\pi R_e^2 = \epsilon_0 E \cdot 4\pi R_e^2$$

$\Rightarrow$ $q = -8.85 \times 10^{-12} \times 150 \times 4 \times 3.14 \times (6.37 \times 10^6)^2$

$\Rightarrow$ $q = -676.55 \times 10^3\,C \simeq -680\,KC$ **Ans. (C)**

Sol. 21

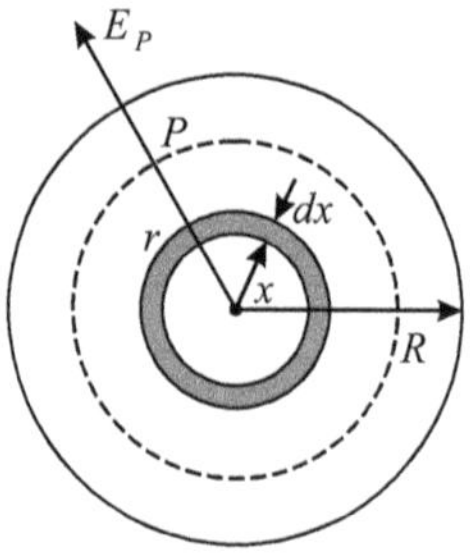

The charge contained in the elemental layer of radius x and width dx is

$$dq = \rho(x) \cdot dV = \rho(x) \cdot 4\pi x^2 dx$$

Total charge in spherical region from $x = 0$ to $x = r$ ($r < R$) is

$$q = \int dq = \int_0^r \rho_0 \left(1 - \frac{x}{R}\right) \cdot 4\pi x^2 dx$$

$\Rightarrow$ $$q = 4\pi\rho_0 \int_0^r \left(x^2 - \frac{x^3}{R}\right) dx$$

$\Rightarrow$ $$q = 4\pi\rho_0 \left[\frac{x^3}{3} - \frac{x^4}{4R}\right]_0^r dx = 4\pi\rho_0 r^3 \left[\frac{1}{3} - \frac{r}{4R}\right]$$

Electric force strength at point P ($x = r$) is given as

$$E_P = \frac{Kq}{r^2}$$

(for spherically symmetric distribution of charge)

$\Rightarrow$ $$E_P = \frac{1}{4\pi \epsilon_0 r^2} \cdot 4\pi\rho_0 r^3 \left[\frac{1}{3} - \frac{r}{4R}\right]$$

$\Rightarrow$ $$E_P = \frac{\rho_0}{\epsilon_0}\left(\frac{r}{3} - \frac{r^2}{4R}\right)$$ **Ans. (B)**

Sol. 22 Due to a complete disc of radius b, electric field at point P is given as

$$E_{disc} = \frac{\sigma}{2\epsilon_0}\left(1 - \frac{h}{\sqrt{b^2 + h^2}}\right) = \frac{\sigma}{2\epsilon_0}\left(1 - \frac{h}{b}\right) \ldots (1)$$

[As $h << b$]

Electric field at point P only due to the circular disc of radius 'a' is given as

$$E_{disc} = \frac{\sigma}{2\epsilon_0}\left(1 - \frac{h}{\sqrt{a^2 + h^2}}\right) = \frac{\sigma}{2\epsilon_0}\left(1 - \frac{h}{b}\right) \ldots (1)$$

[As $h << a$]

Electric force P due to disc with the hole of radius 'a' is given as

$$E_{given\ object} = E_{disc} - E_{hole}$$

$$= \frac{\sigma}{2\epsilon_0}\left[\left(1 - \frac{h}{b}\right) - \left(1 - \frac{h}{a}\right)\right]$$

$$= \frac{\sigma}{2\epsilon_0}\left[\frac{h}{a} - \frac{h}{2a}\right] \qquad (\text{As } b = 2a)$$

$$= \frac{\sigma h}{4\epsilon_0 a} = Ch \Rightarrow C = \frac{\sigma}{4\epsilon_0 a}$$ **Ans. (A)**

Sol. 23

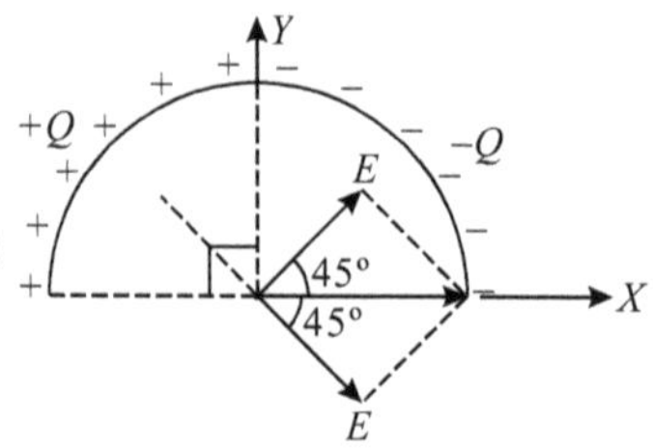

We use $\pi R = L$

$\Rightarrow$ $$R = \frac{20}{\pi}\text{ cm}$$

where $$E = \frac{2kQ\sin\frac{\phi}{2}}{\phi R^2} = \frac{2kQ\sin\left(\frac{\pi}{4}\right)}{\left(\frac{\pi}{2}\right)R^2} \ldots (1)$$

$$E_{net\ at\ O} = \sqrt{2}E \ldots (2)$$

From equation-(1) $$E = \frac{4}{\pi R^2} \cdot \frac{1}{4\pi\epsilon_0} \cdot Q \cdot \frac{1}{\sqrt{2}}$$

$$= \frac{10\epsilon_0}{\pi^2 \epsilon_0 \left(\frac{20}{\pi} \times 10^{-2}\right)^2 \sqrt{2}}$$

$\Rightarrow$ $$E = \frac{1000}{\sqrt{2} \times 400 \times 10^{-4}} = \frac{2.5}{\sqrt{2}} \times 10^{+4}\text{ N/C}$$

From equation-(2), net electric force at O is given as

$$E_{net\ at\ O} = \sqrt{2}E = 2.5 \times 10^4 = 25 \times 10^3\text{ N/C}$$

as $\vec{E}_0 = 25 \times 10^3\,\hat{i}$ N/C **Ans. (D)**

Sol. 24 As net charge enclosed in cavity is $+Q - Q = 0$ so net induced charge on inner surface of shell $Q_1 = 0$.

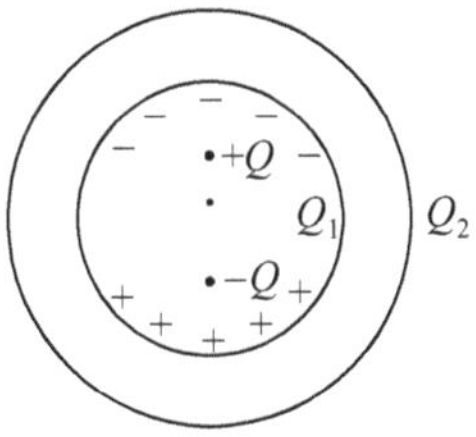

As shown we use $\sigma_1 \neq 0$ and varying with position inner surface of shell = 0

$\Rightarrow$ charge on outer surface

$$Q_2 = 0 \text{ and } \sigma_2 = 0$$ **Ans. (B)**

Sol. 25

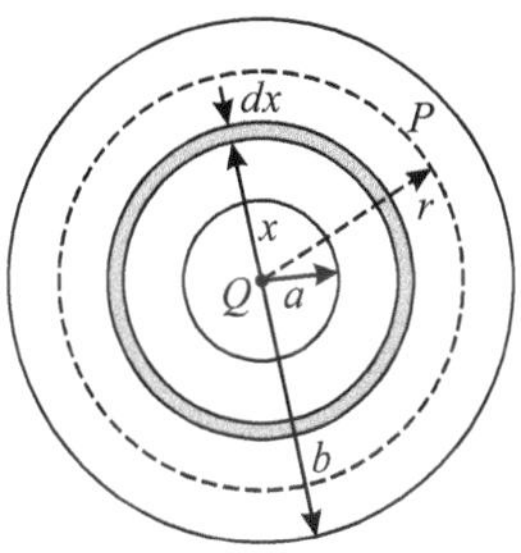

Electric field at point P is given as

$$E_P = \frac{K(Q + q_{en})}{r^2} \quad \ldots(1)$$

Where q_{en} is the charge between charge contained within the elemental layer of radius x and width dx is

$$dq = \rho \cdot dv$$

$$dq = \left(\frac{A}{x}\right) \cdot 4\pi x^2 dx$$

Total charge in region from $x = a$ to $x = r$ is

$$q_{en} = \int dq = 4\pi A \int_{x=a}^{x=r} x\, dx$$

$$\Rightarrow \quad q_{en} = 4\pi A\left[\frac{x^2}{2}\right]_a^r = 2\pi A r^2 - 2\pi A a^2$$

From equation-(1) $E_P = \frac{K}{r^2}(Q + 2\pi A r^2 - 2\pi A a^2) \quad \ldots(2)$

Given that E_p in equation-(2) is a constant (independent of r) if equation-(2)

$$Q = 2\pi A a^2 \Rightarrow E = K(2\pi A) \rightarrow \text{constant}$$

$$\Rightarrow \quad A = \frac{Q}{2\pi a^2}$$ **Ans. (A)**

Sol. 26

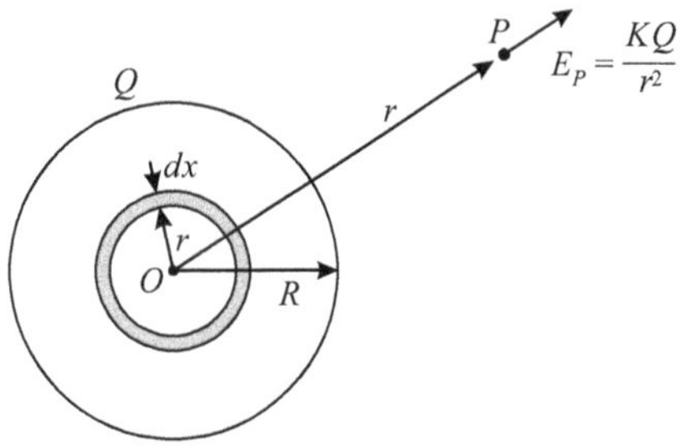

To calculate total charge on ball we consider on elemental shell inside the ball as shown of volume given as

$$dV = 4\pi r^2 dr$$

(For spherically symmetric charge distribution)

charge in the elemental shell thickness is

$$dq = \rho dV = \rho_0 = \left(1 - \frac{r}{R}\right) \cdot 4\pi r^2 dr$$

Total ball's charge is given as

$$Q = \int dq = \int_0^R 4\pi\rho_0\left(1 - \frac{r}{R}\right) \cdot r^2 dr$$

$$\Rightarrow \quad Q = 4\pi\rho_0 \int_0^R \left(r^2 - \frac{r^3}{R}\right) dr$$

$$\Rightarrow \quad Q = 4\pi\rho_0 \left[\frac{r^3}{3} - \frac{r^4}{4R}\right]_0^R$$

$$\Rightarrow \quad Q = 4\pi\rho_0 \left[\frac{R^3}{3} - \frac{R^3}{4}\right]$$

$$\Rightarrow \quad Q = 4\pi\rho_0 \left(\frac{R^3}{12}\right) = \frac{1}{3}\pi\rho_0 R^3$$

Electric field at an outer point of ball is given as

$$E_P = \frac{KQ}{r^2} = \frac{1}{4\pi \in_0} \cdot \frac{\left(\frac{1}{3}\pi\rho_0 R^3\right)}{r^2} = \frac{\rho_0 R^3}{12 \in_0 r^2}$$ **Ans. (D)**

Sol. 27

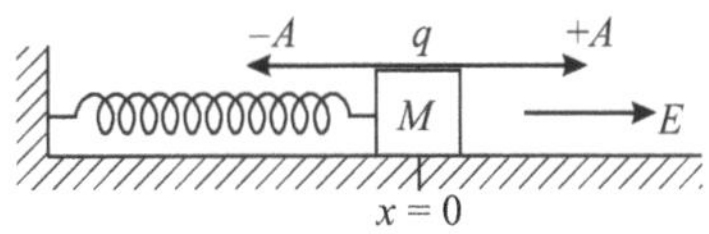

Oscillation energy of system $E_T = \frac{1}{2} m\omega^2 A^2$

PE of oscillations at equilibrium position = 0

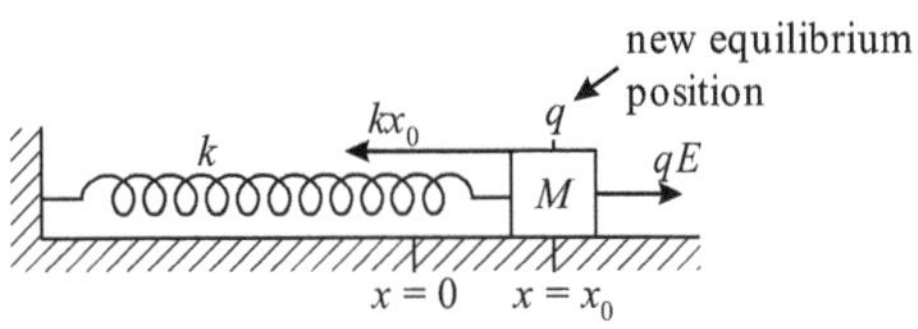

At equilibrium position $kx_0 = qE$

$$\Rightarrow \quad x_0 = \frac{qE}{k}$$

PE of oscillation at equilibrium position $= \frac{1}{2}kx_0^2 = \frac{1}{2}\frac{q^2E^2}{k}$

Total oscillation energy after EF is applied is given as

$$E_T = \frac{1}{2}m\omega^2A^2 + U_{\text{equilibrium}}$$

$$E_T = \frac{1}{2}m\omega^2A^2 + \frac{1}{2}\frac{q^2E^2}{k}$$ **Ans. (A)**

Sol. 28

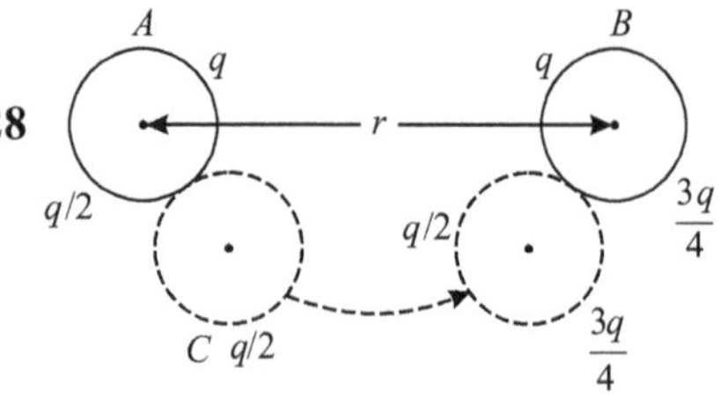

When C is touched to A, charge q is divided equally among A and C

$\Rightarrow$ final charge of A is $q_A = \frac{q}{2}$

Now C is touched to B, charge $\left(q + \frac{q}{2} = \frac{3q}{2}\right)$ is divided equally among B and C

$\Rightarrow$ Final charge on B is $q_B = \frac{3q}{4}$

So final force between A and B is

$$F' = \frac{K\left(\frac{a}{2}\right)\left(\frac{3q}{4}\right)}{r^2} = \frac{3}{8}\frac{Kq^2}{r^2} = \frac{3}{8}F$$ **Ans. (D)**

Sol. 29

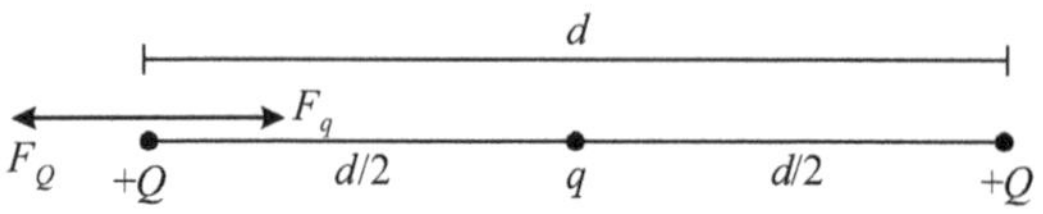

For equilibrium of charge Q, we use

$$\vec{F_Q} + \vec{F_q} = 0$$

$$\vec{F_Q} = -\vec{F_q}$$

$$\frac{kQQ}{d^2} = -\frac{kQq}{(d/2)^2}$$

$$\Rightarrow \quad q = -\frac{Q}{4}$$ **Ans. (C)**

Sol. 30 Electric field on axis of ring at a distance h is given as

$$E = \frac{kQx}{(h^2 + R^2)^{3/2}}$$

For maximum electric field, we use

$$\frac{dE}{dx} = 0$$

Solving we get

$$x = \frac{R}{\sqrt{2}}$$ **Ans. (C)**

Sol. 31

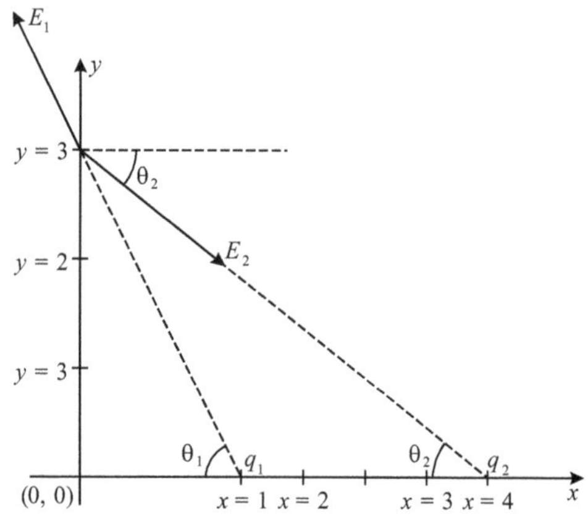

Let $\vec{E_1}$ & $\vec{E_2}$ are the values of electric field due to q_1 & q_2 respectively are magnitude of

$$E_2 = \frac{1}{4\pi\epsilon_0}\frac{q_2}{r^2}$$

$$E_2 = \frac{9\times10^9\times(25)\times10^{-6}}{(4^2+3^2)}\text{ V/m}$$

$$E_2 = 9\times10^3\text{ V/m}$$

$$\Rightarrow \quad \vec{E_2} = 9\times10^3(\cos\theta_2\hat{i} - \sin\theta_2\hat{j})\text{ V/m}$$

Where $\tan\theta_2 = \frac{3}{4}$

$$\Rightarrow \quad \vec{E_2} = 9\times10^3\left(\frac{4}{5}\hat{i} - \frac{3}{5}\hat{j}\right) = (72\hat{i} - 54\hat{j})\times10^2\text{ V/m}$$

Magnitude of

$$E_1 = \frac{1}{4\pi\epsilon_0}\frac{\sqrt{10}\times10^{-6}}{(1^2+3^2)}$$

$$\Rightarrow \quad E_1 = (9\times10^9)\times\sqrt{10}\times10^{-7}\text{ V/m}$$

$$\Rightarrow \quad E_1 = 9\sqrt{10}\times10^2\text{ V/m}$$

$$\Rightarrow \quad \vec{E_1} = 9\sqrt{10}\times10^2\left[\cos\theta_1(-\hat{i}) + \sin\theta_1\hat{j}\right]\text{V/m}$$

Where $\tan\theta_1 = 3$

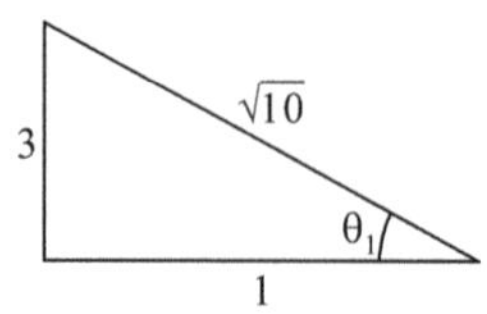

$$\Rightarrow \quad E_1 = 9\times\sqrt{10}\times10^2\left[\frac{1}{\sqrt{10}}(-\hat{i}) + \frac{3}{\sqrt{10}}\hat{j}\right]\text{V/m}$$

$$\Rightarrow \quad E_1 = 9\times10^2\left[-\hat{i} + 3\hat{j}\right] = \left[-9\hat{i} + 27\hat{j}\right]10^2\text{ V/m}$$

$$\Rightarrow \quad \vec{E}_{\text{net}} = \vec{E_1} + \vec{E_2} = (63\hat{i} - 27\hat{j})\times10^2\text{ V/m}$$ **Ans. (C)**

Sol. 32 Total charge in spherical volume is given as

$$Q = \int\rho dV$$

Where dV is the volume of elemental shall shown in figure

$$\Rightarrow \quad Q = \int_0^R \frac{A}{r^2}e^{-2r/a}(4\pi r^2 dr)$$

$$\Rightarrow \quad Q = 4\pi A\int_0^R e^{-2r/a}dr$$

$$\Rightarrow \quad Q = 4\pi A\left(\frac{e^{-2r/a}}{-\frac{2}{a}}\right)_0^R$$

$$\Rightarrow \quad Q = 4\pi A\left(-\frac{a}{2}\right)(e^{-2R/a}-1)$$

$$\Rightarrow \quad Q = 2\pi aA(1-e^{-2R/a})$$

$$\Rightarrow \quad R = \frac{a}{2}\log\left(\frac{1}{1-\frac{Q}{2\pi aA}}\right)$$

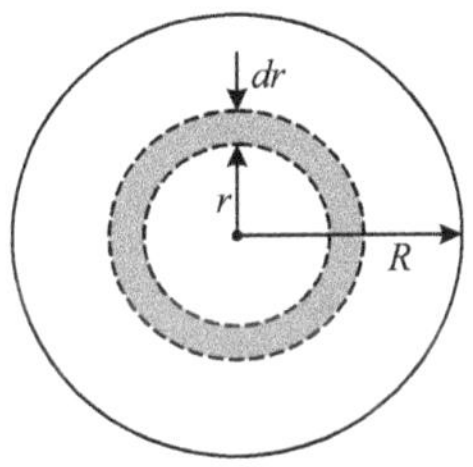

$dV = 4\pi r^2 dr$

Ans. (D)

Sol. 33

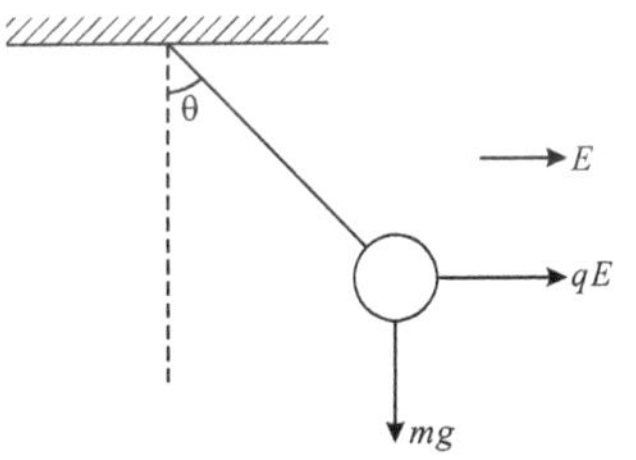

For equilibrium of bob, we use

$$\tan\theta = \frac{qE}{mg} = \frac{5\times10^{-6}\times2000}{2\times10^{-3}\times10}$$

$$\Rightarrow \quad \tan\theta = \frac{1}{2}$$

$$\Rightarrow \quad \theta = \tan^{-1}(0.5)$$

Ans. (C)

Sol. 34

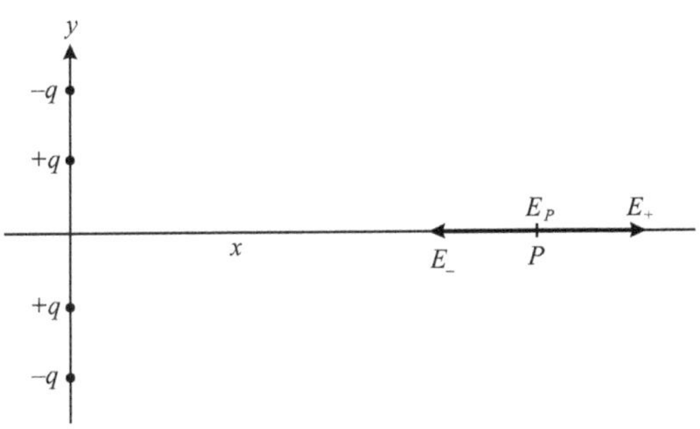

Electric field due to $+q$ charges at point P is

$$E_+ = \frac{K(2q)x}{(x^2+d^2)^{3/2}} = \frac{2kqD}{(D^2+d^2)^{3/2}}$$

Electric field due to $-q$ charges at point P is

$$E_- = \frac{K(2q)x}{(x^2+4d^2)^{3/2}} = \frac{2kqD}{(D^2+4d^2)^{3/2}}$$

Net Electric field at point P is given as

$$E_P = E_+ - E_- = 2kqD\left[\frac{1}{(D^2+d^2)^{3/2}} - \frac{1}{(D^2+4d^2)^{3/2}}\right]$$

For $D >> d$ we use $\Rightarrow E_P = E_P = \frac{2kqD}{D^3}$

$$\left[\left(1+\frac{d^2}{D^2}\right)^{-3/2} - \left(1+\frac{4d^2}{D^2}\right)^{-3/2}\right]$$

$$\Rightarrow \quad E_P = \frac{2kq}{D^2}\left[\left(1-\frac{3d^2}{2D^2}\right)-\left(1-\frac{3}{2}\cdot\frac{4d^2}{D^2}\right)\right]$$

$$\Rightarrow \quad E_P = \frac{2kq}{D^2}\left[1-\frac{3d^2}{2D^2}-1+\frac{6d^2}{D^2}\right] = \frac{9kqd^2}{D^4}$$

$$\Rightarrow \quad E_P \propto \frac{1}{D^4}$$ **Ans. (D)**

Sol. 35

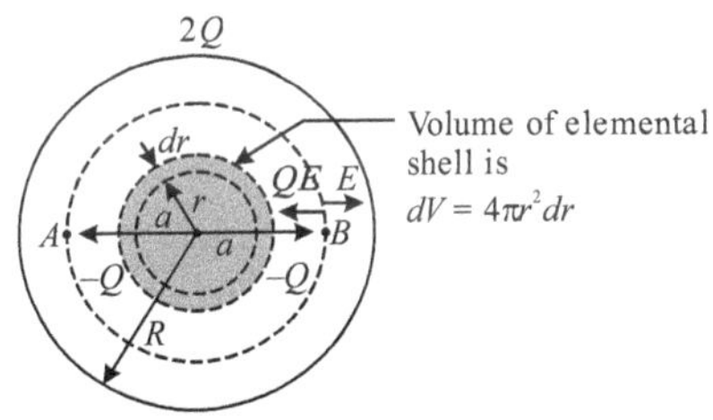

Charge in shell

$$dq = \rho(r)\cdot 4\pi r^2 dr$$

$$\Rightarrow \quad 2Q = \int dq = \int_0^R kr\cdot 4\pi r^2 dr$$

$$\Rightarrow \quad 2Q = 4\pi\frac{R^4}{4} = \pi kR^4$$

$$\Rightarrow \quad k = \frac{2Q}{\pi R^4} \quad \ldots(1)$$

Electric field at a distance a from centre is

$$E_a = \left(\frac{1}{4\pi\epsilon_0}\right)\frac{q_a}{a^2}$$

Where $\quad q_a = \int_0^a kr\cdot 4\pi r^2 dr$

$$\Rightarrow \quad q_a = 4\pi\frac{kq^4}{4}$$

$$\Rightarrow \quad E_a = \frac{1}{4\pi\epsilon_0}\cdot\frac{\pi ka^4}{a^2} = \frac{1}{4\pi\epsilon_0}\cdot\pi ka^2$$

Repulsive force between A and B is given as

$$F = \frac{1}{4\pi \epsilon_0} \cdot \frac{Q^2}{(2a)^2}$$

As net force on A and B is zero, we use

$$\Rightarrow \quad F = QE$$

$$\frac{1}{4\pi \epsilon_0} \cdot \frac{Q^2}{4a^2} = Q \cdot \frac{1}{4\pi \epsilon_0} \cdot \pi k a^2$$

$$\Rightarrow \quad \frac{Q}{4a^2} = \pi\left(\frac{2Q}{\pi R^4}\right) \cdot a^2 \Rightarrow a^4 = \frac{1}{8} R^4$$

$$\Rightarrow \quad a = 8^{-1/4} \cdot R$$

Ans. (C)

15.2 Electric Flux and Gauss's Law

Sol. 1

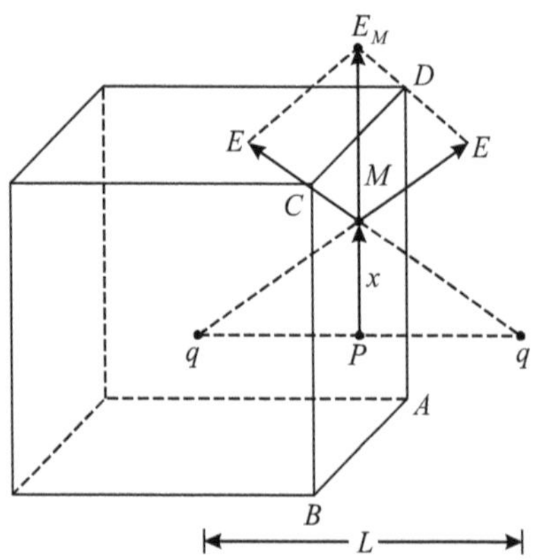

Here $P \rightarrow$ centre of square $ABCD$

$M \rightarrow$ a point in plane of square $ABCD$ at a distance x from P.

Due to the two charges, electric field resultant at any point M in plane of square $ABCD$ is parallel (tangential) to $ABCD$. So there is no electric field normal to surface $ABCD$ hence electric flux through surface $ABCD$ is zero. **Ans. (B)**

Sol. 2 Using Gauss's law, we use

$$\phi_{\text{not outward}} = \frac{q_{\text{end}}}{\epsilon_0}$$

$$\Rightarrow \quad \phi_2 - \phi_1 = \frac{q_{\text{end}}}{\epsilon_0}$$

$$\Rightarrow \quad q_{\text{end}} = \epsilon_0(\phi_2 - \phi_1)$$

Ans. (A)

Sol. 3

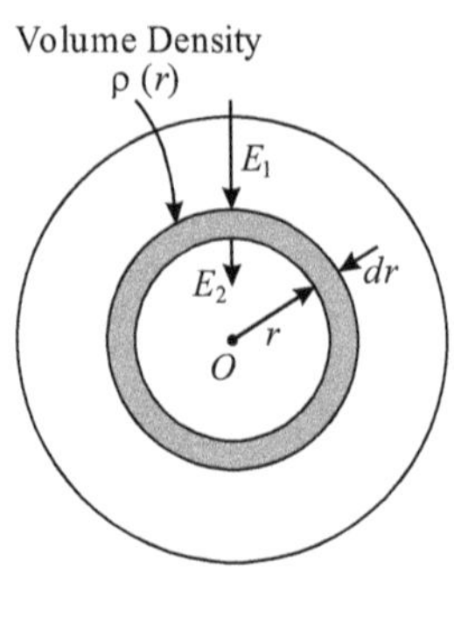

Potential at a distance r from centre is

$$\phi = ar^2 + b$$

$\Rightarrow$ EF at a distance r is given as

$$\vec{E} = -\frac{dv}{dr}\hat{r}$$

$$\Rightarrow \quad \vec{E} = -2ar\hat{r} \qquad \ldots(1)$$

–ve sign implies field is in radially inward direction

$\Rightarrow$ charge on ball is –ve

From equation-(1) we can see that $E \propto r$ so this happens inside a uniformly charged ball and if its volume charge density is ρ then inside EF of this ball is given as

$$E_{\text{in}} = \frac{\rho r}{3\epsilon_0} \qquad \ldots(2)$$

from (1) and (2), we have

$$\frac{\rho r}{3\epsilon_0} = 2ar \Rightarrow \rho = 6a\epsilon_0$$

Ans. (B)

Alternative : If E_1 and E_2 are EF at outer surfaces of the elemental shell as shown then by Gauss's law, we have

$$\phi_{\text{out}} - \phi_{\text{in}} = E_2(4\pi r^2) - E_1(4\pi(r+dr)^2) = \frac{q_{\text{encl}}}{\epsilon_0}$$

$$8\pi ar^3 - 8\pi ar^3\left(1+\frac{dr}{r}\right)^3 = \frac{\rho(r)[4\pi r^2 dr]}{\epsilon_0}$$

$$8\pi ar^3 - 8\pi ar^3\left(1+\frac{3dr}{r}\right) = \frac{\rho(r)[4\pi r^2 dr]}{\epsilon_0}$$

$$\Rightarrow \quad \frac{\rho(r)dr}{\epsilon_0} = -6adr$$

$$\Rightarrow \quad \rho(r) = -6a\epsilon_0$$

Ans. (B)

Sol. 4

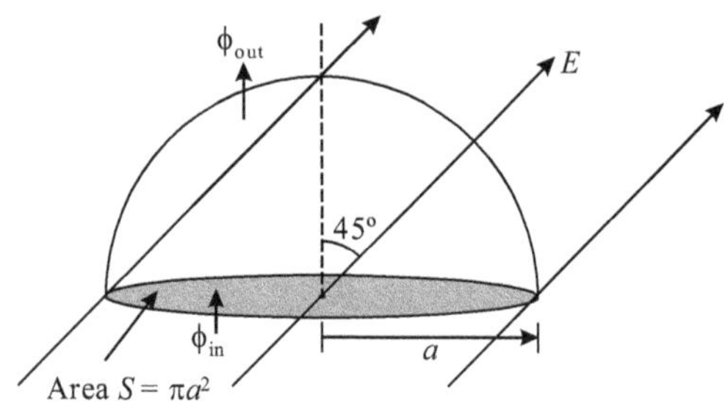

As no charge is enclosed in hemisphere

$$\Rightarrow \quad \phi_{\text{in}} = \phi_{\text{out}}$$

$$\Rightarrow \quad \phi_{\text{out}} = \phi_{\text{in}} = ES\cos 45° = E\pi a^2\left(\frac{1}{\sqrt{2}}\right)$$

Through the curved surface of hemisphere **Ans. (B)**

Sol. 5

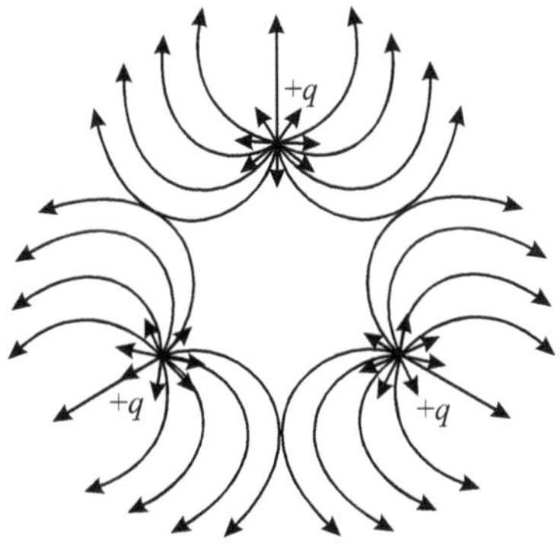

As electric lines originated from a +ve charge are repelled by other neighbouring +ve charges **Ans. (C)**

Sol. 6

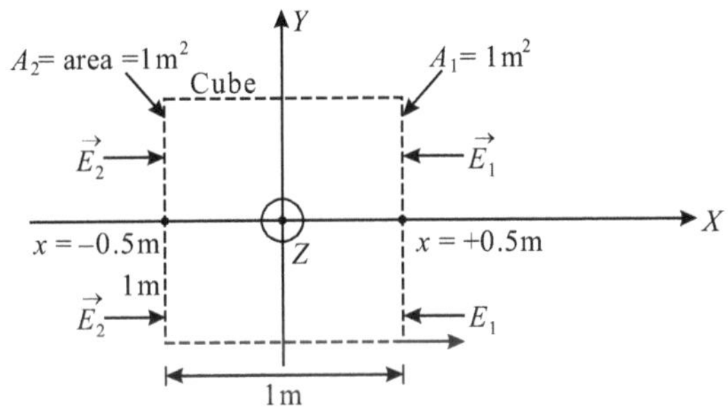

EF in the region is given as

$$\vec{E} = -\frac{dV}{dx}\hat{i}$$

at $x = 0.5\text{ m} \rightarrow \vec{E_1} = -4\hat{i}$

at $x = -0.5\text{ m} \rightarrow \vec{E_2} = 4\hat{i}$

Total flux going into the cubical surface is given as

$$\phi = \vec{E_1}\cdot\vec{A_1} + \vec{E_2}\cdot\vec{A}$$

$\Rightarrow$ $\phi = 4.1 + 4.1 = 8$ units

By Gauss's Law $\phi_{out} = \dfrac{q_{encl}}{\epsilon_0}$

$\Rightarrow$ $-8 = \dfrac{q_{encl}}{\epsilon_0} \Rightarrow q_{encl} = -8\epsilon_0$ **Ans. (D)**

Sol. 7

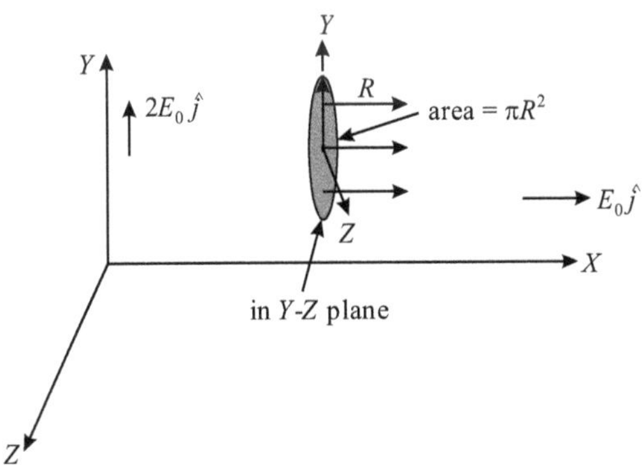

As circular area is perpendicular to X-axis (as in YZ plane) so only the electric flux due to field $E_0\hat{i}$ will pass through so we have

$$\phi = E_0 \times A = E_0 \cdot \pi r^2$$
$$= 100 \times 3.14 \times 4 \times 10^{-4}$$
$$= 12.56 \times 10^{-2}$$
$$= 0.1256\ \text{Nm}^2/\text{C}$$ **Ans. (A)**

Sol. 8

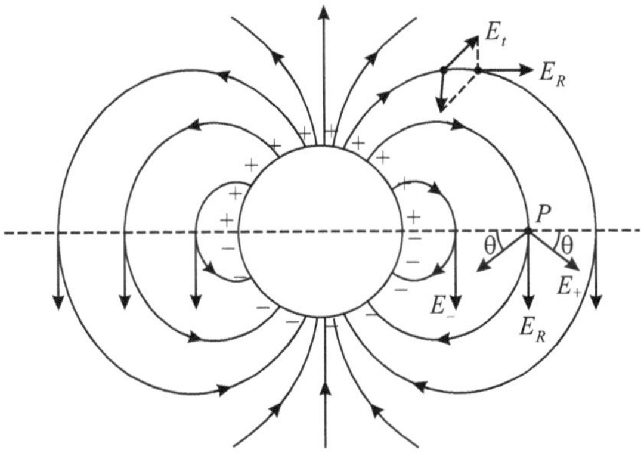

The preference of termination of electric lines is always for a –ve charge (nearly) rather at ∞. **Ans. (A)**

Sol. 9 Using Gauss's law we have electric flux coming out from any closed surface is given as

$$\phi_s = \frac{q_{\text{enclosed by S}}}{\epsilon_0}$$

flux through surfaces S_1, S_2, S_3 and S_4 are

$$\phi_1 = \frac{2q}{\epsilon_0};$$

$$\phi_2 = \frac{3q - q}{\epsilon_0} = \frac{2q}{\epsilon_0};$$

$$\phi_3 = \frac{2q}{\epsilon_0};$$

$$\phi_4 = \frac{8q - 2q - 4q}{\epsilon_0} = \frac{2q}{\epsilon_0}$$

$\Rightarrow$ $\phi_1 = \phi_2 = \phi_3 = \phi_4$ **Ans. (C)**

Sol. 10

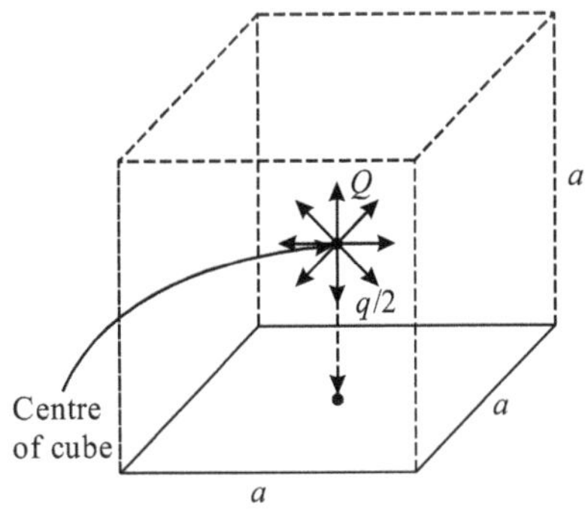

As all six surfaces (squares of side a) of imaginary cube considered are symmetrically located w.r. to charge Q at centre. Total electric flux coming out from cubical surface is

$$\phi_0 = \frac{Q}{\epsilon_0}$$

By symmetry, flux through the given square is

$$\phi = \frac{\phi_0}{6} = \frac{Q}{6\epsilon_0}$$ **Ans. (B)**

16.1 Electric Potential and Dipole

Sol. 1 Work done in moving a charge from one point to another (say A to B) is given as

$$W_{AB} = q \times (V_B - V_A) = \text{charge} \times \text{potential difference}$$

$\Rightarrow$ Potential difference

$$V_B - V_A = \frac{W_{AB}}{q}$$

$$= \frac{2}{20}$$

$$= 0.1\text{ V}$$ **Ans. (A)**

Sol. 2

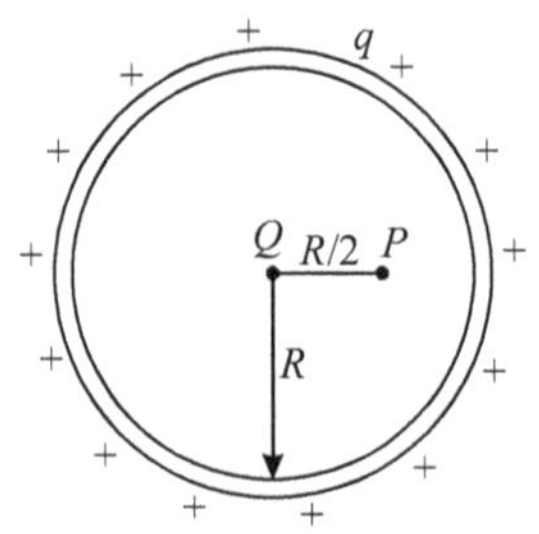

Electric potential at point P is given as

$$V_P = V_{QP} + V_{qP}$$

$$\Rightarrow \quad V_P = \frac{KQ}{\left(\frac{R}{2}\right)} + \frac{Kq}{R}$$

$$\Rightarrow \quad V_P = \frac{2Q}{4\pi\epsilon_0 R} + \frac{q}{4\pi\epsilon_0 R}$$ **Ans. (C)**

Sol. 3 At the point of closest approach, the KE of particle transforms into the PE of system. Thus by conservation of energy, we have

$$\Rightarrow \frac{1}{2}mv^2 + O = O + \frac{KqQ}{r} \Rightarrow r = \frac{2KqQ}{mv^2}$$

$$\Rightarrow r \propto \frac{1}{v^2}$$

If q was given a speed $2v$ then now closest approach distance will be

$$\frac{r_1}{r_2} = \left(\frac{v_2}{v_1}\right)^2 \Rightarrow r_2 = \left(\frac{v_1}{v_2}\right)^2 r_1 = \left(\frac{v_1}{2v}\right)^2 r = \frac{r}{4}$$ **Ans. (D)**

Sol. 4 Electric potential at point C_2 is

$$V_{C_2} = \left(+\frac{Kq}{\sqrt{d^2+R^2}}\right) + \left(-\frac{Kq}{R}\right)$$

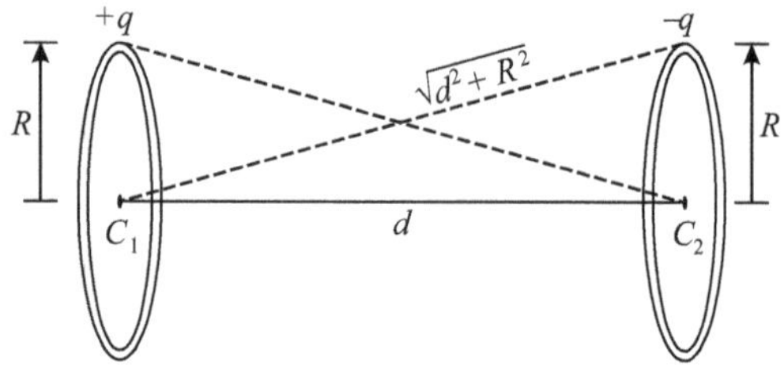

Electric potential at point C_1 is

$$V_{C_1} = V_{+q} + V_{-q}$$

$$\Rightarrow \quad V_{C_1} = \frac{Kq}{R} + \left(-\frac{Kq}{\sqrt{d^2+R^2}}\right)$$

Potential difference between centres of rings is

$$\Delta V = V_{C_1} - V_{C_2} = \frac{q}{2\pi\epsilon_0}\left(\frac{1}{R} - \frac{1}{\sqrt{d^2+R^2}}\right)$$ **Ans. (B)**

Sol. 5

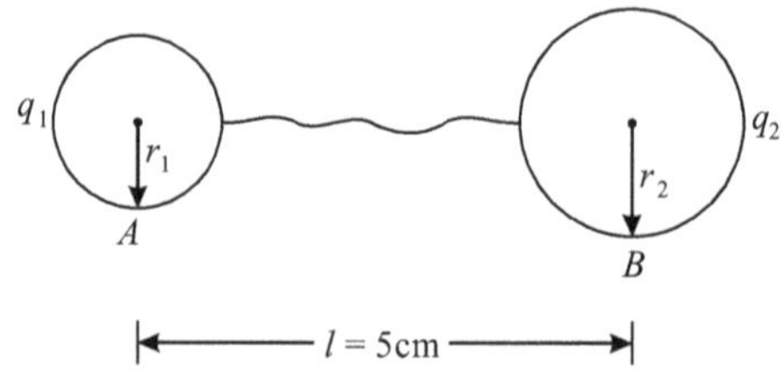

We can ignore induction effects on spheres due to charge on other sphere as separation

$$l >> r_1, r_2$$

If q_1 and q_2 are charges on A and B after connection, we have

$$V_A = V_B$$

$$\frac{Kq_1}{r_1} = \frac{Kq_2}{r_2}$$

Electric field magnitude on surfaces of A and B are

$$E_A = \frac{Kq_1}{r_1^2}$$

and $$E_B = \frac{Kq_2}{r_2}$$

$$\Rightarrow \quad \frac{E_A}{E_B} = \frac{r_2}{r_1} = \frac{2}{1}$$ **Ans. (C)**

Sol. 6 Non-uniform EF

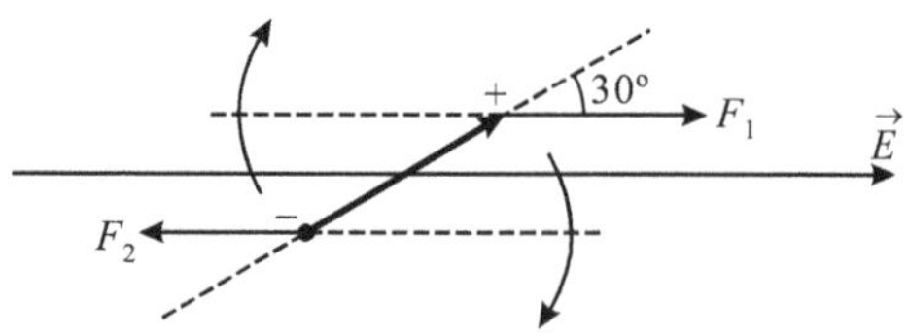

The electric force on a dipole placed in non-uniform EF is given as

$$F = \vec{P}\cdot\frac{\overrightarrow{dE}}{dx} = P\frac{dE}{dx}\cos 30^\circ$$

Force along the direction of EF

As EF is non uniform here $F_1 \neq F_2$

so dipole will experience a translational force

$$|F_1 - F_2| = \left|\vec{P}\cdot\frac{\overrightarrow{dE}}{dx}\right|$$

As F_1 and F_2 are not acting along same lines of action so there will be a torque due to these forces present on dipole. **Ans. (C)**

Sol. 7 We know that EF in space is given as

$$\vec{E} = -\frac{dV}{dx}\hat{i} \quad \text{(for unidirectional EF)}$$

$$\Rightarrow \quad \vec{E} = -\left[20\left(-\frac{1}{(x^2-4)}\times 2x\right)\right]\hat{i}$$

$$\Rightarrow \quad \vec{E} = \frac{40x}{(x^3-4)}\hat{i}\Bigg|_{x=4\mu m} = \frac{40\times 4}{(16-4)^2} = \frac{160}{144}\hat{i}\ \text{V/}\mu\text{m}$$

$$\Rightarrow \quad \vec{E} = \frac{10}{9}\hat{i}\ \text{V/}\mu\text{m} \quad \text{(along + X-direction)}$$ **Ans. (D)**

Sol. 8 Due to symmetry $V_0 = 0$

At O net EF is in direction as shown in figure-1

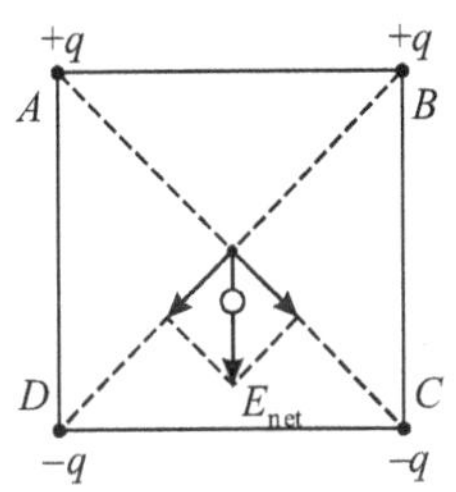

Figure-1

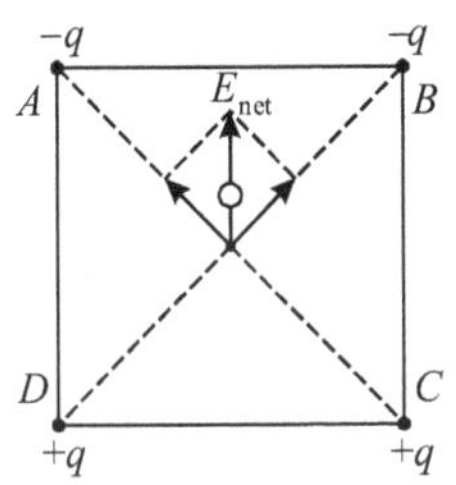

Figure-2

After interchanging charges as given in que Still $V_0 = 0$ (remain unchanged)

As shown in figure-2 direction of $\vec{E}$ is reversed. **Ans. (D)**

Sol. 9 $r_1 = \sqrt{2^2 + 0^2} = 2$

$$r_2 = \sqrt{(\sqrt{2})^2 + (\sqrt{2})^2} = 2$$

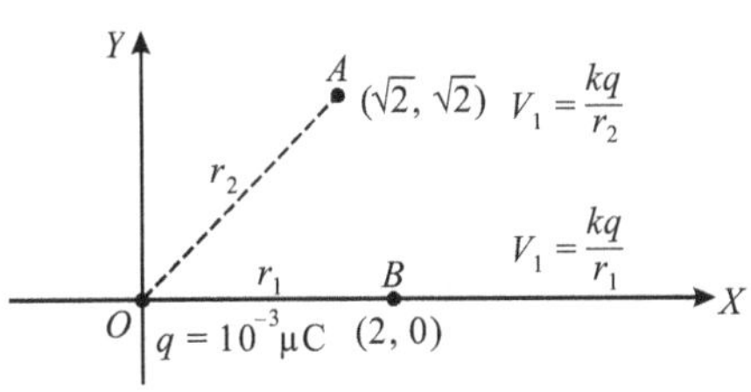

As $r_1 = r_2$ so electric potentials at points A and B are equal

$\Rightarrow \quad V_A = V_B$ **Ans. (B)**

Sol. 10 Work done in moving charge q from point P to Q is

$$W = q(V_Q - V_P)$$

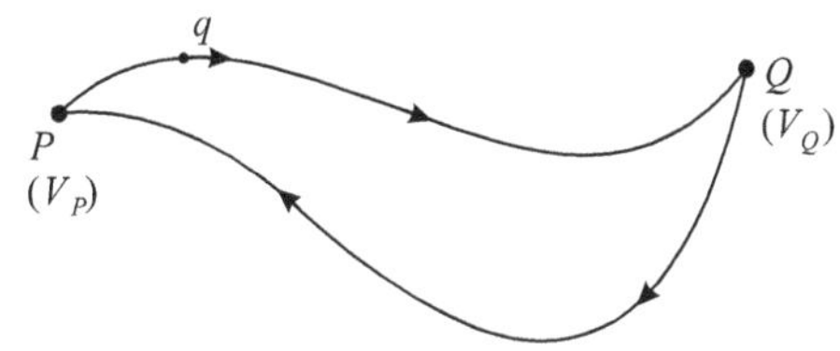

Being EF a conservative field work done does not depend on the path chosen.

Work done in moving 'q' from Q to P is

$$W' = q(V_P - V_Q)$$

$$W_{closed} = W + W' = 0$$ **Ans. (D)**

Sol. 11 Work done in moving a charge q from point P to Q is given as

$$W = q(V_Q - V_P)$$

$$\Rightarrow \quad W = [100\times(-1.6\times 10^{-19})]\times[-4-(10)]$$

$$\Rightarrow \quad W = 2.24\times 10^{-16}\ \text{J}$$ **Ans. (B)**

Sol. 12 KE_{gain} of Q when it moves from X to O is

$$KE_{gain} = Q(V_X - V_O)$$

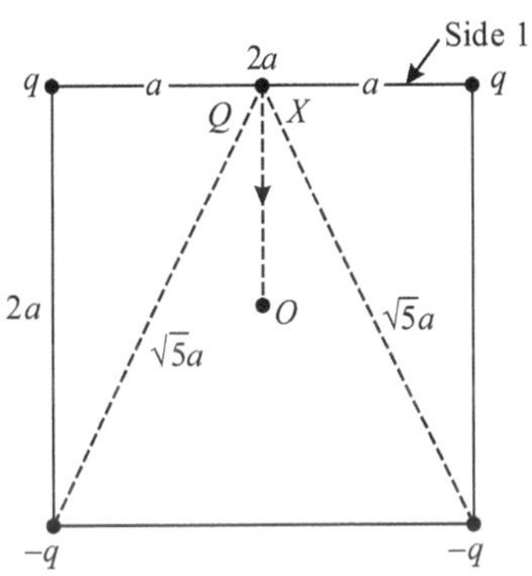

Potential at point X is

$$V_X = \frac{2Kq}{a} - \frac{2Kq}{\sqrt{5}a} = \frac{2Kq}{a}\left(1-\frac{1}{\sqrt{5}}\right)$$

Potential at point centre of square O is

$$V_O = O$$

$$\Rightarrow \quad KE_{gain} = Q(V_n - O) = QV_X$$

$$\Rightarrow \quad KE_{gain} = \frac{2KqQ}{a}\left(1-\frac{1}{\sqrt{5}}\right)$$

$$\Rightarrow \quad KE_{gain} = \frac{1}{4\pi\epsilon_0}\cdot\frac{2qQ}{a}\left(1-\frac{1}{\sqrt{5}}\right)$$ **Ans. (A)**

Sol. 13 Inside a uniformly charged sphere EF is given as

$$E_{in} = \frac{\rho r}{3 \in_0}$$

$\Rightarrow$ S-2 is true

Work done is moving 'q' from centre to surface of sphere is

$$W = \Delta U = q\,(V_S - V_C) = q\left[-\int_0^r E dr\right]$$

$$\Rightarrow \qquad W = -q\int_0^r \frac{\rho r}{3 \in_0}\cdot dr$$

$$\Rightarrow \qquad W = -\frac{q\rho}{3 \in_0}\frac{r^2}{2}$$

$\Rightarrow$ S-1 is false

Alternative : We can see that $\frac{q\rho}{3 \in_0}$ is not having dimensions of energy

$\Rightarrow$ S-1 is false **Ans. (D)**

Sol. 14 As the surface charge densities of the two shells are equal

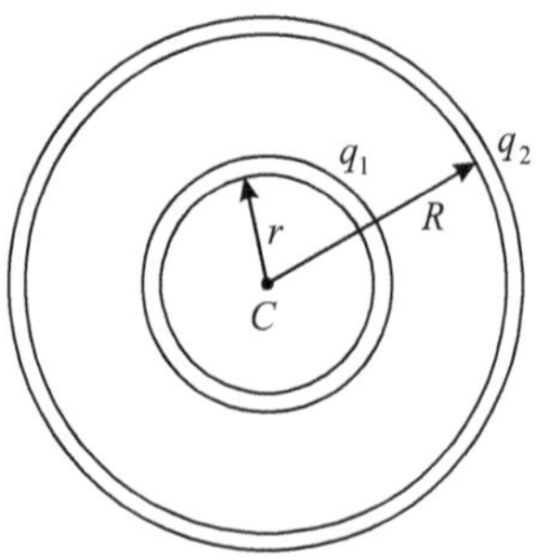

$$\Rightarrow \qquad \frac{q_1}{4\pi r^2} = \frac{q_2}{4\pi R^2} = \frac{Q}{4\pi(r^2 + R^2)}$$

Electric potential at common centre is given as

$$V_C = \frac{Kq_1}{r} + \frac{Kq_2}{R}$$

$$\Rightarrow \qquad V_C = \frac{1}{4\pi \in_0}\left[\frac{Qr^2}{r(r^2+R^2)} + \frac{QR^2}{R(r^2+R^2)}\right]$$

$$\Rightarrow \qquad V_C = \frac{(r+R)Q}{4\pi \in_0 (r^2 + R^2)}$$ **Ans. (C)**

Sol. 15

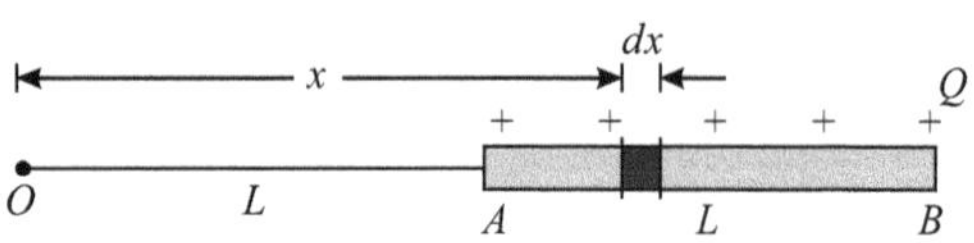

Charge on element

$$dq = \frac{Q}{L}dx$$

Potential at O due to dq is given as

$$dV = \frac{Kdq}{x} = \frac{KQ}{Lx}dx$$

Potential at O due to whole rod AB is

$$V = \int dV = \int_{x=L}^{x=2L}\frac{KQ}{Lx}dx$$

$$\Rightarrow \qquad V = \frac{KQ}{L}\left[\ln x\right]_L^{2L} = \frac{KQ}{L}[\ln 2L \ln L]$$

$$\Rightarrow \qquad V = \frac{KQ}{L}\ln(2) = \frac{Q}{4\pi \in_0 L}\ln(2)$$ **Ans. (D)**

Sol. 16 As the force between dipoles placed 1 to X-axis will be zero so we consider these along X-axis as shown.

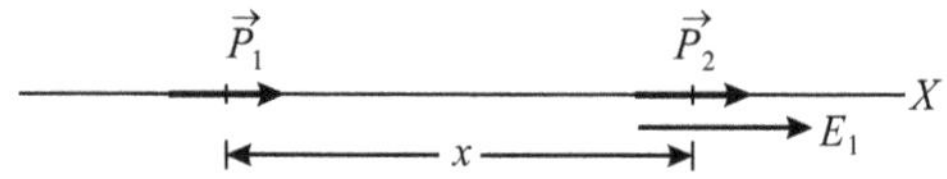

EF due to $\vec{P_1}$ at the location of $\vec{P_2}$ is given as

$$E_1 = \frac{2kP_1}{x^3}$$

In non uniform EF the force on a dipole placed along direction of EF is given as

$$\vec{F} = \vec{P_2}\cdot\frac{\vec{dE_1}}{dx}\hat{i}$$

$$\Rightarrow \qquad \vec{F} = P_2\cdot\frac{d}{dx}\left(\frac{2Kp_1}{x^3}\right)\hat{i} = 2Kp_1p_2\left(-\frac{3}{x^4}\right)\hat{i}$$

$$\Rightarrow \qquad \vec{F} = -\frac{6Kp_1p_2}{x^4}\cdot\hat{i}$$

$$\Rightarrow \qquad \vec{F} = -\frac{1}{4\pi \in_0}\cdot\frac{6p_1p_2}{x^4}\hat{i}$$ **Ans. (C)**

Sol. 17 S-1 is true

Work done in moving a test charge between any two point is

$$W = q_0(V_A - V_B) = O$$

if $V_A = V_B$ (us A and B are on an equipotential surface)

S-2 is false

As at every point of equipotential surface, electric lines are $\perp$ to this surface not mutually $\perp$ to each other. **Ans. (C)**

Sol. 18

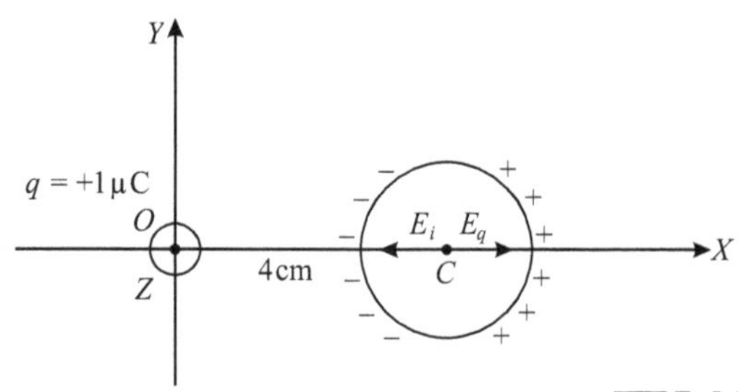

As net EF at C is zero (inside a conductor) we use

$$E_i = E_q = \frac{Kq}{x^2}$$

$$\Rightarrow \quad E_i = \frac{9\times10^9\times10^{-6}}{(4\times10^{-2})^2} = 5.625\times10^6\,\text{V/m}$$

Potential at C is only due to q (As $V_i = 0$) and it is given as

$$V = \frac{Kq}{x} = \frac{9\times10^9\times10^{-6}}{4\times10^{-2}} = 2.25\times10^5\,\text{V} \quad \textbf{Ans. (C)}$$

Sol. 19 As in the direction of EF always potential decreases

$\Rightarrow$ we can use $\quad V_X > V_Y > V_\infty$

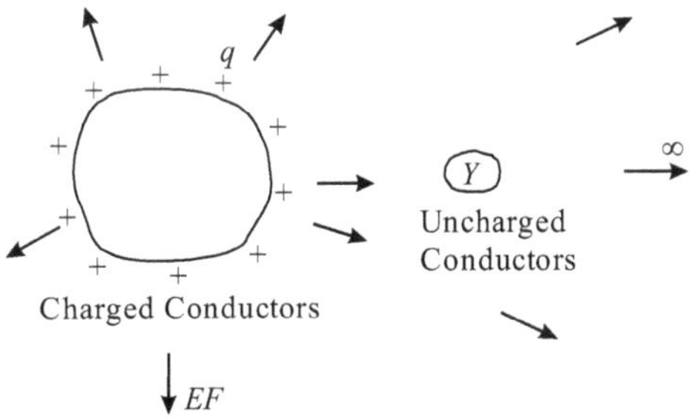

Because uncharged conductor will only affect its neighbourhood EF slightly due to induced charges on it **Ans. (A)**

Sol. 20 At equilibrium of charges, we have

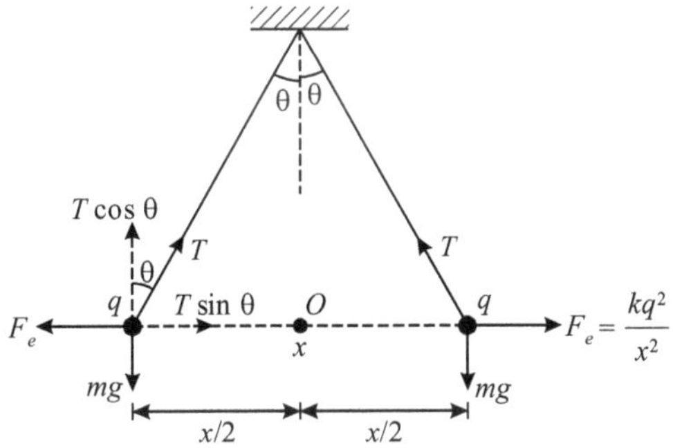

$$\text{T}\sin\theta = \frac{kq^2}{x^2} \qquad \dots(1)$$

$$\text{T}\cos\theta = mg \qquad \dots(2)$$

$$(1)(2) \Rightarrow \quad \tan\theta = \frac{kq^2}{mgx^2}$$

$$\Rightarrow \quad \frac{q}{x} = \sqrt{\frac{mg\tan\theta}{k}}$$

Electric potential at point O is $V_0 = \dfrac{kq}{x/2}\times 2 = \dfrac{4kq}{x}$

$$= 4k\sqrt{\frac{mg\tan\theta}{k}} = 4\sqrt{mgk\tan\theta} \quad \textbf{Ans. (D)}$$

Sol. 21 We use

$$V_A - V_O = -\int_0^2 E\,dx$$

here $\quad V_A < V_O$ So $V_A - V_O \rightarrow -$ve

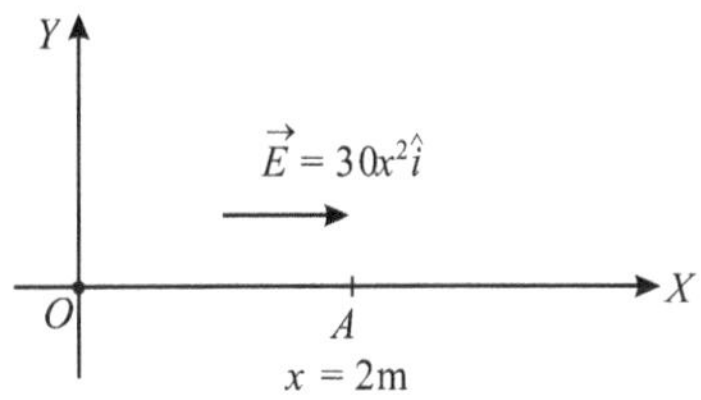

As along direction of EF potential decreases

$$\Rightarrow \quad V_A - V_0 = -\int_0^2 30x^2\,dx$$

$$\Rightarrow \quad V_A - V_0 = -30\left[\frac{x^3}{3}\right]_0^2 = -10[8-0]$$

$$\Rightarrow \quad V_A - V_0 = -80\,\text{V} = -80\,\text{J/C} \qquad \textbf{Ans. (C)}$$

Sol. 22 Potential on surface of a solid uniformly charged sphere is

$$V_O = \frac{KQ}{R}$$

If at $r = R_1$, $V = \dfrac{3V_O}{2}$ $(> V_0 \Rightarrow r < R)$

$$\Rightarrow \quad \frac{5V_O}{4} = \frac{V_O}{2}\left(3 - \frac{r^2}{R^2}\right)$$

$$\Rightarrow \quad 5 = 6 - \frac{2r^2}{R^2} \Rightarrow r = \frac{R}{\sqrt{2}} = R_2$$

If at $r = R_3$, $V = \dfrac{3V_O}{4}$ $(< V_O \Rightarrow r > R)$

$$\Rightarrow \quad \frac{V_O}{4} = \frac{KQ}{r} \Rightarrow \frac{KQ}{4R} = \frac{KQ}{r} \Rightarrow r = 4R = R_4 \quad \text{Ans. (D)}$$

Here $R_4 - R_3 = 4R - \dfrac{4R}{3} = \dfrac{8R}{3} > R_2$ (As $R_2 = \dfrac{R}{\sqrt{2}}$) **Ans. (C)**

Both are correct.

Sol. 23 Along the direction of EF, potential decreases and in uniform EF, potential difference between two points is given as

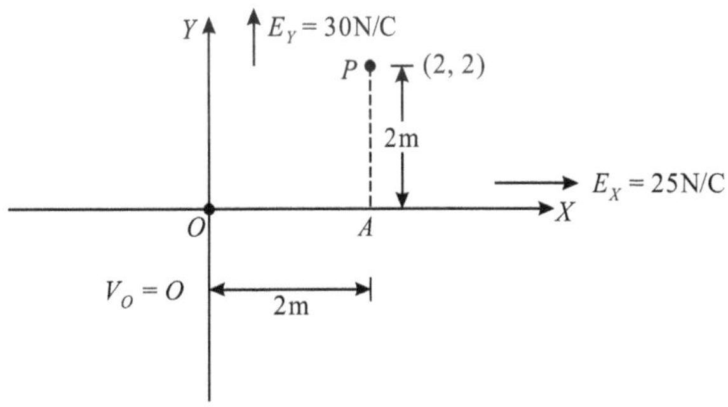

$$V_X - V_Y = Ed$$

In moving from O to P we can write

$$V_O - V_A = 25\times2$$

$$V_A - V_P = 30\times2$$

$\Rightarrow \quad V_O - V_P = 110$

$\Rightarrow \quad V_P = -110\text{ V or } -110\text{ J/C}$ **Ans. (A)**

Sol. 24 For $|Z| \le |m$ $\quad \vec{E} = -\frac{dV}{dZ}\hat{k} = 10|Z|\hat{k}$

$-1\text{m} < Z < 1\text{m}$

and for $|Z| \ge 1\text{m}$ $\quad \vec{E} = -\frac{dV}{dZ}\hat{k} = \pm 10\hat{k}$

$Z > 1\text{m}$ and $Z < -1\text{m}$ constant EF

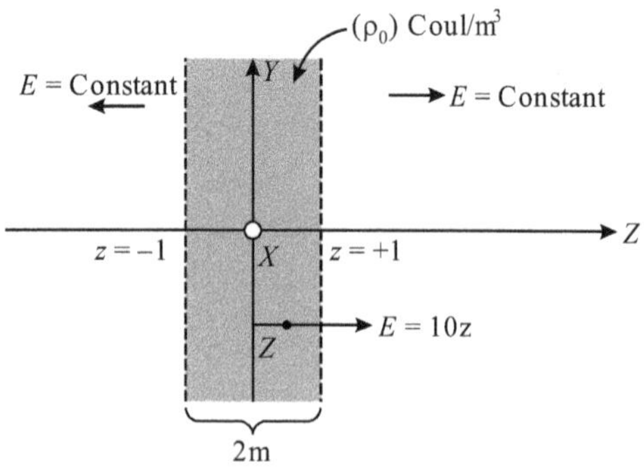

By analyzing the EF we can state that it is produced by a uniform charge distribution in a thick sheet of width 2m as shown.

Due to a thick volume charged sheet, inside at a distance Z from central plane the EF is given as

$$E_{\text{in}} = \frac{\rho Z}{\epsilon_O} = 10z$$

$\Rightarrow \quad \rho_0 = 10\,\epsilon_0 \quad \text{for } |z| \le 1\text{m}$

$\Rightarrow \quad \rho_0 = 0 \quad \text{for } |z| \ge 1\text{m}$ **Ans. (B)**

Sol. 25 $r_2 - r_1 = r_1 - r_0 = r_N - r_{N-1}$

$\Delta V \quad \Delta V \quad \Delta V$ Potential differences

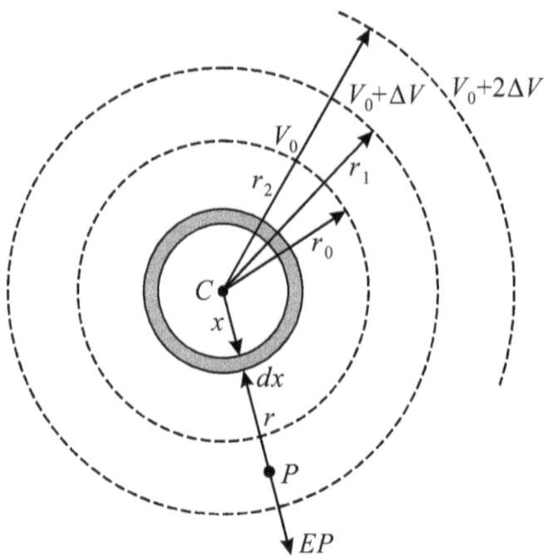

Only in uniform EF the separation between equipotential surface having same ΔV is constant.

To calculate EF at a distance r from centre of charge distribution then it is given by

$$E_P = \frac{kq}{r^2}$$

Where $q \rightarrow$ charge enclosed within sphere of rad $r = 0$ to $r = r$. Charge in sphere of radius r is given as

$$q = \int dq = \int_0^r \rho(x) 4\pi x^2 dx$$

$\Rightarrow$ EF at point P is

$$E_P = \frac{K\int_0^r \rho(x)\cdot 4\pi x^2 dx}{r^2} = C\text{ (constant)}$$

Differentiating we get

$$\frac{d}{dr}\left[k\int_0^r \rho(x)\cdot 4\pi x^2 dx\right] = 2\text{Cr}$$

$$\frac{1}{4\pi\epsilon_0}\cdot[\rho(r)4\pi r^2 - 0] = 2\,\text{Cr}$$

$\Rightarrow \quad \rho(r) = \frac{2C\epsilon_0}{r} \quad \Rightarrow \quad \rho(r) \propto \frac{1}{r}$ **Ans. (C)**

Sol. 26 Here $\quad V_P > V_M > V_B$

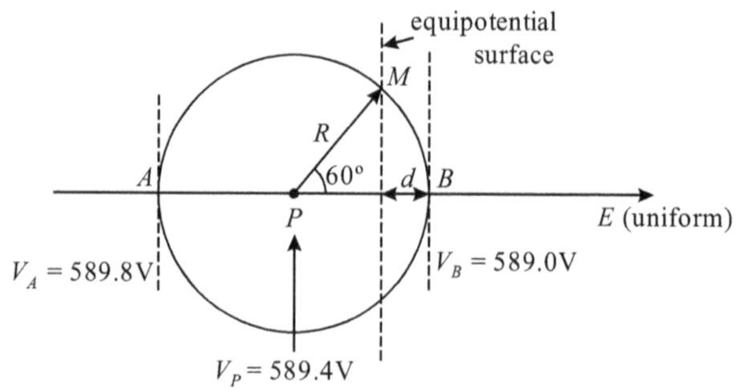

There exist only option (B) which is 589.2 V satisfying the given condition. **Ans. (B)**

Alternative : In given figure shown we use here

$$d = R(1 - \cos 60^\circ) = \frac{R}{2}$$

We use from the given values, we have

$$V_A - V_B = E(2R) = 0.8$$

$\Rightarrow \quad ER = 0.4$

$$V_M - V_B = Ed = \frac{ER}{2}$$

$\Rightarrow \quad V_M = \frac{ER}{2} + V_B$

$\Rightarrow \quad V_M = 0.2 + 589.0 = 589.2\text{ V}$ **Ans. (B)**

Sol. 27 Charges on shells A, B and C are

$$q_A = +\sigma \times 4\pi a^2$$

$$q_B = -\sigma \times 4\pi b^2$$

$$q_C = -\sigma \times 4\pi e^2$$

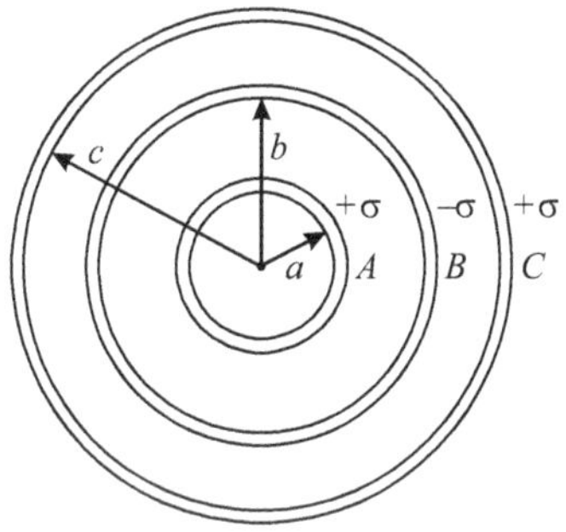

Potential of shell B due to all the shell charges is given as

$$V_B = V_{q_A} + V_{q_B} + V_{q_C}$$

$$\Rightarrow \quad V_B = \frac{Kq_A}{b} + \frac{Kq_B}{b} + \frac{Kq_C}{C}$$

$$\Rightarrow \quad V_B = \frac{1}{4\pi\epsilon_0}\left(\frac{\sigma 4\pi a^2}{b} - \frac{\sigma 4\pi b^2}{b} + \frac{\sigma 4\pi c^2}{c}\right)$$

$$\Rightarrow \quad V_B = \frac{\sigma}{\epsilon_0}\left(\frac{a^2}{b} - b + c\right) = \frac{\sigma}{\epsilon_0}\left(\frac{a^2 - b^2}{b} + c\right)$$

Ans. (B)

Sol. 28

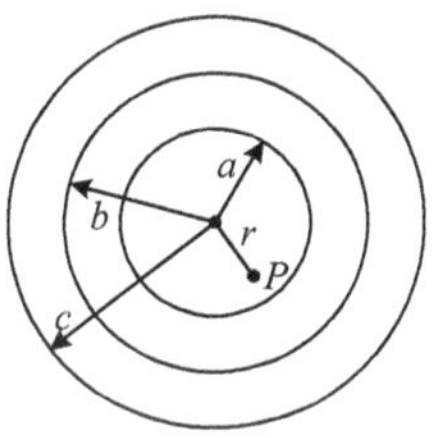

Potential at point P, due to the three shells is given as

$$V = \frac{kQ_a}{a} + \frac{kQ_b}{b} + \frac{kQ_c}{c}$$

As $Q_a : Q_b : Q_c :: a^2 : b^2 : c^2$ [since $\sigma_a = \sigma_b = \sigma_c$]

$$\Rightarrow \quad Q_a = \left[\frac{a^2}{a^2 + b^2 + c^2}\right]Q$$

$$\Rightarrow \quad Q_b = \left[\frac{b^2}{a^2 + b^2 + c^2}\right]Q$$

$$\Rightarrow \quad Q_c = \left[\frac{c^2}{a^2 + b^2 + c^2}\right]Q$$

$$\Rightarrow \quad V = \frac{Q}{4\pi\epsilon_0}\left[\frac{(a+b+c)}{a^2 + b^2 + c^2}\right]$$

Ans. (B)

Sol. 29 At a distance x on right of dipole B, we use

$$V = \frac{4qa}{(R+x)^2} = \frac{2qa}{(x^2)}$$

$$\Rightarrow \quad \sqrt{2}x = R + x$$

$$\Rightarrow \quad x = \frac{R}{\sqrt{2} - 1}$$

Distance of this point from A is

$$x_A = \frac{R}{\sqrt{2} - 1} + R = \frac{\sqrt{2}R}{\sqrt{2} - 1}$$

Ans. (C)

Sol. 30 Electric field at equatorial points of dipole is given as

$$E = \frac{Kp}{r^3}$$

At point P, force on Q is

$$F = \frac{Kp}{r^3}Q \quad \ldots (1)$$

At point P', force on Q is

$$F' = \frac{KpQ}{(r/3)^3} = 27F$$

Ans. (D)

Sol. 31 Figure below shows the given charge configuration

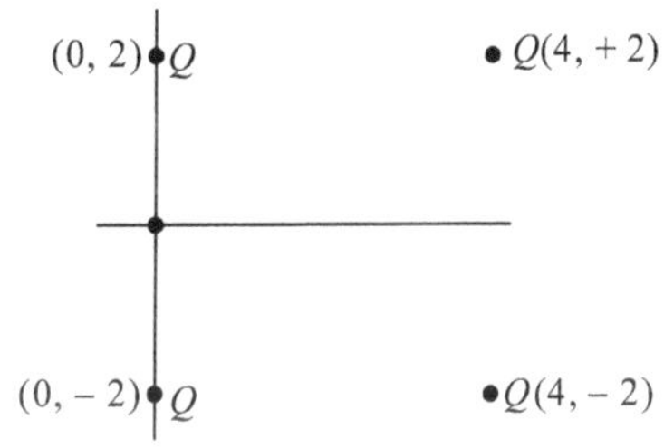

Potential at origin is given as

$$V_0 = \frac{KQ}{2} + \frac{KQ}{2} + \frac{KQ}{\sqrt{20}} + \frac{KQ}{\sqrt{20}}$$

$$\Rightarrow \quad V_0 = KQ\left(1 + \frac{1}{\sqrt{5}}\right) = \frac{Q^2}{4\pi\varepsilon_0}\left(1 + \frac{1}{\sqrt{5}}\right)$$

Work required to put a fifth charge Q at origin is given as

$$W = QV = \frac{Q^2}{4\pi\varepsilon_0}\left(1 + \frac{1}{\sqrt{5}}\right)$$

Ans. (B)

Sol. 32 Total electrostatic energy of system is given as

$$U = \left[\frac{q^2}{a} + \frac{Qq}{a} + \frac{Qq}{a\sqrt{2}}\right] = 0$$

$$\Rightarrow \quad q = -Q\left[1 + \frac{1}{\sqrt{2}}\right]$$

$$\Rightarrow \quad Q = \frac{-q\sqrt{2}}{\sqrt{2} + 1}$$

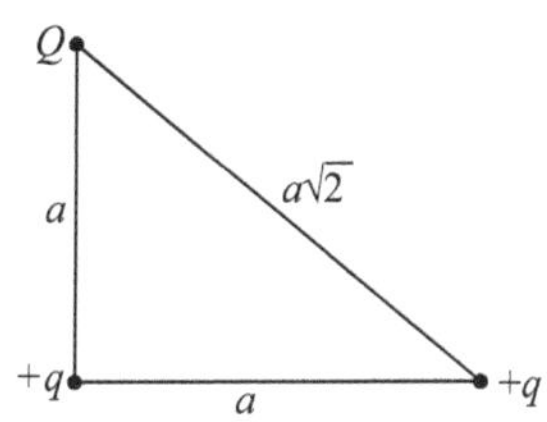

Ans. (A)

Sol. 33 Graph in figure shows that from centre to surface of sphere the physical quantity is constant than at outer points it decreases with distance so it must be electric potential due to a uniformly charged conducting shell. **Ans. (B)**

Sol. 34 Potential energy of a dipole in electric field given as

$$U = -\vec{P}\cdot\vec{E}$$

$\Rightarrow$ $U = -PE\cos\theta$

$\Rightarrow$ $U = -(10^{-29})(10^3)\cos 45^\circ$

$\Rightarrow$ $U = -0.707 \times 10^{-26}$ J

$\Rightarrow$ $U = -7 \times 10^{-27}$ J **Ans. (B)**

Sol. 35

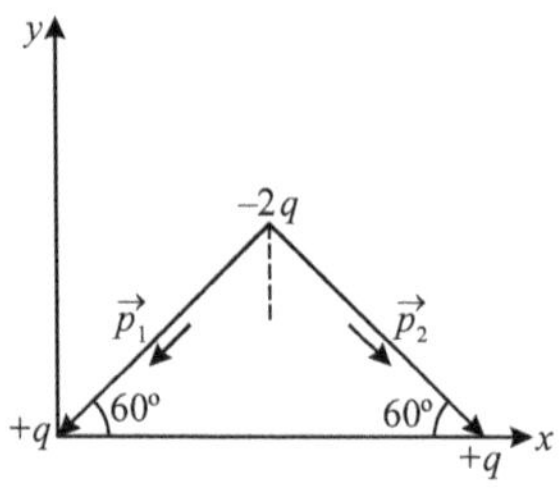

Given system is equivalent to two dipoles as shown above where dipole moments are given as

$$\vec{p}_1 = ql(-\cos 60^\circ\hat{i} - \sin 60^\circ\hat{j}) = \frac{ql}{2}(\hat{i}+\sqrt{3}\hat{j})$$

$$\vec{p}_2 = ql(\cos 60^\circ\hat{i} - \sin 60^\circ\hat{j}) = \frac{ql}{2}(\hat{i}-\sqrt{3}\hat{j})$$

Net dipole moment of system is

$$\vec{p}_{net} = \vec{p}_1 + \vec{p}_2 = -ql\sqrt{3}\hat{j}$$ **Ans. (C)**

Sol. 36 At any instant 't' total energy of charge distribution is constant thus by energy conservation, we use

$$\frac{1}{2}mv^2 + \frac{KQ^2}{2R} = 0 + \frac{KQ^2}{2R_0}$$

$$\Rightarrow \quad \frac{1}{2}mV^2 = \frac{KQ^2}{2R_0} - \frac{KQ^2}{2R}$$

$$\Rightarrow \quad v = \sqrt{\frac{2}{m}\frac{KQ^2}{2}\cdot\left(\frac{1}{R_0} - \frac{1}{R}\right)}$$

$$\Rightarrow \quad v = \sqrt{\frac{KQ^2}{m}\left(\frac{1}{R_0} - \frac{1}{R}\right)}$$

From above expression we can see that with increase in R, v increases and $\frac{dv}{dR}$ decreases. **Ans. (A)**

Sol. 37 As given in the first case, outer shall is given charge $+Q$

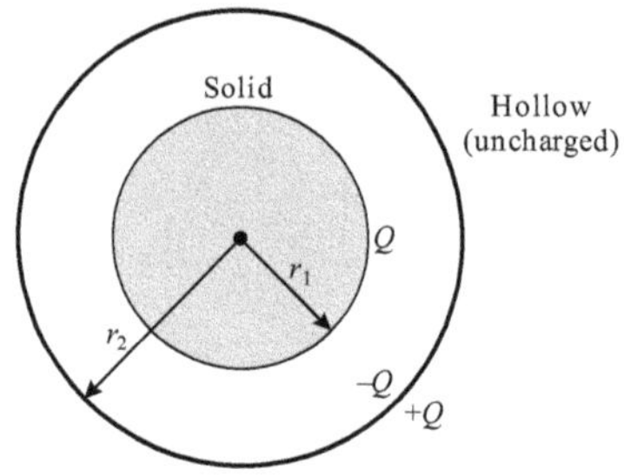

The potential difference of inner and outer spheres is given as

$$V_{in} - V_{out} = \left(\frac{kQ}{r_1}\right) - \left(\frac{kQ}{r_2}\right)$$

$$\Rightarrow \quad V_{in} - V_{out} = kQ\left(\frac{1}{r_1} - \frac{1}{r_2}\right) = V$$

In the second case, outer shall is given charge $-4Q$

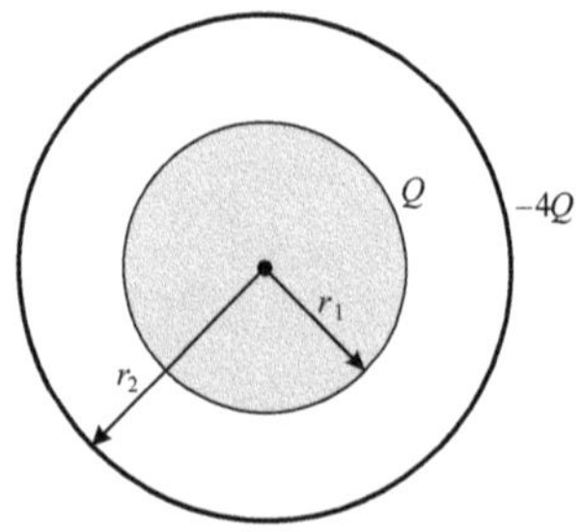

New potential difference is given as

$$V_{in} - V_{out} = \left(\frac{kQ}{r_1} - \frac{4kQ}{r_2}\right) - \left(\frac{kQ}{r_2} - \frac{4kQ}{r_2}\right)$$

$$= \frac{kQ}{r_1} - \frac{kQ}{r_2}$$

$$= kQ\left(\frac{1}{r_1} - \frac{1}{r_2}\right) = V$$ **Ans. (A)**

Sol. 38

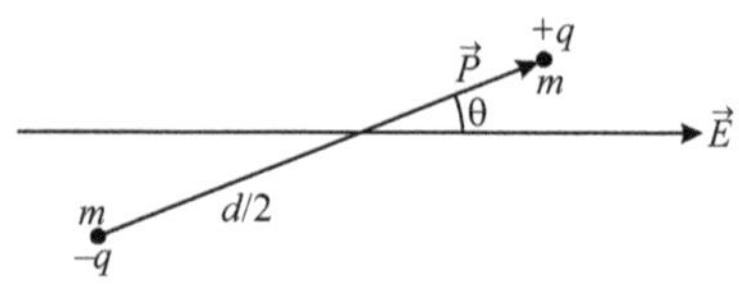

Moment of inertia of charges

$$I = 2 \times m\left(\frac{d}{2}\right)^2 = \frac{md^2}{2}$$

Torque on dipole is

$$\tau = PE\sin\theta = (qd)E\sin\theta = I\alpha$$

$$\Rightarrow \quad \alpha = -\frac{qEd}{I}\cdot\theta \quad \text{[As for small } \theta, \sin\theta \simeq \theta]$$

$-$ sign denotes θ to be in opposite direction to α

$$\Rightarrow \quad \alpha = -\frac{2qE}{md}\cdot\theta$$

Comparing with $\alpha = -\omega^2\theta$ we get $\omega = \sqrt{\frac{2qE}{md}}$ **Ans. (C)**

Sol. 39 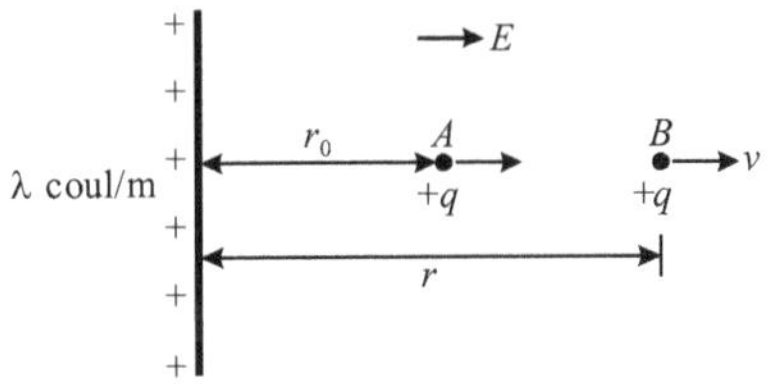

By work energy theorem, we use

$$\frac{1}{2}mv^2 = -q(V_A - V_B) \qquad \dots(1)$$

$$\Rightarrow \quad V_A - V_B = \int_{r_0}^{r} E\cdot dr = \int_{r_0}^{r} \frac{\lambda}{2\pi\epsilon_0 r}dr = \frac{\lambda}{2\pi\epsilon_0}[\ln r]_{r_0}^{r}$$

$$\Rightarrow \quad V_A - V_B = \frac{\lambda}{2\pi\epsilon_0}\ln\left(\frac{r}{r_0}\right)$$

From equation-(1), we use

$$\frac{1}{2}mv^2 = q\left[\frac{\lambda}{2\pi\epsilon_0}\ln\left(\frac{r}{r_0}\right)\right]$$

$$\Rightarrow \quad v \propto \sqrt{\ln\left(\frac{r}{r_0}\right)}$$

Ans. (D)

Sol. 40 Given electric field is

$$\vec{E} = (20x+10)\hat{i}$$

Potential difference is given as

$$V_1 - V_2 = -\int_{-5}^{1}(20x+10)dx$$

$$\Rightarrow \quad V_1 - V_2 = -(10x^2+10x)_{-5}^{1}$$

$$\Rightarrow \quad V_1 - V_2 = 10(25-5-1-1)$$

$$\Rightarrow \quad V_1 - V_2 = 180\,\text{V}$$

Ans. (C)

Sol. 41 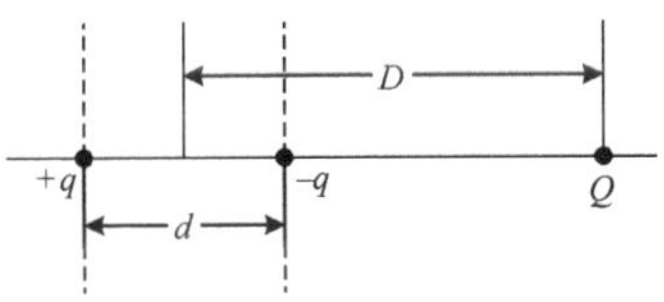

Total potential energy of this system is given as

$$U_{total} = U_{self\ of\ dipole} + U_{interaction}$$

$$\Rightarrow \quad U_{total} = -\frac{kq^2}{d} - \left(\frac{kQ}{D^2}\right)qd$$

$$\Rightarrow \quad U_{total} = -k\left[\frac{q^2}{d} + \frac{qQd}{D^2}\right]$$

Ans. (D)

Sol. 42 By conservation of energy, we have

$$U_i + K_i = U_f + K_f$$

$$\Rightarrow \quad \frac{Kq^2}{\sqrt{16a^2+9a^2}} + \frac{1}{2}mv^2 = \frac{Kq^2}{3a}$$

$$\Rightarrow \quad \frac{1}{2}mv^2 = \frac{Kq^2}{a}\left(\frac{1}{3}-\frac{1}{5}\right) = \frac{2Kq^2}{15a}$$

$$\Rightarrow \quad v = \sqrt{\frac{4Kq^2}{15ma}}$$

Ans. (B)

Sol. 43 By energy conservation, we have

$$\frac{Kq_1q_2}{r_1} - \frac{Kq_1q_2}{r_2} = \frac{1}{2}mv^2$$

$$\Rightarrow \quad \frac{(9\times10^9)\times10^{-12}}{10^{-3}} - \frac{(9\times10^9)\times10^{-12}}{9\times10^{-3}} = \frac{1}{2}\times(4\times10^{-6})v^2$$

$$\Rightarrow \quad v^2 = 4\times10^6$$

$$\Rightarrow \quad v = 2\times10^3\ \text{m/s}$$

Ans. (A)

Sol. 44 Due to inside dipole charge induced on inner surface will be non uniform as shown below

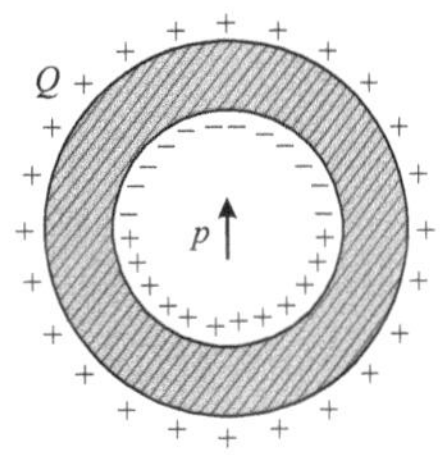

In this case as total charge of dipole is 0, so charge induced on outside surface is 0 due to it. Due to non uniform electric field of dipole, the charge induced on inner surface is non zero and non uniform so, for any observer outside the shell, the resultant electric field is due to Q which is uniformly distributed on outer surface only and it is given as

$$E = \frac{KQ}{r^2} = \frac{1}{4\pi\epsilon_0}\cdot\frac{Q}{r^2}$$

Ans. (A)

Sol. 45 At point M in figure as both charges of dipole are at same distance potential at M is

$$V = 0$$

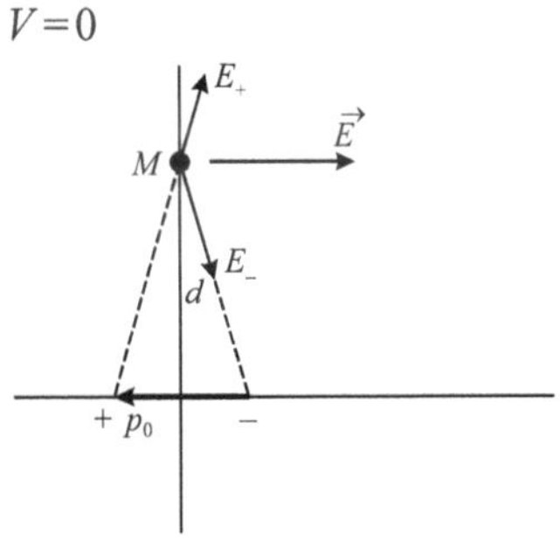

As M is an equatorial point of dipole electric field at M is given as

$$E = \frac{Kp}{d^3}$$

Vectorially, it is written as

$$\vec{E} = -\frac{1}{4\pi\epsilon_0}\frac{p_0}{d^3}$$

Ans. (D)

16.2 Capacitance

Sol. 1 Capacitance of a spherical conductor is

$$C = 4\pi\epsilon_0 R$$

$$\Rightarrow \quad C = \frac{1}{9\times10^9}\times 1$$

$$= 1.1\times10^{-10}\,\text{F}$$ **Ans. (A)**

Sol. 2 For n identical capacitors (C) parallel combination equivalent capacitance is

$$C_{eq} = C + C + C + \ldots n \text{ times} = nC$$

Across a voltage V the energy stored is

$$U = \frac{1}{2}C_{eq}V^2 = \frac{1}{2}nCV^2$$ **Ans. (B)**

Sol. 3 Work done in charging a capacitor of capacitance C with a charge q is

$$W = \frac{q^2}{2C}$$

$$= \frac{1}{2}\left(\frac{8\times10^{-18}}{100\times10^{-6}}\right)^2$$

$$= 32\times10^{-32}\,\text{J}$$

Electrostatic energy stored in capacitor. **Ans. (D)**

Sol. 4

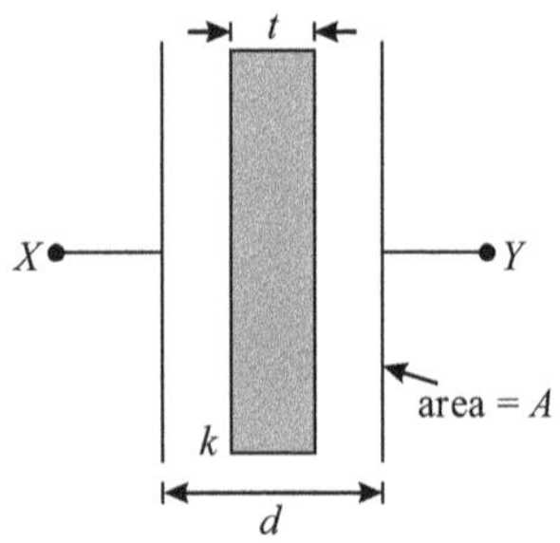

$$C_{XY} = \frac{\epsilon_0 A}{\frac{t}{k} + (d - t)}$$

If $t \to 0 \Rightarrow C_{XY} = \frac{\epsilon_0 A}{d} = C$ **Ans. (B)**

Sol. 5

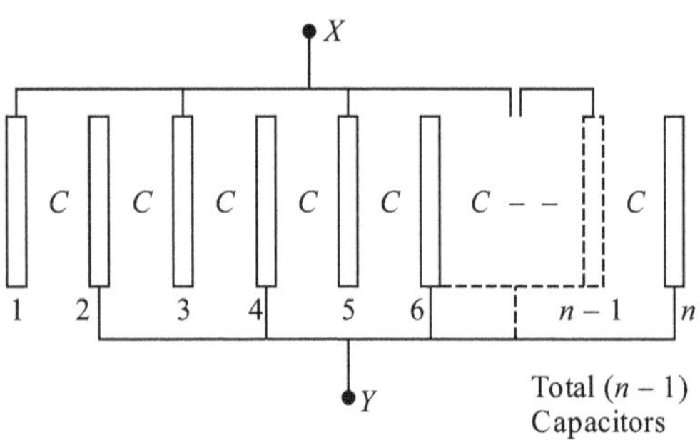

In above system of plates one plate of each capacitor is connected to Y and its other plate is connected to X. Thus all $(n - 1)$ capacitors formed are in parallel combination.

$$\Rightarrow \quad C_{XY} = C + C + \ldots (n-1) \text{ times}$$

$$= (n-1)C$$ **Ans. (A)**

Sol. 6

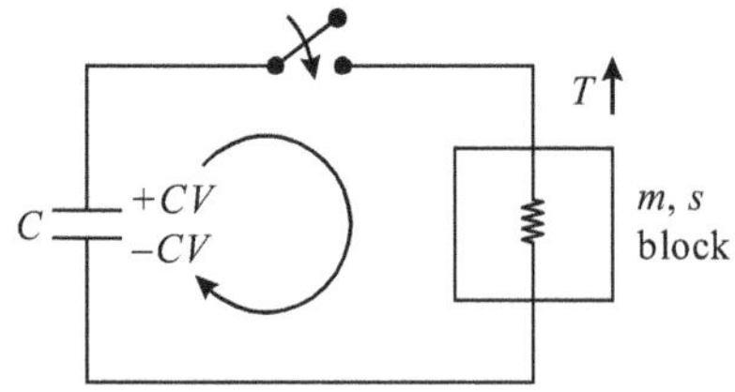

On discharging the capacitor, heat supplied to block is

$$H = U_{cap} = \frac{1}{2}CV^2$$

If temperature of block is raised by ΔT

$$\Rightarrow \quad H = ms\Delta T = \frac{1}{2}CV^2$$

$$\Rightarrow \quad V = \sqrt{\frac{2ms\Delta T}{C}}$$ **Ans. (D)**

Sol. 7

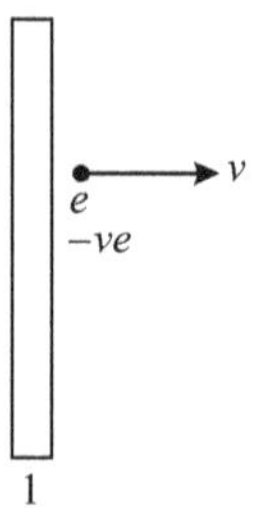

Work done in travelling from plate 1 to 2 on electron by the EF of plates is given as

$$W = e\times(V_2 - V_2) = \Delta K = \frac{1}{2}mv^2$$

$$\Rightarrow \quad W = (-1.6\times10^{-19})\times(-20)$$

$$\Rightarrow \quad W = \frac{1}{2}\times 9.1\times10^{-31}\times v^2$$

$$\Rightarrow \quad v = \sqrt{\frac{2\times1.6\times10^{-19}\times20}{9.1\times10^{-31}}} = 2.65\times10^6\,\text{m/s}$$ **Ans. (A)**

Sol. 8

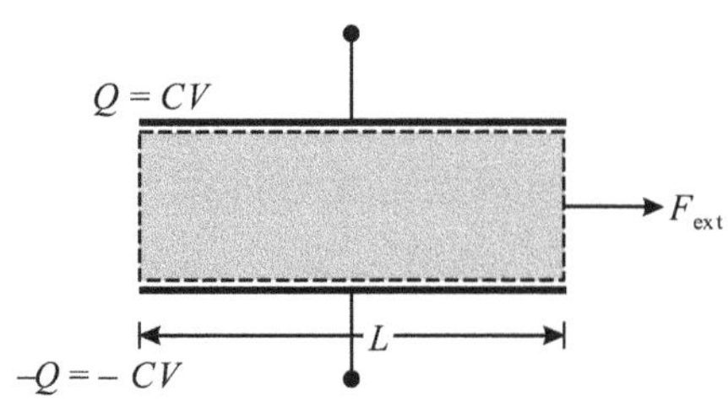

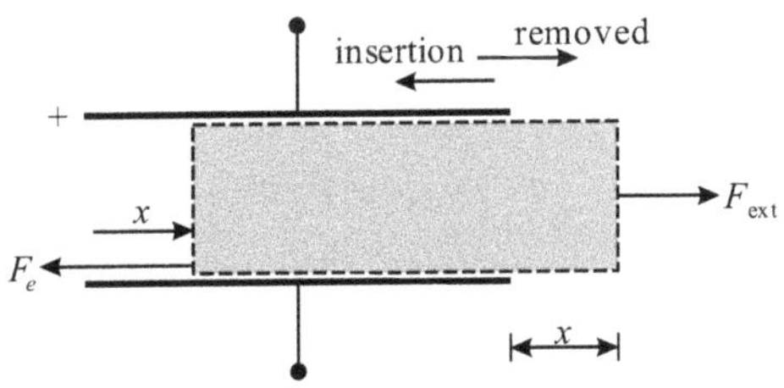

Work done in removing the dielectric slab slowly from capacitor is

$$\text{removal} \rightarrow W_1 = \int_0^l F_{ext} \cdot dx$$

$$\text{insertion} \rightarrow W_2 = \int_0^l F_{ext} \cdot dx$$

For slow insertion or removal of slab $F_{ext} = F_e$

Net work done is $W_1 + W_2 = 0$ **Ans. (D)**

Sol. 9

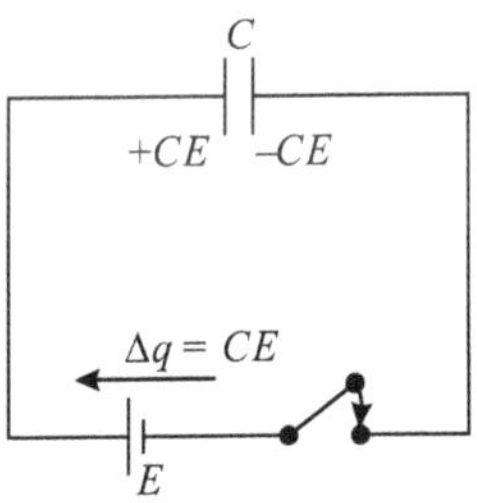

After capacitor is fully charged (when potential difference across plates is equal to that of battery EMF) charge on its plates is equal to

$$q_f = CE \text{ and } q_i = 0$$

Energy stored in capacitor is

$$U = \frac{1}{2} CE^2$$

Work done by battery $W = \Delta qE = CE^2$

$$\Rightarrow \quad \frac{U}{W} = \frac{\frac{1}{2}CE^2}{CE^2} = \frac{1}{2}$$ **Ans. (D)**

Sol. 10

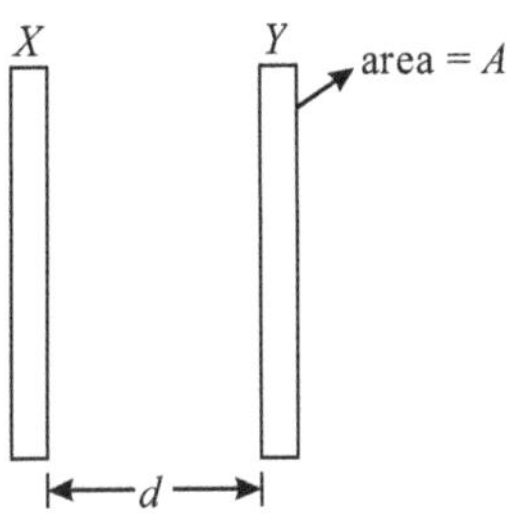

$$C = \frac{\epsilon_0 A}{d} = 9pF \Rightarrow \epsilon_0 A = 9d$$

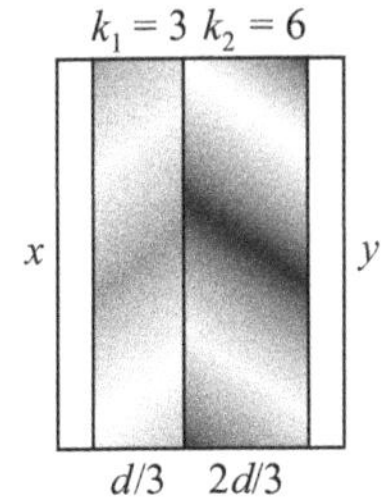

Capacitance of capacitor with multiple dielectric field slabs parallel to plates is given as

$$C' = \frac{\epsilon_0 A}{\frac{t_1}{k_1} + \frac{t_2}{k_2}}$$

$$\Rightarrow \quad C' = \frac{9d}{\frac{d/3}{3} + \frac{2d/3}{6}}$$

$$\Rightarrow \quad C' = \frac{9}{\frac{1}{9} + \frac{1}{9}} = \frac{81}{2} = 40.5 \text{ pF}$$ **Ans. (C)**

Sol. 11 (a) $V_L = \frac{2Q}{3C}$

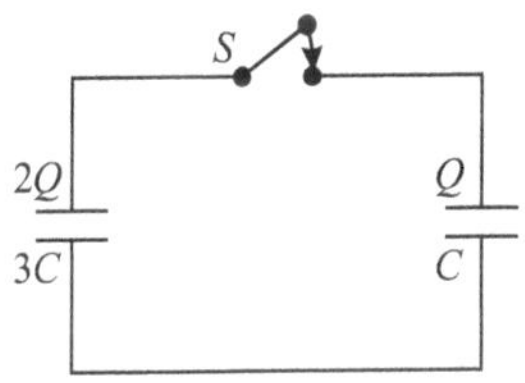

$V_R = \frac{Q}{C}$

As $V_R > V_L$

then on closing S flows from R to L

(b) $V_L = \frac{Q}{C}$

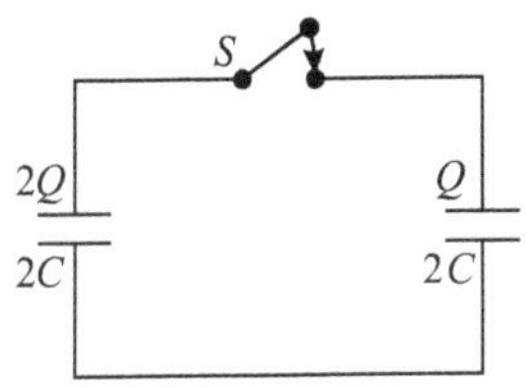

$V_R = \frac{Q}{2C}$

As $V_L > V_R$

then on closing S charge flows from L to R. **Ans. (C)**

Sol. 12 Case-I : n_1 capacitors, each C_1 when connected in series then its equivalent capacitance is given as

$$C_{eq_1} = \frac{C_1}{n_1}$$

$\Rightarrow$ Energy stored with potential difference = 4V is given as

$$U_1 = \frac{1}{2}\left(\frac{C_1}{n_1}\right)(4V)^2 \quad \dots(1)$$

Case-II : n_1 capacitors, each C_2 when connected in parallel then its equation capacitance is given as

$$C_{eq_2} = n_2 C_2$$

$\Rightarrow$ Energy stored with p.d. = V is given as

$$U_2 = \frac{1}{2}(n_2 C_2)V^2 \quad \dots(2)$$

Given that $U_1 = U_2$

$$\Rightarrow \quad \frac{1}{2}\left(\frac{C_1}{n_1}\right)(4V)^2 = \frac{1}{2}(n_2 C_2)V^2$$

$$\Rightarrow \quad 16\frac{C_1}{n_1} = n_2 C_2$$

$$\Rightarrow \quad C_2 = \frac{16C_1}{n_1 n_2}$$

Ans. (D)

Sol. 13 Statement-1 : It is true because the radius of a sphere required for making 1F cap is gives as

$$C = 4\pi\epsilon_0 R$$

$$1 = 4\pi\epsilon_0 R$$

$$\Rightarrow \quad R = \frac{1}{4\pi\epsilon_0} = 9\times 10^9\text{m}$$

Statement-2 : As $6.4\times 10^6\text{m} < 9\times 10^9\text{m}$

$\Rightarrow$ It is not possible with earth

$$C_{\text{earth}} < 1\text{F}$$

Ans. (D)

Sol. 14 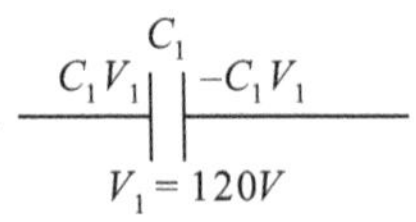

$q_1 = C_1V_1$

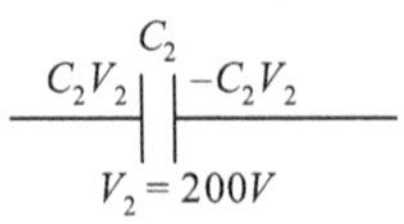

$q_2 = C_2V_2$

If the capacitors are joined in parallel with same polarity plates together then

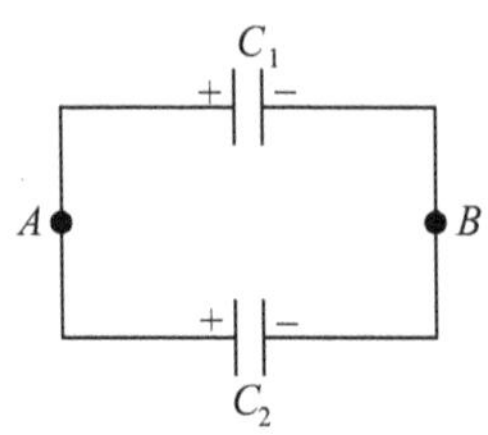

$$V_{AB} = \frac{q_1 + q_2}{C_1 + C_2} \neq 0$$

If the capacitors are joined in parallel with opposite polarity plates together then

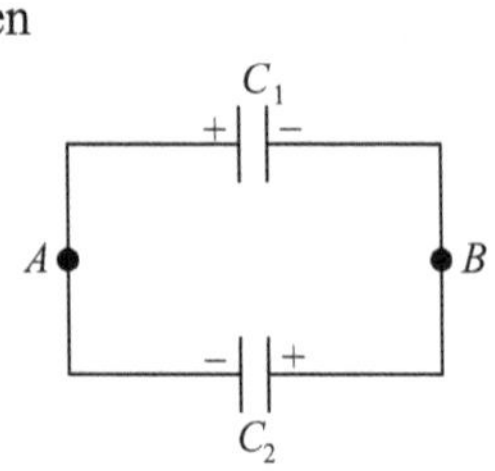

$$V_{AB} = \frac{q_1 + q_2}{C_1 + C_2} \neq 0$$

$$\Rightarrow \quad q_1 = q_2 \quad \Rightarrow \quad C_1V_1 = C_2V_2$$

$$\Rightarrow \quad 120\,C_1 = 200\,C_2$$

$$\Rightarrow \quad 3C_1 = 5C_2$$

Ans. (B)

Sol. 15 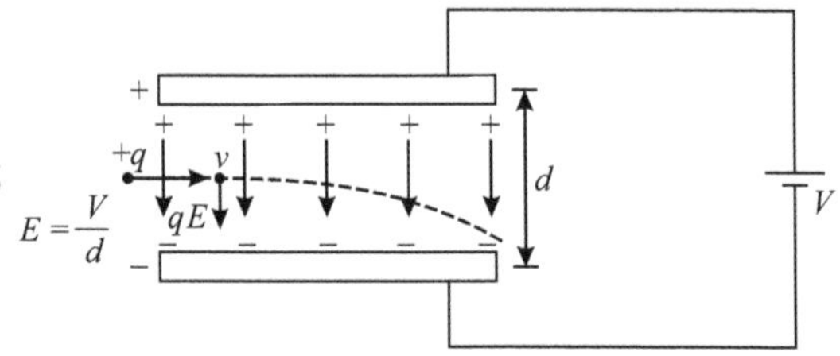

Acceleration of charged particle due to electric force is

$$a = \frac{F}{m} = \frac{qE}{m} \rightarrow \text{constant}$$

at constant acceleration, particle follows projectile motion with

$$g_{eff} = \frac{qE}{m}$$

$\Rightarrow$ path $\rightarrow$ parabolic trajectory

Ans. (C)

Sol. 16 When a capacitor is charged to a potential difference V volt then we have

$$q = CV$$

If V starts varying then q changes and current in connecting wires is given as

$$i = \frac{dq}{dt} = C\frac{dV}{dt}$$

$$\Rightarrow \quad \frac{dq}{dt} = \frac{i}{C} = \frac{2}{1\times 10^{-6}} = 2\times 10^6\text{ V/s}$$

Ans. (C)

Sol. 17 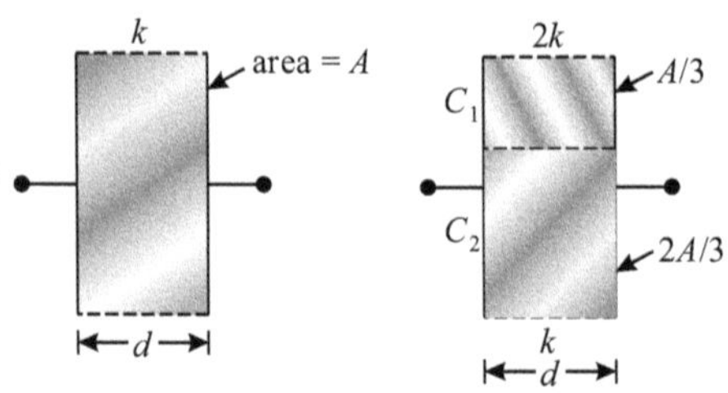

$$C_0 = \frac{K\epsilon_0 A}{d} \qquad \ldots(1)$$

$$C = C_1 + C_2$$

$$\Rightarrow \quad C = \frac{(2k)\epsilon_0 A/3}{d} + \frac{k\epsilon_0 (2A/3)}{d}$$

$$= \frac{4k\epsilon_0 A}{3d} \qquad \ldots(2)$$

From (1) and (2) $\Rightarrow \dfrac{C}{C_0} = \dfrac{4}{3}$

Ans. (B)

Sol. 18 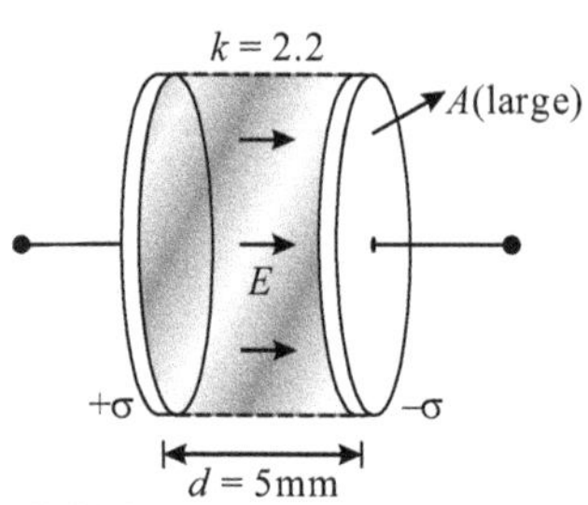

Electric field between plates of a capacitor inside dielectric is given as

$$E = \frac{\sigma}{k\in_0} = 3\times10^4 \text{ V/m}$$

$\Rightarrow \quad \sigma = 3\times10^4\times k\in_0$

$\Rightarrow \quad \sigma = 3\times10^4\times2.2\times8.89\times10^{-12}$

$\Rightarrow \quad \sigma = 5.87\times10^{-7}\text{ C/m}^2$ **Ans. (A)**

Sol. 19 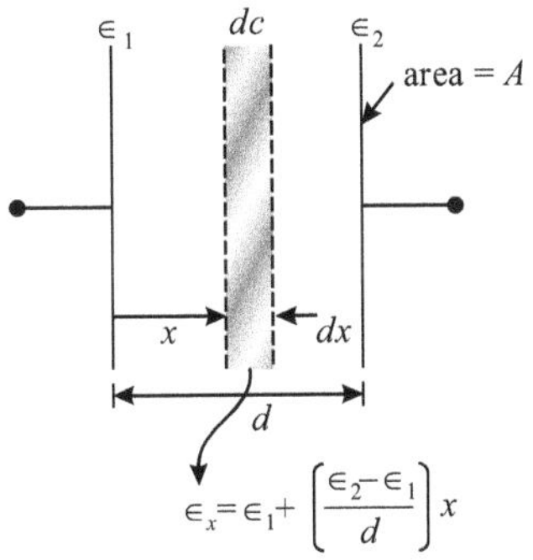

Using $\in_x = mx + C$

at $x = 0, \in = \in_1$

at $x = d, \in = \in_2$

The capacitance of elemental capacitor as shown in

$$dC = \frac{\in_x \in_0 A}{dx}$$

All such dC's are in series combination so we use

$$\frac{1}{C} = \int\frac{1}{dC}$$

$$\Rightarrow \quad \frac{1}{C} = \int_0^d \frac{dx}{\in_0 A\left(\in_1 + \frac{\in_2-\in_1}{d}\cdot x\right)}$$

$$\Rightarrow \quad \frac{1}{C} = \frac{1}{\in_0 A}\frac{d}{(\in_2-\in_1)}\left(\ln\left[\in_1 + \frac{\in_2-\in_1}{d}\cdot x\right]\right)_0^d$$

$$\Rightarrow \quad \frac{1}{C} = \frac{d}{\in_0 A(\in_2-\in_1)}[\ln \in_2 - \ln \in_1]$$

$$\Rightarrow \quad \frac{1}{C} = \frac{d\ln\left(\frac{\in_2}{\in_1}\right)}{\in_0(\in_2-\in_1)A}$$

$$\Rightarrow \quad C = \frac{\in_0(\in_2-\in_1)A}{d\ln\left(\frac{\in_2}{\in_1}\right)}$$ **Ans. (D)**

Sol. 20 Case-I : 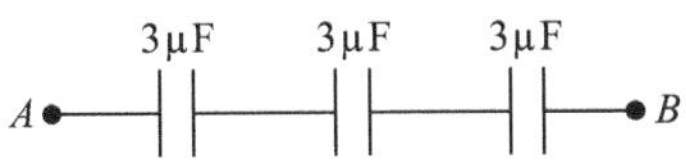

$$\frac{1}{C_{AB}} = \frac{1}{3}+\frac{1}{3}+\frac{1}{3} = 1\mu\text{F} \Rightarrow 1\mu\text{F}$$

Case-II : 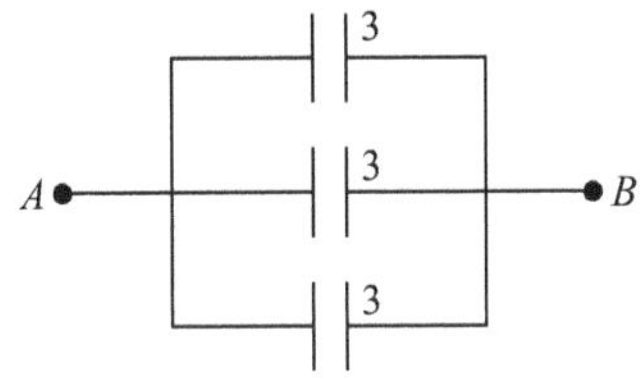

$$C_{AB} = 3+3+3 = 9\mu\text{F}$$

Case-III : 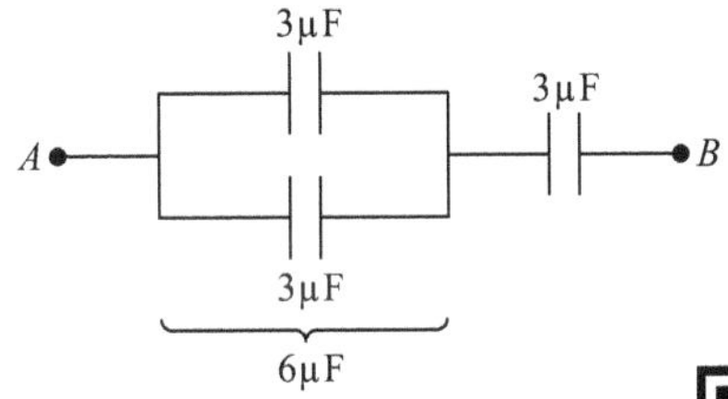

$$C_{AB} = \frac{6\times3}{6+3} = \frac{18}{9} = 2\mu\text{F}$$

Case-IV : 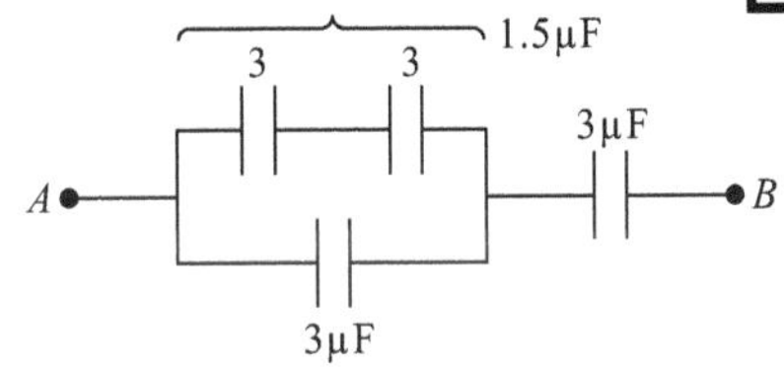

$$C_{AB} = 1.5+3 = 4.5\ \mu\text{F}$$ **Ans. (D)**

Sol. 21 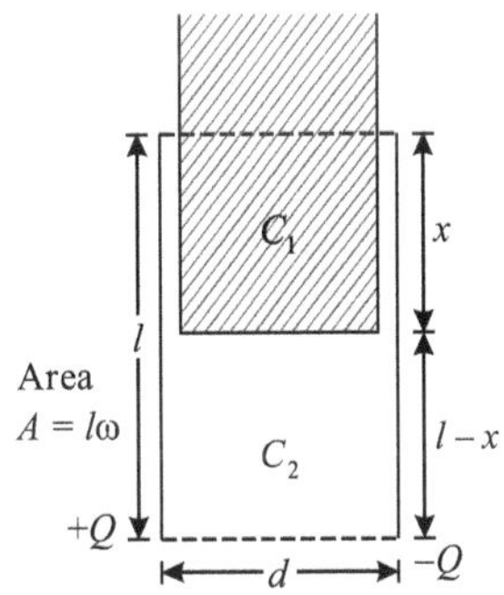

As C_1 and C_2 capacitors are in parallel with values

$$C_1 = \frac{K\in_0(xw)}{d} \quad \text{and} \quad C_2 = \frac{\in_0(l-x)w}{d}$$

$$C_{eq} = C_1 + C_2 = \frac{\in_0 w}{d}[l + x(k-1)]$$

Energy in capacitor with dielectric filled upto a distance x is

$$U = \frac{Q^2}{2C} = \frac{Q^2}{2\frac{\in_0 w}{d}[l+x(k-1)]}$$

$\Rightarrow$ force on dielectric is given as

$$F = -\frac{\partial U}{\partial x}$$

$$\Rightarrow \quad F = \frac{Q^2 d}{2\in_0 w}\left[\frac{1}{[l+x(k-1)]^2}\times(k-1)\right]$$

at $x = 0$ or when dielectric is at the edge, we have

$$F_{x=0} = \frac{Q^2 d(k-1)}{2\epsilon_0 wl^2}$$ **Ans. (C)**

Sol. 22

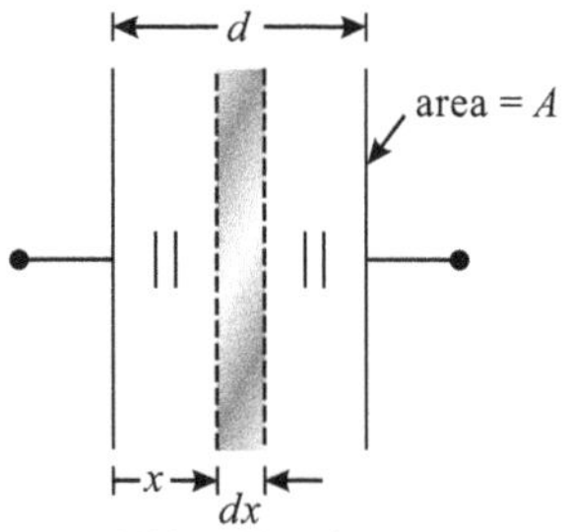

$K(x) = K_0 + \lambda x$

Capacitance with vacuum between plates is given as

$$C_0 = \frac{\epsilon_0 A}{d} \qquad \ldots(1)$$

Capacitance of elemental capacitor as shown in figure is

$$dC = \frac{k(x)\epsilon_0 A}{dx} = \frac{(k_0 + \lambda x)\epsilon_0 A}{dx}$$

Between plates all such dC's are taken in series so equation capacitance with dielectric between plates is given as

$$\frac{1}{C} = \int \frac{1}{dC}$$

$$\Rightarrow \qquad \frac{1}{C} = \int_0^d \frac{1}{\epsilon_0 A} \cdot \frac{dx}{k_0 + \lambda x}$$

$$\Rightarrow \qquad \frac{1}{C} = \frac{1}{\lambda \epsilon_0 A} \cdot [\ln(k_0 + \lambda d) - \ln k_0]$$

$$\Rightarrow \qquad \frac{1}{C} = \frac{\lambda \epsilon_0 A}{\ln\left(1 + \frac{\lambda d}{k_0}\right)} = \frac{\lambda C_0 d}{\ln\left(1 + \frac{\lambda d}{k_0}\right)}$$ **Ans. (C)**

Sol. 23 Using nodal analysis we distribute potential in given circuit as shown in figure.

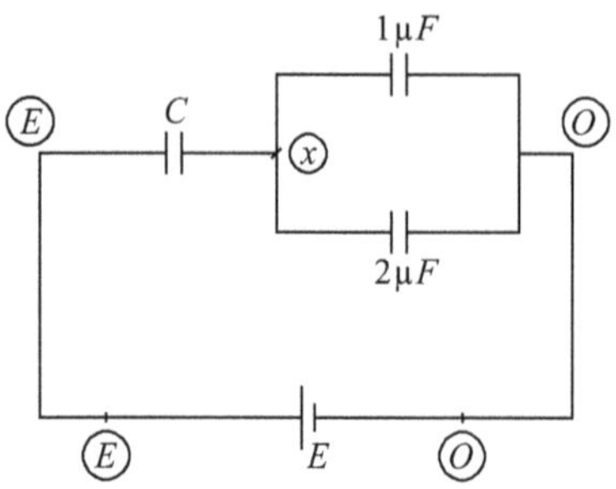

For unknown x, potential x, nodal equation is

$$1(x) + 2(x) + C(x - E) = 0$$

$$\Rightarrow \qquad x = \frac{CE}{3 + C}$$

charge on $2\mu F$ capacitor is

$$q_{2\mu F} = Q_2 = 2x = \frac{2CE}{3+C} \quad \text{(Not linear relation)}$$

When $C = 1\mu F \rightarrow Q_2 = \frac{2E}{4} = \frac{E}{2}$

When $C = 3\mu F \rightarrow Q_2 = \frac{6E}{6} = E$

Slope of Q_2 *vs* C graph is given as

$$\frac{dQ_2}{dC} = 2E\left[\frac{(3+C)1 - C(1)}{(3+C)^2}\right] = \frac{6E}{(3+C)^2}$$

As C increase $\Rightarrow \frac{dQ_2}{dC}$ decreases **Ans. (B)**

Sol. 24

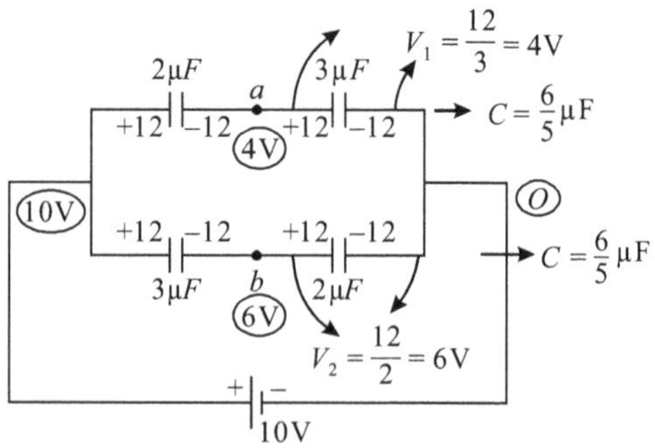

When S is open, charges on each capacitor in circuit is

$$q = CV$$

$$q = \frac{6}{5} \times 10 = 12\mu C$$

When S is closed

Charges flows from b to a

As $\qquad V_B > V_A$

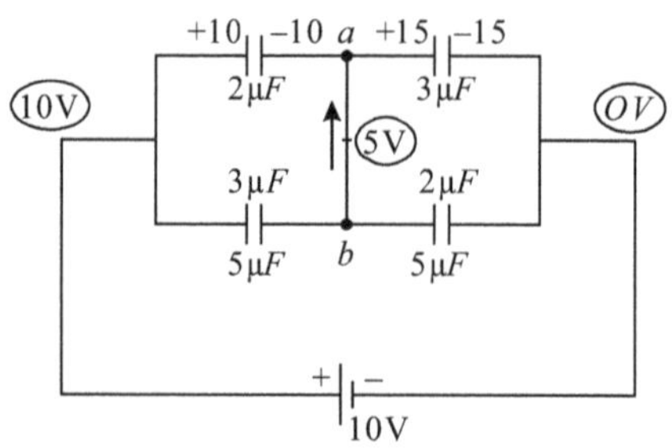

Charge connected to the two plates at point a is

$$q_a = -10 + 15 = +5\mu C$$

Initial charge on these plates

$$q_{a_i} = -12 + 12 = 0$$

On closing the switch $5\mu C$ charge flows from b to a. **Ans. (A)**

Sol. 25

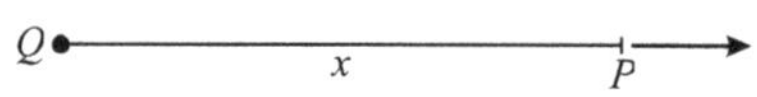

$$E_P = \frac{kQ}{x^2} \qquad \ldots(1)$$

Charges on capacitors in circuit can be calculated by using nodal analysis quickly

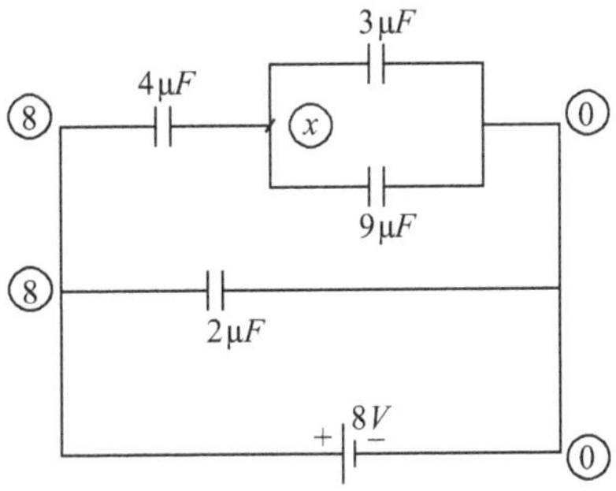

Nodal equation for potential x is given as

$$3x+9x+4(x-8)=0$$

$\Rightarrow$ $12x+4(x-8)=0$

$\Rightarrow$ $4x+8 \Rightarrow x=2V$

$$q_{4\mu F}=4\times 6=24\mu C$$

and $q_{9\mu F}=9\times 2=18\mu C$

$\Rightarrow$ $Q=24\times 18=42\mu C$

From (1) $E_P=\dfrac{9\times 10^9\times 42\times 10^{-6}}{(30)^2}$

$\Rightarrow$ $E_P=420\text{ N/C}$ **Ans. (C)**

Sol. 26 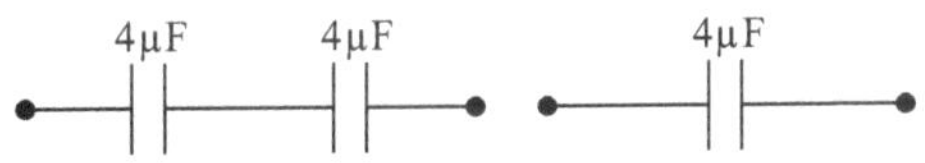

$C_{eq}=\frac{4}{2}=2\mu F$

To get 6μF equivalent

Capacitance circuit combination is as shown in figure

Two in series and one in parallel

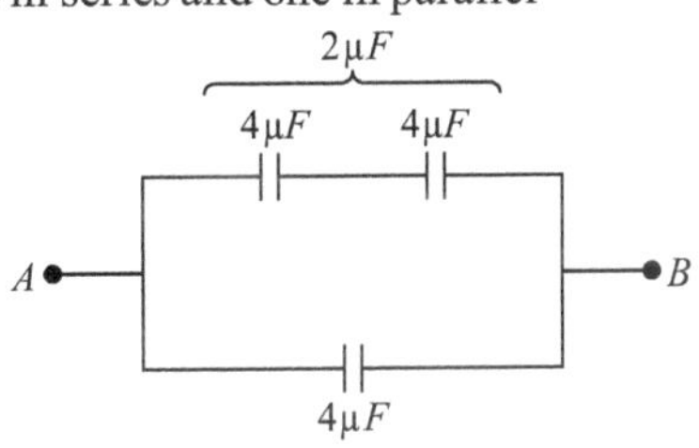

$$C_{AB}=2+4=6\mu F$$ **Ans. (D)**

Sol. 27 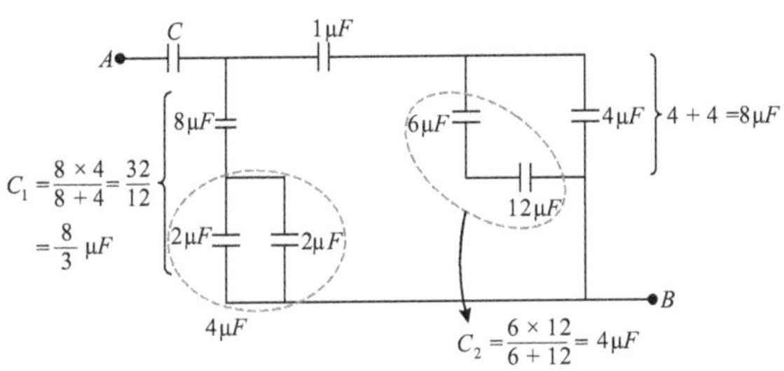

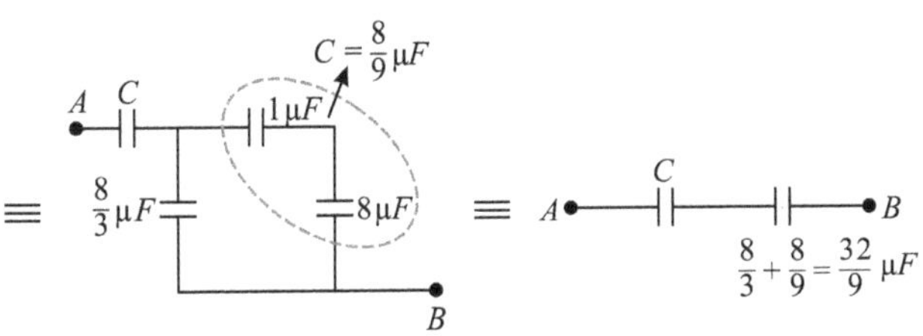

As capacitance across A and B is 1μF

$$\Rightarrow \quad \frac{1}{1}=\frac{1}{C}+\frac{1}{\frac{32}{9}}$$

$$\Rightarrow \quad \frac{1}{C}=1-\frac{9}{32}=\frac{23}{32} \quad \Rightarrow \quad C=\frac{32}{23}\mu F$$ **Ans. (A)**

Sol. 28 Supply voltage = 1000 V

Potential is equally divided in identical capacitors in series so for max 300V p.d. we need 4 capacitors (each 1μF) in series across supply.

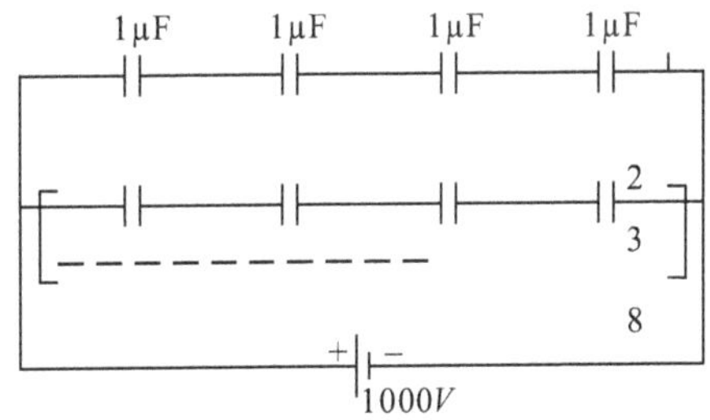

$$C_{eq}=\frac{C}{4}=0.25\mu F$$

To make 2μF capacitance we have to connect 8 such branches (4 in series) in parallel

$\Rightarrow$ $C_F=0.25\times 8=2\mu F$

Thus total capacitors required = $4\times 8=32$ **Ans. (B)**

Sol. 29 The capacitance of an isolated metal sphere is

$$C=4\pi\epsilon_0 R$$

If q is the charge on sphere than energy stored in its EF is

(Self energy of charged sphere)

$$U=\frac{q^2}{2C}=4.5$$

$\Rightarrow$ $(4\times 10^{-6})=4.5\times 2\times 4\pi\epsilon_0 R$

$\Rightarrow$ $R=\dfrac{16\times 10^{-12}}{9}\times 9\times 10^9=16\times 10^{-3}=16\text{ mm}$ **Ans. (D)**

Sol. 30 For a given capacitor (parallel plate)

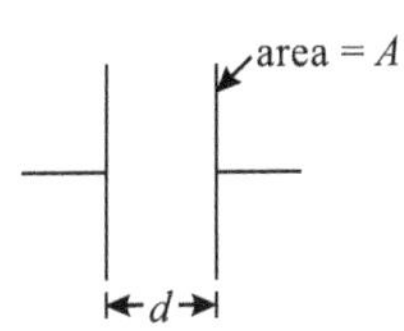

$$C=\frac{\epsilon_0 A}{d}$$

For a given charge to capacitor the potential difference across plates is

$$V=\frac{q}{C} \quad \Rightarrow \quad \text{As } C\uparrow \Rightarrow V\downarrow$$

To maintain the potential difference across capacitors, C must remain constant if d is increased to $d+2.4$ $\Rightarrow$ now capacitance with slab is given as

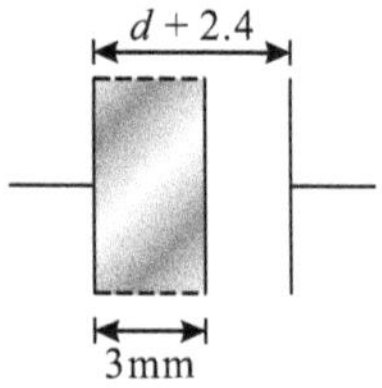

$$C' = \frac{\epsilon_0 A}{d' - t + \frac{t}{k}} \qquad t \rightarrow \text{thickness of slab}$$

To maintain the potential difference we use

$$C' = C$$

$$\frac{\epsilon_0 A}{d' - t + \frac{t}{k}} = \frac{\epsilon_0 A}{d}$$

$$\Rightarrow \quad d + 2.4 - 3 + \frac{3}{k} = d$$

$$\Rightarrow \quad \frac{3}{k} = 0.6 \quad \Rightarrow \quad k = 5$$

Ans. (C)

Sol. 31

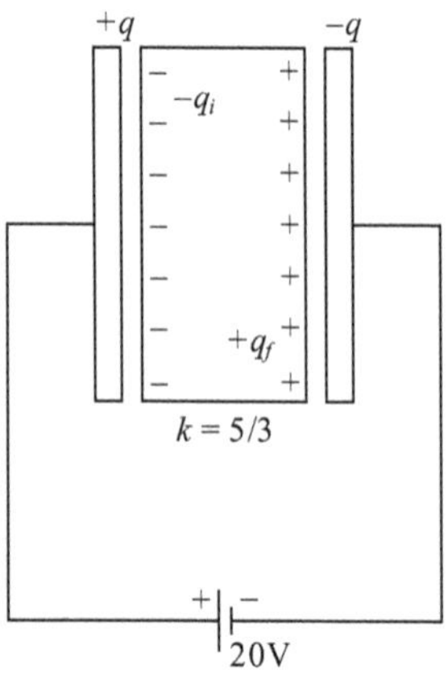

$$C = 90\text{pF} \rightarrow C' = \frac{5}{3} \times 90 = 150\text{ pF}$$

Capacitance with dielectric

In steady state charge on capacitor plates is

$$q = C'V$$

$$q = 3000\text{ pC}$$

Free charge on plates

Induced charges on the surface of dielectric is given as

$$q_i = q\left(1 - \frac{1}{k}\right) = 3000\left(1 - \frac{1}{5/3}\right)$$

$$\Rightarrow \quad q_i = 3000 \times \frac{2}{5} = 1200\text{ pC} = 1.2\text{ nC}$$

Ans. (A)

Sol. 32 If charge on a capacitor plates is q then force of attraction between plates is given as

$$F = \frac{q^2}{2\epsilon_0 A} = \frac{C^2V^2}{2\epsilon_0 A} \qquad (\text{As } q = CV)$$

and $\quad C = \frac{\epsilon_0 A}{d} \quad \Rightarrow \quad F = \frac{\epsilon_0 AV^2}{2d^2}$

$$\Rightarrow \quad V = \sqrt{\frac{2Fd^2}{\epsilon_0 A}}$$

$$= \sqrt{\frac{2 \times 25 \times 10^{-6} \times (1.5 \times 10^{-2})^2}{8.85 \times 10^{-12} \times 200 \times 10^{-4}}} \simeq 250\text{ V}$$

Ans. (C)

Sol. 33 Simplifying the above circuit, it can be redrawn as

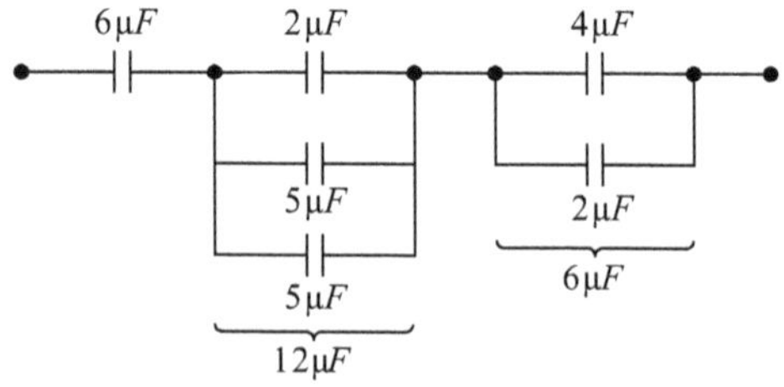

Here equivalent capacitance across AB is given as

$$\frac{1}{C_{AB}} = \frac{1}{6} + \frac{1}{12} + \frac{1}{6}$$

$$\frac{1}{C_{AB}} = \frac{2+1+2}{12} = \frac{5}{12}$$

$$\Rightarrow \quad C_{AB} = \frac{12}{5} = 2.4\ \mu\text{F}$$

Ans. (D)

Sol. 34 When C_1 is charged by battery then its charge will be

$$q_{C_1} = C_1V = 1 \times 60 = 60\ \mu\text{C}$$

Combination of C_2 and C_3 in series is

$$C' = \frac{C_2C_3}{C_2 + C_3} = \frac{3 \times 6}{3 + 6} = 2\mu\text{F}$$

After switch is shifted to position (2), the circuit will be

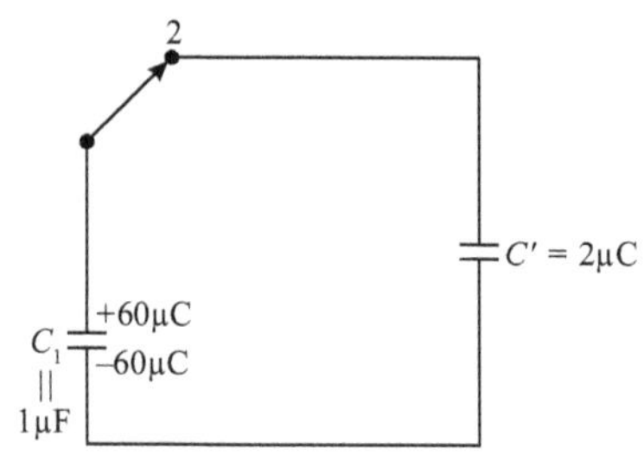

Final voltage across C_1 and C' is

$$V_f = \frac{q}{C_1 + C'} = \frac{60}{1+2} = 20\text{V}$$

$\Rightarrow$ Final charge on C' is

$$q_C = C'V_f = 2 \times 20 = 40\ \mu\text{C}$$

Ans. (D)

Sol. 35

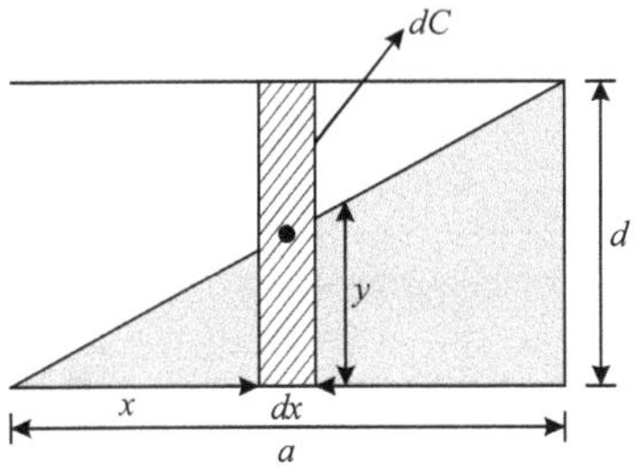

Here we consider elemental capacitor dC as shown in figure for which, we use

$$\frac{y}{x} = \frac{d}{a}$$

$$y = \frac{d}{a}x$$

$$dy = \frac{d}{a}(dx)$$

Capacitance of elemental capacitor is taken as

$$\frac{1}{dC} = \frac{y}{k \in_0 \cdot adx} + \frac{(d-y)}{\in_0 adx}$$

$$\Rightarrow \quad \frac{1}{dC} = \frac{1}{\in_0 adx}\left(\frac{y}{k} + d - y\right)$$

All elemental capacitors are considered in parallel so equivalent capacitance is

$$C_{eq} = \int dC = \int \frac{\in_0 adx}{\frac{y}{k} + d - y}$$

$$\Rightarrow \quad C_{eq} = \in_0 a \cdot \frac{a}{d}\int_0^d \frac{dy}{d + y\left(\frac{1}{k} - 1\right)}$$

$$\Rightarrow \quad C_{eq} = \frac{\in_0 a^2}{\left(\frac{1}{k} - 1\right)d}\left[\ln\left(d + y\left(\frac{1}{k} - 1\right)\right)\right]_0^d$$

$$\Rightarrow \quad C_{eq} = \frac{k \in_0 a^2}{(1-k)d}\ln\left(\frac{d + d\left(\frac{1}{k} - 1\right)}{d}\right)$$

$$\Rightarrow \quad C_{eq} = \frac{k \in_0 a^2}{(1-k)d}\ln\left(\frac{1}{k}\right) = \frac{k \in_0 a^2 \ln k}{(k-1)d}$$ **Ans. (C)**

Sol. 36 Equivalent circuit of given system is shown below

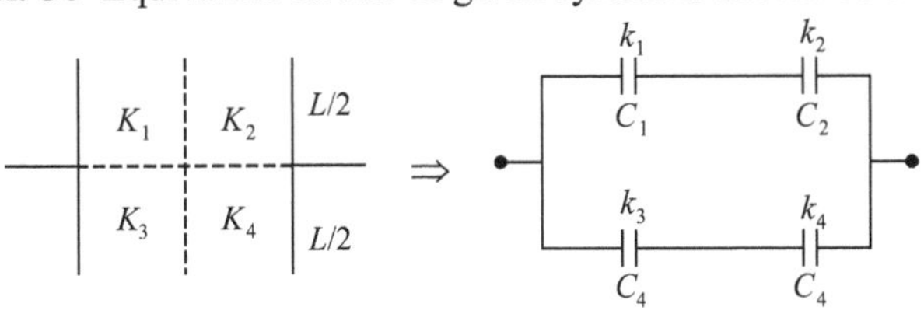

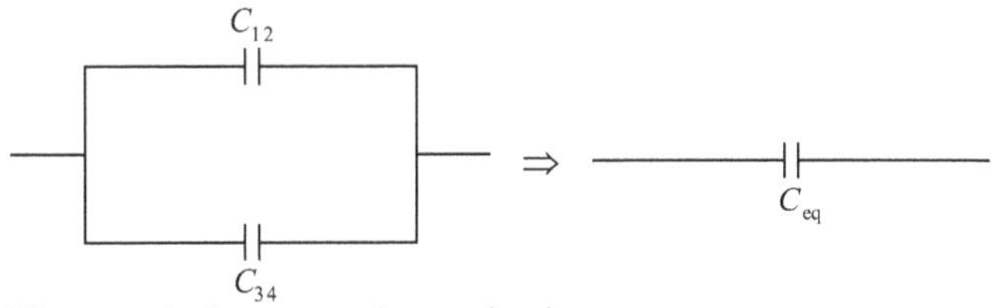

Thus equivalent capacitance is given as

$$C_{12} = \frac{C_1C_2}{C_1 + C_2} = \frac{\frac{k_1 \in_0 \frac{L}{2} \times L}{d/2} \cdot \frac{k_2\left[\in_0 \frac{L}{2} \times L\right]}{d/2}}{(k_1 + k_2)\left[\frac{\in_0 \cdot \frac{L}{2} \times L}{d/2}\right]}$$

$$C_{12} = \frac{k_1k_2}{k_1 + k_2}\frac{\in_0 L^2}{d}$$

In the same way we get,

$$C_{34} = \frac{k_3k_4}{k_3 + k_4}\frac{\in_0 L^2}{d}$$

$$\Rightarrow \quad C_{eq} = C_{12} + C_{34} = \left[\frac{k_1k_2}{k_1 + k_2} + \frac{k_3k_4}{k_3 + k_4}\right]\frac{\in_0 L^2}{d} \qquad \ldots(1)$$

Comparing with

$$C_{eq} = \frac{k \in_0 L^2}{d} \qquad \ldots(2)$$

$$\Rightarrow \quad k_{eq} = \frac{k_1k_2(k_3 + k_4) + k_3k_4(k_1 + k_2)}{(k_1 + k_2)(k_3 + k_4)}$$

This does not match with any of the options, so none of the given options is correct **Ans. (None)**

Sol. 37 Let dielectric constant of material used be K

$$\Rightarrow \quad \frac{10 \in_0 A/3}{d} + \frac{12 \in_0 A/3}{d} + \frac{14 \in_0 A/3}{d} = \frac{K \in_0 A}{d}$$

$\Rightarrow \quad K = 12$ **Ans. (A)**

Sol. 38 Initial energy of capacitor

$$U_i = \frac{1}{2}CV^2 = \frac{1}{2} \times 12 \times 10^{-12} \times (10)^2$$

$$\Rightarrow \quad U_i = 600 \text{ pJ}$$

Since battery is disconnected so charge remain same at 120pc. Final energy of capacitor will be

$$U_f = \frac{q^2}{2C}$$

$$\Rightarrow \quad U_f = \frac{1}{2} \times \frac{120 \times 120}{12 \times 6.5} = 92 \text{ pJ}$$

Work done by capacitor is

$$W = U_i - U_f = 600 - 92 = 508 \text{ pJ}$$ **Ans. (C)**

Sol. 39 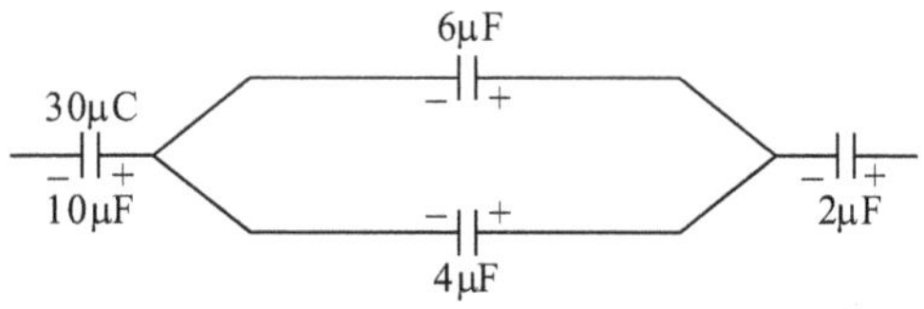

6μF & 4μF are in parallel & total charge on this combination is 30 μC

$\Rightarrow$ Charge on 6μF capacitor

$$= \frac{6}{6+4} \times 30 = 18\ \mu C$$

Thus charge is on right plate of 6 μF capacitor is +18 μC.

Ans. (D)

Sol. 40 Equivalent capacitance is given as

$$C_{eq} = \frac{6}{13}\mu F = \frac{3C}{13} \qquad [\text{As } C = 2\mu F \text{ each}]$$

Therefore three capacitors must be in parallel with 4 in series so, we have

$$\frac{1}{C_{eq}} = \frac{1}{3C} + \frac{1}{C} + \frac{1}{C} + \frac{1}{C} + \frac{1}{C}$$

$$C_{eq} = \frac{3C}{13} = \frac{6}{13}\mu F \qquad \textbf{Ans. (D)}$$

Sol. 41 Initial energy in capacitor is

$$U_i = \frac{Q^2}{2C}$$

After turning the switch to position B, find energy in circuit is

$$U_f = \frac{Q^2}{8C}$$

Heat dissipated is given as

$$H = U_i - U_f = \frac{3Q^2}{8C} \qquad \textbf{Ans. (A)}$$

Sol. 42

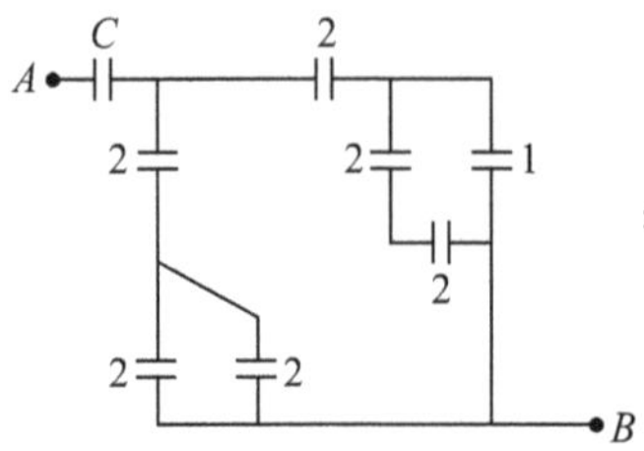

$\Rightarrow$ 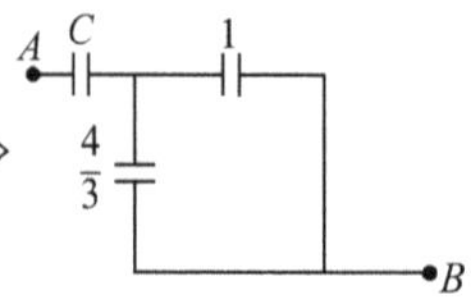

$\Rightarrow$ 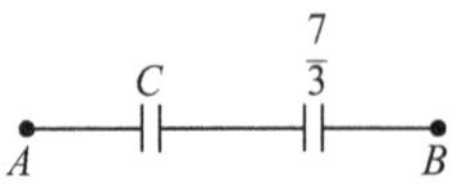

Equivalent capacitance across A and B is given as

$$\Rightarrow \qquad \frac{\frac{7C}{3}}{\frac{7}{3}+C} = \frac{1}{2}$$

$$\Rightarrow \qquad 14\,C = 7 + 3\,C$$

$$\Rightarrow \qquad C = \frac{7}{11}\mu F$$

Ans (B)

Sol. 43 At $t = 4$s, charge on capacitor plates is constant so current in circuit will be zero.

Ans. (C)

Sol. 44 Electric field between plates of a capacitor is given as

$$E = \frac{\sigma}{\epsilon_0} = \frac{Q}{A\epsilon_0}$$

$$\Rightarrow \qquad Q = AE\epsilon_0$$

$$\Rightarrow \qquad Q = (1)(100)(8.85 \times 10^{-12})$$

$$\Rightarrow \qquad Q = 8.85 \times 10^{-10}\text{C}$$

Ans. (D)

Sol. 45 Maximum electric field between plates

$$E_m = 10^6\ \text{V/m}$$

If d is the plate separation, we use

$$E_m = \frac{V_m}{d}$$

$$\Rightarrow \qquad d = \frac{V_m}{E_m} = \frac{500}{10^6} = 5 \times 10^{-4}\text{m}$$

Capacitance $\quad C = \dfrac{k\epsilon_0 A}{d}$

$$\Rightarrow \qquad k = \frac{Cd}{\epsilon_0 A} = \frac{15\times10^{-12}\times5\times10^{-4}}{8.85\times10^{-12}\times10^{-4}}$$

$$= \frac{15\times5}{8.85} = 8.465 \sim 8.5$$

Ans. (D)

Sol. 46 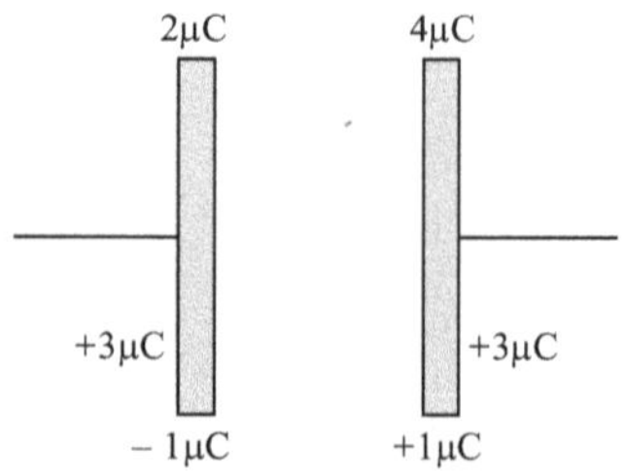

$$q_{cap} = 1\ \mu C$$

$\Rightarrow$ Potential difference across plates

$$V = \frac{q}{C} = \frac{1\mu C}{1\mu F} = 1\text{ V}$$

Ans. (D)

Sol. 47 Work done in pulling plates apart is given as

$$W = U_f - U_i$$

$$\Rightarrow \qquad W = \frac{q^2}{2C_f} - \frac{q^2}{2C_i}$$

$$\Rightarrow \qquad W = \frac{(5\times10^{-6})^2}{2}\left(\frac{1}{2\times10^{-6}} - \frac{1}{5\times10^{-6}}\right)$$

$$\Rightarrow \qquad W = \frac{15}{4} \times 10^{-6}$$

$$\Rightarrow \qquad W = 3.75 \times 10^{-6}\text{ J}$$

Ans. (A)

Sol. 48

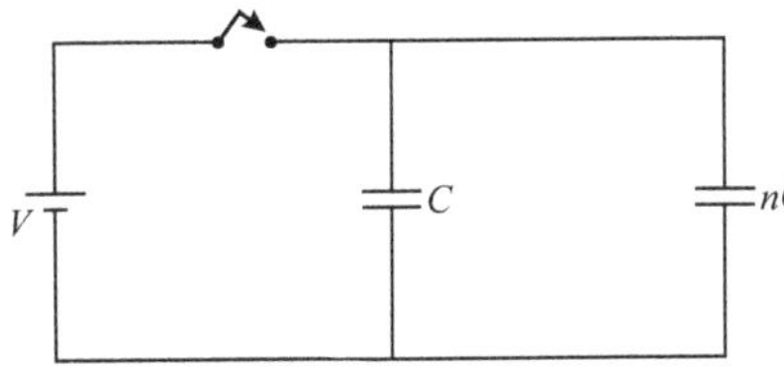

After fully charging, battery is disconnected from circuit so charge on system remain constant. after disconnecting battery system now is shown below

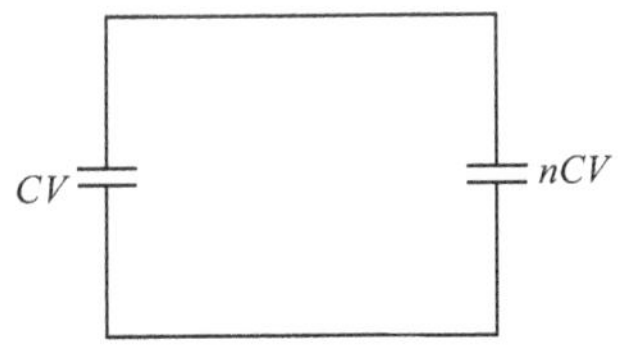

Total charge of the system

$$= CV + nCV$$

$$= (n+1)CV$$

After the insertion of dielectric of constant K, if final potential difference becomes V_f, we use

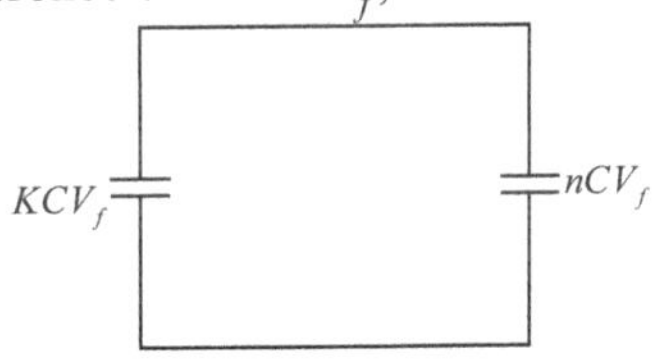

$$V_f = \frac{\text{total charge}}{\text{total capacitance}}$$

$$\Rightarrow \quad V_f = \frac{(n+1)CV}{KC+nC} = \frac{(n+1)V}{K+n}$$ **Ans. (C)**

Sol. 49 As $q = CV$ so the slope of graph will give capacitance. Slope will be more in parallel combination. Hence capacitance in parallel should be 50 μF & in series combination should be 8 μF. From the given option only in option 40 μF & 10 μF it is obtained as

$$C_{\text{parallel}} = 40 + 10 = 50 \mu\text{F}$$

$$C_{\text{series}} = \frac{40 \times 10}{40+10} = 8 \ \mu\text{F}$$ **Ans. (D)**

Sol. 50 In electric field effective gravity for pendulum is given as

$$g_{\text{eff}} = \sqrt{g^2 + \left(\frac{qE}{m}\right)^2}$$

Time period of pendulum is taken as

$$T = 2\pi\sqrt{\frac{l}{g_{\text{eff}}}}$$

$$\Rightarrow \quad T = 2\pi\sqrt{\frac{l}{\sqrt{g^2 + \left(\frac{qE}{m}\right)^2}}}$$ **Ans. (A)**

Sol. 51 In first case we consider the three capacitors of each dielectric are in series.

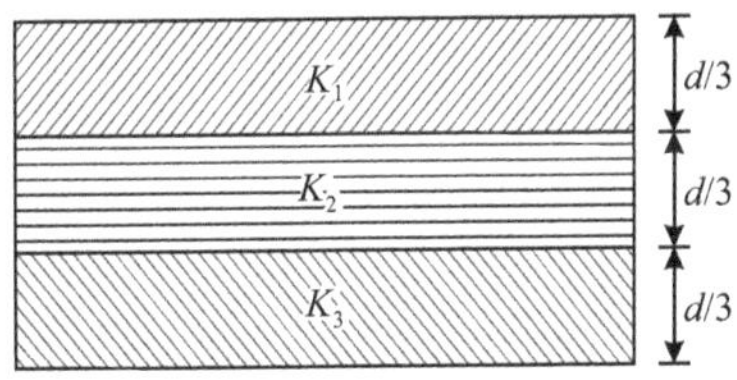

Capacitance for each dielectric is given as

$$C_1 = \frac{3\varepsilon_0 AK_1}{d}$$

$$C_2 = \frac{3\varepsilon_0 AK_2}{d}$$

$$C_3 = \frac{3\varepsilon_0 AK_3}{d}$$

Combination capacitance is taken as

$$\frac{1}{C_{\text{eq}}} = \frac{1}{C_1} + \frac{1}{C_2} + \frac{1}{C_3}$$

$$\Rightarrow \quad C_{\text{eq}} = \frac{3\varepsilon_0 AK_1K_2K_3}{d(K_1K_2 + K_2K_3 + K_3K_1)} \quad \ldots(1)$$

In second case we consider capacitance of each dielectric are in parallel

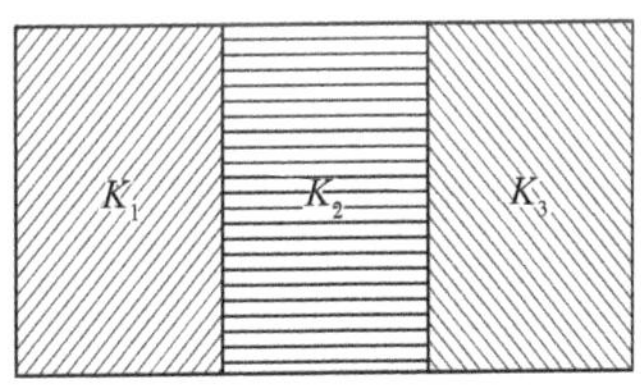

Capacitance of each dielectric is given as

$$C_1 = \frac{\varepsilon_0 K_1 A}{3d}$$

$$C_2 = \frac{\varepsilon_0 K_2 A}{3d}$$

$$C_3 = \frac{\varepsilon_0 K_3 A}{3d}$$

Combination capacitance is taken as

$$C'_{\text{eq}} = C_1 + C_2 + C_3$$

$$C'_{\text{eq}} = \frac{\varepsilon_0 A}{3d}(K_1 + K_2 + K_3) \quad \ldots(2)$$

Ratio of energy stored in two cases is

$$\frac{E_1}{E_2} = \frac{\frac{1}{2}C_{eq}.V^2}{\frac{1}{2}C'_{eq}V^2}$$

$$\Rightarrow \quad \frac{E_1}{E_2} = \frac{9K_1K_2K_3}{(K_1+K_2+K_3)(K_1K_2+K_2K_3+K_3K_1)}$$ **Ans. (A)**

Sol. 52

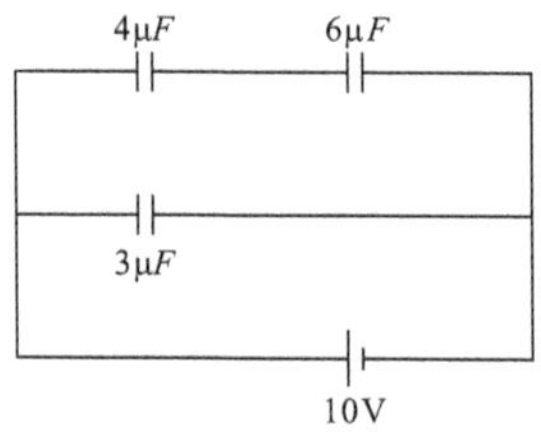

In above circuit we consider 4μF and 6 μF capacitors in series

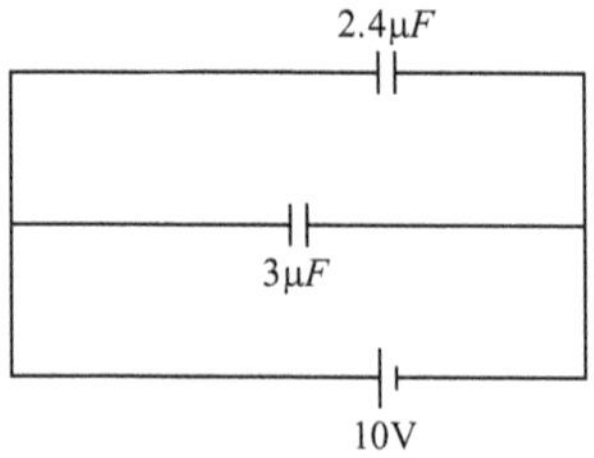

Charge supplied by battery to 2.4 μF capacitor is

$$q = 2.4 \times 10 = 24\ \mu C$$

In series both capacitors will have same charge of 24 μC.

Ans. (B)

17.1 Current Electricity

Sol. 1 If a resistance wire is of length l and cross sectional area A then its resistance is given as

$$R = \rho\frac{l}{A} \quad [\rho \rightarrow \text{wire resistivity}]$$

volume of wire $\quad V = RA$

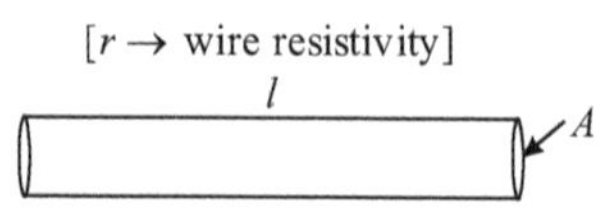

If length of wire is stretched by 100%

$\Rightarrow \quad l_f = 2l$

As $\quad V \rightarrow$ constant $\Rightarrow \quad A_f = \dfrac{A}{2}$

$\Rightarrow$ Final resistance of wire will become

$$R_f = \rho\frac{l_f}{A_f} = \rho\left(\frac{2l}{A/2}\right) = 4\rho\frac{l}{A} = 4\text{R}$$

Change $= R_f - R = 4R - R = 3R$ **Ans. (D)**

Sol. 2 Thermistors are such resistances of which resistance changes rapidly with small change in this temperature. These are used mainly as temperature sensing devices.

Metal oxides which have high temperature coefficients of resistivity are most suitable for making thermistors. **Ans. (C)**

Sol. 3 $\dfrac{l_1}{l_2} = \dfrac{4}{3}$ and $\dfrac{r_1}{r_2} = \dfrac{2}{3} \Rightarrow \dfrac{A_1}{A_2} = \left(\dfrac{2}{3}\right)^2 = \dfrac{4}{9}$

Resistances of the given wires

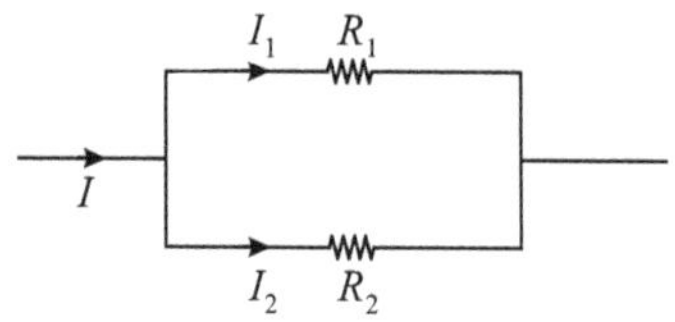

$$R_1 = \rho\frac{l_1}{A_1} \text{ and } R_2 = \rho\frac{l_2}{A_2}$$

$$\Rightarrow \quad \frac{R_1}{R_2} = \frac{l_1}{A_1} \times \frac{A_2}{l_2} = \frac{4}{3} \times \frac{9}{4} = \frac{3}{1}$$

$$\Rightarrow \quad I_1R_1 = I_2R_2$$

$$\Rightarrow \quad \frac{I_1}{I_2} = \frac{R_2}{R_1} = \frac{1}{3}$$

Ans. (B)

Sol. 4 If the given resistances are R_1 and R_2

$$\Rightarrow \quad S = R_1 + R_2 \quad \text{and} \quad P = \frac{R_1R_2}{R_1 + R_2}$$

We use $\quad S = nP$

$$R_1 + R_2 = n\left(\frac{R_1R_2}{R_1 + R_2}\right)$$

$$\Rightarrow \quad n = \frac{(R_1 + R_2)^2}{R_1R_2} = \frac{R_1^2 + R_2^2 + 2R_1R_2}{R_1R_2}$$

$$\Rightarrow \quad n = \frac{R_1}{R_2} + \frac{R_2}{R_1} + 2$$

$$\Rightarrow \quad n = \left(\sqrt{\frac{R_1}{R_2}} - \sqrt{\frac{R_2}{R_1}}\right)^2 + 4$$

here n is minimum when $R_1 = R_2 \Rightarrow n_{min} = 4$ **Ans. (A)**

Sol. 5 We use $\quad R_T = R_0(1 + \alpha T)$

where $\quad R_T \rightarrow$ resistance at T°C

$R_0 \rightarrow$ resistance at 0°C

Initially at 100°C, $R_T = 100\Omega$

$$\Rightarrow \quad 100 = R_0(1 + 100\alpha) \quad \ldots(1)$$

Later at T°C, $\quad R_T = 200\Omega$

$$\Rightarrow \quad 200 = R_0(1 + \alpha T)$$

$$(2) \div (1) \Rightarrow \quad \frac{200}{100} = \frac{1 + \alpha T}{1 + 100\alpha}$$

$$\Rightarrow \quad 2 + 200\alpha = 1 + \alpha T$$

$$\Rightarrow \quad T - 200 = \frac{1}{\alpha} = \frac{1000}{5} = 200$$

$$\Rightarrow \quad T = 200 + 200 + 400°\text{C}$$

Ans. (B)

Sol. 6 $\quad \rho_B = 2\rho_A$

and $\quad r_B = 2r_A \Rightarrow A_B = 4A_A$

As resistances of the two wires are equal

$$\Rightarrow \quad R_A = R_B$$

$$\Rightarrow \quad \rho_A \cdot \frac{l_A}{A_A} = \rho_B \cdot \frac{l_B}{A_B}$$

$$\Rightarrow \quad \frac{l_B}{l_A} = \frac{\rho_A}{\rho_B} \cdot \frac{A_B}{A_A} = \frac{1}{2} \times \frac{4}{1} = 2$$

Ans. (D)

Sol. 7 If resistance at 0°C is considered to be R_0, we use at t°C

$$R_t = R_0(1 + \alpha t)$$

at 50°C, we have

$$5 = R_0(1 + 50\alpha)$$

$$\Rightarrow \quad 5 - R_0 = 50R_0\alpha \qquad \ldots(1)$$

at 100°C, we have

$$6 = R_0(1 + 100\alpha)$$

$$\Rightarrow \quad 6 - R_0 = 100R_0\alpha \qquad \ldots(2)$$

$$\frac{(1)}{(2)} \Rightarrow \frac{5 - R_0}{6 - R_0} = \frac{1}{2}$$

$$\Rightarrow \quad 10 - 2R_0 = 6 - R_0$$

$$\Rightarrow \quad R_0 = 10 - 6 = 4\,\Omega$$

Ans. (D)

Sol. 8 Current density at a distance r from point A is given as

$$J = \frac{I}{2\pi r^2}$$

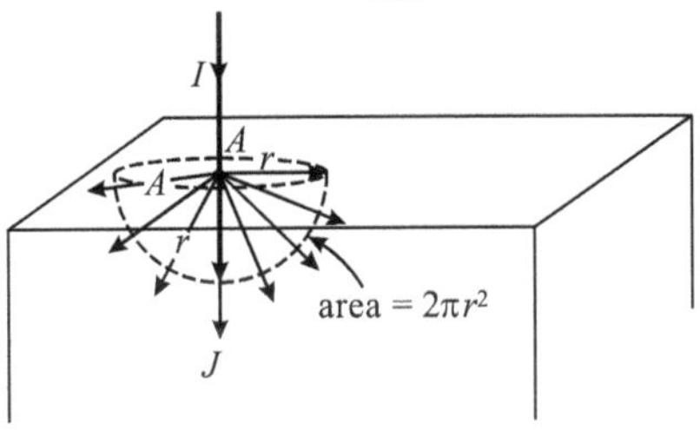

Electric field at a distance r from A is

$$E(r) = \rho J = \frac{\rho I}{2\pi r^2}$$

Ans. (C)

Sol. 9 Here potential difference between points B and C can be given as

$$\Delta V_{BC} = \int_B^C E_{(r)} \cdot dr$$

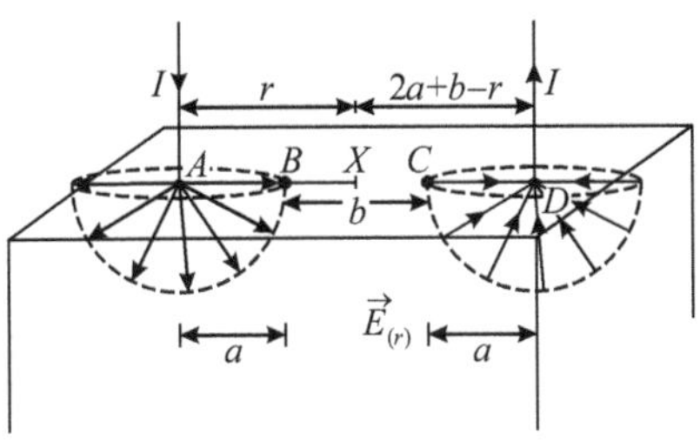

At point X (at a distance r from A) the net EF is given as

$$E_{(r)} = E_{\text{due to } A} + E_{\text{dut to } D}$$

$$E_{(r)} = \frac{\rho I}{2\pi r^2} + \frac{\rho I}{2\pi(2a + b - r)^2}$$

potential difference between points B and C is given as

$$\Delta V_{BC} = \int_a^{a+b} \frac{\rho I}{2\pi}\left(\frac{1}{r^2} + \frac{1}{(2a + b - r)^2}\right) \cdot dr$$

$$\Rightarrow \quad \Delta V_{BC} = \frac{\rho I}{2\pi}\left[-\frac{1}{r} + \frac{1}{2a + b - r}\right]_a^{a+b}$$

$$\Rightarrow \quad \Delta V_{BC} = \frac{\rho I}{2\pi}\left[\left(-\frac{1}{a + b} + \frac{1}{a}\right) - \left(-\frac{1}{a} + \frac{1}{a + b}\right)\right]$$

$$\Rightarrow \quad \Delta V_{BC} = \frac{\rho I}{2\pi}\left[\frac{2}{a} - \frac{2}{a + b}\right]$$

$$\Rightarrow \quad \Delta V_{BC} = \frac{\rho I}{\pi a} - \frac{\rho I}{\pi(a + b)}$$

Ans.

Sol. 10 Using $R = R_0(1 + \alpha\Delta T)$

$$\Rightarrow \quad 100 = R_0(1 + \alpha(27)) \qquad \ldots(1)$$

$$\Rightarrow \quad 150 = R_0(1 + \alpha(227)) \qquad \ldots(2)$$

$$(2) \div (1) \Rightarrow \quad \frac{150}{100} = \frac{1 + \alpha(227)}{1 + \alpha(27)}$$

$$\Rightarrow \quad \frac{3}{2} = (1 + 227\alpha)(1 + 27\alpha)^{-1}$$

$$\text{for } v \text{ small } \alpha \Rightarrow \quad \frac{3}{2} = 1 + 200\alpha$$

$$\Rightarrow \quad \alpha = \frac{1}{400} = 2.5 \times 10^{-3}\ {}^\circ\text{C}^{-1}$$

As per given values for small variations in resistance and small range of temperature variation S-1 is false.

But 27ºC to 227ºC is large variation as α is considered constant.

$\Rightarrow$ for given values S-1 is true if α is considered constant.

By definition S-2 is true. **Ans. (B)**

Sol. 11 For series combination at a temperature T we use

$$R_{eq} = 2R_0(1 + \alpha_S T) \qquad \ldots(1)$$

(equivalent coefficient for series combination)

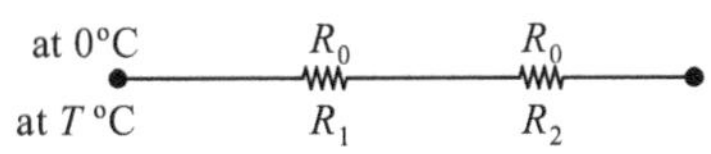

Here we have $R_1 = R_0(1 + \alpha_1 T)$

and $R_2 = R_0(1 + \alpha_2 T)$

Equivalent resistance of series combination as (at TºC)

$$R_{eq} = R_1 + R_2 = R_0(1 + \alpha_1 T) + R_0(1 + \alpha_2 T)$$

$$\Rightarrow \quad R_{eq} = R_0(2 + (\alpha_1 + \alpha_2)T)$$

$$\Rightarrow \qquad R_{eq} = 2R_0\left(1+\left(\frac{\alpha_1+\alpha_2}{2}\right)T\right) \qquad \ldots(2)$$

Comparing (1) and (2) gives

$$\alpha_s = \frac{\alpha_1+\alpha_2}{2}$$

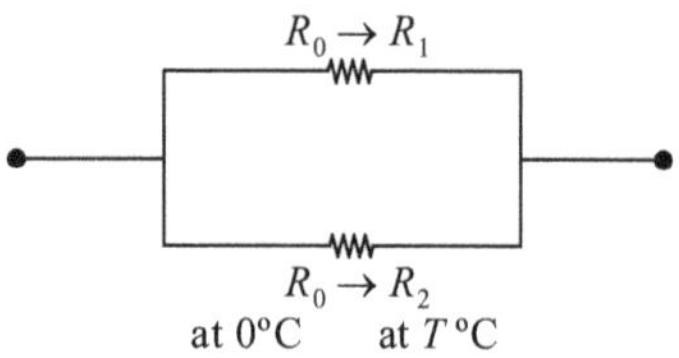

For parallel combination

At 0°C we have $R'_{eq} = \dfrac{R_0}{2}$

At 0°C we have $R''_{eq} = \dfrac{R_0}{2}(1+\alpha_p T) \qquad \ldots(3)$

[$\alpha_p T$ = equal coefficient for parallel combination]

Equal resistance at T°C is given as

$$R''_{eq} = \frac{R_1R_2}{R_1+R_2} = \frac{R_0^2(1+\alpha_1 T)(1+\alpha_2 T)}{R_0(2+(\alpha_1+\alpha_2)T)}$$

[terms of consider $\alpha_1\ \alpha_2 << 1$]

$$\Rightarrow \quad R''_{eq} = \frac{R_0(1+(\alpha_1+\alpha_2)T)}{2\left(1+\left(\frac{\alpha_1+\alpha_2}{2}\right)T\right)} \text{ [term of consider } \alpha_1\ \alpha_2 << 1]$$

$$\Rightarrow \quad R''_{eq} = \frac{R_0}{2}(1+(\alpha_1+\alpha_2)T)\left(1+\left(\frac{\alpha_1+\alpha_2}{2}\right)T\right)^{-1}$$

$$\Rightarrow \quad R''_{eq} = \frac{R_0}{2}(1+(\alpha_1+\alpha_2)T)\left(1-\left(\frac{\alpha_1+\alpha_2}{2}\right)T\right)$$

$$\Rightarrow \quad R''_{eq} = \frac{R_0}{2}\left(1+\left(\frac{\alpha_1+\alpha_2}{2}\right)T\right) \qquad \ldots(4)$$

[As $(\alpha_1+\alpha_2)^2 << 1$]

Comparing (3) and (4) we have

$$\alpha_p = \frac{\alpha_1+\alpha_2}{2}$$ **Ans. (A)**

Sol. 12 Tolerance of a resistance R is given as

$$\frac{dR}{R}\times 100 = 5\% \text{ for } R = 100\,\Omega$$

If we use

$$R_e = R+R+R+R = 4R$$

where $R_e = 400\,\Omega$ and $R = 100\,\Omega$

Tolerance of equivalent resistance

$$\text{Tolerance} = \frac{dR_e}{R_e}\times 100 = \frac{4dR}{4R}\times 100$$

$$= \frac{dR}{R}\times 100 = 5\%$$ **Ans. (B)**

Sol. 13 Resistance of a conductor is given as

$$R = \frac{\rho l}{A}$$

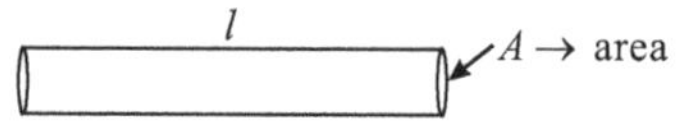

When a conductor is stretched its volume remain constant

$$V = lA = \text{constant}$$

$$A = \frac{V}{l}$$

$$\Rightarrow \quad R = \left(\frac{\rho}{V}\right)l^2$$

$$\Rightarrow \quad R \propto l^2$$

Taking log on both sides

$$\ln R = \ln\left(\frac{\rho}{V}\right) + 2\ln(\rho)$$

differentiating $\dfrac{dR}{R} = \dfrac{2dl}{l}$

$$\Rightarrow \quad \frac{dR}{R}\times 100 = 2\left(\frac{dl}{l}\times 100\right)$$

$$= 2\times 0.1 = 0.2\% \text{ increase}$$ **Ans. (D)**

Sol. 14 In final state the wire AOB behaves like two resistances $\dfrac{R}{2}$ in parallel combination so new resistance of shorter wire is given as

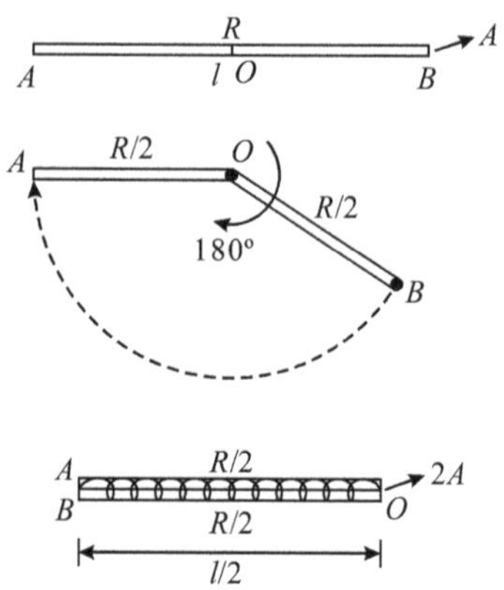

$$R_{new} = \frac{\frac{R}{2}\times\frac{R}{2}}{\frac{R}{2}+\frac{R}{2}} = \frac{R}{4}$$ **Ans. (C)**

Sol. 15 Using ohm's law for a conductor, we have

$$V = IR \quad \text{where } I = neAv_d \text{ and } R = \rho\frac{l}{A}$$

$$\Rightarrow \quad V = neAv_d \times \frac{\rho l}{A}$$

$$\Rightarrow \quad \rho = \frac{V}{nev_d l}$$

$\Rightarrow \quad \rho = \dfrac{5}{8\times10^{28}\times1.6\times10^{-19}\times2.3\times10^{-4}\times0.1}$

$\Rightarrow \quad \rho = \dfrac{2}{12.8}\times10^{-4} = 1.6\times10^{-5}\,\Omega\text{-m}$ **Ans. (D)**

Sol. 16 In a material current density is related to drift speed as

$$J = nev_d$$

If $\quad v_d = C\sqrt{E}$

$\Rightarrow \quad \dfrac{I}{A} = neC\sqrt{E}$

As $\quad E \propto V$

$\Rightarrow \quad I = IneC\sqrt{C_1V}$

$\Rightarrow \quad I \propto \sqrt{V}$ or $I^2 \propto V$ **Ans. (C)**

Sol. 17

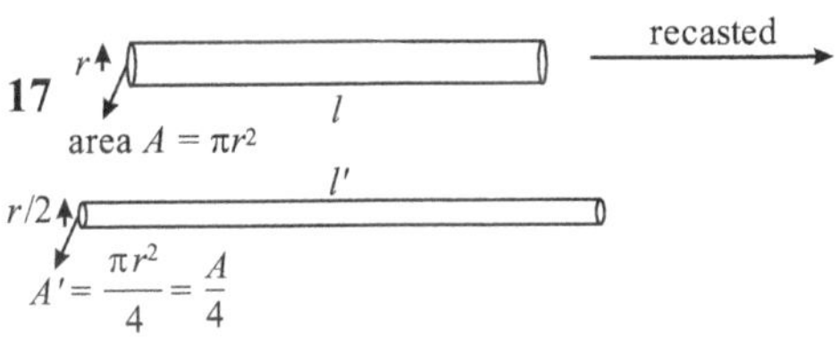

$$R = \rho\frac{l}{A} = 100 \qquad \ldots(1)$$

As volume of material remain same

$\Rightarrow \quad Al = A'l'$

$\Rightarrow \quad l' = 4l$

New resistance of wire will be

$$R' = \rho\frac{l'}{A'} = 16\rho\frac{l}{A}$$

$\Rightarrow \quad R' = 1600\,\Omega$ **Ans. (A)**

Sol. 18

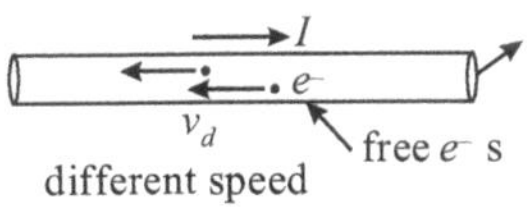

The drift speed of free e^-s (charge carries) is related to current density as

$$v_d = \frac{J}{ne} = \frac{J}{\rho}$$

(As in this case volume charge density (C/m^3) is $\rho = ne$)

$\Rightarrow \quad v_d = \dfrac{I}{\rho A} \quad [\text{as } J = \dfrac{I}{A}]$

Time the charges take in traveling a distance d is

$$t = \frac{d}{v_d} = \frac{\rho A d}{I}$$ **Ans. (C)**

Sol. 19 We use $\quad R_T = R_0(1+\alpha T)$

at room temperature

$$R_T = 100\,\Omega$$

$\Rightarrow \quad 100 = R_0(1+\alpha T_R) \qquad \ldots(1)$

at temperature $T_R + 500°\text{C}$, $R'_T = \dfrac{220}{2} = 110\,\Omega$

$\Rightarrow \quad 110 = R_0(1+\alpha(T_R+500)) \qquad \ldots(2)$

$(2)\div(1) \quad \Rightarrow \quad \dfrac{110}{100} = \dfrac{1+\alpha(T_R+500)}{1+\alpha T_R}$

$$= [1+\alpha(T_R+500)](1+\alpha T_R)^{-1}$$

As α is very small we neglect terms of α^2 which gives

$$\frac{110}{100} = 1 + 500\alpha \qquad [RT_1 = RT_0(1+\alpha\Delta T)]$$

$\Rightarrow \quad \alpha = \dfrac{10}{500\times100}$

$= 2\times10^{-4}\,°\text{C}^{-1}$ **Ans. (C)**

Sol. 20

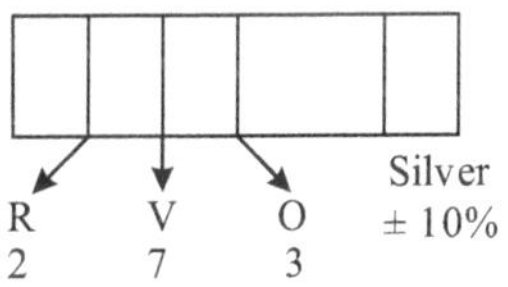

$\Rightarrow \quad R = 27\times10^3\,\Omega \pm 10\%$

$\Rightarrow \quad R = 27\,\text{K}\Omega \pm 10\%$ **Ans. (D)**

Sol. 21 Current in wire is given as

$$I = neAv_d$$

$\Rightarrow \quad v_d = \dfrac{I}{neA} = \dfrac{1.5}{9\times10^{28}\times1.6\times10^{-19}\times5\times10^{-6}}$

$\Rightarrow \quad v_d = 0.02\times10^{-3}$ m/s **Ans. (D)**

Sol. 22

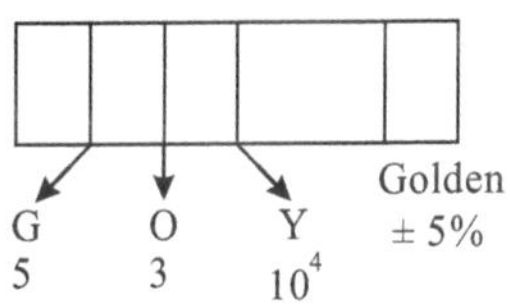

$\Rightarrow \quad R = 53\times10^4 \pm 5\% = 530\,\text{k}\Omega \pm 5\%$ **Ans. (B)**

Sol. 23 $\quad P = i^2R$

For i_{max}, R must be minimum

From color coding

$$R = 50\times10^2\,\Omega$$

Thus we use $\quad 2 = i^2_{max}\times50\times10^2$

$\Rightarrow \quad i^2_{max} = \dfrac{2}{5000} = 4\times10^{-4}$ A

$\Rightarrow \quad i_{max} = 20$ mA **Ans. (D)**

Sol. 24 When red is replaced with green 1st digit changes to 5 so new resistance will be 500 Ω **Ans. (C)**

Sol. 25 $R = 3\Omega = \dfrac{\rho l}{A}$

If $l' \to 2l \;\Rightarrow\; A' = \dfrac{A}{2}$ (As volume is constant)

New resistance of wire will be

$$R' = \frac{\rho l'}{A'} = \frac{\rho(2l)}{A/4} = 4\frac{\rho l}{A} = 4R = 4\times 3 = 12\Omega$$

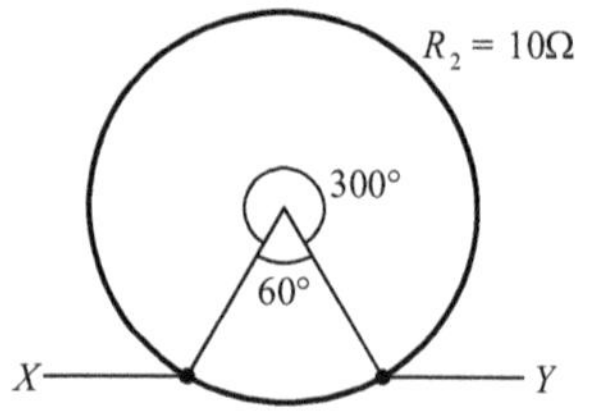

$$R_1 = \frac{12}{360}\times 60 = 2\Omega$$

Here $R_{XY} = \dfrac{R_1R_2}{R_1+R_2} = \dfrac{10\times 2}{10+2}$

$\Rightarrow$ $R_{XY} = \dfrac{20}{12}\Omega = \dfrac{5}{3}\Omega$ **Ans. (B)**

Sol. 26 Resistivity of a conductor is given as

$$\rho = \frac{m}{ne^2\tau}$$

$$= \frac{9.1\times10^{-31}}{8.5\times10^{28}\times(1.6\times10^{-19})\times25\times10^{-15}}$$

$\Rightarrow$ $\rho \cong 1.67\times10^{-8}\,\Omega$-m **Ans. (D)**

Sol. 27 The mobility of free electrons in conductor is given as

$$\mu = \frac{v_d}{E}$$

and $E = \rho J$

$$\Rightarrow \quad \mu = \frac{1.1\times10^{-3}}{1.7\times10^{-8}\times\dfrac{5}{\pi\times25\times10^{-6}}}$$

$$\Rightarrow \quad \mu = \frac{1.1\times10^{-3}\times\pi\times25\times10^{-6}}{1.7\times10^{-8}\times5} \approx 1.01\ \text{m}^2/\text{Vs}$$ **Ans. (D)**

Sol. 28

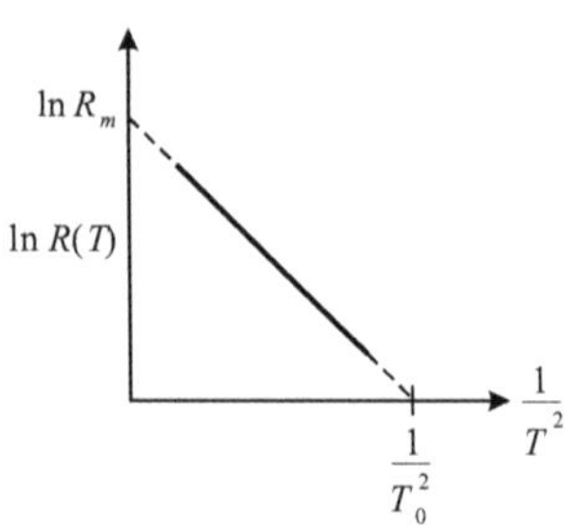

Equation of straight line

$$\ln R(\text{T}) = -\frac{\ln R_m}{(1/T_1^2)}\times\frac{1}{T^2} + \ln R_m$$

$$\Rightarrow \quad \ln R(T) = \ln R_m\left(1 - \frac{T_1^2}{T^2}\right)$$

$$\Rightarrow \quad \ln\left(\frac{R(T)}{R_m}\right) = -\ln \text{R}_\text{m}\left(\frac{T_1^2}{T^2}\right)$$

$$\Rightarrow \quad R(T) = R_m\cdot e^{-T_1^2/T^2(\ln R_m)} = R_0\, e^{-T_0^2/T^2}$$ **Ans. (C)**

Sol. 29 Elemental resistance of shell is

$$dR = \rho.\ \frac{dx}{4\pi x^2}$$

$$dR = \rho\cdot\frac{dx}{4\pi x^2}$$

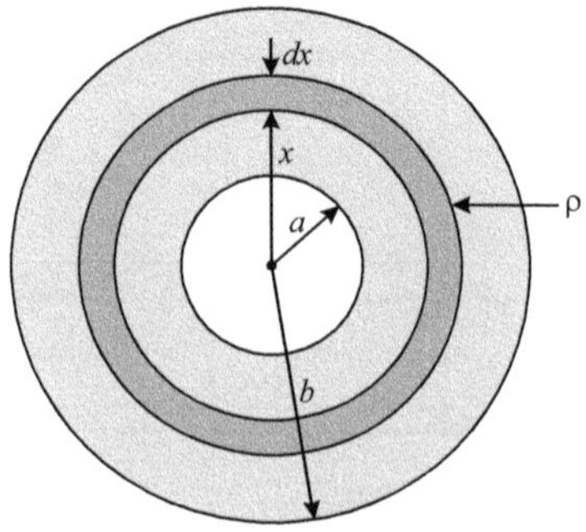

$$R = \int dR = \frac{\rho}{4\pi}\int_a^b\frac{dx}{x^2}$$

$$\Rightarrow \quad R = \frac{\rho}{4\pi}\left[-\frac{1}{x}\right]_a^b$$

$$= \frac{\rho}{4\pi}\left[\frac{1}{a}-\frac{1}{b}\right]$$

$$= \frac{\rho(b-a)}{4\pi ab}$$ **Ans. (B)**

17.2 Circuit Analysis

Sol. 1 $R_{eq} = \dfrac{6\times3}{6+3} = 2\,\Omega$

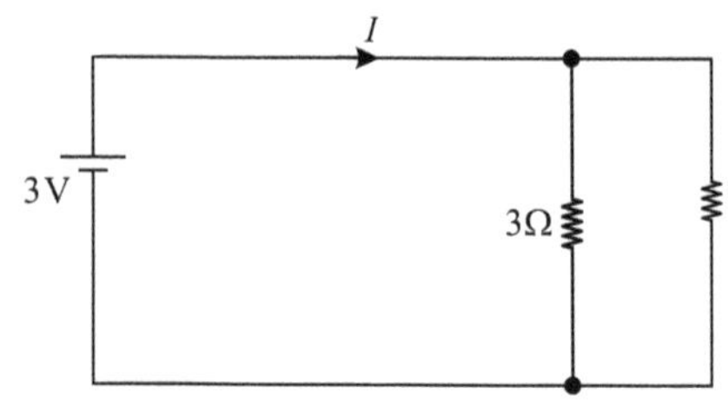

By Ohm's law, circuit current is

$$I = \frac{3}{2} = 1.5\,\text{A}$$ **Ans. (B)**

Sol. 2

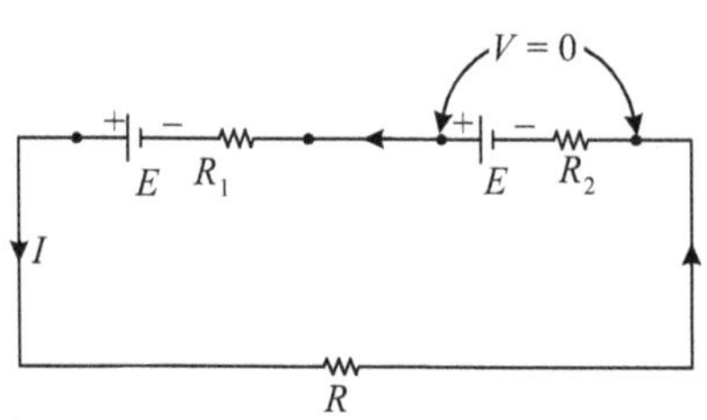

Current in circuit

$$I=\frac{2E}{R_1+R_2+R}$$

Potential difference across second source

$$V=E-IR_2=0$$

$$\Rightarrow \quad E=IR_2$$

$$\Rightarrow \quad E=\frac{2ER_2}{R_1+R_2+R}$$

$$\Rightarrow \quad R_1+R_2+R=2R_2$$

$$\Rightarrow \quad R=R_2-R_1$$ **Ans. (B)**

Sol. 3 As the circuit shown in figure is a balanced wheatstone bridge we can remove the middle resistor

$\Rightarrow$ 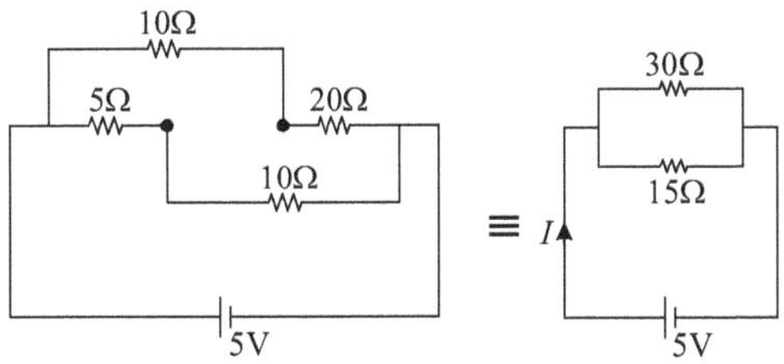

$$R_{eq}=\frac{30\times 15}{45}=10\,\Omega$$

$$\Rightarrow \quad I=\frac{5}{10}$$

$$=0.5\text{ A}$$ **Ans. (B)**

Sol. 4 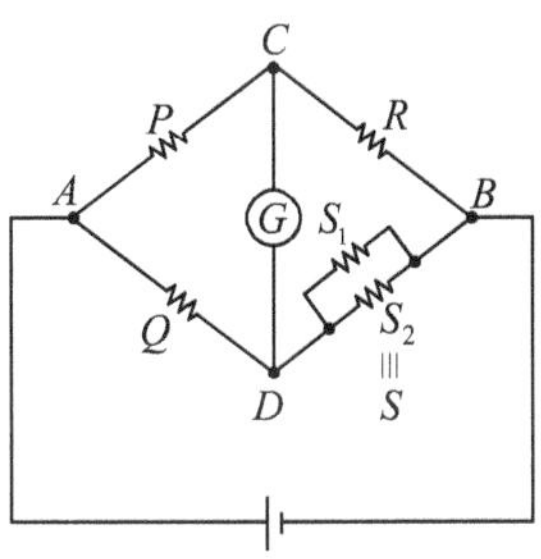

Balancing condition for the given wheatstone bridge is

$$\frac{P}{Q}=\frac{R}{S}$$

here $$S=\frac{S_1S_2}{S_1+S_2}$$

$$\Rightarrow \quad \frac{P}{Q}=\frac{R(S_1+S_2)}{S_1S_2}$$ **Ans. (B)**

Sol. 5 (1) Kirchhoff's first law (KCL) gives

Σi at a function = 0

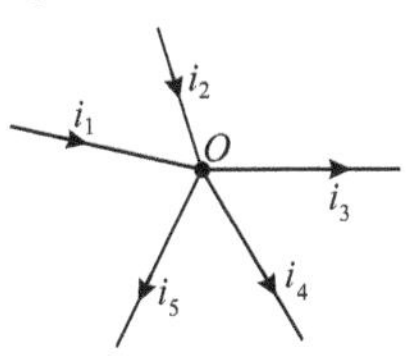

$$i_1+i_2=i_3+i_4+i_5$$

As $i \rightarrow$ charge flow/s

$\Rightarrow$ KCL is based on "Conservation of charge".

(2) Kirchhoff's Second law (KVL)

$$\Sigma iR=\Sigma E$$

(potential drop across all resistances in a loop)

= (*EMF* in that loop)

As *EF* is conservative

$\Rightarrow$ Work done in a closed path = 0 (any loop of circuit)

$\Rightarrow$ Thus KVL is based on "Energy Conservation" **Ans. (D)**

Sol. 6 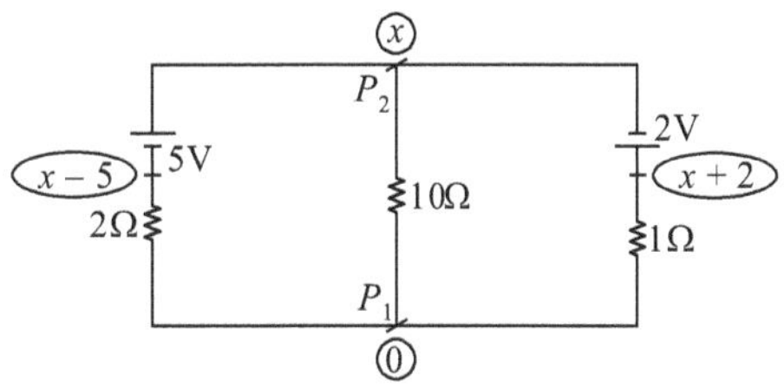

Distributing potentials in circuit and writing KCL equation for x we have

$$\frac{x-0}{10}+\frac{x+2}{1}+\frac{x-5}{2}=0$$

$$\Rightarrow \quad x+10x+20+5x-25=0$$

$$\Rightarrow \quad 16x=5$$

$$\Rightarrow \quad x=\frac{5}{16}\simeq 0.3\text{V}$$

$$I_{10\Omega}=\frac{x-0}{10}=\frac{0.3}{10}=0.03\text{A}$$

As $x>0 \Rightarrow$ current flows from P_2 to P_1 **Ans. (C)**

Sol. 7 If E_2 is removed and potential difference across A and B is equal to $E_1=90$V then on placing E_2 back in circuit no current will flow through E_2.

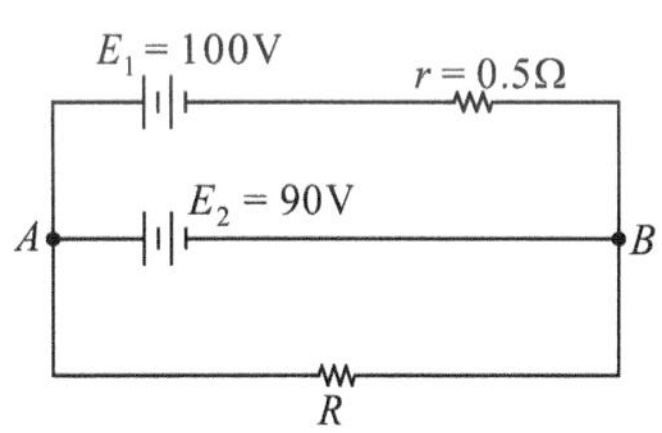

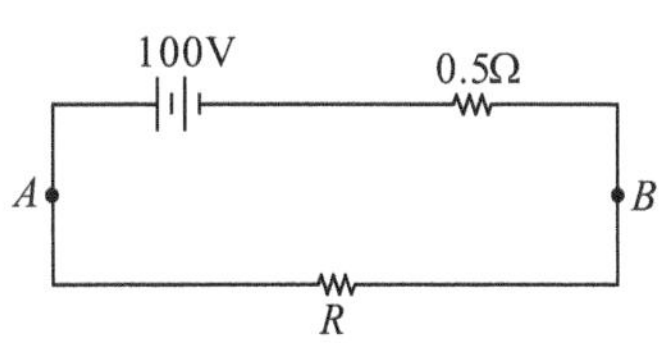

When E_2 is removed the potential difference across A and B is given as

$$V_A - V_B = \frac{100}{R+0.5} \times R = 90$$

$\Rightarrow$ $100R = 90R + 45$

$\Rightarrow$ $10R = 45$

$\Rightarrow$ $R = 4.5\,\Omega$ **Ans. (C)**

Sol. 8

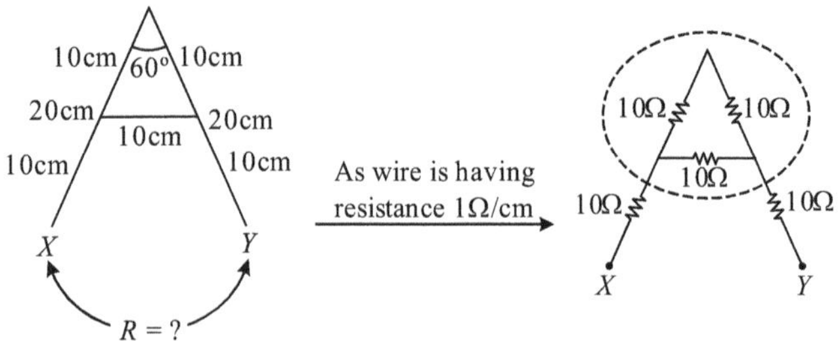

$$r_i = \frac{20 \times 10}{20+10} = \frac{20}{3}\Omega$$

here $R_{xy} = 10 + \frac{20}{3} + 10$

$\Rightarrow$ $R_{xy} = 20 + \frac{20}{3} = \frac{80}{3}\Omega = 26.7\Omega$ **Ans. (D)**

Sol. 9

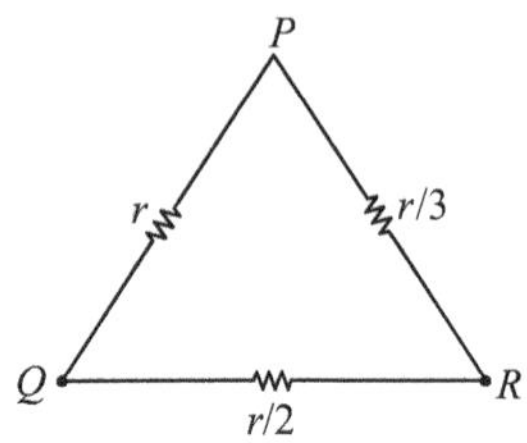

$$r_{PQ} = r \,||\left(\frac{1}{3}+\frac{r}{2}\right)$$

$\Rightarrow$ $$r_{PQ} = \frac{r \times \frac{5r}{6}}{r + \frac{5r}{6}} = \frac{5r}{11}$$

$$r_{QR} = \frac{r}{2} \,||\, \frac{4r}{3} = \frac{\frac{r}{2} \times \frac{4r}{3}}{\frac{r}{2} + \frac{4R}{3}} = \frac{4r}{11}$$

and $$r_{PR} = \frac{r}{3} \,||\, \frac{3r}{2} = \frac{\frac{r}{3} \times \frac{3r}{2}}{\frac{r}{3} + \frac{3r}{2}} = \frac{3r}{11}$$

Highest

$\Rightarrow$ $r_{PQ} > r_{QR} > r_{PR}$ **Ans. (B)**

Sol. 10

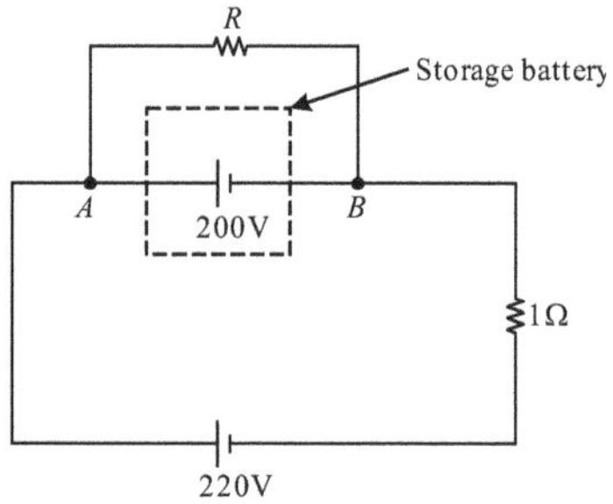

If storage battery is removed and if $V_A - V_B > 200$ V then a current passes through this battery to charge it

$\Rightarrow$ in open circuit condition (without 200 V battery)

$$V_A - V_B = \frac{220}{R+1} \times R$$

For battery to be charge

$$V_A - V_B > 200\,\text{V}$$

$\Rightarrow$ $\frac{220R}{R+1} > 200 \Rightarrow 220R > 200R + 200$

$\Rightarrow$ $R > 10\,\Omega \Rightarrow$ here $11\,\Omega > 10\,\Omega$ **Ans. (C)**

Sol. 11

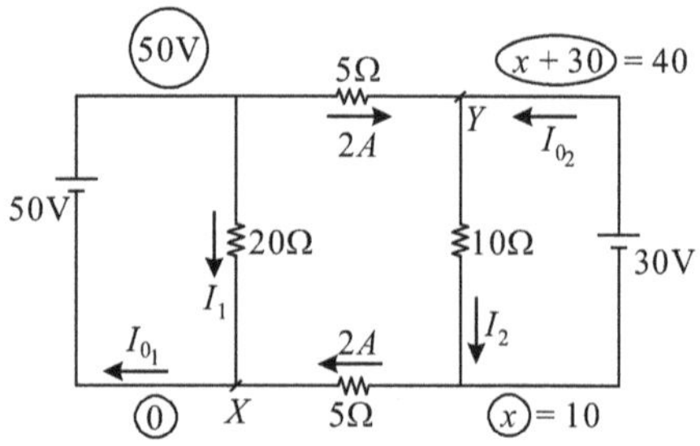

Currents is 20Ω and 10Ω resistances are

$$I_1 = \frac{50}{20} = 2.5\,\text{A}$$

$$I_2 = \frac{30}{10} = 3\,\text{A}$$

KCL equation for (x)

$$\frac{x + 30 - 50}{5} = 0$$

$\Rightarrow$ $2x = 20\,\text{V}$

$\Rightarrow$ $x = 10\,\text{V}$

At junction (x) $I_{0_1} = 2 + 2.5 = 4.5\,\text{A}$

at junction (y) $I_{0_2} = 3 - 2 = 1\,\text{A}$ **Ans. (C)**

Sol. 12 Modified circuit in

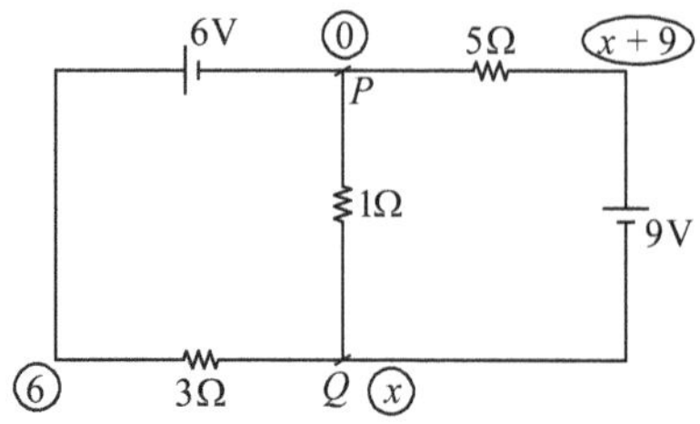

Distributing potentials in the circuit and writing KCL equation for (x), we have

$$\frac{x-6}{3}+\frac{x-0}{1}+\frac{x+9-0}{5}=0$$

$$\Rightarrow \quad 5x-30+15x+3x+27=0$$

$$\Rightarrow \quad 23x=3$$

$$\Rightarrow \quad x=\frac{3}{23}V$$

(As $x>0$ current flows from Q to P)

Current in 1Ω resistance is

$$I=\frac{x}{1}=\frac{3}{23}\simeq 0.13\text{ A}$$ **Ans. (C)**

Sol. 13 We can write equation of potential drop from point D to A along path $DEBA$

$$V_{4\Omega}=0 \text{ as } I_{4\Omega}=0$$

$$V_D-4\text{V}+9\text{V}=V_A$$

$$\Rightarrow \quad V_A-V_D=5\text{V}$$ **Ans. (C)**

Sol. 14

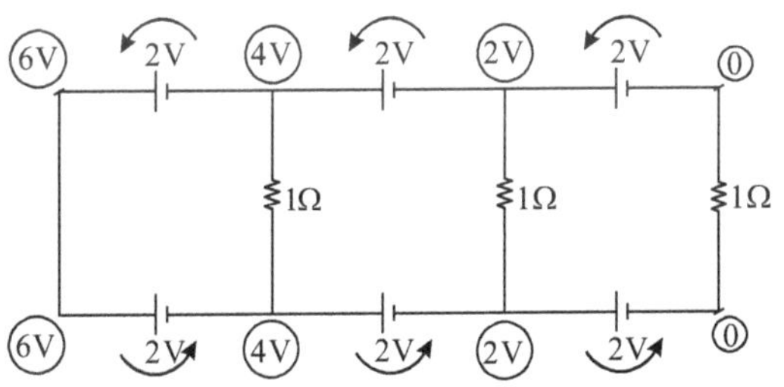

As shown in figure, after distribution of potentials in circuit, potential difference across each resistance is equal to zero.

$$\Rightarrow \quad I_{\text{each }1\Omega}=\frac{V}{R}=0$$ **Ans. (B)**

Sol. 15

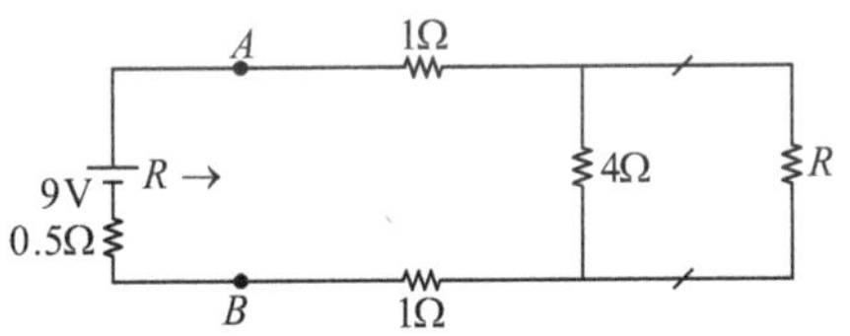

Here R is the equivalent Resistance across terminals A and B

For ∞ grid we use

Across $AB \rightarrow R = 2+\frac{4R}{4+R}$

$$\Rightarrow \quad 4R+R^2=8+6R$$

$$\Rightarrow \quad R^2-2R-8=0$$

$$\Rightarrow \quad R^2-2R-8=0$$

$$\Rightarrow \quad R=\frac{2\pm\sqrt{4+32}}{2}=\frac{2\pm 6}{2}$$

$$\Rightarrow \quad R=\frac{8}{2}=4\,\Omega$$

$\Rightarrow$

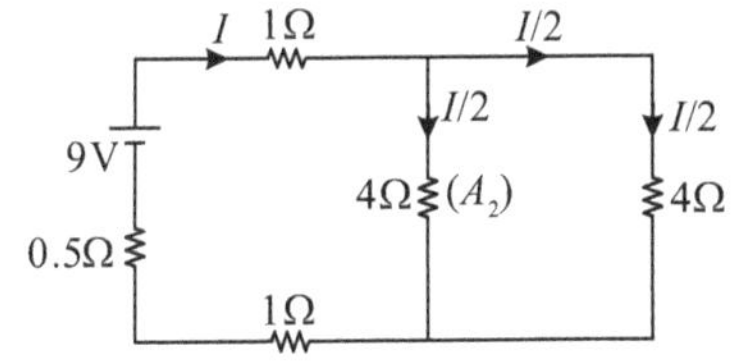

$$I=\frac{9}{4.5}=2\text{ A}$$

Reading of (A_2) is

$$I_2=\frac{2}{2}=1\text{ A}$$

In further part of grid current will divide half in 4 Ω and half to next section

$\Rightarrow$ Reading of (A_3) is

$$\frac{I}{4}=\frac{2}{4}=0.5\text{ A}$$

Reading of voltmeter (V) is

$$=IR=2\times 4=8\text{ V}$$

Here current supplied by battery is

$$I=\frac{9}{0.5+4}=\frac{9}{4.5}=2\text{ A (Reading of } A_1)$$ **Ans. (A)**

Sol. 16

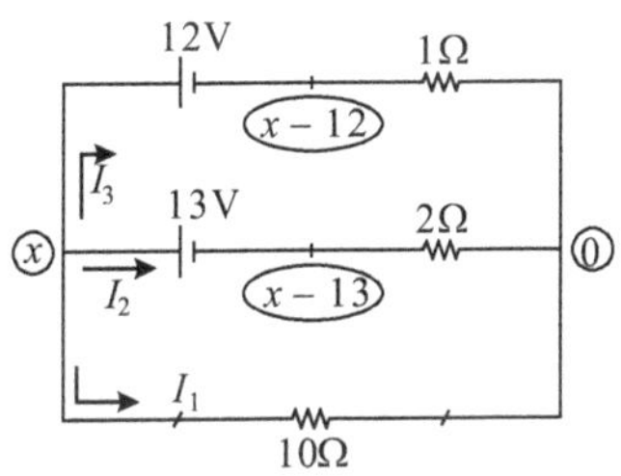

Distributing potentials at junctions of circuit as shown and using KCL for junction with potential x, it gives

$$I_1+I_2+I_3=0$$

$$\Rightarrow \quad \frac{x-0}{10}+\frac{x-13}{2}+\frac{x-12}{1}=0$$

$$\Rightarrow \quad x+5x-65+10x-120=0$$

$$\Rightarrow \quad 16x=185$$

$$\Rightarrow \quad x=11.56\text{ V}$$ **Ans. (B)**

Sol. 17 Equivalent circuit of above situation is

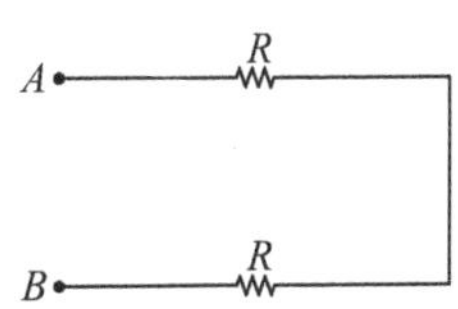

$$\Rightarrow \quad R_{AB}=R+R=2R$$ **Ans. (A)**

Sol. 18 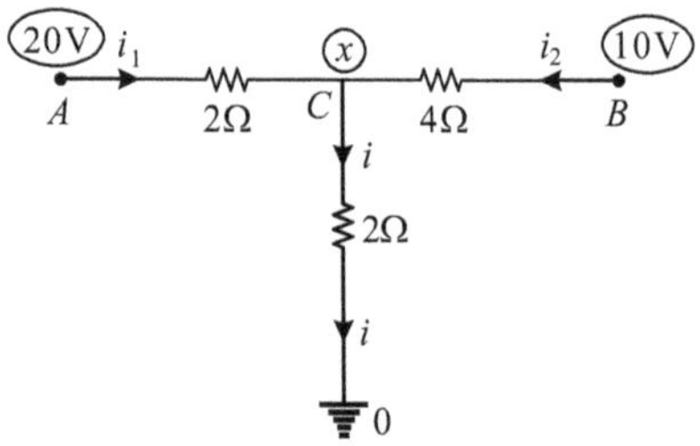

Using KCL equation at C in above circuit, we have

$$\frac{20-x}{2}+\frac{10-x}{4}=\frac{x-0}{2}$$

$\Rightarrow$ $x=10$

$\Rightarrow$ $i=\frac{x}{2}=\frac{10}{2}=5\text{ A}$ **Ans. (B)**

Sol. 19 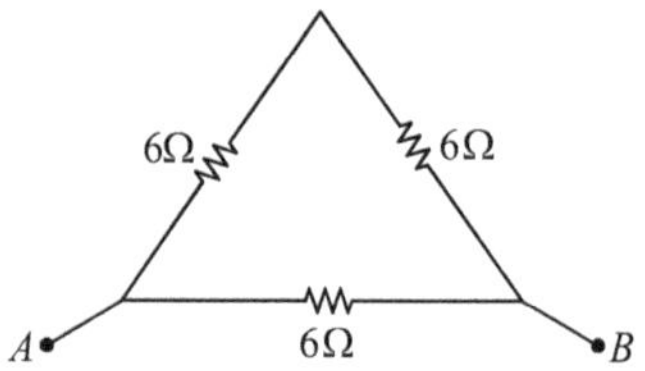

R_{eq} between any two vertex will be

$$\frac{1}{R_{eq}}=\frac{1}{12}+\frac{1}{6}$$

$\Rightarrow$ $R_{eq}=4\ \Omega$ **Ans. (C)**

Sol. 20 From the given circuit, we can write

$$i_1=\frac{10}{20}=0.5\text{ A}$$

$$i_2=\frac{0}{20}=0\text{ A}$$

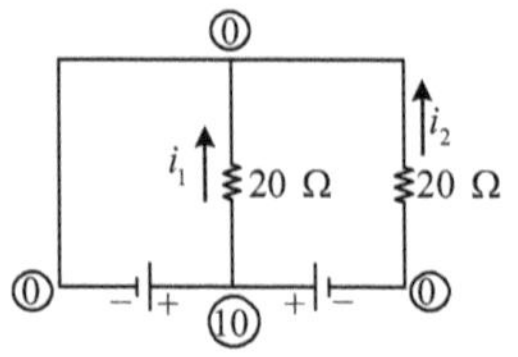

Ans. (D)

Sol. 21 Potential difference across AB will be equal to battery equivalent across CD, given as

$$V_{AB}=V_{CD}=\frac{\frac{E_1}{r_1}+\frac{E_2}{r_2}+\frac{E_3}{r_3}}{\frac{1}{r_1}+\frac{1}{r_2}+\frac{1}{r_3}}=\frac{\frac{1}{1}+\frac{2}{1}+\frac{3}{1}}{\frac{1}{1}+\frac{1}{1}+\frac{1}{1}}$$

$$V_{AB}=\frac{6}{3}=2\text{ V}$$ **Ans. (D)**

Sol. 22 Using KCL, we can write

$$I_3=0.8-0.4=0.4\text{ A}$$

$$I_2=0.4+0.4+0.3$$

$\Rightarrow$ $I_2=1.1\text{ A}$

$I_6=0.4\text{ A}$ **Ans. (A)**

Sol. 23 Here we consider the three batteries in parallel combination so the potential difference across a and b is given as

$$E_{eq}=\frac{\frac{E_1}{2R_1}+\frac{E_2}{R_2}+\frac{E_3}{2R_1}}{\frac{1}{2R_1}+\frac{1}{R_2}+\frac{1}{2R_1}}$$

$\Rightarrow$ $$E_{eq}=\frac{\frac{2}{2}+\frac{4}{2}+\frac{4}{2}}{\frac{1}{2}+\frac{1}{2}+\frac{1}{2}}$$

$\Rightarrow$ $$E_{eq}=\frac{5}{\frac{3}{2}}=\frac{10}{3}=3.3$$ **Ans. (B)**

Sol. 24 The circuit given is reduced as shown below

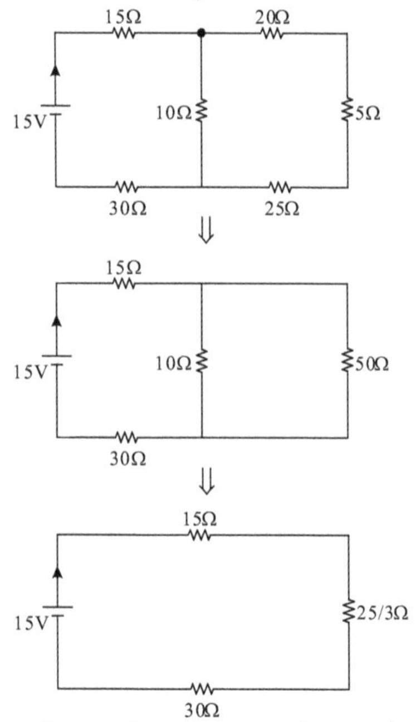

Equivalent resistance across battery is given as

$$R_{eq}=15+\frac{25}{3}+30=\frac{45+25+90}{3}=\frac{160}{3}\Omega$$

Thus current supplied by battery is given as

$$I=\frac{E}{R_{eq}}=\frac{15\times 3}{160}=\frac{9}{32}\text{A}$$ **Ans. (C)**

Sol. 25 The equivalent circuit of the given square is redrawn as shown below

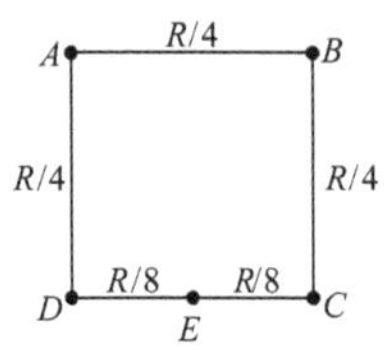

The equivalent resistance across E and C is given as

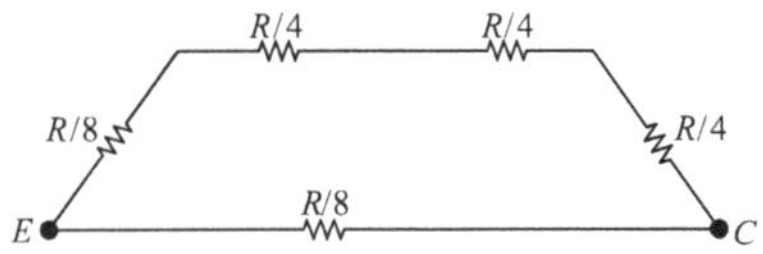

$$\frac{1}{R_{eq}} = \frac{8}{7R} + \frac{8}{R}$$

$$\Rightarrow \quad \frac{1}{R_{eq}} = \frac{8+56}{7R}$$

$$\Rightarrow \quad R_{eq} = \frac{7R}{64}$$ **Ans. (C)**

17.3 Thermal Effects of Current & RC Circuits

Sol. 1 By Faraday's first law of electrolysis

$$m = Zq = ZIt$$

$m \propto It$ **Ans. (B)**

Sol. 2 Initial power dissipation in wire

$$P_1 = \frac{(220)^2}{R} \qquad [R \to \text{resistance of wire}]$$

When wire cut in two equal pieces, each will have resistance $R/2$ and these are connected in parallel

$$\Rightarrow \quad R_{eq} = \frac{R}{4}$$

new power dissipation will be

$$P_2 = \frac{(220)^2}{R/4} = 4\left[\frac{(220)^2}{R}\right] = 4P_1$$

$$\Rightarrow \quad P_2/P_1 = 4$$ **Ans. (B)**

Sol. 3 Here $R||2 = \frac{2R}{2+R}$ resistance is across 15V battery so power dissipation is given as

$$P = \frac{V^2}{R} = \frac{(15)^2}{2R} \times (2+R) = 150$$

$$\Rightarrow \quad 15 \times (2+R) = 10 \times 2R$$

$$\Rightarrow \quad 30 + 15R = 20R$$

$$\Rightarrow \quad R = \frac{30}{5} = 6\,\Omega$$ **Ans. (B)**

Sol. 4 As minimum current that can be detected by galvanometer is

$$I_{min} = 10^{-5}\text{A}$$

If minimum temperature difference which can be measured is $\Delta\theta$ then we use

$$I_{min} = \frac{25 \times 10^{-6} \times \Delta\theta}{R}$$

$$\Rightarrow \quad \Delta\theta = \frac{I_{min} \times R}{25 \times 10^{-6}} = \frac{10^{-5} \times 40}{25 \times 10^{-6}} = \frac{400}{25}$$

$$\Rightarrow \quad \Delta\theta = 16\,°\text{C}$$ **Ans. (A)**

Sol. 5 Mass of metal at negative pole deposited or at positive pole used is given as

$$m = Zq$$

for Zn & Cu electrodes, we use

$$\frac{m_1}{m_2} = \frac{Z_1}{Z_2}$$

$$\Rightarrow \quad m_2 = m_1 \times \frac{Z_2}{Z_1} = 0.13\text{g} \times \frac{31.5}{32.5}$$

$$\Rightarrow \quad m_2 \simeq 0.126\text{g}$$ **Ans. (C)**

Sol. 6 The resistance of bulb is

$$R = \frac{V^2}{P} = \frac{(220)^2}{1000}\,\Omega$$

Power consumed across 110V supply is given as

$$P' = \frac{V'^2}{R} = \frac{(110)^2}{(220)^2/1000} = \frac{1000}{4}$$

$$\Rightarrow \quad P' = 250\text{ W}$$ **Ans. (C)**

Sol. 7 The neutral temperature is the temperature of hot junction when thermo EMF is maximum and cold junction is kept at 0°C.

Here in this case

$$E = a\theta + b\theta^2$$

E is maximum when

$$\frac{dE}{d\theta} = a + 2b\theta = 0$$

$$\Rightarrow \quad \theta = -\frac{a}{2b} = -350°\text{C}$$

We check $\quad \frac{d^2E}{d\theta^2} = 2b\,(+ve)$ value

$\Rightarrow$ E will never attain a maxima.

So no neutral temperature is possible is this case. **Ans. (D)**

Sol. 8 Mass of metal deposited is given as

$$m = Zq$$

charge flown $\quad q = It = 3 \times 2 = 6\text{C}$

mass $\quad m = 3.35 \times 10^{-7} \times 6$

$\quad = 20.10 \times 10^{-7}$ kg **Ans. (C)**

Sol. 9 The heat required to raise the temperature of 1kg water from 10°C to 40°C is

$$\Delta Q = mS_{\omega}\Delta t$$

$$\Rightarrow \quad \Delta Q = 1 \times 4180 \times 30 \text{ J} \qquad [s_{\omega} = 4180 \text{ J/°C kg}]$$

It t is the time required then heat supplied by heater is

$$836t = 4180 \times 30$$

$$t = \frac{300}{2} = 150 \text{ s} \quad \textbf{And. (A)}$$

Sol. 10 Constant current source

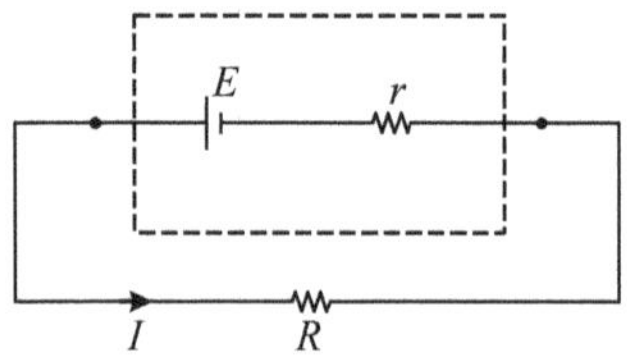

Current supplied

$$I = \frac{E}{r+R}$$

Here if $r >> R$

$$\Rightarrow \quad I = \frac{E}{r} \rightarrow \text{constant for small variations in } R \qquad \textbf{Ans. (B)}$$

Sol. 11 If q_1 and q_2 are charges flown through copper & silver voltameter then we use

$$q = q_1 + q_2 \qquad \ldots(1)$$

[As the two voltmeter are in parallel]

mass of metal deposited is given as

$$m = Zq$$

As for both, mass deposited are same

$$\Rightarrow \quad Z_1q_1 = Z_2q_2$$

$$\Rightarrow \quad q_1 = q_2\left(\frac{Z_2}{Z_1}\right)$$

$$\Rightarrow \quad q = q_2\left(\frac{Z_2}{Z_1} + 1\right)$$

$$\Rightarrow \quad q_2 = \frac{q}{\frac{Z_2}{Z_1} + 1} \qquad \textbf{Ans. (A)}$$

Sol. 12 When a heater coil is placed across a supply of voltage V then heat produced rate is

$$H_1 = \frac{V^2}{R}$$

Each part of coil after cut has a resistance $R/2$ so now heat generation rate is

$$H_2 = \frac{V^2}{R/2} = 2\left(\frac{V^2}{R}\right) = 2H_1 \qquad \textbf{Ans. (B)}$$

Sol. 13 For the given temperature its hot resistance is given as

$$R_{\text{hot}} = \frac{V^2}{P}$$

$$= \frac{(200)^2}{100} = 400\,\Omega$$

Given that $\quad R_{\text{cold}} = \frac{1}{10} \cdot R_{\text{hot}}$

$$= \frac{1}{10} \times 400 = 40\,\Omega \qquad \textbf{Ans. (B)}$$

Sol. 14

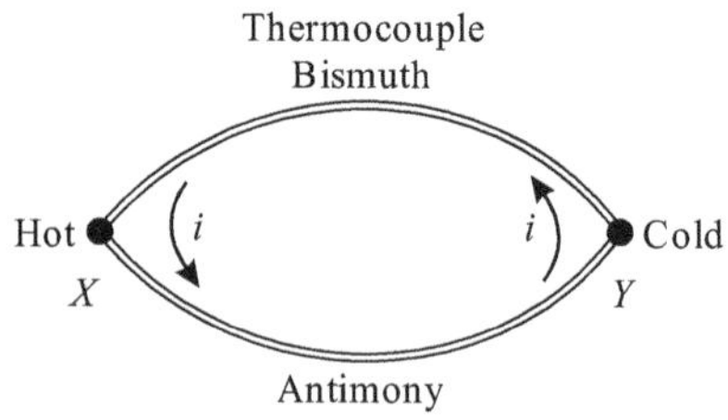

Thermoelectric series

Bismuth

↓

Antimony **Ans. (D)**

Sol. 15 The resistance of bulb is

$$R = \frac{V^2}{P} = \frac{(220)^2}{100}$$

Power consumed when operated at 110V is given as

$$P_C = \frac{V_1^2}{R} = \frac{(110)^2}{(220)^2/100} = \frac{100}{4}$$

$$\Rightarrow \quad P_C = 25 \text{ W} \qquad \textbf{Ans. (C)}$$

Sol. 16 When a capacitor discharges through a resistor, its charge as time function is given as

$$q = Qe^{-t/RC}$$

Where Q is charge at $t = 0$

Energy in capacitor

$$U = \frac{q^2}{2C} = \left(\frac{Q}{2C}\right)^2 e^{-2t/RC} = U_0e^{-2t/RC}$$

Given that at $t = t_1$, $U = \frac{U_0}{2}$

$$\Rightarrow \quad \frac{U_0}{2} = U_0e^{-2t_1/RC}$$

$$\Rightarrow \quad \frac{2t_1}{RC} = \ln(2)$$

$$\Rightarrow \quad t_1 = \frac{RC}{2}\ln(2) \qquad \ldots(1)$$

At $t = t_2$, $\quad q = \dfrac{Q}{4}$

$$\Rightarrow \quad \frac{Q}{4} = Qe^{-t_2/RC}$$

$$\Rightarrow \quad \frac{t_2}{RC} = \ln(4) = 2\ln(2)$$

$$\Rightarrow \quad t_2 = 2RC\ln(2) \qquad \ldots(2)$$

$$\frac{(1)}{(2)} \Rightarrow \quad \frac{t_1}{t_2} = \frac{\frac{RC}{2}\ln(2)}{2RC\ln(2)} = \frac{1}{4} \qquad \textbf{Ans. (D)}$$

Sol. 17 For discharging of capacitors in RC circuit, voltage across capacitor combination is given as

$$V = V_0 e^{-t/RC_{eq}}$$

For parallel combination $C_{eq} = 2C$ and at $t = 10s$, $V = \dfrac{V_0}{2}$

$$\Rightarrow \quad \frac{V_0}{2} = V_0 e^{-10/2RC} \qquad \ldots(1)$$

for series combination

$C_{eq} = C/2$ and at $t = t_1$,

$$V = \frac{V_0}{2}$$

$$\Rightarrow \quad \frac{V_0}{2} = V_0 e^{-2t_1/RC} \qquad \ldots(2)$$

Using (1) and (2) we have

$$\frac{2t_1}{RC} = \frac{10}{2RC}$$

$$\Rightarrow \quad t_1 = \frac{10}{4} = 2.5s \qquad \textbf{Ans. (D)}$$

Sol. 18

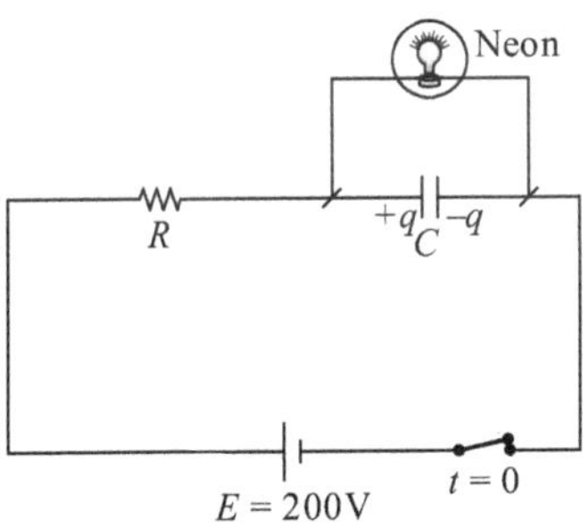

In RC circuit charge on capacitor as a function of time is given as

$$q = CE(1 - e^{-t/RC})$$

p.d. across capacitor as a function of time is given as

$$V = \frac{q}{C} = E(1 - e^{-t/RC})$$

at $t = 5$, $\quad V = 120V$

$$120 = 200(1 - e^{-5/RC})$$

$$\Rightarrow \quad e^{-5/RC} = 1 - \frac{120}{200} = \frac{2}{5}$$

$$\Rightarrow \quad \frac{5}{RC} = \ln(2.5) = 2.303 \times 0.4$$

$$\Rightarrow \quad R = \frac{5}{2\times10^{-6}\times 2.303\times 0.4} = 2.7 \times 10^6\,\Omega \qquad \textbf{Ans. (A)}$$

Sol. 19

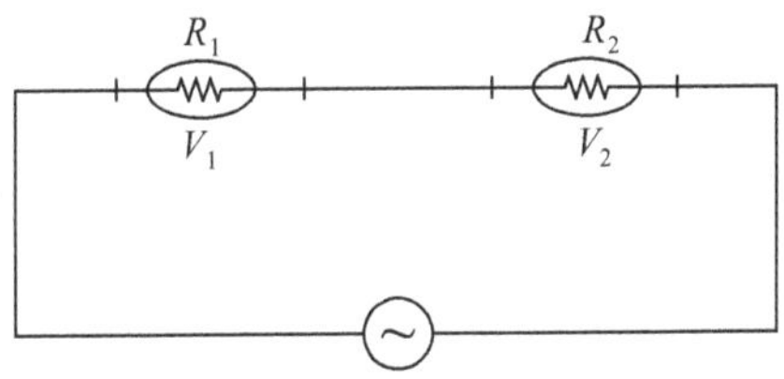

Resistances of bulbs are given as

$$R_1 = \frac{V_1^2}{P_1} = \frac{(220)^2}{25}$$

and $\quad R_2 = \dfrac{V_2^2}{P_2} = \dfrac{(220)^2}{100} \quad \Rightarrow \quad R_1 > R_2$

$$V_1 = \frac{440R_1}{R_1 + R_2} \quad \text{and} \quad V_2 = \frac{440R_2}{R_1 + R_2}$$

As $\quad V_1 + V_2 = 440$

$$\Rightarrow \quad V_1 > V_2$$

$$\Rightarrow \quad V_1 > 220V \quad \text{and} \quad V_2 < 220V$$

$\Rightarrow$ 25 W bulb will fuse **Ans. (D)**

Sol. 20 In RC discharging circuit if at $t = 0$, voltage across capacitor is V_0 then after time $t = t$, it is given as

$$V = V_0 e^{-t/RC} \qquad [\text{from graph at } t = 0,\ V_0 = 25\text{ V}]$$

where time constant $\tau = RC$

$$\Rightarrow \quad V = V_0 e^{-t/\tau} \qquad \ldots(1)$$

at $t = 200s$ from graph we can see that $V = 5$ V

from equation-(1), we use

$$5 = 25\, e^{-200/\tau}$$

$$\Rightarrow \quad \ln 5 = \frac{200}{\tau}$$

$$\Rightarrow \quad \tau = \frac{200}{\ln(5)} = \frac{200}{2.303\log(5)} = \frac{200}{2.303\times 0.7}$$

[As $\log 5 = 0.698$]

$$\Rightarrow \quad \tau \cong \frac{200}{1.6} \simeq 125 \text{ s} \qquad \textbf{Ans. (A)}$$

Sol. 21 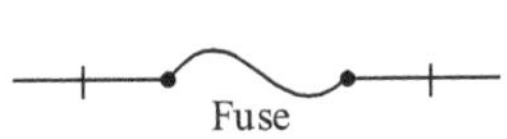

it burns when its temperature exceeds the melting point of resistance material. Heat production rate in fuse

$$H = \frac{V^2}{R}$$

When the current is switched on, R is low as it is cold (at room temperature) and with flow of current R increases and H decreases.

(due to rise in temperature)

$\Rightarrow$ Statement-1 is true.

Statement-2 : When a bulb is lit up, due to heat its resistance increases

$\Rightarrow$ Statement-2 is true & explaining Statement-1 also

$\Rightarrow$ **Ans. (C)**

Sol. 22

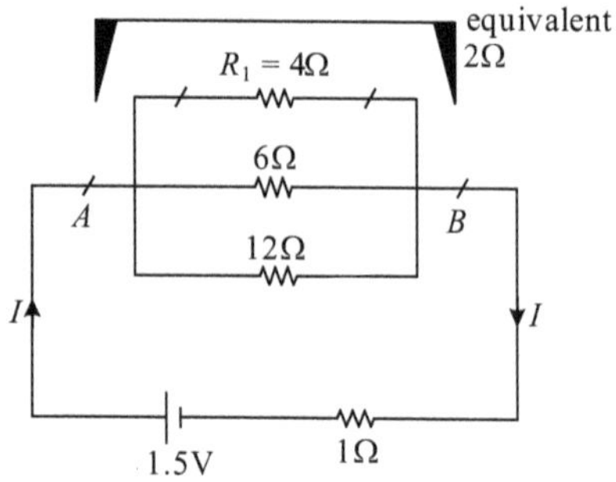

Equivalent resistance of parallel combination is

$$\frac{1}{R} = \frac{1}{4}+\frac{1}{6}+\frac{1}{12} = \frac{3+2+1}{12} = \frac{6}{12} = 0.5$$

$\Rightarrow$ $R = 2\,\Omega$ (Across A and B)

Circuit current

$$I = \frac{1.5}{1+2} = \frac{1.5}{3} = \frac{1}{2} = 0.5\text{ A}$$

Potential difference across A and B is

$$V_{AB} = IR = 0.5 \times 2 = 1\text{ V}$$

Rate of Joule heating in 4Ω resistance is

$$P_{\text{th}} = \frac{V_{AB}^2}{R_1} = \frac{(1)^2}{4} = 0.25\text{ W}$$ **Ans. (C)**

Sol. 23 Bulb brightess α thermal power produced in it

$\Rightarrow$ Brightness $\propto \frac{V^2}{R}$

$\Rightarrow$ Brightness $\propto \frac{1}{R}$ [As V is same for both bulbs]

$$R_1 = 3\,\Omega \quad \text{and} \quad R_2 = 6\,\Omega$$

Bulb-1 will glow brighter **Ans. (B)**

Sol. 24

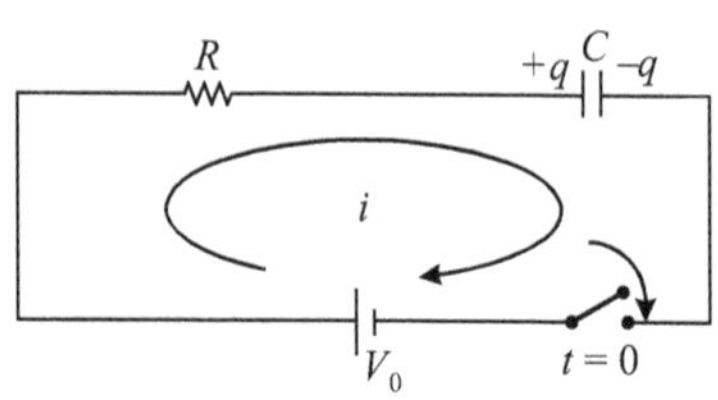

The charge on capacitor as a function of time is

$$q = CV_0(1 - e^{-t/RC})$$

Charge current in circuit is

$$i = \frac{dq}{dt} = \frac{V_0}{R}e^{-t/RC}$$

Voltage across capacitor at time t is

$$V_C = \frac{q}{C} = V_0(1 - e^{-t/RC}) \qquad \ldots(1)$$

and voltage across resistance is

$$V_R = iR = V_0 e^{-t/RC} \qquad \ldots(2)$$

At time t is given that

$$V_C = 7V_R$$

So from (1) and (2)

$$\%(1 - e^{-t/RC}) = 7\% e^{-t/RC}$$

$\Rightarrow$ $1 = 8e^{-t/RC}$

$\Rightarrow$ $\frac{t}{RC} = \ln(8)$

$\Rightarrow$ $t = RC\ln(8) = 3RC\ln(2)$ **Ans. (A)**

Sol. 25

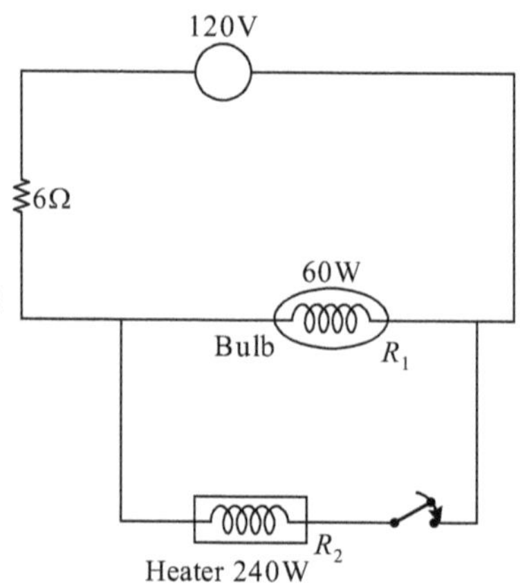

Resistance of heater

$$R_2 = \frac{V^2}{P} = \frac{(120)^2}{240} = 60\,\Omega$$

Resistance of bulb

$$R = \frac{V^2}{P} = \frac{(120)^2}{60} = 240\,\Omega$$

Initially voltage across bulb is

$$V_1 = \frac{120}{246} \times 240$$

$$\simeq 117.07\text{ V}$$

When heater is switched on the equation equivalent of bulb and heater is

$$R_{\text{eq}} = \frac{240 \times 60}{240 + 60} = 48\,\Omega$$

Potential difference across bulb & heater combination is

$$V_2 = \frac{120}{54} \times 48 \simeq 106.6\,\text{V}$$

Change in potential difference across bulb is

$$\Delta V = V_1 - V_2 \simeq 10.04\,\text{V}$$ **Ans. (D)**

Sol. 26

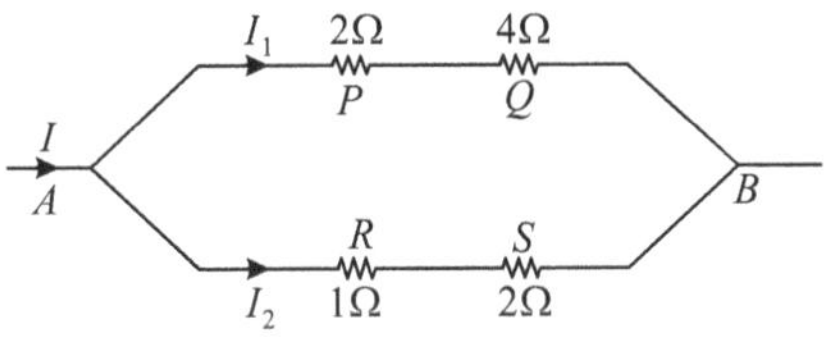

Here I_1 and I_2 are given as

$$I_1 = \frac{I}{3} \text{ and } I_2 = \frac{2I}{3}$$

Thermal power dissipated in resistances are

$$H_P = I_1^2(2) = \frac{2I^2}{9}\,; H_R = I_2^2\,(1) = \frac{4I^2}{9}$$

$$H_Q = I_1^2\,(4) = \frac{4I^2}{9}\,; H_S = I_2^2\,(2) = \frac{8I^2}{9}$$

$\Rightarrow$ $$H_S > (H_R = H_Q) > H_P$$ **Ans. (B)**

Sol. 27 Total power consumption of building at full load is

$$P = 15 \times 40 + 5 \times 100 + 5 \times 80 + 1000$$

$\Rightarrow$ $$P = 600 + 500 + 400 + 1000 = 1400\text{W}$$

The current drawn from mains supply at full load is

$$I = \frac{P}{V} = \frac{2500}{220} = \frac{125}{11} = 11.36\,\text{A}$$

Maximum current supplied at full load

As fuse rating must be kept such that maximum current at full load can sustain. **Ans. (C)**

Sol. 28

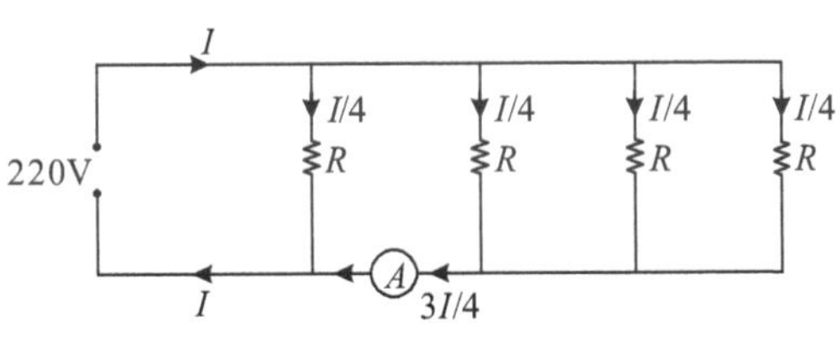

As each bulb is of 100W consumption

$\Rightarrow$ $$\frac{I}{4} = \frac{P}{V} = \frac{100}{220} = \frac{5}{11} = 0.45\,\text{A}$$

The current through ammeter is

$$I_A = \frac{3I}{4} = 3 \times 0.45 = 1.35\,\text{A}$$ **Ans. (C)**

Sol. 29 In steady state no current flows through capacitive branches of circuit

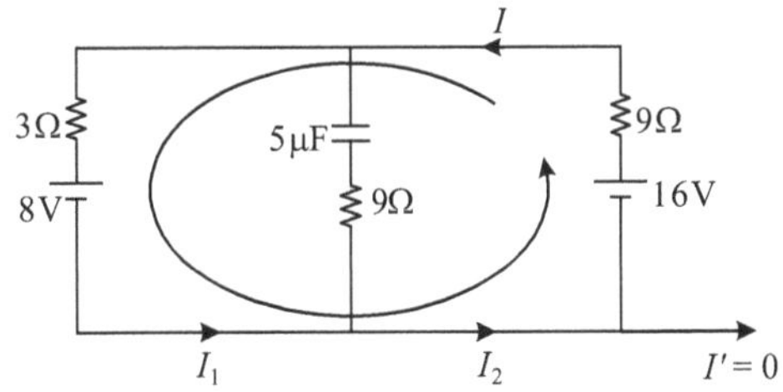

As current flows only in outer loop

$\Rightarrow$ $$I_1 = I_2 = I$$

Current in outer loop is

$$I = \frac{8}{9+3}$$

$$= \frac{8}{12} = \frac{2}{3}A = 0.67\,\text{A}$$ **Ans. (B)**

Sol. 30

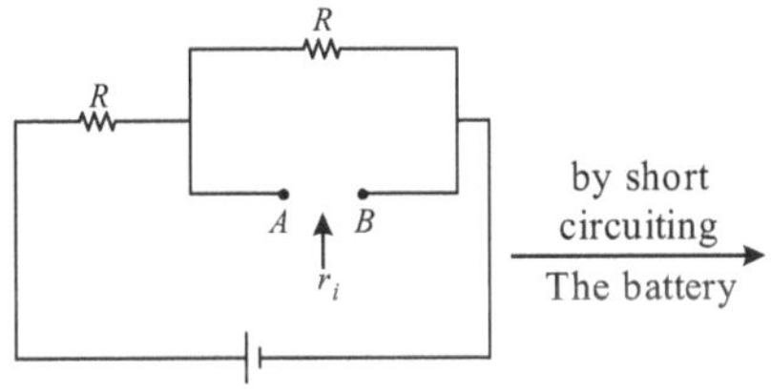

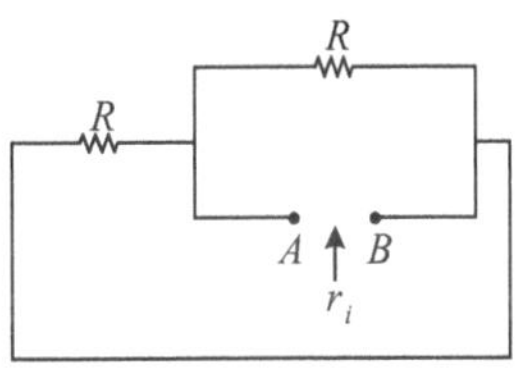

$\Rightarrow$ $$r_i = R \parallel R = R/2$$

According to maximum power transfer theorem, heat generation in r is maximum when

$$r = r_i = R/2 = fR$$

$\Rightarrow$ $$f = \frac{1}{2}$$ **Ans. (A)**

Sol. 31 At 300 K, $R = 100\Omega$

and at 500 K, $R = 120\Omega$

$$120 = 100\,[1 + \alpha(200)]$$

$\Rightarrow$ $$20 = \alpha(20000) \;\Rightarrow\; \alpha = 10^{-3}\,°\text{C}^{-1}$$

When toaster is connected across $V = 200$ volt then thermal power produced is

$$P_{\text{th}} = \frac{V^2}{R} = \frac{(200)^2}{100[1 + \alpha(T - 300)]}$$

at temperature T

$$300\text{K} < T < 500\,\text{K}$$

As T is raised at a constant rate (linear variation) from 300K to 500 K, so time function of temperature is given as

$$T = 300 + \frac{500 - 300}{30} \cdot t = 300 + \frac{20}{3}t$$

Total work done in these 30s of rise in temperature is

$$W = \int_0^{30} P_{th} \cdot dt = \int_0^{30} \frac{(200)^2}{100\left[1+\alpha\left(\frac{20t}{3}\right)\right]} \cdot dt$$

$$\Rightarrow \quad W = 400 \times \frac{300}{20\alpha}\left[\ln\left(1+\frac{200\alpha t}{3}\right)\right]_0^{30}$$

$$\Rightarrow \quad W = \frac{60}{\alpha}[\ln(1+200\alpha) - \ln(1)]$$

$$\Rightarrow \quad W = 60000 \ln(1.2)$$

Ans. (None)

None of the given option is correct.

Sol. 32

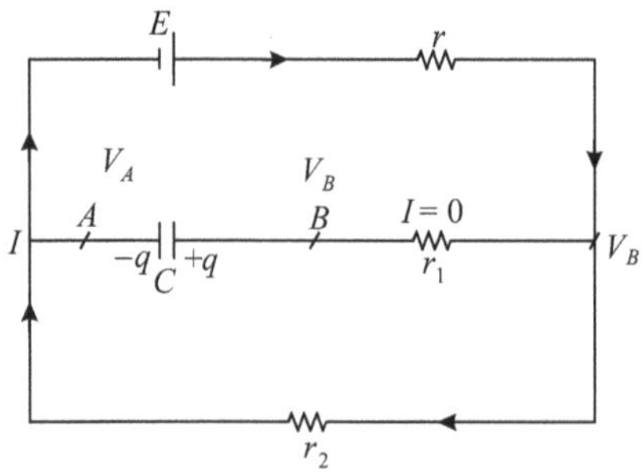

In steady state no current flows through capacitive branch so current only flows in outer loop as shown, and it is given as

$$I = \frac{E}{r + r_2}$$

Potential difference across capacitor is

$$V_B - V_A = Ir_2 = \frac{Er_2}{r+r_2}$$

$\Rightarrow$ Charge on capacitor is

$$q = C(V_B - V_A) = \frac{CEr_2}{r+r_2}$$

Ans. (A)

Sol. 33 For a group of resistances connected across a battery the power dissipated is given as

$$P = \frac{V^2}{R_{eq}}$$

$$R_I = 1\,\Omega \rightarrow P_1 = 9\text{ W}$$

$$R_{II} = 1\,\Omega \,||\, 2\,\Omega \,||\, 2\,\Omega$$

$$P_2 = 9/0.5 = 18\text{ W} = 0.5\,\Omega$$

$$R_{III} = 1.5\,\Omega \rightarrow P_3 = 9/1.5 = 6\text{ W}$$

$$\Rightarrow \quad P_2 > P_1 > P_3$$

Ans. (C)

Sol. 34

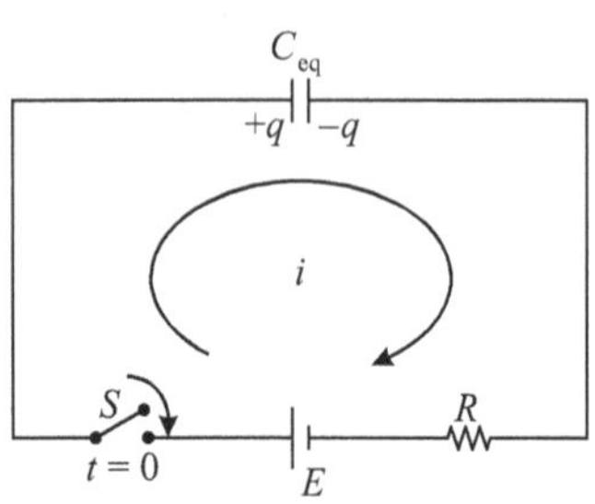

If S is closed at $t = 0$ then charge on capacitor plates as a function of time is given as

$$q = C_q E\,(1 - e^{-t/RC_{eq}})$$

As C_1 and C_2 are in series, charge on their plates must be equal to q only.

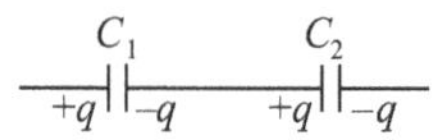

Ans. (A)

Sol. 35

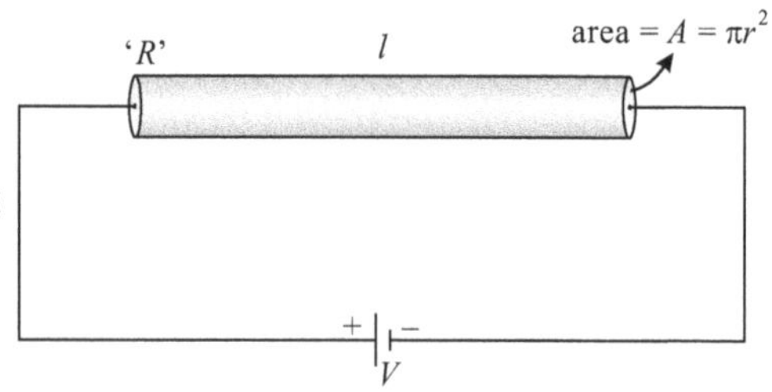

Resistance of wire is

$$R = \frac{\rho l}{A}$$

Thermal power produced in wire

$$P_H = \frac{V^2}{R}$$

when $\quad l \rightarrow l/2$

and $\quad r \rightarrow 2r$

$\Rightarrow \quad A \rightarrow 4A$

$\Rightarrow$ New resistance of wire becomes

$$R' = \frac{\rho(l/2)}{4A} = \frac{1}{8}\frac{\rho l}{A} = \frac{R}{8}$$

Now thermal power produced in wire becomes

$$P'_H = \frac{V^2}{R/8} = \frac{8V^2}{R} = 8P_H$$

Ans. (A)

Sol. 36 Power dissipation in resistance is

$$P = I^2R$$

$$\Rightarrow \quad 4.4 = 4 \times 10^{-6} R$$

$$\Rightarrow \quad R = 1.1 \times 10^6\,\Omega$$

When 11 V power supply is used then

$$P' = \frac{11^2}{R} = \frac{11^2}{1.1} \times 10^{-6} = 11 \times 10^{-5}\text{ W}$$

Ans. (A)

Sol. 37 In series combination, equivalent resistance is 2R thus power consumed is $60\text{W} = \frac{\varepsilon^2}{2\text{R}}$

In parallel combination, equivalent resistance is $R/2$ thus new power is consumed is

$$P' = \frac{\varepsilon^2}{(R/2)} = 4P = 4 \times 60 = 240\text{ W}$$

Ans. (B)

Sol. 38 Resistances of the two bulbs are given as

$$R_1 = \frac{220^2}{25}$$

and $$R_2 = \frac{220^2}{100}$$

When bulbs are connected in series, we use

$$I = \frac{220}{R_1 + R_2}$$

Power consumed by the bulbs are

$$P_1 = i^2 R_1$$

$$\Rightarrow \quad P_1 = \frac{220^2}{\left(\frac{220^2}{25} + \frac{220^2}{100}\right)} \times \frac{220^2}{25}$$

$$\Rightarrow \quad P_1 = \frac{400}{25} = 16 \text{ W}$$

and $$P_2 = i^2 R_2 = \frac{220^2}{\left(\frac{220^2}{25} + \frac{220^2}{100}\right)} \times \frac{220^2}{25} = 4 \text{ W}$$

Ans. (C)

Sol. 39 Rate of magnetic energy storage in inductor is

$$\frac{d}{dt}\left(\frac{1}{2}LI^2\right) = LIdI$$

$$\Rightarrow \quad LIdI = I^2R$$

$$LdI = IR$$

We use current as $$I = \frac{E}{R}(1 - e^{-Rt/2})$$

$$\Rightarrow \quad L \times \frac{E}{10}(-e^{-t/2}) \times \frac{-1}{2} = \frac{E}{10}(1 - e^{-t/2}) \times 10$$

$$\Rightarrow \quad e^{-t/2} = 1 - e^{-t/2}$$

$$\Rightarrow \quad t = 2 \ln 2$$

Ans. (C)

Sol. 40

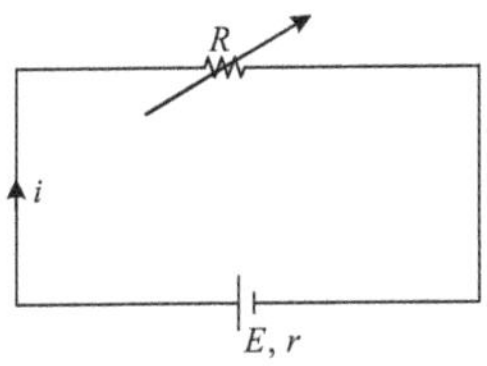

Current in external resistance is given as

$$i = \frac{E}{r+R}$$

Power generated in R is given as

$$P = i^2 R$$

$$P = \frac{E^2 R}{(r+R)^2}$$

For maximum power, we use

$$\frac{dP}{dR} = 0$$

$$\Rightarrow \quad E^2\left[\frac{(r+R)^2 \times 1 - R \times 2(r+R)}{(r+R)^4}\right] = 0$$

$$\Rightarrow \quad r = R$$

Ans. (D)

Sol. 41 Applying point potential method of potential distribution in given circuit as shown below

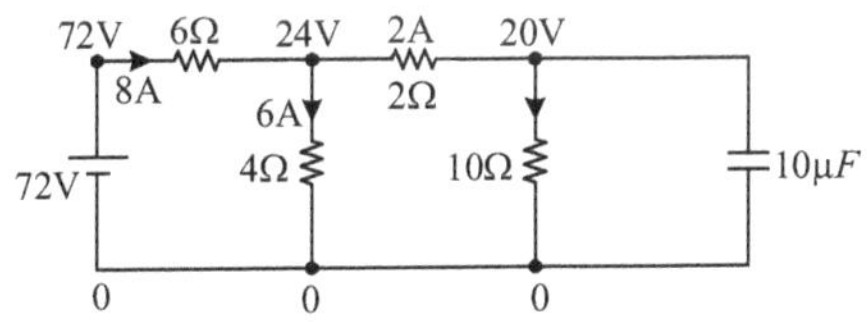

$$q = CV$$

$$q = 10\ \mu F \times 20 = 200\ \mu\text{C}$$

Ans. (C)

Sol. 42

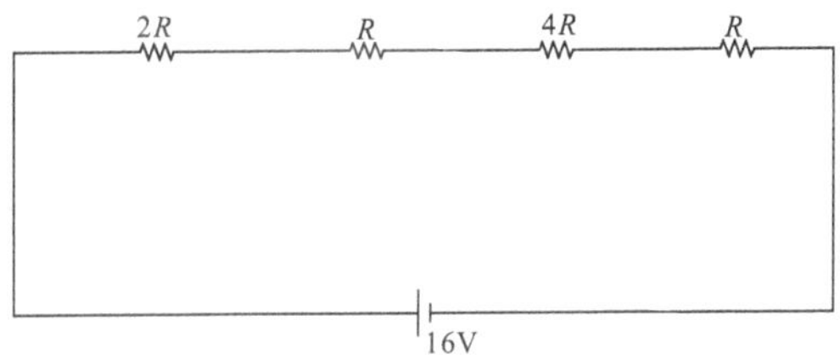

Power consumed by equivalent resistance $8R$ is circuit is

$$P = \frac{16^2}{8R} = 4$$

$$\Rightarrow \quad R = 8\ \Omega \quad \textbf{Ans. (A)}$$

Sol. 43 Total power supplied in time t is

$$Q = P \times t$$

and we use $$Q = mc\Delta T + mL$$

Power supplied by mains is $P = \frac{V_{rms}^2}{R}$

Substituting all values, we have

$$4200 \times 80 + 2260 \times 10^3 = \frac{(200)^2}{20} \times t$$

$$\Rightarrow \quad t = 1298 \text{ s}$$

$$\Rightarrow \quad t \simeq 22 \text{ min}$$

Ans. (B)

17.4 Electrical Measuring Instruments

Sol. 1 Ammeter → its internal resistance is very low

Voltmeter → its internal resistance is very high

When we wish to convert an ammeter into a voltmeter then we connect a high resistance in series. **Ans. (C)**

Sol. 2 To convert a galvanometer/ammeter into a higher range ammeter we need to connect a shunt resistance to bypass the excess current.

The required shunt resistance can be given as

$$(I_R - I_m)R_S = I_m Rg$$

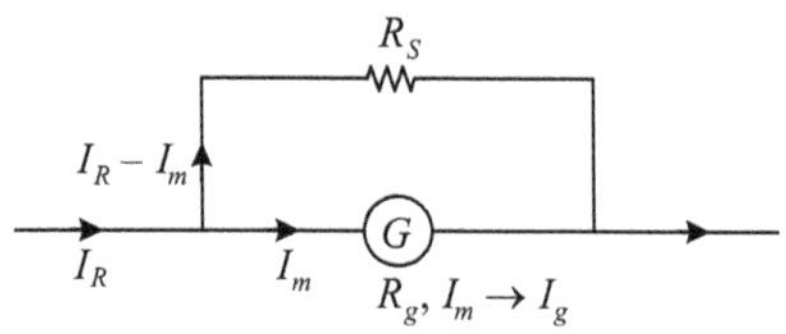

$$\Rightarrow \quad R_S = \frac{I_m R_g}{I_R - I_m} = \frac{1\times 0.81}{10-1} = \frac{0.81}{9}$$

$$\Rightarrow \quad R_S = 0.09\,\Omega$$ **Ans. (D)**

Sol. 3 Across potentiometer wire, potential gradient is

$$\lambda = \frac{E}{L} = \frac{E}{100\text{ cm}}$$

For the given battery if unknown emf is E_{uk} and this balances at l length on wire so we use

$$E_{uk} = \lambda l = \frac{E}{100}\times 30 = \frac{30E}{100}$$ **Ans. (D)**

Sol. 4 Case-I

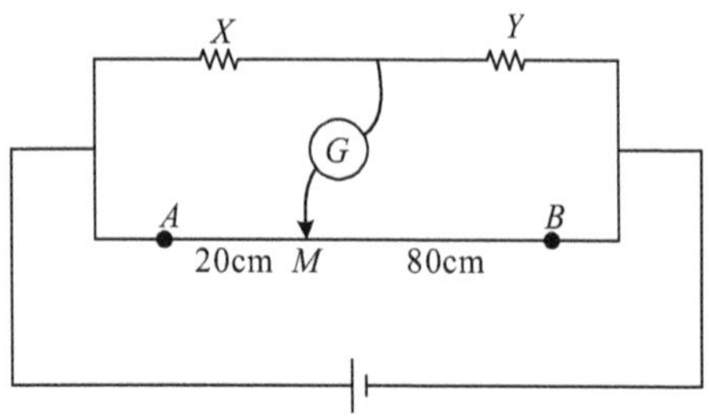

At null point we have bridge balanced

$$\Rightarrow \quad \frac{X}{20} = \frac{Y}{80} \quad \Rightarrow \quad y = 4x$$

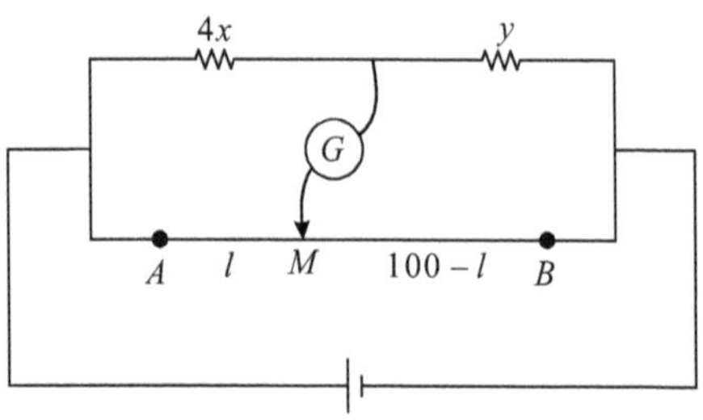

At null point as bridge is balanced

$$\Rightarrow \quad \frac{4X}{l} = \frac{Y}{100-l}$$

$$\Rightarrow \quad 2l = 100$$

$$\Rightarrow \quad l = 50\text{ cm}$$ **Ans. (A)**

Sol. 5 If λ be the potential gradient on potentiometer wire then

Case-I : emf of cell is

$$E = \lambda(240) \qquad \ldots(1)$$

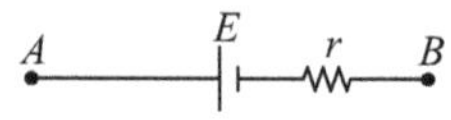

Case-II : potential difference across cells is

$$V_{AB} = \frac{2E}{2+r} = \lambda(120) \qquad \ldots(2)$$

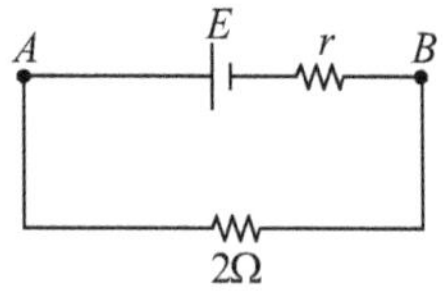

$$V_{AB} = \left(\frac{E}{r+2}\right)\cdot 2$$

$$\frac{(1)}{(2)} \Rightarrow \quad \frac{2+r}{2} = \frac{240}{120} = 2$$

$$\Rightarrow \quad 2 + r = 4$$

$$\Rightarrow \quad r = 2\,\Omega$$ **Ans. (D)**

Alternative : The internal resistance of a cell by potentiometer experiment is given as

$$r = \left(\frac{l_1}{l_2} - 1\right)R$$

$$\Rightarrow \quad r = \left(\frac{240}{120} - 1\right)\times 2 = 2\,\Omega$$ **Ans. (D)**

Sol. 6

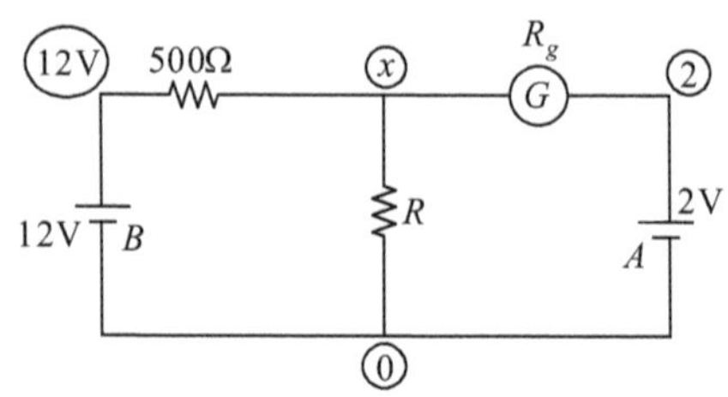

Distributing potentials at different junctions of circuit by considering zero volt reference at bottom junctions.

As current through 'G' is zero

$$\Rightarrow \quad x = 2\text{ V}$$

Writing KCL equation at junction x, we have

$$\frac{x-12}{500} + \frac{x-0}{R} + \frac{x-2}{R_g} = 0$$

$$x = 2\text{ V}$$

$$\Rightarrow \quad -\frac{10}{500} + \frac{2}{R} = 0 \quad \Rightarrow \quad R = 100\,\Omega$$ **Ans. (B)**

Sol. 7 Full deflection current of galvanometer

$$I_{max} = \frac{150}{10} = 15\text{ mA}$$

Full deflection voltage of galvanometer

$$V_{max} = \frac{150}{10} = 75\,\text{mV}$$

at a potential difference of 75mV across galvanometer, a current of 15mA flows through it. So coil resistance is given as

$$R_C = \frac{V_{max}}{I_{max}} = \frac{75}{15} = 5\,\Omega$$

New range of galvanometer to be used as voltmeter is

$$V_R = 150 \times 1 = 150\,\text{V}$$

So if R is external resistance connected in series, we use

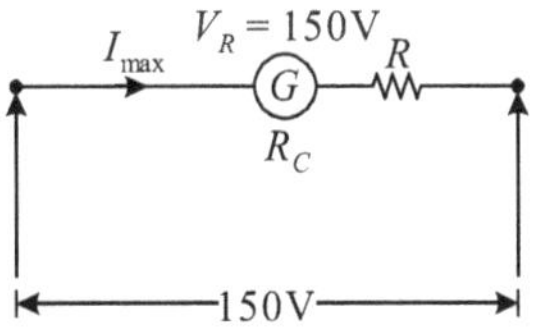

So use $$\frac{150}{R+R_C} = I_{max} = 15\,\text{mA}$$

$$\Rightarrow \quad R + R_C = \frac{150}{15 \times 10^{-3}} = 10^4\,\Omega$$

$$\Rightarrow \quad R = 10^4\,\Omega - 5\,\Omega = 9995\,\Omega$$

Ans. (D)

Sol. 8 For bridge to be balanced we use

$$\frac{55}{20} = \frac{R}{100-20}$$

$$\Rightarrow \quad R = \frac{55}{20} \times 80 = 55 \times 4$$

$$= 220\,\Omega$$

Ans. (B)

Sol. 9

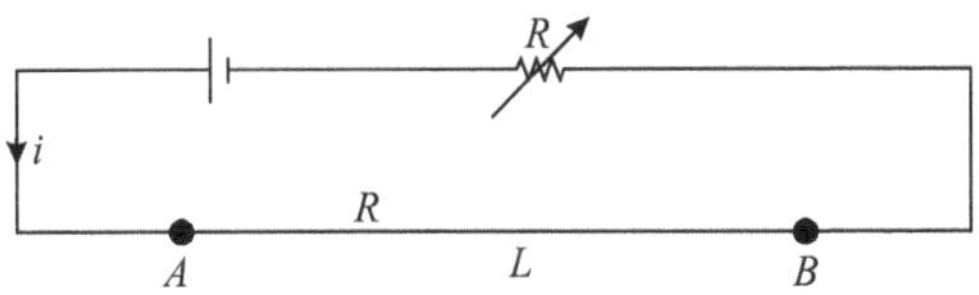

Resistance of potentiometer wire is given as

$$R = \frac{\rho L}{A}$$

Potential gradient on wire AB is given as

$$\lambda = \frac{V_{AB}}{L} = \frac{iR}{L} = \frac{i\rho}{A} = \frac{0.2 \times 4 \times 10^{-7}}{8 \times 10^{-1}}$$

$$\Rightarrow \quad \lambda = 0.1\,\text{V/m}$$

Ans. (D)

Sol. 10 (i) $\rightarrow$ As no current flows through cell, its terminal p.d. is equal to its emf so (i) is correct.

(ii) $\rightarrow$ As potential gradient on potentiometer wire is given as

$$\lambda = \frac{V_{AB}}{L}$$

Measured emf $E = \lambda\, l_{bal}$ [If λ is small then l_{bal} is large]

$\Rightarrow$ Precision will be more closer to actual value

So (ii) is correct

(iii) & (iv) are NOT correct **Ans. (C)**

Sol. 11

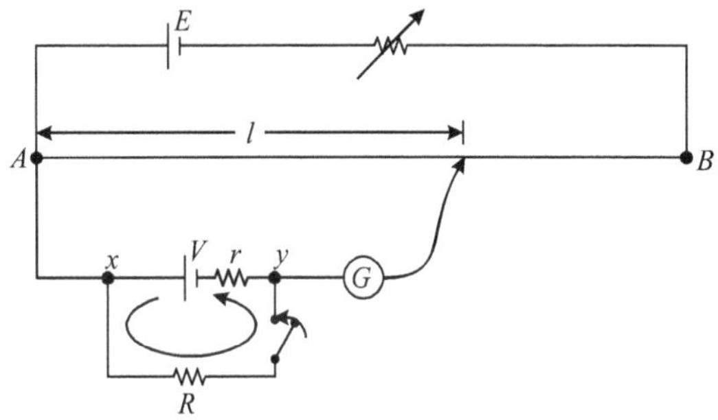

Initially when the cell of EMF V is balanced at length l then we use

$$V = \lambda l \qquad \ldots(1)$$

$$r = R$$

After connecting external resistance, we use

$$V_{xy} = \frac{V}{2R} \cdot R = \frac{V}{2}$$

If after connecting R balancing length is l' then we use

$$\frac{V}{2} = \lambda l' \qquad \ldots(2)$$

from (1) and (2) $\Rightarrow$ $l' = l/2$ **Ans. (C)**

Sol. 12

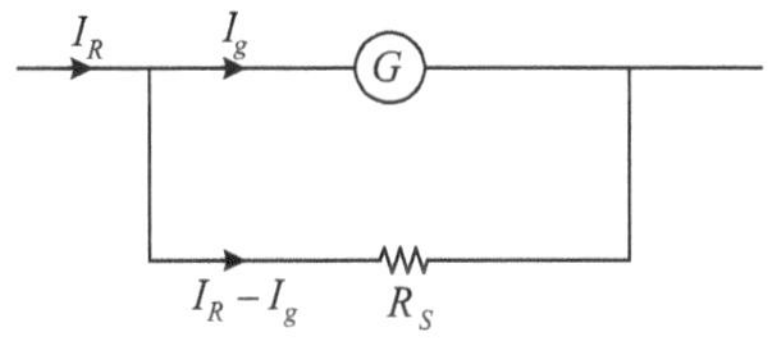

$$R_A = \frac{R_g R_S}{R_g + R_S}$$

Internal resistance of Ammeter.

If I_R increases then more current has to be bypassed through shunt resistance so R_S should be decreased or to decrease it an additional shunt is required across then ammeter.

$\Rightarrow$ Here S-1 is false & S-2 is true. **Ans. (D)**

Sol. 13 To verify Ohm's law we use

$$V = IR$$

By varying R, we measure V across R and I through R & plot the graph in V and I

The graph obtained in s straight line which verifies Ohm's law.

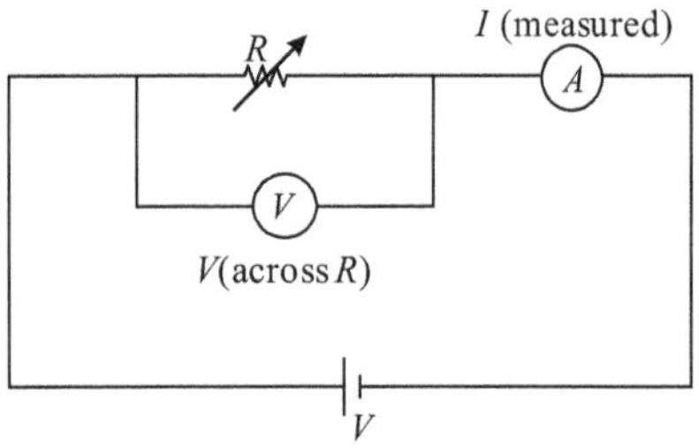

Only option (A) closely matches with above circuit used for experimental verification of Ohm's law.

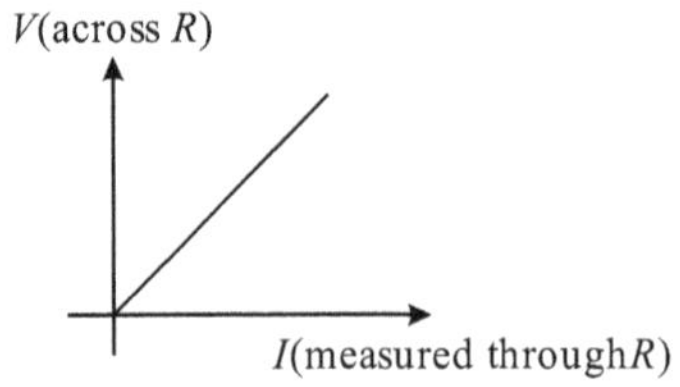

Ans. (A)

Sol. 14 Initially when balancing length is 40 cm, the balancing condition is

$$\frac{X}{40} = \frac{Y}{60} \quad \ldots(1)$$

later when $3x$ is balanced against y, if balancing length is l, then we use

$$\frac{3X}{l} = \frac{Y}{100-l} \quad \ldots(2)$$

$$\frac{(1)}{(2)} \Rightarrow \frac{l}{40\times 3} = \frac{100-l}{60}$$

$$\Rightarrow 6l = 1200 - 12l$$

$$\Rightarrow l = \frac{1200}{18} = \frac{200}{3} = 66.66 \text{ cm} \sim 67 \text{ cm}$$

Ans. (C)

Sol. 15

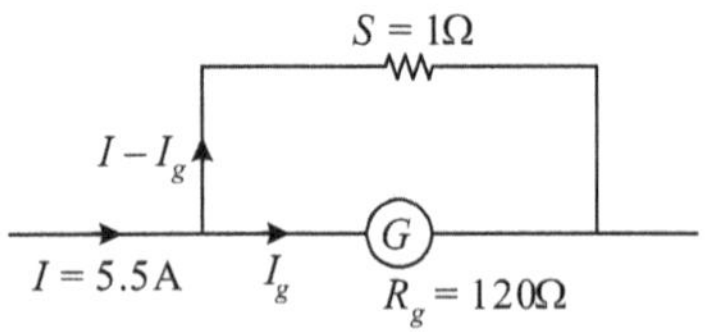

The supplied current of 5.5A will divide in inverse ratio of S and R_g and through "G" its full deflection current I_g flows so, we use

$$I_g = I\times\frac{S}{S+R_g} = \frac{5.5\times 1}{1+120}$$

$$= \frac{5.5}{1210} = \frac{55}{1210}$$

$$\Rightarrow I_g \simeq 0.045 \text{ A}$$

Ans. (D)

Sol. 16 If resistance of galvanometer is G then we use

Case-1 : When $R_3 = 0$, current through galvanometer is

$$I = \frac{E}{R_1 + \dfrac{R_2 G}{R_2+G}} \times \frac{R_2}{R_2+G} \quad \ldots(1)$$

Produces deflection d

Case-2 : $R_3 = 107\,\Omega$, current through galvanometer is $[G' = G + R_3^*]$

$$I' = \frac{E}{\left(R_1 + \dfrac{R_2 G'}{R_2+G'}\right)} \times \frac{R_2}{R_2+G'} \quad \ldots(2)$$

produces deflection $d/2$ for half deflection in galvanometer, we use

$$I' = \frac{I}{2}$$

$$\frac{E}{R_1 + \dfrac{R_2 G'}{R_2+G'}} \times \frac{R_2}{R_2+G'} = \frac{1}{2}\left(\frac{E}{R_1 + \dfrac{R_2 G}{R_2+G}}\right) \times \frac{R_2}{R_2+G}$$

$$\Rightarrow 2R_1R_2 + 2R_1G + 2R_2G = R_1R_2 + R_1G' + R_2G'$$

[here $G' = G + 107$]

$$\Rightarrow 9970\times 30 + 2\times 9970\times G + 60\,G$$

$$= 9970\,G + 9970\times 107 + 30\,G + 30\times 107$$

$$\Rightarrow 10000\,G = 10000\times 107 - 9970\times 30$$

$$\Rightarrow 10000\,G = (10700 - 2991)\times 100 = 7709\times 100$$

$$\Rightarrow G = \frac{7709\times 100}{10000} = 77.09\,\Omega$$

Ans. (D)

Sol. 17 Among given circuits D is most appropriate. When K_1 is closed and K_2 is open then current in galvanometer is

$$I = \frac{E}{R+G} \quad \ldots(1)$$

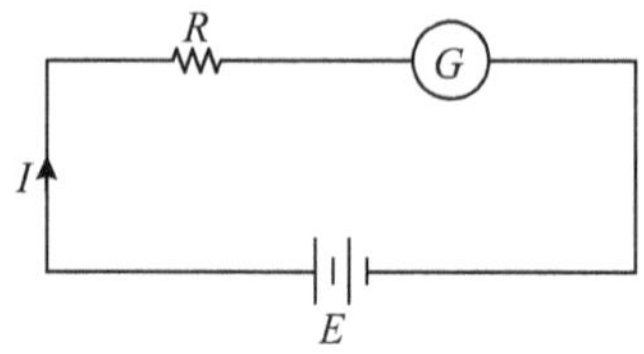

When both K_1 and K_2 are closed current supplied by battery is

$$I' = \frac{E}{R + \dfrac{GS}{G+S}} \quad \ldots(2)$$

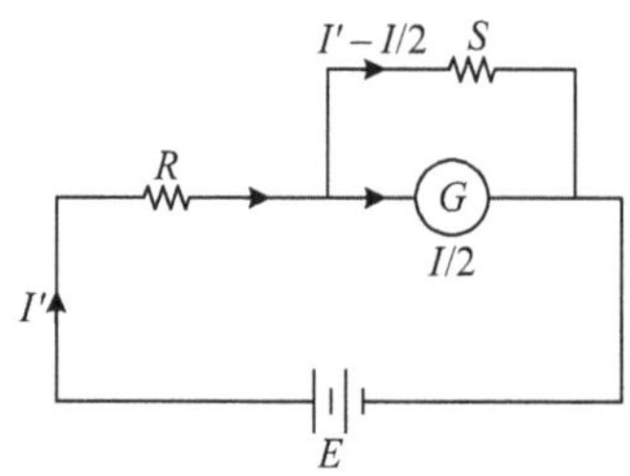

Here I' is divided in inverse ratio of S and G so we use

$$I' \times \frac{S}{S+G} = \frac{I}{2}$$

$$\frac{E.S}{RG+RS+GS} = \frac{E}{2(R+G)}$$

$$2RS+2SG = RG+RS+GS$$

$$\Rightarrow \quad G = \frac{RS}{R-S}$$

Ans. (D)

Sol. 18 When two batteries are connected in parallel then the terminal potential difference of this combination is given as

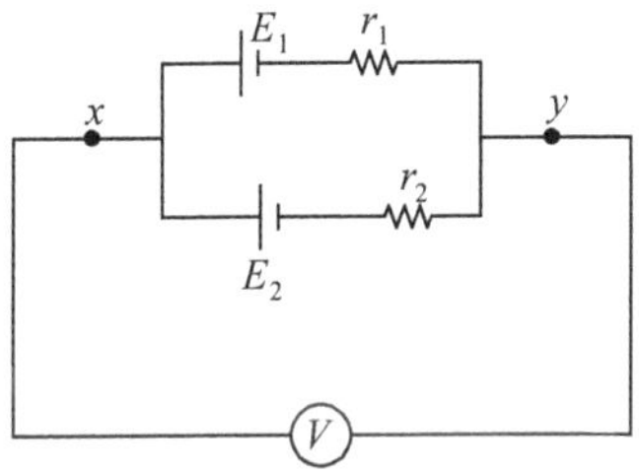

$$V_{xy} = \frac{\dfrac{E_1}{r_1}+\dfrac{E_2}{r_2}}{\dfrac{1}{r_1}+\dfrac{1}{r_2}}$$

[Already discussed in theory]

$$\Rightarrow \quad V_{xy} = \frac{\dfrac{10}{1}+\dfrac{15}{0.6}}{\dfrac{1}{1}+\dfrac{1}{0.6}} = \frac{10+25}{1+\dfrac{5}{3}} = \frac{35}{8/3} = \frac{105}{8}$$

$$\Rightarrow \quad V_{xy} = 13.1\,\text{V}$$

Ans. (C)

Sol. 19

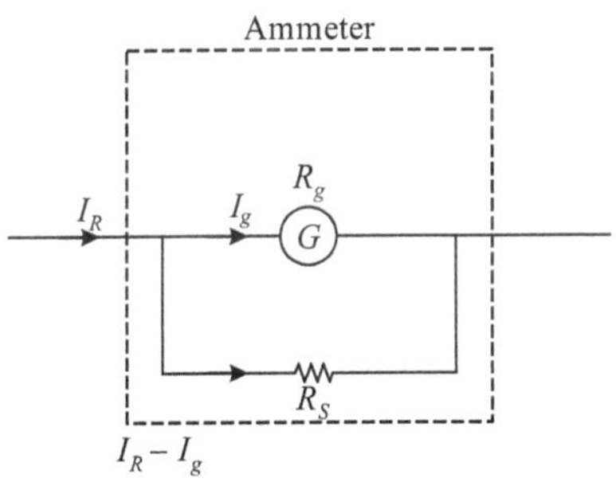

$$R_g = 100\Omega$$
$$I_g = 10^{-3}\,\text{A}$$
$$I_R = 10\,\text{A}$$

As R_g and R_S are in parallel we use

$$I_gR_g = (I_R - I_g)R_S$$

$$\Rightarrow \quad R_S = \frac{I_gR_g}{I_R - I_g}$$

As $I_g << I_R \Rightarrow \quad R_S = \frac{I_gR_g}{I_R} = \frac{10^{-3}\times 100}{10} = 10^{-2}\Omega$

Ans. (A)

Sol. 20 Step-1 :

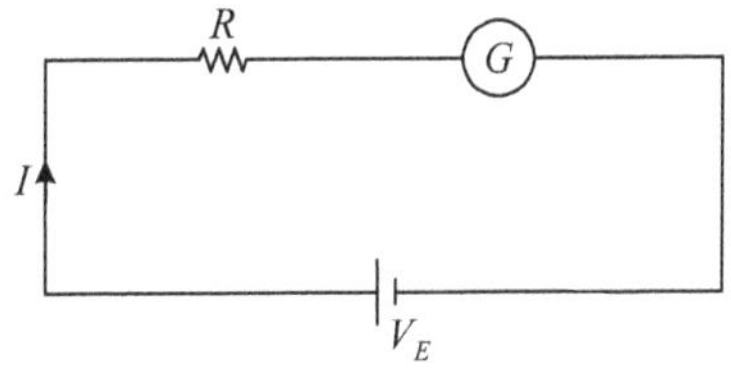

Current through galvanometer is given as

$$I = \frac{V_E}{(R+G)} \qquad \ldots(1)$$

Step-2 :

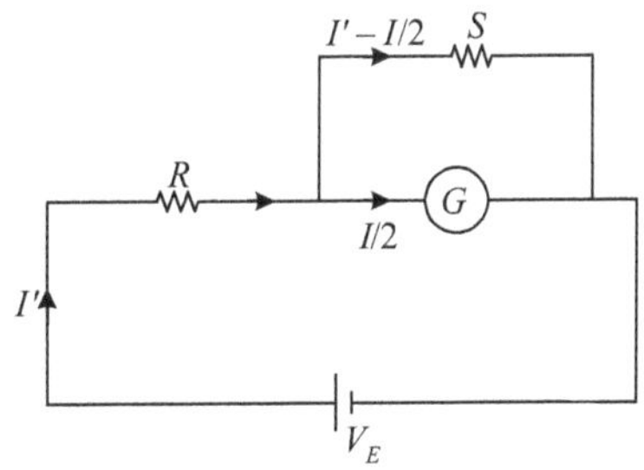

Current supplied by battery

$$I' = \frac{V_E}{R+\dfrac{SG}{S+G}} \qquad \ldots(2)$$

Here I' is distributed in inverse ratio of S and G so we use

$$I/2 = \frac{I'S}{S+G}$$

From (1) and (2) we have

$$\frac{V_E}{2(R+G)} = \frac{V_E}{\left(R+\dfrac{SG}{S+G}\right)} \cdot \frac{S}{(S+G)}$$

$$\Rightarrow \quad \frac{1}{2(R+G)} = \frac{S(S+G)}{[R(S+G)SG](S+G)}$$

$$\Rightarrow \quad RS+RG+SG = 2RS+2SG$$

$$\Rightarrow \quad RG = S(R+G)$$

$\Rightarrow$ **Ans. (A)**

Sol. 21

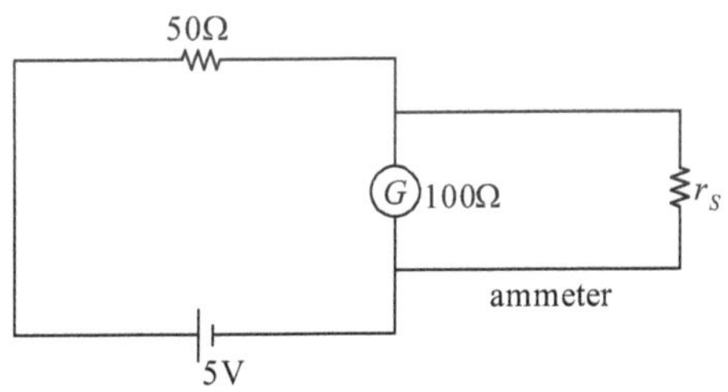

Current in circuit without the ammeter

$$I = \frac{5}{50} = 0.1\,\text{A}$$

When ammeter is there is circuit, the new current will be

$$I' = \frac{5}{50+\dfrac{100r_S}{100+r_S}}$$

As $I' < I \simeq I' = 0.991$ (within 1% initial value I)

$$\Rightarrow \quad \frac{5}{50+\dfrac{100r_S}{100+r_S}} = 0.99 \times 0.1\text{ A}$$

$$\Rightarrow \quad \frac{5}{50+\dfrac{100r_S}{100+r_S}} = 0.099\text{ A}$$

$$\Rightarrow \quad 50+\frac{100r_S}{100+r_S} = \frac{5}{0.099} = 50.50\,\Omega$$

$$\Rightarrow \quad \frac{100r_S}{100+r_S} = 0.50\,\Omega$$

$$\Rightarrow \quad 100\,r_S = 50 + 0.50\,r_S$$

$$\Rightarrow \quad r_S = \frac{50}{99.50} = 0.502\Omega$$

(to be connected in parallel to G)

Ans. (D)

Sol. 22 If resistance of galvanometer is R_g, we have

Case-1 : At 40 div with $R = 2400\,\Omega$ if current sensitivity of "G" is i_d A/div then we use

$$40i_d = \frac{2}{2400+R_g} \qquad \ldots(1)$$

Case-2 : At 20 div with $R = 4900\,\Omega$, we use

$$20i_d = \frac{2}{4900+R_g} \qquad \ldots(2)$$

$$\frac{(1)}{(2)} \Rightarrow \quad 2 = \frac{4900+R_g}{2400+R_g}$$

$$\Rightarrow \quad 4800 + 2R_g = 4900 + R_g$$

$$\Rightarrow \quad R_g = 100\,\Omega$$

from equation-(1)

$$40i_d = \frac{2}{2400+100}$$

$$\Rightarrow \quad i_d = \frac{2}{40\times 2500} = \frac{1}{5}\times 10^{-4}$$

$$= 20\times 10^{-6}\,A/\text{div}$$

$$= 20\,\mu\text{A/div}$$

Case-3 : from 10 div if $R = R$

$$\Rightarrow \quad 10i_d = \frac{2}{R+10}$$

$$\Rightarrow \quad 10\times 20\times 10^{-6} = \frac{2}{R+100}$$

$$\Rightarrow \quad R + 100 = 10000$$

$$\Rightarrow \quad R = 9900\,\Omega$$

As there are total 50 div in "G"

$\Rightarrow$ Range of "G" is

$$I_{max} = 50\times 20\times 10^{-6}$$

$$= 10^{-3}\,A = 1\text{ mA}$$

Ans. (A)

Sol. 23 All options (A), (B), (C) and (D) are covered in the theory option (D) is false.

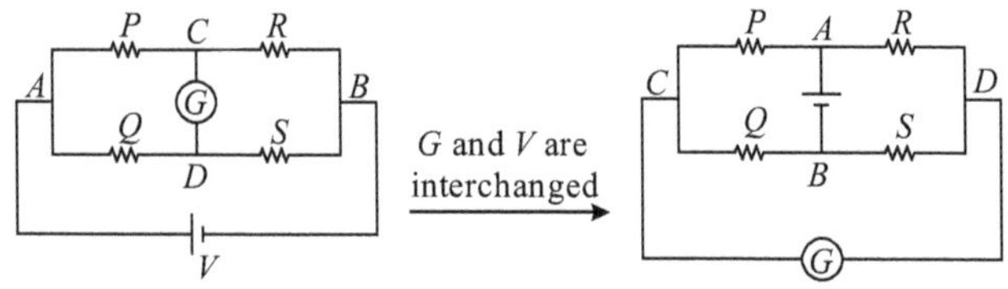

Balancing condition

$$\rightarrow \quad \frac{P}{Q} = \frac{R}{S}$$

Balancing condition

$$\rightarrow \quad \frac{Q}{S} = \frac{P}{R}$$

Option (D) is false. **Ans. (D)**

Sol. 24

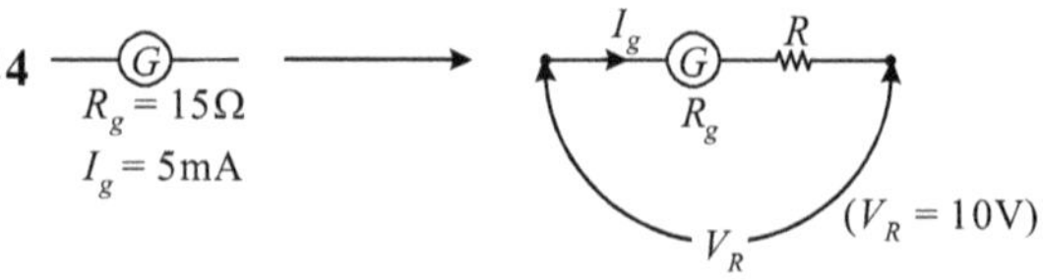

So for full deflection at voltage V_R, we use

$$I_g = \frac{V_R}{R_g + R}$$

$$\Rightarrow \quad R = \frac{V_R}{I_g} - R_g$$

$$\Rightarrow \quad R = \frac{10\times 1000}{0.005} - 15 = 2000 - 15 = 1985\,\Omega$$

Ans. (C)

Sol. 25 In first case we use

$$IR_1 = \lambda l_1 \qquad \ldots(1)$$

where $I \rightarrow$ current flowing through R_1 and R_2 in series

$\lambda \rightarrow$ potential gradient on wire PQ

In second case when K_3 is plugged into points 1 and 3 then R_1 and R_2 in series are balanced across points P and M where

$$PM = l_2$$

$$\Rightarrow \quad I(R_1 + R_2) = \lambda l_2 \qquad \ldots(2)$$

$$\frac{(2)}{(1)} \Rightarrow \quad \frac{R_1+R_2}{R_1} = \frac{l_2}{l_1}$$

$$\Rightarrow \quad 1 + \frac{R_2}{R_1} = \frac{l_2}{l_1}$$

$$\Rightarrow \quad \frac{R_2}{R_1} = \frac{l_2 - l_1}{l_1}$$

$$\Rightarrow \quad \frac{R_1}{R_2} = \frac{l_1}{l_2 - l_1}$$

Ans. (D)

Sol. 26 For meter bridge initially we use

$$AN = 60 \text{ cm}$$

$$NB = 40 \text{ cm}$$

$$\Rightarrow \quad \frac{P}{60} = \frac{Q}{40}$$

$$\Rightarrow \quad \frac{4}{60} = \frac{Q}{40} \quad Q = \frac{8}{3}\,\Omega$$

In next state $P \rightarrow 4 + R$ and

$$AN = 80 \text{ cm}$$

$$NB = 20 \text{ cm}$$

$$\Rightarrow \quad \frac{4+R}{80} = \frac{8/3}{20}$$

$$\Rightarrow \quad 12 + 3R = 32$$

$$\Rightarrow \quad 3R = 20 \quad \Rightarrow \quad R = 20/3\,\Omega$$ **Ans. (D)**

Sol. 27 The internal resistance of cell using potentiometer experiment is given as

$$r = R\left(\frac{l_1}{l_2} - 1\right) = 5 \times \left(\frac{52}{40} - 1\right)$$

$$= 5 \times \frac{12}{40} = \frac{12}{8} = 1.5\,\Omega$$

Alternative : If cell of EMF E is balanced at l_1 length of wire

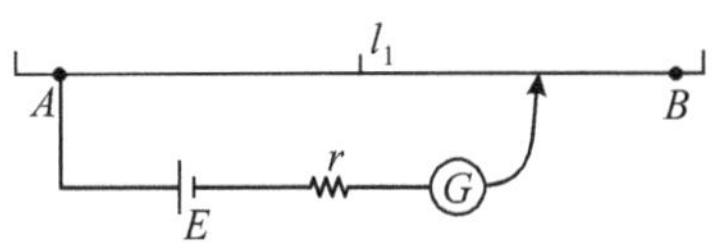

$$\Rightarrow \quad E = \lambda l_1 \ (\lambda = \text{potential gradient}) \qquad \ldots(1)$$

When call is shunted with a resistance R

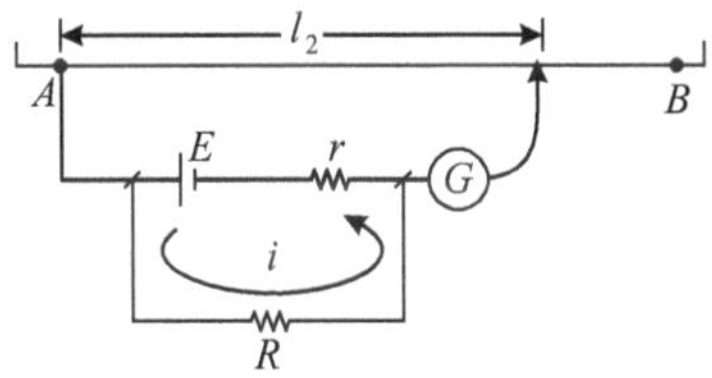

$$i = \frac{E}{r+R}$$

The potential difference across cell terminals after shunting is

$$V_E = iR = \frac{ER}{r+R} = \lambda l_2 \qquad \ldots(2)$$

$$\frac{(2)}{(1)} \quad \Rightarrow \quad \frac{R}{r+R} = \frac{l_2}{l_1}$$

$$\Rightarrow \quad r = R\left(\frac{l_1}{l_2} - 1\right) = 1.5\,\Omega$$ **Ans. (B)**

Sol. 28

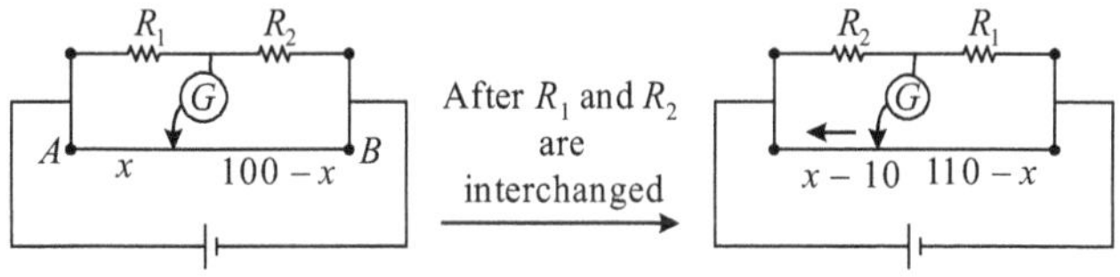

For balanced meter bridge, we use

$$\frac{R_1}{x} = \frac{R_2}{100 - x} \qquad \ldots(1)$$

$$\frac{R_2}{x - 10} = \frac{R_1}{110 - x} \qquad \ldots(2)$$

$$\frac{(1)}{(2)} \quad \Rightarrow \quad \frac{110 - x}{x} = \frac{x - 10}{100 - x}$$

$$\Rightarrow \quad (110 - x)(100 - x) = x(x - 10)$$

$$\Rightarrow \quad 11000 + x^2 - 210x = x^2 - 10x$$

$$\Rightarrow \quad 200x = 11000$$

$$\Rightarrow \quad x = \frac{110}{2} = 55 \text{ cm}$$

From equation-(1) we have

$$\frac{R_1}{55} = \frac{R_2}{45} = \frac{R_1 + R_2}{100} = \frac{1000}{100} = 10$$

[As $R_1 + R_2 = 1k\Omega$]

$$\Rightarrow \quad R_1 = 550\,\Omega$$ **Ans. (C)**

Sol. 29 When bridge is balanced, we have

$$\frac{x}{l_1} = \frac{y}{100 - l_1}$$

$$\Rightarrow \quad \frac{x}{39.5} = \frac{y}{60.5}$$

$$\Rightarrow \quad x = y \times \frac{39.5}{60.5} = 12.5 \times \frac{39.5}{60.5}$$

$$= 8.16\,\Omega$$

When x and y are interchanged then balancing length will be

$$l_2 = 100 - l_1$$

$$\Rightarrow \quad l_2 = 100 - 39.5 = 60.5 \text{ cm}$$ **Ans. (B)**

Sol. 30 Case-I : Without using a shunt resistance

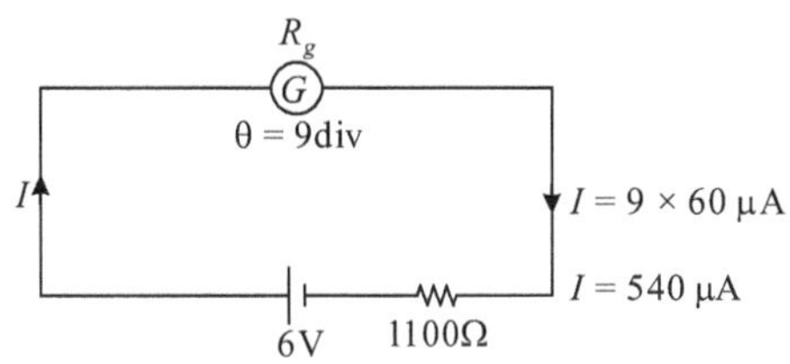

$$\Rightarrow \quad I = \frac{6}{11000 + R_g} = 540 \times 10^{-6}$$

$$\Rightarrow \quad 11000 + R_g = \frac{10^5}{9}$$

$$= 11111.11\,\Omega$$

$$\Rightarrow \quad R_g = 11111.11 - 1000$$

$$= 111.11\,\Omega$$

Case-II : with a shunt resistance

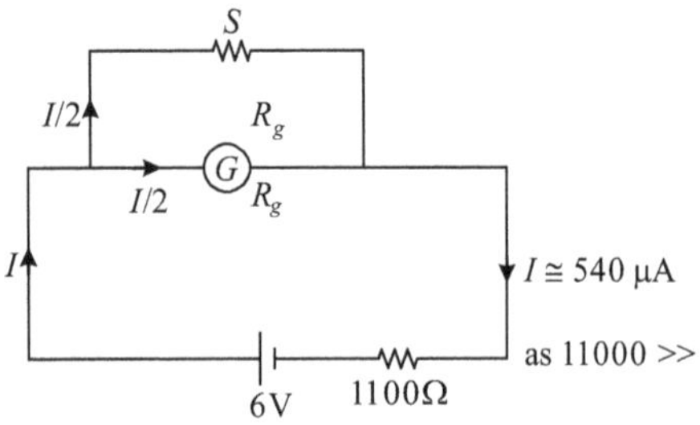

as $11000 >> \dfrac{Sr_g}{S+R_g}$

For half deflection in galvanometer we should use

$$S \simeq R_g = 111.11\,\Omega$$ **Ans. (B)**

Sol. 31

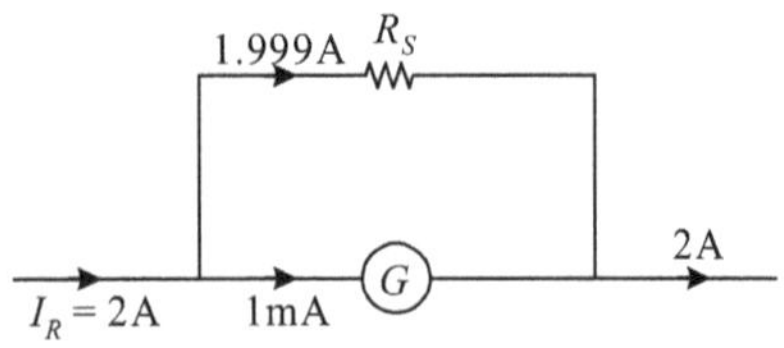

$$R_g = 25\,\Omega$$

$$I_g = 1\text{ mA} = 0.001\text{ A}$$

For making an ammeter of 2A range, we use

$$I_g R_g = (I_R - I_g) R_S$$

$$0.001 \times 25 = 1.999 \times R_S$$

$$\Rightarrow \quad R_S = \frac{25}{1999} \simeq 1.25 \times 10^{-2}\,\Omega$$ **Ans. (D)**

Sol. 32

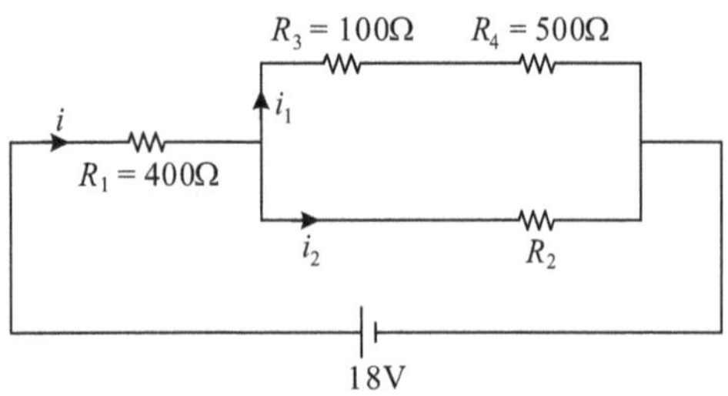

Given that

$$V_4 = 5\text{ V}$$

$$\Rightarrow \quad i_1 = \frac{V_4}{R_4} = 0.01\text{ A}$$

$$\Rightarrow \quad V_3 = i_1 R_3 = 1\text{ V}$$

$$\Rightarrow \quad V_3 + V_4 = 6\text{ V} = V_2$$

$$\Rightarrow \quad V_1 + V_3 + V_4 = 18\text{ V}$$

$$\Rightarrow \quad V_1 = 12\text{ V}$$

$$\Rightarrow \quad i = \frac{V_1}{R_1} = 0.03\text{ A}$$

$$\Rightarrow \quad i - i_1 = 0.02\text{ A}$$

$$\Rightarrow \quad R_2 = \frac{V_2}{i_2} = \frac{6}{0.02} = 300\,\Omega$$ **Ans. (A)**

Sol. 33 Current in primary circuit is

$$i = \frac{\varepsilon}{13r}$$

At no deflection

$$i \times \left(\frac{12r}{L} \times x\right) = \frac{\varepsilon}{2}$$

$$\Rightarrow \quad \frac{\varepsilon}{13r}\left(\frac{12r}{L} \times x\right) = \frac{\varepsilon}{2}$$

$$\Rightarrow \quad x = \frac{13L}{24}$$ **Ans. (D)**

Sol. 34 Using V and I readings measured resistance will be parallel combination of R_V and R, so we use

$$0.95\,R = \frac{R\,R_V}{R + R_V}$$

$$\Rightarrow \quad 0.95\,R^2 + 0.95\,RR_V = RR_V$$

$$\Rightarrow \quad 0.95\,R = 0.05\,R_V$$

$$\Rightarrow \quad R_V = 19 \times 30 = 570\,\Omega$$ **Ans. (B)**

Sol. 35 $R_1 = 32 \times 10 = 320\,\Omega$

For wheat stone bridge, balancing condition is

$$\Rightarrow \quad \frac{R_1}{R_3} = \frac{R_2}{R_4}$$

$$\Rightarrow \quad \frac{320}{R_3} = \frac{80}{40}$$

$$\Rightarrow \quad R_3 = 160\,\Omega$$

Colour code of 160Ω resistance is

(Brown, Blue, Brown) **Ans. (B)**

Sol. 36 Potential gradient with $R_h = 2\Omega$ is

$$\lambda = \left(\frac{6}{2+4}\right) \times \frac{4}{L} = \frac{4}{100}\frac{\text{V}}{\text{cm}}$$

For emf ε_1 let null point be at l cm at J, we use

$$\varepsilon_1 = 0.5\text{ V} = 0.04l \qquad \ldots(1)$$

Now with $R_h = 6\Omega$ new potential gradient is

$$\lambda' = \left(\frac{6}{4+6}\right) \times \frac{4}{L} = \frac{2.4}{100} = 0.024\frac{\text{V}}{\text{cm}}$$

For emf ε_2 null deflection length is same, so we use

$$\varepsilon_2 = 0.024l \qquad \ldots(2)$$

Dividing equation-(1) by (2) we get

$$\frac{0.5}{\varepsilon_2} = \frac{0.04}{0.024}$$

$$\Rightarrow \quad \varepsilon_2 = 0.3 \text{ V} \quad \textbf{Ans. (C)}$$

Sol. 37 At $R = 400\ \Omega$ bridge is balanced so we use

$$\frac{P}{400} = \frac{Q}{X} \quad \ldots(1)$$

On interchanging P and Q for balance $R = 405\ \Omega$. So we use

$$\frac{Q}{405} = \frac{P}{X} \quad \ldots(2)$$

(1) × (2) gives

$$X = \sqrt{400 \times 405} = 402.5\ \Omega \quad \textbf{Ans. (D)}$$

Sol. 38 Given that

$$R_g = 20\ \Omega$$

$$N_{\text{Left}} = N_{\text{Right}} = 30$$

Figure of merit of galvonometer is

$$\text{FOM} = \frac{I}{\phi} = 0.005 \text{ A/Div}$$

$$\text{Current sensitivity} = \left(\frac{1}{0.005}\right) \text{Div/A}$$

$$Ig_{\max} = 0.005 \times 30$$

$$= 15 \times 10^{-2} = 0.15 \text{ A}$$

For this galvanometer to be used as 15 V range voltmeter, we use

$$15 = 0.15\,[20 + R]$$

$$\Rightarrow \quad 100 = 20 + R$$

$$\Rightarrow \quad R = 80\ \Omega \quad \textbf{Ans. (A)}$$

Sol. 39 Initial balancing condition of bridge is

$$\frac{R_1}{R_2} = \frac{2}{3} \quad \ldots(1)$$

When 10 Ω is connected in series with R_1, we use

$$\frac{R_1 + 10}{R_2} = 1$$

$$\Rightarrow \quad R_1 + 10 = R_2 \quad \ldots(2)$$

From (1) and (2)

$$\frac{2R_2}{3} + 10 = R_2$$

$$\Rightarrow \quad 10 = \frac{R_2}{3}$$

$$\Rightarrow \quad R_2 = 30\ \Omega$$

and $\quad R_1 = 20\ \Omega$

If a resistance R is connected in parallel with $R_1 + 10$ then we use

$$\frac{\dfrac{30 \times R}{30 + R}}{30} = \frac{2}{3}$$

$$\Rightarrow \quad R = 60\ \Omega \quad \textbf{Ans. (B)}$$

Sol. 40 In first case galvanometer current is

$$i_g = \frac{E}{220 + R_g} = C\theta_0 \quad \ldots(1)$$

In second case galvanometer current is

$$i_g = \left(\frac{E}{220 + \dfrac{5R_g}{5 + R_g}}\right) \times \frac{5}{(R_g + 5)} = \frac{C\theta_0}{5}$$

$$\Rightarrow \quad \frac{5E}{225R_g + 1100} = \frac{C\theta_0}{5} \quad \ldots(2)$$

(1)/(2) gives

$$\Rightarrow \quad \frac{225R_g + 1100}{1100 + 5R_g} = 5$$

$$\Rightarrow \quad 5500 + 25R_g = 225R_g + 1100$$

$$\Rightarrow \quad 200R_g = 4400$$

$$\Rightarrow \quad R_g = 22\ \Omega \quad \textbf{Ans. (D)}$$

Sol. 41 If resistance of part AP is R_1 and PB is R_2 so for null deflection, we use

$$\frac{R'}{R'} = \frac{R_1}{R_2}$$

$$\Rightarrow \quad R_1 = R_2$$

On the given wire AB the resistance of an elemental length dl is given as

$$dR = \frac{C}{\sqrt{l}}\,dl \qquad (\text{where } C \rightarrow \text{constant})$$

Thus resistance of length AB is given as

$$R_1 = C\int_0^l l^{-1/2}\,dl = C(2, \sqrt{l})$$

Similarly of resistance of length PB is given as

$$R_2 = C\int_l^1 l^{-1/2}\,dl = C(2 - 2\sqrt{l})$$

Using $\quad R_1 = R_2$, we have

$$C_2\sqrt{l} = C(2 - 2\sqrt{l})$$

$$\Rightarrow \quad 2\sqrt{l} = 1$$

$$\Rightarrow \quad \sqrt{l} = \frac{1}{2}$$

$$\Rightarrow \quad l = \frac{1}{4}\text{m} = 0.25 \text{ m} \quad \textbf{Ans. (A)}$$

Sol. 42

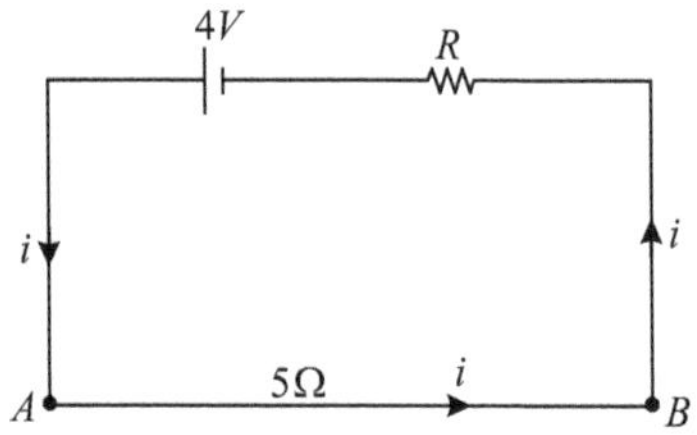

If current flowing in the wire is i, then we use

$$i=\left(\frac{4}{R+5}\right)\text{A}$$

Resistance of 10 cm length of wire is

$$\frac{5}{10}=0.5\,\Omega$$

Potential difference on 10 cm of wire is

$$5\times10^{-3}=\left(\frac{4}{R+5}\right)\times0.5$$

$$\Rightarrow \quad \frac{4}{R+5}=10^{-2}$$

$$\Rightarrow \quad R+5=400\,\Omega$$

$$\Rightarrow \quad R=395\,\Omega$$

Ans. (C)

Sol. 43 Full deflection current of galvanometer is

$$I_g=4\times10^{-4}\times25=10^{-2}\,\text{A}$$

To convert the galvanometer into a voltmeter, we connect a resistance R in series, such that

$$2.5=(50+R)\,10^{-2}$$

$$\Rightarrow \quad R=200\,\Omega$$ **Ans. (C)**

Sol. 44

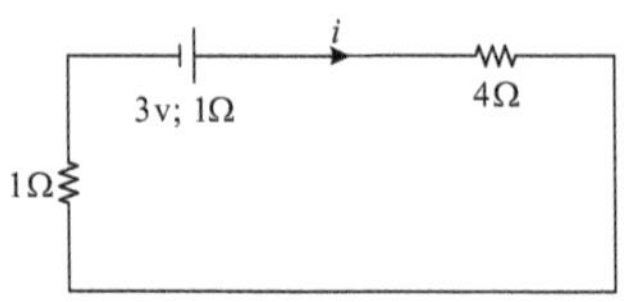

Resistance of wire AB is given as

$$R_{AB}=400\times0.01=4\,\Omega$$

Circuit current is

$$i=\frac{1.5+1.5}{1+1+4}=\frac{3}{6}=0.5\text{A}$$

Voltmeter reading is

$$V_R=i\times R$$

$$\Rightarrow \quad V_R=0.5\times0.01\times50$$

$$\Rightarrow \quad V_R=0.25\,\text{V}$$

Ans. (B)

Sol. 45

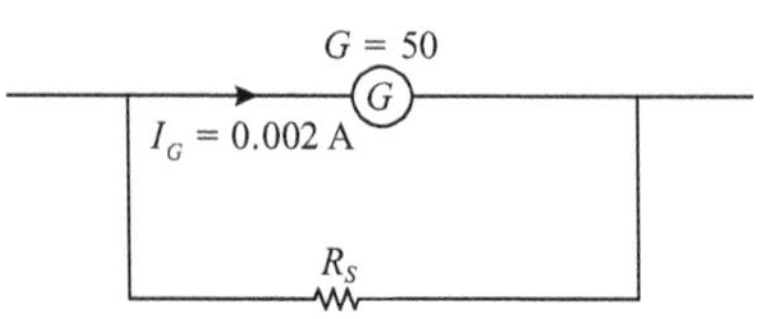

Shunt resistance is calculated as

$$R_S(0.5-0.002)=50\times0.002$$

$$R_S=\frac{50\times0.002}{(0.5-0.002)}=\frac{0.1}{0.498}=0.2$$ **Ans. (A)**

Sol. 46 For balanced state of bridge we use

$$X=\frac{R(100-l)}{l}$$

for (1) $$X=\frac{1000\times(100-60)}{60}\approx666.67\,\Omega$$

for (2) $$X=\frac{100\times(100-13)}{13}\approx669.23\,\Omega$$

for (3) $$X=\frac{10\times(100-1.5)}{1.5}\approx656.67\,\Omega$$

for (4) $$X=\frac{1\times(100-1)}{1}\approx99\,\Omega$$

So option (A) is completely different so inconsistent with given data. **Ans. (A)**

Sol. 47 Equivalent resistance of circuit across battery is

$$R_{\text{eq}}=\frac{15\times10}{25}+2+2r$$

$$=8+2r$$

Current in circuit is

$$i=\frac{3}{8+2r}$$

Potential difference across 10Ω resistance is

$$2=iR_{\text{eq}}=\frac{3}{8+2r}\times6$$

$$\Rightarrow \quad 16+4r=18$$

$$\Rightarrow \quad r=0.5\,\Omega$$

Ans. (D)

Sol. 48 Full deflection current

$$I_g=20\times10^{-6}\times50=10^{-3}\,\text{A}=1\text{ mA}$$

For 2 V range of voltmeter, we use

$$I_g=\frac{2}{100+R_1}\Rightarrow R_1=\frac{2}{10^{-3}}-100=1900\,\Omega$$

For 10 V range of voltmeter, we use

$$I_g=\frac{10}{2000+R_2}\Rightarrow R_2=\frac{2}{10^{-3}}-2000=8000\,\Omega$$

For 20 V range of voltmeter, we use

$$I_g=\frac{20}{10000+R_3}\Rightarrow R_3=\frac{2}{10^{-3}}-10000=10000\,\Omega$$

Ans. (D)

Sol. 49 Reading of voltmeter V and current I in circuit (reading of ammeter) can be related as

$$V = E - Ir$$

at $I = 1A,\ V = V_0 \simeq 0$

$\Rightarrow$ $0 = E - r$

$\Rightarrow$ $E = r$

at $V = 1.5\,\text{V}, I = 0$

$\Rightarrow$ $V = E = 1.5\,\text{V}$

$\Rightarrow$ $r = 1.5\,\Omega$ **Ans. (C)**

18.1 Magnetic Effects of Current

Sol. 1

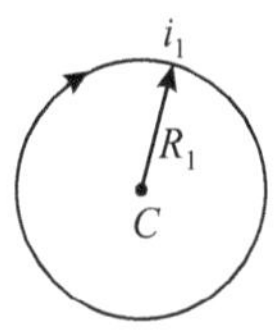

Magnetic induction at C is

$$B_C = \frac{\mu_0 i_1}{2R_1}$$

for coil A, $i_1 = i$ and $R_1 = R \Rightarrow B_A = \frac{\mu_0 i}{2R}$ ...(1)

for coil B, $i_2 = 2i$ and $R_2 = 2R \Rightarrow B_B = \frac{\mu_0 (2i)}{2(2R)} = \frac{\mu_0 i}{2R}$ ...(2)

$\Rightarrow$ (1)/(2) $\Rightarrow$ $\frac{B_A}{B_B} = 1$ **Ans. (A)**

Sol. 2

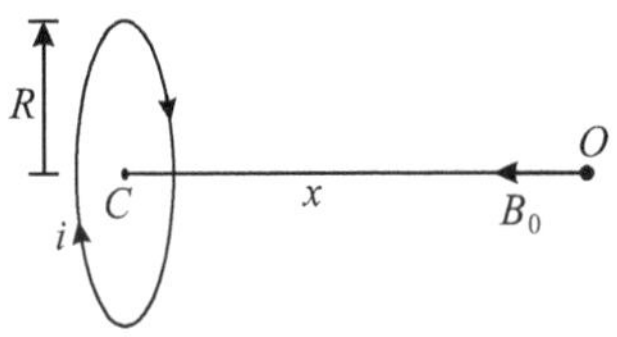

Magnetic field induction at point O is given as

$$B_0 = \frac{\mu_0 i R^2}{2(x^2 + R^2)^{3/2}} = 54\,\mu\text{T} \quad \text{...(1)}$$

Magnetic field induction at point C is given as

$$B_C = \frac{\mu_0 i}{2R} \quad \text{...(2)}$$

$\Rightarrow$ (2)/(1) $\Rightarrow$ $\frac{B_C}{54} = \frac{(x^2 + R^2)^{3/2}}{R^3} = \frac{(4^2 + 3^2)}{3^3} = \frac{125}{27}$

$\Rightarrow$ $B_C = 54 \times \frac{125}{27} = 250\,\mu\text{T}$ **Ans. (A)**

Sol. 3

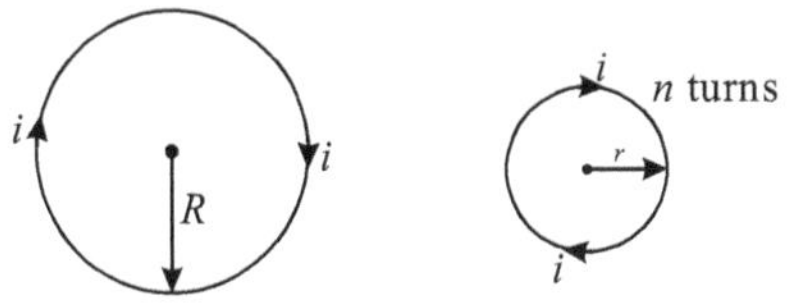

Magnetic induction at centre of one turn loop is given as

$$B = \frac{\mu_0 i}{2R} \quad \text{...(1)}$$

As length of wire is constant, we use

$$2\pi R = 2\pi r \times n \Rightarrow r = R/n$$

Magnetic induction at centre of n turn loop is

$$B' = \frac{\mu_0 i}{2r} \times n = \frac{\mu_0 i n^2}{2R} = n^2 B$$ **Ans. (B)**

Sol. 4 Using Ampere's law on path M, we get

$$\oint_M \vec{B} \cdot \vec{dl} = \mu_0 i_{\text{end}}$$

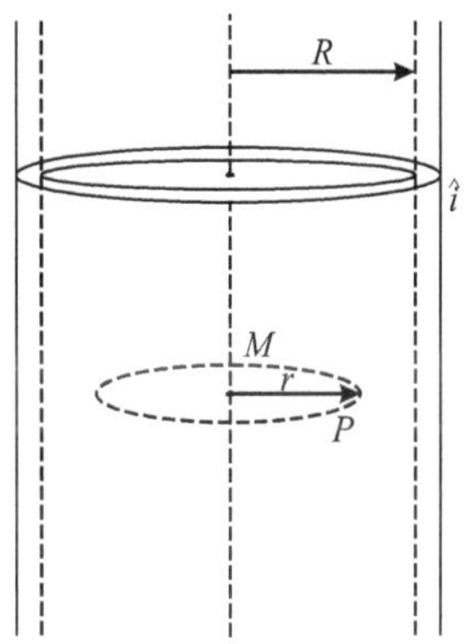

As the conductor is hollow

$\Rightarrow$ $i_{\text{enclosed } M} = 0$

$\Rightarrow$ $\oint_M \vec{B} \cdot \vec{dl} = 0$

$\Rightarrow$ $B_{\text{in } M} = 0$ **Ans. (B)**

Sol. 5

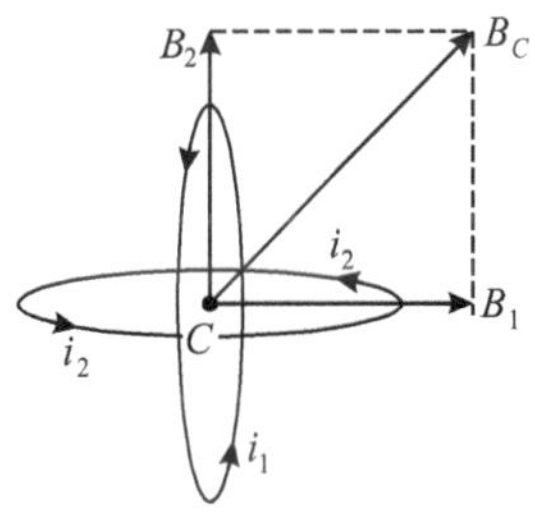

If B_1 and B_2 are magnetic induction at centre of the two coils, we use

$$B_1 = \frac{\mu_0 i_1}{2r_1} \text{ and } B_2 = \frac{\mu_0 i_2}{2r_2}$$

$$r_1 = r_2 = r = 2\pi\,\text{cm}$$

Net magnetic induction at common centre of coils is given as

$$B_C = \sqrt{B_1^2 + B_2^2} = \frac{\mu_0}{2r}\sqrt{i_1^2 + i_2^2}$$

$\Rightarrow$ $B_C = \frac{4\pi \times 10^{-7}}{2 \times 2\pi \times 10^{-2}} \sqrt{(3)^2 + (4)^2}$

$\Rightarrow$ $B_C = 5 \times 10^{-5}\,\text{Wb/m}^2$ **Ans. (C)**

Sol. 6 For a long solenoid having turns per unit length n and current i the magnetic induction as

$$B = \mu_0 n i$$

For case-I: $n = 200$ turns/cm

$$\Rightarrow \quad 6.28 \times 10^{-2} = \mu_0 \times 200 \times 10^2 \times i \quad \ldots (1)$$

For case-II: $n = 100$ turns/cm; $i = i/3$

$$\Rightarrow \quad B = \mu_0 \times 100 \times 10^2 \times i/3 \quad \ldots (2)$$

By (2)/(1) $\Rightarrow$ $$\frac{B}{6.28 \times 10^{-2}} = \frac{1}{3 \times 2}$$

$$\Rightarrow \quad B = \frac{6.28 \times 10^{-2}}{6}$$

$$\Rightarrow \quad B = 1.05 \times 10^{-2}\,\text{T}$$ **Ans. (A)**

Sol. 7

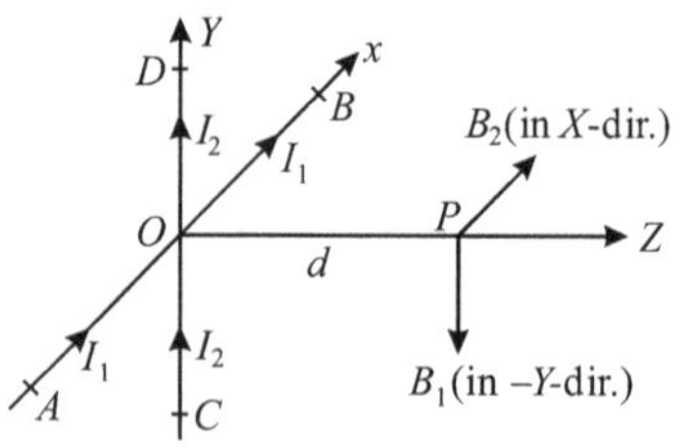

The magnetic inductions at point P due to currents I_1 and I_2 are

$$B_1 = \frac{\mu_0 I_1}{2\pi d} \text{ and } B_2 = \frac{\mu_0 I_2}{2\pi d}$$

As B_1 and B_2 are mutually perpendicular resulting magnetic induction at point P is given as

$$B_P = \sqrt{B_1^2 + B_2^2} = \frac{\mu_0}{2\pi d}\sqrt{I_1^2 + I_2^2}$$ **Ans. (B)**

Sol. 8 Magnetic field at interior point P of wire is given as

$$B_P = \frac{\mu_0 i x}{2\pi a^2} \quad \ldots (1)$$

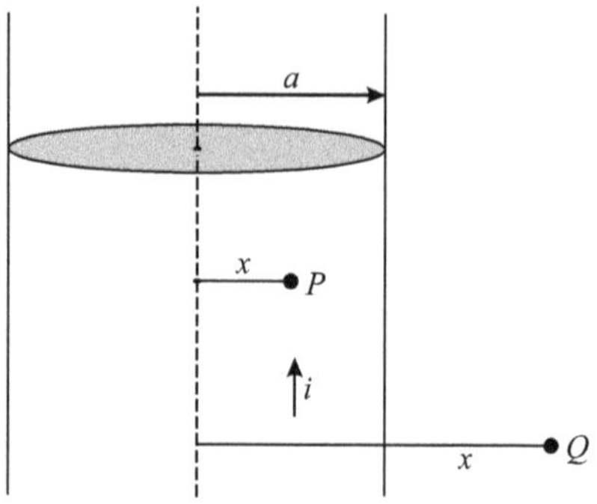

Magnetic field at outer point Q is given as

$$B_Q = \frac{\mu_0 i}{2\pi x} \quad \ldots (2)$$

at $x = \dfrac{a}{2}$

$$B_P = \frac{\mu_0 i\left(\frac{a}{2}\right)}{2\pi a^2}$$

$$\Rightarrow \quad B_P = \frac{\mu_0 i}{4\pi a}$$

at $x = 2a$

$$B_Q = \frac{\mu_0 i}{2\pi(2a)}$$

$$\Rightarrow \quad B_Q = \frac{\mu_0 i}{4\pi a}$$

$$\Rightarrow \quad \frac{B_P}{B_Q} = 1$$ **Ans. (D)**

Sol. 9 If we consider a path M at interior points of pipe of radius x and use Ampere's law, then

$$\oint_M \vec{B} \cdot \vec{dl} = \mu_0 \cdot i_{\text{end}}$$

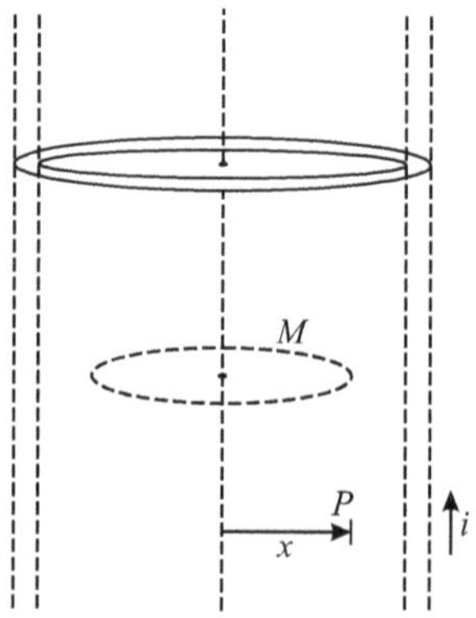

Being a hollow pipe, here

$$i_{\text{enclosed}} = 0$$

$$\Rightarrow \quad \oint_M \vec{B} \cdot \vec{dl} = 0$$

$$\Rightarrow \quad B_{\text{in } M} = 0$$ **Ans. (D)**

Sol. 10

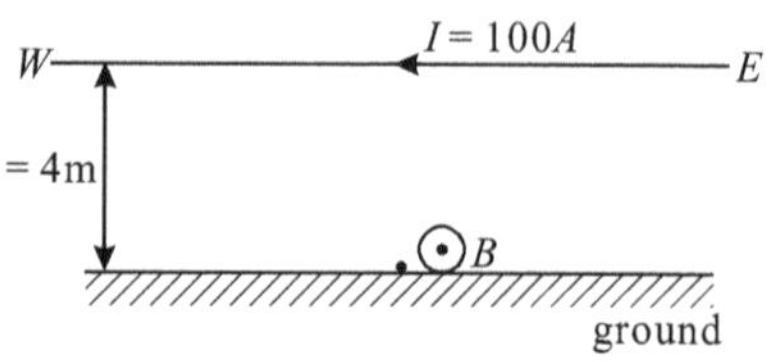

N E W S $\rightarrow$ (horizontal direction)

Outward $\rightarrow$ South

Inward $\rightarrow$ North

Magnetic induction on ground is

$$B = \frac{\mu_0 I}{2\pi h} = \frac{4\pi \times 10^{-7} \times 100}{2\pi \times 4}$$

$$\Rightarrow \quad B = 5 \times 10^{-6}\,\text{T in south}$$ **Ans. (C)**

Sol. 11 The magnetic induction due to a circular arc carrying a current I and subtending angle θ at centre is given as

$$B = \frac{\mu_0 I \theta}{4\pi R} \quad [R \rightarrow \text{radius of arc}]$$

At point O, $B_{AB} = 0$

$$B_{BC} = \frac{\mu_0 I(\pi/6)}{4\pi b} = \frac{\mu_0 I}{24b} \otimes \text{ (inward)}$$

$$B_{CD} = 0$$

$$B_{DA} = \frac{\mu_0 I(\pi/6)}{4\pi a} = \frac{\mu_0 I}{24a} \odot \text{ (outward)}$$

Net magnetic induction at O is given as

$$B_0 = B_{DA} - B_{BC}$$

$$\Rightarrow \quad B_0 = \frac{\mu_0 I}{24a} - \frac{\mu_0 I}{24b} = \frac{\mu_0 I}{24}\left[\frac{1}{a} - \frac{1}{b}\right]$$

$$= \frac{\mu_0 I(b-a)}{24ab}$$

Ans. (D)

Sol. 12

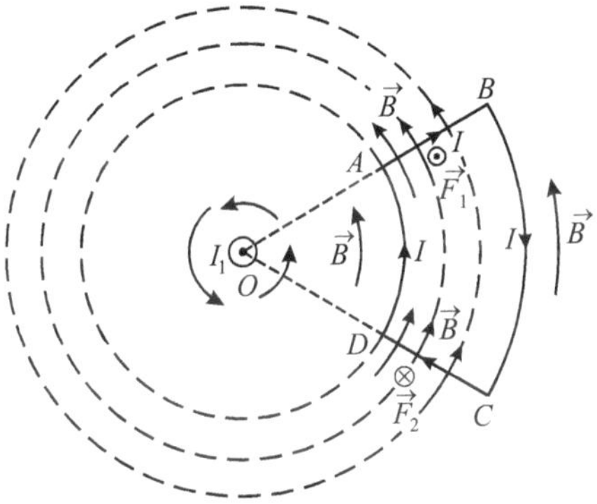

Force on wire AB due to I_1 is given in outward direction $\rightarrow \vec{F_1}$ and Force on wire CD due to I_1 is in inward direction $\rightarrow \vec{F_2}$

by symmetry we have $|\vec{F_1}| = |\vec{F_2}|$

Force on DA and BC are zero as field due to I_1 is parallel to all segments of these arc. **Ans. (D)**

Sol. 13 Due to a long current carrying wire magnetic induction in its surrounding is as shown below for given situation.

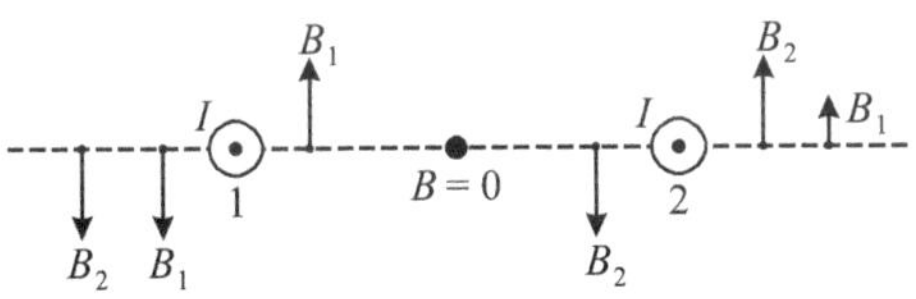

$$B = \frac{\mu_0 i}{2\pi r}$$

Ans. (A)

Sol. 14

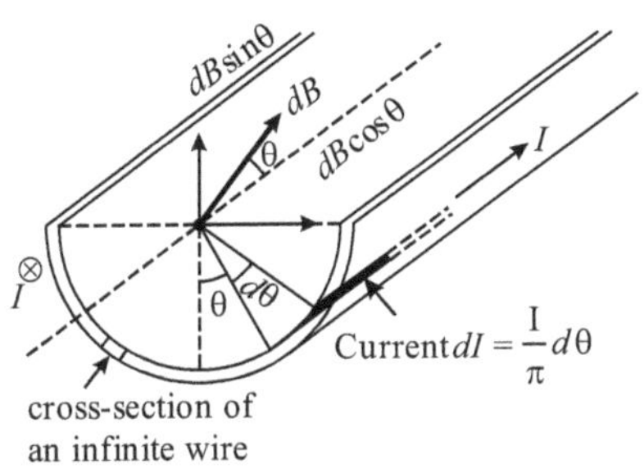

Current flowing in the elemental strip considered on surface of the half cylindrical infinite wire of width '$Rd\theta$'.

The magnetic induction due to this elemental current dI at any point on axis is given as

$$dB = \frac{\mu_0 dI}{2\pi R}$$

Net magnetic induction along axis can be given as $B = \int dB \cos\theta$

$$\Rightarrow \quad B = \int_{-\frac{\pi}{2}}^{+\frac{\pi}{2}} \frac{\mu_0\left(\frac{I}{\pi}d\theta\right)}{2\pi R}\cos\theta$$

$$\Rightarrow \quad B = \frac{\mu_0 I}{2\pi^2 R}\int_{-\frac{\pi}{2}}^{+\frac{\pi}{2}} \cos d\theta$$

$$\Rightarrow \quad B = \frac{\mu_0 I}{2\pi^2 R}[\sin\theta]_{-\pi/2}^{+\pi/2} = \frac{\mu_0 I}{2\pi^2 R}[1-(-1)]$$

$$\Rightarrow \quad B = \frac{\mu_0 I}{\pi^2 R}$$

Ans. (C)

Sol. 15

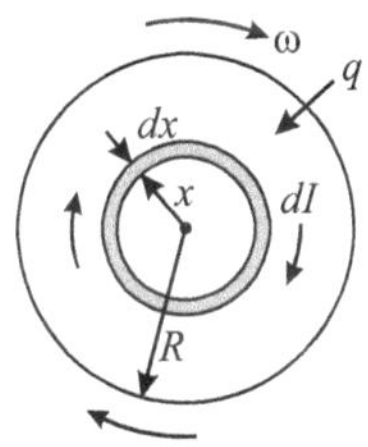

Charge on elemental ring (strip) of radius x and width dx is given as

$$dq = \frac{q}{\pi R^2}2\pi x dx = \frac{2qxdx}{R^2}$$

Current in this elemental ring is

$$dI = \frac{dq\omega}{2\pi} = \frac{qxdx\omega}{\pi R^2}$$

The magnetic induction at centre of disc due to current in the elemental ring is

$$dB = \frac{\mu_0 dI}{2x} = \frac{\mu_0}{2x} \times \frac{qx\omega dx}{\pi R^2} = \frac{\mu_0 q\omega}{2\pi R^2}dx$$

Net magnetic induction at centre of disc is given as

$$B = \int dB = \frac{\mu_0 q\omega}{2\pi R^2}\int_0^R dx$$

$$\Rightarrow \quad B = \frac{\mu_0 q\omega}{2\pi R^2}\cdot R = \frac{\mu_0 q\omega}{2\pi R}$$

$$\Rightarrow \quad B \propto \frac{1}{R}$$

Ans. (B)

Sol. 16 Magnetic induction at centre of a circular coil

$$B_1 = \frac{\mu_0 I}{2R} \qquad \ldots(1)$$

MI at an axial point at a distance x from coil centre is

$$B_2 = \frac{\mu_0 I R^2}{2(R^2 + x^2)^{3/2}}$$

at $x = 2\sqrt{2}\,R$, $$B_2 = \frac{\mu_0 I R^2}{2(R^2 + 8R^2)^{3/2}} = \frac{\mu_0 I}{54R} \quad \ldots(2)$$

from (1) and (2)

$$\Rightarrow \quad \frac{B_1}{B_2} = \frac{\mu_0 I / 2R}{\mu_0 I / 54R} = 27$$ **Ans. (B)**

Sol. 17

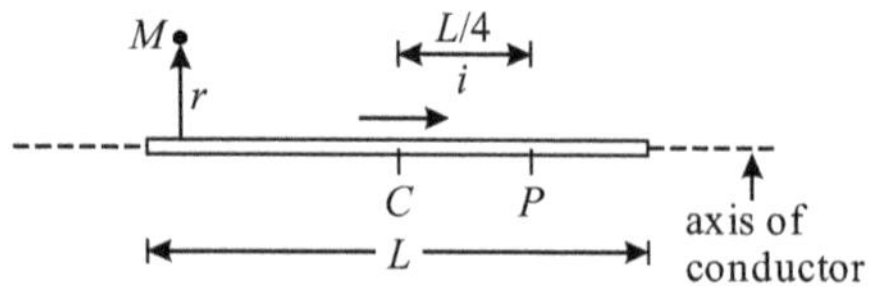

Along the axis of a current carrying conductor magnetic induction is always zero.

$$B_M = \frac{\mu_0 i}{2\pi r}$$

and $$B_{\text{inside}} = \frac{\mu_0 i r}{2\pi r^2} = \frac{1}{2}\mu_0 J r$$

at P as $r = 0$ (inside) $\Rightarrow B_P = 0$ **Ans. (A)**

Sol. 18

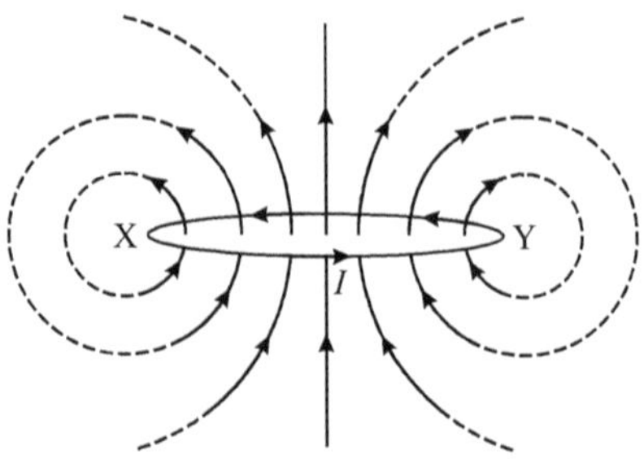

3D-view of magnetic lines configuration due to a circular loop.

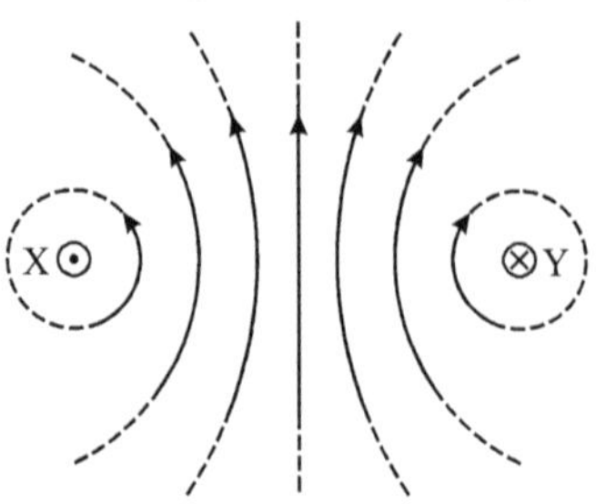

Cross-sectional view **Ans. (A)**

Sol. 19 If length of wires is l then

Case I : $l = 2\pi R_1 \Rightarrow 2R_1 = l/\pi$

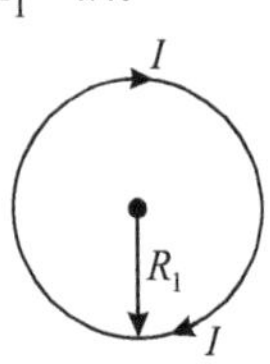

Magnetic Induction at Centre of loop is

$$B_1 = \frac{\mu_0 I}{2R_1} = \frac{\mu_0 I \pi}{l} \quad \ldots(1)$$

Case II : $l = 3 \times 2\pi R_2 \Rightarrow 2R_2 = l/3\pi$

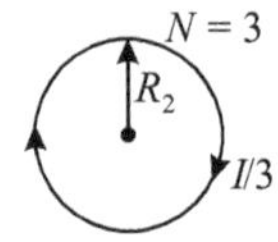

Magnetic Induction at Centre of loop is

$$B_2 = \frac{\mu_0 (I/3)}{2R_2} \times 3 = \frac{3\mu_0 I \pi}{l} \quad \ldots(2)$$

By (1)/(2) $\Rightarrow$ $$\frac{B_1}{B_2} = \frac{1}{3}$$ **Ans. (B)**

Sol. 20

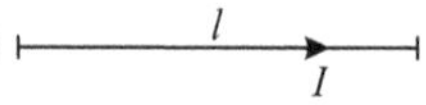

Case-A:

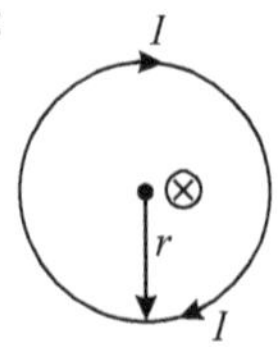

Magnetic induction at the centre of coil is

$$B_A = \frac{\mu_0 I}{2r} = \frac{\mu_0 I \pi}{l} \quad [l = 2\pi r]$$

Case-B:

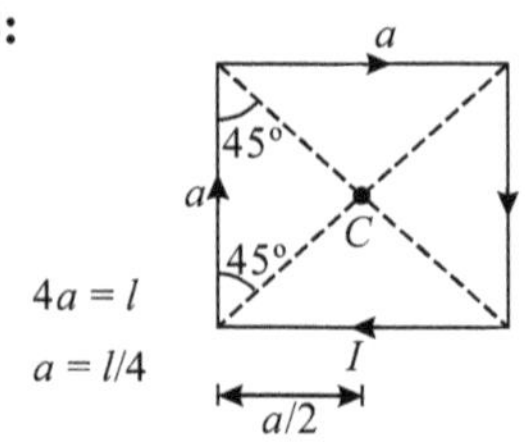

Magnetic induction due to square coil at its centre is given as

$$B_B = 4\left[\frac{\mu_0 I}{4\pi(a/2)}[\cos 45° + \cos 45°]\right]$$

$$\Rightarrow \quad B_B = \frac{2\mu_0 I}{\pi a}\sqrt{2}$$

$$\Rightarrow \quad B_B = \frac{2\sqrt{2}\mu_0 I}{\pi\left(\frac{l}{4}\right)} = \frac{8\sqrt{2}\mu_0 I}{\pi l} \quad \ldots(2)$$

By (1)/(2) $\Rightarrow$ $$\frac{B_A}{B_B} = \frac{\pi^2}{8\sqrt{2}}$$ **Ans. (D)**

Sol. 21

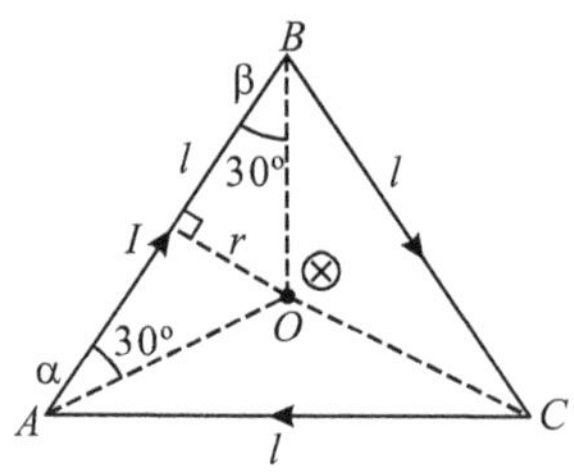

Magnetic induction due to a finite length wire is given as

$$B_0 = \frac{\mu_0 I}{4\pi r}(\cos\alpha + \cos\beta)$$

$$\Rightarrow \quad B_0 = \frac{\mu_0 I}{4\pi\left(\frac{l}{2\sqrt{3}}\right)}(2\cos 30°) \qquad [\because r = \frac{l}{2\sqrt{3}}]$$

$$\Rightarrow \quad B_0 = \frac{\sqrt{3}\times 4\pi\times 10^{-7}\times 1}{2\pi\times 4.5\times 10^{-2}}\times 2\times\frac{\sqrt{3}}{2}$$

Net MI at O is given as

$$B_{net} = 3B_0 = \frac{3\times 3\times 2\times 10^{-5}}{4.5}$$

$$\Rightarrow \quad B_{net} = 4\times 10^{-5}\ \text{Wb/m}^2$$

Ans. (A)

Sol. 22 Magnetic induction due to the two coils at P are in same direction so we have.

$$B = \frac{\mu_0 IR^2}{2(x^2+R^2)^{3/2}}\times 2\Bigg|_{x=R/2} \times N$$

$$\Rightarrow \quad B = \frac{\mu_0 IR^2 N}{\left(\frac{R^2}{4}+R^2\right)^{3/2}} = \frac{\mu_0 IR^2 N}{\left(\frac{5}{4}\right)^{3/2} R^3}$$

$$\Rightarrow \quad B = \frac{8\mu_0 IN}{(5)^{3/2} R}$$

Ans. (B)

Sol. 23 Magnetic field at O is given as

$$\vec{B} = \frac{\mu_0 i}{4\pi}\theta\left[\frac{1}{r_1}-\frac{1}{r_2}\right]$$

$$\Rightarrow \quad \vec{B} = \frac{4\pi\times 10^{-7}}{16}\times 10\times\left[\frac{1}{3\times 10^{-2}}-\frac{1}{5\times 10^{-2}}\right]$$

$$\Rightarrow \quad \vec{B} = \frac{\pi}{3}\times 10^{-5}\ \text{T}$$

Ans. (A)

Sol. 24

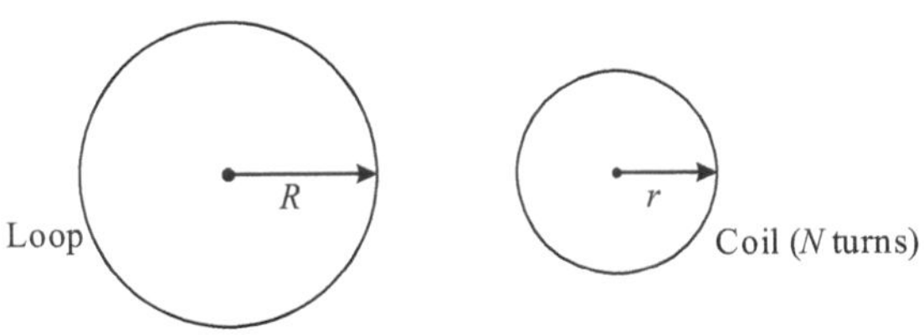

For loop
$L = 2\pi R$

For coil
$L = N\times 2\pi r$

$\Rightarrow \quad R = Nr$

Magnetic field at the centres are given as

$$B_L = \frac{\mu_0 i}{2R}$$

and

$$B_C = \frac{\mu_0 Ni}{2r} = \frac{\mu_0 N^2 i}{2R}$$

$$\Rightarrow \quad \frac{B_L}{B_C} = \frac{1}{N^2}$$

Ans. (C)

Sol. 25 Magnetic field at 'O' will be only due to 'PS' and 'QN' only, so we have

$$B_0 = B_{PS} + B_{QN}$$

Let current in each wire is i so magnetic field at O is given as

$$B_0 = \frac{\mu_0 i}{4\pi d} + \frac{\mu_0 i}{4\pi d}$$

$$\Rightarrow \quad 10^{-4} = \frac{\mu_0 i}{2\pi d} = \frac{2\times 10^{-7}\times i}{4\times 10^{-2}}$$

$$\Rightarrow \quad i = 20\ \text{A}\ \text{(into the page)}$$

Ans. (D)

Sol. 26 Dimension of $\sqrt{\frac{\varepsilon_0}{\mu_0}}$

$$[\varepsilon_0] = [M^{-1}L^{-3}T^4A^2]$$

$$[\mu_0] = [MLT^{-2}A^{-2}]$$

dimensions of

$$\sqrt{\frac{\varepsilon_0}{\mu_0}} = \left[\frac{M^{-1}L^{-3}T^4A^2}{MLT^{-2}A^{-2}}\right]^{\frac{1}{2}}$$

$$= [M^{-2}L^{-4}T^6A^4]^{1/2}$$

$$= [M^{-1}L^{-2}T^3A^2]$$

Ans. (B)

Sol. 27

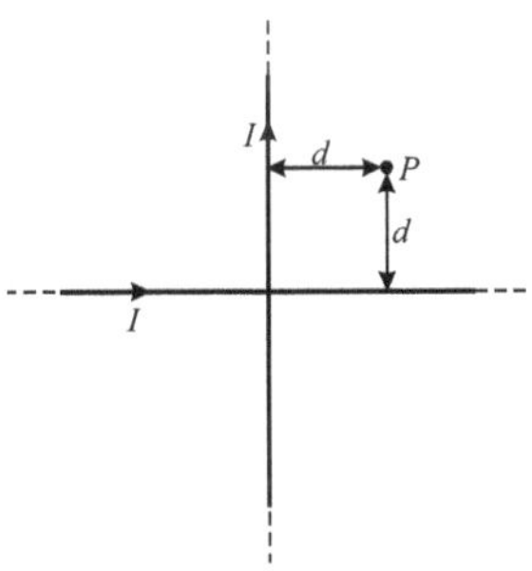

Magnetic field at point P is given as

$$\vec{B}_{net} = \frac{\mu_0 i}{2\pi d}(-\hat{k}) + \frac{\mu_0 i}{2\pi d}(\hat{k}) = 0$$

Ans. (A)

Sol. 28 Magnetic field at centroid O is given as

$$B = 3\left[\frac{\mu_0 i}{4\pi r}(\sin 60° + \sin 60°)\right]$$

Here r is given as

$$r = \frac{a}{2\sqrt{3}} = \frac{1}{2\sqrt{3}}$$

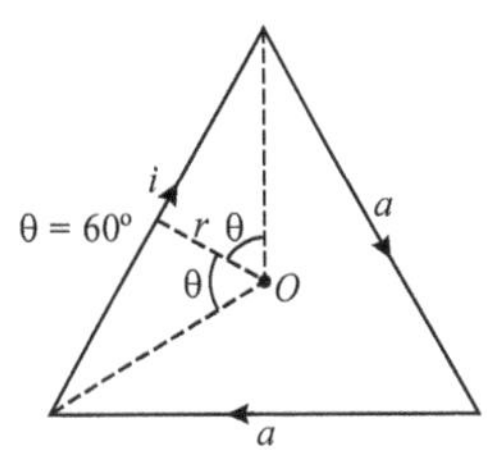

$$\Rightarrow \quad B = 3\left[\frac{4\pi\times10^{-7}\times10\times2\sqrt{3}}{4\pi\times1}\left[\frac{\sqrt{3}}{2}+\frac{\sqrt{3}}{2}\right]\right]$$

$$\Rightarrow \quad B = 18\times10^{-6} = 18\,\mu\text{T}$$ **Ans. (A)**

Sol. 29 Dimensions of given physical quantities are

$$[\varepsilon_0] = M^{-1}L^{-3}T^4A^2$$

$$[\mu_0] = MLT^{-2}A^{-2}$$

$$[R] = ML^2T^{-3}A^{-2}$$

$$\Rightarrow \quad [R] = \left[\sqrt{\frac{\mu_0}{\varepsilon_0}}\right]$$ **Ans. (C)**

Sol. 30 Magnetic field due to a rotating ring is given as

$$B = \frac{\mu_0 i}{2R} = \frac{\mu_0 q\omega}{2R2\pi}$$

$$\Rightarrow \quad q = 3\times10^{-5}\,\text{C}$$ **Ans. (B)**

Sol. 31

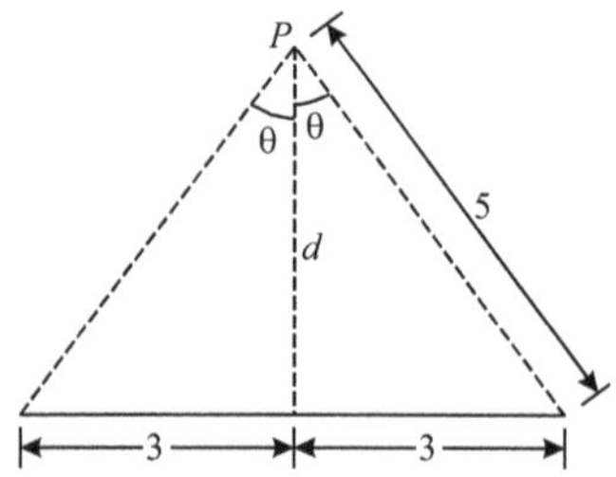

Magnetic field at point P is given as

$$B = \frac{\mu_0 I}{4\pi d}(2\sin\theta)$$

Here $\quad d = 4$ cm and $\sin\theta = \frac{3}{5}$

$$\Rightarrow \quad B = \frac{10^{-7}\times5}{0.04}\times2\times\frac{3}{5} = 1.5\times10^{-5}\,\text{T}$$ **Ans. (D)**

18.2 Magnetic Force on Moving Charges

Sol. 1 If B_1 is the magnetic induction at the location of all dl on wire 2 due to wire 1, than force on dl is given as

$$F = B_1 i_2 dl$$

here $$B_1 = \frac{\mu_0 i_1}{4\pi r}(\cos\theta + \cos\theta)$$

[we use the expression of $B(r)$ due to a finite length wire]

$$\Rightarrow \quad F = \frac{\mu_0 i_1 i_2 dl}{4\pi r}(2\cos\theta)$$

$$\Rightarrow \quad F = \frac{\mu_0 i_1 i_2 dl}{2\pi r}\cos\theta$$ **Ans. (C)**

Sol. 2 In uniform magnetic field the radius of circular motion is given as

$$r = mv/qB$$

$\Rightarrow$ Time period of revolution is given as

$$T = \frac{2\pi r}{v} = \frac{2\pi m}{qB}$$

$\Rightarrow$ T is independent of v. **Ans. (A)**

Sol. 3

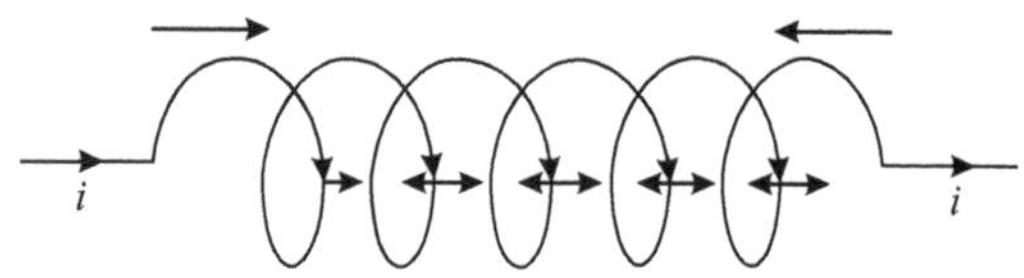

Due to parallel currents in each turn of spring due to same direction of current flow all turns attract each other.

$\Rightarrow$ spring will get compressed. **Ans. (B)**

Sol. 4 When a charge enters in magnetic field normally then it follows a circular path of radius R given as

$$R = \frac{mv}{qB} = \frac{p}{qB} \Rightarrow R \propto \frac{1}{q}$$

As charges of e^- and p^+ are same

$$\Rightarrow \quad R_e = R_p$$ **Ans. (A)**

Sol. 5

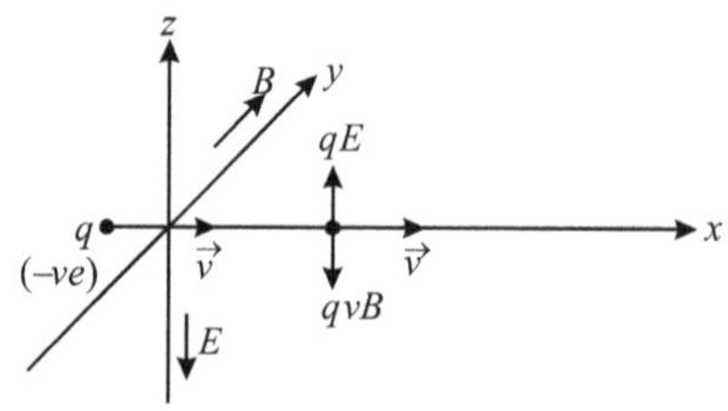

As particle continues motion in x-direction that means electric and magnetic forces on it are balanced.

$$\Rightarrow \quad qE = qvB$$

$$\Rightarrow \quad B = \frac{E}{v} = \frac{10^4}{10} = 10^3\,\text{Wb/m}^2$$ **Ans. (A)**

Sol. 6 Work done in turning a magnetic needle by angle $\theta_1 = 60°$ to $\theta_2 = 60°$ is given as

$$W = MB(\cos\theta_1 - \cos\theta_2)$$

$$W = MB(1 - 1/2) = \frac{1}{2}MB \quad \ldots(1)$$

Torque needed to hold the needle at $\theta_2 = 60°$ to the direction of magnetic field is

$$\tau = MB\sin\theta = MB\left(\frac{\sqrt{3}}{2}\right)$$

$$\Rightarrow \quad \tau = \sqrt{3}W$$ **Ans. (A)**

Sol. 7

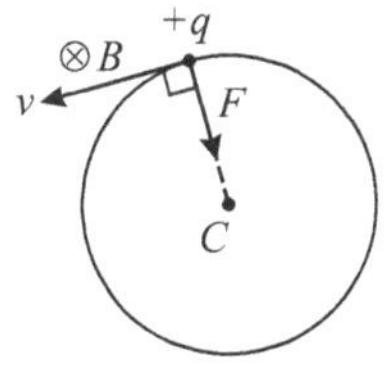

Magnetic force on a moving charge

$$\vec{F} = q(\vec{v} \times \vec{B})$$

as $\vec{F}$ is always ⊥ to $\vec{v}$

⇒ For a freely moving charge in Magnetic force, work done is always equal to zero. **Ans. (B)**

Sol. 8 Case-I :

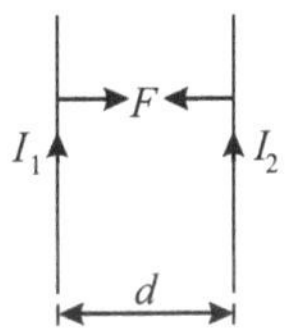

attractive force per unit length between wires is given as

$$F = \frac{\mu_0 I_1 I_2}{4\pi d} \quad \ldots(1)$$

Case-II :

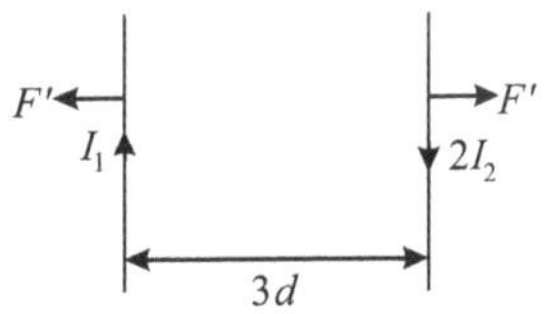

Repulsive force per unit length between wires is given as

$$F' = -\frac{\mu_0 I_1 (2I_2)}{4\pi (3d)} \quad \ldots(2)$$

By (2)/(1) ⇒ $\frac{F'}{F} = -\frac{2}{3} \Rightarrow F' = -\frac{2}{3}F$ **Ans. (C)**

Sol. 9 For perpendicular motion of a charge particle in magnetic field, we use

$$qvB = \frac{mv^2}{r}$$

$$\Rightarrow \quad r = \frac{mv}{qB}$$

Time period of one revolution

$$T = \frac{2\pi r}{v} = \frac{2\pi}{v} \times \frac{mv}{qB} = \frac{2\pi m}{qB}$$ **Ans. (D)**

Sol. 10

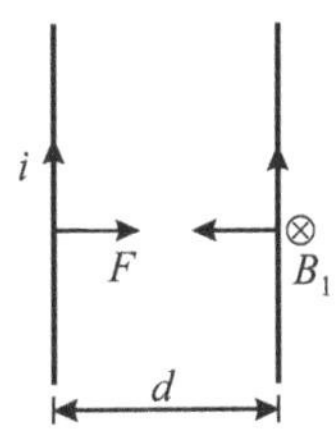

$B_1 \rightarrow$ magnetic induction due to wire 1 at the location of wire 2.
Force per unit length between the wires is given as

$$F = B_1 il = \left(\frac{\mu_0 i}{2\pi d}\right) il$$

$$\Rightarrow \quad \frac{F}{l} = \frac{\mu_0 i^2}{2\pi d}$$ **Ans. (A)**

Sol. 11

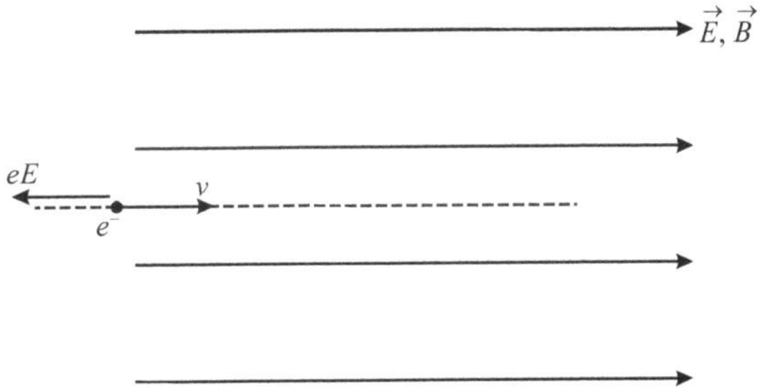

Acceleration of electron

$$a = -\frac{eE}{m}$$

As $\vec{v}$ is in same direction as $\vec{B}$ so no magnetic force will act on electron but electric force due to $\vec{E}$ in direction opposite to $\vec{v}$ so $|\vec{v}|$ will start decreasing. **Ans. (B)**

Sol. 12

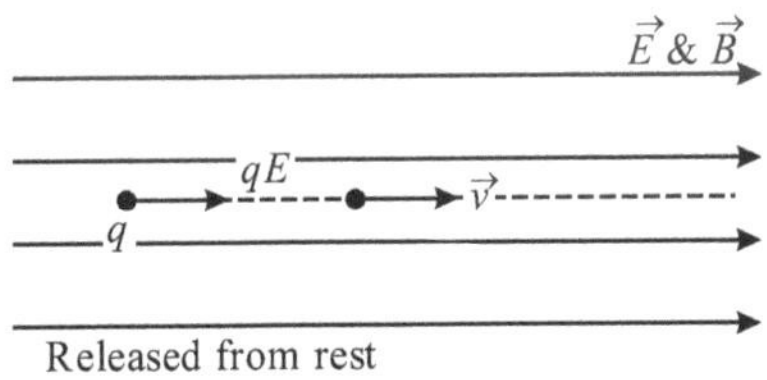

Charge on releasing will start moving due to electric force and it moves along the direction of $\vec{E}$ and $\vec{B}$. At every point the velocity $\vec{v}$ of particle is parallel to magnetic induction $\vec{B}$.

⇒ $F_{magnetic} = 0 \Rightarrow$ path is a straight line **Ans. (B)**

Sol. 13 Whenever a charge moves perpendicular to the direction of magnetic field its path is a circular motion.

⇒ direction of its $\vec{v}$ continuously change and as magnitude force is ⊥ to motion so no work is done on particle.

⇒ $|\vec{v}| \rightarrow$ constant

⇒ KE → constant **Ans. (A)**

Sol. 14

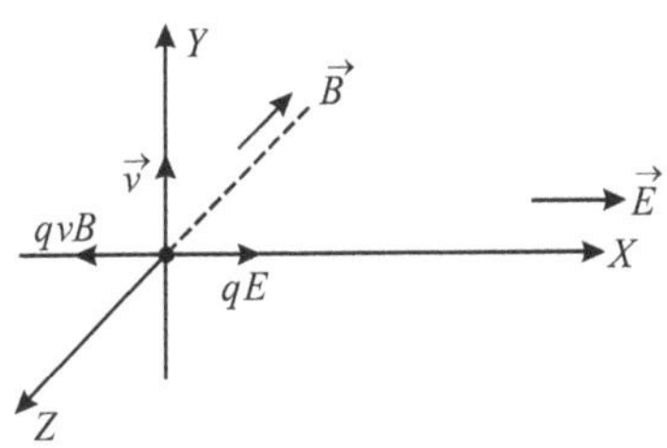

As change comes out without any change in magnitude & direction of $\vec{v}$.

$$\Rightarrow \quad qE = qvB$$

$$\Rightarrow \quad v = \frac{E}{B}$$

velocity vector will be $\vec{v} = \frac{E}{B}\hat{v}$

here we can use $\hat{v} = \frac{(\vec{E}\times\vec{B})}{EB}$

$$\Rightarrow \quad \vec{v} = \frac{E}{B}\cdot\frac{\vec{E}\times\vec{B}}{EB} = \frac{\vec{E}\times\vec{B}}{B^2}$$

Ans. (A)

Sol. 15 For uniformly charged & uniform object when rotates then its ratio of magnetic moment to angular momentum remain constant it is given as

$$\frac{\mu}{L} = \frac{q}{2m}$$

For disc, we use $q = \sigma\times\pi R^2$ & angular momentum $L = \left(\frac{1}{2}mr^2\right)\omega$

$$\Rightarrow \quad \mu = \frac{\sigma(\pi R^2)}{2m}\times\frac{1}{2}mR^2\omega$$

$$\Rightarrow \quad \mu = \frac{1}{4}\sigma\pi R^4\omega$$

Ans. (D)

Sol. 16 By Lorentz force equation, the force experienced by the charge is given as

$$\vec{F} = q\vec{E} + q(\vec{v}\times\vec{B})$$

$$\Rightarrow \vec{F} = 3q\hat{i} + q\hat{j} + 2q\hat{k} + q[(3\hat{i}+4\hat{j}+\hat{k})\times(\hat{i}+\hat{j}-3\hat{k})]$$

$$\Rightarrow \vec{F} = 3q\hat{i} + q\hat{j} + 2q\hat{k} + q[(3\hat{k} - 9(-\hat{j}) + 4(-\hat{k}) - 12\hat{i} + \hat{j} - \hat{i}]$$

$$\Rightarrow \vec{F} = 3q\hat{i} + q\hat{j} + 2q\hat{k} + q[-13\hat{i} + 10\hat{j} - \hat{k}]$$

$$\Rightarrow \vec{F} = -10q\hat{i} + 11q\hat{j} + q\hat{k} \Rightarrow F_y = 11q$$

Ans. (B)

Sol. 17 In magnetic field the radius of circular motion for a charge particle is given as

$$r = \frac{mv}{qB} = \frac{\sqrt{2mk}}{qB}$$

For proton, we use $m = m_p$ & $q = +e \Rightarrow r_p = \frac{\sqrt{2m_pk}}{eB}$

For deutron, we use $m = 2m_p$ & $q = +e$

$$\Rightarrow \quad r_d = \frac{\sqrt{2(2m_p)k}}{eB} = \sqrt{2}r_p$$

for α-particle, we use $m = 4m_p$ & $q = +2e$

$$\Rightarrow \quad r_\alpha = \frac{\sqrt{2(4m_p)k}}{2eB} = r_p$$

$$\Rightarrow \quad r_\alpha = r_p < r_d$$

Ans. (C)

Sol. 18 Initially for equilibrium of wire, we use

$$kx_1 = mg \Rightarrow k\times(0.005) = 0.01\times 10 \quad \ldots(1)$$

where k = force constant of spring

When wire is subjected to the magnetic field springs further stretch by 0.3 cm.

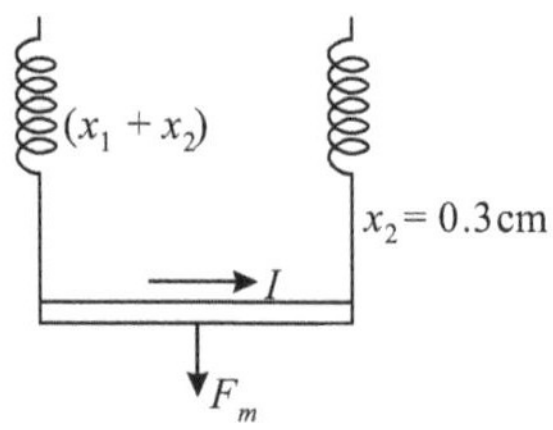

To apply the magnetic force in downward direction the magnetic field must be applied out of the page.

$$F_m = BIl = B\times\left(\frac{24}{12}\right)\times(0.05)$$

The magnetic force is balanced by the spring force due to additional stretch in springs i.e. 0.3 cm.

$$\Rightarrow \quad kx_2 = F_m \Rightarrow k\times 0.003 = B\times 2\times 0.05 \ \ldots(2)$$

$$\text{By (2)/(1)} \Rightarrow \frac{B\times 0.1}{0.1} = \frac{0.003}{0.005} = \frac{3}{5} = 0.6$$

$$\Rightarrow \quad B = 0.6 \text{ T out of the page}$$

Ans. (A)

Sol. 19

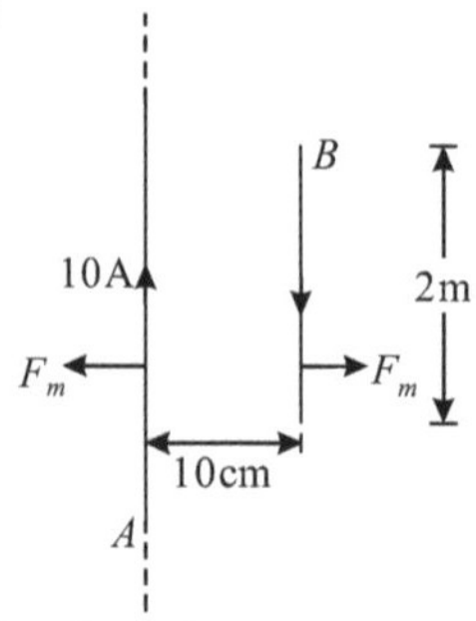

Repulsive force between two opposite and parallel currents is

$$F_m = \frac{\mu_0 I_1 I_2 l}{2\pi d}$$

$$\Rightarrow \quad F_m = \frac{4\pi\times 10^{-7}\times 10\times 2\times 2}{2\pi\times 0.1}$$

$$\Rightarrow \quad F_m = 8\times 10^{-5}\text{N}$$

Ans. (A)

Sol. 20 under the influence of crossed magnetic field & electric field the charge particle if pass undeflected.

$$\Rightarrow \quad \vec{F_e} = -\vec{F_m}$$

$$\Rightarrow \quad |\vec{F_e}| = |\vec{F_m}|$$

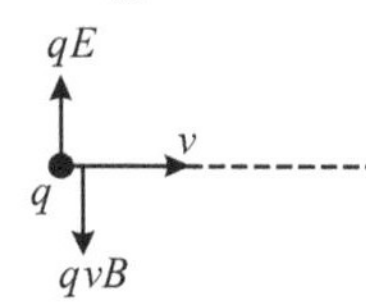

$$qE = qvB$$

$$\Rightarrow \quad v = \frac{E}{B} = \frac{7.7\times10^3}{0.14} = 55\times10^3 \text{ m/s}$$

$$\Rightarrow \quad v = 55 \text{ km/s}$$

Ans. (C)

Sol. 21 By Lorentz force equation, the magnetic force on a moving charge is

$$\vec{F} = q(\vec{v}\times\vec{B})$$

$$\Rightarrow \quad \vec{F} = 2\times10^{-6}[(2\hat{i}+3\hat{j})\times10^6\times2\hat{j}]$$

$$\Rightarrow \quad \vec{F} = 4\times10^{-6}\times10^6[(2\hat{i}+3\hat{j})\times\hat{j}]$$

$$\Rightarrow \quad \vec{F} = 4\times2\hat{k} = 8\hat{k} \text{ N } (+z \text{ direction})$$

Ans. (A)

Sol. 22 In a magnetic field radius of charge particle's circular motion is given as

$$R = \frac{mv}{qB} = \frac{\sqrt{2mk}}{qB}$$

if the particle is accelerated by a potential difference V

$$\Rightarrow \quad K = qV$$

$$\Rightarrow \quad R = \frac{\sqrt{2mqV}}{qB}$$

for proton $q = e, m = m_p \Rightarrow R_p = \dfrac{1}{B}\dfrac{\sqrt{2m_peV}}{q}$

for deutron $q = e, m = 2m_p \Rightarrow R_D = \dfrac{1}{B}\dfrac{\sqrt{2(2m_p)eV}}{q} = R$

$$\Rightarrow \quad \frac{R_p}{R_D} = \frac{1}{\sqrt{2}} \Rightarrow R_p = \frac{R}{\sqrt{2}}$$

Ans. (B)

Sol. 23 S-1: In perpendicular motion to a uniform/Non uniform magnetic field as

$$\vec{F}_{\text{magnetic}} = q(\vec{v}\times\vec{B})$$

$\Rightarrow$ $\vec{F}_{\text{magnetic}}$ is $\perp$ to $\vec{v}$ so magnetic force never do work on a moving charge

$\Rightarrow$ KE $\rightarrow$ constant

$\Rightarrow$ S-1 it true

$\vec{F}_{\text{magnetic}}$ is $\perp$ to $\vec{v}$

$\Rightarrow$ **S-2 :** it is true and explaining **S-1** **Ans. (D)**

Sol. 24

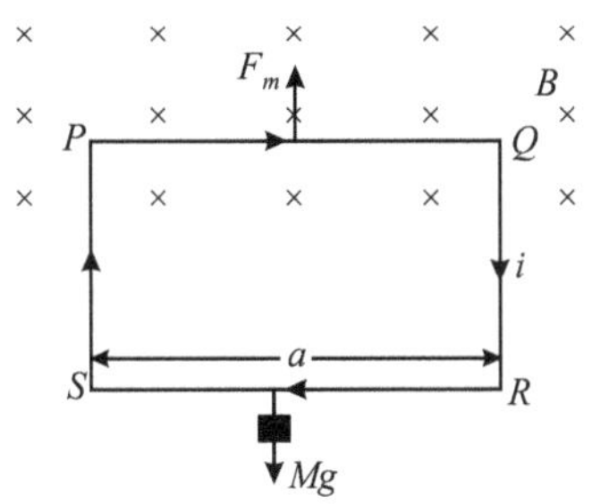

Magnetic force on loop is due to wire PQ, given as

$$F_m = Bia$$

Given that $Bia > mg \Rightarrow$ net force is in upward direction so weight rises up. As net work done by magnetic field is zero and here work is done against gravity. **Ans. (C)**

Sol. 25 As charge is not getting deflected $\Rightarrow$ no net force is acting on it.

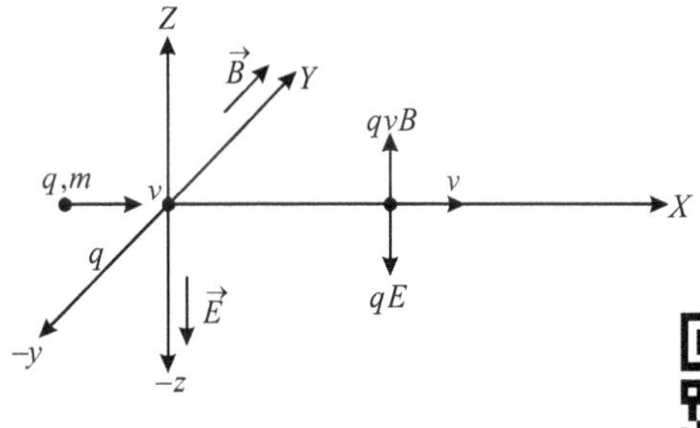

Here we use

$$qE = qvB$$

$$B = \frac{E}{v} = \frac{10^4}{10} = 10^3 \text{ Wb/m}^2$$

Ans. (C)

Sol. 26 Given that $E_y \propto J_xB_z$

$$\Rightarrow \quad E_y = kJ_xB_z$$

$$\Rightarrow \quad k = \frac{E_y}{J_xB_z}$$

$$\Rightarrow \quad U(k) = \frac{U(E_y/B_z)}{U(J_x)} = \frac{\text{m/s}}{\text{A/m}^2} = \frac{\text{m}^3}{\text{As}}$$

Ans. (B)

Sol. 27 Work done in displacing the wire by an elemental displacement dx is

$$dW = BIldx = 3.0\times10^{-4}\times e^{-0.2x}\times10\times3\times dx$$

Total work done in displacing the wire from $x = 0$ to $x = 2$m is

$$W = \int dW = \int_0^{2.0} 9\times10^{-3}\cdot e^{-0.2x}dx$$

$$\Rightarrow \quad W = 9\times10^{-3}\left[\frac{e^{-0.2x}}{-0.2}\right]_0^2$$

$$\Rightarrow \quad W = \frac{9\times10^{-3}}{0.2}(1-e^{-0.4})$$

$$\Rightarrow \quad W = 4.5\times10^{-2}(1-(2.71)^{-0.4}) \text{ J}$$

Power spent is $P = \dfrac{W}{t} = \dfrac{4.5\times10^{-2}\left[1-(2.71)^{-0.4}\right]}{5\times10^{-3}}$

$$\Rightarrow \quad P = 2.96 \text{ W}$$

Ans. (B)

Sol. 28 The force between two parallel current carrying wires is given as

$$F = \frac{\mu_0 I_1 I_2 l}{2\pi d}$$

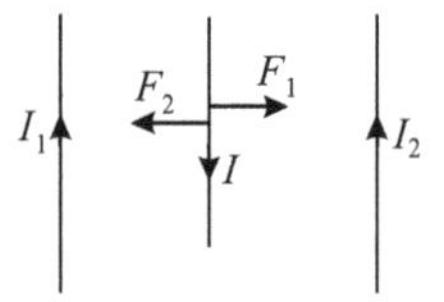

As $F_1 > F_2$

Net force on length l of middle wire is

$$F_{net} = F_1 - F_2 = \frac{\mu_0 I_1 Il}{2\pi(0.03)} - \frac{\mu_0 I_2 Il}{2\pi(0.05)}$$

$$\Rightarrow \quad F_{net} = \frac{\mu_0 Il}{2\pi}\left[\frac{30}{0.03} - \frac{20}{0.05}\right]$$

$$\Rightarrow \quad F_{net} = \frac{4\pi \times 10^{-7} \times 10 \times 0.25}{2\pi} \times (1000 - 400)$$

F_{net} is acting toward right

$$\Rightarrow \quad F_{net} = 2 \times 10^{-7} \times 10 \times 0.25 \times 600$$

$$\Rightarrow \quad F_{net} = 3 \times 10^{-4}\,\text{N}$$ **Ans. (A)**

Sol. 29

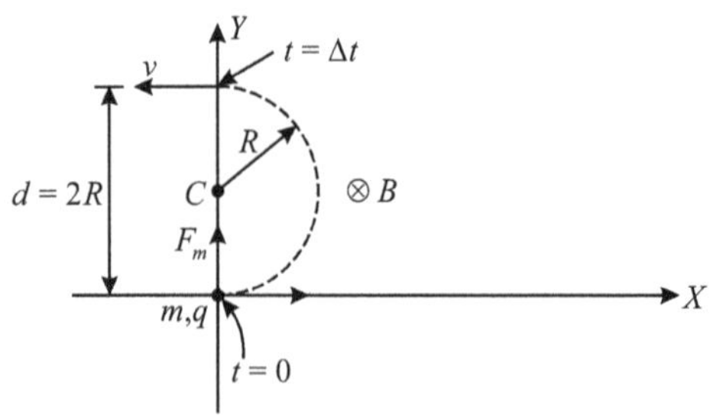

In magnetic field, radius of circular motion of charge is given as

$$R = \frac{mv}{qB} = \frac{d}{2} \quad \ldots(1)$$

Time required to complete half circle is given as

$$\Delta t = \frac{T}{2} = \frac{1}{2}\left(\frac{2\pi m}{qB}\right) = \frac{\pi m}{qB} \quad \left[\text{As } T = \frac{2\pi m}{qB}\right]$$

from equation-(1) $\Delta t = \pi\left(\frac{d}{2v}\right) = \frac{\pi d}{2v}$

also from equation-(1) $B = \frac{2mv}{qd}$ **Ans. (C)**

Sol. 30 For stable equilibrium $U \to$ minimum as $U = -\vec{M}\cdot\vec{B}$

so for $\theta = 0°$ it will be in stable equilibrium

and for unstable equilibrium $U \to \max \Rightarrow$ at $\theta = 180°$

it will be in unstable equilibrium.

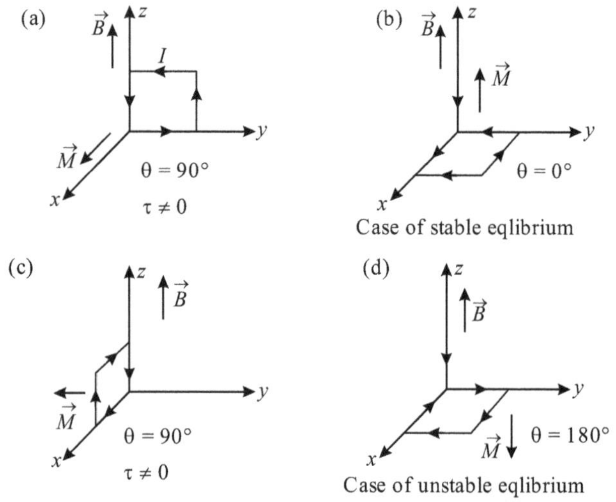

(b) and (d) are cases of stable and unstable equilibrium respectively. **Ans. (C)**

Sol. 31

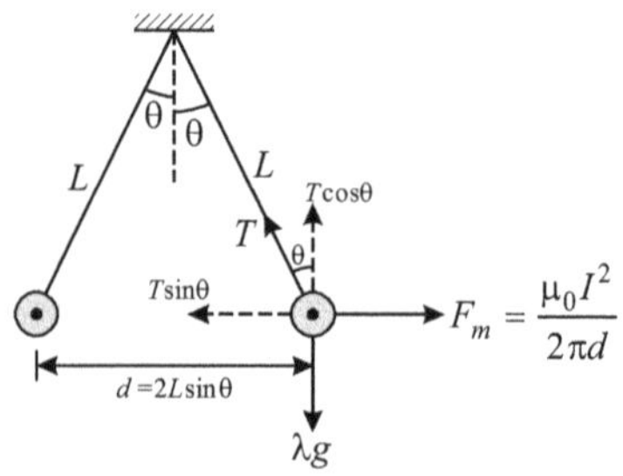

For equilibrium of wires, we have

$$T\sin\theta = \frac{\mu_0 I^2}{2\pi(2L\sin\theta)} \quad \ldots(1)$$

$$T\cos\theta = \lambda g \quad \ldots(2)$$

By (1)/(2) $\Rightarrow \dfrac{\sin^2\theta}{\cos\theta} = \dfrac{\mu_0 I^2}{4\pi L\lambda g}$

$$\Rightarrow \quad I = 2\sin\theta\sqrt{\frac{\pi\lambda g L}{\mu_0\cos\theta}}$$ **Ans. (B)**

Sol. 32 For solenoid having closely spaced turns these behave like closed loop coils and in uniform magnetic field produced by a solenoid the net force on coils (of other solenoid) is equal to zero.

$\Rightarrow \vec{F_1} = \vec{F_2} = 0$ **Ans. (A)**

Sol. 33 $F_m = BIl = 2BIR\sin\theta$ (magnetic force on wire)

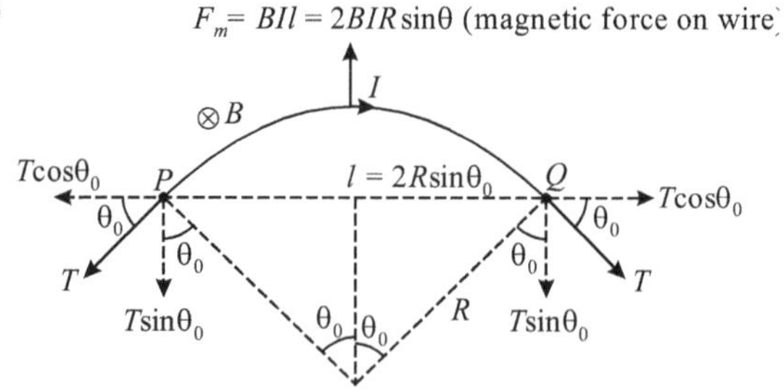

As wire is in equilibrium, we use

$$2T\sin\theta_0 = 2BIR\sin\theta_0$$

$\Rightarrow \quad T = BIR$ **Ans. (C)**

Sol. 34 Force between two parallel current carrying wires is given as

$$F = \frac{\mu_0 I_1 I_2 l}{2\pi d}$$

$\Rightarrow \quad F \propto I_1 I_2$

$\Rightarrow$ If I_1 and I_2 are in same direction (same sign) $\Rightarrow I_1I_2 \rightarrow +$ve then wires attract each other $\Rightarrow F \rightarrow -$ve. And

if I_1 and I_2 are opposite (opposite signs) $\Rightarrow I_1I_2 \rightarrow -$ve then wires repel each other $\Rightarrow F \rightarrow +$ve. **Ans. (A)**

Sol. 35

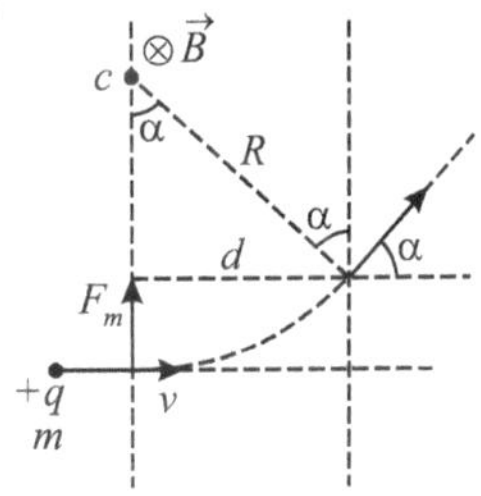

Kinetic energy of proton $K = qV = \frac{1}{2}mv^2$

The radius of proton motion inside the Magnetic Field is

$$R = \frac{mv}{qB} = \frac{\sqrt{2mk}}{qB}$$

from figure $\quad \sin\alpha = \frac{d}{R} = \frac{qBd}{\sqrt{2mK}} = \frac{qBd}{\sqrt{2mqV}}$

$\Rightarrow \quad \sin\alpha = Bd\sqrt{\frac{q}{2mV}}$ **Ans. (D)**

Sol. 36 A loop with a current is in stable equilibrium in a uniform magnetic field when its PE is minimum, which is given as

$$U = -\vec{M}\cdot\vec{B} = -MB\cos\theta$$

Stable equilibrium $\Rightarrow \theta = 0°$ when $\vec{M}$ is parallel to $\vec{B}$. **Ans. (C)**

Sol. 37 By Lorentz force equation, force on a moving charge in EM field is given as

$$\vec{F} = q\vec{E} + q(\vec{v}\times\vec{B}) = 0$$

[As no force is acting on charge]

$$\vec{E} = -(\vec{v}\times\vec{B}) = -v_0(3\hat{i} - \hat{j} + 2\hat{k})\times B_0(\hat{i} + 2\hat{j} - 4\hat{k})$$

$$\vec{E} = -v_0B_0(6\hat{k} + 12\hat{j} + \hat{k} + 4\hat{k} + 2\hat{j} - 4\hat{i})$$

$$\vec{E} = -v_0B_0(14\hat{j} + 7\hat{k})$$ **Ans. (D)**

Sol. 38

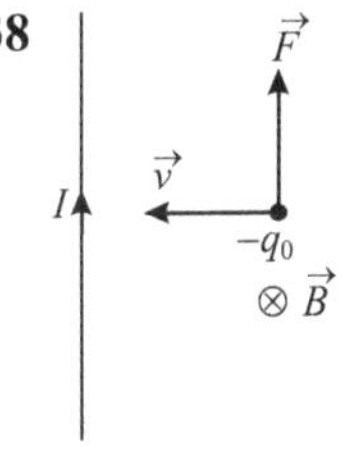

As force $\vec{F}$ and magnetic induction $\vec{B}$ is specified then using Flaming's left hand rule or Right hand Palm Rule we can find the direction of $\vec{v}$ for $-q_0$ charge.

By RHPR fingers point along $\vec{B}$

Palm face vector along force $\vec{F}$

$\Rightarrow$ direction of thumb which is toward right gives the $\vec{v}$ for +ve charge so here as charge is –ve $\Rightarrow \vec{v}$ is leftward. **Ans. (B)**

Sol. 39 Case-I :

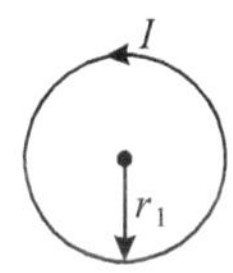

Magnetic moment $\quad m = I\pi r_1^2$

Magnetic induction at centre

$$B_1 = \frac{\mu_0 I}{2r_1} \quad \ldots(1)$$

Case-II : $I \rightarrow$ constant but $m' = 2m$

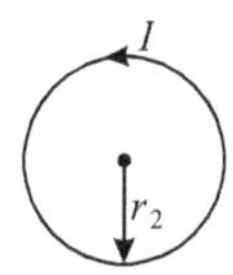

$$m' = I\pi r_2^2 = (I\pi r_1^2)$$

$\Rightarrow \quad r_2 = \sqrt{2}\, r_1$

$$B_2 = \frac{\mu_0 I}{2r_2} = \frac{\mu_0 I}{2\sqrt{2}r_1} \quad \ldots(2)$$

By (1)/(2) $\Rightarrow \quad \frac{B_1}{B_2} = \sqrt{2}$ **Ans. (C)**

Sol. 40 In a magnetic field for a charge particle in circular motion, radius is given as

$$r = \frac{mv}{qB} = \frac{\sqrt{2mK}}{qB}$$

for electron $\quad r_e = \frac{\sqrt{2m_eK}}{eB}$

for proton $\quad r_p = \frac{\sqrt{2m_pK}}{eB}$

for α particle $\quad r_\alpha = \frac{\sqrt{2(4m_p)K}}{(2e)B} = \frac{\sqrt{2m_pK}}{eB} = r_p$

As $m_e < m_p$

$\Rightarrow \quad r_e < r_p = r_\alpha$ **Ans. (B)**

Sol. 41

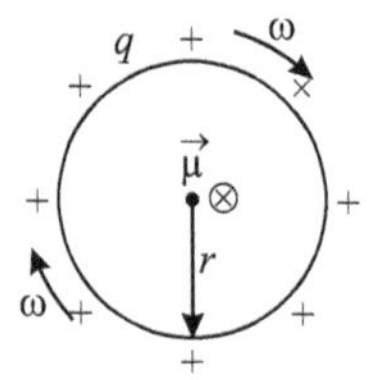

The rotating charged loop behaves like a circular current, which is given as

$$I = \frac{q\omega}{2\pi}$$

$\Rightarrow$ Magnetic moment of this rotating charge is given as

$$\mu = I \times \pi r^2 = \frac{q\omega}{2\pi} \times \pi r^2 = \frac{1}{2} q\omega r^2 \quad \textbf{Ans. (A)}$$

Alternative : We use

$$\frac{\mu}{L} = \frac{q}{2m}$$

$$\Rightarrow \quad \mu = \frac{q}{2m} \times L = \frac{q}{2m} \times mr^2\omega$$

$$\Rightarrow \quad \mu = \frac{1}{2} q\omega r^2 \quad \textbf{Ans. (A)}$$

Sol. 42 Force on loop due to wire is given as

$$|\vec{F}| = m\left|\frac{dB}{dr}\right|$$

Where m is the dipole moment of loop

$$m = i\pi a^2$$

We use

$$\frac{dB}{dx} = \frac{d}{dx}\left(\frac{\mu_0 I}{2\pi x}\right) = -\frac{\mu_0 I}{2\pi x^2}$$

$$\Rightarrow \quad F = \frac{\mu_0 m I}{2\pi x^2} = \frac{\mu_0 i I \pi a^2}{2\pi x^2}$$

$$\Rightarrow \quad F \propto \frac{a^2}{d^2} \quad [\text{As } x = d] \quad \textbf{Ans. (C)}$$

According to Newtons III law same force will be exerted by loop on wire.

Sol. 43 In circular motion under magnetic field, we use

$$\frac{mv^2}{R} = qvB$$

$$mv = qBR \quad \ldots(1)$$

As path becomes straight line by electric field, we use

$$qE = qvB$$

$$\Rightarrow \quad E = vB \quad \ldots(2)$$

From equation-(1) & (2)

$$m = \frac{qB^2R}{E}$$

$$\Rightarrow \quad m = 2.0 \times 10^{-24}\,\text{kg} \quad \textbf{Ans. (A)}$$

Sol. 44 Charge on element of with dx on rod revolving in a circle of radius x is given as

$$dq = \lambda dx$$

its magnetic moment is given as

$$dm = (n)\left(\frac{\rho_0 x}{l}\right) dx \cdot (\pi x^2)$$

Total magnetic moment of rod is

$$M = \int dm = \frac{n\rho_0\pi}{l} \cdot \int_0^l x^3 \cdot dx = \frac{n\rho_0\pi}{l} \cdot \left[\frac{l^4}{4}\right]$$

$$\Rightarrow \quad M = \frac{n\rho_0\pi l^3}{4} \quad \textbf{Ans. (A)}$$

Sol. 45 Radius of circular motion of electron is

$$r = \frac{\sqrt{2mK}}{eB} = \frac{\sqrt{2meV}}{eB}$$

$$\Rightarrow \quad r = \frac{\sqrt{\frac{2m}{e} \cdot V}}{B} = \frac{\sqrt{\frac{2 \times 9.1 \times 10^{-31}}{1.6 \times 10^{-19}}(500)}}{100 \times 10^{-3}}$$

$$\Rightarrow \quad r = \frac{\sqrt{\frac{9.1}{0.16} \times 10^{-10}}}{10^{-1}} = \frac{3}{.4} \times 10^{-4}\,\text{m}$$

$$\Rightarrow \quad r = 7.5 \times 10^{-4}\,\text{m} \quad \textbf{Ans. (A)}$$

Sol. 46

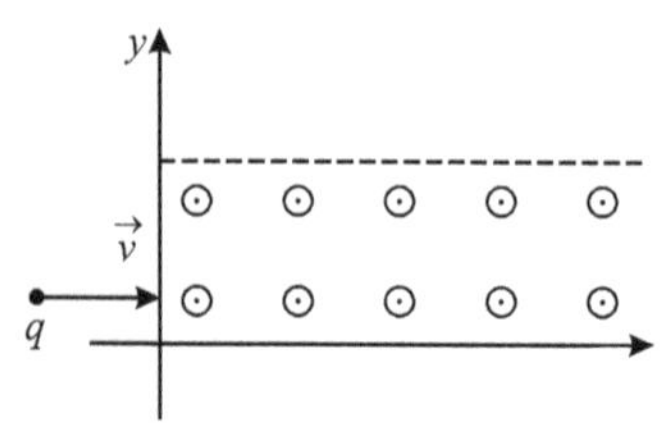

As the initial position of entering particle is not given, acceleration at leaving point cannot be calculated.

Ans. (None)

Sol. 47 Work will only be done against electric force, equal to

$$W = q\vec{E} \cdot \vec{r}$$

$$\Rightarrow \quad W = q(2\hat{i} + 3\hat{j}) \cdot (\hat{i} + \hat{j}) = 5q \quad \textbf{Ans. (D)}$$

Sol. 48 Kinetic energy of a charge q accelerated by a potential difference V is

$$k = qV$$

Radius of circular motion in magnetic field is given as

$$r = \sqrt{\frac{2mk}{q_B}} = \sqrt{\frac{2mkV}{q_B}}$$

$$\Rightarrow \quad r \propto \sqrt{\frac{m}{q}}$$

$$\Rightarrow \quad \frac{r_p}{r_\alpha} = \frac{1}{\sqrt{2}}$$ **Ans. (A)**

Sol. 49

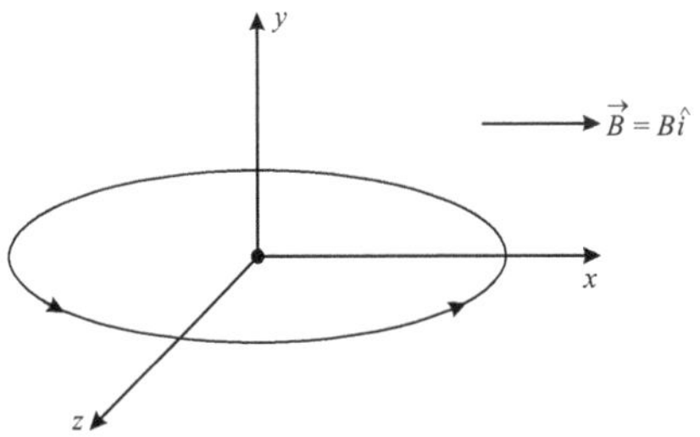

Magnetic moment of coil is given as

$$\overrightarrow{M} = NIA\,\hat{j}$$

$$\vec{\tau} = NI(\pi r^2)\hat{j}$$

Torque on loop (coil)

$$= \overrightarrow{M} \times \overrightarrow{B}$$

$$= NI(\pi r^2)\,B \sin 90° \,(-\hat{k})$$

$$= NI\pi r^2 B(-\hat{k})$$ **Ans. (A)**

Sol. 50 Given that resitance of galvanometer and shunt are

$$R_g = 50\,\Omega$$

$$R_S = 5000\,\Omega$$

Full deflecttion current is

$$i_g = 4 \times 10^{-3}$$

Thus range of voltmeter is given as

$$V_R = i_g(G + S)$$

$$\Rightarrow \quad V_R = 4 \times 10^{-3}(50 + 5000)\,V$$

$$\Rightarrow \quad V_R = 4 \times 10^{-3}(5050)\,V$$

$$\Rightarrow \quad V_R = 20.2\,V$$ **Ans. (B)**

Sol. 51 Force acting on segments of square frame are shown in figure below

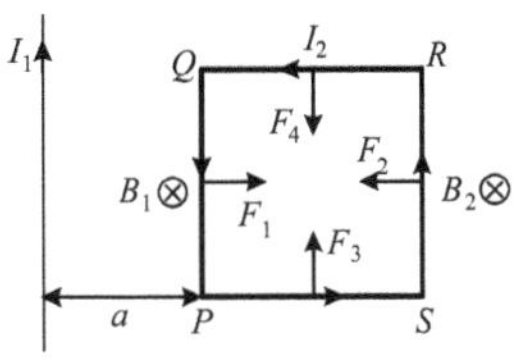

Here F_3 & F_4 cancel each other and force on PQ will be

$$F_1 = I_2 B_1 a$$

$$\Rightarrow \quad F_1 = I_2\left(\frac{\mu_0 I_1}{2\pi a}\right)a$$

$$\Rightarrow \quad F_1 = \frac{\mu_0 I_1 I_2}{2\pi}$$

Force on RS will be

$$F_2 = I_2 B_2 a$$

$$\Rightarrow \quad F_2 = I_2 \frac{\mu_0 I_1}{2\pi(2a)} a$$

$$\Rightarrow \quad F_2 = \frac{\mu_0 I_1 I_2}{4\pi}$$

Net force is given as

$$F_{net} = F_1 - F_2 = \frac{\mu_0 I_1 I_2}{4\pi} \text{ repulsion}$$ **Ans. (B)**

Sol. 52 Torque on coil is given as

$$|\vec{\tau}| = |\overrightarrow{M} \times \overrightarrow{B}|$$

$$\Rightarrow \quad \tau = NI \times A \times B \times \sin 45°$$

$$\Rightarrow \quad \tau = 100 \times 3 \times 12.5 \times 10^{-4} \times 1 \times 0.707$$

$$\Rightarrow \quad \tau = 0.26\,\text{Nm} \simeq 0.27\,\text{Nm}$$ **Ans. (B)**

Sol. 53 Torque on coil is given as

$$\tau = \overrightarrow{M} \times \overrightarrow{B}$$

$$C\theta = i\,N\,A\,B$$

$$\Rightarrow \quad 10^{-6} \times \frac{\pi}{180} = 10^{-3} \times 10^{-4} \times 175 \times B$$

$$\Rightarrow \quad B = 10^{-3}\,\text{T}$$ **Ans. (A)**

Sol. 54 Radius of circular motion in magnetic field is given as

$$r = \frac{mv}{qB} = \frac{\sqrt{2mK}}{qB}$$

$$\Rightarrow \quad r_{He} = r_p > r_e$$ **Ans. (C)**

Sol. 55

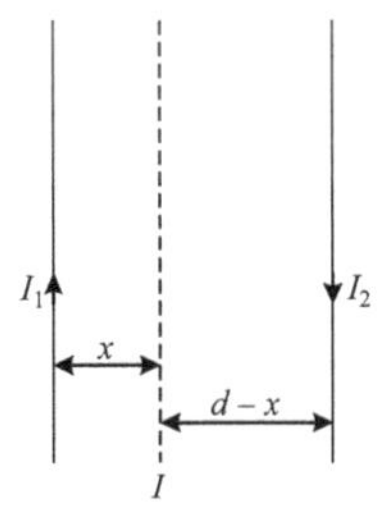

Net force on wire carrying current I per unit length is given as

$$\frac{\mu_0 I_1 I}{2\pi x} + \frac{\mu_0 I_2 I}{2\pi(d - x)} = 0$$

$$\frac{I_1}{x} = \frac{I_2}{x - d}$$

$$\Rightarrow \quad x = \frac{I_1 d}{I_1 - I_2}$$ **Ans. (B)**

Sol. 56 Given that $I_g = 10^{-4}$ A

and if galvanometer resistance is R_g, we use

$$I_g = \frac{V_R}{R_g + R}$$

$$10^{-4} = \frac{5}{R_g + 2\times10^6}$$

$$\Rightarrow \quad R_g = 5\times10^4 - 2\times10^6$$

$$= -195\times10^4\,\Omega$$

Practically not possible

None of the options can be chosen. **Ans. (None)**

Sol. 57 When galvanometer is used as an ammeter shunt is used in parallel with galvanometer as shown in figure below

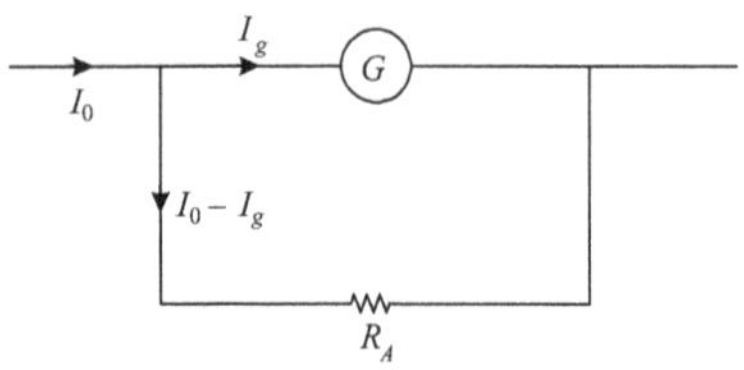

For full deflection, we use

$$I_g G = (I_0 - I_g)R_A$$

$$\Rightarrow \quad R_A = \left(\frac{I_g}{I_0 - I_g}\right)G$$

When galvanometer is used as a voltmeter, resistance is used in series with galvanometer as shown in figure below

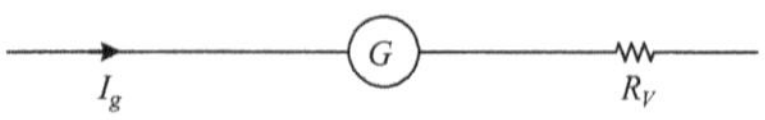

For full deflection, we use

$$I_g(G + R_V) = V = GI_0 \quad (\text{given } V = GI_0)$$

$$\Rightarrow \quad R_V = \frac{(I_0 - I_g)G}{I_g}$$

$$\Rightarrow \quad R_A R_V = G^2$$

and $$\frac{R_A}{R_V} = \left(\frac{I_g}{I_0 - I_g}\right)^2$$ **Ans. (B)**

Sol. 58 Angle of deflection is given as

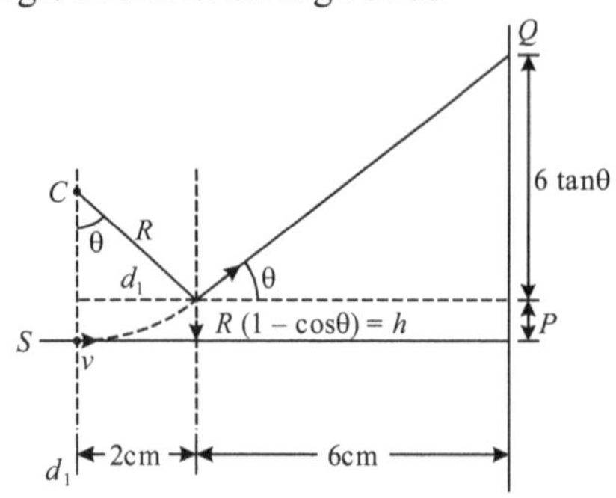

$$\sin\theta = \frac{d_1}{R} = \frac{d_1}{mv/qB} = \frac{qBd_1}{mv} = \frac{qBd_1}{\sqrt{2mE}}$$

$$\Rightarrow \quad \sin\theta = \frac{1.6\times10^{-19}\times1.5\times10^{-3}\times0.02}{\sqrt{2\times9.1\times10^{-31}\times100\times1.6\times10^{-19}}}$$

$$= \frac{4.8\times10^{-24}}{\sqrt{29}\times10^{-24}} = \frac{4.8}{5.4} \simeq 0.89$$

$$\Rightarrow \quad \cos\theta = \sqrt{1-(0.89)^2} = \sqrt{0.2} = 0.45$$

$$\Rightarrow \quad \tan\theta = \frac{0.89}{0.45} \simeq 2$$

$$\Rightarrow \quad d = PQ = 6\tan\theta + h \cong 12 + h\text{ cm}$$ **Ans. (A)**

19.1 Classical and Terrestrial Magnetism

Sol. 1

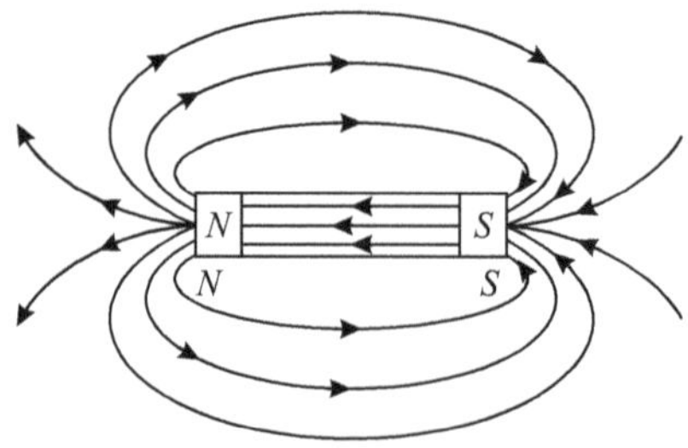

Configuration of magnetic lines in surrounding of a bar magnet. Inside a bar magnet magnetic lines travel from S to N pole.

Ans. (D)

Sol. 2 The oscillation period of a bar magnet in magnetic field is

$$T = 2\pi\sqrt{\frac{I}{MB}}$$

$$I = \frac{ml^2}{12} \quad \text{and} \quad M = m_s.l$$

If magnet is broken in two equal parts of length $l/2$ then its moment of inertia & magnetic moment will become

$$I' = \frac{(m/2)(l/2)^2}{12} = \frac{I}{8}$$

& $$M' = m_s\left(\frac{l}{2}\right) = \frac{M}{2}$$

New time period of oscillation will be

$$\Rightarrow \quad T' = 2\pi\sqrt{\frac{I'}{M'B}} = 2\pi\sqrt{\frac{I/8}{M/2B}} = \frac{T}{2}$$

$$\Rightarrow \quad \frac{T'}{T} = \frac{1}{2}$$ **Ans. (B)**

Sol. 3 In a vibration magnetometer time period of oscillations of a magnet is

$$T = 2\pi\sqrt{\frac{I}{MB}}$$

Here $$I = \frac{ml^2}{12} \ (m = \text{mass})$$

& $$M = m_s l \ (m = \text{pole strength})$$

If magnet is cut in 3 parts each of length $l/3$ then we use new moment of inertia and magnetic moment of combination is

$$I' = 3\times\frac{(m/3)(l/3)^2}{12} = \frac{I}{9}$$

and $M' = m_s\frac{l}{3}\times 3 = M$

New time period will be

$$T' = 2\pi\sqrt{\frac{I'}{M'B}} = 2\pi\sqrt{\frac{I/9}{MB}} = \frac{T}{3} = \frac{2}{3}\text{s}$$ **Ans. (B)**

Sol. 4

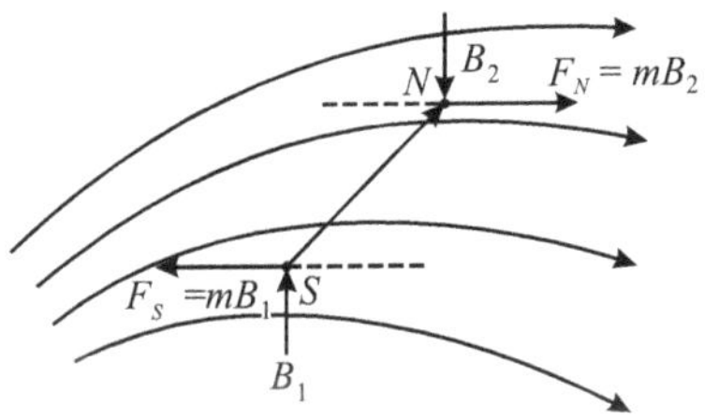

As in non uniform magnetic field $B_1 \neq B_2$

$\Rightarrow$ There will be a net force acting on the needle and as direction F_N and F_S are not along same line.

$\Rightarrow$ There will be a net torque acting as well. **Ans. (C)**

Sol. 5

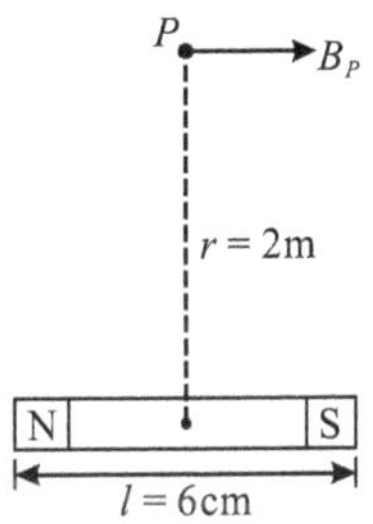

As $r >> l$, here we can use the expression of magnetic induction due to a magnetic dipole, which is given as

for equatorial points $B_P = \frac{kM}{r^3} = \frac{\mu_0}{4\pi}\frac{M}{r^3}$

$$\Rightarrow \quad B_P = 10^{-7}\times\frac{4}{(2)^3} = 0.5\times 10^{-7}\,\text{T}$$

$$\Rightarrow \quad B_P = 5\times 10^{-8}\,\text{T}$$ **Ans. (C)**

Sol. 6

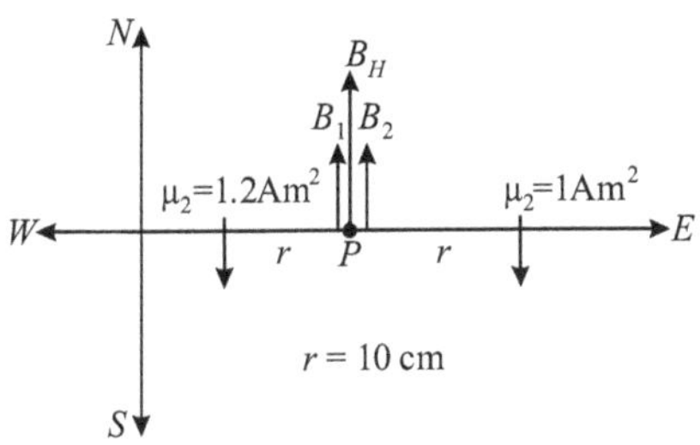

Magnetic moment of the two dipole are μ_1 = 1.20 Am^2 & μ_2 = 1.00 Am^2 and due to these dipoles, magnetic induction at mid point P are given as

$$B_1 = \frac{k\mu_1}{r^3} = \frac{\mu_0}{4\pi}\frac{\mu_1}{r^3} = \frac{10^{-7}\times 1.20}{(0.1)^3}$$

$$\Rightarrow \quad B_1 = 1.2\times 10^{-4}\,\text{Wb/m}^2$$

and $$B_2 = \frac{k\mu_2}{r^3} = \frac{\mu_0}{4\pi}\frac{\mu_2}{r^3} = \frac{10^{-7}\times 1.00}{(0.1)^3}$$

$$\Rightarrow \quad B_2 = 1\times 10^{-4}\,\text{Wb/m}^2$$

$\Rightarrow$ Net magnetic induction at P is

$$B_P = B_H + B_1 + B_2$$

$$\Rightarrow \quad B_P = 3.6\times 10^{-5} + 1.2\times 10^{-4} + 1\times 10^{-4}$$

$$\Rightarrow \quad B_P = 2.56\times 10^{-4}\,\text{Wb/m}^2$$ **Ans. (B)**

Sol. 7

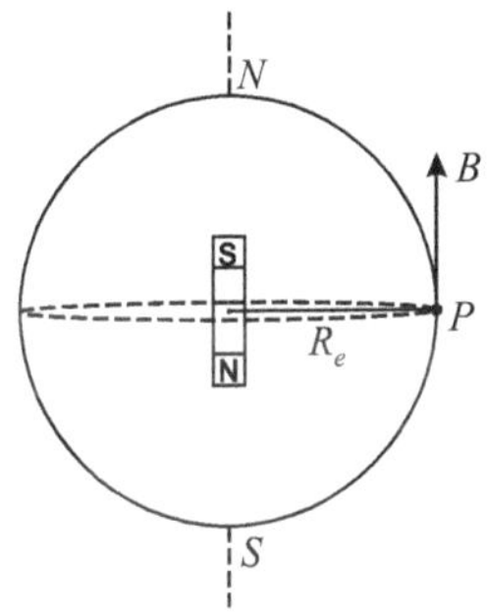

Magnetic induction at equatorial point P due to Earth's magnetic dipole is given as

$$B_P = \frac{kM}{R_e^3}$$

$$\Rightarrow \quad B_P = \frac{10^{-7}\times 8\times 10^{22}}{(6.4\times 10^6)^3}T$$

$$\Rightarrow \quad B_P = 0.302\times 10^{-4}\,\text{T} = 0.302\,\text{G}$$

$$1\,\text{G} = 10^{-4}\,\text{T}$$ **Ans. (D)**

Sol. 8

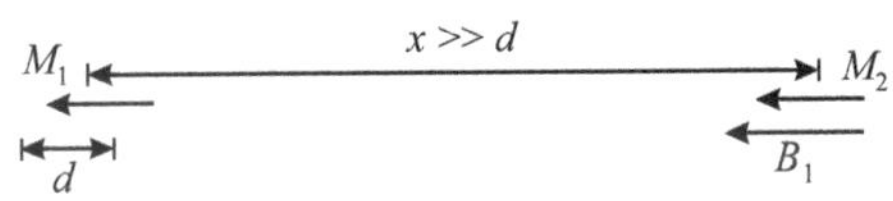

[B_1 is the axial point Magnetic field due to M_1 at a distance x]

Magnetic induction due to M_1 at the location of M_2 is given as

$$B_1 = \frac{2kM_1}{x^3}$$

In non-uniform Magnetic field, force on dipole M_2 is given as

$$\vec{F} = -\vec{M}_2\frac{\overrightarrow{dB_1}}{dx}\hat{i}$$

In magnitude $$|\vec{F}| = |\vec{M}_2\times 2kM_1\frac{d}{dx}(x^{-3})|$$

$$\Rightarrow \quad |\vec{F}| = \frac{6kM_1M_2}{x^4} \Rightarrow F\propto x^{-4}$$ **Ans. (D)**

Sol. 9

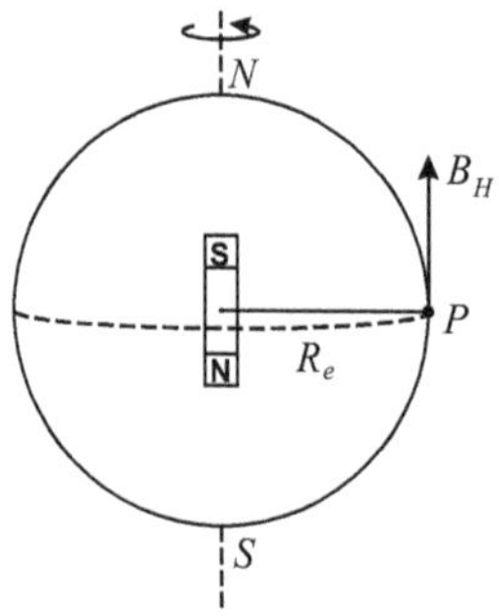

Neglecting declination

For a magnetic dipole at equatorial point, the magnetic induction is given as

$$B_H = \frac{kM}{R_e^3} = 4 \times 10^{-5}$$

$$\Rightarrow \quad M = \frac{4\times10^{-5}\times(6.4\times10^{6})^3}{10^{-7}}$$

$$\Rightarrow \quad M = 1.048 \times 10^{23}\ \text{Am}^2$$

Ans. (A)

Sol. 10

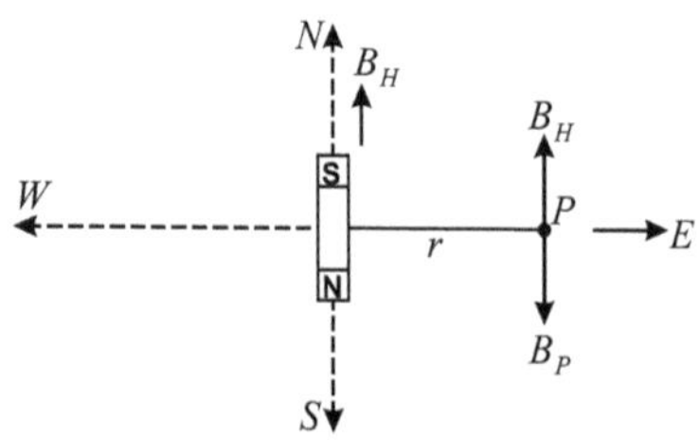

Considering the magnet as a magnetic dipole, magnetic induction at equatorial point P is given as

$$B_P = \frac{KM}{r^3}$$

As at P $\quad B_{net} = 0$ (null point) Þ $B_P = B_H$

$$\Rightarrow \quad \frac{10^{-7}\times M}{(0.3)^3} = 3.6\times10^{-5}$$

$$\Rightarrow \quad M = \frac{3.6\times10^{-5}\times27\times10^{-3}}{10^{-7}} = 9.72$$

Ans. (C)

Sol. 11

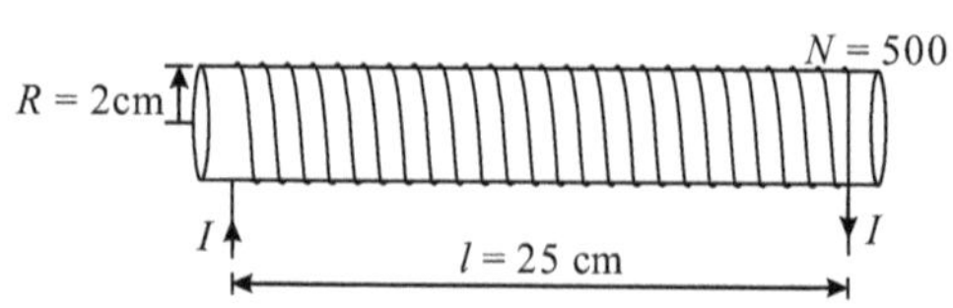

Magnetic moment of solenoid

$$\mu = I(\pi R^2)\times N$$

Volume of solenoid $V = l(\pi R^2)$

Magnetization $\quad |\overrightarrow{M}| = \frac{\mu}{V} = \frac{I\pi R^2 N}{l\pi R^2}$

$$= \frac{IN}{l} = \frac{15\times500}{0.25}$$

$$\Rightarrow \quad |\overrightarrow{M}| = 3\times10^4\ \text{A/m}$$

Ans. (C)

Sol. 12 Stable equilibrium of a dipole in magnetic field exist when angle θ between its dipole moment and net magnetic induction is zero.

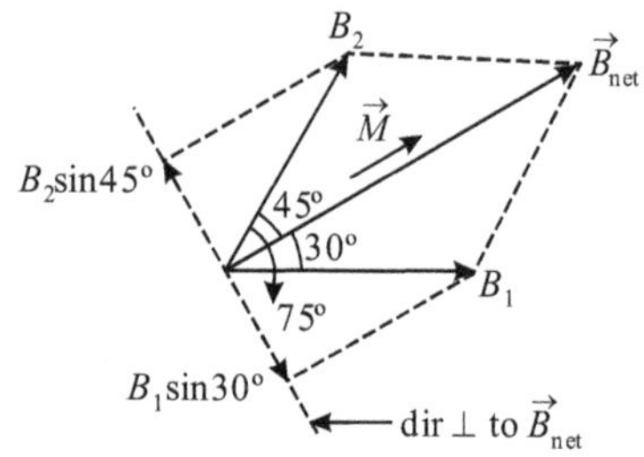

here θ between $\overrightarrow{M}$ and $\vec{B}_{net}$ should be zero. The components of B_1 and B_2 along the direction normal to $\vec{B}_{net}$ must be equal.

$$\Rightarrow \quad B_1\sin30° = B_2\sin45°$$

$$15\times\frac{1}{2} = B_2\times\frac{1}{\sqrt{2}}$$

$$\Rightarrow \quad B_2 = \frac{15}{\sqrt{2}} = 10.6\,\text{mT}$$

Ans. (B)

Sol. 13 For SHM oscillations of magnetic needle must be with small angular amplitude so that restoring torque on it should be given as

$$\tau = MB\sin\theta \approx MB\theta$$

[For small θ we use $\sin\theta \approx \theta$]

⇒ angular acceleration of needle is given as

$$\alpha = -\frac{\tau}{I} = -\frac{MB}{I}\theta$$

[As $\alpha \propto \theta$ ⇒ it executes SHM]

Comparing with $\quad \alpha = -\omega^2\theta$, we use $\omega = \sqrt{\frac{MB}{I}}$

Time period of SHM is

$$T = \frac{2\pi}{\omega} = 2\pi\sqrt{\frac{I}{MB}}$$

$$\Rightarrow \quad T = 2\times3.14\sqrt{\frac{7.5\times10^{-6}}{6.7\times10^{-2}\times0.1}} \approx 0.665\text{s}$$

For 10 oscillations time taken is $= 0.665\times10 = 6.66$s **Ans. (C)**

Sol. 14 Potential energy of a magnetic dipole in constant magnetic field is given as

$$U = -\overrightarrow{M}\cdot\vec{B}$$

$$\Rightarrow \quad U = -MB\cos\theta$$

Maximum PE is for θ = 180° when $U_{max} = -MB(-1) = MB$

Minimum PE is for θ = 0° when $U_{min} = -MB$

Zero PE is for θ = 90° when $U = -MB(0) = 0$

and torque on dipole $\vec{\tau} = \overrightarrow{M}\times\vec{B}$

$$\Rightarrow \quad \vec{\tau} = MB\sin\theta$$

$\Rightarrow \quad \tau = MB \sin 180° = 0$ (Min)

$\Rightarrow \quad \tau = MB \sin 0 = 0$ (Min)

$\Rightarrow \quad \tau = MB \sin 90° = MB$ (Max) **Ans. (C)**

Sol. 15 To keep needle horizontal net torque on it must be zero and net magnetic field at the location of dipole is $\sqrt{2}B$

$$\sqrt{2}B \cdot \mu \sin 45° = F \cdot \frac{l}{2}$$

$$\Rightarrow \quad F = 2mB$$

$$\Rightarrow \quad F = 2 \times 1.8 \times 18 \times 10^{-6}$$

$$\Rightarrow \quad F = 6.48 \times 10^{-5}\,\text{N}$$ **Ans. (B)**

Sol. 16 For oscillations of magnetic dipoles, we use

$$T = 2\pi\sqrt{\frac{I}{\mu B}}$$

For hoop, we use

$$T_h = 2\pi\sqrt{\frac{mR^2}{(2\mu)B}}$$

For cylinder, we use

$$T_C = 2\pi\sqrt{\frac{1/2mR^2}{\mu B}} \simeq T_r$$ **Ans. (D)**

Sol. 17 Magnetic moment of square loop is given as

$$m = NIA = 1 \times I \times a^2$$

Here $\quad a$ = side of square

Now, $\quad 4a = 2\pi r$

$$r = \frac{2a}{\pi}$$

For circular loop, we use

$$m' = 1 \times I \times \pi r^2$$

$$\Rightarrow \quad m' = 1 \times I \times \pi \times \left(\frac{2a}{\pi}\right)^2$$

$$m' = \frac{4m}{\pi}$$ **Ans. (B)**

Sol. 18 For a horizontal compass needle the frequency of small oscillations in earth's magnetic field is given as

$$f = \frac{1}{2\pi}\sqrt{\frac{MB_N}{I}}$$ (horizontal component of earth's field)

When $\quad \theta_D = 45° \rightarrow B_{H_1} = B_1 \cos 45°$

$$f_1 = \frac{1}{2\pi}\sqrt{\frac{MB_1 \cos 45°}{I}} \quad \ldots(1)$$

When $\quad \theta'_D = 30° \rightarrow B_{H_2} = B_2 \cos 30°$

$$f_2 = \frac{1}{2\pi}\sqrt{\frac{MB_2 \cos 30°}{I}} \quad \ldots(2)$$

$$(1)/(2) \Rightarrow \quad \frac{30}{40} = \sqrt{\frac{B_1/\sqrt{2}}{B_2\sqrt{3}/2}}$$

$$\Rightarrow \quad \frac{B_1}{B_2} = \frac{9}{16} \times \sqrt{\frac{3}{2}} \simeq 0.7$$ **Ans. (C)**

19.2 Magnetic Properties of Materials

Sol. 1 Curie temperature is the temperature above which a ferromagnetic material changes to paramagnetic material. **Ans. (A)**

Sol. 2 Materials used for making electromagnets must be quickly magnetized on flow of current & also quickly demagnetize when current is stopped.

$\Rightarrow$ Retentivity and coercivity both must be low. **Ans. (B)**

Sol. 3 Paramagnetic and Diamagnetic materials are weakly magnetized along & opposite to external field.

$\Rightarrow$ N_2 is attracted by external magnet & N_3 is repelled by external magnet weakly Ferromagnetic material strongly magnetized along the direction of external magnetic field.

$\Rightarrow$ N_1 is strongly attracted by magnet. **Ans. (B)**

Sol. 4 For any material always

$$\epsilon_r > 1$$

For a diamagnetic material we use

$$\mu_r < 1$$ **Ans. (B)**

Sol. 5 For demagnetization of magnet we use

$$H_c = \frac{B}{\mu_0} = nI = \frac{N}{l}I$$

$$I = \frac{H_c \times l}{N} = \frac{3 \times 10^3 \times 0.1}{100} = 3\,\text{A}$$ **Ans. (C)**

Sol. 6 Diamagnetic $\rightarrow$ repel external magnetic field $\rightarrow$ Bar *B*.

Paramagnetic $\rightarrow$ magnetize in same direction of external Magnetic field $\rightarrow$ Bar *C*.

Ferromagnetic $\rightarrow$ Highly magnetize in the same direction of external magnetic field $\rightarrow$ Bar *A*. **Ans. (B)**

Sol. 7 In a superconductor it always magnetizes to an extent so that it completely expel external magnetic field and no net magnetic field exist inside a superconducting material.

$\Rightarrow \quad B_{\text{inside}} = 0$

or $\quad B_{\text{sup}} = 0$ **Ans. (B)**

Sol. 8 In cases of generator, transformer & electromagnets we need to use materials which can be quickly magnetized & demagnetized without much loss of energy and we know.

Loss of energy/cycle $\propto$ area of *B-H* curve. **Ans. (D)**

Sol. 9 For demagnetization of ferromagnetic material it is the coercivity which is given as

$$H = 100 \text{ A/m}$$

For a solenoid $\quad H = nI$

$$\Rightarrow \quad I = \frac{H}{n} = \frac{(100 \text{ A/m})}{(10^5 \text{ turns/m})} = 10^{-3} \text{ A}$$

$\Rightarrow \quad I = 1 \text{ mA}$ **Ans. (B)**

Sol. 10 Coercivity of material is given as

$$H_c = \frac{B}{\mu_0}$$

$$\Rightarrow \quad H = ni = \frac{N}{l}i = \frac{100}{0.2} \times 5.2 \text{ A/m}$$

$\Rightarrow \quad H = 2600 \text{ A/m}$ **Ans. (B)**

Sol. 11 Magnetic susceptibility of a material is

$$\chi = \frac{I}{H}$$

Where intensity of magnetization is

$$I = \frac{\text{Magnetic moment}}{\text{Volume}}$$

$$\Rightarrow \quad I = \frac{20 \times 10^{-6}}{10^{-6}} = 20 \text{N/m}^2$$

$$\Rightarrow \quad \chi = \frac{20}{60 \times 10^{+3}} = \frac{1}{3} \times 10^{-3}$$

$\Rightarrow \quad \chi = 0.33 \times 10^{-3} = 3.3 \times 10^{-4}$ **Ans. (C)**

Sol. 12 We know that susceptibility of a material is given as

$$X \alpha \frac{1}{T_C}$$

$$\Rightarrow \quad \frac{X_1}{X_2} = \frac{T_{C_2}}{T_{C_1}}$$

$$\Rightarrow \quad \frac{2.8 \times 10^{-4}}{X_2} = \frac{300}{350}$$

$$\Rightarrow \quad X_2 = \frac{2.8 \times 350 \times 10^{-4}}{300}$$

$\Rightarrow \quad X_2 = 3.266 \times 10^{-4}$ **Ans (C)**

Sol. 13

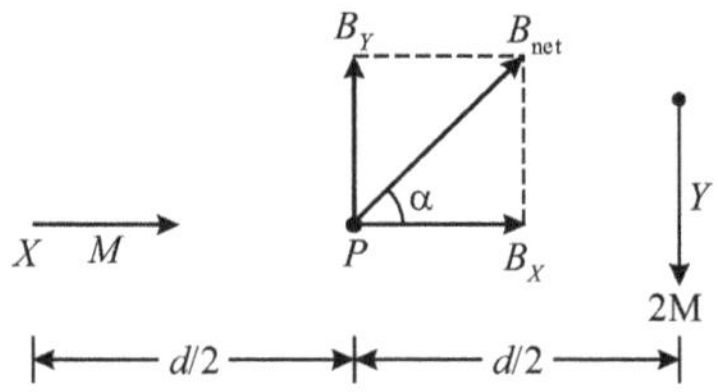

B_X is given as

$$B_X = \frac{2kM_X}{(d/2)^3} = 2\left(\frac{\mu_0}{4\pi}\right) \cdot \frac{M}{(d/2)^3} \qquad (B_X = B_Y)$$

B_Y is given as

$$B_Y = \frac{kM_Y}{(d/2)^3} = \left(\frac{\mu_0}{4\pi}\right) \cdot \frac{2M}{(d/2)^3} \qquad (\alpha = 45°)$$

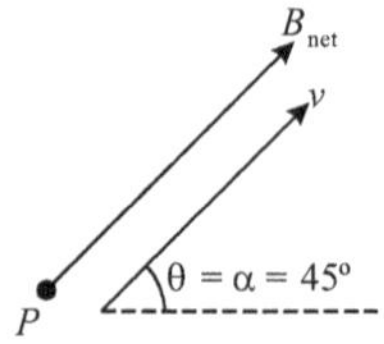

At P, $\vec{B}_{\text{net}}$ is || to $\vec{v}$

$\Rightarrow \quad \vec{F}_{\text{mag}} = q(\vec{v} \times \vec{B}_{\text{net}}) = 0$ **Ans. (D)**

20.1 Electromagnetic Induction

Sol. 1

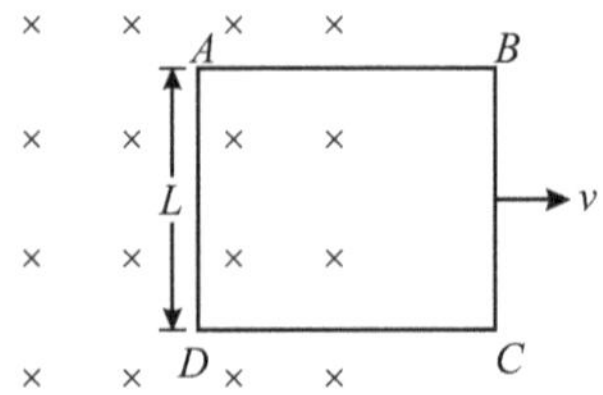

During motion, only wire segment *AD* will cut magnetic flux and motional emf is induced in it, which is given as

$$e_{AD} = BvL$$

$v_A - v_D = BvL$ **Ans. (D)**

Sol. 2

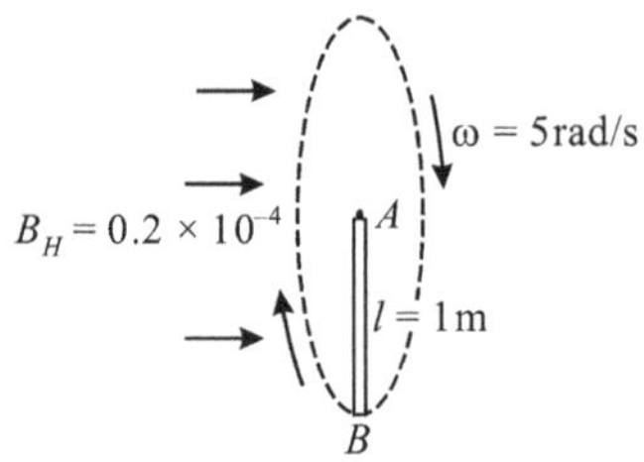

EMF induced across a rotating conductor in uniform and normal magnetic field is given as

$$e = \frac{1}{2} B\omega l^2$$

$$\Rightarrow \qquad e = \frac{1}{2} \times 0.2 \times 10^{-4} \times 5 \times (1)^2$$

$$\Rightarrow \qquad e = 0.5 \times 10^{-4}\,\text{V} = 50\,\mu\text{V}$$ **Ans. (B)**

Sol. 3

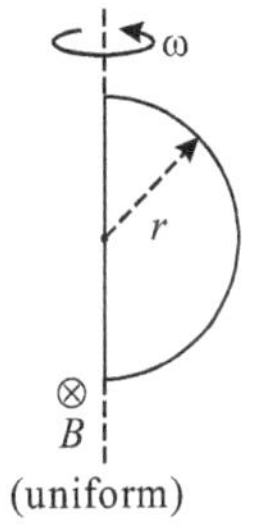

EMF induced in a rotating loop at angular frequency ω is given as

$$e = NBA\sin\omega t = \frac{B\pi\omega r^2}{2}\sin\omega t$$

induced current in loop is $i = e/R$

rate of heat generated (thermal power) in loop is

$$P_{th} = i^2R = \frac{(B\pi\omega r^2)^2}{4R}\sin^2\omega t$$

$$\Rightarrow \text{mean power } \langle P_{th}\rangle = \frac{(B\pi\omega r^2)^2}{4R}\langle\sin^2\omega t\rangle$$

$$\Rightarrow \qquad \langle P_{th}\rangle = \frac{(B\pi\omega r^2)^2}{4R}\times\frac{1}{2} = \frac{(B\pi\omega r^2)^2}{8R}$$ **Ans. (B)**

Sol. 4 For a given n turns coil when it is moved such that the flux changes through it is given as

$$\Delta\phi = n(W_2 - W_1)$$

$\Rightarrow$ emf induced in coil is

$$e = -\frac{\Delta\phi}{\Delta t} = -\frac{n(W_2 - W_1)}{t}$$

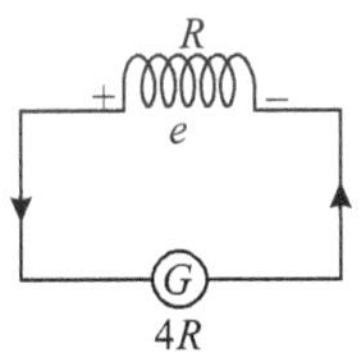

effective resistance of circuit = $5R$

$\Rightarrow$ Current induced in coil will be

$$i = \frac{e}{5R} = -\frac{n(W_2 - W_1)}{5Rt}$$ **Ans. (B)**

Sol. 5

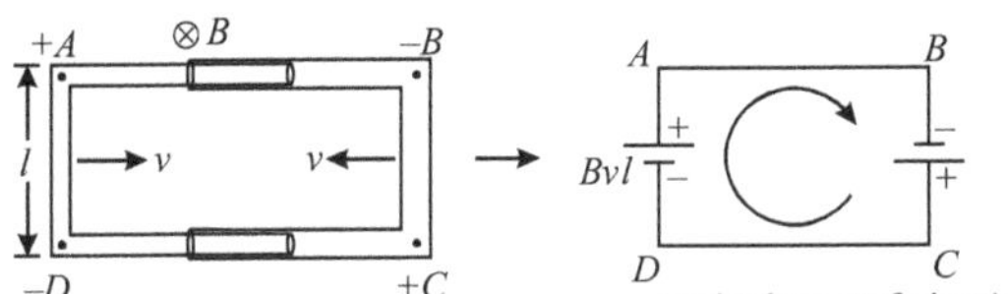

equivalent emf circuit of tubes

As sections AD and BC are moving normal to their lengths so emf is induced only in these sections so motional emf induced is

$$e_{AD} = Bvl \text{ and } e_{BC} = Bvl$$

Total loop emf in circuit

$$e_T = e_{AD} + e_{BC} = 2Bvl$$ **Ans. (D)**

Sol. 6 Motional emf across the conductor PQ is

$$e_{PQ} = Blv$$

$\Rightarrow$ equivalent circuit will be (with emf)

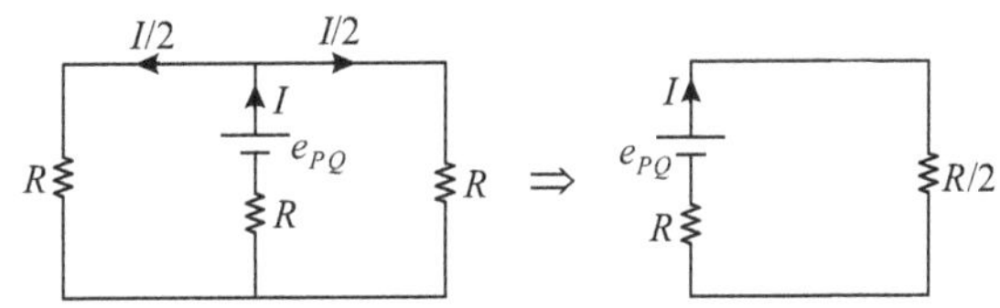

From figure, current supplied by battery is

$$I = \frac{e_{PQ}}{R + \frac{R}{2}} = \frac{Blv}{\frac{3R}{2}} = \frac{2Blv}{3R}$$

Currents through side resistances is

$$I_1 = I_2 = \frac{I}{2} = \frac{Blv}{3R}$$ **Ans. (C)**

Sol. 7 For a straight wire moving perpendicular to the magnetic field, the motional emf is given as

$$e = Bvl$$

$$\Rightarrow \qquad e = 0.3 \times 10^{-4} \times 5 \times 20$$

$$\Rightarrow \qquad e = 3 \times 10^{-3}\,\text{V} = 3\,\text{mV}$$ **Ans. (B)**

Sol. 8 Here $B_H = 5 \times 10^{-5}$ T (As 1T = 1 N/A-m)

$l = 2$m [vertical]

$v = 1.5$ m/s

As vertical aerial cuts horizontal component of earth's field normally so motional emf induced in it is given as

$$e = Blv = 5 \times 10^{-5} \times 2 \times 1.5$$

$$\Rightarrow \qquad e = 1.5 \times 10^{-4} = 0.15\,\text{mV}$$ **Ans. (B)**

Sol. 9

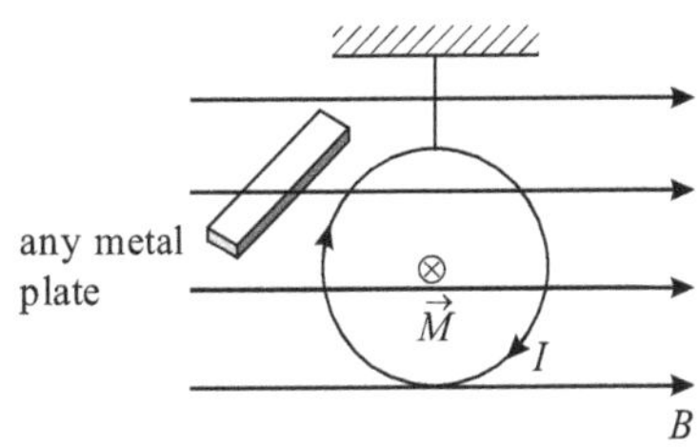

Torque on coil $\qquad \vec{\tau} = \vec{M} \times \vec{B}$

$$\tau = MB\sin\theta$$

$$\tau = MB$$

at $\theta = 0°$, coil is in state of stable equilibrium about which it oscillates.

When an aluminium plate (any metal plate) is kept near the coil then due to oscillation of coil $\vec{B}_C$ due to coil will be time varying and eddy currents are induced in plate and that induction opposes its cause which is oscillation of coil so these are damped out. **Ans. (A)**

Sol. 10 When $|\Delta\phi|$ flux is changed through a coil of resistance R, then charge flown through the coil is given as

$$\Delta q_f = \frac{|\Delta\phi|}{R}$$

$\Rightarrow$ $$|\Delta\phi| = \Delta q_f \times R = \text{area under } i\text{-}t \text{ curve} \times R$$

$\Rightarrow$ $$|\Delta\phi| = \frac{1}{2}\times 4\times 0.1\times 10 = 2\text{ Wb}$$ **Ans. (C)**

Sol. 11

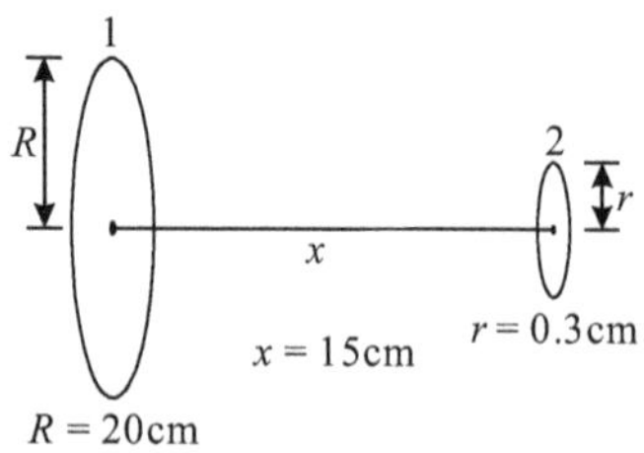

If i_2 current flows in smaller loop then flux linked with bigger loop can be given as

$$\phi_1 = Mi_2 \quad \ldots(1)$$

where $M \rightarrow$ coefficient of mutual inductance between loops 1 and 2.

To calculate M, we supply a current i, through bigger loop & calculate flux linked with smaller loop as

$$\phi_2 = Mi_1 \quad \ldots(2)$$

Magnetic induction at location of loop 2 due to loop 1 is

$$B = \frac{\mu_0 i_1 R^2}{2(R^2+x^2)^{3/2}}$$

flux through loop -2 is $\phi_2 = B.\pi r^2 = \dfrac{\mu_0 \pi r^2 R^2}{2(R^2+x^2)^{3/2}} i_1 \quad \ldots(3)$

Comparing equation-(2) and (3) gives

$$M = \frac{\mu_0 \pi r^2 R^2}{2(R^2+x^2)^{3/2}}$$

Using above value of M in equation-(1), we get

$$\phi_1 = \frac{\mu_0 \pi r^2 R^2}{2(R^2+x^2)^{3/2}} i_2$$

$\Rightarrow$ $$\phi_1 = \frac{4\pi\times10^{-7}\times\pi\times(0.3\times10^{-2})^2\times(0.2)^2\times 2}{2\times((0.2)^2+(0.15)^2)^{3/2}}$$

$\Rightarrow$ $$\phi_1 = \frac{4\pi\times10\times10^{-7}\times9\times10^{-6}\times4\times10^{-2}}{(0.25)^3}$$

$\Rightarrow$ $$\phi_1 = 9.216\times10^{-11}\text{ Wb}$$ **Ans. (A)**

Sol. 12 Motional EMF induced in the element of rod as shown is

$$de = B(x\omega)dx$$

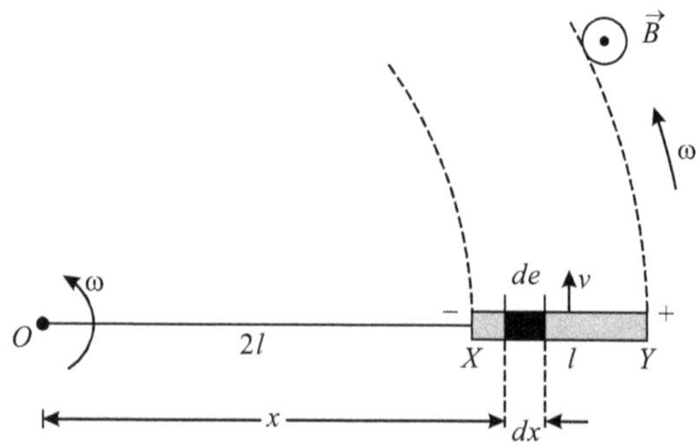

Total EMF induced in rod is

$$e = \int de = B\omega\int_{2l}^{3l} x\,dx = B\omega\left[\frac{x^2}{2}\right]_{2l}^{3l}$$

$\Rightarrow$ $$e = \frac{1}{2}B\omega l^2[9-4] = \frac{5}{2}B\omega l^2$$ **Ans. (D)**

Sol. 13

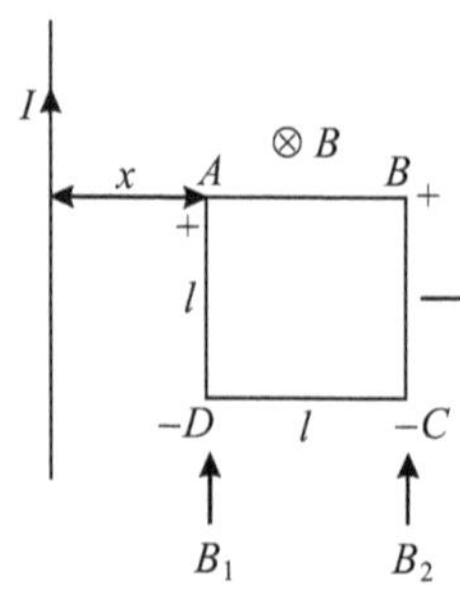

The motional emf will be induced in wire segments BC and AD which are given as

$$e_{AD} = B_1 vl = \frac{\mu_0 I}{2\pi x} vl \quad \ldots(1)$$

and $$e_{BC} = B_2 vl = \frac{\mu_0 I}{2\pi(l+x)} vl \quad \ldots(2)$$

Total emf induced in loop $ABCD$ is given as ($e_{AD} > e_{BC}$)

$$e_T = e_{AD} - e_{BC}$$

from (1) and (2)

$$e_T = \frac{\mu_0 Ivl}{2\pi}\left[\frac{1}{x} - \frac{1}{x+l}\right]$$

$\Rightarrow$ $$e_T = \frac{4\pi\times10^{-7}\times1\times10\times0.1}{2\pi}\left[\frac{1}{0.1} - \frac{1}{0.2}\right]$$

$\Rightarrow \qquad e_T = 2 \times 10^{-7} \times \dfrac{1}{0.2} = 10^{-6}\,\text{V}$

$\Rightarrow \qquad e_T = 1\,\mu\text{V}$ **Ans. (B)**

Sol. 14 emf induced in the coil is

$$e = \left|\frac{d\phi_L}{dt}\right|$$

[ϕ_2 - flux linked with all turns of coil]

here $\qquad \phi_L = N \times \phi = N(BA)$

average emf induced $e = \left|\dfrac{\Delta\phi_L}{\Delta t}\right| = NA\dfrac{\Delta B}{\Delta t}$

$\Rightarrow \qquad e = 1000 \times 4 \times 10^{-4} \times \dfrac{10^{-2}}{0.01}$

$\Rightarrow \qquad e = 0.4\,\text{V} = 400\,\text{mV}$ **Ans. (A)**

Sol. 15 In a cylindrical region TVMF, the induced EF at a distance r from axis of field is given as

for $r < R \rightarrow \qquad E_i = \dfrac{1}{2} r \dfrac{dB}{dt} \Rightarrow E \propto r$

for $r < R \rightarrow \qquad E_0 = \dfrac{1}{2}\dfrac{R^2}{r}\dfrac{dB}{dt} \Rightarrow E \propto \dfrac{1}{r}$ **Ans. (A)**

Sol. 16 Magnetic field components at Delhi
As Delhi is in Northern hemisphere, we use

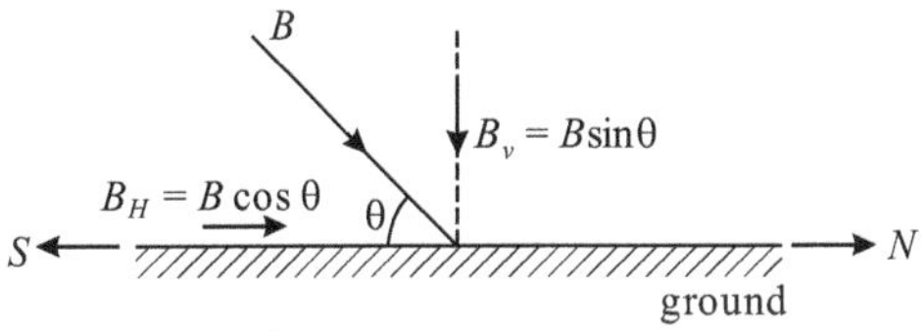

$$\sin\theta = 2/3$$

$$\cos\theta = \sqrt{1-\sin^2\theta} = \frac{\sqrt{5}}{3}$$

Flight of plane

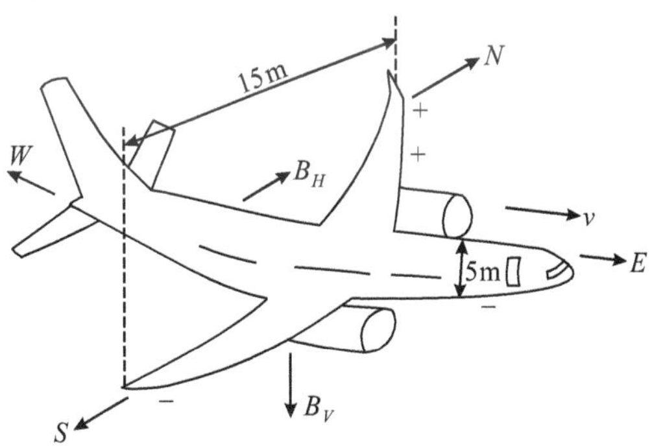

Motional EMF between top and bottom of plane

$$e_{TB} = v_B = Blv = B_H hv = B\cos\theta\, hv$$

$\Rightarrow \qquad v_B = 5 \times 10^{-5}\,T \times \dfrac{\sqrt{5}}{3} \times 5 \times 240$

$\Rightarrow \qquad v_B = 44.7 \times 10^{-3}\,\text{V}$

Motional EMF between the tips of wings is

$$e_{\text{Tips}} = v_w = Blv = B_v wv = B\sin\theta wv$$

$\Rightarrow \qquad v_w = 5 \times 10^{-5} \times \dfrac{2}{3} \times 15 \times 240$

$\Rightarrow \qquad v_w = 120 \times 10^{-3}\,\text{V}$ **Ans. (D)**

Sol. 17

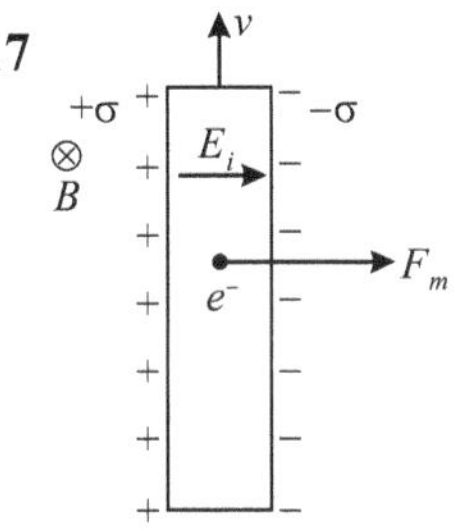

Due to motion of sheet, free electrons in sheet experience force toward right, given as

$$F_m = evB$$

Due to magnetic forces the free e^-'s drift and develops internal EF due to induced charges so that net force on free electron should be zero.

Where $E_i = \sigma/\epsilon_0$ (due to induced charges on surface of sheet)

we use $\qquad eE_i = evB$

$\Rightarrow \qquad \dfrac{\sigma}{\epsilon_0} = vB$

$\Rightarrow \qquad \sigma = \epsilon_0 vB$

$\Rightarrow$ On left and right surfaces of sheet induced charges density will be $+\epsilon_0 vB$ and $-\epsilon_0 vB$. **Ans. (B)**

Sol. 18 Magnetic flux through the loop is

$$\phi = B\pi r^2 = \pi r^2 B_0 e^{-t/\tau}$$

loop emf induced in the loop is $e = \left|\dfrac{d\phi}{dt}\right| = \dfrac{\pi r^2 B_0}{\tau} e^{-t/\tau}$

induced current in the loop is $i = \dfrac{e}{R} = \dfrac{\pi r^2 B_0}{\tau R} e^{-t/\tau}$

Total heat generated in loop is

$$H = \int_0^\infty i^2 R\,dt = \frac{\pi^2 r^4 B_0^2}{\tau^2 R}\int_0^\infty e^{-2t/\tau}\,dt$$

$\Rightarrow \qquad H = \dfrac{\pi^2 r^4 B_0^2}{\tau^2 R}\left[\dfrac{e^{-2t/\tau}}{-2/\tau}\right]_0^\infty$

$\Rightarrow \qquad H = \dfrac{\pi^2 r^4 B_0^2}{2\tau R}\left[e^{-2t/\tau}\right]_\infty^0$

$\Rightarrow \qquad H = \dfrac{\pi^2 r^4 B_0^2}{2\tau R}[1-0] = \dfrac{\pi^2 r^4 B_0^2}{2\tau R}$ **Ans. (B)**

Sol. 19 The charge which flows through a coil of resistance R due to change in magnetic flux $\Delta\phi$ is given as

$$q_f = \frac{\Delta\phi}{R} = \text{area under } i\text{-}t \text{ curve}$$

$$\frac{1}{2} \times 10 \times 0.5 = \frac{\Delta\phi}{100}$$

$\Rightarrow$ $\Delta\phi = 250$ Wb **Ans. (A)**

Sol. 20

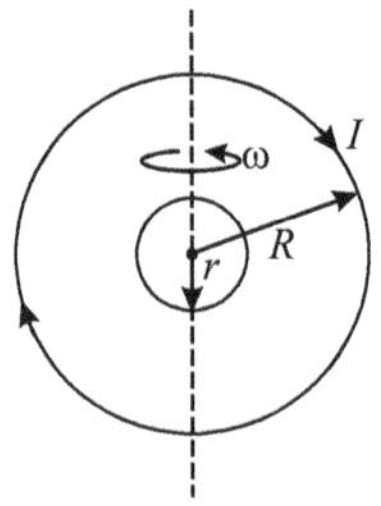

As $r << R$. Magnetic field at centre of larger coil is almost uniform for smaller coil.

The smaller coil is rotating in the magnetic field of larger coil at centre $\left(B_C = \frac{\mu_0 I}{2R}\right)$ so emf induced due to rotation of smaller coil is $e = NBA\omega \sin\omega t$

$\Rightarrow$ $$e = \frac{\mu_0 I}{2R}(\pi r^2)\omega \sin\omega t$$ **Ans. (C)**

Sol. 21

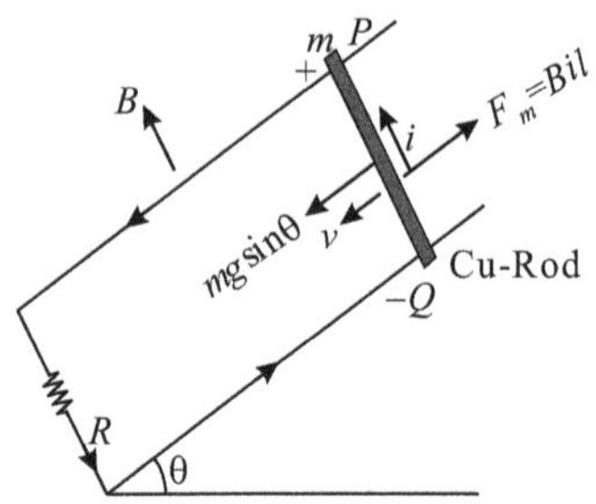

When rod is moving at speed v, the motional emf induced in rod will be

$$e_{PQ} = Bvl$$

$\Rightarrow$ Induced current is

$$i = \frac{e_{PQ}}{R} = \frac{Bvl}{R}$$

Magnetic force on rod due to B is

$$F_m = Bil = \frac{B^2vl^2}{R}$$

At terminal speed $a_{rod} = 0$

$\Rightarrow$ $$mg\sin\theta = F_m$$

$\Rightarrow$ $$mg\sin\theta = \frac{B^2vl^2}{R}$$

$\Rightarrow$ $$v = \frac{mgR\sin\theta}{B^2l^2}$$ **Ans. (B)**

Sol. 22

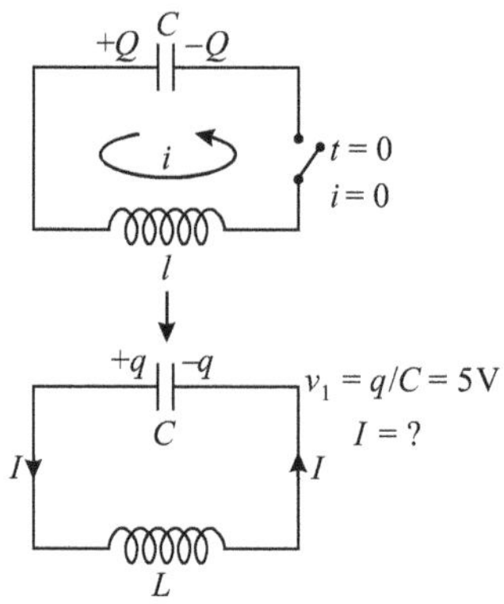

Initial energy in capacitor is

$$E = \frac{1}{2}CV^2 = \frac{1}{2} \times 0.2 \times 10^{-6} \times (10)^2 = 10\ \mu\text{J}$$

Energy of circuit (sum of $E_C + E_L$) when potential difference across capacitor is 5 V, is given as

$$E = \frac{1}{2}CV_1^2 + \frac{1}{2}LI^2$$

$\Rightarrow$ $$10 \times 10^{-6} = \frac{1}{2} \times 0.2 \times 10^{-6} \times (5)^2 + \frac{1}{2} \times 0.5 \times 10^{-3} \times I^2$$

$\Rightarrow$ $$2.5 \times 10^{-4} I^2 = 7.5 \times 10^{-6} \Rightarrow I^2 = 3 \times 10^{-2}$$

$\Rightarrow$ $$I = \frac{\sqrt{3}}{10}\text{A} = 0.1732\ \text{A}$$ **Ans. (A)**

Sol. 23

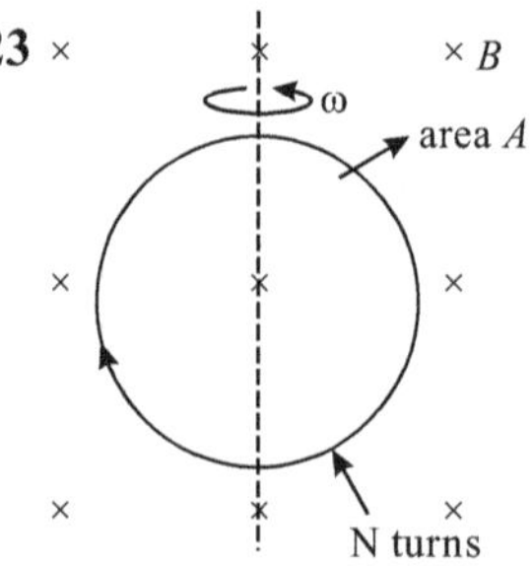

$$\left.\begin{array}{l}\vec{A} \rightarrow \otimes \\ \vec{B} \rightarrow \otimes\end{array}\right] \Rightarrow \theta = 0^\circ$$

Initial flux through coil $\phi = BAN$

after time t angle between $\vec{A}$ and $\vec{B}$ will be

$$\theta = \omega t$$

$\Rightarrow$ Magnetic flux through coil after time t is given as

$$\phi = (\vec{B} \cdot \vec{A})N = NBA\cos\omega t$$

EMF induced in coil is

$$e = -\frac{d\phi}{dt} = NBA\omega \sin\omega t = e_0 \sin\omega t$$

Here $e_0 = NBA\omega \rightarrow$ maximum emf **Ans. (A)**

Sol. 24 Charge flown through the loop is given as

$$Q = \frac{\Delta\phi}{R} = \frac{1}{10}A(B_f - B_i)$$

$$\Rightarrow \quad Q = \frac{1}{10}\times 3.5\times 10^{-3}\left(0.4\sin\frac{\pi}{2} - 0\right)$$

$$\Rightarrow \quad Q = \frac{1}{10}(3.5\times 10^{-3})(0.4-0)$$

$$\Rightarrow \quad Q = 1.4\times 10^{-4} = 0.14\,\text{mC}$$

Ans. (A)

Sol. 25 Potential difference between two faces perpendicular to x-axis will be the motional emf given as

$$e = Bvl = 12\,\text{mV}$$

Ans. (C)

Sol. 26 Work done in reversing direction of current is

$$W = \text{M.B.}(\cos\theta_1 - \cos\theta_2)$$

$$\Rightarrow \quad W = 2\times 10^{-2}\times 1\cos(0.125)$$

$$\Rightarrow \quad W = 0.019\,\text{J}$$

Closest possible option is (B) **Ans. (B)**

Sol. 27 Motional *emf* induced in wire is

$$e = Bvl$$

$$\Rightarrow \quad e = 0.3\times 10^{-4}\times 5\times 10$$

$$\Rightarrow \quad e = 1.5\times 10^{-3}\,\text{V}$$

Ans (D)

Sol. 28 For damped oscillation amplitude is given as

$$A = A_0 e^{-t/r}$$

Where $\gamma = \frac{2m}{b}\left(b = \frac{B^2l^2}{R}\right)$

If at $t = t_1$, $A = \frac{A_0}{e}$

$$\Rightarrow \quad \frac{A_0}{e} = A_0 e^{-t/r}$$

$$\Rightarrow \quad \gamma = \frac{2mR}{B^2l^2}$$

$$\Rightarrow \quad \gamma = \frac{2\times 50\times 10^{-3}\times 10}{(0.1)^2\times(0.1)^2} = 10^4\,\text{s}$$

Time period of oscillations

$$\Rightarrow \quad T_0 = 2\pi\sqrt{\frac{m}{k}} = 2\pi\sqrt{\frac{50\times 10^{-3}}{0.5}} = \frac{2\pi}{\sqrt{10}}\,\text{s}$$

Number of oscillation of strip in t_1 s are

$$N = \frac{t_1}{T_0} = \frac{10^4}{2\pi\sqrt{10}} \sim 5000$$

Ans. (B)

Sol. 29 Resistance of Wheatstone bridge is

$$R_{BF} = \frac{4\times 2}{4+2} = \frac{8}{6} = \frac{4}{3}\Omega$$

Equal resistance of circuit is

$$R_{eq} = 1.33 + 1.7 = 3.03\,\Omega \simeq 3\Omega$$

EMF induced in L is

$$e = Bvl = 1\times 0.05\times 0.01 = 5\times 10^{-4}\,\text{V}$$

Circuit current $i = \frac{e}{R_{eq}} = \frac{5\times 10^{-4}}{3} = 1.66\times 10^{-4}\,\text{A}$

$$= 166\,\mu\text{A} \simeq 170\,\mu\text{A}$$

Ans. (B)

20.2 Self and Mutual Induction

Sol. 1 As in given circuit all three inductors are connected in parallel so we use

$$\frac{1}{L_{eq}} = \frac{1}{L_1} + \frac{1}{L_2} + \frac{1}{L_3}$$

$$= \frac{1}{3} + \frac{1}{3} + \frac{1}{3} = 1$$

$$\Rightarrow \quad L_{eq} = 1\,\text{H}$$

Ans. (D)

Sol. 2 If in a coil of self inductance 'L' current changes by di in time dt then magnitude of emf induced in coil is given as

$$e = L\frac{di}{dt}$$

$$\Rightarrow \quad e = L\frac{\Delta i}{\Delta t}$$

$$\Rightarrow \quad 8 = L\times\frac{4A}{0.05\text{s}}$$

$$\Rightarrow \quad L = \frac{8\times 0.05}{4}$$

$$\Rightarrow \quad L = 0.1\,\text{H}$$

Ans. (D)

Sol. 3 If two coils with self induction coefficients L_1 and L_2 are placed to each other than their mutual induction is given as

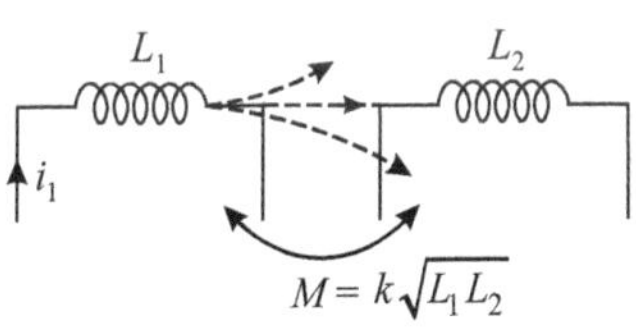

Where $k \to$ Coupling factor [depends upon the fraction of magnetic flux of first coil which in linked with second coil]

Here k indirectly depends upon separation & orientation of the two coil.

for full flux linkage we use $k = 1$

for No flux linkage we use $k = 0$

Either due to large separation or kept at 90° angle to each other. **Ans. (B)**

Sol. 4

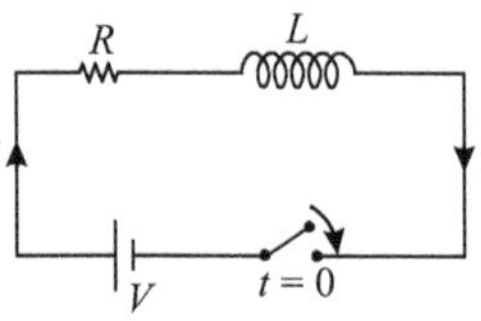

In series *L-R* circuit the current grows according to relation

$$i = \frac{V}{R}(1 - e^{\frac{-Rt}{L}})$$

$V/R \rightarrow$ steady state current

if at $t = t_1$, $i = V/2R$

$$\Rightarrow \quad \frac{V}{2R} = \frac{V}{R}(1 - e^{\frac{-Rt_1}{L}})$$

$$\Rightarrow \quad e^{-Rt_1/L} = \frac{1}{2}$$

$$\Rightarrow \quad \frac{Rt_1}{L} = \ln(2)$$

$$\Rightarrow \quad t_1 = \frac{L}{R}\ln 2 = \frac{0.3}{2} \times 0.693$$

$$\Rightarrow \quad t_1 = 0.104 \text{ s}$$

Ans. (B)

Sol. 5 emf induced in a coil due to time varying magnetic flux is given as

$$e = -\frac{d\phi}{dt}$$

$$\Rightarrow \quad e = -\frac{d}{dt}(10t^2 - 50t + 250)$$

$$\Rightarrow \quad e = -[20t - 50]$$

at $t = 3$s, $\quad e = -[20 \times 3 - 50] = -10$ V

Ans. (B)

Sol. 6 When circuit is in steady state, current through inductor is $\quad I_0 = E/R$

Now battery is disconnected & terminals *A* & *B* are short circuited.

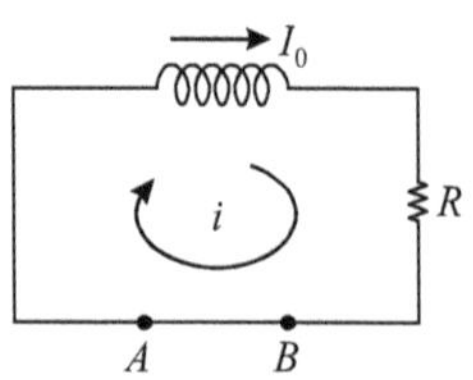

During decay current in circuit is given as

$$i = I_0 e^{-Rt/L} = \frac{E}{R}e^{-Rt/L}$$

$$\Rightarrow \quad i = \frac{100}{100} e^{\frac{-100 \times 10^{-3}}{100 \times 10^{-3}}} = e^{-1}$$

$$\Rightarrow \quad i = \frac{1}{e} \text{ A}$$

Ans. (A)

Sol. 7

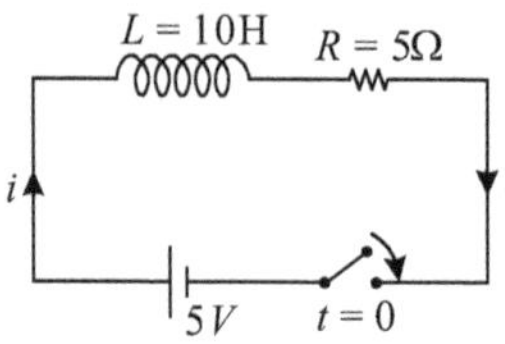

For *R-L* circuit current as a function of time is given as

$$i = \frac{V}{R}(1 - e^{\frac{-Rt}{L}})$$

$$\Rightarrow \quad i = \frac{5}{5}(1 - e^{\frac{-5 \times 2}{10}})$$

$$\Rightarrow \quad i = 1 - e^{-1}$$

Ans. (D)

Sol. 8

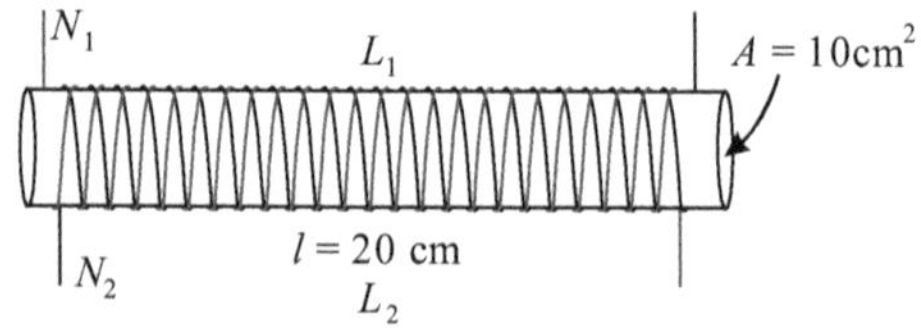

We use $\quad L_1 = \frac{\mu_0 N_1^2 A}{l}$

and $\quad L_2 = \frac{\mu_0 N_2^2 A}{l}$

As the two solenoids are wound over each other due to complete linkage of flux, the mutual induction between the two is given as $\quad M = k\sqrt{L_1 L_2} = \sqrt{L_1 L_2} \quad$ [As $k = 1$]

$$\Rightarrow \quad M = \frac{\mu_0 A}{l} N_1 N_2$$

$$\Rightarrow \quad M = \frac{4\pi \times 10^{-7} \times 10 \times 10^{-4}}{0.2} \times 300 \times 400$$

$$\Rightarrow \quad M = 2.4\pi \times 10^{-4} \text{ H}$$

Ans. (D)

Sol. 9

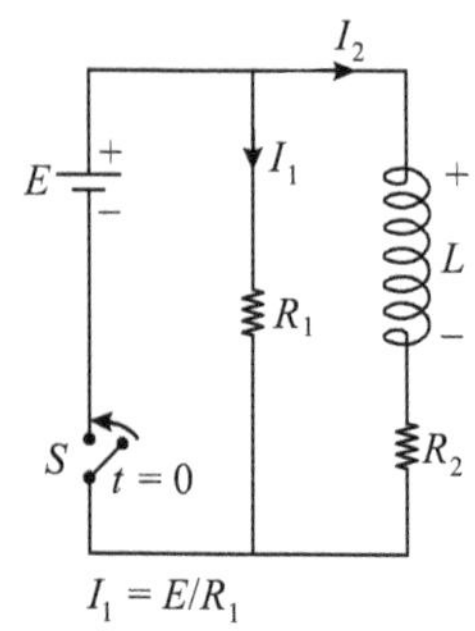

Current I_2 is given as

$$I_2 = \frac{E}{R_2}\left(1 - e^{\frac{-R_2 t}{L}}\right)$$

Potential drop across *L* is given as

$$e_L = L\frac{dI_2}{dt} = \frac{LE}{R_2}(e^{\frac{-R_2 t}{L}} \times \frac{R_2}{L}) = E e^{-R_2 t/L}$$

$$\Rightarrow \quad e_L = 12 e^{\frac{-2t}{0.4}} = 12 e^{-5t} \text{ V}$$

Ans. (B)

Sol. 10 At $t=0$, inductor behaves like an open circuit so we use

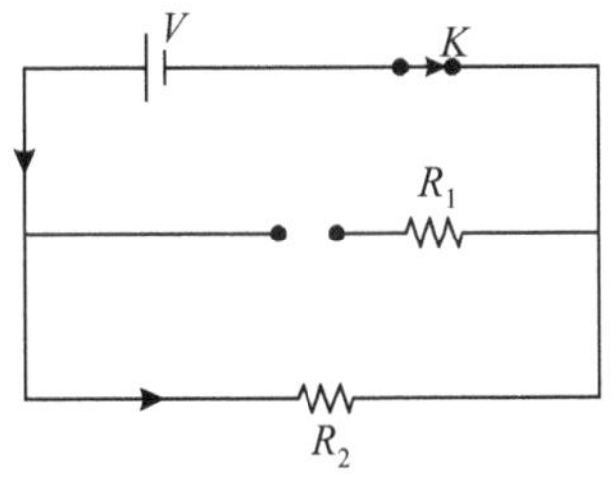

$\Rightarrow$ at $t=0$, $\quad I = V/R_2$

At $t=\infty$, i.e. in steady state inductor behaves like a straight wire so we use

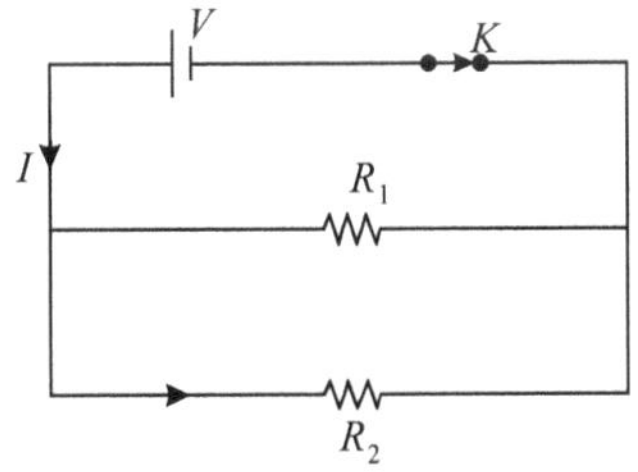

$$R_{eq} = \frac{R_1R_2}{R_1+R_2}$$

$$\Rightarrow \quad I = V/R_{eq} = \frac{V(R_1+R_2)}{R_1R_2}$$ **Ans. (C)**

Sol. 11 Ideally magnetic induction inside a solenoid everywhere remain constant

$$B = \mu_0 nI = \mu_0\left(\frac{N}{L}\right)I$$

Magnetic flux linked with solenoid is

$$\phi = NBA = N\left(\frac{\mu_0 NI}{L}\right)\pi r^2 \qquad [\because A = \pi r^2]$$

$$\Rightarrow \quad \phi = \left(\frac{\mu_0 \pi N^2 r^2}{L}\right)I$$

Here self inductance of the solenoid is

$$L_{sol} = \left(\frac{\mu_0 \pi N^2 r^2}{L}\right)I$$

As magnetic induction on the sides of solenoid is less than $\mu_0 nI$

$$\Rightarrow \quad \phi = L_{sol} I$$

here $\quad L_{sol} < \dfrac{\mu_0 \pi N^2 r^2}{L}$

$\Rightarrow$ **S-1** is true

S-2 is its explanation **Ans. (B)**

Sol. 12

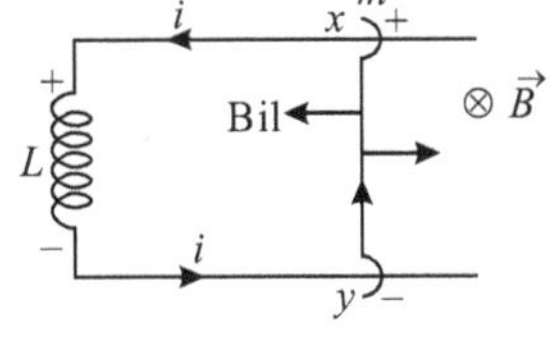

Motional emf induced in connector xy is

$$e_{xy} = Bvl$$

For the inductor as it is in parallel to connector xy, we use

$$Bvl = L\frac{di}{dt} \qquad \ldots(1)$$

If a be the acceleration (–ve) of connector we use

$$a = -\frac{Bil}{m} \qquad \ldots(2)$$

From equation-(1) $\int di = \dfrac{Bl}{L}\int vdt$

$$\Rightarrow \quad i = \frac{Bl}{L}\cdot x$$

[Where $x = \int vdt$ is the displacement of connector]

From equation-(2) $\quad a = -\dfrac{Bl}{m}\left[\dfrac{Bl}{L}\cdot x\right]$

$$\Rightarrow \quad \frac{d^2x}{dt^2} + \frac{B^2l^2}{mL}x = 0$$

This is the differential equation of SHM so connector is executing SHM. **Ans. (D)**

Sol. 13 As S_2 is kept open so the circuit is simple RC transient and charge on capacitor is given at (time function)

$$q = CV(1-e^{-t/RC})$$

$$q = CV(1-e^{-t/\tau})$$

At $t=\tau \rightarrow \quad q = CV(1-e^{-1})$

At $t=2\tau \rightarrow \quad q = CV(1-e^{-2})$ **Ans. (C)**

Sol. 14 for two closely spaced coils, emf induced in one coil due to current variation in other coil is given as

$$V(t) = -M\frac{dI(t)}{dt}$$

[M = mutual inductance between coils]

$\Rightarrow \quad V(t) = -M \times$ slope of $I(t)$ curve

$$\text{slope} = -\frac{V(t)}{M}$$

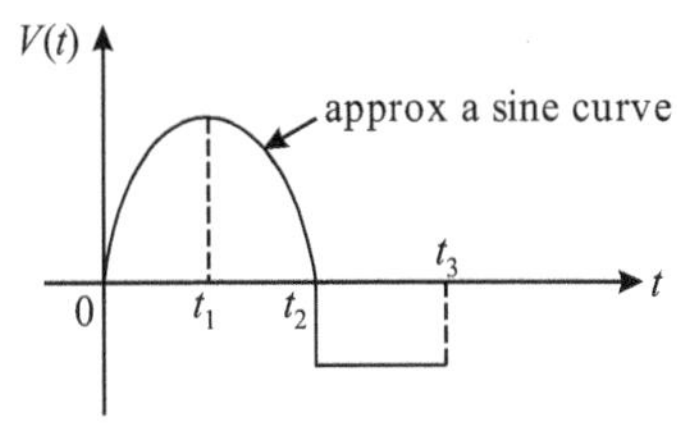

from $\quad 0$ to $t_1 \rightarrow$ Slope of $I(t)$ curve is increasing (+ve)

t_1 to $t_2 \rightarrow$ Slope of $I(t)$ curve is decreasing (+ve)

t_2 to $t_3 \rightarrow$ Slope of $I(t)$ curve is constant (+ve) as $V(t)$ is –ve

$\Rightarrow$ 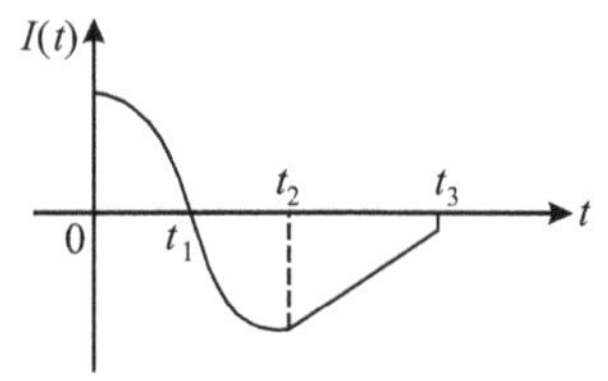

$\Rightarrow$ None of the given options is correct. **Ans. (None)**

Sol. 15 After point C is connected to B then circuit will be

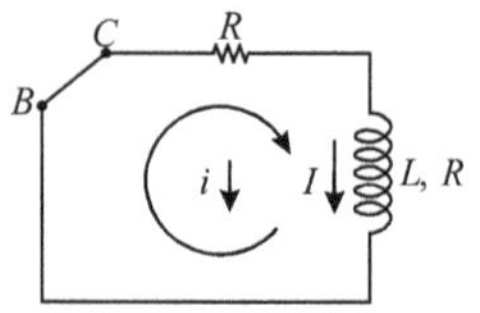

in the given loop we can write

$$V_R + V_L = 0$$

$$\Rightarrow \quad \frac{V_R}{V_L} = -1 \qquad \textbf{Ans. (C)}$$

Sol. 16 frequency of LC oscillation is given as

$$f = \frac{1}{2\pi\sqrt{LC}}$$

for $L_1\ (> L_2)$ we use $f_1 < f_2$ or $T_1 > T_2$

In case of L_1 maximum charge on capacitor Q_{max} will decrease more slowly. **Ans. (A)**

Sol. 17 When K_1 is closed the steady current in circuit is given as-

$$I_0 = \frac{V}{R} = \frac{15}{150} = 0.1\text{ A}$$

when at $t = 0$, K_1 is opened and K_2 is closed then inductor current I_0 will start decaying through R and its time function as

$$i = I_0 e^{-Rt/L} = 0.1 \times e^{-150\times10^{-3}/0.03}$$

$$\Rightarrow \quad i = 0.1 \times e^{-5} = \frac{1}{1500}\text{ A} = 0.67\text{ mA} \qquad \textbf{Ans. (D)}$$

Sol. 18 Emf induced (average) in a coil having self inductance L is given as

$$e = \left|L\frac{\Delta i}{\Delta t}\right|$$

$$\Rightarrow \quad 50 = L\left(\frac{5-2}{0.1}\right)$$

$$\Rightarrow \quad L = \frac{50}{30} = \frac{5}{3} = 1.67\text{ H} \qquad \textbf{Ans. (D)}$$

Sol. 19 In series RC circuit current for charging of capacitor is

$$i_a = \frac{E}{R}e^{-t/RC} \qquad \ldots(a)$$

[RC - Transient circuit]

for series LR circuit growth of current is

$$i_b = \frac{E}{R}(1 - e^{-Rt/L}) \qquad \ldots(b)$$

[RL - Transient circuit] **Ans. (C)**

Sol. 20 Transformer efficiency is given as

$$\eta = \frac{P_{out}}{P_{in}} = \frac{V_S I_S}{V_P I_P}$$

$$\Rightarrow \quad 0.9 = \frac{23 \times I_S}{230 \times 5}$$

$$\Rightarrow \quad I_S = 45\text{ A} \qquad \textbf{Ans. (D)}$$

Sol. 21 Given that

$$L\frac{di}{dt} = 25$$

$$\Rightarrow \quad L \times \frac{15}{1} = 25$$

$$\Rightarrow \quad L = \frac{5}{3}\text{ H}$$

Change in energy of inductance is

$$\Delta E = \frac{1}{2}L(I_2^2 - I_1^2)$$

$$\Rightarrow \quad \Delta E = \frac{1}{2} \times \frac{5}{3} \times (25^2 - 10^2)\text{ J}$$

$$\Rightarrow \quad \Delta E = \frac{5}{6} \times 525 = 437.5\text{ J} \qquad \textbf{Ans. (A)}$$

Sol. 22 From time $t = 0$ to $t = t_0$, current in RL circuit grows and after to current decays.

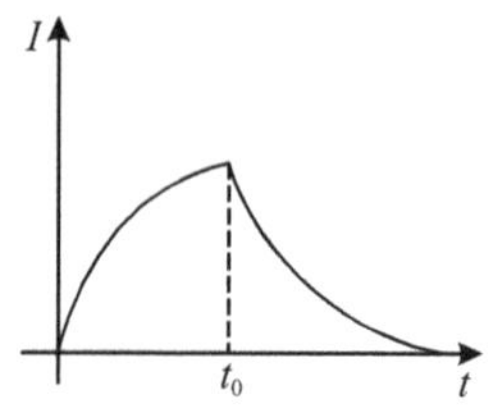

In this case most appropriate option is (B) **Ans. (B)**

Sol. 23 Mutual inductance of the given coils is

$$M = \mu_0 n_1 n_2 \pi r_1^2 l \qquad \ldots(1)$$

Self inductance of inner coils is

$$L = \mu_0 n_1^2 \pi r_1^2 l \qquad \ldots(2)$$

From (1) and (2)

$$\frac{M}{L} = \frac{n_2}{n_1} \qquad \textbf{Ans. (D)}$$

Sol. 24 Self inductance of a solenoid is

$$L=\mu_0 n^2 A l$$

Keeping total length of copper wire same on increasing the frame size to 3 times, the length of solenoid on which terms are wound will become $l/3$ so new self inductance will be

$$L'=\mu_0 n^2(9A)(l/3)=3L$$

Ans. (B)

Sol. 25 Ideal inductor will behave like zero resistance long time after switch is closed so equivalent circuit will be

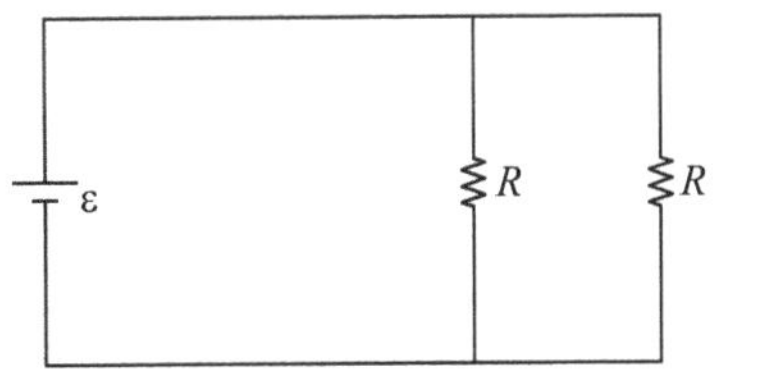

Current through battery is

$$I=\frac{2\varepsilon}{R}=\frac{2\times 15}{5}=6\,\text{A}$$

Ans. (A)

Sol. 26 Flux linked with solenoid is given as

$$\phi=NBA=LI$$

$$\Rightarrow \quad N\mu_0 nI\pi R^2=LI$$

$$\Rightarrow \quad N\mu_0\left(\frac{N}{l}\right)I\pi R^2=LI$$

N & R are to be taken constant, so we use

Self inductance $\quad (L)\propto \dfrac{1}{l}\propto\dfrac{1}{\text{length}}$

Ans. (B)

Sol. 27 If mutual induction between coils P and Q is M, we use

$$\phi_Q=MI_P \quad \ldots(1)$$

and

$$\phi_P=MI_Q \quad \ldots(2)$$

$$\Rightarrow \quad \frac{\phi_Q}{\phi_P}=\frac{I_P}{I_Q}$$

$$\Rightarrow \quad \frac{10^{-3}}{\phi_P}=\frac{3}{2}$$

$$\Rightarrow \quad \phi_P=\frac{2}{3}\times 10^{-3}=0.67\times 10^{-3}\,\text{Wb}$$

Ans. (B)

Sol. 28 Current in solenoid

$$I_{(t)}=kte^{-\alpha t}$$

Flux through solenoid is

$$\phi_{(t)}=B.A=(\mu_0 n I_{(t)})\times\pi R^2$$

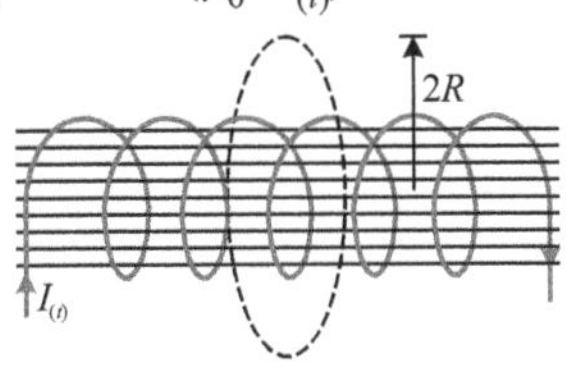

(same flux will pass through coil of radius $2R$)

emf induced in outer coil is

$$e=-\frac{d\phi_{(t)}}{dt}$$

$$\Rightarrow \quad e=-\mu_0 n\pi R^2 k\,[e^{-\alpha t}+t(-\alpha e^{-\alpha t})]$$

$$\Rightarrow \quad e=-Ce^{-\alpha t}(1-\alpha t)$$

Current induced in coil is

$$I_{\text{coil}}=\frac{e}{R_{\text{coil}}}$$

$$\Rightarrow \quad I_{\text{coil}}=-\frac{C}{R}e^{-\alpha t}(1-\alpha t)$$

at $\quad t=0\rightarrow I_{\text{coil}}\rightarrow 0$

at $\quad t=\infty\rightarrow I_{\text{coil}}\rightarrow 0$

$\quad I_{\text{coil}}\rightarrow -\text{ve}$

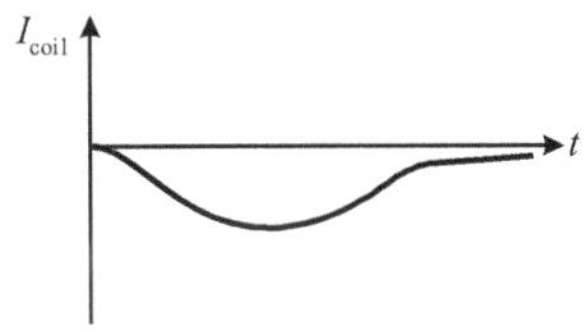

Ans. (B)

Sol. 29

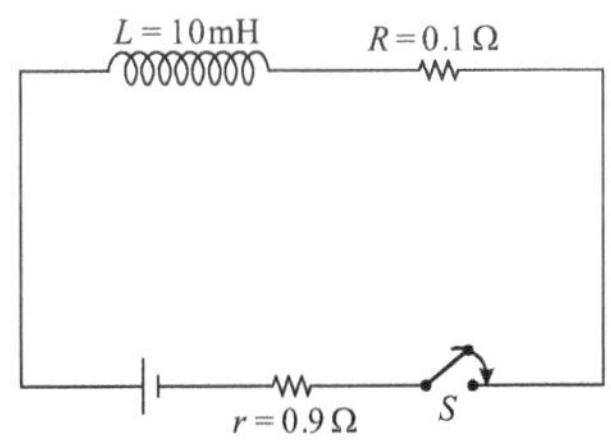

Current in circuit as a function of time is

$$i=i_0(1-e^{-t/\tau})$$

$$\Rightarrow \quad \frac{80}{100}i_0=i_0(1-e^{-t/\tau})$$

$$\Rightarrow \quad 0.8=1-e^{-t/\tau}$$

$$\Rightarrow \quad e^{-t/\tau}=0.2=\frac{1}{5}$$

$$\Rightarrow \quad -\frac{t}{\tau}=\ln\left(\frac{1}{5}\right)$$

$$\Rightarrow \quad -\frac{t}{\tau}=-\ln(5)$$

$$\Rightarrow \quad t=\tau.\ln(5)$$

$$\Rightarrow \quad t=\frac{L}{R_{eq}}\cdot\ln(5)$$

$$\Rightarrow \quad t=\frac{10\times 10^{-3}}{(0.1+0.9)}\times 1.6$$

$$\Rightarrow \quad t=1.6\times 10^{-2}$$

$$\Rightarrow \quad t=0.016\,\text{s}$$

Ans. (B)

Sol. 30 Charge flown through battery is given as

$$q = \int I dt$$

$$\Rightarrow \quad q = \int_0^{L/R} \frac{E}{R}\left[1 - e^{\frac{-Rt}{L}}\right] dt$$

$$\Rightarrow \quad q = \frac{EL}{R^2}\frac{1}{e}$$

$$\Rightarrow \quad q = \frac{EL}{2.7R^2}$$

Ans. (B)

21.1 Alternating Current

Sol. 1 By transformer equation, we have

$$\frac{e_1}{e_2} = \frac{N_1}{N_2} \qquad \ldots(1)$$

and for an ideal transformer we use its power equation for primary and secondary coil is

$$e_1 i_1 = e_2 i_2 \qquad \ldots(2)$$

From (1) and (2) we have

$$\frac{i_2}{i_1} = \frac{N_1}{N_2}$$

$$\Rightarrow \quad i_2 = \frac{N_1}{N_2} i_1 = \frac{140}{280} \times 4$$

$$\Rightarrow \quad i_2 = 2\text{ A}$$

Ans. (B)

Sol. 2

$$X_L = \omega L \angle \pi/2$$

$$R = R \angle 0^\circ$$

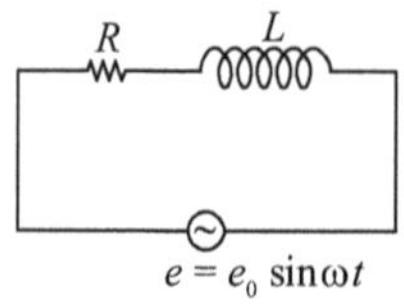

Impedance triangle

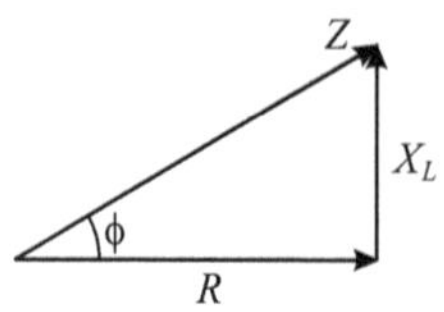

Power factor $\cos\phi = \frac{R}{Z} = \frac{R}{\sqrt{R^2 + \omega^2 L^2}}$ **Ans. (B)**

Sol. 3 The transformer core is laminated to reduce eddy current losses.

Non laminated core :

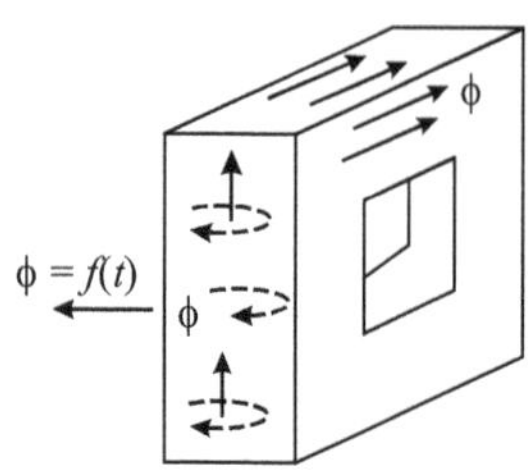

Time varying magnetic flux causes eddy currents to induced over all the cross-section

Laminated core :

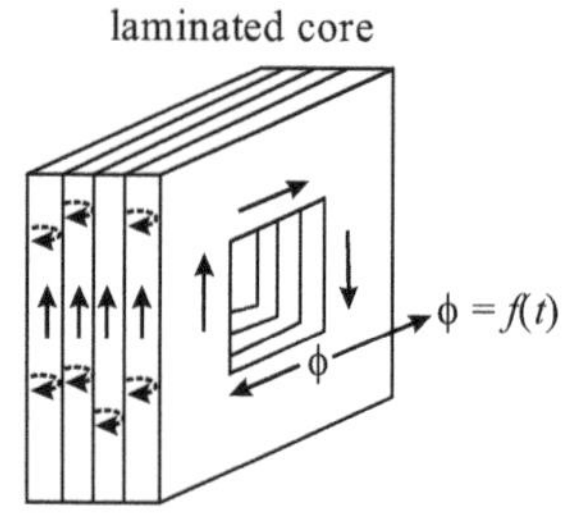

Due to laminated core eddy currents are induced within laminated strips only so losses due these miniature current are reduced. **Ans. (A)**

Sol. 4

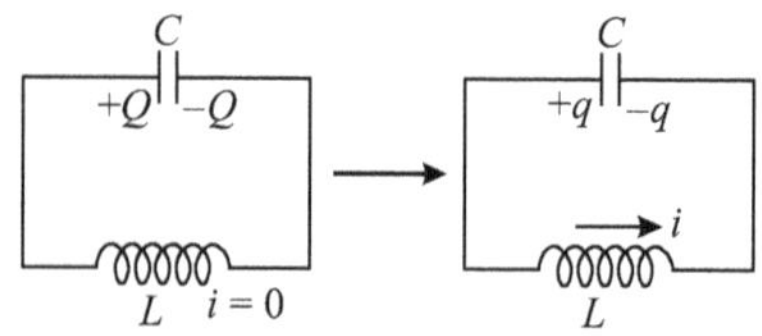

By conservation of energy

$$\frac{Q^2}{2C} = \frac{q^2}{2C} + \frac{1}{2}Li^2$$

When capacitor & inductor have equal amounts of energy stored

we use $\frac{q^2}{2C} = \frac{1}{2}Li^2$

$$\frac{Q^2}{2C} = 2\left(\frac{q^2}{2C}\right)$$

$$\Rightarrow \quad q^2 = \frac{Q^2}{2}$$

$$\Rightarrow \quad q = \frac{Q}{\sqrt{2}}$$

Ans. (C)

Sol. 5

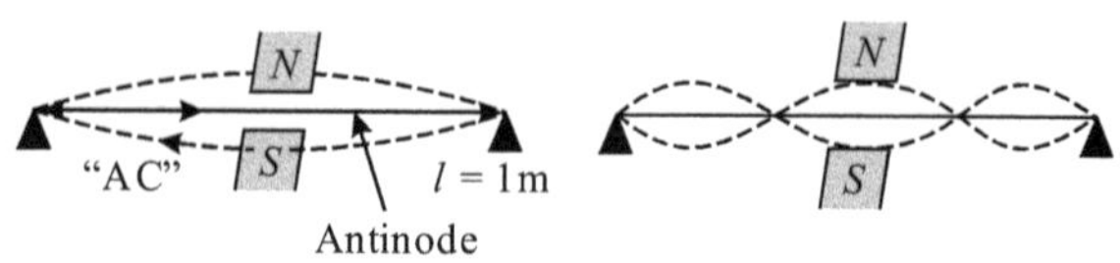

$$T = 10 \times 9.8\text{ N}$$

$$\mu = 9.8 \times 10^{-3}\text{ kg/m}$$

Here the fundamental frequency of oscillation of wire is given as

$$n_0 = \frac{1}{2l}\sqrt{\frac{T}{\mu}} = \frac{1}{2 \times 1}\sqrt{\frac{10 \times 9.8}{9.8 \times 10^{-3}}}$$

$$\Rightarrow \quad n_0 = \frac{1}{2} \times 100 = 50\text{ Hz}$$

So when $n = n_0$ or $3n_0$ or $5n_0$ … the wire will oscillate in resonance.

50 Hz, 150 Hz, 250 Hz.… **Ans. (A)**

Sol. 6 For a given LCR circuit, its resonant frequency is given as

$$f_R = \frac{1}{2\pi\sqrt{LC}}$$

If $C \to 2C$ then to keep f_R same we must use $L \to L/2$

So that product $LC \to$ constant. **Ans. (C)**

Sol. 7

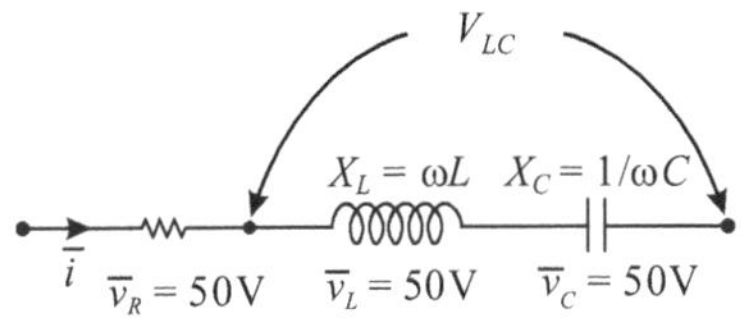

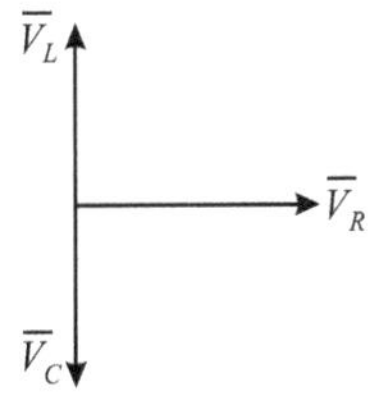

As $\overline{V}_C$ is 90° lagging w.r. to current $\overline{i}$ and $\overline{V}_L$ is 90° ahead w.r. to current $\overline{i}$. So we have $\overline{V}_L$ and $\overline{V}_C$ at phase difference of 180°.

Here $\qquad \overline{V}_{LC} = \overline{V}_L + \overline{V}_C = 0$

As $\qquad |\overline{V}_L| = |\overline{V}_C|$ **Ans. (D)**

Sol. 8 A *DC* ammeter measures the average value of current passing through it.

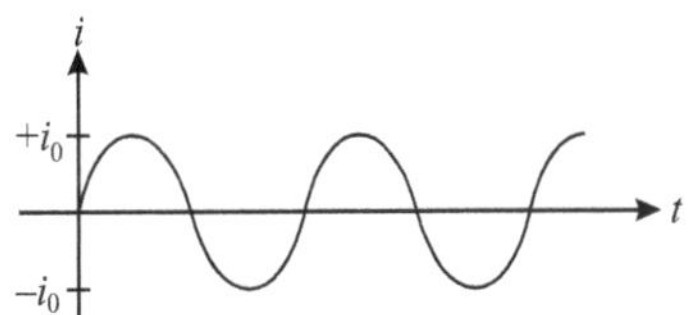

Thus average of a symmetric *AC* is zero. **Ans. (C)**

Sol. 9 Due to an inductor $L \to$ current legs behind emf by $\pi/2$ in phase.

Due to a capacitor $C \to$ current leads emf by $\pi/2$ in phase.

If a resistance R is there in AC circuit then the phase difference between current and emf is between 0 and $\pi/2$.

$\Rightarrow$ Only due to L or C or both L & C phase difference remain $\pi/2$. **Ans. (B)**

Sol. 10 For a given AC circuit power factor is given as

$$\cos\phi = R/Z$$

$$\Rightarrow \qquad \cos\phi = \frac{12}{15} = \frac{4}{5} = 0.8$$ **Ans. (A)**

Sol. 11 Power is maximum at resonant frequency given as

$$f_R = \frac{1}{2\pi\sqrt{LC}}$$

$$\Rightarrow \qquad C = \frac{1}{4\pi^2 L f_R^2}$$

$$\Rightarrow \qquad C = \frac{1}{4\times10\times10\times(50)^2} = 10^{-6}\,\text{F}$$

$$\Rightarrow \qquad C = 1\,\mu\text{F}$$ **Ans. (C)**

Sol. 12 The magnetic flux passing through the coil of area A in magnetic field is given as

$$\phi = \vec{B}\cdot\vec{A}$$

When coil in at an angle $\theta = \omega t$ then total linked flux with coil is

$$\phi_L = N(BA\cos\omega t)$$

emf induced in coil is

$$e = \left|\frac{d\phi_L}{dt}\right| = NBA\sin\omega t = e_0\sin\omega t$$

(AC induced emf in generator)

Maximum emf $\qquad e_0 = NBA\omega$ **Ans. (D)**

Sol. 13

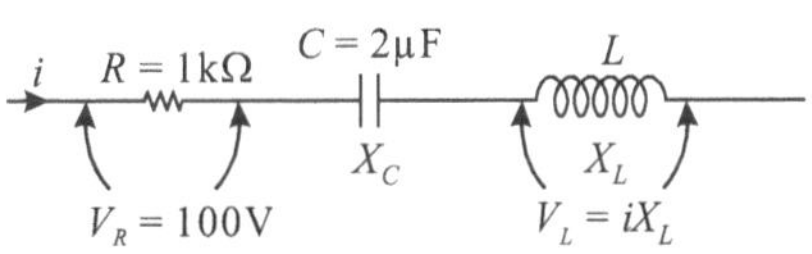

At resonant frequency in LCR circuit

$$X_L = X_C$$

$$X_L = \frac{1}{\omega C}$$

Circuit current $\qquad i = \frac{V_R}{R} = \frac{100}{1000} = 0.1\,\text{A}$

Voltage across inductor is given as

$$V_L = iX_L = i\left(\frac{1}{\omega C}\right) = 0.1\times\frac{1}{200\times2\times10^{-6}}$$

$$\Rightarrow \qquad V_L = \frac{1}{4\times10^{-3}} = \frac{1000}{4} = 250\text{V}$$ **Ans. (C)**

Sol. 14 Average power consumption in an AC circuit is given as

$$P_{avg} = E_{rms} I_{rms}\cos\phi \qquad \ldots(1)$$

Here ϕ is the phase difference between emf and current in circuit

Here $\qquad E = E_0\sin\omega t$

and $\qquad I = I_0\sin(\omega t - \pi/2)$

$\Rightarrow \qquad \phi = \pi/2$

$\Rightarrow$ From equation-(1)

$$P_{avg} = 0$$ **Ans. (B)**

Sol. 15 Due to capacitor and inductor separately, power factor of circuit is same as $\phi = 30°$ in both cases (one lagging & one leading)

$\Rightarrow$ $X_L = X_C$ so this is the condition of 'Resonance'.

In LCR circuit at resonance, power dissipation is given as

$$P_{avg} = \frac{e^2}{R}$$

$$= \frac{(220)^2}{200} = 242 \text{ W} \qquad \textbf{Ans. (A)}$$

Sol. 16

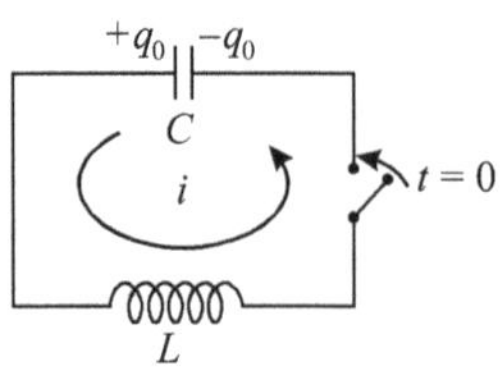

The frequency of LC oscillations is $\omega = \frac{1}{\sqrt{LC}}$

So the charge on capacitor as a function of time can be given as

$$q = q_0 \sin(\omega t + \alpha)$$

at $t = 0$, $q = q_0$

$$\Rightarrow \qquad \alpha = \pi/2$$

$$\Rightarrow \qquad q = q_0 \cos(\omega t) \qquad \ldots(1)$$

If after t, when charge on capacitor is q such that

$$\frac{q_0^2}{2C} = \frac{q^2}{2C} + \frac{1}{2}Li^2 \qquad \left[\text{with } \frac{q^2}{2C} = \frac{1}{2}Li^2\right]$$

$$\Rightarrow \qquad \frac{q_0^2}{2C} = 2 \times \frac{q^2}{2C}$$

$$\Rightarrow \qquad q = \frac{q_0}{\sqrt{2}}$$

[At this charge on cap $U_{EF} = U_{MF}$]

from equestion-(1) we have

$$q = q_0 \cos\omega t$$

$$\Rightarrow \qquad \frac{q_0}{\sqrt{2}} = q_0 \cos(\omega t)$$

$$\Rightarrow \qquad \omega t = \frac{\pi}{4}$$

$$\Rightarrow \qquad \frac{1}{\sqrt{LC}} t = \frac{\pi}{4}$$

$$\Rightarrow \qquad t = \frac{\pi}{4}\sqrt{LC} \qquad \textbf{Ans. (D)}$$

Sol. 17 For a given circuit having an inductor L and a capacitor C, the resonant frequency is given as

$$f_R = \frac{1}{2\pi\sqrt{LC}} = 10000 \text{ Hz} \qquad \ldots(1)$$

When gas of dielectric constant k is filled then $C \to kC$

$\Rightarrow$ new resonant frequency will be

$$f_R' = \frac{1}{2\pi\sqrt{kLC}} = 9950 \text{ Hz} \qquad \ldots(2)$$

$$\text{By (1)/(2)} \Rightarrow \sqrt{k} = \frac{10000}{9950}$$

$$\Rightarrow \qquad k = \left(\frac{1000}{995}\right)^2 \simeq 1.01 \quad \textbf{Ans. (C)}$$

Sol. 18 We can see in figure that $V_L = V_C$ this happens at resonance only

$$\Rightarrow \qquad V_R = e = 220 \text{ V}$$

and circuit current is purely resistivity so

$$i = \frac{e}{R} = \frac{220}{100} = 2.2 \text{ A} \qquad \textbf{Ans. (D)}$$

Sol. 19 In series LCR circuit at resonance circuit current is maximum so voltage across R is also maximum & it happens at resonant frequency, given as

$$\omega_R = \frac{1}{\sqrt{LC}} = \frac{1}{\sqrt{24 \times 2 \times 10^{-6}}}$$

$$\Rightarrow \qquad \omega_R = \frac{1000}{\sqrt{48}} = \frac{1000}{6.9} = 144.5 \text{ Hz}$$

Closest option is (B) **Ans. (B)**

Sol. 20

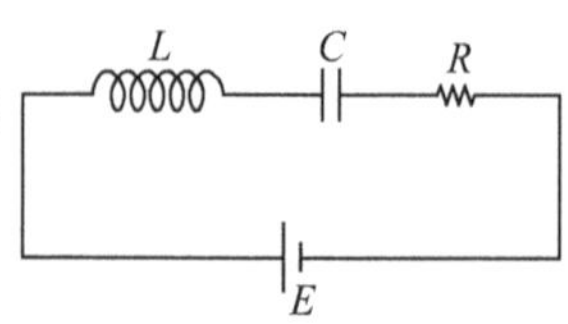

Initially when a DC voltage is applied then in steady state after a long time charge on capacitor is given as

$$q = CE$$

$$\Rightarrow \qquad E = \frac{q}{C} = \frac{10^{-9}}{10^{-11}} = 100 \text{ V}$$

Now DC source is replaced by an AC source of peak value 100 volt.

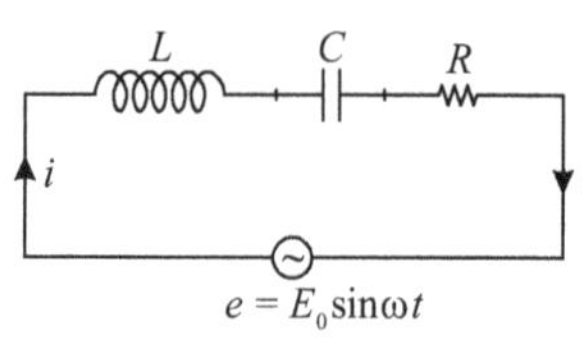

At resonance $\omega = \frac{1}{\sqrt{LC}}$ and circuit current will be

$$i = e/R$$

$$\text{Peak current as} \qquad i_0 = \frac{E_0}{R} = \frac{100}{100} = 1 \text{A}$$

Peak voltage across capacitor at resonance is given as

$$V_C = i_0 X_C = 1 \cdot \left(\frac{1}{\omega C}\right)$$

Peak charge on capacitor is

$$q = CV_C = 1 \cdot \left(\frac{1}{\omega C}\right) = \sqrt{LC}$$

$\Rightarrow \quad q = \sqrt{10^{-5} \times 10^{-11}}$

$\Rightarrow \quad q = 10^{-8}\,\text{C}$ **Ans. (D)**

Sol. 21 For a series RLC circuit quality factor is given as

$$Q = \frac{\omega_0}{\Delta\omega}$$

Here resonant frequency $\omega_0 \rightarrow$ maximum power in circuit

from plot $\quad \omega_0 = 5\,\text{kHz}$

and bandwidth $\Delta\omega \rightarrow$ difference of half power frequencies

$$\Delta\omega = \omega_2 - \omega_1$$

From plot we use $\quad \omega_1 \cong 3.6\,\text{kHz}$ and $\omega_2 \cong 6.4\,\text{kHz}$

$\Rightarrow$ Bandwidth $\quad \Delta\omega \cong 6.4 - 3.6 = 2.8\,\text{kHz}$

So quality factor of circuit is given as

$$Q = \frac{5 \times 10^3}{2.8 \times 10^3} = 1.8$$

Closest option here is (B) **Ans. (B)**

Sol. 22

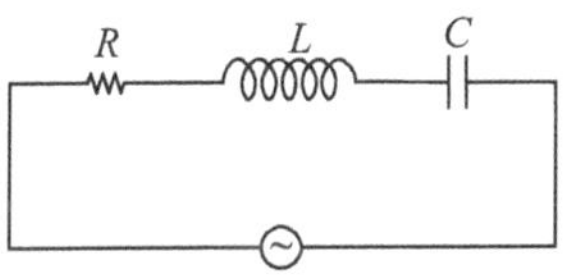

(A) At resonance $X_L = X_C$ so circuit impedance is purely resistive so $\phi = 0°$ and circuit current is in same phase with applied voltage. "It is correct".

(B) At resonant frequency $\omega = \frac{1}{\sqrt{LC}}$ we have $\omega L = \frac{1}{\sqrt{LC}}$

"It is correct".

(C) At resonance circuit current is $i = e/R$

if R is reduced $\Rightarrow$ i increases.

So voltage across capacitor $V_C = iX_C$ increases with i.

"It is correct".

(D) At resonance circuit impedance is $Z = \sqrt{R^2 + (X_L - X_C)^2}$

As $X_L = X_C \Rightarrow Z$ is minimum.

"It is Not correct". **Ans. (D)**

Sol. 23 Case-I :

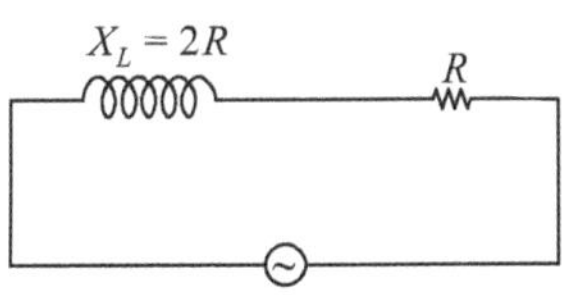

Circuit impedance $\quad Z_1 = \sqrt{R^2 + X_L^2} = \sqrt{5}R$

Power factor $\quad \cos\phi_1 = \frac{R}{Z_1} = \frac{1}{\sqrt{5}}$

Case - II :

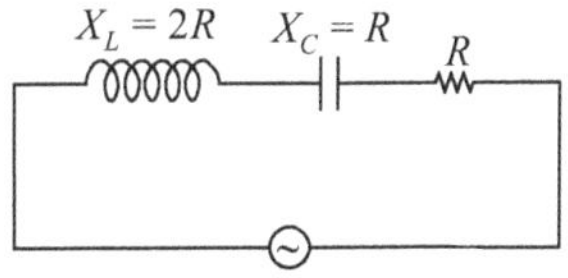

Circuit impedance $\quad Z_2 = \sqrt{R^2 + (X_L - X_C)^2} = \sqrt{2}R$

New power factor, $\cos\phi_2 = \frac{R}{Z_2} = \frac{1}{\sqrt{2}}$

Ratio of new power factor to old power factor is

$$\frac{\cos\phi_2}{\cos\phi_1} = \frac{1/\sqrt{2}}{1/\sqrt{5}} = \sqrt{\frac{5}{2}}$$ **Ans. (D)**

Sol. 24

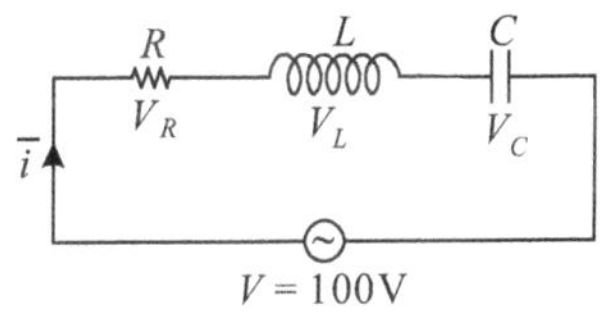

Phasor diagram of voltages is shown below

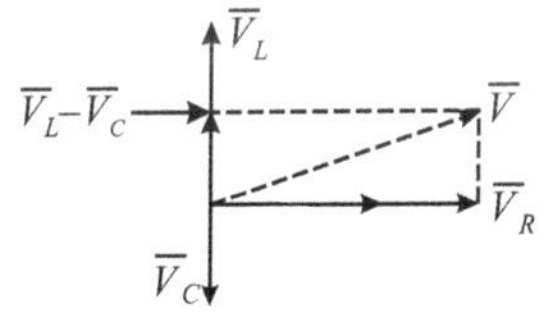

In phasor form $\quad \bar{V} = \bar{V}_R + \bar{V}_L + \bar{V}_C$

in magnitude $\quad V = \sqrt{V_R^2 + (V_L - V_C)^2}$ ...(1)

if $\quad V_L : V_C : V_R = 1 : 2 : 3$

Then we can use in magnitudes $V_L = k$, $V_C = 2k$, $V_R = 3k$

From equation-(1) we have

$$100 = \sqrt{(3k)^2 + (k)^2} = \sqrt{10}k$$

$\Rightarrow \quad k = 10\sqrt{10}$

Here we have $\quad V_R = 3k = 30\sqrt{10} \cong 94.5$

Closest option here is option (C) **Ans. (C)**

Sol. 25

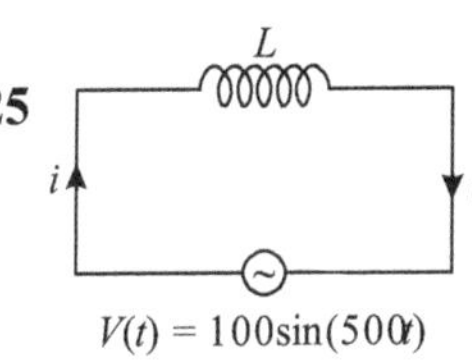

The current through inductor (pure) always lag behind applied emf by 90°

$\Rightarrow \quad i = i_0 \sin(500t - 90°)$

Here $$i_0 = \frac{e_0}{Z} = \frac{e_0}{X_L} = \frac{e_0}{\omega L} = \frac{100}{500 \times 0.02}$$

$$\Rightarrow \quad i_0 = \frac{100}{10} = 10\,\text{A}$$

$$\Rightarrow \quad i = 10 \sin(500t - \pi/2)$$

$$\Rightarrow \quad i = -10 \cos(500\,t)$$ **Ans. (B)**

Sol. 26 As currents & voltages can be measured (RMS value) using Hot wire type meters which use the amount of power dissipation as thermal energy in the meter coil. This does not be affected by change in direction of current.

In devices which work on basis of torque due to magnetic forces will not be useful as due to continuous change in direction of current average torque on coil/magnet = 0. **Ans. (A)**

Sol. 27 Initially current is leading the applied voltage

$$\Rightarrow \quad X_C > X_L \text{ or } \frac{1}{\omega C} > \omega L$$

when C' is connected with C, $X_{CC'} = X_L$

$$\Rightarrow \quad \frac{1}{\omega C_{eq}} < \frac{1}{\omega C} \text{ so as } C_{eq} > C \Rightarrow C' \text{ is connected in parallel}$$

to C, so we have

$$C_{eq} = C + C'$$

After C' is connected we use for resonance

$$\frac{1}{\omega(C + C')} = \omega L$$

$$\Rightarrow \quad C + C' = \frac{1}{\omega^2 L}$$

$$\Rightarrow \quad C' = \frac{1}{\omega^2 L} - C = \frac{1 - \omega^2 LC}{\omega^2 L}$$ **Ans. (C)**

Sol. 28 using direct current 10A current flows at 80V emf

$\Rightarrow$ Resistance of arc lamp $R = \frac{80}{10} = 8\Omega$

$\Rightarrow$ When arc lamp is connected with an inductor across AC supply then 10A current should flow for lamp to function properly.

$$\Rightarrow \quad i = \frac{220}{Z} \Rightarrow 10 = \frac{220}{Z}$$

$$\Rightarrow \quad Z = 22\,\Omega$$

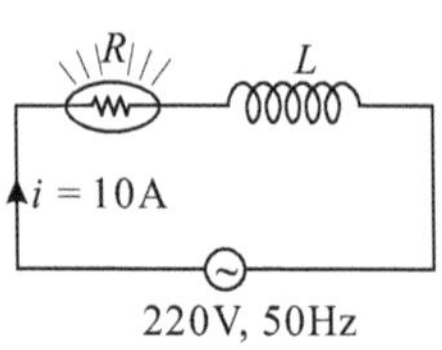

For a series LR circuit impedance is given as

$$Z = \sqrt{R^2 + \omega^2 L^2}$$

[Here $\omega = 2\pi(50) = 100\pi$ rad/s]

$$\Rightarrow \quad 22 = \sqrt{(8)^2 + (100\pi)^2 L^2}$$

$$\Rightarrow \quad 64 + (100\pi)^2 L^2 = (22)^2$$

$$\Rightarrow \quad (100\pi)^2 L^2 = 484 - 64 = 420$$

$$\Rightarrow \quad L = \frac{\sqrt{420}}{100\pi} = 0.0648\,\text{H} \simeq 0.065\,\text{H}$$ **Ans. (D)**

Sol. 29 Across a series LR circuit when AC emf is connected, circuit impedance will be

$$Z = \sqrt{R^2 + \omega^2 L^2}$$

and current in circuit is given as

$$i = i_0 \sin(\omega t - \phi)$$

$$\Rightarrow \quad i = \frac{V_0}{\sqrt{R^2 + \omega^2 L^2}} \sin(\omega t - \phi)$$

Where $\phi = \cos^{-1}\left(\frac{R}{\sqrt{R^2 + \omega^2 L^2}}\right)$ and $i_0 = \frac{e_0}{\sqrt{R^2 + \omega^2 L^2}}$

Ans. (D)

Sol. 30 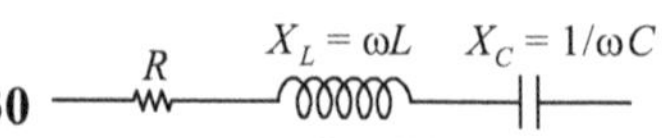

$$R = 5\Omega$$

$$X_L = \omega L = 320 \times 0.025 = 8\Omega$$

$$X_C = \frac{1}{\omega C} = \frac{1}{320 \times 10^{-3}} = 3.1\Omega$$

Impedance triangle (phasor form)

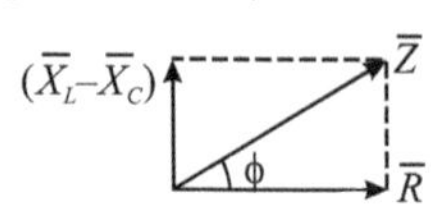

Circuit impedance is given as

$$Z = \sqrt{R^2 + (X_L - X_C)^2}$$

$$Z = \sqrt{(5)^2 + (4.9)^2} = 7\Omega$$

From triangle we have

$$\phi = \tan^{-1}\left(\frac{X_L - X_C}{R}\right) = \tan^{-1}\left(\frac{4.9}{5}\right)$$

$$\Rightarrow \quad \phi \simeq \pi/4$$ **Ans. (B)**

Sol. 31

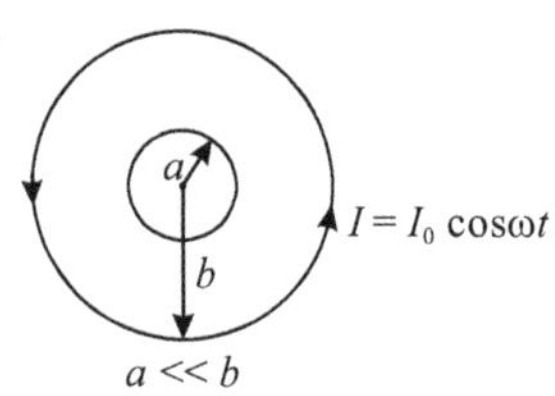

Magnetic flux linked with the smaller loop is

$$\phi = \frac{\mu_0 I}{2b} \times \pi a^2 = \frac{\mu_0 \pi a^2}{2b} I_0 \cos\omega t$$

Emf induced in smaller loop is

$$e = -\frac{d\phi}{dt} = \frac{\mu_0 \pi a^2 I_0}{2b} \omega \sin \omega t \qquad \textbf{Ans. (A)}$$

Sol. 32 The quality factor for a series RLC circuit is given as

$$Q = \frac{\omega_R}{\Delta\omega}$$

$$= \frac{\omega_0 L}{R} \qquad \textbf{Ans. (A)}$$

Sol. 33 Average power consumed by an AC circuit is given as

$$P_{avg} = e_{rms} i_{rms} \cos\phi$$

Here $\quad e_{rms} = \frac{e_0}{\sqrt{2}} = \frac{100}{\sqrt{2}} V$

and $\quad i_{rms} = \frac{i_0}{\sqrt{2}} = \frac{20}{\sqrt{2}} \text{A}; \phi = \pi/4$

$$\Rightarrow \quad P_{avg} = \frac{100}{\sqrt{2}} \times \frac{20}{\sqrt{2}} \times \frac{1}{\sqrt{2}} = \frac{1000}{\sqrt{2}} W$$

Wattless current is given as

$$I_{\omega L} = i_{rms} \sin\phi = \frac{20}{\sqrt{2}} \times \frac{1}{\sqrt{2}} = 10\text{A} \qquad \textbf{Ans.(B)}$$

Sol. 34 Power equation of a transformer is given as

$$e_1 i_1 \times \eta = e_2 i_2$$

$$\Rightarrow \quad 2300 \times 5 \times \frac{90}{100} = 230 \times i_2$$

$$\Rightarrow \quad i_2 = 45\text{A} \qquad \textbf{Ans. (C)}$$

Sol. 35 Capacitive reactance of circuit is given as

$$x_C = \frac{1}{\omega C} = \frac{1}{100\pi \times 120 \times 10^{-6}}$$

$$\Rightarrow \quad x_C = 26.52\,\Omega$$

Inductive Reactance is

$$x_L = \omega L = 100\pi \times 20 \times 10^{-3} = 2\pi\Omega$$

$$\Rightarrow \quad x_C - x_L = 20.24 \approx 20$$

Phasor inpedance triangle

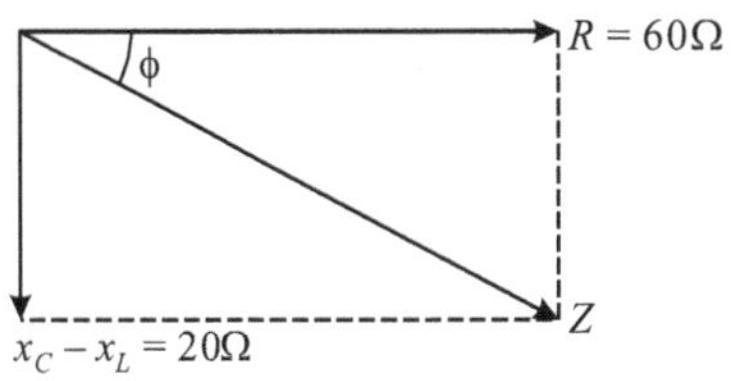

$$\Rightarrow \quad Z = \sqrt{R^2 + (x_C - x_L)^2}$$

$$\Rightarrow \quad Z = 20\sqrt{10}\Omega$$

$$\Rightarrow \quad \cos\phi = \frac{R}{z} = \frac{3}{\sqrt{10}}$$

$$\Rightarrow \quad P_{avg} = \text{VI} \cos\phi \text{ where } I = \frac{V}{Z}$$

$$\Rightarrow \quad P_{avg} = \frac{V^2}{Z} \cos\phi$$

$$\Rightarrow \quad P_{avg} = 8.64 \text{ W}$$

Energy spent is

$$Q = P.t = 8.64 \times 60 = 5.18 \times 10^2 \text{ J} \qquad \textbf{Ans. (D)}$$

Sol. 36 For C-R_2 circuit

$$X_C = \frac{1}{\omega_C} = \frac{2}{10^{-6} \times \sqrt{3} \times 100} = \frac{2 \times 10^4}{\sqrt{3}} \Omega$$

Phase difference of I_2 with AC source emf is given as

$$\tan\theta_1 = \frac{X_C}{R_2} = \frac{2 \times 10^4}{\sqrt{3} \times 20} = \frac{1000}{\sqrt{3}}$$

$\Rightarrow$ θ_1 is close to 90° with current leading emf

For L–R_1 circuit, we use

$$X_L = \omega_L = 100 \times \frac{\sqrt{3}}{10} = 10\sqrt{3}\,\Omega$$

Phase difference of I_1 with AC source emf is given as

$$\tan\theta_2 = \frac{X_L}{R_2} = \frac{10\sqrt{3}}{10} = \sqrt{3}$$

$\Rightarrow \quad \theta_2 = 60°$ with current lagging

Thus phase difference between I_1 and I_2 is given as

$90° + 60° = 150°$ **Ans. (None)**

Sol. 37

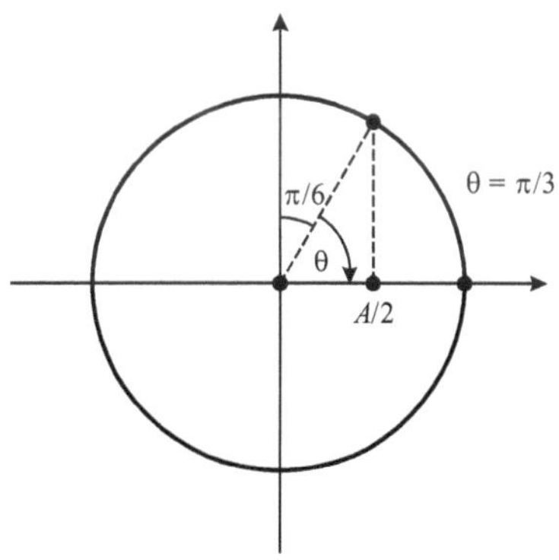

$$V(t) = 220 \sin(100\pi t) \text{ volt}$$

Time taken to change phase angle by $\pi/3$ is

$$t = \frac{\theta}{\omega} = \frac{\frac{\pi}{3}}{100\pi} = \frac{1}{300} \text{ s}$$

$$\Rightarrow \quad t = 3.3 \text{ ms} \qquad \textbf{Ans. (C)}$$

Sol. 38 Given phase difference $\theta = \pi/4$ and $\omega = 100$ rad/s

Phase difference will be 45° when circuit reactance and resitance are equal

$\Rightarrow$ Reactance (X) = Resistance (R) now by checking options

Option (A)

$$R = 1000\,\Omega$$

and $$X_C = \frac{1}{10^{-6}\times 100} = 10^4\Omega$$

Option (B) $R = 10^3\,\Omega$

and $X_L = 10^{-3}\times 100 = 10^{-1}\Omega$

Option (C) $R = 10^3\,\Omega$

and $X_L = 10\times 10^{-3}\times 100 = 1\,\Omega$

Option (D) $R = 10^3\,\Omega$

and $$X_C = \frac{1}{10\times 10^{-6}\times 100} = 10^3\Omega$$

Option (D) matches the given condition. **Ans. (D)**

Sol. 39 Given that for transformer

$$N_P = 300, N_s = 150, P_0 = 2200\text{ W}$$
$$I_s = 10\text{ A}$$

Output power of transformer is

$$P_0 = V_0 I_0$$

$$\Rightarrow \quad 2200 = V_0\times 10$$

$$\Rightarrow \quad V_0 = 220\text{ V}$$

Using transformer equation, we have

$$\frac{V_i}{V_0} = \frac{N_p}{N_s}$$

$$\Rightarrow \quad V_i = 2\times 220 = 440\text{ V}$$

For an ideal transformer

$$V_i I_i = V_O I_O$$

$$\Rightarrow \quad I_i = \frac{2200}{440} = 5\text{ A}$$ **Ans. (B)**

22.1 Electromagnetic Wave

Sol. 1 (A) Polarization $\rightarrow$ Restricting oscillation plane of $\vec{E}$ in EMW $\rightarrow$ happens only in transverse waves.

(B) Interference $\rightarrow$ Phenomenon resulting superposition of two coherent wave $\rightarrow$ happens in both transverse and longitudinal waves.

(C) Reflection $\rightarrow$ Bouncing of wave energy in same medium of incidence $\rightarrow$ happens in both transverse and longitudinal waves.

(D) Diffraction $\rightarrow$ Bending of waves at the boundry of obstacle/objects $\rightarrow$ happens in both transverse and longitudinal waves. **Ans. (A)**

Sol. 2 For an EM wave frequency always remain same and if its wavelength changes from λ to λ' then we use

$$\frac{c}{\lambda} = \frac{v}{\lambda'} \qquad \ldots(1)$$

(Where v is wave speed in other medium)

As nothing is specified, we consider medium as non-magnetic ($\mu_r = 1$).

So wave speed in medium is

$$v = \frac{1}{\sqrt{\mu_0 \epsilon_0 \epsilon_r}} = \frac{c}{\sqrt{4.0}} = \frac{c}{2}$$

From (1) $\Rightarrow \frac{c}{\lambda} = \frac{c}{2\lambda'} \Rightarrow \lambda' = \frac{\lambda}{2}$ **Ans. (A)**

Sol. 3 Total power of source = 4×10^3 W

Energy of one photon in eV

$$E = \frac{4\times 10^3}{10^{20}}\times\frac{1}{1.6\times 10^{-19}}\text{ eV}$$

$$\Rightarrow \quad E = 2.5\times 10^2 = 250\text{ eV}$$

Wavelength of these photons can be given as

$$\lambda = \frac{hc}{E} = \frac{12431}{250}\text{ Å} = 49.72\text{ Å}$$

As X-Ray wavelength range varies from 0.1Å to 100Å **Ans. (B)**

Sol. 4 Polarization direction of EMF is always along the plane of oscillation of electromagnetic force.

$\Rightarrow$ $\vec{X}$ is parallel to $\vec{E}$

direction of propagation of an EM wave is given by poynting vector which gives direction of flow of energy in the EM wave and it is given as

$$\vec{S} = \frac{1}{\mu_0}(\vec{E}\times\vec{B})$$

Where $\vec{E}\times\vec{B} \rightarrow$ direction of EM waves propagation.

$\Rightarrow$ $\vec{k}$ is parallel to $(\vec{E}\times\vec{B})$ **Ans. (B)**

Sol. 5 The low wavelength part of EM spectrum is

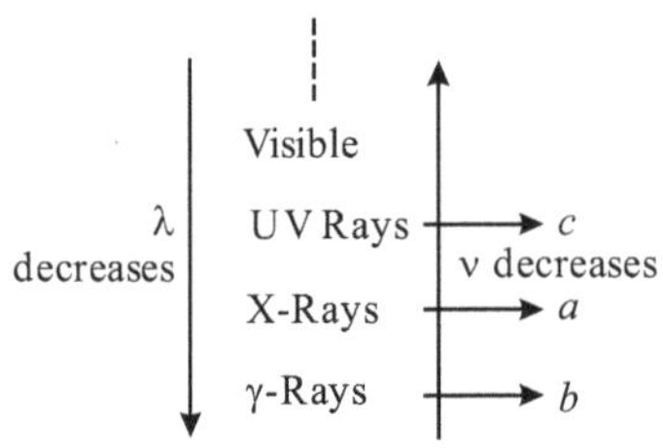

$\Rightarrow$ $b > a > c$ **Ans. (A)**

Sol. 6 Wave propagation in along $\hat{y}$ direction

$\Rightarrow$ $\vec{E}\times\vec{B}$ must be along $\hat{y}$ direction.

$\vec{E}\times\vec{B}$ $\rightarrow$ direction of Poynting vector.
Given that magnetic field is in $+x$ direction i.e. along $\hat{x}$

$\Rightarrow$ $\vec{E}$ must be along $\hat{z}$ direction.

Electric field function is $\vec{E} = E_0\cos(\omega t - \frac{2\pi}{\lambda}y)\,\hat{z}$ **Ans. (C)**

Sol. 7 In EM waves propagation, peak values of electric field E_0 and magnetic field B_0 are related as

$$\frac{E_0}{B_0} = c$$

$$\Rightarrow \quad E_0 = B_0 C$$

$$\Rightarrow \quad E_0 = 20\times10^{-9}\times3\times10^{8}$$

$$\Rightarrow \quad E_0 = 6\text{ V/m}$$ **Ans. (B)**

Sol. 8 The wavelength of given radiation λ can be given as

$$E = \frac{hc}{\lambda} = 11\text{ keV}$$

$$\Rightarrow \quad \lambda = \frac{hc}{E} = \frac{6.63\times10^{-34}\times3\times10^{8}}{11000\times1.6\times10^{-19}}$$

$$\Rightarrow \quad \lambda = 11.3\times10^{-10}\text{ m} = 11.3\text{ Å}$$

As X-rays wavelength range is from 0.1Å to 100Å so this radiation is of X-Rays. **Ans. (A)**

Sol. 9 (A) False - As an EM wave can travel in a medium as well in free space as it does not require a medium for propagation.
(B) False - As EM wave consist of transverse oscillations of EF and MF so these are transverse waves.
(C) False - As EM wave are produced by time variation of EF which can be there by moving charges which accelerates.
(D) True - EM waves carry momentum as well as energy (photons) **Ans. (D)**

Sol. 10 Wave function can be expressed as $f(x - vt)$ by

$$\vec{E} = 4\times10^{-7}\,\vec{E}\left(x - \frac{50t}{4\times10^{-7}}\right)$$

The speed of EM wave in the given medium is given as

$$v = \frac{50}{4\times10^{-7}}\text{ m/s} = 1.25\times10^{8}\text{ m/s}$$

In terms permittivity & permeability of medium, wave speed can be expressed as

$$v = \frac{1}{\sqrt{\mu\in}} = \frac{1}{\sqrt{\mu_0\in_0\in_r}}$$

[As $\mu_r = 1$ for non magnetic medium]

$$\Rightarrow \quad v = \frac{c}{\sqrt{\in_r}}$$

$$\Rightarrow \quad 1.25\times10^{8} = \frac{3\times10^{8}}{\sqrt{\in_r}}$$

$$\Rightarrow \quad \sqrt{\in_r} = \frac{3}{1.25} = 2.4$$

$$\Rightarrow \quad \in_r = (2.4)^2 = 5.76$$ **Ans. (B)**

Sol. 11 In EM wave propagation if E is electric field and B is magnetic field then these are related as

$$\frac{E}{B} = c$$

$$\Rightarrow \quad B = \frac{E}{c}$$

Energy density in electric field wave propagation for electric field and magnetic field are

$$u_E = \frac{1}{2}\in_0 E^2$$

$$u_B = \frac{1}{2}\frac{B^2}{\mu_0} = \frac{1}{2\mu_0}\left(\frac{E}{c}\right)^2 = \frac{1}{2\mu_0}\frac{E^2}{c^2}$$

In free space we use $c = \frac{1}{\sqrt{\mu_0\in_0}}$

$$\Rightarrow \quad u_B = \frac{1}{2\mu_0}(\mu_0\in_0)E^2 = \frac{1}{2\in_0 E^2}$$

$$\Rightarrow \quad u_B = u_E$$ **Ans. (C)**

Sol. 12 (1) $\rightarrow$ γ-rays ~ 10^{-12} m $\rightarrow G$
(2) $\rightarrow$ X-rays ~ 10^{-10} m $\rightarrow X$
(3) $\rightarrow$ UV-rays ~ 10^{-7} m $\rightarrow U$
(4) $\rightarrow$ Visible ~ 10^{-6} m $\rightarrow V$
(5) $\rightarrow$ IR-rays ~ 10^{-4} m $\rightarrow I$
(6) $\rightarrow$ Microwaves ~ 10^{-4} m or more $\rightarrow M$
(7) $\rightarrow$ Radio waves ~ 10^{-3} m to several km $\rightarrow R$
ascending order of wavelength is G, X, U, V, I, M, R **Ans. (C)**

Sol. 13 Average energy density of Electric field in EM waves is given as

$$\langle u_E\rangle = \frac{1}{4}\in_0 E^2$$

$$\Rightarrow \quad \langle u_E\rangle = \frac{1}{4}\times8.8\times10^{-12}\times(4)^2$$

$$\Rightarrow \quad \langle u_E\rangle = 35.2\times10^{-12}\text{ J/m}^3$$ **Ans. (C)**

Sol. 14 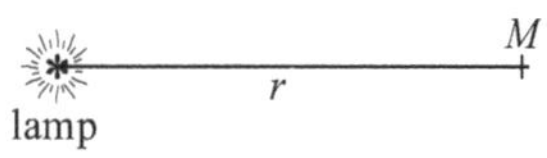

$$P_{EM} = 100\times\frac{3}{100} = 3\text{W}$$

Intensity of EM radiation at $r = 5$m from lamp is given as

$$I = \frac{P}{4\pi r^2} = \frac{3}{4\pi(5)^2} = \frac{3}{100\pi}\ \text{W/m}^2$$

If E_0 is the amplitude of Electric field in wave at point M the we use

$$I = \frac{1}{2}\epsilon_0 E^2 c$$

$$\Rightarrow \quad E_0 = \sqrt{\frac{2I}{\epsilon_0 c}}$$

$$\Rightarrow \quad E_0 = \sqrt{\frac{2\times(3/100\pi)}{\left(\frac{1}{4\pi\times9\times10^9}\right)\times3\times10^8}} = \sqrt{7.2}\ \frac{V}{m}$$

$$\Rightarrow \quad E_0 = 2.68\ \text{V/m}$$

Ans. (B)

Sol. 15 As propagation direction is $+X$ (along $\hat{i}$) so $\vec{B}$ must be along $\hat{j}$ or $\hat{k}$ or in YZ plane.

The amplitude of magnetic field is given as

$$B_0 = \frac{E_0}{c} = \frac{27}{3\times10^8} = 9\times10^{-8}\ \text{T}$$

In a propagating wave general equation is written as

$$B = B_0 \sin 2\pi\left(\frac{x}{\lambda} - \frac{t}{T}\right)$$

Average frequency of wave is $\omega = 2\pi\nu = 2\pi(2\times10^{14})$ Hz

Wavelength of wave is $\lambda = \frac{c}{\nu} = \frac{3\times10^8}{2\times10^{14}} = 1.5\times10^{-6}$ m

$\Rightarrow$ Most appropriate here is option (D), otherwise none is correct. **Ans. (D)**

Sol. 16 For EM wave we use always $\vec{E}\cdot\vec{B} = 0$

and direction of propagation of wave must be along $\vec{E}\times\vec{B}$

here $\vec{E}\times\vec{B}$ must be along $+\hat{k}$ (direction of propagation is z direction).

(A) $(2\hat{i}+3\hat{j}).(\hat{i}+2\hat{j}) = 2+6 = 8 \neq 0$

$\Rightarrow$ Not correct

(B) $(-2\hat{i}-3\hat{j}).(3\hat{i}-2\hat{j}) = -6+6 = 0$

$(-2\hat{i}-3\hat{j})\times(3\hat{i}-2\hat{j}) = 4\hat{k}+(-(-9\hat{k})) = 13\hat{k}$

$\Rightarrow$ Correct.

(C) $(3\hat{i}+4\hat{j}).(4\hat{i}-3\hat{j}) = 12-12 = 0$

$(3\hat{i}+4\hat{j})\times(4\hat{i}-3\hat{j}) = -9\hat{k}-16\hat{k} = -25\hat{k}$ ($-z$ direction)

$\Rightarrow$ Not correct.

(D) $(\hat{i}+2\hat{j})(2\hat{i}-\hat{j}) = 2-2 = 0$

$(\hat{i}+2\hat{j})\times(2\hat{i}-\hat{j}) = -\hat{k}-4\hat{k} = -5\hat{k}$ ($-z$ direction)

$\Rightarrow$ Not correct. **Ans. (B)**

Sol. 17 EM spectrum

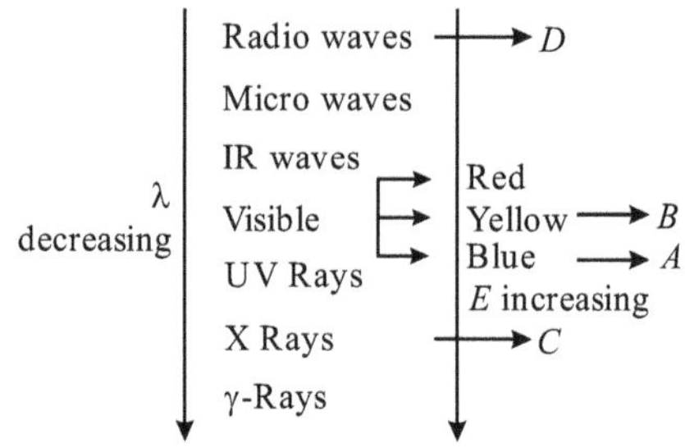

order of increasing energy, $D \to B \to A \to C$ **Ans. (C)**

Sol. 18

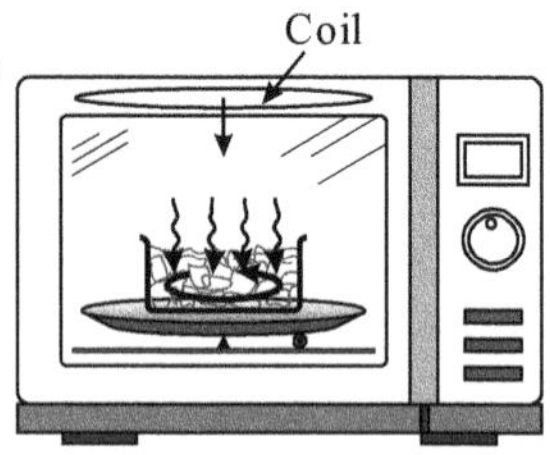

Microwave Oven

Coil of oven produces time varying magnetic field due to which microwaves are produced.

Due to vibrations in ions (water molecules in ionized state or dipoles) localized heat is produced in substance due to which cooking take place. **Ans. (C)**

Sol. 19 (A) Propagation direction $\to +Y$ and $\hat{E} = \hat{z}$ and $\hat{B} = \hat{y}$

This is not possible as $\vec{B}$ is along direction of propagation.

$\Rightarrow$ Not correct.

(B) Propagation direction $\to +Y$ and $\hat{E} = \hat{y}$ and $\hat{B} = \hat{z}$

This is not possible as $\vec{E}$ is along direction of propagation.

$\Rightarrow$ Not correct.

(C) Propagation direction $\to +X$ and $\hat{E} = \left(\frac{\hat{y}+\hat{z}}{\sqrt{2}}\right)$ and $\hat{B} = \left(\frac{\hat{y}+\hat{z}}{\sqrt{2}}\right)$

This is not possible as $\vec{E}$ and $\vec{B}$ are parallel $\Rightarrow$ Not correct.

(D) Propagation direction $\to +X$ and $\hat{E} = \left(\frac{\hat{y}-\hat{z}}{\sqrt{2}}\right)$ and $\hat{B} = \left(\frac{\hat{y}+\hat{z}}{\sqrt{2}}\right)$

here $\vec{E}\cdot\vec{B} = 0$ and both $\vec{E}$ and $\vec{B}$ are $\perp$ to X i.e. in YZ plane and $\vec{E}\times\vec{B}$ direction is along $(\hat{y}-\hat{z})\times(\hat{y}+\hat{z}) = \hat{x}-(-\hat{x}) = 2\hat{x}$ ($+x$ direction). Where $\vec{E}\times\vec{B}$ is direction of energy flow in EM waves. $\Rightarrow$ Correct. **Ans. (D)**

Sol. 20 Electric field amplitude is related to B_0 with speed of EM wave as

$$c = E_0/B_0$$

$$\Rightarrow \quad E_0 = cB_0$$

As wave is travelling along $-X$ direction so vector $\vec{E}\times\vec{B}$ must be along the direction of propagation of EM wave i.e. along $-\hat{i}$ direction. So here Electric field can be given as

$$\vec{E} = E_0 \sin(kx+\omega t)\,\hat{k}$$

$$\Rightarrow \quad \vec{E} = cB_0 \sin(kx+\omega t)\,\hat{k}$$

Ans. (A)

Sol. 21 As we know for an EM wave we use

$$\frac{\partial B}{\partial t} = -\frac{\partial E}{\partial z}$$

$$\Rightarrow \quad \frac{\partial B}{\partial t} = -(-2E_0 k \sin kz \cos\omega t)$$

$$\Rightarrow \quad \partial B = 2E_0 k \sin kz \cos\omega t\,\partial t$$

$$\Rightarrow \quad \int \partial B = 2E_0 k \sin kz \int \cos\omega t\,\partial t$$

$$\Rightarrow \quad B = 2E_0 k \sin kz \left(\frac{\sin\omega t}{\omega}\right)$$

As we use $\frac{\omega}{k} = c$

$$\Rightarrow \quad B = \frac{2E_0}{c}\sin kz \sin\omega t$$

Ans. (C)

Sol. 22 Wave equation are

$$E_1 = E_{01}\cos\left(\frac{\omega z}{c} - \omega t\right) \qquad \ldots(1)$$

[We use $\omega = 2\pi\nu$ and $\omega = kv$ or $v = \omega/k$]

$$E_2 = E_{02}\cos\left(\frac{2\omega z}{c} - \omega t\right) \qquad \ldots(2)$$

$$\Rightarrow \quad E_2 = E_{02}\cos\left(\frac{\omega z}{c/2} - \omega t\right) \qquad [v = c \text{ in air}]$$

In air speed of EM wave is $v_{air} = c$

In medium speed of EM wave is $v_m = c/2$

The speed of EM wave can be written as

$$v = \frac{1}{\sqrt{\mu\in}} = \frac{1}{\sqrt{\mu_0\in_0\mu_r\in_r}}$$

As here medium is non magnetic $\Rightarrow \mu_{r_1} = \mu_{r_2} = 1$

In this case we use

$$\frac{v_{air}}{v_m} = \frac{c}{c/2} = \frac{\frac{1}{\sqrt{\in_{r_1}}}}{\frac{1}{\sqrt{\in_{r_2}}}} = \sqrt{\frac{\in_{r_2}}{\in_{r_1}}}$$

$$\Rightarrow \quad \frac{\in_{r_2}}{\in_{r_1}} = 4 \Rightarrow \frac{\in_{r_1}}{\in_{r_2}} = \frac{1}{4}$$

Ans. (C)

Sol. 23 If $\vec{B}$ is the magnetic field then it should be such that $\vec{E}\times\vec{B}$ is along the direction of propagation of EM wave i.e. parallel to $\left(\frac{\hat{i}+\hat{j}}{\sqrt{2}}\right)$

taking cross product $\hat{k}\times\left(\frac{\hat{i}-\hat{j}}{\sqrt{2}}\right) = \frac{\hat{j}-(-\hat{i})}{\sqrt{2}} = \left(\frac{\hat{i}+\hat{j}}{\sqrt{2}}\right)$

So option (A) and (B) here are resulting $\vec{B}$ such the direction of wave propagation is along the given direction.

here $\vec{r}$ is the wave displacement along direction of propagation

So $\quad \vec{r} = r\hat{r} = r\left(\frac{\hat{i}+\hat{j}}{\sqrt{2}}\right)$

By substituting $\vec{r}$ in (A) and (B) we get

$$\text{(A)} \quad \frac{E_0}{c}\left(\frac{\hat{i}-\hat{j}}{\sqrt{2}}\right)\cos\left[\underbrace{10^4\left(\frac{\hat{i}-\hat{j}}{\sqrt{2}}\right)\cdot r\left(\frac{\hat{i}+\hat{j}}{\sqrt{2}}\right)}_{0} - \overset{\omega=2\pi\nu}{\overbrace{3\times10^{12}t}}\right]$$

So (A) represent on oscillating field not the travelling field.

$$\text{(B)} \quad \frac{E_0}{c}\left(\frac{\hat{i}-\hat{j}}{\sqrt{2}}\right)\cos\left[10^4\left(\frac{\hat{i}-\hat{j}}{\sqrt{2}}\right)\cdot r\left(\frac{\hat{i}+\hat{j}}{\sqrt{2}}\right) - 3\times10^{12}t\right]$$

$$\Rightarrow \quad \frac{E_0}{c}\left(\frac{\hat{i}-\hat{j}}{\sqrt{2}}\right)\cos[10^4 r - 3\times10^{12}t] \text{ travelling wave equation.}$$

Ans. (B)

Sol. 24 For an EM wave $\frac{E}{B} = c \Rightarrow |\vec{B}| = \frac{E}{c}$

If I is intensity then energy density in EM wave is given as

$$\frac{1}{2}\in_0 E_0^2 = \frac{I}{c}$$

$$\Rightarrow \quad E_0 = \sqrt{\frac{2I}{\in_0 c}}$$

We also know that direction of propagation of EM wave is along vector $\vec{E}\times\vec{B}$ which should be along $\hat{j}$ here. **Ans. (C)**

Sol. 25 As z_1 and z_2 are two neighbouring points where EF = 0

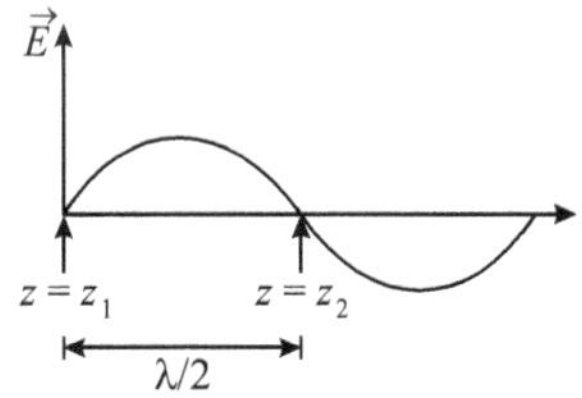

As we have $|(z_2 - z_1)| = \frac{\lambda}{2}$

or $\quad \lambda = c/\nu$

$$\Rightarrow \quad |z_2 - z_1| = \frac{c}{2\nu}$$

$$\Rightarrow \quad \nu = \frac{c}{2|z_2 - z_1|} = \frac{3\times10^8}{2|z_2 - z_1|}$$

$$\Rightarrow \qquad \nu = \frac{1.5\times10^8}{|z_2 - z_1|} \qquad \textbf{Ans. (C)}$$

Sol. 26 In EM wave E, B and C are related as

$$B = \frac{E}{C} = \frac{6.3}{3\times10^8} = 2.1\times10^{-8}\,\text{T}$$

Direction vectors $\hat{E}$, $\hat{B}$ and $\hat{C}$ are related as

$$\Rightarrow \qquad \hat{E}\times\hat{B} = \hat{C}$$

$$\Rightarrow \qquad \hat{j}\times\hat{B} = \hat{i}$$

$$\Rightarrow \qquad \hat{B} = \hat{k} \qquad \textbf{Ans. (C)}$$

Sol. 27 Average energy density of magnetic field,

$u_B = \dfrac{B_0^2}{2\mu_0}$, B_0 is maximum value of magnetic field.

Average energy density of electric field,

$$u_E = \frac{\epsilon_0 E_0^2}{2}$$

Now, $\qquad E_0 = cB_0$, and $c^2 = \dfrac{1}{\mu_0 \epsilon_0}$

$$u_E = \frac{\epsilon_0}{2}\times c^2 B_0^2$$

$$\Rightarrow \qquad u_E = \frac{\epsilon_0}{2}\times\frac{1}{\mu_0\epsilon_0}\times B_0^2 = \frac{B_0^2}{2\mu_0} = u_B$$

$$\Rightarrow \qquad u_E = u_B$$

As energy density of electric & magnetic field is same, energy associated with equal volume will be equal. **Ans. (C)**

Sol. 28 Electric and magnetic field amplitudes are related as

$$E_0 = B_0\times c$$

$$\Rightarrow \qquad E_0 = 100\times10^{-6}\times3\times10^8$$

$$\Rightarrow \qquad E_0 = 3\times10^4\,\text{N/C} \qquad \textbf{Ans. (D)}$$

Sol. 29 The magnetic field magnitude is given as

$$B = \frac{E}{c} = \frac{10}{c}$$

We write EF at $t = 0$ as

$$\vec{E}(x, y) = (10\,\hat{j})\cos[(6\,\hat{i} + 8\,\hat{k})\cdot(x\,\hat{i} + z\,\hat{k})]$$

$$= (10\,\hat{j})\cos[\vec{k}\cdot\vec{r}]$$

Here $\vec{k} = 6\,\hat{i} + 8\,\hat{k}$ is the direction along which wave propagation

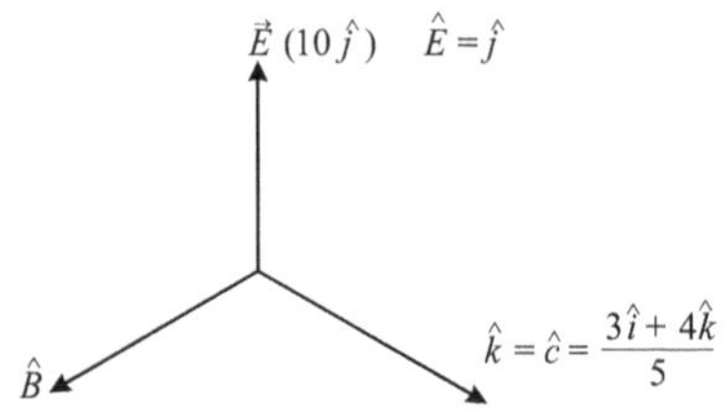

Direction vector of MF is given as

$$\hat{B} = \hat{c}\times\hat{E} = \frac{-4\hat{i} + 3\hat{k}}{5}$$

MF vector is given as

$$\vec{B} = \frac{10}{c}\left(\frac{-4\hat{i} + 3\hat{j}}{5}\right)\cos[\vec{k}\cdot\vec{r} - \omega t]$$

$$\Rightarrow \qquad \vec{B} = \frac{10}{c}\left[\frac{3\hat{i} - 4\hat{i}}{5}\right]\cos[6x + 8z - 10ct] \qquad \textbf{Ans. (B)}$$

Sol. 30 Speed of EM wave in free space is

$$c = \frac{1}{\sqrt{\mu_0 \epsilon_0}}$$

In a medium speed is given as

$$v = \frac{1}{\sqrt{k\,\epsilon_0\,\mu_0}} \qquad [\text{For transparent medium } \mu_r \approx \mu_0]$$

$$\Rightarrow \qquad \frac{c}{v} = \sqrt{k} = n \qquad \text{(refractive index of medium)}$$

EM wave intensity is given as

$$I = \frac{1}{2}\epsilon_0 E_0^2 c = \frac{1}{2}\epsilon_0 kE^2 v$$

$$\Rightarrow \qquad E_0^2 c = kE^2 v$$

$$\Rightarrow \qquad \frac{E_0^2}{E^2} = \frac{kv}{c} = \frac{n^2}{n}$$

$$\Rightarrow \qquad \frac{E_0}{E} = \sqrt{n}$$

For magnetic field we can use

$$\frac{B_0^2 c}{2\mu_0} = \frac{B^2 v}{2\mu_0}$$

$$\Rightarrow \qquad \frac{B_0}{B} = \frac{1}{\sqrt{n}} \qquad \textbf{Ans. (B)}$$

Sol. 31 Intensity of electromagnetic wave is given by

$$I = \frac{\text{Power}}{\text{Area}} = \frac{1}{2}\varepsilon_0 E_0^2\,c$$

$$\Rightarrow \qquad I = \frac{27\times10^{-3}}{10\times10^{-6}} = \frac{1}{2}\times9\times10^{-12}\times E_0^2\times3\times10^8$$

$$\Rightarrow \qquad E_0 = \sqrt{2}\times10^3\,\text{kV/m}$$

$$\Rightarrow \qquad E_0 = 1.4\,\text{kV/m} \qquad \textbf{Ans. (C)}$$

Sol. 32 Power of refracted transmitted light is 96% so, we have

$$P_{\text{refracted}} = \frac{96}{100}P_I$$

$$\Rightarrow \qquad K_2 A_t^2 = \frac{96}{100}K_1 A_i^2$$

$$\Rightarrow \quad r_2 A_t^2 = \frac{96}{100} r_1 A_i^2$$

$$\Rightarrow \quad A_t^2 = \frac{96}{100}\times\frac{1}{\frac{3}{2}}\times(30)^2 \qquad \text{[Given that } A_i = 30\text{ V/m]}$$

$$\Rightarrow \quad A_t = \sqrt{\frac{64}{100}\times(30)^2} = 24\text{ V/m}$$ **Ans. (B)**

Sol. 33 Intensity of electromagnetic wave is given as

$$I = \varepsilon_0 c E_{rms}^2$$

and $\quad E_{rms} = cB_{rms}$

$$\Rightarrow \quad I = \varepsilon_0 c^3 B_{rms}^2$$

$$\Rightarrow \quad B_{rms} = \sqrt{\frac{I}{\epsilon_0 c^3}}$$

$$\Rightarrow \quad B_{rms} \approx 10^{-4}$$ **Ans. (B)**

Sol. 34 The direction of propogation of an *EM* wave is direction of $\vec{E}\times\vec{B}$. Thus from given information

$$\hat{i} = \hat{j}\times\hat{B}$$

$$\Rightarrow \quad \hat{B} = \hat{k}$$

$$c = \frac{E}{B}$$

$$\Rightarrow \quad B = \frac{E}{c} = \frac{6}{3\times10^8}$$

$B = 2\times10^{-8}$ T Along *z* direction **Ans. (C)**

Sol. 35 If we use that direction of light propagation will be along $\vec{E}\times\vec{B}$. Then (D) option is correct

Detailed solution is as following

magnitude of $\quad E = CB$

$$E = 3\times10^8\times1.6\times10^{-6}$$

$$E = 4.8\times10^2$$

$\vec{E}$ and $\vec{B}$ are perpendicular to each other

$$\Rightarrow \quad \vec{E}\cdot\vec{B} = 0$$

⇒ either direction of $\vec{E}$ is $\hat{i}-2\hat{j}$ or $-\hat{i}+2\hat{j}$ from given option

Also wave propagation direction is parallel to $\vec{E}\times\vec{B}$ which is $-\hat{k}$

⇒ $\vec{E}$ is along $(-\hat{i}+2\hat{j})$ **Ans. (D)**

Sol. 36 Maximum Electric field is given as

$$E = (B)(c)$$

$$\vec{E}_0 = (3\times10^{-5})c\,(-\hat{j})$$

$$\vec{E}_1 = (2\times10^{-6})c\,(-\hat{i})$$

Maximum force

$$\vec{F}_{0\max} = q\vec{E} = qc(-3\times10^{-5}\hat{j}-2\times10^{-6}\hat{i})$$

$$|\vec{F}_{0\max}| = 10^{-4}\times3\times10^8\sqrt{(3\times10^{-5})^2+(2\times10^{-6})^2}$$

$$= 0.9\text{ N}$$

$$F_{rms} = \frac{F_{0\max}}{\sqrt{2}} = 0.63\text{ N (approx)}$$ **Ans. (D)**

Sol. 37 Direction of $\vec{E}\times\vec{B}$ is along the propogaion of EM wave

$$\vec{E}\times\vec{B} \parallel \vec{v} \qquad \ldots(1)$$

Given that wave is propagating along positive *z*-axis and $\vec{E}$ along positive *x*-axis. Thus from Maxwell's equation we have

$$\vec{\nabla}\times\vec{E} = -\frac{\partial B}{\partial t}$$

$$\Rightarrow \quad \frac{\partial E}{\partial z} = -\frac{\partial B}{dt}$$

and $\quad B_0 = \frac{E_0}{c}$ **Ans. (B)**

Sol. 38 Electric field wave is given as

$$\vec{E} = E_0\hat{n}\sin[\omega t + (6y - 8z)]$$

Comparing with equation

$$\vec{E} = E_0\hat{n}\sin[\omega t - \vec{s}\cdot\vec{r}]$$

Where $\vec{s}$ is along direction of propagation and

$$\vec{r} = x\hat{i} + y\hat{i} + z\hat{k}$$

$$\vec{s}\cdot\vec{r} = -6y + 8z$$

$$\Rightarrow \quad \vec{s} = -6\hat{i} + 8\hat{k}$$

Unit vector along direction of propagation as

$$\hat{s} = \frac{\vec{s}}{|\vec{s}|} = \frac{-6\hat{j}+8\hat{k}}{10} = \frac{-3\hat{j}+4\hat{k}}{5}$$ **Ans. (C)**

Sol. 39 Magnetic field when electromagnetic wave propagates in +*z* direction so general equation of magnetic field is

$$B = B_0\sin(kz - \omega t)$$

Where $\quad B_0 = \frac{E_0}{C} = \frac{60}{3\times10^8} = 2\times10^{-7}$ T

We use $\quad k = \frac{2\pi}{\lambda} = 0.5\times10^3\text{ m}^{-1}$

and $\quad \omega = 2\pi f = 1.5\times10^{11}$ rad/s **Ans. (C)**

23.1 Geometrical Optics-I Reflection of Light

Sol. 1

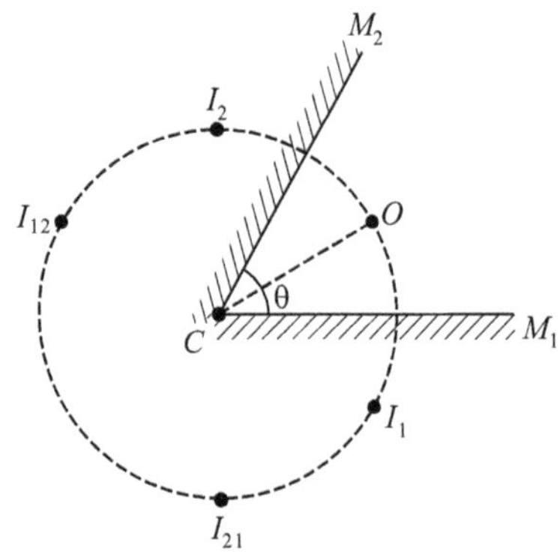

at $\theta = 60°$

$$\frac{180}{60°} = \frac{30}{\text{integer}}$$

If $\frac{180}{\theta} \in I$ then no of images produced are given as

$$N = \frac{360}{\theta} - 1$$

$$N = \frac{360}{60} - 1 = 6 - 1 = 5 \qquad \textbf{Ans. (A)}$$

Sol. 2 If θ is angle between mirrors that no of images obtained are

$$N = \frac{360}{\theta} - 1 = 3$$

$$\Rightarrow \quad \frac{360}{\theta} = 4$$

$$\Rightarrow \quad \theta = \frac{360}{4} = 90° \qquad \textbf{Ans. (B)}$$

Sol. 3

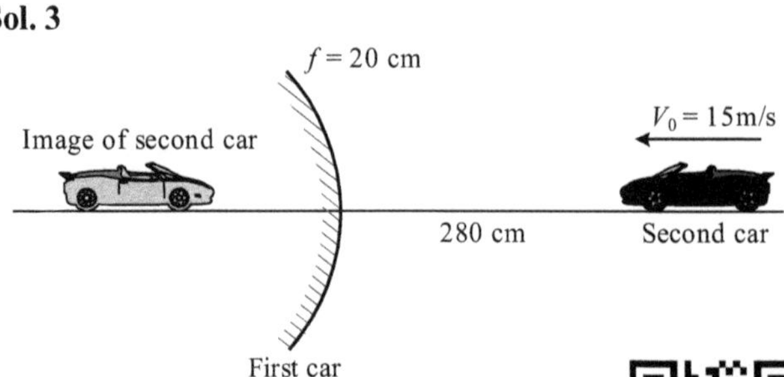

For reflection at convex mirror, we use

$$u = +280 \text{ cm}$$

$$f = -20 \text{ cm}$$

$$\Rightarrow \quad v = \frac{uf}{u-f} = \frac{280 \times (-20)}{300} = -\frac{280}{15} \text{ cm}$$

Image velocity in a mirror is given as

$$v_i = m^2 \cdot V_0 = \left|\frac{v}{u}\right|^2 \times V_0 = \left(\frac{1}{15}\right)^2 \times 15 \text{ [magnification]}$$

$$\Rightarrow \quad v_i = \frac{1}{15} \text{ m/s} \qquad \textbf{Ans. (D)}$$

Sol. 4

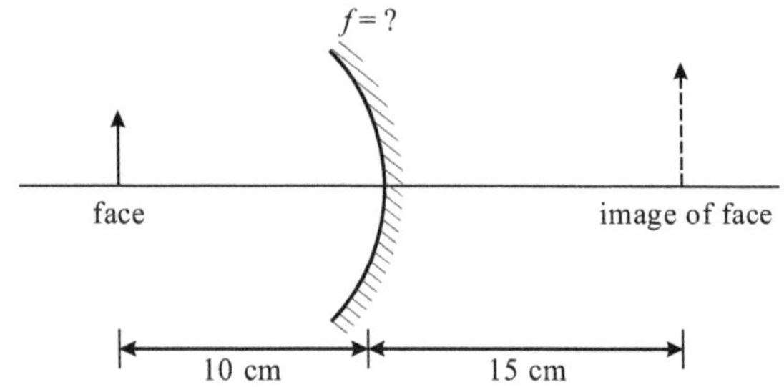

$$R_{\text{mirror}} = 2f$$

For mirror formula we use

$$u = -10 \text{ cm}$$

$$V = +15 \text{ cm}$$

$$\Rightarrow \quad \frac{1}{f} = \frac{1}{u} + \frac{1}{V}$$

$$\Rightarrow \quad \frac{1}{f} = -\frac{1}{10} + \frac{1}{15} = -\frac{5}{150} = -\frac{1}{30}$$

$$\Rightarrow \quad f = -30 \text{ cm}$$

$$\Rightarrow \quad R = 2f = -2 \times 30 = -60 \text{ cm.} \qquad \textbf{Ans. (C)}$$

Sol. 5 As object oscillates $\Rightarrow$ image will also oscillate so (a) is correct.

For the two ext positions of object at $u_1 = 8$ cm and $u_2 = 12$ cm the corresponding image location are – (for $f = 5$ cm)

For left position of object

$$u_1 = 8 \text{ cm } (+ve)$$

$$f = 5 \text{ cm } (+ve)$$

$$\Rightarrow \quad \frac{1}{f} = \frac{1}{v_1} + \frac{1}{u_r}$$

$$\Rightarrow \quad v_1 = \frac{u_1 f}{u_1 - f} = \frac{8 \times 5}{8-5} = +\frac{40}{3} \text{ cm}$$

For right position of object

$$u_2 = 12 \text{ cm}$$

$$f = +5 \text{cm}$$

$$\Rightarrow \quad \frac{1}{f} = \frac{1}{v_2} + \frac{1}{u_2}$$

$$\Rightarrow \quad v_2 = \frac{u_2 f}{u_2 - f} = \frac{12 \times 5}{12-5} + \frac{60}{7} \text{cm}$$

As v_1 and v_2 are not same so the extreme position in oscillation not symmetric about image mean position which is

$$x_0 = 10 \text{ cm } (2f)$$

$\Rightarrow$ Option (c) is correct

The distance between extreme positions of image is

$$\Delta S = v_1 - v_2$$

$$= \frac{40}{3} - \frac{60}{7} = \frac{280-180}{21} = \frac{100}{21}\text{cm}$$

$\Rightarrow$ Option (d) is correct **Ans. (C)**

Sol. 6

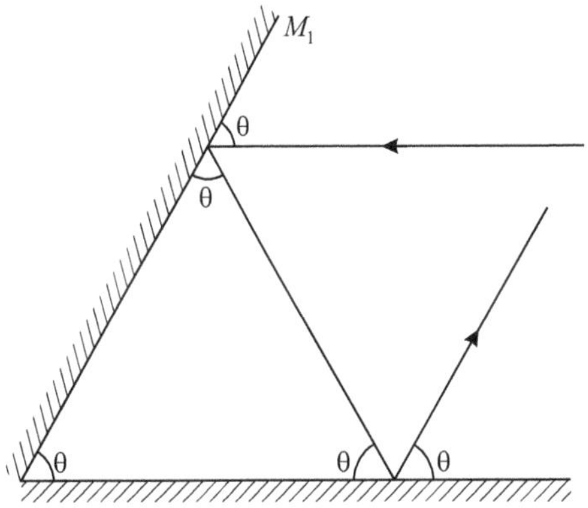

Assuming angles between two mirrors be θ as per geometry, a triangle sum of angles of is

$$3\theta = 180°$$

$$\Rightarrow \quad \theta = 60° \quad \textbf{Ans. (D)}$$

Sol. 7

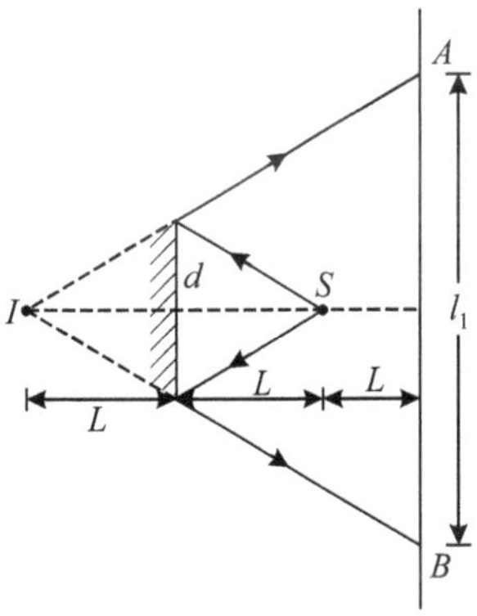

Field of view of image is $l_1 (= AB)$, given as

$$\frac{l_1}{3L} = \frac{d}{L}$$

$$\Rightarrow \quad l_1 = 3d$$ **Ans. (A)**

Sol. 8 Magnification for spherical mirror is given as

$$m = \frac{f}{f-u}$$

$$\Rightarrow \quad 5 = \frac{-40}{-40-u}$$

$$\Rightarrow \quad u = -32\text{ cm}$$ **Ans. (D)**

Sol. 9 Light from particle P will be reflected at mirror.

So for mirror formula, we use

$$u = -5\text{ cm},$$

$$f = -\frac{R}{2} = -20\text{ cm}$$

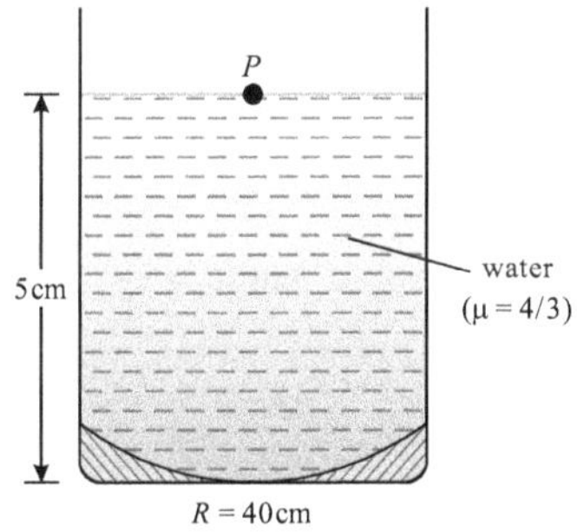

Using mirror formula, we get

$$\frac{1}{v} + \frac{1}{u} = \frac{1}{f}$$

$$\Rightarrow \quad \frac{1}{v} - \frac{1}{5} = -\frac{1}{20}$$

$$\Rightarrow \quad \frac{1}{v} = \frac{1}{5} - \frac{1}{20} = \frac{3}{20}$$

$$\Rightarrow \quad v = +\frac{20}{3}\text{ cm}$$

This image will act as object for light getting refracted at water surface so, object distance is taken as

$$d = 5 + \frac{20}{3} = \frac{35}{3}\text{ cm}$$

Light rays from this object will refract at plane surface (water level) so after refraction, final image will form at

$$d' = d\left(\frac{\mu_2}{\mu_1}\right)$$

$$\Rightarrow \quad d' = \left(\frac{35}{3}\right)\left(\frac{1}{4/3}\right)$$

$$\Rightarrow \quad d' = \frac{35}{4} = 8.75\text{cm}$$

$$\Rightarrow \quad d' \approx 8.8\text{ cm}$$ **Ans. (A)**

23.2 Geometrical Optics-II Refraction of Light

Sol. 1

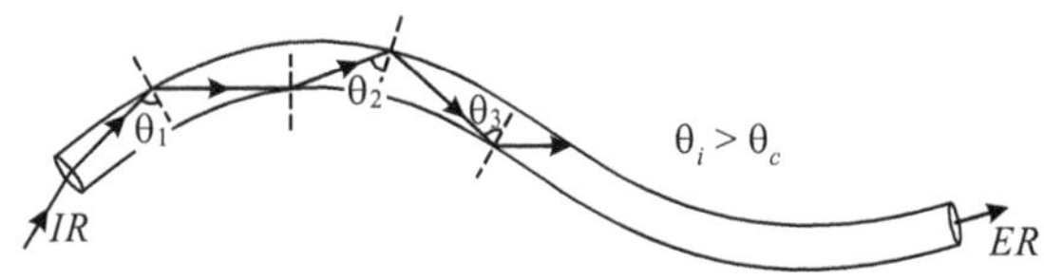

In an optical fibre at every point of inner surface light is internally reflected so light comes out only from the other end of fibre. Total internal reflection. **Ans. (A)**

Sol. 2 As light ray is internally reflected

$$\Rightarrow \quad 45° > \theta_c = \sin^{-1}\left(\frac{1}{n}\right)$$

$$\Rightarrow \quad \frac{1}{n} < \sin 45°$$

$$\Rightarrow \quad n > \sqrt{2}$$ **Ans. (B)**

Sol. 3

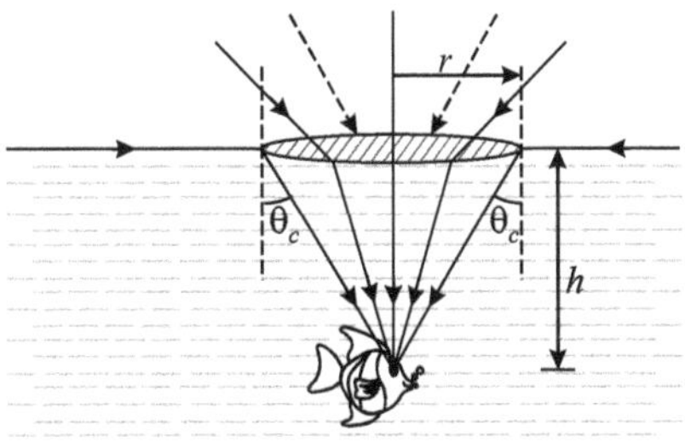

$\theta_c \rightarrow$ critical angle of water air interface

$$\theta_c = \sin^{-1}\left(\frac{1}{\mu_\omega}\right) = \sin^{-1}\left(\frac{3}{4}\right)$$

$$\Rightarrow \quad \sin\theta_c = \frac{3}{4}$$

$$\Rightarrow \quad \tan\theta_c = \frac{3}{\sqrt{7}}$$

In the figure shown, we have

$$\tan\theta_c = \frac{r}{h}$$

$$\Rightarrow \quad r = h\tan\theta_c$$

$$\Rightarrow \quad r = 12\times\frac{3}{\sqrt{7}} = \frac{36}{\sqrt{7}}\text{ cm}$$

Ans. (B)

Sol. 4

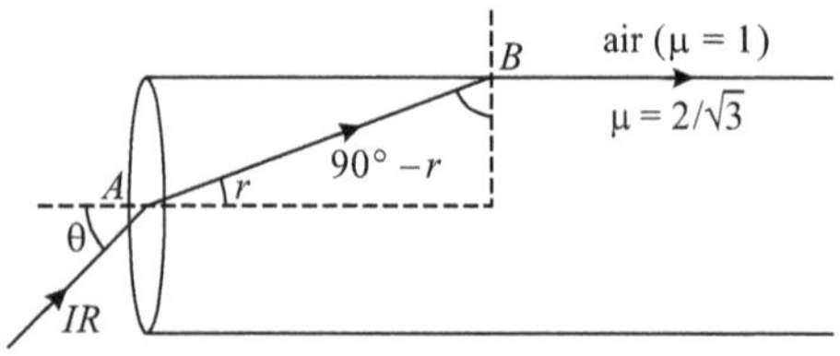

If light ray to graze along the wall of rod

$$\Rightarrow \quad 90^\circ - r = \theta_c = \sin^{-1}\left(\frac{1}{\mu}\right) = \sin^{-1}\left(\frac{\sqrt{3}}{2}\right) = 60^\circ$$

$$\Rightarrow \quad r = 90^\circ - 60^\circ = 30^\circ$$

By Snell's law at point A we have

$$\Rightarrow \quad \sin\theta = \mu\sin r$$

$$= \frac{2}{\sqrt{3}}\times\frac{1}{2} = \frac{1}{\sqrt{3}}$$

$$\Rightarrow \quad \theta = \sin^{-1}\left(\frac{1}{\sqrt{3}}\right)$$

Ans. (B)

Sol. 5

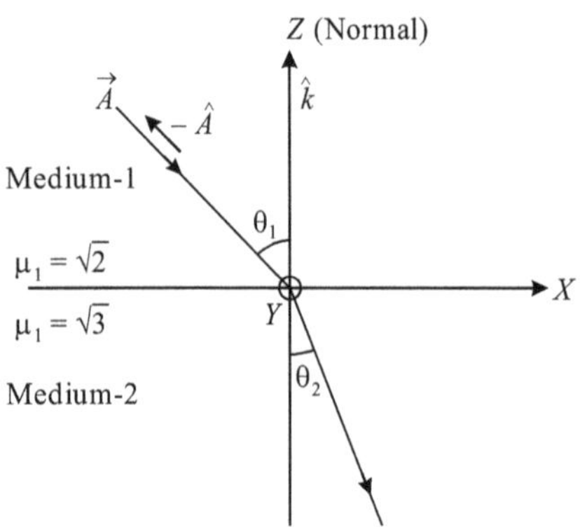

By Snell's law, we use

$$\sqrt{2}\sin\theta_1 = \sqrt{3}\sin\theta_2 \qquad \ldots(1)$$

The angle between $\hat{k}$ and $-\hat{A}$ is given as

$$\cos\theta_1 = -\hat{A}\cdot\hat{k} \qquad \ldots(2)$$

Unit vector along $\vec{A}$ is given as

$$\hat{A} = \frac{\vec{A}}{|\vec{A}|} = \frac{6\sqrt{3}\hat{i}+8\sqrt{3}\hat{j}-10\hat{k}}{\sqrt{(6\sqrt{3})^2+(8\sqrt{3})^2+(10)^2}}$$

$$\Rightarrow \quad \hat{A} = \frac{1}{20}(6\sqrt{3}\hat{i}+8\sqrt{3}\hat{j}-10\hat{k})$$

$$\Rightarrow \quad \hat{A} = \frac{3\sqrt{3}}{10}\hat{i}+\frac{4\sqrt{3}}{10}\hat{j}-\frac{1}{2}\hat{k}$$

From equation (2), we have

$$\cos\theta_1 = -\left(\frac{3\sqrt{3}}{10}\hat{i}+\frac{4\sqrt{3}}{10}\hat{j}-\frac{1}{2}\hat{k}\right)\cdot\hat{k} = \frac{1}{2}$$

$$\Rightarrow \quad \theta_1 = 60^\circ$$

Now from equation-(1), we have

$$\sqrt{2}\times\frac{\sqrt{3}}{2} = \sqrt{3}\sin\theta_2$$

$$\Rightarrow \quad \sin\theta_2 = \frac{1}{\sqrt{2}} \Rightarrow \theta_2 = 45^\circ$$ **Ans. (D)**

Sol. 6

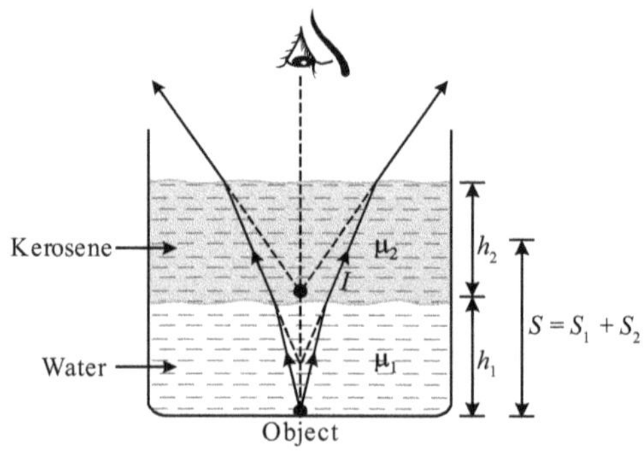

Object shift along normal is given as

$$S_1 = h_1\left(1-\frac{1}{\mu_1}\right)$$

$$S_2 = h_2\left(1-\frac{1}{\mu_2}\right)$$

Total shift of object as seen normally is given as

$$S = h_1\left(1-\frac{1}{\mu_2}\right)+h_2\left(1-\frac{1}{\mu_2}\right)$$ **Ans. (C)**

Sol. 7

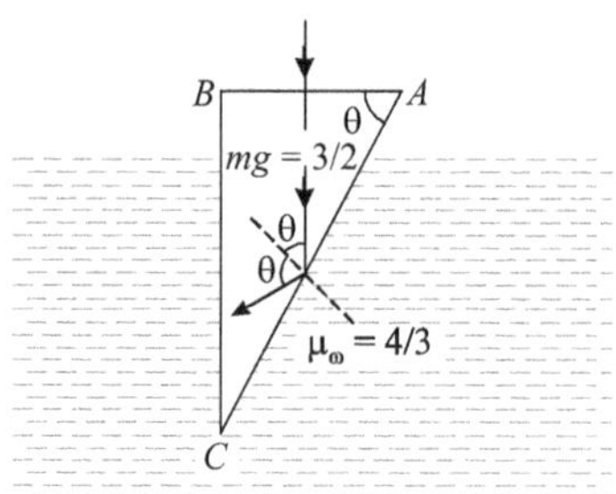

Light will be totally reflected at face AC to reach face BC if

$$\theta > \theta_c$$

[$\theta_c \rightarrow$ critical angle glass-water interface]

$$\Rightarrow \quad \sin\theta > \sin\theta_c$$

$$\Rightarrow \qquad \sin\theta > \frac{1}{{}_{\omega}\mu_g} = \frac{\mu_\omega}{\mu g} = \frac{4/3}{3/2}$$

$$\Rightarrow \qquad \sin\theta > \frac{8}{9}$$ **Ans. (C)**

Sol. 8

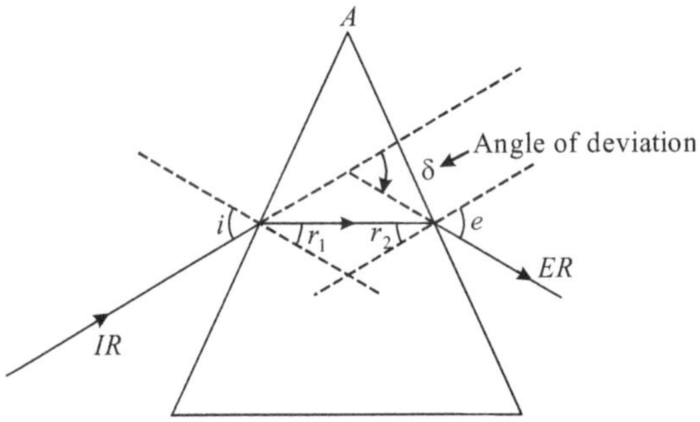

$$\delta = i + e - A$$

When $\quad i = e \rightarrow \delta$ is minimum

$$\Rightarrow \qquad r_1 = r_2 = A/2$$

and light passes through prism symmetrically

$$\sin i = \mu \sin r_1$$

$$\sin e = \mu \sin r_2 = \mu \sin (A - r_1)$$

$$\Rightarrow \qquad e = \sin^{-1}\left[\mu \sin\left(A - \sin^{-1}\left(\frac{\sin i}{\mu}\right)\right)\right]$$

$$\Rightarrow \qquad \delta = i + \sin^{-1}\left[\mu \sin^{-1}\left(A \sin^{-1}\left[\frac{\sin i}{\mu}\right]\right)\right] - A$$

This is a non linear relation. **Ans. (C)**

Sol. 9 As light travels 30 cm in 0.2 ns, the speed of light in medium is

$$v = \frac{3\times10^{-2}}{0.2\times10^{-9}} = 1.5\times10^8 \text{ m/s}$$

Refractive index of medium is

$$\mu = \frac{c}{v} = \frac{3\times10^8}{1.5\times10^8} = 2$$

Critical angle for the medium is

$$\theta_c = \sin^{-1}\left(\frac{1}{\mu}\right) = \sin^{-1}\left(\frac{1}{2}\right) = 30°$$

For all incidence angle $\theta > 30°$, light will suffer TIR

In A & B only for case (B), light will suffer TIR **Ans. (B)**

Sol. 10

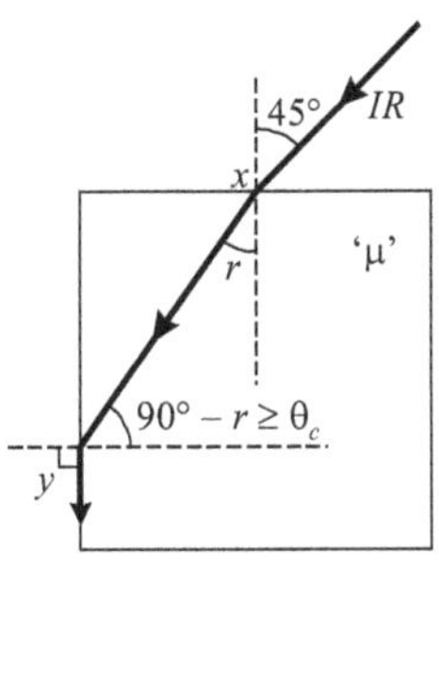

We use Snell's law at x, we have

$$\sin 45° = \mu \sin r \qquad \ldots(1)$$

At point y for TIR here $90° - r = \theta_c$

$$\Rightarrow \qquad \mu \sin(90° - r) = \sin 90°$$

$$\Rightarrow \qquad \mu \cos r = 1 \qquad \ldots(2)$$

$$(1)^2 + (2)^2 \quad \Rightarrow \quad \mu^2 \sin^2 r + \mu^2 \cos^2 r = 1 + \frac{1}{2} = \frac{3}{2}$$

$$\Rightarrow \qquad \mu^2 = \frac{3}{2} \quad \Rightarrow \quad \mu = \sqrt{\frac{3}{2}}$$ **Ans. (D)**

Sol. 11

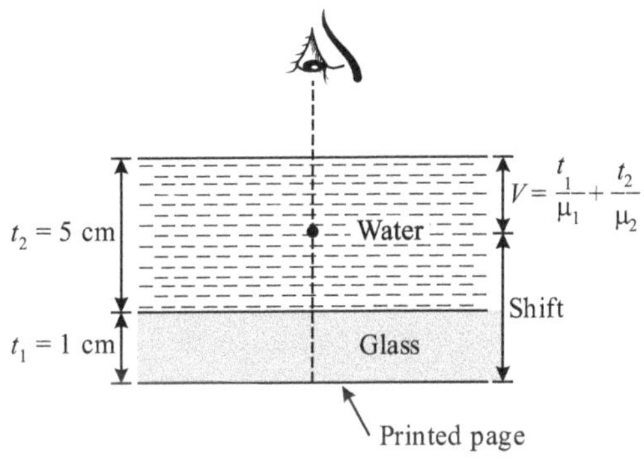

Apparent shift of printed page due to glass and water is given as

$$S = t_1\left(1 - \frac{1}{\mu_1}\right) + t_2\left(1 - \frac{1}{\mu_2}\right)$$

$$\Rightarrow \qquad S = 1\left(1 - \frac{1}{3/2}\right) + 5\left(1 - \frac{1}{4/3}\right)$$

$$\Rightarrow \qquad S = 1\times\frac{1}{3} + 5\times\frac{1}{4} = \frac{1}{3} + \frac{5}{4}$$

$$\Rightarrow \qquad S = \frac{19}{12} \text{ cm}$$

$$\Rightarrow \qquad S \simeq 1.58 \text{ cm}$$ **Ans. (None)**

Sol. 12

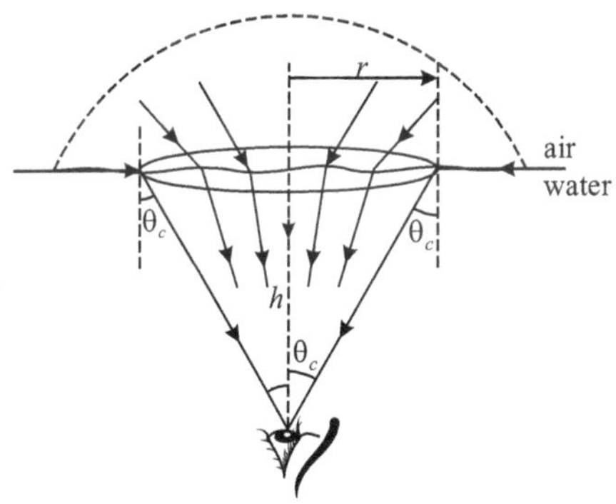

The critical angle at air-water interface is given as

$$\theta_c = \sin^{-1}\left(\frac{1}{\mu_\omega}\right) = \sin^{-1}\left(\frac{3}{4}\right)$$

$$\Rightarrow \qquad \sin\theta_c = \frac{3}{4} \quad \Rightarrow \quad \tan\theta_c = \frac{3}{\sqrt{7}}$$

$$\tan\theta_c = \frac{r}{h}$$

$$\Rightarrow \quad r = h\tan\theta_c$$

$$\Rightarrow \quad h = h\left(\frac{3}{\sqrt{7}}\right) = \frac{15\times 3}{\sqrt{3}}\text{ cm}$$ **Ans. (D)**

Sol. 13

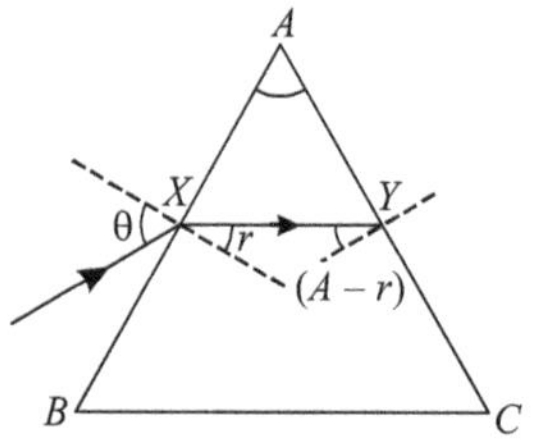

If $(A-r) > \theta_c$ then light will not get not get transmitted through force AC. So in this case we use for transmission of light

$$A - r < \theta_c \qquad \ldots(1)$$

At point x, we use Snell's law

$$\sin\theta = \mu\sin r$$

$$\Rightarrow \quad r = \sin^{-1}\left(\frac{\sin\theta}{\mu}\right)$$

From equation-(1)

$$A - \sin^{-1}\left(\frac{\sin\theta}{\mu}\right) < \sin^{-1}\left(\frac{1}{\mu}\right) \qquad \left[\text{As } \theta_c = \sin^{-1}\left(\frac{1}{\mu}\right)\right]$$

$$\Rightarrow \quad \sin^{-1}\left(\frac{\sin\theta}{\mu}\right) > A - \sin^{-1}\left(\frac{1}{\mu}\right)$$

$$\Rightarrow \quad \frac{\sin\theta}{\mu} > \sin\left(A - \sin^{-1}\left(\frac{1}{\mu}\right)\right)$$

$$\Rightarrow \quad \theta > \sin^{-1}\left[\mu\sin\left(A - \sin^{-1}\left(\frac{1}{\mu}\right)\right)\right]$$ **Ans. (A)**

Sol. 14 We are given with

$$\left.\begin{matrix} i = 35^\circ \\ \delta = 40^\circ \\ e = 79^\circ \end{matrix}\right] \rightarrow \delta = i + e - A$$

Prism angle is given as

$$A = i + e - \delta = 35^\circ + 79^\circ - 40^\circ = 74^\circ$$

We use RI of glass as

$$\mu = \frac{\sin\left(\frac{A+\delta_m}{2}\right)}{\sin\left(\frac{A}{2}\right)}$$

[we use $\delta_m < 40^\circ$ as at 40° $i \neq e$]

$$\Rightarrow \quad \mu = \frac{\sin\left(\frac{A+\delta_m}{2}\right)}{\sin(37^\circ)} \qquad [\text{As } \sin 37^\circ = 3/5]$$

$$\Rightarrow \quad \mu = \frac{\sin\left(\frac{A+\delta_m}{2}\right)}{3/5}$$

$$\Rightarrow \quad \mu = \frac{5}{3}\sin(37^\circ + \theta)$$

Where $\theta = \frac{\delta_m}{2} < 20^\circ$

$$\Rightarrow \quad \mu = \frac{5}{3}\sin[37 + (< 20^\circ)]$$

$$\Rightarrow \quad \mu = \frac{5}{3}\sin[< 57^\circ]$$

$$\Rightarrow \quad \mu < \frac{5}{3}\sin 57^\circ \qquad \left[\text{As } \sin 60^\circ = \frac{\sqrt{3}}{2}\right]$$

$$\Rightarrow \quad \mu < \frac{5}{3}\times 1.67$$

$$\Rightarrow \quad \mu < 1.45$$

Here out of given options closest possible maximum volume can be 1.5. **Ans. (A)**

Sol. 15

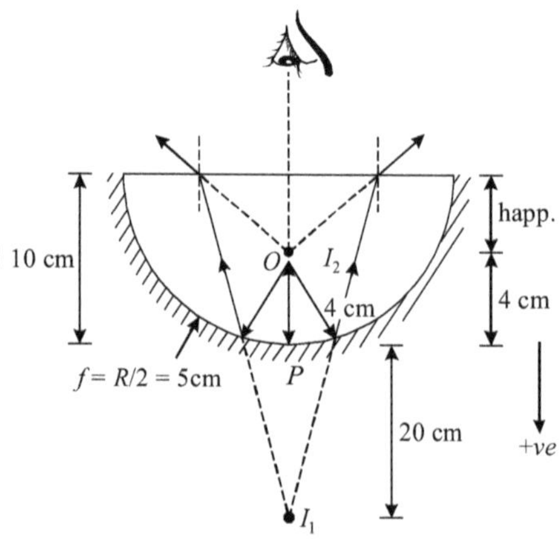

For reflection at spherical mirror we use

$$u = -4\text{ cm}$$

$$f = -5\text{ cm}$$

So from mirror formula, we have

$$\frac{1}{f} = \frac{1}{u} + \frac{1}{v}$$

$$\Rightarrow \quad \frac{1}{v} = \frac{1}{f} - \frac{1}{u} = -\frac{1}{5} + \frac{1}{4} = +\frac{1}{20}$$

$$\Rightarrow \quad v = +20\text{ cm}$$

Due to refraction from plane surface of glass to air final image is seen at

$$h_{ap} = \frac{h}{\mu} = \frac{30}{1.5} = 20\text{ cm}$$ **Ans. (B)**

Sol. 16 We are given that

$$n_{12} = \frac{n_1}{n_2} = \frac{1}{\sin\theta_c} \qquad (n_1 > n_2)$$

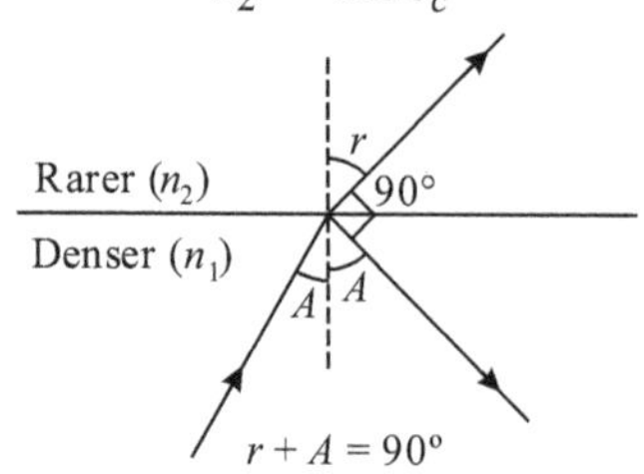

By Snells' law we have

$$n_1 \sin A = n_2 \sin r$$

$$\frac{\sin A}{\sin \theta_c} = \sin(90° - A) = \cos A$$

$\Rightarrow$ $$\tan A = \sin \theta_c$$

$\Rightarrow$ $$A = \tan^{-1}(\sin \theta_c)$$ **Ans. (D)**

Sol. 17

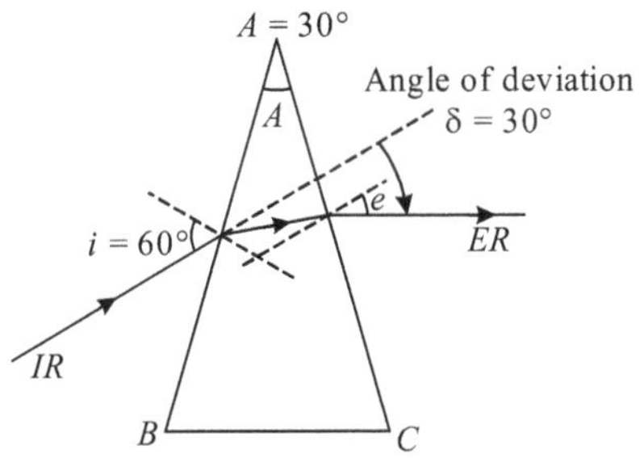

For a trihedral prism we know

$$\delta = i + e - A$$

$\Rightarrow$ $$e = \delta + A - i = 30° + 30° - 60° = 0°$$

$\Rightarrow$ Emergent ray comes out of prism perpendicular to face AC.

Ans. (C)

Sol. 18 Using refraction formula at spherical surface, we have

$$\frac{\mu_2}{v} - \frac{\mu_1}{u} = \frac{\mu_2 - \mu_1}{R}$$

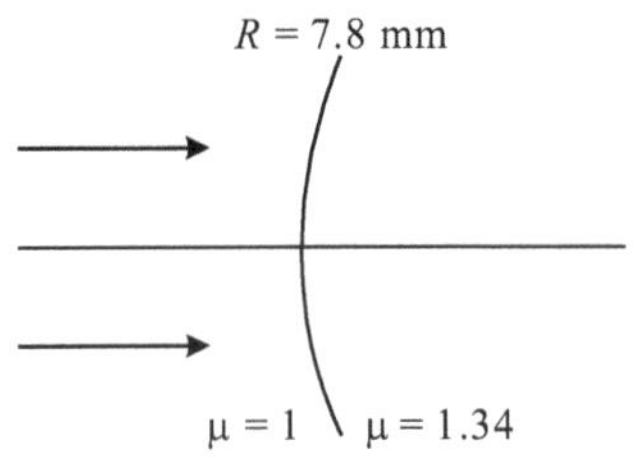

$\Rightarrow$ $$\frac{1.34}{v} - \frac{1}{\infty} = \frac{1.34 - 1}{7.8}$$

$\Rightarrow$ $$v = 30.7\text{mm}$$ **Ans. (C)**

Sol. 19 For minimum deviations by a prism, we use

$$i = e$$

and $$r_1 = r_2 = \frac{A}{2} = 30°$$

by Snell's law, we use

$$\mu_{air} \sin i = \mu_p \sin r$$

$\Rightarrow$ $$1 \times \sin i = \sqrt{3} \times \frac{1}{2} = \frac{\sqrt{3}}{2}$$

$\Rightarrow$ $$i = 60°$$ **Ans. (D)**

Sol. 20 By Snell's law, we use

$$1.\ \sin 40° = 1.31 \sin \theta_2$$

$\Rightarrow$ $$\sin\theta_2 = \frac{0.64}{1.31} = \frac{64}{131}$$

$\Rightarrow$ $$\tan\theta_2 = \frac{64}{\sqrt{(131)^2 - (64)^2}} = \frac{64}{\sqrt{13065}} \simeq 0.49 = \frac{d}{x}$$

$\Rightarrow$ $$x = \frac{d}{0.49}$$

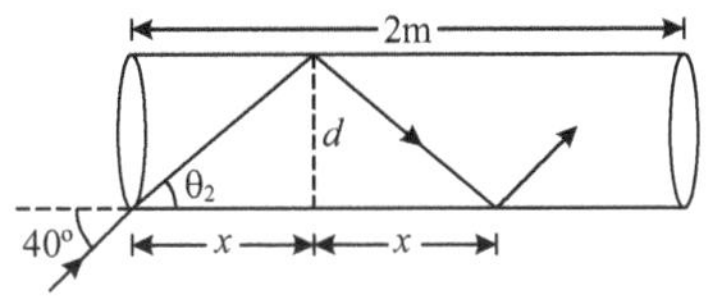

Total number of reflections in fiber are

$$N = \frac{2 \times 0.49}{20 \times 10^{-6}} = 55991$$

Closes to 55000 **Ans. (A)**

Sol. 21 From Snell's law

$$1.\sin 60° = \mu \sin 30°$$

$\Rightarrow$ $$\mu = \sqrt{3}$$

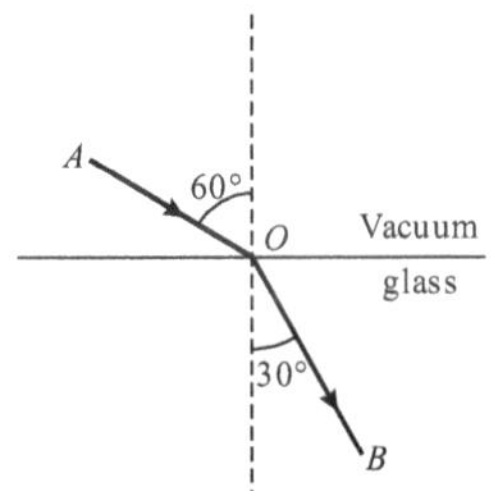

Optical path length of light ray from A to B is

$$OA = \mu(OB)$$

$$= \frac{a}{\cos 60°} + \sqrt{3}\frac{b}{\cos 30°}$$

$$= 2a + 2b$$ **Ans. (A)**

Sol. 22

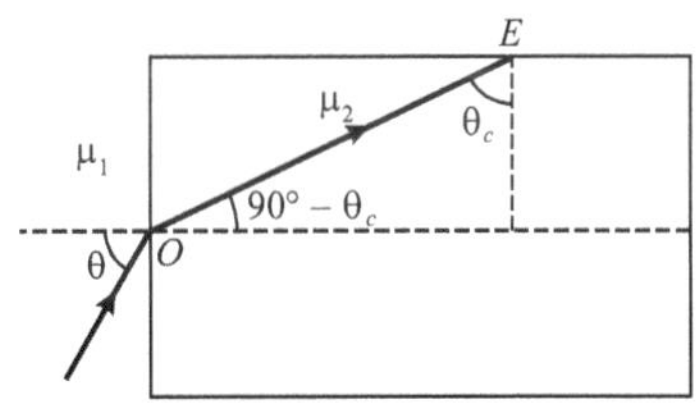

For total internal reflection at point E, we use

$$\sin\theta_c = \frac{\mu_1}{\mu_2}$$

For refraction at point O, we use

$$\mu_1 \sin\theta = \mu_2 \sin(90° - \theta_c)$$

$\Rightarrow$ $$\sin\theta = \frac{\mu_2\sqrt{1 - \frac{\mu_1^2}{\mu_2^2}}}{\mu_1}$$

$$\Rightarrow \qquad \theta = \sin^{-1}\sqrt{\frac{\mu_2^2-\mu_1^2}{\mu_1^2}}$$

For total internal reflection to take place at point E, we use

$$\theta < \sin^{-1}\sqrt{\frac{\mu_2^2}{\mu_1^2}-1}$$ **Ans. (B)**

23.3 Geometrical Optics-III Thin Lenses

Sol. 1

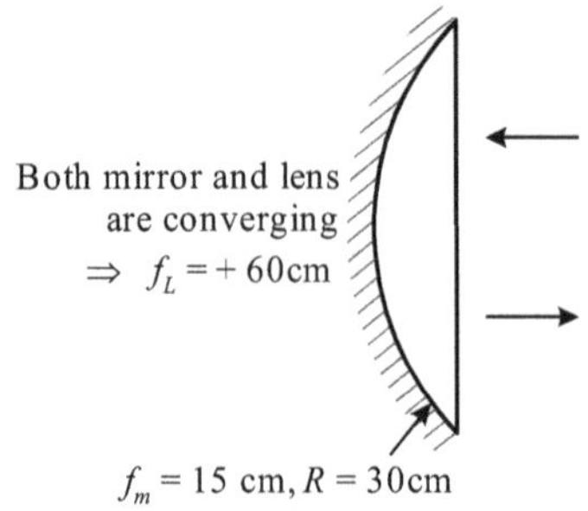

This system acts like a combination of a concave mirror and a plano convex lens

Focal length of mirror is

$$f_m = \frac{R}{2} = 15\,\text{cm}$$

Focal length of lens is

$$f_L = \frac{R}{\mu-1} = \frac{30}{0.5} = 60\,\text{cm}$$

Equivalent focal length of combination is given as

$$f_{eq} = \frac{2}{f_L}+\frac{1}{f_M} = \frac{2}{60}+\frac{1}{15} = \frac{1}{30}+\frac{1}{15} = \frac{1}{10}$$

$\Rightarrow \quad f_{eq} = 10$ cm $\qquad$ $(+ve \Rightarrow$ converging mirror)

This system acts like a concave mirror of focal length 10 cm

A real image of size of object is produced when object is placed at a distance

$$2f_e = 20\,\text{cm}$$ **Ans. (A)**

Sol. 2 Given that $P_{air} = -5$D $\qquad$ [$-ve$ Diverging lens]

$\Rightarrow$ focal length of lens in air

$$f_{air} = \frac{1}{5} = 0.2\,\text{m} = 20\,\text{cm}$$

By lens makers formula we have

$$P = (\mu-1)\left[\frac{1}{R_1}-\frac{1}{R_2}\right]$$

$$-\frac{5}{100} = (1.5-1)\left[\frac{1}{R_1}-\frac{1}{R_2}\right] \qquad \ldots(1)$$

If focal length of lens in liquid medium is f_L then we use

$$\frac{1}{f_L} = \left(\frac{\mu_g}{\mu_L}-1\right)\left(\frac{1}{R_1}-\frac{1}{R_2}\right)$$

$$\Rightarrow \qquad \frac{1}{f_L} = \left(\frac{1.5}{1.6}-1\right)\left[\frac{1}{R_1}-\frac{1}{R_2}\right] \qquad \ldots(2)$$

$$\frac{(1)}{(2)} \Rightarrow \qquad -\frac{f_L}{20} = -\frac{0.5}{0.1}\times 1.6$$

$$\Rightarrow \qquad f_L = 5\times 20\times 1.6 = 160\,\text{cm} = 1.6\,\text{m}$$

Power lens in liquid medium

$$P_{liq} = \frac{\mu_L}{f_L} = \frac{1.6}{1.6} = +1D$$ **Ans. (A)**

Sol. 3 When thin lenses are kept in contact then equivalent power of combination is given as

$$P_{eq} = P_1 + P_2$$
$$= -15D + 5D = 10D$$

equivalent focal length of combination is given as

$$f_{eq} = \frac{1}{P_{eq}} = -\frac{1}{10}$$
$$= -0.1\,\text{m} = -10\,\text{cm}$$ **Ans. (B)**

Sol. 4 Object and image distances from a lens are related by "Lens formula", given as

$$\frac{1}{v}-\frac{1}{u} = \frac{1}{f} \qquad [u \to \text{object distance from lens}]$$

[$v \to$ image distance from lens]

$$\Rightarrow \qquad v = \frac{uf}{u-f}$$

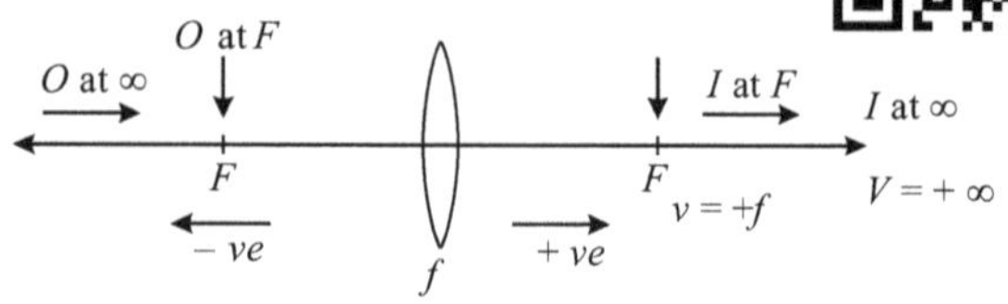

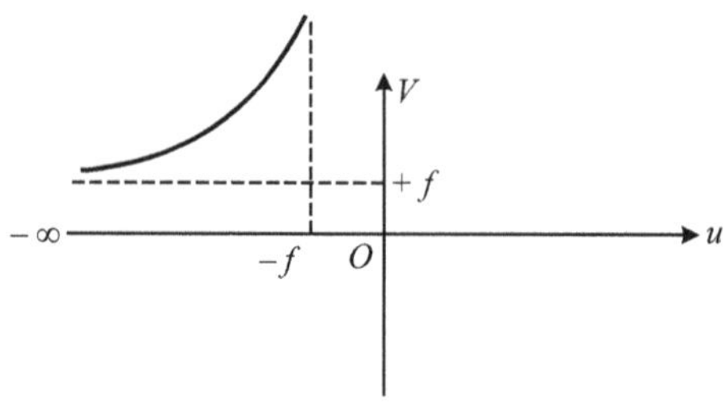

Ans. (C)

Sol. 5 For a convex lens the curve between $|u|$ and $|v|$ is shown as

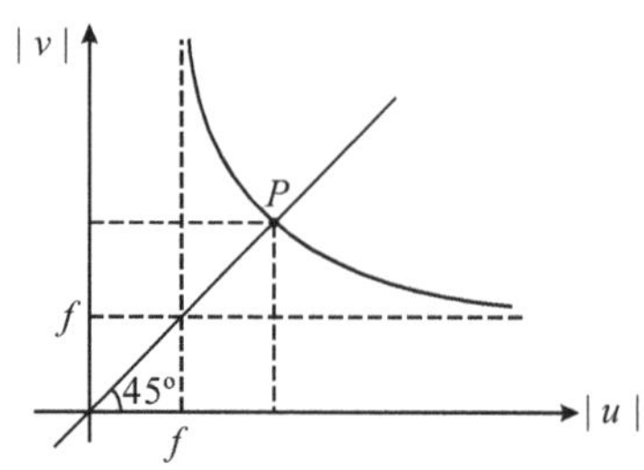

P is graph is a point where object distance and image distance are equal and in case of a convex lens for a real image

$$|u| = |v| = 2f$$

$\Rightarrow \quad P \rightarrow (2f, 2f)$ **Ans. (D)**

Sol. 6 Case-I :

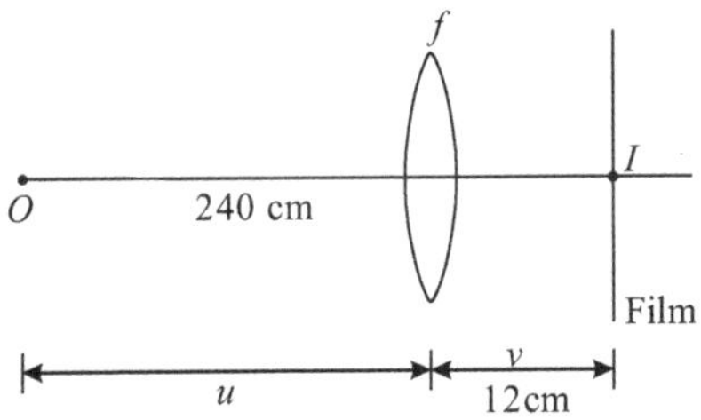

Using lens formula, we have

$$u = -240 \text{ cm}$$

$$v = +12 \text{ cm}$$

$$f = ?$$

$$\frac{1}{v} - \frac{1}{u} = \frac{1}{f}$$

$$\Rightarrow \quad \frac{1}{12} + \frac{1}{240} = \frac{1}{f}$$

$$\Rightarrow \quad f = \frac{12 \times 240}{252} = \frac{240}{21} \text{ cm}$$

Case-II :

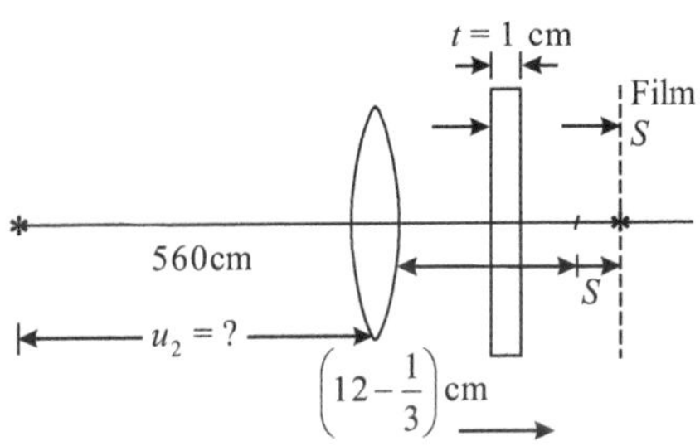

Due to glass plate image produced by lens will be shifted to right by S were

$$S = t\left(1 - \frac{1}{\mu}\right)$$

Shift $$S = 1\left(1 - \frac{1}{3/2}\right) = 1 - \frac{2}{3} = \frac{1}{3} \text{ cm}$$

Lens has to produce the final image at a distance $\left(12 - \frac{1}{3}\right)$ cm so as to get sharp image on screen.

For case-II

$$v_2 = +\left(12 - \frac{1}{3}\right) = +\frac{35}{3} \text{ cm}$$

$$f = +\frac{240}{21} \text{ cm}$$

object distance $\quad u_2 = ?$

By lens formula

$$\frac{1}{v} - \frac{1}{u} = \frac{1}{F}$$

$$\frac{3}{35} - \frac{1}{u} = \frac{21}{240}$$

$$\Rightarrow \quad \frac{1}{u} = \frac{3}{35} - \frac{21}{240}$$

$\Rightarrow \quad u = -560 \text{ cm}$ **Ans. (A)**

Sol. 7 Refractive index of lens material is

$$\mu = \frac{c}{v} = \frac{3 \times 10^8}{2 \times 10^8} = 1.5$$

From the given information, we have (by figure)

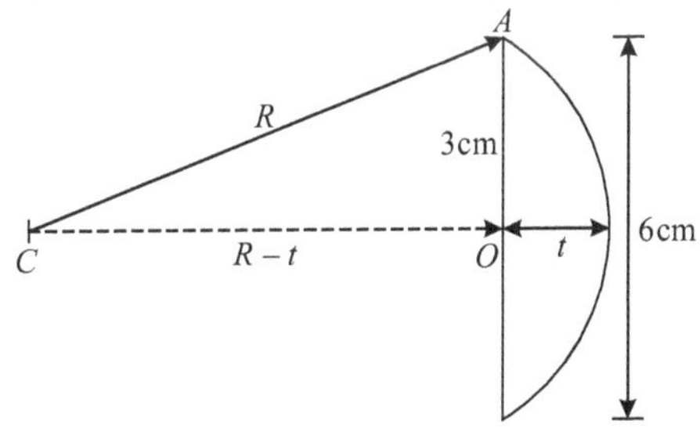

Very thin lines with thickness $t = 3$ mm

In ΔCAO, we use

$$R^2 = (3)^2 + (R - 0.3)^2$$

$$\Rightarrow \quad R^2 = 9 + R^2 + 0.09 - 0.6R$$

$$\Rightarrow \quad 0.6R = 9.09$$

$$\Rightarrow \quad R = \frac{90.9}{6} \simeq 15 \text{ cm}$$

Focal length of a thin lens ($t << R$) is given by lens makes formula as

$f = \dfrac{R}{\mu - 1} = \dfrac{15}{1.5 - 1} = 30 \text{ cm}$ **Ans. (C)**

Sol. 8

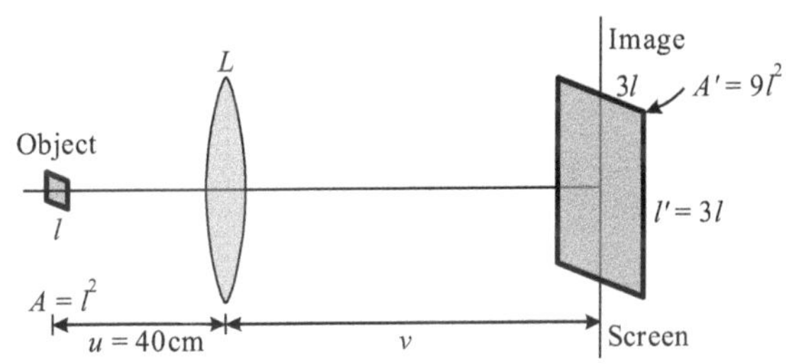

As image size in 3 times the object size

$$\Rightarrow \quad \frac{v}{u} = 3 \Rightarrow v = 3 \times 40$$

$$\Rightarrow \quad = 120 \text{ cm}$$

By lens formula we have

$$u = -40 \text{ cm}$$

$$v = +120 \text{ cm}$$

$$\Rightarrow \quad \frac{1}{f} = \frac{1}{v} - \frac{1}{u} = \frac{1}{120} + \frac{1}{40} = \frac{4}{120} = \frac{1}{30}$$

$\Rightarrow \quad f = +30 \text{ cm}$ **Ans. (D)**

Sol. 9 By lens makes formula

$$\frac{1}{f}=(\mu-1)\underbrace{\left[\frac{1}{R_1}-\frac{1}{R_2}\right]}_{C}$$

$$\Rightarrow \quad f=\frac{1}{C(\mu-1)}=\frac{1}{C(3/2-1)}=\frac{2}{C}$$

When the lens is submerged in a medium of RI μ_m then we use

$$\frac{1}{f}=\left(\frac{\mu}{\mu_m}-1\right)C$$

$$\Rightarrow \quad f=\frac{\mu_m}{C(\mu-\mu_m)}$$

For medium having RI $\mu_1=4/3$, we have focal length

$$f_1=\frac{4/3}{C\left(\frac{3}{2}-\frac{4}{3}\right)}=\frac{4/3}{C\left(\frac{1}{6}\right)}=\frac{8}{C}$$

For medium having RI $\mu_2=5/3$, we have focal length

$$f_2=\frac{5/3}{C\left(\frac{3}{2}-\frac{5}{3}\right)}=\frac{5/3}{C\left(-\frac{1}{6}\right)}=-\frac{10}{C}$$

Ans. (B)

Sol. 10

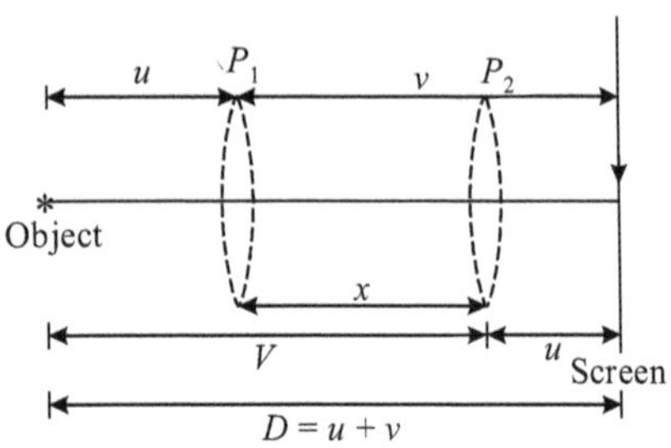

$$I_1=\frac{v}{u}S_0$$

$$I_2=\frac{u}{v}S_0$$

The lens displacement is given as

$$x=v-u=10\text{ cm} \quad \ldots(1)$$

$$\frac{I_1}{I_2}=\frac{v^2}{u^2}=\frac{3}{2} \quad \ldots(2)$$

from (1) and (2), we use

$$v-\sqrt{\frac{2}{3}}v=10$$

$$\Rightarrow \quad v=\frac{\sqrt{3}(10)}{\sqrt{3}-\sqrt{2}}$$

$$\Rightarrow \quad u=\sqrt{\frac{2}{3}}(10) \quad \Rightarrow \quad u=\frac{\sqrt{2}(10)}{\sqrt{3}-\sqrt{2}}$$

Separation between object and screen is given as

$$D=u+v$$

$$\Rightarrow \quad D=10\left(\frac{\sqrt{3}+\sqrt{2}}{\sqrt{3}-\sqrt{2}}\right)\text{cm}$$

$$\Rightarrow \quad D=10\left(\frac{1.732+1.414}{1.732-1.414}\right)=10\times\frac{3.146}{0.318}$$

$$\Rightarrow \quad D=98.9\text{ cm}$$

Ans. (D)

Sol. 11 Method-1 : As we know a lens is submerged in a medium of RI greater then that of lens material then behaviour of lens is changed i.e. convex $\rightarrow$ converging

Here as $\mu_1>\mu$, $\Rightarrow$ Concave lens converges parallel light beam.

Ans. (A)

Method-2 : Using lens makers formula

$$\frac{1}{f}=({}_m\mu_L-1)\left[\frac{1}{R_1}-\frac{1}{R_2}\right]$$

here say $|R_1|=|R_2|$ (we consider an equiconcave lens) and

$${}_m\mu_L=\frac{\mu}{\mu_1}$$

$R_1\rightarrow -ve\,(-R)$

$R_2\rightarrow +ve\,(+R)$

$$\Rightarrow \quad \frac{1}{f}=\underbrace{\left(\frac{\mu}{\mu_1}-1\right)}_{-ve}\underbrace{\left[-\frac{1}{R}-\frac{1}{R}\right]}_{-ve}$$

As $(\mu<\mu_1)$

$\Rightarrow$ $f\rightarrow +ve$

$\Rightarrow$ Lens behave as a converging lens **Ans. (A)**

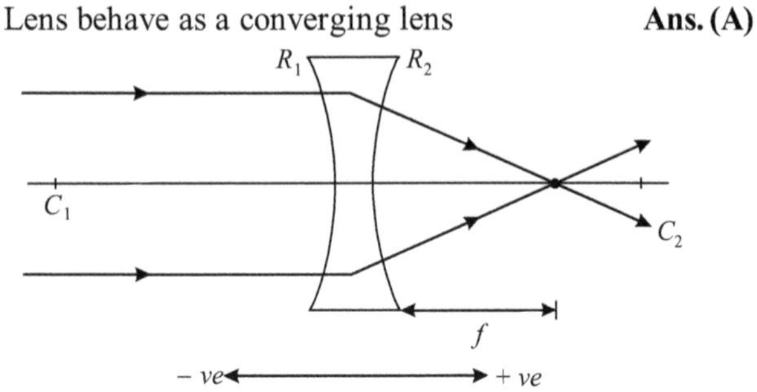

Sol. 12 The lens mirror combination behaves like a spherical mirror of which focal length f_m is given as

$$f_{plain}=\infty$$

$$f_{lens}=f$$

$$\frac{1}{f_m}=\frac{2}{f_{lens}}+\frac{1}{f_{plain}}=\frac{2}{f}$$

$$\Rightarrow \quad f_m=\frac{f}{2}$$

Using mirror formula we have

$$u=+a$$

$$V=+a/3$$

$$\Rightarrow \quad \frac{1}{f_m} = \frac{1}{u} + \frac{1}{V} = \frac{1}{a} + \frac{3}{a} = \frac{4}{a}$$

$$\Rightarrow \quad \frac{a}{4} = \frac{f}{2}$$

$$\Rightarrow \quad a = 2f$$ **Ans. (D)**

Sol. 13

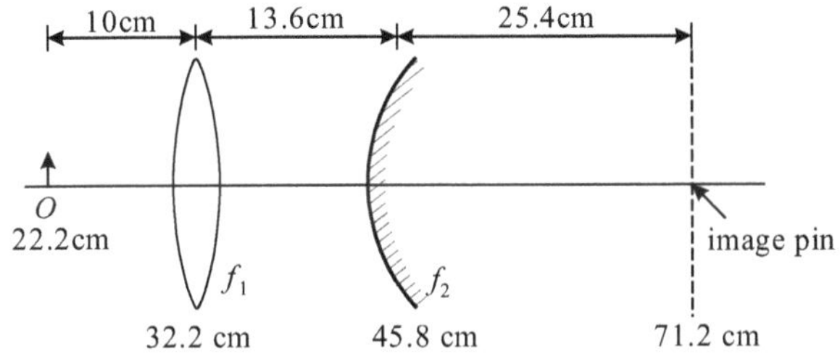

Option (C) and (D) are not possible as $f_2 \neq 25.4$ cm.

In option (A) and (B) $f_2 < 25.4$ cm $\Rightarrow$ Image produced by lens must be on right side of mirror

$\Rightarrow$ for lens $v > 13.6$ cm $\Rightarrow$ Real enlarged image produced by lens

$$\Rightarrow \quad f_1 < 10 \text{ cm}$$

As to produce enlarged real image ($|v| > |u|$) object must be located at

$$|f_1| < u < |2f_1|$$

$\Rightarrow$ Only possible option is (A) where $f_1 = 7.8$

& here (B) is also not possible as $f_1 > 10$ cm **Ans. (A)**

Sol. 14 First refraction at convex lens L_1

$u = -60$ cm lens

$f = +30$ cm formula

$$\frac{1}{f} = \frac{1}{v} - \frac{1}{u}$$

$$\Rightarrow \quad \frac{1}{v} = \frac{1}{30} - \frac{1}{60} = \frac{1}{60}$$

$$\Rightarrow \quad v = +60 \text{ cm}$$

Second refraction at concave lens L_2 for which object is image, produced by L_1

$u = +40$ cm lens

$f = -120$ cm formula

$$\frac{1}{f} = \frac{1}{v} - \frac{1}{u}$$

$$\Rightarrow \quad \frac{1}{v} = -\frac{1}{120} + \frac{1}{40} = \frac{1}{60}$$

$$\Rightarrow \quad v = +60 \text{ cm}$$

Third refraction at L_2 again for which I_3 is the object

$u = -40$ cm

$f = -120$ cm

Using lens formula

$$\frac{1}{f} = \frac{1}{v} - \frac{1}{u}$$

$$\Rightarrow \quad \frac{1}{v} = -\frac{1}{120} - \frac{1}{40} = -\frac{1}{30}$$

$$\Rightarrow \quad v = -30 \text{ cm}$$

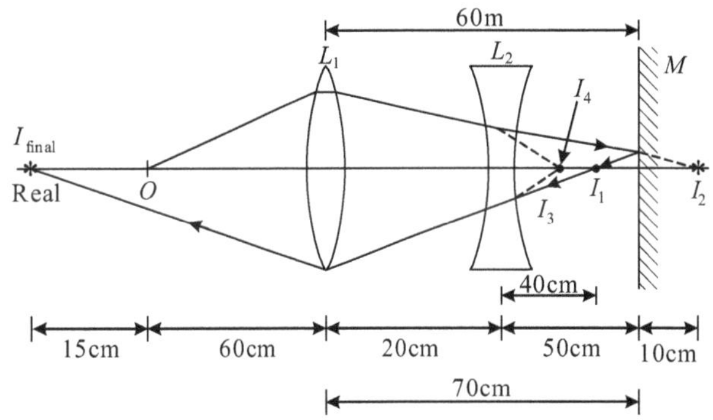

For the refraction at L_1 again for which I_4 is the object

$u = -50$ cm lens

$f = +30$ cm formula

$$\frac{1}{v} = \frac{1}{f} + \frac{1}{u} = \frac{1}{30} - \frac{1}{50} = \frac{5-3}{150} = \frac{1}{75}$$

$$\Rightarrow \quad v = 75 \text{ cm}$$

If whole combination ($L_1 + L_2 + M$) is used once then image produced I_3 is located at a distance 60 cm from the convex lens. So most appropriate answer here is option (A). **Ans. (A)**

Sol. 15

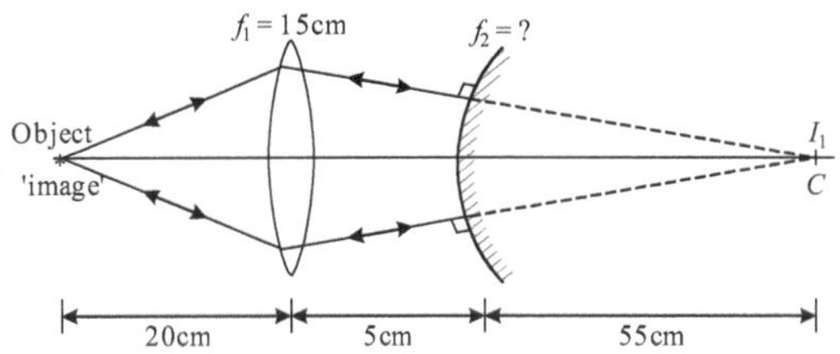

For refraction through lens we use lens formula for which we use

$u = -20$ cm

$f_1 = +15$ cm

$$\Rightarrow \quad \frac{1}{f} = \frac{1}{v} - \frac{1}{u}$$

$$\Rightarrow \quad \frac{1}{v} = \frac{1}{15} - \frac{1}{20} = \frac{1}{60}$$

$$\Rightarrow \quad v = 60 \text{ cm}$$

If image is produced on object then it can happen light rays retrace the path of incident ray after reflection or IR at mirror falls normally. For this I_1 must be at centre of curvature of mirror

$$\Rightarrow \quad R = 55 \text{ cm}$$

$$\Rightarrow \quad f_2 = \frac{R}{2} = 27.5 \text{ cm}$$ **Ans. (A)**

Sol. 16 If μ be the refractive index of lens and R be its radius of curvature then focal length of a thin plano convex lens is given by lens makers formula as

$$f = \frac{R}{(\mu-1)} \qquad \left[\text{As } \frac{1}{f} = (\mu-1)\left[\frac{1}{R_1} - \frac{1}{R_2}\right]\right] \quad (R_1 = R,\ R_2 = \infty)$$

Case-I : (Figure-A)

$$f_{\text{mirror}} = \infty$$

$$f_L = \frac{R}{\mu-1}$$

Effective focal length of lens system is given as

$$\frac{1}{f_{eq}} = \frac{2}{f_L} + \frac{1}{f_M}$$

$$\Rightarrow \quad f_{eq} = \frac{f_L}{2} = \frac{R}{2(\mu-1)}$$

$$\Rightarrow \quad \frac{R}{2(\mu-1)} = 28 \text{ cm} \qquad \ldots(1)$$

Case-II : (Figure-B)

$$f_M = \frac{R}{2}$$

$$f_L = \frac{R}{2(\mu-1)}$$

Equivalent focal length of lens system is given as

$$\frac{1}{f'_{eq}} = \frac{2}{f_C} + \frac{1}{f_M}$$

$$\Rightarrow \quad \frac{1}{f'_{eq}} = \frac{2(\mu-1)}{R} + \frac{2}{R} = \frac{24}{R}$$

$$\Rightarrow \quad f'_{eq} = \frac{R}{2\mu} = 10 \text{ cm} \qquad \ldots(2)$$

We use $\frac{(1)}{(2)}$

$$\Rightarrow \quad \frac{\mu}{\mu-1} = \frac{28}{10} = 2.8$$

$$\Rightarrow \quad \mu = 2.8\,\mu - 2.8$$

$$\Rightarrow \quad \mu = \frac{2.8}{1.8} = \frac{14}{9} = 1.55$$

Ans. (B)

Sol. 17

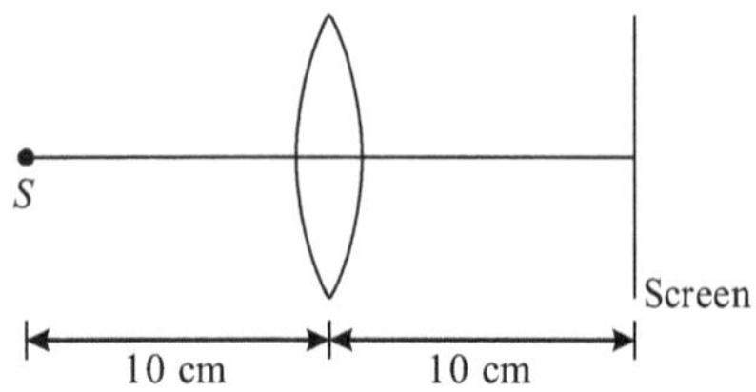

By lens formula, we have

$$\frac{1}{v} - \frac{1}{u} = \frac{1}{f}$$

$$\Rightarrow \quad \frac{1}{10} - \frac{1}{-10} = \frac{1}{f}$$

$$\Rightarrow \quad f = 5 \text{ cm}$$

Shift due to slab $= t\left(1 - \frac{1}{\mu}\right)$ in the direction of incident ray

$$\Rightarrow \quad S = 1.5\left(1 - \frac{2}{3}\right) = 0.5$$

Again using lens formula, we have

$$\frac{1}{v} - \frac{1}{-9.5} = \frac{1}{5}$$

$$\Rightarrow \quad \frac{1}{v} = \frac{1}{5} - \frac{2}{19} = \frac{9}{95}$$

$$\Rightarrow \quad v = \frac{95}{9} = 10.55 \text{ cm}$$

Ans. (A)

Sol. 18 By lens maker's formula, we have

$$\frac{1}{2f_2} = \frac{1}{f_1} = (\mu_1 - 1)\left(\frac{1}{\infty} - \frac{1}{-R}\right)$$

and
$$\frac{1}{f_2} = (\mu_2 - 1)\left(\frac{1}{-R} - \frac{1}{\infty}\right)$$

$$\Rightarrow \quad \frac{(\mu_1 - 1)}{R} = \frac{(\mu_2 - 1)}{2R}$$

$$2\mu_2 - \mu_1 = 1$$

Ans. (B)

Sol. 19 From lens formula, we have

$$\frac{1}{v} - \frac{1}{u} = \frac{1}{f}$$

$$\frac{1}{v} - \frac{1}{(-20)} = \frac{1}{(.3)} = \frac{10}{3}$$

$$\frac{1}{v} = \frac{10}{3} - \frac{1}{20}$$

$$\frac{1}{v} = \frac{197}{60};\ v = \frac{60}{197} \text{ m}$$

Lateral magnification by lens is

$$m = \left(\frac{v}{u}\right) = \frac{\left(\frac{60}{197}\right)}{20} = \frac{3}{197}$$

Velocity of image w.r.t. to lens is given by

$$v_I = m^2 v_O$$

Direction of velocity of image is same as that of object

As $\qquad v_O = 5$ m/s

$$\Rightarrow \quad v_I = \left(\frac{60\times1}{197\times20}\right)^2 (5)$$

$\Rightarrow \quad v_I = 1.16 \times 10^{-3}$ m/s toward lens **Ans. (C)**

Sol. 20 For first lens, using lens formula, we have

$$\frac{1}{v} - \frac{1}{-20} = \frac{1}{5}$$

$$v = \frac{20}{3} \text{ cm}$$

For second lens, using lens formula, we have

$$u_2 = \frac{20}{3} - 2 = \frac{14}{3} \quad \text{(object distance)}$$

$$\Rightarrow \quad \frac{1}{v_2} - \frac{1}{\frac{14}{3}} = \frac{1}{-5}$$

$\Rightarrow \quad v_2 = 70$ cm **Ans. (D)**

Sol. 21 Using lens marker's formula, new focal length of lens will be given as

$$\frac{1}{f} = (\mu_{rel} - 1)\left(\frac{1}{R_1} - \frac{1}{R_2}\right)$$

As focal length of lens will change hence image disappears from the screen. **Ans. (A)**

Sol. 22

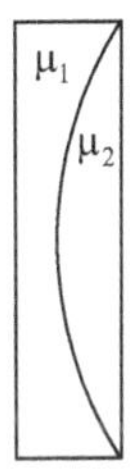

Focal length of this combination is given as

$$\frac{1}{f} = \frac{1}{f_1} + \frac{1}{f_2} = \frac{1-\mu_1}{R} + \frac{\mu_2 - 1}{R} = \frac{\mu_2 - \mu_1}{R}$$ **Ans. (C)**

Sol. 23

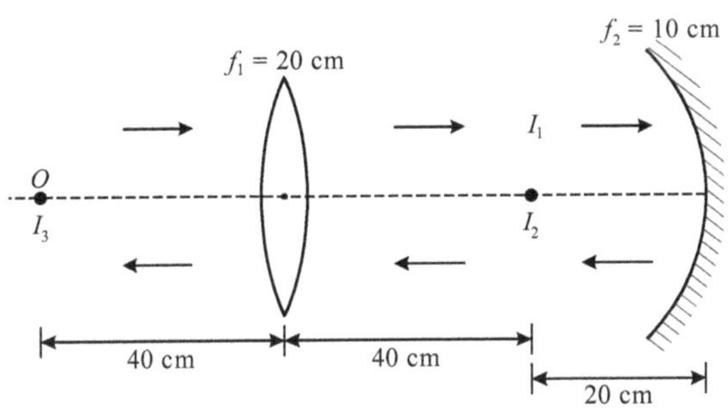

After all refractions and reflection final image is produced at object i.e. 40 cm from lens and of same size.

If I_2 is considered final image after one refraction and one reflection. Then **Ans. (None)**

Safe and most appropriate.

Sol. 24

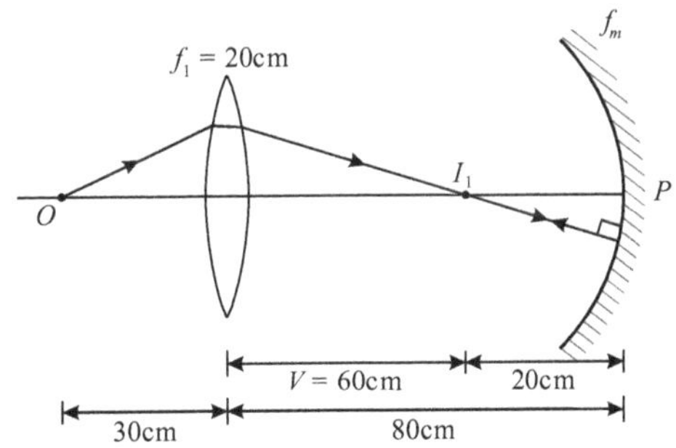

By lens formula we use

$$u = -30 \text{ cm}$$

$$f = +0\,20 \text{ cm}$$

$$\Rightarrow \quad \frac{1}{v} + \frac{1}{30} = \frac{1}{20}$$

$$\Rightarrow \quad V = +60 \text{ cm}$$

I_1 will be producing image at same position in cancave mirror if it is centre of curvature

$$\Rightarrow \quad R_m = 20 \text{ cm}$$

$$f_m = \frac{R_m}{2} = 10 \text{ cm}$$

A cancave mirror produces a virtual image for object to be located from P to F.

$u_{max} = f_m = 10$ cm **Ans. (B)**

Sol. 25 Initially we use

$$f_L = \frac{R}{2(\mu - 1)} R$$

We also use $\frac{1}{f_{eq}} = \frac{1}{f_L} + \frac{1}{\infty} + \frac{1}{f_L} = \frac{2}{f_L}$

$$\Rightarrow \quad f_L = 2f_{eq} = 18 \text{ cm} = R$$

After insertion of liquid, the focal length of plano concave lens

is $f_{L_2} = \frac{R}{\mu_1 - 1} = \frac{18}{\mu_1 - 1}$ (–ve) as diverging lens

equal focal length for the combination now is

$$\frac{1}{f'_{eq}} = \frac{2}{f_L} + \frac{2}{f_{L_2}} \quad \text{[here we use } 2f'_{eq} = 27 \text{ cm]}$$

$$\frac{1}{27/2} = \frac{2}{18} - \frac{2(\mu_1 - 1)}{18}$$

$$\Rightarrow \quad -\frac{\mu_1 - 1}{18} = \frac{1}{27} - \frac{1}{18} = \frac{2-3}{54} = -\frac{1}{54}$$

$$\Rightarrow \quad \mu_1 - 1 = \frac{1}{3}$$

$\Rightarrow \quad \mu_1 = 1 + \frac{1}{3} = \frac{4}{3}$ **Ans. (B)**

Sol. 26 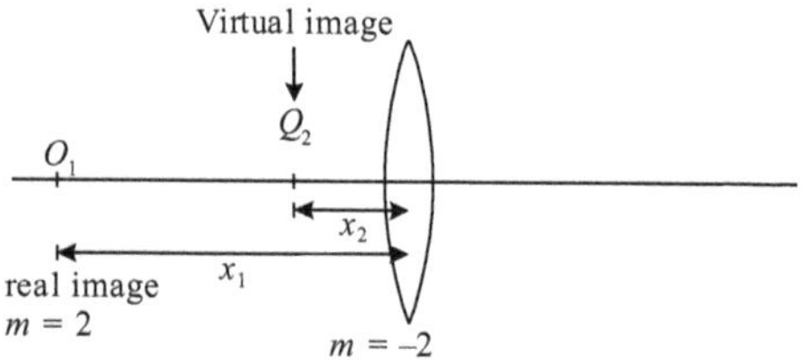

In first case we use

$$m = \frac{f}{x_1 - f} = 2$$

$$\Rightarrow \quad f = 2x_1 - 2f$$

$$\Rightarrow \quad x_1 = \frac{3f}{2} \qquad \ldots(1)$$

In second case we use

$$m = \frac{f}{x_2 - f} = -2$$

$$\Rightarrow \quad f = -2x_2 + 2f$$

$$\Rightarrow \quad x_2 = \frac{f}{2} \qquad \ldots(2)$$

$$(1)/(2) \Rightarrow \quad \frac{x_1}{x_2} = \frac{3}{1}$$

Ans. (D)

Sol. 27 By lens maker's formulas, focal lengths of the two lenses are

$$f_1 = \frac{R}{\mu_1 - 1} \text{ and } f_2 = -\frac{R}{\mu_2 - 1}$$

For combination, we use

$$\frac{1}{f_{eq}} = \frac{1}{f_1} + \frac{1}{f_2} = \frac{\mu_1 - 1}{R} - \frac{\mu_2 - 1}{R} = \frac{\mu_1 - \mu_2}{R}$$

$$\Rightarrow \quad f_{eq} = \frac{R}{\mu_1 - \mu_2}$$

Ans. (D)

Sol. 28 By lens formula, we have

$$\frac{1}{v} - \frac{1}{u} = \frac{1}{f}$$

$$\Rightarrow \quad 1 - \frac{v}{u} = \frac{v}{f}$$

$$\Rightarrow \quad 1 - m = \frac{v}{f}$$

$$\Rightarrow \quad m = 1 - \frac{v}{f}$$

At $v = a$,

$$m_1 = 1 - \frac{a}{f}$$

At $v = a + b$,

$$m_2 = 1 - \frac{a+b}{f}$$

$$m_2 - m_1 = c = \left(1 - \frac{a+b}{f}\right) - \left(1 - \frac{a}{f}\right)$$

In magnitude $c = \dfrac{b}{f}$

$$f = \frac{b}{c}$$

Ans. (D)

23.4 Dispersion of Light

Sol. 1 For a prism of refracting angle A, the refractive index 'μ' and minimum deviation angle D' are related as

$$\mu = \frac{\sin\left(\frac{A+D}{2}\right)}{\sin\left(\frac{a}{2}\right)}$$

As μ increases for a given prism ⇒ D increases

here $\mu_R = 1.520 \rightarrow D_1$

and $\mu_B = 1.525 \rightarrow D_2$

As $\mu_B > \mu_R$

⇒ $D_2 > D_1$ **Ans. (A)**

Sol. 2 We know that refractive index of a material, is related to light wavelength as

$$\mu = A + \frac{B}{\lambda^2}$$

So for blue light $\lambda_B < \lambda_R$

⇒ $\mu_B > \mu_R$

By lens makers formula, we use

$$\frac{1}{f} = (\mu - 1)\left[\frac{1}{R_1} - \frac{1}{R_2}\right]$$

As $\mu_B > \mu_R$

⇒ $f_B < f_R$ **Ans. (B)**

Sol. 3

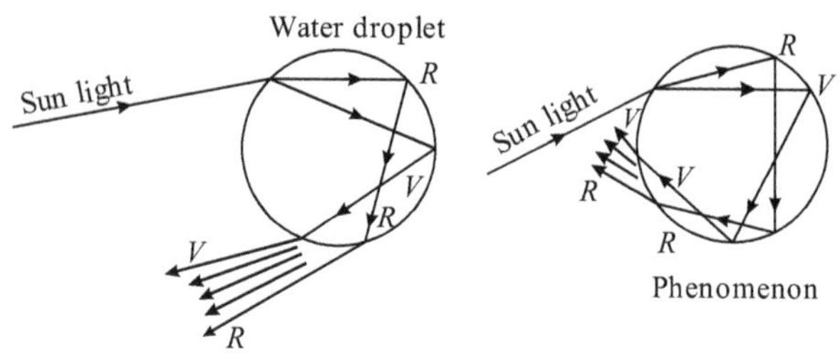

When light enters & emerges out from drop → refraction

At inner surface of drop → Internal reflection

inside the drop → dispersion.

Here internal reflection is not total internal reflection but for the given options most appropriate is option (A) only. **Ans. (A)**

Sol. 4

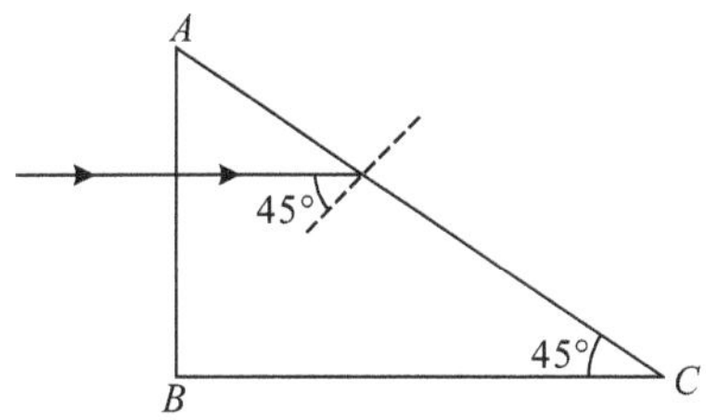

For *TIR* at face *AC*

$$\theta_C < 45°$$

$$\Rightarrow \quad \sin\theta_C < \frac{1}{\sqrt{2}}$$

$$\Rightarrow \quad \frac{1}{\mu} < \frac{1}{\sqrt{2}} \Rightarrow \mu > \sqrt{2}$$

Those colours which have $\mu > 1.414$ will suffer total internal reflection at face *AC* and will not come out.

$\Rightarrow$ Green & Blue colour light will not come out. **Ans. (D)**

Sol. 5 For doublet lenses at a separation d and with focal lengths f_1 and f_2 the equivalent focal length is given as

$$\frac{1}{f_{eq}} = \frac{1}{f_1} + \frac{1}{f_2} - \frac{d}{f_1 f_2}$$

Here $\quad f_{eq} = 10 \text{ cm}$

and $\quad d = 2 \text{ cm}$

$$\Rightarrow \quad \frac{1}{10} = \frac{1}{f_1} + \frac{1}{f_2} - \frac{2}{f_1 f_2} = \left(\frac{f_2 + f_1 - 2}{f_1 f_2}\right)$$

For option (A) $f_1 = 18$ cm & $f_2 = 20$ cm

$$\Rightarrow \quad \frac{1}{10} = \frac{20 + 18 - 2}{18 \times 20} = \frac{36}{18 \times 20} = \frac{1}{10}$$

$\Rightarrow \quad$ LHS = RHS **Ans. (A)**

Sol. 6 As for thin prism, minimum deviation is

$$D_m = (\mu - 1)A$$

So on increasing the wavelength, μ decreases & hence D_m decreases. **Ans. (B)**

23.5 Optical Instruments

Sol. 1 Resolving power of an astronomical telescope is given as

$$R_\lambda = \frac{D}{1.22\lambda}$$

$D \rightarrow$ Aperture diameter of objective lens

$\lambda \rightarrow$ Wavelength of light

An astronomical telescope is use to observe distant objects and heavenly bodies so resolution required is high so keep high resolving power D must be kept large. **Ans. (B)**

Sol. 2 For an optical instrument we know that

Resolving power $\propto \frac{1}{\lambda}$

$$\Rightarrow \quad \frac{R_{\lambda_1}}{R_{\lambda_2}} = \frac{\lambda_2}{\lambda_1} = \frac{5000\text{Å}}{4000\text{Å}} = \frac{5}{4}$$ **Ans. (D)**

Sol. 3 As object is placed in front of objective lens between F and $2F$ points $\Rightarrow$ image produced is real and enlarged. **Ans. (C)**

Sol. 4 Angular width of resolution

$$\theta = \frac{1.22\lambda}{D} = \frac{d}{x}$$

$$\Rightarrow \quad x = \frac{dD}{1.22\lambda} = \frac{10^{-3} \times 3 \times 10^{-3}}{1.22 \times 500 \times 10^{-9}}$$

$$\Rightarrow \quad x = \frac{3}{1.22 \times 0.5} = 4.92 \text{ m} = 5 \text{ m}$$ **Ans. (A)**

Sol. 5 For a travelling microscope readings are taken from meter scale on the vertical mount and vernier scale on the microscope.

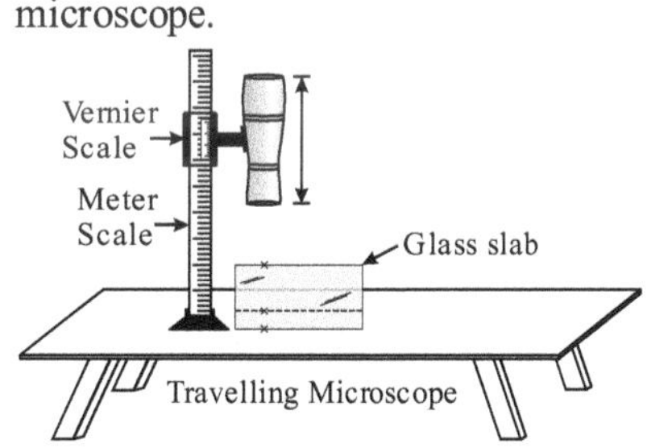

Ans. (A)

Sol. 6 For normal adjustment (to obtain final image at ∞) the image due to objective lens must be obtained at focal point of eyepiece lens as shown, lens formula for objective

$$f_e = +20 \text{ mm}$$

$$u_e = -25 \text{ mm}$$

$$\Rightarrow \quad \frac{1}{f_e} = \frac{1}{v_e} - \frac{1}{u_e}$$

$$\Rightarrow \quad \frac{1}{v_e} = \frac{1}{f_e} + \frac{1}{u_e} = \frac{1}{20} - \frac{1}{25} = \frac{1}{100}$$

$$\Rightarrow \quad v_e = 100 \text{ mm}$$

lens separation $= f_e + v_e$

$= 20 \text{ mm} + 100 \text{ mm} = 120 \text{ mm}$ **Ans. (C)**

Sol. 7 Limit of resolution of given telescope is

$$\theta_L = \frac{1.22\lambda}{D} = \frac{1.22 \times 5.5 \times 10^{-7}}{3 \times 10^{-2}}$$

$$\Rightarrow \quad \theta_L = 2.33 \times 10^{-5} \text{ rad}$$

The angle subtended by the given wire mesh on telescope is

$$\theta = \frac{x}{L} = \frac{2\times10^{-3}}{80} = 2.5\times10^{-5}\text{ rad}$$

As $\theta > \theta_L$ so the mesh spacing will be clearly seen. **Ans. (A)**

Sol. 8 Lens formula for objective lens

$$u_0 = -200\text{ cm}$$

$$f_0 = +50\text{ cm}$$

$$\Rightarrow \quad \frac{1}{f_0} = \frac{1}{v_0} - \frac{1}{u_0}$$

$$\Rightarrow \quad \frac{1}{50} = \frac{1}{v_0} + \frac{1}{200}$$

$$\Rightarrow \quad \frac{V}{v_0} = \frac{1}{50} - \frac{1}{200} = \frac{3}{200}$$

$$\Rightarrow \quad v_0 = \frac{200}{3}\text{ cm}$$

Lens formula for eyepiece lens

$$u_e = -u_e$$

$$v_e = -D = -25\text{ cm}$$

$$\Rightarrow \quad \frac{1}{f_e} = \frac{1}{v_e} - \frac{1}{u_e}$$

$$\Rightarrow \quad \frac{1}{5} = -\frac{1}{25} + \frac{1}{u_e}$$

$$\Rightarrow \quad \frac{1}{u_e} = \frac{1}{5} + \frac{1}{25} = \frac{6}{25}$$

$$u_e = \frac{25}{6}\text{ cm}$$

Magnification produced by objective lens

$$M_0 = \frac{v_0}{u_0} = \frac{+200/3}{-200} = -\frac{1}{3}$$

Magnification produced by objective lens

$$M_e = \frac{v_e}{u_e} = \frac{-25}{-25/6} = +6$$

Total magnification by telescope is

$$M_T = M_0 \times M_e = -\frac{1}{3}\times 6 = -2$$ **Ans. (D)**

Sol. 9 S-1: True as reflecting telescopes are easier to construct due to handling of spherical mirrors. As only one side of mirror is used so on other side the mirror can be mounted where in case of lenses mounting can only be done at thin edges which is practically difficult. Thats why very large size telescopes are made by using mirrors instead of lenses

$\Rightarrow$ **S-2 :** It is true and explaining S-1. **Ans. (C)**

Sol. 10 For standard adjustment (image at D) of a telescope the magnifying power is given as

$$M = \frac{f_0}{f_e}\left(1 - \frac{f_e}{D}\right)$$

$$\Rightarrow \quad M = \frac{30}{3}\left(1 - \frac{3}{25}\right)$$

$$\Rightarrow \quad M = \frac{30}{3}\times\frac{22}{25} = \frac{44}{5} = 8.8$$ **Ans. (D)**

Sol. 11 For normal adjustment (final image at ∞) the magnifying power of microscope is given as

$$M = \left|\frac{V_0}{u_0}\cdot\frac{D}{f_e}\right| \quad \ldots(1)$$

by using lens formula for objective lens, we have

$$\frac{1}{f_0} = \frac{1}{v_0} - \frac{1}{u_0}$$

$$\Rightarrow \quad \frac{1}{1.2} = \frac{1}{v_0} + \frac{1}{1.25}$$

$$\Rightarrow \quad \frac{1}{v_0} = \frac{1}{1.2} - \frac{1}{1.25} = \frac{1}{30}$$

$$\Rightarrow \quad v_0 = 30\text{ cm}$$

From-(1), we have

$$M = \frac{30}{1.25}\times\frac{25}{3} = 200$$ **Ans. (A)**

Sol. 12

$$D = 0.25\times 2$$

$$\Rightarrow \quad D = 0.5\text{ cm}$$

We use for limit of resolution we have

$$\theta = \frac{1.22}{D} = \frac{d}{x}$$

$$\Rightarrow \quad d = \frac{1.22\lambda x}{D} = \frac{1.22\times500\times10^{-9}\times0.25}{0.5\times10^{-2}}$$

$$\Rightarrow \quad d = 30.5\times10^{-6}\text{ m} = 30.5\ \mu\text{m}$$ **Ans. (B)**

Sol. 13 Magnifying power of telescope

$$M = \frac{\theta}{\alpha} = \frac{\text{Angle subtanded by image}}{\text{Angle subtanded by object}}$$

For normal setting we consider final image at ∞

from image $$\frac{\theta}{\alpha} = \frac{f_0}{f_e}$$

$$\Rightarrow \quad \theta = \frac{f_0}{f_e}\cdot\alpha$$

$$\Rightarrow \quad \theta = \frac{150}{5}\times\frac{1}{20}$$

$$\Rightarrow \quad \theta = 1.5\text{ rad}$$

$\Rightarrow \quad \theta = 1.5 \times \dfrac{180^\circ}{\pi} = 86^\circ$

Closest possible option is (C). **Ans. (C)**

Sol. 14 As $\quad m = \dfrac{\text{Image size}}{\text{object size}} = 20$

Angular magnification for normal condition
Angular size

Image size (Apparent size observer is scaling)
$= 20 \times$ object size **Ans. (C)**

Sol. 15 Step-I : A reading of bottom surface = 'a' without slab

Step-II : A reading of bottom surface with the slab = 'b'

Step-III : A reading of top surface of slab = 'c'

$\Rightarrow$ Thickness of slab $t = a - c$ and we use

$$(a-b) = (a-c)\left[1 - \frac{1}{\mu}\right]$$

$$\Rightarrow \quad 1 - \left(\frac{a-b}{a-c}\right) = \frac{1}{\mu}$$

$$\Rightarrow \quad \mu = \left(\frac{a-c}{b-c}\right)$$

Thus readings a, b and c are required. **Ans. (C)**

Sol. 16 Limit of resolution of telescope is given as

$$\theta = \frac{1.22\lambda}{D}$$

$$\Rightarrow \quad \theta = \frac{1.22 \times 500 \times 10^{-9}}{200 \times 10^{-2}}$$

$\Rightarrow \quad \theta = 305 \times 10^{-9}$ rad **Ans. (A)**

Sol. 17 Limit of resolution for objective of telescope is

$$\theta = \frac{1.22\lambda}{d}$$

$$\Rightarrow \quad \theta = \frac{1.22 \times 600 \times 10^{-9}}{250 \times 10^{-2}}$$

$\Rightarrow \quad \theta = 2.9 \times 10^{-7}$ rad **Ans. (C)**

Sol. 18 Numerical aperature of the microscope is given as

$$NA = \frac{0.61\lambda}{d}$$

Where d = minimum sparation between two points to be seen as distinct

$$\Rightarrow \quad d = \frac{0.61\lambda}{NA} = \frac{(0.61) \times (5000 \times 10\text{m}^{-10})}{1.25}$$

$\Rightarrow \quad d = 2.4 \times 10^{-7}$ m

$\Rightarrow \quad d = 0.24\ \mu$m **Ans. (A)**

24.1 Interference of Light

Sol. 1 Interference take place only when two or more coherent waves superpose

Coherence $\rightarrow$ The waves have exactly same frequency and constant (invarient in time) phase difference. **Ans. (D)**

Sol. 2

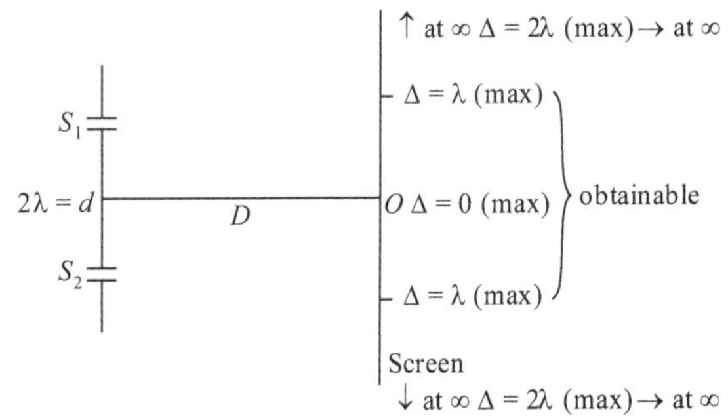

A maxima is obtained whenever path difference in waves from S_1 and S_2 is a multiple of λ, here out of 5 maxima only 3 can be obtained on screen in *YDSE*. **Ans. (C)**

Sol. 3 Case-I : If slits are considered as sources in *YDSE* for a source of light

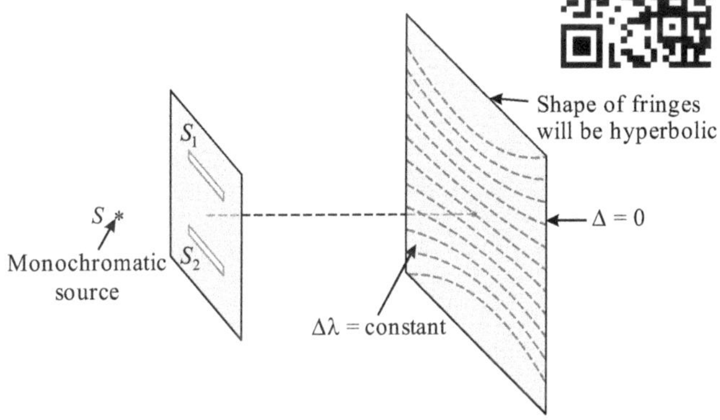

Because the locus of all points on screen which have path difference from S through S_1 and S_2 is constant is a hyperbola.

Case-II : If slits are considered as sources in *YDSE* for a line source

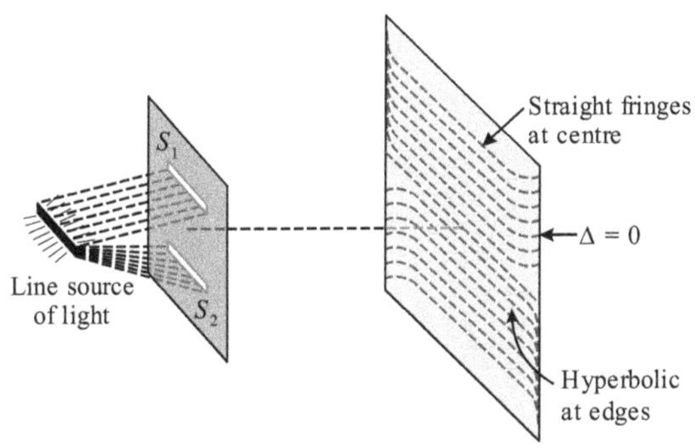

Here most appropriate is option (A). **Ans. (A)**

Sol. 4

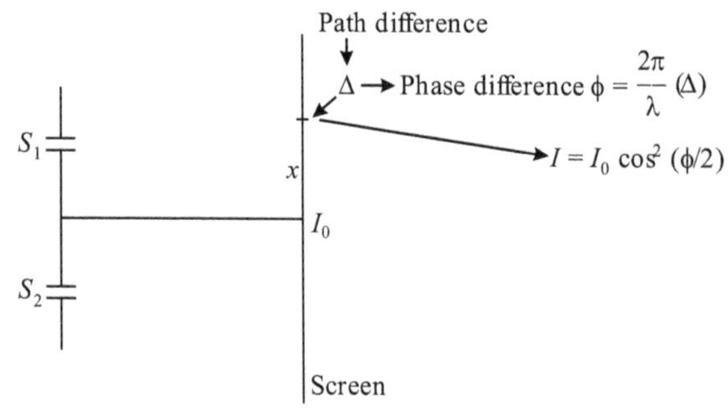

On either side of central maxima on screen the light intensity is given as

$$I = I_0 \cos^2(\phi/2) \quad \ldots(1)$$

Where $\phi \rightarrow$ phase difference between waves coming from S_1 and S_2 to that point for a point having path difference $\lambda/6$, then phase difference is

$$\phi = \frac{2\pi}{\lambda}(\Delta) = \frac{2\pi}{\lambda} \times \frac{\lambda}{6} = \frac{\pi}{3}$$

$\Rightarrow$ from equation-(1)

$$I = I_0 \cos^2\left(\frac{\lambda}{6}\right) = I_0\left(\frac{\sqrt{3}}{2}\right)^2 = \frac{3I_0}{4}$$

$$\Rightarrow \quad \frac{I}{I_0} = \frac{3}{4}$$

Ans. (D)

Sol. 5 For a given wavelength of light used in *YDSE*, the distance of nth bright fringe from central maxima on screen is given as

$$x_{nB} = \frac{n\lambda D}{d}$$

[where $D \rightarrow$ distance between slit plane and screen]
[$d \rightarrow$ slit separation]

For known wavelength (λ_1) and unknown wavelength (λ_2) we use

$$\frac{3\lambda_1 D}{d} = \frac{4\lambda_2 D}{d}$$

$$\Rightarrow \quad \lambda_2 = \frac{3\lambda_1}{4} = \frac{3}{4} \times 590\,\text{nm}$$

$$\Rightarrow \quad \lambda_2 = 442.5\,\text{nm}$$

Ans. (A)

Sol. 6 Initially $\rightarrow$ beam is cylindrical

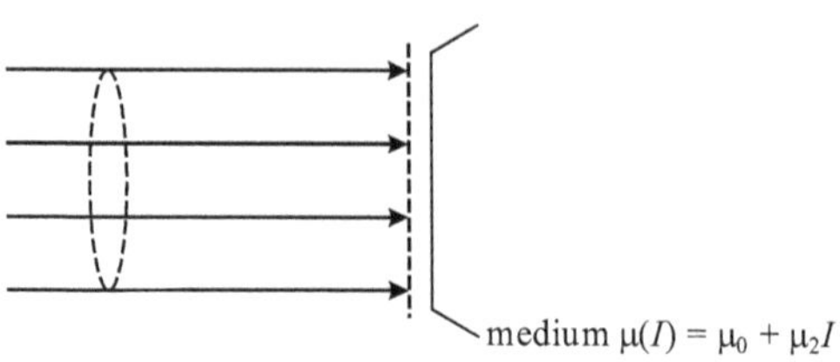

As light travel at same speed at all points of beam so wavefront will be planer. **Ans. (A)**

Sol. 7

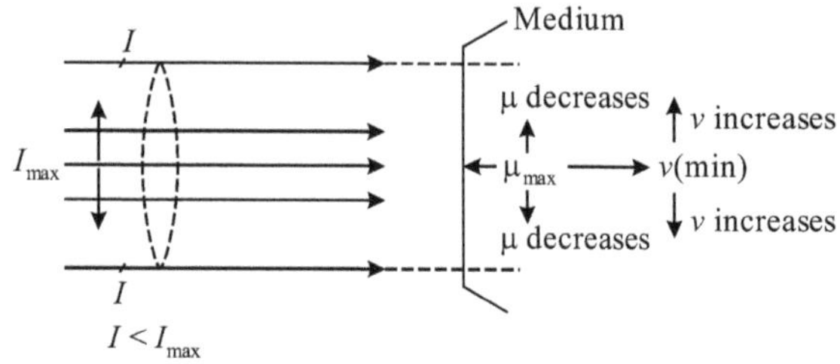

$$\mu(I) = \mu_0 + \mu_2 I \quad \ldots(1)$$

As on the two all sides of axis of beam as intensity of light decreases so from equation-(1) refractive index also decreases

$$\mu = \frac{\text{Speed of light in vaccum}}{\text{Speed in medium}} = \frac{c}{v}$$

$$\Rightarrow \quad v = \frac{c}{\mu}$$ **Ans. (B)**

Sol. 8

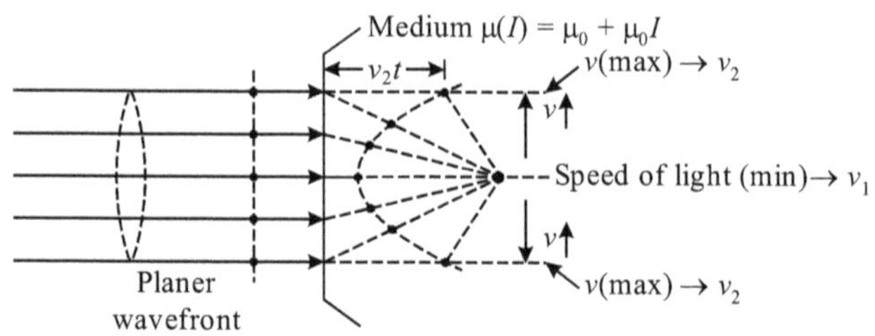

Concave wavefront after beam enters in medium

$\Rightarrow$ beam converges to a point. **Ans. (C)**

Sol. 9

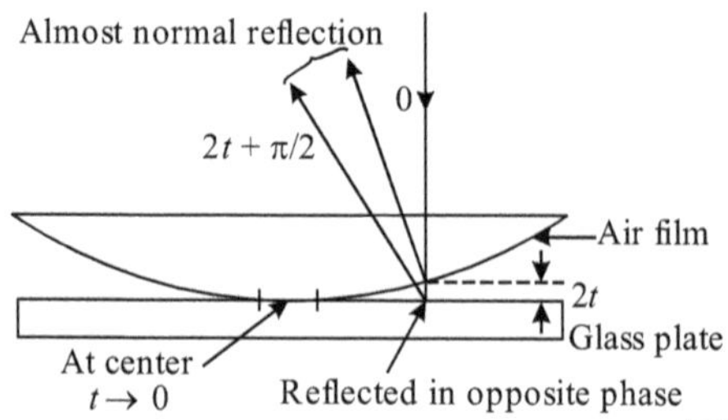

Statement-1 : It is true because a wave whenever reflected from a denser maximum suffers a phase change of π rad or a path delay of $\lambda/2$

$\Rightarrow$ $\Delta = \lambda/2 \rightarrow$ destructive interference $\Rightarrow$ "dark"

$\Rightarrow$ **Statement-2 :** is true & explaining statement-1. **Ans. (D)**

Sol. 10 In first case as sources are coherent
So constructive interference take place at screen centre so the resulting intensity will be given as

$$I_1 = I_{max} = (\sqrt{I_1} + \sqrt{I_2})^2 = 4I_0 \quad \ldots(1)$$

In second case as slits are in coherent sources so no interference take place and intensity will be

$$I_2 = I_R = I_1 + I_2 = 2I_0 \quad [I_0 = kA] \qquad \ldots(2)$$

Intensity due to each slit on screen

$$I_1 = I_2 = I_0$$

$$\frac{I_1}{I_2} = 2 \quad \textbf{Ans. (B)}$$

Sol. 11

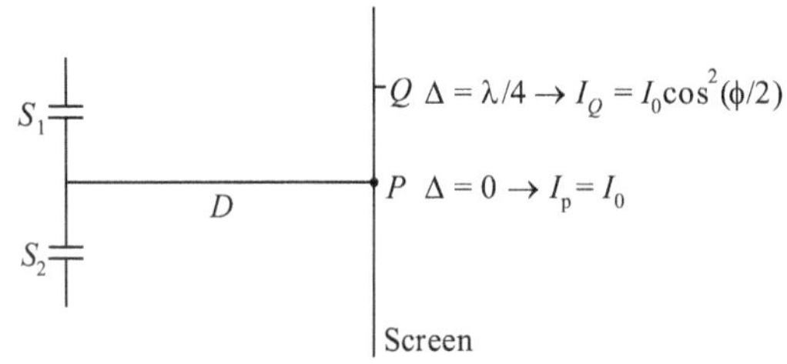

Intensity of light at a point on screen where phase difference between waves coming from S_1 and S_2 is & can be given as

$$I = I_0 \cos^2(\phi/2) \qquad \ldots(1)$$

At point P as $\Delta = 0 \Rightarrow \phi = 0$

So $I_P = I_0$

and at point Q as

$$\Delta = \lambda/4$$

$$\Rightarrow \quad \phi = \frac{2\pi}{\lambda}\cdot\Delta = \frac{2\pi}{\lambda}\left(\frac{\lambda}{4}\right) = \frac{\pi}{2}$$

$$\Rightarrow \quad I_Q = I_0\cos^2(\pi/4) = I_0\left(\frac{1}{\sqrt{2}}\right)^2 = \frac{I_0}{2}$$

$$\Rightarrow \quad \frac{I_P}{I_Q} = \frac{I_0}{I_0/2} = \frac{2}{1} \qquad \textbf{Ans. (B)}$$

Sol. 12 If light amplitude is double from one slit then its intensity will be four times so if light intensities from the two slits are I_1 and I_2 then we use

If $I_1 = I_0 \Rightarrow I_2 = 4I_0$ (from wider slit)

Maximum intensity on screen is given as

$$I_m = (\sqrt{I_1} + \sqrt{I_2})^2 = (\sqrt{I_0} + \sqrt{4I_0})^2 = 9I_0$$

$$\Rightarrow \quad I_0 = \frac{I_m}{9}$$

For a phase difference ϕ, then resulting intensity is given as

$$I_R = I_1 + I_2 + 2\sqrt{I_1 I_2}\cos\phi \qquad \ldots(1)$$

Using I_1 and I_2 in equation (1), we get

$$I_R = I_0 + 4I_0 + 2\sqrt{4I_0^2}\cos\phi$$

$$\Rightarrow \quad I_R = I_0(5 + 4\cos\phi)$$

$$\Rightarrow \quad I_R = \frac{I_m}{9}(5 + 4\cos\phi)$$

$$\Rightarrow \quad I_R = \frac{I_m}{9}(5 + 4(2\cos^2\phi/2 - 1))$$

$$\Rightarrow \quad I_R = \frac{I_m}{9}(1 + 8\cos^2\phi/2) \qquad \textbf{Ans. (A)}$$

Sol. 13

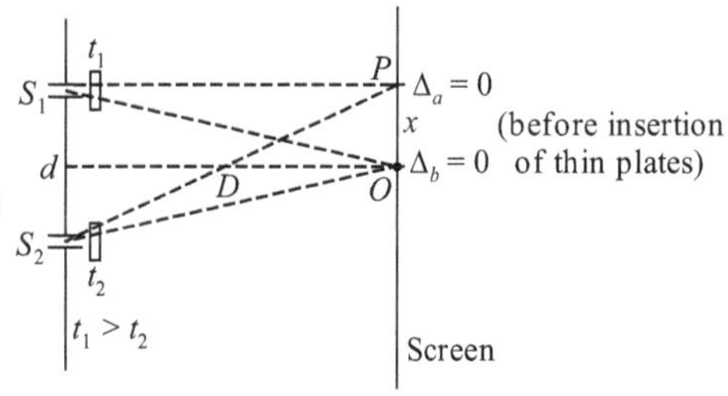

After insertion of thin glass plates in path of beams from S_1 and S_2 as shown, path difference at screen center becomes

$$\Delta_a = t_1(\mu - 1) - \mu_2(\mu - 1)$$

Say after insertion of plates at point P (above O) the physical path difference $\left(\frac{dx}{D}\right)$ will balance the optical path difference due to plates so finally effective path difference at P is zero at point P we use

$$\Delta_{ap} = \frac{dx}{D} - \Delta_a = 0$$

$$\Rightarrow \quad \frac{dx}{D} = t_1(\mu - 1) - t_2(\mu - 1)$$

$$\Rightarrow \quad x = \frac{D}{d}(\mu - 1)(t_1 - t_2) \qquad \ldots(1)$$

In *YDSE* fringe width is given as

$$\beta = \frac{\lambda D}{d} \Rightarrow \frac{D}{d} = \frac{\beta}{\lambda}$$

$$\Rightarrow \text{From (1)} \quad x = \frac{\beta}{\lambda}(\mu - 1)(t_1 - t_2) \qquad \textbf{Ans. (C)}$$

Sol. 14

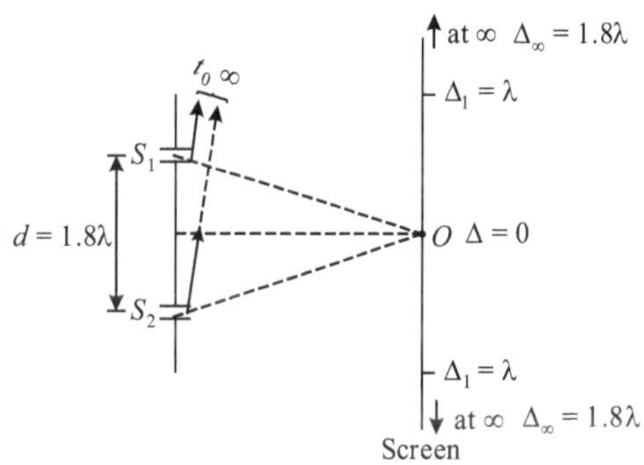

At ∞ on screen path difference in waves coming from states S_1 and S_2 is

$$\Delta_\infty = d = 1.8\lambda$$

between screen centre at ∞ somewhere path difference will be $\Delta_1 = \lambda$ on both sides of screen centre.

So at $\Delta = 0$ and $\Delta_1 = \lambda$ (two sides)

Constructive interference of light takes place

$\Rightarrow$ No of maxima = 3 **Ans. (B)**

Sol. 15 Intensity of light on *YDSE* screen due to a slit is proportional is slit width so we consider light intensities due to the two slits on screen are I_0 and 25 I_0.

The ratio of maximum to minimum intensities in interference pattern is given as

$$\frac{I_{max}}{I_{min}} = \left(\frac{\sqrt{I_1}+\sqrt{I_2}}{\sqrt{I_2}-\sqrt{I_1}}\right)^2$$

$$\Rightarrow \quad \frac{I_{max}}{I_{min}} = \left(\frac{5+1}{5-1}\right)^2 = \left(\frac{6}{4}\right)^2 = \frac{9}{4}$$ **Ans. (C)**

Sol. 16

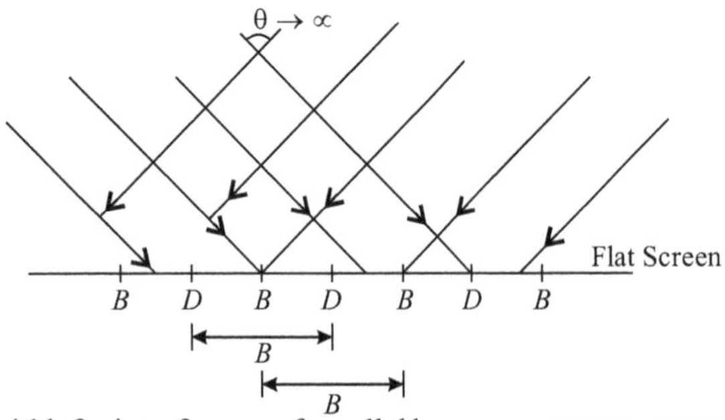

Fringe width for interference of parallel beams on a surface is given as

$$B = \frac{\lambda}{\alpha}$$ **Ans. (B)**

Sol. 17

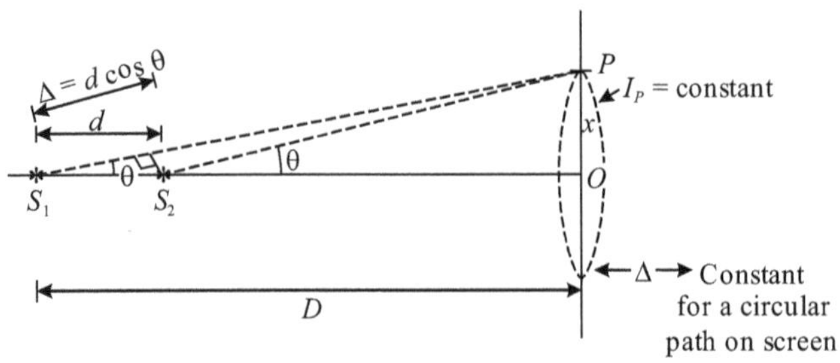

here $d << D$

On screen the locus of points where θ is a constant and at all these points path difference

$\Delta = d\cos\theta$ is a constant, will be a circle.

$\Rightarrow$ Fringes obtained are circular with centre at point O. **Ans. (D)**

Sol. 18 In *YDSE* fringe width is given as

$$\beta = \frac{\lambda D}{d}$$

$$\Rightarrow \quad \beta \propto \lambda$$

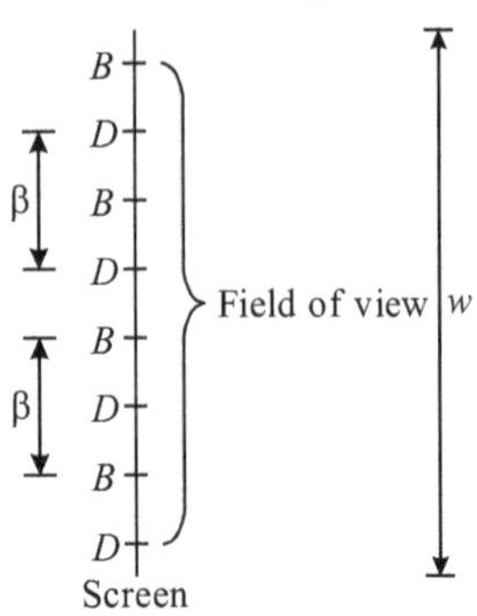

$$\text{No of fringes} = \frac{\text{field of view}}{\beta} = \frac{wd}{\lambda D}$$

$$\Rightarrow \quad \text{No of fringes} \propto \frac{1}{\lambda}$$

$\Rightarrow$ Statement-1 is true & statement-2 is also true and explaining statement-1. **Ans. (A)**

Sol. 19 For interference maxima of wavelength λ_1, the path difference we use bright fringe

$$\Delta = m\lambda_1 \qquad \ldots(1)$$

For interference minima (Dark fringe) of wavelength λ_2, the path difference we use

$$\Delta = (2n+1)\frac{\lambda_2}{2} \qquad \ldots(2)$$

At one point on screen if bright fringe of λ, coincides with dark fringe of λ_2 then we use

$$\Rightarrow \quad \Delta = m\lambda_1 = (2n+1)\frac{\lambda_2}{2}$$

$$\Rightarrow \quad \frac{m}{2n+1} = \frac{\lambda_2}{2\lambda_1} = \frac{600}{2\times 500} = \frac{3}{5}$$

For least distance from central maxima where these fringes coincide will be corresponding if

$$m = 3$$

and $$2n+1 = 5$$

$$\Rightarrow \quad \Delta = 3\lambda_1 = 3\times 500 = 1500\,\text{nm}$$ **Ans. (D)**

20.

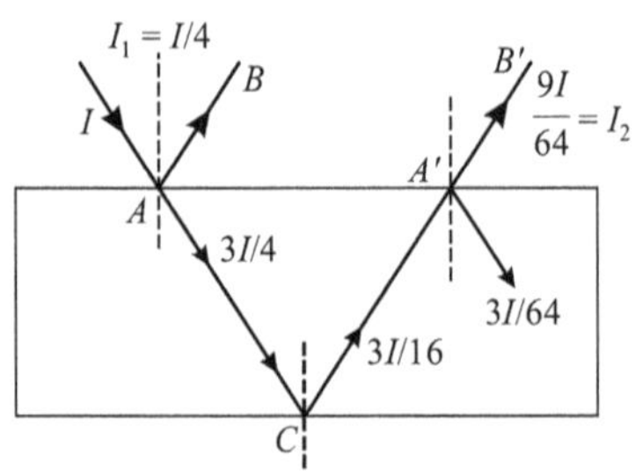

When rays AB and $A'B'$ interfere then the ratio of maximum to minimum intensity in interference pattern is given as

$$\frac{I_{max}}{I_{min}} = \left(\frac{\sqrt{I_1}+\sqrt{I_2}}{\sqrt{I_1}-\sqrt{I_2}}\right)^2 = \left(\frac{\sqrt{\frac{1}{4}}+\sqrt{\frac{9}{64}}}{\sqrt{\frac{1}{4}}-\sqrt{\frac{9}{64}}}\right)^2$$

$$\Rightarrow \quad \frac{I_{max}}{I_{min}} = \left(\frac{\frac{1}{2}+\frac{3}{8}}{\frac{1}{2}-\frac{3}{8}}\right)^2$$

$$\Rightarrow \quad \frac{I_{max}}{I_{min}} = \left(\frac{7}{1}\right)^2 = \frac{49}{1}$$ **Ans. (A)**

Sol. 21 For n coherent waves when, interfere constructively then resulting intensity is given as

$$I_{max} = n^2 I_0 \quad \ldots(1)$$

for n non-coherent waves when superpose then resulting intensity at the point of superposition is given as

$$I'_{max} = nI_0 \quad \ldots(2)$$

$$\Rightarrow \quad \frac{I_{max}}{I_{max}} = \frac{n^2 I_0}{nI_0} = n$$

Ans. (D)

Sol. 22 If I_1 and I_2 are interfering intensities then maxima & minima intensities are given as

$$I_{max} = (\sqrt{I_1} + \sqrt{I_2})^2$$

and $$I_{min} = (\sqrt{I_1} - \sqrt{I_2})^2$$

$$\Rightarrow \quad \frac{I_{max}}{I_{min}} = \left(\frac{4+3}{4-1}\right)^2 = \left(\frac{7}{1}\right)^2 = \frac{49}{1}$$

Ans. (D)

Sol. 23

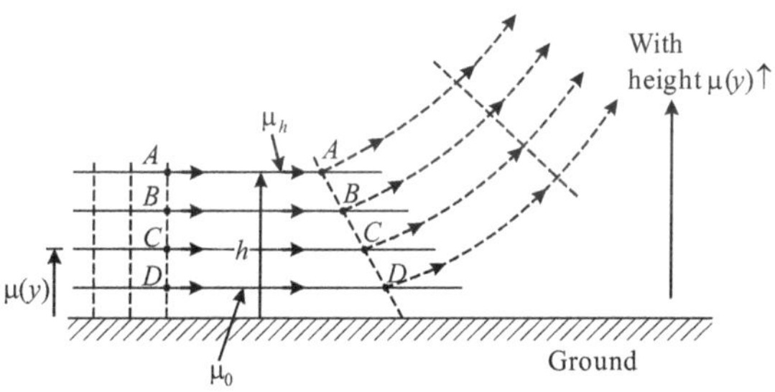

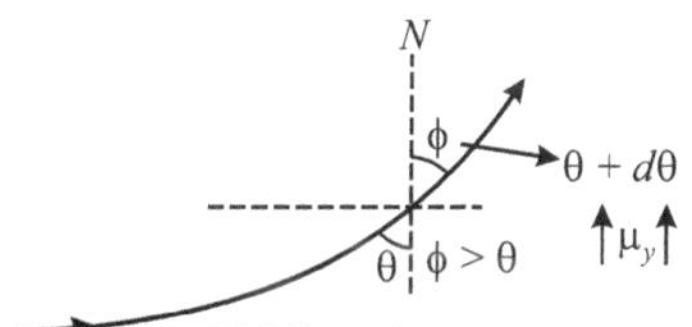

Height trajectory

$\mu_h > \mu_0 \Rightarrow v$ (speed of light) decreases with height so the speed of light at some height is lesser then compared to ground. **Ans. (D)**

Sol. 24

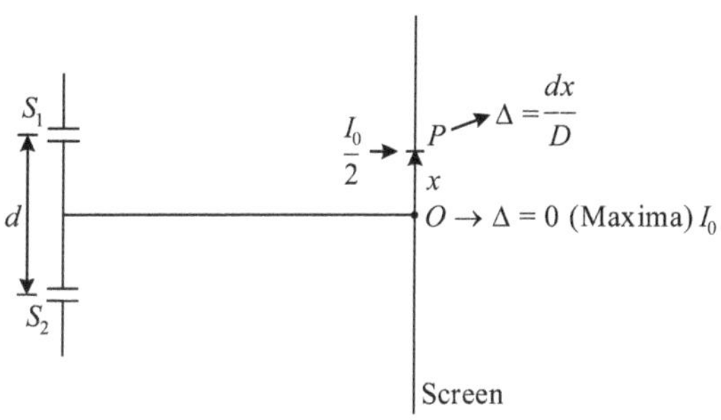

In *YDSE* fringe width is given as

$$\beta = \frac{\lambda D}{d} \quad \ldots(1)$$

The path difference between waves reaching point P at a distance x from screen centre is

$$\Delta_P = \frac{dx}{D}$$

$\Rightarrow$ Phase difference at P

$$\phi_P = \frac{2\pi}{\lambda}\cdot\Delta_P = \frac{2\pi}{\lambda}\cdot\frac{dx}{D} \quad \ldots(2)$$

If phase difference at point P is ϕ_P

$\Rightarrow$ Resulting light intensity at P is given as

$$I_P = I_0 \cos^2\left(\frac{\phi_P}{2}\right) = \frac{I_0}{2}$$

$$\Rightarrow \quad \cos^2\left(\frac{\phi_P}{2}\right) = \frac{1}{\sqrt{2}}$$

$$\Rightarrow \quad \frac{\phi_P}{2} = \frac{\pi}{4} \quad \text{or} \quad \phi_P = \frac{\pi}{2}$$

From equation-(2) we have

$$\phi_P = \frac{2\pi}{\lambda}\frac{dx}{D} = \frac{\pi}{2}$$

$$\Rightarrow \quad x = \frac{D\lambda}{4d} = \frac{\beta}{4}$$

Ans. (C)

Sol. 25 In *YDSE* the distance of nth bright fringe from central maximum is given as

$$x_{nB} = \frac{n\lambda D}{d}$$

Where $D = 1.5\text{m}$

$d = 0.5 \times 10^{-3}\,\text{m}$

If n^{th} bright fringe of λ, coincides with n_2^{th} bright fringe of λ_2 then we use

$$x_{n_1B} = x_{n_2B}$$

$$\Rightarrow \quad \frac{n_1\lambda_1 D}{d} = \frac{n_2\lambda_2 D}{d}$$

$$\Rightarrow \quad \frac{n_1}{n_2} = \frac{\lambda_2}{\lambda_1} = \frac{520}{560} = \frac{4}{5}$$

Last distance from central maxima of coinciding bright fringe is

$$x_{n_1B} = \frac{4 \times 650 \times 10^{-9} \times 1.5}{0.5 \times 10^{-3}}$$

$n_1 = 4$

$n_2 = 5$

$$\Rightarrow \quad x_{n_1B} = 7.8 \times 10^{-3}\text{m}$$

Ans. (D)

Sol. 26 Ratio of maximum to minimum intensities of interference is given as

$$\frac{I_{max}}{I_{min}} = 16$$

If I_1 and I_2 are intensities of compoment waves, we have

$$\Rightarrow \quad \frac{I_{max}}{I_{min}} = \frac{\sqrt{I_1}+\sqrt{I_2}}{\sqrt{I_1}-\sqrt{I_2}} = 4$$

$$\Rightarrow \quad \sqrt{\frac{I_1}{I_2}} = \frac{5}{3}$$

$$\Rightarrow \quad \frac{I_1}{I_2} = \frac{25}{9}$$

Ans. (B)

Sol. 27

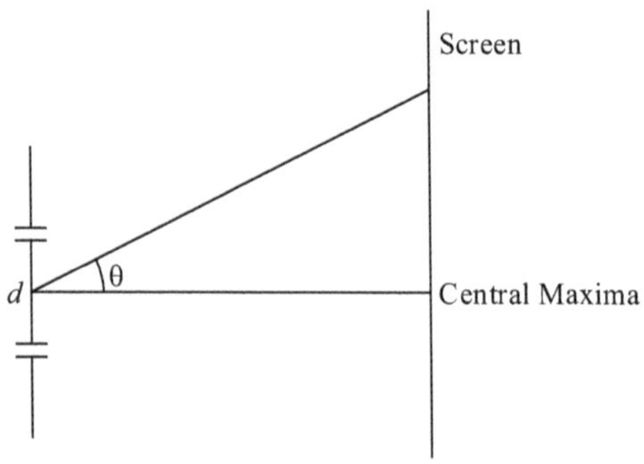

Path difference for angular position θ is

$$d \sin \theta = n\lambda$$

$$0.32 \times 10^{-3} \sin (30°) = n \times 500 \times 10^{-9}$$

$$n = 320$$

Thus including central maxima there will be 641 fringes seen in the given region. **Ans. (B)**

Sol. 28 Path difference

$$\Delta = d \sin\theta \approx d\theta$$

$$\Rightarrow \quad \Delta = 0.1 \times \frac{1}{40} \text{mm} = 2500 \text{ nm}$$

On bright fringe, path difference must be integral multiple of λ.

$$\Rightarrow \quad 2500 = n\lambda_1 = m\lambda_2$$

$$\Rightarrow \quad \lambda_1 = 625,$$

$$\lambda_2 = 500 \text{ (from } m = 5 \text{ and } n = 4)$$

Ans. (B)

Sol. 29 For first minima in front of one slit, path difference at point P is

$$\sqrt{5}d - 2d = \frac{\lambda}{2}$$

$$\Rightarrow \quad d = \frac{\lambda}{2(\sqrt{5}-2)}$$

Ans. (D)

Sol. 30 Given that path difference at the point

$$\Delta x = \frac{\lambda}{8}$$

Phase difference at this point is

$$\Delta\phi = \frac{(2\pi)}{\lambda} \times \frac{\lambda}{8} = \frac{\pi}{4}$$

Intensity at this point is

$$I = I_0 \cos^2\left(\frac{\Delta\phi}{2}\right) = I_0 \cos^2\left(\frac{\pi}{8}\right)$$

$$\Rightarrow \quad \frac{I}{I_0} = \cos^2\left(\frac{\pi}{8}\right) = \frac{1+\cos\left(\frac{\pi}{4}\right)}{2}$$

$$\Rightarrow \quad \frac{I}{I_0} = \frac{1+0.707}{2} = 0.85$$

Ans. (C)

Sol. 31 For diffraction, location of 1st minima

$$y_1 = \frac{D\lambda}{b} = 0.2469\, D\lambda$$

location of 2nd minima is

$$y_2 = \frac{2D\lambda}{b} = 0.4938 D\lambda$$

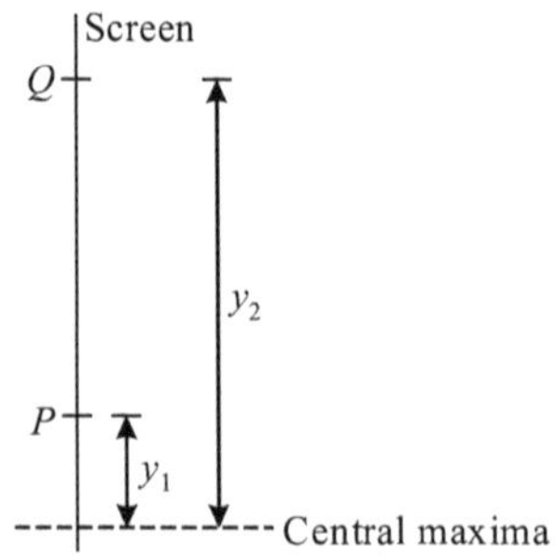

Now for interference

Path difference at point P on screen is

$$\frac{dy}{D} = 4.8\,\lambda$$

Path difference at point Q on screen is

$$\frac{dy}{D} = 9.6\,\lambda$$

So orders of maxima in between P & Q is are 5, 6, 7, 8, 9

Thus 5 bright fringes are present between P & Q. **Ans. (D)**

Sol. 32 Given that ratio of amplitudes is

$$\frac{a_1}{a_2} = \frac{1}{3}$$

Ratio of intensities of the waves can be given as

$$\frac{I_1}{I_2} = \left(\frac{a_1}{a_2}\right)^2 = \frac{1}{9}$$

Ratio of maximum to minimum intensities is given as

$$\frac{I_{max}}{I_{min}} = \left(\frac{\sqrt{I_1}+\sqrt{I_2}}{\sqrt{I_1}-\sqrt{I_2}}\right)^2 = \left(\frac{1+3}{1-3}\right)^2 = 4$$

Ans. (A)

Sol. 33 If we introduce a thin sheet in path of light from one of the slits then path difference introduced is

$$\Delta = t(\mu - 1)$$

If fringe pattern displaces by x, we use

$$\Delta = \frac{xd}{D}\left(\frac{\lambda D}{d}\right) = n\lambda$$

$\Rightarrow \quad t(\mu - 1) = n\lambda$

$\Rightarrow \quad t = \frac{n\lambda}{\mu - 1}$ **Ans. (None)**

Sol. 34 If I_1 and I_2 are light intensities from the two slits, we use

$$I_2 = I_0$$

$\Rightarrow \quad I_1 = 4I_0$

We use $\frac{I_{max}}{I_{min}} = \left[\frac{\sqrt{I_1} + \sqrt{I_2}}{\sqrt{I_1} - \sqrt{I_2}}\right]^2$

$\Rightarrow \quad \frac{I_{max}}{I_{min}} = \left[\frac{2\sqrt{I_0} + \sqrt{I_0}}{2\sqrt{I_0} - \sqrt{I_0}}\right]^2 = \frac{9}{1}$ **Ans. (B)**

Sol. 35

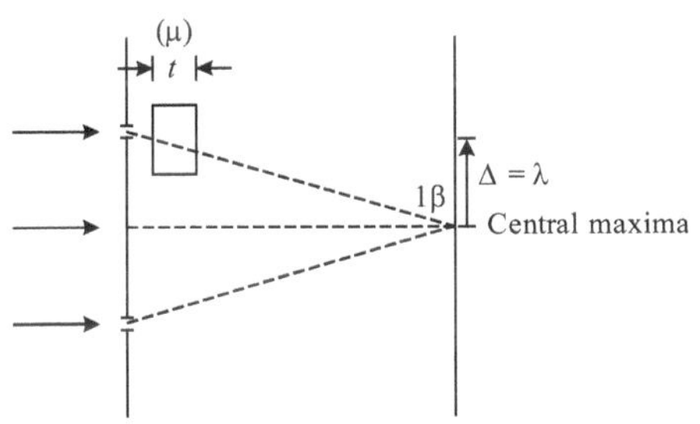

Path difference due to introducing a thin film is given as

$$\Delta = t(\mu - 1)$$

As shift of central maxima is by one fringe width

$\Rightarrow \quad \Delta = \lambda$

$t(\mu - 1) = \lambda$

$\Rightarrow \quad t = \frac{\lambda}{\mu - 1}$ **Ans. (D)**

24.2 Diffraction of Light

Sol. 1

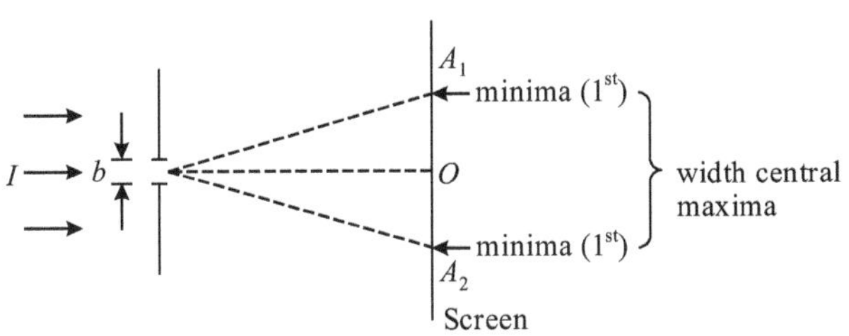

Intensity of central maxima on screen.

$$I_0 = \frac{\text{Power incident on central max}}{\text{area of screen } (A_1 A_2)}$$

If slit width is doubled then power incident on screen will get doubled and we use $\theta = \frac{\lambda}{b}$ (for central maxima)

So for double slit width we have angular separation $\to \theta'$

$$= \frac{\lambda}{2b} = \frac{\theta}{2}$$

$\Rightarrow$ On doubling slit width the area of central maxima will reduce to half

So from equation we use

$$I_{new} = \frac{P \text{ on central max} \times 2}{\text{area } A_1 A_2 \times 1/2} = 4I_0$$ **Ans. (B)**

Sol. 2 If single slit diffraction for n^{th} minima on screen we use

$b \sin\theta = n\lambda$ $[b \to \text{width of slit}]$

For first minima, we use

$$n = 1$$

$\Rightarrow \quad b \sin\theta = \lambda$

$\Rightarrow \quad b = \frac{\lambda}{\sin\theta} = \frac{5 \times 10^{-7}}{0.5} = 10 \times 10^{-7}\ \text{m}$

$\Rightarrow \quad b = 10 \times 10^{-5}\ \text{cm}$ **Ans. (C)**

Sol. 3 **Statement-1 :** As we know wavelength of radio waves is more then that of microwaves so the diffraction possibility is more for radio waves because diffraction is more in case of wavelength of the order of obstacles in surrounding or in path of waves.

$\Rightarrow$ **Statement-1 :** is true

Statement-2 : As both are EM waves so for

$$\lambda_R > \lambda_\mu$$

$\Rightarrow \quad \nu_R < \nu_\mu$

$\Rightarrow$ **Statement-2 :** is false **Ans. (C)**

Sol. 4 The minimum separation between two lines which can be just resolved by a microscope is given as

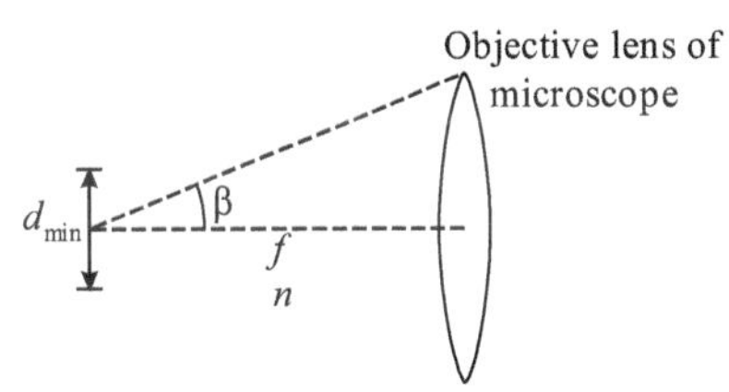

$$d_{min} = \frac{1.22\lambda}{2n\sin\beta}$$

Resolving power of microscope

$$R = \frac{1}{d_{min}} = \frac{2n\sin\beta}{1.22\lambda}$$

$$\sin 2\beta = 2\sin\beta\cos\beta$$

& for small β, we use

$$\cos\beta \simeq 1$$

$\Rightarrow \quad R \propto n \sin 2\beta$ **Ans. (C)**

Sol. 5

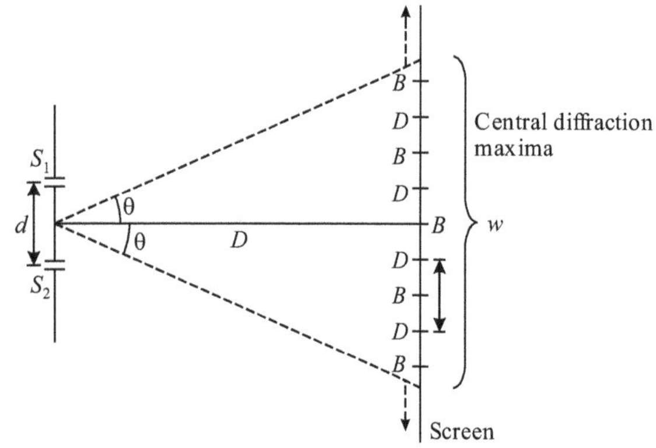

$d = 6.1b$

$b \rightarrow$ slit width

As interference of significant intensity of tight from the two slits will take place in the region of central diffraction maxima as shown in figure

Fringe width of *YDSE* pattern is

$$\beta = \frac{\lambda D}{d} \quad \ldots(1)$$

Angular width of central diffraction maxima is given as

$$2\theta = \frac{2\lambda}{b}$$

Width of central diffraction maxima is

$$w = 2\theta D = \frac{2\lambda D}{b} \quad \ldots(2)$$

No of bright fringes in width of central maxima is

$$N = \frac{w}{\beta} = \frac{2\lambda D}{b} \times \frac{d}{\lambda D}$$

We use $\qquad d = 1b$

$$\Rightarrow \quad N = \frac{2 \times 1.1b}{b} = 12.2$$

$\Rightarrow$ Complete 12 bright fringes are obtained in width of central maxima. **Ans. (C)**

Sol. 6 In a single slit diffraction pattern if θ is angular position of a point on screen from central maxima then we have

For a wavelength λ_1 (red light) at the position of first minimum we use

$$b \sin \theta = \lambda_1 \quad \ldots(1)$$

For one there wavelength λ_1 at the same angle θ for first maximum, we use

$$b \sin \theta = \frac{3\lambda_2}{2} \quad \ldots(2)$$

We use from (1) and (2)

$$\lambda_1 = \frac{3\lambda_2}{2}$$

$$\Rightarrow \quad \lambda_2 = \frac{2\lambda_1}{3} = \frac{2 \times 6600}{3}$$

$$= 4400 \text{ Å}$$

Ans. (B)

Sol. 7

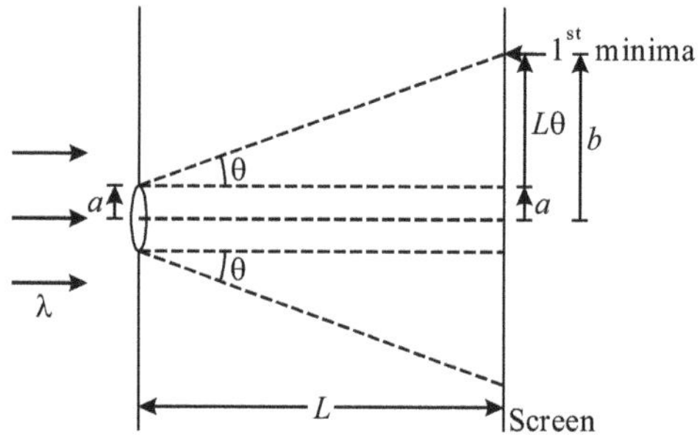

The angular position of first diffraction minima on either side of the hole is

$$a \sin \theta = \lambda \qquad [n = 1]$$

$$\sin \theta = \frac{\lambda}{a}$$

For small θ we use

$$\theta = \frac{\lambda}{a} \qquad [\sin \theta \simeq \theta]$$

The radius of spot can be given as

$$b = L\theta + a$$

Spread due to differaction = Geometric spread

$$\Rightarrow \quad b = \frac{L\lambda}{a} + a$$

b is minimum when

$$\frac{db}{da} = 0 = -\frac{L\lambda}{a^2} + 1$$

$$\Rightarrow \quad a = \sqrt{L\lambda}$$

So $\qquad b_{min} = \frac{L\lambda}{\sqrt{L\lambda}} + \sqrt{L\lambda} = 2\sqrt{L\lambda}$ **Ans. (C)**

Sol. 8

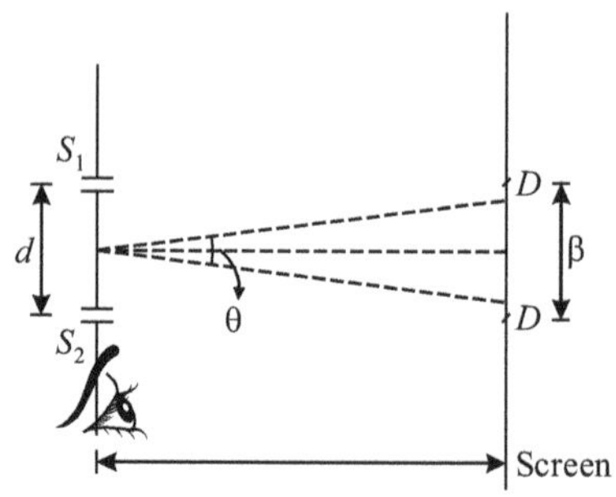

Angular fringe width in *YDSE* is given as

$$\theta = \frac{\beta}{D} = \frac{(\lambda D / d)}{D} = \frac{\lambda}{d}$$

Fringes will not be clearly seen by observer if $\theta < \theta_R$ (resolution limit of eye)

The fringes will just disappear at $d = d_0$

$\Rightarrow$ We use $\theta = \theta_R$ at $d = d_0$

$$\frac{\lambda}{d_0} = \frac{1}{60} \times \frac{\pi}{180}$$

$$\Rightarrow \quad d_0 = \frac{60 \times 180 \times 600 \times 10^{-9}}{3.14}$$

$\Rightarrow \quad d_0 = 2.06 \times 10^{-3}\,\text{m}$

$\Rightarrow \quad d_0 = 2.06\,\text{mm}$ **Ans. (C)**

Sol. 9 The angular resolution limit of a telescope is given as

$$\theta_R = \frac{1.22\lambda}{D}$$

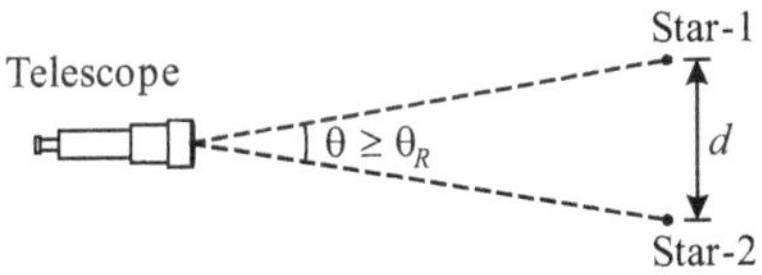

The two starts will be just resolved by the telescope when

$$L = 10 \text{ light yr}$$

$$\theta = \frac{d}{L}$$

$$\theta = \theta_R$$

$$\Rightarrow \quad \frac{d}{L} = \frac{1.22\lambda}{D}$$

$$\Rightarrow \quad d = \frac{1.22\lambda}{D}$$

$$\Rightarrow \quad d = \frac{1.22 \times 600 \times 10^{-9} \times 10 \times 9.46 \times 10^{15}}{0.3}$$

$\Rightarrow \quad d = 2.308 \times 10^{11}\,\text{m}$

$\Rightarrow \quad d = 2.308 \times 10^{8}\,\text{km}$ **Ans. (A)**

Sol. 10

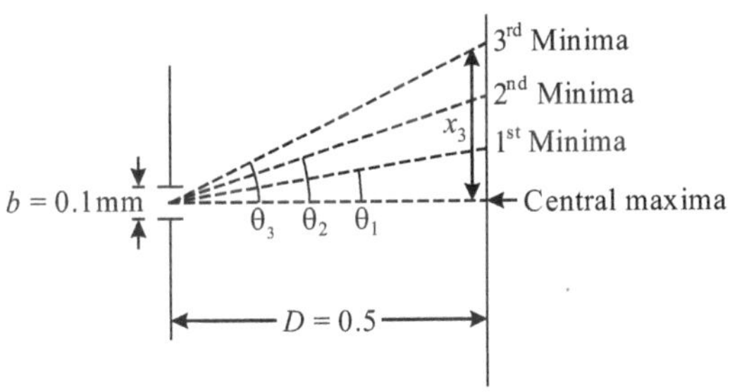

The angular separation of nth minima from screen centre (central maxima) is given as

$$b \sin \theta_n = n\lambda$$

from 3rd dark band ($n = 3$) we have

$$\theta_3 \simeq \sin \theta_3 = \frac{3\lambda}{b} \quad [\text{as } \theta \text{ is } v \text{ small}]$$

distance of 3rd minima from O is

$$x_3 = D\theta_3 = 0.5 \times \frac{3 \times 6000 \times 10^{-10}}{0.1 \times 10^{-3}}$$

$\Rightarrow \quad x_3 = 9 \times 10^{-3}\,\text{m} = 9\,\text{mm}$ **Ans. (B)**

Sol. 11

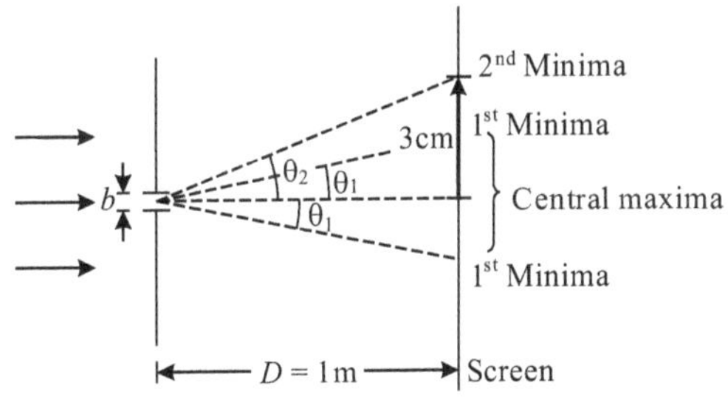

For first minima in diffraction pattern we use

$$b \sin \theta = n\lambda \qquad \ldots(1)\ [n = 1]$$

The angular position of second minimum from screen centre is given as

$$\tan \theta_2 \simeq \sin \theta_2 \simeq \theta_2 = \frac{3\text{cm}}{100\text{cm}}$$

$$\Rightarrow \quad \theta_2 = \frac{3\text{cm}}{1\text{cm}} = 0.03$$

from equation (1), for 2nd minima, we use

$$b \sin \theta_2 = 2\lambda$$

$$\Rightarrow \quad \theta_2 = \frac{2\lambda}{b} = 0.03$$

Angular width of central maxima in diffraction pattern is given as

$$2\theta_1 = \frac{2\lambda}{b} = 0.03$$

width of central maxima is

$$w = 2\theta_1 D = 0.03 \times 1\,\text{m} = 0.03\,\text{m} = 3\,\text{cm}$$

Ans. (B)

Sol. 12

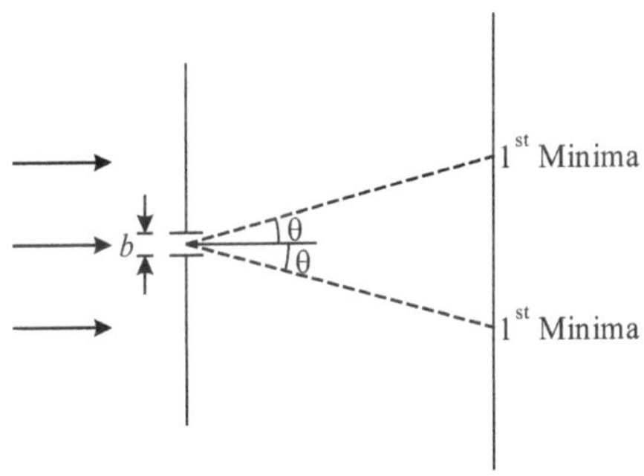

Angular position of first minima

$$b \sin \theta = \lambda$$

$$\Rightarrow \quad \sin \theta = \frac{\lambda}{b}$$

So wavelength of light used

$$\lambda = b \sin 30° = \frac{b}{2}$$

given $\quad 2\theta = 60°$

$\Rightarrow \quad \theta = 30°$

If similar two slits are used in *YDSE* at a separation '*d*' then we have

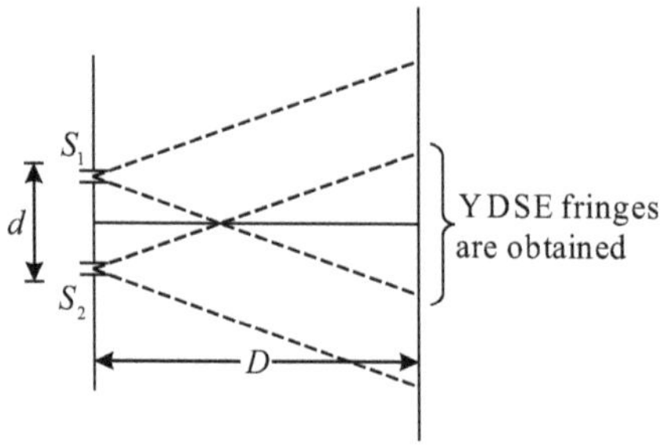

Fringe width in *YDSE* interference pattern is given as

$$b = \frac{\lambda D}{d} = \frac{bD}{2d}$$

⇒ Slit separation

$$d = \frac{bD}{2\beta} = \frac{10^{-6} \times 0.5}{2 \times 0.01} = 25 \times 10^{-6} \text{m}$$

⇒ $d = 25\ \mu\text{m}$ **Ans. (A)**

Sol. 13

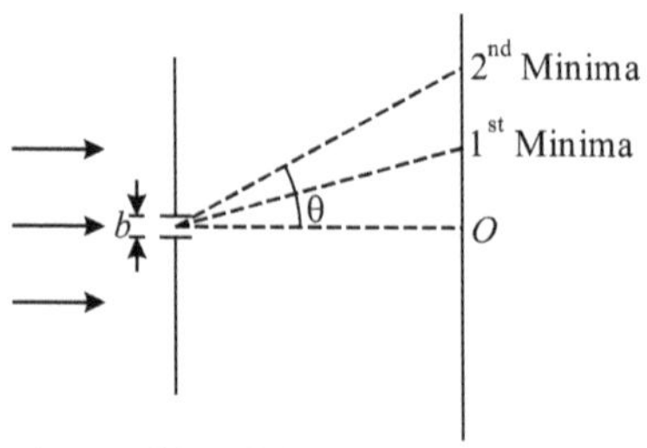

If at angular position θ from central maxima we get n^{th} minima then we use

$$b \sin \theta = n\lambda$$

So for 2nd minima we use $n = 2$

⇒ $b \sin \theta = 2\lambda$

⇒ $\sin \theta = \frac{2\lambda}{b} = \frac{2 \times 550 \times 10^{-9}}{22 \times 10^{-7}} = 50 \times 10^{-2}$

⇒ $\sin \theta = 0.5$

⇒ $\theta = \frac{\pi}{6}$ **Ans. (D)**

24.3 Polarization of Light

Sol. 1 Interference, reflection and diffraction are phenomenon which can be seen for any type of waves but polarization is based upon restricting the plane of oscillation of $\vec{E}$ in *EM* waves only in one plane & plane of oscillations are analyzed only in transverse waves. **Ans. (A)**

Sol. 2 Using Brewster's law, we have reflected light is completely polarized when IR incident at such an angle for which reflected and refracted ray are mutually perpendicular

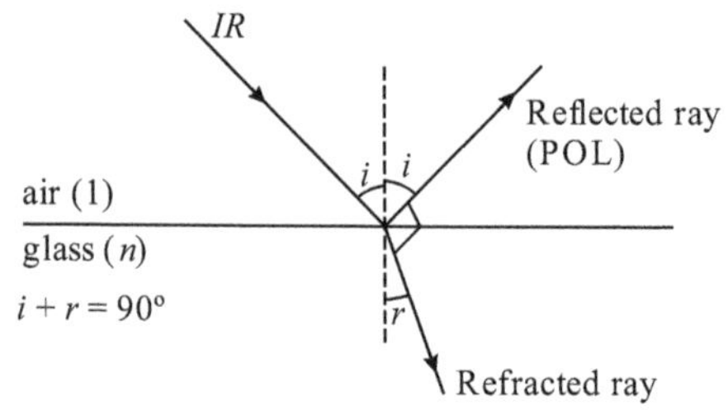

By Snell's law

$$\sin i = n \sin r = n \sin (90 - i)$$

$$\sin i = n \cos i$$

⇒ $\tan i = n$

⇒ $i = \tan^{-1} (n)$ **Ans. (D)**

Sol. 3

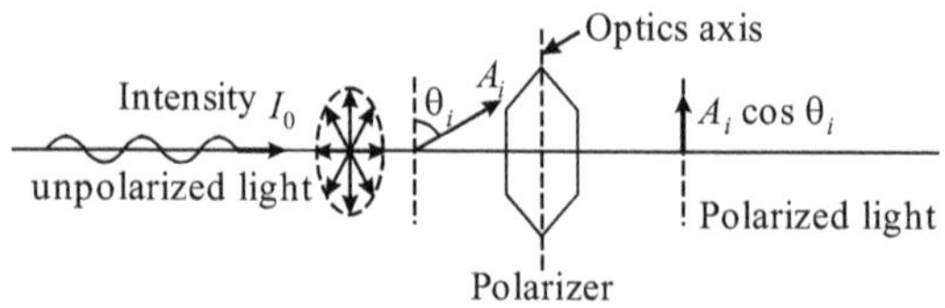

If a specific oscillation is considered before incidence on polarizer which has oscillations at an angle θ to optic axis as shown then in polarized light *A* cos θ component will be transmitted so total intensity of polarized light can be given as

$$I_0 = \Sigma k A_i^2 = kA^2$$

$$I_P = k\Sigma(A_i \cos \theta_i)^2$$

$$I_P = kA_i^2 (\Sigma\cos^2 \theta_i) \text{ (average)}$$

$$I_P = kA_i^2\left(\frac{1}{2}\right) = \frac{1}{2}kA^2 = \frac{1}{2}I_0$$

Here half incident light intensity does not get transmitted.

Ans. (A)

Sol. 4 When sunlight is scattered by air particles in atmosphere

$$\text{Scattering} \propto \frac{1}{\lambda^4}$$

⇒ λ is least for blue/violet light so sky appears blue.

When light is seen at a specific region of atmosphere then polarized light due to scattering also enters the observer's eye so through polarizer (analyzer) light intensity varies on rotating it.

Ans. (C)

Sol. 5

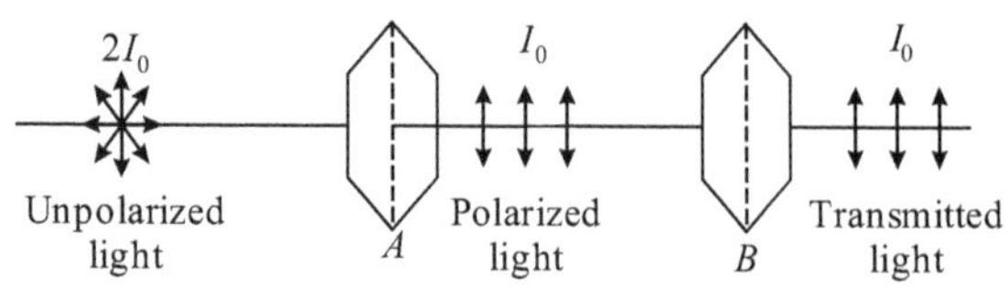

Final situation

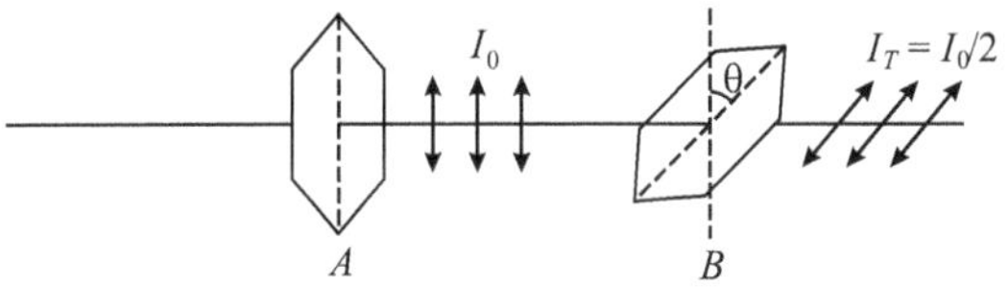

By Malus law the intensity of transmitted beam is

$$I_T = I_0 \cos^2 \theta$$

$$\Rightarrow \quad \frac{I_0}{2} = I_0 \cos^2\theta$$

$$\Rightarrow \quad \cos \theta = \frac{1}{\sqrt{2}}$$

$$\Rightarrow \quad \theta = 45°, 135°, 225°, 315° \qquad \textbf{Ans. (A)}$$

Sol. 6

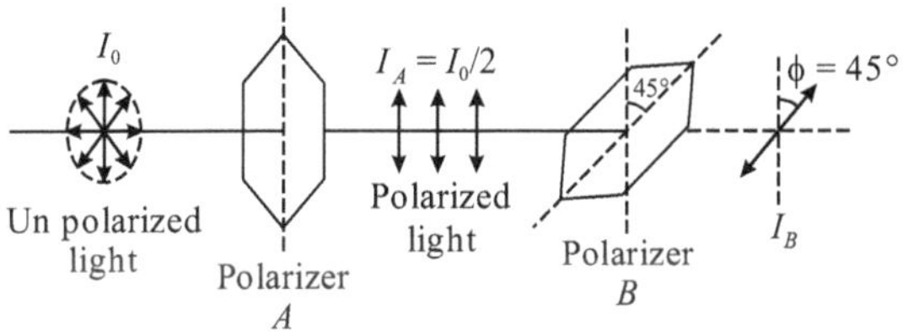

Intensity of light emerging from polarizer B is given by Malus law as

$$I_B = I_A \cos^2 \phi = \frac{I_0}{2} \cos^2 (45°)$$

$$\Rightarrow \quad I_B = \frac{I_0}{2} \times \left(\frac{1}{\sqrt{2}}\right)^2 = \frac{I_0}{4} \qquad \textbf{Ans. (C)}$$

7.

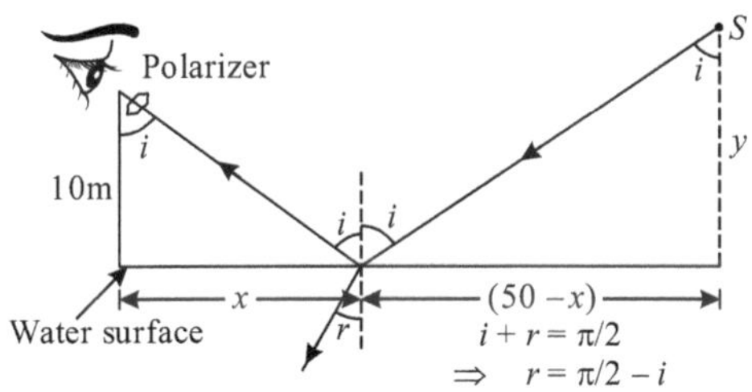

We use Snell's law at the point of reflection

$$\sin i = \mu_\omega \sin r$$

$$\Rightarrow \quad \sin i = \mu_\omega \sin\left(\frac{\pi}{2} - 1\right)$$

$$\Rightarrow \quad \tan i = \mu_\omega = \frac{4}{3}$$

In the figure shown we use

$$\tan i = \frac{x}{10} = \frac{50 - x}{y} = \frac{4}{3} \qquad \ldots(1)$$

$$\Rightarrow \quad y = \frac{10(50 - x)}{x}$$

$$\text{and} \quad x = \frac{40}{3} m \simeq 13.33 \text{ m}$$

$$\Rightarrow \quad y = \frac{10\left(50 - \frac{40}{3}\right)}{\frac{40}{3}} = \frac{3}{4} \times \frac{110}{3} = 27.5 \text{ m} \qquad \textbf{Ans. (B)}$$

Sol. 8

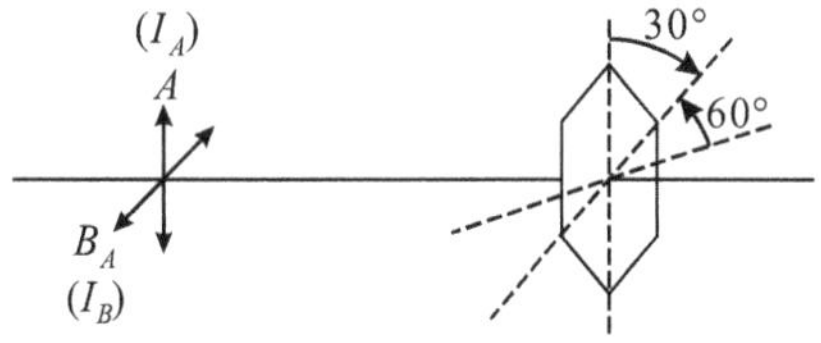

Here the polaroid axis after rotation by 30° from polarization plane of light A it is at an angle 60° from the polarization plane of light B.

By Malus law the transmitting intensities of lights A and B are

$$I_{AT} = I_A \cos^2 30°$$

$$\text{and} \quad I_{BT} = I_B \cos^2 60°$$

$$\text{Given that} \quad I_{AT} = I_{BT}$$

$$\Rightarrow \quad I_A \cos^2 30° = I_B \cos^2 60°$$

$$\Rightarrow \quad \frac{I_A}{I_B} = \frac{\cos^2 60°}{\cos^2 30°} = \frac{\left(\frac{1}{2}\right)^2}{\left(\frac{\sqrt{3}}{2}\right)^2} = \frac{1}{3} \qquad \textbf{Ans. (D)}$$

Sol. 9

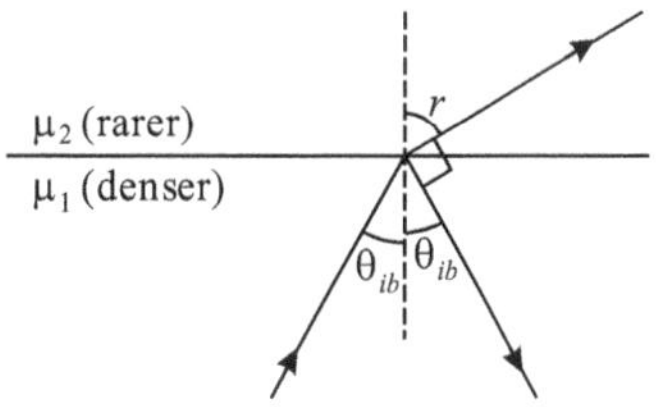

For Brewster's angle of incidence, we use

$$r = \frac{\pi}{2} - \theta_{ib}$$

By Snell's law, we have

$$\mu_1 \sin \theta_{ib} = \mu_2 \sin r = \mu_2 \sin\left(\frac{\pi}{2} - \theta_{ib}\right)$$

$$\text{We use} \quad \sin \theta_{ic} = \frac{\mu_2}{\mu_1} \qquad \ldots(1)$$

$$\Rightarrow \quad \mu_1 \sin \theta_{ib} = \mu_2 \cos \theta_{ib}$$

$$\text{We also use} \quad \frac{\sin \theta_{ic}}{\sin \theta_{ic}} = 1.28 \qquad \ldots(2)$$

$$\Rightarrow \quad \frac{\mu_2}{\mu_1} = \frac{\sin \theta_{ib}}{\cos \theta_{ib}} \qquad \ldots(3)$$

from (1), (2) and (3) we get

$$\sin \theta_{ic} = \frac{\sin \theta_{ib}}{\cos \theta_{ib}}$$

$$\Rightarrow \quad \cos \theta_{ib} = \frac{1}{1.28}$$

So from equation-(3), we have

$$\frac{\mu_2}{\mu_1}=\frac{\sin\theta_{ib}}{\cos\theta_{ib}}=\frac{\sqrt{1-\cos^2\theta_{ib}}}{\cos\theta_{ib}}$$

$$\Rightarrow \quad \frac{\mu_2}{\mu_1}=\sqrt{\left(\frac{1}{\cos\theta_{ib}}\right)^2-1}=\sqrt{(1.28)^2-1}$$

$$\Rightarrow \quad \frac{\mu_2}{\mu_1}=\sqrt{0.64}=0.8$$

Ans. (C)

Sol. 10

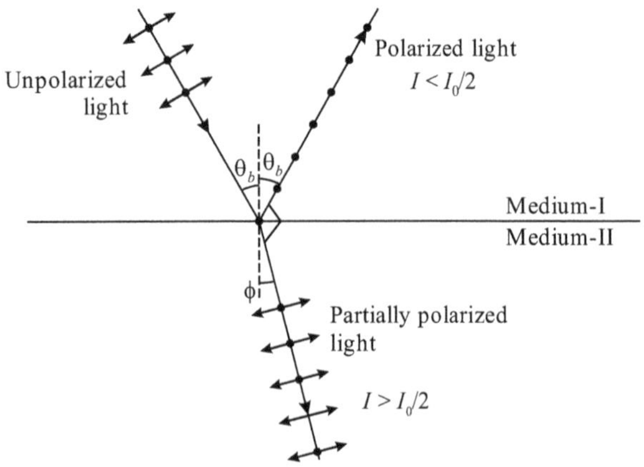

As transmitted light is partially polarized and it also contains the oscillation in the plane parallel to boundary which are there is reflected light.

So intensity of reflected light

$$I_{\text{reflected}}<I_0/2$$

Ans. (A)

Sol. 11

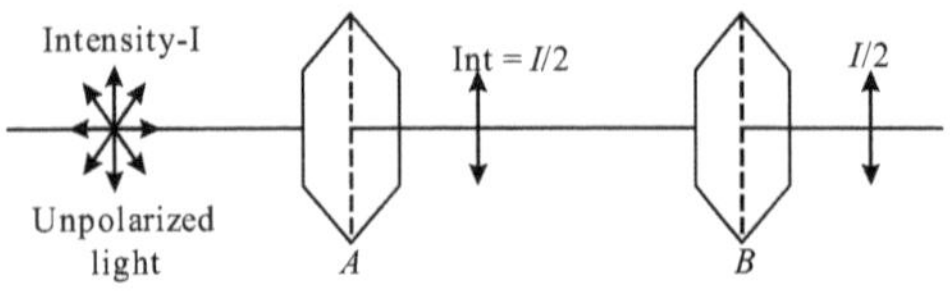

Initially when A and B are present then light intensity does not change on passing through $B \Rightarrow$ Axis of B is parallel to axis of A.

After inserting polarizer C, intensity after B becomes $I/8$

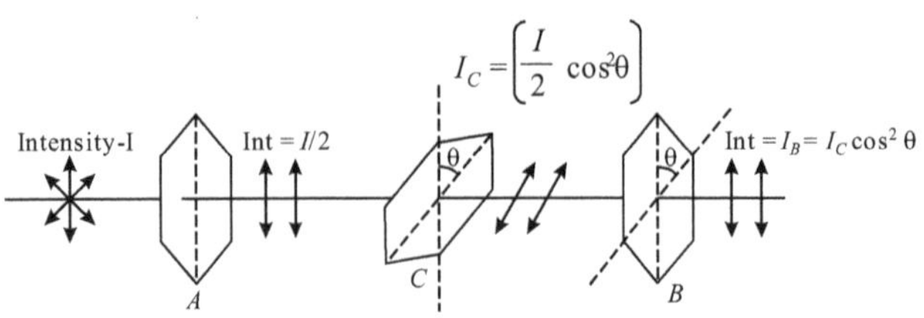

By Malus law $\quad I_C=\frac{I}{2}\cos^2\theta$

and $\quad I_B=I_C\cos^2\theta=\frac{I}{2}\cos^4\theta$

Given that finally $\quad I_B=\frac{I}{8}$

$$\Rightarrow \quad \frac{I}{8}=\frac{I}{2}\cos^4\theta$$

$$\Rightarrow \quad \cos^4\theta=\frac{1}{4}$$

$$\Rightarrow \quad \cos\theta=\frac{1}{\sqrt{2}}$$

$$\Rightarrow \quad \theta=45^\circ$$

Ans. (C)

Sol. 12

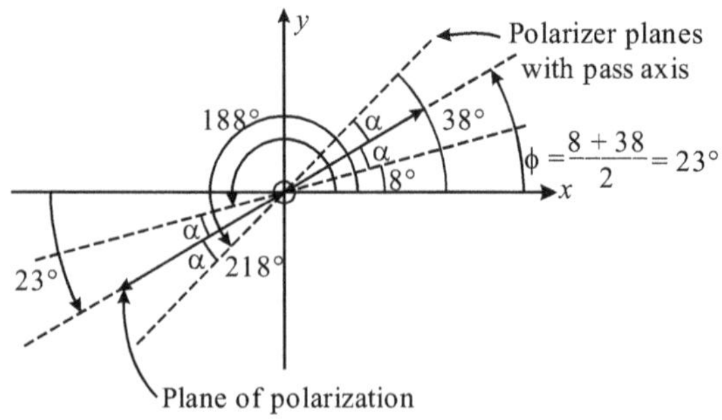

As the light intensity after passing through polarizer is same for the given pass axis (polarization axis)

$\Rightarrow$ Plane of polarization must be at same angle with the given direction of pass axis.

$$I_T=I\cos^2\alpha$$

Values of $\quad \phi=23^\circ$

or $\quad 180+23=203^\circ$

Ans. (A)

Sol. 13 Case-I :

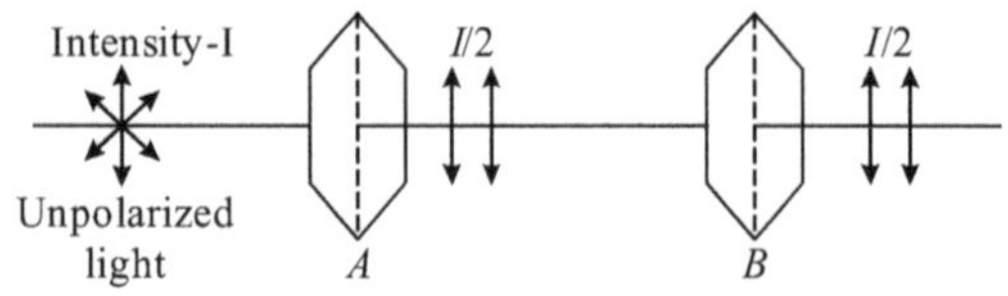

As light intensity does not change after passing through polarizer B

$\Rightarrow$ Axis of B is parallel to axis of A

Case-II :

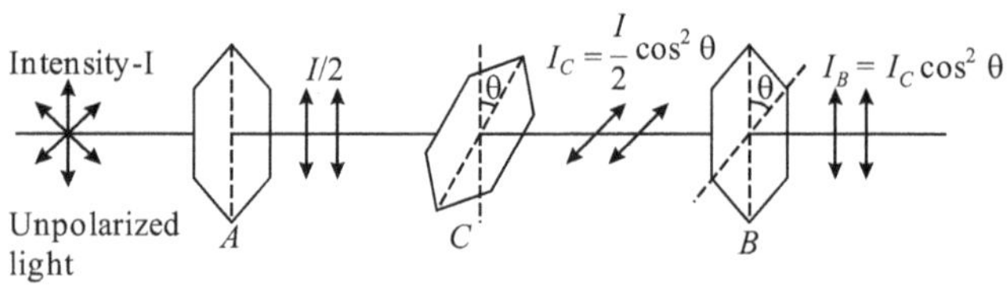

By Malus law have

$$I_C=\frac{I}{2}\cos^2\theta$$

and $\quad I_B=I_C\cos^2\theta=\frac{I}{2}\cos^4\theta$

Given that $I_B = \frac{I}{3}$

$$\Rightarrow \quad \frac{I}{3} = \frac{I}{2}\cos^4\theta$$

$$\Rightarrow \quad \cos\theta = \left(\frac{2}{3}\right)^{1/4} \quad \textbf{Ans. (A)}$$

Sol. 14 For light to never be reflected as polarized, incident angle on liquid glass interface must be less then brewster's angle so we use

$$\theta_{CL} < i_b$$

$$\Rightarrow \quad \sin\theta_c < \sin i_b$$

Here we use $\left[\tan i_b = \frac{1.5}{\mu}\right]$

$$\Rightarrow \quad \sin i_b = \frac{1.5}{\sqrt{\mu^2 + (1.5)^2}}$$

$$\Rightarrow \quad \frac{1}{\mu} < \frac{1.5}{\sqrt{\mu^2 + (1.5)^2}}$$

$$\Rightarrow \quad \sqrt{\mu^2 + (1.5)^2} < 1.5 \times \mu$$

$$\Rightarrow \quad \mu^2 + (1.5)^2 < (\mu \times 1.5)^2$$

$$\Rightarrow \quad \mu > \frac{3}{\sqrt{5}} \quad \textbf{Ans. (A)}$$

Sol. 15 Since unpolarised light falls on P_1, intensity of light transmitted from P_1 is given as

$$I_1 = I_0/2$$

Pass axis of P_2 is at an angle of 30° with P_1 so intensity of light transmitted from P_2 is given by Malus law as

$$I_2 = \frac{I_0}{2}\cos^2 30° = \frac{3I_0}{8}$$

Pass axis of P_3 is at an angle of 60° with P_2 so intensity of light transmitted from P_3 is given as

$$I_3 = \frac{3I_0}{8}\cos^2 60° = \frac{3I_0}{32}$$

$$\Rightarrow \quad \left(\frac{I_0}{I}\right) = \frac{32}{3} = 10.67 \quad \textbf{Ans. (D)}$$

25.1 Photoelectric Effect

Sol. 1 As we know work function ϕ and threshold wavelength λ_{th} are related as

$$\phi = \frac{hc}{\lambda_{th}}$$

$$\Rightarrow \quad \phi \propto \frac{1}{\lambda_{th}}$$

We use $\frac{\phi_{Na}}{\phi_{Cu}} = \frac{\lambda_{Cu}}{\lambda_{Na}} = \frac{2.3}{4.5}$

$$\Rightarrow \quad \frac{\lambda_{Na}}{\lambda_{Cu}} = \frac{45}{23} \simeq \frac{2}{1} \quad \textbf{Ans. (C)}$$

Sol. 2 For photoelectric effect of given light frequencies on photocathods (work function = ϕ) we use

$$hf_1 = \phi + \frac{1}{2}mv_1^2 \quad \ldots(1)$$

and

$$hf_2 = \phi + \frac{1}{2}mv_2^2 \quad \ldots(2)$$

$$(1)-(2) \Rightarrow \quad h(f_1 - f_2) = \frac{1}{2}m(v_1^2 - v_2^2)$$

$$\Rightarrow \quad v_1^2 - v_2^2 = \frac{2h}{m}(f_1 - f_2) \quad \textbf{Ans. (A)}$$

Sol. 3 The largest wavelength (minimum energy) which can cause photoelectric emission is the threshold wavelength and it is given as :

$$\frac{hc}{\lambda_{th}} = \phi$$

$$\Rightarrow \quad \lambda_{th} = \frac{hc}{\phi} = \frac{6.63 \times 10^{-34} \times 3 \times 10^8}{4 \times 1.6 \times 10^{-19}}$$

$$\Rightarrow \quad \lambda_{th} = \frac{12431}{4}\,\text{Å} \simeq 310.8\,\text{nm} \quad \textbf{Ans. (C)}$$

Sol. 4 In photo electric effect for a radiation frequency $\nu\ (\nu > \nu_{th})$ when falls on a metal with work function ϕ then these are related to the $KE_{(max)}$ of ejected electron as

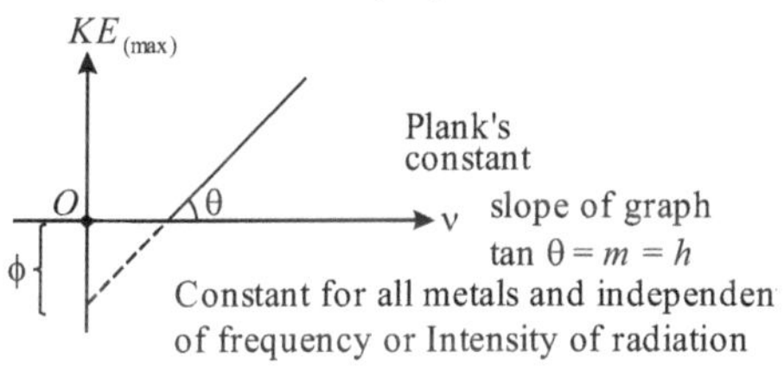

$$h\nu = \phi + KE_{max}$$

$$\Rightarrow \quad KE_{max} = h\nu - \phi$$

$$y = mx + c \quad \textbf{Ans. (D)}$$

Sol. 5

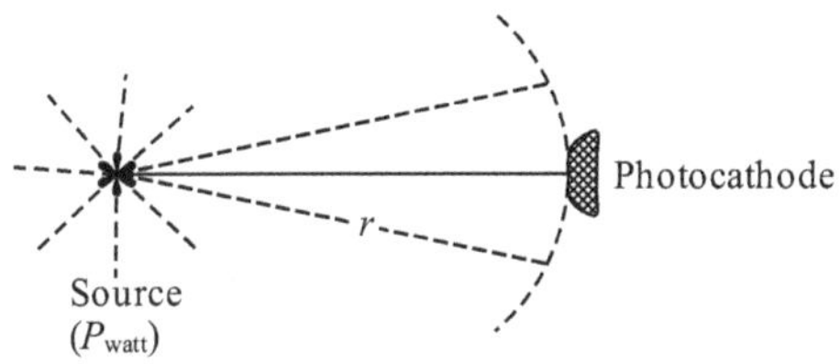

Intensity of light due to source at the location of photocathode is

$$I = \frac{P}{4\pi r^2}$$

Here $I \propto \frac{1}{r^2}$

So when distance r is changed from 1 m to 0.5 m so intensity will become 4 times.

$\Rightarrow$ Number of ejected electrons will also increase by a factor of 4. **Ans. (B)**

Sol. 6

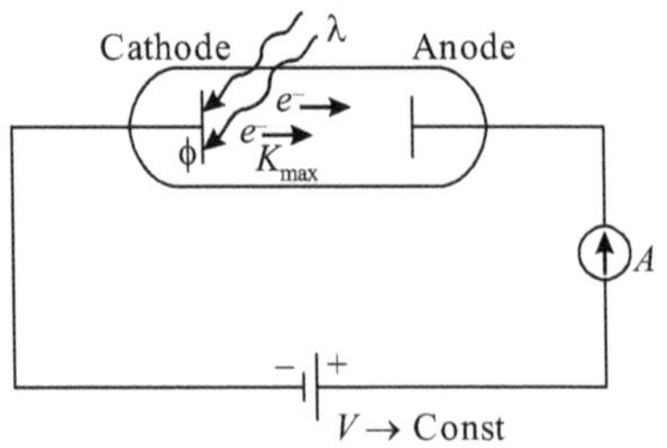

At cathode we use

$$\frac{hc}{\lambda} = \phi = K_{max}$$

As $\lambda \uparrow$ the energy $\frac{hc}{\lambda} \downarrow \Rightarrow K_{max} \downarrow$

As $K_{max} \downarrow \Rightarrow$ less electrons will reach anode $\Rightarrow$ current I decreases. **Ans. (B)**

Sol. 7 As threshold frequency gives the work function of a metal surface

$$\Rightarrow \qquad \phi = h\nu_{th} = 6.2 \text{ eV}$$

Stopping potential for a metal surface gives the maximum kinetic energy for ejected photoelectrons

$$\Rightarrow \qquad KE_{max} = eV_0 = 5 \text{ eV}$$

Energy of incident radiation

$$E_i = \phi = KE_{max} = 6.2 + 5 = 11.2 \text{ eV}$$

Wavelength λ of incident radiation is given as

$$\lambda = \frac{hc}{E_i} = \frac{12431}{11.2} \text{ Å} \simeq 1109.9 \text{ Å}$$

As UV region wavelength exist from ~ 100 Å to 4000 Å

Ans. (A)

Sol. 8 The energy of incident light photon is taken as

$$E = \frac{hc}{\lambda} = \frac{12431}{4000} \text{ eV} = 3.10 \text{ eV}$$

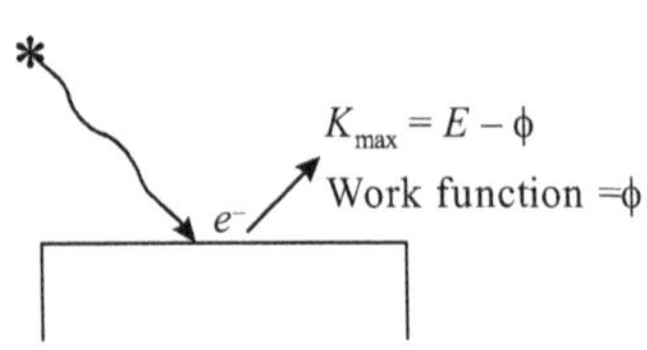

Here we use $KE_{max} = E - \phi$

$$1.68 = 3.10 - \phi$$

$$\Rightarrow \qquad \phi = 3.10 - 1.68 = 1.42 \text{ eV}$$ **Ans. (D)**

Sol. 9 In Statement-1 we use

$$E_i = h\nu = \phi + KE_{max} \qquad \ldots(1)$$

and stopping potential

$$V_0 = \frac{KE_{max}}{e}$$

When UV light is replaced by X-rays. Then we use $E_{i_2} > E_{i_1}$

$$\Rightarrow \qquad KE_{max_2} > KE_{max_1}$$

from equation-(1) as ϕ is a constant KE_{max} increases with E_i

$$\Rightarrow \qquad V_{0_2} > V_{0_1}$$

$\Rightarrow$ Statement-1 is True

In Statement-2 we have

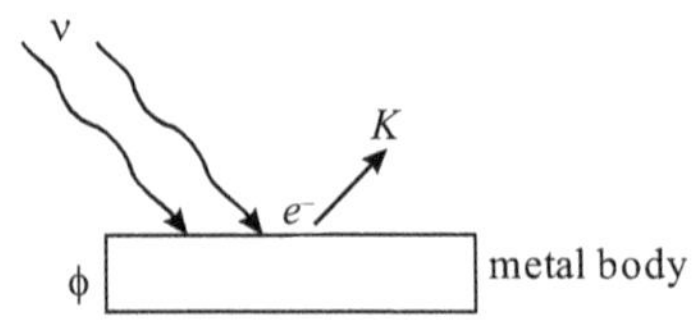

electrons in the process of ejection from metal surface looses some energy in collisions and other phenomenon so ejected KE is less than KE_{max}.

$\Rightarrow$ Statement-2 is False. **Ans. (A)**

Sol. 10 For Statement-1 we know in photoelectric effect

$$h\nu = \phi + KE_{max} \qquad \ldots(1)$$

$$\Rightarrow \qquad h\nu = \phi + eV_0 \qquad \ldots(2)$$

Due to doubling the value of frequency ν, KE_{max} and V_0 will not get doubled.

$\Rightarrow$ Statement-1 is false and Statement-2 is true as by equation (1) and (2) it is clear that KE_{max} and V_0 are linearly dependent on frequency ν. **Ans. (B)**

Sol. 11 Photocurrent $\propto$ photon flux of incident light (Intensity of light)

and $KE_{max} = h\nu - \phi$

if $\nu \leftarrow 2\nu \Rightarrow \quad KE'_{max} = 2h\nu - \phi$

$$\Rightarrow \qquad KE'_{max} > 2KE_{max}$$

$\Rightarrow$ Statement-1 is False

For Statement-2

As $KE_{max} = h\nu - \phi$

linear function

$\Rightarrow$ Statement-2 is True **Ans. (B)**

Sol. 12 (A) Threshold frequency $\nu_{th} = \frac{W}{h}$. So to emit electrons $\nu > \nu_{th}$

$\Rightarrow$ (A) is correct

(B) As there is no time leg in absorption of a photon and emission of electron

$\Rightarrow$ (B) is correct

(C) By photo electric effect equation max kinetic energy of ejected electrons is given as

$$KE_{max} = h\nu - W$$

$\Rightarrow$ (C) is wrong

(D) As KE_{max} depends only on incident frequency

$\Rightarrow$ (D) is correct **Ans. (C)**

Sol. 13

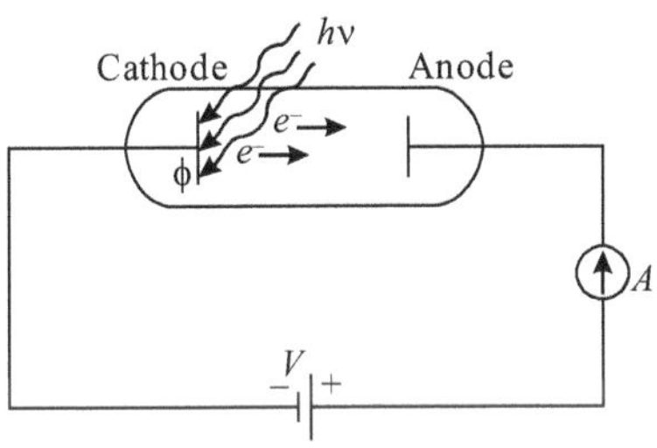

For photoelectric effect at cathode we use

$$h\nu = \frac{hc}{\lambda} = \phi + KE_{max}$$

As $\lambda \uparrow \Rightarrow \frac{hc}{\lambda} \downarrow$ and $KE_{max} \downarrow$

At lesser KE of ejected electrons the circuit current reduces.

Ans. (D)

Sol. 14

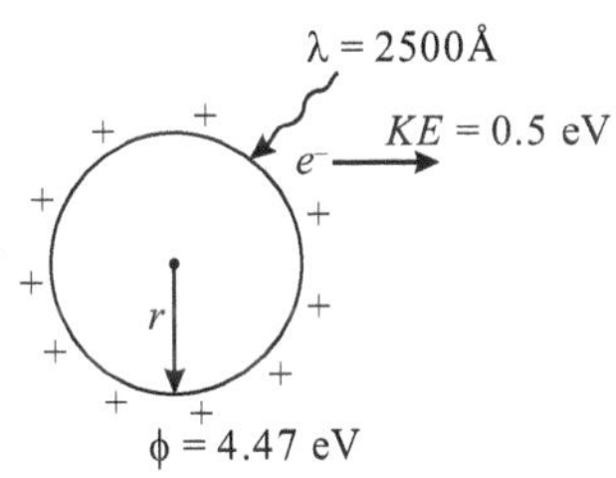

$$r = 1\text{ cm} = 10^{-2}\text{ m}$$

Energy of incident photon is

$$E_i = \frac{hc}{\lambda}$$

$$\Rightarrow \quad E_i = \frac{12431}{2500} = 4.97\text{ eV}$$

Maximum kinetic energy of ejected e^- from ball is

$$KE_{max} = E_i - \phi = 4.97 - 4.47 = 0.5\text{ eV}$$

As positive charge gets accumulated on ball due to ejection of electrons, this positive charge will attract back the ejected electrons, these electrons will not escape from ball if ball's potential will rise to 0.5 V.

If q is the charge acquired by ball upto this time, then we use

$$V = \frac{kq}{r} = \frac{1}{4\pi\varepsilon_0} \cdot \frac{q}{r}$$

$$\Rightarrow \quad 0.5 = \frac{9\times10^{9}\times q}{10^{-2}}$$

$$\Rightarrow \quad q = \frac{5\times10^{-3}}{9\times10^{9}} = 5.5\times10^{-13}\text{ C}$$ **Ans. (A)**

Sol. 15 By Photo electric effect equation we use

$$\frac{hc}{\lambda} = \phi + eV_0$$

$$\Rightarrow \quad V_0 = \left(\frac{hc}{e}\right)\frac{1}{\lambda} - \frac{\phi}{e}$$

Comparing with $y = mx + c$

Slope of V_0 vs $\frac{1}{\lambda}$ curve is

$$\Rightarrow \quad \tan\theta = \frac{hc}{e} \rightarrow \text{constant for all metals}$$

As in graph shown slope for A and B is not same. **Ans. (D)**

Sol. 16 The radiation energy emitted in $3 \rightarrow 2$ transition is given as

$$\Delta E = E_3 - E_2 = (-1.51\text{ eV}) - (-3.4\text{ eV})$$

$$\Rightarrow \quad \Delta E = 1.89\text{ eV}$$

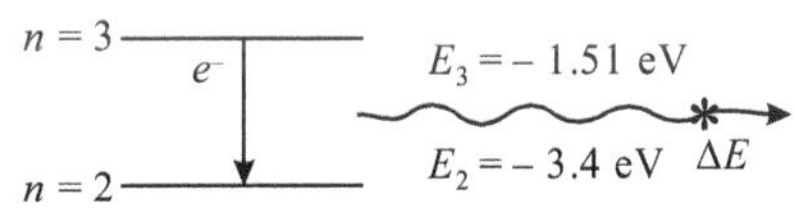

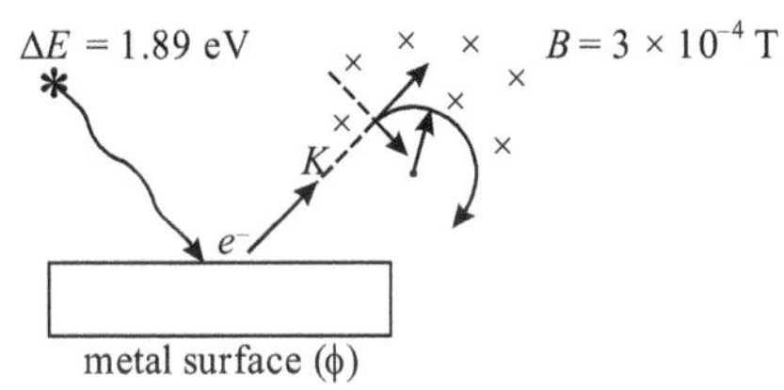

In magnetic field the radius of circular motion is given as

$$r = \frac{mv}{qB}$$

$$\Rightarrow \quad p = mv = qBr$$

The maximum kinetic energy of ejected electron is given as

$$K_{max} = \frac{p^2}{2m} = \frac{q^2B^2r^2}{2m}$$

$$\Rightarrow \quad K_{max} = \frac{(1.6\times10^{-19})^2\times(3\times10^{-4})^2\times(10\times10^{-3})^2}{2\times9.1\times10^{-31}\times(1.6\times10^{-19})}\,eV$$

$$\Rightarrow \quad K_{max} = 0.8\text{ eV}$$

For photoelectric effect at metal surface we use

$$h\nu = \phi + KE_{max}$$

$$\Rightarrow \quad \phi = h\nu - KE_{max} = 1.89 - 0.8 = 1.09\text{ eV}$$

Ans. (B)

Sol. 17 Energy of each photon in the two radiations are

$$E_1 = \frac{hc}{\lambda_1} = \frac{12431}{4972}\text{ eV} \simeq 2.5\text{ eV}$$

and
$$E_2 = \frac{hc}{\lambda_2} = \frac{12431}{6216}\text{ eV} \simeq 1.99\text{ eV} (< 2.3\text{ eV})$$

So second radiation is not capable of ejecting electrons from the metal. The power incident on 1 cm^2 of metal surface in first radiation is

$$P = IA = 1.8\times10^{-3}\times10^{-4} = 1.8\times10^{-7}\text{ W}$$

Number of photons incident on metal/second are

$$N = \frac{P\lambda}{hc}$$

Number of electrons ejected in 2 second are

$$N_e = \frac{P\lambda}{hc}\times 2$$

$$\Rightarrow \quad N_e = \frac{1.8\times10^{-7}\times4972\times10^{-10}}{6.63\times10^{-34}\times3\times10^{8}}\times 2$$

$$\Rightarrow \quad N_e = 9\times10^{11}$$ **Ans. (B)**

Sol. 18 If ϕ be the work function of metal used, then we have

$$\frac{1}{2}mv^2 = \frac{hc}{\lambda} - \phi$$

$$\Rightarrow \quad v = \sqrt{\frac{2hc}{\lambda m} - \frac{2\phi}{m}} \quad \ldots(1)$$

If λ is changed to $\frac{3\lambda}{4}$ so now the speed of fastest electron emitted now will be

$$v' = \sqrt{\frac{2hc}{\lambda m}\left(\frac{4}{3}\right) - \frac{2\phi}{m}} \quad \ldots(2)$$

From equation-(1) $$\sqrt{\frac{4}{3}}\,v = \sqrt{\frac{2hc}{\lambda m}\left(\frac{4}{3}\right) - \frac{2\phi}{m}\left(\frac{4}{3}\right)} \quad \ldots(3)$$

from equation-(2) and (3) comparing, we get

$$v' > \sqrt{\frac{4}{3}}\,v$$ **Ans. (A)**

Sol. 19 Using photo electric effect equation we have

$$\frac{hc}{\lambda_1} = \phi + \text{eV} \quad \ldots(1)$$

$$\frac{hc}{\lambda_2} = \phi + 3\text{ eV} \quad \ldots(2)$$

(1) × 3 – (2) gives

$$\Rightarrow \quad \frac{3hc}{\lambda_1} - \frac{hc}{\lambda_2} = 2\phi \quad \Rightarrow \quad \phi = \frac{3hc}{2\lambda_1} - \frac{hc}{2\lambda_2}$$

When wavelength λ_3 is used

$$\frac{hc}{\lambda_3} = \phi + eV_0$$

$$\Rightarrow \quad V_0 = \frac{1}{e}\left(\frac{hc}{\lambda_3} - \phi\right) = \frac{hc}{e}\left(\frac{1}{\lambda_3} - \frac{3}{2\lambda_1} + \frac{1}{2\lambda_2}\right)$$ **Ans. (D)**

Sol. 20 We use photo electric effect equation which gives

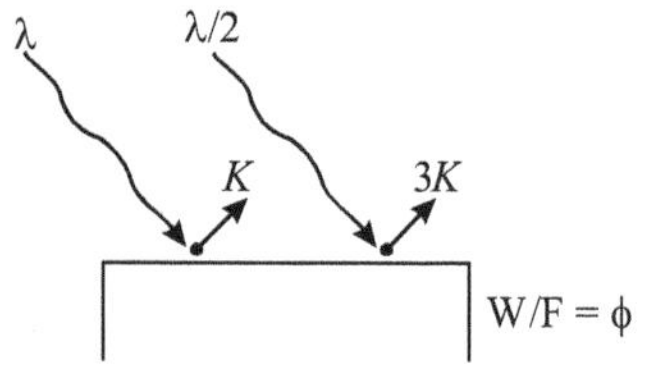

$$\frac{hc}{\lambda} = \phi + k \quad \ldots(1)$$

$$\frac{hc}{(\lambda/2)} = \phi + 3k \quad \ldots(2)$$

Using (1) × 3 – (2) $\Rightarrow \frac{hc}{\lambda} = 2\phi \Rightarrow \phi = \frac{hc}{2\lambda}$ **Ans. (A)**

Sol. 21 By photo electric effect equation we use

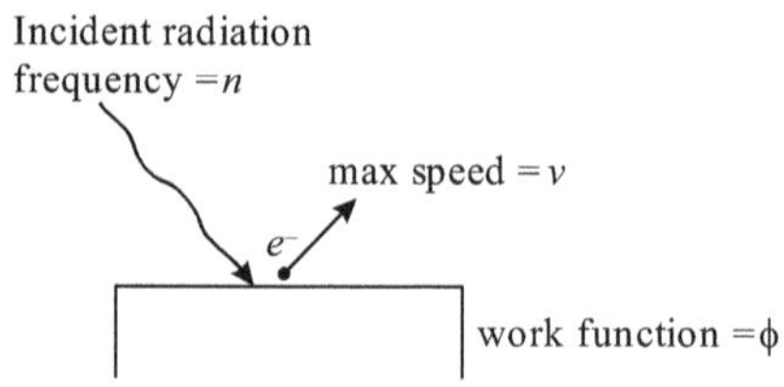

$$hn = \phi + \frac{1}{2}\text{mv}^2$$

$$\Rightarrow \quad v = \sqrt{\frac{2}{m}(hn - \phi)} \quad \ldots(1)$$

If incident radiation frequency is increased to $3n \Rightarrow$ The speed of ejected fastest electron is given as

$$\Rightarrow \quad v' = \sqrt{\frac{2}{m}(3hn - \phi)}$$

$$\Rightarrow \quad v' > \sqrt{3}\,v$$ **Ans. (C)**

Sol. 22

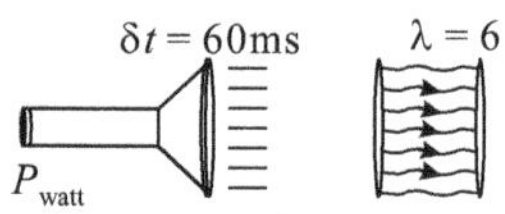

Pulse energy $E = \rho\delta t$

Energy of one photon in pulse $E_{ph} = \frac{hc}{\lambda}$

Number of photons in light pulse are

$$N = \frac{E}{E_{ph}} = \frac{E\lambda}{hc} = \frac{\rho\delta t\lambda}{hc}$$

$$\Rightarrow \quad N = \frac{0.5\times10^{3}\times60\times10^{-3}\times6\times10^{-7}}{6.62\times10^{-34}\times3\times10^{8}}$$

$$\Rightarrow \quad N \simeq \frac{10^{-7}\times 10}{10^{-34}\times 10^{8}} = 10^{20} \qquad \textbf{Ans. (A)}$$

Sol. 23 We use for velocity of ejected photoelectrons

$$\frac{hc}{\lambda_1} = \phi + \frac{1}{2}m(2v)^2 \qquad \ldots(1)$$

and $$\frac{hc}{\lambda_2} = \phi + \frac{1}{2}mv^2 \qquad \ldots(2)$$

$$\frac{(1)}{(2)} \Rightarrow \frac{\frac{hc}{\lambda_1}-\phi}{\frac{hc}{\lambda_2}-\phi} = 4$$

$$\Rightarrow \quad \frac{hc}{\lambda_1} - \phi = \frac{4hc}{\lambda_2} - 4\phi$$

$$\Rightarrow \quad \frac{4hc}{\lambda_2} - \frac{hc}{\lambda_1} = 3\phi$$

$$\Rightarrow \quad \phi = \frac{1}{3}hc\left(\frac{4}{\lambda_2} - \frac{1}{\lambda_1}\right)$$

$$\Rightarrow \quad \phi = \frac{1}{3}\times 1240\left(\frac{4\times 350 - 540}{350\times 540}\right)$$

$$\Rightarrow \quad \phi = 1.8 \text{ eV} \qquad \textbf{Ans. (A)}$$

Sol. 24 $B = B_0\sin(\pi \times 10^7 c)t + B_0\sin(2\pi \times 10^7 c)t$ since there are two *EM* waves with different frequency, to get maximum kinetic energy we take the photon with higher frequency

$$B_1 = B_0\sin(\pi \times 10^7 c)t$$

$$\nu_1 = \frac{10^7}{2}\times c$$

$$B_2 = B_0\sin(2\pi \times 10^7 c)t$$

$$\nu_2 = 10^7 c$$

Where c is speed of light

As $\nu_2 > \nu_1$ so *KE* of photoelectron will be maximum for photon of higher energy $\nu_2 = 10^7 c$ Hz

We use

$$h\nu = \phi + KE_{max} \qquad \ldots(1)$$

Energy of photon incident is

$$E_{ph} = h\nu = 6.6 \times 10^{-34} \times 10^7 \times 3 \times 10^9$$

$$\Rightarrow \quad E_{ph} = 6.6 \times 3 \times 10^{-19} \text{ J}$$

$$\Rightarrow \quad E_{ph} = \frac{6.6\times 3\times 10^{-19}}{1.6\times 10^{-19}} \text{eV} = 12.375 \text{ eV}$$

from (1)

$$KE_{max} = E_{ph} - \phi$$

$$\Rightarrow \quad KE_{max} = 12.375 - 4.7$$

$$= 7.675 \text{ eV} \approx 7.7 \text{ eV} \qquad \textbf{Ans. (A)}$$

Sol. 25 Energy incident on plate per second is

$$E = IA$$

Number of electrons emitted per second are

$$n = \frac{E\times 0.1}{10\text{eV}} = \frac{16\times 10^{-3}\times 1\times 10^{-4}\times 0.1}{10\times 1.6\times 10^{-19}}$$

$$= 10^{11} \text{ electrons/sec}$$

Maximum *KE* of ejected electrons is

$$KE = 10 - 5 = 5 \text{ eV} \qquad \textbf{Ans. (D)}$$

Sol. 26 For photoelectric effect, we use

$$\frac{hc}{\lambda_1} = \phi + eV_1 \qquad \ldots(i)$$

and $$\frac{hc}{\lambda_2} = \phi + eV_2 \qquad \ldots(ii)$$

(i) – (ii)

$$hc\left(\frac{1}{\lambda_1} - \frac{1}{\lambda_2}\right) = e(V_1 - V_2)$$

$$\Rightarrow \quad V_1 - V_2 = \frac{hc}{e}\left(\frac{\lambda_2 - \lambda_1}{\lambda_1 - \lambda_2}\right)$$

$$\Rightarrow \quad \Delta V = (1240 \text{ nm} - V)\frac{100\text{nm}}{300\text{nm}\times 400\text{nm}}$$

$$\Rightarrow \quad \Delta V = 1 \text{ V} \qquad \textbf{Ans. (B)}$$

Sol. 27 By photoelectric effect equation, we use

$$h_\nu = \phi + \frac{eV_0}{2} \qquad \ldots(1)$$

and $$\frac{h\nu}{2} = \phi + eV_0 \qquad \ldots(2)$$

Above equations cannot holds so given data is incorrect.

Ans. (None)

Sol. 28 Frequency of the incident light is

$$\nu = \frac{\omega}{2\pi} = \frac{2\pi\times 6\times 10^{14}}{2\pi} = 6 \times 10^{14} \text{ Hz}$$

Light wavelength is

$$\lambda = \frac{c}{\nu} = \frac{3\times 10^8}{6\times 10^{14}} = 5000 \text{ Å}$$

Photo energy $E = \dfrac{12375}{5000} = 2.475 \text{ eV}$

KE (max) of ejected photoelectrons is

$$KE_{max} = E - \phi = 2.475 - 2 = 0.475 \text{ eV}$$

Stopping potential

$$V = \frac{KE_{max}}{e} = 0.475 \text{ V} \simeq 0.48 \text{ V} \qquad \textbf{Ans. (A)}$$

Sol. 29 Max *KE* of ejected photoelectrons is given as

$$K_{max} = \frac{hc}{\lambda} - \phi$$

$$\Rightarrow \quad K_{max} = \frac{hc}{\lambda} - \frac{hc}{\lambda_0} = \frac{hc}{260\text{ nm}} - \frac{hc}{380\text{ nm}}$$

$$\Rightarrow \quad K_{max} = 1237\left[\frac{1}{260} - \frac{1}{380}\right]$$

$$= 1237\left[\frac{380-260}{3800\times 260}\right]\text{eV} \simeq 1.5\text{ eV}$$ **Ans. (A)**

Sol. 30 V_0 vs ν graph starts at threshold frequency so from graph, we have

$$\Rightarrow \quad \nu_{th} = 4\times 10^{14}\text{ Hz}$$

So work function for this metal is given as

$$\phi = h\nu_{th} = \frac{6.63\times 10^{-34}\times 4\times 10^{14}}{1.6\times 10^{-19}}\text{eV}$$

$$\Rightarrow \quad \phi = 1.657\text{ eV} \simeq 1.66\text{ eV}$$ **Ans. (C)**

25.2 Matter Waves and X-Ray

Sol. 1

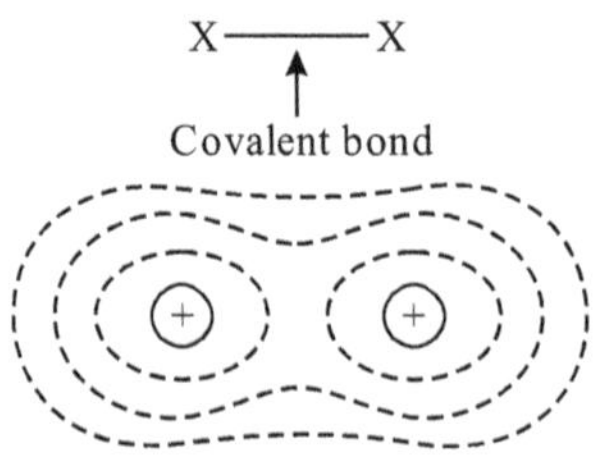

Overlapping region of electron cloud due to Van der waals force between electrons and Nu.

This is a region between two atoms where probability density of existence of electrons now exist. **Ans. (A)**

Sol. 2

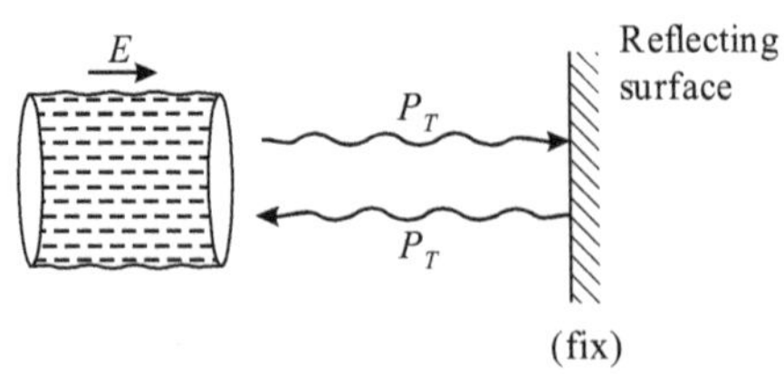

If radiation wavelength is λ, then no of photons in the radiation are $N = \frac{E}{hc/\lambda} = \frac{E\lambda}{hc}$

Each photon in radiation has momentum

$$p = \frac{h}{\lambda}$$

$\Rightarrow$ Total momentum in radiation beam is

$$p_T = \text{N}\times\frac{R}{\lambda} = \frac{E\lambda}{kc}\times\frac{h}{\lambda} = \frac{E}{c}$$

Total momentum change due to reflection is momentum transferred to surface

$$\Delta p_T = p_T - (-p_T) = 2p_T = \frac{2E}{c}$$ **Ans. (B)**

Sol. 3 The momentum of a moving free electron at KE 'K' is given as

$$p = \sqrt{2mK} \quad [\text{As } k = \frac{1}{2}mv^2]$$

De-Broglie wavelength of moving electron can be given as

$$\lambda = \frac{h}{p} = \frac{h}{\sqrt{2mK}} \quad \ldots(1)$$

If K is doubled to $2K$ then new De-Broglie wavelength is given as

$$\lambda' = \frac{h}{\sqrt{2m(2K)}} = \frac{\lambda}{\sqrt{2}}$$ **Ans. (C)**

Sol. 4 As a photon having wavelength λ has momentum associated with it is given by concept of matter wave

$$p = \frac{h}{\lambda} \quad \ldots(1)$$

If ν be the photon frequency we use

$$\lambda = \frac{c}{\nu}$$

from equation-(1) $p = \frac{h}{c/\nu} = \frac{h\nu}{c}$ **Ans. (D)**

Sol. 5 Path difference in the electron beams which are diffracted from points O and O' are

$$\Delta = 2d\cos i$$

For constructive interference in the two beams we use

$$\Delta = 2d\cos i = n\lambda \quad (n \in \text{I})$$

Wavelength of electron wave is given as

$$\lambda = \frac{h}{mv} = \frac{h}{\sqrt{2mK}} = \frac{h}{\sqrt{2meV}} \quad [\text{As } K = eV]$$

Here we have

$$2d\cos i = \frac{nh}{\sqrt{2meV}}$$

$$\Rightarrow \quad V = \frac{n^2h^2}{8med^2\cos^2 i}$$

For least value of V, we take $n = 1$

$$\Rightarrow \quad V = \frac{h^2}{8med^2\cos^2 i}$$

$$\Rightarrow \quad V = \frac{(6.6\times10^{-34})^2}{8\times9.1\times10^{-31}\times1.6\times10^{-16}\times(10^{-10})^2\times\left(\sqrt{3}/2\right)^2}$$

$$\Rightarrow \quad V \simeq 50\,\text{V}$$ **Ans. (B)**

Sol. 6 Path difference in waves AOA' and $BO'B'$ after reflection from atomic layers is given as

$$\Delta = 2d\cos i$$

For constructive interference at angle i we use

$$\Delta = 2d\cos i = n\lambda_{dB} \quad [n \in I]$$ **Ans. (B)**

Sol. 7 Statement-1 : It is true as in Davisson-Germer experiment interference of electrons considering their de-Broglie wavelength was experimentally verified.

Statement-2 : It is true and explaining S-1 also. **Ans. (D)**

Sol. 8 The de-Broglie wavelength for a particle with charge q and mass m which is accelerated at potential difference V is given as

$$\lambda = \frac{h}{\rho} = \frac{h}{\sqrt{2mK}} = \frac{h}{\sqrt{2mqV}} \quad [K = qV]$$

For e^- we use $m = m_e, q = e, V = V$

$$\Rightarrow \quad \lambda_e = \frac{h}{\sqrt{2m_e eV}} \quad \ldots(1)$$

for a proton we use $m = m_p, q = e, V = 4V$

$$\Rightarrow \quad \lambda_p = \frac{h}{\sqrt{2m_p e(4V)}} = \frac{h}{2\sqrt{2m_p eV}} \quad \ldots(2)$$

So we have from (1) and (2)

$$\frac{\lambda_e}{\lambda_p} = \frac{h/\sqrt{2m_e eV}}{h/2\sqrt{2m_p eV}}$$

$$\Rightarrow \quad \frac{\lambda_e}{\lambda_p} = 2\sqrt{\frac{m_p}{m_e}}$$ **Ans. (D)**

Sol. 9 **(1)** Davisson and Germer experiment $\rightarrow$ related to diffraction of electron as a wave. $\rightarrow$(i)

(2) Millikan's oil drop experiment $\rightarrow$ used in determining e/m of an electron. $\rightarrow$(ii)

(3) Rutherfored experiment $\rightarrow$ gold foil α-scattering experiment discovered Nu. $\rightarrow$(iv)

(4) Franck-Hertz experiment $\rightarrow$ energy level quantization. $\rightarrow$(iii)

Ans. (B)

Sol. 10

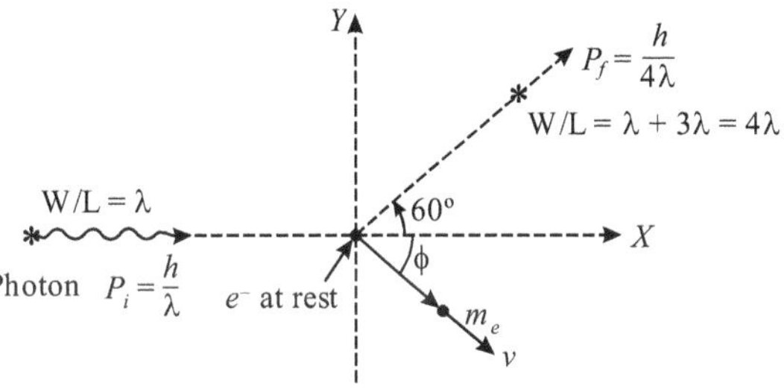

By conservation of linear momentum in above scattering, we use

$$\text{X-dir} \rightarrow \frac{h}{\lambda} = \frac{h}{4\lambda}\cos 60^\circ + m_e v\cos\phi \quad \ldots(1)$$

$$\text{Y-dir} \rightarrow \frac{h}{4\lambda}\sin 60^\circ = m_e v\sin\phi \quad \ldots(2)$$

From (1) and (2) we have

$$m_e v\cos\phi = \frac{h}{\lambda} - \frac{h}{8\lambda} = \frac{7h}{8\lambda} \quad \ldots(1)$$

$$m_e v\sin\phi = \frac{h}{4\lambda}\left(\frac{\sqrt{3}}{2}\right) \quad \ldots(2)$$

$$(2)(1) \Rightarrow \tan\phi = \frac{\sqrt{3}}{7}$$

$$\Rightarrow \quad \tan\phi = \frac{\sqrt{3}}{7} \simeq 0.25$$ **Ans. (C)**

Sol. 11 For a dust particle if we consider its mass

$$m \sim 1\,\mu\text{g} = 10^{-6}\times10^{-3}\,\text{kg}$$
$$= 10^{-9}\,\text{kg}$$

If we consider it to be moving at 1 cm/s ~ 0.01 m/s

Then its de-Broglie wavelength is given as

$$\lambda = \frac{h}{\rho} = \frac{h}{mv} = \frac{6.63\times10^{-34}}{10^{-9}\times10^{-2}} \simeq 6.63\times10^{23}\,\text{m}$$

In such cases where mass of a moving particle is significant than its de-Broglie wavelength is so low that experimentally it cannot be determined. **Ans. (D)**

Sol. 12 For list-1

(A) Frank-Hertz Experiment $\rightarrow$ (q) $\rightarrow$ (ii)

it has given the discrete spectral lines for H or any atom.

It was the first practical experiment which analyzed quantum behaviour of radiation emitted from an atom.

(B) Photo electric effect $\rightarrow (q) \rightarrow$ (i)

Photon energy is absorbed by electron to eject from a metal surface.

(C) Davisson Germer Experiment $\rightarrow (r) \rightarrow$ (iii)

Scattering and interference of electrons in a specific direction by considering de-Broglie wavelength of moving electrons.

Ans. (C)

Sol. 13 For circular motion in magnetic field we use

$$qvB = \frac{mv^2}{r} \qquad \ldots(1)$$

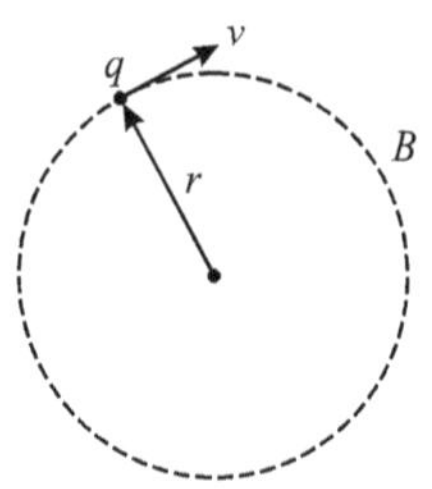

By Bohr model using quantization of angular momentum, we have

$$mvr = \frac{nh}{2\pi} \qquad \ldots(2)$$

$$\Rightarrow \quad r = \frac{nh}{2\pi mv}$$

$$\Rightarrow \quad mv^2 = \frac{qBnh}{2\pi m}$$

from (1) and (2) we use

$$mv = qBr$$

$$\Rightarrow \quad mv = qB\left(\frac{nh}{2\pi mv}\right)$$

TE of charge particle is

$$E = \frac{1}{2}mv^2 = \frac{qBnh}{4\pi m}$$

Ans. (C)

Sol. 14 de-Broglie wavelength of an electron in n^{th} level is given as

$$\lambda = \frac{h}{p_n} = \frac{h}{mv_n} \qquad \ldots(1)$$

As we know $\quad v_n = \dfrac{2\pi kZe^2}{nh}$

$$\Rightarrow \quad v_n \rightarrow \frac{1}{n} \qquad \ldots(2)$$

From (1) and (2) we use

$$\lambda \propto n$$

$\Rightarrow$ For $n = 4$ we use

$$\lambda_{n=4} = 4\lambda_{n=1} \text{ (for ground state)}$$

Ans. (B)

Sol. 15 KE of an electron accelerated by a voltage V is

$$K = eV$$

de-Broglie wavelength of this e^- is given as

$$\lambda = \frac{h}{P} = \frac{h}{\sqrt{2mK}}$$

$$\Rightarrow \quad \lambda = \frac{6.6\times10^{-34}}{\sqrt{2\times9.1\times10^{-31}\times1.6\times10^{-19}\times50}}$$

$$\cong 1.7\times10^{-10}\,\text{m} = 1.7\,\text{Å}$$

Ans. (C)

Sol. 16 After passing through slit the first minima angular position can be given as

$$d\sin\theta = \lambda \qquad \ldots(1)$$

Where λ we consider de-Broglie wavelength of incident electron out of which it is given that one electron acquires momentum p_y in y-div so for this electron we use

$$\lambda = \frac{h}{|P_y|} \qquad \ldots(2)$$

From equation-(1) and (2) we use for all these electrons which gain momentum p_y in y-direction (one electron is given in question)

$$d\sin\theta = \frac{h}{|P_y|}$$

$$\Rightarrow \quad \frac{h}{|P_y|d} = \sin\theta$$

Here for majority of other electrons s we use $\sin\theta < 1$

$$\Rightarrow \quad \frac{h}{|P_y|d} < 1$$

$$\Rightarrow \quad |P_y|d > h$$

Ans. (A)

Sol. 17

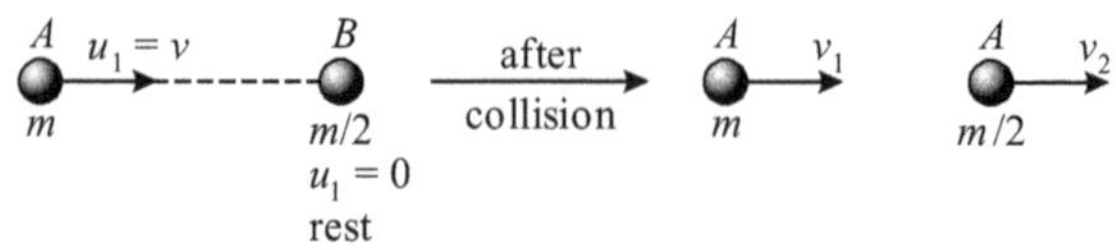

For head-on electric collision we use

$$v_1 = \left(\frac{m_1 - m_2}{m_1 + m_2}\right)u_1 + \left(\frac{2m_2}{m_1 + m_2}\right)u_2$$

$$\Rightarrow \quad v_1 = \frac{m/2}{3m/2}\cdot v + 0 = \frac{v}{3}$$

$$\Rightarrow \quad v_2 = \left(\frac{m_1 - m_1}{m_1 + m_2}\right)u_2 + \left(\frac{2m_1}{m_1 + m_2}\right)u_1$$

$$\Rightarrow \quad v_2 = 0 + \frac{2m}{3m/2}\cdot v = \frac{4v}{3}$$

After collision de-Broglie wavelength of A and B are

$$\Rightarrow \quad \lambda_A = \frac{h}{mv_1} \text{ and } l_B = \frac{h}{\left(\frac{m}{2}\right)v_2} = \frac{3h}{2mv}$$

$$\Rightarrow \quad \frac{\lambda_A}{\lambda_B} = \frac{3h/mv}{3h/2mv} = 2 \qquad \textbf{Ans. (D)}$$

Sol. 18 For continuous X-Ray the cut of wavelength λ_{min} is related to the voltage applied in discharge tube V as

$$\lambda_{min} = \frac{hc}{eV} = \frac{12431}{V} \text{ Å}$$

Taking log on both sides

$$\log \lambda_{min} = \log \frac{12431}{V} = \log(12431) - \log V$$

$$\Rightarrow \quad \log \lambda_{min} = -\log V + \log(12431) \qquad \textbf{Ans. (C)}$$

Sol. 19 It is given that de-Broglie wavelength for proton and α-particle are equal

$$\Rightarrow \quad \lambda_p = \lambda_\alpha$$

$$\Rightarrow \quad \frac{h}{m_\rho v_\rho} = \frac{h}{m_\alpha v_\alpha}$$

$$\Rightarrow \quad \frac{v_\rho}{v_\alpha} = \frac{m_\alpha}{m_\rho} = \frac{4m_\rho}{m_\rho} = \frac{4}{1} \quad [\text{As } m_\alpha = 4m_\rho] \qquad \textbf{Ans. (C)}$$

Sol. 20 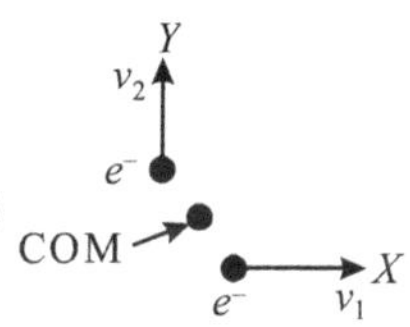

As $\vec{v}_1 = v_1\hat{i} = \dfrac{h}{m\lambda_1}\hat{i}$

$$\vec{v}_2 = v_2\hat{i} = \frac{h}{m\lambda_2}\hat{j}$$

We also use

$$\lambda_1 = \frac{h}{mv_1} \text{ and } \lambda_2 = \frac{h}{mv_2}$$

Vector speed of COM of this two electron system is given as

$$\overrightarrow{V_C} = \frac{m\vec{v}_1 + m\vec{v}_2}{m+m} = \frac{1}{2}(\vec{v}_1 + \vec{v}_2)$$

$$\Rightarrow \quad \overrightarrow{V_C} = \frac{1}{2}\frac{h}{m}\left(\frac{1}{\lambda_1}\hat{i} + \frac{1}{\lambda_2}\hat{j}\right)$$

If reference frame of COM, velocity of first electron travelling in X-direction is given as

$$\vec{V}_{1C} = \vec{v}_1 - \vec{v}_2$$

$$\Rightarrow \quad \vec{V}_{1C} = \frac{h}{m\lambda_1}\hat{i} - \left(\frac{h}{2m\lambda_1}\hat{i} + \frac{h}{2m\lambda_2}\hat{j}\right)$$

$$\Rightarrow \quad \vec{V}_{1C} = \frac{h}{2m\lambda_1}\hat{i} - \frac{h}{2m\lambda_2}\hat{j}$$

$$\Rightarrow \quad |\vec{V}_{1C}| = \frac{h}{2m}\sqrt{\frac{1}{\lambda_1^2 + \frac{1}{\lambda_2^2}}}$$

[speed of first electron in frame of COM]

de-Broglie wavelength of first electron in COM frame is given as

$$\lambda_{1C} = \frac{h}{mv_{1C}} = \frac{h}{m} \times \frac{1}{\frac{h}{2m}\sqrt{\frac{1}{\lambda^2} + \frac{1}{\lambda_2^2}}}$$

$$\Rightarrow \quad \lambda_{1C} = \frac{2\lambda_1\lambda_2}{\sqrt{\lambda_1^2 + \lambda_2^2}} \qquad \textbf{Ans. (C)}$$

Sol. 21 By Bohr's II Postulate we have

$$mvr = \frac{nh}{2\pi} \Rightarrow mv = \frac{nh}{2\pi r}$$

de-Broglie wavelength of electron in second excited state ($n = 3$), we use

$$\lambda = \frac{h}{mv} = \frac{2\pi h}{n} = \frac{2\pi}{n}(0.529\, n^2) \; [\text{as } z = 1]$$

$$\Rightarrow \quad \lambda = 1.06\,\pi n \Rightarrow \lambda \propto n$$

As $n = 3$ for λ_B and $n = 1$ for λ_G so we use

$$\frac{\lambda_B}{\lambda_G} = \frac{3}{1}$$

$$\Rightarrow \quad \lambda_B = 3\lambda_G \qquad \textbf{Ans. (D)}$$

Sol. 22 deBroglie wavelength of electrons is

$$\lambda = \frac{h}{p} \qquad [\lambda = 7.5 \times 10^{-12}]$$

$$\Rightarrow \quad p = \frac{h}{\lambda}$$

Kinetic energy of electrons is

$$KE = \frac{p^2}{2m} = \frac{(h/\lambda)^2}{2m} = \frac{\left\{\frac{6.6\times10^{-34}}{7.5\times10^{-12}}\right\}}{2\times9.1\times10^{-31}} \text{ J}$$

$$\Rightarrow \quad KE = 25 \text{ keV} \qquad \textbf{Ans. (C)}$$

Sol. 23 Given that de-Broglie wavelength of electron is

$$\lambda_e = \frac{h}{mv} = 10^{-3}\left(\frac{3\times10^8}{6\times10^{14}}\right)$$

$$\Rightarrow \quad v = \frac{6.63\times10^{-34}\times6\times10^{14}}{9.1\times10^{-31}\times3\times10^5}$$

$$\Rightarrow \quad v = 1.45 \times 10^6 \text{ m/s} \qquad \textbf{Ans. (A)}$$

Sol. 24 De-Broglie wavelength of an accelerated charge particle is given as

$$\lambda = \frac{h}{p} = \frac{h}{\sqrt{2mK}} = \frac{h}{\sqrt{2mqV}}$$

$$\Rightarrow \quad \frac{\lambda_A}{\lambda_B} = \frac{\sqrt{2m_B q_B V_B}}{\sqrt{2m_A q_A V_A}} \sqrt{\frac{4m \times q \times 2500}{m \times q \times 50}} = 2\sqrt{50}$$

$$\Rightarrow \quad \frac{\lambda_A}{\lambda_B} = 2 \times 7.07 = 14.14$$

Ans. (B)

Sol. 25 Momentum of the two particle are given as

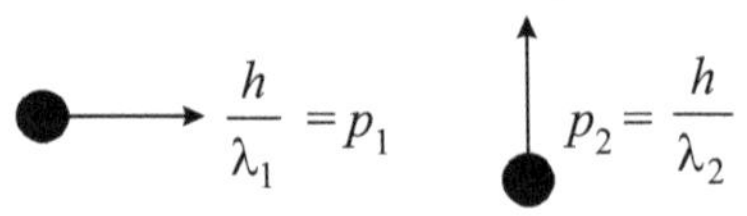

$$\vec{p}_1 = \frac{h}{\lambda_1}\hat{i}$$

and $$\vec{p}_2 = \frac{h}{\lambda_2}\hat{j}$$

Using momentum conservation before and after collision we have

$$\vec{p} = \vec{p}_1 + \vec{p}_2$$

$$\vec{p} = \frac{h}{\lambda_1}\hat{i} + \frac{h}{\lambda_2}\hat{j}$$

$$|\vec{p}| = \sqrt{\left(\frac{h}{\lambda_1}\right)^2 + \left(\frac{h}{\lambda_2}\right)^2}$$

de-Broglie wavelength of final particle λ is given as

$$\frac{h}{\lambda} = \sqrt{\left(\frac{h}{\lambda_1}\right)^2 + \left(\frac{h}{\lambda_2}\right)^2}$$

$$\Rightarrow \quad \frac{1}{\lambda^2} = \frac{1}{\lambda_1^2} + \frac{1}{\lambda_2^2}$$

Ans. (D)

Sol. 26

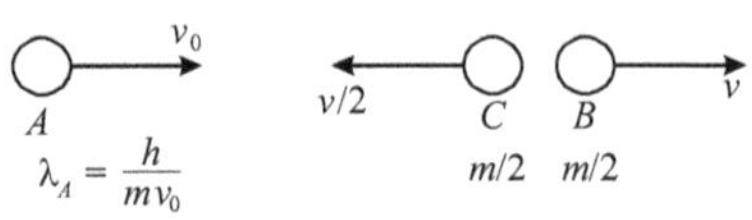

By conservation of momentum

$$mv_0 = \frac{m}{2}v - \frac{mv}{4} = \frac{mv}{4}$$

$$\Rightarrow \quad v = 4v_0$$

De-Broglie wavelength of B is

$$\lambda_B = \frac{2h}{mv} = \frac{2h}{4mv_0} = \frac{\lambda_A}{2}$$

De-Broglie wavelength of C is

$$\lambda_C = \frac{4h}{mv} = \frac{4h}{4mv_0} = \lambda_A$$

Ans. (D)

Sol. 27

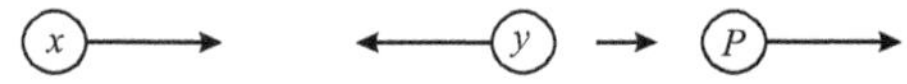

By momentum conservation before and after collision, we have

$$p_x - p_y = p_P$$

$$\Rightarrow \quad \frac{h}{\lambda_x} - \frac{h}{\lambda_y} = \frac{h}{\lambda_P}$$

$$\Rightarrow \quad \lambda_p = \frac{\lambda_x \lambda_y}{|\lambda_y - \lambda_x|}$$

Ans. (C)

Sol. 28

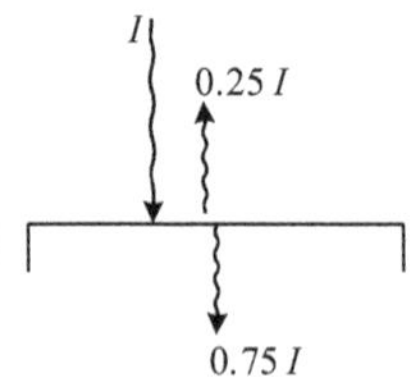

Pressure on surface due to incident, reflected and absorbed radiation is

$$p = 0.25\left[\frac{2I}{c}\right] + 0.75\left[\frac{I}{c}\right]$$

$$= 1.25\frac{I}{c} = 1.25 \times \frac{50}{3\times10^8} = 20.83 \times 10^{-8}\,\text{N}$$

Ans. (D)

Sol. 29 Radiation pressure on surface is

$$p = \frac{I}{c}$$

Momentum transferred in time Δt is

Impulse $$J = F \cdot \Delta t = \frac{I}{c} \times A \times \Delta t$$

$$= \frac{25\times10^4}{3\times10^8} \times 25 \times 10^{-4} \times 40 \times 60$$

$$= 5 \times 10^{-3}\,\text{N-s}$$

Ans. (A)

Sol. 30 From an energy source of power P_{watt}, number of photons emitted/sec are

$$n = \frac{P\lambda}{hc}$$

$$= \frac{2\times10^{-3} \times 500\times10^{-9}}{6.6\times10^{-34} \times 3\times10^8} \simeq 1.5 \times 10^{16}\,\text{ph/s}$$

Ans. (B)

26.1 Atomic Structure

Sol. 1 Ionization energy (to remove on e^-) from n^{th} state of a hydrogenic atom is given as

$$IP_n = \frac{13.6Z^2}{n^2}$$

here from Li^{+2} ($Z = 3$) and $n = 2$ we use

$$IP_2 = \frac{13.6\times3^2}{2^2} = \frac{13.6\times9}{4}$$

$$\Rightarrow \quad IP_2 = 3.4 \times 9 = 30.6\,\text{eV}$$

Ans. (A)

Sol. 2 As H and D have slight variation in spectrum, it is mainly because the nuclei of these two have different masses so if we account for this, we need to use reduced mass of electron while revolving around nucleus at centre of orbit. Here reduced mass of electron are different so energies of different energy levels will be slightly different. **Ans. (C)**

Sol. 3 Ionization of H-atom from $n = 1$ to ∞ requires 13.6 eV So from n^{th} state of H-atom ionization requires the energy

$$IP_n = \frac{13.6Z^2}{n^2}\text{ eV}$$

$$\Rightarrow \quad IP_2 = \frac{13.6\times 1^2}{2^2} = \frac{13.6}{4} = 3.4\text{ eV}$$ **Ans. (C)**

Sol. 4

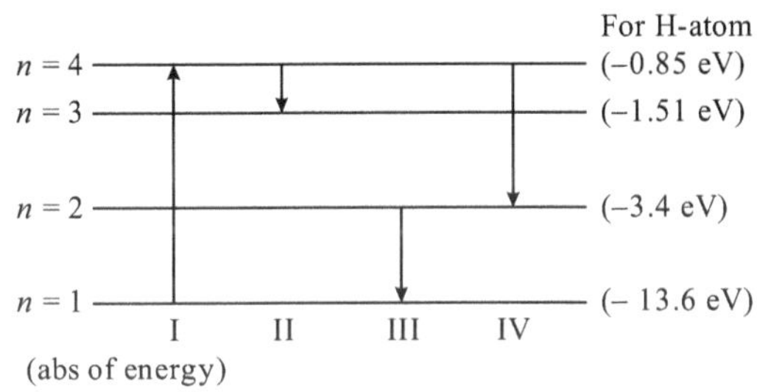

$$E_{\text{II}} = (-0.95\text{ eV}) - (-1.51\text{ eV}) = 0.66\text{ eV}$$

$$E_{\text{III}} = (-3.4\text{ eV}) - (-13.6\text{ eV}) = 10.2\text{ eV}$$

$$E_{\text{IV}} = (-0.85\text{ eV}) - (-3.4\text{ eV}) = 2.55\text{ eV}$$

Ans. (A)

Sol. 5 Options (A) and (D) are absorbing photons

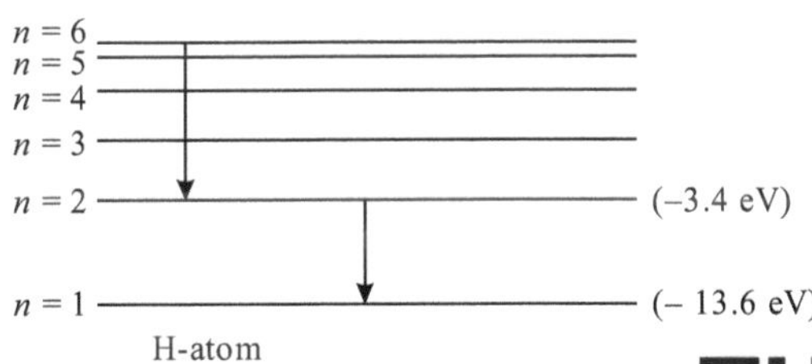

$$E_{6\to 2} < 3.4\text{ eV}$$

$$E_{2\to 1} = (-3.4\text{ eV}) - (-13.6\text{ eV})$$

$$= 10.2\text{ eV}$$ **Ans. (C)**

Sol. 6 In Bohr model for n^{th} orbit we can use for circular motion of electron

$$\frac{mv^2}{r} = \frac{k}{r}$$

$$\Rightarrow \quad mv^2 = k$$

$$v = \sqrt{\frac{k}{m}} \quad \text{(independent of } n\text{)}$$

kE of electron is $T_n = \frac{1}{2}mv_n^2 = \frac{1}{2}k$ (independent of n)

By Bohr's quantization postulate we have

$$mvr = \frac{nh}{2\pi}$$

$$\Rightarrow \quad r_n = \frac{nh}{2\pi mv}$$

$$\Rightarrow \quad r_n \propto n$$

Ans. (B)

Sol. 7 As $\lambda_{\text{IR}} > \lambda_{\text{UV}}$

$\Rightarrow \quad \Delta E_{\text{IR}} < \Delta E_{\text{UV}}$

For H-atom

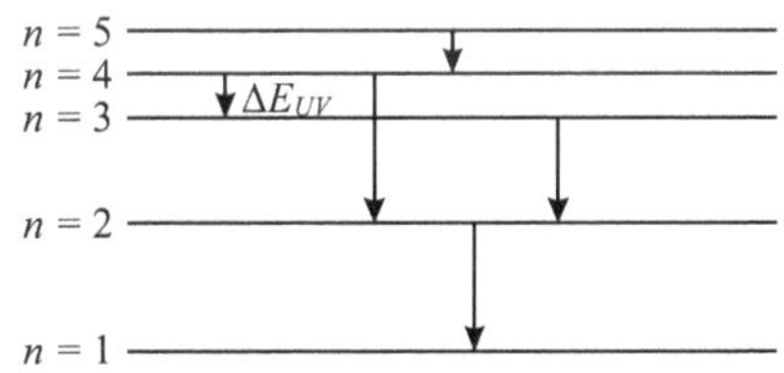

In H-atom IR radiation will be corresponding to a transition which emits energy less than $4 \to 3$ transition

(A) $E_{4\to 2} > E_{4\to 3}$

(B) $E_{5\to 4} < E_{4\to 3}$

(C) $E_{2\to 1} > E_{4\to 3}$

(D) $E_{3\to 2} > E_{4\to 3}$ **Ans. (B)**

Sol. 8 Energy involved in transition from n_1 to n_2 Bohr orbit is given as

$$\Delta E = 13.6\,Z^2\left(\frac{1}{n_1^2} - \frac{1}{n_2^2}\right)\text{ eV}$$

$$\Rightarrow \quad \Delta E = 13.6 \times (3)^2 \times \left[\frac{1}{1^2} - \frac{1}{3^2}\right]\text{ eV}$$

$$\Rightarrow \quad \Delta E = 13.6 \times 8 \times \frac{8}{9} = 108.8\text{ eV}$$ **Ans. (A)**

Sol. 9

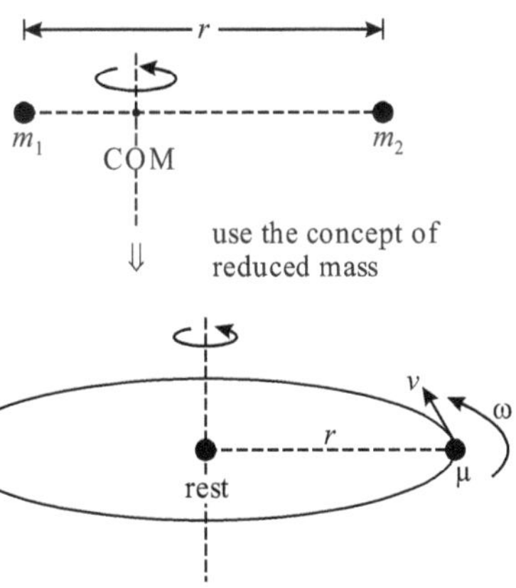

Rotational KE of system is given as

$$E = \frac{1}{2}I\omega^2 = \frac{1}{2}(\mu r^2)\left(\frac{v}{r}\right)^2 = \frac{1}{2}\mu v^2 \ldots(1)$$

Where $\left(\mu = \frac{m_1 m_2}{m_1 + m_2}\right)$

Using Bohr's quantization rule, we have

$$\mu v r = \frac{nh}{2\pi}$$

$$\Rightarrow \qquad v = \frac{nh}{2\pi\mu r} \qquad \ldots(2)$$

From equation (1) and (2) we have

$$E_{RE} = \frac{1}{2}\mu\left(\frac{nh}{2\pi\mu r}\right)^2$$

$$\Rightarrow \qquad E_{RE} = \frac{1}{2}\mu\left(\frac{n^2h^2}{4\pi^2\mu^2r^2}\right)$$

$$\Rightarrow \qquad E_{RE} = \frac{n^2h^2}{8\pi^2\left(\frac{m_1m_2}{m_1+m_2}\right)r^2} = \frac{(m_1+m_2)n^2h^2}{8\pi^2m_1m_2r^2}$$

Most appropriate among given options is option (A). **Ans. (A)**

Sol. 10 If an atom (*H*-atom) the electron is excited to n^{th} state than no of spectral lines in emission spectrum are

$$N = {}^nC_2 = \frac{n(n-1)}{2}$$

$$\Rightarrow \qquad N = \frac{4\times3}{2} = 6$$

Ans. (A)

Sol. 11 By Rydberg's formula we use

$$\frac{1}{\lambda} = RZ^2\left[\frac{1}{n_1^2} - \frac{1}{n_2^2}\right]$$

[We use $n_1 = n$ and $n_2 = n+1$]

$$\Rightarrow \qquad \frac{1}{\lambda} = RZ^2\left[\frac{1}{n^2} - \frac{1}{(n+1)^2}\right]$$

$$\Rightarrow \qquad \frac{1}{\lambda} = RZ^2\left[\frac{(n+1)^2 - n^2}{n^2(n+1)^2}\right]$$

$$\Rightarrow \qquad \frac{1}{\lambda} = R\left[\frac{2n+1}{n^2(n+1)^2}\right]$$

$$\Rightarrow \qquad \lambda = \frac{n^2(n+1)^2}{(2n+1)R}$$

for $n >> 1$

$$\Rightarrow \qquad \lambda = \frac{n^4}{2nR} = \frac{n^3}{2R}$$

Ans. (B)

Sol. 12

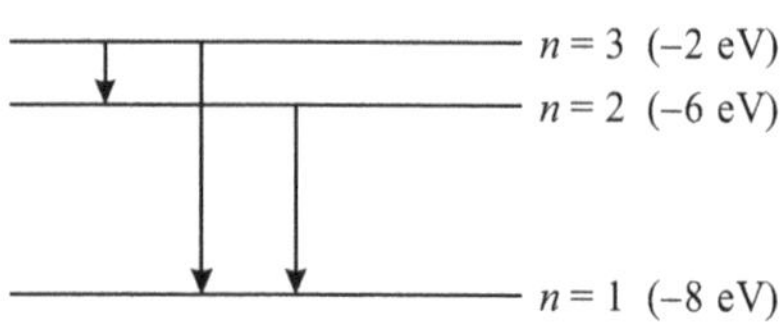

For the three energy levels there will be 3 emission lines in spectrum corresponding to energies

$$\Delta E_{32} = (-2) - (-6) = 4\text{ eV}$$

$$\Delta E_{31} = (-2) - (-8) = 6\text{ eV}$$

$$\Delta E_{21} = (-6) - (-8) = 2\text{ eV}$$

Wavelength emitted corresponding to the energies are

$$\lambda = \frac{hc}{E} = \frac{12431}{E}\text{ Å}$$

$$\lambda_{32} = \frac{12431}{4}\text{ Å} = 3108\text{ Å} \simeq 310\text{ nm}$$

$$\lambda_{31} = \frac{12431}{6}\text{ Å} = 2072\text{ Å} \simeq 207\text{ nm}$$

$$\lambda_{21} = \frac{12431}{2}\text{ Å} = 6215.5\text{ Å} \simeq 621\text{ nm}$$

$\Rightarrow$ 465 nm is not present in the spectrum **Ans. (B)**

Sol. 13 For Li^{+2} $Z = 3$ and by Rydberg's formula these wavelengths are given as :

$$\underset{n_2}{2} \longrightarrow \underset{n_1}{1} \qquad \frac{1}{\lambda_{21}} = R(3)^2\left[\frac{1}{1^2} - \frac{1}{2^2}\right] = 9R \times \frac{3}{4}$$

$$\Rightarrow \qquad \lambda_{21} = \frac{4}{27R}$$

$$\underset{n_2}{3} \longrightarrow \underset{n_1}{1} \qquad \frac{1}{\lambda_{31}} = R(3)^2\left[\frac{1}{1^2} - \frac{1}{3^2}\right] = 9R \times \frac{8}{9}$$

$$\Rightarrow \qquad \lambda_{31} = \frac{9}{72R}$$

$$\underset{n_2}{3} \longrightarrow \underset{n_1}{2} \qquad \frac{1}{\lambda_{32}} = R(3)^2\left[\frac{1}{2^2} - \frac{1}{3^2}\right] = 9R \times \frac{5}{36}$$

$$\Rightarrow \qquad \lambda_{32} = \frac{36}{45R}$$

$$\Rightarrow \qquad \frac{\lambda_{32}}{\lambda_{31}} = \frac{36/45R}{9/42R} = \frac{32}{5} = 6.4$$

$$\frac{\lambda_{21}}{\lambda_{31}} = \frac{4/27R}{9/72R} = \frac{32}{27} = 1.2$$

Ans. (C)

Sol. 14 Speed of electron in n^{th} orbit is given as

$$v_n = \frac{2\pi ke^2}{nh} \qquad (\text{for } z = 1)$$

$$\Rightarrow \qquad v_n \propto \frac{1}{n}$$

Graph-*B* shows the correct case **Ans. (A)**

Sol. 15 As α particles are positively charged so these must be repelled away from the nucleus so here option (C) is not possible.

Ans. (C)

Sol. 16 IP of Li^{+2} which is a hydrogenic atom can be given by Bohr model as energy required to remove third electron from Li^{+2}

$\Rightarrow \quad IP = 13.6\,Z^2 = 13.6\,(3)^2 = 13.6 \times 9 = 122.4\text{ eV}$

Total energy required to remove all 3 electrons from Li atom is

$$E = 5.4 + 75.6\text{ eV} + 122.4\text{ eV}$$

$\Rightarrow \quad E = 203.4\text{ eV}$ **Ans. (C)**

Sol. 17 If ν be the frequency of radiation emitted, we use

$$\Delta E = (IP)\,Z^2\left[\frac{1}{n_1^2} - \frac{1}{n_2^2}\right]$$

(For a transition $n_2 \to n_1$)

$$\Rightarrow \quad \Delta E = h\nu = (IP)\,Z^2\left[\frac{1}{(n-1)^2} - \frac{1}{n^2}\right]$$

$$\Rightarrow \quad \nu = \frac{(IP)Z^2}{h}\left[\frac{2n-1}{n^2(n-1)^2}\right]$$

Here $2n >> 1$ and $n >> 1$

$$\Rightarrow \quad \nu = \frac{(IP)(Z)^2}{h}\left(\frac{2n}{n^4}\right) = \frac{2IP(Z^2)}{h}\left(\frac{1}{n^3}\right)$$

$$\Rightarrow \quad \nu \propto \frac{1}{n^3}$$

Ans. (D)

Sol. 18 Equivalent current due to e^- motion in n^{th} orbit is given as

$$I_n = \frac{e\omega_n}{2\pi} = \frac{ev_n}{2\pi r_n}$$

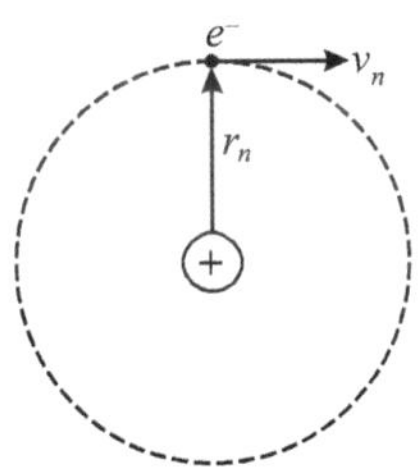

n^{th} orbit

So magnetic moment due to electron current in n^{th} orbit is

$$\mu_n = I_n \times \pi r_n^2 = \frac{ev_n}{2\pi r_n} \times \pi r_n^2$$

$$\mu_n = \frac{1}{2}\,ev_n r_n \qquad \ldots(1)$$

For n^{th} orbit (for $Z = 1$)

$$v_n = \frac{2\pi Ke^2}{nh}$$

and $$r_n = \frac{n^2h^2}{4\pi^2 Ke^2 m}$$

From equation-(1) we use

$$\mu_n = \frac{1}{2}e\left(\frac{2\pi Ke^2}{nh}\right) \times \left(\frac{n^2h^2}{4\pi^2 Ka^2 m}\right)$$

$$\Rightarrow \quad \mu_n = \frac{enh}{4\pi m}$$

Out of given options most appropriate is option (C).
But here for n^{th} excited state $n \leftarrow n+1$. **Ans. (C)**

Sol. 19

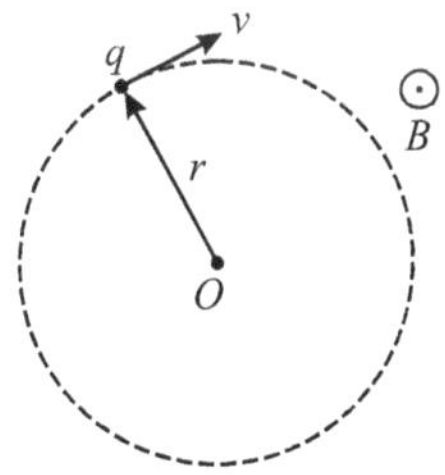

We use $$2\pi r = n\lambda = n\left(\frac{h}{mv}\right)$$

$$\Rightarrow \quad 2\pi r = \frac{nh}{mv} \qquad \ldots(1)$$

For circular motion in magnetic field, we have

$$qvB = \frac{mv^2}{r}$$

$$\Rightarrow \quad mv = qBr \qquad \ldots(2)$$

From (1) and (2) we use

$$2\pi r = \frac{nh}{qBr}$$

$$\Rightarrow \quad 2\pi qBr^2 = nh$$

$\Rightarrow \quad r = \sqrt{n}$ **Ans. (C)**

Sol. 20

We know the excitation energies of H-atom are :

$$\Delta E_{1\to 2} = 10.2\text{ eV}$$
$$\Delta E_{1\to 3} = 12.09\text{ eV}$$
$$\Delta E_{1\to 4} = 12.75\text{ eV}$$

By the collisions with incident electron, the H-atom can get excited max to $n = 3$ level

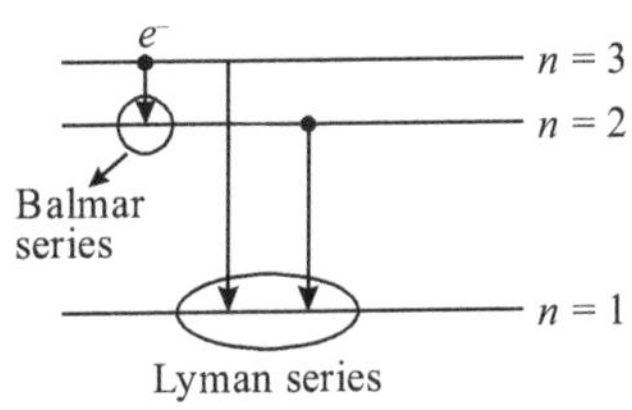

Ans. (A)

Sol. 21 For a revolving electron in speed v_n and radius r_n for n^{th} orbit, we use :

$$F = \frac{mv_n^2}{r_n}$$

$$\Rightarrow \quad \frac{e^2}{4\pi E_0}\left(\frac{1}{r_n^2}+\frac{\beta}{r_n^3}\right) = \frac{mv_n^2}{r_n} \qquad \ldots(1)$$

By using Bohr's quantization postulate we use

$$\Rightarrow \quad mv_n r_n = \frac{nh}{2\pi} \qquad \ldots(2)$$

$$\Rightarrow \quad v_n = \frac{nh}{2\pi m r_n}$$

From equation (1) we have

$$\frac{e^2}{4\pi\varepsilon_0}\left[\frac{1}{r_n}+\frac{\beta}{r_n^2}\right] = m\left[\frac{nh}{2\pi m r_n}\right]^2$$

We multiply r_n^2 in LHS and RHS gives :

$$\frac{e^2}{4\pi\varepsilon_0}[r_n+\beta] = \frac{n^2h^2}{4\pi^2 m}$$

$$\Rightarrow \quad r_n = n^2\frac{h^2\varepsilon_0}{e^2\pi m} - \beta$$

$$\Rightarrow \quad r_n = n^2 a_0 - \beta$$

Where $a_0 = \dfrac{h^2\varepsilon_0}{e^2\pi m}$ **Ans. (C)**

Sol. 22 For a given transition $n_2 \to n_1$, the wavelength emitted can be given as :

$$\frac{1}{\lambda} = RZ^2\left[\frac{1}{n_1^2}-\frac{1}{n_2^2}\right]$$

As $n_2 = 2$ and $n_1 = 1$ for all

$$\Rightarrow \quad \lambda \propto \frac{1}{Z^2}$$

For ${}_1H^1$ and ${}_1H^2$ we take $Z=1$

$$\Rightarrow \quad \lambda_1 \cong \lambda_2$$

and for $({}_2He^4)^+$ $Z=2$

$$\Rightarrow \quad \lambda_3 = \frac{\lambda_1}{4}$$

and for $({}_3Li^6)^{+2}$ $Z=3$

$$\Rightarrow \quad \lambda_4 = \frac{\lambda_1}{9}$$

$$\Rightarrow \quad \lambda_1 = \lambda_2 = 4\lambda_3 = 9\lambda_4$$ **Ans. (C)**

Sol. 23 For a given hydrogenic atom (Li^{+2}) the ionization energy for an electron in n^{th} state is given as :

$$IP_n = 13.6\,\frac{Z^2}{n^2}$$

Here $Z = 3$ and for first excited state we use $n=2$

$$\Rightarrow \quad IP_{n=2} = 13.6 \times \frac{(3)^2}{(2)^2}$$

$$= \frac{13.6\times 9}{4} = 30.3 \text{ eV}$$ **Ans. (B)**

Sol. 24 *KE*, *PE* and *TE* of electron in a hydrogenic atom according to Bohr model is given as :

$$TE = -\frac{KZe^2}{2r_n};\ KE = \frac{KZe^2}{2r_n};\ PE = -\frac{KZe^2}{r_n}$$

When electron makes a transition from excited state to ground state r_n decreases.

As *KE* is positive

$\Rightarrow$ $KE \uparrow$ (+ve); $PE \downarrow$ (–ve); $TE \downarrow$ (–ve) **Ans. (A)**

Sol. 25 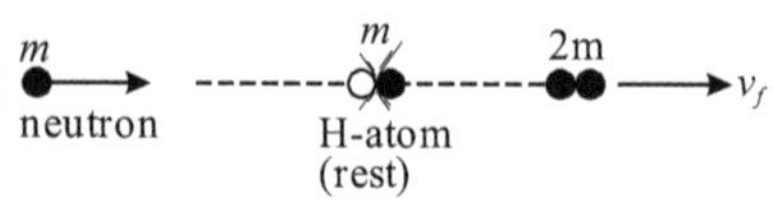

By conservation of momentum if inelastic collision occurs

$$mv = 2mv_f$$

$$\Rightarrow \quad v_f = \frac{v}{2}$$

Loss in *KE* $\quad \Delta E_2 = \frac{1}{2}mv^2 - \frac{1}{2}(2m)\left(\frac{v}{2}\right)^2 = \frac{1}{4}mv^2$

For this collision to happen ΔE_2 must be minimum 10.2 eV (Minimum energy which can be absorbed by *H*-atom in excitation from $n=1$ to $n=2$.)

$$\Rightarrow \quad \frac{1}{4}mv^2 = 10.2 \text{ eV}$$

$$\Rightarrow \quad K_{in} = 20.4 \text{ eV}$$ **Ans. (A)**

Sol. 26 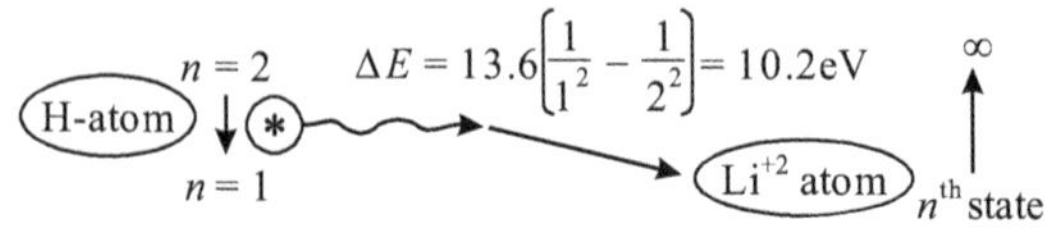

Energy required to completely remove the electron of Li^{+2} in n^{th} state is

$$\Delta E_{IP(n)} = \frac{13.6Z^2}{n^2} = \frac{13.6(3)^2}{n^2}$$

$$\Rightarrow \quad 10.2 \geq \frac{13.6\times 9}{n^2}$$

$$\Rightarrow \quad n^2 \geq \frac{13.6\times 9}{10.2}$$

$$\Rightarrow \quad n^2 \geq 12$$

$$\Rightarrow \quad n \geq 2\sqrt{3} \cong 3.5$$ **Ans. (B)**

Sol. 27 For first transition, we use

$$\frac{hc}{\lambda_1} = -E-(-2E) = E$$

$$\Rightarrow \quad \lambda_1 = \frac{hc}{E} \quad \ldots(1)$$

For second transition, we use

$$\frac{hc}{\lambda_2} = -E-\left(-\frac{4E}{3}\right) = \frac{E}{3}$$

$$\Rightarrow \quad \lambda_2 = \frac{3hc}{E} \quad \ldots(2)$$

From equation-(1) and (2)

$$\frac{\lambda_1}{\lambda_2} = \frac{hc/E}{3hc/E} = \frac{1}{3}$$

Ans. (B)

Sol. 28 Due to motion of electron in n^{th} orbit, the magnetic induction at centre is given by Biot Savart's law as

$$B = \frac{\mu_0}{4\pi}\frac{qv\sin\theta}{r^2} = \frac{\mu_0}{4\pi}\cdot\frac{ev_n}{r_n^2}$$

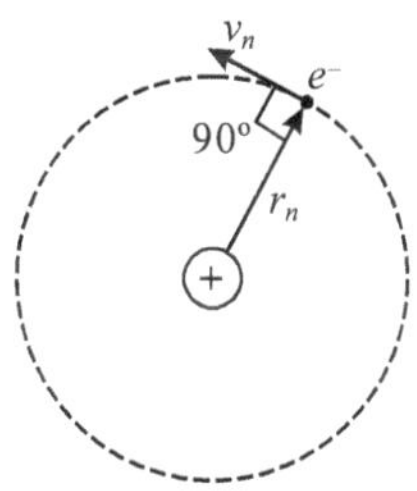

As for n^{th} orbit – for $Z = 1$

$$v_n = \frac{2\pi ke^2}{nh}$$

$$r_n = \frac{n^2h^2}{4\pi^2ke^2m}$$

$$\Rightarrow \quad B_n = \frac{\mu_0}{4\pi}e \times \frac{2\pi ke^2}{nh} \times \left(\frac{4\pi^2ke^2m}{n^2h^2}\right)^2$$

$$\Rightarrow \quad B_n \propto n^{-5}$$

Ans. (B)

Sol. 29 From Bohr's II Postulate we use

$$mvr = \frac{h}{2\pi} \; [n=1] \quad \ldots(1)$$

Acceleration of electron in first orbit of *H*-atom is :

$$a = \frac{v^2}{r} \quad \ldots(2)$$

From equation-(1) we use

$$v = \frac{h}{2\pi mr}$$

From equaton-(2) we have

$$a = \left(\frac{h}{2\pi mr}\right)^2 \times \frac{1}{r} = \frac{h^2}{4\pi^2m^2r^3}$$

Ans. (C)

Sol. 30 Series limit of Lyman series is corresponding to transition $\infty \to 1$

Here by using Rydberg's formula, we have

$$\nu_L = \frac{c}{\lambda_{\infty\to1}} = c\left[RZ^2\left(\frac{1}{1^2}-\frac{1}{\infty^2}\right)\right]$$

$$\Rightarrow \quad \nu_L = RcZ^2 \quad \ldots(1)$$

For Pfund series limit is corresponding to transition $\infty \to 5$

So we use

$$\nu_F = \frac{c}{\lambda_{\infty\to5}} = c\left[RZ^2\left(\frac{1}{5^2}-\frac{1}{\infty^2}\right)\right]$$

$$\Rightarrow \quad \nu_F = RcZ^2\left(\frac{1}{25}\right) = \frac{\nu_L}{25} \quad \ldots(1)$$

Sol. 31 For transition from n^{th} state to ground state emitted wavelength is given as

$$\frac{1}{\Lambda_n} = RZ^2\left[\frac{1}{1^2}-\frac{1}{n^2}\right]$$

As $n >> 1$

$$\Lambda_n = \frac{1}{RZ^2}\left(1-\frac{1}{n^2}\right)^{-1}$$

$$\Lambda_n = \frac{1}{RZ^2}\left(1+\frac{1}{n^2}\right) \quad \ldots(1)$$

[As $\frac{1}{n^2} << 1$ we use $(1+x)^b \simeq (1+xb)$]

For n^{th} state de-Broglie wavelength of e^- is given as :

$$\lambda_n = \frac{h}{p_n} = \frac{h}{mv}$$

By angular momentum quantization postulate of Bohr model, we use

$$mvr = \frac{nh}{2\pi}$$

$$\Rightarrow \quad \lambda_n = \frac{h}{mv}$$

$$= \frac{2\pi}{n} \times (0.529 \times n^2) \simeq 1.06\,\pi n \quad \ldots(2)$$

From equation-(1) we have

$$\Lambda_n = \frac{1}{RZ^2} + \frac{1}{RZ^2n^2}$$

$$\Rightarrow \quad \Lambda_n = \frac{1}{RZ^2} + \frac{(1.06\pi)^2}{RZ^2}\cdot\frac{1}{\lambda_n^2}$$

$$\Rightarrow \quad \Lambda_n = A + \frac{B}{\lambda_n^2}$$

Ans. (A)

Sol. 32 Radius of nth orbit for Muonic atom is

$$r_\mu = \frac{n^2h^2}{4\pi^2kZe^2m_\mu} \simeq \frac{n^2h^2}{4\pi^2kZe^2m_e} \times \frac{1}{200} = \frac{r_e}{200}$$

[As $m_\mu = 200\,m_e$]

$\Rightarrow$ Option (a) is correct.

Speed of muon in n^{th} orbit can be given as

$$v_\mu = \frac{2\pi kZe^2}{nh}$$

[It is independent of mass of orbiting particle]

$$\Rightarrow \quad v_\mu = v_e$$

$\Rightarrow$ Option (b) is NOT correct.

The ionization energy in Bohr model is given as

$$E_\mu = \frac{2\pi^2 k^2 Z^2 e^4 m_\mu}{n^2 h^2}$$

$$\simeq \frac{2\pi^2 k^2 Z^2 e^4 m_e}{n^2 h^2} \times 200 = 200\, E_e$$

$\Rightarrow$ Option (c) is correct.

Momentum of muon in atom is given as :

$$P_\mu = m_\mu v_\mu = 200\, m_e \cdot v_e = 200\, P_e$$

$\Rightarrow$ Option (d) is correct. **Ans. (D)**

Sol. 33 $He \xrightarrow{E_1} He^{+} \xrightarrow{E_2} He^{+2}$

Total energy required to completely ionize He atom is $(E_1 + E_2)$.

As He^+ is a hydrogenic atom (single electron system) so E_2 in the *IP* for He^+ which is given by Bohr model as

$$E_2 = 13.6 Z^2 = 13.6 \times (2)^2 = 13.6 \times 4 = 54.4 \text{ eV}$$

It is given that

$$E_2 = 2.2\, E_1$$

$$\Rightarrow \quad E_1 = \frac{54.4}{2.2} \times \frac{272}{11} = 24.72 \text{ eV}$$

Total required energy is $E_1 + E_2 = 54.4 + 24.72 = 79.12$ eV

Ans. (B)

Sol. 34 Energy of photon $= \dfrac{12500}{980} = 12.75$ eV

So electron will excite to $n = 4$ state

As radius of nth orbit $r_n \propto n^2$ so the radius of atom will be $16a_0$. **Ans. (D)**

Sol. 35 For $M \to L$ shell transition, we use

$$\frac{1}{\lambda} = K\left(\frac{1}{2^2} - \frac{1}{3^2}\right) = \frac{K \times 5}{36} \quad \ldots (1)$$

For $N \to L$ shell transition, we use

$$\frac{1}{\lambda'} = K\left(\frac{1}{2^2} - \frac{1}{4^2}\right) = \frac{K \times 3}{16} \quad \ldots (2)$$

From (1) and (2)

$$\Rightarrow \quad \lambda' = \frac{20}{27}\lambda$$

Ans. (C)

Sol. 36 For the given circular orbit, we use

$$F = \frac{dU}{dr} = kr = \frac{mv^2}{r} \quad \ldots (1)$$

and

$$mvr = \frac{nh}{2\pi} \quad \ldots (2)$$

$$\Rightarrow \quad r^2 \propto n$$

$$\Rightarrow \quad r \propto \sqrt{n}$$

Energy of particle in n^{th} energy level is

$$E = \frac{1}{2}kr^2 + \frac{1}{2}mv^2$$

$$\Rightarrow \quad E = \frac{1}{2}kr^2 + \frac{1}{2}kr^2 = kr^2$$

$$\Rightarrow \quad E \propto n$$

Ans. (D)

Sol. 37 The energy lost by electrons in excitation of mercury atoms is

$$\Delta E = 5.6 - 0.7 = 4.9 \text{ eV}$$

Minimum wavelength (maximum energy) of photons emitted by mercury atoms in de-excitation is given as

$$\lambda = \frac{hc}{\Delta E} = \frac{1240}{4.9} = 253 \text{ nm}$$

Ans. (C)

Sol. 38 Energy released for transition $n = 2$ to $n = 1$ of hydrogen atom is given as

$$E = 13.6\, Z^2\left(\frac{1}{n_1^2} - \frac{1}{n_2^2}\right)$$

For $Z = 1$, $n_1 = 1$, $n_2 = 2$, we get

$$E = 13.6 \times 1 \times \left(\frac{1}{1^2} - \frac{1}{2^2}\right)$$

$$\Rightarrow \quad E = 13.6 \times \frac{3}{4} \text{ eV}$$

For He^+ ion $z = 2$

(A) $n = 1$ to $n = 4$

$$E = 13.6 \times 2^2 \times \left(\frac{1}{1^2} - \frac{1}{4^2}\right) = 13.6 \times \frac{15}{4} \text{ eV}$$

(B) $n = 2$ to $n = 4$

$$E = 13.6 \times 2^2 \times \left(\frac{1}{2^2} - \frac{1}{4^2}\right) = 13.6 \times \frac{3}{4} \text{ eV}$$

(C) $n = 2$ to $n = 5$

$$E = 13.6 \times 2^2 \times \left(\frac{1}{2^2} - \frac{1}{5^2}\right) = 13.6 \times \frac{21}{25} \text{ eV}$$

(D) $n = 2$ to $n = 3$

$$E = 13.6 \times 2^2 \times \left(\frac{1}{2^2} - \frac{1}{3^2}\right) = 13.6 \times \frac{5}{9} \text{ eV}$$

Energy required for transition of He^+ for $n = 2$ to $n = 4$ matches exactly with energy released in transition of H for $n = 2$ to $n = 1$

Ans. (B)

Sol. 39 Using Rydberg's formula for first line of Balmer series, we have

$$\frac{1}{660} = R\left(\frac{1}{2^2} - \frac{1}{3^2}\right) = \frac{5R}{36} \qquad \dots(1)$$

For second line of Balmer series, we have

$$\frac{1}{\lambda} = R\left(\frac{1}{2^2} - \frac{1}{4^2}\right) = \frac{3R}{16} \qquad \dots(2)$$

From equations-(1) and (2), we use

$$\frac{\lambda}{660} = \frac{5 \times 16}{36 \times 3}$$

$$\Rightarrow \quad \lambda = \frac{4400}{9} = 488.88 \simeq 488.9 \text{ nm}$$ **Ans. (C)**

Sol. 40 Energy levels in Hydrogen like atom are given as

$$E = -13.6 \frac{Z^2}{n^2} \text{ eV}$$

As He^+ is 1st excited state, we use

$$z = 2, n = 2$$

$$\Rightarrow \quad E = -13.6 \text{ eV}$$

As total energy of He^+ in 1st excited state is –13.6 eV, its ionisation energy will be +13.6 eV. **Ans. (B)**

Sol. 41 For 6 spectral lines emission, excitation must be upto $n = 4$

$\Rightarrow$ Energy required for transition from $n = 1$ to $n = 4$ is

$$\Delta E = E_4 - E_1$$

$$= (-0.85 z^2) - (-13.6 z^2) \text{ eV}$$

$$= 12.75 z^2 \text{ eV}$$

Wavelength of radiation of this energy

$$12.75 \times 9 = \frac{hc}{\lambda}$$

or $$\lambda = \frac{hc}{9 \times 12.75e} = \frac{12431}{9 \times 12.75} \text{Å} \simeq 108 \text{ Å}$$ **Ans. (C)**

Sol. 42

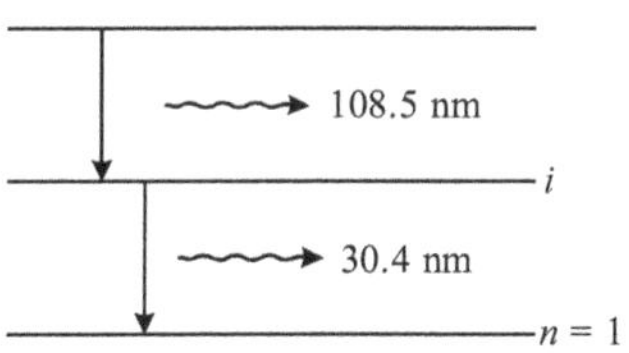

$$\Delta E_1 = \frac{1240}{108.5} \text{ eV} = 11.73 \text{ eV}$$

$$\Delta E_2 = \frac{1240}{30.4} \text{ eV} = 40.79 \text{ eV}$$

$$= 52.22 \text{ eV}$$

$$E_n - E_1 = 52.22 \text{ eV}$$

For He^+, we use $z = 2$

$$\left(-\frac{13.6z^2}{n^2}\right) - (-13.6z^2) = 52.22 \text{ eV}$$

$$54.4 - \frac{54.4}{n^2} = 52.22 \text{ eV}$$

$$\Rightarrow \quad 54.4 - 52.22 = \frac{54.4}{n^2}$$

$$\Rightarrow \quad n^2 = \frac{54.4}{2.18} = 24.95 \simeq 25$$

$$\Rightarrow \quad n = 5$$ **Ans. (A)**

Sol. 43 By Rydberg's formula

$$\frac{1}{\lambda} = R\left(\frac{1}{n_1^2} - \frac{1}{n_2^2}\right)$$

For $n = 4 \rightarrow n = 3$ transition

$$\frac{1}{\lambda_1} = R\left[\frac{1}{3^2} - \frac{1}{4^2}\right] = \frac{16-9}{144} R$$

$$\Rightarrow \quad \lambda_1 = \frac{144}{7R} \qquad \dots(1)$$

For $n = 3 \rightarrow n = 2$ transition

$$\frac{1}{\lambda_2} = R\left[\frac{1}{2^2} - \frac{1}{3^2}\right] = \frac{9-4}{36} R$$

$$\Rightarrow \quad \lambda_2 = \frac{36}{5R} \qquad \dots(2)$$

(1)/(2) gives $$\frac{\lambda_1}{\lambda_2} = \frac{144/7}{36/5} = \frac{20}{7}$$ **Ans. (D)**

Sol. 44 De-Broglie wavelength is given as

$$\lambda = \frac{h}{mv} = \frac{hr}{mvr}$$

By Bohr's II Postulate, we use

$$mvr = \frac{hr}{2\pi}$$

$$\Rightarrow \quad \lambda = \frac{hr}{(hr/2\pi)} = \frac{2\pi r}{n} \qquad [n = 3]$$

$$\Rightarrow \quad \lambda = \frac{2 \times 3.14 \times 4.65 \times 10^{-10}}{3}$$

$$\Rightarrow \quad \lambda \simeq 9.7 \text{ Å}$$ **Ans. (C)**

27.1 Radioactivity

Sol. 1 Only charged particles moving in field experience magnetic force and get deflected

$\Rightarrow$ At the instant emission can contain

(i) e^- (– ve)

(ii) protons (+ ve)

(iii) He^{+2} (+ ve)

Will get deflected in MF **Ans. (A)**

Sol. 2 By Radioactive Decay Law we have

$$N = N_0 2^{-t/T}$$

at $T = 5$ years and $t = 15$ years

$$\Rightarrow \quad N = N_0 (2)^{-15/5}$$

$$\Rightarrow \quad N = \frac{N_0}{2^3} = \frac{N_0}{8}$$ **Ans. (A)**

Sol. 3 Using Radioactive decay law, we have

$$A_c = A_{co} e^{-\lambda t}$$

here at $t = 0, A_{co} = 5000$ dis/min

at $t = 5$ min, $A_c = 1250$ dis/mim

$$\Rightarrow \quad 1250 = 5000\, e^{-\lambda(5)}$$

$$\Rightarrow \quad e^{5\lambda} = \frac{5000}{1250} = 4$$

$$\Rightarrow \quad 5\lambda = \ln(4) = 2\ln(2)$$

$$\Rightarrow \quad \lambda = \frac{2}{5}\ln(2) = 0.4\ln(2)\,\text{min}^{-1}$$ **Ans. (A)**

Sol. 4 Least wavelength $\Rightarrow$ maximum energy

here γ-rays have highest energy

So it will be having least wavelength among other option

"fact based question" **Ans. (A)**

Sol. 5

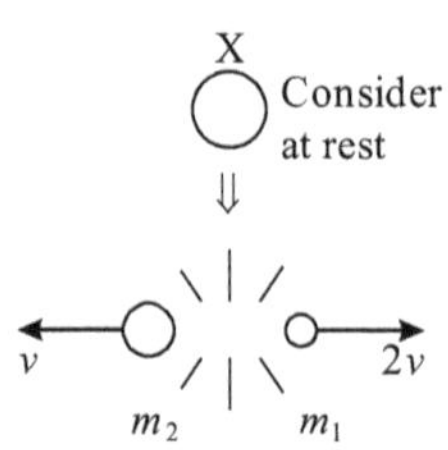

By conservation of linear momentum

$$m_1(2v) = m_2 v$$

$$m_2 = 2m_1$$

$$\Rightarrow \quad A_2 = 2A_1 \quad A_1, A_2 \text{ are molar atomic masses}$$

By concept of fermi radius, the nuclear radius is given as

$$R = R_0 (A)^{1/3}$$

$$\Rightarrow \quad R_1 = R_0 (m_1)^{1/3}$$

and

$$R_2 = R_0 (2m_1)^{1/3}$$

$$\Rightarrow \quad \frac{R_1}{R_2} = \frac{1}{2^{1/3}}$$ **Ans. (D)**

Sol. 6 We use formula of fermi radius of a Nu, given as

$$R = R_0 (A)^{1/3}$$

For ^{27}Al we use

$$3.6 \times 10^{-15} = R_0 (27)^{1/3} \quad \ldots(1)$$

For ^{125}Te we use

$$R = R_0 (125)^{1/3} \quad \ldots(2)$$

(2) and (1) gives $\rightarrow$ $\frac{R}{3.6 \times 10^{-15}} = \left(\frac{125}{27}\right)^{1/3} = \frac{5}{3}$

$$\Rightarrow \quad R = \frac{5}{3} \times 3.6 \times 10^{-15}\,\text{m}$$

$$= 6 \times 10^{-15}\,\text{m} = 6 \text{ fermi}$$ **Ans. (A)**

Sol. 7 Using RA decay equation we have

$$N = N_0 (2)^{-t/T}$$

At $t = 15$ min, $N = \frac{N_0}{8}$ as $\frac{7}{8}N_0$ decay in times 15 min

$$\Rightarrow \quad \frac{N_0}{8} = N_0 (2)^{-15/T}$$

$$\Rightarrow \quad 2^{-3} = 2^{-15/T}$$

$$\Rightarrow \quad \frac{15}{T} = 3$$

$$\Rightarrow \quad T = \frac{15}{3} = 5\,\text{min}$$ **Ans. (C)**

Sol. 8 Half life of X = Mean life of Y

If λ_X and λ_Y are decay constants of elements X and Y, we have

$$\frac{\ln Z}{\lambda_X} = \frac{1}{\lambda_Y}$$

$$\Rightarrow \quad \lambda_X = (\ln 2)\lambda_Y = 0.696\,\lambda_Y$$

$$\Rightarrow \quad \lambda_X < \lambda_Y$$

$\Rightarrow$ For same N_0 we use $\lambda_X N_0 < \lambda_Y N_0$ **Ans. (B)**

Sol. 9 From *RA* decay equation we have

$$N = N_0 (2)^{-t/T} \quad [T = 20 \text{ min}]$$

at $t = t_2$ we use $N = \frac{N_0}{3}$ (as $\frac{2}{3}N_0$ has decayed)

$$\Rightarrow \quad \frac{N_0}{3} = N_0 (2)^{-t_2/20} \quad \Rightarrow \quad \frac{t_2}{20} = \log_2(3)$$

$$\Rightarrow \quad t_2 = 20\log_2(3) \quad \Rightarrow \quad \log_2(3) = \frac{t_2}{20}$$

at $t = t_1$ we use $N = \frac{2N_0}{3}$ (as $\frac{N_0}{3}$ has decayed)

$$\Rightarrow \quad \frac{2N_0}{3} = N_0(2)^{-t_1/20}$$

$$\Rightarrow \quad \frac{t_1}{20} = \log_2\left(\frac{3}{2}\right) = \log_2(3) - 1$$

$$\Rightarrow \quad \frac{t_1}{20} = \frac{t_2}{20} - 1 \quad \Rightarrow \quad t_2 - t_1 = 20 \text{ min}$$ **Ans. (A)**

Sol. 10 By RA decay equation we have $N = N_0(2)^{-t/T}$

For $A \rightarrow N_A = N_0(2)^{-80/20} = \frac{N_0}{2^4} = \frac{N_0}{16}$

Number of nuclei of A decayed in 80 min are

$$N_{AD} = N_0 - N_A = N_0 - \frac{N_0}{16} = \frac{15N_0}{16}$$

For $B \rightarrow N_B = N_0(2)^{-80/20} = \frac{N_0}{2^2} = \frac{N_0}{4}$

Number of nuclei of B decayed in 80 min are

$$N_{BD} = N_0 - N_B = N_0 - \frac{N_0}{4} = \frac{3N_0}{4}$$

$$\Rightarrow \quad \frac{N_{AD}}{N_{BD}} = \frac{15N_0/16}{3N_0/4} = \frac{5}{4}$$ **Ans. (D)**

Sol. 11 Using radioactive decay equation we have

$$A_C = A_{C_0}(2)^{-t/T}$$

at

$t = 0,\ A_{C_0} = 1600$ counts/s

$t = 8\text{s},\ A_C = 100$ counts/s

$$\Rightarrow \quad 100 = 1600(2)^{-8/T}$$

$$\Rightarrow \quad 2^{-8/T} = 2^4 \quad \Rightarrow \quad \frac{8}{T} = 4 \quad \Rightarrow \quad T = 2\text{ s}$$

at $t = 6\text{s}$, A_C is given as

$$A_C = 1600(2)^{-6/2} = \frac{1600}{2^3} = \frac{1600}{8} = 200 \text{ counts/s}$$ **Ans. (D)**

Sol. 12 Using radioactive decay equation at $t = 1$ yr we have ($T \rightarrow$ half life in yrs)

$$N_1 = N_0(2)^{-1/T} \quad [N_0 = 10^{20} \text{ atoms}]$$

at $t = 2$ yr $\quad N_2 = N_0(2)^{-2/T}$

at $t = 3$ yr $\quad N_3 = N_0(2)^{-3/T}$

N_0 of α-particles emitted in 3rd yr

$$\Delta N_3 = N_2 - N_3 = N_0((2)^{-2/T} - (2)^{-3/T}) \quad \ldots(1)$$

N_0 of α-particles emitted in 2nd yr

$$\Delta N_2 = N_1 - N_2 = N_0((2)^{-1/T} - (2)^{-2/T}) \quad \ldots(2)$$

N_0 of α-particles emitted in 1st yr

$$\Delta N_1 = N_0 - N_1 = N_0(1 - (2)^{-1/T}) \quad \ldots(3)$$

From (1) and (2) we have

$$\frac{\Delta N_3}{\Delta N_2} = \frac{N_0(2^{-2/T} - 2^{-3/T})}{N_0(2^{-1/T} - 2^{-2/T})} = 0.3$$

$$\Rightarrow \quad 2^{-2/T} - 2^{-3/T} = 0.32^{-1/T} - 0.32^{-2/T}$$

$$\Rightarrow \quad 1.32^{-2/T} - 2^{-3/T} = -0.32^{-1/T} = 0$$

$$\Rightarrow \quad 1.32^{-1/T} - 2^{-2/T} = -0.3 = 0 \text{ [multiplied the equation with } 2^{1/T}]$$

Using $2^{-1/T} = x \quad \Rightarrow \quad x^2 - 1.3x + 0.3 = 0$

$$\Rightarrow \quad x = \frac{1.3 \pm \sqrt{1.69 - 1.2}}{2} = \frac{1.3 \pm 0.7}{2} = 1 \text{ or } 0.3$$

Here we use $\quad 2^{-1/T} = 0.3$

From equation-(1) we have

$$\Delta N_1 = N_0(1 - 2^{-1/T}) = N_0(1 - 0.3) = 0.7\,N_0$$

$$\Rightarrow \quad \Delta N_1 = 0.7 \times 10^{20} = 7 \times 10^{19}$$ **Ans. (D)**

Sol. 13 (A) ✓ $A_C = \lambda N$ independent of any extra factor fission

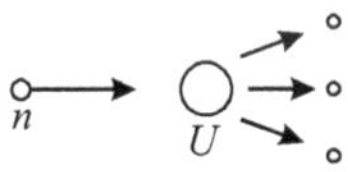

using moderator fission rate can be controlled

(B) × Nuclear forces are independent of charge

(C) × Nuclei having same number of neutrons are called isotones

(D) × For photons momentum $P = \frac{h}{\lambda} \rightarrow$

for matter waves $\lambda = \frac{h}{mv}$ **Ans. (A)**

Sol. 14 In case of simultaneous decay modes of a radioactive element equivalent decay constant is given as

$$\lambda_{eq} = \lambda_\alpha + \lambda_\beta \quad [\lambda_\alpha = \frac{\ln_2}{T_\alpha} \text{ and } \lambda_\beta = \frac{\ln_2}{T_\beta}]$$

Average half life of element is

$$T = \frac{\ln_2}{\lambda_{eq}} = \frac{\ln_2}{\lambda_\alpha + \lambda_\beta}$$

$$\Rightarrow \quad T = \frac{\ln_2}{\frac{\ln_2}{T_\alpha} + \frac{\ln_2}{T_\beta}} = \frac{T_\alpha T_\beta}{T_\alpha + T_\beta}$$ **Ans. (C)**

Sol. 15 Decay rate of a radioactive element is given as $A_C = \lambda N$.
Given that half life of A = mean life of B

$$\Rightarrow \quad \frac{\ln 2}{\lambda_A} = \frac{1}{\lambda_B}$$

$$\Rightarrow \quad \lambda_A = \ln(2) \cdot \lambda_B = 0.693\,\lambda_B$$

$\Rightarrow \quad \lambda_B > \lambda_A$ so B will decay faster than A **Ans. (D)**

Sol. 16 If a radioactive nuclei element is produced at a constant rate r nuclei/s than the rate of accumulation of these nuclei is given as

$$\frac{dN}{dt} = r - \lambda N$$

$$\Rightarrow \quad \int_0^N \frac{dN}{r-\lambda N} = \int_0^t dt$$

$$\Rightarrow \quad -\frac{1}{\lambda}\left[\ln(r-\lambda N)\right]_0^N = t$$

$$\Rightarrow \quad \ln\left(\frac{r-\lambda N}{r}\right) = -\lambda t$$

$$\Rightarrow \quad \ln\frac{r-\lambda N}{r} = e^{-\lambda t}$$

$$\Rightarrow \quad N = \frac{r}{\lambda}(1-e^{-\lambda t}) \qquad \ldots(1)$$

From equation-(1) $N = 50$ at time t gives

$$50 = \frac{1000}{0.5}(1-e^{-0.5t})$$

$$\Rightarrow \quad 50 = 200(1-e^{-0.5t})$$

$$\Rightarrow \quad \frac{1}{4} = 1 - e^{-0.5t}$$

$$\Rightarrow \quad e^{0.5t} = \frac{4}{3}$$

$$\Rightarrow \quad 0.5t = \ln\left(\frac{4}{3}\right)$$

$$\Rightarrow \quad t = 2\ln\left(\frac{4}{3}\right)\text{s}$$ **Ans. (B)**

Sol. 17 Activity at $t = 0$ (living animal) is $A_{C_0} = 16$ dpm/gm
If animal died t yrs ago then activity now is $A_C = 12$ dpm/gm
using radioactive decay equation we have

$$A_C = A_{C_0}(2)^{-t/T}$$

$$\Rightarrow \quad 12 = 16(2)^{-t/T}$$

$$\Rightarrow \quad 2^{-t/5760} = \frac{4}{3}$$

$$\Rightarrow \quad \frac{t}{5760} \times \log 2 = 2\log 2 - \log 3$$

$$\frac{t}{5760} \times 0.301 = 0.602 - 0.477 = 0.125$$

$$\Rightarrow \quad t = \frac{0.125 \times 5760}{0.301} \simeq 2392 \text{ yrs}$$ **Ans. (B)**

Sol. 18 By radioactive decay equation, we have

$$A_C = A_{C_0}(2)^{-t/T} \qquad \ldots(1)$$

From freshly cut tree at $t = 0$ activity is $A_{C_0} = 20$ dpm
If a tree was cut t yrs ago its activity is $A_C = 2$ dpm
from equation-(1) $\quad 2 = 20(2)^{-t/5730}$

$$2^{t/5730} = 10$$

$$\Rightarrow \quad t = \frac{\log 10}{\log 2} \times 5730$$

$$\Rightarrow \quad t = \frac{5730}{0.301} = 19036.55 \text{ yrs.}$$ **Ans. (C)**

Sol. 19 Number of Na^{24} nuclei in beginning (at $t = 0$) are

$$N_0 = \frac{1\times10^{-3}\times6.023\times10^{23}}{24\times10^{-3}}$$

$$\Rightarrow \quad N_0 = \frac{6.023\times10^{23}}{24} \cong 0.25\times10^{23}$$

by radioactive decay equation number of Na^{24} nuclei left at $t = 7.5$ hr are

$$N = N_0(2)^{-t/T} = N_0(2)^{-7.5/15} = \frac{N_0}{\sqrt{2}}$$

Number of decays in 7.5 hr are

= Number of β emissions

$$N_\beta = N_0 - \frac{N_0}{\sqrt{2}} = N_0\left(1-\frac{1}{\sqrt{2}}\right)$$

$$\Rightarrow \quad N_\beta = N_0(1-0.707) \simeq 0.3\,N_0$$

$$\Rightarrow \quad N_\beta = 0.3\times0.25\times10^{23} = 7.5\times10^{21}$$ **Ans. (B)**

Sol. 20 After time t, number of undecayed nuclei of A is given by radioactive decay equation as

$$N_A = N_0(2)^{-t/T} \qquad \ldots(1)$$

[where $N_0 \rightarrow$ number of nuclei of A at $t = 0$]

Number of nuclei of B at time t are

$$N_B = N_0 - N_A = N_0(1-(2)^{-t/T}) \qquad \ldots(2)$$

$$(2)\div(1) \quad \Rightarrow \quad \frac{N_B}{N_A} = \frac{N_0[1-(2)^{-t/T}]}{N_0(2)^{-t/T}} = 0.3$$

$$\Rightarrow \quad 1-(2)^{-t/T} = 0.3(2)^{-t/T}$$

$$\Rightarrow \quad 1.3(2)^{-t/T} = 1$$

$$\Rightarrow \quad 2^{t/T} = 1.3$$

$$\Rightarrow \quad t = \frac{T\log(1.3)}{\log(2)}$$ **Ans. (D)**

Sol. 21 At $t = 0$, Activity of sample is $A_{C_0} = 0.8\ \mu Ci$

$$= 0.8 \times 10^{-6} \times 3.7 \times 10^{10} \text{ dps}$$

At $t = 10$hrs activity/cm^3 is given as AC/cm^3 = 300 dpm

$$= \frac{300}{60} = 5 \text{ dps}$$

If V is the volume of animal's blood then total activity at

$$t = 10 \text{ hr is } A_C = 5V \text{ dps}$$

Using radioactive decay equation, we have

$$A_C = A_{C_0} e^{-\lambda t}$$

$$5V = 0.8 \times 3.7 \times 10^4 \times 0.84$$

$$V = \frac{0.8 \times 3.7 \times 10^4 \times 0.84}{5} \text{ cm}^2$$

$$\Rightarrow \quad V = 4.97 \times 10^{-3} \text{ m}^3 \simeq 5 \text{ ltr}$$

Ans. (D)

Sol. 22 Activity of a sample

$$A_C = \lambda N$$

If S_1 has N_0 nuclei then S_2 will be having $\frac{N_0}{2}$ nuclei at that instant, so we use

$$A_{CS_1} = \lambda_1 N_0 \quad \ldots(1)$$

and

$$A_{CS_2} = \lambda_2 \left(\frac{N_0}{2}\right) \quad \ldots(2)$$

$$(1) \div (2) \Rightarrow \quad \frac{5}{10} = \frac{\lambda_1}{\lambda_2 / 2}$$

$$\Rightarrow \quad \frac{\lambda_1}{\lambda_2} = \frac{1}{4} = \frac{T_2}{T_1}$$

$$\Rightarrow \quad T_1 = 4T_2$$

So half life of S_1 is 4 times that of S_2 **Ans. (D)**

Sol. 23 For *RA* material *A* if there ae $2N_0$ atoms then, we use

$$A_{C_A} = \lambda_A (2N_0) \quad \ldots(1)$$

For material *B*, number of atoms are N_0 so, we use

$$A_{C_B} = \lambda_B (N_0) \quad \ldots(2)$$

$$(1)/(2) \quad \frac{A_{CA}}{A_{CB}} = \frac{2\lambda_A}{\lambda_B} = \frac{10}{20}$$

$$\Rightarrow \quad 4\lambda_A = \lambda_B$$

If half lives of A and B are T_A and T_B

$$\Rightarrow \quad \frac{T_A}{T_B} = \frac{\lambda_B}{\lambda_A} = 4$$

Ans. (A)

Sol. 24 Activities of A and B after time t are given as

$$R_A = A_{COA}\, e^{-\lambda_A t} \left[\text{Here } \lambda_A = \frac{\ln(2)}{T_A} = \frac{\ln(2)}{\ln(2)} = 1\right]$$

$$R_B = A_{COB}\, e^{-\lambda_B t}$$

$$\Rightarrow \quad \frac{R_B}{R_A} = \frac{A_{COB}\, e^{-\lambda_B t}}{A_{COA}\, e^{-\lambda_A t}} = e^{-(\lambda_B - \lambda_A)t} = e^{-3t}$$

$$\Rightarrow \quad \lambda_B - \lambda_A = 3$$

$$\Rightarrow \quad \lambda_B = 3 + 1 = 4$$

$$\Rightarrow \quad T_B = \frac{\ln(2)}{\lambda_B} = \frac{\ln(2)}{4}$$

Ans. (C)

Sol. 25 By radioactive decay law, we use

$$A = A_0 (2)^{-t/T}$$

at $t = 0$, $A_O = 1600$ cps

at $t = 8$s, $A_C = 100$ cps

$$\Rightarrow \quad 100 = 1600\, (2)^{-8/T} \text{ half life}$$

$$\Rightarrow \quad 2^{8/T} = 2^4$$

$$\Rightarrow \quad \frac{8}{T} = 4$$

$$\Rightarrow \quad T = 2\text{s}$$

At $t = 6$s, we use

$$A = 1600\, (2)^{-6/2} = \frac{1600}{2^3} = \frac{1600}{8}$$

$$= 200 \text{ cps}$$

Ans. (C)

Sol. 26 If we write nulcear equation for α and β decay

$$^{232}_{90}\text{Th} \longrightarrow {}^{208}_{78}Y + 6\,^4_2\alpha$$

$$^{208}_{78}Y \longrightarrow {}^{208}_{82}X + 4\,^0_{-1}\beta$$

$$Z = 82 \text{ and } A = 208$$

Ans. (D)

Sol. 27 By using *RA* decay law, we have

For $A \leftarrow N = N_0 e^{-\lambda t} \rightarrow$ For A

$$N_A = N_0 e^{-10\lambda t} \quad \ldots(1)$$

$$N_B = N_0 e^{-\lambda t} \quad \ldots(2)$$

$$(1)/(2) \Rightarrow \quad \frac{N_A}{N_B} = e^{-9\lambda t} = \frac{1}{e}$$

$$\Rightarrow \quad 9\lambda t = 1$$

$$\Rightarrow \quad t = \frac{1}{9\lambda}$$

Ans. (B)

Sol. 28 By *RA* decay equation for *A* and *B*, number of nuclei after time *t* are

$$N_A = N_0\, e^{-5\lambda t}$$

and

$$N_B = N_0\, e^{-\lambda t}$$

$$\Rightarrow \quad \frac{N_A}{N_B} = \frac{e^{-5\lambda t}}{e^{-\lambda t}} = \frac{1}{e^2}$$

$\Rightarrow \quad e^{-4\lambda t} = e^{-2}$

$\Rightarrow \quad 4\lambda t = 2$

$\Rightarrow \quad t = \dfrac{1}{2\lambda}$ **Ans. (C)**

Sol. 29 By radioactive decay equation, we use

For $A \rightarrow \quad N_A = N_0 2^{-t/T_A}$

Number of decayed nuclei

$$N_{DA} = N_0 - N_A$$

$$= N_0 - N_0 2^{-6} = N_0\left(1 - \frac{1}{64}\right) = \frac{63}{64} N_0$$

For $B \rightarrow \quad N_B = N_0 2^{-t/T_B}$

Number of decayed nuclei

$$N_{DB} = N_0 - N_B$$

$$= N_0 - N_0 2^{-3} = N_0\left(1 - \frac{1}{8}\right) = \frac{7}{8} N_0$$

$$\frac{N_{DA}}{N_{DB}} = \frac{63/64 N_0}{7/8 N_0} = \frac{9}{8}$$ **Ans. (A)**

27.2 Nuclear Physics

Sol. 1 Fusion will start when average thermal energy of particles in gaseous state will be equal to the PE of the fusing particles

Average thermal energy $E = \dfrac{3}{2}kT$

$\Rightarrow$ here $\dfrac{3}{2}kT \geq 7.7 \times 10^{-14}$ for fusion to start

$\Rightarrow \quad T \geq \dfrac{7.7\times10^{-14}}{1.5\times1.38\times10^{-23}} \simeq 3.7 \times 10^{9}$ **Ans. (D)**

Sol. 2 We know that in general radio active decay causes α, β and γ emissions

$\alpha \rightarrow He^{+2}$

$\beta \rightarrow e^-(\bar{\nu})$ or $e^+(\nu)$ where $\bar{\nu}$, ν are neutrinos

$\gamma \rightarrow$ EM radioactive with $\lambda < 0.1$ Å

Out of given options in general protons are NOT emitted.

Ans. (A)

Sol. 3 In the decay sequence parent nucleus ($Z = 92$) is emitting

8 α particles $\Rightarrow$ this will decrease Z by $8 \times 2 = 16$ (–ve)

4 β^- particles $\Rightarrow$ this will decrease Z by $4 \times 1 = 4$ (+ve)

2 β^+ particles $\Rightarrow$ this will decrease Z by $2 \times 1 = 2$ (–ve)

Final Z of the resulting end product is

$$Z_{\text{end product}} = 92 - 16 + 4 - 2 = 92 - 14 = 78$$ **Ans. (B)**

Sol. 4 By conservation of linear momentum we use

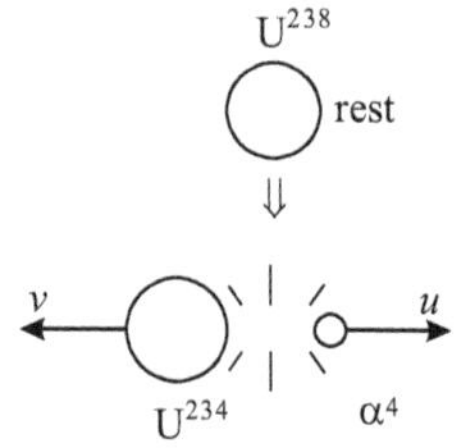

$$234\,v = 4u$$

$\Rightarrow \quad v = \dfrac{4u}{234}$ **Ans. (C)**

Sol. 5 At the point of closest approach the initial KE of α-particle gets transformed into the electrical PE of (α – U) system, so we use

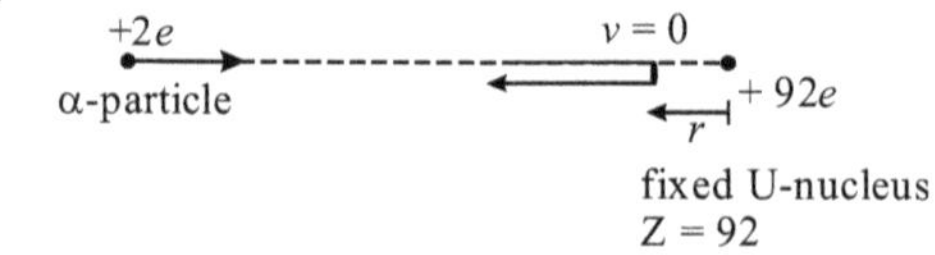

$$KE_i = \frac{Kq_1q_2}{r} \quad [q_1 = +92e \;\&\; q_2 = +2e]$$

$\Rightarrow \quad r = \dfrac{9\times10^{9}\times92\times2\times1.6\times10^{-19}}{5\times10^{6}}$

$\Rightarrow \quad r = 5.3 \times 10^{-14}$ m

$\Rightarrow \quad r = 5.3 \times 10^{-12}$ cm **Ans. (C)**

Sol. 6 The reaction of nuclear fusion is

$${}_1H^2 + {}_1H^2 \longrightarrow {}_2H^2 + \Delta E$$

Here the Q-value of nuclear reaction (energy released) is given as

$$\Delta E = 4\,(BE_N)_{{}_2He^4} - 2 \times 2 \times (BE_N)_{{}_1H^2}$$

$\Rightarrow \quad \Delta E = 4 \times 7 - 4 \times 1.1 = 28 - 4.4 = 23.6$ MeV **Ans. (C)**

Sol. 7 $X(n, \alpha) \rightarrow {}^7_3Li$ can be written as

$${}^1_0n + {}^A_ZX \longrightarrow {}^7_3Li + {}^4_2\alpha$$

In above reaction conserving charge and mass number gives

$1 + A = 7 + 4 \quad \rightarrow \quad A = 11 - 1 = 10$

$0 + Z = 3 + 2 \quad \rightarrow \quad Z = 5$

Element X is ${}^{10}_5B$ **Ans. (B)**

Sol. 8 We use the γ-ray intensity at a penetration depth x, is gives as

$$I = I_0 e^{-\mu x}$$

[$\mu \rightarrow$ absorption coefficient of lead for γ-rays]

$$I_0 = I$$

and $I = \frac{I}{8}$ at $x = 36$ mm

$$\Rightarrow \frac{I}{8} = Ie^{-\mu(36)}$$

$$\Rightarrow 36\mu = \ln 8 = 3\ln 2$$

$$\Rightarrow \mu = \frac{3\ln 2}{36} = \frac{\ln 2}{12}\ \text{mm}^{-1}$$

In next case at $I = \frac{I}{2}$ we use

$$x = r$$

$$\Rightarrow \frac{I}{2} = Ie^{-\mu r}$$

$$\Rightarrow \mu r = \ln(2)$$

$$\Rightarrow r = \frac{\ln(2)}{\mu} = 12\ \text{mm}$$

Ans. (D)

Sol. 9 Fact based

emitted β-particles from a radioactive source spreaded energy spectrum with speeds ranging Option (C) is

most appropriate matching option. **Ans. (C)**

Sol. 10 The nuclear reaction of above given case is written as

$$^{7}_{3}\text{Li} + {}^{1}_{1}\text{p} \longrightarrow {}^{8}_{4}\text{Be} + a$$

In every nuclear reaction the obtained product can be found in excited state which can further transit to ground state and emit gamma radiation.

$$^{7}_{3}\text{Li} + {}^{1}_{1}\text{p} \longrightarrow {}^{8}_{4}\text{Be}^{*}$$

$$^{8}_{4}\text{Be}^{*} \longrightarrow {}^{8}_{4}\text{Be} + \gamma$$

$$\Rightarrow a = \gamma$$ **Ans. (C)**

Sol. 11 At the point of closest approach whole KE transform into electric PE

$$\Rightarrow \frac{1}{2}mv^2 = \frac{Kq_1q_2}{r_m} = \frac{K(2e)(Ze)}{r_m}$$

$$\Rightarrow r_m = \frac{4KZe^2}{mv^2}$$

$$\Rightarrow r_m \propto \frac{1}{m}$$

Ans. (B)

Sol. 12 In the given nuclear reaction

$$p + {}^{7}_{3}\text{Li} \longrightarrow 2\,{}^{4}_{2}\text{He}$$

As some initial energy of proton is required so by conservation of mass-energy in this equation we use

$$E_p + E_{Li} = 2E_{He}$$

$$\Rightarrow E_p + 7 \times 5.60 = 2 \times 4 \times 7.06$$

$$\Rightarrow E_p = 56.48 - 392 = 17.28\ \text{MeV}$$ **Ans. (B)**

Sol. 13 A gamma ray emission takes place only from on excited Nu when it transit to ground state

$$^{A}_{Z}\text{X}^{*} \longrightarrow {}^{A}_{Z}\text{X} + \gamma\text{-photon}$$

Nucleus in excited state — Nucleus in ground state

As Z and A remain same is γ-emission so no change in proton and neutron number. **Ans. (B)**

Sol. 14 $^{17}_{8}$O nucleus contains 8 protons and 9 neutrons

The mass defect in formation of nucleus $^{17}_{8}$O is

$$\Delta m = 8M_P + 9M_n - M_O$$

energy released at the time of formation $^{17}_{8}$O nucleus is

$$\Delta E = \Delta Mc^2 = [8M_P + 9M_n - M_O]c^2$$ **Ans. (B)**

Sol. 15 Nucleus → Neutrons, Protons

Ans. (B)

Sol. 16 In case of fission of heavy nuclei or fusion of light Nu overall $(BE)_N$ increases and energy is released

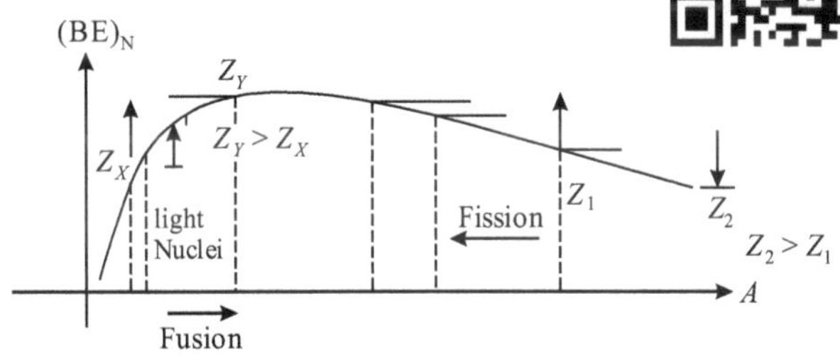

⇒ From above graph we can state that **S-1** is true and **S-2** is false. **Ans. (D)**

Sol. 17 In any reaction if overall BE per nucleon increases

⇒ Energy is released

(i) $A + B \rightarrow C + E$

As $E_A, E_B < E_C$ ⇒ energy is released ⇒ E is +ve

(ii) $C \rightarrow A + B + E$

As $E_C > E_A, E_B \Rightarrow$ energy is absorbed $\Rightarrow$ E is –ve

(iii) $D+E \rightarrow F+E$

As $E_F < E_D, E_E \Rightarrow$ energy is absorbed $\Rightarrow$ E is –ve

(iv) $F \rightarrow D+E+E$

As $E_D, E_E > E_F \Rightarrow$ energy is released $\Rightarrow$ E is +ve

Ans. (C)

Sol. 18 The effective nuclear reaction is

$$ {}^{A}_{Z}X \longrightarrow {}^{A_1}_{Z_1}Y + 3\,{}^{4}_{2}\alpha + 2F_e $$

By conservation of charge, we use

$$Z = Z_1 + 6 + 2 \Rightarrow Z_1 = Z - 8$$

By conservation mass, we use

$$A = A_1 + 12 \Rightarrow A_1 = A - 12$$

In Y ratio of neutrons to protons is $\dfrac{N_n}{N_P} = \dfrac{A_1 - Z_1}{Z_1} = \dfrac{A - Z - 4}{Z - 8}$

Ans. (C)

Sol. 19 As in the given decay process energy is evolved

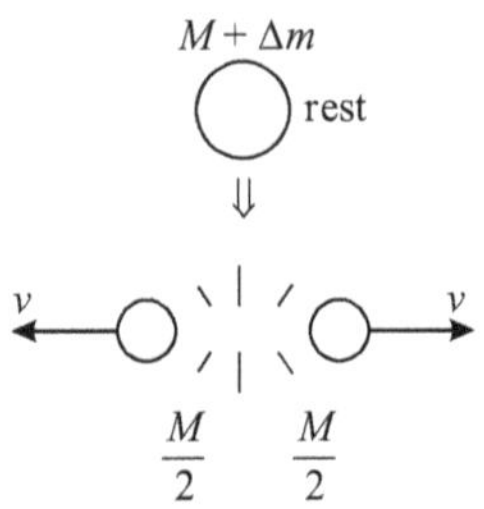

$\Rightarrow$ Final BE per nucleon must be less than initial BE per nucleon

$$\Rightarrow \quad E_2 < E_1$$

Ans. (C)

Sol. 20 Energy evolved $= \Delta mc^2$

due to released energy the daughter nuclei gain KE so we use

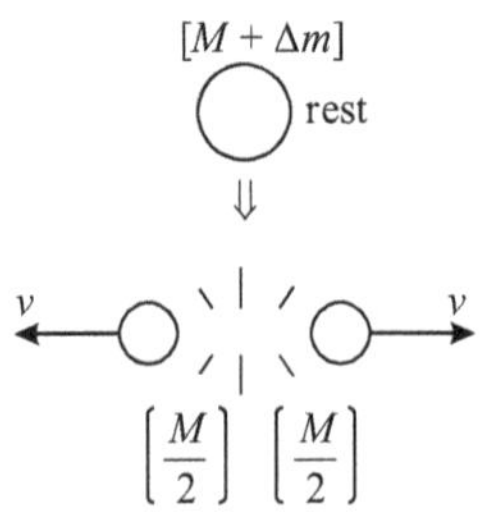

$$\Delta mc^2 = \frac{1}{2}\left(\frac{M}{2}\right)v^2 + \frac{1}{2}\left(\frac{M}{2}\right)v^2 + \frac{M}{2}v^2$$

$$\Rightarrow \quad v = \sqrt{\frac{2\Delta m}{M}} \cdot c$$

Ans. (C)

Sol. 21 $0 \le E_\beta \le E_1 - E_2$

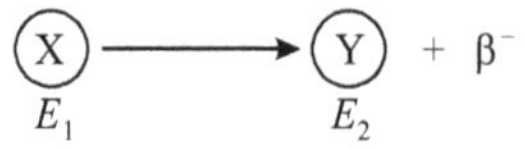

β particles are emitted with varying energy in range from O to $E_1 - E_2$. This is because the energy different is carried by another particle called nutrino.

S-1: It is correct

S-2: By neutrino hypothesis energy and momentum conservation was considered to be valid so it explains S-1. **Ans. (C)**

Sol. 22 As $P_i = 0 \Rightarrow P_f = 0$

$$\Rightarrow \quad P_{m_1} = P_{sm_1}$$

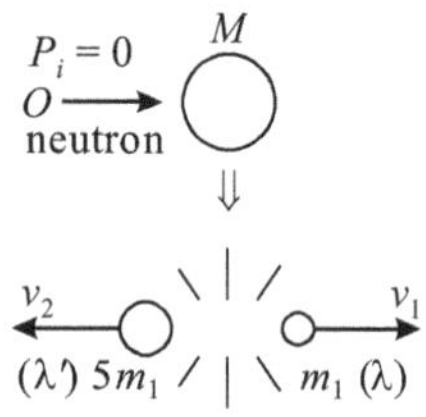

We use de-Broglie wavelength of m_1 is λ

$$\Rightarrow \quad \lambda = \frac{h}{P_{m_1}}$$

For de-Broglie wavelength of $5m_1$ we use

$$\lambda' = \frac{h}{P_{m_2}} = \frac{h}{P_{m_1}} = \lambda$$

Ans. (D)

Sol. 23 If charge and mass of particle X is considered a and b then we use

By conservation of mass $\rightarrow 235 + 1 = 141 + 92 + 3b$

$$\Rightarrow \quad 3b = 236 - 233 = 3$$

$$\Rightarrow \quad b = 1$$

By conservation of charge $\rightarrow 92 + 0 = 56 + 36 + 3a$

$$\Rightarrow \quad 3a = 92 - 92 = 0$$

$$\Rightarrow \quad a = 0$$

$\Rightarrow \; {}^{b}_{a}x \rightarrow {}^{1}_{0}x \Rightarrow {}^{1}_{0}n$ **Ans. (C)**

Sol. 24 The nuclear reaction is

$$ {}^{2}_{1}H + {}^{2}_{1}H \longrightarrow {}^{4}_{2}He + \Delta E$$

The released energy is equal to difference in total binding energies of product in this reaction.

Total BE of the reaction $E_R = 4 \times 1.1 = 4.4$ MeV

Total BE of the reaction $E_P = 7 \times 4 = 28$ MeV

Energy released (Q-value) $\Delta E = E_P - E_R = 28 - 4.4 = 23.6$ MeV

Ans. (C)

Sol. 25 Mass m produces energy $E = mc^2$

Power of reactor $P = 10^9\ W$

Mass of fuel consumed per second is $m' = \frac{P}{c^2} = \frac{10^9}{(3\times10^8)^2}$ kg

Mass of fuel consumed per hour is $m = m' \times 3600$

$$\Rightarrow \quad m = \frac{3600\times10^9}{8\times10^{16}} = 4\times10^{-5}\text{ kg}$$

$$\Rightarrow \quad m = 4\times10^{-2}\text{ gm} \quad \textbf{Ans. (C)}$$

Sol. 26 We consider an element X

$$X^* \longrightarrow X + E_r \qquad (\lambda_N \sim 0.1\text{ Å or less} \approx 0.01\text{Å})$$

(E_r = Energy of the order of some MeV)

For atom X when its e^- transit from $n = 2$ to $n = 1$

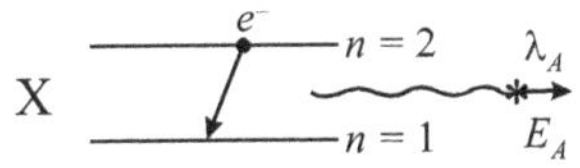

$$\frac{1}{\lambda_A} = R(z-\sigma)^2\left[\frac{1}{1^2} - \frac{1}{2^2}\right]$$

Energy is of the order of few eV only.

For photons we can use

$$E_r = \frac{hc}{\lambda_N} \text{ and } E_A = \frac{hc}{\lambda_A}$$

$$\Rightarrow \frac{\lambda_N}{\lambda_A} = \frac{E_A}{E_r} = \frac{\text{order of } eV}{\text{order of } MeV} \simeq \frac{1}{10^6} = 10^{-6} \qquad \textbf{Ans. (A)}$$

Sol. 27 Formula for fermi radius is $R = R_0(A)^{1/3}$

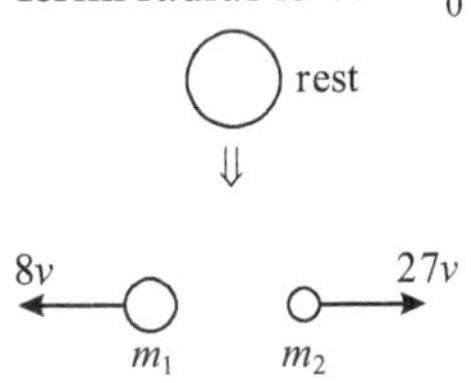

By conservation of linear momentum we use

$$m_1(8v) = m_2(27v)$$

$$8m_1 = 27m_2$$

If R_1 and R_2 are the radii of nuclei of masses m_1 and m_2, we have

$$\frac{R_1}{R_2} = \left(\frac{m_1}{m_2}\right)^{1/3}$$

$$\Rightarrow \quad \frac{R_1}{R_2} = \left(\frac{27}{8}\right)^{1/3} = \frac{3}{2} \quad \textbf{Ans. (C)}$$

Sol. 28 For the given nuclear reaction energy released is

$$E_B = (BE)_{\text{product}} - (BE)_{\text{reactant}}$$

$$= 2\times7.07\times4 + 7.86\times12 - 8.03\times20$$

$$= -9.72\text{ MeV}$$

Thus minimum 9.72 MeV or more energy to be supplied. Thus option (C) is most appropriate **Ans. (C)**

Sol. 29 Mass densities of all nuclei are same so their ratio is 1 **Ans. (A)**

28.1 Semiconductor

Sol. 1 Energy band gap

in case of metals $E \sim 0$ eV [here V.B. and C.B. are overlapped]

in case of Semiconductor $E < 2$ - 2.5 eV

in case of insulators $E > 5$ - 6 eV **Ans. (C)**

Sol. 2 For conductors on increasing temperature mobility of charge carriers decrease and concentration of free charge carriers does not change.

$$\rho_c = \frac{1}{\sigma} = \frac{1}{n_e e\mu_e}$$

$\Rightarrow$ on increasing T, ρ decreases

For semiconductors on increasing temperature the concentration of charge carriers increase due to more e-h pair generation at high temperature

$$\rho_{sc} = \frac{1}{\sigma_{sc}} = \frac{1}{n_e e\mu_e + n_h e\mu_h}$$

$\Rightarrow$ on increasing T, ρ decreases.

Effect of temperature in a semi-conductor on n_e and n_h is high compared to μ_e and μ_h. **Ans. (C)**

Sol. 3 As band gap of Si is 1.14 eV. This energy can be obtained at room temperature by thermal agitation so at room temperature it conducts electricity.

But at $T = 0$ K temperature no energy will be available for generation of e-h pairs

$\Rightarrow$ No free charge carriers will be there is Si lattice so Si behave like an insulator at 0 K temperature. **Ans. (C)**

Sol. 4 As "Copper" is a metal $\Rightarrow$ its resistance increases with rise in temperature, here as temperature is reduced, its resistance "decreases". And as "Germanium" in a semiconductor

$\Rightarrow$ its resistance decreases with rise in temperature here as temperature is reduced, its resistance "increases". **Ans. (C)**

Sol. 5 With temperature increase due to thermal agitation more e - h pairs are generated in a Semi-conductor due to which concentration of charge carriers in electricity conduction increases. **Ans. (B)**

Sol. 6

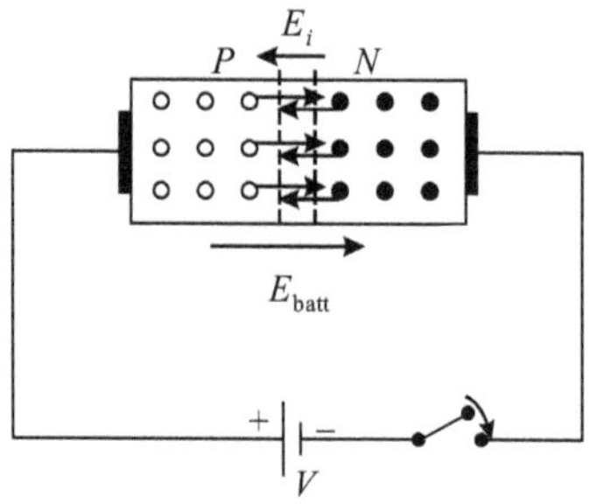

$E_i \rightarrow$ Electric field due to P and N type semi-conductor doping ions.

As external electric field due to battery overcomes the barrier field so barrier height reduces and diffusion of electrons and holes start in depletion layer so it becomes narrower. **Ans. (A)**

Sol. 7 Conductivity increases $\Rightarrow$ more e - h pair generation starts in Semi-conductor.

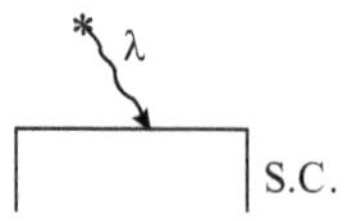

Energy band gap of Semi-conductor is given as

$$\Delta E_g = \frac{hc}{e\lambda} \text{eV}$$

$$\Delta E_g = \frac{12431}{(24800 \text{Å})} \text{eV} \approx 0.5 \text{ eV}$$ **Ans. (D)**

Sol. 8 The currents in the Semiconductor due to electrons and holes can be given as

$$I_e = n_e e A\, v_{d_e}$$

$$I_h = n_e e A\, v_{d_h}$$

here $A \rightarrow$ cross section area of Semi-conductor substrate.

$$\Rightarrow \quad \frac{I_e}{I_h} = \frac{n_e v_{d_e}}{n_n v_{d_h}}$$

$$\Rightarrow \quad \frac{v_{d_e}}{v_{d_h}} = \frac{I_e}{I_h} \times \frac{n_h}{n_e}$$

$$\Rightarrow \quad \frac{v_{d_e}}{v_{d_h}} = \frac{7}{4} \times \frac{5}{7} = \frac{5}{4}$$ **Ans. (C)**

Sol. 9 If conductivity increases with temperature $\Rightarrow$ the substance is a semiconductor (Si or Ge or any other)

Bonding $\rightarrow$ Covalent bonds **Ans. (B)**

Sol. 10 If density of a material increases, the energy levels in atoms become more prominent bands or E_c and E_v increase due to influence by neighbouring atoms.

On decreasing lattice constant E_c and E_v will increase and E_g decreases.

$\Rightarrow$ If lattice constant increase $\Rightarrow E_c$ and E_v will decrease $\Rightarrow E_g$ will increase $\Rightarrow$ density decrease. **Ans. (B)**

Sol. 11 The band energy gap in C, Si and Ge are

$$\Delta E_C(\text{diamond}) \approx 5.5 \text{ eV (insulator)}$$

$$\Delta E_{GE} \cong 0.7 \text{ eV}$$

$$\Delta E_{Si} \cong 1.14 \text{ eV}$$

ΔE_{Ge} and $\Delta E_{Si} < 2$ eV so these gain energy to excite electrons from valence band to conduction band at room temperature

Ans. (A)

Sol. 12 Statement-1 : For a semiconductor as temperature rises, its conductivity increases $\Rightarrow$ Resistance decreases

$\Rightarrow$ its α is negative $\Rightarrow$ True

Statement-2 : Conductivity of a semiconductor is given as

$$\sigma = n_e e \mu_e + n_h e \mu_h$$

[n_e, $n_h \rightarrow e^-$ and hole concentration]

As temperature rises more e - h pair generation take place

$\Rightarrow$ n_e, n_h increases $\Rightarrow \sigma \uparrow$ **Ans. (D)**

Sol. 13 As Cu is a metal we use its resistivity

$$\rho = \rho_0(1 + \alpha T)$$

For Cu with temperature its resistance increase linearly.

For intrinsic Si the carrier concentration is given as

$$n_i = n_0 e^{-\Delta E_g / kT}$$

the conductivity of a substance $\sigma \propto n_i$

$\Rightarrow$ with increase in temperature resistivity of a Semiconductor decreases exponentially. **Ans. (A)**

Sol. 14 Conductivity of a semiconductor is given as

$$\sigma = n_e \mu_e e + n_n \mu_h e$$

$$\Rightarrow \quad \sigma = (n_e \mu_e + n_n \mu_h) e$$

$$\Rightarrow \quad \sigma = [5 \times 10^{18} \times 2.0 + 5 \times 10^{19} \times 0.01] \times 1.6 \times 10^{-19}$$

$$\Rightarrow \quad \sigma = [10^{19} + 0.05 \times 10^{19}] \times 1.6 \times 10^{-19}$$

$$\Rightarrow \quad \sigma = 1.05 \times 1.6 = 1.68\, (\Omega\text{-m})^{-1}$$ **Ans. (A)**

Sol. 15 Resistivity of a semiconductor is given as

$$\rho = \frac{1}{\sigma} = \frac{1}{ne\mu_e}$$

$$\Rightarrow \qquad \rho = \frac{1}{10^{19} \times 1.6 \times 10^{-19} \times 1.6}$$

$$\Rightarrow \qquad \rho = 0.4\,\Omega\text{m} \quad \textbf{Ans. (B)}$$

28.2 PN Junction and Diodes

Sol. 1

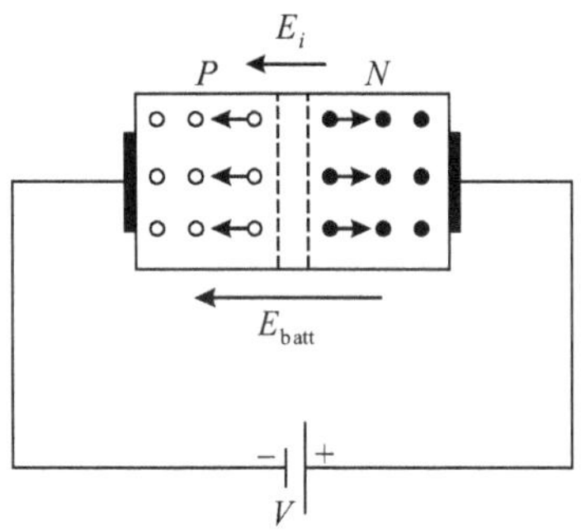

In Reverse Bias mode the electric field by battery adds up to the internal Electric field of the junction.

So barrier height increases $\Rightarrow$ Electric field in depletion layer increase in Reverse Bias condition. **Ans. (C)**

Sol. 2 As "Copper" is a metal $\Rightarrow$ its resistance increases with rise in temperature. Here as temperature is reduced $\Rightarrow$ its resistance "decreases".

And As "Germanium" is a Semiconductor $\Rightarrow$ its resistance decreases with rise in temperature.

Here as temperature is reduced $\Rightarrow$ its resistance "increases". **Ans. (C)**

Sol. 3 In above circuit for the applied battery D_2 is Forward biased and D_1 is reverse biased so the equivalent circuit for this case is

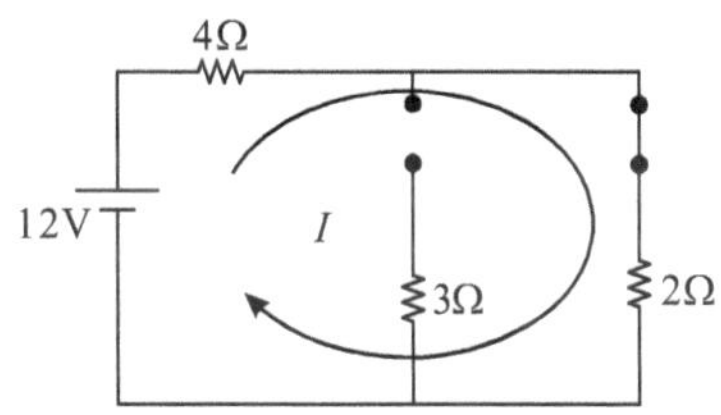

Circuit current $\quad I = \frac{12}{4+2} = \frac{12}{6} = 2\text{A}$ **Ans. (B)**

Sol. 4 In Reverse Bias condition

P side of diode $\rightarrow$ low potential

N side of diode $\rightarrow$ high potential

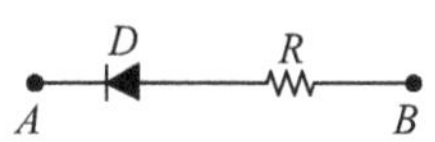

For Reverse Bias we use $V_A > V_B$ **Ans. (D)**

Sol. 5

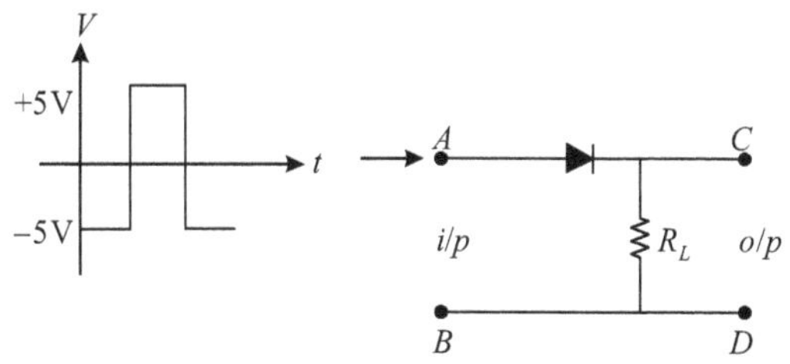

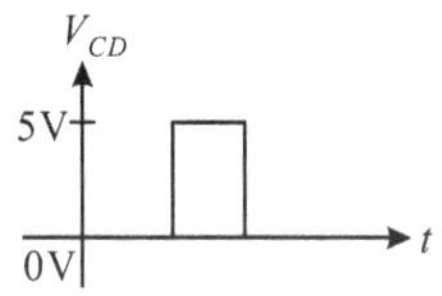

Current will pass through diode only when it is forward biased or when input voltage at *A* is more than 0 *V* so the current through *R* will only be there for $V_A = 0$ to 5 V. **Ans. (A)**

Sol. 6 The given circuit is using "*D*" as a half wave rectifier. **Ans. (B)**

Sol. 7 LED of difference band gaps emit different coloured visible radiations because the band gap is of the order of energy of visible light.

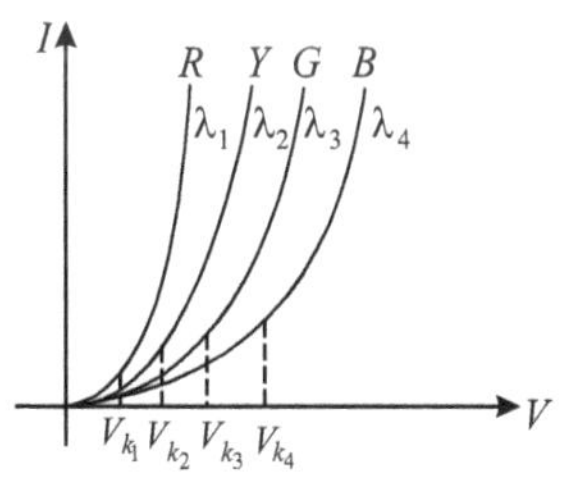

Knee point voltage increases as band gap increase $\Rightarrow$ for higher ΔE_g (for lower λ) V_K is more.

Here $\Delta E_{G_1} < \Delta E_{G_2} < \Delta E_{G_3} < \Delta E_{G_4}$

$\Rightarrow \lambda_1 > \lambda_2 > \lambda_3 > \lambda_4$ **Ans. (A)**

Sol. 8 Here in circuit we can see due to 6V battery diodes D_1 and D_2 are forward biased and D_3 is reverse biased so equivalent circuit of this case is

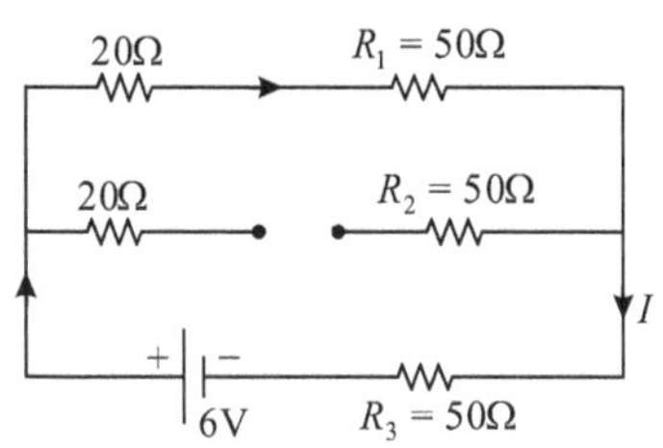

Current through R_3 (in outer loop) is given as

$$I = \frac{6}{120} = \frac{1}{20} = 50\text{ mA} \qquad \textbf{Ans. (A)}$$

Sol. 9 In Forward Bias connection P side of diode is connected to high potential and N side is connected to low potential.

For Forward Bias $V_A > V_B$ **Ans. (A)**

Sol. 10 Equivalent circuit of given situation is shown below

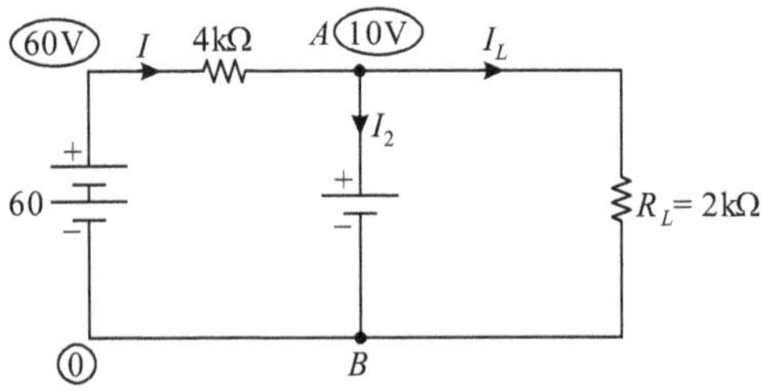

Current I in 4 kΩ resistance is

$$I = \frac{60-10}{4000} = \frac{5}{400} = 12.5 \text{ mA}$$

Current in 2 kΩ resistance is

$$I_L = \frac{10-0}{2000} = 5 \text{ mA}$$

and $$I_Z = I - I_L$$
$$= 12.5 - 5 = 7.5 \text{ mA}$$ **Ans. (C)**

Sol. 11 Light radiation (visible) range is form ~ 4000 Å to 7000 Å for average wavelength of about 6000 Å the corresponding energy should be

$$E\,(\text{in eV}) = \frac{hc}{\lambda e} = \frac{12431}{\lambda(\text{in Å})}$$

$$= \frac{12431}{6000} \approx 2.07 \text{ eV}$$ **Ans. (D)**

Sol. 12

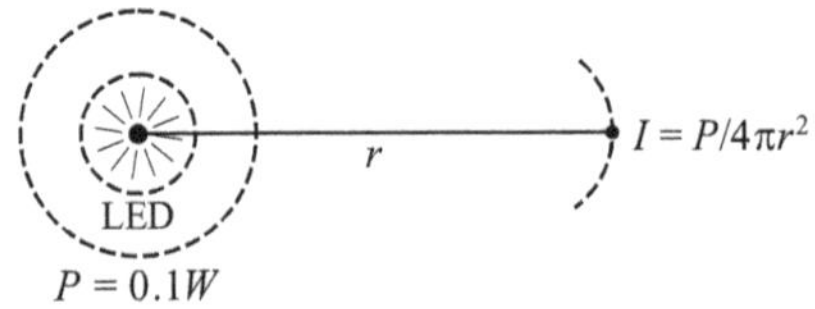

For an EM wave the energy density in light beam is given as

$$u_e = \frac{I}{c} = \frac{1}{2}\epsilon_0 E^2$$

$$\Rightarrow \quad E = \sqrt{\frac{2I}{c\epsilon_0}} = \sqrt{\frac{2P}{4\pi\epsilon_0 cr^2}}$$

$$\Rightarrow \quad E = \sqrt{\frac{2\times 0.1\times 9\times 10^9}{3\times 10^8 \times (1)^2}} = \sqrt{6}$$

$$\Rightarrow \quad E = 2.45 \text{ V/m}$$ **Ans. (D)**

Sol. 13

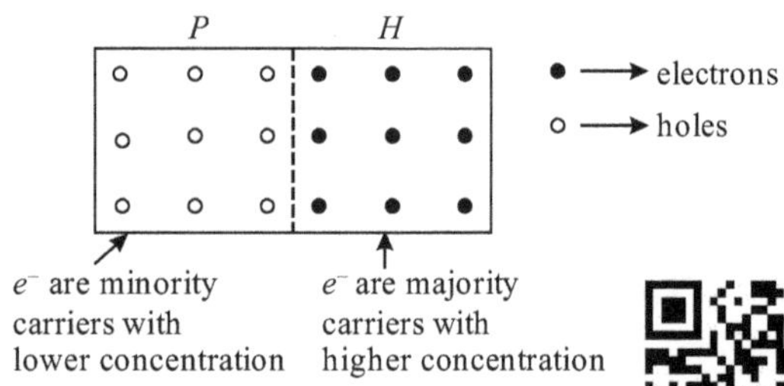

As in n-side concentration of electrons is higher compared to p-side so electron diffusion take place from n side to p side. **Ans. (D)**

Sol. 14 Situation-1 : When +ve terminal of battery is at A, circuit will be

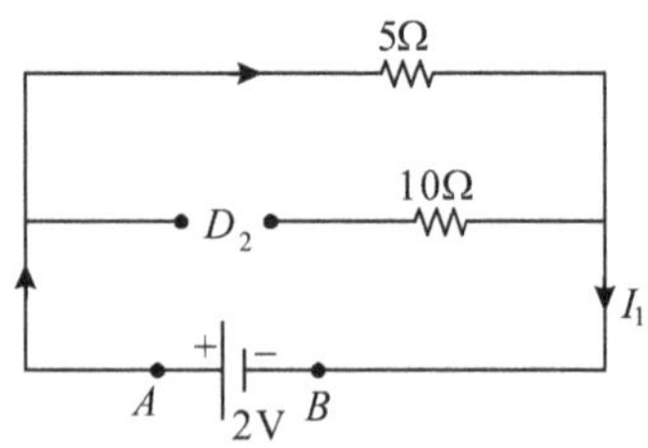

Current through battery is

$$I_1 = \frac{2}{5} = 0.4 \text{ A}$$

Situation-2 : When +ve terminal of battery is at B, circuit will be

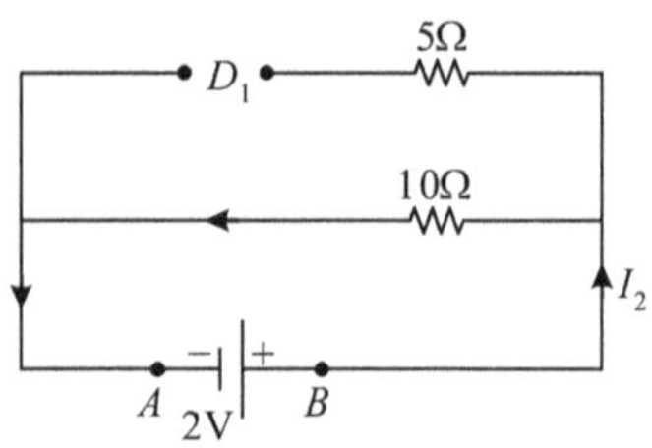

Current through battery is

$$I_2 = \frac{2}{10} = 0.2 \text{ A}$$ **Ans. (A)**

Sol. 15 (i) → Normal diode

(ii) → Zener diode

(iii) → Photo cell/Solar cell (Reverse Bias)

(iv) → Light dependent resistance (Forward Bias) **Ans. (C)**

Sol. 16

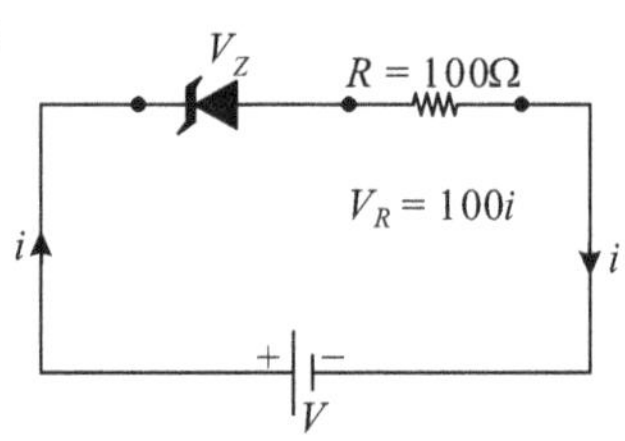

In this circuit potential drop across zener diode is

$$V_Z = V - 100\,i$$

Zener diode has a power dissipation rating of 1 W (max)

$\Rightarrow \quad V_z i = 1$

$\Rightarrow \quad (V - 100i)i = 1$

$\Rightarrow \quad Vi - 100i^2 = 1$

$\Rightarrow \quad 100i^2 - Vi + 1 = 0$

For real values of i we must have

$$D \geq 0$$

$\Rightarrow \quad V^2 - 400 \geq 0$

$\Rightarrow \quad V \geq 20$ volt **Ans. (B)**

Sol. 17 Forward bias resistance is given as

$$r_{FB} = \left.\frac{\Delta V}{\Delta I}\right|_{FB} = \frac{0.8 - 0.7}{(20 - 10) \times 10^{-3}}$$

$$\Rightarrow \quad r_{FB} = \frac{0.1}{10 \times 10^{-3}} = 10\,\Omega$$

Reverse bias resistance is given as

$$r_{RB} = \left.\frac{\Delta V}{\Delta I}\right|_{RB} = \frac{10 - 0}{(1 - 0) \times 10^{-6}} = 10^7\,\Omega$$

$$\Rightarrow \quad \frac{r_{FB}}{r_{RB}} = \frac{10}{10^7} = 10^{-6}$$ **Ans. (B)**

Sol. 18 As silicon is forward biased, it will behave like an emf of 0.7 V which is the barrier potential for a Si diode in Forward bias. So equivalent circuit here is

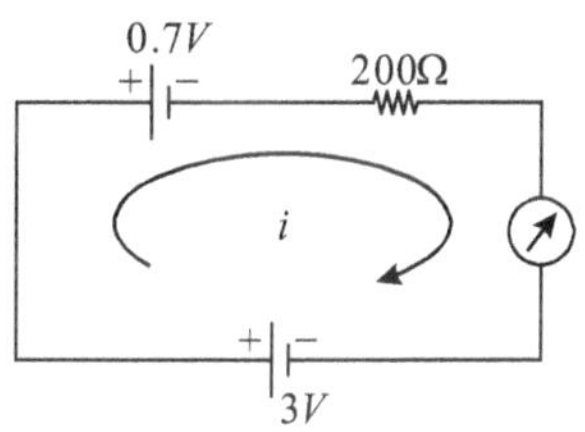

Current in circuit is

$$I = \frac{3 - 0.7}{200} = \frac{2.3}{200} = 0.00115\,\text{A}$$

$\Rightarrow \quad I = 11.5$ mA **Ans. (C)**

Sol. 19 In the given circuit across terminals A and B the potential difference due to 15 V battery (if Zener is not present) is given as

$$V_{AB} = \frac{15}{2000} \times 1500 = \frac{225}{20} = 11.25\,\text{V}$$

So here $V_{AB} > V_Z \Rightarrow$ Zener will operate in breakdown region and maintain V_Z across terminals A and B.

Under the state of breakdown the equilibrium circuit will be

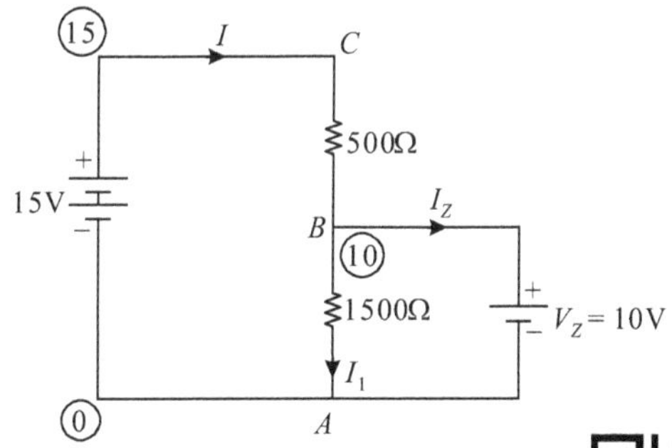

Current in 1500 Ω resistance is

$$I_1 = \frac{10}{1500} = \frac{1}{150}\,\text{A}$$

Current in 500 Ω resistance is

$$I = \frac{15 - 10}{500} = \frac{5}{500} = \frac{1}{100}\,\text{A}$$

Current through zener diode is

$$I_Z = I - I_1 = \frac{1}{100} - \frac{1}{150} = \frac{3 - 2}{300}$$

$$\Rightarrow \quad I_Z = \frac{1}{300} = 3.33\,\text{mA}$$ **Ans. (B)**

Sol. 20 Initially *Ge* is forward biased so current will pass through *Ge* diode with a drop of 0.3V so *Si* will not be forward biased.

If "*Ge*" is reversed then current will flow through "*Si*" diode hence an effective drop of (0.7 – 0.3) = 0.4V is observed. **Ans. (C)**

Sol. 21 Assuming zener diode doesn't breakdown, current in circuit

$$= \frac{120}{15000} = 8\,\text{mA}$$

$\Rightarrow$ The voltage drop across diode = 80 V > 50 V.

$\Rightarrow$ The diode will breakdown and maintain 50 V across 10 kΩ resistance as shown below

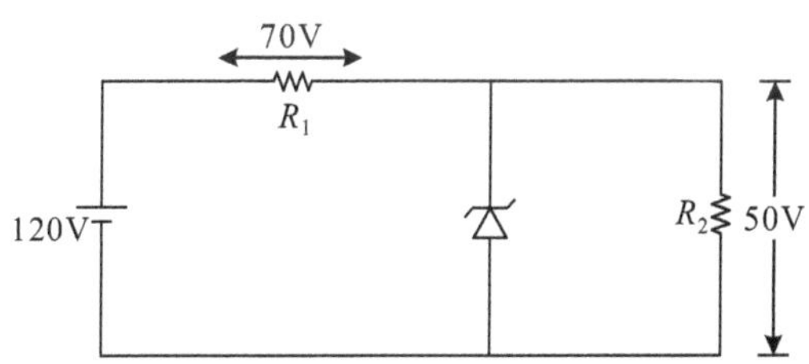

$\Rightarrow$ Current is $R_1 = \dfrac{70}{5000} = 14$ mA

$\Rightarrow$ Current is $R_2 = \dfrac{50}{10000} = 5$ mA

$\Rightarrow$ Current through diode = 9 mA **Ans. (D)**

Sol. 22 As voltage across zener diode must be less than 10V so it will not work in breakdown region, & its resistance will be infinite & current through it = 0. **Ans. (D)**

Sol. 23 Here diode D_2 is reverse biased and D_1 is forward biased so current flows in outer loop only, given as

$$I = \frac{6}{100+150+50} = \frac{6}{300}$$

$\Rightarrow \quad I = 0.02\text{ A}$ **Ans. (B)**

Sol. 24

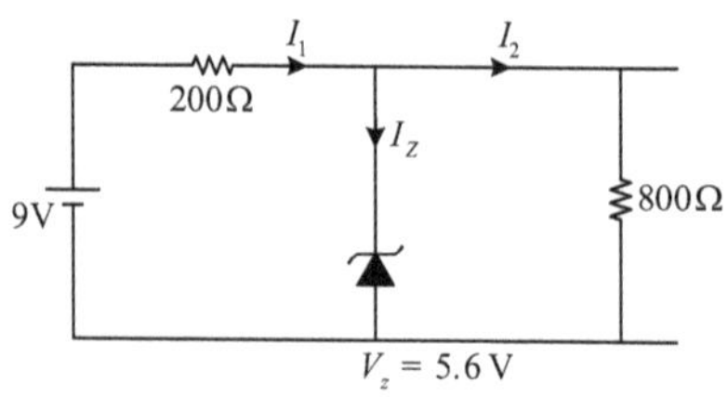

In first loop

$$9V = 200\,I_1 + 5.6$$

$$\Rightarrow \quad I_1 = \frac{3.4}{200} = 17\text{ mA}$$

In second loop

$$I_2 = \frac{5.6}{800} = 7\text{ mA}$$

Here $\quad I_Z = I_1 - I_2$

$\Rightarrow \quad I_Z = 17 - 7 = 10\text{ mA}$ **Ans. (C)**

Sol. 25 At $\quad V = 8\text{V}$

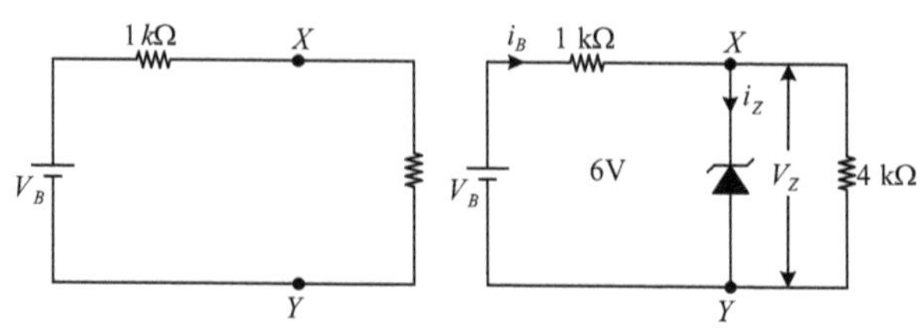

$$V_{XY} = \frac{V_B}{5} \times 4 = \frac{32}{5} = 6.4\text{ V} > \frac{V_Z}{6\text{V}}$$

At $V = 16$ V

$$V_{XY} = \frac{16 \times 4}{5} = \frac{64}{5} = 12.8\text{ V} > 6\text{ V}$$

In breakdown state

$$V_{XY} = 6\text{V}$$

$$\Rightarrow \quad i_L = \frac{6}{4\text{k}} = 1.5 \times 10^{-3}\text{ A}$$

$$\Rightarrow \text{ Battery current } i_B = \frac{V_B - V_Z}{1000}$$

At 8 V $\rightarrow \quad i_B = \dfrac{8-6}{1000} = 2 \times 10^{-3}$ A

$$i_Z = i_B - i_L = (2 - 1.5) \times 10^{-3} = 0.5\text{ mA}$$

At 16 V $\rightarrow \quad i_B' = \dfrac{16-6}{1000} = 10 \times 10^{-3}$ A

$$i_Z = i_B' - i_L = (10\ 1.5) \times 10^{-3} = 8.5\text{ mA}$$ **Ans. (B)**

Sol. 26 Current I_2 will be maximum when input voltage is also maximum at 16 V

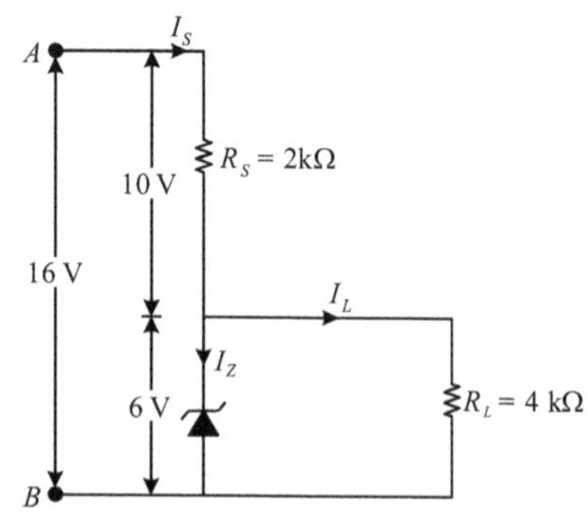

At voltage across R_L is maintained at $6V$

$$\Rightarrow \quad I_L = \frac{6}{4000} = 1.5\text{ mA}$$

Current through R_S is

$$I_S = \frac{10}{2000} = 5\text{ mA}$$

$\Rightarrow$ Zoner current

$$I_Z = I_S - I_L = 5 - 1.5 = 3.5\text{ mA}$$ **Ans. (B)**

28.3 Transistors

Sol. 1 In a transistor emitter is most heavily doped as it supplies majority carriers to base or through base to collector for its functioning.

Base $\rightarrow$ lightly doped

Collector $\rightarrow$ Moderately doped

Fact based. **Ans. (A)**

Sol. 2

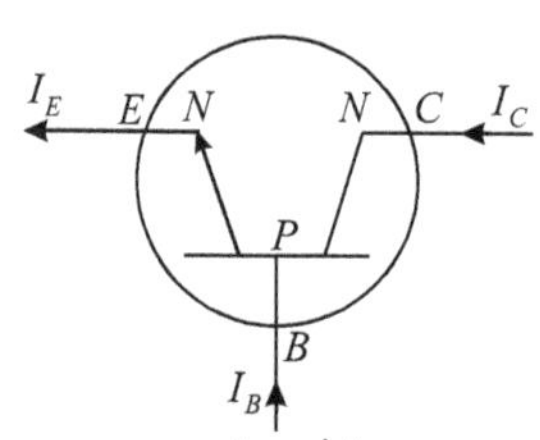

npn transistor

For normal operation of transistor BE junction is forward biased and in this condition electrons are supplied from emitter to Base and holes are supplied from Base to emitter. **Ans. (D)**

Sol. 3 In a common base amplifier input signal is applied at BE junction and output signal is taken at BC junction.

So here input current flows through emitter to collector and come out as output current.

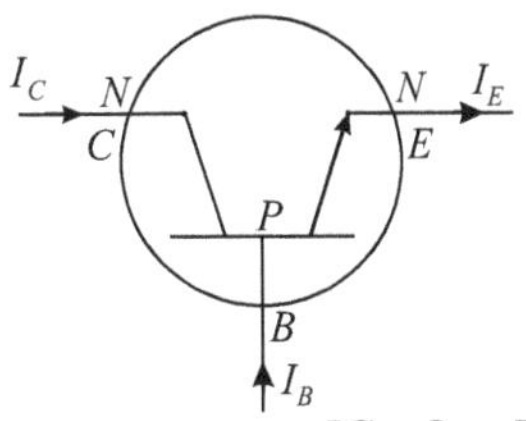

$$IC = I_E - I_B$$

Here I_C = output current and I_E = input current

All time variations pattern in I_E will also be there in I_C so the output signal is in same phase with input signal. (0°) **Ans. (D)**

Sol. 4 Base current in transistor is given as

$$I_B = I_E - I_C = 5.60 - 5.488 = 0.112 \text{ mA}$$

Base current amplification factor is given as

$$\beta = \frac{I_C}{I_B} = \frac{5.488}{0.112} = 49$$ **Ans. (A)**

Sol. 5 As no conduction is found between P and Q

$\Rightarrow$ P and Q are either of E and C of transistor

$\Rightarrow$ R is base

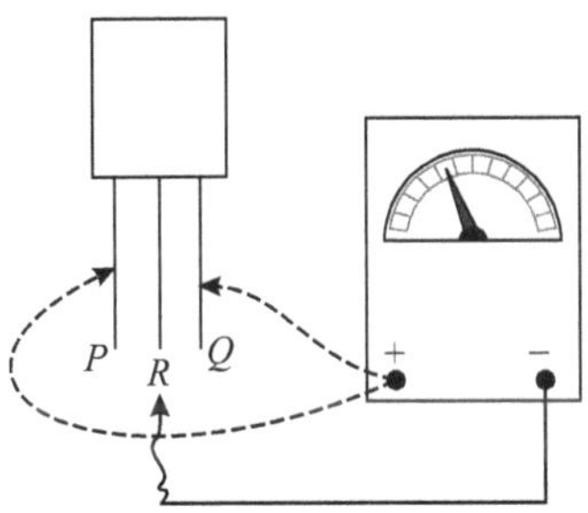

When R is kept –ve (low position) and either of P and Q +ve (high position) then some conduction is there between BE and BC junction which are in FB mode.

$\Rightarrow$ R is n side on transistor and P and Q are p sides.

$\Rightarrow$ *pnp* transistor. **Ans. (B)**

Sol. 6

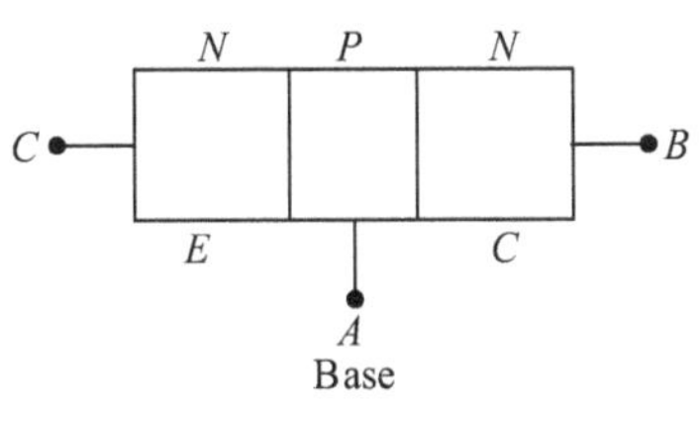

As current is supplied $\Rightarrow$ A must be P side of semiconductor as in a *pn* junction EF exist from N side to P side.

As conductivity of lead wires is more than moist fingers.

$\Rightarrow$ C must be emitter where doping is heavy compared to collector.

$\Rightarrow$ Large deflection (current) **Ans. (C)**

If we consider moist fingers to be equally conducting than.

Ans. (B) or (C)

Sol. 7 For a transistor currents I_E, I_C and I_B through its Emitter, Collector and Base then its current gains are defined as

$$\alpha = \frac{I_C}{I_E} \text{ and } \beta = \frac{I_C}{I_B}$$

and always we use

$$I_E = I_C + I_B$$

$$\Rightarrow \quad \frac{I_C}{\alpha} = I_C + \frac{I_C}{\beta}$$

$$\Rightarrow \quad \frac{1}{\alpha} = 1 + \frac{1}{\beta} \Rightarrow \alpha = \frac{\beta}{1+\beta}$$ **Ans. (B)**

Sol. 8

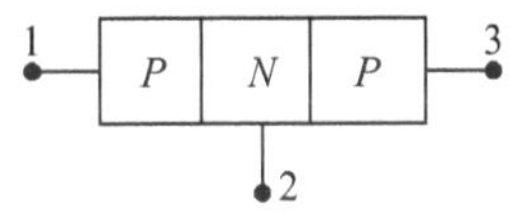

If 2 is +ve and 3 is –ve

$\Rightarrow$ Reverse Biased between 2 & 3 $\Rightarrow R_{\text{high}}$

If 2 is +ve and 1 is –ve

$\Rightarrow$ Reverse Biased between 2 & 1 $\Rightarrow R_{\text{high}}$

If 1 is +ve and 2 is –ve

$\Rightarrow$ Forward Biased between 1 & 2 $\Rightarrow R_{\text{Low}}$

If 3 is +ve and 2 is –ve

$\Rightarrow$ Forward Biased between 3 & 2 $\Rightarrow R_{\text{Low}}$ **Ans. (B)**

Sol. 9 For a CE configuration, input resistance

$$r_i = \frac{\delta V_{BE}}{\delta I_B}$$

Generally BE junction in active region is kept forward biased so upto $V_{BE} = V_{BP} \approx 0.6$ V.

r_i should remain constant and for $V_{BE} > 0.6$V value of r_i should decreased as it happens in Forward bias *p-n* junction. **Ans. (C)**

Sol. 10 Common emitter current gain $\beta = 69$

$$\Rightarrow \quad \alpha = \frac{\beta}{1+\beta} = \frac{69}{70}$$

we use $$\alpha = \frac{I_C}{I_E}$$

$$\Rightarrow \quad I_C = \alpha I_E = \frac{69}{70} \times 7.0 = 6.9 \text{ mA}$$ **Ans. (B)**

Sol. 11

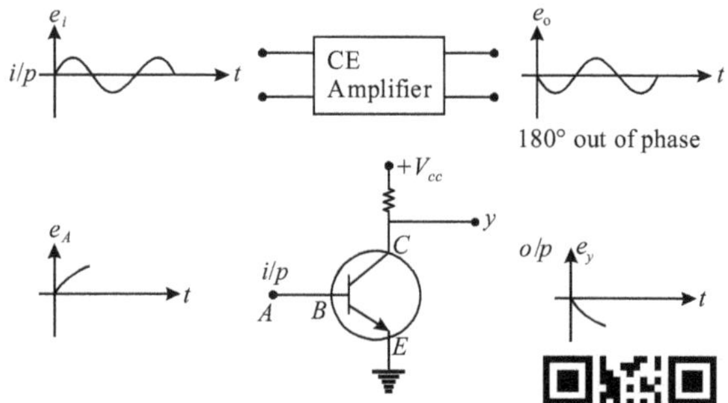

Behaves like a NOT gate (inverter)

Ans. (B)

Sol. 12

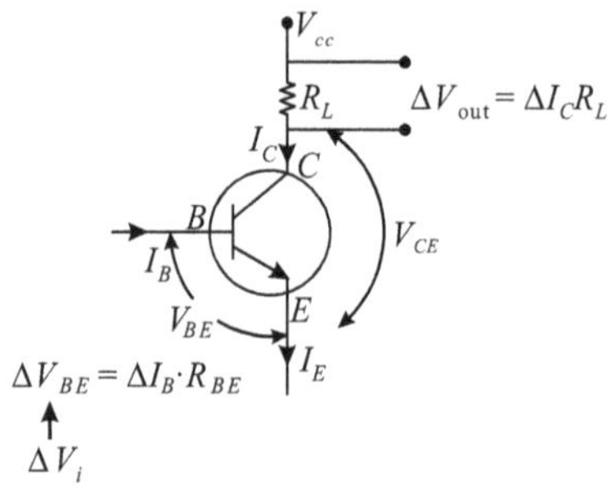

Current gain $\beta = \dfrac{\Delta I_C}{\Delta I_B}$

Voltage gain $A_V = \dfrac{\Delta V_{out}}{\Delta V_{in}} = \dfrac{\Delta I_C R_L}{\Delta I_B R_{BE}}$

$\Rightarrow \quad A_V = \beta \dfrac{R_L}{R_{BE}}$

Power gain = Voltage gain × Current gain

$$A_P = \beta^2 \frac{R_L}{R_{BE}}$$ **Ans. (D)**

Sol. 13 At saturation, *CE* voltage becomes zero, so we have

$$V_{CE} = 0$$

$\Rightarrow \quad V_{CE} = V_{CC} - I_C R_C$

$\Rightarrow \quad I_C = \dfrac{V_{CC}}{R_C} = 5 \times 10^{-3}\,\text{A}$

Given that $\beta_{dc} = 200$

$\Rightarrow \quad \beta_{dc} = \dfrac{I_C}{I_B}$

$\Rightarrow \quad I_B = \dfrac{5 \times 10^{-3}}{200}$

$\Rightarrow \quad I_B = 25\,\mu\text{A}$

At input side of transistor, we use

$$V_{BB} = I_B R_B + V_{BE}$$

$\Rightarrow \quad V_{BB} = (25\mu\text{A})(100k\Omega) + 1V$

$\Rightarrow \quad V_{BB} = 3.5V$ **Ans (B)**

Sol. 14 At saturation

$$V_{CE} = 0$$

$\Rightarrow \quad i_C = \dfrac{V_{CC}}{R_C} = \dfrac{10}{1000} 0.01\,\text{A}$

De current gain of transistor is

$$\beta = \frac{i_C}{i_B}$$

$\Rightarrow \quad i_B = \dfrac{i_C}{\beta} = \dfrac{0.01}{250} = 4 \times 10^{-5}\,\text{A} = 40\,\mu\text{A}$ **Ans. (C)**

Sol. 15 Input current

$$i_b = 15\,\mu\text{A}$$

Output current

$$i_c = 3\,\text{mA}$$

Output resistance

$$R_L = 10\,\text{mV}$$

Input voltage

$$V_{BE} = 10\,\text{mv}$$

Input resistance

$$r_{in} = \frac{V_{BE}}{i_b}$$

$\Rightarrow \quad r_{in} = \dfrac{10 \times 10^{-3}}{15 \times 10^{-6}} = \dfrac{2000}{3} = 0.67\,\text{k}\Omega$

Output voltage

$$v_{out} = i_c R_L = 3\,\text{V}$$

Voltage gain

$$A_V = \frac{V_{out}}{V_{in}} = \frac{3}{10 \times 10^{-3}} = 300$$ **Ans. (D)**

Sol. 16 Power gain in *dB* is given as

$$10 \log_{10} = \left(\frac{P_{out}}{P_{in}}\right) = 60$$

$\Rightarrow \quad \dfrac{P_{out}}{P_{in}} = \text{P}_{gain} = 60^6$

We use $\quad P_{gain} = \beta^2 \times \dfrac{R_{out}}{R_{in}}$

$\Rightarrow \quad 10^6 = \beta^2 \times \dfrac{10^4}{100}$

$\Rightarrow \quad \beta^2 = 10^4$

$\Rightarrow \quad \beta = 100$ **Ans. (D)**

Sol. 17 Voltage gain of a transistor is

$$V_{gain} = \left(\frac{\Delta I_C}{\Delta I_B}\right) \times \frac{R_{out}}{R_{in}}$$

$$= \left(\frac{5\times10^{-3}}{100\times10^{-6}}\right) \times \frac{100\times10^{3}}{100} = 5\times10^{4}$$

Power gain $P_{gain} = \frac{\Delta I_c}{\Delta I_b} \times V_{gain}$

$$= \frac{5\times10^{-3}}{100\times10^{-6}} \times 5\times10^{4} = 2.5\times10^{6}$$ **Ans. (C)**

28.4 Logic Gates

Sol. 1

+5V as 1

0V as 0

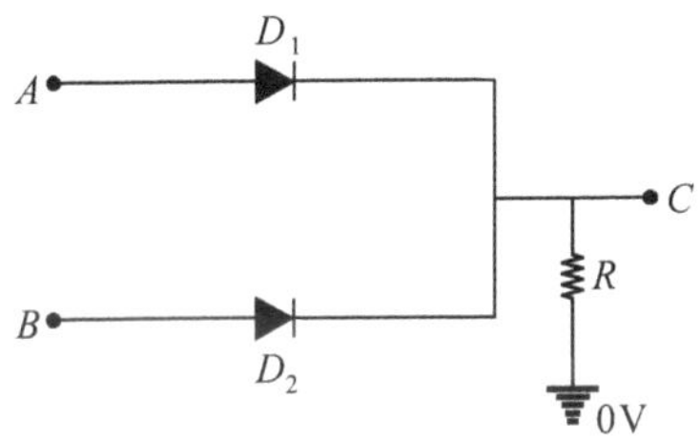

if current passes through $R \Rightarrow C$ is at high potential state (1)
Truth table of given circuit

A	B	C	
0	0	0	→ D_1, D_2 are RB
0	1	1	→ D_1 is RB & D_2 is FB
1	0	1	→ D_1 is FB & D_2 is RB
1	1	1	→ D_1, D_2 are FB

$\Rightarrow$ This is Truth table of OR gate. **Ans. (D)**

Sol. 2

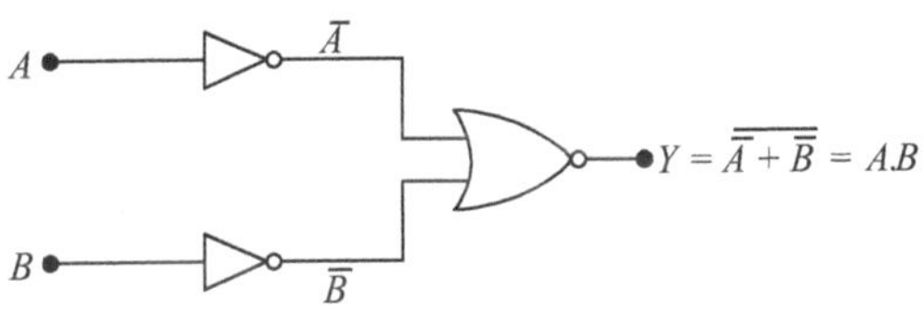

This logic circuit is equivalent to AND gate. The truth table used will be

A	B	Y
0	0	0
0	1	0
1	0	0
1	1	1

Ans. (D)

Sol. 3

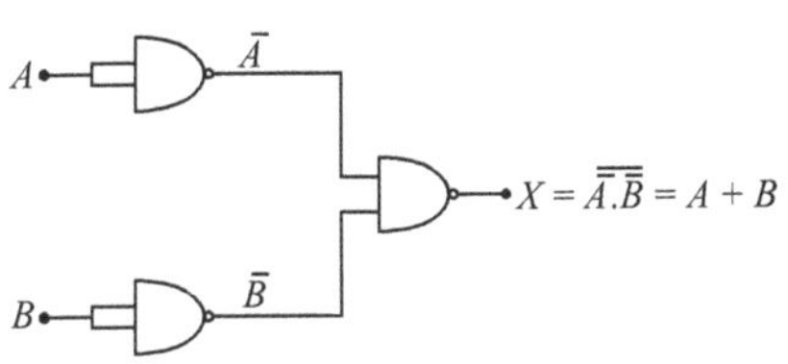

looking at output, this combination represents an OR gate. **Ans. (A)**

Sol. 4

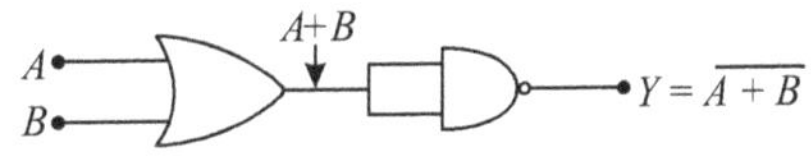

The output of this logic circuit is same as NOR Gate. **Ans. (B)**

Sol. 5

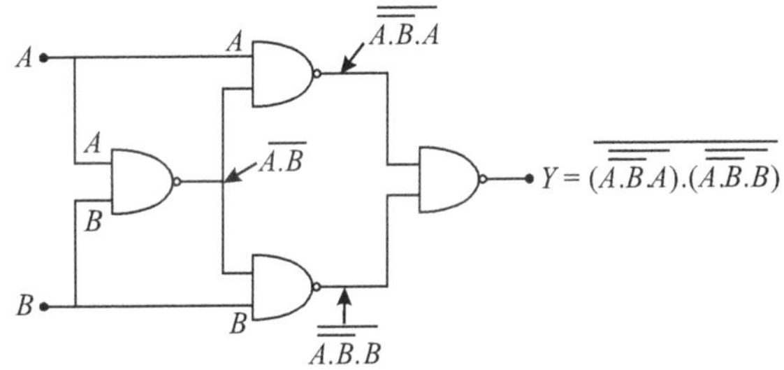

The simplified Boolean expression is

$$Y = \overline{A{\cdot}B}{\cdot}A + \overline{A{\cdot}B}{\cdot}B$$

for $A = 0, B = 0 \Rightarrow Y = 1.0 + 1.0 = 0 + 0 = 0$

for $A = 0, B = 1 \Rightarrow Y = 1.0 + 1.1 = 0 + 1 = 1$

for $A = 1, B = 0 \Rightarrow Y = 1.1 + 1.0 = 1 + 0 = 1$

for $A = 1, B = 1 \Rightarrow Y = 0.1 + 0.1 = 0 + 0 = 0$ **Ans. (A)**

Sol. 6 In the given circuit either of switches A or B is closed will make the lamp glow.

Inputs → if ON state of a switch is taken as '1' and OFF state is taken as '0'

output → lamp glow → '1'

lamp off → '0'

Truth table

A	B	Y
0	0	0
0	1	1
1	0	1
1	1	1

This is truth table of OR gate. **Ans. (B)**

Sol. 7

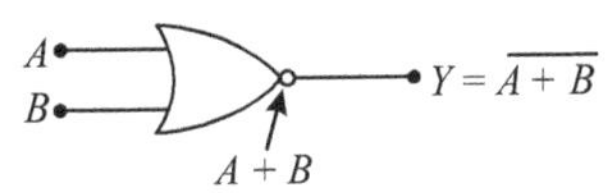

Ans. (A)

Sol. 8 Here the output of given system of gates is given as

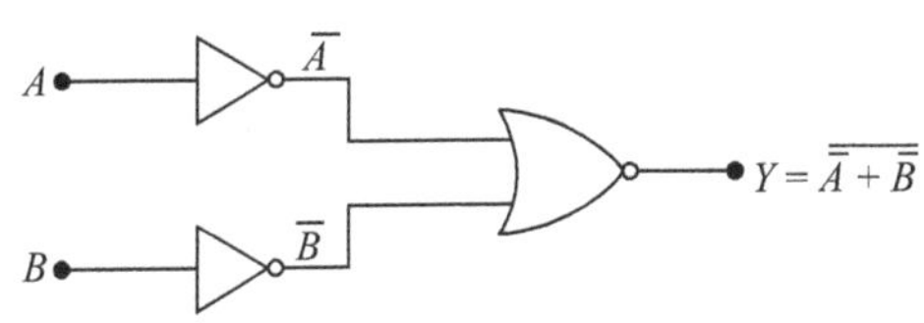

Truth table

A	B	Y
0	0	0
0	1	0
1	0	0
1	1	1

This is the truth table for AND gate. **Ans. (C)**

Sol. 9

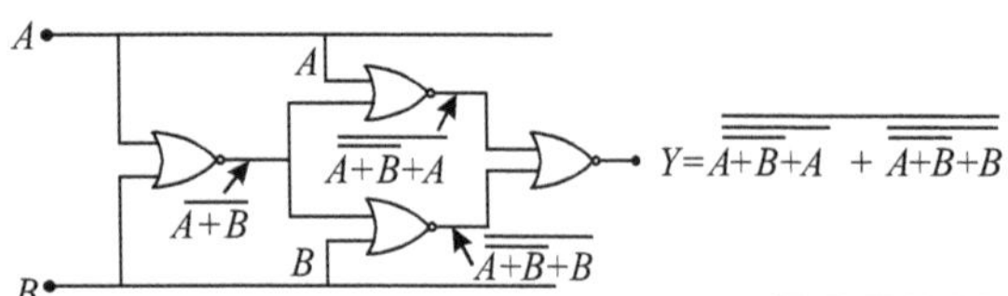

To make the truth table of this system we use

$A=0, 0, 1, 1$

$B=0, 1, 0, 1$

$$Y=\overline{\overline{\overline{A+B}+A}+\overline{\overline{A+B}+B}}$$

At $A=0, B=0 \Rightarrow$ $Y=\overline{\overline{\overline{1+0}+1+0}} = \overline{0+0} = 1$

At $A=0, B=1 \Rightarrow$ $Y=\overline{\overline{0+0}+\overline{0+1}} = \overline{1+0} = 0$

At $A=1, B=0 \Rightarrow$ $Y=\overline{\overline{0+1}+\overline{0+0}} = \overline{0+1} = 0$

At $A=1, B=1 \Rightarrow$ $Y=\overline{\overline{0+1}+\overline{0+1}} = \overline{0+0} = 1$ **Ans. (A)**

Sol. 10 (A) $C=\overline{\overline{A \cdot B}+\overline{A}}$ (B) $C=(A+B)\cdot B$

A	B	C
0	0	0
0	1	0
1	0	1
1	1	0

A	B	C
0	0	1
0	1	0
1	0	1
1	1	1

(C) $C=(A \cdot B)+\overline{B}$ (D) $C=\overline{A.B}$

A	B	C
0	0	0
0	1	1
1	0	0
1	1	1

A	B	C
0	0	1
0	1	1
1	0	1
1	1	0

Ans. (A)

Sol. 11 In the given circuit if any input A or B will be at high potential +5 V (state-1) then transistor will be in ON state and C will get connected to earth (0 V) so C will be at stable 0. Only when A & B are 0 then C will be at high position (state-1)

Truth table of this circuit will be

A	B	Y
0	0	1
0	1	0
1	0	0
1	1	0

This is Truth table of NAND gate. **Ans. (A)**

Sol. 12

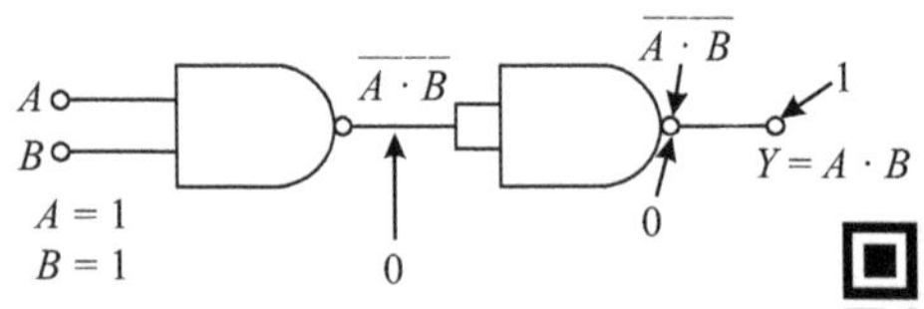

$Y = A.B$ is expression for AND gate.

Ans. (A)

Sol. 13 In given circuit V_{out} terminal is connected directly to $V_{CC} = 6\ V \Rightarrow V_{out}$ is always high at state 1.

So the output is at logic 1 irrespective of input states.

So here it is Not representing any logic gate.

$\Rightarrow$ None of the option is correct. **Ans. (None)**

Sol. 14 For A and B both zero, output Y is zero and if either of A or B is 1, output Y is 1.

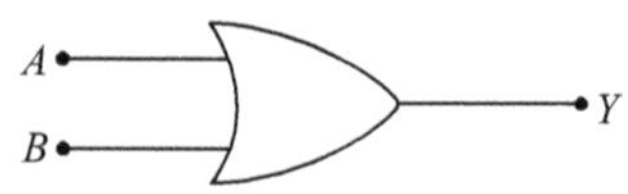

Ans. (A)

Sol. 15 Truth table for circuit

	a	b	c	Y
(A)	0	0	1	0
(B)	1	0	0	0
(C)	1	0	1	1
(D)	0	1	0	0

Boolean expression for the output of this gated circuit is

$$Y=(A+B)C$$ **Ans. (C)**

Sol. 16

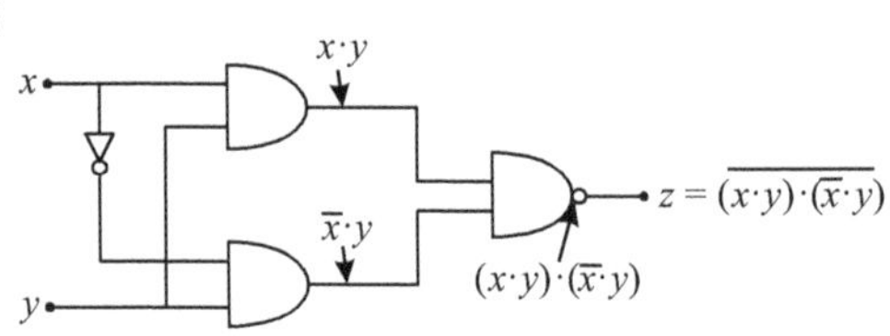

We calculate z for cases

$x = 0, 0, 1, 1$

$y = 0, 1, 0, 1$

output $\quad z = \overline{(x.y).(\bar{x}.y)}$

At $x=0, y=0 \Rightarrow \quad z = \overline{0 \cdot 0} = 1$

At $x=0, y=1 \Rightarrow \quad z = \overline{0 \cdot 1} = 1$

At $x=1, y=0 \Rightarrow \quad z = \overline{0 \cdot 0} = 1$

At $x=1, y=1 \Rightarrow \quad z = \overline{1 \cdot 0} = 1$

Ans. (C)

Sol. 17 $\quad P = \bar{X} + Y$

$$Q = \overline{\bar{Y} \cdot X} = Y + \bar{X}$$

$$\text{Output} = \overline{P + Q}$$

To make output '1' $P + Q$ must be '0'

So, $Y = 0$ and $X = 1$ **Ans. (D)**

Sol. 18 The boolean output of given circuit is shown in figure below

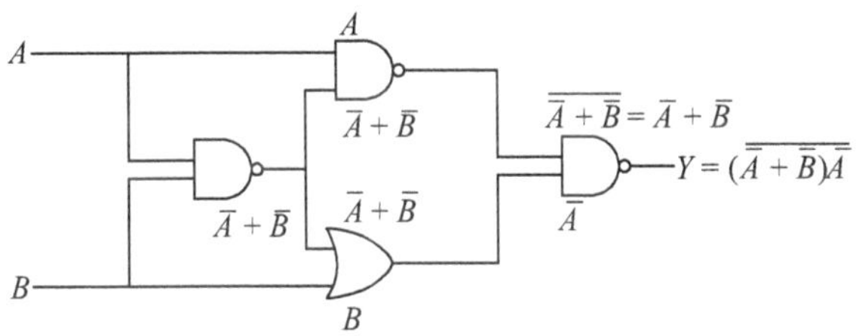

The output is simplified as

$$Y = \overline{(\bar{A} + \bar{B})\bar{A}}$$

$$\Rightarrow \quad Y = \overline{\bar{A} + \bar{A}B}$$

$$\Rightarrow \quad Y = A(\overline{\bar{A}B})$$

$$\Rightarrow \quad Y = A(A + \bar{B})$$

$$\Rightarrow \quad Y = A + A\bar{B} = A\bar{B}$$

Ans. (B)

Sol. 19 Truth table of the given logic circuit is written as

A	B	Equivalent
0	0	0
0	1	1
1	0	1
1	1	1

This is the table for 'OR' gate **Ans. (A)**

Sol. 20

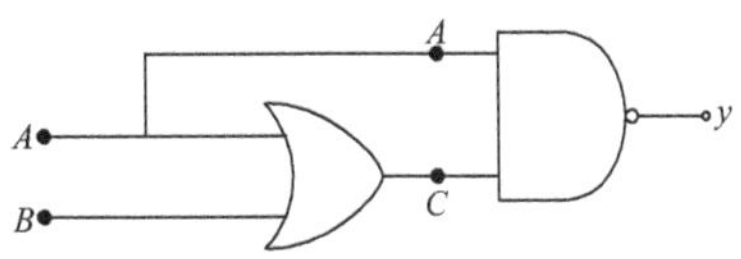

$$C = A + B$$

and $\quad y = \overline{A.C}$

Truth table for this circuit is given as

A	B	$C = (A + B)$	$A.C.$	$y = \overline{A.C}$
0	0	0	0	1
0	1	1	0	1
1	0	1	1	0
1	1	1	1	0

Ans. (A)

29.1 Communication Systems

Sol. 1 (A) True

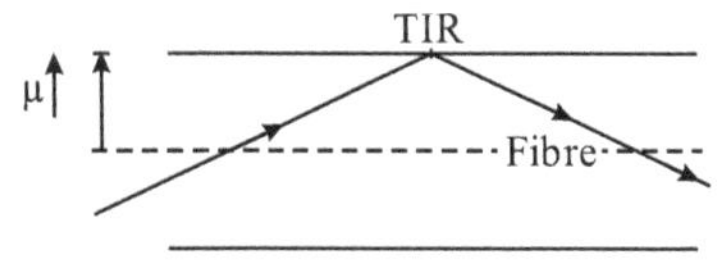

(B) Not true

As only by internal reflection light passes in optical fibre so outside interference is not possible.

(C) True

As absorption or scattering of light in OF is negligible.

(D) True

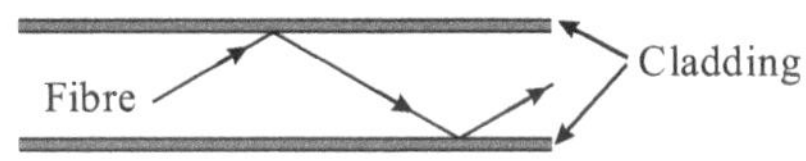

Ans. (B)

Sol. 2 **Statement-1 :** Due to energy absorption by Earth in 'Ground waves' the range of these is less so in sky wave range in relativity vary large as these are reflected from ionospheric layers.

$\Rightarrow$ True

Due to Change in state of ionization in ionosphere with temperature and other physical conditions the signals in sky waves are less stable.

$\Rightarrow$ **Statement-2 :** It is correct & explaining Statement-1 also.

Ans. (A)

Sol. 3 $h = 500\text{ m}$

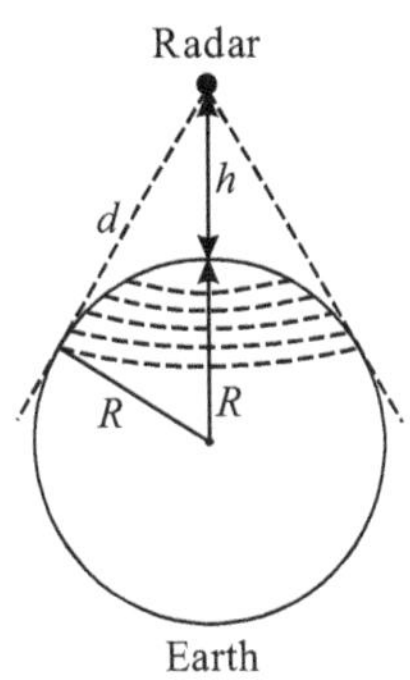

If d is the maximum distance from radar upto which objects can be detected on earth surface, we use

$$(h+R)^2 = d^2 + R^2$$

$$\Rightarrow \quad d = \sqrt{(h+R)^2 - R^2}$$

$$\Rightarrow \quad d = \sqrt{h^2 + 2hR}$$

$$\Rightarrow \quad d = \sqrt{(500)^2 + 2\times 500\times 6400\times 10^3}$$

$$\Rightarrow \quad d = 500\sqrt{1 + 2\times 6.4\times \frac{10^6}{500}}$$

$$\Rightarrow \quad d = 80000\text{ m}$$

Ans. (A)

Sol. 4 If a source of power P watt emits radiation of wavelength λ then number of photons emitted/s from source are

$$N = \frac{P}{hc/\lambda} = \frac{P\lambda}{hc}$$

$$\Rightarrow \quad N = \frac{10\times 10^3 \times 500}{6.62\times 10^{-34}\times 3\times 10^8}$$

$$\Rightarrow \quad N = \frac{5}{6.62\times 3}\times 10^{34-8+6} = 0.25\times 10^{32}$$

$$\Rightarrow \quad N = 2.5\times 10^{31}$$

Ans. (B)

Sol. 5 Wavelength of EM waves transmitted is

$$\lambda = \frac{c}{\nu} = \frac{3\times 10^8}{830\times 10^3} \simeq 361\text{ m}$$

The electric field and magnetic field in EM waves are related as

$$\frac{E}{B} = c$$

$$\Rightarrow \quad E = Bc = 4.82\times 10^{-11}\times 3\times 10^8$$

$$\Rightarrow \quad E = 14.46\times 10^{-3}$$

$$\Rightarrow \quad E = 0.01446\text{ V/m}$$

Ans. (D)

Sol. 6 **Statement-1 :** True

As in sky wave propagation we use frequency upto 30 MHz. (short waves) and long distance communication is possible due to reflection by ionospheric layers.

Statement-2 : Not correct

In ionosphere → plasma. (free ions) The frequency which are above 30 MHz can penetrate the atmosphere → No total Internal reflection occurs because with frequency Refractive Index of ionosphere changes. **Ans. (A)**

Sol. 7 (A) In sky wave signals ionization status of ionosphere changes with time due to thermal agitation change in day, night, seasons & almost every hour. (Correct)

(B) frequency which are more than 30 MHz can penetrate atmospheric layers in ionosphere. (Correct)

(C) ionosphere penetration occurs at $f > 30$ MHz (Not correct)

(D) In night ionospheric ionization is more stable due to lesser temperature variation. (Correct) **Ans. (C)**

Sol. 8 In LOS mode the maximum distance of clear communication is given as

$$d_{max} = \sqrt{2Rh_T} + \sqrt{2Rh_R}$$

$$\Rightarrow \quad d_{max} = \sqrt{2\times 6400\times 10^3\times 32} + \sqrt{2\times 6400\times 10^3\times 50}$$

$$\Rightarrow \quad d_{max} = \sqrt{64\times 64\times 10^5} + \sqrt{6400\times 10^5} = (64+80)\times\sqrt{10^5}$$

$$\Rightarrow \quad d_{max} = 144\times\sqrt{10}\times 10^2$$

$$\Rightarrow \quad d_{max} = 144\times 3.16\times 100$$

$$\Rightarrow \quad d_{max} \simeq 45.5\text{ km}$$

Ans. (B)

Sol. 9 For sky wave propagation of radio waves, frequency range is

$$\Delta f_{SW} \sim 2\text{ MHz} \leftrightarrow 30\text{ MHz}$$

Among options given only option (B) is in this range.

Fact based. **Ans. (B)**

Sol. 10 for sky wave propagation radio waves, frequency range is

$$\Delta f_{SW} \sim 2\text{ MHz} \leftrightarrow 30\text{ MHz}$$

Among the given options only option (B) is in range of this.

Fact based. **Ans. (B)**

Sol. 11 For resonating condition in antena for transmission of EM wave the antenna length should be of the order of λ.

Power transmission in EM waves is inversely proportional to λ^2

$$\Rightarrow \quad P_{eff} \propto \frac{1}{\lambda^2}$$

$$\Rightarrow \quad P_{eff} = K\left(\frac{1}{\lambda}\right)^2$$

Ans. (A)

Sol. 12 Maximum distance upto which signal can be broadcasted is

$$d_{max} = \sqrt{2Rh_T} + \sqrt{2Rh_R}$$

$$\Rightarrow \quad d_{max} = \sqrt{2\times 6.4\times 10^6}\left[\sqrt{140} + \sqrt{40}\right]$$

$$\Rightarrow \quad d_{max} = 65\text{ km}$$

Ans. (D)

Sol. 13 As range of transmission on ground is

$$R \propto \sqrt{h}$$

$\Rightarrow$ To double the range h should be made 4 times. **Ans. (B)**

Sol. 14 Range of transmission in LOS communication is given as

$$\text{Range} = \sqrt{2Rh_T} + \sqrt{2Rh_R}$$

$$\Rightarrow \quad 50 \times 10^3 = \sqrt{2 \times 6400 \times 10^3 \times h_T} + \sqrt{2 \times 6400 \times 10^3 \times 70}$$

$$\Rightarrow \quad h_T = 32\,\text{m}$$

Ans. (C)

Sol. 15 The given question is based on conceptual facts. Optical fibre communication uses infrared light for signal transmission, radar uses radio waves, sonar uses ultrasound and mobile phones use microwaves for their signal transmission. **Ans. (D)**

29.2 Modulation and Demodulation

Sol. 1 Modulation $\rightarrow$ using high frequency carrier frequency for long distance communication of modulation signals because of noise in direct transmission of these signals.

(A) $\rightarrow$ Speed of EM wave remain same even after modulation. Time lag between T_x and R_x is

$$t = \frac{\text{distance}}{c} \rightarrow \text{remain same (Not correct)}$$

(B) $\rightarrow$ After modulation frequency increases $\Rightarrow$ wavelength decreases $\Rightarrow$ size of antenna required reduce. (Correct)

(C) $\rightarrow$ $\Delta\omega = 2\omega_m$ and fractional bandwidth $= \dfrac{\Delta\omega}{\omega_c} = \dfrac{2\omega_m}{\omega_c} \downarrow$ (Correct)

(D) $\rightarrow$ As fractional band width on spectrum decreases $\Rightarrow$ selectivity increases. (Correct) **Ans. (A)**

Sol. 2 $E = E_c \cos\omega_c t + E_m \cos\omega_m t \cos\omega_c t$

$$\Rightarrow \quad E = E_c \cos\omega_c t + \frac{E_m}{2}\cos(\omega_c - \omega_m)t + \frac{E_m}{2}\cos(\omega_c + \omega_m)t$$

where $\omega_c \rightarrow$ Carrier frequency

$(\omega_c - \omega_m) \rightarrow$ lower side band frequency

$(\omega_c + \omega_m) \rightarrow$ upper side band frequency

Ans. (B)

Sol. 3 Noise $\rightarrow$ unwanted signal which is caught from atmosphere and distort the transmitting signal.
It is low frequency (or higher wavelengths)

To avoid (easily extract) noise upto maximum limit the signal frequency is to be kept high that's why modulation is needed.

Tolerance of modulated signal against noise (Noise tolerance) is high for a high frequency wave (short wave). **Ans. (B)**

Sol. 4 During de-modulation the maximum frequency which can be detected is given as

$$f_{max} = \frac{1}{2\pi\mu RC}$$

$$\Rightarrow \quad f_{max} = \frac{1}{2 \times 3.14 \times 0.6 \times 100 \times 10^3 \times 250 \times 10^{-12}}$$

$$\Rightarrow \quad f_{max} = \frac{10^8}{6.28 \times 6 \times 250} = \frac{1000 \times 10^5}{6.28 \times 6 \times 250}$$

$\Rightarrow \quad f_{max} \simeq 10^4\,\text{Hz} \simeq 10\,\text{kHZ}$ **Ans. (B)**

Sol. 5

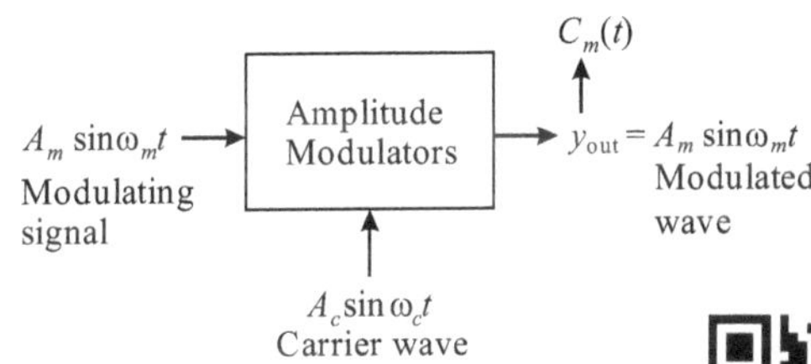

Where $\quad R = A_c + A_m \sin\omega_m t$

$\Rightarrow$ Output as modulated wave is given as

$$C_m(t) = (A_c + A_m \sin\omega_m t)\sin\omega_c t$$
$$= A(1 + \sin\omega_m t)\sin\omega_c t$$

Here modulation index $\mu = \dfrac{A_m}{A_c} = \dfrac{A}{A} = 1$ **Ans. (C)**

Sol. 6 When a modulating signal of frequency ω_m is modulated on a carrier wave of frequency ω_c then the resulting modulated signal will have three frequencies

(1) Carrier frequency $\rightarrow \omega_c \rightarrow 2000$ kHz

(2) lower side band frequency $\rightarrow \omega_c \rightarrow 1995$ kHz

(2) upper side band frequency $\rightarrow \omega_c + \omega_m \rightarrow 2005$ kHz **Ans. (A)**

Sol. 7 Carrier wave is used for long distance propagation of signal using modulation.
here modulating signal $\rightarrow$ audio signal
if carrier wave is $y_c(t) = A_c \sin\omega_c t$
in AM, amplitude of carrier wave is varied in proportion to amplitude of MW (MW = audio signal modulating wave)
in FM, frequency of carrier wave is varied in proportion to amplitude of MW. **Ans. (C)**

Sol. 8 The modulated signal is

$$C_m(t) = 30 \sin 300\,\pi t + 10 \cos 200\,\pi t - 10 \cos 400\,\pi t \quad \ldots(1)$$

If is of the form

$$C_m(t) = A_c \sin\omega_c t + \frac{\mu A_c}{2}\cos(\omega_c - \omega_m)t - \frac{\mu A_c}{2}\cos(\omega_c + \omega_m)t \ldots (2)$$

Comparing-(1) and (2) we get

$$A_c = 30$$

$$\Rightarrow \quad \omega_c = 300\pi = 2\pi f_c$$

$$\Rightarrow \quad f_c = 150 \text{ Hz}$$

$$\omega_c - \omega_m = 200\pi$$

$$\omega_c + \omega_m = 400\pi$$

$$\Rightarrow \quad \omega_m = 100\pi = 2\pi f_w$$

$$f_w = 50 \text{ Hz}$$

$$\Rightarrow \quad \frac{\mu A_c}{2} = 10$$

$$\Rightarrow \quad \frac{\mu(30)}{2} = 10$$

$$\Rightarrow \quad \mu = 2/3$$

Ans. (B)

Sol. 9 Human speech signal range ~ 200 Hz ↔ 2700 Hz

$$(\Delta f_{HS}) = 2500 \text{ Hz}$$

Music signal range ~ 10200 Hz ↔ 15200 Hz

$$\Delta f_{MS} = 5000 \text{ Hz}$$

In case both the signals are transmitted together than overall frequency range. 200 Hz ↔ 15200 Hz

The frequency range for both signals will be $\Delta f_{\text{both}} = 15000\text{Hz}$

$$\frac{\Delta\omega_{\text{both}}}{\Delta\omega_{\text{MS}}} = \frac{\Delta f_{\text{both}}}{\Delta f_{\text{MS}}} = \frac{15000}{2500} = 6$$

Ans. (C)

Sol. 10 If carrier frequency is ω_c than side band frequencies as

lower side band $\quad \omega_1 = \omega_c - \omega_m$

upper side band $\quad \omega_2 = \omega_c + \omega_m$

in transmission of modulated signal all frequencies between $\omega_1 \leftrightarrow \omega_2$ are transmitted.

Out of given option ω_m is NOT contained in modulated wave. **Ans. (C)**

Sol. 11 Modulation index

$$m = \frac{V_{0m}}{V_{0c}} = \frac{5}{25} = \frac{1}{5} = 0.2$$

Modulation frequency $f_m = 20$ kHz

Carrier frequency $\quad f_c = 1.2 \text{ MHz} = 1200 \text{ kHz}$

Side band frequencies as

$$f_{LSB} = f_c - f_m = 1200 - 20 = 1180 \text{ kHz}$$

and $\quad f_{USB} = f_c + f_m = 1200 + 20 = 1220 \text{ kHz}$

Ans. (D)

Sol. 12 Modulation Index

$$m = \frac{V_{0m}}{V_{0c}}$$

$$\Rightarrow \quad V_{0m} = m \times V_{0c}$$

$$\Rightarrow \quad V_{0m} = 0.8 \times 14 = 11.2 \text{ V}$$

Ans. (A)

Sol. 13 Total transmission band width available is

$$= \frac{10}{100} \times 10 \times 10^9 \text{ Hz} = 10^9 \text{ Hz}$$

Each channel requires and individual Band width = 5×10^3 Hz

If the total Band width can accommodate N channels then we use

$$10^9 = N \times 5 \times 10^3$$

$$\Rightarrow \quad N = 2 \times 10^5$$ **Ans. (C)**

Sol. 14 The frequency of carrier wave is given as

$$f_c = \frac{1}{2\pi\sqrt{LC}}$$

$$\Rightarrow \quad f_c = \frac{1}{2\pi\sqrt{49\times10^{-6}\times2.5\times10^{-9}}}$$

$$\Rightarrow \quad f_c = \frac{1}{2\pi\times7\times5\times10^{-8}} = \frac{10^8}{70\pi}$$

$$f_c = 454.54 \text{ kHz}$$

The frequency range used for transmission of this signal is

$$\Delta f = 2f_m$$

$$f_1 \leftrightarrow f_2 = f_c \pm f_m$$

$$\Rightarrow \quad f_1 = f_c - f_m = 454.54 - 12 = 442.54 \text{ kHz}$$

$$\Rightarrow \quad f_2 = f_c + f_m = 454.54 + 12 = 466.54 \text{ kHz}$$

Ans. (C)

Sol. 15 Total bandwidth available for transmission is

$$\Delta BW = 300 \times 10^3 \text{ Hz}$$

Band width of each channel is $CBW = 2f_m = 2 \times 15 \times 10^3$

$$\Rightarrow \quad CBW = 30 \times 10^3 \text{ Hz}$$

If N are the number of channels accommodated for transmission, we have

$$300 \times 10^3 = N \times 30 \times 10^3$$

$$\Rightarrow \quad N = 10$$ **Ans. (B)**

Sol. 16 Frequency of source is given as

$$f = \frac{3\times10^8}{8\times10^{-7}} = \frac{30}{8}\times10^{14} \text{ Hz}$$

$$\Rightarrow \quad f = 3.75 \times 10^{14} \text{ Hz}$$

$$1\% \text{ of } f = 0.0375 \times 10^{14} \text{ Hz}$$

$$\Rightarrow \quad \Delta f = 3.75 \times 10^{12} \text{ Hz} = 3.75 \times 10^6 \text{ MHz}$$

$\Rightarrow$ Number of channels

$$= \frac{3.75\times10^6}{6} = 6.25 \times 10^5$$ **Ans. (D)**

Sol. 17 $f_{\text{carrier}} = \frac{250}{0.1}$ kHz

$\Rightarrow$ Range of signal = 2250 Hz to 2750 Hz

In given options; for 2000 kHz

$$f_{\text{mod}} = 200 \text{ Hz}$$

So its range is 1800 Hz to 2200 Hz

Which is not overlapping with the existing range. **Ans. (B)**

Sol. 18 The given signal can be written as

$$V(t) = 10 \sin B + \frac{3}{2}[2 \cos A \sin B]$$

$$\Rightarrow \quad V(t) = 10 \sin B + 10+\frac{3}{2} \; [\sin (A + B) - \sin (A - B)]$$

$$\Rightarrow \quad V(t) = 10 \sin B\,(5.5 \times 10^5\, t) + \frac{3}{2} \; [\sin (57.2 \times 10^4\, t)]$$

$$- \sin (52.8 \times 10^4\, t)$$

Here $\omega_1 = 57.2 \times 10^4 = 2\pi f_1$

$$\Rightarrow \quad f_1 = \frac{57.2\times10^4}{2\times\left(\frac{22}{7}\right)} = 9.1 \times 10^4$$

$$\Rightarrow \quad f_1 \simeq 91 \text{kHz}$$

and $\omega_2 = 52.8 \times 10^4 t = 2\pi f_2$

$$\Rightarrow \quad f_2 = \frac{52.8\times10^4}{2\times\left(\frac{22}{7}\right)}$$

$$\Rightarrow \quad f_2 \simeq 84 \text{ kHz}$$ **Ans. (C)**

Sol. 19 Analysis of graph shows

(1) Amplitude varies between

8 – 10V or 9 ± 1

(2) Two time period are

100 μs (modulating wave) & 8 μs (carrier wave)

Thus modulated signal is $\left[9\pm1\sin\left(\frac{2\pi t}{T_1}\right)\right]\sin\left(\frac{2\pi t}{T_2}\right)$

$= [9 \pm 1\sin (2\pi \times 10^4 t)] \sin (2.5\pi \times 10^5 t)$ **Ans. (D)**

Sol. 20 Given that by modulation carrier wave amplitude changes maximum upto 160 V, so we use

$$A_m + A_c = 160$$

$$\Rightarrow \quad A_m + 100 = 160$$

$$\Rightarrow \quad A_m = 60$$

Modulation index is

$$\mu = \frac{A_m}{A_c} = \frac{60}{100}$$

$\Rightarrow \quad \mu = 0.6$ **Ans. (A)**

Sol. 21 To minimise attenuation, wavelength of carrier waves is close to 1500 nm **Ans. (D)**

Sol. 22 AM signal contains one carrier frequency and two (ω_0) side band frequencies

$$\text{USB} = \omega_0 + \omega$$

$\text{LSB} = \omega_0 - \omega$ **Ans. (B)**

Sol. 23 The physical size of antenna of reciver and transmitter both directly proportional to the wavelength of carrier wave so both are inversely proportional to carrier frequency. **Ans. (C)**

Sol. 24 Modulation inxex

$$= \frac{V_{0m}}{V_{0c}} = \frac{100}{400} = 0.25$$

USB frequency $= f_c + f_m$

LSB frequency $= f_c - f_m$

USB – LSB $= 2f_m = 2 \times 100 \times 10^6 = 2 \times 10^8$ Hz **Ans. (D)**

Sol. 25 Modulation index

$$m = \frac{A_m}{A_c} = \frac{2}{4} = 0.5$$

Carrier frequency

$$f_c = \frac{\omega_c}{2\pi} = \frac{20000\pi}{2\pi} = 10000 \text{ Hz} = 10 \text{ kHz}$$

Modulating frequency

$$f_m = \frac{\omega_m}{2\pi} = \frac{2000\pi}{2\pi} = 1000 \text{ Hz} = 1 \text{ kHz}$$

LSB frequency $= f_c - f_m = 10 - 1 = 9$ kHz **Ans. (A)**

30.1 Errors & Experiments

Sol. 1 Least count of main scale = 1mm

$$\text{least count of circular scale} = \frac{\text{Pitch}}{\text{No. of CSD}} = \frac{0.5\text{ mm}}{50} = 0.01\text{mm}$$

[As 2 turns of CS gives 1mm displacement on MS]

Diameter of wire = MS + CS – ZE

$$= 3 \text{ mm} + 35 \times 0.01 \text{ mm} - (-0.03 \text{ mm})$$

$= 3 + 0.35 + 0.03 = 3.38$ mm **Ans. (D)**

Sol. 2 Momentum of body is given as

$$p = mv = \underbrace{3.513}_{4SD} \times \underbrace{5.00}_{3SD}$$

$$\Rightarrow \quad p = 17.56500 \text{ kg m/s}$$

Rounding off to 3SD, we get

$$p = 17.6 \text{ kg m/s}$$ **Ans. (A)**

Sol. 3 Given that 29 MSD = 30 VSD

$$\Rightarrow \quad 1\text{VSD} = \frac{29}{30}\text{MSD} \qquad [1\text{ MSD} = 0.5°]$$

Least count of Instrument

$$\text{LC} = 1\text{ MSD} - 1\text{ VSD}$$

$$\Rightarrow \quad \text{LC} = 1\text{ MSD} - \frac{29}{30}\text{MSD}$$

$$\Rightarrow \quad \text{LC} = \frac{1}{30}\text{MSD} = \frac{0.5°}{30} = \frac{1°}{60}$$

$$\Rightarrow \quad \text{LC} = 1'(\text{one minute})$$ **Ans. (D)**

Sol. 4

	No of SD
1st figure $\underset{12\ 345}{23.023}$	5
2nd figure $\underset{1}{0.0003}$	1
2nd figure $\underset{1\ 2}{2.1} \times 10^{-3}$	2

Ans. (A)

Sol. 5 Least count of circular scale

$$= \frac{\text{Pitch}}{\text{no. of division}}$$

$$= \frac{1\text{ mm}}{100} = 0.01 \text{ mm}$$

$$= 0.001 \text{ cm}$$

Diameter of wire = MS + CS

$$= 0 + 52 \times 0.001 = 0.052 \text{ cm}$$ **Ans. (A)**

Sol. 6 Least count of Main Scale = 0.5°

$$\text{least count of Vernier Scale} = \frac{1}{30} \times 0.5° = \frac{1°}{60}$$

$$\text{Prism Angle} = 58.5\,(\text{MS}) + 9(\text{VS}) \times \frac{1}{60}$$

$$= 58.5° + 0.15° = 58.65°$$ **Ans. (C)**

Sol. 7 By Ohm's law we use $V = IR$

$$\Rightarrow \quad R = V/I$$

% (max) error in measurement of R is

$$100 \times \frac{dR}{R} = \left(\frac{dV}{V} + \frac{dI}{I}\right) \times 100$$

$$\Rightarrow \quad \frac{dR}{R} \times 100 = 3 + 3 = 6\%$$ **Ans. (A)**

Sol. 8 (A) 5.3200 cm $\rightarrow$ LC = 0.0001 cm

(B) 5.3 cm $\rightarrow$ LC = 0.1 cm

(C) 5.32 cm $\rightarrow$ LC = 0.01 cm

(D) 5.320 cm $\rightarrow$ LC = 0.001 cm **Ans. (D)**

Sol. 9 Least count = 1 MSD – 1 VSD

If we consider a and b are 1 MSD and 1 VSD lengths then we use

$$Na = (N+1)b \Rightarrow b = \frac{Na}{N+1}$$

$$LC = a - \frac{Na}{N+1} = \frac{a}{N+1}$$ **Ans. (D)**

Sol. 10 The length measured is 3.50 cm.
least count $\rightarrow$ 0.01 cm

(A) Meter scale LC = 0.1 cm

(B) Vernier Calliper $\text{LC} = \frac{1}{10}\text{mm} = 0.1 \text{ mm} = 0.01 \text{ cm}$

(C) Screw Gauge $\text{LC} = \frac{1}{100}\text{mm} = 0.01 \text{ mm} = 0.001 \text{ cm}$

(D) Screw Gauge $\text{LC} = \frac{1}{50}\text{mm} = 0.02 \text{ mm} = 0.002 \text{ cm}$ **Ans. (B)**

Sol. 11 The current is given as

$$I = (e^{1000V/T} - 1)\text{ mA} \rightarrow \text{at } I = 5 \text{ mA we have}$$

$$e^{1000V/T} = 6 \text{ mA}$$

$$dI = \frac{1000}{T} e^{1000V/T}\, dV \text{ mA}$$

$$\Rightarrow \quad dI = \frac{1000}{300} \times 6 \times 0.01 \times 10^{-3}$$

$$\Rightarrow \quad dI = 2 \times 10^{-4} \text{ A} = 0.2 \text{ mA}$$ **Ans. (A)**

Sol. 12 For a simple pendulum time period is given as

$$T = 2\pi\sqrt{\frac{L}{g}}$$

$$\Rightarrow \quad g = \frac{4\pi^2 L}{T^2}$$

Maximum relative error in measurement of g is given as

$$\frac{dg}{g} = \frac{dL}{L} + 2\frac{dT}{T}$$

$$\%\frac{dg}{g} \times 100 = \left(\frac{0.1}{20} + 2 \times \frac{1}{90}\right) \times 100$$

$$= (0.005 + 0.0222) \times 100 = 2.72\%$$ **Ans. (B)**

Sol. 13 If λ be the potential gradient on potentiometer wire than we use $V_s = \lambda l_1$

$$\Rightarrow \quad 1.1 = \lambda(440) \Rightarrow \lambda = \frac{1.1}{440} = \frac{1}{400} \text{ V/cm}$$

If V_R be the potential difference across the resistance we use for balancing length $l_2 = 220$ cm

$$V_R = \lambda l_2$$

$$\Rightarrow \quad V_R = \frac{1}{400} \times 220 = \frac{22}{40} = 0.55 \text{ V}$$

As reading of voltmeter is 0.5 V so error in reading is

$$\varepsilon = 0.5 - 0.55 = -0.05 \text{ V}$$

Ans. (D)

Sol. 14 (1) Rotation period of earth, $T = 24$ Hz $= 86400$ s $\sim 10^5$ s

(2) Revolution period of earth, $T = 365 \times 86400 \sim 10^7$ s

(3) Period of light wave ($\lambda \to 4000 - 8000$ Å)

$$T = \frac{\lambda}{c} = \frac{6\times10^{-7}}{3\times10^8} = 2 \times 10^{-15} \sim 10^{-15} \text{ s}$$

(4) Period of sound wave ($\nu \to 20 - 20000$ Hz)

$$T = \frac{1}{\nu} = \frac{1}{1000} = 10^{-3} \text{ s}$$

Ans. (A)

Sol. 15 Using the expression of time period of simple pendulum, g is calculated as

$$g = \frac{4\pi^2 L}{T^2}$$

% relative error in measurement of g is given as

$$\frac{dg}{g} \times 100 = \frac{dL}{L} \times 100 + 2\frac{dT}{T} \times 100$$

$$\Rightarrow \quad \frac{dg}{g} \times 100 = \frac{0.01}{20} \times 100 + 2 \times \frac{1}{90} \times 100$$

$$= (0.005 + 0.0222) \times 100$$

$$= 0.0272 \times 100 = 2.72\%$$

Ans. (D)

Sol. 16 Least count of VS is $LC_V = \frac{0.1}{10} = 0.01$ cm

The diameter readings are

(1) $0.5 + 8 \times 0.01 - (-0.03) = 0.61$ cm $\quad R_1$

(2) $0.5 + 4 \times 0.01 - (-0.03) = 0.57$ cm $\quad R_2$

(3) $0.5 + 6 \times 0.01 - (-0.03) = 0.59$ cm $\quad R_3$

mean diameter of ball is

$$D = \frac{R_1 + R_2 + R_3}{3} = \frac{0.61 + 0.57 + 0.59}{3}$$

$$\Rightarrow \quad D = 0.59 \text{ cm}$$

Ans. (B)

Sol. 17 Mean value of measurement

$$T_m = \frac{T_1 + T_2 + T_3 + T_4}{4} = \frac{90 + 91 + 95 + 92}{4}$$

$$\Rightarrow \quad T_m = 92 \text{ s}$$

deviation values of measurements from mean value is

$$\Delta T_1 = |T_1 - T_m| = 2\text{s}$$

$$\Delta T_2 = |T_2 - T_m| = 1\text{s}$$

$$\Delta T_3 = |T_3 - T_m| = 3\text{s}$$

$$\Delta T_4 = |T_4 - T_m| = 0\text{s}$$

mean deviation of the data set from mean value is

$$\Delta T_m = \frac{\Delta T_1 + \Delta T_2 + \Delta T_3 + \Delta T_4}{4}$$

$$\Delta T_m = \frac{2 + 1 + 3 + 0}{4} = 1.5 \text{ s}$$

mean time measurement

$$T_m \pm \Delta T_m = 92 \pm 1.5 \text{ s}$$

Ans. (C)

Sol. 18 Main scale least count (1 division) = 0.5 mm

$$\text{Circular scale lest count} = \frac{0.5}{50} \text{ mm} = 0.01 \text{ mm}$$

zero error = 5 division on the –ve side of scale

$$= 5 \times 0.01 \text{ mm} = 0.05 \text{ mm}$$

Thickness of Al scale Reading = 0.5 (MS) mm + 25 (CS) × 0.01

$$= 0.5 + 0.25 = 0.75 \text{ mm}$$

As sheet thickness = 0.75 – (–0.05) = 0.80 mm **Ans. (D)**

Sol. 19 For the given relation of surface tension, the relative % error in its measurement is given as

$$\frac{dT}{T} = \frac{dr}{r} + \frac{dh}{h} = \frac{dD}{D} + \frac{dh}{h} \quad \ldots(1)$$

In measurement of diameter and water rise the least count of readings are

$$dr = \frac{dD}{2} = \frac{0.01\times10^{-2}}{2} \text{ m}$$

and $\quad dh = 0.01 \times 10^{-2}$ m

from equation-(1) we use

$$\%\text{ error} = \frac{dT}{T} \times 100 = \left(\frac{dD}{D} + \frac{dh}{h}\right) \times 100$$

$$\Rightarrow \quad \frac{dT}{T} \times 100 = \left(\frac{0.01\times10^{-2}}{1.25\times10^{-2}} + \frac{0.01\times10^{-2}}{1.45\times10^{-2}}\right) \times 100$$

$$\Rightarrow \quad \frac{dT}{T} \times 100 = \frac{1}{1.25} + \frac{1}{1.45} = 0.8 + 0.69\%$$

$$= 1.49\%$$

Ans. (D)

Sol. 20 From the given expression the measurement of P gives its relative maximum % error as

$$\frac{dP}{P}=\frac{1}{2}\left(\frac{da}{a}\right)+2\left(\frac{db}{b}\right)+3\left(\frac{dc}{c}\right)+4\left(\frac{dd}{d}\right)$$

$$\Rightarrow \quad \% \text{ error in } P=\frac{1}{2}\times 2+2\times 1+3\times 3+4\times 5$$

$$=1+2+9+20=32\%$$

Ans. (C)

Sol. 21 Expression for A is

$$A=\frac{P^3Q^3}{\sqrt{R}S}$$

Maximum relative % error in A is given as

$$\frac{dA}{A}\times 100\%=\left(3\frac{dP}{P}+2\frac{dQ}{Q}+\frac{1}{2}\frac{dR}{R}+\frac{dS}{S}\right)\times 100\%$$

$$\Rightarrow \quad \% \text{ error in } A=3\times 0.5+2\times 1+\frac{1}{2}\times 3+1.5$$

$$=1.5+2+1.5+1.5=6.5\%$$ **Ans. (D)**

Sol. 22 5 revolution displacement of 0.25 cm

$$\Rightarrow \quad 1 \text{ main scale division}=\frac{0.25}{5}=0.05 \text{ cm}$$

$$\text{least count of Screw gauge}=\frac{1 \text{ main scale division}}{\text{no. of circular scale division}}$$

$$=\frac{0.05}{100}=5\times 10^{-4}\text{ cm}$$

Reading for thickness of wire = 4 MS + 30 CS

wire thickness $=4\times 0.05+30\times 5\times 10^{-4}$ cm

$$=0.2+0.015=0.215 \text{ cm}$$ **Ans. (D)**

Sol. 23 If R is radius of a sphere then

its surface area $S=4\pi R^2$

$$\Rightarrow \quad \frac{dS}{S}=\frac{2dR}{R}=\alpha$$

$$\Rightarrow \quad \frac{dR}{R}=\frac{\alpha}{2}$$

its volume $V=\frac{4}{3}\pi R^3$

$$\Rightarrow \quad \frac{dV}{V}=\frac{3dR}{R}=3\left(\frac{\alpha}{2}\right)=\frac{3}{2}\alpha$$ **Ans. (C)**

Sol. 24 Using Kepler's law we have

$$T^2=\frac{4\pi^2}{GM}R^3$$

taking log on both sides and differentiating, gives

$$2\ln T=\ln\left(\frac{4\pi^2}{GM}\cdot R^3\right)$$

$$\Rightarrow \quad 2\ln T=\ln\left(\frac{4\pi^2R^3}{G}\right)-\ln M$$

$$\Rightarrow \quad 2\frac{dT}{T}=\frac{dM}{M}$$

$$\Rightarrow \quad \frac{dM}{M}=2\times 10^{-2}$$ **Ans. (C)**

Sol. 25 Using mass and length of a cube, its density is given as

$$\rho=\frac{\text{Mass}}{\text{Volume}}=\frac{M}{L^3}$$

relative error (maximum) in density calculation is given as

$$\frac{\Delta\rho}{\rho}=\frac{\Delta M}{M}+3\frac{\Delta L}{L}$$

$$\Rightarrow \quad \frac{\Delta\rho}{\rho}=1.5\%+3(1\%)=1.5+3=4.5\%$$ **Ans. (C)**

Sol. 26 Resistance of a conductor is given as

$$R=\frac{\rho l}{A} \text{ and volume } (V)=Al$$

$$R=\frac{\rho l^2}{V}$$

$$\Rightarrow \quad \frac{\Delta R}{R}=\frac{2\Delta l}{l}=1\%$$ **Ans. (C)**

Sol. 27 Least count of screw gauge is given as

$$LC=\frac{\text{Pitch}}{\text{No. of division}}$$

$$\Rightarrow \quad LC=0.5\times 10^{-2}\text{ mm}$$

Zero error is given as

$$z=3\times 0.5\times 10^{-2}\text{ mm } (+ve)$$

$$\Rightarrow \quad z=1.5\times 10^{-2}\text{ mm}=0.015\text{ mm}$$

Thickness of sheet measured by screw gauge is

$$t=MSR+CSR-ZE \;(+ve \text{ zero error})$$

$$\Rightarrow \quad t=5.5\text{ mm}+(48\times 0.5\times 10^{-2})-0.015$$

$$\Rightarrow \quad t=5.5+0.24-0.015=5.725\text{ mm}$$ **Ans. (B)**

Sol. 28 For volume of cylinder, relative error is given as

$$\frac{\Delta V}{V}=2\frac{\Delta d}{d}+\frac{\Delta h}{h}=2\left(\frac{0.1}{12.6}\right)+\frac{0.1}{34.2}\simeq 0.016$$

$$V = 34.2 \times \frac{\pi}{4} \times (12.6)^2$$

$\Rightarrow$ $V = 4262.23 \text{ cm}^3$

and $\Delta V = 0.016 \times 4262.23 = 68.19 \text{ cm}^3$

Volume in 3 SD is given as

$$V = 4260 \pm 68 \text{ cm}^3$$

Out of given options most appropriate is (A). **Ans. (A)**

Sol. 29 Least count of a screw gauge is given as

$$\text{Least count} = \frac{\text{Pitch}}{\text{Number of division on circular scale}}$$

$\Rightarrow$ $5 \times 10^{-6} = \dfrac{10^{-3}}{N}$

$\Rightarrow$ $N = 200$ **Ans. (C)**

Sol. 30 Time of one oscillation measured is

$$T = \frac{30 \text{ sec}}{20}$$

Maximum possible error in measurement (least count) is

$$\Delta T = \frac{1}{20} \text{ sec}$$

Length of pendulum is given as

$$L = 55 \text{ cm}$$

$$\Delta L = 1\text{mm} = 0.1 \text{ cm}$$

Value of g is calculated as

$$g = \frac{4\pi^2 L}{T^2}$$

Percentage error in measurement of g is

$$\frac{\Delta g}{g} \times 100\% = \left(\frac{\Delta L}{L} + \frac{2\Delta T}{T}\right) 100\%$$

$$= \left(\frac{0.1}{55} + \frac{2\left(\frac{1}{20}\right)}{\frac{30}{20}}\right) 100\% \simeq 6.8\%$$ **Ans. (D)**

Sol. 31 Density of cube is given as

$$\rho = \frac{m}{v} \qquad [v = l^3]$$

Percentage error or fractional error in density measurement is

$$\frac{\Delta \rho}{\rho} = \frac{\Delta m}{m} + 3\frac{\Delta l}{l}$$

$\Rightarrow$ $\Delta \rho = \left[\dfrac{0.10}{10.00} + 3\left(\dfrac{0.01}{0.10}\right)\right] \times \dfrac{10.00}{(0.10)^3} \text{ kg/m}^3$

$\Rightarrow$ $\Delta \rho = [0.01 + 0.3] \times 10^4 \text{ kg/m}^3$

$\Rightarrow$ $\Delta \rho = 0.31 \times 10^4 \text{ kg/m}^3$ **Ans. (None)**

Sol. 32 Total Area of seven squares is

$$S = A_1 + A_2 + \ldots A_7$$

$\Rightarrow$ $S = A + A + \ldots 7 \text{ times}$

$\Rightarrow$ $S = 37.03 \text{ cm}^2$

Addition of 7 terms all having 2 terms beyond decimal, so final answer must have 2 terms beyond decimal (as per rules of significant digits.) **Ans. (C)**

* * * * *

CHAPTER-1 UNITS AND DIMENSIONS

1.1 Measurements and Dimensional Analysis

1. A quantity f is given by $f=\sqrt{\frac{hc^5}{G}}$ where c is speed of light, G universal gravitational constant and h is the Planck's constant. Dimension of f is that of: **[9 Jan. 2020 (Morning)]**

(A) Momentum (B) Area

(C) Energy (D) Volume

CHAPTER-2 MOTION IN A STRAIGHT LINE

2.1 Motion in One Dimension

2. A ball is dropped from the top of a 100 m high tower on a planet. In the last $\frac{1}{2}s$ before hitting the ground, it covers a distance of 19 m. Acceleration due to gravity (in ms^{-2}) near the surface on that planet is _______. **[8 Jan. 2020 (Evening)]**

3. The distance x covered by a particle in one dimensional motion varies with time t as $x^2 = at^2 + 2bt + c$. If the acceleration of the particle depends on x as x^{-n}, where n is an integer, the value of n is _______. **[9 Jan. 2020 (Morning)]**

CHAPTER-3 MOTION IN A PLANE

3.1 Motion in Two Dimensions

4. The sum of two forces $\vec{P}$ and $\vec{Q}$ is $\vec{R}$ such that $|\vec{R}| = |\vec{P}|$. The angle θ (in degrees) that the resultant of $2\vec{P}$ and $\vec{Q}$ will make with $\vec{Q}$ is, _______. **[7 Jan. 2020 (Evening)]**

5. A particle is moving along the x-axis with its coordinate with the time 't' given be $x(t) = 10 + 8t - 3t^2$. Another particle is moving the y-axis with its coordinate as a function of time given by $y(t) = 5 - 8t^3$. At $t = 1$s, the speed of the second particle as measured in the frame of the first particle is given as $\sqrt{v}$. Then v (in m/s) is _______. **[8 Jan. 2020 (Morning)]**

6. A particle moves such that its position vector $\vec{r}(t) = \cos\omega t\hat{i} + \sin\omega t\hat{j}$ where ω is a constant and t is time. Then which of the following statements is true for the velocity $\vec{v}(t)$ and acceleration $\vec{a}(t)$ of the particle: **[8 Jan. 2020 (Evening)]**

(A) $\vec{v}$ is perpendicular to $\vec{r}$ and $\vec{a}$ is directed towards the origin

(B) $\vec{v}$ and $\vec{a}$ both are parallel to $\vec{r}$

(C) $\vec{v}$ and $\vec{a}$ both are perpendicular to $\vec{r}$

(D) $\vec{v}$ is perpendicular to $\vec{r}$ and $\vec{a}$ is directed away from the origin

7. A particle starts from the origin at $t = 0$ with an initial velocity of $3.0\hat{i}$ m/s and moves in the x-y plane with a constant acceleration $(6.0\hat{i} + 4.0\hat{j})$ m/s^2. The x-coordinate of the particle at the instant when its y-coordinate is 32 m is D meters. The value of D is: **[9 Jan. 2020 (Evening)]**

(A) 50 (B) 32

(C) 60 (D) 40

CHAPTER-4 LAWS OF MOTION

4.1 Newton's Laws of Motion

8. A mass of 10 kg is suspended by a rope of length 4 m, from the ceiling. A force F is applied horizontally at the mid-point of the rope such that the top half of the rope makes an angle of 45° with the vertical. Then F equals: (Take $g = 10\ ms^{-2}$ and the rope to be massless) : **[7 Jan. 2020 (Evening)]**

(A) 100 N (B) 90 N

(C) 75 N (D) 70 N

CHAPTER-5 WORK ENERGY POWER & APPLICATIONS

5.1 Work Energy Power

9. A 60 HP electric motor lifts an elevator having a maximum total load capacity of 2000 kg. If the frictional force on the elevator is 4000 N, the speed of the elevator at full load is close to: (1 HP = 746 W, $g = 10\ ms^{-2}$) : **[7 Jan. 2020 (Morning)]**

(A) 1.7 ms^{-1} (B) 2.0 ms^{-1}

(C) 1.9 ms^{-1} (D) 1.5 ms^{-1}

10. A particle ($m = 1$ kg) slides down a frictionless track (AOC) starting from rest at a point A (height 2 m). After reaching C, the particle continues to move freely in air as a projectile. When it reaching its highest point P (height 1 m), the kinetic energy of the particle (in J) is : (Figure drawn is schematic and not to scale; take $g = 10\ ms^{-2}$)_____ : **[7 Jan. 2020 (Morning)]**

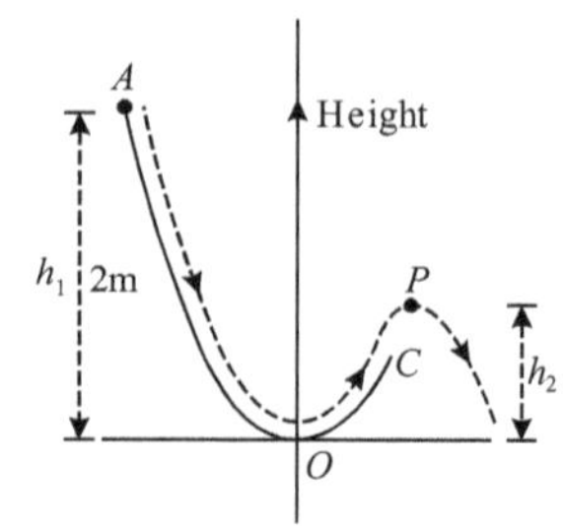

11. An elevator in a building can carry a maximum of 10 persons, with the average mass of each person being 68 kg. The mass of the elevator itself is 920 kg and it moves with a constant speed 3 m/s. The frictional force opposing the motion is 6000 N. If the elevator is moving up with its full capacity, the power delivered by the motor to the elevator ($g = 10$ m/s^2) must be at least :

[7 Jan. 2020 (Evening)]

(A) 56300 W (B) 48000 W

(C) 66000 W (D) 62360 W

12. Consider a force $\vec{F} = -x\hat{i} + y\hat{j}$. The work done by this force in moving a particle from point $A(1, 0)$ to $B(0, 1)$ along the line segment is : (all quantities are in SI units) **[9 Jan. 2020 (Morning)]**

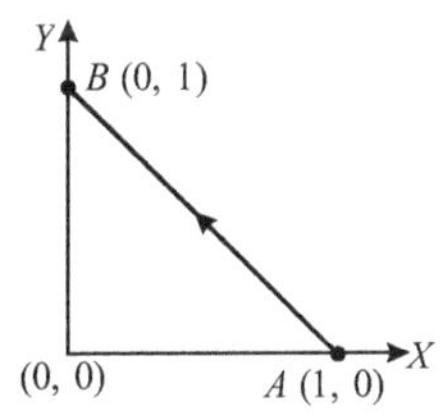

(A) $\frac{3}{2}$ (B) 1

(C) 2 (D) $\frac{1}{2}$

5.2 Circular Motion

13. A particle of mass m is fixed to one end of a light spring having force constant k and unstretched length l. The other end is fixed. The system is given an angular speed ω about the fixed end of the spring such that it rotates in a circle in gravity free space. Then the stretch in the spring is : **[8 Jan. 2020 (Morning)]**

(A) $\frac{ml\omega^2}{k+m\omega^2}$ (B) $\frac{ml\omega^2}{k-m\omega^2}$

(C) $\frac{ml\omega^2}{k-\omega m}$ (D) $\frac{ml\omega^2}{k+m\omega}$

CHAPTER-6 SYSTEM OF PARTICLES & ROTATIONAL MOTION

6.1 System of Particles

14. Three point particles of masses 1.0 kg, 1.5 kg and 2.5 kg are placed at three corners of a right angle triangle of sides 4.0 cm, 3.0 cm and 5.0 cm as shown in the figure. The center of mass of the system is at a point : **[7 Jan. 2020 (Morning)]**

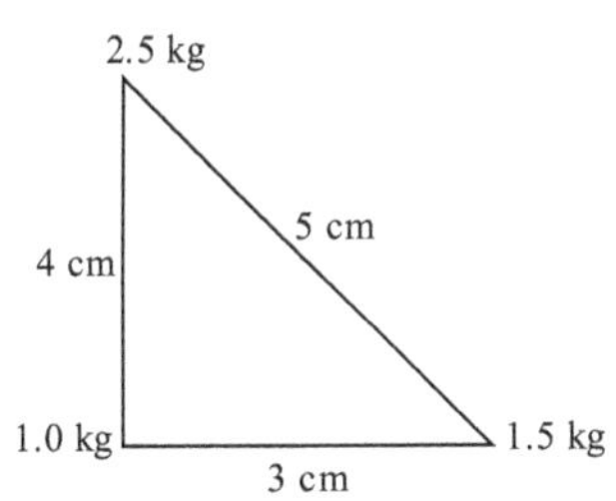

(A) 1.5 cm right and 1.2 cm above 1 kg mass

(B) 0.9 cm right and 2.0 cm above 1 kg mass

(C) 0.6 cm right and 2.0 cm above 1 kg mass

(D) 2.0 cm right and 0.9 cm above 1 kg mass

15. The coordinates of centre of mass of a uniform flag shaped lamina (thin flat plate) of mass 4 kg. (The coordinates of the same are shown in figure) are : **[8 Jan. 2020 (Morning)]**

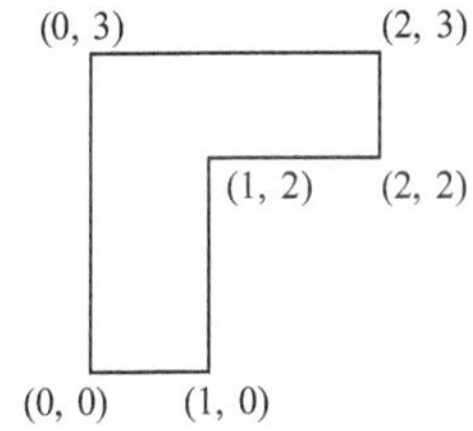

(A) (1.25 m, 1.50 m) (B) (1 m, 1.75 m)

(C) (0.75 m, 0.75 m) (D) (0.75 m, 1.75 m)

16. A body A, of mass m = 0.1 kg has an initial velocity of $3\hat{i}\,\text{ms}^{-1}$. It collides elastically with another body, B of the same mass which has an initial velocity of $5\hat{j}\,\text{ms}^{-1}$. After collision, A moves with a velocity $\vec{v} = 4(\hat{i}+\hat{j})$. The energy of B after collision is written as $\frac{x}{10}J$. The value of x is ________. **[8 Jan. 2020 (Morning)]**

17. As shown in figure, when a spherical cavity (centred at O) of radius 1 is cut out of a uniform sphere of radius R (centred at C), the centre of mass of remaining (shaded) part of sphere is at G, i.e, on the surface of the cavity. R can be determined by the equation : **[8 Jan. 2020 (Evening)]**

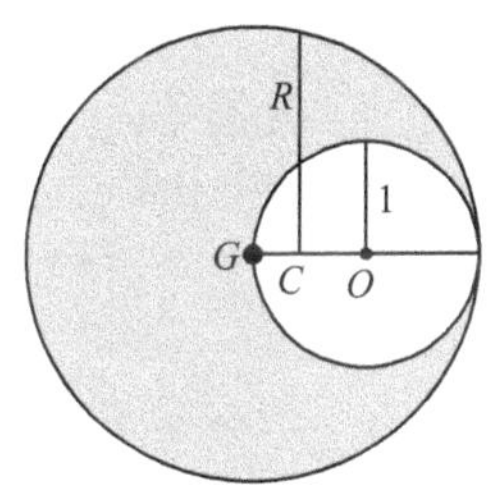

(A) $(R^2-R+1)(2-R)=1$ (B) $(R^2+R-1)(2-R)=1$

(C) $(R^2+R+1)(2-R)=1$ (D) $(R^2-R-1)(2-R)=1$

18. A particle of mass m is dropped from a height h above the ground. At the same time another particle of the same mass is thrown vertically upwards from the ground with a speed of $\sqrt{2gh}$. If they collide head-on completely inelastically, the time taken for the combined mass to reach the ground, in units of $\sqrt{\frac{h}{g}}$ is : **[8 Jan. 2020 (Evening)]**

(A) $\frac{1}{2}$ (B) $\sqrt{\frac{1}{2}}$

(C) $\sqrt{\frac{3}{4}}$ (D) $\sqrt{\frac{3}{2}}$

19. Two particles of equal mass m have respective initial velocities $u\hat{i}$ and $u\left(\frac{\hat{i}+\hat{j}}{2}\right)$. They collide completely inelastically. The energy lost in the process is : **[9 Jan. 2020 (Morning)]**

(A) $\frac{3}{4}mu^2$ (B) $\frac{1}{8}mu^2$

(C) $\sqrt{\frac{2}{3}}mu^2$ (D) $\frac{1}{3}mu^2$

20. A rod of length L has non-uniform linear mass density given by $\rho(x) = a + b\left(\frac{x}{L}\right)^2$, where a and b are constants and $0 \le x \le L$. The value of x for the centre of mass of the rod is at : **[9 Jan. 2020 (Evening)]**

(A) $\frac{4}{3}\left(\frac{a+b}{2a+3b}\right)L$ (B) $\frac{3}{2}\left(\frac{a+b}{2a+b}\right)L$

(C) $\frac{3}{2}\left(\frac{2a+b}{3a+b}\right)L$ (D) $\frac{3}{4}\left(\frac{2a+b}{3a+b}\right)L$

21. A particle of mass m is projected with a speed u from the ground at an angle $\theta = \frac{\pi}{3}$ w.r.t. horizontal (x-axis). When it has reached its maximum height, it collides completely inelastically with another particle of the same mass and velocity $u\hat{i}$. The horizontal distance covered by the combined mass before reaching the ground is : **[9 Jan. 2020 (Evening)]**

(A) $\frac{3\sqrt{2}}{4}\frac{u^2}{g}$ (B) $2\sqrt{2}\frac{u^2}{g}$

(C) $\frac{3\sqrt{3}}{8}\frac{u^2}{g}$ (D) $\frac{5}{8}\frac{u^2}{g}$

6.2 Rigid Body Dynamics

22. As shown in the figure, a bob of mass m is tied by a massless string whose other end portion is wound on a fly wheel (disc) of radius r and mass m. When released from rest the bob starts falling vertically. When it has covered a distance of h, the angular speed of the wheel will be : **[7 Jan. 2020 (Morning)]**

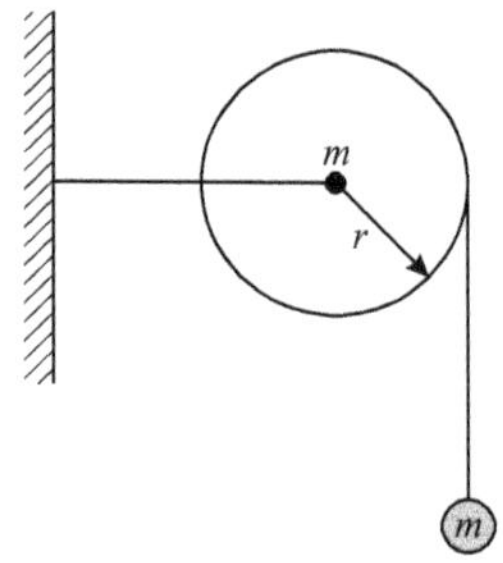

(A) $\frac{1}{r}\sqrt{\frac{2gh}{3}}$ (B) $r\sqrt{\frac{3}{4gh}}$

(C) $\frac{1}{r}\sqrt{\frac{4gh}{3}}$ (D) $r\sqrt{\frac{3}{2gh}}$

23. The radius of gyration of a uniform rod of length l, about an axis passing through a point $\frac{l}{4}$ away from the centre of the rod, and perpendicular to it, is : **[7 Jan. 2020 (Morning)]**

(A) $\frac{1}{8}l$ (B) $\sqrt{\frac{7}{48}}l$

(C) $\sqrt{\frac{3}{8}}l$ (D) $\frac{1}{4}l$

24. Mass per unit area of a circular disc of radius a depends on the distance r from its centre as $\sigma(r) = A + Br$. The moment of inertia of the disc about the axis, perpendicular to the plane and passing through its centre is : **[7 Jan. 2020 (Evening)]**

(A) $2\pi a^4\left(\frac{A}{4}+\frac{aB}{5}\right)$ (B) $\pi a^4\left(\frac{A}{4}+\frac{aB}{5}\right)$

(C) $2\pi a^4\left(\frac{aA}{4}+\frac{B}{5}\right)$ (D) $2\pi a^4\left(\frac{A}{4}+\frac{B}{5}\right)$

25. Consider a uniform cubical box of side a on a rough floor that is to be moved by applying minimum possible force F at a point b above its centre of mass (see figure). If the coefficient of friction is $\mu = 0.4$, the maximum possible value of $100 \times \frac{b}{a}$ for a box not to topple before moving is ________. **[7 Jan. 2020 (Evening)]**

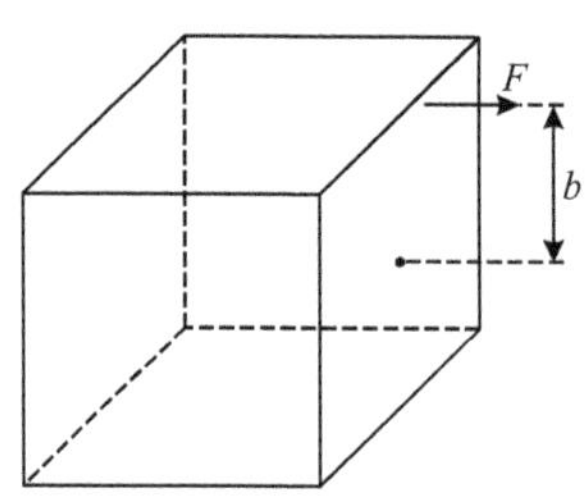

26. Consider a uniform rod of mass $M = 4$ m and length l pivoted about its centre. A mass m moving with velocity v making angle $\theta = \frac{\pi}{4}$ to the rod's long axis collides with one end of the rod and sticks to it. The angular speed of the rod-mass system just after the collision is : **[8 Jan. 2020 (Morning)]**

(A) $\frac{3}{7\sqrt{2}}\frac{v}{l}$ (B) $\frac{3\sqrt{2}}{7}\frac{v}{l}$

(C) $\frac{4}{7}\frac{v}{l}$ (D) $\frac{3}{7}\frac{v}{l}$

27. A uniform sphere of mass 500 g rolls without slipping on a plane horizontal surface with its centre moving at a speed of 5.00 cm/s. Its kinetic energy is : **[8 Jan. 2020 (Evening)]**

(A) 8.75×10^{-4} J (B) 8.75×10^{-3} J

(C) 6.25×10^{-4} J (D) 1.13×10^{-3} J

28. Three solid spheres each of mass m and diameter d are stuck together such that the lines connecting the centres form an equilateral triangle of side of length d. The ratio I_0/I_A of moment of inertia I_0 of the system about an axis passing the centroid and about center of any of the spheres I_A and perpendicular to the plane of the triangle is : **[9 Jan. 2020 (Morning)]**

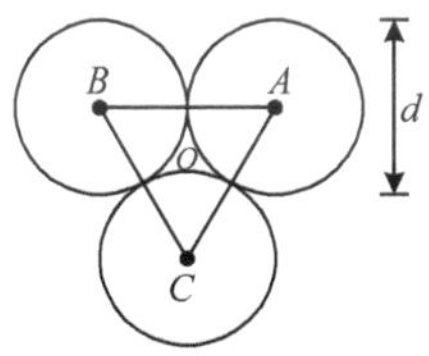

(A) $\frac{13}{23}$ (B) $\frac{15}{13}$

(C) $\frac{23}{13}$ (D) $\frac{13}{15}$

29. One end of a straight uniform 1m long bar is pivoted on horizontal table. It is released from rest when it makes an angle 30° from the horizontal (see figure). Its angular speed when it hits the table is given as $\sqrt{n}\,\text{s}^{-1}$, where n is an integer. The value of n is ________. **[9 Jan. 2020 (Morning)]**

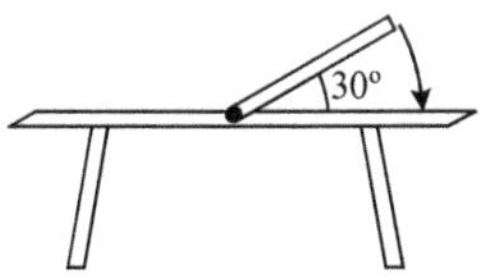

30. A spring mass system (mass m, spring constant k and natural length l) rest in equilibrium on a horizontal disc. The free end of the spring is fixed at the centre of the disc. If the disc together with spring mass system, rotates about it's axis with an angular velocity ω, ($k >> m\omega^2$) the relative change in the length of the spring is best given by the option : **[9 Jan. 2020 (Evening)]**

(A) $\frac{2m\omega^2}{k}$ (B) $\frac{m\omega^2}{3k}$

(C) $\sqrt{\frac{2}{3}}\left(\frac{m\omega^2}{k}\right)$ (D) $\frac{m\omega^2}{k}$

31. A uniformly thick wheel with moment of inertia I and radius R is free to rotate about its centre of mass (see figure). A massless string is wrapped over its rim and two blocks of masses m_1 and m_2 ($m_1 > m_2$) are attached to the ends of the string. The system is released from rest. The angular speed of the wheel when m_1 descents by a distance h is : **[9 Jan. 2020 (Evening)]**

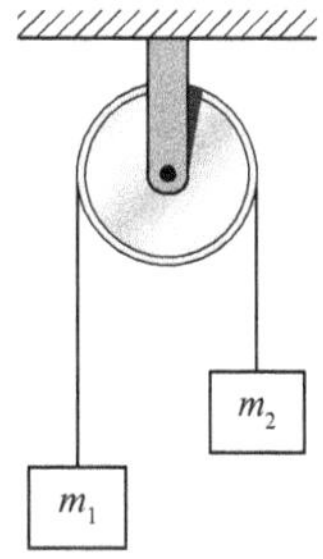

(A) $\left[\frac{m_1 + m_2}{(m_1 + m_2)R^2 + I}\right]^{\frac{1}{2}} gh$ (B) $\left[\frac{2(m_1 - m_2)gh}{(m_1 + m_2)R^2 + I}\right]^{\frac{1}{2}}$

(C) $\left[\frac{2(m_1 + m_2)gh}{(m_1 + m_2)R^2 + I}\right]^{\frac{1}{2}}$ (D) $\left[\frac{(m_1 - m_2)}{(m_1 + m_2)R^2 + I}\right]^{\frac{1}{2}} gh$

CHAPTER-7 GRAVITATION

7.1 Gravitational Field

32. A box weighs 196 N on a spring balance at the north pole. Its weight recorded on the same balance if it is shifted to the equator is close to (Take $g = 10\ \text{ms}^{-2}$ at the north pole and the radius of the earth = 6400 km) : **[7 Jan. 2020 (Evening)]**

(A) 195.66 N (B) 194.66 N

(C) 194.32 N (D) 195.32 N

33. Consider two solid spheres of radii $R_1 = 1$m, $R_2 = 2$m and masses M_1 and M_2, respectively. The gravitational field due to sphere (1) and (2) are shown. The value of $\frac{M_1}{M_2}$ is : **[8 Jan. 2020 (Morning)]**

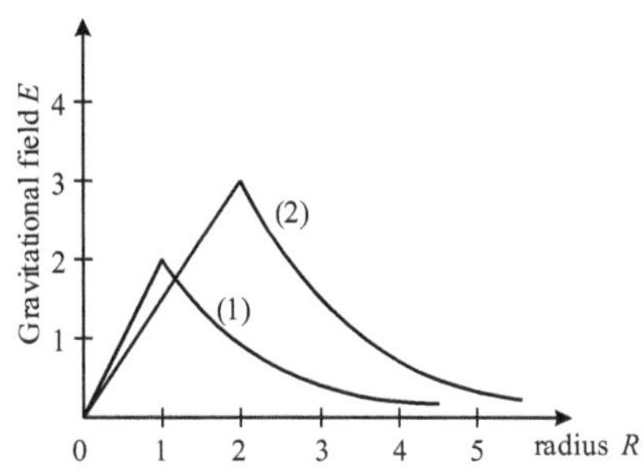

(A) $\frac{1}{2}$ (B) $\frac{2}{3}$

(C) $\frac{1}{3}$ (D) $\frac{1}{6}$

7.2 Satellite Motion

34. A satellite of mass m is launched vertically upwards with an initial speed u from the surface of the earth. After it reaches height R(R = radius of the earth), it ejects a rocket of mass $\frac{m}{10}$ so that subsequently the satellite moves in a circular orbit. The kinetic energy of the rocket is (G is the gravitational constant; M is the mass of the earth) : **[7 Jan. 2020 (Morning)]**

(A) $\frac{m}{20}\left(u-\sqrt{\frac{2GM}{3R}}\right)^2$ (B) $5m\left(u^2-\frac{119}{200}\frac{GM}{R}\right)$

(C) $\frac{3m}{8}\left(u+\sqrt{\frac{5GM}{6R}}\right)^2$ (D) $\frac{m}{20}\left(u^2+\frac{113}{200}\frac{GM}{R}\right)$

35. An asteroid is moving directly towards the centre of the earth. When at a distance of $10R$ (R is the radius of the earth) from the earths centre, it has a speed of 12 km/s. Neglecting the effect of earths atmosphere, what will be the speed of the asteroid when it hits the surface of the earth (escape velocity from the earth is 11.2 km/s) ? Give your answer to the nearest integer in kilometer/s_______. **[8 Jan. 2020 (Evening)]**

36. A body A of mass m is moving in a circular orbit of radius R about a planet. Another body B of mass $\frac{m}{2}$ collides with A with a velocity which is half $\left(\frac{\vec{v}}{2}\right)$ the instantaneous velocity $\vec{v}$ of A. The collision is completely inelastic. Then, the combined body : **[9 Jan. 2020 (Morning)]**

(A) Starts moving in an elliptical orbit around the planet
(B) Continues to move in a circular orbit
(C) Falls vertically downwards towards the planet
(D) Escapes from the Planet's Gravitational field.

37. Planet A has mass M and radius R. Planet B has half the mass and half the radius of Planet A. If the escape velocities from the Planets A and B are v_A and v_B, respectively, then $\frac{v_A}{v_B}=\frac{n}{4}$. The value of n is : **[9 Jan. 2020 (Evening)]**

(A) 4 (B) 1
(C) 2 (D) 3

CHAPTER-8 MECHANICAL PROPERTIES OF SOLIDS

8.1 Elasticity

38. A body of mass m = 10 kg is attached to one end of a wire of length 0.3 m. The maximum angular speed (in rad s^{-1}) with which it can be rotated about its other end in space station is (Breaking stress of wire = 4.8×10^7 Nm^{-2} and area of cross-section of the wire = 10^{-2} cm^2) is : **[9 Jan. 2020 (Morning)]**

39. Two steel wires having same length are suspended from a ceiling under the same load. If the ratio of their energy stored per unit volume is 1 : 4, the ratio of their diameters is : **[9 Jan. 2020 (Evening)]**

(A) $1:\sqrt{2}$ (B) $1:2$
(C) $2:1$ (D) $\sqrt{2}:1$

CHAPTER-9 MECHANICAL PROPERTIES OF FLUIDS

9.1 Fluid Statics

40. Consider a solid sphere of radius R and mass density $\rho(r)=\rho_0\left(1-\frac{r^2}{R^2}\right), 0<r\le R$. The minimum density of a liquid in which it will float is : **[8 Jan. 2020 (Morning)]**

(A) $\frac{\rho_0}{5}$ (B) $\frac{\rho_0}{3}$

(C) $\frac{2\rho_0}{3}$ (D) $\frac{2\rho_0}{5}$

41. Two liquids of densities ρ_1 an $\rho_2(\rho_2=2\rho_1)$ are filled up behind a square wall of side 10 m as shown in figure. Each liquid has a height of 5 m. The ratio of the forces due to these liquids exerted on upper part MN to that at the lower part NO is (Assume that the liquids are not mixing) : **[8 Jan. 2020 (Evening)]**

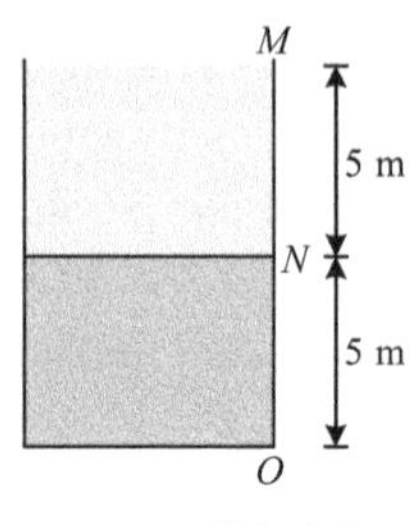

(A) 1/4 (B) 2/3
(C) 1/3 (D) 1/2

9.2 Fluid Dynamics

42. An ideal fluid flows (laminar flow) through a pipe of non-uniform diameter. The maximum and minimum diameters of the pipes are 6.4 cm and 4.8 cm, respectively. The ratio of the minimum and the maximum velocities of fluid in this pipe is :

[7 Jan. 2020 (Evening)]

(A) $\frac{\sqrt{3}}{2}$ (B) $\frac{3}{4}$

(C) $\frac{81}{256}$ (D) $\frac{9}{16}$

43. Water flows in a horizontal tube (see figure). The pressure of water changes by 700 Nm^{-2} between A and B where the area of cross section are 40 cm^2 and 20 cm^2, respectively. Find the rate of flow of water through the tube. (Density of water = 1000 kgm^{-3}) [9 Jan. 2020 (Morning)]

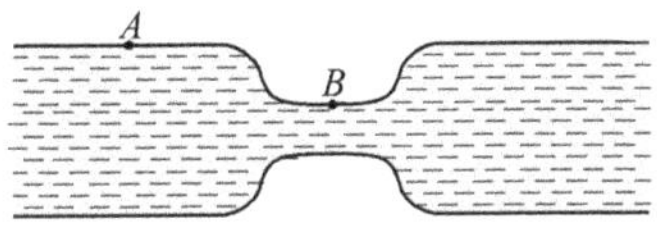

(A) 1810 cm^3/s (B) 3020 cm^3/s
(C) 2720 cm^3/s (D) 2420 cm^3/s

9.4 Surface Tension

44. A small spherical droplet of density d is floating exactly half immersed in a liquid of density ρ and surface tension T. The radius of the droplet is (take note that the surface tension applies an upward force on the droplet) : [9 Jan. 2020 (Evening)]

(A) $r=\sqrt{\frac{2T}{3(d+\rho)g}}$ (B) $r=\sqrt{\frac{3T}{(2d-\rho)g}}$

(C) $r=\sqrt{\frac{T}{(d-\rho)g}}$ (D) $r=\sqrt{\frac{T}{(d+\rho)g}}$

CHAPTER-10 THERMAL PROPERTIES OF MATTER

10.1 Thermal Expansion

45. A non-isotropic solid metal cube has coefficients of linear expansion as :
$5 \times 10^{-5}/°C$ along the x-axis and $5 \times 10^{-6}/°C$ along the y and the z-axis. If the coefficient of volume expansion of the solid is $C \times 10^{-16}/°C$ then the value of C is ______.
[7 Jan. 2020 (Morning)]

46. A leak proof cylinder of length 1 m, made of a metal which has very low coefficient of expansion is floating vertically in water at 0°C such that its height above the water surface is 20 cm. When the temperature of water is increased to 4°C, the height of the cylinder above the water surface becomes 21 cm. The density of water at $T = 4°C$, relative to the density at $T = 0°C$ is close to : [8 Jan. 2020 (Morning)]
(A) 1.01 (B) 1.04
(C) 1.03 (D) 1.26

10.2 Calorimetry

47. M grams of steam at 100°C is mixed with 200 g of ice at its melting point in a thermally insulated container. If it produces liquid water at 40°C [heat of vaporization of water is 540 cal/g and heat of fusion of ice is 80 cal/g], the value of M is______.
[7 Jan. 2020 (Evening)]

48. Three containers C_1, C_2 and C_3 have water at different temperatures. The table below shows the final temperature T when different amounts of water (given in litres) are taken from each containers and mixed (assume no loss of heat during the process). [8 Jan. 2020 (Evening)]

C_1	C_2	C_3	T
1l	2l	–	60°C
–	1l	2l	30°C
2l	–	1l	60°C
1l	1l	1l	θ

The value of θ (in °C to the nearest integer) is ______.

CHAPTER-11 THERMODYNAMICS

11.1 Laws of Thermodynamics

49. A litre of dry air at STP expands adiabatically to a volume of 3 litres. If $\gamma = 1.40$, the work done by air is : $(3^{1.4} = 4.6555)$
[Take air to be an ideal gas] : [7 Jan. 2020 (Morning)]
(A) 90.5 J (B) 48 J
(C) 60.7 J (D) 100.8 J

50. Two moles of an ideal gas with $\frac{C_P}{C_V} = \frac{5}{3}$ are mixed with 3 moles of another ideal gas with $\frac{C_P}{C_V} = \frac{4}{3}$. The value of $\frac{C_P}{C_V}$ for the mixture is : [7 Jan. 2020 (Morning)]
(A) 1.50 (B) 1.42
(C) 1.45 (D) 1.47

51. A Carnot engine operates between two reservoirs of temperatures 900 K and 300 K. The engine performs 1200 J of work per cycle. The heat energy (in J) delivered by the engine to the low temperature reservoir, in a cycle, is______:
[7 Jan. 2020 (Morning)]

52. Two infinite planes each with uniform surface charge density $+\sigma$ are kept in such a way that the angle between them is 30°. The electric field in the region shown between them is given by : [7 Jan. 2020 (Morning)]

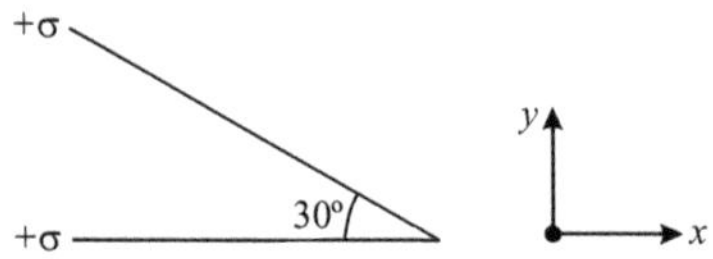

(A) $\dfrac{\sigma}{\varepsilon_0}\left[\left(1+\dfrac{\sqrt{3}}{2}\right)\hat{y}+\dfrac{\hat{x}}{2}\right]$ (B) $\dfrac{\sigma}{2\varepsilon_0}\left[\left(1-\dfrac{\sqrt{3}}{2}\right)\hat{y}-\dfrac{\hat{x}}{2}\right]$

(C) $\dfrac{\sigma}{2\varepsilon_0}\left[(1+\sqrt{3})\hat{y}+\dfrac{\hat{x}}{2}\right]$ (D) $\dfrac{\sigma}{2\varepsilon_0}\left[(1+\sqrt{3})\hat{y}-\dfrac{\hat{x}}{2}\right]$

53. Two ideal Carnot engines operate in cascade (all heat given up by one engine is used by the other engine to produce work) between temperatures, T_1 and T_2. The temperature of the hot reservoir of the first engine is T_1 and the temperature of the cold reservoir of the second engine is T_2. T is temperature of the sink of first engine which is also the source for the second engine. How is T related to T_1 and T_2, if both the engines perform equal amount of work? **[7 Jan. 2020 (Evening)]**

(A) $T=\dfrac{2T_1T_2}{T_1+T_2}$ (B) $T=\sqrt{T_1T_2}$

(C) $T=\dfrac{T_1+T_2}{2}$ (D) $T=0$

54. Under an adiabatic process, the volume of an ideal gas gets doubled. Consequently the mean collision time between the gas molecule changes from τ_1 to τ_2. If $\dfrac{C_p}{C_v}=\gamma$ for this gas then a good estimate for $\dfrac{\tau_2}{\tau_1}$ is given by : **[7 Jan. 2020 (Evening)]**

(A) $\left(\dfrac{1}{2}\right)^{\frac{\gamma+1}{2}}$ (B) 2

(C) $\dfrac{1}{2}$ (D) $\left(\dfrac{1}{2}\right)^{\gamma}$

55. A thermodynamic cycle $xyzx$ is shown on a V-T diagram. The P-V diagram that best describes this cycle is : (Diagrams are schematic and not to scale) : **[8 Jan. 2020 (Morning)]**

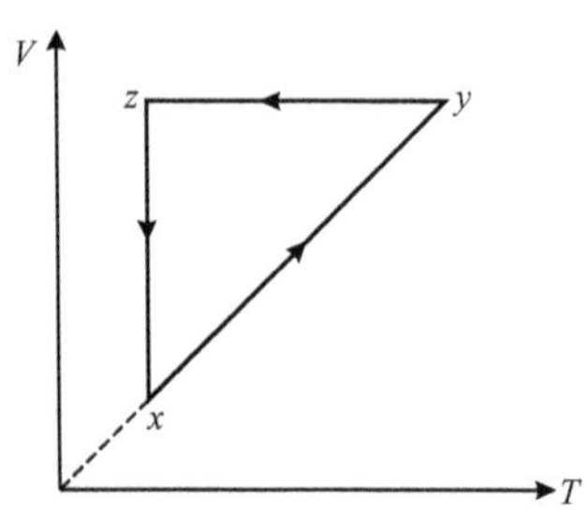

(A)
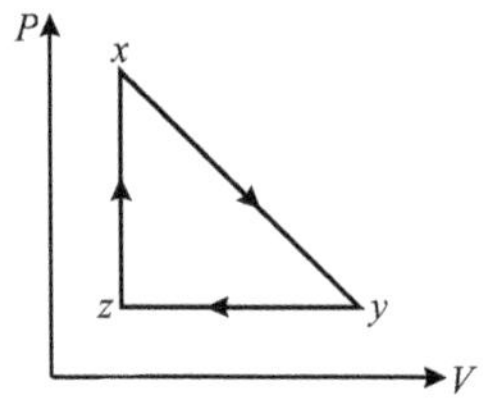

(B)
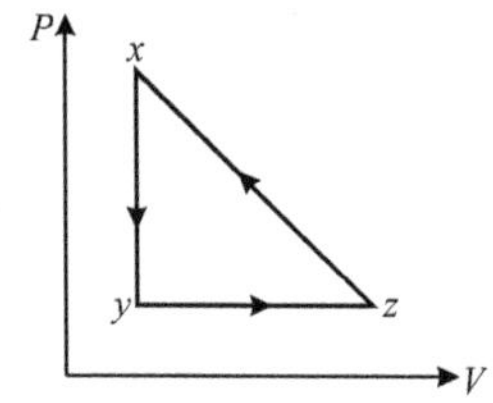

(C)
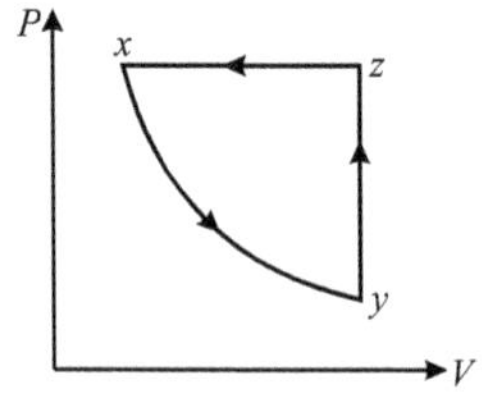

(D)
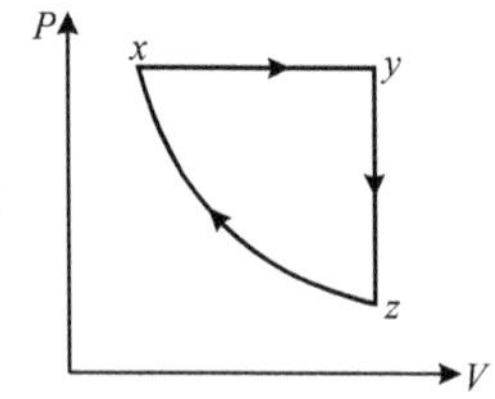

56. A Carnot engine having an efficiency of $\dfrac{1}{10}$ is being used as a refrigerator. If the work done on the refrigerator is 10 J, the amount of heat absorbed from the reservoir at lower temperature is : **[8 Jan. 2020 (Evening)]**

(A) 99 J (B) 100 J

(C) 90 J (D) 1 J

57. Consider a mixture of n moles of helium gas and 2 n moles of oxygen gas (molecules taken to be rigid) as an ideal gas. Its C_P/C_V value will be : **[8 Jan. 2020 (Evening)]**

(A) 67/45 (B) 19/13

(C) 23/15 (D) 40/27

58. Consider two ideal diatomic gases A and B at some temperature T. Molecules of the gas A are rigid, and have a mass m. Molecules of the gas B have an additional vibrational mode, and have a mass $\dfrac{m}{4}$. The ratio of the specific heats (C_V^A and C_V^B) of gas A and B, respectively is : **[9 Jan. 2020 (Morning)]**

(A) 7 : 9 (B) 5 : 7

(C) 3 : 5 (D) 5 : 9

59. Which of the following is an equivalent cyclic process corresponding to the thermodynamic cyclic given in the figure ? Where, 1 → 2 is adiabatic. (Graphs are schematic and are not to scale) **[9 Jan. 2020 (Morning)]**

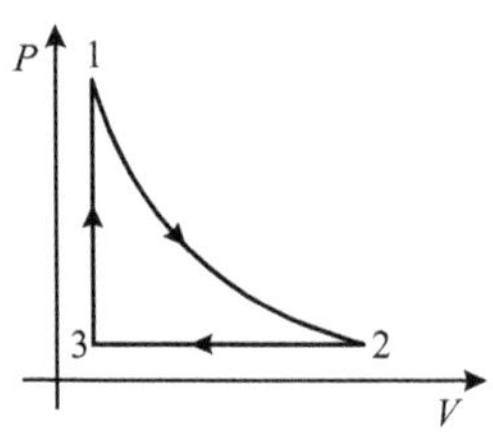

(A) 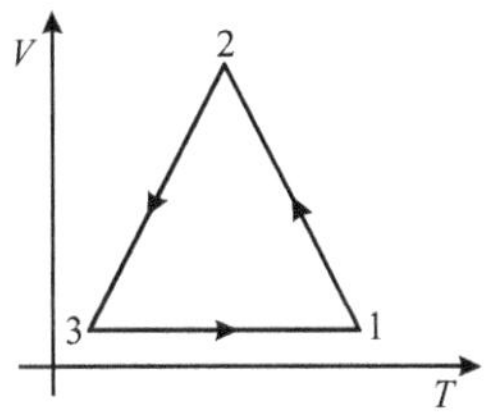

(B) 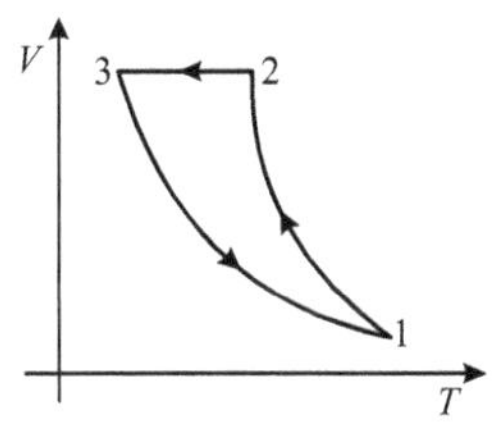

(C) 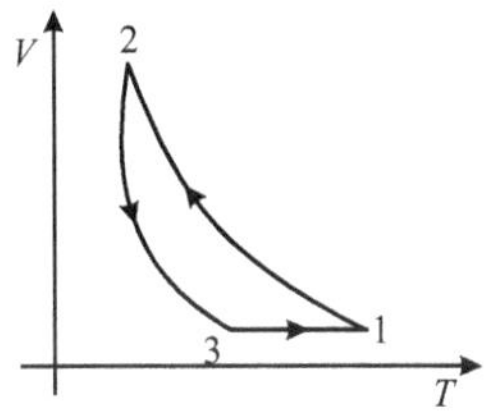

(D) 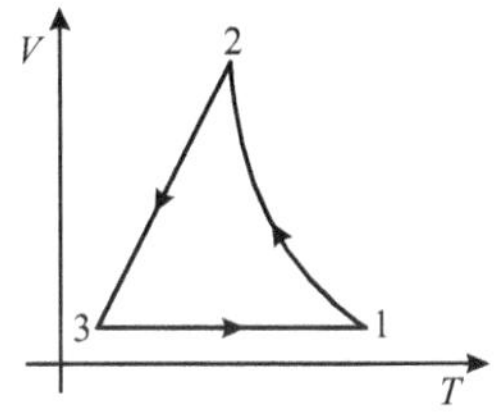

60. Starling at temperature 300 K, one mole of an ideal diatomic gas ($\gamma = 1.4$) is first compressed adiabatically from volume V_1 to $V_2 = \frac{V_1}{16}$. It is then allowed to expand isobarically to volume $2V_2$. If all the processes are the quasi-static then the final temperature of the gas (in °K) is (to the nearest integer) ________. **[9 Jan. 2020 (Evening)]**

CHAPTER-12 KINETIC THEORY

12.1 Kinetic Theory of Gases

61. The plot that depicts the behavior of the mean free time t (time between two successive collisions) for the molecules of an ideal gas, as a function of temperature (T), qualitatively, is : (Graphs are schematic and not drawn to scale) : **[8 Jan. 2020 (Morning)]**

(A) 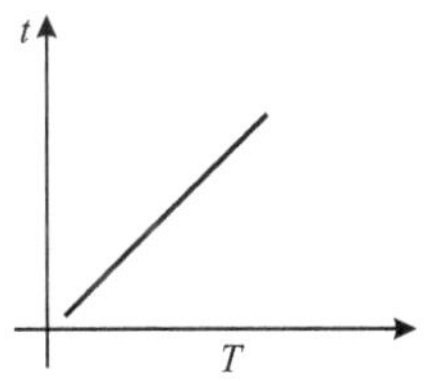

(B) 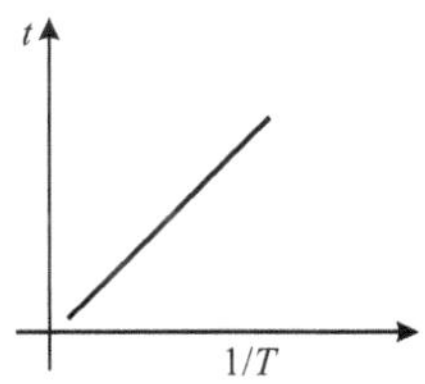

(C) 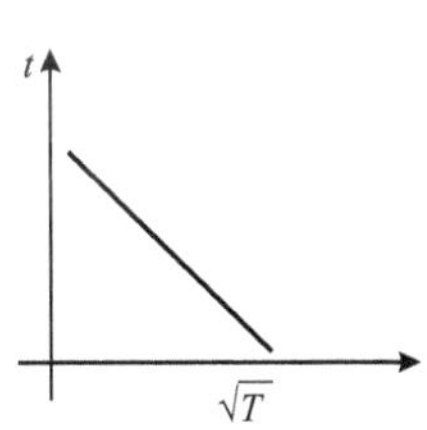

(D) 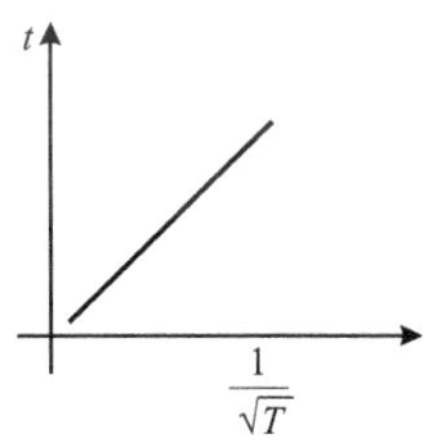

62. Two gases-argon (atomic radius 0.07 nm, atomic weight 40) and xenon (atomic radius 0.1 nm, atomic weight 140) have the same number density and are at the same temperature. The ratio of their respective mean free times is closest to : **[9 Jan. 2020 (Evening)]**

(A) 3.67 (B) 4.67
(C) 1.83 (D) 2.3

CHAPTER-13 OSCILLATIONS

13.1 Simple Harmonic Motion

63. A small circular loop of conducting wire has radius a and carries current I. It is placed in a uniform magnetic field B perpendicular to its plane such that when rotated slightly about its diameter and released, it starts performing simple harmonic motion of time period T. If the mass of the loop is m then : **[9 Jan. 2020 (Evening)]**

(A) $T=\sqrt{\frac{\pi m}{2IB}}$ (B) $T=\sqrt{\frac{2\pi m}{IB}}$

(C) $T=\sqrt{\frac{\pi m}{IB}}$ (D) $T=\sqrt{\frac{2m}{IB}}$

CHAPTER-14 WAVES

14.1 Wave Motion

64. Speed of a transverse wave on a straight wire (mass 6.0 g, length 60 cm and area of cross-section 1.0 mm^2) is 90 ms^{-1}. If the Young's modulus of wire is 16×10^{11} Nm^{-2}, the extension of wire over its natural length is : **[7 Jan. 2020 (Morning)]**

(A) 0.02 mm (B) 0.04 mm
(C) 0.03 mm (D) 0.01 mm

65. A transverse wave travels on a taut steel wire with a velocity of v when tension in it is 2.06×10^4 N. When the tension is changed to T, the velocity changed to $v/2$. The value of T is close to : **[8 Jan. 2020 (Evening)]**

(A) 10.2×10^2 N (B) 5.15×10^3 N
(C) 2.50×10^4 N (D) 30.5×10^4 N

14.2 Stationary Waves and Beats

66. A one metre long (both ends open) organ pipe is kept in a gas that has double the density of air at STP. Assuming the speed of sound in air at STP is 300 m/s, the frequency difference between the fundamental and second harmonic of this pipe is ________ Hz. **[8 Jan. 2020 (Morning)]**

67. A wire of length L and mass per unit length 6.0×10^{-3} kgm^{-1} is put under tension of 540 N. Two consecutive frequencies that it resonates at are 420 Hz and 490 Hz. Then L in meters is : **[9 Jan. 2020 (Evening)]**

(A) 8.1 m (B) 5.1 m
(C) 1.1 m (D) 2.1 m

14.3 Doppler's Effect

68. A stationary observer receives sound from two identical tuning forks, one of which approaches and the other one recedes with the same speed (much less than the speed of sound). The observer hears 2 beats/sec. The oscillation frequency of each tuning fork is $v_0 = 1400$ Hz and the velocity of sound in air is 350 m/s. The speed of each tuning fork is close to : **[7 Jan. 2020 (Evening)]**

(A) $\frac{1}{8}$ m/s (B) $\frac{1}{2}$ m/s

(C) 1 m/s (D) $\frac{1}{4}$ m/s

CHAPTER-15 ELECTRIC CHARGES AND FIELDS

15.1 Electric Force and Electric Field

69. Three charged particle A, B and C with charges $-4q$, $2q$ and $-2q$ are present on the circumference of a circle of radius d. The charged particles A, C and centre O of the circle formed an equilateral triangle as shown in figure. Electric field at O along x-direction is : **[8 Jan. 2020 (Morning)]**

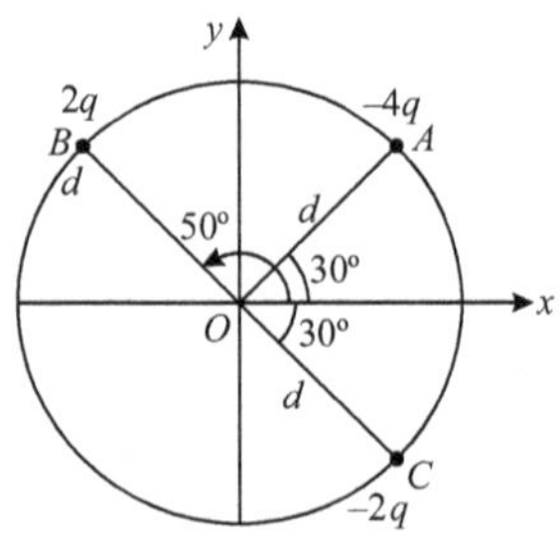

(A) $\frac{2\sqrt{3}q}{\pi\varepsilon_0 d^2}$ (B) $\frac{\sqrt{3}q}{4\pi\varepsilon_0 d^2}$

(C) $\frac{3\sqrt{3}q}{4\pi\varepsilon_0 d^2}$ (D) $\frac{\sqrt{3}q}{\pi\varepsilon_0 d^2}$

70. A particle of mass m and charge q is released from rest in a uniform electric field. If there is no other force on the particle, the dependence of its speed v on the distance x travelled by it is correctly given by (graphs are schematic and not drawn to scale) : **[8 Jan. 2020 (Evening)]**

(A) 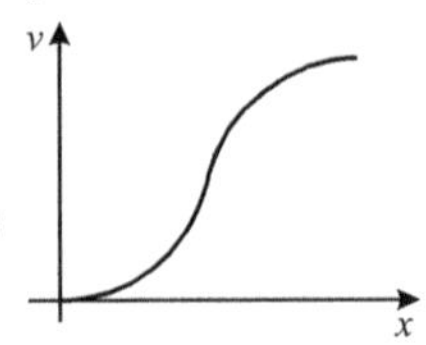

(B) 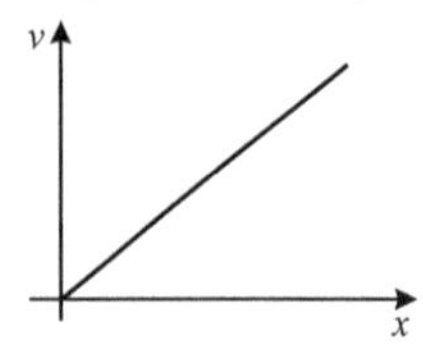

(C) 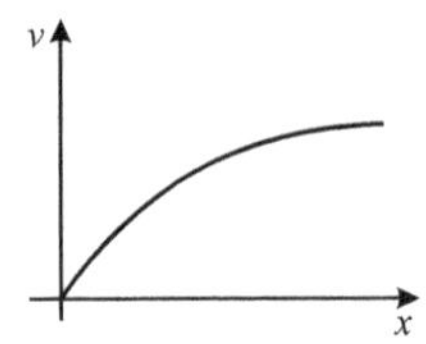

(D) 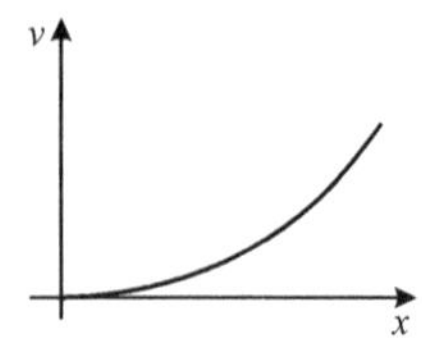

71. Consider a sphere of radius R which carries a uniform charge density ρ. If a sphere of radius $\frac{R}{2}$ is carved out of it, as shown, the ratio $\frac{|\vec{E}_A|}{|\vec{E}_B|}$ of magnitude of electric field $\vec{E}_A$ and $\vec{E}_B$, respectively, at points A and B due to the remaining portion is : **[9 Jan. 2020 (Morning)]**

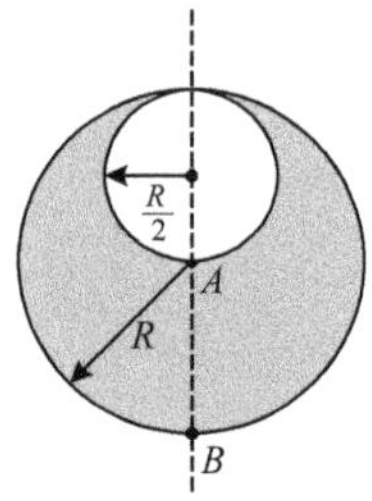

(A) $\frac{18}{54}$ (B) $\frac{21}{34}$

(C) $\frac{17}{54}$ (D) $\frac{18}{34}$

15.2 Electric Flux and Gauss's Law

72. In finding the electric field using Gauss Law the formula $|\vec{E}| = \frac{q_{enc}}{\varepsilon_0 |A|}$ is applicable. In the formula ε_0 is permittivity of free space, A is the area of Gaussian surface and q_{enc} is charge enclosed by the Gaussian surface. The equation can be used in which of the following situation ? **[8 Jan. 2020 (Morning)]**

(A) Only when the Gaussian surface is an equipotential surface

(B) Only when $|\vec{E}|$ = constant on the surface

(C) For any choice of Gaussian surface

(D) Only when the Gaussian surface is an equipotential surface and $|\vec{E}|$ is constant on the surface

73. An electric field $\vec{E} = 4x\hat{i} - (y^2 + 1)\hat{j}$ N/C passes through the box shown in figure. The flux of the electric field through surfaces $ABCD$ and $BCGF$ are marked as ϕ_I and ϕ_{II} respectively. The difference between $(\phi_I - \phi_{II})$ is (in Nm^2/C) ________. **[9 Jan. 2020 (Evening)]**

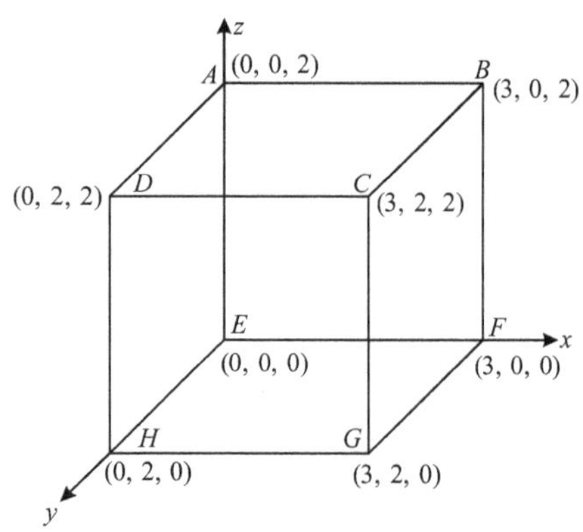

CHAPTER-16 ELECTRIC POTENTIAL AND CAPACITANCE

16.1 Electric Potential and Dipole

74. Consider two charged metallic spheres S_1 and S_2 of radii R_1 and R_2, respectively. The electric fields E_1 (on S_1) and E_2 (on S_2) on their surfaces are such that $E_1/E_2 = R_1/R_2$. Then the ratio V_1(on S_1)/V_2 (on S_2) of the electrostatic potentials on each sphere is : **[8 Jan. 2020 (Evening)]**

(A) (R_2/R_1) (B) $(R_1/R_2)^3$

(C) R_1/R_2 (D) $(R_1/R_2)^2$

75. An electric dipole of moment $\vec{p} = (-\hat{i} - 3\hat{j} + 2\hat{k}) \times 10^{-29}$ C.m is at the origin (0, 0, 0). The electric field due to this dipole at $\vec{r} = +\hat{i} + 3\hat{j} + 5\hat{k}$ (note that $\vec{r} \cdot \vec{p} = 0$) is parallel to : **[9 Jan. 2020 (Morning)]**

(A) $(-\hat{i} + 3\hat{j} - 2\hat{k})$ (B) $(+\hat{i} - 3\hat{j} - 2\hat{k})$

(C) $(+\hat{i} + 3\hat{j} - 2\hat{k})$ (D) $(-\hat{i} - 3\hat{j} + 2\hat{k})$

16.2 Capacitance

76. A parallel plate capacitor has plates of area A separated by distance 'd' between them. It is filled with a dielectric which has a dielectric constant that varies as $k(x) = K(1 + \alpha x)$ where 'x' is the distance measured from one of the plates. If $(\alpha d) << 1$, the total capacitance of the system is best given by the expression : **[7 Jan. 2020 (Morning)]**

(A) $\frac{AK\varepsilon_0}{d}\left(1 + \frac{\alpha d}{2}\right)$ (B) $\frac{A\varepsilon_0 K}{d}\left(1 + \left(\frac{\alpha d}{2}\right)^2\right)$

(C) $\frac{A\varepsilon_0 K}{d}\left(1 + \frac{\alpha^2 d^2}{2}\right)$ (D) $\frac{AK\varepsilon_0}{d}(1 + \alpha d)$

77. A 60 pF capacitor is fully charged by a 20 V supply. It is then disconnected from the supply and is connected to another uncharged 60 pF capacitor in parallel. The electrostatic energy that is lost in this process by the time the charge is redistributed between them is (in nJ) ____ : **[7 Jan. 2020 (Evening)]**

78. Effective capacitance of parallel combination of two capacitors C_1 and C_2 is 10 μF. When these capacitors are individually connected to a voltage source of $1V$, the energy stored in the capacitor C_2 is 4 times that of C_1. If these capacitors are connected in series, their effective capacitance will be : **[8 Jan. 2020 (Morning)]**

(A) 3.2 μF (B) 8.4 μF

(C) 1.6 μF (D) 4.2 μF

79. A capacitor is made of two square plates each of side 'a' making a very small angle α between them, as shown in figure. The capacitance will be close to : **[8 Jan. 2020 (Evening)]**

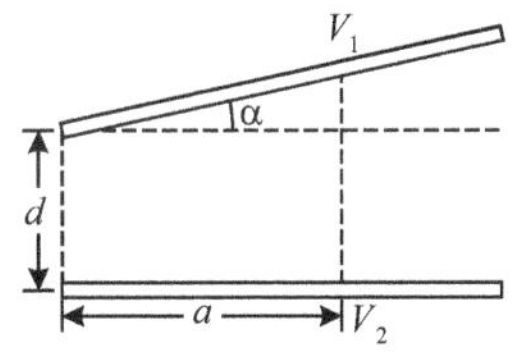

(A) $\frac{\varepsilon_0 a^2}{d}\left(1 - \frac{3\alpha a}{2d}\right)$ (B) $\frac{\varepsilon_0 a^2}{d}\left(1 - \frac{\alpha a}{4d}\right)$

(C) $\frac{\varepsilon_0 a^2}{d}\left(1 + \frac{\alpha a}{d}\right)$ (D) $\frac{\varepsilon_0 a^2}{d}\left(1 - \frac{\alpha a}{2d}\right)$

CHAPTER-17 CURRENT ELECTRICITY

17.2 Circuit Analysis

80. The current I_1(in A) flowing through 1 Ω resistor in the following circuit is : **[7 Jan. 2020 (Morning)]**

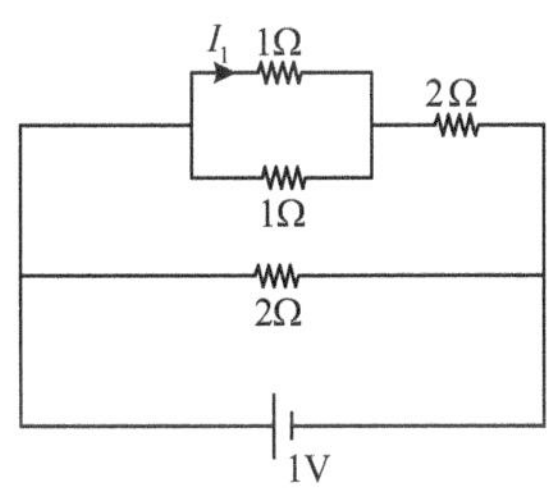

(A) 0.5 (B) 0.2

(C) 0.25 (D) 0.4

81. The series combination of two batteries, both of the same emf 10 V, but different internal resistance of 20Ω and 5Ω, is connected to the parallel combination of two resistors 30 Ω and RΩ. The voltage difference across the battery of internal resistance 20Ω is zero, the value of R (in Ω) is _______. **[8 Jan. 2020 (Evening)]**

82. In the given circuit diagram, a wire is joining points B and D. The current in this wire is : **[9 Jan. 2020 (Morning)]**

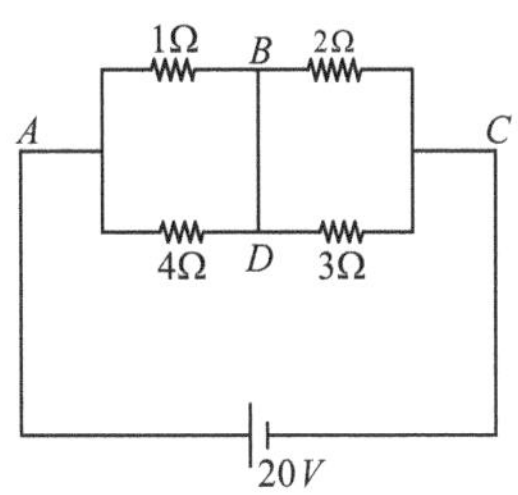

(A) 4A (B) 2A
(C) 0.4A (D) Zero

17.3 Thermal Effects of Current & RC Circuits

83. In a building there are 15 bulbs of 45 W, 15 bulbs of 100 W, 15 small fans of 10 W and 2 heaters of 1 kW. The voltage of electric main is 220 V. The minimum fuse capacity (rated value) of the building will be : **[7 Jan. 2020 (Evening)]**
(A) 10 A (B) 25 A
(C) 15 A (D) 20 A

17.4 Electrical Measuring Instruments

84. The balancing length for a cell is 560 cm in a potentiometer experiment. When an external resistance of 10Ω is connected in parallel to the cell, the balancing length changes by 60cm. If the internal resistance of the cell is $\frac{N}{10}\Omega$, where N is an integer then value of N is_______. **[7 Jan. 2020 (Evening)]**

85. The length of a potentiometer wire is 1200 cm and it carries a current of 60 mA. For a cell of emf 5V and internal resistance of 20Ω, the null point on it is found to be a 1000 cm. The resistance of whole wire is : **[8 Jan. 2020 (Morning)]**
(A) 120 Ω (B) 60 Ω
(C) 80 Ω (D) 100 Ω

86. Four resistances of 15 Ω, 12 Ω, 4 Ω and 10 Ω respectively in cyclic order to form. Wheatstone's network. The resistance that is to be connected in parallel with the resistance of 10 Ω to balance the network is _______ Ω. **[8 Jan. 2020 (Morning)]**

87. A galvanometer having a coil resistance 100 Ω gives a full scale deflection when a current of 1 mA is passed through it. What is the value of the resistance which can convert this galvanometer into a voltmeter giving full scale deflection for a potential difference of 10 V ? **[8 Jan. 2020 (Evening)]**
(A) 9.9 kΩ (B) 8.9 kΩ
(C) 7.9 kΩ (D) 10 kΩ

88. In a meter bridge experiment S is a standard resistance. R is a resistance wire. It Is found that balancing length is $l = 25$ cm. If R is replaced by a wire of half length and half diameter that of R of same material, then the balancing distance l' (in cm) will now be_______. **[9 Jan. 2020 (Evening)]**

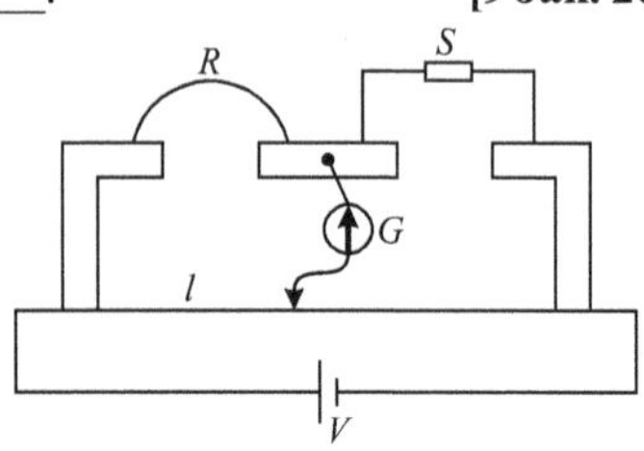

CHAPTER-18 MOVING CHARGES & MAGNETISM

18.1 Magnetic Effects of Current

89. The dimension of $\frac{B^2}{2\mu_0}$, where B is magnetic field and μ_0 is the magnetic permeability of vacuum, is : **[7 Jan. 2020 (Evening)]**
(A) $ML^{-1}T^{-2}$ (B) ML^2T^{-1}
(C) MLT^{-2} (D) ML^2T^{-2}

90. A very long wire $ABDMNDC$ is shown in figure carrying current I. AB and BC parts are straight, long and at right angle. At D wire forms a circular turn $DMND$ of radius R. AB, BC parts are tangential to circular turn at N and D. Magnetic field at the centre of circle is : **[8 Jan. 2020 (Evening)]**

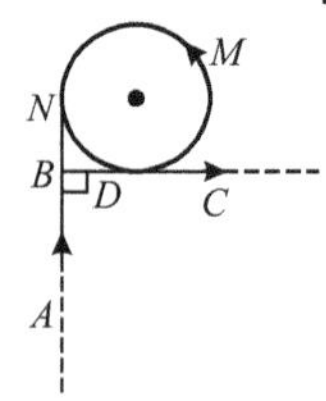

(A) $\frac{\mu_0 I}{2R}$ (B) $\frac{\mu_0 I}{2\pi R}(\pi+1)$
(C) $\frac{\mu_0 I}{2\pi R}\left(\pi+\frac{1}{\sqrt{2}}\right)$ (D) $\frac{\mu_0 I}{2\pi R}\left(\pi-\frac{1}{\sqrt{2}}\right)$

91. A long, straight wire of radius a carries a current distributed uniformly over its cross-section. The ratio of the magnetic fields due to the wire at distance $\frac{a}{3}$ and $2a$, respectively from the axis of the wire is : **[9 Jan. 2020 (Morning)]**
(A) $\frac{2}{3}$ (B) $\frac{3}{2}$
(C) $\frac{1}{2}$ (D) 2

18.2 Magnetic Force on Moving Charges

92. A particle of mass m and charge q has an initial velocity $\vec{v} = -v_0\hat{j}$. If an electric field $\vec{E} = E_0\hat{i}$ and magnetic field $\vec{B} = B_0\hat{i}$ act on the particle, its speed will double after a time : **[7 Jan. 2020 (Evening)]**
(A) $\frac{2mv_0}{qE_0}$ (B) $\frac{3mv_0}{qE_0}$
(C) $\frac{\sqrt{3}mv_0}{qE_0}$ (D) $\frac{\sqrt{2}mv_0}{qE_0}$

93. Protons with kinetic energy of 1 MeV moves from south to north. It gets an acceleration of 10^{12} m/s^2 by an applied magnetic field (west to east). The value of magnetic field : (Rest mass of proton is 1.6×10^{-27} kg) : **[8 Jan. 2020 (Morning)]**

(A) 71 mT (B) 7.1 mT

(C) 0.071 mT (D) 0.71 mT

94. A charged particle of mass 'm' and charge 'q' moving under the influence of uniform electric field $E\vec{i}$ and a uniform magnetic field $B\vec{k}$ follows a trajectory from point P to Q as shown in figure. The velocities at P and Q are respectively, $v\vec{i}$ and $-2v\vec{j}$. Then which of the following statements (a, b, c, d) are the correct ? (Trajectory shown is schematic and not to scale) :

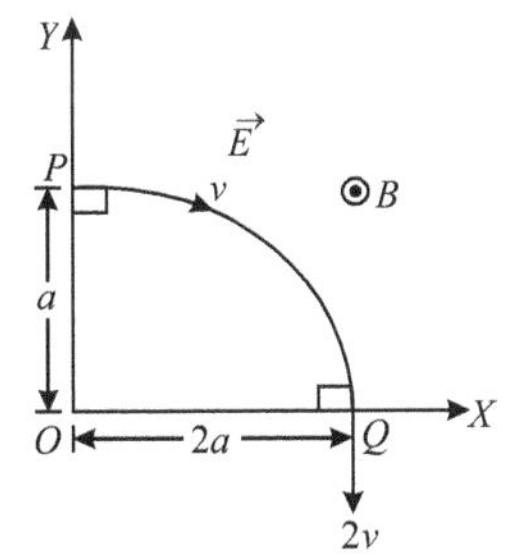

(a) $E = \frac{3}{4}\left(\frac{mv^2}{qa}\right)$

(b) Rate of work done by the electric field at P is $\frac{3}{4}\left(\frac{mv^3}{a}\right)$

(c) Rate of work done by both the fields at Q is zero

(d) The difference between the magnitude of angular momentum of the particle at P and Q is 2 mav.

[9 Jan. 2020 (Morning)]

(A) (a), (b), (c), (d) (B) (a), (b), (c)

(C) (b), (c), (d) (D) (a), (c), (d)

95. An electron gun is placed inside a long solenoid of radius R on its axis. The solenoid has n turns/length and carries a current I. The electron gun shoots an electron along the radius of the solenoid with speed v. If the electron does not hit the surface of the solenoid, maximum possible value of v is (all symbols have their standard meaning) : **[9 Jan. 2020 (Evening)]**

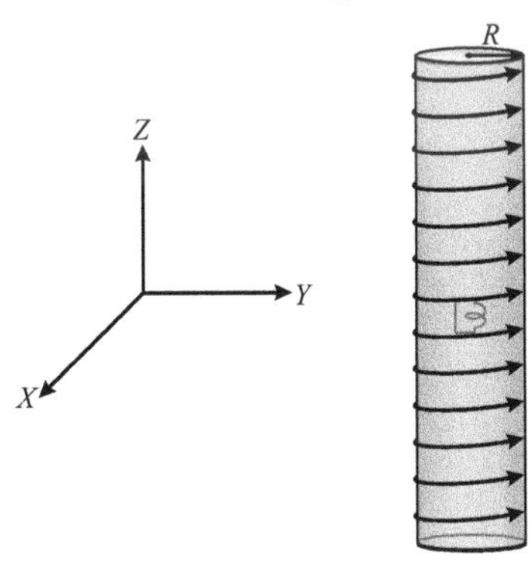

(A) $\frac{e\mu_0 nIR}{m}$ (B) $\frac{e\mu_0 nIR}{2m}$

(C) $\frac{2e\mu_0 nIR}{m}$ (D) $\frac{e\mu_0 nIR}{4m}$

CHAPTER-19 MAGNETISM AND MATTER

19.2 Magnetic Properties of Materials

96. The figure gives experimentally measured B vs H variation in a ferromagnetic material. The retentivity, coercivity and saturation, respectively, of the material are : **[7 Jan. 2020 (Evening)]**

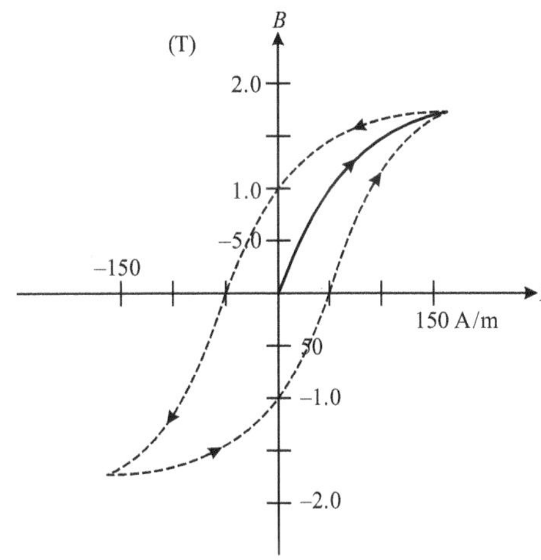

(A) 150 A/m, 1.0 T and 1.5 T (B) 1.0 T, 50 A/m and 1.5 T

(C) 1.5 T, 50 A/m and 1.0 T (D) 1.5 T, 50 A/m and 1.0 T

CHAPTER-20 ELECTROMAGNETIC INDUCTION

20.1 Electromagnetic Induction

97. A long solenoid of radius R carries a time (t)-dependent current $I(t) = I_0 t(1 - t)$. A ring of radius $2R$ is placed coaxially near its middle. During the time interval $0 \le t \le 1$, the induced current (I_R) and the induced EMF(V_R) in the ring change as : **[7 Jan. 2020 (Morning)]**

(A) At $t = 0.5$ direction of I_R reverses and V_R is zero

(B) Direction of I_R remains unchanged and V_R is zero at $t = 0.25$

(C) Direction of I_R remains unchanged and V_R is maximum at $t = 0.5$

(D) At $t = 0.25$ direction of I_R reverses and V_R is maximum

98. Consider a circular coil of wire carrying constant current I, forming a magnetic dipole. The magnetic flux through an infinite plane that contains the circular coil and excluding the circular coil area is given by ϕ_i. The magnetic flux through the area of the circular coil area is given by ϕ_0. Which of the following option is correct ? **[7 Jan. 2020 (Morning)]**

(A) $\phi_i = -\phi_0$ (B) $\phi_i = \phi_0$

(C) $\phi_i < \phi_0$ (D) $\phi_i > \phi_0$

99. A loop $ABCDEFA$ of straight edges has six corner points $A(0,0,0)$, $B(5,0,0)$, $C(5,5,0)$, $D(0, 5, 0)$, $E(0, 5, 5)$ and $F(0, 0, 5)$. The magnetic field in this region is $\vec{B} = (3\hat{i} + 4\hat{k})T$. The quantity of flux through the loop $ABCDEFA$ (in Wb) is ________. **[7 Jan. 2020 (Morning)]**

100. A planar loop of wire rotates in a uniform magnetic field. Initially, at $t = 0$, the plane of the loop is perpendicular to the magnetic field. If it rotates with a period of 10 s about an axis in its plane then the magnitude of induced emf will be maximum and minimum, respectively at : **[7 Jan. 2020 (Evening)]**

(A) 2.5 s and 7.5 s (B) 5.0 s and 7.5s
(C) 5.0 s and 10.0 s (D) 2.5s and 5.0 s

101. At time $t = 0$ magnetic field of 100 Gauss is passing perpendicularly through the area defined by the closed loop shown in the figure. If the magnetic field reduces linearly to 500 Gauss, in the next 5s, then induced EMF in the loop is : **[8 Jan. 2020 (Morning)]**

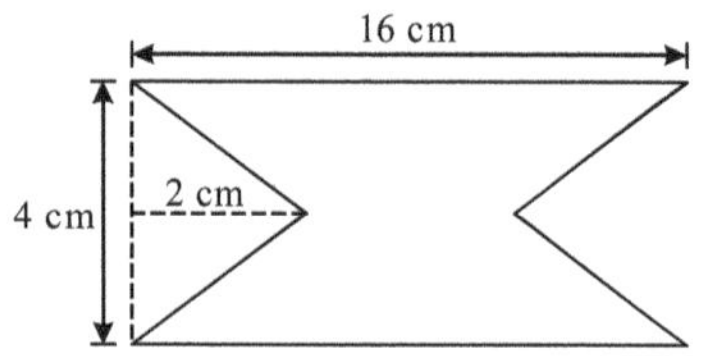

(A) 36 μV (B) 48 μV
(C) 56 μV (D) 28 μV

20.2 Self and Mutual Induction

102. An emf of 20 V is applied at time $t = 0$ to a circuit containing in series 10 mH inductor and 5Ω resistor. The ratio of the currents at time $t = \infty$ and at $t = 40$ s is close to : (Take $e^2 = 7.389$) : **[7 Jan. 2020 (Evening)]**

(A) 1.06 (B) 1.15
(C) 1.46 (D) 0.84

103. A shown in the figure, a battery of emf ε is connected to an inductor L and resistance R in series. The switch is closed at $t = 0$. The total charge that flows from the battery, between $t = 0$ and $t = t_c$ (t_c is the time constant of the circuit) is : **[8 Jan. 2020 (Evening)]**

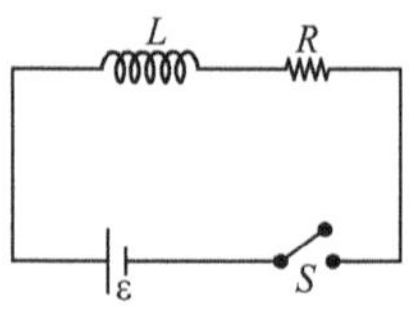

(A) $\frac{\varepsilon L}{R^2}\left(1-\frac{1}{e}\right)$ (B) $\frac{\varepsilon R}{eL^2}$

(C) $\frac{\varepsilon L}{R^2}$ (D) $\frac{\varepsilon L}{eR^2}$

104. In a fluorescent lamp choke (a small transformer) 100 V of reverse voltage is produced when the choke current changes uniformly from 0.25 A to 0 in a duration of 0.025 ms. The self-inductance of the choke (in mH) is estimated to be ________. **[9 Jan. 2020 (Morning)]**

CHAPTER-21 ALTERNATING CURRENT

21.1 Alternating Current

105. A LCR circuit behaves like a damped harmonic oscillator. Comparing it with a physical spring-mass damped oscillator having damping constant 'b', the correct equivalence would be : **[7 Jan. 2020 (Morning)]**

(A) $L \leftrightarrow m, C \leftrightarrow \frac{1}{k}, R \leftrightarrow b$

(B) $L \leftrightarrow \frac{1}{b}, C \leftrightarrow \frac{1}{m}, R \leftrightarrow \frac{1}{k}$

(C) $L \leftrightarrow m, C \leftrightarrow k, R \leftrightarrow b$

(D) $L \leftrightarrow k, C \leftrightarrow b, R \leftrightarrow m$

106. In LC circuit the inductance $L = 40$ mH and capacitance $C = 100$ μF. If a voltage $V(t) = 10 \sin(314\,t)$ is applied to the circuit, the current in the circuit is given as : **[9 Jan. 2020 (Evening)]**

(A) $0.52 \cos 314\,t$ (B) $0.52 \sin 314\,t$
(C) $10 \cos 314\,t$ (D) $5.2 \cos 314\,t$

CHAPTER-22 ELECTROMAGNETIC WAVES

22.1 Electromagnetic Wave

107. If the magnetic field in a plane electromagnetic wave is given by $\vec{B} = 3 \times 10^{-8} \sin(1.6 \times 10^3 x + 48 \times 10^{10} t)\,\hat{j}$ T, then what will be expression for electric field ? **[7 Jan. 2020 (Morning)]**

(A) $\vec{E} = (9 \sin(1.6 \times 10^3 x + 48 \times 10^{10} t)\,\hat{k}$ V/m)

(B) $\vec{E} = (3 \times 10^{-8} \sin(1.6 \times 10^3 x + 48 \times 10^{10} t)\,\hat{i}$ V/m)

(C) $\vec{E} = (60 \sin(1.6 \times 10^3 x + 48 \times 10^{10} t)\,\hat{k}$ V/m)

(D) $\vec{E} = (3 \times 10^{-8} \sin(1.6 \times 10^3 x + 48 \times 10^{10} t)\,\hat{j}$ V/m)

108. The electric field of a plane electromagnetic wave is given by

$$\vec{E} = E_0 \frac{\hat{i}+\hat{j}}{\sqrt{2}} \cos(kz + \omega t)$$

At $t = 0$, a positively charged particle is at the point $(x, y, z) = \left(0, 0, \frac{\pi}{k}\right)$. If its instantaneous velocity at $(t = 0)$ is $\upsilon_0\,\hat{k}$, the force acting on it due to the wave is : **[7 Jan. 2020 (Evening)]**

(A) Zero (B) Parallel to $\frac{\hat{i}+\hat{j}}{\sqrt{2}}$

(C) Antiparallel to $\frac{\hat{i}+\hat{j}}{\sqrt{2}}$ (D) Parallel to $\hat{k}$

109. A plane electromagnetic wave of frequency 25 GHz is propagating in vacuum along the z-direction. At a particular point in space and time, the magnetic field is given by $\vec{B} = 5\times10^{-8}\,\hat{j}T$. The corresponding electric field $\vec{E}$ is (speed of light $c = 3\times10^{8}\ \text{ms}^{-1}$) : **[8 Jan. 2020 (Evening)]**

(A) $1.66\times10^{-16}\,\hat{i}$ V/m (B) $15\hat{i}$ V/m
(C) $-1.66\times10^{-16}\,\hat{i}$ V/m (D) $-15\hat{i}$ V/m

110. The electric fields of two plane electromagnetic plane waves in vacuum are given by

$\vec{E}_1 = E_0\,\hat{j}\cos(\omega t - kx)$ and

$\vec{E}_2 = E_0\,\hat{k}\cos(\omega t - ky)$

At $t = 0$, a particle of charge q is at origin with a velocity $\vec{v} = 0.8c\,\hat{j}$ (c is the speed of light in vacuum). The instantaneous force experienced by the particle is : **[9 Jan. 2020 (Morning)]**

(A) $E_0q(-0.8\hat{i} + \hat{j} + \hat{k})$ (B) $E_0q(0.8\hat{i} - \hat{j} + 0.4\hat{k})$
(C) $E_0q(0.8\hat{i} + \hat{j} + 0.2\hat{k})$ (D) $E_0q(0.4\hat{i} - 3\hat{j} + 0.8\hat{k})$

111. A plane electromagnetic wave is propagating along the direction $\frac{\hat{i}+\hat{j}}{\sqrt{2}}$, with its polarization along the direction $\hat{k}$. The correct form of the magnetic field of the wave would be (here B_0 is an appropriate constant) : **[9 Jan. 2020 (Evening)]**

(A) $B_0\frac{\hat{i}-\hat{j}}{\sqrt{2}}\cos\left(\omega t - k\frac{\hat{i}+\hat{j}}{\sqrt{2}}\right)$

(B) $B_0\frac{\hat{i}+\hat{j}}{\sqrt{2}}\cos\left(\omega t - k\frac{\hat{i}+\hat{j}}{\sqrt{2}}\right)$

(C) $B_0\hat{k}\cos\left(\omega t - k\frac{\hat{i}+\hat{j}}{\sqrt{2}}\right)$

(D) $B_0\frac{\hat{j}-\hat{i}}{\sqrt{2}}\cos\left(\omega t + k\frac{\hat{i}+\hat{j}}{\sqrt{2}}\right)$

CHAPTER-23 RAY OPTICS AND OPTICAL INSTRUMENTS

23.1 Geometrical Optics I - Reflection of Light

112. An object is gradually moving away from the focal point of a concave mirror along the axis of the mirror. The graphical representation of the magnitude of linear magnification (m) versus distance of the object from the mirror (x) is correctly given by : **[8 Jan. 2020 (Evening)]**
(Graphs are drawn schematically and are not to scale)

(A) 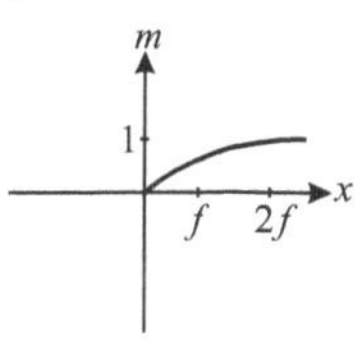

(B) 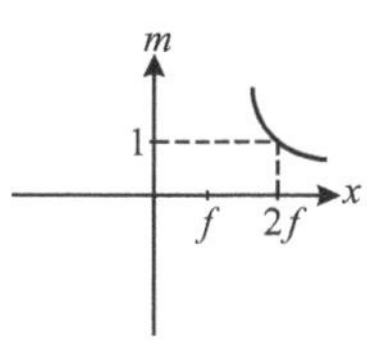

(C) 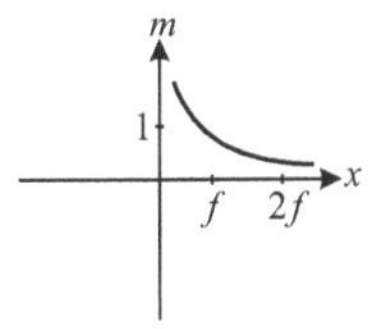

(D) 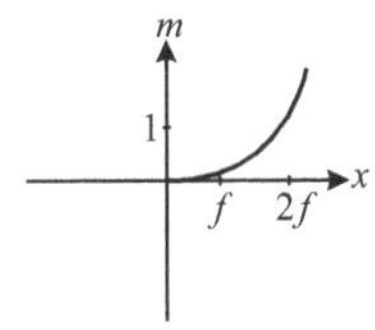

23.2 Geometrical Optics II - Refraction of Light

113. The critical angle of a medium for a specific wavelength, if the medium has relative permittivity 3 and relative permeability $\frac{4}{3}$ for this wavelength, will be : **[8 Jan. 2020 (Morning)]**

(A) 60° (B) 15°
(C) 45° (D) 30°

114. A vessel of depth $2h$ is half filled with a liquid of refractive index $2\sqrt{2}$ and the upper half with another liquid of refractive index $\sqrt{2}$. The liquids are immiscible. The apparent depth of the inner surface of the bottom of vessel will be : **[9 Jan. 2020 (Morning)]**

(A) $\frac{h}{\sqrt{2}}$ (B) $\frac{3}{4}h\sqrt{2}$

(C) $\frac{h}{2(\sqrt{2}+1)}$ (D) $\frac{h}{3\sqrt{2}}$

115. There is a small source of light at some depth below the surface of water (refractive index = 4/3) in a tank of large cross sectional surface area. Neglecting any reflection from the bottom and absorption by water, percentage of light that emerges out of surface is (nearly) :
[Use the fact that surface area of a spherical cap of height h and radius of curvature r is $2\pi rh$] **[9 Jan. 2020 (Evening)]**

(A) 17% (B) 21%
(C) 34% (D) 50%

23.3 Geometrical Optics III - Thin Lenses

116. A thin lens made of glass (refractive index = 1.5) of focal length f = 16 cm is immersed in a liquid of refractive index 1.42. If its focal length in liquid is f_1, then the ratio f_1/f is closest to the integer : **[7 Jan. 2020 (Evening)]**

(A) 1 (B) 5
(C) 9 (D) 17

117. A point object in air is in front of the curved surface of a plano-convex lens. The radius of curvature of the curved surface is 30 cm and the refractive index of the lens material is 1.5, then the focal length of the lens (in cm) is ____________.
[8 Jan. 2020 (Morning)]

23.5 Optical Instruments

118. If we need a magnification of 375 from a compound microscope of tube length 150 mm and an objective of focal length 5 mm, the focal length of the eye-piece, should be close to : **[7 Jan. 2020 (Morning)]**

(A) 22mm (B) 12mm
(C) 33mm (D) 2mm

119. The magnifying power of a telescope with tube 60 cm is 5. What is the focal length of its eye piece ? **[8 Jan. 2020 (Morning)]**

(A) 30 cm (B) 40 cm
(C) 20 cm (D) 10 cm

120. The aperture diameter of a telescope is 5m. The separation between the moon and the earth is 4×10^5 km. With light of wavelength of 5500 Å, the minimum separation between objects on the surface of moon, so that they are just resolved, is close to : **[9 Jan. 2020 (Morning)]**

(A) 20m (B) 600m
(C) 60m (D) 200m

CHAPTER-24 WAVE OPTICS

24.1 Interference of Light

121. In a Young's double slit experiment, the separation between the slits is 0.15 mm. In the experiment, a source of light of wavelength 589 nm is used and the interference pattern is observed on a screen kept 1.5 m away. The separation between the successive bright fringes on the screen is : **[7 Jan. 2020 (Evening)]**

(A) 6.9mm (B) 5.9mm
(C) 4.9mm (D) 3.9mm

122. In a double slit experiment, at a certain point on the screen the path difference between the two interfering waves is $\frac{1}{8}$ th of a wavelength. The ratio of the intensity of light at that point to that at the centre of a bright fringe is : **[8 Jan. 2020 (Evening)]**

(A) 0.568 (B) 0.672
(C) 0.760 (D) 0.853

123. Three harmonic waves having equal frequency ν and same intensity I_0, have phase angles 0, $\frac{\pi}{4}$ and $-\frac{\pi}{4}$ respectively. When they are superimposed the intensity of the resultant wave is close to : **[9 Jan. 2020 (Morning)]**

(A) $5.8\,I_0$ (B) $0.2\,I_0$
(C) I_0 (D) $3\,I_0$

124. In a Young's double slit experiment 15 fringes are observed on a small portion of the screen when light of wavelength 500 nm is used. Ten fringes are observed on the same section of the screen when another light source of wavelength λ is used. Then the value of λ is (in nm) ________. **[9 Jan. 2020 (Evening)]**

24.2 Diffraction of Light

125. Visible light of wavelength 6000×10^{-8} cm falls normally on a single slit and produces a diffraction pattern. It is found that the second diffraction minimum is at 60° from the central maximum. If the first minimum is produced at θ_1, then θ_1 is close to : **[7 Jan. 2020 (Morning)]**

(A) 20° (B) 45°
(C) 30° (D) 25°

24.3 Polarization of Light

126. A polarizer-analyser set is adjusted such that the intensity of light coming out of the analyser is just 10% of the original intensity. Assuming that the polarizer-analyser set does not absorb any light, the angle by which the analyser need to be rotated further to reduce the output intensity to be zero, is : **[7 Jan. 2020 (Morning)]**

(A) 18.4° (B) 71.6°
(C) 90° (D) 45°

CHAPTER-25 DUAL NATURE OF RADIATION, MATTER & X-RAYS

25.1 Photoelectric Effect

127. A beam of electromagnetic radiation of intensity 6.4×10^{-5} W/cm^2 is comprised of wavelength, $\lambda = 310$ nm. It falls normally on a metal (work function $\varphi = 2eV$) of surface area of 1 cm^2. If one in 10^3 photons ejects an electron, total number of electrons ejected in 1 s is 10^x. ($hc = 1240\, eVnm$, $1eV = 1.6 \times 10^{-19}$ J), then x is________. **[7 Jan. 2020 (Morning)]**

128. When photon of energy 4.0 eV strikes the surface of a metal A, the ejected photoelectrons have maximum kinetic energy $T_A eV$ end de-Broglie wavelength λ_A. The maximum kinetic energy of photoelectrons liberated from another metal B by photon of energy 4.50 eV is $T_B = (T_A - 1.5)$ eV. If the de-Broglie wavelength of these photoelectrons $\lambda_B = 2\lambda_A$, then the work function of metal B is : **[8 Jan. 2020 (Morning)]**

(A) 3 eV (B) 2 eV
(C) 4 eV (D) 1.5 eV

129. The dimension of stopping potential V_0 in photoelectric effect in units of Planck's constant 'h', speed of light 'c' and Gravitational constant 'G' and ampere A is : **[8 Jan. 2020 (Morning)]**

(A) $h^2\, G^{3/2}\, c^{1/3}\, A^{-1}$ (B) $h^{-2/3}\, c^{-1/3}\, G^{4/3}\, A^{-1}$
(C) $h^{1/3}\, G^{2/3}\, c^{1/3}\, A^{-1}$ (D) $h^{2/3}\, c^{5/3}\, G^{1/3}\, A^{-1}$

CHAPTER-30 ERRORS AND EXPERIMENTS

30.1 Errors and Experiments

148. A simple pendulum is being used to determine the value of gravitational acceleration g at a certain place. Th length of the pendulum is 25.0 cm and a stop watch with 1s resolution measures the time taken for 40 oscillations to be 50 s. The accuracy in g is : **[8 Jan. 2020 (Evening)]**

(A) 3.40% (B) 5.40%
(C) 4.40% (D) 2.40%

149. If the screw on a screw-gauge is given six rotations, it moves by 3 mm on the main scale. If there are 50 divisions on the circular scale the least count of the screw gauge is : **[9 Jan. 2020 (Morning)]**

(A) 0.001 mm (B) 0.001 cm
(C) 0.02 mm (D) 0.01 cm

150. For the four sets of three measured physical quantities as given below. Which of the following options is correct ?

(i) $A_1 = 24.36, B_1 = 0.0724, C_1 = 256.2$
(ii) $A_2 = 24.44, B_2 = 16.082, C_2 = 240.2$
(iii) $A_3 = 25.2, B_3 = 19.2812, C_3 = 236.183$
(iv) $A_4 = 25, B_4 = 236.191, C_4 = 19.5$ **[9 Jan. 2020 (Evening)]**

(A) $A_4 + B_4 + C_4 < A_1 + B_1 + C_1 < A_3 + B_3 + C_3 < A_2 + B_2 + C_2$
(B) $A_1 + B_1 + C_1 < A_3 + B_3 + C_3 < A_2 + B_2 + C_2 < A_4 + B_4 + C_4$
(C) $A_1 + B_1 + C_1 = A_2 + B_2 + C_2 = A_3 + B_3 + C_3 = A_4 + B_4 + C_4$
(D) $A_4 + B_4 + C_4 < A_1 + B_1 + C_1 = A_2 + B_2 + C_2 = A_3 + B_3 + C_3$

* * * * *

ANSWER KEY

1.1 Measurements and Dimensional Analysis

1	(C)				

2.1 Motion in One Dimension

2	(8)	**3**	(3.00)		

3.1 Motion in Two Dimensions

4	(90°)	**5**	(580.00)	**6**	(A)
7	(C)				

4.1 Newton's Laws of Motion

8	(A)				

5.1 Work Energy Power

9	(C)	**10**	(10)	**11**	(C)
12	(B)				

5.2 Circular Motion

13	(B)				

6.1 System of Particles

14	(B)	**15**	(D)	**16**	(1.00)
17	(C)	**18**	(D)	**19**	(B)
20	(D)	**21**	(C)		

6.2 Rigid Body Dynamics

22	(C)	**23**	(B)	**24**	(A)
25	(75)	**26**	(B)	**27**	(A)
28	(A)	**29**	(15.00)	**30**	(D)
31	(B)				

7.1 Gravitational Field

32	(D)	**33**	(D)		

7.2 Satellite Motion

34	(B)	**35**	(16)	**36**	(A)
37	(A)				

8.1 Elasticity

38	(4.00)	**39**	(D)		

9.1 Fluid Statics

40	(D)	**41**	(A)		

9.2 Fluid Dynamics

42	(D)	**43**	(C)		

9.4 Surface Tension

44	(B)				

10.1 Thermal Expansion

45	(60)	**46**	(A)		

10.2 Calorimetry

47	(40)	**48**	(50)		

11.1 Laws of Thermodynamics

49	(A)	**50**	(B)	**51**	(600)
52	(B)	**53**	(C)	**54**	(A)
55	(D)	**56**	(C)	**57**	(B)
58	(B)	**59**	(D)		
60	(1818.00 to 1819)				

12.1 Kinetic Theory of Gases

61	(D)	**62**	(C)		

13.1 Simple Harmonic Motion

63	(B)				

14.1 Wave Motion

64	(C)	**65**	(B)		

14.2 Stationary Waves and Beats

66	(106.06)	**67**	(D)		

14.3 Doppler's Effect

68	(D)				

15.1 Electric Force and Electric Field

69	(D)	**70**	(C)	**71**	(D)

15.2 Electric Flux and Gauss's Law

72	(D)	**73**	(–48.00)		

16.1 Electric Potential and Dipole

74	(D)	**75**	(C)		

16.2 Capacitance

76 (A) **77** (6) **78** (C)
79 (D)

17.2 Circuit Analysis

80 (B) **81** (30) **82** (B)

17.3 Thermal Effects of Current & RC Circuits

83 (D)

17.4 Electrical Measuring Instruments

84 (12) **85** (D) **86** (10.00)
87 (A) **88** (40.00)

18.1 Magnetic Effects of Current

89 (A) **90** (C) **91** (A)

18.2 Magnetic Force on Moving Charges

92 (C) **93** (D) **94** (B)
95 (B)

19.2 Magnetic Properties of Materials

96 (B)

20.1 Electromagnetic Induction

97 (A) **98** (A) **99** (175)
100 (D) **101** (C)

20.2 Self and Mutual Induction

102 (A) **103** (D) **104** (10.00)

21.1 Alternating Current

105 (A) **106** (A)

22.1 Electromagnetic Wave

107 (A) **108** (C) **109** (B)
110 (C) **111** (A)

23.1 Geometrical Optics I - Reflection of Light

112 (B)

23.2 Geometrical Optics II - Refraction of Light

113 (D) **114** (B) **115** (A)

23.3 Geometrical Optics III - Thin Lenses

116 (C) **117** (60.00)

23.5 Optical Instruments

118 (A) **119** (D) **120** (C)

24.1 Interference of Light

121 (B) **122** (D) **123** (A)
124 (750.00)

24.2 Diffraction of Light

125 (D)

24.3 Polarization of Light

126 (A)

25.1 Photoelectric Effect

127 (11) **128** (C) **129** (Bonus)
130 (B)

25.2 Matter Waves and X-Ray

131 (B) **132** (C) **133** (C)
134 (A)

26.1 Atomic Structure

135 (C) **136** (C) **137** (486)
138 (D)

27.1 Radioactivity

139 (D)

28.2 PN Junction and Diodes

140 (B) **141** (12.00) **142** (C)
143 (A) **144** (40.00)

28.3 Transistors

145 (C)

28.4 Logic Gates

146 (B) **147** (B)

30.1 Errors and Experiments

148 (C) **149** (B) **150** (Bonus)

SOLUTIONS

1.1 Measurements and Dimensional Analysis

Sol. 1 Dimension of h, c and G are

$$[h] = M^1L^2T^{-1}$$
$$[C] = L^1T^{-1}$$
$$[G] = M^{-1}L^3T^{-2}$$

So dimensions of f can be calculated as

$$[f] = \sqrt{\frac{M^1L^2T^{-1}\times L^5T^{-5}}{M^{-1}L^3T^{-2}}} = M^1L^2T^{-2}$$ **Ans. (C)**

2.1 Motion in One Dimension

Sol. 2 Total time taken by ball to travel 81 m is t s and for complete 100 m is $\left(t+\frac{1}{2}\right)$ s, by speed equation, we use

$$\Rightarrow \quad 81 = \frac{1}{2}\times g\times t^2$$
$$\Rightarrow \quad t = 9\sqrt{\frac{2}{g}}$$
$$\Rightarrow \quad 100 = \frac{1}{2}\times g\times\left(t+\frac{1}{2}\right)^2$$
$$\Rightarrow \quad t+\frac{1}{2} = 10\sqrt{\frac{2}{g}}$$
$$\Rightarrow \quad 9\sqrt{\frac{2}{g}}+\frac{1}{2} = 10\sqrt{\frac{2}{g}}$$
$$\Rightarrow \quad \frac{1}{2} = \sqrt{\frac{2}{g}}$$
$$\Rightarrow \quad a = 8 \text{ m/s}^2$$ **Ans. (8)**

Sol. 3 From the given expression, we have

$$x = \sqrt{at^2+2bt+c}$$

Velocity of particle is given as

$$v = \frac{dx}{dt} = \frac{1}{2\sqrt{at^2+2bt+c}}\times(2at+2b)$$
$$\Rightarrow \quad v = \frac{at+b}{x}$$
$$\Rightarrow \quad vx = at+b$$

Differentiating above expression gives

$$vdx + xdv = adt$$
$$\Rightarrow \quad x\frac{dv}{dx}+v = a\times\frac{dt}{dx}$$

Multiplying both side by v, we get

$$\left(v\frac{dv}{dx}\right)x+v^2 = a$$

We use acceleration $f = v\frac{dv}{dx}$ so we have

$$fx = a - v^2$$
$$\Rightarrow \quad fx = a-\left(\frac{at+b}{x}\right)^2$$
$$\Rightarrow \quad fx = \frac{ax^2-(at+b)^2}{x^2}$$
$$\Rightarrow \quad fx = \frac{a(at^2+2bt+c)-(at+b)^2}{x^2}$$
$$\Rightarrow \quad fx = \frac{ac-b^2}{x^2}$$
$$\Rightarrow \quad f = \frac{ac-b^2}{x^3}$$
$$\Rightarrow \quad f\propto\frac{1}{x^3}$$
$$\Rightarrow \quad n = 3$$ **Ans. (3.00)**

3.1 Motion in Two Dimensions

Sol. 4 Figure-(1) shows the initial given condition

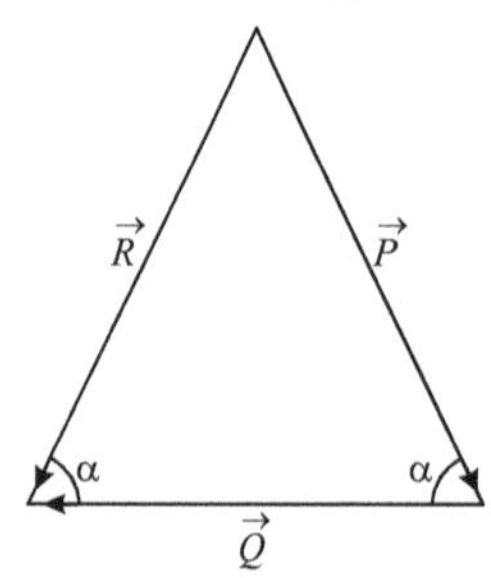

Figure-(1)

After $\vec{P}$ is made twice then the situation is shown in figure-(2) below

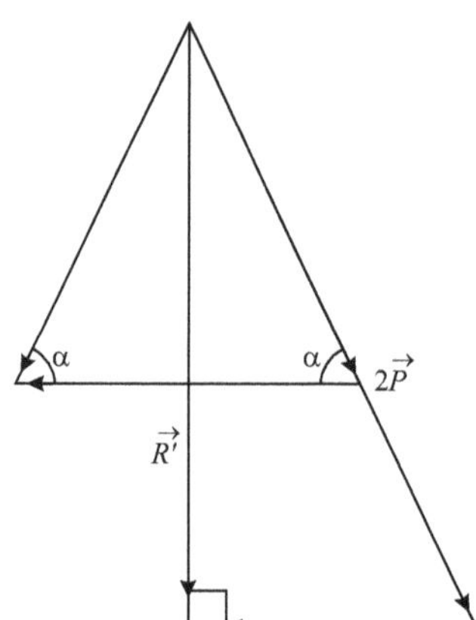

Figure-(2)

Thus angle here is 90°. **Ans. (90°)**

Sol. 5 Position along x-direction is given as

$$x = 10 + 8t - 3t^2$$

$\Rightarrow$ $v_{x_1} = 8 - 6t$

$\Rightarrow$ $(v_{x_1})_{t=1} = 2\hat{i}$

Position along y-direction is given as

$$y = 5 - 8t^3$$

$\Rightarrow$ $v_{y_2} = -24t^2$

$\Rightarrow$ $(v_{y_2})_{t=1} = -24\hat{j}$

Now given that

$$\sqrt{v} = \sqrt{v_{x_1}^2 + v_{y_2}^2}$$

$$\sqrt{v} = \sqrt{(24)^2 + (2)^2} = \sqrt{580}$$

$\Rightarrow$ $v = 580$ m/s **Ans. (580.00)**

Sol. 6 Given that

$$\vec{r}(t) = \cos\omega t\,\hat{i} + \sin\omega t\,\hat{j}$$

Velocity vector is given as

$$\vec{v} = \frac{\overrightarrow{dr}}{dt} = -\omega\sin\omega t\hat{i} + \omega\cos\omega t\hat{j}$$

Acceleration vector is given as

$$\vec{a} = \frac{\overrightarrow{dv}}{dt} = -\omega^2\cos\omega t\,\hat{i} - \omega^2\sin\omega t\,\hat{j} = -\omega^2\vec{r}$$

$\Rightarrow$ $\vec{v}\cdot\vec{r} = 0$ **Ans. (A)**

Sol. 7 For uniform acceleration motion in both x and y direction, we use

$$x = u_x t + \frac{1}{2}a_x t^2 \quad \ldots(1)$$

and $$y = u_y t + \frac{1}{2}a_y t^2 \quad \ldots(2)$$

$\Rightarrow$ $32 = 0 \times t + \frac{1}{2}(4)(t)^2$ [Given that $u_y = 0$]

$\Rightarrow$ $t^2 = 16$

$\Rightarrow$ $t = 4$ s

Along x-direction, we use from equation-(1)

$$x = 3 \times 4 + \frac{1}{2} \times 6 \times 4^2$$

$\Rightarrow$ $x = 12 + 48 = 60$ m **Ans. (C)**

4.1 Newton's Laws of Motion

Sol. 8 FBD of the mass is shown in figure

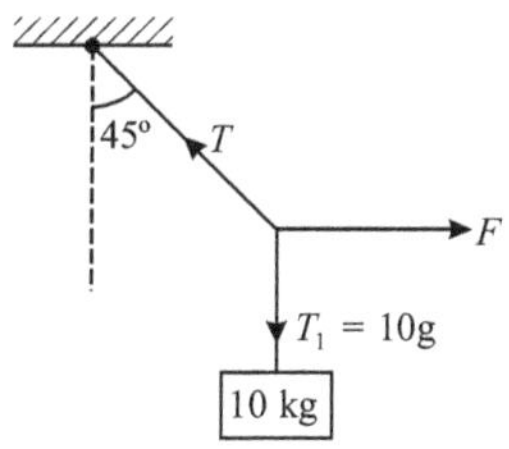

For equilibrium of system, we use

$$T\sin 45° = F \quad \ldots(1)$$

and $$T\cos 45° = 10g \quad \ldots(2)$$

$\Rightarrow$ $F = 10g$

$\Rightarrow$ $F = 100$ N **Ans. (A)**

5.1 Work Energy Power

Sol. 9 Upward force by motor on elevator is

$$F = 2000\,g + f_r = 20000 + 4000$$

$$F = 24000 \text{ N}$$

At full load with speed v, power of motor is

$\Rightarrow$ $60 \times 746 = (24000)\,v$

$\Rightarrow$ $v = \frac{60 \times 746}{24000} = 1.865$

$\Rightarrow$ $v \approx 1.9$ m/s **Ans. (C)**

Sol. 10 Using energy conservation between point A & P, we have

$$mgh_1 = mgh_2 + KE$$

$$mg \times 2 = mg \times 1 + KE$$

$\Rightarrow$ $KE = mg \times 1 = 10$ J **Ans. (10)**

Sol. 11 Free body diagram of elevator is

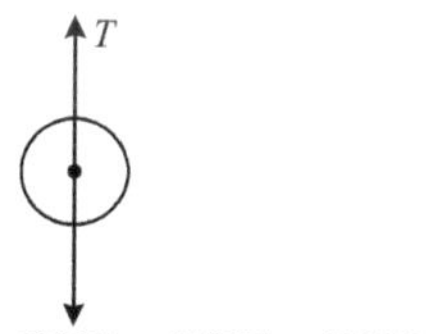

As elevator moving with constant speed, so we use

$$T = 6800 + 9200 + 6000$$

$\Rightarrow$ $T = 22000$ N

Max power is given as

$$P_{max} = T.v = 22000 \times 3$$

$\Rightarrow$ $P_{max} = 66000$ W **Ans. (C)**

Sol. 12 Work done by force is given as

$$W = \int_{\vec{r}_i}^{\vec{r}_g} \vec{F}\cdot\overrightarrow{dr}$$

We use $\vec{F} = -x\hat{i} + y\hat{i}$ and $\overrightarrow{dr} = dx\hat{i} + dy\hat{j}$

$\Rightarrow$ $W = \int_1^0 -x\,dx + \int_0^1 y\,dy$

$\Rightarrow$ $W = \left[\frac{-x^2}{2}\right]_1^0 + \left[\frac{y^2}{2}\right]_0^1$

$\Rightarrow$ $W = -\left(\frac{0^2}{2} - \frac{1^2}{2}\right) + \left(\frac{1^2}{2} - \frac{0^2}{2}\right)$

$\Rightarrow$ $W = 1$ J **Ans. (B)**

5.2 Circular Motion

Sol. 13 FBD of particle is shown in figure

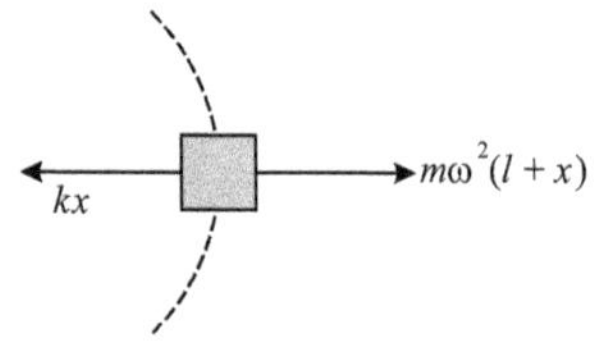

For uniform circular motion, we use

$$kx = ml\omega^2 + mx\omega^2$$

$$\Rightarrow \quad x = \frac{ml\omega^2}{k - m\omega^2}$$ **Ans. (B)**

6.1 System of Particles

Sol. 14 Using coordinate axis in given situation, considering 1 kg mass at origin we use

$$x_{cm} = \frac{1(0)+1.5(3)+2.5(0)}{5} = 0.9 \text{ cm}$$

$$y_{cm} = \frac{1(0)+1.5(0)+2.5(4)}{5} = 2 \text{ cm}$$ **Ans. (B)**

Sol. 15 In figure we consider masses m_1 and m_2 of two sections shown of 3kg and 1kg at centre coordinates (0.5, 1.5) and (1.5, 2.5)

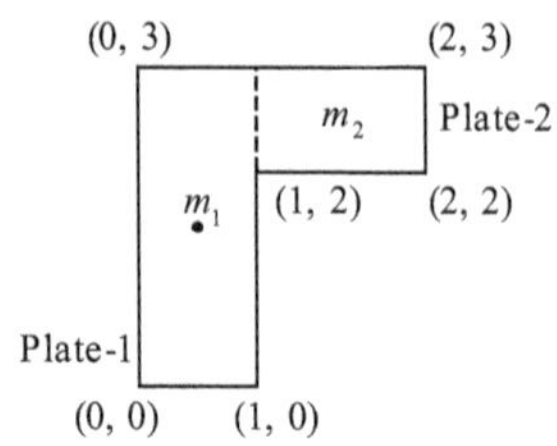

Thus coordinates of centre of mass are given as

$$x_{cm} = \frac{m_1x_1 + m_2x_2}{m_1 + m_2} = \frac{3\times0.5+1\times1.5}{4} = 0.75$$

$$y_{cm} = \frac{m_1y_1 + m_2y_2}{m_1 + m_2} = \frac{3\times1.5+1\times2.5}{4} = 1.75$$ **Ans. (D)**

Sol. 16 Using conservation of momentum, we have

$$(0.1)(3\hat{i}) + (0.1)(5\hat{j}) = (0.1)(4)(\hat{i}+\hat{j}) + (0.1)\vec{v}$$

$$\Rightarrow \quad \vec{v} = -\hat{i} + \hat{j}$$

$$\Rightarrow \quad |\vec{v}| = \sqrt{2}$$

Kinetic energy of B is

$$K = \frac{1}{2}mv^2 = \frac{1}{2}(0.1)(2) = \frac{1}{10}$$

$$\Rightarrow \quad x = 1$$ **Ans. (1.00)**

Sol. 17 Let $m_1 \rightarrow$ mass of body
$m_2 \rightarrow$ mass of cavity

Using mass moment concept of centre of mass, we have

$$m_1R_1 = m_2R_2$$

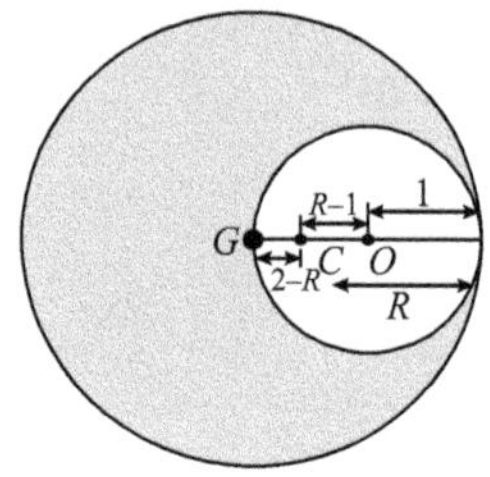

$$\Rightarrow \quad \left(\frac{4}{3}\pi R^3\rho - \frac{4}{3}\pi(1)^3\rho\right)(2-R) = \frac{4}{3}\pi(1)^3\rho \times (R-1)$$

$$\Rightarrow \quad (R^3-1)(2-R) = R-1$$

$$\Rightarrow \quad (R^2+R+1)(2-R) = 1$$ **Ans. (C)**

Sol. 18 Particles will collide in air after time given as

$$t_0 = \frac{h}{\sqrt{2gh}} = \sqrt{\frac{h}{2g}}$$

At the instant of collision velocities of the two particle are

$$v_A = gt_0 = \sqrt{\frac{gh}{2}}$$

$$v_B = u_B - gt_0 = \sqrt{\frac{gh}{2}}$$

$$\Rightarrow \quad v_A = v_B = \sqrt{\frac{gh}{2}}$$

Height at which collision takes place is

$$h_1 = h - \frac{1}{2}gt_0^2 = \frac{3h}{4}$$

As after collision, combined mass comes to rest, time taken to reach ground is given as

$$t_1 = \sqrt{\frac{2\times(3h/4)}{g}} = \sqrt{\frac{3h}{2g}}$$ **Ans. (D)**

Sol. 19 Using momentum conservation, we have

$$mu\hat{i} + mu\left(\frac{\hat{i}+\hat{j}}{2}\right) = (m+m)\vec{v}$$

$$\Rightarrow \quad \vec{v} = \frac{3}{4}u\hat{i} + \frac{u}{4}\hat{j}$$

$$\Rightarrow \quad v = \frac{u}{4}\sqrt{10}$$

After collision final kinetic energy is given as

$$K_i = \frac{1}{2}2\text{m}\left(\frac{u}{4}\sqrt{10}\right)^2 = \frac{5}{8}\text{mu}^2$$

Initial kinetic energy before collision is

$$K_i = \frac{1}{2}\text{mu}^2 + \frac{1}{2}\text{m}\left(\frac{u}{\sqrt{2}}\right)^2 = \frac{6}{8}\text{mu}^2$$

Loss in kinetic energy is given as

$$\Delta K = K_i - K_f = \frac{1}{8}\text{mu}^2$$

Ans. (B)

Sol. 20 Distance of centre of mass of rod from one end is given as

$$x_{cm} = \frac{\int x dm}{\int dm} = \frac{\int (\lambda dx)x}{\int dm}$$

$$\Rightarrow \quad x_{cm} = \frac{\int_0^L \left(a + \frac{bx^2}{L^2}\right) x dx}{\int_0^L \left(a + \frac{bx^2}{L^2}\right) dx}$$

$$\Rightarrow \quad x_{cm} = \frac{\frac{aL^2}{2} + \frac{b}{L^2}\cdot\frac{L^4}{4}}{aL + \frac{b}{L^2}\cdot\frac{L^3}{3}}$$

$$\Rightarrow \quad x_{cm} = \frac{\left(\frac{4a+2b}{8}\right)L}{\frac{(3a+b)}{3}} = \frac{3}{4}\frac{(2a+b)L}{(3a+b)}$$

Ans. (D)

Sol. 21 Velocity of particle at highest point is given as

$$V_x = u\cos 60° = \frac{u}{2}$$

Using momentum conservation at the highest point, final velocity of combined mass is given as

$$\frac{mu}{2} + mu = 2mv'$$

$$\Rightarrow \quad v' = \frac{3v}{4}$$

Range of combined particle after collision is given as

$$R = \frac{3v}{4}\sqrt{\frac{2H}{g}}$$

$$\Rightarrow \quad R = \frac{3v}{4}\sqrt{\frac{2}{g}\left(\frac{u^2\sin^2 60°}{2g}\right)}$$

$$\Rightarrow \quad R = \frac{3}{4}\frac{\sqrt{3}}{2}\cdot\frac{u^2}{g}$$

$$\Rightarrow \quad R = \frac{3\sqrt{3}u^2}{8g}$$

Ans. (C)

6.2 Rigid Body Dynamics

Sol. 22 Using work energy theorem, we use

$$mgh = \frac{1}{2}mv^2 + \frac{1}{2}\times\left(\frac{1}{2}mr^2\right)\times\frac{v^2}{r^2} = \frac{3}{4}mv^2$$

$$\Rightarrow \quad v = \sqrt{\frac{4}{3}gh}$$

Angular speed is given as

$$\omega = \frac{v}{r} = \frac{1}{r}\sqrt{\frac{4gh}{3}}$$

Ans. (C)

Sol. 23 Moment of inertia of given system is taken as

$$m\frac{l^2}{12} + m\frac{l^2}{16} = mk^2$$

$$\Rightarrow \quad \frac{7l^2}{48} = k^2$$

$$\Rightarrow \quad k = \sqrt{\frac{7}{48}}\cdot l$$

Ans. (B)

Sol. 24 We consider an elemental ring in disc as shown

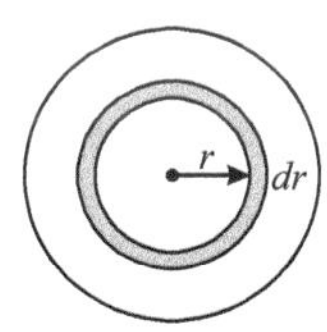

Moment of inertia of this ring is given as

$$dI = dmr^2$$

$$\Rightarrow \quad dI = \sigma 2\pi r\, dr\, r^2$$

$$\Rightarrow \quad dI = 2\pi(A + Br)\cdot r^3\cdot dr$$

Total moment of inertia of disc is

$$I = \int dI = 2\pi\int_0^a (Ar^3 + Br^4)dr$$

$$\Rightarrow \quad I = 2\pi a^4\left(\frac{A}{4} + \frac{aB}{5}\right)$$

Ans. (A)

Sol. 25 Forces are shown in figure below

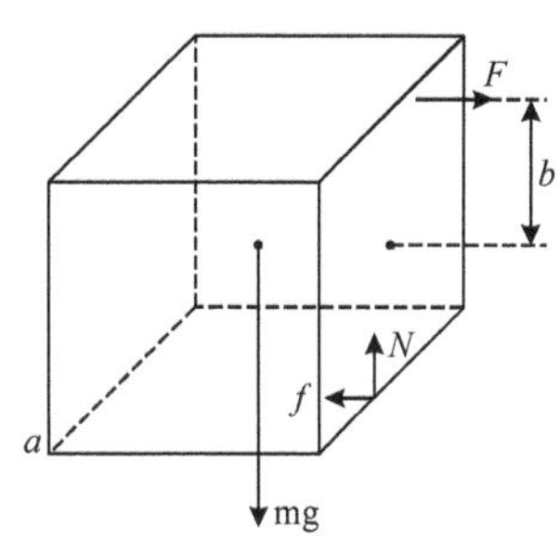

To move the block

$$F = \mu mg \qquad \ldots(1)$$

For bock not to topple, we use

$$F\left(b+\frac{a}{2}\right)=mg\,\frac{a}{2} \qquad \ldots(2)$$

$$\Rightarrow \mu mg\left(b+\frac{a}{2}\right)=mg\times\frac{a}{2}$$

$$\Rightarrow \left(b+\frac{a}{2}\right)\mu=\frac{a}{2}$$

$$\Rightarrow 0.4=\mu=\frac{a}{2b+a}$$

$$\Rightarrow 0.8b+0.4a=a$$

$$\Rightarrow 0.8b=0.6a$$

$$\Rightarrow \frac{b}{a}=\frac{3}{4}$$

Ans. (75)

Sol. 26 Figure below shows the given situation

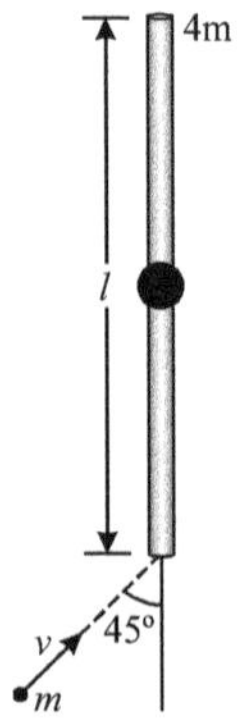

Using by conservation of angular momentum about the hinge, we have

$$m\left(\frac{v}{\sqrt{2}}\right)\left(\frac{l}{2}\right)=\left[\frac{4ml^2}{12}+\frac{ml^2}{4}\right]\omega$$

$$\Rightarrow \omega=\frac{3\sqrt{2}}{7}\left(\frac{v}{l}\right)$$

Ans. (B)

Sol. 27 Kinetic energy of rolling sphere is given as

$$KE=\frac{1}{2}mv^2+\frac{1}{2}I\omega^2$$

$$\Rightarrow KE=\frac{1}{2}mv^2\left(1+\frac{K^2}{R^2}\right)$$

$$\Rightarrow KE=8.75\times10^{-4}\text{ J}$$

Ans. (A)

Sol. 28 Using parallel axis theorem, moment of inertia of system is given as

$$I_0=3\times\left[\frac{2}{5}M\left(\frac{d}{2}\right)^2+M\left(\frac{d}{\sqrt{3}}\right)^2\right]=\frac{13}{10}\text{Md}^2 \qquad \ldots(1)$$

Moment of inertia about axis passing through A is given as

$$I_A=I_0+3M\left(\frac{d}{\sqrt{3}}\right)^2$$

$$\Rightarrow I_A=\frac{13}{10}\text{Md}^2+\text{Md}^2$$

$$\Rightarrow I_A=\frac{23}{10}\text{Md}^2 \qquad \ldots(2)$$

From (1) and (2), we have

$$\frac{I_0}{I_A}=\frac{13}{23}$$

Ans. (A)

Sol. 29 Using work energy theorem, we have

$$\Rightarrow mg\left(\frac{l}{2}\sin30^\circ\right)+\frac{1}{2}I\omega^2$$

$$\Rightarrow mg\times\frac{1}{2}\times\frac{1}{2}\times\frac{1}{2}\left(\frac{m(1)^2}{3}\right)\omega^2$$

$$\Rightarrow \omega^2=\frac{3g}{2} \Rightarrow \omega=\sqrt{15}$$

$$\Rightarrow n=15$$

Ans. (15.00)

Sol. 30 Free body diagram of mass in frame of disc is shown in figure

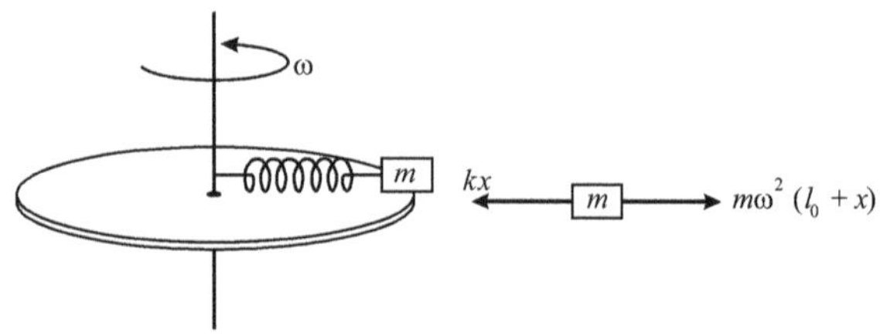

For equilibrium of mass m disc frame, we use

$$kx=m\omega^2(l_0+x)$$

$$x=\frac{m\omega^2 l_0}{k-m\omega^2}\approx\frac{m\omega^2 l_0}{k}$$

Relative change in spring length is

$$\frac{x}{l_0}=\frac{m\omega^2}{k}$$

Ans. (D)

Sol. 31 Using work energy theorem, we have

$$(m_1-m_2)\,gh=\frac{1}{2}(m_1+m_2)\,v^2+\frac{1}{2}I\omega^2$$

$$\Rightarrow (m_1-m_2)\,gh=\frac{1}{2}(m_1+m_2)(\omega R)^2+\frac{1}{2}I\omega^2$$

$$\Rightarrow (m_1-m_2)\,gh=\frac{\omega^2}{2}[(m_1+m_2)R^2+I]$$

$$\Rightarrow \omega=\sqrt{\frac{2(m_1-m_2)gh}{(m_1+m_2)R^2+I}}$$

Ans. (B)

7.1 Gravitational Field

Sol. 32 Weight at equator is given as

$$W_{eq}=W_{pole}-m\omega^2R_e$$

$$\Rightarrow \quad W_{eq} = 196 - \frac{196}{10} \times \left(\frac{2\pi}{86400}\right)^2 \times 6.4 \times 10^5$$
$$= 195.34\,\text{N}$$ **Ans. (D)**

Sol. 33 Gravitational field on the surface of a solid sphere is given as

$$g_s = \frac{GM}{R^2}$$

From the graphs, we have

$$\frac{GM_1}{(1)^2} = 2 \text{ and } \frac{GM_2}{(2)^2} = 3$$

$$\Rightarrow \quad \frac{M_1}{M_2} = \frac{1}{6}$$ **Ans. (D)**

7.2 Satellite Motion

Sol. 34 Using energy conservation, we have

$$K_i + U_i = K_f + U_f$$

$$\frac{1}{2}mu^2 + \left(-\frac{GMm}{R}\right) = \frac{1}{2}mv^2 - \frac{GMm}{2R}$$

$$v = \sqrt{u^2 - \frac{GM}{R}} \quad \ldots(1)$$

At height R rocket must carry all the momentum radial direction and along tangent to orbit it should carry the momentum such that satellite has left only with orbital speed $\sqrt{\frac{GM}{2R}}$ for circular motion. Thus by momentum conservation, we use

Along orbital direction (along tangent)

$$\frac{m}{10}v_{R0} = \frac{9m}{10}\sqrt{\frac{GM}{2R}}$$

$$v_{R0} = 9\sqrt{\frac{GM}{2R}} \quad \ldots(2)$$

Along radial direction

$$\frac{m}{10}v_{RR} = mv$$

$$\Rightarrow \quad \frac{m}{10}v_{RR} = m\sqrt{u^2 - \frac{GM}{R}} \quad \ldots(3)$$

$$\Rightarrow \quad v_{RR} = 10\sqrt{u^2 - \frac{GM}{R}}$$

Now kinetic energy of rocket is given as

$$K = \frac{1}{2}m(v_{R0}^2 + v_{RR}^2)$$

$$\Rightarrow \quad K = \frac{m}{20}\left(81\frac{GM}{2R} + 100u^2 - 100\frac{GM}{R}\right)$$

$$\Rightarrow \quad K = \frac{m}{20}\left(100u^2 - \frac{119GM}{2R}\right)$$

$$\Rightarrow \quad K = 5m\left(u^2 - \frac{119GM}{200R}\right)$$ **Ans. (B)**

Sol. 35 Using conservation of energy, we have

$$P_1 + K_1 = P_2 + K_2$$

$$\Rightarrow \quad -\frac{GM_em}{10R} + \frac{1}{2}mv_0^2 = -\frac{GM_em}{R} + \frac{1}{2}mv^2$$

$$\Rightarrow \quad +\frac{9}{10} \times \frac{GM_em}{R} + \frac{1}{2}mv_0^2 = \frac{1}{2}mv^2$$

$$\Rightarrow \quad \frac{9}{10} \times \frac{1}{2}M \times v_e^2 + \frac{1}{2}mv_0^2 = \frac{1}{2}mv^2$$

$$\Rightarrow \quad v^2 = \frac{9}{10}v_e^2 + v_0^2$$

$$\Rightarrow \quad v^2 = \frac{9}{10} \times (11.2)^2 + (12)^2$$

$$\Rightarrow \quad v^2 = 112.896 + 144$$

$$\Rightarrow \quad v = 16.027\,\text{km/s}$$

$$\Rightarrow \quad v = 16\,\text{km/s}$$ **Ans. (16)**

Sol. 36 The orbital speed of body is given as

$$v_0 = \sqrt{\frac{GM}{R}}$$

After collision, if combined mass moves with speed v_1, we use by conservation of momentum

$$mv_0 + \frac{m}{2}\frac{v_0}{2} = \left(\frac{3m}{2}\right)v_1$$

$$\Rightarrow \quad v_1 = \frac{5v_0}{6}$$

As this speed after collision is less than escape speed $\sqrt{2}v_0$, so the combined body cannot from escape gravitational field. So its motion will be elliptical around the planet. **Ans. (A)**

Sol. 37 Escape velocity from a planet is given as

$$v_e = \sqrt{\frac{2GM}{R}}$$

For planets A and B, we use

$$v_A = \sqrt{\frac{2GM}{R}}$$

and $$v_B = \sqrt{\frac{2G(M/2)}{R/2}} = \sqrt{\frac{2GM}{R}}$$

$$\Rightarrow \quad \frac{v_A}{v_B} = 1 = \frac{n}{4} \Rightarrow n = 4$$ **Ans. (A)**

8.1 Elasticity

Sol. 38 Maximum tension in wire is given as

$$T = m\omega^2 l$$

Breaking stress is wire is given as

$$S = \frac{T}{A} = \frac{m\omega^2 l}{A}$$

$$\Rightarrow \quad \omega^2 = \frac{4.8\times10^7\times(10^{-2}\times10^{-4})}{10\times0.3} = 16$$

$$\Rightarrow \quad \omega = 4$$

Ans. (4.00)

Sol. 39 Elastic energy stored per unit volume is given as

$$u = \frac{1}{2}\times \text{stress} \times \text{strain}$$

$$\Rightarrow \quad u = \frac{(\text{stress})^2}{2Y}$$

Given that

$$\frac{u_1}{u_2} = \frac{1}{4}$$

$$\Rightarrow \quad 4u_1 = u_2$$

$$4\left(\frac{1}{Y}\right)\left[\frac{mg}{\pi d_1^2/4}\right]^2 = \left(\frac{1}{2Y}\right)\left[\frac{mg}{\pi d_2^2/4}\right]^2$$

$$\Rightarrow \quad 4 = \left(\frac{d_1}{d_2}\right)^4$$

$$\Rightarrow \quad \frac{d_1}{d_2} = \sqrt{2}:1$$

Ans. (D)

9.1 Fluid Statics

Sol. 40 For minimum density of liquid, sphere will be floating while completely submerged so the buoyant force must be balancing the weight. Mass of sphere is given as

$$m = \int_0^R \rho(4\pi r^2 dr)$$

$$\Rightarrow \quad m = \rho_0\int_0^R\left(1-\frac{r^2}{R^2}\right)4\pi r^2 dr$$

$$\Rightarrow \quad m = 4\pi\rho_0\left[\frac{R^3}{3}-\frac{R^3}{5}\right] = \frac{8}{15}\pi\rho_0 R^3$$

For floating, we use

$$mg = V\rho_l g$$

$$\Rightarrow \quad \frac{8}{15}4\pi\rho_0 R^3 = \frac{4}{3}\pi R^3\rho_l g$$

$$\Rightarrow \quad \rho_l = \frac{2}{5}\rho_0$$

Ans. (D)

Sol. 41 Average pressure on upper part side wall of liquid is $\frac{\rho gh}{2}$, so force on wall MN is given as

$$F_1 = \frac{\rho gh}{2}\times A$$

Average pressure on lower part side wall of liquid is $\left(\rho gh + \frac{2\rho gh}{2}\right)$, so force on side wall ON is given as

$$F_2 = \left(\rho gh + \frac{2\rho gh}{2}\right)A$$

$$\Rightarrow \quad \frac{F_1}{F_2} = \frac{1}{4}$$

Ans. (A)

9.2 Fluid Dynamics

Sol. 42 By continuity equation, we use

$$A_1v_1 = A_2v_2$$

$$\Rightarrow \quad \frac{v_{\min}}{v_{\max}} = \frac{A_{\min}}{A_{\max}}$$

$$\Rightarrow \quad \frac{v_{\min}}{v_{\max}} = \left(\frac{4.8}{6.4}\right)^2$$

$$\Rightarrow \quad \frac{v_{\min}}{v_{\max}} = \frac{9}{16}$$

Ans. (D)

Sol. 43 Rate of flow of water through the tube is given as

$$Q = A_Av_A = A_Bv_B$$

$$\Rightarrow \quad (40)v_A = (20)v_B$$

$$\Rightarrow \quad v_B = 2v_A \qquad \ldots(1)$$

Applying Bernoulli's theorem at A and B, we have

$$P_A + \frac{1}{2}\rho v_A^2 = P_B + \frac{1}{2}\rho v_B^2$$

$$\Rightarrow \quad P_A - P_B = \frac{1}{2}\rho(v_B^2 - v_A^2)$$

$$\Rightarrow \quad 700 = \frac{1}{2}\times1000(4v_A^2 - v_A^2)$$

$$\Rightarrow \quad v_A = 0.68\text{ m/s} = 68\text{ cm/s}$$

$$\Rightarrow \quad Q = A_Av_A$$

$$\Rightarrow \quad Q = (40)(68) = 2720\text{ cm}^3/\text{s}$$

Ans. (C)

9.4 Surface Tension

Sol. 44 For floatation of droplet, net upward force on droplet balances its weight. Net upward force on droplet is

$$F_{\text{up}} = F_{\text{Buoyancy}} + F_{\text{surface tension}}$$

$$\Rightarrow \quad F_{\text{up}} = \left(\frac{2}{3}\pi R^3\right)\rho g + T(2\pi R)$$

For equilibrium of droplet, we use

$$\left(\frac{2}{3}\pi R^3\right)\rho g + T(2\pi R) = d\left(\frac{4}{3}\pi R^3\right)g$$

$$\Rightarrow \quad T(2\pi R) = \left(\frac{2}{3}\pi R^3\right)g\,[2d-\rho]$$

$$\Rightarrow \quad R = \sqrt{\frac{3T}{(2d-\rho)g}}$$ **Ans. (B)**

Force of surface tension can be upward (as given in question) so in this case liquid does not wet the droplet.

10.1 Thermal Expansion

Sol. 45 Coefficient of volume expansion is given as

$$\gamma = \alpha_x + \alpha_y + \alpha_z$$
$$\Rightarrow \quad \gamma = 5\times 10^{-5} + 5\times 10^{-6} + 5\times 10^{-6}$$
$$\Rightarrow \quad \gamma = (50+5+5)\times 10^{-6}$$
$$\Rightarrow \quad \gamma = 60\times 10^{-6}$$
$$\Rightarrow \quad C = 60$$ **Ans. (60)**

Sol. 46 Initially we use

$$\rho_0 S(80)\,g = mg \qquad \ldots(1)$$

later $$\rho S(79)\,g = mg \qquad \ldots(2)$$

$$\Rightarrow \quad \frac{\rho}{\rho_0} = \frac{80}{79} = 1.012$$

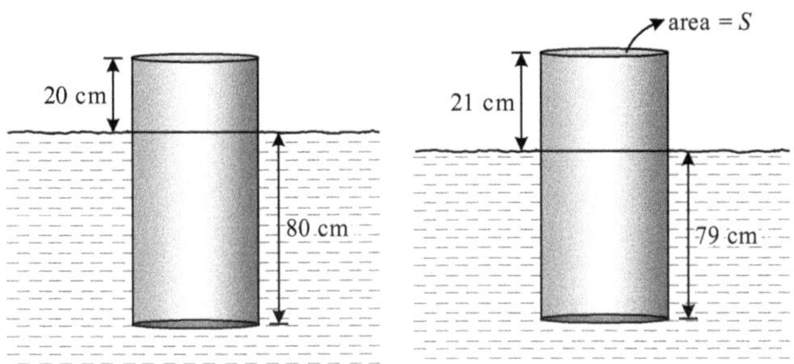

Ans. (A)

10.2 Calorimetry

Sol. 47 For exchange of heat in mixture, we use

Heat Supplied = Heat Gained

$$M\times 540 + M\times 1\times 60 = 200\times 80 + 200\times 1\times 4$$
$$\Rightarrow \quad M = 40\text{ gm}$$ **Ans. (40)**

Sol. 48 According to table using heat exchange between water amounts, we use

Case-I : $1(T_1-60) + 2(T_2-60) = 0$

$$\Rightarrow \quad 1T_1 + 2T_2 = (1+2)60°$$
$$T_1 + 2T_2 = 180 \qquad \ldots(1)$$

Case-II : $2(T_1-60) + 1(T_2-30) = 0$

$$\Rightarrow \quad T_2 + 2T_3 = (1+2)30°$$
$$T_2 + 2T_3 = 90 \qquad \ldots(2)$$

Case-III : $2(T_1-60) + 1(T_3-60) = 0$

$$\Rightarrow \quad 2T_1 + 1T_3 = (1+2)\,60°$$
$$2T_1 + T_3 = 180 \qquad \ldots(3)$$

Adding (1) + (2) + (3) gives

$$3(T_1+T_2+T_3) = 450$$
$$\Rightarrow \quad T_1+T_2+T_3 = 150° \qquad \ldots(4)$$

Case-IV : $(T_1-\theta) + (T_2-\theta) + (T_3-\theta) = 0$

$$\Rightarrow \quad T_1+T_2+T_3 = (1+1+1)\,\theta$$
$$\Rightarrow \quad 150° = 3\theta$$
$$\Rightarrow \quad \theta = 50°\text{C}$$ **Ans. (50)**

11.1 Laws of Thermodynamics

Sol. 49 Work done in adiabatic process is given as

$$W = \frac{nR(T_1-T_2)}{\gamma-1} = \frac{P_1V_1 - P_2V_2}{0.4}$$

$$\Rightarrow \quad W = \frac{100\times 1 - \dfrac{100}{4.6555}\times 3}{0.4} = 88.90$$

Thus closest answer is option (A) **Ans. (A)**

Sol. 50 For equivalent r for mixture we use

$$\frac{n_1+n_2}{r_{eq}-1} = \frac{n_1}{r_1-1} + \frac{n_2}{r_2-1}$$

$$\Rightarrow \quad \frac{5}{r_{eq}-1} = \frac{2}{2/3} + \frac{3}{1/3}$$

$$\Rightarrow \quad r_{eq} = \frac{17}{12} = 1.416$$
$$\simeq 1.42$$ **Ans. (B)**

Sol. 51 For Carnot Cycle, we use

$$\frac{Q_1}{Q_2} = \frac{T_1}{T_2}$$

$$\Rightarrow \quad \frac{Q_2+1200}{Q_2} = \frac{900}{300} \qquad [\text{As } Q_1 = Q_2 + W]$$

$$\Rightarrow \quad Q_2 + 1200 = 3Q$$
$$\Rightarrow \quad Q_2 = 600\text{ J}$$ **Ans. (600)**

Sol. 52 Electric field due to each charged sheet is given as

$$E = \frac{\sigma}{2\varepsilon_0}$$

Which in direction normal to planes as shown in figure

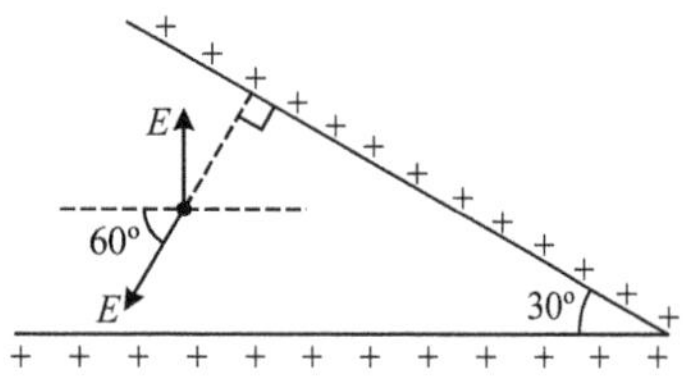

Net electric field between plates is given as

$$\vec{E}_R = E\cos 60° (-\hat{x}) + (E - E\sin 60°)(\hat{y})$$

$$\Rightarrow \quad \vec{E}_R = \frac{\sigma}{2\varepsilon_0}\left[-\frac{\hat{x}}{2} + \left(1 - \frac{\sqrt{3}}{2}\right)\hat{y}\right]$$ **Ans. (B)**

Sol. 53 For the engines, we use

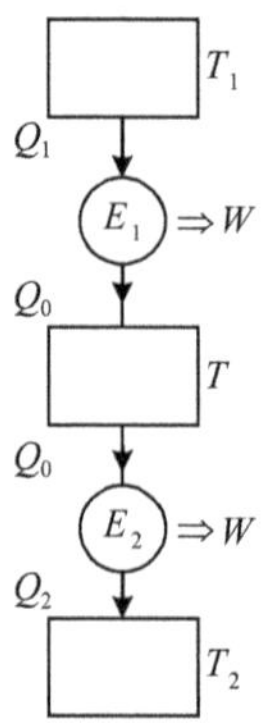

$$\frac{Q}{Q_0} = \frac{T_1}{T}$$

and $$\frac{Q_0}{Q_2} = \frac{T_1}{T}$$

As work is equal by both engines, we use

$$\frac{Q_2 + W}{Q_2} = \frac{T_1}{T}$$

and $$\frac{Q_2}{Q_2 - W} = \frac{T}{T_2}$$

Solving we get

$$T = \frac{T_1 + T_2}{2}$$ **Ans. (C)**

Sol. 54 Mean collision time is taken as

$$t \propto \frac{V}{\sqrt{T}} \qquad \ldots(1)$$

And for adiabatic process, we use

$$TV^{\gamma-1} = \text{constant} \qquad \ldots(2)$$

$$\Rightarrow \quad t \propto V^{\frac{\gamma+1}{2}}$$ **Ans. (A)**

Sol. 55 Given processes are

$x \to y$ Isobaric, $y \to z \Rightarrow$ Isochoric, $z \to x \Rightarrow$ Isothermal

So by slope analysis correct option is (D) **Ans. (D)**

Sol. 56 Efficiency for refrigerator cycle is

$$\eta = \frac{W}{Q_1} = \frac{W}{W + Q_2}$$

$$\frac{1}{10} = \frac{10}{10 + Q_2}$$

$$\Rightarrow \quad Q_2 = 90\text{ J}$$

Heat absorbed from the reservoir at lower temperature is 90 J

Ans. (C)

Sol. 57 C_P/C_V for mixture is given as

$$\gamma_{\text{mixture}} = \frac{n_1 C_{P_1} + n_2 C_{P_2}}{n_1 C_{V_1} + n_2 C_{V_2}}$$

$$\Rightarrow \quad \gamma_{\text{mixture}} = \frac{n \times \left(\frac{5R}{2}\right) + 2n\left(\frac{7R}{2}\right)}{n \times \frac{3R}{2} + 2n\left(\frac{5R}{2}\right)}$$

$$\Rightarrow \quad \gamma_{\text{mixture}} = \frac{19}{13}$$ **Ans. (B)**

Sol. 58 For a diatomic gas molecule there are 5 degrees of freedom if vibration mode is absent and if vibration mode is there then degrees of freedom will be 7 so specific heats in the two cases are $C_V^A = \frac{5}{2}R$ and $C_V^B = \frac{7}{2}R \Rightarrow \frac{C_V^A}{C_V^B} = \frac{5}{7}$.

Ans. (B)

Sol. 59 In the given *PV* diagram process 2 to 3 is isobaric & process 3 to 1 is isochoric. This is shown only in option (D).

Ans. (D)

Sol. 60 For an adiabatic process, we use

$$PV^{\gamma} = \text{constant}$$

and $$TV^{\gamma-1} = \text{C}$$

For the given situation, we use

$$300 \times V^{\frac{7}{5}-1} = T_2\left(\frac{V}{16}\right)^{\frac{7}{5}-1}$$

$$\Rightarrow \quad T_2 = 300 \times 2^{4\times\frac{2}{5}}$$

For isobaric process, we use

$$\frac{V_2}{T_2} = \frac{V_3}{T_3}$$

Using $V_3 = 2V_2$, we get

$$\frac{1}{2} = \frac{T_2}{T_3} \Rightarrow T_3 = 2T_2$$

$$\Rightarrow \quad T_3 = 2 \times 300 \times 2^{\frac{8}{5}} = 1818.85$$

Ans. (1818.00 to 1819)

12.1 Kinetic Theory of Gases

Sol. 61 Mean free time $\propto \frac{1}{\text{Mean speed}}$

$$\Rightarrow \quad t \propto \frac{1}{\sqrt{T}}$$ **Ans. (D)**

Sol. 62 Mean free path is given as

$$\lambda = \frac{1}{\sqrt{2}\pi n_v d^2}$$

Mean time between successive collision is given as

$$\tau = \frac{\lambda}{v_m} = \frac{1}{\sqrt{2}\pi n_v d^2 v_m} = \frac{1}{\sqrt{2}\pi n_v d^2}\sqrt{\frac{\pi M}{8RT}}$$

For two gases, we use

$$\frac{\tau_1}{\tau_2} = \sqrt{\frac{M_1}{M_2}\frac{d_2^2}{d_1^2}}$$

$$\Rightarrow \quad \frac{\tau_1}{\tau_2} = \sqrt{\frac{40}{140}\frac{(0.1)^2}{(0.07)^2}} = 1.09$$

Thus most appropriate answer among given option is (C).

Ans. (C)

13.1 Simple Harmonic Motion

Sol. 63 Figure below shows the given situation

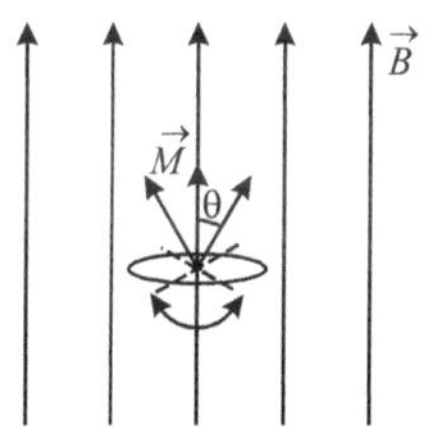

If loop is slightly tilted from equilibrium position ($\theta = 0°$) then restoring torque on it is given as

$$\tau = -MB\sin\theta$$

$$\Rightarrow \quad \tau \simeq -MB\,\theta$$

Thus restoring angular acceleration is given as

$$\alpha = -\left(\frac{MB}{I}\right)\theta$$

Comparing with $\alpha = \omega^2\theta$, we get

$$\Rightarrow \quad \omega = \sqrt{\frac{MB}{I}} = \sqrt{\frac{(i)(\pi R^2)B}{\left(\frac{mR^2}{2}\right)}}$$

$$\Rightarrow \quad \omega = \sqrt{\frac{2i\pi B}{m}}$$

Thus time period of oscillations is

$$T = \frac{2\pi}{\omega} = \sqrt{\frac{2\pi m}{iB}}$$

Ans. (B)

14.1 Wave Motion

Sol. 64 Velocity of waves on wire is given as

$$v = \sqrt{\frac{T}{\mu}}$$

$$\Rightarrow \quad 90 = \sqrt{\frac{\frac{YA}{l}\Delta l}{\frac{m}{l}}} = \sqrt{\frac{16\times10^{11}\times10^{-6}\times\Delta l}{6\times10^{-3}}}$$

$$\Rightarrow \quad \Delta l = \frac{8100\times3}{8}\times10^{-8} \cong 3\times10^{-5}\,\text{m}$$

$$= 0.03\,\text{mm}$$

Ans. (C)

Sol. 65 Velocity of transverse wave in a string is $v \propto \sqrt{T}$. When velocity is changed to $v/2$ then tension changes to $T/4$, so new tension is

$$T' = \frac{2.06\times10^4}{4}$$

$$\Rightarrow \quad T' = 5.15\times10^3\,\text{N}$$

Ans. (B)

14.2 Stationary Waves and Beats

Sol. 66 Speed of sound in air is given as

$$v_s = \sqrt{\frac{\gamma P}{\rho}}$$

$$\Rightarrow \quad \frac{v_{gas}}{v_{air}} = \sqrt{\frac{\rho_{air}}{\rho_{gas}}}$$

$$\Rightarrow \quad \frac{v_{gas}}{300} = \frac{1}{\sqrt{2}}$$

$$\Rightarrow \quad v_{gas} = \frac{300}{\sqrt{2}}\,\text{m/s}$$

$$\Rightarrow \quad v_{gas} = 150\sqrt{2}\,\text{m/s}$$

For open organ pipe, we use

$$n_2 - n_1 = \frac{v_{gas}}{2l} = \frac{150\sqrt{2}}{2(1)} = 75\sqrt{2}\ \text{Hz}$$

$$\Rightarrow \quad \Delta n = 106.06\,\text{Hz}$$

Ans. (106.06)

Sol. 67 For n^{th} harmonic, frequency is given as

$$\frac{nv}{2l} = 420$$

For next $(n+1)^{th}$ harmonic, we use

$$\frac{(n+1)v}{2l} = 490$$

$$\Rightarrow \quad \frac{v}{2l} = 70$$

$$\Rightarrow \quad l = \frac{v}{140} = \frac{1}{140}\sqrt{\frac{540}{6\times10^{-3}}} = \frac{1}{140}\sqrt{90\times10^3}$$

$$\Rightarrow \quad l = \frac{300}{140} = 2.142\,\text{m}$$

Ans. (D)

14.3 Doppler's Effect

Sol. 68 

Frequency from fork A

$$\nu_A = \left(\frac{c}{c-v}\right)\nu_0$$

Frequency from fork B

$$\nu_B = \left(\frac{c}{c+v}\right)\nu_0$$

Beat frequency

$$\Rightarrow \quad f_B = \nu_1 - \nu_2$$

$$\Rightarrow \quad f_B = c\nu_0\left(\frac{1}{c-v} - \frac{1}{c+v}\right)$$

$$\Rightarrow \quad f_B = c\nu_0\left(\frac{c+v-c+v}{c^2-v^2}\right) = \frac{2c\nu_0^2 v}{c^2-v^2}$$

$$\Rightarrow \quad f_B \approx \frac{2c\nu_0 v}{c^2}$$

$$\Rightarrow \quad f_B = \frac{2\nu_0 v}{c}$$

$$\Rightarrow \quad 2 = \frac{2\times 1400 \times v}{350}$$

$$\Rightarrow \quad v = \frac{1}{4}\ \text{m/s}$$

Ans. (D)

15.1 Electric Force and Electric Field

Sol. 69 Summing up all components of electric field along X-direction, we get

$$E_x = \frac{K(4q)}{R^2}\cos 30^\circ + \frac{K(2q)}{R^2}\cos 30^\circ + \frac{K(2q)}{R^2}\cos 30^\circ$$

$$\Rightarrow \quad E_x = (2\sqrt{3}+\sqrt{3}+\sqrt{3})\ \text{kg/R}^2 = \frac{\sqrt{3}q}{\pi\varepsilon_0 d^2}$$

Ans. (D)

Sol. 70 Speed after travelling a distance x is given as

$$v^2 = u^2 + 2ax$$

$$\Rightarrow \quad v^2 = 0 + 2\left(\frac{qE}{m}\right)x$$

$$\Rightarrow \quad v^2 = \frac{2qE}{m}x$$

$$\Rightarrow \quad v \propto \sqrt{x}$$

Thus option (C) is correct. **Ans. (C)**

Sol. 71 If we fill the empty space of cavity with $+\rho$ and $-\rho$ charge density in this region, the electric fields at points A and B are

$$E_A = 0 + \frac{k\rho.\frac{4}{3}\pi\left(\frac{R}{2}\right)^3}{\left(\frac{R}{2}\right)^2} = k\rho\,\frac{4}{3}\pi\left(\frac{R}{2}\right) \qquad \ldots(1)$$

$$\text{and } E_B = \frac{k\rho\cdot\frac{4}{3}\pi R^3}{R^2} - \frac{k\rho\cdot\frac{4}{3}\pi\left(\frac{R}{2}\right)^3}{\left(\frac{3R}{2}\right)^2}$$

$$\Rightarrow \quad E_B = k\rho\frac{4}{3}\pi R - k\rho\frac{4}{3}\pi\frac{R}{18} = k\rho\cdot\frac{4}{3}\pi\left(\frac{17R}{18}\right) \qquad \ldots(2)$$

Using equation-(1) and (2), we have

$$\frac{E_A}{E_B} = \frac{9}{17} = \frac{18}{34}$$

Ans. (D)

15.2 Electric Flux and Gauss's Law

Sol. 72 By Gauss's law we use

$$\oint \vec{E}\cdot\overrightarrow{dA} = \frac{q_{enc}}{\varepsilon_0}$$

If $\vec{E}$ is normal to surface and uniform over the surface then we can use

$$\oint \vec{E}\cdot\overrightarrow{dA} = EA$$

$$\Rightarrow \quad E = \frac{q_{enc}}{\varepsilon_0 A}$$

Thus option-(D) is correct. **Ans. (D)**

Sol. 73 Electric flux through face $ABCD$ is

$$\phi_I = 0 \quad \text{as} \quad E_z = 0$$

Electric flux through face $BCGF$ is

$$\phi_{II} = E_x\cdot S = (4\times 3)\times(2\times 2) = 48\ \text{Nm}^2/\text{C}$$

$$\Rightarrow \quad \phi_I - \phi_{II} = 0 - 48 = -48\ \text{Nm}^2/\text{C}$$

Ans. (–48.00)

16.1 Electric Potential and Dipole

Sol. 74 Electric fields on surfaces of spheres are

$$E_1 = \frac{KQ_1}{R_1^2}$$

$$E_2 = \frac{KQ_2}{R_2^2}$$

Given that ratio of electric fields is

$$\frac{E_1}{E_2} = \frac{R_1}{R_2}$$

$$\Rightarrow \quad \frac{\frac{KQ_1}{R_1^2}}{\frac{KQ_2}{R_2^2}} = \frac{R_1}{R_2}$$

$$\Rightarrow \quad \frac{Q_1}{Q_2} = \frac{R_1^3}{R_2^3}$$

$$\Rightarrow \quad \frac{V_1}{V_2} = \frac{KQ_1/R_1}{KQ_2/R_2} = \frac{R_1^2}{R_2^2}$$

Ans. (D)

Sol. 75 As $\vec{r}$ and $\vec{p}$ are perpendicular to each other so the point is on the equatorial plane and electric field at this point will be in direction opposite to the dipole moment so we use

$$\hat{E} = -\hat{p}$$

Thus $\vec{E}$ is parallel to $-\vec{p}$ **Ans. (C)**

16.2 Capacitance

Sol. 76 Here we use,

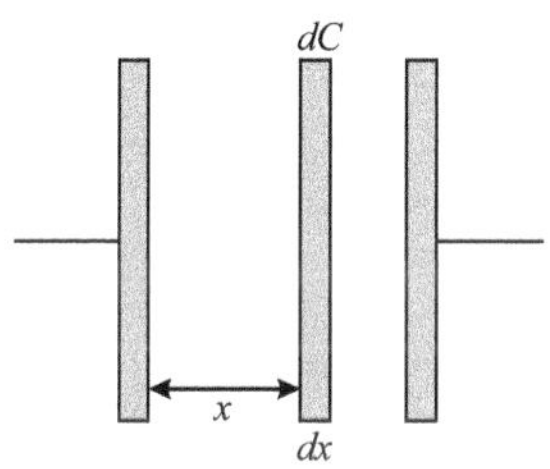

$$dC = \frac{(A)K(1+\alpha x)\varepsilon_0}{dx}$$

Now all such elemental capacitors are is series so we use,

$$\frac{1}{C} = \int \frac{1}{dC} = \int_0^d \frac{dx}{AK\varepsilon_0(1+\alpha x)}$$

$$\Rightarrow \quad \frac{1}{C} = \int_0^d \frac{1}{AK\varepsilon_0} \cdot (1-\alpha x)dx$$

$$\Rightarrow \quad \frac{1}{C} = \frac{1}{AK\varepsilon_0}\left[d - \frac{\alpha d^2}{2}\right] = \frac{d}{AK\varepsilon_0}\left(1-\frac{\alpha d}{2}\right)$$

$$\Rightarrow \quad \frac{1}{C} = \frac{d}{AK\varepsilon_0}\left(1-\frac{\alpha d}{2}\right)$$

$$\Rightarrow \quad C = \frac{AK\varepsilon_0}{d}\left(1+\frac{\alpha d}{2}\right)$$

Ans. (A)

Sol. 77 After second connection charge is equally distributed in both capacitors

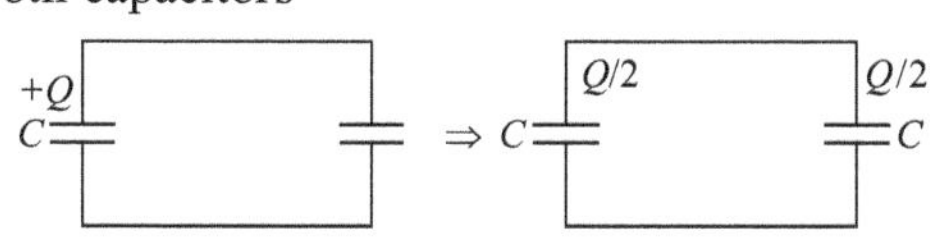

$$Q = CV$$

Energy loss is given as

$$\Delta Q_L = \frac{Q^2}{2C} - \left[\frac{(Q/2)^2}{2C} \times 2\right] = \frac{Q^2}{4C}$$

$$\Rightarrow \quad \Delta Q_L = \frac{1}{4}CV^2$$

$$\Rightarrow \quad \Delta Q_L = \frac{1}{4} \times 60 \times 10^{-12} \times 4 \times 10^2$$

$$\Rightarrow \quad \Delta Q_L = 6 \text{ nJ}$$

Ans. (6)

Sol. 78 In parallel combination, we use

$$C_1 + C_2 = 10 \quad \ldots(1)$$

For energy stored, it is given that

$$\frac{1}{2}C_2V^2 = 4 \times \frac{1}{2}C_1V^2$$

$$\Rightarrow \quad C_2 = 4C_1 \quad \ldots(2)$$

$$\Rightarrow \quad C_1 = 2$$

$$\& \quad C_2 = 8$$

For series combination, we use

$$C_{eq} = \frac{C_1C_2}{C_1+C_2} = 1.6\ \mu\text{F}$$

Ans. (C)

Sol. 79 Assume small element dx at a distance x from left end

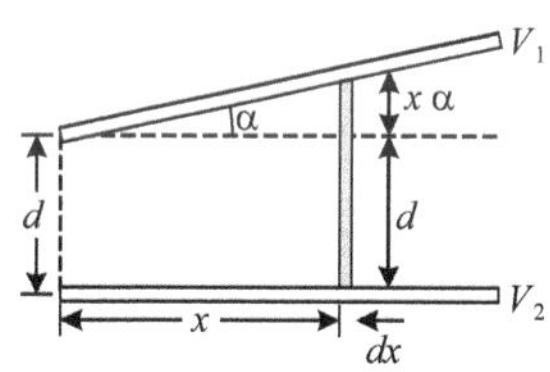

Capacitance of this element is given as

$$dC = \frac{\varepsilon_0(adx)}{d+x\alpha}$$

Thus total capacitance is given as

$$C = \int_0^a \frac{\varepsilon_0 adx}{d+x\alpha}$$

$$\Rightarrow \quad C = \frac{\varepsilon_0 a}{\alpha}\ln\left(\frac{1+\alpha a}{d}\right)\Bigg|_0^a \quad \left(\text{We use } \ln(1+x) \approx x - \frac{x^2}{2}\right)$$

$$\Rightarrow \quad C = \frac{\varepsilon_0 a^2}{d}\left(1-\frac{\alpha a}{2d}\right).$$

Ans. (D)

17.2 Circuit Analysis

Sol. 80 Equivalent resistance of upper branch of circuit is 2.5 Ω so total current in upper branch is 1/2.5 = 0.4A which divides equally in two 1Ω resistances so through each current is 0.2 A.

Ans. (B)

Sol. 81 Figure below shows the circuit for the described situation

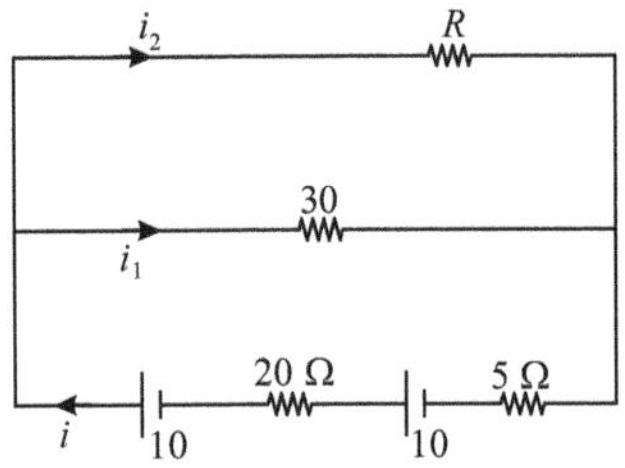

In above circuit, we use potential difference across batteries

$$E_1 = E - ir$$

$\Rightarrow$ $$E_1 = E - ir$$

$\Rightarrow$ $$i = 0.5\,\text{A}$$

and $$E_2 = E - ir$$

$\Rightarrow$ $$E_2 = 10 - 0.5 \times 5$$

$\Rightarrow$ $$E_2 = 7.5\,\text{V}$$

Using KCL at right most junction, we have

$$0.5 = \frac{7.5}{R} + \frac{7.5}{30}$$

$$R = 30\,\Omega$$ **Ans. (30)**

Sol. 82 Considering potentials as shown in circuit below we write *KLL* equation for *x*

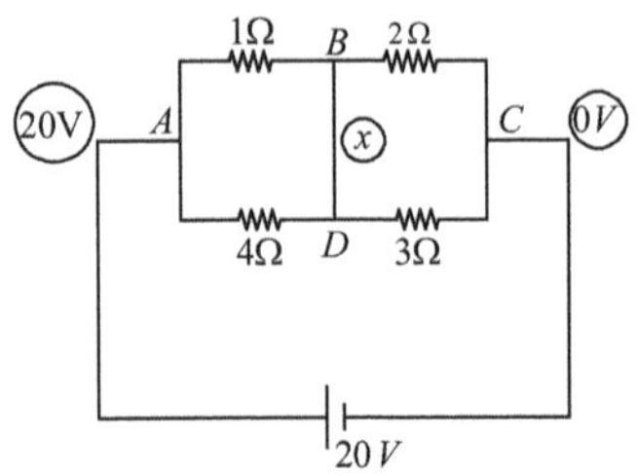

$$\frac{x}{2} + \frac{x}{3} + \frac{x-20}{1} + \frac{x-20}{4} = 0$$

$\Rightarrow$ $$6x + 4x + 12x - 240 + 3x - 60 = 0$$

$\Rightarrow$ $$25x = 300$$

$\Rightarrow$ $$x = 12\text{V}$$

Thus current in 1Ω and 2Ω resistance are

$$I_{1\Omega} = \frac{20-12}{1} = 8A$$

and $$I_{2\Omega} = \frac{12}{2} = 6A$$

Current in branch *BD* is

$$I_{BD} = I_{1\Omega} - I_{2\Omega} = 2A$$ **Ans. (B)**

17.3 Thermal Effects of Current & RC Circuits

Sol. 83 Total power consumption with current *I* at full load of building is

$$220I = P = 15 \times 45 + 15 \times 100 + 15 \times 10 + 2 \times 10^3$$

$\Rightarrow$ $$I = \frac{4325}{220} = 19.66$$

$\Rightarrow$ $$I \approx 20\,\text{A}$$ **Ans. (D)**

17.4 Electrical Measuring Instruments

Sol. 84 Internal resistance of cell by potentiometer is given as

$$r = R\left(\frac{l_1 - l_2}{l_2}\right)$$

$\Rightarrow$ $$r = 10 \times \frac{60}{500} r$$

$\Rightarrow$ $$r = 1.2\,\Omega = \frac{12}{10}\Omega$$ **Ans. (12)**

Sol. 85 If λ is potential gradient of potentiometer wire & *L* is total length of wire, then we use

$$5 = \lambda l$$

$\Rightarrow$ $$5 = \frac{\Delta V}{L} l$$

$\Rightarrow$ $$\Delta V = \frac{5 \times L}{l} = 5 \times \frac{12}{10} = 6\text{V}$$

By Ohm's Law, we use

$$\Delta V = IR$$

$\Rightarrow$ $$R = \frac{\Delta V}{I} = \frac{6}{60 \times 10^{-3}} = 100\,\Omega$$ **Ans. (D)**

Sol. 86 For balancing the wheatstone network as shown in figure, we use

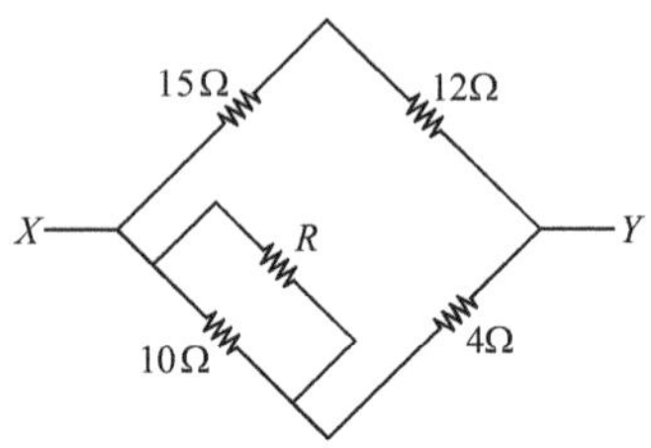

$$15 \times 4 = 12 \times \frac{10R}{10+R}$$

$\Rightarrow$ $$R = 10\,\Omega$$ **Ans. (10.00)**

Sol. 87 For the given galvanometer

$$I_g = 1\,\text{mA},$$

$$R_g = 100\,\Omega$$

To convert it into a voltmeter, we use shunt resistance *R*, so we have

$$V = I_g(R + R_g)$$

$\Rightarrow$ $$10 = 1 \times 10^{-3}(R + 100)$$

$\Rightarrow$ $$R = 9.9\,\text{k}\Omega$$ **Ans. (A)**

Sol. 88 In balancing condition we use

$$\frac{R}{S} = \frac{25}{75}$$

If R is replaced by a wire half length and one fourth cross-section then resistance becomes $2R$, so we use

$$\frac{2R}{S} = \frac{l'}{100-l'}$$

$$\Rightarrow \quad 2\times\frac{1}{3} = \frac{l'}{100-l'} = 3l' = 200-2l'$$

$$\Rightarrow \quad 5l' = 200$$

$$\Rightarrow \quad l' = 40$$

Ans. (40.00)

18.1 Magnetic Effects of Current

Sol. 89 The given expression is of magnetic energy stored per unit volume so its dimensions are $ML^{-1}T^{-2}$. **Ans. (A)**

Sol. 90 The given wire loop can be considered in three parts A, B and C as shown

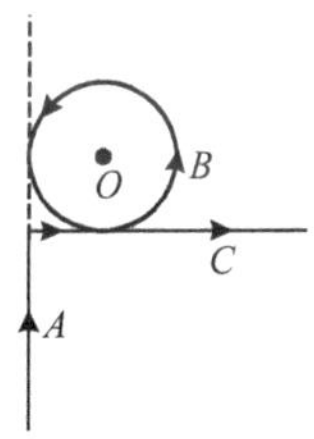

Magnetic field at point O is given as

$$B_0 = B_A + B_B + B_C$$

$$\Rightarrow \quad B_0 = \frac{\mu_0 I}{4\pi R}(\sin 90° - \sin 45°)\otimes + \frac{\mu_0 I}{2R}\odot + \frac{\mu_0 I}{4\pi R}(\sin 45° + \sin 90°)\odot$$

$$\Rightarrow \quad B_0 = \frac{\mu_0 I}{2\pi R}(\sin 45° + \pi)$$

$$\Rightarrow \quad B_0 = \frac{\mu_0 I}{2\pi R}\left(\pi + \frac{1}{\sqrt{2}}\right)$$

Ans. (C)

Sol. 91 If current density in wire is J, then magnitude field due to wire is outside is given as

$$B_1 = \frac{\mu_0 I}{2\pi r} = \frac{\mu_0 (J\pi a^2)}{2\pi(2a)} = \frac{\mu_0 Ja}{4}$$

Due to wire inside field is given as

$$B_2 = \frac{1}{2}\mu_0 Jr = \frac{1}{2}\mu_0 J\left(\frac{a}{3}\right) = \frac{\mu_0 Ja}{6}$$

$$\Rightarrow \quad \frac{B_2}{B_1} = \frac{\mu_0 Ja\times 4}{\mu_0 J6a} = \frac{2}{3}$$

Ans. (A)

18.2 Magnetic Force on Moving Charges

Sol. 92 We use, after time t

$$(2v_0)^2 = v_0^2 + v_x^2$$

$$\Rightarrow \quad v_x = \sqrt{3}\, v_0$$

As particle is only accelerated by electric field in X-direction, we use

$$\sqrt{3}\, v_0 = 0 + \frac{qE_0}{m}t$$

$$\Rightarrow \quad t = \frac{\sqrt{3}v_0 m}{qE_0}$$

Ans. (C)

Sol. 93 Acceleration of protons is

$$a = \frac{qvB}{m}$$

$$\Rightarrow \quad B = \frac{ma}{qv}$$

$$\Rightarrow \quad B = \frac{ma\sqrt{m}}{\sqrt{2k}}$$

$$\Rightarrow \quad B = \frac{m^{3/2}a}{e\sqrt{2k}} = \frac{(1.6\times10^{-27})^{3/2}\times10^{12}}{1.6\times10^{-19}\sqrt{2\times1\times10^6\times1.6\times10^{-19}}}$$

$$\Rightarrow \quad B = 0.71\,\text{mT}$$

Ans. (D)

Sol. 94 As work is only done by EF on charge particle, we use

$$qE(2a) = \frac{1}{2}m(2v)^2 - \frac{1}{2}mv^2$$

$$\Rightarrow \quad qE(2a) = \frac{3}{2}mv^2$$

$$\Rightarrow \quad E = \frac{3}{4}\frac{mv^2}{qa}$$

Thus option (A) is correct.

Rate of work done by electric field is

$$P = \vec{F}\cdot\vec{v} = Fv\cos\theta = Fv$$

$$P = qEv$$

$$\Rightarrow \quad P = q\left(\frac{3}{4}\frac{mV^2}{qa}\right)v$$

$$\Rightarrow \quad P = q\left(\frac{3}{4}\frac{mv^3}{qa}\right)$$

$$\Rightarrow \quad P = \frac{3}{4}\frac{mv^3}{a}$$

Thus option (B) is correct.

The angle between electric force and velocity at point Q is 90°, so rate of work done will be zero at Q.

Thus option (C) is correct.

Initial angular momentum about origin is

$$L_i = mva$$

Final angular momentum about origin is

$$L_f = m(2v)(2a)$$

Change in angular momentum $L_f - L_i = 3mva$

Thus option (D) is NOT correct. **Ans. (B)**

Sol. 95 For not to hit the surface of solenoid the radius of electron motion inside must be less than $R/2$, so we use

$$\frac{R}{2} = \frac{mv}{qB}$$

$$\Rightarrow \quad \frac{R}{2} = \frac{mv_{max}}{e(\mu_0 ni)}$$

$$\Rightarrow \quad v_{max} = \frac{e\mu_0 nIR}{2m}$$

Ans. (B)

19.2 Magnetic Properties of Materials

Sol. 96 By definition and from figure, we get

Retentivity = 1.0 T

Coercivity = 50 A/m

Saturation = 1.5 T **Ans. (B)**

20.1 Electromagnetic Induction

Sol. 97

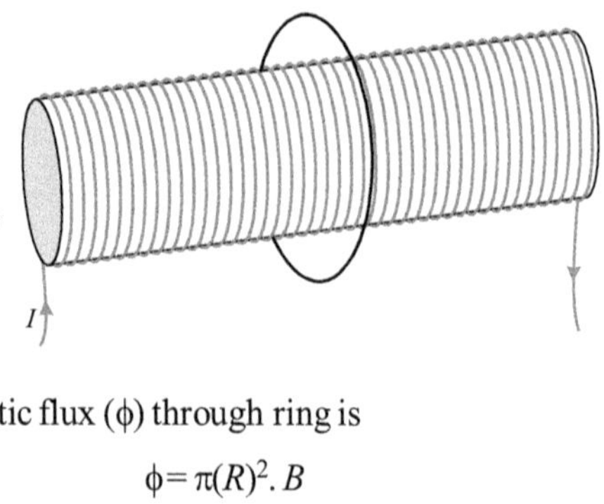

Magnetic flux (ϕ) through ring is

$$\phi = \pi(R)^2 . B$$

$$\Rightarrow \quad \phi = (\pi R^2)(\mu_0 nI)$$

$$\Rightarrow \quad \phi = (\pi R^2 \mu_0 nI_0)(t - t^2)$$

Induced emf is given as

$$e = \frac{d\phi}{dt}$$

$$\Rightarrow \quad e = (\pi R^2 \mu_0 nI_0)(2t - 1)$$

Induced current is given as

$$I_R = \frac{\pi R^2 \mu_0 nI_0 (2t-1)}{R_{ring}}$$

In above expressive, we can see that

e and I_R are zero at $t = \frac{1}{2} = 0.5$ s

and their sign also changes at $t = 0.5$ s **Ans. (A)**

Sol. 98 Magnetic flux which comes out from the area enclosed by coil is same which goes into the surrounding plane of coil from area outside it. **Ans. (A)**

Sol. 99 Figure shows the loop formed by given corner coordinates

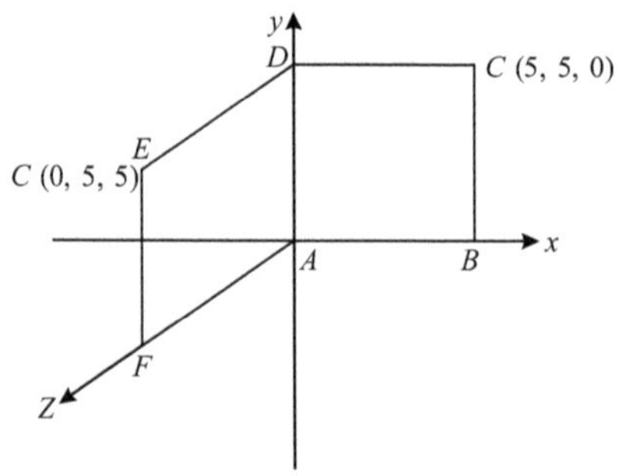

Area vector of loop is given as

$$\vec{A}_{ABCD} = 25\hat{k}$$

$$\vec{A}_{ADEF} = 25\hat{i}$$

$$\vec{A}_{net} = 25\hat{i} + 25\hat{k}$$

Magnetic flux through loop is given as

$$\phi = \vec{B} \cdot \vec{A}$$

$$\Rightarrow \quad \phi = 25 \times 3 + 25 \times 4$$

$$\Rightarrow \quad \phi = 175 \; Wb$$

Ans. (175)

Sol. 100 For a rotating coil in uniform magnetic field. Induced emf is given as $|e| = BA\omega \sin \omega t$

emf will be maximum at $\omega t = \frac{\pi}{2}$

$$\Rightarrow \quad \left(\frac{2\pi}{T}\right)t = \frac{\pi}{2}$$

$$\Rightarrow \quad \left(\frac{2\pi}{10}\right)t = \frac{\pi}{2}$$

$$\Rightarrow \quad t = 2.5 \text{ s}$$

and emf will be minimum at $\omega t = \pi$

$$\Rightarrow \quad \left(\frac{2\pi}{10}\right)t = \pi$$

$$\Rightarrow \quad t = 5 \text{ s}$$

Ans. (D)

Sol. 101 For linear reduction in magnetic field, we use

$$\frac{dB}{dt} = \frac{1000-500}{5} = 100\text{ G/s}$$

$$= 100 \times 10^{-4}\text{ T/s}$$

Area of loop is

$$A = 16 \times 4 - 4 \times 2 = 56\text{ cm}^2$$

EMF induced in loop is

$$\varepsilon = \frac{dB}{dt}A = 100 \times 10^{-4} \times 56 \times 10^{-4} = 56\,\mu\text{V}$$

Ans. (C)

20.2 Self and Mutual Induction

Sol. 102 Growth of current in RL circuit is given as

$$i = i_0(1 - e^{-Rt/L})$$

$$\Rightarrow \quad \frac{i_0}{i} = \frac{1}{1-e^{-2\times10^4}}$$

$$\Rightarrow \quad \frac{i_0}{i} \simeq 1$$

Ans. (A)

Sol. 103 For current growth in LR circuit, we use

$$i = i_0(1 - e^{-Rt/L}) = i_0(1 - e^{-t/T_C})$$

Charge flow through circuit is given as

$$q = \int_0^{T_C} i\,dt$$

$$\Rightarrow \quad q = \int_0^{T_C} \frac{\varepsilon}{R}(1 - e^{-t/T_C})$$

$$\Rightarrow \quad q = \frac{\varepsilon}{R}\left(t - \frac{e^{-t/T_C}}{-1/T_C}\right)\Bigg|_0^{T_C}$$

$$\Rightarrow \quad q = \frac{e}{R}(T_C - T_C e^{-1}) - \frac{e}{R}(0 + T_C)$$

$$\Rightarrow \quad q = \frac{e}{R} \times T_C e^{-1}$$

$$\Rightarrow \quad q = \frac{\varepsilon}{R} \times \frac{L}{R}\frac{1}{e}$$

$$\Rightarrow \quad q = \frac{\varepsilon L}{eR^2}$$

Ans. (D)

Sol. 104 Induced voltage in choke coil is given as

$$V = L\left|\frac{di}{dt}\right|$$

$$\Rightarrow \quad L = \frac{V}{\left|\frac{di}{dt}\right|} = \frac{100}{\frac{0.25}{0.025\times10^{-3}}} = 10\text{ mH}$$

Ans. (10.00)

21.1 Alternating Current

Sol. 105 For LCR circuit by KVL, we use

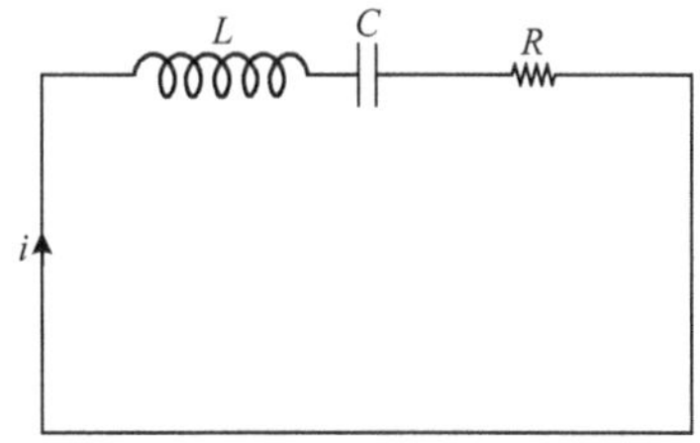

$$-L\frac{di}{dt} - \frac{q}{C} - iR = 0$$

$$L\frac{d^2q}{dt^2} + \frac{1}{C}q + R\frac{dq}{dt} = 0$$

For damped oscillator net force

$$= -kx - bv = ma$$

$$\frac{md^2x}{dt^2} + kx + \frac{bdx}{dt} = 0$$

by comparing; Equivalence is

$$L \to m;\ C \to \frac{1}{k};\ R \to b$$

Ans. (A)

Sol. 106 Figure below shows the circuit for situation described in question

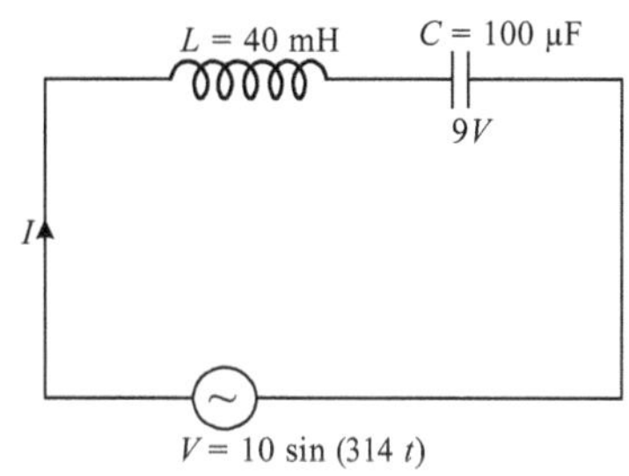

Here reactance of L and C are given as

$$X_L = \omega L = 314 \times 10 \times 10^{-3} = 12.56\,\Omega$$

$$X_C = \frac{1}{\omega C} = \frac{1}{314\times100\times10^{-6}} = \frac{10^4}{314} = 31.84\,\Omega$$

As $\overline{X_C}$ and $\overline{X_L}$ are in opposite phase we use

$$\overline{Z} = \overline{X_C} - \overline{X_L}$$

$$Z = 31.84 - 12.56 = 19.28\,\Omega$$

Peak current in circuit is given as

$$I = \frac{10}{19.28} = 0.52\text{ A}$$

As $X_C > X_L$, in this case current will lease voltage by $\pi/2$ so it is given as

$$I = 0.52\sin\left(314t + \frac{\pi}{2}\right)$$

Ans. (A)

22.1 Electromagnetic Wave

Sol. 107 Given that $\vec{B}$ is along $\hat{j}$ and wave is travelling in X-direction so using

$$\vec{E}\times\vec{B} = \vec{C} = -\hat{i}$$

$\Rightarrow$ $\vec{E}$ is along $\hat{k}$ and its magnitude is given as

$$E = 3\times10^{-8}\times3\times10^{8} = 9\text{ V/m}$$ **Ans. (A)**

Sol. 108 Force on a charge particle is given as

$$\vec{F} = q(\vec{E}+\vec{v}\times\vec{B})$$

Electric field at location of charge is

$$\vec{E} = E_0\left(\frac{\hat{i}+\hat{j}}{\sqrt{2}}\right)\cos\pi = -E_0\left(\frac{\hat{i}+\hat{j}}{\sqrt{2}}\right)$$

For EM wave, we use

$$\vec{E}\times\vec{B} = \vec{c}$$

$$\Rightarrow \quad +E_0\left(\frac{\hat{i}+\hat{j}}{\sqrt{2}}\right)\times\vec{B} = c\hat{k}$$

$$\Rightarrow \quad \vec{B} = -\left(\frac{\hat{i}-\hat{j}}{\sqrt{2}}\right)\frac{E_0}{c}$$

Thus force on charge is given as

$$\vec{F} = q\left(-E_0\left(\frac{(\hat{i}+\hat{j})}{\sqrt{2}}\right) - v_0\hat{k}\times\left(\frac{(\hat{i}-\hat{j})}{\sqrt{2}}\right)\frac{E_0}{c}\right)$$

$$\Rightarrow \quad \vec{F} \simeq -qE_0\left(\frac{\hat{i}+\hat{j}}{\sqrt{2}}\right) \qquad [\text{As } v_0 << c]$$

$\Rightarrow$ F is antiparallel to $\dfrac{\hat{i}+\hat{j}}{\sqrt{2}}$ **Ans. (C)**

Sol. 109 For an electromagnetic wave, we use

$$\vec{E} = \vec{B}\times\vec{c}$$

$$\Rightarrow \quad \vec{E} = (5\times10^{-8}\,\hat{j})\times(3\times10^{8}\,\hat{k})$$

$$\Rightarrow \quad \vec{E} = 15\,\hat{i}\text{ V/m}$$ **Ans. (B)**

Sol. 110 Electric field vector is given as

$$\vec{E}_1 = E_0\,\hat{j}\cos(\omega t - kx)$$

Its corresponding magnetic field will be

$$\vec{B}_1 = \frac{E_0}{c}\,\hat{k}\cos(\omega t - kx)$$

and for second wave corresponding fields are

$$\vec{E}_2 = E_0\,\hat{k}\cos(\omega t - ky)$$

$$\vec{B}_2 = \frac{E_0}{c}\,\hat{i}\cos(\omega t - ky)$$

Net force on charge particle is given as

$$\Rightarrow \vec{F} = q\vec{E}_1 + q\vec{E}_2 + q\vec{v}\times B_1 + q\vec{v}\times B_2$$

$$\Rightarrow \vec{F} = qE_0\,\hat{j} + qE_0\,\hat{k} + q(0.8c\,\hat{j})\times\left(\frac{E_0}{c}\hat{k}\right) + q(0.8c\,\hat{j})\times\left(\frac{E_0}{c}\hat{i}\right)$$

$$\Rightarrow \vec{F} = qE_0\,\hat{j} + qE_0\,\hat{k} + 0.8qE_0\,\hat{i} - 0.8qE_0\,\hat{k}$$

$$\Rightarrow \vec{F} = qE_0\,[0.8\,\hat{i} + 1\,\hat{j} + 0.2\,\hat{k}]$$ **Ans. (C)**

Sol. 111 Direction of polarisation of wave is given as

$$\hat{E} = \hat{k}$$

Direction of propagation

$$\hat{c} = \hat{E}\times\hat{B} = \frac{\hat{i}+\hat{j}}{\sqrt{2}}$$

$$\Rightarrow \quad \hat{B} = \frac{\hat{i}-\hat{j}}{\sqrt{2}}$$ **Ans. (A)**

23.1 Geometrical Optics I - Reflection of Light

Sol. 112 By mirror formula, we have

$$\frac{1}{v}+\frac{1}{u} = \frac{1}{f}$$

At focus $\quad m = \infty$ and $x = f$

At centre $\quad m = -1$ and $x = 2f$

Thus option (B) is most appropriate. **Ans. (B)**

23.2 Geometrical Optics II - Refraction of Light

Sol. 113 Critical angle for a medium is given as

$$\sin\theta_C = \frac{1}{\mu} = \frac{1}{\sqrt{3\times4/3}}$$

[where Refractive index $\mu = \sqrt{\mu_0\varepsilon_0}$]

$$\theta_C = 30°$$ **Ans. (D)**

Sol. 114 Figure shows the situation described in question

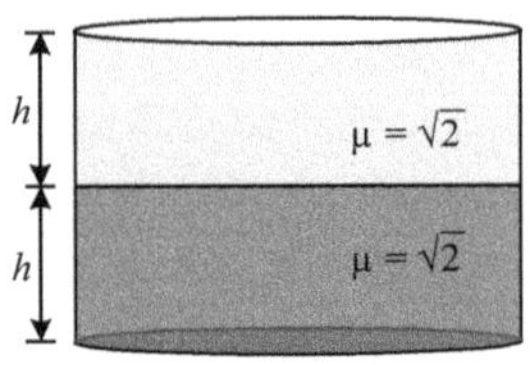

For normal viewing, apparent depth of vessel is given as

$$h_{app} = \frac{h}{\mu_1}+\frac{h}{\mu_2}$$

$$\Rightarrow \quad h_{app} = \frac{h}{\sqrt{2}}+\frac{h}{2\sqrt{2}}$$

$$\Rightarrow \quad h_{app} = \frac{3}{2\sqrt{2}}h = \frac{3}{4}\sqrt{2}\,h$$ **Ans. (B)**

Sol. 115 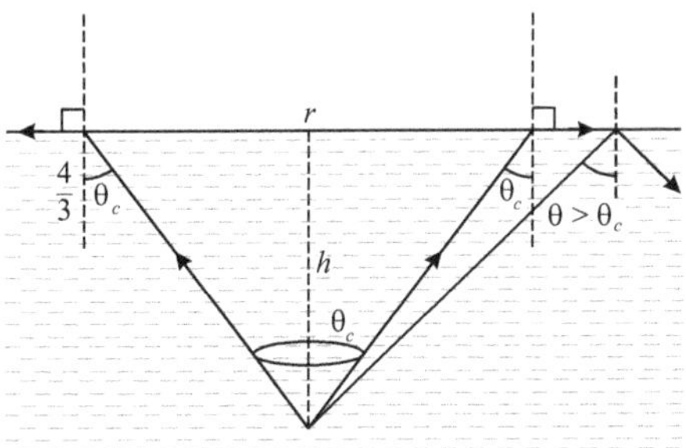

Critical angle for air-water interface is given as

$$\frac{4}{3}\sin\theta_c = 1\sin 90^\circ$$

$$\sin\theta_c = \frac{3}{4}$$

Solid angle of cone from which light escapes out from water surface is

$$\Omega = 2\pi(1-\cos\theta)$$

$$\Rightarrow \quad \Omega = 2\pi\left(1-\frac{\sqrt{7}}{4}\right)$$

Fraction of light energy which escapes out from water is

$$P_{esc} = \frac{P}{4\pi}\times\Omega$$

$$\Rightarrow \quad f = \frac{P_{esc}}{P} = \frac{1}{2}\left(1-\frac{\sqrt{7}}{4}\right)$$

$$\Rightarrow \quad f = \frac{0.33}{2} = 0.17$$

$\Rightarrow$ % fraction = 17% **Ans. (A)**

23.3 Geometrical Optics III - Thin Lenses

Sol. 116 Using Lens Maker's Formula, we have

$$\frac{1}{f} = \left(\frac{\mu_2}{\mu_1}-1\right)\left(\frac{1}{R_1}-\frac{1}{R_2}\right)$$

In air, we use

$$\frac{1}{f} = \left(\frac{1.5}{1}-1\right)\left(\frac{1}{R_1}-\frac{1}{R_2}\right) \quad \ldots(1)$$

In liquid, we use

$$\frac{1}{f_1} = \left(\frac{1.5}{1.42}-1\right)\left(\frac{1}{R_1}-\frac{1}{R_2}\right) \quad \ldots(2)$$

$$\Rightarrow \quad \frac{f_1}{f} = \frac{0.5}{0.056} = 8.93 \approx 9$$ **Ans. (C)**

Sol. 117 By use of Lens-Maker's formula, we have

$$\frac{1}{f} = (\mu-1)\left(\frac{1}{R_1}-\frac{1}{R_2}\right)$$

$$\Rightarrow \quad \frac{1}{f} = (1.5-1)\left(\frac{1}{30}-0\right)$$

$$\Rightarrow \quad f = 60\text{ cm}$$ **Ans. (60.00)**

23.5 Optical Instruments

Sol. 118 Magnification of compound microscope is

$$m = \frac{LD}{f_e\times f_0} = \frac{150\times 250}{f_e\times 25} = 375$$

$$\Rightarrow \quad f_e = 20\text{ mm}$$ **Ans. (A)**

Sol. 119 Tube length is taken as

$$L = f_0 + f_e = 60\text{ cm}$$

and $$M = \frac{f_0}{f_e} = 5$$

$$\Rightarrow \quad f_0 = 5f_e$$

$$\Rightarrow \quad 6f_e = 60\text{ cm}$$

$$\Rightarrow \quad f_e = 10\text{ cm}$$ **Ans. (D)**

Sol. 120 If the distance between objects on moon is l and D is diameter of telescope and d = distance between earth and moon, then we use

$$\theta = \frac{1.22\lambda}{D}$$

$$\Rightarrow \quad \frac{l}{d} = \frac{1.22\lambda}{D}$$

$$\Rightarrow \quad l = \frac{1.22\times(5500\times 10^{-10})\times(4\times 10^{8})}{5} = 53.68\text{ m}$$

Here most appropriate option is (C) **Ans. (C)**

24.1 Interference of Light

Sol. 121 Fringe width in YDSE is given as,

$$\beta = \frac{D\lambda}{d}$$

$$\Rightarrow \quad \beta = \frac{1.5\times 589\times 10^{-9}}{0.15\times 10^{-3}}\text{ m}$$

$$\Rightarrow \quad \beta = 5.9\times 10^{-3}\text{ m}$$

$$\Rightarrow \quad \beta = 5.9\text{ mm}$$ **Ans. (B)**

Sol. 122 Intensity of light on YDSE screen is given as

$$I = I_0\cos^2\left(\frac{\phi}{2}\right)$$

$$\frac{I}{I_0} = \cos^2\left(\frac{\phi}{2}\right)$$

Phase difference for the given point is

$$\phi = \frac{2\pi}{\lambda}\times\frac{\lambda}{8} = \frac{\pi}{4}$$

$$\Rightarrow \quad \frac{I}{I_0} = \cos^2\left(\frac{\pi}{8}\right)$$

$$\Rightarrow \quad \frac{I}{I_0} = 0.853$$ **Ans. (D)**

Sol. 123 If amplitude of each wave is considered as A then for resulting amplitude, we use the figure below

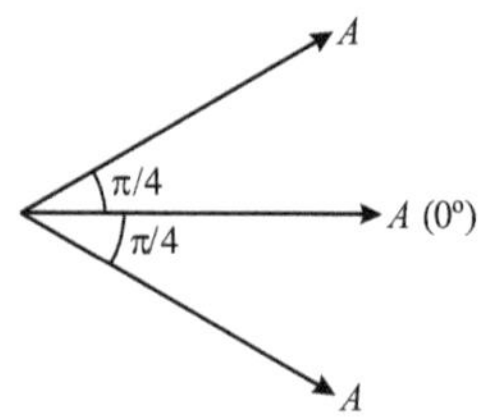

Resulting amplitude is given as

$$R = A + 2\left(\frac{A}{\sqrt{2}}\right) = A(\sqrt{2}+1)$$

As intensity is proportional to square of amplitude, we use

$$I_0 = kA^2 \text{ and } I_R = kR^2$$

$$\Rightarrow \quad \frac{I_R}{I_0} = (\sqrt{2}+1)^2$$

$$\Rightarrow \quad I_R \cong 5.8\, I_0$$ **Ans. (A)**

Sol. 124 The length of portion on screen used for 15 fringes is

$$l = 15B = 15\frac{\lambda_0 D}{d}$$

When wavelength changes to λ, 10 fringes are observed on same position of screen, so we use

$$l = 10\frac{\lambda D}{d} = \frac{15\lambda_0 D}{d}$$

$$\Rightarrow \quad l = \frac{15\lambda_0}{10} = \frac{3}{2}\times 500 = 750\,\text{nm}$$ **Ans. (750.00)**

24.2 Diffraction of Light

Sol. 125 For second minima, we use

$$b \sin\theta_2 = 2\lambda$$

$$\Rightarrow \quad \sin 60° = \frac{2\lambda}{b}$$

$$\Rightarrow \quad \frac{\lambda}{b} = \sqrt{\frac{3}{4}}$$

Now for first minima, we use

$$\sin\theta_1 = \frac{\lambda}{b} = \frac{\sqrt{3}}{4}$$

$$\Rightarrow \quad \theta_1 \simeq 25°$$ **Ans. (D)**

24.3 Polarization of Light

Sol. 126 Initially 10% light comes out so we use

$$\frac{I_0}{10} = \frac{I_0}{2}\times\cos^2\theta$$

$$\Rightarrow \quad \cos\theta = \frac{1}{\sqrt{5}}$$

$$\Rightarrow \quad \theta = 63.44°$$

Thus rotation angle is $\theta_R = 90 - 63.44° = 26.56°$

Out of given options, closest is (A) **Ans. (A)**

25.1 Photoelectric Effect

Sol. 127 Total photons incident on surface are

$$n = \frac{I\lambda}{hc}\times A$$

$$\Rightarrow \quad n = \frac{6.4\times10^{-5}\times1}{\frac{1240}{310}\times1.6\times10^{-19}} = \frac{6.4\times10^{-5}\times310\times1}{1240\times1.6\times10^{-19}}$$

$$\Rightarrow \quad n = 10^{14} \text{ photons/sec}$$

Number of electrons emitted from surface per second are

$$N_e = 10^{14}\times10^{-3} = 10^{11} \text{ per/sec}$$

$$\Rightarrow \quad x = 11$$ **Ans. (11)**

Sol. 128 Given that $\lambda_B = 2\lambda_A$

$$\Rightarrow \quad \frac{h}{\sqrt{2T_B m}} = \frac{2h}{\sqrt{2T_A m}}$$

$$\Rightarrow \quad T_A = 4T_B \qquad \ldots(1)$$

Also given that

$$T_B = (T_A - 1.5)\,\text{eV} \qquad \ldots(2)$$

$$\Rightarrow \quad 3T_B = 1.5\,\text{eV}$$

$$\Rightarrow \quad T_B = 0.5\,\text{eV}$$

and $\quad 0.5\,\text{eV} = 4.5\,\text{eV} - \phi_B$

$$\Rightarrow \quad \phi_B = 4\,\text{eV}$$ **Ans. (C)**

Sol. 129 Using dimensions of given physical quantities, we have

$$[V_0] = [h]^x[c]^y[G]^z[A]^w$$

$$\Rightarrow \quad \frac{\text{ML}^2\text{T}^{-2}}{\text{AT}} = (\text{ML}^2\text{T}^{-1})^x(\text{LT}^{-1})^y(\text{M}^{-1}\text{L}^3\text{T}^{-2})^z A^w$$

Comparing dimensions at LHS and RHS, we have

$\Rightarrow \quad w = -1, x = 0, z = 1$ and $y = 5$

$$\Rightarrow \quad [V_0] = h^0 c^5 G^{-1} A^{-1}$$

So Bonus. **Ans. (Bonus)**

Sol. 130 If the metal surface work function is ϕ then maximum kinetic energy of ejected electrons will be

$$KE_{max} = \frac{hc}{\lambda} - \phi$$

In magnetic field, radius of circular motion will be

$$R = \frac{mv}{qB} = \frac{\sqrt{2mKE_{max}}}{qB} = \frac{\sqrt{2m\left(\frac{hc}{\lambda} - \phi\right)}}{qB}$$

$$\Rightarrow \frac{R^2 q^2 B^2}{2m} = \frac{hc}{\lambda} - \phi$$

$$\Rightarrow \phi = \frac{hc}{\lambda} - \frac{R_{max}^2 q^2 B^2}{2m} = 1.0899 \text{ eV} \approx 1.1\text{eV}$$ **Ans. (B)**

25.2 Matter Waves and X-Ray

Sol. 131 Photon energy is given as

$$E = \frac{hc}{\lambda_{photon}} \quad \ldots(1)$$

and $$\lambda_{electron} = \frac{h}{\sqrt{2mE}} \quad \ldots(2)$$

From (1) and (2) we have

$$\Rightarrow \frac{\lambda_{electron}}{\lambda_{photon}} = \frac{1}{c}\left(\frac{E}{2m}\right)^{1/2}$$ **Ans. (B)**

Sol. 132 Using de-Broglie hypothesis, wavelength of electron is given as

$$\lambda = \frac{h}{mv}$$

$$\Rightarrow \lambda_0 = \frac{h}{m\sqrt{2}v_0} \quad [\text{As } v = \sqrt{2}v_0] \quad \ldots(1)$$

After time t, we use

$$\lambda' = \frac{h}{\sqrt{v_0^2 + v_0^2 + \left(\frac{eE_0 t}{m}\right)^2}}$$

$$\Rightarrow \lambda' = \frac{h}{m\sqrt{2v_0^2 + \frac{e^2 E_0^2 t^2}{m^2}}} \quad \ldots(2)$$

From (1) and (2), we use

$$\Rightarrow \lambda' = \frac{\lambda_0}{\sqrt{1 + \frac{e^2 E_0^2 t^2}{2m^2 v_0^2}}}$$ **Ans. (C)**

Sol. 133 de-Broglie wavelength of particle of kinetic energy E is given as

$$\lambda = \frac{h}{\sqrt{2mE}} \quad \ldots(1)$$

and $$\frac{\lambda}{2} = \frac{h}{\sqrt{2m(E + \Delta E)}} \quad \ldots(2)$$

From (1) and (2), we have

$$\frac{E + \Delta E}{E} = 4$$

$$\Rightarrow \Delta E = 3E$$ **Ans. (C)**

Sol. 134 Acceleration of electron due to electric field is

$$a = \frac{eE}{m}$$

Speed of electron after time t is given as

$$v = u + at = \left(\frac{eE}{m}\right)t$$

de Broglie wavelength of electron is

$$\lambda = \frac{h}{mv} = \frac{h}{eEt}$$

$$\Rightarrow \frac{d\lambda}{dt} = -\frac{h}{|e|Et^2}$$ **Ans. (A)**

26.1 Atomic Structure

Sol. 135 Time period of revolution of electron in n^{th} orbit is given as

$$T \propto \frac{n^3}{Z^2}$$

$$\Rightarrow \frac{T_2}{T_1} = \frac{(2)^3}{(1)^3} = 8$$

$$\Rightarrow T_2 = 8 \times 1.6 \times 10^{-16}\text{s}$$

Frequency in $n = 2$ orbit is

$$f_2 = \frac{1}{T_2} = \frac{10^{16}}{8 \times 1.6}$$

$$\approx 7.8 \times 10^{14}\text{Hz}$$ **Ans. (C)**

Sol. 136 For scattering experiment we use number of α particles at a specific angle are inversely proportional to fourth power of sine of scattering angle. Thus most appropriate option is (C).

Ans. (C)

Sol. 137 Using Rydberg's formula, we use

$$\frac{1}{\lambda} = R_H\left(\frac{1}{2^2} - \frac{1}{n_2^2}\right) \quad \text{[For Balmer Series]}$$

For first and second line of Balmer Series, we have

$$\frac{\lambda_2}{\lambda_1} = \frac{\left(\frac{1}{2^2} - \frac{1}{3^2}\right)}{\left(\frac{1}{2^2} - \frac{1}{4^2}\right)}$$

$$\Rightarrow \quad \frac{\lambda_2}{6561} = \frac{5/36}{3/16}$$

$$\Rightarrow \quad \lambda_2 = \frac{20}{27} \times 6561$$

$$\Rightarrow \quad \lambda_2 = 4860\ \text{Å}$$

$$\Rightarrow \quad \lambda_2 = 486\ \text{nm}$$

Ans. (486)

Sol. 138 Ionization energy of H-like ion with atomic number z is given as

$$IP = 13.6\,Z^2 = 9 \times 13.6\ \text{eV}$$

$$\Rightarrow \quad Z = 3$$

Using Rydberg formula, we use

$$\frac{1}{\lambda} = RZ^2\left(\frac{1}{1^2} - \frac{1}{3^2}\right) = 1.09 \times 10^7 \times 9 \times \frac{8}{9}$$

$$\Rightarrow \quad \lambda = 11.4\ \text{nm}$$

Ans. (D)

27.1 Radioactivity

Sol. 139 From radioactive decay law, we use

$$A = A_0 2^{-t/T}$$

$$\Rightarrow \quad 500 = 700\ 2^{-t/T}$$

$$\Rightarrow \quad 0.7 \simeq 2^{-t/T}$$

$$\Rightarrow \quad 2^{-1/2} \simeq 2^{-t/T}$$

$$\Rightarrow \quad T \simeq 2t = 2 \times 30 = 60\ \text{m}$$

Ans. (D)

28.2 PN Junction and Diodes

Sol. 140 As diode is forword based, we use circuit as

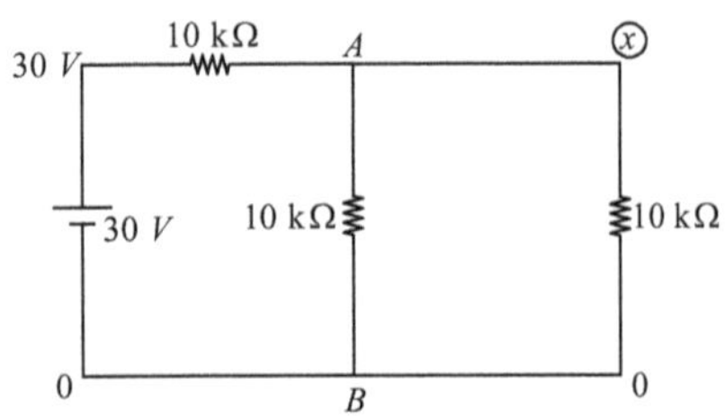

Using KCL equation for x, we have

$$\frac{30-x}{10} + \frac{0-x}{10} + \frac{0-x}{10} = 0$$

$$3 = \frac{3x}{10}$$

$$x = 10\ \text{V}$$

Ans. (B)

Sol. 141 In the figure shown we can if diode D_1 is forward biased potential of point A will become

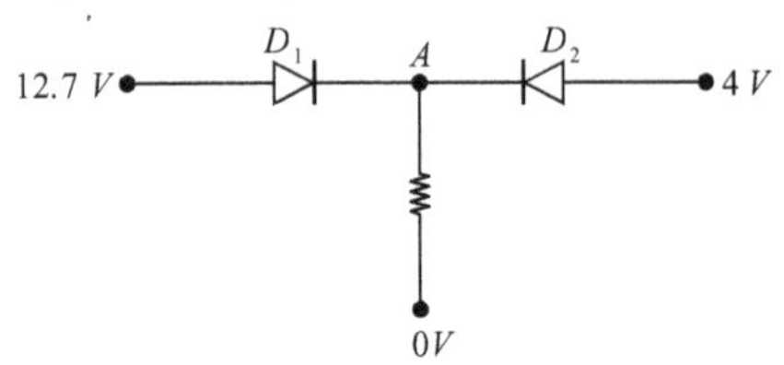

$$V_A = 12.7 - 0.7 = 12\text{V}$$

and diode D_2 will be reverse biased and behave like open circuit.

Ans. (12.00)

Sol. 142 In given circuit both the diodes will be reverse biased so effective circuit is shown below

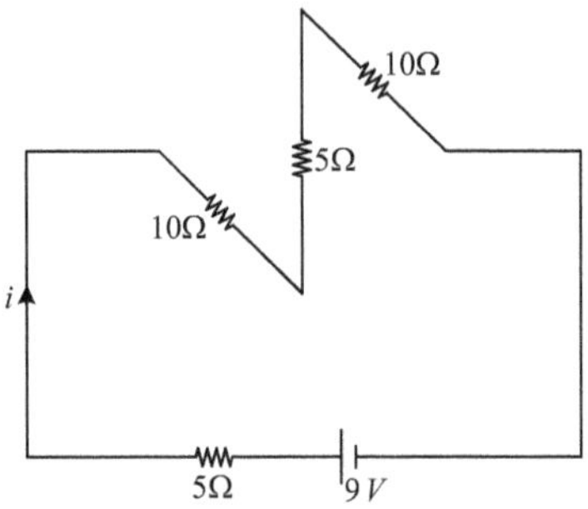

Thus current in circuit is given as

$$i = \frac{9}{(5+10+5+10)} = \frac{9}{30}\text{A}$$

Ans. (C)

Sol. 143 Diode in circuit of capacitor A will remain reverse biased so no current flows. Thus charge on capacitor A will remain constant at

$$Q_A = CV$$

Discharge current flows in circuit of capacitor B so charge capacitor is given as

$$\Rightarrow \quad q = CVe^{-\frac{t}{RC}}$$

at $t = CR$, charge is given as

$$Q_B = CVe^{-1} = \frac{CV}{e}$$

Ans. (A)

Sol. 144 For the given circuit potential difference across 400 Ω resistance is 4V, so current in circuit is

$$I = \frac{4}{400} = \frac{1}{100}A$$

Power dissipated in each diode is given as

$$P_d = \text{VI}$$

$$\Rightarrow \quad P_d = 4 \times \frac{1}{100}W$$

$$\Rightarrow \quad P_d = 40 \times 10^{-3}\ \text{mW}$$

Ans. (40.00)

28.3 Transistors

Sol. 145 In figure shown if any switch is closed base of transistor will become forward biased and output potential difference becomes zero so truth table is as shown below which is for

$$Y = \bar{A} \cdot \bar{B}$$

A	B	Y
0	0	1
1	0	0
0	1	0
1	1	0

Ans. (C)

28.4 Logic Gates

Sol. 146 Reversible operation is only possible in one input, one output systems so only option is (B). **Ans. (B)**

Sol. 147 For the given circuit boolean expression of output Y is given as

$$Y = \overline{\overline{AB}.A}$$

$$\Rightarrow \quad Y = \overline{\overline{AB}} + \bar{A}$$

$$\Rightarrow \quad Y = 0 + 0$$

$$\Rightarrow \quad Y = 0$$

Ans. (B)

30.1 Errors and Experiments

Sol. 148 Time period of a simple pendulum is given as

$$T = 2\pi\sqrt{\frac{l}{g}}$$

$$\Rightarrow \quad g = \frac{4\pi^2 l}{T^2}$$

Fractional error is measurement of g is

$$\frac{\Delta g}{g} = \frac{\Delta l}{l} + \frac{2\Delta T}{T}$$

$$\Rightarrow \quad \frac{\Delta g}{g} = \frac{0.1}{25} + \frac{2\times 1}{50} = 0.044$$

Thus percentage error or accuracy is given as

$$\frac{\Delta g}{g} = 4.4\%$$

Ans. (C)

Sol. 149 As by on six rotation, reading of main scale changes by 3mm, so main scale least count is 0.5 mm. Number of division on circular scale are 50 so least count of the screw gauge will be

$$LC = \frac{0.5}{50}\text{ mm} = 0.001\text{ cm}$$

Ans. (B)

Sol. 150 For the given set of physical quantities we use

$$A_1 + B_1 + C_1 = 24.36 + 0.0724 + 256.2$$
$$= 280.6324$$
$$= 280.6$$
$$A_2 + B_2 + C_2 = 24.44 + 16.082 + 240.2$$
$$= 280.722$$
$$= 280.7$$
$$A_3 + B_3 + C_3 = 25.2 + 19.2812 + 236.183$$
$$= 280.6642$$
$$= 280.7$$
$$A_4 + B_4 + C_4 = 25 + 236.191 + 19.5$$
$$= 280.691$$
$$= 281$$
$$A_4 + B_4 + C_4 > A_3 + B_3 + C_3 = A_2 + B_2 + C_2 > A_1 + B_1 + C_1$$

Out of given options none is correct. **Ans. (Bonus)**

* * * * *

CHAPTER-1 UNITS AND DIMENSIONS

1.1 Measurements and Dimensional Analysis

1. If speed V, area A and force F are chosen as fundamental units, then the dimension of Young's modulus will be :

[2 Sept. 2020 (Morning)]

(A) $FA^{-1}V^{0}$ (B) $FA^{2}V^{-1}$
(C) $FA^{2}V^{-3}$ (D) $FA^{2}V^{-2}$

2. If momentum (P), area (A) and time (T) are taken to be the fundamental quantities then the dimensional formula for energy is : **[2 Sept. 2020 (Evening)]**

(A) $[PA^{-1}T^{-2}]$ (B) $[PA_{1}/2T^{-1}]$
(C) $[P_{2}AT^{-2}]$ (D) $[P_{1}/2AT^{-1}]$

3. Dimensional formula for thermal conductivity is (here K denotes the temperature) : **[4 Sept. 2020 (Morning)]**

(A) $MLT^{-3}K$ (B) $MLT^{-2}K$
(C) $MLT^{-2}K^{-2}$ (D) $MLT^{-3}K^{-1}$

4. A quantity x is given by (IFv^{2}/WL^{4}) in terms of moment of inertia I, force F, velocity v, work W and Length L. The dimensional formula for x is same as that of :

[4 Sept. 2020 (Evening)]

(A) Planck's constant (B) Force constant
(C) Energy density (D) Coefficient of viscosity

5. The quantities $x = \frac{1}{\sqrt{\mu_0 \epsilon_0}}$, $y = \frac{E}{B}$ and $z = \frac{l}{CR}$ are defined where C-capacitance, R-Resistance, l-length, E-Electric field, B-magnetic field and ϵ_0, μ_0-free space permittivity and permeability respectively. Then : **[5 Sept. 2020 (Evening)]**

(A) Only x and y have the same dimension
(B) x, y and z have the same dimension
(C) Only x and z have the same dimension
(D) Only y and z have the same dimension

CHAPTER-2 MOTION IN A STRAIGHT LINE

2.1 Motion in One Dimension

6. The speed verses time graph for a particle is shown in the figure. The distance travelled (in m) by the particle during the time interval $t = 0$ to $t = 5$ s will be ______ :

[4 Sept. 2020 (Evening)]

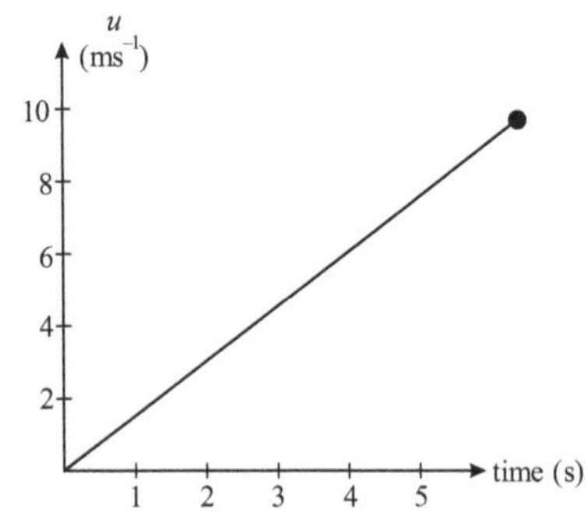

7. A balloon is moving up in air vertically above a point A on the ground. When it is at a height h_1, a girl standing at a distance d (point B) from A (see figure) sees it at an angle 45° with respect to the vertical. When the balloon climbs up a further height h_2, it is seen at an angle 60° with respect to the vertical if the girl moves further by a distance 2.464 d (point C). Then the height h_2 is (given tan 30° = 0.5774) : **[5 Sept. 2020 (Morning)]**

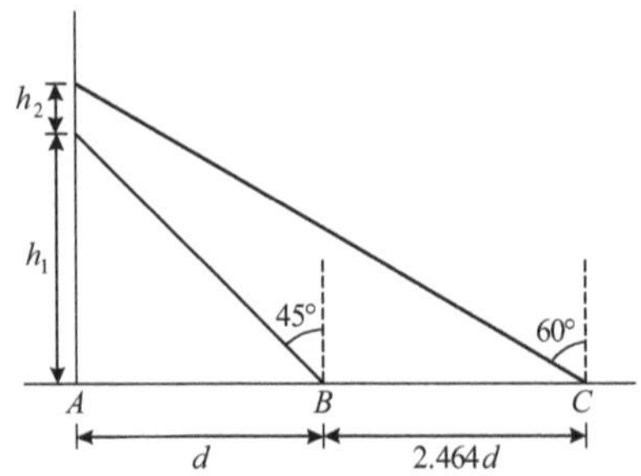

(A) d (B) $0.732d$
(C) $1.464d$ (D) $0.464d$

8. A helicopter rises from rest on the ground vertically upwards with a constant acceleration g. A food packet is dropped from the helicopter when it is a height h. The time taken by the packet to reach the ground is close to [g is the acceleration due to gravity] : **[5 Sept. 2020 (Morning)]**

(A) $t = \sqrt{\frac{2h}{3g}}$ (B) $t = 1.8\sqrt{\frac{h}{g}}$

(C) $t = 3.4\sqrt{\left(\frac{h}{g}\right)}$ (D) $t = \frac{2}{3}\sqrt{\left(\frac{h}{g}\right)}$

9. The velocity (v) and time (t) graph of a body in a straight line motion is shown in the figure. The point S is at 4.333 seconds. The total distance covered by the body in 6s is :

[5 Sept. 2020 (Evening)]

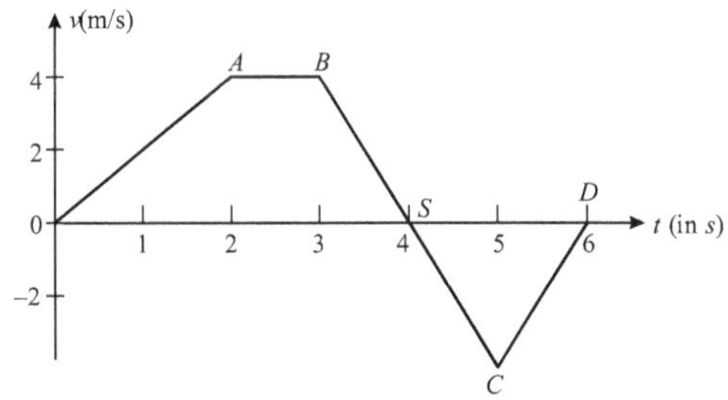

(A) 12m (B) $\frac{49}{4}$ m

(C) 11m (D) $\frac{37}{3}$ m

CHAPTER-3 MOTION IN A PLANE
3.1 Motion in Two Dimensions

10. Starting from the origin at time $t = 0$, with initial velocity $5\hat{j}$ ms^{-1}, a particle moves in the x-y plane with a constant acceleration of $(10\hat{i} + 4\hat{j})$ ms^{-2}. At time t, its coordinates are (20 m, y_0m). The values of t and y_0, are respectively : **[4 Sept. 2020 (Morning)]**

(A) 4 s and 52 m (B) 2 s and 24 m
(C) 2 s and 18 m (D) 5 s and 25 m

3.2 Relative Motion

11. Train A and train B are running on parallel tracks in the opposite directions with speeds of 36 km/hour and 72 km/hour, respectively. A person is walking in train A in the direction opposite to its motion with a speed of 1.8 km/hr. Speed (in ms^{-1}) of this person as observed from train B will be close to : (take the distance between the tracks as negligible) : **[2 Sept. 2020 (Morning)]**

(A) 30.5 ms^{-1} (B) 29.5 ms^{-1}
(C) 31.5 ms^{-1} (D) 28.5 ms^{-1}

12. When a car is at rest, its driver sees rain drops falling on it vertically. When driving the car with speed v, he sees that rain drops are coming at an angle 60° from the horizontal. On further increasing the speed of the car to $(1 + \beta)v$, this angle changes to 45°. The value of b is close to : **[6 Sept. 2020 (Evening)]**

(A) 0.41 (B) 0.50
(C) 0.37 (D) 0.73

CHAPTER-4 LAWS OF MOTION
4.1 Newton's Laws of Motion

13. A small ball of mass is thrown upward with velocity u from the ground. The ball experiences a resistive force mkv^2 where v is its speed. The maximum height attained by the ball is : **[4 Sept. 2020 (Evening)]**

(A) $\frac{1}{2k}\tan^{-1}\frac{ku^2}{g}en$ (B) $\frac{1}{2k}\ln\left(1+\frac{ku^2}{g}\right)$

(C) $\frac{1}{k}\tan^{-1}\frac{ku^2}{2g}$ (D) $\frac{1}{k}\ln\left(1+\frac{ku^2}{2g}\right)$

14. A spaceship in space sweeps stationary interplanetary dust. As a result, its mass increases at a rate $\frac{dM(t)}{dt} = bv^2(t)$, where $v(t)$ is its instantaneous velocity. The instantaneous acceleration of the satellite is : **[5 Sept. 2020 (Evening)]**

(A) $-\frac{2bv^3}{M(t)}$ (B) $-\frac{bv^3}{2M(t)}$

(C) $-bv^3(t)$ (D) $-\frac{bv^3}{M(t)}$

4.2 Friction

15. An insect is at the bottom of a hemispherical ditch of radius 1 m. It crawls up the ditch but starts slipping after it is at height h from the bottom. If the coefficient of friction between the ground and the insect is 0.75, then h is : (g = 10 ms^{-2}) **[6 Sept. 2020 (Morning)]**

(A) 0.80m (B) 0.60m
(C) 0.45m (D) 0.20m

CHAPTER-5 WORK ENERGY POWER & APPLICATIONS
5.1 Work Energy Power

16. A small block starts slipping down from a point B on an inclined plane AB, which is making an angle θ with the horizontal section BC is smooth and the remaining section CA is rough with a coefficient of friction m. It is found that the block comes to rest as it reaches the bottom (point A) of the inclined plane. If $BC = 2AC$, the coefficient of friction is given by $m = k\tan\theta$). The value of k is : **[2 Sept. 2020 (Morning)]**

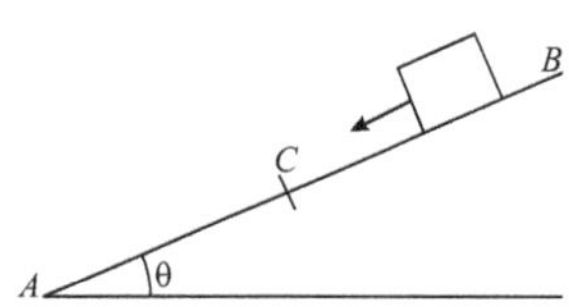

17. A cricket ball of mass 0.15 kg is thrown vertically up by a bowling machine so that it rises to a maximum height of 20 m after leaving the machine. If the part pushing the ball applies a constant force F on the ball and moves horizontally a distance of 0.2 m while launching the ball, the value of F(in N) is ($g = 10\ ms^{-2}$) __________: **[3 Sept. 2020 (Morning)]**

18. A particle is moving unidirectionally on a horizontal plane under the action of a constant power supplying energy source. The displacement (s)-time (t) graph that describes the motion of the particle is (graphs are drawn schematically and are not to scale) : **[3 Sept. 2020 (Evening)]**

(A) 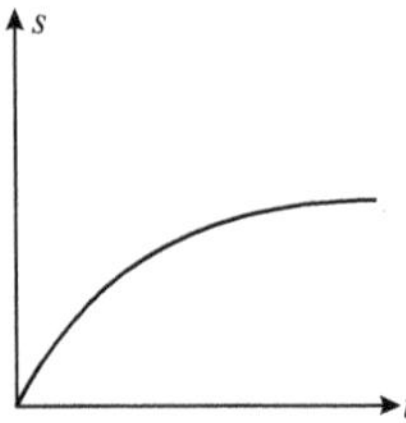

(B) 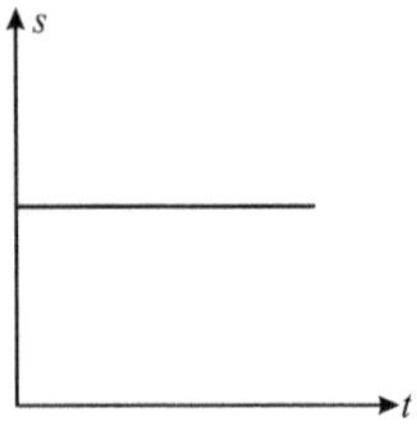

(C) 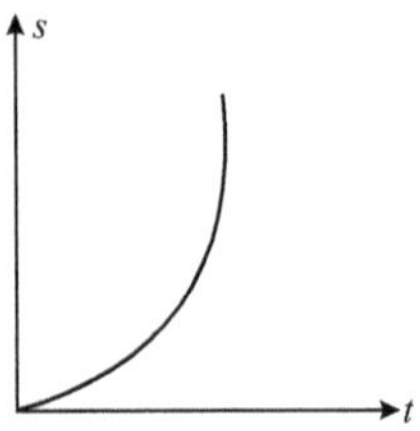

(D) 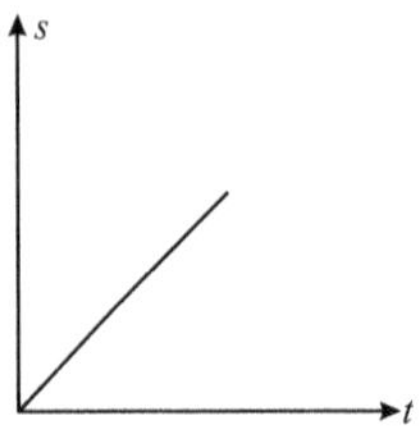

19. A block starts moving up an inclined plane of inclination 30° with an initial velocity of v_0. It comes back to its initial position with velocity $\frac{v_0}{2}$. The value of the coefficient of kinetic friction between the block and the inclined plane is close to $\frac{I}{1000}$. The nearest integer to I is__________: **[3 Sept. 2020 (Evening)]**

20. A person pushes a box on a rough horizontal platform surface. He applies a force of 200 N over a distance of 15 m. Thereafter, he gets progressively tired and his applied force reduces linearly with distance of 100 N. The total distance through which the box has been moved is 30 m. What is the work done by the person during the total movement of the box ? **[4 Sept. 2020 (Evening)]**

(A) 5690 J (B) 5250 J

(C) 3280 J (D) 2780 J

21. A body of mass 2 kg is driven by an engine delivering a constant power 1 J/s. The body starts from rest and moves in a straight line. After 9 seconds, the body has moved a distance (in m)________: **[5 Sept. 2020 (Evening)]**

22. If the potential energy between two molecules is given by $U = \frac{A}{r^6} + \frac{B}{r^{12}}$, then at equilibrium, separation between molecules, and the potential energy are : **[6 Sept. 2020 (Morning)]**

(A) $\left(\frac{B}{A}\right)^{1/6}, 0$ (B) $\left(\frac{B}{2A}\right)^{1/6}, -\frac{A^2}{2B}$

(C) $\left(\frac{2B}{A}\right)^{1/6}, -\frac{A^2}{4B}$ (D) $\left(\frac{2B}{A}\right)^{1/6}, -\frac{A^2}{2B}$

5.2 Circular Motion

23. A bead of mass m stays at point $P(a, b)$ on a wire bent in the shape of a parabola $y = 4Cx^2$ and rotating with angular speed ω (see figure). The value of ω is (neglect friction) : **[2 Sept. 2020 (Morning)]**

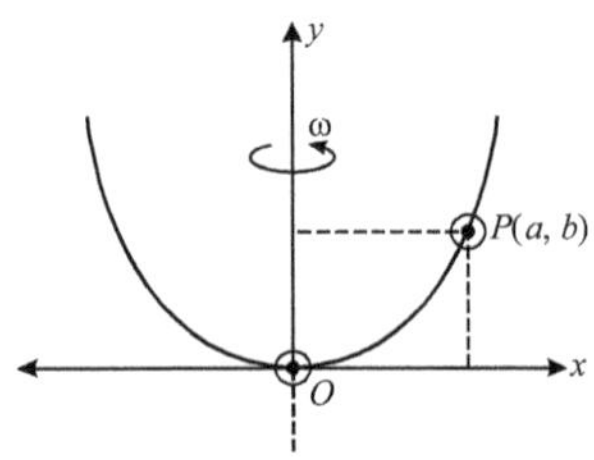

(A) $\sqrt{\frac{2gC}{ab}}$ (B) $2\sqrt{2gC}$

(C) $\sqrt{\frac{2g}{C}}$ (D) $2\sqrt{gC}$

24. A cylindrical vessel containing a liquid is rotated about its axis so that the liquid rises at its sides as shown in the figure. The radius of vessel is 5 cm and the angular speed of rotation is ω rad s^{-1}. The difference in the height, h (in cm) of liquid at the centre of vessel and at the side will be : **[2 Sept. 2020 (Morning)]**

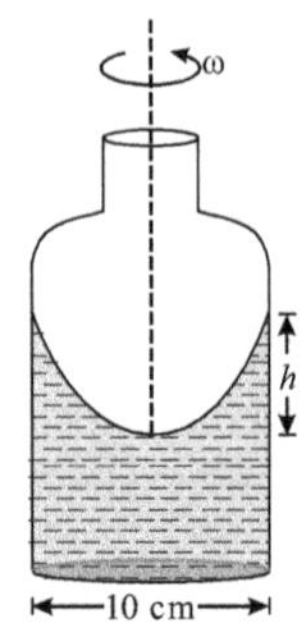

(A) $\frac{25\omega^2}{2g}$ (B) $\frac{2\omega^2}{5g}$

(C) $\frac{5\omega^2}{2g}$ (D) $\frac{2\omega^2}{25g}$

25. A clock has a continuously moving second's hand of 0.1 m length. The average acceleration of the tip of the hand (in units of ms^{-2}) is of the order of : **[6 Sept. 2020 (Morning)]**

(A) 10^{-3} (B) 10^{-2}

(C) 10^{-4} (D) 10^{-1}

CHAPTER-6 SYSTEM OF PARTICLES & ROTATIONAL MOTION

6.1 System of Particles

26. A particle of mass m with an initial velocity $u\hat{i}$ collides perfectly elastically with a mass $3m$ at rest. It moves with a velocity $v\hat{j}$ after collision, then, v is given by : **[2 Sept. 2020 (Morning)]**

(A) $v = \sqrt{\frac{2}{3}}\,u$ (B) $v = \frac{1}{\sqrt{6}}\,u$

(C) $v = \frac{u}{\sqrt{3}}$ (D) $v = \frac{u}{\sqrt{2}}$

27. A particle of mass m is moving along the x-axis with initial velocity $u\hat{i}$. It collides elastically with a particle of mass 10 m at rest and then moves with half its initial kinetic energy (see figure). If $\sin\theta_1 = \sqrt{n}\,\sin\theta_2$ then value of n is : **[2 Sept. 2020 (Evening)]**

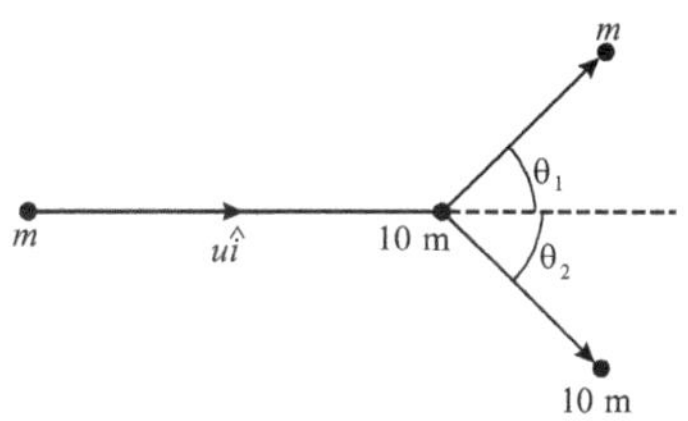

28. A square shaped hole of side $l = \frac{a}{2}$ is carved out at a distance $d = \frac{a}{2}$ from the centre 'O' of a uniform circular disc of radius a. If the distance of the centre of mass of the remaining portion from O is $-\frac{a}{X}$, value of X (to the nearest integer) is : **[2 Sept. 2020 (Evening)]**

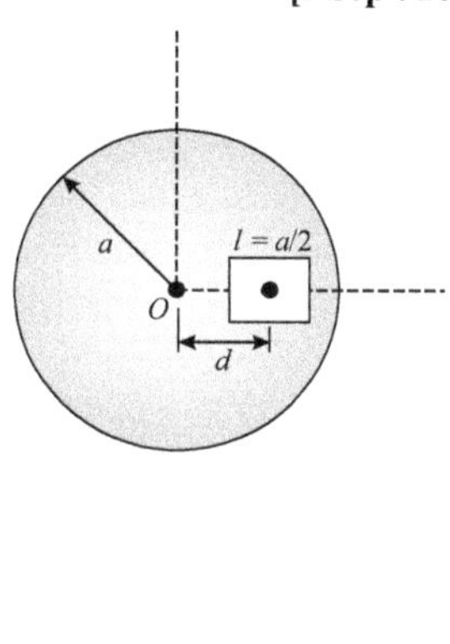

29. A block of mass 1.9 kg is at rest at the edge of a table, of height 1 m. A bullet of mass 0.1 kg collides with the block and sticks to it. If the velocity of the bullet is 20 m/s in the horizontal direction just before the collision then the kinetic energy just before the combined system strikes the floor, is [Take $g = 10$ m/s^2. Assume there is no rotational motion and loss of energy after the collision is negligible.] : **[3 Sept. 2020 (Evening)]**

(A) 21 J (B) 23 J

(C) 19 J (D) 20 J

30. A Tennis ball is released from a height h and after freely falling on a wooden floor it rebounds and reaches height $\frac{h}{2}$. The velocity versus height of the ball during its motion may be represented graphically by. (Graph are drawn schematically and on not to scale) : **[4 Sept. 2020 (Morning)]**

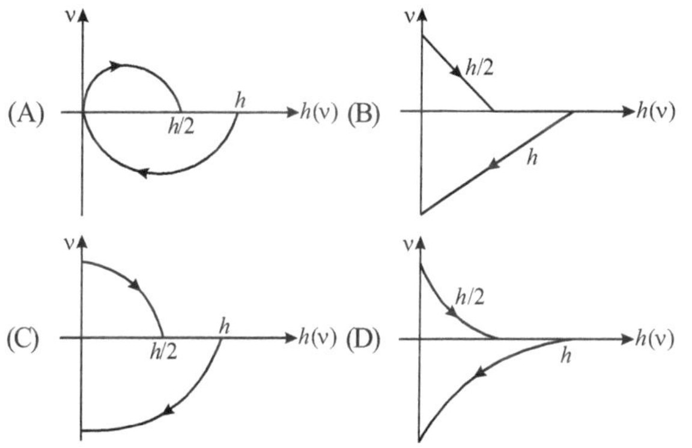

31. Blocks of masses m, $2m$, $4m$ and $8m$ are arranged in a line on a frictionless floor. Another block of mass m, moving with speed v along the same line (see figure) collides with mass m in perfectly inelastic manner. All the subsequent collisions are also perfectly inelastic. By the time the last block of mass $8m$ starts moving the total energy loss is p% of the original energy. Value of 'p' is close to : **[4 Sept. 2020 (Morning)]**

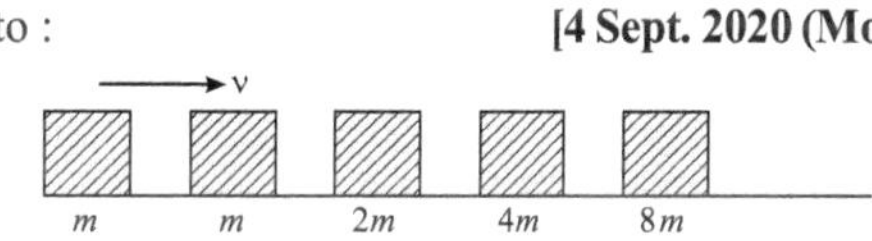

(A) 77 (B) 37

(C) 87 (D) 94

32. Two bodies of the same mass are moving with the same speed, but in different directions in a plane. They have a completely inelastic collision and move together thereafter with a final speed which is half of their initial speed. The angle between the initial velocities of the two bodies (in degree) is : **[6 Sept. 2020 (Morning)]**

33. Particle A of mass m_1 moving with velocity $(\sqrt{3}\hat{i} + \hat{j})\,ms^{-1}$ collides with another particle B of mass m_2 which is at rest initially. Let $\vec{V_1}$ and $\vec{V_2}$ be the velocities of particles A and B

after collision respectively. If $m_1 = 2m_2$ and after collision $\vec{V_1} = (\hat{i} + \sqrt{3}\hat{j})\text{ms}^{-1}$, the angle between $\vec{V_1}$ and $\vec{V_2}$ is : **[6 Sept. 2020 (Evening)]**

(A) 60° (B) 15°
(C) –45° (D) 105°

34. The centre of mass of a solid hemisphere of radius 8 cm is X cm from the centre of the flat surface. Then value of x is : **[6 Sept. 2020 (Evening)]**

6.2 Rigid Body Dynamics

35. Shown in the figure is rigid and uniform one meter long rod AB held in horizontal position by two strings tied to its ends and attached to the ceiling. The rod is of mass 'm' and has another weight of mass $2m$ hung at a distance of 75 cm from A. The tension in the string at A is : **[2 Sept. 2020 (Morning)]**

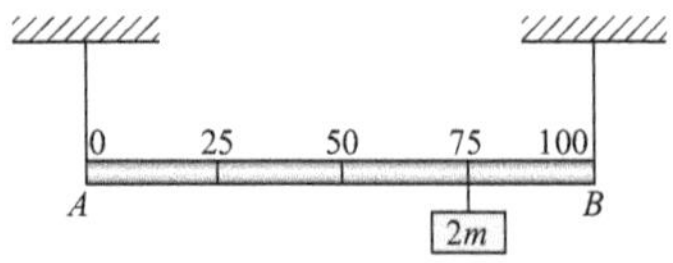

(A) $2mg$ (2) $0.5mg$
(C) $0.75mg$ (D) $1mg$

36. A uniform cylinder of mass M and radius R is to be pulled over a step of height a ($a < R$) by applying a force F at its centre 'O' perpendicular to the plane through the axes of the cylinder on the edge of the step (see figure). The minimum value of F required is : **[2 Sept. 2020 (Morning)]**

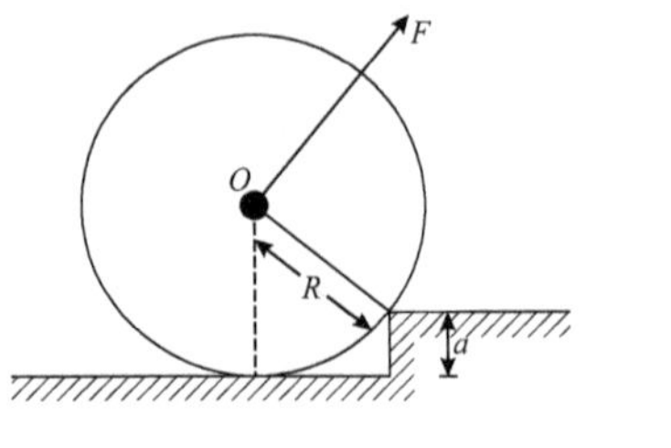

(A) $Mg\sqrt{1-\frac{a^2}{R^2}}$ (B) $Mg\sqrt{\left(\frac{R}{R-a}\right)^2-1}$

(C) $Mg\frac{a}{R}$ (D) $Mg\sqrt{1-\left(\frac{R-a}{R}\right)^2}$

37. Two uniform circular discs are rotating independently in the same direction around their common axis passing through their centres. The moment of inertia and angular velocity of the first disc are 0.1 kg-m^2 and 10 rad s^{-1} respectively while those for the second one are 0.2 kg-m^2 and 5 rad s^{-1} respectively. At some instant they get stuck together and start rotating as a single system about their common axis with some angular speed. The Kinetic energy of the combined system is : **[2 Sept. 2020 (Evening)]**

(A) $\frac{10}{3}$ J (B) $\frac{2}{3}$ J

(C) $\frac{5}{3}$ J (D) $\frac{20}{3}$ J

38. Moment of inertia of a cylinder of mass M, length L and radius R about an axis passing through its centre and perpendicular to the axis of the cylinder is given as

$$I = M\left(\frac{R^2}{4}+\frac{L^2}{12}\right)$$

If such a cylinder is to be made for a given mass of material, the ratio L/R for it to have minimum possible I is : **[3 Sept. 2020 (Morning)]**

(A) $\sqrt{\frac{2}{3}}$ (B) $\frac{3}{2}$

(C) $\sqrt{\frac{3}{2}}$ (D) $\frac{2}{3}$

39. A block of mass $m = 1$ kg slides with velocity $v = 6$ m/s on a frictionless horizontal surface and collides with a uniform vertical rod and sticks to it as shown. The rod is pivoted about O and swings as a result of the collision making angle θ before momentarily coming to rest. If the rod has mass $M = 2$ kg, and length $l = 1$ m, the value of θ is approximately : (Take $g = 10$ m/s^2) : **[3 Sept. 2020 (Morning)]**

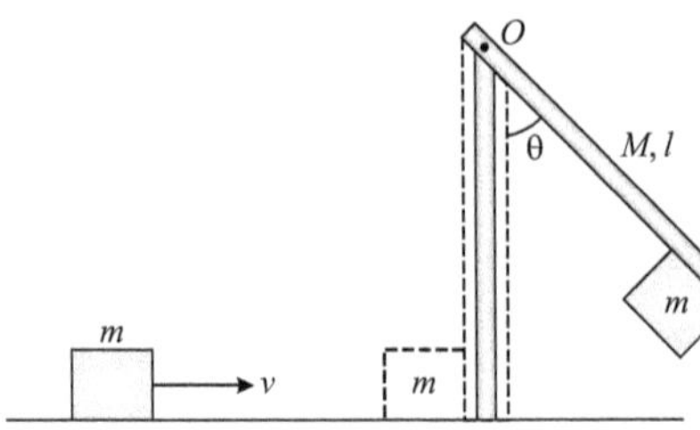

(A) 69° (B) 63°
(C) 55° (D) 49°

40. A person of 80 kg mass is standing on the rim of a circular platform of mass 200 kg rotating about its axis as 5 revolutions per minute (rpm). The person now starts moving towards the centre of the platform. What will be the rotational speed (in rpm) of the platform when the person reaches its centre_______: **[3 Sept. 2020 (Morning)]**

41. A uniform rod of length 'l' is pivoted at one of its ends on a vertical shaft of negligible radius. When the shaft rotates at angular speed ω the rod makes an angle θ with it (see figure). To find θ equate the rate of change of angular momentum (direction going into the paper) $\frac{ml^2}{12}\omega^2 \sin\theta \cos\theta$ about the centre of mass (CM) to the torque provided by the horizontal and vertical forces F_H and F_V about the CM. The value of θ is then such that : **[3 Sept. 2020 (Evening)]**

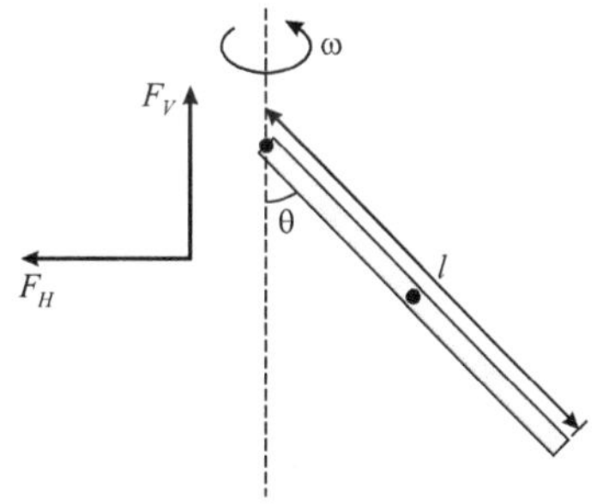

(A) $\cos\theta = \frac{g}{2l\omega^2}$ (B) $\cos\theta = \frac{3g}{2l\omega^2}$

(C) $\cos\theta = \frac{2g}{3l\omega^2}$ (D) $\cos\theta = \frac{g}{l\omega^2}$

42. A massless equilateral triangle EFG of side 'a' (As shown in figure) has three particles of mass m situated at its vertices. The moment of inertia of the system about the line EX perpendicular to EG in the plane of EFG is $\frac{N}{20}ma^2$ where N is an integer. The value of N is__________: **[3 Sept. 2020 (Evening)]**

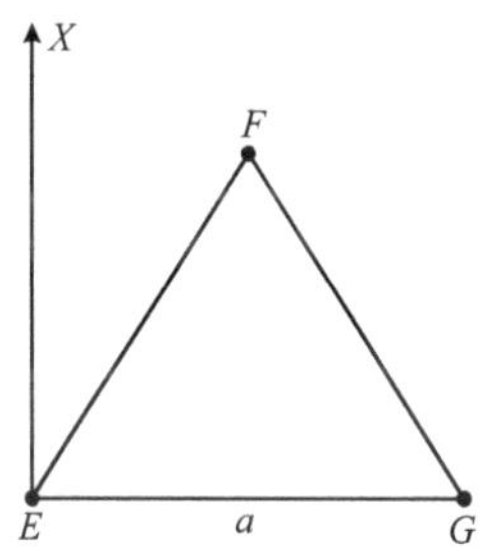

43. ABC is a plane lamina of the shape of an equilateral triangle. D, E are mid points of AB, AC and G is the centroid of the lamina. Moment of inertia of the lamina about an axis passing through G and perpendicular to the plane ABC is I_0. If part ADE is removed, the moment of inertia of the remaining part about the same axis is $\frac{NI_0}{16}$ where N is an integer. Value of N is ____ : **[4 Sept. 2020 (Morning)]**

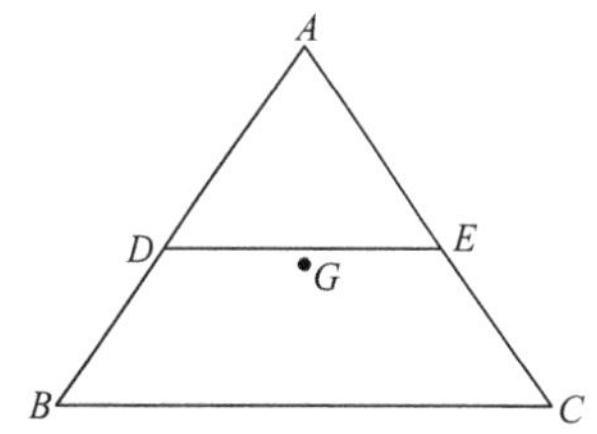

44. A circular disc of mass M and radius R is rotating about its axis with angular speed ω_1. If another stationary disc having radius $\frac{R}{2}$ and same mass M is dropped co-axially on to the rotating disc. Gradually both discs attain constant angular speed ω_2. The energy lost in the process is p% of the initial energy. Value of p is ______ : **[4 Sept. 2020 (Morning)]**

45. Consider two uniform discs of the same thickness and different radii $R_1 = R$ and $R_2 = \alpha R$ made of the same material. If the ratio of their moments of inertia I_1 and I_2, respectively, about their axes is $I_1 : I_2 = 1 : 16$ then the value of α is : **[4 Sept. 2020 (Evening)]**

(A) $\sqrt{2}$ (B) 2

(C) 4 (D) $2\sqrt{2}$

46. For a uniform rectangular sheet shown in the figure, the ratio of moments of inertia about the axes perpendicular to the sheet and passing through O (the centre of mass) and O' (corner point) is : **[4 Sept. 2020 (Evening)]**

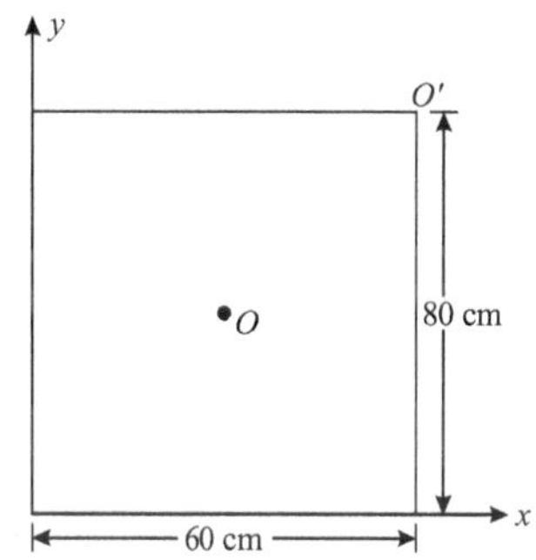

(A) 1/2 (B) 2/3

(C) 1/8 (D) 1/4

47. A wheel is rotating freely with an angular speed ω on a shaft. The moment of inertia of the wheel is I and the moment of inertia of the shaft is negligible. Another wheel of moment of inertia $3I$ initially at rest is suddenly coupled to the same shaft. The resultant fractional loss in the kinetic energy of the system is : **[5 Sept. 2020 (Morning)]**

(A) 0 (B) $\frac{1}{4}$

(C) $\frac{3}{4}$ (D) $\frac{5}{6}$

48. A force $\vec{F} = (\hat{i}+2\hat{j}+3\hat{k})$ N acts at a point $(4\hat{i}+3\hat{j}-\hat{k})m$. Then the magnitude of torque about the point $(\hat{i}+2\hat{j}+\hat{k})m$ will be $\sqrt{x}$ N-m. The value of x is : **[5 Sept. 2020 (Morning)]**

49. A thin rod of mass 0.9 kg and length 1 m is suspended, at rest, from one end so that it can freely oscillate in the vertical plane. A particle of move 0.1 kg moving in a straight line with velocity 80 m/s hits the rod at its bottom most point and sticks to it (see figure). The angular speed (in rad/s) of the rod immediately after the collision will be ______ : **[5 Sept. 2020 (Evening)]**

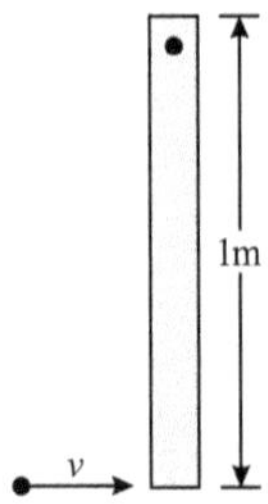

50. Shown in the figure is a hollow icecream cone (it is open at the top). If its mass is M, radius of its top, R and height, H, then its moment of inertia about its axis is : **[6 Sept. 2020 (Morning)]**

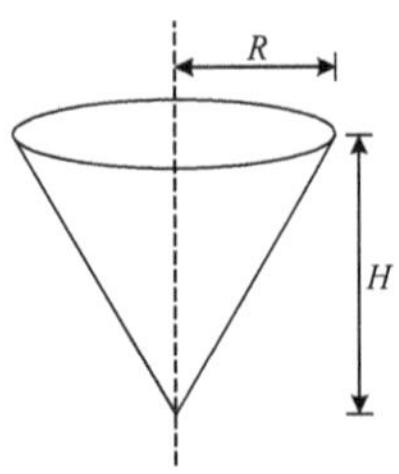

(A) $\frac{MR^2}{2}$ (B) $\frac{MH^2}{3}$

(C) $\frac{MR^2}{3}$ (D) $\frac{M(R^2+H^2)}{4}$

51. Four point masses, each of mass m, are fixed at the corners of a square of side l. The square is rotating with angular frequency ω, about an axis passing through one of the corners of the square and parallel to its diagonal, as shown in the figure. The angular momentum of the square about this axis is : **[6 Sept. 2020 (Morning)]**

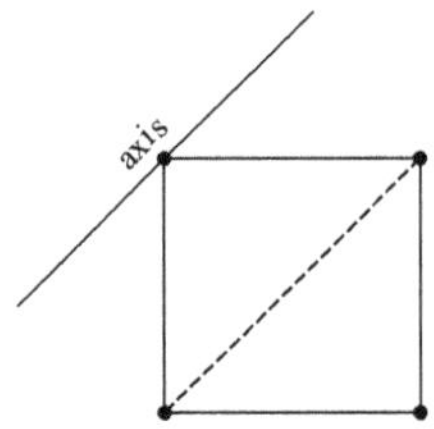

(A) $2\,ml^2\omega$ (B) $3\,ml^2\omega$

(C) $ml^2\omega$ (D) $4\,ml^2\omega$

52. The linear mass density of a thin rod AB of length L varies from A to B as $\lambda(x) = \lambda_0\left(1+\frac{x}{L}\right)$, where x is the distance from A. If M is the mass of the rod then its moment of inertia about an axis passing through A and perpendicular to the rod is : **[6 Sept. 2020 (Evening)]**

(A) $\frac{5}{12}ML^2$ (B) $\frac{3}{7}ML^2$

(C) $\frac{2}{5}ML^2$ (D) $\frac{7}{18}ML^2$

CHAPTER-7 GRAVITATION

7.1 Gravitational Field

53. The height 'h' at which the weight of a body will be the same as that at the same depth 'h' from the surface of the earth is (Radius of the earth is R and effect of the rotation of the earth is neglected) : **[2 Sept. 2020 (Evening)]**

(A) $\frac{\sqrt{5}R-R}{2}$ (B) $\frac{\sqrt{5}}{2}R-R$

(C) $\frac{R}{2}$ (D) $\frac{\sqrt{3}R-R}{2}$

54. The mass density of a planet of radius R varies with the distance r from its centre as $\rho(r) = \rho_0\left(1-\frac{r^2}{R^2}\right)$. Then the gravitational field is maximum at : **[3 Sept. 2020 (Evening)]**

(A) $r=\frac{1}{\sqrt{3}}R$ (B) $r=\sqrt{\frac{5}{9}}R$

(C) $r=\sqrt{\frac{3}{4}}R$ (D) $r=R$

55. On the x-axis and a distance x from the origin, the gravitational field due to a mass distribution is given by $\frac{Ax}{(x^2+a^2)^{3/2}}$ in the x-direction. The magnitude of gravitational potential on the x-axis at a distance x, taking its value to be zero at infinity, is : **[4 Sept. 2020 (Morning)]**

(A) $\frac{A}{(x^2+a^2)^{1/2}}$ (B) $\frac{A}{(x^2+a^2)^{3/2}}$

(C) $A(x^2+a^2)^{3/2}$ (D) $A(x^2+a^2)^{1/2}$

56. The value of the acceleration due to gravity is g_1 at a height $h=\frac{R}{2}$ (R = radius of the earth) from the surface of the earth. It is again equal to g_1 at a depth d below the surface of the earth. The ratio $\left(\frac{d}{R}\right)$ equals : **[5 Sept. 2020 (Morning)]**

(A) $\frac{7}{9}$ (B) $\frac{4}{9}$

(C) $\frac{1}{3}$ (D) $\frac{5}{9}$

57. The acceleration due to gravity on the earth's surface at the poles is g and angular velocity of the earth about the axis passing through the pole is ω. An object is weighed at the equator and at a height h above the poles by using a spring balance. If the weights are found to be same, then h is : ($h \ll R$, where R is the radius of the earth) : **[5 Sept. 2020 (Evening)]**

(A) $\frac{R^2\omega^2}{8g}$ (B) $\frac{R^2\omega^2}{4g}$

(C) $\frac{R^2\omega^2}{g}$ (D) $\frac{R^2\omega^2}{2g}$

7.2 Satellite Motion

58. The mass density of a spherical galaxy varies as $\frac{K}{r}$ over a large distance 'r' from its centre. In that region, a small star is in a circular orbit of radius R. Then the period of revolution, T depends on R as : **[2 Sept. 2020 (Morning)]**

(A) $T \propto R$ (B) $T^2 \propto \frac{1}{R^3}$

(C) $T^2 \propto R$ (D) $T^2 \propto R^3$

59. A satellite is moving in a low nearly circular orbit around the earth. Its radius is roughly equal to that of the earth's radius R_e. By firing rockets attached to it, its speed is instantaneously increased in the direction of its motion so that is become $\sqrt{\frac{3}{2}}$ times larger. Due to this the farthest distance from the centre of the earth that the satellite reaches is R, value of R is : **[3 Sept. 2020 (Morning)]**

(A) $4R_e$ (B) $3R_e$

(C) $2R_e$ (D) $2.5R_e$

60. A body is moving in a low circular orbit about a planet of mass M and radius R. The radius of the orbit can be taken to be R itself. Then the ratio of the speed of this body in the orbit to the escape velocity from the planet is : **[4 Sept. 2020 (Evening)]**

(A) 1 (B) 2

(C) $\frac{1}{\sqrt{2}}$ (D) $\sqrt{2}$

61. A satellite is in an elliptical orbit around a planet P. It is observed that the velocity of the satellite when it is farthest from the planet is 6 times less than that when it is closest to the planet. The ratio of distances between the satellite and the planet at closest and farthest points is : **[6 Sept. 2020 (Morning)]**

(A) 1 : 6 (B) 3 : 4

(C) 1 : 3 (D) 1 : 2

62. Two planets have masses M and $16\,M$ and their radii are a and $2a$, respectively. The separation between the centres of the planets is $10a$. A body of mass m is fired from the surface of the larger planet towards the smaller planet along the line joining their centres. For the body to be able to reach at the surface of smaller planet, the minimum firing speed needed is : **[6 Sept. 2020 (Evening)]**

(A) $\sqrt{\frac{GM^2}{ma}}$ (B) $\frac{3}{2}\sqrt{\frac{5GM}{a}}$

(C) $4\sqrt{\frac{GM}{a}}$ (D) $2\sqrt{\frac{GM}{a}}$

CHAPTER-8 MECHANICAL PROPERTIES OF SOLIDS

8.1 Elasticity

63. A wire of density 9×10^3 kg cm^{-3} is stretched between two clamps 1 m apart. The resulting strain in the wire is 4.9×10^{-4}. The lowest frequency of the transverse vibrations in the wire is (Young's modulus of wire $Y = 9 \times 10^{10}$ Nm^{-2}), (to the nearest integer). **[2 Sept. 2020 (Evening)]**

64. An object of mass m is suspended at the end of a massless wire of length L and area of cross-section, A. Young modulus of the material of the wire is Y. If the mass is pulled down slightly its frequency of oscillation along the vertical direction is : **[6 Sept. 2020 (Morning)]**

(A) $f=\frac{1}{2\pi}\sqrt{\frac{YA}{mL}}$ (B) $f=\frac{1}{2\pi}\sqrt{\frac{YL}{mA}}$

(C) $f=\frac{1}{2\pi}\sqrt{\frac{mA}{YL}}$ (D) $f=\frac{1}{2\pi}\sqrt{\frac{mL}{YA}}$

CHAPTER-9 MECHANICAL PROPERTIES OF FLUIDS

9.1 Fluid Statics

65. A air bubble of radius 1 cm in water has an upward acceleration 9.8 cm s^{-2}. The density of water is 1 gm cm^{-3} and water offers negligible drag force on the bubble. The mass of the bubble is ($g = 980$ cm/s^2) : **[4 Sept. 2020 (Morning)]**

(A) 3.15 gm (B) 4.51 gm
(C) 4.15 gm (D) 1.52 gm

66. Two identical cylindrical vessels are kept on the ground and each contain the same liquid of density d. The area of the base of both vessels is S but the height of liquid in one vessel is x_1 and in the other, x_2. When both cylinders are connected through a pipe of negligible volume very close to the bottom, the liquid flows from one vessel to the other until it comes to equilibrium at a new height. The change in energy of the system in the process is : **[4 Sept. 2020 (Evening)]**

(A) $gdS(x_2 + x_1)^2$ (B) $\frac{3}{4}gdS(x_2 - x_1)^2$
(C) $\frac{1}{4}gdS(x_2 - x_1)^2$ (D) $gdS(x_2^2 + x_1^2)$

67. A hollow spherical shell at outer radius R floats just submerged under the water surface. The inner radius of the shell is r. If the specific gravity of the shell material is $\frac{27}{8}$ w.r.t. water, the value of r is : **[5 Sept. 2020 (Morning)]**

(A) $\frac{4}{9}R$ (B) $\frac{8}{9}R$
(C) $\frac{1}{3}R$ (D) $\frac{2}{3}R$

9.2 Fluid Dynamics

68. A fluid is flowing through a horizontal pipe of varying cross-section, with speed v ms^{-1} at a point where the pressure is P Pascal. At another point where pressure is $\frac{P}{2}$ Pascal its speed is V ms^{-1}. If the density of the fluid is r kgm^{-3} and the flow is streamline, then V is equal to :**[6 Sept. 2020 (Evening)]**

(A) $\sqrt{\frac{P}{2\rho}+v^2}$ (B) $\sqrt{\frac{P}{\rho}+v^2}$
(C) $\sqrt{\frac{2P}{\rho}+v^2}$ (D) $\sqrt{\frac{P}{\rho}+v}$

9.3 Viscosity

69. A cube of metal is subjected to a hydrostatic pressure of $4GPa$. The percentage change in the length of the side of the cube is close to : (Given bulk modulus of metal, $B = 8 \times 10^{10}$ Pa) : **[4 Sept. 2020 (Evening)]**

(A) 0.6 (B) 1.67
(C) 5 (D) 20

70. In an experiment to verify Stokes law, a small spherical ball of radius r and density ρ falls under gravity through a distance h in air before entering a tank of water. If the terminal velocity of the ball inside water is same as its velocity just before entering the water surface, then the value of h is proportional to (ignore viscosity of air) : **[5 Sept. 2020 (Evening)]**

(A) r (B) r^4
(C) r^3 (D) r^2

9.4 Surface Tension

71. A capillary tube made of glass of radius 0.15 mm is dipped vertically in a beaker filled with methylene iodide surface tension $= 0.05$ Nm^{-1}, density $= 667$ kgm^{-3}) which rises to height h in the tube. It is observed that the two tangents drawn from liquid-glass interfaces (from opposite sides of the capillary) make an angle of 60° with one another. Then h is close to ($g = 10$ ms^{-2}) : **[2 Sept. 2020 (Evening)]**

(A) 0.137 m (B) 0.172 m
(C) 0.087 m (D) 0.049 m

72. Pressure inside two soap bubbles are 1.01 and 1.02 atmosphere, respectively. The ratio of their volumes is : **[3 Sept. 2020 (Morning)]**

(A) 8 : 1 (B) 0.8 : 1
(C) 2 : 1 (D) 4 : 1

73. When a long glass capillary tube of radius 0.015 cm is dipped in a liquid, the liquid rises to a height of 15 cm within it. If the contact angle between the liquid and glass to close to 0°, the surface tension of the liquid, in milli Newton m^{-1}, is [$\rho_{(liquid)}$ = 900 kgm^{-3}, g = 10 ms^{-2}] (Give answer in closest integer)_________: **[3 Sept. 2020 (Morning)]**

CHAPTER-10 THERMAL PROPERTIES OF MATTER

10.1 Thermal Expansion

74. When the temperature of a metal wire is increased from 0°C to 10°C, its length increases by 0.02 %. The percentage change in its mass density will be closest to : **[2 Sept. 2020 (Evening)]**

(A) 0.008 (B) 0.06
(C) 0.8 (D) 2.3

75. A bakelite beaker has volume capacity of 500 cc at 30°C. When it is partially filled with V_m volume (at 30°) of mercury, it is found that the unfilled volume of the beaker remains constant as temperature is varied. If $\gamma_{(beaker)} = 6 \times 10^{-6}$ °C^{-1} and

$\gamma_{(mercury)} = 1.5 \times 10^{-4}\,°C^{-1}$, where γ is the coefficient of volume expansion, then V_m (in cc) is close to__________: **[3 Sept. 2020 (Morning)]**

76. Two different wires having lengths L_1 and L_2, and respective temperature coefficient of linear expansion α_1 and α_2, are joined end-to-end. Then the effective temperature coefficient of linear expansion is : **[5 Sept. 2020 (Evening)]**

(A) $4\frac{\alpha_1\alpha_2}{\alpha_1+\alpha_2}\frac{L_2L_1}{(L_2+L_1)^2}$ (B) $2\sqrt{\alpha_1\alpha_2}$

(C) $\frac{\alpha_1+\alpha_2}{2}$ (D) $\frac{\alpha_1L_1+\alpha_2L_2}{L_1+L_2}$

10.2 Calorimetry

77. A calorimeter of water equivalent 20 g contains 180 g of water at 25°C. '*m*' grams of steam at 100°C is mixed in it till the temperature of the mixture is 31°C. The value of '*m*' is close to (Latent heat of water = 540 cal g^{-1}, specific heat of water = 1 cal $g^{-1}\,°C^{-1}$) : **[3 Sept. 2020 (Evening)]**

(A) 2.6 (B) 2
(C) 4 (D) 3.2

78. The specific heat of water = 4200 J kg^{-1} K^{-1} and the latent heat of ice = 3.4×10^5 J kg^{-1}. 100 grams of ice at 0°C is placed in 200 g of water at 25°C. The amount of ice that will melt as the temperature of water reaches 0°C is close to (in grams) : **[4 Sept. 2020 (Morning)]**

(A) 61.7 (B) 63.8
(C) 69.3 (D) 64.6

79. A bullet of mass 5 g, travelling with a speed of 210 m/s, strikes a fixed wooden target. One half of its kinetic energy is converted into heat in the bullet while the other half is converted into heat in the wood. The rise of temperature of the bullet if the specific heat of its material is 0.030 cal/(g –°C) (1 cal = 4.2×10^7 ergs) close to : **[5 Sept. 2020 (Morning)]**

(A) 83.3°C (B) 87.5°C
(C) 119.2°C (D) 38.4°C

10.3 Heat Transfer

80. A metallic sphere cools from 50°C to 40°C in 300 s. If atmospheric temperature around is 20°C, then the sphere's temperature after the next 5 minutes will be close to : **[3 Sept. 2020 (Evening)]**

(A) 33°C (B) 35°C
(C) 31°C (D) 28°C

81. Amount of solar energy received on the earth's surface per unit area per unit time is defined a solar constant. Dimension of solar constant is : **[3 Sept. 2020 (Evening)]**

(A) ML^2T^{-2} (B) MLT^{-2}
(C) $M^2L^0T^{-1}$ (D) ML^0T^{-3}

82. Three rods of identical cross-section and lengths are made of three different materials of thermal conductivity K_1, K_2, and K_3, respectively. They are joined together at their ends to make a long rod (see figure). One end of the long rod is maintained at 100°C and the other at 0°C (see figure). If the joints of the rod are at 70°C and 20°C in steady state and there is no loss of energy from the surface of the rod, the correct relationship between K_1, K_2 and K_3 is : **[6 Sept. 2020 (Evening)]**

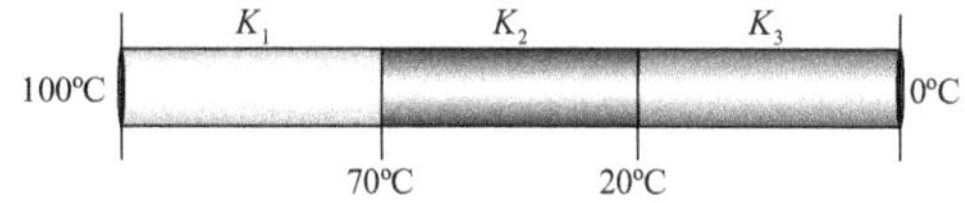

(A) $K_1 : K_3 = 2 : 3$; $K_2 : K_3 = 2 : 5$
(B) $K_1 < K_2 < K_3$
(C) $K_1 : K_2 = 5 : 2$; $K_1 : K_3 = 3 : 5$
(D) $K_1 > K_2 > K_3$

CHAPTER-11 THERMODYNAMICS

11.1 Laws of Thermodynamics

83. A gas mixture consists of 3 moles of oxygen and 5 moles of argon at temperature *T*. Assuming the gases to be ideal and the oxygen bond to be rigid, the total internal energy (in units of *RT*) of the mixture is : **[2 Sept. 2020 (Morning)]**

(A) 11 (B) 15
(C) 20 (D) 13

84. An engine takes in 5 moles of air at 20°C and 1 atm, and compresses it adiabatically to 1/10th of the original volume. Assuming air to be a diatomic ideal gas made up of rigid molecules, the change in its internal energy during this process comes out to be *X* kJ. The value of *X* to the nearest integer is : **[2 Sept. 2020 (Morning)]**

85. A heat engine is involved with exchange of heat of 1915 J, –40 J, +125 J and *QJ*, during one cycle achieving an efficiency of 50.0%. The value of *Q* is : **[2 Sept. 2020 (Evening)]**

(A) 640 J (B) 400 J
(C) 980 J (D) 40 J

86. A balloon filled with helium (32°C and 1.7 atm.) bursts. Immediately afterwards the expansion of helium can be considered as : **[3 Sept. 2020 (Morning)]**

(A) Irreversible isothermal (B) Irreversible adiabatic
(C) Reversible adiabatic (D) Reversible isothermal

87. Consider a gas of triatomic molecules. The molecules are assumed to be triangular and made of massless rigid rods whose vertices are occupied by atoms. The internal energy of a mole of the gas at temperature T is : **[3 Sept. 2020 (Morning)]**

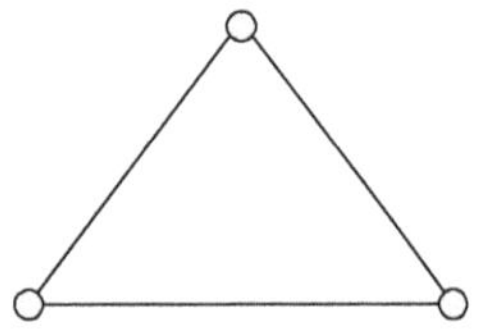

(A) $\frac{9}{2}RT$ (B) $\frac{3}{2}RT$

(C) $\frac{5}{2}RT$ (D) $3RT$

88. If minimum possible work is done by a refrigerator in converting 100 grams of water at 0°C to ice, how much heat (in calories) is released to the surrounding at temperature 27°C (Latent heat of ice = 80 Cal/gram) to the nearest integer ? **[3 Sept. 2020 (Evening)]**

89. A closed vessel contains 0.1 mole of a monoatomic ideal gas at 200 K. If 0.05 mole of the same gas at 400 K is added to it, the final equilibrium temperature (in K) of the gas in the vessel will be closed to ______ : **[4 Sept. 2020 (Morning)]**

90. Match the thermodynamic processes taking place in a system with the correct conditions. In the table, ΔQ is the heat supplied, ΔW is the work done and ΔU is change in internal energy of the system : **[4 Sept. 2020 (Evening)]**

Process	Condition
(I) Adiabatic	(1) $\Delta W = 0$
(II) Isothermal	(2) $\Delta Q = 0$
(III) Isochoric	(3) $\Delta U \neq 0, \Delta W \neq 0, \Delta Q \neq 0$
(IV) Isobaric	(4) $\Delta U = 0$

(A) I-2, II-4, III-1, IV-3 (B) I-2, II-1, III-4, IV-3
(C) I-4, II-1, III-2, IV-3 (D) I-1, II-2, III-3, IV-4

91. The change in the magnitude of the volume of an ideal gas when a small additional pressure ΔP is applied at a constant temperature, is the same as the change when the temperature is reduced by a small quantity ΔT at constant pressure. The initial temperature and pressure of the gas were 300 K and 2 atm respectively. If $|\Delta T| = C|\Delta P|$ then value of C in (K/atm) is ______ : **[4 Sept. 2020 (Evening)]**

92. Three different processes that can occur in an ideal monoatomic gas are shown in the P vs V diagram. The paths are labelled as $A \to B$, $A \to C$ and $A \to D$. The change in internal energies during these process are taken as EAB, EAC and EAD and the workdone as WAB, WAC and WAD. The correct relation between these parameters are : **[5 Sept. 2020 (Morning)]**

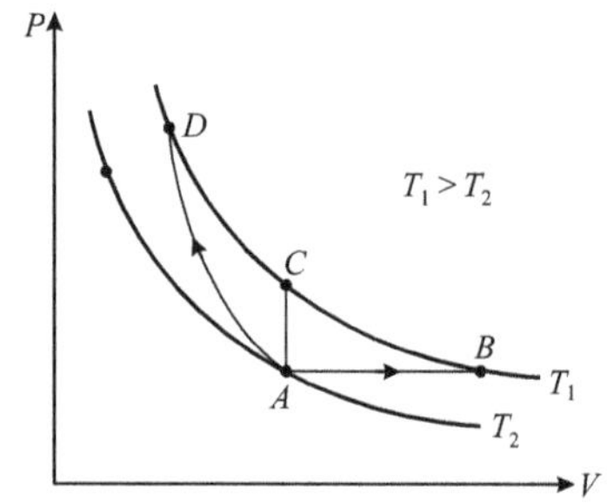

(A) $EAB = EAC = EAD$, $WAB > 0$, $WAC = 0$, $WAD > 0$

(B) $EAB < EAC < EAD$, $WAB > 0$, $WAC > WAD$

(C) $EAB = EAC < EAD$, $WAB > 0$, $WAC = 0$, $WAD < 0$

(D) $EAB > EAC > EAD$, $WAB < WAC < WAD$

93. Number of molecules in a volume of 4 cm^3 of a perfect monoatomic gas at some temperature T and at a pressure of 2 cm of mercury is close to ? (Given, mean kinetic energy of a molecule (at T) is 4×10^{-14} erg, $g = 980$ cm/s^2, density of mercury = 13.6 g/cm^3) : **[5 Sept. 2020 (Morning)]**

(A) 5.8×10^{18} (B) 5.8×10^{16}
(C) 4.0×10^{18} (D) 4.0×10^{16}

94. In an adiabatic process, the density of a diatomic gas becomes 32 times its initial value. The final pressure of the gas is found to be n times the initial pressure. The value of n is : **[5 Sept. 2020 (Evening)]**

(A) 326 (B) $\frac{1}{32}$

(C) 32 (D) 128

95. Molecules of an ideal gas are known to have three translational degrees of freedom and two rotational degrees of freedom. The gas is maintained at a temperature of T. The total internal energy, U of a mole of this gas, and the value of $\gamma\left(=\frac{C_P}{C_v}\right)$ given, respectively, by : **[6 Sept. 2020 (Morning)]**

(A) $U = \frac{5}{2}RT$ and $\gamma = \frac{6}{5}$ (B) $U = 5RT$ and $\gamma = \frac{7}{5}$

(C) $U = 5RT$ and $\gamma = \frac{6}{5}$ (D) $U = \frac{5}{2}RT$ and $\gamma = \frac{7}{5}$

96. An engine operates by taking a monatomic ideal gas through the cycle shown in the figure. The percentage efficiency of the engine is close to : **[6 Sept. 2020 (Evening)]**

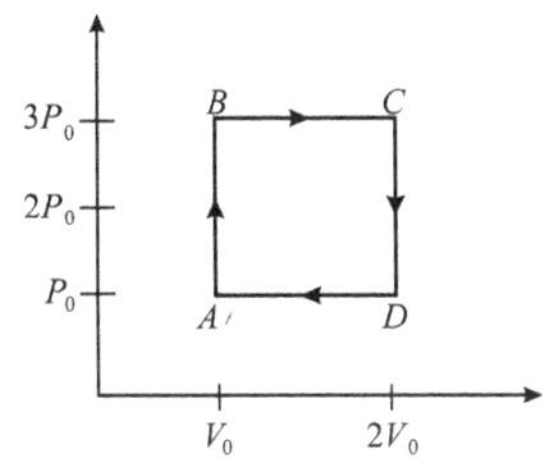

CHAPTER-12 KINETIC THEORY
12.1 Kinetic Theory of Gases

97. An ideal gas in a closed container is slowly heated. As its temperature increases, which of the following statements are true : **[2 Sept. 2020 (Evening)]**

(i) The mean free path of the molecules decreases
(ii) The mean collision time between the molecules decreases
(iii) The mean free path remains unchanged
(iv) The mean collision time remains unchanged

(A) (iii) and (iv) (B) (i) and (ii)
(C) (i) and (iv) (D) (ii) and (iii)

98. To raise the temperature of a certain mass of gas by 50°C at a constant pressure, 160 calories of heat is required. When the same mass of gas is cooled by 100°C at constant volume, 240 calories of heat is released. How many degrees of freedom does each molecule of this gas have (assume gas to be ideal) ? **[3 Sept. 2020 (Evening)]**

(A) 5 (B) 3
(C) 6 (D) 7

99. Match the C_P/C_V ratio for ideal gases with different type of molecules : **[4 Sept. 2020 (Morning)]**

Molecular type	C_P/C_V
(A) Monoatomic	(I) 7/5
(B) Diatomic rigid molecules	(II) 9/7
(C) Diatomic non-rigid molecules	(III) 4/3
(D) Triatomic rigid molecules	(IV) 5/3

(A) A-IV, B-I, C-II, D-III (B) A-IV, B-II, C-I, D-III
(C) A-III, B-IV, C-II, D-I (D) A-II, B-III, C-I, D-IV

100. Nitrogen gas is at 300°C temperature. The temperature (in K) at which the rms speed of a H_2 molecule would be equal to the rms speed of a nitrogen molecule, is _____. (Molar mass of N_2 gas 28g) : **[5 Sept. 2020 (Evening)]**

101. Initially a gas of diatomic molecules is contained in a cylinder of volume V_1 at a pressure P_1 and temperature 250 K. Assuming that 25% of the molecules get dissociated causing a change in number of moles. The pressure of the resulting gas at temperature 2000 K, when contained in a volume $2V_1$ is given by P_2. The ratio P_2/P_1 is : **[6 Sept. 2020 (Morning)]**

102. In a dilute gas at pressure P and temperature T, the mean time between successive collisions of a molecule varies with T as : **[6 Sept. 2020 (Evening)]**

(A) $\sqrt{T}$ (B) $\frac{1}{T}$
(C) $\frac{1}{\sqrt{T}}$ (D) T

CHAPTER-13 OSCILLATIONS
13.1 Simple Harmonic Motion

103. The displacement time graph of a particle executing SHM is given in figure : (sketch is schematic and not to scale)

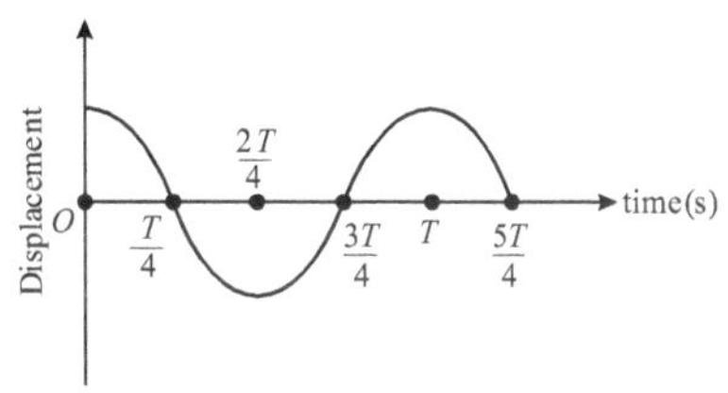

Which of the following statements is/are true for this motion : **[2 sept. 2020 (Evening)]**

(i) The force is zero at $t = \frac{3T}{4}$
(ii) The acceleration is maximum at $t = T$
(iii) The speed is maximum at $t = \frac{T}{4}$
(iv) The potential energy is equal to kinetic energy of the oscillation at $t = \frac{T}{2}$

(A) (i), (ii) and (iv) (B) (ii), (iii) and (iv)
(C) (i) and (iv) (D) (i), (ii) and (iii)

104. A block of mass m attached to massless spring is performing oscillatory motion of amplitude 'A' on a frictionless horizontal plane. If half of the mass of the block breaks off when it is passing through its equilibrium point, the amplitude of oscillation for the remaining system become fA. The value of f is : **[3 Sept. 2020 (Evening)]**

(A) $\frac{1}{2}$ (B) $\sqrt{2}$
(C) 1 (D) $\frac{1}{\sqrt{2}}$

105. A ring is hung on a nail. It can oscillate, without slipping or sliding (i) in its plane with a time period T_1 and, (ii) back and forth in a direction perpendicular to its plane, with a period T_2. The ratio $\frac{T_1}{T_2}$ will be : **[5 Sept. 2020 (Evening)]**

(A) $\frac{2}{\sqrt{3}}$ (B) $\frac{\sqrt{2}}{3}$

(C) $\frac{2}{3}$ (D) $\frac{3}{\sqrt{2}}$

106. When a particle of mass m is attached to a vertical spring of spring constant k and released, its motion is described by $y(t) = y_0 \sin^2 \omega t$, where '$y$' is measured from the lower end of unstretched spring. Then ω is : **[6 Sept. 2020 (Evening)]**

(A) $\sqrt{\frac{g}{y_0}}$ (B) $\sqrt{\frac{g}{2y_0}}$

(C) $\frac{1}{2}\sqrt{\frac{g}{y_0}}$ (D) $\sqrt{\frac{2g}{y_0}}$

CHAPTER-14 WAVES

14.1 Wave Motion

107. A uniform thin rope of length 12 m and mass 6 kg hangs vertically from a rigid support and a block of mass 2 kg is attached to its free end. A transverse short wavetrain of wavelength 6 cm is produced at the lower end of the rope. What is the wavelength of the wavetrain (in cm) when it reaches the top of the rope ? **[3 Sept. 2020 (Morning)]**

(A) 9 (B) 12

(C) 6 (D) 3

108. For a transverse wave travelling along a straight line, the distance between two peaks (crests) is 5 m, while the distance between one crest and one trough is 1.5 m. The possible wavelengths (in m) of the waves are : **[4 Sept. 2020 (Morning)]**

(A) $1, 2, 3, \ldots$ (B) $\frac{1}{2}, \frac{1}{4}, \frac{1}{6}, \ldots$

(C) $1, 3, 5, \ldots$ (D) $\frac{1}{1}, \frac{1}{3}, \frac{1}{5}, \ldots$

109. Assume that the displacement(s) of air is proportional to the pressure difference (Δp) created by a sound wave. Displacement(s) further depends on the speed of sound (v), density of air (r) and the frequency (f). If $\Delta p \sim 10 Pa$, $v \sim 300$ m/s, $p \sim 1$ kg/m^3 and $f \sim 1000$Hz, then s will be the order of (take multiplicative constant to be 1) : **[5 Sept. 2020 (Morning)]**

(A) 10mm (B) $\frac{3}{100}$mm

(C) 1mm (D) $\frac{1}{10}$mm

14.2 Stationary Waves and Beats

110. Two identical strings X and Z made of same material have tension T_X and T_Z in them. It their fundamental frequencies are 450 Hz and 300 Hz, respectively, then the ratio T_X/T_Z is : **[2 Sept. 2020 (Morning)]**

(A) 0.44 (B) 1.5

(C) 2.25 (D) 1.25

111. In a resonance tube experiment when the tube is filled with water up to height of 17.0 cm from bottom, it resonates with a given tuning fork. When the water level is raised the next resonance with the same tuning fork occurs at a height of 24.5 cm. If the velocity of sound in air is 330 m/s, the tuning fork frequency is : **[5 Sept. 2020 (Morning)]**

(A) 1100 Hz (B) 3300 Hz

(C) 2200 Hz (D) 550 Hz

14.3 Doppler's Effect

112. The driver of a bus approaching a big wall notices that the frequency of his bus's horn changes from 420 Hz to 490 Hz, when he hears it after it gets reflected from the wall. Find the speed of the bus if speed of the sound is 330 ms^{-1} : **[4 Sept. 2020 (Evening)]**

(A) 91 kmh^{-1} (B) 71 kmh^{-1}

(C) 81 kmh^{-1} (D) 61 kmh^{-1}

113. A driver in a car, approaching a vertical wall notices that the frequency of his car horn, has changed from 440 Hz to 480 Hz, when it gets reflected from the wall. If the speed of sound in air is 345 m/s, then the speed of the car is : **[5 Sept. 2020 (Evening)]**

(A) 36 km/hr (B) 24 km/hr

(C) 18 km/hr (D) 54 km/hr

114. A sound source S is moving along a straight track with speed v, and is emitting sound of frequency v_0 (see figure). An observer is standing at a finite distance, at the point O, from the track. The time variation of frequency heard by the observer is best represented by :

(t_0 represents the instant when the distance between the source and observer is minimum) **[6 Sept. 2020 (Morning)]**

(A) 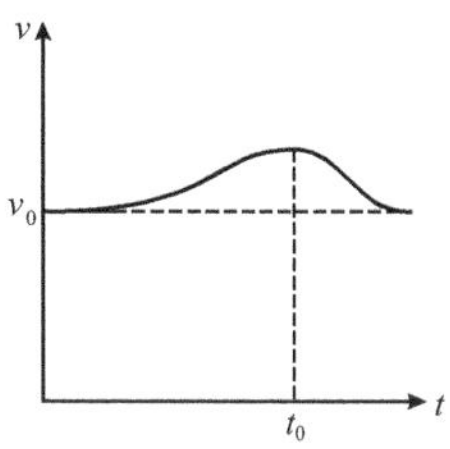

(B) 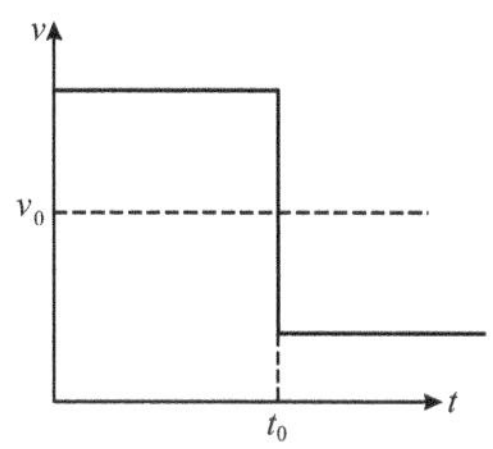

(C) 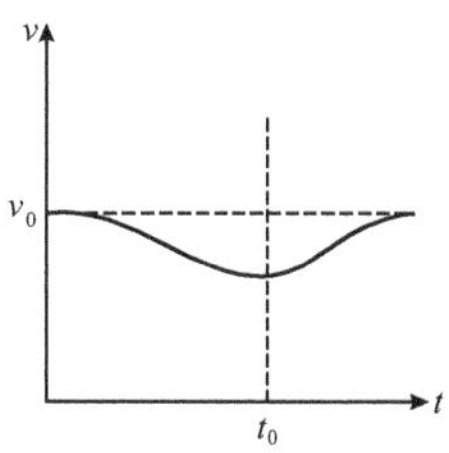

(D) 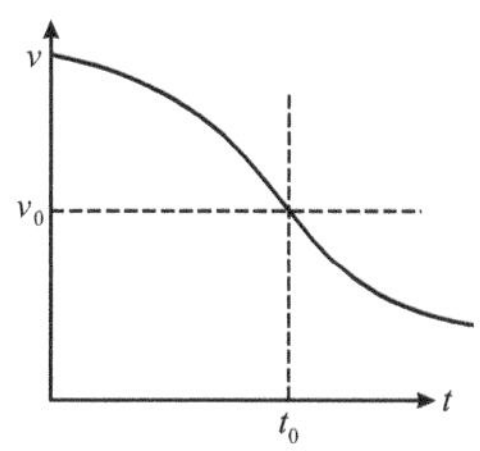

CHAPTER-15 ELECTRIC CHARGES AND FIELDS

15.1 Electric Force and Electric Field

115. A charge Q is distributed over two concentric conducting thin spherical shells radii r and R ($R > r$). If the surface charge densities on the two shells are equal, the electric potential at the common centre is : **[2 Sept. 2020 (Evening)]**

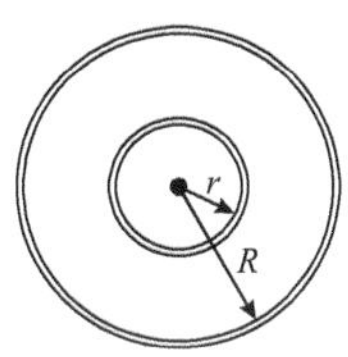

(A) $\frac{1}{4\pi\varepsilon_0}\frac{(R+2r)Q}{2(R^2+r^2)}$ (B) $\frac{1}{4\pi\varepsilon_0}\frac{(R+r)}{2(R^2+r^2)}Q$

(C) $\frac{1}{4\pi\varepsilon_0}\frac{(R+r)}{(R^2+r^2)}Q$ (D) $\frac{1}{4\pi\varepsilon_0}\frac{(2R+r)}{(R^2+r^2)}Q$

116. A small point mass carrying some positive charge on it, is released from the edge of a table. There is a uniform electric field in this region in the horizontal direction. Which of the following options then correctly describe the trajectory of the mass : (Curves are drawn schematically and are not to scale). **[2 Sept. 2020 (Evening)]**

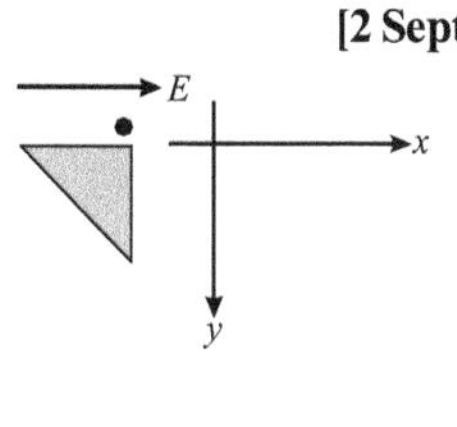

(A) 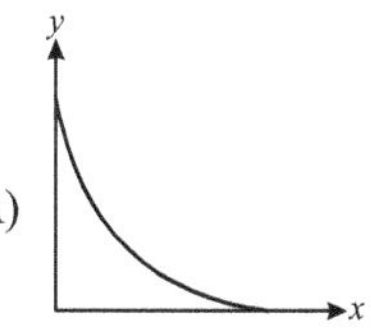

(B) 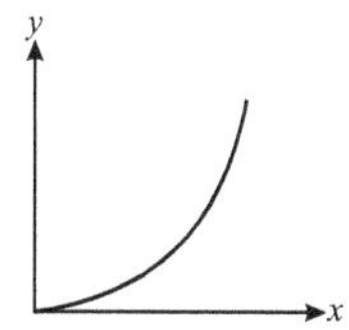

(C) 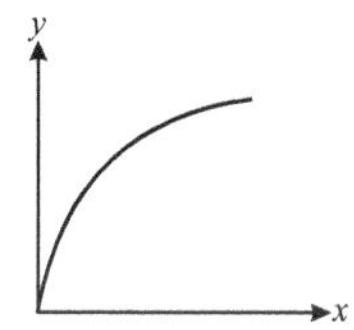

(D) 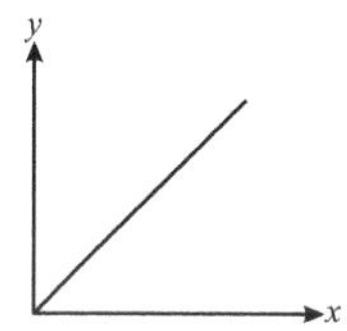

117. Charges Q_1 and Q_2 are at points A and B of a right angle triangle OAB (see figure). The resultant electric field at point O is perpendicular to the hypotenuse, then Q_1/Q_2 is proportional to : **[6 Sept. 2020 (Morning)]**

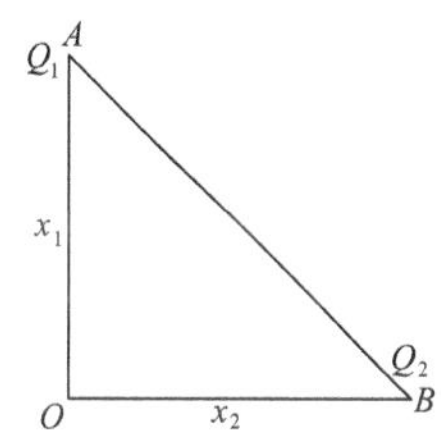

(A) $\frac{x_2^2}{x_1^2}$ (B) $\frac{x_1^3}{x_2^3}$

(C) $\frac{x_1}{x_2}$ (D) $\frac{x_2}{x_1}$

118. Consider the force F on a charge 'q' due to a uniformly charged spherical shell of radius R carrying charge Q distributed uniformly over it. Which one of the following statements is true for F, if 'q' is placed at distance r from the centre of the shell : **[6 Sept. 2020 (Evening)]**

(A) $F = \frac{1}{4\pi\varepsilon_0}\frac{Qq}{r^2}$ for $r > R$

(B) $\frac{1}{4\pi\varepsilon_0}\frac{qQ}{R^2} > F > 0$ for $r < R$

(C) $F = \frac{1}{4\pi\varepsilon_0}\frac{Qq}{r^2}$ for all r

(D) $F = \frac{1}{4\pi\varepsilon_0}\frac{Qq}{R^2}$ for $r < R$

15.2 Electric Flux and Gauss's Law

119. Two charged thin infinite plane sheets of uniform surface charge density σ_+ and σ_- where $|\sigma_+| > |\sigma_-|$ intersect at right angle. Which of the following best represents the electric field lines for this system : **[4 Sept. 2020 (Morning)]**

(A) 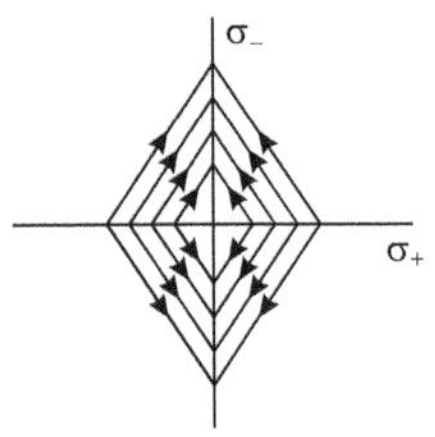

(B) 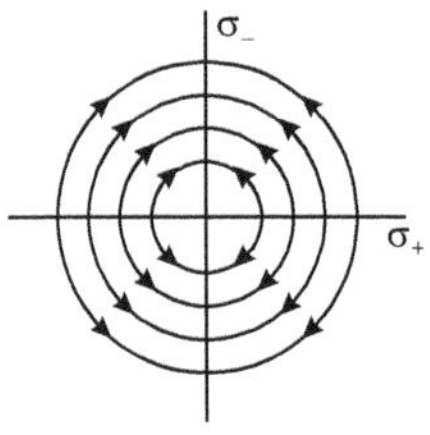

(C) 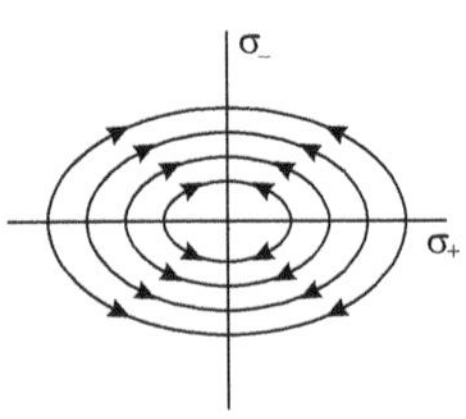

(D) 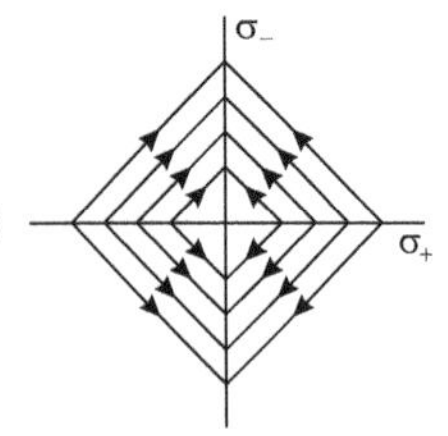

CHAPTER-16 ELECTRIC POTENTIAL AND CAPACITANCE

16.1 Electric Potential and Dipole

120. Two isolated conducting spheres S_1 and S_2 of radius $\frac{2}{3}R$ and $\frac{1}{3}R$ have 12 μC and – 3 μC charges, respectively, and are at a large distance from each other. They are now connected by a conducting wire. A long time after this is done the charges on S_1 and S_2 are respectively : **[3 Sept. 2020 (Morning)]**

(A) 6 μC and 3 μC (B) +4.5 μC and – 4.5 μC

(C) 3 μC and 6 μC (D) 4.5 μC on both

121. Concentric metallic hollow spheres of radii R and $4R$ hold charges Q_1 and Q_2 respectively. Given that surface charge densities of the concentric spheres are equal, the potential difference $V(R) - V(4R)$ is : **[3 Sept. 2020 (Evening)]**

(A) $\frac{3Q_1}{16\pi\varepsilon_0 R}$ (B) $\frac{Q_2}{4\pi\varepsilon_0 R}$

(C) $\frac{3Q_1}{4\pi\varepsilon_0 R}$ (D) $\frac{3Q_2}{4\pi\varepsilon_0 R}$

122. Hydrogen ion and singly ionized helium atom are accelerated, from rest, through the same potential difference. The ratio of final speeds of hydrogen and helium ions is close to : **[3 Sept. 2020 (Evening)]**

(A) 5 : 7 (B) 1 : 2

(C) 10 : 7 (D) 2 : 1

123. A two point charges $4q$ and $-q$ are fixed on the x-axis at $x = -\frac{d}{2}$ and $x = \frac{d}{2}$, respectively. If a third point charge 'q' is taken from the origin to $x = d$ along the semicircle as shown in the figure, the energy of the charge will : **[4 Sept. 2020 (Morning)]**

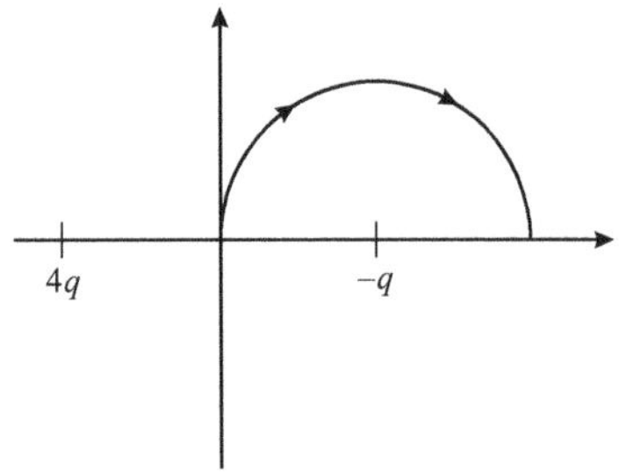

(A) Increase by $\frac{2q^2}{3\pi\varepsilon_0 d}$ (B) Increase by $\frac{3q^2}{4\pi\varepsilon_0 d}$

(C) Decrease by $\frac{4q^2}{3\pi\varepsilon_0 d}$ (D) Decrease by $\frac{q^2}{4\pi\varepsilon_0 d}$

124. A particle of charge q and mass m is subjected to an electric field $E = E_0(1 - ax^2)$ in the x-direction, where a and E_0 are constants. Initially the particle was at rest at $x = 0$. Other than the initial position the kinetic energy of the particle becomes zero when the distance of the particle from the origin is : **[4 Sept. 2020 (Evening)]**

(A) $\sqrt{\frac{2}{a}}$ (B) $\sqrt{\frac{1}{a}}$

(C) a (D) $\sqrt{\frac{3}{a}}$

125. A solid sphere of radius R carries a charge $(Q + q)$ distributed uniformly over its volume. A very small point like piece of it of mass m gets detached from the bottom of the sphere and falls down vertically under gravity. This piece carries charge q. If it acquires a speed v when it has fallen through a vertical height y (see figure), then : (assume the remaining portion to be spherical) : **[5 Sept. 2020 (Morning)]**

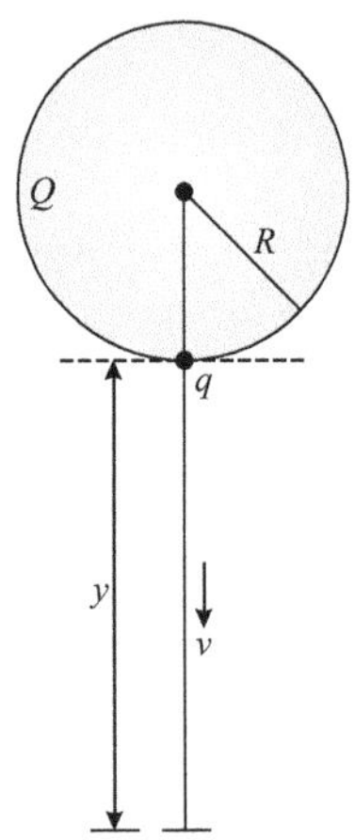

(A) $v^2 = 2y\left[\frac{qQ}{4\pi\epsilon_0 R(R+y)m} + g\right]$

(B) $v^2 = y\left[\frac{qQ}{4\pi\epsilon_0 R^2 ym} + g\right]$

(C) $v^2 = 2y\left[\frac{qQR}{4\pi\epsilon_0 (R+y)^3 m} + g\right]$

(D) $v^2 = y\left[\frac{qQ}{4\pi\epsilon_0 R(R+y)m} + g\right]$

126. Ten charges are placed on the circumference of a circle of radius R with constant angular separation between successive charges. Alternate charges 1, 3, 5, 7, 9 have charge $(+q)$ each, while 2, 4, 6, 8, 10 have charge $(-q)$ each. The potential V and the electric field E at the centre of the circle are respectively : (Take $V = 0$ at infinity) **[5 Sept. 2020 (Evening)]**

(A) $V = \frac{10q}{4\pi\epsilon_0 R}; E = \frac{10q}{4\pi\epsilon_0 R^2}$

(B) $V = 0, E = \frac{10q}{4\pi\epsilon_0 R^2}$

(C) $V = 0, E = 0$

(D) $V = \frac{10q}{4\pi\epsilon_0 R}; E = 0$

127. Two identical electric point dipoles have dipole moments $\vec{p}_1 = p\hat{i}$ and $\vec{p}_2 = -p\hat{i}$ and are held on the x-axis at distance 'a' from each other. When released, they move along the x-axis with the direction of their dipole moments remaining unchanged. If the mass of each dipole is 'm', their speed when they arc infinitely far apart is : **[6 Sept. 2020 (Evening)]**

(A) $\frac{p}{a}\sqrt{\frac{1}{\pi\varepsilon_0 ma}}$ (B) $\frac{p}{a}\sqrt{\frac{3}{2\pi\varepsilon_0 ma}}$

(C) $\frac{p}{a}\sqrt{\frac{1}{2\pi\varepsilon_0 ma}}$ (D) $\frac{p}{a}\sqrt{\frac{2}{\pi\varepsilon_0 ma}}$

16.2 Capacitance

128. A 5 μF capacitor is charged fully by a 220 V supply. It is then disconnected from the supply and is connected in series to another uncharged 2.5 μF capacitor. If the energy change during the charge redistribution is $\frac{X}{100}$ J then value of X to the nearest integer is : **[2 Sept. 2020 (Morning)]**

129. A 10 μF capacitor is fully charged to a potential difference of 50 V. After removing the source voltage it is connected to an uncharged capacitor in parallel. Now the potential difference across them becomes 20 V. The capacitance of the second capacitor is : **[2 Sept. 2020 (Evening)]**

(A) 10 μF (2) 15 μF

(C) 20 μF (4) 30 μF

130. In the circuit shown in the figure, the total charge is 750 μC and the voltage across capacitor C_2 is 20 V. Then the charge on capacitor C_2 is : **[3 Sept. 2020 (Morning)]**

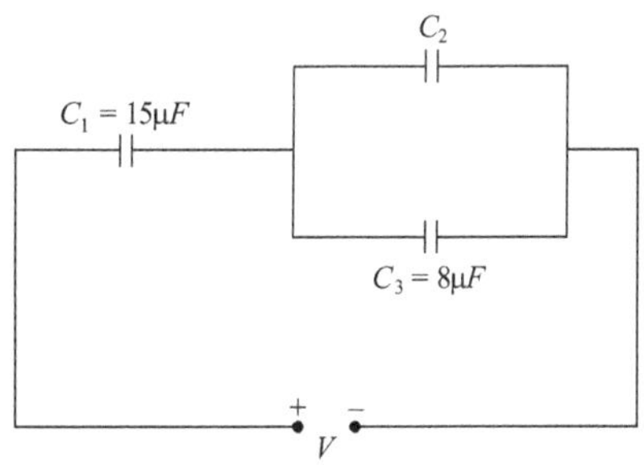

(A) 590 μC (B) 450 μC

(C) 650 μC (D) 160 μC

131. A capacitor C is fully charged with voltage V_0. After disconnecting the voltage source, it is connected in parallel with another uncharged capacitor of capacitance $\frac{C}{2}$. The energy loss in the process after the charge is distributed between the two capacitors is : **[4 Sept. 2020 (Evening)]**

(A) $\frac{1}{6}CV_0^2$ (B) $\frac{1}{2}CV_0^2$

(C) $\frac{1}{3}CV_0^2$ (D) $\frac{1}{4}CV_0^2$

132. Two capacitors of capacitances C and $2C$ are charged to potential differences V and $2V$, respectively. These are then connected in parallel in such a manner that the positive terminal of one is connected to the negative terminal of the other. The final energy of this configuration is : **[5 Sept. 2020 (Morning)]**

(A) $\frac{9}{2}CV^2$ (B) $\frac{25}{6}CV^2$

(C) Zero (D) $\frac{3}{2}CV^2$

133. A parallel plate capacitor has plate of length 'l', width 'w' and separation of plates is 'd'. It is connected to a battery of emf V. A dielectric slab of the same thickness 'd' and of dielectric constant $k = 4$ is being inserted between the plates of the capacitor. At what length of the slab inside plates, will be energy stored in the capacitor be two times the initial energy stored? **[5 Sept. 2020 (Evening)]**

(A) $l/4$ (B) $l/2$

(C) $l/3$ (D) $2l/3$

134. In the circuit shown, charge on the 5 μF capacitor is : **[5 Sept. 2020 (Evening)]**

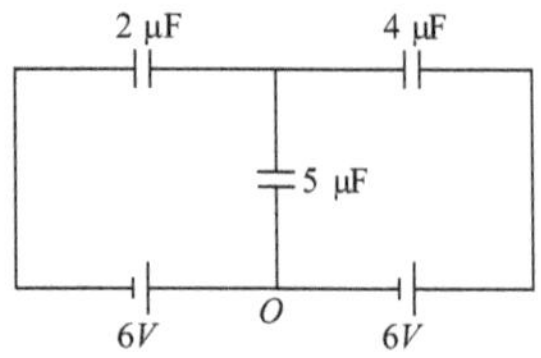

(A) 5.45 μC (B) 16.36 μC
(C) 10.90 μC (D) 18.00 μC

CHAPTER-17 CURRENT ELECTRICITY

17.1 Current Electricity

135. Consider four conducting materials copper, tungsten, mercury and aluminium with resistivity $\rho_C < \rho_T < \rho_M$ and ρ_A respectively. Then : **[2 Sept. 2020 (Morning)]**

(A) $\rho_A > \rho_T > \rho_C$ (B) $\rho_C > \rho_A > \rho_T$
(C) $\rho_A > \rho_M > \rho_C$ (D) $\rho_M > \rho_A > \rho_C$

17.2 Circuit Analysis

136. The value of current i_1 flowing from A to C in the circuit diagram is : **[4 Sept. 2020 (Evening)]**

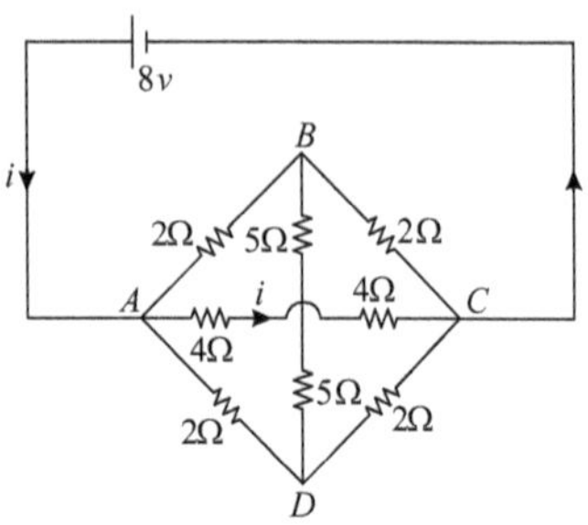

(A) 5 A (B) 2 A
(C) 4 A (D) 1 A

137. Four resistances 40 Ω, 60 Ω, 90 Ω and 110 Ω make the arms of a quadrilateral *ABCD*. Across *AC* is a battery of emf 40 V and internal resistance negligible. The potential difference across *BD* is *V* is _______ : **[4 Sept. 2020 (Evening)]**

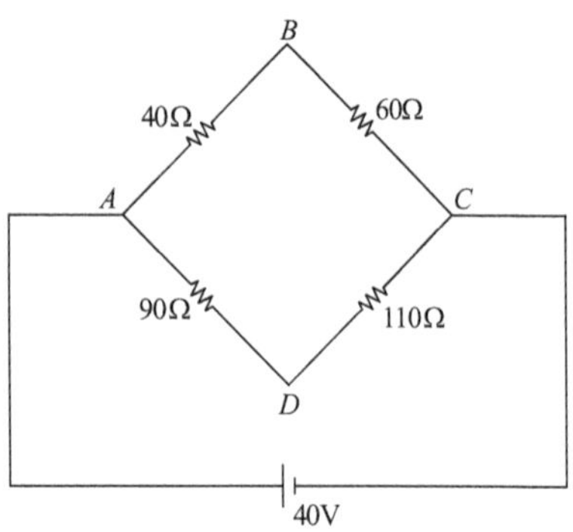

138. In the circuit, given in the figure currents in different branches and value of one resistor are shown. Then potential at point *B* with respect to the point *A* is : **[5 Sept. 2020 (Evening)]**

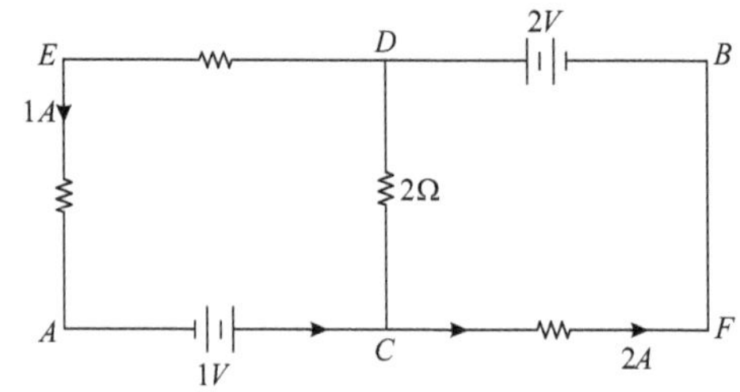

(A) +1 V (B) −1 V
(C) −2 V (D) +2 V

139. In the figure shown, the current in the 10 V battery is close to : **[6 Sept. 2020 (Evening)]**

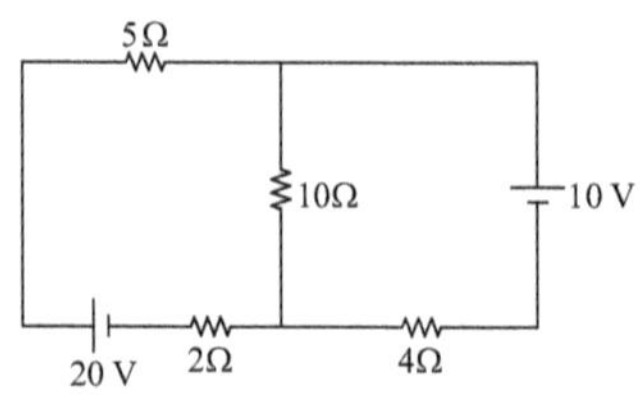

(A) 0.36 A from negative to positive terminal
(B) 0.71 A from positive to negative terminal
(C) 0.21 A from positive to negative terminal
(D) 0.42 A from positive to negative terminal

17.3 Thermal Effects of Current & RC Circuits

140. An ideal cell of emf 10 V is connected in circuit shown in figure. Each resistance is 2 W. The potential difference (in *V*) across the capacitor when it is fully charged is : **[2 Sept. 2020 (Evening)]**

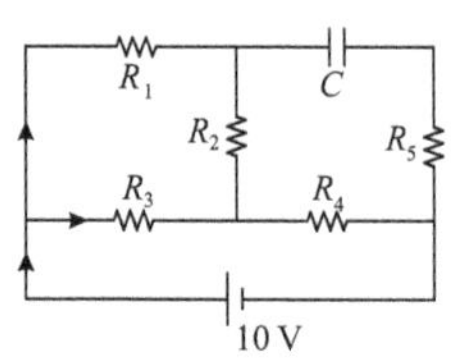

141. Model of a torch battery of length *l* to be made up of a thin cylindrical bar of radius '*a*' and a concentric thin cylindrical shell of radius '*b*' filled in between with an electrolyte of resistivity ρ (see figure). If the battery is connected to a resistance of value R, the maximum Joule heating in R will take place for : **[3 Sept. 2020 (Morning)]**

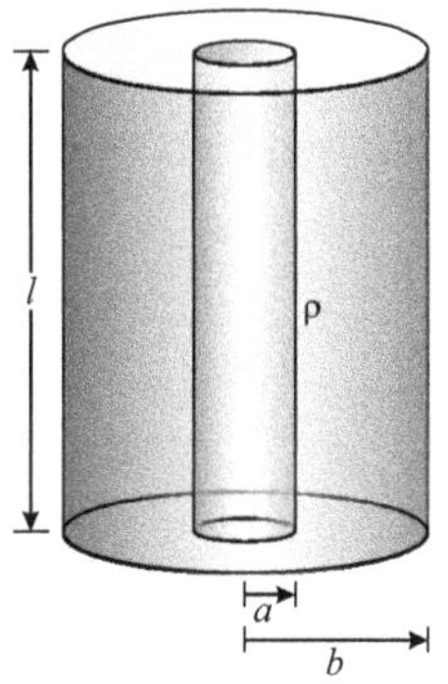

(A) $R=\frac{2\rho}{\pi l}\ln\left(\frac{b}{a}\right)$ (B) $R=\frac{\rho}{\pi l}\ln\left(\frac{b}{a}\right)$

(C) $R=\frac{\rho}{2\pi l}\left(\frac{b}{a}\right)$ (D) $R=\frac{\rho}{2\pi l}\ln\left(\frac{b}{a}\right)$

142. A battery of 3.0 V is connected to a resistor dissipating 0.5 W of power. If the terminal voltage of the battery is 2.5 V, the power dissipated within the internal resistance is : **[4 Sept. 2020 (Morning)]**

(A) 0.50 W (B) 0.125 W
(C) 0.072 W (D) 0.10 W

143. An electrical power line, having a total resistance of 2W, delivers 1 kW at 220 V. The efficiency of the transmission line is approximately: **[5 Sept. 2020 (Morning)]**

(A) 72% (B) 96%
(C) 91% (D) 85%

17.4 Electrical Measuring Instruments

144. A potentiometer wire PQ of 1 m length is connected to a standard cell E_1. Another cell E_2 of emf 1.02 V is connected with a resistance 'r' and switch S (as shown in figure). With switch S open, the null position is obtained at a distance of 49 cm from Q. The potential gradient in the potentiometer wire is : **[2 sept. 2020 (Evening)]**

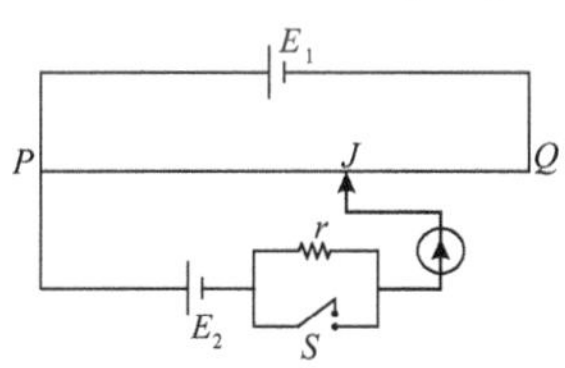

(A) 0.02 V/cm (B) 0.04 V/cm
(C) 0.01 V/cm (D) 0.03 V/cm

145. Which of the following will *NOT* be observed when a multimeter (operating in resistance measuring mode) probes connected across a component, are just reversed ? **[3 Sept. 2020 (Evening)]**

(A) Multimeter shows *NO* deflection in both cases *i.e.* before and after reversing the probes if the chosen component is capacitor

(B) Multimeter shows a deflection, accompanied by a splash of light out of connected component in one direction and *NO* deflection on reversing the probes if the chosen component is *LED*

(C) Multimeter shows *NO* deflection in both cases *i.e.* before and after reversing the probes if the chosen component is metal wire

(D) Multimeter shows an equal deflection in both cases *i.e.* before and after reversing the probes if the chosen component is resistor

146. Two resistors 400Ω and 800Ω are connected in series across a 6 V battery. The potential difference measured by a voltmeter of 10kΩ across 400Ω resistor is close to : **[3 Sept. 2020 (Evening)]**

(A) 2V (B) 1.95 V
(C) 2.05 V (D) 1.8 V

147. A galvanometer of resistance G is converted into a voltmeter of range 0 – 1 V by connecting a resistance R_1 in series with it. The additional resistance that should be connected in series with R_1 to increase the range of the voltmeter to 0 – 2V will be : **[5 Sept. 2020 (Morning)]**

(A) R_1 (B) R_1+G
(C) R_1-G (D) G

148. A galvanometer is used in laboratory for detecting the null point in electrical experiments. If, on passing a current of 6 mA it produces a deflection of 2°, its figure of merit is close to : **[5 Sept. 2020 (Evening)]**

(A) 3×10^{-3} A/div (B) 333° A/div
(C) 6×10^{-3} A/div (D) 666° A/div

149. A circuit to verify Ohm's law uses ammeter and voltmeter in series or parallel connected correctly to the resistor. In the circuit : **[6 Sept. 2020 (Evening)]**

(A) Ammeter is always connected in series and voltmeter in parallel.
(B) Both, ammeter and voltmeter must be connected in series.
(C) Both ammeter and voltmeter must be connected in parallel.
(D) Ammeter is always used in parallel and voltmeter is series.

CHAPTER-18 MOVING CHARGES & MAGNETISM

18.1 Magnetic Effects of Current

150. Magnitude of magnetic field (in SI units) at the centre of a hexagonal shape coil of side 10 cm, 50 turns and carrying current I (Ampere) in units of $\frac{\mu_0 I}{\pi}$ is : **[3 Sept. 2020 (Morning)]**

(A) $250\sqrt{3}$ (B) $5\sqrt{3}$

(C) $500\sqrt{3}$ (D) $50\sqrt{3}$

151. A wire A, bent in the shape of an arc of a circle, carrying a current of $2A$ and having radius 2 cm and another wire B, also bent in the shape of arc of a circle, carrying a current of $3A$ and having radius of 4 cm, are placed as shown in the figure. The ratio of the magnetic fields due to the wires A and B at the common centre O is : **[4 Sept. 2020 (Morning)]**

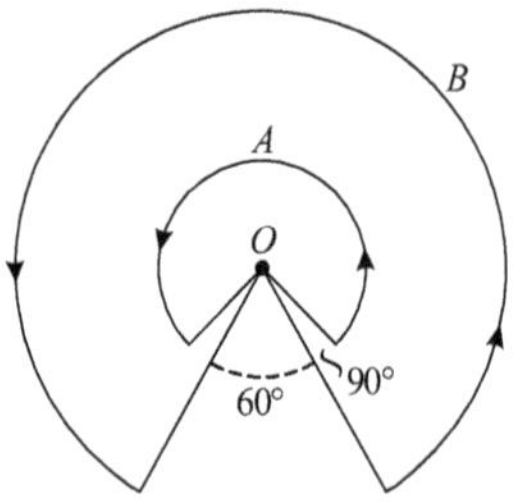

(A) 4 : 6 (B) 6 : 4
(C) 6 : 5 (D) 2 : 5

18.2 Magnetic Force on Moving Charges

152. A beam of protons with speed 4×10^5 ms^{-1} enters a uniform magnetic field of 0.3 T at an angle of 60° to the magnetic field. The pitch of the resulting helical path of protons is close to : (Mass of the proton = 1.67×10^{-27} kg, charge of the proton = 1.69×10^{-19} C) **[2 Sept. 2020 (Morning)]**

(A) 12 cm (B) 4 cm
(C) 5 cm (D) 2 cm

153. A charged particle (mass m and charge q) moves along X-axis with velocity V_0. When it passes through the origin it enters a region having uniform electric field $\vec{E} = -E\hat{j}$ which extends upto $x = d$. Equation of path of electron in the region $x > d$ is : **[2 Sept. 2020 (Morning)]**

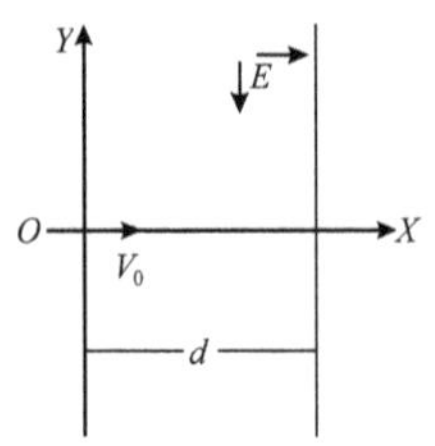

(A) $y = \frac{qEd}{mV_0^2}\left(\frac{d}{2} - x\right)$ (B) $y = \frac{qEd}{mV_0^2}(x - d)$

(C) $y = \frac{qEd}{mV_0^2}x$ (4) $y = \frac{qEd^2}{mV_0^2}x$

154. The figure shows a region of length 'l' with a uniform magnetic field of 0.3 T in it and a proton entering the region with velocity 4×10^5 ms^{-1} making an angle 60° with the field. If the proton completes 10 revolution by the time it cross the region shown, 'l' is close to (mass of proton = 1.67×10^{-27} kg, charge of the proton = 1.6×10^{-19} C) : **[2 sept. 2020 (Evening)]**

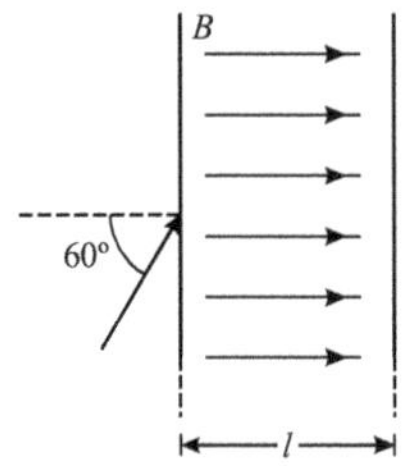

(A) 0.11 m (B) 0.22 m
(C) 0.44 m (D) 0.88 m

155. A wire carrying current I is bent in the shape $ABCDEFA$ as shown, where rectangle $ABCDA$ and $ADEFA$ are perpendicular to each other. If the sides of the rectangles are of lengths a and b, then the magnitude and direction of magnetic moment of the loop $ABCDEFA$ is : **[2 sept. 2020 (Evening)]**

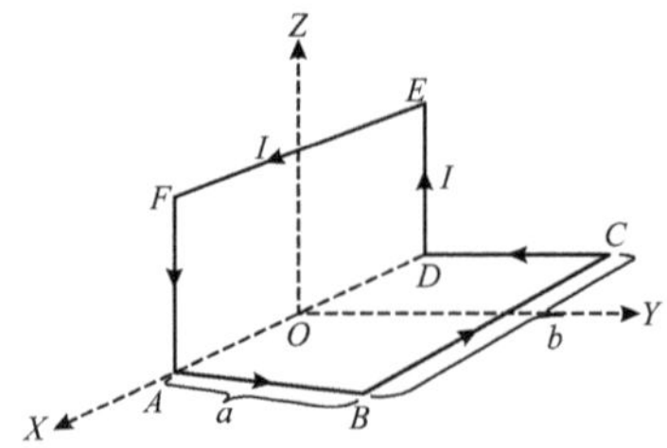

(A) $\sqrt{2}\,abI$, along $\left(\frac{\hat{j}}{\sqrt{2}} + \frac{\hat{k}}{\sqrt{2}}\right)$

(B) $\sqrt{2}\,abI$, along $\left(\frac{\hat{j}}{\sqrt{5}} + \frac{2\hat{k}}{\sqrt{5}}\right)$

(C) abI, along $\left(\frac{\hat{j}}{\sqrt{2}} + \frac{\hat{k}}{\sqrt{2}}\right)$

(D) abI, along $\left(\frac{\hat{j}}{\sqrt{5}} + \frac{2\hat{k}}{\sqrt{5}}\right)$

156. A charged particle carrying charge 1 μC is moving with velocity $(2\hat{i} + 3\hat{j} + 4\hat{k})$ ms^{-1}. If an external magnetic field of $(5\hat{i} + 3\hat{j} - 6\hat{k}) \times 10^{-3}$ T exists in the region where the particle is moving then the force on the particle is $\vec{F} \times 10^{-9}$ N. The vector $\vec{F}$ is : **[3 Sept. 2020 (Morning)]**

(A) $-0.30\hat{i} + 0.32\hat{j} - 0.09\hat{k}$

(B) $-300\hat{i} + 320\hat{j} - 90\hat{k}$

(C) $-30\hat{i} + 32\hat{j} + 9\hat{k}$

(D) $-3.0\hat{i} + 3.2\hat{j} - 0.9\hat{k}$

157. A galvanometer coil has 500 turns and each turn has an average area of 3×10^{-4} m^2. If a torque of 1.5 Nm is required to keep this coil parallel to magnetic field when a current of 0.5 A is flowing through it, the strength of the field (in *T*) is_____: **[3 Sept. 2020 (Evening)]**

158. A circular coil has moment of inertia 0.8 kg m^2 around any diameter and is carrying current to produce a magnetic moment of 20 Am^2. The coil is kept initially in a vertical position and it can rotate freely around a horizontal diameter. When a uniform magnetic field of 4 T is applied along the vertical, it starts rotating around its horizontal diameter. The angular speed the coil acquires after rotating by 60° will be : **[4 Sept. 2020 (Evening)]**

(A) 10 rad s^{-1} (B) 20 π rad s^{-1}
(C) 10 π rad s^{-1} (D) 20 rad s^{-1}

159. A square loop of side $2a$, and carrying current I, is kept in XZ plane with its centre at origin. A long wire carrying the same current I is placed parallel to the z-axis and passing through the point (0, b, 0), ($b >> a$). The magnitude of the torque on the loop about z-axis is given by : **[5 Sept. 2020 (Morning)]**

(A) $\frac{2\mu_0 I^2 a^2}{\pi b}$ (B) $\frac{\mu_0 I^2 a^3}{2\pi b^2}$

(C) $\frac{\mu_0 I^2 a^2}{2\pi b}$ (D) $\frac{2\mu_0 I^2 a^3}{\pi b^2}$

160. A particle of charge q and mass m is moving with a velocity $-v\hat{i}$ ($v \neq 0$) towards a large screen placed in the $Y-Z$ plane at a distance d. If there is a magnetic field $\vec{B} = B_0\hat{k}$, the minimum value of v for which the particle will not hit the screen is : **[6 Sept. 2020 (Morning)]**

(A) $\frac{qdB_0}{2m}$ (B) $\frac{qdB_0}{m}$

(C) $\frac{2qdB_0}{m}$ (D) $\frac{qdB_0}{3m}$

161. An electron is moving along $+x$ direction with a velocity of 6×10^6 ms^{-1}. It enters a region of uniform electric field of 300 V/cm pointing along $+y$ direction. The magnitude and direction of the magnetic field set up in this region such that the electron keeps moving along the x direction will be : **[6 Sept. 2020 (Morning)]**

(A) 5×10^{-3} T, along $+z$ direction
(B) 3×10^{-4} T, along $-z$ direction
(C) 3×10^{-4} T, along $+z$ direction
(D) 5×10^{-3} T, along $-z$ direction

162. A particle moving in the xy plane experiences a velocity dependent force $\vec{F} = k(v_y\hat{i} + v_x\hat{j})$, where v_x and v_y are the x and y components of its velocity $\vec{v}$. If $\vec{a}$ is the acceleration of the particle, then which of the following statements is true for the particle : **[6 Sept. 2020 (Evening)]**

(A) Quantity $\vec{v} \cdot \vec{a}$ is constant in time
(B) Kinetic energy of particle is constant in time
(C) Quantity $\vec{v} \times \vec{a}$ is constant in time
(D) $\vec{F}$ arises due to a magnetic field

163. A charged particle going around in a circle can be considered to be a current loop. A particle of mass m carrying charge q is moving in a plane with speed v under the influence of magnetic field $\vec{B}$. The magnetic moment of this moving particle : **[6 Sept. 2020 (Evening)]**

(A) $-\frac{mv^2\vec{B}}{B^2}$ (B) $-\frac{mv^2\vec{B}}{2\pi B^2}$

(C) $-\frac{mv^2\vec{B}}{3B^2}$ (D) $-\frac{mv^2\vec{B}}{2B^2}$

164. A square loop of side $2a$ and carrying current I is kept in xz plane with its centre at origin. A long wire carrying the same current I is placed parallel to z-axis and passing through point (0, b, 0), ($b >> a$). The magnitude of torque on the loop about z-axis will be : **[6 Sept. 2020 (Evening)]**

(A) $\frac{2\mu_0 I^2 a^2 b}{\pi(a^2+b^2)}$ (B) $\frac{\mu_0 I^2 a^2 b}{2\pi(a^2+b^2)}$

(C) $\frac{\mu_0 I^2 a^2}{2\pi b}$ (D) $\frac{2\mu_0 I^2 a^2}{\pi b}$

CHAPTER-19 MAGNETISM AND MATTER

19.1 Classical and Terrestrial Magnetism

165. A small bar magnet placed with its axis at 30° with an external field of 0.06 T experiences a torque of 0.018 Nm. The minimum work required to rotate it from its stable to unstable equilibrium position is : **[4 Sept. 2020 (Morning)]**

(A) 9.2×10^{-3} J (B) 6.4×10^{-2} J
(C) 11.7×10^{-3} J (D) 7.2×10^{-2} J

166. An iron rod of volume 10^{-3} m^3 and relative permeability 1000 is placed as core in a solenoid with 10 turns/cm. If a current of 0.5A is passed through the solenoid, then the magnetic moment of the rod will be : **[5 Sept. 2020 (Evening)]**

(A) 0.5×10^2 Am^2 (B) 50×10^2 Am^2
(C) 500×10^2 Am^2 (D) 5×10^2 Am^2

19.2 Magnetic Properties of Materials

167. A perfectly diamagnetic sphere has a small spherical cavity at its centre, which is filled with a paramagnetic substance. The whole system is placed in a uniform magnetic field $\vec{B}$. Then the field inside the paramagnetic substance is :

[3 Sept. 2020 (Evening)]

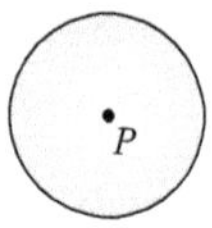

(A) Zero

(B) $\vec{B}$

(C) Much large than $|\vec{B}|$ but opposite to $\vec{B}$

(D) Much large than $|\vec{B}|$ and parallel to $\vec{B}$

168. A paramagnetic sample shows a net magnetisation of 6 A/m when it is placed in an external magnetic field of 0.4 T at a temperature of 4 K. When the sample is placed in an external magnetic field of 0.3 T at a temperature of 24 K, then the magnetisation will be : **[4 Sept. 2020 (Evening)]**

(A) 4 A/m (B) 0.75 A/m

(C) 2.25 A/m (D) 1 A/m

CHAPTER-20 ELECTROMAGNETIC INDUCTION

20.1 Electromagnetic Induction

169. A circular coil of radius 10 cm is placed in a uniform magnetic field of 3.0×10^{-5} T with its plane perpendicular to the field initially. It is rotated at constant angular speed about an axis along the diameter of coil and perpendicular to magnetic field so that it undergoes half of rotation in 0.2s. The maximum value of *EMF* induced (in μV) in the coil will be close to the integer. **[2 Sept. 2020 (Morning)]**

170. An elliptical loop having resistance R, of semi major axis a, and semi minor axis b is placed in a magnetic field as shown in the figure. If the loop is rotated about the x-axis with angular frequency ω, the average power loss in the loop due to Joule heating is : **[3 Sept. 2020 (Morning)]**

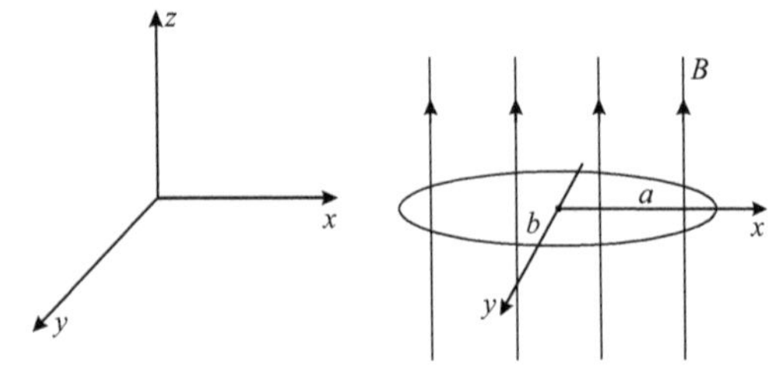

(A) $\frac{\pi^2 a^2 b^2 B^2 \omega^2}{2R}$ (B) Zero

(C) $\frac{\pi^2 a^2 b^2 B^2 \omega^2}{R}$ (D) $\frac{\pi abB\omega}{R}$

171. A uniform magnetic field B exists in a direction perpendicular to the plane of a square loop made of a metal wire. The wire has a diameter of 4 mm and a total length of 30 cm. The magnetic field changes with time at a steady rate $dB/dt = 0.032\ \text{Ts}^{-1}$. The induced current in the loop is close to (Resistivity of the metal wire is 1.23×10^{-8} Wm) :

[3 Sept. 2020 (Evening)]

(A) 0.61 A (B) 0.34 A

(C) 0.43 A (D) 0.53 A

172. A small bar magnet is moved through a coil at constant speed from one end to the other. Which of the following series of observations will be seen on the galvanometer G attached across the coil ? Three positions shown describe : (a) the magnet's entry (b) magnet is completely inside and (c) magnet's exit : **[4 Sept. 2020 (Morning)]**

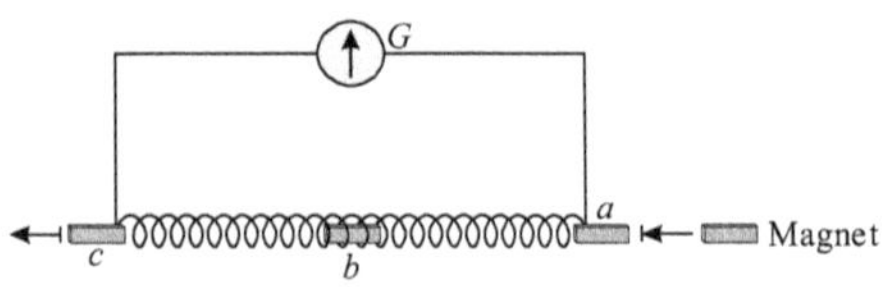

(A)

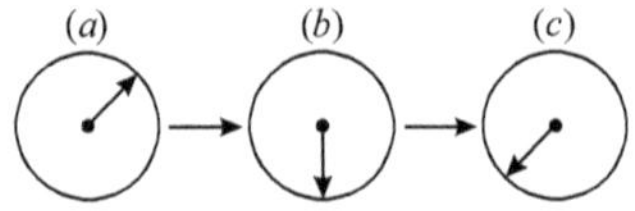

(B)

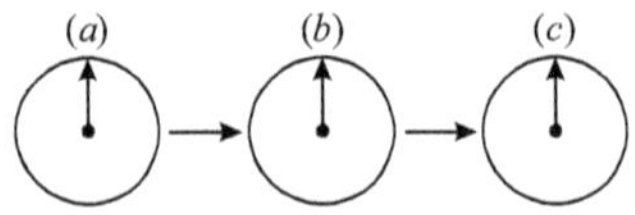

(C)

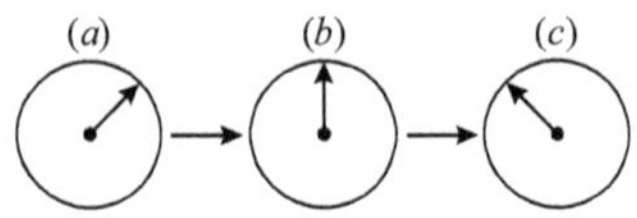

(D)

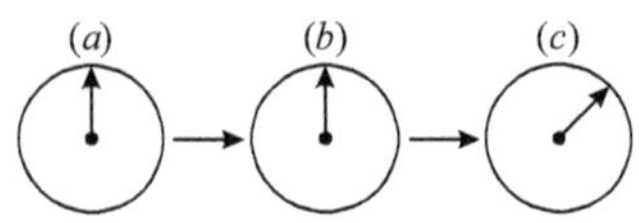

173. Two concentric circular coils, C_1 and C_2, are placed in the XY plane. C_1 has 500 turns, and a radius of 1 cm. C_2 has 200 turns and radius of 20 cm. C_2 carries a time dependent current $I(t) = (5t^2 - 2t + 3)$. A where t is in s. The emf induced in C_1 (in μV), at the instant $t = 1$s is $\frac{4}{x}$. The value of x is :

[5 Sept. 2020 (Morning)]

174. An infinitely long straight wire carrying current I, one side opened rectangular loop and a conductor C with a sliding connector are located in the same plane, as shown in the figure. The connector has length l and resistance R. It slides to the right with a velocity v. The resistance of the conductor and the self inductance of the loop are negligible. The induced current in the loop, as a function of separation r, between the connector and the straight wire is : **[5 Sept. 2020 (Evening)]**

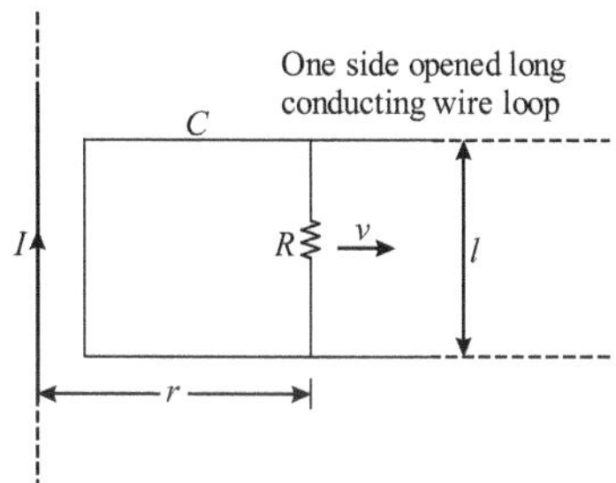

(A) $\frac{\mu_0}{\pi}\frac{Ivl}{Rr}$
(B) $\frac{\mu_0}{2\pi}\frac{Ivl}{Rr}$
(C) $\frac{2\mu_0}{\pi}\frac{Ivl}{Rr}$
(D) $\frac{\mu_0}{4\pi}\frac{Ivl}{Rr}$

20.2 Self and Mutual Induction

175. Magnetic materials used for making permanent magnets (P) and magnets in a transformer (T) have different properties of the following, which property best matches for the type of magnet required : **[2 Sept. 2020 (Morning)]**
(A) T: Large retentivity, small coercivity
(B) P: Small retentivity, large coercivity
(C) T: Large retentivity, large coercivity
(D) P: Large retentivity, large coercivity

176. A series L-R circuit is connected to a battery of *emf* V. If the circuit is switched on at $t = 0$, then the time at which the energy stored in the inductor reaches $(1/n)$ times of its maximum value, is : **[4 Sept. 2020 (Evening)]**

(A) $\frac{L}{R}\ln\left(\frac{\sqrt{n}-1}{\sqrt{n}}\right)$
(B) $\frac{L}{R}\ln\left(\frac{\sqrt{n}}{\sqrt{n}+1}\right)$
(C) $\frac{L}{R}\ln\left(\frac{\sqrt{n}}{\sqrt{n}-1}\right)$
(D) $\frac{L}{R}\ln\left(\frac{\sqrt{n}+1}{\sqrt{n}-1}\right)$

177. A part of a complete circuit is shown in the figure. At some instant, the value of current I is $1A$ and it is decreasing at a rate of $10^2 A\ s^{-1}$. The value of the potential difference $V_P - V_Q$, (in volts) at that instant, is : **[6 Sept. 2020 (Morning)]**

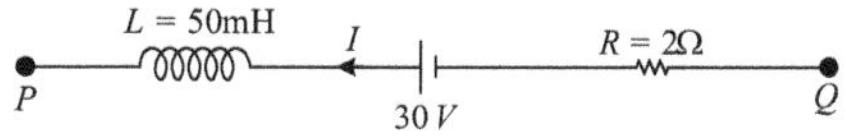

CHAPTER-21 ALTERNATING CURRENT

21.1 Alternating Current

178. An inductance coil has a reactance of 100 W. When an AC signal of frequency 1000 Hz is applied to the coil, the applied voltage leads the current by 45°. The self-inductance of the coil is : **[2 Sept. 2020 (Evening)]**
(A) 1.1×10^{-2} H
(B) 1.1×10^{-1} H
(C) 5.5×10^{-5} H
(D) 6.7×10^{-7} H

179. A 750 Hz, 20 V (r.m.s.) source is connected to a resistance of 100 Ω, an inductance of 0.1803 H and a capacitance of 10 μF all in series. The time in which the resistance (heat capacity 2 J/°C) will get heated by 10°C. (assume no loss of heat to the surroundings) is close to : **[3 Sept. 2020 (Morning)]**
(A) 418 s
(B) 245 s
(C) 348 s
(D) 365 s

180. An AC circuit has $R = 100\ \Omega$, $C = 2\ \mu F$ and $L = 80$ mH, connected in series. The quality factor of the circuit is : **[6 Sept. 2020 (Morning)]**
(A) 0.5
(B) 2
(C) 20
(D) 400

181. For the given input voltage waveform $V_{in}(t)$, the output voltage waveform $V_D(t)$, across the capacitor is correctly depicted by : **[6 Sept. 2020 (Morning)]**

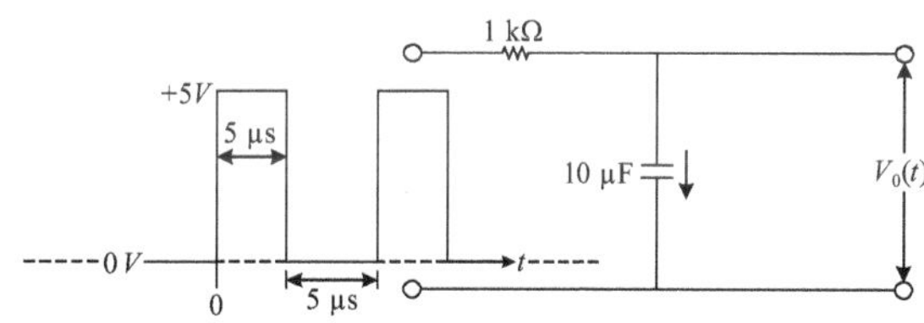

(A)
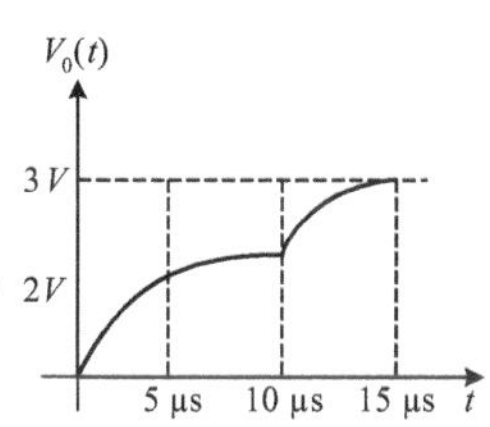

(B)
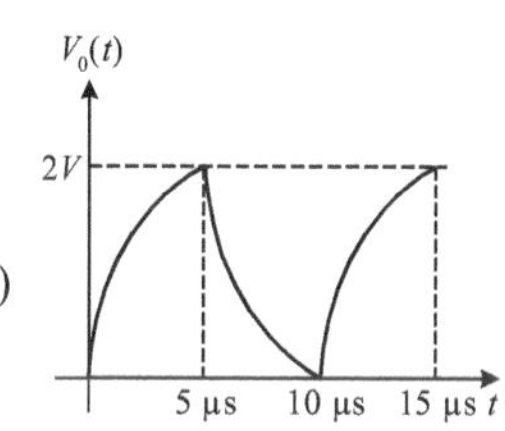

(C)
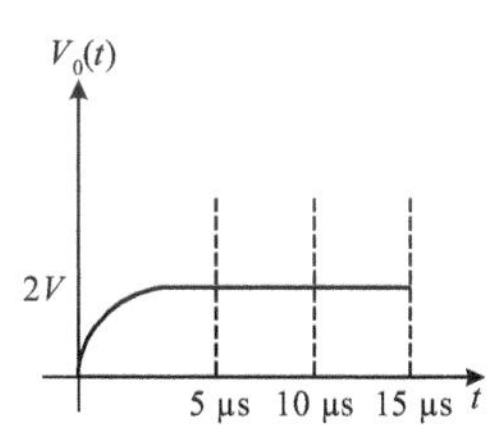

(D)
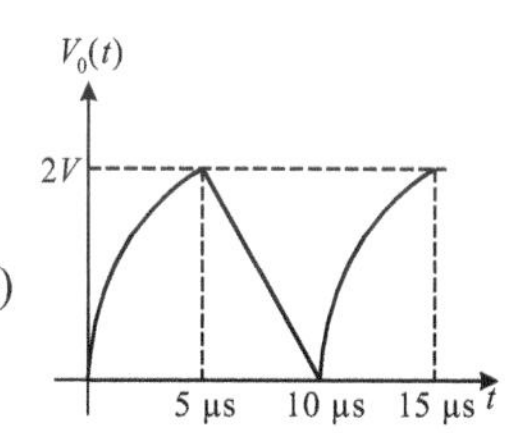

182. In a series LR circuit, power of 400 W is dissipated from a source of 250 V, 50 Hz. The power factor of the circuit is 0.8. In order to bring the power factor to unity, a capacitor of value C is added in series to the L and R. Taking the value of C as $\left(\frac{n}{3\pi}\right)\mu F$, then value of n is : **[6 Sept. 2020 (Evening)]**

CHAPTER-22 ELECTROMAGNETIC WAVES

22.1 Electromagnetic Wave

183. A plane electromagnetic wave, has frequency of 2.0×10^{10} Hz and its energy density is 1.02×10^{-8} J/m^3 in vacuum. The amplitude of the magnetic field of the wave is close to :

[Take: $\frac{1}{4\pi\varepsilon_0} = 9 \times 10^9 \frac{Nm^2}{C^2}$ and speed of light $= 3 \times 10^8$ ms^{-1}]

[2 Sept. 2020 (Morning)]

(A) 180 nT (B) 160 nT
(C) 150 nT (D) 190 nT

184. In a plane electromagnetic wave, the directions of electric field and magnetic field are represented by $\hat{k}$ and $2\hat{i} - 2\hat{j}$, respectively. What is the unit vector along direction of propagation of the wave : **[2 Sept. 2020 (Evening)]**

(A) $\frac{1}{\sqrt{2}}(\hat{i} + \hat{j})$ (B) $\frac{1}{\sqrt{5}}(\hat{i} + 2\hat{j})$

(C) $\frac{1}{\sqrt{5}}(2\hat{i} + \hat{j})$ (D) $\frac{1}{\sqrt{2}}(\hat{j} + \hat{k})$

185. The magnetic field of a plane electromagnetic wave is given as $\vec{B} = 3 \times 10^{-8} \sin[200\pi(y + ct)]\,\hat{i}\ T$. Where $c = 3 \times 10^8$ ms^{-1} is the speed of light. The corresponding electric field is : **[3 Sept. 2020 (Morning)]**

(A) $\vec{E} = 10^{-6} \sin[200\pi(y+ct)]\,\hat{k}$ V/m

(B) $\vec{E} = -9 \sin[200\pi(y+ct)]\,\hat{k}$ V/m

(C) $\vec{E} = 9 \sin[200\pi(y+ct)]\,\hat{k}$ V/m

(D) $\vec{E} = 3 \times 10^{-8} \sin[200\pi(y+ct)]\,\hat{k}$ V/m

186. The electric field of a plane electromagnetic wave propagating along the x direction in vacuum is $\vec{E} = E_0 \hat{j} \cos(\omega t - kx)$. The magnetic field $\vec{B}$, at the moment $t = 0$ is: **[3 Sept. 2020 (Evening)]**

(A) $\vec{B} = E_0\sqrt{\mu_0 \epsilon_0} \cos(kx)\,\hat{j}$

(B) $\vec{B} = \frac{E_0}{\sqrt{\mu_0 \epsilon_0}} \cos(kx)\,\hat{k}$

(C) $\vec{B} = E_0\sqrt{\mu_0 \epsilon_0} \cos(kx)\,\hat{k}$

(D) $\vec{B} = \frac{E_0}{\sqrt{\mu_0 \epsilon_0}} \cos(kx)\,\hat{j}$

187. Choose the correct option relating wavelengths of different parts of electromagnetic wave spectrum :

[4 Sept. 2020 (Morning)]

(A) $\lambda_{x\text{-rays}} < \lambda_{\text{micro waves}} < \lambda_{\text{radio waves}} < \lambda_{\text{visible}}$

(B) $\lambda_{\text{visible}} > \lambda_{x\text{-rays}} > \lambda_{\text{radio waves}} > \lambda_{\text{micro waves}}$

(C) $\lambda_{\text{radio waves}} > \lambda_{\text{micro waves}} > \lambda_{\text{visible}} > \lambda_{x\text{-rays}}$

(D) $\lambda_{\text{visible}} < \lambda_{\text{micro waves}} < \lambda_{\text{radio waves}} < \lambda_{x\text{-rays}}$

188. The electric field of a plane electromagnetic wave is given by $\vec{E} = E_0(\hat{x} + \hat{y})\sin(kz - \omega t)$. Its magnetic field will be given by : **[4 Sept. 2020 (Evening)]**

(A) $\frac{E_0}{c}(\hat{x} - \hat{y})\cos(kz - \omega t)$ (B) $\frac{E_0}{c}(-\hat{x} + \hat{y})\sin(kz - \omega t)$

(C) $\frac{E_0}{c}(\hat{x} - \hat{y})\sin(kz - \omega t)$ (D) $\frac{E_0}{c}(\hat{x} + \hat{y})\sin(kz - \omega t)$

189. An electron is constrained to move along the y-axis with a speed of 0.1 c (c is the speed of light) in the presence of electromagnetic wave, whose electric field is $\vec{E} = 30\hat{j} \sin(1.5 \times 10^7 t - 5 \times 10^{-2} x)$ V/m The maximum magnetic force experienced by the electron will be :
(Given $c = 3 \times 10^8$ ms^{-1} and electron charge $= 1.6 \times 10^{-19}$ C) :

[5 Sept. 2020 (Morning)]

(A) 1.6×10^{-19} N (B) 4.8×10^{-19} N
(C) 3.2×10^{-18} N (D) 2.4×10^{-18} N

190. The correct match between the entries in column-I and column-II are : **[5 Sept. 2020 (Evening)]**

Column-I	Column-II
Radiation	**Wavelength**
(a) Microwave	(i) 100m
(b) Gamma rays	(ii) 10^{-15} m
(c) A.M. radio waves	(iii) 10^{-10} m
(d) X-rays	(iv) 10^{-3} m

(A) (a)-(ii), (b)-(i), (c)-(iv), (d)-(iii)
(B) (a)-(i), (b)-(iii), (c)-(iv), (d)-(ii)
(C) (a)-(iii), (b)-(ii), (c)-(i), (d)-(iv)
(D) (a)-(iv), (b)-(ii), (c)-(i), (d)-(iii)

191. Suppose that intensity of a laser is $\left(\frac{315}{\pi}\right)$ W/m^2. The rms electric field, in units of V/m associated with this source is close to the nearest integer is ($\epsilon_0 = 8.86 \times 10^{-12}$ C^2 Nm^{-2}; $c = 3 \times 10^8$ ms^{-1}): **[6 Sept. 2020 (Morning)]**

192. For a plane electromagnetic wave, the magnetic field at a point x and time t is $\vec{B}(x, t) = [1.2 \times 10^{-7} \sin(0.5 \times 10^3 x + 1.5 \times 10^{11} t)]\,T$. The instantaneous electric field $\vec{E}$ corresponding to $\vec{B}$ is : (speed of light $c = 3 \times 10^8$ ms^{-1}) :

[6 Sept. 2020 (Evening)]

(A) $\vec{E}(x,t) = [36\sin(0.5\times10^3x + 1.5\times10^{11}t)\,\hat{k}]\frac{V}{m}$

(B) $\vec{E}(x,t) = [36\sin(0.5\times10^3x + 1.5\times10^{11}t)\,\hat{j}]\frac{V}{m}$

(C) $\vec{E}(x,t) = [36\sin(1\times10^3x + 0.5\times10^{11}t)\,\hat{j}]\frac{V}{m}$

(D) $\vec{E}(x,t) = [36\sin(1\times10^3x + 1.5\times10^{11}t)\,\hat{j}]\frac{V}{m}$

CHAPTER-23 RAY OPTICS AND OPTICAL INSTRUMENTS

23.1 Geometrical Optics I - Reflection of Light

193. A spherical mirror is obtained as shown in the figure from a hollow glass sphere. If an object is positioned in front of the mirror, what will be the nature and magnification of the image of the object ? (Figure drawn as schematic and not to scale) **[2 Sept. 2020 (Morning)]**

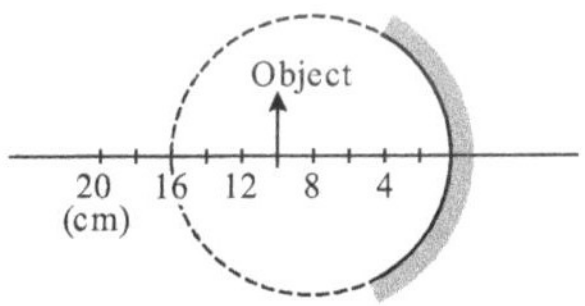

(A) Inverted, real and magnified
(B) Erect, virtual and magnified
(C) Erect, virtual and unmagnified
(D) Inverted, real and unmagnified

194. When an object is kept at a distance of 30 cm from a concave mirror, the image is formed at a distance of 10 cm from the mirror. If the object is moved with a speed of 9 cms^{-1}, the speed (in cms^{-1}) with which image moves at that instant is ______: **[3 Sept. 2020 (Evening)]**

23.2 Geometrical Optics II - Refraction of Light

195. A light ray enters a solid glass sphere of refractive index $\mu = \sqrt{3}$ at an angle of incidence 60°. The ray is both reflected and refracted at the farther surface of the sphere. The angle (in degrees) between the reflected and refracted rays at this surface is: **[2 Sept. 2020 (Evening)]**

196. An observer can see through a small hole on the side of a jar (radius 15 cm) at a point at height of 15 cm from the bottom (see figure). The hole is at a height of 45 cm. When the jar is filled with a liquid up to a height of 30 cm the same observer can see the edge at the bottom of the jar. If the refractive index of the liquid $N/100$, where N is an integer, the value of N is______: **[3 Sept. 2020 (Morning)]**

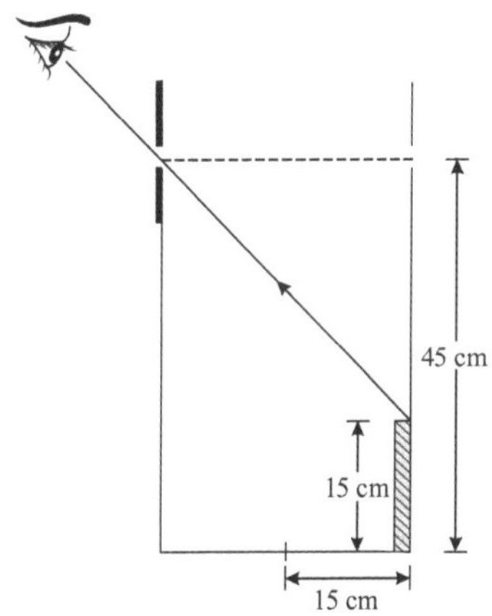

197. A prism of angle $A = 1°$ has a refractive index $\mu = 1.5$. A good estimate for the minimum angle of deviation (in degrees) is close to N/10. Value of N is ______: **[5 Sept. 2020 (Evening)]**

23.3 Geometrical Optics III - Thin Lenses

198. The distance between an object and a screen is 100 cm. A lens can produce real image of the object on the screen for two different positions between the screen and the object. The distance between these two positions is 40 cm. If the power of the lens is close to $\left(\frac{N}{100}\right)$D where N is an integer, the value of N is ______: **[4 Sept. 2020 (Evening)]**

199. For a concave lens of focal length f, the relation between object and image distance u and v, respectively, from its pole can best be represented by ($u = v$ is the reference line) : **[5 Sept. 2020 (Morning)]**

(A)

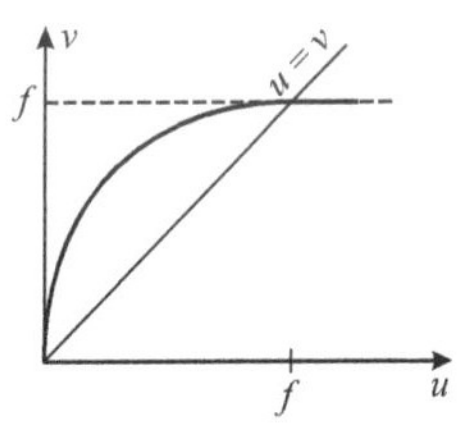

(B)

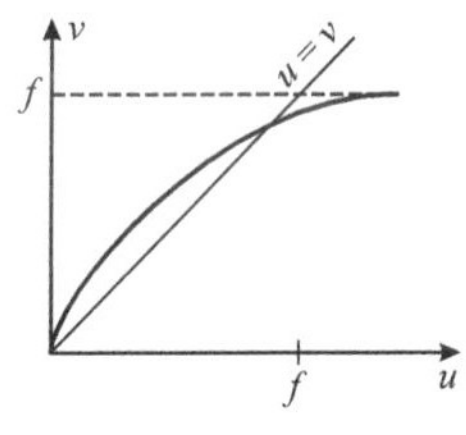

(C)

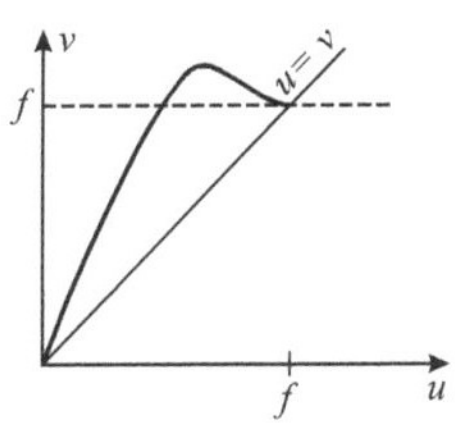

(D)

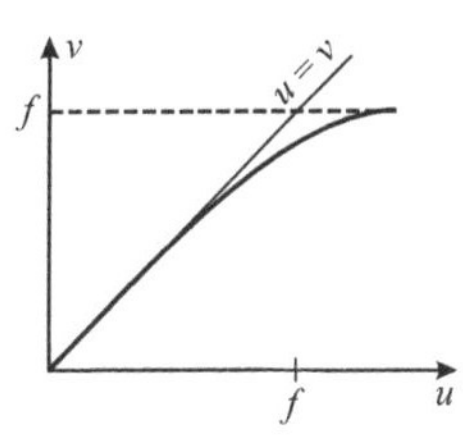

200. A point like object is placed at a distance of 1 m in front of a convex lens of focal length 0.5 m. A plane mirror is placed at a distance of 2 m behind the lens. The position and nature of the final image formed by the system is : **[6 Sept. 2020 (Morning)]**

(A) 1 m from the mirror, virtual
(B) 1 m from the mirror, real
(C) 2.6 m from the mirror, real
(D) 2.6 m from the mirror, virtual

201. A double convex lens has power P and same radii of curvature R of both the surfaces. The radius of curvature of a surface of a plano-convex lens made of the same material with power $1.5\,P$ is : **[6 Sept. 2020 (Evening)]**

(A) $\frac{R}{2}$ (B) $2R$
(C) $\frac{3R}{2}$ (D) $\frac{R}{3}$

23.5 Optical Instruments

202. In a compound microscope, the magnified virtual image is formed at a distance of 25 cm from the eye-piece. The focal length of its objective lens is 1 cm. If the magnification is 100 and the tube length of the microscope is 20 cm, then the focal length of the eye-piece lens (in cm) is ________ : **[4 Sept. 2020 (Morning)]**

203. A compound microscope consists of an objective lens of focal length 1 cm and an eye piece of focal length 5 cm with a separation of 10 cm. The distance between an object and the objective lens, at which the strain on the eye is minimum is $\frac{n}{40}$ cm. The value of n is : **[5 Sept. 2020 (Morning)]**

CHAPTER-24 WAVE OPTICS

24.1 Interference of Light

204. Interference fringes are observed on a screen by illuminating two thin slits 1 mm apart with a light source (l = 632.8 nm). The distance between the screen and the slits is 100 cm. If a bright fringe is observed on a screen at a distance of 1.27 mm from the central bright fringe, then the path difference between the waves, which are reaching this point from the slits is close to : **[2 Sept. 2020 (Morning)]**

(A) 1.27 mm (B) 2 nm
(C) 2.87 nm (D) 2.05 mm

205. In a Young's double slit experiment, 16 fringes are observed in a certain segment of the screen when light of wavelength 700 nm is used. If the wavelength of light is changed to 400 nm, the number of fringes observed in the same segment of the screen would be : **[2 Sept. 2020 (Evening)]**

(A) 28 (B) 24
(C) 18 (D) 30

206. In a Young's double slit experiment, light of 500 nm is used to produce an interference pattern. When the distance between the slits is 0.05 mm, the angular width (in degree) of the fringes formed on the distance screen is close to : **[3 Sept. 2020 (Morning)]**

(A) 0.07° (B) 0.17°
(C) 1.7° (D) 0.57°

207. Two light waves having the same wavelength λ in vacuum are in phase initially. Then the first wave travels a path L_1 through a medium of refractive index n_1 while the second wave travels a path of length L_2 through a medium of refractive index n_2. After this the phase difference between the two waves is : **[3 Sept. 2020 (Evening)]**

(A) $\frac{2\pi}{\lambda}(n_1L_1 - n_2L_2)$ (B) $\frac{2\pi}{\lambda}\left(\frac{L_2}{n_1} - \frac{L_1}{n_2}\right)$

(C) $\frac{2\pi}{\lambda}\left(\frac{L_1}{n_1} - \frac{L_2}{n_2}\right)$ (D) $\frac{2\pi}{\lambda}(n_2L_1 - n_1L_2)$

208. Two coherent sources of sound, S_1 and S_2, produce sound waves of the same wavelength, $\lambda = 1$ m, in phase. S_1 and S_2 are placed 1.5 m apart (see figure) A listener, located at L, directly in front of S_2 finds that the intensity is at a minimum when he is 2 m away from S_2. The listener moves away from S_1, keeping his distance from S_2 fixed. The adjacent maximum of intensity is observed when the listener is at a distance d from S_1. Then, d is : **[5 Sept. 2020 (Evening)]**

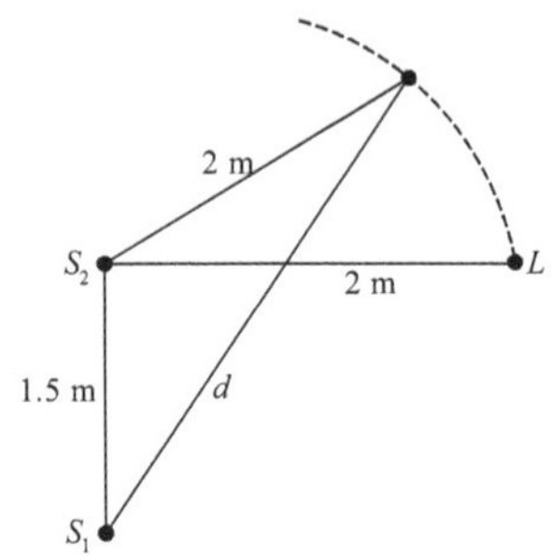

(A) 12m (B) 3m
(C) 5m (D) 2m

209. In the figure below, P and Q are two equally intense coherent sources emitting radiation of wavelength 20 m. The separation between P and Q is 5 m and the phase of P is ahead of that of Q by 90°. A, B and C are three distinct points of observation, each equidistant from the midpoint of PQ. The intensities of radiation at A, B, C will be in the ratio : **[6 Sept. 2020 (Morning)]**

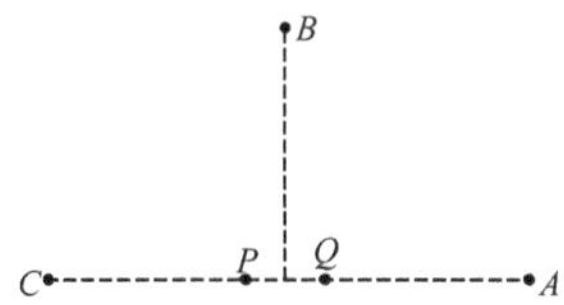

(A) 0 : 1 : 2 (B) 4 : 1 : 0
(C) 0 : 1 : 4 (D) 2 : 1 : 0

210. A Young's double-slit experiment is performed using monochromatic light of wavelength l. The intensity of light at a point on the screen, where the path difference is l, is K units. The intensity of light at a point where the path difference is $\frac{\lambda}{6}$ is given by $\frac{nK}{12}$, where n is an integer. The value of n is : **[6 Sept. 2020 (Evening)]**

24.2 Diffraction of Light

211. Orange light of wavelength 6000×10^{-10} m in illuminates a single slit of width 0.6×10^{-4} m. The maximum possible number of diffraction minima produced on both sides of the central maximum is______: **[4 Sept. 2020 (Evening)]**

212. A beam of electrons of energy E scatters from a target having atomic spacing of 1 Å. The first maximum intensity occurs at $\theta = 60°$. Then E (in eV) is_____.
(Planck constant $h = 6.64 \times 10^{-34}$ Js, 1 eV $= 1.6 \times 10^{-19}$ J, electron mass $m = 9.1 \times 10^{-31}$ kg) : **[5 Sept. 2020 (Morning)]**

24.3 Polarization of Light

213. A beam of plane polarised light of large cross sectional area and uniform intensity of 3.3 Wm^{-2} falls normally on a polariser (cross sectional area 3×10^{-4} m^2) which rotates about its axis with an angular speed of 31.4 rad/s. The energy of light passing through the polariser per revolution, is close to : **[4 Sept. 2020 (Morning)]**

(A) 1.0×10^{-5} J (B) 5.0×10^{-4} J
(C) 1.0×10^{-4} J (D) 1.5×10^{-4} J

CHAPTER-25 DUAL NATURE OF RADIATION, MATTER & X-RAYS

25.1 Photoelectric Effect

214. When radiation of wavelength λ is used to illuminate a metallic surface, the stopping potential is V. When the same surface is illuminated with radiation of wavelength 3λ, the stopping potential is $\frac{V}{4}$. If the threshold wavelength for the metallic surface is $n\lambda$ then value of n will be : **[2 Sept. 2020 (Morning)]**

215. When the wavelength of radiation falling on a metal is changed from 500 nm to 200 nm, the maximum kinetic energy of the photoelectrons becomes three times larger. The work function of the metal is close to : **[3 Sept. 2020 (Morning)]**

(A) 0.61 eV (B) 0.52 eV
(C) 0.81 eV (D) 1.02 eV

216. Given figure shows few data points in a photo electric effect experiment for a certain metal. The minimum energy for ejection of electron from its surface is : (Planck's constant $h = 6.62 \times 10^{-34}$ J.s) : **[4 Sept. 2020 (Morning)]**

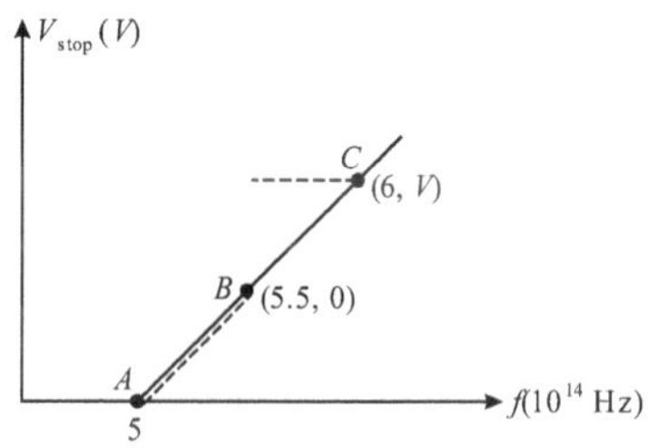

(A) 2.27 eV (B) 2.59 eV
(C) 1.93 eV (D) 2.10 eV

217. In a photoelectric effect experiment, the graph of stopping potential V versus reciprocal of wavelength obtained is shown in the figure. As the intensity of incident radiation is increased : **[4 Sept. 2020 (Evening)]**

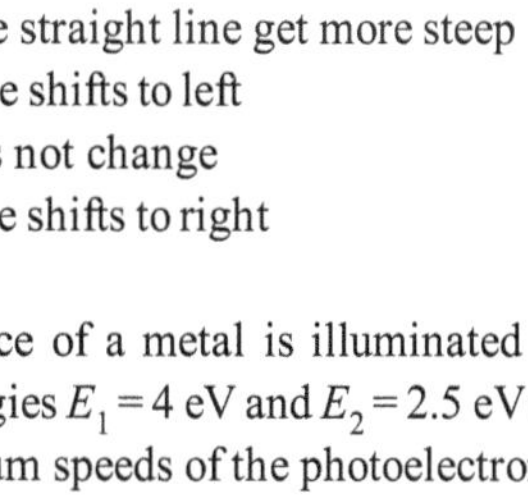

(A) Slope of the straight line get more steep
(B) Straight line shifts to left
(C) Graph does not change
(D) Straight line shifts to right

218. The surface of a metal is illuminated alternately with photons of energies $E_1 = 4$ eV and $E_2 = 2.5$ eV respectively. The ratio of maximum speeds of the photoelectrons emitted in the two cases is 2. The work function of the metal in (eV) is______: **[5 Sept. 2020 (Evening)]**

25.2 Matter Waves and X-Ray

219. A particle is moving 5 times as fast as an electron. The ratio of the de-Broglie wavelength of the particle to that of the electron is 1.878×10^{-4}. The mass of the particle is close to : **[2 Sept. 2020 (Evening)]**

(A) 4.8×10^{-27} kg (B) 1.2×10^{-28} kg

(C) 9.1×10^{-31} kg (D) 9.7×10^{-28} kg

220. Two sources of light emit X-rays of wavelength 1 nm and visible light of wavelength 500 nm, respectively. Both the sources emit light of the same power 200 W. The ratio of the number density of photons of X-rays to the number density of photons of the visible light of the given wavelengths is : **[3 Sept. 2020 (Evening)]**

(A) $\frac{1}{500}$ (B) 500

(C) 250 (D) $\frac{1}{250}$

221. Particle A of mass $m_A = \frac{m}{2}$ moving along the x-axis with velocity v_0 collides elastically with another particle B at rest having mass $m_B = \frac{m}{3}$. If both particles move along the x-axis after the collision, the change $\Delta\lambda$ in de-Broglie wavelength of particle A, in terms of its de-Broglie wavelength (λ_0) before collision is : **[4 Sept. 2020 (Morning)]**

(A) $\Delta\lambda = 4\lambda_0$ (B) $\Delta\lambda = \frac{5}{2}\lambda_0$

(C) $\Delta\lambda = 2\lambda_0$ (D) $\Delta\lambda = \frac{3}{2}\lambda_0$

222. An electron, a doubly ionized helium ion (He^{++}) and a proton are having the same kinetic energy. The relation between their respective de-Broglie wavelengths λ_e, $\lambda_{He^{++}}$ and λ_p is : **[6 Sept. 2020 (Morning)]**

(A) $\lambda_e < \lambda_p < \lambda_{He^{++}}$ (B) $\lambda_e < \lambda_{He^{++}} = \lambda_p$

(C) $\lambda_e > \lambda_{He^{++}} > \lambda_p$ (D) $\lambda_e > \lambda_p > \lambda_{He^{++}}$

223. Assuming the nitrogen molecule is moving with r.m.s. velocity at 400 K, the de-Broglie wavelength of nitrogen molecule is close to :
(Given : Nitrogen molecule weight
4.64×10^{-26} kg, Boltzmann constant
1.38×10^{-23} J/K, Planck constant
6.63×10^{-34} J.s) **[6 Sept. 2020 (Evening)]**

(A) 0.34 Å (B) 0.24 Å

(C) 0.20 Å (D) 0.44 Å

CHAPTER-26 ATOMS

26.1 Atomic Structure

224. In a hydrogen atom the electron makes a transition from $(n+1)^{th}$ level to the nth level. If $n >> l$, the frequency of radiation emitted is proportional to : **[2 Sept. 2020 (Evening)]**

(A) $\frac{1}{n^4}$ (B) $\frac{1}{n^3}$

(C) $\frac{1}{n^2}$ (D) $\frac{1}{n}$

225. In the line spectra of hydrogen atom, difference between the largest and the shortest wavelengths of the Lyman series is 304 Å. The corresponding difference for the Paschen series in Å is : ______ : **[4 Sept. 2020 (Morning)]**

226. A particle of mass 200 MeV/c^2 collides with a hydrogen atom at rest. Soon after the collision the particle comes to rest, and the atom recoils and goes to its first excited state. The initial kinetic energy of the particle (in eV) is $\frac{N}{4}$. The value of N is :
(Given the mass of the hydrogen atom to be 1 GeV/c^2) : **[5 Sept. 2020 (Morning)]**

CHAPTER-27 NUCLEI

27.1 Radioactivity

227. In a radioactive material, fraction of active material remaining after time t is 9/16. The fraction that was remaining after $t/2$ is : **[3 Sept. 2020 (Morning)]**

(A) $\frac{3}{4}$ (B) $\frac{7}{8}$

(C) $\frac{4}{5}$ (D) $\frac{3}{5}$

228. Activities of three radioactive substances A, B and C are represented by the curves A, B and C, in the figure. Then their half-lives $T_{1/2}(A) : T_{1/2}(B) : T_{1/2}(C)$ are in the ratio : **[5 Sept. 2020 (Morning)]**

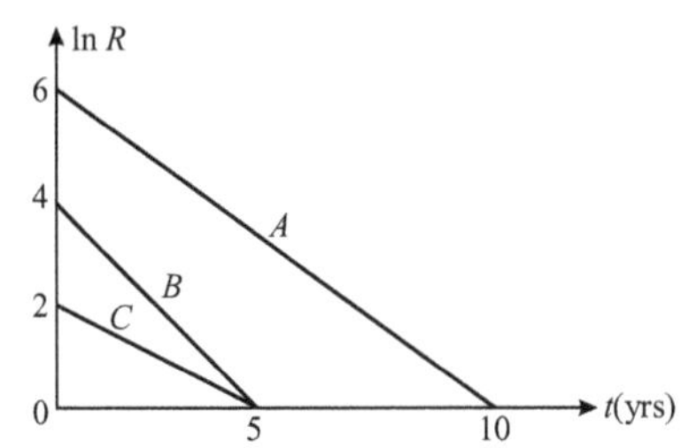

(A) 3 : 2 : 1 (B) 4 : 3 : 1
(C) 2 : 1 : 3 (D) 2 : 1 : 1

229. A radioactive nucleus decays by two different processes. The half life for the first process is 10 s and that for the second is 100 s. The effective half life of the nucleus is close to :

[5 Sept. 2020 (Evening)]

(A) 9 s (B) 55 s
(C) 6 s (D) 12 s

27.2 Nuclear Physics

230. In a reactor, 2 kg of $_{92}U^{235}$ fuel is fully used up in 30 days. The energy released per fission is 200 MeV. Given that the Avogadro number, $N = 6.023 \times 10^{26}$ per kilo mole and $1\ eV = 1.6 \times 10^{-19}$ J. The power output of the reactor is close to :

[2 Sept. 2020 (Morning)]

(A) 125 MW (B) 60 MW
(C) 35 MW (D) 54 MW

231. The radius R of a nucleus of mass number A can be estimated by the formula $R = (1.3 \times 10^{-15})A^{1/3}$ m. It follows that the mass density of a nucleus is of the order of :
$(M_{prot.} \cong M_{neut.} \simeq 1.6 \times 10^{-27}\text{kg})$: **[3 Sept. 2020 (Evening)]**

(A) 10^{24} kg m^{-3} (B) 10^{3} kg m^{-3}
(C) 10^{17} kg m^{-3} (D) 10^{10} kg m^{-3}

232. Find the binding energy per nucleon for $^{120}_{50}Sn$. Mass of proton $m_P = 1.00783U$, mass of neutron $m_n = 1.00867U$ and mass of tin nucleus $m_{Sn} = 119.902199$ U. (Take 1U = 931 MeV) :

[4 Sept. 2020 (Evening)]

(A) 8.5 MeV (B) 7.5 MeV
(C) 8.0 MeV (D) 9.0 MeV

233. You are given that Mass of $^{7}_{3}Li$ = 7.0160 u, Mass of $^{4}_{2}He = 4.0026\ u$ and Mass of $^{1}_{1}H = 1.0079\ u$. When 20 g of $^{7}_{3}Li$ is converted into $^{4}_{2}He$ by proton capture, the energy liberated, (in kWh), is : [Mass of nucleon = 1 GeV/c^2]

[6 Sept. 2020 (Morning)]

(A) 8×10^6 (B) 1.33×10^6
(C) 6.82×10^5 (D) 4.5×10^5

234. Given the masses of various atomic particles $m_p = 1.0072$ u, $m_n = 1.0087$ u, $m_e = 0.000548$ u, $m_{\bar{\nu}} = 0$, $m_d = 2.0141$ u, where p is proton, n is neutron, e is electron, $\bar{\nu}$ is antineutrino and d is deuteron. Which of the following process is allowed by momentum and energy conservation :

[6 Sept. 2020 (Evening)]

(A) $n + p \rightarrow d + g$
(B) $e^{+} + e^{-} \rightarrow g$
(C) $n + n \rightarrow$ deuterium atom (electron bound to the nucleus)
(D) $p \rightarrow n + e^{+} + \bar{\nu}$

CHAPTER-28 SEMICONDUCTORS ELECTRONICS

28.2 PN Junction and Diodes

235. When a diode is forward biased, it has a voltage drop of 0.5 V. The safe limit of current through the diode is 10 mA. If a battery of emf 1.5 V is used in the circuit, the value of minimum resistance to be connected in series with the diode so that the current does not exceed the safe limit is :

[3 Sept. 2020 (Morning)]

(A) 100 Ω (B) 50 Ω
(C) 300 Ω (4) 200 Ω

236. If a semiconductor photodiode can detect a photon with a maximum wavelength of 400 nm, then its band gap energy is :

[3 Sept. 2020 (Evening)]

Planck's constant $h = 6.63 \times 10^{-34}$ Js
Speed of light $c = 3 \times 10^{8}$ m/s

(A) 2.0 eV (B) 1.5 eV
(C) 3.1 eV (D) 1.1 eV

237. Take the breakdown voltage of the zener diode used in the given circuit as 6 V. For the input voltage shown in figure below, the time variation of the output voltage is : (Graphs drawn are schematic and not to scale) : **[4 Sept. 2020 (Morning)]**

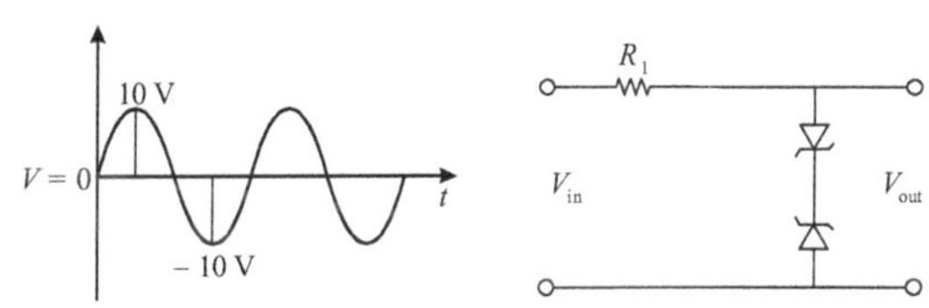

(A)
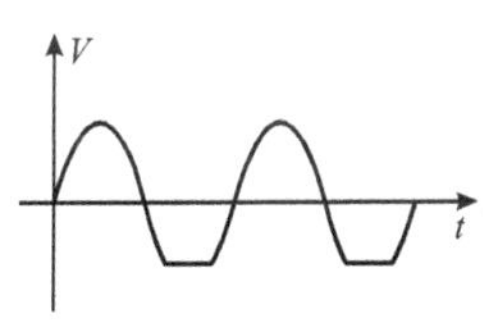

(B)
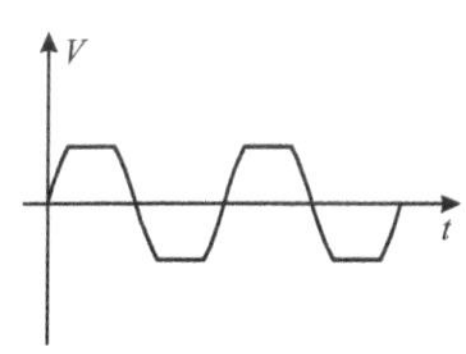

(C)
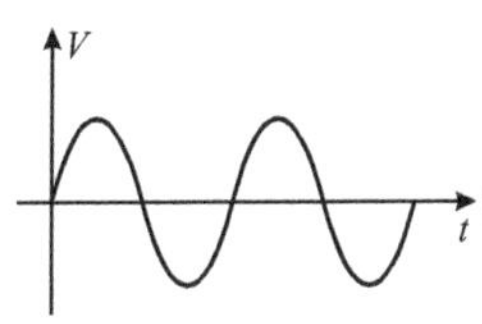

(D)
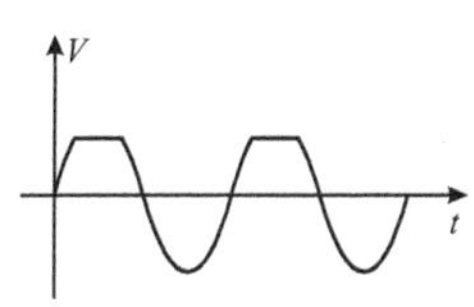

238. With increasing biasing voltage of a photodiode, the photocurrent magnitude : **[5 Sept. 2020 (Morning)]**

(A) Increases initially and saturates finally

(B) Increases initially and after attaining certain value, it decreases

(C) Increases linearly

(D) Remains constant

239. Two Zener diodes (A and B) having breakdown voltages of 6V and 4V respectively, are connected as shown in the circuit below. The output voltage V_O variation with input voltage linearly increasing with time, is given by : ($V_{input} = 0V$ at $t = 0$) (figures are qualitative) : **[5 Sept. 2020 (Evening)]**

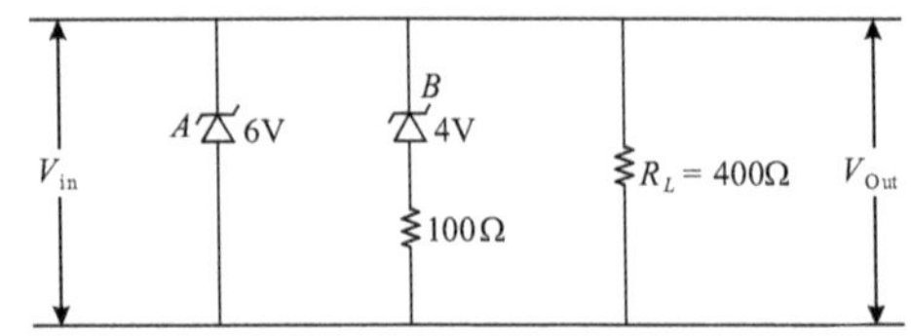

(A)
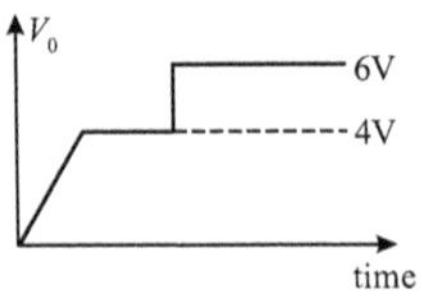

(B)
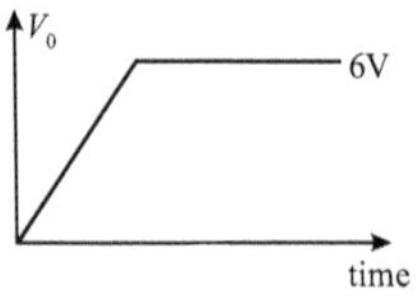

(C)
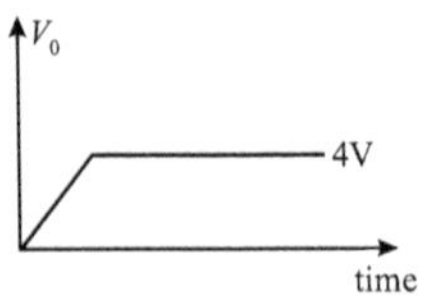

(D)
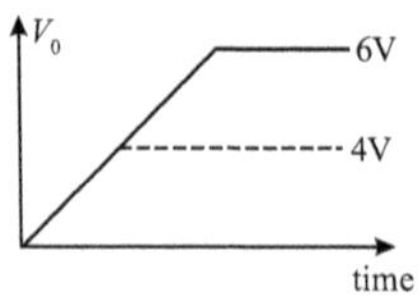

28.3 Transistors

240. The output characteristics of a transistor is shown in the figure. When VCE is 10 V and $1C = 4.0$ μA, then value of back is : **[6 Sept. 2020 (Evening)]**

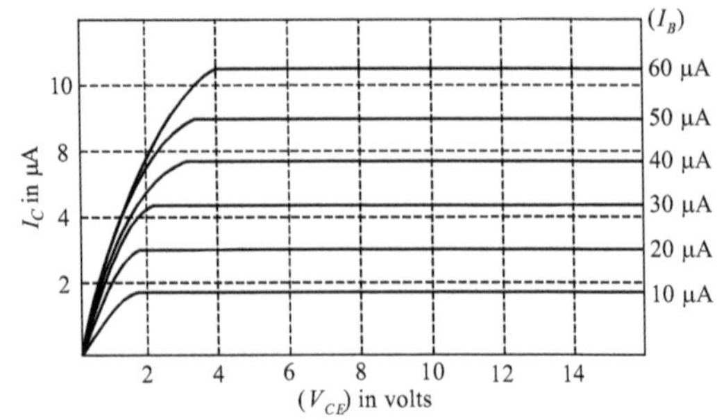

28.4 Logic Gates

241. In the following digital circuit, what will be the output at 'Z', when the input (A, B) are (1, 0), (0, 0), (1, 1), (0, 1) : **[2 Sept. 2020 (Evening)]**

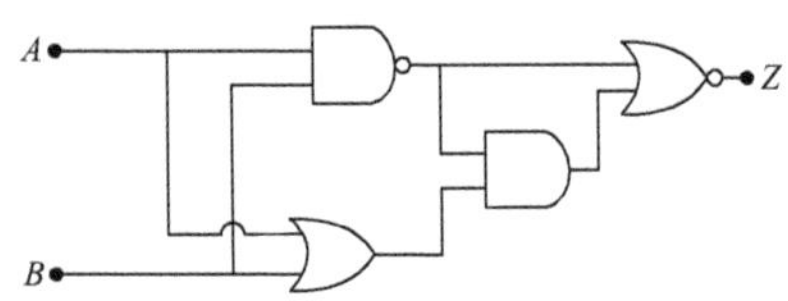

(A) 1, 0, 1, 1 (B) 0, 1, 0, 0

(C) 0, 0, 1, 0 (D) 1, 1, 0, 1

242. Identify the operation performed by the circuit given below : **[4 Sept. 2020 (Evening)]**

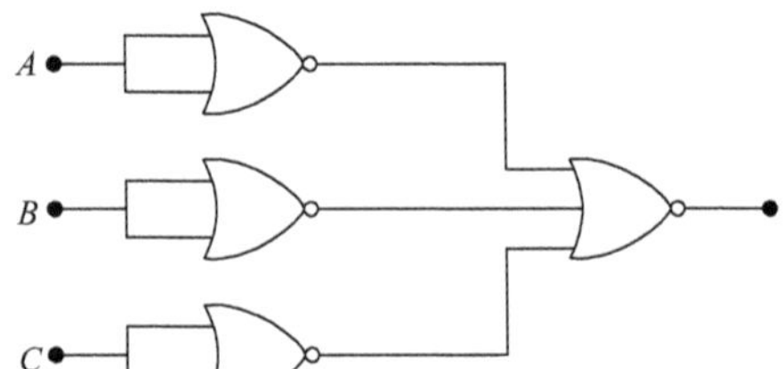

(A) AND (B) NAND

(C) OR (D) NOT

243. Identify the correct output signal Y in the given combination of gates (as shown) for the given inputs A and B : **[6 Sept. 2020 (Morning)]**

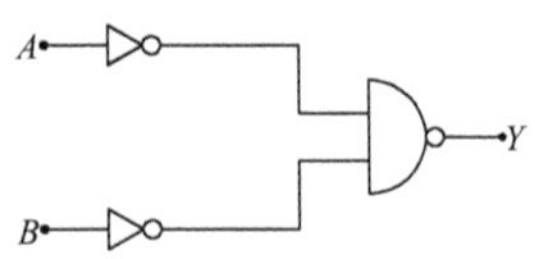

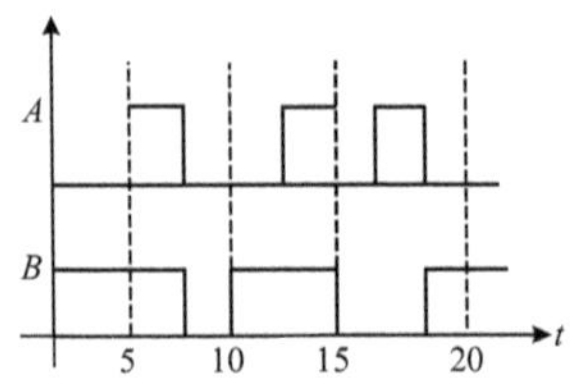

(A)
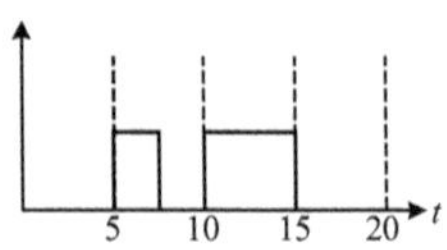

(B)
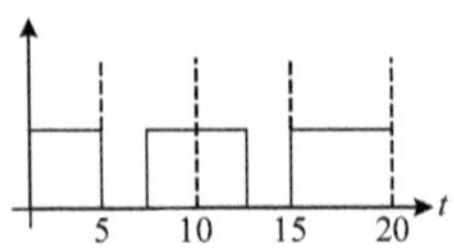

(C)
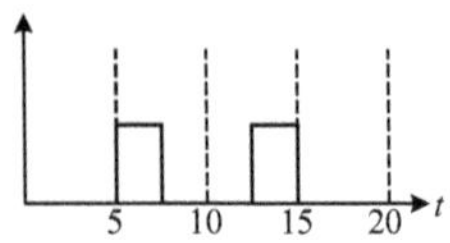

(D)
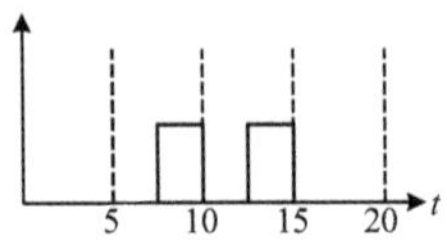

CHAPTER-29 COMMUNICATION SYSTEMS

29.2 Modulation and Demodulation

244. An amplitude modulated wave is represented by the expression $v_m = 5(1 + 0.6 \cos 6280t) \sin(2\pi \times 10^4 t)$ volts. The minimum and maximum amplitudes of the amplitude modulated wave are, respectively : **[2 Sept. 2020 (Morning)]**

(A) 5 V, 8 V (B) $\frac{3}{2}$ V, 5 V

(C) $\frac{5}{2}$ V, 8 V (D) 3 V, 5 V

CHAPTER-30 ERRORS AND EXPERIMENTS

30.1 Errors and Experiments

245. The least count of the main scale of a vernier callipers is 1 mm. Its vernier scale is divided into 10 divisions and coincide with 9 divisions of the main scale. When jaws are touching each other, the 7th division of vernier scale coincides with a division of main scale and the zero of vernier scale is lying right side of the zero of main scale. When this vernier is used to measure length of a cylinder, the zero of the vernier scale is between 3.1 cm and 3.2 cm and 4th *VSD* coincides with a main scale division. The length of the cylinder is : (*VSD* is vernier scale division) : **[2 Sept. 2020 (Morning)]**

(A) 3.21 cm (B) 2.99 cm
(C) 3.2 cm (D) 3.07 cm

246. Using screw gauge of pitch 0.1 cm and 50 divisions on its circular scale, the thickness of an object is measured. It should correctly be recorded as : **[3 Sept. 2020 (Morning)]**

(A) 2.123 cm (B) 2.125 cm
(C) 2.121 cm (D) 2.124 cm

247. A physical quantity z depends on four observables a, b, c and d, as $z = \dfrac{a^2 b^{\frac{2}{3}}}{\sqrt{c}\, d^3}$. The percentage of error in the measurement of a, b, c and d 2%, 1.5%, 4% and 2.5% respectively. The percentage of error in z is : **[5 Sept. 2020 (Morning)]**

(A) 12.25% (B) 14.5%
(C) 16.5% (D) 13.5%

248. A screw gauge has 50 divisions on its circular scale. The circular scale is 4 units ahead of the pitch scale marking, prior to use. Upon one complete rotation of the circular scale, a displacement of 0.5 mm is noticed on the pitch scale. The nature of zero error involved, and the least count of the screw gauge, are respectively : **[6 Sept. 2020 (Morning)]**

(A) Negative, 2 μm (B) Positive, 10 μm
(C) Positive, 0.1 μm (D) Positive, 0.1 mm

249. The density of a solid metal sphere is determined by measuring its mass and its diameter. The maximum error in the density of the sphere is $\left(\dfrac{x}{100}\right)$%. If the relative errors in measuring the mass and the diameter are 6.0% and 1.5% respectively, the value of x is : **[6 Sept. 2020 (Morning)]**

250. A student measuring the diameter of a pencil of circular cross-section with the help of a vernier scale records the following four readings 5.50 mm, 5.55 mm, 5.45 mm; 5.65 mm. The average of these four readings is 5.5375 mm and the standard deviation of the data is 0.07395 mm. The average diameter of the pencil should therefore be recorded as : **[6 Sept. 2020 (Evening)]**

(A) (5.5375 ± 0.0739) mm (B) (5.538 ± 0.074) mm
(C) (5.54 ± 0.07) mm (D) (5.5375 ± 0.0740) mm

* * * * *

ANSWER KEY

1.1 Measurements and Dimensional Analysis

1 (A) **2** (B) **3** (D)
4 (C) **5** (B)

2.1 Motion in One Dimension

6 (20) **7** (A) **8** (C)
9 (D)

3.1 Motion in Two Dimensions

10 (C)

3.2 Relative Motion

11 (B) **12** (D)

4.1 Newton's Laws of Motion

13 (B) **14** (D)

4.2 Friction

15 (D)

5.1 Work Energy Power

16 (3) **17** (150) **18** (C)
19 (346) **20** (B) **21** (18.00)
22 (C)

5.2 Circular Motion

23 (B) **24** (A) **25** (A)

6.1 System of Particles

26 (D) **27** (10.00) **28** (23.00)
29 (A) **30** (C) **31** (D)
32 (120.00) **33** (D) **34** (3.00)

6.2 Rigid Body Dynamics

35 (D) **36** (D) **37** (D)
38 (C) **39** (B) **40** (9)
41 (B) **42** (25) **43** (11)
44 (20) **45** (B) **46** (D)
47 (C) **48** (195) **49** (20.00)
50 (A) **51** (B) **52** (D)

7.1 Gravitational Field

53 (A) **54** (B) **55** (A)
56 (D) **57** (D)

7.2 Satellite Motion

58 (C) **59** (C) **60** (C)
61 (A) **62** (B)

8.1 Elasicity

63 (35.00) **64** (A)

9.1 Fluid Statics

65 (C) **66** (C) **67** (B)

9.2 Fluid Dynamics

68 (B)

9.3 Viscosity

69 (B) **70** (B)

9.4 Surface Tension

71 (C) **72** (A) **73** (101)

10.1 Thermal Expansion

74 (B) **75** (20) **76** (D)

10.2 Calorimetry

77 (B) **78** (A) **79** (B)

10.3 Heat Transfer

80 (A) **81** (D) **82** (A)

11.1 Laws of Thermodynamics

83 (B) **84** (4×10^3 J) **85** (C)
86 (B) **87** (D) **88** (8791)
89 (266) **90** (A) **91** (150)
92 (A) **93** (C) **94** (D)
95 (D) **96** (19.00)

12.1 Kinetic Theory of Gases

97 (D) **98** (C) **99** (A)
100 (41.00) **101** (5.00) **102** (C)

13.1 Simple Harmonic Motion

103 (D) **104** (D) **105** (A)
106 (B)

14.1 Wave Motion

107 (B) 108 (D) 109 (B)

14.2 Stationary Waves and Beats

110 (C) 111 (C)

14.3 Doppler's Effect

112 (A) 113 (D) 114 (D)

15.1 Electric Force and Electric Field

115 (C) 116 (D) 117 (C)
118 (A)

15.2 Electric Flux and Gauss's Law

119 (A)

16.1 Electric Potential and Dipole

120 (A) 121 (A) 122 (D)
123 (C) 124 (D) 125 (A)
126 (C) 127 (C)

16.2 Capacitance

128 (–4.03) 129 (B) 130 (A)
131 (A) 132 (D) 133 (C)
134 (B)

17.1 Current Electricity

135 (D)

17.2 Circuit Analysis

136 (D) 137 (2) 138 (A)
139 (C)

17.3 Thermal Effects of Current & RC Circuits

140 (8.00) 141 (D) 142 (D)
143 (B)

17.4 Electrical Measuring Instruments

144 (A) 145 (A) 146 (B)
147 (B) 148 (A) 149 (A)

18.1 Magnetic Effects of Current

150 (C) 151 (C)

18.2 Magnetic Force on Moving Charges

152 (B) 153 (A) 154 (C)
155 (A) 156 (C) 157 (20)
158 (A) 159 (A) 160 (B)
161 (A) 162 (C) 163 (D)
164 (A)

19.1 Classical and Terrestrial Magnetism

165 (D) 166 (D)

19.2 Magnetic Properties of Materials

167 (A) 168 (B)

20.1 Electromagnetic Induction

169 (15) 170 (A) 171 (A)
172 (C) 173 (5.00) 174 (B)

20.2 Self and Mutual Induction

175 (D) 176 (C) 177 (33.00)

21.1 Alternating Current

178 (A) 179 (C) 180 (B)
181 (A) 182 (400.00)

22.1 Electromagnetic Wave

183 (B) 184 (A) 185 (B)
186 (C) 187 (C) 188 (B)
189 (B) 190 (D) 191 (194.00)
192 (B)

23.1 Geometrical Optics I - Reflection of Light

193 (D) 194 (1)

23.2 Geometrical Optics II - Refraction of Light

195 (90.00) 196 (158) 197 (5.00)

23.3 Geometrical Optics III - Thin Lenses

198 (476) 199 (D) 200 (C)
201 (D)

23.5 Optical Instruments

202 (5) 203 (50.00)

24.1 Interference of Light

204 (A) 205 (A) 206 (D)

207 (A) **208** (B) **209** (D)
210 (9.00)

24.2 Diffraction of Light

211 (198) **212** (50.00)

24.3 Polarization of Light

213 (C)

25.1 Photoelectric Effect

214 (9) **215** (A) **216** (A)
217 (C) **218** (2.00)

25.2 Matter Waves and X-Ray

219 (D) **220** (A) **221** (A)
222 (D) **223** (B)

26.1 Atomic Structure

224 (B) **225** (10553) **226** (51.00)

27.1 Radioactivity

227 (A) **228** (C) **229** (A)

27.2 Nuclear Physics

230 (B) **231** (C) **232** (A)
233 (B) **234** (A)

28.2 PN Junction and Diodes

235 (A) **236** (C) **237** (B)
238 (A) **239** (D)

28.3 Transistors

240 (150.00)

28.4 Logic Gates

241 (C) **242** (A) **243** (None)

29.2 Modulation and Demodulation

244 (None)

30.1 Errors and Experiments

245 (D) **246** (D) **247** (B)
248 (B) **249** (1050.00) **250** (C)

* * * * *

Sol. 12 When speed of car is v, we use

$$\overrightarrow{V_{RC}} = \overrightarrow{V_R} - \overrightarrow{V_C}$$

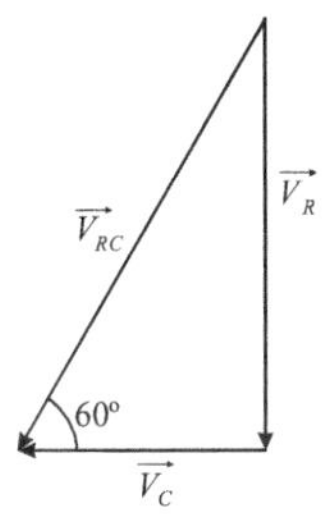

$$\tan 60^\circ = \frac{V_R}{V} = \sqrt{3}$$

$$\Rightarrow \quad V_R = \sqrt{3}\,v \qquad \ldots(1)$$

When speed of car changes to $V_C' = V(1+\beta)$, we use

$$\overrightarrow{V_{RC'}} = \overrightarrow{V_R} - \overrightarrow{V_{C'}}$$

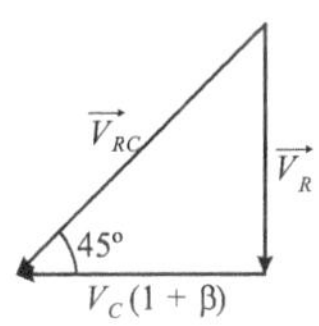

$$\Rightarrow \quad \tan 45^\circ = \frac{V_R}{V(1+\beta)} = 1$$

$$\Rightarrow \quad V_R = v(1+\beta) \qquad \ldots(2)$$

From (1) and (2)

$$1+\beta = \sqrt{3}$$

$$\Rightarrow \quad \beta = 1.732 - 1 = 0.732$$ **Ans. (D)**

4.1 Newton's Laws of Motion

Sol. 13 White going up acceleration of ball is given as

$$\frac{vdv}{dy} = a = -(kv^2 + g)$$

$$\Rightarrow \quad \int_u^0 \frac{vdv}{kv^2+g} = \int_0^H -dy$$

$$\Rightarrow \quad \frac{1}{2k}\ln\left(\frac{ku^2+g}{g}\right) = H$$ **Ans. (B)**

Sol. 14 Force on planet due to increase in mass an be taken as Opposing force on spaceship due to increase in mass is given as

$$F = v\frac{dM_{(t)}}{dt} = -bv^3 [\text{As } F_{ext} = 0]$$

Thus acceleration of spaceship is given as

$$a = \frac{F}{m_{(t)}} = -\frac{bv^3}{m_{(t)}}$$ **Ans. (D)**

4.2 Friction

Sol. 15 At the point of slipping as shown in figure, we use

$$\tan\theta = \mu = \frac{3}{4}$$

$$\Rightarrow \quad \cos\theta = \frac{4}{5}$$

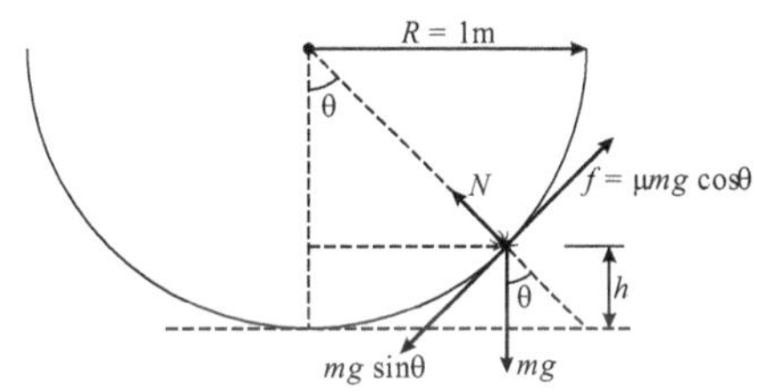

Value of h is given as

$$h = R(1-\cos\theta) = R\left(1-\frac{4}{5}\right) = \frac{R}{5}$$ **Ans. (D)**

5.1 Work Energy Power

Sol. 16 Using work-energy theorem, we use

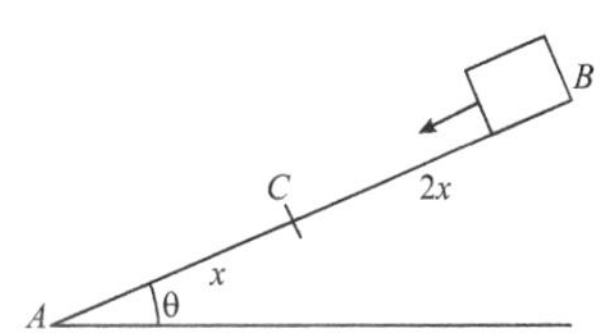

$$mg\sin\theta\,(x+2x) - \mu mg\cos\theta\,x = 0$$

$$\Rightarrow \quad 3\sin\theta - \mu\cos\theta = 0$$

$$\Rightarrow \quad \mu = 3\tan\theta$$

$$\Rightarrow \quad k = 3$$ **Ans. (3)**

Sol. 17 By work energy theorem, we use

$$Fl = mgh$$

$$\Rightarrow \quad F \times 0.2 = 0.15 \times 10 \times 20$$

$$\Rightarrow \quad F = \frac{0.15 \times 200}{0.2} = 150\,\text{N}$$ **Ans. (150)**

Sol. 18 If power is constant, we use

$$K = Pt$$

$$\Rightarrow \quad \frac{1}{2}mv^2 = Pt$$

$$\Rightarrow \quad \frac{ds}{dt} = v = \sqrt{\frac{2Pt}{m}}$$

$$\Rightarrow \quad \int_0^s ds = \int_0^t \sqrt{\frac{2Pt}{m}}\, dt$$

$$\Rightarrow \quad s = \sqrt{\frac{2P}{m}}\left(\frac{2}{3}t^{3/2}\right)$$

$$\Rightarrow \quad s \propto t^{3/2}$$ **Ans. (C)**

Sol. 19 When block goes up its acceleration is

$$a_1 = -g(\sin\theta + \mu\cos\theta)$$

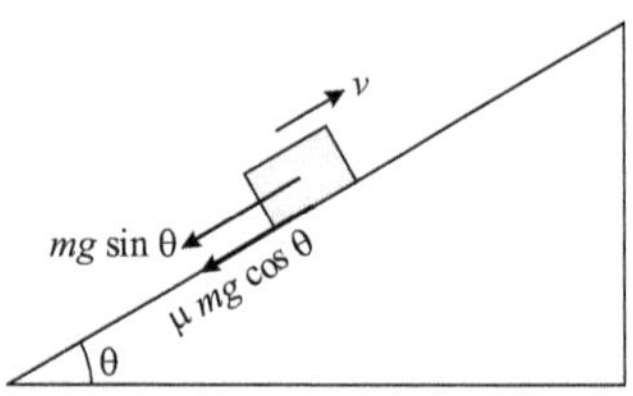

Distance it goes up the plane is given as

$$l = \frac{v_0^2}{2a_1}$$

When it comes down its acceleration is given as

$$a_2 = g(\sin\theta - \mu g\cos\theta)$$

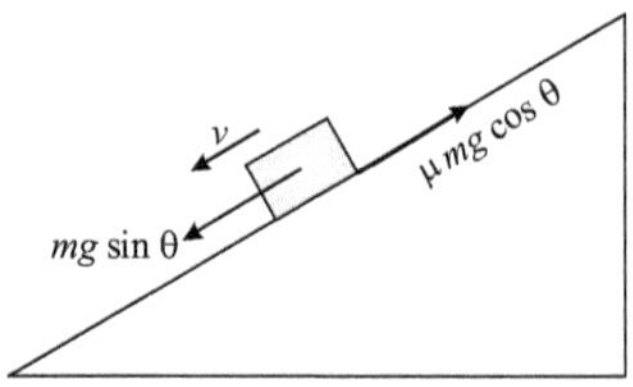

Velocity at bottom point is given as

$$\left(\frac{v_2}{2}\right)^2 = 2a_2 l$$

$$\Rightarrow \quad \frac{v_0^2}{4} = 2g(\sin\theta - \mu\cos\theta)\cdot\frac{v_0^2}{2g(\sin\theta + \mu\cos\theta)}$$

$\Rightarrow$ $\sin\theta + \mu\cos\theta = 4(\sin\theta - \mu\cos\theta)$

At $\theta = 30°$

$$\frac{1}{2} + \frac{\mu\sqrt{3}}{2} = 4\left(\frac{1}{2} - \frac{\mu\sqrt{3}}{2}\right)$$

$$\Rightarrow \quad 1 + \mu\sqrt{3} = 4 - 4\sqrt{3}\mu$$

$$\Rightarrow \quad 5\sqrt{3}\mu = 3$$

$$\Rightarrow \quad \mu = \sqrt{\frac{3}{5}} = 0.3464 = \frac{I}{1000}$$

$$\Rightarrow \quad I = 346$$ **Ans. (346)**

Sol. 20 Given that applied force with position is

$$F = 200\text{ N for } x \le 15\text{ m}$$

$$F = 200 - \frac{100}{15}(x - 15)\Big\}$$

For $15 \le x < 30$

$$\Rightarrow \quad F = 300 - \frac{100}{15}x$$

Total work done by force is

$$W = 200 \times 15 + \int_{15}^{30}\left(30 - \frac{100}{15}x\right)dx$$

$$\Rightarrow \quad W = 300 + 300 \times 15 - \frac{100}{15}\cdot\left[\frac{(30)^2}{2} - \frac{(15)^2}{2}\right]$$

$$\Rightarrow \quad W = 7500 - \frac{100}{30}[900 - 225]$$

$$\Rightarrow \quad W = 7500 - 2250 = 5250\text{ J}$$ **Ans. (B)**

Sol. 21 As power of engine in constant we use

$$P = mva = mv\frac{dv}{dt}$$

$$\Rightarrow \quad \int_0^v v\,dv = \frac{P}{m}\int_0^t dt$$

$$\Rightarrow \quad v = \sqrt{\frac{2Pt}{m}} = \frac{dx}{dt}$$

$$\Rightarrow \quad \int_0^s dx = \sqrt{\frac{2P}{m}}\int_0^t t^{1/2}dt$$

$$\Rightarrow \quad s = \sqrt{\frac{2P}{m}}\cdot\frac{t^{3/2}}{(3/2)} = \frac{2}{3}\sqrt{\frac{2P}{m}}t^{3/2}$$

$$\Rightarrow \quad s = \frac{2}{3}\sqrt{\frac{2\times 1}{2}}\times(9)^{3/2}$$

$$\Rightarrow \quad s = \frac{2}{3}\times 27 = 18\text{ m}$$ **Ans. (18.00)**

Sol. 22 The field force between particles is given as

$$F = \left|\frac{dU}{dr}\right| = -(A(-6r^{-7})) + B(-12r^{-13})$$

$$\Rightarrow \quad F = \frac{6A}{r^7} - \frac{12B}{r^{13}}$$

At equilibrium, we have $F = 0$, so we use

$$0 = \frac{6A}{r^7} - \frac{12B}{r^{13}}$$

$$\Rightarrow \quad \frac{6A}{12B} = \frac{1}{r^6}$$

$$\Rightarrow \quad r = \left(\frac{2B}{A}\right)^{1/6}$$

At equilibrium position, potential energy is given as

$$U_{\left(r=\left(\frac{2B}{A}\right)^{1/6}\right)} = -\frac{A}{2B/A} + \frac{B}{4B^2/A^2}$$

$$\Rightarrow \quad U = \frac{-A^2}{2B} + \frac{A^2}{4B} = \frac{-A^2}{4B}$$ **Ans. (C)**

5.2 Circular Motion

Sol. 23 Figure below shows the forces acting on bead

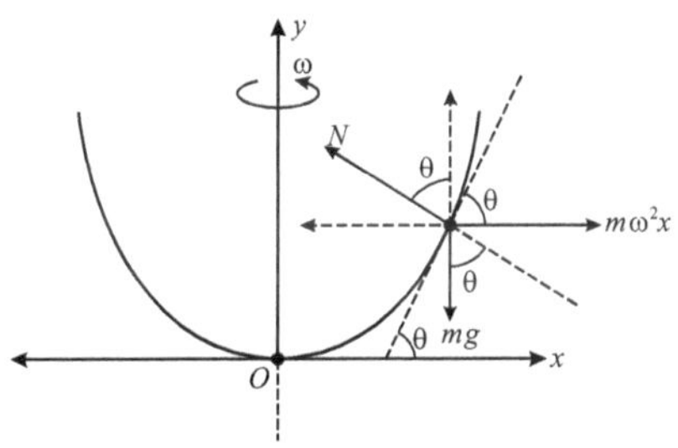

For equilibrium of bead in frame of wire, we use

$$m\omega^2 x \cos\theta = mg\sin\theta$$

$$\Rightarrow \quad \omega^2 x = g\tan\theta$$

Slope of parabola at point P is $\frac{dy}{dx} = \tan\theta$

$$\Rightarrow \quad \frac{dy}{dx} = \frac{\omega^2 x}{g} = 8Cx \qquad [\text{As } y = 4Cx^2]$$

$$\Rightarrow \quad \omega^2 = 8Cg$$

$$\Rightarrow \quad \omega = \sqrt{8Cg}$$ **Ans. (B)**

Sol. 24 As given is theory explanation shape of curved surface is given as

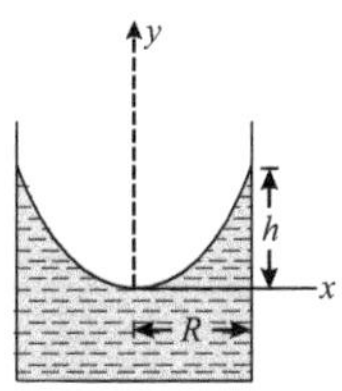

$$y = \frac{\omega^2 x^2}{2g}$$

$\Rightarrow$ at $x = R$, we have

$$h = \frac{\omega^2 R^2}{2g} = \frac{\omega^2 (5)^2}{2g} = \frac{25\omega^2}{2g}$$ **Ans. (A)**

Sol. 25 Angular speed of clock's second's hand is given as

$$\omega = \frac{2\pi}{T} = \frac{2\pi}{60} = 0.105 \text{ rad/sec}$$

Acceleration of tip of hand is

$$a = \omega^2 R$$

$$\Rightarrow \quad a = (0.105)^2 (0.1) \text{ m/s}^2$$

$$\Rightarrow \quad a = 0.0011 \text{ m/s}^2$$

$$\Rightarrow \quad a = 1.1 \times 10^{-3} \text{ m/s}^2$$ **Ans. (A)**

6.1 System of Particles

Sol. 26 By momentum conservation principle, we have

$$m(u\hat{i}) + 3m(0) = mv\hat{j} + 3m\vec{v}_1$$

$$\Rightarrow \quad mu\hat{i} - mv\hat{j} = 3m\vec{v}_1$$

$$\Rightarrow \quad \vec{v}_1 = \frac{u\hat{i} - v\hat{j}}{3}$$

$$\Rightarrow \quad |\vec{v}_1| = \frac{\sqrt{u^2 + v^2}}{3}$$

$$\Rightarrow \quad v_1^2 = \frac{u^2 + v^2}{9} \qquad \ldots(1)$$

As collision is perfectly elastic, we have

$$KE_i = KE_j$$

$$\Rightarrow \quad \frac{1}{2}mu^2 + \frac{1}{2}3m(0)^2 = \frac{1}{2}mv^2 + \frac{1}{2}3mv_1^2$$

$$\Rightarrow \quad u^2 = v^2 + 3v_1^2$$

$$\Rightarrow \quad u^2 = v^2 + 3\frac{(u^2 + v^2)}{9}$$

$$\Rightarrow \quad 3u^2 = 3v^2 + u^2 + v^2$$

$$\Rightarrow \quad 2u^2 = 4v^2$$

$$\Rightarrow \quad v = \frac{u}{\sqrt{2}}$$ **Ans. (D)**

Sol. 27

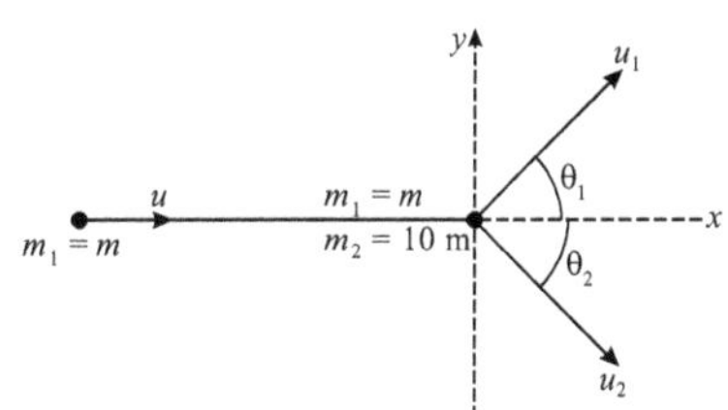

By momentum conservation along y direction, we have

$$m_1 u_1 \sin\theta_1 = m_2 u_2 \sin\theta_2$$

$$\Rightarrow \quad mu_1 \sin\theta_1 = 10mu_2 \sin\theta_2 \qquad \ldots(1)$$

Given that

$$kE_{m_{1(\text{final})}} = \frac{1}{2} kE_{m_{1(\text{initial})}}$$

$$\Rightarrow \quad \frac{1}{2} mu_1^2 = \frac{1}{2} \times \frac{1}{2} mu^2$$

$$\Rightarrow \quad u_1 = \frac{u}{\sqrt{2}} \qquad \ldots(2)$$

As collision is elastic, we use

$$\frac{1}{2} mu^2 = \frac{1}{2} mu_1^2 + \frac{1}{2}(10\,m)u_2^2$$

$$\Rightarrow \quad \frac{1}{2} mu^2 = \frac{1}{2} \times \frac{1}{2} mu^2 + \frac{1}{2} \times 10m \cdot u_2^2$$

$$\Rightarrow \quad \frac{1}{4} mu^2 = \frac{1}{2} \times 10 \times mu_2^2$$

$$\Rightarrow \quad u_2 = \frac{u}{\sqrt{20}} \qquad \ldots(3)$$

From (1), (2) and (3), we have

$$\frac{u}{\sqrt{2}} \sin\theta_1 = 10 \cdot \frac{u}{\sqrt{20}} \sin\theta_2$$

$$\Rightarrow \quad \sin\theta_1 = \sqrt{10} \sin\theta_2$$

$$\Rightarrow \quad n = 10$$

Ans. (10.00)

Sol. 28

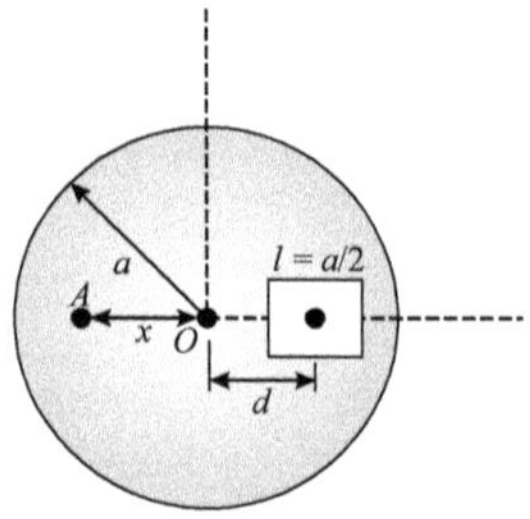

If centre of mass of remaining portion is at point A by mass moment concept, we have

$$m_1 x = m_2\left(\frac{a}{2}\right)$$

Hence m_1 is the mass of complete disc and m_2 is the removed mass from square hole.

As masses are proportional to surface are, we use

$$\left(\pi a^2 - \frac{a^2}{4}\right)x = \frac{a^2}{4}\left(\frac{a}{2}\right)$$

$$\Rightarrow \quad x = \frac{a}{2(4\pi - 1)}$$

$$\Rightarrow \quad X = 2(4\pi - 1)$$

$$\Rightarrow \quad X = 8\pi - 2 = 23.12$$

Thus nearest integral value of X is 23. **Ans. (23.00)**

Sol. 29

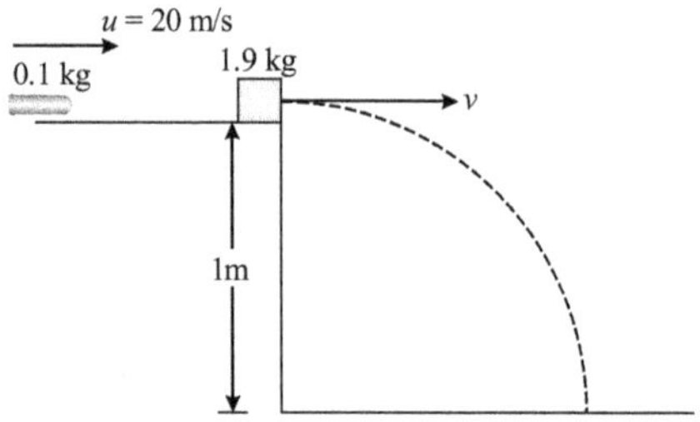

By linear momentum conservation we use

$$0.1 \times 20 = 2v$$

$$\Rightarrow \quad v = 1 \text{ m/s}$$

Final kinetic energy of combined system is given as

$$K_f = mgh + \frac{1}{2} mv^2$$

$$\Rightarrow \quad K_f = 2 \times 10 \times 1 + \frac{1}{2} \times 2\,(1)^2$$

$$\Rightarrow \quad K_f = 20 + 1 = 21 \text{ J}$$

Ans. (A)

Sol. 30 At a height h speed is given as

$$v = \sqrt{2g(H - h)}$$

The graph is parabolic and most appropriate option is (C) which can be analyzed by slope variation. **Ans. (C)**

Sol. 31 By conservation of momentum when last block starts moving

$$mv = 16\,mv'$$

$$\Rightarrow \quad v' = \frac{v}{16}$$

Initial energy $E_1 = \frac{1}{2} mv^2$

Final energy $E_2 = \frac{1}{2}(16m)\left(\frac{v}{16}\right)^2 = \frac{1}{32} mv^2$

Energy loss $\Delta E = E_1 - E_2 = \frac{15}{32} mv^2$

Percentage energy loss is given as

$$\%\text{ loss} = \frac{\frac{15}{32} mv^2}{\frac{1}{2} mv^2} \times 100 = \frac{15}{16}\%$$

Ans. (D)

Sol. 32 Figure below shows the oblique collision in the given situation

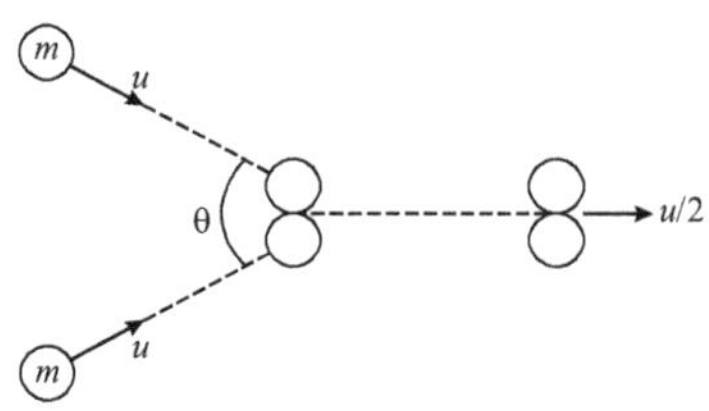

By conservation of momentum, we use

$$2mu\cos\frac{\theta}{2} = 2m\left(\frac{u}{2}\right)$$

$$\Rightarrow \quad \cos\frac{\theta}{2} = \frac{1}{2}$$

$$\Rightarrow \quad \frac{\theta}{2} = 60^\circ$$

$$\Rightarrow \quad \theta = 120^\circ$$

Ans. (120.00)

Sol. 33 Using conservation of linear momentum before and after collision, we have

$$m_1\vec{u} = m_1\vec{v}_1 + m_2\vec{v}_2$$

$$\Rightarrow \quad 2m_2(\sqrt{3}\hat{i}+\hat{j})+0 = 2m_2(\hat{i}+\sqrt{3}\hat{j})+m_2\vec{v}_2$$

$$\Rightarrow \quad \vec{v}_2 = 2(\sqrt{3}\hat{i}+\hat{j})-2(\hat{i}+\sqrt{3}\hat{j})$$

$$\Rightarrow \quad \vec{v}_2 = 2(\sqrt{3}\hat{i}-\hat{j})+2(\hat{i}-\sqrt{3}\hat{j})$$

$$\Rightarrow \quad \vec{v}_2 = 2(\sqrt{3}\hat{i}-\hat{j})(\hat{i}-\hat{j})$$

Angle between $\vec{v}_1$ and $\vec{v}_2$ is given as

$$\cos\theta = \frac{\vec{v}_1\cdot\vec{v}_2}{|\vec{v}||\vec{v}_2|} = \frac{2(\sqrt{3}-1)(1-\sqrt{3})}{2\times 2\sqrt{2}(\sqrt{3}-1)}$$

$$\Rightarrow \quad \cos\theta = \frac{1-\sqrt{3}}{2\sqrt{2}}$$

$$\Rightarrow \quad \theta = 105^\circ$$

Ans. (D)

Sol. 34 Centre of mass of hemisphere from its base centre is given as

$$x = \frac{3R}{8} = 3\text{ cm}$$

$$x = 3$$

Ans. (3.00)

6.2 Rigid Body Dynamics

Sol. 35 Figure below shows the forces acting on rod

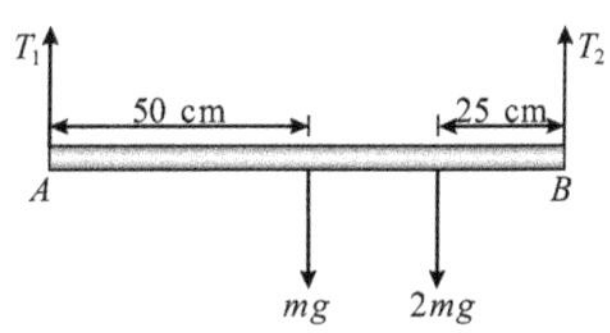

For equilibrium of rod, we use torque about point B

$$\tau_b = 0$$

$$2mg\times 25 + mg\times 50 - T_1\times 100 = 0$$

$$\Rightarrow \quad 100\,T_1 = 100mg$$

$$\Rightarrow \quad T_1 = mg$$

Ans. (D)

Sol. 36 For maximum torque F must be applied normal to radius as shown in figure

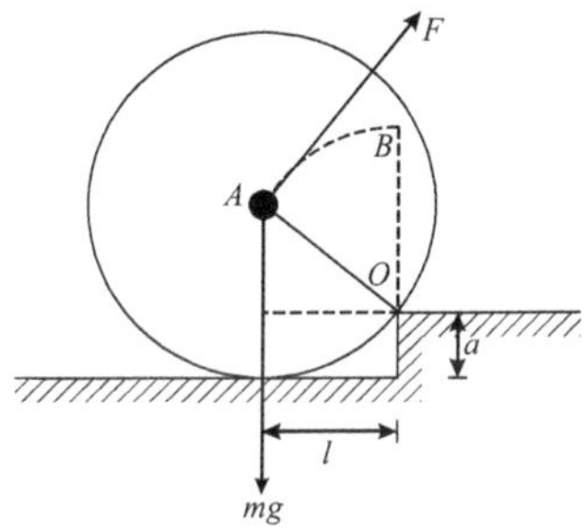

To pull over the edge force should be able to lift cylinder centre from A to B so at point A torque of F must be more than that of gravity about O, so we use

$$FR \geq Mgl$$

$$\Rightarrow \quad F \geq \frac{Mg}{R}\sqrt{R^2-(R-a)^2}$$

$$\Rightarrow \quad F \geq Mg\sqrt{1-\left(\frac{R-a}{R}\right)^2}$$

Ans. (D)

Sol. 37 By angular momentum conservation, we use

$$L_1 + L_2 = L_{\text{final}}$$

$$\Rightarrow \quad I_1\omega_1 + I_2\omega_2 = (I_1+I_2)\omega_f$$

$$\Rightarrow \quad 0.1\times 10 + 0.2\times 5 = (0.1+0.2)\times\omega_f$$

$$\Rightarrow \quad \omega_f = \frac{20}{3}$$

Kinetic energy of combined disc system can be given as

$$\Rightarrow \quad K = \frac{1}{2}(I_1+I_2)\,\omega_f^2$$

$$\Rightarrow \quad K = \frac{1}{2}(0.1+0.2).\left(\frac{20}{3}\right)^2$$

$$\Rightarrow \quad K = \frac{0.3}{2}\times\frac{400}{9} = \frac{120}{18} = \frac{20}{3}\text{ J}$$

Ans. (D)

Sol. 38 For constant mass, volume is also constant, so we use

$$\pi R^2 L = \text{Constant}$$

$$\Rightarrow \quad R^2L = C \qquad \ldots(1)$$

Moment of inertia as given as

$$I = M\left(\frac{R^2}{4}+\frac{L^2}{12}\right)$$

$$\Rightarrow \quad I = M\left(\frac{C}{4L}+\frac{L^2}{12}\right)$$

For I to be minimum, we use

$$\frac{dI}{dL} = M\left[-\frac{C}{4L^2}+\frac{L}{6}\right]=0$$

$$\Rightarrow \quad \frac{C}{4L^2} = \frac{L}{6}$$

$$\Rightarrow \quad L=\left[\frac{3C}{2}\right]^{1/3}$$

From equation-(1), we use

$$R^2 = \frac{C}{L} = \frac{C}{\left[\frac{3}{2}C\right]^{1/3}} = \frac{C^{2/3}}{\left(\frac{3}{2}\right)^{1/3}}$$

$$\Rightarrow \quad R = \frac{C^{1/3}}{\left(\frac{3}{2}\right)^{1/6}}$$

$$\Rightarrow \quad \frac{L}{R} = \frac{\left(\frac{3}{2}\right)^{1/3}\cdot(C)^{1/3}}{C^{1/3}}\times\left(\frac{3}{2}\right)^{1/6}$$

$$\Rightarrow \quad \frac{L}{R} = \left(\frac{3}{2}\right)^{1/2} = \sqrt{\frac{3}{2}}$$

Ans. (C)

Sol. 39 By angular momentum conservation, we have

$$mvl = \frac{Ml^2}{3}\omega \times ml^2\omega$$

$$\Rightarrow \quad \omega = \frac{mvl}{\left(ml^2+\frac{Ml^2}{3}\right)}$$

$$\Rightarrow \quad \omega = \frac{1\times6\times1}{\frac{2}{3}+1} = \frac{18}{5}\text{ rad/s}$$

By energy conservation, we have

$$\frac{1}{2}\left(M\frac{l^2}{3}\right)\omega^2+\frac{1}{2}(ml^2)\omega^2 = mgl(1-\cos\theta)+\frac{Mgl}{2}(1-\cos\theta)$$

$$\Rightarrow \quad \frac{1}{2}\left(\frac{2\times(1)^2}{3}\right)\left(\frac{18}{5}\right)^2+\frac{1}{2}(1\times(1)^2)\left(\frac{18}{5}\right)^2 = (1+1)\times10\times1(1-\cos\theta)$$

$$\Rightarrow \quad \frac{5}{6}\times\left(\frac{18}{5}\right)^2 = 20(1-\cos\theta)$$

$$\Rightarrow \quad 1-\cos\theta = \frac{18}{5}\times\frac{3}{20}$$

$$\Rightarrow \quad \cos\theta = 1-\frac{27}{50}$$

$$\Rightarrow \quad \theta \simeq 63^\circ$$

Ans. (B)

Sol. 40 By conservation of angular momentum, we use

$$\left(80R^2+\frac{200R^2}{2}\right)\omega_i = \left(0+\frac{200R^2}{2}\right)\omega_f$$

$$\Rightarrow \quad 180\,\omega_i = 100\,\omega_f$$

$$\Rightarrow \quad \omega_f = 1.8\,\omega_i = 1.8\times5 = 9\text{ rev/min}$$

Ans. (9)

Sol. 41 Horizontal and vertical forces given in system are shown in below figure are

$$F_H = m\omega^2\frac{l}{2}\sin\theta$$

$$F_V = mg$$

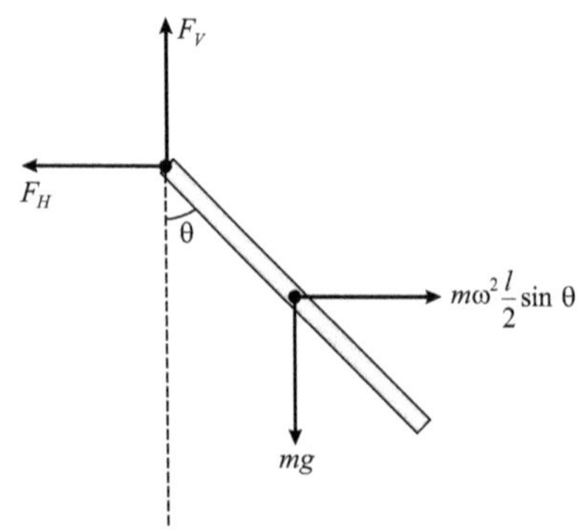

As stated in question, we use

$$mg\times\frac{l}{2}\sin\theta - m\omega^2\frac{l}{2}\sin\theta\cdot\frac{l}{2}\cos\theta = \frac{ml^2}{12}\omega^2\sin\theta\cos\theta$$

$$\Rightarrow \quad \cos\theta = \frac{3g}{2\omega^2 l}$$

Ans. (B)

Sol. 42 Moment of inertia about axis EX is given as

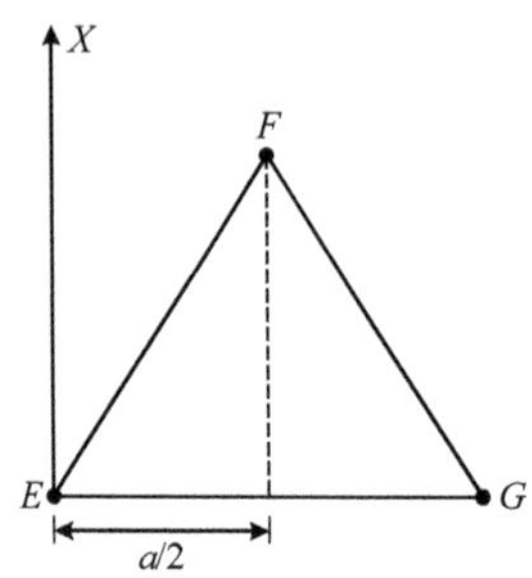

$$I = I_E+I_F+I_G$$

$$\Rightarrow \quad I = 0+m\left(\frac{a}{2}\right)^2+ma^2$$

$$\Rightarrow \quad I = \frac{5}{4}ma^2 = \frac{25}{20}ma^2$$

$$\Rightarrow \quad N = 25$$

Ans. (25)

Sol. 43 Moment of inertia of an equilateral triangle about its centroid is

$$I_0 = \frac{ma^2}{12}$$

ΔADE is also an equilateral triangle of side $\frac{a}{2}$ so its moment of inertia about its centroid is given as

$$I_{ADE} = \frac{m_1\left(\frac{a}{2}\right)^2}{12} = \frac{m_1 a^2}{48} = \frac{ma^2}{192} \qquad \left(m_1 = \frac{m}{4}\right)$$

MI of system about G after removing ΔADE is given as

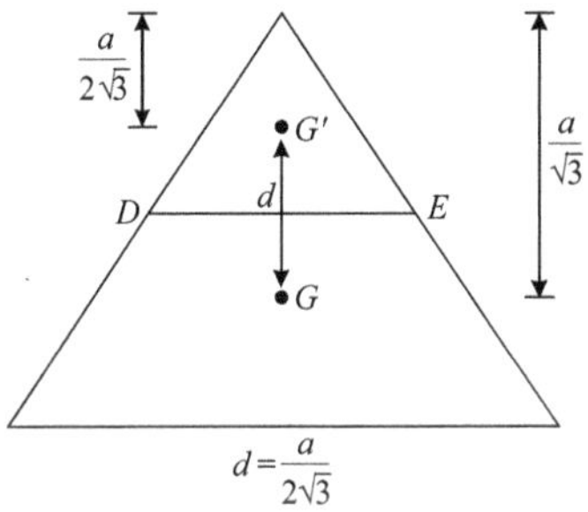

$$I' = I_0 - \left(I_{ADE} + \frac{m}{4}d^2\right)$$

$$\Rightarrow \quad I' = \frac{ma^2}{12} - \left(\frac{ma^2}{192} + \frac{m}{4}\cdot\left(\frac{a^2}{12}\right)\right)$$

$$\Rightarrow \quad I' = ma^2\left[\frac{1}{12} - \frac{1}{192} - \frac{1}{48}\right]$$

$$\Rightarrow \quad I' = \frac{11}{192}ma^2 = \frac{11}{16}I_0 = \frac{NI_0}{16}$$

Thus $\quad N = 11$ **Ans. (11)**

Sol. 44 By angular momentum conservation, we use

$$\frac{1}{2}MR^2\omega_1 = \left[\frac{1}{2}MR^2 + \frac{1}{2}M\left(\frac{R}{2}\right)^2\right]\omega_2$$

$$\Rightarrow \quad \omega_2 = \frac{4\omega_1}{5}$$

Initial kinetic energy

$$K_1 = \frac{1}{2}\left(\frac{1}{2}MR^2\right)\omega_1^2$$

Final kinetic energy

$$K_2 = \frac{1}{2}\left(\frac{5}{8}MR^2\right)\left(\frac{4\omega_1}{5}\right)^2$$

$$\Rightarrow \quad K_2 = \frac{1}{2}\left(\frac{2}{5}MR^2\right)\omega_1^2$$

Kinetic energy loss

$$\Delta K = K_1 - K_2 = \frac{1}{2}\left(\frac{1}{10}MR^2\right)\omega^2$$

Percentage loss P is given as

$$P = \frac{\Delta K}{K_1}\times 100 = \frac{\frac{1}{10}MR^2}{\frac{1}{2}MR^2}\times 100$$

$$\Rightarrow \quad P = 20\%$$ **Ans. (20)**

Sol. 45 Moment of inertia of the two discs are given as

$$I_1 = \frac{1}{2}MR^2 = \frac{1}{2}(\rho\pi R^2)R^2$$

and $$I_2 = \frac{1}{2}M'(\alpha R)^2 = \frac{1}{2}(\rho\pi\alpha^2 R^2)\alpha^2 R^2$$

Where ρ is surface mass density of discs

Given that $$\frac{I_1}{I_2} = \frac{1}{\alpha^4} = \frac{1}{16}$$

$$\Rightarrow \quad \alpha = 2$$ **Ans. (B)**

Sol. 46 Moment of inertia about axis passing through O is

$$I_0 = \frac{M}{12}(l^2 + b^2) = \frac{M}{12}((80)^2 + (60)^2)$$

MI about axis passing through O' is

$$I_0' = I_0 + Md^2$$

$$\Rightarrow \quad I_{0'} = I_0 + M(50)^2$$

$$\Rightarrow \quad I_{0'} = \frac{M}{12}((80)^2 + (60)^2) + M(50)^2$$

$$\Rightarrow \quad \frac{I_0}{I_{0'}} = \frac{\frac{M}{12}((80)^2 + (60)^2)}{\frac{M}{12}((80)^2 + (60)^2) + M(50)^2} = \frac{1}{4}$$

Ans. (D)

Sol. 47 On coupling of another wheel we use conservation of angular momentum as

$$I\omega = (I + 3I)\omega_1$$

$$\Rightarrow \quad \omega_1 = \frac{\omega}{4}$$

Initial kinetic energy of wheel

$$K_1 = \frac{1}{2}I\omega^2$$

Final kinetic energy of coupled wheels

$$K_2 = \frac{1}{2}(4I)\left(\frac{\omega}{4}\right)^2 = \frac{1}{8}I\omega^2$$

Fractional loss in kinetic energy

$$f=\frac{K_1-K_2}{K_1}=\frac{\frac{1}{2}-\frac{1}{8}}{\frac{1}{2}}=\frac{3}{4}$$ **Ans. (C)**

Sol. 48 Torque of force is given as

$$\vec{\tau}=\vec{r}\times\vec{F}$$

Here $\vec{r}=\vec{r}_2-\vec{r}_1=(3\hat{i}+\hat{j}-2\hat{k})$

$\Rightarrow$ $\vec{\tau}=(3\hat{i}+\hat{j}-2\hat{k})\times(\hat{i}+2\hat{j}+3\hat{k})$

$\Rightarrow$ $\vec{\tau}=(7\hat{i}-11\hat{j}+5\hat{k})$

$\Rightarrow$ $|\vec{\tau}|=\sqrt{49+121+25}=\sqrt{195}$ N-m **Ans. (195)**

Sol. 49 By conservation of angular momentum, we use

$$L_i=L_f$$

$$mvl=\left(ml^2+\frac{ML^2}{3}\right)\omega$$

$\Rightarrow$ $0.1\times 80\times 1=\left(0.1\times(1)^2+\frac{0.9\times(1)^2}{3}\right)\omega^2$

$\Rightarrow$ $\omega^2=\frac{8}{0.4}=20$ rad/s **Ans. (20.00)**

Sol. 50 Moment of inertia of a hollow cone is $\frac{1}{2}MR^2$ as already derived in theory. **Ans. (A)**

Sol. 51 Moment of inertia of all point masses fixed on square about given axis of rotation is given as

$$I=0+2\left(m\left(\frac{l}{\sqrt{2}}\right)^2\right)+m(\sqrt{2}\,l)^2$$

$\Rightarrow$ $I=ml^2+2ml^2=3ml^2$

Angular momentum of system is given as

$$L=I\omega=3ml^2\omega$$ **Ans. (B)**

Sol. 52 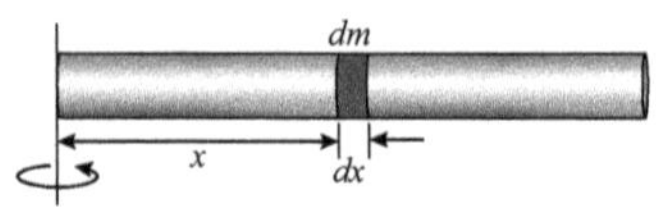

We consider an element as shown in figure. The moment of inertia of element is given as

$$I=\int dmx^2=\int\lambda dx\cdot x^2$$

$\Rightarrow$ $I=\int_0^L x^2\lambda_0\left(1+\frac{x}{L}\right)dx$

$\Rightarrow$ $I=\lambda_0\int_0^L\left(x^2+\frac{x^3}{L}\right)dx$

$\Rightarrow$ $I=\lambda\left[\frac{L^3}{3}+\frac{L^3}{4}\right]$

$\Rightarrow$ $I=\frac{7L^3\lambda_0}{12}$ ...(1)

Mass of rod is given as

$$M=\int_0^L\lambda dx=\int_0^L\lambda_0\left(1+\frac{x}{L}\right)dx$$

$\Rightarrow$ $M=\lambda_0\left(L+\frac{L}{2}\right)=\frac{3\lambda_0 L}{2}$ ...(2)

From (1) & (2), we use

$$I=\frac{7}{12}L^3\left(\frac{2M}{3L}\right)$$

$\Rightarrow$ $I=\frac{7}{18}ML^2$ **Ans. (D)**

7.1 Gravitational Field

Sol. 53 Value of g at a height h above earth surface is

$$g_{out}=\frac{GM}{(R+h)^2}$$

Value of g at a depth h below the earth surface is

$$g_{in}=\frac{GM(R-h)}{R^3}$$

Given that

$$g_{out}=g_{in}$$

$\Rightarrow$ $\frac{GM}{(R+h)^2}=\frac{GM(R-h)}{R^3}$

$\Rightarrow$ $(R-h)(R+h)^2=R^3$

Simplifying above equation gives

$$h^2+Rh-R^2=0$$

$\Rightarrow$ $h=\frac{-R+\sqrt{R^2+4R^2}}{2}$

$\Rightarrow$ $h=\frac{(\sqrt{5}-1)R}{2}$ **Ans. (A)**

Sol. 54 Mass contained in a region of radius r is

$$m=\int_0^r\rho(r)\cdot 4\pi r^2 dr$$

$$m=\int_0^r\rho_0\left(1-\frac{r^2}{R^2}\right)4\pi r^2 dr$$

$\Rightarrow$ $m=4\pi\rho_0\left[\frac{r^3}{3}-\frac{r^5}{5R^2}\right]$

Gravitational field due to this expression of EF, we use

$$g = \frac{Gm}{r^2}$$

$$\Rightarrow \quad g = \frac{G}{r^2}\left(4\pi\rho_0\left(\frac{r^3}{3} - \frac{r^5}{5R^2}\right)\right)$$

$$\Rightarrow \quad g = 4\pi G\rho_0\left(\frac{r}{3} - \frac{r^3}{5R^2}\right)$$

g is maximum when

$$\frac{dg}{dr} = 0$$

$$\Rightarrow \quad \frac{dg}{dr} = 4\pi G\rho_0\left(\frac{1}{3} - \frac{3r^2}{5R^2}\right) = 0$$

$$\Rightarrow \quad r = \sqrt{\frac{5}{9}}R$$

Ans. (B)

Sol. 55 Potential on X-axis is given as

$$V = -\int_\infty^x E_g \cdot dx = -\int_\infty^x \frac{Ax}{(x^2+a^2)^{3/2}}dx$$

$$\Rightarrow \quad V = \left[\frac{A}{(x^2+a^2)^{1/2}}\right]_\infty^x = \frac{A}{\sqrt{x^2+a^2}}$$

Ans. (A)

Sol. 56 Gravity at a height $h = \frac{R}{2}$ above earth surface is given as

$$g_{h\text{ up}} = \frac{GM}{\left(R + \frac{R}{2}\right)^2}$$

gravity at a depth d below earth surface is given as

$$g_{d\text{ in}} = \frac{GM(R-d)}{R^3}$$

Given that

$$g_{h\text{ up}} = g_{d\text{ in}}$$

$$\Rightarrow \quad \frac{1}{\left(\frac{3R}{2}\right)^2} = \frac{R-d}{R^3}$$

$$\Rightarrow \quad \frac{4R}{9} = R - d$$

$$\Rightarrow \quad \frac{d}{R} = \frac{5}{9}$$

Ans. (D)

Sol. 57 Gravity on equator of earth is given as

$$g_e = g_s - \omega^2 R$$

Gravity at a height h above earth surface is

$$g_h = g_s\left(1 - \frac{2h}{R}\right) \quad [\text{For } h << R]$$

Given that $g_e = g_h$

$$\Rightarrow \quad g_s - \omega^2 R = g_s - \frac{2g_s h}{R}$$

$$\Rightarrow \quad h = \frac{\omega^2 R^2}{2g}$$

Ans. (D)

7.2 Satellite Motion

Sol. 58 Mass of galaxy inside radius R is given as

$$M = \int_0^R \rho_{(r)} \cdot 4\pi r^2 dr$$

$$\Rightarrow \quad M = \int_0^R \frac{K}{r} \cdot 4\pi r^2 dr$$

$$\Rightarrow \quad M = 4\pi K\left[\frac{R^2}{2}\right] = 2\pi KR^2$$

For orbital motion of star of mass m, we use

$$\frac{GMm}{R^2} = \frac{mv^2}{R}$$

$$\Rightarrow \quad v = \sqrt{\frac{GM}{R}} = \sqrt{2\pi KGR}$$

Time period of revolution is given as

$$T = \frac{2\pi R}{v} = \frac{2\pi R}{\sqrt{2\pi KGR}}$$

$$\Rightarrow \quad T = \sqrt{\frac{2\pi R}{KG}}$$

$$\Rightarrow \quad T \propto \sqrt{R}$$

$$\Rightarrow \quad T^2 \propto R$$

Ans. (C)

Sol. 59 Figure below shows the trajectory of satellite after firing rockets

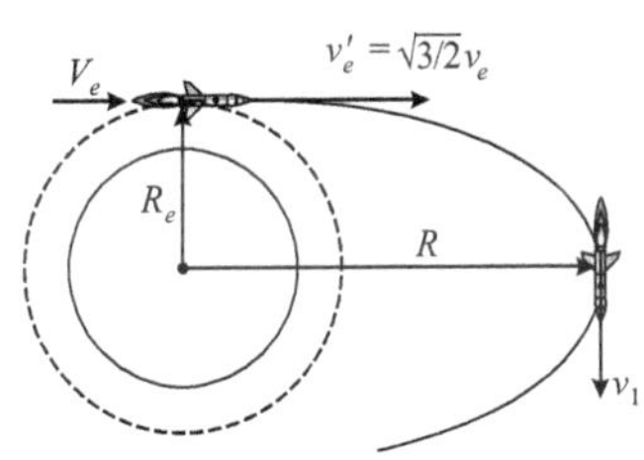

$$v_e = \sqrt{\frac{GM_e}{R_e}}$$

By conservation of angular momentum, we have

$$v_e' \cdot R_e = v_1 R$$

$$\Rightarrow \quad \sqrt{\frac{3}{2}\frac{GM_e}{R_e}} \cdot R_e = v_1 R$$

$$\Rightarrow \quad v_1 = \sqrt{\frac{3}{2}\frac{GM_e m}{R_e}\cdot\frac{R_e}{R}}$$

By conservation of energy we use

$$-\frac{GM_e m}{R_e} + \frac{3}{4}\frac{GM_e m}{R_e} = -\frac{GM_e m}{R} + \frac{1}{2}mv_1^2$$

$$\Rightarrow \quad -\frac{GM_e m}{4R_e} = -\frac{GM_e m}{R} + \frac{1}{2}m\left(\frac{3}{2}\frac{GM_e m}{R_e}\right)\cdot\frac{R_e^2}{R^2}$$

$$\Rightarrow \quad \frac{3}{4}\frac{R_e}{R^2} - \frac{1}{R} + \frac{1}{4R_e} = 0$$

$$\Rightarrow \quad \frac{1}{R} = \frac{1 \pm \sqrt{1-4\left(\frac{3}{4}\right)\left(\frac{1}{4}\right)}}{2\times\frac{3}{4}R_e}$$

$$\Rightarrow \quad \frac{1}{R} = \frac{1 \pm \frac{1}{2}}{\frac{3}{2}R_e}$$

$$\Rightarrow \quad R = 3R_e$$

Here R_e is not possible as $R > R_e$ **Ans. (B)**

Sol. 60 Orbital speed very close to planet surface is given as

$$v_0 = \sqrt{\frac{GM}{R}}$$

escape velocity from the planet surface is given as

$$v_e = \sqrt{\frac{2GM}{R}}$$

$$\Rightarrow \quad \frac{v_0}{v_e} = \frac{1}{\sqrt{2}}$$ **Ans. (C)**

Sol. 61 By angular momentum conservation, we use

$$v_{max} r_{min} = v_{min} r_{max}$$

$$\Rightarrow \quad \frac{r_{min}}{r_{max}} = \frac{v_{min}}{v_{max}} = \frac{1}{6}$$ **Ans. (A)**

Sol. 62

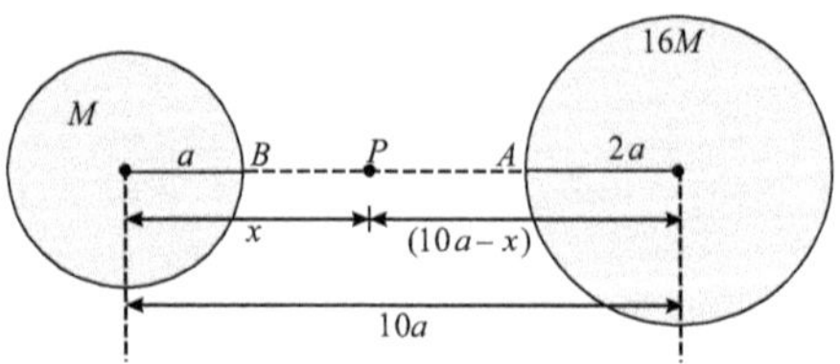

Null point P can be obtained as

$$\frac{GM}{x^2} = \frac{G(16M)}{(10a-x)^2}$$

$$\Rightarrow \quad \frac{1}{x} = \frac{4}{(10a-x)}$$

$$\Rightarrow \quad 4x = 10a - x$$

$$\Rightarrow \quad x = 2a \qquad \ldots(1)$$

For body fired from point A at speed v, it should cross null point P in order to reach point B so by energy conservation, we use

$$-\frac{GMm}{8a} - \frac{G(16M)m}{2a} + \frac{1}{2}mv^2$$

$$= -\frac{GMm}{2a} - \frac{G(16M)m}{8a}$$

$$\Rightarrow \quad v^2 = 2GM\left[\frac{1}{8a} + \frac{16}{2a} - \frac{1}{2a} - \frac{16}{8a}\right]$$

$$\Rightarrow \quad v^2 = 2GM\left[\frac{1+64-4-16}{8a}\right]$$

$$\Rightarrow \quad v^2 = 2GM\left[\frac{45}{8a}\right]$$

$$\Rightarrow \quad v = \sqrt{\frac{90GM}{8a}}$$

$$\Rightarrow \quad v = \frac{3}{2}\sqrt{\frac{5GM}{a}}$$ **Ans. (B)**

8.1 Elasticity

Sol. 63 Strain in wire is 4.9×10^{-4}

Stress in wire is given as

$$\text{Stress} = Y \times \text{Strain}$$

$$= 9 \times 10^{10} \times 4.9 \times 10^{-4}\ \text{N/m}^2$$

Lowest vibration frequency of clamped wire is

$$n_f = \frac{1}{2l}\sqrt{\frac{T}{\mu}} = \frac{1}{2l}\sqrt{\frac{T}{\rho A}} = \frac{1}{2l}\sqrt{\frac{\text{Stress}}{\rho}}$$

$$\Rightarrow \quad n_f = \frac{1}{2\times 1} \times \sqrt{\frac{9\times10^{10}\times4.9\times10^{-4}}{9\times10^3}}$$

$$\Rightarrow \quad n_f = \frac{1}{2} \times \sqrt{4.9\times10^3}$$

$$\Rightarrow \quad n_f = \frac{7}{2} \times 10 = 35\ \text{Hz}$$ **Ans. (35.00)**

Sol. 64 An elastic wire can be considered as a spring of force constant k given as

$$k = \frac{YA}{l}$$

Thus time period of oscillations of object can be given as

$$T = 2\pi\sqrt{\frac{m}{k}}$$

So frequency of oscillations is given as

$$f = \frac{1}{2\pi}\sqrt{\frac{k}{m}} = \frac{1}{2\pi}\sqrt{\frac{YA}{ml}} \quad \textbf{Ans. (A)}$$

9.1 Fluid Statics

Sol. 65 For bubble, we use

$$F_B - mg = ma$$

$$\Rightarrow \quad V\rho_\omega g - mg = ma$$

$$\Rightarrow \quad m = \frac{V\rho_\omega g}{g+a} = \frac{\frac{4}{3}\times 3.14\times(1)^3\times 1\times 980}{980+9.8}$$

$$\Rightarrow \quad m \simeq 4.15 \text{ gm} \quad \textbf{Ans. (C)}$$

Sol. 66 Initial energy of liquid in the two vessels is

$$U_1 = (\rho S x_1)g\left(\frac{x_1}{2}\right) + (\rho S x_2)g\left(\frac{x_2}{2}\right)$$

If after equilibrium, final level becomes x_0, we use

$$U_2 = 2(\rho S x_0)g\left(\frac{x_0}{2}\right)$$

As volume of liquid is constant we use

$$x_0 = \frac{x_1 + x_2}{2}$$

Change in energy is given as

$$\Delta U = U_1 - U_2 = \rho S g\left[\left(\frac{x_1^2}{2}+\frac{x_2^2}{2}\right) - x_0^2\right]$$

$$\Rightarrow \quad \Delta U = \rho S g\left[\frac{x_1^2}{2}+\frac{x_2^2}{2} - \left(\frac{x_1+x_2}{2}\right)^2\right]$$

$$\Rightarrow \quad \Delta U = \rho S g(x_1 - x_2)^2 \quad \textbf{Ans. (C)}$$

Sol. 67 For floating of shell, we use

$$\frac{4}{3}\pi(R^3 - r^3)\rho_s g = \frac{4}{3}\pi R^3 \rho_w g$$

$$\Rightarrow \quad 1-\left(\frac{r}{R}\right)^3 = \frac{8}{27}$$

$$\Rightarrow \quad \frac{r}{R} = \left(\frac{19}{27}\right)^{1/3} = \frac{19^{1/3}}{3}$$

$$\Rightarrow \quad r = 0.88 \simeq \frac{8}{9}R \quad \textbf{Ans. (B)}$$

9.2 Fluid Dynamics

Sol. 68 Using Bernoulli's Equation at the two points, we have

$$P + \frac{1}{2}\rho v^2 = \frac{P}{2} + \frac{1}{2}\rho V^2$$

$$\Rightarrow \quad \frac{2P}{2\rho} + \frac{1}{2}\frac{\rho v^2}{\rho} \times 2 = V^2$$

$$\Rightarrow \quad V = \sqrt{\frac{P}{\rho} + v^2} \quad \textbf{Ans. (B)}$$

9.3 Viscosity

Sol. 69 Bulk modulus of material is given as

$$B = \frac{\Delta P}{\left|\frac{\Delta V}{V}\right|}$$

$$\Rightarrow \quad \left|\frac{\Delta V}{V}\right| = \frac{\Delta P}{B} = \frac{4\times 10^9}{8\times 10^{10}} = \frac{1}{20}$$

$$\Rightarrow \quad \frac{\Delta l}{l} = \frac{1}{3}\left|\frac{DV}{V}\right| = \frac{1}{3}\times\frac{1}{20} = \frac{1}{60}$$

Percentage change in length of cube is given as

$$\frac{\Delta l}{l}\times 100 = \frac{1}{60}\times 100 = \frac{5}{3} = 1.67\% \quad \textbf{Ans. (B)}$$

Sol. 70 After falling through a height h as the velocity of ball is equal to terminal velocity, we use

$$\sqrt{2gh} = \frac{2}{9}\frac{r^2 g}{\eta}(\rho_l - \rho)$$

$$\Rightarrow \quad h = \frac{2}{81}\frac{r^4 g\,(\rho_l - \rho)^2}{\eta^2}$$

$$\Rightarrow \quad h \propto r^4 \quad \textbf{Ans. (B)}$$

9.4 Surface Tension

Sol. 71 Figure below shows that angle of contact is $\theta = 30°$

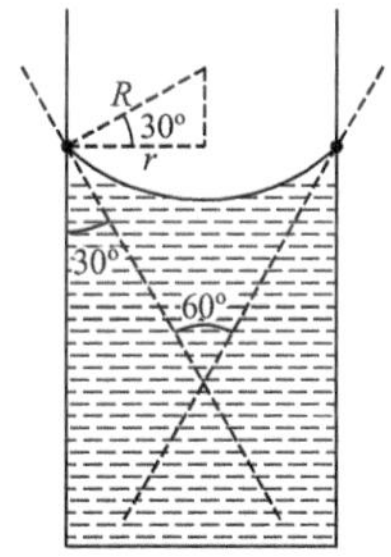

Capillary height of liquid in tube is given as

$$h = \frac{2T}{\rho g r\cos\theta}$$

$$\Rightarrow \quad h = \frac{2\times 0.05}{667\times 10\times\left(\frac{0.15\times 10^{-3}\times 2}{\sqrt{3}}\right)}$$

$$\Rightarrow \quad h = 0.087 \text{ m} \quad \textbf{Ans. (C)}$$

Sol. 72 By excess pressure formula, we use

$$0.01 = \frac{4T}{R_1}$$

and $$0.02 = \frac{4T}{R_2}$$

$$\Rightarrow \quad \frac{R_1}{R_2} = 2$$

$$\Rightarrow \quad R_1 = 2R_2$$

Ratio of volumes is given as

$$\frac{V_1}{V_2} = \frac{R_1^3}{R_2^3} = 8$$ **Ans. (A)**

Sol. 73 Capillary rise tube is given as

$$h = \frac{2T\cos\theta}{r\rho g}$$

$$\Rightarrow \quad T = \frac{\rho g r h}{2\cos\theta}$$

$$\Rightarrow \quad T = \frac{(900)(10)(15\times10^{-5})(15\times10^{-2})}{2}$$

$$\Rightarrow \quad = 101.25\times10^{-3}\,\text{N/m}$$

$$\Rightarrow \quad = 101.25\,m\text{N/m}$$ **Ans. (101)**

10.1 Thermal Expansion

Sol. 74 It is given that by thermal expansion

$$\frac{\Delta l}{l} = \frac{0.02}{100} = \propto \Delta T$$

$$\Rightarrow \quad \propto = \frac{2\times10^{-4}}{\Delta T}$$

Thus volume expansion coefficient is

$$\gamma = 3\alpha = \frac{6\times10^{-4}}{\Delta T}$$

Density variation is given as

$$\Delta\rho = -\rho\gamma\Delta T$$

$$\Rightarrow \quad \frac{\Delta\rho}{\rho} = -\gamma\Delta T = -6\times10^{-4} = -0.06\%$$ **Ans. (B)**

Sol. 75 The unfitted volume of beaker is given as

$$\Delta V = (V_0 - V_m)$$

After increasing temperature by ΔT, we use

$$\Delta V' = (V_0' - V_m')$$

As $\Delta V' = \Delta V$, we use

$$V_0 - V_m = V_0(1+\gamma_b\Delta T) - V_m(1+\gamma_M\Delta T)$$

$$\Rightarrow \quad V_0\gamma_b = V_m\gamma_m$$

$$\Rightarrow \quad V_m = \frac{V_0\gamma_b}{\gamma_m} = \frac{(500)(6\times10^{-6})}{(1.5\times10^{-4})}$$

$$\Rightarrow \quad V_m = 20\,\text{cm}^3$$ **Ans. (20)**

Sol. 76 At T°C, Total length of combination s

$$L = L_1 + L_2$$

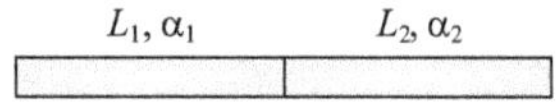

At $T + \Delta T$, Total length becomes

$$L'_{eq} = L'_1 + L'_2$$

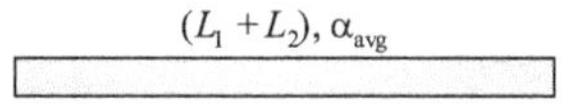

There we us $$L'_1 = L_1(1+\alpha_1\Delta T)$$

and $$L'_2 = L_2(1+\alpha_2\Delta T)$$

$$\Rightarrow \quad L'_{eq} = (L_1+L_2)(1+\alpha_{avg}\Delta T)$$

$$\Rightarrow \quad (L_1+L_2)(1+\alpha_{avg}\Delta T) = L_1 + L_2 + L_1\alpha_1\Delta T + L_2\alpha_2\Delta T$$

$$\Rightarrow \quad (L_1+L_2)\alpha_{avg} = L_1\alpha_1 + L_2\alpha_2$$

$$\Rightarrow \quad \alpha_{avg} = \frac{L_1\alpha_1 + L_2\alpha_2}{L_1+L_2}$$ **Ans. (D)**

10.2 Calorimetry

Sol. 77 Given mixture can be taken as

Mixture = 200 g water at 25°C + m gram steam at 100°C

Heat supplied by steam = $m \times 540 + m \times 1 \times 69 = 609m$

Heat absorbed by water = $200 \times 1 \times 6 = 1200$ cal

by conservation of energy, we use

$$1200 = 609m$$

$$\Rightarrow \quad m = \frac{1200}{609} \simeq 2\,\text{g}$$ **Ans. (B)**

Sol. 78 Here the water will supply heat to ice for melting

$$m_w s_w \Delta\theta = m_{ice} L_{ice}$$

$$m_{ice} = \frac{0.2\times4200\times25}{3.4\times10^5}$$

$$\Rightarrow \quad m_{ice} = 0.0617\,\text{kg}$$

$$\Rightarrow \quad m_{ice} = 61.7\,\text{gm}$$

Remaining ice will remain un-melted so correct answer is (A). **Ans. (A)**

Sol. 79 Given that half of KE is converted to heat, we use

$$\frac{1}{2}\times\frac{1}{2}mv^2 = ms\Delta T$$

$$\Rightarrow \quad \Delta T = \frac{(210)^2}{4\times0.03\times4200} \simeq 87.5°\text{C}$$ **Ans. (B)**

10.3 Heat Transfer

Sol. 80 Using average form of Newton's law of cooling it gives

$$\frac{50-40}{300} = k\left(\frac{50+40}{2}-20\right) \quad \ldots(1)$$

and $$\frac{40-T}{300} = k\left(\frac{40+T}{2}-20\right) \quad \ldots(2)$$

Dividing (1) and (2), it gives

$$\frac{10}{40-T} = \frac{25}{T/2}$$

$\Rightarrow$ $5T = 100 - 25T$

$\Rightarrow$ $30T = 1000$

$\Rightarrow$ $T = \frac{100}{3}$°C **Ans. (A)**

Sol. 81 Solar constant has units of Joule/s-m^2 thus its dimensional formula is given as

$$\left[\frac{\text{Joule}}{\text{sec}-m^2}\right] \equiv \frac{[ML^2T^{-2}]}{[T][L^2]} = MT^{-3}$$ **Ans. (D)**

Sol. 82

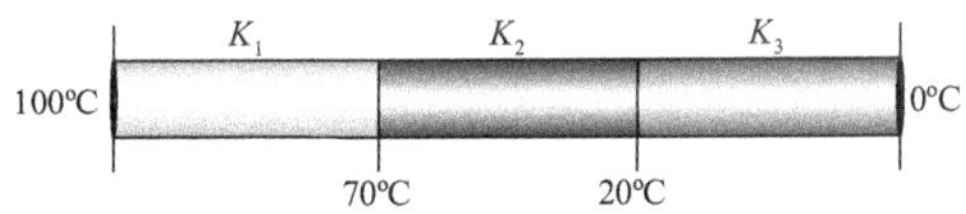

We consider rods have same length l and area of cross-section A so for steady state conductor, we use

$$\left(\frac{\Delta Q}{\Delta t}\right)_{AB} = \left(\frac{\Delta Q}{\Delta t}\right)_{BC} = \left(\frac{\Delta Q}{\Delta t}\right)_{CD}$$

$$\frac{(100-70)K_1A}{l} = \frac{(70-20)K_2A}{l} = \frac{(20-0)K_3A}{l}$$

$\Rightarrow$ $30K_1 = 50K_2 = 20K_3$

$\Rightarrow$ $3K_1 = 2K_3$

$\Rightarrow$ $\frac{K_1}{K_3} = \frac{2}{3}$

and $\frac{K_2}{K_3} = \frac{2}{5} = 2:5$ **Ans. (A)**

11.1 Laws of Thermodynamics

Sol. 83 Total internal energy of mixture is given as

$$U = \frac{f_1}{2}n_1RT + \frac{f_2}{2}n_2RT$$

$\Rightarrow$ $U = \frac{5}{2}\times 3RT + \frac{3}{2}\times 5RT = 15RT$ **Ans. (B)**

Sol. 84 As volume compresses to $V/10$, we use

$TV^{\gamma-1}$ = Constant [r = 7/5 For air]

$\Rightarrow$ $T_1V_1^{2/5} = T_2\left(\frac{V}{10}\right)^{2/5}$

$\Rightarrow$ $T_2 = T(10)^{2/5}$

Change in internal energy in adiabatic process is

$$\Delta U = \frac{R(T_2-T_1)}{\gamma-1}$$

$\Rightarrow$ $$\Delta U = \frac{5\times 8.31\times 293\times(10^{2/5}-1)}{\frac{7}{5}-1}$$

$\cong 4\times 10^3$ J **Ans. (4 × 10³ J)**

Sol. 85 Efficiency of heat engine is given as

$$\eta = \frac{\text{Work done}}{\text{Heat supplied}}$$

$\Rightarrow$ $\eta = \frac{1915-40+125-Q}{1915+125} = 0.5$

$\Rightarrow$ $\frac{2000-Q}{2040} = 0.5$

$\Rightarrow$ $2000 - Q = 1020$

$\Rightarrow$ $Q = 980$ J **Ans. (C)**

Sol. 86 As all sudden process are considered irreversible and in these processes no time is available for heat exchange so considered adiabatic. **Ans. (B)**

Sol. 87 For triangular molecule degrees of freedom are given translational $dof = 3$

Rotational, $dof = 3$

Thus internal energy of gas is given as

$$U = \frac{f}{2}nRT = 3RT$$ **Ans. (D)**

Sol. 88 For minimum work, we use Carnot cycle so, coefficient of performance is given as

$$COP = \frac{Q_{in}}{W} = \frac{Q_{in}}{Q_{out}-Q_{in}} = \frac{273}{300-273} = \frac{273}{27}$$

$\Rightarrow$ $W = \frac{27}{273}\times Q_{in} = \frac{27}{273}\times 100\times 80$ cal

Heat rejected by refrigerator is given as

$Q_{out} = W + Q_1$

$\Rightarrow$ $Q_{out} = \frac{27}{273}\times 100\times 80 + 100\times 80$ cal

$\Rightarrow$ $Q_{out} = 8791.2$ cal **Ans. (8791)**

Sol. 89 For mixture of gases, equilibrium temperature is given as

$$T_{eq} = \frac{n_1T_1 + n_2T_2}{n_1 + n_2} = \frac{0.1\times 200 + 0.05\times 400}{0.1+0.05}$$

$$\Rightarrow \quad T_{eq} = \frac{20+20}{0.15} = \frac{40}{0.15} = \frac{800}{3}$$

$$\Rightarrow \quad T_{eq} = 266.66\text{ K}$$ **Ans. (266)**

Sol. 90 (I) In Adiabatic process no heat exchange take place so option (B) is correct match

(II) In Isothermal process no change internal energy so option (D) is correct match

(III) In Isochoric process V = constant. So $\Delta\omega = 0$

(IV) In isobaric process as P = constant option (C) is correct match. **Ans. (A)**

Sol. 91 By gas law we use

$$PV = nRT$$

At constant pressure

$$P\Delta V = nR\Delta T$$

$$\Rightarrow \quad \Delta T = \left(\frac{P}{nR}\right)\Delta V$$

At constant temperature

$$P\Delta V + V\Delta P = 0$$

$$\Rightarrow \quad \Delta P = -\frac{P\Delta V}{V}$$

Value of C is given as

$$C = \left|\frac{\Delta T}{\Delta P}\right| = \frac{V}{nR} = \frac{T}{P} = \frac{300}{2}$$

$$\Rightarrow \quad C = 150\text{ K/atm}$$ **Ans. (150)**

Sol. 92 From the given indicter diagram, we can see for all processes as ΔT is same

$$\Delta U = nC_v\Delta T = \text{same}$$

$AB \rightarrow$ volume is increasing $\Rightarrow$ $W > 0$

$AD \rightarrow$ volume is decreasing $\Rightarrow$ $W < 0$

$AC \rightarrow$ volume is constant $\Rightarrow$ $W = 0$ **Ans. (A)**

Sol. 93 Using gas law, we use

$$n = \frac{PV}{RT}$$

Mean kinetic energy of a molecule is given as

$$\frac{3}{2}kT = 4\times 10^{-14}$$

$$\Rightarrow \quad RT = \frac{8}{3}\times 10^{-14}\times N_{AV} \qquad \left[\text{As } k = \frac{R}{N_{AV}}\right]$$

Total molecules in gas are given as

$$N = \frac{PV}{RT}\times N_{AV}$$

$$\Rightarrow \quad N = \frac{2\times 13.6\times 980\times 4}{\frac{8}{3}\times 10^{-14}} = 3.99\times 10^{18}$$ **Ans. (C)**

Sol. 94 In adiabatic process we use

$$PV^{\gamma} = \text{constant}$$

$$\Rightarrow \quad P\left(\frac{m}{\rho}\right)^{\gamma} = \text{constant}$$

$$\Rightarrow \quad P \propto \rho^{\gamma}$$

$$\Rightarrow \quad \frac{P_f}{P_i} = \left(\frac{\rho_f}{\rho_i}\right)^{\gamma} = (32)^{7/5} = 2^7 = 128$$ **Ans. (D)**

Sol. 95 Internal energy of gas for f degrees of freedom is given as

$$U = \frac{f}{2}nRT = \frac{5}{2}RT$$

Ratio of specific heats is given as

$$\gamma = \frac{C_P}{C_V}$$

$$\Rightarrow \quad \gamma = \frac{f+2}{f} = \frac{5+2}{5} = \frac{7}{5}$$ **Ans. (D)**

Sol. 96 Work done in cycle $ABCDA$ is given as

$$W_{ABCDA} = 2P_0V_0$$

Total heat input in cycle is given as

$$Q_{in} = Q_{AB} + Q_{BC}$$

Here, we use in process AB

$$Q_{AB} = nC(T_B - T_A)$$

$$\Rightarrow \quad Q_{AB} = \frac{n3R}{2}(T_B - T_A)$$

$$\Rightarrow \quad Q_{AB} = \frac{3}{2}(P_BV_B - P_AV_A)$$

$$\Rightarrow \quad Q_{AB} = \frac{3}{2}(3P_BV_0 = P_0V_0) = 3P_0V_0$$

For process BC, we use

$$Q_{BC} = nC_P(T_C - T_B)$$

$$\Rightarrow \quad Q_{BC} = \frac{n5R}{2}(T_C - T_B)$$

$$\Rightarrow \quad Q_{BC}=\frac{5}{2}(P_CV_C-P_BV_B)$$

$$\Rightarrow \quad Q_{BC}=\frac{5}{2}(6P_0V_0-3P_0V_0)=\frac{15}{2}P_0V_0$$

Efficiency of cycle is given as

$$\eta=\frac{W}{Q_{in}}\times 100$$

$$\Rightarrow \quad \eta=\frac{2P_0V_0}{3P_0V_0+\frac{15}{2}P_0V_0}\times 100$$

$$\Rightarrow \quad \eta=\frac{400}{21}=19.04\approx 19$$ **Ans. (19.00)**

12.1 Kinetic Theory of Gases

Sol. 97 The mean free path of molecules of an ideal gas is given as

$$\lambda=\frac{V}{\sqrt{2}\pi d^2N}$$

Due to rise in temperature speed of molecules increases so mean collision time between molecules decreases and mean free path is independent of temperature. **Ans. (D)**

Sol. 98 Given that at constant pressure

$$\Delta Q=nC_P\Delta T$$

$$\Rightarrow \quad 160=n\times C_P(50) \quad \ldots(1)$$

At constant volume, we use

$$240=nC_V(100) \quad \ldots(2)$$

From (1) & (2)

$$\gamma=\frac{C_P}{C_V}=\frac{160/50}{240/100}=\frac{4}{3}$$

As gas has $\gamma=\frac{4}{3}$

$$\Rightarrow \quad \gamma=\frac{f+2}{f}$$

$$\Rightarrow \quad f=6$$ **Ans. (C)**

Sol. 99 For monoatomic molecule $\to f=3$

$$\Rightarrow \quad \gamma=\frac{f+2}{f}=\frac{5}{3}$$

For Diatomic rigid molecules $\to f=5$

$$\Rightarrow \quad \gamma=\frac{7}{5}$$

For Diatomic non rigid molecules $\to f=7$

$$\Rightarrow \quad \gamma=\frac{9}{7}$$

For Triatomic rigid molecules $\to f=6$

$$\Rightarrow \quad \gamma=\frac{8}{6}=\frac{4}{3}$$ **Ans. (A)**

Sol. 100 RMS speed of gas molecule is given as

$$V_{rms}=\sqrt{\frac{3RT}{M}}$$

Given that

$$V_{N_2}=V_{H_2}$$

$$\Rightarrow \quad \sqrt{\frac{3RT_{N_2}}{M_{N_2}}}=\sqrt{\frac{3RT_{H_2}}{M_{H_2}}}$$

$$\Rightarrow \quad \frac{573}{28}=\frac{T_{H_2}}{2}$$

$$\Rightarrow \quad T_{H_2}=40.928\text{ K}$$ **Ans. (41.00)**

Sol. 101 By gas law we use

$$P_1V_1=nR(250) \quad \ldots(1)$$

If $\frac{n}{4}$ mole molecules gets dissociated then final number of moles will become $n+\frac{n}{4}=\frac{5n}{4}$ so we use again by gas law

$$P_2(2V_1)=\frac{5n}{4}R(2000) \quad \ldots(2)$$

From (1) and (2), we have

$$\frac{P_1}{2P_2}=\frac{4\times 250}{5\times 2000}$$

$$\Rightarrow \quad \frac{P_1}{P_2}=\frac{1}{5} \text{ or } \frac{P_2}{P_1}=5$$ **Ans. (5.00)**

Sol. 102 As mean speed of molecules is $v_m\propto\sqrt{T}$ mean time between collisions is given as

$$t_m=\frac{\lambda}{v_m}$$

$$\Rightarrow \quad t_m\propto\frac{1}{\sqrt{T}}$$ **Ans. (C)**

13.1 Simple Harmonic Motion

Sol. 103 (i) At $t=\frac{3T}{4}$ displacement is zero where velocity is maximum and acceleration force is zero.

(ii) At $t = T$, displacement is maximum so acceleration is maximum.

(iii) At $t = \frac{T}{4}$, displacement is zero so velocity is maximum.

(iv) at $t = \frac{T}{2}$ it shows displacement is maximum where kinetic energy = 0 and potential energy is maximum.

Hence (i), (ii) and (iii) are correct. **Ans. (D)**

Sol. 104 Angular frequency of SHM of block is given as

Initial $\omega_1 = \sqrt{\frac{K}{m}}$

Initial $\omega_2 = \sqrt{\frac{K}{m/2}}$

At mean position mass breaks off at maximum speed so we use

$$\omega_1 A_1 = \omega_2 A_2$$

$\Rightarrow$ $A_2 = \frac{\omega_1 A_1}{\omega_2} = \frac{A_1}{\sqrt{2}}$ **Ans. (D)**

Sol. 105 The ring acts like a compound pendulum. In its oscillations in plane, time period is given as

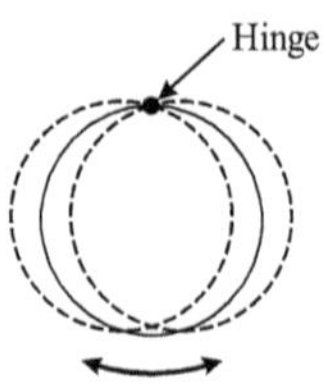

$$T_1 = 2\pi\sqrt{\frac{2MR^2}{MgR}} = 2\pi\sqrt{\frac{2R}{g}}$$

In its oscillations back and forth. Perpendicular to vertical plane, its time period is given as

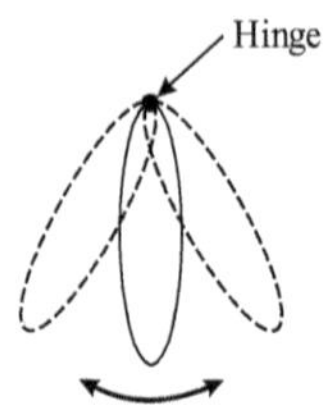

$$T_2 = 2\pi\sqrt{\frac{\frac{3}{2}MR^2}{MgR}} = 2\pi\sqrt{\frac{3M}{2g}}$$

$\Rightarrow$ $\frac{T_1}{T_2} = \frac{\sqrt{2}}{\sqrt{3/2}} = \frac{2}{\sqrt{3}}$ **Ans. (A)**

Sol. 106 Given motion equation is

$$y = y_0 \sin^2 \omega t = \frac{y_0}{2}(1 - \cos 2\omega t)$$

$\Rightarrow$ $y - \frac{y_0}{2} = -\frac{y_0}{2}\cos 2\omega t$

As y is measured from lower end of unstretched spring here $\frac{y_0}{2}$ is the amplitude of SHM. Thus we use

$$\frac{y_0}{2} = \frac{mg}{k}$$

SHM frequency is given as

$$2\omega = \sqrt{\frac{k}{m}} = \sqrt{\frac{2g}{y_0}}$$

$\Rightarrow$ $\omega = \sqrt{\frac{g}{2y_0}}$ **Ans. (B)**

14.1 Wave Motion

Sol. 107 Tension in repeat bottom is

$$T_1 = 2g = 20\text{ N}$$

tension in rope at top is

$$T_2 = 8g = 80\text{ N}$$

Wave speed in rope is given as

$$v = \sqrt{\frac{T}{\mu}}$$

As frequency of pulse is same at bottom and top, we use

$$\frac{v_1}{\lambda_1} = \frac{v_2}{\lambda_2}$$

$\Rightarrow$ $\lambda_2 = \frac{v_2}{v_1} \times \lambda_1 = \sqrt{\frac{T_2}{T_1}} \times \lambda_1$

$\Rightarrow$ $\lambda_2 = \sqrt{\frac{80}{20}} \times 6 = 12\text{ cm}$ **Ans. (B)**

Sol. 108 Given that

$$5\text{m} = N_2\lambda$$

and $1.5\text{m} = (2N_1 + 1)\lambda/2$

$\Rightarrow$ $\frac{5}{1.5} = \frac{2N_2}{2N_1 + 1}$

$\Rightarrow$ $10N_1 + 5 = 3N_2$

Thus for $N_1 = 1 \Rightarrow N_2 = 5 \Rightarrow \lambda = 1$

and $N_1 = 4 \Rightarrow N_2 = 15 \Rightarrow \lambda = 1/3$

and $N_1 = 7 \Rightarrow N_2 = 25 \Rightarrow \lambda = 1/5$ **Ans. (D)**

Sol. 109 Pressure amplitude is given as

$$\Delta P_0 = \frac{2\pi}{\lambda} BA$$

$$\Rightarrow \quad \Delta P_0 = \frac{2\pi}{\lambda} \times \rho v^2 \times A = \frac{w}{v} \times \rho v^2 \times A$$

$$\Rightarrow \quad A = \frac{\Delta P_0}{\rho v \omega} = \frac{10}{1 \times 300 \times 2 \times 3.14 \times 1000}$$

$$\Rightarrow \quad A = \frac{10}{300 \times 6.28} \times 10^{-3}\,\text{m}$$

Among all given options, (B) is correct. **Ans. (B)**

14.2 Stationary Waves and Beats

Sol. 110 Fundamental frequency of oscillation for a clamped string is given as

$$f = \frac{l}{2l}\sqrt{\frac{T}{\mu}}$$

For identical string l and m will be same

$$\Rightarrow \quad f \propto \sqrt{T}$$

$$\Rightarrow \quad \frac{450}{300} = \sqrt{\frac{T_x}{T_y}}$$

$$\Rightarrow \quad \frac{T_x}{T_y} = \frac{9}{4} = 2.25$$

Ans. (C)

Sol. 111 For two successive resonance lengths in tube, we use

$$\Rightarrow \quad \lambda = 2(l_2 - l_1)$$

$$\Rightarrow \quad \lambda = 2 \times (24.5 - 17)$$

$$\Rightarrow \quad \lambda = 2 \times 7.5 = 15\,\text{cm}$$

As $v = f\lambda$, we use

$$\Rightarrow \quad 330 = f \times 15 \times 10^{-2}$$

$$\Rightarrow \quad f = \frac{330}{15} \times 100\ \text{Hz}$$

$$\Rightarrow \quad f = \frac{110 \times 100}{5}\ \text{Hz}$$

$$\Rightarrow \quad f = 2200\,\text{Hz}$$

Ans. (C)

14.3 Doppler's Effect

Sol. 112 Apparent frequency of bus as received by wall is

$$f_w = 420 \times \left(\frac{330}{330 - v_B}\right)$$

Apparent frequency as received by driver is

$$f_D = 420 \times \left(\frac{330 + v_0}{330}\right)\left(\frac{330}{330 - v_B}\right)$$

$$\Rightarrow \quad 490 = 420 \times \left(\frac{330 + v_B}{330 - v_B}\right)$$

$$\Rightarrow \quad \frac{7}{6} = \frac{330 + v_B}{330 - v_B}$$

$$\Rightarrow \quad v_B = \frac{330}{13}\,\text{m/s}$$

$$\Rightarrow \quad v_B = \frac{330}{13} \times \frac{18}{5}$$

$$\approx 91\,\text{km/hr}$$

Ans. (A)

Sol. 113 Frequency received by wall is

$$h_\omega = h_0\left[\frac{C}{C - V_C}\right]$$

reflected frequency received by driver is

$$h_d = h_\omega\left[\frac{C + V_C}{C}\right] = n_0\left(\frac{C + V_C}{C - V_C}\right)$$

$$\Rightarrow \quad 480 = 440\frac{(C + V_C)}{(C - V_C)}$$

$$\Rightarrow \quad 12C - 12V_C = 11C + 11V_C$$

$$\Rightarrow \quad 23V_C = C$$

$$\Rightarrow \quad V_C = \frac{C}{23} = \frac{345}{23} = 15\,\text{m/s}$$

$$\Rightarrow \quad V_C = 54\,\text{kph}$$

Ans. (D)

Sol. 114 Observer is located at finite distance from track as shown in figure.

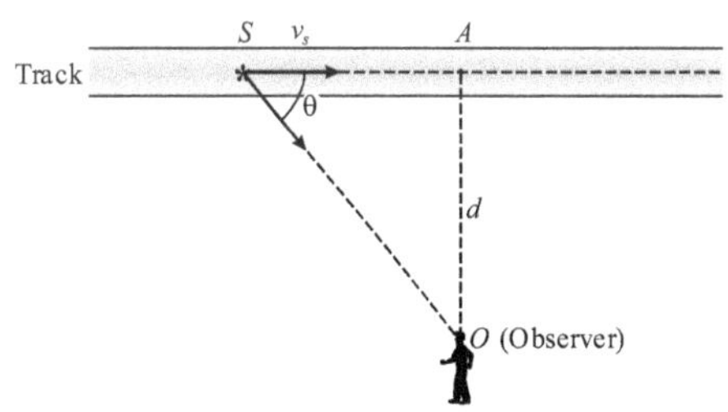

When source crosses point A the apparent frequency which was higher then v_0 will drop below v_0 as it started receding the observer. There correct option can be (B) or (D) but (B) is possible only when observer is located on track so in this case

(D) is correct answer. This can also be mathematically verified with expression of apparent frequency, given as

$$f_{\text{app}} = \nu_0\left[\frac{v_{\text{sound}}}{v_{\text{sound}} - v_s\cos\theta}\right]$$ **Ans. (D)**

15.1 Electric Force and Electric Field

Sol. 115 Let the charges on inner and outer spheres are q_1 and q_2.

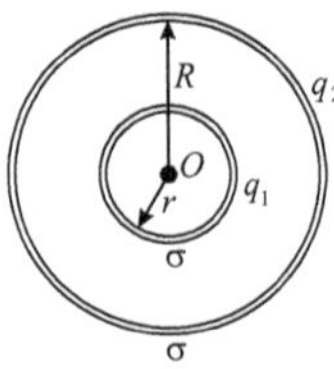

As charge density is same on both shells, we use

$$\sigma = \frac{q_1}{4\pi r^2} = \frac{q_2}{4\pi R^2}$$

$$\Rightarrow \quad \frac{q_1}{q_2} = \frac{r^2}{R^2}$$

$$q_1 + q_2 = Q$$

$$\Rightarrow \quad \frac{q_2 r^2}{R^2} + q_2 = Q$$

$$\Rightarrow \quad q_2 = \frac{qR^2}{(r^2+R^2)}$$

$$\Rightarrow \quad q_1 = \frac{r^2}{R^2}\cdot\frac{qR^2}{(R^2+r^2)} = \frac{qr^2}{(R^2+r^2)}$$

Potential at common centre is given as

$$V_C = \frac{kQ_1}{r} + \frac{kQ_2}{R}$$

$$\Rightarrow \quad V_C = K\left[\frac{Qr^2}{r(R^2+r^2)} + \frac{QR^2}{R(R^2+r^2)}\right]$$

$$\Rightarrow \quad V_C = \frac{KQ(r+R)}{(R^2+r^2)}$$

$$\Rightarrow \quad V_C = \frac{1}{4\pi\epsilon_0}\frac{(R+r)}{(R^2+r^2)}Q$$ **Ans. (C)**

Sol. 116 As initial speed of particle is zero, particle starts moving in the direction of acceleration which is given as

$$a = \sqrt{g^2 + \left(\frac{qE}{m}\right)^2}$$

at angle $\theta = \tan^{-1}\left(\frac{qE}{mg}\right)$ from Y direction.

Thus particle follows a straight line path and acceleration along the direction of acceleration. **Ans. (D)**

Sol. 117 Electric field at O due to Q_1 and Q_2 are

$$\vec{E}_1 = -\frac{kQ_1}{x_1^2}\hat{i}$$

and $$\vec{E}_2 = -\frac{kQ_2}{x_2^2}\hat{i}$$

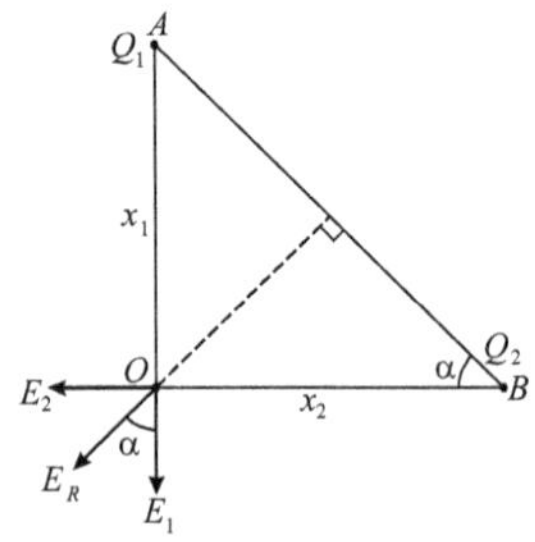

For $\overrightarrow{E_R}$ to be normal to line AB, we use

$$\tan\alpha = \frac{E_2}{E_1} = \frac{x_1}{x_2}$$

$$\Rightarrow \quad \frac{Q_2}{Q_1} = \frac{x_2}{x_1}$$ **Ans. (C)**

Sol. 118 For a uniformly charged body electric field is always in radial direction for all outside points so force on charge q (outside) is given as

$$F = \frac{1}{4\pi\epsilon_0}\cdot\frac{qQ}{r^2}$$

and at all interior points as

$$E = 0,\ F = 0$$ **Ans. (A)**

15.2 Electric Flux and Gauss's Law

Sol. 119

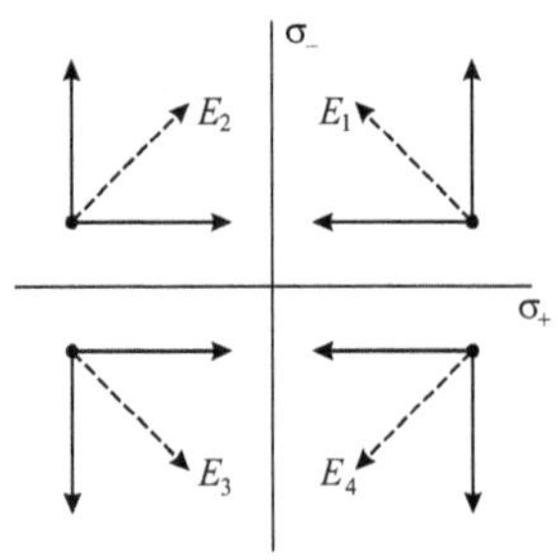

In figure above E_1, E_2, E_3 and E_4 represent the resultants electric field in region due to the two sheets. As $|\sigma_+| > |\sigma_-|$ option (A) is more appropriate. **Ans. (A)**

16.1 Electric Potential and Dipole

Sol. 120 As total charge remain conserved, if final charges are q_1' and q_2', we use

$$q_1 + q_2 = q_1' + q_2' = 12\ \mu C - 3\ \mu C = 9\ \mu C$$

As spheres are connected their potential must be same

$$\Rightarrow \quad V_1 = V_2$$

$$\Rightarrow \quad \frac{Kq_1'}{\left(\frac{2R}{3}\right)} = \frac{Kq_2'}{\left(\frac{R}{3}\right)}$$

$$\Rightarrow \quad q_1' = 2q_2'$$

$$\Rightarrow \quad 2q_2' + q_2' = 9\ \mu C$$

$$\Rightarrow \quad q_2' = 3\ \mu C$$

and $\quad q_1' = 6\ \mu C$ **Ans. (A)**

Sol. 121 Potential of inner sphere is given as

$$V_i = \frac{KQ_1}{R} + \frac{KQ_2}{4R} \quad \ldots(1)$$

and that of outer sphere is given as

$$V_O = \frac{KQ_1}{4R} + \frac{KQ_2}{4R} \quad \ldots(2)$$

As surface density of charges are equal, we use

$$\frac{Q_1}{4\pi R_1^2} = \frac{Q_2}{4\pi R_2^2}$$

$$\Rightarrow \quad \frac{Q_1}{R} = \frac{Q_2}{4R}$$

$$\Rightarrow \quad Q_2 = 4Q_1$$

From (1) & (2) we use

$$V_i - V_O = \frac{3}{4}\frac{KQ_1}{R} = \frac{3Q_1}{16\pi \epsilon_0 R}$$ **Ans. (A)**

Sol. 122 For a charge, kinetic energy attained due to electric potential difference is given as

$$q\Delta V = \frac{1}{2}mv^2$$

$$\Rightarrow \quad v = \sqrt{\frac{2q\Delta V}{m}}$$

$$\Rightarrow \quad \frac{v_1}{v_2} = \sqrt{\frac{q \times m_0}{q \times m_1}} = 2$$ **Ans. (D)**

Sol. 123 Due to $-\frac{q}{2}$ there is no charge in energy of q so energy change is only due to $4q$, which is given as

$$\Delta U = 4kq^2\left[\frac{1}{\frac{3d}{2}} - \frac{1}{\frac{d}{2}}\right]$$

$$\Rightarrow \quad \Delta U = \frac{8kq^2}{d}\left[\frac{1}{3} - 1\right] = -\frac{16kq^2}{3d}$$

$$\Rightarrow \quad \Delta U = -\frac{4q^2}{3\pi \epsilon_0 d} \text{ (Decreased)}$$ **Ans. (C)**

Sol. 124 Particle given kinetic energy due to work done by electric field so kinetic energy becomes zero again when

$$W = \int_0^{x_1} qEdx = \int_0^{x_1} qE_0(1 - ax^2)dx = 0$$

$$\Rightarrow \quad qE_0\left[x_1 - \frac{ax_1^3}{3}\right] = 0$$

$$\Rightarrow \quad x_1 = \sqrt{\frac{3}{a}}$$ **Ans. (D)**

Sol. 125 By work energy theorem for mass m, we use

$$\frac{kQq}{R} + mgy = \frac{kQq}{R+y} + \frac{1}{2}mv^2$$

$$\Rightarrow \quad v^2 = 2gy + \frac{2kQqy}{mR(R+y)}$$ **Ans. (A)**

Sol. 126 By symmetry of charges $E_C = 0$ and sum of total charges is zero, so at centre $V_C = 0$. **Ans. (C)**

Sol. 127 Initial interaction energy of dipoles is given as

$$U = -\vec{p_1} \cdot \vec{E_2}$$

$$\Rightarrow \quad U = -p_1\left(\frac{2Kp_2}{a^3}\right)\cos 180^\circ$$

$$= \frac{2Kp_1p_2}{a^3} = \frac{1}{2\pi \epsilon_0} \cdot \frac{p^2}{a^3}$$

At infinite separation interaction energy becomes zero so by conservation of energy, we have

$$\frac{p^2}{2\pi \epsilon_0 a^3} = 2 \times \frac{1}{2}mv^2$$

$$\Rightarrow \quad v = \sqrt{\frac{p^2}{2\pi \epsilon_0 ma^3}} = \frac{p}{a}\sqrt{\frac{1}{2\pi \epsilon_0 ma}}$$ **Ans. (C)**

16.2 Capacitance

Sol. 128 Initial energy stored in capacitor is

$$U_i = \frac{1}{2} \times 5 \times 10^{-6} (220)^2$$

Final common potential of capacitors is given as

$$V = \frac{220 \times 5 + 0 \times 2.5}{5 + 2.5} = 220 \times \frac{2}{3}$$

Final energy stored in capacitors is

$$U_f = \frac{1}{2} (5 + 2.5) \times 10^{-6} \left(220 \times \frac{2}{3}\right)^2$$

$$\Delta U = U_f - U_i$$

$$\Rightarrow \quad \Delta U = -403.33 \times 10^{-4}$$

$$\Rightarrow \quad -403.33 \times 10^{-4} = \frac{X}{100}$$

$$X = -4.03$$

Ans. (–4.03)

Sol. 129 Initial charges on capacitors are

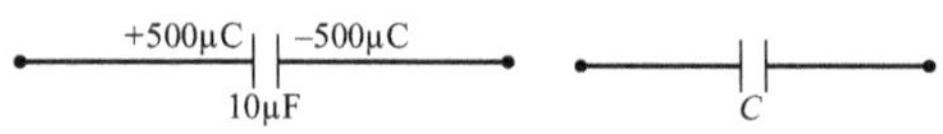

After connection

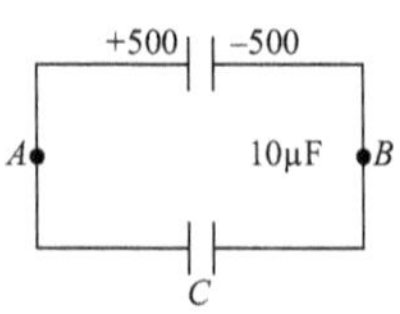

Potential difference across A and B is

$$V_{AB} = \frac{q_{\text{total}}}{q_{\text{total}}} = \frac{500}{C + 10} = 20$$

$$\Rightarrow \quad 20C + 200 = 500$$

$$\Rightarrow \quad 20C = 300$$

$$\Rightarrow \quad C = 15\ \mu\text{F}$$

Ans. (B)

Sol. 130 If total charge is 750 μC then charge on C_2 is given as

$$q_{C_2} = 750 - q_{C_3} = 750 - 8 \times 20 = 590\ \mu\text{C}$$

Ans. (A)

Sol. 131 Initial energy in capacitor is

$$U_1 = \frac{1}{2} C V_0^2$$

After connection with $\frac{C}{2}$ final potential of system is given as

$$V_f = \frac{CV_0}{C + \frac{C}{2}} = \frac{2V_0}{3}$$

Final energy in system is

$$U_2 = \frac{1}{2}\left(\frac{3C}{2}\right)\left(\frac{2V_0}{3}\right)^2 = \frac{1}{2} C V_0^2$$

Energy loss is given as

$$\Delta U_L = U_1 - U_2 = \left(\frac{1}{2} - \frac{1}{3}\right) C V_0^2 = \frac{1}{6} C V_0^2$$

Ans. (A)

Sol. 132 Initial charges on capacitors are given as

$$q_1 = CV$$

$$q_2 = 2C \times 2V = 4CV$$

When capacitors are connected with opposite polarity, find change on combination is

$$q_f = 4CV - CV = 3CV$$

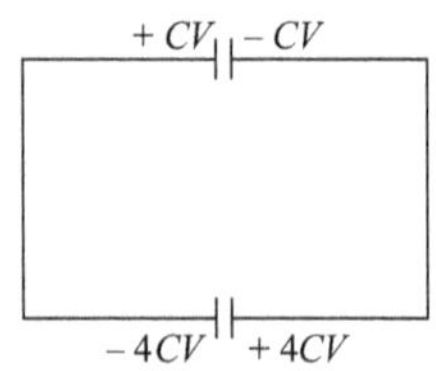

Thus final energy in combination is

$$U_f = \frac{q_f^2}{2(3C)} = \frac{(3CV)^2}{6C} = \frac{3}{2} C V^2$$

Ans. (D)

Sol. 133

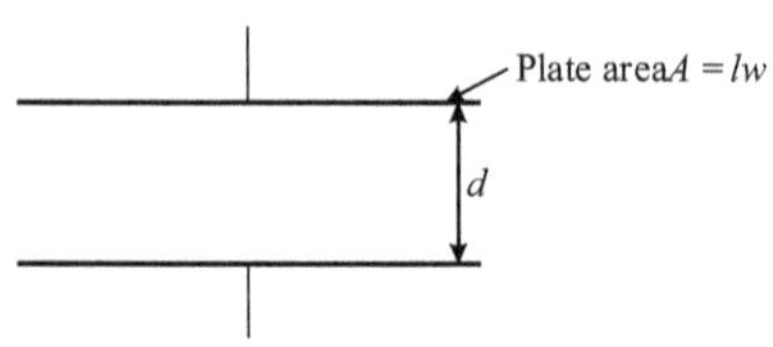

Capacitance of empty capacitor shown above is given as

$$C = \frac{\epsilon_0 lw}{d}$$

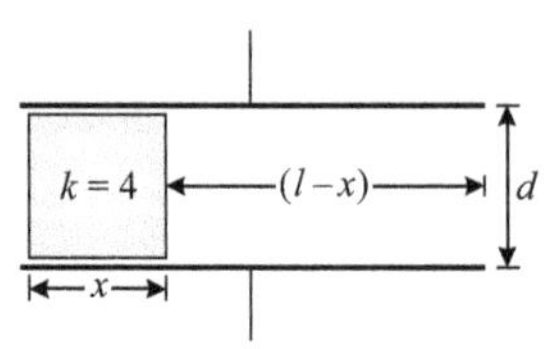

Capacitance of capacitor with dielectric shown above is given as

$$C_f = \frac{\epsilon_0 (4)xw}{d} + \frac{\epsilon_0 (l - x)w}{d}$$

For given applied potential difference final energy be two times if capacitance is two times, so we use

$$C_f = 2C$$

$$\frac{4\epsilon_0 xw}{d} + \frac{3\epsilon_0 lw}{d} - \frac{\epsilon_0 xw}{d} = 2\frac{\epsilon_0 lw}{d}$$

$$\Rightarrow \quad \frac{3\epsilon_0 xw}{d} = \frac{\epsilon_0 lw}{d}$$

$$\Rightarrow \quad x = \frac{l}{3}$$ **Ans. (C)**

Sol. 134 Figure below shows the potential distribution on different points of circuit by

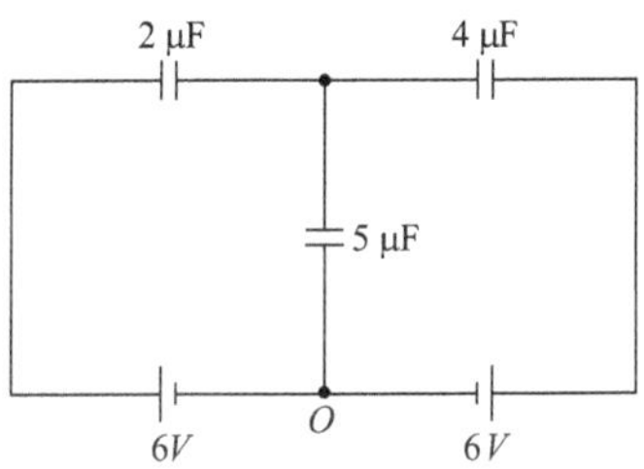

Taking potential of point O at 0

Writing equation for x, we have

$$q_1 + q_2 + q_3 = 0$$

$$2(x-6) + 4(x-6) + 5(x) = 0$$

$$\Rightarrow \quad x = \frac{36}{11}$$

$$\Rightarrow \quad q_3 = \frac{36(5)}{11} = \frac{180}{11}$$

$$q_3 = 16.36\,\mu\text{C}$$ **Ans. (B)**

17.1 Current Electricity

Sol. 135 Resistivity of different commonly used metals are given in increasing order below

Silver
Copper
Gold
Aluminium
Magnetism
Tungsten
Zinc
Nickel
Iron
Lead
Mercury
Germanium
Silicon

$\Rightarrow \quad \rho_M > \rho_A > \rho_C$ **Ans. (D)**

17.2 Circuit Analysis

Sol. 136 Current in branch of 8Ω resistance is given as

$$i = \frac{V}{R} = \frac{8}{8} = 1\text{ A}$$ **Ans. (D)**

Sol. 137 Current in upper & power branches are given as

$$i_1 = \frac{40}{100} = 0.4\text{ A}$$

and $$i_2 = \frac{40}{200} = 0.2\text{ A}$$

We write equation of potential drop from B to D as

$$V_B - 60 \times 0.4 + 110 \times 0.2 = V_D$$

$$\Rightarrow \quad V_B - V_D = 24 - 22 = 2\text{ V}$$ **Ans. (2)**

Sol. 138 Writing equation of potential drop A to B, we have,

$$V_A + 1 + 1 \times 2 - 2 = V_B$$

(Current in 2 Ω resistance is 1 A downward)

$$\Rightarrow \quad V_B - V_A = 1\text{ V}$$ **Ans. (A)**

Sol. 139 By principle of superposition current in 4Ω resistance is given as

$$I_{10V} = I_{4\Omega} = I_{20V} - I_{10V}$$

$$\Rightarrow \quad I_{10V} = \frac{20}{7 + \frac{20}{7}} \times \frac{10}{14} - \frac{10}{4 + \frac{70}{17}}$$

$$\Rightarrow \quad I_{10V} = \frac{140}{69} \times \frac{10}{14} - \frac{170}{138}$$

$$\Rightarrow \quad I_{10V} = \frac{100}{69} - \frac{170}{138} = \frac{30}{138} \simeq 0.22\text{ A}$$ **Ans. (C)**

NOTE : This question can also be solved using KCL or KVL or by Thevenin's analysis. Students are advised to solve this question by all these methods for conceptual clarity.

17.3 Thermal Effects of Current & RC Circuits

Sol. 140 Considering loop currents i_1 and i_2 in circuit as shown in figure

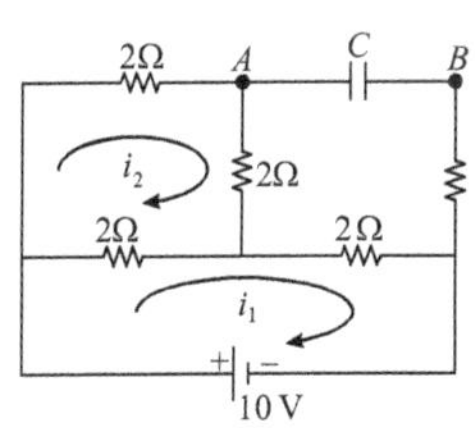

Using KVL equation for i_1 and i_2 gives

$$4i_1 - 2i_2 = 10 \quad \ldots(1)$$

$$6i_2 - 2i_1 = 0 \quad \ldots(2)$$

Solving (1) and (2) gives

$$12i_2 - 2i_2 = 10$$

$\Rightarrow$ $$i_2 = 1A$$

and $$i_1 = 3A$$

Potential difference between A and B is given as

$$V_A - 2\times 1 - 2\times 3 = V_B$$

$\Rightarrow$ $$V_A - V_B = 8V$$ **Ans. (8.00)**

Sol. 141 Resistance of an elemental shell of radius x and width dx inside the battery for radial flow of current as shown in figure is given as

$$dr = \frac{\rho dx}{2\pi xl}$$

$\Rightarrow$ $$r = \int dr = \int_a^b \frac{\rho}{2\pi l}\frac{dx}{x} = \frac{\rho}{2\pi l}\ln\left(\frac{b}{a}\right)$$

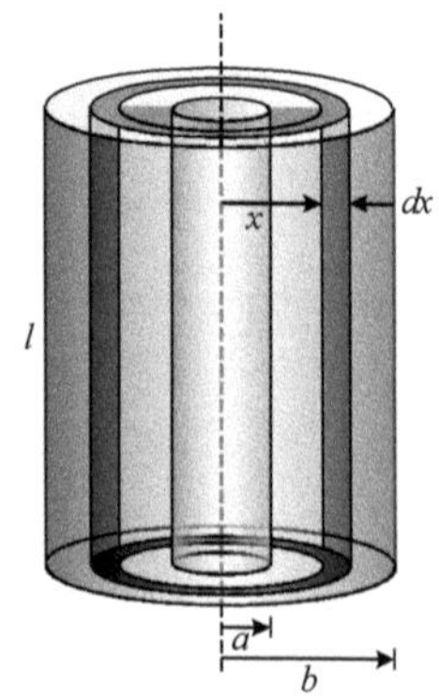

When battery is connected to external resistance R, maximum heat is produced in external resistance when

$$R = r = \frac{\rho}{2\pi l}\ln\left(\frac{b}{a}\right)$$ **Ans. (D)**

Sol. 142 If circuit current is i, we use

$$i^2R = 0.5$$

and $$iR = 2.5$$

Terminal voltage of battery is given as

$$V = E - ir$$

$\Rightarrow$ $$2.5 = 3 - ir$$

$\Rightarrow$ $$ir = 0.5$$

$\Rightarrow$ $$\frac{R}{r} = 5$$

Power dissipated in r is given as

$$P_r = i^2r$$

Power dissipated in external R is given as

$$P_R = i^2R$$

$\Rightarrow$ $$\frac{P_r}{P_R} = \frac{r}{R} = \frac{1}{5}$$

$\Rightarrow$ $$P_r = \frac{1}{5}\times 0.5 = 0.1\text{ W}$$ **Ans. (D)**

Sol. 143 Power delivered by transmission line is

$$P = vi = 1000\text{ W}$$

$\Rightarrow$ $$i = \frac{1000}{220} = \frac{50}{11}\text{A}$$

Power loss in line is

$$P_L = i^2R = \left(\frac{50}{11}\right)^2 \times 2 = \frac{5000}{121}\text{ W}$$

line efficiency is given as

$$\eta = \frac{1000}{1000 + \frac{5000}{121}}\times 100$$

$\Rightarrow$ $$\eta = \frac{121}{126}\times 100 \simeq 96\%$$ **Ans. (B)**

17.4 Electrical Measuring Instruments

Sol. 144 If λ be the potential gradient then for balancing length $l = 100 - 49 = 51$ cm, we use

$$E_2 = \lambda l$$

$\Rightarrow$ $$1.05 = \lambda \times 51$$

$\Rightarrow$ $$\lambda = \frac{1.02}{51}$$

$\Rightarrow$ $$\lambda = 0.02\text{ V/cm}$$ **Ans. (A)**

Sol. 145 (A) Is incorrect as there is no connectivity between plates of capacitor so infinite resistance

(B) Is correct because *LED* works only in forward based mode

(C) For metal wire if considered as a perfect conductor, this is correct

(D) For a resistance both sides resistance value is equal so this is correct. **Ans. (A)**

Sol. 146 Effective circuit is shown below

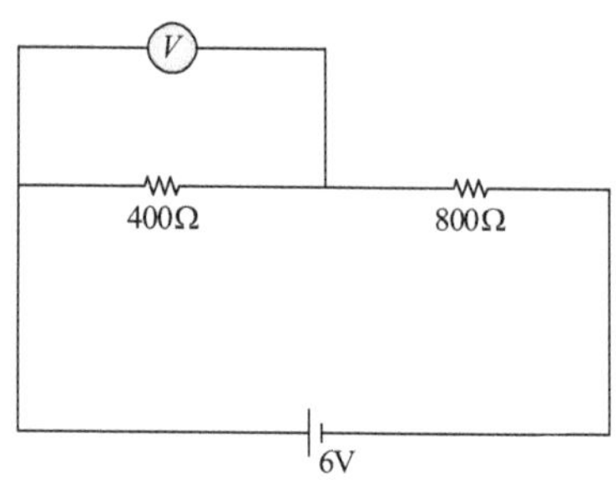

Equivalent resistance across battery is

$$R_{eq} = \frac{400\times 10000}{10400} + 800 = 1185\ \Omega$$

Potential difference measured is

$$V_{400\Omega} = \frac{6}{1185}\times 400 \cong 1.95\text{ V}$$ **Ans. (B)**

Sol. 147 In first case, we use

$$1 = I_g(G+R_1) \quad \ldots(1)$$

In second case, we use

$$2 = I_g(G+R_1+R_2) \quad \ldots(2)$$

from (1) and (2), we use

$$\frac{G+R_1+R_2}{G+R_1} = 1$$

$$\Rightarrow \quad 1+\frac{R_2}{G+R_1} = 2$$

$$\Rightarrow \quad R_2 = G+R_1$$ **Ans. (B)**

Sol. 148 Figure of Merit of galvanometer is given as

$$C = \frac{i}{\theta}$$

$$\Rightarrow \quad C = \frac{6\times 10^{-3}}{2} = 3\times 10^{-3}\text{ Am}^2$$ **Ans. (A)**

Sol. 149 For normal operation of circuit ammeter is to be connected in series because it has low resistance and voltmeter is to be connected in parallel as it has high resistance.

Ans. (A)

18.1 Magnetic Effects of Current

Sol. 150 In figure shown below, magnetic field due to coil at centre is given as

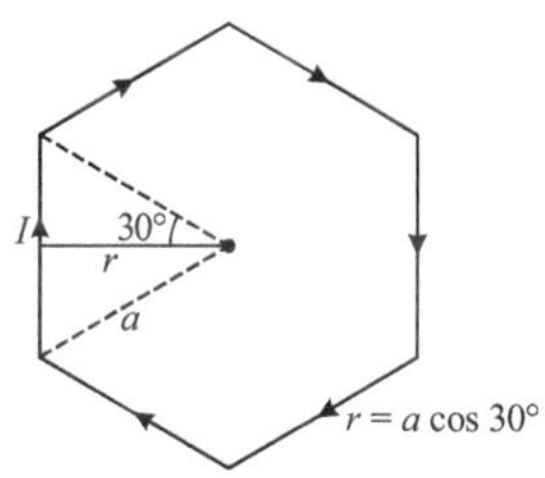

$$B = \frac{6\mu_0 I}{4\pi r}(2\sin 30^\circ)\times 50$$

$$\Rightarrow \quad B = \frac{6\mu_0 I}{4\pi a\cos 30^\circ}\times 2\sin 30^\circ \times 50$$

$$\Rightarrow \quad B = \frac{\mu_0 I}{\pi}\frac{150}{\sqrt{3}a} = \frac{50\sqrt{3}}{0.1}\frac{\mu_0 I}{\pi}$$

$$\Rightarrow \quad B = 500\sqrt{3}\frac{\mu_0 I}{\pi}$$ **Ans. (C)**

Sol. 151 In given figure magnetic fields due to wires A and B are

$$B_A = \frac{\mu_0 I\left(\frac{3\pi}{2}\right)}{4\pi R_A}$$

and $$B_B = \frac{\mu_0 I\left(\frac{5\pi}{3}\right)}{4\pi R_B}$$

$$\Rightarrow \quad \frac{B_A}{B_B} = \frac{I_A(3/2)}{R_A}\times\frac{R_B}{I_B(5/6)}$$

$$\Rightarrow \quad \frac{B_A}{B_B} = \frac{2(3/2)\times 4}{2\times 3\times(5/3)} = \frac{12}{10} = \frac{6}{5}$$ **Ans. (C)**

18.2 Magnetic Force on Moving Charges

Sol. 152 Pitch of helix in magnetic field is given as

$$p_H = \frac{2\pi m}{qB}v\cos\theta$$

$$\Rightarrow \quad p_H = \frac{2(3.14)(1.67\times 10^{-27})\times 4\times 10^5\times\cos 60^\circ}{(1.69\times 10^{-19})(0.3)}$$

$$\Rightarrow \quad p_H = 0.04\text{ m} = 4\text{ cm}$$ **Ans. (B)**

Sol. 153 Acceleration of particle is given as

$$\vec{a} = \frac{-qE}{m}\hat{j}$$

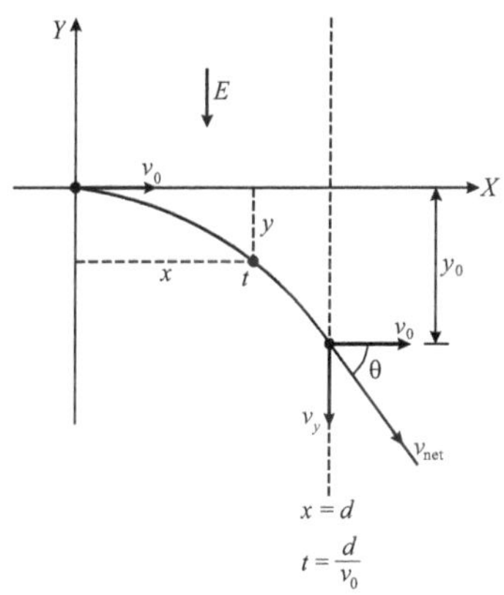

x & y coordinate of particle at time $t = t$ is

$$x = v_0 t \quad \ldots(1)$$

$$y = \frac{1}{2}\left(-\frac{qE}{m}\right)t^2 \quad \ldots(2)$$

From (1) and (2)

$$y = -\frac{qEx^2}{2mv_0^2}$$

At $x = d, t = \frac{d}{v}$ $\Rightarrow$ $v_x = v_0$ and $v_y = \frac{qE}{m}\left(\frac{d}{v_0}\right)$

$$\Rightarrow \tan\theta = \frac{v_y}{v_0} = \frac{qEd}{mv_0^2};\quad y_0 = \frac{-qEd^2}{2mv_0^2}$$

Equation of straight line motion of particle after $x = d$ is

$$y = -(x-d)\tan\theta - y_0$$

$$\Rightarrow \quad y = -x\left(\frac{qEd}{mv_0^2}\right) + \frac{qEd^2}{mv_0^2} - \frac{qEd^2}{2mv_0^2}$$

$$\Rightarrow \quad y = \frac{qEd}{mv_0^2}\left(\frac{d}{2} - x\right)$$ **Ans. (A)**

Sol. 154 For helical motion of proton, time period is given as

$$T = \frac{2\pi m}{qB}$$

Total time taken by proton is

$$t = 10\,T$$

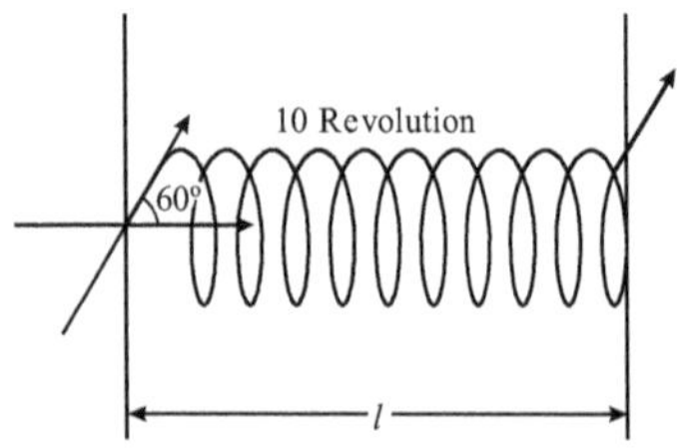

Velocity of proton along axis is

$$v_x = v\cos 60° = \frac{v}{2}$$

Thus distance travelled along axis is

$$l = \frac{v}{2} \times t = \frac{v}{2} \times 10 \times \frac{2\pi m}{qB}$$

$$\Rightarrow \quad l = 4 \times 10^5 \times 10 \times \frac{3.14 \times 1.67 \times 10^{-27}}{1.6 \times 10^{-19} \times 0.3}\,\text{m}$$

$$\Rightarrow \quad l = 0.439\,\text{m}$$ **Ans. (C)**

Sol. 155 Magnetic moment of loop *ABCDA* is

$$\vec{M}_1 = abI\,\hat{k}$$

Magnetic moment of loop *ADEFA* is

$$\vec{M}_2 = abI\,\hat{j}$$

Total magnetic moment of loop *ABCDEAA* is

$$\vec{M} = \vec{M}_1 + \vec{M}_2$$

$$\Rightarrow \quad \vec{M} = abI(\hat{k} + \hat{j})$$

$$\Rightarrow \quad \vec{M} = \sqrt{2}\,abI\left(\frac{\hat{k}}{\sqrt{2}} + \frac{\hat{j}}{\sqrt{2}}\right)$$ **Ans. (A)**

Sol. 156 Force on particle is given as

$$\vec{F} = q(\vec{v} \times \vec{B})$$

$$\Rightarrow \quad \vec{F} = 1(2\hat{i} + 3\hat{j} + 4\hat{k}) \times (5\hat{i} + 3\hat{j} - 6\hat{k}) \times 10^{-3} \times 10^{-6}\,\text{N}$$

$$\Rightarrow \quad \vec{F} = (6\hat{k} + 12\hat{j} - 15\hat{k} - 18\hat{i} + 20\hat{j} - 12\hat{i}) \times 10^{-9}\,\text{N}$$

$$\Rightarrow \quad \vec{F} = (-30\hat{i} + 32\hat{j} - 9\hat{k}) \times 10^{-9}\,\text{N}$$ **Ans. (C)**

Sol. 157 Magnetic torque on coil when it is parallel to magnetic field is

$$\tau = BINA = \tau_{ext}$$

$$\Rightarrow \quad B = \frac{\tau_{ext}}{INA} = \frac{1.5}{0.5 \times 500 \times 3 \times 10^{-4}}$$

$$\Rightarrow \quad B = \frac{100}{5} = 20\,\text{T}$$ **Ans. (20)**

Sol. 158 Work done is rotation of coil by 60° is given as

$$\omega = MB(\cos 30° - \cos 90°) = \frac{1}{2}I\omega^2$$

$$\Rightarrow \quad \omega = \sqrt{\frac{2MB\left(\frac{\sqrt{3}}{2}\right)}{I}} = \sqrt{\frac{20 \times 4 \times \sqrt{3}}{0.8}}$$

$$\Rightarrow \quad \omega = \sqrt{100\sqrt{3}} = 10(3)^{1/4}\,\text{rad/s}$$

$$\Rightarrow \quad \omega \simeq 13.16\,\text{rad/s}$$

Out of given options closest is option (A). **Ans. (A)**

Sol. 159 Torque on loop is given as

$$\tau = MB = 4a^2I \times \frac{\mu_0 I}{2\pi b}$$

$$\Rightarrow \quad \tau = \frac{2\mu_0 I^2 a^2}{\pi b}$$ **Ans. (A)**

Sol. 160 Particle will not hit the screen if its radius of circular motion is less then d, so we use

$$R < d$$

$$\Rightarrow \quad \frac{mv}{qB_0} < d$$

$$\Rightarrow \quad v = \frac{qB_0 d}{m}$$ **Ans. (B)**

Sol. 161 Force on electron due to electric field is given as

$$F_e = eE = 1.6 \times 10^{-19} \times 300 \times 100 \text{ (Along } -\text{Y-direction)}$$

For electron to keep moving along X-direction, net force on it must be zero thus magnetic force must balance electric force so act along +Y-direction so for electron moving in +X-direction, magnetic force is given as

$$\vec{F_m} = -e(\vec{v}\times\vec{B})$$

For $\vec{F_m}$ along +Y-direction $\vec{B}$ must be along +Z-direction such that

$$evB = eE$$

$$\Rightarrow \quad B = \frac{E}{v} = \frac{300\times 100}{6\times 10^6} = 5\times 10^{-3}\,\text{T}$$ **Ans. (A)**

Sol. 162 Velocity vector is given as

$$\vec{v} = v_x\hat{i} + v_y\hat{j} \quad \ldots(1)$$

Here force on particle is given as

$$\vec{F} = k(v_y\hat{i} + v_x\hat{j}) \quad \ldots(2)$$

From (1) and (2), if mass of particle is m, we use

$$m = \frac{dv_x}{dt} = kv_y$$

and $$m\frac{dv_y}{dt} = kv_x$$

$$\Rightarrow \quad \frac{dv_x}{dv_y} = \frac{v_y}{v_x}$$

$$\Rightarrow \quad v_x dv_x - v_y dv_y = 0$$

$$\Rightarrow \quad v_x^2 - v_y^2 = \text{constant}$$

In option (C) $\vec{v}\times\vec{a} = (v_x\hat{i} + v_y\hat{j})\times\frac{k}{m}(v_y\hat{i} + v_x\hat{j})$

$$= (v_x^2 - v_y^2)\times\frac{k}{m}\hat{k} = \text{Constant}$$ **Ans. (C)**

Sol. 163 Radius of circular motion of particle is given as

$$r = \frac{mv}{qB}$$

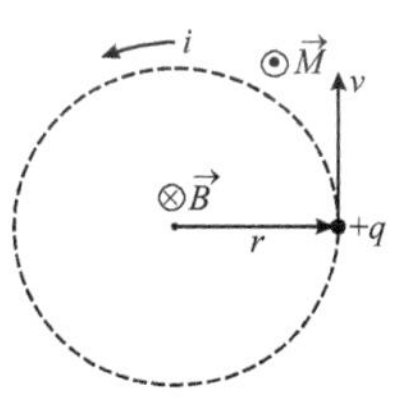

Angular speed of particle is given as

$$\omega = \frac{v}{r} = \frac{qB}{m}$$

Magnetic moment of revolving charge is given as

$$M = i\cdot\pi r^2$$

$$\Rightarrow \quad M = \frac{q\omega}{2\pi}\cdot\pi\left(\frac{mv}{qB}\right)^2 = \frac{1}{2}q\left(\frac{qB}{m}\right)\left(\frac{m^2v^2}{q^2B^2}\right)$$

$$\Rightarrow \quad M = \frac{mv^2}{2B}$$

From figure it is clear that direction of $\vec{M}$ is opposite to $\vec{B}$ so we write

$$\vec{M} = -\frac{mv^2}{2B^2}\vec{B}$$ **Ans. (D)**

Sol. 164 Figure shows the forces acting on two side wires of square loop due to the long wire placed at (0, b, 0) which is given as

$$F = \frac{\mu_0 I^2(2a)}{2\pi\sqrt{b^2+a^2}}$$

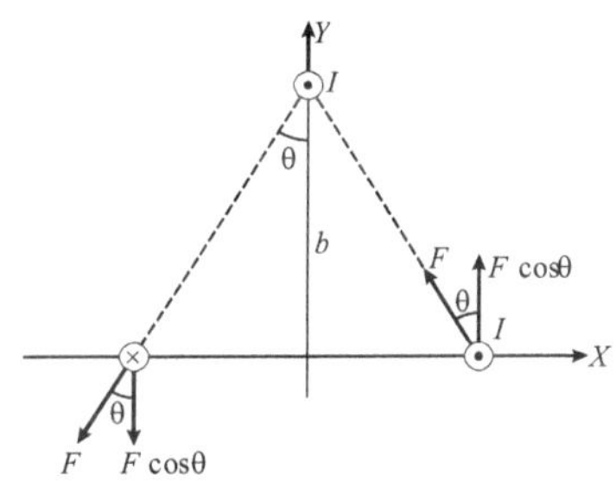

Torque on loop is given as

$$\tau = F\cos\theta\times 2a$$

$$\Rightarrow \quad \tau = \frac{\mu_0 I^2(2a)}{2\pi\sqrt{b^2+a^2}}\times\frac{b}{\sqrt{b^2+a^2}}\times 2a$$

$$\Rightarrow \quad \tau = \frac{2\mu_0 I^2 a^2 b}{\pi(b^2+a^2)} \simeq \frac{2\mu_0 I^2 a^2}{\pi b}\,(b >> a)$$

For all the option given (A) is most appropriate. **Ans. (A)**

19.1 Classical and Terrestrial Magnetism

Sol. 165 Stable equilibrium position is at $\theta = 0°$ and unstable equilibrium position is at $\theta = 180°$

Torque on magnet is given as

$$\tau = MB\sin\theta$$

With given values, we use

$$0.018 = M\times 0.06\times(10.5)$$

$$\Rightarrow \quad M = \frac{0.018}{0.06\times 0.5} = 0.6\,\text{A-m}^2$$

Work done in rotation of magnet from 0° to 180° is given as

$$W = MB(\cos 0° - \cos 180°)$$

$$\Rightarrow \quad W = 2MB = 2 \times 0.6 \times 0.06$$

$$\Rightarrow \quad W = 7.2 \times 10^{-2}\,\text{J}$$ **Ans. (D)**

Sol. 166 Magnetic moment of electromagnet is given as

$$M = \mu_r NiA$$

$$\Rightarrow \quad M = \mu_r (nl) iA$$

$$\Rightarrow \quad M = \mu_r niV$$

$$\Rightarrow \quad M = 1000 \times 10 \times 100 \times 0.5 \times 10^{-3}$$

$$\Rightarrow \quad M = 500\,\text{Am}^2$$ **Ans. (D)**

19.2 Magnetic Properties of Materials

Sol. 167 The perfect diamagnetic material does not allow external field to enter into it so material inside will not get magnetised. **Ans. (A)**

Sol. 168 By Curies law we use

$$\chi \propto \frac{1}{T}$$

$$\Rightarrow \quad \chi_1 T_1 = \chi_2 T_2$$

$$\Rightarrow \quad \frac{6}{0.4} \times 4 = \frac{I}{0.3} \times 24$$

$$\Rightarrow \quad I = \frac{0.3}{0.4} = 0.75\,\text{A/m}$$ **Ans. (B)**

20.1 Electromagnetic Induction

Sol. 169 *EMF* induced in a rotating coil in magnetic field is

$$e = BA\omega \sin\omega t$$

Maximum *EMF* is $e_{max} = BA\omega = BA\left(\frac{2\pi}{T}\right)$

$$\Rightarrow \quad e_{max} = 3 \times 10^{-5} \times 3.14 \times (0.1)^2 \times \frac{2 \times 3.14}{0.4}$$

$$\Rightarrow \quad e_{max} = \frac{6 \times 10}{4} \times 10^{-6}$$

$$\Rightarrow \quad e_{max} = 15\,\mu\text{V}$$ **Ans. (15)**

Sol. 170 *emf* induced in rotating loop is

$$e = NBA\omega \sin \omega t$$

Here $N = 1$

$$\Rightarrow \quad e = BA\omega \sin \omega t$$

Average power developed in loop is

$$P_{avg} = \left|\frac{e^2_{avg}}{R}\right| = \frac{B^2 A^2 \omega^2 (\sin^2 \omega t)_{avg}}{R}$$

$$\Rightarrow \quad P_{avg} = \frac{B^2 A^2 \omega^2}{2R} \qquad \left[\text{As } (\sin^2 \omega t)_{avg} = \frac{1}{2}\right]$$

$$\Rightarrow \quad P_{avg} = \frac{B^2 \omega^2}{2R} (\pi ab)^2 = \frac{\pi^2 a^2 b^2 B^2 \omega^2}{2R}$$ **Ans. (A)**

Sol. 171 Resistance of wire is

$$R = \frac{\rho l}{A} = \frac{1.23 \times 10^{-8} \times 30 \times 10^{-2}}{\pi (2 \times 10^{-3})^2}$$

EMF induced in loop is given as

$$e = \frac{AdB}{dt} = a^2 \frac{dB}{dt} = (7.5 \times 10^{-2})^2 \times 0.032$$

Induced current in loop is

$$i = \frac{e}{R} = \frac{(7.5 \times 10^{-2})^2 \times 0.032}{1.23 \times 10^{-8} \times 30 \times 10^{-2}} \times 3.14 \times (2 \times 10^{-3})^2$$

$$\Rightarrow \quad i \simeq 0.61\,\text{A}$$ **Ans. (A)**

Sol. 172 At the time of entering of magnet into coil and at the time of leaving the coil due to flux change in opposite direction emf is induced in opposite sense and when magnet is completely inside flux will not change so reading of galvanometer remain zero. Correct match is with option (C). **Ans. (C)**

Sol. 173 Magnetic field at centre of C_2 is

$$B = \frac{\mu_0 I}{2R_2} \cdot N_2$$

flux linked through coil C_1 due to magnetic field of C_2 is

$$\phi = B.\pi R_1^2 N_1 = \frac{\mu_0 \pi R_1^2 N_1 N_2}{2R_2} . (5t^2 - 2t + 3)$$

emf induced in C_1 is given as

$$e = \frac{d\phi}{dt} = \frac{\mu_0 \pi R_1^2 N_1 N_2}{2R_2} \times (10t - 2)$$

$$\Rightarrow \quad e = \frac{4\pi \times 10^{-7} \times (0.01)^2 \times \pi \times 200 \times 500}{2 \times 0.2} \times (10 - 2)$$

$$\Rightarrow \quad e = 8 \times 10^{-4}\,\text{V} = 0.8\,\text{mV}$$ **Ans. (5.00)**

Sol. 174 Emf induced in sliding conductor is given as

$$e = Bvl$$

$$\Rightarrow \quad e = \frac{\mu_0 I}{2\pi r} vl$$

Induced current in loop is calculated as

$$i = \frac{e}{R} = \frac{\mu_0 I v l}{2\pi r R}$$ **Ans. (B)**

20.2 Self and Mutual Induction

Sol. 175 For making of permanent magnets, materials are preferred which have large retentivity and large coercivity.
Ans. (D)

Sol. 176 Maximum energy stored in inductor is

$$U_{max} = \frac{1}{2}LI_{max}^2$$

In series L-R circuit growth of current as a function of time is given as

$$i = I_{max}(1 - e^{-Rt/L})$$

When circuit current becomes $\frac{I_{max}}{\sqrt{n}}$ then energy in inductor becomes $\left(\frac{1}{n}\right)$ times maximum value. So we use

$$\frac{I_{max}}{\sqrt{n}} = I_{max}(1 - e^{-Rt/L})$$

$$\Rightarrow \quad e^{-Rt/L} = 1 - \frac{1}{\sqrt{n}} = \frac{\sqrt{n}-1}{\sqrt{n}}$$

$$\Rightarrow \quad -\frac{Rt}{L} = \ln\left(\frac{\sqrt{n}-1}{\sqrt{n}}\right)$$

$$\Rightarrow \quad t = \frac{L}{R}\ln\left(\frac{\sqrt{n}}{\sqrt{n}-1}\right)$$

Ans. (C)

Sol. 177 In the given circuit writing equation of potential drop from P to Q, we get

$$V_P - L\frac{di}{dt} - 30 + iR = V_Q$$

$$\Rightarrow \quad V_P - V_Q = 50 \times 10^{-3} \times 10^2 + 30 - 1 \times 2 = 33\text{ V}$$

NOTE : As current is decreasing in circuit self induced emf in inductor will have high potential end at left side. **Ans. (33.00)**

21.1 Alternating Current

Sol. 178 As due to inductance voltage lead current by 45° that means resistance of coil is equal to that of inductive reactance as shown in phasor diagram below

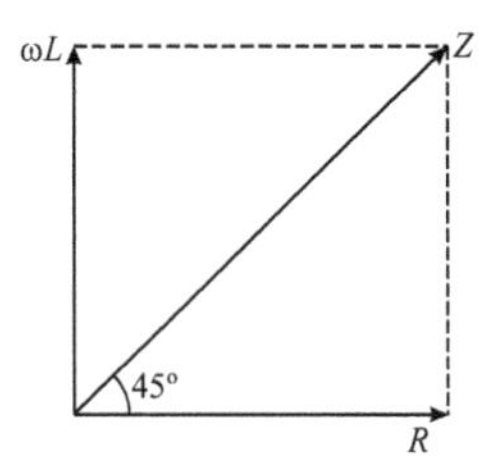

We use

$$R = X_L = \omega L \text{ and } Z = \sqrt{R^2 + \omega^2 L^2}$$

$$\Rightarrow \quad L = \frac{R}{\omega} = \frac{100}{2000\pi} = \frac{1}{20 \times 3.14}$$

$$\Rightarrow \quad L = 1.6 \times 10^{-2}\text{ H}$$

Most appropriate, among given answers is (A) **Ans. (A)**

Sol. 179 Given circuit impedance is

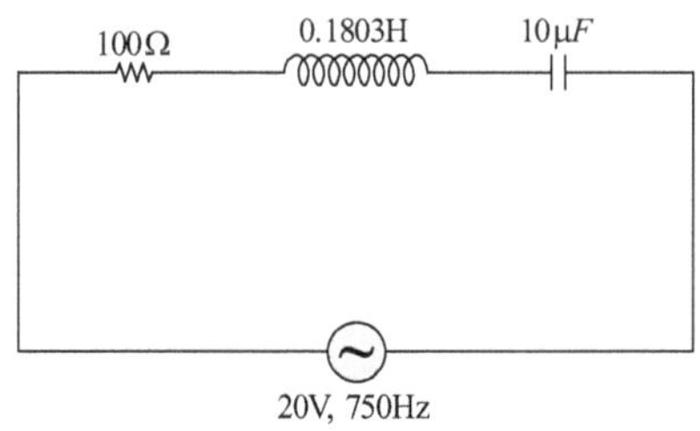

$$Z = \sqrt{R^2 + (X_L - X_C)^2}$$

$$\Rightarrow \quad Z = \sqrt{(100)^2 + \left(2\pi nL - \frac{1}{2\pi nL}\right)^2}$$

$$\Rightarrow \quad Z = \sqrt{(100)^2 + \left[2\times3.14\times750\times0.1803 - \frac{1}{2\times3.14\times750\times10\times10^{-6}}\right]^2}$$

$$\Rightarrow \quad Z \simeq 834\,\Omega$$

Current in circuit is given as

$$i = \frac{e}{Z} = \frac{20}{834}A$$

Thermal power produced in resistance is

$$P_{th} = i^2R = \left(\frac{20}{834}\right)^2 \times 100\text{W}$$

In time t if temperature of resistance is raised by 10°C, we use

$$\theta = ms\Delta\theta$$

$$\Rightarrow \quad P_{th}t = ms\Delta\theta$$

$$\Rightarrow \quad t = \frac{2\times10}{\left(\frac{20}{834}\right)^2\times100} = \frac{(834)^2}{20\times100}$$

$$\Rightarrow \quad t \cong 348\text{ s}$$

Ans. (C)

Sol. 180 Q-factor of a series RLC circuit is given by

$$Q = \frac{1}{R}\sqrt{\frac{L}{C}} = \frac{1}{100}\sqrt{\frac{80\times10^{-3}}{2\times10^{-6}}}$$

$$\Rightarrow \quad Q = \frac{1}{100}\sqrt{40\times10^3}$$

$$\Rightarrow \quad Q = \frac{200}{100} = 2$$ **Ans. (B)**

Sol. 181 During $t = 0$ to $t = 5\ \mu C$ charge on capacitor grows and becomes constant for $t = 5\ \mu C$ to $t = 10\ \mu C$ then again grows with time so most appropriate option is (A) **Ans. (A)**

Sol. 182 Power dissipated in circuit is given as

$$P = \left(\frac{e^2}{Z}\right)\cos\phi$$

$$\Rightarrow \quad 400 = \frac{(250)^2}{Z}\times 0.8$$

$$\Rightarrow \quad Z = 125\ \Omega$$

Capacitance required to change power factor to unity is such that

$$\frac{1}{\omega C} = \omega L = Z\sin\phi$$

$$\Rightarrow \quad C = \frac{1}{\omega Z\sin\phi}$$

$$\Rightarrow \quad C = \frac{1}{2\pi(50)\times125\times0.6}$$

$$\Rightarrow \quad C = \frac{\left(\frac{1}{2500}\right)}{3\pi}F = \frac{n}{3\pi}\mu F$$

$$\Rightarrow \quad n = \frac{10^6}{2500} = 400$$ **Ans. (400.00)**

22.1 Electromagnetic Wave

Sol. 183 Energy density of electromagnetic waves is given as

$$u_m = \frac{B_0^2}{2\mu_0}$$

$$\Rightarrow \quad B_0 = \sqrt{2\mu_0 u_m} = \sqrt{2\times4\pi\times10^{-7}\times1.02\times10^{-8}}$$

$$\cong \sqrt{256\times10^{-16}}$$

$$= 16\times10^{-8}\,\text{T} = 160\,\text{nT}$$ **Ans. (B)**

Sol. 184 Unit vectors along electric and magnetic fields are

$$\hat{E} = \hat{k}$$

$$\hat{B} = \frac{\vec{B}}{|B|} = \frac{2\hat{i}-2\hat{j}}{2\sqrt{2}}$$

$$\Rightarrow \quad \hat{B} = \frac{1}{\sqrt{2}}(\hat{i}-\hat{j})$$

Direction of wave propagation is given as

$$\hat{c} = \hat{E}\times\hat{B}$$

$$\Rightarrow \quad \hat{c} = \hat{k}\times\left[\frac{1}{\sqrt{2}}(\hat{i}-\hat{j})\right]$$

$$\Rightarrow \quad \hat{c} = \frac{1}{\sqrt{2}}(\hat{k}\times\hat{i} - \hat{k}\times\hat{j})$$

$$\Rightarrow \quad \hat{c} = \frac{1}{\sqrt{2}}(\hat{i}+\hat{j})$$ **Ans. (A)**

Sol. 185 Direction of speed in an electromagnetic wave is given as

$$\hat{c} = \hat{E}\times\hat{B}$$

$\hat{c}$: $-\hat{k}$ (analysed); $\hat{E}$: (given) $-\hat{j}$; $\hat{B}$: $\hat{i}$ (given)

Amplitudes of electric field and magnetic fields are related as

$$E_0 = cB_0 = 3\times10^8\times3\times10^{-8} = 9\ \text{V/m}$$

As wave is propagating in $-y$-direction direction of $\vec{E}$ must be along $-\hat{k}$ so we have

$$\vec{E} = -9\sin(200\pi(y+ct))\hat{k}$$ **Ans. (B)**

Sol. 186 At any instance magnetic field magnitude in electromagnetic wave is given as

$$B_0 = \frac{E_0}{c} = E_0\sqrt{\mu_0\in_0}$$

Direction of wave propagation is given by Poynting vector

$$\vec{S} = \frac{1}{\mu_0}(\vec{E}\times\vec{B})$$

$$\hat{S} = \hat{E}\times\hat{B}$$

$\hat{i}$ (given) $\quad \hat{j}$ (given) $\quad \Rightarrow \hat{k}$ (analysed)

Thus magnetic field is given as

$$B = B_0\cos(\omega t - kx)\hat{k}$$

at $t = 0$ $\quad B = E_0\sqrt{\mu_0\in_0}\cos kx\hat{k}$ **Ans. (C)**

Sol. 187 Electromagnetic spectrum is given as

$\lambda_{\text{increasing}}$ ↑			
	Radio waves		few cm to few m
	Micro waves		0.01 mm to few cm
	IR	~	8000 Å – 10^5 Å
	Visible	~	4000 Å – 8000 Å
	UV	~	100 Å – 4000 Å
	X-Rays	~	0.1 Å – 100 Å

Ans. (C)

Sol. 188 From wave equation, we can see the direction of wave propagation is $+\hat{k}$ so we use

$$\hat{k} = \hat{E}\times\hat{B}$$

Unit vector along $\vec{E}$ is given as

$$\hat{E} = \frac{1}{\sqrt{2}}(\hat{i}+\hat{j})$$

$$\Rightarrow \quad \hat{k} = \frac{1}{2}(\hat{i}+\hat{j})\times\hat{B}$$

$$\Rightarrow \quad \hat{B} = \frac{1}{\sqrt{2}}(-\hat{i}+\hat{j})$$

Equation of magnetic field wave is given as

$$\vec{B} = \frac{E_0}{C}(-\hat{x}+\hat{y})\ \sin(kz-\omega t)$$ **Ans. (B)**

Sol. 189 Magnetic field amplitude in *EM* wave is given as

$$B_0 = \frac{E_0}{C} = \frac{30}{1.5\times10^7}\times 5\times10^{-2} = 10^{-7}\,\text{T}$$

Maximum force on electron is given as

$$F_{max} = evB_0 = 1.6\times10^{-19}\times0.1\times3\times10^8\times10^{-7}$$

$$\Rightarrow \quad F_{max} = 4.8\times10^{-19}\,\text{N}$$ **Ans. (B)**

Sol. 190 *EM* radiation spectrum is given below in decreasing order of wavelength

$\lambda_{decreasing}$ ↓

Radio waves	–	few cm to several meters
Micro waves	–	$10^{-2} - 10^{-5}$ m
IR	–	8000 Å $- 10^{-5}$ m
Visible	–	4000 – 8000 A
UV	–	100 – 4000 Å
X-Rays	–	0.01 – 100 Å
Gama rays	–	10^{-5} to 10^{-3} Å

Ans. (D)

Sol. 191 Intensity of EM wave is given as

$$I = \epsilon_0 E^2 c$$

$$\Rightarrow \quad E^2 = \frac{I}{\epsilon_0\, c}$$

$$\Rightarrow \quad E^2 = \frac{315}{\pi\,\epsilon_0}\times\frac{1}{3\times10^8}$$

$$\Rightarrow \quad E^2 = \frac{315\times4\times9\times10^9}{3\times10^8} = 315\times120$$

$$\Rightarrow \quad E = \sqrt{315\times120} \simeq 194.4$$ **Ans. (194.00)**

Sol. 192 For given electromagnetic wave electric field amplitude is given as

$$E_0 = cB_0 = 3\times10^8\times1.2\times10^{-7} = 36\ \text{V/m}$$

As wave is travelling in X-direction, we use $\hat{c} = -\hat{i}$

From Poynting vector direction, we use

$$\hat{c} = \hat{E}\times\vec{B}$$

$$-\hat{i} = \hat{E}\times\hat{k}$$

Thus $\hat{E} = -\hat{j}$ so instantaneous electric field is given as

$$\vec{E} = -36\sin(0.5\times10^3x + 1.5\times10^{11}t)\,\hat{j}\ \text{V/m}$$

Ans. (B)

23.1 Geometrical Optics I - Reflection of Light

Sol. 193 Using mirror formula, we have

$$f = -4\ \text{cm}$$

$$u = -10\ \text{cm}$$

Substituting values in formula

$$\frac{1}{v}+\frac{1}{u} = \frac{1}{f}$$

$$\Rightarrow \quad \frac{1}{v}+\left(\frac{1}{-10}\right) = \frac{1}{-4}$$

$$\Rightarrow \quad \frac{1}{v} = \frac{1}{10}-\frac{1}{4}$$

$$\Rightarrow \quad \frac{1}{v} = \frac{4-10}{40}$$

$$\Rightarrow \quad v = \frac{40}{-6}\ \text{cm}$$

$$\Rightarrow \quad v = \frac{-20}{3}\ \text{cm}$$

$$m = \frac{-v}{u} = \frac{-\left(-\frac{20}{3}\right)}{-10} = \frac{-2}{3}$$

$\Rightarrow$ Image will be real, inverted and diminished. **Ans. (D)**

Sol. 194 For the given situation, object and image distances are given as

$$u = 30\ \text{cm}$$

$$v = 10\ \text{cm}$$

Object speed = 9 cm/s

Considering object motion along axis, we use

$$\text{Image speed} = \left|\frac{v}{u}\right|^2\times\text{Object speed}$$

$$v_{image} = \left(\frac{1}{3}\right)^2\times 9 = 1\ \text{cm/s}$$ **Ans. (1)**

23.2 Geometrical Optics II - Refraction of Light

Sol. 195

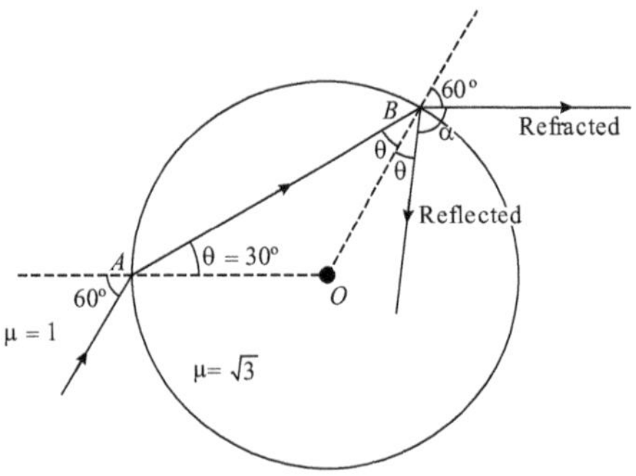

Using Snell's law at point A, we have

$$1 \times \sin 60^\circ = \sqrt{3} \times \sin\theta$$

$$\Rightarrow \quad \frac{\sqrt{3}}{2} = \sqrt{3} \times \sin\theta$$

$$\Rightarrow \quad \sin\theta = \frac{1}{2}$$

$$\Rightarrow \quad \theta = 30^\circ$$

At point B, we have

$$\theta + 60^\circ + \alpha = 180^\circ$$

$$\Rightarrow \quad 30^\circ + 60^\circ + \alpha = 180^\circ$$

$$\Rightarrow \quad \alpha = 90^\circ$$

Ans. (90.00)

Sol. 196 From figure, we have

$$\tan r = \frac{15}{30} = \frac{1}{2}$$

$$\Rightarrow \quad \sin r = \frac{1}{\sqrt{5}}$$

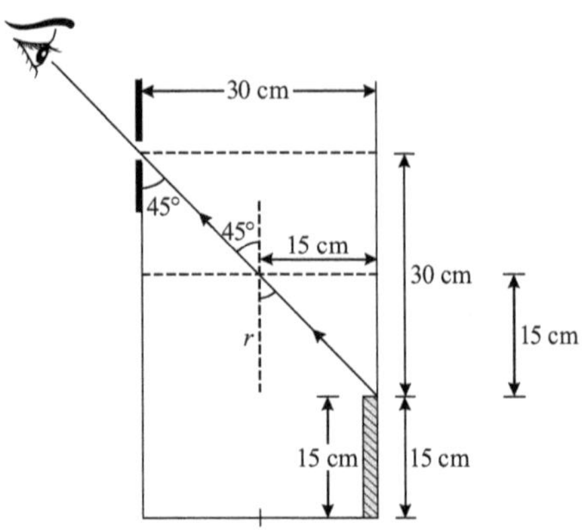

By smell's law, we have

$$\mu \sin r = \sin 45^\circ$$

$$\Rightarrow \quad \mu = \frac{\frac{1}{\sqrt{2}}}{\frac{1}{\sqrt{5}}} = \sqrt{\frac{5}{2}} = 1.581$$

Given value of

$$\mu = \frac{N}{100} = 1.581$$

$$\Rightarrow \quad N = 158.1$$

Ans. (158)

Sol. 197 Angle of minimum deviation for small angled prism is given as

$$\delta_{min} = (\mu - 1)A$$

$$= (1.5 - 1)1$$

$$= 0.5^\circ$$

$$\delta_{min} = \frac{N}{10} = 0.5^\circ$$

$$\Rightarrow \quad N = 5$$

Ans. (5.00)

23.3 Geometrical Optics III - Thin Lenses

Sol. 198 Using displacement method on optical bench we have

$$f = \frac{D^2 - d^2}{4D}$$

$$\Rightarrow \quad f = \frac{100^2 - 40^2}{4(100)} = 21 \text{ cm}$$

Power of lens is given as

$$P = \frac{1}{f} = \frac{100}{21} D = \left(\frac{N}{100}\right)_D$$

$$\Rightarrow \quad N = \frac{10000}{21} \simeq 476$$

Ans. (476)

Sol. 199 By lens formula, we have

$$\frac{1}{v} - \frac{1}{u} = \frac{1}{f}$$

For $\quad v = u$ (both positive)

This happens only when object is at pole

For + *ve* u, *ve* is given as

$$\frac{1}{v} = \frac{1}{f} + \frac{1}{u}$$

Thus $|v| < |u|$ so the $u - v$ curve must lie below $u = v$ reference line and at $u \to \infty$, $v = f$ thus option (D) is correct. **Ans. (D)**

Sol. 200 For lens refraction, we use its image produced is I_1 at position v from lens, given as

$$u = -1 \text{ m}$$

$$f = +0.5 \text{ m}$$

$$\Rightarrow \quad v = \frac{uf}{u+f} = \frac{-1 \times 0.5}{-0.5} = 1 \text{ m}$$

Image I_1 produced by convex lens above will act like object for plane mirror so in mirror its image I_2 is produced at a distance 1 m behind the mirror. (Refer ray diagram given below)

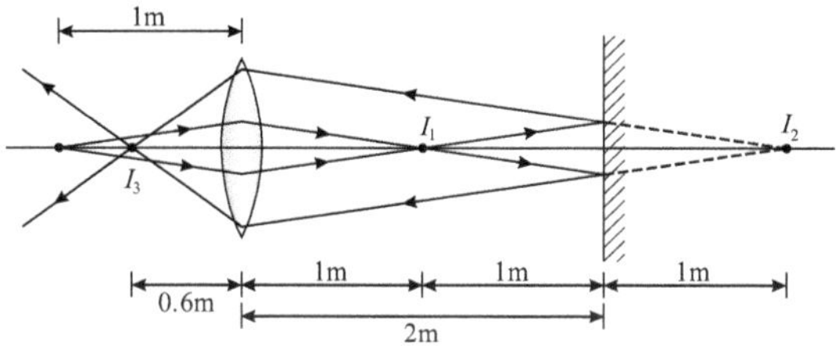

For lens this image I_2 will act as object so again for lens refraction we use

$$u = +3\,\text{m}$$

$$f = -0.5\,\text{m}$$

$$\Rightarrow \quad v = \frac{uf}{u+f} = \frac{3\times -0.5}{2.5} = -\frac{3}{5} = -0.6\,\text{m}$$

This final image is produced to the left of lens at 0.6 m and real. So correct answer is option-(C) **Ans. (C)**

Sol. 201 By lens makers formula, we use

$$P = (\mu - 1)\left[\frac{2}{R}\right] \quad \ldots(1)$$

For another plano convex lens with power $1.5P$, we use

$$1.5P = (\mu - 1)\left[\frac{1}{R'}\right] \quad \ldots(2)$$

From (1) and (2), we have

$$1.5 = \frac{1/R'}{2/R}$$

$$\Rightarrow \quad R' = \frac{R}{2\times 1.5} = \frac{R}{3}$$ **Ans. (D)**

23.5 Optical Instruments

Sol. 202 If object is placed at a distance x from objective, its image is produced at for objective

$$\frac{1}{v} + \frac{1}{x} = 1$$

$$\Rightarrow \quad v = \frac{x}{x-1}$$

and $$m_0 = \left|\frac{v}{u}\right| = \frac{1}{x-1}$$

For eye-piece

$$\frac{1}{-25} - \frac{1}{-\left(20 - \frac{x}{x-1}\right)} = \frac{1}{f_e}$$

and $$m_e = \frac{25}{20 - \frac{x}{x-1}}$$

Given that

$$\left(\frac{1}{x-1}\right)\times\left(\frac{25}{20 - \frac{x}{x-1}}\right) = 100$$

Solving we get $$x = \frac{81}{76}\,\text{cm}$$

For eye piece lens formula, we use

$$-\frac{1}{25} - \frac{1}{20 - \frac{81/76}{81/76 - 1}} = \frac{1}{f_e}$$

Solving we get

$$f_e \cong 4.5\,\text{cm}$$

Answer can be 4 or 5. **Ans. (5)**

Sol. 203 Given focal lengths of objective and eyepiece are

$$f_0 = 1\,\text{cm}$$

$$f_e = 5\,\text{cm}$$

$$l = 10\,\text{cm (Tube length)}$$

For final image at infinity (for least strain on eye), image produced by objective must be at focal point of eyepiece i.e. 5 cm from it. For objective we use

$$f_0 = +1\,\text{cm}$$

$$V_0 = +5\,\text{cm}$$

$$\Rightarrow \quad u = \frac{V_0 f_0}{V_0 - f_0} = \frac{5\times 1}{4} = 1.25\,\text{cm} = \frac{N}{40}$$

$$\Rightarrow \quad N = 50$$ **Ans. (50.00)**

24.1 Interference of Light

Sol. 204 Distance of nth bright fringe from centre is

$$Y = \frac{nD\lambda}{d}$$

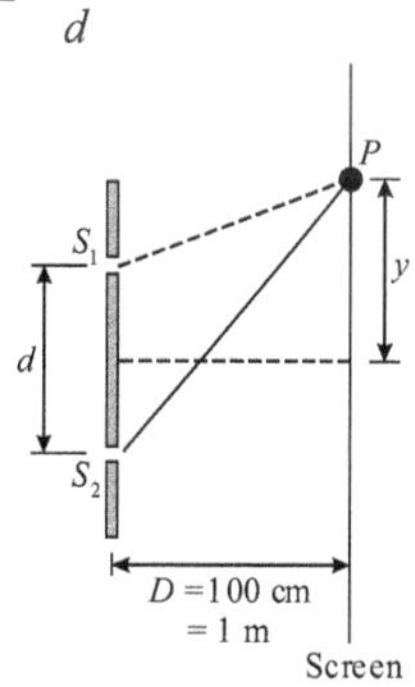

$$\Rightarrow \quad n = \frac{Yd}{D\lambda} = \frac{1.27\times 10^{-3}\times 10^{-3}}{1\times 632.8\times 10^{-9}} = 2$$

For bright fringe path difference is given as

$$\Delta x = n\lambda$$

$$\Rightarrow \quad \Delta x = 2 \times 632.8\,\text{nm}$$

$$\Rightarrow \quad \Delta x = 1265.6\,\text{nm}$$

$$\Rightarrow \quad \Delta x = 1.27\,\mu\text{m}$$

Ans. (A)

Sol. 205 If N fringes are there in screen segment of length l and fringe width is β, we use

$$N\beta = l$$

$$\Rightarrow \quad N\left(\frac{\lambda D}{d}\right) = l$$

For first wavelength $\lambda_1 = 700$ nm, we use

$$\frac{N_1\lambda_1 D}{d} = l$$

For second wavelength $\lambda_2 = 400$ nm, we use

$$\frac{N_2\lambda_2 D}{d} = l$$

$$\Rightarrow \quad N_1\lambda_1 = N_2\lambda_2$$

$$\Rightarrow \quad 16 \times 700 = N_2 \times 400$$

$$\Rightarrow \quad N_2 = 28$$

Ans. (A)

Sol. 206 In YDSE setup given data is

$$d = 0.05 \times 10^{-3}\,\text{m}$$

and $\quad \lambda = 5 \times 10^{-7}\,\text{m}$

Angular fringe width is given as

$$\theta = \frac{\beta}{D} = \frac{\lambda D/d}{D} = \frac{\lambda}{d}\ \text{rad}$$

In degrees it is given as

$$\theta = \frac{\lambda}{d} \times \frac{180}{\pi}\ \text{degree}$$

$$\Rightarrow \quad \theta = \frac{5\times10^{-7}\times180}{5\times10^{-5}\times3.14} \simeq 0.57^\circ$$

Ans. (D)

Sol. 207 Optical path difference between two waves is given as

$$\Delta x = n_1L_1 - n_2L_2$$

Phase difference is given as

$$\Delta\phi = \frac{2\pi}{\lambda}\Delta k = \frac{2\pi}{\lambda}(n_1L_1 - n_2L_2)$$

Ans. (A)

Sol. 208 At initial position

$$S_1L = \sqrt{2^2 + \left(\frac{3}{2}\right)^2} = \frac{5}{2}\,\text{m}$$

and $\quad S_2L = 2\,\text{m}$

Path difference at initial position is given as

$$\Delta = S_1L - S_2L = 2.5 - 2 = 0.5\,\text{m}$$

Here $\quad \Delta = \dfrac{\lambda}{2}$

Thus intensity is minimum. When L moves away to new position L', next maximum is obtained when

$$\Delta = \lambda = 1\,\text{m}$$

$$\Rightarrow \quad d = S_1L' = S_2L' + 1 = 3\,\text{m}$$

Ans. (B)

Sol. 209 As P is 90° ahead in phase so path lead of P w.r. to Q is

$$\Delta = \frac{\lambda}{2\pi} \times \frac{\pi}{2} = \frac{\lambda}{4} = \frac{20}{4} = 5\,\text{m}$$

For point A, path difference in waves from P and Q is

$$\Delta_A = 5 - 5 = 0$$

Thus at A intensity is maximum I_0

For point B, path difference in waves from P and Q is

$$\Delta = 5\,\text{m}$$

$$\Rightarrow \quad \text{Phase difference } \phi = \frac{2\pi}{\lambda}\Delta = \frac{2\pi}{20} \times 5 = \frac{\pi}{2}$$

Thus intensity at B is

$$I_B = I_0\cos^2\frac{\phi}{2} = I_0\cos^2\frac{\phi}{4} = \frac{I_0}{2}$$

For point C, path difference in waves from P and Q is

$$\Delta = 5 + 5 = 10\,\text{m}$$

As $\Delta = \dfrac{\lambda}{2}$ so intensity will be minimum at C due to distractive interference, so we have

$$\Rightarrow \quad I_A : I_B : I_C = I_0 : \frac{I_0}{2} : 0 = 2 : 1 : 0$$

Ans. (D)

Sol. 210 The point on screen where path difference is λ is the point of bright fringe so at a general point resulting light intensity is given as

$$I_R = K\cdot\cos^2\frac{\phi}{2}$$

At the point where path difference is $\dfrac{\lambda}{6}$, phase difference is given as

$$\phi = \frac{2\pi}{\lambda} \times \frac{\lambda}{6} = \frac{\pi}{3}$$

Thus intensity at this point is given as

$$\Rightarrow \quad I = \frac{3K}{4} = \frac{nK}{12}$$

$$\Rightarrow \quad n = 9$$

Ans. (9.00)

24.2 Diffraction of Light

Sol. 211 For n^{th} diffraction minima, we use

$$b \sin \theta = n\lambda$$

$$\Rightarrow \quad \sin \theta = \frac{n\lambda}{b} \leq 1 \quad \text{(for maximum limit)}$$

$$\Rightarrow \quad n \leq \frac{b}{\lambda} = \frac{6\times10^{-5}}{6\times10^{-7}} = 100$$

Thus maximum number of minima on either side are 99 so total minima on screen are

$$99 \times 2 = 198$$ **Ans. (198)**

Sol. 212 First intensity maximum occurs at an angle θ, given as

$$2d \sin \theta = \lambda = \frac{h}{\sqrt{2mE}}$$

$$\Rightarrow \quad E = \frac{h^2}{8md^2 \sin^2 \theta} \times \frac{1}{e} eV$$

$$\Rightarrow \quad E = \frac{(6.64\times10^{-34})^2}{8\times9.1\times10^{-31}\times(10^{-10})^2\times\frac{3}{4}\times1.6\times10^{-19}}$$

$$\Rightarrow \quad E \cong 50.47$$ **Ans. (50.00)**

24.3 Polarization of Light

Sol. 213 Total power of light incident on polarizer is

$$P = IA = 3.3 \times 3 \times 10^{-4} = 9.9 \times 10^{-4} \text{W}$$

Average light intensity passes through polarizer is $I/2$ so we use amount of light energy passing through it in one revolution is

$$E = \frac{I}{2} \times A \times t = \frac{1}{2} \times 9.9 \times 10^{-4} \times \frac{2\pi}{31.4}$$

$$\Rightarrow \quad E = 4.95 \times 10^{-4} \times 0.2$$

$$\Rightarrow \quad E = 0.99 \times 10^{-4}$$

$$\Rightarrow \quad E \simeq 1 \times 10^{-4} \text{J}$$ **Ans. (C)**

25.1 Photoelectric Effect

Sol. 214 We use for given data

$$\frac{hc}{\lambda} = \frac{hc}{\lambda_{th}} + eV \quad \ldots(1)$$

and
$$\frac{hc}{3\lambda} = \frac{hc}{\lambda_{th}} + \frac{eV}{4} \quad \ldots(2)$$

$4 \times (2) - (1)$ gives

$$\frac{1}{3}\frac{hc}{\lambda} = \frac{3hc}{\lambda_{th}}$$

$$\Rightarrow \quad \lambda_{th} = 9\lambda$$

$$\Rightarrow \quad n = 9$$ **Ans. (9)**

Sol. 215 By photoelectric effect equation, we use

$$\frac{hc}{\lambda} = \phi + k_m$$

We use
$$\frac{hc}{500 \text{ nm}} = \phi + k_m \quad \ldots(1)$$

and
$$\frac{hc}{200 \text{ nm}} = \phi + 3k_m \quad \ldots(2)$$

from (1) and (2) we get

$$3 = \frac{\frac{hc}{200 \text{ nm}} - \phi}{\frac{hc}{500 \text{ nm}} - \phi}$$

We use $\quad hc = 1243.1 \text{ eV} - \text{nm}$

$$\Rightarrow \quad 3\times\frac{1243.1}{500} - 3\phi = \frac{1243.1}{200} - \phi$$

$$\Rightarrow \quad 2\phi = 1243.1\left[\frac{3}{500} - \frac{1}{200}\right]$$

$$\Rightarrow \quad \phi = \frac{1243.1}{1000}[6-5]$$

$$\Rightarrow \quad \phi = \frac{1.243}{2} = 0.612 \text{ eV}$$ **Ans. (A)**

Sol. 216 From the given graph, we use

$$h\nu = \phi + eV_0$$

At point B, $\quad V_0 = 0,$

$$\nu = 5.5 \times 10^{14} \text{ Hz}$$

$$\Rightarrow \quad h(5.5 \times 10^{14}) = \phi$$

$$\Rightarrow \quad \phi = \frac{6.62\times10^{-34}\times5.5\times10^{14}}{1.6\times10^{-19}}$$

$$\Rightarrow \quad \phi = 2.27 \text{ eV}$$ **Ans. (A)**

Sol. 217 With variation in intensity we consider change in photon flux at constant frequency so the graph does not change. **Ans. (C)**

Sol. 218 For emitted photoelectrons, we use

$$E_1 = \phi + k_1$$

and
$$E_2 = \phi + k_2$$

As
$$\frac{\nu_1}{\nu_2} = 2$$

$$\Rightarrow \quad \frac{k_1}{k_2} = \frac{v_1^2}{v_2^2} = 4$$

Given that $\frac{k_1}{k_2} = 4 = \frac{4-\phi}{2.5-\phi}$

$$\Rightarrow \quad 10 - 4\phi = 4 - \phi$$

$$\Rightarrow \quad \phi = 2\text{ eV}$$

Ans. (2.00)

25.2 Matter Waves and X-Ray

Sol. 219 deBroglie wavelength of particle is given as

$$\lambda_d = \frac{h}{p} = \frac{h}{mv}$$

$$\Rightarrow \quad \lambda_{dp} = \frac{h}{m(5v)} \quad \ldots(1)$$

$$\Rightarrow \quad \lambda_{de} = \frac{h}{m_e v} \quad \ldots(2)$$

Given that

$$\frac{\lambda_{dp}}{\lambda_{de}} = 1.878 \times 10^{-4}$$

$$\Rightarrow \quad m = \frac{m_e}{5 \times 1.878 \times 10^{-4}}$$

$$\Rightarrow \quad m = \frac{9.1 \times 10^{-31}}{5 \times 1.878 \times 10^{-4}}$$

$$\Rightarrow \quad m = 9.7 \times 10^{-28}\text{ kg}$$

Ans. (D)

Sol. 220 Number of photons emitted per second are given as

$$N_1 = \frac{P\lambda_1}{hc}$$

and

$$N_2 = \frac{P\lambda_2}{hc}$$

$$\Rightarrow \quad \frac{N_1}{N_2} = \frac{\lambda_1}{\lambda_2} = \frac{1}{500}$$

Ans. (A)

Sol. 221 Before Collision

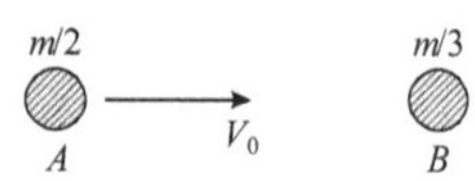

After Collision

By momentum conservation, we use

$$\frac{m}{2}V_0 = \frac{m}{2}V_A + \frac{m}{3}V_B$$

$$\Rightarrow \quad V_0 = V_A + \frac{2V_B}{3} \quad \ldots(1)$$

As collision is elastic

$$V_0 = V_B - V_A \quad \ldots(2)$$

Solving (1) & (2), we get

$$V_A = \frac{V_0}{5} \text{ and } V_B = \frac{6V_0}{5}$$

De-Broglie wavelength of A before collision

$$\lambda_{DA} = \frac{2h}{mV_0}$$

De-Broglie wavelength A after collision

$$\lambda'_{DA} = \frac{h}{mV_A} = \frac{10h}{mV_0}$$

$$\Rightarrow \quad \Delta\lambda = \lambda'_{DA} - \lambda_{DA} = \frac{8h}{mv_0} = 4\lambda_{DA} = 4\lambda_0$$

Ans. (A)

Sol. 222 For a particle of kinetic energy E, de-Broglie wavelength is given as

$$\lambda = \frac{h}{p} = \frac{h}{\sqrt{2mE}}$$

$$\Rightarrow \quad \lambda \propto \frac{1}{\sqrt{m}}$$

As $m_{He^{++}} > m_p > m_e$

$$\Rightarrow \quad \lambda_{He^{++}} < \lambda_p < \lambda_e$$

Ans. (D)

Sol. 223 RMS speed of molecules is given as

$$v_{rms} = \sqrt{\frac{3RT}{M}} = \sqrt{\frac{3KT}{m}}$$

Here $M = m\,N_{AV}$ where m is mass of one molecule and $R = k \cdot N_{AV}$

de-Broglie wavelength of molecules is given as

$$\lambda = \frac{h}{mv}$$

$$\Rightarrow \quad \lambda = \frac{h}{m\sqrt{\frac{3KT}{m}}}$$

$$\Rightarrow \quad \lambda = \frac{h}{\sqrt{3KTm}}$$

$$\Rightarrow \quad \lambda = \frac{6.63 \times 10^{-34}}{\sqrt{3 \times 1.38 \times 10^{-23} \times 400 \times \left(\frac{28 \times 10^{-3}}{6.023 \times 10^{-23}}\right)}}$$

$$\Rightarrow \quad \lambda = \frac{6.63 \times 10^{-11}}{2.77} = 2.39 \times 10^{-11}\text{ m}$$

$$\Rightarrow \quad \lambda \simeq 0.24\text{ Å}$$

Ans. (B)

26.1 Atomic Structure

Sol. 224 For very high value of n, transition between adjoining orbits causes emission of radiation of frequency equal to revolution frequency of electron in n^{th} orbit which is given as

$$f_n = \frac{v_n}{2\pi r_n} = \frac{\dfrac{2\pi kZe^2}{nh}}{\dfrac{n^2h^2}{2\pi kZe^2 m}}$$

$$\Rightarrow \quad f_n \propto \frac{Z^2}{n^3}$$

$$\Rightarrow \quad f_n \propto \frac{1}{n^3}$$

Ans. (B)

Sol. 225 For lyman series first & last time wavelengths are given as

$$\frac{1}{\lambda_1} = R\left[\frac{1}{1^2}-\frac{1}{2^2}\right] = \frac{3R}{4}$$

$$\frac{1}{\lambda_L} = R\left[\frac{1}{1^2}-\frac{1}{\infty^2}\right] = R$$

$$\Delta\lambda_L = \lambda_1 - \lambda_L = \frac{4}{3R}-\frac{1}{R} = \frac{1}{3R} = 304\ \text{Å}$$

For paschen series we use

$$\frac{1}{\lambda_1} = R\left[\frac{1}{3^2}-\frac{1}{4^2}\right] = \frac{7R}{144}$$

$$\Rightarrow \quad \frac{1}{\lambda_L} = R\left[\frac{1}{3^2}-\frac{1}{\infty^2}\right] = \frac{R}{9}$$

$$\Rightarrow \quad \Delta\lambda_p = \lambda_1 - \lambda_L = \frac{144}{7R}-\frac{9}{R} = \frac{81}{7R}$$

$$\Rightarrow \quad \Delta\lambda_p = \frac{81}{7}\times 304\times 3 = 10553.14\ \text{Å}$$

Ans. (10553)

Sol. 226 If initial KE of particle is E_1 and KE of H-atom after collision is E_2 we use

$$E_1 - 10.2\ \text{eV} = E_2$$

By conservation of momentum we use

$$\sqrt{2m_pE_1} = \sqrt{2m_HE_2}$$

$$m_pE_1 = m_H(E_1 - 10.2)$$

$$\Rightarrow \quad E_1 = \frac{10.2\times m_H}{m_H - m_P}eV$$

$$\Rightarrow \quad E_1 = \frac{10.2\times 10^9}{10^9 - 2\times 10^8}eV$$

$$\Rightarrow \quad E_1 = \frac{10.2}{0.8} = 12.75\ eV = \frac{N}{4}$$

$$\Rightarrow \quad N = 51$$

Ans. (51.00)

27.1 Radioactivity

Sol. 227 By radiative decay equation, we use

$$N = N_0e^{-\lambda t}$$

Given that $\dfrac{N}{N_0} = \dfrac{9}{16} = e^{-\lambda t}$

After time $t/2$, fraction remaining is

$$\left(\frac{N}{N_0}\right)_{t/2} = e^{-\lambda t/2} = \sqrt{e^{-\lambda t}} = \sqrt{\frac{9}{16}} = \frac{3}{4}$$

Ans. (A)

Sol. 228 By radioactive decay equation, we have

$$R = R_0\,e^{-\lambda t}$$

$$\Rightarrow \quad \ln R = \ln R_0 - \lambda t$$

For element A we use

$$\lambda_A = \frac{6}{10}$$

$$\Rightarrow \quad T_A = \frac{10}{6}\ln 2$$

For element B we use

$$\lambda_B = \frac{6}{5}$$

$$\Rightarrow \quad T_B = \frac{5\ln 2}{6}$$

For element C we use

$$\lambda_C = \frac{2}{5}$$

$$\Rightarrow \quad T_C = \frac{5\ln 2}{2}$$

$$\Rightarrow \quad T_A : T_B : T_C = \frac{10}{6}:\frac{5}{6}:\frac{15}{6} :: 2:1:3$$

Ans. (C)

Sol. 229 For simultaneous decay modes of nucleus, effective decay constant is given as

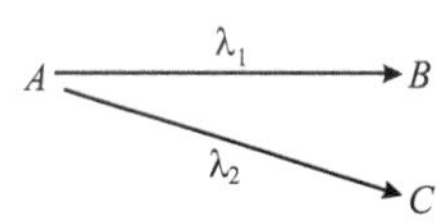

$$\lambda_{eff} = \lambda_1 + \lambda_2$$

$$\Rightarrow \quad \frac{\ln 2}{T_{eff}} = \frac{\ln 2}{T_1} + \frac{\ln 2}{T_2}$$

$$\Rightarrow \quad T_{eff} = \frac{T_1T_2}{T_1+T_2} = \frac{100\times 10}{100+10} = \frac{100}{11} = 9.09\ \text{s}$$

Ans. (A)

27.2 Nuclear Physics

Sol. 230 Power output is energy released per second. Total energy released in 30 days is given as

$$E = 200\,\text{MeV} \times \left[\frac{2\times 6.03\times 10^{23}}{235\times 10^{-3}}\right]$$

Power output of reactor is give.n as

$$\Rightarrow \quad P = \frac{E}{T} = \frac{200\times 10^{6}\times 1.6\times 10^{-19}}{30\times 24\times 86400} \times \frac{2\times 6.023\times 10^{23}}{235\times 10^{-3}}$$

$$\cong 63.2\times 10^{6}\,\text{W}$$ **Ans. (B)**

Sol. 231 Mass density of nucleus is given as

$$\rho = \frac{m}{V}$$

$$\Rightarrow \quad \rho = \frac{m}{V} = \frac{A}{\frac{4}{3}\pi R^{3}} = \frac{3}{4\pi r^{3}}$$

$$\Rightarrow \quad \rho = 2.3\times 10^{17}\,\text{kg/m}^3$$ **Ans. (C)**

Sol. 232 Binding energy of ${}^{120}_{50}S_n$ given as

$$\Delta E = \Delta mc^2$$

Where $\Delta m = 50(1.00783) + 70(1.00867) - 119.902199$

$$\Rightarrow \quad \Delta E = \Delta m \times 931 = 1020.56\,\text{MeV}$$

BE per nucleon is given as

$$(BE)_n = \frac{1020.56}{120} = 8.5\,\text{MeV}$$ **Ans. (A)**

Sol. 233 Mass defect of proton capture reaction is given as

$${}^{7}_{3}\text{Li} + {}^{1}_{1}\text{H} \rightarrow 2\,{}^{4}_{2}\text{He}$$

$$\Delta m = m_{\text{Li}} + m_{\text{H}} - 2M_{\text{He}}$$

$$\Rightarrow \quad \Delta m = [(7.016 + 1.0079) - 2\times 4.0026] = 0.0187\,u$$

Energy released by 20 gm of Li is given as

$$E = \frac{\Delta mc^2}{m_{\text{Li}}} \times 20$$

$$\Rightarrow \quad E = \frac{0.0187\times 1.6\times 10^{-19}\times 10^{9}}{7.016\times 1.6\times 10^{-24}} \times 20\,\text{J}$$

$$\Rightarrow \quad E = 5.33\times 10^{12}\,\text{J}$$

$$\Rightarrow \quad E = \frac{5.33\times 10^{12}}{1000\times 3600} = 1.48\times 10^{6}\,\text{kWh}$$ **Ans. (B)**

Sol. 234 Only in option-(A) masses of reactants is more than that of product so this is only possible reaction. **Ans. (A)**

28.2 PN Junction and Diodes

Sol. 235 Circuit of described situation is shown below

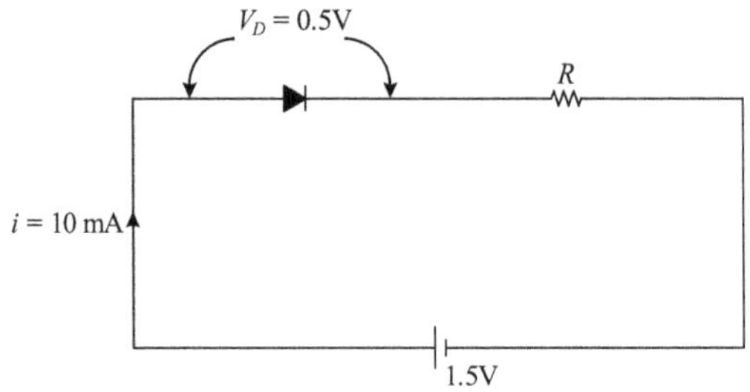

Using KVL equation in loop gives

$$1.5 - 0.5 - iR = 0$$

$$\Rightarrow \quad R = \frac{1}{i} \times \frac{1}{10\times 10^{-3}} = 100\,\Omega$$ **Ans. (A)**

Sol. 236 Energy band gap is given as

$$\Delta E = \frac{hc}{\lambda} = \frac{19.89\times 10^{-26}}{400\times 10^{-9}\times 1.6\times 10^{-19}}$$

$$\Rightarrow \quad \Delta E = 3.107\,\text{eV}$$ **Ans. (C)**

Sol. 237 As there are two zener diodes in opposite sense in both cycles one will breakdown and cut off voltage at 6 V so option (B) is correct. **Ans. (B)**

Sol. 238 For photodiode on increasing reverse voltage, current increases then saturation arises at a specific intensity of light falling on it.

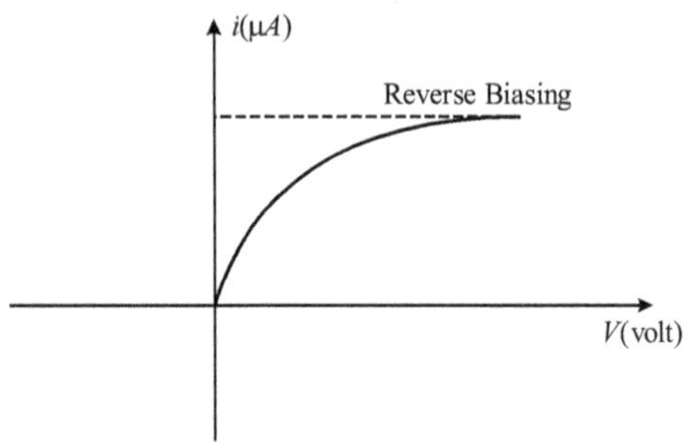

Ans. (A)

Sol. 239 When input voltage increases from 0, breakdown in 4V zener take place when $V_{\text{in}} = 4$V but when $V_{\text{in}} > 4$V then also V_{out} increases because of potential drop across 100Ω resistance and when $V_{\text{in}} = 6$V or more breakdown in 6V zener occurs and V_{out} remain constant at 6V now on lord. **Ans. (D)**

28.3 Transistors

Sol. 240 From the given graph at $I_C = 4$ mA, we calculate ΔI_B and ΔI_C as

$$\Delta I_B = (30 - 20) = 10\ \mu A$$

$$\Delta I_C = (4.5 - 3)\ mA = 1.5\ mA$$

β_{ac} is given as

$$\beta_{ac} = \frac{\Delta I_C}{\Delta I_B} = \frac{1.5 mA}{10 \mu A} = 150$$

$\Rightarrow$ $\beta_{ac} = 150$ **Ans. (150.00)**

28.4 Logic Gates

Sol. 241 In below figure intermediate outputs of gates are mentioned as C, D and E

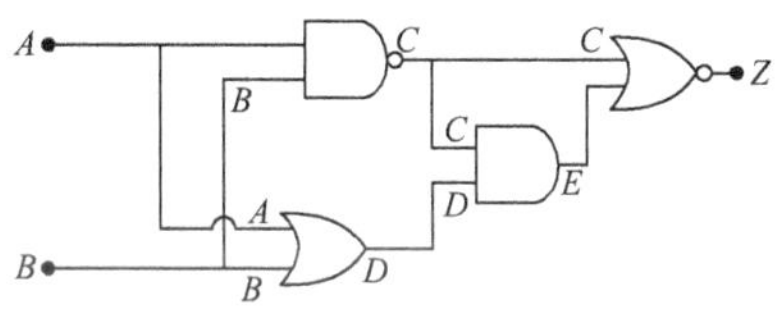

Relation between output z and inputs A and B are deduced as

$$Z = \overline{(C + E)}$$

$\Rightarrow$ $Z = \overline{(C + CD)}$

$\Rightarrow$ $Z = \overline{(C(1 + D))} = \overline{C}$

$\Rightarrow$ $Z = \overline{\overline{AB}} = AB$

Thus truth table for given circuit is made as shown in figure

A	B	Z
0	0	0
0	1	0
1	0	0
1	1	1

Ans. (C)

Sol. 242 The truth table of given circuit is shown below

A	B	C	
0	0	0	0
1	0	0	0
0	1	0	0
0	0	1	0
1	1	0	0
1	0	1	0
0	1	1	0
1	1	1	1

This corresponds to AND gate. **Ans. (A)**

Sol. 243 Truth table of given logic circuit is given as

A	B	Y
0	0	0
0	1	1
1	0	1
1	1	1

For the given input wavelength correct output wavelength is given as

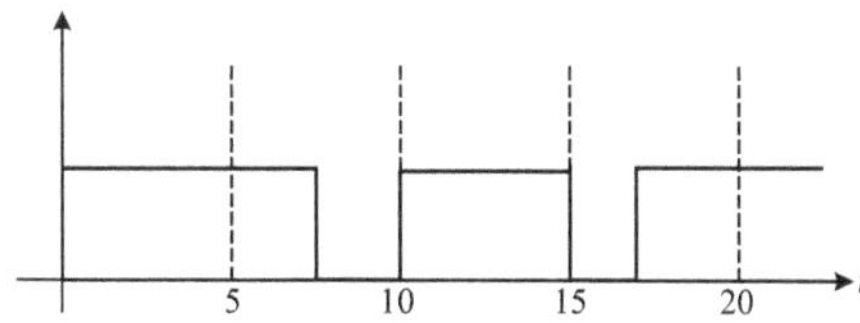

None of given option is correct. **Ans. (None)**

29.2 Modulation and Demodulation

Sol. 244 Modulated signed voltage is given as

$$V_m = 5\,(1 + 0.6 \cos(6280t) \sin(2\pi \times 10^4 t)$$

$$V_m = 5 \sin(2\pi \times 10^4 t) + 3 \cos(6280t) \sin(2\pi \times 10^4 t)$$

In above wave maximum amplitude is

$$V_{max} = 5 + 3 = 8\ V$$

$$V_{min} = 5 - 3 = 2\ V$$

Ans. (None)

30.1 Errors and Experiments

Sol. 245 Zero error in scale is

$$Z = 0 + 0.01 \times 7 = 0.07\ cm$$

Reading of vernier caliper for length of cylinder is

$$R = 3.1 + 0.01 \times 4 - 0.07 = 3.07\ cm$$

Ans. (D)

Sol. 246 Pitch of screw gauge $p = 0.1$ cm

No of divisions on circular scale

$$N = 50$$

Least count of screw gauge is

$$L = \frac{p}{N} = 0.002\ cm$$

The result of measurement by this screw gauge must be in multiples of L, so here result would be 2.124 cm. **Ans. (D)**

Sol. 247 Percentage error in measurement of z is given as

$$\frac{\Delta z}{z} = \frac{2\Delta a}{a} + \frac{2}{3}\frac{\Delta b}{b} + \frac{1}{2}\frac{\Delta c}{c} + \frac{3\Delta d}{d}$$

$\Rightarrow$ $\frac{z}{\Delta z} = 2 \times 2 + \frac{2}{3} \times 15 + \frac{1}{2} \times 4 + 3 \times 2.5\%$

$\Rightarrow$ $\frac{\Delta z}{z} = 14.5\%$ **Ans. (B)**

Sol. 248 Pitch of circular scale = 0.5 mm

Number of divisions of CS = 50

$$\text{Least count} = \frac{0.5}{50}\text{ mm}$$

$$= 0.01\text{ mm}$$

$$= 0.01 \times 10^{-3}\text{ m}$$

$$= 10 \times 10^{-6}\text{ m} = 40\ \mu\text{m}$$

Zero error $z = 4 \times 0.01 = 0.04$ mm (+ve) **Ans. (B)**

Sol. 249 Fractional error in density measurement is given as

$$\frac{d\rho}{\rho} = \frac{dm}{m} + \frac{3dD}{D} \qquad \left[\text{As } \rho = \frac{m}{\frac{4}{3}\pi\left(\frac{D}{2}\right)^3}\right]$$

We use

$$\frac{d\rho}{\rho} = \times 100 = 6 + 3 \times 1.5 = 10.5\% = \frac{x}{100}$$

$\Rightarrow$ $x = 1050$ **Ans. (1050.00)**

Sol. 250 As all readings are given in 3 significant digits, final results must be in 3 significant digits only so option (C) can be only correct answer. **Ans. (C)**

* * * * *

www.ingramcontent.com/pod-product-compliance
Ingram Content Group UK Ltd.
Pitfield, Milton Keynes, MK11 3LW, UK
UKHW061548190726
13845UKWH00030B/1075

9 789390 187744